Dietary Reference Intakes: Recommended Intakes for Individuals: Minerals

Life Stage Group	Calcium (mg/day)	Chromium (µg/day)	Copper (µg/day)	Fluoride (mg/day)	Iodine (µg/day)	Iron (mg/day)	Magnesium (mg/day)	Manganese (mg/day)	Molybdenum (µg/day)	Phosphorus (mg/day)	Selenium (µg/day)	Zinc (mg/day)	Sodium (g/day)	Chloride (g/day)	Potassium (g/day)
Infants															
0–6 mo	200	0.2*	200*	0.01*	110*	0.27*	30*	0.003*	2*	100*	15*	2*	0.12*	0.18*	0.4*
7–12 mo	260	5.5*	220*	0.5*	130*	**11**	75*	0.6*	3*	275*	20*	**3**	0.37*	0.57*	0.7*
Children															
1–3 y	700	11*	**340**	0.7*	**90**	**7**	**80**	1.2*	**17**	**460**	**20**	**3**	1.0*	1.5*	3.0*
4–8 y	1,000	15*	**440**	1*	**90**	**10**	**130**	1.5*	**22**	**500**	**30**	**5**	1.2*	1.9*	3.8*
Males															
9–13 y	**1,300**	25*	**700**	2*	**120**	**8**	**240**	1.9*	**34**	**1,250**	**40**	**8**	1.5*	2.3*	4.5*
14–18 y	**1,300**	35*	**890**	3*	**150**	**11**	**410**	2.2*	**43**	**1,250**	**55**	**11**	1.5*	2.3*	4.7*
19–30 y	**1,000**	35*	**900**	4*	**150**	**8**	**400**	2.3*	**45**	**700**	**55**	**11**	1.5*	2.3*	4.7*
31–50 y	**1,000**	35*	**900**	4*	**150**	**8**	**420**	2.3*	**45**	**700**	**55**	**11**	1.5*	2.3*	4.7*
51–70 y	**1,000**	30*	**900**	4*	**150**	**8**	**420**	2.3*	**45**	**700**	**55**	**11**	1.3*	2.0*	4.7*
> 70 y	**1,200**	30*	**900**	4*	**150**	**8**	**420**	2.3*	**45**	**700**	**55**	**11**	1.2*	1.8*	4.7*
Females															
9–13 y	**1,300**	21*	**700**	2*	**120**	**8**	**240**	1.6*	**34**	**1,250**	**40**	**8**	1.5*	2.3*	4.5*
14–18 y	**1,300**	24*	**890**	3*	**150**	**15**	**360**	1.6*	**43**	**1,250**	**55**	**9**	1.5*	2.3*	4.7*
19–30 y	**1,000**	25*	**900**	3*	**150**	**18**	**310**	1.8*	**45**	**700**	**55**	**8**	1.5*	2.3*	4.7*
31–50 y	**1,000**	25*	**900**	3*	**150**	**18**	**320**	1.8*	**45**	**700**	**55**	**8**	1.5*	2.3*	4.7*
51–70 y	**1,200**	20*	**900**	3*	**150**	**8**	**320**	1.8*	**45**	**700**	**55**	**8**	1.3*	2.0*	4.7*
> 70 y	**1,200**	20*	**900**	3*	**150**	**8**	**320**	1.8*	**45**	**700**	**55**	**8**	1.2*	1.8*	4.7*
Pregnancy															
≤ 18 y	**1,300**	29*	**1,000**	3*	**220**	**27**	**400**	2.0*	**50**	**1,250**	**60**	**12**	1.5*	2.3*	4.7*
19–30 y	**1,000**	30*	**1,000**	3*	**220**	**27**	**350**	2.0*	**50**	**700**	**60**	**11**	1.5*	2.3*	4.7*
31–50 y	**1,000**	30*	**1,000**	3*	**220**	**27**	**360**	2.0*	**50**	**700**	**60**	**11**	1.5*	2.3*	4.7*
Lactation															
≤ 18 y	**1,300**	44*	**1,300**	3*	**290**	**10**	**360**	2.6*	**50**	**1,250**	**70**	**13**	1.5*	2.3*	[illegible]
19–30 y	**1,000**	45*	**1,300**	3*	**290**	**9**	**310**	2.6*	**50**	**700**	**70**	**12**	1.5*	2.3*	[illegible]
31–50 y	**1,000**	45*	**1,300**	3*	**290**	**9**	**320**	2.6*	**50**	**700**	**70**	**12**	1.5*	2.3*	[illegible]

Note: This table (taken from the DRI reports, see www.nap.edu) presents Recommended Dietary Allowances (RDAs) in **bold** type and Adequate Intakes (AIs) in ordinary type followed by an aste RDAs and AIs may both be used as goals for individual intakes. RDAs are set up to meet the needs of almost all (97–98%) individuals in a group. For healthy breastfed infants, the AI is the mean The AI for all other life stage and gender groups is believed to cover needs of all individuals in the group, but lack of data or uncertainty in the data prevent being able to specify with confidence percentage of individuals covered by this intake.

Source: Dietary Reference Intake Tables: The Complete Set. Institute of Medicine, National Academy of Sciences. Available online at www.nap.edu.

Acceptable Macronutrient Distribution Ranges (AMDR) for Healthy Diets as a Percent of Energy						
Age	**Carbohydrate**	**Added Sugars**	**Total Fat**	**Linoleic Acid**	**α-Linolenic Acid**	**Protein**
1–3 y	45–65	≤ 25	30–40	5–10	0.6–1.2	5–20
4–18 y	45–65	≤ 25	25–35	5–10	0.6–1.2	10–30
≥ 19 y	45–65	≤ 25	20–35	5–10	0.6–1.2	10–35

Source: Institute of Medicine, Food and Nutrition Board. "Dietary Reference Intakes for Energy, Carbohydrates, Fiber, Fat, Protein, and Amino Acids." Washington, D.C.: National Academy Press, 2002.

Dietary Reference Intakes: Recommended Intakes for Individuals: Carbohydrates, Fibre, Fat, Fatty Acids, Protein, and Water								
Life Stage Group	**Carbohydrate (g/day)**	**Fibre (g/day)**	**Fat (g/day)**	**Linoleic Acid (g/day)**	**α-Linolenic Acid (g/day)**	**Protein**		**Water (litres)**
						(g/kg/day)	**(g/day)**	
Infants								
0–6 mo	60*	ND	31*	4.4*†	0.5*‡	1.52*	9.1*	0.7*
7–12 mo	95*	ND	30*	4.6*†	0.5*‡	**1.2**	**11.0**	0.8*
Children								
1–3 y	**130**	19*	**ND**	7*	0.7*	**1.05**	**13**	1.3*
4–8 y	**130**	25*	**ND**	10*	0.9*	**0.95**	**19**	1.7*
Males								
9–13 y	**130**	31*	**ND**	12*	1.2*	**0.95**	**34**	2.4*
14–18 y	**130**	38*	**ND**	16*	1.6*	**0.85**	**52**	3.3*
19–30 y	**130**	38*	**ND**	17*	1.6*	**0.80**	**56**	3.7*
31–50 y	**130**	38*	**ND**	17*	1.6*	**0.80**	**56**	3.7*
51–70 y	**130**	30*	**ND**	14*	1.6*	**0.80**	**56**	3.7*
> 70 y	**130**	30*	**ND**	14*	1.6*	**0.80**	**56**	3.7*
Females								
9–13 y	**130**	26*	**ND**	10*	1.0*	**0.95**	**34**	2.1*
14–18 y	**130**	26*	**ND**	11*	1.1*	**0.85**	**46**	2.3*
19–30 y	**130**	25*	**ND**	12*	1.1*	**0.80**	**46**	2.7*
31–50 y	**130**	25*	**ND**	12*	1.1*	**0.80**	**46**	2.7*
51–70 y	**130**	21*	**ND**	11*	1.1*	**0.80**	**46**	2.7*
> 70 y	**130**	21*	**ND**	11*	1.1*	**0.80**	**46**	2.7*
Pregnancy	**175**	28*	**ND**	13*	1.4*	**1.1**	**RDA + 25g**	3.0*
Lactation	**210**	29*	**ND**	13*	1.3*	**1.3**	**RDA + 25g**	3.8*

ND = not determined *Values are AI (Adequate Intakes) † Refers to all ω-6 polyunsaturated fatty acids ‡ Refers to all ω-3 polyunsaturated fatty acids

Source: Institute of Medicine, Food and Nutrition Board. "Dietary Reference Intakes for Energy, Carbohydrates, Fiber, Fat, Fatty Acids, and Protein." Washington, D.C.: National Academy Press, 2002.

ALL THE HELP, RESOURCES, AND PERSONAL SUPPORT YOU AND YOUR STUDENTS NEED!

www.wileyplus.com/resources

2-Minute Tutorials and all of the resources you and your students need to get started

Student support from an experienced student user

Collaborate with your colleagues, find a mentor, attend virtual and live events, and view resources
www.WhereFacultyConnect.com

Pre-loaded, ready-to-use assignments and presentations created by subject matter experts

Technical Support 24/7
FAQs, online chat,
and phone support
www.wileyplus.com/support

Your *WileyPLUS* Account Manager, providing personal training and support

Nutrition: Science and Applications

Canadian Edition

Lori A. Smolin, Ph.D.
University of Connecticut

Mary B. Grosvenor, M.S., R.D.

Debbie Gurfinkel, Ph.D.
University of Toronto

John Wiley & Sons Canada, Ltd.

Library and Archives Canada Cataloguing in Publication
Smolin, Lori A.
Nutrition : science and application / Lori A. Smolin, Mary B. Grosvenor, Debbie Gurfinkel. — Canadian ed.
Includes bibliographical references and index.
ISBN 978-0-470-67634-9
1. Nutrition. I. Grosvenor, Mary B. II. Gurfinkel, Debbie, 1955- III. Title.
QP141.S66 2011 613.2 C2011-904629-6

Production Credits

Acquisitions Editor	Rodney Burke
Vice President & Publisher	Veronica Visentin
Senior Marketing Manager	Anne-Marie Seymour
Editorial Manager	Karen Staudinger
Production Manager	Tegan Wallace
Developmental Editor	Andrea Grzybowski
Media Editor	Channade Fenandoe
Editorial Assistant	Laura Hwee
Photo Research & Permissions Coordinator	Luisa Begani
Design	Mike Chan
Typesetting	Thomson Digital
Cover Design	Joanna Vieira
Cover Image	© iStockphoto/Vetta Collection
Printing and binding	Quad

Printed and bound in the United States.

1 2 3 4 QD 15 14 13 12

John Wiley & Sons Canada, Ltd.
6045 Freemont Blvd.
Mississauga, Ontario L5R 4J3
Visit our website at: www.wiley.ca

About the Authors

Lori A. Smolin received a Bachelor of Science degree from Cornell University, where she studied human nutrition and food science. She received a doctorate from the University of Wisconsin at Madison, where her doctoral research focused on B vitamins, homocysteine accumulation, and genetic defects in homocysteine metabolism. She completed postdoctoral training both at the Harbor–UCLA Medical Center, where she studied human obesity, and at the University of California—San Diego, where she studied genetic defects in amino acid metabolism. She has published articles in these areas in peer-reviewed journals. Dr. Smolin is currently at the University of Connecticut, where she teaches in the Department of Nutritional Science. Courses she has taught include introductory nutrition, life cycle nutrition, food preparation, nutritional biochemistry, general biochemistry, and introductory biology.

Mary B. Grosvenor holds a bachelor of arts in English and a master of science in Nutrition Science, affording her an ideal background for nutrition writing. She is a registered dietitian and has worked in clinical as well as research nutrition, in hospitals and communities large and small in the western United States. She teaches at the community college level and has published articles in peer-reviewed journals in nutritional assessment and nutrition and cancer. Her training and experience provide practical insights into the application and presentation of the science in this text.

Debbie Gurfinkel earned her undergraduate degree at the University of Toronto, specializing in food chemistry. She earned her MSc from the University of Toronto's Department of Nutritional Sciences. Her thesis work was on the chemistry of dietary fibre in wheat bran. Her doctoral research, in the same department, investigated the anti-carcinogenic activity of soybean constituents. Her interest in cancer research continued at the post-doctoral level, where she worked in the area of cancer drug discovery at the Ontario Cancer Institute, Princess Margaret Hospital in Toronto.

The use of education technologies in the classroom and the development of online courses has been a continuing interest. She has been involved in authoring an online course on nutrition for pharmacists and holds a Certificate in E-learning from the University of Toronto's School of Continuing Studies.

Since 2004, she has lectured at the undergraduate level at the University of Toronto in the Department of Nutritional Sciences. Her teaching duties have included a wide range of courses including basic human nutrition and food chemistry, as well as vitamin and mineral metabolism, nutritional genomics, and the metabolic aspects of nutrition and human disease. Before joining the University of Toronto, she taught several food science courses at Ryerson University's School of Nutrition. To date she has taught over 3,000 students

Dedication

To my sons, Zachary and Max, and to my husband, David Knecht, who help me maintain perspective and recognize and appreciate the important things in life.

LAS

To my boys, in appreciation for the time this work takes from them and the inspiration (and in recent years, editorial assistance) they give to me.

MBG

To Jack, Steven, and Jeffrey, for their unwavering support and endless patience while I worked on this project.

DG

Preface

Nutrition: Science and Applications, Canadian Edition is an adaptation of the American textbook ***Nutrition: Science and Applications, Second Edition***. This new Canadian edition has as its foundation the nutrition fundamentals and outstanding features of the original American edition. The integration of Canadian content, with this excellent core material, results in a textbook that is both comprehensive and relevant to the Canadian student. It is intended as an introductory nutrition textbook for science-oriented students at the university or college level. The scientific aspects of nutrient function are detailed using the basic principles of biology, physiology, and biochemistry.

Integrated Canadian Content

The Canadian content of the textbook has several recurring themes. Canada's Food Guide is described in detail early in the text and its usefulness as a tool for making nutritious food choices is emphasized throughout. Data from the Canadian Community Health Survey is presented throughout, often in the form of critical thinking exercises in which students are asked to interpret results and evaluate their implications. Canadian regulations such as those related to food labelling, natural health products, and food safety and the activities of food regulatory agencies such as Health Canada and the Canadian Food Inspection Agency are widely discussed. Results of research on Canadian populations, often conducted at Canadian universities, is included in many chapters.

Critical Thinking Enhances Problem-Solving Skills

Nutrition: Science and Applications, Canadian Edition, takes a critical thinking approach to understanding and applying human nutrition. Like other introductory texts it offers students the basics of nutrition by exploring the nutrients, their functions in the body, and sources in the diet. But its unique critical thinking approach gives students an understanding of the "whys" and the "hows" behind nutrition processes and recommendations. In each chapter, Critical Thinking exercises introduce nutrition-related problems and lead students through the logical questions and thought processes needed to find a solution. The critical thinking exercises included in the textbook fall into two categories:

- Critical thinking exercises related to the health issues and food choices of the individual
- Critical thinking exercises related to public health issues arising from the results of nutritional research studies, including the Canadian Community Health Survey

Critical Thinking Exercises Consider the Food Choices of the Individual

- Can I help my mom manage her blood cholesterol?
- Why have I gained 10 pounds?
- What should I eat to reduce my risk of cancer?
- How can I change my diet to better support my athletic training?

These are some of the questions students want answered when they enroll in nutrition classes. To answer these and other health-related questions and to continuously fuel student interest, discussions of the relationships among nutrition, health, and disease are integrated throughout the text. Almost every chapter contains a critical thinking exercise in which an individual faces a relevant health issue, often presented with a clinical flavour (e.g., a middle-aged man trying to lower his serum cholesterol levels because of his family history of heart disease). Students are challenged to analyze the individual's situation and food intake and use their knowledge to make dietary recommendations. "Applications" at the end of each chapter then ask students to use this same process of logical scientific inquiry, along with the information in the chapter, to assess their own diets and modify them to promote health and to reduce the risk of nutrient deficiencies and nutrition-related chronic diseases. This critical thinking approach gives students the tools they need to bring nutrition out of the classroom and apply the logic of science to their own nutrition concerns.

Critical Thinking Exercises Encourage an Understanding of Nutrition Research

Nutrition: Science and Applications, Canadian Edition introduces nutrition research methodology in the first chapter of the text by explaining the strengths and limitations of both observational studies and intervention trials. In an approach that is relatively unique in introductory textbooks, this introductory information is reinforced with at least one critical thinking exercise in each chapter that presents the results of a nutrition research study and challenges the student to interpret the results and their implications. Students' attention is often drawn to the link between nutritional research findings and public health policy (for instance, research on the relationship between fish consumption and heart disease and the recommendation in Canada's Food Guide to consume at least two servings of fish each week). These exercises teach students to evaluate the nutrition information they encounter in scientific literature, which is essential to their role as informed consumers and as future scientists and/or health professionals.

Integrated Metabolism Reinforces Understanding

Nutrition: Science and Applications, Canadian Edition is distinctive in its integrated approach to the presentation of nutrient metabolism. Metabolism is one of the most challenging topics for students, but it is a topic that applies to each nutrient and is intimately linked to its function and its impact on human health. The text includes a comprehensive discussion of metabolism as it applies to each of the energy-yielding nutrients, it shows how the micronutrients are involved, and then ties everything together in discussions of energy metabolism during weight management and physical activity. Integrating discussions of metabolism throughout appropriate chapters in the text makes metabolism more manageable and memorable for students because it presents this challenging material in smaller segments and illustrates its relevance to the nutrient being discussed. It also reinforces understanding of metabolic processes by revisiting key concepts with each nutrient and adding relevant new information. For example, the text introduces a simple overview of metabolism in Chapter 3 and then builds on this base with a more complex discussion of carbohydrate metabolism in Chapter 4. This information is added to and reviewed in chapters covering lipids, proteins, micronutrients, energy balance, and exercise. Each discussion of metabolism is highlighted by a metabolism icon. To tie the concepts together, the illustrations use the same basic diagram with new information superimposed over familiar portions to demonstrate how each nutrient fits into the process. The nutrients and steps of metabolism are also colour coded for easier recognition.

Two Approaches to the Presentation of Micronutrient Function Provide Flexibility

Nutrition: Science and Applications, Canadian Edition presents two approaches to the discussion of micronutrient function. In one approach, the micronutrients are described chapter by chapter, where the multiple functions of each vitamin and mineral are described in a single section devoted to that nutrient. A second approach, often referred to as a "functional" approach, is also presented. A feature entitled Focus On Integrating Nutrient Function in the Body, unique to the Canadian edition, describes the role that multiple micronutrients play in several body systems beginning with a discussion of cell metabolism and a description of nutrients that play important roles in gene expression, energy metabolism, antioxidant activity, and fluid balance. Next, the central nervous system and the role that nutrients play in nerve and brain health and development is described. The complex role of nutrients in the immune system is considered next. How nutrients function in the formation of erthyrocytes, the control of blood clotting, as well as the maintenance of healthy blood vessels is also presented in a section on the circulatory system. Finally, the nutrients essential to bone health are discussed. The overall impact of this Focus On feature, which appears after Chapter 12, is to emphasize to students that micronutrients function throughout the body. Furthermore, where appropriate, the roles of amino acids and fatty acids are included, which broadens the picture of integrated function. The presence of these two approaches to the presentation of micronutrient function affords instructors maximum flexibility in their course design.

Focusing on the Total Diet Illustrates the Best Way to Evaluate Food Intake

Nutrition: Science and Applications, Canadian Edition presents the message that each nutrient and food choice makes up only a small part of a person's total diet and that it is the overall dietary pattern that determines the healthfulness of the diet. For example, choosing whole-wheat bread provides a more nutrient-dense source of carbohydrate than choosing a slice of chocolate cake. However, this does not mean people should never have chocolate cake. No one food choice is bad as long as the sum of the food choices over a period of days or weeks makes up a healthy overall diet. The concept of dietary patterns, such as those described in *Eating Well with Canada's Food Guide*, the Mediterranean Diet, and the CDA's *Just the Basics* and *Beyond the Basics* are introduced early in the text. Healthy eating indices are described and research data showing the relationship between index scores and disease risk is presented. Assessments of the quality of the overall Canadian diet using the Canadian Healthy Eating Index are also described.

Distinctive Features

This text includes a number of features that both spark student interest and help them learn the basics of nutrition.

Case Study

Each chapter begins with a short case study. These health- and disease-oriented vignettes help spur student curiosity and provide an introduction to some of the concepts that will be explained in the chapter. For example, the case study for Chapter 6 discusses the experiences of a Canadian International Development Agency (CIDA) intern observing protein-energy malnutrition for the first time and wondering why Songe, a malnourished child, would have a swollen abdomen. Chapter 9 examines the case of a toddler in the Canadian North who has rickets, and Chapter 10 recounts the story of an athlete who experienced dehydration while running the Montreal marathon. These intriguing stories link the material in the chapter with everyday nutrition issues.

9 The Fat-Soluble Vitamins

Case Study

The pediatrician in an Iqaluit clinic in Nunavut is concerned about a young patient of hers. The patient, a little girl named Eva, is 14 months old, but has only one tooth, slightly bowed legs, and bumps on her ribs, classic symptoms of the vitamin D deficiency disease rickets. Vitamin D is needed for proper formation and maintenance of bones and teeth. Without sufficient vitamin D, a child's legs bow under the weight of standing, and bony bumps appear on each of the ribs. The poorly formed bones can break easily, and teeth erupt late and are very prone to decay. Eva is showing these symptoms.

The doctor recently read a scientific paper that reported cases of rickets in Canada, especially among babies and young children who are breastfed and receive little sunlight.[1] Breast milk, while being an excellent food for growing infants and toddlers, contains little vitamin D. The ultraviolet light of the sun can be used by the skin to biosynthesize vitamin D, but in the remote north of Nunavut, sunlight is too limited for this process to occur. The doctor learns from Eva's parents that she has been breastfed all her life and her mother is unaware that breastfed babies need vitamin D supplements. The pediatrician's diagnosis of rickets is confirmed by a blood test that finds very low levels of vitamin D in Eva's blood as well as indicators of abnormal bone breakdown.

Like Eva's parents, most people in Canada are not familiar with rickets, but rare cases of the deficiency disease still occur in young children, especially in circumstances similar to Eva's.

In this chapter, you will learn about the important functions of vitamin D as well as the other fat-soluble vitamins, vitamins A, E, and K.

(Peter Griffith/Masterfile)

(©iStockphoto)

(Rita Maas Studio/StockFood America)

Chapter Outline

9.1 Fat-Soluble Vitamins in the Canadian Diet

9.2 Vitamin A
- Vitamin A in the Diet
- Vitamin A in the Digestive Tract
- Vitamin A in the Body
- Recommended Vitamin A Intake
- Vitamin A Deficiency: A World Health Problem
- Vitamin A Toxicity and Supplements

9.3 Vitamin D
- Vitamin D in the Diet
- Vitamin D in the Body
- Recommended Vitamin D Intake
- Vitamin D Deficiency
- Vitamin D Supplements
- Vitamin D Toxicity

9.4 Vitamin E
- Vitamin E in the Diet
- Vitamin E in the Body
- Recommended Vitamin E Intake
- Vitamin E Deficiency
- Vitamin E Supplements and Toxicity

9.5 Vitamin K
- Vitamin K in the Diet
- Vitamin K in the Body
- Recommended Vitamin K Intake
- Vitamin K Deficiency
- Vitamin K Toxicity and Supplements

Chapter Outline

This brief outline of the chapter's content provides students and instructors with an overview of the major topics presented in the chapter.

Case Study Outcome

Each chapter ends with a "case study outcome," which completes the case study stories begun in the chapter introduction. For example, the outcome at the end of Chapter 6 describes how young Songe's protein deficiency symptoms improve when his diet is supplemented with protein, and discusses the CIDA worker's efforts to improve protein nutrition by educating the people in the community. These "outcomes" review concepts covered in the chapter and illustrate the application of nutrition knowledge to clinical situations.

Learning Objectives

Each chapter section begins with one or more learning objectives. These present brief clear descriptions of what the students should be able to do to demonstrate mastery of the material once they have read the section.

7.2 Exploring Energy Balance

Learning Objectives

- Explain the principle of energy balance.
- Describe the processes involved in generating ATP from food.
- Describe the components of energy expenditure.
- Indicate how excess dietary energy is stored in the body.

The principle of **energy balance** states that when energy consumption equals energy expenditure, body weight remains constant. Energy balance can be achieved at any weight—fat, thin, or in between (**Figure 7.2**); it simply means that body weight is not changing. If, however, less energy is taken in than expended, energy balance is negative and weight will be lost. On the other hand, if the amount of energy taken in exceeds the amount expended, energy balance is positive and the extra energy will be stored in the body, causing weight to increase.

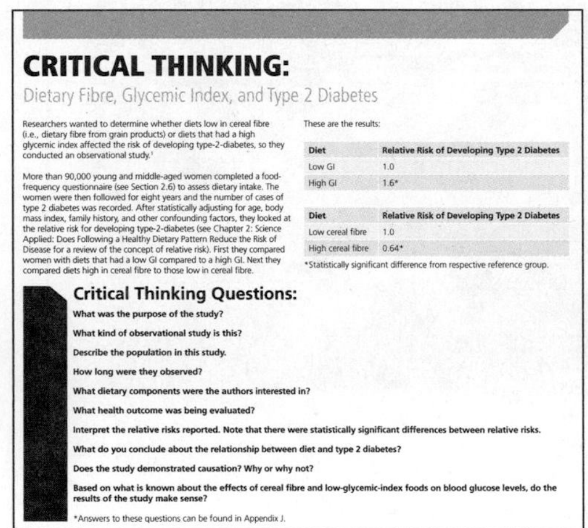

CRITICAL THINKING:

Dietary Fibre, Glycemic Index, and Type 2 Diabetes

Researchers wanted to determine whether diets low in cereal fibre (i.e., dietary fibre from grain products) or diets that had a high glycemic index affected the risk of developing type-2-diabetes, so they conducted an observational study.[1]

More than 90,000 young and middle-aged women completed a food-frequency questionnaire (see Section 2.6) to assess dietary intake. The women were then followed for eight years and the number of cases of type 2 diabetes was recorded. After statistically adjusting for age, body mass index, family history, and other confounding factors, they looked at the relative risk for developing type-2-diabetes (see Chapter 2: Science Applied: Does Following a Healthy Dietary Pattern Reduce the Risk of Disease for a review of the concept of relative risk). First they compared women with diets that had a low GI compared to a high GI. Next they compared diets high in cereal fibre to those low in cereal fibre.

These are the results:

Diet	Relative Risk of Developing Type 2 Diabetes
Low GI	1.0
High GI	1.6*

Diet	Relative Risk of Developing Type 2 Diabetes
Low cereal fibre	1.0
High cereal fibre	0.64*

*Statistically significant difference from respective reference group.

Critical Thinking Questions:

What was the purpose of the study?

What kind of observational study is this?

Describe the population in this study.

How long were they observed?

What dietary components were the authors interested in?

What health outcome was being evaluated?

Interpret the relative risks reported. Note that there were statistically significant differences between relative risks.

What do you conclude about the relationship between diet and type 2 diabetes?

Does the study demonstrated causation? Why or why not?

Based on what is known about the effects of cereal fibre and low-glycemic-index foods on blood glucose levels, do the results of the study make sense?

*Answers to these questions can be found in Appendix J.

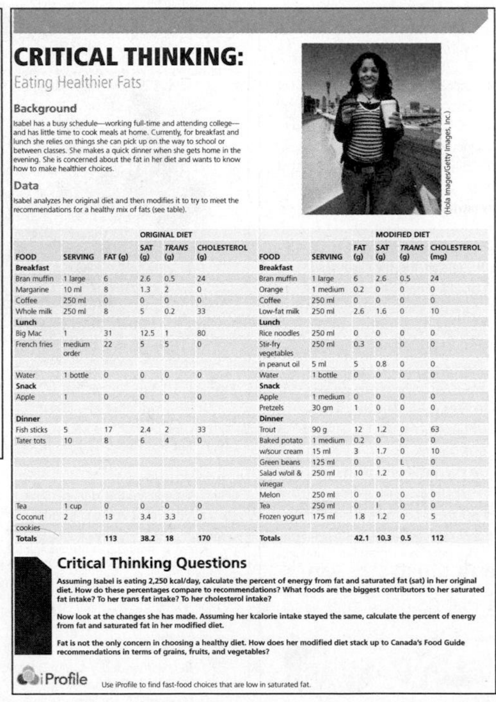

CRITICAL THINKING:

Eating Healthier Fats

Background

Isabel has a busy schedule—working full-time and attending college—and has little time to cook meals at home. Currently, for breakfast and lunch she relies on things she can pick up on the way to school or between classes. She makes a quick dinner when she gets home in the evening. She is concerned about the fat in her diet and wants to know how to make healthier choices.

(Hola Images/Getty Images, Inc.)

Data

Isabel analyzes her original diet and then modifies it to try to meet the recommendations for a healthy mix of fats (see table).

			ORIGINAL DIET					MODIFIED DIET			
FOOD	SERVING	FAT (g)	SAT (g)	*TRANS* (g)	CHOLESTEROL (g)	FOOD	SERVING	FAT (g)	SAT (g)	*TRANS* (g)	CHOLESTEROL (mg)
Breakfast						**Breakfast**					
Bran muffin	1 large	6	2.6	0.5	24	Bran muffin	1 large	6	2.6	0.5	24
Margarine	10 ml	8	1.3	2	0	Orange	1 medium	0.2	0	0	0
Coffee	250 ml	0	0	0	0	Coffee	250 ml	0	0	0	0
Whole milk	250 ml	8	5	0.2	33	Low-fat milk	250 ml	2.6	1.6	0	10
Lunch						**Lunch**					
Big Mac	1	31	12.5	1	80	Rice noodles	250 ml	0	0	0	0
French fries	medium order	22	5	5	0	Stir-fry vegetables	250 ml	0.3	0	0	0
						in peanut oil	5 ml	5	0.8	0	0
Water	1 bottle	0	0	0	0	Water	1 bottle	0	0	0	0
Snack						**Snack**					
Apple	1	0	0	0	0	Apple	1 medium	0	0	0	0
						Pretzels	30 gm	1	0	0	0
Dinner						**Dinner**					
Fish sticks	5	17	2.4	2	33	Trout	90 g	12	1.2	0	63
Tater tots	10	8	6	4	0	Baked potato	1 medium	0.2	0	0	0
						w/sour cream	15 ml	3	1.7	0	10
						Green beans	125 ml	0	0	L	0
						Salad w/oil & vinegar	250 ml	10	1.2	0	0
						Melon	250 ml	0	0	0	0
Tea	1 cup	0	0	0	0	Tea	250 ml	0	I	0	0
Coconut cookies	2	13	3.4	3.3	0	Frozen yogurt	175 ml	1.8	1.2	0	5
Totals		**113**	**38.2**	**18**	**170**	**Totals**		**42.1**	**10.3**	**0.5**	**112**

Critical Thinking Questions

Assuming Isabel is eating 2,250 kcal/day, calculate the percent of energy from fat and saturated fat (sat) in her original diet. How do these percentages compare to recommendations? What foods are the biggest contributors to her saturated fat intake? To her trans fat intake? To her cholesterol intake?

Now look at the changes she has made. Assuming her kcalorie intake stayed the same, calculate the percent of energy from fat and saturated fat in her modified diet.

Fat is not the only concern in choosing a healthy diet. How does her modified diet stack up to Canada's Food Guide recommendations in terms of grains, fruits, and vegetables?

iProfile Use iProfile to find fast-food choices that are low in saturated fat.

Critical Thinking

These exercises, which appear in each chapter, use a critical thinking approach to making decisions and solving problems regarding nutrition. Two types of critical thinking exercises are included in each chapter: one that focuses on individuals and one that presents nutrition research. The critical thinking exercises that focus on individuals help students apply their nutrition knowledge to everyday situations by presenting a nutrition-related problem and then asking a series of questions that lead the student through the logical progression of thought processes needed to solve the problem. Many of these exercises focus on modifying an individual's diet to reduce disease risk or maintain health. For example, one of the exercises in Chapter 5 takes the student through the process of evaluating an individual's risk factors for developing heart disease and recommending dietary changes. The exercise in Chapter 6 shows students how to use complementary proteins to plan a vegetarian diet that is based on traditional Indian cuisine.

The critical thinking exercises that focus on nutrition research introduce students to the interpretation of data. In Chapter 4, the critical thinking exercise that focuses on research challenges students to interpret the results of an observational study that sought to determine the relationship between glycemic index and the risk of developing type two diabetes. In Chapter 6, students evaluate the relationship between vegetarian diets and the risk of death from cardiovascular disease.

Your Choice

These boxed features included in each chapter provide a critical review of products and issues that are the focus of individual nutrition-related consumer choices. For example, the Your Choice box in Chapter 1 discusses how grabbing breakfast on the run can affect your waistline and your wallet. The Your Choice box in Chapter 4 discusses the challenge of choosing whole-grain products and the box in Chapter 5 discusses foods that can lower your blood cholesterol.

YOUR CHOICE

(©iStockphoto)

Natural Health Products in Food Format

Vitamin and mineral supplements are examples of natural health products (see Section 2.5). We tend to think of these supplements as coming in the form of pills or drops. But consider the following: an energy bar that contains soy protein and 23 vitamins and minerals; bottled water with vitamins added to it; a carbonated drink containing B-vitamins and caffeine. These are also examples of the natural health products that have appeared on the Canadian market. They are products that are described by Health Canada's Natural Health Products Directorate as natural health products in "food format."[1]

The main difference between natural health products and food is that natural health products are defined as having a therapeutic purpose, for example: "the diagnosis, treatment, mitigation or prevention of a disease, disorder, or abnormal physical state or its symptoms in humans; restoring or correcting organic functions in humans; or modifying organic functions in humans, such as modifying those functions in a manner that maintains or promotes health." They do not require a prescription.[1]

Food, on the other hand, is recognized by its traditional use as a food to "provide nourishment, nutrition or hydration, or to satisfy hunger, thirst or a desire for taste, texture, or flavour."[1]

There is concern that natural health products in food format may confuse the consumer. All natural health products are labelled with a maximum daily dose. In the case of vitamin and mineral supplements, these are intended to ensure that consumers do not exceed the upper tolerable intake levels (UL). Dietitians of Canada have issued a position paper opposing the sale of natural health products in food format because consumers may perceive them as a food, be unaware that there are maximum doses, and exceed the ULs for some nutrients, with potential adverse effects.[2] For example, the overconsumption of vitamin water during very hot summer days could result in excessive intake of some vitamins.

An additional criticism of natural health products in food format made by Dietitians of Canada is that there is no nutrition facts label on these products. A natural health product in beverage form, for example, may contain added sugar, as well as vitamins, but consumers would not know the caloric content of the product. Health Canada is evaluating the issues related to natural health products in food format.[1]

If you are considering using a natural health product, whether in tablet or food format, you have to thoughtfully consider its suitability as part of your self-care.[3] Remember that many nutrients are readily available from the diet and you may not need to use a natural health product. Dietary modifications may be a more effective way to improve your health as they will provide not only the selected nutrients available in a supplement, but macronutrients, fibre, as well as additional nutrients and phytochemicals.

If you are considering their use, though, the labels of natural health products, although different from food labels, will provide you with important information that will allow you to make a safer and more informed choice. The Natural Health Products Directorate evaluates each natural health product for quality, safety, and efficacy (e.g., whether it does what it claims to do) and when a product meets the directorate's standards, it is awarded a natural products number (NPN). Therefore, the first thing you should look for on the label is the NPN, as absence of a number means that the product has not been evaluated by Health Canada.

Additional information that appears on the label includes the recommended dose and directions on how to take the product (e.g., orally, with or without food). The label will also indicate the health benefit of taking the product, for example, a vitamin C supplement might be described as "a factor in the maintenance of bones, cartilage, teeth and gums." Also listed on the label are the medicinal (e.g., amount of vitamin C) and non-medicinal ingredients (e.g., any other ingredients used to formulate the supplement); if it is a beverage with added sugar, sugar would be listed here.

You will also see information intended to protect you from harm. This includes, where applicable, warnings against using of the product if done in combination with other natural health products or prescription drugs, or if you have certain medical conditions, such as diabetes or heart disease, or if you are pregnant or breast feeding. Any known adverse reactions will also be described.[4]

When in doubt, also consider consulting with a health professional such as a dietitian, doctor, a pharmacist, a nurse or naturopath prior to using the product.[4] When considering a natural health product, such as a vitamin or mineral supplement, be sure to use all the information available to help you to make your choice.

[1]Health Canada. Classification of products at the food-natural health products interface: products in food formats. Available online at. http://www.hc-sc.gc.ca/dhp-mps/prodnatur/legislation/docs/food-nhp-aliments-psn-guide-eng.php. Accessed March 27, 2011.

[2]Dietitians of Canada. Natural health products in food format. Available online at http://www.dietitians.ca/Dietitians-View/Natural-Health-Products-in-Food-Format.aspx. Accessed March 27, 2011.

[3]Health Canada. Informing You about Natural Health Products. Available online at http://www.hc-sc.gc.ca/dhp-mps/prodnatur/fiche_info_sheets-eng.php. Accessed March 27, 2011.

[4]Health Canada. Informing yourself about natural health products-information sheet #5- for consumers- informing yourself. Available online at http://www.hc-sc.gc.ca/dhp-mps/prodnatur/fiche_info_sheets_5-eng.php. Accessed March 27, 2011.

Label Literacy

"Label Literacy" boxes present in-depth information on food and natural health product labels. Label Literacy is designed to show students how to use labels to make wise choices. For example, the box in Chapter 4 shows how food labels can be used to avoid foods that are high in added sugars. Chapter 6 has a Label Literacy feature that discusses how someone with food allergies can use food labels to avoid ingesting particular ingredients and illustrates some resources available from Health Canada to assist them. In Chapter 14, Label Literacy discusses the label of a prenatal supplement, a natural health product.

LABEL LITERACY

How Much Vitamin C Is in Your Orange Juice?

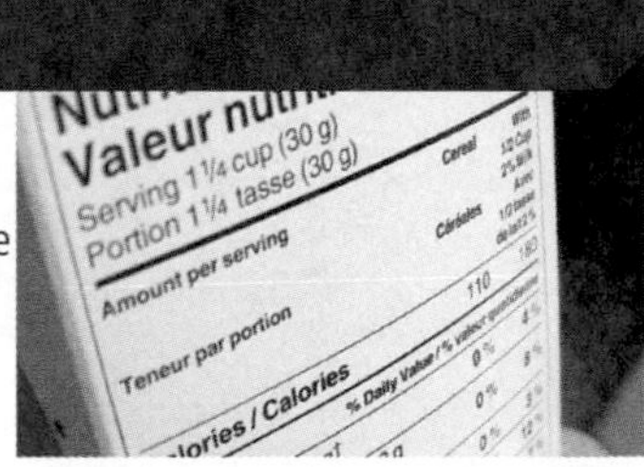

How much folate is in your breakfast cereal? And how much iron is in a box of raisins? It can be difficult to tell from the Nutrition Facts section of a food label exactly how much of a micronutrient is in a food. Food labels are required to provide the % Daily Values for vitamin A, vitamin C, iron, and calcium, but not the actual amount (See Section 2.5). To determine the amount of one of these nutrients in a serving of food, you need to know its Daily Value (see table). Once you know the Daily Value, you can multiply it by the % Daily Value on the label to determine the amount in a serving of the food. Follow these steps to find out how much vitamin C is in a 250 ml (1 cup) of orange juice:

1. Look up the Daily Value:

Vitamin	Daily Value
Vitamin A	1,000 RE (retinol equivalents)
Vitamin D	5 μg
Vitamin E	10 mg
Vitamin K	80 μg
Biotin	30 μg
Pantothenic acid	7 mg
Vitamin C	60 mg
Thiamin	1.3 mg
Riboflavin	1.6 mg
Niacin	23 mg
Vitamin B_6	1.8 mg

2. Find the % Daily Value (%DV) on the food label (see figure): % DV for vitamin C in orange juice = 120%. 3. Multiply the % Daily Value by the Daily Value to find out how much is in a serving: 60 mg X 120% DV = 60 X 1.2 = 72 mg vitamin C.

You can next compare the amount of the nutrient in your food to your RDA. For example, the RDA for vitamin C for a 21-year-old woman is 75 mg. So 250 ml of juice contains 72/75 = 96% of a young woman's RDA. For a young man of the same age, the RDA = 90 mg, so 250 ml of juice contains 80% of a young man's RDA.

Even if you don't look up the Daily Value and do calculations, the % Daily Value on the food label helps you judge how much the food contains. As a general guideline, if the % Daily Value of a nutrient is 5% or less, the food is a poor source; if it is 15% or higher, the food is a good source. Whether you are converting a Daily Value into the amount of a vitamin or are just looking at the Daily Value, be sure to consider how many servings you plan to eat. Remember that doubling the serving doubles the nutrients and kcalories.

Orange Juice

SCIENCE APPLIED

Leptin: Discovery of an Obesity Gene

A discovery made by Dr. Jeffrey Friedman and colleagues in 1994 brought hope to millions of people. Perhaps the cause of obesity had been found and a cure might be close behind. Was relief in sight for those who suffer from the physical and social consequences of obesity?

Dr. Friedman's work began with a strain of mice called *ob* for obese. *Ob* mice become grossly obese, gaining up to three times the normal body weight. The *ob* strain arose spontaneously in 1950 in the mouse colony at the Jackson Laboratory in Bar Harbor, Maine. Friedman and colleagues unraveled the cause for the obesity in this strain of mice when they identified and cloned the gene that was responsible.[1]

Researchers used a series of breeding experiments to localize the gene to a particular stretch of DNA. They then looked to see if any of the genes in this stretch of DNA were expressed in adipose tissue. The search yielded a single gene. Evidence that this gene was involved in the regulation of body weight was obtained by examining the gene and the protein it codes for in the *ob* mice. Researchers found that this protein, which they named leptin, was either not produced or produced in an inactive form in the obese mice. Soon afterward, a similar gene was identified in humans.

Optimism about the role of the protein hormone leptin in human obesity was so great that a biotechnology firm (Amgen) paid $25 million for the commercial rights to leptin in the hope that it could be used to treat human obesity. Those hopes grew even higher when Friedman and colleagues were able to demonstrate that injections of the hormone could restore the genetically obese mice to normal weight.[2,3] Unfortunately, the role of leptin in human obesity has not lived up to expectations. Mutations in this gene are not responsible for most human obesity.[4] In fact, obese humans generally have high blood leptin levels.[5] High doses of leptin administered to obese humans produces only modest weight loss.[6]

The leptin receptor—a protein in the brain to which leptin must bind to produce weight reduction—was also identified.[7] The fact that obese humans have high levels of leptin suggested that the cause of human obesity might involve an abnormality in leptin receptors. If leptin receptors were defective, the leptin produced would have no place to bind and would not be able to signal mechanisms to promote weight reduction. Thus far, however, defective leptin receptors have not been found to be an important cause of human obesity.[8]

A mouse with a defect in the leptin gene (*ob*) may weigh three times as much as a normal mouse. Both of these mice have defective *ob* genes but the one on the right was treated with leptin injections. (© AP/Wide World Photos)

Continued study of the role of leptin in obesity has confirmed that it is an important signal involved in the long-term regulation of body fat, but it does not act alone. There are many steps, involving many genes, that occur between the production of leptin and alterations in food intake and energy expenditure. Researchers have discovered about a dozen molecules that interact with leptin in the brain to control appetite.[9] For example, neuropeptide Y and melanin-concentrating hormone boost appetite, whereas alpha-melanocyte stimulating hormone blunts appetite, and a protein called SOCS3 reduces the sensitivity of leptin receptors.

Despite the fact that the identification of leptin has not produced a cure for human obesity, its discovery lit up the field. This research was an important advance in our understanding of the genetics of body weight regulation. Continued work will someday answer the questions that remain about why some of us are obese and some of us are lean.

[1] Zhang, Y., Proenca, R., Maffei, M. et al. Positional cloning of the mouse obese gene and its human homologue. *Nature* 372:425–432, 1994.

[2] Halaas, J. L., Gajiwala, K. S., Maffei, M. et al. Weight-reducing effects of the plasma protein encoded by the obese gene. *Science* 269:543–546, 1995.

[3] Pelleymounter, M. A., Cullen, M. J., Baker, M. B. et al. Effects of the obese gene product on body weight regulation in *ob/ob* mice. *Science* 269:540–543, 1995.

[4] Montague, C. T., Farooqi, I. S., Whitehead, J. P. et al. Congenital leptin deficiency is associated with severe early onset obesity in children. *Nature* 387:903–908, 1997.

[5] Considine, R. V., Sinha, M. K., Heiman, M. L. et al. Serum immunoreactive-leptin concentrations in normal weight and obese humans. *N. Engl. J. Med.* 334:292–295, 1996.

[6] Gura, T. Obesity research: Leptin not impressive in clinical trial. *Science* 286:881–882, 1999.

[7] Tartaglia, L. A., Dembski, M., Weng X. et al. Identification and expression cloning of a leptin receptor, OB-R. *Cell* 83:1263–1271, 1995.

[8] Tsigos, C., Kyrou, I., and Raptis, S. A. Monogenic forms of obesity and diabetes mellitus. *J. Pediatr. Endocrinol. Metab.* 15:241–253, 2002.

[9] Gura,T. Tracing leptin's partners in regulating body weight. *Science* 287:1738–1741, 2000.

Science Applied

These boxed features included in every chapter highlight nutrition research studies that have led to discoveries key to our current understanding of nutrition. Often written with a historical perspective, the features help students appreciate the research process. For example, in Chapter 1 the Seven Countries Study is discussed. This classic study, begun in 1958, identified the relationship between saturated fat intake and heart disease. Other science applied topics include how weightlessness in space is used to study bone loss, how the obesity gene leptin was identified, how pellagra was determined to be a nutritional deficiency, and how LDL receptors (which help remove LDL cholesterol from the blood) were discovered.

Process Diagrams

Sometimes the hardest part of understanding an illustration is knowing where to start. To help students start at the beginning and understand the information presented in the more complex line art, the steps in the processes are numbered and include a narrative describing what happens at each step. Process diagrams for metabolism help students reduce this often intimidating topic to a series of easy-to-follow steps.

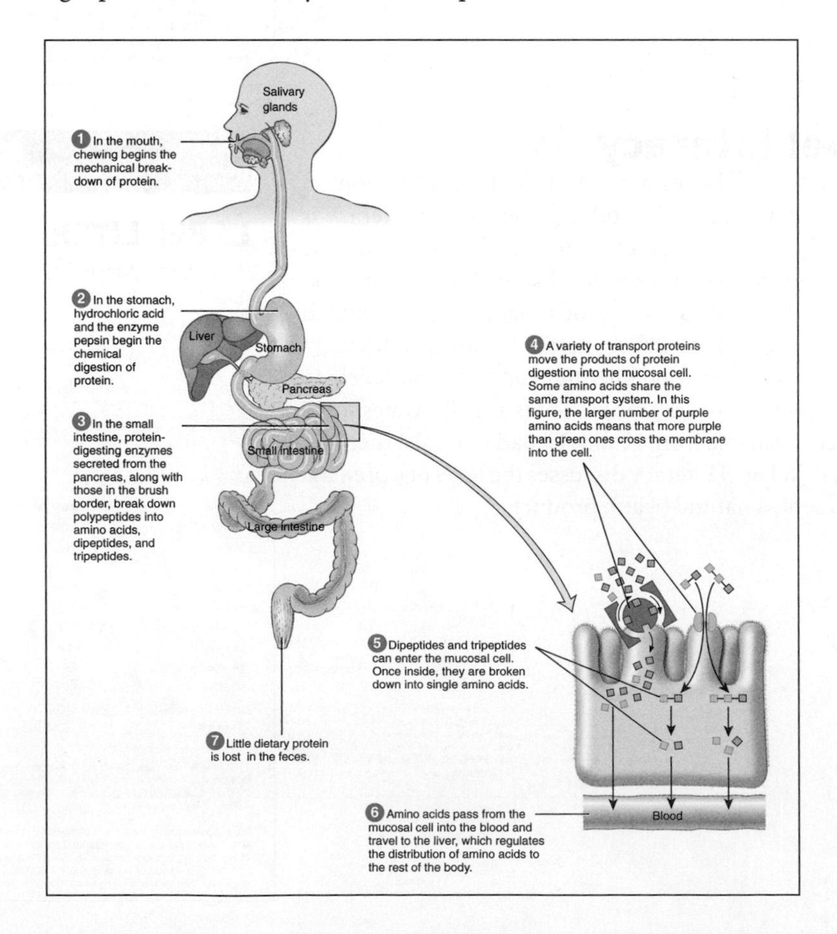

Applications

These exercises, which appear at the end of each chapter, are divided into two parts: one that focuses on the student's personal diet and nutrition concerns and a second that relates to more general nutrition issues. Both require the student to think critically and apply key nutrition concepts. Some of these exercises feature clinical applications and therefore also help reinforce the importance of nutrition in health promotion and disease prevention. Some of these can be done as collaborative learning exercises, which encourage students to work together and learn from each other to solve a problem.

Chapter Summary

Each chapter ends with a summary that highlights important concepts addressed in each section of the chapter.

Review Questions

A set of questions appears at the end of each chapter to test students' understanding of the key points covered. It can be used by students as a study tool to test their understanding of the important information in the chapter.

Focus On

Several topics have been selected for more detailed discussion because of their relevance to nutrition. To provide adequate coverage of these important topics, they are discussed in sections called Focus On. There are 7 of these throughout the book: Focus on Alcohol; Focus on Obesity, Metabolism, and Disease Risk; Focus on Phytochemicals; Focus on Nonvitamin/Nonmineral Supplements; Focus on Integrating Nutrition Function in the Body; Focus on Eating Disorders; and Focus on Biotechnology.

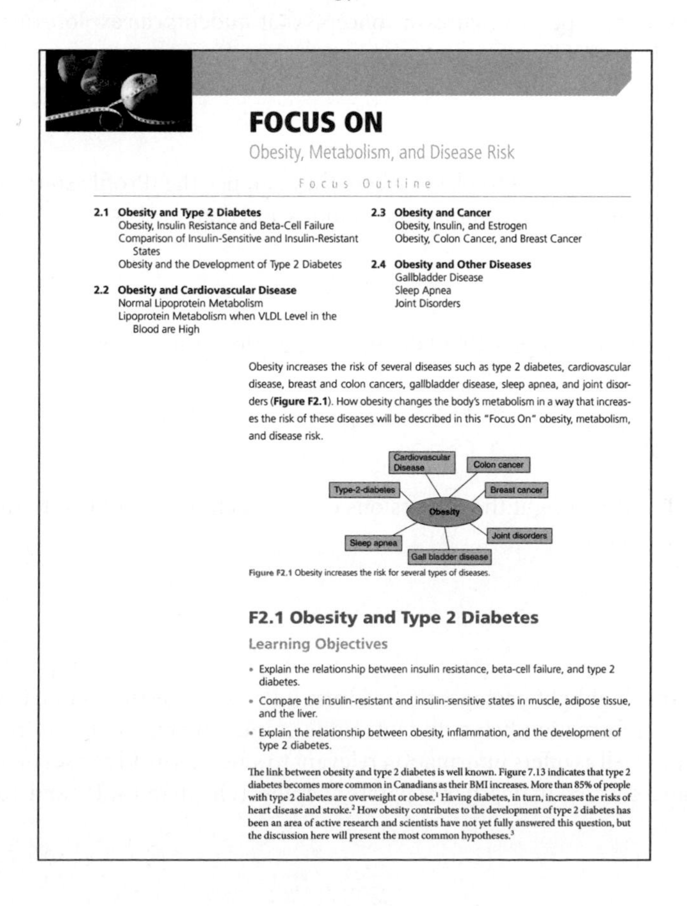

FOCUS ON

Obesity, Metabolism, and Disease Risk

Focus Outline

2.1 Obesity and Type 2 Diabetes
Obesity, Insulin Resistance and Beta-Cell Failure
Comparison of Insulin-Sensitive and Insulin-Resistant States
Obesity and the Development of Type 2 Diabetes

2.2 Obesity and Cardiovascular Disease
Normal Lipoprotein Metabolism
Lipoprotein Metabolism when VLDL Level in the Blood are High

2.3 Obesity and Cancer
Obesity, Insulin, and Estrogen
Obesity, Colon Cancer, and Breast Cancer

2.4 Obesity and Other Diseases
Gallbladder Disease
Sleep Apnea
Joint Disorders

Obesity increases the risk of several diseases such as type 2 diabetes, cardiovascular disease, breast and colon cancers, gallbladder disease, sleep apnea, and joint disorders (**Figure F2.1**). How obesity changes the body's metabolism in a way that increases the risk of these diseases will be described in this "Focus On" obesity, metabolism, and disease risk.

Figure F2.1 Obesity increases the risk for several types of diseases.

F2.1 Obesity and Type 2 Diabetes

Learning Objectives

- Explain the relationship between insulin resistance, beta-cell failure, and type 2 diabetes.
- Compare the insulin-resistant and insulin-sensitive states in muscle, adipose tissue, and the liver.
- Explain the relationship between obesity, inflammation, and the development of type 2 diabetes.

The link between obesity and type 2 diabetes is well known. Figure 7.13 indicates that type 2 diabetes becomes more common in Canadians as their BMI increases. More than 85% of people with type 2 diabetes are overweight or obese.[1] Having diabetes, in turn, increases the risks of heart disease and stroke.[2] How obesity contributes to the development of type 2 diabetes has been an area of active research and scientists have not yet fully answered this question, but the discussion here will present the most common hypotheses.[3]

Marginal Features

The text includes a number of marginal features that aid student comprehension, enhance interest, and point out where particular types of information can be found.

Definitions of New Terms

New terms are highlighted in bold and defined in the margin. This provides easy access to new terms as they appear in the text. These and other terms are included in an extensive glossary at the back of the text.

> **The Scientific Method**
>
> Advances in nutrition are made using the **scientific method**. The scientific method offers a systematic, unbiased approach to evaluating the relationships among food, nutrients, and health. The first step of the scientific method is to make an observation and ask questions about the observation. The next step is to propose a **hypothesis**, or explanation for the observation. Once a hypothesis has been proposed, experiments can be designed to test it. The experiments must provide objective results that can be measured and repeated. The results of one experiment are generally not considered sufficient proof that a hypothesis is correct. If, however,
>
> **scientific method** The general approach of science that is used to explain observations about the world around us.
>
> **hypothesis** An educated guess made to explain an observation or to answer a question.

Canadian Content Icons highlight sections of the text that contain significant Canadian content.

How It Works Icons appear by figures or concepts that students can explore further through animations in *WileyPLUS*.

iProfile icons point students to places where they can use the iProfile software to answer questions about the nutrient composition of foods and diets.

Video Icons appear by topics that students can explore further by watching videos in *WileyPLUS*.

Metabolism Icons highlight the discussions of how each nutrient fits into the metabolic processes of nutrition.

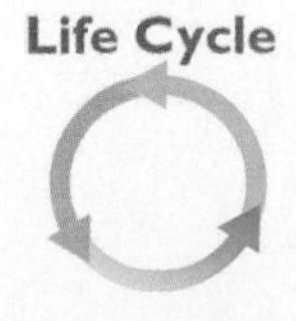

Life Cycle Icons highlight passages that address how life stage affects nutrient needs and concerns. This information helps students understand why nutrient requirements differ with life stage as well as offers information relevant to students in all phases of life. Life cycle nutrition is also covered in depth in separate chapters (chapters 14, 15, and 16).

Chapter-by-Chapter Overview

CHAPTER 1, "Nutrition: Food for Health," begins by discussing the modern diet and how healthy it is. It emphasizes that food choices affect current and future health. This chapter provides an overview of the nutrients and their roles in the body, and defines the basic principles of balance, variety, and moderation that are key to a healthy diet. It also introduces the scientific method, including a detailed description of observational studies and intervention trials, and the steps students need to follow to sort accurate from inaccurate nutrition information.

Canadian content in this chapter includes a discussion of the quality of the Canadian diet, based on CCHS data, and a discussion of reliable Canadian sources of nutrition information.

CHAPTER 2, "Nutrition Guidelines: Applying the Science of Nutrition," defines the Dietary Reference Intakes (DRIs) and discusses how they are used to access the nutritional status of individuals and populations, including the use of the EAR cutpoint method. *Eating Well with Canada's Food Guide* and food labels are also discussed. The final section of this chapter details how to assess the nutritional health of populations and individuals. The concept of dietary patterns is defined and the discussion of nutritional research introduces the concept of relative risk, which recurs in subsequent presentations of research data.

Canadian content in this chapter includes a detailed page-by-page description of Canada's Food Guide, including a discussion of how it was developed. Also discussed are the nutrition facts table, permitted Canadian health claims, and the labelling of natural health products.

CHAPTER 3, "Digestion, Absorption, and Metabolism," discusses how food is digested; how nutrients from foods are absorbed into the body and transported to the cells where they are broken down to provide energy or used to synthesize structural or regulatory molecules, and finally how wastes are removed. This chapter provides an overview of metabolism that serves as a launching pad for the more in-depth metabolism information presented in subsequent chapters.

Canadian content in this chapter includes a discussion of the labelling of prebiotic and probiotic yogurt.

CHAPTERS 4, 5, AND 6 feature the energy-yielding nutrients carbohydrates, lipids, and proteins. Each begins with a discussion of the respective macronutrient in the modern diet. The body of each chapter discusses dietary sources of the nutrient and its functions, digestion and absorption, and metabolism. Each chapter ends with a discussion of how to choose a diet that meets recommendations. Emphasis is placed on the types and proportions of these that are optimal for health. Chapter 4, "Carbohydrates: Sugars, Starches, and Fibre," discusses the health impact of refined grains and foods high in added sugar versus whole grains and unrefined sources of sugars. The concept of glycemic index is explained and research on the relationship between GI and type 2 diabetes is presented. It includes the sources of carbohydrates in the Canadian diet, CCHS data on the fibre intake of Canadians, Canadian whole grain stamps, and a discussion of the Canadian Diabetes Association's *Just the Basics* and *Beyond the Basics* meal planning systems.

Chapter 5, "Lipids," emphasizes the importance of the type of fat consumed to human health. Research data are presented on the relationship between different fats (e.g. vegetable oils, tub and stick margarines, and butter) and cardiovascular disease, as well as a discussion of the effects of fish consumption on disease risk. This chapter includes CCHS data on fat consumption, a discussion on the decline of *trans* fat in the Canadian food supply, and permitted fat replacers in Canada. Chapter 6, "Proteins and Amino Acids," discusses animal and plant sources of protein and points out that either plant or animal proteins can meet protein needs, but these protein sources bring with them different combinations of nutrients. In addition to discussing how to meet protein needs, this chapter includes information on how to plan a healthy vegetarian diet. The Canadian content of Chapter 6 includes a discussion of the sources of protein in the Canadian diet, the information on food allergies available from Health Canada, and how protein rating is calculated for labelling purposes.

Focus on Alcohol discusses alcohol metabolism and the health risks associated with excessive alcohol consumption and the cardiovascular benefits associated with moderate intake.

CHAPTER 7, "Energy Balance and Weight Management," introduces energy balance and shows how small changes in diet and behaviour can alter long-term energy balance. It discusses the obesity epidemic and the effect that excess body fat has on health. Research data on the relationship between BMI and disease risk is presented. The most up-to-date information on how body weight is regulated is discussed as well as the role of genetic versus environmental factors in determining body fatness. The chapter includes recommendations for healthy body weight and composition and equations for determining energy needs. It also discusses weight loss options that range from simple energy restriction to risky surgical approaches.

Canadian content in Chapter 7 includes CCHS data on the prevalence of overweight and obesity in Canada and its relationship to comorbidities, such type 2 diabetes, heart disease, and hypertension. The Canadian Medical Association clinical practice guidelines for the treatment of obesity are also discussed.

Focus On Obesity, Metabolism, and Disease Risk explains the metabolic changes that occur in the obese body that contribute to the development of cardiovascular disease, type 2 diabetes, and cancer as well as other diseases.

CHAPTER 8, "The Water-Soluble Vitamins," begins with a general overview of the vitamins—where they are found in the diet, factors affecting their bioavailability, and how they function. Each of the B vitamins and vitamin C are then discussed individually, providing information on sources in the diet, functions in the body, impact on health, recommended intakes, use as dietary supplements, and potential for toxicity. A detailed description of the homocysteine hypothesis is given and the discussion of vitamins B6, folate, and B12 emphasize the interrelationships between the three nutrients. The section on vitamin C introduces the concept of oxidative stress and the role of antioxidants. This chapter also discusses choline.

Canadian content in Chapter 8 includes CCHS data on the intake of water-soluble vitamins, discussion of natural health products, and Canadian data on the effectiveness of folate fortification in reducing neural tube defects.

CHAPTER 9, "The Fat-Soluble Vitamins," introduces the fat-soluble vitamins within the context of the modern diet and then presents each one with a discussion of their sources in the diet, functions in the body, impact on health, recommended intakes, use as dietary supplements, and potential for toxicity. The controversies associated with the recent changes to the DRIs for vitamin D are discussed.

Canadian content in Chapter 9 includes a discussion of rickets in Canada and research on populations in Canada vulnerable to vitamin D deficiency. CCHS data on vitamin A intake is presented and natural health products are discussed.

Focus on Phytochemicals discusses the role of phytochemicals in nutrition and health. These substances are not dietary essentials but can positively impact health. Different categories of phytochemicals are presented, along with a discussion of how to maximize their intake.

CHAPTER 10, "Water and the Electrolytes," addresses water, a nutrient often overlooked, and the electrolytes because they help regulate the distribution of body water. This chapter presents information on where these nutrients are found in the diet and discusses their functions in the body and their relationship to health and disease. A discussion of hypertension and salt sensitivity illustrates the importance of sodium, potassium, and other minerals and dietary components in blood pressure regulation. Advances in our understanding of how dietary patterns affect hypertension is stressed by the research data on the DASH diet, a dietary pattern that has been shown to lower blood pressure.

Canadian content in Chapter 10 includes a discussion of Canadian regulations for the labelling of bottled water, a description of the Sodium Reduction Strategy for Canada, and the CCHS data on sodium and potassium intake.

CHAPTER 11, "Major Minerals and Bone Health," discusses the remaining major minerals, calcium, phosphorus, magnesium, and sulfur. This chapter discusses their functions in the body and availability in the diet as well as their relationship to health and disease. Because most of these play an important role in bone health, this chapter also includes a section on the relationship between nutrition and the development of osteoporosis. Research data is presented on the influence of inulin (a dietary fibre) on the bioavailability of calcium and changes in bone density.

Canadian content in Chapter 11 includes CCHS data on the intake of calcium, magnesium, and phosphorus; data on Canadian intake of milk; and discussion of CaMOS studies.

Focus on Nonvitamin/Nonmineral Supplements targets natural health products that contain substances other than micronutrients. Micronutrient supplements are discussed in chapters 8 through 12 with the appropriate nutrient. This focus will help students evaluate the benefits and risks associated with these products.

CHAPTER 12, "The Trace Elements," discusses the trace elements in a format similar to that used for other micronutrients. An emphasis is placed on the unique roles of some minerals as well as the similarities that some have in their functions and the interactions that exist among them. Discussions of the health issues related to these nutrients help create interest, as do discussions of the pros and cons of trace element supplements.

Canadian content in chapter 12 includes CCHS data on iron and zinc intake, discussion of zinc fortification in Canada, and regulations related to iodized salt.

Focus on Integrating Nutrient Function in the Body Now that all nutrients have been discussed individually, this feature shows how many nutrients function together in many systems in the body. Nutrients that have important functions in each of these systems are described: gene expression, energy metabolism, antioxidant activity, fluid balance, the central nervous system, the immune system, circulatory system and blood, and bone. This feature emphasizes how nutrients function together in all parts of the body, to maintain health.

CHAPTER 13, "Nutrition and Physical Activity," discusses the relationships among physical activity, nutrition, and health. It emphasizes the importance of exercise for health maintenance as well as the impact nutrition can have on exercise performance. Because nutrients fuel activity, this chapter serves as a review of energy metabolism. By this point in the text students have studied all the essential nutrients, so a complete discussion of the macronutrients and micronutrients needed for energy metabolism can be included. An expanded discussion of ergogenic aids directs students to use a risk-benefit analysis of these products before deciding whether or not to use them.

Canadian content in Chapter 13 includes a discussion of the health impact of inactivity in Canada and CCHS data on activity levels among Canadians. The Canadian Physical Activity Guidelines for adults are included and a Canadian research on interval training is presented.

CHAPTER 14, "Nutrition During Pregnancy and Lactation," addresses the role of nutrition in human development by discussing the nutritional needs of women during pregnancy and lactation as well as the nutritional needs of infants. The benefits of breastfeeding are discussed.

Canadian content in Chapter 14 includes data on the weight gain of Canadian women during pregnancy, a history of how folic acid research in neural tube defects led to the fortification of the Canadian food supply, and a discussion of Health Canada's assessment of the mercury risk from fish. The chapter also discusses the Canadian Prenatal Nutrition Program as well as statistics on breastfeeding in Canada.

CHAPTER 15, "Nutrition from Infancy to Adolescence," begins with a discussion of the rising rates of childhood obesity and other chronic diseases and the importance of learning healthy eating habits early in life. The chapter discusses nutrient needs and how they change from infancy through adolescence.

Canadian content in Chapter 15 includes CCHS data on the quality of diets of children and adolescents, CCHS data on the relationship between the body weight of parent and child, research on the quality of food in school cafeterias, the regulations on the labelling of food directed to children and the Physical Activity Guidelines for children and adolescents.

Focus on Eating Disorders includes a comprehensive discussion of the different types of eating disorders, many of which are diagnosed during adolescence, their causes, consequences, and treatment. This focus also addresses the sociocultural factors that influence body ideal as well as what to do if you have a friend or relative you suspect has an eating disorder.

CHAPTER 16, "Nutrition and Aging: The Adult Years," addresses how nutrition affects health and how the physiological changes that occur with aging affect nutritional needs and the ability to meet them. The impact that chronic disease, medications, and socioeconomic changes have on the risk of malnutrition is discussed.

Canadian content in Chapter 16 includes discussion of life expectancy, the Canadian Physical Activity Guidelines for seniors, the SCREEN program for detecting nutritional problems in the elderly, and research on the factors that contribute to health.

CHAPTER 17, "Food Safety," discusses the risks and benefits associated with the Canadian food supply and includes information on the impact of microbial hazards, chemical toxins, food additives, irradiation, and food packaging. The use of HACCP (Hazard Analysis Critical Control Point) to ensure safe food is described as well as the advanced technology used to help prevent and track foodborne illness.

Canadian content in Chapter 17 includes a description of the role of the Canadian Food Inspection Agency in tracking foodborne illness, the precautions implemented in Canada to prevent mad cow disease, the history of pasteurization, and the Canadian regulation of organic foods.

Focus on Biotechnology explains how plants are genetically modified and addresses the potential benefits and risks associated with this expanding technology. The role of Health Canada and the Canadian Food Inspection Agency in accessing the safety of novel plants is discussed.

CHAPTER 18, "World Hunger and Malnutrition," discusses the coexistence of hunger and malnutrition along with obesity in both developed and developing nations around the world. It examines the causes of world hunger and solutions that can impact the amounts and types of food and nutrients that are available.

Canadian content in Chapter 18 includes a discussion of the Micronutrient Initiative and the Sprinkles Global Health Initiative. The causes, consequences, and responses to food insecurity in Canada are also described.

Student Resources in WileyPLUS

WileyPLUS

WileyPLUS is an innovative, research-based online environment for effective teaching and learning. *WileyPLUS* builds students' confidence because it takes the guesswork out of studying by providing students with a clear roadmap: what to do, how to do it, if they did it right. Students will take more initiative so you'll have greater impact on their achievement in the classroom and beyond.

Videos

Videos produced by current news sources connect students to nutrition in the world around them by allowing students to explore relevant topics in nutrition through the lens of current events.

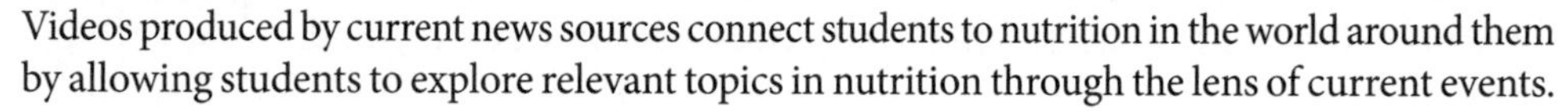

How It Works Animations

Wiley has developed a new set of animations for nutrition students and professors. We surveyed professors across North America to find out which topics are the most difficult to teach and learn and which processes are most essential to the introductory nutrition course. After much research and reviewing, we developed animations on these topics to make these difficult processes easier for students to learn and bring the process diagrams from the book to life.

Absorption of Nutrients
Flow of Blood During Absorption
Digestion and Absorption of Carbohydrates
Glucose Metabolism
Regulation of Glucose Metabolism
Maintaining Normal Blood Glucose Levels
Blood Glucose Regulation
Digestion and Absorption of Lipids
Lipid Transport
Metabolism of Lipids
Digestion and Absorption of Proteins
Metabolism of Proteins
Role of B Vitamins in Metabolism
Action of Antioxidants Against Free Radicals
Action of Vitamins as Coenzymes
Acid-Base Balance

These animations and accompanying quiz questions are available in *WileyPLUS*.

iProfile: Assessing Your Diet and Energy Balance

Updated with Canadian nutritional data and food brands, this online dynamic assessment software contains nutrient values for a wide range of foods, including fast food and brand name products (both Canadian and American), as well as ethnic and cultural choices. If a food is not in the database, iProfile allows users to add foods to the database to keep pace with the ever-growing market of available products. To make calculating energy expenditure easier, iProfile includes an easy-to-use section on physical activity. In addition to the ability to track and analyze diets and exercise, some distinctive features of the software include serving size animations, a quick self-quiz, single nutrient reports, menu planning, and a user-friendly design.

Nutrition News Finder

This searchable resource provides links to nutrition-related news articles from major publications from around the world, updated every 15 minutes.

iProfile Sample Profiles and Assignments

A collection of sample profiles that students and instructors can import into their own iProfile program is now available in *WileyPLUS*. Each sample profile includes a profile, food journal, and activity journal for a fictional person. *WileyPLUS* contains automatically gradable assessment questions about each sample profile. Since every student report is different, using reports from these sample profiles provide a standard way to teach and learn the concepts needed to analyze the reports from food and activity journals.

iProfile Essay Questions in WileyPLUS

Essay questions that ask about students' individual reports based on their food and activity journals are available in *WileyPLUS*. These essay questions ask students to reflect on their own diets and activities. For example, students are asked to look at their fibre intakes, compare them to the recommendations for fibre, and suggest foods that could increase the amount of fibre in their diets.

Resources for Instructors

Instructor Companion Website

www.wiley.com/go/smolin

A dedicated companion website for instructors provides many resources for preparing and presenting lectures. Included are:

- **Additional Critical Thinking exercises**
- **PowerPoint Lectures** that combine important images with major concepts from each chapter
- Questions for use with **Clicker Systems**
- **Test bank**—Available online, the test bank includes multiple-choice questions as well as short answer questions.
- **Computerized Test Bank**—This computerized version of the test bank makes preparing clear, concise tests quick and easy. It is available in both Windows and Mac formats.

To the Student

Nutrition is a subject that all of you have a personal interest in, whether you are concerned about your own nutritional health, a parent with diabetes, or a friend with an eating disorder. You may enroll in a nutrition course to learn what to eat and how to choose healthy foods and then be surprised when the course talks about protein synthesis, lipid transport, and anaerobic metabolism. A good course and textbook should do both.

As authors, our goal is to provide you with tools that can be used throughout your life. We could tell you what makes a healthy breakfast, but if you didn't understand why the foods in the breakfast were healthy choices you would not be able to make healthy choices from a different set of breakfast foods or use the same principles to choose a healthy dinner. On the other hand, for example, if you understand why saturated fat affects blood cholesterol you will not forget how to choose a heart-healthy diet.

The critical thinking approach we have used in this text will help you understand the science of nutrition and give you the decision-making skills you need to navigate the scores of choices you face when deciding what to eat and which of the latest nutrition headlines to believe. By becoming a knowledgeable consumer, you will be able to make informed choices about diet and lifestyle, whether you use this information to improve your own health or to pursue a career in nutrition.

Acknowledgements

Many thanks to our reviewers, whose comments helped to improve this textbook:

Olasunkanmi Adegoke, York University
Nooshin Alizadeh-Pasdar, University of British Columbia
Teresa Bosse, Athabasca University
Sébastien Boyas, University of Ottawa
Sukhinder K. Cheema, Memorial University
Julia Ewaschuk, University of Alberta
Catherine Field, University of Alberta
Debra Fingold, Seneca College
Mary Flesher, University of British Columbia
Kristin Hildahl-Shawn, University of Manitoba
Paul LeBlanc, Brock University
Janet Le Patourel, Langara College
Karen B. McLaren, Canadore College
Tien Nguyen, University of Ottawa
Csilla Reszegi
Yaseer Shakur, George Brown College
Elizabeth Strachan, University of Windsor
Norman Temple, Athabasca University
Christine Wellington, University of Windsor
Zach Weston, Wilfrid Laurier University

I would like to acknowledge the authors of the American edition of ***Nutrition: Science and Applications,*** Lori Smolin and Mary Grosvenor, for providing me with an excellent foundation upon which to create this Canadian edition.

Others who provided invaluable assistance with the writing of this text include contributing author, Sharon Parker, who wrote the Canadian content for Chapter 18; Melanie Byland, who wrote Appendix E on the CDA's Beyond the Basics meal planning system; and Lauryn Choleva, who worked on Appendices B, C, and G.

There are many people to thank at Wiley beginning with publisher Veronica Visentin and acquisitions editor Rodney Burke who first approached me with the idea of creating this textbook. Special thanks to developmental editor Andrea Grzybowski for her patient and steady guidance through the many steps involved in creating a finished manuscript. Other members of the Wiley team whose hard work must be acknowledged include copyeditor Sophie Vitkovitsky, proofreader Emma Cole, photo researcher and permissions coordinator Luisa Begani, and indexer Belle Wong. Thanks to Laserwords for their excellent artwork.

Debbie Gurfinkel
Toronto, October 2011

Brief Contents

Contents

1

Nutrition:

FOOD FOR HEALTH

Case Study

Kaitlyn knew the potato chips in the dorm vending machine weren't a good choice—but aside from candy, they were her only option. Kaitlyn had been in classes and at work until late in the evening. When she finally sat down to study, she realized that she had missed dinner. She needed to eat something to keep her going until she finished the first chapter of her nutrition textbook, but her residence offered few food choices late at night. So she opted for the chips and fruit punch from the vending machine—they seemed to be the healthiest choices available.

As a first-year university student who has never been away from home, Kaitlyn has gained a few pounds and is beginning to be concerned about her weight. Her father recently had cardiac bypass surgery, and her mother takes medication for high blood pressure. Kaitlyn knows that because of this family history, her diet is a particularly important part of her future health. The residence cafeteria, although not great, does offer a variety of choices. The problem is that Kaitlyn doesn't know how to choose a healthy diet. She tries to keep some healthy snacks in her room, but her options are limited because she doesn't have a refrigerator. Several of her friends have started taking supplements like Mega B to give them more energy and ginkgo biloba to improve their memory. Kaitlyn is tempted to start taking them herself but remembers that her high school soccer coach told her that some supplements can be dangerous. To optimize her health, then, Kaitlyn needs to learn the basics of nutrition science and perfect the art of making nutritionally sound decisions and healthy food choices—a goal that is a little overwhelming at first.

By reading the first chapter of her nutrition textbook, Kaitlyn will begin to understand how to achieve this goal, as will you as you read along with her.

(PhotoAlto/Sigrid Olsson/Getty Images, Inc.)

(Susie M. Elsing Food Photography/StockFood America)

Chapter Outline

1.1 Nutrition and the Canadian Diet

CANADIAN CONTENT

Learning Objectives

- Discuss how eating habits have changed over the past 50 years.
- Compare Canadians' intake of milk and alternatives, vegetables and fruit, grain products, meat and alternatives, and total fat to recommendations for a healthy diet.

nutrition A science that studies the interactions that occur between living organisms and food.

nutrients Chemical substances in foods that provide energy and structure and help regulate body processes.

Nutrition is a science that studies all the interactions that occur between living organisms and food. Food provides **nutrients** and energy, which are needed to keep us alive and healthy, to support growth, and to allow reproduction. Sometimes, however, fast-paced lifestyles and food choices made available through modern technology contribute to a diet that contains too much or too little of some of the nutrients we need (**Figure 1.1**).

The Modern Canadian Food Supply

For much of human history, in order to get enough to eat people needed to spend most of their day obtaining food ingredients and preparing meals. Even 100 years ago, the time spent for meal preparation was measured in hours—hours spent peeling, chopping, baking, roasting, stewing, and then cleaning. Today, a microwaveable dinner that includes meat, rice, vegetables, and dessert can be ready in 5 minutes.

Figure 1.1 The availability of fast food has changed the Canadian diet, but at what price? (© Ilene MacDonald/Alamy)

The modern Canadian food supply includes an endless assortment of eating options. Many of these choices are foods that have been part of the human diet for centuries—fresh fruits and vegetables, meats, and grains. But others are newer additions—frozen vegetables, canned soups, packaged meats, frozen prepared meals, and snack foods. Fifty years ago, people ate most of their meals at home, with their families, at a leisurely pace. Today, more single-parent households and families with both parents working means that a parent is no longer at home in the afternoon preparing meals and waiting for the kids to come home from school. After-school activities often impinge on family meal times. Fewer young adults than in the past know how to prepare a full meal and shoppers of all ages are choosing to buy more convenient, **processed foods** that can be boiled or microwaved rather than raw ingredients that need to be chopped, seasoned, and cooked. People opt for frozen french fries that just need to be warmed in the oven instead of raw potatoes that need to be peeled, cut, and then prepared by frying. The increase in the availability and variety of processed foods has made it easier and quicker to get something to eat. Snack foods such as potato chips, nachos, sugar-sweetened beverages, and other snack foods are readily available in fast food outlets, vending machines, school and workplace cafeterias, gasoline stations, etc. For example, on any given day, about one-quarter of Canadians eat food prepared in a fast-food restaurant.[1]

processed foods Foods that have been specially treated or changed from their natural state.

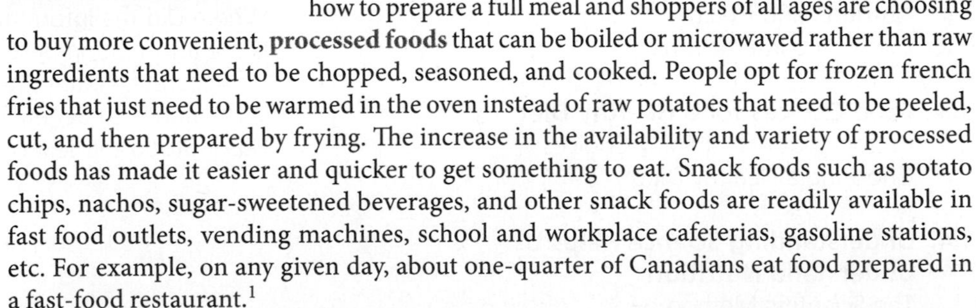

These changes in the Canadian food supply have made it easier to obtain a meal or snack but they have not necessarily improved our nutritional health (see Your Choice: Convenience Has Its Costs). The easy availability of these foods can lead to overeating. Over the past century, the major nutrition concerns in Canada have shifted from providing enough nutrients to meet people's needs to limiting overconsumption. The excess intake of energy and certain nutrients increases the risk of obesity and chronic diseases, such as heart disease, hypertension, and diabetes, that are common among Canadians.

Canadian Community Health Survey This is a comprehensive survey of health-related issues, including the eating habits of 35,000 Canadians, that was begun in 2000 and continues to collect data annually. Results of this survey will be presented throughout this textbook.

How Healthy Is the Canadian Diet?

The Canadian diet is not as healthy as it could be. Nutritious foods are divided into four major groups: vegetables and fruits, grain products, milk and alternatives, and meat and alternatives. A recent survey of Canadians' health, including eating habits, called the **Canadian Community Health Survey (CCHS),** found that many Canadians are not consuming as many nutritious

YOUR CHOICE

(©iStockphoto)

Convenience Has Its Costs

It's Monday morning, and you're exhausted. As you lie in bed, you decide that you'll stop for a muffin and coffee on your way to school to save a couple of minutes. You won't have to turn on your coffee pot, wait for the toaster, or clean up after you eat. But what is the cost of this convenience in terms of dollars and energy content? The examples given here show how much more expensive convenience foods are in dollar cost, and how much higher in energy content they are, compared to similar foods consumed at home. Scientists measure the energy content of food in kilocalories, which is exactly the same as the popular term "calorie." Eating more kilocalories than we expend during daily activities will result in weight gain.

Consider that morning coffee and muffin. At home you might pour yourself an 8-ounce mug of coffee with whole milk and sugar, which would cost about 20 cents and provide about 50 kilocalories of energy along with a little protein and calcium. Toast or an English muffin with butter would add about 30 cents to the cost, along with 150 kilocalories and some B vitamins and iron. If instead you stop for a 16-ounce mocha and a healthy-looking bran muffin at the corner coffee shop, you will spend around $4.00 and ingest about 850 kilocalories, and not get many more nutrients than the toast and coffee at home would provide. The impact of stopping for coffee and a muffin once in a while when you are running late is minimal, but making it an everyday habit is expensive and can break your kilocalorie "budget" as well.

A homemade turkey sandwich for lunch, as an example, would cost about $2.00 and provide approximately 320 kilocalories. A 12-inch sub from the shop on the corner, in comparison, costs two or three times as much and may contain an additional 500 kilocalories—even more if you add a soft drink and cookie. Dropping $4.00 into the vending machine for an afternoon snack will get you a soft drink and a bag of chips, for a total of more than 300 kilocalories with few vitamins or minerals. Instead, a snack of fruit and baby carrots from home costs only about 50 cents and provides an assortment of essential nutrients for only about 100 kilocalories. Although the modern Canadian lifestyle tempts us with convenience, we need to consider the costs to both our wallets and our waistlines.

(P. Mittongtare/FoodPix/Jupiter Images Corp.)

vegetables, fruits, milk and alternatives, and grain products as they should. As **Figure 1.2a** indicates, about half of adult Canadians do not consume even five servings of vegetables and fruit per day, while current recommendations for adults are for 7 to 10 servings per day. About two-thirds of Canadians over age 30 do not consume the recommended servings of milk and milk alternatives. Many Canadians also consume fewer grain products (which include bread, cereals, pasta, and rice) than recommended. Only for the meat and alternative group were the percentage of Canadians consuming below recommended levels relatively low.[2]

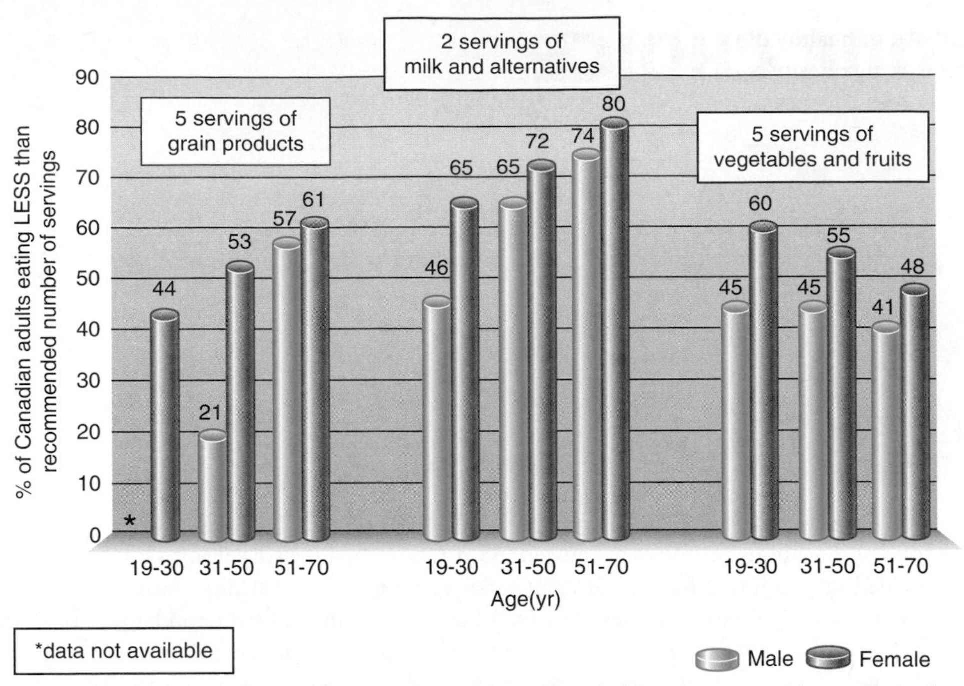

Figure 1.2a Many Canadians eat fewer grain products, milk and alternatives, vegetables and fruits than recommended.

On the other hand, between 17%-28% of adults (aged 19-70) consume higher than recommended levels of fat (**Figure 1.2b**). High-fat food choices most popular among Canadians include pizza, hamburgers, hot dogs, cookies, donuts, and muffins. Furthermore, a high proportion of total daily calories comes from snack foods. While snacks can certainly include nutritious foods, the CCHS found that Canadians were not always making the best choices. High-fat foods, such as those listed above, and high-sugar foods, such as regular soft drinks, were often selected instead.[2]

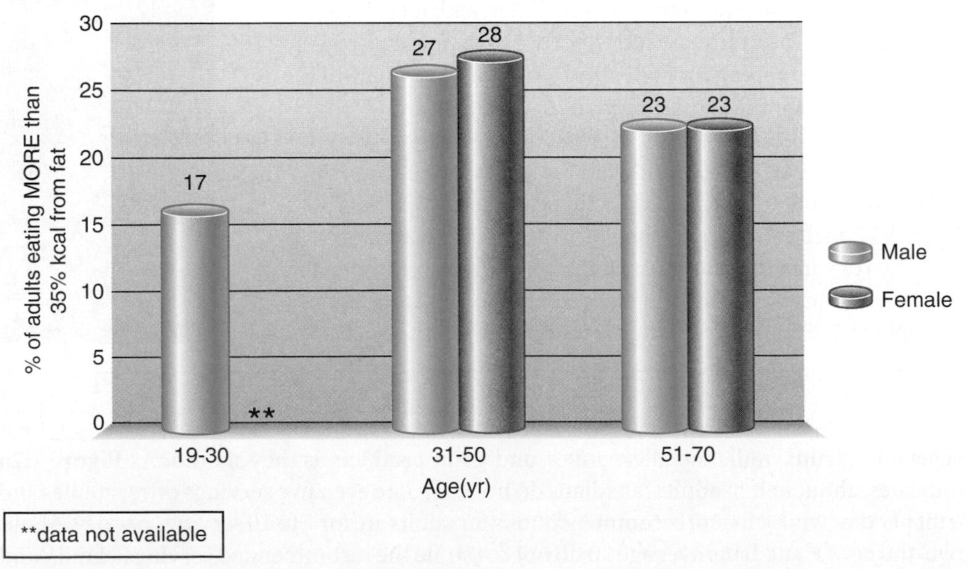

Figure 1.2b The maximum recommended level of intake of fat is 35% of our daily kcal. Many Canadians eat more fat than recommended.
Source: Garriguet, D. 2004 Overview of Canadians' Eating Habits. Statistics Canada no: 82-620-MIE-No 2. Available online at www.statcan.gc.ca/pub/82-620-m/82-620-m2006002-eng.pdf. Accessed Jan 28, 2010

An unhealthy dietary pattern along with a lack of physical activity increases the risk of developing obesity and chronic diseases such as heart disease and stroke, diabetes, and certain types of cancer. As indicated in **Figure 1.3**, these chronic diseases are the major causes of death among Canadians.

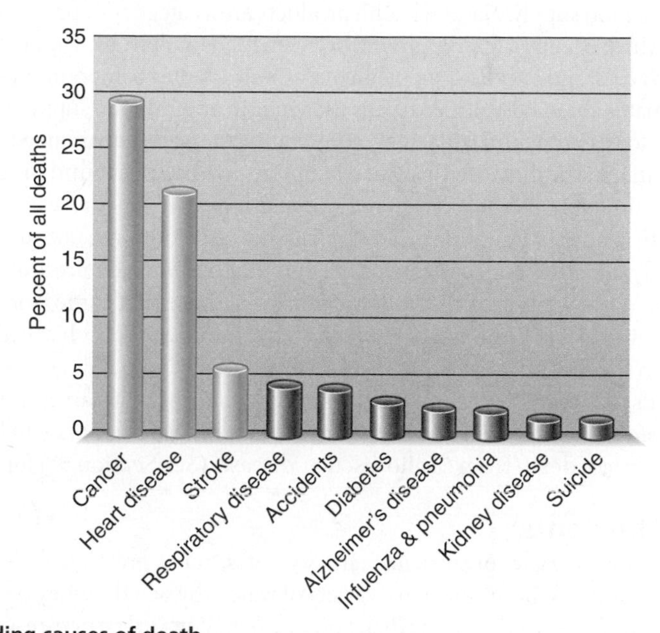

Figure 1.3 **Leading causes of death**
Of the 10 leading causes of death in Canada, the top three are nutrition-related.
Source: Statistics Canada: Leading Causes of Death, by sex: www40.statcan.gc.ca/l01/cst01/hlth36a-eng.htm.

As you study nutrition, you will learn more about the relationship between diet and disease and how to make wise food choices. To choose a healthy diet that provides the right amounts of energy and each nutrient, we need to understand how our bodies obtain nutrients from food, which nutrients are essential, how much we need, and which foods provide healthy sources of nutrients. We also need to recognize which nutrition information to believe.

1.2 Food Provides Nutrients

Learning Objectives

- Define the term *essential nutrient* and list the six classes of nutrients.
- Define fortified foods.
- Define natural health products.
- Describe the three general functions of nutrients.
- Define malnutrition.
- Define diet-gene interaction.

essential nutrients Nutrients that must be provided in the diet because the body either cannot make them or cannot make them in sufficient quantities to satisfy its needs.

To date, approximately 45 nutrients have been determined to be essential to human life. **Essential nutrients** must be supplied in the diet to support life; they either cannot be made by the body or cannot be made in sufficient quantities to meet needs. For example, our bodies cannot synthesize vitamin C, but we need it to stay healthy. If we do not consume vitamin C in the foods we eat, we will begin to show signs of vitamin C deficiency. If vitamin C is not added back to the diet, the deficiency will eventually be fatal.

Our intake of essential nutrients is determined by our food choices. Some foods are naturally high in nutrients and some contain nutrients added during processing. Foods to which nutrients

fortified foods Foods to which one or more nutrients have been added, typically to replace nutrient losses during processing or to prevent known inadequacies in the Canadian diet.

natural health products Natural health products are a category of products regulated by Health Canada that include vitamin and mineral supplements, amino acids, fatty acids, probiotics, herbal remedies, and homeopathic and other traditional medicines.They occupy a middle ground between food and drugs.

phytochemicals Substances found in plant foods (*phyto* means plant) that are not essential nutrients but may have health-promoting properties.

zoochemicals Substances found in animal foods (*zoo* means animal) that are not essential nutrients but may have health-promoting properties.

energy-yielding nutrients Nutrients that can be metabolized to provide energy in the body.

macronutrients Nutrients needed by the body in large amounts. These include water and the energy-yielding nutrients: carbohydrates, lipids, and proteins.

micronutrients Nutrients needed by the body in small amounts. These include vitamins and minerals.

organic molecules Those containing carbon bonded to hydrogen.

inorganic molecules Those containing no carbon–hydrogen bonds.

kilocalorie (kcalorie, kcal) The unit of heat that is used to express the amount of energy provided by foods. It is the amount of heat required to raise the temperature of 1 kilogram of water 1 degree Celsius (1 kcalorie = 4.18 kjoules).

kilojoule (kjoule, kJ) A unit of work that can be used to express energy intake and energy output. It is the amount of work required to move an object weighing one kilogram a distance of 1 metre under the force of gravity (4.18 kjoules = 1 kcalorie).

legumes The starchy seeds of plants belonging to the pea family; includes peas, peanuts, beans, soybeans, and lentils.

have been added are called **fortified foods**. In Canada, vitamins and minerals can be added to a number of foods such as white flour, breakfast cereals, milk, orange juice, infant formula, and plant-based beverages such as soy, rice, and almond milk. The type and amount of nutrients that can be added are regulated and are intended to restore nutrient losses caused by processing or to prevent known inadequacies in the Canadian diet. **Natural health products** are another source of nutrients in the food supply. Natural health products are a category of products regulated by Health Canada which occupy a middle ground between foods and drugs. They include vitamins, minerals, amino acids, and fatty acid supplements, as well as other compounds (see definition). The CCHS estimates about 40% of Canadians use vitamin and mineral supplements.[3]

In addition to essential nutrients, food contains substances that are needed by the body but are not essential in the diet. Lecithin, for example, is a substance found in egg yolks that is needed for nerve function. It is not considered an essential nutrient because it can be manufactured in the body in adequate amounts. The typical diet also contains substances that are not made by the body and are not necessary for life, but that have health-promoting properties. Those that come from plants are called **phytochemicals**; those that come from animal foods are called **zoochemicals**. For example, a phytochemical found in broccoli called sulforaphane is not essential in the diet but has effects in the body that may help reduce the risk of cancer. Certain fatty acids found in fish oils are examples of zoochemicals; they are not essential as they can be synthesized in the body, but additional dietary intake has been linked to health benefits such as reduced risk of death from cardiovascular disease. (See Section 5.6 for more details.)

Classes of Nutrients

Chemically, there are six classes of nutrients: carbohydrates, lipids, proteins, water, vitamins, and minerals. These classes can be grouped in a variety of ways—by whether they provide energy to the body, by how much is needed in the diet, and by their chemical structure. Carbohydrates, lipids, and proteins provide energy and thus are referred to as **energy-yielding nutrients**. Alcohol also provides energy but is not considered a nutrient because it is not needed to support life (see Focus on Alcohol). Along with water, the energy-yielding nutrients constitute the major portion of most foods and are required in relatively large amounts by the body. Therefore, they are referred to as **macronutrients** (*macro* means large). Their requirements are measured in kilograms (kg) or grams (g) (see Appendix K). Vitamins and minerals are classified as **micronutrients**, because they are needed in small amounts in the diet (*micro* means small). The amounts required are expressed in milligrams (1 mg = 1/1000 g) or micrograms (1 μg = 1/1,000,000 g). Structurally, carbohydrates, proteins, lipids, and vitamins are **organic molecules** so they are referred to as organic nutrients. Minerals and water are **inorganic molecules** so they are referred to as inorganic nutrients.

The Energy-Yielding Nutrients The energy provided by carbohydrates, lipids, and proteins is measured in **kilocalories** (abbreviated as kcalories or kcals) or in **kilojoules** (abbreviated as kjoules or kJs). The more common term, "calorie," is technically 1/1000 of a kilocalorie, but when it is spelled with a capital "C," it indicates kilocalories. For instance, the term "Calories" on food labels actually refers to kilocalories (**Figure 1.4**). However, in the popular press, the term "calorie" (small "c") is often used to express the kcalorie content of a food or diet.

Carbohydrates provide a readily available source of energy to the body. They contain four kcalories per gram (**Table 1.1**). Carbohydrates include sugars such as those in table sugar, fruit, and milk, and starches such as those in vegetables and grains. Sugars are the simplest form of carbohydrate, and starches are more complex carbohydrates made of many sugars linked together (**Figure 1.5**). Most fibre is also carbohydrate, and cannot be digested, and therefore provides very little energy. However, it is important for gastrointestinal health. Fibre is found in vegetables, fruits, **legumes**, and whole grains. Carbohydrates is discussed in detail in Chapter 4.

Table 1.1 Energy Provided by Macronutrients and Alcohol

	Kcalories/Gram	Kjoules/Gram
Carbohydrate	4	16.7
Lipid	9	37.6
Protein	4	16.7
Alcohol	7	29.3

Lipids, commonly called fats and oils, provide 9 kcalories per gram. They are a concentrated source of energy in food and a lightweight storage form of energy in the body. There are several types of lipids that are important in nutrition. Triglycerides are the type that is most abundant in foods and in the body. The fat on the outside of a steak, the butter and oil that is added to food during cooking, and the layer of fat under a person's skin are all comprised almost entirely of triglycerides. Triglycerides are made up of fatty acids (see Figure 1.5). Different types of fatty acids have different health effects. Diets high in saturated fatty acids increase the risk of heart disease whereas those high in monounsaturated and polyunsaturated fatty acids may reduce risks. Cholesterol is another type of lipid; high levels in the blood can increase heart disease risk. Lipids are discussed in detail in Chapter 5.

Protein is needed for growth and maintenance of body structures and regulation of body processes. It can also be used to provide energy—4 kcalories per gram. Meat, fish, poultry, milk, grains, vegetables, and legumes all provide protein. Like carbohydrate and lipid, protein is not a single substance. There are thousands of different proteins in the human body and in the diet. All of these are made up of units called amino acids. Different combinations of amino acids are linked together to form different types of proteins (see Figure 1.5). Some amino acids can be made by the body, and others are essential in the diet. The proteins in animal products better match our need for amino acids than do plant proteins but, both plant and animal proteins can provide all the amino acids we need. Proteins are discussed in detail in Chapter 6.

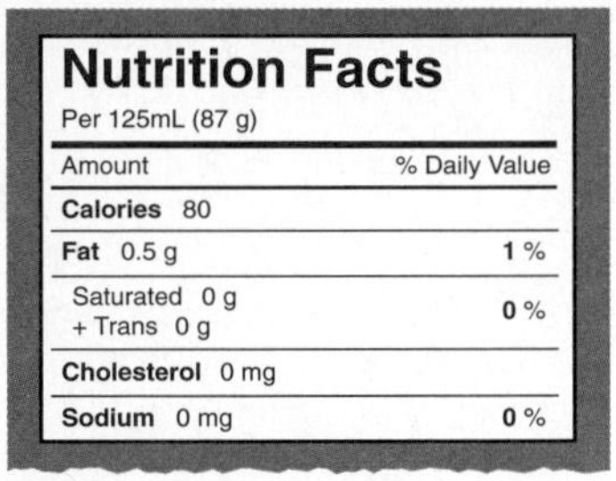

Water Water is a nutrient in a class by itself. It is a macronutrient that does not provide energy. Water makes up about 60% of the human body by weight and is required in kilogram amounts in the daily diet. Water serves many functions in the body, including acting as a lubricant, a transport fluid, and a regulator of body temperature. Water can be obtained from the beverages we drink and also from food, as many foods contain large amounts of water. About 60% of the weight of raw beef is water and fruits and vegetables can be anywhere from 70%-98% water.

Nutrition Facts
Per 125mL (87 g)

Amount	% Daily Value
Calories 80	
Fat 0.5 g	1 %
Saturated 0 g + Trans 0 g	0 %
Cholesterol 0 mg	
Sodium 0 mg	0 %

Figure 1.4 Kilocalories on food labels
The "Calories" listed near the top of the Nutrition Facts panel of a food label technically refer to the number of kilocalories (kcalories) in a serving.

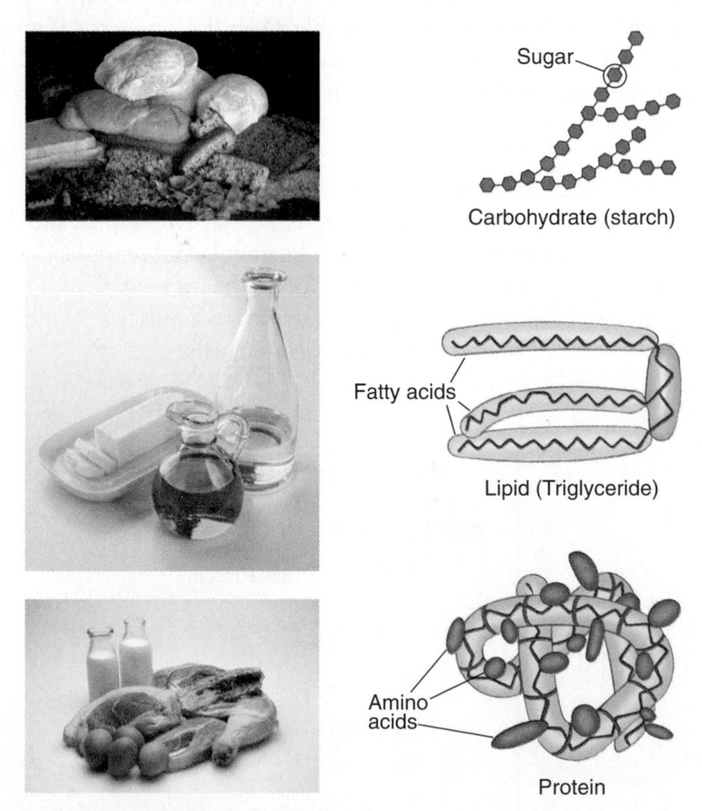

Figure 1.5 Carbohydrates, lipids, and proteins
Starches are a type of carbohydrate made of sugars linked together; most lipids, such as the triglyceride shown here, contain fatty acids; proteins are made of amino acids linked together. (Photographs, © © © © istockphoto.com/Rosette Jordaan; Tetra Images/Getty Images; © Imagemore Co., Ltd./Corbis)

Micronutrients Vitamins and minerals are needed in small amounts. Vitamins are organic molecules that do not provide energy, but are needed to regulate body processes. Thirteen substances have been identified as vitamins. Each has a unique structure and provides a unique function in the body. Many are involved in helping the body use the energy from carbohydrates, lipids, and proteins; others function in processes such as bone growth, vision, blood clotting, oxygen transport, and tissue growth and development. Vitamins are discussed in detail in Chapters 8 and 9.

Minerals are inorganic molecules. Like vitamins they do not provide energy. Many have regulatory roles and some are important structurally. They are needed for bone strength, the transport of oxygen, the transmission of nerve impulses, and numerous other functions. Requirements have been established for many of the minerals, but some are required in such small amounts that their role in maintaining health is still not fully understood. Minerals are discussed in detail in Chapters 10, 11, and 12.

Vitamins and minerals are found in most foods. Fresh foods are a good, natural source of vitamins and minerals, and many processed foods are fortified with micronutrients. Food processing and preparation can also cause vitamin losses because some are destroyed by exposure to light, heat, and oxygen. Minerals are more stable but can still be lost along with vitamins in the water used in cooking and processing. Nevertheless, frozen, canned, and otherwise processed foods can still be good sources of vitamins and minerals.

Functions of Nutrients

Together, the macronutrients and micronutrients provide energy, structure, and regulation, which are needed for growth, maintenance and repair, and reproduction. Each nutrient provides one or more of these functions, but all nutrients together are needed to maintain health.

Providing Energy Inside the body, biochemical reactions release the energy contained in carbohydrates, lipids, and proteins. Some of this energy is used to synthesize new compounds and maintain basic body functions, some is used to fuel physical activity, and some is lost as heat. When the energy in the carbohydrates, lipids, and proteins consumed in the diet is not needed immediately, it can be stored, primarily as fat. These stores can provide energy when dietary sources are unavailable. Over the long term, if more energy is consumed than is needed, body stores get larger, and body weight increases. If less energy is consumed than is needed, the body will burn its stores to meet its energy needs, and body weight will decrease.

Forming Structures Most of the weight of the human body is due to water, protein, and fat (**Figure 1.6**). These nutrients, along with the minerals, are needed to form and maintain the shape and structure of the body. Proteins form the ligaments and tendons that hold bones together and attach muscles to bones. Protein also forms the framework of bones and teeth that is hardened by mineral deposits, and the overall structure of muscles. At the cellular level, lipids and proteins make up the membranes that surround cells.

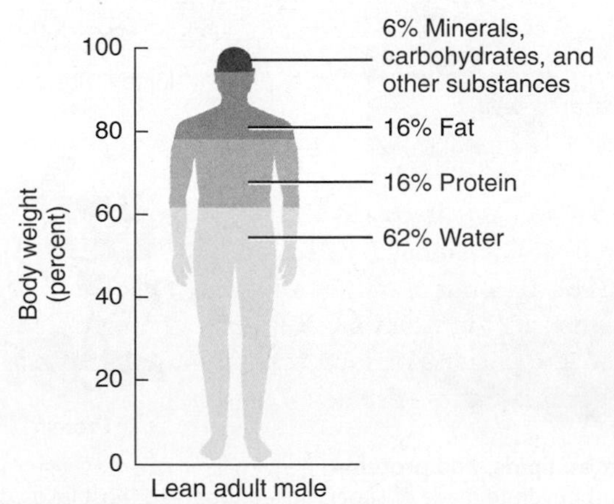

Figure 1.6 **Composition of the human body**
Water, protein, and fat are the most abundant nutrients in the human body.

Regulating Body Processes Together all of the reactions that occur in the body are referred to as **metabolism**. Metabolic processes must be regulated to maintain a stable environment inside the body, referred to as **homeostasis**. All six classes of nutrients have important regulatory roles (**Table 1.2**). The enzymes that catalyze the chemical reactions of metabolism are made up of proteins. These proteins combine with vitamins and minerals to speed up or slow down the reactions as needed to maintain homeostasis. Because water is the solvent for metabolism, most of these reactions occur in the watery component of the cells. Water also helps to regulate body temperature. When body temperature increases, water lost through sweat helps to cool the body.

metabolism The sum of all the chemical reactions that take place in a living organism.

homeostasis A physiological state in which a stable internal body environment is maintained.

Table 1.2 Examples of Nutrient Functions in the Body

Function	Nutrient	Example
Energy	Carbohydrate	Glucose is a carbohydrate that provides energy to body cells.
	Lipid	Fat is the most plentiful source of stored fuel in the body.
	Protein	Protein consumed in excess of protein needs will be used for energy.
Structure	Lipid	Lipids are the principal component of the membranes that surround each cell.
	Protein	Protein in connective tissue holds bones together and holds muscles to bones. Protein in muscles defines their shape.
	Minerals	Calcium and phosphorus are minerals that harden teeth and bones.
Regulation	Lipid	Estrogen is a lipid hormone that helps regulate the female reproductive cycle.
	Protein	Leptin is a protein that helps regulate the size of body fat stores.
	Carbohydrate	Sugar chains attached to proteins circulating in the blood signal whether the protein should remain in the blood or be removed by the liver.
	Water	Water in sweat helps cool the body to regulate body temperature.
	Vitamins	B vitamins regulate the use of macronutrients for energy.
	Minerals	Sodium is a mineral that helps regulate blood volume.

Nutrition and Health

What we eat has an enormous impact on how much we weigh, how healthy we are now, and how likely we are to develop chronic diseases like heart disease and diabetes in the future. Consuming either too little or too much of one or more nutrients or energy can cause **malnutrition**. Malnutrition can affect our health today and can impact on our health 20, 30, or 40 years from now.

malnutrition Any condition resulting from an energy or nutrient intake either above or below that which is optimal.

Dietary Deficiencies **Undernutrition** is a form of malnutrition caused by a deficiency of energy or nutrients. It may be caused by a deficient intake, increased requirements, or an inability to absorb or use nutrients. Starvation, the most severe form of undernutrition, is a deficiency of energy that causes weight loss, poor growth, the inability to reproduce, and if severe enough, death (**Figure 1.7a**). Deficiencies of individual nutrients can also cause serious health problems. These health problems often reflect the body functions that rely on the deficient nutrient. For example, vitamin A is necessary for vision; a deficiency of vitamin A can result in blindness. Vitamin B_{12} is needed for normal nerve function. A deficiency of this vitamin, which is common in older adults because absorption often decreases with age, causes changes in mental status.

undernutrition Any condition resulting from an energy or nutrient intake below that which meets nutritional needs.

Some nutrient deficiencies cause symptoms quickly. In only a matter of hours an athlete exercising in hot weather may become dehydrated due to a deficiency of water. Drinking water relieves the headache, fatigue, and dizziness caused by dehydration almost as rapidly as these symptoms appeared. Other nutritional deficiencies may take weeks, months, or even years to become apparent. The symptoms of the vitamin C deficiency disease scurvy do not occur until the diet has been deficient in vitamin C for weeks or months. Too little calcium in the teenage years causes no immediate symptoms but can cause bones to be weak and break too easily when people reach their fifties or sixties.

Dietary Excesses **Overnutrition**, an excess of nutrients, is also a form of malnutrition. When excesses of specific nutrients are consumed, an adverse or toxic reaction may occur. For example, a single excessive dose of iron can cause liver failure, and too much vitamin B_6 over a

overnutrition Poor nutritional status resulting from an energy or nutrient intake in excess of that which is optimal for health.

few weeks or months can cause nerve damage. Most nutrient toxicities are due to excessive intakes of vitamin and mineral supplements. Foods generally do not contain high enough concentrations of nutrients to cause toxic reactions.

The kinds of overnutrition that are most common in the Canada today do not have immediate toxic effects but contribute to the development of chronic diseases, such as heart disease, type-2 diabetes and some types of cancer, in the long term. Diets providing more energy than needed have resulted in a Canadian population where 60% of adults are overweight or obese (**Figure 1.7b**).[4] Diets high in salt contribute to high blood pressure, those high in saturated fat and cholesterol play a role in heart disease, and those high in red meats and saturated fats and low in fruits and vegetables and fibre may increase the risk of colon cancer.[5]

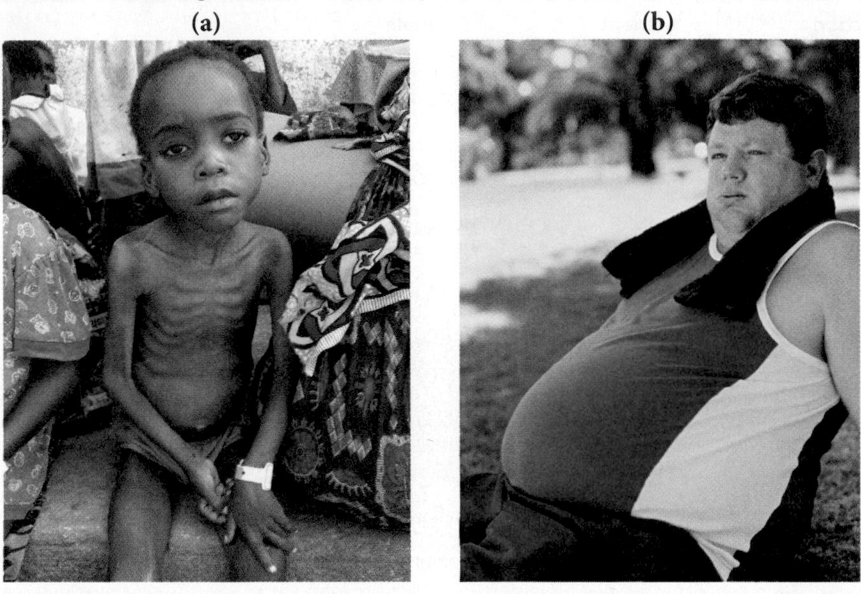

Figure 1.7 Malnutrition includes both undernutrition (a) and overnutrition (b). (Tom Koene/Peter Arnold, Inc.; Stockbyte/Getty Images, Inc.)

Diet–Gene Interactions

genes Units of a larger molecule called DNA that are responsible for inherited traits.

Diet affects your health, but diet alone does not determine whether you will develop a particular disease. Each of us inherits a unique combination of **genes**. Genes are composed of DNA which contain the information that a cell needs to synthesize specific protein. Some of these genes affect your risk of developing chronic diseases such as heart disease, cancer, high blood pressure, and diabetes, but their impact is affected by what you eat (**Figure 1.8**). Your genetic makeup determines the impact a certain nutrient will have on you. For example, some people inherit a combination of genes that results in a tendency to have high blood pressure, possibly because the proteins produced by these genes are less effective in excreting sodium. When

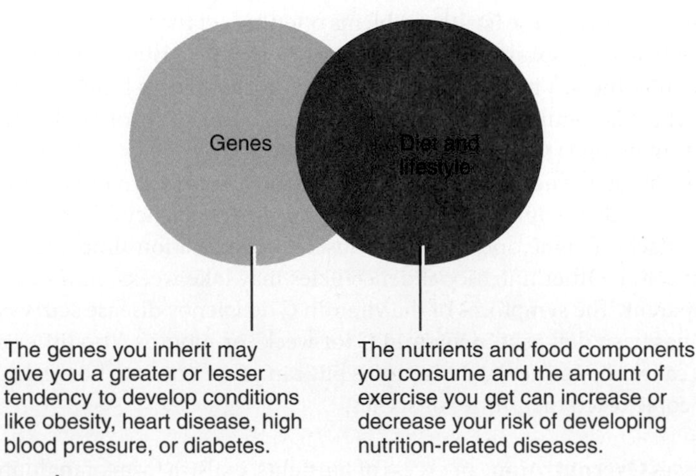

Figure 1.8 Diet and genes
Your actual risk of disease results from the interplay between the genes you inherit and the diet and lifestyle choices you make.

these individuals consume large amounts of sodium, their blood pressure increases (discussed further in Section 10.3). Others inherit genes that allow them to consume a high-sodium diet without a rise in blood pressure, possibly because they are able to excrete sodium more readily than most people. Those whose genes dictate a rise in blood pressure with a high-sodium diet can reduce their blood pressure, and the complications associated with high blood pressure, by eating a diet that is low in sodium.

Our increasing understanding of human genetics has given rise to the discipline of **nutritional genomics** or **nutrigenomics**, which explores the interaction between genetic variation and nutrition.[6] This research has led to the development of the concept of "personalized nutrition," the idea that a diet based on the genes an individual has inherited can be used to prevent, moderate, or cure chronic disease. Although today we do not know enough to take a sample of your DNA and use it to tell you what to eat to optimize your health, we do know that certain dietary patterns can reduce the risk of many chronic diseases.

nutritional genomics or **nutrigenomics** The study of how diet affects our genes and how individual genetic variation can affect the impact of nutrients or other food components on health.

1.3 Food Choices for a Healthy Diet

Learning Objectives

- List factors other than nutrition that affect food choices.
- Define nutrient density.
- Explain the importance of variety, balance, moderation, and kcalorie control in selecting a healthy diet.

Each of the food choices we make contributes to our diet as a whole. This diet must provide enough energy to fuel the body and all the essential nutrients and other food components—in the right proportions—to prevent deficiencies, promote health, and protect against chronic disease. No single food choice is good or bad in and of itself, but all of our choices combined make up a dietary pattern that is either healthy or not so healthy.

Figure 1.9 What factors influence the foods you purchase at the supermarket? (© istockphoto.com/ Steve Jacobs)

Factors that Affect Food Choices

There are hundreds of food choices to make and hundreds of reasons for making them (**Figure 1.9**). Even though the foods we eat provide the nutrients and energy necessary to maintain health, the foods we choose are not necessarily determined by the nutrients these foods contain. Our food choices and food intake are affected not only by nutrient needs but also by what is available to us, where we live, what is within our budget and compatible with our lifestyle, what we like, what is culturally acceptable, what our emotional and psychological needs are, what we think we should eat, and how we are influenced by media.

Availability The food available to an individual or a population is affected by geography, socioeconomics, and health status. Geography is important in developing parts of the world, where dietary choices are often limited to foods produced locally. Nutrients that are lacking in local foods will be lacking in the population's diet. This is less of a factor in more developed countries, because the ability to store, transport, and process food allows year-round access to seasonal foods and foods grown and produced at distant locations (**Figure 1.10**).

Figure 1.10 The fresh produce you buy in your local store may come from across the street or across the ocean. (David Frazier/ Stone/Getty Images)

Even if foods are available in the store, it does not mean that they are available to all individuals. Socioeconomic factors such as income level, living conditions, and lifestyle as well as education affect the types and amounts of foods that are available. Individuals with limited incomes can choose only the types and amounts of foods that they can afford. Individuals who do not own cars can only purchase what they can carry home. Those without refrigerators or stoves are limited in what foods can be prepared at home. And those who cannot or do not have time to cook are limited to prepared foods and restaurant meals.

Health status also affects the availability of food. People who cannot carry heavy packages are limited in what they can purchase. People with food allergies, digestive problems, and dental disease are limited in the foods that are safe and comfortable for them to eat. People consuming special diets to manage disease conditions are limited to foods that meet their dietary prescriptions.

Figure 1.11 A plate of silkworms, such as these being sold in a Vietnamese market, may not seem very appetizing to you, but insects are a part of the diet in many parts of the world. (AFP/Getty Images)

Cultural and Family Background Food preferences and eating habits are learned as part of each individual's family, cultural, national, and social background. They are among the oldest and most entrenched features of every culture. In Japan, rice is the focus of the meal, whereas in Italy, pasta is commonly consumed. Curries characterize Indian cuisine and we expect refried beans and tortillas when we go out for Mexican food. The foods we are exposed to as children influence what foods we buy and cook as adults. If your mother never served artichokes or Swiss chard, you may not consider eating them as an adult. If you grew up in Asia or Africa, you might consider grasshoppers or termites an acceptable food choice, but in Canada, insects are considered food contaminants (**Figure 1.11**). If you did not grow up in a culture that eats insects, you may be unwilling to try them now.

What would a birthday be without a cake, or Thanksgiving without a turkey? Each of us associates holidays such as Christmas, Passover, Chinese New Year, Diwali, and Eid with specific foods that are traditional in our family, religion, and culture. Seventh-Day Adventists are vegetarians; Jews and Muslims do not eat pork; Sikhs and Hindus do not eat beef. Even for those who choose not to observe religious dietary rules, habit may dictate many mealtime decisions. Jewish kosher laws prohibit the consumption of meat and milk in the same meal. Often Jews who, as adults, do not follow kosher law may choose not to serve milk at dinners that also include meat, simply because they never had it as children.

Social Acceptability In addition to being part of our cultural heritage, food is the centrepiece of many of our everyday social interactions. We get together with friends for a meal or for a cup of coffee and dessert. The dinner table is often the focal point for communication within the family—a place where the experiences of the day are shared. Social events dictate our food choices in a number of ways. When invited to a friend's house for dinner, we may eat foods we do not like out of politeness to our hosts. We sometimes alter our food choices because of peer pressure. For example, an adolescent may feel that stopping for a cheeseburger or taco after school is an important part of being accepted by his or her peers.

Personal Preference We eat what we like. Tradition, religion, and social values may dictate what foods we consider appropriate, but personal preferences for taste, smell, appearance, and texture affect which foods we actually consume. How would you feel about giving up your favorite foods? Probably not too good, and you are not alone. Even though most people understand that nutrition is important to their health, many do not choose a healthy diet because they do not want to give up their favourite foods and they do not want to eat foods they do not like.[7] Personal convictions also affect food choices; a vegetarian would not choose a meal that contains meat, and an environmentalist may not buy foods packaged in non-recyclable containers.

Figure 1.12 What foods make you feel better when you are sad, tired, or lonely? (White Packert/Photonica/Getty Images, Inc.)

Psychological and Emotional Factors Food represents comfort, love, and security. We learn to associate food with these feelings as infants suckling while cradled in our mothers' arms. As children and as adults, comfort foods such as hot tea and chicken soup help us to feel better when we are sick (**Figure 1.12**). We use food as a reward when we are good—An excellent report card is celebrated with an ice cream cone. We sometimes take away food as punishment—a child who misbehaves is sent to bed without dinner. We consider ourselves good when we eat healthy foods and bad when we order a decadent dessert. We celebrate milestones and reward life's accomplishments with food. Food may also be an expression and a moderator of mood and emotional states. When we are upset, some of us turn to chocolate or overeat in general while others eat less or stop eating altogether.

Health Concerns An individual's perceptions of what makes a healthy diet affect their food and nutrition choices. For example, someone may choose low-carbohydrate foods if they believe that these choices will help them lose weight. They may limit red meat intake to reduce their risk of heart disease, or they may purchase organically produced foods if they believe that reducing pesticide exposure will prevent illness. In a recent survey of attitudes about diet and nutrition, 43% of respondents said that they have made changes in their eating behaviour to achieve a healthful, nutritious diet and 38% felt they knew what healthy eating was but for one reason or another were not able to do it.[7]

Media Food choices are often influenced by the messages someone receives from the media. Food advertisers try to promote a particular food product hoping to influence food purchases and preferences. Magazines, newspapers, books, television, and the Internet make a vast amount of information on food, nutrition, and health available to the individual. Sometimes this information is accurate and useful; other times it is inaccurate and confusing. (See Sorting Out Nutrition Information: Section 1.6).

Choosing a Healthy Diet

Adequacy and Nutrient Density A healthy or adequate diet is one that provides the right amount of energy to keep weight in the desirable range; the proper types and amounts of carbohydrates, proteins, and fats; plenty of water; and sufficient but not excessive amounts of essential vitamins and minerals. To ensure **adequacy**, a diet must be rich in nutrient-dense foods. **Nutrient density** is a measure of the nutrients a food provides compared to its energy content. Nutrient-dense foods contain substantial amounts of nutrients per kcalorie. For example, broccoli is more nutrient-dense than french fries (**Figure 1.13**). The broccoli is a good source of calcium, vitamin C, vitamin A, and folate and only contributes about 20 kcalories per 125 ml (1/2 cup). The french fries provide little vitamin A and much smaller amounts of vitamin C, calcium, and folate and contribute about 80 kcalories per 125 ml. The french fries also provide less fibre and more fat than the broccoli. This does not mean you should never have french fries, but it does mean that if many of your choices throughout the day are foods that are low in nutrient density, such as soft drinks, snack foods, and baked goods, it will be hard to meet your nutrient needs. On the other hand, if you know how to choose nutrient-dense foods, you can meet all your nutrient needs and have kcalories left over for occasional treats that are low in nutrients and high in kcalories (**Table 1.3**). An adequate diet is based on variety, balance, moderation, and calorie control. Using these principles you can develop a personal strategy for making better choices and maintain your health for the long term.

adequacy A state in which there is a sufficient amount of a nutrient or nutrients in the diet to maintain health.

nutrient density An evaluation of the nutrient content of a food in comparison to the kcalories it provides.

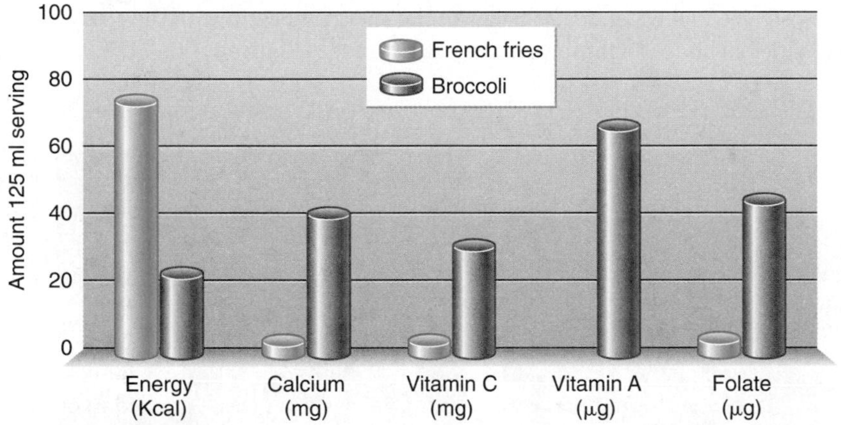

Figure 1.13 Nutrient density
Choosing broccoli instead of french fries will provide fewer kcalories and more calcium, vitamin C, vitamin A, and folate.

Table 1.3 Choices to Boost Nutrient Density

Lower Nutrient Density Choice	Higher Nutrient Density Choice
Instead of this . . .	**Have this . . .**
Soft drink	Low-fat milk
Chocolate candies	Fruit and nut trail mix
Apple pie	Fresh apple
Potato chips and sour cream dip	Baked tortilla chips and salsa
Triple fudge brownie	Oatmeal raisin cookie
Fried chicken	Roast chicken without skin
French fries	Oven-baked potato wedges

Eat a Variety of Foods No one food can provide all the nutrients the body needs for optimal health. Eating a variety of foods, however, helps ensure an adequate nutrient intake. Variety means including grains, vegetables, fruits, milk and alternatives, and meat and alternatives in the diet. Some of these foods are rich in protein and minerals, others in vitamins and phytochemicals. All are important. Variety also means choosing many different foods from within each food group. For instance, if you choose three servings of vegetables a day and they are all carrots, it is unlikely that you will meet your nutrient needs. Carrots provide fibre and vitamin A but are a poor source of vitamin C. If instead you have carrots, peppers, and broccoli, you will be getting vitamin C along with more vitamin A, vitamin K, fibre, and phytochemicals than carrots alone would provide. Likewise, if you always choose red meat from the high-protein choices, you will be missing out on the fibre in beans and the healthy fats in nuts and fish. Variety comes not only from choosing different foods each day, but also each week, and each season. If you had apples and grapes today, have blueberries and cantaloupe tomorrow. If tomatoes do not look appetizing in the winter, replace them with a winter vegetable like squash.

Choosing a varied diet is also important because there are interactions between different foods and nutrients. These interactions may be positive, enhancing nutrient utilization, or negative, inhibiting nutrient use. For example, consuming iron with orange juice enhances iron absorption, while consuming iron with milk may reduce its absorption. In a varied diet, these interactions balance out. In addition, some foods may contain natural toxins, residues of pesticides and fertilizers, or other contaminants. Choosing a variety of foods avoids an excess of any one of these substances.

Balance Your Choices Balance involves mixing and matching foods in proportions that allow you to get enough of the nutrients you need and not too much of the ones that might harm your health. A balanced diet provides plenty of whole grains, vegetables, and fruits. It contains enough but not too much of each of the vitamins and minerals, as well as protein, carbohydrate, fat, and water. It also balances the energy taken in with the energy used up in daily activities so body weight stays in the healthy range (**Figure 1.14**).

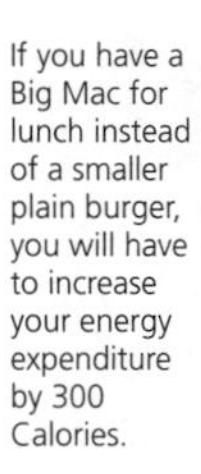

If you have a Big Mac for lunch instead of a smaller plain burger, you will have to increase your energy expenditure by 300 Calories.

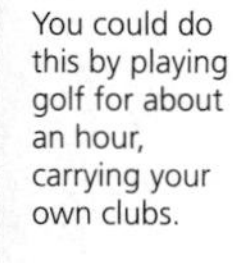

You could do this by playing golf for about an hour, carrying your own clubs.

If you have a grande mocha frappuccino instead of a regular iced coffee, you will have to increase your energy expenditure by 370 Calories.

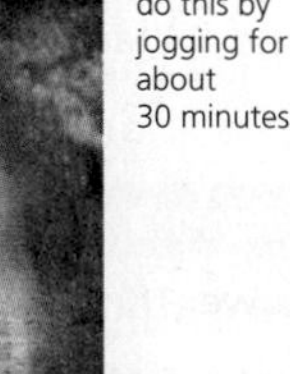

You could do this by jogging for about 30 minutes.

Figure 1.14 Balance kcalories in with kcalories out
Extra kcalories you consume during the day must be balanced by increasing the kcalories you burn in physical activity to avoid weight gain. (Top left, Andy Washnik; top right, Picturenet/Blend Images/Getty Images, Inc.; bottom left, Andy Washnik; bottom right, Kate Thompson/NG Image Collection)

Everything in Moderation Moderation means everything is okay, as long as you do not overdo it. If you like burgers or potato chips, they can be included in your diet, but you have to watch the size of your portions and how frequently you consume these foods. As indicated in Figure 1.14, choosing larger portions sizes or high-calorie foods can result in excess kcalories. If these kcalories are not expended through physical activity, they can result in weight gain. Have you ever sat down in front of the TV with a bag of chips and before you knew it, half the bag was gone? If you have, then you know how easy it is to let portion sizes get out of control. Moderation means not consuming too much energy, too much fat, too much sugar, too much salt, or too much alcohol. Choosing moderately will help you maintain a healthy weight and help prevent chronic diseases like heart disease, cancer, and type-2 diabetes that compromise the quality of life of many Canadians. The fact that more Canadians are obese than ever before demonstrates that we have not been practicing moderation when it comes to energy intake. Moderation will make it easier to balance your diet and will allow you to enjoy a greater variety of foods.

One factor that may make moderation difficult is a phenomenon called **portion distortion**.[8] Researchers have found that over the last 40 years, portion sizes in restaurants and the sizes of single portion snack foods have increased. For example, in the 1960s, cola drinks were sold in 250-ml-sized bottles (110 calories for a sugar-containing drink), while today's single portion bottle contains almost 600 ml and 260 kcalories. Similarly, hamburgers, muffins, and bagels have increased in size. Many more examples can be found at Portion Distortion, a website created by the US Dept of Heath as part of their Obesity Education Initiative (find it by searching NHLBI portion distortion using any search engine).

portion distortion The increase in portion sizes for typical restaurant and snack foods, observed over the last 40 years.

Ensuring Kcalorie Control Kcalorie control refers to the specific aspects of balance and moderation that are related to energy intake. Ensuring that energy intake from foods balances energy expended in daily activities will ensure kcalorie control as will moderation in food choices, so that too much energy is not consumed.

1.4 Understanding Science Helps Us Understand Nutrition

Learning Objectives

- List the steps of the scientific method.
- Describe the features of a good experiment.

Nutrition, like all science, continues to develop as new discoveries provide clues to the right combination of nutrients needed for optimal health. As knowledge and technology advance, new nutrition principles are developed. Sometimes, established beliefs and concepts must give way to new ideas, and recommendations change. Today more and more consumers are seeking information about nutrition and how to improve their diets. But they may find this frustrating because the experts seem to change their minds so often. One day, consumers are told margarine is better for them than butter; the next day, a report says that it is just as bad. Developing an understanding of the process of science and how it is used to study the relationship between nutrition and health can help consumers make wise nutrition decisions, whether they involve what to have for breakfast or determining whether a headline about vitamin E supplements is true.

The Scientific Method

Advances in nutrition are made using the **scientific method**. The scientific method offers a systematic, unbiased approach to evaluating the relationships among food, nutrients, and health. The first step of the scientific method is to make an observation and ask questions about the observation. The next step is to propose a **hypothesis**, or explanation for the observation. Once a hypothesis has been proposed, experiments can be designed to test it. The experiments must provide objective results that can be measured and repeated. The results of one experiment are generally not considered sufficient proof that a hypothesis is correct. If, however,

scientific method The general approach of science that is used to explain observations about the world around us.

hypothesis An educated guess made to explain an observation or to answer a question.

theory An explanation based on scientific study and reasoning.

the hypothesis is confirmed in a number of studies, that is, if the results are reproduced, then the hypothesis is considered strong enough for a **theory**, or a scientific explanation, to be established. Scientific theories are accepted only as long as they cannot be disproved and continue to be supported by all new evidence that accumulates. Even a theory that has been accepted by the scientific community for years can be proved wrong.

The discovery of the relationship between nutrition and pellagra, a disease now known to be caused by a deficiency of the vitamin niacin, is an example of how the scientific method has been used to study nutrition (**Figure 1.15**). In the early 1900s, pellegra was a common disease in the southeastern United States, where corn was a major component of the diet, particularly among the poor. Scientists observed that in institutions such as hospitals, orphanages, and prisons, residents suffered from pellagra, but the staff did not. If pellagra were an infectious disease, both populations would be affected. The hypothesis proposed was that pellagra was due to a dietary deficiency. To test this hypothesis, nutritious foods such as fresh meats and vegetables were added to the residents' diet. The symptoms of pellagra disappeared, supporting the hypothesis that pellagra is due to a deficiency of something in the diet. This experiment and others, shown in the flow chart below, led to the theory that pellagra is caused by a dietary deficiency. This theory, which still holds today. was strengthened by the discovery of the vitamin niacin, and the observation that corn, as it was used in the southeastern US at the turn of the last century, was found to be a poor source of niacin. (see Chapter 8: Science Applied: Pellagra: Infectious Disease or Dietary Deficiency?).

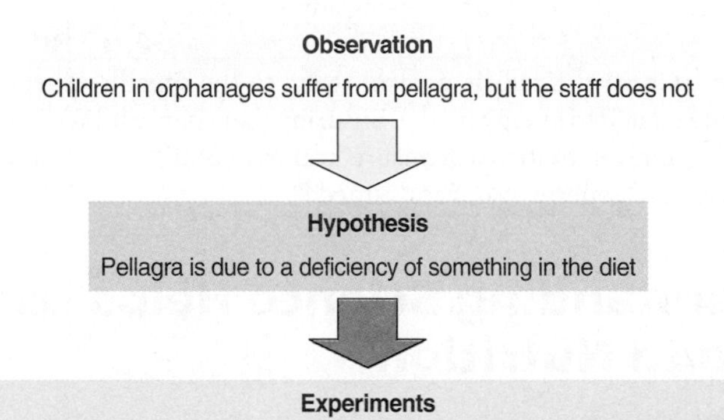

Experiments

Experimental design: Nutritious foods including meat, milk, and vegetables are added to the diets of children in two orphanages.

Results: Those consuming the healthier diets recover from pellagra. Those without the disease who eat the new diet do not contract pellagra, supporting the hypothesis that it is caused by dietary deficiency.

Experimental design: Eleven volunteers are fed a diet believed to be lacking in the dietary substance that prevents pellagra.

Results: Six of the eleven develop symptoms of pellagra after five months of consuming the experimental diet, supporting the hypothesis that it is caused by a dietary deficiency.

Continued experiments: Various studies involving both animals and humans eventually reveal that nicotinic acid, better known as the B vitamin niacin, cures pellagra.

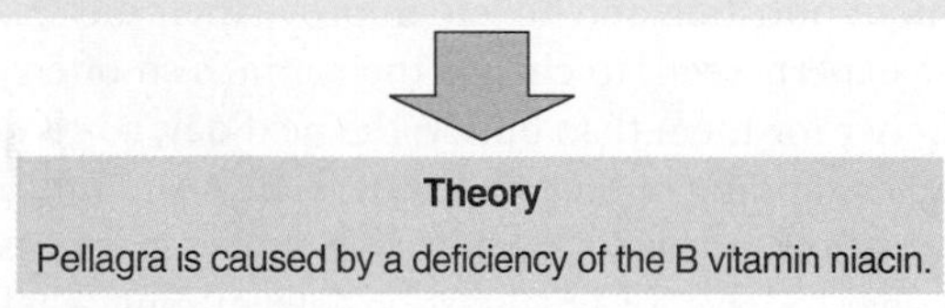

Figure 1.15 The scientific method
The scientific method is a process used to ask and answer scientific questions through observation and experimentation.

What Makes a Good Experiment?

For the scientific method to generate reliable theories, the experiments done to test hypotheses must generate reliable results and be interpreted accurately. Features of a well-designed nutrition experiment include (a) quantifiable data, (b) an appropriate experimental population, (c) appropriate number of subjects, (d) suitable study duration, (e) statistical analysis of results, and (f) publication after a peer-review process.

Quantifiable Data Scientific experiments are designed to provide quantifiable data. This means that the data can be measured in a way that provides numerical results and uses methods that can be repeated by others. Examples of measurements that are typically included in nutrition studies are sex, age, dietary intake (using methods discussed in Section 2.6), weight, blood pressure, levels of nutrients in the blood or urine, other biological indicators of future disease development, or the actual presence of disease. These indicators of future disease development are often called **biomarkers**. For example, the risk of getting **cardiovascular disease** is related to the levels of cholesterol in the blood. People who eventually develop cardiovascular disease tend to have higher levels of cholesterol in their blood for some time before they get the disease, than people who do not develop cardiovascular disease. Therefore cholesterol levels would be considered a biomarker of cardiovascular disease The advantage of using biomarkers is that they often respond quickly to a dietary treatment. For example, some diets can lower cholesterol levels in a few weeks, while cardiovascular disease may take decades to develop. So the use of biomarkers can be very useful in quickly determining the effectiveness of a dietary treatment. On the other hand, some studies do measure the actual presence of disease. These studies often last many years. For example, a study might compare the number of heart attacks (the major outcome of cardiovascular disease) in a group of individuals given a vitamin supplement for five to ten years, compared to a group that did not get the supplement.

biomarker A biological measurement that is an indicator of future disease development.

cardiovascular disease A disease that results from damage to blood vessels, such as the coronary arteries of the heart, which can cause heart attack, or the blood vessels of the brain, which can result in stroke.

Feelings or impressions are more difficult to assess than measurements of body weight and blood pressure, but not impossible. In order to be useful in science, feelings and opinions must be quantified using standardized questionnaires and compared, where possible, to a quantifiable biomarker. For example, if a person taking a supplement to build muscle reports that they feel stronger after taking it, this alone is considered anecdotal information that has little value on its own. However, if perceptions of strength are assessed with a questionnaire which has a quantitative component (e.g., asking people to rate their strength on a numerical scale) and changes in muscle mass are documented by measuring muscle mass, these are objective measurements that can be quantified and repeated.

Appropriate Experimental Population For an experiment to produce useful results, the right experimental population must be studied. For example, a food or supplement that claims to improve performance in trained athletes must be tested using trained athletes. Scientists who are interested in studying how diet might help prevent heart disease generally select older adults, in whom the development of heart disease is common. In this population, it is easy to determine if diet has increased or decreased the number of adults who develop heart disease. On the other hand, young adults in their 20s and 30s get so little heart disease to begin with that diet might not have a measurable effect. They would not be the most suitable subjects.

Appropriate Number of Subjects The number of subjects included in a study is also important. Statistical methods are used to determine how many subjects are needed to demonstrate the effect of the experimental treatment. The number of subjects will depend on the type of study and the effect being tested. The fewer **variables** included in a study, the fewer experimental subjects needed to demonstrate an effect. To be successful, an experiment must show that the treatment being tested causes a result to occur more frequently than it would occur by chance. Fewer subjects are needed to demonstrate an effect that rarely occurs by chance. For example, if only one person in a million can increase his muscle mass by weight training for 4 weeks, then an experiment to see if a supplement increases muscle mass in athletes weight training for four weeks would require only a few subjects to demonstrate an effect. However, if 1 in 4 athletes can improve his muscle mass by weight training for 4 weeks, then many more subjects are needed to show that the supplement further increases muscle mass.

variable A factor or condition that is changed in an experimental setting.

Suitable Study Duration It is important for scientists to design their study to last long enough to see an effect. For example, some biomarkers such as blood cholesterol levels can respond to diet very quickly and results can be seen in a few weeks or months. Diseases such as heart disease or cancer take decades to develop, so studies investigating disease outcomes often last many years. In the example above, the young adults (in their 20s and 30s) would take 20-40 years to develop heart disease, and that is why they would not typically be used as subjects for a study on heart disease and diet, unless researchers are able to conduct a study of this duration. One such study, the Framingham Study, has examined the health of three generations; the first generation was enrolled in the study in 1948, the second in 1971, and the third in 2002.[9–11]

Statistical Analysis of Results Statistical methods are essential for the analysis of the results. When differences between dietary intakes are found to be associated with disease risk, it is essential to determine whether the difference is due to chance alone, or represents a true difference due to dietary intake. This is determined, using statistical methods, by calculating the probability that the difference is due to chance. If the probability of a chance difference is low, then the differences are described as being *statistically significant* and this is generally a desirable outcome for most experiments. After the analysis is complete, scientists interpret their results and draw conclusions about how they can be applied. Generally if the probability of chance is less than 5% or $p < 0.05$, then results are considered statistically significant.

Publication of Results After Peer Review The sharing of experimental results is essential to the progress of science, so after completing an experiment, scientists publish their results in a scientific journal. Scientific experimentation, however, is a very complex process and to ensure that only well-conducted and properly interpreted studies are published, these journals require that prior to publication, two or three experts (who did not take part in the research that is being evaluated) agree that the experiment under review was well conducted and that the results were interpreted fairly. This process is called **peer review**. Nutrition articles can be found in peer-reviewed journals such as the *American Journal of Clinical Nutrition*, the *Journal of Nutrition*, the *Journal of the American Dietetic Association*, the *New England Journal of Medicine*, and the *International Journal of Sport Nutrition*. Journals published by Canadian organizations include the *Canadian Journal of Dietetic Practice and Research* (published by Dietitians of Canada), *CMAJ* (Canadian Medical Association Journal) *Health Reports*, published by Statistics Canada, and *Biochemistry and Cell Biology* and *Applied Physiology, Nutrition, and Metabolism*, both published by the National Science and Engineering Research Council of Canada.

peer review A process by which the quality of a science experiment is reviewed by experts. Experts must agree that the experiment is of good quality before it can be published.

1.5 Nutrition Research

Learning Objectives

- Distinguish between correlation (association) and causation.
- Distinguish between a prospective cohort study and a case-control study.
- Distinguish between observational (epidemiological) studies and intervention trials.
- Describe experimental controls, including control groups, placebos, and blinded studies.
- Explain the purpose of balance studies and depletion-repletion studies.
- Discuss how science monitors the ethics of human and animal studies.
- Describe the components of a research paper.

Nutrition research studies are done to determine nutrient requirements, to learn more about the metabolism of nutrients, and to understand the role of nutrition in health and disease. Perfect tools do not exist for addressing all these questions. However, there are many types of research studies that can be useful in understanding the relationship between people and their diets.

epidemiology The study of the interrelationships between health and disease and other factors in the environment or lifestyle of different populations.

observational study An epidemiological study that looks for associations between health and disease and environmental or lifestyle factors. There is no intervention or attempt to alter the behaviour or lifestyle of the study participants. Information about the subjects' lifestyle and health is collected and analyzed.

Types of Nutrition Studies

Observational Studies Some of our nutrition knowledge has been obtained by observing relationships between diet and health in different populations throughout the world. This study of diet, health, and disease patterns is called **epidemiology**. Epidemiological studies are also called **observational studies** because the dietary intake and the health of the population being studied are being observed. There is no attempt to intervene or alter the dietary intake of the population. From these observations, patterns or associations are identified between diet and disease. For instance, diets of individuals with different intakes of vegetables can be compared with the incidence of cancer. When scientists conducted such studies, they generally found that as vegetable intake increased, the risk of getting cancer decreased.[12] Another way of saying this is that scientists found an **association**, or **correlation**, between diets high in vegetables and a lower incidence of cancer. This does not prove that vegetables *cause* less cancer, that is, it does not prove **causation**; only that there is an association between the two variables. The correlation can be direct (positive) or inverse (negative). A direct or positive relationship is observed when increased nutrient intake increases disease risk; it is called an inverse or negative relationship when increased nutrient intake decreases disease risk.

correlation or **association** Two or more factors occurring together. The correlation can be direct (positive) or inverse (negative). A direct or positive relationship is observed when increased nutrient intake increases disease risk; an inverse or negative relationship is observed when decreased nutrient intake increases disease risk. Correlations do not prove causation.

One of the limitations of observational studies is the presence of **confounding factors**, which are factors associated with both the dietary intake of interest and the disease. For example, in the cancer study, it may have been that the people who ate more vegetables were by chance younger than the people who did not eat vegetables. We know that cancer is less common in younger people, so perhaps the observed relationship between vegetables and cancer is really due to differences in age. This would be an example of age being a confounding factor, as it is associated with both vegetable intake and cancer incidence. Complex statistical methods can eliminate the effect of some confounding factors. In the example above, mathematical adjustment for age can compare the effect of high and low vegetable intake on cancer incidence per year of aging (so age is no longer a contributing factor). Age is a very common confounding factor and virtually all observational studies are age-adjusted. Sex is another common confounding factor, so the results of men and women are often presented separately. Statisticians try to identify and correct for all confounding factors but it is not possible to completely eliminate their effect. So when scientists observe that a group of individuals with a high vegetable intake have less cancer than a group with a low vegetable intake they cannot be 100% certain that vegetable intake is the only difference between the two groups. Other factors may be contributing to the observation. That is why we describe the results as indicating an association or correlation, but not proving causation, that is, not proving that a high vegetable intake causes less cancer.

causation A relationship between two factors where one factor causes the second factor to occur.

confounding factor In scientific studies, a factor that is related to both the outcome being investigated (e.g., disease) and a factor that might influence outcome (dietary intake).

There are several types of observational studies; two important ones are the **prospective cohort study** and the **case-control study**. In a prospective cohort study, the dietary intake of a healthy population is recorded and the health of this population is followed for a number of years, that is, it is prospective or moving forward in time. The Science Applied section describes one such study, the Seven Countries Study, in which a population of almost 13,000 men participated. As noted above, selection of an appropriate population is essential in any experiment. For the Seven Countries Study, a population of men aged 40-59 were selected because this is a population which would be expected to develop heart disease over the 10-year duration of the study. (Note: One limitation of the Seven Countries Study is that it was conducted only on men; scientists doing such a study today would include both men and women.) This population was followed for many years and the number of men who developed heart disease over the duration of the study was recorded. Scientists were interested in finding out whether men who consumed a certain type of diet were more or less likely to get heart disease. Researchers were able to show an association between saturated fat intake and heart disease (**Figure 1.16**). The association is a direct or positive one, meaning that as saturated fat intake increased, the incidence of heart disease increased.

prospective cohort study An observational study in which dietary intake information is collected by researchers and the health of the study participants is observed, usually for several years. At the end of the study, scientists determine whether there are any correlations between dietary intake and the incidence of disease.

case-control study A type of observational study that compares individuals with a particular condition under study with individuals of the same age, sex, and background who do not have the condition.

Case-control studies compare individuals with a particular condition to similar individuals without the condition. For example, a case-control study of cancer might include a comparison of the dietary intake of a 45-year-old man with cancer to a man of the same age and ethnic background who is free of the disease. This matching would be done for all participants of the study. If a pattern emerged, such as a lower vegetable intake among the cancer patients (i.e., the cases) compared to individuals without cancer (i.e., the controls), then a hypothesis can be proposed that there is an inverse association between cancer risk and vegetable intake.

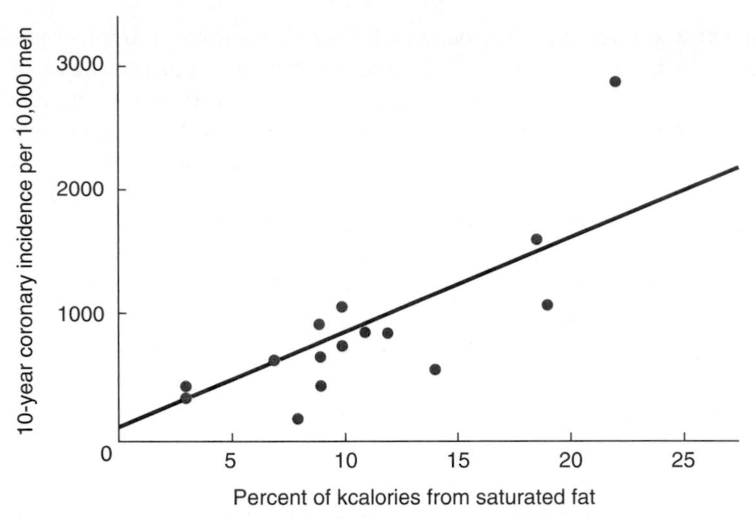

Figure 1.16 Association between saturated fat intake and heart disease
This graph shows that populations with high intakes of saturated fat have a higher incidence of coronary heart disease, but it does not tell us if the high incidence of heart disease is caused by the high intake of saturated fat. (Seven Countries Study)

human intervention trial or **clinical trial** A study of a population in which there is an experimental manipulation of some members of the population; observations and measurements are made to determine the effects of this manipulation, compared to members who did not undergo the manipulation.

control group A group of participants in an experiment that is identical to the experimental group except that no experimental treatment is used. It is used as a basis of comparison.

treatment group A group of participants in an experiment who are receiving an experimental treatment. The effects of the treatment are compared to the control group.

randomization The process by which participants in an intervention trial are assigned to either a treatment group or a control group entirely by chance.

placebo A fake medicine or supplement that is indistinguishable in appearance from the real thing. It is used to disguise the control from the experimental groups in an experiment.

single-blind study An experiment in which either the study participants or the researchers are unaware of which subjects are in the control or experimental group.

Human Intervention Trials The observations and hypotheses that arise from epidemiological studies can be tested using human intervention trials or studies, also called clinical trials or clinical studies. Unlike the observational study, this type of experiment actively intervenes in the lives of a population and examines the effect of this intervention. Nutrition intervention studies generally explore the effects that result from altering people's diets. Often intervention trials are conducted to test the hypotheses generated by observational studies. An intervention trial that would follow naturally from the Seven Countries Study's findings (see Science Applied: From Seven Countries to the Mediterranean Diet) would be to compare the number of cases of heart disease that develop, over the long term, for two groups of individuals—a **control group**, in which participants are asked to consume a typical Western diet, which tends to be high in saturated fat; and a **treatment group,** in which the diet is low in saturated fat and high in polyunsaturated fats. Such studies have been conducted and have found a reduced risk of cardiovascular disease when polyunsaturated fats are consumed.[13] When setting up an intervention trial, scientists recruit suitable participants for the study. They then randomize the population into the control and intervention groups. This means that whether an individual gets a high saturated fat diet or a high polyunsaturated fat diet is due totally to chance. This **randomization** is the strength of the intervention study. In observational studies, it is not possible to completely eliminate the effects of confounding factors. Because of randomization, however, any confounding factors, including ones that remain unidentified, are equally distributed between the two groups; for example, the individuals in the two groups will have the same average age, weight, blood pressure, same distribution of men and women, etc. This means that the only difference between the two groups is the intervention, or differences in diet. Because *only* the dietary intake differs, if there is a reduction in the incidence of heart disease in the treatment group, then scientists can conclude that unsaturated fat causes less heart disease. They have shown causation. This is the major difference between an observational study and an intervention trial.

Another important element of the intervention trial is the use of the **placebo**, in order to make the control and experimental groups indistinguishable. A placebo is identical in appearance to the actual treatment but has no therapeutic value. Consider, for example, an intervention trial to test whether a protein drink improves performance in trained athletes. Trained athletes would be used as the subjects and would be randomized into a control and treatment group.

If the experimental group is consuming a protein drink, an appropriate placebo for the control group would be a drink that looks and tastes just like the protein drink but does not contribute any nutrients. Using a placebo prevents participants in the experiment from knowing whether or not they are receiving the actual supplement. When the subjects do not know which treatment they are receiving, the study is called a **single-blind study**. Using a placebo in a single-blind study helps to prevent the expectations of subjects from biasing the

SCIENCE APPLIED

From "Seven Countries" to the Mediterranean Diet

In 1916, Dutch physician C. D. de Langen hypothesized that a cholesterol-rich diet was associated with high blood cholesterol levels and coronary heart disease. But it was not until the 1950s that the associations among diet, blood cholesterol, and coronary heart disease were investigated. In 1958, Professor Ancel Keys initiated the Seven Countries Study, designed to test the hypothesis that coronary heart disease was a nutritional problem related to fat intake. Between 1958 and 1964, the study enrolled 12,763 men from 40 to 59 years of age from 16 different regions within seven countries on three continents.

The study evaluated health status, dietary intake, body weight, blood pressure, blood cholesterol level, and other health-related parameters at regular intervals. Patterns began to emerge. In northern European countries, the diet was high in dairy products; in the United States, it was high in meat; in southern Europe, it was high in vegetables, legumes, fish, and wine; and in Japan, it was high in cereals, soy products, and fish.[1] After 10 years, 1512 of the study participants were dead—413 of them from coronary heart disease[2]—but mortality differed strikingly with location. The Isle of Crete had only one coronary death out of 686 men studied, whereas eastern Finland had 78 coronary deaths among 817 participants.[3]

The coronary death rate was correlated with the average percentage of kcalories from saturated fat. In countries such as Japan, where the diet was low in total and saturated fat, blood cholesterol levels were lower, as was the risk of dying from heart disease. In countries such as Finland, where the diet was rich in saturated fat and cholesterol, blood cholesterol levels were higher, as was the incidence of heart disease. The results showed that blood cholesterol was strongly correlated with coronary heart disease deaths both for populations and for individuals.

A closer examination of the data revealed some paradoxes. In different populations, the same blood cholesterol level represented different degrees of cardiovascular risk. For example, someone living in northern Europe with a blood cholesterol level of 200 had five times the risk of dying of heart disease as someone with a blood cholesterol level of 200 living in Mediterranean southern Europe.[4] The data suggested that blood cholesterol was not the only risk factor in heart disease. When other dietary and lifestyle factors were examined, patterns related to the risk of coronary heart disease emerged. In southern Europe around the Mediterranean Sea—where the incidence of coronary heart disease was low—the diet was plentiful in vegetables and whole grains, and olive oil was the source of fat. In addition, the lifestyle was not stressful and wine was routinely, but not excessively, consumed with meals. This dietary pattern became known as the "Mediterranean diet."

Many features of the Mediterranean diet have been included in current population-wide recommendations for reducing the risk of coronary heart disease (see figure). The standard prescription for reducing heart disease risk includes consuming olive oil and other unsaturated oils as the source of dietary fat, including plenty of whole grains and fruits and vegetables, exercising, not smoking, and reducing stress. You will read more about the Mediterranean diet in Section 2.4

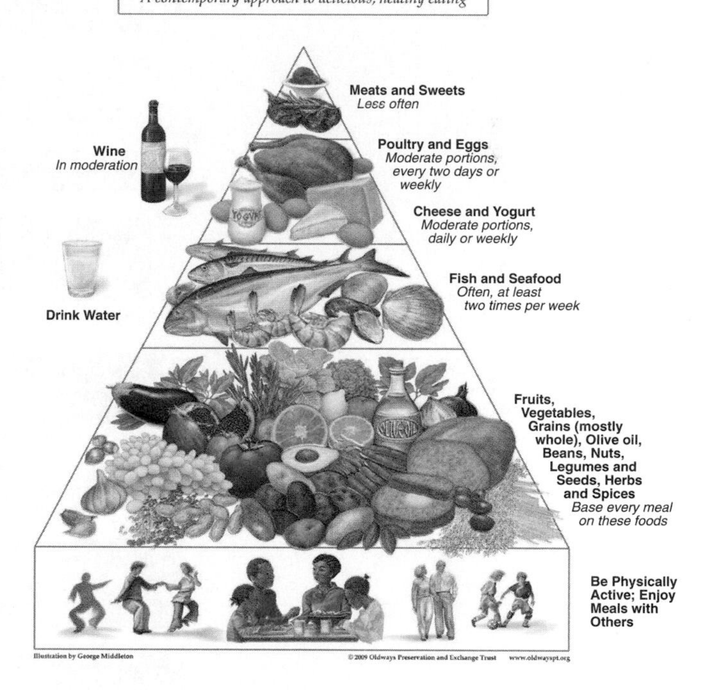

After more than 50 years, subjects of the Seven Countries Study are still being studied and epidemiological data are being collected, analyzed, and used as the basis for countless laboratory studies and intervention trials. As more and more pieces of the coronary heart disease puzzle fall into place, scientists and government agencies no doubt will continue to modify their recommendations.

[1] Menotti, A., Kromhout, D., Blackburn, H., et al. Food intake patterns and 25-year mortality from coronary heart disease: Cross-cultural correlations in the Seven Countries Study. The Seven Countries Research Group. *Eur. J. Epidemiol.* 15:507–515, 1999.

[2] The diet and all-causes death rate in the Seven Countries Study. *Lancet* 2:58–61, 1981.

[3] Keys, A. *Seven Countries: A Multivariate Analysis of Death and Coronary Heart Disease.* A Commonwealth Fund book. Cambridge, MA: Harvard University Press, 1980.

[4] Kromhout, D. Serum cholesterol in cross-cultural perspective: The Seven Countries Study. *Acta. Cardiol.* 45:155–158, 1999.

results. For example, if the athletes think they are taking the protein supplement, they may be convinced that they are getting stronger and as a result work harder in their training and develop bigger muscles even without the supplement. Errors can also occur if investigators' expectations bias the results or the interpretation of the data. This type of error can be avoided by designing a **double-blind study** in which neither the subjects nor the investigators know who is in which group until after the results have been analyzed.

double-blind study An experiment in which neither the study participants nor the researchers know who is in the control or the experimental group.

Placebos can be developed when dietary supplements are being used in an experiment, but when entire diets are altered, a placebo may not be possible. The participants may be able to discern which group they are in although researchers may try to avoid disclosing to participants some details about the diets so as to reduce bias from expectations. The investigators can still be blinded even when participants are not.

While intervention trials may sound very attractive, they have their limitations. The impact of a dietary treatment may not be evident in a short experiment and may require may years to become apparent. Yet an intervention trial is difficult to conduct for long periods of time, especially if you are asking participants to sustain major changes to their diet. For this reason, in nutritional research, both observational studies, which demonstrate association but can be sustained for many years, and intervention trials, which demonstrate causation, but are harder to maintain for long durations, are important in understanding the impact of diet on human health. See **Figure 1.17** for a comparison of how the same nutritional problem can be studied by both an observational study and an intervention trial. Also see Critical Thinking: Using the Scientific Method to read how both intervention trials and observational studies are used to test a hypothesis.

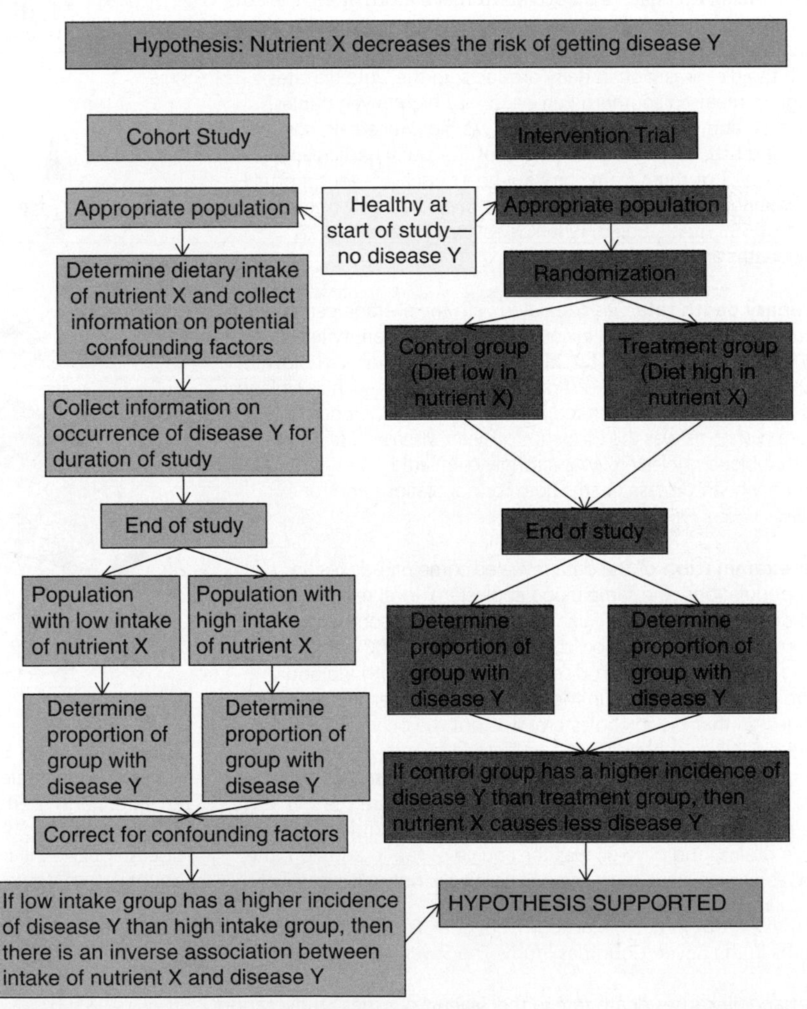

Figure 1.17 Comparison of an observational study (prospective cohort study) and an intervention trial, addressing the same hypothesis.

Laboratory Studies Laboratory studies are conducted in research facilities such as hospitals and universities. They are used to learn more about how nutrients function and to evaluate the relationships among nutrient intake, levels of nutrients in the body, and health. They may study nutrient requirements and functions in whole organisms—either humans or animals—or they may focus on nutrient functions at the cellular, biochemical, or molecular levels.

Studies Using Whole Organisms Many nutrition studies are done by feeding a specific diet to a person or animal and monitoring the physiological effects of that diet. There are many approaches to determining nutrient requirements, that is, how much of a nutrient someone needs to maintain health. Two approaches that will be described here include **depletion-repletion** and **balance studies.**

depletion-repletion study A study that feeds subjects a diet devoid of a nutrient until signs of deficiency appear, and then adds the nutrient back to the diet to a level at which symptoms disappear and health is restored.

Depletion-repletion studies are a classic method for studying the functions of nutrients and estimating the requirement for a particular nutrient. This type of study involves depleting a nutrient by feeding experimental subjects a diet devoid of that nutrient. After a period of time, if the nutrient is essential, symptoms of a deficiency will develop. The symptoms provide information on how the nutrient functions in the body. The nutrient is then added back to the diet, or repleted, until the symptoms are reversed. The requirement for that nutrient is the amount needed to reverse the deficiency symptoms and restore health. An example of a depletion-repletion study will be discussed in Section 2.2.

Another method for determining nutrient functions and requirements is to compare the intake of a nutrient with its excretion. This is known as a **balance study**. If more of a nutrient is consumed than is excreted, balance is positive and it is assumed that the nutrient is being used or stored by the body. If more of the nutrient is excreted than is consumed, balance is negative, indicating that some is being lost from the body. When the amount consumed equals the amount lost, the body is neither gaining nor losing that nutrient and is said to be in a steady state or in balance (**Figure 1.18**). By varying the amount of a nutrient consumed and then measuring the amount excreted, it is possible to determine the minimum amount of that nutrient needed to replace body losses. This type of study can be used to determine protein requirements because protein is not stored in the body and will be discussed in more detail in Section 6.6. It is not useful for determining the requirements for nutrients such as fat and iron that are stored when an excess is available.

balance study A study that compares the total amount of a nutrient that enters the body with the total amount that leaves the body.

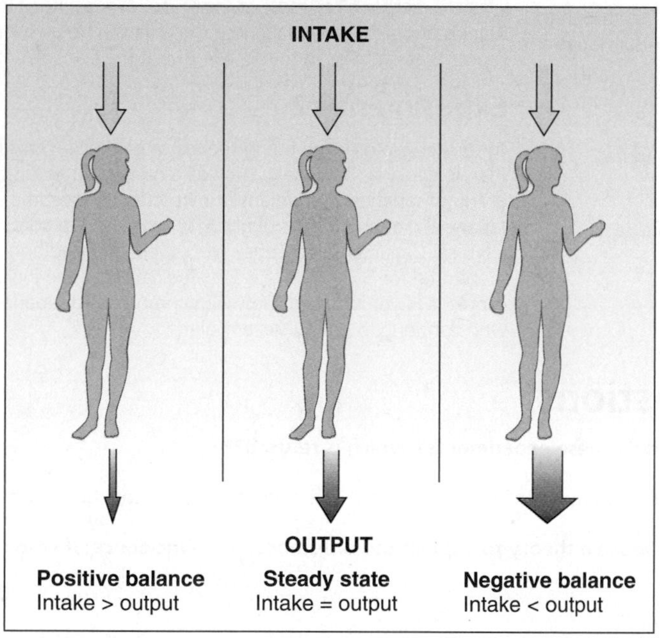

Figure 1.18 Nutrient balance
Nutrient balance studies compare the amount of a nutrient that is consumed with the amount excreted to determine whether the body is in positive balance, steady state, or negative balance.

Using Animals to Study Human Nutrition Ideally, studies of human nutrition would be done in humans. However, because studying humans is costly, time-consuming, inconvenient for subjects, and in some cases impossible for ethical reasons, many studies are done using experimental animals.

An ideal animal model is one with metabolic and digestive processes similar to humans. For example, cows are rarely used in human nutrition research because they digest their food in four stomach-like chambers as opposed to a single stomach. Pigs, on the other hand, are a good model because they digest food in a manner similar to that of humans. In addition to digestion and metabolism, factors such as cost and time must be considered. Pigs and other large animals are expensive to use, and they take a long time to develop nutrient deficiencies. Smaller laboratory animals, such as rats and mice, are therefore the most common experimental animals. They are inexpensive, have short life spans, reproduce quickly, and the effects

CRITICAL THINKING:
Using the Scientific Method

(Lori Smolin)

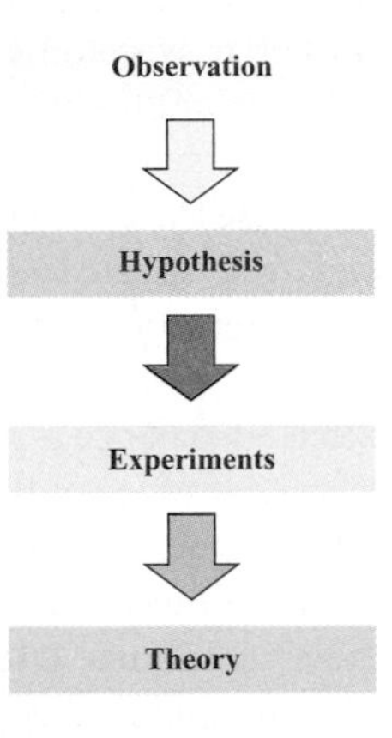

Observation

While analyzing some epidemiological data, Dr. Tanaka notices that the incidence of colon cancer is much greater in Canada than it is in Japan. He proposes two hypotheses to explain this observation.

Hypotheses

1. The difference may occur because the Japanese people inherit genetic factors that protect against colon cancer.

2. The difference may be due to differences between the diet in Canada and in Japan.

Experiment #1

To generate data to support or refute each of his hypotheses, Dr. Tanaka reviews the medical records and dietary intakes of second-generation Japanese Canadians. These individuals are not different genetically from Japanese living in Japan, but they have been consuming a typical Western diet for all or most of their lives. The results of his study indicate that second-generation Japanese Canadians have an incidence of colon cancer that is similar to that of the rest of the Canadian population.

Experiment #2

To further explore why the incidence of colon cancer is lower in Japan than in Canada, Dr. Tanaka designs a human intervention trial. One group of Japanese Canadians is instructed to consume a traditional Japanese diet and a control group is instructed to continue eating a typical Canadian diet. Those who adopt the traditional diet are monitored for 5 years. The results indicate that the incidence of colon cancer is lower in this group when compared to Japanese Canadians who continue to eat a Western diet.

Critical Thinking Questions

Which of the two hypotheses is supported by these experiments? Which is refuted?*

Could you propose other experiments to obtain more information about the causes of colon cancer in Canada?

Based on the evidence presented here, propose a theory to explain the difference in the incidence of colon cancer in Canada and Japan.

*The answers to these questions can be found in Appendix J.

iProfile To see one example of how Western and Japanese diets might differ, use iProfile to compare the fat content of a fast food burger, commonly consumed in the Western world, and sushi, a traditional Japanese food.

---

of nutritional changes develop rapidly. Their food intake can be easily controlled, and their excretions can be measured accurately using special cages. Even when using small animals, the species of animal must be carefully chosen. For example, rats are more resistant to heart disease than are humans, so they are not a good model for studying the effect of diet on heart disease. Rabbits, on the other hand, do develop heart disease and can be used to study diet–heart disease relationships. Both rabbits and rats, however, are poor choices for a study of vitamin C requirements because they can synthesize this vitamin in their bodies. Guinea pigs would be a better choice because the guinea pig is one of the few animals, other than humans, that cannot make vitamin C in its body (**Figure 1.19**). Even the best animal model is not the same as a human, and care must be taken when extrapolating the results to the human population.

For example, a study that uses rats to show that a calcium supplement increases bone density can hypothesize, but not conclude, that the supplement will have the same effect in humans.

Figure 1.19 What are the advantages and disadvantages of using guinea pigs and other laboratory animals to study human nutrition? (Lori Smolin)

Studies Using Cells Another alternative to conducting studies in humans is to study cells either extracted from humans or animals or grown in the laboratory (**Figure 1.20**). Biochemistry can be used to study how nutrients are used to provide energy and how they regulate biochemical reactions in cells. Molecular biology can be used to study how genes regulate cell functions. The types and amounts of nutrients available to cells can affect the action of genes. For example, vitamin A can directly activate or inactivate certain genes. Knowledge gained from biochemical and molecular biological research can be used to study nutrition-related conditions that affect the entire organism. Molecular biology can help us understand the hereditary basis of diseases like heart disease, cancer, diabetes, and obesity and is helping us to identify individuals who have a genetic susceptibility to specific diseases so that intervention can begin early.

Ethical Concerns in Scientific Study

Ethical issues are often raised in the process of conducting nutrition research. Whenever possible, researchers use alternatives to human subjects or experimental animals. For example, studying body fluids, cells grown in the laboratory, and even computer models can help predict how changes in nutrient intake affect body processes. However, such alternatives cannot always be used; human and animal experimentation is still necessary to answer many questions. To avoid harm and protect the rights of humans and animals used in experimental research, government guidelines have been developed.

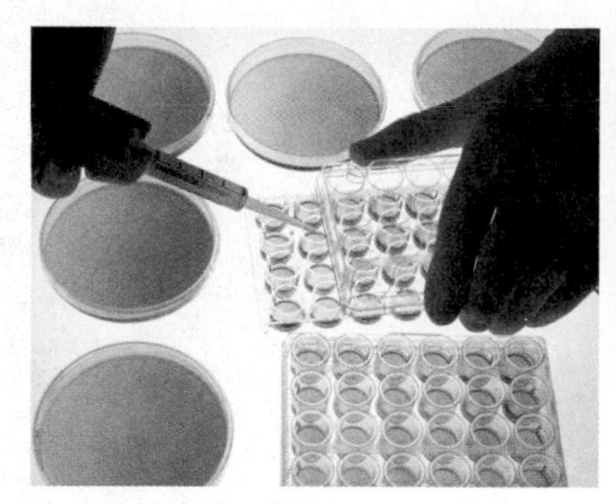

Figure 1.20 The ability to grow cells in the laboratory allows scientists to study nutrients without using whole organisms. (SPL/Photo Researchers, Inc.)

Human Subjects Before an experiment involving human subjects can be conducted, the study must first be reviewed by a committee of scientists and nonscientists to ensure that the rights of the subjects are respected and that the risk of physical, social, and psychological injury is balanced against the potential benefit of the research. Before subjects participate in a study, the researchers must explain to them the purpose of the research, the procedures used, and the possible risks and benefits. In addition to an oral explanation of the study, each subject must be given a written description of the study and its risks and benefits. Those who choose to participate must then sign a consent form stating exactly what they have agreed to do. Signing a consent form does not mean a subject must complete the study if it turns out to be difficult to do so—subjects can leave a study at any time. This informed consent process is part of the strict safety and ethical regulations that must be followed when research involves human subjects. This protects subjects, but limits the type of study that can be done on humans. For example, much of what we know today about the effects of starvation in humans was determined during World War II by conducting depletion-repletion studies using conscientious objectors as experimental subjects. These subjects were monitored physically and psychologically while they were starved and then re-fed. These individuals experienced some level of suffering during the trials and risked longer-lasting physical and psychological harm. This type of study would not be approved if researchers wanted to repeat it today.

Animal Studies As with experiments involving humans, the federal government mandates that panels of scientists review experiments that propose to use animals. These panels consider whether the need for animals is justified and whether all precautions will be taken to avoid pain and suffering. Animal housing and handling are strictly regulated, and a violation of these guidelines can close a research facility.

Genetic Modification The development of new techniques in the field of molecular biology has given rise to ethical issues regarding the manipulation of genes. Guidelines for manipulating genes have been developed and are constantly being revised to stay abreast of advances in this field.

Accessing Nutrition Research Information

As noted above, research scientists publish in peer-reviewed journals. Research articles are typically composed of the following parts:

1. Abstract: A short paragraph that summarizes the experiment and its main findings.
2. Introduction: Describes the purpose of the experiment and summarizes the current scientific knowledge on the subject of the paper.

3. Materials and Methods: Describes the methods used, e.g., type of experiment (e.g., case-control study, clinical study); the population used (e.g., men, women, age range, health status, etc.); the duration of the study; the biomarkers measured and the statistical analysis conducted.
4. Results: Describes the results of the study.
5. Discussion and conclusion: Interprets the results and explains their usefulness.
6. Bibliography: A list of publications referred to in the paper.

The abstract is useful because it can be read quickly and provides the reader with the most important information about a paper. Scientific databases store these abstracts and can be searched by keyword, so anyone interested in learning about a nutrition topic can read the abstracts on a topic and get a quick overview of current scientific knowledge. One popular database that can be accessed by anyone through the Internet is Pubmed, which is supported by the National Institutes of Health in the United States (www.ncbi.nlm.nih.gov/sites/entrez?db=pubmed). **Figure 1.21** shows the different parts of a Pubmed abstract.

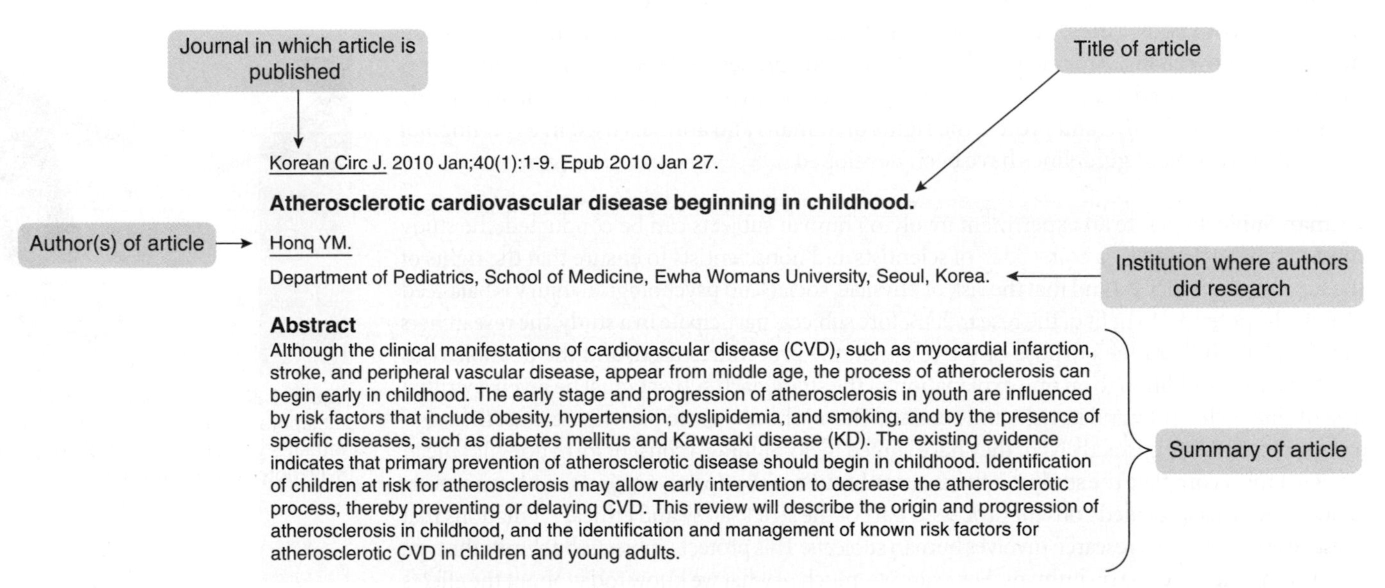

Figure 1.21 Anatomy of an Abstract: An abstract is a summary of a scientific paper.
Source: PubMed, US National Library of Medicine National Institutes of Health. Accessed Oct 20. 2011. www.ncbi.nlm.nih.gov/pubmed.

1.6 Sorting Out Nutrition Information

Learning Objectives

- Name seven questions you should ask yourself when evaluating the reliability of nutritional information.
- Discuss why individual testimonies are not considered reliable sources of information.

We are bombarded with nutrition information. Some of what we hear is accurate and based on science, and some of it is incorrect or exaggerated to sell products or make news headlines more enticing: oat bran lowers cholesterol, antioxidants prevent cancer, low-carbohydrate diets promote weight loss, vitamin C cures the common cold, vitamin E slows aging. Sifting through this information and distinguishing the useful from the useless can be overwhelming. Just as scientists use the scientific method to expand their understanding of the world around us, each of us can use an understanding of how science is done to evaluate nutrition claims by asking the questions discussed below and summarized in **Table 1.4**.

Table 1.4 Ask These Questions BEFORE You Believe it
• Does the claim presented make sense?
• If it is too incredible to believe, disregard it.
• Where did the information come from?
• If it is based on personal opinions, be aware that one person's perception does not make something true.
• Information from government agencies, universities, and nonprofit organizations is generally sound.
• Check the credentials of the person providing the information. If they do not have a legitimate degree in nutrition or medicine, view the material with skepticism. If no credentials are listed, there is no way to determine if they are qualified to give this information.
• Was the information based on well-designed experiments?
• Reliable information is based on scientific studies that use proper controls, include enough experimental subjects to get reliable results, and collect quantifiable data.
• Studies published in peer-reviewed journals have been evaluated for accuracy before they can be published.
• Were the experimental results interpreted accurately?
• Compare news reports to study results to see if the importance of the study has been exaggerated to make the headline more attractive.
• If the study was done on animals, consider carefully whether the results will also apply to humans.
• Who stands to benefit?
• If the information is helping to sell a product, it may be biased toward that product.
• If the information is making a magazine cover or newspaper headline more appealing, the claims may be exaggerated to promote sales.
• Has it stood the test of time?
• If the study is the first to support a particular finding, wait before changing your diet based on the result.
• If the finding has been shown repeatedly in different studies over a period of years, it will become the basis for reliable nutrition recommendations.
• Does it pose a risk?
• Be sure the expected benefit of the product is worth the risk associated with using it.

Does the Information Make Sense?

The first question to ask yourself when evaluating nutrition information is does the claim being made make sense? Some claims are too incredible to be true. For example, the hypothetical advertisement for WellShield illustrated in **Figure 1.22** states that this product will make illness a thing of the past. This is certainly appealing, but it is hard to believe, and common sense should tell you that it is too good to be true. The claim that WellShield will reduce cold symptoms, however, is not so outrageous.

Where Did the Information Come From?

If the claim seems reasonable, look to see where it came from. Was it a personal testimony, a government recommendation, or advice from a health professional? Was it the result of a research study? Is it in a news story or an advertising promotion? Is it on television, in a magazine, or on a Web page?

Although health professionals such as dietitians are sources of reliable nutrition information, Canadians are also exposed to many messages about food and nutrition information in the mass media. Mass media are very powerful tools in promoting health and nutrition messages. Information that would take individual health-care workers years to disseminate can reach millions of individuals in a matter of hours or days. Much of this information is reliable, but some can be misleading. The motivation for news stories is often to sell subscriptions or improve ratings, not to promote the nutritional health of the population. Some nutrition and health information originates from food manufacturers, and is usually in the form of marketing and advertising designed to sell existing products or target new ones. This promotional information can be confusing to the consumer, who may not know how to

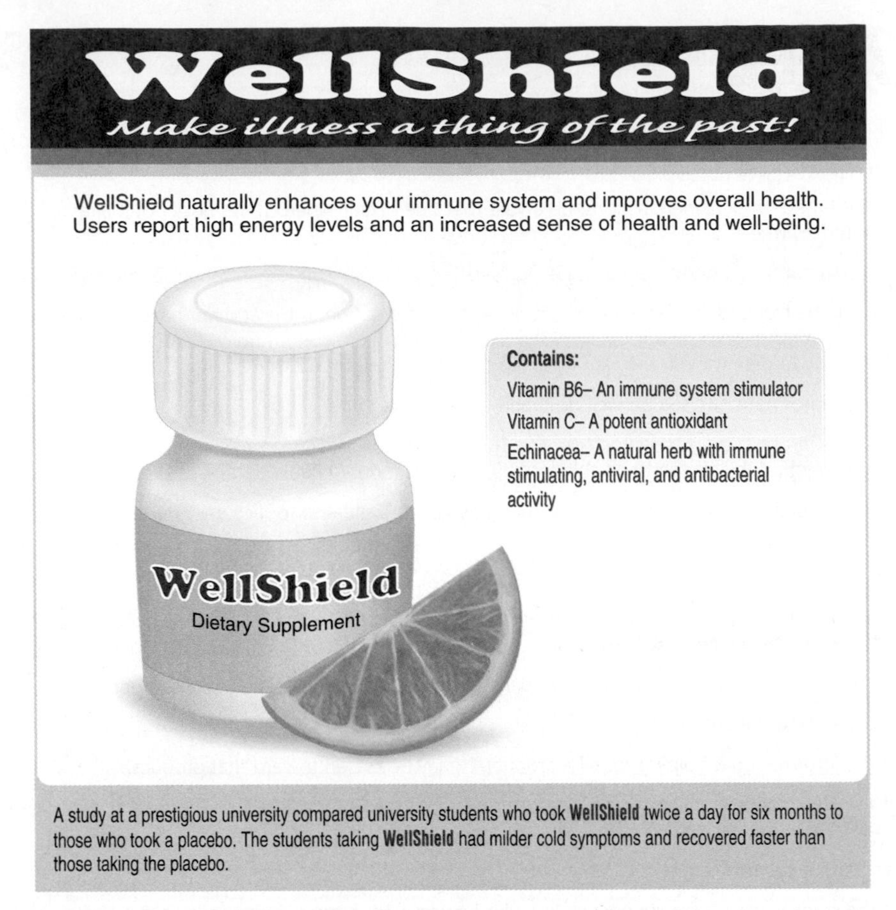

Figure 1.22 This hypothetical supplement advertisement illustrates the types of nutrition claims that consumers must be prepared to evaluate.

interpret it. For instance, in the 1990s, the recommendation to reduce fat intake to protect against heart disease and help maintain a healthy weight created a demand for products low in fat. Food manufacturers responded by creating fat-free, low-fat, and reduced-fat products at an astonishing rate. Weight- and health-conscious consumers responded by increasing their consumption of these foods, but waistlines continued to increase. The message that reducing fat intake promotes health was received, but consumers did not understand that fat-free foods are not kcalorie-free and simply adding low-fat foods to the diet was not a prescription for good health. Knowing what information to believe and how to use this information to choose a diet can be challenging (see Label Literacy: Look Beyond the Banner on the Label).

Individual Testimonies Claims that come from individual testimonies have not been tested by experimentation. For example, the claim in Figure 1.22 that WellShield improves energy levels and overall sense of health and well-being is based on the comments of supplement users and is therefore anecdotal, and not based on measured parameters.

CANADIAN CONTENT

Information From Government, Nonprofit, and Educational Institutions The three most important government agencies in Canada that provide health and dietary information are Health Canada, the Canadian Food Inspection Agency, and the Public Health Agency of Canada. Health Canada is the government agency that deals with many health-related matters including the development of recommendations regarding healthy dietary practices. These recommendations are developed by committees of scientists who interpret the latest well-conducted research studies and use their conclusions to provide guidelines for the population as a whole. For example, Health Canada developed *Eating Well with Canada's Food Guide* and, in collaboration with American agencies, the Dietary Reference Intakes, both of which are discussed in Sections 2.3 and 2.2, respectively. Health Canada provides

LABEL LITERACY

CANADIAN CONTENT

Look Beyond the Banner on the Label

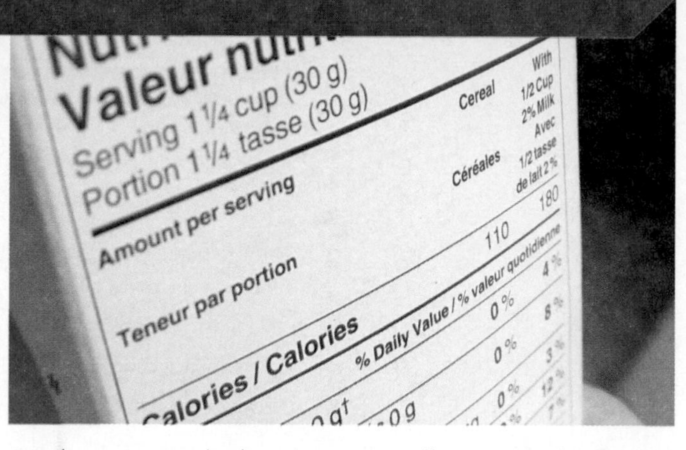

(©Tim Pohl/iStockphoto)

Extra-lean!; Reduced in energy!; Fat-free! Claims like these on a food package stand out and catch your eye, but what do these terms mean? Food labels provide a lot of nutritional information but the interpretation of this information is not always straightforward. Food labels must conform to Canadian Food and Drugs regulations, which are described in the Canadian Food Inspection Agency's Guide to Food Labelling and Advertising[1] and use standard definitions, but these definitions may not necessarily make sense to the consumer. Extra-lean, for example, can be used only to describe meat products and means that the product is less than 7.5% fat. Reduced in energy means that a product contains at least 25% less than a similar reference food. A fat-free cookie means that it contains less than 0.5 grams of fat per reference serving (which is also defined by regulations).

Furthermore, many food label descriptors highlight individual nutrients. But because no single nutrient makes a food good or bad for you, you must look beyond these descriptors to understand the overall contribution that the food makes to your diet. For example, chocolate cookies labelled "fat free" may not be your best choice if you are trying to reduce your sugar intake or increase your fibre consumption. Product names, which also comply with regulations of the Food and Drug Act, can still be confusing. Vegetable stew with beef must contain at least 12% beef, while beef stew must contain at least 20% beef. Weiners with beans contains at least 25% wieners, while beans with wieners contains at least 10% wieners.[2] To get the whole picture, you need to look beyond the healthy-sounding descriptors and the product name. Read the label, which must include the food's nutrient content and information to see how it fits into the diet as a whole. Section 2.5 and Label Literacy boxes throughout this book provide more information on how to read food labels.

[1] Canadian Food Inspection Agency. Canadian Food Inspection Agency's Guide to Food Labelling and Advertising. Available online at http://www.inspection.gc.ca/english/fssa/labeti/guide/toce.shtml. Accessed January 28, 2010.

[2] Justice Canada Food & Drugs Regulations-Division 14: Meat B.14.070:Meat Specialities. Available online at laws-lois.justice.gc.ca/eng/regulations/C.R.C.%2C_c._870/page-207.html#h-104. Accessed on March 20, 2011.

information about food safety and recommendations on food choices and the amounts of specific nutrients needed to avoid nutrient deficiencies and excesses, and to prevent chronic diseases. These recommendations are used to develop food-labelling regulations and are the basis for public health policies and programs. Health Canada publishes very useful information, on many health-related topics, on their website at www.hc-sc.gc.ca. The Canadian Food Inspection Agency (www.inspection.gc.ca) enforces Canadian regulations to ensure the safety of the Canadian food supply, for example, the prevention of food-borne illness, the contamination of food with toxic chemicals, and the mislabelling of food. The Public Health Agency of Canada has the goal of preventing disease and improving the health of Canadians and, like Health Canada, also provides useful nutrition-related information on their website at www.phac-aspc.gc.ca. Another useful website for Canadians to access is www.healthycanadians.ca, which provides information to Canadians on a variety of important issues including food and nutrition.

Nonprofit organizations such as Dietitians of Canada, the Canadian Diabetes Association, the Canadian Medical Association, the Heart and Stroke Foundation of Canada, and the Canadian Cancer Society are also good sources of nutrition information. See **Table 1.5** for web addresses of these organizations. The purpose of the information they provide to the public is to improve health. Reports that come from universities are supported by research and are also a reliable place to look for information. Many universities provide information that targets the general public and university research studies are usually published in peer-reviewed journals and are well scrutinized. The WellShield ad in Figure 1.22 cites a university research study that demonstrated reduced cold symptoms. If the study was published in a peer-reviewed journal, the information is probably reliable.

Table 1.5 Non-profit organizations in Canada that provide reliable nutrition information

Organization	Web Address
Dietitians of Canada	www.dietitians.ca
Canadian Diabetes Association	www.diabetes.ca
Canadian Medical Association	www.cma.ca/
Heart and Stroke Foundation of Canada	www.heartandstroke.com
Canadian Cancer Society	www.cancer.ca
Osteoporosis Canada	www.osteoporosis.ca
Hypertension Canada	www.hypertension.ca

Qualified Individuals Knowing who is providing the information can help you decide whether or not to believe it. What are the credentials of the individual providing the information? Where do they work? Do they have a degree in nutrition or medicine? If you are looking at an article or a website, check the credentials of the author. Care must be taken even when obtaining information from nutritionists. Although "nutritionists" and "nutrition counselors" may provide accurate information, these terms are not legally defined and are used by a wide range of people from college professors with doctoral degrees from reputable universities to health food store clerks with no formal training. One reliable source of nutrition information is the registered dietitian. Registered dietitians (RD or RDt) are nutrition professionals who have completed a 4-year college degree in a nutrition-related field and who have met established criteria to certify them in providing nutrition counseling. In addition to the term registered dietitian, the terms registered dietitian nutritionist (RDN), professional dietitian (PDt) or dietetiste professionnelle (Dt. P) are also used in Canada to denote professionals with the same training.

Is the Information Based on Well-Designed, Accurately Interpreted Research Studies?

If the source of the information seems reliable and cites a research study, ask if the study was well designed and if the results were interpreted accurately. Even well-designed, carefully executed, peer-reviewed experiments can be a source of misinformation if the experimental results are interpreted incorrectly or if the implications of the results are exaggerated. For example, the headline in Figure 1.22 states that WellShield improves overall health. However, the study investigates only cold symptoms; overall health is not addressed.

For some nutrition claims, not enough information is given to evaluate the validity of the studies on which they are based. Others, however, do provide the details of how a study was done. For example, the university study described in the WellShield ad is an intervention trial that compares the severity of cold symptoms and the cold recovery time (quantifiable parameters) of college students who took the WellShield supplement (experimental group) with the same parameters in a similar group of students who took a placebo (control group). The results indicate that the experimental group had milder cold symptoms and shorter recovery times than the control or placebo group. This supports the claim that the product can reduce cold symptoms. The ad, on the other hand, has not provided consumers with enough information to find the study or its abstract on Pubmed, so they can judge the claims for themselves.

Who Stands to Benefit?

Another important question to ask when judging nutrition claims is who stands to benefit from the information? If a person or company will profit from the information presented, you should be wary. Information presented in newspapers and magazines and on television may be biased or exaggerated because it helps sell magazines or boost ratings. Consider whether the claim is making a magazine cover or newspaper headline more appealing. Claims that are part of an advertisement should be viewed skeptically because advertisements are designed to increase product sales and the company stands to profit from your belief in that claim. For example, in an advertisement for a vitamin E supplement, a company may cite a study that shows that rats fed a diet high in vitamin E, that is, the vitamin E comes from food sources only, live longer than those consuming less vitamin E. However, the fact that a diet high in vitamin E increased longevity does not mean that Vitamin E supplements will have the same effect. In addition, this study was done in rats. Can the result be extrapolated to human health? Just because rats consuming diets high in vitamin E live longer does not mean that vitamin E supplements will extend human life. Information on the Internet is also likely to be biased toward a product or service if it comes from a site where you can buy the product.

Has It Stood the Test of Time?

Often the results of a new scientific study are on the morning news the same day they are published in a peer-reviewed journal. Sometimes this information is accurate, but a single study is never enough to develop a reliable theory. Results need to be reproducible before they can be used to make dietary recommendations. Headlines based on a single study should therefore be viewed skeptically. The information may be accurate, but there is no way to know because there has not been time to repeat the work and reaffirm the conclusions. If, for example, someone has found the secret to easy weight loss, the information will undoubtedly appear again if the finding is valid. If it is not, it will fade away with all the other weight loss cures that have come and gone.

CASE STUDY OUTCOME

Kaitlyn found it hard to eat a healthy diet living in residence for the first time. Her choices were limited to the meals in the cafeteria and it was difficult to figure out what a healthy diet was. Now that she's read this chapter, she knows that a healthy diet includes a variety of foods. She aims for an assortment of choices from within each of the food groups. She recognizes that eating a healthy diet involves using moderation, so she doesn't consume too much of any one food and avoids calorie-rich foods high in fat and sugar. Kaitlyn now knows that she can occasionally have study snacks like potato chips, as long as she selects nutrient-dense foods most of the time. She is gaining confidence in her ability to balance her food intake with her activity. In addition, her understanding of the scientific method and how to evaluate nutrition information has given her the tools she needs to make decisions about following fads and using dietary supplements. Now, in the middle of her second semester and armed with a better understanding of nutrition, Kaitlyn is maintaining a healthy weight and using the principles of adequacy, variety, balance, kcalorie control, and moderation to choose a nutrient-dense diet that minimizes her risks of developing heart disease and high blood pressure.

APPLICATIONS

iProfile

Personal Nutrition

1. How healthy is your diet?

   a. How many different vegetables and fruits did you eat today? How about this week? If you average fewer than five a day, make some suggestions that would increase the fruit and vegetable variety in your diet.

   b. How often do you eat treats such as a doughnut or an extravagant dessert? Are you exercising regularly?

   c. Do you order large portions? Use iProfile to look up how many more kcalories are in a large burger, fries, and drink than in a medium-sized order.

   d. Do you ever eat foods right out of the package? If you do, it is hard to tell how much you really ate. Suggest some things you could do to control your portion size.

2. What factors affect your food choices?

   a. List four food items you ate today or yesterday.

   b. For each food listed, indicate the factor or factors that influenced your selection of that particular food. For example, if you ate a candy bar before your noon class, did you choose it because it was available in the vending machine outside the lecture hall, because you did not have enough money for anything else, because you just like candy bars, because you were depressed, because all of your friends were eating them, because you believe it is good for you, or for some other reason?

   c. For each food, indicate what information you used in making the selection. For example, did you read the label on the product, or consider something you had read or heard recently in the news media?

   d. List three factors that commonly influence your food choices.

   e. List three types of information you regularly use to make your food choices.

General Nutrition Issues

1. **The Good Heart Study is evaluating the relationship between a vitamin supplement and heart disease. The study involves 200 participants, who are randomized into two equal groups. Pills identical in appearance are administered to all participants. Group 1 receives two tablets per day of the supplement and group 2 receives two placebo tablets per day. After 1 month, 50 subjects in group 1 have blood cholesterol levels at or below the recommended level. In group 2, 47 of the subjects have blood cholesterol levels at or below the recommended level. Statistical analysis indicated that there was no statistically significant difference between the two groups.**

   a. Is the study blinded?

   b. What does randomization mean and why is it important?

   c. Does the vitamin supplement lower cholesterol?

2. **Does this nutritional supplement live up to its claims?**

   a. Examine a nutritional supplement ad provided by your instructor or select one from a health- or fitness-related magazine or website. What is the claim made about this product, and is it believable?

   b. Does the ad refer to any research studies? If so, do they seem well controlled? Were the results based on objective measurements? Were the conclusions consistent with the results obtained? Were they published in peer-reviewed journals? Can you find the abstract for the study in Pubmed?

   c. Were claims based on anecdotal reports of individual users?

   d. Who stands to benefit if you spend money on this product?

   e. Based on this ad alone, would you choose to take this supplement? Why or why not?

SUMMARY

1.1 Nutrition and the Canadian Diet

- Nutrition is a science that encompasses all the interactions that occur between living organisms and food. Canadians today are eating more fast food, processed foods, and prepared foods and spending less time preparing meals and eating at home than 50 years ago. This is affecting the healthfulness of the diet.
- Many Canadians are not eating foods that meet the recommendations for a healthy diet. This contributes to the incidence of chronic diseases such as diabetes, obesity, and heart disease.

1.2 Food Provides Nutrients

- About 45 nutrients are essential to human life. Nutrients consumed come from those naturally present in foods, those added to fortified foods, and those contained in natural health products, such as vitamin and mineral supplements. In addition to nutrients, food provides phytochemicals and zoochemicals, and nonessential substances, which may provide health benefits. There are six classes of nutrients: carbohydrates, lipids, proteins, water, vitamins, and minerals.
- Food contains nutrients that are needed by the body for growth, maintenance and repair, and reproduction. Carbohydrates, lipids, and proteins are energy-yielding nutrients. The energy they provide to the body is measured in kcalories or kjoules. Carbohydrates, lipids, protein, water, and minerals provide structure to the body, and all nutrient classes help regulate the biochemical reactions of metabolism to maintain homeostasis.
- When energy or one or more nutrients are deficient or excessive in the diet, malnutrition may result. Malnutrition includes both

undernutrition and overnutrition. Undernutrition is caused by a deficiency of energy or nutrients. Overnutrition may be caused by a toxic dose of a nutrient or chronic over-consumption of energy or of nutrients that increases the risk of chronic disease. Depending on the cause, the symptoms of malnutrition can occur in the short term or over the course of many weeks, months, or even years.

- The diet you consume can affect your genetic predisposition for developing a variety of chronic diseases.

1.3 Food Choices for a Healthy Diet

- Food choices are affected by food availability, sociocultural influences, personal tastes, emotional factors, and what we think we should eat to stay healthy. No one food choice is good or bad, and no one choice can make a diet healthy or unhealthy—each choice contributes to the diet as a whole.
- A healthy diet includes a variety of nutrient-dense foods from each food group as well as a variety of foods from within each group. It balances energy and nutrient intake with needs and moderates choices to keep intakes of energy, fat, sugar, salt, and alcohol within reason.

- Go to WileyPLUS to view a video clip on the local food movement.

1.4 Understanding Science Helps Us Understand Nutrition

- The science of nutrition uses the scientific method to determine the relationships between food and the nutrient needs and health of the body. The scientific method involves making observations of natural events, formulating hypotheses to explain these events, designing and performing experiments to test the hypotheses, and developing theories that explain the observed phenomena based on the experimental results.
- To be valid, nutrition information should be based on experiments that use quantifiable measurements, the right type and number of experimental subjects, are of appropriate duration, are carefully analyzed using statistical methods, have results that are correctly interpreted, and are of sufficient quality to be published in a peer-reviewed journal.

1.5 Nutrition Research

- The science of nutrition uses many different types of experimental approaches to determine nutrient functions and requirements. Observational studies identify relationships between diet and health. Two types of observational studies are the case control study and the prospective cohort study. Intervention trials can test hypotheses, many developed from the results of observational studies. Laboratory studies use biochemical and molecular methods to study whole organisms or cells.
- Ethical guidelines protect humans and animals involved in research studies, but limit the type of experiments that can be done.
- Pubmed is an internet database that can be used to access abstracts of research papers on nutrition.

1.6 Sorting Out Nutrition Information

- When judging nutrition claims, first consider whether the information makes sense and whether it comes from a reliable source, such as educational institutions, government and nonprofit organizations. Individual testimonies cannot be trusted because they have not been tested by experimentation.
- If information is based on experimentation, determine if the studies were well designed and accurately interpreted.
- Information that promotes a product or in any other way benefits the person or organization providing it should be viewed with skepticism.
- Accurate information will be supported by more than a single research study.

REVIEW QUESTIONS

1. How does the typical Canadian diet compare to recommendations for a healthy diet?
2. What does the science of nutrition study?
3. What is an essential nutrient?
4. List six classes of nutrients and indicate which provide energy.
5. List three functions provided by nutrients.
6. List three ways in which what you eat today can affect your health.
7. What is malnutrition?
8. List three factors other than biological need that influence what we eat.
9. Why is it important to choose a variety of foods?
10. How does moderation help maintain a healthy weight?
11. List the steps of the scientific method.
12. What is a control group?
13. What is a placebo?
14. What is a double-blind study?
15. What are the characteristics of an observational study?
16. What are the characteristics of a prospective cohort study?
17. What are the characteristics of an intervention trial?
18. What is a confounding factor? How are they dealt with in an observational study? In an intervention trial?
19. Why are animals used to study human nutrition?
20. What factors should be considered when judging nutrition claims reported in the media?

REFERENCES

1. Garriguet, D. Canadians' Eating Habits *Health Reports* Vol 18 (2). Statistics Canada Cat 82-003 May 22, 2007 pp. 17-32. Available online at http://www.statcan.gc.ca/pub/82-003-x/2006004/article/habit/9609-eng.pdf. Accessed April 21, 2011.
2. Garriguet, D. 2004. Overview of Canadians' Eating Habits. Statistics Canada no: 82-620-MIE-No 2. Available online at http://www.statcan.gc.ca/pub/82-620-m2006002-eng.pdf. Accessed April 21, 2011.
3. Guo, G., Willows, N., Kuhle, S., et al. Use of vitamin and mineral supplements among Canadian adults. *Can. J. Public Health* 100(5): 357-360, 2008.
4. Tjepkema, M. Measured Obesity: Adult obesity in Canada: Measured height and weight. Nutrition findings from the Canadian Community Health Survey 2-620-MWE Issue No 1. Component of Statistics Canada Catalogue #: 82-620-MWE 2005001. Available

online at http://www.statcan.gc.ca/pub/82-620-m/2005001/article/adults-adultes/8060-eng.htm. Accessed April 21, 2011.

5. Diet, Nutrition and the Prevention of Chronic Diseases. Report of a Joint WHO/FAO Expert Consultation. Available online at http://whqlibdoc.who.int/trs/WHO_TRS_916.pdf. Accessed May 14, 2009.
6. Kaput, J. Nutrigenomics–2006 update. *Clin. Chem. Lab. Med.* 45 :279–287, 2007.
7. American Dietetic Association. Nutrition and You: Trends 2008. Available online at www.eatright.org/trends/2008. Accessed May 14, 2009.
8. Portion Distortion. U.S. Department of Health and Human Sciences. Available online at http://hp2010.nhlbihin.net/portion/. Accessed March 5, 2011.
9. Dawber, T.R., Meadors, G.F., Moore, F.E., Jr. Epidemiological approaches to heart disease: the Framingham Study *Am J Public Health Nations Health*. 41(3):279-81, 1951.
10. Feinleib, M., Kannel, W.B., Garrison, R,J., et al. The Framingham Offspring Study. Design and preliminary data. *Prev Med*. 4(4): 518-25, 1975.
11. Parikh, N.I., Hwang, S.J., Larson, M.G., et al. Parental occurrence of premature cardiovascular disease predicts increased coronary artery and abdominal aortic calcification in the Framingham Offspring and Third Generation cohorts. *Circulation*. 116(13):1473-81, 2007.
12. World Cancer Research Fund/American Institute for Cancer Research. *Food, Nutrition, Physical Activity and the Prevention of Cancer: a Global Perspective* . Washington DC. AICR 2007. Available online at www.dietandcancerreport.org. Accessed Jan 28, 2010.
13. Mensink, R.P., Zock, P.L., Kester, A.D., et al. Effects of dietary fatty acids and carbohydrates on the ratio of serum total to HDL cholesterol and on serum lipids and apolipoproteins: a meta-analysis of 60 controlled trials. *Am. J. Clin. Nutr.* 77(5):1146-55, 2005.

2 Nutrition Guidelines:

APPLYING THE SCIENCE OF NUTRITION

Case Study

Li was in his first year at university. Away from home and having to make his own food choices at the dorm cafeteria, Li was overwhelmed. Should breakfast be oatmeal, doughnuts, an omelette, or just fruit? He liked burgers and fries for lunch, but maybe a salad and a cold sandwich were better options. Dinner offered even more choices—there was always some type of meat dish, a pasta choice, fish, and a vegetarian entrée. For dessert there were cakes, pies, and big vats of scoop-your-own ice cream to tempt him. Li wondered whether there were guidelines that could help him decide how much and what types of food he should be eating, so he decided to ask a professor in the university's department of nutritional sciences.

The professor suggested that he visit the Health Canada website (www.hc-sc.gc.ca) to find out how to choose a healthy diet. There he could learn about *Eating Well with Canada's Food Guide* and discover how many servings of vegetables and fruit, meat and meat alternatives, grains products, and milk and milk alternatives he would need to consume to meet his needs based on his age and sex. The professor also advised him that making healthy choices from each food group and choosing a variety of foods were just as important as the amounts of food he ate. If he chose foods that were high in fat and sugar, he might exceed his energy needs even if he ate the right number of servings from each group. The professor explained which foods Li should emphasize to maximize his nutrient intake. He also showed him how to use food labels to choose foods to meet his goals.

This chapter will detail how you can use nutrition guidelines and food labels to make nutritious food choices.

(© Zhang Bo/IstockPhoto)

(Burke/Triolo Productions/FoodPix)

Chapter Outline

2.1 Nutrition Recommendations for the Canadian Diet

Learning Objectives

- List some reasons why population-wide nutritional recommendations are developed.
- Describe the two approaches that are taken to formulate nutrition recommendations.

People need to eat to survive, but health-conscious individuals want to do more than survive. They want to choose diets that will optimize their health. An optimal diet would contain just the right amount of each nutrient to prevent deficiencies and maintain health, and, for some people, to maintain a healthy pregnancy or to allow for growth. What is optimal for each individual also depends on a variety of other factors, such as their genetic background, their activity level, and the other nutrients in their diet (**Figure 2.1**).

Figure 2.1 An individual's nutritional needs depend on age, gender, and genetic makeup. (Bob Thomas/Stone/Getty Images)

The science of nutrition has determined which nutrients are necessary to keep humans alive and how the amounts needed vary at different stages of life, such as pregnancy or infancy. But it is not currently possible to determine the optimal amount of each nutrient that should be included in the diet of each individual. Instead, the methods of science have been used to establish general recommendations for the types and amounts of nutrients that will maintain the health of individuals and populations. To be useful to health-conscious consumers, these amounts have been translated into recommendations about food choices. These recommendations are also used as a standard of comparison to assess whether populations and individuals are consuming diets that promote health.

Early Nutrition Recommendations

Some of the first nutrition recommendations were made in England in the 1860s, when the Industrial Revolution caused a rise in urban populations, leading to large numbers of homeless and hungry people. The government wanted to know the least expensive way to keep these people alive and maintain the workforce. As a result, a dietary standard was established based on what the average working person ate in a typical day. This method of estimating nutrient needs was used until World War I, when the British Royal Society made specific recommendations about foods that not only would sustain life but also would be protective of health. They recommended that fruits and green vegetables be included in a healthy diet, and that milk be included in the diets of all children. Since then, the governments of many countries have established their own sets of dietary standards based on the nutritional problems and dietary patterns specific to their populations and the interpretations of their scientists. Generally, the differences between guidelines from country to country are small. The World Health Organization and the Food and Agriculture Organization of the United Nations, organizations concerned with international health, publish a set of dietary standards to apply worldwide[1] (see Appendix F).

CANADIAN CONTENT

Nutrition Recommendations in Canada

dietary pattern A description of a way of eating that includes the types and amounts of recommended foods and food groups, rather than individual nutrients.

There are two major approaches to formulating nutrition recommendations. One way is a nutrient-based approach, which describes the amounts of individual nutrients that are needed (e.g., how much vitamin C a person needs to consume). The other approach is a food-based approach in which a **dietary pattern** is recommended. A dietary pattern describes the amounts and types of food to eat to ensure an adequate intake of nutrients; for example, it describes the number of servings of vegetables and fruits a person should eat. This ensures adequate intake of many nutrients, such as vitamin C, potassium, and fibre, which are all abundant in vegetables and fruits. The food-based approach also makes recommendations on foods to eat (and foods to avoid) to reduce the risk of chronic disease (e.g., avoid foods high in sugar, salt, or fat).

All recommendations, whether food- or nutrient-based, are updated periodically as the science of nutrition evolves and new information becomes available.

The first nutrient-based nutrition standards in Canada were published in 1939 and were periodically updated. In 1983, they were named the Recommended Nutrient Intakes (RNIs). A similar process took place in the United States over the same time period. The American standards were called Recommended Dietary Allowances (RDAs) and were first published in 1943. Recommendations were made for the intake of kcalories and nutrients at risk for deficiency—protein, vitamins, and minerals. Levels of intake were based on amounts that would prevent nutrient deficiencies.[2] Over the years since these first standards were developed, our knowledge of nutrient needs has increased and patterns of dietary intake and disease have changed. Overt nutrient deficiencies are now rare in both Canada and the United States, but the incidence of diet-related chronic diseases, such as heart disease, cancer, and obesity, has increased. In response to these changes in our diet and health, a process began in the early 1990s in which scientists from both Canada and United States collaborated to develop new nutrition standards to replace both the RDAs in the United States and the RNIs in Canada. This new set of energy and nutrient intake recommendations is called the **Dietary Reference Intakes, or DRIs**. These are designed to promote health as well as prevent nutrient deficiencies. This joint American–Canadian effort has helped to standardize recommendations for North America.[3]

dietary reference intakes (DRIs) A set of reference values for the intake of energy, nutrients, and food components that can be used for planning and assessing the diets of healthy people in the United States and Canada.

In addition to the DRIs, Canada's Food Guide, a food-based dietary pattern, has been a central tool in the promotion of nutritious eating. The first guide (**Figure 2.2**) was called *Canada's Official Food Rules* and was developed in 1942. This was during wartime, when food rationing was in place, and the government wanted to optimize the nutrient intake of Canadians during a time of food shortages. In 1961, Canada's Food Rules became Canada's Food Guide and the guide has been updated several times since then. Each modification incorporated new nutritional knowledge about the relationship between food choices and health (see Appendix D). The most current revision took place in 2007. *Eating Well with Canada's Food Guide*, the full title of the latest version of Canada's Food Guide, promotes a food-based dietary pattern, that is, its intent is to show people how to obtain all of their nutrients and promote good health through the proper selection of a variety of foods.[4] Canada's Food Guide represents one dietary pattern. We will discuss another dietary pattern, the Mediterranean diet (see Chapter 1: Science Applied: From "Seven Countries to the Mediterranean Diet"), later in this chapter.

CANADA'S OFFICIAL FOOD RULES

These are the Health-Protective Foods

Be sure you eat them every day in at least these amounts.

(Use more if you can)

MILK—Adults—½ pint. Children-more than 1 pint. And some CHEESE, as available.

FRUITS—One serving of tomatoes daily, or of a citrus fruit, or of tomato or citrus fruit juices, and one serving of other fruits, fresh, canned or dried.

VEGETABLES (In addition to potatoes of which you need one serving daily)—Two servings daily of vegetables, preferably leafy green, or yellow, and frequently raw.

CEREALS AND BREAD—One serving of a whole-grain cereal and 4 to 6 slices of Canada Approved Bread, brown or white.

MEAT, FISH, etc.—One serving a day of meat, fish, or meat substitutes. Liver, heart or kidney once a week.

EGGS—At least 3 or 4 eggs weekly.

Eat these foods first, then add these and other foods you wish.

Some source of Vitamin D such as fish liver oils, is essential for children, and may be advisable for adults.

Figure 2.2 Canada's Official Food Rules were developed in 1942 to promote nutritious eating during difficult wartime rationing.
Source: Health Canada. *Canada's Food Guide from 1942 to 1992.* Available online at www.hc-sc.gc.ca/fn-an/food-guide-aliment/context/fg_history-histoire_ga-eng.php#1942. Accessed April 22, 2011.

Canada's Food Guide also advises Canadians to use food labels to compare foods and make better food choices. We will see how food labels show consumers which nutrients are provided by packaged foods and how the amounts contained in a serving compare with the recommendations for an overall healthy diet.

2.2 Dietary Reference Intakes

Learning Objectives

- Describe the types of nutrient intake recommendations included in the Dietary Reference Intakes (DRIs) and explain the purpose of each.
- Explain how a requirement distribution can be used to determine the probability that an individual is meeting his or her requirement.
- Explain how the EAR cutpoint method can be used to determine the proportion of a population that is meeting its requirements.
- List the variables used to calculate the EERs.
- Describe some applications of DRIs.

Life Cycle

The DRIs are designed to be used for planning and assessing the diets of healthy people. They do not apply to people who are ill. They include recommendations for energy, carbohydrate, fat, protein, and micronutrients, including fat-soluble vitamins A, D, E, and K, water-soluble vitamins such as the B vitamins and vitamin C, and minerals such as calcium, phosphorus, magnesium, fluoride, selenium, iron, zinc, copper, sodium, and potassium.[5-7]

The DRIs include values for different **life-stage groups**. These have been established to account for the physiological differences among infants, children, adolescents, adults, older adults, and pregnant and lactating women. The pregnancy and lactation recommendations are divided into age categories to distinguish the unique nutritional needs of pregnancy and lactation in teenagers and older mothers (see inside cover). Differences between males and females are also taken into account.

life-stage groups Groupings of individuals based on stages of growth and development, pregnancy, and lactation, that have similar nutrient needs.

recommended dietary allowances (RDAs) Intakes that are sufficient to meet the nutrient needs of almost all healthy people in a specific life-stage and gender group.

adequate intakes (AIs) Intakes that should be used as a goal when no RDA exists. These values are an approximation of the average nutrient intake that appears to sustain a desired indicator of health.

tolerable upper intake levels (ULs) Maximum daily intakes that are unlikely to pose a risk of adverse health effects to almost all individuals in the specified life-stage and gender group.

Nutrient Recommendations

The DRIs for macronutrients and micronutrients include four different sets of reference values:[5]

1. Estimated average requirement (EAR)
2. Recommended dietary allowance (RDA)
3. Adequate intake
4. Tolerable upper intake level (UL)

These values are set for each life-stage group. Two sets of values, the **Recommended Dietary Allowances (RDAs)** and the **Adequate Intakes (AIs)**, can be used to set goals for individual intake and can be used to plan or evaluate individual diets, while the **Tolerable Upper Intake Levels (ULs)** help individuals prevent nutrient toxicities. The **estimated average requirement (EAR)** is used to determine the RDA (this will be explained immediately below) and is used to evaluate the adequacy of nutrient intakes for groups of people or populations.

estimated average requirements (EARs) Intakes that meet the estimated nutrient needs of 50% of individuals in a gender and life-stage group.

criterion of adequacy A functional indicator, such as the level of a nutrient in the blood, that can be measured to determine the biological effect of a level of nutrient intake.

Establishing Estimated Average Requirements (EARs) In order to understand how these DRIs are used, it is important to understand a bit about how they are determined. An EAR is the amount of a nutrient that is estimated to meet the needs of 50% of people in the same sex and life-stage group. To set an EAR, scientists must establish a measurable marker of adequacy, based on an understanding of the function of the nutrient in the body. This **criterion of adequacy** is typically the activity of an enzyme, the amount of nutrient stored in the body, the amount of nutrient excreted in the urine, or the level of a nutrient or metabolite in the blood, which can be evaluated to determine the biological effect of a level of nutrient intake. Appropriate criteria of adequate intake must be established for each nutrient in each life-stage and sex group.

Let us consider a hypothetical experiment to determine the EAR for a nutrient we will call vitamin X. The criterion of adequacy for vitamin X is saturation of the vitamin in the blood. Saturation in the blood means that as someone's dietary intake of vitamin X increases, the blood level of vitamin X increases up to a certain point and then levels off. Once saturation is reached, increasing the dietary intake of vitamin X will not result in a further increase of the amount of the vitamin in the blood. This saturation occurs because the body has all the vitamin X that it requires to maintain health and any additional vitamin X is being excreted or eliminated from the body. So the intake of vitamin X, at which saturation in the blood occurs, represents the requirement for vitamin X.

In order to determine the EAR for vitamin X, scientists can perform a depletion-repletion experiment (this was described briefly in Section 1.5). A group of people, at the same life stage, are fed a vitamin-X-free diet until there is virtually no vitamin X detectable in their blood (this is the depletion phase of the experiment). As soon as this happens, the subjects are re-fed (the repletion phase of the experiment) and the blood levels of vitamin X increase until saturation is reached. (**Figure 2.3** shows both phases of the experiment.)

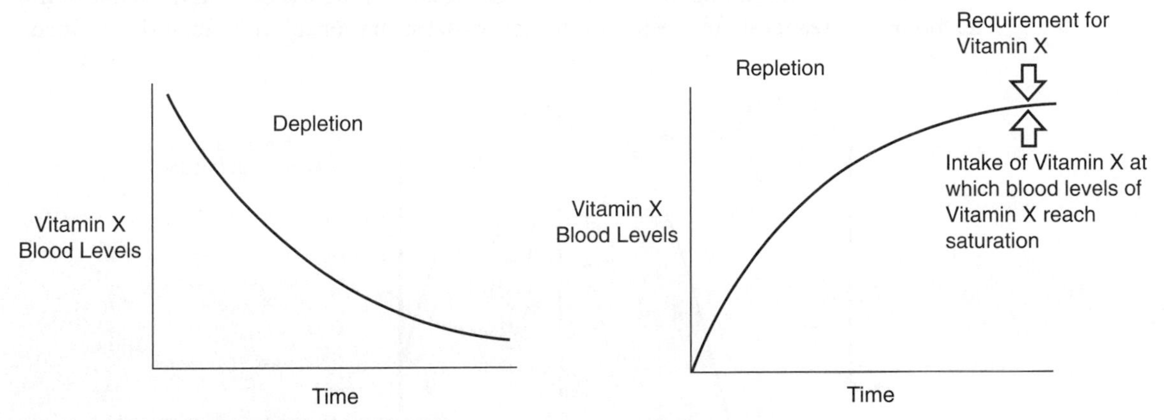

Figure 2.3 Depletion-repletion experiment
A depletion-repletion experiment can be conducted to determine nutrient requirements. See text for more details.

For each person in the experiment, the vitamin X intake at which their blood levels reach saturation is the requirement of vitamin X for that person. Because of genetic and biological differences between people, no two individuals in the depletion-repletion experiment will have exactly the same requirement. When you plot the requirements, you typically get a **requirement distribution** like that shown in **Figure 2.4**. The shape of the distribution is called a bell curve or a normal (or binomial) distribution. The vertical line marking the centre of the requirement distribution divides the group in half. Fifty percent of people in the repletion-depletion experiment have requirements below this value (as shown by the area with lines in figure 2.4) and 50% have requirements above this value. This centre line is the estimated average requirement (EAR), that is, the nutrient intake that meets the nutrient requirement for 50% of the population in the life stage measured in the experiment.

requirement distribution A plot of the nutrient requirements for a group of individuals in the same life stage. Typically, the plot has the shape of a bell curve, i.e., a normal or binomial distribution.

How can the results of this experiment be applied to the general population? Obviously, most people do not participate in depletion-repletion experiments and never know their exact

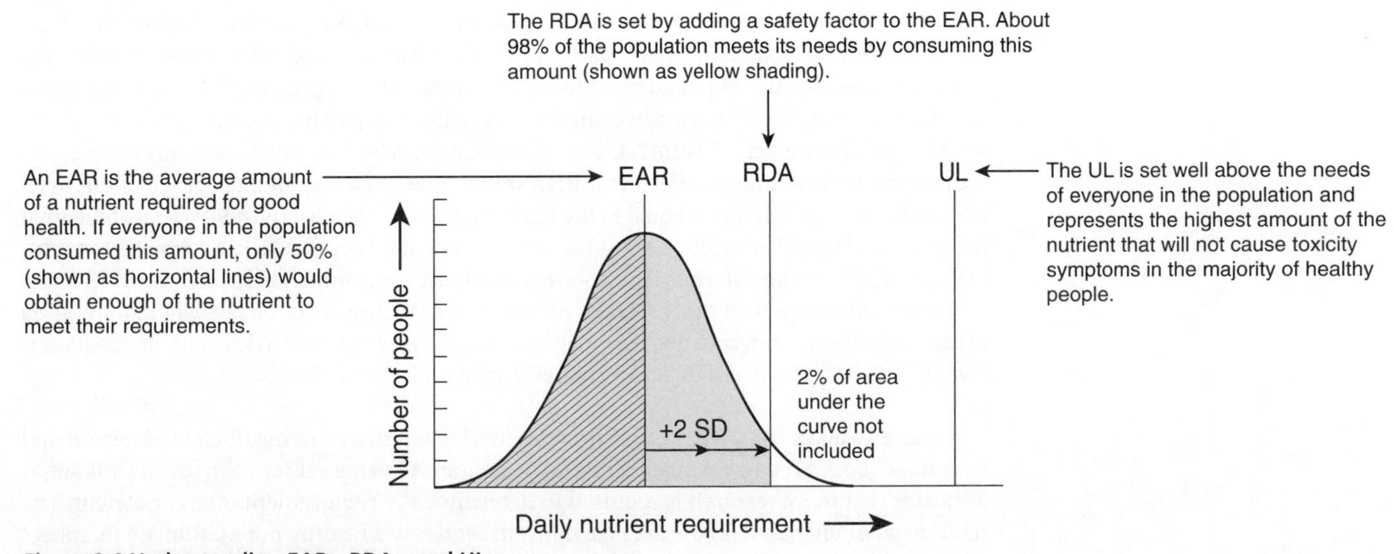

Figure 2.4 Understanding EARs, RDAs, and ULs
EARs and RDAs are determined by assessing the different amounts of nutrients required by different individuals in a population group and plotting the resulting values. Because a few individuals in the group need only a small amount, a few need a large amount, and most need an amount that falls between the extremes, the result is a bell-shaped curve like the one shown here.

nutrient requirement. What the requirement distribution does allow an individual to determine is the *probability* that they are meeting their requirement. For example, if an individual has an *intake* equal to the EAR, there is a 50% *probability* that they are meeting their *requirement.*

The probability of meeting a nutrient requirement can be calculated for any nutrient intake, as long as you have a requirement distribution. For example, in **figure 2.5**, if an individual has a usual intake equal to the vertical line, then the probability that they are meeting their nutrient requirements is equal to the shaded area under the requirement distribution for that nutrient. In the example shown in the figure, this is equal to 84%. (Note that the mathematics involved in determining that the area under the curve is 84% is beyond the scope of this textbook. The reader should recognize, however, that such probabilities can be determined.)

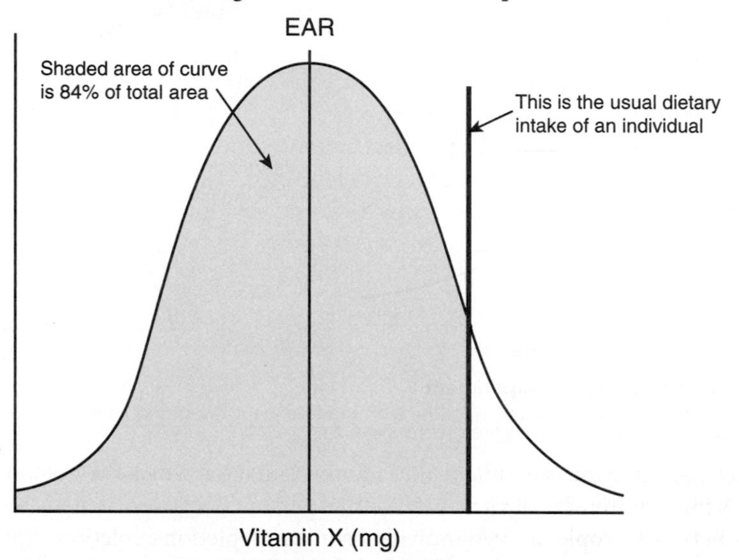

Figure 2.5 Determining the probability that an individual is meeting their requirement
Given someone's usual intake for a nutrient, the probability that they are meeting their requirement can be calculated. This is the requirement distribution for vitamin X. The red line represents the usual dietary intake of vitamin X for an individual. The probability that this person's intake is meeting their requirement is indicated by the proportion of the curve to the left of the red line or in the example here, 84%.

Establishing Recommended Dietary Allowances (RDAs) From the discussion of nutrient requirements, it should be apparent that the best way that individuals can optimize their health is to maximize the probability that they are meeting their requirement for any particular nutrient. The Recommended Dietary Allowance or RDA is the dietary reference intake that allows individuals to do this. This is because the RDAs are intakes that meet the nutrient requirements, not of 50% of healthy individuals as is the case with the EAR, but almost 98% of individuals in a population (the exact value is between 97%-98% of the population). An RDA is therefore a higher value than the EAR. An RDA is determined by starting with the EAR value and using a statistic called the standard deviation (SD). Standard deviation is a measure of the range or width of the requirement distribution curve. The RDA is mathematically equal to the EAR plus two standard deviations (RDA = EAR + 2SD) as shown in Figure 2.4. The value of almost 98% also corresponds, in Figure 2.4, to the area shaded in yellow to the left of the RDA vertical line on the requirement distribution curve. If an individual had an intake equal to the RDA, there would be an almost 98% probability that the individual would be meeting his or her requirement. It is because the RDA represents such a high probability for meeting nutrient requirements that it is recommended as an individual target intake. An intake less than the RDA does not necessarily indicate that the nutrient requirements of that particular person have not been met; however, the probability of a deficiency inadequate is lowest if intake meets the RDA, and increases as intake falls below the RDA.[7]

Adequate Intakes (AIs) These are estimates used when there are insufficient experimental data to set an EAR and calculate an RDA. When an AI value rather than an RDA is set, it indicates that more research is required to determine the requirement of that nutrient. The AIs are generally based on the average nutrient *intake* by a healthy population, all members of which are meeting nutrient requirements, by an established criteria of adequacy. For example, the AI for calcium intake for young infants is based on the average calcium intake of infants that are exclusively breastfed. It is assumed that breast milk provides adequate levels of calcium for the developing infant. A healthy individual whose intake of a specific nutrient is at or above the AI is unlikely to have an inadequate intake of that nutrient, therefore, like the RDA, the AI is a useful individual goal. If the intake is below the AI, it may or may not be adequate, but since the true requirement distribution of the nutrient is unknown, it is not possible to calculate any probabilities of adequacy as can be done when an EAR is available.

Tolerable Upper Intake Levels (ULs) The fourth set of values, the Tolerable Upper Intake Levels, represent the maximum level of daily intake of a nutrient that is unlikely to pose a risk of adverse health effects to almost all individuals in the specified group. These are not recommended levels, but levels of intake that can probably be tolerated (see Figure 2.4). ULs are used as a guide for limiting intake when planning diets and evaluating the possibility of overconsumption. The exact level of intake that will cause an adverse effect cannot be known with certainty for each individual, but if a person's intake is below the UL, there is good assurance that an adverse effect will not occur.

To establish a UL, a specific adverse effect or indicator of excess is considered. The lowest level of intake that causes the adverse effect is determined, and the UL is set far enough below this level that even the most sensitive people in the population are unlikely to be affected. If adverse effects have been associated only with intake from supplements, the UL is based only on this source. Therefore, for some nutrients, these values represent intake from supplements alone; for some, intake from supplements and fortified foods; and for others, total intake from all sources, that is, food, fortified food, water, nonfood sources, and supplements. For many nutrients, data are insufficient to establish a UL value.

Using the EAR to Determine the Prevalence of Adequate Intake Within a Group The descriptions given above demonstrate how the DRIs can be used to set individual targets and to determine the probability that an individual's intake is adequate. While this is useful for an individual making personal food choices, nutrition researchers want to have information, not on individuals, but on how well large groups or populations are meeting their nutrient requirements. Fortunately, it is possible to determine the adequacy of a population's intake using a method called the **EAR cutpoint method.**[7] Although the mathematical validity of the method requires a complicated proof, the application of the method is straightforward. When evaluating a population, the nutrient intake of many individuals in that population is determined and a distribution of intake is plotted. This **intake distribution** has a median intake (see **Figure 2.6**). Most intake distributions are similar to but not exactly normal distributions.

EAR cutpoint method A method that indicates the proportion of a population that is not meeting its requirements, indicated by the proportion of the population with intakes below the EAR.

intake distribution A plot of the intakes of a specific nutrient in a population.

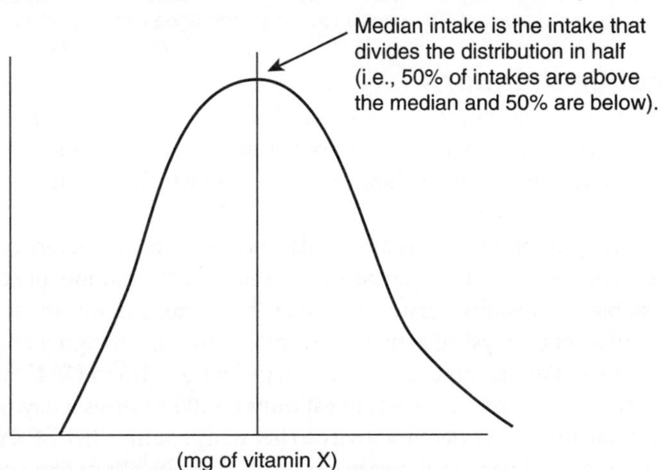

Figure 2.6 An intake distribution. The usual intake of vitamin X is measured for a large number of people and these intakes are then plotted to produce the distribution shown.

To use the EAR cutpoint method, a vertical line equivalent to the EAR for the nutrient is drawn on the *intake distribution*. The area under the curve to the left of the EAR line, or the proportion of the population whose intake is *less* than the EAR, is the proportion of the population that is *not* meeting its requirement. Note that we cannot specifically identify which individuals are not meeting their requirements, we can only describe the proportion of the population that has an adequate or inadequate intake. **Figure 2.7**, which describes the EAR cutpoint method, shows two intake distributions. In (a), the population is poorly nourished with respect to vitamin X, with 78% not meeting their requirement. When considering populations, a desirable goal is to have a median intake that ensures that 90%, or more, of the population is meeting their requirement, that is, when using the EAR cutpoint method, no more than 10% of the curve is to the left of the EAR vertical line, as shown in (b).

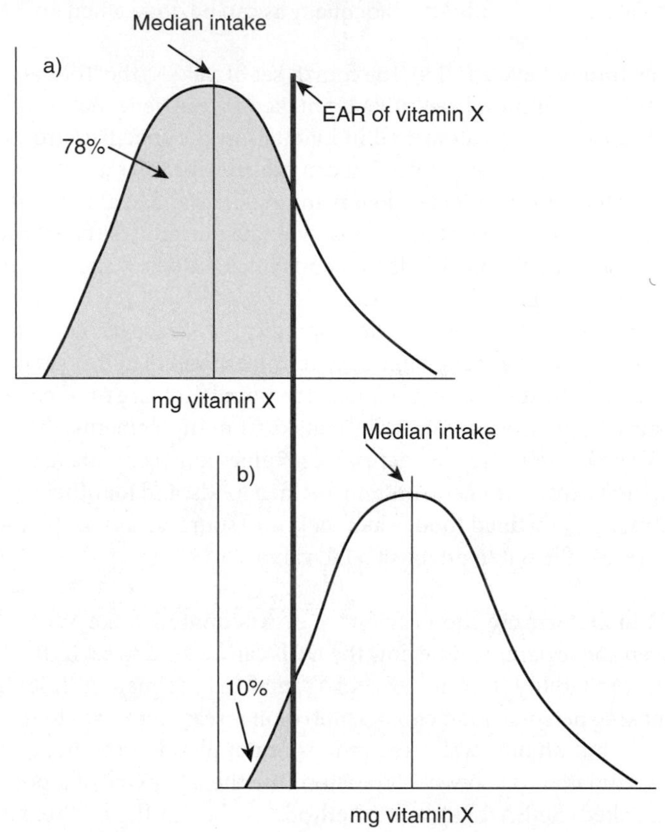

Figure 2.7 EAR cutpoint method
Two intake distributions are shown (a) and (b). The red line indicates the position of the EAR (mg of Vitamin X). (a) In this intake distribution, the grey-shaded area represents the proportion of the population not meeting their requirement, or 78%. This population is poorly-nourished with respect to vitamin X. (b) In this distribution, only 10% of the population is not meeting their requirement or conversely 90% of the population is meeting their requirement. This population is well-nourished with respect to vitamin X.

Energy Recommendations

The DRIs make two types of recommendations regarding energy intake. One provides an estimate of how much energy is needed to maintain body weight and the other provides information about the proportion of each of the energy-yielding nutrients from which this energy should come.

estimated energy requirements (EERs) Average energy intakes predicted to maintain body weight in healthy individuals.

Estimated Energy Requirements The recommendations for energy intake are called **Estimated Energy Requirements (EERs)**.[8] They can be used to calculate the number of kcalories needed to keep weight stable in a healthy person. Variables in the calculations include age, gender, weight, height, and level of physical activity (see inside cover). Changing any of these variables changes the EER. For example, a 17-year-old girl who is 1.6 m (5′ 4″) tall and weighs 57 kg (127 lbs) and gets no exercise needs to eat only 1,730 kcalories a day to maintain her weight. If she adds an hour of moderate activity to her daily routine, her EER will increase to 2,380 kcalories and she will need to increase her food intake by about 650 kcalories per day to maintain her weight. If she grows taller or gains weight, these factors will also increase her energy needs. (The EERs are discussed in more depth in Section 7.3.)

acceptable macronutrient distribution ranges (AMDRs) Ranges of intake for energy-yielding nutrients, expressed as a percentage of total energy intake, that are associated with reduced risk of chronic disease while providing adequate intakes of essential nutrients.

Acceptable Macronutrient Distribution Ranges (AMDRs) The proportion of each of the energy-yielding nutrients in the diet is just as important as the total amount of energy consumed. Therefore, the DRIs make recommendations for the proportions of carbohydrate, fat, and protein that make up a healthy diet. These are called **Acceptable Macronutrient Distribution Ranges (AMDRs)**. These recommendations are expressed as ranges because healthy diets can contain many different combinations of carbohydrate, protein, and fat.[8] According to the AMDRs, a healthy diet for an adult can contain from 45% to 65% of kcalories from carbohydrate, 20% to 35% from fat, and 10% to 35% from protein. When kcalorie intake stays the same, changing the proportion of one of these will change the proportion of the others as well. So, for example, in a diet that provides 2,000 kcalories with 50% of kcalories from carbohydrate, the other 50% will come from protein and fat. If the kcalories are kept the same but carbohydrate intake is decreased, the percentage of fat and/or protein will increase (**Figure 2.8**). The AMDRs allow flexibility in food choices based on individual preferences while still providing a diet that minimizes disease risk. AMDR values have also been set for specific amino acids and fatty acids (Appendix A).

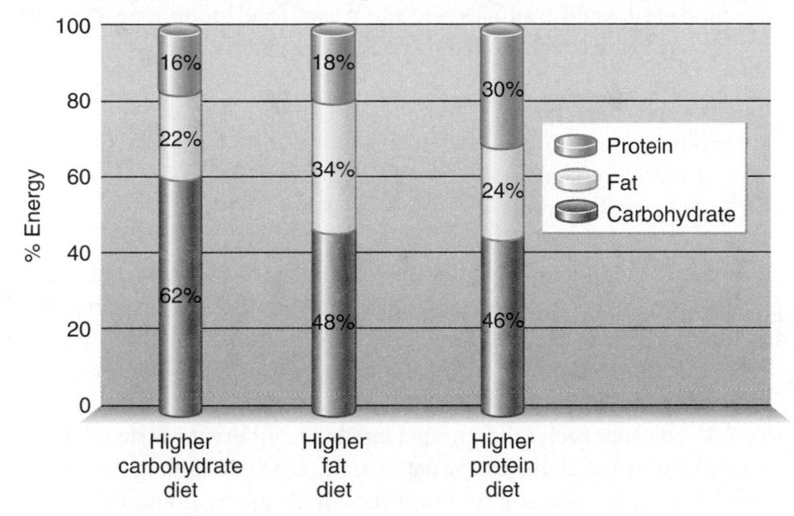

Figure 2.8 Proportions of energy-yielding nutrients
Changing the proportion of one of the energy-yielding nutrients in the diet changes the proportions of the others.

Applications of the Dietary Reference Intakes

The DRIs have many uses. They provide a set of standards that can be used to plan diets, to assess the adequacy of diets, and to make judgements about excessive intakes for individuals and populations.[7]

For example, they can be used as a standard for meals prepared for schools, hospitals, and other health-care facilities. They can be used to determine standards for food labelling and to develop practical tools for diet planning, such as food group systems. They can also be used to interpret information gathered about the food consumed by a population to help identify potential nutritional inadequacies that may be of public health concern.

Despite their many uses, dietary standards cannot be used to identify with certainty whether a specific person has a nutritional deficiency or excess. At best, only probabilities of adequacy can be determined. To evaluate an individual's nutritional status, dietary, clinical, biochemical, and body-size measurements are needed, as discussed in Section 2.6.

2.3 *Eating Well with Canada's Food Guide*

Learning Objectives

- List the four food groups in Canada's Food Guide.
- Give some examples of a Canada's Food Guide serving.
- Explain the purpose of the additional statements found in Canada's Food Guide.
- Describe the process used to ensure that Canada's Food Guide meets the nutrient requirements of most Canadians.
- Describe the online services available to Canadians to personalize Canada's Food Guide and assess their food intake.

While the DRIs establish the amounts of each nutrient that an individual should consume to ensure a high probability of nutritional adequacy, in order to obtain these nutrients we must consume food. Health Canada developed the document *Eating Well with Canada's Food Guide* to assist Canadians in choosing a healthy diet. It is a guide to a dietary pattern that has the characteristics of the healthy diet described in Section 1.3, that is, varied, balanced, moderate, kcalorie-controlled, and comprised of nutrient-dense foods.

The Guide: Page by Page

Page 1 *Eating Well with Canada's Food Guide* can be recognized by its rainbow graphic, which appears on the first page of this 6-page guide (**Figure 2.9**). Each arc of the rainbow represents one of the four food groups and examples of foods in each food group appear on each arc. The four food groups, which are colour-coded throughout the guide, are:

a. vegetables and fruits (green)
b. grain products (yellow)
c. milk and alternatives (blue)
d. meat and alternatives (red)

Figure 2.9 Canada's Food Guide Page 1 The first page of *Eating Well with Canada's Food Guide* is distinguished by its unique rainbow graphic.
Source: Health Canada. *Eating Well with Canada's Food Guide.* Available online at www.hc-sc.gc.ca/fn-an/food-guide-aliment/index-eng.php. Accessed Mar 5, 2011.

The size of the arc roughly corresponds to the number of servings recommended in each food group, so that the graphic indicates that Canadians should be eating primarily vegetables and fruit and grain products, and lesser amounts of milk and alternatives and meat and alternatives. For a nutritious diet, food from all four food groups should be consumed, since no single food group is a source of all nutrients.

As **Table 2.1** indicates, each food group contributes different nutrients which in combination can contribute to a healthy eating pattern. For example, among the most important nutrients provided by vegetables and fruits are vitamin A, folate, vitamin C, and potassium, while grain products, particularly whole grains, provide fibre, iron, B vitamins, and magnesium. Vegetable and fruit consumption, as well as high-fibre diets, have also been associated with reduced risk of several chronic diseases, such as cardiovascular disease and some types of cancer.[9,10]

Milk and alternatives are important for nutrients such as calcium and vitamin D, which reduce the risk of the bone disease, osteoporosis. Meat and alternatives supply protein as well as iron, zinc and some B vitamins.

Page 2 The second page of Canada's Food Guide charts the number of servings required by males and females at various ages (**Figure 2.10**). It also states the purpose of Canada's Food Guide, which includes meeting the nutrient needs of Canadians and reducing the risk of obesity and chronic diseases such as type 2 diabetes, heart disease, certain types of cancer, and osteoporosis.

Table 2.1 Nutrient Contribution of Each Food Group

	SOME IMPORTANT NUTRIENTS IN THE FOOD GROUPS			
Key Nutrient	***Vegetables and Fruit***	***Grain Products***	***Milk and Alternatives***	***Meat and Alternatives***
Protein			✓	✓
Fat			✓	✓
Carbohydrate	✓	✓	✓	
Fibre	✓	✓		
Thiamin		✓		✓
Riboflavin		✓	✓	✓
Niacin		✓		✓
Folate	✓	✓		
Vitamin B6	✓			✓
Vitamin B12			✓	✓
Vitamin C	✓			
Vitamin A	✓		✓	
Vitamin D			✓	

Calcium			✓	
Iron		✓		✓
Zinc		✓	✓	✓
Magnesium	✓	✓	✓	✓
Potassium	✓	✓	✓	✓

Source: *Eating Well with Canada's Food Guide*: A Resource for Educators and Communicators. Available online at www.hc-sc.gc.ca/fn-an/alt_formats/hpfb-dgpsa/pdf/pubs/res-educat-eng.pdf. © Her Majesty the Queen in Right of Canada, represented by the Minister of Health (2011).

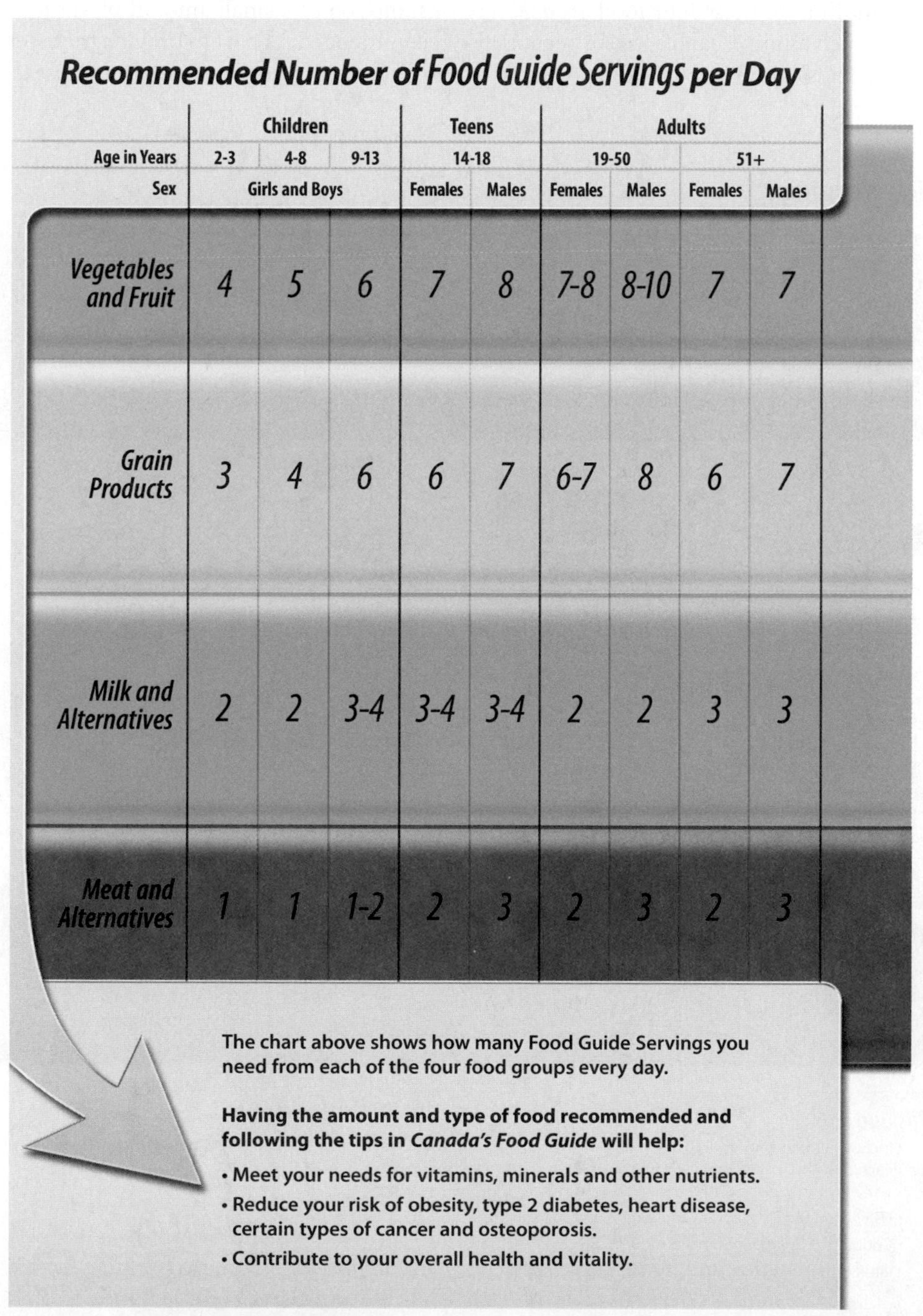

Recommended Number of Food Guide Servings per Day

	Children			Teens		Adults			
Age in Years	2-3	4-8	9-13	14-18		19-50		51+	
Sex	Girls and Boys			Females	Males	Females	Males	Females	Males
Vegetables and Fruit	4	5	6	7	8	7-8	8-10	7	7
Grain Products	3	4	6	6	7	6-7	8	6	7
Milk and Alternatives	2	2	3-4	3-4	3-4	2	2	3	3
Meat and Alternatives	1	1	1-2	2	3	2	3	2	3

The chart above shows how many Food Guide Servings you need from each of the four food groups every day.

Having the amount and type of food recommended and following the tips in *Canada's Food Guide* will help:

- **Meet your needs for vitamins, minerals and other nutrients.**
- **Reduce your risk of obesity, type 2 diabetes, heart disease, certain types of cancer and osteoporosis.**
- **Contribute to your overall health and vitality.**

Figure 2.10 Canada's Food Guide Page 2.
This page shows the number of food guide servings recommended daily.
Source: Health Canada. *Eating Well with Canada's Food Guide.* Available online at www.hc-sc.gc.ca/fn-an/food-guide-aliment/index-eng.php. Accessed March 5, 2011. © Her Majesty the Queen in Right of Canada, represented by the Minister of Health (2011).

Page 3 The third page of the food guide defines a Canada's Food Guide serving (**Figure 2.11**). What is very important to recognize from the examples given on this page is that a Food Guide serving is a relatively small amount of food . For example, 125 ml (1/2 cup) of orange juice is one Canada's Food Guide serving. One slice of bread is one serving, while a small bagel or 250 ml (1 cup) of rice are two servings. A glass of milk (250 ml) is one serving of milk and alternative, while a modest 175 ml (3/4 cup) of yogurt also counts as a serving. The guide's servings of meat and alternatives are also relatively small. A serving of meat weighs about 75 g, roughly the size of a deck of playing cards. **Table 2.2** shows how common objects and body parts, i.e. your hand, can be used to estimate serving sizes and help manage the amount of food you eat. Canada's Food Guide dietary pattern is one in which small amounts of a variety of foods are advocated. The importance of choosing a variety is emphasized in a separate panel on page 4, while on page 5, a panel on how to count Canada's Food Guide servings in a meal is included. In addition to these four food groups, the consumption of a small amount of unsaturated oils, such as olive, canola, or soybean, is also recommended, in part to provide the essential fatty acids (discussed in Section 5.6) and also to reduce the risk of cardiovascular disease.[11]

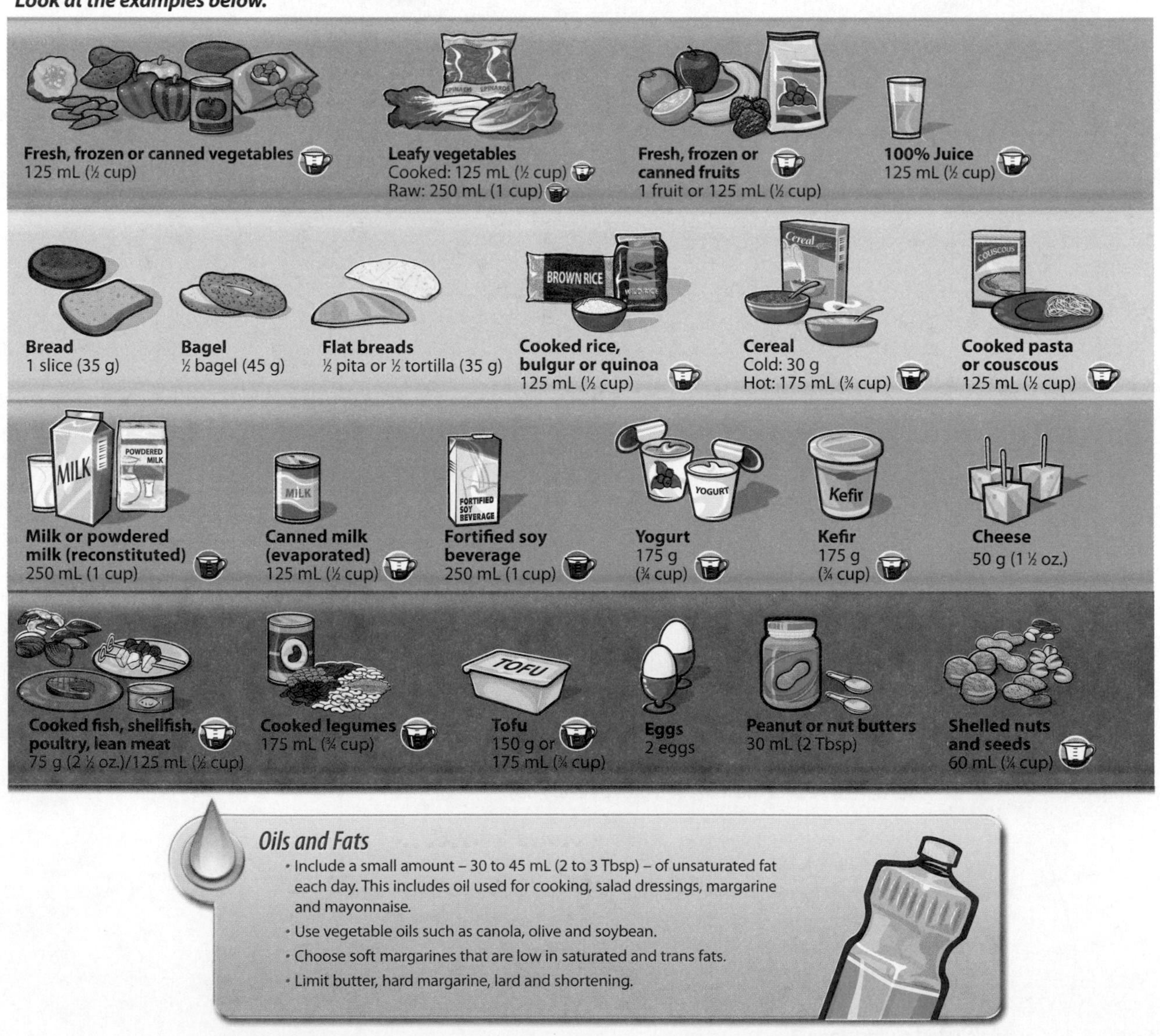

Figure 2.11 Canada's Food Guide Page 3:
This page illustrates the meaning of a Canada Food Guide serving.
Source: Health Canada. *Eating Well with Canada's Food Guide*. Available online at www.hc-sc.gc.ca/fn-an/food-guide-aliment/index-eng.php. Accessed March 3, 2011. © Her Majesty the Queen in Right of Canada, represented by the Minister of Health (2011).

Table 2.2 How Common Objects and Body Parts Can Be Used to Estimate Serving Sizes

The following object or body part:		Is approximately:	Which is an example of Canada's Food Guide serving of:
Tennis ball OR ice cream scoop		125 ml (1/2 cup)	Cooked vegetables Orange juice
1 pair of rolled-up socks OR Fist		250 ml (1 cup)	Leafy green vegetables Milk
Two thumbs OR 2 nine-volt batteries		Volume taken up by 50 g cheese	Cheese
Deck of cards OR Palm of hand OR Computer mouse		Volume taken up by 75 g of meat	Meat, fish
Golf ball		30 ml	Peanut butter
Nuts covering palm of hand		60 ml	Nuts
Thumb tip OR Die		5 ml	5 ml is not a CFG serving, but it can be used to estimate use of margarine and other solid fats.

(Photos by Luisa Begani)

Page 4 In order to optimize the quality the food chosen, page 4 lists a number of additional statements about food choices (**Figure 2.12**). Four statements, one for each food group, encourage the consumption of low-fat, low-sugar, and low-salt choices. Choosing foods low in fat and sugar increases nutrient density and reduces the number of calories consumed, which in turn reduces the risk of obesity. Obesity is associated with many chronic diseases, such as type 2 diabetes, cardiovascular disease, and some types of cancer, so reducing obesity contributes to reduced risk of these diseases. Choosing low-salt foods reduces sodium intake, which in turn reduces the risk of developing high blood pressure, or hypertension, which also contributes to the development of cardiovascular disease (see Section 10.3).

Canadians are also advised to consume at least one Canada's Food Guide serving each of dark green and orange vegetables. These vegetables are particularly rich in folate and vitamin A (see **Table 2.3**).

To ensure an adequate intake of fibre, Canadians are advised to ensure that half of the grain-product servings are whole grains (whole grains are discussed in more detail in Section 4.1). The consumption of 500 ml of low-fat milk is advised to ensure adequate intakes of both vitamin D and calcium. While vitamin D can be biosynthesized in the skin when skin is exposed to ultraviolet light, in Canada, during the winter months, sunlight is not direct enough for biosynthesis to take place. A dietary source of vitamin D is required, so Canadian milk producers are required, by law, to add vitamin D to fluid milk. For those who cannot

Make each Food Guide Serving count…
wherever you are – at home, at school, at work or when eating out!

- **Eat at least one dark green and one orange vegetable each day.**
  - Go for dark green vegetables such as broccoli, romaine lettuce and spinach.
  - Go for orange vegetables such as carrots, sweet potatoes and winter squash.
- **Choose vegetables and fruit prepared with little or no added fat, sugar or salt.**
  - Enjoy vegetables steamed, baked or stir-fried instead of deep-fried.
- **Have vegetables and fruit more often than juice.**

- **Make at least half of your grain products whole grain each day.**
  - Eat a variety of whole grains such as barley, brown rice, oats, quinoa and wild rice.
  - Enjoy whole grain breads, oatmeal or whole wheat pasta.
- **Choose grain products that are lower in fat, sugar or salt.**
  - Compare the Nutrition Facts table on labels to make wise choices.
  - Enjoy the true taste of grain products. When adding sauces or spreads, use small amounts.

- **Drink skim, 1%, or 2% milk each day.**
  - Have 500 mL (2 cups) of milk every day for adequate vitamin D.
  - Drink fortified soy beverages if you do not drink milk.
- **Select lower fat milk alternatives.**
  - Compare the Nutrition Facts table on yogurts or cheeses to make wise choices.

- **Have meat alternatives such as beans, lentils and tofu often.**
- **Eat at least two Food Guide Servings of fish each week.***
  - Choose fish such as char, herring, mackerel, salmon, sardines and trout.
- **Select lean meat and alternatives prepared with little or no added fat or salt.**
  - Trim the visible fat from meats. Remove the skin on poultry.
  - Use cooking methods such as roasting, baking or poaching that require little or no added fat.
  - If you eat luncheon meats, sausages or prepackaged meats, choose those lower in salt (sodium) and fat.

Enjoy a variety off oods from the four food groups.

Satisfy your thirst with water!

Drink water regularly. It's a calorie-free way to quench your thirst. Drink more water in hot weather or when you are very active.

* Health Canada provides advice for limiting exposure to mercury from certain types of fish. Refer to www.healthcanada.gc.ca for the latest information.

Figure 2.12 Canada's Food Guide Page 4
This page lists additional statements to help make every serving count.
Source: Health Canada. *Eating Well with Canada's Food Guide*. Available online at www.hc-sc.gc.ca/fn-an/food-guide-aliment/index-eng.php. Accessed March 3, 2011.

Table 2.3: Vegetables Vitamin A and folate content of selected orange and leafy green vegetables (Based on data from Canadian Nutrient File)

Vegetable (One Canada Food Guide serving)	Vitamin A content (% RDA for 20-year-old male)	Folate content (% RDA for 20-year-old male)
Cooked chopped broccoli (125 ml)	7	22
Romaine lettuce (250 ml fresh, chopped)	29	20
Boiled spinach (125 ml)	55	35
Raw carrot (1 medium)	57	3
Sweet potato, baked (125 ml)	112	7
Winter squash, all varieties, baked (125 ml)	31	6

consume milk because of lactose intolerance or allergies, soy milk fortified with vitamin D and calcium to the same levels as those found in milk is a useful alternative. Also, although not mandatory, some manufacturers are adding vitamin D to yogurt and cheese. By reading food labels carefully, consumers can identify which of these products are also sources of vitamin D (see more on food labels in Section 2.5).

With respect to meat and alternatives, Canadians are advised to consume plant sources of protein, such as beans, lentils, and tofu often. Diets high in plant-based proteins have been associated with reduced risk of chronic disease.[12] As **Table 2.4** indicates, in addition to protein, beans and lentils also contain large amounts of fibre and folate, not found in animal sources of protein. Fish consumption has been associated with reduced risk of death from cardiovascular disease, so at least two servings of fish per week are suggested.[13] These beneficial effects are believed to be due to the presence of omega-3 fatty acids, which are discussed in Chapter 5: Critical Thinking: Fish Consumption and Heart Disease.

Table 2.4 Comparison of nutrient content of vegetable and animal proteins (Based on data from Canadian Nutrient File)

Meat and alternatives (one Food Guide serving)	Protein (g)	Fibre (g)	Fat (g)	Folate (% RDA for 20-year-old male)
Lentils (cooked, 175 ml)	13	6	0.5	67
Navy beans (cooked, 175 ml)	11	9	1	47
Beef burger, extra lean (75 g)	23	–	8	1

The importance of eating a variety of foods is emphasized in a separate box on page 4. This page also advises Canadians to drink water, as it is a calorie-free way to quench thirst. The consumption of sugar-containing soft drinks, particularly among children, is believed to be a contributing factor in the development of obesity.[14]

Taken together, the Canada's Food Guide serving numbers and portion sizes (on pages 2 and 3) define the amount of food to be consumed in each food group, and the additional statements ensure that the best type of food within each food group is selected. In this way, as indicated at the top of page 4, Canadians are able to make every Canada's Food Guide serving "count" toward the promotion of nutritious eating, and as illustrated on page 5, know how to "count" Canada's Food Guide servings (see Critical Thinking: Canada's Food Guide: Additional Statements).

Page 5 Because nutrient requirements vary with stages in the life cycle, specific recommendations are made for children, women of childbearing age, and Canadians over 50 on the fifth page of the guide (**Figure 2.13**). Life cycle nutrition is discussed in more detail in Chapters 14, 15, and 16.

Page 6 The last page of the guide includes a number of suggestions to Canadians about how to improve their diet and how to become more physically active, adding at least 30-60 minutes of moderate activity daily, in at least 10-minute periods (**Figure 2.14**). The reader is also directed to Canada's Physical Activity Guide for more details (more about physical activity guidelines in Section 13.2). Modest increases in physical activity like the ones advised here are associated with improved health.[15] The "take a step today panel" lists small changes that people can make toward a healthier lifestyle. For example, research has shown that eating breakfast regularly helps those trying to lose weight, possibly by reducing appetite later in the day.[16]

Also included on page 6 is a long list of "foods to limit," e.g., cakes, pastries, cookies, potato chips, etc. These foods are very high in either sugars, salts, and/or unhealthy saturated and *trans* fats. Excessive consumption of these foods increases the risk of obesity and contributes to the development of chronic diseases such as cardiovascular disease and type-2 diabetes. Another important piece of advice given is for Canadians to use nutrition labels to compare foods and help them make better food choices. Nutrition labelling will be discussed in detail in Section 2.5.

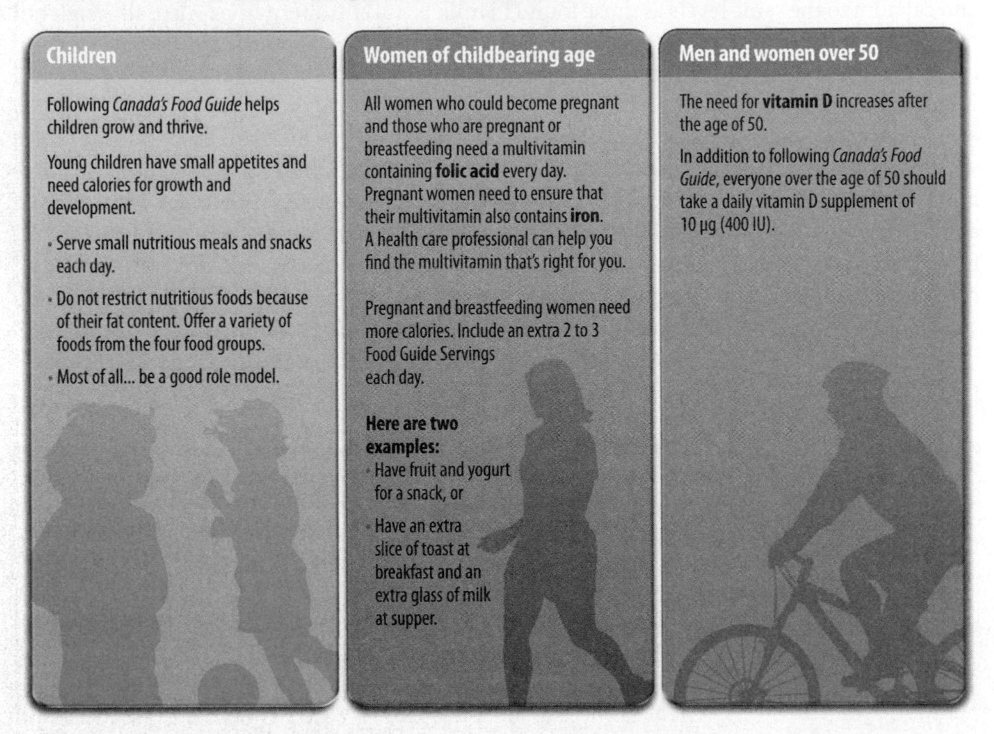

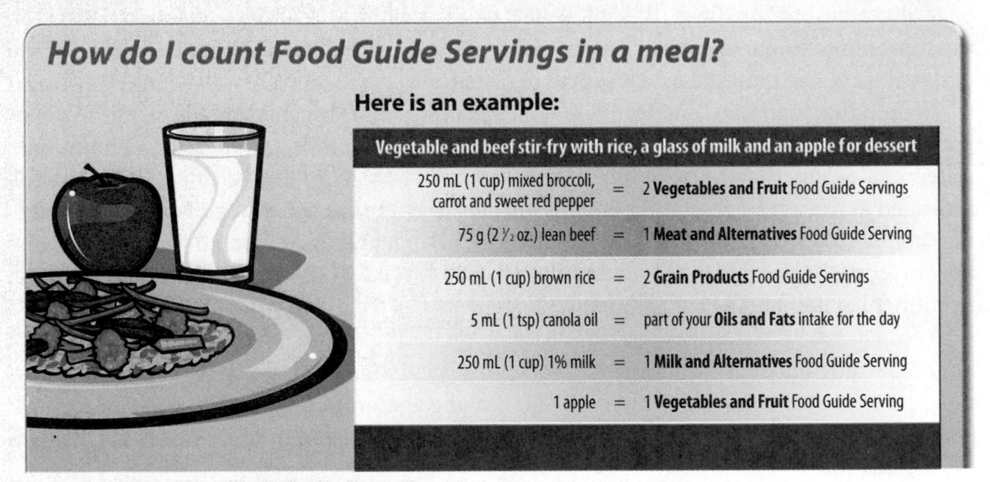

Vegetable and beef stir-fry with rice, a glass of milk and an apple for dessert		
250 mL (1 cup) mixed broccoli, carrot and sweet red pepper	=	2 **Vegetables and Fruit** Food Guide Servings
75 g (2 ½ oz.) lean beef	=	1 **Meat and Alternatives** Food Guide Serving
250 mL (1 cup) brown rice	=	2 **Grain Products** Food Guide Servings
5 mL (1 tsp) canola oil	=	part of your **Oils and Fats** intake for the day
250 mL (1 cup) 1% milk	=	1 **Milk and Alternatives** Food Guide Serving
1 apple	=	1 **Vegetables and Fruit** Food Guide Serving

Figure 2.13 Canada's Food Guide Page 5
This page provides advice for various life stages.
Source: Health Canada. *Eating Well with Canada's Food Guide*. Available online at www.hc-sc.gc.ca/fn-an/food-guide-aliment/index-eng.php. Accessed March 3, 2011. © Her Majesty the Queen in Right of Canada, represented by the Minister of Health (2011).

How the Current Version of Canada's Food Guide was Developed

When the current 2007 Canada's Food Guide was being created, it went through several versions before being finalized, because developers wanted to ensure that 90% of the population was meeting their requirements for most nutrients when following the guide. In order to check whether this 90% target was being met, developers created 500 different menus or dietary simulations, based on foods commonly consumed in Canada. These simulations conformed to the first version of the new, 2007 Canada's Food Guide and the nutrient content of these menus was analyzed to determine whether the 90% target had been achieved. In fact, these first tests indicated inadequate intakes for some nutrients, so the guide was modified and a second version created. For example, an early test revealed low intakes of vitamin A and folate, so the additional statement on the daily consumption of orange and green leafy vegetables, which are high in both nutrients, was added. When the menus were

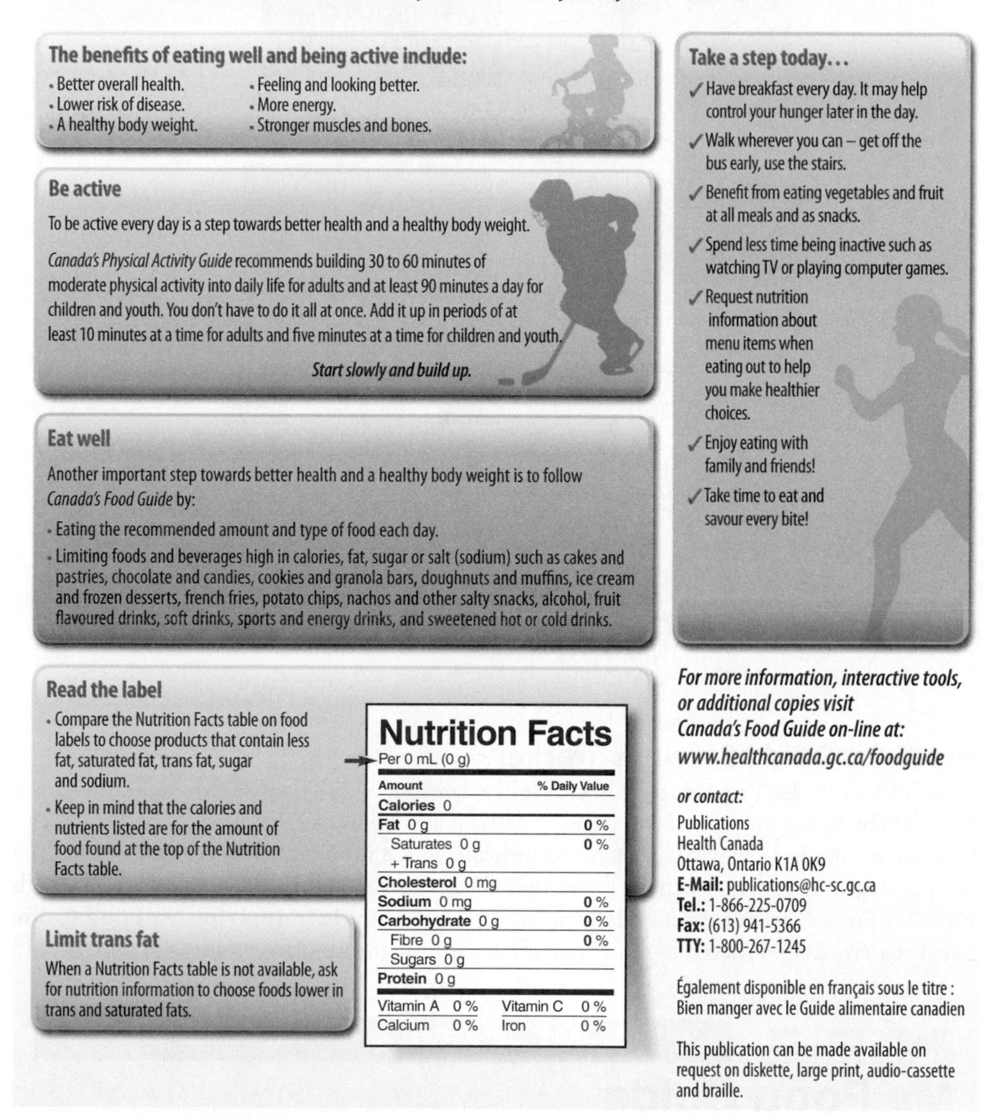

Eat well and be active today and every day!

The benefits of eating well and being active include:
- Better overall health.
- Lower risk of disease.
- A healthy body weight.
- Feeling and looking better.
- More energy.
- Stronger muscles and bones.

Be active

To be active every day is a step towards better health and a healthy body weight.

Canada's Physical Activity Guide recommends building 30 to 60 minutes of moderate physical activity into daily life for adults and at least 90 minutes a day for children and youth. You don't have to do it all at once. Add it up in periods of at least 10 minutes at a time for adults and five minutes at a time for children and youth.

Start slowly and build up.

Eat well

Another important step towards better health and a healthy body weight is to follow *Canada's Food Guide* by:
- Eating the recommended amount and type of food each day.
- Limiting foods and beverages high in calories, fat, sugar or salt (sodium) such as cakes and pastries, chocolate and candies, cookies and granola bars, doughnuts and muffins, ice cream and frozen desserts, french fries, potato chips, nachos and other salty snacks, alcohol, fruit flavoured drinks, soft drinks, sports and energy drinks, and sweetened hot or cold drinks.

Read the label
- Compare the Nutrition Facts table on food labels to choose products that contain less fat, saturated fat, trans fat, sugar and sodium.
- Keep in mind that the calories and nutrients listed are for the amount of food found at the top of the Nutrition Facts table.

Nutrition Facts
Per 0 mL (0 g)

Amount	% Daily Value
Calories 0	
Fat 0 g	0 %
Saturates 0 g	0 %
+ Trans 0 g	
Cholesterol 0 mg	
Sodium 0 mg	0 %
Carbohydrate 0 g	0 %
Fibre 0 g	0 %
Sugars 0 g	
Protein 0 g	
Vitamin A 0 %	Vitamin C 0 %
Calcium 0 %	Iron 0 %

Limit trans fat

When a Nutrition Facts table is not available, ask for nutrition information to choose foods lower in trans and saturated fats.

Take a step today...
- ✓ Have breakfast every day. It may help control your hunger later in the day.
- ✓ Walk wherever you can – get off the bus early, use the stairs.
- ✓ Benefit from eating vegetables and fruit at all meals and as snacks.
- ✓ Spend less time being inactive such as watching TV or playing computer games.
- ✓ Request nutrition information about menu items when eating out to help you make healthier choices.
- ✓ Enjoy eating with family and friends!
- ✓ Take time to eat and savour every bite!

For more information, interactive tools, or additional copies visit Canada's Food Guide on-line at: www.healthcanada.gc.ca/foodguide

or contact:

Publications
Health Canada
Ottawa, Ontario K1A 0K9
E-Mail: publications@hc-sc.gc.ca
Tel.: 1-866-225-0709
Fax: (613) 941-5366
TTY: 1-800-267-1245

Également disponible en français sous le titre : Bien manger avec le Guide alimentaire canadien

This publication can be made available on request on diskette, large print, audio-cassette and braille.

Figure 2.14 Canada's Food Guide Page 6
This page includes more advice on how to eat well and be active.
Source: Health Canada. *Eating Well with Canada's Food Guide*. Available online at www.hc-sc.gc.ca/fn-an/food-guide-aliment/index-eng.php. Accessed March 3, 2011. © Her Majesty the Queen in Right of Canada, represented by the Minister of Health (2011).

modified to conform to this additional statement, vitamin A and folate intakes increased to satisfactory levels. This process was repeated until the objective of the 90% target was largely met (see **Figure 2.15**). This was achieved only after 50 versions of Canada's Food Guide were tested. One requirement that could not be realistically addressed by any version of the guide was the vitamin D requirement for individuals over 50 (which is higher than for younger people). As a result, on page 5 of the guide, a vitamin D supplement is recommended for everyone over 50.[17]

My Food Guide

Health Canada provides additional online interactive tools to assist Canadians in making nutritious food choices based on Canada's Food Guide. These include My Food Guide (found on the Health Canada website under My Food Guide), which allows an individual to create a personalized food guide that lists the number of recommended servings for each food group and provides pictures of a Canada's Food Guide serving of personal food favourites which you can select from an online list (see **Figure 2.16**). Recognizing the cultural diversity of Canada, a number of ethnic foods are included on these lists.

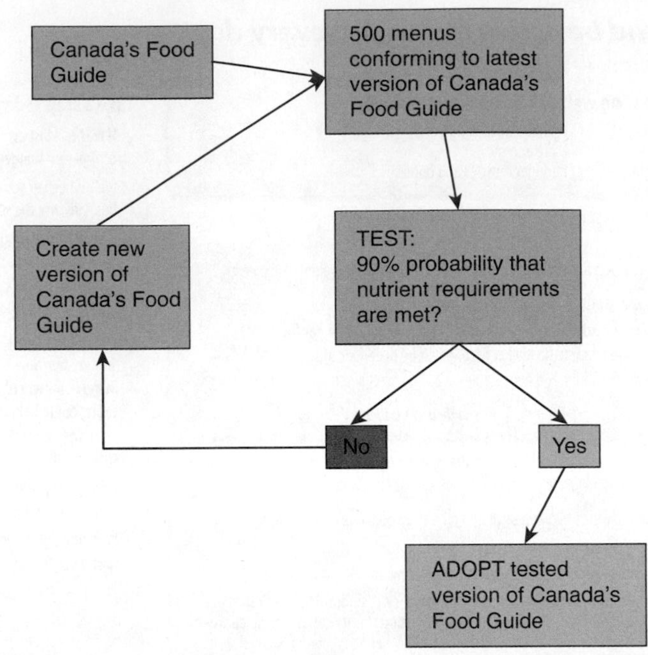

Figure 2.15 The development of Canada's Food Guide

My Food Guide Servings Tracker

Health Canada also provides a Food Guide Tracker that allows the user to record his or her food intake. It is composed of two pages; the first is a Food Intake Record on which users can write down the food they have eaten and indicate which food group each item is from. The second page has a series of check boxes that allow users to compare their food intake with Canada's Food Guide recommendations (see **Critical Thinking: Applying My Food Guide Servings Tracker**). When recording food intake, ideally the weight or volume of food should

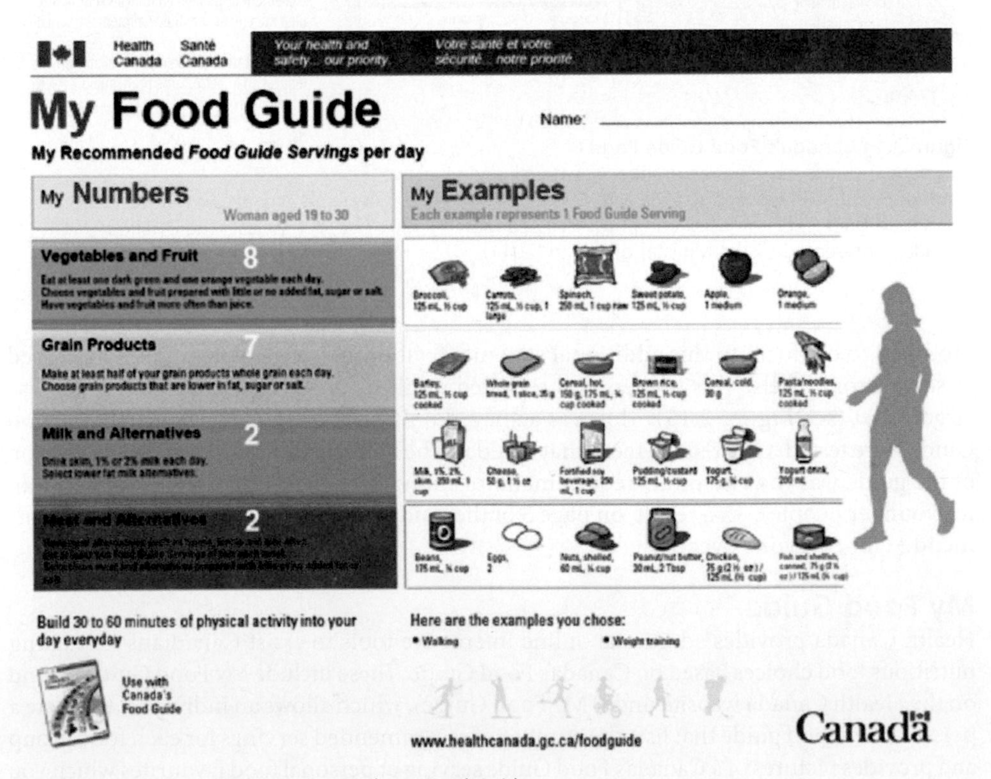

Figure 2.16 Health Canada's Online tool, My Food Guide.
Source: Health Canada. My Food Guide. Available online at www.healthcanada.gc.ca/foodguide. Accessed April 14, 2011.

CRITICAL THINKING:

Canada's Food Guide: Additional Statements

CANADIAN CONTENT

The additional statements that appear on page 4 (Figure 2.12) of Canada's Food Guide are very important for the selection of a nutritious diet. More advice is included on page 5 (Figure 2.13) and 6 (Figure 2.14) of Canada's Food Guide to further help Canadians plan their diet.

Critical Thinking Questions

Test your understanding of these additional statements and advice with the following questions. Answers to all Critical Thinking questions can be found in Appendix J.

An elderly woman requires seven servings of vegetables and fruits daily. Is she following Canada's Food Guide if she consumes three servings of orange juice and four servings of mashed potatoes?

A sedentary young man consumes the appropriate number of servings of foods from all four food groups. Is he following Canada's Food Guide if he regularly enjoys his food deep-fried or served with creamy sauces?

A young woman consumes fortified soy milk instead of cow's milk. Is she following Canada's Food Guide, with respect to milk and alternatives?

A young man eats a well-balanced nutritious diet but has french fries at a fast-food outlet once every three weeks. Is he following the recommendations of Canada's Food Guide?

Is a middle-aged woman who regularly skips breakfast following Canada's Food Guide?

An elderly woman, interested in increasing her intake of vegetables, does so by consuming canned soups regularly. Is she following Canada's Food Guide? (Additional information: One serving of soup contains 75% of the woman's Adequate Intake for sodium).

A young woman wants to increase her intake of fish, but is concerned about mercury contamination. Does Canada's Food Guide direct her to information on this subject?

A middle-aged man routinely drinks tap water because it is inexpensive. Is this the reason why Canada's Food Guide recommends drinking water?

A 58-year old woman does not take any supplements because she believes that all nutrients should come from food. Has she interepreted Canada's Food Guide correctly?

---

be measured accurately. As this is often impractical, the comparisons shown in Table 2.2 allows reasonable estimates of portion size to be made.

EaTracker.ca

Dietitians of Canada maintains a website (www.eatracker.ca) that provides the public with a wealth of nutritional information (**Figure 2.17**). This includes the ability to analyze the nutrient content of the food that the user recorded on his or her food tracker sheet. One enters the foods recorded on the sheet into the eaTracker database and a profile to the nutrient intakes is generated along with suggestions on how to modify intake to improve diet quality.

Meal Planning with Canada's Food Guide The example below indicates how one can use Canada's Food Guide to plan meals, using the example of a 16-year-old male.

Step 1: Determine the required numbers of Canada Food Guide servings from page two of the guide (Figure 2.10) for the individual. The results for the adolescent male is shown in **Table 2.5**.

Table 2.5: Recommended Canada Food Guide servings for 16-year-old male

Food Group	No. of Food Guide servings recommended
Vegetables and fruit	8
Grain products	7
Milk and alternatives	3-4
Meat and alternatives	3
Unsaturated vegetable oil	30-45 ml

CRITICAL THINKING:

Applying My Food Guide Servings Tracker

Background

For the first time in his life, Jarad is living on his own. He has gained a bit of weight, and is beginning to realize that he needs to pay more attention to the kinds of foods he eats. Jarad is 22 years old. He spends less than 30 minutes walking from class to class each day and gets little other exercise. To find out how well his current diet compares to the recommendations of Canada's Food Guide, he visits the Health Canada Web site and prints the My Food Guide Servings Tracker sheet for his age and sex. He then keeps a record of everything he consumes for one day by completing the Food Intake Record. His results are shown below.

MEAL		VEGETABLES AND FRUIT	GRAIN PRODUCTS	MILK AND ALTERNATIVES	MEAT AND ALTERNATIVES	OILS AND FATS	FOODS TO LIMIT
							Limit foods and beverages high in calories, fat, sugar, or salt.
BREAKFAST	Corn flakes (30 g)		1				
	Whole milk (250 ml)			1			
	Orange juice (250 ml)	2					
	Coffee						
	Cream						X
	Sugar						X
LUNCH	Hamburger, regular beef (75 g)				1		
	Bun, white (45 g)		1				
	Onions, tomato, lettuce (125 ml)	1					
	French fries (250 ml)	2					
	Milk shake						
	Whole milk (250 ml)			1			
	Ice cream (125 ml)			0.5			
DINNER	Lasagna						
	Noodles, not whole-grain (250 ml)		2				
	Tomato sauce (125 ml)	1					
	Regular ground beef (75 g)				1		
	Mozzarella cheese, whole milk (25 g)			0.5			
	Ricotta cheese, whole milk (50 g)			1			
	Olive oil (10 ml)					10 ml	
	Cola, regular (1 can)						X
SNACKS	Whole grain crackers (15 g)		0.5				
	Cheddar cheese (25 g)			0.5			
	Cola, regular (1 can)						X
TOTAL NUMBER OF FOOD GUIDE SERVINGS		6	4.5	4.5	2	10 ml	

Critical Thinking Questions

Obtain the tracking sheet for a male Jarad's age by going to www.healthcanada.gc.ca/foodguide and selecting My Food Guide Servings Tracker from the bar on the left. Using the check boxes, how well does Jarad's diet compare to the recommendations for each food group for a man of his age?

Is he following the recommendations of Canada's Food Guide's additional statements?

Why do you think he is gaining weight?

What changes do you think Jarad should make to his diet and activity levels?

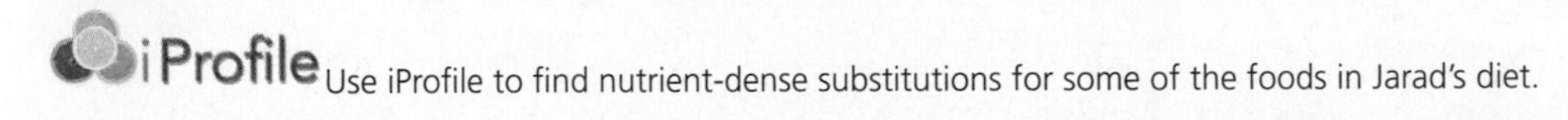

Use iProfile to find nutrient-dense substitutions for some of the foods in Jarad's diet.

Figure 2.17 EaTracker
EaTracker allows you to enter the foods that you have consumed for one day and provides feedback.

Step 2: Create a template of meals and snacks suiting the preferences of the individual. In our example (see **Table 2.6**), the adolescent likes to snack, so a template was created that included 3 meals and 2 snacks. The appropriate number of servings from each food group are distributed across these meals and snacks as shown below. The recommendations for unsaturated oils are also included.

Table 2.6 Serving Template

Meal	Oil (ml)	Vegetables & Fruit (Canada's Food Guide Servings)	Grain Products (Canada's Food Guide Servings)	Milk & Alternatives (Canada's Food Guide Servings)	Meat & Alternatives (Canada's Food Guide Servings)
Breakfast		2	1	1	
Snack					1
Lunch	15-30	3	4		1
Snack		1		1	
Supper	15	2	2	1	1
Total	**30-45**	**8**	**7**	**3**	**3**
Recommended	**30-45**	**8**	**7**	**3-4**	**3**

Step 3: Once the template is complete, different menus can be created that conform to Canada's Food Guide by replacing the food group servings with specific types of food as shown below **(Table 2.7).** In selecting these foods, it is essential to ensure that the additional statements are also taken into account, for example, be sure to include a dark green and orange vegetable, select foods low in fat, sugar, and salt, etc. (The checklist provided on page 1 of the My Food Guide Servings Tracker is useful for this purpose.)

Table 2.7: Example of daily menu generated using the serving template in table 2.6

	Number of Canada Food Guide Servings				
Meals	**Oil (ml)**	**Vegetables & Fruit**	**Grain products**	**Milk & alternatives**	**Meat & Alternatives**
Breakfast:					
Orange juice (125 ml)		1			
Fresh strawberries (125 ml)		1			
Hot oatmeal (175 ml)			1		
Yogurt (175 ml)				1	
Snack:					
Nuts (60 ml)					1
Lunch:					
Salad with spinach, some cherry tomatoes, red pepper (500 ml)		2			

Table 2.7: ***(Continued)***

		Number of Canada Food Guide Servings			
Meals	**Oil (ml)**	**Vegetables & Fruit**	**Grain products**	**Milk & alternatives**	**Meat & Alternatives**
Canola-oil-based salad dressing	15				
Vegetable soup (250 ml)		1			
Salmon sandwiches consisting of:					
four slices whole-wheat bread			4		
canned salmon with bones (75 g)					1
Mayonnaise-type dressing made with canola oil	10				
Snack:					
Apple (1 medium)		1			
Cheddar cheese (50 g)				1	
Supper:					
Baked beans (175 ml)					1
Sweet potato (125 ml)		1			
Mixed cooked vegetables (125 ml) stir-fry		1			
Canola oil	15				
Brown rice (250 ml)			2		
1% milk (250 ml)				1	
Total Servings	**40**	**8**	**7**	**3**	**3**
Recommended servings	**30-45**	**8**	**7**	**3-4**	**3**

Canada's Food Guide for First Nations, Inuit, and Métis

Eating Well with Canada's Food Guide: First Nations, Inuit, and Métis is a food guide that recognizes the cultural values of the aboriginal peoples and the contribution made by both store-bought and traditional foods, such as meats and wild game, fish, and shellfish, bannock (an indigenous bread), wild plants and berries, to the diets of these peoples (see **Figure 2.18**).

Figure 2.18 First Nations, Inuit, and Métis Food Guide
Source: Health Canada: *Eating Well with Canada's Food Guide* for First Nations, Inuit, and Métis. Available online at www.hc-sc.gc.ca/fn-an/food-guide-aliment/fnim-pnim/index-eng.php. Accessed March 3, 2011. © Her Majesty the Queen in Right of Canada, represented by the Minister of Health (2011).

2.4 Other Food Guides and Dietary Patterns

Learning Objectives

- Describe similarities and differences between various national food guides and Canada's Food Guide.
- Describe the purpose of the Canadian Diabetes Association's meal-planning tools, Just the Basics and Beyond the Basics.

Other National Food Guides

The Canadian rainbow graphic is unique among national food guides, while food guides developed in other countries have adopted different shapes to emphasize the proportions of choices that should come from different food groups. Korea and China use a pagoda shape, Mexico, Australia, and most European countries use a pie or plate shape, Japan uses a spinning top (see **Figure 2.19**). The American food guide recently adopted the MyPlate icon (Figure 2.21) after many years of using a pyramid shape. The U.S. guide contains five food groups rather than the four used in *Eating Well with Canada's Food Guide,* but the recommended number of servings and the overall dietary advice are very similar (see Appendix G). This is true for most national guides. The Mediterranean diet represented by a pyramid (see Chapter 1: Science Applied: From "Seven Countries" to the Mediterranean

Diet) is also a popular dietary pattern. Canada's Food Guide and the Mediterranean diet are compared in Critical Thinking: Should Canadians Eat According to Canada's Food Guide or the Mediterranean Diet?

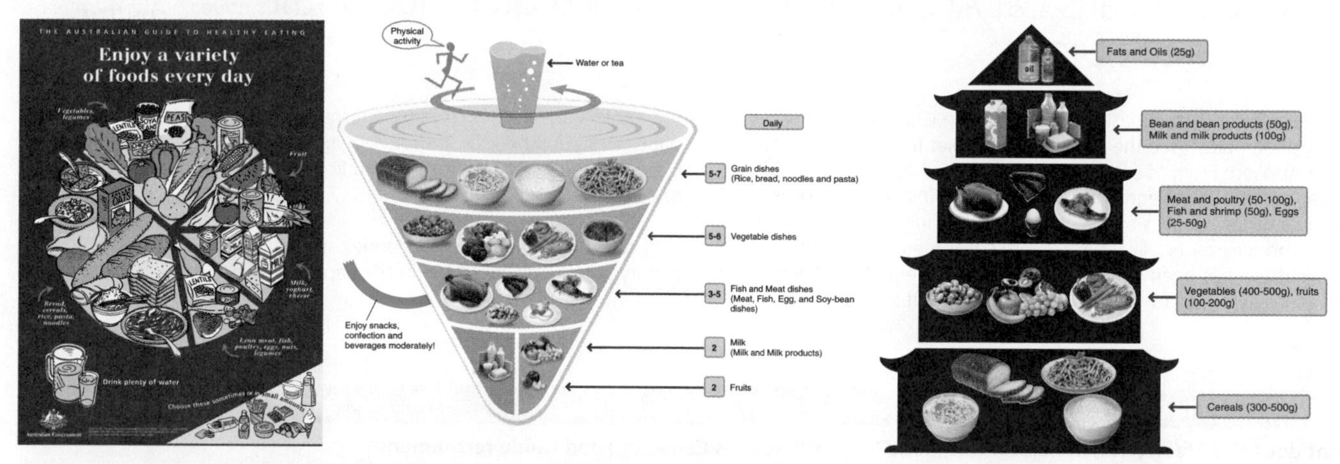

Figure 2.19 Various national food guides (from left to right), the Australian Food Guide, and renditions of the Japanese Food Guide Spinning Top and the Dietary Pagoda from the Chinese Nutrition Society.
Sources: Australian Government Department of Health and Aging, and Healthy Eating; Japanese Ministry of Health, Labour, and Welfare; Chinese Nutrition Society.

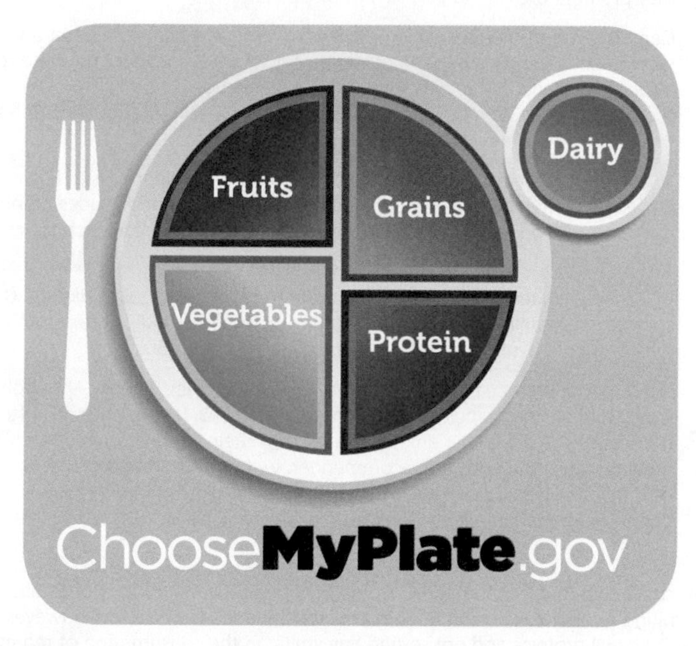

Figure 2.20 MyPlate
MyPlate is a food-group system designed to help Americans plan their individual diets.
Source: United States Department of Agriculture. Available online at www.choosemyplate.gov.

Canadian Diabetes Association: Just the Basics and Beyond the Basics

Canada's Food Guide is not the only meal-planning tool. The Canadian Diabetes Association has developed meal-planning tools for individuals with diabetes, a simple tool called Just the Basics and a more complex system called Beyond the Basics.[18,19] These tools place more emphasis on the carbohydrate content of food than does Canada's Food Guide and are intended to help those with diabetes manage their food intake to successfully control their blood glucose levels, which rise in uncontrolled diabetes. These meal-planning systems will be discussed in Section 4.5 and are similar to the Exchange Lists developed by the American Diabetes Association.

CRITICAL THINKING:

CANADIAN CONTENT

Should Canadians Eat According to Canada's Food Guide or the Mediterranean Diet?

One very popular dietary pattern, introduced in Chapter 1, is the Mediterranean diet[1] (see the Mediterranean Diet Pyramid in Chapter 1: Science Applied: From "Seven Countries" to the Mediterranean Diet). This diet is plentiful in vegetables and whole grains and olive oil is the source of fat. Wine is consumed with meals in moderation, while animal products, especially red meat, are consumed infrequently. This dietary pattern has been studied extensively, and repeatedly been found to be associated with reduced disease risk. It is often considered the "gold standard" of dietary patterns. Several years ago, a scientific paper was published, by authors Downs and Willows, entitled "Should Canadians eat according to the traditional Mediterranean diet pyramid or Canada's Food Guide?"[2]

Let's consider some of the similarities and differences between the two dietary patterns, highlighted in the paper, and listed here.

Comparison of Mediterranean diet with Canada's Food Guide recommendations

What does the Mediterranean diet recommend?	What does Canada's Food Guide recommend?
The Mediterranean diet recommends the moderate consumption of wine, which is defined as no more than 1-2 glasses/day.	Canada's Food Guide lists alcohol as one of the foods to limit, in part because of its calorie content and also because of the concerns about alcoholism. The Mediterranean diet recommends the consumption of wine with meals, but Canadians tend to drink beer and spirits on weekends—a very different and potentially harmful consumption pattern. While the moderate consumption of alcohol appears to be beneficial for preventing cardiovascular disease, the paper also notes concerns that it may promote breast and other types of cancer.
Mediterranean diet recommends the consumption of olive oil.	Canada's Food Guide similarly promotes olive oil as a healthy oil, but also recommends the consumption of canola oil, which is cheaper and more readily available in Canada. Canola oil, which is compositionally similar to olive oil, appears to provide the same benefits and Downs and Willows (2007) also cite studies that show canola may be somewhat more beneficial.
Mediterranean diet recommends the consumption of fresh local produce.	While many Canadians strive to eat locally when possible, this is difficult during the Canadian winter. Canada's Food Guide recognizes this and includes graphics of frozen and canned foods as well as fresh, indicating that these are acceptable choices as well.
Mediterranean diet makes no recommendations about vitamin D—in the sunny Mediterranean, few dietary sources of vitamin D are required.	In Canada, limited sunlight exposure during the winter requires dietary vitamin D sources, so vitamin D sources such as milk or fortified soymilk are recommended, as well as supplementation for Canadians over the age of 50.
Mediterranean diet emphasizes the daily consumption of plant proteins with the consumption of fish, eggs, and poultry a few times weekly and red meat sparingly.	Like the Mediterranean diet, fish is recommended twice weekly. Although Canada's Food Guide recommends the consumption of beans, lentils, and tofu often, it does not emphasize plant over animal sources of protein. Animal sources such as red meat, poultry, and eggs are recommended without any limitations. This is a criticism of Canada's Food Guide, given the evidence that diets high in plant proteins are beneficial to human health.

Compared to the traditional Mediterranean Diet, the paper concluded that Canada's Food Guide may be better suited for Canadians because it recognizes the importance of vitamin D and recommends limiting alcohol consumption. They criticize Canada's Food Guide, however, for grouping plant proteins (e.g., nuts, seeds, beans, lentils, tofu, etc.) with animal proteins and not setting any limits on the consumption of red meat. They suggest that the best option is for Canadians to follow Canada's Food Guide, but to choose vegetarian options for the meat and alternative category frequently.

Critical Thinking Questions

So what do you think? Which pattern is more appropriate? Should Canadians eat according to the traditional Mediterranean diet pyramid or Canada's Food Guide? Do you agree with Downs and Willows' conclusions?

[1] Oldways. *What is the Mediterranean Diet Pyramid?* Available online at www.oldwayspt.org/mediterranean-diet-pyramid. Accessed Feb 22, 2011.

[2] Downs SM, Willows ND. *Should Canadians eat according to the traditional Mediterranean diet pyramid or Canada's Food Guide?* Appl Physiol Nutr Metab. 2008 Jun;33(3):527-35.

2.5 Food and Natural Health Product Labels

Learning Objectives

- List the components that must appear on a food label.
- Use the Nutrition Facts on a food label to select a food that is considered low in saturated fat.
- Use a food label to distinguish which ingredients in the product are present in the greatest and least amounts.
- Describe how Canada's Food Guide recommends that food labels be used.
- Describe what is meant by the % Daily Value.
- Describe three health claims that are permitted on food labels.
- Describe how food labelling differs from the labelling of natural health products.

Food Labels

All packaged foods in Canada are required, by law, to have the following components:

a. the name of the product; the net contents or weight; the date by which the product should be sold, if perishable; and the name and place of business of the manufacturer, packager, or distributor.
b. a list of ingredients listed in descending order, by weight, that is, the most common ingredient in the food is listed first.
c. a nutrition facts table.

Restaurants are not required to provide nutrition information on menus (see Your Choice: How Can You Eat Out and Still Eat Well?)

Ingredient List The ingredients section of the label lists the contents of the product in order of their prominence by weight. For example, many juice drinks are mostly water and sugar. The ingredient list shown in **Figure 2.21** indicates that water and the sweetener high-fructose corn syrup are the first two ingredients and thus are the most abundant ingredients by weight. An ingredient list is required on all products containing more than one ingredient. Food additives, including food colours and flavourings, must be listed among the ingredients (see Label Literacy: Using Food Labels to Choose Wisely.)

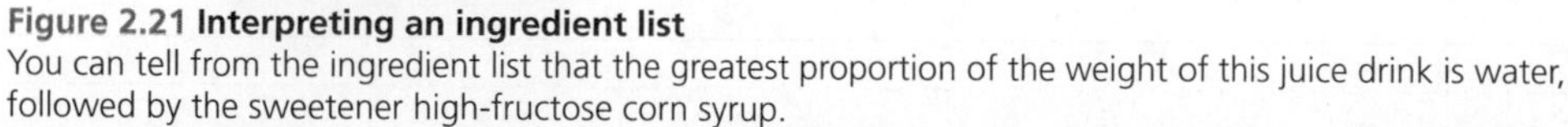

Figure 2.21 Interpreting an ingredient list
You can tell from the ingredient list that the greatest proportion of the weight of this juice drink is water, followed by the sweetener high-fructose corn syrup.

YOUR CHOICE

(©iStockphoto)

How Can You Eat Out and Still Eat Well?

Restaurant meals are typically higher in kcalories, saturated fat, cholesterol, and salt, and lower in fibre, vitamin A, and other micronutrients than the meals we prepare at home.[1] Our eat-on-the-run lifestyle has made choosing healthy foods from restaurant menus an important skill—but it can be a challenge.

Some healthy choices are easy. Skip the fried fish and have it broiled instead. Minimize sauces and spreads that add fat, sugar, and kcalories. Ask that salad dressing be served on the side. Take some of the meal home for tomorrow's lunch. Other restaurant choices are more difficult to make. Items that sound like part of a healthy diet are not always what they seem. What's in that house special turkey tetrazzini, beef lo mein, or fajita wrap?

Many restaurants have responded to consumer concern about healthy diets by highlighting healthy items on their menus and providing nutrition information on the premises or on corporate websites and consumers should check these sources when making decisions about what and where to eat. These activities, however, are voluntary. There are no mandatory labelling requirements for restaurants except when they are making nutrient-content claims, nutrient-function claims, or disease-risk-reduction claims about their foods. Then, they must provide the additional nutritional information supporting the claim. Some people believe that Canada should change its regulations to require that nutrition information be made available at point of purchase. This would make nutrition information visible to patrons before they purchased their food, when it could potentially most influence food choices. Many advocate for this change.[2]

When such information is available, studies suggest that people will reduce the number of calories consumed. A recent study, for example, surveyed more than 7,000 customers from 275 restaurants, representing many popular fast-food outlets. Researchers found that the average purchase contained 827 kcalories and 34% of purchases exceeded 1,000 kcalories. In one outlet, nutrition information was made visible at point of purchase, so the nutrition information was clearly visible to customers before they bought their food. Compared to patrons in other outlets, where nutrition information was available but not at point of purchase, many more customers in this outlet indicated that they saw the nutrition information (34% vs. 4%). Furthermore, patrons who saw the information purchased, on average, 52 fewer calories than patrons who didn't see the information.[3]

(© Blue Jean Images / Alamy)

Until nutritional information in restaurants becomes more readily available, look for menu items that fit into an overall healthy diet. Choose foods you like, and remember that a high-fat or high-kcalorie meal now and then doesn't make your overall diet unhealthy. If you eat out often, though, choose carefully, because these meals make up a greater part of your overall diet. Make use of nutrition information if it is available.

[1] Frazao, E. (Ed.). America's Eating Habits: Changes and Consequences. USDA/Economic Research Service, Agricultural Information Bulletin No. AIB750, May 1999. Available online at www.ers.usda.gov/publications/aib750/. Accessed May 16, 2009.

[2] Dietitians of Canada.Current Issues: The Inside Story. Does menu labelling make a difference to consumer choices? Available online at www.dietitians.ca/Downloadable-Content/Public/menulabelling-position-paper.aspx. Accessed Mar 5, 2011.

[3] Bassett, M.T., Dumanovsky, T., Huang, C., et al. Purchasing behavior and calorie information at fast-food chains in New York City, 2007. *American Journal of Public Health*. 98(8) (August 2008), pp.1457-9.

Nutrition Facts Table This part of the label (**Figure 2.22**) provides information about serving size, total kcalories (on food labels, the term "Calorie" is used to represent kcalories), and the amounts of nutrients per serving. Canada's Food Guide encourages people to compare the Nutrition Facts label on food products to ensure that foods with less fat, saturated fat, *trans* fat, sugar, and sodium are being selected (See Label Literacy: Using Food Labels Wisely.)

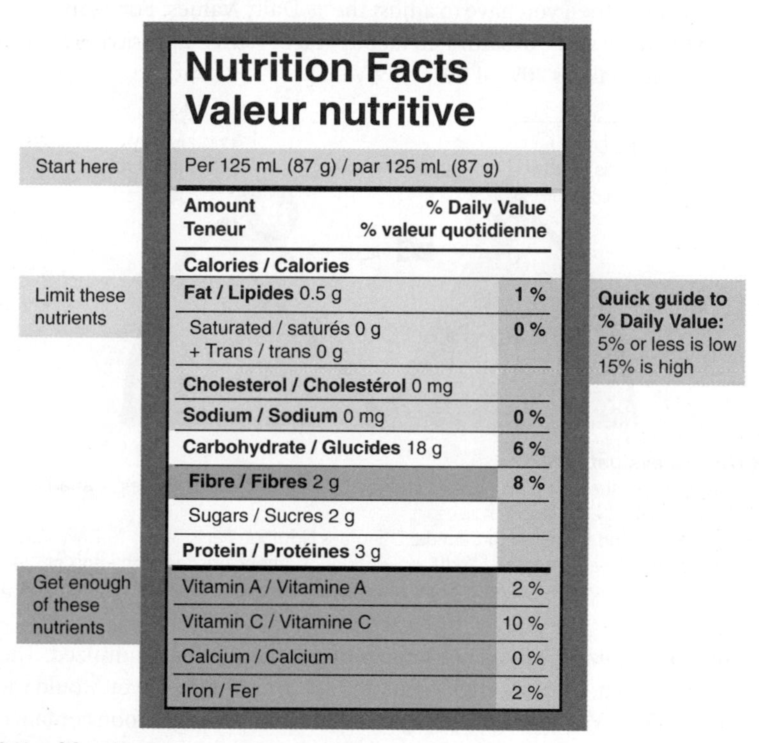

Figure 2.22 Nutrition Facts
This Nutrition Facts panel provides information that can be used to make wise food choices.

Serving Size The serving size of a food product is given in common household and metric measures. Serving size is generally based on a standard list of serving sizes.[20] If, however, a food is sold as a single portion, for example, a small bag of potato chips, then the serving size is the size of the entire package. It is always important to check the definition of the serving size when making a comparison of the nutrition content of different foods, and to compare these servings to the amount of food you would typically eat. If, for example, you usually eat the equivalent of two servings, as described on the label, you are consuming twice as many kcalories and twice the amount of nutrients. It is also important to note that the serving size on labels is not necessarily the same size as a Canada's Food Guide serving.

Daily Values Food labels must list information about the following nutrients on their labels: total fat, saturated fat, *trans* fat, cholesterol, sodium, total carbohydrate, dietary fibre, and protein, as well as vitamin A, vitamin C, calcium, and iron. The label format most commonly used in Canada is shown in Figure 2.22. Information can be presented as a single bilingual label, or as two separate but equal-sized panels in French and English.

Most of the information is presented as a percentage (%) of a standard called the **Daily Value (DV)**, which was developed specifically for food labels. Daily Values are a set of numbers intended to approximate the nutritional needs of a person consuming 2,000 kcalories daily, a good estimate for most adults. These values are best used to indicate whether there is a lot or a little of a nutrient in a food and to help consumers make comparisons between foods. They should not be used in place of appropriate DRIs to assess the adequacy of nutrient intakes. The

daily value A nutrient reference value used on food labels to help consumers make comparisons between foods and select more nutritious food.

actual Daily Values used are shown in **Table 2.7**. The nutrient content of food is expressed as a percentage of these values to assist consumers. A person may not be able to judge whether a food containing 330 mg of calcium per serving contains a lot or a little calcium, but when the calcium content is presented as 30% of a DV, it becomes clear that a substantial amount of the mineral is present in the food.

As a general rule, a Daily Value of 5% or less indicates that the food is low in that nutrient, and a Daily Value of 15% or more indicates that it is high (see **Figure 2.23**). It is important to recognize that the Daily Values are based on the stated serving on the label. If you consume more or less of the food then you have to adjust the % Daily Values. For example if a product contains 10% of the daily value of saturated fat but you eat twice the portion indicated on the label then you actual intake is 20% of the Daily Value.

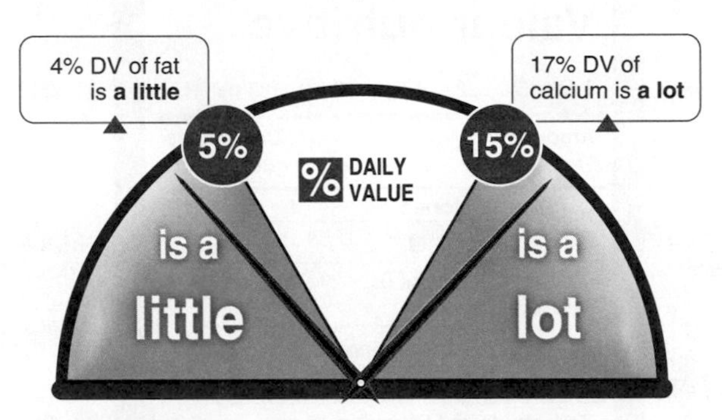

Figure 2.23 How to evaluate DV
This graphic is from a fact sheet on Daily Values created by Health Canada to help Canadians use nutrition labelling to make better food choices.
Source: Used with permission from Health Canada. Using the Nutrition Facts Table: % Daily Value. Available online at www.hc-sc.gc.ca/fn-an/alt_formats/pdf/label-etiquet/nutrition/cons/fact-fiche-eng.pdf. Accessed March 13, 2011. © Her Majesty the Queen in Right of Canada, represented by the Minister of Health (2011).

It is important to recognize the intake of some nutrients should be minimized. These include total calories, saturated fat, *trans* fat, cholesterol, and sodium. For these, you would want to select foods with a low % Daily Value. So, for example, if the stated serving of food contains 20% of the Daily Value for saturated and *trans* fat, then it contains 20% of the maximum amount recommended for a 2,000-kcalorie diet. This would be considered a lot and you might want to pick a food with less saturated and *trans* fat or reduce the amount of this food you will eat. For other nutrients, for example, dietary fibre, protein, vitamins A and C, calcium, and iron, the objective is to maximize intake so a food that contains 20% of the DV for calcium would be considered a good choice. In addition to the nutrients listed on the standard label, if a food processor adds a nutrient to the food, for example, fortifies the food with vitamin D or folate, these nutrients must also be included on the label. Finally, the food processor is free to add information about additional nutrients to the label. For example, many cereal products list information about B vitamins and minerals, as they are present in substantial amounts in these products.

recommended daily intakes (RDIs) Reference values established for vitamins and minerals in Canada in the 1980s and 1990s.

reference standards Reference values established for other several nutrients. The values are based on dietary recommendations for reducing the risk of chronic disease.

Daily Values are based on two sets of standards, the **Recommended Daily Intakes (RDIs)** and the **Reference Standards** (Table 2.7). These numbers predate the development of the DRIs and differ from them for some nutrients. To avoid confusion, only the term "Daily Value" appears on food labels.

Table 2.7 Standards that Make Up the Daily Values

Nutrient	Reference Standards	Amount in 2,000-Kcalorie Diet
Total fat	<30% of kcalories	65 g
Saturated & *trans* fat	<10% of kcalories	20 g
Total carbohydrate	60% of kcalories	300 g
Dietary fibre	11.5 g/1,000 kcalories	25 g
Cholesterol	<300 mg	<300 mg
Sodium	<2,400 mg	<2,400 mg
Potassium	3,500 mg	3,500 mg

Nutrient	Recommended Daily Intake	Nutrient	Recommended Daily Intake
Vitamin A	1,000 RE	Vitamin E	10 mg
Biotin	30 μg	Riboflavin	1.7 mg
Vitamin C	60 mg	Niacin	23 NE
Vitamin B_{66}	1.8 mg	Vitamin B_{12}	2 ug
Thiamin	1.3 mg	Chromium	120 μg
Folic acid	220 μg	Phosphorus	1,100 mg
Pantothenic acid	7 mg	Selenium	50 μg
Vitamin K	80 μg	Calcium	1,100 mg
Iodide	160 μg	Magnesium	250 mg
Molybdenum	75 μg	Manganese	2 mg
Iron	14 mg	Zinc	9 mg
Vitamin D	5 ug	Chloride	3,400 mg
Copper	2 mg		

Source: Canadian Food Inspection Agency: Guide to Food Labelling and Advertising. Chapter 6.1. Available online at www.inspection.gc.ca/english/fssa/labeti/guide/ch6e.shtml#a6_2. Accessed March 13, 2011.

Health Claims

In addition to the mandatory labelling requirements, there are three additional claims that may appear on food labels. These include (a) nutrient-content claims, (b) disease-risk-reduction claims, and (c) nutrient-function claims. These three items are collectively referred to as health claims.

Nutrient-Content Claims Food labels often highlight the level of a nutrient or dietary substance in a product that might be of interest to the consumer, such as "low calorie" or "high fibre." Definitions for nutrient-content descriptors such as "free," "high," and "reduced," are regulated by the Canadian Food Inspection Agency, to ensure that they conform to strict legal definitions. In selecting a product labelled with a descriptor such as "fat free," consumers can be assured that the food meets the defined criteria; in this case, that the product contains less than half a gram of fat per serving, or that the claim an "excellent source of vitamin C" means that a serving of the product must contain at least 50% of the Daily Value for the vitamin (see **Table 2.8**).

Disease-Risk-Reduction Claims Food labels are also permitted to include a number of health claims if they are relevant to the product (see **Figure 2.24**). Disease-risk reduction claims refer to a relationship between a food or nutrient and the risk of a disease or health-related condition. They can help consumers choose products that will meet their dietary needs or health goals. For example, low-fat milk, a good source of calcium, might include on the package label a statement indicating that a diet high in calcium may reduce the risk of developing osteoporosis.

Figure 2.24 Health claims on food labels This package of frozen vegetables displays one of Canada's permitted disease-risk-reduction claims.

LABEL LITERACY

Using Food Labels to Choose Wisely

(©iStockphoto)

Food labels cannot help you include more vegetables and fruits in your diet each day, or ensure that you select a varied diet, but they do provide you with a good source of nutrition information. For example, for breakfast, you might choose granola with whole milk or oatmeal made with nonfat milk. Which is the best choice if you want your diet to be moderate in sugar and fat, and high in nutrient density? You can see on the label that a cup of whole milk provides 150 kcalories and 5 grams of saturated fat. This 5 grams of saturated fat represents 25% of the Daily Value—that is, 25% of the total amount of saturated and *trans* fat recommended per day (20 g) for a 2,000-kcalorie diet. A cup of nonfat milk, in contrast, contains no saturated fat and only 90 kcalories. Both are good sources of calcium and vitamins A and D, and both are in the milk and alternatives group. But nonfat milk better meets the additional statement recommendation to choose a diet low in saturated fat and cholesterol.

Granola and oatmeal each provide a serving from the grain product group—both contain whole grains—but the amounts of fat and added sugars differ. A serving of granola provides 230 kcalories and 9 grams of fat, whereas a serving of oatmeal has only 150 kcalories and 3 grams of fat. The ingredient list reveals that oatmeal contains only rolled oats; no sugars are added. In contrast, the granola contains rolled oats plus brown sugar and honey, in addition to other ingredients. The oatmeal is lower in both fat and added sugar so it is higher in nutrient density.

Knowing how to interpret the information on food labels can help you choose a diet that meets the recommendations of *Eating Well with Canada's Food Guide*. However, this doesn't mean you can never have a doughnut for breakfast because the label identifies it as high in fat and kcalories. Even a high-fat food choice can be part of a healthy diet as long as intake is infrequent. Remember, it is your total diet—not each choice—that counts.

CANADIAN CONTENT

Table 2.8 Descriptors Commonly Used on Food Labels	
Free	Product contains no amount of, or a trivial amount of fat, saturated fat, *trans* fat, cholesterol, sodium, sugars, kcalories, etc. For example, "sugar free" and "fat free" both mean less than 0.5 g per serving. *Trans* fat free means less than 0.2 g of *trans* fat and less than 2 g saturated fat per serving. Synonyms for "free" include "without," "no," and "zero."
Low	Can be used to describe the amount of fat, saturated fat, cholesterol, sodium, kcalories, and other nutrients. Specific definitions have been established for each of these nutrients. For example, "low fat" means that the food contains 3 g or less per serving; "low cholesterol" means that the food contains less than 20 mg of cholesterol (and less than 2 g saturated fat) per serving; "low sodium" means less than 140 mg sodium/100 g of food. Synonyms for "low" include "little," "few," and "low source of."
Lean and Extra Lean	Used to describe the fat content of meat, poultry, seafood, and game meats. "Lean" means that the food contains less than 10 g fat per 100 g. "Extra lean" means that the food contains less than 7.5 g fat per 100 g.
Source of	Foods contain greater than 5% of the daily value of the stated nutrient, e.g., source of vitamin A.
Good Source of	Food contains greater than 15% of the Daily Value for a particular nutrient per serving, except vitamin C, for which foods contain > 30%, e.g., good source of fibre.
Excellent source of	Used for foods that contains 25% or more of the Daily Value for a particular nutrient (except vitamin C, which contains 50% or more). Synonyms include "high" and "rich in," e.g., excellent source of calcium.
Reduced	Nutritionally altered product contains 25% less of a nutrient or of energy than the regular or reference product. Synonyms include "less", "lower" and "light", e.g., reduced in fat.
Light	Used in different ways. See "reduced" above. "Lightly salted" refers to a food in which sodium has been reduced by 50%. The term "light" can also be used to describe properties such as texture and colour, as long as the label explains the intent—for example, "light and fluffy."

Source: Canadian Food Inspection Agency. Available online at www.inspection.gc.ca/english/fssa/labeti/guide/tab7e.shtml. Accessed April 22, 2011.

Food manufacturers who want to include a health claim on their product must apply to Health Canada to have their claim evaluated and approved. Health Canada reviews the submissions and claims are authorized only if an extensive review of the scientific evidence indicates strong evidence in support of the claim. At present, six disease-risk-reduction claims are permitted in Canada (**Table 2.9**), but applications from food manufacturers for additional health claims are continually being considered, so more claims may be added in the future. Disease-risk-reduction claims are generally combined with nutrient-content claims; for example, to make a claim that osteoporosis risk is reduced, the food must also be a good source of calcium.

Table 2.9 Health Claims Used on Food Labels	
Calcium and osteoporosis	"A healthy diet with adequate calcium and vitamin D, and regular physical activity, help to achieve strong bones and may reduce the risk of osteoporosis. [Naming the food] is an excellent source of calcium and vitamin D."OR " [Naming the food] is a good source of calcium."
Sodium and high blood pressure	"A healthy diet containing foods high in potassium and low in sodium may reduce the risk of high blood pressure, a risk factor for stroke and heart disease. [Naming the food] is low in sodium."
Saturated fat and *trans* fat and risk of coronary heart disease	"A healthy diet low in saturated and *trans* fats may reduce the risk of heart disease. [Naming the food] is free of [or low in] saturated and *trans* fats."
Fruits and vegetables and cancer	"A healthy diet rich in a variety of vegetables and fruit may help reduce the risk of some types of cancer."

Table 2.9 *(Continued)*	
Foods low in starch or fermentable sugars and dental caries (cavities)	"Won't cause cavities." "Does not promote tooth decay." "Does not promote dental caries." "Non-cariogenic."
Plant sterols and cholesterol	"A serving (from the Nutrition Facts table) of (naming the food) contains X% of the daily amount of plant sterol to help reduce/lower cholesterol in adults." Daily amount = 2 g; X% must be > 10%

Sources: CFIA Health Claims Section 8.4.5 Summary of Disease Risk Reduction Claims. Available online at www.inspection.gc.ca/english/fssa/labeti/guide/ch8e.shtml. Accessed April 22, 2011.
Health Canada. Plant Sterols and Blood Cholesterol Lowering. Available online at www.inspection.gc.ca/english/fssa/labeti/guide/ch5ae/shtml#a5_7. Accessed April 22, 2011.

Nutrient-Function Claims Nutrient-function claims describe the role of a nutrient or dietary ingredient in maintaining normal structure or function in humans. For example, a structure/function claim about calcium may state that "calcium aids in the formation and maintenance of bones and teeth." These claims may also describe the general well-being that arises from consumption of a nutrient such as "fibre promotes laxation." These claims can be used on foods that contain at least 5% of the Daily Value of that nutrient per serving and must include mention of both the food and nutrient. For example, "milk is an excellent source of calcium which aids in the formation and maintenance of bones and teeth." Canadian regulations list a number of acceptable nutrient-function claims, some of which are shown in **Table 2.10.** If food manufacturers wish to label a food with a nutrient-function claim not on the list, they are encouraged to consult with Health Canada and are expected to have on hand scientific evidence supporting the claim, if this is requested.

Table 2.10 Selected Nutrient-Function Claims (For complete list, see source.)

Nutrient	Function Claim
Vitamin A	Aids normal bone and tooth development Aids in the development and maintenance of night vision Aids in maintaining the health of the skin and membranes
Vitamin D	Factor in the formation and maintenance of bones and teeth Enhances calcium and phosphorus absorption and utilization
Vitamin E	A dietary antioxidant A dietary antioxidant that protects the fat in body tissues from oxidation
Vitamin C	A factor in the development and maintenance of bones, cartilage, teeth, and gums A dietary antioxidant A dietary antioxidant that significantly decreases the adverse effects of free radicals on normal physiological functions A dietary antioxidant that helps to reduce free radicals and lipid oxidation in body tissues
Calcium	Aids in the formation and maintenance of bones and teeth
Phosphorus	Factor in the formation and maintenance of bones and teeth
Magnesium	Factor in energy metabolism, tissue formation, and bone development
Iron	Factor in red blood cell formation

Source: Canadian Food Inspection Agency: Guide to Food labelling and advertising. Ch 8: Health Claims. Available online at www.inspection.gc.ca/english/fssa/labeti/guide/ch8e.shtml#a8_6. Accessed March 20, 2011.

Interactive Tools to Help Canadians Understand Nutrition Labelling

Health Canada has developed several interactive tools to help Canadians better understand nutrition labelling. Health Canada's *Interactive Tools: Nutrition Labelling* is available on Health Canada's website. It helps Canadians to better interpret the meaning of % DV, to recognize the importance of the amount of food indicated as a serving, and to make decisions about which foods to select, based on a comparison of labels. An Interactive Nutrition Label and Quiz is also available on Health Canada's website and, an excellent way to test nutrition labelling skills.

Natural Health Products Labelling

The labelling of vitamin and mineral supplements falls under a different set of regulations than food labelling. Vitamin and mineral supplements, as well as essential fatty acids, amino acids, and probiotics, are considered natural health products and are regulated by the Natural Health Products Directorate of Health Canada.[21] Natural Health Products also include herbal remedies, traditional medicines, and homeopathic products, but these products will not be discussed in detail in this textbook.

The Natural Health Products Directorate was formed in 2004 and its regulations are being phased in over several years. The intention of the directorate is to assess a natural health product and to issue a licence for those products that are judged to be safe and effective under the recommended conditions of use. Natural Health Products, unlike most drugs, do not require a prescription and are selected by consumers for their personal care. Consumers can identify licensed natural health products by looking for the eight-digit Natural Product Number (NPN) (see **Figure 2.25**).

The labelling of natural health products differs from food labelling. There is no nutrition facts table. Instead the label of any natural health product includes the following:

- Product name
- Product licence holder
- Natural Product Number (NPN) or Homeopathic Medicine Number (DIN-HM)
- Product's medicinal ingredients
- Product's nonmedicinal ingredients
- Product's dosage form
- Product's recommended use or purpose (i.e., its health claim or indication)
- Risk information associated with the product's use (i.e., cautions, warnings, contra-indications, and known adverse reactions)

The directorate requires licence applicants to submit evidence of the safety and effectiveness of their product, based on scientific evidence that is similar to that required of food processors who want to include a health claim on their product.[22]

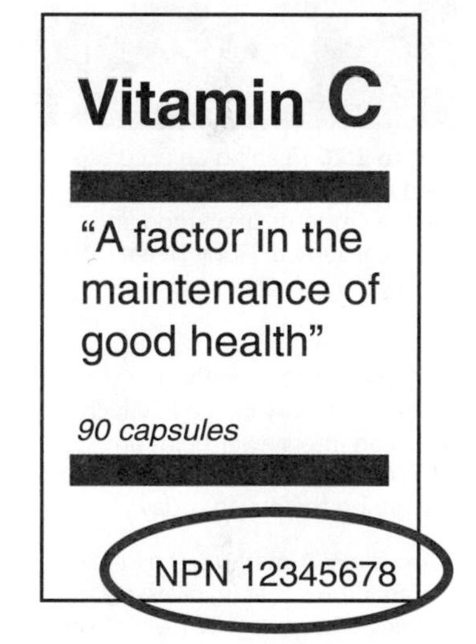

Figure 2.25 All natural health products have a natural health product number (NPN).
Source: Health Canada: Drugs & Health Products: Welcome to the Licensed NHPs Database. Available online at www.hc-sc.gc.ca/dhp-mps/prodnatur/applications/licen-prod/lnhpd-bdpsnh-eng.php. © Her majesty the Queen in Right of Canada, represented by the Minister of Health (2011).

2.6 Assessing Nutritional Health

Learning Objectives

- Name three types of information used in assessing nutritional status.
- Discuss the pros and cons of a dietary recall versus a dietary record for assessing food intake.
- Explain the types of tools used to assess the nutritional health of populations.

To be healthy, people need to consume combinations of foods that provide appropriate amounts of nutrients. Scientists have developed standards for the amounts of nutrients needed and tools for planning diets to meet these needs. But how do we know if the nutritional needs of an individual or the population are being met? Evaluating the **nutritional status** of individuals and populations can identify nutritional needs and be used to plan diets to meet these needs.

nutritional status State of health as it is influenced by the intake and utilization of nutrients.

Individual Nutritional Health

What is your nutritional status? Are you losing weight? Gaining weight? Do you have a history of heart disease in your family? Are you at risk for a nutrient deficiency because you cannot get to the store, cannot afford to buy healthy foods, or you do not know what to eat or how

nutritional assessment An evaluation used to determine the nutritional status of individuals or groups for the purpose of identifying nutritional needs and planning personal healthcare or community programs to meet those needs.

to cook? An individual **nutritional assessment** helps to determine if a person has a nutrient deficiency or excess, or is at risk of one, or if that individual is at risk for chronic diseases that are affected by diet. It requires a review of past and present dietary intake, a clinical assessment that evaluates body size and includes a medical history and physical exam, and laboratory measurements. Even with all these tools, diagnosing a nutritional deficiency or excess is not trivial. Estimates of dietary intake are not always accurate, and symptoms may be indistinguishable from other medical conditions.

Estimating Dietary Intake A good place to start when evaluating nutritional status is to determine what a person typically eats. This can be done by observing all the food and drink consumed by the individual for a specified period of time or by asking the subject to record or recall his or her intake. Neither option is ideal. Being observed can affect an individual's intake, and recording and recalling intake are imprecise measures because these methods rely on the memory and reliability of the consumer. For instance, a person who is attempting to lose weight may tend to report smaller portions than were actually eaten.[23] Despite this problem, the commonly used methods described here are the best tools available for evaluating dietary intake to predict nutrient deficiencies or excesses (see Science Applied: The Art of Assessing Food Intake).

Figure 2.26 Keeping an accurate food diary requires recording the precise amounts of all food and drink consumed. (Tony Freeman/PhotoEdit)

24-hour recall A method of assessing dietary intake in which a trained interviewer helps an individual remember what he or she ate during the previous day.

24-Hour Recall The most common method of assessing dietary intake is a **24-hour recall** in which a trained interviewer asks a person to recall exactly what he or she ate during the preceding 24-hour period. A detailed description of all food and drink, including descriptions of cooking methods and brand names of products, is recorded. Since food intake varies from day to day, repeated 24-hour recalls on the same individual provide a more accurate estimate of typical intake.

food diary A method of assessing dietary intake that involves an individual keeping a written record of all food and drink consumed during a defined period.

Food Diary* or *Food Intake Record Food intake information can also be gathered by having a consumer keep a **food diary**, or record, of all the food and drink consumed for a set period of time. Typically, this is done for two to seven days, including at least one weekend day, since most people eat differently on weekends than during the school or work week. Foods may be weighed or measured (**Figure 2.26**), or portion sizes just estimated (Table 2.2). The record should be as complete as possible, including all beverages, condiments, and the brand names and preparation methods. The tedious nature of this type of record can be a disadvantage, because in some cases, it may cause the consumer to change his or her intake rather than record certain items.

food frequency questionnaire A method of assessing dietary intake that gathers information from individuals about how often certain categories of food are consumed.

Food Frequency A **food frequency questionnaire** lists a variety of foods, and the consumer is asked to estimate the frequency with which he or she consumes each item or food group. The consumer may be asked, "How often do you drink milk?" or, "How many times a week do you eat red meat?" This method cannot be used to itemize a specific day's intake, but it can give a general picture of a person's typical pattern of food intake (**Figure 2.27**).

diet history Information about dietary habits and patterns. It may include a 24-hour recall, a food record, or a food frequency questionnaire to provide information about current intake patterns.

Diet History A **diet history** collects information about dietary patterns. It may review eating habits; for example, people may be asked whether they cook their own meals or skip lunch. It may also include a combination of methods to assess food intake, such as a 24-hour recall along with a food frequency questionnaire. The combination of two or more methods often provides more complete information than one method alone. For instance, if an individual's 24-hour recall does not include milk, but a food frequency questionnaire suggests that the individual usually drinks milk five days a week, the two can be combined to provide a more accurate picture of this individual's typical intake.

Analyzing Nutrient Intake Once information on food intake has been obtained, the nutrient content of the diet can be compared to recommended intakes. This can be done in a number of ways. To get a general picture of dietary intake, an individual's food record can be compared with a guide for diet planning, such as *Eating Well with Canada's Food Guide.* For example, does the individual consume the recommended amount of milk each day? A more precise and extensive analysis of dietary intake can be done by totalling the nutrients contributed by each food item.

SCIENCE APPLIED

The Art of Assessing Food Intake

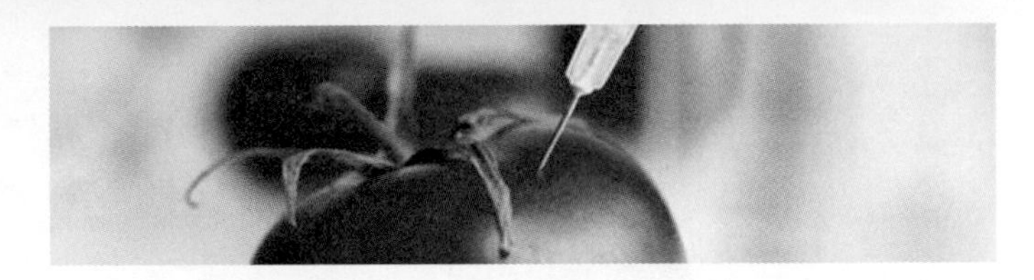

What did you have for dinner last night? How about last week? To study nutrient requirements and functions, scientists need to know exactly how much of each nutrient their subjects are consuming. To identify possible nutrient deficiencies or excesses, a dietitian needs to know a client's typical diet. To make nutrition recommendations for a population, public health officials need to know what the population is eating.

The most accurate way to know what people in a research study are eating is to control it. Subjects typically are housed in a research facility where all the food they consume is specially prepared and served to them. The nutrient content of the diet is calculated to meet the study needs; the food is weighed before it is served; and everything that is served must be consumed. This method, though, is only practical for small numbers of subjects. Larger studies can be managed if a research facility prepares all food and subjects consume it in their homes or workplaces—but only if the subjects are compliant with the diet, resist urges for a candy bar or bag of chips not included in the diet, and are honest about what they consume. Also, these techniques do not provide information on what people typically eat.

Food frequencies, food records, and 24-hour recalls can be used to evaluate what individuals typically eat, to assess a population's nutritional health, or to study the relationships between nutrient intake and disease. The tool used depends on the type of information needed and the available time and cost. The goal is to collect the most reliable data in the fastest and most cost-effective manner. For instance, a study of fruit and vegetable consumption could use a food frequency questionnaire, which is easy for participants and relatively inexpensive for researchers to analyze. Food frequencies are often used in large epidemiological studies of food intake patterns. When more specific information is required, food frequencies may not be the best tool.[1]

Food records and **24-hour recalls** can provide more detailed information. Food records created as subjects consume their meals can be very reliable. However, the act of recording can affect intake. Subjects may decide to skip the handful of chips they would have eaten rather than record that they ate it. In addition, food records are time-consuming for subjects and can be costly for researchers to collect and analyze. The 24-hour recall method, in contrast, can survey large numbers of people in a short time and can be conducted by telephone.[2] The process is easy for the subjects; also, it can provide information that is more comprehensive than a food frequency and more accurate than a food record because subjects are less likely to change their intake.

All these methods involve the possibility for error, as they depend on the memories and reliability of study subjects. The most common error in food intake data is underreporting of intake. Subjects may not remember all the foods they have consumed or choose not to report them all. Portion sizes can be hard to assess accurately, and it can be difficult to know what ingredients are included in foods prepared away from home.

To recognize these biases, improve accuracy, and decrease or control for error in estimates of intake, researchers have studied methods of evaluating food intake and have come up with unique ways of validating accuracy. People can be placed in settings where they are allowed to choose the types and amounts of foods they wish to eat and their choices secretly observed and recorded. This type of observational study can generate general estimates on how well individuals report their intake. More recent methods to test the accuracy of food intake data rely on measuring energy balance (see Section 7.3). In people who are not losing or gaining weight, the energy they eat will equal the energy they expend. A comparison of an individual's energy expenditure, which can be accurately measured, with his or her energy intake can be used to evaluate the reliability of that person's reported intake.[3] These types of investigations have found that underreporting of food intake is common. It occurs more often in women than in men and more in persons who are older, overweight, or trying to lose weight.[4]

Assessments of what people typically eat are used to make and evaluate nutrition recommendations and plan the food supply. Inaccurate dietary intake data can give a misleading view of the public's nutritional status or of the effectiveness of nutritional guidelines. For example, the incidence of childhood obesity has increased over the last few decades. If food intake data indicate that children have not increased their energy intake over this time, scientists might conclude that the increase in obesity is due to a decrease in physical activity and recommend that they increase their activity level. How might this conclusion and recommendation change if it were determined that the children were underreporting their food intake?

[1] Schaefer, E. J., Augustin, J. L., Schaefer, M. M., et al. Lack of efficacy of food frequency questionnaire in assessing dietary macronutrient intakes in subjects consuming diets of known composition. *Am. J. Clin. Nutr.* 71:746–751, 2000.

[2] Godwin, S. L., Chambers, E., 4th, and Cleveland, L. Accuracy of reporting dietary intake using various portion-size aids in-person and via telephone. *J. Am. Diet. Assoc.* 104:585–594, 2004.

[3] Black, A. E., Welch, A. A., and Bingham, S. A. Validation of dietary intakes measured by diet history against 24 h urinary nitrogen excretion and energy expenditure measured by the doubly-labeled water method in middle-aged women. *Br. J. Nutr.* 83:341–354, 2000.

[4] Becker, W., and Welten, D. Underreporting in dietary surveys–implications for development of food-based dietary guidelines. *Public Health Nutr.* 4:683–687, 2003.

Information on the nutrient composition of foods is available on food labels, in published food composition tables, and in computer databases. Food labels provide information only for some nutrients, and they are not available for all foods. Food composition tables generated by government and industry laboratories can provide more extensive information on food composition. Health Canada has compiled the Canadian Nutrient File, available on their website while in the United States, the USDA Nutrient Database for Standard Reference is available at www.

Food Frequency Questionnaire

On the following pages, please check the appropriate column indicating how often you consume each food.

	Once a day	Twice or more a day	Once a week	Twice or more a week	Once a month	Twice or more a month
Milk						
Whole						
Reduced fat						✓
Nonfat	✓					
Yogurt						
Whole						
Reduced fat			✓			
Nonfat						
Cheese						
Hard					✓	
Soft						✓
Reduced fat						
Ice cream						
Regular						
Reduced fat	✓					

Figure 2.27 Food Frequency
This section of a sample food frequency questionnaire can be used to obtain information about dairy product consumption patterns.

nal.usda.gov/fnic/foodcomp/search/. Computer programs with food composition databases are available for professionals and for home use (such as eaTracker.ca), and allow easy-to-use online nutrient analysis. To analyze nutrient intake correctly using a computer program, each food and the exact portion consumed must be entered into the program. If a food is not found in the computer database, an appropriate substitute can be used or the food can be broken down into its individual ingredients. For example, homemade vegetable soup could be entered as generic vegetable soup, or as vegetable broth, carrots, green beans, rice, and so on. If a new product has come on the market, the information from the food label can be added to the database. The advantage of computer diet analysis is that it is fast and accurate. A program can calculate the nutrients for each day or average them over several days. It can also compare nutrient intake to recommended amounts. However, the information generated by computer diet analysis is only useful if it is entered correctly and interpreted appropriately. Also, a nutrient intake that is below the recommended amount does not always indicate a serious deficiency, and intake that meets recommendations does not ensure adequate nutritional status. **Figure 2.28** illustrates how iProfile diet analysis software compares the nutrients in a diet with recommendations.

Nutrient	DRI	Intake	Percent of Recommendation (0% – 50% – 100%)
Vitamin A (RAE)	700 µg	525	75%
Vitamin C	75 mg	86	115%
Iron	18 mg	9.7	54%
Calcium	1000 mg	750	75%
Saturated fat	< 23.8 g	31.9	134%

Figure 2.28 Computerized diet analysis
In this example of an iProfile printout, which shows only a few nutrients, intake of vitamin C and saturated fat is above the recommended amounts and intake of vitamin A, calcium, and iron is below the recommended amounts.

anthropometric measurements
External measurements of the body, such as height, weight, limb circumference, and skinfold thickness.

Anthropometric Measurements Evaluating nutritional health also involves an assessment of an individual's height, weight, and body size. These **anthropometric measurements** can be compared with population standards (see Appendix B) or used to monitor changes in an individual over time (**Figure 2.29**). If an individual's measurements differ significantly from standards, it could indicate a nutritional deficiency or excess; however, this information should be evaluated only within the context of that person's personal and family history. For example, children who are small for their age may have a nutritional deficiency or may simply have inherited their small body size. Individuals who weigh less than the standard may be adequately nourished if they have never weighed more than their current weight and are otherwise healthy.

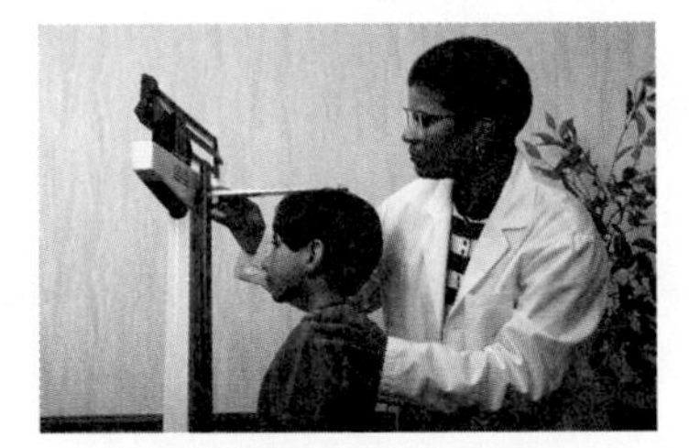

Figure 2.29 Height, weight, and body circumference are examples of anthropometric measurements. (Blair Seitz/Photo Researchers)

Medical History and Physical Exam A medical history is an important component of a nutritional assessment because dietary needs depend on genetic background, life stage, and health status. Family history is important because the risk of developing some nutrition-related diseases is affected by an individual's genes. If your mother died of a heart attack at age 50, you have a higher than average risk of developing heart disease. If you have a family history of diabetes, you have an increased risk of developing this disease. If both of your parents are overweight, it increases the chances that you, too, will have a weight problem.

Life stage is important because nutrient needs vary at different stages. Pregnant women need more of some nutrients and energy to support the development of a healthy newborn. Young infants have higher energy and protein needs per unit body weight than at any other time of life. The needs of older adults change as their body composition changes and the ability to digest and absorb certain nutrients declines.

Existing health conditions also affect dietary needs. Some conditions, such as arthritis, affect the ability to acquire and prepare food. Others affect the kinds of foods that should be consumed or the way nutrients are handled by the body. For example, gastrointestinal disorders may decrease the ability to digest foods and absorb nutrients. Kidney disease alters the ability to excrete nitrogen, a component of protein, and so affects the amount of protein that should be consumed.

In conjunction with personal and family medical history, a careful physical exam can detect the symptoms of, and risk factors for, nutrition-related diseases. In a physical exam, all areas of the body, including the mouth, skin, hair, eyes, and fingernails are examined for indications of poor nutritional status. Symptoms such as dry skin, cracked lips, or lethargy may indicate a nutritional deficiency, but these types of symptoms are nonspecific and may be due to factors unrelated to nutritional status. Determining whether the symptoms noted in a physical exam are due to malnutrition or another disease requires that they be evaluated not only within the context of each individual's medical history, but in conjunction with the results of laboratory measurements.

Laboratory Measurements Measures of nutrients or their by-products in body cells or fluids, such as blood and urine, can be used to detect nutrient deficiencies and excesses (see Appendix C). For instance, levels of various blood proteins are often used to assess protein status. Newly absorbed nutrients are carried to the cells of the body in the blood, therefore, the amounts of some nutrients in the blood may reflect the amount in the current diet rather than the total body status of the nutrient. To assess the status of these nutrients, it may be necessary to measure nutrient functions rather than just nutrient amounts. For example, vitamin B_6 is needed for chemical reactions involved in amino acid metabolism. Measuring the rates of chemical reactions that require vitamin B6 can be used to assess B6 status.

Laboratory data can also be used to evaluate risk for nutrition-related chronic diseases. For instance, heart disease risk can be assessed by measuring cholesterol levels in the blood. Measuring the amount of glucose in the blood can be used to diagnose diabetes. More sophisticated medical tests can be used to obtain additional information about the risk and progression of nutrition-related diseases. For example, procedures are available to determine the extent of coronary artery blockage in an individual with heart disease or to assess bone density in someone at risk for osteoporosis.

Stages of Nutrient Deficiency A nutritional deficiency usually takes time to develop. For example, an individual who is not meeting the requirement for protein may not suffer any physical signs of protein deficiency for months. Deficiencies generally progress through a number of stages. Appropriate nutritional assessment tools can identify deficiencies at any of these stages and allow intervention to restore nutritional health (**Figure 2.30**).

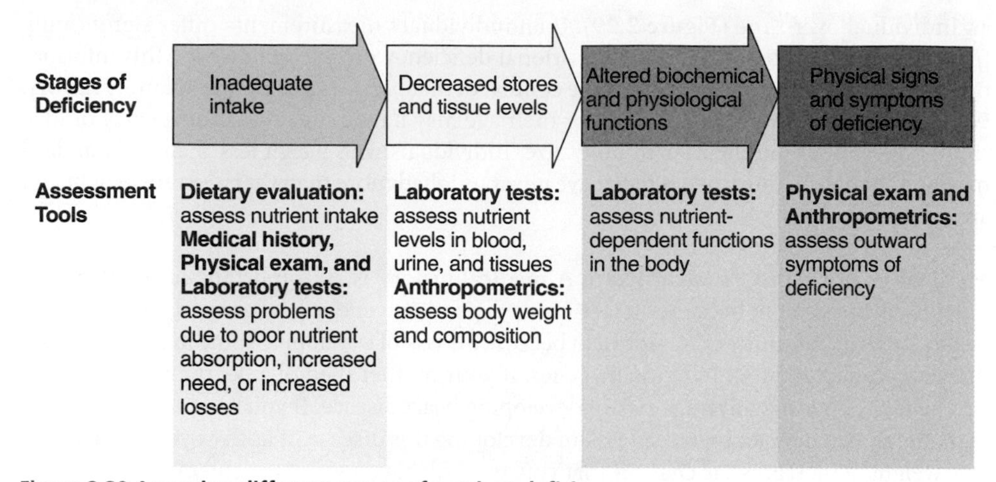

Figure 2.30 **Assessing different stages of nutrient deficiency**
Different assessment tools yield different types of information that can be used to evaluate the severity of a nutrient deficiency.

The earliest stage of nutrient deficiency is intake that is inadequate to meet needs. This may occur due to a deficient diet, poor absorption, increased need, or increased losses from the body. Assessment of dietary intake can help identify low levels in the diet and a physical exam and medical history can identify conditions that might reduce absorption or increase need or losses. The next stage of deficiency is declining nutrient stores in the body. Anthropometric measures that assess body weight and body fat can be used to monitor energy stores. Laboratory tests that measure levels of nutrients and their by-products in blood and tissues can be used to detect decreases in nutrient stores. For instance, measuring levels of the iron-containing protein ferritin in the blood can be used to detect low iron stores. The next stage of deficiency is altered biochemical or physiological functions, such as low enzyme activities or reduced amounts of regulatory or structural molecules. Finally, if a deficiency persists, function is disrupted enough that physical signs and symptoms, referred to as clinical symptoms, become apparent. For instance, a deficiency of iron causes fatigue, weakness, and decreased work capacity (see Critical Thinking: Assessing Nutritional Health).

Nutritional Health of the Population

We know that there is enough food available in Canada to meet the needs of the population. We also know that poor nutritional choices from this food supply result in diets high in some nutrients and low in others. This kind of information is obtained by monitoring what foods are available and what is consumed.

food disappearance surveys
Surveys that estimate the food use of a population by monitoring the amount of food that leaves the marketplace.

Monitoring the Food Supply The food available to a population is estimated using **food disappearance surveys**. The food supply includes all that is grown, manufactured, or imported for sale in the country. Food use, or "disappearance," is estimated by measuring what food is sold. These types of surveys are used to estimate what is available to the population, provide year-to-year comparisons, and identify trends in the diet; but they tend to overestimate actual intake because they do not consider losses that occur during processing, marketing, and home use. But assuming that losses remain relatively constant over time, the data can highlight changes in the types of food Canadians eat. For example, **Figure 2.31** illustrates the food disappearance data on milk consumption since 1990, using data collected by Statistics Canada. It shows that the consumption of whole milk, which is high in fat, has declined since the 1990s and the consumption of lower-fat milks has modestly increased. From this it can be concluded that fat intake from milk has declined. But the graph also indicates that total milk consumption has been declining. This may alert the government that calcium intake from milk has decreased and that there may be a risk for low calcium intake in the population. A limitation of disappearance data is that they do not give any information about how milk consumption is distributed throughout the population or who is at risk of inadequate calcium intake.

CANADIAN CONTENT

CRITICAL THINKING:

Assessing Nutritional Health

Background

Darra is a 23-year-old university student. Recently, she has been feeling tired and has had difficulty concentrating in class. She goes to the health clinic where she is weighed and measured. A physician does a physical exam and asks about her medical history. She suspects that Darra is anemic, so she orders a blood sample for laboratory analysis. Darra is referred to a dietitian to assess her dietary intake.

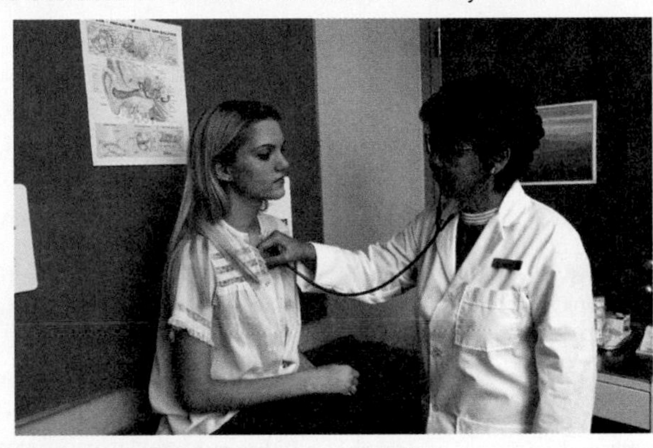

(Michael Newman/PhotoEdit)

Clinical assessment

The physician notes that Darra appears thin and pale. Anthropometric measurements of height and weight tell us that she is1.6 m (5'4") tall and weighs 52 kg (114 lbs.) She recalls that a year ago, she weighed 54.5 kg (120 lbs) and hasn't been trying to lose weight. Although her body weight is in the normal range, her unintentional weight loss is a concern.

Dietary assessment

Darra tells the dietitian that she stopped eating red meat last year. Using information from a 24-hour recall, the dietitian enters her diet into a computer program. A portion of the analysis is shown here.

NUTRIENT	VALUE	% OF RECOMMENDATION
Kcalories	1,500	68%
Protein	46 g	100%
Vitamin C	110 mg	146%
Vitamin A	1,028 mg	147%
Iron	6 mg	33%
Calcium	1,300 mg	130%

Laboratory assessment

The results of her blood test indicate that her blood hemoglobin level is 112 g per L (litre) of blood and that her hematocrit, which measures the ratio of the volume of blood cells to the total volume of blood, is 0.340. When iron intake is low, blood hemoglobin and hematocrit decline.

Critical Thinking Questions

Do you think Darra has iron deficiency anemia? Evaluate this by looking up the normal values for hemoglobin and hematocrit in Appendix C.

What about her iron intake? Compare her iron intake to the recommendations for a woman of her age.

Should Darra be concerned about the nutrients she is consuming in excess of the recommended amount? Use the DRI tables to determine if they are likely to pose a risk.

iProfile Use iProfile to find foods that are good sources of iron.

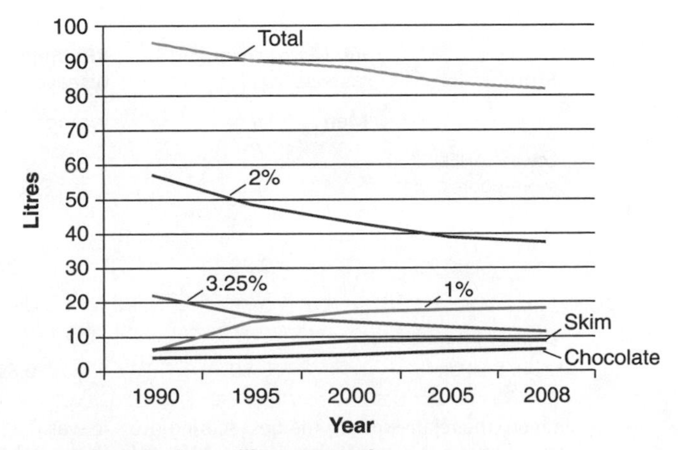

Figure 2.31 Trends in Canadian milk consumption
Food disappearance data illustrate that there was an increase in the consumption of lower-fat milks and a decrease in whole-milk consumption since 1990.
Source: Canadian Dairy Commission: The Industry: Per Consumption of Milk & Cream: Available online at www.dairyinfo.gc.ca/pdf/camilkcream.pdf. Accessed March 15, 2011.

SCIENCE APPLIED

Does Following a Healthy Dietary Pattern Reduce the Risk of Disease?

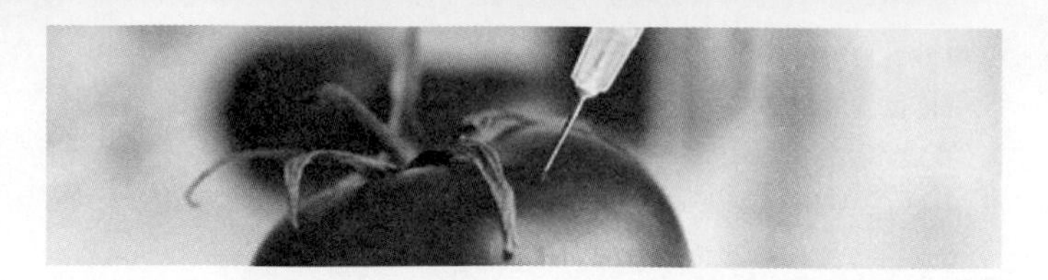

Healthy eating indices such as the CHEI, have been used to assess diet quality. Also, dietary patterns such as Canada's Food Guide have been developed with the intention of reducing the risk of chronic disease. One interesting question to ask, then, is whether any scientists have determined directly whether there is an association between overall dietary pattern and disease risk or, put simply, are people who eat an overall nutritious diet healthier than people who don't? While no such comprehensive study has yet to be conducted using the Canadian Healthy Eating Index, the AHEI (the U.S. version of this index) has been used in a study to answer exactly this question. In order to understand the results of this study, however, it is necessary to understand the concept of relative risk.

Consider this very simple example illustrated in the diagram below. There are two groups of 10 people. The two groups differ in that one group has a high intake of nutrient X and the other has a low intake of nutrient X. The health of these two groups is watched for several years and the number of people in each group who develop disease A is recorded. At the end of the experiment, two people in the high-intake group have disease A, while four in the low-intake group have disease A. The frequency of disease A in the low-intake group is 0.4 (i.e., 4 /10), while in the-high intake group, it is 0.2 (i.e., 2/10). The ratio of the two frequencies is a measure of relative risk.

Low nutrient intake

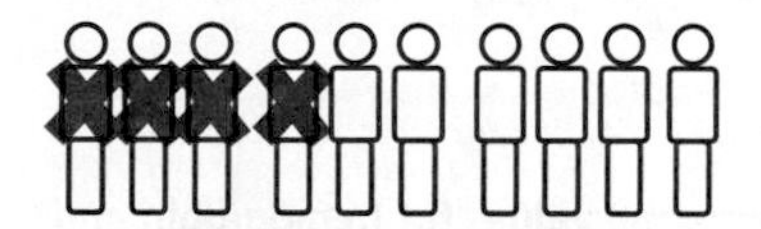

N = 10

Number with Disease A = 4

Frequency of Disease A = 0.4

Relative Risk:

$$\frac{\text{Frequency of Disease A (low intake)}}{\text{Frequency of Disease A (low intake)}} = \frac{0.4}{0.4} = 1$$

High nutrient intake

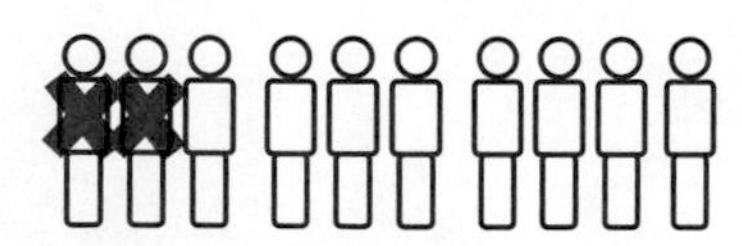

N = 10

Number with Disease A = 2

Frequency of Disease A = 0.2

Relative Risk:

$$\frac{\text{Frequency of Disease A (high intake)}}{\text{Frequency of Disease A (low intake)}} = \frac{0.2}{0.4} = 0.5$$

When determining relative risk data, researchers select a reference group to which the other groups are compared (i.e., the denominator in the relative risk calculation) and it is assigned a relative risk of 1. In our example above, the reference group is the low-intake group. Therefore, compared to the low-nutrient group, the high-nutrient group has a relative risk of 0.5. This means that compared to the low-intake group, the high-intake group has a 50% reduction of risk of getting disease A. This is summarized in the table below.

Nutrient Intake	Relative Risk
Low intake of nutrient	1
High intake of nutrient	0.5

A relative risk (RR) greater than 1 means an increased risk compared to the reference group, while as in the example here, a relative risk less than 1 means reduced risk. In addition to calculating the RR, additional analysis is typically conducted to determine whether the difference between groups is statistically significant.

Several years ago, a study entitled "Evaluating adherence to recommended diets in adults: the Alternate Healthy Eating Index" was published.[1] This paper describes an observational study in which the dietary intake for 100,000 people was scored using the AHEI, after which their health was assessed for chronic disease over several years.

The AHEI described in this study gave high scores to diets that had high intakes of fruits and vegetables (excluding potatoes), fish and poultry, plant proteins such as beans, lentils, and tofu, cereal fibre, and polyunsaturated fatty acids. A person following Canada's Food Guide would score very well on this index.

The participants of the study were divided into five equal groups called quintiles in order of their score. The first quintile was made up of people with the lowest scores (lowest 20%); the next quintiles had progressively higher scores, and the top quintile (top 20%) had the highest scores. Researchers next determined the number of cases of major chronic diseases that occurred over the duration of the study. "Major chronic disease" was defined as any occurrence of cardiovascular disease, cancer, or death by any natural causes. Relative risk for each quintile relative to the lowest quintile was calculated. These were the results:

Score	RR: Major chronic disease	RR: Major chronic disease
	Men	women
Lowest AHEI score-poorest diet	1.0	1.0
	0.96	0.97
	0.88	0.92
	0.79	0.95
Highest AHEI score-best diet	0.80	0.89

In men, the relative risk in the best scoring quintile was 0.80; another way of saying this is that there was a 20% reduction in risk. In women, there was also a reduction, but of only 11%. These results were statistically significant, that is, they were not likely to be due to chance.

Score	RR: CVD Men	RR: CVD Women
Lowest AHEI score-poorest diet	1.0	1.0
	0.85	0.95
	0.79	0.80
	0.67	0.75
Highest score AHEI-best score	0.61	0.72

The experiment was repeated looking only at cases of cardiovascular disease and the risk reduction was even greater in this case, reflecting a strong link between cardiovascular disease and diet.

From these results, the authors concluded that a healthy dietary pattern, such as Canada's Food Guide, is associated with reduced disease risk.

(McCullough, M.L., Willett, W.C. **Evaluating adherence to recommended diets in adults: the Alternate Healthy Eating Index**.Public Health Nutr. 9(1A):152-7, 2006.)

Monitoring Nutritional Status The nutritional status of the population is monitored by examining and comparing trends in food intake and health. This is done by interviewing individuals within the population to determine what food is actually consumed, and collecting information on health and nutritional status. The Canadian Community Health Survey began surveying Canadians on health related issues in 2001, and continues to collect data. The nutrient intake of thousands of Canadians has been compiled and the results of this survey will be referred to throughout this text.

Healthy Eating Indices In order to measure the overall quality of a dietary pattern, scientists have developed a scoring system, or index. One such index used to evaluate the American diet was called the Healthy Eating Index (HEI) and a second Alternative Healthy Eating Index (AHEI) was also developed that scored more dietary factors.[24] Recently, a similar index was developed to evaluate the adequacy of the diet of Canadians, called the Canadian Healthy Eating Index (CHEI). This provides a score of overall diet quality by measuring various components of the diet for its conformance to Canada's Food Guide (e.g., points are awarded for consuming orange and green leafy vegetables, whole grain products, etc.). An individual who carefully follows all the recommendations of Canada's Food Guide scores 100. In a recent analysis of the quality of the Canadian diet using the CHEI, the average score for Canadians was 58.8, with only 0.5% of Canadians scoring above 80 and 16% scoring below 50, suggesting that many Canadians can make improvements in their diet.[25] (See Science Applied: Does Following a Healthy Dietary Pattern reduce the Risk of Disease?)

The Integrated Pan-Canadian Healthy Living Strategy The results of population studies help governments develop strategies to improve the health of Canadians. In 2002, a meeting of federal, provincial, and territorial ministers of health identified the need for such an initiative. As a result, after consultation, a healthy-living strategy was developed, and the Co-ordinating Committee for the Healthy Living Network was formed to promote collaboration between all levels of government, non-governmental agencies, aboriginal organizations, and the private sector, to implement this strategy. It was called the Integrated Pan-Canadian Health Living Strategy,[26] and its goals are to improve overall health outcomes, focusing specifically on healthy eating, physical activity, and healthy body weight, and to reduce regional and income-related health disparities.

The Healthy Living Strategy identified the following targets:

- By 2015, increase by 20% the proportion of Canadians who make healthy food choices.
  - According to the 2003 CCHS, the proportion of the adult population consuming fruits and vegetables at least five times per day was 39%. A 20% increase would mean that 46.8% of people would be doing so.
- By 2015, increase by 20% the proportion of Canadians who participate in at least 30 minutes of moderate to vigorous activity daily.
  - According to the 2003 CCHS, 50.4% of Canadians engaged in physical activity as described above. A 20% increase would mean that 60.5% of people would be doing so.
- By 2015, increase by 20% the proportion of Canadians at a normal body weight, defined as a BMI (body mass index) of 18.5 to 24.9.

- According to the 2003 CCHS, 46.7% of Canadian fell into the normal body weight category. A 20% increase would mean the 56% of Canadians would be in the normal category.
- The Healthy Living Strategy also intends to address health disparities by targeting programming to low-income Canadians, who tend to be less healthy than their high-income counterparts, and First Nations people, to who tend to be less healthy than other Canadians.

CASE STUDY OUTCOME

Li wanted to make sure that he was making nutritious food choices. The dorm cafeteria offered a seemingly endless variety of selections, and he didn't know which foods to choose. He was tempted by all those great desserts but knew that he would regret the extra weight. To plan his diet, he learned to use Canada's Food Guide, which promotes a diet that ensures that he will meet his nutrient requirements and reduces his risk of chronic disease. By comparing his diet to the recommendations of Canada's Food Guide, Li figured out whether he was consuming some foods in excess as well as the types of foods he needed to add to his diet. When he bought snacks, he used the information on food labels to determine nutritional benefits. He tried to go for regular walks, even when he had a lot of studying to do. As a result of his planning, he kept his weight where he wanted it, found extra energy for school work, and was able to ace his final exams!

APPLICATIONS

Personal Nutrition

1. **What do you eat? To find out, use a My Food Guide Servings Tracker sheet for your sex-age group by accessing the My Food Guide Servings Tracker on the Health Canada website. Visit www.healthcanada.gc.ca/foodguide and click on My Food Guide Servings Tracker. Print three copies of the tracker for your gender and age group. Use the Food Intake Record to keep a diary of everything you eat for three days (or alternatively, use a form provided by your instructor). Since you may eat differently on weekends, record for two weekdays and one weekend day. To make sure you don't forget anything, carry your record with you and record food as it is consumed. This food record will be used in Applications throughout this book to focus on particular nutrients. List the foods and amounts that you eat for each day in the leftmost column of the Tracker sheet. Make the record as complete as possible by using the following tips (and also refer to the sample food record in this section):**

   a. Include all food and drink, and be as specific as possible. For example, did you eat a chicken breast or thigh?

   b. Measure or estimate as carefully as possible the portion size that you ate, for example: 125 ml of rice, 10 potato chips, 175 ml of tofu, and 250 ml of milk. See Table 2.2 to help you determine serving sizes.

   c. Record the preparation or cooking method. For example, was your potato peeled? Was your chicken skinless? Was it baked or fried?

   d. Include anything added to your food, for instance, butter, ketchup, or salad dressing.

   e. Don't forget snacks, beverages, and desserts.

   f. If the food is from a fast-food chain, list the name.

   g. You may have to break down mixed dishes into their ingredients. For example, a tuna sandwich can be listed as two slices of whole-wheat bread, 15 ml of mayonnaise, and 75 g of tuna packed in water.

SAMPLE FOOD RECORD

FOOD OR BEVERAGE	KIND/HOW PREPARED	AMOUNT
Chicken salad sandwich:		
wheat bread		2 slices
chicken	skinless breast	75 g
mayonnaise	low-fat	15 ml
diet cola		1 can (355 ml)

2. **Does your diet meet Canada's Food Guide recommendations of your individual tracking sheet? To find out, complete the remaining columns of the Food Intake Record as shown in Critical Thinking: Applying My Food Guide Servings Tracker.**

   a. List the foods and amounts from one day of your food intake record on the left-side column of the form.

   b. Indicate the number of Canada's Food Guide servings for each food item in the correct food-group category. See Section 2.3, Eating Well with Canada's Food Guide: The Guide: Page by Page: Page 3, for examples of Canada's Food Guide servings.

   c. Total up the number of servings consumed in each Food Guide food group for each of the three days and then determine the average number of servings in each food group for the three days. How do they compare to the recommended numbers in the check boxes?

   d. Complete the additional statements check list. How do the types of food you consumed compare to the recommendations?

   e. How many of your choices fell into the "foods to limit" category? Determine more nutrient-dense substitutes for these foods. Are there any food groups and food you need to eat more or less of?

   f. Use iProfile to analyze the nutrient intake of your diet. How well do you conform to the DRIs?

   g. How does your nutritional status, based on conformance to the DRIs, determined in (f), compare to the results of your Canada's Food Guide analysis?

3. **What's in your packaged foods? Select three packaged foods and check the labels.**

   a. How many kcalories are in each serving?

   b. How much total carbohydrate, total fat, and fibre are in a serving of each?

   c. How does each of these foods fit into your overall daily diet with regard to total carbohydrate? Total fat? Dietary fibre?

   d. If you consumed a serving of each of these three foods, how much more fat could you consume during the day without exceeding the Daily Value? How much more total carbohydrate and fibre should you consume that day to meet the Daily Value for a 2,000-kcalorie diet?

General Nutrition Issues

1. **To encourage healthy eating in her family, a nutrition student hangs a copy of *Eating Well with Canada's Food Guide* on the refrigerator. Her family is enthusiastic, and can see from the guide the number of Canada's Food Guide servings from each food group they need to consume. They don't, however, know how to determine the size of a serving. What simple ways can our nutrition student suggest for determining a Canada's Food Guide serving? (Hint: see Table 2.2)**

2. **Review the important aspects of nutrition labelling by visiting the Health Canada interactive tools on the Health Canada website (visit www.healthcanada.gc.ca and search nutrition labelling interactive tools).**

3. **Are "health foods" different than standard products?**

   a. Compare the label from a product such as cereal, crackers, or cookies purchased at the grocery store to the label from a comparable product from a "health" or "natural" food store.

b. Which is higher in total fat? Saturated fat? *Trans* fat? Cholesterol?

c. Which has more kcalories per serving?

d. Which has more sugars?

e. How do the ingredients differ?

f. What other differences or similarities do you notice?

4. Food intake data from population surveys provided the following information about the intake of milk and soft drink in 1977–1978 and 1994–1996:

POPULATION	YEAR	MILK CONSUMPTION (ML PER DAY)	SOFT DRINK CONSUMPTION (ML PER DAY)
Teenage boys	1977–1978	480	210
(ages 12–19)	1994–1996	300	600
Teenage girls	1977–1978	300	210
(ages 12–19)	1994–1996	180	420

a. How might these trends have affected the number of kcalories from added sugars in the teen diet?

b. How might these trends affect calcium intake in teenagers?

SUMMARY

2.1 Nutrition Recommendations for the Canadian Diet

- Nutrition recommendations made to the public for health promotion and disease prevention are based on available scientific knowledge.
- Dietary standards such as the Dietary Reference Intakes (DRIs) provide recommendations for intakes of nutrients and other food components that can be used to plan and assess the diets of individuals and populations. Intakes at these levels will avoid deficiencies and excesses and prevent chronic diseases in the majority of healthy persons.
- *Eating Well with Canada's Food Guide* describes a food-based dietary pattern.

2.2 Dietary Reference Intakes

- The DRIs include four sets of nutrient intake recommendations. The Estimated Average Requirement (EAR) is the amount of a nutrient that is estimated to meet the needs of half of the people in a particular gender and life-stage group. The Recommended Dietary Allowances (RDAs), which are based on the EARs, are recommendations calculated to meet the needs of nearly all healthy individuals (97%–98%) in a specific group. Adequate Intakes (AIs) estimate nutrient needs based on average intakes by healthy populations when there is insufficient scientific evidence to calculate an EAR and RDA. Tolerable Upper Intake Levels (ULs) provide a guide for a safe upper limit of intake.
- Energy recommendations of the DRIs include Estimated Energy Requirements (EERs), which provide a recommendation for energy intakes that will maintain body weight, and Acceptable Macronutrient Distribution Ranges (AMDRs), which recommend the proportions of energy intake that should come from carbohydrate, protein, and fat.
- The Dietary Reference Intakes can be used to evaluate and plan nutrient intakes for populations, as a guide for individual intake, and to make judgements about excessive intakes for individuals and populations.

2.3 Eating Well with Canada's Food Guide

- *Eating Well with Canada's Food Guide* assists Canadians in selecting a nutritious diet. The guide recommends the amounts and types of foods that ensure that nutrient requirements are met and that the risk of chronic disease is reduced. Foods from four food groups are recommended, with the number of servings dependent on age and sex. The four food groups are vegetables and fruit, grain products, milk and alternatives, and meat and alternatives. Additional statements in the guide help to ensure nutritious choices are made.
- Internet resources such as My Food Guide, My Food Guide Servings Tracker, and Eatracker.ca, assist Canadians in evaluating and planning their diets.

2.4 Other Food Guides

- Other countries have food guides, such as the U.S. MyPlate, but they are all very similar in their recommendations.

2.5 Food and Supplement Labels

- All food labels in Canada must contain the product name, the manufacturer, an ingredients list, and a nutrition facts table. The nutrition facts table follows a standard format and is designed to provide consumers with the information they need to compare foods. The % Daily Values allow consumers to make comparisons between foods, and thus make more nutritious choices.
- Additional information that may appear on a food label includes a nutrient-content claim, a nutrient-function claim, and a disease-risk-reduction claim. The claims that are permitted are strictly regulated by Health Canada and the Canadian Food Inspection Agency.
- Supplements, such as vitamins, minerals, essential fatty acids, and amino acids, are considered natural health products and have different labelling requirements than does food.

2.6 Assessing Nutritional Health

- Individual nutritional status is assessed by evaluating dietary intake, examining clinical parameters such as body size, and interpreting laboratory values within the context of an individual's medical history.
- The nutritional status of populations is monitored by measuring what foods are available, what foods are consumed, and how nutrient intake is related to overall health.

REVIEW QUESTIONS

1. Which types of DRI standards can be used as a goal for individual intake?
2. What type of DRI standard can be used to evaluate the adequacy of nutrient intake in a population?
3. How is the EAR cutpoint method used?
4. What is a dietary pattern?
5. What is the purpose of the arcs of the rainbow in the Canada's Food Guide graphic?
6. Why must food from all food groups be consumed?
7. Why are the additional statements in Canada's Food Guide important?
8. How much food from each group does a 56-year-old man need to meet his nutritional needs?
9. How do the Daily Values help consumers compare foods and make good food choices?
10. What kinds of health claims can appear on a food label?
11. What determines the order in which food ingredients are listed on a label?
12. What is the purpose of the Beyond the Basics meal planning system?
13. What is nutritional status?
14. List the components of individual nutritional assessment.
15. How do food disappearance data help monitor the nutritional health of populations?
16. What is the CCHS and why is it important?
17. What is the Pan-Canadian Healthy Living Strategy?

REFERENCES

1. Report of a Joint WHO/FAO Expert Consultation. Diet, nutrition and the prevention of chronic diseases. Available online at http://whqlibdoc.who.int/trs/WHO_TRS_916.pdf. Accessed May 20, 2009.
2. National Research Council, Food and Nutrition Board. *Recommended Dietary Allowances*. 10th ed. Washington DC National Academies Press, 1989.
3. Health Canada. The development of the Dietary Reference Intakes. Available online at http://www.hc-sc.gc.ca/fn-an/nutrition/reference/dri_dev-elab_anref-eng.php. Accessed January 10, 2010.
4. Health Canada. *Eating Well with Canada's Food Guide.* Available online at http://www.hc-sc.gc.ca/fn-an/food-guide-aliment/index-eng.php. Accessed March 5, 2011.
5. Institute of Medicine, Food and Nutrition Board. *Dietary Reference Intakes for Calcium, Phosphorus, Magnesium, Vitamin D, and Fluoride*. Washington, DC: National Academies Press, 1997.
6. Institute of Medicine, Food and Nutrition Board. *Dietary Reference Intakes for Thiamin, Riboflavin, Niacin, Vitamin B-6, Folate, Vitamin B-12, Pantothenic Acid, Biotin, and Choline*. Washington, DC: National Academies Press, 1998.
7. Institute of Medicine, Food and Nutrition Board. Dietary Reference intakes, Applications in Dietary Planning. Washington DC: National Academies Press, 2003.
8. Institute of Medicine, Food and Nutrition Board. *Dietary Reference Intakes for Energy, Carbohydrates, Fiber, Fat, Protein, and Amino Acids*. Washington, DC: National Academies Press, 2002.
9. Anderson, J.W., Baird, P., Davis, R.H., Jr., et al. "Health benefits of dietary fiber." *Nutr Rev* 67(4):188-205, 2009.
10. WCRF/AICR Expert Report. *Food, Nutrition, Physical Activity and the Prevention of Cancer: a Global Perspective.* Available online at http://www.dietandcancerreport.org. Accessed Jan 28, 2010.
11. Mensink, R.P., Zock, P.L., Kester, A.D., et al. Effects of dietary fatty acids and carbohydrates on the ratio of serum total to HDL cholesterol and on serum lipids and apolipoproteins: a meta-analysis of 60 controlled trials. *Am J Clin Nutr.* 77(5):1146-55, 2003.
12. Hu, F.B. "Plant-based foods and prevention of cardiovascular disease: an overview." *Am J Clin Nutr.* 78(3 Suppl), 544S-551S, 2003.
13. He, K., Song, Y., Daviglus, M.L., et al. Accumulated evidence on fish consumption and coronary heart disease mortality: a meta-analysis of cohort studies.*Circulation* 109(22):2705-11, 2004.
14. Malik, V.S., Schulze, M.B., Hu, F.B. Intake of sugar-sweetened beverages and weight gain: a systematic review. *Am J Clin Nutr.* 84(2):274-88, 2006.
15. Haskell, W.L., Lee, I.M., Pate, R.R., et al. Physical activity and public health: updated recommendation for adults from the American College of Sports Medicine and the American Heart Association. *Med Sci Sports Exerc.* 39(8):1423-34, 2007.
16. Hill, J.O., Wyatt, H., Phelan, S., et al. The National Weight Control Registry: is it useful in helping deal with our obesity epidemic? *J Nutr Educ Behav.* 37(4):206-10, 2005.
17. Katamay, S.W., Esslinger, K.A., Vigneault, M., et al. *Eating Well with Canada's Food Guide* (2007): development of the food intake pattern. *Nutr Rev.* 65(4):155-66, 2007.
18. Canadian Diabetes Association. Just the Basics. Available online at http://www.diabetes.ca/files/jtb17x_11_cpgo3_1103.pdf. Accessed Mar 5, 2011.
19. Canadian Diabetes Association. Beyond the Basics. Available online at http://www.diabetes.ca/for-professionals/resources/nutrition/beyond-basics/. Accessed Mar 5, 2011.
20. Canadian Food Inspection Agency: Guide to Food Labelling and Advertising: Chapter 6. Available online at http://www.inspection.gc.ca/english/fssa/labeti/guide/ch6e.shtml#a6_2. Accessed Feb 1, 2010.
21. Health Canada. Natural Health Products Directorate. 2011. Available online at http://www.hc-sc.gc.ca/ahc-asc/branch-dirgen/hpfb-dgpsa/nhpd-dpsn/index-eng.php. Accessed March 5, 2011.
22. Health Canada. Evidence for the safety and efficacy of finished natural health products. Available online at http://www.hc-sc.gc.ca/dhp-mps/prodnatur/legislation/docs/efe-paie-eng.php#13. Accessed April 22, 2011.
23. Johansson, L., Solvoll, K., Bjørneboe, G-E. A., et al. Under- and overreporting of energy intake related to weight status and lifestyle in a nationwide sample. *Am. J. Clin Nutr.* 68:226–274, 1998.
24. McCullough, M.L., Willett, W.C. Evaluating adherence to recommended diets in adults: the Alternate Healthy Eating Index. *Public Health Nutr.* 9(1A):152-7, 2006.
25. Garriguet, D. Diet Quality in Canada. *Health Rep* 20(3): 41-52, 2009.
26. Public Health Agency of Canada. The Integrated Pan-Canadian Healthy Living Strategy. Available online at http://www.phac-aspc.gc.ca/hp-ps/hl-mvs/ipchls-spimmvs-eng.php. Accessed Feb 1, 2010.

3 Digestion, Absorption, and Metabolism

Case Study

Sheila was worried as she left the doctor's office. She had just learned that now, in addition to having high blood pressure and high cholesterol, her blood sugar levels were elevated. Her weight was seriously affecting her health, and the doctor had suggested gastric bypass surgery.

Sheila had been overweight since she was 18, and now at 38, she was heavier than she had ever been—131 kg (289 lb) on her 1.8 m (5′10″) frame. She felt fat and miserable and even simple things in her life were difficult. She hated squeezing into restaurant booths and airplane seats, and shopping for clothes was depressing. She had tried every diet, but when she did manage to lose weight, she would gain it back—and then some. Exercise was challenging because she was too embarrassed to put on a bathing suit and her knees hurt too much to walk very far.

The surgical procedure Sheila was considering would alter her digestive tract in two ways. First, it would reduce the size of her stomach so it could hold less food; second, it would shorten the length of her intestines so her body could not absorb as much of what she ate. She knew it would be a drastic move—the surgical procedure itself carries risks and would permanently alter the amounts and types of foods she could eat. She would have to eat very small meals throughout the day. Overeating or choosing the wrong foods could result in pain, sweats, chills, and nausea. Even if she were careful, diarrhea could be a constant problem. Bypassing part of her intestines would also put her at risk for vitamin deficiencies. On the plus side, though, she would feel less hungry, eat less, and lose weight.

Should she choose a procedure such as this that permanently changes her digestive tract?

This chapter describes the digestive tract that gastric bypass surgery so dramatically alters.

(Digital Vision/Getty Images, Inc.)

(© istockphoto.com/Robyn Mackenzie)

Chapter Outline

3.1 Food Becomes Us

Learning Objective

- Describe the organization of life from atoms to organisms.

The old adage that you are what you eat is not literally true, but biochemically it is a fact. The food we eat provides all of the energy we need to stay alive and active and all the raw materials we need to build and maintain our body structures and synthesize regulatory molecules.

atoms The smallest units of an element that still retain the properties of that element.

elements Substances that cannot be broken down into products with different properties.

chemical bonds Forces that hold atoms together.

molecules Units of two or more atoms of the same or different elements bonded together.

Atoms and Molecules

To be useful to us, the food we eat must be broken into smaller components, absorbed into the body, and then converted into forms that can be used. Our bodies and the food we eat, like all matter on Earth, are made of units called **atoms**. Atoms cannot be further broken down by chemical means. Atoms of different **elements** have different characteristics. Carbon, hydrogen, oxygen, and nitrogen are the most abundant elements in our bodies and in the foods we eat. These atoms can be linked by forces called **chemical bonds** to form **molecules** (**Figure 3.1** and Appendix I). The chemistry of all life on Earth is based on organic molecules, which are those that contain carbon bonded to hydrogen. As discussed in Chapter 1, carbohydrates, lipids, proteins, and vitamins are nutrient classes that are made up of organic molecules, whereas water and minerals are inorganic nutrients because they do not contain carbon-hydrogen bonds.

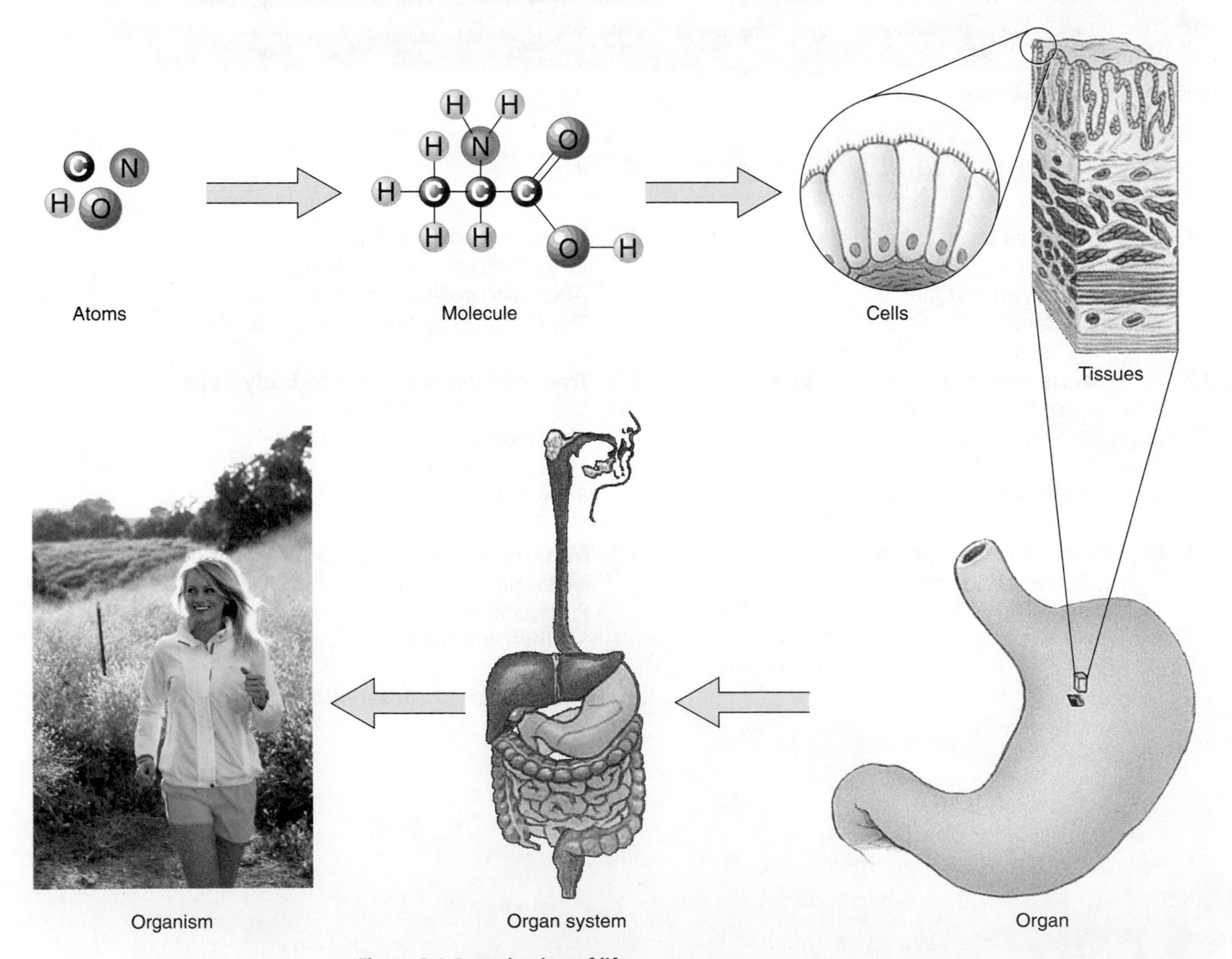

Figure 3.1 Organization of life
The organization of life begins with atoms that form molecules, which are then organized into cells to form tissues, organs, organ systems, and whole organisms. (Photo: Fuse/Getty Images, Inc.)

Cells, Tissues, and Organs

In any living system, molecules are organized into structures that form **cells**, the smallest unit of life. Cells of similar structure and function are organized into tissues. The human body contains four types of tissue: muscle, nerve, epithelial, and connective. These tissues are organized in varying combinations into **organs**, which are discrete structures that perform specialized functions in the body (see Figure 3.1). The stomach is an example of an organ; it contains all four types of tissue.

cells The basic structural and functional units of plant and animal life.

organs Discrete structures composed of more than one tissue that perform a specialized function.

Organ Systems

Most organs do not function alone but are part of a group of co-operative organs called an organ system. The organ systems in humans include the nervous system, respiratory system (lungs), urinary system (kidneys and bladder), reproductive system, cardiovascular system (heart and blood vessels), lymphatic/immune system, muscular system, skeletal system, endocrine system (hormones), integumentary system (skin and body linings), and digestive system (**Table 3.1**). An organ may be part of more than one organ system. For example, the pancreas is part of the endocrine system as well as the digestive system. Organ systems work together to support the entire organism.

Table 3.1 Organ Systems and Their Functions

Organ System	What It Includes	What It Does
Nervous	Brain, spinal cord, and associated nerves	Responds to stimuli from the external and internal environments; conducts impulses to activate muscles and glands; integrates activities of other systems.
Respiratory	Lungs, trachea, and air passageways	Supplies the blood with oxygen and removes carbon dioxide.
Urinary	Kidneys and their associated structures	Eliminates wastes and regulates the balance of water, electrolytes, and acid in the blood.
Reproductive	Testes, ovaries, and their associated structures	Produces offspring.
Cardiovascular	Heart and blood vessels	Transports blood, which carries oxygen, nutrients, and wastes.
Lymphatic/Immune	Lymph and lymph structures, white blood cells	Defends against foreign invaders; picks up fluid leaked from blood vessels; transports fat-soluble nutrients.
Muscular	Skeletal muscles	Provides movement and structure.
Skeletal	Bones and joints	Protects and supports the body; provides a framework for the muscles to use for movement.
Endocrine	Pituitary, adrenal, thyroid, pancreas, and other ductless glands	Secretes hormones that regulate processes such as growth, reproduction, and nutrient use.
Integumentary	Skin, hair, nails, and sweat glands	Covers and protects the body; helps control body temperature.
Digestive	Mouth, pharynx, esophagus, stomach, intestines, pancreas, liver, and gallbladder	Ingests and digests food; absorbs nutrients into the blood; eliminates unabsorbed food residues.

Source: Adapted from Marieb, E. N., and Hoehn, K. *Human Anatomy and Physiology*, 7th ed. Menlo Park, CA: Benjamin Cummings, 2007.

The digestive system is the organ system primarily responsible for the movement of nutrients into the body; however, several other organ systems are also important in the process of using these nutrients. The endocrine system secretes chemical messengers that help regulate food intake and absorption. The nervous system aids in digestion by sending nerve signals that help control the passage of food through the digestive tract. Once absorbed, nutrients are transported to individual cells by the cardiovascular system. The body's urinary, respiratory, and integumentary systems allow for the elimination of metabolic waste products.

3.2 An Overview of the Digestive System

Learning Objectives

- Define "digestion" and "absorption."
- Explain the roles of mucus, enzymes, nerves, and hormones in the digestive tract.
- Describe the function of phagocytes, lymphocytes, and antibodies.

digestion The process of breaking food into components small enough to be absorbed into the body.

absorption The process of taking substances into the interior of the body.

The digestive system provides two major functions: **digestion** and **absorption**. Most food must be digested in order for the nutrients it contains to be absorbed into the body. For example, a slice of whole-wheat bread does not travel through the digestive system intact. First it is broken apart, releasing its carbohydrate, protein, and fat. Most of the carbohydrate is digested to sugars, the protein to amino acids, and the fat to fatty acids. Sugars, amino acids, and fatty acids can be absorbed. The fibre from the bread is carbohydrate that cannot be digested and therefore is not absorbed into the body. Fibre and other unabsorbed substances pass through the digestive tract and are excreted in the **feces**.

feces Body waste, including unabsorbed food residue, bacteria, mucus, and dead cells, which is excreted from the gastrointestinal tract by passing through the anus.

Structure of the Gastrointestinal Tract

The main part of the digestive system is the **gastrointestinal tract**. It is also referred to as the GI tract, gut, digestive tract, intestinal tract, or alimentary canal. It can be thought of as a hollow tube, about 10 metres in length, that runs from the mouth to the anus. The organs of the gastrointestinal tract include the mouth, pharynx, esophagus, stomach, small intestine, large intestine, and anus (**Figure 3.2a**). The inside of the tube that these organs form is called the lumen (**Figure 3.2b**). Food within the lumen of the gastrointestinal tract has not been absorbed and is therefore technically still outside the body. Therefore, if you swallow something that cannot be digested, such as an apple seed, it will pass through your digestive tract and exit in the feces without ever being broken down or entering your blood or cells. Only after food is transferred into the cells that line the intestine by the process of absorption is it actually "inside" the body.

gastrointestinal tract A hollow tube consisting of the mouth, pharynx, esophagus, stomach, small intestine, large intestine, and anus, in which digestion and absorption of nutrients occur.

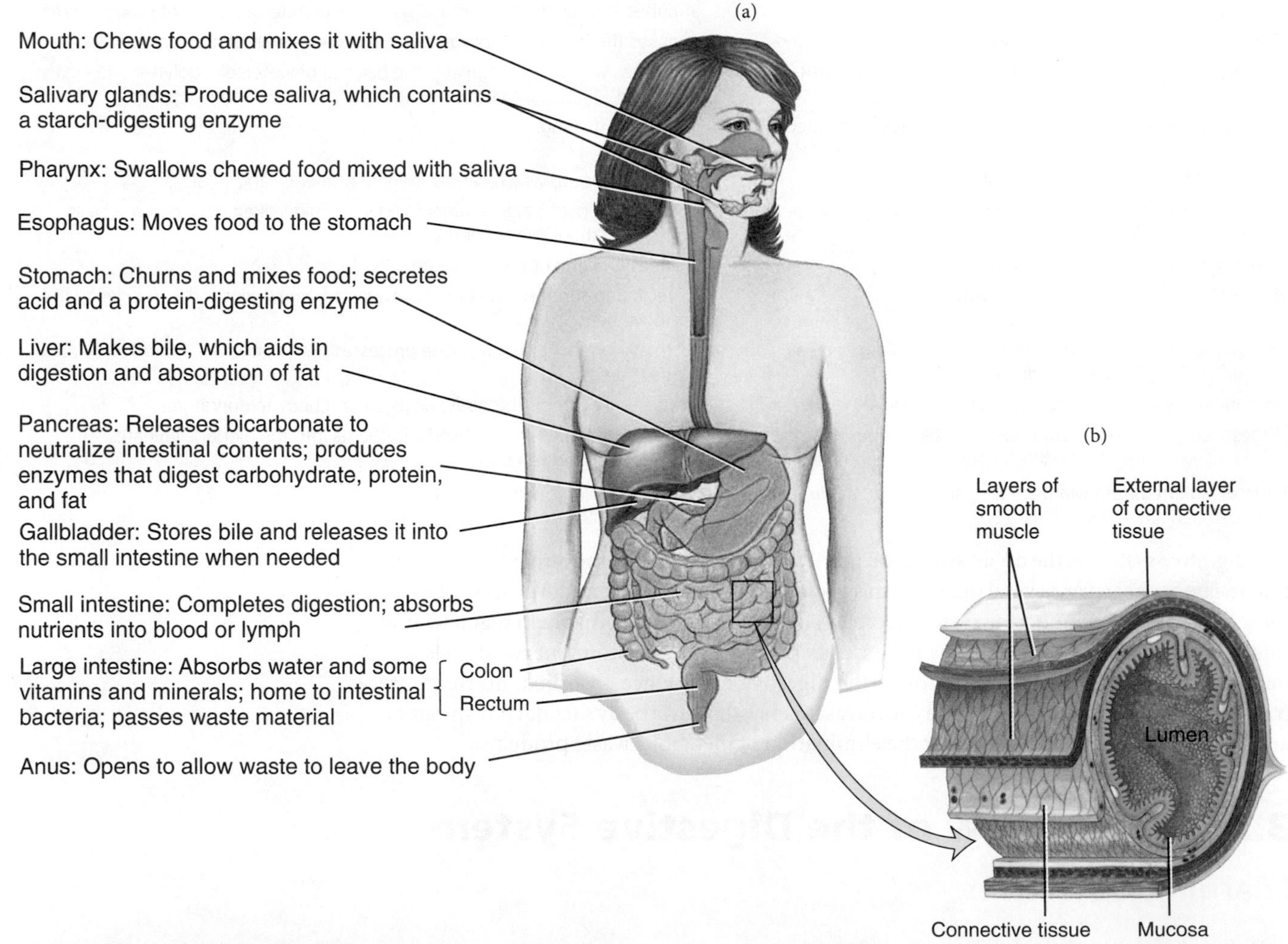

Figure 3.2 Structure of the digestive system
(a) The digestive system consists of the organs of the gastrointestinal tract, as well accessory organs that include the salivary glands, liver, gallbladder, and pancreas.
(b) A cross-section through the wall of the small intestine shows the four tissue layers: mucosa, connective tissue, smooth muscle layers, and outer connective tissue layer. (**Source:** Adapted from Solomon E. P., Schmidt R. R., and Adragna P. J., *Human Anatomy and Physiology*, 2nd ed. Philadelphia: Saunders College Publishing, 1990.)

Transit Time The amount of time it takes for food to pass the length of the GI tract from mouth to anus is referred to as **transit time**. In a healthy adult, transit time is about 24-72 hours. It is affected by the composition of the diet, physical activity, emotions, medications, and illnesses. To measure transit time, researchers add a nonabsorbable dye to a meal and measure the time between consumption of the dye and its appearance in the feces. The shorter the transit time, the more rapid the passage through the digestive tract.

transit time The time between the ingestion of food and the elimination of the solid waste from that food.

Structure of the Gut Wall The wall of the GI tract contains four layers of tissue (see Figure 3.2b). Lining the lumen is the **mucosa**, a layer of mucosal cells that serves as a protective layer and is responsible for the absorption of the end products of digestion. The cells of the mucosa are in direct contact with churning food and harsh digestive secretions. Therefore, these cells have a short lifespan—only about 2-5 days. When these cells die, they are sloughed off into the lumen, where some components are digested and absorbed and the remainder are excreted in the feces. Because mucosal cells reproduce rapidly, the mucosa has high nutrient requirements and is therefore one of the first parts of the body to be affected by nutrient deficiencies. Surrounding the mucosa is a layer of connective tissue containing nerves and blood vessels. This layer provides support, delivers nutrients to the mucosa, and provides the nerve signals that control secretions and muscle contractions. Layers of smooth muscle—the type over which we do not have voluntary control—surround the connective tissue. The contraction of smooth muscles mixes food, breaks it into smaller particles, and propels it through the digestive tract. The final, external layer is also made up of connective tissue and provides support and protection.

mucosa The layer of tissue lining the GI tract and other body cavities.

Digestive Secretions

Digestion inside the lumen of the GI tract is aided by digestive secretions. One of these substances is **mucus**, a viscous material produced by cells in the mucosal lining of the gut. Mucus moistens, lubricates, and protects the digestive tract. **Enzymes**, another component of digestive system secretions, are protein molecules that speed up chemical reactions without themselves being consumed or changed by the reactions (**Figure 3.3**). In digestion, enzymes accelerate the breakdown of nutrients. Different enzymes are needed for the breakdown of different nutrients. For example, an enzyme that digests carbohydrate has no effect on fat, and one that digests fat has no effect on carbohydrate. Digestive enzymes and their actions are summarized in **Table 3.2.**

mucus A viscous fluid secreted by glands in the GI tract and other parts of the body, which acts to lubricate, moisten, and protect cells from harsh environments.

enzymes Protein molecules that accelerate the rate of specific chemical reactions without being changed themselves.

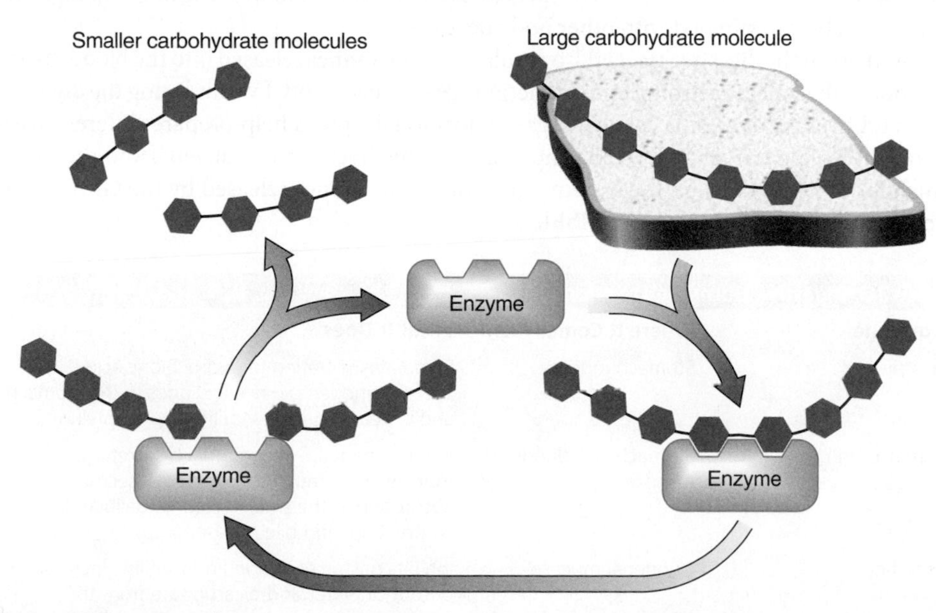

Figure 3.3 Enzyme action
Enzymes speed up chemical reactions without themselves being altered by the reaction. In this example, the enzyme amylase breaks a large carbohydrate molecule into two smaller ones.

Table 3.2 Enzyme Functions

Enzyme	Where It Is Found	What It Does
Salivary amylase	Mouth	Breaks starch (a large carbohydrate molecule) into smaller carbohydrate molecules.
Rennin	Stomach	Causes the milk protein casein to curdle.
Pepsin		Breaks proteins into polypeptides and amino acids.
Trypsin	Pancreas	Breaks proteins and polypeptides into shorter polypeptides.
Chymotrypsin		Breaks proteins and polypeptides into shorter polypeptides.
Carboxypeptidase		Breaks polypeptides into amino acids.
Pancreatic Lipase		Breaks triglycerides into monoglycerides, fatty acids, and glycerol.
Pancreatic amylase		Breaks starch into shorter glucose chains and maltose.
Carboxypeptidase, aminopeptidase, and dipeptidase	Small intestine	Breaks polypeptides into amino acids.
Lipase		Breaks monoglycerides into fatty acids and glycerol.
Sucrase		Breaks sucrose into glucose and fructose.
Lactase		Breaks lactose into glucose and galactose.
Maltase		Breaks maltose into glucose.
Dextrinase		Breaks short chains of glucose into individual glucose molecules.

Regulation of Gastrointestinal Function

Nerve signals help regulate activity in the GI tract. The sight and smell of food, as well as the presence of food in the gut, stimulate nerves throughout the GI tract. For example, food in the mouth can trigger a nerve impulse that signals the stomach to prepare itself for the arrival of food. Nerve signals cause muscle contractions that churn, mix, and propel food through the gut at a rate that allows for optimal absorption of nutrients. Nerve signals also stimulate or inhibit digestive secretions. For example, food in the mouth stimulates digestive secretions in the stomach. After food has passed through a section of the digestive tract, digestive secretions decrease and muscular activity slows to conserve energy and resources for other body processes. The nerves in the GI tract can also communicate with the brain so digestive activity can be coordinated with other body needs.

hormones Chemical messengers that are produced in one location, released into the blood, and elicit responses at other locations in the body.

Activity in the digestive tract is also regulated by **hormones** released into the bloodstream. Hormones that affect gastrointestinal function are produced both by cells lining the digestive tract and by a number of accessory organs. Hormonal signals help prepare different parts of the gut for the arrival of food and thus regulate the digestion of nutrients and the rate at which food moves through the system. Some of the hormones released by the GI tract and their functions are summarized in **Table 3.3.**

Table 3.3 Digestive Hormone Functions

Hormone	Where It Comes From	What It Does
Gastrin	Stomach mucosa	Stimulates secretion of hydrochloric acid (HCl) and pepsinogen by gastric glands in the stomach and increases gastric motility and emptying.
Somatostatin	Stomach and duodenal mucosa	Inhibits the following: stomach secretion, motility, and emptying; pancreatic secretion; absorption in the small intestine; gallbladder contraction; and bile release.
Secretin	Duodenal mucosa	Inhibits gastric secretion and motility; increases output of water and bicarbonate from the pancreas; increases bile output from the liver.
Cholecystokinin (CCK)		Stimulates contraction of the gallbladder to expel bile; increases output of enzyme-rich pancreatic juice.
Gastric inhibitory peptide		Inhibits gastric secretion and motility.

The Gastrointestinal Tract and Immune Function

The GI tract is an important part of the immune system, which protects the body from infection by foreign invaders. The GI tract limits the absorption of toxins and disease-causing organisms, but if an invading substance, or **antigen**, does get past the intestinal cells, the immune system has a number of ways to destroy it before it spreads to other parts of the body. The cells of the immune system consist of the different types of white blood cells **(Figure 3.4)**. Because many of these reside in the GI tract, a foreign substance or harmful organism that enters the body through the mucosa can be destroyed quickly. The first types of cells to come to the body's defense are called phagocytes. They target any invader, engulf it, and destroy it by breaking it up so that antigens are "presented" at the surface of the phagocyte **(Figure 3.4a)** These antigens are then detected by lymphocytes, which react by producing and secreting protein molecules called **antibodies**. Antibodies bind to invading antigens and help to destroy them, in part by making it easier for phagocytes to engulf them **(Figure 3.4b)**. Each antibody is designed to fight off only one specific antigen. Once the body has made antibodies to a specific antigen, it remembers and can rapidly produce these antibodies to fight that antigen any time it enters the body. Other lymphocytes bind directly to infected cells and destroy them **(Figure 3.4c)**. This type of lymphocyte helps eliminate cells that have been infected by viruses and bacteria and can also destroy cancer cells and foreign tissue.

antigen A foreign substance (almost always a protein) that, when introduced into the body, stimulates an immune response.

antibodies Proteins produced by cells of the immune system that destroy or inactivate foreign substances in the body.

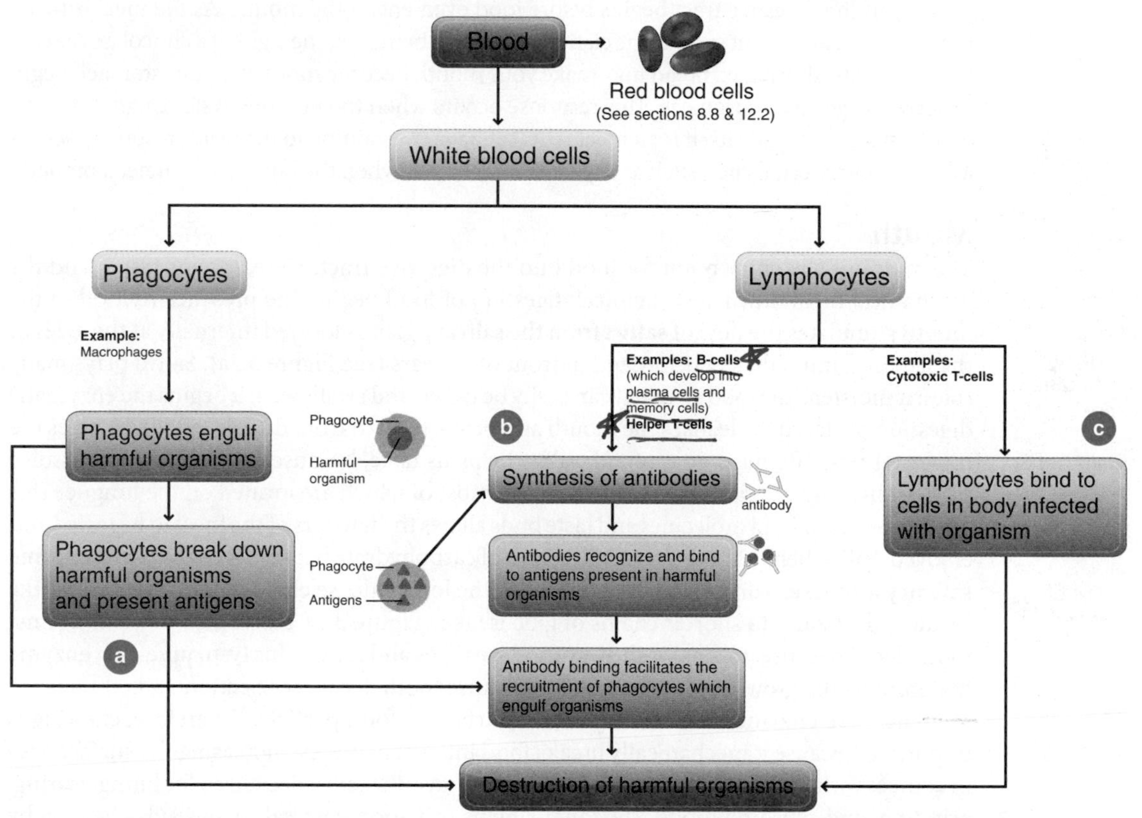

Figure 3.4 The cells of the immune system. The immune system can destroy harmful organisms in several ways: a) the action of the phagocytes; b) the synthesis of antibodies, which bind to organisms, and c) the binding, by lymphocytes, to infected cells. These three steps all ultimately lead to the destruction of harmful organisms.

The immune system protects us from many invaders without our even being aware that the battle is going on. Unfortunately, the response of the immune system to a foreign substance is also responsible for allergic reactions. An allergic reaction occurs when the immune system produces antibodies to a substance, called an **allergen**, that is present in our diet or environment. For example, a food allergy occurs when proteins absorbed from food are seen as foreign and trigger an immune response. The immune response causes symptoms that range from hives to life-threatening breathing difficulties (see Section 6.3). Food allergies affect less than 2% of adults and up to 6% of children under 3 years of age. They are responsible for 150 deaths each year. Foods that most commonly cause allergies in adults include seafood, peanuts, tree nuts, fish, and eggs.[1]

allergen A substance that causes an allergic reaction.

3.3 Digestion and Absorption

Learning Objectives

- Describe what happens in each of the organs of the gastrointestinal tract.
- Discuss factors that influence how quickly food moves through the gastrointestinal tract.
- Explain how the structure of the small intestine enhances its function.
- Distinguish simple diffusion, facilitated diffusion, and active transport.

To be used by the body, food must be eaten and digested, and the nutrients must be absorbed and transported to the cells of the body. Because most foods we consume are mixtures of carbohydrate, fat, and protein, the physiology of the digestive tract is designed to allow the digestion of all of these components without competition among them.

Sights, Sounds, and Smells

Activity in the digestive tract begins before food even enters the mouth. As the meal is being prepared, sensory input such as the clatter of the table being set, the sight of a chocolate cake, or the smell of freshly baked bread may make your mouth become moist and your stomach begin to secrete digestive substances. This response occurs when the nervous system signals the digestive system to ready itself for a meal. This cephalic (pertaining to the head) response occurs as a result of external cues, such as sight and smell, even when the body is not in need of food.

Mouth

The mouth is the entry point for food into the digestive tract. Here, food is tasted and the mechanical breakdown and chemical digestion of food begin. The presence of food in the mouth stimulates the flow of **saliva** from the salivary glands located internally at the sides of the face and immediately below and in front of the ears (see Figure 3.2a). Saliva plays many roles: it moistens the food so that it can easily be tasted and swallowed; it begins the enzymatic digestion of starch; it cleanses the mouth and protects teeth from decay; and it lubricates the upper GI tract. By moistening food, saliva helps us taste, because food molecules dissolve in the saliva and are carried to the taste buds, most of which are located on the tongue. This contact between food molecules and taste buds allows the flavours of the meal to be tasted and enjoyed. Saliva begins the chemical digestion of carbohydrate because it contains the enzyme **salivary amylase**. Salivary amylase can break the long glucose chains of starch in foods like bread and cereal into shorter chains of glucose (see Figure 3.3). Saliva also protects against tooth decay because it helps wash away food particles and it contains **lysozyme**—an enzyme that inhibits the growth of bacteria that may cause tooth decay (see Section 4.6).

saliva A watery fluid produced and secreted into the mouth by the salivary glands. It contains lubricants, enzymes, and other substances.

salivary amylase An enzyme secreted by the salivary glands that breaks down starch.

lysozyme An enzyme in saliva, tears, and sweat that is capable of destroying certain types of bacteria.

Digestive enzymes can act only on the surface of food particles. Therefore, chewing is important because it mechanically breaks food into small pieces, increasing the surface area in contact with digestive enzymes. Adult humans have 32 teeth, specialized for biting, tearing, grinding, and crushing foods. The tongue helps mix food with saliva and aids chewing by constantly repositioning food between the teeth. Chewing also breaks apart fibre that traps nutrients in some foods. If the fibre is not broken apart, some nutrients cannot be absorbed. For example, unless chewed thoroughly, a raisin or a kernel of corn will travel intact through the digestive tract and the vitamins and minerals in the fibrous skin and the nutrients within the raisin or corn kernel will be less available.

Pharynx

The tongue initiates swallowing by moving the bolus of chewed food mixed with saliva back toward the **pharynx**. The pharynx is shared by the digestive tract and the respiratory tract: food and liquid pass through the pharynx on their way to the stomach, and air passes here on its way to and from the lungs. We are able to start the muscular contractions of swallowing by choice, but once initiated, swallowing becomes involuntary and proceeds under the control of nerves. During swallowing, the air passages are blocked by a valve-like flap of tissue called the **epiglottis**, which ensures that the bolus of food passes to the stomach, not the lungs

pharynx A funnel-shaped opening that connects the nasal passages and mouth to the respiratory passages and esophagus. It is a common passageway for food and air and is responsible for swallowing.

epiglottis A piece of elastic connective tissue at the back of the throat that covers the opening of the passageway to the lungs during swallowing.

(**Figure 3.5a**). Sometimes food can pass into an upper air passageway. It is usually dislodged with a cough, but if it becomes stuck, it can block the flow of air and cause choking. A quick response is required to save the life of a person whose airway is completely blocked. The Heimlich manoeuvre, which forces air out of the lungs by using a sudden application of pressure to the upper abdomen, can blow an object out of the blocked air passage (see **Figure 3.5b**).

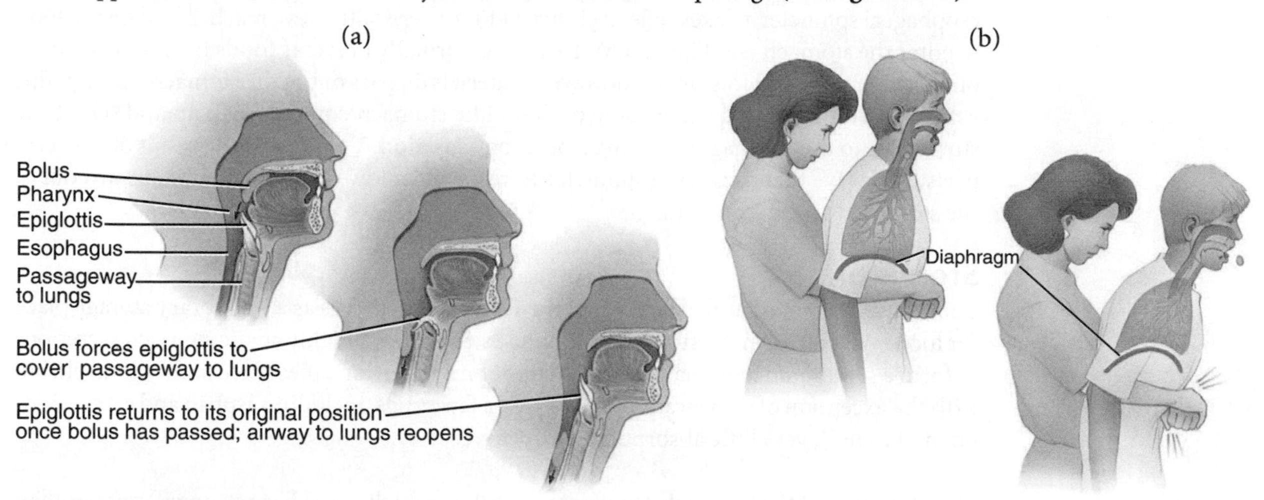

Figure 3.5 Swallowing
(a) When a bolus of food is swallowed, it pushes the epiglottis down over the opening to the air passageways. (b) If food becomes lodged in the airways, it can be dislodged by the Heimlich manoeuvre.

Esophagus

The **esophagus** is a tube that passes through the diaphragm, a muscular wall separating the abdomen from the chest cavity where the lungs are located, to connect the pharynx and stomach. In the esophagus the bolus of food is moved along by rhythmic contractions of the smooth muscles, called **peristalsis** (**Figure 3.6**). Peristalsis is like an ocean wave that moves through the muscle, producing a narrowing in the lumen that pushes food and fluid in front of it. It takes only about 4-8 seconds for solid food to move from the pharynx down the esophagus to the stomach and even less time for liquids to make the trip. The peristaltic waves in the esophagus are so powerful that food and fluids will reach your stomach even if you are upside down. This contractile

esophagus A portion of the GI tract that extends from the pharynx to the stomach.

peristalsis Coordinated muscular contractions that move food through the GI tract.

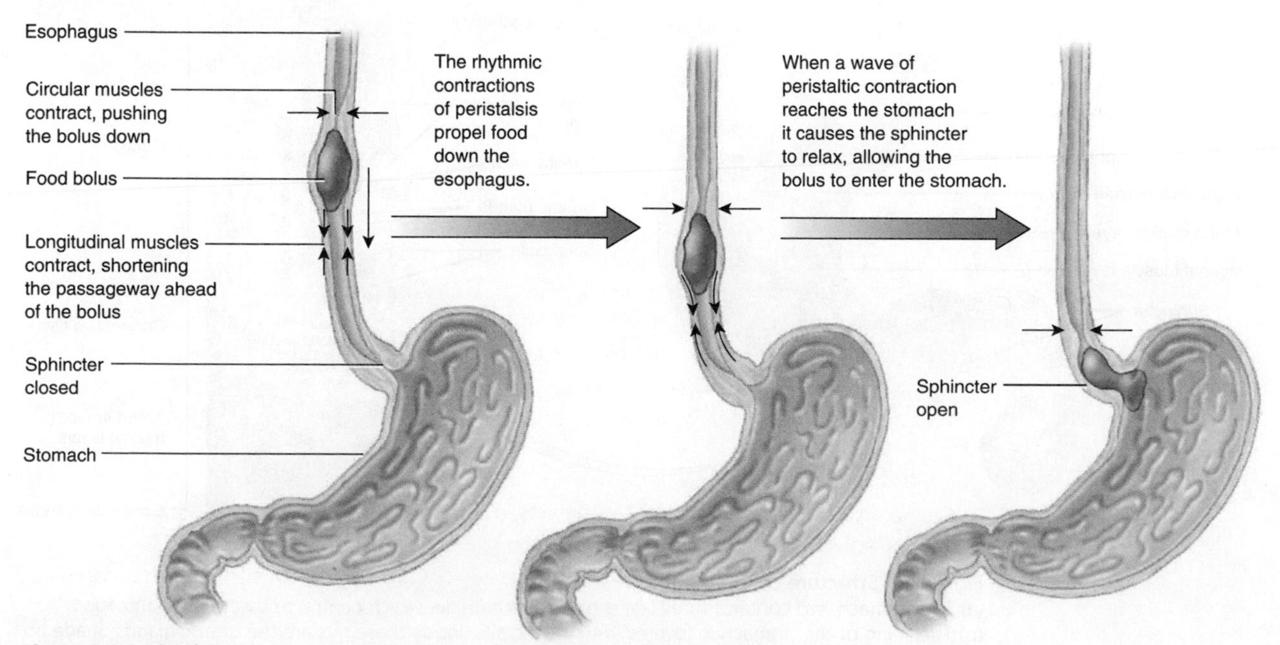

Figure 3.6 Peristalsis
The rhythmic contractions of peristalsis propel food down the esophagus. When a peristaltic wave reaches the stomach, the gastroesophageal sphincter relaxes to allow the food bolus to enter.

movement, which is controlled automatically by the nervous system, occurs throughout the GI tract, pushing the food bolus along from the pharynx through the large intestine.

sphincter A muscular valve that helps control the flow of materials in the GI tract.

To move from the esophagus into the stomach, food must pass through a **sphincter**, a muscle that encircles the tube of the digestive tract and acts as a valve. When the muscle contracts, the valve is closed. The gastroesophageal sphincter, also called the cardiac or lower esophageal sphincter, relaxes reflexively just before a peristaltic wave reaches it, allowing food to enter the stomach (see Figure 3.6). This valve normally prevents foods from moving back out of the stomach. Occasionally, however, materials do pass out of the stomach through this valve. Heartburn occurs when some of the acidic stomach contents leak up and out of the stomach into the esophagus, causing a burning sensation. Vomiting is the result of a reverse peristaltic wave that causes the sphincter to relax and allow the food to pass upward out of the stomach toward the mouth.

Stomach

The stomach is an expanded portion of the GI tract that serves as a temporary storage place for food. While held in the stomach, the bolus is mixed with highly acidic stomach secretions to form a semiliquid food mass called **chyme**. Some digestion takes place in the stomach, but with the exception of some water, alcohol, and a few drugs, including aspirin and acetaminophen (Tylenol), very little absorption occurs here.

chyme A mixture of partially digested food and stomach secretions.

The Structure of the Stomach The stomach walls are thicker and have stronger muscles than other segments of the GI tract. Two layers of muscle, one running longitudinally down the tract and one running around it, surround most of the GI tract. The stomach has an additional layer that circles it diagonally, allowing for powerful contractions that thoroughly churn and mix the stomach contents (**Figure 3.7a**).

The surface of the stomach mucosa is covered with cells that produce large amounts of protective mucus. This surface is interrupted by millions of tiny openings called gastric pits. These pits lead to gastric glands, which contain several different types of cells that secrete substances into the stomach (**Figure 3.7b**). These stomach secretions are collectively referred to as gastric juice. Gastric glands also contain cells that secrete a variety of hormones and hormone-like compounds into the blood.

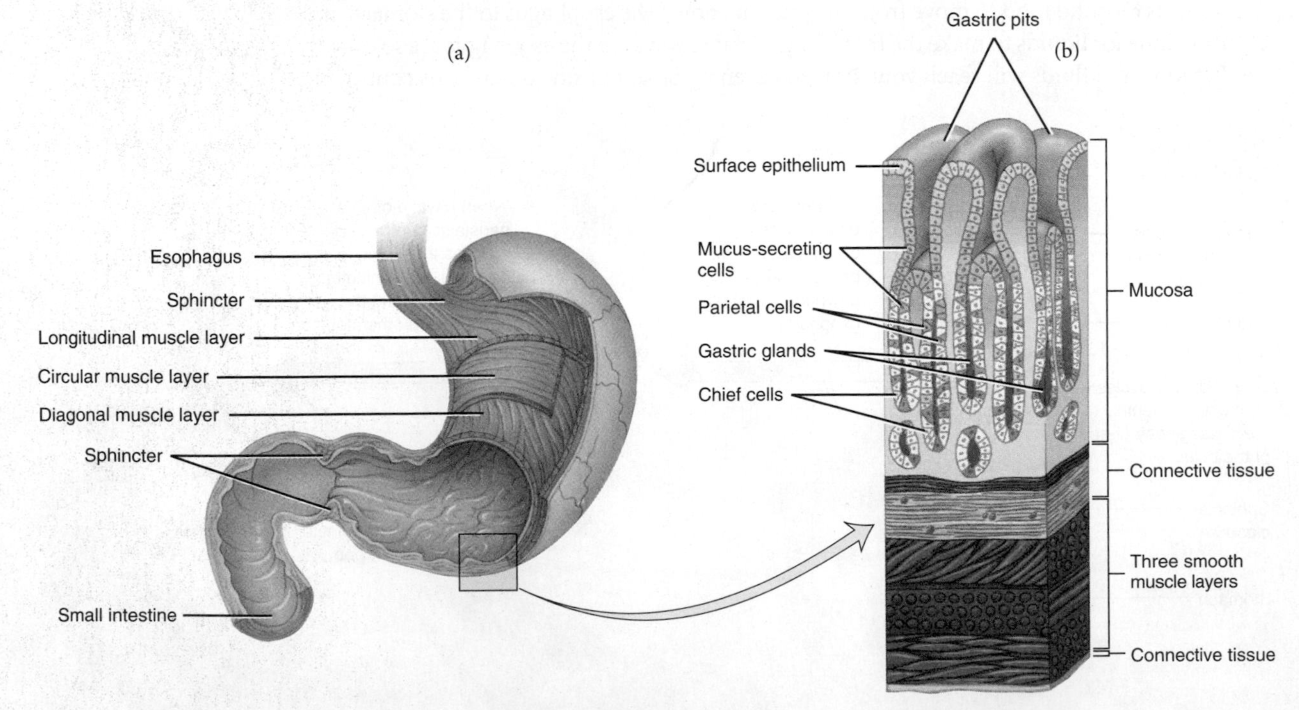

Figure 3.7 Structure of the stomach
(a) The stomach wall contains three layers of smooth muscle, which contract powerfully to mix food.
(b) The lining of the stomach is covered with gastric pits. Inside these pits are the gastric glands, made up of different types of cells that produce the components of gastric juice.

Composition of Gastric Juice Gastric juice consists of a number of substances, one of which is hydrochloric acid, produced by **parietal cells**. Hydrochloric acid acidifies the stomach contents and as a result kills most bacteria present in food. Parietal cells also produce intrinsic factor, which is needed for the absorption of vitamin B_{12} (see Section 8.9). Gastric juice also contains **pepsinogen**, produced by chief cells. Pepsinogen is an inactive form of the protein-digesting enzyme **pepsin**. Pepsin breaks proteins into shorter chains of amino acids called polypeptides. In children, the stomach glands also produce the enzyme rennin, which acts on the milk protein casein to convert it into a curdy substance resembling sour milk.

Pepsin functions best in the acidic environment of the stomach. This acidic environment stops the function of salivary amylase. Therefore, the digestion of starch from foods such as bread and potatoes temporarily stops in the stomach, although it starts again in the small intestine. On the other hand, digestion of protein from foods such as meat, milk, and legumes, begins in the stomach. The protein of the stomach wall is protected from the acid and pepsin by a thick layer of mucus. If the mucus layer is penetrated, pepsin and acid can damage the underlying tissues and cause a **peptic ulcer**, an erosion of the stomach wall or some other region of the GI tract. One of the leading causes of ulcers is acid-resistant bacteria called *Helicobacter pylori* that infect the lining of the stomach, destroying the protective mucosal layer and causing damage to the stomach wall.[2,3] (See Science Applied: Discovering What Causes Ulcers.)

parietal cells Cells in the stomach lining that make hydrochloric acid and intrinsic factor in response to nervous or hormonal stimulation.

pepsinogen An inactive protein-digesting enzyme produced by gastric glands and activated to pepsin by acid in the stomach.

pepsin A protein-digesting enzyme produced by the gastric glands. It is secreted in the gastric juice in an inactive form and activated by acid in the stomach.

peptic ulcer An open sore in the lining of the stomach, esophagus, or small intestine.

Regulation by Nerves and Hormones How much your stomach churns and how much gastric juice is released is regulated by signals from both nerves and hormones. These signals originate from three different sites—the brain, the stomach, and the small intestine. The thought, smell, sight, or taste of food causes the brain to send nerve signals that stimulate gastric motility and secretion, preparing the stomach to receive food (**Figure 3.8**). Food entering the stomach stimulates the release of gastric secretions and an increase in motility. It does this by stretching local nerves, sending signals to the brain, and promoting the secretion of the hormone **gastrin**. Gastrin then triggers the release of gastric juice and increases stomach motility. Chyme moving out of the stomach must pass through the pyloric sphincter, which helps regulate the rate at which food empties from the stomach. Food entering the small intestine triggers hormonal and nervous signals that can decrease stomach motility and secretions and slow stomach emptying. This ensures that the amount of chyme entering the small intestine does

gastrin A hormone secreted by the stomach mucosa that stimulates the secretion of gastric juice.

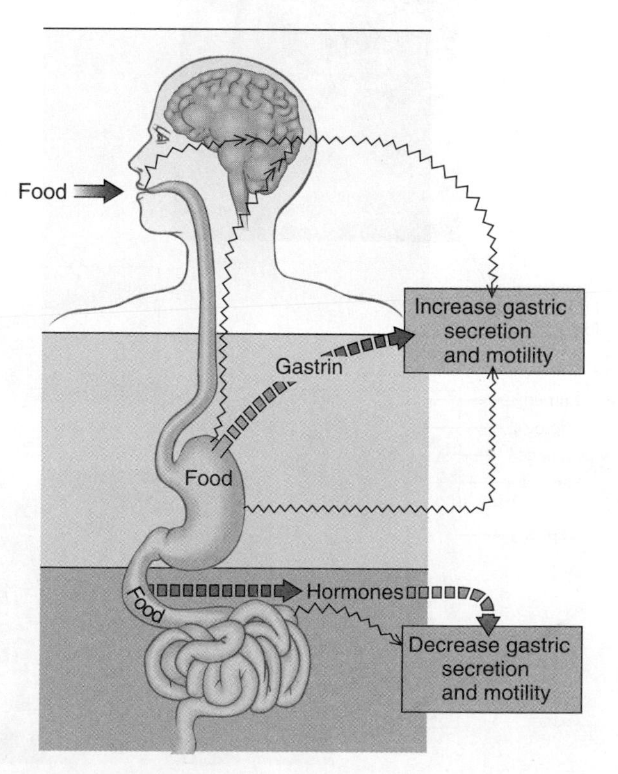

Figure 3.8 Regulation of stomach motility and secretion
Nerve signals (zigzag arrows) and hormonal signals (dashed arrows) from the brain, stomach, and small intestine regulate stomach activity.

not exceed the ability of the intestine to process it. Emotional factors can also affect gastric emptying. For example, sadness and fear tend to slow emptying, while aggression tends to increase gastric motility and speed emptying.

The Rate of Stomach Emptying Chyme normally leaves the stomach in 2-6 hours, but this rate is determined by the size and composition of the meal. A large meal will take longer to leave the stomach than a small meal, and a solid meal will leave the stomach more slowly than a liquid meal. The nutritional composition of a meal also affects how long it stays in the stomach. A meal that is partly solid and partly liquid and contains about 25% of energy as protein, 45% as carbohydrate, and 30% as fat will be in the stomach for an average amount of time, about 4 hours. A high-fat meal will stay in the stomach the longest because fat entering the small intestine causes the release of hormones that slow GI motility, thus slowing stomach emptying. A meal that is primarily protein will leave more quickly, and a meal of mostly carbohydrate will leave the fastest. The reason you are often ready to eat again soon after a dinner of vegetables and rice is that this high-carbohydrate, low-fat meal leaves the stomach rapidly. What you choose for breakfast can also affect when you become hungry for lunch. Toast and coffee will leave your stomach far more quickly than a larger meal with more protein and some fat, such as a bowl of cereal with low-fat milk, toast with peanut butter, and a glass of juice.

Small Intestine

The small intestine is the main site of digestion of food and absorption of nutrients. It is a narrow tube about 6 m in length, and divided into three segments. The first 30 cm are the duodenum, the next 2.4 m are the jejunum, and the last 3.3 m are the ileum. The structure of the small intestine is specialized to allow maximal absorption of the nutrients. In addition to its length, the small intestine has three other features that increase the area of its absorptive surface (**Figure 3.9**). First, the intestinal walls are arranged in large circular folds, which increase the surface area in contact with nutrients. Second, its entire inner surface is covered with finger-like projections called **villi** (singular, villus). And finally, each of these villi is covered with tiny **microvilli**, often referred to as the **brush border**. Together these features provide a

villi (villus) Finger-like protrusions of the lining of the small intestine that participate in the digestion and absorption of nutrients.

microvilli or **brush border** Minute, brush-like projections on the mucosal cell membrane that increase the absorptive surface area in the small intestine.

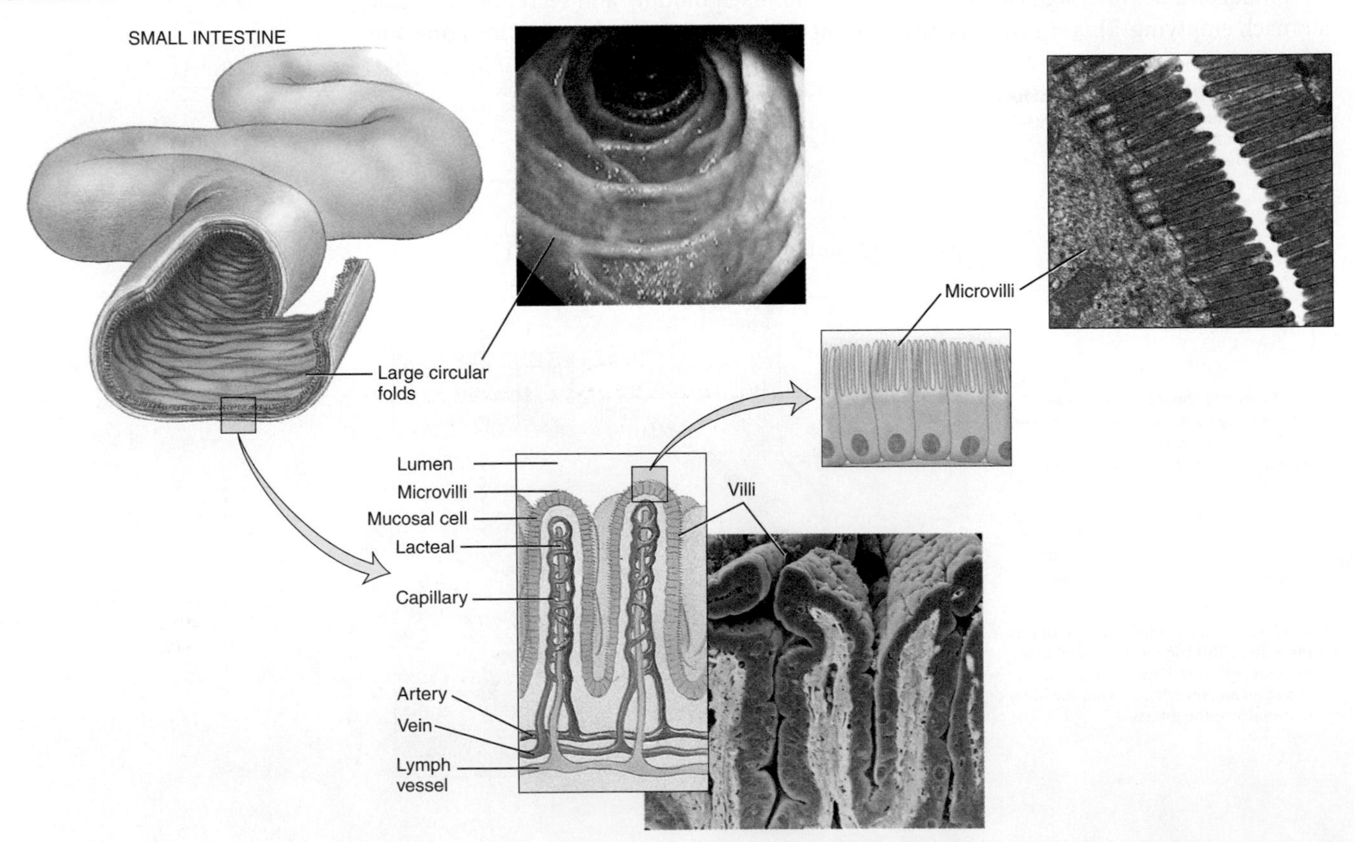

Figure 3.9 Structure of the small intestine
The small intestine contains large circular folds, villi, and microvilli, all of which increase the absorptive surface area. (top/center: David M. Martin/Photo Researchers; right: STEVE GSCHMEISSNER/SPL/Getty Images; bottom: SPL/Photo Researchers, Inc.)

surface area that is about the size of a tennis court (250 m^2 or 2,700 ft^2). Each villus contains a blood vessel and a lymph vessel or **lacteal**, which are located only one cell layer away from the nutrients in the intestinal lumen. Nutrients must cross the mucosal cell layer to reach the bloodstream or lymphatic system for delivery to the tissues of the body.

Chyme is propelled through the small intestine by peristalsis, and the mixing of chyme with digestive secretions is aided in the small intestine by rhythmic local constrictions called **segmentation** (**Figure 3.10**). Segmentation also enhances absorption by repeatedly moving chyme over the surface of the intestinal mucosa.

lacteal A tubular component of the lymphatic system that carries fluid away from body tissues. Lymph vessels in the intestine are known as lacteals and can transport large particles such as the products of fat digestion.

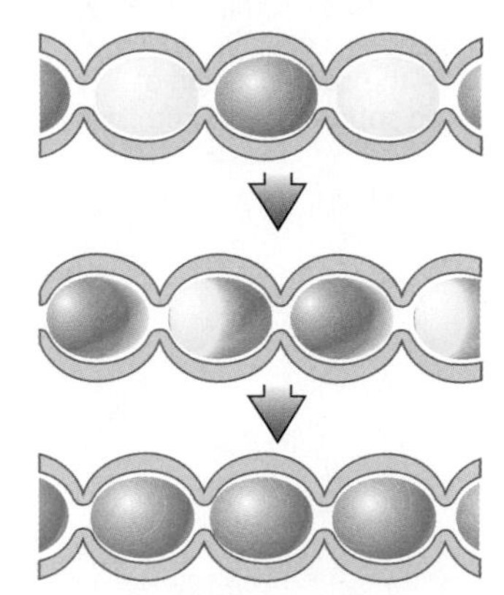

Figure 3.10 Segmentation
The alternating contraction and relaxation of segments of the small intestine move food forward and backward, mixing it rather than propelling it forward.

Enzymes and Secretions in the Small Intestine The cells of the small intestine produce some digestive enzymes as well as a watery, mucus-containing fluid called intestinal juice, that aids in absorption. However, normal digestion and absorption in the small intestine also require secretions from the **pancreas** and **gallbladder**. The pancreas secretes pancreatic juice, which contains both bicarbonate ions and digestive enzymes. The bicarbonate ions neutralize the acid in chyme, making the environment in the small intestine neutral rather than acidic as it is in the stomach. This neutrality allows enzymes from the pancreas and small intestine to function. The enzyme pancreatic amylase continues the job of breaking starch into sugars that was started in the mouth by salivary amylase. Pancreatic protein-digesting enzymes, including trypsin and chymotrypsin (see Table 3.2), continue to break protein into shorter and shorter chains of amino acids, and pancreatic fat-digesting enzymes, called **lipases**, break triglycerides into fatty acids. Intestinal digestive enzymes, found attached to or inside the cells lining the small intestine, are involved in the digestion of sugars into single sugar units and the digestion of small polypeptides into amino acids.

The gallbladder secretes **bile**, a substance produced in the liver and stored in the gallbladder that is necessary for fat digestion and absorption. Bile secreted into the small intestine mixes with fat and emulsifies it, or breaks it into smaller droplets, allowing lipases to more efficiently access the fat. Bile then helps form tiny droplets of digested fats, which move up against the mucosal lining, facilitating fat absorption (see Section 5.3).

segmentation Rhythmic local constrictions of the intestine that mix food with digestive juices and speed absorption by repeatedly moving the food mass over the intestinal wall.

Hormonal Control of Secretions The release of bile and pancreatic juice into the small intestine is controlled by two hormones secreted by the mucosal lining of the duodenum (see Table 3.3). **Secretin** signals the pancreas to secrete pancreatic juice rich in bicarbonate ions and stimulates the liver to secrete bile into the gallbladder. **Cholecystokinin (CCK)** signals the pancreas to secrete digestive enzymes and causes the gallbladder to contract and release bile into the duodenum (**Figure 3.11**).

pancreas An organ that secretes digestive enzymes and bicarbonate ions into the small intestine during digestion.

gallbladder An organ of the digestive system that stores bile, which is produced by the liver.

lipases Fat-digesting enzymes.

bile A substance made in the liver and stored in the gallbladder, which is released into the small intestine to aid in fat digestion and absorption.

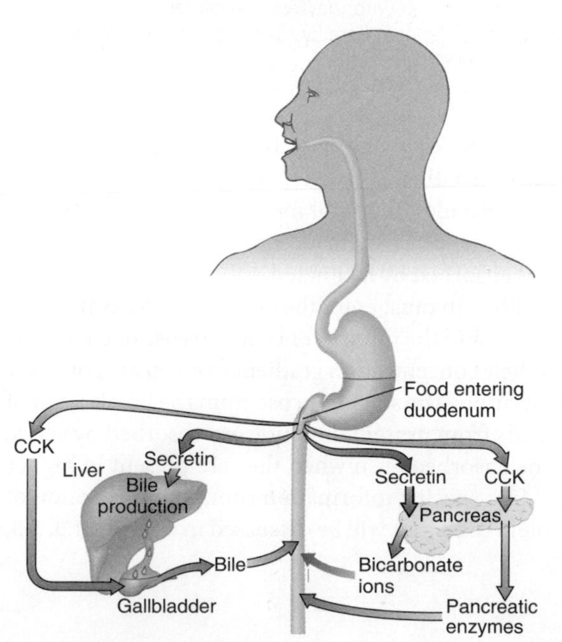

Figure 3.11 Hormonal control of secretions into the small intestine
Food entering the duodenum triggers the release of the hormones secretin and cholecystokinin (CCK). These stimulate the secretion of bicarbonate ions, digestive enzymes, and bile into the small intestine.

secretin A hormone released by the duodenum that signals the release of pancreatic juice rich in bicarbonate ions and stimulates the liver to secrete bile into the gallbladder.

cholecystokinin (CCK) A hormone released by the duodenum that stimulates the release of pancreatic juice rich in digestive enzymes and causes the gallbladder to contract and release bile into the duodenum.

simple diffusion The movement of substances from an area of higher concentration to an area of lower concentration. No energy is required.

osmosis The passive movement of water across a semipermeable membrane in a direction that will equalize the concentration of dissolved substances on both sides.

How Are Nutrients Absorbed? The small intestine is the primary site of absorption for water, vitamins, minerals, and the products of carbohydrate, fat, and protein digestion. To be absorbed, these nutrients must pass from the lumen of the GI tract into the mucosal cells lining the tract and then into the blood or lymph. Several different mechanisms are involved (**Figure 3.12**). Some molecules are absorbed by diffusion—the process by which a substance moves from an area of higher concentration to an area of lower concentration. Substances that move from higher to lower concentrations are said to move down their concentration gradient. When a concentration gradient exists and the nutrient can pass freely from the lumen of the GI tract across the cell membrane into the mucosal cell, the process is called **simple diffusion**, and it does not require the input of energy. Vitamin E and fatty acids are absorbed by simple diffusion. Water is also absorbed by diffusion, but the diffusion of water is called **osmosis**. In osmosis, there is a net movement of water in a direction that will balance the concentration of dissolved substances on either side of a membrane. For example, if there is a high concentration of sugar in the lumen of the intestine, water will actually move from the mucosal cells into the lumen. As the sugar is absorbed, and the concentration of sugar in the lumen decreases, water will move back into the mucosal cells by osmosis.

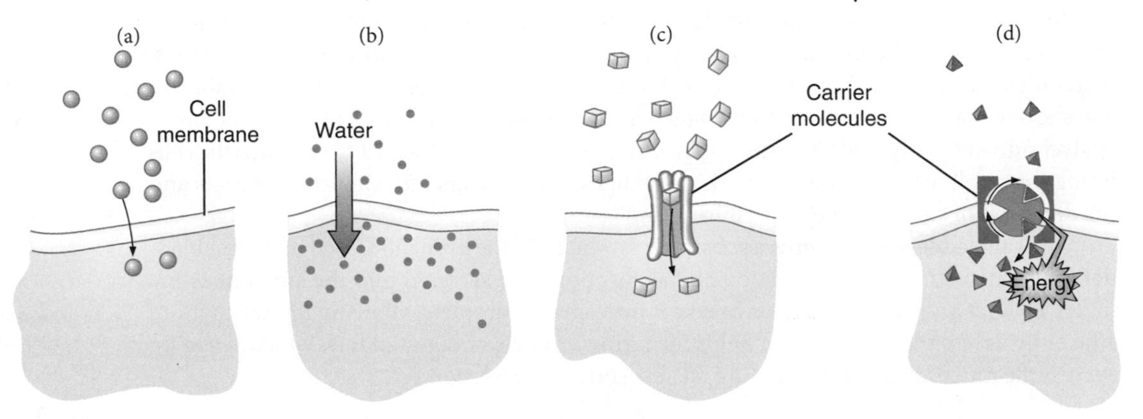

Figure 3.12 Absorption mechanisms
(a) Simple diffusion, which requires no energy, is shown here by the purple balls that move from an area of higher concentration to an area of lower concentration.
(b) Osmosis is the diffusion of water. The water molecules represented by the purple arrow move from an area with a low concentration of dissolved substances (blue dots) to an area with a high concentration of dissolved substances.
(c) Facilitated diffusion, which requires no energy, is shown here by the yellow cubes that move from an area of higher concentration to an area of lower concentration with the help of a carrier.
(d) Active transport, which requires energy and a carrier, is shown here by the red pyramids that move from an area of lower concentration to an area of higher concentration.

facilitated diffusion The movement of substances across a cell membrane from an area of higher concentration to an area of lower concentration with the aid of a carrier molecule. No energy is required.

active transport The transport of substances across a cell membrane with the aid of a carrier molecule and the expenditure of energy. This may occur against a concentration gradient.

Many nutrients, however, cannot pass freely across cell membranes; they must be helped across by carrier molecules in a process called **facilitated diffusion**. Even though these nutrients are carried across the cell membrane by other molecules, they still move down a concentration gradient from an area of higher concentration to one of lower concentration without requiring energy; the sugar fructose, found in fruit, is absorbed by facilitated diffusion.

Substances unable to be absorbed by diffusion must enter the body by **active transport**, a process that requires both a carrier molecule and the input of energy. This use of energy allows substances to be transported against their concentration gradient from an area of lower concentration to an area of higher concentration. The sugar glucose from the breakdown of the starch in a slice of bread and amino acids from protein digestion are absorbed by active transport. This allows these nutrients to be absorbed even when they are present in higher concentrations inside the mucosal cells. More specific information about the absorption of the products of carbohydrate, fat, and protein digestion will be discussed in sections 4.3, 5.3, and 6.3, respectively.

colon The largest portion of the large intestine.

rectum The portion of the large intestine that connects the colon and anus.

Large Intestine

Components of chyme that are not absorbed in the small intestine pass through the ileocecal valve to the large intestine, which includes the **colon** and **rectum**. Although most absorption occurs in the small intestine, water and some vitamins and minerals are also absorbed in the colon. Peristalsis here is slower than in the small intestine. Water, nutrients, and fecal matter may spend 24 hours in the large intestine, in contrast to the 3-5 hours it takes for chyme to move through the small intestine. This slow movement favours the growth of bacteria, referred to as the **intestinal microflora**. These bacteria are permanent, beneficial residents of this part of the GI tract (see Your Choice: Should You Feed Your Flora? and Label Literacy: Choosing Your Yogurt). The microflora act on unabsorbed portions of food, such as the fibre, producing nutrients that the bacteria themselves can use or, in some cases, that can be absorbed into the body. For example, the microflora synthesize small amounts of fatty acids, some B vitamins, and vitamin K, some of which can be absorbed. One additional by-product of bacterial metabolism is gas, which causes flatulence. In normal adult humans, between 200-2,000 mL of intestinal gas is produced per day.

intestinal microflora Microorganisms that inhabit the large intestine.

Materials not absorbed in the colon are excreted as waste products in the feces. The feces are a mixture of undigested, unabsorbed matter, dead cells, secretions from the GI tract, water, and bacteria. The amount of bacteria in the feces varies but can make up more than half the weight of the feces. The amount of water in the feces is affected by fibre and fluid intake. Fibre retains water, so when adequate fibre and fluid are consumed, feces have a high water content and are easily passed. When inadequate fibre or fluid is consumed, feces are hard and dry, and difficult to eliminate.

The end of the colon is connected to the rectum, where feces are stored prior to defecation. The rectum is connected to the **anus**, the external opening of the digestive tract. The rectum works with the colon to prepare the feces for elimination. Defecation is regulated by a sphincter that is under voluntary control. It allows the feces to be eliminated at convenient and appropriate times. The digestion and absorption of carbohydrate, fat, and protein are summarized in **Figure 3.13**.

anus The outlet of the rectum through which feces are expelled.

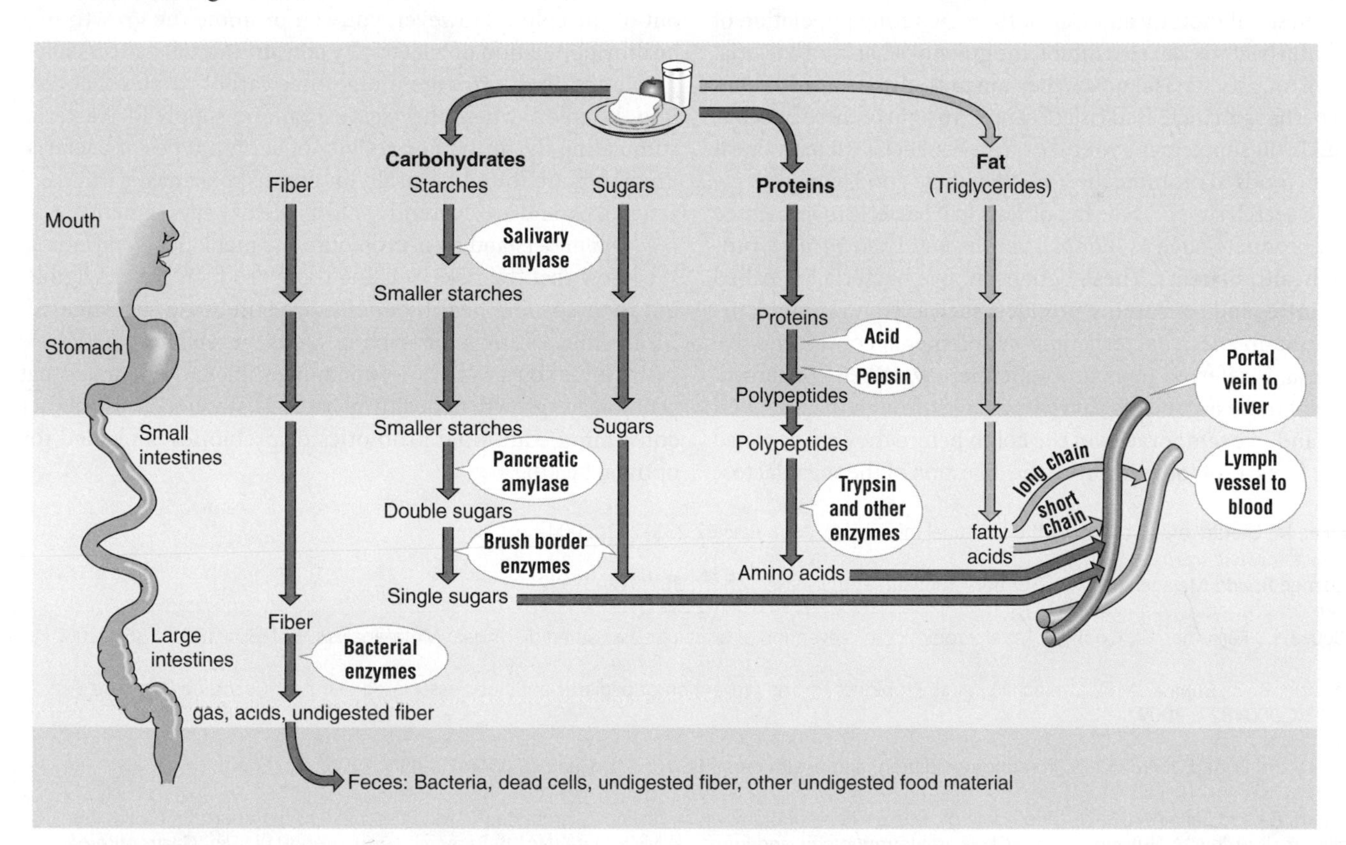

Figure 3.13 An overview of digestion and absorption
Some carbohydrate digestion occurs in the mouth and some protein digestion occurs in the stomach, but the majority of macronutrient digestion and the absorption of nutrients occur in the small intestine.

YOUR CHOICE

(©iStockphoto)

Should You Feed Your Flora?

The large intestine of a healthy adult is home to 300-500 different species of bacteria, and the gut has about 10 times more bacterial cells than the total number of cells in your entire body.[1] These bacteria are usually beneficial. They break down indigestible dietary substances, improve the digestion and absorption of essential nutrients, synthesize some vitamins, and metabolize harmful substances. They are important for intestinal immune function, proper growth and development of cells in the colon, and optimal intestinal motility and transit time.[2] A strong population of healthful bacteria can also inhibit the growth of harmful bacteria. If the wrong bacteria take over, they can cause diarrhea, infections, and perhaps an increased risk of cancer. In light of these benefits, should you supplement your diet with beneficial bacteria or eat certain foods to promote the growth of these good bugs?

Research supports the hypothesis that bacteria in fermented dairy products, such as *Bifidobacterium* and *Lactobacillus*, provide health benefits. These beneficial, live bacteria are called **probiotics** and consuming products such as yogurt that naturally contain these bacteria, and supplements containing live bacteria, is referred to as probiotic therapy. When consumed, some of these organisms survive passage through the upper GI tract and live temporarily in the colon before they are excreted in the feces. Probiotics improve the digestion of the sugar lactose in lactose-intolerant people (see Section 4.3), prevent diarrhea associated with antibiotic use,[3] reduce the duration of diarrhea due to intestinal infections and other causes,[4] and have beneficial effects on immune function in the intestine.[5] There is also evidence that a healthy microflora may relieve constipation, reduce allergy symptoms, modify the risk of colon cancer, and affect body weight.[6–8] One problem with probiotics is that when they are no longer consumed, the added bacteria are rapidly washed out of the colon. However, you can promote the growth of a healthy population of bacteria by consuming substances called prebiotics. **Prebiotics** are indigestible carbohydrates that pass into the colon, where they serve as a food supply for bacteria, stimulating the growth or activity of certain types of bacteria. Prebiotics are found naturally in onions, bananas, garlic, and artichokes, and are currently sold as dietary supplements.

Our understanding of probiotics and prebiotics is expanding. We know that the risks of using these products are negligible, but their specific health benefits are still being investigated. Meanwhile, eating a diet rich in fruits, vegetables, and whole grains gives you a wide variety of indigestible carbohydrates that promote the growth of healthful bacteria. It is not clear whether consuming additional probiotics or prebiotics is needed for optimal health.[9]

[1] Volker, M. Dietary modification of the intestinal microbiota. *Nutr. Rev.* 62:235–242, 2004.

[2] Guarner, F., and Malagelada, J. R. Gut flora in health and disease. *Lancet.* 361:512–519, 2003.

[3] D'Souza, L., Rajkumar, C., Cooke, J., et al. Probiotics in prevention of antibiotic associated diarrhoea: Meta-analysis. *BMJ* 324:1361–1366, 2002.

[4] Johnston, B. C., Supina, A. L., Ospina, M., et al. Probiotics for the prevention of pediatric antibiotic-associated diarrhea. *Cochrane Database Syst. Rev.* 18:CD004827, 2007.

[5] Gill, H., and Prasad, J. Probiotics, immunomodulation, and health benefits. *Adv Exp Med Biol* 606:423–454, 2008.

[6] Isolauri, E., and Salminen, S. Nutrition, Allergy, Mucosal Immunology, and Intestinal Microbiota (NAMI) Research Group Report: Probiotics: Use in allergic disorders: A Nutrition, Allergy, Mucosal Immunology, and Intestinal Microbiota (NAMI) Research Group Report. *J. Clin. Gastroenterol.* 42(Suppl 2):S91–S96, 2008.

[7] Rescigno, M. The pathogenic role of intestinal flora in IBD and colon cancer. *Curr. Drug. Targets* 9:395–403, 2008.

[8] Bäckhed, F., Ding, H., Wang, T. et al. The gut microbiota as an environmental factor that regulates fat storage. *Proc. Natl. Acad. Sci. USA* 101:18–23, 2004.

[9] Blaut, M. Relationship of prebiotics and food to intestinal microflora. *Eur. J. Nut.* 41(Suppl. 1):111–116, 2002.

3.4 Digestion and Health

Learning Objectives

- Describe the cause of heartburn and GERD.
- Discuss when alternative feeding methods are needed to provide nutrients.
- Explain how changes in the digestive system throughout life affect digestion and absorption.

Probiotics Specific types of live bacteria found in foods that are believed to have beneficial effects on human health.

Prebiotics Indigestible carbohydrates that pass into the colon, where they serve as a food supply for bacteria, stimulating the growth or activity of certain types of beneficial bacteria.

The digestive system is adaptable and able to handle a wide variety of foods. However, minor problems related to the digestive tract are common and almost everyone experiences some type of GI distress at one time or another. The type of foods that can be consumed and the function of the digestive tract are also affected by certain stages of life. If gastrointestinal problems limit the ability to obtain adequate energy and nutrients, alternative methods must be employed to provide the nutrients necessary for life.

Common Digestive Problems

Minor digestive problems, such as heartburn, constipation, and diarrhea, are common and, in most cases, have little effect on nutritional status. However, more long-term or severe problems can have serious consequences for nutrition and overall health. Some digestive problems and their causes, consequences, and solutions are given in **Table 3.4**.

Table 3.4 Digestive Problems and Nutritional Consequences

Problem	Causes	Consequences	Treatment/Management
Dry mouth	Disease, medications	Decreased food intake due to changes in taste, difficulty chewing and swallowing, increased tooth decay, and gum disease.	Change medications, use artificial saliva.
Dental pain and loss of teeth	Tooth decay, gum disease	Reduced food intake due to impaired ability to chew, reduced nutrient absorption due to incomplete digestion.	Change consistency of foods consumed.
Heartburn, Gastroesophageal reflux disease (GERD)	Stomach acid leaking into esophagus due to overeating, anxiety, stress, pregnancy, hiatal hernia, or disease processes	Pain and discomfort after eating, ulcers, increased cancer risk.	Reduce meal size, avoid high-fat foods, consume liquids between rather than with meals, remain upright after eating, take antacids and other medications.
Hiatal hernia	Pressure on the abdomen from persistent or severe coughing or vomiting, pregnancy, straining while defecating, or lifting heavy objects	Heartburn, belching, GERD, and chest pain	Reduce meal size, avoid high-fat foods, consume liquids between rather than with meals, remain upright after eating, take antacids and other medications, lose weight.
Ulcers	Infection of stomach by *H. pylori*, acid-resistant bacteria that penetrate the mucous layer and damage the epithelial lining; chronic use of drugs such as aspirin and ibuprofen that erode the mucosa; GERD	Pain, bleeding, and possible abdominal infection	Antibiotics to treat infection, antacids to reduce acid, change in medications.
Vomiting	Bacterial and viral infections, medications, other illnesses, eating disorders, pregnancy, food allergies	Dehydration and electrolyte imbalance; if chronic, can damage the mouth, gums, esophagus, and teeth.	Medications to treat infection, fluid and electrolyte replacement.
Diarrhea	Bacterial and viral infections, medications, food intolerance	Dehydration and electrolyte imbalance	Medications to treat infection, fluid and electrolyte replacement.
Constipation	Low fibre intake, low fluid intake, high fibre in combination with low fluid intake, weak intestinal muscles	Discomfort, intestinal blockage, formation of outpouchings in the intestinal wall (diverticula) (see Chapter 4)	High-fibre, high-fluid diet, exercise, medications.
Irritable bowel syndrome	When the muscle contractions in the intestines are stronger and last longer than normal, or are slower and weaker than normal.	Abdominal pain or cramping and changes in bowel function—including bloating, gas, diarrhea, and constipation.	Manage stress and make changes in diet and lifestyle, take fibre supplements, antidiarrheal medications, other medications.
Pancreatic disease	Cystic fibrosis or pancreatitis	Malabsorption of fat, fat-soluble vitamins, and vitamin B_{12} due to reduced availability of pancreatic enzymes and bicarbonate.	Oral supplements of digestive enzymes.
Gallstones	Deposits of cholesterol, bile pigments, and calcium in the gallbladder or bile duct	Pain and poor fat digestion and absorption.	Low-fat diet, surgical removal of the gallbladder.

Heartburn Heartburn is one of the most common digestive complaints. It occurs when the acidic stomach contents leak back into the esophagus, causing a burning sensation in the chest or throat (**Figure 3.14**). The more technical term for the leakage of stomach contents back into the esophagus is gastroesophageal reflux. Occasional heartburn from reflux is common, but if it occurs more than twice a week, it may indicate a condition called **gastroesophageal reflux disease (GERD)**. If left untreated, GERD can eventually lead to more serious health problems such as bleeding, ulcers, and cancer. Whether you have occasional heartburn or GERD, it can cause discomfort after eating a large meal and limit the types and amounts of food you can consume. Symptoms can be reduced by eating small meals and avoiding spicy foods, fatty and fried foods, citrus fruits, chocolate, caffeinated beverages, tomato-based foods, garlic, onions, and mint. Remaining upright after eating, wearing loose clothing, avoiding smoking and alcohol, and losing weight may also help relieve symptoms. For many people, medications are needed to manage symptoms.

gastroesophageal reflux disease (GERD) A chronic condition in which acidic stomach contents leak back up into the esophagus, causing pain and damaging the esophagus.

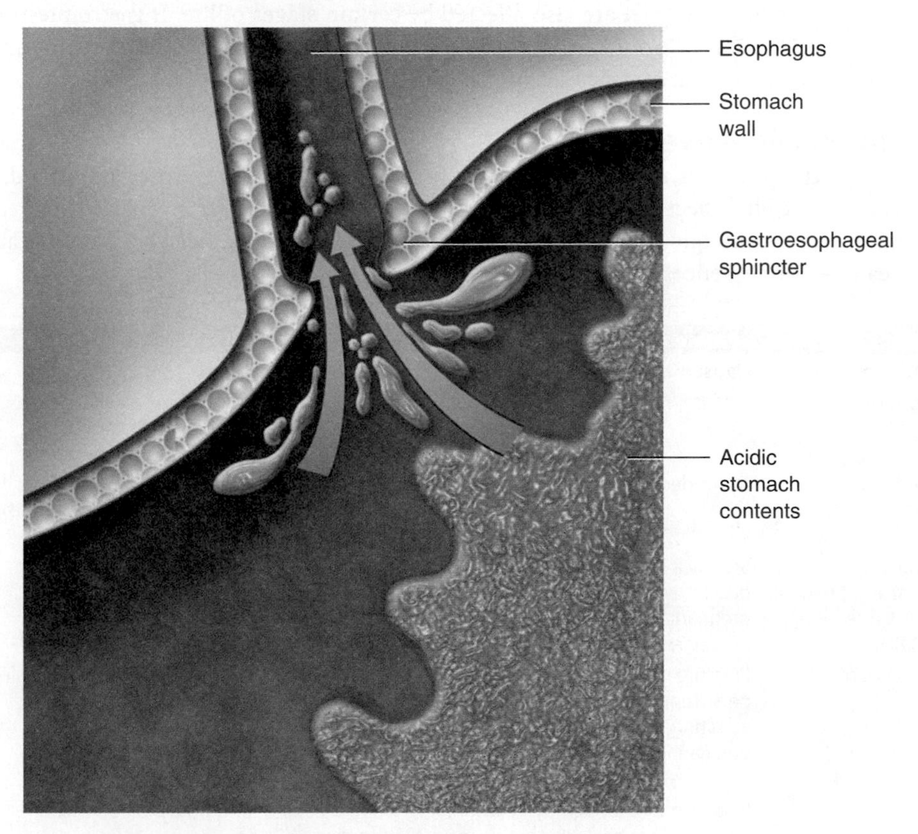

Figure 3.14 Gastroesophageal reflux
Heartburn and GERD occur when stomach acid leaks through the gastroesophageal sphincter, which separates the esophagus from the stomach, and irritates the lining of the esophagus.

Heartburn and GERD are sometimes caused by a hiatal hernia. The opening in the diaphragm that the esophagus passes through is called the hiatus. A hiatal hernia occurs when the upper part of the stomach bulges through this opening into the chest cavity. Hiatal hernias are common, occurring in about one-quarter of people older than 50. They're more common in women, in smokers, and in people who are overweight. Most hiatal hernias cause no signs or symptoms, but larger ones may allow food and acid to back up into the esophagus, causing heartburn and chest pain. A hiatal hernia can usually be treated by heartburn medications, but in severe cases, surgery may be needed.

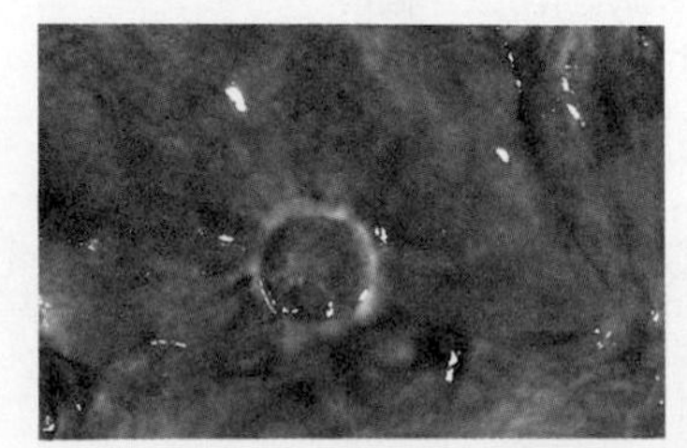

Figure 3.15
Ulcers, such as this one in the lining of the stomach, are most often due to *Helicobacter Pylori* infection. (CNRI/SPL/Photo Researchers)

Ulcers Ulcers can arise both in the esophagus and the stomach. They occur when the mucosa is eroded away, exposing the underlying tissues to the gastric juices (**Figure 3.15**). If the damage reaches the nerve layer, it causes pain, and if capillaries are damaged, gastrointestinal bleeding can occur. If the ulcer perforates the wall of the GI tract, a serious abdominal infection can occur. Ulcers can result from GERD or the chronic use of drugs such as aspirin and ibuprofen that erode the mucosa of the GI tract, but, as previously discussed, ulcers

LABEL LITERACY

Choosing Your Yogurt

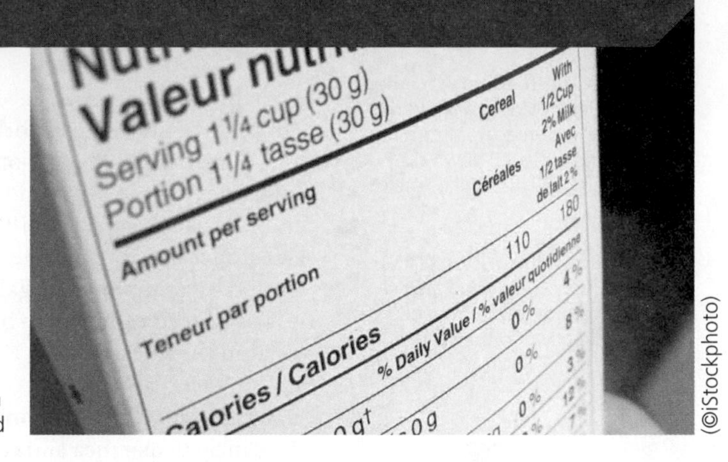

(©iStockphoto)

CANADIAN CONTENT

In recent years, there has been considerable interest in the health benefits of probiotic bacteria (see Your Choice: Should You Feed Your Flora?), and probiotic products, most notably yogurt and yogurt-based drinks, have entered the Canadian food market. A careful review of the yogurt label will help you distinguish between a probiotic yogurt and a regular yogurt. The differences are highlighted below.

Note, as highlighted, that the probiotic yogurt is clearly labelled as such, and lists, as required by Health Canada, the specific Latin names of the bacterial cultures used. Health Canada also permits nutrient-function claims for probiotics that can include one of the following phrases:[4]

1) probiotic that naturally forms part of the gut flora.

2) provides live micro-organisms that naturally form part of the gut flora.

3) probiotic that contributes to healthy gut flora.

4) provides live micro-organisms that contribute to healthy gut flora.

Note that both yogurts were made from vitamin-D-fortified milk and are a source of both calcium and vitamin D.

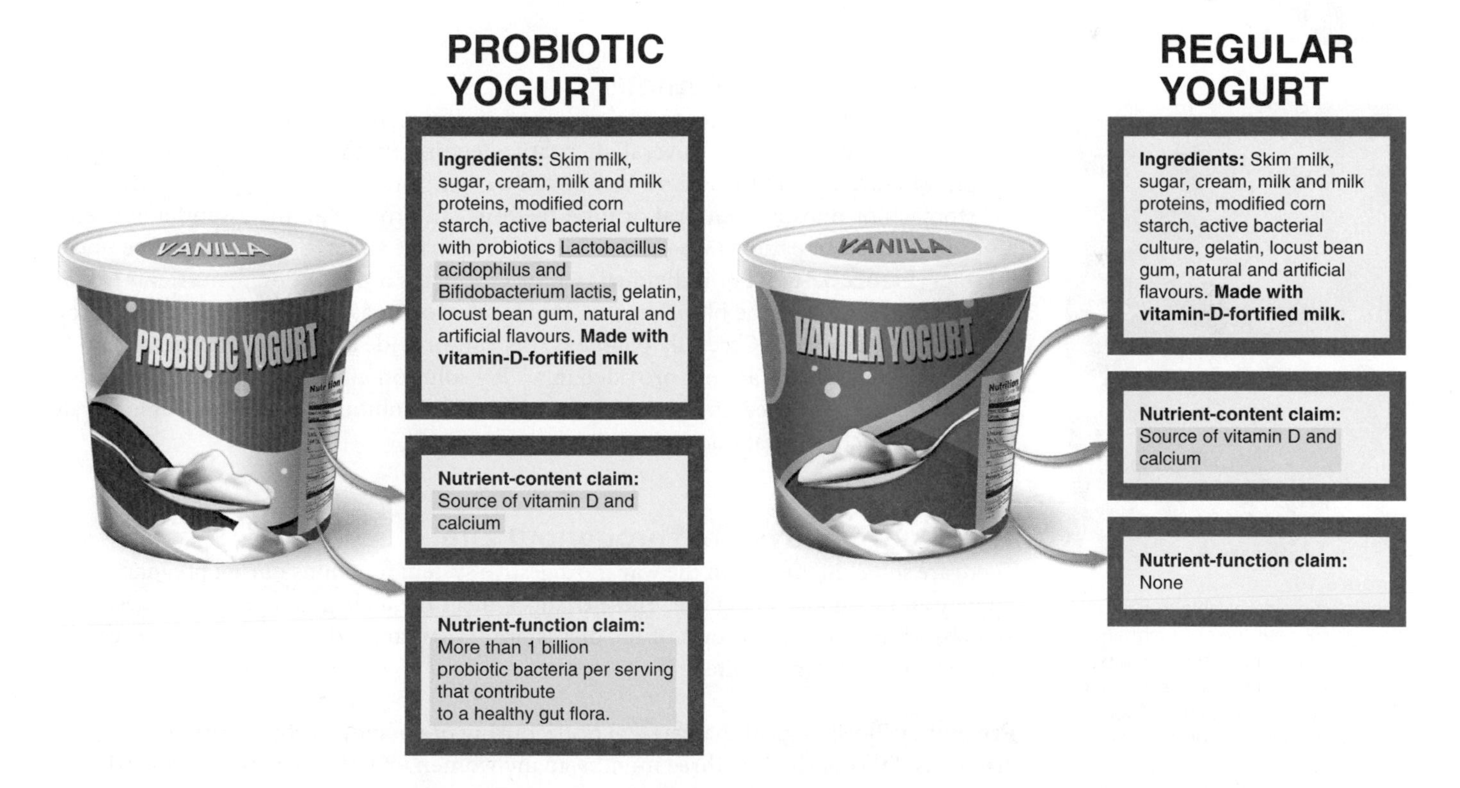

[1] Health Canada. Accepted claims about the nature of probiotic micro-organisms in food. Available online at http://www.hc-sc.gc.ca/fn-an/label-etiquet/claims-reclam/probiotics_claims-allegations_probiotiques-eng.php. Accessed Feb 4, 2010.

are more often caused by a bacterial infection in the stomach. Mild ulcers can affect nutrient intake by limiting food choices to those that do not cause discomfort, but more severe ulcers can cause bleeding that can be life-threatening (see Science Applied: Discovering What Causes Ulcers).

Pancreatic and Gallbladder Problems Changes in accessory organs of the GI tract can affect digestion and cause discomfort. If the pancreas is not functioning normally, the

enteral or **tube-feeding** A method of feeding by providing a liquid diet directly to the stomach or intestine through a tube placed down the throat or through the wall of the GI tract.

total parenteral nutrition (TPN) A technique for nourishing an individual by providing all needed nutrients directly into the circulatory system.

availability of enzymes needed to digest carbohydrate, fat, and protein may be reduced, limiting the ability to absorb these essential nutrients. If the gallbladder is not releasing bile, it can impair fat absorption. A common condition that affects the gallbladder is gallstones. These are clumps of solid material that form in the gallbladder or bile duct and can cause pain when the gallbladder contracts in response to fat in the intestine. Gallstones can interfere with bile secretion and reduce fat absorption. They are usually treated by removing the gallbladder. In this case, bile simply drips into the intestine as it is produced rather than being stored and squeezed out in larger amounts in response to fat in the intestine.

Diarrhea and Constipation Common discomforts that are related to problems in the intestines include diarrhea and constipation. Diarrhea refers to frequent watery stools. It occurs when material moves through the colon too quickly for sufficient water to be absorbed or when water is drawn into the lumen from cells lining the intestinal tract. Diarrhea may occur when the lining of the small intestine is inflamed so nutrients and water are not absorbed. The inflammation may be caused by infection with a micro-organism or by other irritants. Constipation refers to hard, dry stools that are difficult to pass, and can be caused by a diet that contains insufficient fluid or fibre, lack of exercise, or a weakening of the muscles of the large intestine. A condition that may cause either diarrhea or constipation is irritable bowel syndrome. Its cause is unknown, but it often results in abnormal contractions of the muscles in the intestinal wall. Symptoms include abdominal pain or cramping, bloating, gas, diarrhea, and constipation.

(a)

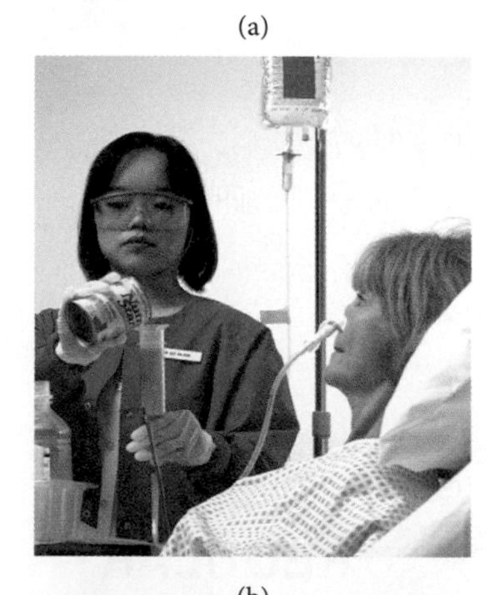

(b)

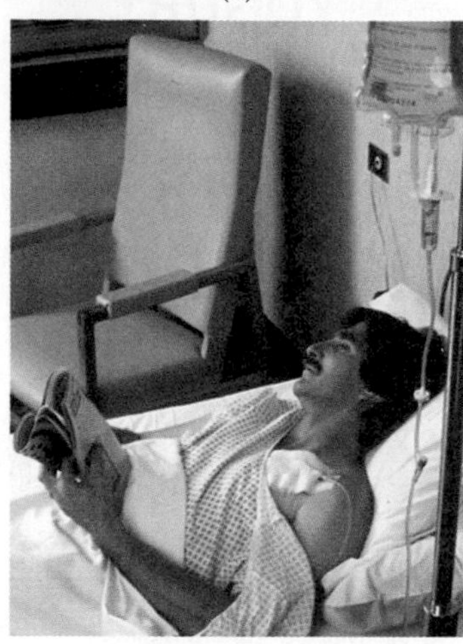

Figure 3.16
People who are not able to eat enough to meet their nutrient and energy needs can be nourished with (a) enteral feeding (tube-feeding) if their GI tract is able to digest food and absorb nutrients or (b) total parenteral nutrition if their gut is not functional. (Ed Eckstein/ Phototake; L. Steinmark/Custom Medical Stock Photo, Inc.)

Alternate Feeding Methods

For individuals who are unable to consume food or digest and absorb the nutrients needed to meet their requirements, several alternative feeding methods have been developed. People who are unable to swallow can be fed a liquid diet through a tube inserted into the stomach or intestine. **Enteral** or **tube-feeding** can provide all the essential nutrients and can be used in patients who are unconscious or have suffered an injury to the upper GI tract (**Figure 3.16a**). For individuals whose GI tract is not functional, nutrients can be provided directly into the bloodstream. This is referred to as **total parenteral nutrition (TPN)** (**Figure 3.16b**). Carefully planned TPN can provide all the nutrients essential to life. When all nutrients are not provided in a TPN solution, nutrient deficiencies develop quickly. Inadvertently feeding patients incomplete TPN solutions has helped demonstrate the essentiality of several trace minerals.

The Digestive System Throughout Life

There are some differences in the way the digestive system functions during pregnancy and infancy and with advancing age. These changes affect the ability to ingest and digest food and absorb nutrients. However, if the diet is properly managed, nutritional status can be maintained at all stages of life.

Life Cycle

Pregnancy Physiological changes that occur during pregnancy may cause gastrointestinal problems. During the first three months, many women experience nausea, referred to as morning sickness. This term is a misnomer, because the nausea can occur at any time of the day. Morning sickness is believed to be due to pregnancy-related hormonal changes. In most cases, it can be dealt with by eating frequent small meals and avoiding foods and smells that cause nausea. Eating dry crackers or cereal may also help. In severe cases where uncontrollable vomiting occurs, nutrients may need to be given intravenously to maintain nutritional health.

Later in pregnancy, the enlarged uterus puts pressure on the stomach and intestines, which can make it difficult to consume large meals. In addition, the placenta produces the hormone progesterone, which causes the smooth muscles of the digestive tract to relax. The muscle-relaxing effects of progesterone may relax the gastroesophageal sphincter enough to allow the stomach contents to move back into the esophagus, causing heartburn.

SCIENCE APPLIED

Discovering What Causes Ulcers

Peptic ulcers are lesions in the wall of the stomach or other part of the GI tract. They cause pain and, sometimes, bleeding. For many years, they were thought to be caused by excess stomach acid. Treatments included milk to coat the GI lining, a bland diet, stress reduction, and drugs to neutralize acid or reduce its secretion. These treatments usually reduced irritation and decreased symptoms, but ulcers were a chronic condition even with treatment. Today, due to the observations and perseverance of Drs. B. J. Marshall and J. R. Warren, many ulcers can be cured.

Dr. Warren is a pathologist who observed curvy-shaped bacteria in some of the stomach biopsies he examined under the microscope. He noted that the bacteria were always present in tissue that was inflamed and that the number of organisms correlated with the degree of inflammation.[1] In 1982, he and Dr. Marshall isolated and grew these bacteria, later named *Helicobacter pylori (H. pylori)*, in the laboratory.[2] They believed that the bacteria were the cause of peptic ulcers.

The hypothesis that ulcers were caused by bacteria, however, was not immediately embraced by the scientific community. At the time, the accepted theory was that ulcers were caused by too much stomach acid and it was widely accepted that bacteria could not survive in the strong acid of the stomach. To persuade other scientists that the hypothesis was correct, stronger evidence demonstrating that the bacteria were the cause of ulcer symptoms, such as gastric inflammation, was needed. Therefore, Marshall decided to use himself as an experimental subject to gather evidence to support his theory.

Marshall's experiment involved first having a small sample of his gastric mucosa examined to confirm that it was not infected with bacteria. He then drank a vial containing *H. pylori* that had been isolated from a patient with chronic gastric inflammation. Ten days later, he developed symptoms of gastric inflammation. A follow-up sample of his gastric mucosa confirmed that his stomach lining was now inflamed and *H. pylori* bacteria could be seen attached to the mucosa (see figure).[3] Fortunately, his symptoms resolved quickly with antibiotic therapy. This confirmed the connection between *H. pylori* and stomach inflammation, but since Marshall did not develop an ulcer, that link was still unproven. The connection between *H. pylori* and ulcers was eventually supported by epidemiological studies that showed an increased incidence of ulcers in persons infected with the bacteria.[1]

As our understanding of how this organism survives in the acid environment of the stomach grew, the idea that bacteria caused ulcers became more universally accepted. In the stomach, *H. pylori* use a tail-like structure to swim through the protective mucus layer and adhere to the mucosal cells. Once attached to the mucosal cells, they use an enzyme to produce substances that neutralize the acid immediately around them.[4] The bacteria damage mucosal cells and increase the release of gastrin, which subsequently increases stomach acid secretion.[5] The presence of bacteria causes inflammation and other immune responses that lead to additional tissue damage. About 30%-50% of the world's population is infected with *H. pylori*.[1] In Canada, about 35% of the adult population has been estimated to be infected.[6] Most people who are infected with *H. pylori* do not have symptoms and fewer than 20% go on to develop an ulcer, but infection with this organism is now known to also be associated with cancers of the stomach.[1, 6]

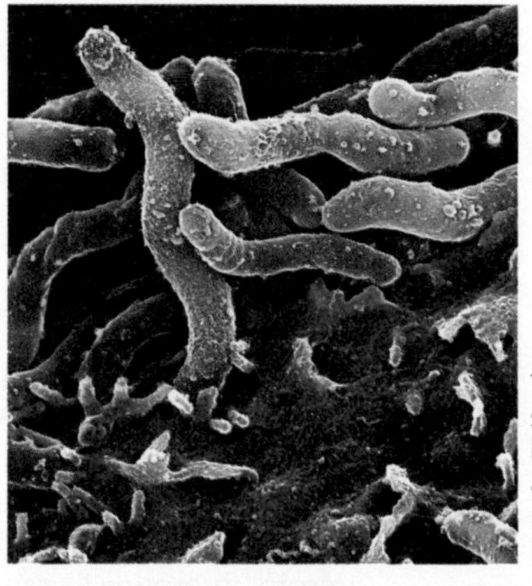

This electron micrograph shows the helical-shaped bacteria *Helicobacter pylori* attached to the gastric mucosa. (© Science Photo Library/Photo Researchers)

Because of the observations and persistence of Drs. Warren and Marshall, a patient diagnosed with an ulcer today is given antibiotics rather than a bland diet and antacids. Successful antibiotic therapy combined with acid-suppression therapy can allow the ulcer to heal, eliminate the bacteria that caused the disease, and cure the patient. A hypothesis that was at first rejected by the scientific community is now a theory supported by scientific evidence.

[1] Lynch, N. A. *Helicobacter pylori*: A paradigm revised. Available online at www.faseb.org/opa/pylori/pylori.html/. Accessed January 6, 2005.

[2] Marshall, B. J., and Warren, J. R. Unidentified curved bacilli on gastric epithelium in active chronic gastritis. *Lancet* I:1311–1315, 1984.

[3] Marshall, B. J., Armstrong, J. A., McGechie, D. B., et al. Attempt to fulfill Koch's postulates for pyloric. *Campylobacter. Med. J. Aust.* 142:436–439, 1985.

[4] McGee, D. J., and Mobley, H. L.T. Mechanisms of *Helicobacter pylori* infection: Bacterial factors. *Curr. Top. Microbiol. Immunol.* 241:156–180, 1999.

[5] Joseph, I. M., and Kirschner, D. A. Model for the study of *Helicobacter pylori* interaction with human gastric acid secretion. *J. Theor. Biol.* 228:55–80, 2004.

[6] Pérez-Pérez G.I., Bhat N., Gaensbauer, J., et al. Country-specific constancy by age in cagA+ proportion of *Heliobacter pylori* infections. *Int J Cancer.* 1997 Jul 29;72(3):453-6.

In the large intestine, relaxed muscles and the pressure of the uterus cause less efficient peristaltic movements and may result in constipation. Increasing water intake, eating a diet high in fibre, and exercising regularly can help prevent and relieve constipation (see sections 14.1 and 14.2).

Infancy The digestive system is one of the last to fully mature in developing humans. At birth, the digestive tract is functional, but a newborn is not ready to consume an adult diet. The most obvious difference between the infant and adult digestive tracts is that newborns are not able to chew and swallow solid food. They are born with a suckling reflex that allows them to consume liquids from a nipple placed toward the back of the mouth. A protrusion reflex causes anything placed in the front of the mouth to be pushed out by the tongue. As head control increases, this reflex disappears, making spoon-feeding possible.

Digestion and absorption also differ between infants and adults. In infants, the digestion of milk protein is aided by rennin, an enzyme produced in the infant stomach that is not found in adults.[4] The stomachs of newborns also produce the enzyme gastric lipase. This enzyme is present in adults but plays a more important role in infants, where it begins the digestion of the fats in human milk. The absorption of fat from the infant's small intestine is inefficient. Low levels of pancreatic enzymes in infants limit starch digestion; however, enzymes at the brush border of the small intestine allow the milk sugar lactose to be digested and absorbed.

The ability to absorb intact proteins is greater in infants than that in adults. The absorption of whole proteins can cause food allergies (see Section 15.2), but it also allows infants to absorb immune factors from their mothers' milk. These proteins provide temporary immunity to certain diseases.

The bacteria in the large intestine of infants are also different from those in adults because of the all-milk diet infants consume, which is the reason that the feces of breastfed babies are almost odorless. Another feature of the infant digestive tract is the lack of voluntary control of elimination. Between the ages of 2 and 3, this ability develops and toilet training is possible.

Aging Although there are few dramatic changes in nutrient requirements with aging, changes in the digestive tract and other systems may affect the palatability of food and the ability to obtain proper nutrition. The senses of smell and taste are often diminished or even lost with age, reducing the appeal of food. A reduction in the amount of saliva may make swallowing difficult, decrease the taste of food, and also promote tooth decay. Loss of teeth and improperly fitting dentures may limit food choices to soft and liquid foods or cause solid foods to be poorly chewed. Gastrointestinal secretions may also be reduced, but this rarely impairs absorption, because the levels secreted in healthy elderly adults are still sufficient to break down food into forms that can be absorbed. A condition called **atrophic gastritis** that causes a reduction in the secretion of stomach acid is also common in the elderly. This may decrease the absorption of several vitamins and minerals and may allow bacterial growth to increase (see Sections 8.9 and 16.4). Constipation is a common complaint among the elderly. It may be caused by decreased motility and elasticity in the colon, weakened abdominal and pelvic muscles, and a decrease in sensory perception (see Critical Thinking: Gastrointestinal Problems Can Affect Nutrition).

atrophic gastritis An inflammation of the stomach lining that causes a reduction in stomach acid and allows bacterial overgrowth.

3.5 Transporting Nutrients to Body Cells

Learning Objectives

- Explain why the cardiovascular system is important in nutrition.
- Compare the path of an amino acid and a large fatty acid from absorption to delivery to a cell.

Nutrients absorbed into the mucosal cells of the intestine enter the blood circulation by either the **hepatic portal circulation** or the **lymphatic system**. The hepatic portal circulation is part of the cardiovascular system, which consists of the heart and blood vessels. Amino acids from protein digestion, simple sugars from carbohydrate digestion, and the water-soluble products of fat digestion are absorbed into **capillaries**, which are part of the hepatic portal circulation. The products of fat digestion that are not water-soluble are taken into lacteals, which are small vessels of the lymphatic system, before entering the blood.

hepatic portal circulation The system of blood vessels that collects nutrient-laden blood from the digestive organs and delivers it to the liver.

lymphatic system The system of vessels, organs, and tissues that drains excess fluid from the spaces between cells, transports fat-soluble substances from the digestive tract, and contributes to immune function.

capillaries Small, thin-walled blood vessels where the exchange of gases and nutrients between blood and cells occurs.

The Cardiovascular System

The cardiovascular system is a closed network of tubules through which blood is pumped. Blood carries nutrients and oxygen to the cells of all the organs and tissues of the body and removes waste products from these same cells. Blood also carries other substances, such as hormones, from one part of the body to another.

The heart is the workhorse of the cardiovascular system. It is a muscular pump with two circulatory loops—one that delivers blood to the lungs and one that delivers blood to the rest of the body (**Figure 3.17**). The blood vessels that transport blood and dissolved substances toward the heart are called **veins**, and those that transport blood and dissolved substances away from the heart are called **arteries**. As arteries carry blood away from the heart, they branch many times to form smaller and smaller blood vessels. The smallest arteries are called arterioles. Arterioles then branch to form capillaries. Capillaries are thin-walled vessels that are just large enough to allow one red blood cell to pass at a time. From the capillaries, oxygen and nutrients carried by the blood pass into the cells, and waste products pass from the cells into the blood. In the capillaries of the lungs, blood releases carbon dioxide to be exhaled and

veins Vessels that carry blood toward the heart.

arteries Vessels that carry blood away from the heart.

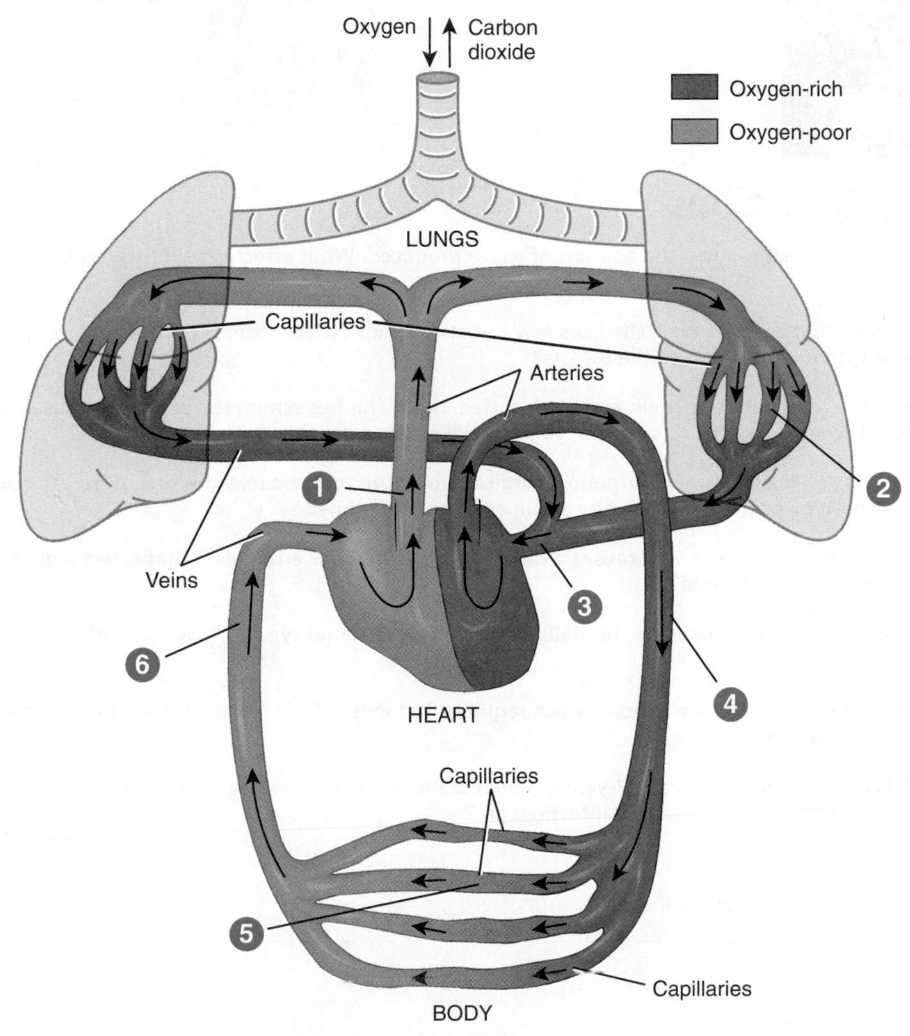

Figure 3.17 Cardiovascular system
Blood pumped to the lungs picks up oxygen and delivers nutrients. Blood pumped to the rest of the body delivers oxygen and nutrients.

1. Oxygen-poor blood that reaches the heart from the rest of the body is pumped through the arteries to the capillaries of the lungs.
2. In the capillaries of the lungs, oxygen from inhaled air is picked up by the blood, and carbon dioxide is released into the lungs and exhaled.
3. Oxygen-rich blood returns to the heart from the lungs via veins.
4. Oxygen-rich blood is pumped out of the heart into the arteries that lead to the rest of the body.
5. In the capillaries of the body, nutrients and oxygen move from the blood to the body's tissues, and carbon dioxide and other waste products move from the tissues to the blood, to be carried away.
6. Oxygen-poor blood returns to the heart via veins.

CRITICAL THINKING:

Gastrointestinal Problems Can Affect Nutrition

Many factors affect how well we digest food and absorb nutrients. For each situation described below, think about how digestion and absorption are affected and what the consequences are for nutritional status.

© istockphoto.com/James Driscoll

(©iStockphoto)

Critical Thinking Questions

A 50-year-old man is taking medication that reduces the amount of saliva produced. What effect would this have on his nutrition and health?

An 80-year-old woman has dentures that don't fit well; she likes raw carrots but can't chew them thoroughly. How will this affect the digestion and absorption of nutrients in the carrots?

For breakfast, a college student has cereal with skim milk and black coffee. His friend has scrambled eggs with sausage, biscuits with butter, and a glass of whole milk. Which person's stomach will empty faster? Why?

A 40-year-old woman weighing 160 kg (352 lbs) has undergone a surgical procedure that reduced the size of her stomach. How will this affect the amount and type of food that can be consumed at any one time?

A 63-year-old woman has a disease of the pancreas that causes a deficiency of pancreatic enzymes. What effect would this have on her ability to digest and absorb protein?

A 56-year-old man has gallstones, which cause pain when his gallbladder contracts. What type of foods should be avoided and why?

A 47-year-old woman undergoes treatment for colon cancer, which requires that most of her large intestine be surgically removed. How would this affect fluid needs?

After reading about the benefits of a high-fibre diet, an 18-year-old man dramatically increases the amount of fibre he consumes. How might this affect the feces? The amount of intestinal gas?

Use iProfile to find out how much fibre is in a cup of raw carrots.

picks up oxygen to be delivered to the cells. Blood in capillaries in the villi of the small intestine picks up water-soluble nutrients absorbed from the diet. Blood then flows from capillaries into the smallest veins, the venules, which converge to form larger and larger veins for return to the heart. Therefore, blood starting in the heart is pumped through the arteries to the capillaries of the lungs where it picks up oxygen. It then returns to the heart via the veins and is pumped out again into the arteries that lead to the rest of the body. In the capillaries of the body, blood delivers oxygen and nutrients and removes wastes before returning to the heart via the veins.

The volume of blood flow, and hence the amounts of nutrients and oxygen that are delivered to an organ or tissue, depends on the need. When a person is resting, about 25% of the blood goes to the digestive system, about 20% to the skeletal muscles, and the rest to the heart, kidneys, brain, skin, and other organs.[4] After a large meal, a greater proportion will go to the intestines to support digestion and absorption and to transport nutrients. When a person engages in strenuous exercise, about 85% of blood flow will be directed to the skeletal muscles to deliver nutrients and oxygen and remove carbon dioxide and waste products. Attempting to exercise after a large meal creates a conflict: the body cannot supply both the intestines and the muscles with enough blood to support their respective activities. The muscles win, and food remains in the intestines, often resulting in cramps.

The Hepatic Portal Circulation

In the small intestine, water-soluble molecules, including amino acids, sugars, water-soluble vitamins, and water-soluble products of fat digestion, cross the mucosal cells of the villi and enter capillaries. These capillaries merge to form venules at the base of the villi. The venules then merge to form larger and larger veins, which eventually form the **hepatic portal vein**. The hepatic portal vein transports blood directly to the liver, where absorbed nutrients are processed before they enter the general circulation (**Figure 3.18**).

hepatic portal vein The vein that transports blood from the GI tract to the liver.

The liver acts as a gatekeeper between substances absorbed from the intestine and the rest of the body. Depending on the immediate needs of the body, some nutrients are stored in the liver, some are broken down or changed into different forms, and others are allowed to pass through unchanged for delivery to other body cells. For example, the liver, with the help of hormones from the pancreas, keeps the concentration of sugar in the blood constant. The liver modulates blood glucose by removing absorbed glucose from the blood and storing it, by sending absorbed glucose on to the tissues of the body, or by releasing liver glucose (from stores or synthesis) into the blood. The liver also plays an important role in the synthesis and breakdown of amino acids, proteins, and fats. It modifies the products of protein breakdown to form molecules that can be safely transported to the kidneys for excretion. The liver also contains enzyme systems that protect the body from toxins that are absorbed by the GI tract.

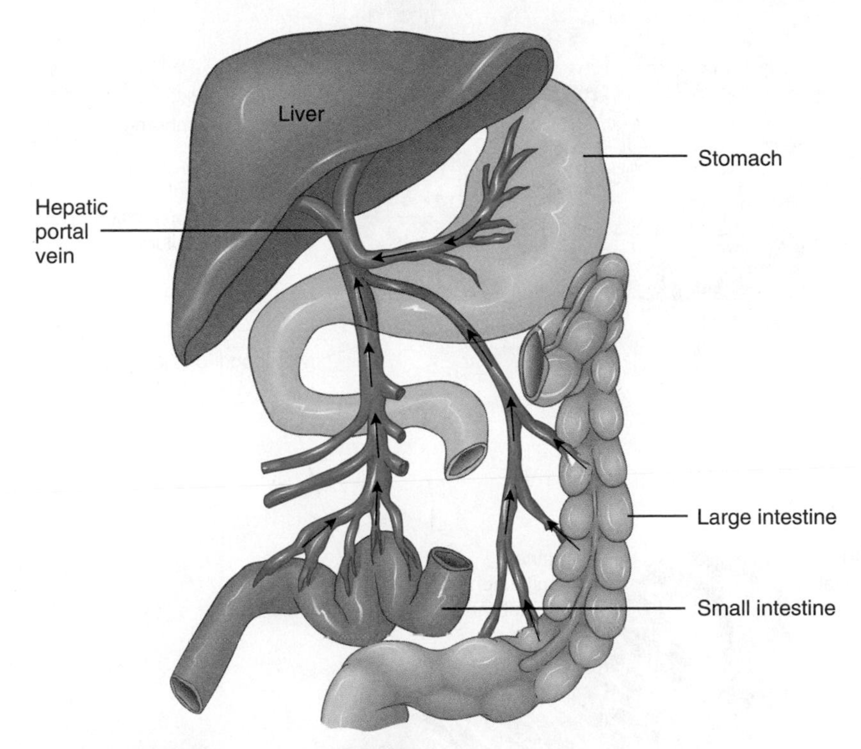

Figure 3.18 Hepatic portal circulation
The hepatic portal circulation carries blood from the stomach and intestines to the hepatic portal vein and then to the liver.

The Lymphatic System

The lymphatic system consists of a network of tubules (lymph vessels), structures, and organs that contain infection-fighting cells. Fluid that has accumulated in tissues drains into the lymphatic system, where it is filtered past a collection of infection-fighting lymphocytes and phagocytes. The cleansed fluid is then returned to the bloodstream. If the fluid contains a foreign substance, it may trigger an immune response. By draining the excess fluid and any disease-causing agents it contains away from the spaces between cells, the lymphatic system provides immunity and prevents the accumulation of fluid from causing swelling.

In the intestine, fat-soluble materials such as triglycerides, cholesterol, and fat-soluble vitamins are incorporated into particles, called chylomicrons, that are too large to enter the intestinal capillaries (see Section 5.3). These pass from the intestinal mucosa into the lacteals, the smallest of the lymph vessels, which drain into larger lymph vessels. Lymph vessels from the intestine and most other organs of the body drain into the thoracic duct, which empties into the bloodstream near the neck. Therefore, substances that are absorbed via the lymphatic system do not pass through the liver before entering the general blood circulation.

cell membrane The membrane that surrounds the cell contents.

selectively permeable Describes a membrane or barrier that will allow some substances to pass freely but will restrict the passage of others.

cytosol The liquid found within cells.

organelles Cellular organs that carry out specific metabolic functions.

mitochondrion (mitochondria) Cellular organelle responsible for providing energy in the form of ATP for cellular activities.

Body Cells

In order for nutrients to be used by the body, they must enter the cells. To enter a cell, substances must first cross the **cell membrane**. The cell membrane maintains homeostasis in the cell by controlling what enters and what exits. It is **selectively permeable**, allowing some substances, such as water, to pass freely back and forth, while limiting the passage of others. Nutrients and other substances are transported from the bloodstream into cells by simple and facilitated diffusion and active transport. Inside the cell membrane are the **cytosol**, or cell fluid, and **organelles** that perform functions necessary for cell survival. The largest organelle is the nucleus, which contains the cell's genetic material (**Figure 3.19**). The organelle where metabolic reactions that provide energy occur is called the **mitochondrion**.

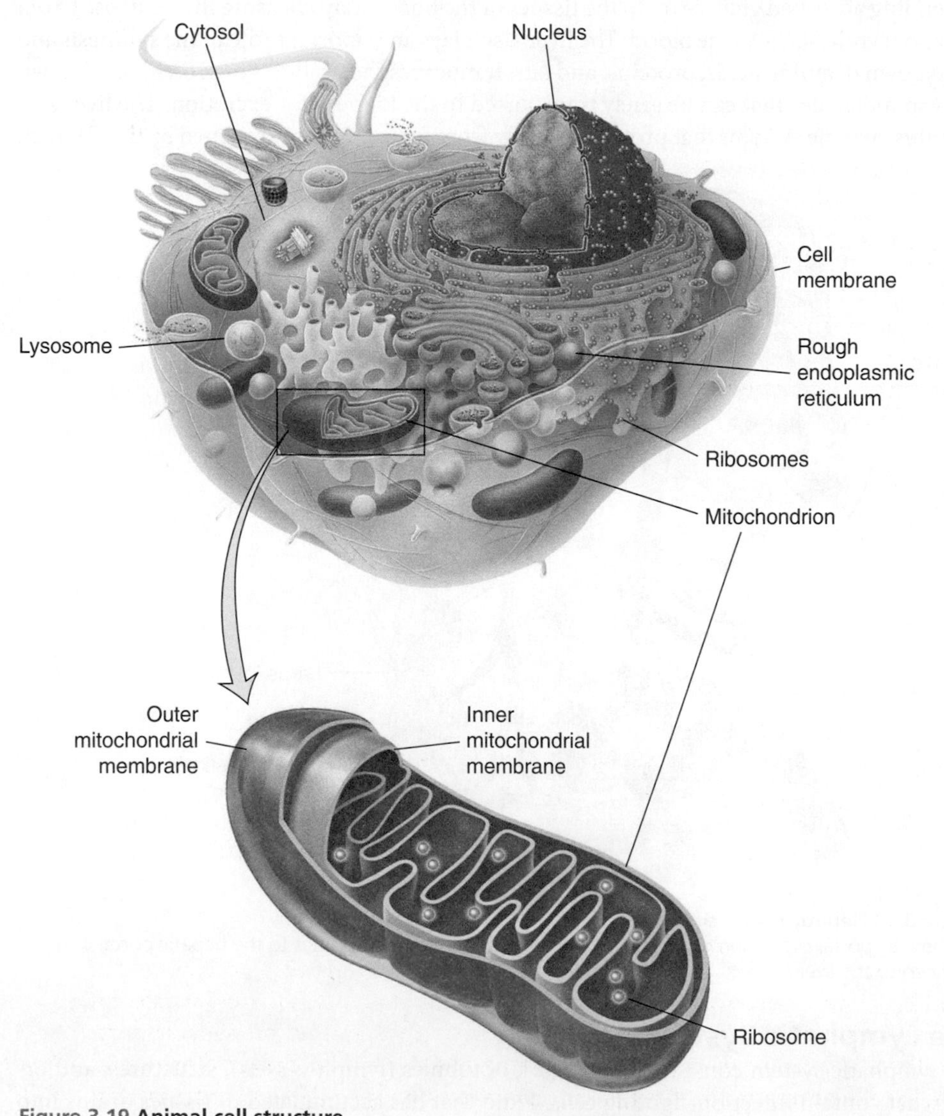

Figure 3.19 Animal cell structure
All human cells are surrounded by a cell membrane and most contain a nucleus, mitochondria, lysosomes, endoplasmic reticulum, and ribosomes in their cytosol.

3.6 Metabolism of Nutrients: An Overview

Learning Objectives

- Explain how the respiratory system is related to cellular respiration.
- Name the dietary fuel sources used to produce ATP.

By the mechanisms described thus far, foods are digested and the products of digestion absorbed, transported, and delivered to body cells. Each nutrient plays a unique role in metabolism. If the proper amounts and types of nutrients are not delivered to cells, the metabolic reactions cannot proceed optimally, resulting in poor health. The following discussion provides only a brief overview of how nutrients are used by the cells. Details about the metabolism of each nutrient will be discussed in appropriate chapters throughout this text.

metabolic pathway A series of chemical reactions inside of a living organism that result in the transformation of one molecule into another.

coenzyme A small, organic molecule (not a protein but sometimes a vitamin) that is necessary for the proper functioning of many enzymes.

catabolic The processes by which substances are broken down into simpler molecules, releasing energy.

ATP (adenosine triphosphate) The high-energy molecule used by the body to perform energy-requiring activities.

anabolic Energy-requiring processes in which simpler molecules are combined to form more complex substances.

cellular respiration The reactions that break down carbohydrates, fats, and proteins in the presence of oxygen to produce carbon dioxide, water, and energy in the form of ATP.

Metabolic Pathways

Depending on the body's needs, the glucose, fatty acids, and amino acids absorbed from the diet will be broken down to provide energy, be used to synthesize essential structural or regulatory molecules, or be transformed into energy-storage molecules. The conversion of one molecule into another often involves a series of reactions. The series of biochemical reactions needed to go from a raw material to the final product is called a **metabolic pathway** (see Appendix I). For each of the reactions of a metabolic pathway to proceed at an appropriate rate, an enzyme is required. These enzymes often need help from **coenzymes**. The B vitamins are important coenzymes in metabolism (See Section 8.1).

Some metabolic pathways break large molecules into smaller ones. These **catabolic** pathways release energy trapped in the chemical bonds that hold molecules together. Some of this energy is lost as heat, but some is converted into a form that can be used by the body called **adenosine triphosphate (ATP)** (**Figure 3.20**). ATP can be thought of as the energy currency of the cell. The chemical bonds of ATP are very high in energy, and when they break, the energy is released and can be used to power body processes, such as circulating blood or conducting nerve impulses—or it can be used to synthesize new molecules needed to maintain and repair body tissues. Metabolic pathways that use energy from ATP to build body compounds are referred to as **anabolic** pathways. The anabolic and catabolic pathways of metabolism occur in the body continually and simultaneously (**Figure 3.21**).

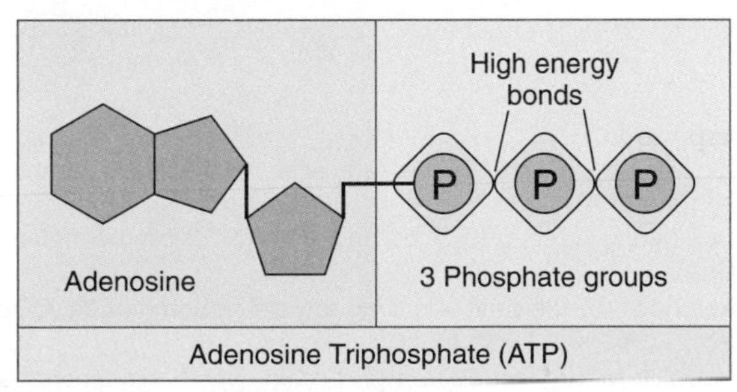

Figure 3.20 Structure of ATP
ATP consists of an adenosine molecule attached to three phosphate groups. The bonds between the phosphate groups are very high in energy, which is released when the bonds are broken.

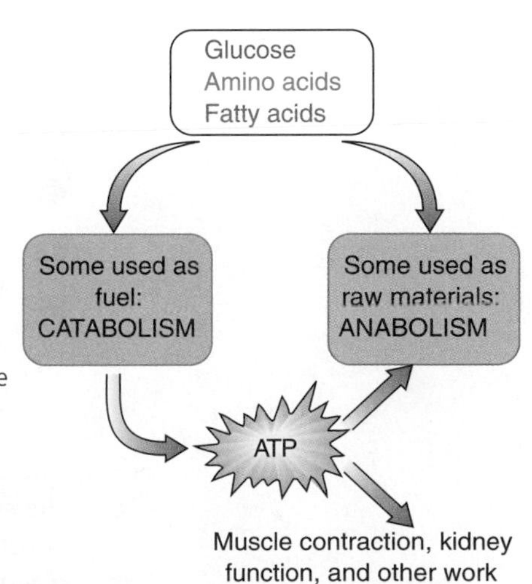

Figure 3.21 Anabolism and catabolism
Nutrients delivered to body cells can be used either in catabolic reactions to produce ATP or as raw materials in anabolic reactions that use ATP to synthesize molecules needed by the body.

Producing ATP

Carbohydrate, fat, and protein, both from the diet and from body stores, can be used to produce ATP via a catabolic pathway called **cellular respiration**. First carbohydrate, fat, and protein must be broken down into glucose, fatty acids, and amino acids, respectively. Then they can be metabolized, producing ATP along with carbon dioxide and water. In cellular respiration, oxygen brought into the body by the respiratory system and delivered to cells by the circulatory system is used and carbon dioxide is released. This carbon dioxide is then transported to the lungs where it is exhaled.

acetyl-CoA A metabolic intermediate formed during the breakdown of glucose, fatty acids, and amino acids. It is a 2-carbon compound attached to a molecule of CoA.

citric acid cycle Also known as the Krebs cycle or the tricarboxylic acid cycle, this is the stage of cellular respiration in which two carbons from acetyl-CoA are oxidized, producing two molecules of carbon dioxide.

electron High-energy particle carrying a negative charge that orbits the nucleus of an atom.

electron transport chain The final stage of cellular respiration in which electrons are passed down a chain of molecules to oxygen to form water and produce ATP.

The reactions of cellular respiration are central to all energy-yielding processes in the body. Without available oxygen, only glucose can be used to produce ATP (see Section 13.3). When oxygen is available, glucose, fatty acids, and amino acids can all be broken down to yield 2-carbon units that form a molecule called **acetyl-CoA** (**Figure 3.22**). To form acetyl-CoA, glucose must first be split in half by a pathway called glycolysis (see Section 4.4). To form acetyl-CoA, the carbon chains that make up fatty acids are broken into 2-carbon units by a pathway called beta-oxidation (see Section 5.5). Amino acids vary in structure, but after the amino group is removed by deamination, they can be broken down into units that can form acetyl-CoA (see Section 6.4). Acetyl-CoA from all of these sources is broken down inside the mitochondria via the metabolic pathway known as the **citric acid cycle**. In this pathway, the two carbons of acetyl-CoA are removed one at a time, forming carbon dioxide molecules, releasing **electrons**, and generating a small amount of ATP. The electrons, which are high in energy, are passed to shuttling molecules for transport to the **electron transport chain**.

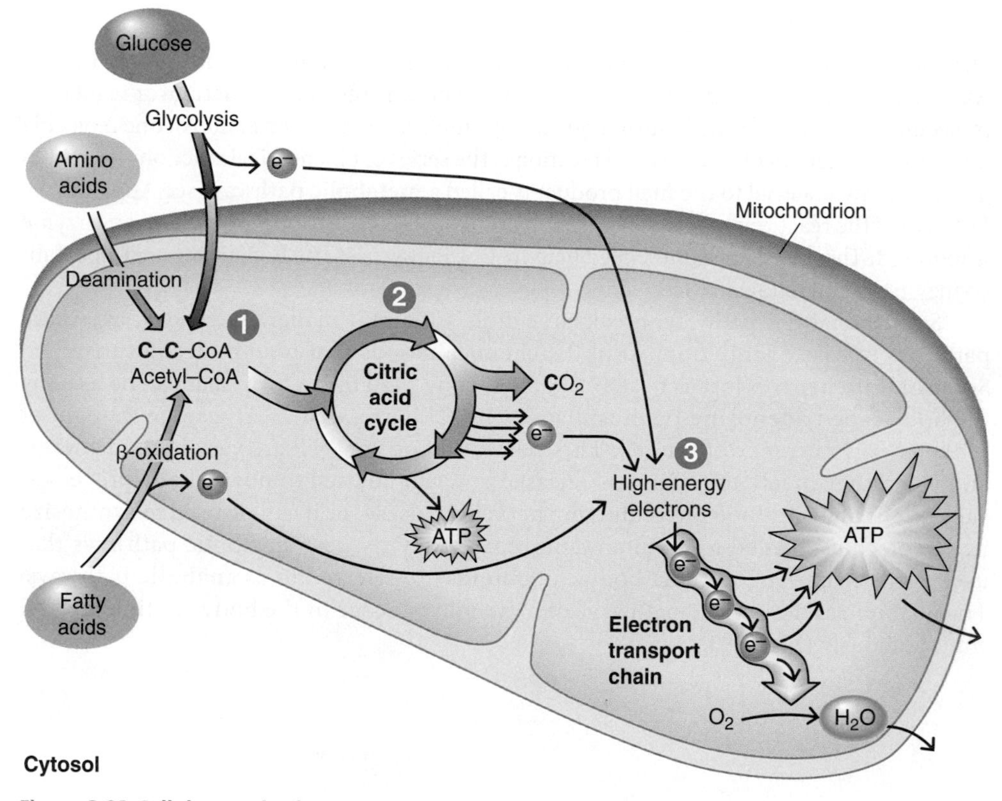

Figure 3.22 Cellular respiration
Cellular respiration uses oxygen to convert glucose, fatty acids, and amino acids into carbon dioxide, water, and energy, in the form of ATP.

1. In the presence of oxygen, glucose, fatty acids, and amino acids can be metabolized to produce acetyl-CoA (C-C-CoA).
2. Acetyl-CoA is broken down by the citric acid cycle to yield carbon dioxide (CO_2) and high-energy electrons.
3. The electrons are shuttled to the electron transport chain, where their energy is used to generate ATP and they are combined with oxygen and hydrogen to form water.

oxidized Refers to a compound that has lost an electron or undergone a chemical reaction with oxygen.

reduced Refers to a compound that has gained an electron.

The electron transport chain consists of a series of molecules that accept electrons from the shuttling molecules and pass them from one to another down the chain. When one substance loses an electron, another must pick up that electron. A substance that loses an electron is said to be **oxidized** and one that gains an electron is said to be **reduced**. Reactions that transfer electrons are called oxidation-reduction reactions and are very important in energy metabolism. As electrons are passed along the electron transport chain, their energy is released and used to make ATP. The final molecule to accept electrons in the electron transport chain is oxygen. When oxygen accepts electrons, it is reduced and forms a molecule of water (see Figure 3.22).

Synthesizing New Molecules

Glucose, fatty acids, and amino acids that are not broken down for energy are used in anabolic pathways to synthesize structural, regulatory, or storage molecules. Glucose molecules can be used to synthesize glycogen, a storage form of carbohydrate. If the body has enough glycogen, glucose can also be used to synthesize fatty acids. Fatty acids can be used to synthesize triglycerides that are stored as body fat. Amino acids can be used to synthesize the various proteins that the body needs, such as muscle proteins, enzymes, protein hormones, and blood proteins. Excess amino acids can be converted into fatty acids and stored as body fat.

3.7 Elimination of Metabolic Wastes

Learning Objective

- List four routes for eliminating waste products from the body.

Substances that cannot be absorbed by the body, such as fibre, are excreted from the GI tract in feces. The waste products generated by the metabolism of absorbed substances, such as carbon dioxide, nitrogen, and water, must also be removed from the body. These are eliminated by the lungs, the skin, and the kidneys (**Figure 3.23**).

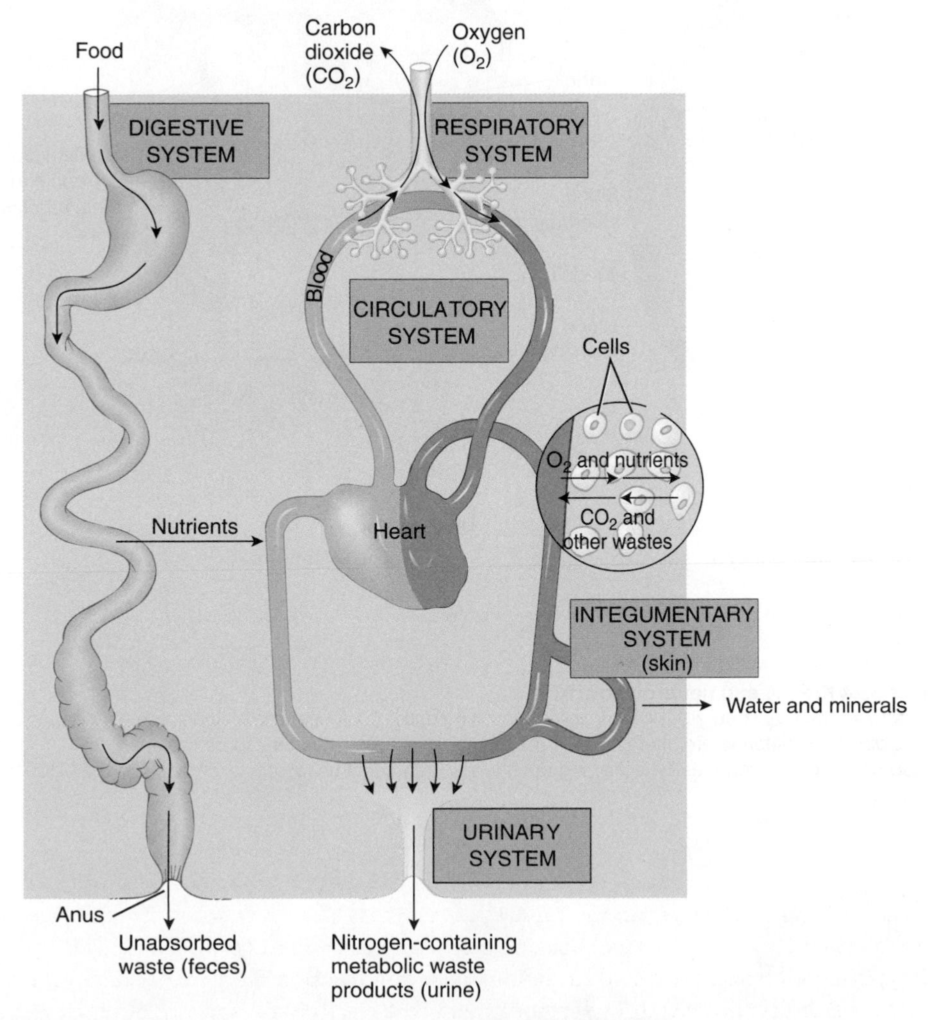

Figure 3.23 Taking in nutrients and oxygen and eliminating wastes
The digestive system takes in nutrients and the respiratory system takes in oxygen, which are then distributed to all body cells by the circulatory system. The urinary, respiratory, and integumentary systems transfer metabolic wastes to the external environment.

Lungs and Skin

Carbon dioxide produced by cellular respiration leaves the cells and is transported to the lungs by red blood cells. At the lungs, red blood cells release their load of carbon dioxide, which is then exhaled. In addition to carbon dioxide, a significant amount of water is lost from the lungs by evaporation. Water, along with protein breakdown products and minerals, is also lost through the skin in perspiration, or sweat.

Kidneys

nephron The functional unit of the kidney which performs the job of filtering the blood and maintaining fluid balance.

glomerulus A ball of capillaries in the nephron that filters blood during urine formation.

The kidneys are the primary site for the excretion of water, metabolic waste products, and excess minerals. Each kidney consists of about 1 million **nephrons**. The nephrons consist of a **glomerulus** where the blood is filtered and a series of tubules where molecules that have been filtered out of the blood can be reabsorbed (**Figure 3.24**). As blood flows through the glomerulus, most of the small, dissolved molecules are filtered out. Protein molecules and blood cells are too large to be removed by the glomerulus. Filtered substances that are needed are then reabsorbed back into the blood. Components that are not needed are not reabsorbed but are passed down the ureters to the bladder and excreted in the urine. The amounts of water and other substances excreted in the urine are regulated so that homeostasis is maintained (see Section 10.1).

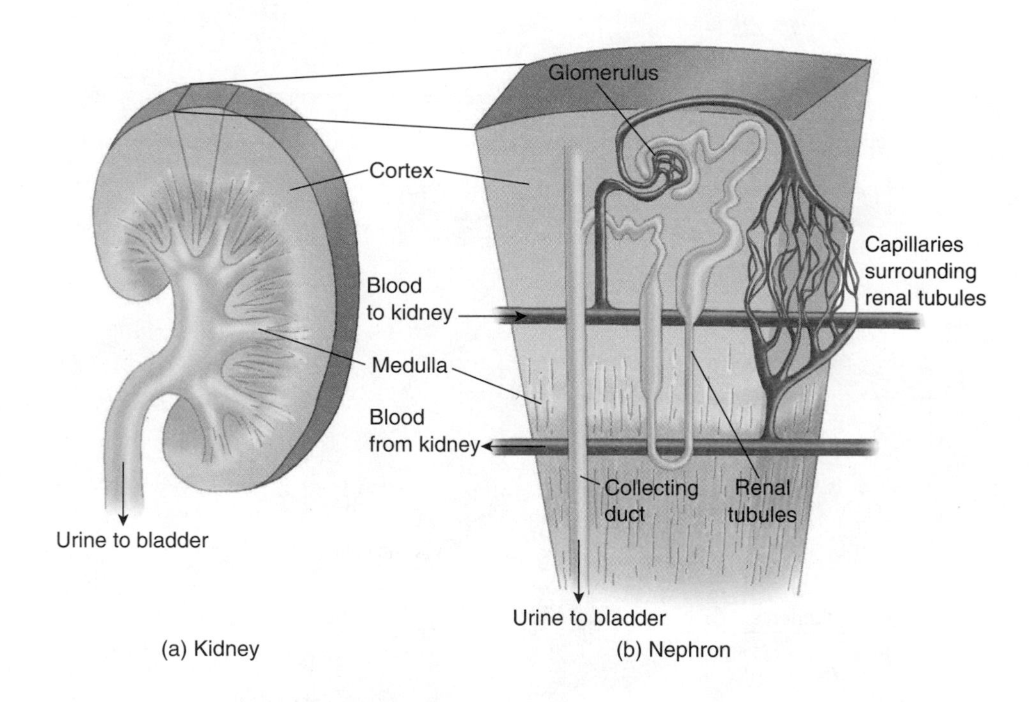

Figure 3.24 Kidney and nephron structure
(a) A kidney consists of an outer cortex and a central medulla; (b) A nephron includes the glomerulus, where dissolved materials are filtered out of the blood; renal tubules, where some materials are reabsorbed into the blood; and a collecting duct, which transports unabsorbed materials to the bladder.

CASE STUDY OUTCOME

It was a difficult decision but Sheila decided that the benefits of bariatric surgery outweighed the risk. The surgery itself went well and, in the past year, Sheila and has been able to lose 45 kg (100 pounds). The changes in Sheila's GI tract have caused some problems, though. First, she must carefully limit what and how much she eats, because alterations in her stomach have bypassed the gastroesophageal sphincter, which regulates the rate at which material leaves her stomach. When she eats too much or chooses too many high-carbohydrate foods, her meal "dumps" into her intestines and draws in water, resulting in dizziness, sweating, and diarrhea. Second, her small intestine has less surface area for nutrient absorption now that a portion of it has been bypassed. To prevent vitamin and mineral deficiencies, Sheila must take supplements and get monthly vitamin B_{12} injections. Although she often longs to sit down to a large meal and not have to worry about the consequences of overeating, she believes the decision to have the procedure was the right one for her. Her blood pressure, blood cholesterol, and blood sugar have all decreased. She can now bend to tie her own shoes, and when she travels, she can fit into a single airplane seat. By sticking to her diet and exercise program, she plans to lose another 18 kg (40 pounds) and maintain the loss.

APPLICATIONS

Personal Nutrition

iProfile

1. **Select one meal from the food record you kept in Chapter 2 and using iProfile, review the protein, carbohydrate, and fat content of each food.**
   a. List foods that would not begin chemical digestion until they have left the mouth.
   b. List foods that might require bile for digestion and explain why.
   c. List foods that might begin their digestion in your stomach and explain why.
2. **Imagine you wake up on a Sunday morning and join some friends for a large breakfast consisting of a cheese omelette and sausage (foods high in fat and protein), a croissant with butter (which contains carbohydrate but is also very high in fat), and a small glass of orange juice. After the meal, you remember that you have plans to play basketball with a friend in just an hour.**
   a. If you keep your plans and play basketball, what problems might you experience while exercising?
   b. Had you remembered your plans for strenuous exercise before you had breakfast, what type of meal might you have selected to ensure that your stomach would empty more quickly?

General Nutrition Issues

1. **There are hundreds of products available to aid digestion. Go to the drug store, or search the Internet, and select a product claiming to aid digestion.**
   a. List the claims made for the product.
   b. Use the information in Chapter 1 on judging nutritional claims to analyze the information given.
   c. What nutrients, if any, does the product provide?
   d. What risks, if any, does it carry?
   e. Would you take it if you were experiencing digestive problems? Why or why not?
2. **Plan a lunch menu for a seniors centre. It should provide food from each food group of *Eating Well with Canada's Food Guide*. Assume that many of the diners have difficulty chewing.**
   a. Evaluate whether your menu would be appropriate for someone with gallstones who needs to limit their intake of fat.
   b. Evaluate whether your menu would be appropriate for someone who is trying to increase their intake of fibre and fluid to prevent constipation.
   c. Evaluate whether your menu would be appropriate for someone who must restrict their salt intake.

SUMMARY

3.1. Food Becomes Us

- Our bodies and the foods we eat are all made from the same building blocks—atoms. Atoms are linked by chemical bonds to form molecules.
- Molecules can form cells, which are the smallest units of life. Cells of similar structure and function are organized into tissues, and tissues into organs and organ systems.

3.2. An Overview of the Digestive System

- The digestive system is the organ system primarily responsible for the movement of nutrients into the body. It provides two major functions: digestion and absorption. Digestion is the process by which food is broken down into units that are small enough to be absorbed. Absorption is the process by which nutrients are transported into the body.
- The gastrointestinal (GI) tract consists of a hollow tube that begins at the mouth and continues through the pharynx, esophagus, stomach, small intestine, and large intestine.
- The digestion of food and the absorption of nutrients in the lumen of the GI tract is aided by the secretion of mucus and enzymes.
- The passage of food and the secretion of digestive substances is regulated by nervous and hormonal signals.
- Immune system cells and tissues located in the gastrointestinal tract help eliminate disease-causing organisms or chemicals.

3.3. Digestion and Absorption

- The processes involved in digestion begin in response to the smell or sight of food and continue as food enters the digestive tract at the mouth.
- In the mouth, food is broken into smaller pieces by the teeth and mixed with saliva. Carbohydrate digestion is begun in the mouth by salivary amylase.
- From the mouth, food passes through the pharynx and into the esophagus. The rhythmic contractions of peristalsis propel it down the esophagus to the stomach.
- The stomach acts as a temporary storage tank for food. The muscles of the stomach mix the food into a semiliquid mass called chyme, and gastric juice containing hydrochloric acid and pepsin begins protein digestion. Stomach emptying is regulated by the amount and composition of food consumed and by nervous and hormonal signals from the stomach and small intestine.
- The small intestine is the primary site of nutrient digestion and absorption. In the small intestine, bicarbonate from the pancreas neutralizes stomach acid, and pancreatic and intestinal enzymes digest carbohydrate, fat, and protein. The digestion of fat in the small intestine is aided by bile from the gallbladder. Bile helps make fat available to fat-digesting enzymes by breaking it into small droplets; it also facilitates fat absorption. Secretions from the pancreas and liver are regulated by the hormones secretin and cholecystokinin (CCK), produced by the duodenum.
- The absorption of food across the intestinal mucosa occurs by several different processes. Simple diffusion, osmosis, and

facilitated diffusion do not require energy but depend on a concentration gradient. Active transport requires energy but can transport substances against a concentration gradient. The absorptive surface area of the small intestine is increased by folds, finger-like projections called villi, and tiny projections called microvilli, which cover the surface of the villi.

- Components of chyme that are not absorbed in the small intestine pass on to the large intestine, where some water and nutrients are absorbed. The large intestine is populated by bacteria that digest some of these unabsorbed materials, such as fibre, producing small amounts of nutrients and gas. The remaining unabsorbed materials are excreted in feces.

3.4. Digestion and Health

- Heartburn and GERD are common digestive problems that are caused by the leakage of stomach contents into the esophagus. Ulcers are caused by infection with *Helicobacter pylori*, GERD, or medications that damage the mucosa. Gallstones can cause pain and interfere with fat digestion.
- Tube-feeding can nourish a patient who is unable to ingest or chew food on his or her own. Total parenteral nutrition is necessary when the gut is not able to digest and/or absorb nutrients.
- Digestive system function is affected by life stage. During pregnancy, physiological changes cause morning sickness, heartburn, and constipation. During infancy, the immaturity of the GI tract limits what foods can be ingested and digested. In older adults, changes in the digestive tract may decrease the appeal of food and the ability to digest and absorb nutrients.

3.5. Transporting Nutrients to Body Cells

- Absorbed nutrients are delivered to the cells of the body by the cardiovascular system. The heart pumps blood to the lungs to pick up oxygen and eliminate carbon dioxide. From the lungs, blood returns to the heart and is pumped to the rest of the body to deliver oxygen and nutrients and remove carbon dioxide and other wastes before returning to the heart. Blood is pumped away from the heart in arteries and returned to the heart in veins. Exchange of nutrients and gases occurs at the smallest blood vessels, the capillaries.
- The products of carbohydrate and protein digestion and the water-soluble products of fat digestion enter capillaries in the intestinal villi and are transported to the liver via the hepatic portal circulation. The liver serves as a processing centre, removing the absorbed substances for storage, converting them into other forms, or allowing them to pass unaltered. The liver also protects the body from toxic substances that may have been otherwise absorbed.
- The fat-soluble products of digestion enter lacteals in the intestinal villi. Lacteals join larger lymph vessels. The nutrients absorbed via the lymphatic system enter the blood circulation without first passing through the liver.
- Cells are the final destination of absorbed nutrients. To enter the cells, nutrients must be transported across cell membranes.

3.6. Metabolism of Nutrients: An Overview

- Within the cells, glucose, fatty acids, and amino acids absorbed from the diet can be broken down to provide energy in the form of ATP, used to power body activities and synthesize essential structural or regulatory molecules, or be transformed into energy-storage molecules. The sum of all the chemical reactions of the body is called metabolism.
- The reactions that completely break down macronutrients in the presence of oxygen to produce water, carbon dioxide, and ATP are referred to as cellular respiration. Glucose, fatty acids, and amino acids can all be broken down into 2-carbon molecules that form acetyl-CoA. The reactions of the citric acid cycle and the electron transport chain complete the breakdown of acetyl-CoA to form carbon dioxide and water and generate ATP. Dietary glucose, fatty acids, and amino acids are used to synthesize structural, regulatory, or storage molecules.

3.7. Elimination of Metabolic Wastes

- Unabsorbed materials are excreted in the feces. Carbon dioxide is eliminated in exhaled air. Water is lost via the lungs and skin.
- Water, metabolic waste products, and excess minerals are excreted by the kidneys.

REVIEW QUESTIONS

1. What is the smallest unit of plant and animal life?
2. List three organ systems involved in the digestion and absorption of food.
3. How do teeth function in digestion?
4. What is peristalsis? What is segmentation?
5. List two functions of the stomach.
6. How is the movement of material through the digestive tract regulated?
7. List three mechanisms by which nutrients are absorbed.
8. Where does most digestion and absorption occur?
9. How does the structure of the small intestine aid absorption?
10. What products of digestion are transported by the lymphatic system?
11. What path does an amino acid follow from absorption to delivery to the cell? Compare this to the path a large fatty acid would follow from absorption to delivery to the cell.
12. What is the form of energy used by cells?
13. Explain what occurs during the citric acid cycle and the electron transport chain.
14. What happens to material that is not absorbed in the small intestine?
15. How do the lungs and kidneys help eliminate metabolic waste products?

REFERENCES

1. Formanek, R. Food allergies: When food becomes the enemy. *FDA Consumer*, July/Aug, 2001.
2. Konturek, S. J., Konturek, P. C., Konturek, J. W., et al. *Helicobacter pylori* and its involvement in gastritis and peptic ulcer formation. *J. Physiol. Pharmacol. 57* (Suppl– 3):29–50, 2006.
3. Lynch, N. A. *Helicobacter pylori* and ulcers: A paradigm revisited. Available online at http://opa.faseb.org/pdf/pylori.pdf. Accessed December 20, 2008.
4. Marieb, E. N., and Hoehn, K. *Human Anatomy and Physiology*, 7th ed. Menlo Park, CA: Benjamin Cummings, 2007.

4 Carbohydrates:

SUGARS, STARCHES, AND FIBRE

Case Study

Pearl's enthusiasm for her low-carbohydrate diet was flagging, and the 7 kg (15 pounds) she had recently lost was slowly creeping back onto her 1.6 m (5'6") frame. She had always been heavier than she liked and had been delighted to lose some weight. Now, though, the 4.5 kg (10 pounds) she regained had brought her up to 69 kg (153 pounds). This weight was still in the healthy range, so she decided to forget the low-carb weight-loss diet approach and just focus on eating healthy foods. Pearl looked up *Eating Well with Canada's Food Guide* on the Health Canada website and was surprised to discover that many carbohydrate foods were recommended. She learned that Canada's Food Guide recommended that she eat six Canada Food Guide servings of grain products daily and that at least half of them should be whole grains. She also saw that she needed to increase her intake of fruits and vegetables to seven Canada Food Guide servings daily. These foods, too, contain carbohydrate and are high in fibre.

The first step Pearl took to improve her diet was to keep a bag of cut-up raw vegetables and a bowl of fruit salad in her refrigerator, ready for snacking or adding to a meal. To increase her intake of whole grains, she began to make smarter choices at the grocery store. She switched from white rice to brown rice and chose multi-grain breads and healthy-sounding cereals. A few weeks later, she had her diet analyzed at a health fair. She was dismayed to see that even after all the changes she had made in her diet, her fibre intake was still below the recommended 25 g per day. When she got home, she took a look at the food labels in her cupboard. She found that her seven-grain bread didn't actually contain any whole grains. Her breakfast cereal did contain whole wheat, but provided only 2 g of fibre per cup—and it also contained 18 g of sugars she hadn't intended to consume. Choosing healthy carbohydrates was turning out to be almost as difficult as eliminating them.

This chapter will discuss the differences between the various types of carbohydrates found in foods and will help you to identify the most nutritious choices.

Cole Group/Photodisc/Getty Images, Inc

Chapter Outline

4.1 Carbohydrates in the Canadian Diet

Learning Objectives

- List three ways by which carbohydrates can be classified.
- Discuss the difference between refined and unrefined carbohydrates.
- Explain why added sugars are considered empty kcalories.

Carbohydrates are the basis of our diet. Of the three macronutrients, we consume more carbohydrates than either fat or protein. There are several ways to classify carbohydrates. One way is based on their chemical composition. Another way is by its physiological impact on blood glucose using a measure called the glycemic index. Both of these classifications will be discussed in Section 4.5. Carbohydrates can also be classified by the degree to which they have been processed. They are found in foods as diverse as whole-wheat bread, chocolate cake, fresh fruit, milk, legumes (e.g., beans, lentils), and carbonated soft drinks. Carbohydrates are a readily available source of energy; they supply 4 kcal per gram. However, the additional nutritional impact they deliver varies, depending on whether the carbohydrate is **refined** or in its natural state. The carbohydrates in the whole-wheat bread, the fresh fruit, and the milk are considered unrefined or whole-food sources of carbohydrate, because they have not been altered from their natural state by processing. These foods contain vitamins, minerals, dietary fibre (an indigestible carbohydrate) and other health-promoting substances, as well as digestible carbohydrates. The cake and the soft drink provide only digestible carbohydrates. Processing, or refining, reduces the amount of the vitamins, minerals, and/or dietary fibre present in the whole food and sometimes adds high amounts of sugar, fat, or salt (**Table 4.1**).

refined Refers to foods that have undergone processes that change or remove various components of the original food.

Table 4.1 More and Less Refined Carbohydrate Food Choices

	Less Refined	More Refined	High in Added Sugars
Cereals	Oatmeal, shredded wheat	Corn flakes, Rice Puffs, Cheerios	Lucky Charms, Frosted Shredded Wheat
Breads	Whole-wheat bread, whole-wheat bagel	White bread, English muffin, white bagel	Doughnut, Danish pastry
Grains	Whole-wheat pasta, brown rice, bulgur wheat, barley, quinoa	White pasta, white rice, rice cakes	Rice pudding, Rice Krispie treats, sweetened rice cakes
Fruit	Raspberries, apple, orange	Canned fruit, dried fruit, orange juice	Canned fruit in heavy syrup, fruit pies, sweetened dried fruit, candied fruit, fruit punch
Vegetables	Baked potato, zucchini	French fries, fried zucchini	Candied yams, carrot cake, pumpkin pie
Legumes (e.g., Beans & lentils)	Boiled raw beans	Beans canned in brine (contains salt)	Canned beans in maple syrup

Figure 4.1a indicates the intake of carbohydrates in the Canadian diet, while **Figure 4.1b** indicates various carbohydrate sources. These numbers are based on disappearance data collected by Statistics Canada. Figure 4.1a shows that carbohydrate intake has increased since the 1970s. The major sources of carbohydrates in the Canadian diet are cereals, which include breakfast foods, wheat flour, and other grain products such rice, corn, oats, rye, and barley (Figure 4.1b). As we shall see later in the chapter, Canadians have a very low intake of dietary fibre, suggesting that many of the cereal products that Canadians consume are refined and are low in fibre. The next largest source of carbohydrate is from sugar and syrups; this list includes sucrose (cane and beet sugar), maple syrup, and honey. Another source of sugar comes from soft drinks. These three items make up about two-thirds of the carbohydrates

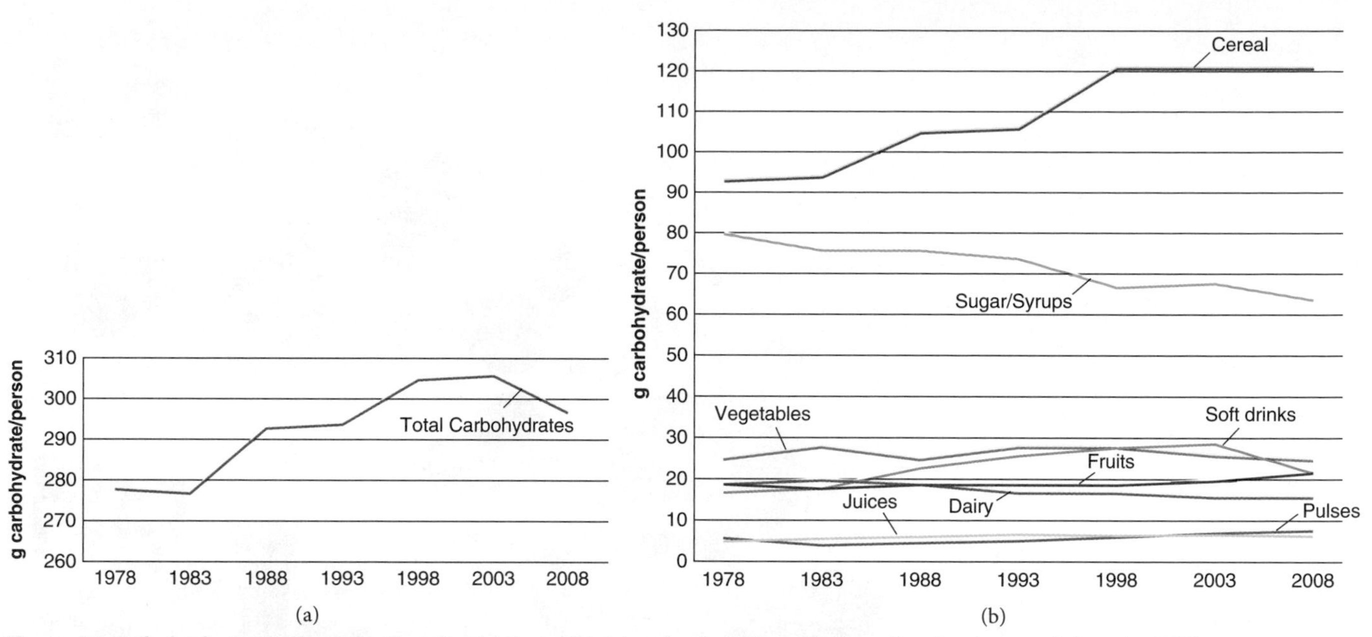

Figure 4.1 Carbohydrates in the Canadian diet (a) the total intake of carbohydrates by Canadians has increased since the 1970s. (b) Carbohydrates in the Canadian diet come from cereals such as wheat and rice, and sugars and syrups, which include sucrose (cane and beet sugar, maple syrup, and honey). The amount of sugars and syrups in the diet has declined over the last two decades. This is believed to be due to an increase in the use of high-fructose corn syrup (HFCS). Statistics Canada does not collect specific information on the use of HFCS, but its usage is inferred from the carbohydrate contribution of soft drinks. These are the sources of HFCS in the Canadian diet, and the intake of this sweetener has generally increased since the 1970s.
Source: Statistics Canada. CANSIM Table 003-0080. Nutrients in the food supply, by source of nutritional equivalent and commodity, annual. Available online at http://cansim2.statcan.gc.ca/cgi-win/cnsmcgi.exe?Lang=E&CNSM-Fi=CII/CII_1-eng.htm. Accessed Feb 2, 2010.

consumed by Canadians. The carbohydrates in today's diet are more refined than they were 50–100 years ago.

Canada's Food Guide recommends that we eat more unrefined and nutrient-dense sources of carbohydrates such as whole grains, vegetables, legumes, and fruits, and fewer foods high in refined carbohydrates and **added sugars**, such as baked goods and soft drinks (**Figure 4.2**).

added sugars Sugars and syrups that have been added to foods during processing or preparation.

Refined Carbohydrates

Unrefined food sources of carbohydrate such as whole grains, legumes, vegetables, fruit, and milk contain a variety of nutrients in addition to carbohydrates. Whole grains, legumes, and vegetables provide B vitamins, some minerals, and fibre. Fruits provide vitamins A and C along with fibre. Milk contains the sugar lactose, but is a good source of the B vitamin riboflavin and the mineral calcium. In contrast, refined sources of carbohydrate, such as the corn flakes you may have had for breakfast, are made from corn that has been ground, sieved, washed, cooked, extruded, and dried. During these refining steps, many of the nutrients and other healthful components of the corn kernel are lost. When we eat the entire kernel or seed of a grain, such as corn or wheat, we are eating an unrefined or **whole-grain** product. The whole-grain kernel includes three parts (**Figure 4.3**). The outermost **bran** layers contain most of the fibre and are also a good source of vitamins. The **germ**, which lies at the base of the kernel, is the plant embryo where sprouting occurs. This germ is the source from which we obtain the commonly used vegetable oils such as corn or safflower oil. It is also is rich in vitamin E and contains protein, fibre, and the B vitamins riboflavin, thiamin, and vitamin B_6. The remainder of the kernel is the **endosperm**, which is primarily starch but also contains most of the protein and some vitamins and minerals. During the milling of grain into flour, the grinding detaches the germ and bran from the endosperm. Whole-grain flours such as whole-wheat flour include most of the bran, germ, and endosperm (see Your Choice: Choosing Whole Grains). White flour, however, is produced from just the endosperm, so fibre and some vitamins, minerals, and phytochemicals naturally found in the whole grain are therefore lost. In order to restore

whole-grain The entire kernel of grain, including the bran layers, the germ, and the endosperm.

bran The protective outer layers of whole grains. It is a concentrated source of dietary fibre.

germ The embryo or sprouting portion of a kernel of grain, which contains vegetable oil, protein, fibre, and vitamins.

endosperm The largest portion of a kernel of grain, which is primarily starch and serves as a food supply for the sprouting seed.

Figure 4.2 **Carbohydrate recommendations for a healthy diet**
Choose more unrefined sources of carbohydrates while limiting foods high in refined carbohydrates and added refined sugars. (© Andy Washnik)

Figure 4.3 **Parts of a whole-grain kernel**
A kernel of wheat contains outer layers of bran, the plant embryo (germ), and a carbohydrate-rich endosperm. (Kevin Morris/Stone/ © Getty Images)

some of the lost nutrients, refined grains sold in Canada are fortified with some, but not all, of the nutrients lost in processing. **Fortified or enriched grains** contain added thiamin, riboflavin, niacin, and iron, and are fortified with folate. However, they do not contain added vitamin E, magnesium, vitamin B_6, or a number of other nutrients that are also removed by milling.

fortified or **enriched grains** Grains to which specific amounts of thiamin, riboflavin, niacin, and iron have been added. Since 1998, folic acid has also been added to enriched grains.

Added Sugars

If you sprinkle some sugar on your corn flakes, you are adding another refined source of carbohydrate. This added sugar was most likely extracted from a sugar beet, boiled, bleached, and purified. It adds kcalories without adding any nutrients other than carbohydrate and reduces the nutrient density of your breakfast. But the sugar you add to food isn't the only source of added sugars in the diet—much of the added sugar we consume comes from desserts, beverages, and snacks that we purchase already prepared. Refined added sugar is estimated to make up about 13% of the kcalories in the Canadian diet.[1] Added sugars are not nutritionally or chemically different from sugars occurring naturally in foods. The only difference is that they have been separated from their plant sources and therefore are not consumed with all of the fibre, vitamins, minerals, and other substances found in the original plant. Because added sugars provide few nutrients for the number of kcalories they contain, they have a low nutrient density and are thought of as **empty kcalories**. Unrefined or whole food sources of sugar, such as fruit, provide vitamins, minerals, and phytochemicals as well as kcalories. For example, a 355 ml soft drink contains about 140 kcal but almost no nutrients other than sugar. Three kiwis also have about 140 kcal but contribute vitamin C, folate, potassium, and some calcium as well as fibre (**Figure 4.4**).

empty kcalories Refers to foods that contribute energy but few other nutrients.

Nutrient	Soft Drink (355 ml)	Kiwi (3 Medium)
Vitamin A (μg)	0	7
Vitamin C (mg)	0	171
Folate (μg DFE)	0	87
Potassium (mg)	7	757
Calcium (mg)	7	60
Protein (g)	0	2
Fibre (g)	0	8
Carbohydrate (g)	35	34
Sugars (g)	33	26
Energy (kcals)	136	139

Figure 4.4 Three kiwis and 355 ml of a soft drink provide about the same amounts of energy and carbohydrate, but the kiwis also contain fibre and a variety of micronutrients. (© Luisa Begani)

4.2 The Chemistry of Carbohydrates

Learning Objectives

- Name the major monosaccharides and disaccharides in the Canadian diet.
- Distinguish between glycogen and starch.
- Distinguish between soluble and insoluble fibre and name food sources of each.

Chemically, carbohydrates are compounds that contain carbon (carbo), as well as hydrogen and oxygen in the same proportion, as in water (hydrate). They are typically divided into monosaccharides, disaccharides, and polysaccharides. The monosaccharides and disaccharides are often referred to as **simple carbohydrates**, also known as sugars, and polysaccharides,

simple carbohydrates Carbohydrates known as sugars that include monosaccharides and disaccharides.

complex carbohydrates Carbohydrates composed of monosaccharide molecules linked together in straight or branching chains. They include glycogen, starches, and fibres.

which include starches and dietary fibre, are referred to as **complex carbohydrates.** Both can provide a source of energy to fuel the body.

Monosaccharides and Disaccharides

The basic unit of carbohydrate is a single sugar unit, a **monosaccharide** (*mono* means one). When two sugar units combine, they form a **disaccharide** (*di* means two). Monosaccharides and disaccharides are known as simple sugars or simple carbohydrates. Fruits, vegetables, and milk are sources of simple carbohydrates. The sugars we add to food such as white table sugar, brown sugar, molasses, and confectioner's sugar are also simple carbohydrates. These are produced by refining the sugar from plants such as sugar cane and sugar beets.

monosaccharide A single sugar unit, such as glucose.

disaccharide A sugar formed by linking two monosaccharides.

glucose A monosaccharide that is the primary form of carbohydrate used to provide energy in the body. It is the sugar referred to as blood sugar.

galactose A monosaccharide that combines with glucose to form lactose or milk sugar.

fructose A monosaccharide that is the primary form of carbohydrate found in fruit.

sucrose A disaccharide that is formed by linking fructose and glucose. It is commonly known as table sugar or white sugar.

lactose A disaccharide that is formed by linking galactose and glucose. It is commonly known as milk sugar.

maltose A disaccharide made up of 2 molecules of glucose. It is formed in the intestines during starch digestion.

Monosaccharides The three most common monosaccharides in the diet are glucose, galactose, and fructose. Each contains 6 carbon, 12 hydrogen, and 6 oxygen atoms, but differ in their arrangement (**Figure 4.5**). **Glucose** circulates in the blood and is the most important carbohydrate fuel for the body. It is produced in plants by the process of photosynthesis, which uses energy from the sun to combine carbon dioxide and water (**Figure 4.6**). Glucose occurs as a monosaccharide in honey but most often is found as part of a disaccharide or starch. **Galactose** is also rarely present as a monosaccharide in the food supply, but occurs most often as a part of lactose, the disaccharide in milk.

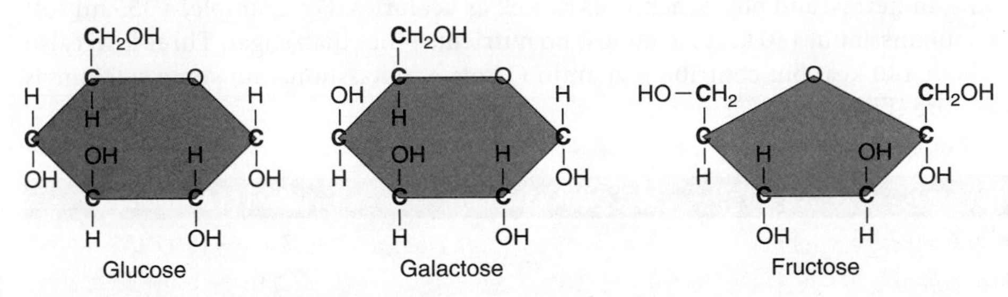

Figure 4.5 Structures of common monosaccharides
Glucose, galactose, and fructose have the same chemical formulas, but the atoms are arranged differently.

Fructose is a monosaccharide that tastes sweeter than glucose; it is found in fruits and vegetables and makes up more than half the sugar in honey. Because fructose does not cause as great a rise in blood glucose as other sugars, it is sometimes used in products for people with diabetes. However, because fructose causes an increase in blood lipids, its use should be limited. Too much fructose consumed in fruits or juices can also cause diarrhea in children. Most of the fructose in our diet comes from high-fructose corn syrup. This sweetener is produced by modifying starch extracted from corn to produce a syrup that is approximately half glucose and half fructose. High-fructose corn syrup is sweeter and less expensive than table sugar and is used in most soft drinks. The increase in the use of high-fructose corn sweeteners that has occurred since its introduction in the 1970s has been suggested to be related to the increased incidence of diabetes and obesity.[2]

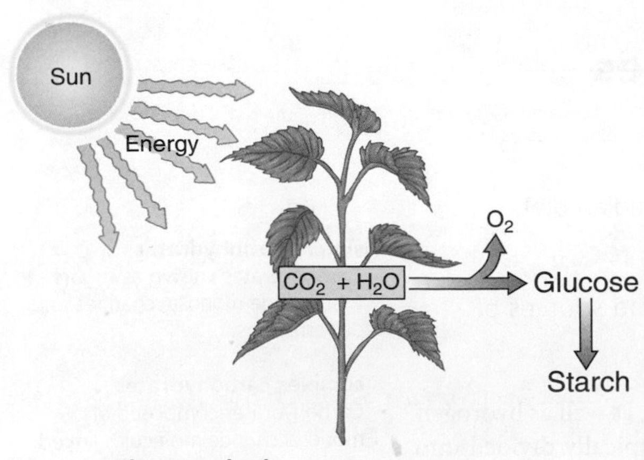

Figure 4.6 Photosynthesis
Photosynthesis uses energy from the sun to convert carbon dioxide and water into glucose, which can be stored as starch.

Disaccharides Disaccharides are simple carbohydrates made up of two monosaccharides linked together (**Figure 4.7**). **Sucrose**, or common white table sugar, is the disaccharide formed by linking glucose to fructose. It is found in sugar cane, sugar beets, honey, and maple syrup. **Lactose**, or milk sugar, is glucose linked to galactose. Lactose is the only sugar found naturally in animal foods. It contributes about 30% of the energy in whole cow's milk and about 40% of the energy in human milk. **Maltose** is a disaccharide consisting of two molecules of glucose. This sugar is made whenever starch is broken down. For example, it is responsible for the slightly sweet taste experienced when bread is held in the mouth for a few minutes. As salivary amylase begins digesting the starch, some sweeter-tasting maltose is formed.

YOUR CHOICE

Choosing Whole Grains

CANADIAN CONTENT

You know you should eat more whole-grain products, but how can you spot them at the store? Do you just put brown bread instead of white into your shopping cart? Unfortunately, it's not that easy. Bread may be brown because of ingredients such as molasses, not necessarily because it is made from whole-grain ingredients. Product names can also be deceptive. Healthy-sounding terms like "multi-grain" or "seven-grain" simply mean the product contains more than one type of grain, not that these grains are necessarily in their unrefined state. "Wheat" refers to the type of grain, not how refined it is; "stone ground" refers to how the grain was processed, not whether the bran and germ are included; and terms like "bran" and "oat" may refer to ingredients added to the product, not whether the product is made predominantly from a whole grain.

In Canada during the milling process, the germ and bran portion of the kernel can be separated and then recombined to make various types of flour, such as whole grain, whole wheat, white cake and pastry flour, and all-purpose white flour. If all the parts of the kernel are used, then the flour is considered whole grain.[1]

Canadian regulations allow up to 5% of the kernel in whole-wheat flour to be removed, to help reduce rancidity. Rancidity refers to the off-odours that develop when fats in the kernel's germ react with oxygen in the air, so whole-wheat flour, as it is sold in Canada, will have much of the germ reduced. This means that 100% whole-wheat flour is not the same as whole-grain flour. It is, however, a nutritious choice, as much of the bran, and its dietary fibre content, is still present.[1]

To determine if a product is whole grain, you have to look for the words "whole grain" on the label and the ingredient list, for example, whole-grain wheat, whole-grain rolled oats, whole-grain rye, etc. Also, foods to which wheat bran or oat bran has been added to increase the fibre content provide the benefits of unrefined grain ingredients but are not necessarily whole-grain products. Don't forget to look at the rest of the ingredient list, too. For example, Lucky Charms cereal is made with whole grains, but marshmallows are the second ingredient, making this choice high in sugar.

The Whole Grains Council (www.wholegrainscouncil.org) is a consumer advocacy group, with food-industry sponsors, that was formed in the United States to promote the consumption of whole grains. One of their projects is the whole-grain stamp to help consumers identify whole-grain products. The stamp, shown below, indicates the number of grams of whole grain a food portion contains.

In 2008, the stamp appeared in Canada. To be able to use the stamp in Canada, a food portion must contain at least 8 g of whole grain per portion. To use the 100% stamp (shown below on the right) all the ingredients must be whole grain, so this might appear on a bag of brown rice, whole grain flour, or oatmeal that contain no other ingredients.

Eating Well with Canada's Food Guide recommends that at least half of your recommended servings of grain products be whole grains. To help identify these whole grains in your food, look for the whole-grain stamp and read the ingredients list carefully. Don't base your decision on the colour and name of the product.

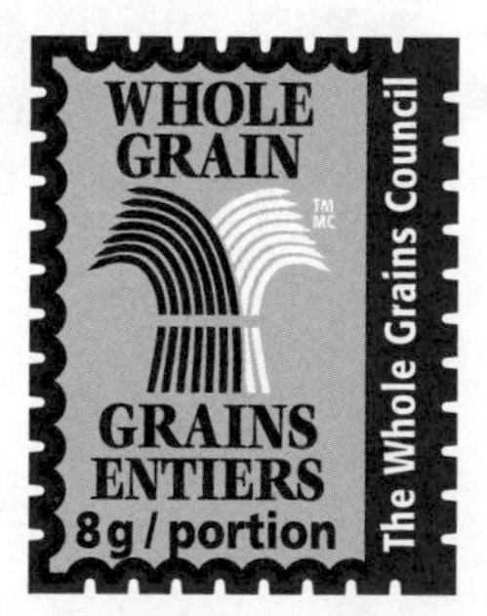

THE BASIC STAMP
Product may contain some refined grain

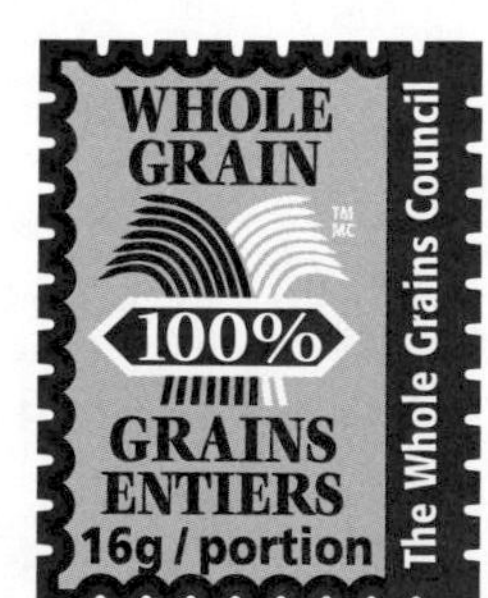

THE 100% STAMP
All grain ingredients are whole grains

These whole-grain stamps, developed by the Whole Grains Council, help Canadian consumers identify whole-grain products.
Source: Courtesy Oldways and the Whole Grains Council, wholegrainscouncil.org

[1] Health Canada Whole Grains-Get the Facts. Available online at http://www.hc-sc.gc.ca/fn-an/nutrition/whole-grain-entiers-eng.php. Accessed February 2. 2010.

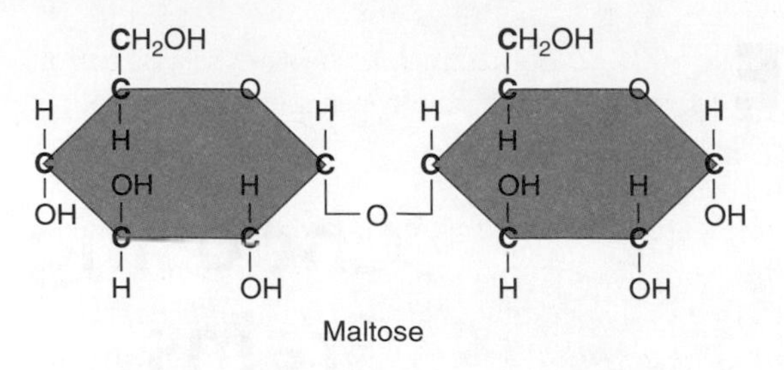

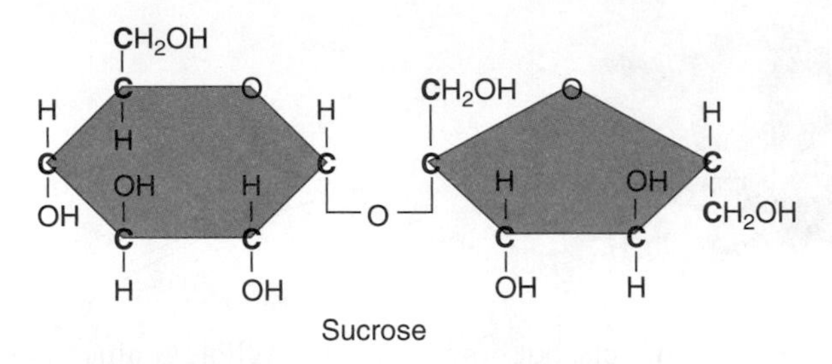

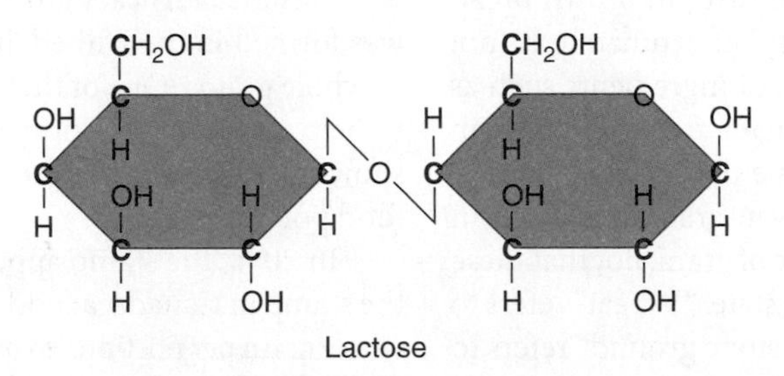

Figure 4.7 Structures of common disaccharides
Maltose, sucrose, and lactose are made up of different pairs of monosaccharides.

hydrolysis reaction A type of chemical reaction in which a large molecule is broken into two smaller molecules by the addition of water.

Making and Breaking Sugar Chains The chemical reaction that breaks the bonds between sugar molecules is called a **hydrolysis reaction** (**Figure 4.8**). Hydrolysis reactions use water to add a hydroxyl group (OH) to one sugar and a hydrogen atom (H) to the other. The reaction that links

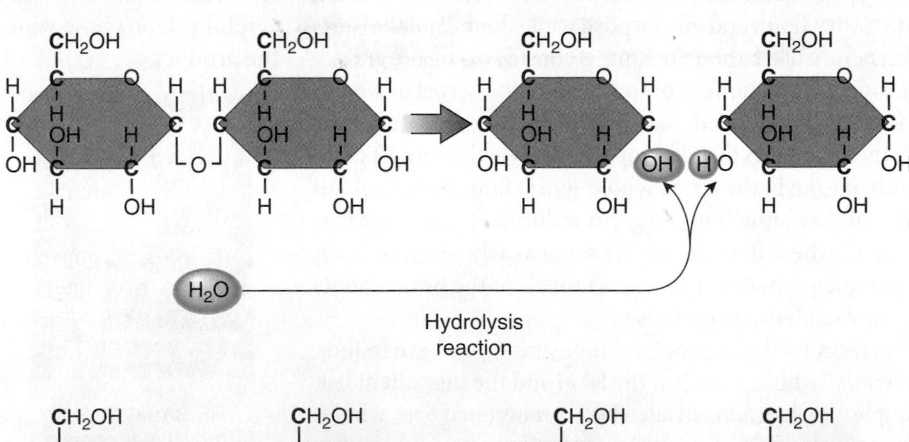

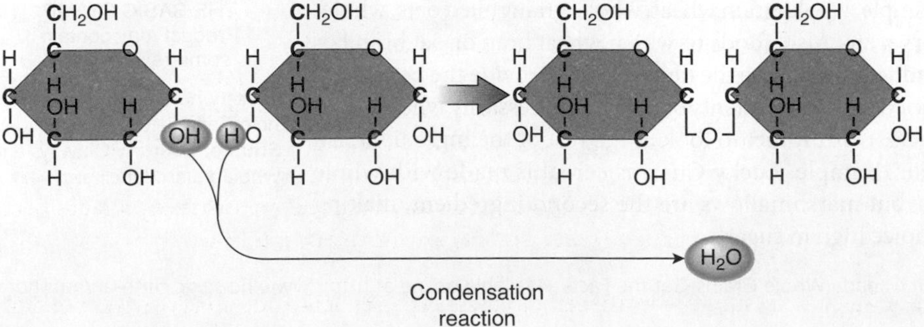

Figure 4.8 Hydrolysis and condensation reactions
In this hydrolysis reaction, a molecule of maltose is broken into its component glucose molecules; two glucose molecules are joined to form the disaccharide maltose in a condensation reaction.

two sugars together is called a **condensation reaction**. Condensation reactions release a molecule of water by taking a hydroxyl group from one sugar and a hydrogen atom from the other.

condensation reaction Type of chemical reaction in which two molecules are joined to form a larger molecule and water is released.

Polysaccharides

Polysaccharides, or complex carbohydrates, are made up of many monosaccharides linked together in chains. They are generally not sweet to the taste the way simple carbohydrates are. Short chains of 3-10 monosaccharides are called **oligosaccharides**, and longer chains are called **polysaccharides** (*poly* means many). The polysaccharides include **glycogen** in animals, and **starch** and fibre in plants (**Figure 4.9**).

oligosaccharides Short-chain carbohydrates containing 3-10 sugar units.

polysaccharides Carbohydrates containing many monosaccharides units linked together.

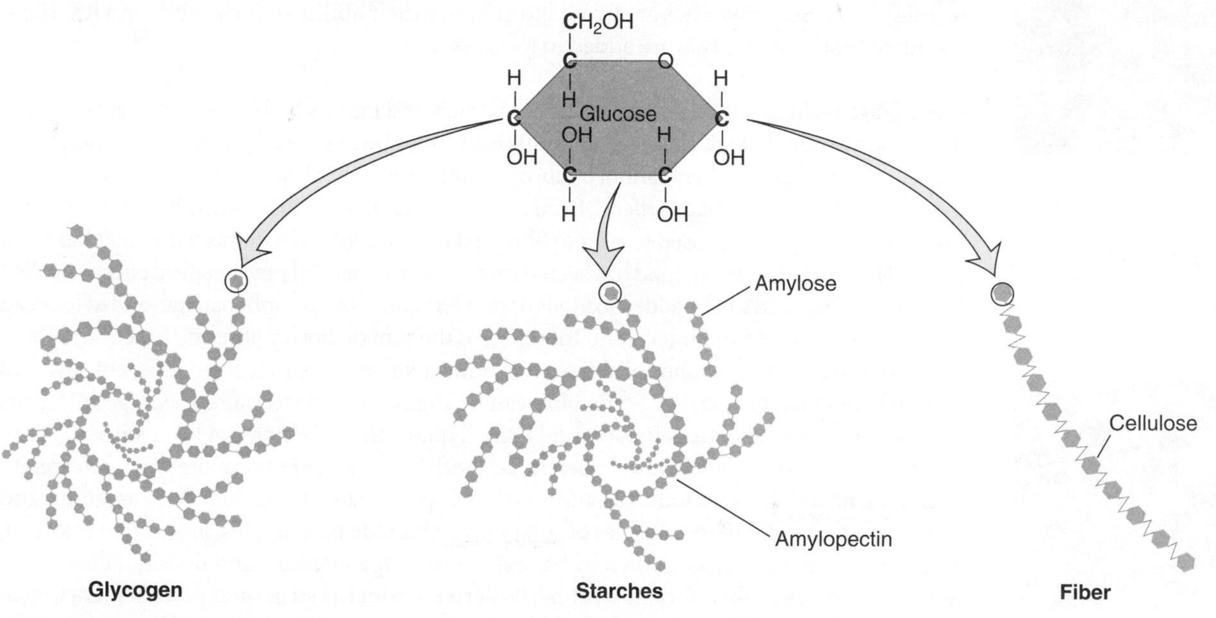

Figure 4.9 Complex carbohydrates
Glycogen, starches, and the fibre cellulose are made up of straight or branching chains of glucose.

Oligosaccharides Some oligosaccharides are formed in the gut during the breakdown of polysaccharides. These are then further digested to monosaccharides. Other oligosaccharides are found naturally in foods such as beans and other legumes, onions, bananas, garlic, and artichokes. Many of these are not digested by human enzymes in the digestive tract and pass into the colon, where they are broken down by the intestinal microflora, which means they can affect the types of bacteria that grow in the colon and have beneficial effects on gastrointestinal (GI) health. Oligosaccharides present in human milk make the infant stool easier to pass, help promote the growth of a healthy intestinal microflora, and may protect the infant from infections that cause diarrhea.[3]

glycogen A carbohydrate made of many glucose molecules linked together in a highly branched structure. It is the storage form of carbohydrate in animals.

starch A carbohydrate made of many glucose molecules linked in straight or branching chains. The bonds that hold the glucose molecules together can be broken by human digestive enzymes.

Glycogen Glycogen is the storage form of carbohydrate in animals. It is a polysaccharide made up of highly branched chains of glucose molecules (see Figure 4.9), a structure that allows it to be broken down quickly when glucose is needed. In humans, glycogen is stored in the muscles and in the liver. Muscle glycogen provides glucose to the muscle as a source of energy during activity; liver glycogen releases glucose into the bloodstream for delivery to cells throughout the body. We do not consume glycogen in our food because glycogen present in animal muscles is broken down soon after slaughter and so is not present when the meat is consumed.

The amount of glycogen in the body is relatively small—about 200-500 g. The amount of glycogen stored in muscle can be temporarily increased by a regimen called carbohydrate loading or glycogen supercompensation. This regimen is often used by endurance athletes to build up glycogen stores before an event. Extra glycogen can mean the difference between running only 30 km or finishing a full marathon (42.2 km) before exhaustion takes over. Glycogen supercompensation is discussed in more detail in Section 13.6.

Starches Starch is the storage form of carbohydrate in plants. It is made up of two types of molecules: amylose, which consists of long, straight chains of glucose molecules, and amylopectin, which consists of branched chains of glucose molecules (see Figure 4.9). Starch accumulates in roots and tubers (the underground energy-storage organ of some plants) where it provides

legumes Plants in the pea or bean family, which produce an elongated pod containing large starchy seeds. Examples include green peas, lentils, kidney beans, and peanuts.

Figure 4.10
This starchy root vegetable known as cassava or manioc is a dietary staple in some parts of Africa. (Magar-StockFood Munich/© StockFood America)

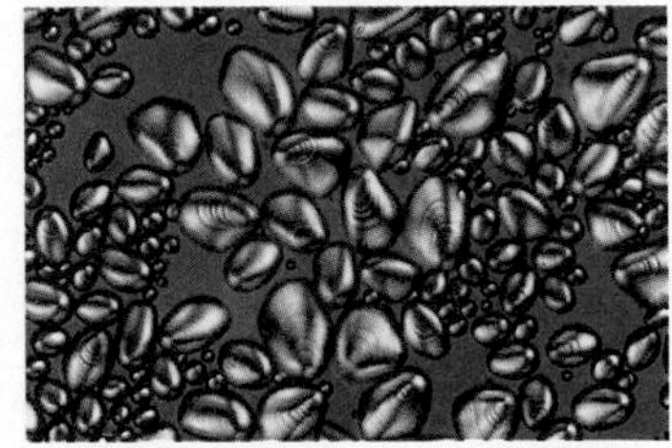

Figure 4.11
These starch granules inside a potato cell, like starch granules in all plant cells, have a unique size, shape, and organization that accounts for the properties of the starch during cooking. (© Eric V. Grave/Photo Researchers, Inc.)

dietary fibre A mixture of indigestible carbohydrates and lignin that is found intact in plants.

functional fibre Isolated indigestible carbohydrates that have been shown to have beneficial physiological effects in humans.

total fibre The sum of dietary fibre and functional fibre.

soluble fibre Fibre that dissolves in water or absorbs water to form viscous solutions and can be broken down by the intestinal microflora. It includes pectins, gums, and some hemicelluloses.

insoluble fibre Fibre that, for the most part, does not dissolve in water and cannot be broken down by bacteria in the large intestine. It includes cellulose, some hemicelluloses, and lignin.

energy for the growth and reproduction of the plant. We consume starch in roots and tubers such as potatoes, sweet potatoes, beets, turnips, and cassava (**Figure 4.10**). Starch accumulates in seeds as well, as an energy source for the developing plant embryo, and we consume this sort of starch, too, in grain seeds such as wheat, barley, and rye. We also consume starch in **legumes**, such as lentils, soybeans, and pinto and kidney beans.

In addition to the starch naturally present in foods, our diets also contain refined starch such as cornstarch, which is added to thicken foods such as sauces, puddings, and gravies. Starch can be used for this purpose because the granules swell when heated in water (**Figure 4.11**). As a starch-thickened mixture cools, high-amylose starches form bonds between the molecules, forming a gel. Some starches are treated to enhance their ability to form a gel, and it is these modified food starches that are added to foods as thickeners.

Fibre Fibre includes certain complex carbohydrates and lignins (substances in plants that are not carbohydrates but are classified as fibre) that cannot be digested by human enzymes. Since they cannot be digested, they cannot be absorbed into the body. However, fibre consumed in the diet can have beneficial health effects, from reducing constipation to lowering blood cholesterol. The term **dietary fibre** is used to refer to fibre that is found intact in plants. Fibre that has been isolated from its plant source and has been shown to have beneficial physiological effects is called **functional fibre**, and can be added to foods or supplements. For example, oat bran added to bread would be considered functional fibre. **Total fibre** is the sum of dietary fibre and functional fibre.[4]

Fibre includes a number of different chemical substances that have different physical and physiological properties. Some fibre can be digested by bacteria in the large intestine, producing gas and short-chain fatty acids, small quantities of which can be absorbed. These fibres also form viscous solutions when mixed with water and are therefore often referred to as **soluble fibres**. They are found around and inside plant cells, and include pectins, gums, and some hemicelluloses. Food sources of soluble fibre include oats, apples, beans, and seaweed. Fibre that cannot be broken down by bacteria in the large intestine and does not dissolve in water is called **insoluble fibre**. It is primarily derived from the structural parts of plants, such as the cell walls, and includes cellulose, some hemicelluloses, and lignins. Food sources of insoluble fibre include wheat bran and rye bran, which are mostly hemicellulose and cellulose, and vegetables such as broccoli, which contain woody fibre composed partly of lignins. Most foods of plant origin contain mixtures of soluble and insoluble fibre.

In addition to the soluble and insoluble fibre found in whole grains, fruits, and vegetables, our diet contains fibre that is added to foods during processing. Pectin is a soluble fibre found in fruits and vegetables that forms a gel when sugar and acid are added. It is used to thicken yogurt and to form jams and jellies. Carbohydrate gums such as xanthan gum and locust bean gum are also sources of soluble fibre. They combine with water and are used to keep solutions from separating. Gravies, puddings, reduced-fat salad dressings, and frozen desserts are examples of products that contain carbohydrate gums. Pectins and gums are also used in reduced-fat products to mimic the texture of fat (see Section 5.7). Inulin is a functional fibre, derived from chicory root, that is added to food as a prebiotic (see Chapter 3: Label Literacy). Insoluble fibre sources such as wheat bran are added to foods like breads and muffins to reduce kcalorie content and meet consumer demand for high-fibre foods.

4.3 Carbohydrates in the Digestive Tract

Learning Objectives

- Describe the steps of carbohydrate digestion.
- Define lactose intolerance and explain why it causes gas and bloating when milk is consumed.
- Discuss the effects of soluble and insoluble dietary fibre and other indigestible carbohydrates on gastrointestinal function and health.

Disaccharides and complex carbohydrates must be digested to monosaccharides to be absorbed. Some people are unable to digest the disaccharide lactose. It spills into the colon, causing uncomfortable side effects. All humans lack the digestive enzymes needed to completely break down a variety of oligosaccharides, certain forms of starch, and fibre. These indigestible carbohydrates have important effects on the health and function of the digestive system and the body as a whole.

Digestible Carbohydrates

Digestion of starch begins in the mouth, where the enzyme salivary amylase starts breaking it into shorter polysaccharides. The majority of starch and disaccharide digestion occurs in the small intestine. Here, pancreatic amylases complete the job of breaking down starch into monosaccharides, disaccharides, and oligosaccharides. The digestion of disaccharides and oligosaccharides is completed by enzymes attached to the brush border of the villi in the small intestine (**Figure 4.12**). At the brush border, maltose is broken down into two glucose molecules by the enzyme maltase, sucrose is broken down by sucrase to yield glucose and fructose, and lactose is broken down by **lactase** to form glucose and galactose. The resulting monosaccharides—glucose, galactose, and fructose—are then absorbed and transported to the liver via the hepatic portal vein.

lactase An enzyme located in the brush border of the small intestine that breaks the disaccharide lactose into glucose and galactose.

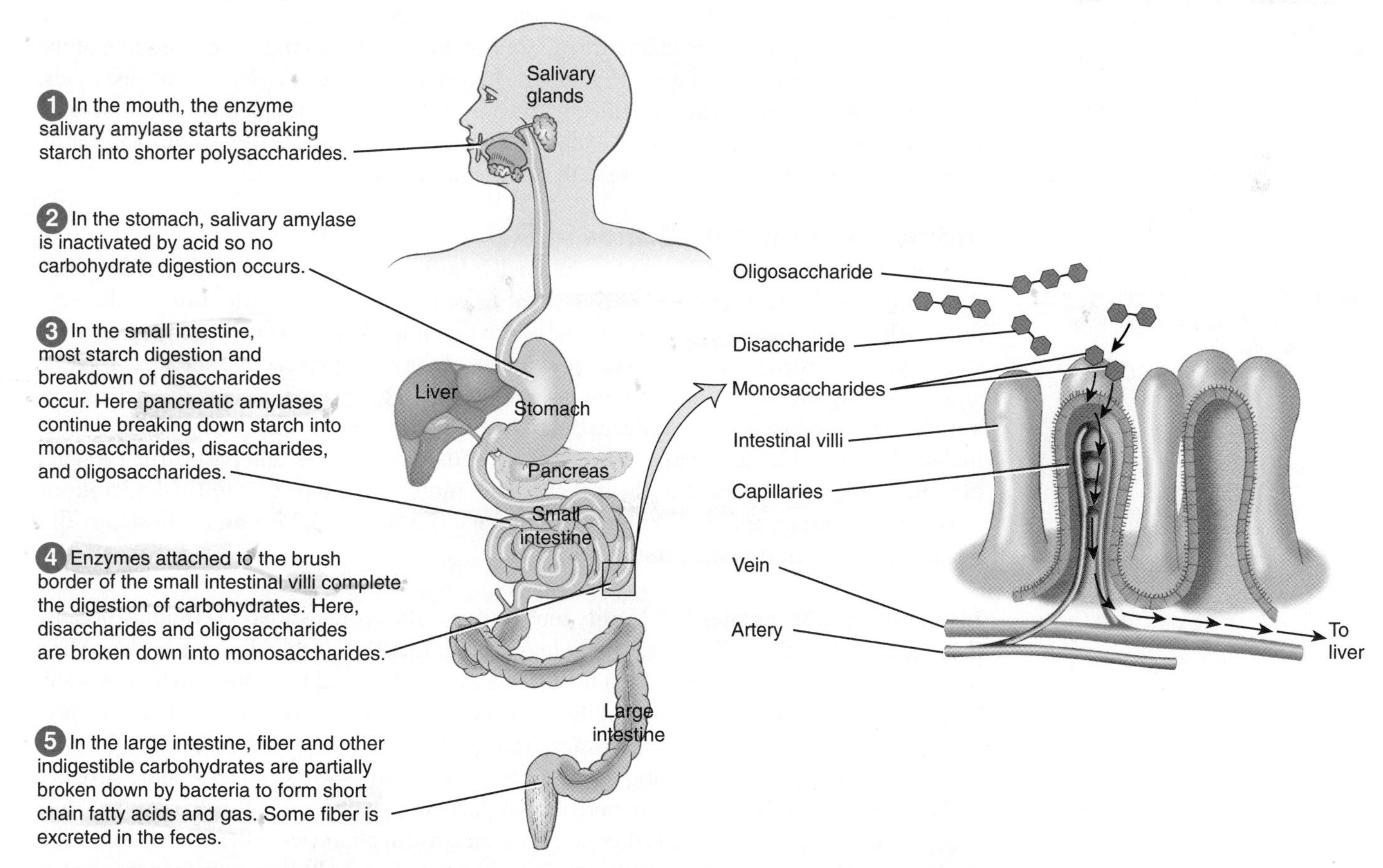

Figure 4.12 Overview of carbohydrate digestion and absorption
During digestion, enzymes break starches and sugars into monosaccharides, which are absorbed. Most of the fibre and other indigestible carbohydrates are excreted in the feces.

Lactose Intolerance

Lactose intolerance is a condition in which there is not enough of the enzyme lactase in the small intestine to digest the milk sugar lactose. When this occurs, the undigested lactose passes into the large intestine, where it draws in water and is metabolized by bacteria producing acids and gas. This causes symptoms that include abdominal distension, flatulence, cramping, and diarrhea. Human infants normally produce enough of the enzyme lactase to digest the lactose in their all-milk diet. Enzyme activity may begin to decrease at two years of age but the symptoms of lactose intolerance usually do not become apparent until after the age of six, and may not be evident until adulthood. Whether or not an individual retains the ability to digest lactose into adulthood depends on the genes they inherit.[5] Lactose intolerance may also occur as a result of an intestinal infection or other disease. It

lactose intolerance The inability to digest lactose because of a reduction in the levels of the enzyme lactase. It causes symptoms including intestinal gas and bloating after dairy products are consumed.

is then referred to as secondary lactose intolerance and may disappear when the primary condition is resolved.

Incidence of Lactose Intolerance The incidence of lactose intolerance varies enormously depending on ethnic background. Lactose intolerance is more common in Asian, African, Native American, and Mediterranean populations than it is among northern and western Europeans. Nearly 100% of adults in Asian populations are lactose intolerant as opposed to just 5% or fewer adults in northwestern European populations.[6]

Figure 4.13 These foods are good natural sources of calcium that are low in lactose. (© Luisa Begani)

Meeting Calcium Needs In Canada, milk is an important source of calcium. Canada's Food Guide recommend 2-3 servings of milk or milk alternatives each day. Because the degree of lactose intolerance varies, some individuals can consume small amounts of dairy products without symptoms and can meet their calcium needs by dividing the three cups into many smaller portions. Those who cannot tolerate any lactose can meet their calcium needs with foods like tofu, fish, and vegetables (**Figure 4.13**). These foods provide dietary calcium in cultures where lactose intolerance is common. For example, in Asia, tofu and fish consumed with bones supply calcium, and in the Middle East, cheese and yogurt provide much of the calcium. These fermented products are more easily tolerated than milk because some of the lactose originally present is digested by bacteria or lost in processing. Calcium-fortified foods, such as soy milk, which is also fortified with vitamin D, calcium supplements, milk treated with the enzyme lactase, and lactase tablets, which can be consumed with or before milk products, are also available for those with lactose intolerance.

Indigestible Carbohydrates

resistant starch Starch that escapes digestion in the small intestine of healthy people.

Carbohydrates that are not digested in the small intestine include soluble and insoluble fibre, some oligosaccharides, and **resistant starch**. Fibre and oligosaccharides are not digested because human enzymes cannot break the bonds that hold their subunits together. Resistant starch is not digested because the natural structure of the grain protects it or because cooking and processing alter its digestibility. For instance, heating makes potato starch more digestible but cooling the cooked potato reduces the starch's digestibility. Foods high in resistant starch include legumes, unripe bananas, and cold, cooked potatoes, rice, and pasta. The presence of indigestible carbohydrates in the diet affects GI motility, the type of intestinal microflora, nutrient absorption, and the amount of intestinal gas. Soluble and insoluble fibres have differing effects on the digestive system.

Insoluble Fibre Stimulates GI Motility Indigestible carbohydrates affect GI motility because they increase the volume of material in the lumen of the intestine. Insoluble fibres, such as wheat bran, increase the bulk of material in the feces. Soluble fibres and resistant starch draw water into the intestine. The combination of the increased bulk and additional water allow for easier evacuation of the stool. Insoluble fibres also promote healthy bowel function because the extra bulk stimulates peristalsis, causing the muscles of the colon to work more, become stronger, and function better. The increase in peristalsis reduces transit time—the time it takes food and fecal matter to move through the digestive tract. In African countries, where the diet contains 40-150 g of fibre per day, the transit time is 36 hours or less, half the average transit time of British citizens who consume a low-fibre diet. (See Science Applied: Cereal Fibres and Health.)

Soluble Fibres, Resistant Starch, and Oligosaccharides Promote a Healthy Microflora When soluble fibres, resistant starch, prebiotics such as inulin, and oligosaccharides reach the colon, they serve as a food source for the microflora that reside there. Diets high in these substances promote the maintenance of beneficial species of bacteria in the colon. When these carbohydrates are broken down, it results in the production of short-chain fatty acids. These fatty acids serve as a fuel source for cells in the colon as well as other body tissues and may play a role in regulating cellular processes. They also inhibit the growth of undesirable bacteria and favour the growth of *Lactobacilli* and *Bifidobacteria*, which are well adapted to acid conditions.[7] In addition to inhibiting the growth of disease-causing bacteria, these short-chain fatty acids may help prevent and treat inflammation in the bowel, which causes diarrhea, as well as protect against colon cancer.[8,9] (See Chapter 3: Your Choice: Should You Feed Your Flora?)

Soluble Fibres Slow Nutrient Absorption Soluble fibre increases the volume of the intestinal contents by absorbing water and forming viscous solutions. These effects slow nutrient absorption by slowing the passage of food through the GI tract, by decreasing the amount of contact between nutrients and the absorptive surface of the small intestine, and by reducing contact between digestive enzymes and food. In the stomach, soluble fibre causes distension and slows emptying. In the small intestine, the added volume and viscosity slow the absorption of sugars and other nutrients (**Figure 4.14**). This can be beneficial because it slows the absorption of glucose and thereby reduces fluctuations in blood glucose, which would be undesirable especially in people with type 2 diabetes (Section 4.5). Soluble fibre also binds cholesterol and bile, which is made from cholesterol, reducing their absorption. This is beneficial because it can lower blood cholesterol and help reduce the risk of heart disease.

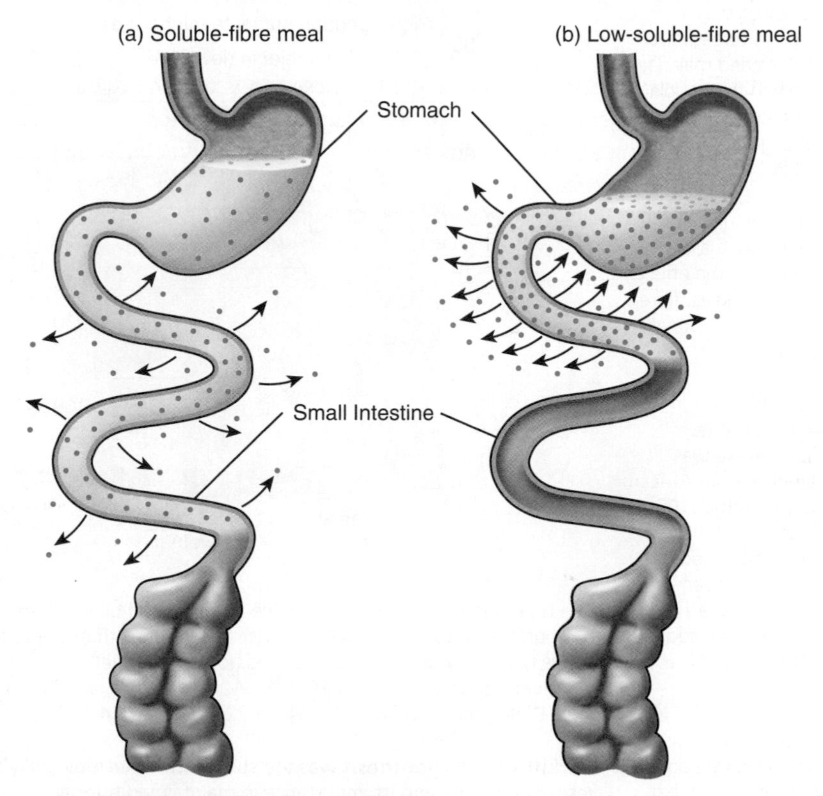

Figure 4.14 Effect of fibre on digestion and absorption
(a) A soluble-fibre-rich meal dilutes the stomach and small intestinal contents and slows the digestion and absorption of nutrients (green dots). (b) Nutrients from a low-soluble-fibre meal are more concentrated in the gastrointestinal tract, resulting in more rapid digestion and absorption.

Fibre also binds certain minerals, preventing their absorption. For instance, wheat-bran fibre binds the minerals zinc, calcium, magnesium, and iron. Too much fibre can reduce the absorption of these essential minerals. However, when mineral intake meets recommendations, a reasonable intake of high-fibre foods does not compromise mineral status.

A high-fibre diet also increases the volume of food needed to meet energy requirements. This is beneficial for someone who is trying to lose weight because he or she feels satiated after eating fewer kcalories. A high-fibre diet may be a disadvantage for someone with a small stomach capacity because that person may satisfy his or her hunger before nutrient requirements are met. Generally, this is a problem only when the diet is low in protein or micronutrients or when high-fibre diets are consumed by young children, whose small stomachs limit the amount of food they can eat.

Life Cycle

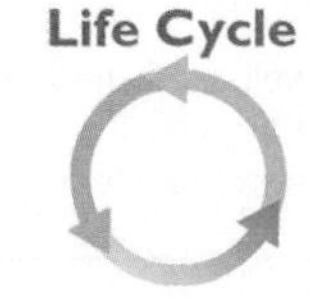

Indigestible Carbohydrates Increase Intestinal Gas Anyone who has ever eaten beans knows of their potentially embarrassing side effect of flatulence. The reason beans cause gas is that they are particularly high in the oligosaccharides raffinose and stachyose, neither of which can be digested by enzymes in the human stomach and small intestine. They pass into the large intestine where the bacteria that live there digest them, producing gas and other

SCIENCE APPLIED

Cereal Fibres and Health

One hundred and fifty years ago, self-proclaimed health advocates Sylvester Graham, John Harvey Kellogg, and Charles W. Post promoted cereal foods as health tonics. These pioneers of what are now Kellogg's and Post cereals were not scientists. Often the health information they promoted was outlandish. Graham preached that food should never be eaten hot, that water should not be consumed with a meal, and that lewdness, along with chicken pie, was the cause of cholera. Kellogg told his patients that coffee could cripple the liver and that bouillon was a solution of poisons. And Post advertised that his whole-grain cereal, called Grape Nuts, tightened up loose teeth and cured tuberculosis and malaria.[1] While these ideas have not held up over time, the suggestion that whole grains are healthy is part of current nutrition wisdom. Canada's Food Guide recommends a diet that includes whole grains to promote gastrointestinal health and reduce the risks of heart disease, cancer, and diabetes.

Scientific support for the role of unrefined cereal grain consumption in health started to accumulate in the 1940s, when scientists such as A. R. P. Walker and Denis Burkitt began observing and investigating the effects of high-fibre foods on health. At the time they began their studies, fibre was referred to as roughage and many regarded it as a gastrointestinal irritant rather than a dietary component important for health.

Walker and colleagues began to relate a population's dietary pattern with their disease pattern—emphasizing the role of fibre.[2] They observed that in Western populations, where fibre intake was between 15-30 g per day, feces were smaller and harder than in African populations consuming diets containing from 70 g to more than 100 g of fibre per day. They hypothesized that fibre increased stool weight and decreased transit time. This hypothesis was supported by studies that compared intestinal transit time and stool weight in Ugandan subjects, who ate a high-fibre diet, to British subjects, who ate a lower-fibre diet (see figure). To further test this hypothesis, they added unprocessed bran to the diet of the British subjects. The added fibre reduced transit time and increased stool weight.[3]

In 1956, Denis Burkitt, traveling in Africa, noticed that many diseases that were common among whites in Europe and Africa were rare among black Africans. With these observations in mind, Burkitt proposed that a variety of conditions common in industrialized society, including diabetes, obesity, heart disease, constipation, diverticular disease, hemorrhoids, and varicose veins, were caused by the overconsumption of refined carbohydrates. He postulated that three manufactured foods—refined sugar, white flour, and white rice—caused virtually all the diseases of civilization.[2] Burkitt hypothesized that a deficiency of dietary fibre may underlie the development of these diseases.[3,4] This became known as the fibre hypothesis. Burkitt was often quoted as saying that the health of a country's people could be determined by the size of their stools and whether the stools floated or sank.[5]

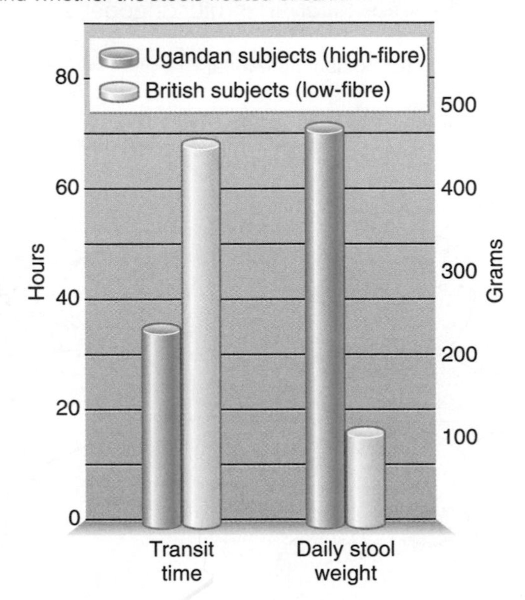

The Ugandan subjects, who consume an unrefined diet high in fibre, have greater stool weights and shorter transit times than the British subjects who consume a more refined, lower-fibre diet.
(**Source:** Adapted from Burkitt, D. P., Walker, A. R. P., and Painter, N. S. Dietary fibre and disease. *JAMA* 229:1068–1074, 1974.)

Burkitt's fibre hypothesis was the stimulus for much of today's research on fibre and its importance in maintaining normal gastrointestinal function and reducing the incidence of chronic disease. Today, the whole-grain cereals promoted by Graham, Kellogg, and Post over a century ago are still considered a sort of health tonic, but this time, scientific data are available to support the benefits.

[1] Deutsch, R. M. The New Nuts Among the Berries. Palo Alto, CA: Bull Publishing Company, 1977.

[2] Trowell, H. Dietary fibre: A paradigm. In: Trowell, H., Burkitt, D., and Heaton, K., eds. *Dietary Fibre, Fibre-Depleted Foods and Disease*. London: Academic Press, 1985, 1–20.

[3] Burkitt, D. P., Walker, A. R. P., and Painter, N. S. Dietary fibre and disease. *JAMA* 229:1068–1074, 1974.

[4] Burkitt, D. P., and Trowell, H. C. *Refined Carbohydrate and Disease: Some Implications of Dietary Fibre*. London: Academic Press, 1975.

[5] Story, J.A., and Kritchevsky, D. Denis Parsons Burkitt. *J. Nutr.* 124:1551–1554, 1994.

by-products. This gas can cause abdominal discomfort and flatulence. To alleviate the problem, over-the-counter enzyme tablets and solutions (such as Beano®) can be consumed to break down oligosaccharides before they reach the intestinal bacteria, thereby reducing the amount of gas produced.

As with oligosaccharides, intestinal gas is a by-product of the bacterial breakdown of soluble fibre and resistant starch. A sudden increase in the fibre content of the diet can cause abdominal discomfort, gas, and diarrhea. Constipation can also be a problem if fibre intake is increased without an increase in fluid intake. To avoid these problems, the fibre content of the diet should be increased gradually and fluid intake should also be increased.

4.4 Carbohydrates in the Body

Learning Objectives

- Describe the steps involved in metabolizing glucose to produce ATP.
- Describe the steps involved in gluconeogenesis.
- Discuss how carbohydrate intake is related to ketone production.

Carbohydrates are central to energy production in the body and also provide other essential functions. The monosaccharide galactose is an important molecule in nervous tissue. It also combines with glucose to make lactose in women who are producing breast milk. Two other monosaccharides that are of great importance to the body are deoxyribose and ribose. These sugars are components of DNA and RNA (ribonucleic acid), respectively, which contain the genetic information for the synthesis of proteins. Deoxyribose and ribose can be synthesized by the body and are not found in significant amounts in the diet. Ribose is also a component of the vitamin riboflavin. Oligosaccharides are also important in our bodies. They are found attached to proteins or lipids on the surface of cells, where they help to signal information about the cells. Another type of carbohydrate that is important in the body is mucopolysaccharides, which are a type of polysaccharide that functions with proteins in body secretions and structures. Mucopolysaccharides give mucus its viscous consistency and provide cushioning and lubrication in connective tissue.

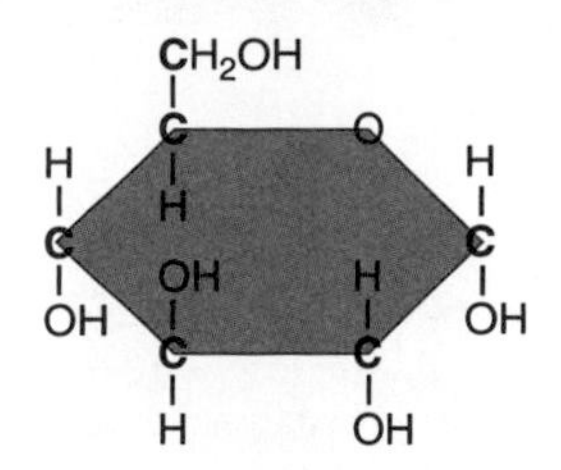

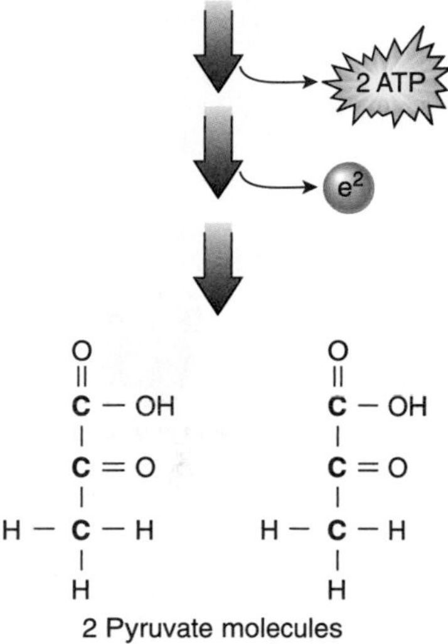

Figure 4.15 Glycolysis
In the cytosol of the cell, glycolysis breaks glucose into two molecules of pyruvate, electrons are released, and two ATP molecules are produced.

Using Carbohydrate to Provide Energy

After absorption, monosaccharides travel to the liver. The monosaccharides fructose and galactose are metabolized for energy. Glucose may also be broken down to provide energy, or passed into the bloodstream for delivery to other body tissues that can use it to provide energy. It may also be stored in the liver as glycogen and, to a lesser extent, used to synthesize fat.

To generate ATP, glucose is metabolized through **cellular respiration**. Cellular respiration uses 6 molecules of oxygen to convert 1 molecule of glucose into 6 molecules of carbon dioxide, 6 molecules of water, and about 38 molecules of ATP:

cellular respiration The reactions that break down carbohydrates, fats, and proteins in the presence of oxygen to produce carbon dioxide, water, and ATP.

$$\underset{\text{glucose}}{C_6H_{12}O_6} + \underset{\text{oxygen}}{6O_2} \Rightarrow \underset{\text{carbon dioxide}}{6CO_2} + \underset{\text{water}}{6H_2O} + ATP$$

The carbon dioxide produced by cellular respiration is transported to the lungs, where it is eliminated in exhaled air. Providing energy through cellular respiration involves four interconnected stages, glycolysis, acetyl CoA formation, citric acid cycle and electron transport chain, (see Appendix I).

Glycolysis The first stage of cellular respiration takes place in the cytosol of the cell and is called **glycolysis**, meaning glucose breakdown. Because oxygen isn't needed for this reaction, glycolysis is sometimes called **anaerobic metabolism**. In glycolysis, the 6-carbon sugar glucose is broken into two, 3-carbon molecules called pyruvate (**Figure 4.15**, **Figure 4.16**). The reactions generate two molecules of ATP for each molecule of glucose and release high-energy electrons that are passed to shuttling molecules, which can transport them to the last stage of cellular respiration. When oxygen is limited, no further metabolism of glucose or production of ATP occurs.

glycolysis (also called **anaerobic metabolism**) Metabolic reactions in the cytosol of the cell that split glucose into two, 3-carbon pyruvate molecules, yielding two ATP molecules.

Acetyl-CoA Formation When oxygen is present, **aerobic metabolism** can proceed. In the mitochondria, one carbon is removed from pyruvate and released as CO_2. The remaining 2-carbon compound combines with a molecule of coenzyme A (CoA) to form acetyl-CoA

aerobic metabolism Metabolism in the presence of oxygen, which can completely break down glucose to yield carbon dioxide, water, and as many as 38 ATP molecules.

(Figure 4.16). High-energy electrons are released and passed to shuttling molecules for transport to the last stage of cellular respiration. Acetyl-CoA then enters the third stage of breakdown, the citric acid cycle.

How it Works

View in your WileyPLUS
www.wileyplus.com

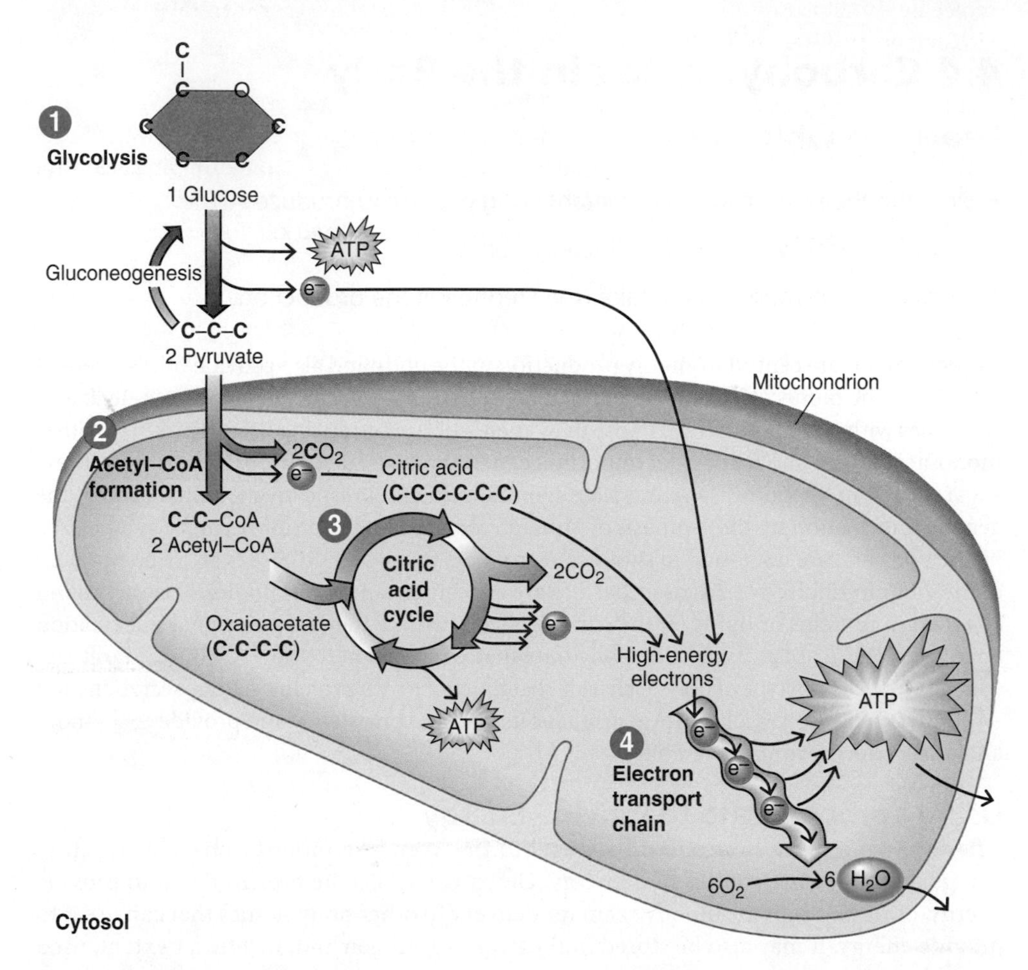

Figure 4.16 Glucose Metabolism
The reactions of cellular respiration split the bonds between carbon atoms in glucose, releasing energy that is used to synthesize ATP. ATP is used to power the energy-requiring processes in the body.

1. Glycolysis splits glucose, a 6-carbon molecule, into two molecules of pyruvate, a 3-carbon molecule (C-C-C). This step produces high-energy electrons (e^-) and a small amount of ATP. Each pyruvate is then either broken down to produce more ATP or used to make glucose via gluconeogenesis.
2. When oxygen is available, pyruvate can be used to produce more ATP. The first step is to remove one carbon as carbon dioxide from each pyruvate. This produces a 2-carbon molecule that combines with coenzyme A to form acetyl-CoA (C-C-CoA) and releases high-energy electrons.
3. Each acetyl-CoA enters the citric acid cycle, reacting with 4-carbon compound, oxaloacetate, to produce citric acid. From citric acid 2 carbons are lost as carbon dioxide, high-energy electrons are released, and a small amount of ATP is produced.
4. In the final step of cellular respiration, the electron transport chain accepts the high-energy electrons released in previous steps and uses the energy to synthesize ATP. The electrons are combined with oxygen and hydrogen to form water.

Citric Acid Cycle In the third stage, acetyl-CoA combines with oxaloacetate, a 4-carbon molecule derived from carbohydrate, to form a 6-carbon molecule called citric acid and begin the citric acid cycle (Figure 4.16). The reactions of the citric acid cycle then remove one carbon at a time, to produce carbon dioxide. After two carbons have been removed in this manner, a 4-carbon oxaloacetate molecule is re-formed and the cycle can begin again. These chemical reactions produce two ATP molecules per glucose molecule and also remove electrons, which are passed to shuttling molecules for transport to the fourth and last stage of cellular respiration, the electron transport chain.

Electron Transport Chain The electron transport chain consists of a series of molecules, most of which are proteins, associated with the inner membrane of the mitochondria. These molecules accept electrons from the shuttling molecules and pass them from one to another down the chain until they are finally combined with oxygen to form water (Figure 4.16). As the electrons are passed along, their energy is trapped and used to make more than 30 ATP molecules per glucose molecule. The reactions of cellular respiration are central to all energy-yielding processes in the body.

Carbohydrate and Protein Breakdown

When carbohydrate intake is low, some glucose can be obtained from the breakdown of glycogen. This glucose is released into the blood to prevent blood glucose from dropping below the normal range. Glucose is also supplied by a metabolic pathway called **gluconeogenesis** (production of new glucose). Gluconeogenesis, which occurs in liver and kidney cells, is an energy-requiring process that forms glucose from 3-carbon molecules, which come primarily from amino acids derived from protein breakdown. Some amino acids, referred to as glucogenic amino acids, can form pyruvate and oxaloacetate. These can then be used to make glucose (**Figure 4.17**, Figure 4.16). Fatty acids and ketogenic amino acids cannot be used to make glucose because the reactions that break them down produce primarily 2-carbon molecules that form acetyl-CoA, and acetyl-CoA cannot be converted to pyruvate (Figure 4.17). Gluconeogenesis is essential for meeting the body's immediate need for glucose, particularly when carbohydrate intake is very low, but it uses amino acids from proteins that could be used for other essential functions, such as growth and maintenance of muscle tissue. Since adequate dietary carbohydrate eliminates the need to use amino acids from protein to synthesize glucose, carbohydrate is said to spare protein.

gluconeogenesis The synthesis of glucose from simple, noncarbohydrate molecules. Amino acids from protein are the primary source of carbons for glucose synthesis.

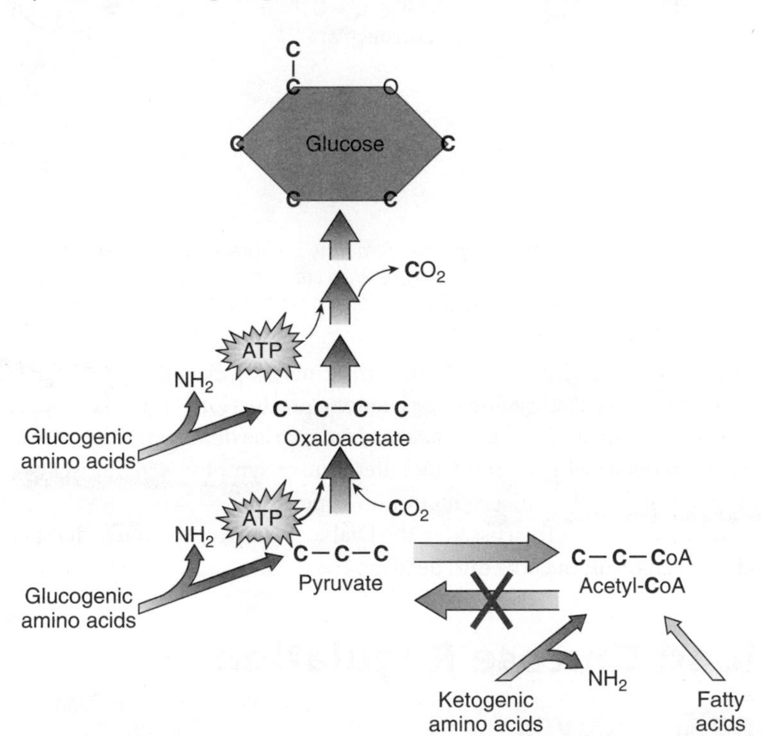

Figure 4.17 Gluconeogenesis
Gluconeogenesis uses 3 carbon molecules and energy from ATP to synthesize glucose; 2-carbon compounds, such as acetyl-CoA, cannot be used to make glucose.

Carbohydrate and Fat Breakdown

Carbohydrate is also needed for the metabolism of fat, so when the supply of carbohydrate is limited, fat cannot be completely broken down. This is because fatty acids are broken into molecules of acetyl-CoA. Acetyl-CoA cannot be metabolized via the citric acid cycle unless it can combine with a 4-carbon oxaloacetate molecule derived from carbohydrate metabolism. When carbohydrate is in short supply, oxaloacetate is limited so acetyl-CoA cannot be broken down to form carbon dioxide and water and produce ATP. Instead, the liver converts it into compounds known as **ketones** or **ketone bodies**, which are released into the blood (**Figure 4.18**). Ketones can be used as an energy source by tissues, such as those in the heart,

ketones or **ketone bodies** Molecules formed in the liver when there is not sufficient carbohydrate to completely metabolize the 2-carbon units produced from fat breakdown.

muscle, and kidney. Ketone production is a normal response to starvation or to a diet very low in carbohydrate. Even the brain, which requires glucose, can adapt to obtain a portion of its energy from ketones.

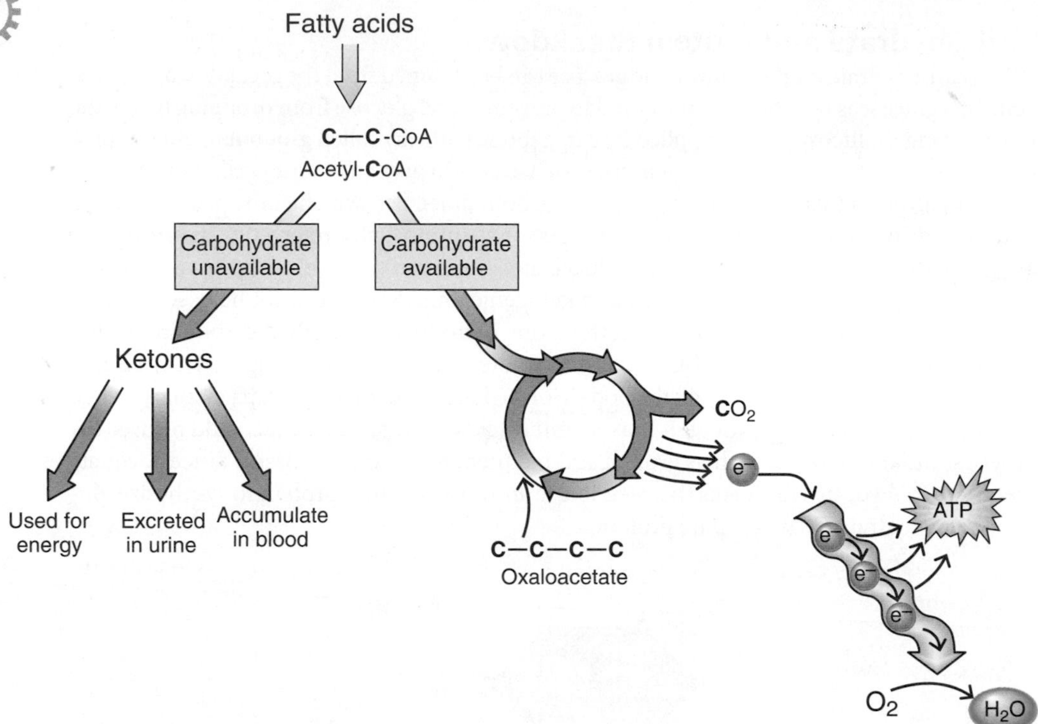

Figure 4.18 Ketone formation
When carbohydrate is available (right), acetyl-CoA from fatty acid breakdown can combine with oxaloacetate and enter the citric acid cycle and no ketones are formed; when carbohydrate (and thus oxaloacetate) are in short supply (left), acetyl-CoA molecules cannot enter the citric acid cycle, and the liver converts them to ketones.

Excess ketones are excreted by the kidney in urine. However, if fluid intake is too low to produce enough urine to excrete ketones, or if ketone production is high, ketones can build up in the blood, causing ketosis. Mild ketosis, which may arise during moderate energy restriction, such as might occur with a weight-loss diet, causes symptoms including headache, dry mouth, foul-smelling breath, and a reduction in appetite. High ketone levels, such as might occur with untreated diabetes (discussed in the Diabetes Mellitus section), increase the acidity of the blood and can result in coma and death.

4.5 Blood-Glucose Regulation

Learning Objectives

- Explain how insulin and glucagon are involved in regulating blood glucose.
- Explain how the glycemic index is measured.
- Distinguish between glycemic index and glycemic load.
- Compare the causes and consequences of type 1 and type 2 diabetes.
- Describe the treatment for diabetes.

diabetes mellitus A disease caused by either insufficient insulin production or decreased sensitivity of cells to insulin. It results in elevated blood-glucose levels.

Blood-glucose levels are normally tightly controlled by the liver and hormones secreted by the pancreas. If these hormones are not produced normally or if the body does not respond to them normally, blood-glucose levels can rise too high or drop too low. **Diabetes mellitus** is a disease in which blood-glucose levels are consistently above the normal range, while **hypoglycemia** is a condition in which blood-glucose levels drop below the normal range.

hypoglycemia A low blood-glucose level, usually below 2.2 to 2.8 mmol/L of blood plasma.

Regulating Blood Glucose

Normally, fasting blood glucose, or more precisely, **plasma** glucose, measured after an 8- to 12-hour overnight fast, is maintained at about 3.3 to 5.5 mmol/L. Maintaining this level ensures adequate glucose will be available to body cells. A steady supply of glucose is particularly important for nerve cells, including those in the brain and red blood cells, because these cells rely almost exclusively on glucose as an energy source.

plasma The liquid portion of the blood that remains when the blood cells are removed.

Glycemic Response The carbohydrate consumed in food is digested and absorbed and enters the bloodstream, causing blood glucose to rise. How quickly and how high blood glucose rises after carbohydrate is consumed is referred to as **glycemic response**, which is affected by both the amount and type of carbohydrate eaten and the amount of fat and protein in that food or meal. Because carbohydrate must be digested and absorbed to enter the blood, how quickly a food leaves the stomach and how fast it is digested and absorbed in the small intestine all affect how long it takes glucose to get into the blood. **Figure 4.19** illustrates a **blood-glucose response curve.** Figure 4.19 shows the change in blood glucose that occurs after a meal. Any food containing glucose or a digestible carbohydrate that breaks down to glucose during digestion will cause a rise in blood glucose, but this rise is temporary, as glucose is taken from the blood by the body's cells. It shows that the high-soluble-fibre meal does not cause as high a rise in blood glucose as does a low-soluble-fibre meal. This is because the viscosity of the soluble fibre slows the rate at which glucose leaves the intestine and enters the blood. The rise in blood-glucose levels after a meal is thus more gradual (Figure 4.14).

glycemic response The rate, magnitude, and duration of the rise in blood glucose that occurs after a particular food or meal is consumed.

blood-glucose response curve A curve that illustrates the change in blood glucose that occurs after consuming food.

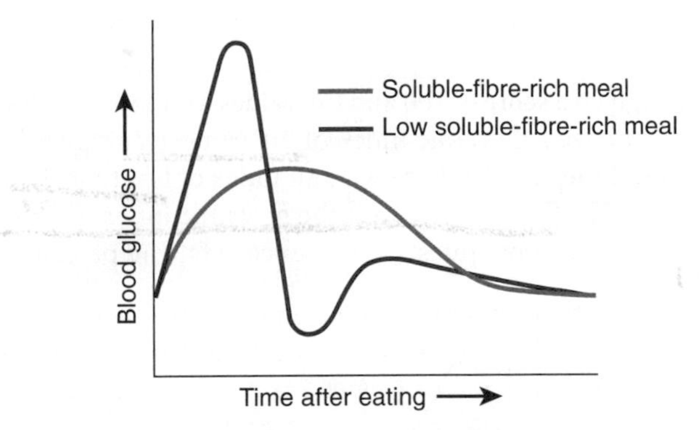

Figure 4.19 Effect of soluble fibre on blood-glucose response
Blood glucose rises rapidly after a high-carbohydrate, low-soluble-fibre meal, whereas the rise in blood glucose is delayed and blunted after a high-carbohydrate meal that is rich in soluble fibre, even though the amount of carbohydrate is the same in both meals.

Different foods have different glycemic responses, that is, they produce different-shaped blood-glucose response curves, so the **glycemic index** was developed to quantify these differences.[10] Figure 4.14, described earlier with respect to soluble fibre, similarly describes the difference between low-glycemic-index food (A) and high-glycemic-index food (B) (**Figure 4.20**). Low-glycemic-index food slows the rate at which glucose is absorbed.

glycemic index A ranking of the effect on blood glucose of a food of a certain carbohydrate content relative to an equal amount of carbohydrate from a reference food such as white bread or glucose.

Although many high-fibre foods have a low glycemic index, it is important to recognize that high-fibre and low-glycemic-index foods do not have exactly the same physiological properties. For example, yogurt, a low-glycemic index food, is also low in fibre, while some high-fibre cereals also have a high glycemic index, possibly because of processing or the type of fibre present in the food (e.g., insoluble fibre rather than soluble fibre).

When multiple carbohydrate-containing foods are consumed in a meal, the overall glycemic index of a meal can be calculated as a weighted average of the glycemic indices of each of the foods.[11]

Glycemic index is measured by comparing the ratio of the area under a blood-glucose response curve for a portion of food containing 50 g of digestible carbohydrate (also referred to as available carbohydrate) compared to the response curve of 50 g of glucose

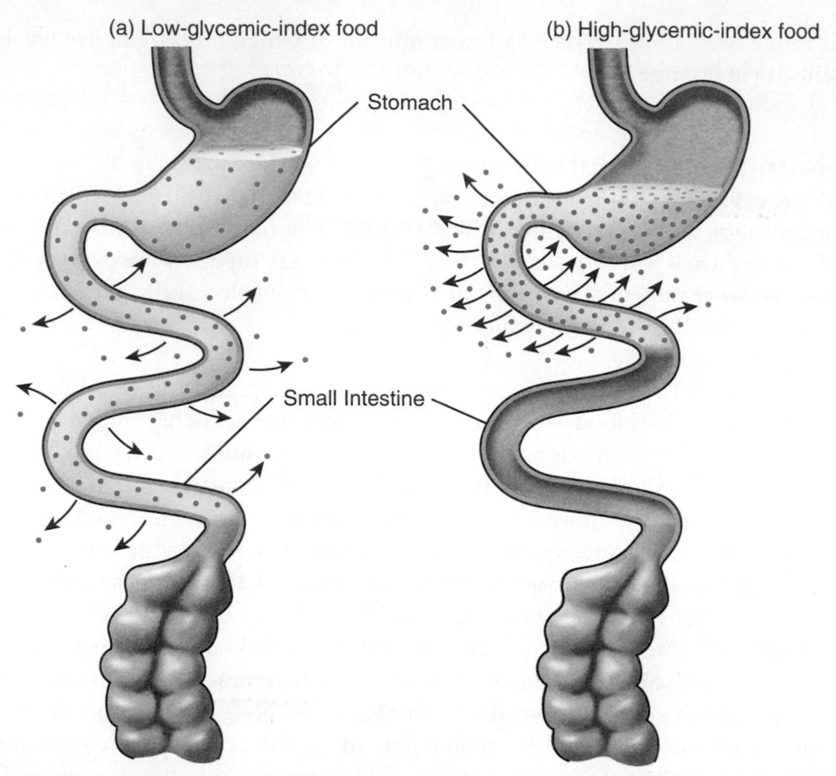

Figure 4.20 Effect of GI on nutrient absorption
With low-glycemic-index food (a), the digestion and absorption of glucose is slower than a high-glycemic-index food (b).

(**Figure 4.21**). Glucose is given a score of 100 and the values for food samples are expressed relative to this. Foods that have a glycemic index of 70 or more compared to glucose are considered high-glycemic-index foods; those with an index of less than 55 are considered low-glycemic-index foods. The factors that affect the glycemic index of a food include the chemical composition of the carbohydrates, the presence of fat and protein, its processing, and its fibre content.[12]

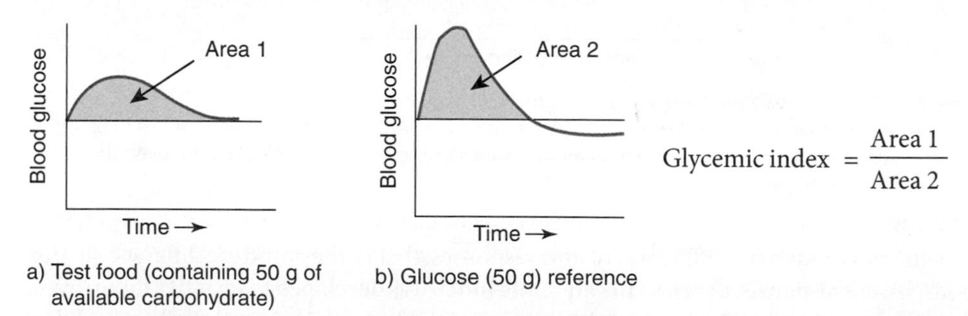

Figure 4.21 Measuring glycemic index Glycemic index is the ratio of the area under the glucose response curve of the test food (a) compared to the area under the curve for 50 g of glucose (b).

glycemic load An index of the glycemic response that occurs after eating specific foods. It is calculated by multiplying a food's glycemic index by the amount of available carbohydrate in a serving of the food.

Because the glycemic index is based on 50 g of available carbohydrate and this is often more than is typically found in a portion of food, the **glycemic load** was developed to take into account both the glycemic index of the food and the amount of carbohydrate in a typical portion.[13] To calculate glycemic load, the grams of available carbohydrate in a serving of food are multiplied by that food's glycemic index, expressed as a percent. A glycemic load of 20 or more is considered high, whereas a value of less than 11 is considered low.

For example, a portion of food with a glycemic index of 80 and 10 g available carbohydrate has a glycemic load of 8 (10 g × 0.80). A food with a glycemic index of 60 and 20 g of available carbohydrate has a glycemic load of 12 (20 × 0.60). The first food will result in a smaller rise in blood glucose than the second food, when consumed.

Insulin A rise in blood glucose triggers the pancreas to secrete the hormone **insulin**, which allows glucose to be taken into the cells of the body. Insulin is secreted by a specific type of pancreatic cell called the beta cell. In the liver, insulin promotes the storage of glucose as glycogen and, to a lesser extent, fat. In muscle, insulin stimulates the uptake of glucose for energy production and the synthesis of muscle glycogen for energy storage (**Figure 4.22**). It also stimulates protein synthesis and, in fat-storing cells, it increases glucose uptake from the blood and stimulates lipid synthesis. These actions remove glucose from the blood, decreasing levels.

insulin A hormone made in the pancreas that allows the uptake of glucose by body cells and has other metabolic effects such as stimulating protein and fat synthesis and the synthesis of glycogen in liver and muscle.

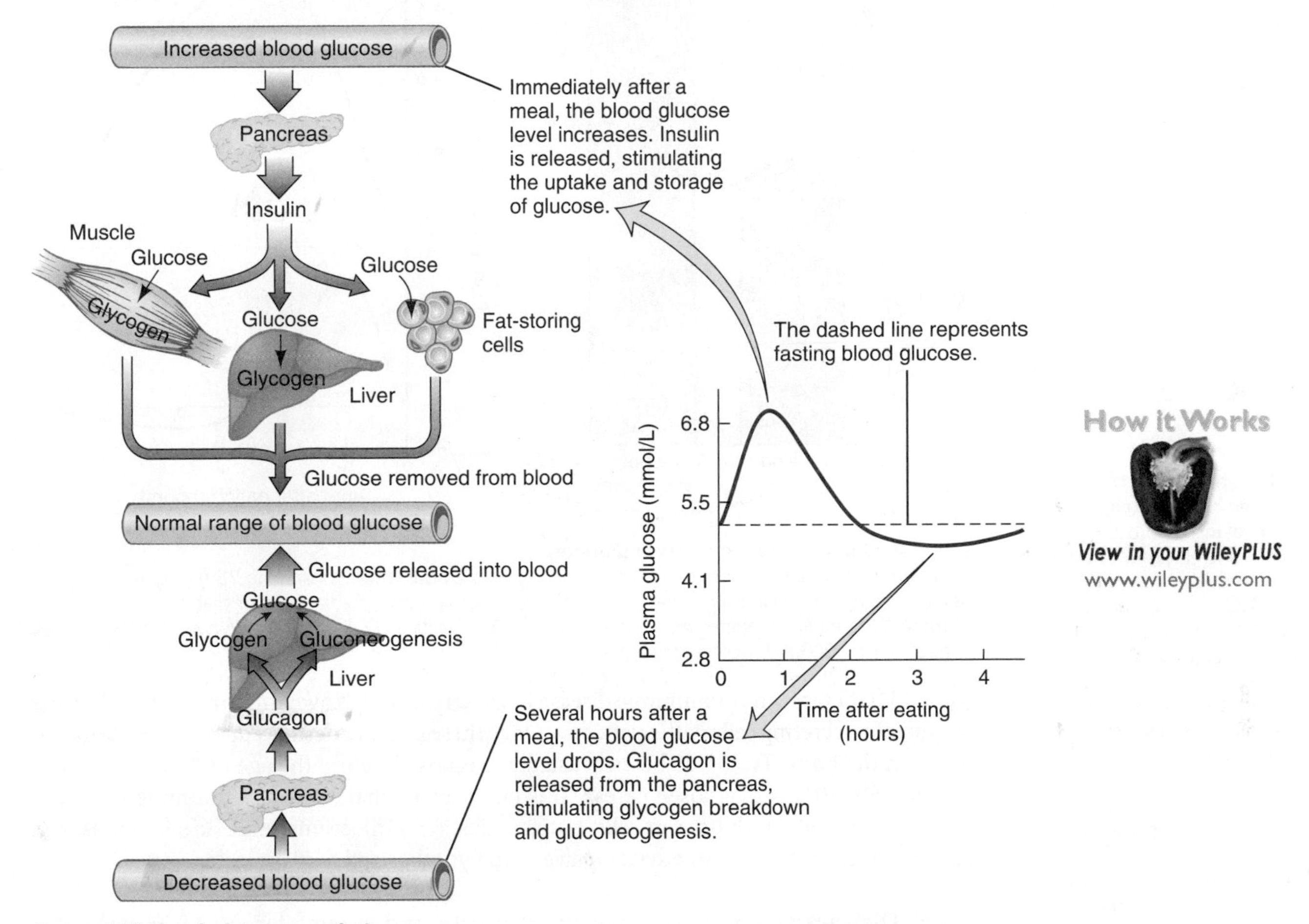

Figure 4.22 Blood-glucose regulation
Blood-glucose levels are regulated by the hormones insulin and glucagon, secreted by the pancreas

How it Works
View in your WileyPLUS
www.wileyplus.com

Glucagon When no carbohydrate has been consumed for a few hours, the glucose level in the blood—and consequently the glucose available to the cells—begins to decrease. This triggers another group of cells in the pancreas, the alpha cells, to secrete the hormone **glucagon**, which signals liver cells to break down glycogen into glucose, which is released into the bloodstream. Glucagon also stimulates the liver to synthesize new glucose molecules by gluconeogenesis (see Figure 4.22). Newly synthesized glucose is released into the blood to prevent blood glucose from dropping below the normal range. Gluconeogenesis can also be stimulated by the hormone epinephrine, also known as adrenaline. This hormone, which is released in response to dangerous or stressful situations, enables the body to respond to emergencies. It causes a rapid release of glucose into the blood to supply the energy needed for action.

glucagon A hormone made in the pancreas that stimulates the breakdown of liver glycogen and the synthesis of glucose to increase blood sugar.

Diabetes Mellitus

Diabetes mellitus, commonly called diabetes, is a major public health problem in Canada. According to the Canadian Diabetes Association, about three million Canadians have diabetes and the number is expect to rise to 3.7 million by 2020. Diabetes is characterized by high blood-glucose levels due to either a lack of insulin or an unresponsiveness or resistance to

insulin (**Figure 4.23**). The elevated glucose causes damage to the large blood vessels, leading to an increased risk of heart disease and stroke. It also causes changes in small blood vessels and nerves. Diabetes is the leading cause of blindness in adults, causes kidney failure, and is the major reason for nontraumatic lower-limb amputations.[14] There are three main types of diabetes: type 1, type 2, and gestational diabetes, which occurs during pregnancy.

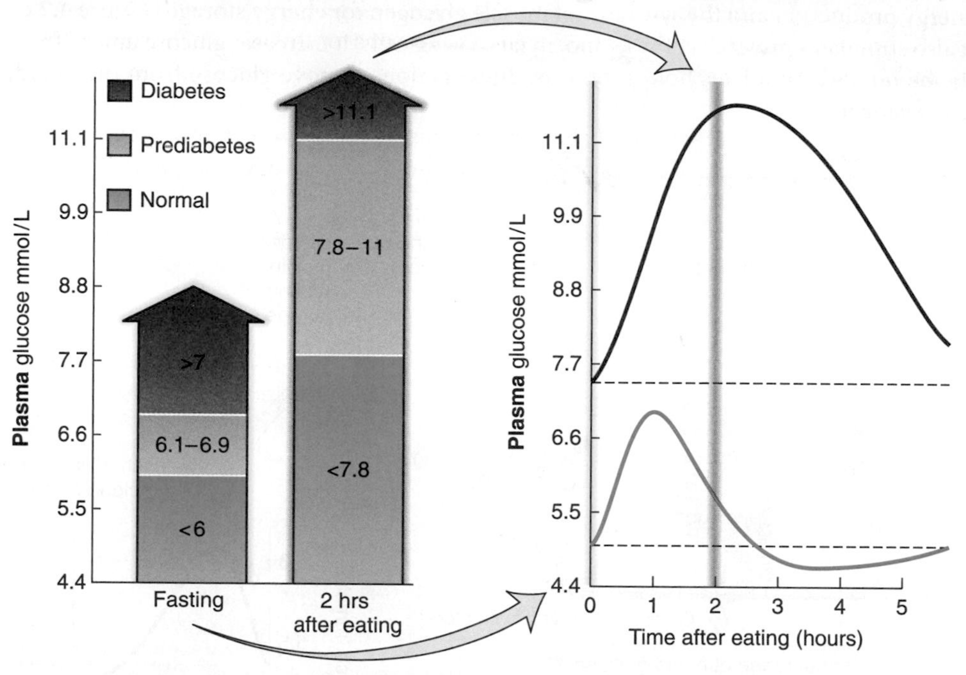

Figure 4.23 Blood-glucose levels in diabetes
Plasma-glucose levels measured after an 8-hour fast and 2 hours after consuming 75 g of glucose determine whether an individual has normal plasma-glucose levels, prediabetes, or diabetes.
(**Source:** © Canadian Diabetes Association 2008 Clinical Practice Guidelines. http://www.diabetes.ca/files/cpg2008/cpg-2008.pdf. Accessed Feb 2, 2010.

type 1 diabetes A form of diabetes that is caused by the autoimmune destruction of insulin-producing cells in the pancreas, usually leading to absolute insulin deficiency; previously known as insulin-dependent diabetes mellitus or juvenile-onset diabetes.

type 2 diabetes A form of diabetes that is characterized by insulin resistance and relative insulin deficiency; previously known as noninsulin-dependent diabetes mellitus or adult-onset diabetes.

insulin resistance A situation when tissues become less responsive to insulin and do not take up glucose as readily. As a result glucose levels in the blood rise.

metabolic syndrome A collection of health risks, including excess fat in the abdominal region, high blood pressure, elevated blood triglycerides, low high-density lipoprotein (HDL) cholesterol, and high blood glucose that increases the chance of developing heart disease, stroke, and diabetes. The condition is also known by other names including Syndrome X, insulin resistance syndrome, and dysmetabolic syndrome.

pre-diabetes or **impaired glucose tolerance** A fasting blood-glucose level above the normal range but not high enough to be classified as diabetes.

Type 1 Diabetes is an autoimmune disease in which the body's own immune system destroys the insulin-secreting cells of the pancreas. Once these cells are destroyed, insulin is no longer made in the body. Type 1 diabetes is usually diagnosed before the age of 30 and accounts for only 5%-10% of diagnosed cases.[15] It is not known what causes the immune system to malfunction and attack its own cells, but genetics, viral infections, exposure to toxins, and abnormalities in the immune system have been hypothesized to play a role.

Type 2 Diabetes is the more common form of diabetes, and accounts for about 95% of all cases. It is often the result of a decrease in the sensitivity of cells to insulin, called **insulin resistance**; tissues that normally respond to insulin, such as the muscle, liver and adipose tissue, become less responsive to insulin and do not take up gluocse as readily. As a result, only limited amounts of glucose can enter the cells and blood levels of glucose rise. Large amounts of insulin are therefore required to allow cells to take up enough glucose to meet their energy needs. Type 2 diabetes is believed to be due to a combination of genetic and lifestyle factors. Risk of developing this disease is increased in people with a family history of diabetes, in those who are overweight or obese, particularly if they carry their extra body fat in the abdominal region, and those who have a sedentary lifestyle. In Chapter 7: Focus On: The Relationship between Obesity and Disease, we will examine more closely the relationship between obesity and diabetes. Type 2 diabetes may occur as part of a combination of conditions called **metabolic syndrome**, which includes obesity, elevated blood pressure, altered blood-lipid levels, and insulin insensitivity.

Type 2 diabetes is often preceded by a condition called **pre-diabetes** or **impaired glucose tolerance** in which blood-glucose levels are above normal but not high enough to be diagnosed as diabetes (see Figure 4.23). Progression to diabetes among those with pre-diabetes is not inevitable. Weight loss and increased physical activity among people with pre-diabetes can prevent or delay diabetes and may return blood-glucose levels to normal.[15]

Type 2 diabetes has typically been diagnosed in persons over the age of 40, but its incidence is increasing among younger individuals. This change is thought to be due to the increasing incidence of obesity and overweight in younger age groups. According to the Canadian Diabetes Association, about 95% of children diagnosed with type 2 diabetes are overweight.

First Nations people appear to develop type 2 diabetes more frequently than non-First Nations people. In a recent study, the prevalence of diabetes among First Nations adults was 20% for women and 16% for men, compared to 5.5% and 6.2% respectively for non-First Nations.[16] According to the Canadian Diabetes Association, type 2 diabetes is being diagnosed in First Nations children as young as age eight. The reasons for this are not fully understood but are likely due to a combination of genetic and environmental factors.

Life Cycle

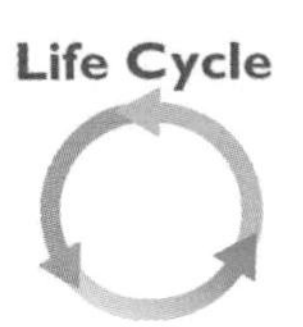

Gestational Diabetes is a form of diabetes that occurs in women during pregnancy. It may be caused by the hormonal changes of pregnancy. The high levels of glucose in the mother's blood increase the risk of complications for the fetus (see Section 14.1). Gestational diabetes usually disappears once the pregnancy is complete and hormones return to nonpregnant levels. However, individuals who have had gestational diabetes have an increased risk for developing type 2 diabetes later in life.

gestational diabetes A form of diabetes that occurs during pregnancy and resolves after the baby is born.

Diabetes Symptoms The symptoms of diabetes result from the fact that without sufficient insulin, glucose cannot be used normally. Cells that require insulin for glucose uptake are starved for glucose, and cells that can use glucose without insulin are exposed to damagingly high levels.

Immediate Symptoms The immediate symptoms of diabetes may include excessive thirst, frequent urination, blurred vision, and weight loss. Excessive thirst and frequent urination occur because blood-glucose levels rise so high that the kidneys excrete glucose, which draws fluid with it, increasing the volume of urine. Blurred vision occurs when excess glucose enters the lens of the eye, drawing in water and causing the lens to swell. Most notably in type 1 diabetes, weight loss in adults and impaired growth in children occur because glucose cannot enter cells to be used for energy, so the body responds as it does in starvation, breaking down fat and protein to supply fuel. With limited carbohydrate for fatty acid metabolism, ketones are formed and released into the blood. Some ketones are used as fuel by muscle and adipose (fat) tissue, but in type 1 diabetes, they are produced more rapidly than they can be used and thus accumulate in the blood. This elevation of ketones causes an increase in the acidity of the blood called ketoacidosis. In type 2 diabetes, ketoacidosis usually does not develop because there is enough insulin to allow some glucose to be used, so fewer ketones are produced.

Long-Term Complications The long-term complications of diabetes include damage to the heart, blood vessels, kidneys, eyes, and nerves. This damage is thought to be a result of prolonged exposure to high levels of blood glucose. When glucose is high, it can bind to proteins contributing to blood-vessel damage and abnormalities in blood-cell function. Damage to the large blood vessels leads to an increased risk of heart disease and stroke; in fact, heart disease is a major complication and the leading cause of premature death among people with diabetes. Changes in small blood vessels and nerves lead to kidney failure, blindness, and nerve dysfunction. For example, accumulation of glucose in the eye damages small vessels in the retina, leading to blindness (**Figure 4.24**). High glucose levels cause kidney failure by damaging kidney cells and small blood vessels in the kidney. Exposure to high glucose also affects the function of peripheral nerves, often causing numbness and tingling in the feet. In addition to these problems, infections are more common in diabetes because high blood-glucose levels favor microbial growth; infections are usually the cause of amputations of the toes, feet, and legs.

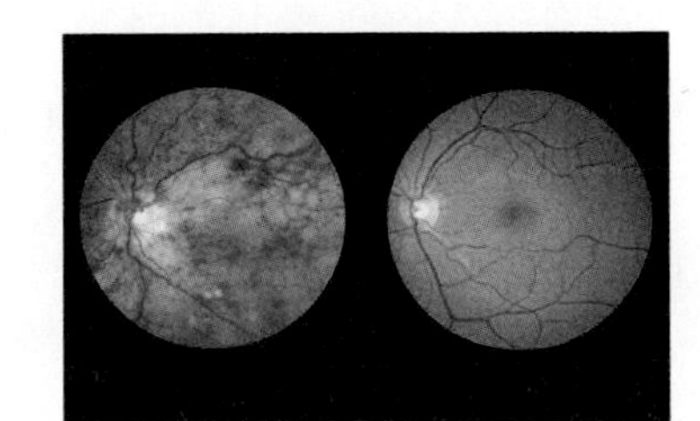

Figure 4.24
(Left) Damaged retinal blood vessels caused by diabetes. (Right) Normal retinal blood vessels. (© SBHA/Stone/Getty Images)

Diabetes Treatment The goal of diabetes treatment is to keep blood-glucose levels within the normal range. This involves diet, exercise, and, in many cases, medication. Blood-glucose levels should be monitored frequently to assure that levels are staying in the healthy range. Adherence to this type of treatment regimen can reduce the incidence of elevated blood-glucose levels and the complications they cause.

CANADIAN CONTENT

Diet Plans: Canadian Diabetes Association Just the Basics and Beyond the Basics To help control blood-glucose levels, carbohydrate intake should be distributed throughout the day. The Canadian

The Canadian Diabetes Association recommends a simple dietary plan, *Just the Basics,* for the management of diabetes. This plan helps the user choose the correct proportion of the right foods, by visualizing the plate divided into three parts, and filling each part appropriately (see **Figure 4.25**).

• Eat more vegetables. These are very high in nutrients and low in calories.

• Choose starchy foods such as whole grain breads and cereals, rice, noodles, or potatoes at every meal. Starchy foods are broken down into glucose, which your body needs for energy.

• Include fish, lean meats, low-fat cheeses, eggs, or vegetarian protein choices as part of your meal.

Figure 4.25 The *Just the Basics* plan for healthy eating from the Canadian Diabetes Association. This simple plan helps to control the amount and type of food consumed by dividing a plate into 3 parts and filling each part with specific foods.
Source: Just the basics. Canadian Diabetes Association. Available online at http://www.diabetes.ca/files/jtb17x_11_cpgo3_1103.pdf. Accessed May 13, 2011.

The Canadian Diabetes Association also has a more sophisticated program called *Beyond the Basics: Meal Planning for Healthy Eating, Diabetes Prevention and Management.* This system allows individuals to plan meals based on a combination of foods high in carbohydrates and foods low in carbohydrates. *Beyond the Basics* is discussed in more detail in Appendix E. The appendix also compares Canada's Food Guide and Beyond the Basics; although both dietary patterns are nutritious, there are interesting differences between them.

The Canadian Diabetes Association also recommends the replacement of high-glycemic-index foods with low-glycemic-index foods in its clinical practice guidelines[17] and provides consumers with a table of the glycemic index of common foods, suggesting that a low-glycemic food be selected at each meal (**Figure 4.26**).

Both the *Just the Basics* and *Beyond the Basics* plans result in a diet that is adequate in energy, protein, and micronutrients. Overweight individuals may need to restrict energy intake to promote weight loss, which can be beneficial for maintaining blood-glucose levels in the normal range.

Exercise Exercise is an important component of diabetes management because exercise increases the sensitivity of body cells to insulin. Therefore, more glucose can enter the cells with less insulin. It also promotes weight loss, which further reduces insulin resistance. Individuals with diabetes are encouraged to maintain regular exercise patterns. A change in the amount of exercise an individual participates in may change the amount of food and medication required to keep blood glucose in the normal range.

Medication When diet and exercise cannot keep blood glucose in the normal range, drug treatments are needed. In type 1 diabetes, insulin production is absent, so insulin must be injected. Insulin cannot be taken orally because it is a protein that would be broken down in the GI tract, losing its ability to function. Type 2 diabetes can often be treated with medications that increase pancreatic insulin production, decrease glucose production by the liver, enhance insulin action, or slow carbohydrate digestion to keep blood glucose in

A lot of starchy foods have a high Glycemic index (GI). Choose Medium and low GI foods more often.

Low GI (55 or Less)*† choose most often √√√	**Medium GI** (56-69) *† choose more often √√	**High GI** (70 or more)* † choose less often √
BREADS: 100% stone ground whole wheat Heavy mixed grain Pumpernickel	**BREADS:** Whole wheat Rye Pita	**BREADS:** White bread Kaiser roll Bagel, white
CEREAL: All Bran™ Bran Buds with Psyllium™ Oatmeal Oat Bran™	**CEREAL:** Grapenuts™ Shredded Wheat™ Quick oats	**CEREAL:** Branflakes Cornflakes Rice Krisples™ Cheerios™
GRAINS: Parboiled or converted rlce Barley Bulgar Pasta/Noodles	**GRAINS:** Basmati rice Brown rice Couscous	**GRAINS:** Short grain rice
OTHER: Sweet potato Yam Legumes Lentils Chickpeas Kidney beans Split peas Soy beans Baked beans	**OTHER:** Potato, New/White Sweet corn Popcorn Stoned Wheat Thins™ Ryvita™ (rye crisps) Black bean soup Green pea soup	**OTHER:** Potato, baking (Russet) French fries Pretzels Rice cakes Soda Crackers

Figure 4.26 The glycemic index of various foods. The Canadian Diabetes Association. recommends the consumption of low-and medium-glycemic-index foods more often.
* Expressed as a percentage of the value for glucose † Canadian values where available
Source: Canadian Diabetes Association. Glycemic Index. Avaiable online at http://www.diabetes.ca/Files/diabetes_gl_final2_cpg03.pdf. Accessed May 13, 2011.

the normal range. In some cases of type 2 diabetes, injected insulin is needed to achieve normal blood-glucose levels.

Hypoglycemia

Hypoglycemia is a condition in which blood sugar drops low enough to cause symptoms including irritability, nervousness, sweating, shakiness, anxiety, rapid heartbeat, headache, hunger, weakness, and sometimes seizure and coma. It can occur in people with diabetes as a result of over-medication or an imbalance between insulin level and carbohydrate intake. People with diabetes must learn to recognize the symptoms of hypoglycemia and immediately treat them by consuming a source of quickly absorbed carbohydrate, such as juice or hard candy. Following this, a meal should be consumed within about 30 minutes to keep glucose in the healthy range.

In individuals without diabetes, hypoglycemia can result from abnormalities in the production of or response to insulin or other hormones involved in blood sugar regulation. There are two forms of hypoglycemia. The first, reactive hypoglycemia occurs in response to the consumption of high-carbohydrate foods. The rise in blood glucose from the carbohydrate stimulates insulin release. However, too much insulin is secreted, resulting in a rapid fall in blood glucose to an abnormally low level. The treatment for reactive hypoglycemia is a diet that prevents rapid changes in blood glucose. Small, frequent meals low in simple carbohydrates and high in protein and fibre are recommended. A second form of hypoglycemia, fasting hypoglycemia, is not related to food intake. In this disorder, abnormal insulin secretion results in episodes of low blood-glucose levels. This condition is often caused by pancreatic tumours.

4.6 Carbohydrates and Health

Learning Objectives

- Explain the health risks and benefits of diets high in unrefined carbohydrates and diets high in refined carbohydrates, respectively.
- Discuss the role of carbohydrates in weight management.
- Discuss how different types of carbohydrates affect the risk of getting diabetes.
- Describe how soluble fibre can lower blood-cholesterol levels.
- Describe the process of cancer development.

Carbohydrate-rich foods are the basis of healthy diets around the world.[4] They provide about half of the kcalories in the Canadian diet and as much as two-thirds in developing countries. Nonetheless, the consumption of carbohydrates has been blamed for a host of chronic health problems, from dental caries and hyperactivity to obesity and heart disease. The incongruity relates to the health effects of different forms and sources of dietary carbohydrates: a dietary pattern that is high in unrefined carbohydrates, such as whole grains, fruits, and vegetables, has been associated with a lower incidence of a variety of chronic diseases, whereas diets high in refined carbohydrates, such as added sugars and white flour, may contribute to chronic disease risk.[18]

dental caries The decay and deterioration of teeth caused by acid produced when bacteria on the teeth metabolize carbohydrate.

Dental Caries

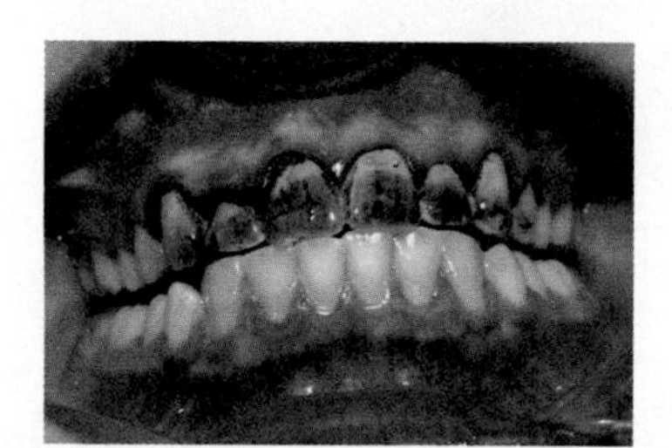

Figure 4.27 The regions on these teeth that are stained brown indicate the presence of dental plaque. The main component of dental plaque is bacterial colonies. (SPL/Photo Researchers, Inc.)

The most well-documented health problem associated with a diet high in carbohydrates is **dental caries**, or tooth cavities. Cavities are caused when bacteria that live in the mouth form colonies, known as plaque, on the tooth surface (**Figure 4.27**). If the plaque is not brushed, flossed, or scraped away, the bacteria metabolize carbohydrate from the food we eat, producing acid, which can dissolve the enamel and underlying structure of the teeth, forming cavities. Bacteria can metabolize both naturally occurring and added refined sugars and starches. Some types of food are more cavity-causing than others. Simple carbohydrate, particularly sucrose, is the most rapidly used food source for bacteria and therefore easily produces tooth-damaging acids. But starchy foods that stick to the teeth can also promote tooth decay. Foods such as gummy candies, cereals, crackers, cookies, and raisins and other sticky dried fruits tend to remain on the teeth longer, providing a continuous supply of nutrients to decay-causing bacteria. Other foods, such as chocolate, ice cream, and bananas, are rapidly washed away from the teeth and therefore are less likely to promote cavities. Frequent snacking, sucking on hard candy, or slowly sipping a soft drink can also increase the risk of cavities by providing a continuous food supply for the bacteria. Limiting sugar intake can help prevent dental caries, but other dietary factors and proper dental hygiene are important even if the diet is low in sugar. Dairy products, sugarless gums (sweetened with sugar alcohols), and fluoride reduce caries formation. Brushing teeth after eating reduces cavity risk no matter what food is consumed.

Life Cycle

Does Sugar Cause Hyperactivity?

The consumption of sugary foods has been suggested as a cause of hyperactivity in children (see Section 15.3). The rise in blood glucose following a meal high in simple carbohydrates has been hypothesized to provide the energy for the excessive activity of a hyperactive child. However, a review of the research on sugar intake and behaviour failed to support the hypothesis that sugar contributes to behavioural changes in most children.[19] Hyperactive behaviour that is observed after sugar consumption is likely the result of other circumstances, for example, the excitement of a birthday party rather than the cake is more likely the cause of such behaviour. Hyperactivity might also be caused by lack of sleep, overstimulation, caffeine consumption, the desire for more attention, or lack of physical activity.

Do Carbohydrates Affect Body Weight?

Carbohydrates are often referred to as "fattening." They provide 4 kcal per gram compared with 9 kcal per gram provided by fat. In fact, it is the fats that we often add to our high-carbohydrate foods that increase their kcalorie tally. A medium-sized baked potato provides about 160 kcal, but the 2 tablespoons of sour cream you add brings the total to 225 kcal (**Figure 4.28**). A plate of plain pasta has about 200 kcal, but with a high-fat sauce, the kcalories rise to 300; add sausage and the meal is now 450 kcal. No single macronutrient, be it carbohydrates, fat, or protein, is "fattening" in isolation; it is the total number of calories consumed and whether that total exceeds energy expenditure that determine weight gain. Nonetheless, some types of carbohydrates have been implicated in weight gain.

Figure 4.28
High-carbohydrate foods like baked potatoes are not high in kcalories, but the toppings used on them can easily double their kcalorie count. (© istockphoto.com/Joe Gough)

High-Fructose Corn Syrup and Body Weight High-fructose corn syrup, widely used as a sweetener, particularly of soft drinks, has been implicated in weight gain because its introduction into the food supply in the 1970s correlates with the rise in obesity. Fructose is handled by the body differently than glucose. It does not stimulate insulin production and has different effects on other hormonal signals that regulate food intake and body weight.[2] Fructose metabolism in the liver favours fat synthesis. Studies in mice indicate that dietary fructose produces a greater increase in body fat than the same amount of sucrose.[20] So, is high-fructose corn syrup making us fat? It certainly is adding kcalories to our diets, but high-fructose corn syrup provides almost the same amount of fructose as does sucrose, which is broken down to glucose and fructose before it is absorbed. High-fructose corn syrup alone cannot account for the rise in obesity. The increase in total energy intake and reduction in physical activity remain the most significant factors contributing to the increased incidence of obesity.[21]

Low-Carbohydrate Weight-Loss Diets The rationale behind consuming a low-carbohydrate diet for weight loss is that foods high in carbohydrate stimulate the release of insulin, which is a hormone that promotes energy storage. It is suggested that the more insulin you release, the more fat you will store. High-glycemic-index foods, which increase blood sugar and consequently stimulate insulin release, are therefore hypothesized to shift metabolism toward fat storage. In contrast, a low-carbohydrate diet causes less of a rise in insulin and therefore is suggested to promote fat loss. Weight loss while consuming a low-carbohydrate diet may also be affected by ketone levels and the amount of protein in the diet. Ketones help suppress appetite and the high protein content of a low-carbohydrate diet can be satiating, so both help the dieter eat less. Studies on the effectiveness of low-carbohydrate diets for weight loss indicate that they result in greater short-term weight loss (6 months) than low-fat diets.[22] The weight loss on these diets, as with any weight-loss diet, is caused by consuming less energy than is expended.[23-25] Regrettably, most people who diet tend to regain their weight after a year, and this is true regardless of whether a low-carbohydrate or low-fat diet is followed.

Unrefined Carbohydrates and Weight Management Research has suggest that diets with a low glycemic index may be helpful in weight loss.[26] Diets including low-glycemic-index foods may make weight loss easier because they slow the absorption of energy-yielding nutrients and enhance the feeling of fullness while the dieter is consuming less food.

Refined Carbohydrates and Diabetes

Evidence is accumulating that the type of dietary carbohydrate one eats plays a role not only in the treatment of diabetes, but in the development of type 2 diabetes in susceptible individuals.[27] The risk of developing type 2 diabetes is lower in populations that consume diets high in fibre and low in glycemic index ratings than in populations that eat a diet high in refined starches and added sugars (see Critical Thinking: Dietary Fibre, Glycemic Index, and Type 2 Diabetes).[28-30] In particular, a high intake of sugar-sweetened beverages has been associated with an increased risk for development of type 2 diabetes.[31] One mechanism that may explain this relationship is that a dietary pattern that is high in refined starches and added sugars causes a greater glycemic response and therefore increases the amount of insulin needed to maintain normal blood-glucose levels. Over the long term, in susceptible individuals, the high demand for insulin eventually may wear out the insulin-producing beta cells in the pancreas. When there are more low-glycemic-index foods in the diet, there is a more gradual rise in blood sugar and therefore a lower insulin demand.

CRITICAL THINKING:

Dietary Fibre, Glycemic Index, and Type 2 Diabetes

Researchers wanted to determine whether diets low in cereal fibre (i.e., dietary fibre from grain products) or diets that had a high glycemic index affected the risk of developing type-2-diabetes, so they conducted an observational study.[1]

More than 90,000 young and middle-aged women completed a food-frequency questionnaire (see Section 2.6) to assess dietary intake. The women were then followed for eight years and the number of cases of type 2 diabetes was recorded. After statistically adjusting for age, body mass index, family history, and other confounding factors, they looked at the relative risk for developing type-2-diabetes (see Chapter 2: Science Applied: Does Following a Healthy Dietary Pattern Reduce the Risk of Disease for a review of the concept of relative risk). First they compared women with diets that had a low GI compared to a high GI. Next they compared diets high in cereal fibre to those low in cereal fibre.

These are the results:

Diet	Relative Risk of Developing Type 2 Diabetes
Low GI	1.0
High GI	1.6*

Diet	Relative Risk of Developing Type 2 Diabetes
Low cereal fibre	1.0
High cereal fibre	0.64*

*Statistically significant difference from respective reference group.

Critical Thinking Questions:

What was the purpose of the study?*

What kind of observational study is this?

Describe the population in this study.

How long were they observed?

What dietary components were the authors interested in?

What health outcome was being evaluated?

Interpret the relative risks reported. Note that there were statistically significant differences between relative risks.

What do you conclude about the relationship between diet and type 2 diabetes?

Does the study demonstrated causation? Why or why not?

Based on what is known about the effects of cereal fibre and low-glycemic-index foods on blood glucose levels, do the results of the study make sense?

*Answers to these questions can be found in Appendix J.

[1] Schulze, M. B., Liu, S., Rimm, E. B., et al. Glycemic index, glycemic load, and dietary fiber intake and incidence of type 2 diabetes in younger and middle-aged women. *Am. J. Clin. Nutr.* 80:348-356, 2004.

Carbohydrates and Heart Disease

Just as with weight management and diabetes, when considering heart disease, some carbohydrates may be protective while others may increase risk. Evidence shows that a diet high in sugar can raise blood lipid levels and thereby increase the risk of heart disease.[32] On the other hand, diets high in whole grains have been found to reduce the risk of heart disease.[33-36] In an analysis of more than 150,000 people, those with the highest dietary fibre intake had a 29% lower risk of coronary heart disease than those with the lowest intake.[37] In general, people with the highest intake of whole grains—about three servings a day—have a 20-30% lower risk of heart disease than those consuming the fewest whole grains. Whole grains provide fibre, resistant starch, oligosaccharides, omega-3 fatty acids, vitamins, minerals, antioxidants, and other phytochemicals that may be protective against heart disease (see Section 5.6).

One of the ways a diet high in whole grains and other unrefined carbohydrates may reduce the risk of heart disease is by reducing blood-cholesterol levels. Soluble fibre binds cholesterol and bile acids, which are made from cholesterol, in the digestive tract. Normally, bile acids secreted into the GI tract are absorbed and reused. When bound to fibre, they are excreted in the feces rather than being absorbed (**Figure 4.29**). The liver must then use cholesterol from the blood to synthesize new bile acids. This provides a mechanism for eliminating cholesterol from the body and reducing blood-cholesterol levels. Soluble fibres from legumes, oats, guar gum, pectin, flax seed, and psyllium (a grain used in bulk-forming laxatives such as Metamucil) are effective at reducing cholesterol, but insoluble fibres such as wheat bran or cellulose are not.[38] Soluble fibre may also reduce blood cholesterol by inhibiting cholesterol synthesis in the liver or by increasing the removal of cholesterol from the blood.[39]

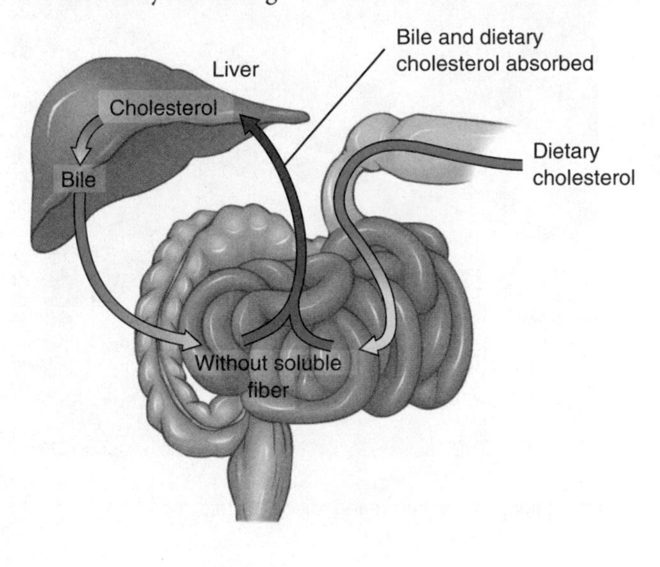

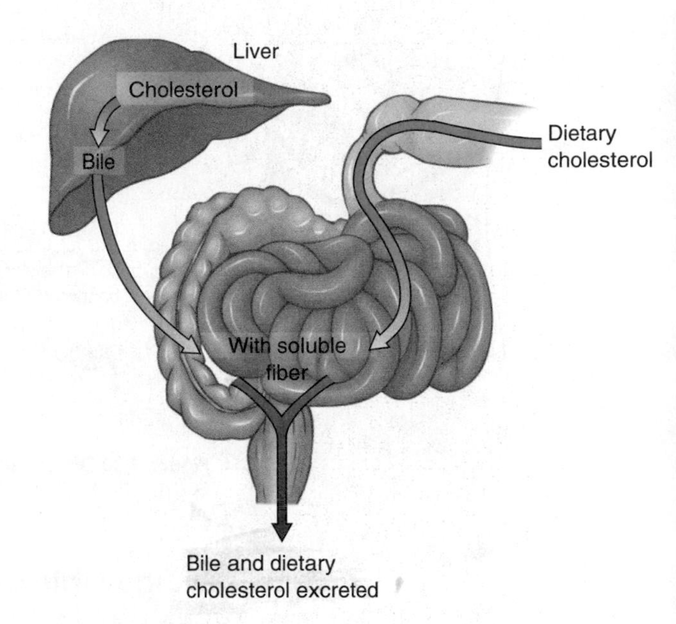

Figure 4.29 Effect of soluble fibre on cholesterol absorption
(a) Bile, which contains cholesterol and bile acids made from cholesterol, is absorbed and returned to the liver when soluble fibre is not present in the digestive tract. (b) When soluble fibre is present, the fibre binds cholesterol and bile acids so that they are excreted rather than absorbed.

In addition to lowering blood-cholesterol levels, high fibre foods help lower blood pressure, normalize blood-glucose levels, prevent obesity, and affect a number of other parameters, all of which help reduce the risk of heart disease.[38]

Indigestible Carbohydrates and Bowel Disorders

Whole grains, fruits, and vegetables are good sources of fibre and also contain resistant starch and oligosaccharides. Diets high in these indigestible carbohydrates can relieve or prevent certain bowel disorders that are caused by pressure in the lumen of the colon.[40] As discussed earlier, a mixture of soluble and insoluble fibres in the colon adds bulk and absorbs water. This increases stool weight, speeds transit, and makes the feces larger and softer, thus reducing the amount of pressure needed for defecation. The presence of these fibres and other indigestible carbohydrates helps to reduce the incidence of constipation, which makes muscles strain to move stool that is too hard and is the main cause of increased pressure in the colon. This excess pressure contributes to the formation of **hemorrhoids**, the swelling of veins in the rectal or anal area. Excess pressure is also believed to cause weak spots in the colon to bulge out and become **diverticula** (**Figure 4.30**). When these outpouchings form, the condition is called **diverticulosis**. A high-fibre diet reduces pressure in the lumen of the colon and therefore reduces the possibility of developing diverticulosis.[41,42] If diverticulosis does develop, fecal matter may occasionally accumulate in these outpouchings, causing irritation, pain, inflammation, and infection, a condition known as **diverticulitis**. Treatment of diverticulitis usually includes antibiotics to reduce bacterial growth and a temporary decrease in fibre intake to prevent irritation of the inflamed tissues. Once the inflammation is resolved, however, a high-fibre intake is recommended to increase fecal bulk, decrease transit time, ease stool elimination, and reduce future attacks of diverticulitis.

hemorrhoids Swollen veins in the anal or rectal area.

diverticula Sacs or pouches that protrude from the wall of the large intestine in the disease **diverticulosis**. When these become inflamed, the condition is called **diverticulitis**.

Although fibre helps soften stools and prevent constipation, if fibre is consumed without sufficient fluid, it can also cause constipation. The more fibre there is in the diet, the more water is needed to keep the stool soft. When too little fluid is consumed, the stool becomes hard and difficult to eliminate. In severe cases when fibre intake is excessive and fluid intake is low, intestinal blockage can occur.

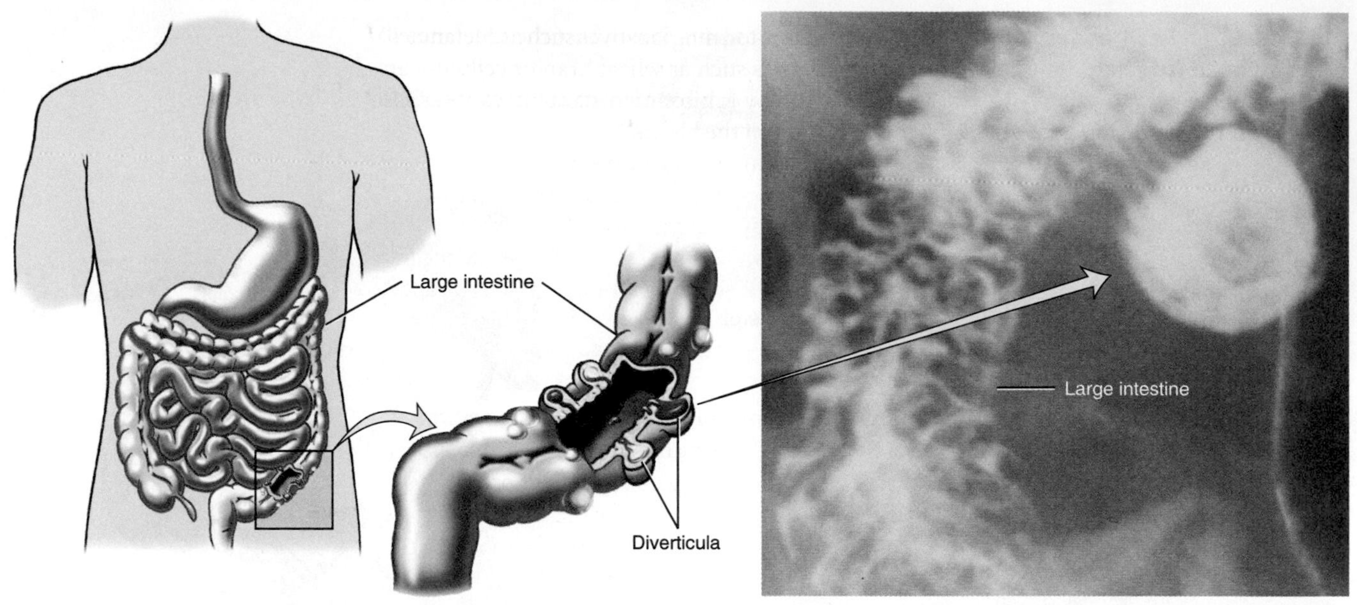

Figure 4.30 Diverticula
Diverticula (the singular is *diverticulum*) in the colon. (© L.V. Bergman/Project Masters, Inc.)

Indigestible Carbohydrates and Colon Cancer

Cancer is a disease that affects the way cells behave. Different cancers originate in different parts of the body and have different causes and effects. The type of cancer depends on the type of cell that is originally affected—for example, lung, breast, or colon—and on how the genetic material has been altered. Some people are more susceptible to cancer due to a genetic predisposition, but the development of most cancers is also believed to be influenced by environmental **carcinogens** from the diet, tobacco smoke, or air pollution. In the case of colon cancer, substances consumed in the diet or produced in the GI tract that come into contact with mucosal cells may contribute to cancer development.

carcinogens Cancer-causing substances.

Characteristics of Cancer Cells Cells become cancerous as a result of **mutations** in their genetic material that allow them to reproduce without restraint and grow in abnormal locations. Normal body cells reproduce only to replace lost cells or to accommodate normal growth, but cancer cells divide continuously, forming enlarged cell masses known as tumours. Further mutations allow them to invade and colonize areas reserved for other cells, referred to as **malignancy** or **metastasis** (**Figure 4.31**). The cancer cells eventually crowd out the normal cells, robbing them of nourishment, and preventing them from functioning properly. Some carcinogens act by damaging DNA and inducing mutations. These are usually referred to as tumour initiators since the induction of mutations in key genes is thought to be the initiating event in cancer development. Other carcinogens contribute to cancer development by stimulating cells to divide. Such compounds are referred to as tumour promoters, since the increased cell division they induce enlarges the population of mutated cells, which is necessary for cancer to progress.

mutations Changes in DNA caused by chemical or physical agents.

malignancy or **metastasis** A mass of cells showing uncontrolled growth, a tendency to invade and damage surrounding tissues, and an ability to seed daughter growths to sites remote from the original growth.

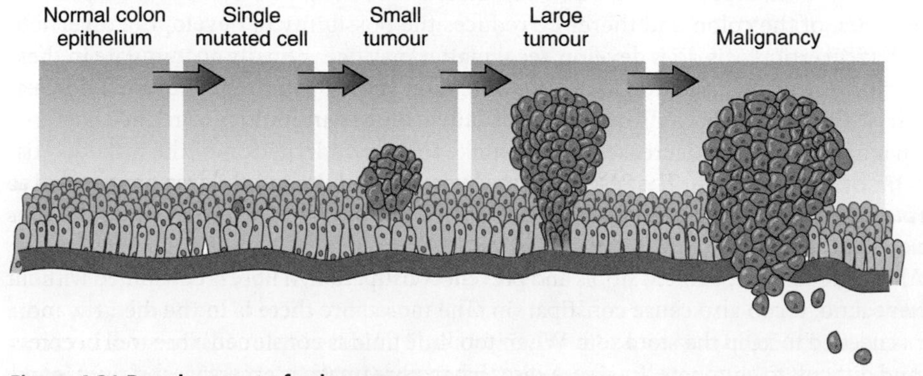

Figure 4.31 Development of colon cancer
Colon cancer, like other cancers, progresses from a single mutated cell to a tumour to a malignancy.

Fibre and Cancer Development Epidemiological studies have shown that the incidence of colon cancer is lower in populations consuming diets high in fibre.[41,43,44] Several hypotheses have been suggested to explain how fibre might affect the development of colon cancer. One is related to its ability to decrease contact between the mucosal cells of the large intestine and the fecal contents, which may contain tumour initiators or tumour promoters. Fibre decreases contact by increasing fecal bulk, diluting the colon contents, and speeding transit. Another theory relates to the effect of fibre on the intestinal microflora and the by-products of microbial metabolism, such as fatty acids, that accumulate there. These by-products may directly affect colon cells or may cause changes in the environment of the colon that can affect the development of colon cancer. It has also been hypothesized that high-fibre diets protect against colon cancer because of the antioxidant vitamins and phytochemicals that are present along with fibre in plant foods.

Recent intervention studies have not supported the epidemiological observations of a connection between fibre intake and colon cancer.[45–47] A number of reasons have been suggested for this discrepancy—the interventions were not long enough, the fibre dose was not high enough, the type of cancer monitored was not appropriate, or the fibre itself is not really protective, but rather some other component in the diet of the low-cancer populations may have a protective effect. Despite these results, the scientific consensus is still that there is enough evidence that diets high in fibre protect against colon cancer to recommend an increase in fibre intake.[41]

4.7 Meeting Carbohydrate Recommendations

Learning Objectives

- Modify a diet so it meets current recommendations for the types and amounts of carbohydrate.
- Discuss the role of alternative sweeteners in weight-loss.

The average Canadian diet provides plenty of carbohydrate, but whether or not this carbohydrate promotes or harms our health depends on the food sources and types of carbohydrates we choose. A healthy diet is high in complex carbohydrates from whole grains, legumes, fruits, and vegetables, and simple carbohydrates from unrefined foods such as fresh fruit and low-fat dairy products. This diet is high in fibre, micronutrients, and phytochemicals, and low in saturated fat and cholesterol. Unfortunately, the typical Canadian diet is lower in whole grains, fruits, and vegetables and higher in added sugars than is recommended.

Types of Carbohydrate Recommendations

A small amount of carbohydrate is needed to fuel the brain. Additional carbohydrate provides an important source of energy in the diet and adequate fibre offers many health benefits. Therefore, the DRIs make several kinds of recommendations for carbohydrate intake: an RDA for total carbohydrate, a range of acceptable carbohydrate intakes called the Acceptable Macronutrient Distribution Range (AMDR), and an AI for fibre. Because no specific toxicity is associated with high intakes of carbohydrate in general or of different types of carbohydrates, no UL has been established for total carbohydrate, added sugars, or fibre.

The RDA for Carbohydrate The RDA for carbohydrate for adults and children has been set at 130 g per day, based on the average minimum amount of glucose used by the brain.[5] In a diet that meets energy needs, this amount will provide adequate glucose and prevent ketosis. This amount of carbohydrate provides only 420 kcal; that's about 25% of the energy in a 2,000-kcal diet. It is equivalent to a breakfast of 250 ml (1 cup) of juice, two slices of toast with jam, and a bowl of cereal with half a banana and milk (**Figure 4.32**). Most people consume well in excess of this amount over the course of a day. A diet that includes only this much carbohydrate and meets kcalorie needs would be very high in protein and fat.

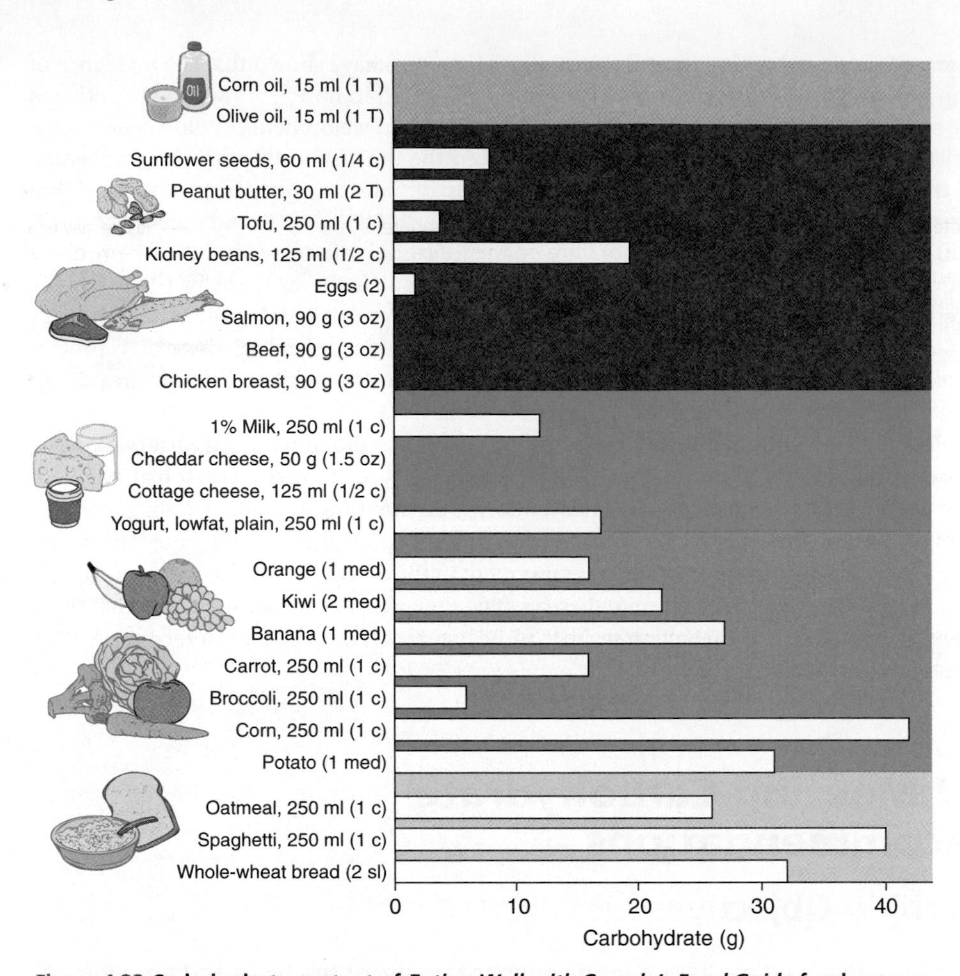

Figure 4.32 Carbohydrate content of *Eating Well with Canada's Food Guide* food groups
Grains, vegetables, and legumes are the best sources of complex carbohydrates; fruits and milk are good sources of naturally occurring simple carbohydrates. Plant-based meat alternatives are also good sources of carbohydrates.

The Range of Healthy Carbohydrate Intakes No single ratio of macronutrients defines a perfect diet. Healthy diets can be made up of many different combinations of carbohydrate, protein, and fat. The AMDR for carbohydrate intake for a healthy diet has been set at 45-65% of energy. Choosing a diet in this range will allow you to meet your energy needs without consuming excessive amounts of protein or fat. The CCHS reports that children and teens get about 55% of their calories from carbohydrates and adults get about 50% of their calories from carbohydrates, right in the middle of the acceptable range.[48] The sources of carbohydrate are more important than the absolute amount. In order to promote a diet that meets the needs for all nutrients, the DRIs recommend that most carbohydrate should come from unrefined food sources; no more than 25% of energy should come from added refined sugars.[49] This percentage is based on the amount of added sugar that can be consumed without reducing the nutrient density of the diet so much that essential nutrient needs cannot be met. This is higher than the World Health Organization, which recommends no more than 10% of kcalories from sugar.[50] *Canada's Food Guide* recommends that Canadians choose foods with little or no added sugar. Unfortunately, the CCHS did not collect data on the amount of added sugar comsumed by Canadians, but it has been estimated at about 13% of kcalories for Canadians.[1] In the United States, government dietary surveys do measure added sugar and American intake is 16%,[4] close to the Canadian value of 13%, suggesting this estimate is probably sound.

The AI for Fibre For fibre, an AI has been set at 38 g and 25 g per day for young adult men and women, respectively, based on the amount of fibre needed to reduce heart disease risk

CRITICAL THINKING:

How are Canadians Doing with Respect to Fibre Intake?

As noted in Section 4.7, fibre intake is evaluated using an Adequate Intake. As discussed in Section 2.2, an AI is estimated when there are insufficient scientific data to establish an EAR. Instead an AI is established by assessing the median intake of a population where everyone is known to be meeting their requirements, by an established criteria (Section 2.2). When assessing whether a population is meeting its requirement for a nutrient that has an AI, the intakes are interpreted in the following way: when 50% or more of a population consumes more than the AI, it can be concluded that the population, as a whole, is meeting its requirements.[1] If, however, less than 50% of the population has intakes above the AI, then the population may or may not be consuming adequate amounts of fibre. This may not seem logical, but it is a consequence of the limitations of the AI and an explanation of these limitations is beyond the scope of this textbook. What you should know is that if less than 50% of a population has intakes above the AI, then it is not possible to draw any firm conclusions about the adequacy of intake. This graph shows the CCHS results for fibre intake of Canadian adults. The green line marks the 50% cutoff.

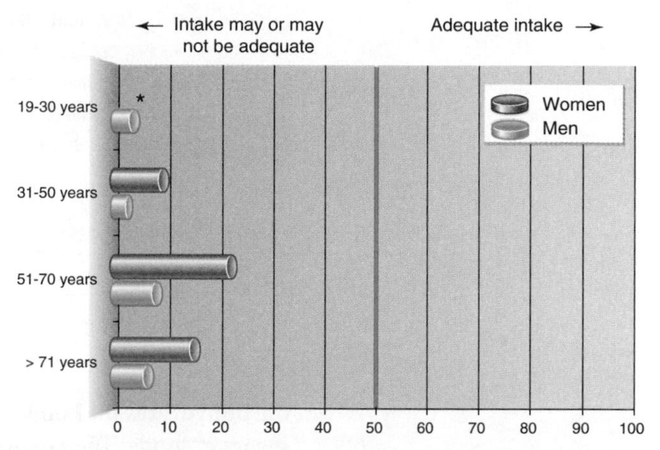

Source: Health Canada and Statistics Canada: Canadian Community Health Survey Cycle 2.2. Nutrition (2004) Nutrient Intakes from Food Cat No. 978-0-662-06542-5.
*Data for women 19-30 years is not reported.

Critical Thinking Questions

What would you conclude about the fibre intake of Canadians?

[1] Health Canada. Canadian Community Health Survey cycle 2.2. Nutrition (2004) A Guide to Assessing and Interpreting the Data. Available online at www.hc-sc.ca/fn-an/surveill/nutrition/commun/cchs_guide_escc_2-eng.php#2.2.3. Accessed Feb 28, 2011.

(see Critical Thinking: How are Canadians Doing with Respect to Fibre Intake?).[4] Eating a bowl of raisin bran with a half-cup of strawberries for breakfast, a sandwich on whole wheat bread with lettuce and tomatoes and an apple for lunch, eggplant parmesan for dinner, and popcorn for a snack will provide about 25 g of fibre. Specific AIs for fibre have been set for different life-stage groups (see inside cover).

Tools for Assessing Carbohydrate Intake

How does your diet compare to the recommendation of getting 45-65% of your kcalories from carbohydrate? **Table 4.2** illustrates how to calculate carbohydrate intake as a percent of energy. This same calculation can be used to determine the percent of energy from carbohydrate in individual foods. To calculate the percent of energy from carbohydrate, you need to know the amount of carbohydrate in a food or in the diet. This can be estimated from the food tables used in *Beyond the Basics* (see Appendix E), using values from food labels, or using food composition tables.

Table 4.2 Calculating Percent Energy from Carbohydrate
Determine
• The total energy (kcalorie) intake for the day
• The grams of carbohydrate in the day's diet
Calculate Energy from Carbohydrate
• Carbohydrate provides 4 kcalories per gram

Table 4.2 *(Continued)*
• Multiply grams of carbohydrate by 4 kcalories per gram
Energy (kcalories) from carbohydrate = grams carbohydrate × 4 kcalories/gram carbohydrate
Calculate % Energy from Carbohydrate
• Divide energy from carbohydrate by total energy and multiply by 100 to express as a percent
$\text{Percent of energy from carbohydrate} = \frac{\text{kcalories from carbohydrate}}{\text{Total kcalories}} \times 100$
For example:
A diet contains 2,500 kcalories and 350 g of carbohydrate
350 g of carbohydrate × 4 kcal/g = 1,400 kcal of carbohydrate
$\frac{\text{1,400 kcal of Carbohydrate}}{\text{2,500 kcal}} \times 100 = 56\%$ of energy (kcal) from Carbohydrate

Carbohydrates on Food Labels Food labels list the grams of total carbohydrate, fibre, and sugars in foods. The amounts of soluble and insoluble fibre may be listed if manufacturers choose to include them, but are not required. Total carbohydrate and fibre are also listed as a percent of the Daily Value. The Daily Value for total carbohydrate is calculated as 60% of the energy. For a 2,000-kcal diet, this represents 300g of carbohydrate ([2,000 kcal × 0.6]/ 4 kcal/g of carbohydrate = 300 g). The Daily Value for fibre is 25 g in a 2,000-kcal diet. A slice of whole-wheat bread contains 2 g of fibre per serving, which is 8% of the Daily Value. No Daily Value has been established for sugars, but labels can help identify products that are high in sugars (see Label Literacy: The Scoop on Sugar). Descriptors such as "high fibre" and "a good source of fibre" can help you find high-fibre products (see **Table 4.3**).

Table 4.3 Sugar and Fibre on Food Labels	
Sugar-free	Product contains no amount, or a trivial amount, of sugars (less than 0.5 g per serving). Synonyms for "free" include "without," "no," and "zero."
Reduced sugar	Nutritionally altered product contains 25% less sugar than the regular or reference product.
Lower in sugar	Whether altered or not, a food contains 25% less sugar than the reference food. Also "less sugar," "lower sugar."
No added sugars or without added sugars	No sugar or sugar-containing ingredient is added during processing.
Source of fibre	Food contains at least 2 g/serving.
High source of fibre	Food contains at least 4 g/serving.
Very high source of fibre	Food contains at least 6 g/serving.

Source: Canadian Food Inspection Agency. Guide to Food Labelling and Advertising Chapter 7: Nutrient Content Claims. Available online at www.inspection.gc.ca/english/fssa/labeti/guide/ch7be.shtml#a7_23. Accessed Oct 2, 2011.

Translating Recommendations into Healthy Diets

In order to promote a healthy balance of carbohydrates, *Eating Well with Canada's Food Guide* recommends that Canadians consume whole grains, vegetables, and fruits, and limit their intake of added sugars from foods such as bakery products, candy, and soft drinks.

Eat More Vegetables, Fruits,Whole Grains and Legumes Canada's Food Guide emphasizes the consumption of vegetables and fruits. To maximize fibre intake, most fruit choices should be whole fruits rather than juices. An apple provides about 80 kcal-90 kcal and 2.7 g of fibre, whereas 250 ml (1 cup) of apple juice provides the same amount of energy but almost no fibre (0.2 g). Canada's Food Guide also recommends that you choose at least half of your grain products from whole-grain sources (see Your Choice: Choosing Whole Grains). Whole grains are good sources of fibre as well as micronutrients and phytochemicals. Choosing legumes as a meat alternative will also increase fibre intake—serving (175 ml) of cooked black beans has about 10 g of fibre.

LABEL LITERACY

The Scoop on Sugar

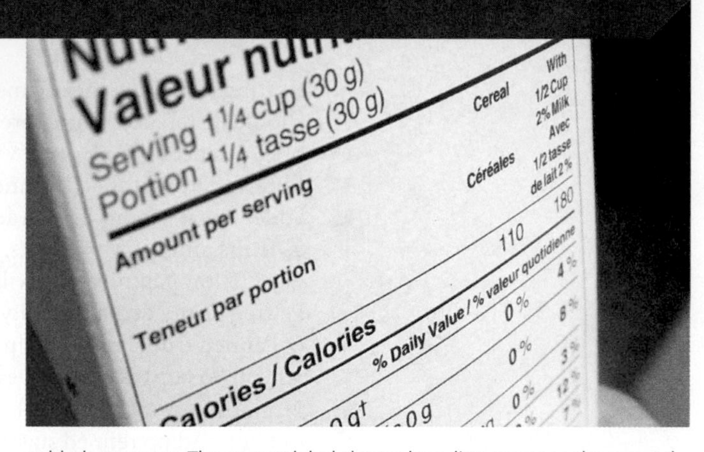

Cutting down on the sugar you add to your morning coffee and cereal will reduce your sugar intake. You might be surprised to learn, though, that this strategy isn't cutting out the biggest sources of sugar in your diet. Most of the sugar you consume comes from added sugars in prepared foods like soft drinks, cookies, and snacks. Recognizing foods high in added sugars is an important step toward cutting down on these empty kcalories.

Identifying packaged foods that are high in added refined sugar isn't always easy because food labels don't differentiate between added and natural sugar. As you can see in the strawberry yogurt label shown here, the Nutrition Facts lists the total grams of sugars (monosaccharides and disaccharides)—28 g in this case. This amount includes both the sugar found naturally in the strawberries and milk and the sugar added for additional sweetness. Some food labels make it easier to identify foods that have not had sugar added in processing by including a nutrition claim such as "no added sugar" or "without added sugar." For products that do not contain descriptors such as these, you can sort out their sugar sources by reading the ingredient list.

To catch foods high in added sugar, check labels for sweeteners listed early in the ingredient list. The sooner they appear, the more the product contains by weight. Remember, though, that the sum of the added sweeteners hiding farther down in the ingredient list may be considerable. Recognizing added sugars can be a challenge. The only sweetener that can be called "sugar" on the ingredient list is sucrose; this may come in the form of brown, powdered, granulated, or raw sugar. However, there are many other sugars added to foods; for example, invert sugar, dextrose, glucose, maltose, lactose, and fructose are sugars added in dry form. Corn syrup, honey, molasses, malt syrup, sugar syrup, fruit-juice concentrates, and high-fructose corn syrup are added as syrups. The yogurt label shown here lists sugar as the second ingredient and high-fructose corn syrup as the fourth. Together these two contribute much of the sugar in this food.

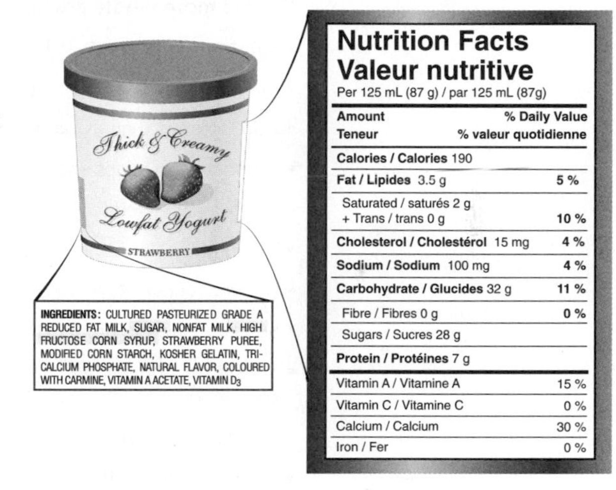

Limit Added Sugars Foods that contain the most added sugars in the Canadian diet include soft drinks, candy, cakes, cookies, pies, fruit drinks, and dairy desserts, but added sugars are also found in thousands of other processed products (see **Table 4.4**). These are the foods that *Canada's Food Guide* suggest we limit in our diet. Sugars are also added at home when we sprinkle sugar on our cereal or spoon honey into our tea (1 tsp (5 ml) sugar weighs 4 grams). The greater your consumption of foods high in added sugars, the harder it is to consume enough nutrients without gaining weight. Added sugars provide kcalories with little if any of the essential nutrients. People who eat foods and beverages high in added sugars tend to eat more kcalories and fewer micronutrients.[51]

Table 4.4 How Much Added Sugar Do You Eat?

Food	Added Sugar (g)
Doughnut, 7.5 cm diameter	10 g
Cookies, 2 medium chocolate chip	15
Frosted corn flakes, 30 g	12
Cake, frosted, 1 piece	24
Pie, fruit, 2 crust, 1 slice	24
Fruit, canned in heavy syrup, 250 ml	16
Chocolate milk, 2%, 250 ml	12
Low-fat fruit yogurt, 250 ml	28
Ice cream, vanilla, 250 ml	12
Chocolate bar, 30 g	20
Fruit drink, 350 ml	48
Cola, not diet, 350 ml	36
Cola, not diet, 600 ml	60

Putting It All Together To meet the recommendations for a healthy diet, refined carbohydrates should be replaced with unrefined sources (**Table 4.5**). For example, choosing a stir-fry meal of 75 g of beef and plenty of vegetables on brown rice can provide the same kcalories but more fibre and less fat than a dinner of steak, white rice, and a small salad with dressing. To limit added sugars, foods high in added sugar should be replaced with natural sources of sugar such as fruits and dairy products. If instead of a 600-ml bottle of a cola drink you have a 250-ml glass of low-fat milk, you will consume 140 fewer kcalories, no added sugar, and you will be getting plenty of high-quality protein, calcium, and other micronutrients. Using fresh instead of canned fruit can also help increase fibre and decrease added refined sugars. For example, 125 ml (½ cup) of pear halves canned in heavy syrup provides 90 kcal, 1 g of fibre, and almost 20 g of sugar, most of which is added in the syrup. One large fresh pear provides 90 kcal, 4 g of fibre, and no refined sugar (see Critical Thinking: Becoming Less Refined).

Table 4.5 How to Choose Carbohydrates Wisely
Choose more whole grains
• Substitute brown rice, wild rice, bulgur, or quinoa for white rice.
• Make sandwiches with whole wheat rather than white bread.
• Add legumes such as kidney, black, and pinto beans to casseroles and salads.
• Add barley to soups and stews.
• Choose packaged foods that contain 10% or more of the Daily Value for fibre.
• When baking at home, substitute whole-wheat flour for white or unbleached flour.
• Eat whole-grain breakfast cereals, such as Wheaties, Shredded Wheat, Grape Nuts, muesli, and oatmeal.
• Substitute whole-grain rolls, tortillas, and crackers for those made from refined grains.
• Substitute whole-wheat pasta or pasta made from 50% whole wheat and 50% white flour for conventional pastas.
Limit added sugars
• When cooking at home, use less sugar; try adding one-fourth less sugar than called for in the recipe.
• Use less added sugar in coffee and tea and on cereals and pancakes.
• Eat fewer high-sugar prepared foods such as cookies, cakes, and candies.
• Snack on fruit. If fresh fruits are not available, choose frozen or canned fruits without added sugar.
• Read food labels to choose foods low in added sugars.

Alternative Sweeteners

A love of sweets and the bad press surrounding sugar have driven the technological development of an increasing number of alternative or artificial sweeteners. These sugar substitutes, which provide little or no energy, are added to a host of low-kcalorie and "light" foods such as yogurts, ice creams, and soft drinks. Although many sugar substitutes are technically not carbohydrates, they were developed to replace simple sugars in food products or as an alternative for table sugar at home. Alternative sweeteners consumed in reasonable amounts are generally safe for healthy people;[52] however, to assure that they are not misused, Health Canada has defined acceptable daily intakes (ADIs)—levels that should not be exceeded when using these products. The ADI is an estimate of the amount of the sweetener per kilogram of body weight that an individual can safely consume every day over a lifetime with minimal risk.

The Role of Alternative Sweeteners in the Diet Replacing foods high in added sugars with foods sweetened with sugar substitutes will cut down on kcalories and decrease sugar intake, but it will not increase the intake of whole grains or fresh fruits and vegetables—key components of a healthy diet. Because foods that are high in added sugar tend to be nutrient-poor choices, replacing them with artificially sweetened alternatives does not increase the nutrient density of the diet. However, these products can be used in moderation if they are part of a healthy diet based on whole grains, vegetables, and fruits.

Alternative sweeteners have been shown to reduce the incidence of dental caries and can be helpful for managing blood sugar levels in diabetes, but their usefulness for weight loss has been more controversial. The rising consumption of sugared beverages has been blamed for

CRITICAL THINKING:

Becoming Less Refined

Background

Emma thinks that a good diet is important, and is concerned that she eats too much added sugar and not enough fibre. She records her food intake for a day and calculates its nutrient content. Her fibre and sugar intake are shown below.

Diet Analysis

Emma's diet analysis indicates that her daily diet provides about 2,340 kcal, 67 g of protein, 80 g of fat, 350 g of carbohydrate of which 200 g is sugars, and 19.6 g of fibre.

CURRENT DIET

CURRENT DIET

FOOD	SERVING	FIBRE (G)	SUGARS (G)
Breakfast			
White bread with jelly and margarine	2 slices	1.2	2
	15 ml	0.1	6
	5 ml	0	0
Fruit punch	250 ml	0	22
Lunch			
Macaroni and cheese	250 ml	1.4	8
Milk	250 ml	0	12
Apple	1 medium	3.7	18
Snack			
Soft drink	600 ml	0	63
3 Musketeers bar	1 regular	1	34
Dinner			
Roast beef	75 g	0	0
Flour tortillas	1,20-cm size	1.4	1
Pinto beans	250 ml	10	1
Snack			
Ice cream	175 ml	0.8	17
Ice cream with cherry syrup	30 ml	0	16
Total		**19.6**	**200**

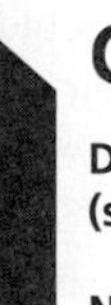

Critical Thinking Questions

Determine whether Emma eats enough carbohydrate by calculating the percent of kcalories from carbohydrate in her diet (see Table 4.2).

Now consider her fibre intake. Does it meet recommendations? If not, list specific changes she could make in her diet to increase her fibre intake to the recommended level.

What about added refined sugars? Which food adds the most sugar to her diet? Identify other foods in her diet that are high in added sugar. Suggest foods she could substitute for these to reduce her added sugar intake.

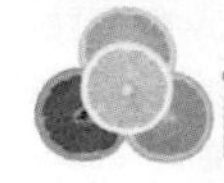

iProfile Use iProfile to find substitutions for her high-sugar, low-fibre snacks.

the increasing prevalence of obesity[2] so if individuals trying to lose weight replace sugar and high-sugar foods, such as soft drinks, with artificially sweetened products, they will lower their kcalorie intake. Short-term studies show that this lower energy intake can promote a reduction in body weight, but it is not possible to draw conclusions about the benefits for long-term weight maintenance.[53] Although alternative sweeteners can help reduce kcalorie intake, on their own they are not the solution to the obesity epidemic.

CANADIAN CONTENT

Types of Alternative Sweeteners The main artificial sweeteners which can be added to foods by manufacturers in Canada are aspartame, sucralose, and acesulfame K (acesulfame

Figure 4.33
Consumers often recognize their favourite sugar substitutes by the distinctive colours of the packaging. (© Andy Washnik)

potassium) (**Figure 4.33**). These are used alone or in combination to sweeten a variety of foods.

Stevia, a sweetener 300 times sweeter than sugar, and relatively new on the market, is made from the stevia plant and is regulated in Canada as a natural health product. It is not approved for use as a sweetener which can be added to food. The consumer, however, can purchase it as a herbal product in health food stores.[54] Two artificial sweeteners, cyclamate and saccharin, were implicated in the late 1960s and early 1970s as causing cancer in animal experiments; the animals received extremely high doses. The Canadian government responded by banning the use of these two sweeteners in food, but because of the lack of any alternative noncaloric sweeteners at the time, permitted their limited sale. Cyclamate could be purchased as a tabletop sweetener, while saccharin could be purchased in pharmacies. Recently, Health Canada began a re-evaluation of the scientific evidence of saccharin's safety with the possible intention of reinstating it as a food additive. This was considered because the early cancer studies on animals were subsequently judged to be irrelevant to humans.[55]

Aspartame In 1965, James Schlatter, at the pharmaceutical company G. D. Searle, was working with a chemical made up of two amino acids, aspartic acid and phenylalanine, when he spilled some of the chemical on his fingers. Shortly afterward, he licked his finger to pick up a piece of paper and discovered an intensely sweet taste. This accidental discovery led to the development of the artificial sweetener aspartame. Since aspartame is made of amino acids, the building blocks of protein, it is not a carbohydrate. Approved for some uses in the 1980s, aspartame is used in chewing gum, breakfast cereals, fruit spreads, yogurt, and beverages. Because aspartame breaks down when heated, it works best in products that are not cooked. Common trade names for this sweetener include NutraSweet, Equal, and NutriTaste. Each gram of aspartame contains 4 kcal, but since it is about 200 times as sweet as sugar, only 1/200th as much of it is needed to achieve the same level of sweetness.

As with other artificial sweeteners, safety concerns have been raised about aspartame. It contains the amino acid phenylalanine and, therefore, can be dangerous to individuals with a genetic disorder called phenylketonuria (PKU). These individuals have an abnormality that affects the metabolism of phenylalanine, and must restrict their intake of this amino acid to prevent brain damage (see Section 6.5). For this reason the warning "Aspartame contains phenylalanine" must appear on the packaging of foods that contain aspartame **(Figure 4.34).** There is also a concern that consuming aspartame might cause dangerously high blood phenylalanine levels in the general public. Phenylalanine occurs naturally in protein. A 120 g (4-ounce) hamburger has 12 times more phenylalanine than a 355 ml aspartame-sweetened soft drink. However, when phenylalanine is ingested without the other amino acids found in high-protein foods, blood and brain levels increase to a greater extent. Headaches, dizziness, seizures, nausea, allergic reactions, and other side effects have been reported following ingestion of aspartame; however, double-blind placebo-controlled studies have not been able to reproduce these symptoms.[52] There has also been concern that the use of aspartame might be associated with an increased risk of brain cancer in children, but controlled studies found no evidence that aspartame is a carcinogen or that there is a correlation between aspartame use and brain cancer incidence.[56] Overall, the consensus of the scientific community is that aspartame is safe for most people.

Each piece contains: SORBITOL, MALTITOL, GUM BASE, XYLITOL, CALCIUM CARBONATE, MANNITOL, VEGETABLE DERIVED GLYCERIN, GUM ARABIC, ASPARTAME (7.3 mg), ACESULFAME-POTASSIUM (3.5 mg), SOY LECITHIN, CANDELILLA WAX, TALC, NATURAL AND ARTIFICIAL FLAVOURS, COLOUR.
ASPARTAME CONTAINS PHENYLALANINE.

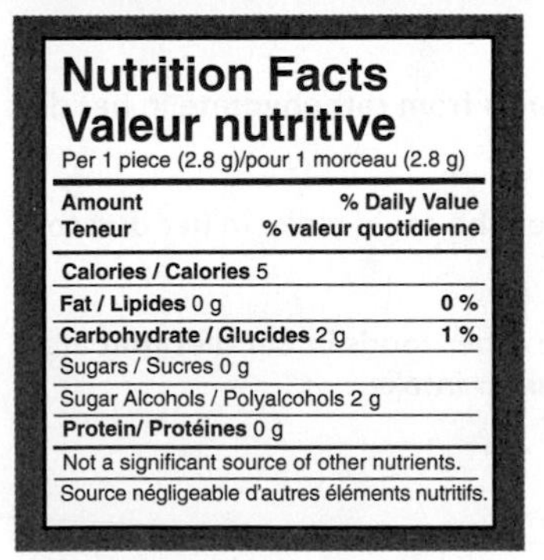

Nutrition Facts
Valeur nutritive
Per 1 piece (2.8 g)/pour 1 morceau (2.8 g)

Amount Teneur	% Daily Value % valeur quotidienne
Calories / Calories 5	
Fat / Lipides 0 g	0 %
Carbohydrate / Glucides 2 g	1 %
Sugars / Sucres 0 g	
Sugar Alcohols / Polyalcohols 2 g	
Protein/ Protéines 0 g	
Not a significant source of other nutrients. Source négligeable d'autres éléments nutritifs.	

Figure 4.34 Sugar alcohols on food labels
The Nutrition Facts panel on a label from sugar-free gum shows the total amount of *sugar alcohols* in each piece of gum. The ingredients include the three sugar alcohols used (highlighted in yellow). Note that the amounts of artificial sweeteners aspartame and acesulfame-potassium are shown (highlighted in green) as well as the warning: "Aspartame contains phenylalanine."

Health Canada has set an ADI of 40 mg of aspartame per kg of body weight. A packet of sweetener contains about 37 mg of aspartame.[57] A 355-ml soft drink sweetened with aspartame contains about 225 mg. To exceed the ADI, a 70-kg adult would have to consume almost 12 cans of aspartame-sweetened soft drinks a day and a 35-kg child would have to consume almost 6 cans of soft drinks.[52]

Sucralose Sucralose (trichlorogalactosucrose) was discovered in 1976 and is the only nonkcaloric sweetener made from sugar. To make this sweetener, the sugar molecule is modified so it cannot be digested and passes through the digestive tract unchanged. Approved for use in Canada, it is about 600 times sweeter than sucrose. It is sold as Splenda and can be used as a tabletop sweetener that is added directly to foods, and since it is heat stable, it can be used in baked goods, too.[52] Sucralose is used in beverages, chewing gum, frozen desserts, puddings, jams and jellies, syrups, and many other products, and has been extensively tested and found to be safe, even for children and pregnant and lactating women.[58] The ADI is 9 mg per kg of body weight; one packet contains 12 mg of sucralose.[57]

Acesulfame K Acesulfame potassium, or acesulfame K, is 200 times as sweet as sugar and provides no energy. It is used in chewing gum, powdered drink mixes, gelatins, puddings, soft drinks, and nondairy creamers. It is heat stable, so it can be used in baking. The ADI has been set at 15 mg per kg of body weight[57]; a packet of sweetener contains about 50 mg of acesulfame K.[52]

Neotame Neotame is a sugar substitute that is 7,000-13,000 times sweeter than sucrose. Like aspartame, it is made from the amino acids aspartic acid and phenylalanine, but the bond between the amino acids is harder to break than the bond in aspartame, so it is more stable. Since it cannot be broken down, releasing phenylalanine, it is not a problem for people with PKU. Health Canada approved the use of neotame as a general-purpose sweetener in 2007. Because it does not break down when heated, it can be used in both cooking and baking applications as well as in foods and beverages that are not heated.

Sugar Alcohols **Sugar alcohols**, also called polyols, such as sorbitol, mannitol, lactitol, and xylitol, are chemical derivatives of sugar that are used as low-kcalorie sweeteners. Because they are not digested, absorbed, or metabolized to the same extent as monosaccharides and disaccharides, they generally provide less energy than sucrose. Maltitol provides 3 kcal per gram, lactitol 2 kcal per gram, and erythritol only 0.2 kcal per gram.

sugar alcohols Sweeteners that are structurally related to sugars but provide less energy than monosaccharides and disaccharides because they are not well absorbed.

Sugar alcohols are not monosaccharides or disaccharides, so they can be used in products labeled "sugar free." The grams of sugar alcohols in a serving must be listed in the Nutrition Facts portion of the food label under carbohydrates (Figure 4.34). Products sweetened with sugar alcohols, such as chewing gums, candies, ice creams, and baked goods, may carry the health claim statement that they do not promote tooth decay. These products are less likely to promote tooth decay because the bacteria in the mouth cannot metabolize sugar alcohols as rapidly as sucrose. Consumption of large amounts of sugar alcohols (more than 50 g of sorbitol or 20 g of mannitol per day) can cause diarrhea.

CASE STUDY OUTCOME

Although Pearl lost weight when she consumed a low-carbohydrate diet, she wasn't able to stick with it for very long—a common complaint. Low-carb diet plans are based on the premise that high-carbohydrate foods cause a sharp rise in blood sugar and subsequently a rise in insulin, thereby promoting the storage of body fat. The restrictions of her diet eliminated virtually all choices from the grains and fruits food groups. She missed snacking on fruit, having sandwiches for lunch, eating pasta and rice at dinner, and enjoying cakes and cookies for dessert. She was always lightheaded and craved sweets.

When Pearl stopped her low-carb diet and went back to her old ways of eating, she began to regain weight. Fortunately, she started using *Eating Well with Canada's Food Guide* to choose the right types and amounts of carbohydrates. In addition to recognizing that she needed to consume more whole-grain products, Pearl soon learned that she also had to take note of the additional statements in Canada's Food Guide. These advised her to read food labels carefully and to choose foods lower in sugar (as well as fat and salt). As a result, she was able to identify, and add to her diet, high-fibre cereals that did not contain large amounts of added sugar. She learned that among the foods in the meat and alternatives group, beans and lentils were also high in fibre. She also made a conscious effort to regularly include low-glycemic-index foods, such as yogurt, barley, oatmeal, sweet potato, and beans, in her diet. These foods limit the rate of glucose absorption and hence dampen the rise in blood sugar and insulin. In combination with fruits and vegetables which Pearl had also begun to eat regularly, her new diet was flavourful and varied, so she was able to more easily stick with it in the long term. By watching portion sizes, Pearl lost 4 pounds in 2 months.

APPLICATIONS

Personal Nutrition

iProfile

1. **How much carbohydrate is in your diet?**
   a. Use iProfile and the 3-day diet record you kept in Chapter 2 to calculate your average carbohydrate and energy intake.
   b. What is the percent of energy from carbohydrate in your diet?
   c. How does your percentage of carbohydrate intake compare with the recommended 45-65% of energy from carbohydrate?
2. **Look at the sources of carbohydrate in your diet.**
   a. Are most of your carbohydrate choices from unrefined or refined sources?
   b. Suggest some changes that would increase your intake of unrefined carbohydrates.
   c. List some foods in your diet that are high in added, refined sugars.
   d. Suggest some changes that would reduce the amount of added sugar in your diet.
3. **How much fibre is in your diet?**
   a. Use iProfile to calculate the grams of fibre in your original diet.
   b. Does your intake meet the AI for fibre for someone of your age and gender?
   c. How do the changes you suggested in question 2b affect the fibre content of your diet? If your modified diet still does not meet fibre recommendations, how might you further increase your intake?

General Nutrition Issues

1. **Cheryl is trying to increase her fibre intake. For lunch, she typically orders a sandwich from a local deli. Her favourite is turkey on a Kaiser roll with lettuce and mayo. Below, column A lists the types of bread available, column B lists the fillings, and column C lists the vegetables that can be added to the sandwiches. Can you suggest three different sandwiches for Cheryl that would provide more fibre than her typical sandwich?**

COLUMN A	COLUMN B	COLUMN C
Kaiser roll	Sliced turkey	Lettuce
Sourdough bread	Soyburger	Tomatoes
Whole-wheat bread	Ham	Green and red peppers
Rye bread	Hummus	Onions
Sesame bagel	Falafel	Cucumber
Pumpernickel bread	Tuna salad	Olives
Pita bread	Peanut butter	Pickles
Oat bread	Grilled eggplant	Alfalfa sprouts

2. **Bob weighs about 14 kg (30 pounds) more than he wants to weigh, so he decides to try to shed weight quickly with a low-carbohydrate weight-loss diet. The diet allows an unlimited amount of beef, chicken, and fish as well as limited fruits and vegetables; breads, grains, and cereals are not allowed. Bob is overjoyed with his initial rapid weight loss, but after about a week, his weight loss slows down and he begins to feel tired and light-headed. He is having headaches and notices a funny smell on his breath. A nutritional assessment suggests that Bob needs about 2,500 kcal a day to maintain his weight. His weight-loss diet provides about 1,000 kcal, 25 g of carbohydrate, 125 g of protein, and 44 g of fat per day. He consumes about 750 ml of fluid daily.**
   a. Explain why Bob is tired, light-headed, and has headaches and an unusual odour on his breath.
   b. What recommendations do you have to reduce these symptoms?
3. **Go to a bookstore or the library and look up a sample 1-day menu from a diet book that advocates a low-carbohydrate intake. Enter these foods into the iProfile diet analysis computer program.**
   a. How does this diet compare to the recommended intakes for saturated fat?
   b. For calcium?
   c. For fibre?
4. **Using the Canadian Diabetes Association Just the Basics as a guide, list the foods you might choose to fill your plate for a breakfast, lunch, and supper. Be sure that you choose one low-glycemic-index food for each meal.**

SUMMARY

4.1 Carbohydrates in the Canadian Diet

- Some carbohydrates in our diets are from unrefined whole foods such as whole grains, fruits, and vegetables. Others are from refined grain products like white breads and baked goods and added sugars such as those found in candies and soft drinks.
- Whole grains include the entire grain kernel, which is rich in fibre, micronutrients, and phytochemicals.
- Added refined sugars provide energy but few nutrients, so they reduce the nutrient density of the diet.

4.2 Chemistry of Carbohydrates

- Carbohydrates are chemical compounds that contain carbon, hydrogen, and oxygen.
- Simple carbohydrates include monosaccharides and disaccharides and are found in foods such as table sugar, honey, milk, and fruit.
- Complex carbohydrates include oligosaccharides and polysaccharides. Glycogen is a polysaccharide found in animals, and starch and soluble and insoluble fibre are polysaccharides found in plants.

4.3 Carbohydrates in the Digestive Tract

- Sugars and starches consumed in food are broken down in the digestive tract to monosaccharides, which can be absorbed into the blood stream.
- Lactose intolerance occurs when the enzyme lactase is not available in sufficient quantities to digest lactose. Undigested lactose passes into the colon where it draws in water and is metabolized by bacteria, producing gas and acids and causing abdominal distension, gas, cramping, and diarrhea.
- Fibre, some oligosaccharides, and resistant starch are carbohydrates that are not broken down by human digestive enzymes in the stomach and small intestine and therefore pass into the colon. These indigestible carbohydrates benefit health by increasing the amount of water and bulk in the intestine, which stimulates gastrointestinal motility; promoting the growth of a healthy microflora; and slowing nutrient absorption.

4.4 Carbohydrates in the Body

- In the body, carbohydrate provides a source of energy—4 kcal per gram. It also provides other essential roles in cell communication and as part of RNA and DNA.
- Glucose is metabolized through cellular respiration, which begins with glycolysis or anaerobic metabolism. Glycolysis breaks each 6-carbon glucose molecule into 2 3-carbon pyruvate molecules, producing ATP even when oxygen in unavailable. When oxygen is available, aerobic metabolism can proceed. Pyruvate loses a carbon as carbon dioxide to form acetyl-CoA, which is then broken down by the citric acid cycle to form 2 carbon dioxide molecules. Electrons released at each step pass to the electron transport chain, where their energy is used to generate ATP and water is formed.
- Several tissues, including the brain and red blood cells, require glucose as an energy source. When glucose is not available, it can be obtained from the breakdown of glycogen or synthesized from amino acids by the process of gluconeogenesis.
- Carbohydrate is important for fatty acid metabolism because it is needed for acetyl-CoA to enter the citric acid cycle. When carbohydrate is limited, acetyl-CoA is used to make ketones.

4.5 Blood-Glucose Regulation

- Blood-glucose levels are maintained within normal limits by the hormones insulin and glucagon. When blood glucose rises, insulin is released from the pancreas to allow body cells to take up the glucose. When blood glucose falls, glucagon is released to increase blood glucose.
- Blood glucose rises after eating. How quickly and how high blood glucose rises is referred to as glycemic response. Glycemic response can be quantified by measuring glycemic index and glycemic load.
- Diabetes is an abnormality in blood sugar regulation resulting in high blood-glucose levels, which damage tissues and cause complications including heart disease, kidney failure, blindness, and the need for amputations. This occurs either because insufficient insulin is produced or because there is a decrease in the sensitivity of body cells to insulin. Treatment to maintain glucose in the normal range includes diet, exercise, and medication.
- Hypoglycemia is a condition in which blood glucose falls to abnormally low levels, causing symptoms such as sweating, headaches, and rapid heartbeat.

4.6 Carbohydrates and Health

- When carbohydrates—particularly simple carbohydrates—remain in contact with the teeth, they increase the risk of dental caries.
- Consumption of low-glycemic-index foods can reduce glycemic response, enhance satiety, and help maintain a healthy weight. When consumed in excess of needs, carbohydrates contribute to weight gain. Low-carbohydrate diets cause ketosis and have been found to cause weight loss in the short term but compliance and health risks may limit their effectiveness in the long term.

- Carbohydrate-containing foods, with a high glycemic index, cause a larger glycemic response than food, with a low glycemic index, and are associated with an increased risk of type 2 diabetes.
- High-sugar diets can increase heart disease risk by raising blood lipids. Foods containing soluble fibre may help to lower cholesterol levels.
- Indigestible carbohydrates make the stool larger and softer. This reduces the pressure needed to move material through the colon, lowering the risk of hemorrhoids and diverticular disease.
- Cancer cells differ from normal cells because they divide without restraint and are able to grow in areas reserved for other cells. Cells in the colon may be exposed to carcinogens in the colon contents. Fibre may help reduce the risk of colon cancer by decreasing the amount of contact between the cells lining the colon and these carcinogenic substances.

4.7 Meeting Carbohydrate Recommendations

- Guidelines for healthy diets recommend 45-65% of energy from carbohydrates and a fibre intake of 38 g/day for men and 25 g/day for women. Food labels list Daily Value recommendations based on the amounts needed in a 2,000 kcal diet: 300 g of carbohydrate and 25 g of fibre.
- To meet carbohydrate recommendations foods such as whole grains, legumes, fruits, and vegetables, and milk can be consumed. Foods high in added sugars, such as baked goods, candy, and soft drinks, should be limited.
- Alternative sweeteners can be used to reduce the amount of added sugar in the diet. They do not contribute to tooth decay and can help keep blood sugar in the normal range. They can also be used by dieters to reduce the energy content of the diet.

REVIEW QUESTIONS

1. What foods are good sources of unrefined complex carbohydrates? Unrefined simple carbohydrates?
2. What is the basic unit of carbohydrate?
3. List three common simple carbohydrates. Where are they found in the diet or in the body?
4. Describe three types of complex carbohydrates.
5. Why is added sugar considered a source of empty kcalories?
6. How much energy is provided by a gram of carbohydrate?
7. Explain how fibre affects gastrointestinal health.
8. Describe what happens during the process of glycolysis.
9. What are the end products of cellular respiration? During which step is each produced?
10. Explain why carbohydrate is said to spare protein.
11. What is the main function of glucose in the body?
12. What is diabetes and what are the long-term complications of this disease?
13. Why is ketoacidosis a problem only in type 1 diabetes?
14. What health benefits are associated with a diet high in unrefined carbohydrates?
15. How can you use the information on food labels to help you identify foods that are high in added sugars? In fibre?
16. What are the risks and benefits of alternative sweeteners?

REFERENCES

1. Canadian Sugar Institute: Estimates of Sugars Consumption in Canada. Available online at http://www.sugar.ca/english/health professionals/sugarsconsumption.cfm#b1. Accessed Feb 1, 2010.
2. Bray, G. A., Nielsen, S. J., and Popkin, B. M. Consumption of high-fructose corn syrup in beverages may play a role in the epidemic of obesity. *Am. J. Clin. Nutr*. 79:537–543, 2004.
3. Newburg, D. S., Ruiz-Palacios, G. M., Morrow, A. L. Human milk glycans protect infants against enteric pathogens. *Annu. Rev. Nutr*. 25:37–58, 2005.
4. Institute of Medicine. Food and Nutrition Board. Dietary Reference Intakes for Energy, Carbohydrates, Fiber, Fat, Protein and Amino Acids. Washington, DC: National Academies Press, 2002.
5. Swallow, D. M. Genetics of lactase persistence and lactose intolerance. *Ann. Rev. Genetics*. 37:197–219, 2003.
6. The National Digestive Diseases Information Clearinghouse (NDDIC), National Institute of Diabetes and Digestive and Kidney Diseases (NIDDK), National Institutes of Health of the U.S. Department of Health and Human Services. Lactose intolerance. Available online at http://digestive.niddk.nih.gov/ddiseases/pubs/ lactoseintolerance. Accessed March 18, 2008.
7. Blaut, M. Relationship of prebiotics and food to intestinal microflora. *Eur J. Nutr*. 41(Suppl 1):II1-II6, 2002.
8. Topping, D. L., Fukushima, M., and Bird, A. R. Resistant starch as a prebiotic and synbiotic: State of the art. *Proc Nutr. Soc*. 62:171–176, 2003.
9. Andoh, A., Tsujikawa, T., and Fujiyama, Y. Role of dietary fibre and short-chain fatty acids in the colon. *Curr. Pharm. Des*. 9:347–358, 2003.
10. Jenkins D. J., Wolever, T. M., Taylor, R. H.. Glycemic index of foods: a physiological basis for carbohydrate exchange. *Am. J. Clin. Nutr.* 34(3):362-6,1981.
11. Wolever, T. M., Yang, M., Zeng, X. Y., et al. Food glycemic index, as given in glycemic index tables, is a significant determinant of glycemic responses elicited by composite breakfast meals. *Am. J. Clin. Nutr.* 83(6):1306–12, 2006.
12. Wolever, T. M. 1998. Chapter 4: The role of glycemic index in food choice. Carbohydrates in human nutrition . FAO Food and Nutrition Paper 66. Available online at http://www.fao.org/docrep/W8079E/ w8079e0a.htm#definition of glycemic index (gi). Accessed May 7, 2011.
13. Foster-Powel, K., Holt, S. H., and Brand-Miller, J. C. International table of glycemic index and glycemic load values: 2002. *Am. J. Clin. Nutr*. 76:5–56, 2002.
14. The National Institute of Diabetes and Digestive and Kidney Diseases NIDDK National Diabetes Statistics, 2007. Available online at http:// diabetes.niddk.nih.gov/ dm/pubs/statistics/#allages. Accessed April 9, 2009.
15. National Diabetes Education Program. About diabetes and pre-diabetes. Available online at ndep.nih.gov/ diabetes/diabetes.htm. Accessed February 15, 2005.

16. Dyck, R., Osgood, N., Lin, T.H., Gao, et al. Epidemiology of diabetes mellitus among First Nations and non-First Nations adults.*CMAJ*. 182:249–562, 2010.

17. Canadian Diabetes Association Clinical Practice Guidelines. Available online at http://www.diabetes.ca/for-professionals/resources/2008-cpg/. Accessed March 14, 2011.

18. Walker, C., and Reamy, B. V. Diets for cardiovascular disease prevention: What is the evidence? *Am. Fam. Physician*. 79:571–578, 2009.

19. National Institutes of Mental Health. Attention Deficit Hyperactivity Disorder. NIH Publication No. 3572. Available online at http:// www.nimh.nih.gov/Publicat/ ADHD.cfm/. Accessed January 24, 2006.

20. Jurgens, H., Haass, W., Castenada, T. R., et al. Consuming fructose beverages increases body adiposity in mice. *Obes. Res*. 13:1146–1156, 2005.

21. Office of the Surgeon General. The Surgeon General's Call to Action to Prevent and Decrease Overweight and Obesity. Rockville, MD: U.S. Dept. of Health and Human Services, 2001. Available online at www.surgeongeneral. gov/topics/obesity/. Accessed March 4, 2009.

22. Foster, G. D., Wyatt, H. R., Hill, J. O., et al. A randomized trial of a low-carbohydrate diet for obesity. *N. Engl. J. Med*. 348:2082–2090, 2003.

23. Sacks, F. M., Bray, G. A., Carey, V. J., et al. Comparison of weight-loss diets with different compositions of fat, protein, and carbohydrates. *N. Engl. J. Med*. 360:859–873, 2009.

24. Buchholz, A. C., and Schoeller, D. A. Is a calorie a calorie? *Am. J. Cl in. Nutr*. 79:899S–906S, 2004.

25. Wilkinson, D. L., and McCargar, L. Is there an optimal macronutrient mix for weight loss and weight maintenance? *Best Pract. Res. Gastroenterol*. 18:1031–1047, 2004.

26. Thomas, D. E., Elliott, E. J., Baur, L. Low glycaemic index or low glycaemic load diets for overweight and obesity. *Cochrane Database Syst Rev.* 18;(3):CD005105, 2007.

27. Gross, L. S., Li, L., Ford, E. S. Increased consumption of refined carbohydrates and the epidemic of type 2 diabetes in the United States: an ecologic assessment. *Am. J. Clin. Nutr*. 79:774–779, 2004.

28. Van Dam, R. M., Rimm, E. B., Willett, W. C., et al. Dietary patterns and risk for type 2 diabetes mellitus in US men. *Ann. Intern. Med*. 136:201–109, 2002.

29. Willett, W., Manson, J., and Liu, S. Glycemic index, glycemic load, and risk of type 2 diabetes. *Am. J. Clin. Nutr*. 76:274S–280S, 2002.

30. Schulze, M. B., Liu, S., Rimm, E. B., et al. Glycemic index, glycemic load, and dietary fiber intake and incidence of type 2 diabetes in younger and middle-aged women. *Am. J. Clin. Nutr*. 80:348–356, 2004.

31. Schulze, M. B., Manson, J. E., Ludwig, D. S., et al. Sugar-sweetened beverages, weight gain, and incidence of type 2 diabetes in young and middle-aged women. *JAMA* 292:927–934, 2004.

32. Fried, S. K., and Rao, S. P. Sugars, hypertriglyceridemia, and cardiovascular disease. *Am. J. Clin. Nutr*. 78:873S– 880S, 2003.

33. Jacobs, D. R., Meyer, K. A., Kushi, L. H., et al. Whole-grain intake may reduce the risk of ischemic heart disease death in postmenopausal women: The Iowa Women's Health Study. *Am. J. Clin. Nutr*. 68:248–257, 1998.

34. Kushi, L. H., Meyer, K. A., and Jacobs, D. R., Jr. Cereals, legumes, and chronic disease risk reduction: Evidence from epidemiologic studies. *Am. J. Clin. Nutr*. 70(Suppl): 451S–458S, 1999.

35. Steffen, L. M., Jacobs, D. R. Jr., Stevens, J., et al. Associations of whole-grain, refined-grain, and fruit and vegetable consumption with risks of all-cause mortality and incident coronary artery disease and ischemic stroke: The Atherosclerosis Risk in Communities (ARIC) Study. *Am. J. Clin. Nutr*. 78:383–390, 2003.

36. Anderson, J. W. Whole grains protect against atherosclerotic cardiovascular disease. *Proc. Nutr. Soc*. 62:135–142, 2003.

37. Anderson, J. W., Hanna, T. J., Peng, X., et al. Whole grain foods and heart disease risk. *J. Am. Coll. Nutr*. 19(3 Suppl):291S–299S, 2000.

38. Institute of Medicine Food and Nutrition Board, Dietary Reference Intakes: Proposed Definition of Dietary Fibre. Washington, DC: National Academies Press, 2001.

39. Marlett, J. A. Sites and mechanisms for the hypocholesterolemic actions of soluble dietary fibre sources. In: Kritevsky, D., and Bonfield, C., eds. *Fibre in Human Health and Disease*. New York: Plenum Press, 1997, pp. 109–121.

40. Higgins, J. A. Resistant starch: Metabolic effects and potential health benefits. *J. AOAC Int*. 87:761–768, 2004.

41. Marlett, J. A., McBurney, M. I., and Slavin, J. L. Position of the American Dietetic Association: Health implications of dietary fibre. *J. Am. Diet. Assoc*. 102:993–1000, 2002.

42. National Digestive Diseases Information Clearinghouse. NIDDK, NIH. Diverticulosis and Diverticulitis. Available online at digestive.niddk.nih.gov/ddiseases/pubs/ diverticulosis/. Accessed October 27, 2005.

43. Peters, U. N., Sinha, R., Chatterjee, N., et. al. Dietary fibre and colorectal adenoma in a colorectal cancer early detection programme. *Lancet*. 361:1491–1495, 2003.

44. Bingham, S. A., Day, N. E., Luben, R., et al. Dietary fibre in food and protection against colorectal cancer in the European Prospective Investigation into Cancer and Nutrition (EPIC): An observational study. *Lancet* 361:1496–1501, 2003.

45. Alberts, D. S., Martinez, M. E., Roe, D. J., et al. Lack of effect of a high-fibre cereal supplement on the occurrence of colorectal adenomas. *N. Engl. J. Med*. 342:1156–1162, 2000.

46. Schatzkin, A., Lanza, E., Corle, D., et al. Lack of effect of a low-fat, high-fibre diet on the recurrence of colorectal adenomas. *N. Engl. J. Med*. 342:1149–1155, 2000.

47. Park, Y., Hunter, D. J., Spiegelman, D., et al. Dietary fiber intake and colorectal cancer: A pooled analysis of prospective cohort studies. *JAMA* 294:2849–2857, 2005.

48. Garriguet D. Canadians' eating habits. *Health Rep*.18:17–32, 2007.

49. Health Canada. Dietary Reference Intake Tables. Available online at http://www.hc-sc.gc.ca/fn-an/nutrition/reference/table/index-eng.php. Accessed February 28, 2011.

50. WHO. Diet and Chronic Disease Prevention Report of the WHO FAO Expert Commission Geneva 2002, Available online at http://whqlibdoc.who.int/trs/WHO_TRS_916.pdf. Accessed April 10, 2009.

51. US Dept. of Health and Human Services, US Dept. of Agriculture. *Dietary Guidelines for Americans*, 6th ed. Washington, DC, 2005.

52. American Dietetic Association. Position of the American Dietetic Association: Use of nutritive and non-nutritive sweeteners. *J. Am. Diet. Assoc*.104:255–275, 2004.

53. Vermunt, S. H., Pasman, W. J., Schaafsma, G., et al. Effects of sugar intake on body weight: A review. *Obes Rev*. 4:91–99, 2003.

54. Health Canada. Revised Guidelines for the use of stevia in natural health products. Available online at http://www.hc-sc.gc.ca/dhp-mps/prodnatur/legislation/docs/notice-avis-stevia-eng.php. Accessed February 3, 2010.

55. Health Canada. Summary of input received on Health Canada's consultation to reinstate saccharin as a food additive. Available online at http://www.hc-sc.gc.ca/fn-an/securit/addit/sweeten-edulcor/saccharin-summary-eng.php. Accessed February 3, 2010.

56. Gurney, J. G., Pogoda, J. M., Holly, E. A., et al. Aspartame consumption in relation to childhood brain tumour risk: Results from a case-control study. *J. Natl. Cancer Inst*. 89:1072–1074, 1997.

57. Canadian Diabetes Association. Sugar and Sweeteners. Available online at http://www.diabetes.ca/files/en_sweeteners_final.pdf. Accessed Oct 2, 2011.

58. Calorie Control Council. Low calorie sweeteners: Sucralose. Available online at www.caloriecontrol.org/ sucralos.html/. Accessed June 2, 2004.

5 Lipids:

Case Study

Sam's grandfather died of a heart attack at age 50. Sam is a 20-year-old university student. He recently completed a heart disease risk assessment at a health fair and found that he was about 10 kg (25 lb) overweight, his percent body fat was higher than recommended, and his blood cholesterol was slightly elevated at 5.4 mmol/L (normal levels are less than 5.2 mmol/L). He was told that he should see a physician to evaluate his risk for a heart attack. How could a 20-year-old be at risk for heart disease, and how can he lower that risk?

Sam has begun to think about changes he could make in his diet and lifestyle. He eats a lot of red meat and has only one or two servings of fruits and vegetables a day. He drinks whole milk and snacks on ice cream every night. His only exercise is a Tuesday afternoon of Frisbee with friends and lifting weights on Friday nights.

When he tells his friends and family about his health concerns, everyone is quick to offer advice. His girlfriend, who is a vegetarian, recommends that he eliminate meat from his diet. His biology lab partner tells him to cut out all fat. His sister tells him about the Mediterranean diet and recommends he eat pasta with plenty of olive oil every night. His roommate tells him to eat more fish. His mother says he should stop using margarine because of all the *trans* fat it contains. Whose advice should Sam follow?

In this chapter, you will learn about the different types of fats found in the diet and how they affect human health. At the end of the chapter, you will be able to identify the most nutritious choices.

(Karen Pearson/Getty Images, Inc.)

(©iStockphoto)

(Peter Rees/StockFood America)

Chapter Outline

5.1 Fat in the Canadian Diet

Learning Objectives

- Name two qualities that fat adds to foods.
- List which foods are the most common sources of fat in the Canadian diet.
- Discuss how fat intake affects the risk of chronic disease.

lipids A group of organic molecules, most of which do not dissolve in water. They include fatty acids, triglycerides, phospholipids, and sterols.

Lipids, the chemical term for what we commonly call fats, contribute to the texture, flavour, and aroma of food. It's fat that gives ice cream, cheesecake, and cream cheese a smooth texture and rich taste. Sesame oil gives Chinese food its distinctive aroma and olive oil imparts a unique flavour to salads. In addition to texture, flavour, and aroma, fats add kcalories to our food and the amount and type we eat can affect our health. Fats and oils provide 9 kcal/g—more than twice as much as a gram of carbohydrate or protein. Because of the high-kcalorie content of fat, a high-fat diet can make it easier to consume excess energy and make it more difficult to keep weight in the healthy range.

The typical Canadian diet today contains fat from a variety of sources; some are more visible than others. The pat of butter melting on a hot baked potato or the layer of fat around the edge of a sirloin steak is obvious. Less obvious is the fat in baked goods; these help make cookies and crackers crisp, muffins soft, and piecrusts flaky. It is also difficult to recognize the fats that are naturally abundant in whole milk, avocados, and nuts, or the oil that stays with your potatoes after they are removed from the fryer (**Figure 5.1**).

CANADIAN CONTENT

Canada's Fat Intake

The Canadian Community Health Survey (CCHS) provided a detailed look at the fat intake of Canadians. It suggested that the amount of fat consumed by Canadians (as percent of kcalories) has declined over the last 25 years. In 1978, Canadians consumed about 40% of their kcalories from fat; by 2004, this had declined to an average of 31%. While this average was within the AMDR recommendations of 20%-35% kcal from fat, substantial numbers of Canadians were, nonetheless, consuming more than 35% kcal from fat and potentially increasing their risk of chronic disease. Intakes above 35% kcal were most common in the 35-50-year age group in both men and women (**Figure 5.2**).[1]

Figure 5.1 Potatoes absorb oil during frying.
A baked potato has no fat, but a medium order of fries provides about 17 g of fat—the equivalent of about 4 pats of butter. (Paul Poplis/FoodPix/Jupiter Images Corp.)

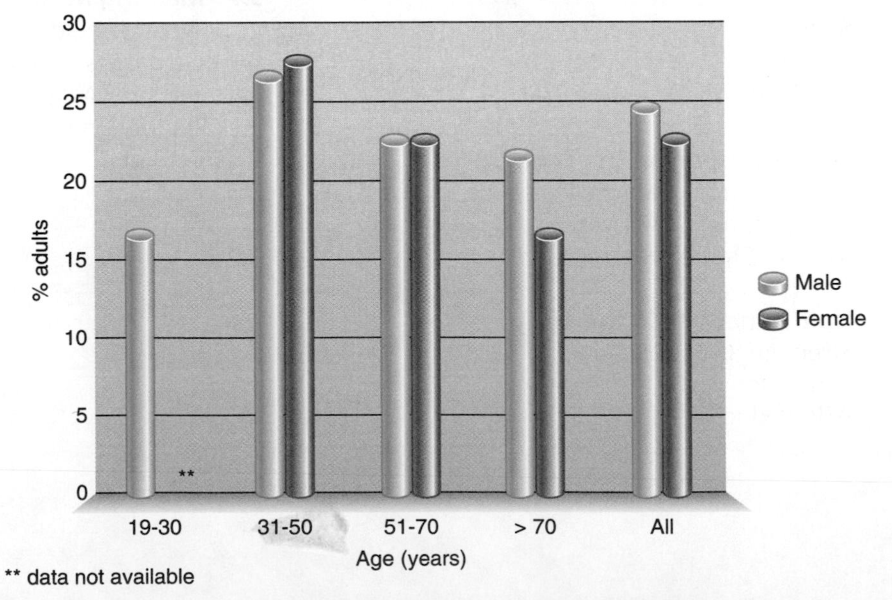

Figure 5.2 Proportion of Canadian adults consuming more that 35% kcal from fat.
Many Canadians exceed the maximum recommended intake of fat, which is 35% kcal.

Source: Health Canada Canadian Community Health Survey, Cycle 2.2, Nutrition (2004) Tables and figures. Articles on Canadians' Nutrient Intakes from Food. Available online at http://www.hc-sc.gc.ca/fn-an/surveill/nutrition/commun/art-nutr-table-eng.php. Accessed June 10, 2011.

Foods that most contribute to fat intake may not be the healthiest choices for Canadians. In terms of food groups, the meat and alternatives and milk and alternatives were the largest contributors, accounting for almost half of the fat intake of Canadian adults; they accounted for 31.6% and 17.9%, respectively, of total fat. A limited number of food items, such as pizzas, sandwiches, submarines, hamburgers, and hot dogs, as well as sweet baked goods, such as muffins, donuts, and cookies, accounted for almost a quarter of the total fat intake. These foods are commonly served in fast-food restaurants and as snacks. It is perhaps not surprising, then, that the CCHS also found that a quarter of Canadians reported that, on the day before their survey interview, they had consumed an item purchased in a fast-food outlet and that Canadians consumed more kcalories as snacks than they did for breakfast.[1]

How Fat Intake Affects Health

Dietary recommendations today recognize that the type of fat consumed is as important, or possibly more important, than the total amount of fat, when considering the risk of chronic diseases such as heart disease and cancer. Populations like Canada's, which have high intakes of fat from meats and dairy products, foods that are high in saturated fat and cholesterol, tend to have a higher incidence of heart disease and certain types of cancer than populations where intake of saturated fat and cholesterol is low. Diets high in processed fats used in margarine and in shortening, a fat commonly used in baked products, also increase risks, because they may contain large amounts of *trans* fats. On the other hand, populations where most of the fat in the diet is comprised of unsaturated fats from foods like fish, nuts, soybean, canola, flaxseed and olive oils seem to be at lower risk for disease. Eating a healthy diet means shifting the type of fats consumed to include more fish and plant-derived sources. As we shall see, these food sources contain more unsaturated fat, which has been linked to health benefits, and less saturated and *trans* fat, which have been linked to increased risk of human disease (**Figure 5.3**). Combining these foods with high-fibre whole grains, fruits, and vegetables promotes health (see Your Choice: Eating to Lower Cholesterol).

Figure 5.3 Fat recommendations for a healthy diet
Choose more fish, vegetable oils, and nuts; limit solid fats from shortening and animal products such as butter, ice cream, and fatty meats.

YOUR CHOICE

Eating to Lower Cholesterol Levels

Many Canadians take medications to reduce their blood cholesterol levels and thus their heart disease risk, but as you will learn in this chapter, diet strongly influences both cholesterol levels in the blood and the risk of heart disease. For example, a dietary pattern that includes soluble fibre, plant sterols, soy protein, and nuts has been found to be as effective at lowering blood cholesterol levels in some people as medication.[1] They are a dietary prescription for reducing the risk of heart disease.

When cholesterol is measured in the blood (or serum), total cholesterol is measured (i.e., all forms of cholesterols), as well as the cholesterol found in low density lipoproteins (LDL cholesterol) which are made in the liver and increase the risk of heart disease, and the cholesterol found in high density lipoproteins (HDL cholesterol), also made in the liver, which reduce the risk of cardiovascular disease. LDL cholesterol and HDL cholesterol are often referred to as "bad" and "good" cholesterol, respectively (Lipoproteins are discussed in more detail in Section 5.4).

Soluble Fibre, such as that in oatmeal and oat bran, lowers total and unhealthy LDL cholesterol levels.[2] The soluble fibre psyllium also lowers total and LDL cholesterol without affecting healthy HDL cholesterol.[3] So lower your cholesterol by eating oatmeal in the morning, apples at lunch, and beans and barley in your stew at dinner.

Flaxseed can lower LDL cholesterol without lowering HDL cholesterol. Bake these seeds in breads, muffins, and cookies, or grind them up and sprinkle them on cereal or yogurt. They provide soluble fibre and omega-3 fatty acids, both of which help protect against high blood lipids and cardiovascular disease.[4]

Soy Protein, when it replaces animal protein in the diet, lowers LDL cholesterol and either increases or has no effect on HDL cholesterol.[5] Although the decrease in LDL cholesterol is small, soy may also protect against heart disease because it is high in polyunsaturated fat, fibre, vitamins, and minerals, and low in saturated fat.[6] Try tofu in your stir-fry and soy milk on your cereal. Foods that are low in saturated and *trans* fat can include a health claim stating that the food may help reduce the risk of heart disease (see Section 2.5).

Plant Sterols and Stanols resemble cholesterol chemically, making it difficult for the digestive tract to distinguish them from cholesterol. They lower blood cholesterol by reducing cholesterol absorption.[7] You can get small quantities of these compounds in many fruits, vegetables, nuts, seeds, cereals, legumes, vegetable oils, and other plant foods. In May 2010, Health Canada approved the addition of plant sterols to food in Canada.[8] Look for a health claim on the label about the relationship between plant sterols, lowering blood cholesterol levels, and reduced risk of heart disease.

Nuts are high in monounsaturated fat, omega-3 fatty acids, antioxidants, and fibre, all of which provide cardiovascular protection. Epidemiological studies have shown that diets rich in nuts are associated with a lower incidence of cardiovascular disease,[4] so grab some nuts for a snack, bake them into breads, sprinkle them on salads, and mix them into stir-fries.

Fish, particularly fatty fish, such as mackerel, lake trout, herring, sardines, char, and salmon, can help reduce your risk of heart disease and protect against sudden cardiac death.[4] Eating fish affects blood lipids positively, stabilizes the heart electrically, decreases atherosclerotic plaque growth, and lowers blood pressure.[9] For this reason, Canada's Food Guide recommends 2 servings of fatty fish a week. The beneficial effects of fish are believed to be due to the presence of omega-3 fatty acids, eicosapentaenoic acid (EPA) and docosahexaenoic acid (DHA). (For more about fish, see Critical Thinking: Fish Consumption and Heart Disease.)

(© istockphoto.com/GMVozd)

Mercury contamination is a concern with fish consumption because the element can accumulate in fish muscle. Health Canada reports that most fish commonly consumed in Canada have very low mercury levels. The exceptions are fresh or frozen tuna, shark, swordfish, marlin, orange roughy, and escolar, and Canadians are advised to limit their intakes to the amounts posted on the Health Canada website.[10] Although fresh and frozen tuna are mentioned, canned light tuna, a commonly consumed fish has low levels of mercury and is not of concern. Canned albacore tuna, a more expensive form of tuna, has sufficient mercury that Health Canada recommends limits on intake for children and women who are planning to get pregnant, are pregnant or are breastfeeding.[10]

[1] Jenkins, D. J. A., Kendall, C. W. C., Marchie, A. et al. Effects of a dietary portfolio of cholesterol-lowering foods vs. lovastatin on serum lipids and C-reactive protein. *JAMA* 290:502–510, 2003.

[2] Brown, L., Rosner, B., Willett, W. W., and Sacks, F. M. Cholesterol-lowering effects of dietary fiber: A meta-analysis. *Am. J. Clin. Nutr.* 69:30–42, 1999.

[3] Anderson, J. W., Allgood, L. D., Lawrence, A. et al. Cholesterol-lowering effects of psyllium intake adjunctive to diet therapy in men and women with hypercholesterolemia: Meta-analysis of 8 controlled trials. *Am. J. Clin. Nutr.* 71:472–479, 2000.

[4] Mente, A., de Koning, L., Shannon, H. S., and Anand, S. S. A systematic review of the evidence supporting a causal link between dietary factors and coronary heart disease. *Arch. Intern. Med.* 169:659–669, 2009.

[5] Anthony, M. S. Soy and cardiovascular disease: Cholesterol lowering and beyond. *J. Nutr.* 130:662S–663S, 2000.

[6] Sacks, F. M., Lichtenstein, A., Van Horn, L. et al. Soy protein, isoflavones, and cardiovascular health: An American Heart Association Science Advisory for professionals from the Nutrition Committee. *Circulation* 113:1034–1044, 2006.

[7] Law, M. Plant sterol and stanol margarines and health. *BMJ* 320:861–864, 2000.

[8] Health Canada. Plant sterols and blood cholesterol lowering. Available online at http://www.hc-sc.gc.ca/fn-an/label-etiquet/claims-reclam/assess-evalu/phytosterols-eng.php. Accessed July 19, 2010.

[9] Holub, D. J., and Holub, B. J. Omega-3 fatty acids from fish oils and cardiovascular disease. *Mol. Cell. Biochem.* 263:217–225, 2004.

[10] Health Canada Consumption Advice: Marking informed choices about fish Available online at http://www.hc-sc.gc.ca/fn-an/securit/chem-chim/environ/mercur/cons-adv-etud-eng.php. Accessed Mar 19, 2011.

5.2 Types of Lipid Molecules

Learning Objectives

- Describe the functions of triglycerides, phospholipids, and cholesterol.
- Compare the structures of saturated, monounsaturated, polyunsaturated, omega-6, omega-3, and *trans* fatty acids.
- Name foods that are sources of saturated, monounsaturated, polyunsaturated, omega-6, omega-3, and *trans* fatty acids.

The primary type of lipid in our food and in our bodies, and what we are typically referring to when we use the word fat, is **triglycerides**. Each triglyceride includes three **fatty acids**. These fatty acids determine the physical properties and health effects of the triglycerides we consume. **Phospholipids** and **sterols** are two other types of lipid that are important in nutrition. The different structures of these lipids affect their function in the body and the properties they give to food.

triglycerides (Triacylglycerols) The major form of lipid in food and in the body. They consist of three fatty acids attached to a glycerol molecule.

fatty acids Organic molecules made up of a chain of carbons linked to hydrogen atoms with an acid group at one end.

phospholipids Types of lipids containing phosphorous. The most common are the phosphoglycerides, which are composed of a glycerol backbone with two fatty acids and a phosphate group attached.

sterols Types of lipids with a structure composed of multiple chemical rings.

Triglycerides and Fatty Acids

Triglycerides, also known as triacylglycerols, consist of a backbone of glycerol with three fatty acids attached, as shown in **Figure 5.4**. If only one fatty acid is attached to the glycerol,

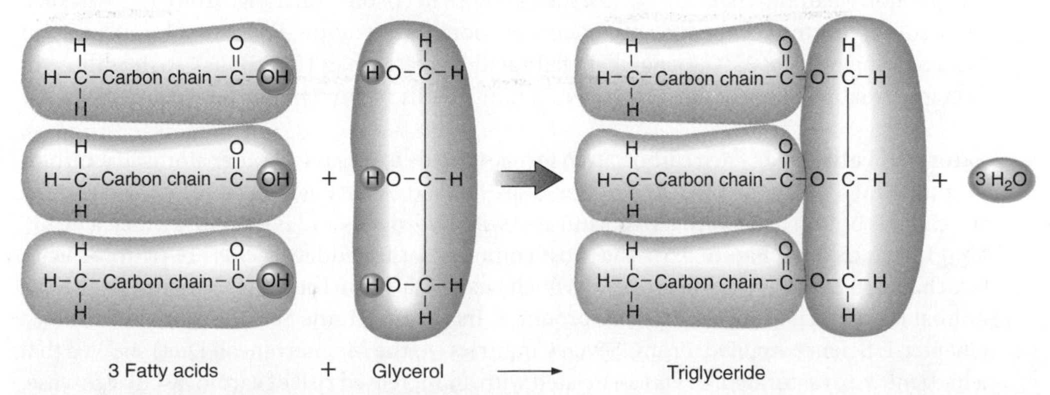

Figure 5.4 Formation and structure of triglycerides
A triglyceride (triacylglycerol) is formed when 3 fatty acids bind to a molecule of glycerol. As each bond is formed, a hydrogen atom (H) from the glycerol and a hydroxyl group (OH) from the acid end of the fatty acid combine to form a molecule of water.

the molecule is called a monoglyceride or monoacylglycerol, and when two fatty acids are attached, it is a diglyceride or diacylglycerol.

Structurally, a fatty acid is a chain of carbon atoms with an acid group (COOH) at one end. The other end of the carbon chain is called the omega or methyl end, and consists of a carbon atom attached to 3 hydrogen atoms (CH_3). Each of the carbons between is attached to its 2 neighbouring carbons and up to 2 hydrogen atoms (**Figure 5.5**). The physical properties of a fatty acid depend on the length of the carbon chain and the type and location of the bonds between the carbon atoms.

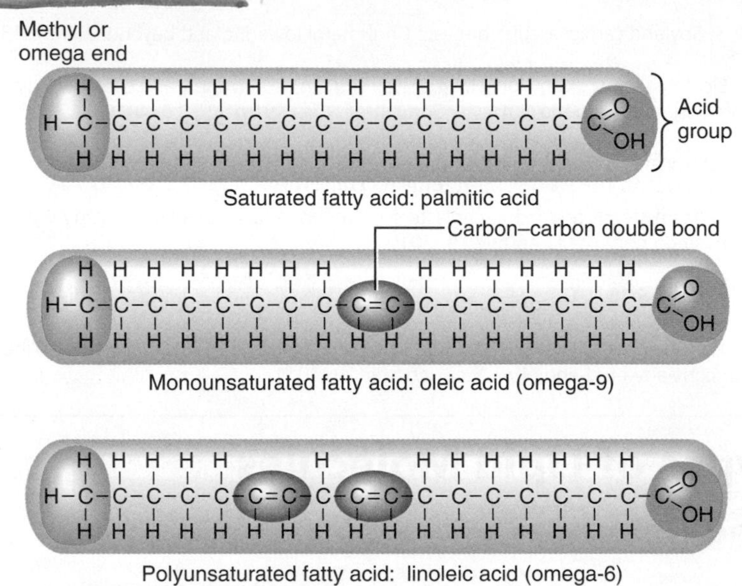

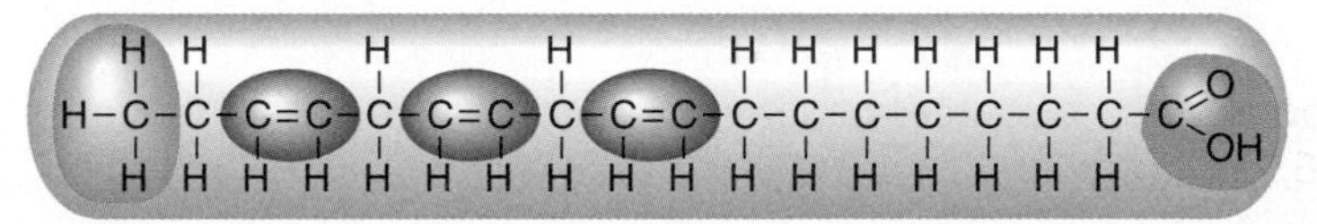

Figure 5.5 Fatty acid structure
These fatty acids, commonly found in our foods, illustrate the structural differences between saturated, monounsaturated, and polyunsaturated fatty acids.

Fatty Acid Chain Length The carbon chains of fatty acids vary in length from a few to 20 or more carbons. Most of the fatty acids found in nature contain an even number of carbon atoms. Most fatty acids in plants and animals, including humans, contain between 14 and 22 carbons. Short-chain fatty acids range from 4-7 carbons in length. They remain liquid at colder temperatures. For example, the short-chain fatty acids in whole milk remain liquid even in the refrigerator. Medium-chain fatty acids, such as those in coconut oil, range from 8-12 carbons. They solidify in the refrigerator but liquefy at room temperature. For example, coconut oil has a melting point of 25°C. Long-chain fatty acids (greater than 12 carbons), such as those in beef fat, usually remain solid at room temperature, with melting points between 50 to 70°C.

Saturated Fatty Acids Each carbon atom forms 4 bonds to link it to 4 other atoms. If a carbon is not bound to 4 other atoms, double bonds are formed. A fatty acid in which each carbon in the chain is bound to 2 hydrogens is saturated with hydrogens and is therefore called a **saturated fatty acid** (see Figure 5.5). The most common saturated fatty acids are palmitic acid, which has 16 carbons, and stearic acid, which has 18 carbons. These are found most often in animal foods such as meat and dairy products. In Chapter 1, the Seven Country study (see Chapter 1: Science Applied: From "Seven Countries" to the Mediterranean Diet) showed that a high intake of saturated fat was associated with an increased risk of cardiovascular disease.

saturated fatty acid A fatty acid in which the carbon atoms are bound to as many hydrogens as possible and which, therefore, contains no carbon-carbon double bonds.

tropical oils A term used in the popular press to refer to the saturated oils—coconut, palm, and palm kernel oil—that are derived from plants grown in tropical regions.

Plant sources of saturated fatty acids include palm oil, palm kernel oil, and coconut oil. These are often called **tropical oils** because they are found in plants common in tropical climates. Tropical oils are rarely added to foods at home but, because they are more resistant to rancidity and have longer shelf lives than oils containing unsaturated fats, they have been used by the food industry in cereals, crackers, salad dressings, and cookies. However, concern about the health risks associated with saturated fats led many of the large food manufacturers to reformulate their products to avoid the use of tropical oils.

Unsaturated Fatty Acids Unsaturated fatty acids contain some carbons that are not saturated with hydrogens. The carbons within the chain that are bound to only 1 hydrogen form carbon-carbon double bonds (see Figure 5.5). A fatty acid containing 1 double bond in its carbon chain is called a **monounsaturated fatty acid**. In our diets, the most common monounsaturated fatty acid is oleic acid (Figure 5.5), which is prevalent in olive and canola oils. A fatty acid with more than one double bond in its carbon chain is said to be a **polyunsaturated fatty acid**. The most common polyunsaturated fatty acid is linoleic acid (Figure 5.5), found in corn, safflower, and soybean oils. Unsaturated fatty acids melt at cooler temperatures than saturated fatty acids of the same chain length. Therefore, the more unsaturated bonds a fatty acid contains, the more likely it is to be liquid at room temperature.

monounsaturated fatty acid A fatty acid that contains 1 carbon-carbon double bond.

polyunsaturated fatty acid A fatty acid that contains 2 or more carbon-carbon double bonds.

Omega-3 and Omega-6 Fatty Acids There are different categories of unsaturated fatty acids, depending on the location of the first double bond in the carbon chain. If the first double bond occurs between the third and fourth carbons, counting from the omega end (CH_3) of the chain, the fat is said to be an **omega-3(ω-3) fatty acid** (see Figure 5.5). Alpha-linolenic acid, found in vegetable oils, and eicosapentaenoic acid (EPA) and docosahexaenoic acid (DHA), found in fish oils, are omega-3 fatty acids (**Table 5.1**). If the first double bond occurs between the sixth and seventh carbons (from the omega end), the fatty acid is called an **omega-6 (ω -6) fatty acid**. Linoleic acid, found in corn and safflower oils, and arachidonic acid, found in meat and fish, are omega-6 fatty acids. Linoleic acid is the major omega-6 fatty acid in the North American diet. Both omega-3 and omega-6 fatty acids are used to synthesize regulatory molecules in the body and the biological effect of the molecule synthesized depends on the structure of the fatty acid from which it is made. Therefore, the ratio of omega-3 to omega-6 fatty acids in the diet is important in processes such as blood pressure regulation, blood clotting, and immune function. Replacing saturated fatty acids in the diet with polyunsaturated fatty acids has been associated with a reduced risk of cardiovascular disease.[2] Furthermore, increasing the intake of omega-3 fatty acids, especially EPA and DHA, found in fish, has also been associated with reduced deaths from cardiovascular disease.[3] (See Critical Thinking: Fish Consumption and Heart Disease.)

omega-3 (ω-3) fatty acid A fatty acid containing a carbon-carbon double bond between the third and fourth carbons from the omega end.

omega-6 (ω -6) fatty acid A fatty acid containing a carbon-carbon double bond between the sixth and seventh carbons from the omega end.

Table 5.1 Sources of Omega-3 and Omega-6 Fatty Acids. The omega-3 fatty acids found in fish are mainly EPA and DHA. The omega-3 fatty acid in plant food is mainly alpha-linolenic acid.

Food	Portion	Omega-3 (g)		Omega-6 (g)
Fish Products				
Swordfish, salmon, trout, cooked	90 g	1.5	Mainly EPA & DHA	0.5
Sole, cod, cooked	90 g	0.13		4.3
Shrimp, cooked	90 g	0.27		0.08
Mussels, clams, cooked	90 g	0.7		0.15
Tuna, canned	90 g	0.23		0.04
Plant Products				
Flax seed	15 ml	2.0	Mainly alpha-linolenic acid	0
Canola oil	15 ml	1.27		2.77
Walnuts	60 ml	2.72		11.4
Peanuts	60 ml	0.09		4.01
Almonds	60 ml	0.13		4.3
Sunflower seeds	60 ml	0.02		10.5
Spinach, cooked	125 ml	0.08		0.01
Mustard greens, cooked	125 ml	0.02		0.02

***Cis* Versus *Trans* Double Bonds** The position of the hydrogen atoms around a double bond also affects the properties of unsaturated fatty acids. Most unsaturated fatty acids found in nature have both hydrogen atoms on the same side of the double bond, called the *cis* configuration. When the hydrogen atoms are on opposite sides of the double bond, called the *trans* configuration, the fatty acid is a ***trans* fatty acid** (**Figure 5.6**). A *trans* fatty acid has a higher melting point than the same fatty acid in the *cis* configuration.

***trans* fatty acid** An unsaturated fatty acid in which the hydrogen atoms are on opposite sides of the double bond.

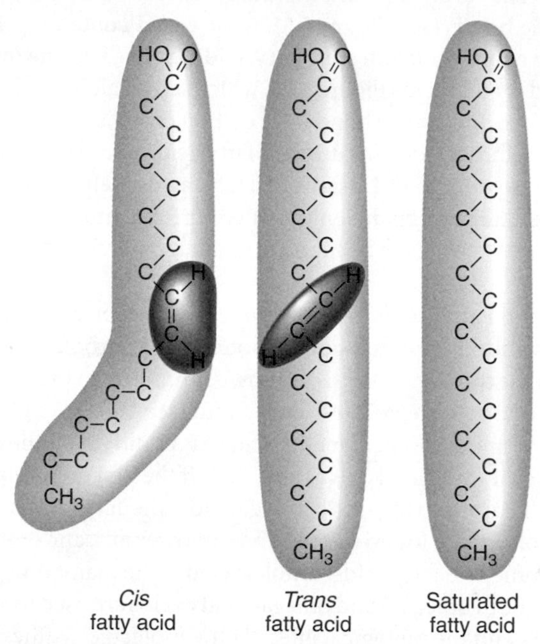

Figure 5.6 *Cis* versus *trans* fatty acids
In *cis* fatty acids, the orientation of hydrogen atoms around the double bond causes a bend in the carbon chain. In *trans* fatty acids, the orientation of hydrogen atoms does not cause a bend, so the carbon chain is straighter, and resembles a saturated fatty acid.

hydrogenation The process whereby hydrogen atoms are added to the carbon-carbon double bonds of unsaturated fatty acids, making them more saturated.

Most of the *trans* fat we eat comes from products that have undergone **hydrogenation.** Hydrogenation bubbles hydrogen gas into liquid oil, which causes some of the double bonds in the oil to accept hydrogen atoms and become saturated. The resulting fat has more of the properties of a saturated fat, such as increased stability against rancidity and a higher melting point. However, during hydrogenation, only some of the bonds become saturated. Some of those that remain unsaturated are altered, converting them from the *cis* to the *trans* configuration. The resulting product therefore, contains more *trans* fatty acids than the original oil. Hydrogenated or partially hydrogenated vegetable oils are a primary ingredient in stick margarine and vegetable shortening because they raise the melting point of the products, making them more solid at room temperature. They are also used to lengthen shelf life in other processed foods such as cookies, crackers, breakfast cereals, and potato chips. When consumed in the diet, *trans* fats raise blood-cholesterol levels and increase the risk of heart disease. Because of their relationship to heart disease risk, information about *trans* fat content is included on food labels and many manufacturers have reformulated their products to reduce or eliminate *trans* fatty acids. For example, unlike stick margarines which are firm and solid at room temperature, most margarines now available in Canadian markets are tub margarines, which contain little or no *trans* fat and more monounsaturated and polyunsaturated fatty acids. As a result, they are softer and less solid than stick margarines and thus sold in tubs. Consumers can easily determine how much saturated and *trans* fat are present in a product by checking the nutrition facts table; manufacturers can voluntarily include information about polyunsaturated and monounsaturated fat content as well (**Figure 5.7**).

Some *trans* fats occur naturally in very small amounts in milk and dairy products. These fatty acids differ in composition from hydrogenated fats, and include conjugated linoleic acid (CLA), which is different from the essential fatty acid, linoleic acid. There is some scientific evidence that CLA may have some health benefits, but detrimental effects have also been reported. More research is needed on the health impact of natural *trans* fats.[4]

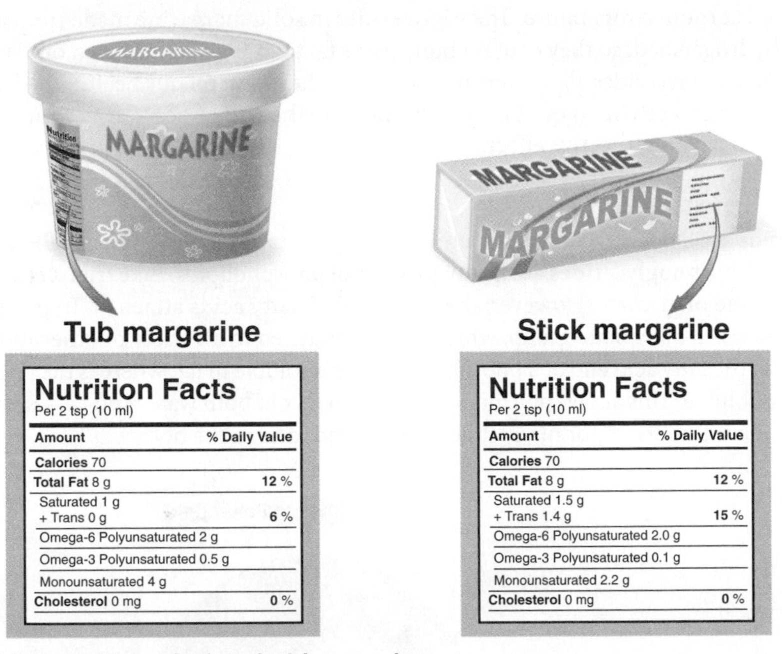

Tub margarine

Nutrition Facts
Per 2 tsp (10 ml)

Amount	% Daily Value
Calories 70	
Total Fat 8 g	12 %
Saturated 1 g + Trans 0 g	6 %
Omega-6 Polyunsaturated 2 g	
Omega-3 Polyunsaturated 0.5 g	
Monounsaturated 4 g	
Cholesterol 0 mg	0 %

Stick margarine

Nutrition Facts
Per 2 tsp (10 ml)

Amount	% Daily Value
Calories 70	
Total Fat 8 g	12 %
Saturated 1.5 g + Trans 1.4 g	15 %
Omega-6 Polyunsaturated 2.0 g	
Omega-3 Polyunsaturated 0.1 g	
Monounsaturated 2.2 g	
Cholesterol 0 mg	0 %

Figure 5.7 Comparison of tub and stick margarines
Food processors are required to indicate the total amount of saturated and *trans* fats in a product. They can also voluntarily provide additional information on the monounsaturated and polyunsaturated fat in the product, as well. The two labels shown here are examples of labels with this additional information. The higher saturated and *trans* fat content of the stick margarine (DV 15% vs. 6%) is evident as is the higher polyunsaturated content of the tub margarine (6.5 g vs 4.3 g/10 ml).

Fatty Acids and the Properties of Triglycerides Triglycerides may contain any combination of fatty acids: long, medium, or short chain, saturated or unsaturated, *cis* or *trans* (**Figure 5.8**). The types of fatty acids in triglycerides determine their texture, taste, and physical characteristics. For example, the amounts and types of fatty acids in chocolate allow it to remain solid at room temperature, snap when bitten into, and then melt quickly and smoothly in the mouth. The triglycerides in red meat contain predominantly long-chain, saturated fatty acids, so the fat on a piece of steak is solid at room temperature. The triglycerides in olive oil contain predominantly monounsaturated fatty acids, whereas those in corn oil are mostly polyunsaturated. These

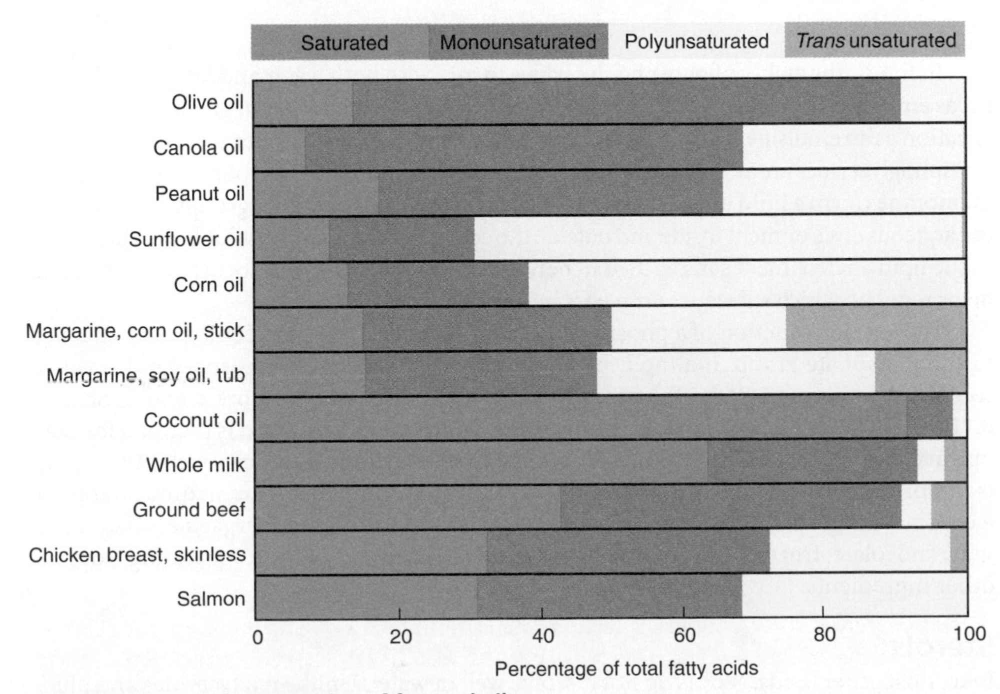

Figure 5.8 Fatty acid composition of fats and oils
The fats and oils in our diets contain combinations of saturated, monounsaturated, polyunsaturated, and *trans* fatty acids; shown here as a percentage of the total amount of fat in the item. (From USDA)

fats are liquid at room temperature. The triglycerides in solid margarine made from corn oil have been hydrogenated, so they contain more *trans* fat than the original corn oil. The types of fatty acids in triglycerides also determine the effect they have on our health. As discussed later in the chapter, certain types of fatty acids increase the risk of heart disease and cancer, whereas others have a protective effect.

Phospholipids

phosphoglycerides A class of phospholipid consisting of a glycerol molecule, 2 fatty acids, and a phosphate group.

Phospholipids are lipids attached to a chemical group containing phosphorus called a phosphate group. **Phosphoglycerides** are the major class of phospholipids. Like triglycerides, they have a backbone of glycerol. However, they have only 2 fatty acids attached. In place of the third fatty acid is a phosphate group, which is then attached to a variety of other molecules (**Figure 5.9**). The fatty acid end of a phosphoglyceride is soluble in fat, whereas the phosphate end is water-soluble. This allows phosphoglycerides to mix in both water and fat—a property that makes them important for many functions in foods and in the body.

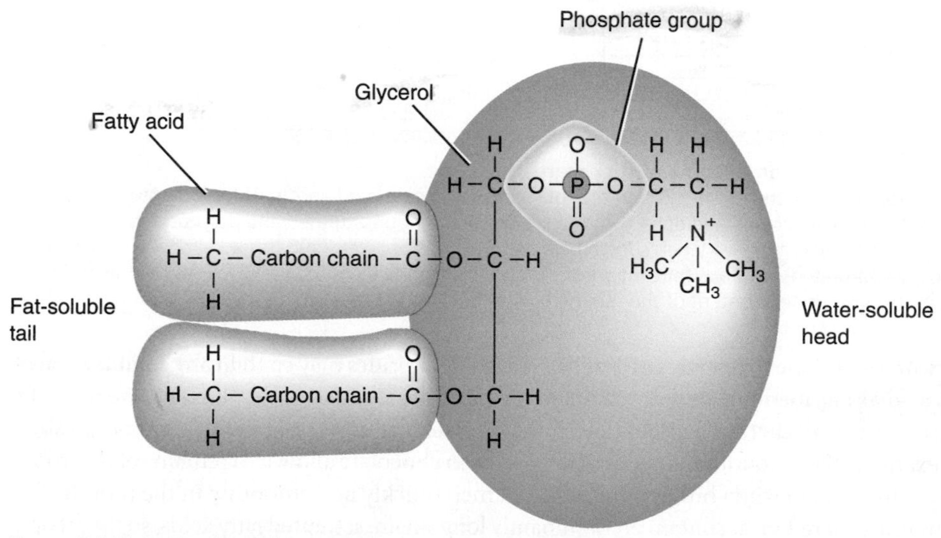

Figure 5.9 Phosphoglyceride structure
Phosphoglycerides, such as the lecithin molecule shown here, consist of a water-soluble head containing a phosphate group and a lipid-soluble tail of fatty acids.

emulsifiers Substances that allow water and fat to mix by breaking large fat globules into smaller ones.

lipid bilayer Two layers of phosphoglyceride molecules oriented so that the fat-soluble fatty acid tails are sandwiched between the water-soluble phosphate-containing heads.

lecithin A phosphoglyceride composed of a glycerol backbone, two fatty acids, a phosphate group, and a molecule of choline.

In foods, the ability of phosphoglycerides to mix with both water and fat allows them to act as **emulsifiers (Figure 5.10a).** For example, egg yolks, which contain phosphoglycerides, function as an emulsifier in cake batter, where they allow the oil and water to mix. In the body, phosphoglycerides are an important component of cell membranes. The phosphoglycerides in membranes form a **lipid bilayer**, in which the water-soluble phosphate groups orient toward the aqueous environment inside and outside the cell, while the water-insoluble fatty acids stay in the lipid environment sandwiched in between (**Figure 5.10b**). This forms the barrier that helps regulate which substances can pass into and out of the cell.

The specific function of a phosphoglyceride depends on the molecule that is attached to the phosphate group. If a molecule of choline is attached, the phospholipid is called **lecithin** (see Figure 5.9; the functions of choline are discussed in more detail in Section 8.11). In the body, lecithin is a major constituent of cell membranes and is required for their optimal function. It is also used to synthesize the neurotransmitter acetylcholine, which is important in the memory centre of the brain. Eggs and soybeans are natural sources of lecithin. Lecithin is also used by the food industry as an additive in margarine, salad dressings, chocolate, frozen desserts, and baked goods to keep the oil from separating from the other ingredients.

Sterols

cholesterol A lipid that consists of multiple chemical rings and is made only by animal cells.

Like most other lipids, sterols do not dissolve well in water. Unlike triglycerides and phosphoglycerides, their structure consists of multiple chemical rings. Sterols are found in both plants and animals. **Cholesterol**, probably the best-known sterol, is found only in animals

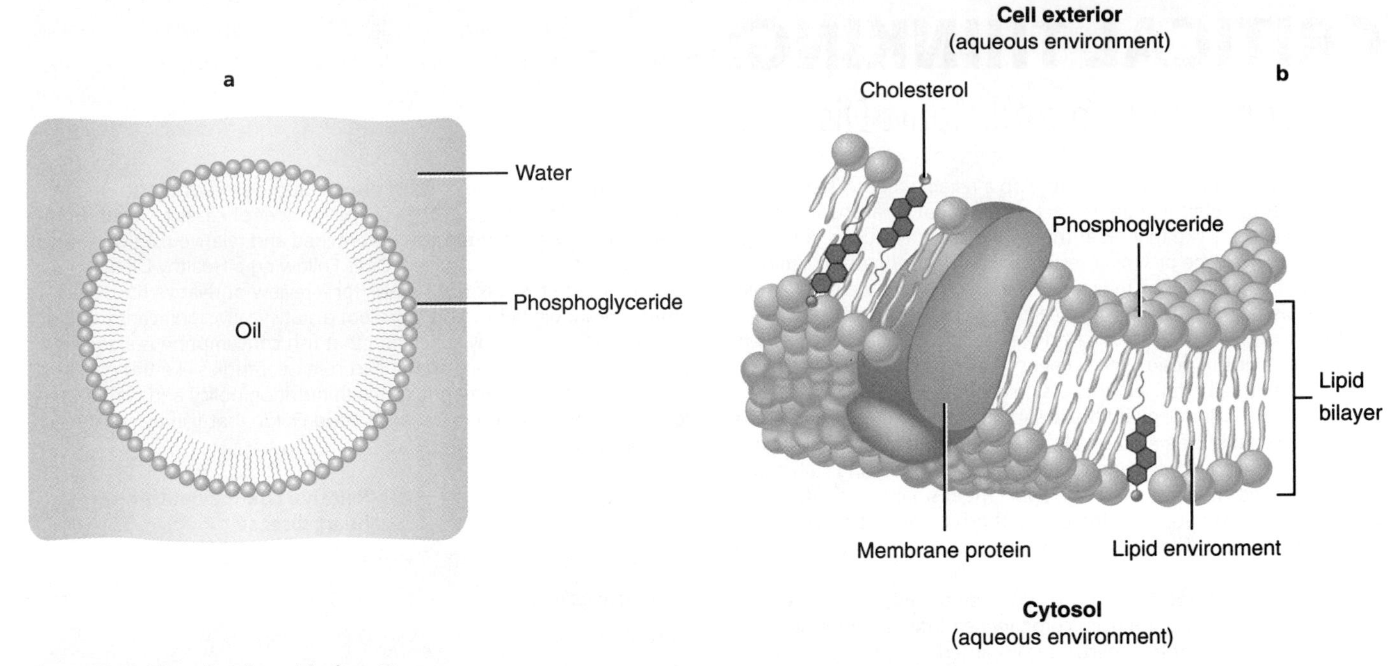

Figure 5.10 Phosphoglyceride functions
(a) Phosphoglycerides act as emulsifiers in foods because they can surround droplets of oil, allowing them to remain suspended in a watery environment. (b) In cell membranes, phosphoglycerides form a lipid bilayer by orienting the water-soluble phosphate-containing heads toward the watery environment inside and outside of the cell and the fatty acid tails toward the interior of the membrane.

(**Figure 5.11**). Cholesterol is necessary in the body, but because the liver manufactures it, it is not essential in the diet. More than 90% of the cholesterol in the body is found in cell membranes (see Figure 5.10b). It is also part of myelin, the coating on many nerve cells. Cholesterol is needed to synthesize vitamin D in the skin; cholic acid, a component of bile; some hormones, such as testosterone and estrogen, which promote growth and the development of sex characteristics; and cortisol, which promotes glucose synthesis in the liver. When cholesterol circulates in the blood at high levels, it may increase the risk of heart disease.

Although cholesterol is not an essential nutrient, it is present in foods from animal sources and some people consume large amounts of dietary cholesterol. Egg yolks and organ meats such as liver and kidney are high in cholesterol. One egg yolk contains about 213 mg of cholesterol, while organ meats contain about 300 mg/75 g serving. Lean red meats and skinless chicken contain about 90 mg/75 g, whereas fish contains 50 mg/75 g. Diets high in cholesterol may increase the risk of heart disease for some cholesterol-sensitive individuals, because this cholesterol is absorbed and converted in the body to undesirable forms of cholesterol; for example, LDL-cholesterol, so called "bad" cholesterol .

Plant foods contain other sterols, but do not contain cholesterol unless animal products are combined with them in cooking or processing. Plant sterols can help reduce cholesterol levels in the body by decreasing cholesterol absorption from the diet and Health Canada permits food manufacturers to add plant sterols to foods because of this cholesterol lowering effect.[5] (See Your Choice: Eating to Lower Cholesterol Levels.)

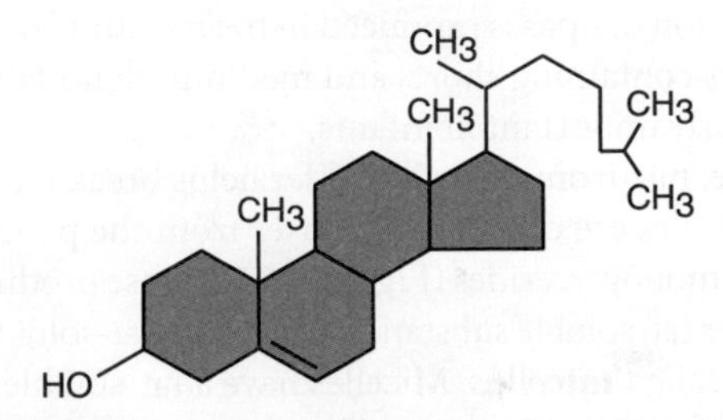

Figure 5.11 Cholesterol structure
The four coloured rings indicate the backbone structure common to all sterols.

CRITICAL THINKING:

Fish Consumption and Heart Disease

A single study is never enough to develop a reliable theory (see Section 1.4). In order to determine the impact on health of a particular nutrient or diet, researchers like to look at the results of many studies. If multiple studies have the same or similar results and a trend or pattern can be discerned, this strengthens the conclusion made about the relationship between diet and health. When researchers look at the results of multiple studies, this process is called "reviewing the literature." When studies are similar in design the numerical results of multiple studies can be pooled, that is, combined, into a single study, with a much larger sample size and a single quantitative set of results. This is called a meta-analysis. The results of a meta-analysis can provide strong evidence for a relationship between diet and disease risk, often revealing statistically significant trends that would not be apparent in smaller, individual studies. In order to determine the effect of fish consumption on the risk of death from heart disease, scientists performed such a meta-analysis.[1]

The results of 11 prospective cohort studies on the relationship between fish consumption and death from coronary heart disease (diseases of the coronary arteries of the heart, usually caused by atherosclerosis, discussed in Section 5.6) were combined, representing 222,364 individuals. As is the case in prospective cohort studies, the population's dietary intake, especially of fish, was determined using methods such as food-frequency questionnaires (described in Section 2.6). The populations were all free of heart disease at the start of the study and were followed for almost 12 years. The number of cases of death from heart disease that occurred was determined and relative risks calculated (see Chapter 2: Science Applied: Does Following a Healthy Dietary Pattern Reduce the Risk of Disease? for a review of relative risk). The results are shown below and represent a statistically significant trend. The authors of the study concluded that fish consumption is associated with reduced risk of death from heart disease. Studies like this meta-analysis help to formulate public health nutrition policy and contributed to the recommendation in Canada's Food Guide that fish be consumed twice a week.

Fish consumption	Relative risk for death from heart disease
Less than 1 serving/month	1
1-3 servings/month	0.89
1 serving/week	0.85
2-4 servings/week	0.77
5 or more servings/week	0.62

Critical Thinking Questions

Describe one strength and one limitation of the study.*

Based on the results presented here, do you agree or disagree with the authors' conclusions? Explain your answer by interpreting the results of the table above.

*Answers to all Critical Thinking questions can be found in Appendix J.

[1]He, K., Song, Y., Daviglus, M. L. et al. Accumulated evidence on fish consumption and coronary heart disease mortality: a meta-analysis of cohort studies. *Circulation.* 109(22):2705–11, 2004.

5.3 Lipids in the Digestive Tract

Learning Objectives

- Describe the steps involved in the digestion of triglycerides.
- Explain how micelles facilitate lipid absorption.

In healthy adults, most of the digestion of dietary fat takes place in the small intestine due to the action of lipid-digesting enzymes called lipases. Some triglyceride digestion also occurs in the stomach due to the action of lipases produced in the mouth and stomach. These enzymes work best on triglycerides containing short- and medium-chain fatty acids such as those in milk, and so are particularly important in infants.[6]

In the small intestine, bile from the gallbladder helps break fat into small globules. The triglycerides in these globules are digested by lipases from the pancreas, which break them down into fatty acids and monoglycerides (**Figure 5.12**). These products of triglyceride digestion, cholesterol, and other fat-soluble substances including fat-soluble vitamins mix with bile to form smaller droplets called **micelles**. Micelles have a fat-soluble centre surrounded by a coating of bile acids. They facilitate the absorption of lipids into the mucosal cells of the small intestine by allowing these substances to get close enough to the brush border to diffuse across

micelles Particles formed in the small intestine when the products of fat digestion are surrounded by bile acids. They facilitate the absorption of fat.

into the mucosal cells. Most of the bile acids in micelles are also absorbed and returned to the liver to be reused. Because fat-soluble vitamins and other fat-soluble molecules present in foods must be incorporated into micelles to be absorbed, their absorption depends on the presence of dietary fat. Once inside the mucosal cell, long-chain fatty acids, cholesterol, and other fat-soluble substances require further processing before they can be transported in the blood.

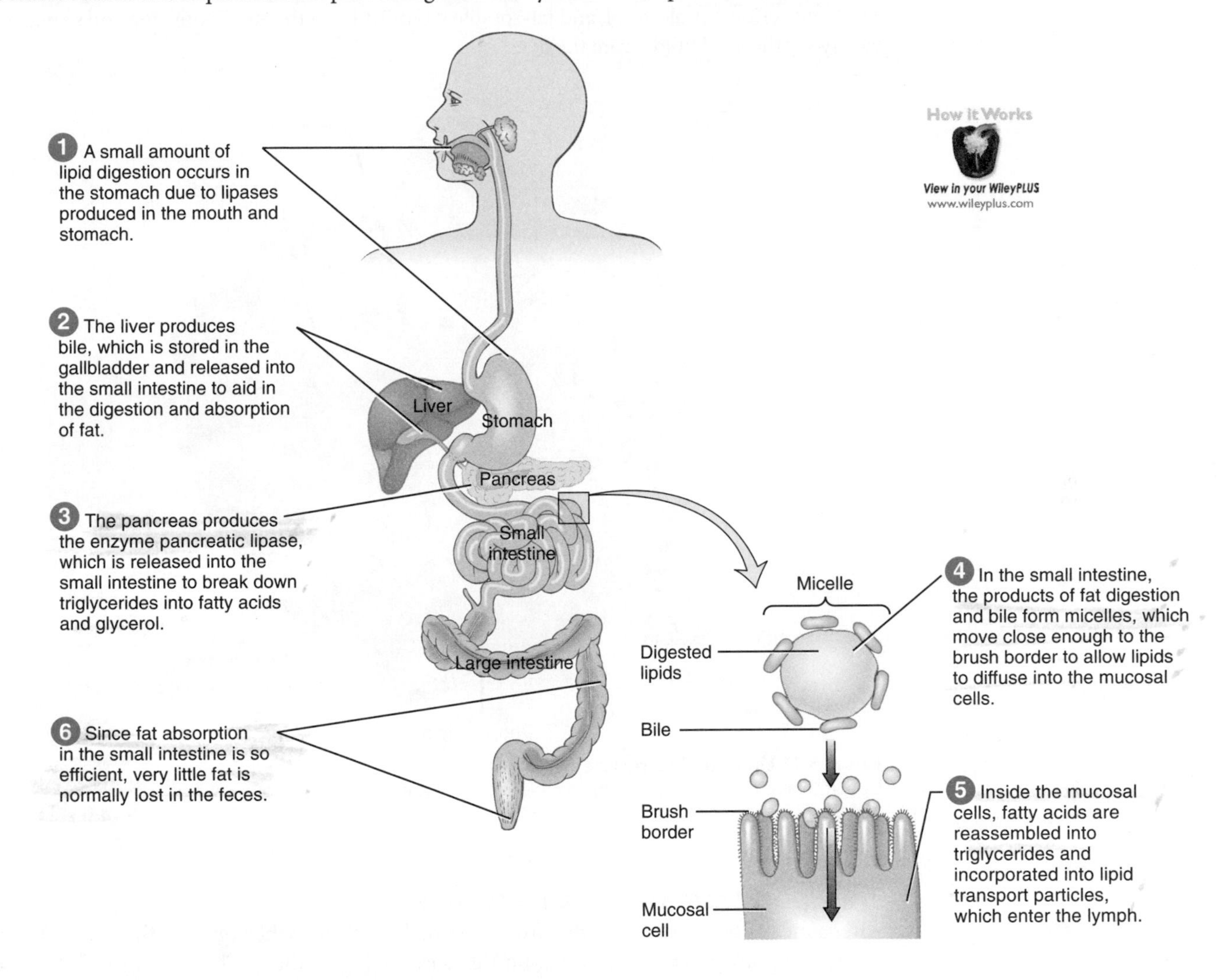

Figure 5.12 Lipid digestion
The bulk of our dietary lipids is triglycerides, which need to be digested before they can be absorbed. The diet also contains smaller amounts of phospholipids, which are partially digested, and cholesterol and fat-soluble vitamins, which are absorbed without digestion.

5.4 Lipid Transport in the Body

Learning Objectives

- Describe how lipids are transported from the small intestine and delivered to cells.
- Describe the role of the liver in the formation of VLDL.
- Describe the relationship between VLDL and LDL.
- Describe how LDL is taken up by tissue.
- Compare and contrast the functions of LDLs and HDLs.

Just as water-insoluble lipids require special mechanisms to be absorbed into the body, they require special transport particles to travel throughout the body in the blood. These transport

lipoproteins Particles containing a core of triglycerides and cholesterol surrounded by a shell of protein, phospholipids, and cholesterol that transport lipids in blood and lymph.

particles, called **lipoproteins**, are created by combining water-insoluble lipids with phospholipids and proteins (**Figure 5.13**). Phospholipids orient with their fat-soluble tails toward the interior and their water-soluble heads toward the outside. The resulting particle consists of a fat-soluble core surrounded by a water-soluble envelope, allowing water-insoluble lipids to be transported in the aqueous environment of the blood. Lipoproteins help transport dietary triglycerides, cholesterol, and fat-soluble vitamins from the small intestine and stored or newly synthesized lipids from the liver.

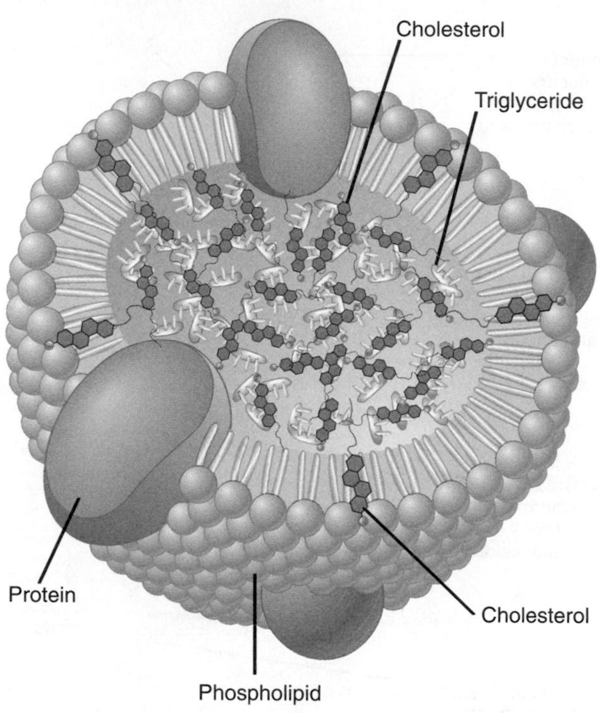

Figure 5.13 Lipoprotein structure
Lipoproteins consist of a core of triglycerides and cholesterol surrounded by a shell of proteins, phospholipids, and cholesterol.

Transport from the Small Intestine

post-prandial state The time following a meal when nutrients from the meal are being absorbed.

chylomicrons Lipoproteins that transport lipids from the mucosal cells of the small intestine and deliver triglycerides to other body cells.

lipoprotein lipase An enzyme that breaks down triglycerides into fatty acids and glycerol; attached to the cell membranes of cells that line the blood vessels.

After a meal, lipids are transported from the small intestine to the organs in the body for use as fuel. The time after a meal, when nutrients eaten during the meal are being absorbed, is referred to as the absorptive or **post-prandial state**. How lipids are transported from the small intestine depends on their solubility in water. Short- and medium-chain fatty acids, which are water-soluble, can be transported from the small intestine in the blood and delivered to cells throughout the body. Lipids that are not soluble in water, such as long-chain fatty acids and cholesterol, cannot enter the bloodstream directly, so they must be incorporated into lipoproteins. For this to occur, long-chain fatty acids and monoglycerides are first assembled into triglycerides by the mucosal cell. These triglycerides are then combined with cholesterol, phospholipids, and a small amount of protein to form lipoproteins called **chylomicrons (Figure 5.14).** Chylomicrons are transferred into the lymphatic system and then enter the bloodstream without first passing through the liver.

As chylomicrons circulate in the blood, the enzyme **lipoprotein lipase**, present on the surface of the cells lining the blood vessels, breaks the triglycerides down into fatty acids and glycerol, which enter the surrounding cells. The fatty acids can be either used as fuel or re-synthesized into triglycerides for storage. What remains of the chylomicron is a chylomicron remnant composed mostly of cholesterol and protein. This goes to the liver and is disassembled. This process is shown in steps 1, 2, and 3 of Figure 5.14.

Transport from the Liver

The liver is the major lipid-producing organ in the body. Here, in the post-prandial state, chylomicron remnants are disassembled and triglycerides from the remnants recovered. Excess

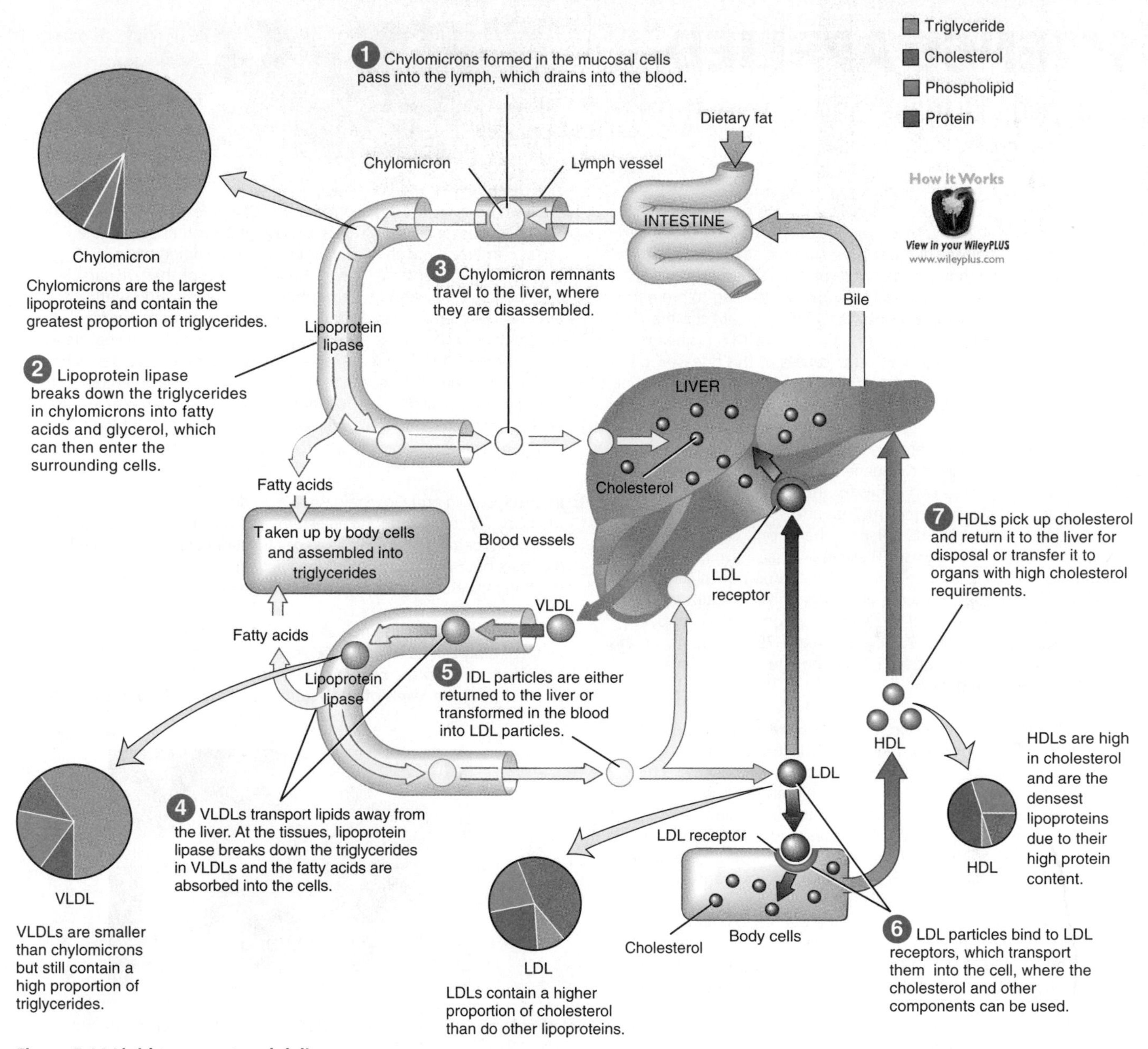

Figure 5.14 Lipid transport and delivery
Chylomicrons and very-low-density lipoproteins transport triglycerides and deliver them to body cells. Low-density lipoproteins transport and deliver cholesterol, while high-density lipoproteins help return cholesterol to the liver for reuse or elimination.

protein, carbohydrate, or alcohol can be broken down as well, and used to make triglycerides or cholesterol. Triglycerides in the liver are then incorporated into lipoprotein particles called **very-low-density lipoproteins (VLDLs)**. Cholesterol synthesized in the liver or delivered in chylomicron remnants can also be incorporated into VLDLs or can be used to make bile. VLDLs transport lipids out of the liver and deliver triglycerides to body cells (see Figure 5.14, step 4). As with chylomicrons, the enzyme lipoprotein lipase breaks down the triglycerides in VLDLs so that the fatty acids can be taken up by surrounding cells. Once the triglycerides are removed from the VLDLs, a denser, smaller, intermediate-density lipoprotein (IDL) remains (see Figure 5.14, step 5). About two-thirds of the IDLs are returned to the liver, and the rest are transformed in the blood into **low-density lipoproteins (LDLs)**. LDLs contain less triglyceride and therefore proportionally more cholesterol than VLDLs and are the primary cholesterol delivery system for cells (see Figure 5.14, step 6).

very-low-density lipoproteins (VLDLs) Lipoproteins assembled by the liver that carry lipids from the liver and deliver triglycerides to body cells.

low-density lipoproteins (LDLs) Lipoproteins that transport cholesterol to cells. Elevated LDL cholesterol increases the risk of cardiovascular disease.

Cholesterol Delivery For LDLs to be taken up by cells, a protein on the surface of the LDL particle called apoprotein B (apo B) must bind to a receptor protein on the cell membrane,

SCIENCE APPLIED

A Genetic Disease and Cholesterol Regulation

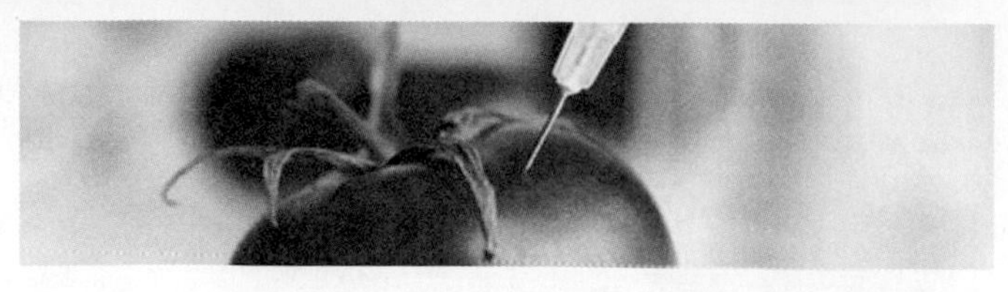

Children with a rare form of the inherited disease familial hypercholesterolemia have blood cholesterol levels that range from 17-26 mmol/L—five times the normal level. Cholesterol is so high in their blood that it deposits in the tissues, forming soft, raised bumps on the skin called xanthomas (see figure of hand later in this box). The elevated cholesterol damages blood vessels, leading to premature atherosclerosis. Chest pain, heart attacks, and death are common before the age of 15 and it is rare for individuals with this disease to survive past age 30.[1,2] Research into the causes of this disease led Drs. Michael Brown and Joseph Goldstein to a discovery that is key to our understanding of how blood cholesterol is regulated.

The severity of familial hypercholesterolemia depends on whether the individual inherits one or two genes for the disease. The children described above inherited two genes for this disease, causing them to have a rare and more severe form that affects only about one in a million people. About 1 in 500 persons inherits only one gene and has a less severe form of the disease. These individuals have moderately elevated blood cholesterol levels from birth and do not usually develop clinical symptoms of heart disease until after age 30. Adults with this disease have blood cholesterol levels ranging from 7.7-14 mmol/L (normal levels are less than 5.2 mmol/L). About 75% of men with this condition will have a heart attack before the age of 60, compared to only 15% of unaffected men.[3]

To study familial hypercholesterolemia, Brown and Goldstein grew cells in culture. They observed that when LDL cholesterol was added to the culture medium, the rate of cholesterol synthesis in cells from normal subjects decreased. But in cells from individuals with familial hypercholesterolemia, the presence of LDL in the medium did not cause a decrease in cholesterol synthesis.[4] By using radioactively labelled LDL particles, Brown and Goldstein were able to demonstrate that in cells from normal individuals, LDL particles bind to the cells and are removed from the surrounding media. In cells from individuals with familial hypercholesterolemia, the LDL particles are unable to bind to the cells and therefore cholesterol cannot be removed from the surrounding media.[5] The binding of LDL to the cell was found to be due to a protein on the surface of the cell membrane that was named the LDL receptor.

The discovery of the LDL receptor led to an understanding of how the body controls the concentration of LDL cholesterol in the blood. The LDL receptor is a cell surface protein; it is synthesized by the cell and inserted into the cell membrane. The binding of an LDL particle to the receptor results in the movement of the LDL particle into the cell. Cholesterol entering the cell suppresses the synthesis of new cholesterol. An individual who inherits one gene for familial hypercholesterolemia has about half the number of LDL receptors as someone without the disease. Someone who inherits two genes has no LDL receptors. This means that LDL cholesterol is not removed from the blood and the liver continues to synthesize large amounts of cholesterol, causing LDL particles to accumulate in the blood. The elevated cholesterol leads to atherosclerosis and eventually heart attacks.

In 1985, Brown and Goldstein were awarded the Nobel Prize in Medicine for their work on cholesterol and LDL receptors. Although their work began by looking at a specific genetic defect, reduced numbers of LDL receptors also explained elevated blood cholesterol levels in individuals without this genetic defect.

A very effective class of cholesterol-lowering drugs, the statins, acts by increasing the number of LDL receptors on the cell membranes, thereby increasing the number of LDL particles taken up by cells and reducing the levels of cholesterol-containing LDL particles in the blood.

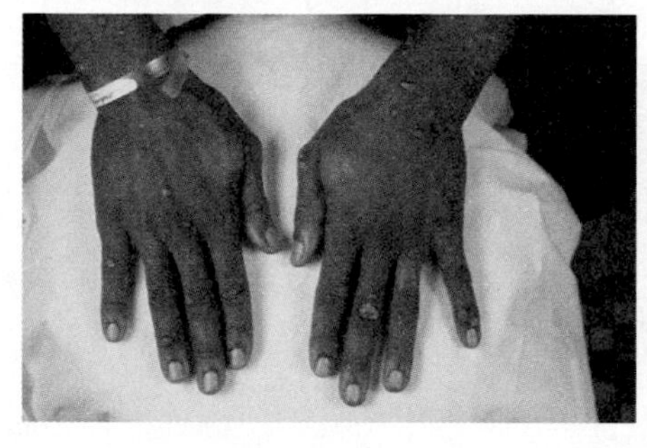

O.J. Staats MD/Custom Medical Stock Photo, Inc.

[1] Fredrickson, D. S., Goldstein, J. L., and Brown, M. S. The familial hypercholesterolemias. In *The Metabolic Basis of Inherited Disease*, 4th ed. J. B. Stanbury, J. B. Wyngaarden, and D. S. Fredrickson, eds. New York: McGraw-Hill, 1974, pp. 604–655.

[2] Goldstein, J. L., and Brown, M. S. The LDL receptor locus and the genetics of familial hypercholesterolemia. *Ann. Rev. Genet.* 13:259–289, 1979.

[3] Stone, N. J., Levy, R. I., Fredrickson, D. S., et al. Coronary artery disease in 116 kindred with familial type II hyperlipoproteinemia. *Circulation* 49:476–478, 1974.

[4] Goldstein, J. L., and Brown, M. S. The low-density lipoprotein pathway and its relation to atherosclerosis. *Ann. Rev. Biochem.* 46:897–930, 1977.

[5] Brown, M. S., and Goldstein, J. L. Familial hypercholesterolemia: Defective binding of lipoproteins to cultured fibroblasts associated with impaired regulation of 3-hydroxy-3-methylglutaryl coenzyme A reductase activity. *Proc. Nat. Acad. Sci.* 71:788–792, 1974.

LDL receptor A protein on the surface of cells that binds to LDL particles and allows their contents to be taken up for use by the cell.

called an **LDL receptor**. This binding allows LDLs to be removed from the blood circulation and enter cells where their cholesterol and other components can be used (see Figure 5.14, step 6). If the amount of LDL cholesterol in the blood exceeds the amount that can be taken up by cells—due to either too much LDL cholesterol or too few LDL receptors—the result is a high level of LDL cholesterol.[7] High levels of LDL particles in the blood have been associated with an increased risk for heart disease (see Science Applied: A Genetic Disease and Cholesterol Regulation). For this reason, LDL cholesterol is often referred to as "bad" cholesterol.

Reverse Cholesterol Transport

Because most body cells have no system for breaking down cholesterol, it must be returned to the liver to be eliminated from the body. This reverse cholesterol transport is accomplished by the densest of the lipoprotein particles, called **high-density lipoproteins (HDLs)** (see Figure 5.14, step 7). These particles originate from the intestinal tract and liver and circulate in the blood, picking up cholesterol from other lipoproteins and body cells. They function as a temporary storage site for lipid. Some of the cholesterol in HDLs is taken directly to the liver for disposal, and some is transferred to organs that have a high requirement for cholesterol, such as those involved in steroid-hormone synthesis. High levels of HDL in the blood help prevent cholesterol from depositing in the artery walls and are associated with a reduction in heart disease risk. For this reason, HDL cholesterol is referred to as "good" cholesterol.

high-density lipoproteins (HDLs) Lipoproteins that pick up cholesterol from cells and transport it to the liver so that it can be eliminated from the body. A high level of HDL decreases the risk of cardiovascular disease.

5.5 Lipid Functions in the Body

Learning Objectives

- Describe the relationship between cholesterol and hormones.
- Describe the relationship between essential fatty acids and inflammation.
- Discuss how fatty acids are used to generate ATP.
- Describe how adipose tissue can provide energy to tissue when energy intake is too low to meet needs.
- Explain what is meant by the glucose-fatty acid cycle.

After lipids have been delivered to cells, they can be used to make structural and regulatory molecules, stored as an energy reserve, or broken down via cellular respiration to produce carbon dioxide, water, and energy in the form of ATP.

Structure and Lubrication

Most of the lipids in the human body are triglycerides stored in **adipose tissue**, which lies under the skin and around internal organs. Deposits of adipose tissue help define our body shape and contours (**Figure 5.15**). In addition to providing stored energy, adipose tissue insulates the body from changes in temperature and provides a cushion to protect internal

adipose tissue Tissue found under the skin and around body organs that is composed of fat-storing cells.

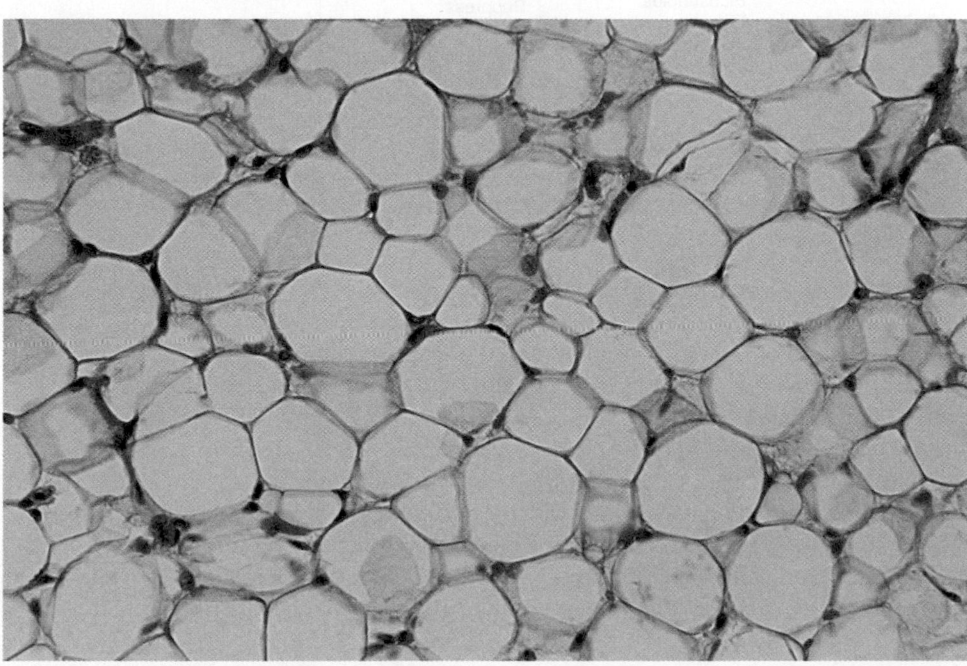

Figure 5.15 Adipose tissue cells contain large droplets of triglyceride that push the other cell components to the perimeter of the cell. (Ed Reschke/Peter Arnold, Inc.)

organs against shock. Lipids are also important structural components at the cellular level, particularly in the brain and nervous system, where they form an insulating coating around nerves. As components of all cell membranes, lipids define the boundaries of cells and partition off their organelles. Lipids are also important for lubricating body surfaces. For example, glands in the skin and mucous membranes of the eyes release oils that lubricate these tissues.

Regulation

Cholesterol and fatty acids are both used to synthesize regulatory molecules in the body. Cholesterol, either consumed in the diet or made in the liver, is used to make a number of hormones, including the sex hormones estrogen and testosterone and the stress hormone cortisol. Polyunsaturated fatty acids are used to make hormone-like molecules that help regulate blood pressure, blood clotting, and other functions such as gene expression (Gene expression is discussed in Section 6.4). However, unlike cholesterol, which is synthesized by the liver, the body cannot make all the fatty acids it needs so certain types must be consumed in the diet.

essential fatty acids Fatty acids that must be consumed in the diet because they cannot be made by the body or cannot be made in sufficient quantities to meet needs.

Essential Fatty Acids The human body is capable of synthesizing most of the fatty acids it needs from glucose or other sources of carbon, hydrogen, and oxygen. Humans, however, are not able to synthesize double bonds in the omega-6 and omega-3 positions. Therefore, the fatty acids linoleic acid (an 18-carbon omega-6 fatty acid) and alpha-linolenic acid (an 18-carbon omega-3 fatty acid) are **essential fatty acids.** Essential fatty acids are important for the formation of the phospholipids that give cell membranes their structure and functional properties. Omega-6 fatty acids are important for growth, skin integrity, fertility, and maintaining red blood cell structure. Arachidonic acid is a 20-carbon omega-6 fatty acid synthesized from linoleic acid (**Figure 5.16**). Linoleic acid is very plentiful in the Canadian diet.

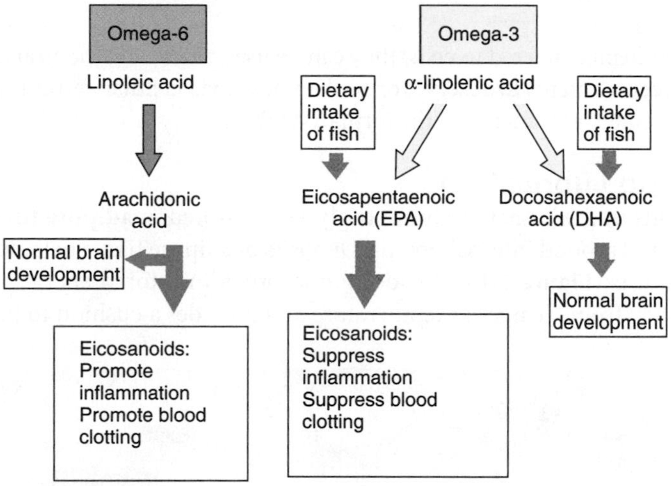

Figure 5.16 Essential fatty acids
If the diet contains enough of the essential fatty acids linoleic acid and alpha-linolenic acid, the longer chain omega-6 and omega-3 fatty acids, arachionic acid, EPA, and DHA can be synthesized. EPA and DHA can also be obtained directly through the consumption of fish.

Alpha-linolenic acid is less plentiful in the Canadian diet. Good dietary sources of alpha-linolenic acid include walnuts, flaxseed, leafy green vegetables, and canola oil (see Table 5.1). EPA and DHA are omega-3 fatty acids that can be synthesized from alpha-linolenic acid. The conversion of alpha-linolenic acid to EPA and DHA, however, is inefficient in humans, and many scientists believe that, for optimal health benefits, EPA and DHA should be obtained directly from the diet, through the consumption of fish, which is the major dietary source of these fatty acids (see Table 5.1).[8] As discussed in Critical Thinking: Fish Consumption and Heart Disease, fish consumption is associated with the reduction of risk of death from cardiovascular disease, and for this reason, Canada's Food Guide recommends 2 servings of fish per week.

Both arachidonic acid and DHA are necessary for normal brain development in infants and young children. Both omega-6 and omega-3 fatty acids, and compounds derived from these fatty acids, also serve as regulators of glucose and fatty-acid metabolism through roles in gene expression.

eicosanoids Regulatory molecules, including prostaglandins and related compounds, that can be synthesized from omega-3 and omega-6 fatty acids.

Eicosanoid Synthesis and Function Both omega-6 and omega-3 fatty acids are used to make hormone-like molecules called **eicosanoids**. Eicosanoids help regulate blood clotting, blood pressure, immune function, and other body processes. The two types of fatty acids tend to have opposing functions. For example, when the omega-6 fatty acid arachidonic acid is the starting material, the eicosanoid synthesized increases blood clotting; whereas when the eicosanoid is made from the omega-3 fatty acid EPA, it decreases blood clotting. Eicosanoids derived from omega-3 fatty acids, such as EPA, have anti-inflammatory properties, while those derived from omega-6 fatty acids (e.g., arachidonic acid) promote inflammation (see Figure 5.16). Since inflammation plays a role in the progression of heart disease, some of the protective effects of omega-3 fatty acids against heart disease and stroke may be due in part to their anti-inflammatory effects.[9] The anti-inflammatory properties of omega-3 fatty acid have also been shown to be of benefit in other inflammatory and autoimmune diseases such as rheumatoid arthritis, inflammatory bowel disorders, and multiple sclerosis.[10]

The ratio of dietary omega-6 to omega-3 essential fatty acids affects the balance of these fatty acids in the tissues and therefore the ratio of the types of eicosanoids made from them. In order to maintain a healthy balance in the body, a dietary ratio of linoleic to alpha-linolenic acid of 5:1-10:1 is recommended. To provide this ratio, a diet that contains 20 g of linoleic acid would need to include 2-4 g of alpha-linolenic acid. However, if the diet contains plenty of arachidonic acid, EPA, and DHA, which are made from linoleic and alpha-linolenic acid, the actual ratio of linoleic to alpha-linolenic is less of a concern.[11] The Canadian diet contains plenty of omega-6 fatty acids, so to get a healthier mix of omega-6s and omega-3s, Canadians should increase their intake of omega-3s from foods such as fish, walnuts, flaxseed, leafy green vegetables, and canola oil (see Table 5.1).[12]

ATP Production

Lipids consumed in the diet can be used as an immediate source of energy or stored in adipose tissue for future use. Throughout the day, triglycerides are continuously stored and then broken down, depending on the immediate energy needs of the body. For example, after a meal, some triglycerides will be stored; then, between meals, some of the stored triglycerides will be broken down to provide energy. When the energy in the diet equals the body's energy requirements, the net amount of stored triglyceride in the body does not change.

Using Triglycerides to Provide Energy In muscle and other tissues, fatty acids and glycerol from triglycerides can be used to produce ATP. In the first step of this process, called **beta-oxidation**, the carbon chain of fatty acids is broken into 2-carbon units that form acetyl-CoA and release high-energy electrons **(Figure 5.17)**.

beta-oxidation The first step in the production of ATP from fatty acids. This pathway breaks the carbon chain of fatty acids into 2-carbon units that form acetyl-CoA and releases high-energy electrons that are passed to the electron transport chain.

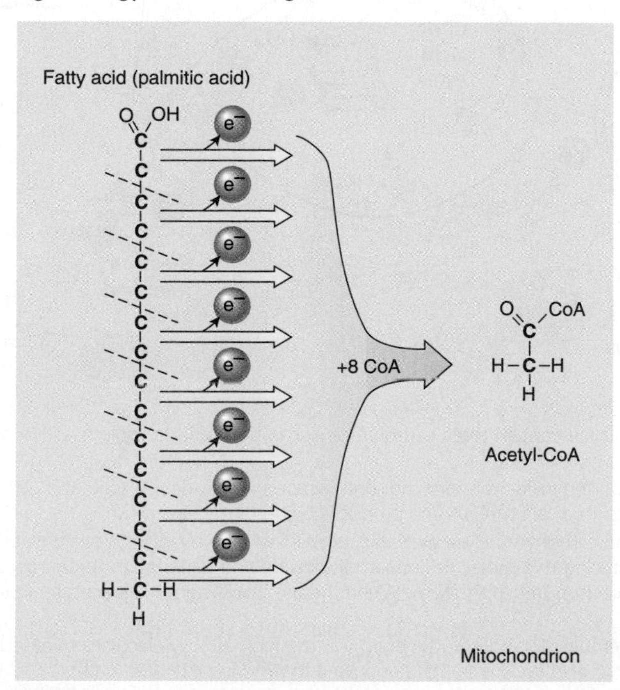

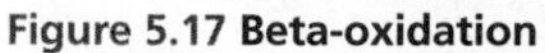

Figure 5.17 Beta-oxidation
Beta-oxidation of a molecule of palmitic acid, which contains 16 carbons, produces 8 molecules of acetyl-CoA and 8 high-energy electrons that are passed to the electron transport chain to generate ATP.

To enter the citric acid cycle, acetyl-CoA combines with oxaloacetate, a 4-carbon molecule derived from carbohydrate, to form a 6-carbon molecule called citric acid. The reactions of the citric acid cycle then remove one carbon at a time to produce carbon dioxide (see **Figure 5.18**). After 2 carbons have been removed in this manner, a 4-carbon oxaloacetate molecule is re-formed and the cycle can begin again. The high-energy electrons released in beta-oxidation and the citric acid cycle are shuttled to the last stage of cellular respiration, the electron transport chain. Molecules in this chain accept electrons and pass them down the chain until they are finally combined with oxygen to form water. The energy in the electrons is used to generate ATP. When there is insufficient oxaloacetate, fatty acids cannot enter the citric acid cycle and instead ketone bodies form (**Figure 5.19**). The glycerol from triglyceride breakdown can also be used to produce ATP or to make glucose via gluconeogenesis. Glycerol makes up only a small proportion of the carbon in a triglyceride molecule, so the amount of glucose that can result from triglyceride breakdown is small.

Storing Fat in Adipose Tissue When energy is ingested in excess of needs, as in when we feast, the excess can be stored in adipose tissue. Excess energy consumed as fat can be transported directly from the intestines to the adipose tissue in chylomicrons. Excess energy consumed

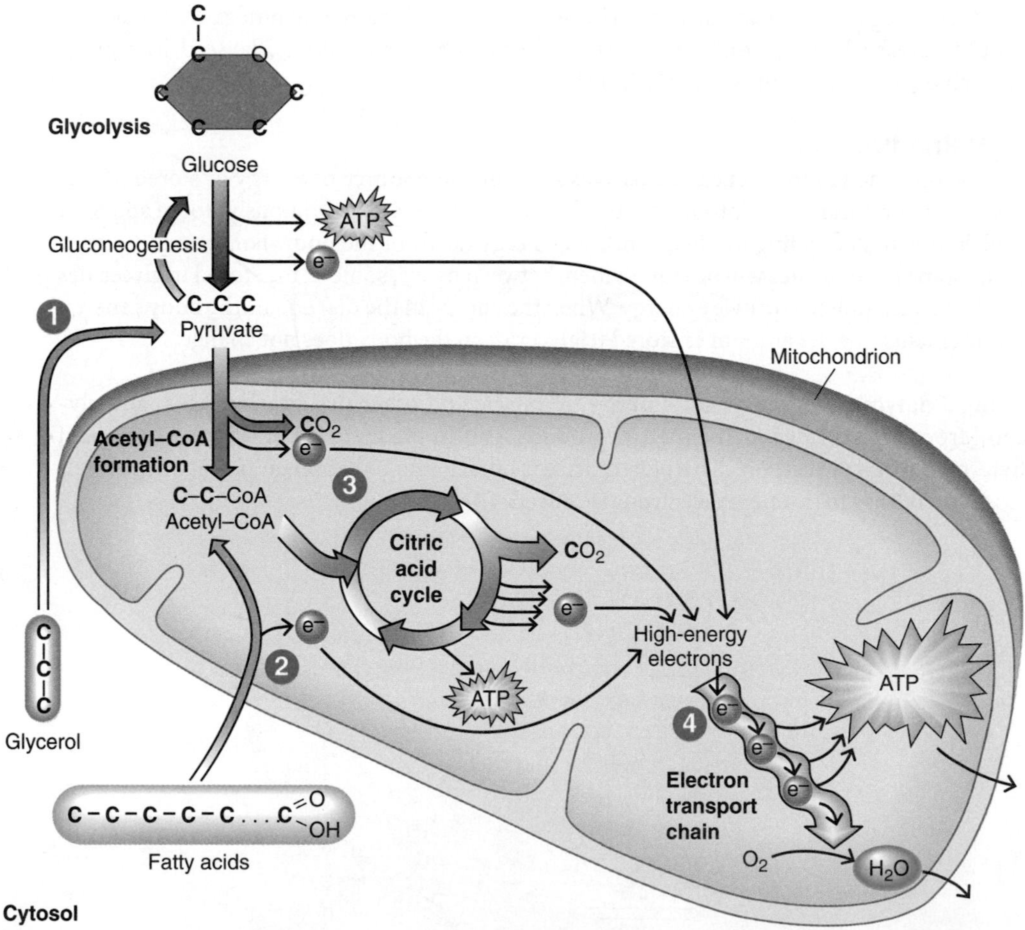

1. Glycerol molecules, which contain three carbon atoms, can be used to produce ATP or small amounts of glucose.
2. Fatty acids are transported inside the mitochondria, where beta-oxidation splits the carbon chains into two-carbon units that form acetyl-CoA and produces high-energy electrons.
3. If oxygen and enough carbohydrate are available, acetyl-CoA combines with oxaloacetate to enter the citric acid cycle, producing two molecules of carbon dioxide and releasing high-energy electrons that are shuttled to the electron transport chain. When there is insufficient oxaloacetate, ketone bodies form (Figure 5.19).
4. In the final step of aerobic metabolism, the energy in the high-energy electrons released from beta-oxidation and the citric acid cycle is trapped and used to produce ATP and water.

Figure 5.18 Triglyceride metabolism
The breakdown of triglycerides yields fatty acids and a small amount of glycerol. Fatty acids provide most of the energy stored in a triglyceride molecule.

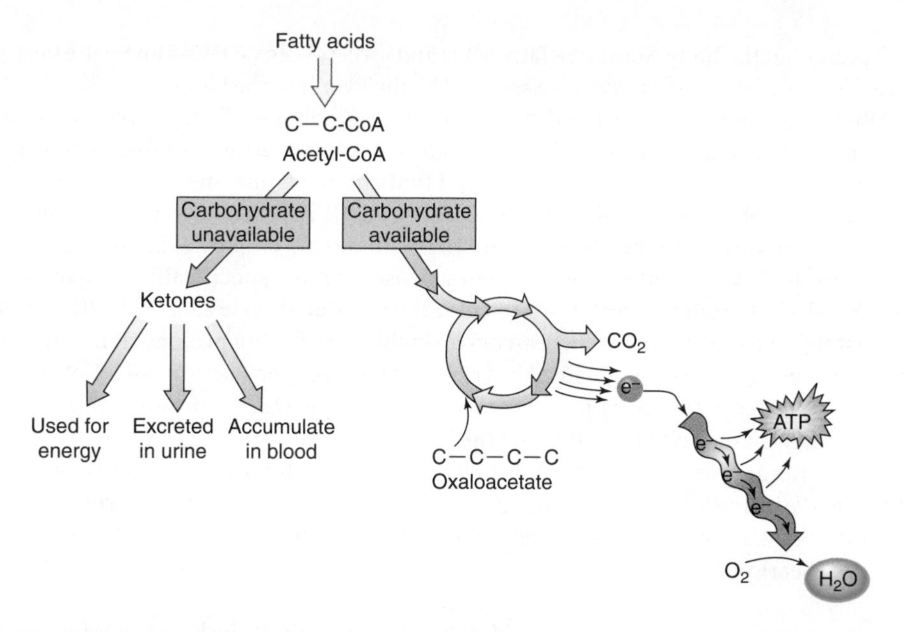

Figure 5.19 Ketone body formation
When glucose, and as a consequence, oxaloacetate levels in the cell are low, fatty acids will be converted to ketone bodies.

as carbohydrate or protein must first go to the liver, where the carbohydrate and protein can be used, although inefficiently, to synthesize fatty acids; these fatty acids are then assembled into triglycerides, which are transported to the adipose tissue in VLDLs. Lipoprotein lipase at the membrane of cells lining the blood vessels breaks down the triglycerides from both chylomicrons and VLDLs so that the fatty acids can enter the fat cells, where they are reassembled into triglycerides for storage (**Figure 5.20**).

The ability of the body to store fat is substantial. Fat cells can increase in weight by about 50 times, and new fat cells can be synthesized when existing cells reach their maximum size (see Section 7.1). Because each gram of fat provides 9 kcal, compared with only 4 kcal/g from carbohydrate or protein, a large amount of energy can be stored in the body as fat without a great increase in size or weight. Even a lean man, whose body fat is only about 10% of his weight, stores more than 50,000 kcal of energy as fat.

Using Stored Fat Adipose tissue releases fatty acids into the blood to use as a source of energy. These fatty acids are referred to as "free" fatty acids to emphasize that they are not in the form of triglycerides. The amount of free fatty acids released is proportional to the amount of adipose

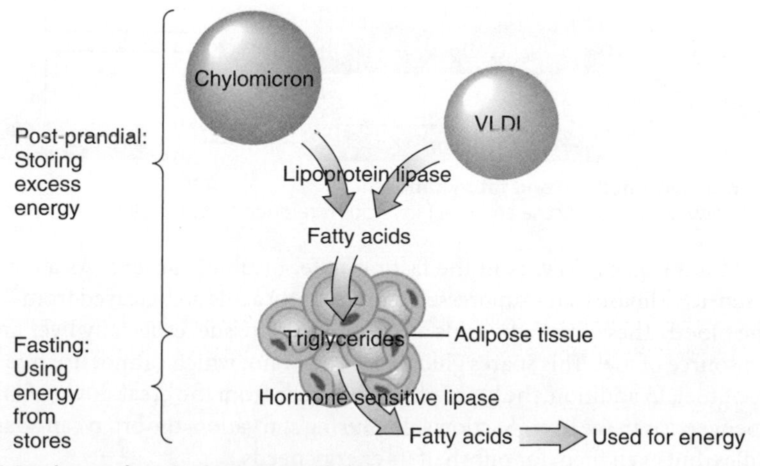

Figure 5.20 Storing and retrieving energy in fat
The enzymes lipoprotein lipase and hormone-sensitive lipase mediate the storage and removal of triglycerides in adipose tissue according to energy intake and energy needs.

tissue present in the body. Some free fatty acids and glycerol can be taken up by the liver and converted to VLDL, which in turn are secreted by the liver into the blood.

hormone-sensitive lipase An enzyme present in adipose cells that responds to chemical signals by breaking down triglycerides into free fatty acids and glycerol for release into the bloodstream.

When less energy is consumed than is needed, as when we fast, the release of these fatty acids is especially stimulated. In this situation, the enzyme **hormone-sensitive lipase** inside the fat cells receives a hormonal signal that enhances enzyme activity, promoting the breakdown of stored triglycerides (see Figure 5.20). The free fatty acids and glycerol are released directly into the blood, where they can be taken up by cells throughout the body to produce ATP. If there is not enough glucose, or more specifically oxaloacetate, to allow acetyl-CoA from fat breakdown to enter the citric acid cycle (Figure 5.19), it will be used instead to make ketones. This can occur in the liver during prolonged fasting when glycogen stores (a source of glucose) have been depleted (see Section 4.4). Ketones can be used as an energy source by muscle and adipose tissue. During prolonged starvation, meaning after many days without food , the brain can adapt to use ketones to meet about half of its energy needs. For the other half, it must use glucose. Fatty acids cannot be used to make glucose, and only a small amount of glucose can be made from glycerol. The body breaks down muscle protein and converts it to glucose in order to fuel the brain (See Section 6.4).

Integration of Fat and Carbohydrate Metabolism The role of the hormones insulin and glucagon on the regulation of blood glucose is discussed in Section 4.5. In this chapter, you have learned about the transport of lipids after a meal (Figure 5.14) and the role of adipose tissue in storing and retrieving energy (Figure 5.20). While the relationship between insulin and glucose is well known, insulin also plays a central role in the regulation of lipid metabolism.

As shown in **Figure 5.21**, when blood-glucose levels are high, such as in the post-prandial state, insulin stimulates the uptake of glucose by the liver and its conversion to the storage polysaccharide, glycogen. Insulin also stimulates the activity of the enzyme lipoprotein lipase, which promotes the uptake of triglycerides from chylomicrons, especially by adipose tissue. There is also suppression by insulin of the activity of hormone-sensitive lipase, which in turn suppresses of the release of free fatty acids from adipose tissue into the blood.

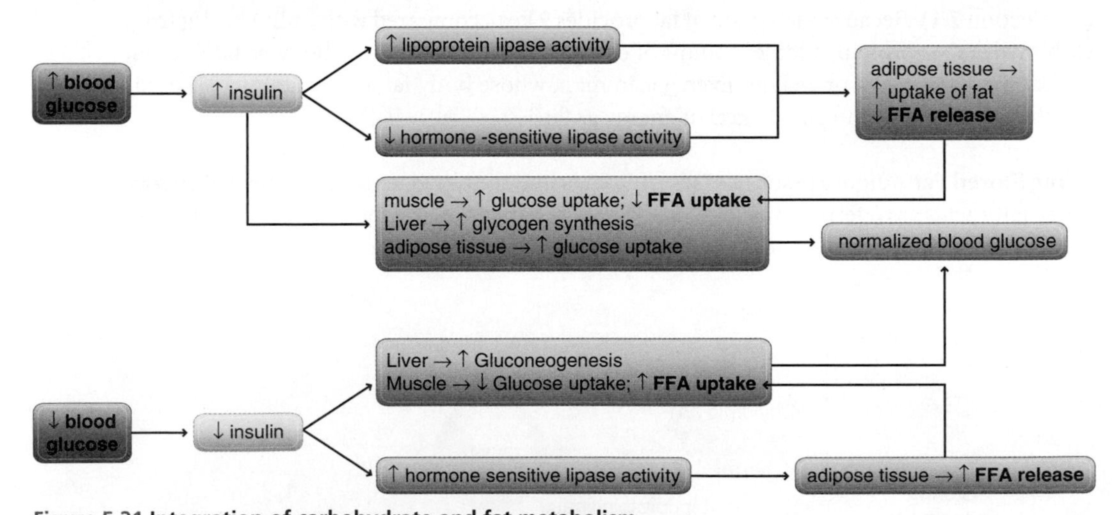

Figure 5.21 Integration of carbohydrate and fat metabolism
The inverse relationship between blood glucose and free fatty acids is referred to as the glucose-fatty acid cycle.

glucose/fatty-acid-cycle The relationship between blood glucose and free fatty acids. When blood glucose levels are high, as in the post-prandial state, free-fatty-acid levels are low. In the fasting state, when blood glucose levels decline, free-fatty-acid levels increase.

When blood sugar is low, as in the fasting state, insulin is absent. As a consequence, hormone-sensitive lipase is not suppressed and free fatty acids are released from adipose tissue into the blood. These free fatty acids are taken up by tissue, especially liver and muscle, and act as a source of fuel. This spares glucose for the brain, which cannot use free fatty acids as a source of fuel. In addition, the brain obtains glucose from the breakdown of amino acids and gluconeogenesis in the liver (Section 4.4). During starvation, the brain can adapt to using ketone bodies, but even then, for only half its energy needs.

This inverse relationship between blood glucose and free fatty acids, as highlighted in Figure 5.21, is referred to as the **glucose/fatty acid cycle.**

5.6 Lipids and Health

Learning Objectives

- Describe the development of atherosclerosis.
- Distinguish dietary lipids that increase the risk of heart disease from those that decrease the risk.
- Discuss the relationship between dietary fat and cancer.
- Explain whether dietary fat intake is related to obesity in the Canadian population.

Adequate amounts of essential fatty acids must be consumed in the diet to maintain health. However, diets high in fat, particularly certain types of fats, are associated with an increased risk for many chronic diseases. The development of **cardiovascular disease** has been linked to diets high in cholesterol, saturated fat, and *trans* fat.[11] The risk of certain types of cancer, including that of the breast, colon, and prostate, has been associated with a high-fat intake. Obesity is also associated with diets high in fat because these diets are usually high in energy and promote storage of body fat. Excess body fat in turn is associated with an increased risk of diabetes, cardiovascular disease, and high blood pressure.

cardiovascular disease Any disease affecting the heart and blood vessels.

Essential Fatty Acid Deficiency

If adequate amounts of linoleic and alpha-linolenic acid are not consumed, an **essential fatty acid deficiency** will result. Symptoms include scaly, dry skin, liver abnormalities, poor healing of wounds, growth failure in infants, and impaired vision and hearing. Essential fatty acid deficiency is rare because the requirement for essential fatty acids is well below the typical intake. Deficiencies have been seen in infants and young children fed low-fat diets and in individuals who are unable to absorb lipids.

essential fatty acid deficiency A condition characterized by dry, scaly skin and poor growth that results when the diet does not supply sufficient amounts of the essential fatty acids.

Cardiovascular Disease

In 2007, cardiovascular disease killed 73,000 Canadians, making it the leading cause of death and accounting for almost 30% of all deaths in Canada that year.[13] Epidemiological studies have shown that diet and lifestyle both affect the risk of developing heart disease. The relationship between dietary fat and heart disease risk is one that has been extensively studied. In general, populations that consume high-fat diets have a higher incidence of heart disease, but this does not always hold true. Populations that consume a diet high in omega-3 fatty acids, such as the Inuit in Greenland, have a low incidence of heart disease.[14] In Mediterranean countries, where the diet is high in monounsaturated fat as well as grains and vegetables, death from heart disease is also less frequent.[15] It is the type of fat that determines the risk of cardiovascular disease (see Science Applied: Fat and Serum Cholesterol Levels).

How Does Atherosclerosis Develop? **Atherosclerosis** is a type of cardiovascular disease in which lipids and fibrous material are deposited in the artery walls, reducing their elasticity and eventually blocking the flow of blood. Although the deposition of lipids within the artery wall is an important component of the process, we now know that inflammation, which is the process whereby the body responds to injury, is what drives the formation of **atherosclerotic plaque.** An inflammatory response is initiated by an injury, such as when you cut yourself. White blood cells, which are part of the immune system, rush to the injured area, blood clots form, and new tissue grows to heal the wound. Similar inflammatory responses occur when an artery is injured, but instead of resulting in healing, they lead to the development of atherosclerotic plaque (**Figure 5.22**).

atherosclerosis A type of cardiovascular disease that involves the buildup of fatty material in the artery walls.

atherosclerotic plaque The cholesterol-rich material that is deposited in the arteries of individuals with atherosclerosis. It consists of cholesterol, smooth-muscle cells, fibrous tissue, and eventually calcium.

The atherosclerotic process begins as a response to an injury that causes changes in the lining of the artery wall. The exact cause of the injury is not known but may be related to elevated blood levels of LDL cholesterol, glucose, or the amino acid homocysteine; high blood pressure; chemicals derived from cigarette smoking; diabetes; genetic alterations; or infectious micro-organisms.[16] The specific cause may be different in different people.

Once the initial injury has occurred, the lining of the artery becomes more permeable to LDL particles, which migrate into the artery wall (see Figure 5.22). In an uninjured blood vessel, the LDL particle would not enter the artery wall, but continue to the capillaries

SCIENCE APPLIED

Dietary Fat and Serum Cholesterol Levels

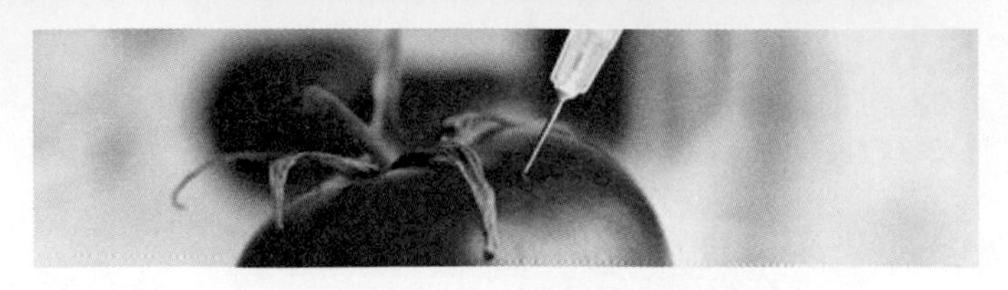

One of the problems that researchers face when studying the effect of diet on disease, such as cardiovascular disease, is that the disease may take decades to develop. If you wanted to determine whether a diet reduced the incidence of cardiovascular disease using an intervention trial, you would theoretically have to feed test subjects an experimental diet for many years to see a result, a very complex and costly process. Fortunately, to determine whether a diet may impact the risk of cardiovascular disease, it is not necessary for the disease itself to be the endpoint or main outcome of a study. Because there have been many studies that have established serum (blood) cholesterol levels as a risk factor for cardiovascular disease, a study that measures the effect of diet on serum cholesterol is sufficient. The good news is that diet can often alter cholesterol levels in 4-8 weeks, greatly simplifying the ability to determine the impact of diet on the risk of cardiovascular disease. Factors such as serum cholesterol and other similar biochemical measurements are often referred to as biomarkers.

It has been stated that the risk of cardiovascular disease increases with increasing levels in the blood of LDL cholesterol and decreases with increased HDL cholesterol. Scientists combine the effect of these two biomarkers by calculating the ratio of total cholesterol to HDL cholesterol. This ratio has been found to be a strong predictor of cardiovascular risk. A decrease in the ratio is associated with a decreased risk of cardiovascular disease.

The graph in this section is taken from a review article in which the authors conducted a meta-analysis of 60 intervention trials to measure the effect of changing the fat composition of a typical American diet on the ratio of total cholesterol to HDL cholesterol, and by inference, on the risk of cardiovascular disease.(See Critical Thinking: Fish Consumption and Heart Disease for an explanation of meta-analysis)

It illustrates clearly why Canada's Food Guide (Section 2.3) advises Canadians to limit their intake of butter, high in saturated fat, and hard margarines and shortening, which contain both saturated and *trans* fats. The ratio of total cholesterol to HDL cholesterol increases when these fats are added to the diet, increasing the risk of cardiovascular disease. Conversely, the replacement of some of the fats with unsaturated oils, for example, soybean, olive, and rapeseed (i.e., canola) decreases the risk of cardiovascular disease.

Effect of fat on a biomarker of cardiovascular-disease risk

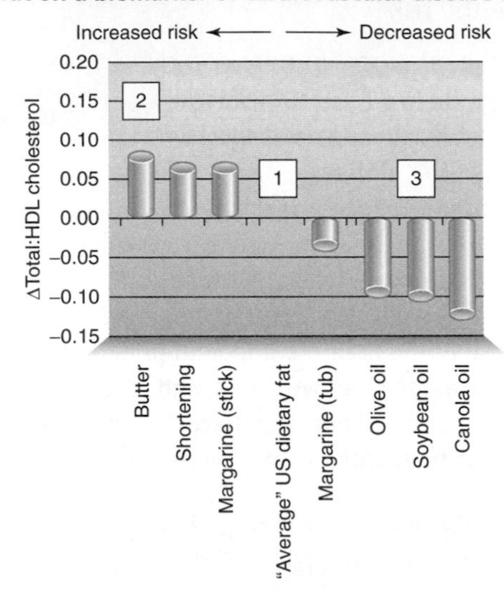

The numbers (1), (2), and (3) below refer to the regions on the graph:

(1) This represents the composition of the typical American diet (the Canadian diet is similar). The bar graphs to the right and left of this typical diet illustrates how the ratio of total cholesterol to HDL cholesterol changes when a fixed portion of the fat in the typical diet is replaced with the shown fat.

(2) This bar shows the effect of replacing part of the fat in the typical diet with butter. The ratio of total cholesterol to HDL cholesterol goes up. Butter is high in saturated fat, so the effect of the replacement is to increase overall saturated fat content of the diet, which increases the risk of cardiovascular disease. Not surprisingly, similar effects are observed for stick (hard) margarine and shortening, which contain large amounts of *trans* fats.

(3) This bar shows the effect of replacing part of the fat in the typical diet with soybean oil, which is high in unsaturated fat. The ratio of total cholesterol to HDL cholesterol goes down, consistent with a reduction in the risk of cardiovascular disease. Similar effects are indicated for other unsaturated oils and also for soft margarine, which is low in *trans* fat and is typically a mixture of 75%-80% unsaturated fats and 20%-25% saturated fats.

Source: Mensink, R. P., Zock, P. L., Kester, A. D., et al. Effects of dietary fatty acids and carbohydrates on the ratio of serum total to HDL cholesterol and on serum lipids and apolipoproteins : a meta-analysis of 60 controlled trials. *Am J. of Clin Nutr* 77(5):1146-1155, 2003.

---

oxidized LDL cholesterol A substance formed when the cholesterol in LDL particles is oxidized by reactive oxygen molecules. It is key in the development of atherosclerosis because it contributes to the inflammatory process.

scavenger receptors Proteins on the surface of macrophages that bind to oxidized LDL cholesterol and allow it to be taken up by the cell.

and then enter nearby tissue. Once inside the blood vessel wall, the LDL particles are modified chemically, often by oxidation to form **oxidized LDL cholesterol.** Oxidized LDL cholesterol is harmful and its presence promotes inflammation in a number of ways. It triggers the production and release of substances that cause immune system cells to stick to the lining of the artery and then to migrate into the artery wall. Once inside, these cells become large white blood cells called macrophages, which have **scavenger receptors** on their surface. Just as LDL receptors bind to LDL cholesterol, scavenger receptors bind to and transport oxidized LDL cholesterol into the interior of the cell. As macrophages fill with more and more oxidized LDL cholesterol, they are transformed into cholesterol-filled foam cells (named because of their foamy appearance under a microscope). Foam cells accumulate in the artery wall and then burst, depositing cholesterol to form a fatty streak (see Figure 5.22).[16,17]

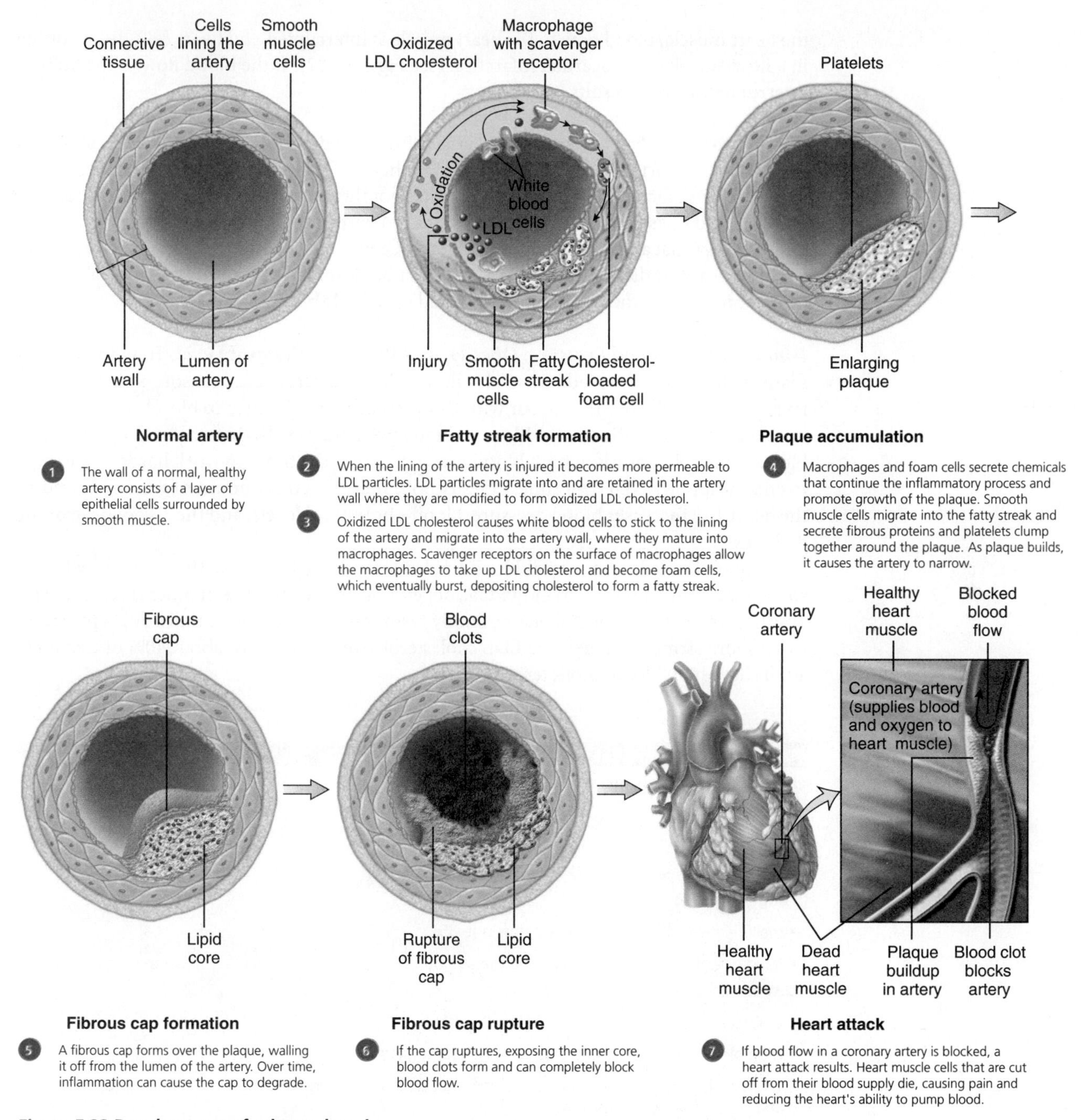

Figure 5.22 Development of atherosclerosis
The inflammation that occurs in response to an injury to the artery wall precipitates the development of atherosclerotic plaque. The buildup of plaque can eventually lead to a heart attack or stroke.

Macrophages and foam cells secrete growth factors and other chemicals that continue the inflammatory response and promote growth of the plaque. The release of growth factors signals smooth-muscle cells from the wall of the artery to migrate into the fatty streak and secrete fibrous proteins. Platelets, which are cell fragments involved in blood clotting, become sticky and clump together around the lesion. As the lesion enlarges, it causes the artery to narrow and lose its elasticity, hampering blood flow. As the process progresses, a fibrous cap of smooth-muscle cells and fibrous proteins forms over the mixture of white blood cells, lipids, and debris, walling it off from the lumen of the artery. The formation of the cap is a way of healing the injury, but if the inflammation continues, substances secreted by immune system cells can degrade this cap.[17] If the cap becomes too thin and ruptures, the material leaks out and causes blood clots to form. The clots can completely block the artery at that spot, or break loose and block an artery elsewhere. If this occurs in blood vessels that supply the heart muscle, blood flow to the heart muscle is interrupted and heart cells die, resulting in a heart attack or myocardial infarction (see Figure 5.22). If the blood flow to the brain is interrupted, a stroke results.

Risk Factors for Heart Disease It is evident from the description of the process above, in which the LDL particle plays a critical role, why high levels of LDL or "bad" cholesterol in the blood are a risk factor for heart disease. But it is not the only risk factor. High blood pressure, diabetes, and obesity, are also strongly associated with the risk of developing heart disease. Other factors that affect risk include age, gender, genetics, and lifestyle factors such as smoking, exercise, and diet. These may affect risk or act indirectly by altering blood cholesterol levels or increasing the risk of injury to blood vessels (**Table 5.2).**

Diabetes, High Blood Pressure, Obesity, and Blood-Cholesterol Levels Individuals with diabetes have an increased risk of developing heart disease. One reason is that the high levels of blood glucose that occur with this disease cause damage to blood vessels, which initiates atherosclerotic events. Elevated blood pressure can also increase risk by damaging blood vessels. In addition, high blood pressure forces the heart to work harder, causing it to enlarge and weaken over time. Obesity both increases the amount of work required by the heart and increases blood pressure, blood-cholesterol levels, and the risk of developing diabetes.

The desirable level for total blood lipids is shown in Table 5.2. These values will vary somewhat from individual to individual, depending on the presence of other risk factors; for example, the presence of diabetes or family history might prompt an individual's physician to set more stringent targets for LDL cholesterol levels. Currently, about 40% of Canadian adults have high blood-cholesterol levels.[18]

Table 5.2 Risk Factors for Heart Disease
Heart disease risk is increased by the following:
Age:
• men $\geq$ 45 years
• women $\geq$ 55 years
Family History:
• male relative with heart disease before age 55
• female relative with heart disease before age 65
Disease Factors:
• Diabetes: fasting blood sugar $\geq$ 7 mmol/L
• High blood pressure: blood pressure $\geq$ 140/90 mm Hg†
• Overweight: body mass index $>$ 25 kg/m^2* and waist circumference $>$ 88 cm (35 inches) for women and $>$ 102 cm (40 inches) for men
Altered Blood Lipid Levels (these levels may vary depending on the presence of other risk factors—see text):**
• LDL cholesterol $\geq$ 3.5 mmol/L
• HDL cholesterol $<$ 1.0 mmol/L
• Total cholesterol $\geq$ 5.2 mmol/L
• Triglycerides $\geq$ 1.7 mmol/L
• Total cholesterol/HDL-cholesterol $>$ 5.0 mmol/L
Lifestyle:
• Cigarette smoking
• A sedentary lifestyle
• A diet high in saturated fat, *trans* fat, and cholesterol and low in fibre fruits, and vegetables

*See Section 7.5 and Appendix B for information about body mass index and how it can be calculated.
†See Section 10.3 for information on hypertension and the measurement of blood pressure
** Source: McPherson R, Frohlich J, Fodor G, Genest J, Canadian Cardiovascular Society. Canadian Cardiovascular Society position statement--recommendations for the diagnosis and treatment of dyslipidemia and prevention of cardiovascular disease. *Can J Cardiol.* 22(11):913-27, 2006.

Age, Gender, Genetics, and Lifestyle Factors The risk of developing heart disease increases with age, as illustrated in **Figure 5.23.** Men and women are both at risk, but men are generally affected a decade earlier than women. This is due in part to the protective effect of the hormone estrogen in women. As women age, the effects of menopause—including the decline in estrogen level and gain in weight—increase heart-disease risk. Although we tend to think of heart disease as a man's disease, cardiovascular disease is the number one killer of both Canadian women and men. To raise awareness of cardiovascular disease, the website The Heart Truth, specifically aimed at women, was created by the Heart and Stroke Foundation (www.thehearttruth.ca).

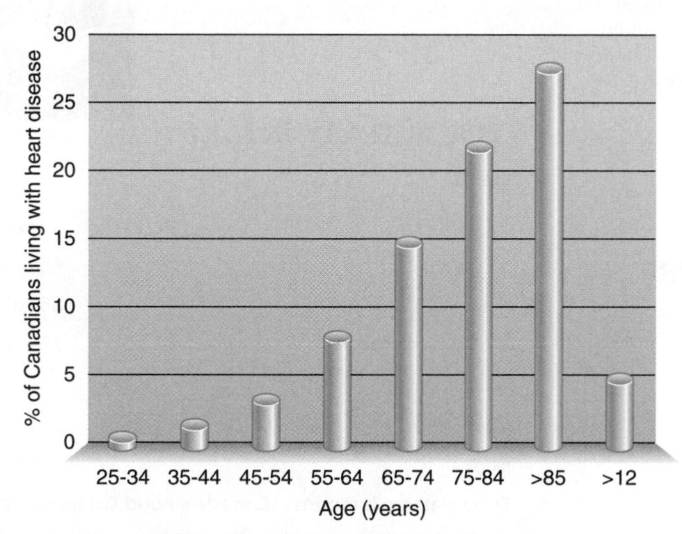

Figure 5.23 The percentage of Canadians living with heart disease.
As Canadians get older more are living with heart disease.
Source: Public Health Agency of Canada. 2009 Tracking Health Disease and Stroke. Available online at http://www.phac-aspc.gc.ca/publicat/2009/cvd-avc/pdf/cvd-avs-2009-eng.pdf. Accessed June 10, 2011.

Lifestyle factors have a significant impact on the risk of heart disease. These factors include activity level, smoking, and diet. An inactive lifestyle and cigarette smoking both increase the risk of heart disease. On the other hand, regular exercise decreases risk by promoting the maintenance of a healthy body weight, reducing the risk of diabetes, increasing HDL cholesterol, and reducing blood pressure. A number of dietary factors, including high intakes of saturated fat, *trans* fat, and cholesterol, increase risk. Other dietary factors, including adequate intake of fibre, fruits and vegetables, unsaturated fats, and antioxidants, may offer a protective effect.

Genetics also affect risk. Individuals with a male family member who exhibited heart disease before the age of 55 or a female family member who exhibited heart disease before the age of 65 are considered to be at greater risk.

Ethnic background may also be a risk factor. Studies suggest that Canadians of South Asian origin (i.e., from India, Pakistan, Bangladesh, or Sri Lanka) have a higher risk for heart disease than Canadians with a European background, who in turn have a higher risk than Chinese Canadians.[19] Research also indicates that First Nations people are also at higher risk for cardiovascular disease, compared to Canadians of European origin.[20] These differences in ethnic groups may reflect both genetic differences between populations, and differences in lifestyle factors, for example, rates of smoking, diets high in saturated fats, that may exist among populations. You can't change your age, gender, or genetic background, but you can control these lifestyle factors (see Critical Thinking: Lowering the Risk of Heart Disease).

Dietary Lipids that Promote Heart Disease Excessive intake of cholesterol, saturated fat, and *trans* fatty acids can increase the risk of cardiovascular disease. Some or all of the effects of these lipids are due to their impact on blood-cholesterol levels.

Dietary Cholesterol The extent to which cholesterol intake affects blood levels depends on an individual's genes. Cholesterol in the blood comes from cholesterol both consumed in the diet

CRITICAL THINKING:

Lowering the Risk of Heart Disease

Background

Rafael is a financial advisor. He spends much of his day at his computer and when he does get out, he is in his car or taking a client out to lunch. When he is home with his family, he enjoys watching his children play soccer and basketball but rarely finds time to exercise himself. His mother died of a heart attack at age 60. Rafael is worried about his own heart-disease risk, so he makes an appointment with his physician. He fills out a questionnaire about his medical history and lifestyle and has blood drawn for cholesterol analysis.

Data

The table summarizes some of the information that may affect Rafael's risk of developing heart disease:

HEART DISEASE RISK QUESTIONNAIRE

Sex:	Male
Age:	35
Family history:	Mother had heart attack at age 60
Height/weight:	1.7 metres/72 kg
BMI	24.9
Blood pressure:	120/70
Smoker:	Yes
Activity level:	Sedentary
Blood values:	
Total cholesterol	5.4 mmol/L
LDL cholesterol	4.3 mmolL
HDL cholesterol	0.9 mmolL
Triglycerides	1.3 mmolL
Total/HDL cholesterol	6 mmol/L

Rafael's physician recommends he meet with a dietitian to evaluate his diet. The following information is obtained from a 24-hour diet recall and a food-frequency questionnaire.

(Brand X Pictures/Getty Images, Inc.)

TYPICAL DAILY INTAKE

Lipids	
Total fat	33% of kcalories
Saturated fat	18% of kcalories
Polyunsaturated fat	8% of kcalories
Monounsaturated fat	7% of kcalories
Trans fat	7% of kcalories
Cholesterol	350 mg/day
Fibre	22 g/day
Food pattern/amounts (Canada's Food Guide servings)	
Whole grains	2
Vegetables and fruits	6
Whole-fat dairy products	3
Red meat	2
Solid fats (butter)	15 ml/day
Nuts and seeds	0
Legumes	0

Critical Thinking Questions

Is Rafael at risk for a heart attack? Review his risk factor questionnaire and list the factors that increase his risk (see Table 5.2).

What about his diet? Based on the information from his typical intake, how does his fat intake compare with recommendations? Are there any other food choices he makes that increase his risk?

Suggest some diet and lifestyle changes that would reduce Rafael's risk of developing heart disease.

Use iProfile to find dinner entrées that are low in saturated fat and cholesterol.

and made by the liver. Generally, about 3-4 times more cholesterol is made by the body than is consumed in the diet. In most individuals, as dietary cholesterol increases, liver cholesterol synthesis decreases so that blood levels do not change.[21] In others, however, liver synthesis does not decrease in response to an increase in dietary cholesterol, so blood-cholesterol levels rise.

Since we cannot easily identify which individuals are especially sensitive to dietary cholesterol, it is prudent for all individuals to avoid excessive intake.

Saturated Fat When the diet is high in saturated fatty acids, liver production of cholesterol-carrying lipoproteins increases and the activity of LDL receptors in the liver is reduced, so that LDL cholesterol cannot be removed from the blood.[22] Therefore, diets high in saturated fat increase LDL cholesterol in the blood. Increased LDL cholesterol then increases the risk of atherosclerosis. When the diet is low in saturated fats and high in unsaturated fat, lipoprotein production decreases and the number of LDL receptors increases, allowing more cholesterol to be removed from the bloodstream. This lowers LDL cholesterol levels and the risk of heart disease (see **Figure 5.24**). Some saturated fatty acids, such as stearic acid found in chocolate and beef, do not increase blood-cholesterol levels. However, these may contribute to heart disease by lowering HDL cholesterol, and by affecting blood platelets and blood clotting, both of which are involved in the development of atherosclerosis.[23]

Trans Fatty Acids Both clinical and epidemiological studies provide evidence that a high *trans* fatty acid intake increases the risk of heart disease. Many studies have found that diets high in *trans* fatty acids cause a greater increase in heart disease risk than those high in saturated fatty acids.[24]Some of the increase in risk is due to the effect of *trans* fatty acids on blood cholesterol levels. *Trans* fatty acids promote the synthesis of cholesterol in the liver.[25] This in turn may cause the observed increases in LDL cholesterol levels. At high intakes, *trans* fatty acids also lower HDL cholesterol (see Figure 5.24). It has also been hypothesized that *trans* fatty acids may increase the risk of heart disease by increasing inflammation, an important risk factor in the development of atherosclerosis.[26]

Dietary Lipids that Protect Against Heart Disease Dietary omega-6 and omega-3 polyunsaturated fatty acids as well as monounsaturated fatty acids tend to decrease the risk of heart disease. Some of this protection is due to a reduction in LDL cholesterol and an increase in HDL cholesterol, but other mechanisms also play a role.

Omega-6 and Omega-3 Polyunsaturated Fat When saturated fat in the diet is replaced by any type of polyunsaturated fat (except *trans*), blood levels of LDL cholesterol decrease.[27] This

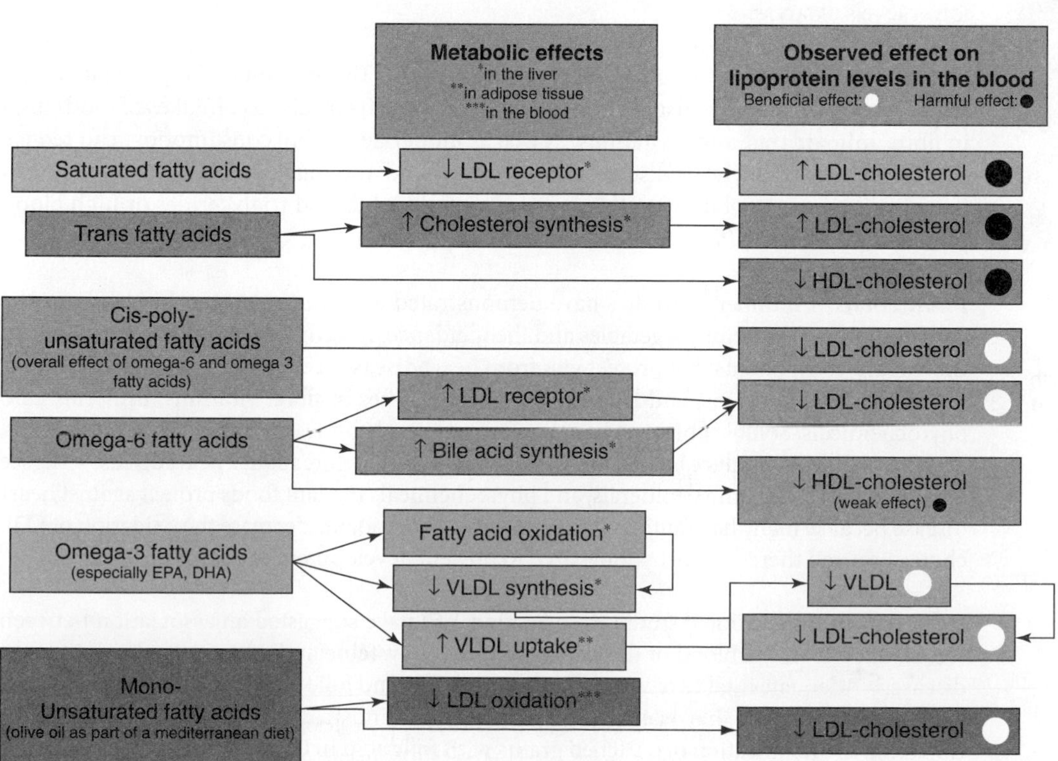

Figure 5.24 Effect of fatty acids on cholesterol levels in the blood and some metabolic effects that may contribute to these observed changes.

may be due to the effects of omega-6 fatty acids on gene expression, specially an increase in the expression of the LDL receptor. This receptor takes up LDL particles into the liver and other tissue, removing them from the blood. Omega-6 fatty acids may also promote bile acid synthesis, increasing the elimination of cholesterol in the feces.[25] However, a high intake of omega-6 polyunsaturated fatty acids may also cause a slight decrease in HDL cholesterol, which is undesirable in terms of heart disease risk (see Figure 5.24).

Increasing omega-3 fatty acid intake, specifically EPA and DHA from fish, lowers levels of LDL cholesterol but not HDL cholesterol.[27] These omega-3 fatty acids also influence gene expression in a way that results in increased fatty acid oxidation in the liver. Increased fatty acid oxidation reduces the fatty acids available for VLDL synthesis and secretion into the blood. Reducing VLDL levels in the blood reduces the level of atherogenic LDL particles (which are derived from VLDLs, see Figure 5.14). Similarly, omega-3 fatty acids act on gene expression in a way that results in an increase in uptake of VLDL particles by adipose tissue, further reducing their levels in the blood[25] (see Figure 5.24). EPA and DHA may also reduce arrhythmias (irregular heartbeat) which reduces the risk of death from cardiovascular disease[28]. As noted, in Section 5.5, eicosanoids derived from EPA reduce blood clotting, which may prevent heart attack (Figure 5.22, steps 6 and 7) and reduce inflammation, which may contribute to the development of atherosclerosis. Because of the beneficial effects of EPA and DHA, Canada's Food Guide recommends consuming fish twice a week to increase the intake of omega-3 fatty acids (**Figure 5.25**).

Figure 5.25 Fish, walnuts, flaxseed, and canola oil are good sources of omega-3 fatty acids. (Andy Washnik)

Monounsaturated Fat Olive oil is very high in monounsaturated fats and is the major source of fat in the traditional Mediterranean diet (see Section 2.4). Populations that consume the traditional Mediterranean diet have a mortality rate from heart disease that is half of that in Canada and the United States (**Figure 5.26**). This is true even when total fat intake provides 40% or more of energy intake.[29] Substituting monounsaturated fat for saturated fat reduces unhealthy LDL cholesterol without decreasing healthy HDL cholesterol and makes LDL cholesterol less susceptible to oxidation[27] (see Figure 5.25). However, the type of fat in the diet is unlikely to be the only factor responsible for the difference in the incidence of heart disease between the Mediterranean countries and Canada. The traditional Mediterranean diet is higher in fruits and vegetables, lower in animal products, includes more wine, and is consumed in countries where the lifestyle includes more day-to-day physical activity and lower levels of stress.

Figure 5.26 The high proportion of monounsaturated fatty acids in olive oil, which is used liberally in the Mediterranean diet, is one factor believed to contribute to the health of the people in the Mediterranean region. (Frank Wieder/StockFood Munich/StockFoodAmerica)

Other Dietary Factors that Affect Heart Disease Risk The amount and type of fat are not the only dietary factors that affect the risk of developing heart disease. Intakes of foods high in fibre, antioxidants, and B vitamins, as well as moderate alcohol consumption, can reduce the risk of developing heart disease. On the other hand, too much added sugar and salt can increase risk by contributing to the development of high blood triglycerides or high blood pressure, respectively (see sections 4.6 and 10.3).

Plant Foods A number of studies have demonstrated an inverse relationship between the consumption of fruits and vegetables and the incidence of cardiovascular disease.[30] Many of the dietary components that protect you from heart disease are found in plant foods. Fruits, vegetables, whole grains, and legumes are good sources of fibre, vitamins, minerals, and phytochemicals. Soluble fibres, such as those in oat bran, legumes, psyllium, pectin, and gums, have been shown to reduce blood cholesterol levels and therefore reduce heart disease risk (see Section 4.6). The vitamins, minerals, and phytochemicals in plant foods protect against heart disease because many have antioxidant functions. Antioxidants decrease the oxidation of LDL cholesterol and therefore are hypothesized to prevent development of plaque in artery walls.[30]

B Vitamins Observational studies (see Section 1.5) have suggested an association between elevated levels in the blood of the amino acid homocysteine and an increased risk of heart disease.[31] Adequate intakes of vitamin B_6, vitamin B_{12}, and folic acid may help protect against heart disease because they keep blood levels of the amino acid homocysteine low (see Section 8.7). The fortification of enriched grains with folic acid in Canada and the United States in 1998 increased folic acid intake. This has resulted in a reduction in blood homocysteine levels in the American population.[32] Although comprehensive studies on homocysteine levels in the Canadian population have not been done, the same is probably true in Canada. There is, however, insufficient evidence from intervention trials (see Section 1.5) to support the use of vitamin supplements for the prevention of heart disease. (See the Homocysteine Hypothesis, Section 8.7).

Niacin is another B vitamin that may affect heart disease risk. When consumed in extremely high doses, the nicotinic acid form of niacin can be used to lower blood cholesterol. Nicotinic acid is inexpensive and widely available without a prescription, but at the high doses needed to lower cholesterol, it is really a drug, not a nutrient. Because of the potential side effects, an individual using nicotinic acid as a drug to lower cholesterol should be monitored by a physician.

Life Cycle

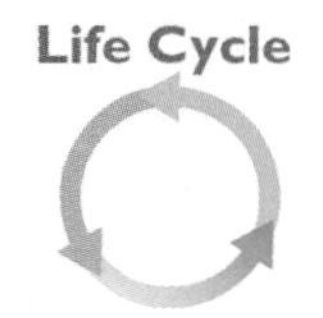

Moderate Alcohol Consumption Scientific research has shown that moderate alcohol consumption may reduce stress, raise levels of HDL cholesterol, and reduce blood clotting and thus reduce the risk of cardiovascular disease.[33] These effects are greater when red wine is consumed, as it commonly is in France, Italy, and Greece. Red wine is high in phytochemicals called polyphenols, which are antioxidants that protect against LDL oxidation, and may have other effects that protect against the development of atherosclerosis.[34] Moderate drinking means no more than 1 drink per day for women and 2 drinks per day for men. One drink is defined as 350 ml of beer, 150 ml of wine, or 50 ml of 80-proof distilled spirits. Greater intake of alcohol increases the risk of accidental deaths, heart disease, cancer, birth defects, and drug interactions and should be avoided. Furthermore, alcohol consumption is not recommended for children or adolescents, pregnant women, individuals who cannot restrict their drinking to a moderate amount, or individuals who plan to drive or perform other activities that require concentration (see Focus on Alcohol). Alcohol can be a dangerous drug, so despite its benefits, Canada's Food Guide lists alcohol among the foods that should be limited.

Overall Dietary Pattern and Canada's Food Guide The question was asked in Chapter 2, "Does following a healthy dietary pattern reduce the risk of disease?" (See Chapter 2, Science Applied: Does Following a Healthy Dietary Pattern Reduce the Risk of Disease?) and the impact of total diet on the risk of cardiovascular disease, using the Alternative Healthy Eating Index to measure diet quality, was discussed. A person following Canada's Food Guide would have a high score using this index and as the results showed, there was a very dramatic decline in overall risk of cardiovascular disease as diet quality increased (**Table 5.3**). Nutritional research can identify specific foods and nutrients that have a beneficial or harmful effect on cardiovascular health, as described above, but it is overall dietary patterns, such as those described in Canada's Food Guide, that ultimately determine health.

Table 5.3 The Risk of Cardiovascular Disease (CVD) Declines as Diet Quality Increases

Score	RR: CVD*	RR: CVD
	Men	Women
Lowest score-poorest diet	1.0	1.0
	0.85	0.95
	0.79	0.80
	0.67	0.75
Highest score-best diet	0.61	0.72

*See Chapter 2: Science Applied: Does Following a Healthy Dietary Pattern Reduce the Risk of Disease? for a review of relative risk (RR)

Source: McCullough, M. L., Willett, W. C. Evaluating adherence to recommended diets in adults: the Alternate Healthy Eating Index. *Public Health Nutr.* 9(1A):152–7, 2006.

Dietary Fat and Cancer

Cancer is the second-leading cause of death in Canada. As with cardiovascular disease, epidemiology suggests that diet and lifestyle affect cancer risk. It is estimated that 30%-40% of cancers are directly linked to dietary and lifestyle choices.[35] Populations consuming diets high in fruits and vegetables tend to have a lower cancer risk. These foods are high in fibre and provide antioxidant vitamins and phytochemicals. In contrast, populations that consume diets high in fat, particularly animal fats, have a higher cancer incidence. The mechanism whereby a high intake of dietary fat increases the incidence of various cancers is less well understood than the relationship between dietary fat and cardiovascular disease; however, dietary fat has been suggested to be both a tumour promoter and tumour initiator, depending on the type of cancer. The good news is that for the most part, the same type of diet that protects you from cardiovascular disease will also protect you from certain forms of cancer.

Dietary Fat and Breast Cancer Breast cancer is the leading form of cancer in women worldwide. About 20,000 Canadian women were diagnosed annually with the disease.[36] Breast cancer is more common in post-menopausal women, in women who have had no children or who had children later in life, and in women with a family history of the disease.[37]

The type of fat in the diet may affect the risk of breast cancer. For example, the incidence of breast cancer in Mediterranean women who rely on olive oil, high in monounsaturated fat, as a source of dietary fat is low despite a total fat intake similar to that in Canada.[38] Epidemiology also supports a protective effect from an increased intake of omega-3 fatty acids from fish.[39] *Trans* fatty acids, on the other hand, have been suggested to increase the risk of breast cancer.[40]

The mechanism by which diet affects breast cancer has been studied in laboratory animals. Since most laboratory animals do not get breast cancer tumours, studies are conducted by implanting breast tumours and examining how diet affects their growth. These experiments demonstrate that dietary fat is a tumour promoter; the tumours are more likely to grow in mice fed a high-fat diet than in those fed a low-fat diet. The type of fat also affects tumour growth; in animals fed diets high in linoleic acid, which is found in polyunsaturated vegetable oils, the tumours grow faster than in rats fed diets high in saturated fatty acids or omega-3 fatty acids. However, these animal studies have not been supported by human trials.[41]

Dietary Fat and Colon Cancer Epidemiology has correlated the incidence of colon cancer with high-fat, low-fibre diets. Diets high in saturated fats from red meat are associated with a higher incidence of colon and prostate cancer.[42] Diets that are high in omega-3 fatty acids from fish are associated with a lower incidence of colon cancer.[43] The connection between dietary fat and colon cancer may be related to the breakdown of fat in the large intestine. Here, bacteria metabolize dietary fat and bile, producing substances that may act as tumour initiators. A high intake of fibre tends to dilute these carcinogens by increasing the volume of feces. High-fibre diets also decrease transit time. Both of these effects reduce the exposure of the intestinal mucosa to hazardous substances in the colon contents (see Section 4.6).

Dietary Fat and Obesity

In addition to affecting the risk of heart disease and cancer, dietary fat has been postulated to contribute to weight gain and obesity. One reason is that fat has 9 kcal/g—almost twice as much as either carbohydrate or protein. Therefore, a high-fat meal contains more kcalories in the same volume as a lower-fat meal. And, because people have a tendency to eat a certain weight or volume of food, if that food is high in fat, it will contain more kcalories and contribute to weight gain.[44] High-fat diets also tend to promote overconsumption because energy from fat is less satiating than energy from carbohydrate, so when eating a high-fat meal, you will eat more kcalories before you feel full.[45] The way fat is metabolized may also contribute to weight gain. It takes less energy for the body to use fat as an energy source and excess dietary fat is stored very efficiently as body fat (see Section 7.2).

CANADIAN CONTENT

Despite evidence that fat is "fattening," the fat content of the diet is not the main reason for obesity in Canada. A recent study comparing the diet composition of obese and non-obese Canadian adults found that the risk of obesity increased with higher intakes of total kcalories in both men and women. In men, a higher fibre intake was found to reduce the risk of obesity, but this relationship was not observed in women. In both men and women, there was no relationship between obesity and the relative amount of carbohydrates, fat, and protein consumed. In short, weight gain occurs when more energy is consumed than expended, regardless of whether the extra energy comes from fat, carbohydrate, or protein.[46]

5.7 Meeting Recommendations for Fat Intake

Learning Objectives

- List the recommendations for fat and cholesterol intake.
- Use food labels to choose foods that provide healthy fats.
- Review a diet and suggest modifications that would help it meet current recommendations for the types and amounts of fat.

About 31% of the energy in the typical Canadian diet comes from fat.[1] This amount will easily meet the minimum requirements for essential fatty acids. However, a healthy diet must also provide the right types of fat and include plenty of whole grains, legumes, vegetables, and fruits, which are high in fibre, micronutrients, and phytochemicals.

Types of Lipid Recommendations

Fat is needed in the diet to provide essential fatty acids, to allow the absorption of fat-soluble vitamins and phytochemicals, and to provide energy. The amounts needed for this are small but a diet that provides only the minimum amount of fat would be very high in carbohydrate, not very palatable, and not necessarily any healthier than diets with more fat. Therefore, the DRI recommendations regarding fat include AIs for essential fatty acids as well as Acceptable Macronutrient Distribution Ranges (AMDR) for essential fatty acids and total fat intake. The DRIs also suggest that since there is no specific dietary need for saturated fats, cholesterol, or *trans* fat, that these be reduced as much as possible.

Recommendations for Essential Fatty Acids The amounts of the essential fatty acids recommended by the DRIs are based on the amounts consumed by a healthy population. The AI for linoleic acid is 12 g/day for women and 17 g/day for men. You can meet your requirement by consuming 125 ml (½ cup) of walnuts or 30 ml (2 tablespoons) of safflower oil. For alpha-linolenic acid, the AI is 1.1 g/day for women and 1.6 g/day for men. Your requirement can be met by eating 60 ml (¼ cup) of walnuts or a 15 ml (1 tablespoon) of canola oil or ground flaxseeds. Consuming these amounts provides the recommended ratio of linoleic to alpha-linolenic acid of between 5:1 and 10:1.[11] AMDRs of 5%-10% of energy from linoleic acid and 0.6%-1.2% of energy from alpha-linolenic acid (with 10% or less of this as EPA and DHA) have been set.[11]

A Healthy Range for Fat Intake The DRIs recommend a total fat intake of 20%-35% of kcalories for adults; this range allows for a variety of individual preferences in terms of food choices. Fat intakes above this range generally result in a higher intake of saturated fat and make it more difficult to avoid consuming excess kcalories. Intakes below 20% increase the probability that vitamin E and essential fatty acid intakes will be low and may contribute to unfavourable changes in HDL and triglyceride levels.

Life Cycle

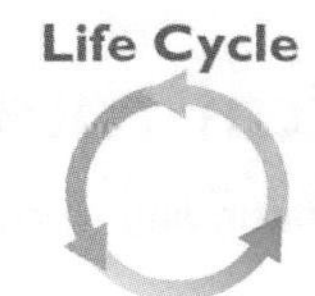

Because children and teens need more total fat in their diets to meet their needs for growth, the AMDRs for fat are higher for these groups: 30%-40% of energy for ages 1-3 years, and 25%-35% of energy for ages 4-18 years. These amounts meet the needs for growth and are unlikely to increase the risk of chronic disease.

The AMDRs for fat intake are not increased during pregnancy or lactation, but the AIs for essential fatty acids are slightly higher than those for nonpregnant women. Recommendations are not different for older adults. In this population, fat intake must be carefully balanced with other nutrients to reduce the risk of malnutrition (see Section 16.3).

Limit Cholesterol, Saturated, and *Trans* Fat The DRIs have not set specific guidelines for cholesterol, saturated fat, or *trans* fat, but they recommend that intake of these be kept to a minimum because the risk of heart disease increases as intakes rise. Daily Values on food labels give more specific recommendations: less than 10% of energy (20 g of saturated fat for a 2,000-kcal diet) as saturated fat and no more than 300 mg of cholesterol per day.[47] Canada's

CRITICAL THINKING:

Eliminating *Trans* Fat from the Canadian Food Supply

CANADIAN CONTENT

It has been calculated that the deaths of 3,000 Canadians annually from cardiovascular disease would be prevented if the amount of *trans* fats in the Canadian food supply could be reduced to 1% of total energy.[1]

Given the strong scientific evidence for the deleterious effects of *trans* fats on human health, in 2003, Canada became the first country to mandate the labelling of *trans* fats on packaged foods, that is, requiring manufacturers to declare on their labels the quantity of *trans* fat present. If a manufacturer wanted to label its product "*trans* fat free," the product had to contain less than 0.2 g of *trans* fat and less than 2 g of saturated fat per serving.[2] This regulation, limiting saturated as well as *trans* fat content, ensures that the manufacturer does not reduce *trans* fat levels by replacement with saturated fats, which, like *trans* fats, also increase the risk of cardiovascular disease.

The impact of these labelling regulations has been positive. Consumer awareness of *trans* fats increased from 45% of Canadians in 1999 to 79% in 2004.[3] Because of this awareness, many food manufacturers were motivated to reduce *trans* fat levels. For example, most breads and salad dressings are now *trans* fat free and there have been reductions in the *trans* fat content of store-bought french fries and snack foods.[3]

But this progress was not sufficient to reduce the level of *trans* fat in the Canadian food supply to desirable levels. A *Trans* Fat Task Force was formed to provide recommendations on how to reduce *trans* fat levels to the lowest levels possible in the Canadian food supply. It was co-chaired by Health Canada and the Heart and Stroke Foundation of Canada and consulted with food-industry and health experts.

This task force made the following recommendations:

1) *Trans* fat content in vegetable oils and soft, spreadable (tub-type) margarine should be limited, by regulation, to $\leq$ 2% of total fat content.

2) In all other foods *trans* fat content should be limited, by regulation, to $\leq$ 5% of total fat content.

These regulations would apply to manufactured foods purchased in grocery stores, as well as foods prepared in restaurants, food service operations, and grocery store bakeries. About 22% of the *trans* fat consumed by Canadians comes from restaurant food. The task force felt these recommendations were reasonable, as alternatives to *trans* fats exist. They conceded that manufacturers of some food products, especially baked goods, faced a more difficult reformulation challenge. *Trans* fats are used in baked products because they enhance flavour and shelf life and alternatives cannot fully duplicate their properties.

The task force submitted its recommendations to the minister of health, who chose not to impose regulations immediately. Instead, food companies were given 2 years to voluntarily reduce *trans* fats or face mandatory regulation. Over the 2-year period (2007-2009), Health Canada monitored the levels of *trans* fat in a large variety of foods to determine whether the food industry was voluntarily complying with the task force recommendations.

The final results of the monitoring program were reported in December 2009, and results were mixed.[4] In April 2010, the health minister conceded that, although considerable progress had been made in reducing *trans* fat in Canada's food supply, many products still contained higher-than-desirable amounts. Only 25% of croissants, 36% of doughnuts, 36% of pies, 45% of brownies, 58% of popcorn, 59% of fast-food chicken products, and 65% of packaged cookies met the recommended targets. The baked-food-products sector was not eliminating *trans* fats from their formulations as quickly as other food sectors, because of the impact of removing *trans* fats on food quality.[5]

In response to the release of these data, the Heart and Stroke Foundation of Canada called on the government to immediately enact regulations making *trans* fat reduction mandatory, noting that 84% of Canadians support a *trans* fat ban.[1]

On a more positive note, the intake of *trans* fats by Canadians dropped from 8.4 g/day in the mid-1990s to 3.4 g/day in 2009. This is still higher than the World Health Organization recommendation of 2 g/day, but it's a dramatic decline that demonstrates the positive impact of the voluntary program[6].

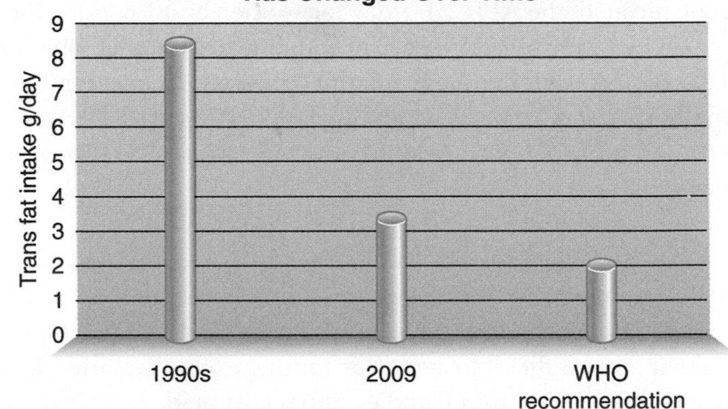

Critical Thinking Questions

In this continuing story, what do you think the government of Canada should do next?

Why do you think the voluntary compliance was so low for baked foods?

Using *trans* fat as an example, suggest general steps that a government might follow when trying to improve the nutrient composition of the food supply.

[1] Heart and Stroke Foundation: Canada's *trans* fat scorecard: Few brownie points in latest survey. Available online at http://www.newswire.ca/en/releases/archive/February2009/12/c8060.html. Accessed May 1, 2010.

[2] Canadian Food Inspection Agency. Guide to Food Labelling and Advertising. 7.18: *Trans* Fatty Acid Claims. Available online at http://inspection.gc.ca/english/fssa/labeti/guide/ch7ae.shtml#a7_18. Accessed May 1, 2010.

[3] Transforming the Food Supply: Report of the *Trans* Fat Task Force June 2006. Available online at http://www.hc-sc.gc.ca/fn-an/alt_formats/hpfb-dgpsa/pdf/nutrition/tf-gt_rep-rap-eng.pdf. Accessed May 1, 2010.

[4] Health Canada. *Trans* Fat Monitoring Program. Available online at http://www.hc-sc.gc.ca/fn-an/nutrition/gras-*trans*-fats/tfa-age_tc-tm-eng.php. Accessed May 1, 2010.

[5] AHN News Staff. Canada's voluntary *trans* fat reduction efforts fail. Available online at http://www.allheadlinenews.com/articles/7018487648. Accessed May 1, 2010.

[6] Ratnayake, W. M., L'Abbé, M. R., Farnworth, S. et al. *Trans* fatty acids: current contents in Canadian foods and estimated intake levels for the Canadian population. *J AOAC Int.* 92(5):1258-1276. 2009.

Food Guide was also designed to achieve intakes of less than 10% of kcalories from saturated fat.[48] No Daily Value has been established for *trans* fat, but the World Health Organization recommends that *trans* fat make up less than 1% of energy or less than 2 g per day for a 2,000-kcal diet.[49] (See Critical Thinking: Eliminating *Trans* Fat from the Canadian Food Supply.)

Guidelines for Prevention of Specific Diseases

In addition to these general recommendations for fat intake, some dietary recommendations target populations at risk for specific diseases. The Canadian Cancer Society recommends that Canadians follow Canada's Food Guide for the prevention of cancer, and provides more detail (**Table 5.4**) in their publication *Eat well, be active.*[50]

CANADIAN CONTENT

Table 5.4 Recommendations to Reduce Cancer Risks
Eat a variety of vegetables and fruits every day. Go for high-fibre foods. Limit how much alcohol you drink. Limit red meat and processed meats. Use less salt and sugar. Cook and prepare food with care.

Source: Canadian Cancer Society, 2004. *Eat Well, Be Active: What you can do.* Available online at http://www.cancer.ca/~/media/CCS/Canada%20wide/Files%20List/English%20files%20heading/Library%20PDFs%20-%20English/Eat-well-Be-active_2011.ashx. Accessed June 1, 2010.

Figure 5.27 The Health Check™ logo on a food product indicates it has met criteria developed by the Heart and Stroke Foundation's registered dietitians based on recommendations in Canada's Food Guide. healthcheck.org (tm) The Health Check logo and word mark are trademarks of the Heart and Stroke Foundation of Canada, used under license.
Source: http://www.heartandstroke.on.ca/site/c.pvI3IeNWJwE/b.3581713/k.7AD9/Health_Check.htm

The Heart and Stroke Foundation of Canada similarly recommends that Canadians follow Canada's Food Guide, also emphasizing increased consumption of fruits, vegetables, and whole grains.[51]

The Health Check™ program of the Heart and Stroke Foundation has developed criteria for grocery and restaurant menu items. Criteria include total fat, saturated fat, trans fat, fibre, sodium, sugar, protein, and certain vitamins and minerals. Items that meet the criteria earn the right to display the logo (see **Figure 5.27**).

The Canadian Cardiovascular Society has published recommendations for the diagnosis (Figure 5.27) and treatment of dyslipidemia (abnormal blood lipids, for example, high LDL cholesterol, high triglycerides) and the prevention of cardiovascular disease.[52] Some of their recommendations are comprised of health-behaviour interventions that include nutritional changes (**Table 5.5**).

Table 5.5 Major Health Behaviour Interventions for the Prevention of Cardiovascular Disease
Major health behaviour interventions for the prevention of cardiovascular disease include:
Smoking cessation, including the use of pharmacological therapy as required.
A diet low in sodium and simple sugars, with substitution of unsaturated fats for saturated fat and *trans* fats, as well as increased consumption of fruits and vegetables.
Caloric restriction to achieve and maintain ideal body weight.
Moderate to vigorous exercise for 30-60 minutes, most (preferably all) days of the week.
Psychological stress management.
Alcohol consumption in moderation is not contraindicated if there are no metabolic or clinical contraindications.

Source: Genest, J. McPherson, R. Frohlich, J. et al., 2009. Canadian Cardiovascular Society/Canadian guidelines for the diagnosis and treatment of dyslipidemia and prevention of cardiovascular disease in the adult-2009 recommendations. *Can. J Cardiol.* 25(10): 567–579, 2009.

The Canadian Cardiovascular Society also recommends drug therapy for individuals with extremely high cholesterol levels or for those for whom lifestyle changes are not effective. The most common drugs used to treat elevated blood cholesterol are the statins; these work by blocking cholesterol synthesis in the liver and by increasing the number of LDL receptors in the liver to remove cholesterol from the blood (see Science Applied: A Genetic Disease and Cholesterol Regulation). Other cholesterol-lowering drugs act in the gastrointestinal tract by preventing cholesterol and bile absorption.

A comparison of Table 5.4 and Table 5.5 illustrates that there is very little difference in the dietary recommendations for disease prevention between cardiovascular disease and cancer. Although it appears that the advice for alcohol consumption varies, careful examination shows that it is in fact similar. Because alcohol consumption is linked to increased risk of some cancers (see Focus on Alcohol), the Canadian Cancer Society emphasizes limiting alcohol intake to less than 1 drink a day for women and fewer than 2 drinks for men; on the other hand, for the prevention of heart disease, in the absence of contraindications, moderate consumption of alcohol is not discouraged by the Canadian Cardiovascular Society. Moderation is defined as no more than 1 drink a day for women and 2 drinks a day for men.

Tools for Assessing Fat Intake

How does your diet compare with the recommendation of 20%-35% of energy from fat and less than 10% of energy as saturated fat? **Table 5.6** illustrates how to calculate fat intake as a percent of energy. This same calculation can be used to determine the percent of energy as fat or saturated fat in individual foods. To calculate the percent of energy as fat, you need to know the amount of fat in a food or in the diet. This can be estimated by using values from food labels or food composition tables or databases.

Table 5.6 Calculating Percent of Energy from Fat

Determine

- The total energy (kcalorie) intake for the day
- The grams of fat in the day's diet

Calculate Energy from Fat

- Fat provides 9 kcal/g
- Multiply grams of fat by 9 kcal/g

$$\text{Energy (kcal) from fat} = \text{grams fat} \times 9\ \text{kcal/g fat}$$

Calculate % Energy from Fat

- Divide energy from fat by total energy and multiply by 100 to express as a percent

$$\text{Percent of energy from fat} = \frac{\text{kcal from fat}}{\text{Total kcal}} \times 100$$

For example:

A diet contains 2,000 kcal and 75 g of fat

75 g of fat × 9 kcal/g = 675 kcal from fat

$$\frac{675\ \text{kcal from fat}}{2{,}000\ \text{kcal}} \times 100 = 34\%\ \text{of energy (kcals) from fat}$$

The Beyond the Basics meal-planning system from the Canadian Diabetes Association (see Section 4.5) also provides information on fat content. The plan divides foods into high- and low-carbohydrate content but also lists the amount of fat present in various food portions (see Appendix E).

Fats on Food Labels Food labels provide an accessible source of information on the fat content of packaged foods. Understanding how to use this information can help you make more informed choices about the foods you include in your diet. Unfortunately, food labels are not always available on fresh meats, which are one of the main contributors of fat in our diets (see Label Literacy: Choosing Lean Meat).

The Nutrition Facts section provides the total number of kcalories, the number of grams of total fat, saturated fat, and *trans* fat, and the number of milligrams of cholesterol in a serving.[53] These are also presented as a percentage of the Daily Value (**Figure 5.28**). This information allows consumers to tell at a glance how one food will fit into the recommendations for fat intake for the day. For example, if a serving provides 50% of the Daily Value for fat—that is, half the recommended maximum daily intake for a 2,000 kcal diet—the rest of the day's foods will have to be carefully selected to not exceed the recommended maximum. To choose foods low in saturated fat and cholesterol, use the general rule that 5% of the Daily Value or less of these in a serving is low and 15% or more is high.

The amount of monounsaturated and polyunsaturated fat is voluntarily included on the labels of some products. For example, in addition to listing the 2 g of saturated fat, the label on a bottle of olive oil may indicate that it contains 2 g of polyunsaturated fat and 10 g of monounsaturated fat per 15 ml (1 tablespoon). There are no Daily Values for polyunsaturated and monounsaturated fat (see Figure 5.7).

Cheese Crackers

Nutrition Facts

Per 4 crackers (43 g)

Amount			% Daily Value
Calories 230			
Fat 14 g			22 %
Saturated 3.5 g + Trans 4 g			18 %
Cholesterol 5 mg			
Sodium 410 mg			17 %
Carbohydrate 23 g			8 %
Fibre 0 g			0 %
Sugars 6 g			
Protein 3 g			
Vitamin A	0 %	Vitamin C	0 %
Calcium	6 %	Iron	6 %

INGREDIENTS: ENRICHED FLOUR (WHEAT FLOUR, NIACIN, REDUCED IRON, THIAMINE MONONITRATE {VITAMIN B1}, RIBOFLAVIN {VITAMIN B2}, FOLIC ACID), PARTIALLY HYDROGENATED SOYBEAN OIL, WHEY (FROM MILK), SUGAR, HIGH FRUCTOSE CORN SYRUP, BUTTER (PASTEURIZED CREAM, SALT, ANNATTO COLOR), CHEDDAR CHEESE (CULTURED MILK, SALT, ENZYMES, AND ANNATTO EXTRACT COLOR), BUTTERMILK SOLIDS, SALT, LEAVENING (BAKING SODA, CALCIUM PHOSPHATE), DISODIUM PHOSPHATE (STABILIZER), NATURAL FLAVOR, SOY LECITHIN (EMULSIFIER), COLOR ADDED (INCLUDES YELLOW 6), MALTED BARLEY FLOUR, PEANUTS.

Figure 5.28 **Fat and cholesterol on food labels**
This label from cheese crackers tells us that they are low in cholesterol but high in total fat, and they add saturated and *trans* fat to the diet. (© istockphoto.com/Karen Mower)

If you want to know the source of fat in a packaged food, you can check the ingredient list. This will show you if a food contains, for example, corn oil, soybean oil, coconut oil, or hydrogenated vegetable oil (see Figure 5.28). Labels may also include terms such as "low-fat," "fat-free," and "low cholesterol" that describe their fat content. Food-labelling regulations have developed standard definitions for these terms and they can be used only in ways that do not confuse consumers (**Table 5.7**). For instance, a food that contains no cholesterol but is high in saturated fat, such as crackers containing coconut oil, cannot be labelled "cholesterol free" because saturated fat in the diet raises blood cholesterol. The "cholesterol-free" statement is permitted only for foods that contain less than 2 mg of cholesterol and 2 g or less of saturated and *trans* fat per serving (Table 5.7). Food labels may also include health claims related

LABEL LITERACY

Choosing Lean Meat

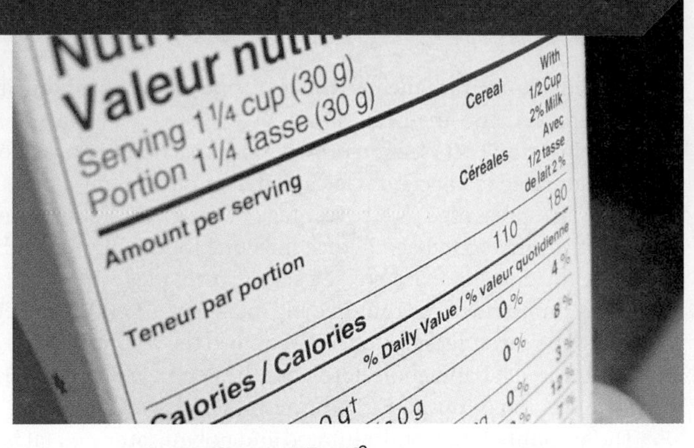

(© istockphoto.com/Tim Pohl)

Meat is a source of saturated fats in the diet, so one way to reduce your saturated fat content is by looking for lean meat such as lean cuts of beef, poultry, and turkey.

The Canadian Food and Drug regulations define "lean" as meat, poultry, turkey, etc, that has not been ground and contains less than 10% fat by weight, and "extra lean" as containing no more than 7.5% fat by weight. Ground meats, such as ground beef, pork, turkey or chicken, follow a different standard. Ground meats can be labelled "lean" if no more than 17% of its weight is fat, while "extra lean" beef contains no more than 10% by weight of fat. All ground meat in Canada must have a nutrition facts label, like the one shown here.[1] Labels are not required for fresh cuts of meat, for example, steaks, roasts, and so on.

In the label shown, 75 g of lean ground beef, or one Canada's Food Guide serving, contains 11 g of fat, and a modest 160 kcal/serving. The saturated and trans fat levels, however, are 27% of DV, indicating that the ground beef has high amounts of these fats. A %DV greater than 15% is considered a lot (see Section 2.5).

So, if lean ground beef has high levels of saturated and trans fats, should you stop eating ground beef altogether? Not necessarily. Beef is an excellent source of protein and micronutrients such as iron, zinc, and vitamin B_{12} and when used in moderation can be a nutritious contribution to a balanced diet. Balance this higher fat choice by eating foods that are low in fat and saturated fat, such as having your beef with several CFG servings of steamed vegetables.[1]

Lean Ground Beef[2]

Nutrition Facts
Per 3/8 cup (75 g)

Amount			% Daily Value
Calories 160			
Fat 11 g			17 %
Saturated 4.5 g + Trans 0.4 g			27 %
Cholesterol 45 mg			
Sodium 50 mg			2 %
Carbohydrate 0 g			0 %
Fibre 0 g			0 %
Sugars 0 g			
Protein 15 g			
Vitamin A	0 %	Vitamin C	0 %
Calcium	0 %	Iron	11 %

[1] Canadian Food Inspection Agency. Ground Beef Standards. Available online at http://www.inspection.gc.ca/english/fssa/labeti/retdet/bulletins/meavia/grohace.shtml. Accessed on Oct 2, 2011.

[2] Beef Information Centre. Nutrition Labelling Mandatory for Ground Beef. Available online at http://www.canadianbeef.info/ca/en/rt/resources/Download.aspx?FileName=150911-EN.PDF&masterNo=150911&itemType=PDF&itemUnits=EACH&catalogCode=RES_RETAIL&subCatalogCode=RES_RETAIL_TRADEEDUCATIONTECH&Language=LAN_ENGLISH&ProfileNo=1869263. Accessed June 10, 2011.

---

to their fat content. For example, foods low in saturated fat and cholesterol may make the health claim that that they help to reduce the risk of coronary heart disease (see Section 2.5 for more information on health claims). Foods to which plant sterols have been added can make similar claims.

Table 5.7 Fat and Cholesterol on Food Labels

Fat-free	Contains less than 0.5 g of fat per serving.
Low-fat	Contains 3 g or less of fat per serving.
Reduced or less fat	Contains at least 25% less fat per serving than the regular or reference product.
Saturated fat-free	Contains less than 0.2 g of saturated fat per serving and less than 0.2 g *trans* fatty acids per serving.
Low saturated fat	Contains 2 g or less of saturated and *trans* fat and not more than 15% of kcalories from saturated fat per serving.
Reduced or less saturated fat	Contains at least 25% less saturated fat than the regular or reference product (without increasing the *trans* fat content).
Free of trans fat	Contains less than 0.2 g of *trans* fatty acids per serving and less than 2 g saturated fat.
Cholesterol-free	Contains less than 2 mg of cholesterol and 2 g or less of saturated and *trans* fat per serving.
Low cholesterol	Contains 20 mg or less of cholesterol and 2 g or less of saturated and *trans* fat per serving.

Reduced or less cholesterol	Contains at least 25% less cholesterol than the regular or reference product and 2 g or less of saturated fat and *trans* per serving.
Lean	Meat or poultry that has not been ground, marine or fresh-water animals, or a product of any of these that contains less than 10 g of fat per 100 g.
Extra lean	Meat or poultry that has not been ground, marine or fresh-water animals, or a product of any of these that contains less than 7.5 g of fat per 100 g.
Source of omega-3 polyunsaturated fatty acids	Contains 0.3 g or more of omega-3 polyunsaturated fatty acids per serving.
Source of omega-6 polyunsaturated fatty acids	Contains 2 g or more of omega-6 polyunsaturated fatty acids per serving.

Source: Canadian Food Inspection Agency. Guide to Food Labelling and Advertising. Nutrient Content Claims. Available on at http://inspection.gc.ca/english/fssa/labeti/guide/ch7e.shtml. Accessed Oct 2, 2011.

Translating Recommendations into Healthy Diets

The typical Canadian diet meets some but not all of the recommendations for fat intake. Typical intake of total fat is within the recommended 20%-35% of kcal for many, but not all Canadians. *Trans* fat levels are high (3.4 g/day) but have dropped from levels of 8.4 g /day in the 1990s.[54] The CCHS found that half of Canadians consumed more than 10% of their calories from saturated fat.[55] In order to improve their diets, Canadians need to get more of their fats from foods like fish, nuts, and vegetable oils, which are sources of polyunsaturated and monounsaturated fatty acids. Using the recommendations of Canada's Food Guide can help consumers choose healthy diets. **Table 5.8** gives some suggestions on how to make healthy fat choices.

Table 5.8 How to Choose Fats Wisely

Limit cholesterol, *trans* fat, and saturated fat

- Choose lower-fat cuts of meat
- Opt for chicken (skinless) and fish
- Try a vegetarian meal once a week—beans and other vegetarian sources of protein are low in *trans* and saturated fat and don't contain cholesterol
- Use low-fat milk, yogurt, and cheese
- Choose a margarine with no *trans* fat
- Cut down on packaged foods that contain *trans* fats

Increase mono- and polyunsaturated fats

- Choose olive, peanut, or canola oil (high in monounsaturated fat) for cooking and salad dressing
- Use corn, sunflower, or safflower oil (high in polyunsaturated fat) for baking
- Snack on nuts and seeds
- Add olives, avocados, nuts, and seeds to your salad

Get enough omega-3s

- Sprinkle flaxseed on your cereal or yogurt or bake it into muffins
- Add another serving of fish to your weekly menu
- Have a leafy green vegetable with dinner
- Put walnuts in your salad

Watch your total fat

- Instead of frying, bake, broil, barbecue, roast, steam, or microwave
- Have a smaller serving of ice cream
- Try oven-baked rather than deep-fried potatoes
- Use half your usual amount of butter

Choose Added Fats Carefully The most concentrated sources of fat—oils, butter, margarine, sauces, and salad dressings—should be chosen carefully to keep the amount and type of fat in the diet healthy (**Figure 5.29**). Limiting the amount of fat added to food at the table and in cooking will reduce total fat consumption. Avoiding solid fats like butter and cream cheese will reduce saturated-fat intake. Using vegetable oils such as canola and olive oil that are high in monounsaturated fat, or corn and soybean oils that are high in polyunsaturated fat instead of butter, will increase the proportion of unsaturated fats. Soybean oil and canola oil are also good plant sources of omega-3 unsaturated fatty acids. Choosing margarines and other spreads that are low in *trans* fat can reduce *trans* fat intake (see Critical Thinking: Eating Healthier Fats).

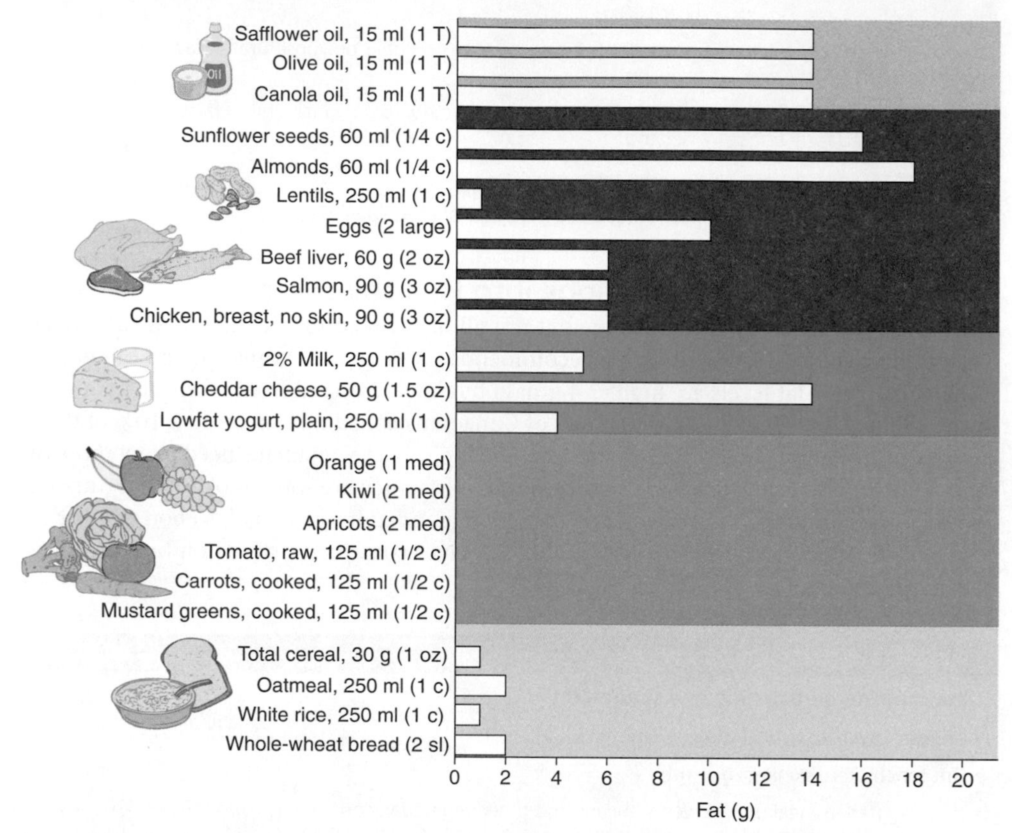

Figure 5.29 The fat content of Canada's Food Guide food groups
Oils, nuts, and seeds are high in fat, but the fat they provide is rich in mono- and polyunsaturated fat. Animal products contain saturated fat and cholesterol.

Choose Protein Sources Wisely Meats and dairy products are the major sources of saturated fat and the only sources of cholesterol in our diets. Trimming the fat from meat, choosing leaner cuts, and removing the skin from poultry can reduce the amount of total fat and saturated fat (**Table 5.9**). Likewise, using nonfat or low-fat milk and milk products will cut both total and saturated fat as well as cholesterol. Choosing fish and shellfish will provide a meal with less saturated fat and more heart-protective, long-chain omega-3 fatty acids. Levels of omega-3 fatty acids are higher in oilier fish such as salmon, trout, and herring (see Table 5.1). Using vegetable sources of protein such as legumes will limit total fat, saturated fat, and cholesterol, and choosing nuts and seeds will add a source of omega-3 and monounsaturated fat.

Table 5.9 Making Choices that Lower Saturated Fat Intake

Higher-Fat Food	Saturated Fat (g)	Total Fat (g)	Energy (kcal)	Lower-Fat Equivalent	Saturated Fat (g)	Total Fat (g)	Energy (kcal)
Steak, fat not trimmed, 90 g cooked	4.8	12.7	211	Steak, fat trimmed, 90 g cooked	1.4	4.2	153
Regular ground beef (25% fat), 90 g cooked	7.5	19	260	Extra-lean ground beef (4% fat), 90 g	1.9	3.3	122

Fried-chicken breast (with skin), 90 g	3.2	12	240	Broiled-chicken breast (no skin), 90 g	1	3.9	149
Fish (fried), 90 g	1.3	5.5	149	Fish (baked), 90 g	0.3	1.4	118
Whole milk, 250 ml	4.6	8	146	Low-fat milk, 250 ml	1.5	4	102
Butter, 5 ml	2.4	3.8	34	Soft margarine with zero *trans* fat, 5 ml	0.5	3.8	34
Regular ice cream, 125 ml	4.5	7.3	133	Low-fat frozen yogurt, 125 ml	0.7	1	98
Potato, baked with sour cream, 1 medium	3.5	5.6	169	Potato, baked plain, 1 medium	0	0.2	113
Doughnut, glazed, 1 large	8.4	22	412	Bagel, 1 large	0.2	1	195
Frozen broccoli with cheese sauce, 250 ml	10.8	17.3	256	Fresh broccoli, 250 ml	0	0.1	27
Ramen noodles, 250 ml	4.4	15.6	426	Egg noodles, 250 ml	0.5	2.4	213
Spaghetti with Alfredo sauce, 250 ml	4.5	20	368	Spaghetti with tomato sauce, 250 ml	0.7	5	250

Watch Processed Foods Generally, whole grains, fruits, and vegetables are low in total fat and saturated fat and have no cholesterol. However, choices from these groups need to be made with care to limit fats that are added in processing or preparation. Processed foods are a major source of fat in the diet. From the grains group, frozen pizzas, macaroni and cheese, flavoured rice dishes, and baked goods such as doughnuts, cookies, and muffins, are sources of dietary fat. In fact, most baked goods are so high in added fats and sugars that about half of their kcalories are considered discretionary kcalories.

Within the fruits group, fresh fruits are a low-fat choice while fruits that are baked into pies and tarts add total fat, *trans* fat, and refined sugar. Olives and avocados are naturally high in fat, but provide monounsaturated fats. Most fresh vegetables have little or no fat. However, fats are often added to vegetables in cooking and processing; french fries and breaded fried vegetables, such as onion rings and mushrooms, are high in fat and energy, and depending on the fat used for frying, can be a source of saturated fat or *trans* fat (see Table 5.8). Careful selection from these food groups can help keep the amount and type of fat in the diet in the recommended range.

Reduced-Fat Foods

There are many reduced-fat and fat-free foods available. Some, such as low-fat milk, are made by simply removing the fat. In other foods, the fat is replaced with ingredients that mimic the taste and texture of fat. Some of these fat substitutes are lower in kcalories because they are carbohydrate- or protein-based. Others are lipids, but contribute fewer kcalories because they are poorly digested and absorbed (**Figure 5.30**).

Figure 5.30 The carbohydrate-based fat replacers used in the muffins and ice cream shown here add kcalories, but fewer than come from the fat they replace, because they are made from carbohydrate. Olestra is a fat replacer used in the United States which replaces the fat in these chips. It is not absorbed so it adds no kcalories to the diet. Olestra is not approved for use in Canada. (Andy Washnik)

The Role of Reduced-Fat Foods in a Healthy Diet To reduce fat intake, people often choose foods that have been modified to reduce their fat content. Some of these foods can make an important contribution to a healthy diet. For example, low-fat dairy products are recommended because they provide all the essential nutrients contained in the full-fat versions but have fewer kcalories and less saturated fat and cholesterol. Using these products increases the nutrient density of the diet as a whole. However, not all reduced-fat foods are nutrient dense and using them does not necessarily transform a poor diet into a healthy one or improve overall diet quality. Some reduced-fat foods are lower-fat versions of nutrient-poor choices such as baked goods and chips. If these reduced-fat desserts and snack foods replace whole grains, fruits, and vegetables, the resulting diet could be low in fat but also low in fibre, vitamins, minerals, and

CRITICAL THINKING:

Eating Healthier Fats

(Hola Images/Getty Images, Inc.)

Background

Isabel has a busy schedule—working full-time and attending college—and has little time to cook meals at home. Currently, for breakfast and lunch she relies on things she can pick up on the way to school or between classes. She makes a quick dinner when she gets home in the evening. She is concerned about the fat in her diet and wants to know how to make healthier choices.

Data

Isabel analyzes her original diet and then modifies it to try to meet the recommendations for a healthy mix of fats (see table).

ORIGINAL DIET

FOOD	SERVING	FAT (g)	SAT (g)	*TRANS* (g)	CHOLESTEROL (g)
Breakfast					
Bran muffin	1 large	6	2.6	0.5	24
Margarine	10 ml	8	1.3	2	0
Coffee	250 ml	0	0	0	0
Whole milk	250 ml	8	5	0.2	33
Lunch					
Big Mac	1	31	12.5	1	80
French fries	medium order	22	5	5	0
Water	1 bottle	0	0	0	0
Snack					
Apple	1	0	0	0	0
Dinner					
Fish sticks	5	17	2.4	2	33
Tater tots	10	8	6	4	0
Tea	1 cup	0	0	0	0
Coconut Cookies	2	13	3.4	3.3	0
Totals		**113**	**38.2**	**18**	**170**

MODIFIED DIET

FOOD	SERVING	FAT (g)	SAT (g)	*TRANS* (g)	CHOLESTEROL (mg)
Breakfast					
Bran muffin	1 large	6	2.6	0.5	24
Orange	1 medium	0.2	0	0	0
Coffee	250 ml	0	0	0	0
Low-fat milk	250 ml	2.6	1.6	0	10
Lunch					
Rice noodles	250 ml	0	0	0	0
Stir-fry vegetables	250 ml	0.3	0	0	0
in peanut oil	5 ml	5	0.8	0	0
Water	1 bottle	0	0	0	0
Snack					
Apple	1 medium	0	0	0	0
Pretzels	30 gm	1	0	0	0
Dinner					
Trout	90 g	12	1.2	0	63
Baked potato	1 medium	0.2	0	0	0
w/sour cream	15 ml	3	1.7	0	10
Green beans	125 ml	0	0	L	0
Salad w/oil & vinegar	250 ml	10	1.2	0	0
Melon	250 ml	0	0	0	0
Tea	250 ml	0	I	0	0
Frozen yogurt	175 ml	1.8	1.2	0	5
Totals		**42.1**	**10.3**	**0.5**	**112**

Critical Thinking Questions

Assuming Isabel is eating 2,250 kcal/day, calculate the percent of energy from fat and saturated fat (sat) in her original diet. How do these percentages compare to recommendations? What foods are the biggest contributors to her saturated fat intake? To her trans fat intake? To her cholesterol intake?

Now look at the changes she has made. Assuming her kcalorie intake stayed the same, calculate the percent of energy from fat and saturated fat in her modified diet.

Fat is not the only concern in choosing a healthy diet. How does her modified diet stack up to Canada's Food Guide recommendations in terms of grains, fruits, and vegetables?

Use iProfile to find fast-food choices that are low in saturated fat.

phytochemicals—a dietary pattern that could increase the risk of chronic disease. However, if used appropriately, fat-modified foods can be part of a healthy diet. For example, if a low-fat salad dressing replaces a full-fat version, it allows consumers to enhance the appeal of a nutrient-rich salad without the added fat or kcalories from the dressing. Replacing regular potato chips with fat-free ones will not add any other nutrients to the diet, but will reduce fat and energy intake. If the extra kcalories are consumed as fruits, vegetables, or whole grains, the overall diet has been improved. If the extra kcalories are not replaced, this substitution could be helpful for weight loss. However, those using low-fat products to aid weight loss should be aware that these foods are not kcalorie-free and cannot be consumed liberally without adding kcalories to the diet and contributing to weight gain. If you are using low-fat products to reduce your kcalorie intake, check the label—don't assume they are low in kcalories.

Fat Substitutes One way to replicate the texture of fat in foods is to replace some of the fat with carbohydrates such as cellulose, dextrins, pectins, gums, or modified starch. These thicken foods and mimic the slippery texture that fat provides. Since they are carbohydrates, they can provide up to 4 kcal/g, but because they are often mixed with water, they typically provide only 1-2 kcal/g.[56]

Protein has also been used to simulate fat. The fat replacer Simplesse is made from egg white and milk proteins that are modified by heating, filtering, and high-speed mixing. The resulting protein consists of millions of microscopic balls that slip and slide over each other to give it the creamy texture of fat. Because the protein is mixed with water, Simplesse contains only 1.3 kcal/g. It is used in frozen desserts, cheese foods, and other products, but it cannot be used for cooking because heat causes it to break down.

Some fat substitutes are made from fats that have been modified to reduce how well they can be digested or absorbed. One way to do this is to use triglycerides made up of poorly absorbed fatty acids. These are digested like other triglycerides, but the fatty acids are only partially absorbed, so they provide fewer kcalories per gram.

In the United States, the artificial fat Olestra (sucrose polyester) is used in foods. It provides no kcalories because it is not digested by either the human enzymes or the bacterial enzymes in the gastrointestinal tract. It is, therefore, excreted in the feces without being absorbed. It is made by attaching fatty acids to a molecule of sucrose. One of the problems with Olestra is that it reduces the absorption of other fat-soluble substances, including the fat-soluble vitamins A, D, E, and K. To avoid depleting these vitamins, Olestra has been fortified with them. However, it is not fortified with beta-carotene and other fat-soluble substances that may be important for health. Another potential problem with Olestra is that it can cause gastrointestinal irritation, bloating, and diarrhea in some individuals because it passes into the colon without being digested. Because of these concerns, Olestra is not approved for use in Canada.

Another modification that changes the health impact of a fat is to increase the proportion of diglycerides. Only about 10% of the glycerides naturally present in plant oils are diglycerides. However, food manufacturers are able to modify these oils to create oils that contain 70% diglycerides. These high-diglyceride oils (marketed as Enova oil) are not lower in kcalories, but when they replace similar oils containing triglycerides, blood lipid levels are lower after eating and body weight and fat accumulation in the abdominal region is reduced.[57] Enova oil was approved for use in Canada, but was discontinued by the manufacturer in 2009 due to concerns about potentially toxic compounds present in the oil.[58]

CASE STUDY OUTCOME

Sam's friends and relatives all offered advice about how to reduce his risk of heart disease, but their solutions weren't things that he could or should follow to the letter. The best approach is for Sam to make a number of small changes in his diet and lifestyle that he can stick with for the long term. To eat a heart-healthy diet he doesn't need to become a vegetarian, as his girlfriend recommended, but he should eat more fruits and vegetables and not make meat the focus of every meal. Despite his lab partner's suggestions, it is not necessary—or even healthy—to eliminate all fat from the diet. Instead, Sam should change the types of foods he eats to reduce his intake of saturated fat, cholesterol, and *trans* fat and increase his monounsaturated and polyunsaturated fat intake. Cutting out stick margarine, as his mother advised, will lower his *trans* fat intake, but he can still enjoy margarine by using food labels to select a *trans* fat-free brand. Adding fish to his diet, as his roommate recommended, will increase his intake of heart-healthy omega-3 fatty acids. And using olive oil, which is plentiful in the Mediterranean diet his sister proposed, will boost his intake of beneficial monounsaturated fatty acids.

With the help of the information in this chapter, Sam incorporated many of the suggestions from his friends and relatives. He eats more fruits, vegetables, and fish; chooses healthy fats; and has started exercising regularly. Over the past 6 months, he has lost 5 kg (10 lb) and lowered his total cholesterol to 4.7 mmol/L.

APPLICATIONS

Personal Nutrition

1. **How do the fats in your diet compare to the recommendations?**

   a. Use iProfile to calculate your average total fat, saturated fat, and cholesterol intake using the 3-day food record you kept in Chapter 2.

   b. How does your fat intake compare with the recommendation of 20%-35% of energy from total fat?

   c. How does your percent of energy from saturated fat compare to the Daily Value recommendation of less than 10% of kcalories?

   d. What foods do you typically consume that are high in saturated fat? Suggest food substitutions that will decrease the amount of saturated fat in your diet.

   e. How does your cholesterol intake compare to the Daily Value recommendation of less than 300 mg/day? If it is greater than 300 mg, suggest some substitutions that will decrease your cholesterol intake.

   f. List some foods in your diet that are high in *trans* fat. What substitutions could you make to decrease your *trans* fat intake?

2. **Do you make healthy choices from all the food groups?**

   a. List the dairy products in your diet and indicate if they are full fat or reduced fat.

   b. List the grain products you typically consume. How many of them are baked goods with added fats? How many of them are eaten with an added high-fat spread or sauce? Suggest changes you could make to reduce the fats added to your carbohydrates.

   c. List the vegetables in your diet. Underline those that are cooked or prepared in a way that increases their fat intake. For example, are they fried or topped with butter or salad dressing?

   d. List the high-protein foods in your diet. Which ones are naturally high in fat? What types of fats do these provide? Which, if any, are high in fat because fat is added in cooking or in sauces or gravies? Which are low in fat?

3. **Are you getting your omega-3s?**

   a. Review all 3 days of the food record you kept in Chapter 2 and list foods you eat that are good sources of omega-3 fatty acids.

   b. If your diet does not contain good sources of omega-3 fatty acids, suggest some substitutions that would boost your omega-3 intake.

4. **How much fat is in your packaged foods?**

   a. Select four packaged foods that you routinely consume.

   b. Examine the food labels. If you consumed only this food for an entire day, how many servings could you eat before exceeding the Daily Value for:

      1. Total fat?

      2. Saturated fat?

      3. Cholesterol?

General Nutrition Issues

1. **The percent of kcalories from fat in the typical Canadian diet has decreased over the last 30 years. There are thousands of reduced-fat products on the market. Nevertheless, there has been an increase in the incidence of obesity in Canada over this same period. Discuss how it is possible that people are cutting down on the percent fat in their diet and still gaining weight.**

2. **Ka Ming is 54 years old and has lived in Canada since 1964. An annual physical reveals that his total blood cholesterol is 7.2 mmol/L and his HDL cholesterol is 0.65 mmol/L. He is of normal weight and does not smoke. He works as a laboratory technician and so spends part of the day on his feet but gets little other exercise. A medical history reveals that none of Ka Ming's relatives in China have had cardiovascular disease.**

   a. Is Ka Ming at risk for developing cardiovascular disease?

   b. Why might the lack of cardiovascular disease in his family history not be a true indication of Ka Ming's risk?

   c. A diet analysis reveals that Ka Ming consumes a mixture of Canadian foods and traditional Chinese foods. He likes a cooked breakfast and typically buys fast food for lunch. Dinner is often Chinese-style food that he cooks himself at home or buys from a family-owned Chinese-Canadian restaurant near his home. His diet contains approximately 40% of its energy from fat, and 15% is from saturated fat. It contains about 350 mg of cholesterol a day. Below is a list of foods that Ka Ming routinely consumes. What modifications, selection suggestions, or cooking tips would decrease his intake of saturated fat and cholesterol?

Breakfast foods	Lunch foods
Scrambled eggs	Ham and cheese sub
Pancakes	Meatball sandwich
Cheese omelette	Fried chicken
Sausages	Pepperoni pizza
Dinner foods	
Crispy fried beef over white rice	
Sweet and sour pork over white rice	
Chicken and broccoli stir-fry over white rice	
Pork fried rice	
Egg rolls	

SUMMARY

5.1 Fat in the Canadian Diet

- According to the CCHS, almost half of the fat consumed by Canadian adults comes from meat and alternatives and milk and alternatives, accounting for 31.6% and 17.9%, respectively, of total fat.
- Lipids add kcalories, texture, and flavour to our foods. Some of the fats we eat are visible, but others are less obvious.
- Changes in the types of fats in the Canadian diet can help reduce the risk of chronic disease.

5.2 Types of Lipid Molecules

- Lipids are a diverse group of organic compounds, most of which do not dissolve in water. Triglycerides, commonly referred to as fat, are the most abundant lipid in our diet and our bodies. They consist of a backbone of glycerol with 3 fatty acids attached. The physical properties and health effects of triglycerides depend on the fatty acids they contain.
- Fatty acids consist of a carbon chain with an acid group at one end. Fatty acids that are saturated with hydrogen atoms are saturated fatty acids and those that contain carbon-carbon double bonds are unsaturated. The length of the carbon chain and the number and position of the carbon-carbon double bonds, as well as the configuration of the double bonds, determine the physical properties and health effects of fatty acids.
- Phosphoglycerides are a type of phospholipid that consist of a backbone of glycerol, 2 fatty acids, and a phosphate group. Phosphoglycerides allow water and oil to mix. They are used as emulsifiers in the food industry and are an important component of cell membranes and lipoproteins.
- Sterols, of which cholesterol is the best known, are made up of multiple chemical rings. Cholesterol is made by the body and consumed in animal foods in the diet. In the body, it is a component of cell membranes and is used to synthesize vitamin D, bile acids, and a number of hormones.

5.3 Lipids in the Digestive Tract

- In the small intestine, muscular churning mixes chyme with bile from the gallbladder to break fat into small globules. This allows pancreatic lipase to access these fats for digestion.
- The products of fat digestion (primarily fatty acids and monoglycerides), cholesterol, phospholipids, and other fat-soluble substances combine with bile to form micelles, which facilitate the absorption of these materials into the cells of the small intestine.

5.4 Lipid Transport in the Body

- In body fluids, water-insoluble lipids are transported as lipoproteins. Lipids absorbed from the intestine are incorporated into chylomicrons, which enter the lymphatic system before entering the blood. The triglycerides in chylomicrons are broken down by lipoprotein lipase on the surface of cells lining the blood vessels. The fatty acids released are taken up by surrounding cells and the chylomicron remnants that remain are taken up by the liver.
- Very-low-density lipoproteins (VLDLs) are synthesized by the liver. With the help of lipoprotein lipase, they deliver triglycerides to body cells. Once the triglycerides have been removed, intermediate-density lipoproteins (IDLs) remain. These are transformed in the blood into low-density lipoproteins (LDLs). LDLs deliver cholesterol to tissues by binding to LDL receptors on the cell surface.
- High-density lipoproteins (HDLs) are made by the liver and small intestine. They help remove cholesterol from cells and transport it to the liver for disposal.

5.5 Lipid Functions in the Body

- Triglycerides provide a concentrated source of energy, After eating, chylomicrons and VLDLs deliver triglycerides to cells for energy or storage. During fasting, triglycerides stored in adipose cells are broken down by hormone-sensitive lipase, and the fatty acids and glycerol are released into the blood. To generate ATP from fatty acids, beta-oxidation first breaks fatty acids into 2 carbon units that form acetyl-CoA, which can then be broken down by the citric-acid cycle.

- Lipids in adipose tissue insulate against shock and temperature changes. Oils lubricate body surfaces and phosphoglycerides and cholesterol contribute structure to cell membranes.
- Hormones synthesized from cholesterol and eicosanoids synthesized from fatty acids have important regulatory roles.

5.6 Lipids and Health

Go to Wiley PLUS to view a video clip on the Mediterranean Diet.

Video

- Linoleic acid (omega-6) and alpha-linolenic acid (omega-3) are considered essential fatty acids because they cannot be synthesized by the body. The proper ratio of omega-6 to omega-3 fatty acids is essential for optimal health.
- Atherosclerosis is a disease characterized by deposits of lipids and fibrous material in the artery wall. It is begun by an injury to the artery wall that triggers an inflammatory response that leads to plaque formation. A key event in plaque formation is the oxidation of LDL cholesterol in the artery wall. Oxidized LDL cholesterol is taken up by macrophages and deposited in the artery wall. Oxidized LDL cholesterol also contributes to plaque formation by promoting inflammation and sending signals that lead to fibrous deposits and blood-clot formation.
- High blood levels of total and LDL cholesterol are a risk factor for heart disease. High blood HDL cholesterol protects against heart disease. A low ratio of total cholesterol:HDL cholesterol is associated with reduced heart disease risk. The risk of heart disease is also increased by diabetes, high blood pressure, and obesity.
- Diets high in saturated fat, *trans* fatty acids, and cholesterol increase the risk of heart disease. Diets high in omega-6 and omega-3 polyunsaturated fatty acids, monounsaturated fatty acids, certain B vitamins, and plant foods containing fibre, antioxidants, and phytochemicals reduce the risk of heart disease. The total dietary and lifestyle pattern is more important than any individual dietary factor in reducing heart-disease risk.
- Diets high in fat are associated with an increased incidence of certain types of cancer. In some types of cancer, such as breast cancer, fat may act as a tumour promoter, increasing the rate of tumour growth. In the case of colon cancer, dietary fat in the colon may act as a tumour initiator by forming compounds that cause mutations.
- Fat contains 9 kcal/g. A high-fat diet therefore increases the likelihood of weight gain, but it is not the primary cause of obesity. Consuming more energy than expended leads to weight gain regardless of whether the energy is from fat, carbohydrate, or protein.

5.7 Meeting Recommendations for Fat Intake

- The DRI recommendations regarding fat include AIs and AMDRs for essential fatty acids and an AMDR for total fat intake of 20%-35% of energy for adults. The DRIs also advise keeping *trans* fats, saturated fats, and cholesterol to a minimum to reduce the risk of heart disease. The Daily Values recommend that total fat account for no more than 30% of energy, that saturated fat account for no more than 10% of energy, and that dietary cholesterol be no more than 300 mg/day.
- To keep the amount and type of fat in the diet healthy, added fats, protein sources, and processed foods must be chosen carefully. Limiting animal fats from meat and dairy products reduces saturated fat intake. Choosing fish increases intake of omega-3 fatty acids. Eating nuts and seeds increase both monounsaturated fats and of omega-3 fatty acids. Processed foods can be high in saturated and *trans* fat. A diet based on whole grains, fruits, vegetables, and lean meats and low-fat dairy products will meet the recommendations for fat intake.
- Fat substitutes are used to create reduced-fat products with taste and texture similar to the original. Some low-fat products are made by using mixtures of carbohydrates or proteins to simulate the properties of fat, and some use lipids that are modified to reduce absorption. Products containing fat substitutes can help reduce fat and energy intake when used in moderation as part of a balanced diet.

REVIEW QUESTIONS

1. According the CCHS, which food groups and individual foods most contribute to the fat intake of Canadian adults?
2. Name two functions of fat in foods.
3. What is a lipid?
4. Name the types of lipids found in the body.
5. What distinguishes a saturated fat from a monounsaturated fat? From a polyunsaturated fat?
6. What type of processing increases the amounts of *trans* fatty acids?
7. List three functions of fat in the body.
8. What is the advantage of storing energy as body fat rather than as carbohydrate?
9. What is the function of bile in fat digestion and absorption?
10. How do HDLs differ from LDLs?
11. How are blood levels of LDLs and HDLs related to the risk of cardiovascular disease?
12. Is essential-fatty-acid deficiency common in developed countries? Why or why not?
13. What types of foods contain cholesterol?
14. How does an atherosclerotic plaque form?
15. What is the AMDR for total dietary fat intake?
16. What information about dietary fat is included on food labels?
17. List two foods that are sources of monounsaturated fatty acids, two that are sources of omega-3 fatty acids, and two that are sources of cholesterol.
18. Describe the role of insulin in lipid metabolism.

REFERENCES

1. Garriguet, D. Canadians' Eating Habits. Health Reports 2007 (May) 18:2:17-32. Statistics Canada Catalogue 82-003. Available online at http://www.statcan.gc.ca/bsolc/olc-cel/olc-cel?catno=82-003-X&artid=9609&lang=eng. Accessed June 14, 2010.
2. Siri-Tarino, P. W., Sun, Q., Hi, F. B. et al. Saturated fatty acids and risk of coronary heart disease: Modulation by replacement nutrients. *Curr. Atheroscler. Rep.* 12:384–390, 2010.

3. Russo, G. L. Dietary n-6 and n-3 polyunsaturated fatty acids: From biochemistry to clinical implications in cardiovascular prevention. *Biochemical Pharmacology* 77:937–946, 2009.

4. Benjamin, S., Spener, F. Conjugated linoleic acids as functional food: an insight into their health benefits. *Nutr. Metab. (Lond).* 6:36, 2009.

5. Health Canada. Plant sterols and blood cholesterol lowering. 2010. Available online at http://www.hc-sc.gc.ca/fn-an/label-etiquet/claims-reclam/assess-evalu/phytosterols-eng.php. Accessed July 19, 2010.

6. Hamosh, M., Iverson, S. J., Kirk, C. L. et al. Milk lipids and neonatal fat digestion: Relationship between fatty acid composition, endogenous and exogenous digestive enzymes and digestion of milk fat. *World Rev. Nutr. Diet.* 75:86–91, 1994.

7. Brown, M. S., and Goldstein, J. L. How LDL receptors influence cholesterol and atherosclerosis. *Sci. Am.* 251:58–66, 1984.

8. Arterburn, L. M., Hall, E. B., Oken, H. Distribution, interconversion, and dose response of n-3 fatty acids in humans. *Am. .J. Clin. Nutr.* 83 (6 Suppl):1467S-1476S, 2006.

9. Mori, T. A., Beilin, L. J. Omega-3 fatty acids and inflammation. *Curr. Atheroscler. Rep.* 6:461–467, 2004.

10. Simopoulos, A. P. Omega-3 fatty acids in inflammation and autoimmune diseases. *J. Am. Coll. Nutr.* 21:495–505, 2002.

11. Institute of Medicine, Food and Nutrition Board. *Dietary Reference Intakes for Energy, Carbohydrates, Fiber, Fat, Protein and Amino Acids.* Washington, DC: National Academies Press, 2002.

12. American Heart Association. Diet and Lifestyle Recommendations. Available online at www.americanheart.org/presenter.jhtml?identifier=851. Accessed February 16, 2009.

13. Statistics Canada: Leading Causes of Death, by sex. Available online at http://www40.statcan.gc.ca/l01/cst01/hlth36a-eng.htm. Accessed February 15, 2011.

14. Ascherio, A., Rimm, E. B., Stampfer, M. J. et al. Marine ω-3 fatty acids, fish intake, and the risk of coronary disease among men. *N. Engl. J. Med.* 332:977–982, 1995.

15. Sofi, F., Abbate, R., Gensini, G.F., Casini, A. Accruing evidence on benefits of adherence to the Mediterranean diet on health: an updated systematic review and meta-analysis. Am J Clin Nutr 92(5):1189-96, 2010.

16. Libby, P. Inflammation and cardiovascular disease mechanisms. *Am. J. Clin. Nutr.* 83:456S–460S, 2006.

17. Kher, N., and Marsh, J. D. Pathobiology of atherosclerosis—A brief review. *Semin. Thromb. Hemost.* 30:665–672, 2004.

18. Heart and Stroke Foundation of Canada. Statistics. Available online at http://www.heartandstroke.on.ca/site/c.pvI3IeNWJwE/b.3581729/k.359A/Statistics.htm. Accessed June 21, 2010.

19. Sheth, T., Nair, C., Nargundkar, M. et al. Cardiovascular and cancer mortality among Canadians of European, South Asian, and Chinese origin from 1979 to 1993: an analysis of 1.2 million deaths. *Canadian Medical Association Journal.* 161(2):132–138, 1999.

20. Anand, S. S., Yusuf, S., Jacobs, R. et al. Risk factors, atherosclerosis, and cardiovascular disease among Aboriginal people in Canada: the Study of Health Assessment and Risk Evaluation in Aboriginal Peoples (SHARE-AP). *Lancet.* 358(9288):1147–1153, 2001.

21. Denke, M. A. Review of human studies evaluating individual dietary responsiveness in patients with hypercholesterolemia. *Am. J. Clin. Nutr.* 62 (suppl): 471S–477S, 1995.

22. Ginsberg, H. N., and Karmally, W. Nutrition, lipids, and cardiovascular disease. In *Biochemical and Physiological Aspects of Human Nutrition.* M. H. Stipanuk, ed. Philadelphia: W. B. Saunders Company, 2000, pp. 917–944.

23. Connor, W. E. Harbingers of coronary heart disease: Dietary saturated fatty acids and cholesterol. Is chocolate benign because of its stearic acid content? *Am. J. Clin. Nutr.* 70:951–952, 1999.

24. Van Horn, L., McCoin, M., Kris-Etherton, P. M. et al. The evidence for dietary prevention and treatment of cardiovascular disease. *J. Am. Diet. Assoc.* 108:287–331, 2008.

25. Fernandez, M. L., West, K. L. Mechanisms by which dietary fatty acids modulate plasma lipids. *J. Nutr.* 135(9):2075-8), 2005.

26. Lopez-Garcia, E., Schulze, M. B., Meigs, J. B. et al. Consumption of *trans* fatty acids is related to plasma biomarkers of inflammation and endothelial dysfunction. *J. Nutr.* 135:562–566, 2005.

27. National Cholesterol Education Program. Adult Treatment Panel III Report, 2001, 2004. Available online at www. nhlbi.nih.gov/guidelines/cholesterol/atp3udp04.htm/. Accessed May 26, 2009.

28. Russo, G. L. Dietary n-6 and n-3 polyunsaturated fatty acids: From biochemistry to clinical implications in cardiovascular prevention. *Biochemical Pharmacology* 77:937–946, 2009.

29. Willett, W. C., Sacks, F., Trichopouluo, A., et al. Mediterranean diet pyramid: A cultural model for healthy eating. *Am. J. Clin. Nutr.* 61(suppl):1402S–1406S, 1995.

30. Keaney, J. F., and Vita, J. A. Vascular oxidative stress and antioxidant protection in atherosclerosis. *J. Cardiopulmonary Rehabil.* 22:225–233, 2002.

31. Humphrey, L. L., Fu, R., Rogers, K. et al. Homocysteine level and coronary heart disease incidence: A systematic review and meta-analysis. *Mayo Clin. Proc.* 83:1203–1212, 2008.

32. Selhub, J., Jacques, P. F., Bostom, A. G. et al. Relationship between plasma homocysteine and vitamin status in the Framingham study population. Impact of folic acid fortification. *Public Health Rev.* 28:117–1145, 2000.

33. Lucas, D. L., Brown, R. A., Wassef, M. et al. Alcohol and the cardiovascular system: Research challenges and opportunities. *J. Am. Coll. Cardiol.* 45:1916–1924, 2005.

34. Zern, T. L., and Fernandez, M. L. Cardioprotective effects of dietary polyphenols. *J. Nutr.* 135:2291–2294, 2005.

35. American Institute for Cancer Research. The Diet and Cancer Link: Dietary Choices Play an Important Role in Reducing Cancer. Available online at www.aicr.org/diet.html/. Accessed June 28, 2004.

36. Statistics Canada. Cancer, new cases, by selected primary site of cancer, by sex. Available online at http://www40.statcan.gc.ca/l01/cst01/hlth61-eng.htm. Accessed June 21, 2010.

37. National Cancer Institute. Breast Cancer Prevention. Available online at http://www.cancer.gov/cancertopics/pdq/prevention/breast/HealthProfessional/page2. Accessed August 16, 2009.

38. Schultz, M., Hoffmann, K., Weikert, C. et al. Identification of a dietary pattern characterized by high-fat food choices associated with increased risk of breast cancer: The European Prospective Investigation into Cancer and Nutrition (EPIC)-Potsdam Study, *Br. J. Nutr.* 100:942–946, 2008.

39. Gago-Dominguez, M., Yuan, J. M., Sun, C. L. et al. Opposing effects of dietary ω-3 and ω-6 fatty acids on mammary carcinogenesis: The Singapore Chinese Health Study. *Br. J. Cancer.* 89:1686–1692, 2003.

40. Voorrips, L. E., Brants, H. A., Kardinaal, A. F. et al. Intake of conjugated linoleic acid, fat and other fatty acids in relation to post menopausal breast cancer: The Netherlands Study on Diet and Cancer. *Am. J. Clin. Nutr.* 76:873–882, 2002.

41. Zock, P. L., Katan, M. B. Linoleic acid intake and cancer risk: A review and meta-analysis. *Am. J. Clin. Nutr.* 68:142–153, 1998.

42. Kushi, L., Giovannucci, E. Dietary fat and cancer. *Am. J. Med.*113(suppl 9B):63S–70S, 2002.

43. Hall, M. N., Chavarro, J. E., Lee, I. M. et al. A 22-year prospective study of fish, n-3 fatty acid intake, and colorectal cancer risk in men. *Cancer Epidemiol. Biomarkers Prev.* 17:1136–1143, 2008.

44. Devitt, A. A., Mattes, R. D. Effects of food unit size and energy density on intake in humans. *Appetite* 42:213–220, 2004.

45. Ledikwe, J. H., Blanck, H. M., Kettel Khan, L. et al. Dietary energy density is associated with energy intake and weight status in US adults. *Am. J. Clin. Nutr.* 83:1362–1368, 2006.

46. Langlois, K., Garriguet, D., Findlay, L. Diet composition and obesity among Canadian adults. Health Reports 20(4): 1-10. Available online at http://www.statcan.gc.ca/bsolc/olc-cel/olc-cel?catno=82-003-X200900410933&lang=eng. Accessed June 21, 2010.

47. Canadian Food Inspection Agency. Nutritional Labelling:Chapter 6: The elements within the nutrition facts table. Available online at http://www.inspection.gc.ca/english/fssa/labeti/guide/ch6e.shtml#a6_3. Accessed July 6, 2011.

48. Katamay, S. W., Esslinger, K. A., Vigneault, M. et al. Eating well with Canada's Food Guide: development of the food intake pattern. *Nutr. Rev.* 65(4):155–66, 2007.

49. WHO: Diet, nutrition, and the prevention of chronic disease. Available online at http://www.who.int/dietphysicalactivity/publications/trs916/download/en/index.html. Accessed June 13, 2011.

50. Canadian Cancer Society. Eat Well, Be Active. What you can do. Available online at http://www.cancer.ca/Canada-wide/Publications/Alphabetical%20list%20of%20publications/Eat%20well%20%20be%20active%20What%20you%20can%20do.aspx?sc_lang=en. Accessed July 19, 2010.

51. Heart and Stroke Foundation. Healthy Living. Canada's Food Guide. http://www.heartandstroke.com/site/c.ikIQLcMWJtE/b.3484315/k.D9C8/Healthy_living__Eating_Well_with_Canadas_Food_Guide.htm. Accessed June 11, 2011.

52. Genest, J., McPherson, R., Frohlich, J. et al. Canadian Cardiovascular Society/Canadian guidelines for the diagnosis and treatment of dyslipidemia and prevention of cardiovascular disease in the adult-2009 recommendations. *Can. J. Cardiol.* 25 (10): 567-579, 2009.

53. Canadian Food Inspection Agency. Nutrition Labelling. Available online at http://inspection.gc.ca/english/fssa/labeti/nutrition-pagee.shtml. Accessed June 24, 2010.

54. Ratnayake, W. M., L'Abbé, M. R., Farnworth, S. et al. *Trans* fatty acids : current contents in Canadian foods and estimated intake levels for the Canadian population. *JAOAC Int.* 92(5):1258–1276, 2009.

55. Health Canada 2004. Canada Community Health Survey Cycle 2.2., Nutrition. Available online at http://www.hc-sc.gc.ca/fn-an/surveill/nutrition/commun/index-eng.php. Accessed July 17, 2010.

56. American Dietetic Association. Position of the American Dietetic Association: Fat replacers. *J. Am. Diet. Assoc.* 105:266–275, 2005.

57. Tada, N., Yoshida, H. Diacylglycerol on lipid metabolism. *Curr. Opin. Lipidol.* 14:29–33, 2003.

58. Martin Frid. 2010. Kurashi-News from Japan 2010: The end of econa oil. Available online at http://martinjapan.blogspot.com/2009/10/end-of-econa-oil.html. Accessed June 10, 2011.

(©iStockphoto)

FOCUS ON
Alcohol

Focus Outline

Almost every human culture since the dawn of civilization has produced and consumed some type of alcoholic beverage. These intoxicating beverages are part of religious ceremonies, social traditions, and even medical prescriptions. Sumerian clay tablets dating back to 2100 BCE record physicians' prescriptions for beer, and in ancient Egypt, both beer and wine were prescribed as part of medical treatment.

Depending on the times and the culture, alcohol use has been touted, casually accepted, denounced, and even outlawed. Today, some people refrain from its use for religious, cultural, personal, and medical reasons, while others enjoy the relaxing effects afforded by drinking these beverages. Whether alcohol consumption represents a risk or provides some benefits depends on who is drinking and how much is consumed. In some groups, moderate alcohol consumption provides health advantages, but excessive alcohol consumption always has medical and social consequences that negatively impact drinkers and their families. It can reduce nutrient intake and affect the storage, mobilization, activation, and metabolism of nutrients. Its breakdown produces toxic compounds that damage tissues, particularly the liver.

F1.1 What's in Alcoholic Beverages?

Learning Objective

- Explain the difference between alcohol and ethanol.

ethanol The type of alcohol in alcoholic beverages. It is produced by yeast fermentation of sugar.

Chemically, any molecule that contains a hydroxyl group (—OH) is an alcohol. Although there are many molecules present in our diet and our bodies that can be classified as alcohols, the term alcohol refers almost always to **ethanol**, also known as grain alcohol, and often to any beverage that contains ethanol (**Figure F1.1**). Alcoholic beverages, whether beer, wine, or distilled liquor, consist primarily of water, ethanol, and varying amounts of sugars. The amounts of other nutrients such as protein, vitamins, and minerals are almost negligible. Carbohydrate and alcohol provide the kcalories in these beverages. An average drink, defined as about 150 ml of wine, 360 ml of beer, or 50 ml of distilled spirits, contains about 12-14 g (about 15 ml) of alcohol, which contributes about 90 kcal (7 kcal/g). The amount of energy contributed by carbohydrate depends on the type of beverage (**Table F1.1**). Canada's Food Guide lists alcohol among the foods to limit.

Figure F1.1 Ethanol, the type of alcohol in all alcoholic beverages, is a small water-soluble molecule. (Andy Washnik)

Table F1.1 Energy, Carbohydrate, and Alcohol Content of Alcoholic Beverages

Beverage	Typical Serving (ml)	Energy (kcals)	Carbohydrate (g)	Alcohol (g)
Long Island iced tea	210 ml	170	24.3	17.8
Gin and tonic	300 ml	114	10.5	10.7
Wine cooler	360 ml	170	20.2	13.2
Beer	360 ml	146	13.2	12.8
Light beer	360 ml	100	4.6	11.3
White wine	150 ml	100	1.2	13.7
Red wine	150 ml	106	2.5	13.7
Bourbon	45 ml	96	0	13.9
Whisky	45 ml	96	0	13.9
Vodka	45 ml	96	0	13.9

F1.2 Absorption, Transport, and Excretion of Alcohol

Learning Objectives

- Explain how food in the stomach affects alcohol absorption.
- Discuss how alcohol is eliminated from the body.

Ingested alcohol is absorbed by simple diffusion along the entire gastrointestinal tract. Only a small amount is absorbed in the mouth and esophagus. Larger amounts are absorbed in the stomach, and the majority of absorption occurs in the duodenum and jejunum of the small intestine. Because alcohol is absorbed rapidly and significant amounts can be absorbed directly from the stomach, the effects of alcohol consumption are almost immediate, especially if it is consumed on an empty stomach. If there is food in the stomach, absorption is slowed because the stomach contents dilute the alcohol, reducing the amount in direct contact with the stomach wall. Food in the stomach also slows absorption because it slows stomach emptying and therefore decreases the rate at which alcohol enters the small intestine, where absorption is the most rapid.

Once absorbed, alcohol enters the bloodstream. It is a small, water-soluble molecule, and therefore is rapidly distributed throughout the body. It crosses cell membranes by diffusion so the amount that enters a cell depends on the concentration gradient across the cell membrane. Blood alcohol level therefore reflects the amount of alcohol in the body and is dependent on the difference between the rates of alcohol absorption and elimination. Peak blood concentrations are attained approximately 1 hour after ingestion (**Figure F1.2**). Many variables, including the kind and quantity of alcoholic beverage consumed, the speed at which the beverage is drunk, the food consumed with it, the weight and gender of the consumer, and the activity of alcohol-metabolizing enzymes in the body, determine blood alcohol level (**Table F1.2**).

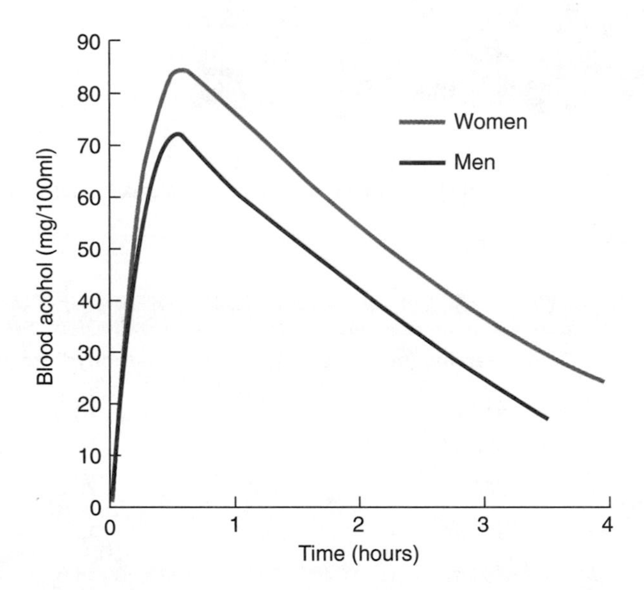

Figure F1.2 Blood alcohol concentrations in men and women
Consuming an equivalent amount of alcohol (0.5 g/kg of body weight) causes higher blood alcohol concentrations in women than in men.

Table F1.2 Factors Affecting Blood Alcohol Level

Factor	Effect
Weight	The more people weigh, the more body water they have, so the more dilute the alcohol in their blood is after they consume a given amount.
Gender	Men have more body water and more stomach alcohol dehydrogenase (ADH) activity and thus have a lower blood alcohol level after consuming a standard amount of alcohol than women of the same size.
Food	Food in the stomach slows alcohol absorption so the more food people eat before drinking, the lower their blood alcohol level will be.
Drinking Rate	The body metabolizes alcohol slowly. As the number of drinks per hour increases, blood alcohol level steadily rises.
The Type of Drink	The amount of alcohol in the drink affects how fast the blood alcohol level rises. When carbonated mixers (such as tonic water or club soda) are used, the body absorbs alcohol more quickly.

Because alcohol is a toxin and cannot be stored in the body, it must be eliminated quickly. Absorbed alcohol travels to the liver via the portal circulation. In the liver it is given metabolic priority and is therefore broken down before carbohydrate, protein, and fat. About 90% of the alcohol consumed is metabolized by the liver, about 5% is excreted into the urine, and the remainder is eliminated via the lungs during exhalation. The alcohol that reaches the kidney acts as a diuretic, increasing water excretion. Therefore, excessive alcohol intake can cause dehydration. The amount lost through the lungs is predictable and reliable enough to be used to estimate blood alcohol level from a measure of breath alcohol (**Figure F1.3**).

Figure F1.3 The amount of alcohol lost in exhaled breath is proportional to the amount of alcohol in the blood. Therefore, a measurement of breath alcohol can be used to estimate blood alcohol level and determine if an individual is driving under the influence of alcohol. (Science Photo Library/Photo Researchers, Inc.)

F1.3 Alcohol Metabolism

Learning Objectives

- Compare the 2 enzyme pathways that metabolize alcohol.
- Explain why alcohol intake increases fatty acid synthesis.

There are two primary pathways for alcohol metabolism: the **alcohol dehydrogenase (ADH)** pathway located in the cytosol of the cell and the **microsomal ethanol-oxidizing system (MEOS)** located in small vesicles called microsomes that form in the cell when they split off from an organelle called the smooth endoplasmic reticulum.[1]

alcohol dehydrogenase (ADH) An enzyme found primarily in the liver and stomach that helps break down alcohol into acetaldehyde, which is then converted to acetyl-CoA.

microsomal ethanol-oxidizing system (MEOS) A liver enzyme system located in microsomes that converts alcohol to acetaldehyde. Activity increases with increases in alcohol consumption.

Alcohol Dehydrogenase

In people who consume moderate amounts of alcohol and/or only consume alcohol occasionally, most of the alcohol is broken down via the ADH pathway. Although liver cells have the highest levels of ADH activity, this enzyme has also been found in all parts of the gastrointestinal tract, with the greatest amounts in the stomach.[2] The amount of alcohol broken down in the stomach may be significant when small amounts of alcohol are consumed. One hypothesis as to why women become intoxicated after consuming less alcohol than men is that women have lower activities of this stomach enzyme.[3,4] Another explanation for why women have higher blood alcohol concentrations than men following a standard amount of alcohol is the fact that they have a higher proportion of body fat and thus less body water than men. Thus, the alcohol they do consume is distributed in a smaller amount of body water (see Figure F1.2).

ADH converts alcohol to acetaldehyde. Acetaldehyde is a toxic compound that is further degraded by the mitochondrial enzyme aldehyde dehydrogenase to a 2-carbon molecule called acetate that forms acetyl-CoA (**Figure F1.4**). These reactions release electrons and hydrogen ions. Although these processes produce ATP, they also slow the citric acid cycle, preventing acetyl-CoA from being further broken down. Instead, the acetyl-CoA generated by alcohol breakdown, as well as acetyl-CoA from carbohydrate or fat metabolism, is used to synthesize fatty acids that accumulate in the liver. Fat accumulation can be seen in the liver after only a single bout of heavy drinking.

Microsomal Ethanol-Oxidizing System

Alcohol can also be metabolized in the liver by a second pathway called the microsomal ethanol-oxidizing system (MEOS). This system is particularly important when greater amounts of alcohol are consumed. MEOS converts alcohol to acetaldehyde, which is then broken down by aldehyde dehydrogenase in the mitochondria (see Figure F1.4). The MEOS system requires oxygen and the input of energy to break down alcohol. In addition to forming acetaldehyde and water, reactive oxygen molecules are generated. Reactive oxygen molecules are compounds that react with components in cells, such as protein, DNA, and lipids in a way that causes them to be oxidized. This oxidation can impair the functioning of the cell by damaging cell membranes, changing the activity of enzymes, and introducing mutations into DNA (see section 8.10 for a more detailed discussion of reactive oxygen molecules). When this happens in the liver it can contribute to liver disease. The rate that ADH breaks down alcohol is fairly constant but MEOS activity increases when more alcohol is consumed. The MEOS also metabolizes other drugs, so as activity increases in response to high alcohol intake, the metabolism of other drugs may be altered.

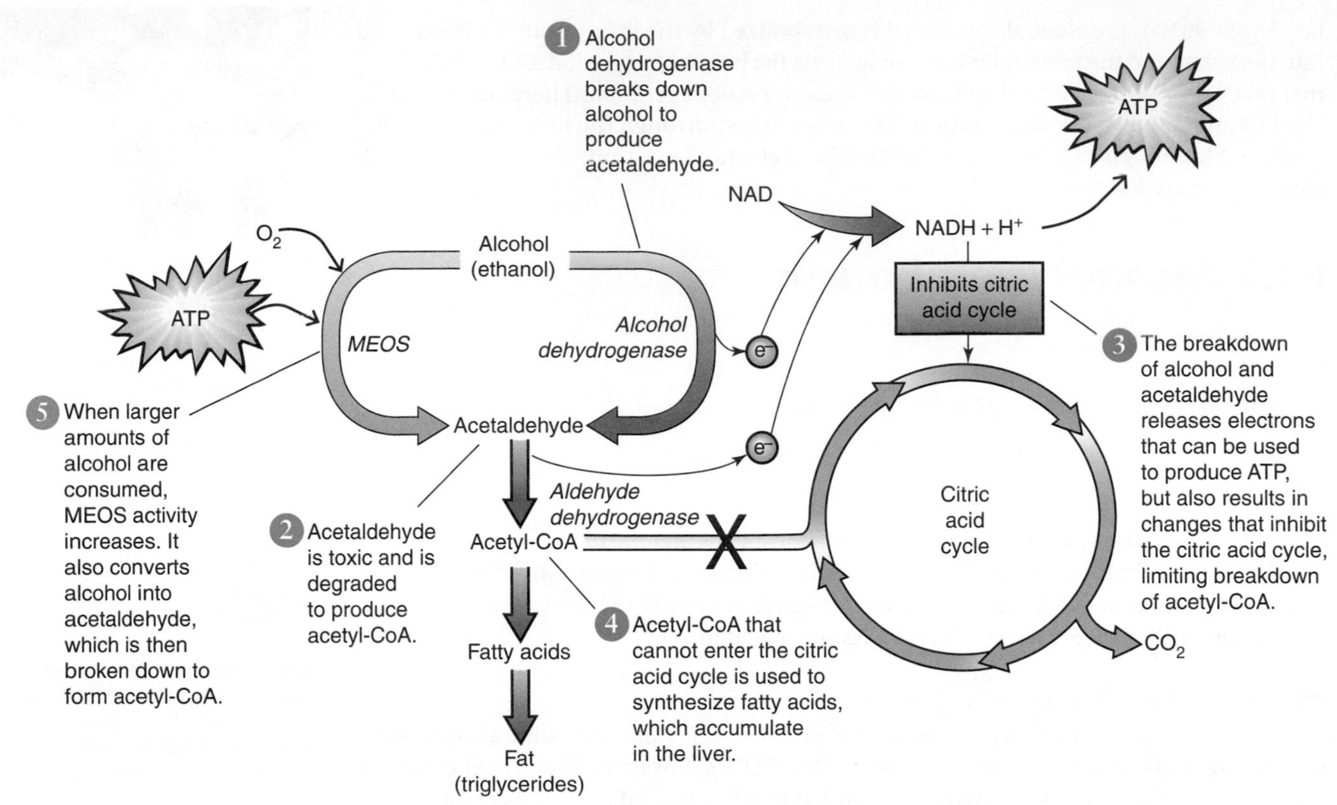

Figure F1.4 Alcohol metabolism
Alcohol can be metabolized by two pathways. The alcohol dehydrogenase pathway predominates when small amounts of alcohol are consumed and the MEOS pathway becomes important when larger amounts are consumed.

Alcohol Metabolism in the Colon

Although alcohol does not reach the colon through the digestive tract, alcohol is transported from the blood into the lumen of the colon. The alcohol that enters the large intestine is then metabolized by bacterial ADH to yield acetaldehyde. The capacity to metabolize alcohol to acetaldehyde by bacteria is greater than the ability to convert acetaldehyde to acetate. Therefore, toxic acetaldehyde accumulates in the colon, where it may contribute to mucosal injury and colon cancer. Acetaldehyde absorbed back into the blood contributes to liver damage.

F1.4 Adverse Effects of Alcohol Consumption

Learning Objectives

- Describe the short-term symptoms of alcohol intoxication.
- Explain why chronic excessive alcohol consumption can lead to malnutrition.
- Describe the long-term effects of alcoholism on the liver.

The consumption of alcohol has short-term effects that interfere with organ function for several hours after ingestion. It also has long-term effects that result from chronic alcohol consumption. Chronic alcohol consumption can cause disease both because it interferes with nutritional status and because it produces toxic compounds during its breakdown.

The effects of alcohol vary with life stage. When consumed during pregnancy, alcohol can cause abnormal brain development and other birth defects in the fetus (see section 14.3). When consumed during childhood and adolescence, when the brain is still developing and changing, alcohol can cause permanent reductions in learning and memory.[5] In everyone, alcohol either directly or indirectly affects every organ in the body and increases the risk of malnutrition and many chronic diseases (**Table F1.3**).

Table F1.3 Health Effects of Chronic Alcohol Use

Health Effect	Role of Alcohol
Birth defects	Increases the risk of fetal alcohol syndrome and alcohol-related birth defects and neurodevelopmental disorders when consumed during pregnancy.
Gastrointestinal problems	Damages the lining of the stomach and small intestine, and contributes to the development of pancreatitis.
Liver disease	Causes fatty liver, alcoholic hepatitis, and cirrhosis.
Malnutrition	Decreases nutrient absorption and alters the storage, metabolism, and excretion of some vitamins and minerals. Associated with a poor diet because alcohol replaces more nutrient-dense energy sources in the diet.
Neurological disorders	Contributes to impaired memory, dementia, and peripheral neuropathy.
Cardiovascular disorders	Associated with cardiovascular diseases such as cardiomyopathy, hypertension, arrhythmias, and stroke.
Blood disorders	Increases the risk of anemia and infection.
Immune function	Depresses the immune system and results in a predisposition to infectious diseases, including respiratory infections, pneumonia, and tuberculosis.
Cancer	Increases the risk for cancer, particularly of the upper digestive tract—including the esophagus, mouth, pharynx, and larynx—and of the liver, pancreas, breast, and colon.
Sexual dysfunction	Can lead to inadequate functioning of the testes and ovaries, resulting in hormonal deficiencies, sexual dysfunction, and infertility. It is also related to a higher rate of early menopause and a higher frequency of menstrual irregularities (duration, flow, or both).
Psychological disturbances	Causes depression, anxiety, and insomnia, and is associated with a higher incidence of suicide.
Mortality	About 4,000 Canadians under the age of 70 die annually from alcohol-related causes.[6]

Acute Effects of Alcohol Consumption

CANADIAN CONTENT

Depending on body size, amount of previous drinking, food intake, and general health, the liver can break down about 15 ml of alcohol per hour. This is the amount of alcohol in one drink (150 ml of wine, 360 ml of beer, or 50 ml of distilled liquor). When alcohol intake exceeds the ability of the liver to break it down, the excess accumulates in the bloodstream until the liver enzymes can metabolize it. The circulating alcohol affects the brain, resulting in impaired mental and physical abilities. In the brain, alcohol acts as a depressant, slowing the rate at which neurological signals are received. First, it affects reasoning; if drinking continues, the vision and speech centres of the brain are affected. Next, large-muscle control becomes impaired, causing lack of coordination. Finally, if alcohol consumption continues, it can result in **alcohol poisoning**, a serious condition that can slow breathing, heart rate, and the gag reflex, leading to loss of consciousness, choking, coma, and even death (**Table F1.4**). This occurs most frequently with **binge drinking**. Binge drinking is most common among young adults. Statistics Canada found that 44% of men and 29% of women aged 18 to 19 years consume five or more drinks per occasion at least 12 times a year. The number drops to 39% of men and 20% of women ages 20 to 34 years, and continues to decline as people get older.[7]

alcohol poisoning When the quantity of alcohol consumed exceeds an individual's tolerance for alcohol and impairs mental and physical abilities.

binge drinking When 5 or more drinks are consumed at one time.

Actual blood alcohol values depend on the amount of alcohol in the beverage, the rate of consumption, foods consumed with the alcohol, gender, and body weight.

The effects of alcohol on the central nervous system are what make driving while under the influence of alcohol so dangerous. Alcohol affects reaction time, eye-hand coordination, accuracy, and balance. Not only does alcohol impair one's ability to operate a motor vehicle, but it also impairs one's judgment in the decision to drive. Even if the individual does not lose consciousness, excess drinking may still result in memory loss. Drinking enough alcohol to cause amnesia is called **blackout drinking**. This puts people at risk because they have no memory of events that occurred during the blackout. During blackouts, individuals may engage in risky behaviours such as unprotected sexual intercourse, property vandalism, or driving a car—and have no memory of it afterward.

blackout drinking Amnesia following a period of excess alcohol consumption.

Table F1.4 Effects of Alcohol on the Central Nervous System

Number of Drinks[a]	Blood Alcohol[b] (%)	Effect on Central Nervous System
2	0.05	Impaired judgment, altered mood, relaxed inhibitions and tensions, increased heart rate
4	0.10	Impaired coordination, delayed reaction time, impaired peripheral vision
6	0.15	Unrestrained behaviour, slurred speech, blurred vision, staggered gait
8	0.20	Double vision, inability to walk, lethargy
12	0.30	Stupor, confusion, coma
≥14	0.35–0.60	Unconsciousness, shock, coma, death

[a]Each drink contains 15 ml of ethanol and is equivalent to 150 ml of wine, 360 ml of beer, or 50 ml of distilled liquor.
[b]Values represent blood alcohol within approximately 1 hour after consumption for a 70 kg individual.

Alcoholism: Chronic Effects of Alcohol Use

alcoholism A chronic disorder characterized by dependence on alcohol and development of withdrawal symptoms when alcohol intake is reduced.

One risk associated with regular alcohol consumption is the possibility of addiction, which is referred to as **alcoholism**. Alcohol addiction, like any other drug addiction, is a physiological problem that needs treatment. Alcoholism is believed to have a genetic component that makes some people more likely to become addicted, but environment also plays a significant role.[8] Thus, someone with a genetic predisposition toward alcoholism whose family and peers do not consume alcohol is much less likely to become addicted than someone with the same genes who drinks regularly with friends.

Alcohol and Malnutrition One of the complications of long-term excessive alcohol consumption is malnutrition. Alcoholic beverages contributes energy but few nutrients; they may replace more nutrient-dense energy sources in the diet, thereby reducing overall nutrient intake. As the percent of kcalories from alcohol increases, the risk of nutrient deficiencies rises. With more moderate alcohol intakes, the drinker typically substitutes alcohol for carbohydrate in the diet and total energy intake increases slightly. When intake of alcohol exceeds 30% of kcalories, protein and fat intake as well as carbohydrate intake decrease and consumption of essential micronutrients such as thiamin and vitamins A and C may fall below recommended amounts.[1] Therefore, a diet that is high in alcohol causes primary malnutrition because nutrient-dense energy sources in the diet are replaced by alcoholic beverages, decreasing overall nutrient intake.

In addition to decreasing nutrient intake, alcohol can contribute to a secondary malnutrition by interfering with nutrient absorption, even when adequate amounts of nutrients are consumed. Alcohol causes inflammation of the stomach, pancreas, and intestine, which impairs the digestion of food and absorption of nutrients into the blood. Alcohol damage to the lining of the small intestine decreases the absorption of several B vitamins and vitamin C.[2] Deficiency of the B vitamin thiamin is a particular concern with chronic alcohol consumption (see Section 8.2). The mucosal damage caused by alcohol also increases the ability of large molecules to cross the mucosa. This allows toxins from the gut lumen to enter the portal blood, increasing the liver‘s exposure to these toxins and, consequently, the risk of liver injury. Alcohol also contributes to malnutrition by altering the storage, metabolism, and excretion of other vitamins and some minerals.

In addition to contributing to undernutrition, alcohol consumption may be related to obesity. Kcalories consumed as alcohol are more likely to be deposited as fat in the abdominal region; excess abdominal fat increases the risk of high blood pressure, heart disease, and diabetes. An analysis of alcohol consumption patterns and body weight showed that individuals who consumed a small amount of alcohol frequently (one drink per day, 3-7 days/week) had the lowest BMI, while those who consumed large amounts infrequently had the highest BMI.[9] Alcohol may contribute to weight gain because liquids are less satiating than solid food and drinking may stimulate appetite, promoting consumption of additional energy sources.

Toxic Effects of Alcohol Chronic alcohol consumption is associated with hypertension, heart disease, and stroke, but the most significant physiological effects occur in the liver. In

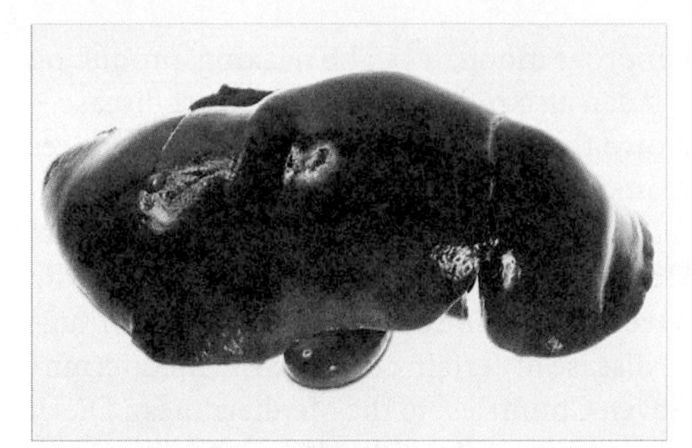

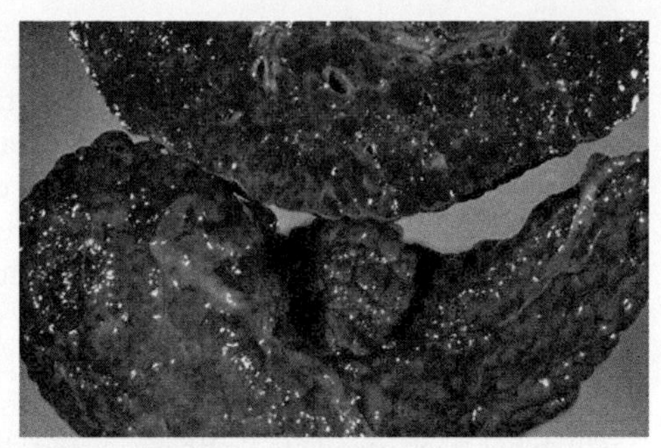

Figure F1.5 Chronic alcohol consumption can cause permanent liver damage. (Martin/Custom Medical Stock Photo, Inc.; Science Heritage/Custom Medical Stock Photo, Inc.)

addition to causing liver damage from malnutrition, alcohol causes liver damage through its toxic effects. Metabolism via ADH produces excess amounts of the electron carrier NADH. NADH inhibits the citric acid cycle and affects the metabolism of lipids, carbohydrates, and proteins (see Figure F1.4; also see section 8.4 for a more detailed discussion of the molecule NADH). High NADH favours fat synthesis and inhibits fatty acid breakdown, leading to fat accumulation in the liver (**Figure F1.5**). Metabolism by MEOS generates free reactive oxygen molecules, which damage cellular components. Whether broken down by ADH or MEOS, toxic acetaldehyde is formed. Acetaldehyde exerts its toxic effects by binding to proteins and by inhibiting reactions and functions of the mitochondria. This decreases the metabolism of acetaldehyde to acetic acid allowing more acetaldehyde to accumulate, causing further liver damage.

Alcoholic liver disease progresses in a number of phases. The first phase is **fatty liver**, a condition that occurs when alcohol consumption increases the synthesis and deposition of fat in the liver. The second phase, **alcoholic hepatitis**, is an inflammation of the liver. Both of these conditions are reversible if alcohol consumption is stopped and good nutritional and health practices are followed. If alcohol consumption continues, **cirrhosis** may develop. This is an irreversible condition in which fibrous deposits scar the liver and interfere with its function (see Figure F1.5). Because the liver is the primary site of many metabolic reactions, cirrhosis is often fatal.

fatty liver The accumulation of fat in the liver.

alcoholic hepatitis Inflammation of the liver caused by alcohol consumption.

cirrhosis Chronic liver disease characterized by the loss of functioning liver cells and the accumulation of fibrous connective tissue.

Alcohol and Cancer In addition to causing liver disease, heavy drinking is associated with certain types of cancer. A meta-analysis of alcohol consumption and cancer risk found an association between alcohol consumption and cancer of the oral cavity, pharynx, esophagus, larynx, breast, liver, colon, rectum, and stomach.[10] Epidemiology shows that alcohol consumption is associated with an increased risk of breast cancer and that risk is directly related to the amount of alcohol consumed. How alcohol promotes breast cancer is still unclear but it has been hypothesized to affect cancer risk by altering estrogen metabolism through the damaging effects that acetaldehyde can have on DNA, by causing oxidative damage, or by interfering with the metabolism of folate, a B vitamin discussed in Section 8.8.[11] There is some evidence that adequate dietary folate protects against the increased risk of breast cancer associated with alcohol consumption.[12]

F1.5 Benefits of Alcohol Consumption and Safe Drinking

Learning Objectives

- Describe how moderate alcohol intake reduces the risk of cardiovascular disease.
- List 3 things you could do to reduce the chances of becoming intoxicated while drinking alcoholic beverages.

For some people, moderate alcohol consumption (defined as no more than 1 drink per day for women and 2 drinks per day for men) may be beneficial. Consuming alcoholic beverages before or with meals can stimulate appetite and improve mood. It can be relaxing, producing a euphoria that can enhance social interactions. It can also reduce the risk of heart disease.

Epidemiological and clinical studies have shown that light to moderate drinking reduces the risk for heart disease and stroke when compared to not consuming alcohol at all.[13] This reduction in heart disease risk results in a reduced mortality in middle-aged and older adults who consume moderate amounts of alcohol.[14] However, at high levels of consumption, the mortality from heart disease is increased[15] (**Figure F1.6**). Wine consumption has been suggested as a reason for the lower incidence of heart disease in certain cultures. The Mediterranean diet (see Chapter 1 Science Applied: From "Seven Countries" to the Mediterranean Diet), which has been associated with a reduced risk of heart disease, includes daily consumption of wine in moderation. And one explanation for the French paradox—the fact that the French eat a diet that is as high or higher in fat than the North American diet but suffer from far less heart disease—is the glass of wine they drink with meals. The particular benefit of red wine is likely due not only to the alcohol, but also to the phytochemicals, called polyphenols, it contains[16,17] (**Figure F1.7**).

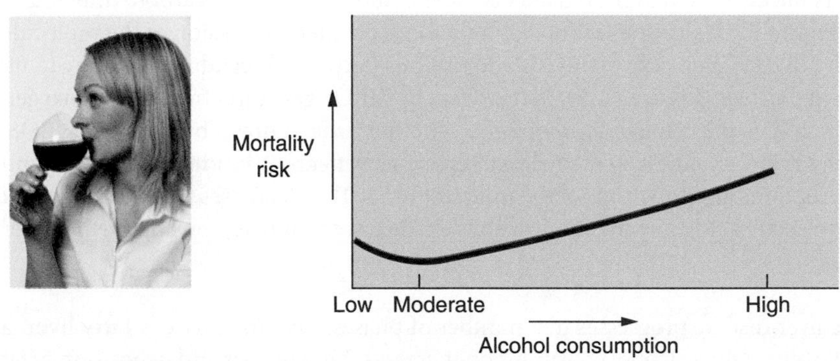

Figure F1.6 Alcohol consumption and mortality
The risk of mortality plotted against the amount of alcohol consumed generally results in a J-shaped curve with lowest mortality at the level that corresponds to moderate alcohol consumption. The actual shape of the curve varies with age and gender. (Westend61/The Canadian Press)

Figure F1.7 Alcohol has a number of effects that reduce the risk of heart disease. The phytochemicals in red wine may add to this beneficial effect. (©Rayes/Digital Vision/Age Fotostock America, Inc.)

There are a number of ways that moderate alcohol consumption reduces the risk of heart disease[15,18] (**Figure F1.8**). The most significant is its effect on HDL cholesterol. Moderate alcohol consumption can increase HDL cholesterol by 30% and is believed to account for about half of alcohol's protective effect. Alcohol also reduces the risk of heart disease by reducing levels of the blood-clotting protein fibrinogen. These lower levels reduce blood clots that can block blood flow to the heart, resulting in a heart attack. Smaller beneficial effects have been attributed to alcohol's effect on insulin sensitivity, inflammatory molecules, and platelets.[18]

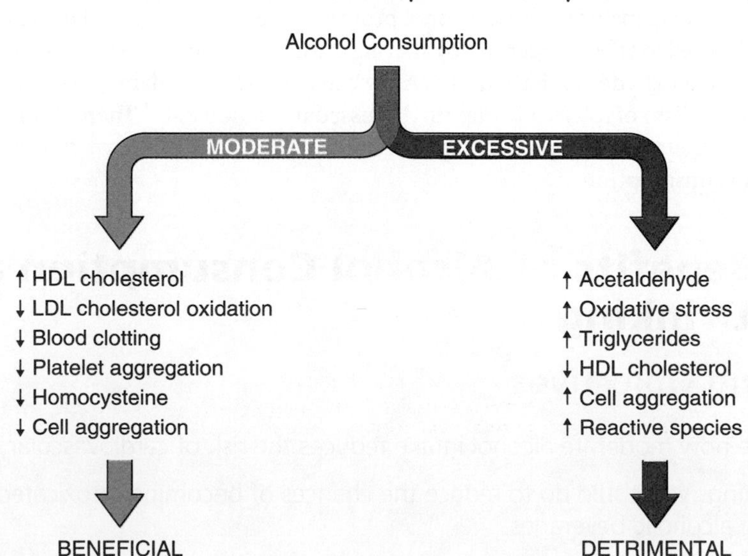

Figure F1.8 Benefits and risks of alcohol consumption
Moderate alcohol consumption may reduce the risk of heart disease, but greater intakes contribute to cardiovascular disease risk.

Although all of these effects have been documented, the benefit any one individual will gain from moderate alcohol consumption depends on their genetic background, health status, and overall lifestyle.

Alcohol, like other drugs, has risks as well as benefits. The risks posed by alcohol depend on the consumer and the amount consumed. Some people should not consume any alcohol. For instance, women who are pregnant or trying to conceive should not consume alcohol because it can damage the fetus (see section 14.3). Children and adolescents should not consume alcohol because they are more likely to suffer its toxic effects—drunkenness and poisoning leading to seizures, coma, and death.[5] Individuals who plan to drive or operate machinery should not consume alcohol because it can impair coordination and reflexes. Alcoholics should avoid alcohol because they cannot restrict their drinking to moderate levels. Finally, individuals taking medications that can interact with alcohol should avoid alcohol. For everyone, the risks of excess alcohol consumption outweigh the benefits.

Figure F1.9 When a moderate amount of alcohol is consumed with food, the risk of intoxication is reduced and the potential benefits are enhanced. (RonTech2000/the Agency Collection/Getty Images)

Whether the benefits of alcohol consumption outweigh the risks, drinking is a personal decision that must take into account medical and social considerations. However, those who choose to drink should do so in moderation. Alcohol should be consumed slowly, no more than 1 drink every 1.5 hours. Sipping, not gulping, allows the liver time to break down the alcohol that has already been consumed. Alternating nonalcoholic and alcoholic drinks will also slow down the rate of alcohol intake and prevent dehydration. Alcohol absorption is most rapid on an empty stomach. Consuming alcohol with meals slows its absorption and may also enhance its protective effects on the cardiovascular system (**Figure F1.9**). Also, the effect of alcohol on HDL is believed to be greater when the liver is processing nutrients from a meal.

Life Cycle

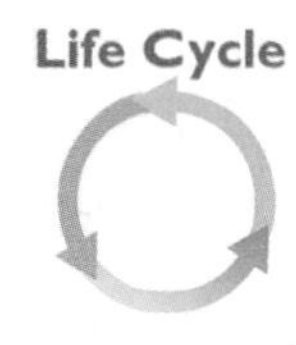

REFERENCES

1. Lieber, C. S. Relationships between nutrition, alcohol use, and liver disease. *Alcohol Res. Health* 27:220–231, 2003.
2. Bode, C., and Bode, J. C. Effect of alcohol consumption on the gut. *Best Pract. Res. Clin. Gastroenterol.* 17:575–592, 2003.
3. Baraona, E., Abittan, C. S., Dohmen, K. et al. Gender differences in pharmacokinetics of alcohol. *Alcohol Clin. Exp. Res.* 25:502–507, 2001.
4. Lai, C. L., Chao, Y. C., Chen, Y. C. et al. No sex and age influence on the expression pattern and activities of human gastric alcohol and aldehyde dehydrogenases. *Alcohol Clin. Exp. Res.* 24:1625–1632, 2000.
5. American Medical Association. Harmful Consequences of Alcohol Use on the Brains of Children, Adolescents and College Students, 2002. Fact Sheet. Available online at www.ama-assn.org/ama1/pub/upload/mm/388/harmful_consequences.pdf. Accessed September 11, 2009.
6. Rehm, J., Patra, J. and Popova, S. Alcohol-attributable mortality and potential years of life lost in Canada 2001: implications for prevention and policy. *Addiction*. 101(3):373-84, 2006.
7. Statistics Canada. Heaving Drinking 2008. Available online at http://www.statcan.gc.ca/pub/82-625-x/2010001/article/11103-eng.htm. Accessed July 30, 2011.
8. Whitfield, J. B. Alcohol and gene interactions. *Clin. Chem. Lab. Med.* 43:480–487, 2005.
9. Breslow, R. A., and Smothers, B. A. Drinking patterns and body mass index in never smokers: National Health Interview Survey, 1997–2001. *Am. J. Epidemiol.* 161:368–376, 2005.
10. Bagnardi, V., Blangiardo, M., La Vecchia, C, et al. A meta-analysis of alcohol drinking and cancer risk. *Br. J. Cancer* 85:1700–1705, 2001.
11. Dumitrescu, R. G., and Shields, P. G. The etiology of alcohol-induced breast cancer. *Alcohol* 35:213–225, 2005.
12. Baglietto, L., English, D. R., Gertig, D. M. et al. Does dietary folate intake modify effect of alcohol consumption on breast cancer risk? Prospective cohort study. *BMJ* 331:807, 2005.
13. Djousse, L., Ellison, R. C., Beiser, A. et al. Alcohol consumption and the risk of ischemic stroke: The Framingham Study. *Stroke* 4:907–912, 2002.
14. Meister, K. A., Whelan, E. M., and Kava, R. The health effects of moderate alcohol intake in humans: An epidemiologic review. *Crit. Rev. Clin. Lab. Sci.* 37:261–296, 2000.
15. Lucas, D. L., Brown, R. A., Wassef, M., et al. Alcohol and the cardiovascular system: Research challenges and opportunities. *J. Am. Coll. Cardiol.* 45:1916–1924, 2005.
16. Gronbaek, M., Becker, U., Johansen, D. et al. Type of alcohol consumed and mortality from all causes, coronary heart disease, and cancer. *Ann. Intern. Med.* 133:411–419, 2000.
17. Goldfinger, T. M. Beyond the French paradox: The impact of moderate beverage alcohol and wine consumption in the prevention of cardiovascular disease. *Cardiol. Clin.* 21:449–457, 2003.
18. Klatsky, A. L. Alcohol and cardiovascular disease. *Expert Rev. Cardiovasc. Ther.* 7:499–506, 2009.

6 Proteins and Amino Acids

Case Study

Teresa, an intern with the Canadian International Development Agency (CIDA), recently arrived in Somalia to help distribute emergency relief supplies. In a rural health clinic, a small 2-year-old boy named Songe catches her eye. His hair is an odd colour, his legs are scrawny, and his belly is so large he looks pregnant. Songe and his new baby brother live with their mother and three other siblings in a refugee camp. His mother says that his symptoms began shortly after his younger brother was born. His gaunt look suggests that he isn't getting enough to eat, but why is his belly so large?

The clinic nurse explains to Teresa that Songe is suffering from protein-energy malnutrition. When his brother was born, Songe stopped nursing at his mother's breast and started eating the camp diet, which is high in starch and fibre. He no longer consumes enough protein to support the needs of his growing body. His belly is bloated because his abdomen is filling with fluid and his liver is filling with fat. There is not enough protein in his blood to hold the fluid there, so it seeps into his abdomen; nor is there enough protein to transport fat, so it accumulates in his liver. Songe is also experiencing health problems that are less visible. Without adequate protein, his immune system cannot function normally, increasing his susceptibility to infections and making the immunizations provided in the camp less effective.

Teresa asks the nurse why Songe's older siblings, who consume the same high-carbohydrate diet, don't have large bellies too. The nurse explains that because Songe's siblings are older, they can eat a larger volume of food. Also, their protein needs per kcalorie are less than Songe's, so they are able to eat enough of the camp diet to meet their protein needs. To treat his malnutrition, the clinic provides Songe with a special high-protein drink to consume along with his regular diet.

In this chapter, you will learn about the important role that protein plays in human nutrition.

(Digital Vision/Getty Images, Inc.)

(©iStockphoto)

(© istockphoto.com/Luca Manieri)

Chapter Outline

6.1 Protein in the Canadian Diet

Learning Objectives

- Indicate which types of foods provide the most concentrated sources of protein.
- Describe how the protein source affects the other nutrients it provides.

Protein is essential to life. The human body is about 15% protein and our muscles, organs, and tissue are comprised of significant amounts of protein. Fortunately, protein deficiency is rare in Canada. The CCHS found that more than 97% of Canadians consumed protein above the AMDR of 10% kcal from protein, the minimum recommended amount. The average intake was typically about 15% kcal from protein over all age groups. This translated into an average protein intake of about 99 g/day for adult men and 73 g/day for adult women, amounts that more than meet the requirements of most Canadians.[1]

Sources of Dietary Protein

In Canada, disappearance data suggest about 62% of dietary protein comes from meat, poultry, fish, eggs, dairy products, and eggs(**Figure 6.1**). These animal products are the most concentrated sources of protein. Bread, rice, and pasta provide small amounts of protein—about 2 g-3 g per Canada's Food Guide serving, but Canadians consume so many cereal products that they represent more than 20% of Canadians' protein intake (**Figure 6.2**). Legumes or pulses, such as lentils, soybeans, peanuts, fava beans, black-eyed peas, chickpeas, and dried beans, as well as nuts and seeds, are also important sources.

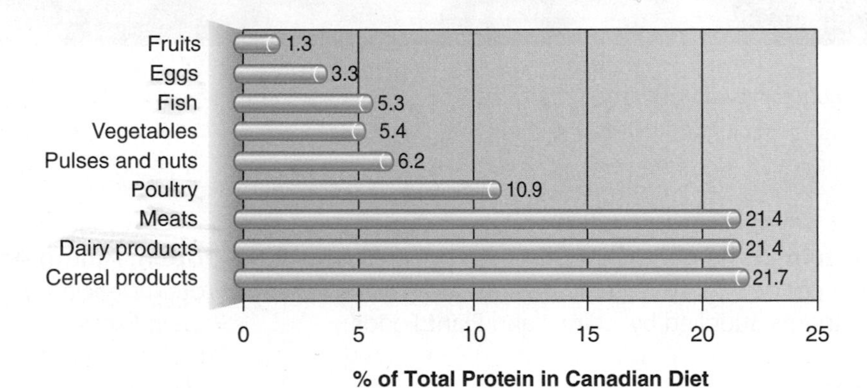

Figure 6.1: Sources of protein in the Canadian diet.
Disappearance data shown here indicates that animal sources, such as eggs, fish, poultry, meat, and dairy products contribute 62% of the protein in the Canadian diet.

Source: Statistics Canada. Table 003-0080-Nutrients in the food supply, by source of nutritional equivalent and commodity, annual, CANSIM (database): http://cansim2.statcan.gc.ca/cgi-win/cnsmcgi.exe?Lang=E&CNSM-Fi=CII/CII_1-eng.htm.

Nutrients Supplied by Animal and Plant Foods

Figure 6.2 Both plant and animal foods provide good sources of protein in the Canadian diet. (Amy Etra/PhotoEdit)

The source of the protein in the diet determines what other nutrients are consumed along with it. Animal products provide an excellent source of B vitamins and minerals such as iron, zinc, and calcium, but they are low in fibre and are often high in saturated fat and cholesterol—a nutrient mix that increases the risk of heart disease. Plant sources of protein are a good source of most, but not all, B vitamins and they also supply iron, zinc, and calcium, but in less-absorbable forms. Foods that provide good sources of plant proteins also contain fibre, phytochemicals, and unsaturated fats—dietary components that should be increased to promote health. Canada's Food Guide recommends whole-grain products, vegetables, and fruits, and smaller amounts of lean meats and low-fat milk products. This dietary pattern provides plenty of protein, without an over-reliance on animal protein, as well as many other nutrients.

6.2 Protein Molecules

Learning Objectives

- Describe the general structure of an amino acid, a polypeptide, and a protein.
- Distinguish between essential and nonessential amino acids.
- Discuss how the order of amino acids in a polypeptide chain affects protein structure and function.

Like carbohydrates and lipids, protein molecules contain the elements carbon, hydrogen, and oxygen, but proteins are distinguished from the other energy-yielding nutrients by the presence of the element nitrogen in their structure.

A protein molecule, whether found in a steak, a kidney bean, or a part of the human body, is constructed of one or more folded, chain-like strands of **amino acids**. The amino-acid chain of each type of protein molecule contains a characteristic number and proportion of amino acids that are bound together in a precise order. This chain folds into a specific orientation, giving each protein a unique, three-dimensional shape that is essential to its particular function. Variations in the number, proportion, and order of amino acids allow for an infinite number of different protein structures.

amino acids The building blocks of proteins. Each contains a central carbon atom bound to a hydrogen atom, an amino group, an acid group, and a side chain.

Amino Acid Structure

Approximately 20 amino acids are commonly found in proteins. Each amino acid consists of a carbon atom bound to 4 chemical groups: a hydrogen atom; an amino group, which contains nitrogen; an acid group; and a fourth group or side chain that varies in length and structure (**Figure 6.3**). Different side chains give specific properties to individual amino acids.

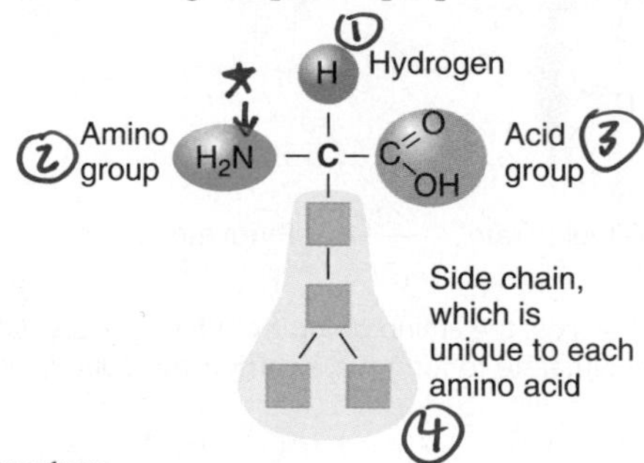

Figure 6.3 Amino acid structure
All amino acids have a similar structure, but each has a different side chain.

Of the 20 amino acids commonly found in protein, 9 cannot be made by the adult human body. These amino acids, called **essential or indispensable amino acids**, must be consumed in the diet (**Table 6.1**). If the diet is deficient in one or more of these amino acids, new proteins containing them cannot be made without breaking down other body proteins to provide them. The 11 **nonessential** or **dispensable amino acids** can be made by the human body and are not required in the diet. When a nonessential amino acid needed for protein synthesis is not available from the diet, it can be made in the body. Most of the nonessential amino acids can be made by the process of **transamination**, in which an amino group from one amino acid, often an essential amino acid, is transferred to a carbon-containing molecule to form a different amino acid (**Figure 6.4**).

Some amino acids are **conditionally essential**, meaning they are essential only under certain conditions. For example, the conditionally essential amino acid tyrosine can be made in the body from the essential amino acid phenylalanine. If phenylalanine is in short supply, tyrosine cannot be made and becomes essential in the diet. Likewise, the amino acid cysteine is only essential when the essential amino acid methionine is in short supply or cannot be converted to cysteine. Other amino acids may be essential at certain times of life, such as premature infancy, or due to certain conditions, such as metabolic abnormalities or physical stress.

essential or **indispensable amino acids** Amino acids that cannot be synthesized by the human body in sufficient amounts to meet needs and therefore must be included in the diet.

nonessential or **dispensable amino acids** Amino acids that can be synthesized by the human body in sufficient amounts to meet needs.

transamination The process by which an amino group from one amino acid is transferred to a carbon compound to form a new amino acid.

conditionally essential amino acids Amino acids that are essential in the diet only under certain conditions or at certain times of life.

Table 6.1 Essential and Nonessential Amino Acids

Essential Amino Acids	Nonessential Amino Acids
Histidine	Alanine
Isoleucine	Arginine*
Leucine	Asparagine
Lysine	Aspartic acid (aspartate)
Methionine	Cysteine (cystine)*
Phenylalanine	Glutamic acid (glutamate)
Threonine	Glutamine*
Tryptophan	Glycine*
Valine	Proline*
	Serine
	Tyrosine*

*These amino acids are considered conditionally essential by the Institute of Medicine, Food and Nutrition Board (*Dietary Reference Intakes for Energy Carbohydrates, Fiber,* Fat, Protein and Amino Acids, Washington, DC: National Academies Press, 2002). See text for explanation.

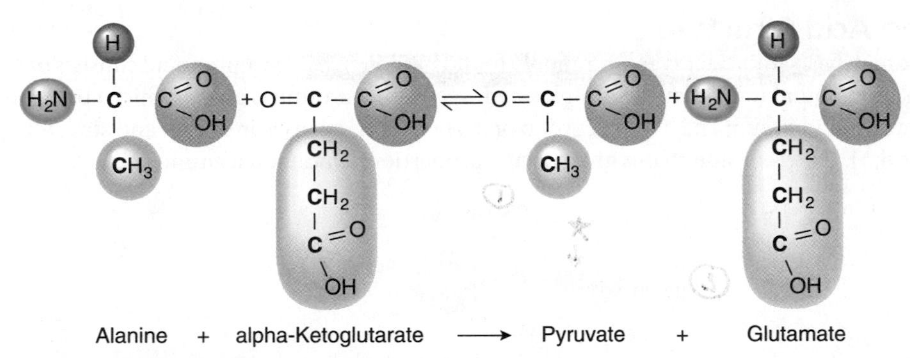

Figure 6.4 Transamination
In this example, transamination transfers the amino group from the nonessential amino acid alanine to the carbon compound alpha-ketoglutarate to form the 3-carbon compound pyruvate and the amino acid glutamate.

Protein Structure

To form proteins, amino acids are linked together by a unique type of chemical bond called a peptide bond. This bond is formed between the acid group of 1 amino acid and the nitrogen atom of the next amino acid (**Figure 6.5**). When 2 amino acids are linked with a peptide bond, it is called a **dipeptide**; when 3 amino acids are linked, they form a **tripeptide**. Many amino acids bonded together constitute a **polypeptide**.

dipeptide Two amino acids linked by a peptide bond. A **tripeptide** is 3 amino acids linked by peptide bonds, and a **polypeptide** is a chain of 3 or more amino acids linked by peptide bonds.

A protein is made of one or more polypeptide chains folded into a complex, three-dimensional shape (see Figure 6.5). The order and chemical properties of the amino acids in a polypeptide chain determine the three-dimensional shape of the protein. Folds and bends in the chain occur when some of the amino acids attract each other and other amino acids repel each other. For example, amino acids at various places along the chain that are attracted to each other may cause segments of the chain to form a coil. Amino acids that are attracted to water will orient to the outside of the structure to be in contact with body fluids, whereas amino acids that repel water will fold to the inside to be away from body fluids. After polypeptide chains have folded, several chains may bind together to form the final protein.

It is the shape of the final protein that determines its function. For example, the elongated shape of the connective-tissue proteins, collagen and alpha-keratin, allows them to give strength to fingernails and ligaments. The oxygen-carrying protein hemoglobin has a spherical shape,

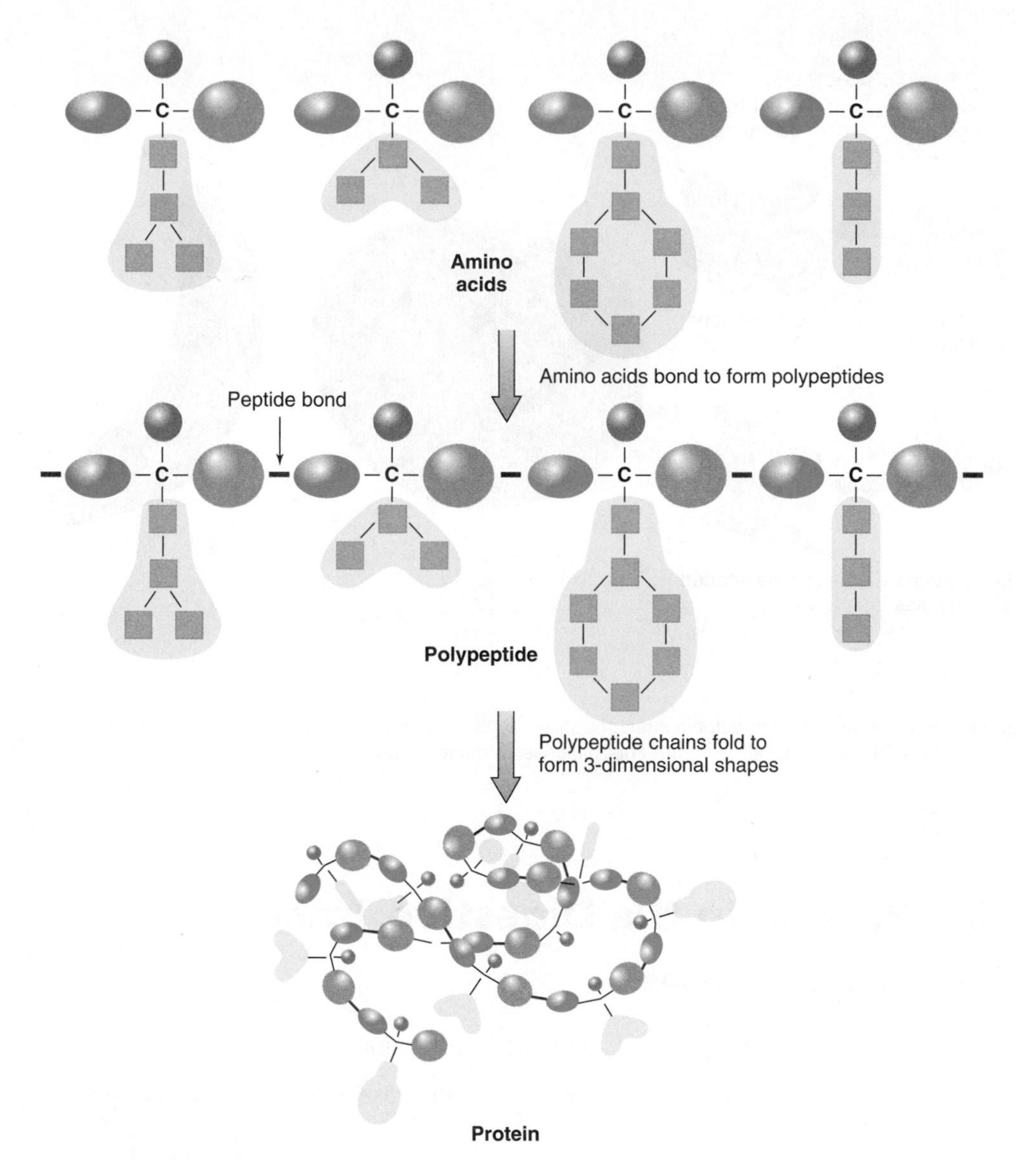

Figure 6.5 Protein structure
When many amino acids are linked by peptide bonds, they form a polypeptide. Folding of the polypeptide chains creates the three-dimensional structure of the protein.

which allows proper functioning of the red blood cells. If the shape of a protein is altered, its function may be disrupted. For example, in the genetic disease sickle cell anemia, a single amino acid in the hemoglobin molecule is altered, causing the protein molecules to bind together in long chains. Thus, rather than having the disc-shaped characteristic of normal red blood cells, red blood cells containing these rope-like strands of sickle cell hemoglobin have a distorted shape that resembles a crescent or sickle (**Figure 6.6**). Sickle-shaped red blood cells can block capillaries, causing inflammation and pain. They also rupture easily, leading to anemia from a shortage of red blood cells.

Changes in protein structure can also be caused by changes in the physical environment of the protein, such as an increase in temperature or a change in pH. Such changes cause protein **denaturation**. In food, the heat of cooking denatures protein, thereby changing its shape and physical properties. For example, a raw egg white is clear and liquid, but once it has been denatured by cooking, it becomes white and firm (**Figure 6.7**). This change in shape may also assist in the digestion of proteins by making it easier for digestive enzymes to break down protein (see Section 6.3).

denaturation The alteration of a protein's three-dimensional structure.

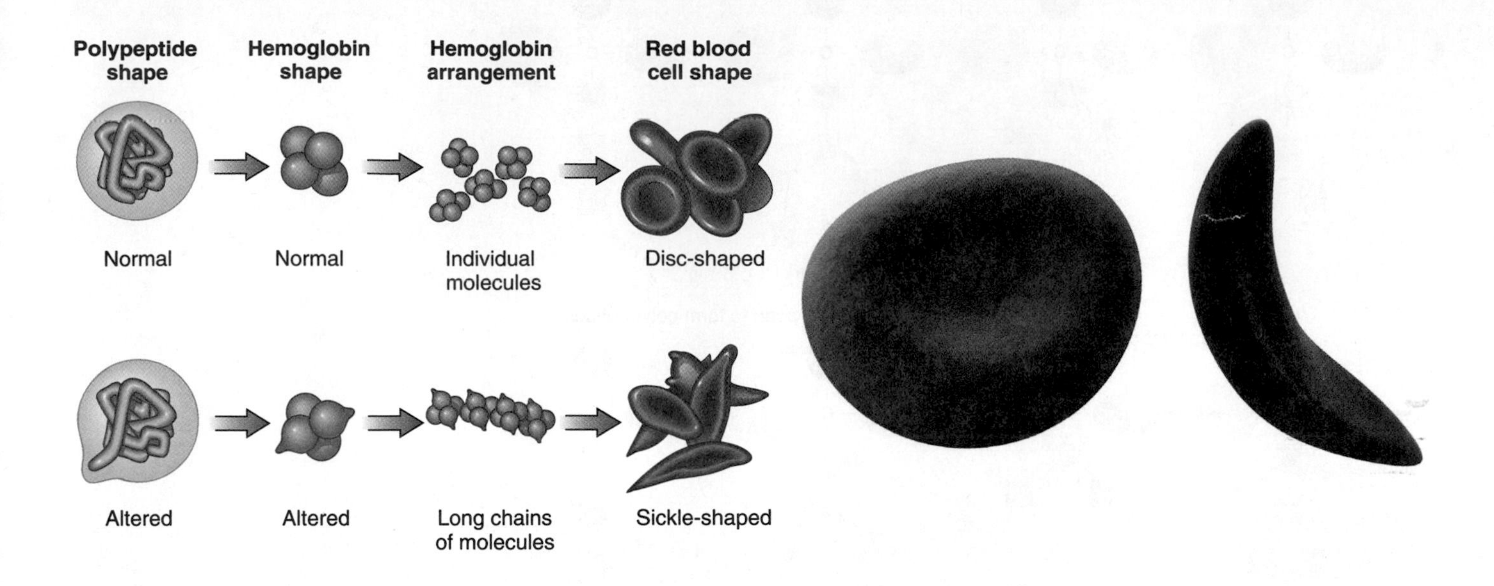

Figure 6.6 Sickle cell anemia
In sickle cell anemia, a change in the sequence of amino acids in hemoglobin alters the shape and function of the protein molecule. Sickle cell hemoglobin forms long chains that distort the shape of red blood cells. (Ingram Publishing/Getty Images, Ltd.)

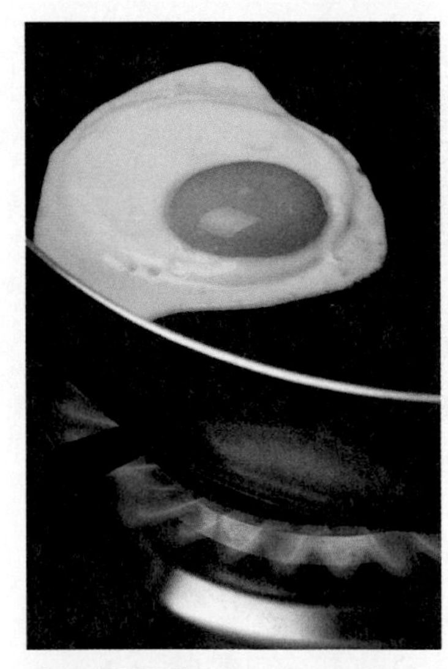

Figure 6.7 Why does the protein in egg white change from clear and liquid to white and firm when it is cooked? (© istockphoto.com/MorePixels)

6.3 Protein in the Digestive Tract

Learning Objectives

- Describe protein digestion and amino acid absorption.
- Explain why an excess of one amino acid could cause a deficiency of another amino acid.
- Discuss how protein digestion is related to food allergies.

Protein enters the digestive tract from food, from digestive secretions, and from sloughed gastrointestinal cells. No matter what the source, the protein must be broken down into amino acids before entering the bloodstream.

Protein Digestion

The chemical digestion of protein begins in the stomach, where hydrochloric acid denatures proteins, opening up their folded structure to make them more accessible to enzyme attack (**Figure 6.8**). The acid also activates the protein-digesting enzyme pepsin, which breaks proteins into polypeptides and amino acids. When the polypeptides enter the small intestine, they are broken into smaller polypeptides, tripeptides, dipeptides, and amino acids by pancreatic protein-digesting enzymes such as trypsin and chymotrypsin. Protein-digesting enzymes in the brush border of the small intestine further break down the small polypeptides. Single amino acids, dipeptides, and tripeptides can be absorbed into the mucosal cells of the small intestine. Once inside, they are broken into single amino acids (see Figure 6.8).

Amino Acid Absorption

Amino acids and di- and tripeptides enter the mucosal cells of the small intestine using one of several energy-requiring active transport systems. Amino acids with similar structures share the same transport system and therefore compete for absorption. For instance, the amino acids leucine, isoleucine, and valine, referred to as branched-chain amino acids because their carbon side chains

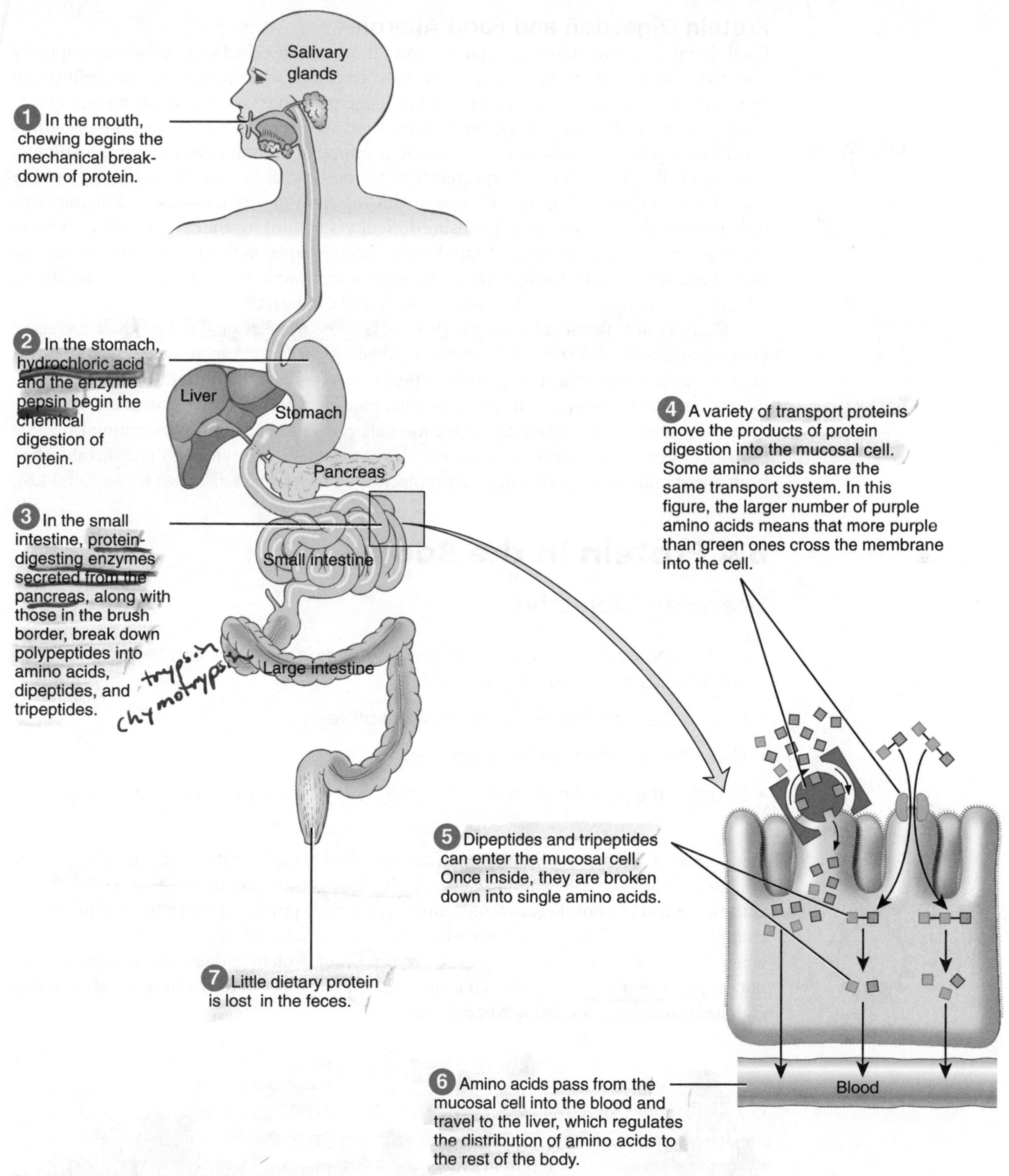

Figure 6.8 Protein digestion and absorption
Protein must be broken down into small peptides and amino acids before it can be absorbed into the mucosal cells.

have a branching structure (see Appendix I), share the same transport system. If there is an excess of any one of the amino acids sharing a transport system, more of it will be absorbed, slowing the absorption of the other competing amino acids. This is generally not a problem when amino acids are consumed in foods because foods contain a variety of amino acids without disproportionately large amounts of any one. However, taking a supplement of one amino acid can provide enough of it to impair absorption of other amino acids that share the same transport system (see Figure 6.8). For example, weight lifters often take supplements of the amino acid arginine, which shares the same transport system as lysine. When large doses of arginine are ingested, the absorption of lysine will be reduced. This could result in a lysine deficiency.

Protein Digestion and Food Allergies

Food allergies are triggered when a protein from the diet is absorbed without being completely digested. The proteins from milk, eggs, peanuts, tree nuts, wheat, soy, fish, and shellfish are common causes of food allergies. The first time the protein is consumed and a piece of it is absorbed intact, the immune system is stimulated (see Section 3.2). When the protein is consumed again, the immune system sees it as a foreign substance and mounts an attack, causing an allergic reaction. Allergic reactions to food can cause symptoms throughout the body. These can involve the digestive system, causing vomiting or diarrhea; the skin, causing a rash or hives; the respiratory tract, causing difficulty breathing; or the cardiovascular system, causing a drop in blood pressure. A rapid, severe allergic reaction that involves more than one part of the body is called **anaphylaxis.** A severe anaphylactic reaction can cause breathing difficulty or a dangerous drop in blood pressure and can be fatal.

anaphylaxis An immediate and severe allergic reaction to a substance (e.g., food or drugs). Symptoms include breathing difficulty, loss of consciousness, and a drop in blood pressure and can be fatal.

Allergies are common in people with gastrointestinal disease, because their damaged intestine allows the absorption of incompletely digested proteins. Allergies are also common in infants, because their immature gastrointestinal tracts are more likely to allow larger polypeptides to be absorbed. Once an infant's intestinal mucosa matures, absorption of incompletely digested proteins is less likely and some food allergies disappear. The absorption of whole proteins by very young infants, however, can also be of benefit, since antibody proteins absorbed from breast milk can provide temporary protection against certain diseases (see Section 14.6).

6.4 Protein in the Body

Learning Objectives

- List the sources of amino acids entering the amino acid pool and the uses for amino acids drawn out of the pool.
- Discuss the steps involved in synthesizing proteins.
- Name four functions of body proteins.
- Describe the conditions under which the body uses protein to provide energy.

amino acid pool All of the amino acids in body tissues and fluids that are available for use by the body.

Once dietary proteins have been digested and absorbed, their constituent amino acids become available to the body. The amino acids in body tissues and fluids are referred to collectively as the **amino acid pool (Figure 6.9).** Amino acids in this pool can be metabolized to provide energy—4 kcal/g. This occurs both when the diet contains protein in excess of needs and when the diet is low in energy. When dietary intake of protein and energy are adequate but not excessive, most amino acids in the amino acid pool are used to synthesize body proteins and other nitrogen-containing compounds.

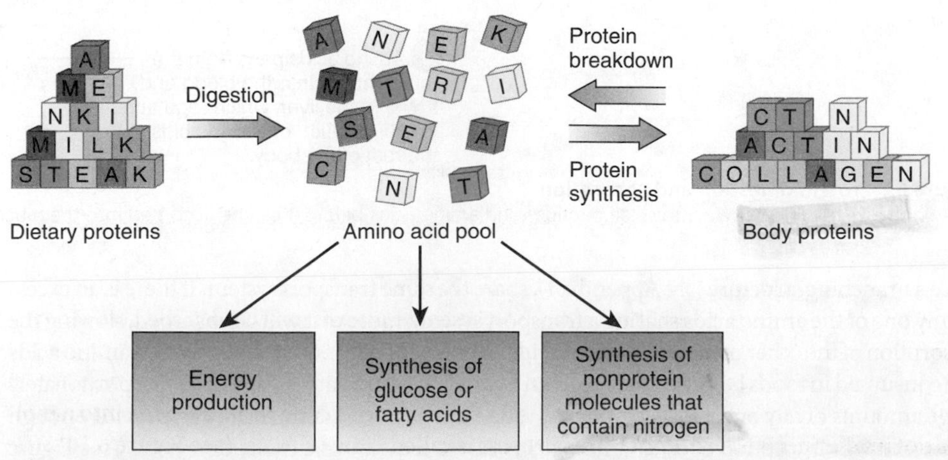

Figure 6.9 Amino acid pool
Amino acids in the available pool come from the diet and from the breakdown of body proteins. They are used to synthesize body proteins and nonprotein molecules, and to generate ATP, or to synthesize glucose or fatty acids.

Protein Turnover

Body proteins are not static but rather are continuously broken down and resynthesized. This process, referred to as **protein turnover**, is necessary for normal growth and maintenance of body tissues and for adaptation to changing situations. The rate at which proteins are made and degraded varies with the protein and is related to its function. Proteins whose concentration must be regulated and proteins that act as chemical signals in the body tend to have high rates of turnover—that is, synthesis and degradation. For example, the level of insulin in the blood can be rapidly increased by increasing its synthesis and decreasing the rate at which it is degraded. To decrease insulin levels, breakdown can be increased and synthesis slowed. Altering levels of synthesis and degradation allows protein levels to quickly change to maintain homeostasis as body conditions change. Structural proteins, such as collagen in connective tissue, have slower rates of turnover. The total amount of body protein broken down each day is large; twice as many of the amino acids in the amino acid pool come from recycled proteins as from dietary proteins.

protein turnover The continuous synthesis and breakdown of body proteins.

Protein Synthesis

The amino acids used to synthesize body proteins come from the amino acid pool. The instructions that dictate which amino acids are needed, and in what order they should be combined, are contained in stretches of DNA called **genes**. DNA is chemically composed of 4 compounds called nucleotide bases. These bases are adenine, abbreviated (A), thymine (T), cytosine (C), and guanine (G). These compounds form complementary pairs (A-T and C-G) on the characteristic double helix structures of DNA (**Figure 6.10**). The order of the bases, abbreviated as TACTTACCG and so on, is referred to as the DNA sequence and it is this sequence that determines the order and type of amino acids that make up a protein. Although there are only 4 bases, they can be combined into sequences for the thousands of proteins in the human body.

gene A length of DNA containing the information needed to synthesize RNA or a polypeptide chain.

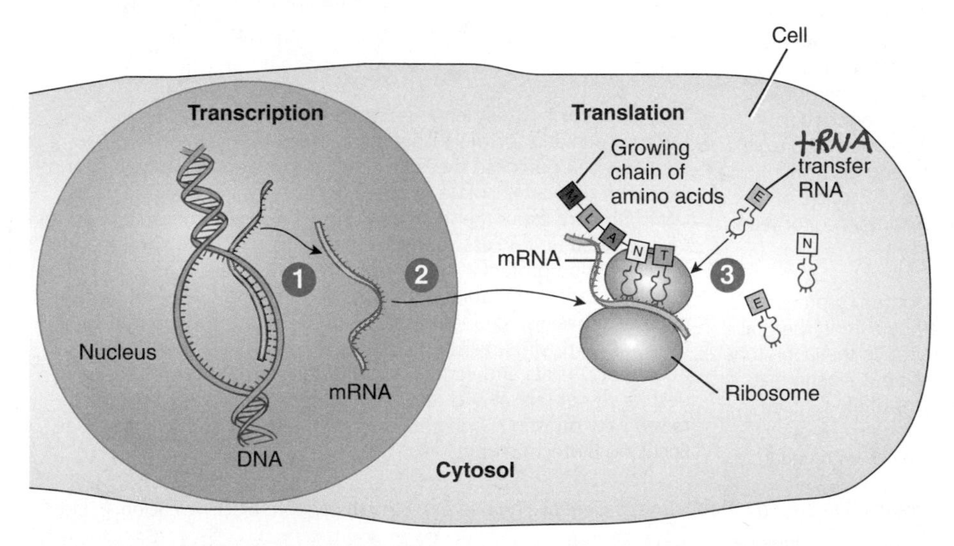

1. In the nucleus, the blueprint, or code, for the protein is copied or transcribed from the DNA gene into a molecule of messenger RNA (mRNA).
2. The mRNA takes the genetic information from the nucleus to structures called ribosomes in the cytosol, where proteins are made.
3. In the cytosol, transfer RNA (tRNA) reads the genetic code and delivers the needed amino acids to the ribosome to form a polypeptide chain.

Figure 6.10 Transcription and translation
DNA in cell nuclei provides the information needed to assemble proteins (transcription and translation).

When a protein is needed, the process of protein synthesis is turned on. The first step in protein synthesis involves copying, or transcribing, the DNA code from the gene into a molecule of messenger RNA (mRNA). This process is called **transcription** (see Figure 6.10). The mRNA then takes this information from the nucleus of the cell to ribosomes in the cytosol where proteins are made. Here, the information in mRNA is translated through another type of RNA, called transfer RNA (tRNA). Transfer RNA reads the code and delivers the needed

transcription The process of copying the information in DNA to a molecule of mRNA.

SCIENCE APPLIED

Discovering How to Manipulate Genes

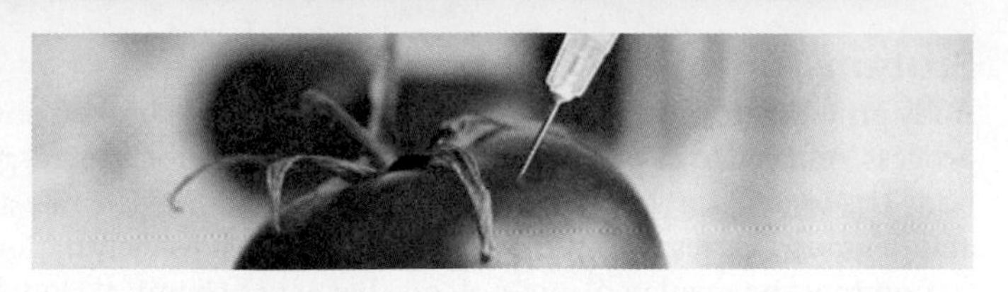

Genetic engineering is the process of manipulating the genetic makeup of plants, animals, and micro-organisms. By modifying DNA, scientists can change the proteins that a cell or organism can make. This technology, developed over the past 40 years, has allowed researchers to create bacteria that make medicines for humans, plants that are disease-resistant, and foods that provide a healthier mix of nutrients.

To alter the genetic composition of a cell, a specific piece of DNA, or gene, from one type of cell must be clipped out and pasted into the DNA in another cell. The new DNA, called recombinant DNA, can then provide the blueprint for new proteins. For example, if scientists take a piece of human DNA containing the gene for the hormone insulin and paste it into the DNA of a bacterium, the bacterium can then make human insulin. Today, more than 80% of people with diabetes who require insulin use genetically engineered insulin. Before genetically engineered insulin was available, the only source of insulin was the pancreases of slaughtered animals such as pigs and sheep.

Before 1970, genetic engineering was not possible because there was no way to cut DNA. The ability to cut DNA emerged from basic laboratory research that studied bacteria. In the 1950s, it was recognized that some strains of bacteria were able to slice viral DNA into pieces.[1] This ability was found to be due to bacterial enzymes, called restriction enzymes, which cut DNA in specific places. By the late 1960s and early 1970s, restriction enzymes were being isolated and characterized.[2, 3] At the same time, bacterial enzymes that repair breaks in DNA, called DNA ligases, were also being studied. DNA ligases had the ability to paste together two strands of DNA. At the time of their discovery, restriction enzymes and DNA ligases were recognized as interesting in the field of microbiology but it was impossible to predict the impact they would have on the fields of biology and medicine. These seemingly obscure bacterial enzymes changed the course of technology and will continue to impact many fields of science for years to come.

Restriction enzymes are present in bacteria as a form of protection. If a virus invades the bacterium, it can defend itself by cutting the viral DNA into little pieces before the virus can cause harm. In the laboratory, these enzymes act like precision scissors that can clip out a gene that produces a specific protein. Because DNA in all forms of life is made of the same building blocks, an enzyme from a bacterium can cut the DNA from a cow, a soybean plant, and a human cell with equal efficiency. When DNA from one cell is combined with that of another, the resulting cells can produce new proteins and provide new functions to the host. For example, a corn plant can be given a gene that makes a protein that is toxic to the corn borer, an insect that attacks corn plants. The corn plant is then resistant to that insect pest.

The discovery of restriction enzymes and DNA ligases provided the tools needed for the techniques of molecular biology. Today, scientists can purchase these enzymes from scientific suppliers, and use them to locate, isolate, prepare, and study small segments of DNA. Obscure experiments done 30 years ago with oddball enzymes have created an endless supply of human insulin for diabetics and blood-clotting proteins for hemophiliacs. They have led to new, more powerful cancer therapies and vaccines to prevent disease. Genetic engineering is also changing the foods we eat. It has created insect-resistant corn, virus-resistant papaya, grains with higher protein quality, and fruits and vegetables with longer shelf life. However, despite the potential and realized benefits, this technology has raised new questions and concerns, both ethical and scientific (see Focus on Biotechnology).

[1] Old, R. W., and Primrose, S. B. *Principles of Gene Manipulation: An Introduction to Genetic Engineering*, 5th ed. London: Blackwell Science, Ltd., 1994.

[2] Meselson, M., and Yuan, R. DNA restriction enzyme from *E. coli*. *Nature* 217:1110–1114, 1968.

[3] Smith, H. O., and Wilcox, K. W. A restriction enzyme from *Hemophilus influenzae*. I. Purification and general properties. *J. Molecular Biol*. 51:379–391, 1970.

---

translation The process of translating the mRNA code into the amino acid sequence of a polypeptide chain.

amino acids to form a polypeptide chain. This process is called **translation**. After translation, polypeptides typically undergo further chemical modifications before achieving their final protein structure and function (see Science Applied: Discovering How to Manipulate Genes).

Limiting Amino Acids During the synthesis of a protein, a shortage of one needed amino acid can stop the process. Just as on an assembly line, if one part is missing, the line stops—a different part cannot be substituted. If the missing amino acid is a nonessential amino acid, it can be synthesized in the body and protein synthesis can continue. If the missing amino acid is an essential amino acid, the body can break down some of its own proteins to obtain this amino acid. If an amino acid cannot be supplied, protein synthesis will stop. The essential amino acid present in shortest supply relative to need is called the **limiting amino acid**, because lack of this amino acid limits the ability to make protein (**Figure 6.11**). If all amino acids are present in adequate amounts at the time of synthesis, proteins will be completed and released for further processing by the cell. Animal foods are generally better sources of protein because animal proteins provide enough of all of the amino acids needed to build human proteins. Plant sources of protein are generally low in one or more of the essential amino acids so are used less efficiently to synthesize body protein. An exception is soy protein, which is as good as animal protein when it comes to supplying essential amino acids (see Your Choice: Should You Increase Your Soy Intake?).

limiting amino acid The essential amino acid that is available in the lowest concentration in relation to the body's needs.

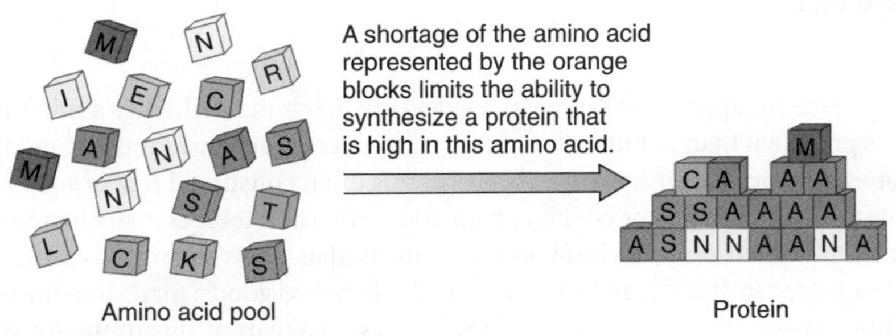

Figure 6.11 Limiting amino acids
The amino acids needed for protein synthesis come from the amino acid pool. If the protein to be made contains more of a particular amino acid than is available, that amino acid limits protein synthesis and is said to be the limiting amino acid.

Regulation of Gene Expression **Gene expression** refers to the process whereby the information coded in a gene is used to produce a product that functions in the body. This means that when a gene is expressed, the product for which it codes is synthesized. Sometimes the final product is a molecule of RNA and sometimes it is a protein. Which gene products are made and when they are made is important to the health of the cell and the organism, therefore, gene expression is carefully regulated. Not all genes are expressed in all cells or at all times. For example, the hormone glucagon is a protein that is made in pancreatic cells. Glucagon is not made by other body cells because the gene is not expressed in cells other than those in the pancreas. The expression of some genes changes depending on the need for the proteins for which they code. For example, when zinc intake is high, the expression of a gene that codes for metallothioneine, a metal-binding protein, is turned on. This allows more of this protein to be synthesized, increasing the capacity to bind zinc and other metal ions. The levels of nutrients in the body also affect the expression of other genes. For example, vitamin A levels affect the expression of genes involved in the maturation and specialization of cells and vitamin D affects the expression of genes that code for calcium-transport proteins (see Section 9.3).

gene expression The events of protein synthesis in which the information coded in a gene is used to synthesize a product, either a protein or a molecule of RNA.

Synthesis of Nonprotein Molecules

Amino acids are needed for the synthesis of a variety of nonprotein molecules that contain nitrogen. **Neurotransmitters** are one such class of molecules. They function to transfer signals between the cells of the nervous system and can stimulate or inhibit a signal. For example, the amino acid tryptophan is used to synthesize the neurotransmitter serotonin, which promotes relaxation. Amino acids are needed for the synthesis of the nitrogen-containing compounds that are the building blocks of DNA and RNA. Other molecules synthesized from amino acids include the skin pigment melanin, the vitamin niacin, creatine needed for muscle contraction, and histamine, which causes blood vessels to dilate.

Neurotransmitters. Molecules that function to transfer signals between the cells of the nervous system and can stimulate or inhibit a signal.

ATP Production

Although carbohydrate and fat are more efficient energy sources, amino acids from the diet and from body proteins are also used to provide energy. Before this can occur, the nitrogen-containing amino group must be removed from the amino acids in a process called **deamination**

deamination The removal of the amino group from an amino acid.

YOUR CHOICE

Should You Increase Your Soy Intake?

Soy products have long been a major protein source in Japan and China, countries where heart disease is less prevalent than it is in Canada. Both soy protein and isoflavones, phytochemicals found in soy, have been hypothesized to be responsible for the lower incidence of heart disease in these Asian countries. Isoflavones are estrogen-like compounds. Their estrogen-like effects have led some to speculate that in addition to affecting heart disease risk, they may reduce the symptoms of menopause (including hot flashes) and prevent bone loss. Soy has also been suggested to have beneficial effects in preventing and treating certain forms of cancer.

Some of the excitement about the health benefits of soy may have been premature. A review of the effect of soy on the risk of heart disease concluded that soy has only a small LDL-cholesterol-lowering effect and this occurs only when large amounts are consumed.[1] Soy has not been found to affect HDL cholesterol, triglyceride levels, or blood pressure. Research has not shown soy to reduce the symptoms of menopause and results are mixed with regard to its effect on postmenopausal bone loss. Little evidence has been found to support the effectiveness of soy in the prevention or treatment of cancers of the breast, uterus, and prostate.[1]

Despite the disappointing news with regard to blood-lipid levels, soy may still be good for your heart and overall health. Soybeans and products made from them provide plant protein equivalent in quality to animal proteins, so choosing soy can help you meet your protein needs by adding high-quality protein to your diet. Soy products are also high in polyunsaturated fat, fibre, vitamins, and minerals, and low in saturated fat. Therefore, replacing foods high in animal protein, like hamburgers, with soy products, like soy burgers, is likely to benefit your health.

Soy-based foods are available in many forms. Soybeans can be eaten boiled or roasted. Cooked soybean sprouts can be served as a side dish. Soy butter, which is similar to peanut butter, can be spread on crackers and sandwiches. Tofu, also known as bean curd, is often consumed raw in soups or salads or cooked in stir fries. Miso and tempeh, which are fermented soybean products, are used in soups and mixed dishes. Soy flour can be incorporated into baked goods; made into the texturized soy protein (TSP) used to make meat substitutes for vegetarian burgers, hot dogs, meatballs, and chicken; or added to animal protein as an extender or filler.

Tofu, also known as bean curd, is a soft, cheese-like product made by curdling fresh, hot soymilk. © Tetra Images

Remember, though, that simply including soy-based foods—or any single food, for that matter—in your diet is not the answer to good health. Replacing some of the animal sources of protein with soy protein may help protect your health, but other dietary and lifestyle factors also influence your overall risk of developing heart disease.

[1] Sacks, F. M., Lichtenstein, A., Van Horn, L., et al. Soy protein, isoflavones, and cardiovascular health: An American Heart Association Science Advisory for professionals from the Nutrition Committee. Circulation 113:1034–1044, 2006.

(**Figure 6.12**). The carbon compounds remaining after the amino group is removed can be used in a number of ways, depending on the needs of the body. If glucose is in short supply, amino acids that break down to form 3-carbon compounds can be used by the liver to synthesize glucose via gluconeogenesis. If energy is needed, the carbon structure of the amino acids can be converted into acetyl-CoA or compounds that directly enter the citric-acid cycle

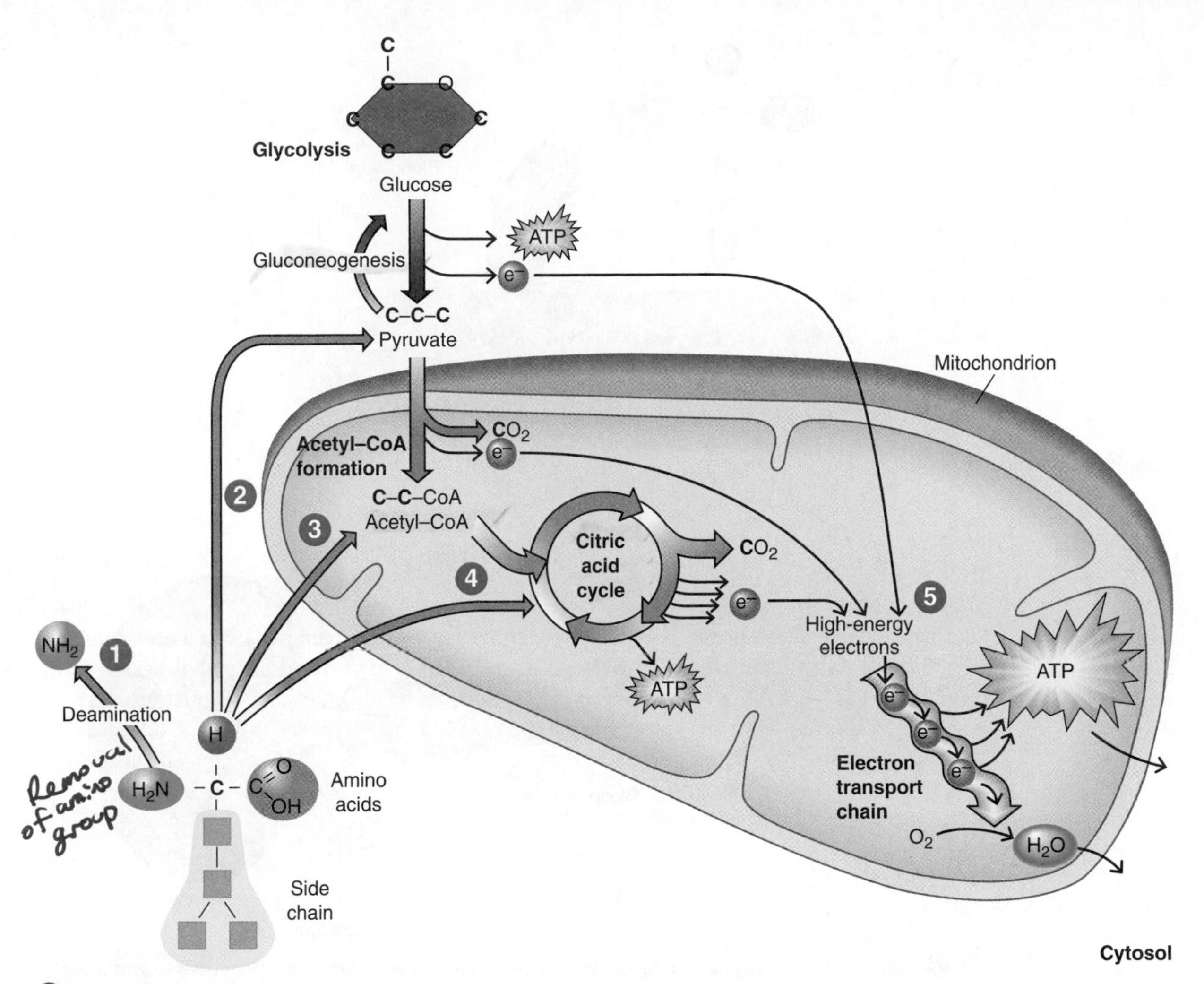

1. The amino group is removed by deamination.
2. Deamination of some amino acids produces 3-carbon molecules that can be used to synthesize glucose, via gluconeogenesis.
3. Deamination of some amino acids results in 2-carbon molecules that form acetyl-CoA, which can enter the citric-acid cycle or be used to synthesize fatty acids.
4. Deamination of some amino acids forms molecules that are intermediates in the citric-acid cycle.
5. High-energy electrons from the breakdown of amino acids are transferred to the electron transport chain where the energy is trapped and used to produce ATP and water.

Figure 6.12 Amino acid metabolism
In order for the body to use amino acids as an energy source, the nitrogen-containing amino group must be removed. The compounds remaining after the amino group has been removed are composed of carbon, hydrogen, and oxygen and can be broken down to produce ATP or used to make glucose or fatty acids.

to produce ATP (see Figure 6.12). The use of amino acids as an energy source increases both when the diet does not provide enough total energy to meet needs, as in starvation, and when protein is consumed in excess of needs. When both protein and energy are plentiful, amino acids can be converted into acetyl-CoA and used to synthesize fat for storage. The nitrogen released from amino acids by deamination forms ammonia, a toxic waste product. High levels of ammonia in the blood can be fatal. To protect the body, the liver combines ammonia with carbon dioxide to form a less toxic waste product called **urea** (**Figure 6.13**). Urea can be eliminated from the body by the kidneys.

urea A nitrogen-containing waste product formed from the breakdown of amino acids that is excreted in the urine.

When Energy Intake Is Low When energy is deficient, body proteins, such as enzymes and muscle proteins, are broken down into amino acids that can then be used to generate ATP or synthesize glucose. This provides energy in times of need, but it also robs the body of functional proteins. The most dispensable proteins are broken down first, conserving others for the numerous critical roles they play, but if the energy deficit continues, more critical proteins, such as those that make up the heart and other internal organs, will also be degraded. A loss of more than 30% of the body's protein reduces the strength of the muscles required for breathing

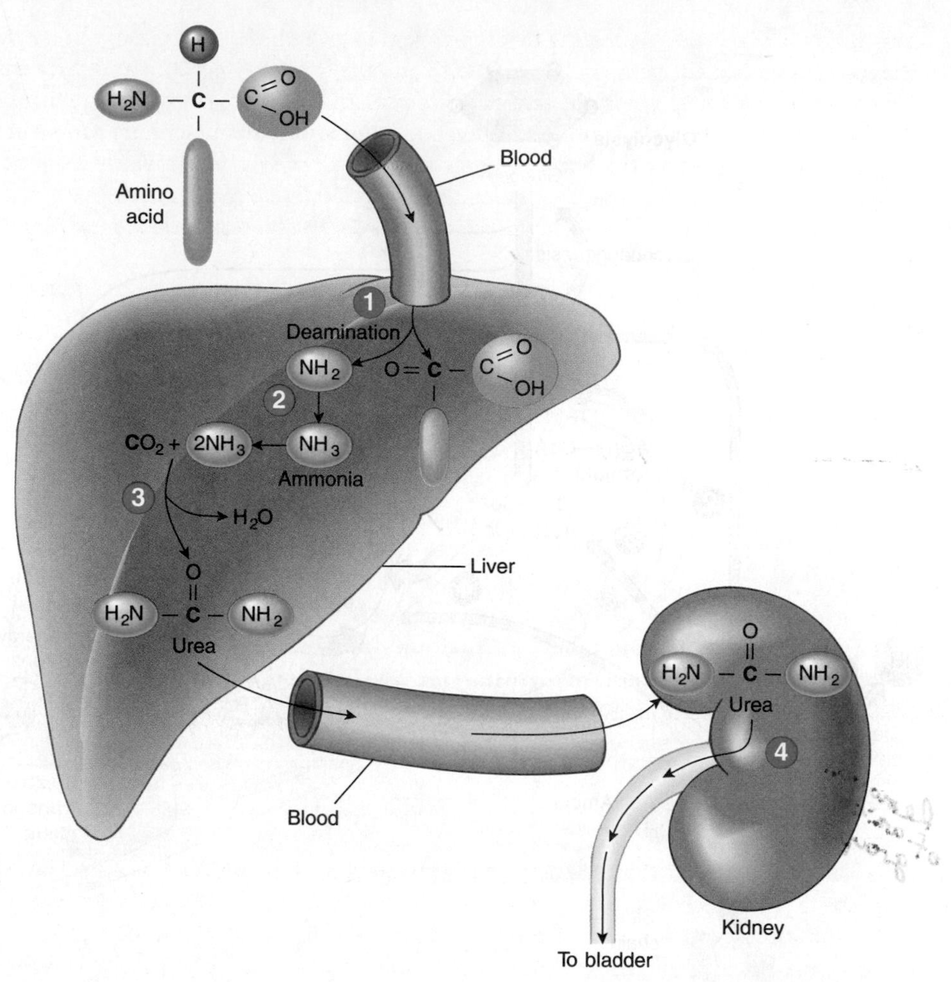

1. Amino acids are deaminated before they can be metabolized to produce ATP or used to synthesize glucose or fat.
2. The amino group forms toxic ammonia.
3. Ammonia is converted into urea in the liver.
4. Urea can safely travel in the blood and is filtered out of the blood by the kidney and eliminated from the body in the urine.

Figure 6.13 Urea Synthesis
The nitrogen from amino acid breakdown is eliminated from the body as urea.

and heart function, depresses immune function, and causes a general loss of organ function that is great enough to cause death.

When Protein Intake Exceeds Needs Amino acids are used for energy when protein intake exceeds protein needs. If the diet is adequate in energy and high in protein, the amino acids from the excess protein are deaminated and used to produce ATP. If both energy and protein exceed needs, the extra amino acids are converted into fatty acids, stored as triglycerides in adipose tissue, and can contribute to weight gain.

Protein Functions

About 15% of body weight is protein. Much of this is due to muscle proteins but there are also many other body proteins, each with a unique function. Some have important structural roles and others help regulate body processes.

Structural Proteins Proteins provide structure to individual cells and to the body as a whole. In cells, proteins are an integral part of the cell membrane, the cytosol, and the organelles. Skin, hair, and muscle are body structures that are composed largely of protein. The most abundant  protein in the body is collagen; it holds cells together and forms the protein framework of bones and teeth. It also forms tendons and ligaments, strengthens artery walls, and is a major constituent of scar tissue. When the diet is deficient in protein, these structures break down. The muscles shrink, the skin loses its elasticity, and the hair becomes thin and can easily be pulled out by the roots. These outward signs of protein deficiency have become marketing strategies for cosmetic companies. Shampoo and hand lotion manufacturers add protein to their products, suggesting that protein applied to the hair or skin will improve its structure. However, the proteins that make up hair and skin can only be made inside the body, so a healthy diet will do more for hair and skin quality than expensive protein shampoos or lotions.

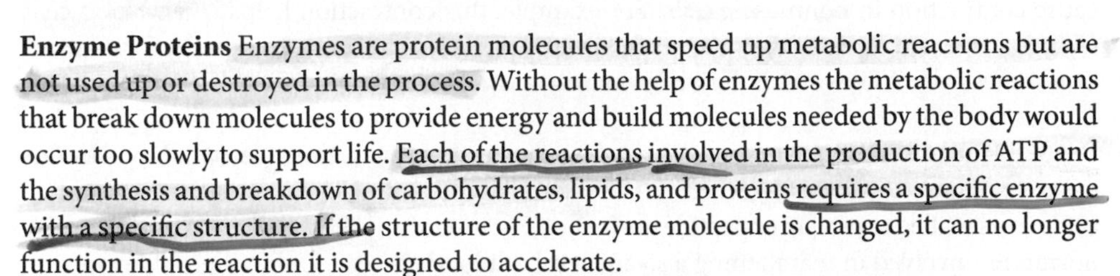

Enzyme Proteins Enzymes are protein molecules that speed up metabolic reactions but are not used up or destroyed in the process. Without the help of enzymes the metabolic reactions that break down molecules to provide energy and build molecules needed by the body would occur too slowly to support life. Each of the reactions involved in the production of ATP and the synthesis and breakdown of carbohydrates, lipids, and proteins requires a specific enzyme with a specific structure. If the structure of the enzyme molecule is changed, it can no longer function in the reaction it is designed to accelerate.

Enzymes that function in the body are made by the body and therefore do not need to be consumed in the diet. Enzymes present in raw foods are denatured by the cooking process and are no longer functional when the cooked food is eaten. When foods are eaten raw, the enzymes present are broken down during digestion and are absorbed from the gastrointestinal tract as amino acids. Purified enzymes sold as dietary supplements are also broken down in the gut. Some of these may provide function while in the gut; for example, lactase, taken by individuals with lactose intolerance, remains functional long enough to break down lactose consumed at the same time. Eventually, these enzymes are also digested and absorbed as amino acids.

Figure 6.14 The protein hemoglobin, which gives red blood cells their red colour, shuttles oxygen to body cells and carries away carbon dioxide. (© istockphoto.com/jpa1999)

Transport Proteins Proteins transport substances into and out of individual cells and throughout the body. At the cellular level, transport proteins present in cell membranes help move substances such as glucose and amino acids across the cell membrane into and out of cells. For example, transport proteins in the intestinal mucosa are necessary to absorb amino acids from the intestinal lumen into the mucosal cells. Transport proteins in the blood carry substances from one organ to another. For example, hemoglobin, the protein in red blood cells, picks up oxygen in the lungs and transports it to other organs of the body (**Figure 6.14**). The proteins in lipoproteins are needed to transport lipids from the intestines and liver to body cells. Some nutrients must be bound to a specific protein to be transported in the blood. When protein is deficient, the nutrients that require specific proteins for transport cannot travel to the cells. For this reason, a protein deficiency can cause a vitamin A deficiency; even if vitamin A is consumed in the diet, without protein, it cannot be transported to the cells where it is needed.

antibodies Proteins produced by the body's immune system that recognize foreign substances in the body and help destroy them.

Proteins that Provide Protection Proteins play an important role in protecting the body from injury and invasion by foreign substances. Skin, which is made up primarily of protein, is the first barrier against infection and injury. Foreign particles such as dirt or bacteria that are on the skin cannot enter the body and can be washed away. If the skin is broken and blood vessels are injured, blood-clotting proteins, including fibrin and thrombin, help prevent too much blood from being lost. If a foreign particle such as a virus or bacterium enters the body, the immune system fights it off by synthesizing proteins called **antibodies**. Each antibody has a unique structure that allows it to attach to a specific invader. When an antibody binds to an invading substance, the production of more antibodies is stimulated, and other parts of the immune system are signaled to help destroy the invading substance. The next time the same type of invading bacterium or virus enters the body, the immune system is already primed to produce specific antibodies to fight off that particular invader. Immunizations against diseases, such as measles, work in a similar way. A small amount of dead or inactivated virus is injected into the body; the injected material does not cause disease, but it does stimulate the immune system to produce antibodies to the virus (**Figure 6.15**). When the body comes in contact with the live virus, the immune system is already primed, so a large-scale immune system attack is mounted and the infection is prevented. When the immune system malfunctions as a result of protein deficiency, human immunodeficiency virus (HIV) infection, or other causes, the ability to protect the body from infection is compromised.

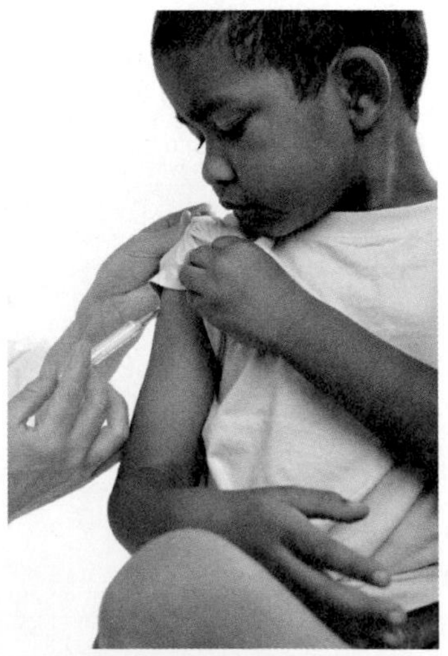

Figure 6.15 How will this immunization prevent the child from contracting measles or mumps? (© istockphoto.com/Alexander Raths)

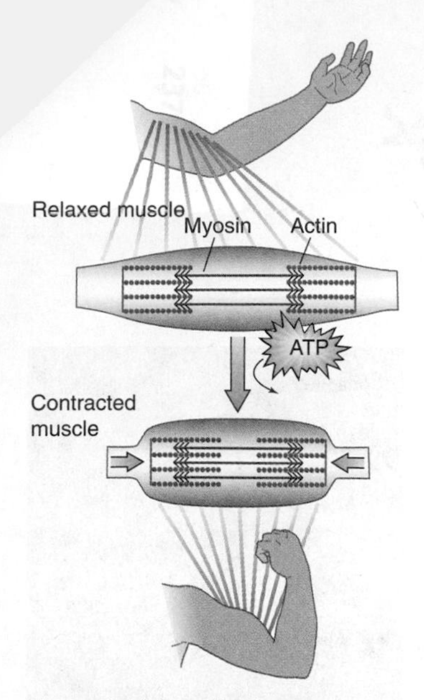

Figure 6.16 Proteins in muscle contraction
During muscle contraction, the proteins actin and myosin slide past each other, causing the muscle fibres to shorten.

Contractile Proteins The proteins in muscles allow us to move. When you climb a flight of stairs, walk across the room, or run around the block, you are relying on the muscle proteins actin and myosin, which cause the contraction of muscles. Contraction occurs when these two proteins slide past each other, shortening the muscle. For example, when you do a biceps curl, the muscles in your biceps shorten as the alternating actin and myosin protein fibres slide past one another (**Figure 6.16**). A similar process causes contraction in the heart muscle and in the muscles that cause constriction in the digestive tract, blood vessels, and glands. Actin and myosin can also cause contraction in nonmuscle cells. For example, this contraction helps white blood cells change shape and move so they can reach infected tissues in the body. The energy for contraction comes from ATP, which is derived primarily from the metabolism of carbohydrate and fat.

Protein Hormones Hormones are chemical messengers that are secreted into the blood by one tissue or organ and act on target cells in other parts of the body. Hormones made from amino acids are classified as protein or peptide hormones. For instance, insulin and glucagon are peptide hormones involved in maintaining a steady level of blood glucose. Unlike steroid hormones, which are made from cholesterol and can diffuse across the cell membrane and enter the cell, peptide hormones act by binding to protein receptors on the surface of the cell membrane.

Proteins that Regulate Fluid Balance The distribution of fluid among body cells, the bloodstream, and the spaces between cells is important for homeostasis. Fluid moves back and forth across membranes to maintain appropriate concentrations of particles and fluids inside and outside cells and tissues (see Section 10.2). Proteins help regulate this fluid balance in two ways. First, protein pumps located in cell membranes transport substances from one side of a membrane to the other. Second, large protein molecules present in the blood hold fluid in the blood by contributing to the osmotic load in the bloodstream. In cases of protein malnutrition, the concentration of these large proteins in the blood decreases, so fluid is no longer held in the blood, and it accumulates in tissues and in the abdomen.

pH A measure of acidity.

Proteins that Regulate Acid-Base Balance The chemical reactions of metabolism require a specific level of acidity, or **pH**, to function properly. In the gastrointestinal tract, acidity levels vary widely. The digestive enzyme pepsin works best in the acid environment of the stomach, whereas the pancreatic enzymes operate best in the more neutral environment of the small intestine. Inside the body, the range of optimal pH is much tighter. The acids and bases produced by metabolic reactions must be neutralized in order to prevent changes in pH, which in turn can prevent life-sustaining metabolic reactions from proceeding normally. The lungs and kidneys help maintain a normal pH by eliminating some of these waste products. Proteins both in the blood and within the cells act as buffers to prevent changes in pH. They function by attracting or releasing hydrogen ions. For instance, the protein hemoglobin in red blood cells helps neutralize acid produced when carbon dioxide reacts with water. Untreated type 1 diabetes is an example of what happens when the amount of acid produced exceeds the ability of the body's proteins and other systems to neutralize it. In type 1 diabetes, the inability to get glucose into cells results in the breakdown of fats and the buildup of ketones, which are acidic. As ketones accumulate, they cause a drop in pH called ketoacidosis (see Section 4.4). The acidic pH denatures proteins and they are unable to perform their functions, resulting in coma and eventually, if not treated, death.

6.5 Protein, Amino Acids, and Health

Learning Objectives

- Compare kwashiorkor with marasmus.
- Explain why protein-energy malnutrition develops more rapidly in young children than in adults.
- Discuss the potential risks associated with a high-protein diet.

A diet adequate in protein is essential to health. Dietary protein is needed for growth and to replace body protein that is broken down and lost each day. If too little protein is consumed, the consequences can be dramatic and devastating. Too much protein, particularly if it is derived primarily from animal sources, may also have negative health effects. In addition. some people are sensitive to specific proteins and amino acids.

protein-energy malnutrition (PEM) A condition characterized by wasting and an increased susceptibility to infection that results from the long-term consumption of insufficient amounts of energy and protein to meet needs.

kwashiorkor A form of protein-energy malnutrition in which only protein is deficient.

marasmus A form of protein-energy malnutrition in which a deficiency of energy in the diet causes severe body wasting.

Protein Deficiency

Because of the availability and variety of foods in developed countries, protein deficiency is uncommon. However, in developing nations, concerns about inadequate protein are very real. Diets deficient in protein are most often deficient in energy as well, but a pure protein deficiency can occur when food choices are extremely limited and the staple food of a population is very low in protein. The term **protein-energy malnutrition (PEM)** is used to refer to the continuum of protein-deficiency conditions ranging from pure protein deficiency, called **kwashiorkor**, to overall energy deficiency, called **marasmus.** Most protein-energy malnutrition is a combination of the two.

Kwashiorkor Kwashiorkor is typically a disease of children. The word "kwashiorkor" comes from the Ga people of the African Gold Coast. It means the disease that the first child gets when a second child is born.[2] When the new baby is born, the older child is no longer breastfed. Rather than receiving protein-rich breast milk, the young child is fed a watered-down version of the diet eaten by the rest of the family. This diet is low in protein and is often high in fibre and difficult to digest. The child, even if able to get adequate energy, is not able to eat a large enough quantity to get adequate protein. Because children are growing, their protein needs per unit of body weight are higher than those of adults, and the effects of a deficiency become evident much more quickly.

The symptoms of kwashiorkor can be explained by examining the roles that proteins play in the body. Because protein is needed for the synthesis of new tissue, growth in height and weight is hampered in children. Because proteins are important in immune function, there is an increased susceptibility to infection. There are changes in hair colour because the skin pigment melanin is not made; the skin flakes because structural proteins are not available to provide elasticity and support. Cells lining the digestive tract die and cannot be replaced, so nutrient absorption is impaired. The bloated belly typical of this condition is a result of both fat accumulating in the liver, because there is not enough protein to transport it to other tisues, and fluid accumulating in the abdomen, because there is not enough protein in the blood to keep water from diffusing out of the blood vessels (Figure 6.17a).

Kwashiorkor occurs most commonly in Africa, South and Central America, the Near East, and the Far East. It has also been reported in poverty-stricken areas in the United States. Although kwashiorkor is thought of as a disease of children, it is seen in hospitalized adults who have high-protein needs due to infection or trauma and a low-protein intake because they are unable to eat.

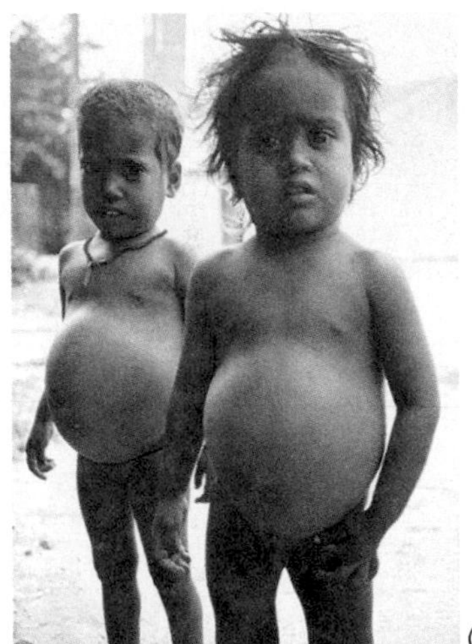

(a)

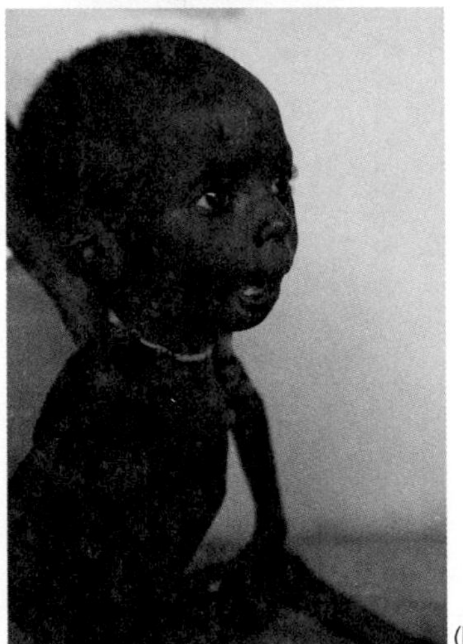

(b)

Figure 6.17 (a) Kwashiorkor is characterized by a bloated belly, whereas (b) marasmus presents as severe wasting. (©FAO/G. Kent Marie Ange Cortellino: ©AP/Wide World Photos)

Marasmus At the other end of the continuum of protein-energy malnutrition is marasmus, meaning to waste away. Marasmus is due to a deficiency of energy, but protein and other nutrients are usually also insufficient to meet needs. Marasmus may have some of the same symptoms as kwashiorkor, but there are also differ ences. In kwashiorkor, some fat stores are retained, since energy intake is adequate. In contrast, marasmic individuals appear emaciated because their body fat stores have been used to provide energy **(Figure 6.17b)**. Since fat is a major energy source and carbohydrate is limited, ketosis may occur in marasmus. This is not so in kwashiorkor because carbohydrate intake is adequate—only protein is deficient.

Marasmus occurs in individuals of all ages but is most devastating in infants and children because adequate energy is essential for growth. Most brain growth takes place in the first year of life, so malnutrition early in life causes a decrease in intelligence and learning ability that persists throughout life. Marasmus often occurs in children who are fed diluted infant formula prepared by caregivers trying to stretch limited supplies. Marasmus occurs less often in breastfed infants. Marasmus is the form of malnutrition that occurs with eating disorders (See Chapter 15: Focus on Eating Disorders).

Protein Excess

Adequate protein intake is absolutely essential to life, but too much protein has been hypothesized to affect the health of the kidneys and bones, and to impact on the healthfulness of the overall diet. For a healthy person, there are no short-term problems associated with consuming a diet very high in protein, but we are still investigating whether the same is true in the long term.

Hydration and Kidney Function As protein intake increases above the amount needed, so does the production of protein breakdown products, such as urea, which must be eliminated from the body by the kidneys. To do this, more water must be excreted in the urine, increasing water losses. Although not a concern for most people, this can be a problem if the kidneys are not able to concentrate urine. For example, the immature kidneys of newborn infants are not able to concentrate urine and therefore they need to excrete more water than adults to eliminate the same amount of urea. Feeding a newborn infant formula that is too high in protein can increase fluid losses and lead to dehydration. High-protein diets are also a risk for people with kidney disease. The increased wastes produced on a high-protein diet may speed the progression of renal failure in these individuals.[3] Despite this, there is little evidence that a high-protein diet increases the risk of kidney disease in healthy people.[4]

Bone Health It has been suggested that the amount and source of protein in the diet affect calcium status and bone health.[5] For healthy bone, intakes of both calcium and protein must be adequate. There is general agreement that diets that are moderate in protein (1.0 to 1.5 g/kg/day) are associated with normal calcium metabolism and do not alter bone metabolism.[5,6] However, a high protein intake may increase the amount of calcium lost in the urine. Some studies suggest that the amount of calcium lost in the urine is greater when protein comes from animal rather than vegetable sources.[7] These findings have contributed to a widely held belief that high-protein diets (especially diets that are high in animal protein) result in bone loss. However, clinical studies do not support the idea that animal protein has a detrimental effect on bone health or that vegetable-based proteins are better for bone health.[6] In fact, when calcium intake is adequate, higher-protein diets are associated with greater bone mass and fewer fractures.[5] This is likely the case because in healthy adults, a high protein intake increases intestinal calcium absorption as well as urinary excretion, so the increase in the amount of calcium lost in the urine does not cause an overall loss of body calcium.

Kidney Stones The increase in urinary calcium excretion associated with high-protein diets has led to speculation that a high protein intake may increase the risk of kidney stones, which are deposits of calcium and other substances in the kidneys and urinary tract. Higher concentrations of calcium and acid in the urine increase the likelihood that the calcium will be deposited, forming these stones. Epidemiological studies suggest that diets that are rich in animal protein and low in fluid contribute to the formation of kidney stones.[8]

Heart Disease and Cancer Risk Another concern with high-protein diets is related more to the dietary components that accompany animal versus plant proteins. Typically, high-protein diets are also high in animal products; this dietary pattern is high in saturated fat and cholesterol and low in fibre and therefore increases the risk of heart disease. These diets are also typically low in grains, vegetables, and fruits, a pattern associated with a greater risk of cancer.[9] Such diets are also usually high in energy and total fat, which may promote excess weight gain.

Proteins and Amino Acids that May Cause Harm

Unlike lipids and carbohydrates, people don't think of protein as contributing to health problems, but for some people, the wrong protein can be harmful. In some cases, this is because a protein in food is recognized and targeted by the immune system, causing an allergic reaction. There are no cures for food allergies, so to avoid symptoms, allergic individuals need to avoid eating foods that contain proteins that cause an allergic reaction (see Label Literacy: Is it Safe for You?). Not all adverse reactions to proteins and amino acids are due to allergies; some are due to food intolerances. These reactions do not involve the immune system. The symptoms of a food intolerance can range from minor discomfort, such as the abdominal distress some people feel after eating raw onions, to more severe reactions.

Aspartame and Phenylketonuria Aspartame is a sugar substitute composed of two amino acids, aspartic acid and phenylalanine. Aspartame is used in a wide variety of foods, including carbonated beverages, gelatin desserts, and chewing gum. Digestion breaks aspartame into aspartic acid, phenylalanine, and an alcohol called methanol. Because the phenylalanine released from aspartame digestion is absorbed into the blood, food products containing this alternative sweetener must be avoided by individuals with a genetic disorder called **phenylketonuria**.

phenylketonuria (PKU) An inherited disease in which the body cannot metabolize the amino acid phenylalanine. If the disease is untreated, toxic by-products called phenylketones accumulate in the blood and interfere with brain development.

Individuals with PKU inherit a defective gene for an enzyme called phenylalanine hydroxylase, which is needed to metabolize phenylalanine. The enzyme does not function properly in those with this faulty gene, and they are unable to convert the essential amino acid phenylalanine to the amino acid tyrosine. Instead, phenylalanine is converted to compounds called phenylketones, which build up in the blood (**Figure 6.18**). In infants and children, high phenylketone levels can interfere with brain development, causing intellectual disability. Pregnant women with PKU must be especially careful to consume a low-phenylalanine diet in order to protect the fetus from high phenylketone levels, which can cause brain abnormalities and other birth defects.[10]

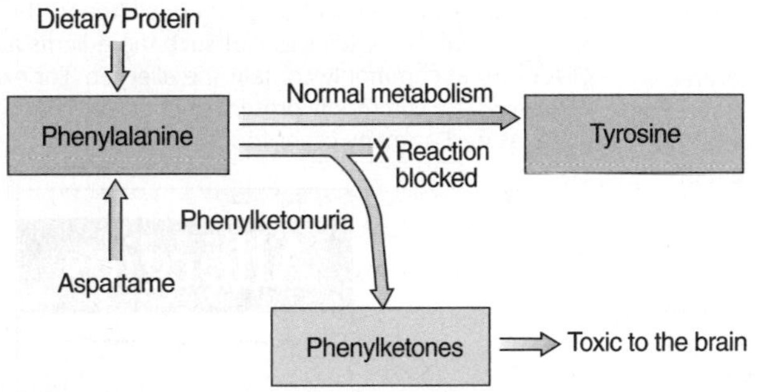

Figure 6.18 Phenylketonuria
In individuals with phenylketonuria, phenylalanine accumulates and is converted to phenylketones, which can interfere with brain development.

PKU afflicts about 1 in 12,000 newborn infants.[11] Infants are tested for this disorder at birth because brain damage can be avoided by a special, low-phenylalanine diet. The diet must provide just enough phenylalanine to meet the body's need for protein synthesis but not so much that the buildup of phenylketones occurs. The diet must also provide sufficient tyrosine, because the disease prevents conversion of phenylalanine to tyrosine, causing it to become an essential amino acid. All proteins naturally contain phenylalanine, so a low-phenylalanine diet must carefully regulate overall protein intake. Special low-phenylalanine or phenylalanine-free formulas are manufactured for infants with this disease. Because foods like diet soft drinks and other artificially sweetened beverages do not contain protein, individuals with this disease might not expect them to contain phenylalanine. Therefore, warnings for individuals with PKU are included on the labels of all products containing aspartame, indicating it contains phenylalanine and showing the actual amount of the sweetener in the product (**Figure 6.19**).

Celiac Disease Gluten intolerance, also called **celiac disease,** celiac sprue, or gluten-sensitive enteropathy, is another form of food intolerance. Individuals with celiac disease cannot tolerate gluten, a protein found in wheat, rye, barley, and other grains. Celiac disease is an autoimmune disease in which gluten causes the body to attack the villi in the small intestine, causing symptoms such as diarrhea, abdominal bloating and cramps, weight loss, and anemia. It occurs in 1:133 individuals.[12] The only treatment is to avoid gluten by eliminating all products containing gluten from the diet.

Monosodium Glutamate Sensitivity Monosodium glutamate (MSG) is a flavour enhancer best known for its use in Chinese cooking. It is added to a variety of packaged foods such as potato chips, canned soups, cured meats, and packaged entrees. It is also sold as a seasoning directly to consumers. MSG consists of the amino acid glutamic acid (or glutamate) bound to sodium and is present in hydrolyzed vegetable proteins. Some people report adverse reactions, including a flushed face, tingling or burning sensations, headache, rapid heartbeat, chest pain, and general weakness after consuming MSG.[13] These symptoms are referred to as MSG symptom complex, and are most likely to occur within an hour after eating about 3 g or more of MSG on an empty

celiac disease A disorder that causes damage to the intestines when the protein gluten is eaten.

LABEL LITERACY

CANADIAN CONTENT

Is it Safe for You?

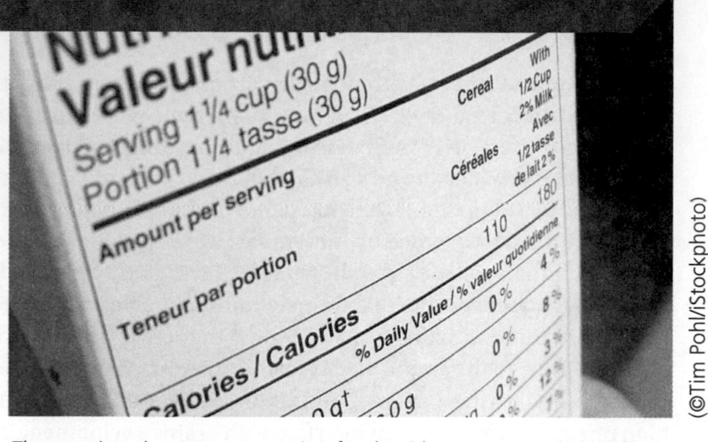

(©Tim Pohl/iStockphoto)

"What is food to one, is to others bitter poison," said the ancient Roman philosopher and scientist Lucretius. Today, fortunately, the information in ingredient lists can help those with food allergies tell the difference. Health Canada has identified the following priority food allergens: peanuts; tree nuts (almonds, Brazil nuts, cashews, hazelnuts, macadamia nuts, pecans, pine nuts, pistachios, walnuts); sesame seeds; milk; eggs; fish (including crustaceans, for example, crab, crayfish, lobster, shrimp) and shellfish (for example, clams, mussels, oysters, scallops); soy; wheat; and sulphites. These products are responsible for 90% of reported adverse reactions to food.[1]

Food manufacturers are required to clearly state if a product contains any of the priority food allergens. One of the components of a food label is a list of ingredients and manufacturers must indicate the presence of a priority allergen in this list.[2] This can be done in several ways. One is to simply list it clearly in the ingredient list. For example, milk can be listed with other ingredients. A second way is to use a parenthetical statement. This is helpful when milk is a component of another ingredient, for example, batter (water, modified cornstarch, wheat flour, salt, sodium bicarbonate, *milk ingredients*, guar gum).

A third way food labels may indicate the presence of allergens is to use the word "contains" followed by the name of the major food allergen, printed at the end of the ingredient list or next to it. For example: *Contains: Milk*.

Finally, products that do not intentionally contain proteins that cause allergies may provide a precautionary warning if there is a potential for cross-contamination from equipment or foods processed in the same facility: for example, *May contain peanuts*. This is a potentially life-saving warning for individuals with peanut allergy, who can have a severe reaction from minute amounts of peanut protein.

Health Canada has published a booklet entitled *Common Food Allergies: A Consumer's Guide to Managing the Risks*.[3] (see figure). In this booklet, Canadians with allergies are advised to read labels carefully, to not eat foods that contain ingredients the names of which they do not recognize, and to avoid foods with precautionary statements about ingredients to which they are allergic.

The warning about not consuming foods with unrecognized ingredients is important, as there are many food ingredients that are derived from foods to which someone can be allergic. Those derived from milk, for example, have names such as caseinate, lactalbumin, or whey protein concentrate, and would not be immediately recognizable as milk products. For each of the priority allergens, the *Common Food Allergies* booklet contains lists of such ingredients as well as the names of foods that commonly contain the allergen. For example, milk or milk derivatives are commonly present in chocolate bars, margarine, and pizza, as well as many other foods.

Health Canada's *Common Food Allergies* booklet. Each allergen is discussed in detail. The discussion of milk allergy, for example, includes other names for milk products as well as lists of foods that contain milk or milk derived products. © Her Majesty the Queen in Right of Canada, represented by the Minister of Health (2011).

[1] Health Canada. Food Allergies. Available online at http://www.hc-sc.gc.ca/hl-vs/iyh-vsv/food-aliment/allerg-eng.php. Accessed July 20, 2010.

[2] Health Canada Health Canada's modifications to regulatory project 1220-Enhanced labeling for food allergens, gluten sources, and added sulphites. Available online at http://www.hc-sc.gc.ca/fn-an/label-etiquet/allergen/proj1220-modifications-eng.php. Accessed July 20, 2010.

[3] Health Canada Available online at http://www.inspection.gc.ca/english/fssa/labeti/allerg/allerge.pdf. Accessed June 29, 2010.

Figure 6.19 Diet soft drinks sweetened with aspartame carry a warning to alert individuals with PKU that the soft drink is a source of phenylalanine.

stomach. However, a typical serving of foods containing MSG includes only about 0.5 mg of MSG. Controlled studies have not been able to consistently document reactions to MSG.[14]

Because glutamate is a neurotransmitter, some brain researchers have expressed concern that very high dietary intakes of glutamate could be toxic to nerves in humans. However, a review of scientific data has found no evidence that dietary MSG causes brain lesions or damages nerve cells in humans. Health Canada, based on a review of the scientific literature, considers MSG to be safe, but advises Canadians that have a sensitivity to the product to avoid it by carefully reading the ingredient list on food labels for monosodium glutamate, potassium glutamate, or hydrolyzed vegetable protein.[15] If the label states that the product contains "no MSG" or "no added MSG" then neither MSG nor hydrolyzed protein containing glutamate has been added to the food. When eating out, you can avoid this additive by asking the restaurant to prepare your food without added MSG.

CANADIAN CONTENT

6.6 Meeting Protein Needs

Learning Objectives

- Discuss how protein needs are determined.
- Explain what is meant by protein quality.
- Review a diet and substitute complementary plant proteins for the animal proteins it contains.

Protein consumed in the diet must supply amino acids to replace losses that occur during protein turnover, to repair damaged tissues, and to synthesize new body proteins for growth. In a typical Canadian diet, protein provides about 15% of the energy, typically resulting in intakes twice the RDA.[1] Canada's Food Guide recommends the consumption of whole grains, fruits, and vegetables, with smaller amounts of lean meats and low-fat milk products, as well as the consumption of beans, lentils, and tofu often. This is a dietary pattern that provides a healthy balance between animal and plant proteins.

Determining Protein Requirements

Historically, recommendations for protein intake were estimated from the amount of protein consumed by healthy, working men in the general population. These protein levels were often as high as 150 g/day. Current recommendations are generally lower than this and are based on **nitrogen balance** studies (**Figure 6.20**). Since protein is the only macronutrient that contains nitrogen, the amount of protein used by the body can be estimated by comparing nitrogen

nitrogen balance The amount of nitrogen consumed in the diet compared with the amount excreted by the body over a given period.

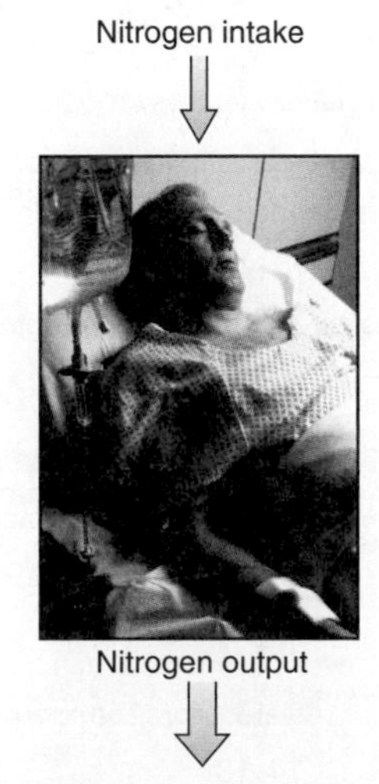

Figure 6.20 Nitrogen balance
Nitrogen balance indicates whether the amount of protein in the body is remaining constant, decreasing, or increasing. (Cameron Lawson/NG Image Collection; Brian Yarvin/Photo Researchers; Roy Toft/NG Image Collection)

intake with nitrogen loss. Nitrogen intake is calculated from dietary protein intake. Nitrogen loss or output is measured by totalling the amounts of nitrogen excreted in urine and feces and that lost from skin, sweat, hair, and nails. The majority of the nitrogen lost is excreted in the urine as urea. Comparing the amount of nitrogen consumed with the amount lost provides information about the amount of protein being synthesized and broken down within the body. An individual who is consuming enough protein to meet body needs is in protein or nitrogen balance. This means the individual is consuming enough nitrogen or protein to replace the amount that is lost from the body. The protein requirement is the smallest amount of dietary protein that will maintain balance when energy needs are met by carbohydrate and fat.

If the body breaks down more protein than it synthesizes, then nitrogen balance is negative; this means more nitrogen is lost than ingested (see Figure 6.20). This indicates that body protein is being lost. Negative nitrogen balance can occur when intake is too low or when the amount of protein breakdown has been increased by a stress such as injury, illness, or surgery.

If the body is synthesizing more protein than it breaks down, nitrogen balance is positive; this indicates that the body is using dietary protein for the synthesis of new body proteins (see Figure 6.20). Positive nitrogen balance occurs when new tissue is synthesized, such as during growth, pregnancy, wound healing, or muscle building (see Critical Thinking: What Does Nitrogen Balance Tell Us?).

The protein requirement of a specific individual can be determined by doing a nitrogen balance study for that individual. Because this procedure cannot be done for everyone, the protein needs of populations must be estimated from balance study data. Recommendations for protein intake for the general public are actually higher than the requirements determined by nitrogen balance studies for individuals. This is to allow a margin of safety that will ensure that the needs of the majority of the population are met.

The RDA for Protein

The RDA for protein for adults is 0.8 g/kg of body weight per day.[4] For a person weighing 70 kg (154 lb), the recommended intake would be 56 g of protein per day (**Table 6.2**).

Table 6.2 Calculating Protein Needs

To determine protein requirement:

• Determine body weight. If weight is measured in pounds, convert it to kilograms by dividing by 2.2;

$$\frac{\text{Weight in lbs}}{\text{2.2 lbs/kg}} = \text{weight in kg}$$

For example:

$$\frac{\text{150 lbs}}{\text{2.2 lbs/kg}} = \text{68 kg}$$

• Determine the grams of protein required per day. Multiply weight in kilograms by the grams of protein per kilogram recommended for the specific gender and life-stage group (see below).

For example:

A 23-year-old woman weighing 68 kg would require 0.8 g/kg/day × 68 kg = 54.4 grams of protein/day.

Group	Age (yrs)	RDA (g/kg/day)
Infants	0–0.5	1.52*
	0.5–1	1.5
Children	1–3	1.1
	4–8	0.95
Adolescents	9–13	0.95
	14–18	0.85
Adults	19 and older	0.8
Pregnancy		Nonpregnant RDA + 25 g/day
Lactation	First 6 months	Nonpregnant RDA + 25 g/day

*This value is an AI.

CRITICAL THINKING:
What Does Nitrogen Balance Tell Us?

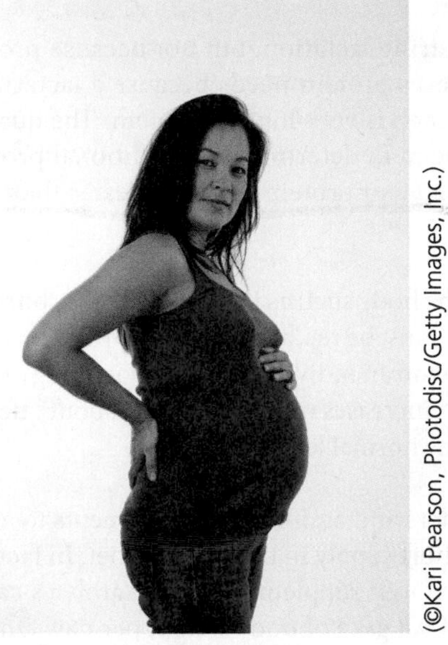
(©Kari Pearson, Photodisc/Getty Images, Inc.)

Background

The Amecht Company wants to include nitrogen balance studies in the assays it performs in its clinical laboratory. To test their methodology, they analyze nitrogen balance in three individuals. The technicians are given information about the daily nitrogen intake of these subjects and analyze samples of urine and feces to determine daily nitrogen losses. Thr formula for nitrogen balance is shown below:

Nitrogen balance = Nitrogen intake – nitrogen output.

Data

Subject A consumed 6.4 g of nitrogen. The laboratory determines that she lost 8.0 g of nitrogen in her urine and feces. The nitrogen balance equation for subject A is

$$6.4\text{ g} - 8.0\text{ g} = -1.6\text{ g}$$

Subject B is a healthy 29-year-old man who weighs 82 kg and consumes an adequate diet providing 2,700 kcal and 70 g of protein a day. His nitrogen values are:

	Nitrogen In	Nitrogen Out
Subject B	11.2	11.2

Subject C is a 31-year-old pregnant woman of average pre-pregnancy weight who is consuming 2,500 kcal and 80 g of protein a day. Her nitrogen values are:

	Nitrogen In	Nitrogen Out
Subject C	12.8	10.4

Critical Thinking Questions

Subject A is in negative nitrogen balance. List some possible reasons that would explain this.*

Calculate Subject B's nitrogen balance. What can you determine about subject B based on your answer?

What would you predict about Subject C's nitrogen balance based on the fact that she is pregnant? Calculate her nitrogen balance. Explain whether it supports your prediction.

*Answers to all Critical Thinking Questions can be found in Appendix J.

iProfile Use iProfile to find a snack that provides the additional 25 g of protein needed during pregnancy.

Protein needs are increased if protein is being deposited in the body as it is during growth, or when protein losses are increased, such as during lactation or when the body is injured.

RDAs have also been established for each of the essential amino acids (see Appendix A); these are not a concern in a typical diet but are important when developing intravenous feeding solutions.

Infants and Children For an individual to grow, new body proteins must be synthesized. During the first year of life, a large amount of protein is required to support the rapid growth rate. Thus, an AI for the first 6 months of life has been set at 1.52 g/kg of body weight per day; for the second 6 months, the RDA is 1.5 g/kg.[4] As the growth rate slows, requirements per unit of body weight decrease but continue to be greater than adult requirements until 19 years of age (**Figure 6.21**).

Figure 6.21 Young children have high protein needs because of their rapid rate of growth. (Cheryl Maeder/Taxi/Getty Images)

Pregnancy and Lactation During pregnancy, both the mother and the fetus are growing. The mother's diet must supply enough protein to provide for the expansion of her blood volume, enlargement of her uterus and breasts, development of the placenta, and growth and development of the fetus. The RDA for pregnant women is 25 g of protein per day above the nonpregnant recommendation. Most women in North America already consume this much protein in their typical diets.

Protein needs are also increased during lactation, but not because protein is being deposited in the body. Lactation increases protein needs because a lactating woman is producing and secreting breast milk, which is very high in protein. The quantity of milk produced and the protein content of the milk determine the additional protein needs of lactation. The RDA during lactation is 25 g of protein per day greater than the RDA for nonlactating women.

Illness and Injury Extreme stresses on the body such as infections, fevers, burns, or surgery increase protein breakdown. These losses must be replaced by dietary protein. Requirements for these types of stresses must be assessed on an individual basis, depending on the extent of the losses. For example, a severe infection increases requirements by about one-third. Burns can increase requirements to 2-4 times the normal level.

Exercise The marketing of protein powders and amino acid supplements to athletes might lead people to believe that protein is in short supply in the athlete's diet. In fact, athletes can obtain plenty of protein in their diets without supplements. Most athletes can meet their protein needs by consuming the RDA of 0.8 g/kg of body weight per day. Only endurance athletes and strength athletes, such as triathletes and body builders, require more than the RDA. The reason endurance athletes need more is that some protein is used for energy and to maintain blood glucose during endurance events, such as ultramarathons and long-distance cycling. Athletes participating in endurance events such as these may benefit from 1.2 g-1.4 g of protein per kg/day. Strength athletes, such as body builders, need extra protein because it provides the raw materials needed for building their large muscles; 1.2-1.7 g/kg/day is recommended.[16] The higher protein needs of endurance and strength athletes can be met without protein supplements, as long as the diet provides adequate kcalories. For example, if a 200-lb (91-kg) man consumes 3,600 kcal/day, 15% of which is from protein (approximately the amount contained in a typical North American diet), he will consume 135 g of protein. This equals about 1.5 g of protein per kg of body weight. Consuming additional protein as food or supplements is unlikely to enhance performance. The protein needs of athletes are also discussed in Section 13.4.

A Range of Healthy Protein Intakes

In addition to the RDA, the DRIs give a recommendation for protein intake as a percentage of kcalories; the Acceptable Macronutrient Distribution Range (AMDR) is 10%-35% of kcalories from protein. This range allows for different food preferences and eating patterns. A protein intake in this range will meet needs, balance with carbohydrate and fat kcalories, and not increase health risks. A diet that provides 10% of kcalories from protein will meet the RDA but is a relatively low-protein diet based on typical eating patterns in Canada.[4] Most people consume about 15% of their kcalories from protein. Very few people consume more protein than the upper end of the healthy range—35% of kcalories.[1] If the proportion of protein goes higher than this, the diet will likely be higher in fat and lower in carbohydrate than is recommended. As discussed previously, the health concerns associated with high-protein diets relate more to the relative amounts of energy-yielding nutrients included in these diets. For a certain level of energy intake, as the protein content of the diet increases, fat also increases because most of the high-protein foods in our diets are animal products that contain fat along with the protein. Therefore, a diet that contains 35% protein would tend to have more fat, saturated fat, and cholesterol and less carbohydrate than a diet with the same number of kcalories that has only 10% protein.

Estimating Protein Intake

To determine how much protein is in your diet, you can use food composition tables or databases to look up the number of grams of protein in individual foods. The Canadian Diabetes

Association's *Beyond the Basics: Meal Planning for Healthy Eating, Diabetes Prevention, and Management* includes food lists that contain the protein content of common food portions (see Appendix E). Food labels provide a ready source of information about the protein content of packaged foods; however, since the labelling of raw meats and fish is not mandatory, many of the greatest sources of protein in the diet do not carry food labels. The Nutrition Facts section lists the number of grams of protein per serving. The ingredient list provides information on the protein and amino acid-containing ingredients in the food. This information can be important for people with allergies and those trying to avoid certain additives (see Label Literacy: Is it Safe for You?).

Considering Protein Quality

Most Canadians eat plenty of protein, but to evaluate protein intake, it is important to consider both the amount and the quality of the protein. **Protein quality** is a measure of how good the protein in a food is at providing the essential amino acids needed by the body. The RDA for protein is calculated assuming that the diet contains a mixture of plant and animal proteins and therefore is of mixed quality. Protein needs can be met with lower-quality proteins, but more protein is needed to supply enough of all the essential amino acids. Choosing a diet that includes a variety of protein-containing foods will provide enough protein and enough of each of the essential amino acids to meet needs.

protein quality A measure of how efficiently a protein in the diet can be used to make body proteins.

Complete and Incomplete Protein Because we are animals, it makes sense that animal proteins generally contain mixtures of amino acids more similar to those we need. This is why animal proteins are considered higher quality than proteins from plant sources. Animal proteins also tend to be digested more easily than plant proteins; only protein that is digested can contribute amino acids to meet requirements.[17] Because they are easily digested and supply essential amino acids in the proper proportions for human use, foods of animal origin are sources of **complete dietary protein**. Plant proteins are usually more difficult to digest and are low in one or more of the essential amino acids and are therefore referred to as **incomplete dietary protein.** As mentioned earlier soy protein, which is a high-quality plant protein, is an exception to this generalization.

complete dietary protein Protein that provides essential amino acids in the proportions needed to support protein synthesis.

incomplete dietary protein Protein that is deficient in one or more essential amino acids relative to body needs.

Measuring Protein Quality Being able to evaluate the protein quality of a food or a diet is valuable when assessing the adequacy of human diets throughout the world. For example, knowing the quality of the protein in cassava is extremely important in determining the adequacy of the diet in countries where it is a staple and both food and protein are scarce.

Protein quality is evaluated experimentally in a number of ways (see **Table 6.3**). One way is to compare the amino acid pattern of the food protein of interest with that found in a reference protein known to be of high quality. A **chemical** or **amino acid score** is calculated by comparing the amount of the limiting amino acid in the test protein with the amount of that amino acid in egg protein. In this analysis, proteins with the most desirable proportions of amino acids will have the highest scores. Amino acid score is a useful measure, but

chemical or **amino acid score** A measure of protein quality determined by comparing the essential amino acid content of the protein in a food with that in a reference protein. The lowest amino acid ratio calculated is the chemical score.

Table 6.3 Measures of Protein Quality

Measure	Formula
Chemical or Amino Acid Score =	$\frac{\text{mg of limiting amino acid per g of test protein}}{\text{mg of limiting amino acid per g of reference protein}} \times 100$
Protein Digestibility–Corrected Amino Acid Score (PDCAAS) =	amino acid score × digestibility factor
Protein Efficiency Ratio (PER) =	$\frac{\text{wt gain when fed test protein}}{\text{wt gain when fed reference protein}}$
Net Protein Utilization (NPU) =	$\frac{\text{nitrogen retained}}{\text{nitrogen consumed}} \times 100$
Biological Value (BV) =	$\frac{\text{nitrogen retained}}{\text{nitrogen absorbed}} \times 100$

protein digestibility–corrected amino acid score (PDCAAS) A measure of protein quality that reflects a protein's digestibility as well as the proportions of amino acids it provides.

it does not consider digestibility. A measurement that considers both amino acid composition and digestibility is the **protein digestibility–corrected amino acid score (PDCAAS) (Figure 6.22).** PDCAAS measures the quality of a protein by comparing its amino acid composition to a reference amino acid pattern which is the amino acid requirements of a 2–5-year-old child (the age group with the highest protein needs relative to size), and then adjusting this for digestibility. A higher PDCAAS means that less of the protein is needed to provide all the needed amino acids.

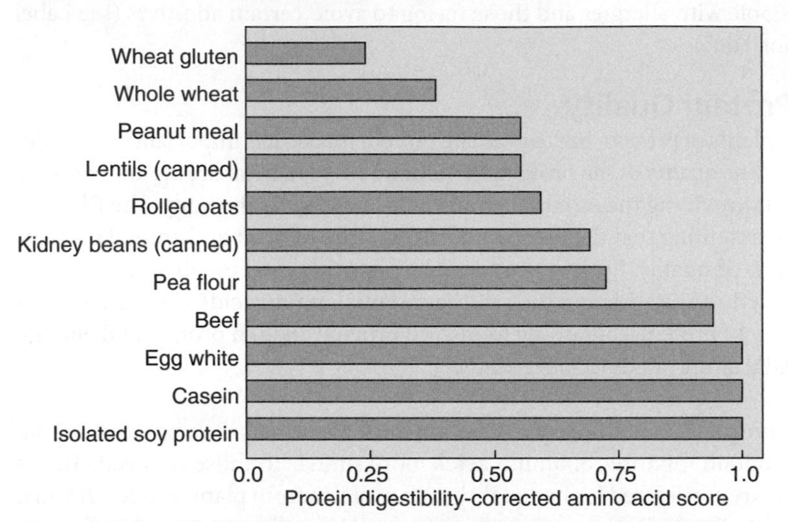

Figure 6.22 Protein digestibility-corrected amino acid score (PDCAAS)
The PDCAAS of plant proteins is generally lower than that of animal proteins. An exception is soy protein.
Source: Protein Quality Evaluation, Report of the Joint FAO/WHO Expert Consultation, FAO/WHO, 1989.

The calculation in **Table 6.4** shows that the limiting amino acids are methionine and cysteine, which are grouped together because they are structurally similar. Methionine is an amino acid commonly low in non-animal proteins. Lysine is another amino acid often found lacking in plant proteins.

Table 6.4 Determining the limiting amino acid in a test protein using the reference amino acid pattern

AMINO ACID	TEST PROTEIN (mg/g protein)	REFERENCE AMINO ACID PATTERN (mg/g protein)	AMINO ACID SCORE (mg amino acid in test protein/ mg amino acid in reference)
Isoleucine	4.1	5.9	4.1/5.9 = 0.69
Leucine	7.3	9.0	7.3/9.0 = 0.77
Lysine	3.5	7.2	3.5/7.2 = 0.49
Methionine + Cysteine	2.0	6.3	2.0/6.3 = 0.32 ← Limiting amino acid
Phenylalanine + Tyrosine	10.3	10.3	10.3/10.3 = 1.0
Threonine	3.5	5.0	3.5/5.0 = 0.7
Tryptophan	1.0	1.3	1.0/1.3 = 0.77
Valine	5.0	6.7	5/6.7 = 0.75

The digestibility of the protein is determined in an animal-feeding experiment in which the digestibility of the test protein is compared to a fully digestible protein. For the example above, the digestibility was found to be 0.8.

Therefore, the PDCAAS = 0.32 × 0.8 = 0.26. The maximum PDCAAS is set at 1.0. Other methods of evaluating protein quality include the **protein efficiency ratio**, which measures how well a protein promotes growth in animals, **net protein utilization**, and **biological value**, which measure how well a protein is used by the body for growth (Table 6.3).

protein efficiency ratio A measure of protein quality determined by comparing the weight gain of a laboratory animal fed a test protein with the weight gain of an animal fed a reference protein.

net protein utilization A measure of protein quality determined by comparing the amount of nitrogen retained in the body with the amount eaten in the diet.

biological value A measure of protein quality determined by comparing the amount of nitrogen retained in the body with the amount absorbed from the diet.

A score, called the protein rating, is used by the Canadian Food Inspection Agency to define protein labelling claims, such as which food is a source of protein and which is an excellent source of protein (**Table 6.5**). The higher the score the better the quality of the protein. The protein rating is calculated by multiplying the quantity of protein present in a Reasonable Daily Intake of the food by the quality of the protein, as determined by the Protein Efficiency Ratio (PER):

CANADIAN CONTENT

$$\text{Protein Rating} = \text{Protein in a Reasonable Daily Intake} \times \text{Protein Efficiency Ratio (PER)}.$$

The Reasonable Daily Intakes of many common foods have been determined and the PER for many common food proteins, which are determined by rat feeding studies, have been established.[18] The calculations shown in **Table 6.6** demonstrate the superior quality of egg protein compared to the protein in white bread.[19]

Table 6.5 Protein Labelling Claims

Claim	Conditions to meet claim
"Source of protein"	Protein rating >20
"Excellent source of protein"	Protein rating >40

Source: Canadian Food Inspection Agency. Nutritional Labelling: Protein claims: http://inspection.gc.ca/english/fssa/labeti/guide/ch7ae.shtml#a7_15.

Table 6.6 Comparing the protein rating of white bread and whole eggs

Calculating the Protein Rating of White Bread

Percent (%) Protein = 8.4
Reasonable Daily Intake = 150 g (5 slices)
Protein in a Reasonable Daily Intake = 0.084 X 150 g = 12.6 g
PER = 1.0
Protein Rating = 12.6 X 1.0 = 12.6

Calculating the Protein Rating of Whole Egg

Percent (%) Protein = 12.8
Reasonable Daily Intake = 100 g (2 eggs)
Protein in a Reasonable Daily Intake = 0.128 X 100 g = 12.8 g
PER = 3.1
Protein Rating = 12.8 X 3.1 = 39.68

Source: Canadian Food Inspection Agency: Nutritional Labelling: Chapter 6. Available online at http://www.inspection.gc.ca/english/fssa/labeti/guide/ch6ae.shtml. Accessed June 30, 2010.

Protein Complementation Assessing protein quality is important in planning diets for regions of the world where protein is scarce. It is less important in industrialized countries, where high-quality protein is readily available to most people. A more appropriate way of evaluating protein quality in an individual diet is to look at the sources of the protein. As discussed earlier in this section, foods of animal origin are sources of complete dietary protein, whereas most plant foods contain proteins that are incomplete. If the protein in a diet comes from both animal and plant sources, it most likely contains adequate amounts of all the essential amino acids needed for protein synthesis. If the protein in a diet comes only from incomplete plant sources, a technique called **protein complementation** can be used to meet protein needs.

Protein complementation. The process of combining proteins from different sources so that they collectively provide the proportions of amino acids required to meet needs.

Protein complementation combines foods containing proteins with different limiting amino acids in order to improve the protein quality of the diet as a whole. By eating plant proteins with complementary amino acid patterns, essential amino acid requirements can be met without consuming any animal proteins. The amino acids that are most often limited in plant proteins are lysine, methionine, cysteine, and tryptophan. As a general rule, legumes are deficient in methionine and cysteine but high in lysine. Grains, nuts, and seeds are deficient in lysine but high in methionine and cysteine. Corn is deficient in lysine and tryptophan but is a good source of methionine. Combining foods with complementary proteins provides all of the essential amino acids. For example, consuming rice, which is limited in the amino acid lysine but high in methionine and cysteine, with beans, which are high in lysine but limited in methionine and cysteine, provides enough of all of the amino acids needed by the body (**Figure 6.23**).

Figure 6.23 Protein complementation
A meal that contains both rice and beans contains enough methionine (met), cysteine (cys), and lysine (lys) to meet the body's need for essential amino acids. (StockFood/Getty Images, Inc.)

Common combinations of grains and legumes that have become cultural staples include beans and rice or beans and wheat or corn tortillas in Central and South America; rice and tofu in China and Japan; rice and lentils in India; rice and black-eyed peas in the southern United States; and peanut butter (peanuts are legumes) and bread throughout Canada and the United States. Plant proteins can also be complemented with animal proteins in order to meet the need for essential amino acids. For example, in Asia, rice is often flavoured with a small amount of spiced beef, chicken, or fish. Although it is not necessary to consume complementary proteins at each meal, the entire day's diet should include proteins from complementary sources in order to satisfy the daily need for amino acids.[20]

Translating Recommendations into Healthy Diets

Following the recommendations of Canada's Food Guide will result in a diet that includes both animal and plant sources of protein. A food guide serving of milk provides about 8 g of protein, and a serving of beans can add another 7-10 g, while a CFG serving of meat will add about 14-20 g. The Canada Food Guide food groups of milk and alternatives and meat and alternatives contain the foods that are the highest in protein, and most choices provide high-quality animal protein (**Figure 6.24**). Each serving of grains and vegetables provides another 1-3 g. But just getting enough protein does not ensure a healthy diet. The sources of dietary protein also affect the healthfulness of the diet, particularly how well it meets the recommendations for fat and fibre intake.

Choose Fish, Lean Meats, and Low-Fat Milk Products Animal foods high in protein are the major source of saturated fat and the only source of cholesterol in the diet, but this doesn't mean you need to become a vegetarian. Choosing wisely can keep your diet healthy. For example, to benefit from the iron, zinc, and B vitamins in meats without adding too much saturated fat, select lean cuts of meat and remove the skin from poultry. Choosing fish adds iron and zinc and also contributes heart-healthy omega-3 fatty acids. Nonfat or low-fat milk and milk products provide high-quality protein and calcium without as much saturated fat as whole milk.

Get More of Your Protein From Plants Plant sources of protein bring with them poly- and monounsaturated fats and dietary fibre. Legumes, for example, provide about 15 g of fibre per 250 ml (1 cup). Much of this is soluble fibre, which helps lower blood cholesterol. Choosing nuts and seeds increases intake of heart-healthy monounsaturated fats as well as fibre. Whole grains and vegetables add fibre, phytochemicals, vitamins, and minerals, and because we eat a larger number of servings from these groups, they make an important contribution to our protein intake. Depending on how they are processed and prepared, grains and vegetables are generally low in saturated fat and contribute no cholesterol.

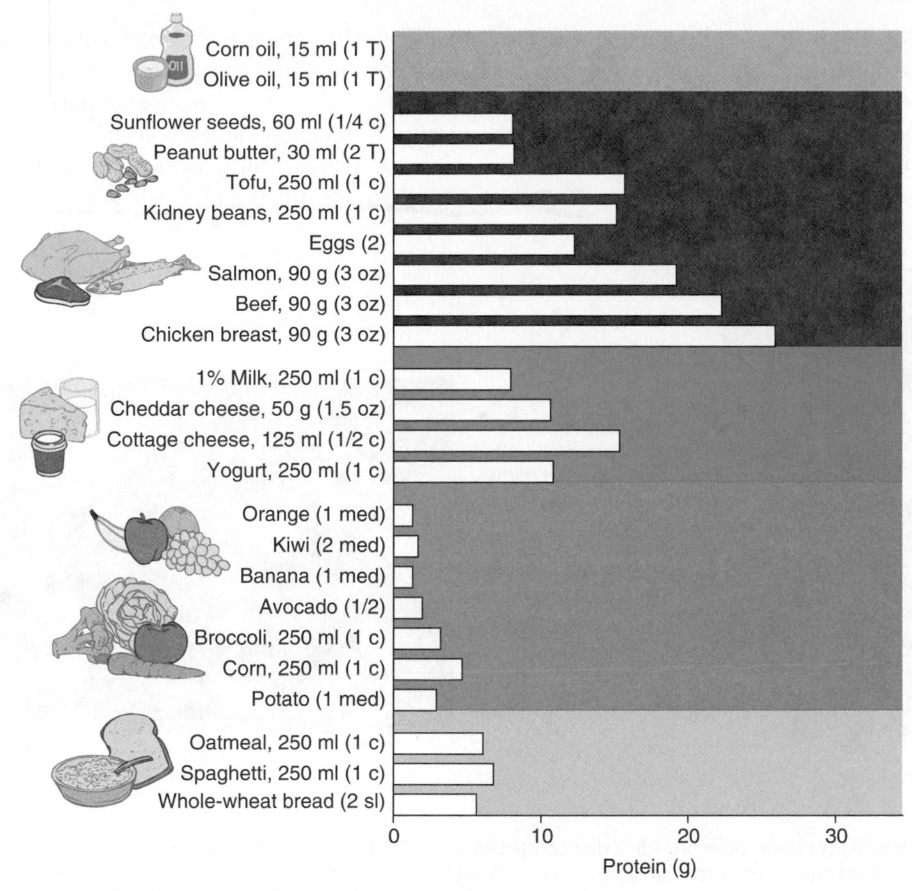

Figure 6.24 Protein content of Canada's Food Guide food groups
Meats, legumes, eggs, and milk products contain the most protein per serving.

Protein and Amino-Acid Supplements Even though protein is plentiful in the Canadian diet, protein and amino acid supplements remain popular among some segments of the population. Protein is needed for proper immune function, healthy hair, and muscle growth, but supplements will impact these only if the diet is deficient in protein in the first place. Increasing protein intake above the requirement does not protect you from disease, make your hair shine, or stimulate muscle growth. Although protein supplements are not harmful for most people, they are an expensive and unnecessary way to increase protein intake. A typical protein drink provides 10-35 g of protein per serving (**Figure 6.25**). It can add about 100-250 kcal to the diet and thus can contribute to weight gain. If consumed consistently, a high intake of protein from supplements or from foods increases water loss in the urine and may contribute to dehydration.

Amino acid supplements are frequently marketed to athletes with claims that they will reduce fatigue, enhance performance, and help build muscle. In general, most claims are not supported by the scientific evidence[21] although some amino acid supplements may provide modest benefits (see Focus on Non-vitamin/Non mineral supplements for a more detailed discussion). Because some amino acids are absorbed using the same transport systems, supplementing one amino acid can cause a deficiency of others that share the same transport system (see Figure 6.8).

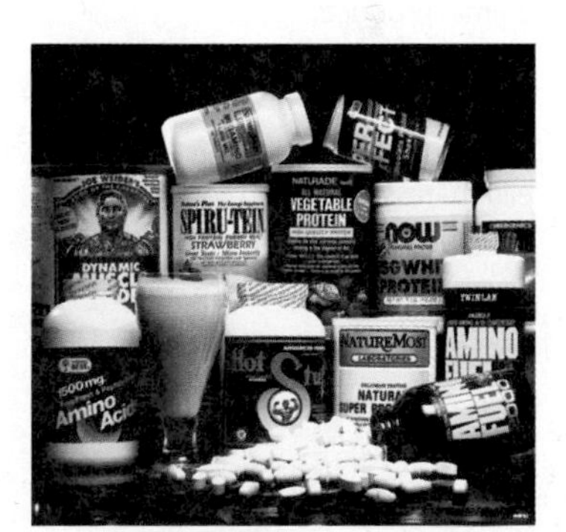

Figure 6.25 Supplements marketed to increase the intake of protein or specific amino acids are expensive and usually unnecessary. (Charles D. Winters/Photo Researchers, Inc.)

6.7 Meeting Needs with a Vegetarian Diet

Learning Objectives

- Describe the benefits associated with vegetarian diets.
- Discuss the nutrients at risk of deficiency in vegetarian diets.

Animal foods are not necessary for health. Combining plant proteins containing different limiting amino acids can provide the right combinations of amino acids to meet protein needs (Figure 6.26). And making wise food choices and in some cases, using fortified foods or supplements, can meet the need for other nutrients. In many parts of the world, plant-protein-based or vegetarian diets have evolved mostly out of necessity because animal sources of protein are unavailable physically or economically. In affluent societies, vegetarian diets are followed for a variety of reasons other than economics, such as health, religion, personal ethics, or environmental awareness. About 4% of Canadians identify themselves as vegetarians.[22]

Figure 6.26 Combining Plant Proteins
Combining complementary sources of incomplete plant proteins can provide a diet containing enough of all the essential amino acids. (Jean Paul Chassenet/Photo Researchers; Corbis Images; © istockphoto.com/Luca Manieri)

Types of Vegetarian Diets

vegetarianism A pattern of food intake that eliminates some or all animal products.

vegan A pattern of food intake that eliminates all animal products.

Traditionally, **vegetarianism** is defined as abstinence from meat, fish, and fowl, with **vegans** consuming the most restrictive vegetarian diets. The term vegetarian has come to include a wide variety of eating patterns, depending on the degree of abstinence from animal products. Semivegetarians are those who avoid only certain types of red meat, fish, or poultry—for example, individuals who avoid all red meat but continue to consume poultry and fish. Lacto-ovo vegetarians are those who eat no animal flesh but do eat eggs and dairy products such as milk and cheese; lacto vegetarians are those who avoid animal flesh and eggs but do consume dairy products. Pescetarians exclude all animal flesh except for fish.

Benefits of Vegetarian Diets

A vegetarian diet can be a healthy, low-cost alternative to a meat-and-potatoes diet. Vegetarians have been shown to have lower body weight and a reduced incidence of obesity and of other chronic diseases, such as diabetes, cardiovascular disease, high blood pressure, and some types of cancer (see Critical Thinking: Scientific Evidence for the Benefits of a Vegetarian Diet).[20,23] The lower body weight of vegetarians is a result of lower energy intake, primarily due to higher intake of fibre, which makes the diet more filling. The reductions in the risk of other chronic diseases may be due to lower body weight and to the fact that these diets are typically lower in saturated fat and cholesterol, which increase disease risk. Or it could be that vegetarian diets are higher in grains, legumes, vegetables, and fruits, which add fibre, vitamins, minerals, antioxidants, and phytochemicals—substances that have been shown to lower disease risk. It is likely that the total dietary pattern, rather than a single factor alone, is responsible for the health-promoting effects of vegetarian diets.

In addition to reducing disease risks, diets that rely more heavily on plant protein than on animal protein are more economical. For example, a vegetarian stir-fry over rice costs about half as much as a meal of steak and potatoes. Yet both meals provide a significant portion of the day's protein requirement.

CRITICAL THINKING:

Scientific Evidence for the Benefits of a Vegetarian Diet

A meta-analysis (see Chapter 5: Critical Thinking: Fish Consumption and Heart Disease for a more detailed discussion of meta-analysis), combines the data from multiple studies to measure the relationship between diet and disease in a very large group of people. This pooling of data often reveals statistically significant trends that would not be apparent in smaller studies.

When researchers, for example, wanted to determine whether a vegetarian diet reduced the risk of death from cardiovascular disease, they combined the results of five prospective cohort studies into one large study of 76,000 individuals, including almost 28,000 vegetarians.[1] They obtained the following results:

	Relative risk of death from cardiovascular disease*
Non-vegetarian	1
Vegetarian	0.76**

* See Chapter 2: Science Applied: Does Having a Healthy Dietary Pattern Reduce the Risk of Disease? for a review of relative risk
**statistically significant difference from non-vegetarian

When the researchers broke down their data based on the age of the individuals, they found the following trend:

	Relative risk of death from cardiovascular disease < 65 years	Relative risk of death from cardiovascular disease 65-79 years	Relative risk of death from cardiovascular disease 80-89 years
Non-vegetarian	1	1	1
Vegetarian	0.55*	0.69*	0.92

*statistically significant difference from non-vegetarian

Finally, the researchers broke down the non-vegetarians into frequent and non-frequent meat eaters.

	Relative risk of death from cardiovascular disease
Frequent meat eater	1.0
Non-frequent meat eater	0.78*
Vegetarian	0.66*

*statistically significant difference from frequent meat eater

Critical Thinking Questions

What do you conclude about disease risk from the first table comparing non-vegetarians and vegetarians?

What effect does age have on disease risk?

What can you conclude from the results shown in the third table comparing frequent and non-frequent meat eaters?

[1] Key, T.J., Fraser, G.E., Thorogood, M., et al. Mortality in vegetarians and non-vegetarians: a collaborative analysis of 8300 deaths among 76,000 men and women in five prospective studies. *Public Health Nutr.* 1(1):33-41, 1998.

Nutrients at Risk in Vegetarian Diets

Despite the health and economic benefits, nutrient deficiencies can be a problem for people consuming unsupplemented vegetarian diets, particularly vegan diets. Most people can easily meet their protein needs with lacto and lacto-ovo vegetarian diets. These diets contain high-quality animal proteins from eggs or milk, which complement the limiting amino acids in the plant proteins. Protein deficiency is a potential risk when vegan diets, which contain little high-quality protein, are consumed by small children and adults with increased protein needs, such as pregnant women and those recovering from illness or injury. These individuals must consume carefully planned diets to meet their protein needs.

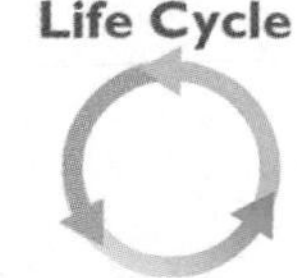

Deficiencies of particular vitamins and minerals are a greater risk for vegetarians than protein deficiency. Of primary concern to vegans is vitamin B_{12}. Because this B vitamin is found almost exclusively in animal products, to meet needs, people following vegan diets must consume vitamin B_{12} supplements or foods fortified with vitamin B_{12}. Another nutrient of concern is calcium. Milk products are the major source of calcium in the North American diet, so diets that eliminate these foods must contain plant sources of calcium or calcium-fortified foods to meet needs. Likewise, since most dietary vitamin D comes from fortified milk products, this vitamin must be obtained from sunlight (see Section 9.3) or consumed in other sources, such as fortified soy milk. Iron and zinc may be deficient in vegetarian diets because the best sources of these minerals are red meats and the iron and zinc present in

CRITICAL THINKING:

Choosing a Healthy Vegetarian Diet

(© istockphoto.com/Joe Gough)

Background

A year ago, Ajay decided to stop eating meat. Now that he is studying protein in his nutrition class, he has become concerned that his vegetarian diet isn't meeting his needs.

Data

Ajay is 26 years old and weighs 70 kg (154 lbs). He records his food intake for one day and then uses iProfile to calculate his protein intake.

FOOD	SERVING	PROTEIN (G)
Breakfast		
Grape Nuts	125 ml	7.2
Milk, low-fat	125 ml	4
Orange juice	175 ml	0.8
Toast, wheat	2 slices	5
Peanut butter	15 ml	4
Coffee	250 ml	0
Lunch		
Dahl (lentil soup)	250 ml	9
Rice	250 ml	6
Banana	1 medium	1
Apple juice	250 ml	0
Dinner		
Green salad with	250 ml	1
dressing	15 ml	0
Rice	250 ml	6
Curried potatoes and	125 ml	1.5
chickpeas	125 ml	5
Yogurt, plain	250 ml	13
Poori (fried bread)	2 pieces	5
Ice cream	125 ml	2
Total		**70.5**

Critical Thinking Questions

Does Ajay get enough protein? Compare his intake with the RDA for someone his age and size.

Does his diet contain complementary proteins? List the protein sources in his diet and explain how they complement each other.

If Ajay decided to become a vegan, what could he substitute for his dairy foods in order to meet his protein needs?

How could he make sure his vegan diet meets his need for calcium and vitamin D (see Table 6.7)?

Use iProfile to find the protein content of your favourite vegetarian entrée.

plant foods is poorly absorbed. Since iron and zinc are low in milk products, lacto-ovo and lacto vegetarians as well as vegans are at risk for deficiencies. Low intakes of omega-3 fatty acids, including EPA and DHA (see Section 5.5) are also a concern in vegan diets.[20] Diets that do not include fish, eggs, or large amounts of sea vegetables will not provide preformed EPA and DHA and therefore higher levels of alpha-linolenic acid are needed. By including flaxseed or flax and canola oils, which are good sources of alpha-linolenic acid, vegetarians can have a healthy ratio of omega-3 to omega-6 fatty acids and be able to synthesize enough of the longer-chain omega-3 fatty acids, EPA and DHA, without consuming animal products.[21] Vegetarian sources of these nutrients are listed in **Table 6.7** (see Critical Thinking: Choosing a Healthy Vegetarian Diet).

Table 6.7 Meeting Nutrient Needs with a Vegan Diet

Nutrient at Risk	Sources in Vegan Diets
Protein	Soy-based products, legumes, seeds, nuts, grains, and vegetables
Vitamin B_{12}	Products fortified with B_{12}, such as soy beverages and cereals, nutritional yeast, vitamin supplements
Calcium	Tofu processed with calcium, broccoli, kale, bok choy, legumes, and products fortified with calcium, such as soy beverages, grain products, and orange juice
Vitamin D	Sunshine, products fortified with vitamin D, such as soy beverages, margarine, and orange juice
Iron	Legumes, tofu, dark green leafy vegetables, dried fruit, whole grains, iron-fortified cereals and breads (absorption is improved by vitamin C, found in citrus fruit, tomatoes, strawberries, and dark green vegetables)
Zinc	Whole grains, wheat germ, legumes, nuts, tofu, and simulated meat products
Omega-3 fatty acids	Canola oil, flaxseed and flaxseed oil, soybean oil, walnuts, sea vegetables (seaweed), and DHA-rich microalgae

Vegetarian Food Groups

Life Cycle

Well-planned vegetarian diets, including vegan diets, are appropriate for all stages of the life cycle, including pregnancy, lactation, infancy, childhood, and adolescence.[21] One way to plan a healthy vegetarian diet is to modify the selections from Canada's Food Guide. The food choices and recommended amounts from the grain products and vegetables and fruit food groups should stay the same for vegetarians. Among the vegetables and fruit selections, including a Canada Food Guide serving (250 ml) of dark green leafy vegetables daily will help meet iron and calcium needs. The foods in the milk and alternatives and the meat and alternatives groups include foods of animal origin. Vegetarians who consume eggs and milk can still choose these foods. Those who avoid all animal foods can choose dry beans, nuts, and seeds from the meat and alternatives group. Fortified soy milk can be substituted for the dairy foods. To obtain adequate vitamin B_{12}, vegans must take supplements, nutritional yeasts, or use products fortified with vitamin B_{12}.

CANADIAN CONTENT

CASE STUDY OUTCOME

When Teresa started her internship at CIDA, she quickly learned the huge effect malnutrition has on the health of children around the world. Songe introduced her to the impact of protein-energy malnutrition (PEM). When his mother stopped nursing him and he no longer consumed adequate protein, Songe developed kwashiorkor; his belly became bloated because fat and fluid accumulated there. After a few months of consuming the high-protein supplement he was given at the clinic, his belly had shrunk and his arms and legs began filling out with muscle. Songe is now 3 years old, and although small for his age, he is a healthy, playful child.

At the clinic, Teresa helps educate families about the right combinations of foods needed to prevent PEM. She is learning all she can about the locally available foods so she can recommend meals that will provide the complementary proteins needed to meet amino acid needs. Teresa knows that with the scarcity of certain types of food in the typical local diet, she will likely see many other young children with symptoms similar to Songe's. Those who don't get early intervention will not be as fortunate as he was. Many children who have PEM die from infections because their immune systems cannot function optimally. Others suffer from permanent mental or physical ailments.

APPLICATIONS

Personal Nutrition

1. How much protein do you eat?

a. Use iProfile to calculate your average daily protein intake using the 3-day food record you kept in Chapter 2.

b. How does your protein intake compare to the RDA for protein for someone of your weight, age, and life stage? How does your protein intake compare to the recommended percent of kcalories from protein of between 10% and 35%? If you consumed just the RDA for protein, what percent of kcalories would this represent?

c. Is your protein intake greater than the RDA? If so, do you think you should decrease your protein intake? Why or why not?

d. Is your protein less than the RDA? If so, modify 1 day of your diet to meet your protein needs.

2. Are high-protein foods also high in saturated fat?

a. Using the 3-day record you kept in Chapter 2, record your total protein and saturated fat intake for each day in the table below.

	PROTEIN (g)	SATURATED FAT (g)
Day 1	________	________
Day 2	________	________
Day 3	________	________

b. What is the relationship between the amount of saturated fat and protein in your diet?

c. For each day, list the 3 foods that contribute the most protein to your diet. Are they animal or plant foods?

d. What percentage of your total saturated fat for that day do these 3 foods provide?

3. What changes would make your diet vegetarian?

a. Make a list of the nondairy animal foods in your diet and then list plant foods you could substitute for these.

b. If you made these substitutions, how much protein would the diet provide? Would it meet your RDA for protein?

iProfile

c. Convert this lacto-vegetarian diet into a vegan diet by substituting plant sources of protein for dairy products. Use protein complementation (see Figure 6.26) to be sure that you meet your need for essential amino acids. Make sure the diet includes at least 8 servings of calcium-rich foods.

General Nutrition Issues

1. A friend of yours is a weight lifter. He has read that if he eats a high-protein diet he will build muscle more quickly. He is 173 cm (5′ 8″) tall and weighs 72.5 kg (160 lbs). He drinks 2 or 3 protein shakes daily and always has 2 eggs for breakfast, a 120 g (4-ounce) hamburger for lunch, and a 180 g (6-ounce) steak for dinner.

a. Use iProfile to determine how much protein the eggs, hamburger, and steak contribute to his diet.

b. Use the Internet to determine how much protein a typical protein shake contains.

c. How does the protein he consumes from these sources compare to his requirement? (Remember that protein needs may be slightly higher for weight lifters.)

d. Does he need the protein shakes to meet his protein needs?

2. What food proteins complement each other? For each food in column A, select one or more in column B that could be combined with it to provide a meal of high-quality protein.

COLUMN A	COLUMN B
Rice	Tofu
Wheat bread	Peanut butter
Corn tortilla	Cheese
Pasta	Kidney beans
Tofu	Cashews
Peanut butter	Corn tortilla
Corn bread	Wheat bread
Soybeans	Chickpeas
Black-eyed peas	Chicken

3. One of the most common examples of protein complementation is a peanut butter sandwich. The table here gives the grams of each essential amino acid in 100 g of peanut butter protein and 100 g of wheat bread protein and in a reference amino acid pattern.

AMINO ACID	PEANUT BUTTER	WHEAT BREAD	REFERENCE AMINO ACID PATTERN
Isoleucine	4.0	3.4	5.9
Leucine	7.7	6.2	9.0
Lysine	3.9	1.7	7.2
Methionine + Cysteine	2.4	3.6	6.3
Phenylalanine + Tyrosine	10.8	6.4	10.3
Threonine	3.0	2.4	5.0
Tryptophan	1.2	1.0	1.3
Valine	4.6	3.8	6.7

a. For both peanut butter and wheat bread, calculate the percentage of each amino acid supplied relative to the amino acid reference pattern.

b. Which is the limiting amino acid in peanut butter (the amino acid present in the smallest amount relative to the reference pattern)? Which is the limiting amino acid in wheat bread? Which of the two food proteins has the higher amino acid score?

SUMMARY

6.1 Protein in the Canadian Diet

- Most of the protein in the Canadian diet comes from animal foods, but protein comes from both animal and plant sources. In developing countries, most dietary protein comes from plant sources.
- Whether protein is from a plant or animal source affects the amounts of fibre, saturated and unsaturated fat, cholesterol, and micronutrients in the diet.

6.2 Protein Molecules

- Amino acids consist of a carbon atom with a hydrogen atom, a nitrogen-containing group, an acid group, and a unique side chain attached. The amino acids that the body is unable to make in sufficient amounts are essential amino acids and must be consumed in the diet.
- Proteins are made of amino acid chains that fold over on themselves to create unique three-dimensional structures. The shape of a protein determines its function.

6.3 Protein in the Digestive Tract

- Digestion breaks dietary protein into small peptides and amino acids, which are absorbed into the mucosal cell using one of several active transport systems.
- Undigested protein fragments that are absorbed can trigger a food allergy.

6.4 Protein in the Body

- The amino acids in body tissues and fluids that are available for the synthesis of protein and other nitrogen-containing molecules or ATP production are known as the amino acid pool; they come from both dietary protein and the degradation of body proteins. The continuous breakdown and resynthesis of body proteins is referred to as protein turnover and is necessary for growth, maintenance, and regulation.
- DNA in the nucleus of cells contains the information needed to make body proteins. In transcription, this information is copied into a molecule of mRNA, which carries it to the cytosol. In translation, tRNA translates the mRNA code into a sequence of amino acids. Which genes are expressed determines which proteins are made.
- Amino acids are used to make nonprotein molecules that contain nitrogen.
- Amino acids can be used to provide energy when the diet doesn't meet energy needs or when protein intake exceeds needs. Amino acids that are used for energy are first deaminated. The amino groups are converted into urea, which can safely be excreted. The carbon compounds that remain can be broken down to generate ATP, or be used to synthesize glucose or fatty acids, depending on the needs of the body.
- In the body, proteins provide structure, regulate body functions as enzymes and hormones, transport molecules in the blood and into and out of cells, function in the immune system, and aid in muscle contraction, fluid balance, and acid balance.

6.5 Protein, Amino Acids, and Health

- Protein-energy malnutrition (PEM) is a concern, primarily in developing countries. Kwashiorkor is a form of PEM that occurs when the protein content of the diet is deficient but energy intake is adequate. It is most common in children. Marasmus is a form of PEM that occurs when total energy intake is deficient.
- High-protein diets increase the production of urea and other waste products that must be excreted in the urine and therefore can increase water losses. High-protein intakes increase urinary calcium losses, but when calcium intake is adequate, high-protein diets are associated with greater bone mass and fewer fractures. Diets high in animal proteins and low in fluid are associated with an increased risk of kidney stones. High-protein diets can be high in saturated fat and cholesterol.
- People with food allergies must avoid certain protein sources. Some proteins and amino acids trigger food intolerances. Those with the genetic disease phenylketonuria and sensitivities to gluten or monosodium glutamate (MSG) must avoid specific foods.

6.6 Meeting Protein Needs

- Nitrogen balance compares the amount of nitrogen consumed in the diet with the amount excreted.
- Nitrogen balance studies have been used to determine an RDA for protein for healthy adults of 0.8 g/kg of body weight per day.
- Healthy diets can include 10%-35% of energy from protein. Growth, pregnancy, lactation, illness, and injury can increase requirements. Certain types of physical activity may also increase protein needs.
- Animal proteins contain a pattern of amino acids that matches the needs of the human body more closely than the pattern of amino acids in plant proteins, and are therefore said to be of higher quality than plant proteins. Diets that include little or no animal protein can provide adequate protein if the sources of protein are complemented to supply enough of all the essential amino acids.
- Recommendations for a healthy diet suggest that we get more of our protein from plant sources. These foods are low in saturated fat and cholesterol and are good sources of monounsaturated and polyunsaturated fats, fibre, micronutrients, and phytochemicals.

6.7 Meeting Needs with a Vegetarian Diet

- Many vegetarian diets include some animal products. Vegan diets exclude all foods of animal origin.
- Vegetarian diets are associated with a lower risk for obesity, diabetes, cardiovascular disease, high blood pressure, and some types of cancer.
- Vegetarian diets can easily meet protein needs, but care must be taken to include enough iron and zinc in lacto-ovo vegetarian diets. Vegan diets must be well planned to meet the needs for calcium, vitamin D, iron, zinc, and omega-3 fatty acids and must include supplements or fortified foods that provide vitamin B_{12}.
- Canada's Food Guide can be used to plan vegetarian diets by choosing plant sources of protein from the meat and alternatives group and substituting fortified soymilk for milk.

REVIEW QUESTIONS

1. List some plant sources of protein.
2. What are amino acids?
3. Describe the general structure of a protein.
4. What is an essential amino acid?
5. What is the amino acid pool and where do these amino acids come from?
6. List 6 functions of proteins in the body.
7. Explain how proteins are synthesized.
8. Why is protein deficiency most common in infants and children?
9. Compare and contrast the causes and symptoms of kwashiorkor and marasmus.
10. How does the typical protein intake in Canada compare to recommendations?
11. What health problems are associated with a diet high in animal proteins?
12. What effect does moderate exercise have on protein needs?
13. What does nitrogen balance suggest about the balance between protein synthesis and protein breakdown in the body?
14. What is protein quality?
15. What is protein complementation?
16. What nutrients are at risk of deficiency in vegan diets?

REFERENCES

1. Health Canada. CCHS cycle 2.2 Nutrition: Nutrient Intakes from Food. Volumes 1, 2 & 3: Cat. No. 978-0-662-06542-5, 2004.
2. Williams, C. D. Kwashiorkor: Nutritional disease of children associated with maize diet. *Lancet* 2:1151–1154, 1935.
3. Alebiosu, C. O. An update on 'progression promoters' in renal diseases. *J. Natl. Med. Assoc.* 95:30–42, 2003.
4. Institute of Medicine, Food and Nutrition Board. Dietary Reference Intakes for Energy, Carbohydrates, Fiber, Fat, Protein and Amino Acids. Washington, DC: National Academies Press, 2002.
5. Heaney, R. P., and Layman, D. K. Amount and type of protein influences bone health. *Am. J. Clin. Nutr.* 87:1567S–1570S, 2008.
6. Kerstetter, J., O'Brien, K., and Insogna, K. Dietary protein, calcium metabolism, and skeletal homeostasis revisited. *Am. J. Clin. Nutr.* 78(Suppl):S584–S592, 2003.
7. Weikert, C., Walter, D., Hoffmann, K. et al. The relation between dietary protein, calcium and bone health in women: Results from the EPIC-Potsdam cohort. *Ann. Nutr. Metab.* 49:312–318, 2005.
8. Siener, R. Impact of dietary habits on stone incidence. *Urol. Res.* 34:131–133, 2006.
9. Van't Veer, P., Jansen, M. C., Klerk, M. et al. Fruits and vegetables in the prevention of cancer and cardiovascular disease. *Public Health Nutr.* 3:103–107, 2000.
10. Brown, A. S., Fernhoff, P. M., Waisbren, S. E. et al. Barriers to successful dietary control among pregnant women with phenylketonuria. *Genet. Med.* 4:84–89, 2002.
11. Seymour, C. A., Cockburn, F., Thomason, M. J. et al. Newborn screening for inborn errors of metabolism: A systematic review. *Health Technol. Assess.* I:1–95, 1997.
12. Fasano, A., Berti, I., Gerarduzzi, T. et al. Prevalence of celiac disease in at-risk and not-at-risk groups in the United States. *Arch. Intern. Med.* 163:268–292, 2003.
13. Walker, R., and Lupien, J. R. The safety evaluation of monosodium glutamate. *J. Nutr.* 130:1049S–1052S, 2000.
14. Geha, R. S., Beiser, A., Ren, C. et al. Review of alleged reaction to monosodium glutamate and outcome of a multicenter double-blind placebo-controlled study. *J. Nutr.* 130:1058S–1062S, 2000.
15. Health Canada. Monosodium glutamate: questions and answers. Available online at http://www.hc-sc.gc.ca/fn-an/securit/addit/msg_qa-qr-eng.php. Accessed June 29, 2010.
16. American Dietetic Association. Position of the American Dietetic Association, Dietitians of Canada, and the American College of Sports Medicine: Nutrition and athletic performance. *J. Am. Diet. Assoc.* 109:509–527, 2009.
17. Stipanuk, M. H. Protein and amino requirements. In *Biochemical and Physiological Aspects of Human Nutrition*. M. H. Stipanuk, ed. St. Louis: Saunders Elsevier, 2006, pp. 419–448.
18. Canadian Food Inspection Agency. Nutritional Labelling: Protein claims. Available online at http://inspection.gc.ca/english/fssa/labeti/guide/ch7ae.shtml#a7_15. Accessed June 30, 2010.
19. Canadian Food Inspection Agency: Nutritional Labelling: Chapter 6. Available online at http://www.inspection.gc.ca/english/fssa/labeti/guide/ch6ae.shtml. Accessed June 30, 2010.
20. American Dietetic Association. Position of the American Dietetic Association: Vegetarian diets. *J. Am. Diet. Assoc.* 109:1266–1282, 2009.
21. Williams, M. H. Facts and fallacies of purported ergogenic amino acid supplements. *Clin. Sports Med.* 18:633–649, 1999.
22. Canadian Council of Food and Nutrition. Consumer Trends: Tracking Nutrition Trends 1989-1994-1997-2001. Available online at http://www.ccfn.ca/pdfs/rap-vol17-1.pdf. Accessed July 20, 2010.
23. Leitzmann, C. Vegetarian diets: What are the advantages? *Forum Nutr.* 57:147–156, 2005.

7 Energy Balance and Weight Management

Case Study

Erin had always been "the chubby one" in her family. She and her sister, Sara, had the same parents and were only 3 years apart in age, yet they were very different. Sara had brown hair, long legs, and a slender build. Why had Erin ended up with freckles, red hair, and chunky thighs? She was always the one who could stand to lose a little weight. She would rather read than ride her bike and often spent the afternoon munching chips while she studied.

Then one day, in her first year of university, Erin decided she didn't want to be "the chubby one" anymore. She changed her diet and started to exercise. At first, it was just a walk after dinner, but soon, exercise became part of her daily routine. She covered a few kilometres every day, whether at the gym, on her bike, or jogging. The extra weight disappeared. Sara, in contrast, spent the fall finishing law school and snacking as she studied for the bar exam. She passed the bar 15 kg heavier than she had ever been. At the family reunion that summer, Sara heard a relative say to her sister, "Didn't you used to be the heavy one?" What happened? How could two women with the same parents have such different builds? And how could they change so much in one year?

Sara and Erin inherited different combinations of genes from their parents. These genes determine their body builds, but lifestyle factors also affect how much body fat they accumulate. Sara inherited her slender build, but eating too much and exercising too little while she studied for the bar exam added unwanted weight. Erin, in contrast, may never have thin thighs, but when she exercises regularly and watches what she eats, she can keep her weight in a healthy range.

In this chapter, you will learn more about the factors, both personal and environmental, that influence body weight.

(Photodisc/Getty Images, Inc.)

(©iStockphoto)

(Robert Daily/Getty Images, Inc.)

Chapter Outline

7.1 The Obesity Epidemic

Learning Objective

- Discuss the incidence of obesity in Canada and around the world.

overweight Being too heavy for one's height. It is defined as having a body mass index (BMI—a ratio of weight to height squared) of 25-29.9 kg/m^2.

obesity A condition characterized by excess body fat. It is defined as a body mass index of 30 kg/m^2 or greater.

In 2004, the Canadian Community Health Survey found that about 60% of adults in Canada were **overweight** or **obese**[1] **(Figure 7.1).** This represents a dramatic upward shift in the numbers. Surveys done between 1978-1979 and between 1986-1992 reported that only 14% and 15% of Canadian were obese during those periods, respectively, while in 2004, the number rose to 23%.[2]

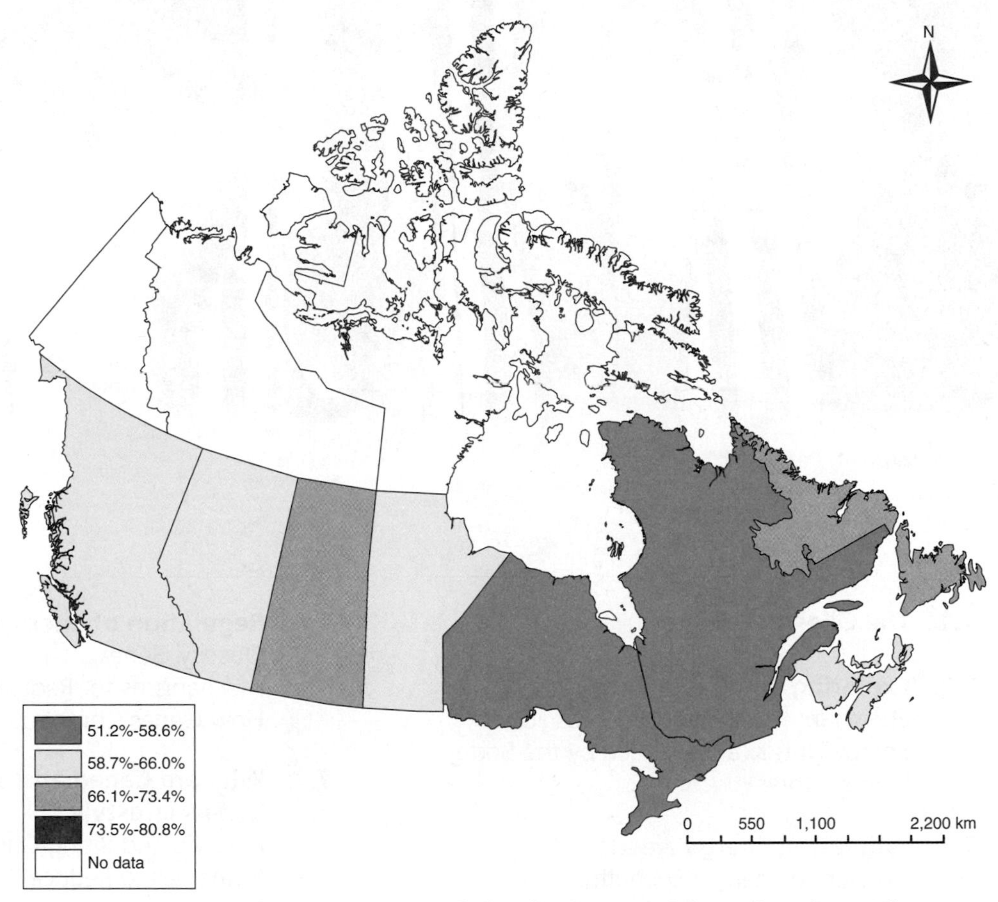

Figure 7.1 Proportion of the Canadian population that is overweight or obese (BMI>25). The national average is 59.3%.

Source: Health Canada. Map of Overweight or Obese According to Measured Body Mass Index (BMI) in Adults in Canada (both males and females) Available online at http://www.hc-sc.gc.ca/fn-an/surveill/atlas/map-carte/mass_adult_over_obes_mf-hf-eng.php. Accessed July 9, 2010.

The problem is not limited to adults. In 2004, 18% of Canadian children and adolescents aged 2-17 were overweight and 8% were obese, up from 12% and 3% respectively in 1978-1979.[3]

The reason Canadians are getting fatter is that changes in the nation's food supply and lifestyle have created an energy imbalance; we are eating more and exercising less, leading to weight gain. The repercussions of this rise in obesity have led public health officials to call it an epidemic. Carrying excess body fat increases the risk of a host of chronic health problems, including high blood pressure, heart disease, high blood cholesterol, diabetes, stroke, gallbladder disease, arthritis, sleep disorders, respiratory problems, and cancers of the breast, uterus, prostate, and colon.

Lifestyle transitions similar to those in Canada are occurring worldwide and the incidence of obesity is following suit. It is such an important trend that the word "globesity" has been coined to reflect the escalation of global obesity and overweight. Around the world, approximately 1.6 billion adults are overweight, and 400 million are obese. The World Health Organization projects that by 2015, approximately 2.3 billion adults will be overweight and more than 700 million will be obese.[4] Once considered a problem only in high-income countries, overweight and obesity are now on the rise in low- and middle-income countries, particularly in urban settings.

7.2 Exploring Energy Balance

Learning Objectives

- Explain the principle of energy balance.
- Describe the processes involved in generating ATP from food.
- Describe the components of energy expenditure.
- Indicate how excess dietary energy is stored in the body.

energy balance The amount of energy consumed in the diet compared with the amount expended by the body over a given period.

The principle of **energy balance** states that when energy consumption equals energy expenditure, body weight remains constant. Energy balance can be achieved at any weight—fat, thin, or in between (**Figure 7.2**); it simply means that body weight is not changing. If, however, less energy is taken in than expended, energy balance is negative and weight will be lost. On the other hand, if the amount of energy taken in exceeds the amount expended, energy balance is positive and the extra energy will be stored in the body, causing weight to increase.

Figure 7.2 Energy balance What you weigh is a balance between how much energy you consume and how much energy you expend.

Energy is defined as the ability to do work. In nutrition, energy is measured in kilocalories (kcalories, kcal), which are units of heat or kilojoules (kjoules, kJ), which are units of work. In Canada and the United States, kcalories are the standard measure of energy in food and the body, while in Europe, kjoules are the measure most commonly used. Technically, a kcalorie is the amount of heat required to raise the temperature of 1 kg of water 1°C. In practical terms, a kcalorie is a measure of the amount of energy that is supplied to or expended by the body.

Energy In: Kcalories Taken in as Food

The energy taken into the body comes from the energy-yielding nutrients (carbohydrates, lipids, and proteins) and alcohol consumed in food and beverages. Individuals who struggle with weight loss often think of the kcalories in food as an enemy—something to be avoided. However, food and the energy it provides are essential for life. Just as gasoline is necessary to run an engine, kcalories are necessary to run the body. The amount of energy (number of kcalories) taken in depends on the total amount of food consumed and the nutrient composition of that food. The energy content of food can be measured precisely in the laboratory or estimated from its nutrient composition.

Determining the Amount of Energy in Food The amount of energy in a food or a mixture of foods can be determined in the laboratory using a **bomb calorimeter.** A bomb calorimeter consists of a chamber surrounded by a jacket of water (**Figure 7.3**). Food is dried, placed in the chamber, and burned. As the food combusts, heat is released, raising the temperature of the water. The increase in water temperature can be used to calculate the amount of energy in the food based on the fact that 1 kcal is the amount of heat needed to increase the temperature of 1 kg of water by 1°C.

bomb calorimeter An instrument used to determine the energy content of food. It measures the heat energy released when a dried food is combusted.

Combusting a food in a bomb calorimeter determines the total amount of energy contained in that food. However, because the body cannot completely digest, absorb, and utilize all of the substances in a food, bomb calorimeter values are slightly higher than the amount of energy the body can obtain from that food. To correct for this difference, feeding experiments have been done to measure the amount of energy that is not available to the body, such as that lost in urine and feces. Subtracting this unavailable energy from the values determined by the bomb calorimeter gives a more accurate estimate of the energy obtained from food. These types of experiments were used to determine the amount of energy provided by the carbohydrate, fat, protein, and alcohol in a mixed diet.

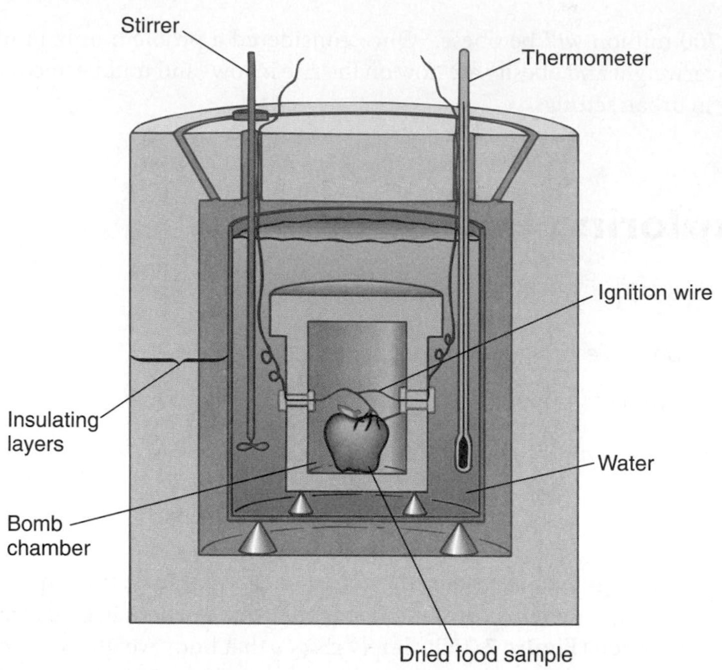

Figure 7.3 Bomb calorimeter
When dried food is combusted inside the chamber of a bomb calorimeter, the rise in temperature of the surrounding water can be used to determine the energy content of the food.

Figure 7.4 You can calculate the kcalories in a slice of pizza if you know how much carbohydrate, protein, and fat it provides. (© istockphoto.com/ Burwell and Burwell Photography)

When the nutrient composition of a food is known, totalling the energy from the carbohydrate, fat, and protein in the food can approximate the energy content (**Table 7.1**). Vitamins, minerals, and water, though essential nutrients, do not provide energy to the body. Carbohydrate and protein provide about 4 kcal/g, so 5 g of sugar, which is almost pure carbohydrate, contains about 20 kcal (5 g × 4 kcal/g) and 5 g of dried egg white, which is almost pure protein, also provides about 20 kcal. Fat, the most concentrated source of energy, provides 9 kcal/g. Five grams of corn oil, which is almost pure fat, contains about 45 kcal (5 g × 9 kcal/g). Alcohol provides 7 kcal/g. Most foods are of mixed composition; for instance, a slice of pizza contains about 15 g of protein, 50 g of carbohydrate, and 10 g of fat (**Figure 7.4**). Its energy content is therefore: (4 kcal/g × 15 g protein) + (4 kcal/g × 50 g carbohydrate) + (9 kcal/g × 10 g fat) = 350 kcal.

Table 7.1 Estimating the Energy Content of Food
Determine: The number of grams of carbohydrate, protein, fat, and alcohol in a food or meal
Calculate the energy provided by each: grams of carbohydrate × 4 kcal/g = kcalories from carbohydrate grams of protein × 4 kcal/g = kcalories from protein grams of fat × 9 kcal/g = kcalories from fat grams of alcohol × 7 kcal/g = kcalories from alcohol
Calculate the total energy: Total energy (kcal) = (kcal from carbohydrate) + (kcal from protein) + (kcal from fat) + (kcal from alcohol) **For example:** A 100 g (125 ml) serving of macaroni and cheese contains 8 g of protein, 20 g of carbohydrate, and 11 g of fat: 20 g of carbohydrate × 4 kcal/g = 80 kcal 8 g of protein × 4 kcal/g = 32 kcal 11 g of fat × 9 kcal/g = 99 kcal **Total energy** = 80 kcal + 32 kcal + 99 kcal = 211 kcal

LABEL LITERACY

CANADIAN CONTENT

How Many Kcalories in That Box, Bowl, or Bottle?

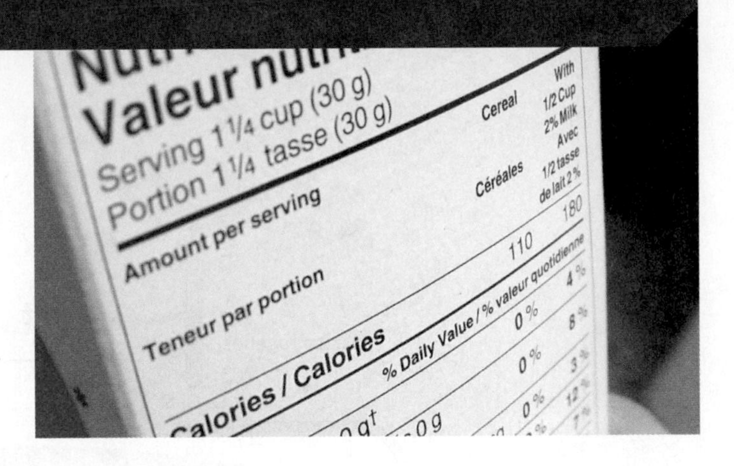

Are you watching your kcalorie intake? The Nutrition Facts portion of a food label can help, if you read it carefully. It provides information about kcalories per serving and serving size. Make sure you check both.

People tend to eat in units—one bottle of juice, one can of iced tea, one bag of potato chips—but the serving size on the label doesn't always reflect the kcalorie count for those units. For example, the label on the iced tea bottle shown here says that a serving has only 100 kcal. But the serving size is 250 ml, and the bottle holds 600 ml. So if you finish the bottle, you will be getting 240 kcal, mostly from added sugars.

The discrepancy between the serving sizes listed on packages and the portion sizes we consume is one of the reasons Canadians are getting fatter. We choose portions that are generally larger than recommended. For example, the label on your ice cream shows that it provides about 140 kcal per serving and that a serving is only 125 ml, a portion the size of a tennis ball. If you scoop 250 ml of ice cream onto your cone or into your bowl, you will be consuming about 280 kcal. Likewise, if you pour yourself 250 ml of granola for breakfast, which according to its nutrition facts table contain 100kcal per 60 ml serving, you are having more than 400 kcal.

Knowing what a serving is can help you keep your kcalories under control, but it isn't always easy. For example, pasta portions are particularly tricky, as the serving size is usually given as dry pasta. What does 60 g of dry spaghetti—for 200 kcal—look like once it is cooked? The answer is about 250 ml, so if you pile 500 ml onto your plate, you are getting 400 kcal rather than 200 kcal. Some product labels list nutritional information both before and after the product is prepared; this can help you figure out what you are choosing. So read carefully, and use all the information on the label to determine how many kcalories are in your portions.

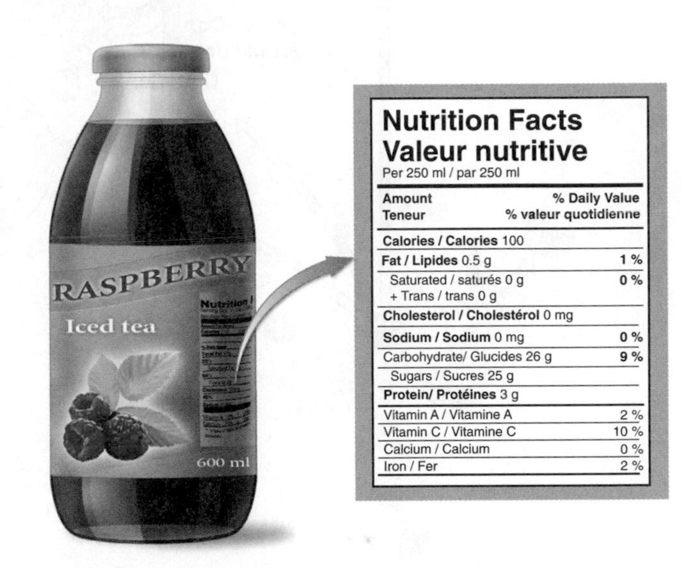

Nutrition Facts
Valeur nutritive
Per 250 ml / par 250 ml

Amount / Teneur	% Daily Value / % valeur quotidienne
Calories / Calories 100	
Fat / Lipides 0.5 g	1 %
Saturated / saturés 0 g + Trans / trans 0 g	0 %
Cholesterol / Cholestérol 0 mg	
Sodium / Sodium 0 mg	0 %
Carbohydrate/ Glucides 26 g	9 %
Sugars / Sucres 25 g	
Protein/ Protéines 3 g	
Vitamin A / Vitamine A	2 %
Vitamin C / Vitamine C	10 %
Calcium / Calcium	0 %
Iron / Fer	2 %

Information on the energy content of foods can be found on food labels and in food composition tables and databases such as iProfile. The Nutrition Facts portion of food labels lists the total kcalories in a serving of food (Label Literacy: How Many Kcalories in that Bowl, Box, or Bottle?). The energy content of foods in a diet can also be estimated from the food group lists in CDA's Beyond the Basics meal planning system (See Appendix E).

Converting Food Energy into ATP Just as the energy in flowing water can be converted into electrical energy, which can then be converted into the light energy emitted by a light bulb, the energy stored in the chemical bonds of carbohydrates, fats, and proteins can be converted into ATP, which can be used to keep you alive and moving. To generate ATP, metabolic reactions break down or oxidize carbohydrate, fat, and protein (**Figure 7.5** and sections 4.4, 5.5, and 6.4). All three can be converted into the common intermediate acetyl-CoA. Glycolysis converts glucose from carbohydrate into pyruvate, which then loses a carbon to form acetyl-CoA. Triglycerides are broken into fatty acids and glycerol. Beta-oxidation breaks fatty acids into 2-carbon units that form acetyl-CoA. Amino acids from protein are deaminated and their carbon skeletons used to make acetyl-CoA or other intermediates. Acetyl-CoA can then enter the citric acid cycle. The high-energy electrons released at various metabolic steps are passed to the electron transport chain where their energy is trapped and used to generate ATP. The ATP is then used to fuel metabolic reactions that build and maintain body components and to power other cellular and body activities. Much of the energy consumed in food is also converted to and lost from the body as heat.

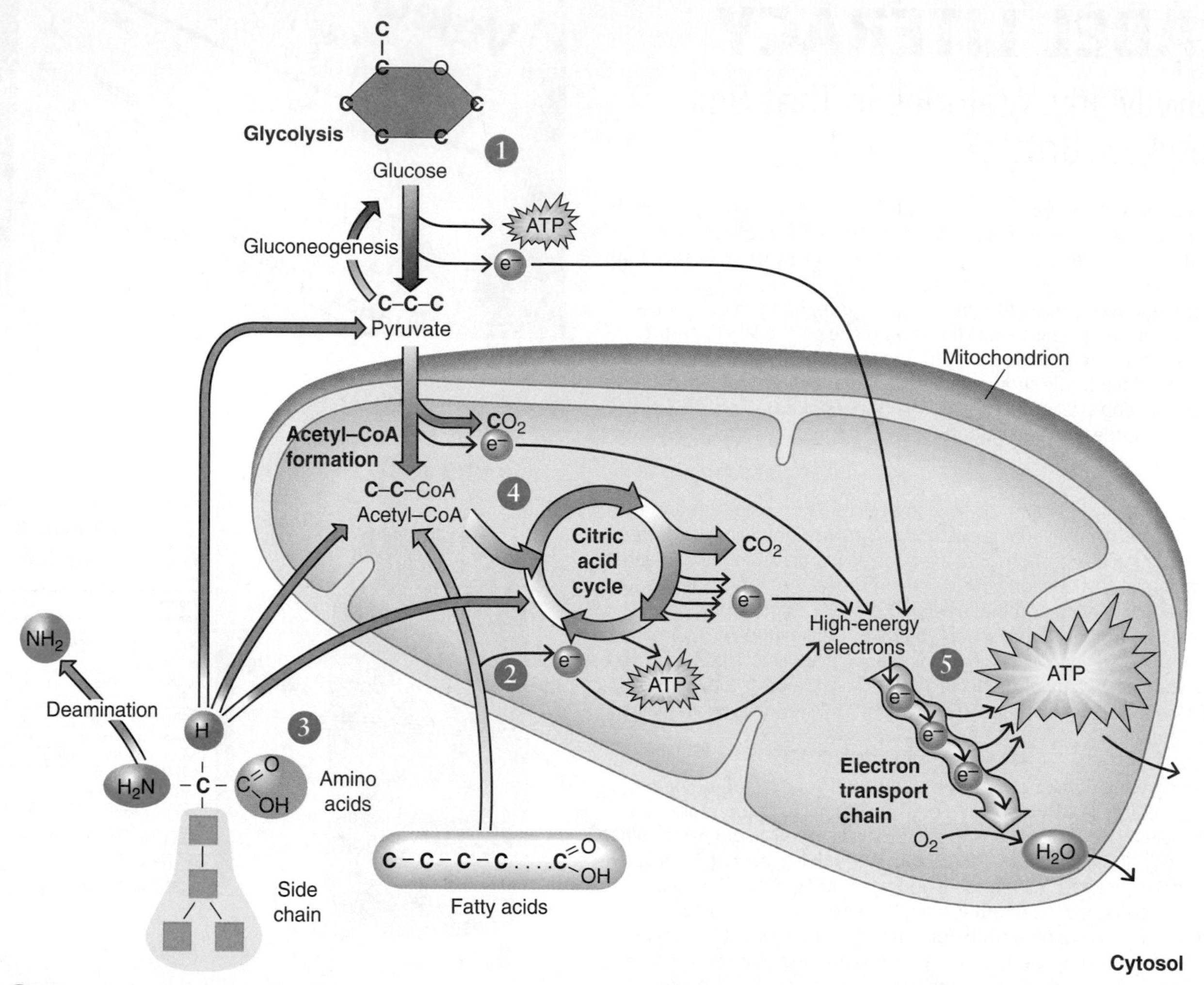

1. Glycolysis breaks glucose in half to yield pyruvate, which is then converted to acetyl-CoA.
2. Beta-oxidation breaks fatty acids into 2-carbon units that form acetyl-CoA.
3. Amino acids are deaminated and can break down to form acetyl-CoA, pyruvate, or other intermediates.
4. The acetyl-CoA from glucose, fatty acid, and amino acid breakdown can enter the citric acid cycle.
5. The electrons released are passed to the electron transport chain and their energy is used to make ATP.

Figure 7.5 Producing ATP from glucose, fatty acids, and amino acids
Glucose, fatty acids, and amino acids can be broken down by the reactions of cellular respiration to yield carbon dioxide, water, and energy in the form of ATP.

total energy expenditure (TEE) The sum of the energy used for basal metabolism, activity, processing food, deposition of new tissue, and production of milk.

basal energy expenditure (BEE) The energy expended to maintain an awake resting body that is not digesting food.

basal metabolic rate (BMR) The rate of energy expenditure under resting conditions. BMR measurements are performed in a warm room in the morning before the subject rises, and at least 12 hours after the last food or activity.

Energy Out: Kcalories Used by the Body

The total amount of energy used by the body each day or **total energy expenditure (TEE)** includes the energy needed to maintain basic bodily functions such as the beating of your heart, as well as that needed to fuel activity and process food. In individuals who are growing or pregnant, total energy expenditure also includes the energy used to deposit new tissues. In women who are lactating, it includes the energy used to produce milk. A small amount of energy is also used to maintain body temperature in a cold environment.

Basal Metabolism For most people, about 60%-75% of the body's total energy expenditure is used for basal metabolism. **Basal energy expenditure (BEE)** includes all of the involuntary things your body does to stay alive such as breathing, circulating blood, regulating body temperature, synthesizing tissues, removing waste products, and sending nerve signals. The rate at which energy is used for these basic functions is called **basal metabolic rate (BMR)** and is often expressed in kcalories per hour. Basal needs include the energy necessary for essential metabolic reactions and life-sustaining functions but do not include the energy needed for physical activity or for the digestion of food and the absorption of nutrients. Therefore, to minimize residual energy expenditure for activity or processing food, BMR is measured in the morning, in a warm room before the subject rises, and at least 12 hours after food intake or activity (**Figure 7.6**). Because of the difficulty of achieving these conditions, measures are often made after about 5-6 hours without food or exercise. When done under these conditions, it is reported as **resting energy expenditure (REE)** or **resting metabolic rate (RMR)**. RMR values are about 10%-20% higher than BMR values.[5]

resting energy expenditure (REE) or **resting metabolic rate (RMR)** Terms used when an estimate of basal metabolism is determined by measuring energy utilization after 5-6 hours without food or exercise.

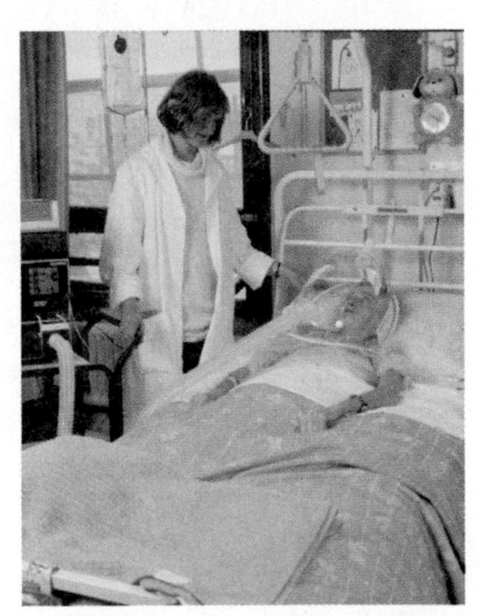

Figure 7.6 To assess BMR, expired gases can be collected and measured by having the subject breathe into a hood. (St. Bartholomew's Hospital/Science/Custom Medical Stock Photo)

Basal needs are affected by factors such as body weight, gender, growth rate, and age. BMR increases with increasing body weight, so it is higher in heavier individuals. It also rises with increasing **lean body mass**; thus, BMR is generally higher in men than in women because men have more lean tissue. BMR increases during periods of rapid growth because energy is required to produce new body tissue. It decreases with age, partly due to the decrease in lean body mass that usually occurs in older adults.

lean body mass Body mass attributed to nonfat body components such as bone, muscle, and internal organs. It is also called fat-free mass.

Basal needs can be altered by certain abnormal conditions. An elevation in body temperature increases BMR. It is estimated that for every 1°C above normal body temperature, there is a 14% increase in BMR. This extra energy use explains why a fever can cause weight loss. Abnormal levels of thyroid hormones can also affect BMR. Individuals who overproduce these hormones burn more energy; in fact, a symptom used to diagnose thyroid hormone excess is unexplained weight loss. Individuals with an underproduction of thyroid hormones require less energy. The fact that hormones produced by the thyroid gland affect energy expenditure is the reason that obesity was once explained as a glandular problem. It is now known that obesity due to a thyroid hormone deficiency is rare.

Metabolic rate may also be affected by low-energy diets. Energy intake below needs may depress resting metabolic rate by 10%-20%, or the equivalent of 100-400 kcal/day.[6] This drop in basal needs decreases the amount of energy needed to maintain weight. It is a beneficial adaptation in starvation, but it makes intentional weight loss more difficult.

Physical Activity Physical activity is the second major component of energy expenditure. It represents the metabolic cost of external work, which includes the energy needed for planned exercise as well as for the activities of daily life, such as cooking, gardening, and house cleaning.

The energy expended for unintentional exercise is called **nonexercise activity thermogenesis (NEAT)**. NEAT accounts for the majority of the energy expended for activity and varies enormously, depending on an individual's occupation and daily movements.[7] For most people, physical activity accounts for 15%-30% of energy requirements, but this varies greatly (**Figure 7.7**). An endurance athlete who trains 5 or 6 hours a day may expend more energy in activity than for basal metabolism. A person's occupation can also have a great effect on the energy expended for activity. For example, a construction worker who spends 8 hours a day doing physical labour uses a great deal more energy in his daily activities than does an office worker who spends most of his day sitting at a desk (**Figure 7.8**).

nonexercise activity thermogenesis (NEAT) The energy expended for everything we do other than sleeping, eating, or sports-like exercise.

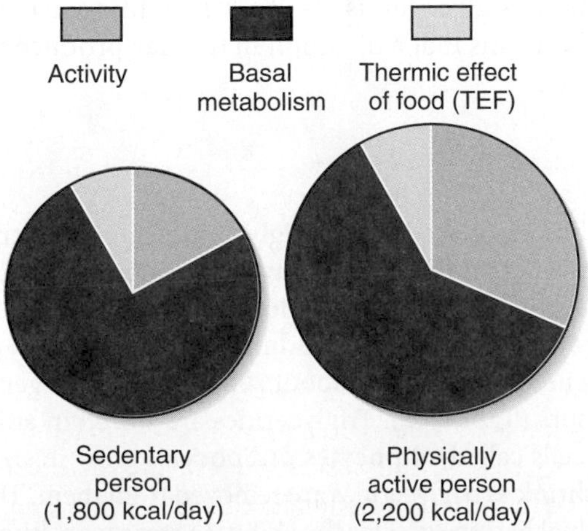

Activity
Basal metabolism
Thermic effect of food (TEF)

Sedentary person (1,800 kcal/day)
Physically active person (2,200 kcal/day)

Figure 7.7 Activity as a percentage of total energy requirement
The percentage of energy expended in physical activity increases from only about 15% in a sedentary person to 30% or more in an active individual. Increasing activity increases total energy expenditure.

The energy required for an activity depends on how strenuous the activity is and the length of time it is performed. For example, walking at a speed of 5-6.5 km per hour requires a moderate degree of exertion and uses about 300 kcal/h for a 70 kg man. The energy required increases progressively as the walking speed and length of time the activity continues increase, and as the body weight of the exerciser rises. In addition to the energy expended during exercise, there is a small increase in energy expenditure for a period of time after exercise has been completed. The energy expended for activity is the one component of our total energy needs over which we have control. To maintain health, people today need to consciously increase their physical activity. This does not mean they have to run marathons. Choosing to take the stairs rather than the elevator, walking rather than taking the bus, and riding a bike rather than driving to the store all increase activity. The energy costs of specific activities are listed in Appendix H.

Figure 7.8 A carpenter may burn about twice as many kcalories in a workday as someone with a desk job. (© Huntstock/Age Fotostock; © istockphoto.com/Chris Schmidt)

thermic effect of food (TEF) or **diet-induced thermogenesis** The energy required for the digestion of food and the absorption, metabolism, and storage of nutrients. It is equal to approximately 10% of daily energy intake.

Thermic Effect of Food Our energy comes from food, but we also need energy to digest food and to absorb, metabolize, and store the nutrients from this food. The energy used for these processes is called the **thermic effect of food** (TEF) or **diet-induced thermogenesis.** This increase in energy expenditure causes body temperature to rise slightly for several hours after eating. The energy required for TEF is estimated to be about 10% of energy intake but can vary depending on the amounts and types of nutrients consumed. Because it takes energy to store nutrients, TEF increases with the size of the meal. A meal that is high in fat has a lower TEF than a meal high in carbohydrate or protein, because dietary fat can be used or stored more efficiently than either protein or carbohydrate. The metabolic cost of either oxidizing or storing dietary fat is only 2%-3% of the energy consumed, whereas the cost of using amino acids by either oxidizing them or incorporating them into proteins is 15%-30% of the energy consumed, and the cost of breaking down carbohydrate or storing it as glycogen is 6%-8%.[8] The difference in the cost of storing different nutrients as fat means that a diet high in fat may produce more body fat than a diet high in carbohydrate.[9]

Energy Stores

Energy is stored in the body as glycogen and triglycerides. Stored energy can be used when intake is less than needs, whether this occurs between meals or over a longer period such as during starvation, fasting, or dieting for weight loss.

Glycogen stores are located in the liver and muscle and fill when dietary carbohydrate is adequate. The body generally stores only about 200-500 g of glycogen—enough to provide glucose for about 24 hours (**Table 7.2**). Triglycerides are stored in adipose tissue, which is made up of fat-storing cells called **adipocytes.** Adipocytes grow in size as they accumulate more triglycerides and shrink as triglycerides are removed from them. The greater the number of adipocytes an individual has, the greater the ability to store fat. Although most adipocytes are formed between infancy and adolescence, excessive weight gain can cause the production of new fat-storing cells in adults.

adipocytes Fat-storing cells.

Table 7.2 Sources of Stored Energy in the Body

Energy Source	Primary Location	Energy (kcal)[a]
Glycogen	Liver and muscle	1,400
Glucose or free fatty acid	Body fluids	100
Triglyceride	Adipose tissue	115,000
Protein	Muscle	25,000

[a]Values represent the approximate amounts in a 70-kg male.

Source: Cahill, G. F. Starvation in man. *N. Engl. J. Med.* 282:668–675, 1970; and Frayn, K. *Metabolic Regulation*: A Human Perspective. London: Portland Press, 1996, pp. 78–102.

Using Body Energy Stores: Weight Loss To function normally, the body needs a steady supply of energy. Some of this energy must come from glucose, which is needed to fuel the brain and several other types of body cells. After eating, energy is supplied by absorbed nutrients. Between meals, the breakdown of glycogen provides glucose, and the breakdown of stored fat meets other energy needs. Typically, these stores are then replaced by energy consumed in the next meal so that there is no net change in the amount of stored energy. However, if energy stores are not replenished, the amount of stored energy—and hence, body weight—will decrease (**Figure 7.9**).

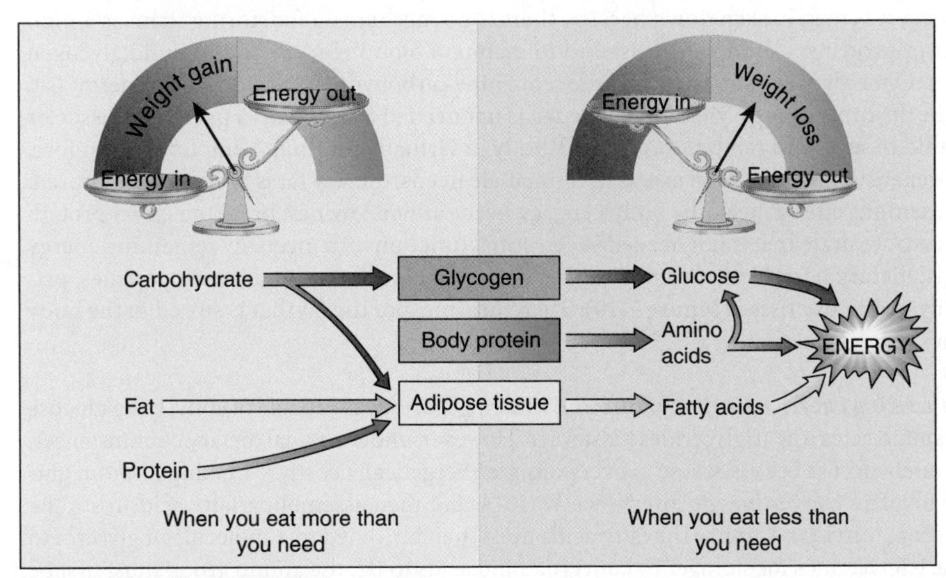

Figure 7.9 Feasting and fasting
When you eat more than you need at that time, some energy is put into body stores. When you haven't eaten in a while, you retrieve energy from these stores. A small amount of body protein is also broken down to amino acids to make glucose and provide energy.

If no food is eaten for more than several hours, the body must shift the way it uses energy to ensure that glucose continues to be available to cells that need it. Glycogen stores can provide glucose, but are limited, so glucose is also supplied by the breakdown of small amounts of body protein, primarily muscle protein, to yield amino acids. Amino acids can then be used to synthesize glucose via gluconeogenesis (see sections 4.4 and 6.4). Once glycogen stores are depleted, all of the glucose must come from gluconeogenesis. Because protein is not stored in the body, the breakdown of protein to provide energy and glucose results in the loss of functional body proteins.

Energy for tissues that don't require glucose is provided by the breakdown of stored fat. If the supply of glucose is limited, such as during starvation, fatty acids delivered to the liver cannot be completely oxidized, so ketones are produced (see Section 4.4). Ketones can be used as an energy source by many tissues. After about 3 days of starvation, even the brain adapts to meet some of its energy needs from ketones (see Section 5.5). This reduces the amount of glucose needed and thus slows the rate of protein breakdown.

If energy intake is restricted for a prolonged period (fasting), substantial amounts of fat are used to provide energy and protein is degraded to provide glucose. This results in weight loss (see Figure 7.9). The magnitude of the weight loss depends on the degree of energy deficit and the length of time over which it occurs. It is estimated that an energy deficit of about 3,500 kcal will result in the loss of 0.5 kg (1 lb) of adipose tissue.

Building Body Stores: Weight Gain People typically eat 3-6 times during the day. For weight to remain stable, the sum of the energy in all of these meals and snacks must equal daily energy expenditure, but at each meal or snack more energy is likely to be consumed than is needed at that moment in time. When excess energy is consumed (feasting), we generally say this excess is stored as fat, but this is an oversimplification of a complex situation. After eating, the body prioritizes how nutrients are used, based on body needs, which nutrients can be stored, and how efficiently they can be stored. Nonetheless, regardless of the composition of the diet, when excess energy is consumed over the long term, fat stores will enlarge and weight will increase (see Figure 7.9).

Hierarchy of Nutrient Use There is a metabolic hierarchy of how fuels are used by the body. Alcohol, although not a nutrient, does supply energy. Because it is toxic and cannot be stored in the body, it is rapidly oxidized. Amino acids from dietary protein are next in the hierarchy. They are first used to synthesize needed body proteins and non-protein molecules; any excess is then broken down because there is no mechanism for storing them as amino acids or proteins. Carbohydrate is used to maintain blood glucose and to build glycogen stores. Once glycogen stores are full, the remaining carbohydrate is oxidized for energy. Fat, unlike the other energy-yielding nutrients, is not needed as a fuel for a particular tissue or to build tissues, and can be stored in the body in virtually unlimited amounts. Therefore, if the energy consumed is in excess of immediate needs, dietary fat is preferentially stored. For example, after a meal, the body's energy needs are met by first breaking down protein and carbohydrate that is not needed for essential functions. To meet any remaining energy needs, dietary fat is oxidized and any dietary fat that is left is stored as triglycerides, primarily in adipose tissue (**Figure 7.10**). Therefore, most of the fat that is stored in the body comes from dietary fat.

Synthesizing Fat from Carbohydrate and Protein The body is capable of converting glucose and amino acids into triglycerides for storage. However, under normal dietary circumstances, this rarely occurs because these conversions are energetically costly.[10] Making fat from glucose involves converting glucose to acetyl Co-A and then assembling fatty acids from the 2-carbon acetyl-CoA units. The fatty acids must then be joined to a molecule of glycerol to make triglycerides for storage. To convert amino acids to fat, the amino group must first be removed and then the carbon skeleton must be broken down to yield acetyl-CoA that can be used for fatty acid synthesis. In contrast, the conversion of dietary fat to body fat requires only the removal and reattachment of fatty acids from the glycerol backbone (see Section 5.4); it takes only 2%-3% of the energy in the fat to store the fat in adipose tissue (**Figure 7.11**). The low metabolic cost of converting dietary fat to stored body fat makes it more efficient for the body to use dietary fat to make body fat and to oxidize carbohydrate and protein to meet immediate energy needs rather than convert it to fat. The conversion of carbohydrate to fat becomes important only when the diet is composed primarily of carbohydrate and energy intake exceeds expenditure.[10]

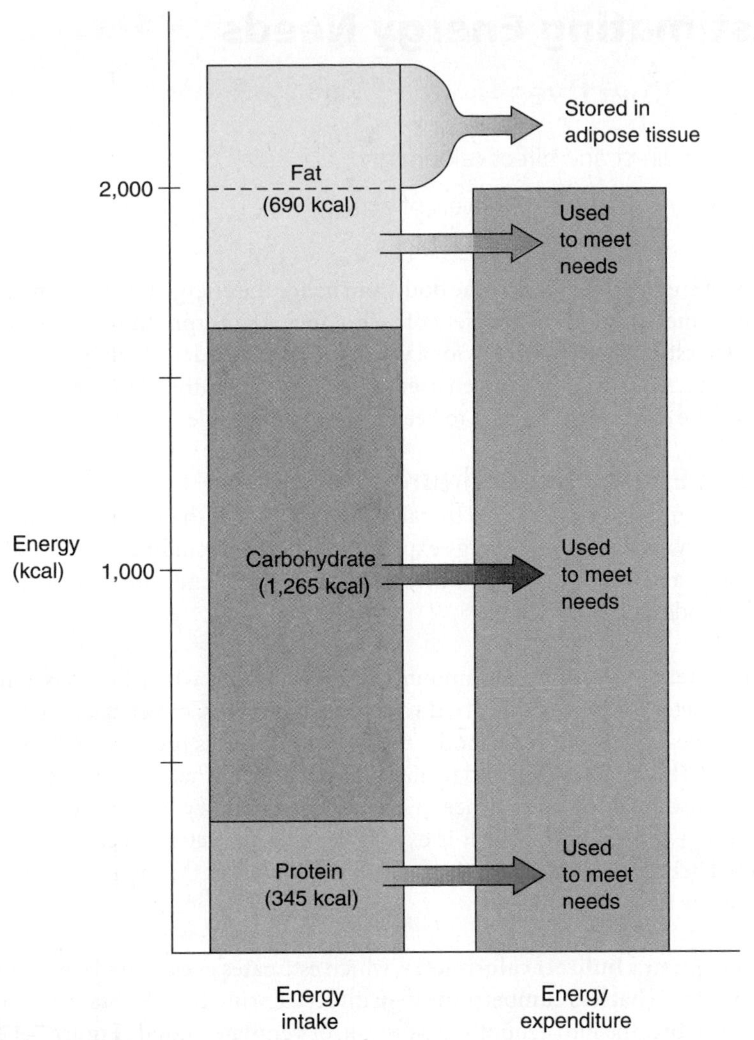

Figure 7.10 Nutrient usage
If energy intake exceeds expenditure, dietary fat is stored in adipose tissue. In this example, all the kcalories from carbohydrate and protein are used to meet body needs, but only some of the kcalories from fat are used. The remaining kcalories are stored in adipose tissue.

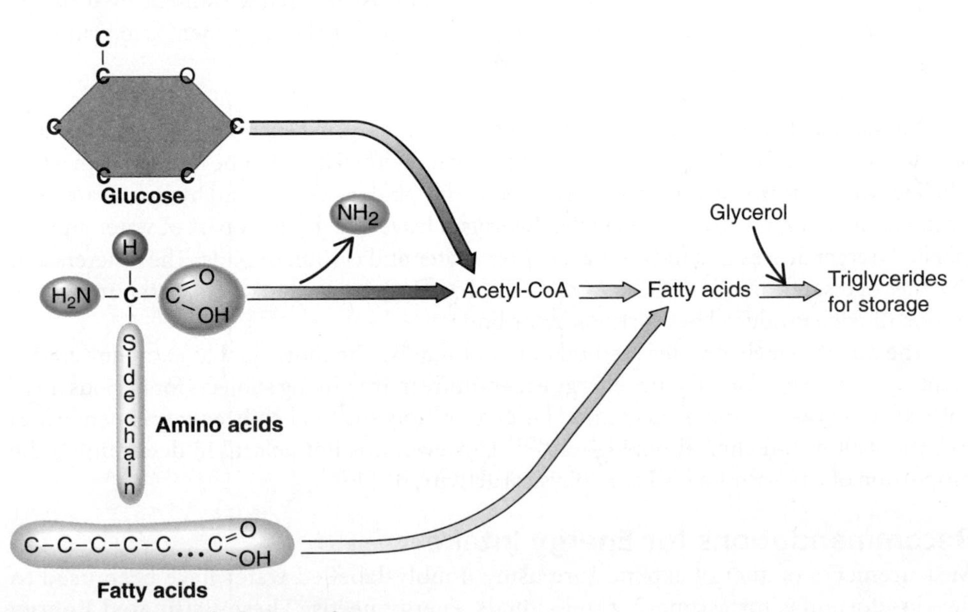

Figure 7.11 Storing kcalories as fat
It is metabolically much simpler and less costly to convert dietary fatty acids to triglycerides for storage than to convert glucose or amino acids to triglycerides.

7.3 Estimating Energy Needs

Learning Objectives

- Distinguish indirect and direct calorimetry.
- Calculate your EER at various levels of activity.

The amount of energy expended by the body, and hence the energy needed to maintain body weight, can be measured using a variety of techniques. Data from these measurements can then be used to estimate energy needs for a variety of people under a variety of circumstances. Calculations of energy needs have been used to generate recommendations about how much energy should be consumed in order to keep body weight stable.

Measuring Energy Expenditure

Energy expenditure can be measured by calorimetry, which is the science of measuring heat flow. Calorimetry can determine energy expenditure directly or indirectly. The use of doubly-labelled water is a newer technique for measuring energy expenditure that can be used over prolonged periods.

direct calorimetry A method of determining energy use that measures the amount of heat produced.

Direct Calorimetry Measuring the amount of heat produced when a food is combusted in a bomb calorimeter is a type of **direct calorimetry**. In humans, direct calorimetry measures the amount of heat given off by the body; the heat produced is proportional to the amount of energy used. This heat is generated by metabolic reactions that both convert food energy into ATP and use ATP for body processes. Direct calorimetry is an accurate method for measuring energy expenditure, but it is expensive and impractical because it requires that the individual being assessed remain in an insulated chamber throughout the procedure in order to measure the heat produced.

indirect calorimetry A method of estimating energy use that compares the amount of oxygen consumed to the amount of carbon dioxide exhaled.

Indirect Calorimetry **Indirect calorimetry**, which estimates energy use by assessing oxygen utilization, is somewhat less cumbersome than direct calorimetry. To obtain a measurement, the subject must breathe into a mouthpiece, mask, or ventilated hood (**Figure 7.12**). Oxygen use and carbon dioxide production are measured by analyzing the difference between the composition of inhaled and exhaled air. The body's energy use can be calculated from these values because the burning of fuels by the body in cellular respiration uses oxygen and produces carbon dioxide. This method can measure the energy used for individual components of expenditure, such as physical activity or BMR. It can also be used to estimate total energy needs, but it is not practical in free-living individuals because the equipment is too cumbersome for long-term use.

doubly-labelled water technique A method for measuring energy expenditure based on measuring the disappearance of heavy (but not radioactive) isotopes of hydrogen and oxygen in body fluids after consumption of a defined amount of water labelled with both isotopes.

isotopes Alternative forms of an element that have different atomic masses, which may or may not be radioactive.

Doubly-Labelled Water A more practical method for measuring energy needs is the **doubly-labelled water technique.** This involves having the individual ingest or be injected with water labelled with **isotopes** of oxygen and hydrogen. The labelled oxygen and hydrogen are used by the body in metabolism. The labelled hydrogen leaves the body as part of water and the labelled oxygen leaves the body as part of both water and carbon dioxide. The difference in the rates of disappearance of these 2 types of isotopes can be used to calculate the amount of carbon dioxide produced by reactions in the body.

The doubly-labelled water method does not require the individual to carry any equipment and can be used to measure energy expenditure in free-living subjects for periods up to 2 weeks. It is now the preferred method for determining the total daily energy expenditures of both healthy and clinical populations.[5,11] However, it is not helpful in determining the proportion of energy used for BMR, physical activity, or TEF.

estimated energy requirements (EER) The amount of energy recommended by the DRIs to maintain body weight in a healthy person based on age, gender, size, and activity level.

Recommendations for Energy Intake

Measurements of energy expenditure using doubly-labelled water have been used to develop formulas for estimating individuals' energy needs. These **Estimated Energy Requirements (EER)**, established by the DRIs, are the current recommendations for energy intake in Canada and the United States.[5] An EER is the amount of energy predicted to maintain energy balance in a healthy person of a defined age, gender, weight, height, and level of physical activity. EERs predict energy expenditure in normal-weight individuals. Equations that predict the amount of energy needed for weight maintenance for overweight and obese individuals are given in Appendix A. No specific RDAs or ULs for energy have been established.

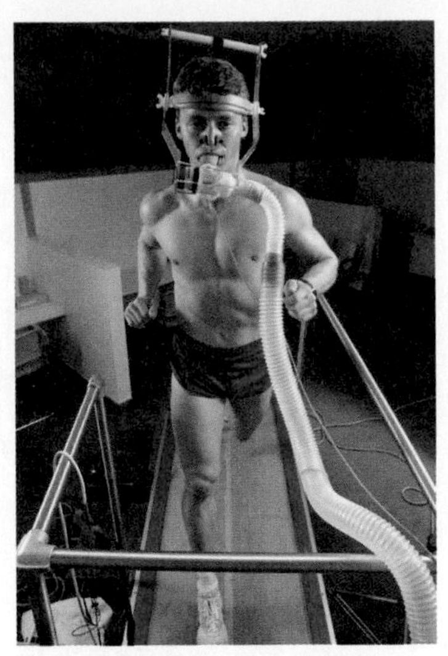

Figure 7.12 Indirect calorimetry can be used to assess the energy expended for specific activities. Subjects breathe into a mouthpiece or through a mask that is placed over their nose and mouth. The amount of O_2 and CO_2 in the inhaled and exhaled air is measured. (Stockbyte/Getty Images, Inc.)

Determining Physical Activity Level In order to calculate an individual's energy needs using the EER equations, his or her activity level must be estimated. The DRIs have defined four physical activity levels: sedentary, low active, active, and very active.[5] A "sedentary" individual is one who does not participate in any activity beyond that required for daily independent living, such as housework, homework, yard work, and gardening. To be in the "low active" category, an adult weighing 70 kg would need to expend an amount of energy equivalent to walking 1 km at a rate of 5 to 6.5 km per hour in addition to the activities of daily living. To be "active," an individual would need to perform daily exercise equivalent to 11 km at a rate of 5 to 6.5 km per hour, and to be "very active," an individual would need to perform the equivalent of walking 27 km at this rate in addition to the activities of daily living. In order to maintain a healthy weight and reduce the risk of chronic disease, physical activity at the "active" level is recommended. Not everyone has the time to walk for 1 ¾ hours every day, but if you engage in more vigorous activities, you will expend the same number of kcalories in less time and still be in the "active" category.

Activity level can be estimated by keeping a daily log of your activities and recording the amount of time spent in each. Activities can then be categorized as activities of daily living, moderate, or vigorous (**Table 7.3**). As seen in **Table 7.4**, your physical activity level is determined based on the time you spend in each category of activity. Each physical activity level is assigned a numerical **PA (physical activity) value** that can then be used in the EER calculation. It is important to carefully estimate activity level because it has a significant effect on energy needs. For example, a 30-year-old woman who is 5' 5" tall and weighs 130 lb. needs about 1,900 kcal/day if she is at the "sedentary" activity level. If she increases her activity to "active," the level recommended by the DRIs, her energy needs increase to 2,370 kcals/day.

PA (physical activity) value A numeric value associated with activity level that is a variable in the EER equations used to calculate energy needs.

Table 7.3 Categorizing Activities

Activities of Daily Living[a]	Moderate Activities[b]	Vigorous Activities[c]
Gardening (no lifting)	Bicycling (leisurely, < 16 km/h)	Aerobics (moderate to heavy)
Watering plants	Calisthenics (light, no weights)	Basketball (vigorous)
Raking leaves	Dancing	Bicycling (>16 km/h)
Mowing the lawn	Gardening/yard work (light)	Climbing (hills or mountains)
Household tasks	Golf (walking and carrying clubs)	Jogging (8 km/hr or faster)
Mopping	Hiking	Rope jumping
Vacuuming	Skating leisurely	Skating vigorously
Doing laundry	Swimming (slow)	Skiing (water, downhill, or cross-country)
Washing dishes	Walking (5.6 km/h, 10-11 min/km)	Swimming (freestyle laps)
Walking from the house to car or bus	Water aerobics	Tennis
Loading/unloading the car	Weight lifting (light workout)	Walking (7.2 km/h)
Walking the dog	Yoga	Weight lifting (vigorous effort)
		Yard work (heavy, chopping wood)

[a]It is assumed that we spend about 2.5 h/day in these types of activities.
[b]Activities that expend about 210-420 kcal/h for a 70-kg (154-lb) individual.
[c]Activities that expend more than 420 kcal/h for a 70-kg (154-lb) individual.

Table 7.4 Determining Physical Activity (PA) Values

	PA values			
	3–18 years		≥19 years	
Physical Activity Level	**Boys**	**Girls**	**Men**	**Women**
Sedentary: Engages in only the activities of daily living and no moderate or vigorous activities.	1.00	1.00	1.00	1.00
Low active: Daily activity equivalent to at least 30 minutes of moderate activity and a minimum of 15-30 minutes of vigorous activity, depending on the intensity of the activity.	1.13	1.16	1.11	1.12
Active: Engages in at least 60 minutes of moderate activities or a minimum of 30-60 minutes of vigorous activity depending, on the intensity of the activity.	1.26	1.31	1.25	1.27
Very active: Engages in at least 2.5 hours of moderate activity or a minimum of 1.0-1.75 hours of vigorous activity, depending on the intensity of the activity.	1.42	1.56	1.48	1.45

Other Factors Affecting Energy Needs Energy needs are affected by gender, height, weight, life stage, and age as well as level of physical activity. These factors are all taken into consideration in the EER equations (**Table 7.5**, or inside cover). Separate equations for men and women reflect gender differences in energy requirements. Height and weight are variables in the equations and when larger numbers are entered, calculated results reflect the higher energy needs of taller, heavier individuals. For example, an active 25-year-old man who is 180 cm (5' 11") tall and weighs 77 kg (170 lb.), requires 3,175 kcalories to maintain his weight. If the same man weighed 100 kg (220 lb.), he would need about 400 kcal/day more to maintain his weight.

Table 7.5 Calculating Your EER

- **Find your weight in kilograms (kg) and your height in metres (m):**

  Weight in kilograms = weight in pounds ÷ 2.2 lb./kg

  Height in metres = height in inches × 0.0254 in./m

  For example: 160 pounds = 160 lbs ÷ 2.2 lb./kg = 72.7 kg

  5' 9" = 69 in × 0.0254 in./m = 1.75 m

- **Estimate the amount of physical activity you get per day and use Table 7.4 to find the PA value for someone your age, gender, and activity level.**

For example, if you are a 19-year-old male who performs 40 minutes of vigorous activity a day, you are in the active category and have a PA of 1.25.

- **Choose the appropriate EER prediction equation below and calculate your EER:**

For example: if you are an active 19-year-old male:

EER = 662 – (9.53 × Age in yrs) + PA [(15.91 × Weight in kg) + (539.6 × Height in m)]

Where age = 19 yrs, weight = 72.7 kg, height = 1.75 m, Active PA value = 1.25

EER = 662 – (9.53 × 19) × 1.25([15.91 × 72.7] + [539.6 × 1.75]) = 3,107 kcal/day

Life Stage EER Prediction Equation[a]
Boys 9–18 yrs EER = 88.5 – (61.9 × Age in yrs) + PA [(26.7 × Weight in kg) + (903 × Height in m)] + 25
Girls 9–18 yrs EER = 135.3 – (30.8 × Age in yrs) × PA [(10.0 × Weight in kg) + (934 × Height in m)] + 25
Men ≥ 19 yrs EER = 662 – (9.53 × Age in yrs) + PA [(15.91 × Weight in kg) + (539.6 × Height in m)]
Women ≥ 19 yrs EER = 354 – (6.91 × Age in yrs) + PA [(9.36 × Weight in kg) + (726 × Height in m)]

[a]These equations are appropriate for determining EERs in normal-weight individuals. Equations that predict the amount of energy needed for weight maintenance in overweight and obese individuals are also available (see Appendix A).

The EER values for infants, children, and adolescents include the energy used to deposit tissues associated with growth. Beginning at age 3, there are separate EER equations for boys and girls because of differences in growth and physical activity. The EER for pregnancy is determined as the sum of the total energy expenditure of a nonpregnant woman plus the energy needed to maintain pregnancy and deposit maternal and fetal tissue. During lactation, EER is the sum of the total energy expenditure of nonlactating women and the energy in the milk produced, minus the energy mobilized from maternal tissue stores.

The impact of age on the energy needs of adults can be striking. As people get older, their EER declines. So, the 25-year-old man described above who needs 3,175 kcalories to maintain his weight when he is 25 years old, will need only 2,935 kcalories when he turns 50. If he decreases his level of activity, as many people do when they age, the number of kcalories he can eat without gaining weight drops even more—he will require only 2,385 kcalories to maintain his 77 kg (170-lb). weight at age 50 if he is sedentary. If people do not decrease their kcalorie intake to compensate for getting older and becoming less active, body weight will increase.

7.4 Body Weight and Health

Learning Objectives

- Name some health problems that are more common in overweight than in normal-weight individuals.
- List the psychological and social consequences of obesity.

Some body fat is essential for health; it provides an energy store, cushions internal organs, and insulates against changes in temperature. Too much body fat, however, can increase the risk of disease and create psychological and social problems. Individuals who have little stored fat also have a greater risk for early death than individuals whose body fat is within the normal range.[12]

Excess Body Fat and Disease Risk

Heart disease, high blood cholesterol, high blood pressure, stroke, diabetes, gallbladder disease, sleep apnea, respiratory problems, arthritis, gout, and certain types of cancers all occur more frequently in obese individuals (**Table 7.6**). These diseases or conditions are often described as **comorbidities**, that is, both obesity and a second disease state are present together.

Comorbidity Two disease states or health conditions that occur together, such as obesity and type 2 diabetes.

Table 7.6 Health Risks Associated with Excess Body Weight
Cardiovascular disease is more likely when body weight is elevated.
• Blood pressure increases as body weight increases.
• Triglyceride levels increase as body weight increases.
• LDL cholesterol increases as body weight increases.
• HDL cholesterol falls as body weight increases.
Type 2 diabetes risk increases with body weight.
• Fasting blood sugar increases with increasing body weight.
• 85% of people with type 2 diabetes are obese.
• Incidence increases as much as 30-fold with a BMI >35.
Respiratory problems are more common in overweight people.
• Sleep apnea is more common in overweight people.
• The workload of muscles used for breathing increases.
• Asthma is worse.
Gallbladder disease is more common in overweight people.
Osteoarthritis and degenerative joint disease increase with increasing weight.
Menstrual irregularities are increased in overweight women.
Cancer risk is higher in overweight people.
• Obese women are at increased risk for cancers of the endometrium, breast, cervix, and ovaries.
• Obese men are at increased risk for colorectal cancer and don't respond as well to prostate cancer treatment.
A sedentary lifestyle further increases risk.
• Obese individuals who are inactive have higher risks of illness and death.
• Inactivity increases the likelihood of developing diabetes and heart disease.

Life Cycle

In addition, the presence of these diseases increases the risk of illness and premature death that is associated with being obese. Obesity also increases the incidence and severity of infectious disease and has been linked to poor wound healing and surgical complications.[13] The magnitude of the health risks increases as the amount of excess fat rises (**Figure 7.13**).

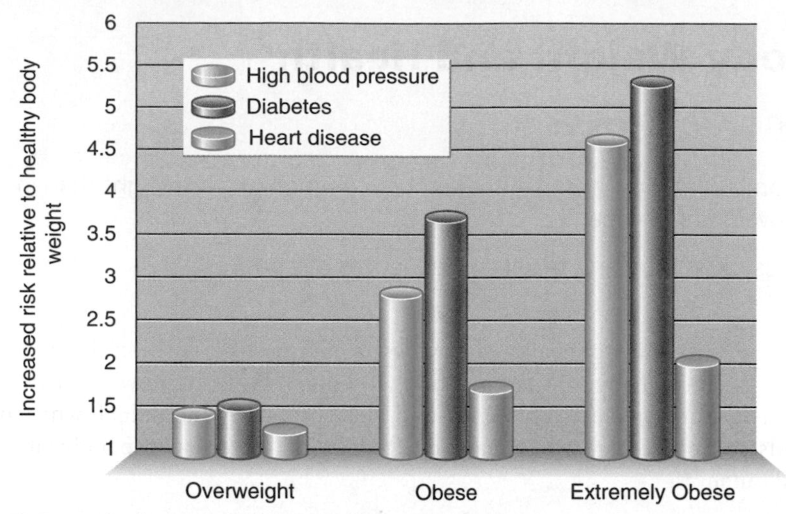

Figure 7.13 **Excess body fat and disease risk among Canadians**
The more excess body fat an individual carries, the greater the risk of a number of chronic diseases. A value of 1.0 represents the risk of having these disorders if your weight is in the healthy range; a value of 2 indicates that the risk is doubled.
Source: Statistics Canada. Nutrition: Findings from the Canadian Community Health Survey. Measured Obesity: Adult Obesity in Canada: Measured Height and Weight: 2005. Available online at http://www.statcan.gc.ca/pub/82-620-m/2005001/article/adults-adultes/8060-eng.htm#5. Accessed July 12, 2010.

Being overweight causes problems throughout life. During pregnancy, carrying excess body fat increases risks both for the mother and the fetus (see Section 14.3).[14] In children and adolescents, being overweight contributes to the development of high blood cholesterol levels, high blood pressure, and elevated blood glucose (see Section 15.3). The high rate of childhood obesity is a major threat to this generation because the longer a person is overweight, the greater the risks; those who gain excess weight at a young age and remain overweight throughout life have the greatest health risks.

The metabolic aspects of obesity and how obesity contributes to increasing the risk of disease are detailed in Focus on Obesity, Metabolism, and Disease Risk.

Psychological and Social Consequences of Obesity

Carrying excess body fat has psychological and social consequences. Our society puts a high value on physical appearance. Being thin is considered attractive and being fat is not. Those who do not conform to standards may pay a high psychological and social price. For example, overweight children are often teased and ostracized. This teasing about body weight is associated with low body satisfaction, low self-esteem, depression, social isolation, and thinking about and attempting suicide.[15,16] If obese children grow into obese adolescents and adults, and most of them do, they may be discriminated against in college admissions, in the job market, in the workplace, and even on public transportation. Obese individuals of every age are more likely to experience depression, a negative self-image, and feelings of inadequacy.[17] The physical health consequences of obesity may not manifest themselves as disease for years, but the psychological and social problems experienced by the obese are felt every day.

Health Implications of Being Underweight

Some people are naturally lean and this reduces their health risks. Research has suggested that having a low level of body fat may reduce the risk of diabetes and other chronic diseases and may even increase longevity.[18] (see Chapter 16, Science Applied: Eat Less—Live Longer?). But it is not good to be too thin. Body fat is needed for cushioning, as an insulator, and as a reserve for periods of illness. People with little energy reserves have a disadvantage during a famine or when battling a medical condition such as cancer that causes wasting and malnutrition. Therefore, despite the lower incidence of certain chronic diseases in people with low body weights, when compared to normal weight, being too lean is associated with an increased risk of early death.[12]

Life Cycle

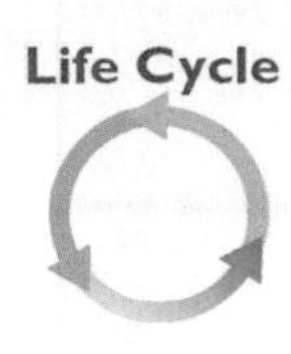

When leanness is due to intentional or forced restriction of food intake, rather than a genetic tendency to be lean, it can create severe health problems. Substantial reductions in body weight due to starvation or eating disorders reduce body fat and muscle mass, affect electrolyte balance, and decrease the ability of the immune system to fight disease. Too little body fat can cause problems at all stages of life. During adolescence, it can delay sexual development. During pregnancy, too little weight gain increases the risk that the baby will have health complications, and in the elderly, too little body fat increases the risk of malnutrition. In developed countries, socioeconomic conditions may create isolated pockets of undernutrition, but severe cases of wasting are usually a result either of self-starvation due to an eating disorder, such as anorexia nervosa (see Chapter 15: Focus on Eating Disorders), or of a disease process, such as AIDS or cancer.

7.5 Guidelines for a Healthy Body Weight

Learning Objectives

- Describe three methods used to estimate percent body fat.
- Calculate your BMI and determine if it is in the healthy range.
- Compare the health implications of excess visceral versus subcutaneous fat.

Guidelines for a healthy body weight are based on the weight at which the risk of illness and death are lowest. These risks are associated not only with body weight, but also with the amount and location of body fat, therefore, assessment of a healthy body weight must consider body composition. Despite this, body composition measures are generally not used clinically to assess health. Instead, **body mass index (BMI)**, which is calculated from body weight and height, is the most common measure used to assess the healthfulness of body weight.

body mass index (BMI) A measure of body weight in relation to height that is used to compare body size with a standard.

Lean Versus Fat Tissue

The human body is composed of lean tissue and body fat. Lean tissue, referred to as lean body mass or fat-free mass, includes bones, muscles, and all tissue except fat tissue. Body fat, or adipose tissue, lies under the skin and around internal organs. The amount of fat an individual carries and where that fat is deposited are affected by age, gender, and genetics, as well as by energy balance.

Life Cycle

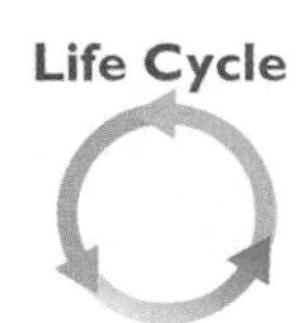

At birth, about 12% of a baby's weight is fat. This percentage increases in the first year of life and then, during childhood, as muscle mass increases, the percentage of body fat decreases. During adolescence, girls gain proportionately more fat and boys gain more muscle mass. As adults, women have more stored body fat than men so the level that is healthy for women is somewhat higher. A healthy level of body fat for young adult women is between 21%-32% of total weight; for young adult men, it is between 8%-19%.[19] There is an increase in body fat during pregnancy to provide energy stores for the mother and fetus. With aging, lean body mass decreases; between the ages of 20-60, body fat typically doubles even if body weight remains the same. This occurs regardless of energy intake. Some of this loss of lean body mass can be prevented by strength training.[20]

Assessing Body Composition

A number of techniques are available for assessing body composition. Many require expensive equipment and must be performed in a research setting by trained technicians. Others are more portable, so they are more appropriate for use in a clinic, office, or health club.

Bioelectric Impedance Analysis **Bioelectric impedance analysis** is the most popular way to measure body composition. It estimates body fat by directing a painless, low-energy electrical current through the body. The difference between the current applied to the first electrode and the current that reaches the second electrode is used to determine the resistance, or the amount of current a substance will stop. Because fat is a poor conductor of electricity, it offers

bioelectric impedance analysis A technique for estimating body composition that measures body fat by directing a low-energy electric current through the body and calculating resistance to flow.

Figure 7.14 Hand-held bioelectric impedance devices such as this one are available for home use. (Yoav Levy/Phototake/Alamy)

skinfold thickness A measurement of subcutaneous fat used to estimate total body fat.
subcutaneous fat Adipose tissue that is located under the skin.

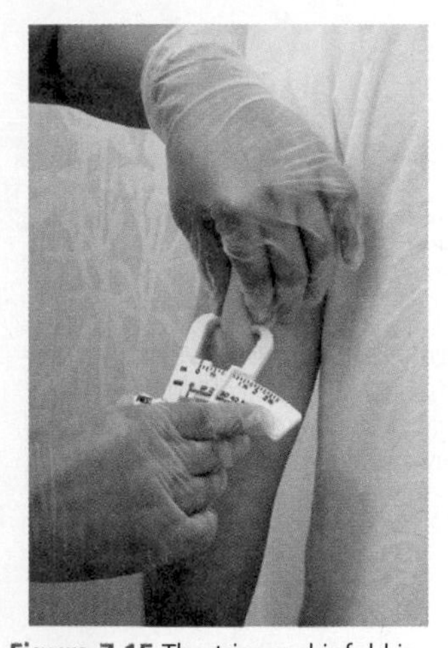

Figure 7.15 The triceps skinfold is measured at the midpoint of the back of the arm. This measure of the thickness of the fat layer under the skin can be used to estimate body fat. (3660 Group Inc./Getty Images)

underwater weighing A technique that uses the difference between body weight underwater and body weight on land to estimate body density and calculate body composition.

Figure 7.16 Underwater weighing is an accurate technique for determining body composition that relies on measurements of body weight on land and under water. (Ultimate Test Labs)

resistance to the current. Thus, the amount of resistance to current flow is proportional to the amount of body fat. The equipment needed for impedance measurements is inexpensive and the process is quick and painless (**Figure 7.14**). However, because impedance is affected by the amount of water in the body, measurements must be performed when the gastrointestinal tract and bladder are empty and body hydration is normal. Measurements performed within 24 hours of strenuous exercise are not accurate because body water is low due to losses in sweat.

Skinfold Thickness Measurements of **skinfold thickness** at various locations on the body can also be used to assess body composition. Skinfold thickness is measured with calipers and is used to assess the amount of **subcutaneous fat** (**Figure 7.15**). Measurements are taken from one or more standard locations. The most common sites are the triceps (the area over the muscles on the back of the upper arm) and the subscapular area (just below the shoulder blade). Mathematical equations are then used to estimate percent body fat from these measurements. Skinfold measurements are noninvasive and, when performed by a trained individual, can accurately predict body fat in normal-weight individuals. These measures are more difficult to perform and less accurate in obese and elderly subjects.

Underwater Weighing An accurate, noninvasive technique for assessing body composition is **underwater weighing**, which involves weighing an individual both on land and in the water. The difference between these two weights can be used to determine body volume and body density, which is proportional to fat-free mass. The percentage of body fat can then be determined using standardized equations. To measure underwater weight, subjects must sit on a scale, expel the air from their lungs, and be lowered into a tank of water (**Figure 7.16**). Although this method is accurate, it requires special equipment and cannot be used for some groups such as small children or frail adults. A newer method for estimating body composition measures air displacement rather than water displacement to determine body volume. The individual is placed in an air-filled chamber (known as the BOD POD) rather than in water. It is accurate and more convenient than underwater weighing.[21]

Dilution Methods Body fat can also be assessed by using the principle of dilution. Because water is present primarily in lean tissue and not in fat, a water-soluble isotope can be ingested or injected into the bloodstream and allowed to mix with the water throughout the body. The concentration of the isotope in a sample of body fluid, such as blood, can then be measured. The extent to which the isotope has been diluted can be used to calculate the amount of lean tissue in the body, and body fat can then be calculated by subtracting lean weight from total body weight. Another technique measures a naturally occurring isotope of potassium. Because potassium is found primarily in lean tissue, a measure of the amount of this isotope in the body can be used to determine the total amount of body potassium, which can then be used to estimate the amount of lean tissue. Dilution techniques are expensive and invasive, usually requiring injections. They are used primarily for research purposes.

Radiologic Methods A variety of radiologic technologies has been used to assess body composition. These are less invasive but more expensive than dilution methods. Computerized tomography (CT), generally employed as a diagnostic technique, can be used to visualize fat and lean tissue. CT is more accurate than underwater weighing, skinfold measures, and total body potassium for evaluation of body composition, and is particularly useful for measuring the amount of **visceral fat** (**Figure 7.17**).[22] Dual-energy X-ray absorptiometry (DXA) is another method that uses low-energy X-rays for assessing body composition. A single investigation can accurately determine total body mass, bone mineral mass, and the amount and percentage of body fat. Another method, magnetic resonance imaging (MRI), uses magnetic fields to create an internal body image. MRI can be used to accurately estimate the amount of abdominal fat, which is associated with the risk of heart disease and other chronic diseases.

Body Mass Index

The current standard for evaluating body weight is body mass index (BMI). Although BMI does not directly assess percent body fat, BMI values correlate well with body fat in most

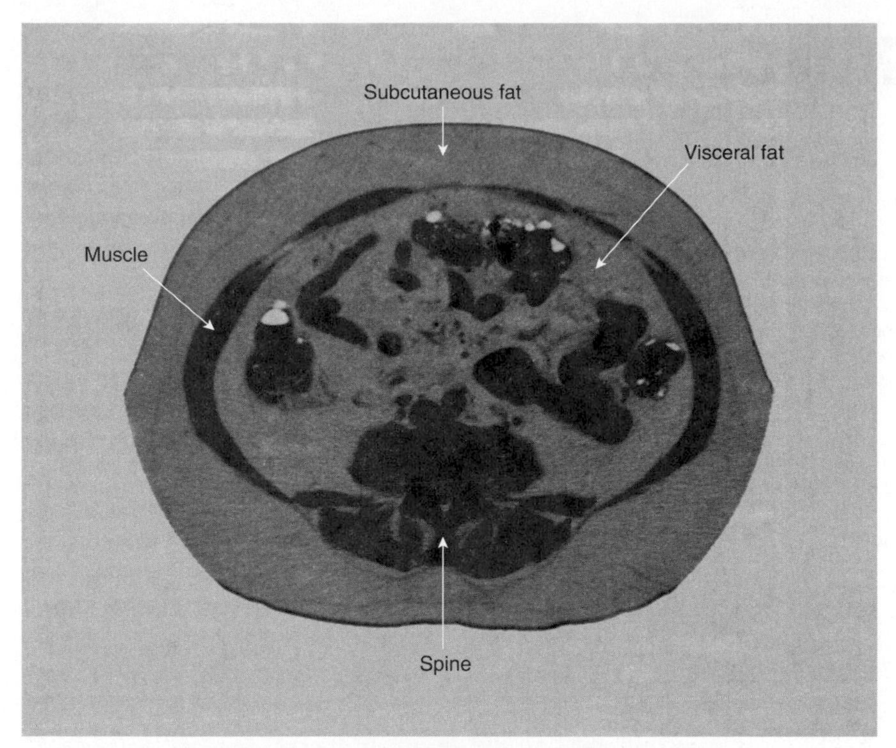

Figure 7.17 Visceral and subcutaneous body fat A CT scan taken at waist level in an overweight female shows that fat in the abdominal region is located both under the skin (subcutaneous) and around the internal organs (visceral). (Courtesy Michael F. Smolin)

visceral fat Adipose tissue that is located in the abdomen around the body's internal organs.

people.[23,24] BMI is calculated from a ratio of weight to height according to either of the following equations:

$$\text{BMI} = \text{weight in kg}/(\text{height in m})^2$$

or,

$$\text{BMI} = \text{weight in lb.}/(\text{height in inches})^2 \times 703^*$$

For example, someone who is 6 feet (72 in or 1.83 m) tall and weighs 180 lb. (81.8 kg) has a BMI of 24.5 kg/m^2. BMI calculators are also available online (See Appendix B).

What Is a Healthy BMI? A healthy body weight is defined as a BMI of 18.5-24.9 kg/m^2. In general, people with a BMI within this range have the lowest health risks. Underweight is defined as a BMI of less than 18.5 kg/m^2, overweight is identified as a BMI of 25-29.9 kg/m^2, and obese as a BMI of 30 kg/m^2 or greater.[24] A BMI of 40 or over is classified as extreme or morbid obesity. See Appendix B.

Limitations of BMI Even though BMI correlates well with the amount of body fat, it is not a perfect tool for evaluating the health risks associated with obesity. This is particularly true in athletes who have highly developed muscles; their BMI may be high because they have an unusually large amount of lean body mass. In these individuals, BMI is high, but body fat and hence disease risk are low (**Figure 7.18**). BMI is also not suitable for evaluating weight in pregnant and lactating women because of their rapidly changing weight and body composition (see Section 14.1). It is also less accurate in individuals who have lost muscle, such as many older adults. Because of these limitations, BMI should not be the only measure used to determine nutritional health and fitness. For example, someone who is in the overweight category based on BMI but consumes a healthy diet and exercises regularly may be more fit and have a lower risk of chronic disease than someone with a BMI in the healthy range who is sedentary and eats a poor diet. While BMI has some limitations, it is nonetheless, an important tool for researchers interested in determining relationships between body weight and disease risk. (See Critical Thinking: Obesity, BMI, and Cancer: Establishing a Relationship).

Figure 7.18 Although wrestler/actor Hulk Hogan has a BMI of 30.3 kg/m^2, which falls into the category of obese, it probably does not indicate that he has excess body fat or an increased risk of disease. (Hubert Boesl/dpa/Landov)

* Different constants are used in the scientific literature (700, 704.5) but the differences are insignificant.

CRITICAL THINKING:

Obesity, BMI, and Cancer: Establishing a Relationship

BMI is an important measure of obesity. To answer the question of whether there was an association between obesity, as measured by BMI, and cancer, scientists conducted prospective cohort studies (see Section 1.5 and Chapter 2: Science Applied: Does Following a Healthy Dietary Pattern Reduce the Risk of Disease? for a review of observational studies and relative risk.). Studies like the two described below helped to establish a link between BMI and two types of cancer, colon cancer and breast cancer. (Also see Focus on Obesity, Metabolism, and Disease Risk).

One such study examined the link between death from colon cancer and body mass index.[1] More than 496,000 middle-aged women and 379,000 middle-aged men were enrolled in the study. They were cancer-free at the start of the study and provided important health information to the researchers, including body weight and height (from which body mass index can be calculated). The participants were followed for 12 years, during which time the deaths from colon cancer were recorded. The relative risk of death from colon cancer was determined and the following results obtained:

BMI	Relative Risk: Men	Relative Risk: Women
<25	1.00	1.00
25- 29.9	1.34*	1.08
>30	1.75*	1.25*
*statistically significant difference from RR = 1.0		

In a second cohort study, researchers enrolled almost 39,000 women, ages 55-74 years.[2] They were cancer-free at the start of the study and were followed for 10 years, during which time the numbers of breast cancer cases were recorded. The following results with obtained:

BMI	Relative Risk:
<25	1.00
25-29.9	1.3*
>30	1.3*
* statistically significant difference from RR = 1.0	

Critical Thinking Questions

What do you conclude about the relationship between obesity and death from colon cancer?*

How does the risk between men and women differ?

The results shown were age-adjusted (see Section 1.5). Why is this important?

What do you conclude about the relationship between obesity and risk for breast cancer?

*Answers to all Critical Thinking questions can be found in Appendix J.

[1]Murphy, T.K., Calle, E.E., Rodriguez, C., et al. Body mass index and colon cancer mortality in a large prospective study. *Am. J. Epidemiol.* 152(9):847-54, 2000.

[2]Lahmann, P.H., Hoffmann, K., Allen N., et al. Body size and breast cancer risk: findings from the European Prospective Investigation into Cancer and Nutrition (EPIC). *Int. J. Cancer.* 111(5):762-71, 2004.

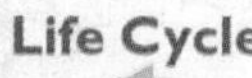

Location of Body Fat

Where body fat is stored affects the health risks associated with having too much. Subcutaneous fat carries less risk than visceral fat, which is deposited around the organs in the abdomen. An increase in visceral fat is associated with a higher incidence of heart disease, high blood pressure, stroke, and diabetes.[25–27] This is because free fatty acids are released from visceral fat more readily than subcutaneous fat and visceral fat is more prone to becoming insulin resistant.[28] Generally, fat in the hips and lower body is subcutaneous, whereas fat deposited around the waist in the abdominal region is primarily visceral fat. Therefore, people who carry their excess fat around and above the waist have more visceral fat. Those who carry their extra fat below the waist in the hips and thighs have more subcutaneous fat. In the popular literature, these body types have been dubbed apples and pears, respectively (**Figure 7.19**).

Where your extra fat is deposited is determined primarily by your genes.[29] Visceral fat storage is more common in men than women. After menopause, visceral fat increases in women. Other factors that affect the amount of visceral fat include stress, tobacco use, and alcohol consumption, all of which predispose people to visceral fat deposition; physical activity is a factor that reduces it.

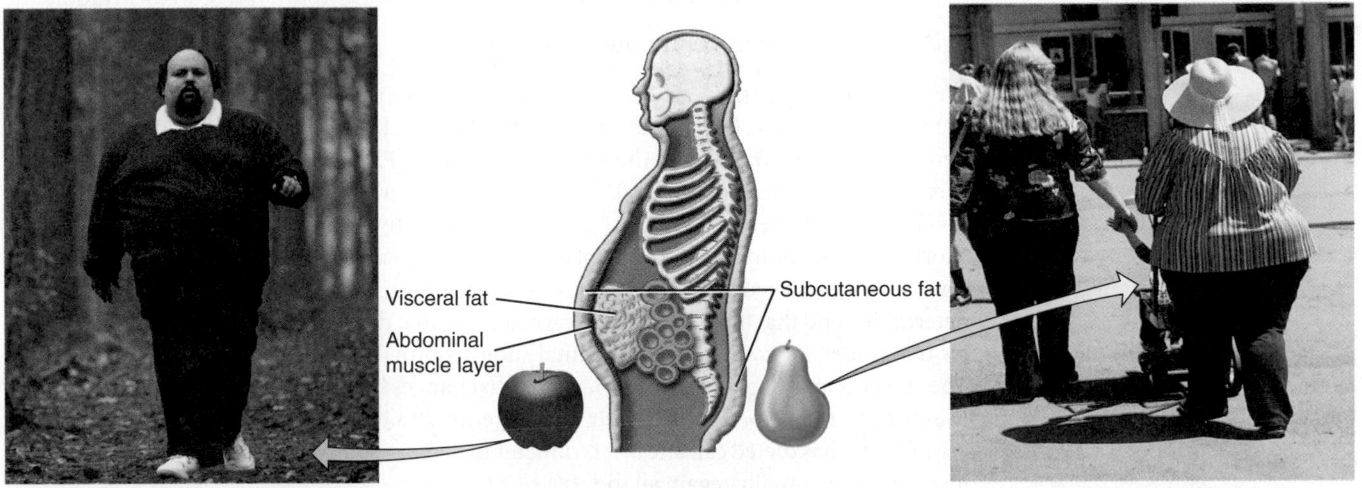

Figure 7.19 The location of body fat can influence health risk
Overweight individuals with apple-shaped body types deposit more fat in the abdominal region. They are at greater risk of developing heart disease and diabetes than are those with pear-shaped body types, who deposit more fat in the hips and thighs, where it is primarily subcutaneous. (©Corbis; Tom McHugh/Photo Researchers, Inc.)

Distinguishing the relative amounts of visceral and subcutaneous fat requires sophisticated imaging techniques. However, the risk associated with visceral fat deposition can be estimated by measuring waist circumference. A waist circumference greater than 102 cm for men and 88 cm for women is associated with an increased risk, and these cutoffs, based on studies of white populations, have been used by Health Canada.[30] In 2006, Canadian Clinical Practice Guidelines on the Management and Prevention of Obesity in Adults and Children were published.[31] This report recommended different waist circumference cutoffs **(Table 7.7)** that vary with ethnic origin and are used in combination with other medical indicators (e.g., cholesterol levels, blood pressure, etc.) to assess risk.

Table 7.7 Waist Circumference Cutoffs Based on Ethnic Origin

	Waist circumference, in cm	
Country or ethnic group	**Men**	**Women**
European; Sub-Saharan Africa*; Eastern and Middle East (Arab)*	≥ 94	≥ 80
South Asian, Chinese; South and Central American**	≥ 90	≥ 80
Japanese	≥ 85	≥ 90

*More specific data for this ethnic group is currently unavailable. The Canadian Medical Association (CMA) recommends this ethnic group use European cutoff points until the data becomes available.

**More specific data for this ethnic group is currenly unavailable. The CMA recommends this ethnic group use South Asian cutoff points until the data becomes available.

Source: Obesity Canada Clinical Practice Guidelines Expert Panel. Canadian clinical practice guidelines on the management and prevention of obesity in adults and children.[31]

7.6 Regulation of Energy Balance

Learning Objectives

- Describe what is meant by a set point for body weight.
- List 4 physiological signals that determine whether you feel hungry or full.
- Explain why leptin levels might be higher in an obese than in a lean individual.
- Discuss how genes contribute to obesity.

We inherit our body shape and characteristics from our parents. Some of us inherit tall, slender bodies with long, thin bones. Others inherit stocky bodies with short, wide bones. Some people have broad hips and others broad shoulders (**Figure 7.20**). Some people naturally carry a larger amount of body fat than others. These inherited characteristics don't change much over time. Even body weight tends to remain relatively constant for long periods despite short-term fluctuations in the amount of exercise we get and the amount of food we consume. This can be explained by the **set-point theory** which suggests that body weight is genetically determined and that there are internal mechanisms that defend against weight change.[32] Studies that under- or overfeed experimental subjects provide support for the set-point theory. The subjects lose or gain weight in response to changes in intake, but when they are allowed to return to their normal diet, their weight returns to its original "set" level.[33] Likewise, as anyone who has dieted can attest, it is difficult to decrease body weight, and most people who lose weight eventually regain all they have lost.

set-point theory The theory that when people finish growing, their weight remains relatively stable for long periods despite periodic changes in energy intake or output.

Figure 7.20 The genes we inherit from our parents are important determinants of our body size and shape. (© istockphoto.com/ Troels Graugaard; Alan Powdrill/Getty Images, Inc.)

If body weight is regulated at a particular set-point, then why are so many of us getting fatter? Despite experimental support for the existence of a set-point, it appears that the mechanisms that defend body weight are not absolute. Changes in physiological, psychological, and environmental circumstances do cause the level at which body weight is regulated to change, usually increasing it over time. For example, body weight increases in most adults between the ages of 30-60 years, and after child-bearing, most women return to a weight that is 0.5-1 kg higher than their prepregnancy weight. This suggests that the mechanisms that defend against weight loss are stronger than those that prevent weight gain.

Obesity Genes

Genes involved in regulating body fatness have been called **obesity genes** because an abnormality in one or more of these could result in obesity. More than 300 genes and regions of human chromosomes have been linked to body weight regulation and, hence, obesity.[34] These genes are responsible for the production of proteins that affect how much food people

obesity genes Genes that code for proteins involved in the regulation of food intake, energy expenditure, or the deposition of body fat. When they are abnormal, the result is abnormal amounts of body fat.

eat, how much energy they expend, and how efficiently their body fat is stored. The combined effects of all these genes help to determine and regulate what people weigh and how much fat they carry. The influence of genes on body weight was demonstrated dramatically by a study done with identical twins at Laval University. During the study, the pairs of identical twins were overfed to the same extent. Each set of twins tended to gain the same amount of weight and to deposit fat in the same parts of their bodies. In contrast, large differences were seen between sets of twins—some of the twin sets gained only 3.2 kg, whereas others gained as much as 13.2 kg[29] (**Figure 7.21**).

Figure 7.21 Identical twins inherit the same genes and thus tend to gain the same amount of weight and deposit fat in the same locations. (Dennis MacDonald/Age Fotostock America, Inc.)

Mechanisms for Regulating Body Weight

To regulate weight and fatness at a constant level, the body must be able to respond both to changes in food intake that occur over a short time frame as well as to more long-term changes in the amount of stored body fat. Signals related to food intake affect **hunger** and **satiety** over a short period of time—from meal to meal—whereas signals from the adipose tissue trigger the brain to adjust both food intake and energy expenditure for long-term regulation.

hunger Internal signals that stimulate one to acquire and consume food.

satiety The feeling of fullness and satisfaction, caused by food consumption, that eliminates the desire to eat.

Short-Term: Regulating Food Intake from Meal-to-Meal How do you know how much to eat for breakfast, or when it is time to eat lunch? To some extent, your level of hunger or satiety determines how much you eat at each meal. These physical sensations that tell people to eat or stop eating are triggered by signals from the GI tract, levels of circulating nutrients, and messages from the brain.[35] Some signals are sent before food is eaten, some are sent while food is in the GI tract, and some occur once nutrients are circulating in the bloodstream.

The simplest type of signal about how much food has been eaten comes from local nerves in the walls of the stomach and small intestine that sense the volume or pressure of food and send a message to the brain to either start or stop eating. Once food is consumed, the presence of nutrients in the GI tract sends information directly to the brain and triggers the release of gastrointestinal hormones that signal eating to stop. Once nutrients have been absorbed, circulating levels of nutrients, including glucose, amino acids, ketones, and fatty acids, are monitored by the brain and may trigger signals to eat or not to eat.[36] Nutrients that are taken up by the brain may affect neurotransmitter concentrations, which then affect the amount and type of nutrients consumed. For example, some studies suggest that when brain levels of the neurotransmitter serotonin are low, carbohydrate is craved, but when it is high, protein is preferred.[36] Absorbed nutrients also affect metabolism in the liver because absorbed water-soluble nutrients go there directly. Changes in liver metabolism, in particular the amount of ATP, are believed to be involved in regulating food intake.

There are many different hormonal signals that regulate different aspects of food intake. For instance, the hormone insulin is released by the pancreas in response to the intake of carbohydrate. Insulin allows glucose to be taken up by cells, thereby reducing circulating levels of glucose and increasing hunger. The hormone **ghrelin** may be the reason people typically feel hungry around lunchtime regardless of when and how much they had for breakfast. It is produced by the stomach and is believed to stimulate the desire to eat meals at usual times. Levels rise an hour or two before a meal and drop to very low levels after a meal. Levels have been found to rise in people who have lost weight, increasing their desire to eat more.[37] Cholecystokinin, released when chyme enters the small intestine, causes us to stop eating by inducing satiety. Another hormone that causes a reduction in appetite is peptide PYY. It is released from the GI tract after a meal and the amount released is proportional to the kcalorie content of a meal.[38]

ghrelin A hormone produced by the stomach that stimulates food intake.

Psychological factors can also affect hunger and satiety. For example, some people eat for comfort and to relieve stress. Others may lose their appetite when these same emotions are felt. Psychological distress can alter the mechanisms that regulate food intake.

Long-Term: Regulating the Amount of Body Fat Short-term regulators of energy balance affect the size and timing of individual meals, but if a change in input is sustained over a long period it can affect long-term energy balance and, hence, body weight and fatness. To regulate the amount of fat at a set level, the body must be able to monitor how much fat is present. This information is believed to come from hormones, such as insulin and **leptin**, which are secreted in proportion to the amount of body fat.[39] Insulin is secreted from the pancreas when blood glucose levels rise; its circulating concentration is proportional to the amount

leptin A protein hormone produced by adipocytes that signals information about the amount of body fat.

of body fat. Insulin interacts with the hypothalamus to reduce food intake and body weight, and insulin levels are believed to affect the amount of leptin produced and secreted. Leptin is a hormone that is produced by the adipocytes and acts in the hypothalamus. The amount of leptin produced is proportional to the size of adipocytes—more leptin is released as fat stores increase. Leptin exerts its effect on food intake and energy expenditure by binding to leptin receptors present in the hypothalamus. This triggers mechanisms that affect energy intake and expenditure. When leptin levels are high, mechanisms that increase energy expenditure and decrease food intake are stimulated, and pathways that promote food intake and hence weight gain are inhibited. When fat stores shrink, less leptin is released. Low leptin levels in the brain allow pathways that decrease energy expenditure and increase food intake to become active.[39] Unfortunately, leptin regulation, like other regulatory mechanisms, is much better at preventing weight loss than at defending against weight gain. Obese individuals generally have high levels of leptin, but these levels are not effective at reducing calorie intake and increasing energy expenditure.[40] This lack of responsiveness to leptin is referred to as **leptin resistance** and is analogous to insulin resistance. (**Figure 7.22**; see Science Applied: Leptin: Discovery of an Obesity Gene).

Leptin resistance A lack of responsiveness to the hormone leptin; characterized, in obesity, by high levels of leptin in the blood but the lack of response to the action of leptin, which is to decrease energy intake and increase energy expenditure.

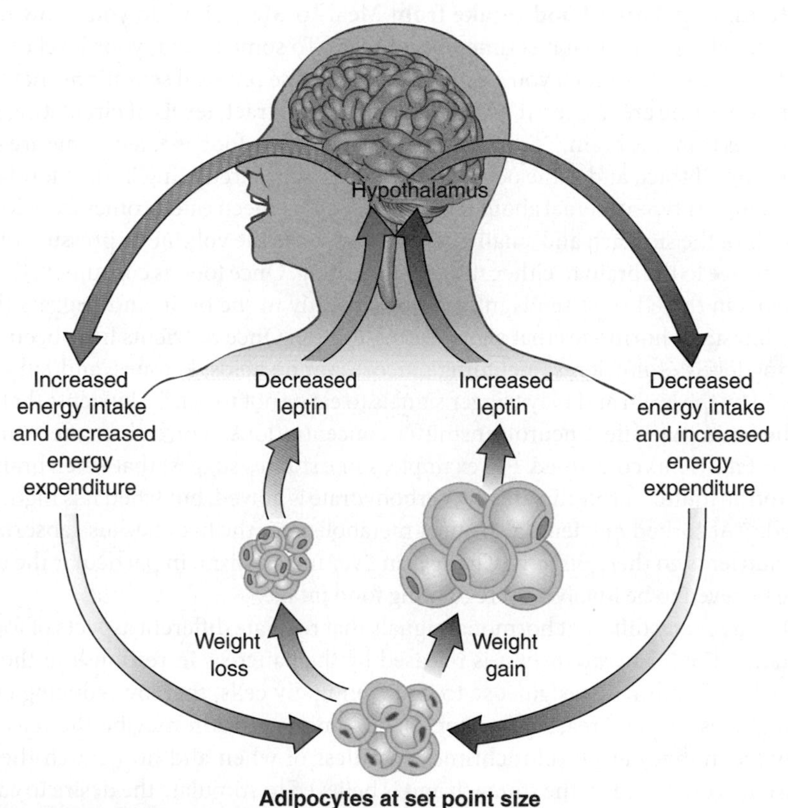

Figure 7.22 Leptin regulation of body fat Changes in the size of the adipocytes affect the amount of leptin released. The amount of leptin reaching the hypothalamus determines the response and helps return body fat stores to a set level.

Hormonal signals involved in the long-term regulation of body weight act in the brain not only to favor shifts in energy balance, but also to affect the sensitivity of the brain to short-term signals of energy balance. For example, during weight loss, low levels of these hormones are hypothesized to decrease the efficacy of satiety signals, suppress pathways that cause weight loss, and activate pathways that contribute to weight gain.[41]

How Genes Contribute to Obesity

When a gene is defective, the protein it codes for is not made or is made incorrectly. When an obesity gene, such as the gene for leptin, is defective, the signals to decrease food intake and/or increase energy expenditure are not received, and weight gain results. A few cases of human obesity have been linked directly to defects in the genes for leptin and leptin receptors,[42] but mutations in single genes such as these are not responsible for most human obesity. Rather,

SCIENCE APPLIED

Leptin: Discovery of an Obesity Gene

A discovery made by Dr. Jeffrey Friedman and colleagues in 1994 brought hope to millions of people. Perhaps the cause of obesity had been found and a cure might be close behind. Was relief in sight for those who suffer from the physical and social consequences of obesity?

Dr. Friedman's work began with a strain of mice called *ob* for obese. *Ob* mice become grossly obese, gaining up to three times the normal body weight. The *ob* strain arose spontaneously in 1950 in the mouse colony at the Jackson Laboratory in Bar Harbor, Maine. Friedman and colleagues unraveled the cause for the obesity in this strain of mice when they identified and cloned the gene that was responsible.[1]

Researchers used a series of breeding experiments to localize the gene to a particular stretch of DNA. They then looked to see if any of the genes in this stretch of DNA were expressed in adipose tissue. The search yielded a single gene. Evidence that this gene was involved in the regulation of body weight was obtained by examining the gene and the protein it codes for in the *ob* mice. Researchers found that this protein, which they named leptin, was either not produced or produced in an inactive form in the obese mice. Soon afterward, a similar gene was identified in humans.

Optimism about the role of the protein hormone leptin in human obesity was so great that a biotechnology firm (Amgen) paid $25 million for the commercial rights to leptin in the hope that it could be used to treat human obesity. Those hopes grew even higher when Friedman and colleagues were able to demonstrate that injections of the hormone could restore the genetically obese mice to normal weight.[2,3] Unfortunately, the role of leptin in human obesity has not lived up to expectations. Mutations in this gene are not responsible for most human obesity.[4] In fact, obese humans generally have high blood leptin levels.[5] High doses of leptin administered to obese humans produces only modest weight loss.[6]

The leptin receptor—a protein in the brain to which leptin must bind to produce weight reduction—was also identified.[7] The fact that obese humans have high levels of leptin suggested that the cause of human obesity might involve an abnormality in leptin receptors. If leptin receptors were defective, the leptin produced would have no place to bind and would not be able to signal mechanisms to promote weight reduction. Thus far, however, defective leptin receptors have not been found to be an important cause of human obesity.[8]

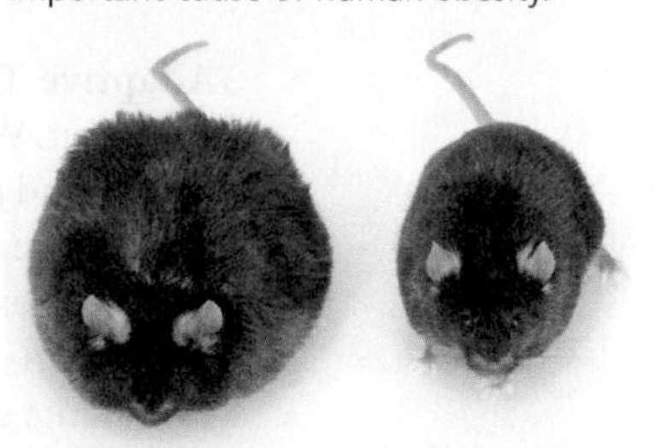

A mouse with a defect in the leptin gene (*ob*) may weigh three times as much as a normal mouse. Both of these mice have defective *ob* genes but the one on the right was treated with leptin injections. (© AP/Wide World Photos)

Continued study of the role of leptin in obesity has confirmed that it is an important signal involved in the long-term regulation of body fat, but it does not act alone. There are many steps, involving many genes, that occur between the production of leptin and alterations in food intake and energy expenditure. Researchers have discovered about a dozen molecules that interact with leptin in the brain to control appetite.[9] For example, neuropeptide Y and melanin-concentrating hormone boost appetite, whereas alpha-melanocyte stimulating hormone blunts appetite, and a protein called SOCS3 reduces the sensitivity of leptin receptors.

Despite the fact that the identification of leptin has not produced a cure for human obesity, its discovery lit up the field. This research was an important advance in our understanding of the genetics of body weight regulation. Continued work will someday answer the questions that remain about why some of us are obese and some of us are lean.

[1] Zhang, Y., Proenca, R., Maffei, M. et al. Positional cloning of the mouse obese gene and its human homologue. *Nature* 372:425–432, 1994.

[2] Halaas, J. L., Gajiwala, K. S., Maffei, M. et al. Weight-reducing effects of the plasma protein encoded by the obese gene. *Science* 269:543–546, 1995.

[3] Pelleymounter, M. A., Cullen, M. J., Baker, M. B. et al. Effects of the obese gene product on body weight regulation in *ob/ob* mice. *Science* 269:540–543, 1995.

[4] Montague, C. T., Farooqi, I. S., Whitehead, J. P. et al. Congenital leptin deficiency is associated with severe early onset obesity in children. *Nature* 387:903–908, 1997.

[5] Considine, R. V., Sinha, M. K., Heiman, M. L. et al. Serum immunoreactive-leptin concentrations in normal weight and obese humans. *N. Engl. J. Med.* 334:292–295, 1996.

[6] Gura, T. Obesity research: Leptin not impressive in clinical trial. *Science* 286:881–882, 1999.

[7] Tartaglia, L. A., Dembski, M., Weng X. et al. Identification and expression cloning of a leptin receptor, OB-R. *Cell* 83:1263–1271, 1995.

[8] Tsigos, C., Kyrou, I., and Raptis, S. A. Monogenic forms of obesity and diabetes mellitus. *J. Pediatr. Endocrinol. Metab.* 15:241–253, 2002.

[9] Gura,T. Tracing leptin's partners in regulating body weight. *Science* 287:1738–1741, 2000.

variations in many genes interact with one another and affect metabolic rate, food intake, fat storage, and activity level. These in turn affect overall body shape and size.

Thrifty Metabolism Many overweight people contend that they eat very few kcalories and yet continue to gain weight. This would imply that their energy expenditure is less than in normal-weight individuals. One possible explanation for this is that overweight individuals inherited a thrifty metabolism. An individual with a thrifty metabolism theoretically uses energy very efficiently so that more of the energy they consume is converted into ATP or deposited in energy stores than in someone with a less efficient metabolism. They would therefore need to eat less to maintain their body weight. Throughout human history, starvation has threatened survival. Over time, the human body has evolved ways to conserve body fat stores and prevent weight loss. Individuals with the "thriftiest" metabolism would have been more likely to survive. In Canada today, however, food is abundant, so people who inherited these "thrifty genes" are more likely to be obese.

Adaptive Thermogenesis The body has mechanisms that help keep weight at a specific set-point. When people overeat occasionally, their metabolism speeds up to burn the extra energy and prevent weight gain.[43–45] Conversely, when body weight is reduced by restricting food intake, energy expenditure decreases to conserve energy.[33] These changes in the amount of energy expended in response to changes in circumstance, such as over- or undereating, changes in environmental temperature, or trauma, are referred to as **adaptive thermogenesis**. Some studies found the drop in BMR seen with weight reduction was greater in obese than in lean subjects, and the increase in BMR seen with weight gain was less in obese than in lean subjects.[33] It has been proposed that these differences in the adaptive responses of lean versus obese subjects may help explain why some people gain weight more easily.

adaptive thermogenesis The change in energy expenditure induced by factors such as changes in ambient temperature and food intake.

Futile Cycling Several biochemical mechanisms have been proposed to explain adaptive thermogenesis. The first is substrate cycling or futile cycling, which wastes energy by allowing opposing biochemical reactions to occur simultaneously. For example, a molecule is formed, consuming ATP, and then is quickly broken down again. Energy is consumed but there is no net change in the number of molecules in the body, and therefore no storage of energy as fat.

Brown Adipose Tissue A second way that excess energy might be dissipated is by separating or uncoupling the electron transport chain from the production of ATP. When this occurs, energy is lost as heat rather than being used to produce ATP. For example, the increase in energy expenditure that occurs when mice are injected with leptin is believed to be due to the stimulation of receptors on a specialized type of adipose tissue called **brown adipose tissue**. Brown adipose tissue can waste energy as heat. This tissue contains many more mitochondria than regular white adipose tissue, and these mitochondria can be uncoupled from the electron transport chain allowing the energy in food to be released as heat. In rats, brown adipose tissue generates heat to prevent weight gain during overfeeding and to provide warmth when the ambient temperature is low. Significant amounts of brown adipose tissue are present in human infants, but until recently it was believed that adults did not have enough brown adipose tissue for it to be relevant physiologically. Newer technology now allows researchers to measure the amount of brown adipose tissue in adults. In a study of young men, brown adipose tissue activity was found to be lower in overweight and obese men, suggesting that it may play a role in human obesity.[46]

brown adipose tissue A type of fat tissue that has a greater number of mitochondria than the more common white adipose tissue. It can waste energy by producing heat.

Level of Activity Activity burns kcalories; this is true whether the activity is planned exercise, or NEAT activities such as housework, walking between classes, fidgeting, and moving to maintain posture. How active you are is affected by your genes and your personal choices. Some people may gain weight more easily because they inherit a tendency to expend less energy on activity.[45] Even if they spend the same amount of time exercising as a lean person, their total energy expenditure may be lower because they expend less energy for NEAT activities (**Figure 7.23**). The impact of NEAT on weight gain was demonstrated by a study that overfed normal-weight individuals. There was a 10-fold variation in the amount of fat they gained. Some subjects were able to increase energy expenditure to a greater extent and so gained less fat. About two-thirds of the increase in energy expenditure that occurred with over-feeding was found to be due to an increase in unplanned exercise.[47] Those who gained the least weight had the greatest levels of involuntary exercise. The mechanisms that cause some people to respond to excess energy intake by becoming restless and increasing NEAT activity while others remain lethargic are still not understood.

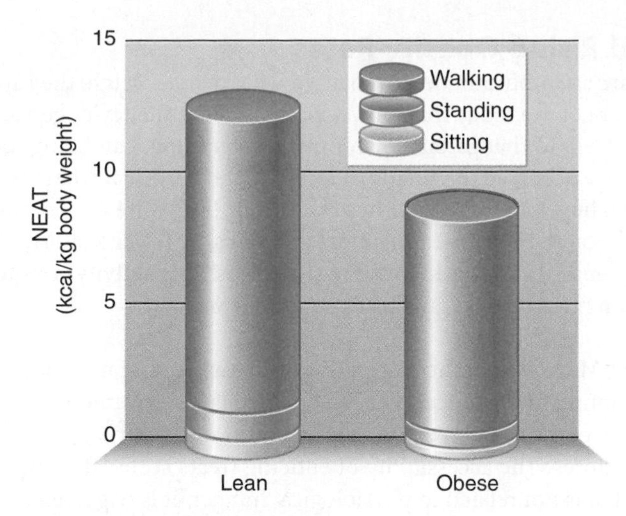

Figure 7.23 NEAT in obese versus lean individuals
Sedentary obese individuals were found to stand and walk less than sedentary lean individuals and therefore expended about 350 kcal/day fewer than their lean counterparts.

Source: Levine, J. A., Lanningham-Foster L. M, McCrady S. K., et al. Interindividual variation in posture allocation: Possible role in human obesity. Science 307:584–586, 2005.

7.7 Why Are Canadians Getting Fatter?: Genes Versus Lifestyle

Learning Objectives

- Compare the impact of genetics and lifestyle on the obesity epidemic.
- Discuss strategies to reduce the incidence of obesity.

The genes you inherit are an important determinant of what you weigh. If one or both of your parents is obese, your risk of becoming obese is increased. Individuals with a family history of obesity are two to three times more likely to be obese, and the risk increases with the magnitude of the obesity.[48] But even if you inherit genes that predispose you to being overweight, your actual weight is determined by the balance between the genes you inherit and the lifestyle choices you make. By studying identical twins, researchers have been able to determine that about 75% of the variation in BMI between individuals can be attributed to genes.[49] This means that the remaining 25% is determined by the environment in which you live and the lifestyle choices you make. Someone with a genetic predisposition to obesity who carefully monitors his or her diet and exercises regularly may never be obese, but someone with no genetic tendency toward obesity who consumes a high-energy diet and gets little exercise may end up overweight. When genetically susceptible individuals find themselves in an environment where food is appealing and plentiful and physical activity is easily avoided, obesity is a likely outcome.

An example of human obesity that clearly demonstrates the interaction of a genetic predisposition and an environment that is conducive to obesity is the Pima Indians living in Arizona. The incidence of obesity in this population is much higher than in the general U.S. population. Studies have identified a number of genes that may be responsible for this group's tendency to store more body fat. Their current lifestyle demands little physical activity and includes many high-kcalorie foods. The outcome is the strikingly high incidence of obesity. In contrast, there is a genetically similar group of Pima Indians living in Mexico, but they are farmers who work in the fields and consume the food they grow. They still have higher rates of obesity than would be predicted from their diet and exercise patterns, suggesting genes that favour high body weight. However, they are significantly less obese than the Arizona Pima Indians.[50]

Lifestyle and Rising Obesity Rates

Although genes are an important determinant in what people weigh, they are not the reason more Canadians are obese today than 30 years ago. The frequency of genes in a population takes many generations to change, but environmental conditions can change quickly. Environmental changes have occurred in Canada in recent years that affect what Canadians eat, how much they eat, and how much exercise they get. Simply put, more Canadians are overweight than ever before because they are eating more and burning fewer kcalories than they did in the past. Food is plentiful and continuously available and little activity is required in our daily lives.[51] This is often referred to as an **obesogenic environment.**

Obesogenic environment An environment that promotes weight gain by encouraging overeating and physical inactivity.

appetite The desire to consume specific foods that is independent of hunger.

People are Eating More Part of the reason Canadians are eating more is the increase in the availability of tempting food choices. Palatable, affordable, convenient food is readily available to the majority of the population 24 hours a day in supermarkets, fast-food restaurants, and convenience stores. The accessibility of enticing treats stimulates **appetite**. Appetite is the desire to eat that is not related to physiological hunger. It is triggered or inhibited by external factors such as the sight, taste, and smell of food, as well as the time of day, emotions, and cultural and social conventions. It is usually appetite and not hunger that makes us stop for an ice cream cone on a summer afternoon or buy chocolate chip cookies and cinnamon rolls while strolling through the mall. We are constantly bombarded with cues to eat—we see and hear about tasty, inexpensive foods in TV ads and we see and smell tempting food in convenience stores, food courts, and vending machines.

Not only have the variety, visibility, and convenience of tempting foods increased in the past 50 years, but so have the portion sizes.[52] (briefly discussed in section 1.3). This phenomenon is called portion distortion, as studies of how much people eat demonstrate that when more food is put in front of people, they eat more.[53] So if you increase the amount of pasta on your plate or cereal in your bowl, you will most likely consume more kcalories. People tend to eat in units, such as one cookie, one sandwich, or one bag of chips, regardless of the size of the unit. When presented with larger units, therefore, such as bigger muffins, burgers, or bottles of soft drinks, people still eat or drink the whole thing. As carbonated beverage standards went from a 355 ml can to a 591 ml bottle, beverage intake increased; energy intake from sweetened beverages alone increased 135% between 1977 and 2001.[54] One of the best examples of how portion size has increased is the food served at fast-food restaurants; the french fry portions and hamburgers served today are 2-5 times bigger than when fast food sold 40 years ago.[55] (**Figure 7.24**). You probably wouldn't order an extra hamburger but you will finish the one you order, even if it is twice as big.

According to a survey by the American Dietetic Association, individuals overestimate the recommended serving size for many foods; fewer than half of respondents accurately estimated the recommended serving sizes for pasta, meat, and vegetables.[56] The serving size is listed on food labels, but people often assume that the kcalories listed are for the entire container, even when the package contains 2 or more servings (see Label Literacy: How Many Kcalories in that Bowl, Box, or Bottle?).

Social changes that have occurred over the last few decades have also contributed to the increase in the number of kcalories Canadians consume. The increasing number of single-parent households and households with two working parents means that the time families have to prepare meals at home is limited. Fifty years ago, hours of time and effort were invested in preparing meals. The main course might cook all afternoon and dessert was baked and served at the conclusion of the family meal. Today, a fast-food meal can be in hand in a matter of minutes and getting a snack just involves opening a package. As a result of our fast-paced lifestyles, prepackaged, convenience, and fast food have become mainstays. These foods are typically higher in energy than meals prepared at home. These factors make overeating much easier.

People are Exercising Less Along with overeating, a decrease in the amount of energy expended, both at work and play, further contributes to weight gain. Fewer Canadian adults today work in jobs that require physical labour. People drive to work, rather than walk or bike, take elevators instead of stairs, use dryers rather than hang clothes outside, and cut the lawn with riding rather than push mowers. All of these simple changes reduce the amount of energy expended to perform the tasks of daily living. A typical office worker today walks only about 3,000-5,000 steps in his or her daily activities. In contrast, in the Amish community where driving automobiles and using electrical appliances and other modern conveniences are not allowed, a typical adult takes 14,000–18,000 steps a day. The overall incidence of obesity in this group is only 4%.[57]

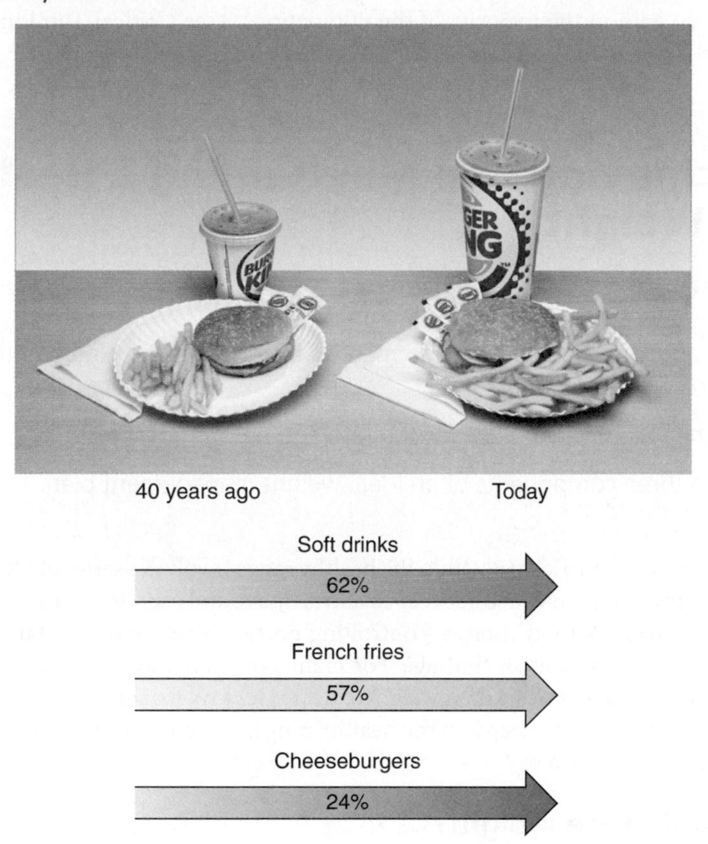

Figure 7.24 Larger portion sizes
The relative size of these two fast-food meals clearly illustrates the dramatic increase in portion sizes since the 1970s.
Source: Adapted from Nielsen, S. J. and Popkin, B. M. Patterns and trends in food portion sizes, 1977–1998. *JAMA* 289:450–453, 2003. (Andy Washnik)

In addition to less-active jobs, busy schedules, long days at work, and commuting make people feel they have no time for active recreation. Instead, at the end of the day, Canadians sit in front of televisions, electronic games, and computers—all sedentary ways to spend leisure time.

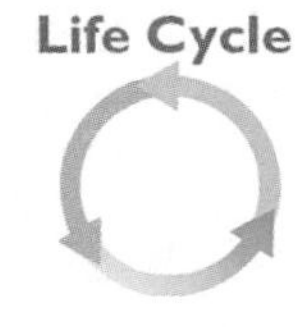

The reduction in physical activity is not restricted to adults. Many schools have reduced or even eliminated physical education programs to save money and find more time for academics. Social changes have increased crime, forcing children to stay inside after school. In the 1960s, children spent their after school hours outdoors with bikes, balls, and friends. Today they are more likely to spend it indoors with video games and computers. The end result is that they burn fewer kcalories and have more opportunities to snack, and consequently they gain weight.

Strategies to Reduce the Incidence of Obesity

To become a thinner country, we need strategies that can help everyone improve their food choices, reduce serving sizes, and increase their physical activity.[58] Although successful weight management ultimately depends on an individual's choices, food manufacturers and restaurants can help us cut kcalories by offering healthier foods and packaging or serving foods in smaller portions. Communities can help increase activity by providing parks, bike paths, and other recreational facilities for people of all ages. Businesses and schools can contribute by offering more opportunities for physical activity at the workplace and during the school day.

Even small changes, if they are consistent, can arrest the increase in obesity in the population. It has been estimated that a population-wide shift in energy balance of only 100 kcalories a day, the equivalent of walking a mile or cutting out a scoop of ice cream, would prevent further weight gain in the majority of the population[51] (see Critical Thinking: Balancing Energy: Genetics and Lifestyle).

7.8 Achieving and Maintaining a Healthy Body Weight

Learning Objectives

- Evaluate an individual's weight and medical history to determine whether weight loss is recommended.
- Discuss the recommendations for the rate and amount of weight loss.
- Name the three components of an ideal weight management plan.

Managing body weight to keep it within the healthy range involves a series of lifestyle choices. It requires maintaining a balance between kcalorie intake and exercise. For some people, this means making healthy food choices, controlling portion sizes, and maintaining an active lifestyle to avoid weight gain as they age. For many others, it may mean developing a meal and exercise plan that will allow their weight to decrease. And for some it may mean working to increase weight and then keep it in the healthy range. The goal for everyone is to achieve a healthy weight and stay there.

Who Should Lose Weight?

CANADIAN CONTENT

In 2006, the Canadian Medical Association published Clinical Practice Guidelines for the management and prevention of obesity in adults and children. They recommend that overweight and obesity in adults be accessed by measuring BMI and waist circumference and assessed in combination with other health indicators. A threshold at which a program in weight management should be considered was a BMI greater than 25 and/or a waist circumference cutoff as indicated in Table 7.7. **Figure 7.25** illustrates a flow chart that describes the steps that can be taken to manage overweight and obesity.[31]

Medical Risk Factors Because obesity, particularly abdominal obesity as indicated by waist circumference, increases the risk for prediabetes, type 2 diabetes, and cardiovascular disease (see Table 7.6), blood pressure, blood glucose, and blood cholesterol and triglyceride levels should be measured to determine an individual's risk for these comorbidities, and if necessary, treatment for these comorbidities should be provided (see Figure 7.25). Also recommended is an assessment for depression, as well as for eating and mood disorders, as these conditions are often associated with obesity and interfere with successful weight loss.

Life Cycle

Weight-Loss Goals and Guidelines

As indicated in Figure 7.25, the medical goal for weight loss in an overweight person is to reduce the health risks associated with being overweight. For most people, a relatively modest weight loss will significantly reduce disease risk. The initial goal of weight loss should therefore be to reduce body weight by about 10% over a period of about 6 months. A slow loss of 10% of body weight is considered achievable for most people. To ensure that most of what is lost is fat and not lean tissue, the weight should be lost slowly at a rate of between 0.5-1 kg/week. If weight is lost more rapidly, the loss is less likely to be maintained, and the additional loss will be from fluid, glycogen, and muscle protein. Most people who lose large amounts of weight or lose weight rapidly eventually regain all that they have lost. Repeated cycles of weight loss and regain, referred to as **weight cycling** or **yo-yo dieting**, decrease the likelihood that future attempts at weight loss will be successful (**Figure 7.26**).[59]

weight cycling or **yo-yo dieting** The repeated loss and regain of body weight.

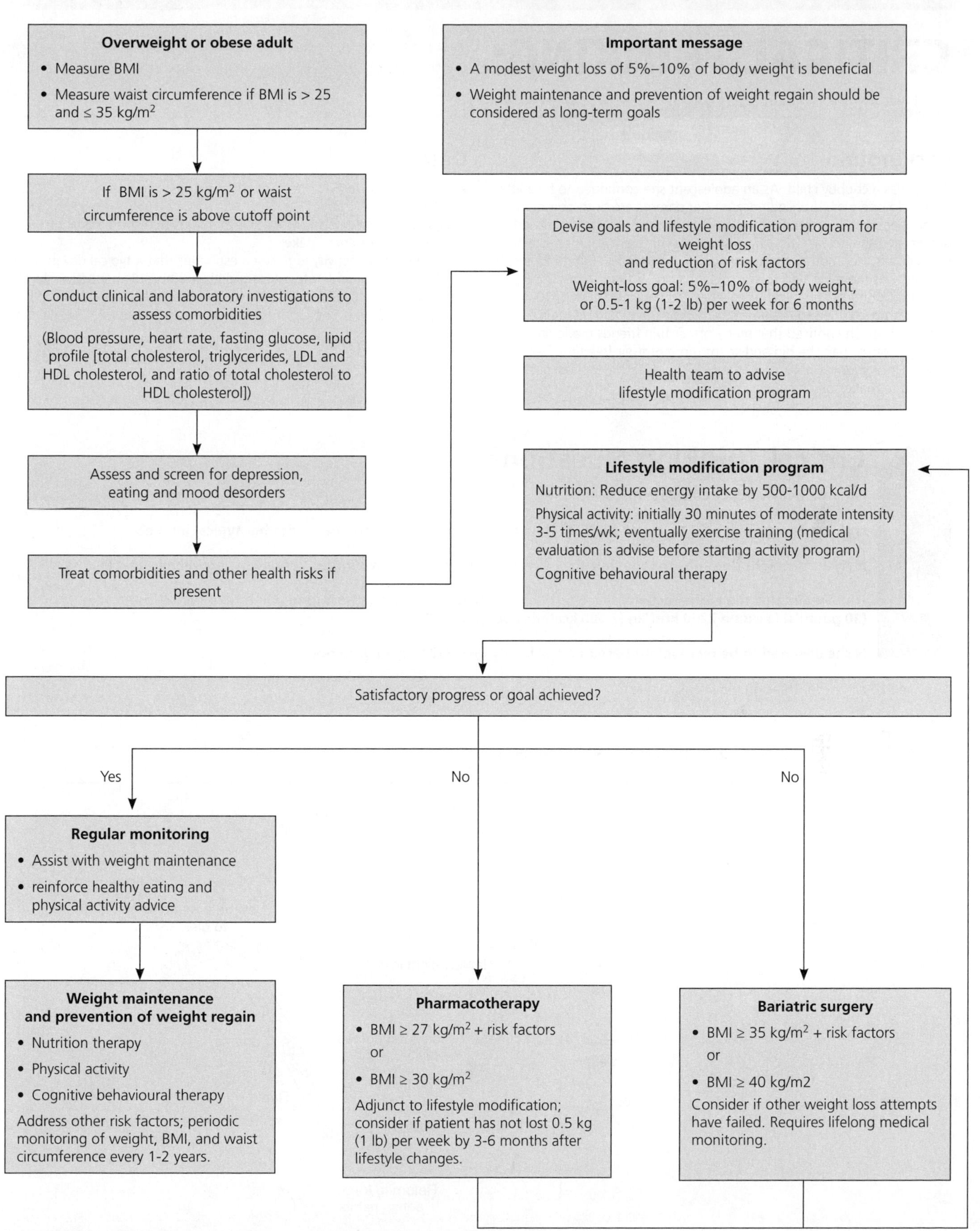

Figure 7.25 Approaches to the management of the overweight or obese adult.
Source: Obesity Canada Clinical Practice Guidelines Expert Panel. Canadian clinical practice guidelines on the management and prevention of obesity in adults and children. 2006.[31]

CRITICAL THINKING:

Balancing Energy: Genetics and Lifestyle

Background

Aysha was a chubby child. As an adolescent she continued to be a little heavy. No one was surprised because her parents are both obese. At home, her mother served healthy meals, but the family never missed dessert and the house was always stocked with plenty of high-kcalorie snack foods. Her mother spent her spare time sewing or knitting and her dad was more inclined to watch sports on TV than to participate in them. During her first year at college, Aysha gained 4.5 kg (10 lb) and became resigned to the inevitability that she would be fat like her parents. Then she noticed that many of her thin friends made different dietary choices than she did and spent more of their free time being active. Since she was now in charge of all her lifestyle choices, she decided to make some changes.

Data

- Aysha is 163 cm (5'4") tall, 23 years old, and currently weighs 70 kg (155 lb).
- Aysha snacks while studying. She estimates that this adds about 350 kcal/day to her intake.
- By keeping an activity log, Aysha estimates that a typical day includes 30 minutes of moderate-intensity activity. This puts her activity level in the "low-active" category. By recording and analyzing her food intake for 3 days, she determines that she eats about 2,450 kcal/day.

Critical Thinking Questions

Is Aysha overweight? Calculate her BMI.*

Is she in energy balance? Calculate her EER. How do her energy needs compare with her typical intake?

If Aysha's weight does not change, but she increases her activity to 60 minutes per day, what will her new EER be?

If she maintains the higher activity level, but doesn't change her intake, how long will it take her to lose 4.5 kg (10 pounds) (assume 7700 kcal/kg [3,500 kcalories per pound])?

Is she destined to be overweight based on her family history? Why or why not?

*Answers to all Critical Thinking questions can be found in Appendix J.

Use iProfile to find snacks that provide less than 100 kcalories.

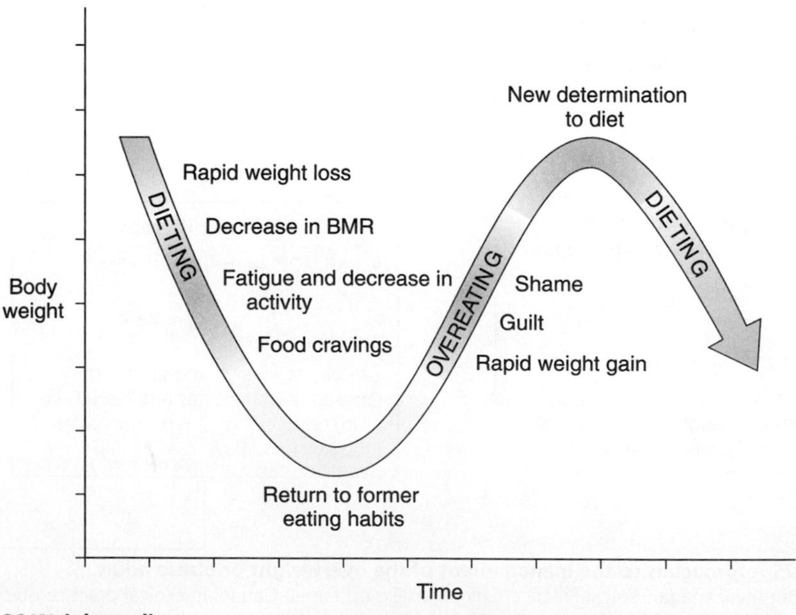

Figure 7.26 Weight cycling
Weight cycling, the repeated loss and regain of body weight, makes future attempts at weight loss less likely to succeed.

The key feature of managing weight loss is a lifestyle modification program (Figure 7.25) comprised of three components: nutrition, physical activity, and cognitive behavioural therapy.

Nutrition In order to promote weight loss without compromising nutrient intake, an individual's diet must be low in energy but provide for all the body's nutrient needs. Nutrient density becomes more important as energy intake is reduced. Canada's Food Guide suggests reducing kcalories from added sugars, solid fats, and alcohol, which provide kcalories but few essential nutrients. Losing weight requires tipping the energy balance scale: eating less or exercising more. A kilogram of body fat provides about 7,700 kcal (3,500 kcal/lb). Therefore, to lose 0.5 kg, you need to decrease your intake or increase your expenditure by this amount. To lose 0.5 kg in a week, you would need to tip your energy balance by about 500 kcal/day. This could mean adding 500 kcal of exercise, subtracting 500 kcal from your food intake, or some combination of the two. The arithmetic is simple, but achieving and maintaining weight loss is not easy. It can be achieved by small but consistent changes to eating habits (**Table 7.8**).

Table 7.8 Tips for Shifting Energy Balance Toward Weight Loss
Watch your serving size
• Pour one serving of chips into a bowl rather than eating right from the big bag.
• Fill your plate once, skip the seconds.
• Check labels to see if your serving size matches the label.
• Don't super size—choose a small drink and a small order of fries.
• Have a plain old burger, not one with a special sauce or extra large patty.
• Don't overeat when you eat out—share an entrée with a friend or take it home for lunch the next day.
Cut down on high-kcalorie foods
• Have one scoop of ice cream rather than two.
• Use less butter or margarine on your toast.
• Bake, roast, or grill, rather than fry your food.
• Skip the baked goods and have fruit for dessert.
• Bring your own lunch instead of buying lunch in the cafeteria.
• Have an apple with lunch instead of a candy bar.
• Have water instead of soft drinks.
• Switch to low-fat milk.
Don't get too hungry
• Eat breakfast, you'll eat less later in the day.
• Fill up on high-fibre foods.
• Increase your vegetable servings.
• Plan nutrient-dense snacks.
• Keep cut-up veggies and fruit available for snacks.
• Cut down on high-fat and high-sugar choices.
If you want to eat more—exercise more
• Go for a bike ride.
• Try bowling instead of watching TV on Friday night.
• Take a walk during your lunch break or after dinner.
• Play tennis—you don't have to be good to get plenty of exercise
• Shoot some hoops.
• Get off the bus one stop early.

Even when choosing nutrient-dense foods, it is difficult to meet nutrient needs with intakes of fewer than 1,200 kcal/day, so dieters consuming less than this should take a multivitamin and mineral supplement. Medical supervision is recommended if intake is 800 kcal/day or less.

Physical Activity Physical activity is an important component of any weight-management program. Exercise promotes fat loss and weight maintenance. It increases energy expenditure, so if intake remains the same, energy stored as fat is used for fuel. An increase in activity of 200 kcal five times a week will result in the loss of 0.5 kg in about 3½ weeks. In addition to increasing energy expenditure, exercise also promotes muscle development. This is important for promoting weight loss because muscle is metabolically active tissue. Increasing muscle mass helps to prevent the drop in metabolic rate that occurs as body weight decreases. Weight loss is better maintained when physical activity is included. In addition, physical activity improves overall fitness and relieves boredom and stress. The Canadian Physical Activity Guidelines for adults (19-64 years) recommends 150 minutes per week of moderate to vigorous activity.[60] The benefits of exercise are discussed further in Section 13.1.

Cognitive Behavioural Therapy People tend to think of weight loss as something they can accomplish by going on a diet. When the weight is lost, they go off the diet. The problem is that when their eating patterns return to what they were previously, they regain the weight. This "on a diet, off a diet" pattern may allow you to look good for the prom but it isn't what is needed for long-term weight management. To manage your weight at a healthy level, you need to establish a pattern of food intake and exercise that allows you to enjoy foods and activities you like without your weight climbing. It should be a pattern that you can comfortably adopt for life.

Changing food consumption and exercise patterns requires identifying the old patterns that led to weight gain and replacing them with new ones to promote and maintain weight loss. This can be accomplished through a process called **behaviour modification**, which is based on the theory that behaviours involve (1) antecedents or cues that lead to the behaviour, (2) the behaviour itself, and (3) consequences of the behaviour. These are referred to as the ABCs of behaviour modification. Changes to behaviour are achieved by helping individuals to understand why they behave as they do.

behaviour modification A process used to gradually and permanently change habitual behaviours.

The first step in a behaviour modification program is to identify cues that lead to eating. You can do this by keeping a log of everything you eat or drink, where you were when you ate, what else you were doing at the time, and what motivated you to eat at that time. Then by analyzing this log, you can see what prompted you to eat excessive amounts or high-kcalorie foods. For instance, sitting in front of the television and mindlessly demolishing a bag of potato chips may cause you to overeat and then feel bad because you consumed the extra kcalories (**Figure 7.27**). In this case, the antecedent is watching TV, the behaviour is mindlessly eating the chips, and the consequence is feeling remorse and gaining weight. The key to modifying this behaviour is to recognize the antecedent, change the behaviour, and replace the negative consequence with a positive one. In this example, not taking food with you to the television, or taking only the portion of food you want to consume, eliminates the antecedent and the behaviour. The consequence is that you have consumed only the food you planned, you do not gain weight, and you feel a sense of accomplishment. Applying behaviour modification techniques to change eating behaviours has been shown to improve long-term weight maintenance.

Suggestions for Weight Gain

As difficult as weight loss is for some people, weight gain can be equally elusive for underweight individuals. The first step toward weight gain is a clinical evaluation to rule out medical reasons for low body weight. This is particularly important when weight loss occurs unexpectedly. If the low body weight is due to low intake or high expenditure, gradually increasing consumption of energy-dense foods is suggested. More frequent meals and high-kcalorie snacks such as nuts, peanut butter, or milkshakes between meals can help increase energy intake. Replacing

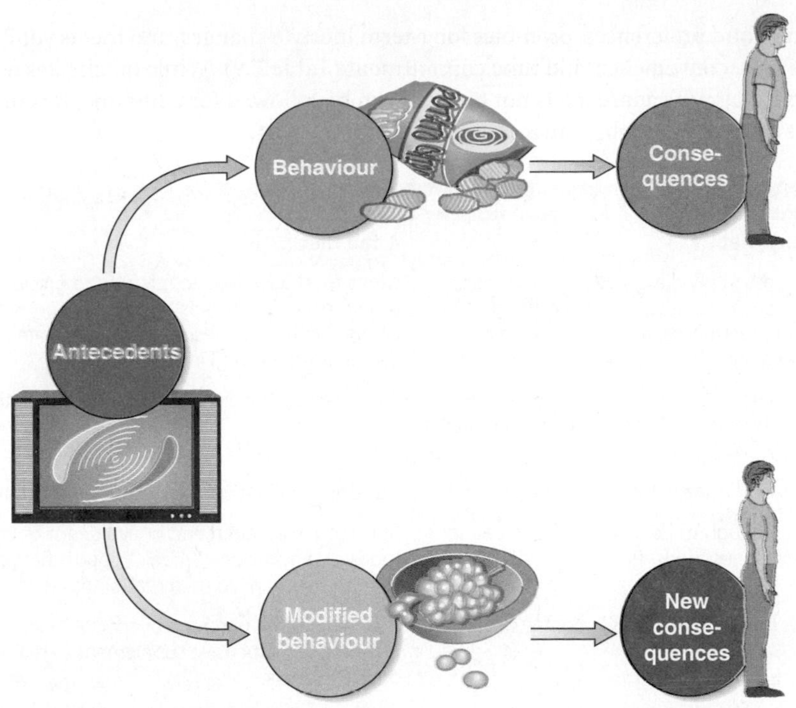

Figure 7.27 Behaviour modification
Behaviour modification can be used to identify and change undesirable behaviours.

low-kcalorie fluids like water and diet beverages with fruit juices and milk may also help. To encourage a gain in muscle rather than fat, strength-training exercise should be a component of any weight-gain program. This approach requires extra kcalories to fuel the activity needed to build muscles. These recommendations apply to individuals who are naturally thin and have trouble gaining weight on the recommended energy intake. This dietary approach may not promote weight gain for those who limit intake because of an eating disorder (see Focus on Eating Disorders).

7.9 Diets, Diets, Everywhere

Learning Objectives

- Distinguish between a healthy weight management program and a fad diet.
- Discuss the importance of the proportions of carbohydrate, protein, and fat in a weight-loss plan.

Want to lose 5 kg (10 lbs) in just five days? What dieter wouldn't? People desperate to lose weight are prey to all sorts of diets that promise quick fixes. They willingly eat a single food for days at a time, select foods based on special fat-burning qualities, and consume odd combinations at specific times of the day. Most diets, no matter how outlandish, will promote weight loss because they reduce energy intake. Even diets that focus on modifying fat or carbohydrate intake or promise to allow unlimited amounts of certain foods work because intake is reduced. The true test of the effectiveness of a weight-loss diet is whether those who follow it maintain their weight loss over the long term.

An ideal diet is one that is part of a weight-management program that promotes weight loss and then maintenance of that loss over the long term. To be successful, the program needs to encourage changes in the lifestyle patterns that led to weight gain. When selecting a program, look for one that is based on sound nutrition and exercise principles, suits your

individual food preferences, promotes long-term lifestyle changes, and meets your needs in terms of cost, convenience, and time commitment (**Table 7.9**). While quick fixes are tempting, if the program's approach is not one that can be followed for a lifetime, it is unlikely to promote successful weight management.

Table 7.9 Distinguishing Between Healthy Diets and Fad Diets

A healthy diet ...	A fad diet ...
Promotes a healthy dietary pattern that meets nutrient needs, includes a variety of foods, suits food preferences, and can be maintained throughout life.	Limits food selections to a few food groups or promotes rituals such as eating only specific food combinations. As a result, it may be limited in certain nutrients and in variety.
Promotes a reasonable weight loss of 0.5 to 1 kg per week and does not restrict kcalories to less than 1,200 a day.	Promotes rapid weight loss of much more than 1 kg/week.
Promotes or includes physical activity.	Advertises weight loss without the need to exercise.
Is flexible enough to be followed when eating out and includes foods that are easily obtained	May require a rigid menu or avoidance of certain foods or may include "magic" foods that promise to burn fat or speed up metabolism.
Does not require costly supplements.	May require the purchase of special foods, weight-loss patches, expensive supplements, creams, or other products.
Promotes a change in behaviour. Teaches new eating habits. Provides social support.	Does not recommend changes in activity and eating habits, recommends an eating pattern that is difficult to follow for life, provides little social support.
Is based on sound scientific principles and may include monitoring by qualified health professionals.	Makes outlandish and unscientific claims, does not support claims that it is clinically tested or scientifically proven, claims that it is new and improved or based on some new scientific discovery, or relies on testimonials from celebrities or connects the diet to trendy places such as Beverly Hills.

The following sections discuss some of the more common methods for reducing kcalorie intake. The advantages and disadvantages of a number of popular diets and commercial weight-management programs are given in **Table 7.10**.

Prepared Meals and Drinks

It is easier to eat less when someone else decides what you're having and puts appropriate portions of that food in front of you (**Figure 7.28**). This is the idea behind diet plans that sell prepackaged meals designed to replace some or all of your usual meals. These diets are easy to follow as long as you are not travelling or eating out, but they can be expensive and are not practical in the long term. Because all meals are provided, they do not teach the food selection skills needed to make a long-term lifestyle change.

Figure 7.28 Frozen prepared meals are popular with dieters because the food is portioned and no decisions or measuring are required. (Rob Melnychuk/Getty Images)

Table 7.10 Pros and Cons of Some Commercial Weight-Loss Diets

Diet	Approach	Pros	Cons
Atkins™' Diet	Very low carbohydrate	Inexpensive; rapid initial weight loss	Difficult to follow in the long term; no social support
Cabbage soup diet	Unlimited amounts of cabbage soup, fruit, coffee, and tea	Rapid weight loss	No social support; does not promote long-term behaviour change; lacks variety
Dukan Diet	High protein and vegetables	Rapid weight loss; provides maintenance diet	Low in fat and carbohydrates; bad breath, constipation
Dieting with the Duchess	Simple nutrition and exercise tips	Inexpensive, flexible	No social support
Eating Thin for Life	Moderation—written as weight-loss success stories, recipes, and menu ideas	Inexpensive	No social support
Fit or Fat	Increased exercise	Safe, inexpensive	No social support
Grapefruit diet	Some foods have special qualities that burn fat	Inexpensive	Based on unsound principles
Jenny Craig	Low energy	Safe, convenient	Expensive; relies on purchase of special foods
Optifast	Very-low-kcalorie formula	Rapid weight loss	Expensive; dangerous if does not include medical supervision
SlimFast	Low energy	Safe	Does not promote long-term behaviour change
South Beach Diet	Initially very low carbohydrate, then more healthy carbohydrates are allowed	Safe, inexpensive, heart-healthy	Initial weight loss is mostly water; no social support
Sugar Busters	Eliminates sugar; low kcalorie—1,200 kcal per /day	Inexpensive	No social support; based on unsound principles; insufficient carbohydrate
The New Beverly Hills Diet	Specific timing and combinations of foods	Inexpensive	Based on unusual principles; does not promote long-term behaviour change; nutritionally unsound
The Zone (and Mastering the Zone) Diet	Low carbohydrate (40% of energy)	Inexpensive, flexible	Based on questionable principles; no social support
Volumetrics Weight Control Plan	Emphasizes foods high in water, fibre and air to promote fullness with few kcalories	Safe, inexpensive	No social support or exercise component
Weight Watchers	Low energy, social support	Safe, inexpensive, flexible	Requires group participation for optimal results

Liquid Meal Replacements Rather than a prepackaged meal, many diet plans replace some or all meals with special beverages. They can make reducing intake easy because they eliminate the problem of choosing appropriate portions of low-kcalorie foods. Many of the liquid weight-loss diets that are available over-the-counter recommend a combination of food and the liquid formula to provide about 800-1,200 kcal/day. These formula plans promote weight loss as long as the foods eaten with them are low in kcalories. They are easy to use and relatively inexpensive but they do little to change eating habits for life. They are not recommended without medical supervision.

very-low-kcalorie diet A weight-loss diet that provides fewer than 800 kcal/day.

protein-sparing modified fast A very-low-kcalorie diet with a high proportion of protein, designed to maximize the loss of fat and minimize the loss of protein from the body.

Very-Low-Kcalorie Diets A **very-low-kcalorie diet** is defined as one with fewer than 800 kcal/day and should only be undertaken while under medical supervision and for no more than 16 weeks.[61] These diets are generally a variation of the **protein-sparing modified fast**, which is a diet providing a high proportion of protein, but little energy. The concept behind this is that the protein in the diet will be used to meet the body's protein needs and will, therefore, prevent excessive loss of body protein. Frequently, very-low-kcalorie diets are offered as a liquid formula. These formulas provide from 300-800 kcal and 50-100 g of protein per day and meet all other nutrient needs.

These diets will cause rapid weight loss; initial weight loss is 1.5-2.5 kg (3-5 lb.) per week. This can provide a psychological boost and motivate the dieter to continue losing weight; however, in most cases, much of this initial weight loss is from water loss. Once the initial water loss ends, weight loss slows. The dieter's basal metabolism slows to conserve energy, and physical activity decreases because the dieter often does not have the energy to continue his or her typical level of activity.

Very-low-kcalorie diets are no more effective than other methods of weight loss in the long term and carry more risks. At these low-energy intakes, body protein is broken down and potassium is excreted. Depletion of potassium can result in an irregular heartbeat and is potentially deadly. Other side effects include gallstones, fatigue, nausea, cold-intolerance, light-headedness, nervousness, constipation or diarrhea, anemia, hair loss, dry skin, and menstrual irregularities.

Low-Fat Diets

Low-fat weight-loss diets have been popular for decades. Fat is high in energy: 9 kcal/g—almost twice as much as either carbohydrate or protein. Low-fat diets therefore tend to provide a greater volume of food for less energy than a diet with more fat. Because people have a tendency to eat a certain weight or volume of food, if that food is low in fat, it will contribute fewer kcalories.[62] Low-fat diets also satisfy hunger after less energy is consumed. Differences in the way dietary fat and dietary carbohydrate are used by the body also explain why low-fat diets are more effective for weight loss. Excess kcalories from dietary fat are stored more efficiently than excess kcalories from carbohydrate, so consuming excess energy from fat leads to a greater accumulation of body fat than consuming excess energy as carbohydrate. Short-term clinical trials demonstrate that a reduction in fat without intentional energy restriction leads to a decrease in energy intake and a modest reduction in body weight.[63,64] Although there are no long-term clinical trials on the effect of a low-fat diet on body weight, they are believed to be as effective as other diets for long-term weight loss.

Problems with low-fat diets occur when people eat large quantities of low-fat foods without considering that these foods are not necessarily low in kcalories. Even a diet low in fat will result in weight gain if energy intake exceeds energy output. In the 1990s, the food industry flooded the market with reduced-fat cookies and cakes. These foods were low in fat, but not in energy, so when consumed in large amounts they caused weight gain, not weight loss.

Low-Carbohydrate Diets

If you cut out the pasta and potatoes you will lose weight, right? The Atkins' Diet, South Beach Diet, Sugar-Busters, Calories Don't Count, the Scarsdale Diet, the Zone, and the Dukan Diet are just a few of the low-carbohydrate, weight-loss diets that have been promoted over the past 50 years. In addition to promising weight loss, these diets claim to improve athletic performance and promote overall health. Low-carbohydrate diets are all based on the premise that a high-carbohydrate intake causes an increase in insulin levels, which promotes the storage of body fat. Restricting carbohydrate intake is hypothesized to reduce insulin, thereby reducing fat storage and promoting fat loss.

Very-Low-Carbohydrate Diets Very restrictive low-carbohydrate diets prohibit foods such as breads, grains, and fruits, and limit vegetable intake while allowing unlimited quantities of meat and high-fat foods that are low in carbohydrate. These diets cause a rapid initial weight loss, most of which is water. This occurs because when carbohydrate intake is low, glycogen stores, along with the water they hold, are lost quickly. Ketones are produced because fat is not completely broken down in the absence of carbohydrate. Excretion of these ketones causes further water loss.

CRITICAL THINKING:

Choosing a Weight-Loss Plan that Works for You

Background

Rose has gained 20 kg over the past few years. She wants to lose weight. She researches her options and narrows her choice of weight-loss programs to three. To decide which weight-loss program is best for her, Rose considers the advantages and disadvantages of each plan in terms of the nutritional soundness of the diet, variety, ease, expense, and how it will affect her activity level.

(© istockphoto.com/ Carolyn Woodcock)

Data

Rose's Information

BMI: 31 kg/m^2
Waist circumference: 92 cm
Blood pressure: 160/92
Total cholesterol: 6.07 mmol/L
LDL cholesterol: 4.2 mmol/L
Activity: plays golf once a month using a golf cart

Weight-loss Options

Low-carbohydrate plan This plan limits carbohydrate intake to 30 g/day—this means she cannot eat any grains, milk, or fruit. Many of her friends have used this to shed pounds and say they were never hungry.
Liquid formula plan This plan uses cans of formula to replace 2 meals a day. A 1-week supply costs $30.00. It is easy because she doesn't need to do much meal planning and she can still eat dinner with her family.
Low-calorie diet This plan is run through her community centre. It includes a walking group that meets 5 days a week and weekly meetings for weigh-ins and nutrition lectures. The cost is $25.00 per week.

Critical Thinking Questions

Would Rose benefit from weight loss based on the data given about her in the table? Check Appendix C to compare Rose's data with normal values.

Evaluate the programs she is considering: are the diets nutritionally sound? Do they recommend activity and include social support? What about cost?

Which plan do you think would benefit Rose most in the long term?

iProfile Use iProfile to look up the kcalorie content of a liquid weight-loss product.

Weight loss continues on low-carbohydrate diets because total energy intake is reduced. Elevated blood ketones and reduced insulin levels suppress appetite, making it easier to reduce food intake. In addition, these diets limit food choices to such an extent that the monotony results in a spontaneous reduction in energy intake.[65] The availability of carbohydrate-modified products such as low-carbohydrate bread and pasta makes these diets more palatable, but may also make them less effective. As with special low-fat products, low-carbohydrate foods are not necessarily low in kcalories and cannot be eaten liberally without affecting energy balance (see Critical Thinking: Choosing a Weight Loss Plan that Works for You).

Although these diets do promote weight loss, more research is needed to determine the health consequences of consuming very low-carbohydrate diets for long periods. These diets are higher in fat and protein than is recommended. High intakes of saturated fat and cholesterol increase the risk of high blood cholesterol and gallbladder disease. They are also lower in fruits, vegetables, whole grains, and milk than is recommended. Low intakes of these foods will reduce intakes of essential nutrients, phytochemicals, and fibre.

Including Unrefined Carbohydrate Not all low-carbohydrate diets severely restrict carbohydrate and most allow the dieter to increase carbohydrate intake over time. The carbohydrates that are included come from unrefined sources such as vegetables and whole grains. These foods are high in fibre and often have a low glycemic index (see section 4.5). Including whole grains and vegetables in the diet increases the intake of fibre, phytochemicals, and micronutrients and may help to regulate blood glucose levels.

Scientific studies have been conducted to compare the effects of these various diets on weight loss. To test whether they promote long-term weight loss, their effect after one year is typically evaluated and the reality is that regardless of the diet, within one year, most people regain all or part of their weight loss. No one diet is substantially better than any other, although low-calorie diets that reduce energy intake by 500-1,000 kcal tend to be more common among successful weight losers.[61]

Canadian Clinical Practice Guidelines for the Management of Obesity suggest that the optimal dietary plan for achieving a healthy body weight is best developed between a health professional, preferably a registered dietitian, the individual, and his or her family. This diet should reduce energy intake but be nutritionally balanced.[31]

7.10 Weight-Loss Drugs and Surgery

Learning Objectives

- Explain how surgery can promote weight loss.
- Discuss when weight-loss drugs or surgery might be considered appropriate.

When lifestyle modifications alone are unsuccessful in achieving weight-loss goals and especially when additional risk factors for cardiovascular disease and type 2 diabetes are present, pharmacotherapy and bariatric surgery are additional therapies.

Drugs and Supplements

Prescription Drugs Few drug therapies are available to promote weight loss.. An ideal drug for the treatment of obesity would permit an individual to lose weight and maintain the loss, be safe when used for long periods of time, have no side effects, and not be addictive. Many attempts have been made to develop such a drug, but weight-loss drugs still carry risks that have resulted in the withdrawal of some from the market . For example, in the United States in 1996, approximately 18 million prescriptions were written for a drug combination called fen-phen (fenfluramine and phentermine) that was being used to reduce food intake. The following year, it was linked to serious heart-valve damage. As a result, fenfluramine and the related drug dexfenfluramine were withdrawn from the market.[66] Although the fen-phen drug combination was never approved for use in Canada, the drugs fenfluramine and phentermine were available by prescription separately and there was concern that they were being used in combination. Health Canada banned fenfluramine in 1997,[67] and in 2008, phentermine was discontinued by the Canadian manufacturer.[68]

CANADIAN CONTENT

Another popular prescription used in weight loss, sibutramine, was withdrawn from the Canadian marketplace in 2010 after reports of numerous adverse effects related to cardiovascular disease.[69]

One commonly prescribed drug is orlistat (brand name Xenical). Orlistat acts by disabling the enzyme lipase, preventing triglycerides from being digested into fatty acids and glycerol. The undigested triglycerides continue through the intestines and are eliminated in the feces **(Figure 7.29)**. This cuts the number of kcalories available to the body, but the fat in the stool may cause gas, diarrhea, and more frequent and hard-to-control bowel movements. Because of these potential side effects, drug therapy is recommended only for those whose health is seriously compromised by their body weight, that is, those with a BMI greater than 30 kg/m^2 or those with a BMI greater than or equal to 27 kg/m^2 who have obesity-related risk factor or diseases.[24]

Weight-Loss Supplements Alternative treatments for obesity, such as dietary supplements, were reviewed by the Obesity Canada Clinical Practice Guidelines Expert Committee. However, there were few studies that lasted more than 6 months, so the long-term impact of these products could not be evaluated. The committee concluded that there was insufficient evidence favouring the use of weight-loss supplements and noted that some have serious adverse effects.[31] Some of these are listed in **Table 7.11**. For more detailed information also see Your Choice: Can a Weight-Loss Supplement Help You Trim the Fat?

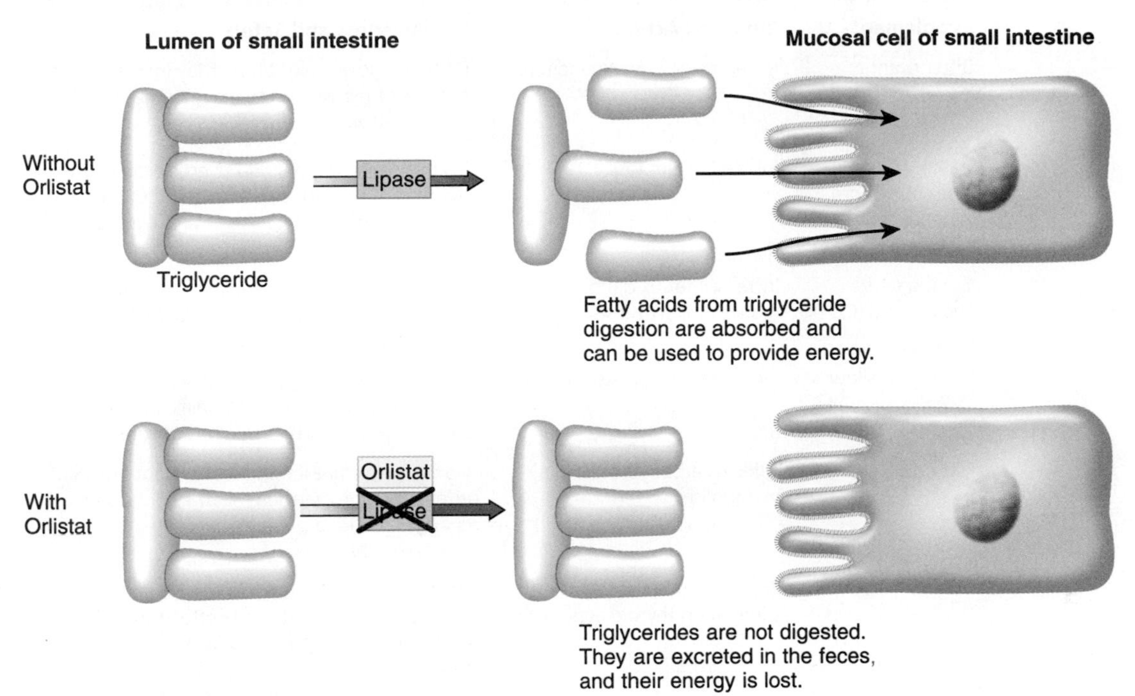

Figure 7.29 Orlistat and fat digestion
Orlistat is a prescription drug that acts by preventing triglyceride digestion.

Bariatric Surgery

In extreme cases, that is, people with a BMI greater than or equal to 40 kg/m^2 (extreme obesity) and those with a BMI between 35-40 kg/m^2 who have other life-threatening conditions that could be remedied by weight loss, surgical procedures are considered. These bariatric surgeries cause weight loss because they alter the GI tract to reduce food intake and nutrient absorption. Such surgical approaches are risky and are advised only for individuals in whom the risk of dying from obesity and its complications is great. Each case must be evaluated individually to assess the potential risks and benefits of the surgery, but it is usually recommended only for those who have tried other methods and failed and whose weight severely limits their quality of life and ability to perform daily activities. To be successful, the individual must understand the procedure and its risks and be aware of how his or her life may change after the operation. Even after surgery, success requires a lifelong behavioural commitment that includes well-balanced eating and physical activity.

Types of Surgical Procedures Weight-loss surgery may restrict the amount of food that can be consumed, limit the amount that can be absorbed, or both. **Gastric banding** is a procedure that just restricts food intake. It uses an adjustable band to create a small pouch at the upper end of the stomach (**Figure 7.30**). This new, smaller stomach, which typically holds about 30 ml, limits the amount of food that can be comfortably consumed at one time and slows the rate at which food leaves the stomach. Another procedure, called **gastric bypass**, bypasses part of the stomach and small intestine. It is currently the most common and most successful type of bariatric surgery. Gastric bypass connects the intestine to the upper, smaller portion of the stomach. The smaller stomach limits intake and the shorter intestine reduces absorption (**Figure 7.31**). Individuals who have any of these procedures must make permanent changes in their eating habits and experience permanent changes in their bowel habits. Some weight regain is common after 2-5 years.

gastric banding A surgical procedure in which an adjustable band is placed around the upper portion of the stomach to limit the volume that the stomach can hold and the rate of stomach emptying.

gastric bypass A surgical procedure to treat morbid obesity that both reduces the size of the stomach and bypasses a portion of the small intestine.

Table 7.11 Common Weight-Loss Supplements

Products with serious side effects

For important additional information also see: Your Choice: Can a Weight-Loss Supplement Help You Trim the Fat?

Supplement	Proposed Action	Effectiveness and Safety
Bitter orange	Increases energy expenditure	Does promote weight loss. May increase the risk of hypertension, arrhythmias, heart attacks, strokes, and seizure.
Cascara, senna, aloe, buckthorn, rhubarb root, and castor oil	Increase water loss	Do not cause fat loss and overuse can cause diarrhea, vomiting, stomach cramps, chronic constipation, fainting, and severe electrolyte imbalances.
Conjugated linoleic acid (CLA)	Increases fat oxidation or reduces fat synthesis	No evidence that it is effective for weight loss in humans. It causes gastrointestinal symptoms.
Country mallow	Increases energy expenditure	Does promote weight loss. May increase the risk of hypertension, arrhythmias, heart attacks, strokes, and seizure. Contains ephedra.
Ginseng	Affects carbohydrate metabolism	No evidence that it enhances weight loss. Side effects include diarrhea, headache, insomnia, changes in blood pressure, and altered bleeding time.
Guggul	Boosts metabolism by stimulating thyroid activity	No human studies support the claim that it boosts thyroid activity. Side effects include gastrointestinal upset, headache, nausea, and hiccups.
Guarana	Suppresses appetite	It contains caffeine (about twice as much as coffee beans) and has been shown to be effective for short-term weight loss when used in combination with ephedra. Side effects include anxiety, nervousness, and difficulty sleeping.
Hydroxycitric acid (extract from *Garcinia cambogia*)	Increases fat oxidation or reduces fat synthesis	Not found to be effective. Side effects include a laxative effect, abdominal pain, and vomiting.
L-carnitine	Increases fat oxidation	Not found to significantly affect total body mass or fat mass. Side effects include nausea, vomiting, abdominal cramps, diarrhea, and body odour. **Cannot be obtained in Canada without a prescription**
Licorice	Increases fat oxidation or reduces fat synthesis	Has been shown to reduce body fat in normal-weight subjects but causes high blood pressure and low blood potassium.
St. John's wort	Enhances mood	No evidence it promotes weight loss. Contains similar ingredients to the antidepressant drug fluoxetine (Prozac) and should not be used by people taking antidepressants.
Yerba maté	Increases energy expenditure	No studies supporting its benefit for weight loss. Side effects include nervousness, dehydration, and nausea.
Yohimbine	Suppresses appetite and the body's ability to store fat	No evidence that it aids weight loss. Potential side effects include anxiety, elevated blood pressure, a feeling of queasiness, insomnia, rapid heartbeat, tremors, and vomiting. **Cannot be obtained in Canada without a prescription**

Table 7.11 Continued

Products that are generally considered safe		
Supplement	**Proposed Action**	**Effectiveness and Safety**
Apple cider vinegar	Increases energy expenditure, reduces hunger and food cravings	Safe, but no evidence that it has any affect on body weight.
Chitosan	Blocks fat absorption	Has not been shown to enhance weight loss. Short-term human trials have not reported any severe adverse effect. No long-term studies have been done.
Chromium	Affects carbohydrate metabolism	Human trials have not demonstrated an effect on body composition or body weight. No reports of adverse effects in humans.
Dandelion	Increases water loss	Does not cause fat loss. No side effects other than rare allergic reactions.
Glucomannan	Increases satiety	Safe, but has not been shown to promote weight loss.
Green tea	Increases fat oxidation or reduces fat synthesis	Safe, but no evidence from controlled clinical trials that tea or tea extracts promote weight loss.
Guar gum	Increases satiety	Safe, but has not been shown to promote weight loss.
Psyllium	Increases satiety	Safe, but has not been shown to promote weight loss.
Pyruvate	Increases fat oxidation or reduces fat synthesis	Some studies have shown a weight-loss benefit in individuals on weight-loss diets, but the dosage used in the studies was very high. No known side effects.
Spirulina (blue-green algae)	Suppresses appetite	Safe, but no evidence that it aids weight loss.

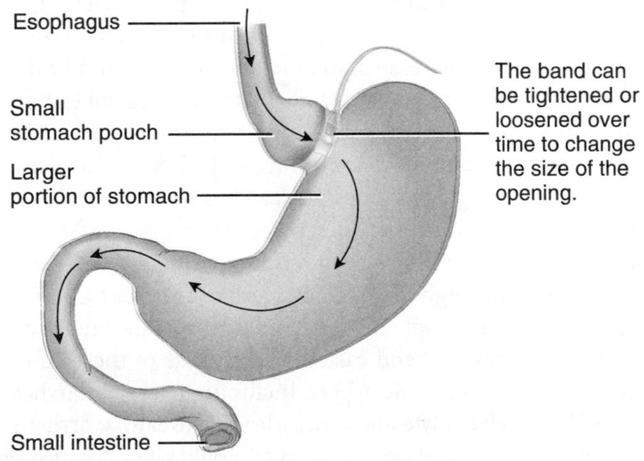

Figure 7.30 Gastric banding
Gastric banding involves surgically placing an adjustable band around the upper part of the stomach, creating a small pouch.

Liposuction is another surgical procedure, but it is primarily a cosmetic one. It involves inserting a large hollow needle under the skin into a localized fat deposit and literally vacuuming out the fat. It is often advertised as a way to remove cellulite, which is just fat that has a lumpy appearance because of the presence of connections to the tissue layers below. The risks of liposuction include those associated with general anesthesia and the possibility of infection. The procedure can reduce the amount of fat in a specific location, but will not significantly

liposuction A procedure that suctions out adipose tissue from under the skin; used to decrease the size of local fat deposits such as on the abdomen or hips.

YOUR CHOICE

Can a Weight-Loss Supplement Help You Trim the Fat?

CANADIAN CONTENT

Losing weight is not easy, and keeping it off is even harder. Wouldn't it be nice if there were a pill that you could take to help lose the weight and keep it off? A walk through any supermarket, pharmacy, or supplement store will reveal an assortment of supplements that advertise that they do just this. But are they safe and do they work?

Some weight-loss supplements claim to increase metabolism. These "fat burners" promise not only to boost metabolism, but also to prevent the loss of lean muscle tissue, suppress appetite, and increase the burning of stored fat. Although they are probably the most effective over-the-counter weight-loss supplements, they are also the most dangerous. One of the most popular and controversial herbal fat burners is ephedra, a stimulant that increases blood pressure and heart rate and constricts blood vessels. Use of ephedra-containing products has been associated with an increased risk of psychiatric, nervous, and gastrointestinal symptoms, heart palpitations, hypertension, arrhythmias, heart attacks, strokes, and seizure.[1] Due to these safety concerns, the use of ephedra is limited in Canada, only for nasal decongestion, and to products containing no more than 8 mg/dose or a daily dose of 32 mg. The products must not contain stimulants, such as caffeine, which aggravate ephedra's adverse effects.[2]

Ephedra-containing products were available under many names, for example Ma Huang, Chinese Ephedra, Ma Huang extract, Ephedra, Ephedra Sinica, Ephedra extract, Ephedra herb powder, *Sida Cordifolia* or epitonin. Sources of caffeine often added to epedra included green tea, guarana, yerba maté, cola nut and yohimbe.[3] After ephedra was restricted, bitter orange, a stimulant with similar side effects as ephedra, appeared in the marketplace.[4] Health Canada advised Canadians that bitter orange was not authorized for use in Canada and should not be used.[5] Green tea extract is another popular supplement used to boost metabolism and aid weight loss. It appears to be safe if used in appropriate amounts, but studies have not shown it to enhance weight loss.[4]

Some weight-loss supplements contain soluble fibre, which absorbs water so your stomach fills up after consuming fewer kcalories. Common sources of fibre include glucomannan, guar gum, and psyllium. Although these fibres are safe to consume, there is little evidence that they promote weight loss.[6,7]

A few products promise to prevent fat synthesis and deposition. Hydroxycitric acid supplements are marketed to block fat synthesis. Although few side effects have been reported, evidence for its effectiveness for weight loss is inconsistent.[7] Conjugated linoleic acid has been shown to reduce body weight and/or fat deposition in animal models.[8] However, there is no evidence that it is effective for weight loss in humans, and it causes gastrointestinal symptoms. Chromium picolinate is marketed to decrease body fat and increase the proportion of lean tissue. Human trials, however, have not supported these claims. There are no reports of adverse effects in humans, but laboratory studies show that the picolinate form may cause oxidative damage to DNA and lipids.[9]

Supplements containing chitosan, which is derived from a molecule found in the shells of crustaceans, claim to block fat absorption. Results of most clinical trials have not shown any increase in weight loss, however, and healthy people taking chitosan do not show increased fecal fat excretion.[10]

Several supplements cause weight loss by increasing the amount of water lost from the body. Water loss decreases body weight but does not cause a decrease in body fat. Dandelion is a diuretic, so it increases water excretion by the kidneys. Other herbal products induce diarrhea, which causes water loss through the GI tract. Herbal laxatives found in weight-loss teas and supplements include senna, aloe, buckthorn, rhubarb root, cascara, and castor oil. Overuse of these substances can have serious side effects, including nausea, diarrhea, vomiting, and electrolyte imbalances, leading to cardiac arrhythmia and death.[11]

In short, if a secret pill could safely help people eat less or burn more kcalories without effort, it would not be a secret for long. The best and safest way to lose weight is to eat less and exercise more.

[1] Shecell, P. G., Hardy, M. L., Morton, S. C. et al. Efficacy and safety of ephedra and ephedrine for weight loss and athletic performance: A meta-analysis. *JAMA* 289: 1537-1545, 2003.

[2] Health Canada. 2003: Health Canada reminds Canadians of the dangers of Ephedra. Available online at http://hc-sc.gc.ca/ahc-asc/media/advisories-avis/_2003/2003_43-eng.ph. Accessed July 12, 2010.

[3] Health Canada. 2008. Health Canada reminds Canadians not to use Ephedra. Available online at http://www.hc-sc.gc.ca/ahc-asc/media/advisories-avis/_2008/2008_41-eng.php. Accessed July 12,2010.

[4] Sarma, D. N., Barrett, M. L., Chavez, M. L. et al. Safety of green tea extracts: A systematic review by the U.S. Pharmacopeia. *Drug Saf.* 31:469–484, 2008.

[5] Health Canada. 2004. Canadian Adverse Reaction Newsletters. October, 2004. Available online at http://www.hc-sc.gc.ca/dhp-mps/medeff/bulletin/carn-bcei_v14n4-eng.php#a.3. Accessed July 12, 2010.

[6] Pittler, M. H., and Ernst, E. Dietary supplements for body-weight reduction: A systematic review. *Am. J. Clin. Nutr.* 79:529–536, 2004.

[7] Saper, R. B., Eisenberg, D. M., and Phillips, R. S. Common dietary supplements for weight loss. *Am. Fam. Physician* 70:1731–1738, 2004.

[8] Li, J. J., Huang, C. J., and Xie, D. Anti-obesity effects of conjugated linoleic acid, docosahexaenoic acid, and eicosapentaenoic acid. *Mol. Nutr. Food Res.* 52:631–645, 2008.

[9] Vincent, J. B. The potential value and toxicity of chromium picolinate as a nutritional supplement, weight loss agent and muscle development agent. *Sports Med.* 33:213–230, 2003.

[10] Gades, M. D., and Stern, J. S. Chitosan supplementation and fecal fat excretion in men. *Obes. Res.* 11:683–688, 2003.

[11] Kurtzweil, P. Dieter's brews make tea time a dangerous affair. *FDA Consumer* 31: July–August, 1997. Available online at www.fda.gov/fdac/features/1997/597_tea.html/. Accessed February 18, 2001.

---

reduce overall body weight. Liposuction has not been found to affect obesity-related metabolic abnormalities, and therefore, unlike overall weight loss, it does not reduce the risk of heart disease and diabetes that is associated with excess body fat.[70]

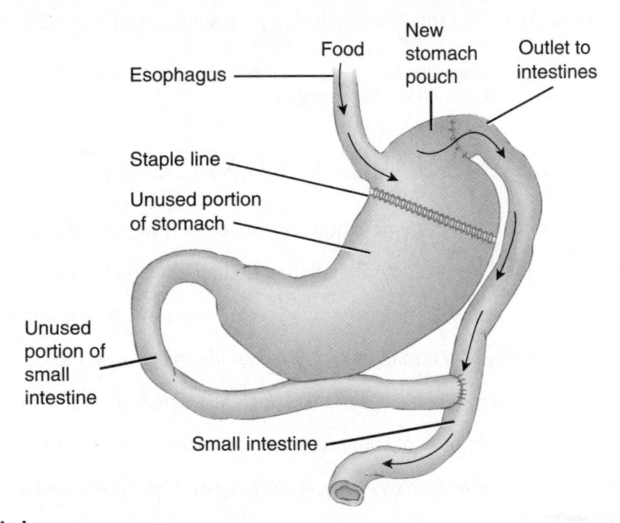

Figure 7.31 Gastric bypass surgery
Gastric bypass surgery reduces the size of the stomach by placing staples across the top of the stomach. It reduces absorption by attaching the stomach pouch to the small intestine, bypassing a portion of the intestine.

Weighing the Benefits and the Risks The most obvious benefit of bariatric surgery is weight loss (**Figure 7.32**). Ninety percent of individuals have a significant (20%-25% of body weight) weight loss and 50%-80% maintain this loss for at least 5 years.[71] The weight loss occurs quickly and continues for about 2 years after surgery. With this weight loss comes a reduction in the presence and risk of diseases related to obesity such as diabetes, high blood pressure, and heart disease. Other benefits commonly reported include improved mobility and stamina, better mood, and increased self-esteem and interpersonal effectiveness. These benefits, however, do not come without a price.

Gastric banding and gastric bypass have short-term surgical risks and complications. Complications that require follow-up surgeries include abdominal hernias, the breakdown of the staple line, and stretched stomach outlets. More than one-third of obese patients who have weight-loss surgery develop gallstones; the risk is increased during rapid or substantial weight loss. Many patients develop "dumping syndrome"—a condition in which food moves too rapidly into the small intestine, causing nausea, weakness, sweating, faintness, and diarrhea after eating. To avoid dumping syndrome, patients must eat small, frequent meals and avoid foods that cause problems. Another risk of bariatric surgery is nutrient deficiencies, particularly of vitamin B_{12}, folate, calcium, and iron.[72] Almost 30% of patients experience deficiency symptoms such as anemia, osteoporosis, and metabolic bone disease. To prevent these deficiencies, patients who have these surgeries must take nutritional supplements. Because rapid weight loss and nutritional deficiencies can harm a developing fetus, women who have weight-loss surgery should not become pregnant until their weight is stable.

Figure 7.32 Acclaimed Canadian soprano, Measha Brueggergosman, shed over 65 kg after gastric bypass surgery in 2005.
Source: Terauds, J. *Heart surgery saves soprano*. The Toronto Star. Available online at: http://www.thestar.com/article/656319. Accessed June 30, 2011. (Oleniuk/Toronto Star, Jon Von Tiedemann Lucas).

CASE STUDY OUTCOME

Many factors interact to determine how much a person weighs. The genes Sara and Erin inherited from their parents determined their body shape and how easily they put on weight. But what they actually weigh is determined by how much they eat and how much they exercise. Erin, who had been heavy all her life, has learned to control her tendency to gain weight by eating a healthy diet and exercising regularly. What started as a plan to lose weight has now become part of her lifestyle. She monitors her weight and food intake and enjoys daily activity. Sara, who never needed to think about her weight, experienced lifestyle changes that promoted weight gain. But she is inspired by the changes her sister has made and has re-evaluated her own lifestyle. After starting a job with a law firm, Sara has begun watching what she eats and has started riding her bike after work to reduce stress and keep her weight under control. In the past 6 months, she has lost about 7 kg. At this year's family picnic, both sisters have a BMI in the healthy range.

APPLICATIONS

Personal Nutrition

1. **Are you in the healthy weight range?**

   a. Use Appendix B to determine if your BMI falls in the healthy range.

   b. If your BMI is greater than 25, measure your waist circumference. Does it indicate you are at increased risk due to visceral fat storage?

   c. Even if you are not overweight, answer the following questions to see how many factors you have that may increase your risk of obesity-related complications if you become overweight.

      1. Do you have a personal or family history of heart disease?
      2. Are you a male older than 45 years?
      3. Are you a postmenopausal female?
      4. Do you smoke cigarettes?
      5. Do you have a sedentary lifestyle?
      6. Do you have high blood pressure?
      7. Do you have high LDL cholesterol, low HDL cholesterol, or high triglycerides?
      8. Do you have diabetes?

2. **Are you in energy balance?**

   iProfile

   a. Use iProfile to calculate your average daily energy intake from the 3-day food record you kept in Chapter 2.

   b. Keep an activity log for several days. Use Table 7.3 and 7.4 to determine your physical activity level and PA value. Calculate your EER (see Table 7.5).

   c. How does your EER compare with your energy intake?

   d. If you consumed and expended this amount of energy every day, would your weight increase, decrease, or stay the same?

   e. If your intake does not equal your EER, how much would you gain or lose in a month? (Assume that 1 kg of fat is equal to 7700 kcalories.)

   f. If your energy intake does not equal your EER, list some specific changes you could make in your diet or the amount of activity you get to make the two balance.

3. **Are genetic or environmental factors a larger influence on your energy balance? Answer the following questions to help you decide.**

   a. How has your weight changed over the past year?

   b. How much have your parents' weights changed since they were 21?

   c. Do you eat more servings of snack foods such as chips and candy bars, or of fruits and vegetables daily? Has this changed over the past year?

   d. How has your activity changed over the past year?

   e. What patterns do you see emerging that can predict if your weight will change over the next year? The next 10 years? The next 20 years? What are your predictions?

4. **What are the goals and guidelines for weight loss?**

   a. Imagine you have a BMI of 30 kg/m^2. Use your own height and the equation for BMI (BMI = weight in kg/(height in m)^2 or BMI = weight in lb./(height in inches)^2 $\times$ 703) to determine about how much weight you would need to lose to get your BMI into the healthy range.

   b. How much weight should you lose per week?

   c. Choose a weight-loss plan that best suits your individual needs and explain why you chose this plan.

   d. Review the criteria for recognizing a healthy weight-loss diet listed in Table 7.9. Does your plan meet each of these guidelines?

   e. Should you consider surgery or drugs? Why or why not?

General Nutrition Issues

1. **What weight patterns and influences are present in your class?**

   a. Do a class survey by collecting everyone's answers to question 3 above. Tabulate the patterns that you see.

   b. Is weight generally increasing or decreasing?

   c. Is exercise increasing or decreasing?

   d. Which are the more popular snacks—prepackaged snack foods or fruits and vegetables?

   e. What percentage of your classmates has one parent whose weight increased by 9 kg (20 lb.) or more since they were 21?

   f. What percentage of your classmates has two parents whose weight increased by 9 kg (20 lb.) or more since they were 21?

2. **How useful are low-kcalorie products?**

   a. Go to the grocery store and select 5-10 products labelled "light," "reduced calorie," "low calorie," or "calorie free." Record the number of kcalories per serving.

   b. Would any of these products be useful for someone on a weight-loss diet?

   c. Can they be consumed in unlimited amounts without significantly increasing energy intake?

   d. Rate these foods in terms of their nutrient density.

3. **Several strains of mice with mutations in genes that regulate body weight have been identified. For each of the following, predict whether the mouse will be over- or underweight and explain why.**

   a. A mouse that makes excess leptin.

   b. A mouse that makes leptin normally, but the leptin receptor in the brain is defective, so it always acts as if large amounts of leptin are bound to it.

   c. A mouse that makes more leptin than normal, but the leptin molecule made is defective and cannot bind to receptors in the brain.

   d. A mouse that makes too much ghrelin.

4. Roger just celebrated his 40th birthday. He is about 18 kg (40 lb.) heavier than he was on his 30th birthday. At his current rate of weight gain, he will be 36 kg (80 lb.) overweight when he reaches age 50. Roger is 173 cm (5′ 8″) tall and weighs 86 kg (190 lb.). He has three young children at home and works full-time as a salesperson. His day starts at about 6 a.m. when he gets up and has breakfast with his wife and children. He spends most of the morning in his car travelling to visit his clients. He usually has at least one doughnut-and-coffee break. Lunch is fast food if he is alone or a restaurant meal if he is with clients. Most evenings he is home for dinner. By the time his children are in bed, it is about 8 p.m. He sits down with his wife for a glass of wine or a bowl of ice cream.

a. Design a weight-loss and exercise regimen for Roger. Assume that his wife and children are supportive regarding dietary changes at home.

b. Does the program you designed fit in with Roger's schedule? Why or why not?

c. If he follows your suggestions, how long will it take him to lose 18 kg (40 lb.)?

SUMMARY

7.1 The Obesity Epidemic

- The rapid rise in the incidence of overweight and obesity in Canada has been called epidemic. It is also a growing problem around the world.

7.2 Exploring Energy Balance

- The principle of energy balance states that if energy intake equals energy needs, body weight will remain constant. A kcalorie is the amount of heat needed to raise the temperature of 1 kg of water 1 degree Celsius.
- Energy is provided to the body by carbohydrate (4 kcal/g), fat (9 kcal/g), protein (4 kcal/g), and alcohol (7 kcal/g). To power the body, these need to be broken down and their energy converted into ATP. The ATP then fuels metabolic reactions that build and maintain body components and provides energy for other cellular and body activities.
- Total energy expenditure (TEE) includes the energy required for basal metabolism, physical activity, the thermic effect of food, growth, milk production in lactating women, and maintenance of body temperature in a cold environment. The largest component of energy expenditure in most people is basal energy expenditure (BEE), which varies depending on body size, body composition, age, and gender. The energy needed for activity, which includes planned exercise and nonexercise activity thermogenesis (NEAT), typically accounts for 15%-30% of energy expenditure but varies greatly depending on the individual. The thermic effect of food (TEF) is the energy required for the digestion of food and the absorption, metabolism, and storage of nutrients. It is equal to about 10% of energy consumed.
- When the diet does not meet needs, body stores are used. Glucose is provided by glycogen stores or synthesized from amino acids by gluconeogenesis. Energy for tissues that don't require glucose is provided by the breakdown of stored fat. When the diet contains excess energy, the extra energy is stored for later use, primarily as fat.

7.3 Estimating Energy Needs

- The amount of energy expended by the body can be determined by direct calorimetry, which measures the heat produced by the body, and indirect calorimetry, which measures oxygen utilization. The doubly-labelled water method estimates energy expenditure by using water labelled with isotopes of hydrogen and oxygen to calculate carbon dioxide production. Current recommendations for energy needs were determined using the doubly-labelled water method.
- An individual's energy needs can be predicted by calculating estimated energy requirements (EER). EER calculations take into account age, gender, life stage, height, weight, and activity level. As physical activity increases, more kcalories need to be consumed to maintain body weight.

7.4 Body Weight and Health

- Some body fat is essential for health but too much increases the risk of developing chronic diseases such as diabetes, heart disease, high blood pressure, gallbladder disease, sleep apnea, arthritis, and certain types of cancer.
- Excess body fat can create psychological and social problems.
- Being naturally lean decreases health risks, but if low body weight is due to starvation or eating disorders, it can affect electrolyte balance and immune function and increase the risk of early death.

7.5 Guidelines for a Healthy Body Weight

- A healthy body weight is a weight at which the risk of illness and death are lowest. Health risks increase when too little or too much body fat is stored, and when the excess fat is visceral—stored around the internal organs.
- The amount of body fat can be assessed using techniques such as bioelectric impedance, skinfold thickness, underwater weighing, isotope dilution techniques, and imaging.
- The most common way to evaluate the healthfulness of body weight is body mass index (BMI), which is calculated from a ratio of weight to height.
- An increase in visceral fat is associated with a greater incidence of heart disease, high blood pressure, stroke, and diabetes. Waist measurement can be used to assess the presence of too much visceral fat.

7.6 Regulation of Energy Balance

- The genes people inherit determine their body shape and characteristics. Genes involved in regulating body fatness are referred to as obesity genes.
- Signals from the GI tract, hormones, and levels of circulating nutrients regulate body weight in the short term by affecting hunger and satiety. Signals that relay information about the size of body fat stores, such as the release of leptin from adipocytes, regulate long-term energy intake and expenditure. Although body weight appears to be regulated around a set-point, changes in physiological, psychological, and environmental circumstances do cause the level at which body weight is regulated to change, usually increasing it over time.
- Defects in one or more of the genes that regulate body weight could lead to the storage of too much body fat. Variations in the genes that regulate metabolic rate, the ability to dissipate excess kcalories, and levels of NEAT have been hypothesized to contribute to obesity.

7.7 Why Are Canadians Getting Fatter?: Genes versus Environment

- People become overweight or obese because they are taking in more kcalories and expending fewer. They eat more today because they are exposed to large portions of a wide variety of tasty, convenient foods that stimulate our appetite. They move less due to modern conveniences, busy lifestyles, and the availability of sedentary ways to spend their leisure time.

- To reduce the incidence of obesity, Canadians need to move more and eat less. Even a shift in energy balance of 100 kcal/day would prevent further weight gain in most people.

7.8 Achieving and Maintaining a Healthy Body Weight

- Whether a person needs to lose weight depends on how much body fat he or she has, where the fat is located, and what the person's health risks are. Risk typically increases with a BMI >25 or with increasing waist circumference.
- Weight management involves adjusting energy intake and expenditure and modifying long-term behaviours. To lose a kilogram of fat, expenditure must exceed intake by approximately 7700 kcalories. A slow, steady weight loss of 0.5-1 kg/week is more likely to be maintained than rapid weight loss.
- A modest weight loss of 5-10% of body weight at a rate of 0.5-1 kg/week for 6 months is sufficient to produce health benefits.
- If underweight is not due to a medical condition, weight gain can be accomplished by increasing energy intake and lifting weights to increase muscle mass.

7.9 Diets, Diets, Everywhere

- There are many diets that vary in composition—some are low-carbohydrate, some are low-fat, others include prepared foods or liquid meal replacements. After one year, weight loss is usually modest and does not vary greatly between diet types.
- A good weight-loss diet is one that allows a wide range of food choices, does not require the purchase and consumption of special foods or combinations of foods, and can be followed for life.
- *Go to WileyPLUS to view a video clip on evaluating diets.*

Video

7.10 Weight-Loss Drugs and Surgery

- Prescription weight-loss drugs are only recommended for individuals who are significantly overweight or have accompanying health risks. Those currently available act by suppressing appetite or blocking fat absorption. Some herbal weight-loss supplements have caused serious side effects.
- The use of surgery to promote weight loss is a drastic measure that is considered only for those whose health is seriously at risk because of their obesity. It causes permanent changes in the GI tract that affect the amount of food that can be consumed and the absorption of nutrients. Even after surgery, weight loss requires changes in eating patterns and behaviour.

REVIEW QUESTIONS

1. What is a kcalorie?
2. Explain how energy balance is related to body weight.
3. Which nutrients provide energy? How much does each provide?
4. What is basal metabolic rate?
5. What is NEAT? How does it affect energy balance?
6. What is the thermic effect of food?
7. Explain why the energy in dietary fat is stored in body fat more efficiently than the energy in dietary carbohydrate.
8. Describe three methods for measuring energy expenditure.
9. What is EER and what variables are used in its calculation?
10. Explain what is meant by a healthy body weight.
11. List 5 health problems that are associated with excess body fat.
12. How does the distribution of body fat affect the risks of excess body fat?
13. List some methods for determining the amount of body fat.
14. How is BMI calculated and why is it commonly used to assess body weight?
15. Explain 3 mechanisms that make you stop eating when you have eaten enough at a meal.
16. Discuss the role of leptin in regulating body weight.
17. List some social and environmental factors that affect energy balance and discuss how these mightinteract with an individual's genetic predisposition to a particular body weight.
18. Why is waist circumference included in the assessment of obesity?
19. How many kcalories must be expended to lose 1 kg of fat?
20. What is the best approach to weight management? Why?
21. Describe how orlistat causes weight loss.
22. How does weight-loss surgery cause negative energy balance?

REFERENCES

1. Canadian Community Health Survey, 2004. Available online at http://www.hc-sc.gc.ca/fn-an/surveill/nutrition/commun/cchs_guide_escc_3_tab1-eng.php. Accessed July 9, 2010.
2. Shields, M. and Tjepkema, M. Trends in adult obesity. Health Reports 17 (3):53–59. Available online at http: //www.statcan.gc.ca/cgi-bin/af-fdr.cgi?l=eng&loc=http://www.statcan.gc.ca/studies-etudes/82-003/archive/2006/9279-eng.pdf&t=Trends in adult obesity, 2006. Accessed July 9, 2010.
3. Shields, M. Overweight and obesity among children and youth. Health Reports 17 (3):27-42, 2006. Available online at http://www.statcan.gc.ca/cgi-bin/af-fdr.cgi?l=eng&loc=http://www.statcan.gc.ca/studies-etudes/82-003/archive/2006/9277-eng.pdf&t=Overweight and obesity among children and youth. Accessed July 9, 2010.
4. World Health Organization. *Obesity and Overweight*, September 2006. Fact sheet no. 311. Available online at www.who.int/mediacentre/factsheets/fs311/en/print.html. Accessed July 2, 2008.
5. Institute of Medicine, Food and Nutrition Board. *Dietary Reference Intakes for Energy, Carbohydrate, Fiber, Fat, Protein and Amino Acids*. Washington, DC: National Academies Press, 2002.
6. Weinsier, R. L., Nagy, T. R., Hunter, G. R. et al. Do adaptive changes in metabolic rate favor weight regain in weight-reduced individuals? An examination of the set-point theory. *Am. J. Clin. Nutr.* 72:1088–1094, 2000.
7. J. A. Levine, C. M. Kotz, NEAT—non-exercise activity thermogenesis—egocentric & geocentric environmental factors vs. biological regulation. *Acta. Physiologica. Scandinavica* 184: 309–318, 2005.
8. Kriketos, A. D., Peters, J. C., and Hill, J. O. Cellular and whole-animal energetics. In: *Biochemical and Physiological Aspects of Human Nutrition*. M. H. Stipanuk, ed. Philadelphia: W.B. Saunders Company, 2000, pp. 411–424.

9. Horton, T. S., Drougas, H., Brachey, A. et al. Fat and carbohydrate overfeeding in humans: Different effects on energy storage. *Am. J. Clin. Nutr*. 62:19–29, 1995.

10. Hellerstein, M. K. De novo lipogenesis in humans: Metabolic and regulatory aspects. *Eur. J. Clin. Nutr*. 53(suppl 1):S53–S65, 2000.

11. Schoeller, D. A. Recent advances from application of doubly-labeled water to measurement of human energy expenditure. *J. Nutr*. 129:1765–1768, 1999.

12. Flegal, K. M., Graubard, B. I., Williamson, D. F., et al. Excess deaths associated with underweight, overweight, and obesity. *JAMA*. 293:1861–1867, 2005.

13. Marti, A., Marcos, A., and Martínez, J. A. Obesity and immune function relationships. *Obes. Rev.* 2:131–140, 2001.

14. Committee to Reexamine IOM Pregnancy Weight Guidelines, Institute of Medicine, National Research Council *Weight Gain During Pregnancy: Reexamining the Guidelines*, Washington, DC: National Academies Press, 2009.

15. Eisenberg, M. E., Neumark-Sztainer, D., and Story, M. Associations of weight-based teasing and emotional well-being among adolescents. *Arch. Pediatr. Adolesc. Med*. 157:733–738, 2003.

16. Strauss, R. S., and Pollack, H. A. Social marginalization of overweight children. *Arch. Pediatr. Adolesc. Med*. 157:746–752, 2003.

17. Onyike, C. U., Crum, R. M., Lee, H. B. et. al. Is obesity associated with major depression? Results from the Third NHANES. *Am. J. Epidemiol*. 158:1139–1147, 2003.

18. Allison, D. B., Zhu, S. K., Plankey, M. et al. Differential associations of body mass index and adiposity with all-cause mortality among men in the first and second National Health and Nutrition Examination Surveys (NHANES I and NHANES II) follow-up studies. *Int. J. Obes. Relat. Metab. Disord*. 26:410–416, 2002.

19. Gallagher, D., Heymsfield, S., Heo, M. et al. Healthy percentage body fat ranges: An approach for developing guidelines based on body mass index. *Am. J. Clin. Nutr*. 72:694–701, 2000.

20. Evans, W. J. Protein nutrition, exercise and aging. *J. Am. Coll. Nutr*. 23:601S–609S, 2004.

21. Fields, D. A., Hunter, G. R., and Goran. M. I.Validation of the BOD POD with hydrostatic weighing: Influence of body clothing. *Int. J. Obes. Relat. Metab. Disord*. 24:200–205, 2000.

22. Goodpaster, B.H. Measuring body fat distribution and content in humans. *Curr. Opin. Clin. Nutr. Metab. Care*:481–487, 2002.

23. U.S. Department of Health and Human Services, U.S. Department of Agriculture. *Dietary Guidelines for Americans*, 6th ed. Washington, DC, 2005.

24. National Institutes of Health. NHLBI. The Practical Guide to Identification, Evaluation and Treatment of Overweight and Obesity in Adults. NIH Publication No. 02-4084. Bethesda, MD: National Institutes of Health, 2002. Available online at http://www.nhlbi.nih.gov/guidelines/obesity/practgde.htm. Accessed June 11, 2009.

25. National Cholesterol Education Program. Adult Treatment Panel III Report, 2001, 2004. Available online at www.nhlbi.nih.gov/guidelines/cholesterol/atp3udp04.htm/. Accessed June 11, 2009.

26. Nicklas, B. J., Penninx, B. W., Cesari M. et al. Association of visceral adipose tissue with incident myocardial infarction in older men and women: The Health, Aging and Body Composition Study. *Obstet. Gynecol. Surv*. 60:173–175, 2005.

27. Redinger, R. N. The physiology of adiposity. *J. Ky. Med. Assoc.* 106:53–62, 2008.

28. Chan, D. C., Barrett, H. P. R., Watts, G. F. 2004. Dyslipidemia in visceral obesity. *Am. J. Cardiovasc. Drugs* 4(40: 227-246, 2004.

29. Bouchard, C., Tremblay, A., Després, J.-P. et al. The response to long term feeding in identical twins. *N. Engl. J. Med.* 322:1477–1482, 1990.

30. Health Canada: Canadian Guidelines for the Classification of body weight in adults. Available online at http://www.hc-sc.gc.ca/fn-an/nutrition/weights-poids/guide-ld-adult/cg_quick_ref-ldc_rapide_ref-table2-eng.php. Accessed July 11, 2010.

31. Obesity Canada Clinical Practice Guidelines Expert Panel. 2006 Canadian clinical practice guidelines on the management and prevention of obesity in adults and children *CMA*. 176 (8) S1-13, 2007.

32. Major, G. C., Doucet, E., Trayhurn, P. et al. Clinical significance of adaptive thermogenesis. *Int. J. Obes. (Lond.)* 31:204–212, 2007.

33. Leibel, R. L., Rosenbaum, M., and Hirsch, J. Changes in energy expenditure resulting from altered body weight. *N. Engl. J. Med.* 332:622–628, 1995.

34. Chagnon, Y. C., Rankinen, T., Snyder, E. E., et al. The human obesity gene map: The 2002 update. *Obes. Res.* 11:313–367, 2003.

35. Wynne, K., Stanley, S., McGowan, B., and Bloom, S. Appetite control. *J. Endocrinol.* 184:291–318, 2005.

36. Smith, G. P. Controls of food intake. In: *Modern Nutrition in Health and Disease*, 10th ed. M. E. Shils, Shike, M., A. C. Ross et al., eds. Philadelphia: Lippincott Williams & Wilkins, 2006, pp. 751–770.

37. Cummings, D. E., Weigle, D. S., Frayo, R. S. et al. Plasma ghrelin levels after diet-induced weight loss or gastric bypass surgery. *N. Engl. J. Med*. 346:1623–1630, 2002.

38. Batterham, R. L., Cowley, M. A., Small, C. J. et al. Gut hormone PYY(3-36) physiologically inhibits food intake. *Nature* 418(6898):650–654, 2002.

39. Friedman, J. M. The alphabet of weight control. *Nature* 385:119–120, 1997.

40. Enriori, P. J., Evans, A. E., Sinnayah, P., et al. Leptin resistance and obesity. *Obesity* 14:254S–258S, 2006.

41. Peters, J. C. Control of energy balance. In Stipanuk, M. H. (ed.). *Biochemical and Physiological Aspects of Human Nutrition*, 2nd ed. Philadelphia: W.B. Saunders, 2006, pp. 618–639.

42. Tsigos, C., Kyrou, I., and Raptis, S. A. Monogenic forms of obesity and diabetes mellitus. *J. Pediatr. Endocrinol. Metab*. 15:241–253, 2002.

43. Tremblay, A., Després, J-P., Thriault, G. et al. Overfeeding and energy expenditure in humans. *Am. J. Clin. Nutr*. 56:857–862, 1992.

44. Diaz, E. O., Prentice, A. M., Goldberg, G. R. et al. Metabolic response to experimental overfeeding in lean and overweight healthy volunteers. *Am. J. Clin. Nutr*. 56:641–655, 1992.

45. Levine, J. A., Lanningham-Foster, L. M, McCrady, S. K. et al. Interindividual variation in posture allocation: Possible role in human obesity. *Science* 307:584–586, 2005.

46. van Marken Lichtenbelt, W. D., Vanhommerig, J.W., Smulders, N. M. et al. Cold-activated brown adipose tissue in healthy men. *N. Engl. J. Med.* 360:1500–1508, 2009.

47. Levine, J. A., Eberhardt, N. L., and Jensen, M. D. Role of nonexercise activity thermogenesis in resistance to fat gain in humans. *Science* 283:212–214, 1999.

48. Bouchard, C. Genetics of human obesity: Recent results from linkage studies. *J. Nutr*. 127:1887S–1890S, 1997.

49. Wardle, J., Carnell, S., Haworth, C. M., et al. Evidence for a strong genetic influence on childhood adiposity despite the force of the obesogenic environment. *Am. J. Clin. Nutr.* 87:398–404, 2008.

50. Schultz, L. O., Bennett, P. H., Ravussin, E. et al. Effects of traditional and western environments on prevalence of type 2 diabetes in Pima Indians in Mexico and the U.S. *Diabetes Care* 29:1866–1871, 2006.

51. Hill, J. O. Can a small-changes approach help address the obesity epidemic? A report of the Joint Task Force of the American Society for Nutrition, Institute of Food Technologists, and International Food Information Council. *J. Clin. Nutr.* 89:477–484, 2009.

52. Young, L. R., and Nestle, M. The contribution of expanding portion size to the obesity epidemic. *Am. J. Public Health* 92:246–249, 2002.

53. Rolls, B. J., Morris, E. L., and Roe, L. S. Portion size of food affects energy intake in normal-weight and overweight men and women. *Am. J. Clin. Nutr*. 76:1207–1213, 2002.

54. Nielsen, S. J,. and Popkin, B. M. Changes in beverage intake between 1977 and 2001. *Am. J. Prev. Med*. 27:205–210, 2004.

55. Nielsen, S. J., and Popkin, B. M. Patterns and trends in food portion sizes, 1977–1998. *JAMA*. 289:450–453, 2003.

56. American Dietetic Association. Nutrition and You: Trends 2002. Available online at www.eatright.org/ada/files/trends02findings.pdf/ Accessed March 23, 2009.

57. Bassett, D. R., Schneider, P. L., and Huntington, G. E. Physical activity in an Old Order Amish community. *Med. Sci. Sports Exerc*. 36:79–85, 2004.

58. Office of the Surgeon General. The Surgeon General's Call to Action to Prevent and Decrease Overweight and Obesity. U.S. Department of Health and Human Services. Rockville, MD, 2001. Available online at www.surgeongeneral.gov/topics/obesity/. Accessed March 4, 2004.

59. Kroke, A., Liese, A. D., Schulz, M. et al. Recent weight changes and weight cycling as predictors of subsequent two year weight change in a middle-aged cohort. *Int. J. Obes. Relat. Metab. Disord*. 26:403–409, 2002.

60. Canadian Society of Exercise Physiologists. Physical Activity Guidelines Information Sheet adults 18-64 years. Available online at http://www.csep.ca/CMFiles/Guidelines/CSEP-InfoSheets-adults-ENG.pdf. Accessed October 16, 2011..

61. Strychar, I. Diet in the management of weight loss. *CMAJ* 174 (1):56-63, 2006.

62. American Dietetic Association. Position of the American Dietetic Association: Fat replacers. *J. Am. Diet. Assoc*. 105:266–275, 2005.

63. Hill, J. O., Melanson, E. L., and Wyatt, H. T. Dietary fat intake and regulation of energy balance: Implications for obesity. *J. Nutr*. 130:284S–288S, 2000.

64. Astrup, A., Astrup, A., Buemann, B. et al. Low-fat diets and energy balance: How does the evidence stand in 2002? *Proc. Nutr. Soc*. 61:299–309, 2002.

65. Westman, E. C., Feinman, R. D., Mavropoulos, J. C., et al. Low-carbohydrate nutrition and metabolism. *Am. J. Clin. Nutr.* 86:276–284, 2007.

66. Frackelmann, K. Diet drug debacle: How two federally approved weight-loss drugs crashed. *Science News* 152:252–253, 1997.

67. Health Canada. Cardiac adverse reaction in patients following the use of fen-phen (a combination of fenfluramine and phentermine), 1997. Available online at http://www.hc-sc.gc.ca/dhp-mps/medeff/advisories-avis/prof/_1997/fenphen_hpc-cps-eng.php. Accessed July 12, 2010.

68. Canadian Pharmacists Association. Compendium of Pharmaceuticals and Specialties Online version (e-cps): Available online at https://www-e-therapeutics-ca.myaccess.library.utoronto.ca/cps.showMonograph.action?simpleMonographId=m266200. Accessed July 12, 2010.

69. Health Canada. Abbott Laboratories voluntarily withdraws weight loss drug sibutramine (Meridia) from the Canadian market. Available online at http://www.hc-sc.gc.ca/ahc-asc/media/advisories-avis/_2010/2010_169-eng.php. Accessed April 10, 2011.

70. Klein, S., Fontana, L., Young, V. L. et al. Absence of an effect of liposuction on insulin action and risk factors for coronary heart disease. *N. Engl. J. Med*. 350:2549–2557, 2004.

71. American Dietetic Association. Position of the American Dietetic Association: Weight management. *J. Am. Diet. Assoc*. 1059:330–346, 2009.

72. WIN, National Institutes of Health. NIDDK. Gastrointestinal Surgery for Severe Obesity. Available online at win.niddk.nih.gov/publications/gastric.htm/. Accessed September 17, 2009.

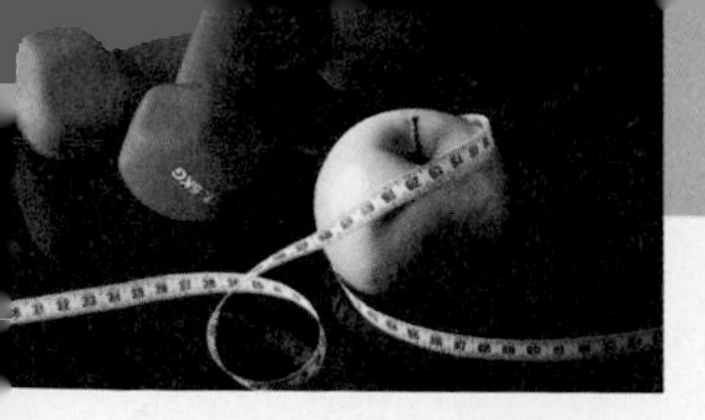

FOCUS ON

Obesity, Metabolism, and Disease Risk

Focus Outline

Obesity increases the risk of several diseases such as type 2 diabetes, cardiovascular disease, breast and colon cancers, gallbladder disease, sleep apnea, and joint disorders (**Figure F2.1**). How obesity changes the body's metabolism in a way that increases the risk of these diseases will be described in this "Focus On" obesity, metabolism, and disease risk.

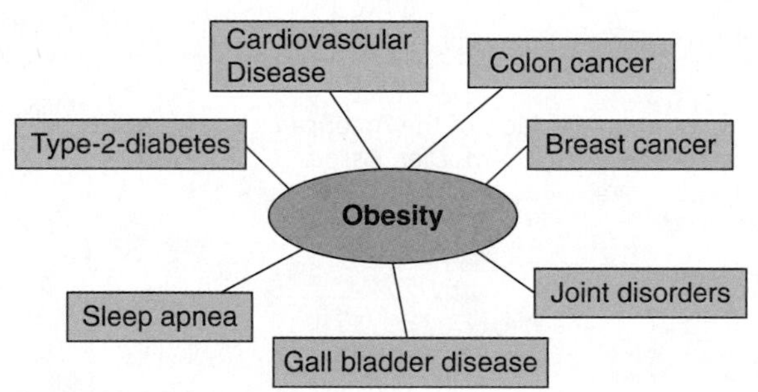

Figure F2.1 Obesity increases the risk for several types of diseases.

F2.1 Obesity and Type 2 Diabetes

Learning Objectives

- Explain the relationship between insulin resistance, beta-cell failure, and type 2 diabetes.
- Compare the insulin-resistant and insulin-sensitive states in muscle, adipose tissue, and the liver.
- Explain the relationship between obesity, inflammation, and the development of type 2 diabetes.

The link between obesity and type 2 diabetes is well known. Figure 7.13 indicates that type 2 diabetes becomes more common in Canadians as their BMI increases. More than 85% of people with type 2 diabetes are overweight or obese.[1] Having diabetes, in turn, increases the risks of heart disease and stroke.[2] How obesity contributes to the development of type 2 diabetes has been an area of active research and scientists have not yet fully answered this question, but the discussion here will present the most common hypotheses.[3]

Obesity, Insulin Resistance and Beta-Cell Failure

In order to understand the relationship between obesity and type 2 diabetes, it is first necessary to understand insulin resistance (see Section 4.5). and its role in the development of type 2 diabetes. Insulin resistance refers to a situation when tissues that normally respond to insulin, such as muscle and adipose tissue, become less responsive to insulin and do not take up glucose as readily. As a result, glucose remains in circulation and blood glucose levels rise. This results first in pre-diabetes and, finally, in type 2 diabetes. The sequence of events that are believed to cause this progression from normal blood glucose to diabetes is illustrated in **Figure F2.2**. The events are related to how well the pancreas is able to increase its secretion of insulin to compensate for the elevated blood glucose. When this compensation is complete, normal glucose levels are restored, but there are high levels of insulin in the blood, a situation called hyperinsulinemia (see Figure F2.2, step 1). If insulin resistance persists, the levels of insulin secreted may not fully compensate for increases in blood glucose. In this case, blood glucose levels rise, initially to pre-diabetic levels (see Figure F2.2, step 2) and finally to levels that are indicative of type 2 diabetes (see Figure F2.2, step 3).

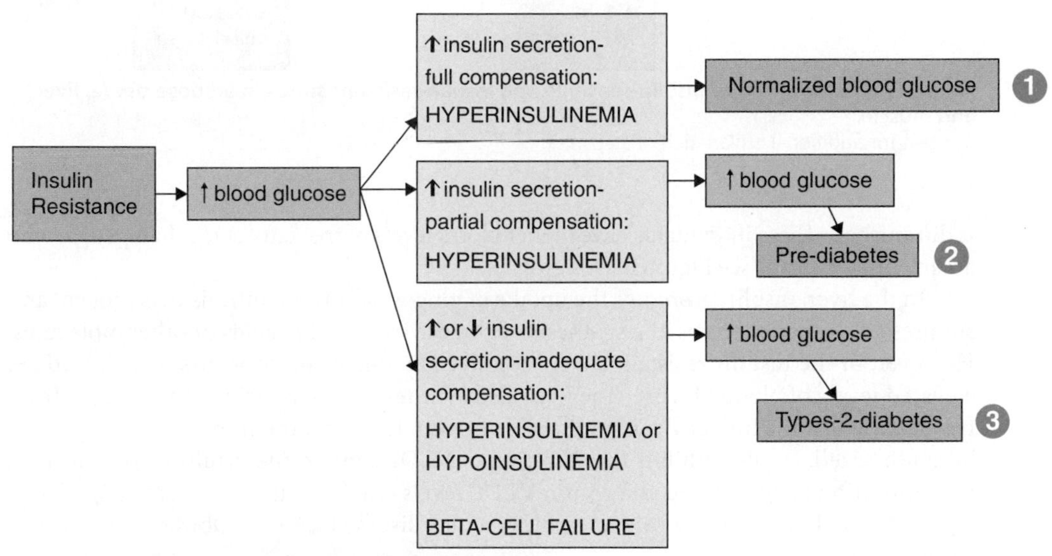

Figure F2.2 Development of type 2 diabetes.
Type 2 diabetes develops as a result of insulin resistance and beta-cell failure. See text for additional explanation of steps 1-3.

The cells of the pancreas that secrete insulin are called the beta-cells. Type 2 diabetes is believed to develop when both insulin resistance is present and some of the beta-cells of the pancreas cease to function properly, which is referred to as **beta-cell failure.** In some cases, there are enough functioning cells to continue to secrete insulin, that is, hyperinsulinemia continues; in other cases, there are too few cells to maintain insulin secretion and insulin secretion falls below normal levels (hypoinsulinemia).

beta-cell failure The malfunctioning of beta-cells of the pancreas, resulting in impaired secretion of insulin, an important factor in the development of type 2 diabetes.

Comparison of Insulin-Sensitive and Insulin-Resistant States

The elevation of blood glucose levels that we see during the development of type 2 diabetes (see Figure F2.2) is the result of changes in metabolism in the liver, adipose, and muscle tissue. Section 5.5 describes how insulin regulates not only glucose metabolism, but lipid metabolism as well. **Figure F2.3** illustrates the differences in carbohydrate and lipid metabolism that occur in each of these organs, when an individual is responsive to insulin (insulin-sensitive), or not (insulin-resistant).

Adipose tissue releases free fatty acids (FFA) into the blood (see Section 5.5), in a process called **lipolysis**; this process is the result of the action of the enzyme hormone-sensitive lipase. Insulin suppresses the effect of this hormone and promotes the uptake of triglycerides from chylomicrons, which come from the small intestine, and very-low-density lipoproteins (VLDLs), which are produced by the liver, by stimulating the enzyme lipoprotein lipase (See Chapter 5). With insulin-resistant individuals, this suppressive effect

lipolysis Breakdown of triglycerides to free fatty acids and glycerol.

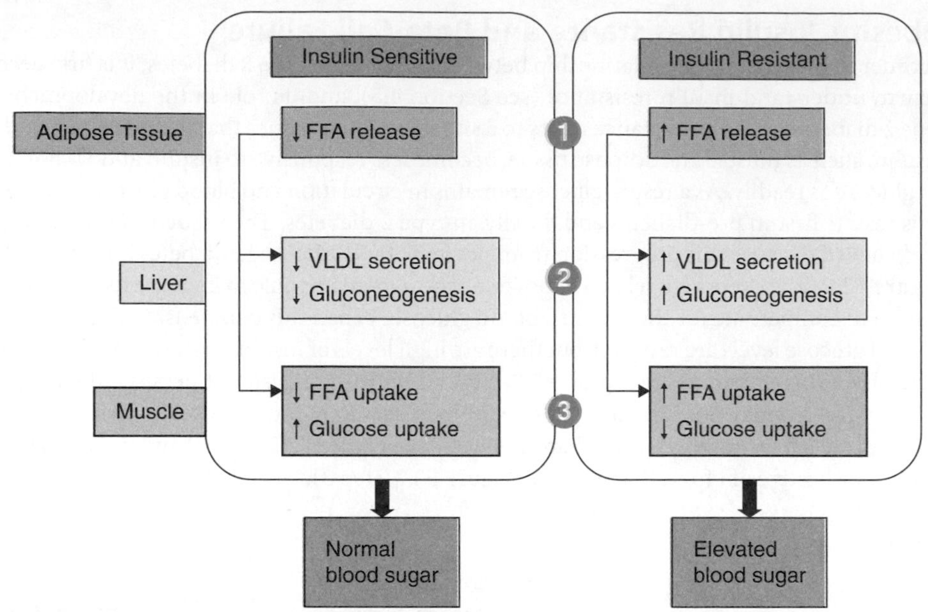

Figure F2.3 Comparison of insulin-sensitive and insulin-resistant states in adipose tissue, liver, and muscle.
See text for additional explanation of steps 1-3.

is diminished. These individuals have higher blood levels of free fatty acids than people who are insulin sensitive (see Figure F2.3, step 1).

In the liver, insulin promotes the uptake of glucose, for the synthesis of glycogen, and suppresses gluconeogenesis, the synthesis of glucose from amino acids or other molecules. However, in the insulin-resistant state, excessive gluconeogenesis occurs, which leads to elevated levels of blood glucose. The liver also synthesizes triglycerides and VLDLs from free fatty acids; the higher the blood levels of free fatty acids, the more triglycerides will be synthesized. Insulin inhibits the secretion of VLDLs, but in the insulin-resistant state, this inhibition is diminished and serum VLDL levels rise (see Figure F2.3, step 2).[4] These elevated VLDL levels play a role in cardiovascular disease (see F2.2 Obesity and Cardiovascular Disease).

Muscle takes up both glucose and fatty acids. In the fasting state, when insulin levels are low, muscle uses free fatty acids as fuel. After a meal, when insulin is secreted in response to glucose, the muscle of an insulin-sensitive individual switches from the uptake of fatty acids to the uptake of glucose as fuel. In the case of an insulin-resistant individual, in part due to the greater availability of free fatty acids, this switch does not occur and there is a diminished uptake of glucose. This also contributes to elevated levels of blood glucose (see Figure F2.3, step 3).

Figure F2.2 describes how insulin resistance can develop into type 2 diabetes. Figure F2.3 shows the metabolic changes that take place in the liver, adipose tissue, and muscle during insulin resistance that elevate blood glucose. With this background, let us consider the role of obesity in the development of type 2 diabetes.

Obesity and the Development of Type 2 Diabetes

Obesity is believed to contribute to the development of type 2 diabetes by promoting both insulin resistance and beta-cell failure. This is due, in part, to changes that occur in the metabolism of adipose tissue in the obese state (**Figure F2.4**).[5,6] Obesity induces an increase in the size of the adipocyte (fat-containing cell). The excessive enlargement of adipocytes, by complex mechanisms that will not be discussed here, causes macrophages to infiltrate adipose tissue. Macrophages are cells of the immune system and they induce inflammation in the adipose tissue (this is analogous to the inflammation induced by macrophages in the development of atherosclerosis; see Figure 5.22). The macrophages also secrete compounds that enter the blood and cause inflammation in other parts of the body. Most people think of inflammation as a process that takes place around an injury, a swelling induced by the immune system to limit the spread of infection. This is acute inflammation, a process that does not last long and helps to promote healing. In the body, however, chronic or long-term inflammation can also occur, such as in adipose tissue. There is no infection, but the macrophages secretes inflammatory compounds, nonetheless, that can alter metabolism throughout the body. Chronic inflammation, for example, interferes with insulin function, inducing insulin resistance (see Figure F2.4, step 1).

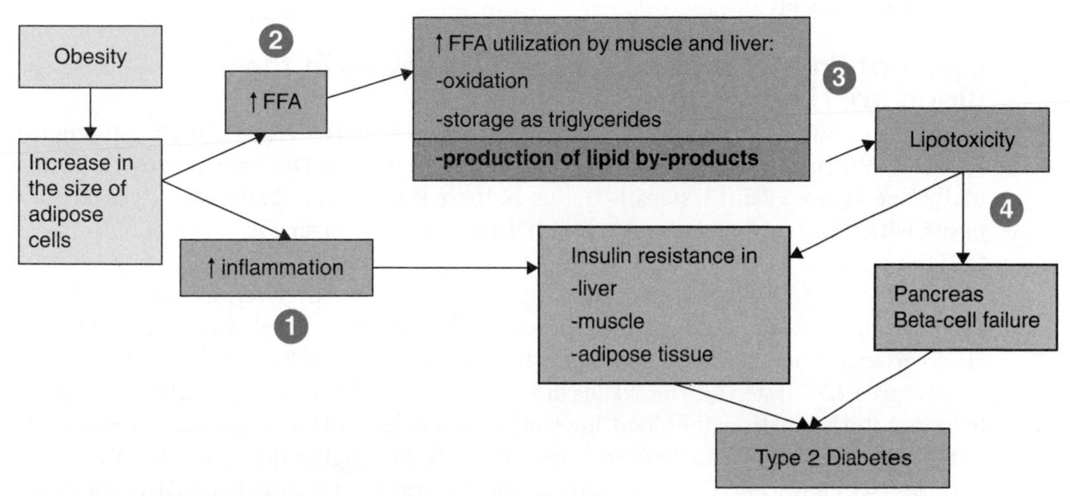

Figure 2.4 Obesity and the development of type 2 diabetes.
Obesity promotes both insulin resistance and beta-cell failure. See text for explanation of steps 1-4.
Source: Adapted from Schenk S, Saberi M, Olefsky, JM. 2008. Insulin sensitivity: modulation by nutrients and inflammation. *J Clin Invest 118*: 2992-3002.

In addition to inflammatory effect, the greater the amount of adipose tissue an individual has, the greater the level of free fatty acids in the blood. These free fatty acids are taken up by muscle (see Figure F2.4, step 2). Some of these fatty acids are oxidized, as a source of energy, and some are stored as triglycerides, but if the amount of free fatty acids taken up is excessive (as might be the case for an inactive obese individual), some are diverted into other metabolic pathways and produce products that induce insulin resistance, an effect called **lipotoxicity** (see Figure F2.4, step 3). A similar lipotoxic effect occurs in the liver, which like muscle, also takes up fatty acids. Lipotoxicity also occurs in the beta-cell of the pancreas, as it too is exposed to high levels of free fatty acids, and this contributes to beta-cell failure (see Figure F2.4, step 4). Obesity, therefore, contributes to the two elements present in type 2 diabetes, insulin resistance in adipose, muscle, and liver, along with beta-cell failure.[7]

lipotoxicity A toxic effect produced by the products of fatty acid metabolism by pathways other than fatty acid oxidation and triglyceride synthesis.

F2.2 Obesity and Cardiovascular Disease

Learning Objective

- Explain the relationship between obesity, the lipid triad, and the development of cardiovascular disease.

Obesity increases the risk of cardiovascular disease by altering the levels of lipoproteins that circulate in the blood; the greater the adipose tissue mass, the greater the release of free fatty acids. These fatty acids are taken up by the liver, where they are synthesized into triglycerides. which are packaged into VLDLs and secreted. As a result, an obese person will tend to have higher levels of VLDLs in the blood (measured as higher serum triglycerides), compared to a lean person (see Section 5.4 for a discussion of lipoproteins). The elevated levels of VLDLs alter lipoprotein transport in a way that promotes the development of atherosclerosis (see Section 5.6 for a discussion of atherosclerosis).

Normal Lipoprotein Metabolism

To understand the development of cardiovascular disease, first let us discuss normal lipoprotein metabolism (section 5.4). When VLDLs, produced in the liver, are taken up by tissue, the enzyme lipoprotein lipase (LPL) breaks down the triglycerides into glycerol and fatty acids, which are taken up by tissue. Cholesterol, which is also present in the VLDL particle, is released at the same time, and taken up by the HDL for transport to the liver process called reverse cholesterol transport, (**Figure F2.5 (a),** step 1). After triglycerides are removed from the VLDL, this particle is converted into IDL and LDL (see Figure F2.5(a), steps 2 and 3). The LDL transports cholesterol to tissue, including the liver (see Figure F2.5(a), step 4).

Lipoprotein Metabolism when VLDL Levels in the Blood are High

Now let us consider lipoprotein metabolism when VLDL levels are high (see Figure F2.5 (b)). When levels of VLDL are high, lipoprotein metabolism, in part, proceeds normally (see Figure F2.5 (b), steps 1-3), that is, there is the usual uptake of triglycerides by tissue, with transfer of cholesterol to the HDL particle and conversion of the VLDL particle to IDL and LDL.

But there are also some atypical transfers of cholesterol and triglycerides between VLDLs and other lipoproteins as well, because the levels of VLDL are very high. Cholesterol from the HDL particle, rather than being taken up by the liver, is transferred back to the VLDL particles (see Figure F2.5 (b) step 4). This results in a cholesterol-depleted HDL particle, which tends to be rapidly cleared from the blood, lowering HDL-cholesterol levels. In addition, as a result of the high levels of VLDL, there are atypical transfers of triglycerides from VLDL to LDL particles (see Figure F2.5 (b), step 5). As a result, this triglyceride-enriched LDL particle (see Figure F2.5 (b), step 6), rather than being directly taken up by the liver, interacts instead with an enzyme called hepatic lipase (which acts like lipoprotein lipase but is found in the liver) to remove the triglycerides, leaving, in the blood, an LDL particle that is smaller in size than the original LDL particle and enriched in cholesterol. This is called a small dense LDL (sd-LDL) (see Figure (b), step 7).

lipid triad or **atherogenic phenotype** A characteristic pattern of lipoproteins in the blood that increases the risk of cardiovascular disease: high serum triglycerids (due to high VLDL), low HDL-cholesterol, and high levels of small dense LDL.

Because of their small size, sd-LDL particles tend to get trapped in blood vessel walls and the cholesterol in the particle is more readily oxidized. Both properties promote the development of atherosclerosis and cardiovascular disease (see section 5.6). The combination of high serum triglycerides (due to high VLDL), low HDL-cholesterol, and high levels of sd-LDL is referred to as the **lipid triad** or the **atherogenic phenotype.**[8]

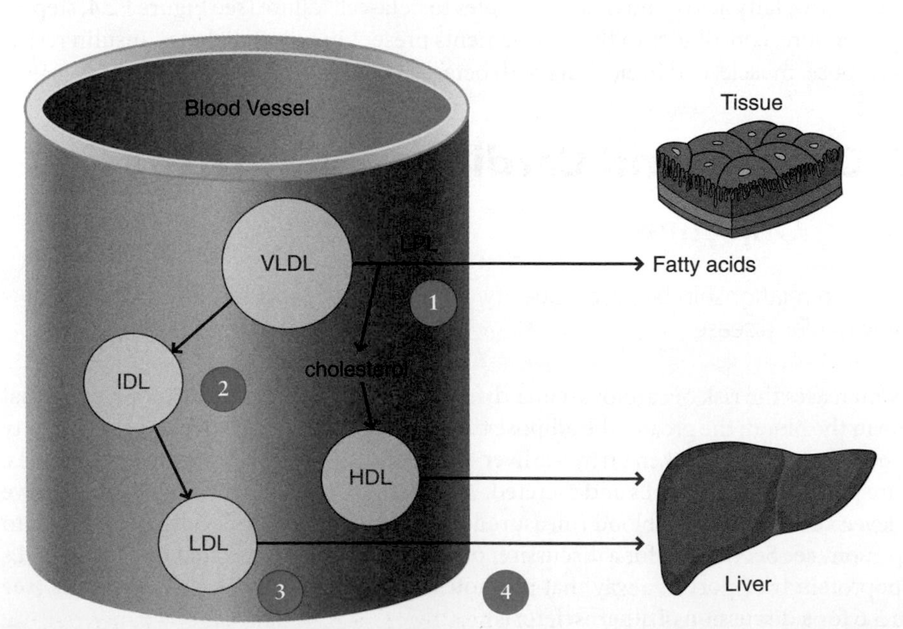

Figure F2.5a The development of the lipid triad
Normal lipoprotein transport. See text for explanation of steps 1-4.

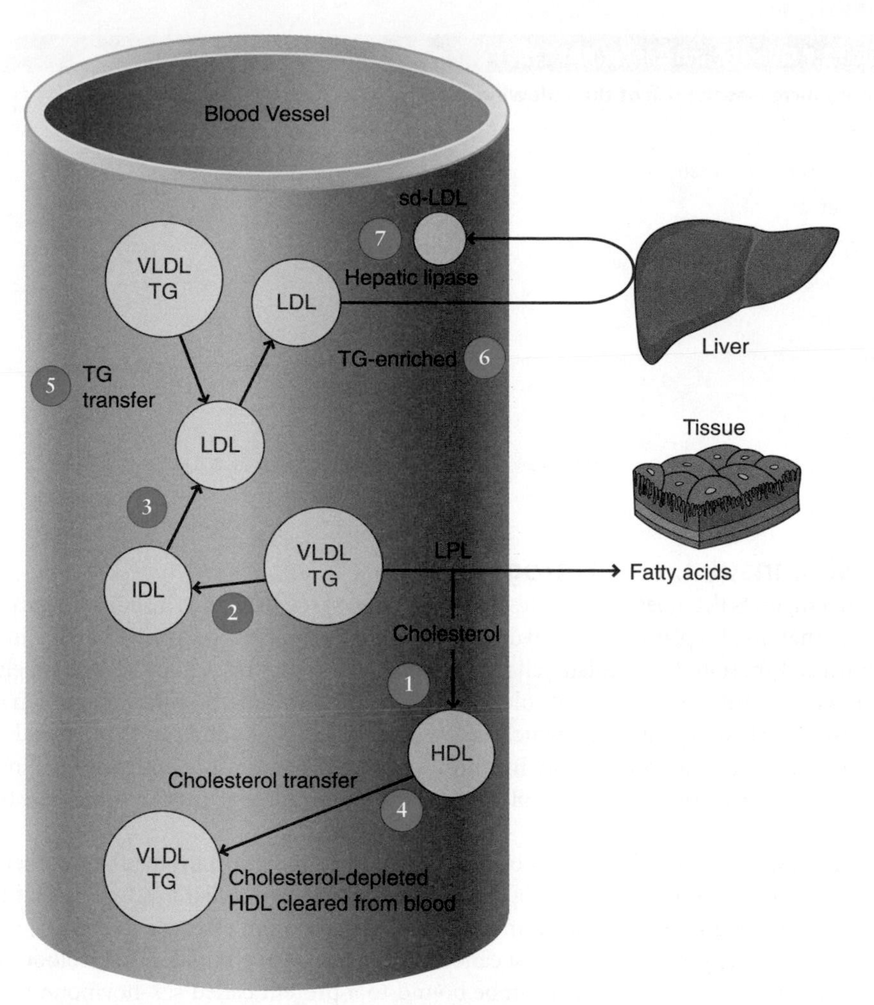

Figure F2.5b Lipoprotein transport when VLDL levels are high.
High levels of VLDL cause atypical transfers between lipoproteins (steps 4 and 5), resulting in the reduction of HDL cholesterol and the formation of sd-LDL (step 7), a combination that promotes the atherogenic process. See text for more explanation. Note: LPL = lipoprotein lipase; TG = triglycerides

F2.3 Obesity and Cancer

Learning Objectives

- Explain the relationship between obesity, insulin and estrogen.
- Explain the relationship between obesity, colon cancer, and breast cancer.

Obesity is associated with several types of cancer as shown in **Table F2.1.** The most widely studied cancers are colorectal and breast cancers, so the discussion will focus on these. Cancer is a disease characterized by a series of genetic mutations that change the way a cell functions and grows (see section 4.6). These mutations cause a cell to grow uncontrollably, resulting in a tumour. This growth is referred to as cell proliferation. A healthy cell has the ability to repair mutated DNA, or if the repair is not possible, to undergo programmed cell death, also called apoptosis. Cancer cells, on the other hand, become resistant to apoptosis. Obesity is linked to cancer because it tends to promote proliferation and suppress apoptosis, by raising the levels of growth factors in the blood. As suggested by their name, growth factors stimulate cell proliferation. If the growth of a healthy cell is stimulated, this is beneficial, but if a cell with cancer-causing mutations is stimulated, than a tumour may develop.

Table F2.1 Obesity and Cancer Risk

Obesity increases the risk of the following cancers:
Colorectal
Breast (postmenopausal)
Endometrial
Kidney
Esophageal
Pancreatic
Liver
Gallbladder
Gastric

Source: Calle, E. E. and Kaaks, R. Overweight, obesity, and cancer: Epidemiological evidence and proposed mechanisms. Nature Reviews Cancer 4: 579-591, 2004.

Obesity, Insulin, and Estrogen

Research suggests that obesity promotes the action of two growth factors: insulin and estrogen.[9] The role that insulin plays in carbohydrate and lipid metabolism is explained in Section 4.5. Insulin also functions to stimulate cell proliferation and to suppress apoptosis and is believed to play a role in the development of colorectal cancer. This stimulating effect on growth is not diminished by insulin resistance, which affects primarily carbohydrate and lipid metabolism in liver, muscle, and adipose tissue. Insulin resistance instead results in hyperinsulinemia, increasing the levels of insulin in the blood available to stimulate growth in vulnerable tissue, such as the colon.

Estrogen stimulates the growth of breast tissue and is important for the maintenance of healthy breast tissue. In postmenopausal women, however, research suggests that it can also promote the growth of breast tumours. The suppression of estrogen activity is, in fact, a successful strategy for treating breast cancer. When estrogen circulates in the blood, it can be present as "free" estrogen, or it can be bound to a protein called sex-hormone-binding globulin (SHBG). When estrogen is bound to SHBG, its bioavailability decreases, meaning it is less able to stimulate tumour growth.

Obesity, Colon Cancer, and Breast Cancer

The link between obesity, insulin, estrogen, and cancer is illustrated in **Figure F2.6**. As discussed, obesity promotes insulin resistance, resulting in an increase in insulin levels in the blood (see Figure F2.6, step 1). The hormone directly stimulates the growth of cells, such as colon cells, increasing the risk of tumour development (see Figure F2.6, step 2). Insulin also impacts estrogen metabolism by suppressing the synthesis of SHBG in the liver, increasing

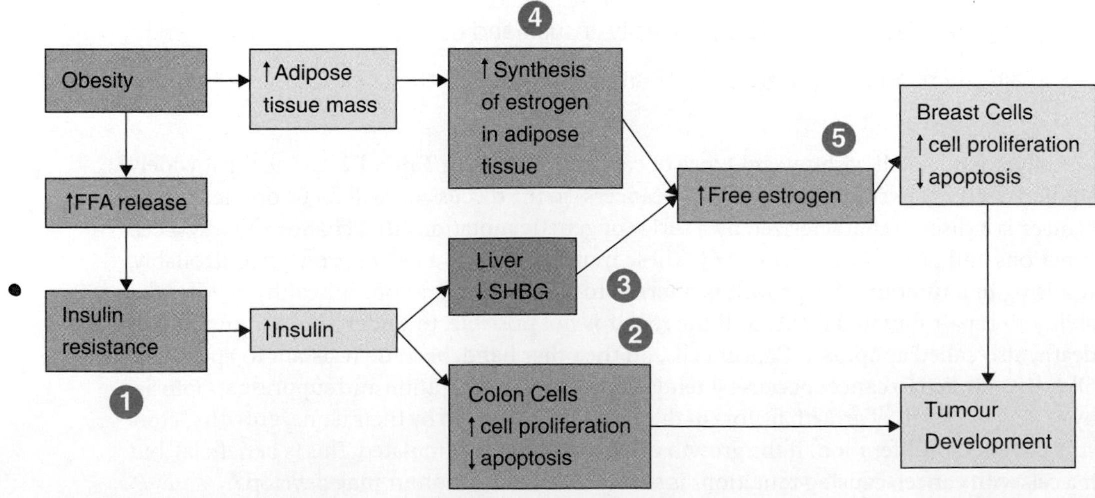

Figure F2.6 Relationship between obesity and cancer.
See text for explanation of steps 1-5. Note: SHBG=sex hormone binding globulin.
Source: Adapted from Calle, E. E. and Kaaks, R. Overweight, obesity, and cancer: Epidemiological evidence and proposed mechanism.[9]

the amount of free estrogen (see Figure F2.6, step 3). Estrogen is also synthesized in adipose tissue, which further contributes to increased levels of free estrogen (see Figure F2.6, step 4). This increased amount of free estrogen can stimulate the proliferation of breast cells, increasing the risk of breast cancer (see Figure F2.6, step 5).

F2.4 Obesity and Other Diseases

Learning Objective

- Describe the relationship between obesity and gallbladder disease, sleep apnea, and joint disorders.

Gallbladder Disease

Obesity is associated with an increase in gallstone formation (**Figure F2.7**). Gallstones are typically composed mostly of cholesterol and may form as a single large stone or many small ones. They often cause no symptoms, but if they lodge in the bile ducts, gallstones can cause pain and cramps. If the passage of bile is blocked, lipid absorption in the small intestine is impaired and the gallbladder can become inflamed.

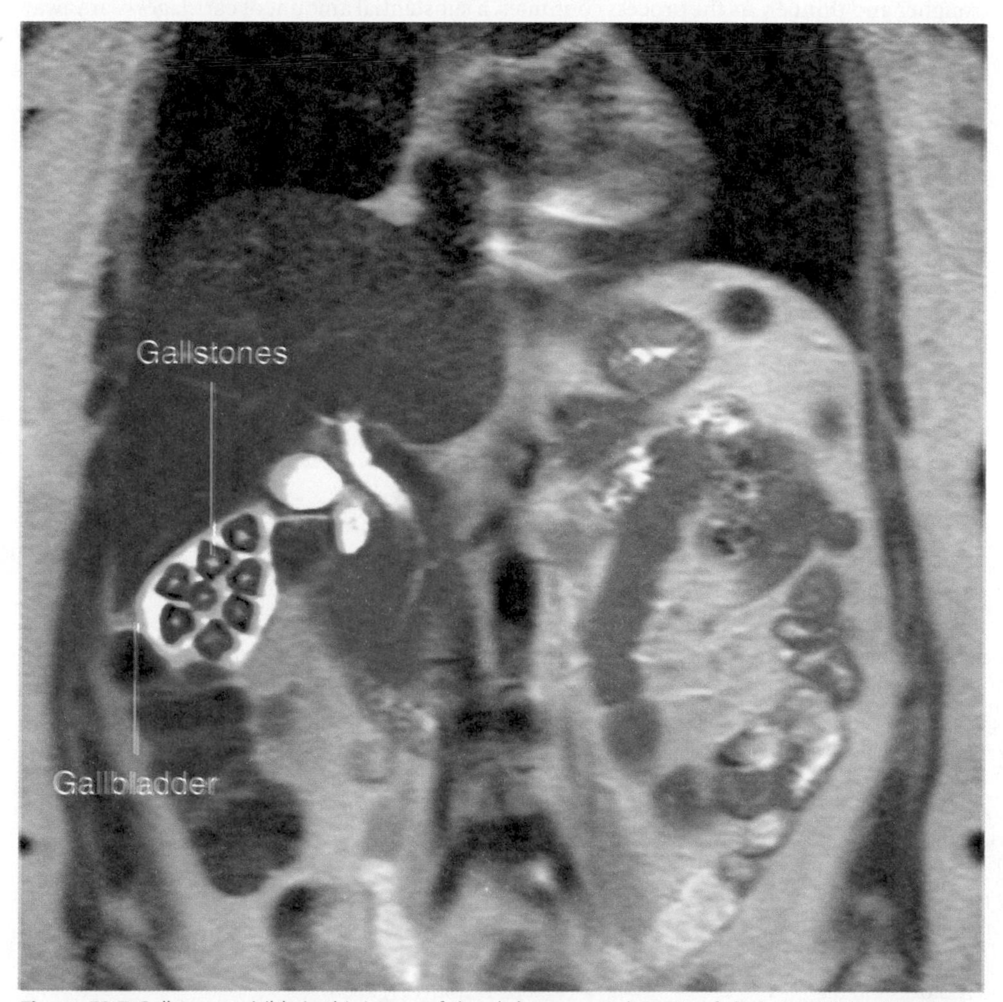

Figure F2.7 Gallstones, visible in this image of the abdomen, are deposits of cholesterol, bile pigments, and calcium in the gallbladder or bile duct. They may form as a single large stone or many small ones. (Simon Fraser/Photo Researchers, Inc.)

The more obese a person is, the greater his or her risk is of developing gallstones. Women who are obese have about 3-7 times the risk of developing gallstones as women at a healthy body weight.[10] The reason that obesity increases the risk of gallstones is unclear, but researchers believe that in obese people, the liver produces too much cholesterol, which deposits in the gallbladder and forms stones. Although the risk of gallstones decreases with a lower body weight, weight loss, particularly rapid weight loss, increases the risk of gallbladder disease because as lipids are released from body stores, cholesterol synthesis increases, increasing the tendency for cholesterol stone formation. Gallstones are one of the most medically important complications of voluntary weight loss.[10]

Sleep Apnea

Sleep apnea is a serious, potentially life-threatening condition characterized by brief interruptions of breathing during sleep. These short stops in breathing can happen up to 400 times every night and therefore patients with sleep apnea sleep very poorly and wake up in the morning still feeling tired, and remain tired throughout the day. Sleep apnea has also been associated with cardiovascular conditions such as high blood pressure, heart attack, stroke, impotence, and irregular heartbeat.[11] It is common in obesity because fatty tissue in the pharynx and neck compress the airway and block airflow. In obese individuals, weight loss may decrease both the frequency and severity of sleep apnea symptoms.[1]

Joint Disorders

Excess weight and fat can also increase the risk of developing osteoarthritis, a type of arthritis that occurs when the cartilage cushioning the joints breaks down and gradually becomes rougher and thinner. As the process continues, a substantial amount of cartilage wears away, so the bones in the joint rub against each other, causing pain and reducing movement. Being overweight is the most common cause of excess pressure on the joints and it can speed the rate at which the cartilage wears down. Losing weight reduces the pressure and strain on the joints and slows the wear and tear of cartilage.[1] In individuals suffering from osteoarthritis, weight loss can help reduce pain and stiffness in the affected joints, especially those in the hips, knees, back, and feet.[12]

Gout is a joint disease caused by high levels of uric acid in the blood. Uric acid is a nitrogenous waste product resulting from the breakdown of DNA and similar types of molecules. It sometimes forms into solid stones or crystal masses that become deposited in the joints, causing pain. The synthesis of uric acid is complex, but has been found to increase with insulin resistance. Not surprisingly then, gout is more common in overweight people, because of their greater risk of insulin resistance, and the risk of developing the disorder increases with higher body weight.

REFERENCES

1. NIDDK Weight Control Information Network. Do You Know the Risks of Being Overweight? Available online at win.niddk.nih.gov/publications/health_risks.htm#sleep/. Accessed June 11, 2009.
2. National Cholesterol Education Program. Adult Treatment Panel III Report, 2001, 2004. Available online at www.nhlbi.nih.gov/guidelines/cholesterol/atp3udp04.htm/. Accessed June 11, 2009.
3. McGarry, J. D. Dysregulation of fatty acid metabolism in the etiology of type 2 diabetes. *Diabetes* 51:7-18, 2002.
4. Lewis, G. F, Carpentier, A. , Adeli, K. et al. Disordered fat storage and mobilization in the pathogenesis of insulin resistance and type 2 Diabetes. *Endocrine Reviews* 23(2):201-229, 2002.
5. Schenk, S., Saberi, M., Olefsky, J. M. Insulin sensitivity: modulation by nutrients and inflammation. *J. Clin. Invest.* 118:2992-3002, 2008..
6. Cusi, K. The role of adipose tissue and lipotoxicity in the pathogenesis of type 2 diabetes. *Curr. Diab. Report* 10:306-315, 2010.
7. Muoio, D. M. and Newgard, C. B. Molecular and metabolic mechanisms of insulin resistance and beta-cell failure in type 2 diabetes. *Nat Rev: Mol Cell Biol* 9:193-205, 2008
8. Frayn, K. N. Metabolic Regulation. In: *A Human Perspective.* 3rd Edition.West Sussex, Wiley Blackwell. pp. 299-303.
9. Calle, E.E., and Kaaks, R. Overweight, obesity, and cancer: Epidemiological evidence and proposed mechanism. *Nat Rev Cancer* 4:579-591, 2004.
10. NIDDK Weight Control Information Network. Dieting and Gallstones. Available online at win.niddk.nih.gov/publications/gallstones.htm/. Accessed June 11, 2009.
11. Phillips, B. Sleep-disordered breathing and cardiovascular disease. *Sleep Med. Rev.* 9:131-140, 2005.
12. Lementowski, P. W. and Zelicof, S. B. Obesity and Osteoarthritis, *Am. J. Orthop.* 37:148-151, 2008.

8 The Water-Soluble Vitamins

Case Study

Chen is 65 years old and has always been healthy. He and his wife exercise regularly, watch their weight, and eat a healthy diet. Recently, Chen experienced a few episodes of forgetfulness, but he laughed it off as old age. Then he began feeling tired and having tingling in his hands and feet, difficulty walking, and diarrhea. Fearing the worst, he made an appointment to see his doctor.

Laboratory tests showed that Chen had low levels of vitamin B_{12}. A diet history revealed that his diet didn't provide much of the vitamin. He eats lots of grains, fruits, and vegetables, which do not provide vitamin B_{12}, and very little meat, which is a source of this vitamin. The doctor explained that Chen's symptoms were due to a vitamin B_{12} deficiency. The deficiency was likely caused by an inflammation of his stomach that reduced his ability to absorb the small amounts of vitamin B_{12} provided by his diet. Chen's doctor gave him an injection of vitamin B_{12} and recommended that he start taking a daily supplement containing the vitamin.

(Hiroshi Yagi/Getty Images, Inc.)

Because of our plentiful food supply and the availability of vitamin supplements, Canadians tend to think of vitamin deficiencies as a thing of the past, or perhaps as conditions that afflict people only in the developing world. For the most part that is true. But, as Chen's case illustrates, some segments of Canada's population still don't get enough of some vitamins due to low intakes, increased needs, or conditions that interfere with absorption or utilization.

In this chapter, you will learn about the water-soluble vitamins and their important functions in the body.

(©iStockphoto)

(Lew Robertson/FoodPix/Jupiter Images)

Chapter Outline

8.1 What Are Vitamins?

Learning Objectives

- Name the sources of vitamins in the Canadian diet.
- Describe how bioavailability affects vitamin requirements.
- Explain the role of coenzymes.

vitamins Organic compounds needed in the diet in small amounts to promote and regulate the chemical reactions and processes needed for growth, reproduction, and maintenance of health.

water-soluble vitamins Vitamins that dissolve in water.

fat-soluble vitamins Vitamins that dissolve in fat.

Vitamins are organic compounds that are essential in the diet in small amounts to promote and regulate the processes necessary for growth, reproduction, and the maintenance of health. When a vitamin is lacking in the diet, deficiency symptoms occur. When the vitamin is restored to the diet, the symptoms resolve. Vitamins have traditionally been grouped based on their solubility in water or fat. This chemical characteristic allows generalizations to be made about how they are absorbed, transported, excreted, and stored in the body. The **water-soluble vitamins** include the B vitamins and vitamin C. The **fat-soluble vitamins** include vitamins A, D, E, and K (**Table 8.1**). The vitamins were initially named alphabetically in approximately the order in which they were identified: A, B, C, D, and E. The B vitamins were first thought to be one chemical substance but were later found to be many different substances, so the alphabetical name was broken down by numbers; thiamin, riboflavin, and niacin were originally referred to as vitamin B_1, B_2, and B_3, respectively. Vitamins B_6 and B_{12} are the only ones that are still commonly referred to by their numbers.

Table 8.1 The Vitamins

Water-Soluble Vitamins	Fat-Soluble Vitamins
B vitamins	Vitamin A
Thiamin (B_1)	Vitamin D
Riboflavin (B_2)	Vitamin E
Niacin (B_3)	Vitamin K
Biotin	
Pantothenic acid	
Vitamin B_6	
Folate	
Vitamin B_{12}	
Vitamin C	

Vitamins in the Canadian Diet

The last of the 13 compounds recognized as vitamins today was characterized in 1948. The ability to isolate and purify vitamins has allowed them to be added to the food supply and incorporated into pills. As a result, the modern diet includes not only vitamins that are naturally present in food but also those that have been added to foods and those consumed as natural health products. Despite the variety of options for obtaining vitamins, it is still possible to consume too little of some vitamins and the likelihood of consuming too much of others is increasing.

Natural Sources of Vitamins Almost all foods contain some vitamins (**Figure 8.1**). Grains are good sources of most of the B vitamins. Leafy green vegetables provide folate, vitamin A, vitamin E, and vitamin K; citrus fruit provides vitamin C. Meat and fish are good sources of all of the B vitamins and milk provides riboflavin and vitamins A and D. Even oils provide vitamins; vegetable oils are high in vitamin E. How much of each of these vitamins remains in a food when it reaches the table depends on how the food is handled. Cooking and storage methods can cause vitamin losses. Processing can cause vitamin losses but can also add vitamins and other nutrients to food.

GRAIN PRODUCTS	VEGETABLES & FRUIT	OILS	MILK & ALTERNATIVES	MEAT & ALTERNATIVES
Thiamin	Riboflavin	Vitamin E	Riboflavin	Thiamin
Riboflavin	Niacin		Vitamin A	Riboflavin
Niacin	Vitamin B_6		Vitamin D	Niacin
Pantothenic acid	Folate		Vitamin B_{12}	Biotin
Vitamin B_6	Vitamin C			Pantothenic acid
Folate	Vitamin A			Folate
	Vitamin E			Vitamin B_{12}
	Vitamin K			Vitamin A
				Vitamin D
				Vitamin K

Figure 8.1 Vitamins in Canada's Food Guide food groups
Vitamins are found in foods from all groups, as well as in oils, but some groups are lacking in specific vitamins.

fortification A term used generally to describe the process of adding nutrients to foods, such as the addition of vitamin D to milk.

enrichment Refers to a food that has had nutrients added to restore those lost in processing to a level equal to or higher than originally present.

CANADIAN CONTENT

Fortified Foods The addition of nutrients to foods is called **fortification**. Consuming fortified foods increases nutrient intake. This can be beneficial if the added nutrients are deficient in the diet but it can also increase the risk of toxicity.

Government-mandated fortification programs have been used to increase nutrient intake and reduce deficiency diseases in populations. Adding nutrients to food is an effective way to supplement nutrients that are deficient in the population's diet without having to rely on consumers to alter their food choices or to take nutrient supplements. Which foods are fortified, which nutrients are added, and how much of a nutrient is added depends on the food supply, the needs of the population, and public health policies. Health Canada regulates which foods must be fortified and which nutrients should be added. The mandatory fortification of table salt with iodine, milk with vitamin D, and the **enrichment** of grains with thiamin, riboflavin, niacin, and iron as well as fortification with folic acid helps to prevent micronutrient deficiencies in the Canadian population (**Figure 8.2**). Today, fortification programs are used throughout the world to increase the intake of nutrients likely to be lacking (see Section 18.4). The levels of nutrients added are based on an amount that is high enough to benefit those who need to increase their intake but not so high as to increase the risk of excessive intakes in others.

ENRICHED PASTA

Nutrition Facts
Valeur nutritive
Per about 1/5 box (85 g)
par environ 1/5 boite (85 g)

Amount / **Teneur**	**% Daily Value** / **% valeur quotidienne**
Calories / Calories 310	
Fat / Lipides 1.5 g	**2 %**
Saturated / saturés 0 g + Trans / trans 0 g	**0 %**
Cholesterol / Cholestérol 0 mg	
Sodium / Sodium 0 mg	**0 %**
Carbohydrate / Glucides 66 g	**22 %**
Fibre / Fibres 8 g	**32 %**
Sugars / Sucres 3 g	
Protein / Protéines 11 g	
Vitamin A / Vitamine A	0 %
Vitamin C / Vitamine C	0 %
Calcium / Calcium	2 %
Iron / Fer	25 %
Thiamine / Thiamine	40 %
Riboflavin / Riboflavine	6 %
Niacin / Niacine	20 %
Folate / Folate	80 %

Figure 8.2 Nutrients in enriched pasta
Health Canada requires the fortification of pasta with folic acid, niacin, thiamin, riboflavin, and iron. All nutrients that are added to the product must be listed on the nutrition facts label, which is shown here in bilingual format.

Natural health products Natural health products are another source of vitamins in the modern diet (see Section 2.5). These products include vitamin and mineral supplements as well as amino acids, fatty acids, probiotics, and traditional herbs and other forms of traditional medicine. Natural health products are regulated by Health Canada's Natural Health Products Directorate to ensure the quality, safety, and efficacy of the products Canadians buy (see Your Choice: Natural Health Products in Food Format). While supplements provide specific nutrients and can help some people meet their nutrient needs, they do not provide all the benefits of a diet containing a wide variety of foods.[1] A varied diet provides phytochemicals and other substances that are not nutrients but that have health-promoting properties as well as energy, water, protein, minerals, or fibre (**Figure 8.3**). Epidemiological studies show that people who eat more fruits and vegetables have a lower incidence of a host of chronic diseases. These same benefits are not duplicated by taking supplements of nutrients found in these foods. Scientists have not yet identified all the substances contained in foods, nor have they determined all of their effects on human health. What is clear is that a wholesome, varied diet is important for optimal health. If chosen with care, supplements are unlikely to be harmful, but consumers should not rely heavily on them to meet their needs.

Understanding Vitamin Needs

Today, in Canada and other industrialized countries, an understanding of the sources and functions of the vitamins, a varied food supply, and the ability to fortify foods and supplement nutrients have helped to eliminate severe vitamin deficiencies as a public health problem (see Critical Thinking: How are Canadians Doing with Respect to Their Intake from Food of

Figure 8.3 Vitamin supplements cannot take the place of a balanced diet. (© istockphoto.com/Woraput Chawalitphon)

Water-Soluble Vitamins?). For example, niacin deficiency, which was common in the southern United States in the early 1900s, is now almost unheard of in both Canada and the United States; severe vitamin deficiencies, such as vitamin C deficiency, which killed countless sailors and soldiers throughout history, is now a rarity; and vitamin A deficiency, which remains a major public health concern worldwide, rarely occurs in developed countries. However, despite all of our knowledge, our varied diet, and shelves of vitamin supplements, not everyone gets enough of every vitamin all the time. Certain segments of the population, such as children, pregnant women, and the elderly, are at particular risk for deficiency. Some vitamin deficiencies are on the rise because of changes in dietary patterns. In addition, marginal deficiencies, which may have been present in the population for a long time, are now being recognized, and their detrimental effects better understood. In order to make recommendations about how much of each vitamin is needed to optimize health, it is important to understand how vitamins are absorbed, transported, used, stored, and excreted.

Bioavailability of Vitamins Whether vitamins come from foods, fortified foods, or supplements, they must be absorbed into the body to perform their functions. About 40%-90% of the vitamins in food are absorbed, primarily in the small intestine (**Figure 8.4**). The composition of the diet and conditions in the body, however, may influence how much of a vitamin is available in the body. **Bioavailability** considers the amount of a nutrient that can be absorbed and utilized by the body.

bioavailability A general term that refers to how well a nutrient can be absorbed and used by the body.

One of the key factors affecting bioavailability is whether the vitamin is soluble in fat or water. Fat-soluble vitamins require fat in the diet for absorption and are poorly absorbed when the diet is very low in fat. The water-soluble vitamins do not require fat for absorption but many depend on energy-requiring transport systems or must be bound to specific molecules in the gastrointestinal tract in order to be absorbed. For example, thiamin and vitamin C are absorbed by energy-requiring transport systems, riboflavin and niacin require carrier proteins for absorption, and vitamin B_{12} must be bound to a protein produced in the stomach before it can be absorbed in the intestine.

Once absorbed into the blood, vitamins must be transported to the cells. Most of the water-soluble vitamins are bound to blood proteins for transport. Fat-soluble vitamins must be incorporated into lipoproteins or bound to transport proteins in order to be transported in the aqueous environment of the blood. For example, vitamins A, D, E, and K are all incorporated into chylomicrons for transport from the intestine. Vitamin A is stored in the liver, but it must be bound to a specific transport protein to be transported in the blood to other tissues; therefore, the amount delivered to the tissues depends on the availability of the transport protein.

Some vitamins are absorbed in inactive **provitamin** or **vitamin precursor** forms that must be converted into active vitamin forms once inside the body. How much of each provitamin can be converted into the active vitamin and the rate at which this occurs affect the amount of a vitamin available to function in the body.

provitamin or **vitamin precursor** A compound that can be converted into the active form of a vitamin in the body.

Storage and Excretion The ability to store and excrete vitamins helps to regulate the amount present in the body. With the exception of vitamin B_{12}, the water-soluble vitamins are easily excreted from the body in the urine. Because they are not stored to any great extent, supplies of water-soluble vitamins are rapidly depleted and they must be consumed regularly in the diet. Nevertheless, it takes more than a few days to develop deficiency symptoms, even when these vitamins are completely absent from the diet. Fat-soluble vitamins, on the other hand, are stored in the liver and fatty tissues and cannot be excreted in the urine. In general, because they are stored to a larger extent, it takes longer to develop a deficiency of fat-soluble vitamins when they are no longer provided by the diet.

Recommended Intakes Recommendations for vitamin intake for healthy populations in the United States and Canada are made by the DRIs (see Section 2.2). The DRIs provide either an RDA value, when sufficient information is available to establish an EAR, or an AI, when a recommendation is estimated from population data. These values are used as a goal for dietary intake by individuals. The DRIs also establish Tolerable Upper Intake Levels (ULs) as a guide to the maximum amount of a nutrient that is unlikely to cause adverse health effects (see inside cover). Meeting vitamin needs without exceeding a safe level of intake requires careful attention to the kinds of foods chosen as well as knowledge of the nutrients added to foods and those consumed in supplements.

YOUR CHOICE

(©iStockphoto)

Natural Health Products in Food Format

Vitamin and mineral supplements are examples of natural health products (see Section 2.5). We tend to think of these supplements as coming in the form of pills or drops. But consider the following: an energy bar that contains soy protein and 23 vitamins and minerals; bottled water with vitamins added to it; a carbonated drink containing B-vitamins and caffeine. These are also examples of the natural health products that have appeared on the Canadian market. They are products that are described by Health Canada's Natural Health Products Directorate as natural health products in "food format."[1]

The main difference between natural health products and food is that natural health products are defined as having a therapeutic purpose, for example: "the diagnosis, treatment, mitigation or prevention of a disease, disorder, or abnormal physical state or its symptoms in humans; restoring or correcting organic functions in humans; or modifying organic functions in humans, such as modifying those functions in a manner that maintains or promotes health." They do not require a prescription.[1]

Food, on the other hand, is recognized by its traditional use as a food to "provide nourishment, nutrition or hydration, or to satisfy hunger, thirst or a desire for taste, texture, or flavour."[1]

There is concern that natural health products in food format may confuse the consumer. All natural health products are labelled with a maximum daily dose. In the case of vitamin and mineral supplements, these are intended to ensure that consumers do not exceed the upper tolerable intake levels (UL). Dietitians of Canada have issued a position paper opposing the sale of natural health products in food format because consumers may perceive them as a food, be unaware that there are maximum doses, and exceed the ULs for some nutrients, with potential adverse effects.[2] For example, the overconsumption of vitamin water during very hot summer days could result in excessive intake of some vitamins.

An additional criticism of natural health products in food format made by Dietitians of Canada is that there is no nutrition facts label on these products. A natural health product in beverage form, for example, may contain added sugar, as well as vitamins, but consumers would not know the caloric content of the product. Health Canada is evaluating the issues related to natural health products in food format.[1]

If you are considering using a natural health product, whether in tablet or food format, you have to thoughtfully consider its suitability as part of your self-care.[3] Remember that many nutrients are readily available from the diet and you may not need to use a natural health product. Dietary modifications may be a more effective way to improve your health as they will provide not only the selected nutrients available in a supplement, but macronutrients, fibre, as well as additional nutrients and phytochemicals.

If you are considering their use, though, the labels of natural health products, although different from food labels, will provide you with important information that will allow you to make a safer and more informed choice. The Natural Health Products Directorate evaluates each natural health product for quality, safety, and efficacy (e.g., whether it does what it claims to do) and when a product meets the directorate's standards, it is awarded a natural products number (NPN). Therefore, the first thing you should look for on the label is the NPN, as absence of a number means that the product has not been evaluated by Health Canada.

Additional information that appears on the label includes the recommended dose and directions on how to take the product (e.g., orally, with or without food). The label will also indicate the health benefit of taking the product, for example, a vitamin C supplement might be described as "a factor in the maintenance of bones, cartilage, teeth and gums." Also listed on the label are the medicinal (e.g., amount of vitamin C) and non-medicinal ingredients (e.g., any other ingredients used to formulate the supplement); if it is a beverage with added sugar, sugar would be listed here.

You will also see information intended to protect you from harm. This includes, where applicable, warnings against using of the product if done in combination with other natural health products or prescription drugs, or if you have certain medical conditions, such as diabetes or heart disease, or if you are pregnant or breast feeding. Any known adverse reactions will also be described.[4]

When in doubt, also consider consulting with a health professional such as a dietitian, doctor, a pharmacist, a nurse or naturopath prior to using the product.[4] When considering a natural health product, such as a vitamin or mineral supplement, be sure to use all the information available to help you to make your choice.

[1]Health Canada. Classification of products at the food-natural health products interface: products in food formats. Available online at. http://www.hc-sc.gc.ca/dhp-mps/prodnatur/legislation/docs/food-nhp-aliments-psn-guide-eng.php. Accessed March 27, 2011.

[2]Dietitians of Canada. Natural health products in food format. Available online at http://www.dietitians.ca/Dietitians-View/Natural-Health-Products-in-Food-Format.aspx. Accessed March 27, 2011.

[3]Health Canada. Informing You about Natural Health Products. Available online at http://www.hc-sc.gc.ca/dhp-mps/prodnatur/fiche_info_sheets-eng.php. Accessed March 27, 2011.

[4]Health Canada. Informing yourself about natural health products-information sheet #5- for consumers- informing yourself. Available online at http://www.hc-sc.gc.ca/dhp-mps/prodnatur/fiche_info_sheets_5-eng.php. Accessed March 27, 2011.

CRITICAL THINKING:

How are Canadians Doing with Respect to Their Intake, from Food, of Water-Soluble Vitamins?

CANADIAN CONTENT

In 2004, Canadian Community Health Survey determined the intake of vitamins Canadians get from food. The proportion of the population with an intake below the EAR indicates the proportion of the population that is not meeting their requirements for nutrients. This analysis is called the EAR cutpoint method (see Section 2.2). Health Canada considers a low prevalence of inadequate intake to be <10%; that is, when the proportion of Canadians with intakes below the EAR of a nutrient is less than 10%, then there are no concerns about the intake of that nutrient in the overall population. This 10% cutoff is indicated by the green line in the bar graph. Nutrients, where the proportion of the population with intakes below the EAR is greater than 10%, that is, to the right of the green line, are of concern.

In addition Health Canada has created a Nutrition and Health Atlas where regional differences in nutrient intakes can be illustrated, such as the map shown for vitamin C.

Regional differences in vitamin C intake across Canada showing the proportion (%) of adults with intakes below the EAR.

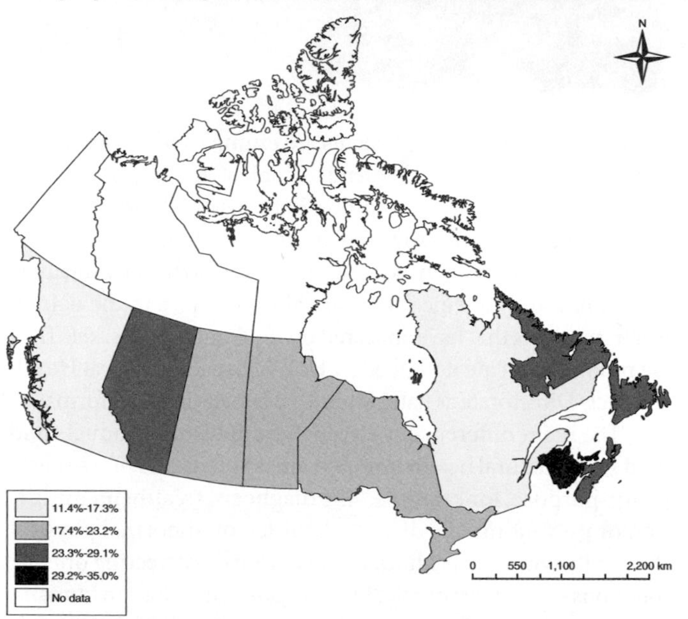

Source: Health Canada. Canada's Nutrition and Health Atlas. Available online at http://www.hc-sc.gc.ca/fn-an/surveill/atlas/map-carte/nutri_vitc_mf-hf-eng.php. Accessed July 22, 2010.

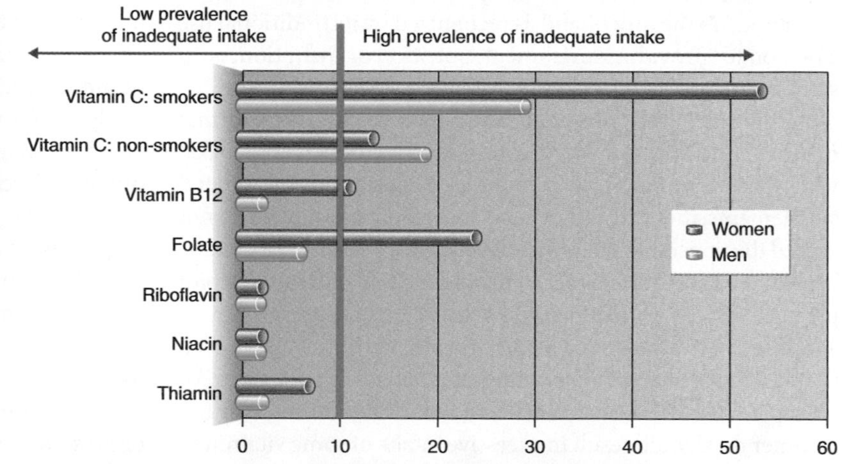

Source: Health Canada and Statistics Canada. 2004. Canada Community Health Survey Cycle 2.2., Nutrient Intakes from Food. Data files on CD. Cat No. 978-0-662-06542-510.

Photolink/Photodisc/Getty Images

Critical Thinking Questions

Based on the data in the bar graph, identify areas of concern; that is, identify the vitamins and the population groups with low intakes.

What Canada's Food Guide food groups would you recommend to Canadians to improve their intake of these vitamins? (Hint: Check the figures in this chapter that illustrate the vitamin content of Canada's Food Guide food groups.)

The data in the bar graph and map are based on the intake of vitamins from food alone. No data on vitamin intake from supplements are presented. Speculate on how the numbers in the bar graph might change if intake form supplements is included.

The map shows the regional differences in vitamin C intake across Canada. Where in Canada is the intake of vitamin C most adequate? Inadequate? Why do you think these differences exist?

*Answers to all Critical Thinking questions can be found in Appendix J.

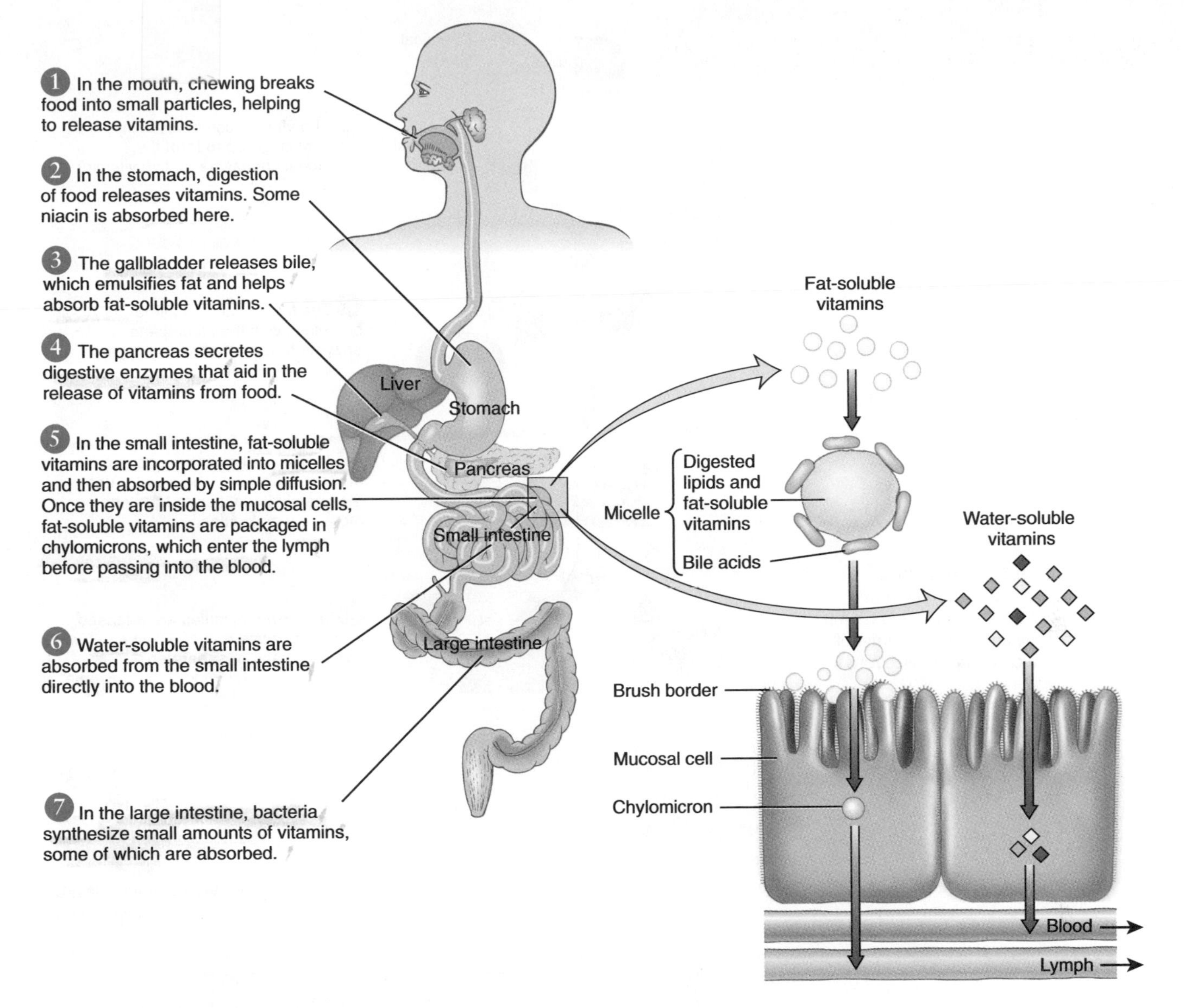

Figure 8.4 Vitamins in the digestive tract
Most vitamin absorption takes place in the small intestine. The mechanism by which vitamins are absorbed and transported affects their bioavailability.

Understanding Vitamin Functions

Vitamins promote and regulate body activities. Each vitamin may have unique roles, but vitamins also function together with other vitamins and nutrients in important metabolic pathways in the body. The B vitamins, for example, all function as components of **coenzymes.** Coenzymes are organic compounds that bind to enzymes and are essential to the function of the enzymes to which they bind (**Figure 8.5**). If someone has an inadequate intake of a vitamin, then the coenzyme cannot form and enzyme activity is impaired.

coenzymes Small nonprotein organic molecules that act as carriers of electrons or atoms in metabolic reactions and are necessary for the proper functioning of many enzymes.

Many B vitamins function together in energy metabolism. Thiamin, riboflavin, niacin, pantothenic acid, and biotin all serve as coenzymes for reactions that release energy from carbohydrate, fat, and protein as well as alcohol (**Figure 8.6**).

Vitamin B_6, folate, and B_{12} function together. Vitamin B_6, folate, and vitamin B_{12} function together in metabolic pathways related to single carbon metabolism. These pathways support fetal development, and the formation of cell blood cells, and the health of nerve tissue and are described in sections 8.7, 8.8, and 8.9.

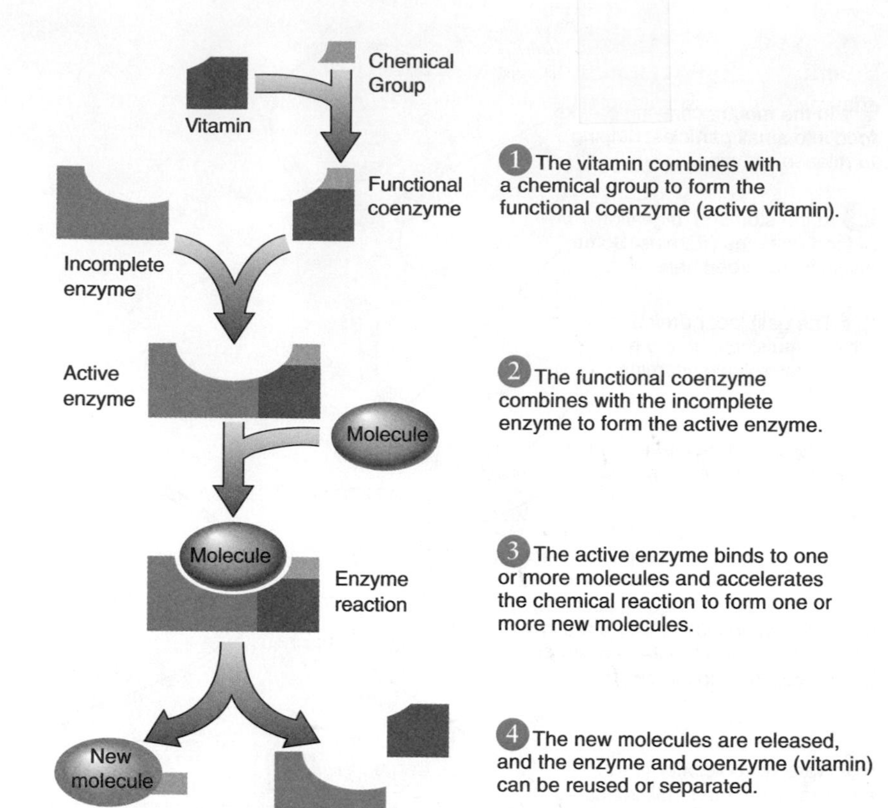

Figure 8.5 Coenzymes
The active coenzyme form of a vitamin is necessary for enzyme activity and acts as a carrier of chemical groups or electrons in the reaction.

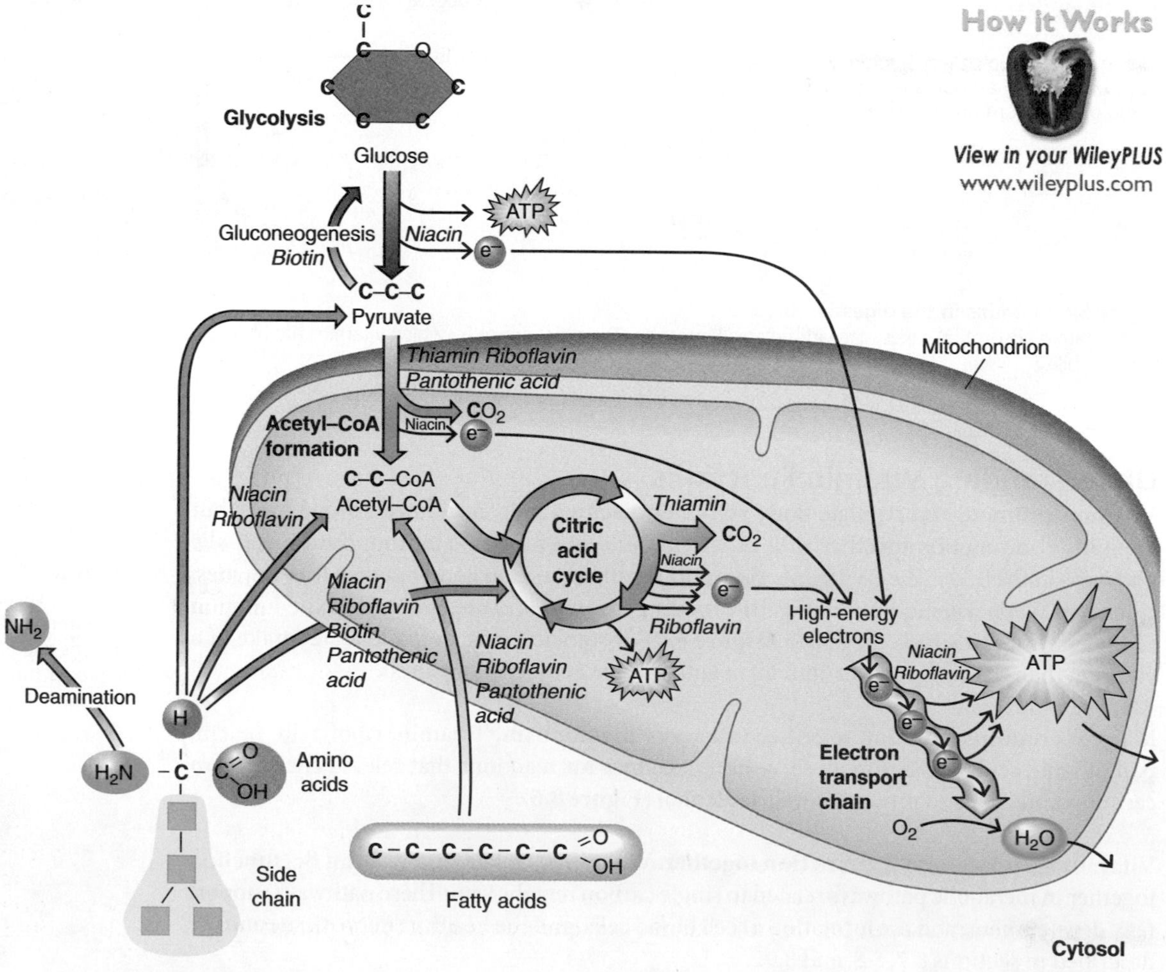

How it Works

View in your WileyPLUS
www.wileyplus.com

Glycolysis
Glucose
Gluconeogenesis
Biotin
Niacin
ATP
e^-
C–C–C
Pyruvate
Mitochondrion
Thiamin Riboflavin
Pantothenic acid
Acetyl–CoA formation
Niacin
CO_2
e^-
C–C–CoA
Acetyl–CoA
Niacin
Riboflavin
Citric acid cycle
Thiamin
CO_2
Niacin
e^-
Riboflavin
ATP
Niacin
Riboflavin
Biotin
Pantothenic acid
Niacin
Riboflavin
Pantothenic acid
High-energy electrons
Niacin
Riboflavin
ATP
Electron transport chain
O_2
H_2O
NH_2
Deamination
H
H_2N
Amino acids
Side chain
C–C–C–C–C–C
Fatty acids
Cytosol

Figure 8.6
B vitamins and energy metabolism
Reactions that require thiamin, riboflavin, niacin, biotin, or pantothenic acid as coenzymes are particularly important in the production of ATP from glucose, fatty acids, and amino acids.

Vitamin C and Vitamin E function together. Like the B vitamins, Vitamin C is also a coenzyme that is essential for the synthesis of neurotransmitters, hormones, and collagen, a protein vital to the structure of connective tissue. It also works with vitamin E to protect the body from oxidative damage.

In this chapter, we will discuss each individual nutrient as well as the ways in which they function together.

8.2 Thiamin

Learning Objectives

- Discuss the role of thiamin in providing energy.
- Explain why a thiamin deficiency causes neurological symptoms.

Thiamin was the first of the B vitamins to be identified and is therefore sometimes called vitamin B_1. The disease that results from a deficiency of this vitamin, **beriberi**, has been present in East Asia for more than 1,000 years and came to the attention of Western medicine in colonial Asia in the nineteenth century. Beriberi became such a problem that the Dutch East India Company sent a team of scientists to find its cause. What they were expecting to find was a micro-organism like those that caused cholera and rabies. What they found for a long time was nothing. For more than 10 years, a young physician named Christian Eijkman tried to induce beriberi in chickens. His success came as a twist of fate. He ran out of food for his experimental chickens and instead of the usual brown rice, he fed them white rice. Shortly thereafter, the chickens came down with beriberi-like symptoms. When he fed them brown rice again, they got well. This provided evidence that the cause of beriberi was not a poison or a micro-organism, but rather something missing from the chickens' diet.

beriberi The disease resulting from a deficiency of thiamin.

Just as a diet of white rice was the cause of beriberi in Eijkman's chickens, a diet consisting primarily of white or polished rice was also the reason the incidence of beriberi in East Asia increased dramatically in the 1800s. Polished or white rice is produced by polishing off the bran layer of brown rice, creating a more uniform product. However, polishing off the bran also removes the thiamin-rich portion of the grain (**Figure 8.7**). Therefore, in populations where white rice was the staple of the diet, beriberi became a common health problem.

Figure 8.7 Why might thiamin deficiency be more common in cultures where unenriched white rice is a dietary staple? (© istockphoto.com/Stepan Popov)

Thiamin in the Diet

Thiamin is widely distributed in foods (**Figure 8.8**). A large proportion of the thiamin consumed comes from enriched grains used in foods such as breakfast cereals, breads, and other baked goods. Pork, whole grains, legumes, nuts, seeds, and organ meats (liver, kidney, heart) are also good sources.

The thiamin in foods may be destroyed during cooking or storage because it is sensitive to heat, oxygen, and low-acid conditions. Thiamin bioavailability is also affected by the presence of antithiamin factors that destroy the vitamin. For instance, there are enzymes in raw shellfish and freshwater fish that degrade thiamin during food storage and preparation and during passage through the gastrointestinal tract. These enzymes are destroyed by cooking so they are only a concern in foods consumed raw. Other anti-thiamin factors that are not inactivated by cooking are found in tea, coffee, betel nuts, blueberries, and red cabbage. Because these make thiamin unavailable to the body, habitual consumption of foods containing anti-thiamin factors increases the risk of thiamin deficiency.[2]

Thiamin in the Body

Thiamin is a vitamin so it does not provide energy, but it is important in the energy-yielding reactions in the body. The active form, **thiamin pyrophosphate**, is a coenzyme for reactions in which carbon dioxide is lost from larger molecules (see Appendix I). For instance, the reaction that forms acetyl-CoA from pyruvate and one of the reactions of the citric acid cycle require thiamin pyrophosphate (see Figure 8.6). Thiamin is therefore essential to the production of ATP from glucose.

thiamin pyrophosphate The active coenzyme form of thiamin. It is the predominant form found inside cells, where it aids reactions in which a carbon-containing group is lost as CO_2.

Thiamin is also needed for the metabolism of other sugars and certain amino acids; the synthesis of the neurotransmitter acetylcholine; and the production of the sugar ribose, which is needed to synthesize RNA (ribonucleic acid).

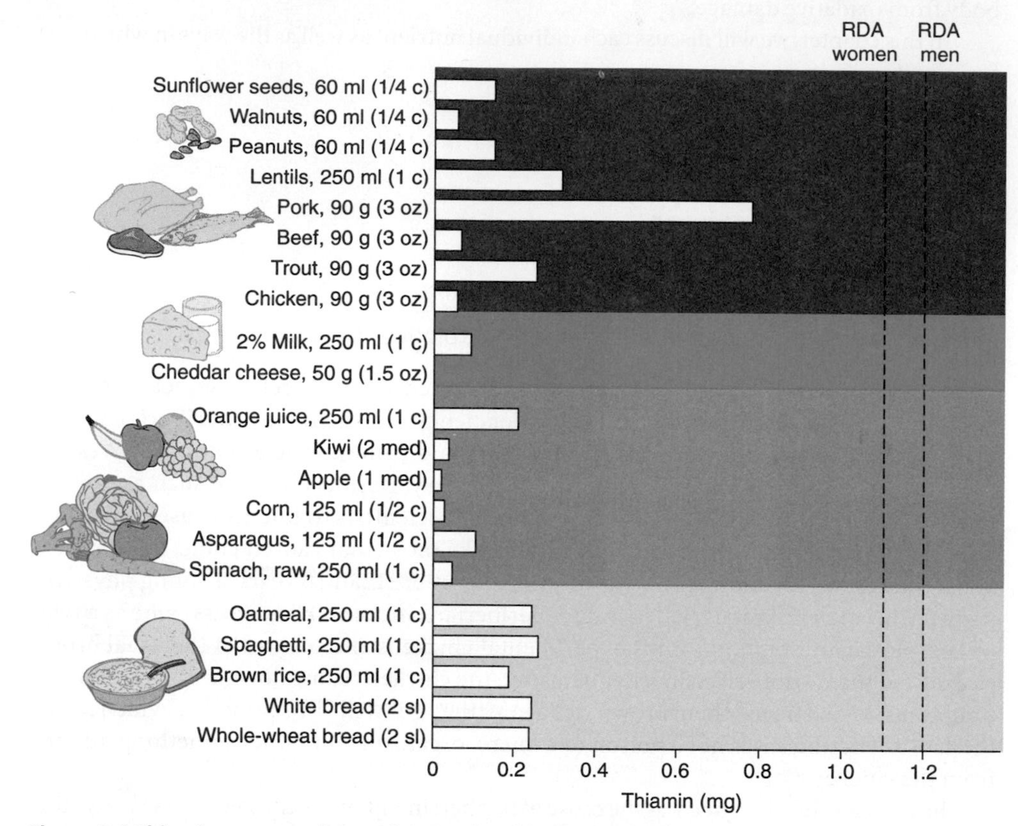

Figure 8.8 Thiamin content of Canada's Food Guide food groups
A combination of grains and foods from the meats and alternatives group can easily supply the RDA of thiamin (dashed lines).

Some, but not all, of the symptoms of beriberi can be explained by the roles of thiamin in glucose metabolism and in the synthesis of the neurotransmitter acetylcholine. The earliest symptoms, depression and weakness, which occur after only about 10 days on a thiamin-free diet, are probably related to the inability to completely use glucose. Since brain and nerve tissue rely on glucose for energy, the inability to form acetyl-CoA rapidly affects nervous system activity. Poor coordination, tingling in the arms and legs, and paralysis may also be caused by the lack of acetylcholine. The reason thiamin deficiency causes cardiovascular symptoms is not well understood.

Recommended Thiamin Intake

The RDA for thiamin for adult men age 19 and older is set at 1.2 mg/day, and for adult women 19 and older, at 1.1 mg/day. The RDA is based on the amount of thiamin needed to achieve and maintain normal activity of a thiamin-dependent enzyme found in red blood cells and normal urinary thiamin excretion.[3] For an average adult, half of the RDA can be obtained from 125 g of pork or 60 ml of shelled sunflower seeds.

The requirement for thiamin is increased during pregnancy to accommodate the needs of growth and energy utilization, and during lactation, to meet the need for increased energy for milk production and to replace the thiamin secreted in milk. There is not enough information to establish an RDA for infants, so an AI has been set based on the thiamin intake of infants fed human milk. A summary of the sources, recommended intakes, functions, deficiencies, and toxicities of thiamin and other water-soluble vitamins is provided in **Table 8.2**.

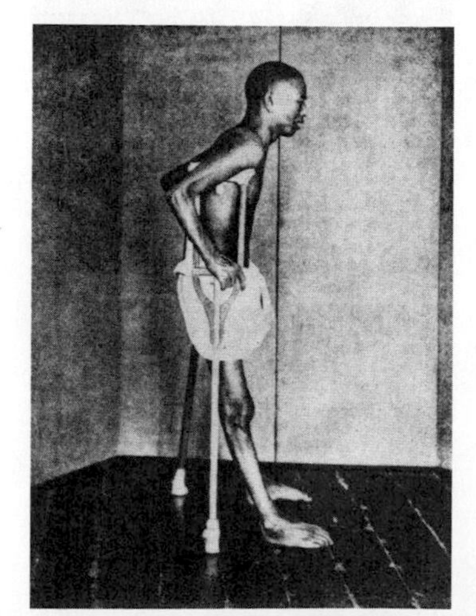

Figure 8.9 In Sri Lanka, the word *beriberi* means "I cannot," referring to the extreme weakness and depression that are the earliest symptoms of the disease. (U.S. Library of Medicine/Photo Researchers, Inc.)

Thiamin Deficiency

Thiamin deficiency results in the disease beriberi, which causes lethargy, fatigue, and other neurological symptoms (**Figure 8.9**). It can also cause cardiovascular problems such as rapid heartbeat, enlargement of the heart, and congestive heart failure.

Table 8.2 A Summary of the Water-Soluble Vitamins and Choline

Vitamin	Sources	Recommended Intake for Adults	Major Functions	Deficiency Diseases and Symptoms	Groups at Risk of Deficiency	Toxicity	UL
Thiamin (vitamin B_1, thiamin mononitrate)	Pork, whole and enriched grains, seeds, nuts, legumes	1.1–1.2 mg/d	Coenzyme in acetyl-CoA formation and citric acid cycle; acetylcholine synthesis; nerve function	Beriberi: weakness, apathy, irritability, nerve tingling, poor co-ordination, paralysis, heart changes	Alcoholics, those living in poverty	None reported	ND
Riboflavin (vitamin B_2)	Dairy products, whole and enriched grains, leafy green vegetables, meats	1.1–1.3 mg/d	Coenzyme in citric acid cycle, lipid metabolism, and electron transport chain	Ariboflavinosis: inflammation of mouth and tongue, cracks at corners of the mouth	None	None reported	ND
Niacin (nicotinamide, nicotinic acid, vitamin B_3)	Beef, chicken, fish, peanuts, legumes, whole and enriched grains. Can be made from tryptophan.	14–16 mg NE/d	Coenzyme in glycolysis, citric acid cycle, electron transport chain, and lipid synthesis and breakdown	Pellagra: diarrhea, dermatitis on areas exposed to sun, dementia	Those consuming a limited diet based on corn, alcoholics	Flushing, nausea, rash, tingling extremities	35 mg/d from fortified foods and supplements
Biotin	Liver, egg yolks, synthesized by bacteria in the gut	30 μg/d[a]	Coenzyme in glucose and fatty acid synthesis and amino acid metabolism	Dermatitis, nausea, depression, hallucinations	Those consuming large amounts of raw egg whites, alcoholics	None reported	ND
Pantothenic acid (calcium pantothenate)	Meat, legumes, whole grains, widespread in foods	5 mg/d[a]	Coenzyme in citric acid cycle and lipid synthesis and breakdown	Fatigue, rash	Alcoholics	None reported	ND
Vitamin B_6 (Pyridoxine, pyridoxal phosphate, pyridoxamine)	Meat, fish, poultry, liver, legumes, whole grains, nuts and seeds	1.3–1.7 mg/d	Coenzyme in protein and amino acid metabolism, neurotransmitter and hemoglobin synthesis	Headache, convulsions, other neurological symptoms, decreased immune function, poor growth, anemia	Alcoholics	Numbness, nerve damage	100 mg/d
Folate (folic acid, folacin, pteroylglutamic acid)	Leafy green vegetables, legumes, nuts, seeds, enriched grains, oranges, liver	400 μg DFE/d	Coenzyme in DNA synthesis and amino acid metabolism	Macrocytic anemia, inflammation of tongue, diarrhea, poor growth, neural tube defects	Pregnant women, premature infants, alcoholics	Masks B_{12} deficiency	1,000 μg/d from fortified food and supplements
Vitamin B_{12} (Cobalamin, cyanocobalamin)	Animal products	2.4 μg/d	Coenzyme in folate and fatty acid metabolism; nerve function	Pernicious anemia, macrocytic anemia, nerve damage	Vegans, elderly, those with stomach or intestinal disease	None reported	ND

Table 8.2 Continued

Vitamin C (ascorbic acid, ascorbate)	Citrus fruits, broccoli, strawberries, greens, peppers, potatoes	75–90 mg/d	Coenzyme in collagen synthesis, hormone and neurotransmitter synthesis; antioxidant	Scurvy: poor wound healing, bleeding gums, loose teeth, bone fragility, joint pain, pinpoint hemorrhages	Alcoholics, elderly people	GI distress, diarrhea	2,000 mg/d
Choline[b]	Egg yolks, organ meats, leafy greens, nuts, body synthesis	425–550 mg/d[a]	Synthesis of cell membranes and neurotransmitters	Liver dysfunction	None	Sweating, low blood pressure, liver damage	3,500 mg/d

[a]Adequate Intake (AI).

[b]Choline may not be an essential vitamin at all life stages but recommendations have been made for its intake.

UL, Tolerable Upper Intake Level; NE, niacin equivalent; DFE, dietary folate equivalent; ND, insufficient data to determine a UL

Overt beriberi is rare in North America today, but thiamin deficiency does occur in alcoholics. They are particularly vulnerable because thiamin absorption is decreased due to the effect of alcohol on the GI tract. In addition, the liver damage that occurs with chronic alcohol consumption reduces conversion of thiamin to active coenzyme forms; thiamin intake also may be low due to a diet high in alcohol and low in nutrient-dense foods. Thiamin-deficient alcoholics may develop a neurological condition known as the **Wernicke-Korsakoff syndrome**. It is characterized by mental confusion, psychosis, memory disturbances, and coma.

Wernicke-Korsakoff syndrome A form of thiamin deficiency associated with alcohol abuse that is characterized by mental confusion, disorientation, loss of memory, and a staggering gait.

Thiamin Toxicity

Since no toxicity has been reported when excess thiamin is consumed from either food or supplements, a UL for thiamin intake has not been established.[3] This does not mean that high intakes are necessarily safe. Intakes of thiamin above the RDA have not been shown to provide health benefits.

Thiamin Supplements

Thiamin supplements containing up to 50 mg/day are widely available and are marketed with the promise that they will provide "more energy." Although thiamin is needed to produce ATP, it does not increase energy levels. Unless thiamin is deficient, increasing thiamin intake does not increase the ability to produce ATP. Because thiamin deficiency causes mental confusion and damages the heart, supplements often promise to improve mental function and prevent heart disease. However, in the absence of a deficiency, supplements do not have these effects. Thiamin is also included in supplements referred to as B-complex supplements (**Table 8.3**).

Table 8.3 Benefits and Risks of Water-Soluble-Vitamin Supplements

Supplement	Claim	Actual Benefits or Risks
B-complex (thiamin, riboflavin, niacin, pantothenic acid, biotin, vitamin B_6, vitamin B_{12})	Increases energy, needed during stress	Needed for energy metabolism but does not provide energy. Low risk of toxicity except for vitamin B_6.
Niacin (nicotinic acid form)	Lowers cholesterol	Medicinal doses may reduce cholesterol levels. Causes flushing, tingling and potentially liver damage. Doses above 35 mg should only be taken under medical supervision.
Vitamin B_6	Prevents heart disease; relieves carpal tunnel syndrome (CTS), and PMS; enhances immune function	Adequate amounts needed to maintain immune function and normal homocysteine levels, which may reduce heart disease risk*—excess provides no additional benefit; higher doses may have a slight benefit in some people with CTS or PMS. Levels above the UL may cause tingling, numbness, and muscle weakness.

Table 8.3 Continued		
Folate (folic acid)	Prevents birth defects, protects against heart disease and cancer	Adequate amounts needed to keep homocysteine levels normal, which may reduce heart disease risk.* Low folate may increase cancer risk. Supplemental sources reduce the risk of birth defects and are recommended for women of childbearing age. May mask a vitamin B_{12} deficiency at high intakes.
Vitamin B_{12}	Prevents heart disease, prevents dementia, reduces fatigue	Adequate amounts needed for nerve function, red blood cell synthesis, and to keep homocysteine levels low, which may reduce heart disease risk.* Supplemental sources recommended for older adults and vegans. No benefit of excess. Low risk of toxicity.
Vitamin C	Prevents colds, reduces cold symptoms, enhances immunity, protects against heart disease and cancer, enhances antioxidant protection	May reduce duration of colds. Important antioxidant but extra as supplements has not been shown to provide additional benefits. Too much can cause GI distress, damage teeth, promote kidney stone formation, and interfere with anticoagulant medications.

*Read "Vitamin B_6, B_{12}, folate and the homocysteine hypothesis," in Section 8.7, Vitamin B_6.

8.3 Riboflavin

Learning Objectives

- Describe the function shared by thiamin and riboflavin.
- Explain why milk sold in clear bottles might be low in riboflavin.

While searching for a cure for beriberi, scientists isolated riboflavin and several other B vitamins in addition to thiamin. This occurred because the extracts they made from vegetables and grains could be separated into two components: one contained thiamin, the antiberiberi factor they sought, and cured beriberi; the other was a mix of B vitamins that was later determined to contain riboflavin along with vitamin B_6, niacin, and pantothenic acid.

Riboflavin in the Diet

Milk is the best source of riboflavin in the Canadian diet. Other important sources include liver, red meat, poultry, fish, and whole and enriched grain products. Vegetable sources include asparagus, broccoli, mushrooms, and leafy green vegetables such as spinach (**Figure 8.10**). Because riboflavin is destroyed by exposure to light, poor handling decreases a food's riboflavin content. This is a problem when milk is stored in clear containers and exposed to light. Cloudy plastic milk bottles block some light, partially protecting the riboflavin, but cardboard or opaque plastic milk containers are even better at preventing losses[4] (**Figure 8.11**).

Riboflavin in the Body

Riboflavin forms the active coenzymes **flavin adenine dinucleotide** (FAD) and **flavin mononucleotide** (FMN) (see Appendix I). FAD functions in the citric acid cycle and is important for the breakdown of fatty acids. Both FMN and FAD function as electron carriers in the electron transport chain (see Figure 8.6). Therefore, adequate riboflavin is crucial in providing energy from carbohydrate, fat, and protein. Riboflavin is also involved directly or indirectly in converting a number of other vitamins, including folate, niacin, vitamin B_6, and vitamin K, into their active forms.

flavin adenine dinucleotide (FAD) and flavin mononucleotide (FMN) The active coenzyme forms of riboflavin. The structure of these molecules allows them to pick up and donate hydrogens and electrons in chemical reactions.

Recommended Riboflavin Intake

The RDA for riboflavin for adult men age 19 and older is 1.3 mg/day and for adult women 19 and older, 1.1 mg/day. This recommendation is based on the amount of riboflavin needed to maintain normal activity of a riboflavin-dependent enzyme in red blood cells and normal riboflavin excretion in the urine.[3] Half a litre provides about half the amount of riboflavin recommended for a typical adult. The recommended intake can be met without milk if the daily diet includes 2-3 servings of meat and 4-5 servings of enriched grain products and high-riboflavin vegetables, such as spinach.

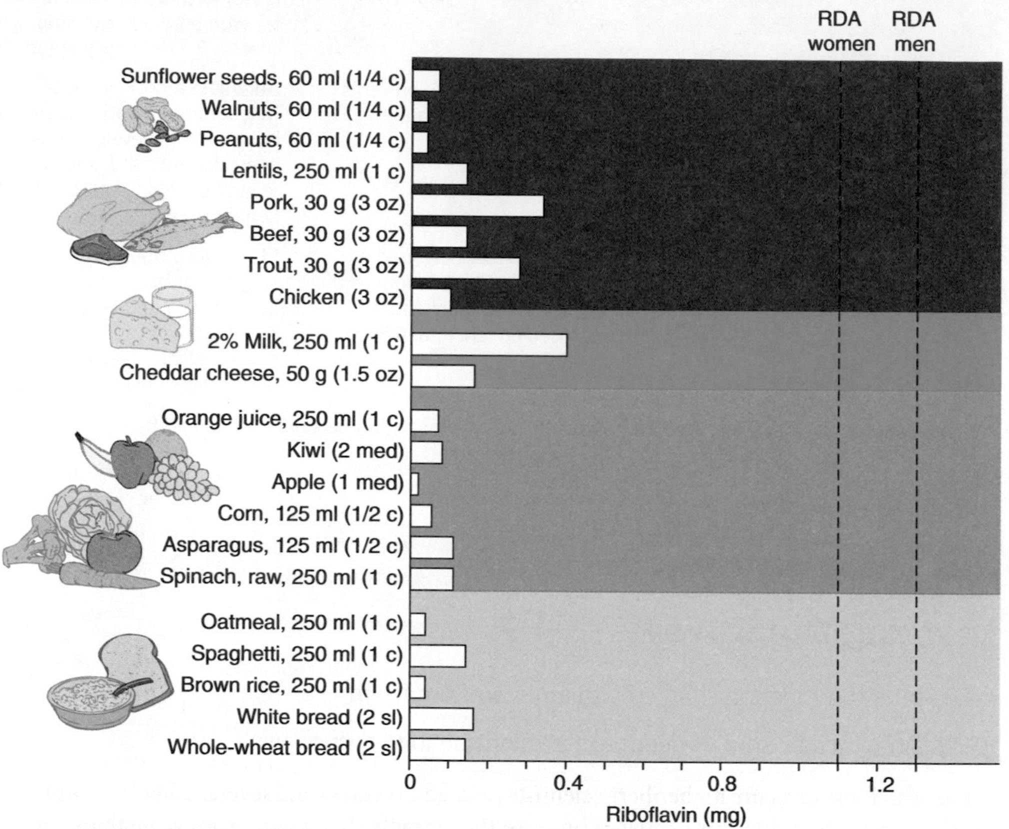

Figure 8.10 Riboflavin content of Canada's Food Guide food groups
Milk and fortified cereals are exceptionally good sources of riboflavin, but a combination of foods is needed to supply the RDA (dashed lines).

Figure 8.11 Why is milk often supplied in opaque or cardboard containers? (SEALTEST is used under licence by Agropur)

Additional riboflavin is recommended during pregnancy to support growth and increased energy utilization, and during lactation to allow for the riboflavin secreted in milk. There is not enough information to establish an RDA for infants, so an AI has been set based on the amount of riboflavin consumed by infants fed human milk.

Riboflavin Deficiency

When riboflavin is deficient, injuries heal poorly because new cells cannot grow to replace the damaged ones. Tissues that grow most rapidly, such as the skin and the linings of the eyes, mouth, and tongue, are the first to be affected by a deficiency. Symptoms of riboflavin deficiency, called **ariboflavinosis**, include inflammation of the eyes, lips, mouth, and tongue; scaly, greasy skin eruptions; cracking of the tissue at the corners of the mouth; and confusion. Deficiency symptoms may develop after approximately 2 months on a riboflavin-poor diet.

Life Cycle

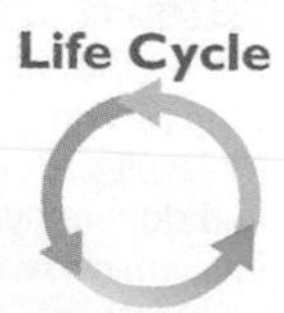

A deficiency of riboflavin is rarely seen alone. It usually occurs in conjunction with deficiencies of other B vitamins. One reason is that the food sources of B vitamins are similar (see Table 8.2). Therefore, a deficiency of riboflavin due to poor diet will likely lead to multiple vitamin deficiencies. Because riboflavin is needed to convert other vitamins into their active forms, some of the symptoms seen with riboflavin deficiency are actually due to deficiencies of these other nutrients.

Riboflavin Toxicity

No adverse effects have been reported from overconsumption of riboflavin from foods or supplements, and there are not sufficient data to establish a UL for this vitamin. Large doses of riboflavin are not well absorbed, and it is readily excreted in the urine. A harmless side effect of high riboflavin intake, such as may be obtained from over-the-counter supplements, is bright yellow urine.

ariboflavinosis The condition resulting from a deficiency of riboflavin.

Riboflavin Supplements

As with thiamin, the role of riboflavin in energy metabolism has led to claims that supplements containing riboflavin, such as B-complex supplements, will provide an energy boost (see Table 8.3). Although riboflavin is needed for energy metabolism, it does not provide energy. Since a deficiency causes skin and eye symptoms, riboflavin has also been suggested as a cure for eye diseases and skin disorders. However, in the absence of a deficiency, supplementation does not affect the eyes or skin.

8.4 Niacin

Learning Objectives

- Discuss why a niacin deficiency is more likely in someone consuming a diet that is based on corn.
- List the 3 D's of pellagra.

A deficiency of niacin results in a disease called **pellagra**, which causes progressive physical and mental deterioration. It was first observed in Europe in the eighteenth century, and in the early twentieth century it became endemic in the southeastern United States (see Science Applied: Pellagra: Infectious Disease or Dietary Deficiency?). The emergence of pellagra can be traced to the cultivation of corn as a dietary staple.[5] It primarily affects the poor who cannot afford a varied diet.

pellagra The disease resulting from a deficiency of niacin.

Niacin in the Diet

Meat and fish are good sources of niacin (**Figure 8.12**). Other sources include legumes, mushrooms, wheat bran, asparagus, and peanuts. Niacin added to enriched flours used in baked goods provides much of the usable niacin in the North American diet. Niacin can also be synthesized in the body from the essential amino acid tryptophan (**Figure 8.13**). In a diet that contains high-protein foods such as milk and eggs, which are poor sources of niacin but good sources of tryptophan, much of the need for niacin can be met by tryptophan. Tryptophan, however, is only used to make niacin if enough is available to first meet the needs of protein synthesis. When the diet is low in tryptophan, it is not used to synthesize niacin. Food composition tables and databases list only preformed niacin in a food, not the amount of niacin that can be made from tryptophan contained within the food.

The association between niacin deficiency and a limited diet based on corn and low in animal products has been attributed to the low-tryptophan content of corn and the fact that the niacin found naturally in corn (and to a lesser extent in other cereal grains) is bound to other molecules and therefore not well absorbed. The treatment of corn with lime water (water and calcium hydroxide), as is done in Mexico and Central America during the making of tortillas, enhances the availability of niacin (**Figure 8.14**). The diet in these regions also contains legumes, which provide both niacin and a source of tryptophan for the synthesis of niacin. As a result, despite their corn-based diet, populations in these regions have not suffered from pellagra. Today, pellagra remains common in India and parts of China and Africa.[6] Efforts to eradicate this deficiency include the development of new varieties of corn that provide more available niacin and more tryptophan than traditional varieties.

Niacin in the Body

Niacin is important in the production of ATP from the energy-yielding nutrients as well as in reactions that synthesize other molecules. There are two forms of niacin: nicotinic acid and nicotinamide (see Figure 8.13). Either form can be used by the body to make the two active coenzymes **nicotinamide adenine dinucleotide (NAD) and nicotinamide adenine dinucleotide phosphate (NADP)**. NAD functions in glycolysis and the citric acid cycle, accepting released electrons and passing them on to the electron transport chain where ATP is formed (see Figure 8.6). NADP acts as an electron carrier in reactions that synthesize fatty acids and cholesterol. The need for niacin is so widespread in metabolism that a deficiency causes damage throughout the body.

nicotinamide adenine dinucleotide (NAD) and nicotinamide adenine dinucleotide phosphate (NADP) The active coenzyme forms of niacin that are able to pick up and donate hydrogens and electrons. They are important in the transfer of electrons to oxygen in cellular respiration and in many synthetic reactions.

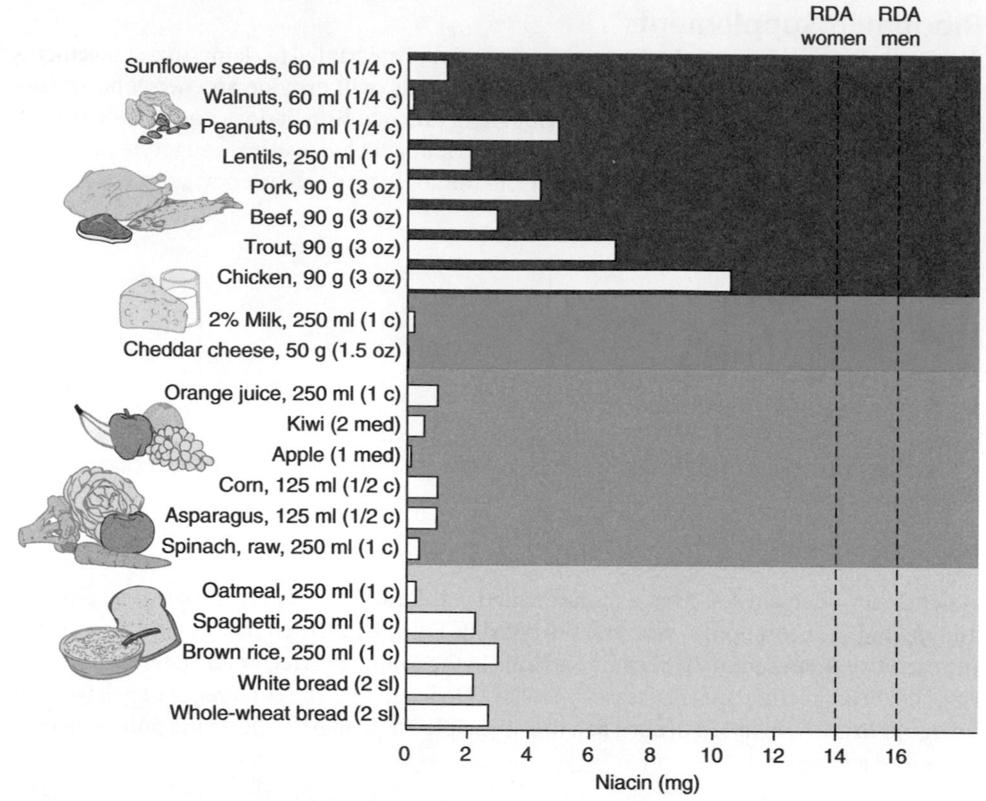

Figure 8.12 Niacin content of Canada's Food Guide food groups
Meat, legumes, and whole and enriched grains are good sources of niacin; the dashed lines represent the RDAs for adult men and women.

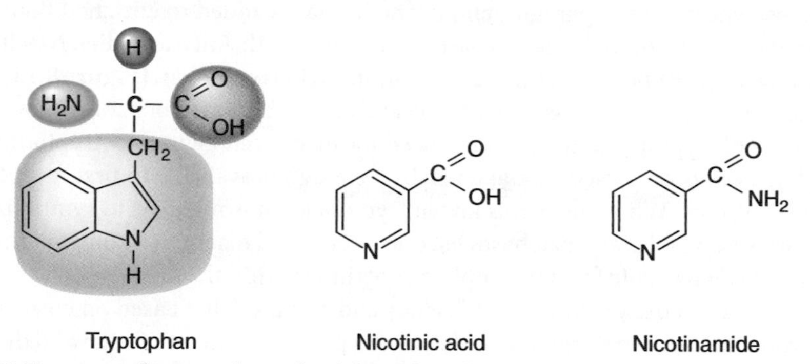

Figure 8.13 Structure of tryptophan, nicotinic acid, and nicotinamide
The amino acid tryptophan can be used to synthesize the two forms of niacin, nicotinic acid, and nicotinamide. These can be converted into the active coenzyme forms of niacin.

Figure 8.14 The treatment of corn with lime water during the preparation of tortillas improves niacin bioavailability and has helped prevent pellagra in Mexico and other Latin American countries. (Jeff Greenberg/Photo Researchers, Inc.)

Recommended Niacin Intake

The RDA for niacin is expressed as **niacin equivalents (NEs)**. One NE is equal to 1 mg of niacin or 60 mg of tryptophan. This allows for the fact that some of the requirement for niacin can be met by the synthesis of niacin from tryptophan. Approximately 60 mg of tryptophan is needed to make 1 mg of niacin. To estimate the niacin contributed by high-protein foods, protein is assumed to be about 1% tryptophan. The criterion used to estimate the average niacin requirement is urinary excretion of niacin metabolites. The RDA for adult men and women of all ages is 16 and 14 mg NE/day, respectively.[3] A meal containing a medium chicken breast and a cup of steamed asparagus provides this amount.

niacin equivalents (NEs) The measure used to express the amount of niacin present in food, including that which can be made from its precursor, tryptophan. One NE is equal to 1 mg of niacin or 60 mg of tryptophan.

Niacin needs are increased during pregnancy to account for the increase in energy expenditure, and during lactation to account for both the increase in energy expenditure and the niacin secreted in milk. There is not enough information to establish an RDA for infants, so an AI has been set based on the amount of niacin found in human milk.

Life Cycle

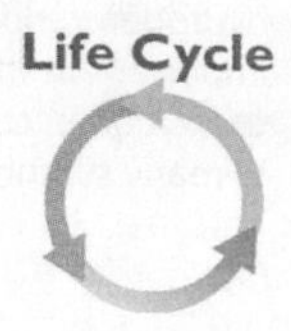

Niacin Deficiency

The early symptoms of pellagra include fatigue, decreased appetite, and indigestion, followed by the three D's: dermatitis, diarrhea, and dementia. If left untreated, niacin deficiency results in a fourth D—death. The dermatitis resembles sunburn and strikes parts of the body exposed to sunlight, heat, or injury (**Figure 8.15**). Gastrointestinal symptoms include a bright-red tongue and may include vomiting, constipation, or diarrhea. Mental symptoms begin with irritability, headaches, loss of memory, insomnia, and emotional instability and progress to psychosis and acute delirium.

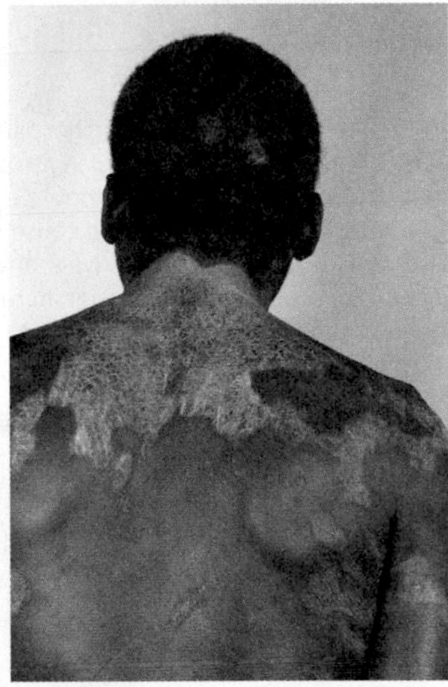

Figure 8.15 The cracked, inflamed skin characteristic of pellagra most commonly appears on areas exposed to sunlight or other stresses. (Dr. M. A. Ansary/SPL/Photo Researchers, Inc.)

Niacin Toxicity

There is no evidence of any adverse effects from consumption of niacin naturally occurring in foods, but supplements can be toxic. The adverse effects of high intakes of niacin include flushing of the skin, a tingling sensation in the hands and feet, a red skin rash, nausea, vomiting, diarrhea, high blood sugar levels, abnormalities in liver function, and blurred vision. Since flushing is the first toxicity symptom to appear as the dose is increased, the UL for adults has been set at 35 mg, the highest level that is unlikely to cause flushing in the majority of healthy people. This value applies to the forms of niacin contained in supplements and fortified foods, but does not include niacin naturally occurring in foods.

Niacin Supplements

Niacin deficiency is not a public health concern in Canada. Despite this, niacin is a commonly used vitamin supplement. It is included in multivitamins as well as in B-complex vitamin supplements marketed to give you more energy. High-dose supplements of this vitamin are also used to treat elevated blood cholesterol (see Table 8.3). When vitamin supplements are taken in large doses to treat or prevent diseases that are not due to vitamin deficiencies, they are really being used as drugs rather than vitamins. Doses of 50 mg/day or greater of the nicotinic acid form of niacin have been found to decrease blood levels of LDL cholesterol and triglycerides and increase HDL cholesterol. These high-dose niacin supplements are also associated with a reduction in recurrent heart attacks and deaths in individuals with cardiovascular disease.[7] Since the UL for niacin is only 35 mg, many people are unable to take niacin as a cholesterol-lowering drug because they experience side effects due to niacin toxicity. Because niacin supplements are available over-the-counter, people may try to treat themselves for high cholesterol, but high doses of vitamins are as dangerous as drugs and should be used only with medical supervision. Supplements of the nicotinamide form of niacin have been investigated for their benefit in the prevention of another disease—type 1 diabetes. A recent intervention trial has not found it to be effective for this purpose.[8]

8.5 Biotin

Learning Objective

- Explain why consuming raw eggs might cause a biotin deficiency.

Biotin was discovered when rats fed protein derived from raw egg white developed a syndrome of hair loss, dermatitis, and neuromuscular dysfunction. The symptoms were due to a deficiency of biotin. The deficiency was caused by a protein in raw egg white, called avidin, which tightly binds biotin and prevents its absorption.

Biotin in the Diet

Good dietary sources of biotin include liver, egg yolks, yogurt, and nuts, while fruit and meat are poor sources. Foods containing raw egg whites should be avoided not only because avidin binds biotin and prevents its absorption, but because raw eggs also may be contaminated with bacteria that can cause food-borne illness (**Figure 8.16**). Thoroughly cooking eggs destroys bacteria and denatures avidin so that it cannot bind biotin.

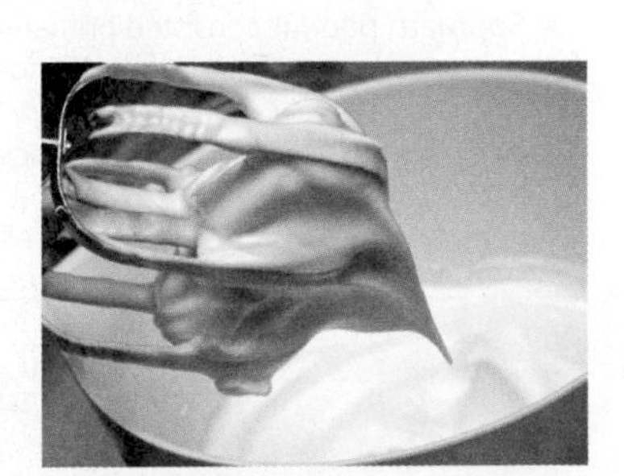

Figure 8.16 Why is the consumption of raw eggs not the best way to meet your biotin needs? (©Paula Thomas/Getty Images)

Biotin in the Body

Biotin is a coenzyme for a group of enzymes that add the acid group COOH to molecules. It functions in energy metabolism because it is needed to make a 4-carbon molecule necessary in the citric acid cycle and in glucose synthesis. It is also important in the metabolism of fatty acids and amino acids (see Figure 8.6).

SCIENCE APPLIED

Pellagra: Infectious Disease or Dietary Deficiency?

In the early 1900s, psychiatric hospitals in the southeastern United States were filled with patients with dementia due to a disease called pellagra. Although we now know pellagra is due to niacin deficiency, at the time, it was thought to be caused by an infectious agent or toxin. As many as 100,000 people were affected by pellagra and upward of 10,000 deaths resulted per year. In response to this epidemic, the U.S. government set up the Thompson-McFadden Pellagra Commission and the U.S. Public Health Service sent Dr. Joseph Goldberger to investigate.

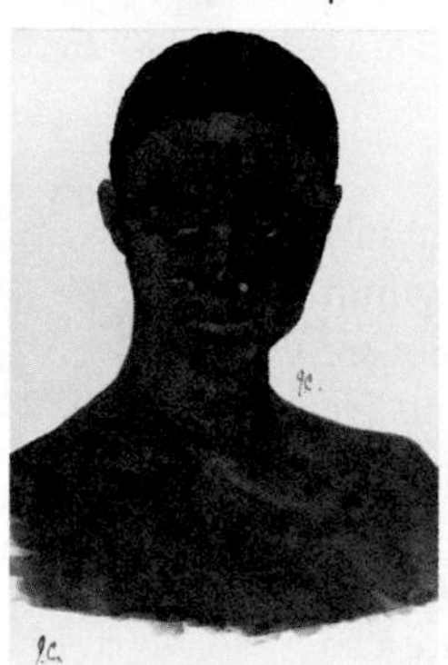

During the summer of 1919, a young artist named John Carroll was assigned to work with Joseph Goldberger's pellagra study. He produced 41 drawings of pellagra patients. The drawing shown here depicts a female pellagra patient in the Georgia State Sanitarium in 1919. (Eskind Biomedical Library, Vanderbilt University Medical Center.)

Between 1912 and 1916, investigators with the Pellagra Commission conducted epidemiological and bacteriological studies. Monkeys and baboons were injected with blood, urine, and other extracts from patients with pellagra; they were fed feces and skin scrapings. None of the animals developed the disease. Although these results supported the view that pellagra was not due to an infectious agent, many questioned whether animal studies could be applied to humans. Perhaps monkeys and baboons were not susceptible to pellagra? The commission still believed that pellagra could be transmitted in some way from a pellagrous to a nonpellagrous person.[1]

Goldberger spent 2 years studying the communities and institutions where pellagra was common. He noticed that pellagra was prevalent among people in institutions but that the attendants and nurses never contracted the disease—a fact that did not support an infectious nature. While looking for factors that distinguished the patients from the staff, Goldberger noted that the patients were fed a diet typical of the Southern poor; it consisted primarily of corn meal, molasses, and fatback or salt pork. The staff ate a more varied diet, so Goldberger began experimenting with diet. In an orphanage and a state hospital, he was able to cure pellagra and prevent recurrences by adding milk, eggs, and more meat to the diet. The next experimental step needed to support his hypothesis was to produce pellagra in healthy people by feeding them a diet similar to the institutional diet. In 1915, at a work farm affiliated with the Mississippi State Penitentiary, 12 convicts volunteered to participate in the experiment in exchange for pardons. They were fed a diet consisting of corn meal, grits, cornstarch, white wheat flour, white rice, cane syrup, sugar, sweet potatoes, small amounts of turnip greens, cabbage, and collards, and a liberal amount of pork fat. After about 6 months, 6 men had developed pellagra. Goldberger concluded that the cause was a deficiency of an amino acid, a mineral, a fat-soluble vitamin, or some as yet unknown vitamin factor.

To provide the final proof that pellagra was not due to an infectious agent, Goldberger and 15 of his colleagues voluntarily injected themselves with blood, swabbed their throats with nasal secretions, and swallowed urine, feces, and skin cells from patients who were severely ill with pellagra (later in the experiment, they put the feces and other materials in capsules). After 6 months, none had become ill. Goldberger had proven that pellagra was not an infectious disease but he continued to search for the dietary cause.

In 1922, a diet comparable to that used to produce pellagra in Goldberger's experiment caused a similar disease in dogs, called black tongue. Subsequent food experiments found that yeast and liver contained a pellagra-preventative factor. These foods could be used to prevent pellagra and cure mild cases. However, despite Goldberger's efforts, the epidemic raged on. Even though it had been demonstrated that yeast and liver could cure pellagra, nothing had been done to change the Southern diet that produced the disease. In addition, in serious cases of the disease, providing food sources of niacin was often ineffective because inflammation of the GI tract, lack of appetite, and vomiting made it difficult for patients to ingest and absorb enough of the pellagra-preventative factor to cure the deficiency. Joseph Goldberger died in 1929, the year the epidemic reached its peak.

In 1937, another research team identified the pellagra-preventative factor as nicotinic acid. With this form of the vitamin isolated, it could be given intravenously, bypassing the digestive tract, and saving the lives of those suffering from pellagra. Despite this advance in treatment, pellagra remained a problem among the Southern poor. Poor dietary habits, poverty, and chronic malnutrition made it a difficult problem to address.[2] Finally, the economic boom created by World War II, combined with a federally sponsored enrichment program, added enough niacin to the diet to end the pellagra epidemic in the United States.

[1] Roe, D. A. *A Plague of Corn: The Social History of Pellagra*. Ithaca, NY: Cornell University Press, 1973.

[2] Syndenstricker, V. P. The history of pellagra, its recognition as a disorder of nutrition and its conquest. *Am. J. Clin. Nutr.* 6:409–441, 1958.

Recommended Biotin Intake

A dietary requirement for biotin has been difficult to estimate because some biotin is produced by bacteria in the gastrointestinal tract and absorbed into the body. Therefore no RDA could be determined but an AI of 30 μg/day has been established for adult men and women based on the amount of biotin found in a typical North American diet.[3]

No additional biotin is recommended for pregnancy, but the AI is increased during lactation to account for the amount secreted in milk. The AI for infants is based on the amount of biotin consumed by infants fed human milk.

Biotin Deficiency and Toxicity

Although biotin deficiency is uncommon, it has been observed in people with malabsorption or protein-energy malnutrition, those receiving tube feedings or total parenteral nutrition without biotin, those taking anticonvulsant drugs for long periods, and those frequently consuming raw egg whites.[3] When biotin intake is deficient, symptoms including nausea, thinning hair, loss of hair colour, a red skin rash, depression, lethargy, hallucinations, and tingling of the extremities gradually appear.

No toxicity has been reported in patients given 200 mg/day of biotin to treat various disease states, and sufficient data are not available to establish a UL.

8.6 Pantothenic Acid

Learning Objective

- Discuss why pantothenic acid deficiency is rare.

Pantothenic acid, which gets its name from the Greek word *pantothen* (meaning "from everywhere"), is widely distributed in foods.

Pantothenic Acid in the Diet

Pantothenic acid is particularly abundant in meat, eggs, whole grains, and legumes. It is found in lesser amounts in milk, vegetables, and fruits (**Figure 8.17**). Pantothenic acid is susceptible to damage by exposure to heat and low- or high-acid conditions.

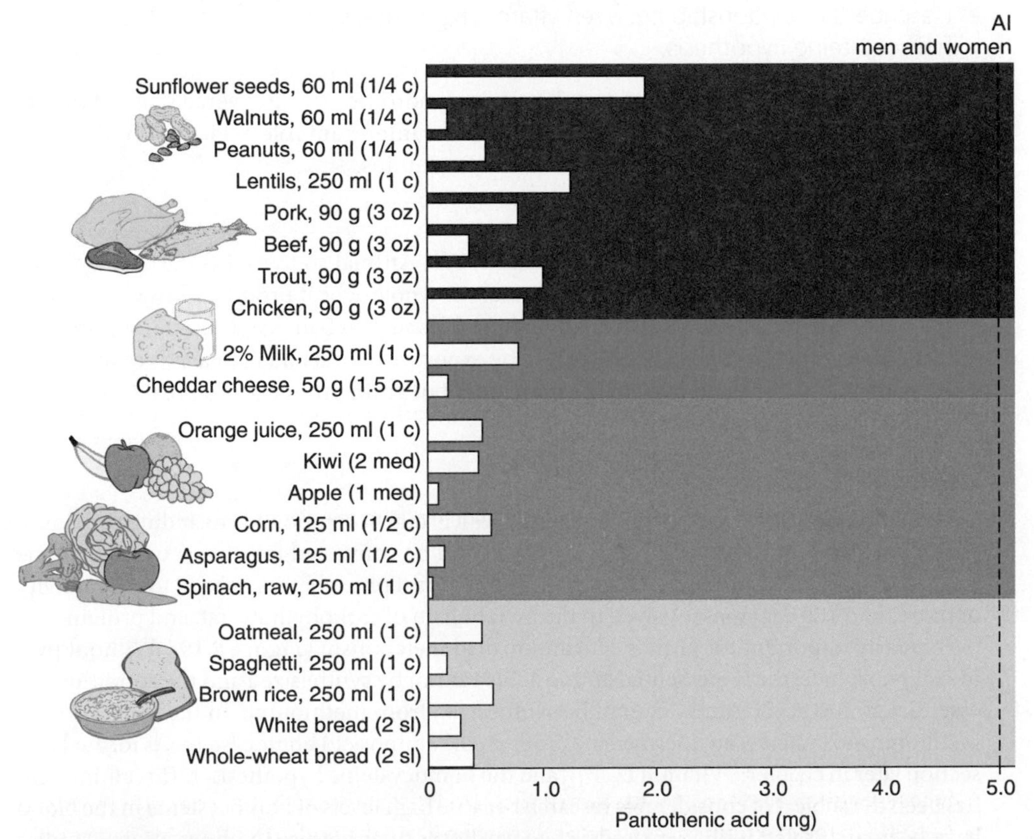

Figure 8.17 Pantothenic acid content of Canada's Food Guide food groups
All CFG groups contain good sources of pantothenic acid; the dashed line represents the AI for adult men and women.

Pantothenic Acid in the Body

In the body, pantothenic acid is part of the coenzyme A (CoA) molecule, which is part of acetyl-CoA, a molecule formed during the breakdown of carbohydrates, fatty acids, and amino acids (see Appendix I). Pantothenic acid is also needed to produce the acyl carrier protein needed for the synthesis of cholesterol and fatty acids (see Figure 8.6).

Recommended Pantothenic Acid Intake

Life Cycle

There is no RDA for pantothenic acid, but an AI of 5 mg/day has been recommended for adult men and women.[3] This value is based on the intake of pantothenic acid sufficient to replace urinary losses. The AI is increased to 6 and 7 mg/day to meet the needs of pregnancy and lactation, respectively.

Pantothenic Acid Deficiency and Toxicity

The wide distribution of pantothenic acid in foods makes deficiency rare in humans. A deficiency of this vitamin alone has not been reported, but it may occur as part of a multiple B-vitamin deficiency resulting from malnutrition or chronic alcoholism.

Pantothenic acid is relatively nontoxic. No toxic symptoms were reported in a study that fed young men 10 g of pantothenic acid per day for 6 weeks. Another study found that doses of 10-20 g/day may result in diarrhea and water retention.[3] Data are not sufficient to establish a UL for pantothenic acid.

8.7 Vitamin B_6

Learning Objectives

- Explain the role of vitamin B_6 in amino acid metabolism.
- Describe the relationship between vitamin B_6, folate, vitamin B_{12}, and the homocysteine hypothesis.

Vitamin B_6 was identified only when a deficiency syndrome was discovered that did not respond to thiamin or riboflavin supplementation. The important role of vitamin B_6 in amino acid metabolism distinguishes it from the other B vitamins.

Vitamin B_6 in the Diet

Vitamin B_6 is found in both animal and plant foods. Animal sources include chicken, fish, pork, and organ meats. Good plant sources include whole-wheat products, brown rice, soybeans, sunflower seeds, and some fruits and vegetables such as bananas, broccoli, and spinach (**Figure 8.18**). Vitamin B_6 is easily destroyed by exposure to heat and light and is easily lost in processing. It can be added back in the enrichment of grain products and breakfast cereals, but its addition is not mandatory. [9]

Vitamin B_6 in the Body

pyridoxine The chemical term for vitamin B_6.

pyridoxal phosphate The major coenzyme form of vitamin B_6 that functions in more than 100 enzymatic reactions, many of which involve amino acid metabolism.

Vitamin B_6, also known as **pyridoxine**, comprises a group of compounds including pyridoxal, pyridoxine, and pyridoxamine. All three forms can be converted into the active coenzyme form, **pyridoxal phosphate** (see Appendix I). Pyridoxal phosphate is needed for the activity of more than 100 enzymes involved in the metabolism of carbohydrate, fat, and protein. It is particularly important for protein and amino acid metabolism (**Figure 8.19**). Without pyridoxal phosphate, the nonessential amino acids cannot be synthesized and the conditionally essential amino acid cysteine cannot be synthesized from methionine. In the conversion of methionine to cysteine, an intermediate product, the amino acid homocysteine, is formed (see section later in chapter: "Vitamin B_6, B_{12}, and the homocysteine hypothesis"). This amino acid has been the subject of considerable research because high levels of homocysteine in the blood have been associated with increased risk to cardiovascular disease in observational studies (See Section 1.5 for a discussion of observational studies). Vitamin B_6, along with folate and vitamin B_{12}, play an important role in the metabolism of the amino acid homocysteine, which will be discussed later in the chapter.

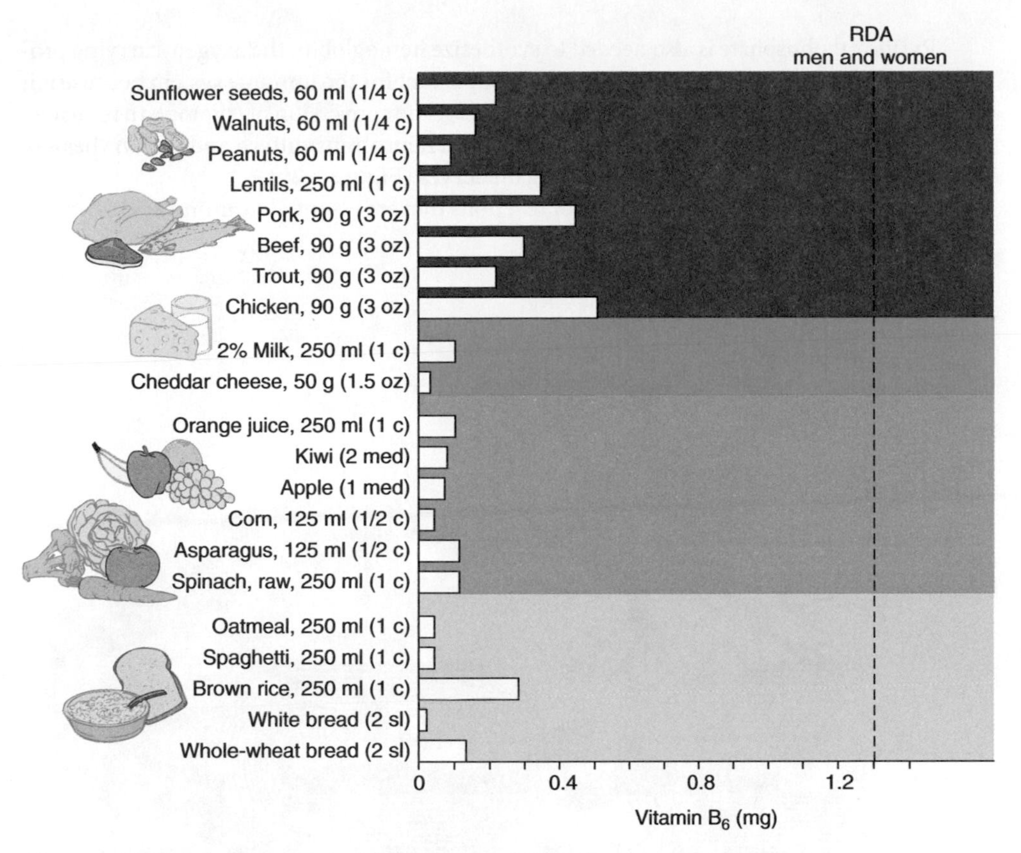

Figure 8.18 Vitamin B_6 content of Canada's Food Guide food groups
Meats, legumes, and whole grains are the best sources of vitamin B_6; the dashed line represents the RDA for adults up to 50 years of age.

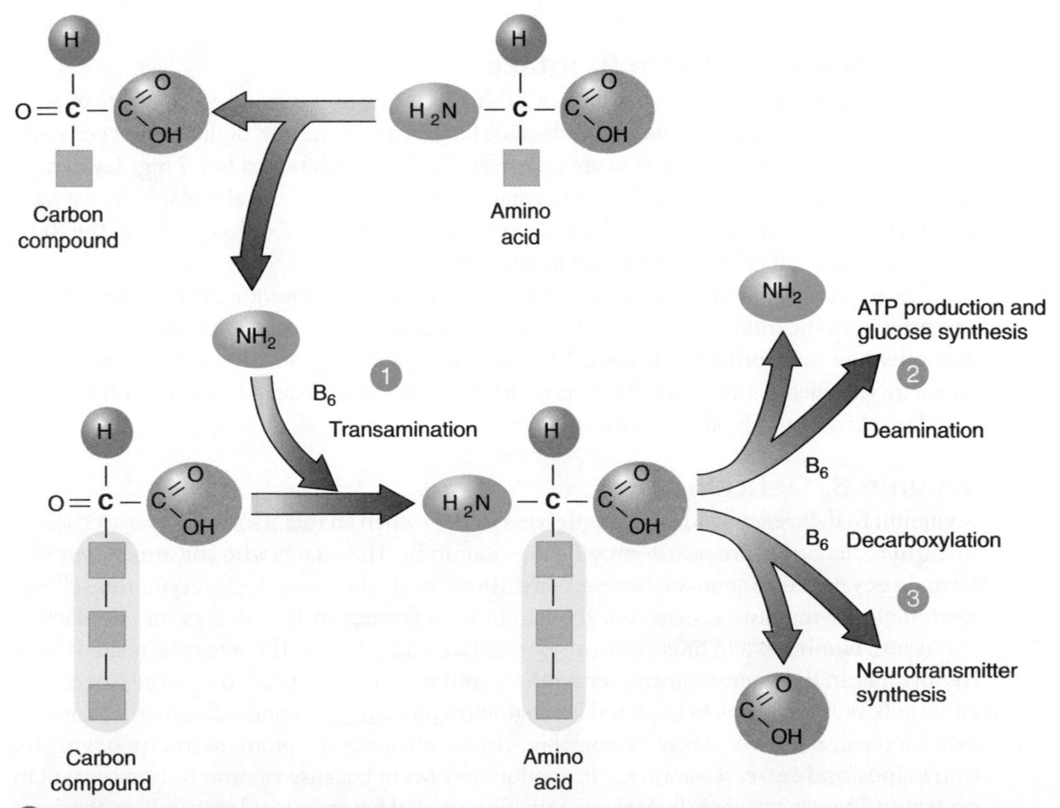

1. Vitamin B_6 is needed to synthesize nonessential amino acids by transamination.
2. Vitamin B_6 is needed for deamination so amino acids can be used to produce ATP or to synthesize glucose.
3. Vitamin B_6 is needed for removes the COOH group from amino acids s they can be used to synthesize neurotransmitters.

Figure 8.19 Functions of vitamin B_6

Pyridoxal phosphate is also needed to synthesize hemoglobin, the oxygen-carrying protein in red blood cells. Pyridoxal phosphate is important for the immune system because it is needed to form white blood cells. It is also needed for the conversion of tryptophan to niacin, the metabolism of glycogen, the synthesis of certain neurotransmitters, and the synthesis of the lipids that are part of the myelin coating on nerves (**Figure 8.20**).

Vitamin B_6 is needed for a number of reactions that are essential to amino acid synthesis and breakdown.

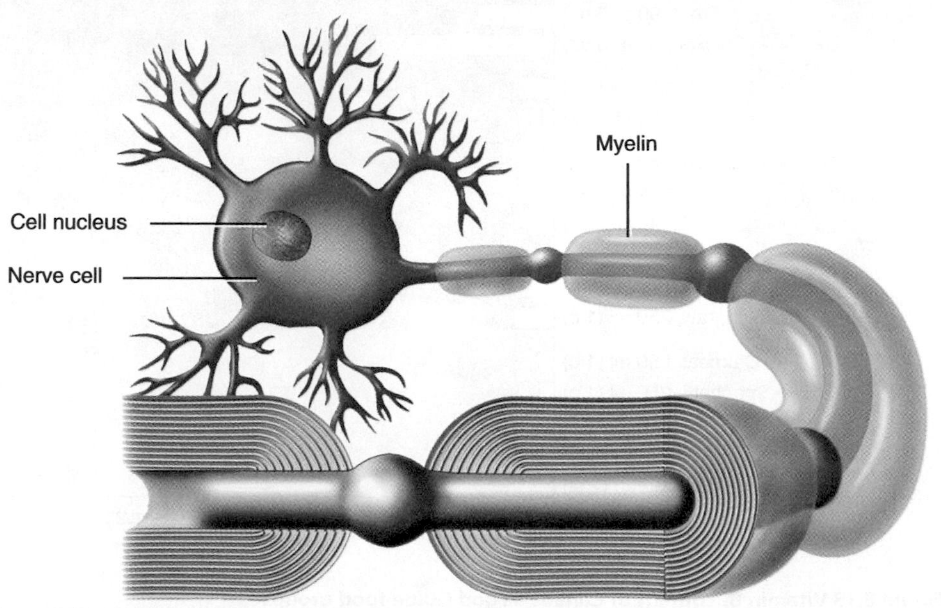

Figure 8.20 Myelin
Both vitamin B_6 and vitamin B_{12} are needed to synthesize and maintain the mylein coating on nerve cells, which is essential for normal nerve transmission.

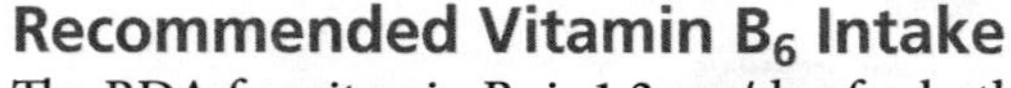

Recommended Vitamin B_6 Intake

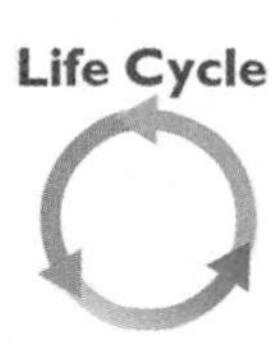

The RDA for vitamin B_6 is 1.3 mg/day for both adult men and women 19-50 years of age.[3] This is the amount needed to maintain adequate blood concentrations of the active coenzyme pyridoxal phosphate. In adults 51 years and older, the RDA is increased to 1.7 mg/day in men and 1.5 mg/day in women to maintain normal blood levels of pyridoxal phosphate. An 85-g serving of chicken, fish, or pork, or half a baked potato, provides about a quarter of the RDA for an average adult; a banana provides about a third.

The RDA for vitamin B_6 is increased during pregnancy to provide for metabolic needs and growth of the mother and fetus. Because the vitamin B_6 concentration in breast milk is dependent on the mother's intake, the RDA is increased during lactation to assure adequate levels are supplied to the infant. There is no RDA for infants, but an AI has been established based on the vitamin B_6 content of human milk.

Vitamin B_6 Deficiency

A vitamin B_6 deficiency syndrome was defined in 1954 when an infant formula was overheated in the manufacturing process, destroying the vitamin B_6. The infants who consumed only this formula developed abdominal distress, convulsions, and other neurological symptoms.[10] The neurological symptoms associated with vitamin B_6 deficiency include depression, headaches, confusion, numbness and tingling in the extremities, and seizures. These may be related to the role of vitamin B_6 in neurotransmitter synthesis and myelin formation. Anemia also occurs in vitamin B_6 deficiency due to impaired hemoglobin synthesis; red blood cells are small (microcytic) and pale due to the lack of hemoglobin. Other deficiency symptoms such as poor growth, skin lesions, and decreased antibody formation may occur because vitamin B_6 is important in protein and energy metabolism. Since vitamin B_6 is needed for amino acid metabolism, the onset of a deficiency can be hastened by a diet that is low in vitamin B_6 but high in protein.

Vitamin B_6 status in the body can be affected by a number of drugs, including alcohol and oral contraceptives. Alcohol decreases the formation of the active coenzyme pyridoxal phosphate and makes it more susceptible to breakdown. Oral contraceptive use has been associated with small decreases in blood levels of pyridoxal phosphate, but vitamin B_6 supplements are not routinely recommended for women taking oral contraceptives.[3]

Vitamin B_6, B_{12}, folate, and the homocysteine hypothesis

Vitamin B_6, vitamin B_{12}, and folate function together to keep the levels of homocysteine in the blood low. High blood levels of homocysteine have been associated, in many observational studies, with an increased risk of cardiovascular disease,[11] as these levels are hypothesized to damage blood vessels and thus promote atherosclerosis (see Section 5.6 for description of atherosclerosis).[12] This has lead to the development of the homocysteine hypothesis, which states that homocysteine is a risk factor for cardiovascular disease and that lowering homocysteine levels in the blood would reduce disease risk. Vitamin B_6, vitamin B_{12}, and different coenzyme forms of folate function in metabolic pathways involved in **single carbon metabolism,** which refers to the transfer of a chemical group containing a single carbon, such as a methyl group, from one compound to another. These transfers are important in the metabolism of homocysteine.

single carbon metabolism The transfer of single carbon groups, such as but not limited to methyl groups ($-CH_3$), between compounds.

Figure 8.21 illustrates the interrelationship between the three vitamins. Vitamin B_6 helps to maintain lower homocycsteine levels by acting as a coenzyme for enzymes that convert homocysteine to the amino acid cysteine (see Figure 8.21, step 1). Folate and vitamin B_{12} function together to maintain lower levels of homocysteine by acting together to convert it to the amino acid methionine. Specifically, vitamin B_{12} interacts with a coenzyme form of folate called methyl folate, converting it to folate and picking up its methyl group to form methyl-vitamin B_{12} (see Figure 8.21, step 2). Next, methyl-vitamin B_{12} transfers its methyl group to homocysteine, converting it to methionine (see Figure 8.21, step 3). Steps 2 and 3 are examples of the transfer of methyl groups that are characteristic of single carbon metabolism. Finally, in order for these reactions to occur again, folate is converted to methyl folate by a series of reactions, one of which requires vitamin B_6 (Figure 8.21, step 4).

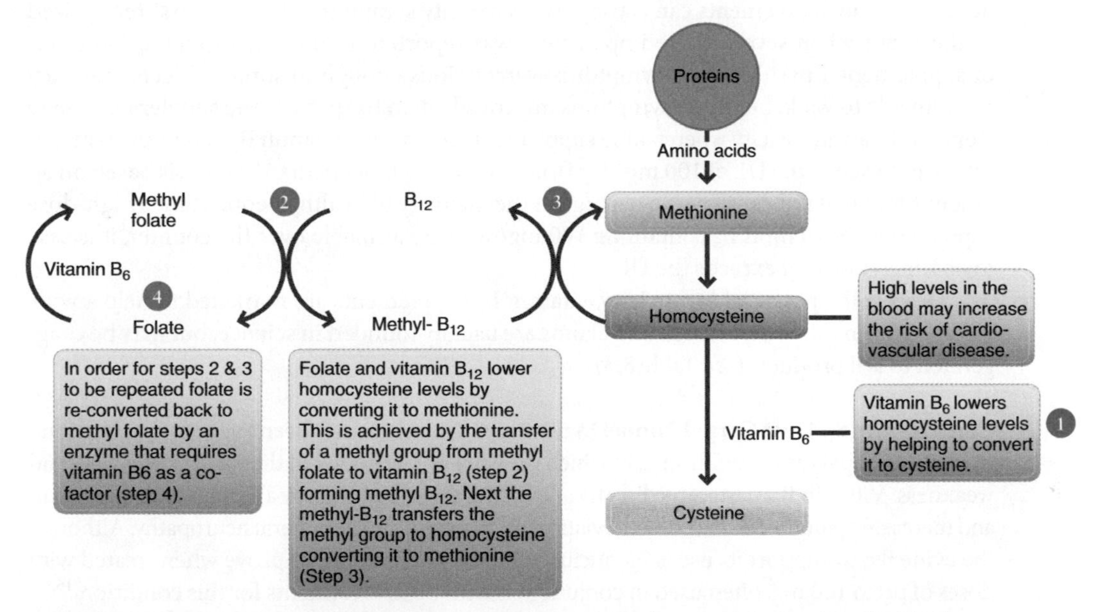

Figure 8.21 Vitamins B_6, B_{12}, and folate function together to lower homocysteine levels in the blood.

Multiple intervention trials (see Section 1.5) have shown that folate supplementation, in combination with vitamins B_{12} and B_6, decreases blood homocysteine levels.[11] Furthermore, there is strong evidence, from multiple observational studies, that increased folate intake and/or decreased blood levels of homocysteine are associated with reduced risk of cardiovascular disease.[13] It would seem reasonable, then, to hypothesize that intervention trials that increase folate intake, particularly in combination with vitamin B_6 and B_{12}, would reduce the risk of cardiovascular disease. Surprisingly, however, several large intervention trials have failed to show any benefit from combined supplementation of folate, vitamin B_{12}, and B_6 on the risk of cardiovascular disease.[14] These relationships are illustrated in **Figure 8.22.**

What does it mean, then, when results of observational trials disagree with the results of intervention trials? Intervention trials demonstrate causation, while observational studies prove only that there is an association between two variables. This means that evidence from intervention trials is generally considered stronger. There are, however, important differences between the intervention trials and observational studies that looked at the homocysteine hypothesis, such as the types of populations studied, the duration of the studies and the vitamin doses consumed. These differences may explain why observational studies and intervention trials have resulted in different conclusions. Clearly, more study is needed to fully understand the role of B vitamins in cardiovascular disease.

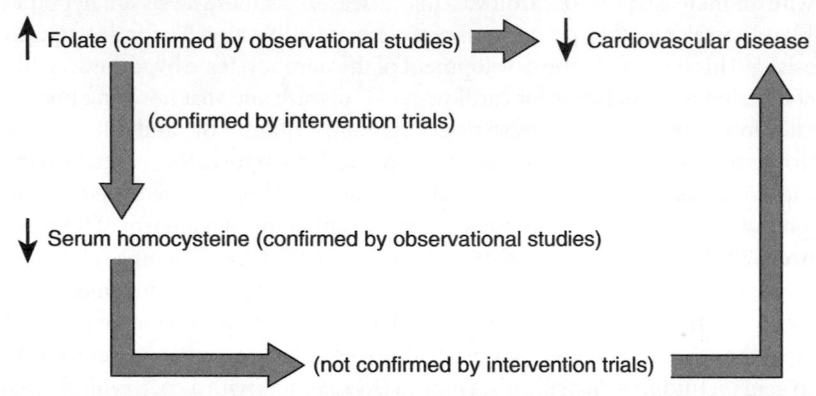

Figure 8.22 **The interrelationships between folate, homocysteine, and cardiovascular disease.** Intervention trials, using folate supplements, have not confirmed observational studies which demonstrate that increased folate intake is associated with reduced risk of cardiovascular disease.

Vitamin B_6 Supplements and Toxicity

No adverse effects have been associated with high intakes of vitamin B_6 from foods, but large doses found in supplements can cause serious toxicity symptoms. This was first recognized in the 1980s when severe nerve impairment was reported in individuals taking 2-6 g/day of supplemental pyridoxine.[15] Symptoms were serious enough in some subjects that they were unable to walk, but these symptoms improved when the pyridoxine supplements were stopped. To avoid toxicity when taking supplements containing vitamin B_6, it is important that intake not exceed the UL of 100 mg/day from food and supplements.[3] The UL is based on an amount that will not cause nerve damage in the majority of healthy people. Since high-dose supplements of vitamin B_6 containing 100 mg/dose are available over the counter, it is easy to obtain a dose that exceeds the UL.

Despite the potential for toxicity, vitamin B_6 supplements are marketed to help a wide variety of ailments. These marketing claims are usually founded in science, but may be exaggerated to sell products (see Table 8.3).

Can Vitamin B_6 Treat Carpal Tunnel Syndrome? Vitamin B_6 has been suggested to be useful in treating carpal tunnel syndrome, in which pressure on the nerves in the hand causes pain and weakness. Vitamin B_6 may act by directly alleviating the symptoms, by altering pain perception and increasing pain threshold, or by alleviating an unrecognized peripheral neuropathy. Although the evidence to support its use is inconclusive, some patients will improve when treated with doses of up to 100 mg, often used in conjunction with other treatments for this condition.[16]

Does Vitamin B_6 Prevent Premenstrual Syndrome? Premenstrual syndrome (PMS) is a collection of physical and emotional symptoms that some women experience prior to menstruation. It causes mood swings, food cravings, bloating, tension, depression, headaches, acne, breast tenderness, anxiety, temper outbursts, and more than 100 other symptoms. The proposed connection between these symptoms and vitamin B_6 is the fact that the vitamin is needed for the synthesis of the neurotransmitters serotonin and dopamine. Insufficient vitamin B_6 has been suggested to reduce levels of these neurotransmitters and cause the anxiety, irritability, and depression associated with PMS. However, there is little evidence that supplementing vitamin B_6 will provide significant benefit for women with PMS.[17]

Will Vitamin B_6 Boost Immunity? Immune function can be impaired by a deficiency of any nutrient that hinders cell growth and division. Therefore, one of the most common claims for vitamin supplements in general is that they improve immune function. Vitamin B_6 is no exception and there are data to support the claim. Vitamin B_6 supplements have been found to improve immune function in older adults.[3] However, since elderly individuals frequently have low intakes of vitamin B_6, it is unclear whether the beneficial effects of supplements are due to an improvement in vitamin B_6 status or immune system stimulation.

8.8 Folate or Folic Acid

Learning Objectives

- Explain how folate deficiency causes anemia.
- Discuss why folic acid supplementation is recommended for women of child-bearing age.

It has been known for more than 100 years that anemia often occurs during pregnancy. In 1937, anemia in a pregnant woman was successfully treated with a yeast preparation named *Wills Factor*, after Dr. Lucy Wills, who treated this patient. The Wills Factor was later isolated from spinach and named *folate*, after the Latin word for foliage. **Folate** and **folacin** are general terms for compounds that have chemical structures and nutritional properties similar to those of **folic acid** (**Figure 8.23**). The chemical name for folate is pteroylglutamic acid.

folate and **folacin** General terms for the many forms of this vitamin.

folic acid The monoglutamate form of folate, which is present in the diet in fortified foods and vitamin supplements.

(a)

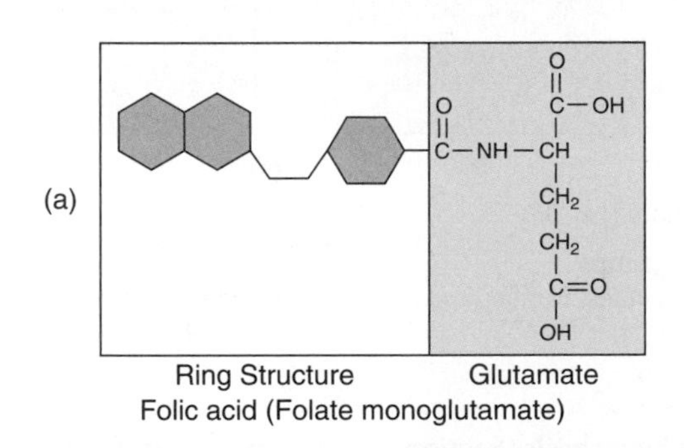

Ring Structure Glutamate
Folic acid (Folate monoglutamate)

(b)

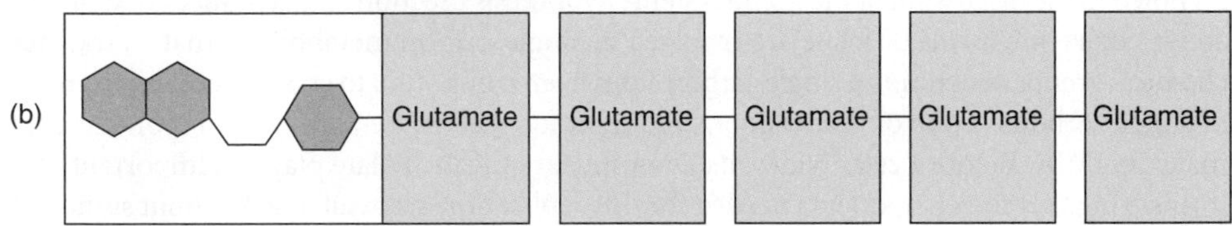

Folate polyglutamate

Figure 8.23 Structure of folate
(a) Folic acid is the monoglutamate form of folate. It is the form found in fortified foods and supplements. (b) Folate polyglutamate is the form found naturally in foods; it includes many glutamate molecules attached to form a chain.

Folate in the Diet and the Digestive Tract

Excellent dietary sources of folate include liver, yeast, asparagus, oranges, legumes, and fortified grain products. Fair sources include vegetables such as corn, green beans, mustard greens, and broccoli, as well as some nuts and seeds. Small amounts are found in meats, cheese, and milk (**Figure 8.24**). Most folate found naturally in food contains a string of glutamate molecules (see Figure 8.23). Glutamate is an amino acid, and folate bound to many glutamates is referred to as folate polyglutamate. Before this form can be absorbed, all but one of the glutamate molecules must be removed by enzymes in the brush border of the small intestine to yield folic acid, the monoglutamate form. It is estimated that about 50% of the folate in food is absorbed.[3] The folic acid form rarely occurs naturally in food but is used in supplements and fortified foods. It is more easily absorbed because it does not require the enzymatic removal of a string of glutamate molecules. The bioavailability of the synthetic folic acid added to grain products and used in supplements is about twice that of the folate naturally found in food. In Canada, since 1998, enriched white flour, corn meal, and pasta have been fortified with folic acid.

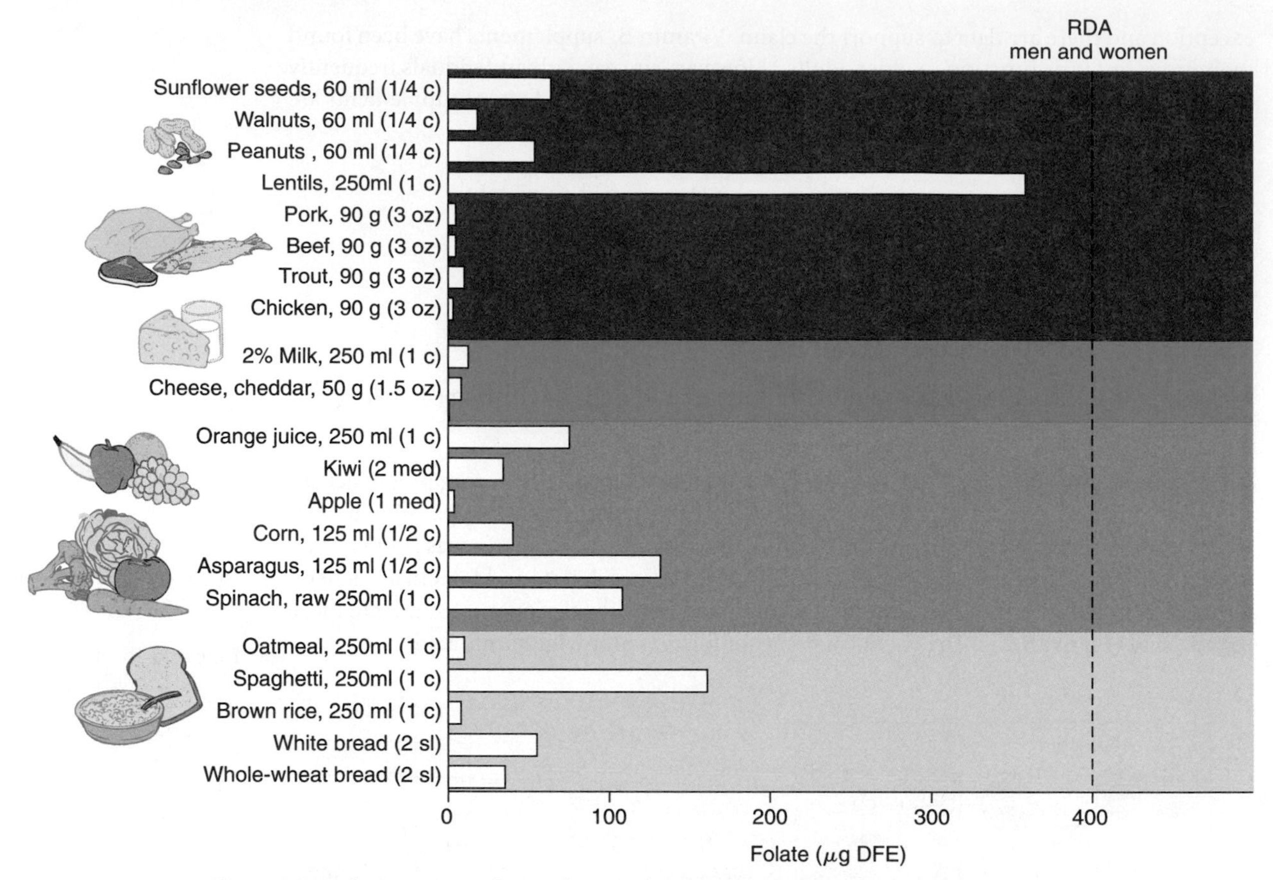

Figure 8.24 Folate content of Canada's Food Guide food groups
Adults can obtain the RDA for folate (dashed line) by eating legumes, fortified foods, and fruits and vegetables.

Folate in the Body

As noted in the discussion of the homocysteine hypothesis (Section 8.7), a number of different active coenzyme forms of folate are involved in single carbon metabolism, that is, transfer chemical groups containing a single carbon atom from compound to compound. This transfer of single carbons is one of the reactions required for the synthesis of the components that make up DNA. Before a cell divides, its DNA must replicate. Folate plays an important role in ensuring that the components needed for this replication are available. Without sufficient folate, the synthesis of DNA is impaired. The role of folate in DNA synthesis is particularly important in tissues where cells are rapidly dividing, such as bone marrow, where red blood cells are made; intestines; skin; and during prenatal development. As discussed in the section on folate deficiency, anemia develops when a person is deficient in folate because DNA synthesis is impaired, preventing the proper formation of red blood cells.

Folate also has a second DNA-related function, a role in gene expression (see Section 6.4). Folate is involved in the transfer of a single carbon atom in the form of methyl groups (-CH_3) to DNA. High levels of DNA methylation result in the silencing of genes, that is, genes in a region of DNA that is methylated are generally not expressed. Low folate intake may alter the levels of DNA methylation and hence which genes are or are not expressed in a cell (**Figure 8.25**). The careful switching on and off of genes is believed to be critical in prenatal development. Babies born to mothers that are folate deficient are more likely to develop birth defects called **neural tube defects.** The neural tube is the part of the embryo that ultimately develops into the spinal cord and brain. While the exact way in which folate affects neural tube development is not fully understood, DNA methylation is believed to play an important role.[18] When the expression of the genes involved in this developmental process is altered because of low intakes of folate, and hence low DNA methylation, serious and often fatal brain and spinal cord defects result (see Section 14.2). This role of folate in gene expression is referred to as an epigenetic role, that is, a role in **epigenetics**, a study of genetics that is unrelated to the sequence of DNA, but to DNA's modification due to reactions such as methylation (see Figure 8.25).

neural tube defects Abnormalities in the brain or spinal cord that result from errors that occur during prenatal development. Defects in the brain are fatal, while those of the spinal cord often result in paralysis.

epigenetics The study of genetics unrelated to changes in the sequence of DNA but to its chemical modification due to reactions such as methylation.

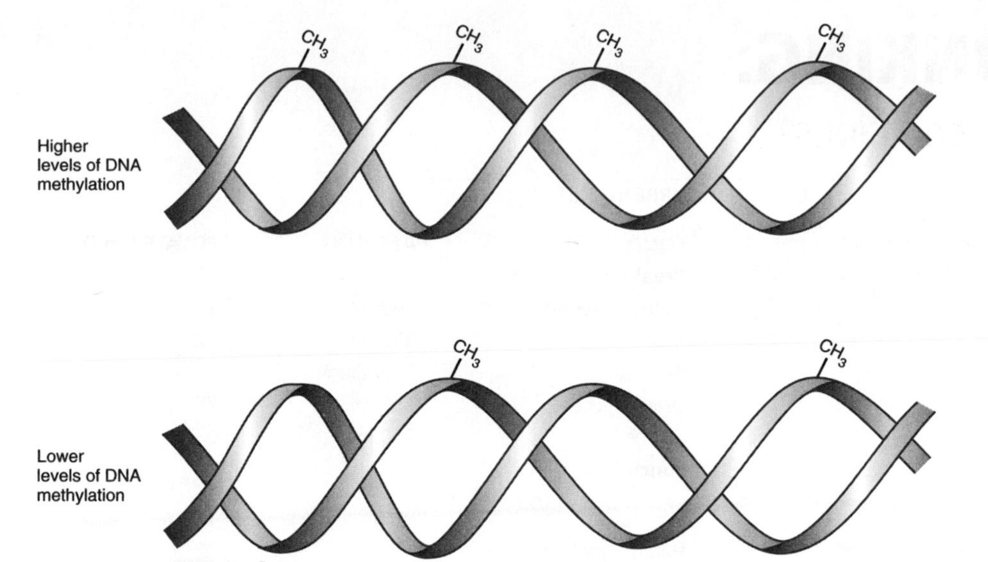

Figure 8.25 DNA methylation
Variations in DNA methylation alter the number and types of genes expressed in a cell. Folate functions in the transfer of methyl groups to the DNA molecule.

Recommended Folate Intake

Life Cycle

The RDA for folate is set at 400 μg **dietary folate equivalents (DFEs)** per day for adult men and women. One DFE is equal to 1 μg of food folate, that is, naturally occurring folate, or 0.6 μg of synthetic folic acid from fortified food or supplements consumed with food. Synthetic folic acid is more readily absorbed so you can consume less synthetic folic acid to obtain the same biological effect, compared to naturally occurring folate. The concept of DFE takes this difference into account (**Table 8.4**).

dietary folate equivalents (DFEs) The unit used to express the amount of folate present in food. One DFE is equivalent to 1 μg of folate naturally occurring in food, 0.6 μg of synthetic folic acid from fortified food or supplements consumed with food, or 0.5 μg of synthetic folic acid consumed on an empty stomach.

Table 8.4 Calculating Dietary Folate Equivalents in Fortified Foods Containing Both Naturally Occurring Folate and Synthetic Folic Acid
Synthetic folic acid is the form of the vitamin added to fortified foods, such as white flour, cornmeal, and pasta, and is more bioavailable than natural forms of folate, so the unit dietary folate equivalent (DFE) has been developed to compare the two forms of the vitamin.
1 μg DFE is equal to 0.6 μg of synthetic folic acid, or conversely, 1 μg of synthetic folic acid is equal to 1.7 μg DFE.
Some fortified foods also contain some naturally occurring folate, so to determine the total amount of μg DFE in some foods, both sources have to be considered.
For example,
100 g of fortified white flour contains the following:
Folic acid, synthetic form 150 μg
Folate, naturally occurring 29 μg
To convert to DFE multiply the μg folic acid by 1.7
Folic acid, synthetic form 150 μg X 1.7 = 255 μg DFE
To find total DFE add together the DFE from synthetic and natural forms:
Folic acid, synthetic form 150 μg X 1.7 = 255 μg DFE
Folate, naturally occurring 29 μg X 1.0 = 29 μg DFE
Total: 284 μg DFE

As noted, since 1998, foods such as enriched white flour, pasta, and cornmeal, have been fortified with folic acid, in order to improve the overall intake of folate among women of child-bearing age and reduce the risk of neural tube defects. In order to further reduce the risk of neural tube defects, Canada's Food Guide and Health Canada also recommends that women of child-bearing age take a multivitamin that contains at least 400 μg of folic acid as well as vitamin B_{12}. This is because the neural tube defects develop 2-3 weeks after conception, a time when many women do not yet know that they are pregnant and also because

CRITICAL THINKING:

Meeting Folate Recommendations

Background

Mercedes would like to have a baby but before she tries to conceive, she wants to be sure she is in the best possible health. She consults her physician who gives her a clean bill of health but suggests she make sure she is getting enough folic acid.

(© istockphoto.com/svariophoto)

Data

Mercedes records her food intake for 1 day to determine her folate intake.

CURRENT DIET

FOOD	AMOUNT	FOLATE (μg DFE)
Breakfast		
Oatmeal, regular	250 ml	2
Milk	250 ml	12
Banana	1 medium	22
Orange juice	250 ml	75
Coffee	250 ml	0
Lunch		
Hamburger	90 g	11
Hamburger bun	1	32
French fries	20 pieces	24
Coke	360 ml	0
Apple	1 medium	4
Dinner		
Chicken	90 g	4
Refried beans	125 ml	106
Macaroni, enriched	125 ml	80
Flour tortilla	1	60
Salad	250 ml	64
Salad dressing	15 ml	1
Milk	250 ml	12
White cake	1 piece	32
Total		**541**

Critical Thinking Questions

Why is folate a concern for Mercedes when planning a pregnancy?

Look at Mercedes's diet. Which foods are highest in folate? Of these, which are fortified with folic acid?

Mercedes has oatmeal for breakfast because it is a whole grain, but why is it low in folate?

If she replaced her enriched macaroni, tortilla, and hamburger bun with whole wheat products, how would this affect her total folate intake?

Would you recommend Mercedes take a folic acid supplement?

Use iProfile to find out how much folate is in the folate-fortified grain products you consume each day.

many pregnancies are unplanned. Starting to take folate after a pregnancy is confirmed is too late to reduce the risk of neural tube defects, although it is beneficial for other aspects of fetal development and maternal health. A well-balanced diet that follows Canada's Food Guide and includes fortified food together with this supplement has been found to provide enough folic acid to measurably reduce the risk of neural tube defects.[19] It is important to note that women of child-bearing age following these recommendations would have a folate intake that exceeds the RDA of 400 μg DFE (see Critical Thinking: Meeting Folate Recommendations).

The RDA for folate during pregnancy is increased to 600 μg DFE/day, from 400 μg DFE for adults, to provide for the increase in cell division. As this amount is difficult to obtain solely from intake of food even when fortified foods are consumed, folate is typically supplemented during pregnancy. The RDA during lactation is 500 μg DFE to account for folate secretion in milk. Needs per unit of body weight are higher for infants and children than for adults because of their rapid growth. Human and cow milk provide enough folate to meet infant needs, but goat milk does not. Infants and children given goat milk may not receive adequate folate unless it is provided from other sources.

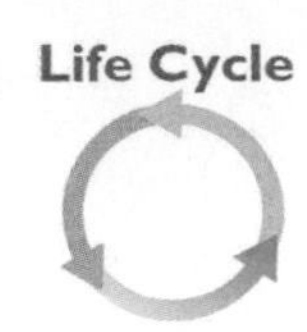

Folate Deficiency

A deficiency of folate leads to a drop in blood folate levels and a rise in blood homocysteine (Figure 8.21), followed by changes that affect rapidly dividing cells. Deficiency symptoms include poor growth, problems in nerve development and function, diarrhea, inflammation of the tongue, and anemia. Anemia results when folate is deficient because cells in the bone marrow that develop into red blood cells cannot duplicate their DNA and so cannot divide. Instead, they just grow bigger. These large immature cells are known as **megaloblasts** and can be converted into large red blood cells called **macrocytes.** The result is that fewer mature red blood cells are produced, and the oxygen-carrying capacity of the blood is reduced. This condition is called **megaloblastic** or **macrocytic anemia** (**Figure 8.26**). Groups at risk of folate deficiency include pregnant women and premature infants because of their rapid rate of cell division and growth; the elderly because of their limited intake of foods high in folate; alcoholics because alcohol inhibits folate absorption; and tobacco smokers because smoke inactivates folate in the cells lining the lungs.[3]

megaloblasts Large, immature red blood cells that are formed when developing red blood cells are unable to divide normally.

macrocytes Larger-than-normal mature red blood cells that have a shortened life span.

megaloblastic or **macrocytic anemia** A condition in which there are abnormally large immature and mature red blood cells in the bloodstream and a reduction in the total number of red blood cells and the oxygen-carrying capacity of the blood.

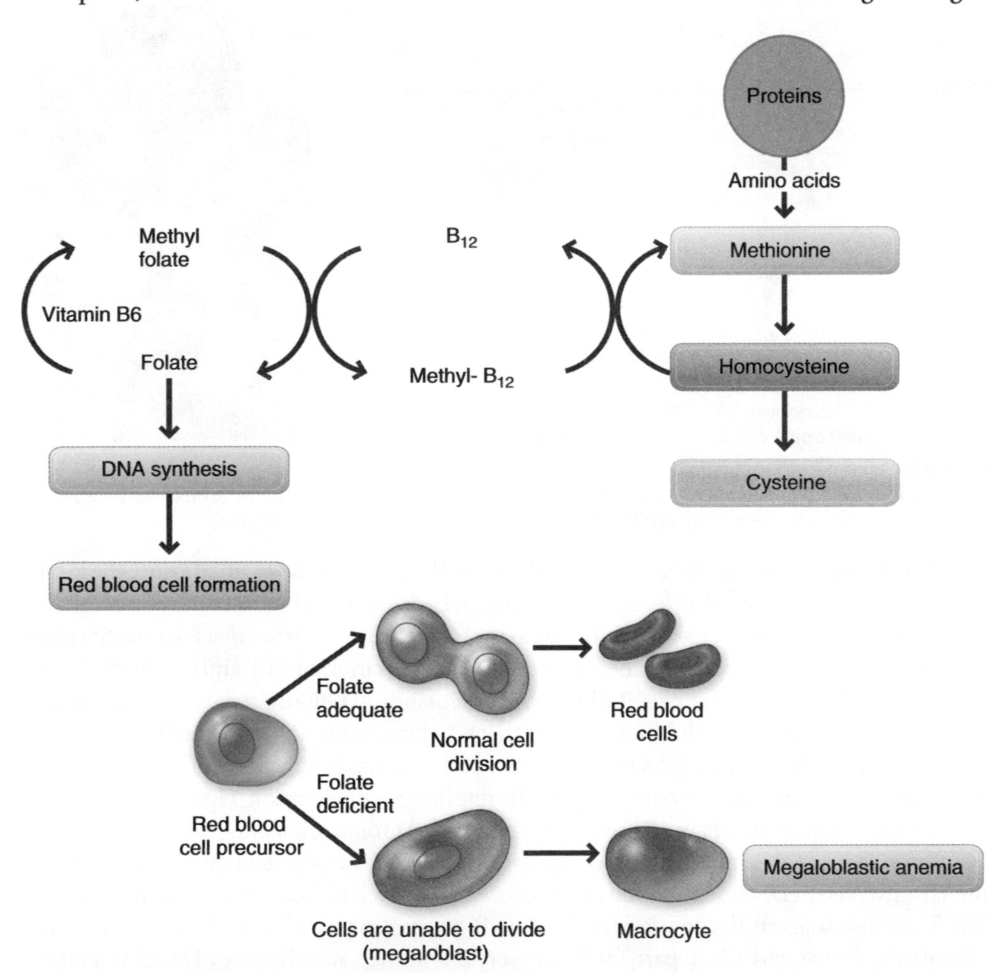

Figure 8.26 Role of folate in macrocytic anemia
Macrocytic anemia occurs when red blood cell precursors are unable to divide, resulting in the formation of abnormally large red blood cells (macrocytes).

It is easy from Figure 8.26 to see how a folate deficiency can impair red blood cell formation. The figure also suggests that a vitamin B_{12} deficiency might also cause the same anemia, and this, as is discussed in the section on vitamin B_{12}, is indeed the case. The absence of vitamin B_{12} prevents the conversion of methyl folate to folate, which is required for DNA synthesis to proceed.

spina bifida A birth defect resulting from the incorrect development of the spinal cord that can leave the spinal cord exposed. This can result in paralysis.

Folate and the Risk of Neural Tube Defects

Neural tube defects, such as **spina bifida, anencephaly**, and other birth defects that affect the brain and spinal cord, are not true folate-deficiency symptoms because not every pregnant woman with inadequate folate levels will give birth to a child with a neural tube defect. Instead,

anencephaly A birth defect due to failure of the neural tube to close that results in the absence of a major portion of the brain, skull, and scalp. This is a fatal defect.

neural tube defects are probably due to a combination of factors that include low folate levels and a genetic predisposition. Folate is necessary for a critical step called *neural tube closure.* (see Section 14.2). When neural tube closure does not occur normally, portions of the brain or spinal cord are not adequately protected (**Figure 8.27**). Neural tube closure is complete by 28 days after conception, therefore, folate status should be adequate even before a pregnancy begins to assure an adequate supply during early development. Studies in which supplemental folic acid was given to women before and during early pregnancy showed that 360-800 μg/day of synthetic folic acid in addition to food folate was associated with a reduced incidence of neural tube defects.[3] The folate fortification program in Canada, which mandated fortification of white flour, cornmeal, and pasta with folic acid, along with advice to take folic acid supplements, has helped to increase the amount of folate in the blood of many Canadian women to levels that are known to be protective against neural tube defects (see Critical Thinking: Folate Intake in Canada: Too Little or Too Much?). A recent survey by the Public Health Agency of Canada found that 58% of women took supplements prior to conception and that 77% of women knew that folic acid supplements protect against neural tube defects.[20]

CANADIAN CONTENT

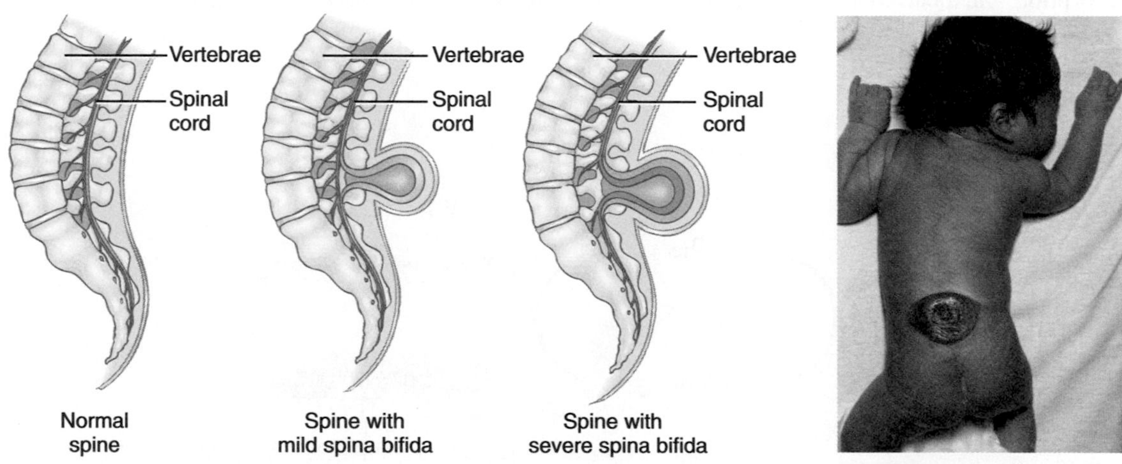

Figure 8.27 **Spina bifida**
The neural tube develops into the brain and spinal cord. If folate is inadequate during neural tube closure, neural tube defects such as spina bifida occur more frequently. (Wellcome Trust/Custom Medical Stock Photo, Inc.)

The incidence of neural tube defects has decreased by almost 50% in Canada since the inception of food fortification.[21] In Newfoundland, which had one of the highest rates of neural tube defects in North America, rates dropped by almost 80%.[22] There is evidence that folate fortification has also resulted in a decrease in other birth defects. Studies in Alberta [23] and Quebec[24] show a reduction in the number of congenital heart defects. A study in Ontario reported a reduction in neuroblastoma, a common childhood cancer that is believe to originate prenatally.[25]

The way in which folate protects against neural tube defects is the subject of considerable research, but is believed to be related to its epigenetic role in DNA methylation. **Figure 8.28** illustrates the metabolic pathways believed to be involved. A methyl group, originating with methyl folate, is transferred through a multi-step process to a derivative of the amino acid methionine which is directly involved in DNA methylation (**Figure 8.25**). Low levels of folate will impair this process. The figure also suggests that low levels of vitamin B_{12} may also play a role in the development of neural tube defects and it is, in part, for this reason that women are advised by Health Canada to take a folic acid supplement that also contains vitamin B_{12}. The role of vitamin B_{12} in neural tube defects is an active area of research. Its importance is emphasized by a recent study that found 5% of pregnant Canadian women may be deficient in vitamin B_{12} during the time of neural tube closure.[26]

Life Cycle

Folate and Cancer

Numerous observational studies have demonstrated that low folate status increases the risk of developing cancer of the ovary, breast, pancreas, and colon.[27–29] Although folate deficiency does not cause cancer, it has been hypothesized that low folate intake enhances an underlying predisposition to cancer. Higher plasma levels of folate are associated with a reduced risk of developing breast cancer, particularly in women at high risk due to alcohol consumption.[30] Alcohol consumption greatly increases the cancer risk associated with a low folate diet.[31]

Higher dietary folate intake and blood folate levels are associated with a lower risk of colorectal cancer. Studies suggest an approximately 40% reduction in the risk of colorectal cancer in individuals with the highest dietary folate intake compared with those with the lowest intake.[32]

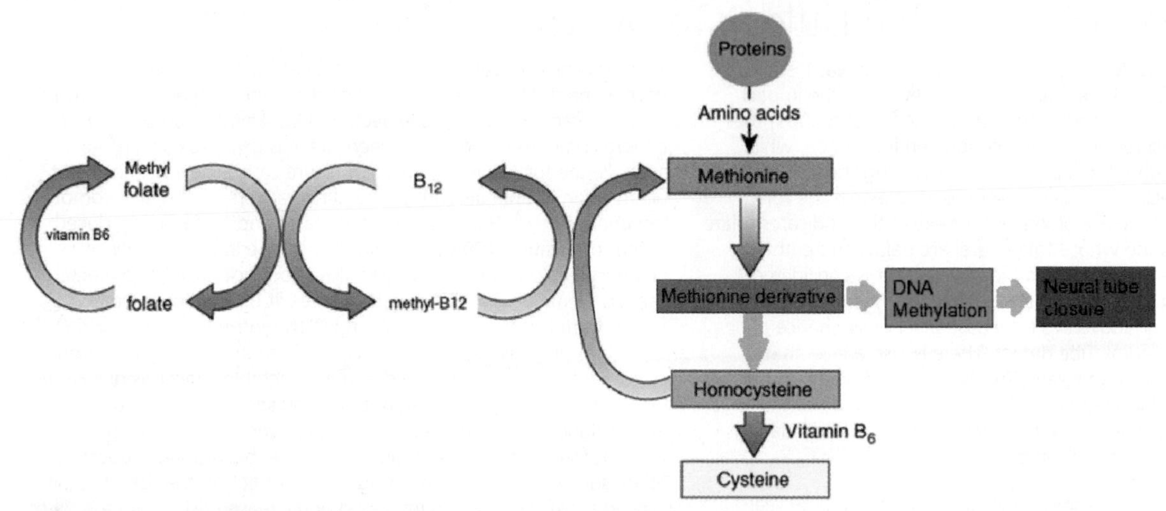

Figure 8.28 The role of folate and vitamin B12 in neural tube closure.
Low levels of folate and/or vitamin B_{12} may impair DNA methylation reactions needed for proper neural tube closure.

While observational studies have shown that folate intake is associated with reduced risk of cancer, intervention trials with folate supplementation have not confirmed these results, while animal studies suggest that there is a more complex relationship between folate and cancer (See Critical Thinking: Folate Intake in Canada: Too Little or Too Much?). This is an area where more research is needed.

Folate Toxicity

Although there is no known folate toxicity, a high intake may mask the early symptoms of vitamin B_{12} deficiency, allowing it to go untreated and resulting in irreversible nerve damage (see Critical Thinking, Folate Intake in Canada: Too Little or Too Much?). The UL for folic acid for adults is set at 1,000 µg per day from supplements and/or fortified foods. This value was determined based on the progression of neurological symptoms seen in patients who are deficient in vitamin B_{12} and taking folic acid supplements.

8.9 Vitamin B_{12}

Learning Objectives

- Name foods that are good sources and poor sources of vitamin B_{12}.
- List the steps involved in vitamin B_{12} absorption.
- Compare the functions of folate and vitamin B_{12}.

Pernicious anemia is a form of anemia that does not respond to iron supplementation. At the time it was first described in 1820, it could not be treated and was fatal. Pernicious anemia is caused by an inability to absorb sufficient vitamin B_{12}.

pernicious anemia An anemia resulting from vitamin B_{12} deficiency that occurs when dietary vitamin B_{12} cannot be absorbed due to a lack of intrinsic factor.

Drs. Minot and Murphy were awarded the Nobel Prize for curing the disease with a diet containing large quantities of liver, which is a good source of vitamin B_{12}. Today, concern focuses on the effects of marginal deficiencies of this vitamin and the potential masking of B_{12} deficiency by high intakes of folic acid.

Vitamin B_{12} in the Diet

Vitamin B_{12} is found almost exclusively in animal products (**Figure 8.29**). It can be made by bacteria, fungi, and algae but not by plants and animals. It accumulates in animal tissue

CRITICAL THINKING:

Folate Intake in Canada: Too Little or Too Much?

Folate fortification of white flour, pasta, and corn meal was instituted in Canada in 1998. Since then, many studies to evaluate the impact of this program have been conducted. Between 2007 and 2009, the Canadian Health Measures Survey was conducted in Canada, which measured various health-related parameters, including the levels of folate in red blood cells.[1] These levels were used to assess the folate status of Canadians. There is a blood level or cutoff that indicates folate deficiency, that is, anyone whose folate level are below this cutoff is folate deficient. There is also a folate blood level that is considered optimal for protecting against neural tube defects, meaning that women with levels at or above this level have a very low chance of having a baby with a neural tube defect. There is also a level that indicates a very high intake of folate. This is not based on any adverse effects but reflects a blood level that was higher than 97% of the population in a recent American survey (1999-2004) of a population that consumes a folate-fortified foods.[2]

The folate levels of Canadian women of child-bearing age (20-39 years) are shown in the figure describing red blood cell folate levels. The researchers found that less than 1% of Canadian women of child-bearing age had a folate deficiency (#1 on figure). Similar results were also observed for Canadian women in other age groups and all Canadian men. About 22% of women, however, did not have levels considered optimal to protect their pregnancy from neural tube defects (#2 on figure). These numbers suggest that Canada's folate fortification program has not fully eliminated the risk of neural tube defects caused by inadequate folate intake. It has been suggested, in response to these results, that the amount of folate added to food should be increased to further reduce the risk.[2]

On the other hand, about 40% of Canadian women of child-bearing had folate levels that were considered high (#3) and this was also observed in Canadian men and women in other age groups. There are some concerns about very high intakes of folate. One is that excessive intakes may mask vitamin B_{12} deficiencies (how this happens is described in Section 8.9). Although cases of this happening are not widely documented, it forms the basis for setting the UL for folate. The population most likely to be affected by masking are the elderly, who often suffer reduced ability to absorb vitamin B_{12}.[3]

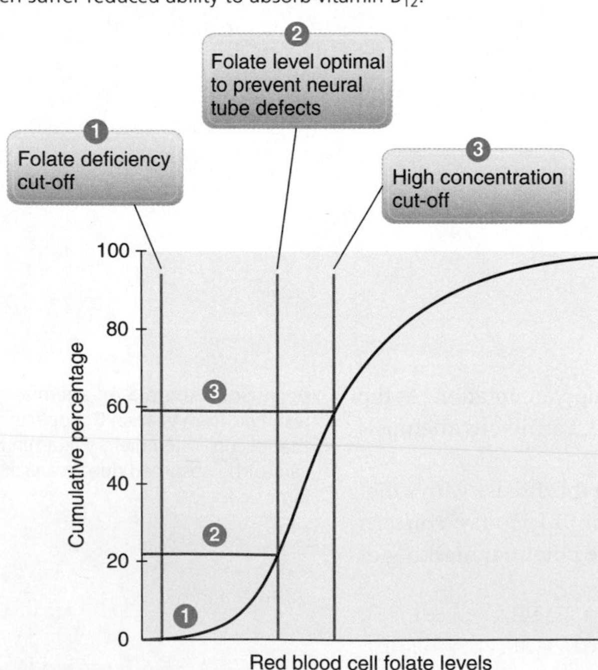

Source: Adapted from Colapinto, C. K., O'Connor, D. L. , Tremblay, M. S. Folate status of the population in the Canadian Health Measures Survey.

There is also a concern about the role that folate plays in cancer development. Many observational studies suggest that folate may be protective against cancer (see Section 8.8). Animal studies suggest a more complex relationship: when folate is given to cancer-free animals, the folate protects against future cancer development, but when given to animals with pre-existing cancers, the folate promotes tumour growth.[4] When both United States and Canada introduced folate fortification 1998, there was a brief rise in the number of diagnosed colon cancer cases (see figure on incidence of colorectal cancer). Some believe this may be the result of additional dietary folate, from fortification, stimulating DNA synthesis and promoting the growth of pre-existing cancerous cells. These pre-existing cancerous cells might not have developed into a detectable cancer as quickly, if higher amounts of folate had not been present in the diet. But this interpretation is controversial, as there are very few human intervention trials on folate and cancer. Studies that have been done suggest that folate supplementation has no significant effect on the risk of colon cancer; there is neither an increased or decreased risk of disease with supplementation.[5]

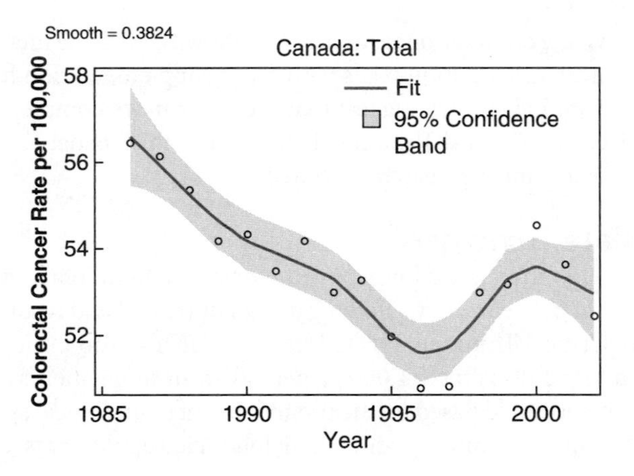

Incidence of colorectal cancer in Canada from 1986-2002; note the increase in cancer cases around 1998, coinciding with the introduction of folate fortification; a similar trend was also observed in the U.S.

Source: Mason, J. B., Dickstein, A., Jacques, P. F. et al. A temporal association between folic acid fortification and an increase in colorectal cancer rates may be illuminating important biological principles: a hypothesis. *Cancer Epidemiol. Biomarkers Prev.*;16(7):1325-9, 2007.

What the data described above suggest is that when a food fortification program that benefits one segment of the population, for example, child-bearing women and their children, is implemented, one must be careful to avoid doing potential harm to another segment of the population, such as perhaps the elderly, who have a greater risk of cancer due to age and also may suffer reduced absorption of vitamin B_{12}.

Critical Thinking Questions

Given this information, do you agree with the recommendation that the amount of folate added to food in Canada should be increased to reduce the number of women of child-bearing age that have folate blood levels below the optimum to reduce neural tube defects?

[1]Health Canada Canadian Health Measures Survey: Questions & Answers. Available online at http://www.hc-sc.gc.ca/ewh-semt/contaminants/human-humaine/chms-ecms_faq-eng.php. Accessed April 2, 2011.

[2]Colapinto, C. K., O'Connor, D. L., Tremblay, M. S. Folate status of the population in the Canadian Health Measures Survey. *CMAJ*. 183(2):E100-6, 2011.

[3]Cuskelly, G., J., Mooney, K. M., Young, I. S. Folate and vitamin B_{12}: friendly or enemy nutrients for the elderly. *Proc. Nutr. Soc.*;66(4):548-58, 2007.

[4]Kim, Y. I.. Folic acid fortification and supplementation--good for some but not so good for others. *Nutr. Rev.* 65(11):504-11, 2007.

[5]Figueiredo, J. C., Mott, L. A., Giovannucci, E, et al. Folic acid and prevention of colorectal adenomas: A combined analysis of randomized clinical trials. *Int J. Cancer.* 129(1):192-203, 2011.

from the diet or from synthesis by bacterial microflora. Bacteria in the human colon produce vitamin B_{12}, but it cannot be absorbed. Vitamin B_{12} is not found in plant products unless they have been contaminated with bacteria, soil, insects, or other sources of vitamin B_{12}, or have been fortified with it. Diets that do not include animal products must include supplements or foods fortified with vitamin B_{12} in order to meet needs.[33]

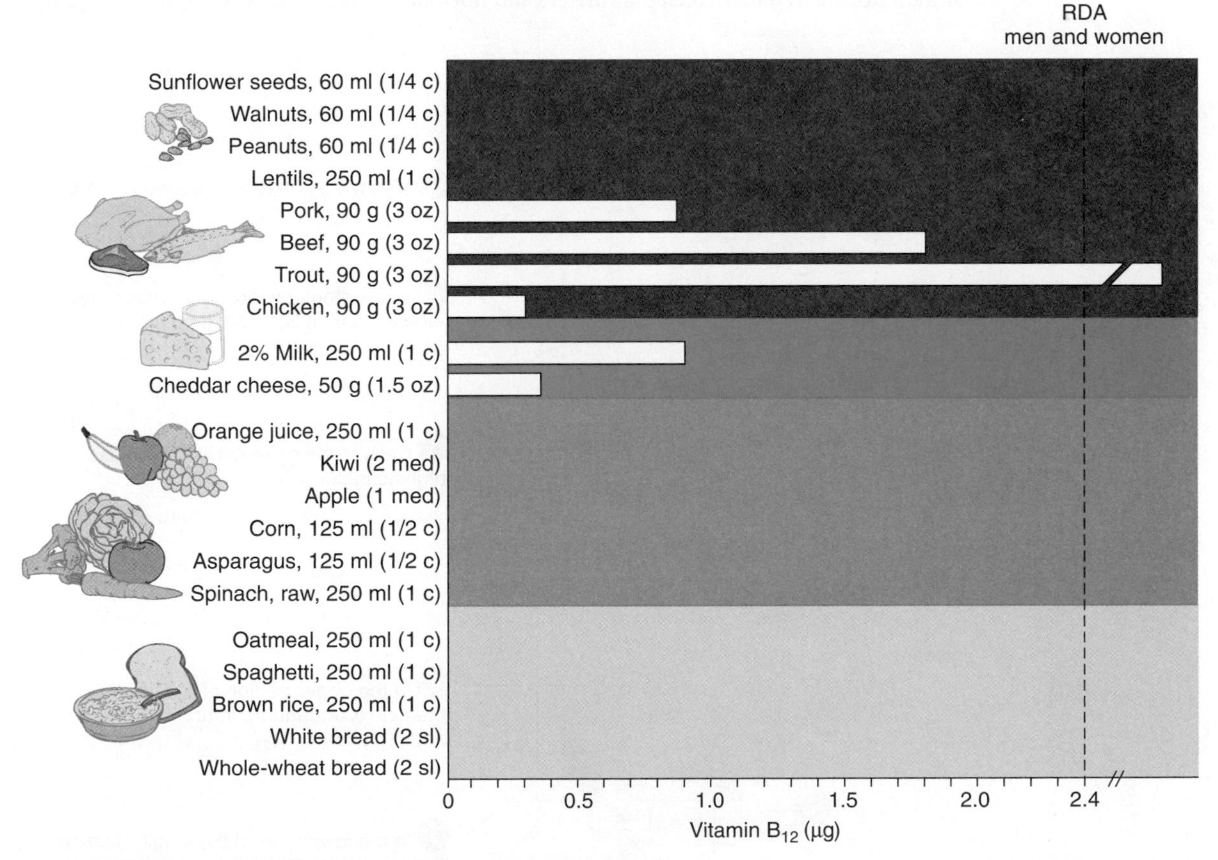

Figure 8.29 Vitamin B_{12} content of Canada's Food Guide food groups
Vitamin B_{12} is found only in foods of animal origin or foods that have been fortified with the vitamin; the dashed line represents the RDA for adult men and women.

Vitamin B_{12} in the Digestive Tract

Naturally occurring vitamin B_{12} is bound to proteins in food and must be released before it can be absorbed. It is released in the stomach by acid and the protein-digesting enzyme pepsin. In the small intestine, vitamin B_{12} binds to **intrinsic factor**, which is a protein secreted by the **parietal cells** in the lining of the stomach. The intrinsic factor-vitamin B_{12} complex

parietal cells Large cells in the stomach lining that produce and secrete intrinsic factor and hydrochloric acid.

intrinsic factor A protein produced in the stomach that is needed for the absorption of adequate amounts of vitamin B_{12}.

binds to receptor proteins in the ileum of the small intestine, allowing the vitamin to be absorbed (**Figure 8.30**). Only a small amount of vitamin B_{12} can be absorbed when intrinsic factor is absent. Vitamin B_{12} absorption is also reduced by low stomach acid and insufficient pancreatic secretions.

Vitamin B_{12} is secreted in bile, but most of this is reabsorbed, rather than being lost in the feces. Because of this efficient recycling, it can take many years of a deficient diet before the symptoms of vitamin B_{12} deficiency appear.

Vitamin B_{12} in the Body

cobalamin The chemical term for vitamin B_{12}.

The terms vitamin B_{12} and **cobalamin** refer to members of a group of cobalt-containing compounds. Vitamin B_{12} is necessary for the maintenance of myelin, which insulates nerves and is essential for normal nerve transmission (see Figure 8.20). Vitamin B_{12} can be converted into either of 2 active cobalamin coenzyme forms, methylcobalamin and adenosylcobalamin (see Appendix I). The function of methylcobalamin is illustrated in figures 8.21, 8.26, and 8.28, where it is more simply designated "methyl-B_{12}." It functions in single carbon metabolism to transfer methyl groups between folate and the amino acid homocysteine. Adenosylcobalamin rearranges carbon atoms so that the breakdown products of fatty acids, with an odd number of carbons (which are rare), can enter the citric acid cycle and be used to provide energy.

Recommended Vitamin B_{12} Intake

The RDA for adults of all ages for vitamin B_{12} is 2.4 μg/day.[3] This is the amount needed to maintain normal red blood cell parameters and normal blood concentrations of vitamin B_{12}.

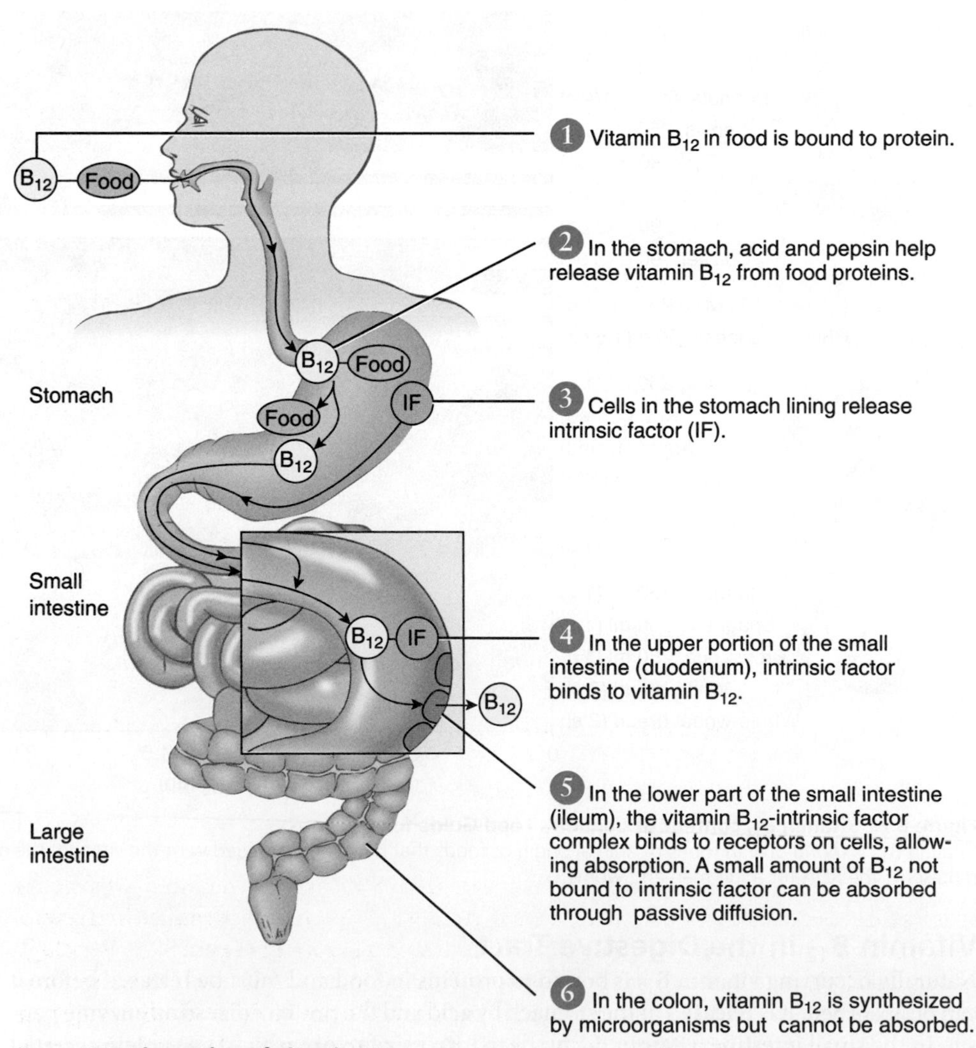

Figure 8.30 Absorption of vitamin B12
Normal absorption of vitamin B_{12} requires a healthy stomach, pancreas, and small intestine.

The RDA for vitamin B_{12} is increased during pregnancy, even though absorption is increased. The RDA during lactation is increased to account for the amount secreted in milk. Pregnant and lactating vegans, like anyone who does not eat animal products, are advised to take a supplement or consume fortified foods to obtain the recommended intake for vitamin B_{12}.[33]

Life Cycle

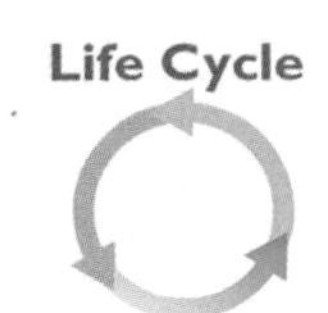

Vitamin B_{12} Deficiency

Symptoms of vitamin B_{12} deficiency include an increase in blood homocysteine levels and a macrocytic, megaloblastic anemia that is indistinguishable from that seen in folate deficiency. This anemia occurs because vitamin B_{12} is needed to convert methyl folate back to the folate form that is active for DNA synthesis (see Figure 8.26). Lack of vitamin B_{12} causes a secondary folate deficiency and, consequently, megaloblastic anemia. Vitamin B_{12} deficiency also causes neurological symptoms, which include numbness and tingling, abnormalities in gait, memory loss, and disorientation. This is because vitamin B_{12} is required for the synthesis of a derivative of the amino acid methionine, which, in turn, is required for the transfer of methyl groups to constituents of the myelin sheath. When these methyl group transfers are impaired by vitamin B_{12} deficiency, degeneration of the myelin that coats the nerves, spinal cord, and brain results. If not treated, this eventually causes paralysis and death.

Blatant deficiencies of vitamin B_{12} are rare because the body stores and reuses it. However, marginal vitamin B_{12} status is of public health concern, particularly for older adults and vegetarians who consume no animal products. Because of the efficient recycling of vitamin B_{12}, it can take many years of a deficient diet before the symptoms of deficiency appear. However, when absorption is impaired, neither dietary vitamin B_{12} nor the vitamin B_{12} secreted in the bile is absorbed, so deficiency symptoms appear more rapidly. This occurs in both individuals with pernicious anemia, an autoimmune disease in which the parietal cells that produce intrinsic factor are destroyed, and in those with **atrophic gastritis**, an inflammation of the stomach lining that results in a reduction in stomach acid and bacterial overgrowth.

atrophic gastritis An inflammation of the stomach lining that results in reduced secretion stomach acid and bacterial overgrowth.

Pernicious Anemia The autoimmune disease pernicious anemia is the major cause of severe vitamin B_{12} deficiency. Without intrinsic factor, vitamin B_{12} cannot be absorbed normally. This anemia can be treated with injections of the vitamin, with a B_{12}-containing nasal gel, or with oral megadoses. The injections, which deliver vitamin B_{12} to the subcutaneous fat or muscle, and the nasal gel, which allows the vitamin to enter the bloodstream through the nasal mucosa, bypass the gastrointestinal tract, and thus the need for intrinsic factor. Megadoses can treat pernicious anemia because they allow adequate amounts of vitamin B_{12} to be absorbed by passive diffusion, which does not require intrinsic factor.

Atrophic Gastritis About 10%-30% of individuals over 50 years of age are unable to absorb food-bound vitamin B_{12} normally because they have atrophic gastritis. When stomach acid is reduced, the enzymes that release protein-bound vitamin B_{12} cannot function properly and the bound vitamin B_{12} cannot be released and absorbed. Atrophic gastritis also causes microbial overgrowth in the intestine. These microbes reduce vitamin B_{12} absorption by competing for available vitamin B_{12}. In severe cases of atrophic gastritis, the production of intrinsic factor is also reduced, further impairing vitamin B_{12} absorption. It is recommended that individuals over the age of 50 meet their RDA by taking a vitamin B_{12}-containing supplement. Because the vitamin B_{12} in these products is not bound to proteins, it is absorbed even when stomach acid is low.

Life Cycle

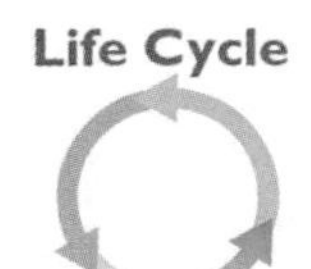

Vegan Diets Vitamin B_{12} deficiency is also a concern among vegans since vitamin B_{12} is only found in foods of animal origin. Severe deficiency has been observed in breastfed infants of vegan women, but marginal deficiency is a concern for all vegans if supplements or fortified foods are not included in the diet.

Vitamin B_{12} Deficiency and Supplemental Folate If individuals with vitamin B_{12} deficiency consume enough folate, they will not develop anemia, which is an easily identified and reversible symptom of vitamin B_{12} deficiency. How this occurs is shown in **Figure 8.31**. Very high folate intakes bypass the need for the conversion of methyl folate to folate, which is a reaction that requires vitamin B_{12}. Without this symptom, diagnosis can be delayed, allowing more serious and irreversible symptoms, such as nerve damage, to progress. Although the fortification of grain products with folic acid has raised concern that additional folate in the food supply could delay diagnosis of vitamin B_{12} deficiency, the amount consumed from a typical diet is unlikely to be high enough to cause problems.

Vitamin B_{12} Toxicity and Supplements

No toxic effects have been reported with excess vitamin B_{12} intakes of up to 100 μg/day from food or supplements. There are not sufficient data to establish a UL for vitamin B_{12}.

Supplements of vitamin B_{12} are available as cyanocobalamin in both oral and injectable forms. Because vitamin B_{12} deficiency causes anemia, supplements of the vitamin, particularly as injections, have been promoted as a pick-me-up for tired, run-down individuals. However, there are no proven benefits of vitamin B_{12} supplementation in individuals who are not vitamin B_{12} deficient. Oral supplements may be of benefit for those at risk for vitamin B_{12} deficiency, such as vegans and individuals over 50 years of age.[3]

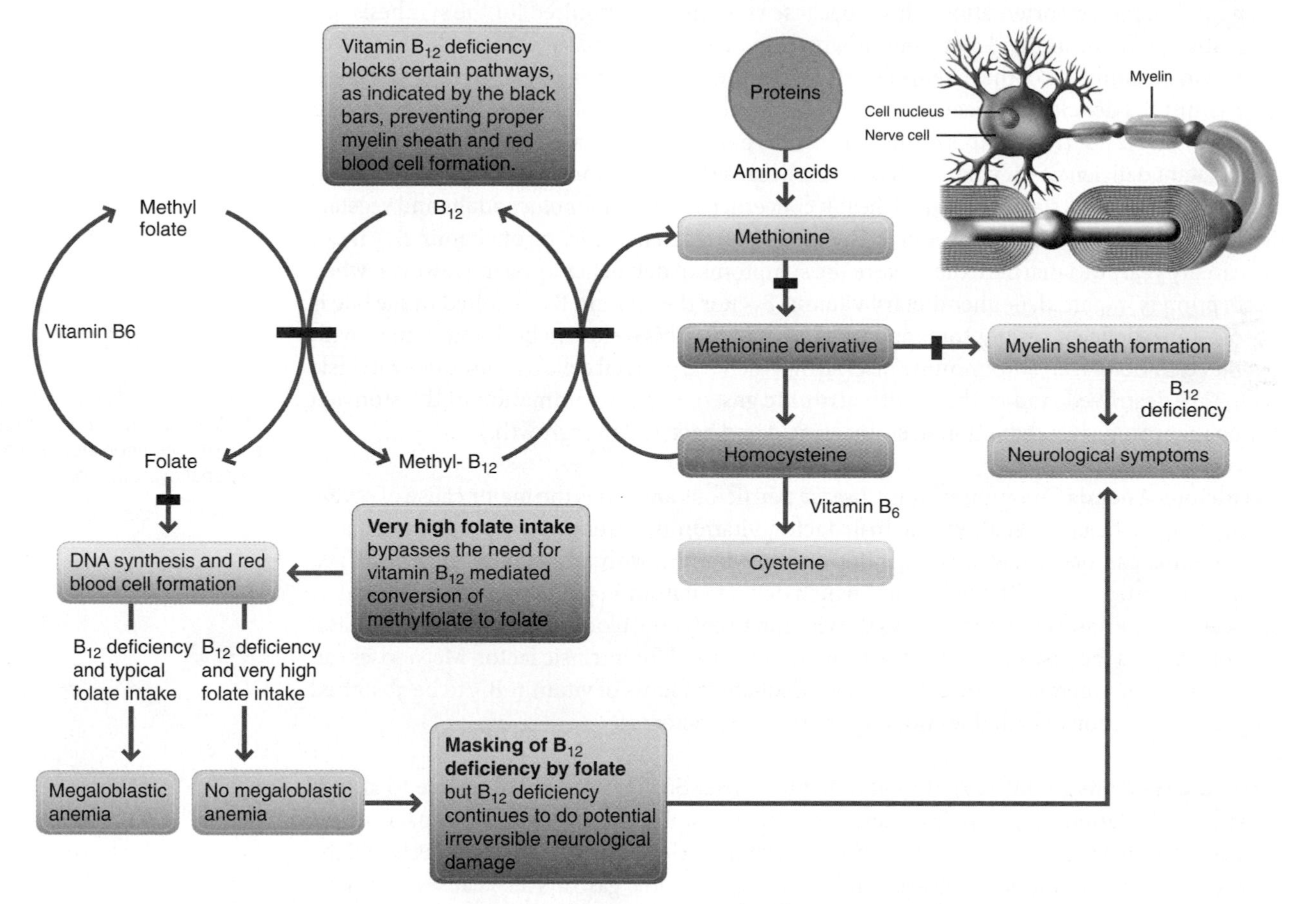

Figure 8.31: Masking of vitamin B12 deficiency by very high intakes of folate.

8.10 Vitamin C

Learning Objectives

- Name foods that are good sources and poor sources of vitamin C.
- Relate the role of vitamin C in the body to the symptoms of scurvy.
- Explain how vitamin C reduces oxidative stress.

scurvy The vitamin C deficiency disease.

ascorbic acid or **ascorbate** The chemical term for vitamin C.

Scurvy has been the scourge of armies, navies, and explorers throughout history. This vitamin C deficiency disease was described by ancient Greeks, Egyptians, and Romans. The reason obtaining enough vitamin C, also known as **ascorbic acid** or **ascorbate**, has been a particular problem for armies and explorers is that fresh fruits and vegetables are its main sources; these foods spoil quickly and don't transport well on long voyages. In the mid-1500s, the indigenous peoples of eastern Canada knew that an extract from white cedar needles would cure the disease. In 1594, after a voyage to the South Seas, Sir Richard Hawkins recommended that this sickness be treated by including citrus fruit in the diet. Despite his recommendation, 10,000 British sailors died of scurvy that same year. More than 100 years later, James Lind, a Scottish physician serving in the British navy, tested various agents for their effectiveness at curing scurvy and reported that two patients given citrus fruits recovered within 6 days. However, it was another 48 years before it was required that lime or lemon juice be included in the rations of the mercantile service, earning British sailors the name *limeys*. Unfortunately, the rest of the world did not heed the lesson of the limeys. In the mid-nineteenth century, during the U.S. Civil War, scurvy was rampant.

Vitamin C in the Diet

Citrus fruits, such as oranges, lemons, and limes, are an excellent source of vitamin C. Other fruits that are high in vitamin C include strawberries and cantaloupe. Vegetables in the cabbage family, such as broccoli, cauliflower, bok choy, and brussels sprouts, as well as green leafy vegetables, green and red peppers, okra, tomatoes, and potatoes, are also good sources (**Figure 8.32**). Meat, fish, poultry, eggs, dairy products, and grains are poor sources. The amount of vitamin C in packaged foods must be listed on food labels as a percentage of the Daily Value. This information can be used to identify packaged foods, such as frozen strawberries and orange juice, that are good sources of vitamin C. (See Label Literacy: How Much Vitamin C is in your orange juice?). Fresh fruits and vegetables, which are the best sources of vitamin C, do not carry food labels.

Vitamin C is unstable and is destroyed by oxygen, light, and heat, so it is readily lost in cooking. This loss is accelerated by contact with copper or iron cooking utensils and by low-acid conditions.

Vitamin C in the Body

Vitamin C is a water-soluble vitamin that donates electrons in biochemical reactions, including those needed for the synthesis and maintenance of connective tissue. Because it can donate electrons, vitamin C also has a more general role as an antioxidant that, along with other antioxidants, protects the body from reactive oxygen molecules (**Figure 8.33**). It also helps maintain the immune system and aids in the absorption of iron.

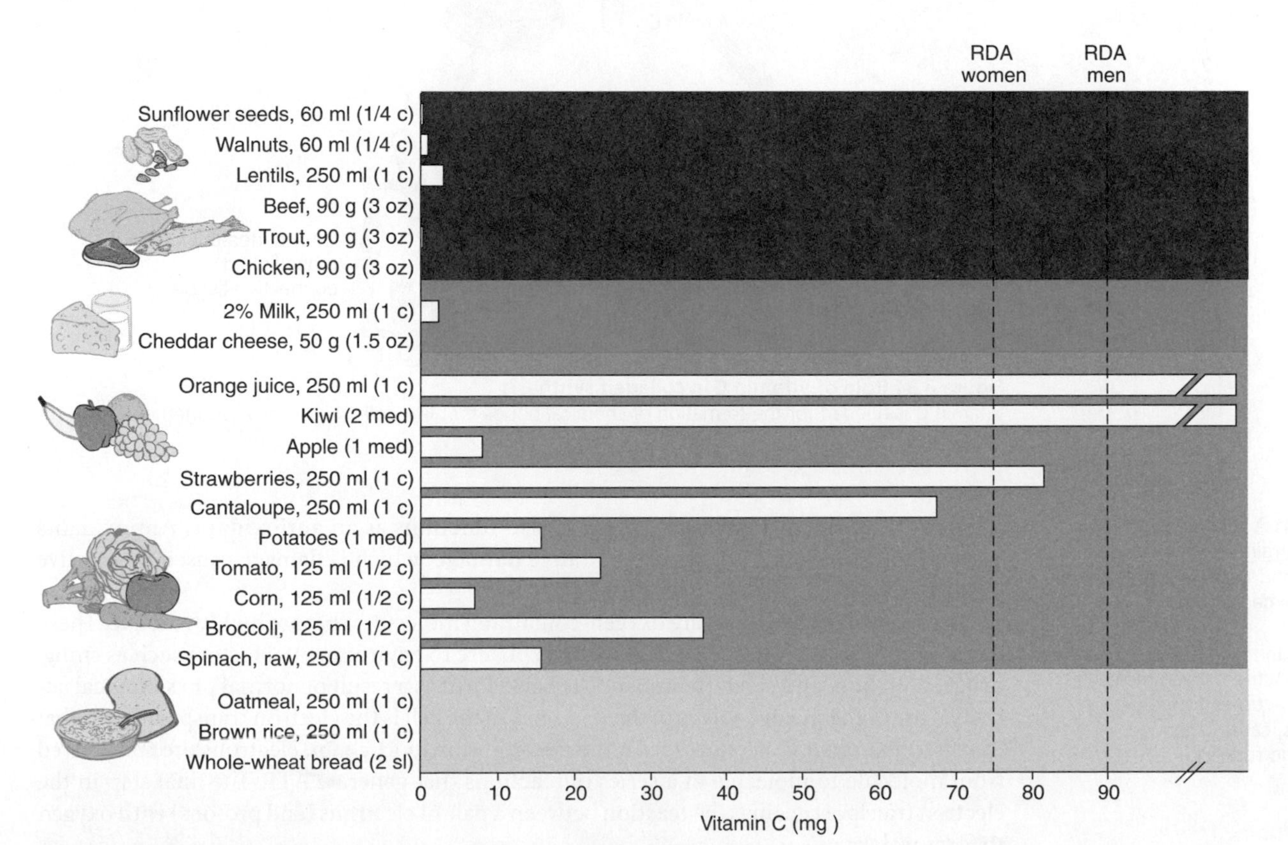

Figure 8.32 Vitamin C content of Canada's Food Guide food groups
Fruits and vegetables are the best sources of vitamin C; the dashed lines represent the RDAs for adult men and women.

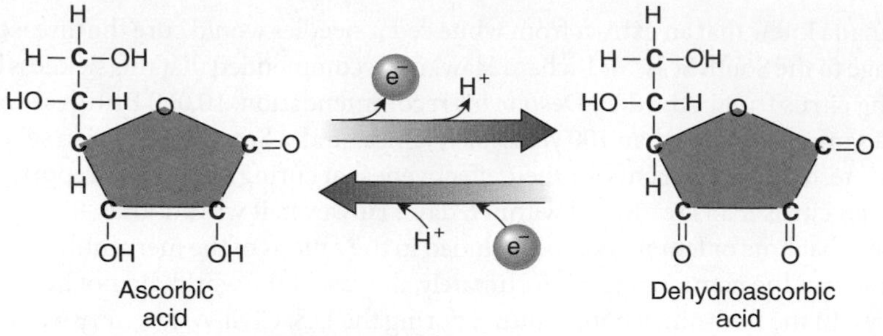

Ascorbic acid

Dehydroascorbic acid

Figure 8.33 Oxidation and reduction of vitamin C
When ascorbic acid donates electrons and hydrogens in chemical reactions, the structure changes to form a molecule of dehydroascorbic acid. This reaction is reversible and vitamin C can be restored when electrons and hydrogens are provided by other antioxidants such as glutathione.

collagen The major protein in connective tissue.

Vitamin C as a Coenzyme Many of the biochemical reactions requiring vitamin C add a hydroxyl group (OH) to other molecules. Two such reactions are essential for the formation of **collagen**, the protein that forms the base of all connective tissue in the body. Vitamin C is needed for activity of the enzymes that add the hydroxyl groups to the amino acids proline and lysine to form hydroxyproline and hydroxylysine, respectively. These hydroxyl groups are necessary for the formation of chemical bonds that cross-link the polypeptide strands of collagen to give it strength (**Figure 8.34**). Vitamin C also serves as a coenzyme in reactions needed for the synthesis of other cell compounds, including neurotransmitters, hormones such as the thyroid and steroid hormones, bile acids, and carnitine needed for fatty acid breakdown.

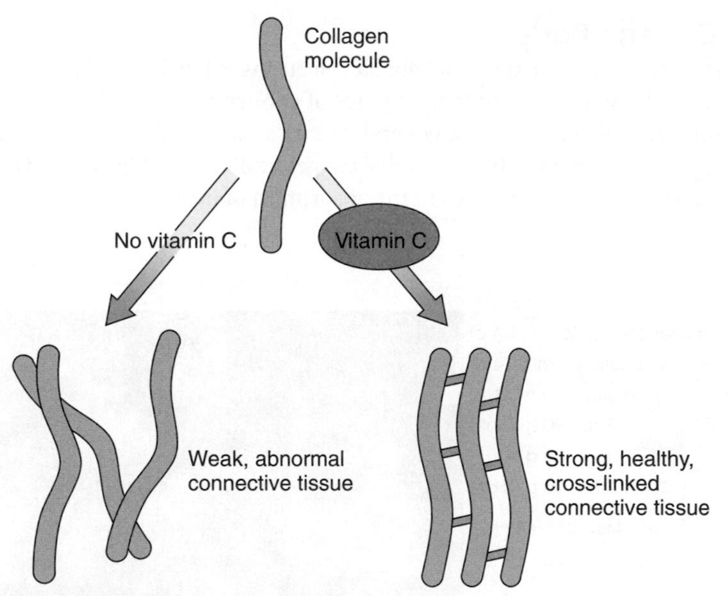

Figure 8.34 Role of vitamin C in collagen synthesis
Vitamin C is needed for the formation of chemical bonds that link collagen molecules together to give connective tissue strength and stability.

antioxidant A substance that is able to neutralize reactive oxygen molecules and thereby reduce oxidative damage.

oxidative damage Damage caused by highly reactive oxygen molecules that steal electrons from other compounds, causing changes in structure and function.

Vitamin C as an Antioxidant Vitamin C also functions as an **antioxidant.** Antioxidants are substances that protect against **oxidative damage**, which is damage caused by reactive oxygen molecules.

Reactive oxygen species are oxygen-containing molecules that are highly reactive. These compounds form in the body as a result of exposure to environmental factors such as smog, drugs, radiation, and cigarette smoke. They also form as a result of normal physiological activity. One of the major sources of these species in the cell is the electron transport chain that was first discussed in Section 3.6. In the electron transport chain, electrons are transferred from molecule to molecule in a series of reactions that generate ATP. The final step in the electron transport chain is the reaction between a pair of electrons (and protons) with oxygen to form water:

$$\frac{1}{2} O_2 + 2 H^+ + 2 e^- \rightarrow H_2O$$

Sometimes, this electron transfer goes wrong and instead of 2 electrons transferring to oxygen to form water, only 1 electron transfers, and instead, a reactive oxygen species called superoxide (O_2-) forms:

$O_2 + e^- + H^+ \rightarrow O_2^- + H^+$

Superoxide molecules are extremely reactive because they carry an odd number of electrons (stable molecules have an even number of electrons). Reactive oxygen species that have an odd number of electrons are also called **free radicals.** In the cell, superoxide will react with oxygen to generate several different types of reactive oxygen species, one of which is hydrogen peroxide. Superoxide, hydrogen peroxide, and other reactive oxygen species react with important molecules in the cells, such as proteins, lipids in cell membranes, and DNA. This results in changes in the structure and function of these molecules. DNA damage is hypothesized to be a major reason for the increase in cancer incidence that occurs with age. Free radical damage to lipoproteins, that is, the oxidation of LDL cholesterol, and lipids in membranes is implicated in the development of atherosclerosis (See Section 5.5).[34] When there is a serious imbalance between the amounts of reactive oxygen molecules and the availability of antioxidant defenses, an organism is said to be experiencing **oxidative stress.** Agents that can induce oxidative stress by causing an increase in reactive oxygen molecules, a decrease in antioxidant defenses, or an increase in oxidative damage are called **pro-oxidants.**

Antioxidants act by destroying reactive oxygen molecules before they can do excessive cellular damage. Some antioxidants are produced in the body; others are consumed in the diet. Vitamin C, vitamin E, and the mineral selenium have been classified as **dietary antioxidants.**[35]

Vitamin C is able to neutralize superoxide molecules by donating electrons to them, that is, by creating an even number of electrons again, a process called scavenging (**Figure 8.35**). Vitamin C has been shown to scavenge reactive oxygen molecules in white blood cells, the lungs, and the stomach mucosa.[35]

free radical One type of highly reactive molecule that causes oxidative damage.

oxidative stress A condition that occurs when there are more reactive oxygen molecules than can be neutralized by available antioxidant defenses. It occurs either because excessive amounts of reactive oxygen molecules are generated or because antioxidant defenses are deficient.

pro-oxidant A substance that promotes oxidative damage.

dietary antioxidant A substance in food that significantly decreases the adverse effects of reactive species on normal physiological function in humans.

How it Works

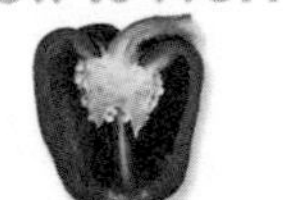

View in your WileyPLUS
www.wileyplus.com

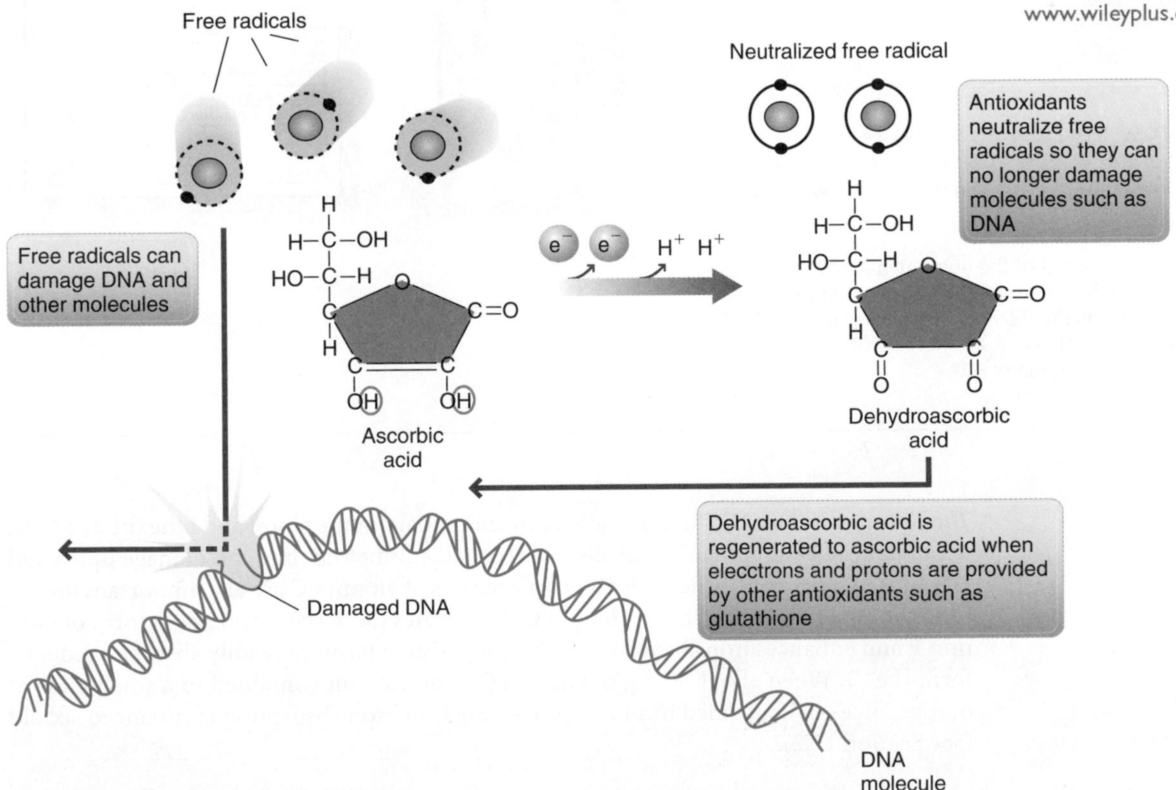

Figure 8.35 Antioxidant role of vitamin C
Vitamin C functions as an antioxidant that donates electrons to neutralize free radicals so that they are no longer damaging.

LABEL LITERACY

How Much Vitamin C Is in Your Orange Juice?

How much folate is in your breakfast cereal? And how much iron is in a box of raisins? It can be difficult to tell from the Nutrition Facts section of a food label exactly how much of a micronutrient is in a food. Food labels are required to provide the % Daily Values for vitamin A, vitamin C, iron, and calcium, but not the actual amount (See Section 2.5). To determine the amount of one of these nutrients in a serving of food, you need to know its Daily Value (see table). Once you know the Daily Value, you can multiply it by the % Daily Value on the label to determine the amount in a serving of the food. Follow these steps to find out how much vitamin C is in a 250 ml (1 cup) of orange juice:

1. Look up the Daily Value:

Vitamin	Daily Value
Vitamin A	1,000 RE (retinol equivalents)
Vitamin D	5 µg
Vitamin E	10 mg
Vitamin K	80 µg
Biotin	30 µg
Pantothenic acid	7 mg
Vitamin C	60 mg
Thiamin	1.3 mg
Riboflavin	1.6 mg
Niacin	23 mg
Vitamin B_6	1.8 mg

2. Find the % Daily Value (%DV) on the food label (see figure): % DV for vitamin C in orange juice = 120%. 3. Multiply the % Daily Value by the Daily Value to find out how much is in a serving: 60 mg X 120% DV = 60 X 1.2 = 72 mg vitamin C.

You can next compare the amount of the nutrient in your food to your RDA. For example, the RDA for vitamin C for a 21-year-old woman is 75 mg. So 250 ml of juice contains 72/75 = 96% of a young woman's RDA. For a young man of the same age, the RDA = 90 mg, so 250 ml of juice contains 80% of a young man's RDA.

Even if you don't look up the Daily Value and do calculations, the % Daily Value on the food label helps you judge how much the food contains. As a general guideline, if the % Daily Value of a nutrient is 5% or less, the food is a poor source; if it is 15% or higher, the food is a good source. Whether you are converting a Daily Value into the amount of a vitamin or are just looking at the Daily Value, be sure to consider how many servings you plan to eat. Remember that doubling the serving doubles the nutrients and kcalories.

Orange Juice

Nutrition Facts
Valeur nutritive
Per 250 ml / par 250 ml

Amount / **Teneur**	**% Daily Value** / **% valeur quotidienne**
Calories / Calories 110	
Fat / Lipides 0 g	0 %
Saturated / saturés 0 g + Trans / trans 0 g	0 %
Cholesterol / Cholestérol 0 mg	
Sodium / Sodium 0 mg	0 %
Potassium/ Potassium 470 mg	13 %
Carbohydrate/ Glucides 27 g	9 %
Sugars / Sucres 23 g	
Protein/ Protéines 2 g	
Vitamin A / Vitamine A	0 %
Vitamin C / Vitamine C	120 %
Calcium / Calcium	2 %
Iron / Fer	0 %
Folate / Folate	25 %

The Role of Vitamin C Vitamin C acts as an antioxidant in the blood and other body fluids. It can neutralize superoxide radicals and free radicals before they can damage lipids and DNA (see Figure 8.35). The antioxidant properties of vitamin C are also important for the functioning of other nutrients. Vitamin C regenerates the active antioxidant form of vitamin E and enhances iron absorption by keeping iron in its more readily absorbed reduced form (Fe^{+2}). When about 50 mg of vitamin C—the amount contained in a small glass of orange juice—is consumed in a meal containing iron, iron absorption is enhanced sixfold (see Section 12.2).

Vitamin C may also act as a pro-oxidant by converting iron and copper to reduced forms that can then generate free radicals. There is some evidence that vitamin C supplements could lead to oxidative damage to DNA. However, studies of the pro-oxidant effects of vitamin C are inconsistent.[36] More research is needed to determine what factors influence whether the antioxidant or pro-oxidant properties of vitamin C predominate.

Recommended Vitamin C Intake

Humans are one of only a few animal species that require vitamin C in the diet, as most animals can synthesize vitamin C in their bodies. For example, a pig makes 8 g of the vitamin each day. The recommendations for vitamin C intake are based on the amount needed to maximize concentrations in neutrophils, a type of white blood cell, with minimal excretion of the vitamin in the urine. The RDA is 90 mg/day for men and 75 mg/day for women. This amount is easily obtained by drinking 250 ml glass of orange juice.

The RDA for vitamin C is increased during pregnancy and lactation. The recommendation for infants is based on the vitamin C content of human milk. Cigarette smoking increases the requirement for vitamin C because vitamin C is used to break down compounds in cigarette smoke. It is recommended that cigarette smokers consume an extra 35 mg of vitamin C daily[35]—the amount in 125 ml of broccoli. Exercise and mental and emotional stress have not been found to affect the need for vitamin C.

Life Cycle

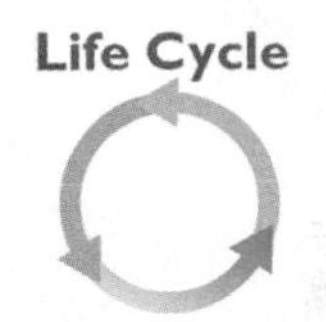

Vitamin C Deficiency

When vitamin C intake is below 10 mg/day, the symptoms of scurvy may appear. These symptoms reflect the role of vitamin C in the maintenance of collagen. Without vitamin C, the bonds holding adjacent collagen molecules together cannot be formed, so healthy collagen cannot be synthesized and maintained, resulting in symptoms such as poor wound healing, the reopening of previously healed wounds, bone and joint aches, bone fractures, and improperly formed and loose teeth. Connective tissue is also important for blood vessel integrity. A vitamin C deficiency, therefore, causes weakened blood vessels and ruptured capillaries, which leads to symptoms such as tiny bleeds around the hair follicles, bleeding gums, and easy bruising (**Figure 8.36**). Iron absorption is reduced when vitamin C is deficient, so anemia may also occur. The psychological manifestations of scurvy include depression and hysteria.

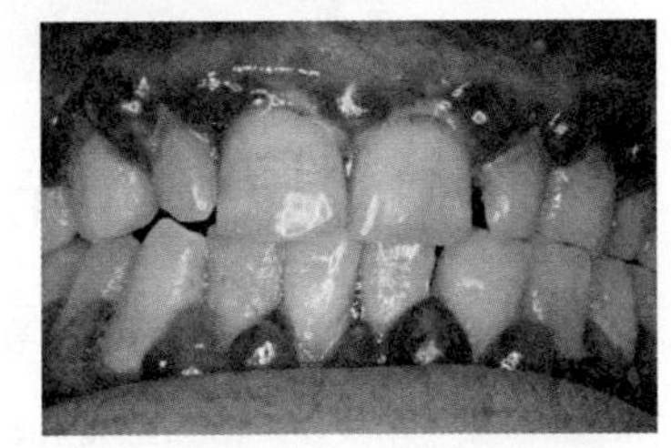

Figure 8.36 When vitamin C is deficient, the symptoms of scurvy begin to appear. The gums become inflamed, swell, and bleed. The teeth loosen and eventually fall out. (Science Photo Library/Photo Researchers, Inc.)

Throughout history, a lack of fruits and vegetables containing vitamin C has caused the death of thousands of sailors, soldiers, and explorers. Today, severe vitamin C deficiency leading to scurvy is rare, but marginal vitamin C deficiency is a concern for individuals who consume few fruits and vegetables. Scurvy has been reported in infants fed diets consisting exclusively of cow's milk and in alcoholics and elderly individuals consuming nutrient-poor diets.

Vitamin C Toxicity

Vitamin C is generally considered nontoxic. Large increases in intake do not cause large increases in the amount of vitamin C in body fluids. This is because the percentage of the dose absorbed decreases as the size of the dose increases and vitamin C absorbed in excess of need is excreted by the kidneys. The most common symptoms that occur with consumption of vitamin C doses of 2 g or more are diarrhea, nausea, and abdominal cramps. These are caused when unabsorbed vitamin C draws water into the intestine. The UL for vitamin C is based on an amount that is unlikely to cause gastrointestinal symptoms in healthy individuals and has been set at 2,000 mg/day from food and supplements. (See Critical Thinking: Natural Health Product Choices).

A high intake of vitamin C should be avoided by individuals prone to kidney stones because it can increase stone formation, by individuals who are unable to regulate iron absorption because it increases absorption, and by those with sickle cell anemia because it can worsen symptoms. Other potential problems associated with vitamin C intakes greater than 3 g/day include interference with drugs prescribed to slow blood clotting and, because the structure of vitamin C is similar to that of glucose, interference with urine tests used to monitor glucose levels. Another concern with high doses of vitamin C taken as supplements is damage to tooth enamel. Vitamin C is an acid that is strong enough to dissolve tooth enamel when vitamin C tablets are chewed.

CRITICAL THINKING:

Natural Health Product Choices

Background:

Hazel is suffering through her third cold of the winter. She is tired of being sick! When she complains at a local health food store, the clerk recommends several natural health products. These include a vitamin C supplement, a B-complex with vitamin C supplement, and zinc lozenges. Anxious to get better as soon as possible, Hazel buys all three products and takes the maximum dose recommended for each product. Several days later, Hazel is suffering from abdominal cramps, nausea and diarrhea. Instead of getting better, she is feeling even worse. Her friend, Amanda, who is a nutrition student, takes a careful look at the contents of each supplement and immediately recognizes Hazel's problem.

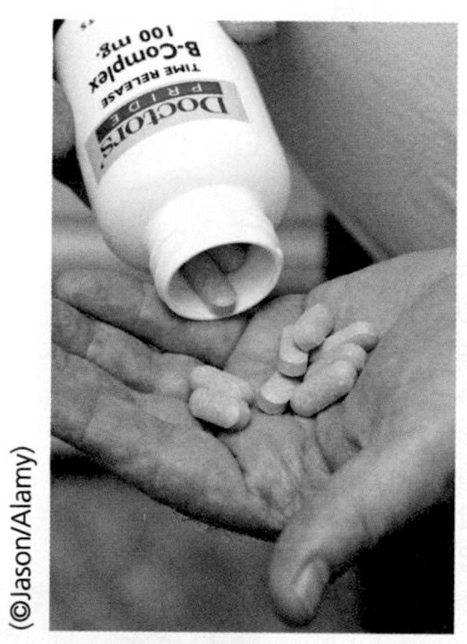

(©Jason/Alamy)

Supplement	Ingredients	Amount/ capsule	Dose
Vitamin C	Vitamin C	1,000 mg	1-2 caps/day
B-complex with C	Thiamin	5 mg	2 caps/day
	Niacin	5 mg	
	Vitamin B_6	6 mg	
	Riboflavin	5 mg	
	Biotin	10 μg	
	Pantothenic Acid	25 mg	
	Folic Acid	50 μg	
	Vitamin B_{12}	10 μg	
	Vitamin C	500 mg	
Zinc Lozenges	Zinc	15 mg	5 lozenges/day (maximum 7 days)
	Vitamin C	250 mg	

Critical Thinking Questions

Review the ingredients in Hazel's supplements. If she takes all these products at the frequency recommended in the table, will she be exceeding the UL for any nutrients? (See the inside cover of the textbook, for a list of ULs). List them.

What is responsible for Hazel's symptoms?

Use iProfile to compare the vitamin C content of an orange and a multivitamin supplement.

How it Works

View in your WileyPLUS
www.wileyplus.com

Vitamin C Supplements

Many people take supplements of vitamin C in the hope that they will prevent or reduce symptoms of the common cold. More recently, however, the role of vitamin C as an antioxidant has been used to promote vitamin C supplements as protection against cardiovascular disease and cancer.

Vitamin C and the Common Cold Studies examining the relationship between vitamin C and the common cold date back to the 1930s. A review of placebo-controlled trials that supplemented diets with vitamin C failed to find that routine vitamin C supplementation reduced the incidence of colds in the healthy population.[37] The incidence of respiratory infections was reduced somewhat in those exposed to periods of severe physical exercise and/or cold environments.[38,39] Regular supplementation with vitamin C did, however, reduce the duration and severity of colds (**Figure 8.37**). To have this effect, the supplements needed to be introduced before the onset of the cold symptoms. The effect of vitamin C on cold symptoms may be due to its direct antiviral effect, its antioxidant effect, its role in stimulating various aspects of immune function, its ability to increase the breakdown of histamine (a molecule that causes inflammation), or a combination of these.[39,40]

Figure 8.37 Vitamin C supplements won't prevent you from catching a cold, but they may help you recover faster. (© SCIENCE PHOTO LIBRARY/ Age Fotostock)

Vitamin C and Cardiovascular Disease Vitamin C supplements have been suggested to reduce the risk of cardiovascular disease by lowering blood pressure, preventing the oxidation of LDL cholesterol, and reducing blood cholesterol levels. Several studies have suggested that blood pressure is inversely related to vitamin C status; however, the data are not conclusive.[41] Vitamin C is suggested to prevent LDL oxidation because of its antioxidant function. Vitamin C may reduce blood cholesterol because it is involved in the synthesis of bile acids from cholesterol in the liver. Adequate vitamin C allows cholesterol to be used for bile synthesis and therefore may reduce the amount of cholesterol in the blood. Despite these roles of vitamin C in protecting LDL cholesterol from oxidation and modulating blood cholesterol levels, data thus far from epidemiology and human intervention trials have provided little evidence to support the use of vitamin C supplements in preventing atherosclerosis in humans.[42]

Vitamin C and Cancer It has been suggested that high doses of vitamin C both treat and prevent cancer. Although controlled trials have not found any benefits of vitamin C in the treatment of patients with advanced cancer,[43] there is evidence supporting a role for vitamin C in cancer prevention. Epidemiological studies have found inverse relationships between the intake of fruits and vegetables, foods high in vitamin C, and cancer incidence.[44] Higher plasma vitamin C levels are associated with a lower cancer mortality in men.[45] As an antioxidant, vitamin C may protect against cancers caused by oxidative damage. In the case of gastrointestinal cancers, vitamin C may prevent cancer by inhibiting the formation of carcinogenic nitrosamines in the gut (see Section 17.5). Despite the association between higher intakes of vitamin C and a lower incidence of various cancers, a low risk of cancer is more closely linked to a dietary pattern that is rich in whole-food sources of antioxidants than to any individual antioxidant. Factors other than vitamin C found in fruits and vegetables are likely to contribute to the overall protective effect.[44]

8.11 Choline

Learning Objective

- Describe the role of choline in the body.

Choline is needed to synthesize a number of important molecules, including a phospholipid found in cell membranes, the neurotransmitter acetylcholine, and the methyl donor betaine. It is also an important source of carbon atoms in biochemical reactions. Choline can be synthesized to a limited extent by humans. However, there is evidence that it is essential in healthy men and women and an AI of 550 mg/day for men and 425 mg/day for women has been established based on the amount needed to prevent liver damage. There are few data to assess whether dietary choline is needed at all stages of life and it is not yet clear whether it is also essential in the diets of infants and children.[3,46] Some studies suggest that during pregnancy, some women may not have an adequate intake.[47]

Choline deficiency causes liver abnormalities. Deficiency is unlikely in healthy humans, but it has been observed in individuals fed a choline-deficient diet and in those receiving total parenteral nutrition without choline.[3] Choline is widely distributed in foods; particularly good sources include egg yolks, organ meats, spinach, nuts, and wheat germ.

Intakes of choline that are much higher than can be obtained from foods can cause body odour, sweating, reduced growth rate, low blood pressure, and liver damage. A UL for adults of 3.5 g/day has been set based on the occurrence of low blood pressure.

CASE STUDY OUTCOME

Chen and his wife work hard to maintain a healthy diet and lifestyle. They were shocked, then, to find that the fruits, vegetables, and whole grains in their diet were not providing all of the nutrients Chen needs. Because of Chen's low-meat diet, he was consuming little vitamin B_{12}. In addition, he had developed a condition common in older adults called atrophic gastritis, which reduces the amount of stomach acid. Stomach acid is essential for releasing vitamin B_{12} from the proteins it is bound to in food and allowing it to be absorbed. Without sufficient vitamin B_{12}, the myelin coating on nerves cannot be maintained, and nerve function is disrupted. To boost Chen's vitamin B_{12} status, his doctor gave him an injection and then instructed him to start taking a supplement containing vitamin B_{12}. The vitamin B_{12} in supplements is not bound to proteins, so it is more easily absorbed when stomach acid is low. Chen is now back to his old self. He continues his healthy diet and lifestyle, but has added a supplement to increase his vitamin B_{12} intake. He is grateful he discovered his condition early, because an untreated vitamin B_{12} deficiency might have resulted in permanent neurological damage.

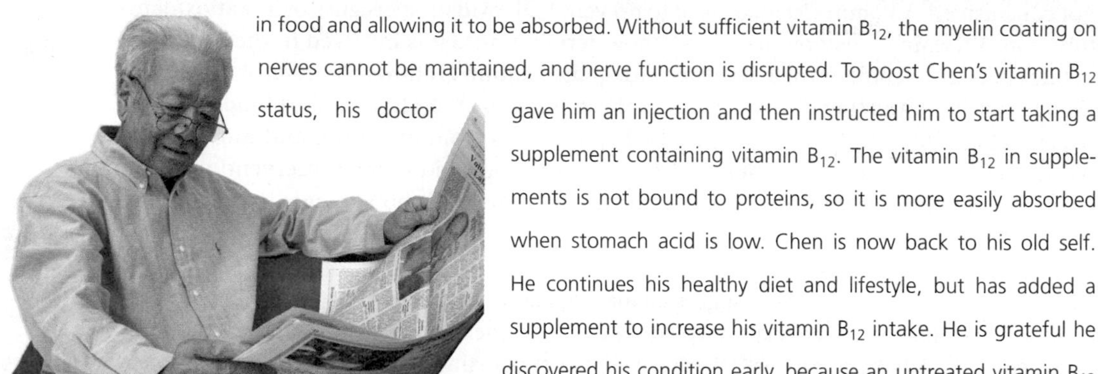

APPLICATIONS

Personal Nutrition

1. **Do you get enough folate?**
   a. Use iProfile and your food intake record from Chapter 2 to determine how much folate your diet contains.
   b. How does your intake compare with the RDA?
   c. If your diet doesn't meet recommendations, suggest some dietary modifications that will help you meet the RDA for folate.
   d. List several natural sources of folate in your diet and several foods that are fortified with folic acid.
2. **Do you get enough vitamin B_{12}?**
   a. Use iProfile and your food intake record from Chapter 2 to determine how much vitamin B_{12} your diet contains.
   b. How does your intake compare with the RDA?
   c. If your diet doesn't meet recommendations, suggest some dietary modifications that will help you meet the RDA for vitamin B_{12}.
   d. If you eliminated meat, fish, poultry, and eggs from your diet, would you be getting enough vitamin B_{12}? List the foods remaining in your diet that contribute vitamin B_{12}.
   e. If you were 60 years old, would your diet meet the recommendation for vitamin B_{12} intake?
3. **How much vitamin C is in your fast-food favourites?**

iProfile

a. Use iProfile or information from company websites to determine the amount of vitamin C in 5 items you commonly order from fast-food restaurants.

b. What percentage of your RDA for vitamin C does each provide? How does this compare with the percentage of your daily energy needs that each provides?

c. What fast-food items are highest in vitamin C? Why?

General Nutrition Issues

1. Evaluate each of the following supplements:

Supplement	Ingredients	Dose
Pyridoxine	100 mg pyridoxine	2 tablets daily
Stress tab	35 mg pyridoxine	3 times daily
	1 mg thiamin	
	1.1 mg riboflavin	
	30 mg niacin	
	500 mg choline	
Folic acid	800 µg folic acid	Once daily

a. Do any of these supplements create a risk for toxicity when taken at the recommended dosage? Which ones and why?

b. Would you recommend them for everyone? For a specific life-stage group? Why or why not?

2. Take a look at the Nutrition Facts label on a box of breakfast cereal.

a. What percent of the Daily Value is included in a serving for each of the water-soluble vitamins?

b. What percent of the Daily Value would you consume for each water-soluble vitamin if you consumed the portion of cereal you typically have for breakfast?

c. Does the amount in your portion exceed the UL for any of these vitamins? Which ones?

SUMMARY

8.1 What Are Vitamins?

- Vitamins are essential organic nutrients that do not provide energy and are required in small quantities in the diet. Canadians consume vitamins that are naturally present in foods, added to foods by fortification, and supplied by supplements. Some foods are fortified with nutrients according to government guidelines to promote public health. Others are fortified according to manufacturers' perceptions of what will sell in the marketplace.
- Vitamins are needed to promote and regulate body processes needed for growth, reproduction, and tissue maintenance. Vitamin deficiencies remain a major health problem worldwide. The amount of a vitamin that is available to the body is regulated by vitamin absorption, transport, activation, storage, and excretion. Recommended intakes for vitamins are expressed as RDAs or AIs. UL values estimate the highest dose that is unlikely to cause toxicity.

Video

Go to WileyPLUS to view a video clip on vitamin supplements.

8.2 Thiamin

- The best food sources of thiamin are lean pork, legumes, and whole or enriched grains.
- Thiamin is required for the formation of acetyl-CoA from pyruvate and for a reaction in the citric acid cycle and is therefore particularly important for the production of ATP from glucose. It is also needed for the synthesis of the neurotransmitter acetylcholine.
- Thiamin deficiency, or beriberi, causes nervous system abnormalities. Deficiencies are common in alcoholics. No toxicity has been identified.

8.3 Riboflavin

- Milk, meat, and enriched grain products are the best food sources of riboflavin.
- Riboflavin coenzymes are needed for the generation of ATP from carbohydrate, fat, and protein.
- Riboflavin deficiency is rarely seen alone because food sources of riboflavin are also sources of other B vitamins and because riboflavin is needed for the utilization of several other vitamins. No toxicity has been identified.

8.4 Niacin

- Beef, chicken, turkey, fish, and enriched grain products are the best food sources of niacin. The amino acid tryptophan can be converted into niacin, so tryptophan from dietary protein can meet some of the niacin requirement.
- Niacin coenzymes are important in the breakdown of carbohydrate, fat, and protein to provide energy and in the synthesis of fatty acids and sterols.
- Niacin deficiency results in pellagra, which is characterized by dermatitis, diarrhea, dementia, and finally, if untreated, death.
- High-dose supplements of the nicotinic acid, a form of niacin, can lower elevated blood cholesterol but frequently cause toxicity symptoms such as flushing, tingling sensations, nausea, and a red skin rash.

8.5 Biotin

- Liver and egg yolks are good sources of biotin.
- Biotin is needed for the synthesis of glucose and fatty acids, and the metabolism of certain amino acids.
- An RDA has not been established because some of our biotin need is met by bacterial synthesis in the GI tract. However, an AI has been set. Toxicity has not been reported.

8.6 Pantothenic Acid

- Pantothenic acid is abundant in the food supply and deficiency is rare.
- Pantothenic acid is part of coenzyme A (CoA), which is required for the production of ATP from carbohydrate, fat, and protein and the synthesis of cholesterol and fatty acids. There is no RDA, but an AI has been established.

8.7 Vitamin B_6

- Food sources of vitamin B_6 include chicken, fish, liver, and whole grains.
- Pyridoxal phosphate, the coenzyme form of vitamin B_6, is needed for the activity of more than 100 enzymes involved in the metabolism of carbohydrate, fat, and protein. Vitamin B_6 is

a coenzyme for transamination and deamination reactions and is therefore particularly important for amino acid metabolism.

- Vitamin B_6 deficiency causes neurological symptoms, anemia due to impaired hemoglobin synthesis, poor immune function, and elevated levels of homocysteine, which may increase the risk of heart disease. High intakes from supplements can cause nervous system abnormalities.

8.8 Folate or Folic Acid

- Food sources of folate include liver, legumes, oranges, leafy green vegetables, and fortified grains.
- Folate is necessary for the synthesis of DNA, so it is especially important for rapidly dividing cells.
- It is recommended that women of child-bearing age consume a supplement of at least 400 μg of folic acid from fortified foods in addition to the folate found in a varied diet to reduce the risk of neural tube defects in the developing fetus.
- Folate deficiency results in macrocytic anemia and can cause an increase in homocysteine levels. Low levels of folate before and during early pregnancy are associated with an increased incidence of neural tube defects in the offspring. A high intake of folate can mask the early symptoms of vitamin B_{12} deficiency.

8.9 Vitamin B12

- Vitamin B_{12} is found almost exclusively in animal products.
- The absorption of vitamin B_{12} from food requires adequate levels of stomach acid, intrinsic factor, and pancreatic secretions.
- Vitamin B_{12} is needed for the metabolism of folate and fatty acids and to maintain the insulating layer of myelin surrounding nerves.
- Vitamin B_{12} deficiency causes homocysteine levels to increase and can result in anemia and permanent nerve damage. In pernicious anemia, severe deficiency occurs due to an absence of intrinsic factor. Vitamin B_{12} deficiency may also occur in vegans, who consume no animal products, and in older individuals with low stomach acid due to atrophic gastritis.

8.10 Vitamin C

- The best food sources of vitamin C are citrus fruits.
- Vitamin C is necessary for the synthesis and maintenance of connective tissue and for the synthesis of hormones and neurotransmitters. Vitamin C is also a water-soluble antioxidant. Antioxidants protect the body from reactive oxygen molecules such as free radicals. These molecules are generated from normal body reactions and come from the environment. They cause damage by stealing electrons from DNA, proteins, carbohydrates, and unsaturated fatty acids.
- Vitamin C deficiency, called scurvy, is characterized by poor wound healing, bleeding, and other symptoms related to the improper formation and maintenance of collagen.
- Vitamin C supplements are the most commonly taken vitamin supplements and are usually used to reduce the symptoms of the common cold.

8.11 Choline

- Choline is a substance necessary for metabolism. It may be required in the diet at certain stages of life, so an AI has been established.

REVIEW QUESTIONS

1. What is a vitamin?
2. List four factors that affect how much of a vitamin is available to the body.
3. Define coenzyme and describe the coenzyme functions of 5 vitamins.
4. Why is thiamin deficiency common in alcoholics?
5. Why should milk be packaged in opaque containers?
6. What is pellagra?
7. How is vitamin B_6 involved in amino acid and protein metabolism?
8. Why is low folate intake of particular concern for women of child-bearing age?
9. Why would someone who has had his stomach removed (or had gastric bypass surgery) need to receive injections of vitamin B_{12} to meet his needs?
10. Why are vegans at risk for vitamin B_{12} deficiency? Why are the elderly at risk?
11. Explain why a deficiency of vitamin B_6, folate, or vitamin B_{12} can all cause an increase in homocysteine levels.
12. Why does vitamin C deficiency cause poor wound healing?
13. What are reactive oxygen molecules and how do they cause damage?
14. What is the role of antioxidants and pro-oxidants in oxidative stress?

REFERENCES

1. American Dietetic Association. Position of the American Dietetic Association: Fortification and nutritional supplements. *J. Am. Diet. Assoc.* 105:1300–1311, 2005.
2. Butterworth, R. F. In: *Modern Nutrition in Health and Disease*, 10th ed. M. E. Shils, M. Shike, A. C. Ross et al., eds. Philadelphia: Lippincott Williams & Wilkins, 2006, pp. 426–433.
3. Institute of Medicine, Food and Nutrition Board. *Dietary Reference Intakes for Thiamin, Riboflavin, Niacin, Vitamin B-6, Folate, Vitamin B-12, Pantothenic Acid, Biotin, and Choline*. Washington, DC: National Academies Press, 1998.
4. Saffert, A., Pieper, G., and Jetten, J. Effect of package light transmittance on vitamin content of milk. *Technology and Science* 21:47–55, 2008.
5. Roe, D. A. *A Plague of Corn: The Social History of Pellagra*. Ithaca, NY: Cornell University Press, 1973.
6. Seal, A. J., Creeke, P. I., Dibari, F. et al. Low and deficient niacin status and pellagra are endemic in postwar Angola. *Am. J. Clin. Nutr.* 85:218–242, 2007.
7. Guyton, J. R., Blazing, M. A., Hagar, J. et al. Extended-release niacin vs. gemfibrozil for the treatment of low levels of high-density

lipoprotein cholesterol. NiaspanGemfibrozil Study Group. *Arch. Intern. Med.* 160:1177–1184, 2000.

8. Gale, E. A., Bingley, P. J., Emmett, C. L., et al. European Nicotinamide Diabetes Intervention Trial (ENDIT) Group. European Nicotinamide Diabetes Intervention Trial (ENDIT): A randomized controlled trial of intervention before the onset of type 1 diabetes. *Lancet* 363:925–931, 2004.
9. Department of Justice. Food and Drugs Regulations Available online at http://laws-lois.justice.gc.ca/eng/regulations/C.R.C.%2C_c._870/. Accessed April 11, 2011.
10. Bessey, O. A., Adam, D. J., and Hansen, A. E. Intake of vitamin B_6 and infantile convulsions: A first approximation of requirements of pyridoxine in infants. *Pediatrics* 20:33–44, 1957.
11. Ntaios, G., Savopoulos, C., Grekas, D. et al. The controversial role of B-vitamins in cardiovascular disease risk: an update. *Arch Cardiovasc Dis.* 102:847-854, 2009.
12. McCully, K. S. Chemical pathology of homocysteine. IV. Excitotoxicity, oxidative stress, endothelial dysfunction, and inflammation. *Ann. Clin. Lab. Sci.* 39(3):219-32, 2009.
13. Mente, A., de Koning, L., Shannon, H. S. et al. A systematic review of the evidence supporting a causal link between dietary factors and coronary heart disease. *Arch. Intern. Med.* 169(7):659-669, 2009.
14. Clarke, R., Halsey, J., Bennett, D. et al. Homocysteine and vascular disease: review of published results of the homocysteine-lowering trials. *J. Inherit. Metab. Dis.* 34(1):83-91, 2011.
15. Schaumburg, H., Kaplan, J., Windebank, A. et al. Sensory neuropathy from pyridoxine abuse. *N. Engl. J. Med.* 309:445–448, 1983.
16. Aufiero, E., Stitik, T. P., Foye, P. M., et al. Pyridoxine hydrochloride treatment of carpal tunnel syndrome: A review. *Nutr. Rev.* 62:96–104, 2004.
17. Bendich, A. The potential for dietary supplements to reduce premenstrual syndrome (PMS) symptoms. *J. Am. Coll. Nutr.* 19:3–12, 2000.
18. Blom, H.J. Folic acid, methylation and neural tube closure in humans. *Birth Defects Res. A Clin. Mo.l Teratol.* 85(4):295-302. 2009.
19. Health Canada. Prenatal Guidelines for Health Professionals-Folate contributes to a healthy pregnancy. Available online at http://www.hc-sc.gc.ca/fn-an/pubs/nutrition/folate-eng.php. Accessed July 17, 2010.
20. Public Health Agency of Canada. What Mothers Say: The Canadian Maternity Experiences Survey. Available online at http://www.phac-aspc.gc.ca/rhs-ssg/pdf/survey-eng.pdf. 2009. Accessed Mar 5, 2011.
21. De Wals, P., Tiarou, F., Van Allen, M. I. et al. Reduction in neural-tube defects after folic acid fortification in Canada. *N. Engl. J. Med.* 357:135–142, 2007.
22. House, J. D., March, S. B., Ratnam, M. S. et al. Improvements in the status of folate and cobalamin in pregnant Newfoundland women are consistent with observed reductions in the incidence of neural tube defects.*Can. J. Public Health.* 97(2):132-135, 2006.
23. Godwin, K. A., Sibbald, B., Bedard, T. et al. Changes in frequencies of select congenital anomalies since the onset of folic acid fortification in a Canadian birth defect registry. *Can. J. Public Health* 99(4):271-275, 2008.
24. Ionescu-Ittu, R., Marelli, A. J., Mackie, A.S. et al. Prevalence of severe congenital heart disease after folic acid fortification of grain products: time trend analysis in Quebec, Canada. *BMJ.* 338:b1673. doi: 10.1136/bmj.b1673, 2009.
25. French, A. E., Grant, R., Weitzman, S. et al. Folic acid food fortification is associated with a decline in neuroblastoma.*Clin. Pharmacol. Ther.* 74(3):288-94, 2003.
26. Ray, J. G., Goodman, J., O'Mahoney, P. R. et al. High rate of maternal vitamin B^{12} deficiency nearly a decade after Canadian folic acid flour fortification. *QJM.* 101(6):475-7, 2008.
27. Hubner, R. A., and Houlston, R. S. Folate and colorectal cancer prevention. *Br. J. Cancer* 100:233–239, 2009.
28. Larsson, S. C., Bergkvist, L., and Wolk, A. Folate intake and risk of breast cancer by estrogen and progesterone receptor status in a Swedish cohort. *Cancer Epidemiol. Biomarkers. Prev.* 17:3444–3449, 2008.
29. Larsson, S. C., Håkansson, N., Giovannucci, E., et al. Folate intake and pancreatic cancer incidence: A prospective study of Swedish women and men. *J. Natl. Cancer. Inst.* 98:407–413, 2006.
30. Zhang, S. M., Willett, W. C., Selhub, J. et al. Plasma folate, vitamin B_6, vitamin B_{12}, homocysteine, and risk of breast cancer. *J. Natl. Cancer Inst.* 95:373–380, 2003.
31. Su, L. J., and Arab, L. Nutritional status of folate and colon cancer risk: Evidence from NHANES I epidemiologic follow-up study. *Ann. Epidemiol.* 11:65–72, 2001.
32. Kim, Y. I. Role of folate in colon cancer development and progression. *J. Nutr.* 133:3731S–3739S, 2003.
33. American Dietetic Association. Position of the American Dietetic Association and Dietitians of Canada: Vegetarian diets. *J. Am. Diet. Assoc.* 103:748–765, 2003.
34. Stocker, R., and Keaney, J. F., Jr. Role of oxidative modifications in atherosclerosis. *Physiol. Rev.* 84:1381–1478, 2004.
35. Food and Nutrition Board, Institute of Medicine. *Dietary Reference Intakes for Vitamin C, Vitamin E, Selenium, and Carotenoids.* Washington, DC: National Academies Press, 2000.
36. Carr, A., and Frei, B. Does vitamin C act as a pro-oxidant under physiological conditions? *FASEB J.* 13:1007–1024, 1999.
37. Douglas, R. M., Hemila, H., D'Souza, R. et al. Vitamin C for preventing and treating the common cold. *Cochrane Database System Review* 18:CD000980, 2004.
38. Hemila, H. Vitamin C supplementation and respiratory infections: A systematic review. *Mil. Med.* 169:920–925, 2004.
39. Jari, R. J., and Harakeh, S. Antiviral and immunomodulatory activities of ascorbic acid. In: *Subcellular Biochemistry*, Vol. 25: Ascorbic Acid: Biochemistry and Biomedical Cell Biology. J. R. Harris, ed. New York: Plenum Press, 1996, pp. 215–231.
40. Johnston, C. S. The antihistamine action of ascorbic acid. In *Subcellular Biochemistry*, Vol. 25: Ascorbic Acid: Biochemistry and Biomedical Cell Biology. J. R. Harris, ed. New York: Plenum Press, 1996, pp. 189–213.
41. Wilburn, A. J., King, D. S., Glisson, J. et al. The natural treatment of hypertension. *J. Clin. Hypertens. (Greenwich)* 6:242–248, 2004.
42. Blomhoff, R. Dietary antioxidants and cardiovascular disease. *Curr. Opin. Lipidol.* 16:47–54, 2005.
43. Shekelle, P., Hardy, M. L., Coulter, I. et al. Effect of the supplemental use of antioxidants vitamin C, vitamin E, and coenzyme Q10 for the prevention and treatment of cancer. *Evid. Rep. Technol. Assess.* (Summ). 75:1–3, 2003.
44. Lee, K. W., Lee, H. J., Surh, Y. J., et al. Vitamin C and cancer chemoprevention: reappraisal. *Am. J. Clin. Nutr.* 78:1074–1078, 2003.
45. Khaw, K. T., Bingham, S., Welch, A. et al. Relation between plasma ascorbic acid and mortality in men and women in EPIC-Norfolk prospective study: A prospective population study. European Prospective Investigation into Cancer and Nutrition. *Lancet* 357:657–663, 2001.
46. Fischer, L. M., daCasta, K. A., Kwack, L. et al. Sex and menopausal status influence human dietary requirements for the nutrient choline. *Am. J. Clin. Nutr.* 85:1275–1285, 2007.
47. Zeisel, S. H., da Costa, K. A. Choline: an essential nutrient for public health. *Nutr Rev.*67(11):615–23, 2009.

9 The Fat-Soluble Vitamins

Case Study

The pediatrician in an Iqaluit clinic in Nunavut is concerned about a young patient of hers. The patient, a little girl named Eva, is 14 months old, but has only one tooth, slightly bowed legs, and bumps on her ribs, classic symptoms of the vitamin D deficiency disease rickets. Vitamin D is needed for proper formation and maintenance of bones and teeth. Without sufficient vitamin D, a child's legs bow under the weight of standing, and bony bumps appear on each of the ribs. The poorly formed bones can break easily, and teeth erupt late and are very prone to decay. Eva is showing these symptoms.

The doctor recently read a scientific paper that reported cases of rickets in Canada, especially among babies and young children who are breastfed and receive little sunlight.[1] Breast milk, while being an excellent food for growing infants and toddlers, contains little vitamin D. The ultraviolet light of the sun can be used by the skin to biosynthesize vitamin D, but in the remote north of Nunavut, sunlight is too limited for this process to occur. The doctor learns from Eva's parents that she has been breastfed all her life and her mother is unaware that breastfed babies need vitamin D supplements. The pediatrician's diagnosis of rickets is confirmed by a blood test that finds very low levels of vitamin D in Eva's blood as well as indicators of abnormal bone breakdown.

Like Eva's parents, most people in Canada are not familiar with rickets, but rare cases of the deficiency disease still occur in young children, especially in circumstances similar to Eva's.

In this chapter, you will learn about the important functions of vitamin D as well as the other fat-soluble vitamins, vitamins A, E, and K.

(Peter Griffith/Masterfile)

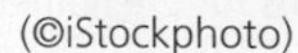

(©iStockphoto)

(Rita Maas Studio/StockFood America)

Chapter Outline

9.1 Fat-Soluble Vitamins in the Canadian Diet

Learning Objectives

- Name the fat-soluble vitamins.
- Discuss factors that impact fat-soluble vitamin status.

Vitamins A, D, E, and K are grouped together due to their solubility in fat (see Section 8.1). Because of this, they require special handling for absorption into and transport through the body. Fat-soluble vitamins require bile and dietary fat for absorption. Once absorbed, they are transported with fats through the lymphatic system in chylomicrons before entering the blood (see Section 5.4). Since excesses of these vitamins can be stored in the body and retrieved as needed, intakes can vary without a risk of deficiency as long as average intake over a period of weeks or months meets needs. Solubility in fat, however, limits their routes of excretion and therefore increases the risk of toxicity.

Despite the body's ability to store the fat-soluble vitamins, deficiencies of vitamins A and D are common in the developing world. Elimination of these deficiencies through supplementation and fortification is a focus of public-health programs. In Canada and other developed countries, severe deficiencies of these nutrients are rare, but trends in the modern diet have affected the amounts of fat-soluble vitamins people consume. Our rising reliance on fast food has reduced our intake of fruits and vegetables, particularly leafy greens, reducing our intake of vitamins A, E, and K (**Figure 9.1**). We work at indoor jobs, exercise at indoor gyms, and protect ourselves from the sun by using sunscreens when we go out, all of which limit our ability to get adequate vitamin D from sunshine. We limit our fat intake to reduce our waistlines and risk of chronic disease. But animal fats are good sources of vitamin D and preformed vitamin A, and vegetable fats provide much of our vitamin E, as well as fat-soluble phytochemicals, some of which serve as precursors to vitamin A. Medications that limit fat absorption also limit fat-soluble-vitamin absorption. Even the low-carbohydrate diet craze has had an impact on the amounts of fat-soluble vitamins consumed from whole foods because eliminating grains, fruits, and many vegetables can reduce intake of vitamin E and vitamin A precursors, as well as vitamin K. To avoid the risks of limited intakes, we take supplements of these vitamins and fortify our food with them. This then increases the risks of consuming toxic amounts. It is unclear whether these new patterns will have a long-term impact on our fat-soluble-vitamin status.

Figure 9.1 Even though vitamins A, E, and K are soluble in fat, leafy green vegetables such as this swiss chard, which are very low in fat, are good sources of vitamin A precursors, as well as vitamin E and vitamin K. (Maximillian Stock/StockFood America)

9.2 Vitamin A

Learning Objectives

- Compare the sources, functions, and potential toxicity of preformed vitamin A and provitamin A.
- Describe the role of vitamin A in night vision.
- Discuss how vitamin A affects gene expression.
- Explain why a vitamin A deficiency can cause eye infections and blindness.

Are carrots really good for your eyes? Carrots are high in provitamin A, and vitamin A is important for vision. This connection between vision and foods that we now know are high in vitamin A has been recognized for centuries. In ancient times, the Egyptians knew that eating liver could improve night vision in those who had difficulty in adjusting from bright light to dim light. In 1968, George Wald earned the Nobel Prize in medicine for identifying the mechanism by which vitamin A is involved in vision. Although this is a key function of vitamin A, attention today is focused more on how vitamin A interacts with genes to regulate growth and **cell differentiation**. Despite our expanding understanding of the functions of vitamin A, deficiency remains a world health problem.

cell differentiation Structural and functional changes that cause cells to mature into specialized cells.

Vitamin A in the Diet

Vitamin A is found preformed and in precursor or provitamin forms in our diet. Preformed vitamin A is found primarily in animal foods, and the provitamin A forms are found in plants (**Figure 9.2**). Both sources can be used to meet vitamin A needs in the body.

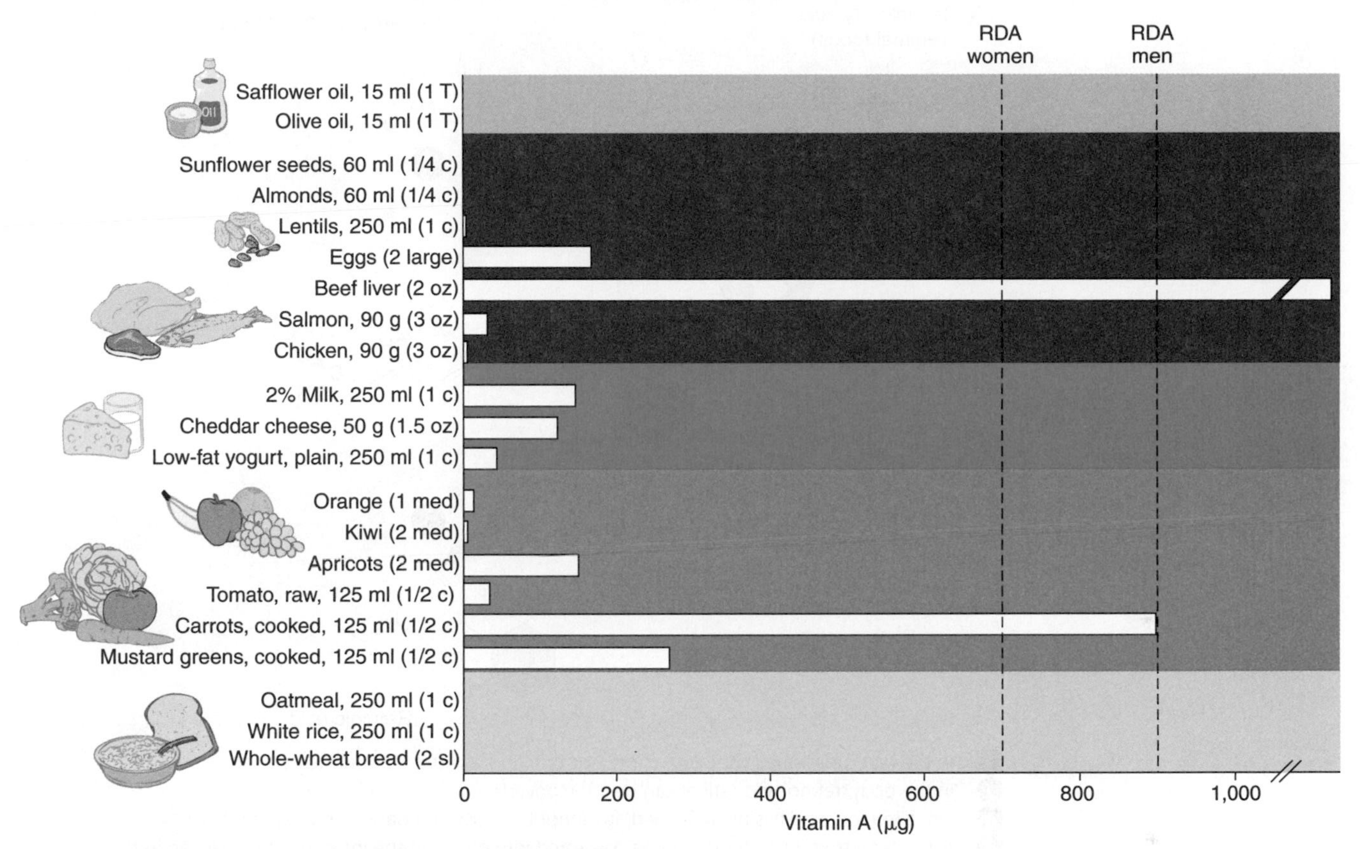

Figure 9.2 Vitamin A content of Canada's Food Guide food groups
Both plant and animal foods are good sources of vitamin A; the dashed lines show the RDAs for men and women.

Preformed vitamin A compounds are known as **retinoids**. The retinoids include retinol, retinal, and retinoic acid. Animal foods such as liver, fish, egg yolks, and milk and alternative products provide preformed vitamin A, primarily as retinol, or retinol attached to a fatty acid. Margarine and nonfat and reduced-fat milk are fortified with retinol because they are often consumed in place of butter and whole milk, which are good sources of this vitamin. Retinal and retinoic acid can be formed in the body from retinol (**Figure 9.3**).

retinoids The chemical forms of preformed vitamin A: retinol, retinal, and retinoic acid.

Plants contain provitamin A compounds called **carotenoids**. Carotenoids are yellow, orange, and red pigments that give these colours to fruits and vegetables. About 50 of the 600 carotenoids that have been isolated provide vitamin A activity. **Beta-carotene (β-carotene),** the most potent precursor, is plentiful in dark orange fruits and vegetables such as mangoes, apricots, cantaloupe, carrots, red peppers, pumpkins, and sweet potatoes as well as in leafy greens, where the orange colour is masked by the green colour of chlorophyll. Other carotenoids that provide some provitamin A activity include alpha-carotene (α-carotene) found in carrots, leafy green vegetables, and winter squash, and beta-cryptoxanthin (β-cryptoxanthin), found in mangoes, papayas, winter squash, and sweet red peppers.[2] Lutein, lycopene, and zeaxanthin are carotenoids with no vitamin A activity (see Chapter 9: Focus on Phytochemicals).

carotenoids Natural pigments synthesized by plants and many micro-organisms. They give yellow and red-orange fruits and vegetables their colour.

beta-carotene (β-carotene) A carotenoid that has more provitamin A activity than other carotenoids. It also acts as an antioxidant.

To help consumers identify food sources of vitamin A, labels on packaged foods must list the vitamin A content as a percentage of the Daily Value. All forms of vitamin A in the diet are fairly stable when heated but may be destroyed by exposure to light and oxygen (see Your Choice: Fresh, Frozen, or Canned—Does It Matter?).

Vitamin A in the Digestive Tract

Both preformed vitamin A and carotenoids are bound to proteins in foods. To be absorbed, they must be released from the protein by pepsin and other protein-digesting enzymes. In the small intestine, the released retinol and carotenoids combine with bile acids and other

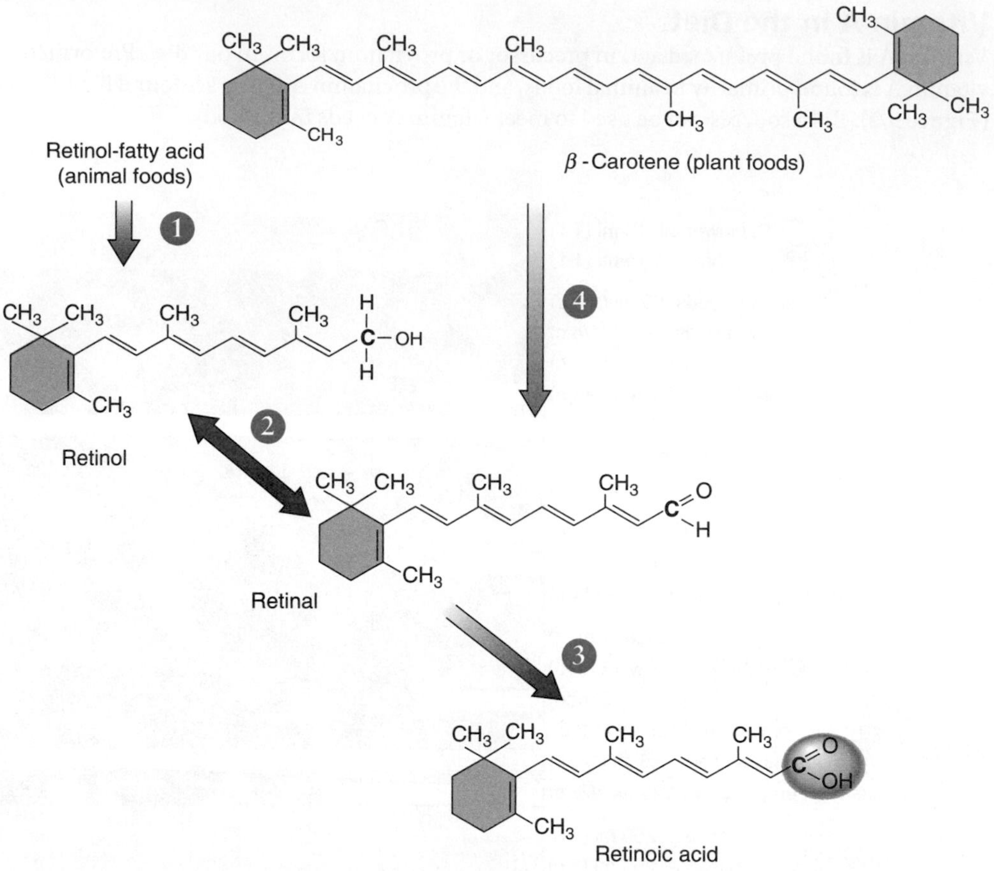

1. In the diet, preformed vitamin A is present primarily as retinol bound to fatty acids.
2. In the body, retinol and retinal can be interconverted.
3. Once retinoic acid has been formed, it cannot be converted back to retinal or retinol.
4. β-carotene from plant foods can be converted into retinal in the intestinal mucosa and in the liver. Cleaving a molecule of β-carotene in half theoretically yields two molecules of retinal; however, because β-carotene is not as well absorbed as preformed vitamin A and may not be efficiently converted to retinal, it takes about 12 mg of dietary β-carotene to yield 1mg of retinol.

Figure 9.3 Forms of Vitamin A
Both preformed and provitamin A from the diet can be used to obtain retinol, retinal, and retinoic acid.

fat-soluble food components to form micelles, which facilitate their diffusion into mucosal cells. Absorption of preformed vitamin A is efficient—70%-90% of what is consumed. The provitamin carotenoids are less well absorbed, and absorption decreases as intake increases, so large amounts are not well absorbed.[3] Once inside the mucosal cells, much of the β-carotene is converted to retinoids (see Figure 9.3).

The fat content of the diet and the ability to absorb fat can affect the amount of vitamin A that is absorbed. A diet that is very low in fat (less than 10 g/day) can reduce vitamin A absorption. This is rarely a problem in industrialized countries, where typical fat intake ranges from 50-100 g/day. However, in populations with low dietary fat intakes, vitamin A deficiency may occur due to poor absorption. Diseases that cause fat malabsorption, as well as some medications, can also interfere with vitamin A absorption and cause a deficiency.

Vitamin A in the Body

Preformed vitamin A and carotenoids absorbed from the diet are transported from the intestine in chylomicrons. These lipoproteins deliver the preformed vitamin A and carotenoids to body tissues such as bone marrow, blood cells, spleen, muscles, kidney, and liver. In the liver, some carotenoids can be converted into retinol. To move from liver stores to the tissues, retinol must be bound to **retinol-binding protein**. There is no specific blood transport protein for carotenoids, but since they are fat soluble, they are incorporated into lipoproteins to travel in the bloodstream.

retinol-binding protein A protein that is necessary to transport vitamin A from the liver to other tissues.

YOUR CHOICE

(©iStockphoto)

Fresh, Frozen, or Canned—Does It Matter?

Vegetables and fruits are excellent sources of many vitamins and minerals. A commonly asked question is which is the best choice, nutritionally, fresh, frozen, or canned? Fresh fruits and vegetables seem healthiest. Frozen are handy, but canned are the least expensive. Does it matter which you choose?

Heat, light, air, and time cause the loss of nutrients from food. Therefore, how a food has been handled is an important consideration when trying to make the most nutritious choice. Fresh produce would seem to be the best, but if the "fresh" vegetable has actually spent a week in a truck travelling to your store, several days on the shelf, and then another week in your refrigerator, it may not be the most nutritious choice. A frozen version may actually supply more vitamins. Although freezing itself can cause some losses, much produce is frozen right in the field, so nutrients are not lost due to time and exposure after picking.

What about canned vegetables or fruits? The canning process uses high temperatures, which reduce the vitamin content. In addition, canned fruit is often high in added sugar, and canned vegetables are high in salt. Despite these disadvantages, canned foods keep a long time, do not require refrigeration, and are typically less expensive than fresh or frozen.

Both frozen and canned vegetables and fruits can be more convenient, requiring less preparation time than fresh produce. If you are trying to increase your consumption of vegetables and fruit and don't have a lot of time for food preparation, these products can be a very convenient starting point. If you do like to prepare your food from scratch, how you handle your produce at home can also affect its nutrient content. Because exposure to oxygen, light, and heat can destroy vitamins, it is best to store food away from heat and light, and, if practical, begin preparation as close to serving time as possible. Cutting vegetables and fruits increases the surface area exposed to light and oxygen, so avoid storing chopped or cut produce for long periods of time. Cooking at higher temperatures and for longer periods causes greater vitamin losses, so a microwave oven, which cooks foods quickly, can help to reduce nutrient losses. Water-soluble vitamins can be washed away in cooking water, so roast, grill, stir-fry, or bake when possible. When foods are cooked in water, use the cooking water to make soups and sauces, so you retrieve some of the nutrients.

© Helen Sessions/Alamy

So, which should you choose? The answer is whichever helps you get the recommended amounts of vegetables and fruits daily. Choose what works for your lifestyle— even a vegetable that has lost some of its vitamins is still a good source of nutrients as well as fibre and phytochemicals. Canada's Food Guide illustrates both fresh, frozen, and canned vegetables and fruits on its rainbow because all can play a role in a balanced diet. Remember that consuming a variety of foods, whether canned, fresh or frozen, is the key to good nutrition.

The different forms of vitamin A have different functions. The body can make the retinal and retinoic acid forms from the retinol and carotenoids in the diet (see Figure 9.3). Retinol is the form that circulates in the blood. Retinal is the form that is important for vision. Retinol and retinal can be interconverted from one to the other. Retinoic acid, which is made from retinol or retinal, cannot be used in the visual cycle (see The Visual Cycle) but is the form that affects gene expression and is responsible for vitamin A's role in cell differentiation, growth, and reproduction.[4] Carotenoids that are not converted to retinoids may act as antioxidants or provide other biological functions.

The Visual Cycle Vitamin A is involved in the perception of light (**Figure 9.4**). In the eye, the retinal form of the vitamin combines with the protein opsin to form the visual pigment **rhodopsin**. Rhodopsin helps transform the energy from light into a nerve impulse that is sent to the brain. This nerve impulse allows us to see.

rhodopsin A light-absorbing compound found in the retina of the eye that is composed of the protein opsin loosely bound to retinal.

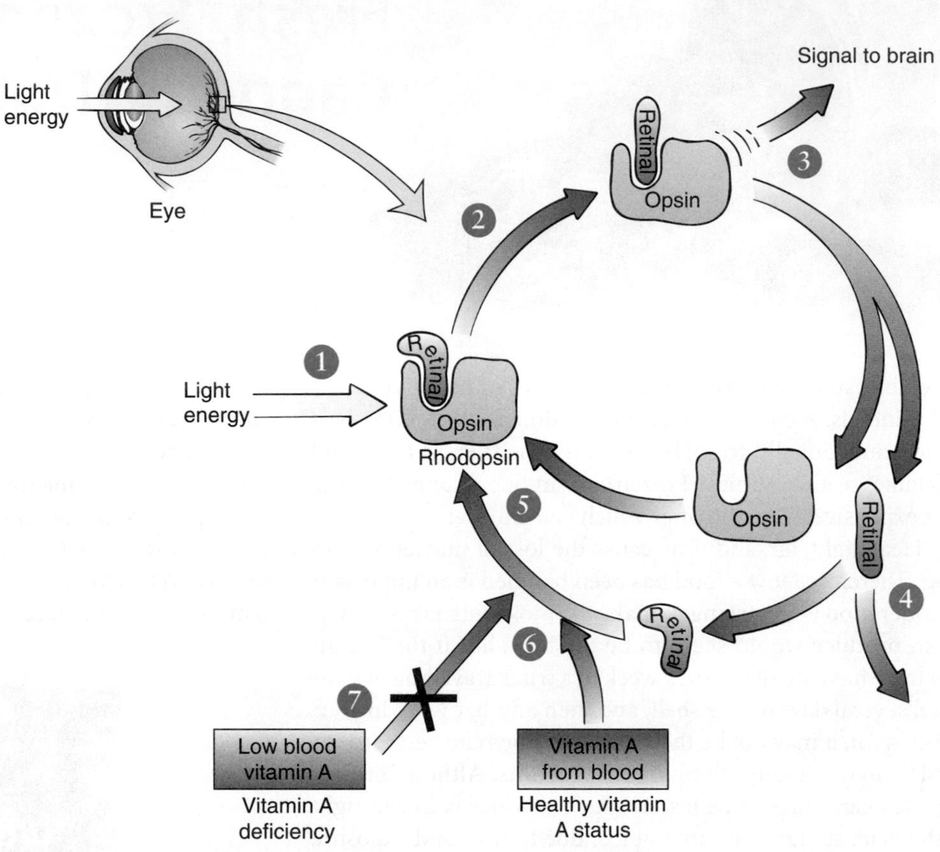

1. Light strikes the visual pigment rhodopsin, which is formed by combining retinal with the protein opsin.
2. The retinal molecule changes from a bent (*cis*) to a straight (*trans*) configuration.
3. A nerve signal is sent to the brain, telling us that there is light, and retinal is released from opsin.
4. Some retinal is lost from the cycle.
5. Some retinal returns to its original *cis* configuration and binds opsin to begin the cycle again.
6. When vitamin A status is normal, vitamin A from the blood replaces any retinal lost from the cycle.
7. When vitamin A is deficient, little vitamin A is available in the blood, and the regeneration of rhodopsin is delayed. Until it is reformed, light cannot be perceived.

Figure 9.4 The visual cycle
Looking into the bright headlights of an approaching car at night is temporarily blinding for all of us, but for someone with vitamin A deficiency, the blindness lasts much longer.

The visual cycle begins when light passes into the eye and strikes rhodopsin. The light changes the retinal in rhodopsin from a curved molecule to a straight one by converting a *cis* double bond in retinal to a *trans* double bond. This change in shape initiates a series of events causing a nerve signal to be sent to the brain and retinal to be released from opsin. After the light stimulus has passed, the *trans* retinal is converted back to its original *cis* form and recombined with opsin to regenerate rhodopsin. Each time this cycle occurs, some retinal is lost and must be replaced by retinol from the blood. The retinol is converted into retinal in the eye. When vitamin A is deficient, there is a delay in the regeneration of rhodopsin, which causes difficulty seeing in dim light, particularly after exposure to a bright light—a condition called **night blindness**. Night blindness is one of the first and more easily reversible symptoms of vitamin A deficiency.

night blindness The inability of the eye to adapt to reduced light, causing poor vision in dim light.

Regulating Gene Expression: Cell Differentiation Cell differentiation is the process whereby cells change in structure and function to become specialized. For instance, in the bone marrow, some cells differentiate into various white blood cells, whereas others differentiate to form red blood cells. Vitamin A affects cell differentiation through its effect on gene expression. This means that it can turn on or turn off the production of certain proteins that regulate functions within cells and throughout the body. By affecting gene expression, vitamin A can also determine what type of cell an undifferentiated cell will become.

In order to affect gene expression, the retinoic acid form of vitamin A enters the nucleus of specific target cells, where it binds to protein receptors; this retinoic acid-protein receptor complex then binds to a regulatory region of DNA (**Figure 9.5**). This binding changes the amount of messenger RNA (mRNA) that is made by the gene. The change in mRNA changes the amount of the protein that is produced. This turning on (or turning off) of the gene increases (or decreases) the production of proteins and thereby affects various cellular functions. For example, vitamin A turns on a gene that makes an enzyme in liver cells. This enzyme enables the liver to make glucose through the process of gluconeogenesis.

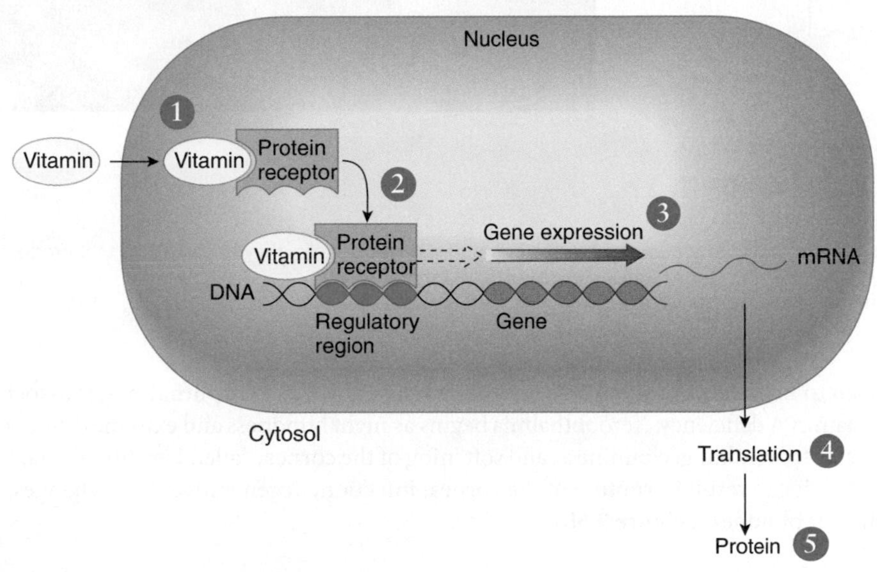

1. Retinoic acid enters the nucleus and binds to a protein receptor.
2. The vitamin-protein receptor complex binds to a regulatory region of DNA.
3. Transcription of the gene is turned on, increasing the amount of mRNA made.
4. mRNA directs translation, increasing the synthesis of the protein coded by this gene.
5. There is an increase in the amount of the protein and hence the cellular functions and body processes affected by this protein.

Figure 9.5 Vitamin A and gene expression
The retinoic acid form of vitamin A (shown as the yellow vitamin) affects cell function by changing gene expression. The steps illustrate what happens when vitamin A turns on gene expression.

Maintenance of Epithelial Tissue Vitamin A's role in cell differentiation is necessary for the maintenance of epithelial tissue. This type of tissue covers external body surfaces and lines internal cavities and tubes. It includes the skin and the linings of the eyes, intestines, lungs, vagina, and bladder. When vitamin A is deficient, epithelial cells do not differentiate normally because vitamin A is not there to turn on or turn off the production of particular proteins. For example, the epithelial tissue on many body surfaces contains cells that produce mucus for lubrication. When mucus-secreting cells die, new cells differentiate into mucus-secreting cells to replace them. When vitamin A is deficient, the new cells do not differentiate properly and instead become cells that produce a protein called **keratin** (**Figure 9.6a**). Keratin is the hard protein that makes up hair and fingernails. As the mucus-secreting cells die and are replaced by keratin-producing cells, the epithelial surface becomes hard and dry. This process is known as keratinization. The hard, dry surface does not have the protective capabilities of normal epithelium and so the likelihood of infection is increased. The risk of infection is compounded by the fact that vitamin A deficiency also decreases immune function.

keratin A hard protein that makes up hair and nails.

All epithelial tissues are affected by vitamin A deficiency, but the eye is particularly susceptible to damage. The mucus in the eye normally provides lubrication, washes away dirt and other particles, and also contains a protein that helps destroy bacteria. When vitamin A is deficient, the lack of mucus and the buildup of keratin cause the cornea to dry and leave the

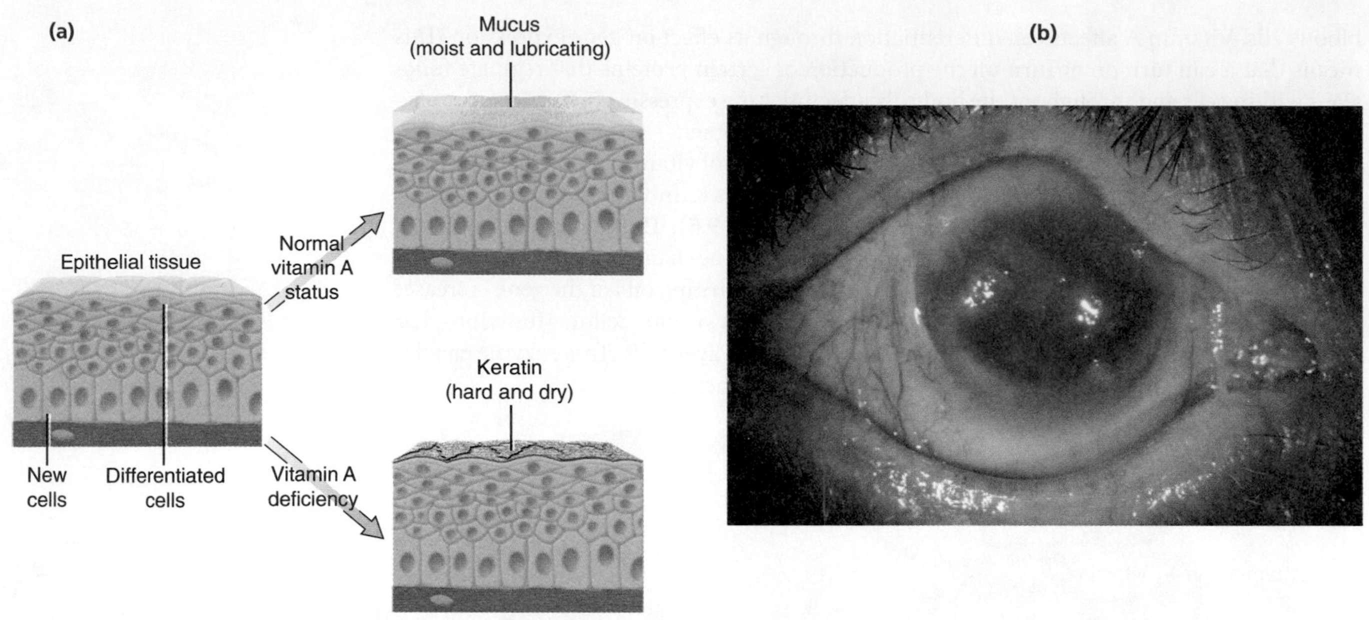

Figure 9.6 Vitamin A deficiency
(a) When vitamin A is deficient, immature cells can't differentiate normally, and instead of mucus-secreting cells, they become cells that produce keratin.
(b) As xerophthalmia progresses, the drying of the cornea results in ulceration, infection, and ultimately blindness. (ISM/Phototake)

xerophthalmia A spectrum of eye conditions resulting from vitamin A deficiency that may lead to blindness. An early symptom is night blindness, and as deficiency worsens, lack of mucus leaves the eye dry and vulnerable to cracking and infection.

keratomalacia Softening and drying and ulceration of the cornea resulting from vitamin A deficiency.

eye open to infection. A spectrum of eye disorders, known as **xerophthalmia,** is associated with vitamin A deficiency. Xerophthalmia begins as night blindness and extreme dryness and progresses to wrinkling, cloudiness, and softening of the cornea, called **keratomalacia.** If left untreated, it can result in rupture of the cornea, infection, degenerative tissue changes, and permanent blindness (**Figure 9.6b**).

The *Immune System* Vitamin A controls the gene expression that induces the differentiation of immature immune system cells into mature white blood cells such as phagocytes and lymphocytes, including antibody-producing cells (see Section 3.2; Figure 3.4).[5] The combined effect of vitamin A deficiency on both epithelial cell function and the cells of the immune system dramatically increases the risk of infection in vulnerable populations such as young children (see Vitamin A Deficiency: A World Health Problem and Chapter 18: Science Applied: Vitamin A: The Anti-Infective Vitamin).

Reproduction and Growth The ability of vitamin A to regulate the differentiation of cells also makes it essential for normal reproduction and growth. Lack of vitamin A during embryonic development results in abnormalities and death. Vitamin A is important for the formation of the heart and circulatory system, the nervous system, respiratory system, and skeleton.[6] In children, poor overall growth is an early sign of vitamin A deficiency. Vitamin A affects the activity of cells that form and break down bone, and a deficiency early in life can cause abnormal jawbone growth, resulting in crooked teeth and poor dental health.

Beta-Carotene: A Vitamin A Precursor and an Antioxidant Some carotenoids, particularly β-carotene, can be converted to vitamin A in the intestinal mucosa and liver. Unconverted carotenoids also circulate in the blood and reach tissues where they may function as antioxidants, a role independent of any conversion to vitamin A. Beta-carotene and other carotenoids are fat-soluble antioxidants that may play a role in protecting cell membranes from damage by free radicals. The antioxidant properties of carotenoids have stimulated interest in their ability to protect against diseases in which oxidative processes play a role, such as cancer, heart disease, and impaired vision due to macular degeneration and cataracts.

Life Cycle

Recommended Vitamin A Intake

The recommended intake for vitamin A is based on the amount needed to maintain normal body stores. The RDA is set at 900 μg/day of vitamin A for men and 700 μg/day for women[3]

(**Table 9.1**). The RDA is increased in pregnancy to account for the vitamin A that is transferred to the fetus, and during lactation to account for the vitamin A lost in milk. The RDA for children is set lower than that for adults based on their smaller body size. For infants, an AI has been set based on the amount of vitamin A consumed by an average, healthy, breastfed infant.

Table 9.1 A Summary of the Fat-Soluble Vitamins

Vitamin	Sources	Recommended Intake for Adults	Major Functions	Deficiency Diseases and Symptoms	Groups at Risk of Deficiency	Toxicity	UL
Vitamin A (retinol, retinal, retinoic acid, vitamin A acetate, vitamin A palmitate, retinyl palmitate, provitamin A, carotene, β-carotene, carotenoids)	Retinol: liver, fish, fortified milk and margarine, butter, eggs Carotenoids: carrots, leafy greens, sweet potatoes, broccoli, apricots, cantaloupe	700–900 μg/d	Vision, health of cornea and other epithelial tissue, cell differentiation, reproduction, immune function	Night blindness, xerophthalmia, poor growth, dry skin, impaired immunity	Those with a limited diet (particularly children and pregnant women), those consuming very low-fat or low-protein diets	Headache, vomiting, hair loss, liver damage, skin changes, bone and muscle pain, fractures, birth defects	3,000 μg/d of preformed vitamin A
Vitamin D (calciferol, cholecalciferol, ergocalciferol, dihydroxy vitamin D)	Egg yolk, liver, fish oils, tuna, salmon, fortified milk, synthesis from sunlight	15-20 μg/d	Absorption of calcium and phosphorus, maintenance of bone	Rickets in children: abnormal growth, misshaped bones, bowed legs, soft bones; Osteomalacia in adults: weak bones and bone and muscle pain	Some breastfed infants, children and elderly (especially those with dark skin and little sun exposure), people with kidney disease	Calcium deposits in soft tissues, growth retardation, kidney damage	100 μg/d
Vitamin E (tocopherol, α-tocopherol)	Vegetable oils, leafy greens, seeds, nuts, peanuts	15 mg/d	Antioxidant, protects cell membranes	Broken red blood cells, nerve damage	Those with poor fat absorption, premature infants	Inhibition of vitamin K activity	1,000 mg/d from supplemental sources
Vitamin K (phylloquinones, menaquinone)	Vegetable oils, leafy greens, synthesis by intestinal bacteria	90–120 μg/d[a]	Coenzyme for synthesis of blood clotting proteins and proteins in bone	Hemorrhage	Newborns (especially premature), people on long-term antibiotics	Anemia and brain damage in infants	ND

[a]Adequate Intake (AI).
UL, Tolerable Upper Intake Level; ND, insufficient evidence to set a UL.

Recommendations for vitamin A intake are expressed in micrograms (μg) of retinol. Retinol can be supplied by both preformed vitamin A and carotenoids in the diet. No quantitative recommendations have been made for intakes of β-carotene or other carotenoids. Their intake is considered only with regard to the amount of retinol they provide. Because carotenoids are less well absorbed and not completely converted to vitamin A, a correction factor, referred to as **retinol activity equivalents (RAE)**, must be applied to carotenoids to determine the amount of usable vitamin A they provide. Twelve micrograms of β-carotene provide 1 RAE of vitamin A, and 24 μg of α-carotene or β-cryptoxanthin provide 1 RAE.[3]

retinol activity equivalent (RAE) The amount of retinol, β-carotene, α-carotene, or β-cryptoxanthin that provides vitamin A activity equal to 1 μg of retinol.

As our understanding of vitamin A has increased, the units in which recommended intakes have been expressed have changed. Prior to 1980, vitamin A was expressed in international units (IUs). In the 1980s a unit called retinol equivalents (REs) was developed to account for differences in absorption between preformed vitamin A and carotenoids. This was later replaced by the more accurate RAE, but these older units are still found in some food composition databases and tables. Values for converting REs and IUs to micrograms of retinol are given in **Table 9.2.**

CRITICAL THINKING:

How Are Canadians Doing with Respect to Their Intake of Vitamin A?

The CCHS collected information on the intake of Vitamin A, from food, and the data are shown in the bar graph.

Health Canada considers the prevalence of inadequate intake to be low when less than 10% of the population has intakes below the EAR, as this means that less than 10% of the population is *not* meeting its nutrient requirement, or, conversely, 90% or more of the population *is* meeting its requirement. The analysis is based on the EAR cutpoint method (see Section 2.2). The 10% cutoff is indicated by the green line in the bar graph. If more than 10% of the population has intakes below the EAR, then this is considered an area of concern. Vitamin A intakes in different parts of Canada are also illustrated in the map from Health Canada's Nutrition and Health Atlas.

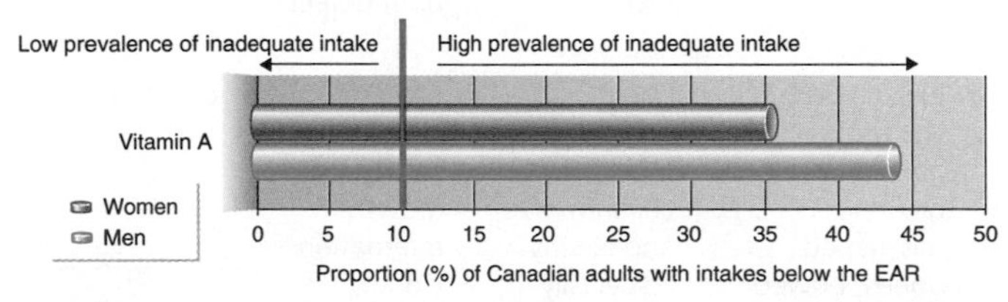

Source: Health Canada. Canadian Community Health Survey, Cycle 2.2, Nutrition (2004) - Nutrient Intakes from Food: Provincial, Regional and National Summary Data Tables, Volume 1, 2 and 3. Cat No. 978-0-662-06542-5

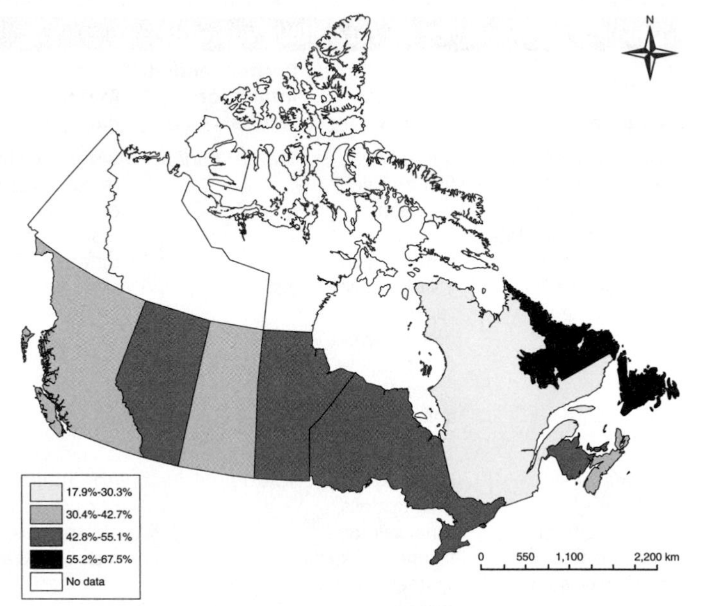

The proportion (%) of adults with vitamin A intakes below the EAR.

Source: Health Canada: Canada's Nutrition and Health Atlas. Available online at http://www.hc-sc.gc.ca/fn-an/surveill/atlas/map-carte/nutri_vita_mf-hf-eng.php. July 31, 2011.

Critical Thinking Questions

What would you conclude about the intake of vitamin A by Canadians?

What does Canada's Food Guide recommend to ensure adequate intake of vitamin A?

The data reflect vitamin A intake from food alone. What impact would vitamin supplements have on overall intake?

Consider the map. Are there regional differences in the vitamin A intake of Canadian adults? Why might these differences exist?

*Answers to all Critical Thinking questions can be found in Appendix J.

Table 9.2 Converting Vitamin A Units

Form and Source	Amount Equal to 1 μg Retinol
Preformed vitamin A in food or supplements	1 μg
	1 RAE
	1 μg RE
	3.3 IU
β-carotene in food[a]	12 μg
	1 RAE
	2 μg RE
	20 IU
α-carotene or β-cryptoxanthin in food	24 μg
	1 RAE
	2 μg RE
	40 IU

[a]Beta-carotene in supplements may be better absorbed than β-carotene in food and so provides more vitamin A activity. It is estimated that 2 μg of β-carotene dissolved in oil provides 1 μg of vitamin A activity.

Vitamin A Deficiency: A World Health Problem

Vitamin A deficiency is a threat to the health, sight, and lives of millions of children in the developing world. Children deficient in vitamin A have poor appetites, are anemic, have an increased susceptibility to infections, including measles, and are more likely to die in childhood. It is estimated that more than 250 million children worldwide are vitamin A deficient and that 250,000-500,000 children go blind annually due to vitamin A deficiency.[7] It is most common in India, Africa, Latin America, and the Caribbean.

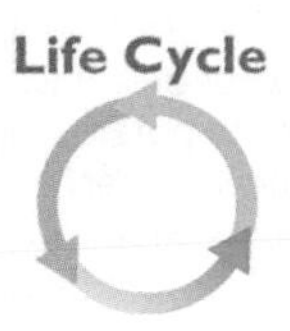

Vitamin A deficiency can be caused by insufficient intakes of vitamin A, fat, protein, or the mineral zinc. As discussed, without fat, vitamin A cannot be absorbed, so a diet very low in fat can cause a deficiency by reducing vitamin A absorption. Protein deficiency can cause vitamin A deficiency because the retinol-binding protein needed to transport vitamin A from the liver cannot be made in sufficient quantities. The importance of zinc for vitamin A utilization is believed to be due to its role in protein synthesis. When zinc is deficient, the proteins needed for vitamin A transport and metabolism are lacking.

Vitamin A deficiency is not common in developed countries, but the CCHS found that some Canadians do not meet the recommendations for this vitamin.[8] (See also: Critical Thinking: How are Canadians Doing with Respect to Their Intake, from Food, of Vitamin A?) Inadequate intakes can be caused by poor food choices even when the food supply is plentiful. In Canada, the intake of vegetables and fruit, many of which are excellent sources of provitamin A, does not meet recommendations.[9] For example, a typical fast-food meal of a hamburger and french fries provides almost no vitamin A (see Critical Thinking: How Much Vitamin A is in Your Fast-Food Meal?).

Vitamin A Toxicity and Supplements

Preformed vitamin A can be toxic at extremely high doses. Acute toxicity has been reported in Arctic explorers who consumed polar bear liver, which contains about 100,000 μg of vitamin A in just 90 g. Although polar bear liver is not a common dish at most dinner tables, supplements of preformed vitamin A also have the potential to deliver a toxic dose. Signs of acute toxicity include nausea, vomiting, headache, dizziness, blurred vision, and a lack of muscle coordination. Chronic toxicity occurs when preformed vitamin A doses as low as 10 times the RDA are consumed for a period of months to years. The symptoms of chronic toxicity include weight loss, muscle and joint pain, liver damage, bone abnormalities, visual defects, dry scaling lips, and skin rashes. Excess vitamin A is a particular concern for pregnant women because it may contribute to birth defects.[3,10] High intakes of vitamin A have also been found to increase the incidence of bone fractures.[11,12] The UL is set at 2,800 μg/day of preformed vitamin A for 14- to 18-year-olds and 3,000 μg/day for adults.

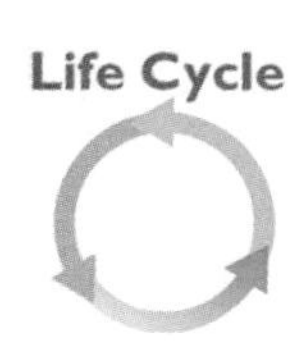

Vitamin A as a Drug Derivatives of vitamin A are currently used as drugs. One derivative of retinoic acid, marketed as Retin A, is used topically to treat acne and to reduce wrinkles due to sun damage. It acts by increasing the turnover of cells. In patients with acne, new cells replace the cells of existing pimples and the rapid turnover of cells prevents new pimples from forming. By a similar mechanism, Retin A can reduce wrinkles and diminish areas of darkened skin and rough skin. Another vitamin A derivative, 13-*cis*-retinoic acid, marketed as Accutane, is taken orally to treat acne. This medication can have serious side effects, including dry, itchy skin and chapped lips, irritated eyes, joint and muscle pain, decreased night vision, depression, and increases in blood lipid levels. In pregnant women, it can cause severe birth defects, including brain damage and physical malformations. Although Retin A and Accutane are derivatives of vitamin A, it is important to remember that they are drugs and should only be taken while under the care of a physician.

Carotenoid Toxicity Because of the toxicity of preformed vitamin A, most supplements provide some or all of their vitamin A as carotenoids (**Table 9.3**). Carotenoids are not toxic because their absorption from the diet decreases at high doses, and once in the body, their conversion to retinoids is limited. Large daily intakes of carotenoids—usually from carrot juice or β-carotene supplements—do, however, lead to a condition known as **hypercarotenemia.** In this condition, large amounts of carotenoids stored in the adipose tissue give the skin a yellow-orange colour (**Figure 9.7**). This is particularly apparent on the palms of the hands

hypercarotenemia A condition in which carotenoids accumulate in the adipose tissue, causing the skin to appear yellow-orange, especially the palms of the hands and the soles of the feet.

CRITICAL THINKING:

How Much Vitamin A is in Your Fast-Food Meal?

Background

John lives on his own, goes to school, and works part-time. He eats Frosted Flakes and milk for breakfast at home and usually brings a ham and cheese sandwich with potato chips and a cola for lunch; dinner is always a fast-food meal. He recently heard a report indicating that fast-food was low in some vitamins, particularly vitamin A. To explore this issue, John uses iProfile to look up the nutrient composition of his favourite fast-food meals.

(©TNT Magazine/Alamy)

Data

FOOD		VITAMIN A (μg)	% DAILY VALUE
McDonalds	Big Mac	89	10
	Fries	2	0.2
Pizza Hut	Pepperoni pizza	358	40
KFC	Chicken leg and breast	24	2.6
	Mashed potatoes and gravy	14	1.5

Critical Thinking Questions

Looking at his fast-food options, John notices that pizza has lots of vitamin A. What ingredients in pizza make it higher in vitamin A than the other choices?

Should John be concerned about his vitamin A intake? Use iProfile to estimate how much vitamin A he is getting in his breakfast and lunch. How much vitamin C is provided by these meals?

How could John's breakfast and lunch be modified to increase his intake of vitamin A and vitamin C?

Use iProfile to find out how much vitamin A is in your favourite fast food meal.

Table 9.3 Benefits and Risks of Fat-Soluble Vitamin Supplements

Supplement	Marketing Claim	Actual Benefits or Risks
Vitamin A (retinoids)	Improves vision, prevents skin disorders, enhances immunity	Needed for vision and eye health, growth, reproduction, and immunity but extra as supplements does not provide additional benefits. Toxic at high doses, can cause birth defects and bone loss.
Carotenoids	Needed for vision, prevents skin disorders, antioxidant	Can provide all functions of vitamin A and is an antioxidant but supplements do not provide additional benefits. High doses can cause orange-coloured skin, especially on the palms of hands and soles of feet, and increase lung cancer risk in smokers.
Vitamin D	Bone health, prevents multiple sclerosis	Needed for calcium absorption and bone maintenance. There is evidence that supplements may reduce the risk of autoimmune diseases and cancer. High doses cause heart and kidney damage.
Vitamin E	Prevents heart disease, improves symptoms of fibrocystic breast disease, promotes immune function, reduces scar formation	Antioxidant, protects cell membranes, little evidence that oral supplements reduce risk of heart disease or that topical application reduces scar formation. High doses interfere with anticoagulant medications.

and the soles of the feet. It is not known to be dangerous, and when intake decreases, the skin returns to its normal colour.

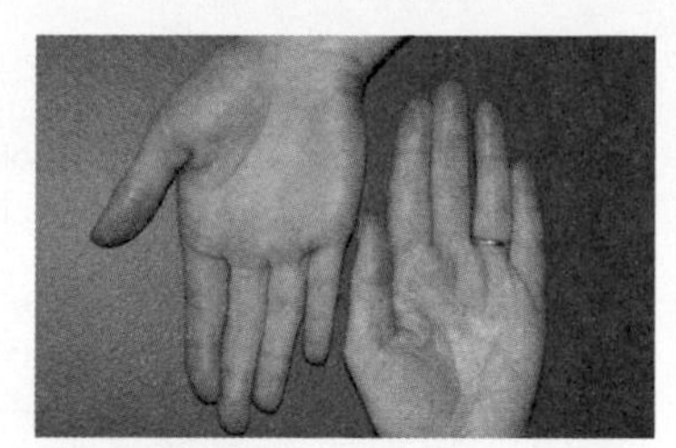

Figure 9.7 This photo compares normal hand colour (right) with more orange colour seen on the hand of a patient with hypercarotenemia (left). (New Zealand Dermatological Society Incorporated, DermNetNZ.org)

Although not considered toxic, carotenoid supplements may be harmful to cigarette smokers. Two clinical research trials found an increased incidence of lung cancer in cigarette smokers who took β-carotene supplements.[13,14] Even though other trials have not shown this effect, until more information is available, smokers are advised to avoid β-carotene supplements and to rely on food sources to obtain carotenoids in their diet. The small amounts found in standard-strength multivitamin supplements are not likely to be harmful for any group. No UL has been determined for carotenoid intake.

9.3 Vitamin D

Learning Objectives

- Explain why vitamin D is known as the "sunshine vitamin."
- Relate the functions of vitamin D to the symptoms that occur when it is deficient.
- Discuss why vitamin D deficiency is on the rise.

Vitamin D is known as the "sunshine vitamin" because it can be produced in the skin by exposure to ultraviolet light. Because vitamin D can be made in the body, there is a long-standing debate as to whether vitamin D is a vitamin or a hormone. Vitamin D acts like a hormone because it is produced in one organ, the skin, and affects other organs, primarily the intestine, bone, and kidney. By definition, vitamins are dietary essentials. However, vitamin D can be formed in the skin, so it is only essential in the diet when exposure to sunlight is limited or the body's ability to synthesize the vitamin is reduced. In a northern country like Canada, where sunlight can be very limited, especially in the winter months, dietary sources of vitamin D are important.

Vitamin D in the Diet

The major source of vitamin D for most humans is exposure to sunlight.[15] Only a few foods are natural sources of vitamin D. These include liver; fatty fish such as salmon, mackerel, and sardines; cod liver oil; and egg yolks (**Figure 9.8**). These foods contain **cholecalciferol**, also known as vitamin D_3. Cholecalciferol is the form of vitamin D that is made in the skin of animals by the action of sunlight on a compound made from cholesterol, called 7-dehydrocholesterol (**Figure 9.9**). Fortified beverages, such as milk, soy milk, and orange juice are important sources of vitamin D in Canada. These may contain vitamin D_3 or another active form of the vitamin called vitamin D_2. It is because of both their vitamin D and calcium content that Canada's Food Guide recommends the consumption of 500 ml of milk or fortified soy milk daily (see Section 2.3).

cholecalciferol The chemical name for vitamin D_3. It can be formed in the skin of animals by the action of sunlight on a form of cholesterol called 7-dehydrocholesterol.

Vitamin D in the Body

Vitamin D from the diet and from synthesis in the skin is inactive until it is chemically altered in the liver and then the kidney. In the liver, a hydroxyl group (OH) is added to vitamin D to form 25-hydroxy vitamin D_3. This is the form of the vitamin that circulates in the blood and is monitored in the blood to indicate the vitamin D status of patients. However, 25-hydroxy vitamin D_3 is inactive and must be modified by the kidney where another hydroxyl group is added to make the active form of vitamin D: 1,25-dihydroxy vitamin D_3 (**Figure 9.10**).

Vitamin D and Bone Health The principal function of vitamin D is to maintain levels of calcium and phosphorus in the blood. Calcium levels in the blood must be carefully regulated because of the important role that calcium plays in muscle contraction. When blood calcium levels drop too low, the parathyroid gland releases **parathyroid hormone (PTH)**. PTH release stimulates enzymes in the kidney to convert 25-hydroxy vitamin D_3 to the active form of the vitamin. Active vitamin D functions by binding to a protein called the **vitamin D receptor** in target tissues and affecting gene expression in a manner that is similar to the action of retinoic acid, shown in Figure 9.5. Its effect at three different target tissues helps to increase blood calcium levels.[16] One target is intestinal cells, where vitamin D increases the

parathyroid hormone (PTH) A hormone released by the parathyroid gland that acts to increase blood calcium levels.

vitamin D receptor (VDR) A protein to which vitamin D binds. This receptor-vitamin complex is then able to bind to DNA and alter gene expression.

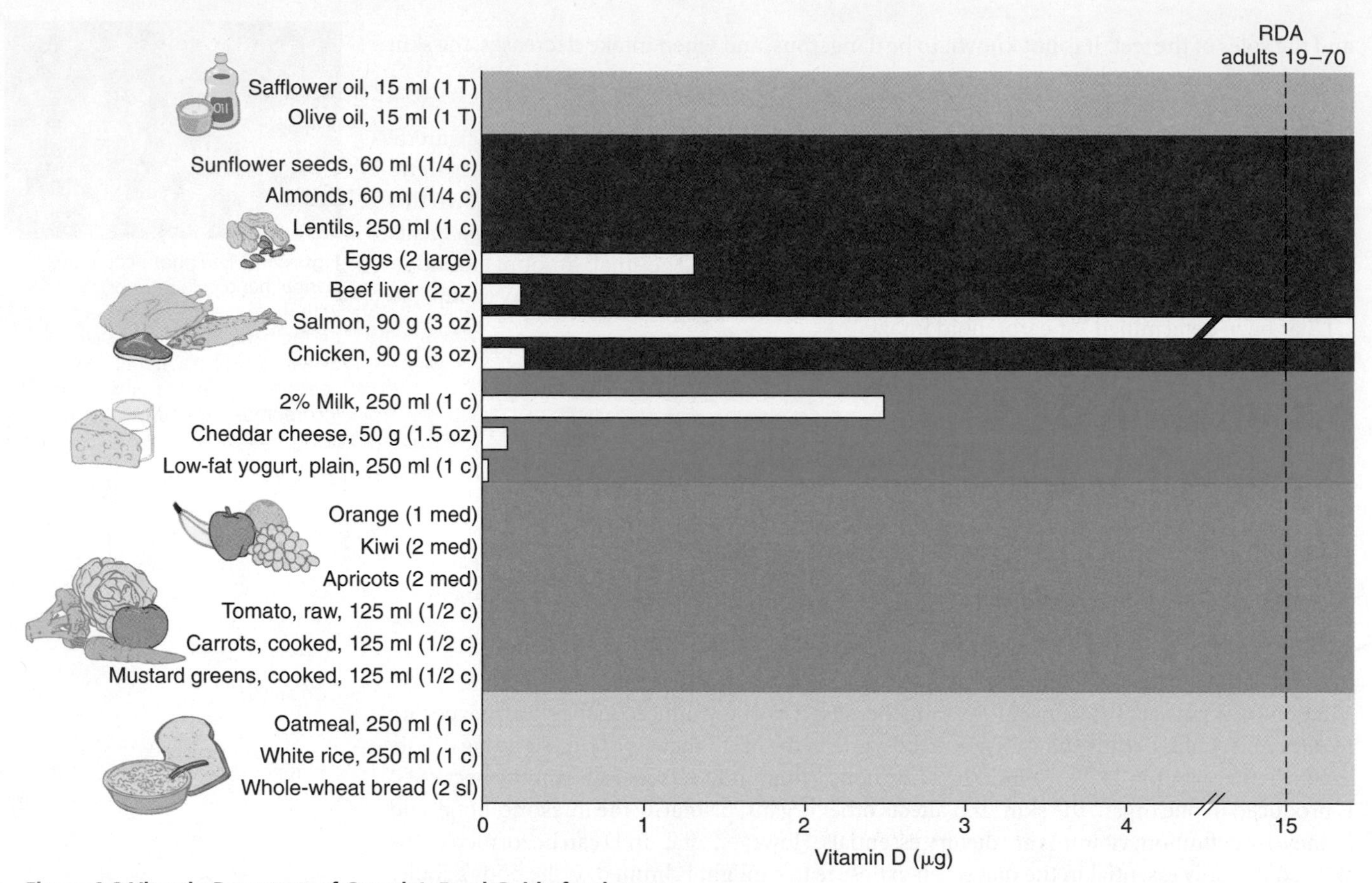

Figure 9.8 Vitamin D content of Canada's Food Guide food groups
Only a few foods are natural sources of vitamin D; the dashed line shows the RDA for adults 19 through 70 years of age.

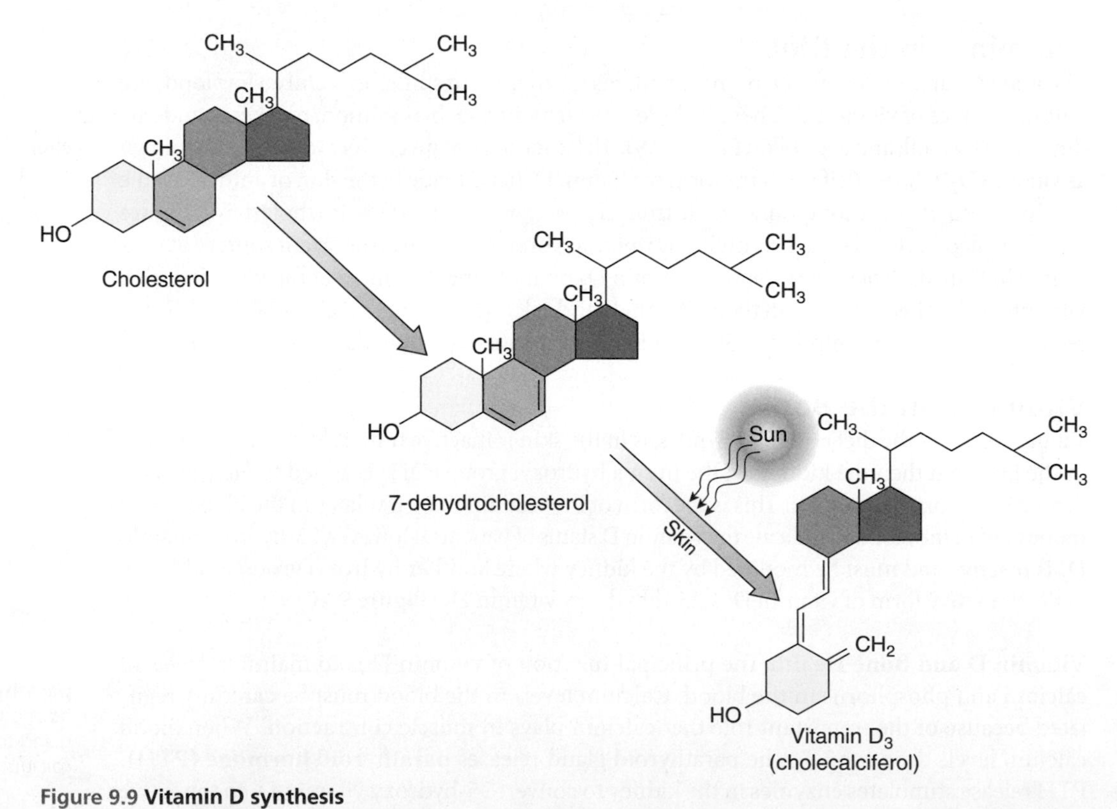

Figure 9.9 Vitamin D synthesis
In the body, 7-dehydrocholesterol can be made from cholesterol. Vitamin D_3 (cholecalciferol) can then be formed by the action of sunlight on 7-deyhydrocholesterol in the skin.

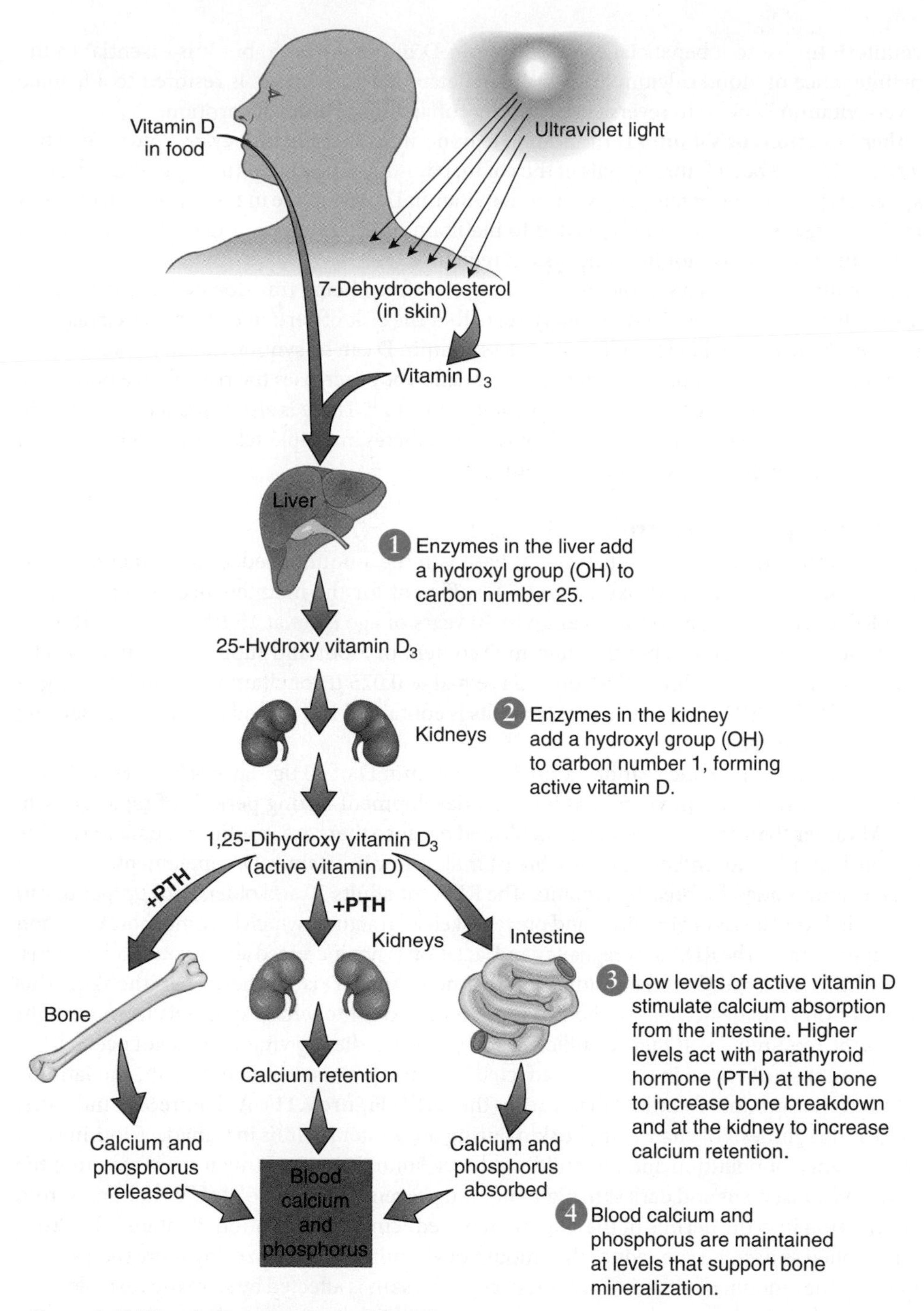

Figure 9.10 Vitamin D activation and functions
In order to function, vitamin D from food and from synthesis in the skin must be activated.

expression of genes that code for the production of intestinal calcium transport proteins. This enhances the active transport of dietary calcium from the intestinal lumen into the body, both to increase blood calcium levels, and also to increase the amounts of calcium available to build bone (see Figure 9.10). It is because of its role in increasing calcium absorption that vitamin D is considered important in the maintenance of bone health, which will be discussed in more detail in Section 11.3.

If dietary calcium is unavailable, higher levels of vitamin D act at the bone and kidney in conjunction with PTH to return blood calcium to normal (see Figure 9.10). At the kidney, vitamin D acts with PTH to increase the amount of calcium retained by the kidneys.[16] At the bone, vitamin D causes precursor cells to differentiate into cells that break down bone. Bone breakdown releases calcium and phosphorus into the blood. This function may seem counterintuitive to a beneficial role for vitamin D to bone health, but it is essential to the maintenance of blood calcium levels. When dietary calcium intake is restored to adequate levels, vitamin D helps to reverse bone loss by enhancing calcium absorption.

Other Functions of Vitamin D In addition to bone, intestine, and kidney, receptors for active vitamin D have been found in cells of the colon, parathyroid gland, pituitary gland, immune system, reproductive organs, and skin; active vitamin D plays a role in the expression of genes in these organs.[16] Vitamin D is needed to maintain normal functioning of the parathyroid gland and is an important immune system regulator.

Vitamin D also plays a role in preventing cells from being transformed into cancerous cells.[17] It has been recognized for many years that the risk of certain cancers is increased in people who live at higher latitudes, where less vitamin D can be synthesized in the skin. Studies now support the hypothesis that vitamin D deficiency increases the risk of developing and dying from colon, breast, ovarian, and prostate cancers.[18] There is also evidence that vitamin D may play a role in the increased risk of type 1 diabetes, multiple sclerosis, and high blood pressure in people who live at higher latitudes.[19]

Recommended Vitamin D Intake

The recommended intake of vitamin D is based on the amount needed in the diet to maintain blood levels of 25-hydroxy vitamin D_3 sufficient for the maintenance of bone health. The RDA for adult men and women up to 70 years of age is set at 15 μg/day.[20] The RDA is expressed in micrograms, but the vitamin D content of foods and supplements may also be given as International Units (IUs); one IU is equal to 0.025 μg of vitamin D_3 (40 IU = 1 μg of vitamin D). The RDA for vitamin D for adults is contained in a Canada Food Guide serving of salmon (see Table 9.1).

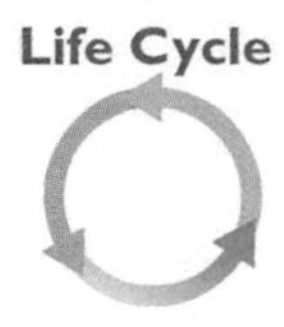

For infants up to one year of age an AI for vitamin D of 10 μg/day has been established. This is to allow sufficient vitamin D for bone development during periods of rapid growth. An AI rather than an RDA has been established because there are insufficient data to reliably establish an RDA for infants. Because breast milk is low in vitamin D, supplemental vitamin D is recommended for breastfed infants. The RDA for adults 70 and older, is 20 μg/per day to maintain blood levels of vitamin D and prevent skeletal fractures which become more common as our bones age. The RDA for pregnancy and lactation is not increased above young adult levels.

The RDA is based on the assumption that no vitamin D is synthesized in the skin. This assumption is made because of the variation in the extent to which synthesis from sunlight meets the requirement. If there is sufficient sun exposure, dietary vitamin D is not needed, but the amount synthesized in the skin is affected by many factors. Climate, season, and latitude all affect the amount of sunlight that reaches the earth (**Figure 9.11**). As Figure 9.11 indicates, there is no synthesis of vitamin in the skin during the winter months in Canada. Clothing and the presence of pollution and tall buildings block sunlight, preventing it from reaching the skin, and sunscreens and dark skin pigmentation prevent the ultraviolet (UV) light rays from penetrating into the dermis of the skin, thereby reducing the formation of vitamin D. Properly applied sunscreen can reduce the amount of vitamin D synthesized by more than 95%.[19] Because the amount of vitamin D synthesized in the skin is affected by so many variables, it is not possible to make a single recommendation regarding the amount of time a person needs to spend in the sun to meet their vitamin needs. For example, during the spring, summer, and fall, light-skinned individuals may need to spend only 5-10 minutes, 2-3 times a week, outdoors with their faces, hands, arms, and legs exposed to meet their vitamin D requirement, whereas very dark-skinned (never sunburn) individuals may need 10-50 times more sun exposure to produce the same amount of vitamin D.[21,22] In the summer, children and active adults usually spend enough time outdoors without sunscreens to provide for their vitamin D requirement. It has been estimated that more than 90% of the vitamin D requirement for most people comes from casual exposure to sunlight.[23] In the absence of adequate exposure to sunlight, careful attention must be paid to dietary sources of vitamin D. The dietary reference intakes for vitamin D, first established in 1997, were updated in 2010. The amount of vitamin D required for optimum health is an area of ongoing research (see Science Applied: New Scientific Knowledge Changes Dietary Reference Intakes).

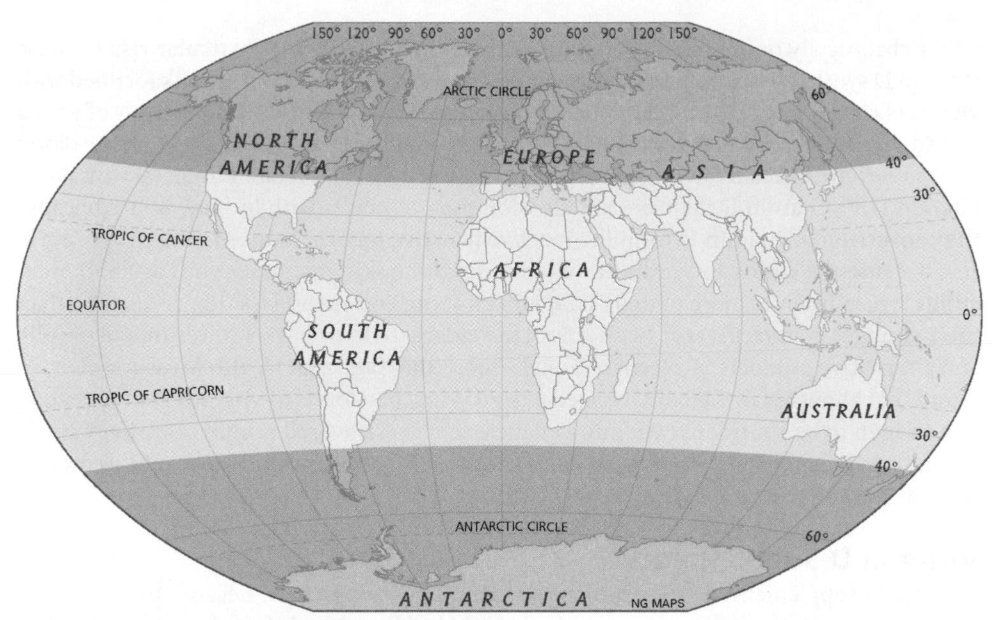

Figure 9.11 Effect of latitude on sun exposure
During the winter at latitudes greater than about 40° north or south, there is not enough UV radiation to synthesize adequate amounts of vitamin D.

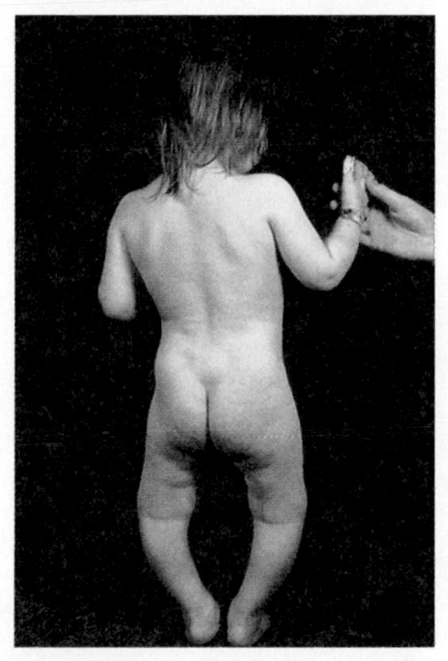

Figure 9.12 Bowed legs are characteristic of rickets. (Biophoto Associates/Photo Researchers)

Vitamin D Deficiency

When we consume foods containing calcium, only about 25%-30% of the calcium we eat is typically absorbed. When vitamin D is deficient, this drops to about 10%-15% of the calcium in the diet. As a result, calcium is not available for proper bone mineralization and abnormalities in bone structure occur.

In children who are deficient in vitamin D, bones are weak because they do not contain enough calcium and phosphorus. This syndrome, called **rickets**, is characterized by bone deformities such as narrow rib cages, known as pigeon breasts, and bowed legs (**Figure 9.12**). The legs bow because the bones are too weak to support the weight of the body. Vitamin D deficiency also prevents children from reaching their genetically programmed height and reduces bone mass and causes muscle weakness. Rickets, first recognized in the 1600s, was common during the Industrial Revolution when large numbers of poorly nourished children lived under a layer of smog in the newly industrialized cities. Tall buildings and smog-filled air reduced children's exposure to sunlight. The fortification of milk with vitamin D has helped to greatly reduce rickets in most developed countries, but it is still a problem in infants and young children with dark skin and those who are breastfed.[24] Rickets is also seen in children with disorders that affect fat absorption and in children, who for any variety of reasons, do not drink milk. Cases of rickets have been reported in Canada, typically found among children living in northern Canada, who are breastfed, and are not receiving a vitamin D supplement, although these cases are very rare (2.9 cases per 100,000 children).[1] A recent study of 16-year-olds in Quebec found that 10% of girls and 13% of boys had blood levels of vitamin D low enough to be classified as Vitamin D deficient.[25]

Life Cycle

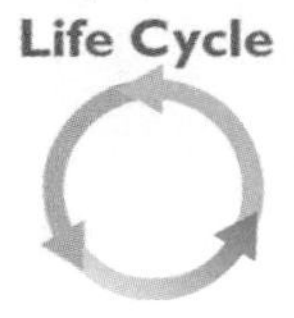

rickets A vitamin D deficiency disease in children that is characterized by poor bone development because of inadequate calcium absorption.

In adults, the vitamin D deficiency disease comparable to rickets is called **osteomalacia**. Because bone growth is complete in adults, osteomalacia does not cause bone deformities, but bones are weakened because not enough calcium is available to form the mineral deposits needed to maintain healthy bone. Insufficient bone mineralization leads to fractures of the weight-bearing bones such as those in the hips and spine. It can precipitate or exacerbate osteoporosis, which is a loss of total bone mass, not just minerals (see Section 11.3). Osteomalacia also causes bone pain and muscle aches and weakness. It is estimated that over half of African Americans in the United States are at risk of vitamin D deficiency

osteomalacia A vitamin D deficiency disease in adults characterized by a loss of minerals from bones. It causes bone pain, muscle aches, and an increase in bone fractures.

either chronically or during the winter months.[23] This group is at particular risk because vitamin D synthesis is low due to dark pigmentation and consumption of milk fortified with vitamin D is low due to the high frequency of lactose intolerance. Similarly, a study of young Canadian adults at the University of Toronto found that the levels of vitamin D in the blood were inversely associated with skin pigmentation and directly related to dietary intake of vitamin D,[26] Vitamin D deficiency is also common in adults with kidney failure because the conversion of vitamin D from the inactive to active form is reduced. The elderly are at risk because the ability to synthesize vitamin D in the skin decreases with age and older adults typically cover more of their skin with clothing and spend less time in the sun than their younger counterparts.[23] In addition, the elderly tend to have a lower intake of milk and alternative products. A recent national survey, the Canadian Health Measures Survey, found that about 4% of Canadians, aged 6 to 79 years, had levels of vitamin D in the blood low enough to be considered vitamin D deficient.[27] These overall numbers are very small, but they do mean that health professionals need to be vigilant for the impact of low intakes of vitamin D or poor sunlight exposure.

Life Cycle

CANADIAN CONTENT

Vitamin D Supplements

Vitamin D supplements are recommended to a number of groups. Because breast milk is low in vitamin D, Health Canada recommends that all breastfed infants be given 10 μg/day of supplemental vitamin D and that the supplement be continued until they are consuming this amount of dietary vitamin D from other sources.[28] The Canadian Pediatric Society recommends a higher dose, 20 μg, during the winter (October to April) for babies in northern Canada (north of 55° latitude, which is about the level of Edmonton).[29] Canada's Food Guide recommends that all Canadians over age 50 take a Vitamin D supplement containing 10 μg of vitamin D. Vitamin D requirements are higher for older Canadians and it is extremely difficult to meet these requirements from food sources alone.

Vitamin D Toxicity

Too much vitamin D in the body can cause high calcium concentrations in the blood and urine, deposition of calcium in soft tissues such as the blood vessels and kidneys, and cardiovascular damage. However, consumption of unfortified foods does not cause vitamin D toxicity, nor does synthesis of vitamin D from exposure to sunlight. Oversupplementation and overfortification can pose a risk. One case of accidental overfortification of milk resulted in the hospitalization of 56 individuals and the death of 2.[30] A UL for adults for vitamin D has been set at 100 μg (4,000 IU) per day.[20] However, based on the fact that sunshine can provide an adult with vitamin D in an amount equivalent to daily oral consumption of 250 μg (10,000 IU) per day, it is now believed that intakes this high will not cause adverse reactions in the majority of healthy people.[31]

9.4 Vitamin E

Learning Objectives

- List two food sources of vitamin E.
- Discuss the function of vitamin E.

Vitamin E is a fat-soluble vitamin with an antioxidant function. It was first identified as a fat-soluble component of grains that was necessary for fertility in laboratory rats. It took almost 30 years to isolate this vitamin and to determine that it is also necessary for reproduction in humans. The chemical name for one form of vitamin E, **tocopherol**, is from the Greek *tos*, meaning childbirth, and *phero*, meaning to bring forth. Although vitamin E has been promoted to slow aging, cure infertility, reduce scarring, and protect against air pollution, research has not shown it to be useful for these purposes. Today we continue to explore the role of this antioxidant in protecting us from chronic disease.

tocopherol The chemical name for vitamin E.

SCIENCE APPLIED

New Scientific Knowledge Changes Dietary Reference Intakes

DRIs were first established for vitamin D in 1997. At that time, less was known about the role of vitamin D in human health than is today, and scientific data were insufficient to establish an EAR; instead, an AI was determined (see Section 2.2). Since 1997, there has been considerable research on the role of vitamin D, both in its role in the maintenance of bone health, and its potential in preventing cancer, diabetes, and enhancing immune function. As a result of this new scientific knowledge, many scientists came to believe that the 1997 AIs for vitamin D were too low and to reap the full benefits of vitamin D, people needed to consume considerably higher levels. In response, in 2007, the Institute of Medicine, the organization that first developed the DRIs, formed a new committee to review this new literature on vitamin D and to determine if new DRIs for the vitamin should be established.

The Institute of Medicine thoroughly reviewed the current scientific literature and came to the following conclusions:

- While observational studies do indicate a possible protective effect by vitamin D against cancer, the results of intervention trials are inconsistent; some studies show a beneficial effect and some do not. Since observational studies demonstrate association, not causation (see Section 1.5), the committee decided the cancer-related evidence was weak. They drew similar conclusions about the evidence on vitamin D's role in diabetes, immune function, and several other health-related conditions.
- On the other hand, there were substantial numbers of well-designed intervention trials on the role of vitamin D in bone health that could be used to establish an EAR.
- Based on the experiments conducted, adequate bone health was achieved when median blood levels of 40 nmol/L-50 nmol/L of vitamin D were maintained.[1]

Next, the committee used scientific data to determine what dietary intake is needed to ensure that vitamin D would be maintained in the blood at levels of 40 nmol/L in each life-stage group and this became the new EAR (10 μg/day for most life stages). Note that this number assumes that there is no exposure to sunlight. Once the EAR is established, an RDA (which is the EAR plus a "safety margin" of 2 standard deviations—see Section 2.2) is calculated and this represents an intake (15 μg for most life stages), in the absence of sunlight, that guarantees that requirements can be met by most of the population.[1] The figure compares the adult 1997 AI to the new 2010 RDA. Since both the AI and RDA represent personal intake targets that ensure a high probability of meeting requirements, comparing them is useful. As the bar graph indicates, the IOM concluded that higher intakes of vitamin D were required at each life stage. The IOM also reviewed a recent survey of the serum vitamin D levels of Canadians and concluded that almost 90% of adult Canadians had serum vitamin levels above the 40 nmol/L required to maintain bone health.[1]

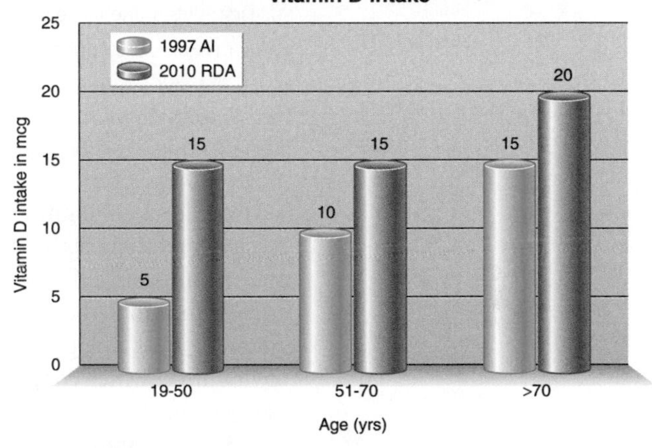

As is often the case in science, these new DRIs have generated controversy. A response by Canadian scientists to these new standards called them "a step in the right direction" but noted that several pieces of evidence, including that serum levels of up to 80 nmol/L improved calcium absorption, were not considered.[2] Many scientists believe that the committee was also being too conservative in dismissing the cancer-related data and other analyses that suggest that vitamin D blood levels of 75-100 nmol/L are required to optimize the benefits of the nutrient. EARs of at least 20-25 μg /day would be required to maintain such blood levels, higher than the new recommendations.[3]

So what happens next? Research into the role of Vitamin D in human health will continue and perhaps new knowledge will emerge that will again result in changes to recommendations. This continuing story illustrates the dynamic nature of science, the challenges of interpreting scientific data, and the relationship between research and recommendations.

[1] Food and Nutrition Board, Institute of Medicine. Dietary Reference Intakes for Calcium and Vitamin D. Washington DC: National Academies Press, 2011.

[2] Schwalfenberg, G. K. and Whiting, S. J. A Canadian response to the 2010 Institute of Medicine vitamin D and calcium guidelines. *Public Health Nutr.* 14(4):746-8, 2011.

[3] Bischoff-Ferrari, H. A., Giovannucci, E., Willett, W. C. et al. Estimation of optimal serum concentrations of 25-hydroxyvitamin D for multiple health outcomes. *Am. J. Clin. Nutr.* 84(1):18-28, 2006.

Vitamin E in the Diet

Eight naturally occurring forms of vitamin E are found in foods. These fall into two groups: the tocopherols and the tocotrienols. There are 4 tocopherol forms (alpha, beta, gamma, and delta) and similarly 4 tocotrienols forms (alpha, beta, gamma, and delta). Of the eight vitamin E forms, **alpha-tocopherol (α-tocopherol)** is found in the highest concentrations in human blood, bound to an α-tocopherol transfer protein, a liver protein that helps distribute vitamin E to body tissues, and is believed to be the biologically most important form of the vitamin. For this reason, natural or synthetic α-tocopherol is the form of vitamin E

alpha-tocopherol (α-tocopherol) The most common form of vitamin E in human blood.

isomers Molecules with the same molecular formula but a different arrangement of the atoms.

used in supplements. The synthetic form of α-tocopherol found in dietary supplements is composed of 8 different **isomers**—only half of these are active in the body. Therefore, synthetic α-tocopherol provides half of the biological activity of natural α-tocopherol; 10 mg of synthetic α-tocopherol provides the function of 5 mg of natural α-tocopherol. Vitamin E content can be expressed in mg, or older designations such as α-tocopherol equivalents (TE) or International Units (IUs). **Table 9.4** illustrates how to interconvert between IU and the synthetic and natural forms of vitamin E.

Table 9.4 Converting Vitamin E Units
To estimate the α-tocopherol intake from foods:
• If values are given as mg α-TEs: mg α-TE × 0.8 = mg α-tocopherol
• If values are given as IUs: First, determine if the source of the α-tocopherol is natural or synthetic.
• For natural α-tocopherol: IU of natural α-tocopherol × 0.67 = mg α-tocopherol
• For synthetic α-tocopherol (dl-α-tocopherol): IU of synthetic α-tocopherol × 0.45 = mg α-tocopherol

While most vitamin E research has focused on the biological effects of α-tocopherol, in recent years, scientists have become interested in the role that the other forms of vitamin E might play in human health. Most notably, researchers have looked at the effect of the tocotrienols[32] and gamma-tocopherol, a form of vitamin E widespread in the diet.[33] These other forms of vitamin E have biological functions that may differ from those of α-tocopherol. Dietary sources of vitamin E, which contain all forms of the vitamin, include nuts and peanuts; plant oils, such as soybean, corn, and sunflower oils; leafy green vegetables; and wheat germ (**Figure 9.13**).

Because vitamin E is sensitive to destruction by oxygen, metals, light, and heat, some is lost during food processing, cooking, and storage. Although it is relatively stable at normal cooking temperatures, the vitamin E in cooking oils may be destroyed if the oil is repeatedly heated to the high temperatures used for deep-fat frying.

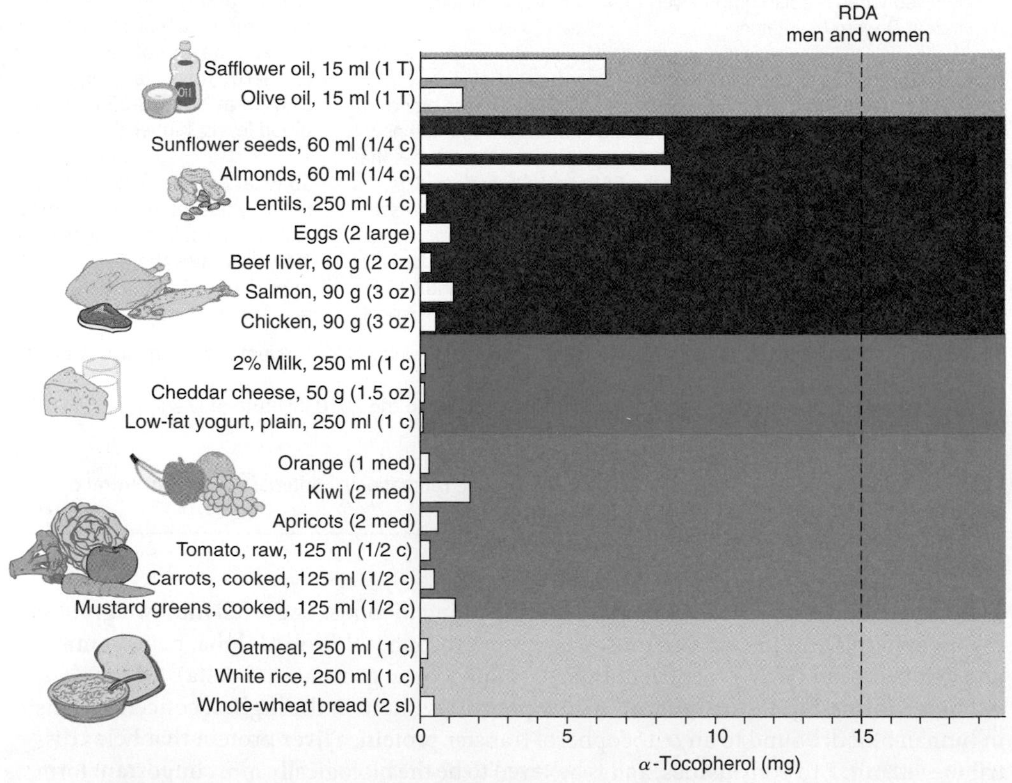

Figure 9.13 Vitamin E content of Canada's Food Guide food groups
Adults can obtain their RDA for vitamin E (dashed line) by consuming plant oils, nuts and seeds, and leafy green vegetables.

Vitamin E in the Body

Vitamin E absorption depends on normal fat absorption. Once absorbed, vitamin E is incorporated into chylomicrons. As chylomicrons are broken down, some vitamin E is distributed to other lipoproteins and delivered to tissues, but most is taken to the liver where, with the help of α-tocopherol transfer protein, the α-tocopherol form is incorporated into very-low-density lipoproteins (VLDLs).[34] The α-tocopherol in VLDLs is distributed to other plasma lipoproteins and delivered to cells.

Vitamin E functions primarily as a fat-soluble antioxidant. It neutralizes reactive oxygen compounds before they damage unsaturated fatty acids in cell membranes. After vitamin E is used to eliminate free radicals, its antioxidant function can be restored by vitamin C (**Figure 9.14**). Because polyunsaturated fats are particularly susceptible to oxidative damage, vitamin E needs increase as polyunsaturated fat intake increases.

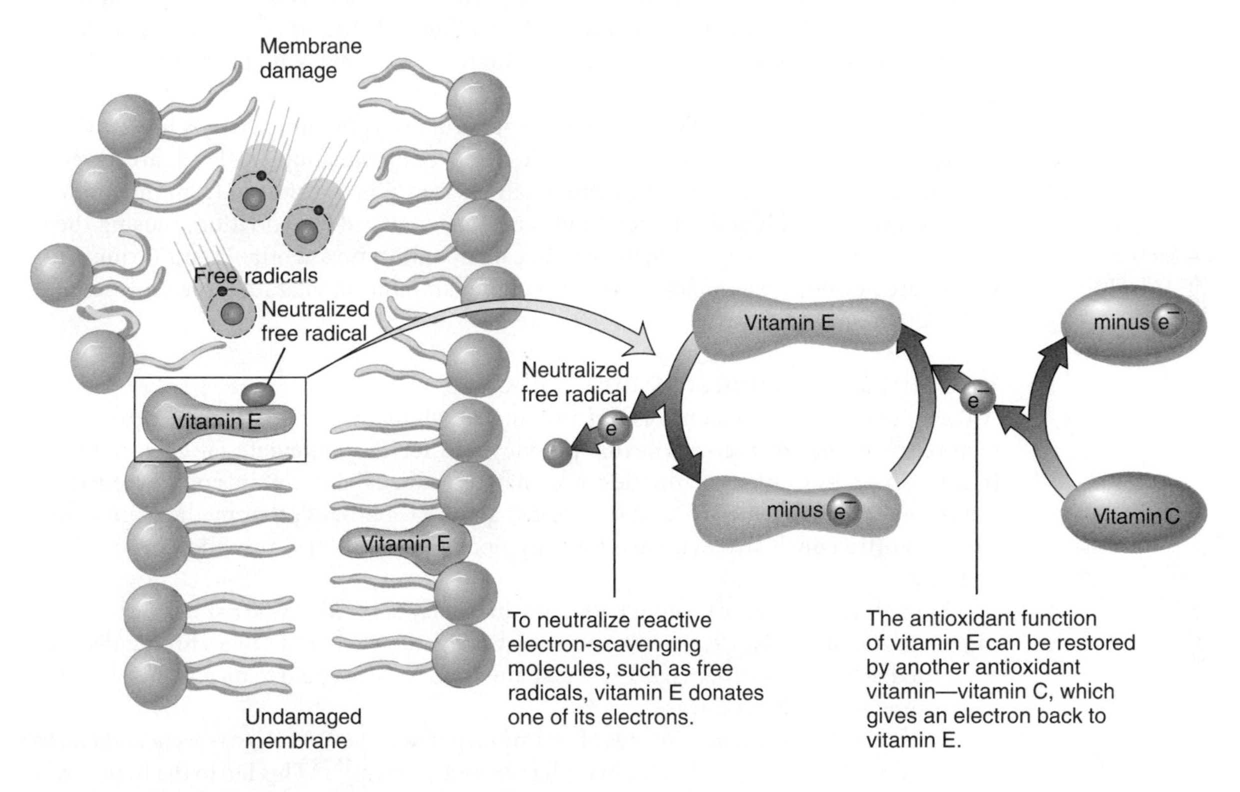

Figure 9.14 Antioxidant mechanism of vitamin E
By neutralizing free radicals, vitamin E guards not only cell membranes, as shown here, but also body proteins, DNA, and cholesterol. Vitamin E must be regenerated by vitamin C to restore it to the form that can act as an antioxidant.

By protecting cell membranes, vitamin E is important in maintaining the integrity of red blood cells, cells in nervous tissue, cells of the immune system, and lung cells, where it is particularly important because oxygen concentrations in those cells are high.[35] Vitamin E can also defend cells from damage by heavy metals, such as lead and mercury, and toxins, such as carbon tetrachloride, benzene, and a variety of drugs. It also protects against some environmental pollutants such as ozone. A number of vitamin E's roles are hypothesized to reduce the risk of heart disease (see Section 5.6). As an antioxidant, it helps protect low-density lipoprotein (LDL) cholesterol from oxidation. Studies done in the laboratory indicate that it also inhibits other events critical to the development of atherosclerosis, such as the formation of blood clots.[36]

Recommended Vitamin E Intake

Life Cycle

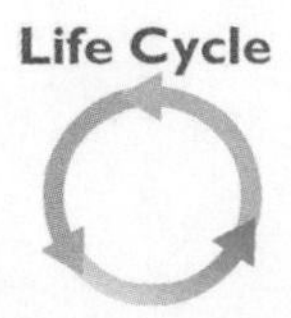

The recommendation for vitamin E intake is based on the amount needed to maintain plasma concentrations of α-tocopherol that protect red blood cell membranes from rupturing. The RDA for adult men and women is set at 15 mg/day of α-tocopherol.[36]

For infants, an AI for vitamin E has been set based on the amount consumed by infants fed principally with human milk. EARs and RDAs for children and adolescents have been estimated from adult values. The RDA for pregnancy is not increased above nonpregnant levels. To estimate the requirement for lactation, the amount secreted in human milk is added to the requirement for nonlactating women.[36]

Vitamin E Deficiency

Vitamin E protects membranes; therefore a deficiency can cause membrane changes. Nerve tissue and red blood cells are particularly susceptible. Vitamin E deficiency is usually characterized by neurological problems associated with nerve degeneration. Symptoms observed in humans include poor muscle coordination, weakness, and impaired vision. Because vitamin E is plentiful in the food supply and is stored in many of the body's tissues, vitamin E deficiency is rare, occurring only in those unable to absorb the vitamin due to fat malabsorption, those with inherited abnormalities in vitamin E metabolism, those with protein-energy malnutrition, and premature infants. For example, in individuals with cystic fibrosis, an inherited condition that reduces fat absorption, deficiency can develop rapidly, causing serious neurological problems, which, if untreated, can become permanent.

Life Cycle

All newborn infants have low blood tocopherol levels because there is little transfer of vitamin E from mother to fetus until the last weeks of pregnancy. The levels are lower in premature infants, who are born before much vitamin E is transferred from the mother. In these infants, red blood cell membranes may be damaged by oxidation, causing them to rupture. This results in a type of anemia called **hemolytic anemia**. Infant formula for premature newborns is supplemented with higher amounts of vitamin E than formula for full-term infants.

hemolytic anemia Anemia that results when red blood cells break open.

Vitamin E Supplements and Toxicity

Although vitamin E deficiency is uncommon, supplements are promoted to grow hair; restore, maintain, or increase sexual potency and fertility; alleviate fatigue; maintain immune function; enhance athletic performance; reduce the symptoms of premenstrual syndrome (PMS) and menopause; slow aging; and treat a host of other medical problems. There is little conclusive evidence that supplemental vitamin E provides any of these benefits.

The antioxidant role of vitamin E suggests that it may help reduce the risk of heart disease, cancer, Alzheimer's disease, macular degeneration, and a variety of other chronic diseases associated with oxidative damage. Particular attention has been paid to its potential benefits in guarding against heart disease.

In observational studies, intakes of vitamin E greater than 100 IU/day were associated with a reduced risk of heart disease in both men and women.[37,38] This led to the hypothesis that the risk of heart disease could be reduced by increasing vitamin E intake with supplements. Although studies on the biological effects of vitamin E supplements indicate that they decrease LDL oxidation, decrease platelet stickiness, and have an anti-inflammatory effect[39] and individuals with the highest vitamin E levels in their blood have been found to have the lowest risk of death,[40] intervention trials examining the effects of vitamin E supplements have not found them to be beneficial in preventing heart disease or other chronic diseases, and some studies have shown a slight increase in the risk of death.[40,41] Therefore, there is not sufficient evidence to recommend supplemental vitamin E to protect against heart disease.

Like other antioxidants, supplements of vitamin E have been suggested to prevent oxidative damage that could lead to cancer development. Vitamin E supplements have also been hypothesized to prevent cancer by boosting immune function and preventing the formation of carcinogenic nitrosamines in the digestive tract. Currently, the evidence supporting a benefit of vitamin E supplements for cancer prevention is inconsistent and limited, so supplements are not recommended for this purpose.

Vitamin E is relatively nontoxic. There is no evidence of adverse effects from consuming large amounts from food. The UL is 1,000/day from supplemental sources. Vitamin E supplements should not be taken by individuals taking blood-thinning medications because it reduces blood clotting and interferes with the action of vitamin K (see Label Literacy: Think Before You Supplement).

9.5 Vitamin K

Learning Objectives

- Describe how vitamin K is involved in blood clotting.
- Explain why newborns and people taking antibiotics are at risk for vitamin K deficiency.

Vitamin K is one of the few vitamins about which extravagant claims are not made. Like the other fat-soluble vitamins, it was discovered inadvertently by feeding animals a fat-free diet. In this case, researchers in Denmark noted that chicks fed this diet developed a bleeding disorder that was cured by feeding them a fat-soluble extract from green plants. Vitamin K was named for *koagulation*, the Danish word for **coagulation**, or blood clotting.

coagulation The process of blood clotting.

Vitamin K in the Diet

As with other fat-soluble vitamins, vitamin K is found in several forms. **Phylloquinone** is the form found in plants and the primary form in the diet. A group of vitamin K compounds, called menaquinones, are found in fish oils and meats and are synthesized by bacteria, including those in the human intestine. **Menaquinones** are the form found in supplements. Dietary deficiencies of vitamin K are very rare.[3] The best dietary sources are liver and leafy green vegetables such as spinach, broccoli, brussels sprouts, kale, and turnip greens. These leafy greens provide about half of the vitamin K in a typical North American diet.[42] Some vegetable oils are also good sources (**Figure 9.15**). Some of the vitamin K produced by bacteria in the human gastrointestinal tract is also absorbed. Vitamin K is destroyed by exposure to light and low- or high-acid conditions.

phylloquinone The form of vitamin K found in plants.

menaquinones The forms of vitamin K synthesized by bacteria and found in animals.

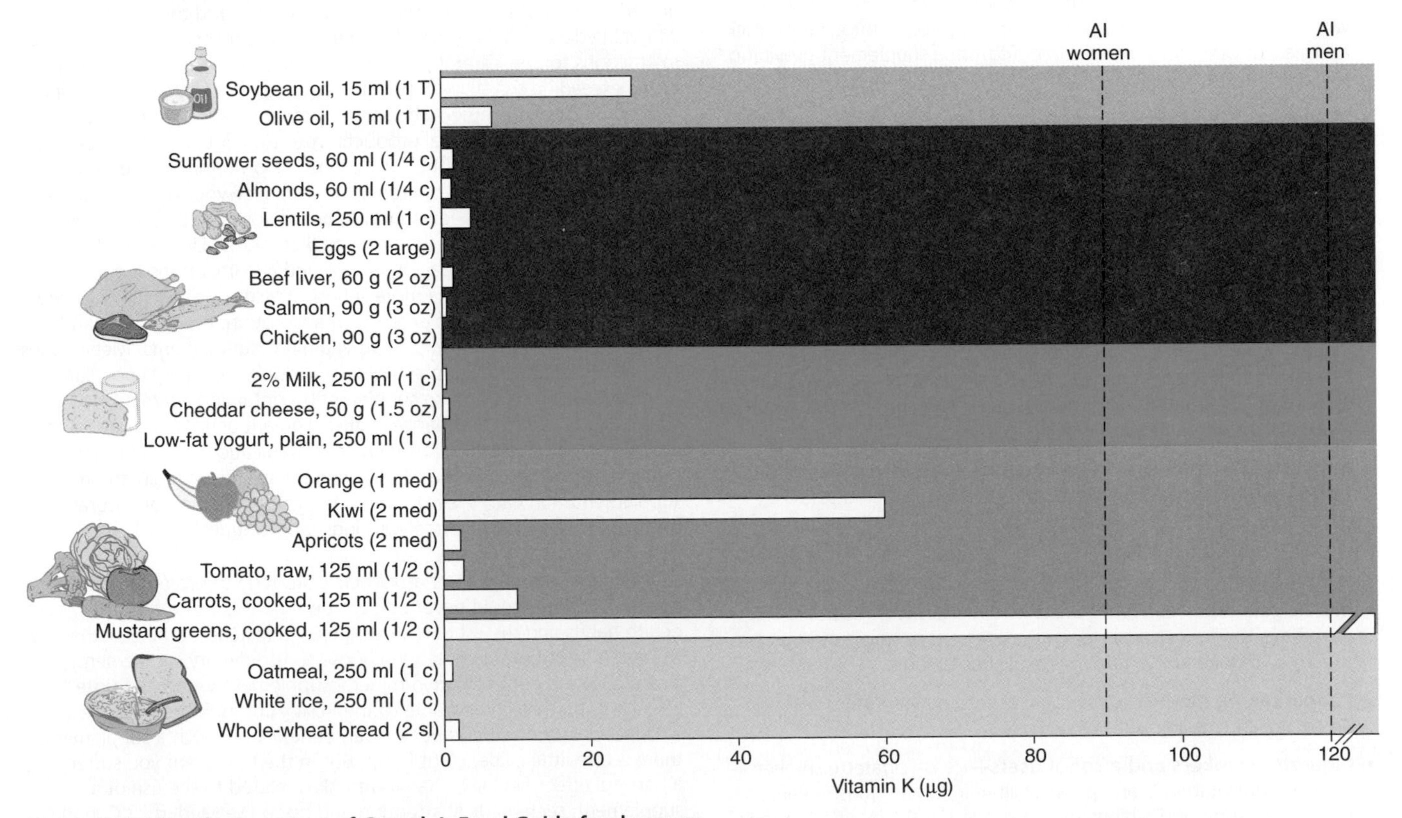

Figure 9.15 Vitamin K content of Canada's Food Guide food groups
The best sources of vitamin K are leafy green vegetables and some plant oils; the dashed lines represent the AIs for adult men and women.

Vitamin K in the Body

Vitamin K is a coenzyme needed for the production of the blood-clotting protein **prothrombin** and other specific blood-clotting factors. Blood-clotting factors are proteins that circulate in the blood in an inactive form. When activated, they lead to the formation of fibrin, the protein

prothrombin A blood protein required for blood clotting.

LABEL LITERACY

Think Before You Supplement

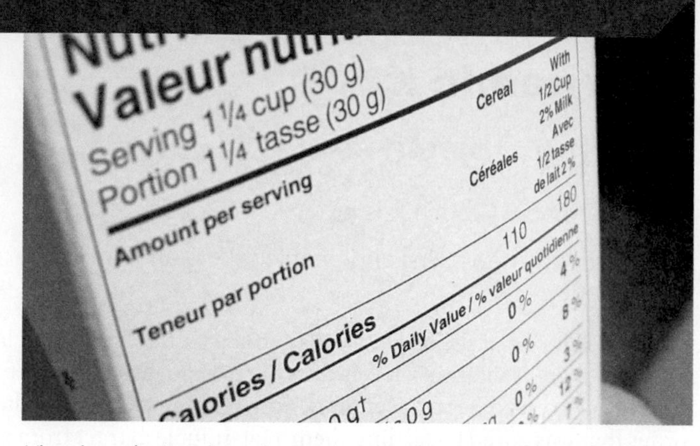

About 40% of Canadian adults use vitamin and mineral supplements.[1]

People take them to energize themselves, to protect themselves from disease, to cure their illnesses, to enhance what they get in food, and simply to ensure against deficiencies. You can purchase almost any nutrient as an individual supplement or you can choose from a surfeit of combinations of nutrients, herbs, and other components. Some claim to be "all natural;" others entice you with terms like "mega," "advanced formula," "high potency," and "ultra." Do you need a supplement? How can you decide which is best?

Eating a variety of foods is the best way to meet nutrient needs, and most healthy adults who consume a reasonably good diet do not need supplements. For some people, however, supplements may be the only way to meet needs because they have low intakes, increased needs, or excess losses. Groups for whom vitamin and mineral supplements are typically recommended include the following:[2]

- **Dieters**—People who consume fewer than 1,600 kcal/day should take a multivitamin-mineral supplement.[2]
- **Vegans and those who eliminate all dairy foods**—To obtain adequate vitamin B_{12}, those who do not eat animal products need to take supplements or consume B_{12}-fortified foods. Because dairy products are an important source of calcium and vitamin D, those who do not consume dairy products due to lactose intolerance, milk allergies, or other reasons may benefit from a supplement providing calcium and vitamin D.
- **Infants and children**—Supplemental fluoride, vitamin D, and iron are recommended under certain circumstances.
- **Young women and pregnant women**—Women of child-bearing age should consume 400 μg of folic acid daily from either fortified foods or supplements (see Sections 8.8 and 14.2). Supplements of iron and folic acid are recommended for pregnant women, and multivitamin and mineral supplements are usually prescribed.
- **Older adults—**Because of the high incidence of atrophic gastritis in adults over 50, vitamin B_{12} supplements or fortified foods are recommended (see Section 8.9). Meeting the AI for vitamin D and calcium may also be difficult for older adults so supplements of these nutrients are often recommended.
- **Individuals with dark skin pigmentation or who cover their bodies when outdoors**—Those with dark skin or who totally cover their bodies when outdoors may not be able to synthesize enough vitamin D to meet needs and may therefore benefit from supplementation.[2]
- **Individuals with restricted diets**—Individuals with health conditions that affect what foods they can eat or how nutrients are used may require vitamin and mineral supplements.
- **People taking medications**—Medications may interfere with the body's use of certain nutrients.
- **Cigarette smokers and alcohol users**—Heavy cigarette smokers require more vitamin C and possibly vitamin E than nonsmokers.[3,4] Alcohol consumption inhibits the absorption of B vitamins and may interfere with their metabolism.

Despite the benefits of supplements to some individuals, they can also carry risks. If you choose to take a vitamin or mineral supplement, whether to ensure an adequate nutrient intake, prevent disease, or optimize health, use some common sense along with the information on the supplement label to assure that the supplement provides the nutrients you need to satisfy your individual concerns, but does not contain ingredients or amounts that could cause adverse effects.

Vitamin and mineral supplements are considered natural health products in Canada and must be labelled in accordance with natural health product regulations (see Section 2.5). The information included on supplement labels can be used to determine if a supplement includes the nutrients you want and whether they are present in appropriate amounts. If you want to increase your calcium intake, check to see if it provides the amount of calcium that you want. If you have low iron stores, check to see if it provides iron. If you are over 50, check to see if it provides enough vitamin D to meet your needs (see figure). Special supplement formulas for men, seniors, and women are available, but may not necessarily provide what you need even if you fit into the group they target. Remember that natural health products do not contain a nutrition facts table so you will have to look up the DRIs and compare it to the stated dose on the bottle.

Health Canada regulates the minimum and maximum dose of vitamins and minerals that are permitted in a supplement and the maximum allowable dose is usually the UL,[5] so it is important to follow the instructions for use carefully; taking more than instructed may result in doses exceeding the UL with possible adverse effects. If you are taking more than one supplement, be sure to compare labels. If the same nutrient is present in several products, your total intake may exceed the UL. One nutrient to be particularly careful of is vitamin A; too much increases the risk of bone fractures.[6] To minimize your risk, do not take supplements of preformed vitamin A, and if you take a multivitamin, look for one that contains vitamin A as β-carotene. Even when ULs are not exceeded, individual conditions and circumstances require consideration. Iron can be a more serious toxicity concern in those with an abnormality that causes excess iron absorption. People who tend to develop kidney stones should avoid vitamin C supplements. Medications may also interact with supplements. Individuals taking anticoagulant medications should not take supplements containing vitamin E or vitamin K, as vitamin E enhances anti-coagulant activity and vitamin K inhibits it. Individuals who routinely take medications should discuss nutrient–drug interactions and the need for specific vitamin and mineral supplementation with their doctor or pharmacist. Labels of natural health products often include warnings and cautions.

Supplements can be part of an effective strategy to promote good health, but they should never be considered a substitute for other good health habits and they should never be used instead of medical therapy to treat a health problem. If you choose to use dietary supplements, a safe choice is a multivitamin/mineral supplement where the dose of each nutrient is comparable to the RDA. Although there is little evidence that the average person can benefit from such a supplement, there is also little evidence of harm. But in the rare event you suffer a harmful effect or illness that you think is related to the use of a supplement, seek medical attention and go to the MedEffect Canada website to report an adverse reaction.

Considering Supplements

Before you choose to take a dietary supplement, consider the following:

- **Why do you want a supplement?** If you are taking it for insurance, does it provide both vitamins and minerals? If you want to supplement specific nutrients, are they contained in the product?

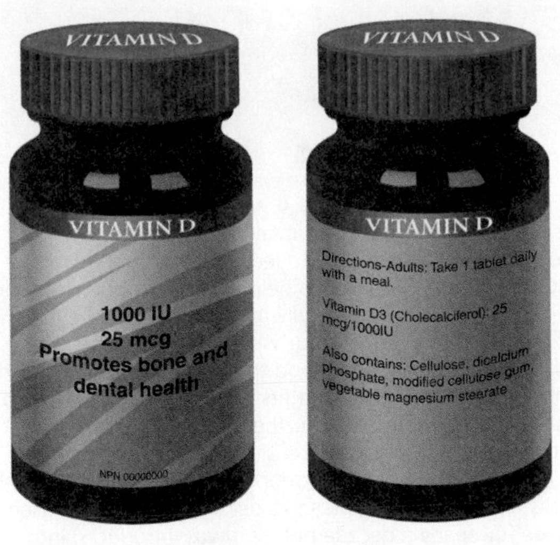

If taken as directed the total vitamin D intake from this supplement is 25 mcg or 25% of the UL which is 100 mcg.

- **Does it have an NPN?** An NPN (natural product number) indicates that the product has been licensed for use by Health Canada and a review of its safety and effectiveness has been made.
- **Does it contain potentially toxic levels of any nutrient?** Compare the dose to the RDA and UL.
- **Does it contain any nonvitamin/nonmineral ingredients?** If so, have you ever had an adverse reactions to any of them?
- **Do you have a medical condition that recommends against certain nutrients or other ingredients?** Are you a smoker? Smoking increases the need for vitamin C, but also increases the risks associated with taking β-carotene.
- **Are you taking prescription medication with which an ingredient in the supplement may interact?** Check with your physician, dietitian, or pharmacist to help identify these interactions.
- **How much does it cost?** Compare product costs and ingredients before you buy. Just as more isn't always better, more expensive is not always better, either.
- **What is the expiration date?** Some nutrients degrade over time so expired products will have less than is on the label.

[1] Guo, X., Willows, N., Kuhle, S. et al. Use of vitamin and mineral supplements among Canadian adults. *Can. J. Public Health* 100(5):357-60, 2009.

[2] American Dietetic Association. Position of the American Dietetic Association: Fortification and nutritional supplements. *J. Am. Diet. Assoc.* 105:1300–1311, 2005.

[3] Food and Nutrition Board, Institute of Medicine. *Dietary Reference Intakes for Vitamin C, Vitamin E, Selenium, and Carotenoids*. Washington, DC: National Academies Press, 2000.

[4] Bruno, R. S., and Traber, M. G. Cigarette smoke alters human vitamin E requirements. *J. Nutr.* 135:671–674, 2005.

[5] Health Canada. Multi vitamin/mineral supplements. Available online at http://www.hc-sc.gc.ca/dhp-mps/prodnatur/applications/licen-prod/monograph/multi_vitmin_suppl-eng.php#app2. Accessed July 31, 2011.

[6] Michaelsson, K., Lithell, H., Vessby, B. et al. Serum retinol levels and the risk of fracture. *N. Engl. J. Med.* 348:287–294, 2003.

---

that forms the structure of a blood clot (**Figure 9.16**). Injuries, as well as the normal wear and tear of daily living, produce small tears in blood vessels. To prevent blood loss, these tears must be repaired with blood clots.

There are other roles for vitamin K in the body but they are less well understood. For example, there are several vitamin-K-dependent proteins in bone that may be involved in bone mineralization and demineralization. Several studies have demonstrated that low-dietary vitamin K intake is associated with low bone-mineral density or increased fractures.[43] Vitamin-K-dependent proteins in blood vessels may be involved in preventing calcification of blood vessels and loss of elasticity.[44]

Recommended Vitamin K Intake

Unlike other fat-soluble vitamins, the body uses vitamin K rapidly, so a constant supply is necessary. The vitamin K produced by bacteria in the gastrointestinal tract contributes to

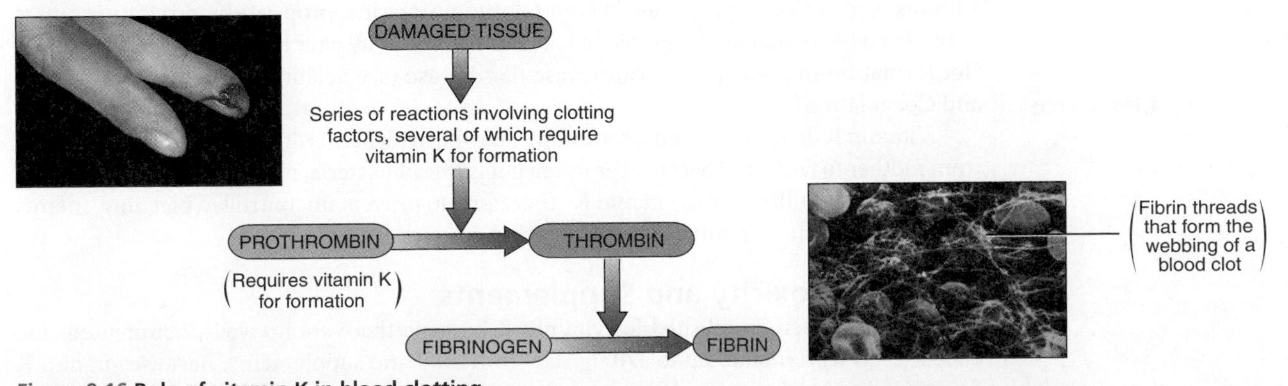

Figure 9.16 Role of vitamin K in blood clotting
Several clotting factors, including prothrombin, require vitamin K for synthesis, These are needed for fibrinogen to be converted into its active form, fibrin, which is a structural component of blood clots. (Amethyst/Custom Medical Stock Photo, Inc.; Science Photo Library/Custom Medical Stock Photo, Inc.)

SCIENCE APPLIED
Cows, Clover, and Coagulation

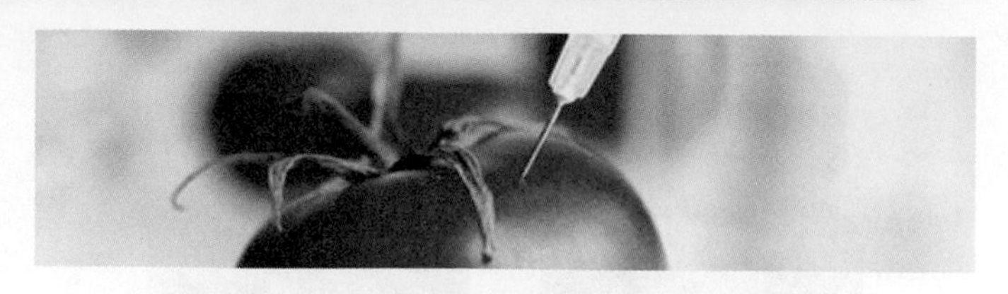

On a snowy night in 1933, a disgruntled farmer delivered a bale of moldy clover hay, a pail of unclotted blood, and a dead cow to the laboratory of Dr. Carl Link at the University of Wisconsin. Link, the university's first professor of biochemistry, was already making a name for himself in the scientific community when he was presented with the farmer's challenge to speed the typically deliberate pace of research. Why were the farmer's cows dying? What could Link do to stop the bleeding disease that was killing them? Link had just begun to research hemorrhagic sweet clover disease, the condition that was killing the cows, but at the time, the only advice he could offer was to find alternative feed and try blood transfusions to save the other animals. Link and his colleagues began a line of inquiry that ultimately led to the development of an anti-blood-clotting factor that today saves the lives of hundreds of thousands of people.

In the 1930s, hemorrhagic sweet clover disease was killing cows across the midwestern prairies of the United States and Canada. It occurred in cattle that were fed moldy, spoiled, sweet clover hay. These animals died because their blood did not clot. Even a minor scratch from a barbed wire fence could be fatal; once bleeding began, it did not stop. Six years after the farmer's challenge, Link and colleagues had isolated the anticoagulant dicumarol from moldy clover. Dicumarol is a derivative of coumarin, which gives clover its sweet scent; mold converts coumarin to dicumarol. Cows fed moldy clover consume dicumarol, which interferes with vitamin K activity and consequently prevents normal blood clotting.

The discovery of dicumarol enhanced our understanding of the blood-clotting mechanism and led to the development of anticoagulant drugs. These drugs help eliminate blood clots and prevent their formation. In carefully regulated doses, they are used to treat heart attacks, which occur when blood clots block one or more of the vessels supplying blood to the heart muscle. Dicumarol, first synthesized in 1940, was the first anticoagulant that could be administered orally to humans. Further work with dicumarol led Link to propose the use of a more potent derivative, called warfarin, as rat poison. When rats consume the odorless, colourless warfarin, their blood fails to clot, and they bleed to death. Warfarin was used as a rodenticide for nearly a decade before it was introduced into clinical medicine in 1954. Sodium warfarin, also known by the brand name Coumadin, soon became the most widely prescribed anticoagulant drug.

Sodium warfarin and dicumarol have been administered to millions of patients to prevent blood clots, which can cause heart attacks and strokes. People taking such medication must consume similar amounts of vitamin K daily to stabilize the activity of their medication. It is ironic that a substance that killed hundreds of cattle across the Great Plains and that efficiently kills rodents in our homes has saved so many human lives and does so by interfering with the activity of a vitamin.

---

needs, but this is not well absorbed and alone is not enough to meet needs. An AI for dietary vitamin K has been set at about 120 μg/day for men and 90 μg/day for women (see Table 9.1). The AI is not increased for pregnancy or lactation. An AI for infants was set based on the amount typically consumed in breast milk.

Vitamin K Deficiency

Abnormal blood coagulation is the major symptom of vitamin K deficiency. When vitamin K is deficient, the blood does not clot to seal ruptured arteries or veins, and blood loss goes unchecked. If the deficiency is severe, it can eventually cause death from blood loss. A deficiency is very rare in the healthy adult population, but it may result from fat malabsorption syndromes or the long-term use of antibiotics, which kill the bacteria in the gastrointestinal tract that are a source of the vitamin. In combination with an illness that reduces the dietary intake of vitamin K, this may precipitate a deficiency. Injections of vitamin K are typically administered before surgery to aid in blood clotting. Since inappropriate blood clotting causes strokes and heart attacks, drugs that block vitamin K activity have been used to reduce blood clot formation in patients with cardiovascular disease (see Science Applied: Cows, Clover, and Coagulation).

Life Cycle

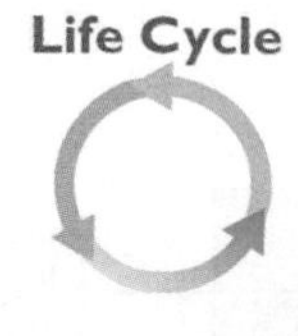

Vitamin K deficiency is most common in newborns. There is little transfer of this vitamin from mother to fetus, and because the infant gut is free of bacteria, no vitamin K is made there. Further, breast milk is low in vitamin K. Therefore, to prevent uncontrolled bleeding, infants are typically given a vitamin K injection within 6 hours of birth.

Vitamin K Toxicity and Supplements

A UL has not been established for vitamin K because there are no well-documented side effects, even with intakes up to 370 μg/day from food and supplements. Because vitamin K functions in blood clotting, high doses can interfere with anticoagulant drugs. Therefore, individuals prescribed these medications should consult with their physicians before taking supplements containing vitamin K.

CASE STUDY OUTCOME

Eva is now a happy, healthy 2-year-old. You would never know that a year ago, she had been diagnosed with the vitamin D deficiency disease rickets. She developed rickets because her diet was low in vitamin D and she was not exposed to enough sunlight to synthesize adequate amounts of the vitamin to meet her needs. Without adequate vitamin D, she was unable to absorb sufficient calcium to support bone growth and health. After diagnosing Eva with rickets, her pediatrician prescribed oral supplements of both vitamin D and calcium and recommended that Eva consume a calcium-rich diet. After only a week of treatment, Eva's laboratory values had improved, and X-rays showed that her bones were healing. After a few months, the bony protrusions on her ribs had disappeared, the slight bowing in her legs was corrected, and new teeth began to emerge. If Eva had not been treated when she was young and still growing, her skeletal deformities would have been permanent. Her parents now make sure her diet contains lots of sources of calcium and vitamin D, including vitamin D from traditional Inuit sources like seal oil and fatty fish, such as salmon.

Furthermore, Eva's mother is now pregnant with Eva's sibling. To ensure that her new baby has a healthy start, she is taking vitamin supplements that contain vitamin D during her pregnancy, and will give Eva's new brother or sister a vitamin D supplement while breastfeeding.

APPLICATIONS

Personal Nutrition

iProfile

1. **Do you get enough vitamin A?**
   a. Use iProfile and your food intake record from Chapter 2 to calculate your average daily intake of vitamin A.
   b. How does your vitamin A intake compare to the RDA for someone of your age and sex?
   c. What are 3 major food sources of vitamin A in your diet?
   d. Do the major food sources of vitamin A in your diet contain preformed vitamin A or carotenoids?
2. **How much vitamin E is in your diet?**
   a. Use iProfile and the food intake record you kept in Chapter 2 to calculate your average intake of vitamin E.
   b. How does your intake of vitamin E compare with the RDA?
   c. If your diet does not meet the RDA, suggest modifications that will add enough vitamin E to your diet to meet the RDA without increasing your energy intake.

General Nutrition Issues

1. **Who is at risk of vitamin D deficiency in your town?**
   a. Assume you have a friend or relative living in a nursing home. Find out about how much time nursing-home residents spend

outdoors without sunscreen or clothing covering most of their skin. Do you think they are at risk for vitamin D deficiency? Why or why not?

b. At higher northern and southern latitudes, little vitamin D is synthesized in the winter months. Is your community located at a latitude greater than 40° north or south?

c. How much time do the school children at your local elementary school spend outdoors during recess? Do you think they are at risk for vitamin D deficiency? Why or why not?

d. Do a survey to determine the percentage of people who apply sunscreen daily. Why might sunscreen affect vitamin D status?

2. **What supplements are your friends taking? Are they safe?**

a. Do a supplement survey of 10 people. Record all of the vitamin and mineral supplements they take including the dose and the number of doses taken per day as well as the reason they chose to take each supplement.

b. Tabulate the total amount of each vitamin and mineral each person takes.

c. Are any of the survey subjects consuming nutrients in excess of recommendations (RDA or AI)?

d. Are any nutrients consumed in excess of the UL?

e. Do you think these supplements will fulfill the expectations of the consumers? Why or why not?

SUMMARY

9.1 Fat-Soluble Vitamins in the Canadian Diet

- Vitamins A, D, E, and K are soluble in fat, which affects how they are absorbed, transported, stored, and excreted. Deficiencies of vitamins A and D are common in the developing world. In Canada, the risk of deficiencies of fat-soluble vitamins is increasing due to low intakes of fruits and vegetables and limited sun exposure.

9.2 Vitamin A

- Vitamin A is found both preformed as retinoids and in precursor forms called carotenoids. The major food sources of preformed vitamin A include liver, eggs, fish, and milk and alternatives. Carotenoids are found in plant foods such as yellow-, orange-, and red-coloured vegetables and fruit, and leafy green vegetables. Some carotenoids are precursors of vitamin A. The most potent is β-carotene.
- In the body, the retinoids, which include retinol, retinal, and retinoic acid, are needed for vision and for the growth and differentiation of cells. Retinol is transported in the blood and can be converted into retinal or retinoic acid. Retinal binds to opsin in the eye to form rhodopsin. When light strikes rhodopsin to begin the visual cycle, it causes a nerve impulse to be sent to the brain so that light is perceived. Retinoic acid affects cell differentiation by altering gene expression. It is needed for healthy epithelial tissue and normal reproduction and immune function. Beta-carotene functions as an antioxidant, a role that is independent of its conversion to vitamin A.
- Vitamin A deficiency is a world health problem that increases the frequency of infectious disease and causes blindness and death in millions of children.
- Preformed vitamin A can be toxic at doses as low as 10 times the RDA and can increase the risk of bone fractures and birth defects. Carotenoids are not toxic, but a high intake can give the skin, especially the palms of the hands and soles of feet, an orange appearance.

9.3 Vitamin D

- Vitamin D can be made in the skin by exposure to sunlight, so dietary needs vary depending on the amount synthesized. Vitamin D is found in fish oils and fortified milk and soy milk.
- Dietary vitamin D as well as vitamin D synthesized in the skin must be modified by the liver and then the kidney to form active vitamin D. Active vitamin D promotes calcium and phosphorus absorption from the intestines and acts with parathyroid hormone to cause the release of calcium from bone and calcium retention by the kidney. These roles are essential for maintaining proper levels of calcium and phosphorus in the body. Adequate vitamin D may also protect against autoimmune diseases and cancer.
- Vitamin D deficiency in children results in a condition called rickets; in adults, vitamin D deficiency causes osteomalacia.
- In 2010, new dietary reference intakes were established for vitamin D to replace the original values first developed in 1997. Debate continues, however, about the appropriate intake to optimize health.

9.4 Vitamin E

- Vitamin E is found in nuts, plant oils, green vegetables, and fortified cereals.
- Vitamin E functions primarily as a fat-soluble antioxidant. It is necessary for reproduction and protects cell membranes from oxidative damage. It also has anti-coagulant properties.
- Vitamin E deficiency can cause hemolytic anemia and neurological problems.
- Many people take vitamin E supplements. There is little risk of toxicity, but there is also little documented evidence of any benefit from supplements.

9.5 Vitamin K

- Vitamin K is found in plants and is synthesized by bacteria in the gastrointestinal tract.
- Vitamin K is a coenzyme essential for the formation of blood clotting factors as well as proteins needed for normal bone mineralization.
- Deficiency causes bleeding and low bone density. Since vitamin K deficiency is a problem in newborns, they are routinely given vitamin K injections at birth.

REVIEW QUESTIONS

1. List two food sources of preformed vitamin A and two of provitamin A.
2. How is vitamin A involved in the perception of light?
3. How does vitamin A affect the proteins made by a cell?
4. Why does a deficiency of vitamin A cause night blindness? Dry eyes?
5. What is β-carotene?
6. Explain why β-carotene is not toxic, but preformed vitamin A is.

7. Why is vitamin D called the "sunshine vitamin?"
8. Name two sources of vitamin D in the diet.
9. What is the primary function of vitamin D?
10. Describe the symptoms of vitamin D deficiency in children.
11. Explain how vitamin D's effect on gene expression alters calcium absorption.
12. What is the function of vitamin E?
13. Name two sources of vitamin E in the diet.
14. What is the main function of vitamin K?
15. What are the symptoms of vitamin K deficiency?

REFERENCES

1. Ward, L. M., Gaboury, I., Ladhani, M. et al. Vitamin D–deficiency rickets among children in Canada. *CMAJ* 177(2):161–166, 2007.
2. Holden, J. M., Eldridge, A. L., Beecher, G. R. et al. Carotenoid content of U.S. foods: An update of the database. *J. Food Composition and Analysis* 12:169–196, 1999.
3. Food and Nutrition Board, Institute of Medicine. *Dietary Reference Intakes: Vitamin A, Vitamin K, Arsenic, Boron, Chromium, Copper, Iodine, Iron, Manganese, Molybdenum, Nickel, Silicon, Vanadium, and Zinc*. Washington, DC: National Academies Press, 2001.
4. Ross, A. C. Vitamin A and carotenoids. In *Modern Nutrition in Health and Disease*, 10th ed. M. E. Shils, M. Shike, A. C. Ross et. al. eds. Philadelphia: Lippincott Williams & Wilkins, 2006, pp. 351–375.
5. Wintergerst, E. S., Maggini, S., and Hornig, D. H. Contribution of selected vitamins and trace elements to immune function. *Ann. Nutr. Metab.* 51:301–323, 2007.
6. Clagett-Dame, M., and DeLuca, H. F. The role of vitamin A in mammalian reproduction and embryonic development. *Annu. Rev. Nutr.* 22:347–381, 2002.
7. World Health Organization. *Micronutrient Deficiencies. Vitamin A Deficiency: The Challenge.* Available online at www.who.int/nutrition/topics/vad/en/index.html. Accessed April 23, 2009.
8. Health Canada. Do Canadian adults meet their nutrient requirements through food intake alone. 2009. Available online at http://www.hc-sc.gc.ca/fn-an/surveill/nutrition/commun/art-nutr-adult-eng.php. Accessed July 18, 2010.
9. Garriguet, D. Canadians' eating habits. *Health Reports* 18(2):17–32, 2007.
10. McCaffery, P. J., Adams, J., Maden, M., et al. Too much of a good thing: Retinoic acid as an endogenous regulator of neural differentiation and exogenous teratogen. *Eur. J. Neurosci.* 18:457–472, 2003.
11. Michaelsson, K., Lithell, H., Vessby, B., et al. Serum retinol levels and the risk of fracture. *N. Engl. J. Med*. 348:287–294, 2003.
12. Feskanich, D., Singh, V., Willett, W. C., et al. Vitamin A intake and hip fractures among postmenopausal women. *JAMA*. 287:47–54, 2002.
13. Cooper, D. A., Eldridge, A. L., and Peters, J. C. Dietary carotenoids and lung cancer: A review of recent research. *Nutr. Rev.* 57:133–145, 1999.
14. Pryor, W. A., Stahl, W., and Rock, C. L. Beta-carotene: From biochemistry to clinical trials. *Nutr. Rev*. 58:39–53, 2000.
15. Holick, M. F. Resurrection of vitamin D deficiency and rickets. *J. Clin. Invest.* 116:2062–2072, 2006.
16. DeLuca, H. F. Overview of the general physiologic features and functions of vitamin D. *Am. J. Clin. Nutr.* 80(Suppl):1689S–1696S, 2004
17. Nagpal, S., Na, S., and Rathnachalam, R. Noncalcemic actions of vitamin D receptor ligands. *Endocr. Rev.* 26:62–68, 2005.
18. Garland, C. F., Garland, F. C., Gorham, E. D. et al. The role of vitamin D in cancer prevention. *Am. J. Public Health* 96:252–261, 2006.
19. Holick, M. F., and Chen, T. C. Vitamin D deficiency: A worldwide problem with health consequences. *Am. J. Clin. Nutr.* 87(Suppl.): 1080S–1086S, 2008.
20. Food and Nutrition Board, Institute of Medicine. Dietary Reference Intakes for Calcium and Vitamin D. Washington DC: National Academies Press, 2011.
21. Holick, M. F. Vitamin D deficiency: What a pain it is. *Mayo Clin. Proc.* 78(12):1457–1459, 2003.
22. Holick, M. F. Vitamin D: Importance in the prevention of cancers, type 1 diabetes, heart disease, and osteoporosis. *Am. J. Clin. Nutr.* 79:362–371, 2004.
23. Holick, M. F. Sunlight and vitamin D for bone health and prevention of autoimmune diseases, cancers, and cardiovascular disease. *Am. J. Clin. Nutr.* 80(Suppl):1678S–1688S, 2004.
24. Weisberg, P., Scanlon, K. S., Li, R. et al. Nutritional rickets among children in the United States: Review of cases reported between 1986 and 2003. *Am. J. Clin. Nutr.*. 80(Suppl): 1697S–1705S, 2004.
25. Mark, S., Gray-Donald, K., Delvin, E. E., et al..Low vitamin D status in a representative sample of youth from Québec, Canada. *Clin Chem*. 54(8):1283–9, 2008.
26. Gozdzik, A., Bartz, J. L., Wu, H. et al. Low wintertime vitamin D levels in a sample of healthy young adults of diverse ancestry living in the Toronto area: associations with vitamin D intake and skin pigmentation. *BMC Public Health* 8:336–344, 2008.
27. Langlois, K., Greene-Finestone, L., Little, J. et al. Vitamin D status of Canadians as measured in the 2007 to 2009 Canadian Health Measures Survey. *Health Rep*. 2010 (1):47–55, 2010.
28. Health Canada. Vitamin D: Recommendations and Health Status. Available online at http://www.hc-sc.gc.ca/fn-an/nutrition/vitamin/vita-d-eng.php. Accessed July 18, 2010.
29. Canadian Pediatric Society. Pregnancy and babies. Vitamin D. Available online at http://www.caringforkids.cps.ca/pregnancybabies/VitaminD.htm. Accessed July 18, 2010.
30. Blank, S., Scanlon, K. S., Sinks, T. H. et al. An outbreak of hypervitaminosis D associated with the overfortification of milk from a home-delivery dairy. *Am. J. Public Health* 85:656–659, 1995.
31. Vieth, R. Vitamin D toxicity, policy, and science. *J. Bone Miner. Res.* 22(Suppl. 2):V64–V68, 2007.
32. Sen, C. K. , Khanna, S., Rink, C. et al. Tocotrienols: the emerging face of natural vitamin E. *Vitam Horm*. 76:203–61, 2007.
33. Jiang, Q., Christen, S., Shigenaga, M. K. et al. gamma-tocopherol, the major form of vitamin E in the U.S. diet, deserves more attention. *Am. J. Clin. Nutr.* 74(6):714–22, 2001.
34. Traber, M. G., Burton, G. W., and Hamilton, R. L. Vitamin E trafficking. *Ann. N.Y. Acad. Sci*. 1031:1–12, 2004.
35. Sabat, R., Guthmann, F., and Rüstow, B. Formation of reactive oxygen species in lung alveolar cells: Effect of vitamin E deficiency. *Lung* 186(2):115–122, 2008.
36. Food and Nutrition Board, Institute of Medicine. *Dietary Reference Intakes for Vitamin C, Vitamin E, Selenium, and Carotenoids*. Washington, DC: National Academies Press, 2000.

37. Rimm, E. B., Stampfer, M. J., Ascherio, A. et al. Vitamin E consumption and the risk of heart disease in men. *N. Engl. J. Med.* 328:1450–1456, 1993.

38. Stampfer, M. J., Hennekens, C. H., Manson, J. E. et al. Vitamin E consumption and the risk of heart disease in women. *N. Engl. J. Med.* 328:1487–1489, 1993.

39. Harris, A., Devaraj, S., and Jailal, I. Oxidative stress, alpha-tocopherol therapy, and atherosclerosis. *Curr. Atheroscler. Rep.* 4:373–380, 2002.

40. Traber, M. G., Frei, B., and Beckman, J. S. Vitamin E revisited: Do new data validate benefits for chronic disease prevention? *Curr. Opin. Lipidol.* 19:30–38, 2008.

41. Bjelakovic, G., Nikolova, D., Gluud, L. L. et al. Mortality in randomized trials of antioxidant supplements for primary and secondary prevention: Systematic review and meta-analysis. *JAMA* 297:842–857, 2007.

42. Booth, S. L., Pennington, J. A. T., and Sadowski, J. A. Food sources and dietary intakes of vitamin K_1 (phylloquinone) in the American diet: Data from the FDA Total Diet Study. *J. Am. Diet. Assoc.* 96:149–154, 1996.

43. Bugel, S. Vitamin K and bone health. *Proc. Nutr. Soc.* 62:839–843, 2003.

44. Vermeer, C., Shearer, M. J., Zittermann, A. et al. Beyond deficiency: Potential benefits of increased intakes of vitamin K for bone and vascular health. *Eur. J. Nutr.* 43:325–335, 2004.

(©iStockphoto)

FOCUS ON
Phytochemicals

Focus Outline

Food presents an unlimited array of tastes, textures, colours, and aromas. With this gastronomic variety and delight come a myriad of nutrient combinations. In addition, food contains substances that have not been identified as nutrients but may promote health and reduce disease risk. Foods have been used in folk medicine for centuries. Today, researchers continue to discover beneficial effects of various food components and explore the relationships among the consumption of specific foods, typical dietary patterns, and health. Foods that provide health benefits beyond basic nutrition are called **functional foods.**[1] **Table F3.1** provides examples of functional foods and their potential benefits. Health-promoting substances in plant foods are called **phytochemicals**, while those found in animal foods are called **zoochemicals**.

functional foods As defined by Health Canada: "A *functional food* is similar in appearance to, or may be, a conventional food, is consumed as part of a usual diet, and is demonstrated to have physiological benefits and/or reduce the risk of chronic disease beyond basic nutritional functions."[2]

phytochemicals Substances found in plant foods (*phyto* means plant) that are not essential nutrients but may have health-promoting properties.

zoochemicals Substances found in animal foods (*zoo* means animal) that are not essential nutrients but may have health-promoting properties.

Table F3.1 Benefits of Functional Foods

Functional Food	Key Components	Potential Benefit
Whole-grain products	Fibre, lignans, phytoestrogens	Reduce the risk of cancer and heart disease.
Oatmeal	β-glucan, soluble fibre	Reduces blood cholesterol.
Grape juice	Polyphenols	Improves cardiovascular health.
Green or black tea	Polyphenols such as tannins, catechins	Reduces the risk of certain types of cancer.
Fatty fish	Omega-3 fatty acids	Reduces the risk of heart disease.
Soy	Phytoestrogens, soy protein	Reduces the risk of cancer and heart disease, reduces symptoms of menopause.
Garlic	Organic sulfur compounds	Reduces the risk of cancer and heart disease.
Spinach, kale, collard greens	Lutein, zeaxanthin	Reduce the risk of age-related blindness (macular degeneration).
Nuts	Monounsaturated fatty acids	Reduce the risk of heart disease.

F3.1 Phytochemicals in the Canadian Diet

Learning Objectives

- Distinguish phytochemicals from essential nutrients.
- Discuss how the colour of fruits and vegetables is related to their phytochemical content.
- Give some examples of the health benefits of phytochemicals.

Phytochemicals include the hundreds, perhaps thousands, of biologically active non-nutritive chemicals found in plants. In the plants themselves, phytochemicals provide biological functions. For example, compounds in onions and garlic are natural pesticides that protect plants from insects. Gardeners often plant these near more vulnerable plant varieties to protect them from insect infestation. Some plant chemicals can be toxic. The chemicals in rhubarb leaves, for example, can cause symptoms ranging from burning in the mouth and throat, abdominal pain, nausea, vomiting, and diarrhea to convulsions, coma, and death from cardiovascular collapse (**Figure F3.1**). On the other hand, many chemicals in plants have been identified that may have a beneficial effect on human health. For example, compounds in soy and tea have been shown to inhibit tumour growth.[3] Despite the variety of effects these chemicals can have, the term "phytochemical" is generally used to refer to those substances found in plants that have health-promoting properties. Many phytochemicals are listed in **Table F3.2.** Phytochemicals are responsible for the health-promoting properties afforded by a variety of natural functional foods.

The phytochemicals in our diet protect our health in a variety of ways. Some, such as carotenoids (see Section 9.2), are **antioxidants** (see Sections 8.10 and 9.4 for a general discussion of the actions of antioxidants). Others provide benefits because they mimic the structure or function of natural substances in the body. Phytoestrogens, such as those in soy, for instance, have structures similar to the hormone estrogen and affect us by blocking or mimicking estrogen action. Phytosterols resemble cholesterol in structure and thus compete with cholesterol for absorption from the gastrointestinal tract. Their presence reduces cholesterol absorption and helps lower blood cholesterol—a major risk factor for cardiovascular disease. Some phytochemicals stimulate the body's natural defenses. For example, indoles and isothiocyanates found in broccoli stimulate the activity of enzymes that help deactivate **carcinogens**. Other phytochemicals are health-promoting due to their ability to alter the way in which cells communicate, to affect DNA repair mechanisms, or to influence other cell processes that may affect cancer development.[4] The phytochemicals described in this chapter are just a few of the ones that have been identified in our food. Many others have been studied and still more have yet to be isolated.

antioxidants Substances that are able to neutralize reactive oxygen molecules and thereby reduce oxidative damage.

carcinogen A substance that causes cancer.

Figure F3.1 Not all of the chemicals in plants benefit our health. The leaves of the rhubarb plant are poisonous. The stems have lower levels of the toxin and are safe to eat. (Image Source/Getty Images, Inc.)

Carotenoids

Carrots, sweet potatoes, acorn squash, apricots, and mangoes get their yellow-orange colour from the carotenoids they contain. These and other yellow-orange and red fruits and vegetables, as well as leafy greens, in which the colour of the carotenoids is masked by the green colour of chlorophyll, are the major sources of dietary carotenoids. People who eat more of these have higher blood carotenoid levels. A high intake of carotenoid-containing fruits and vegetables has been associated with a reduced risk of certain cancers, cardiovascular disease, and age-related eye diseases.[5]

Carotenoids are phytochemicals that have antioxidant properties; some also have vitamin A activity[6] (see Table F3.2). The most prevalent carotenoids in the North American diet include β-carotene, α-carotene, β-cryptoxanthin, lycopene, lutein, and zeaxanthin. Beta-carotene is the best known and it provides the most vitamin A activity, but it may be a less effective antioxidant than others. Lycopene, the carotenoid that gives tomatoes their red colour, does not provide vitamin A activity, but it is a more potent antioxidant than other dietary carotenoids[7] (**Figure F3.2**). The carotenoids lutein and zeaxanthin are antioxidants that accumulate in the macula, which is the central portion of the retina of the eye. High intakes of these are associated with reduced risk of **macular degeneration**, the leading cause of blindness in older adults.

macular degeneration An incurable eye disorder that is caused by deterioration of the retina of the eye. It is the leading cause of blindness in adults over the age of 55 years.

Table F3.2 Phytochemicals in Foods

Phytochemical Name or Class	Food Sources	Biological Activities and Possible Effects
Carotenoids: α-carotene, β-carotene, β-cryptoxanthin, lutein, zeaxanthin, lycopene	Apricots, carrots, cantaloupe, tomatoes, peppers, sweet potatoes, squash, broccoli, spinach, and other leafy greens	Some are converted to vitamin A, provide antioxidant protection; some decrease the risk of macular degeneration.
Flavonoids: flavonols (quercetin, kaempferol, myricetin), flavones (apigenin), flavanols (catechins), anthocyanidins (cyanidin, delphinidin)	Berries, citrus fruits, onions, purple grapes, green tea, red wine, and chocolate	Make capillary blood vessels stronger, block carcinogens, and slow the growth of cancer cells.

Figure F3.2 The red colour of tomatoes is due to the carotenoid lycopene, a potent antioxidant. Processing of tomatoes into tomato paste and sauce increases lycopene concentration and bioavailability. (© Datacraft/Age Fotostock)

Figure F3.3 Cruciferous vegetables are excellent sources of phytochemicals that may protect against cancer. (©iStockphoto)

Table F3.2 *continued*

Phytoestrogens including lignins and isoflavones such as genistein, biochanin A, and daidzein	Tofu, soy milk, soybeans, flax seed, and rye bread	Mimic effect of estrogen, induce cancer cell death, slow the growth of cancer cells, reduce blood cholesterol, may reduce risk of osteoporosis.
Phytosterols: β-sitosterol, stigmasterol, and campesterol	Nuts, seeds, and legumes	Decrease cholesterol absorption, reduce the risk of colon cancer by slowing growth of colon cells.
Capsaicin	Hot peppers	Modulates blood clotting.
Glucosinolates, isothiocyanates, indoles	Broccoli, brussels sprouts, and cabbage	Increase the activity of enzymes that deactivate carcinogens, alter estrogen metabolism, affect the regulation of gene expression.
Sulfides and allium compounds	Onions, garlic, leeks, and chives	Deactivate carcinogens, kill bacteria, protect against heart disease.
Inositol	Sesame seeds and soybeans	Protects against free radicals, protects against cancer.
Saponins	Beans and herbs	Decrease cholesterol absorption, decrease cancer risk, antioxidant.
Ellagic acid	Nuts, grapes, and strawberries	Anticancer properties, prevents the formation of carcinogens in the stomach.
Tannins, catechins	Tea and red wine	Antioxidants, cancer protection.
Curcumin	Turmeric and mustard	Reduces carcinogen formation, antioxidant, anti-inflammatory.
Sulforaphane	Broccoli and other cruciferous vegetables	Detoxifies carcinogens, shown to protect animals from breast cancer.
Limonene	Citrus fruit peels	Inhibits cancer cell growth.

Figure F3.4 Garlic and onions provide sulfur-containing phytochemicals that may help protect against cancer and heart disease. (Ulrich Kerth/StockFood/Getty Images)

Flavonoids

Like carotenoids, flavonoids are plant pigments that add colour to your plate. They are found in fruits, vegetables, wine, grape juice, chocolate, and tea. One of the most abundant types of flavonoids is the anthocyanidins, which give the blue and red colours to blueberries, raspberries, and red cabbage. Other types of flavonoids give the pale yellow colour to potatoes, onions, and orange rinds. These compounds are strong antioxidants that protect against cancer and cardiovascular disease. Citrus fruits contain about 60 flavonoids that inhibit blood clotting and have antioxidant, anti-inflammatory, and anticancer properties[8] (see Your Choice: Chocolate: High-Fat Treat or Functional Food?).

Indoles, Isothiocyanates, and Alliums

Cruciferous vegetables such as broccoli, cauliflower, brussels sprouts, cabbage, bok choy, and greens such as mustard and collards are particularly good sources of sulfur-containing phytochemicals (**Figure F3.3**). These stimulate the activity of enzymes that detoxify carcinogens. Sulforaphane, a phytochemical in broccoli, is particularly effective at boosting the activity of these enzyme systems and has been shown to protect animals from breast cancer.[9] Crucifers also contain phytochemicals called indoles, which inactivate the hormone estrogen. Exposure to estrogen is believed to increase the risk of cancer; therefore, because indoles reduce estrogen exposure, they protect against cancer.

Sulfur compounds called alliums are found in garlic, onions, leeks, chives, and shallots (**Figure F3.4**). These phytochemicals boost the activity of cancer-destroying enzyme systems, protect against oxidative damage, and defend against heart disease by lowering blood cholesterol, blood pressure, and platelet activity.[10] In addition, these compounds prevent bacteria in the gut from converting nitrates into nitrites, which can form carcinogens.

cruciferous A group of vegetables (also called crucifers) named for the cross shape of their four-petal flowers. They include broccoli, brussels sprouts, cabbage, bok choy, cauliflower, kale, kohlrabi, mustard greens, rutabagas, and turnips. Their consumption is linked with lower rates of cancer.

Phytoestrogens and Other Plant Hormones

Human hormones help regulate body processes and maintain homeostasis. Plants have hormones, too, which, when ingested, can affect human health. Phytoestrogens are plant

YOUR CHOICE

(©iStockphoto)

Chocolate: High-Fat Treat or Functional Food?

Few foods provide as many mixed messages as chocolate. It is offered as a reward or treat and is consumed for comfort. It has been hailed as everything from an antidepressant to an aphrodisiac and been accused of causing acne, weight gain, and tooth decay. Yet we consume it with vigour. Is it simply the seductive flavour of chocolate that lures us into consuming this high-kcalorie, high-fat treat, or does it provide other benefits that contribute to our attraction to this confection?

Chocolate is made from the seeds, commonly known as cocoa beans, of the cacao tree. Cocoa beans are processed into cocoa solids, or powder, and cocoa butter, the vegetable fat derived from the bean. The benefits of the chocolate are associated with the phytochemicals found in the cocoa solids.

One of the reasons we crave chocolate is because it makes us feel good. Cocoa contains several chemical compounds that may be responsible for the mood boost it induces. One study found that consuming cocoa causes the brain to produce natural opiates, which dull pain and increase feelings of well-being.[1] Other studies found that a compound called anandamide can mimic the effects of tetrahydrocannabinol (THC)—the active chemical in marijuana—causing a chocolate "high." Phenylethylamine, an amphetamine-like compound that raises blood pressure and blood sugar, may increase alertness and contentment. Chocolate also contains caffeine and related compounds that make us feel more alert.

In addition to making us feel good, cocoa provides small amounts of copper, magnesium, iron, and zinc. Cocoa and dark chocolate, made from at least 70% cocoa, are also very rich in antioxidants, which may protect against heart disease and other health problems. In fact, dark chocolate provides more antioxidants than red wine or blueberries.[2] Dark chocolate, which is rich in flavonoids, has been shown to improve arterial function, inhibit platelet activities that lead to heart disease, lower blood pressure, and improve insulin sensitivity.[3]

Unfortunately, dark chocolate has a bitter taste. To improve the taste, most of the chocolate we typically consume has added sugar. Also to improve the texture of chocolate products, cocoa butter is added. Cocoa butter has a melting point of 37°C (body temperature) which is why chocolate has that creamy "melt in your mouth" feel. This added sugar and cocoa butter, unfortunately, dilutes the concentration of important cocoa phytochemicals, lightens the colour of the chocolate, and increases its caloric content. Half of the fat in cocoa butter is saturated fat, which may also increase the risk of cardiovascular disease. Because of these additions, light chocolate does not contain as many phytochemicals as dark chocolate, while white chocolate, made primarily from cocoa butter and sugar, contains no cocoa at all, and hence, no phytochemicals.

Dark chocolate, therefore, made from 70% or more cocoa, is an ideal example of a functional food, providing essential minerals and antioxidants that may reduce disease risk. More typical forms of lighter chocolate contain much lowerlevels of phytochemicals, and are treats, best enjoyed in moderation.

(Mary Ellen Bartley/Foodpix/Jupiter Images Corp.)

[1]Kuwana, E. Discovering the sweet mysteries of chocolate. Available online at faculty.washington.edu/chudler/choco.html/. Accessed June 30, 2005.

[2]Halvorsen, B. L., Carlsen, M. H., Phillips, K. M. et al. Content of redox-active compounds (i.e., antioxidants) in foods consumed in the United States. *Am. J. Clin. Nutr.* 84(1):95–135, 2006.

[3]Corti, R., Flammer, A. J., Hollenberg, N. K., et al. Cocoa and cardiovascular health. *Circulation* 119:1433–41, 2009.

hormones that are believed to interrupt cancer development and affect health by interfering with the action of the human hormone estrogen. Phytoestrogens include isoflavones and lignins. These molecules are modified by the microflora in the intestines to form compounds that are structurally similar to estrogen. They are believed to block estrogen function by tying up estrogen receptors on cells. Isoflavones are found in soybeans, flaxseed, and barley; they may protect against some types of cancers as well as osteoporosis.[11,12] They are also thought to decrease hot flashes and other symptoms of menopause, although studies have shown this effect to be minimal[13] (see Chapter 6: Your Choice: Should You Increase Your Soy Intake?). Flaxseed is also rich in lignin, the metabolites of which are structurally similar to estrogen and have been shown to inhibit the growth of estrogen-stimulated breast cancer cells.[14]

CANADIAN CONTENT

F3.2 Choosing a Phytochemical-Rich Diet

Learning Objectives

- List foods that are high in phytochemicals.
- Compare the health benefits of taking phytochemical supplements with those of a diet rich in unrefined plant foods.

It has not been possible to make quantitative recommendations for the intake of specific phytochemicals. Many of these substances have not been identified or classified, and many function differently depending on the form in which they are consumed. Their effects may vary if they are removed from the foods in which they are found. Therefore, to benefit the most from phytochemicals, choose a diet based on plant foods. The impact of the total diet is more significant than that of any single phytochemical. For example, in some people, a diet high in plant sterols, soy, almonds, and foods such as oats, barley, psyllium, okra, and eggplant that are high in soluble fibres has been shown to be as effective at lowering cholesterol as prescription medications.[15] Likewise, it has been hypothesized that a diet rich in flaxseed, cruciferous vegetables, and fruits and vegetables in general could significantly reduce the risk of breast, colon, prostate, lung, and other cancers.[16] **Table F3.3** includes some suggestions for increasing the intake of foods that provide a variety of phytochemicals.

Table F3.3 Tips to Increase Phytochemicals
• Choose 5 different colours of fruits and vegetables each day.
• Try a new fruit or vegetable each week.
• Spice up your food—herbs and spices are a great source of phytochemicals.
• Add vegetables to your favourite entrees such as spaghetti sauces and casseroles.
• Try fresh, baked, or dried fruit for dessert.
• Double your typical serving of vegetables.
• Add pesto, spinach, artichokes, or asparagus to pizza.
• Keep spices such as garlic, ginger, and basil on hand to make it easy to add more of these to your cooking.
• Snack on whole-grain crackers.
• Add barley or bulgur to casseroles or stews.
• Switch to whole-wheat bread, brown rice, and whole-wheat pasta.
• Add fruit to your cereal or vegetables to your eggs.
• Dice up some tofu and add it to your stir fry.
• Include nuts in stir fries and baked goods.
• Sprinkle flaxseed in your oatmeal.

Eat More Vegetables and Fruit

Canada's Food Guide recommends a diet with plenty of vegetables and fruit, including a specific recommendation to eat green and orange vegetables daily. In fact, if the vegetables and fruit in your diet include a rainbow of colours—dark green-coloured vegetables as well as yellow-orange, pale yellow, and deep red and purple fruits and vegetables—you are getting an abundance of carotenoids, flavonoids, and other phytochemicals (**Figure F3.5**).

Unfortunately most people do not follow these recommendations. The CCHS indicates that 50% of Canadians eat less than 5 daily servings of fruits and vegetables.[17] In addition, white potatoes, typically consumed as high-fat french fries, represent a disproportionate share of the total vegetable intake, A variety of vegetables that includes the occasional baked potato along with many colourful phytochemical- and nutrient-rich vegetables is a much more nutritious eating pattern.[18]

Figure F3.5 Choosing a rainbow of fruits and vegetables is a good way to include a variety of phytochemicals in your diet. (© iStockphoto.com/Kelly Cline)

Make Half Your Grains Whole

Fruits and vegetables aren't the only source of dietary phytochemicals. In fact, whole grains deliver as many if not more phytochemicals and antioxidants than do fruits and vegetables.[19] Epidemiological studies have found that whole-grain intake is associated with a lower risk of cancer, cardiovascular disease, diabetes, and obesity. These health-promoting properties are believed to be due to the synergistic effects of the wide variety of nutrients and phytochemicals found in whole grains.[20]

In addition to fibre, whole grains are rich in antioxidant phytochemicals called polyphenols as well as phytate, phytoestrogens such as lignan, and plant stanols and sterols. The bran and germ portions of whole-wheat flour contribute more than half of the total polyphenols, flavonoids, lutein, and zeaxanthin, as well as more than 80% of the water-soluble and fat-soluble antioxidant activity.[21] These phytochemicals and vitamins are lost when the bran and germ are removed to make white flour. Canada's Food Guide recommends Canadians consume half of their grain product servings as whole grains.

Choose Plant Proteins

The amounts of health-promoting substances in the diet can further be enhanced by choosing plant sources of protein. Phytochemical-rich soybeans and flaxseed are good sources of protein and, other legumes, nuts, and seeds are high-protein foods that make important fibre and phytochemical contributions. Canada's Food Guide encourages the consumption of these high-phytochemical protein sources by recommending that Canadians choose beans, peas, nuts, and seeds more often.

What about Added Phytochemicals?

In addition to foods that are natural sources of phytochemicals, phytochemical supplements are available and many foods are fortified with phytochemicals. For example, some margarines contain added plant sterols to help lower blood cholesterol, some cereal brands combine oatmeal with soy and flaxseed, and green tea extracts are added to some yogurts. These modified foods are functional foods because they are designed to provide specific health benefits (see Table F3.1)

While phytochemical supplements and phytochemical-fortified foods may offer some specific advantages, they do not provide all the benefits obtained from a diet high in natural sources of these compounds. One reason is that most contain only a few of the many phytochemicals contained in foods. In addition, the benefits provided by many foods are believed to be due to the interactions among a variety of phytochemicals and nutrients and so cannot be replicated by a supplement that contains only one or a few of these substances. Finally, the amounts that can be added to supplements or foods may be too small to have an impact on overall health. For example, a multivitamin advertising that it provides lycopene may include about 0.3 mg per dose, whereas a 125 ml (half-cup) serving of spaghetti sauce will give you about 20 mg.

REFERENCES

1. American Dietetic Association. Position of the American Dietetic Association: Functional foods. *J. Am. Diet. Assoc.* 104:814–826, 2004.
2. Health Canada. Policy Paper-Nutraceuticals/functional foods and Health Claims on foods. Available online at: http://www.hc-sc.gc.ca/fn-an/label-etiquet/claims-reclam/nutra-funct_foods-nutra-fonct_aliment-eng.php. Accessed July 30, 2011.
3. Zhou, J. R., Yu, L., Mai, Z., et al. Combined inhibition of estrogen-dependent human breast carcinoma by soy and tea bioactive components in mice. *Intl. J. Cancer*. 108:8–14, 2004.
4. Zuzino, S. How Plants Protect Us. *Agriculture Research Magazine*, March 2008, pp 9–11. Available online at www.ars.usda.gov/News/docs.htm?docid=18510. Accessed August 29, 2009.
5. Krinsky N. I., Johnson E. J. Carotenoid actions and their relation to health and disease. *Mol. Aspects Med.* 6:459–516, 2005.
6. Rao, A. V., and Rao, L. G. Carotenoids and human health. *Pharmucol. Res.* 5:207–216, 2007.
7. Agarwal, S., Rao, A. V. Tomato lycopene and its role in human health and chronic diseases. *CMAJ*. 163:739–744, 2000.
8. Erdman, J. W., Balentine, D., and Arab, L. Flavonoids and heart health: Proceedings of the ILSI North America Flavonoids Workshop, May 31–June 1, 2005, Washington, DC. *J. Nutr.* 137:718S–737S, 2007.
9. Stan, S. D., Kar, S., Stoner, G. D., et al. Bioactive food components and cancer risk reduction. *J. Cell. Biochem.* 104:339–356, 2008.
10. Butt, M. S., Sultan, M. T., Butt, M. S., et al. Garlic: Nature's Protection against Physiological Threats. *Critical Reviews in Food Science and Nutrition*. 49:538–551, 2009.
11. Messina, M., Gardner, C., and Barnes, S. Gaining insight into the health effects of soy but a long way still to go: Commentary on the Fourth International Symposium on the Role of Soy in Preventing and Treating Chronic Disease. *J. Nutr.* 132:547S–551S, 2002.
12. Messina, M., and Messina, V. Soyfoods, soybean isoflavones, and bone health: A brief overview. *J. Ren. Nutr.* 1:63–68, 2000.
13. Vincent, A., and Fitzpatrick, L. A. Soy isoflavones: Are they useful in menopause? *Mayo Clin. Proc.* 75:1174–1184, 2000.
14. Rice, S. and Whitehead, S. A. Phytoestrogens oestrogen synthesis and breast cancer. *J. Steroid Biochem. Mol. Biol.* 108:186–195, 2008.
15. Jenkins, D. J., Kendall, C. W., Marchie, A. et al. Direct comparison of a dietary portfolio of cholesterol-lowering foods with a statin in hypercholesterolemic participants. *Am. J. Clin. Nutr.* 81:380–387, 2005.
16. Donaldson, M. S. Nutrition and cancer: A review of the evidence for an anti-cancer diet. *Nutr. J.* 3:19–30, 2004.
17. Garriguet, D. Canadians' eating habits. Health Reports 18(2):17-32, 2007.
18. Krebs-Smith, A. M., and Kantor, L. S. Choose a variety of fruits and vegetables daily: Understanding the complexities. *J. Nutr.* 131:487S–501S, 2001.
19. Jones, J. M., Reicks, M., Adams, J. et al. Becoming proactive with the whole-grains message. *Nutr. Today* 39:10–17, 2004.
20. Slavin, J. Why whole grains are protective: Biological mechanisms. *Proc. Nutr. Soc.* 62:129-134, 2003.
21. Adom, K. K., Sorrells, M. E., and Liu, R. H. Phytochemicals and antioxidant activity of milled fractions of different wheat varieties. *J. Agric. Food Chem.* 53:2297–2306, 2005.

10 Water and the Electrolytes

Case Study

The twentieth running of the Montreal Marathon on September 5, 2010, proved to be a memorable event.[1] A record 21,000 people ran the race, which began at the Jacques Cartier Bridge and wound its way through the hills of scenic downtown Montreal before ending at the Olympic Stadium. There, spectators witnessed both triumph and heartbreak, as Kenya's Choge Julius Kirwa won the race by a margin more commonly seen in sprints than marathons. He beat the second-place finisher, Columbian William Naranjo, by three-tenths of a second.

William Naranjo was an experienced runner, a South American champion, and a participant in the Pan American Games. But despite this experience, the last seven kilometres of the race had been gruelling and he was fighting fatigue when he entered the Olympic Stadium after running for more than two hours. Although the finish line was within his sight, his body became less and less responsive as he made his way more and more slowly toward the winner's tape. He was almost at the finish when Kirwa passed him, to take first place. While photographers and reporters surrounded Kirwa, Naranjo crossed the finish line and then collapsed. He was quickly wheeled away on a stretcher to the first aid area of the Olympic Stadium. He had been felled by a case of dehydration, a potent reminder that water is the most essential of nutrients.

(Phil Carpenter, The Gazette)

In this chapter, you will learn about water and the minerals dissolved in it. They make up the "internal sea" in which the reactions that maintain life occur. Water and mineral imbalances can come on rapidly and be devastating, but they can also be alleviated faster than other nutrient deficiencies.

(©iStockphoto)

(Eiichi Onodera/Dex Image/Getty Images, Inc.)

Chapter Outline

10.1 Water: The Internal Sea

Learning Objectives

- Discuss the forces that move water back and forth across cell membranes.
- Describe 5 functions of water in the body.
- List the sources of body water and describe the routes by which water is lost.
- Discuss the effects of dehydration.

The complex molecules necessary for the emergence of life were forged in the Earth's first seas. These primordial seas supported life because they were rich in inorganic minerals as well as organic substances. As organisms grew in complexity, the water and chemicals critical to their survival were incorporated into an internal sea. Just as the right amounts of water, organic molecules, and minerals were necessary for the beginning of life, the right combination is necessary in the body for the maintenance of life. This internal sea allows the reactions necessary for life to proceed.

Functions of Water in the Body

Water is an essential nutrient that we must consume in our diet to survive. Without water, death occurs in only a few days. In the body, water serves as a medium in which chemical reactions take place; it also transports nutrients and wastes, provides protection, helps regulate temperature, participates in chemical reactions, and helps maintain acid-base balance **(Table 10.1)**.

Table 10.1 A Summary of Water and the Electrolytes

Nutrient	Sources	Recommended Intake for Adults	Major Functions	Deficiency Diseases and Symptoms	Groups at Risk of Deficiency	Toxicity	UL[a]
Water	Drinking water, other beverages, and food	2.7–3.7 L/d	Solvent, reactant, protector, transporter, regulator of temperature and pH	Thirst, weakness, poor endurance, confusion, disorientation	Infants, those with fever and diarrhea, elderly, athletes	Confusion, coma, convulsions	ND
Sodium	Table salt, processed foods	<2,300 mg/d, ideally 1,500 mg/d	Major positive extracellular ion, nerve transmission, muscle contraction, fluid balance	Muscle cramps	Those consuming a severely sodium-restricted diet, excessive sweating	Contributes to high blood pressure	2,300 mg/d
Potassium	Fresh fruits and vegetables, legumes, whole grains, milk, and meat	4,700 mg/d or greater	Major positive intracellular ion, nerve transmission, muscle contraction, fluid balance	Irregular heartbeat, fatigue, muscle cramps	Those consuming poor diets high in processed foods, those taking thiazide diuretics	Abnormal heartbeat	ND
Chloride	Table salt, processed foods	<3,600 mg/d, ideally 2,300 mg/d	Major negative extracellular ion, fluid balance	Unlikely	None	None likely	3,600 mg/d

[a]Recommended intakes for water, sodium, potassium, and chloride are Adequate Intakes (AIs); UL, Tolerable Upper Intake Level; ND, insufficient data to determine a UL.

solvent A fluid in which one or more substances dissolve.

solutes Dissolved substances.

Solvent One of the key functions of water in the body is as a **solvent.** A solvent is a fluid in which **solutes** can dissolve to form a solution. Water is an ideal solvent for some substances because it is **polar**; that is, the 2 sides or poles of the water molecule have opposite electrical charges. The polar nature of water comes from its structure, which consists of 2 hydrogen atoms and 1 oxygen atom. These atoms, like all atoms, are made up of a positively charged central core, or nucleus, with negatively charged **electrons** orbiting around it. To form a water molecule, the 2 hydrogen atoms move close enough to share their electrons with an atom of oxygen. But the sharing is not equal. The shared electrons spend more time around the oxygen atom than around the hydrogen atoms, giving the oxygen side of the molecule a slightly negative charge and the hydrogen side a slightly positive charge. This polar nature of water allows it to surround other charged molecules and disperse them. Table salt, which dissolves in water, consists of a positively charged sodium **ion** bound to a negatively charged chloride ion. When placed in water, the sodium and chloride ions move apart, or **dissociate**, because the positively charged sodium ion is attracted to the negative pole of the water molecule and the negatively charged chloride ion is attracted to the positive pole **(Figure 10.1)**.

polar Used to describe a molecule that has a positive charge at one end and a negative charge at the other.

electrons Negatively charged particles.

ion An atom or group of atoms that carries an electrical charge.

dissociate To separate two charged ions.

Transport Blood, which is 90% water, transports oxygen and nutrients to cells. It then carries carbon dioxide and other waste products away from the cells. Blood also distributes hormones and other regulatory molecules throughout the body so they can reach target cells. Water in urine helps to carry waste products, such as urea and ketones, out of the body.

Lubrication and Protection Water functions as a lubricant and cleanser. Watery tears lubricate the eyes and wash away dirt, synovial fluid lubricates the joints, and saliva lubricates the mouth, making it easier to chew, taste, and swallow food. Water inside the eyeballs and spinal cord acts as a cushion against shock. Similarly, during pregnancy, water in the amniotic fluid provides a protective cushion around the fetus.

Regulation of Body Temperature Body temperature is closely regulated to maintain a normal level of around 37°C. If the body temperature rises above 42°C or falls below 27°C, death is likely. The fact that water holds heat and changes temperature slowly helps keep body temperature constant when the outside temperature fluctuates, but water is also more actively involved in temperature regulation.

The water in blood helps regulate body temperature by increasing or decreasing the amount of heat lost from the surface of the body. When body temperature starts to rise, the blood vessels in the skin dilate, causing blood to flow close to the surface of the body where it can release some of the heat to the environment. This occurs with fevers as well as when environmental temperature rises. In a cold environment, blood vessels in the skin constrict, restricting the flow of blood near the surface and conserving body heat.

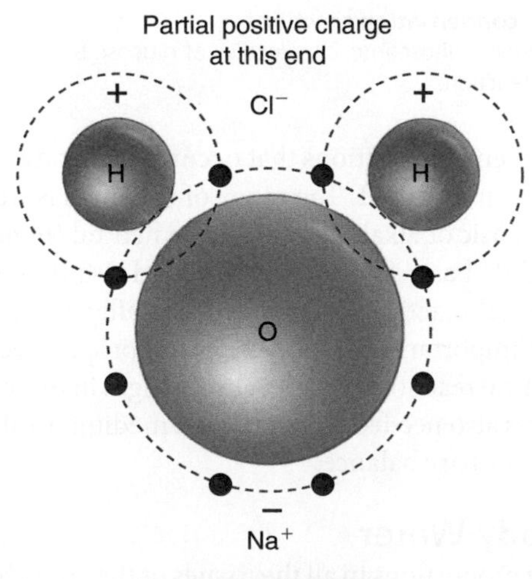

Figure 10.1 The polar nature of water
Electrons in a water molecule, shown as dots, spend more time around the oxygen atom, giving the oxygen side of the molecule a slightly negative charge, and the hydrogen side a slightly positive charge. When sodium chloride is added to water, the positive sodium ion is attracted to the negative pole of the water molecule, and the negative chloride ion is attracted to the positive pole.

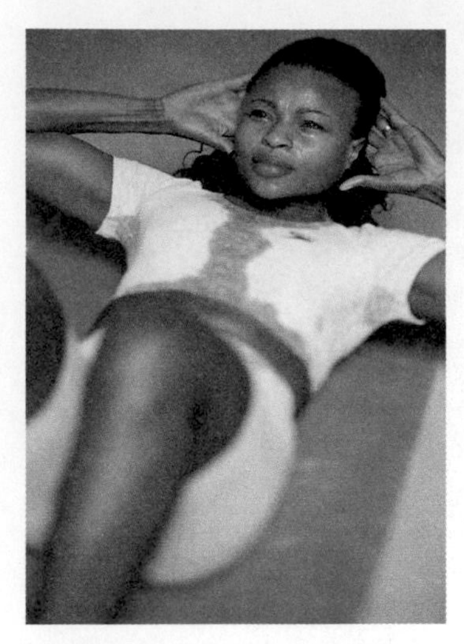

Figure 10.2 Exercising in the heat can dramatically increase water losses from sweat.
(Lori Adamski Peek/Getty Images, Inc.)

Water also helps regulate body temperature through the evaporation of sweat. When body temperature increases, the brain triggers the sweat glands in the skin to produce sweat, which is mostly water. As the sweat evaporates from the skin, heat is lost, cooling the body (**Figure 10.2**).

Chemical Reactions Water is involved in chemical reactions in the body. A **hydrolysis reaction** breaks large molecules into smaller ones by adding water. For example, water is added in the reaction that breaks 1 molecule of maltose into 2 glucose molecules (**Figure 10.3**). Water is also involved in reactions that join 2 molecules. This type of reaction is referred to as a **condensation reaction**. The formation of the disaccharide maltose from 2 glucose molecules is a condensation reaction and therefore requires the removal of a water molecule.

hydrolysis reaction Chemical reaction that breaks large molecules into smaller ones by adding water.

condensation reaction Chemical reaction that joins 2 molecules together. Hydrogen and oxygen are lost from the 2 molecules to form water.

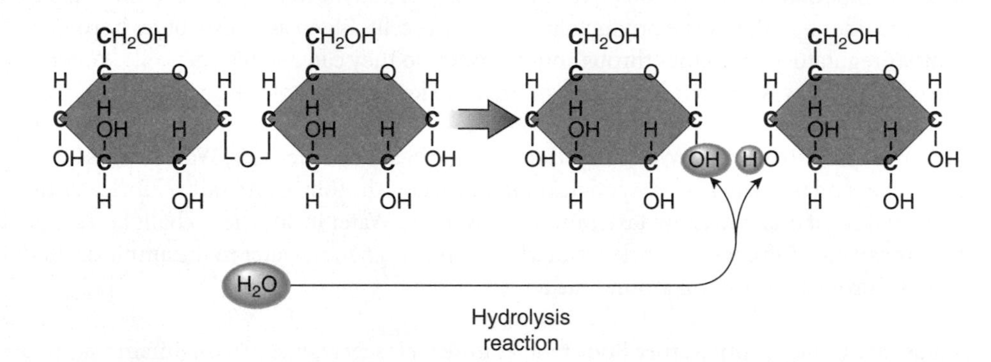

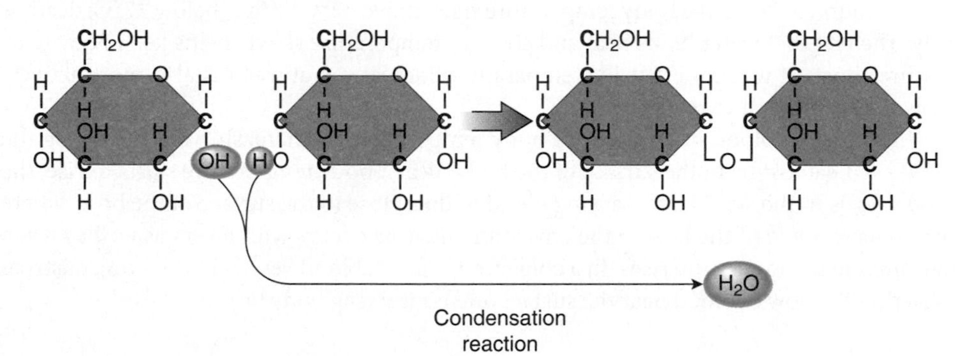

Figure 10.3 **Hydrolysis and condensation reactions**
The cleavage of the disaccharide maltose into 2 molecules of glucose is a hydrolysis reaction; disaccharide formation is a condensation reaction.

pH A measure of the level of acidity or alkalinity of a solution.

Acid-Base Balance The chemical reactions that occur in the body are very sensitive to acidity. Acidity is expressed in units of **pH**. The range of pH units is from 1 to 14, with 1 being very acidic, 14 being very basic or alkaline, and 7 being neutral (**Figure 10.4**). Most reactions in the body occur in slightly basic solutions, around pH 7.4. If body solutions become too acidic or too basic, chemical reactions cannot proceed efficiently. Water and the dissolved substances it contains are important for maintaining the proper level of acidity. Water serves as a medium for the chemical reactions that prevent changes in pH and it participates in some of these reactions. Water is also needed as a transport medium to allow the respiratory tract and kidneys to regulate acid-base balance.

Distribution of Body Water

Water is found in varying proportions in all the tissues of the body; blood is about 90% water, muscle about 75%, and bone about 25%. Adipocytes have a low water content—only about 10%. In adults, about 60% of total body weight is water but this varies with age and other factors that affect body composition. The percentage of water is higher in infants than in adults and decreases as adults age, primarily due to increases in body fat and a loss of muscle mass. Because women typically have a higher percentage of body fat than men, they have less body water. Obese individuals have a lower percentage of body weight as water and a higher percentage as fat than their lean counterparts.

Life Cycle

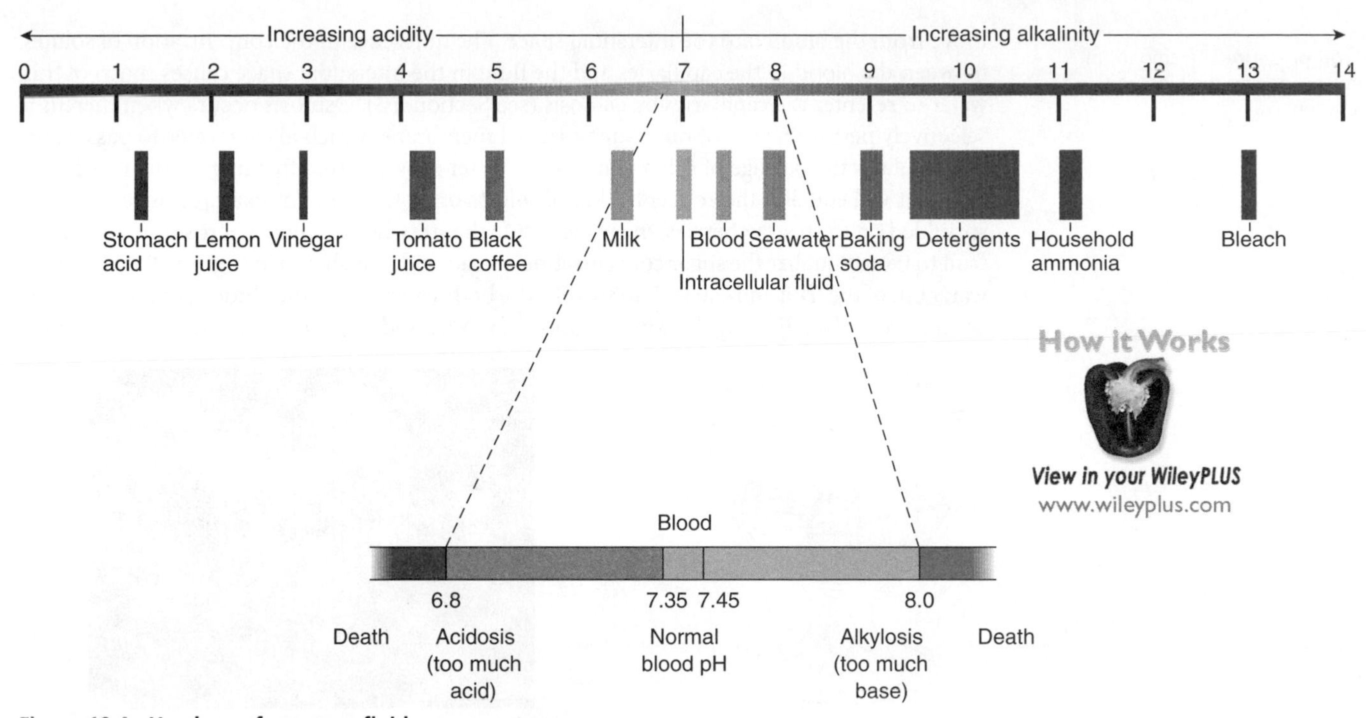

How it Works

View in your WileyPLUS
www.wileyplus.com

Figure 10.4 pH values of common fluids
Fluctuations from the normal blood pH range of 7.35-7.45 leads to acidosis (too much acid) or alkalosis (too much base) and, if severe, can be fatal.

About two-thirds of body water is found inside cells; this is known as **intracellular fluid**. The remaining one-third is outside cells, as **extracellular fluid** (**Figure 10.5a**). Extracellular fluid includes primarily blood plasma, lymph, and the fluid between cells, called **interstitial fluid**. Other extracellular fluids include fluid secreted by glands such as saliva and other digestive secretions, and fluid in the eyes, joints, and spinal cord. The concentration of substances dissolved in body water varies among these body compartments. The concentration of protein is highest in intracellular fluid, lower in extracellular fluid, and even lower in interstitial fluid. Extracellular fluid has a higher concentration of sodium and chloride and a lower concentration of potassium, and intracellular fluid is higher in potassium and lower in sodium and chloride.

intracellular fluid The fluid located inside cells.

extracellular fluid The fluid located outside cells. It includes fluid found in the blood, lymph, gastrointestinal tract, spinal column, eyes, joints, and that found between cells and tissues.

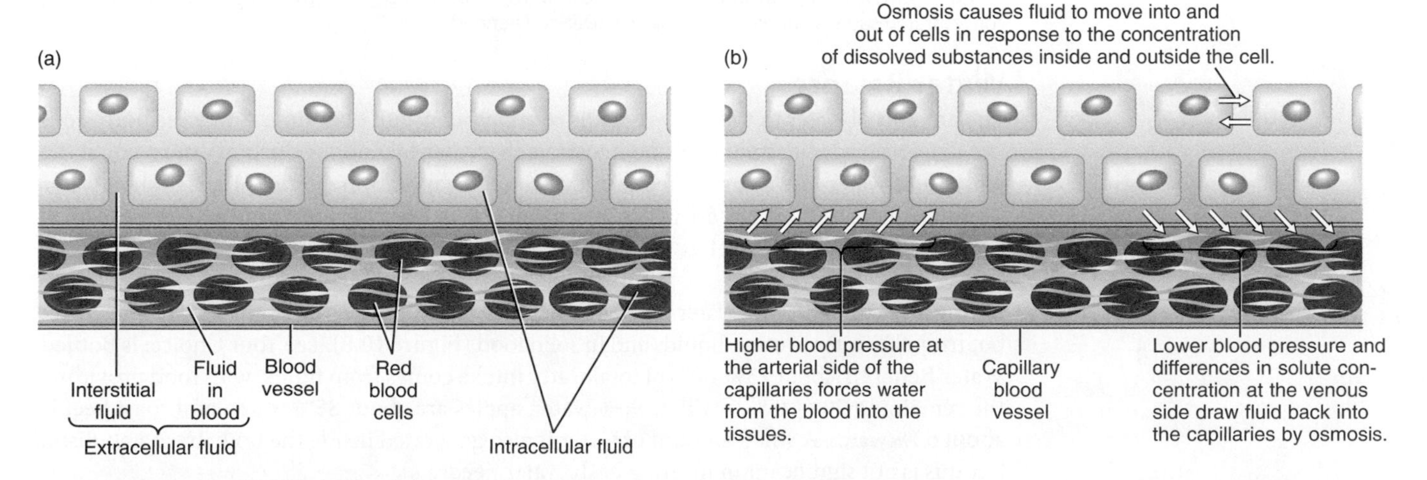

Figure 10.5 Forces that determine the distribution of body water
(a) Body water is distributed between intracellular and extracellular spaces. (b) The amount of water in blood and tissues is determined by blood pressure and the force generated by osmosis.

The movement of water from one compartment to another is affected by fluid pressure and osmosis, which depends on the concentration of solutes in each compartment (**Figure 10.5b**). The fluid pressure of blood against the blood vessel walls, or **blood pressure**, causes water to

interstitial fluid The portion of the extracellular fluid located in the spaces between the cells of body tissues.

blood pressure The amount of force exerted by the blood against the artery walls.

move from the blood into the interstitial space. The difference in the concentration of solutes between the blood in the capillaries and the fluid in the interstitial space causes much of this water to re-enter the capillaries by osmosis (see Section 3.3). Osmosis occurs when there is a selectively permeable membrane, such as a cell membrane, which allows water to pass freely but regulates the passage of other substances. Water moves across this membrane in a direction that will equalize the concentration of solutes on both sides. For example, when sugar is sprinkled on fresh strawberries, the water inside the strawberries moves across the skin of the fruit to try to equalize the sugar concentration on each side. As shown in **Figure 10.6**, this pulls water out of the fruit and causes it to shrink. The body can regulate the amount of water in each compartment by adjusting the concentration of solutes and relying on osmosis to move water.

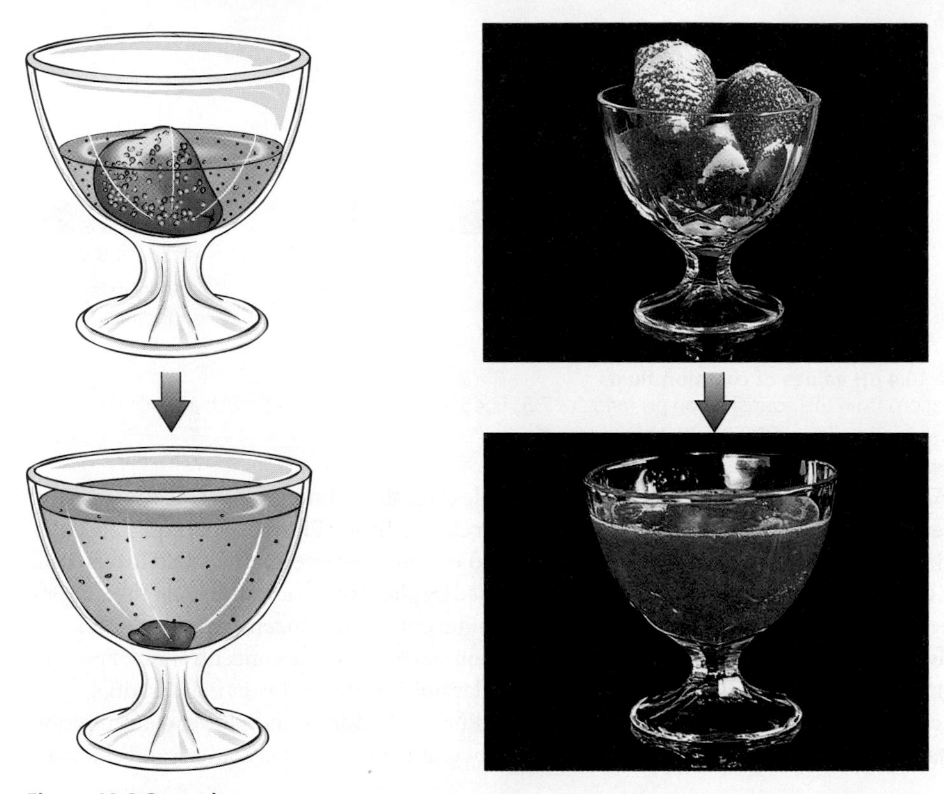

Figure 10.6 Osmosis
When sugar is sprinkled on strawberries, osmosis draws water out of the strawberries to dilute the concentrated sugar solution on the surface. (Dennis Drenner)

Water Balance

The amount of water in the body remains relatively constant over time. Since the body does not store water, to maintain this homeostasis the water taken into the body must equal the amount of water lost in urine, feces, and through evaporation **(Figure 10.7)**. When water losses are increased, as they are in hot weather and with exercise, intake must increase to keep body water at a healthy level.

Water Intake Most of the water in the body comes from the diet—not only as water we drink but from the water in other liquids and in solid food **(Figure 10.8)** (see Your Choice: Is Bottled Water Better?). About 75%-80% of total water intake comes from fluids, with food providing the remaining 20%-25%.[2] Milk is 90% water, apples are about 85% water, and roast beef is about 50% water. A small amount of water is also generated inside the body by metabolism, but this is not significant in meeting body water needs.

Water consumed in the diet is absorbed from the gastrointestinal tract by osmosis. This is possible because of the concentration gradient created by the absorption of nutrients from the lumen into the blood. As nutrients are absorbed, the solute concentration in the lumen decreases. Water therefore moves with the absorbed nutrients toward the area with the highest solute concentration. The rate of water absorption is affected by the volume of water and the concentration of nutrients consumed with it. Consuming a large volume of water increases the rate of water absorption; increasing the nutrients and other solutes it contains decreases the rate of absorption.

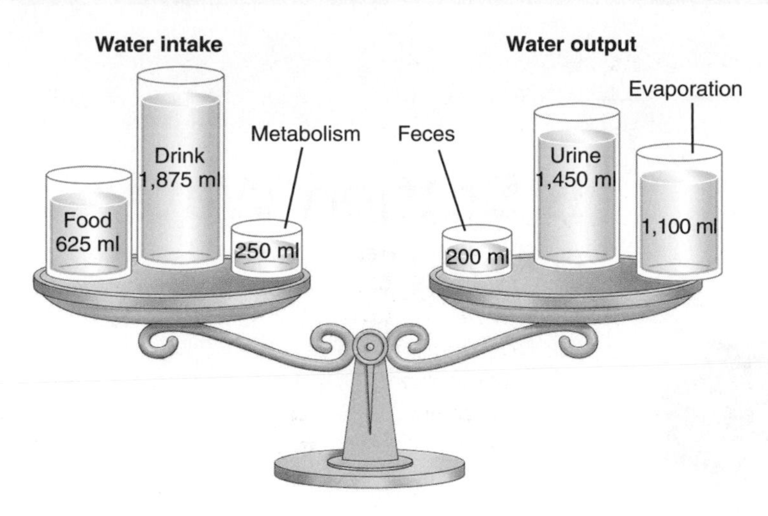

Figure 10.7 Water balance
To maintain water balance, intake must equal output. This figure approximates the amounts of water from different sources that make up water intake and the amounts of lost water in an adult who is not sweating.

Water Losses Water is lost from the body in urine, in feces, through evaporation from the lungs and skin, and in sweat. A typical adult who is not sweating loses about 2.75 L of water daily (see Figure 10.7).

Urinary Losses Typical urine output is 1-2 L/day, but this varies depending on the amount of fluid consumed and the amount of waste to be excreted. The waste products that must be excreted in urine include urea and other nitrogen-containing products from protein breakdown, ketones from fat breakdown, phosphates, sulfates, and other minerals. The amount of urea that must be excreted is increased when dietary or body protein breakdown is increased. Ketone excretion is increased when body fat is broken down. In both cases, the need for water increases in order to produce more urine to excrete the extra wastes.

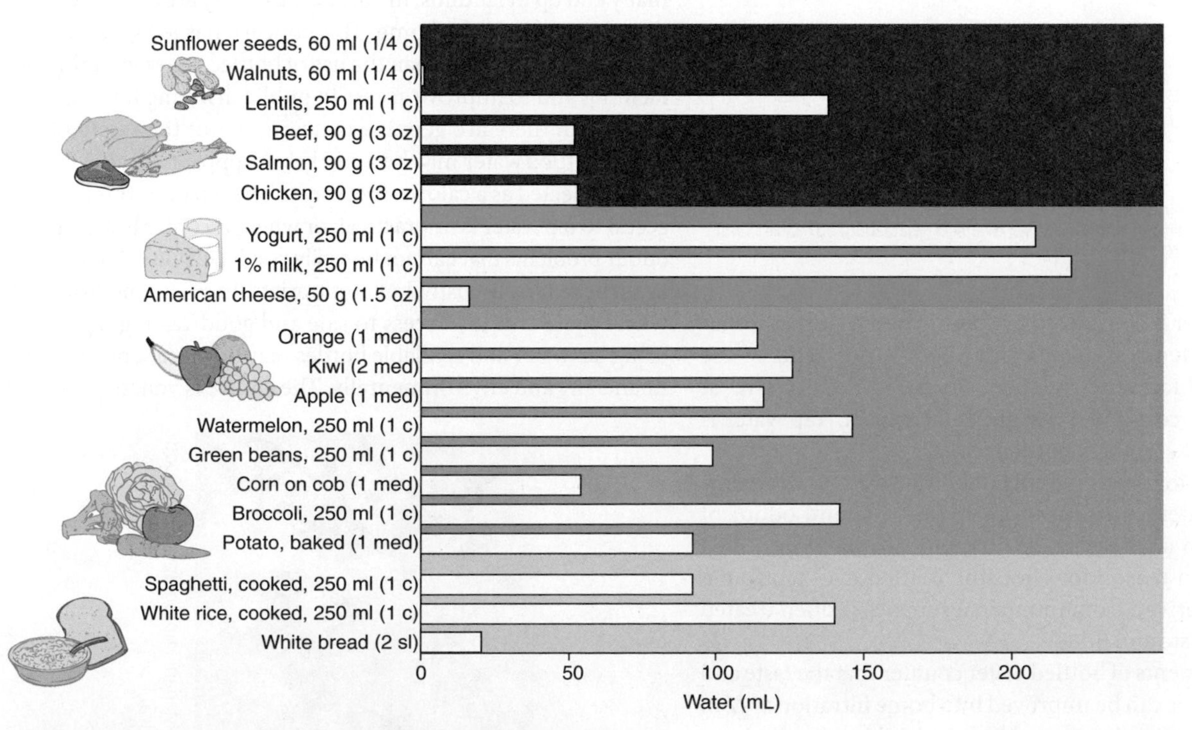

Figure 10.8 Sources of water in Canada's Food Guide food groups
Fruits and vegetables, as well as many choices from other food groups, have a high water content.

YOUR CHOICE

(©iStockphoto)

Is Bottled Water Better?

Health Canada defines bottled water as water intended for human consumption, that is packed in a sealed container. Bottled water can be derived from natural sources such as springs and aquifers or from municipal water supplies.[1]

The per-capita consumption of bottled water in Canada has increased steadily since 1999 (see graph). Statistics Canada recently reported that about a quarter to a third of Canadian households use bottled water as their main source of drinking water.[2]

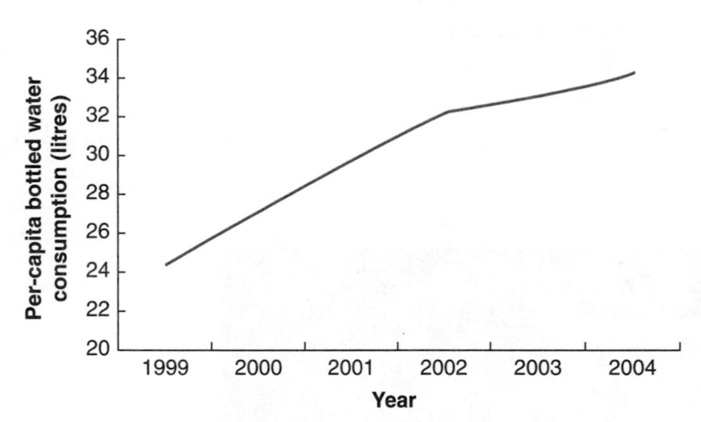

Source: Table 13: Pacific Institute. Per Capita Bottled Water Consumption by Country, 1999-2004. Available online at http://www.worldwater.org/data20062007/Table13.pdf. Accessed August 14, 2010.

Bottled water is considered by some to be a healthy alternative to sugar-laden soft drinks and juices. Opponents of the product say that tap water is an equally healthy alternative, at a fraction of the cost and environmental impact. Tap water is estimated to cost 1/1000 of bottled water.[3]

One reason for the popularity of bottled water is that compared to tap water, consumers prefer the taste and odour of bottled water. In a recent study, 70% of Canadians identified taste as the main reason for choosing bottled over tap water. Bottled water derived from municipal supplies is often treated to improve its taste and odour.[3,4]

Many opponents of bottled water counter that the taste and odour of tap water can be improved by a home filtration system or by simply allowing it to stand before drinking (to dissipate the chlorine and other odour compounds).[5]

Bottled water is also perceived as safer and healthier than tap water; approximately 25% of Canadians identified this as the reason for choosing bottled water.[4] However, the standards applied to both bottled and tap water, by federal, provincial and/or municipal agencies, are very similar. For example, federal Food and Drug regulations that govern the safety of bottled water require the product to be free of pathogenic (disease-causing) bacteria. Similarly, provincial and municipal standards that govern the quality of tap water also require that it be free of pathogenic organisms. Both types of water will contain harmless bacteria.[1]

Some voice concerns about the chemicals present in plastic bottles, although Health Canada has found no evidence that chemicals leach into the water at harmful levels.[1]

The impact on the environment of plastic bottles has also raised concerns. Compared to filling reusable bottles with tap water, plastic bottles consume considerably more energy in production and disposal. Although plastic bottles are recyclable, many end up in landfills, in part because they are often used and discarded outside the home. This concern has led several Canadian municipalities to ban the use of bottled water in municipal facilities and to improve access to public drinking fountains.[6]

When there are genuine concerns about the safety of tap water, bottled water may be a costly but appropriate alternative. When selected as a calorie-free way to quench one's thirst, when access to tap water is impractical, or when tap water has taste and odour problems that cannot be easily resolved, bottled water may again be relatively costly, but appropriate. In situations, however, where there is ready access to safe and good-tasting tap water, using tap water and refillable bottles may be more sensible, both financially and environmentally. The choice is yours.

(Rick Mariani Photography/StockFood America)

[1] Health Canada. Frequently asked questions about bottled water. Available online at http://www.hc-sc.gc.ca/fn-an/securit/facts-faits/faqs_bottle_water-eau_embouteillee-eng.php. Accessed August 14, 2010.

[2] Statistics Canada. Which households drink bottled water? Available online at http://www41.statcan.ca/2009/1762/cybac1762_002-eng.htm. Accessed August 15, 2010.

[3] Windsor Utilities Commission. Tap vs. bottled water. Available online at http://www.wuc.on.ca/conservation/tap_vs_bottled.cfm. Accessed August 16, 2010.

[4] Doria, M. F. Bottled water versus tap water: understanding consumer preferences. *Journal of Water and Health* 4(2):271-275, 2006. Available online at http://www.iwaponline.com/jwh/004/0271/0040271.pdf. Accessed August 14, 2010.

[5] Sierra Club. Bottled Water: Learning the facts and taking action, 2008. Available online at http://www.sierraclub.org/committees/cac/water/bottled_water/bottled_water.pdf. Accessed July 31, 2011.

[6] Polaris Institute. Toronto bans bottled water, 2008.Available online at http://www.polarisinstitute.org/toronto_bans_bottled_water. Accessed July 31, 2011.

Fecal Losses The amount of water lost in the feces is usually small, only about 200 mL/ day (less than a cup). This is remarkable because every day about 9 litres of fluid enter the gastrointestinal tract via food, water, and secretions. Under normal conditions, more than 95% is reabsorbed before the feces are eliminated. However, in cases of severe diarrhea, large amounts of water can be lost through the gastrointestinal tract.

Insensible Losses Water loss due to evaporation from the skin and lungs occurs without the individual being aware that it is occurring; such losses are therefore referred to as **insensible losses.** An inactive person at room temperature loses about 1 L/day through insensible losses, but the amount varies depending on body size, environmental temperature and humidity, and physical activity. For example, a very dry environment, such as that found in an airplane or in the desert, increases evaporative losses.

insensible losses Fluid losses that are not perceived by the senses, such as evaporation of water through the skin and lungs.

Sweat Losses Water lost in sweat is distinct from insensible losses because it is a detectable loss. The amount of water lost through sweat is extremely variable depending on the individual, his or her activity level, and the environmental conditions. An individual doing light work at a temperature of about 30°C will lose about 2-3 L of sweat per day. Strenuous exercise in a hot environment can cause water losses in sweat to be as high as 2-4 L/hour.[3] Adequate water intake is essential to compensate for these losses. Athletes can estimate water loss by weighing themselves before and after exercise. Lost water can then be replaced by consuming at least the equivalent weight in fluids. For instance, an athlete who loses 1 kg during a workout should consume an extra 1-2 L of fluid (1.0 kg = 1 L) (see Section 13.5).

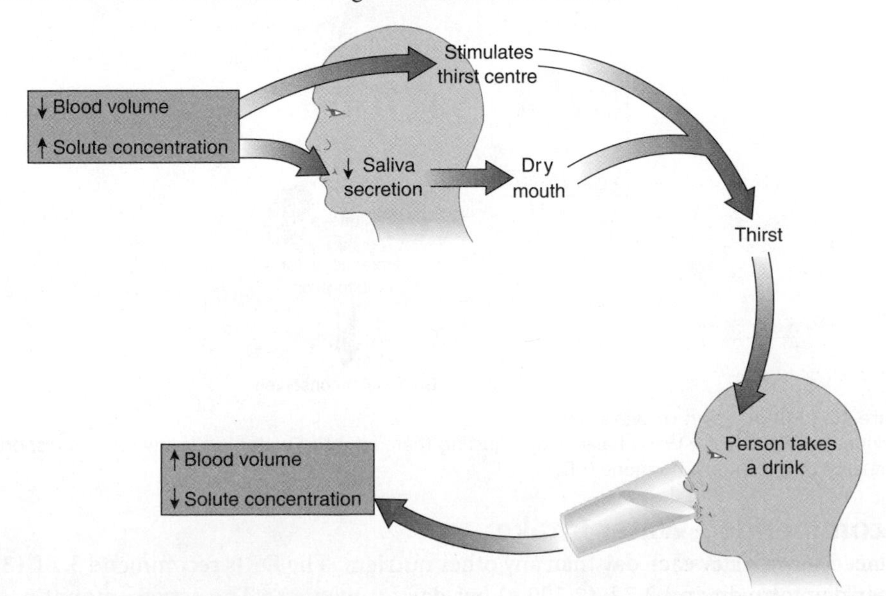

Figure 10.9 Regulation of water intake
The sensation of thirst helps motivate fluid intake in order to restore water balance.

Regulation of Water Intake The desire to drink, or thirst, is triggered by a decrease in the amount of water in the blood, which is sensed by the thirst centre in the hypothalamus of the brain. A decrease in saliva secretion, which causes a dry mouth, also stimulates thirst **(Figure 10.9).**

Over the course of days and weeks, fluid intake from beverages typically consumed during meals, along with fluid intake driven by thirst, is adequate to maintain water homeostasis. However, this is not always the case in the short term. People don't or can't always drink when they are thirsty and thirst is quenched almost as soon as fluid is consumed and long before short-term water balance is restored. Also, the sensation of thirst often lags behind the need for water. For example, athletes exercising in hot weather lose water rapidly but do not experience intense thirst until they have lost so much body water that their physical performance is compromised.[4,5] A person with fever, vomiting, or diarrhea may also be losing water rapidly and thirst mechanisms may not be adequate to replace the fluid. Thirst is a powerful urge, but to maintain adequate levels of water, the body must also regulate water excretion.

Regulation of Water Losses The amount of water lost by the body is regulated by increasing or decreasing the amount of water excreted in the urine. The kidneys serve as a filtering system that regulates the amount of water and dissolved substances retained in the blood and excreted in urine. As blood flows through the kidneys, water and small molecules are filtered out of the blood vessels (see Section 3.7). Some of the water and molecules are reabsorbed into the blood and the rest are excreted in the urine. The amount of water that is reabsorbed depends on conditions in the body. When the concentration of solutes in the blood is high, **antidiuretic hormone (ADH),** which is secreted from the pituitary gland, signals the kidneys to reabsorb water to reduce the amount lost in the urine. This reabsorbed water is returned to the blood, preventing the solute concentration in the blood from increasing further **(Figure 10.10)**. When the solute concentration in the blood is low, ADH levels decrease so less water is reabsorbed and more is excreted in the urine, allowing blood solute concentration to increase to normal. The amount of sodium in the blood, blood volume, and blood pressure also play a role in regulating body water.

antidiuretic hormone (ADH) A hormone secreted by the pituitary gland that increases the amount of water reabsorbed by the kidney and therefore retained in the body.

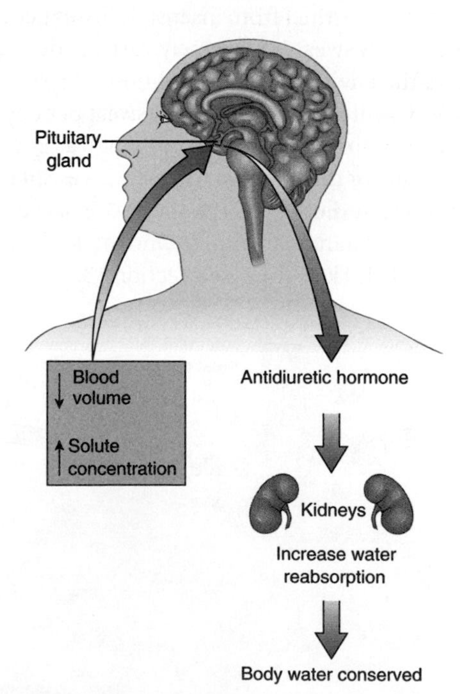

Figure 10.10 **Regulation of water loss**
The kidneys help regulate water balance by adjusting the amount of water lost in the urine in response to the release of antidiuretic hormone (ADH).

Recommended Water Intake

We need more water each day than any other nutrient. The DRIs recommend 3.7 L (3,700 g) per day for men and 2.7 L (2,700 g) per day for women.[2] The actual amount needed may vary depending on activity, heat and humidity, and diet. Activity increases water needs because it increases the amount of water lost in sweat; the amount is greater in hot and humid environments. The composition and adequacy of the diet also affect water needs. A low-kcalorie diet increases water needs because, as body fat and protein are broken down to fuel the body, ketones and urea are produced and extra water is needed for them to be excreted in the urine. A high-protein diet increases the amount of urea and other nitrogenous waste that must be excreted. A high-sodium diet increases water losses because the excess salt must be excreted in the urine. A high-fibre diet increases water needs because more water is held in the intestines and lost in the feces. Despite variations in our fluid needs, most people consume adequate fluids on a day-to-day basis to maintain body water at a normal level.[2]

Life Cycle

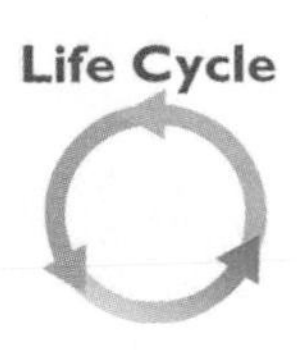

Water needs increase during pregnancy and lactation. During pregnancy, water is needed to increase blood volume, produce amniotic fluid, and nourish the fetus. The AI for pregnancy of 3 L/day is based on the median total water intake during pregnancy. During lactation, the fluid secreted in milk, about 750 mL or 3 cups/day, must be replaced by the mother's fluid intake. Based on the median fluid intake during lactation, the AI is set at 3.8 L/day.[2]

The fluid requirements per unit of body weight are higher for infants than for adults. One reason is that the infant's kidneys cannot concentrate urine as efficiently as adult kidneys, so water loss is greater. Moreover, insensible losses are proportionally greater in infants and children because body surface area relative to body weight is much greater than in adults. In addition to having greater water needs, infants are susceptible to **dehydration** because they cannot ask for a drink when they are thirsty. The AI for infants 0 to 6 months of age is 0.7 L/day and is based on the amount of water consumed in human milk.[2]

dehydration A condition that results when not enough water is present to meet the body's needs.

Water Deficiency: Dehydration

Even minor changes in the amount and distribution of body water can be life-threatening. Without food, an average individual can live for about 8 weeks, but a lack of water reduces survival to only a few days. A deficiency of water causes symptoms more rapidly than any other nutrient deficiency. Likewise, health can be restored in a matter of minutes or hours when fluid is replaced.

Dehydration occurs when the drop in body water is great enough for blood volume to decrease, thereby reducing the ability to deliver oxygen and nutrients to cells and remove waste products. Even mild dehydration—a body water loss of 1%-2% of body weight—can impair physical and cognitive performance.[2] Early symptoms of dehydration include headache, fatigue, loss of appetite, dry eyes and mouth, and dark-coloured urine **(Figure 10.11)**. A loss of 5% body weight as water can cause nausea and difficulty concentrating. Confusion and disorientation can occur when water loss approaches 7% of body weight. A loss of about 10%-20% can result in death.

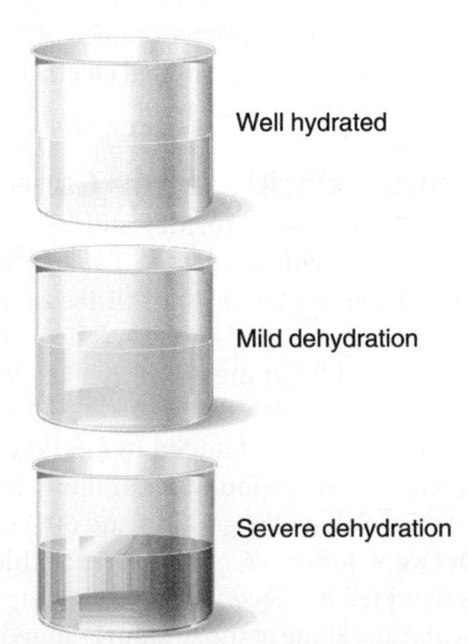

Figure 10.11 Urine colour and hydration status
The darker an individual's urine colour, the greater the level of dehydration. Pale yellow urine indicates good hydration.

Young athletes involved in sports with weight classes, such as wrestling and boxing, sometimes use dehydration to reduce their body weight so they can compete in a lower weight class. Being at the high end of the lower weight class is thought to provide an advantage over smaller opponents in that class.[6] However, when this is accomplished through even mild dehydration, exercise performance can be impaired (see Section 13.5).

Water Intoxication: Overhydration

An excess of water can affect the distribution of water among body compartments and can be just as dangerous as dehydration. However, it is difficult to consume too much water under normal circumstances; overhydration or **water intoxication** may occur with illness or in certain situations during exercise. When there is too much water relative to the amount of sodium in the body, the concentration of sodium in the blood drops, a condition called **hyponatremia.** When this occurs, water moves out of the blood vessels into the tissues by osmosis, causing them to swell. Swelling in the brain can cause disorientation, convulsions, coma, and death.

water intoxication A condition that occurs when a person drinks enough water to significantly lower the concentration of sodium in the blood.

hyponatremia Low blood sodium concentration.

The early symptoms of water intoxication may be similar to dehydration: nausea, muscle cramps, disorientation, slurred speech, and confusion. It is important to determine if the problem is dehydration or water intoxication because if you assume the symptoms are from dehydration and drink plain water, the symptoms will worsen and can result in seizure, coma, or death. To help avoid water intoxication when exercising for more than an hour, beverages such as sports drinks, containing dilute solutions of sodium as well as sugar, should be used to replace water losses (see Section 13.5).

10.2 Electrolytes: Salts of the Internal Sea

Learning Objectives

- Define the term "electrolyte" and explain the functions of electrolytes in the body.
- Contrast the dietary sources of sodium and potassium.
- Explain how blood pressure is regulated.

The water in the body—the internal sea—contains a variety of mineral salts. The right amounts and combinations of these are necessary for the maintenance of life. The distribution of these minerals affects the distribution of water in different body compartments. The properties of these minerals, including electrical charge, affect nerve and muscle functions in the body. Mineral salts dissociate in water to form ions. Positively and negatively charged ions are called **electrolytes.** Although there are many electrolytes in the body, in nutrition, the term is typically used to refer to the three principal electrolytes in body fluids: sodium, potassium, and chloride.

electrolytes Positively and negatively charged ions that conduct an electrical current in solution. Commonly refers to sodium, potassium, and chloride.

Sodium, Potassium, and Chloride in the Canadian Diet

The modern diet is high in salt (sodium chloride) and low in potassium (See Critical Thinking: How are Canadians Doing with Respect to Their Intake, from Food, of Sodium and Potassium?). The reason is that we eat a lot of processed foods, which are high in sodium and generally low in potassium, and too few fresh, unprocessed foods, such as fruits, vegetables, whole grains, and fresh meats, which are low in sodium and high in potassium (**Figure 10.12**).

About 77% of salt consumed is from that added to food during processing and manufacturing. Only about 12% comes from salt found naturally in food, while 11% is from salt added in cooking and at the table.[7] Salt is 40% sodium and 60% chloride by weight, so 9 g of salt contains 3.6 g of sodium (9 g × 40% = 3.6 g) and 5.4 g of chloride. Some of the sodium in processed foods is from salt added for flavouring; potato chips, lunchmeats, and canned soups are all high in sodium chloride. Some of the sodium is added as a preservative because it inhibits bacterial growth. In addition to sodium chloride, other sodium salts, such as sodium bicarbonate, sodium citrate, and sodium glutamate are also used as preservatives. Less than 1% of the salt we consume is from tap water.[2] Softened water or mineral water is often higher

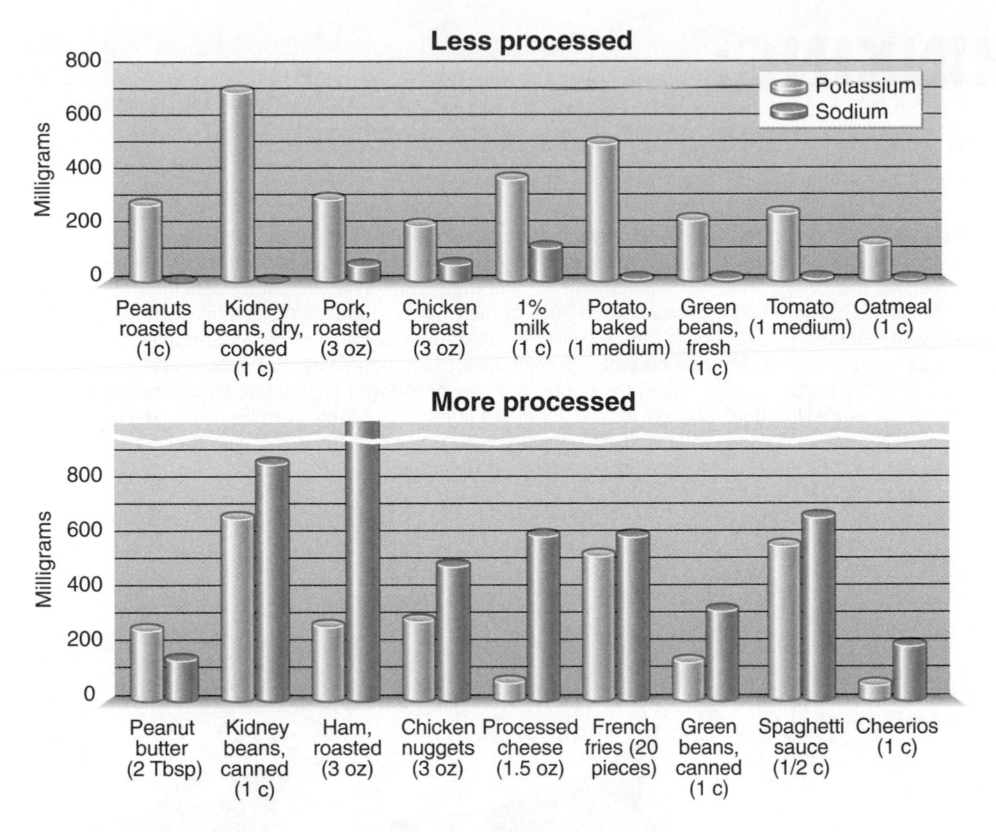

Figure 10.12 Effect of food processing on sodium and potassium
Less processed foods tend to be low in sodium and good sources of potassium, whereas more processed foods are generally higher in sodium and may also be lower in potassium.

in sodium than tap water and, if consumed in large quantities, can contribute significantly to daily sodium intake.

The human diet has not always been high in salt. Prehistoric diets consisted of plant foods such as nuts, berries, roots, and greens. These foods are high in potassium and low in salt. Fresh animal foods, such as meat and milk, are also low in sodium and high in potassium. Because of its value as a food preservative and flavour enhancer, salt was highly prized by ancient cultures in Asia, Africa, and Europe, where it was used in rituals as well as in the preservation of food. Roman soldiers were paid in *sal*, the Latin word for salt from which we get our word "salary." Today, rather than a prized commodity, salt is a substance we attempt to limit in the diet. The reason for restricting salt is that diets high in salt have been implicated as a risk factor for **hypertension.**

hypertension Blood pressure that is consistently elevated to 140/90 mm Hg or greater.

Sodium, Potassium, and Chloride in the Body

Almost all of the sodium, chloride, and potassium consumed in the diet is absorbed. Despite large variations in dietary intake, homeostatic mechanisms act to control the concentrations of these electrolytes in the body, where they help regulate fluid balance and are important for nerve conduction and muscle contraction.

Regulation of Fluid Balance The distribution of fluid among body compartments depends on the concentration of electrolytes and other solutes. Water moves by osmosis in response to solute concentration. All body fluids are in osmotic balance. So, for example, a change in the concentration of solutes in the blood causes a shift in water that affects blood volume as well as interstitial and intracellular fluid volumes. The concentration of specific electrolytes in body compartments differs dramatically. Potassium is the principal positively charged ion inside cells, where it is 30 times more concentrated than outside the cell. Sodium is the most abundant positively charged electrolyte in the extracellular fluid and chloride is the principal negatively charged extracellular ion.

CRITICAL THINKING:

How are Canadians Doing with Respect to Their Intake, from Food, of Sodium and Potassium?

The Canadian Community Health Survey collected information on the intake of sodium and potassium. The results for sodium were expressed as the percentage of the population consuming above the UL. In Section 2.2, the UL was defined as "the maximum level of daily intake of a nutrient that is unlikely to pose a risk of adverse health effects." As discussed in Section 10.3, excessive sodium intake is associated with increased risk of hypertension. Sodium intakes in different parts of Canada are also illustrated in the map from Health Canada's Nutrition and Health Atlas.

Finally data on potassium is presented. For potassium an AI is established. When assessing whether a population is meeting requirements for a nutrient with an AI, the intakes are interpreted this way: when 50% or more of the population has intakes above the AI, it can be concluded that the population is meeting requirements. If, however, less than 50% of the population has intakes above the AI, then the population may or may not be meeting requirements; no firm conclusions about the adequacy of intake can be made (See Chapter 4: Critical Thinking: How are Canadians Doing with Respect to Fibre Intake? for additional details).

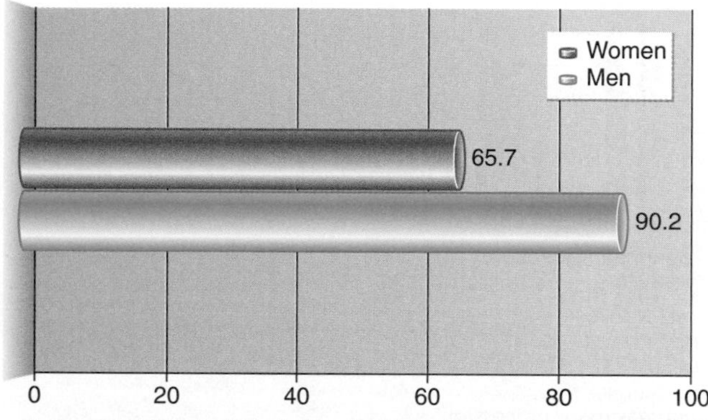

Proportion (%) of adults (+19 yr) with sodium intake above the UL

Source: Health Canada and Statistics Canada. 2004. Canada Community Health Survey Cycle 2.2., Nutrient Intakes from Food. Data files on CD. Cat No. 978-0-662-06542-5

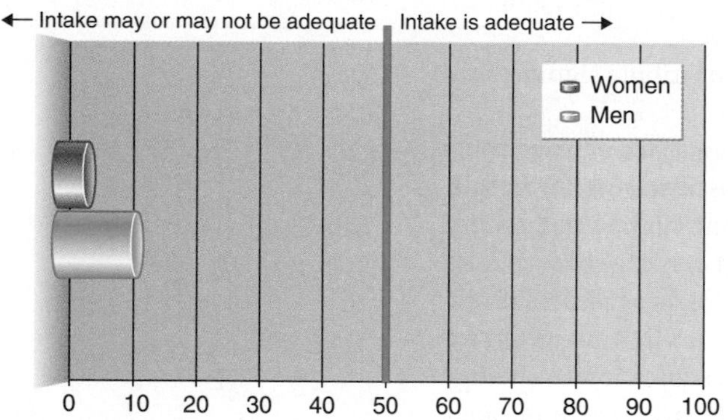

Proportion (%) of adults (+19 yr) with potassium intakes above the AI

Source: Health Canada. Canadian Community Health Survey, Cycle 2.2, Nutrition (2004), Nutrient Intakes from Food: Provincial, Regional and National Summary Data Tables, Volume 1,2 and 3. Cat No. 978-0-662-06542-5

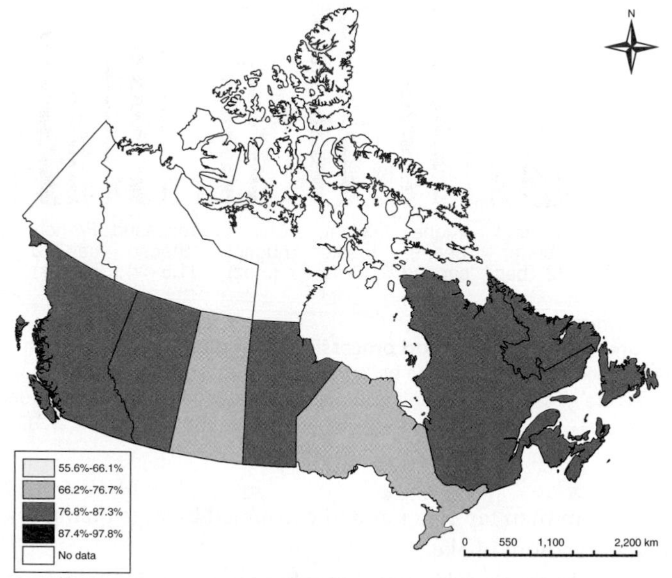

Proportion (%) of adults with sodium intakes above the UL.

Source: Health Canada: Nutrition and Health Atlas. Available online at http://www.hc-sc.gc.ca/fn-an/surveill/atlas/map-carte/nutri_sod_mf-hf-eng.php. Accessed July 31, 2011.

Critical Thinking Questions

Based on the graph showing the proportion of Canadian adults with sodium intakes above the UL, what would you conclude about Canadians' exposure to adverse health effects arising from their sodium intake?*

What does the map suggest about regional differences in sodium intake across Canada?

What does Canada's Food Guide recommend to lower sodium intake?

Can anything be definitively concluded about the potassium intake of Canadian adults?

Which of Canada's Food Guide food groups should Canadians consume to increase their potassium intake (see Figure 10.8)? What is the impact of processing on potassium content of food (see Figure 10.12)?

*Answers to Critical Thinking Questions can be found in Appendix J.

[1] Health Canada. Do Canadian adults meet their nutrient requirements through food intake alone? 2009. Available online at http://www.hc-sc.gc.ca/fn-an/surveill/nutrition/commun/art-nutr-adult-eng.php. Accessed July 18, 2010.

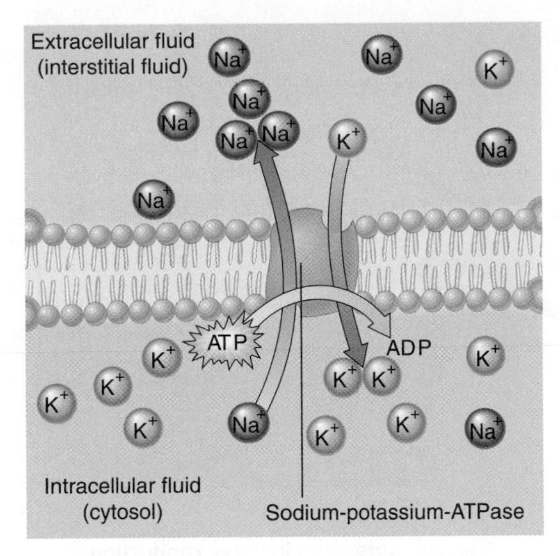

Figure 10.13 Sodium-potassium-ATPase
Each cycle of sodium-potassium-ATPase uses the energy in ATP to pump 2 potassium ions into the cell and expel 3 sodium ions.

The different intracelluar and extracellular concentrations of these electrolytes is maintained by cell membranes and an active transport system called the sodium-potassium-ATPase. Cell membranes keep the majority of sodium outside the cell and potassium inside, but some does leak across the membrane. The sodium-potassium-ATPase maintains the concentration gradient by pumping sodium out of the cell and pumping potassium into the cell (**Figure 10.13**). Maintaining this concentration gradient of electrolytes is important for nerve conduction and muscle contraction, but requires a great deal of energy: it is estimated that the sodium-potassium-ATPase pump accounts for 20%-40% of resting energy expenditure in an adult. The pumping of sodium ions across cell membranes is also linked to nutrient transport. For example, glucose and amino acids are transported by systems that depend on the movement of sodium ions across cell membranes.

Nerve Conduction and Muscle Contraction The concentration gradient of sodium and potassium across nerve cell membranes is important for the conduction of nerve impulses (see Table 10.1). An electrical charge, or membrane potential, exists across nerve cell membranes because the number of negative ions just inside the cell membrane is greater than the number outside. This occurs because the cell membrane allows more positively charged ions to leak out of the cell than to leak into the cell. Nerve impulses are created by a change in the electrical charge across cell membranes. Stimuli, such as touch or the presence of neurotransmitters, change the cell membrane's permeability to sodium, allowing it to rush into the cell (**Figure 10.14**). This reverses, or depolarizes, the charge of the cell membrane at that location, and an electrical current is generated. The nerve impulse travels along the nerve cell as an electrical current. Once the nerve impulse passes, the original membrane potential is rapidly restored by another change in cell membrane permeability; then the original distribution of sodium and potassium ions across the cell membrane is restored by the sodium-potassium-ATPase pump in the cell membrane. A similar mechanism causes the depolarization of the muscle cell membranes, leading to muscle contraction.

Regulation of Electrolyte Balance In northern China, typical sodium chloride intake is greater than 13.9 g/day; in the Kalahari Desert, it is less than 1.7 g/day; and among the Yanomani Indians of Brazil, salt consumption may be less than 0.06 g/day.[2] Despite these variations in intake, homeostatic mechanisms ensure that blood levels of sodium are not significantly different among these groups.

Sodium and chloride homeostasis is regulated to some extent by the intake of both water and salt. When salt intake is high, thirst is stimulated to increase water intake. When salt intake is very low, a salt appetite causes the individual to seek out the mineral. These mechanisms

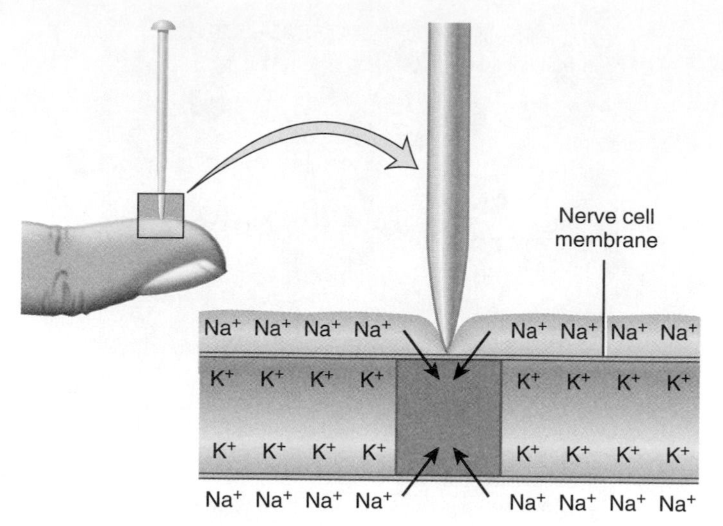

Figure 10.14 The role of sodium and potassium in nerve conduction
You feel a pinprick because it stimulates nerves beneath the surface of the skin. This stimulation increases the permeability of the nerve cell membrane to sodium, which rushes into the cell (shown here), initiating a nerve impulse.

help ensure that appropriate proportions of salt and water are taken in. The kidneys are the primary regulator of sodium, chloride, and potassium balance in the body. Excretion of these electrolytes in the urine is decreased when intake is low and increased when intake is high. Because water follows sodium by osmosis, the ability of the kidneys to conserve sodium provides a mechanism to conserve body water.

Sodium plays a pivotal role in regulating extracellular fluid volume. When the concentration of sodium in the blood increases, water follows, causing an increase in blood volume. An increase in blood volume typically increases blood pressure, an important factor in hypertension (Section 10.3). A decrease in blood volume typically decreases blood pressure. Changes in blood pressure trigger the production and release of proteins and hormones that affect the amount of sodium, and hence water, retained by the kidneys. For example, when blood pressure decreases, the kidneys release the enzyme **renin**, beginning a series of events leading to the production of **angiotensin II** (**Figure 10.15**). Angiotensin II increases blood pressure both by causing the blood vessel walls to constrict and by stimulating the release of the hormone **aldosterone**, which acts on the kidneys to increase sodium (and chloride) reabsorption. Water follows the reabsorbed sodium, helping to maintain blood volume and, consequently, blood pressure. As blood pressure increases, it inhibits the release of renin and aldosterone so that blood pressure does not continue to rise. This system of renin-angiotensin-aldosterone, which increases blood pressure by sodium retention, acts in combination with ADH, which increases blood pressure by water reasborption (Figure 10.10). Some cases of hypertension may be due to malfunctions in the renin-angiotensin-aldosterone system.

renin An enzyme produced by the kidneys that converts angiotensin to angiotensin I.

angiotensin II A compound that causes blood vessel walls to constrict and stimulates the release of the hormone aldosterone.

aldosterone A hormone that increases sodium reabsorption by the kidney and therefore enhances water retention.

The amount of potassium in the body is also tightly regulated. If blood levels begin to rise, mechanisms are activated to stimulate the cellular uptake of potassium. This short-term regulation prevents the amount of potassium in the extracellular fluid from getting lethally high. The long-term regulation of potassium balance depends on aldosterone release, which causes the kidneys to excrete potassium and retain sodium. Aldosterone release is stimulated by high-blood potassium, low-blood sodium, or angiotensin II.

Recommended Sodium, Chloride, and Potassium Intakes

The AI for sodium for adults ages 19-50 years is 1,500 mg/day and for chloride, 2,300 mg/day, which is equivalent to 3.8 g of salt per day[2] (see Table 10.1). This amount ensures an adequate intake of other nutrients and accounts for sodium losses in sweat. The UL is set at 2,300 mg/day

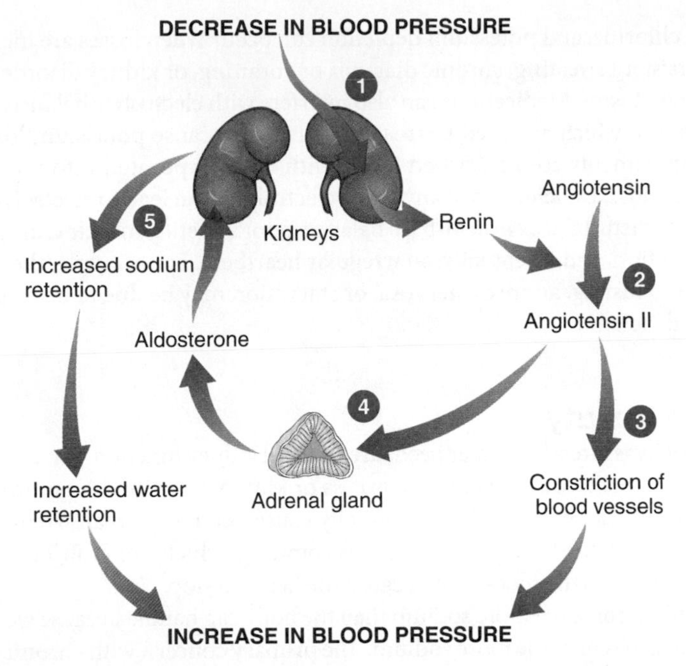

1. A decrease in blood pressure triggers the kidney to release the enzyme renin.
2. Renin converts angiotensin into angiotensin I, which is activated to angiotensin II.
3. Angiotensin II increases blood pressure by constricting the walls of blood vessels.
4. Angiotensin II stimulates release of aldosterone from the adrenal gland.
5. Aldosterone increases sodium reabsorption by the kidneys. Water follows the sodium, helping to maintain blood volume and blood pressure.

Figure 10.15 Regulation of blood pressure
A drop in blood pressure triggers events that cause blood vessels to constrict and the kidneys to retain water. An increase in blood pressure inhibits these events so that blood pressure does not continue to rise.

of sodium and 3,600 mg/day of chloride. This is equivalent to 5.8 g of salt, or about 1 teaspoon per day. The UL is based on the adverse effects of higher levels of sodium on blood pressure.

The Daily Value for sodium of no more than 2,400 mg/day (6 g) is close to the UL for sodium.

The AI for potassium is set at 4,700 mg/day, a level that will lower blood pressure and reduce the adverse effects of sodium on blood pressure. The Daily Value is at least 3,500 mg/day for adults. No UL has been set for potassium because potassium intake from foods is not a risk in healthy people with normal kidney function.

Life Cycle

There is no evidence that sodium and chloride requirements differ during pregnancy so pregnant women are advised to follow the recommendation for the general population for sodium intake[2] (see Section 14.1). At one time, a dietary salt restriction was common during pregnancy to prevent a syndrome known as pregnancy-induced hypertension. The cause of pregnancy-induced hypertension is not known, but salt restriction is no longer recommended. Potassium recommendations are not increased during pregnancy, but during lactation, the AI is higher to account for the potassium lost in milk. In infants, sodium, chloride, and potassium needs are estimated from the amount consumed in human milk, which contains more chloride than sodium. This same chloride-to-sodium ratio has been recommended for infant formulas.

Electrolyte Deficiency

The electrolytes are found in plentiful amounts in the diet, and the kidneys of a healthy individual are efficient at regulating amounts in the body. However, illness and extreme conditions can increase electrolyte losses and affect overall health.

Sodium, chloride, and potassium depletion can occur when losses are increased due to heavy and persistent sweating, chronic diarrhea or vomiting, or kidney disorders that lead to excessive urinary losses. Medications can also interfere with electrolyte balance. For example, thiazide diuretics, which are used to treat hypertension, cause potassium loss. Generally, potassium supplements are prescribed along with or incorporated into medications that cause potassium loss. Deficiencies of any of the electrolytes can lead to electrolyte imbalance, which can cause disturbances in acid-base balance, poor appetite, muscle cramps, confusion, apathy, constipation, and, eventually, an irregular heartbeat. For example, the sudden death that can occur in fasting, anorexia nervosa, or starvation may be due to heart failure caused by potassium deficiency.

Electrolyte Toxicity

Electrolyte toxicity is rare when water needs are met and kidney function is normal. If, however, potassium supplements are consumed in excess or kidney function is compromised, blood levels of potassium can increase and potentially cause death due to an irregular heartbeat. A high oral dose of potassium generally causes vomiting, which limits absorption, but if too much potassium enters the blood, it can cause the heart to stop.

It is difficult to consume more sodium than the body can handle because we usually drink more water when we consume more sodium. The primary concern with chronic high sodium intake is elevated blood pressure. Though rare, elevation of blood sodium can result from dehydration or from massive ingestion of salt, such as may occur from drinking seawater or consuming salt tablets. The symptoms of high blood sodium, or hypernatremia, are similar to those of dehydration. High sodium intake also increases calcium excretion and has been implicated as a risk factor for osteoporosis.[8]

Other Electrolytes

While sodium, potassium and chloride are the principal electrolytes in body fluid, other nutrients play are also present as electrolytes. These nutrients include the cations calcium and magnesium and the anions phosphorus, in the form of phosphate, and sulfur, in the form of sulfur-containing amino acids which are discussed in Chapter 11. Bicarbonate, another anion, is a dissolved form of carbon dioxide.

10.3 Hypertension

Learning Objectives

- Define "hypertension" and list its symptoms and consequences.
- Discuss the effect of dietary salt intake on blood pressure.
- Describe the DASH diet and how it affects blood pressure.

CANADIAN CONTENT

A certain level of blood pressure is necessary to ensure that blood is delivered to all the body tissues. An optimal blood pressure is less than 120/80 mm of mercury (mm Hg). The higher number is systolic pressure, the maximum pressure in the artery. The lower number is diastolic pressure, the minimum pressure in the artery. Blood pressure from 120/80-139/89 is referred to as prehypertension. Prehypertension indicates an increased risk for developing hypertension.[9] Blood pressure that is consistently 140/90 mm of mercury or greater indicates hypertension (see Appendix C).[10] High blood pressure damages artery walls, increases the workload of the heart, and can cause blood vessel walls to weaken or rupture. Hypertension is associated with increased risk of cardiovascular diseases such as atherosclerosis, heart attack, and stroke.[10] It also increases the risks for kidney disease and early death. Hypertension is a serious public health concern in Canada; about 20% of adult Canadians have hypertension and another 20% are prehypertensive. It is also estimated that 17% of Canadians don't know they have hypertension[11] (**Figure 10.16**).

What Causes Hypertension?

Most people with high blood pressure have essential hypertension—hypertension with no obvious external cause. It is a complex disorder, most likely resulting from disturbances in one or more of the mechanisms that control body fluid and electrolyte balance. High blood pressure that occurs as a result of other disorders is referred to as secondary hypertension. For example, if atherosclerosis causes a reduction in blood flow to the kidneys, the kidneys respond by releasing renin. This triggers events that lead to an increase in blood volume and a constriction of blood vessels. This raises blood pressure to the kidneys, but also has the undesirable effect of raising blood pressure throughout the body.

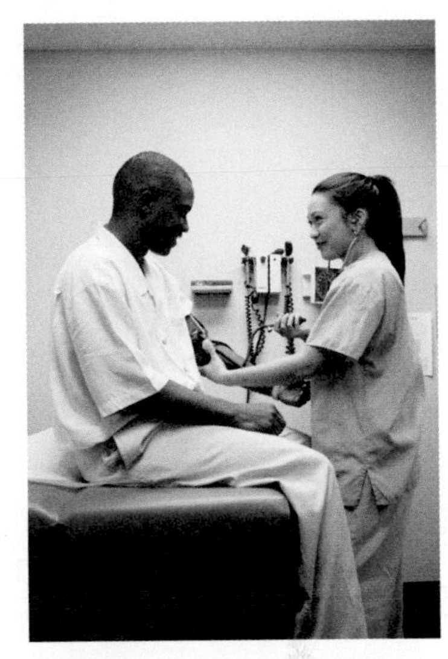

Figure 10.16 Everyone should have regular blood pressure monitoring, because hypertension has no outward symptoms.(Andersen Ross/ Getty Images, Inc.)

Factors that Affect the Risk of Hypertension

The risk of developing high blood pressure is affected by genetics, age, existing disease conditions, and lifestyle factors. A family history of high blood pressure increases your risk of developing this disorder. The genetic basis of hypertension is also illustrated by the fact that it is more common in people of certain ethnic and racial backgrounds. For example, studies of ethnic groups in Canada suggest that South Asians (people from India, Pakistan, Sri Lanka, and Bangledesh), blacks, and Aboriginal populations are at increased risk of hypertension.[12,13] In the United States, African Americans, Puerto Ricans, and Cuban and Mexican Americans have a higher incidence of hypertension than non-Hispanic whites.[14] The increased incidence among African Americans is reflected in their 1.8 times greater rate of death from stroke, 1.5 times greater rate of death from heart disease, and 4.2 times greater rate of hypertension-related kidney failure compared to whites.[15]

The risk of hypertension increases with age. One reason is that as people age, the arteries lose their elasticity. Diabetes also increases the risk of high blood pressure. Kidney damage is generally the cause of high blood pressure in type 1 diabetes. In type 2 diabetes, the higher incidence of hypertension may also be due to the effect of high insulin levels on sodium retention in the kidneys as well as the presence of excess body fat. A lack of physical activity, heavy alcohol consumption, smoking, stress, and a number of dietary factors can also increase blood pressure.[16] Obesity, particularly abdominal obesity, increases the risk of hypertension. Excess adipose tissue adds miles of capillaries through which blood must be pumped. Weight loss can prevent or delay the onset of hypertension in obese individuals.

Diet and Blood Pressure

The connection between sodium and blood pressure is well known. On average, as sodium intake goes up, so does blood pressure, but sodium is not the only mineral that affects blood pressure. Diets high in potassium, calcium, and magnesium are associated with a lower incidence of hypertension.

Salt Intake and Blood Pressure The relationship between salt intake and blood pressure was first identified by examining hypertension in populations with different average dietary salt intakes. It was found that in populations consuming less than 4.5 g of salt per day, average blood pressure was low and hypertension was rare or absent. In populations consuming 5.8 g or more of salt per day, blood pressure increased with sodium intake.[17] More recent intervention trials have examined the effect of different levels of sodium intake on blood pressure. It was found that the lower the amount of sodium in the diet. the lower the blood pressure.[18] When compared to an intake of 3,400 mg of sodium (the average level consumed by Canadians) an intake of 2,400 mg of sodium reduced blood pressure in those with and without hypertension, and even more significant reductions were seen when sodium intake was reduced to 1,500 mg of sodium per day. These studies and others are the basis for the recommendation for Canadians to reduce their sodium intake to less than 2,300 mg/day (see: Sodium Reduction Strategy for Canada). Despite the general effect of sodium intake on blood pressure, not everyone who consumes more than the recommended daily amount of sodium will develop hypertension. Individuals with hypertension, diabetes, and chronic kidney disease, as well as older individuals and certain ethnic groups, tend to be more sensitive to salt intake.[2] It is believed that people are salt sensitive because of a defect in the excretion of sodium by the kidneys.[19]

The consequences of this is shown in **Figure 10.17.** Increased sodium intake, by raising the levels of sodium in the blood, increases fluid retention, resulting in an increase in both blood volume and pressure. In most healthy individuals, kidneys respond by excreting sodium and normal blood pressure is restored. In a salt-sensitive individual, the excretion of sodium is impaired. This may be due to genetic predisposition or disease-induced kidney damage. Scientific evidence also suggests that high sodium levels further exacerbate the problem by causing the constriction of blood vessels.[19] The combination of these two effects results in the persistence of high blood pressure.

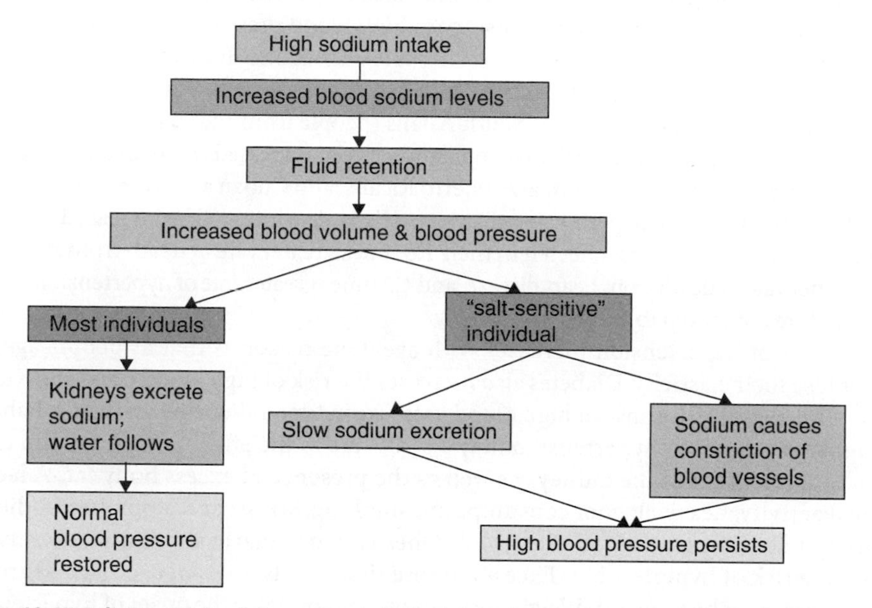

Figure 10.17 Salt Sensitivity Salt-sensitive individuals have impaired sodium excretion, which in combination with a high sodium intake, increases blood pressure.

Potassium, Calcium, and Magnesium Intake and Blood Pressure Dietary patterns with high intakes of fibre and the minerals potassium, magnesium, and calcium are associated with lower blood pressure.[20] For example, populations and individuals consuming vegetarian diets, which are high in these nutrients, generally have lower blood pressure than nonvegetarians.[21] Population surveys have shown that a dietary pattern low in calcium, potassium, and magnesium is associated with hypertension in adults.[22] Despite these data, studies that have explored the impact of individual nutrients on blood pressure are often inconclusive. This may be because the impact of each individual nutrient is small and that other dietary components are also important in blood pressure regulation.

The first intervention trial to look at the effect of a dietary pattern high in potassium, calcium, and magnesium on blood pressure was the DASH trial, which stands for Dietary Approaches to Stop Hypertension[23] (see Science Applied: A Total Dietary Approach to Reducing Blood Pressure). In this trial, the amount of dietary sodium was kept constant and different dietary patterns were compared in terms of their effect on blood pressure. The greatest reduction in blood pressure was found with a dietary pattern that was high in fruits and vegetables and included low-fat milk and alternative products, and lean meat, fish, and poultry. This diet was higher in potassium, magnesium, calcium, and fibre, and lower in fat, saturated fat, and cholesterol than the typical North American diet. The results of the DASH trial demonstrate that changing the dietary pattern can lower blood pressure. In addition, the DASH dietary pattern may also reduce cancer risk, prevent osteoporosis, and protect against heart disease (see Critical Thinking: A Diet for Health). The DASH dietary pattern or DASH Eating Plan is very similar to Canada's Food Guide.

Sodium Reduction Strategy for Canada

In response to the health concerns associated with sodium intake, in 2007, the Canadian minister of health announced the formation of the Sodium Working Group to develop a strategy for the reduction of sodium intake by Canadians.

In July 2010, the working group submitted its report that outlined a three-pronged strategy to for sodium reduction.[24] The objective of the strategy is to reduce the average intake of sodium by Canadians from its current level of 3,400 mg/day to the UL of 2,300 mg/day by 2016 and subsequently to further reduce intake so that at least 95% of Canadians are consuming less than 2,300 mg/day. This would correspond to an average intake of about 1,500 mg/day.

The strategy had three major components:

1. The reduction of sodium levels in processed foods and restaurant foods through the development of sodium reduction targets and voluntary product reformulation by the food industry to meet these targets. A recommendation was also made to change the Daily Value for sodium from 2,400 mg to 1,500 mg.
2. Increasing education and awareness about sodium reduction in the food industry and health care professions. Programs would also be directed to the general population. These would include campaigns to help the consumer understand the importance of sodium reduction and the use of the Nutrition Facts Tables to select low-sodium food products (see Label Literacy: Pass on the Salt).
3. Increasing research in three major areas:
   a. health effects of sodium
   b. food reformulation; given that sodium plays an important role in the taste and preservation of foods, sodium reduction is a significant technological challenge
   c. Policy implementation; how best to translate knowledge into action, and barriers to action.

Finally, all three components listed above are to be monitored and evaluated for effectiveness and progress reports are to be issued regularly.

If the recommendations of the working group are fully implemented, a reduction in the amount of sodium in the food supply is likely to follow. It has been estimated that reducing sodium intake to target levels would reduce hypertension by 30% and cardiovascular disease by 13% and save $1.4 billion in health care costs.[24]

Choosing a Diet and Lifestyle to Prevent Hypertension

Diet and lifestyle are both involved in regulating blood pressure. Maintaining body weight in a healthy range, staying active, and limiting alcohol consumption will help keep blood pressure in the normal range (**Table 10.2**). Consuming a diet described in Canada's Food Guide or the DASH eating plan, rich in fresh fruits, vegetables, legumes, nuts and seeds, whole grains, lean meats, and low-fat milk and alternative products is also important in keeping blood pressure normal (**Table 10.3**).

Table 10.2 Lifestyle Choices to Keep Blood Pressure in the Normal Range
• Eat plenty of fruits and vegetables—they are naturally low in salt and kcalories and rich in potassium.
• Choose and prepare foods with less salt. Limit fast foods, canned foods, and highly processed foods as they tend to be high in salt.
• Aim for a healthy weight—blood pressure increases with increases in body weight and decreases when excess weight is reduced.
• Increase physical activity—it helps lower blood pressure, reduce the risk of other chronic diseases, and manage weight.
• If you drink alcoholic beverages, do so in moderation. Excessive alcohol consumption has been associated with high blood pressure.
• Quit smoking.
• Lower stress.

Source: Hypertension Canada, 2010. Public Recommendations. Available online at http://hypertension.ca/bpc/wp-content/uploads/2010/03/PublicRec2010_120310.pdf. Accessed April 27, 2011.

LABEL LITERACY

Pass on the Salt

CANADIAN CONTENT

Most of the sodium in the Canadian diet comes from processed foods. Sodium chloride is the most common form in which sodium is added to foods; other sodium-containing ingredients include sodium hydroxide, sodium salts such as baking soda and baking powder, and monosodium glutamate (MSG).

Food labels list the sodium-containing ingredients in the ingredient list, and the Nutrition Facts panel gives the total amount of sodium in milligrams and as a percentage of Daily Value to help you make comparisons between food products and assess the amount of sodium present. For example, the Nutrition Facts label shown here indicates that a serving of this spaghetti sauce contains 120 mg of sodium, or 5% of the Daily Value. Hypertension Canada, an organization dedicated to reducing hypertension, advises Canadians that foods that contain less than 200 mg of sodium per serving can be consumed readily but that they should avoid foods that contain more than 400 mg (slightly over 15% of the DV) of sodium per serving.[1]

Food labels can also include nutrient content claims relating to salt or sodium content (see table). The spaghetti sauce shown here can be labelled "Low in Sodium," because it contains less than 140 mg of sodium. Additionally, the following health claim could be made on foods, such as the can of spaghetti sauce shown below, that contain more than 350 mg of potassium and are low in sodium or sodium free: "A healthy diet high in potassium and low in sodium reduces the risk of high blood pressure, which is a risk factor for stroke and heart disease."

One must always be thorough when reading food labels. For example, a food can be labelled "reduced in sodium" when it contains 25% less sodium than a reference food. So a soup manufacturer can label a product that contains 450 mg of sodium "reduced in sodium" if its regular product contains 600 mg. Although all reductions in sodium in the food supply are beneficial, the reduced-sodium soup still exceeds the recommendations of Hypertension Canada (more than 400 mg/serving) and contains 19% of the DV, a high-sodium product.

Nutrient Content Claims for Salt and Sodium on Food Labels	
Sodium-free	Contains less than 5 mg of sodium per serving.
Salt-free	Must meet criterion for "sodium-free."
Low sodium	Contains 140 mg or less of sodium per serving.
Reduced in sodium or salt; less sodium or salt; lower in sodium or salt	Contains at least 25% less sodium per serving than a reference food.
No salt added, without added salt, and unsalted	No salt added during processing, and the food it resembles and for which it substitutes is normally processed with salt.
Lightly salted	Contains at least 50% less added sodium per serving than that added to a reference food.

Nutrition Facts
Valeur nutritive
Per 125 ml / par 125 ml

Amount / Teneur	% Daily Value / % valeur quotidienne
Calories / Calories 50	
Fat / Lipides 1 g	**2 %**
Saturated / saturés 0 g + Trans / trans 0 g	**0 %**
Cholesterol / Cholestérol 0 mg	
Sodium / Sodium 120 mg	**5 %**
Potassium / Potassium 530 g	**15 %**
Carbohydrate / Glucides 9 g	**3 %**
Fibre / Fibres 1 g	**4 %**
Sugars / Sucres 7 g	
Protein / Protéines 3 g	
Vitamin A / Vitamine A	10 %
Vitamin C / Vitamine C	25 %
Calcium / Calcium	2 %
Iron / Fer	10 %

[1] Hypertension Canada. A short summary about dietary sodium for public awareness. Available online at http://www.lowersodium.ca/en/public/tools. Accessed August 15, 2010.

SCIENCE APPLIED

A Total Dietary Approach to Reducing Blood Pressure

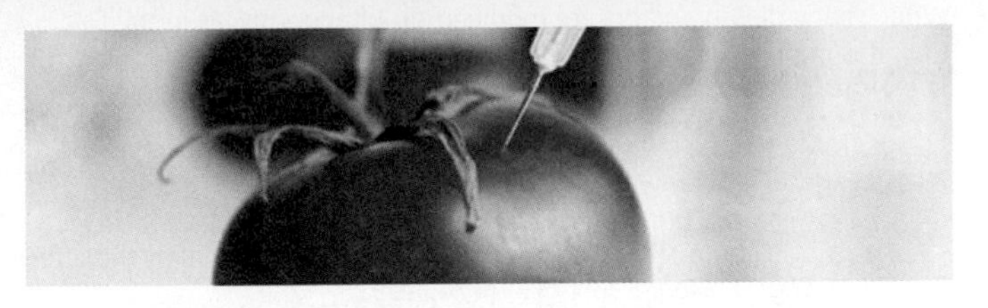

For over 30 years, an American organization, the National Heart, Lung, and Blood Institute (NHLBI) of the National Institutes of Health (NIH) had been making recommendations for the treatment of high blood pressure through modifications of individual dietary components such as sodium and other minerals as well as fat, cholesterol, fibre, and protein. A review of these studies revealed enticing but conflicting and equivocal results. This led the NHLBI to pursue a study that would evaluate the effects of dietary patterns rather than individual dietary components on blood pressure.[1] As a result, an intervention trial called the DASH (Dietary Approaches to Stop Hypertension) trial was conducted, finally providing conclusive evidence that diet does affect blood pressure.[2]

Designing the DASH trial presented several challenges. The study required large numbers of subjects to consume a specific diet for a fairly long period. To accrue enough subjects, the study needed to be run simultaneously from several different study centres in the United States. Subjects' diets had to be standardized across these centres.[1] Controlling dietary intake also presented challenges because the subjects lived at home but had to consume one meal a day, five days a week at the research study centres; the remaining meals were prepared by the research centres and sent home with the subjects to be consumed at home. To control for extra intake, subjects were also required to keep a daily diary of any non-study items consumed.

The DASH study population included 459 adults with slightly elevated blood pressure who were not taking blood pressure medications. Study subjects consumed a control diet for the first 3 weeks of the study, and their blood pressure was monitored. After 3 weeks, they were randomly assigned to one of three dietary groups for 8 weeks. During this intervention period, blood pressure was measured by individuals who did not know to which dietary groups the participants had been assigned.

The study diets, summarized in the table, included a "control" diet, a "fruits and vegetables" diet, and a "combination" diet, the DASH eating plan. Each diet provided about 3,000 mg of sodium—slightly lower than the typical U.S. intake—and enough kcalories to maintain body weight. The "control" diet was a typical American dietary pattern—low in fibre and high in fat and saturated fat. It matched the average U.S. nutrient intake except for potassium, magnesium, and calcium, which were set below average. The "fruits and vegetables" diet increased servings of fruits and vegetables and decreased sweets but was otherwise similar to the control diet. This diet was higher in fibre than the control, and the potassium and magnesium were above the typical American intake. The "combination" diet increased fruits, vegetables, and low-fat milk and alternative products and reduced red meat, sweets, and sugar-sweetened drinks; it was higher in potassium, calcium, magnesium, and fibre than the control diet and was lower in total fat, saturated fat, and cholesterol.

Diet	Sodium Intake
Control diet:	3,000 mg
Low in fibre	
High in total fat and saturated fat	
Fruits and vegetables diet:	3,000 mg
Compared to the control:	
More fruits and vegetables	
DASH eating plan:	*3,000 mg*
Compared to the control diet:	
***More** fruits and vegetables*	
***More** low-fat milk products*	
***Less** red meat, sweets, sugar*	
***More** calcium, magnesium, and fibre*	
***Less** total fats, saturated fats, and cholesterol*	

After only 2 weeks on the diet, participants consuming the "fruits and vegetables" diet had a blood pressure reduction of 2.8 mm Hg systolic and 1.1 mm Hg diastolic. Those consuming the "combination" diet had a reduction in blood pressure of 5.5 mm Hg systolic and 3 mm Hg diastolic compared with the control group.[2] These numbers may seem very small but estimates suggest that if the population adopted the DASH combination diet, the incidence of coronary heart disease would be reduced by 15% and stroke by 27%.

A second trial, DASH-Sodium, compared the effect of the DASH Eating Plan with a control diet (typical American diet) at three different levels of sodium intake, as shown in the table below.[3]

	Control Diet			**DASH diet**		
Sodium Level	**High**	**Intermediate**	**Low**	**High**	**Intermediate**	**Low**
*Sodium intake (mg) **	*3,450*	*2,300*	*1,150*	*3,450*	*2,300*	*1,150*

*Based on 2,100 kcal/day

The study included 412 participants with mild hypertension or prehypertension. In both the DASH diet and the control diet, lowering the sodium lowered blood pressure (see bar graph). The combination of the DASH diet and the lowest sodium intake reduced blood pressure more than either the DASH diet alone or low sodium alone. Compared with the control diet with high sodium, the DASH diet with the lowest sodium lowered systolic blood pressure by 7.1 mm Hg in subjects without hypertension and by 11.5 mm Hg in those with hypertension—an effect equal to or greater than what would be expected from treatment with a single hypertension medication.

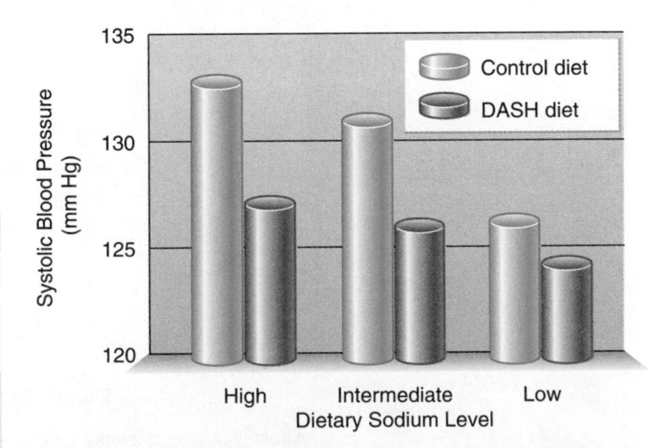

A decrease in sodium intake lowered blood pressure more in the control diet than in the DASH diet. This is likely because blood pressures were already lower in the DASH diet group. The lowest blood pressure resulted from a combination of DASH and low sodium.

When many dietary factors are modified simultaneously, as in the DASH trial, it is impossible to determine which ones cause the effects seen. Therefore, the DASH trial taught scientists less about the physiology of hypertension than do studies that look at individual factors, but it provided more information about a daily dietary pattern that can prevent or reduce hypertension.[1]

[1] Vogt, T. M., Appel, L. J., Obarzanek, E. et al. Dietary Approaches to Stop Hypertension: Rationale, design and methods. *J. Am. Diet. Assoc.* 99:12s–18s, 1999.

[2] Appel, L. J., Moore, T. J., Obarzanek, E. et al. A clinical trial of the effects of dietary patterns on blood pressure. DASH Collaborative Research Group. *N. Engl. J. Med.* 336:1117–1124, 1997.

[3] Sacks, F. M., Svetkey, L. P., Vollmer, W.M. et al. Effects on blood pressure of reduced dietary sodium and the Dietary Approaches to Stop Hypertension (DASH) diet. DASH-Sodium Collaborative Research Group. *N. Engl. J. Med.* 344:3–10, 2001.

Table 10.3 The DASH Eating Plan

		Daily Servings per Kcalorie Level			
Food Group	**Serving Sizes**	**1,600**	**2,000**	**2,600**	**3,100**
Grains[a]	1 slice bread, 30 g (1 oz dry) cereal, 125 ml (1/2 cup) cooked rice, pasta, or cereal	6	7–8	10–11	12–13
Vegetables	250 ml (1 cup) raw leafy vegetables, 125 ml (1/2 cup) cooked vegetables, 175 ml (6 oz) vegetable juice	3–4	4–5	5–6	6
Fruits	175 ml (6 oz) fruit juice, 1 medium fruit, 60 ml (1/4 cup) dried fruit, 125 ml (1/2 cup) fresh, frozen, or canned fruit	4	4–5	5–6	6
Low-fat milk	250 ml (8 oz) milk, 250 ml (1 cup) yogurt, 50 g (1½ oz) cheese	2–3	2–3	3	3–4
Meats, fish, poultry	90 g (3 oz) cooked meat, poultry, or fish	1–2	2 or less	2	2–3
Beans, nuts, and seeds	80 ml (1/3 cup) nuts, 30 ml (2 Tbsp) seeds, 125 ml (1/2 cup) cooked dry beans or peas	3 per week	4–5 per week	1	m
Fat and oils[b]	5 ml (1 tsp) soft margarine, 15 ml (1 Tbsp) low-fat mayonnaise, 30 ml (2 Tbsp) light salad dressing, 5 ml (1 tsp) vegetable oil	2	2–3	3	m
Sweets	15 ml (1 Tbsp) sugar, 15 ml (1 Tbsp) jelly or jam, 15 g (1/2 oz) jelly beans, 250 ml (8 oz) lemonade	0	5 per week	2	2

[a]Whole grains are recommended for most servings to meet fibre recommendations.

[b]Fat content changes the number of servings for fats and oils; 15 ml (One Tbsp) regular salad dressing equals one serving, 15 ml (1 Tbsp) low-fat dressing equals 1/2 serving, and 15 ml (1 Tbsp) of fat-free dressing equals 0 servings.

Source: Dietary Guidelines for Americans, 2005.

CRITICAL THINKING:

A Diet for Health

(©iStockphoto)

Background

Rashamel's father died of a stroke at the age of 54 as a result of undiagnosed and untreated high blood pressure. Because he wants to live to see his grandchildren, Rashamel exercises for about 30 minutes on most days of the week, doesn't smoke, and watches his weight and salt intake. Despite these efforts, at his recent physical his blood pressure was 138/87, in the prehypertension category. Rashamel's doctor suggests he see a dietitian to help him manage his blood pressure with diet.

Data

Rashamel is 43 years old and weighs 80 kg (175 lbs). The dietitian recommends he follow the DASH Eating Plan to reduce his blood pressure. Rashamel's current diet is shown here.

CURRENT DIET

Breakfast	
Orange juice	125 ml (1/2 cup)
1% low-fat milk	250 ml (1 cup)
Mini-Wheats™ w/1 tsp sugar	30 g (1 oz))
Whole-wheat bread w/jelly	1 slice
Margarine	5 ml (1 tsp)
Lunch	
Tuna, canned	75 g (2.5 oz)
Wheat bread	2 slices
Chips	30 g (1 oz)
Cola	1 can (355 ml)
Dinner	
Baked chicken	90 g (3 oz)
White rice	250 ml (1 cup)
Salad	250 ml (1 cup)
Light salad dressing	15 ml (1 Tbsp)
Large dinner roll	1
Margarine	10 ml (2 tsp)
Cantaloupe	125 ml (1/2 cup)
Iced tea (sweetened)	360 ml (12 oz)
Snacks	
Cookies	1 large
Dried apricots	5
Mars bar	1
Cola	1 can (355 ml)

Critical Thinking Questions

How healthy is Rashamel's current diet? Does it meet the recommendations of Canada's Food Guide for grains, fruits, and vegetables?

Now compare his diet to the recommendations of the DASH Eating Plan for someone who eats 2,600 kcal/day (see Table 10.3). Does he meet these recommendations? How many additional servings from grains, fruits, and vegetables would he need to add?

Rashamel's wife is concerned about her risk for osteoporosis. Use Table 10.3 and iProfile to plan a day's meals that follow a 1,600-kcal DASH Eating Plan and meet her calcium needs.

Use iProfile to find out how much sodium is provided by your favourite snack foods.

Limiting sodium intake is also important for preventing hypertension. To reduce sodium intake, Canadians need to limit processed foods and added salt and increase their intake of fresh foods. **Table 10.4** provides some tips for reducing sodium intake. Food labels can be helpful in selecting lower-sodium foods (see Label Literacy: Pass on the Salt). Foods with less salt may taste bland at first, but the preference for highly salted food is acquired. After consuming lower-salt foods for a period of time, the desire for salt decreases. Adding other flavourings such as onions, garlic, lemon juice, vinegar, black pepper, parsley, and other herbs to food may also help satisfy your taste for flavourful food without adding sodium.

Table 10.4 Tips for Reducing Sodium Intake

When shopping: • Use food labels to select foods low in sodium. • Choose unprocessed foods—they have less sodium than processed foods. • Choose fresh or frozen vegetables rather than canned. • Choose fresh or frozen fish, shellfish, poultry, and meat more often than canned or processed forms.
When cooking: • Prepare meals from scratch so you control the amount of salt added. • Do not add salt to the water when cooking rice, pasta, and cereals. • Flavour foods with ingredients such as lemon juice, onion or garlic powder (not salt), pepper, curry, dill, basil, oregano, or thyme rather than salt.
When eating: • Limit salt use at the table. • Limit salted snack foods like potato chips, salted popcorn, and crackers, and replace them with fresh fruits and vegetables. • Limit cured, salted, or smoked meats such as bologna, corned beef, hot dogs, and smoked turkey to a few servings a week or less. Substitute sliced roasted turkey, chicken, or beef. • Limit salty or smoked fish such as sardines, anchovies, or smoked salmon (lox). • Limit foods prepared in salt brine such as pickles, olives, and sauerkraut. • Cut down on cheeses, especially processed cheeses. • Limit the amounts of soy sauce, Worcestershire sauce, barbecue sauce, ketchup, and mustard you add to food.
When eating out: • Choose foods without sauces, or ask for them to be served on the side. • Ask that food be prepared without added salt. • Limit your fast food intake—it is usually very high in salt. • Reduce the salt in your diet gradually so that you learn to enjoy the unsalted flavours in foods.

The average Canadian diet contains only about 2,500-3,500 mg of potassium per day—well below the recommendation of 4,700 mg/day.[25] This is because most Canadians do not consume the recommended amounts of fruits and vegetables—the best sources of potassium per kcalorie (**Figure 10.18**). Meat, milk, and cereal products also provide potassium. Canada's

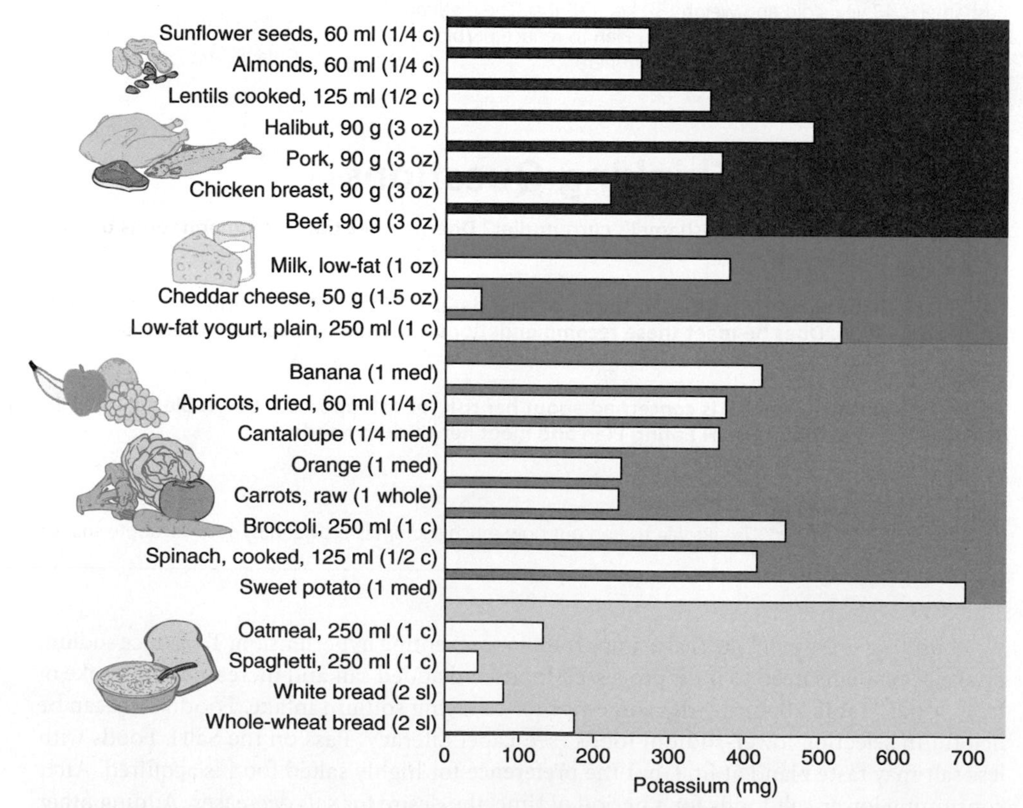

Figure 10.18 Sources of potassium in Canada's Food Guide food groups
Foods from all Canada's Food Guide food groups are good sources of potassium, but whole unprocessed foods are the best.

Food Guide and DASH Diet Plan recommendations for fruit and vegetable intake easily meet the potassium recommendation (**Figure 10.19**). When diets are high in fruits and vegetables, intakes of 8,000-1,000 mg/day are not uncommon.[2]

Figure 10.19 The amounts in the measuring cups shown here (about 500 ml (2 cups) of fruit and 625 ml (2.5 cups) of cooked vegetables) represent the amounts recommended by the DASH Eating Plan for a 2,000-kcal diet. (Andy Washnik)

CASE STUDY OUTCOME

After collapsing at the finish line of the 2010 Montreal Marathon, William Naranjo was taken to the first aid area of the Olympic Stadium, where he began to vomit, one of the many symptoms of dehydration, which also include headache, thirst, dizziness, muscle cramps, chills, and fatigue. In faster runners, like Naranjo, thirst often lags behind dehydration and so runners have to estimate their water losses during a race and ensure that they drink enough while running to compensate. The hilly streets of Montréal proved more challenging than expected for Naranjo and he underestimated how much water he needed to drink. The amount of water in his body began to decline, his blood volume decreased, and sufficient oxygen and nutrients could not be delivered to his exercising muscles. By the time he entered the stadium for his final lap, the lack of blood to his muscles slowed him down just enough to deprive him of the win.

Fortunately, Naranjo recovered quickly after receiving fluids. Despite the disappointment of finishing second, he summed up his Montréal experience positively: "I am happy I came. Everyone treated me well. I hope to come back next year!"

APPLICATIONS

Personal Nutrition

1. **Do you drink enough fluid?**
   a. Keep a log of all the fluids you consume in one day. Calculate your intake by totalling the volume of water, beverages, as well as foods that are liquid at room temperature.
   b. How does your intake on this day compare with the AI?
   c. How should your intake change if you added an hour of jogging or basketball to your day?
2. **How does your diet compare to the DASH Eating Plan?**
   a. Use one day of the food record you kept in Chapter 2 to compare the number of servings you ate from each of the food groups to the number of servings recommended by the DASH Eating Plan for your energy intake as shown in Table 10.3.
   b. Suggest modifications to your diet that would allow it to meet the DASH guidelines.
   c. What difficulties or inconveniences do you see with following this dietary pattern?
   d. What other dietary or lifestyle changes might you make if you are at high risk of hypertension?
3. **How much sodium do processed foods add to your diet?**

iProfile

   a. Use iProfile or food labels to estimate the amount of sodium you consume from processed foods each day.
   b. Make a list of processed foods you commonly eat that contain more than 10% of the Daily Value for sodium (2,400 mg) per serving.
   c. What less-processed choices might you substitute for these?

General Nutrition Issues

1. Virginia's mother has high blood pressure and has had several strokes. A recent physical exam indicates that Virginia, who is 40 years old, also has high blood pressure. Her physician prescribes medication to lower her blood pressure, but he believes that with some changes in diet and lifestyle, Virginia's blood pressure can be brought into the normal range without drugs. Virginia works at a desk job. The only exercise she gets is when she takes care of her nieces and nephews one weekend a month. Her typical diet includes a breakfast of cereal, tomato juice, and coffee. She has a snack of doughnuts and coffee at work, and for lunch she joins co-workers for a fast-food cheeseburger, fries, and milkshake. When she gets home she has a soda and snacks on peanuts or chips. Dinner is usually a frozen dinner with milk.

   a. What dietary changes would you recommend for Virginia?
   b. What lifestyle changes would you recommend?

2. The recommendations of Canada's Food Guide and the DASH Eating Plan are similar.

   a. What recommendations does Canada's Food Guide make in terms of servings for each food group?
   b. Review the recommendations for the DASH Eating Plan for 2,000 kcal (Table 10.3).
   c. Make a chart that compares the amounts of food from each food group recommended by Canada's Food Guide and the DASH Eating Plan.
   d. Which diet do you think would be easier to consume? Why?

SUMMARY

10.1 Water: The Internal Sea

- Water is an essential nutrient that provides many functions in the body. The polar structure of water allows it to act as a solvent for the molecules and chemical reactions involved in metabolism. Water helps to transport other nutrients and waste products within the body and to excrete wastes from the body. It also helps to protect the body, regulate body temperature, and lubricate areas such as the eyes and the joints.
- The adult human body is about 60% water by weight. Body water is distributed between intracellular and extracellular compartments. The amount in each compartment depends largely on the concentration of solutes. Since water will diffuse by osmosis from a compartment with a lower concentration of solutes to one with a higher concentration, the body regulates the distribution of water by adjusting the concentration of electrolytes and other solutes in each compartment.

How it Works

View in your WileyPLUS
www.wileyplus.com

- Water cannot be stored, so intake must balance losses to maintain homeostasis. Water is lost from the body in urine and feces, through evaporation from the skin and lungs, and in sweat. The kidney is the primary regulator of water output. If water intake is low, antidiuretic hormone will cause the kidney to conserve water. If water intake is high, more water will be excreted in the urine.
- The recommended intake of water is 2.7 L/day for women and 3.7 L/day for men; needs vary depending on environmental conditions and activity level. Fluid intake is stimulated by the sensation of thirst, which occurs in response to a decrease in blood volume and an increase in the concentration of solutes.
- Dehydration can occur if water intake is too low or output is excessive. Mild dehydration can cause headache, fatigue, loss of appetite, dry eyes and mouth, and dark-coloured urine. Water intoxication causes hyponatremia, which can result in abnormal fluid accumulation in body tissues.

10.2 Electrolytes: Salts of the Internal Sea

- The Canadian diet is abundant in sodium and chloride from processed foods and table salt but generally low in potassium, which is high in unprocessed foods such as fresh fruits and vegetables. Recommendations for health suggest that we increase our intake of potassium and consume less sodium.

- The minerals sodium, chloride, and potassium are electrolytes that are important in the maintenance of fluid balance and the formation of membrane potentials essential for nerve transmission and muscle contraction.
- The recommended salt intake is 3.8 g/day for adults ages 19-50. Because salt is 40% sodium and 60% chloride by weight, this represents 1,500 mg of sodium and 2,300 mg of chloride. The Daily Value for sodium is somewhat higher—no more than 2,400 mg/day. The UL for sodium, which is based on the increase in blood pressure seen with higher sodium intakes, is only 2,300 mg/day. The DRIs recommend a potassium intake of 4,700 mg/day; the Daily Value is at least 3,500 mg/day for adults. This amount is significantly higher than the typical 2,500-3,500 mg consumed by most Canadian adults. No UL has been set for potassium.
- Electrolyte and fluid homeostasis is regulated primarily by the kidneys. A decrease in blood pressure or blood volume signals the release of the enzyme renin, which helps form angiotensin II. Angiotensin II causes blood vessels to constrict and the hormone aldosterone to be released. Aldosterone causes the kidneys to reabsorb sodium and hence water, thereby preventing any further loss in blood volume. Failure of these regulatory mechanisms may be a cause of hypertension.

10.3 Hypertension

- A healthy blood pressure is less than 120/80 mm of mercury. Blood pressure from 120/80-139/89 is referred to as prehypertension, and blood pressure that is consistently 140/90 mm of mercury or greater indicates hypertension.
- Hypertension is common in Canada. A diet high in sodium increases blood pressure especially in salt-sensitive individuals. High intakes of the minerals potassium, magnesium, and calcium help lower blood pressure. Maintaining a healthy weight and exercise program helps prevent hypertension.
- Diets high in sodium and low in potassium are associated with an increased risk of hypertension. The DASH diet—a dietary pattern moderate in sodium; high in potassium, magnesium, calcium, and fibre; and low in fat, saturated fat, and cholesterol—lowers blood pressure.
- To reduce the risk of hypertension, public health recommendations suggest a dietary pattern that is high in fruits, vegetables, whole grains, legumes, nuts and seeds, and provides low-fat milk and alternative products and lean meat, fish, and poultry. This dietary pattern contains less sodium and more potassium than the typical North American diet. Blood pressure management also requires maintenance of a healthy weight, an active lifestyle, and a limit on alcohol consumption.

REVIEW QUESTIONS

1. Describe the functions of water in the body.
2. How is the total amount of water in the body regulated?
3. How is the amount of water in each body compartment regulated?
4. What is the recommended water intake for adults?
5. List 3 factors that increase water needs.
6. Define electrolyte.
7. How do sodium, potassium, and chloride function in the body?
8. Explain how a drop in blood pressure is returned to normal.
9. What are the consequences of untreated hypertension?
10. What types of foods contribute the most sodium to the North American diet?
11. What types of foods are good sources of potassium?
12. What is the relationship between dietary sodium and blood pressure?
13. What is the DASH diet, and how does it affect blood pressure?

REFERENCES

1. Basu, A. Heartbreak, triumph at Montreal Marathon. Available online at http://www.faceoff.com/Heartbreak+triumph+Montreal+Marathon/3489538/story.html. Accessed April 28, 2011.
2. Institute of Medicine, Food and Nutrition Board. *Dietary Reference Intakes for Water, Salt and Potassium*. Washington, DC: National Academies Press, 2004.
3. Shen, H-P. Body fluids and water balance. In *Biochemical and Physiological Aspects of Human Nutrition*. 2nd ed. M. Stipanuk, ed. St. Louis: Saunders Elsevier, 2006, pp. 973–1000.
4. Murray, B. Hydration and physical performance. *J. Am. Coll. Nutr.* 26: 542S–548S, 2007.
5. Coyle, E. F. Fluid and fuel intake during exercise. *J. Sports Sci.* 22:39–55, 2004.
6. Oppliger, R. A., Case, H. S., Horswill, C. A. et al. American College of Sports Medicine position statement: Weight loss in wrestlers. *Med. Sci. Sports Exerc.* 28:ix–xii, 1996.
7. U.S. Department of Health and Human Services; U.S. Department of Agriculture. Dietary Guidelines for Americans, 2005. Available online at www.healthierus.gov/dietaryguidelines/. Accessed May 19, 2009.
8. Teucher, B., Fairweather-Tait, S. Dietary sodium as a risk factor for osteoporosis: Where is the evidence? *Proc. Nutr. Soc.* 62:859–866, 2003.
9. U.S. Department of Health and Human Services. The Seventh Report of the Joint National Committee on Prevention, Detection, Evaluation, and Treatment of Blood Pressure. NIH Publication No. 04–5230, 2004.
10. Chobanian, A. V., Bakris, G. L., Black, H. R. et al. Seventh report of the Joint National Committee on Prevention, Detection, Evaluation, and Treatment of High Blood Pressure. Joint National Committee on Prevention, Detection, Evaluation, and Treatment of High Blood Pressure. National Heart, Lung, and Blood Institute. *Hypertension* 42:1206–1252, 2003.
11. Wilkins, K., Campbell, N. R. C., Joffres, M. R. et al. Blood pressure in Canadian adults. *Health Reports* 21(1): 1-9. Available online at http://www.statcan.gc.ca/pub/82-003-x/82-003-x2010001-eng.htm. Accessed August 12, 2010.
12. Veenstra, G. Racialized identity and health in Canada: Results from a nationally representative survey. *Social Science and Medicine* 69: 538-542, 2009.
13. Chiu, M., Austin, P. C., Manuel, D. G. et al. Comparison of cardiovascular risk profiles among ethnic groups using population health surveys between 1996 and 2007. *CMAJ* 182(8):E301-310, 2010.
14. Ostchega, Y., Yoon, S. S., Hughes, J., et al. *Hypertension Awareness, Treatment, and Control—Continued Disparities in Adults: United States*, 2005–2006. Available online at www.cdc.gov/nchs/data/databriefs/db03.pdf. Accessed August 25, 2008.

15. American Heart Association. *Heart Disease and Stroke Statistics—2009 Update* (At-a-Glance Version). Available online at www.americanheart.org/presenter.jhtml?identifier=3037327. Accessed October 8, 2009.

16. American Heart Association. Am I at Risk? Factors That Contribute to High Blood Pressure. Available online at www.americanheart.org/presenter.jhtml?identifier=2142. Accessed August 25, 2008.

17. Carvalho, J. J., Baruzzi, R. G., Howard, P. F. et al. Blood pressure in four remote populations in the Intersalt study. *Hypertension* 14:238–246, 1989.

18. Sacks, F. M., Svetkey, L. P., Vollmer, W. M. et al. Effects on blood pressure of reduced dietary sodium and the Dietary Approaches to Stop Hypertension (DASH) diet. DASH–Sodium Collaborative Research Group. *N. Engl. J. Med*. 344:3–10, 2001.

19. Rodriguez-Iturbe, B., Romero, F., Johnson, R.J. Pathophysiological mechanisms of salt-dependent hypertension. *Am. J. Kidney Dis.* 20150(4):655-72, 2007.

20. Houston, M. C., and Harper, K. J. Potassium, magnesium, and calcium: Their role in both the cause and treatment of hypertension. *J. Clin. Hypertens. (Greenwich)* 10(7 Suppl. 2):3–11, 2008.

21. Craig, W. J., Mangels, A. R. American Dietetic Association. Position of the American Dietetic Association: Vegetarian diets. *J. Am. Diet. Assoc.* 109:1266–1282, 2009.

22. Townsend, M. S., Fulgoni, V. L. 3rd, Stern, J. S. et al. Low mineral intake is associated with high systolic blood pressure in the Third and Fourth National Health and Nutrition Examination Surveys: Could we all be right? *Am. J. Hypertens*. 18:261–269, 2005.

23. Appel, L. J., Moore, T. J., Obarzanek, E. et al. A clinical trial of the effects of dietary patterns on blood pressure. *N. Engl. J. Med*. 336:1117–1124, 1997.

24. Sodium Working Group: Health Canada. Sodium Reduction Strategy for Canada: 2010. Available online at http://www.hc-sc.gc.ca/fn-an/nutrition/sodium/strateg/index-eng.php. Accessed August 14, 2010.

25. Health Canada and Statistics Canada. 2004. Canadian Community Health Survey. Cycle 2.2. Nutrient Intakes from Food. Data on CD Cat No. 978-0-662-06542-5.

11 Major Minerals and Bone Health

Case Study

Margie felt a searing pain in her hip when her foot struck the pavement, and the next thing she knew, she was lying in the street.

Her daughter and son-in-law would be arriving soon with her new grandson, and she had been rushing back from the corner store with some last-minute groceries. Stepping off the curb to cross the street had fractured her hip. The visit with her grandson would take place in the hospital rather than her home.

Margie is 75 years old, is 1.5 m (5′ 2″) tall, and weighs 52 kg (115 pounds). She just retired from her engineering job and lives alone in her apartment. She is always busy taking care of her home and doing volunteer work at the library. Although she is generally healthy and has never felt the need for regular checkups, she does have a history of broken bones. Six months ago, she broke her wrist after slipping and falling on an icy patch on the sidewalk. Five years, ago she suffered a stress fracture in her foot while sightseeing on Cape Breton Island. How could stepping off a curb cause her hip to break? How would it affect her future?

A health history reveals that Margie has never got much exercise, and her dietary calcium intake is about a quarter of the recommended 1,200 mg/day. Her hip broke because the bone had been weakened by osteoporosis, a disease characterized by low bone density. Bone density is higher in people who exercise and consume adequate calcium. Margie's lifelong low-calcium intake and limited exercise increased her risk of developing this disease. Her hip fracture will require hospitalization, surgery, and a long period of rehabilitation. It may impair her ability to walk unassisted and cause prolonged or permanent disability. On a positive note, Margie is still in good health, so rehabilitation, diet, and medications can help her recover from the injury.

Many older adults who experience a hip fracture due to osteoporosis require long-term, nursing-home care and never regain their independence.

In this chapter, you will learn about minerals such as calcium and how they function in bone health.

(Barbara Penoyar/Photodisc/Getty Images, Inc.)

(©iStockphoto)

(PhotoDisc. Inc./Getty Images)

Chapter Outline

11.1 What Are Minerals?

Learning Objectives

- Define "mineral" in terms of nutrition.
- Describe how interactions among minerals and other dietary components affect mineral bioavailability.

minerals In nutrition, elements needed by the body in small amounts for structure and to regulate chemical reactions and body processes.

major minerals Minerals needed in the diet in amounts greater than 100 mg/day or present in the body in amounts greater than 0.01% of body weight.

trace elements or **trace minerals** Minerals required in the diet in amounts of 100 mg or less per day or present in the body in amounts of 0.01% of body weight or less.

Minerals are inorganic elements needed by the body as structural components and regulators of body processes. Minerals may combine with other elements in the body, but they retain their chemical identity. Unlike vitamins, they are not destroyed by heat, oxygen, or acid. The ash that remains after a food is combusted in a bomb calorimeter contains the minerals that were present in that food.

Minerals have traditionally been categorized based on the amounts needed in the diet or present in the body. The **major minerals** include those needed in the diet in amounts greater than 100 mg/day or present in the body in amounts greater than 0.01% of body weight. The **trace elements** or **trace minerals** are minerals required by the body in an amount of 100 mg or less per day or present in the body in an amount of 0.01% or less of body weight (**Figure 11.1**). The major minerals include the electrolytes sodium, chloride, and potassium, which are discussed in Chapter 10. Calcium, phosphorus, and magnesium, discussed in this chapter, are major minerals that play a role in bone health, and sulfur is a major mineral that functions in association with other molecules, such as vitamins and amino acids. The trace elements are discussed in Chapter 12.

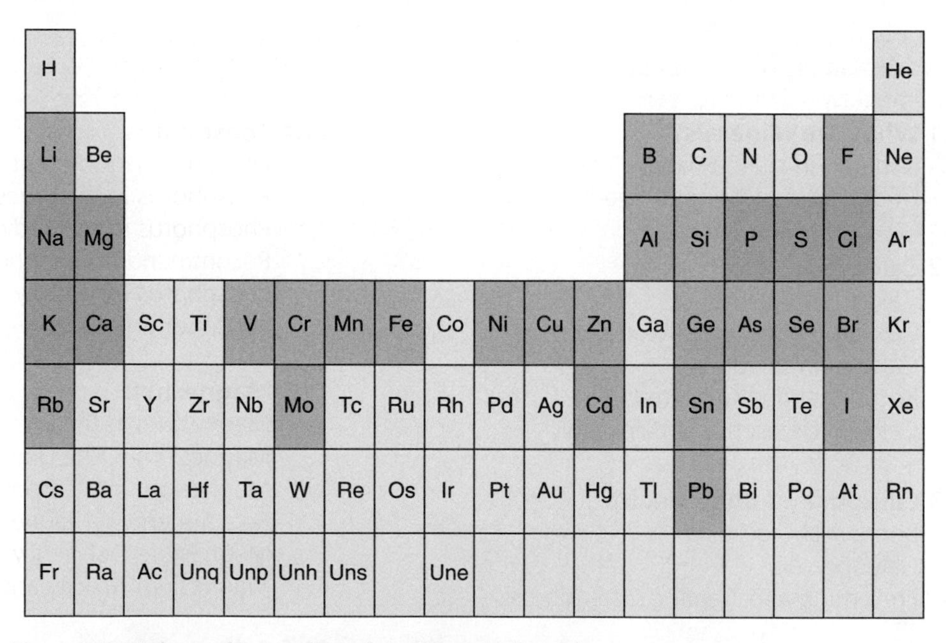

Figure 11.1 Major and trace minerals in the periodic table
The major minerals are shown in purple, and the trace elements are shown in blue.

Minerals in the Canadian Diet

Minerals are found in foods from all groups of Canada's Food Guide (**Figure 11.2**). Most foods naturally contain minerals and some foods provide minerals that are added intentionally by fortification or accidentally through contamination. The mineral content of the diet can be maximized by eating a variety of foods, including many unprocessed foods such as fresh fruits, vegetables, whole grains, lean meats, and low-fat milk and alternative products, as well as fortified foods (**Figure 11.3**). Natural health products are also a source of minerals.

GRAIN PRODUCTS	VEGETABLES AND FRUIT	MILK AND ALTERNATIVES	MEAT AND ALTERNATIVES
Iron, Zinc, Selenium, Copper, Magnesium, Chromium, Sulfur, Manganese, Sodium, Potassium, Phosphorus	Iron, Calcium, Potassium, Magnesium, Molybdenum	Calcium, Zinc, Phosphorus, Potassium, Iodine, Molybdenum	Iron, Zinc, Magnesium, Potassium, Chromium, Sulfur, Iodine, Selenium, Phosphorus, Copper, Manganese, Fluoride

Figure 11.2 Minerals in Canada's Food Guide food groups
Minerals are found in all food groups, but some groups are particularly good sources of specific ones.

Natural Sources In some foods, the amounts of minerals naturally present are predictable because the minerals are regulated components of the plant or animal. For instance, calcium is a component of milk; therefore, drinking a glass of milk reliably provides a known amount. Magnesium is a component of chlorophyll, so it is found in consistent amounts in leafy greens. For some minerals, the amounts in food vary depending on the mineral concentration in the soil and water at the food's source. For example, the soil content of iodine is high near the ocean but usually quite low in inland areas. Therefore, foods grown near the ocean are better sources of iodine than those grown inland. In developed countries, modern transportation systems make foods produced in many locations available, so the diet is unlikely to be deficient in minerals that vary in concentration depending on where the food is produced. In countries where the diet consists predominantly of locally grown foods, individual trace element deficiencies and excesses are more likely to occur.

Figure 11.3 The mineral content of the diet can be maximized by eating a variety of nutrient-dense foods.

Processed Foods Food processing and refining can affect the mineral content of foods. Processing does not destroy minerals, but it can still cause losses. For example, as the structure of a food is broken down, potassium is lost, and when the skins of fruits and vegetables and the bran and germ of grains are detached, minerals such as magnesium, iron, selenium, zinc, and copper are removed. Processing also adds minerals to foods. Sodium is frequently added for flavour or as a preservative. The enrichment of grains has been adding iron to our breads, baked goods, and rice since the 1940s and today, orange juice and soy milk are fortified with calcium. Some minerals enter the food supply inadvertently through contamination. For example, the iodine content of milk products is increased by contamination from the cleaning solutions used on milking machines.

Natural Health Products Natural health products such as mineral supplements are also a source of single or multiple minerals. Some, such as iron and calcium, are recommended for certain groups to meet their needs. Others, like chromium, zinc, and selenium, are taken in the hopes that they will enhance athletic performance, stimulate immune function, or reduce cancer risk. High doses of minerals from supplements can be toxic. Because of the complex interactions among minerals, taking high doses of one can compromise the bioavailability of others, creating a mineral imbalance that can interfere with functions essential to human health. The body's regulatory mechanisms control the absorption and excretion of minerals but have evolved to deal with the amounts of these elements that occur naturally in the diet. Large doses of mineral supplements may override this regulation, causing toxicity.

Understanding Mineral Needs

To maintain health, enough of each mineral must be consumed and the overall diet must contain all the minerals in the correct proportions. The wrong amounts or combinations can cause a deficiency or a toxic reaction. For some minerals, too much or too little causes obvious symptoms that impact short-term health. For example, an inadequate iron intake can cause a decrease in the number and size of red blood cells, reducing the blood's capacity to deliver oxygen and causing fatigue. These symptoms develop over a period of several months. Deficiencies of other minerals cause symptoms only in the long term. For example, a low calcium intake has no short-term consequences, but over the long term reduces bone density, increasing the risk of fractures later in life. Deficiencies of iron, iodine, and calcium are world health problems. Deficiencies of other minerals are rare, occurring only when the food supply is particularly limited or other factors affect mineral absorption or utilization.

Mineral toxicities occur most often as a result of environmental pollution or excessive use of supplements. For example, lead is toxic. Chronic exposure to lead from old lead paint, lead pipes, and soil and air contamination can cause growth retardation and learning disabilities in children (see Section 15.3). Even minerals such as iron that are essential in small doses can be toxic when consumed in excess.

Bioavailability of Minerals The body's ability to absorb specific minerals as well as the composition of the diet and the nutritional status and life stage of the consumer all affect bioavailability. Almost 100% of the sodium consumed in the diet is absorbed, whereas iron absorption may be as low as 5%. Calcium absorption is typically about 25% but during times of life when calcium needs are high, such as pregnancy and infancy, the proportion of dietary calcium absorbed is higher. For some minerals, the amount absorbed depends primarily on the amount that is consumed. For others, the consumption of other minerals carrying the same charge affects absorption. Calcium, magnesium, zinc, copper, and iron all carry a 2^+ charge, and a high intake of one may reduce the absorption of others. For example, a high intake of calcium in a meal containing iron may reduce the absorption of the iron.

phytic acid or **phytate** A phosphorus-containing storage compound found in seeds and grains that can bind minerals and decrease their absorption.

tannins Substances found in tea and some grains that can bind minerals and decrease their absorption.

oxalates Organic acids found in spinach and other leafy green vegetables that can bind minerals and decrease their absorption.

Other substances that are in the gastrointestinal tract at the same time as the mineral also affect bioavailability. Some substances enhance absorption. For example, when iron is consumed with acidic foods, the low pH helps to keep it in its more absorbable chemical form. Substances that bind minerals in the gastrointestinal tract can reduce absorption. **Phytic acid** or **phytate**—an organic compound containing phosphorus that is found in whole grains, bran, and soy products—binds calcium, zinc, iron, and magnesium, limiting their absorption. Phytic acid can be broken down by yeast, so the bioavailability of minerals is increased in yeast-leavened foods such as breads. **Tannins**, found in tea and some grains, can interfere with iron absorption, and **oxalates**, which are organic acids found in spinach, rhubarb, beet greens, and chocolate, have been found to interfere with calcium and iron absorption. Dietary fibre also interferes with mineral absorption. Although Canadians generally do not consume enough of any of these components to cause mineral deficiencies, problems may occur in developing countries when the diet is high in cereal grains and marginal in certain minerals.

Life Cycle

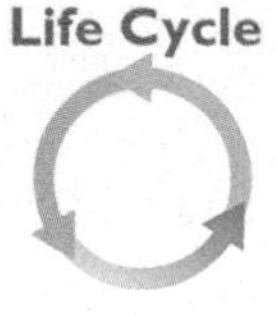

The ability to transport minerals from intestinal mucosal cells to the rest of the body also affects bioavailability. Some minerals must bind to plasma proteins or specific transport proteins to be transported in the blood. This binding helps regulate their absorption and prevents reactive minerals from forming free radicals that could cause oxidative damage. Nutritional status and nutrient intake can affect mineral transport in the body. For instance, when protein intake is deficient, transport proteins (and proteins in general) cannot be synthesized. Therefore, even if a mineral is adequate in the diet, it cannot be transported to the cells where it is needed. Sometimes the intake of a specific mineral can affect how much of another is able to leave the mucosal cell and bind to a transport protein in the plasma.

Mineral Functions Minerals serve a wide range of vital structural and regulatory roles in the body. For example, calcium, phosphorus, magnesium, and fluoride affect the structure and strength of bones. Iodine is a component of the thyroid hormones, which regulate metabolic rate; chromium plays a role in regulating blood glucose levels; and zinc plays an important role in gene expression. Many of the minerals serve as **cofactors** necessary for enzyme activity **(Figure 11.4)**. For instance, selenium is a cofactor for an antioxidant enzyme system.

cofactor An inorganic ion or coenzyme required for enzyme activity.

CRITICAL THINKING:

How are Canadians Doing with Respect to Their Intake, from Food, of Calcium, Magnesium, and Phosphorus?

The CCHS collected information on the intake, from food, of 3 nutrients discussed in this chapter, calcium, phosphorus and magnesium.

Health Canada considers the prevalence of inadequate intake to be low when less than <10% of the population has intakes below the EAR, as this means that less than 10% of the population is *not* meeting their nutrient requirement, or, conversely, 90% of the population *is* meeting its requirement. This 10% cutoff is indicated by the green line in the graph. If more than 10% of the population has intakes below the EAR, then this is considered an area of concern.

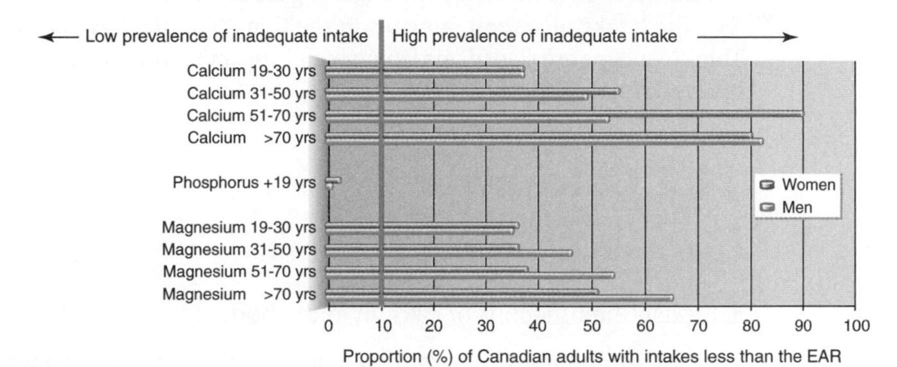

Source: Health Canada. Canadian Community Health Survey, Cycle 2.2, Nutrition (2004) - Nutrient Intakes from Food: Provincial, Regional and National Summary Data Tables, Volume 1,2 and 3. Cat No. 978-0-662-06542-5

Note: At press time, Health Canada had not yet modified CCHS calcium data to reflect 2010 DRIs for calcium; values shown here were estimated by the author (D.G) and should be considered no more than an approximate assessment of Canadian intakes.

Critical Thinking Questions

What would you conclude about Canadians' intake, from food, of calcium, phosphorus, and magnesium?*

How and why does age affect the adequacy of intake?

The data above measure intake from food only. With respect to calcium, it is known that calcium supplements are popular among older women. How would that affect the prevalence of inadequate intake?

* Answers to all Critical Thinking questions can be found in Appendix J.

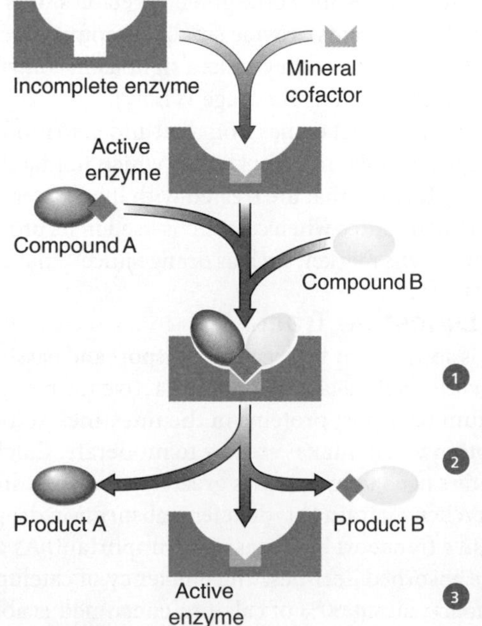

1. The mineral cofactor combines with the incomplete enzyme to form the active enzyme.
2. The active enzyme binds to the molecules involved in the chemical reaction (compounds A and B) and accelerates their transformation into the final products (products A and B).
3. The final products are released, while the enzyme remains unchanged.

Figure 11.4 Minerals as cofactors
The binding of a cofactor to an enzyme activates the enzyme.

Recommended Mineral Intakes As with other nutrients, the DRI recommendations for mineral intakes provide either an RDA value, when sufficient information is available to establish an EAR, or an AI, when population data are used to estimate needs. To help avoid toxic amounts of minerals, ULs have been established when adequate data are available. Recommendations are based on evidence from many types of research studies, ranging from laboratory studies done with animals and clinical trials using human subjects to epidemiological observations. In addition to planned experiments, information about trace element needs has come from the study of inherited diseases affecting trace element utilization and from the study of deficiency symptoms in individuals fed for long periods of time solely by total parenteral nutrition (TPN) solutions that were inadvertently deficient in specific essential minerals. As with other nutrients, when no other data are available, mineral needs can be estimated by evaluating the intake in a healthy population.

The intake of several minerals in the Canadian diet has been evaluated (see Critical Thinking: How are Canadians Doing with Respect to Their Intake, from Food, of Calcium, Magnesium, and Phosphorus?)

11.2 Calcium

Learning Objectives

- Explain the functions of calcium in the body.
- Compare the roles of parathyroid hormone, calcitonin, and vitamin D in the regulation of blood calcium levels.
- List foods that are good sources of calcium.

Calcium is the most abundant mineral in the body.[1] It provides structure to bones and teeth and has essential regulatory roles. Because of the importance of calcium in regulation, blood levels of this mineral are strictly controlled. However, this occurs at the expense of bone calcium; calcium is released from bone when blood levels drop. As a result, if the diet is not sufficient in calcium, over the long term, bone calcium is decreased and the risk of bone fractures due to **osteoporosis** is increased.

osteoporosis A bone disorder characterized by a reduction in bone mass, increased bone fragility, and an increased risk of fractures.

Calcium in the Diet

The main source of calcium in the Canadian diet are milk, cheese, and yogurt. Fish, such as sardines and canned salmon, that are consumed in their entirety, including the bones, are also a good source, as are legumes and some green vegetables such as broccoli, Chinese cabbage, and kale **(Figure 11.5)**. Grains provide smaller amounts of calcium, but because they are consumed in such large quantities they make a significant contribution to dietary calcium intake (see Your Choice: Choose Your Beverage Wisely).

Some of the calcium in the diet comes from that added to foods during food processing. Baked goods such as breads, rolls, and crackers, to which nonfat dry milk powder has been added, provide calcium. Tortillas that are treated with lime water (calcium hydroxide) provide calcium. Tofu is a good source when calcium is used in its processing. In addition, there are numerous products on the market, such as orange juices, that are fortified with calcium.

Calcium in the Digestive Tract

Absorption Calcium is absorbed by both active transport and passive diffusion **(Figure 11.6)**. Active transport depends on the availability of the active form of vitamin D, which induces the synthesis of calcium transport proteins in the intestine. Active transport accounts for most calcium absorption when intakes are low to moderate. Calcium absorption is higher at times when the body's need for calcium is greater. In young adults, about 25% of dietary calcium is absorbed. When vitamin D is deficient, absorption drops to about 10%.[1] At high calcium intakes, passive transport becomes more important. As calcium intake increases, the percentage that is absorbed declines. The efficiency of calcium absorption varies with life stage. During infancy, about 60% of calcium consumed is absorbed. In young adults, absorption is about 25%. In older adults, absorption declines due to a decrease in blood levels of the active form of vitamin D, or a decrease in responsiveness to vitamin D.[2] An

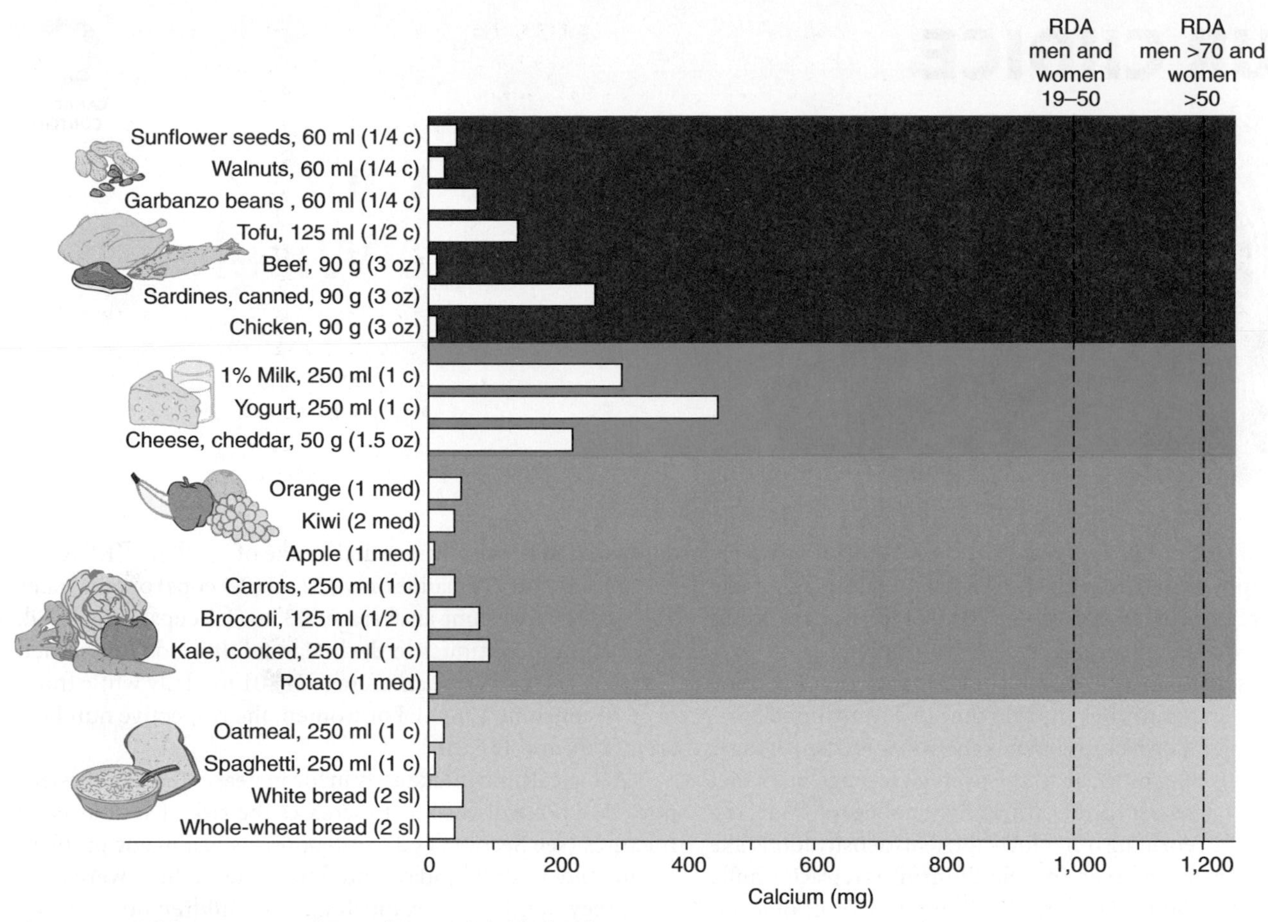

Figure 11.5 Calcium content of Canada's Food Guide food groups
Calcium is provided by many different foods, but milk and alternative products and fish consumed with bones are the best sources; the dashed lines represent the RDA for adults ages 19-50 and for adults over age 50.

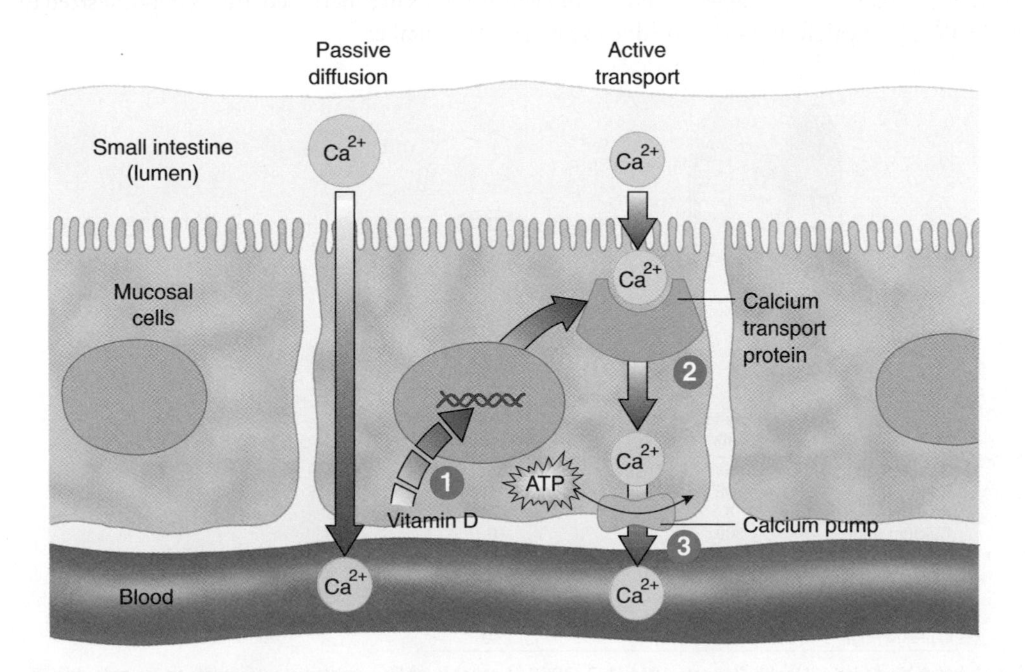

Figure 11.6 Calcium absorption
Some calcium can be absorbed by passive diffusion, particularly when calcium concentrations are high; at lower calcium concentrations, absorption occurs primarily by an active transport mechanism that requires vitamin D.

1. Vitamin D turns on the synthesis of calcium transport proteins.
2. Calcium transport proteins shuttle calcium across the mucosal cell.
3. A calcium pump that requires energy moves calcium from the mucosal cells to the bloodstream.

YOUR CHOICE

CANADIAN CONTENT

(©iStockphoto)

Choose Your Beverage Wisely

We think of soft drinks as thirst quenchers, snacks, or beverages to gulp with our meals. The Canadian adolescent male (14-18 years) consumes an average 376 ml of regular soft drinks daily, while adolescent girls average 179 ml daily.[1]

There is no doubt that carbonated beverages have become part of popular culture, but what do they offer nutritionally?

A 355-ml soft drink can contains about 40 g (10 tsp) of sugar and thus, adds 40 g of sugar to the average teenage boy's diet and 20 g to a teenage girl's diet. This might not be too bad if the soft drink were replacing other low-nutrient-density foods like cookies or cakes, but for many people, the drink is replacing milk.

Replacing 250 ml (1 cup) of milk with 355 ml of a soft drink increases the amount of added sugar in the diet by about 40 g and reduces protein, calcium, vitamin A, vitamin D, and riboflavin intake (see table). Canadian teenage girls, age 14-18 years, consume an average of 917 mg of calcium daily compared to the recommended intake of 1,300 mg (RDA), in part because they consume only 222 ml (0.9 cups) of milk daily. Teenage boys consume an average 323 ml (1.3 cups) daily.[1] Milk consumption continues to decline in adulthood. Young men (19-30 years) consume an average of 201 ml daily while those over 70 consume 136 ml. For women, the respective numbers are 178 ml and 136 ml.[2]

A low-calcium intake early in life increases the risk of osteoporosis, a bone disease that increases the risk of serious bone fractures (see Section 11.3). Osteoporosis is a major problem among older adults today, and when these adults were children, they drank twice as much milk as children do today. By substituting soft drinks for milk, children and teens are putting their bones at risk. In fact, the number of bone fractures among children and young adults has increased; this is hypothesized to be due to low-calcium intakes.[3]

	Low-fat milk	Cola soft drink
Serving size (ml)	250	355
Energy (kcal)	102	150
Protein (g)	8	0
Calcium (mg)	300	0
Phosphorus (mg)	235	45
Riboflavin (mg)	0.4	0
Vitamin A (μg)	144	0
Vitamin D (μg)	2.5	0
Caffeine (mg)	0	40

[1] Garriguet, D. Beverage consumption of children and teens. Health Reports 19(4):1-7, 2008. Available online at http://www.statcan.gc.ca/pub/82-003-x/2008004/article/6500820-eng.pdf. Accessed August 21, 2010.

[2] Garriguet, D. Beverage consumption of Canadian adults. Health Report 19(4):23-29, 2008. Available online at http://www.statcan.gc.ca/cgi-bin/af-fdr.cgi?l=eng&loc=2008004/article/6500821-eng.pdf. Accessed Sept 23, 2010.

[3] NIH; NICHD. Calcium Crisis Affects American Youth, December 10, 2001. Available online at www.nichd.nih.gov/new/releases_bak_20040224/calcium_crisis.cfm. Accessed September 4, 2004.

additional decrease in calcium absorption occurs in women after menopause due to the decrease in estrogen. The DRIs for calcium take into account the limited absorption of calcium. For example, an RDA of 1,000 mg assumes about 25%-30% calcium absorption, meaning that if you ingest 1,000 mg of calcium from food, this will ensure that sufficient calcium is absorbed to maintain bone health.

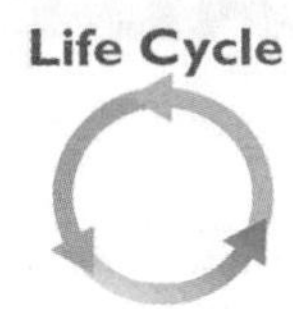

During pregnancy, when calcium is needed for formation of the fetal skeleton, elevated estrogen helps increase calcium absorption. Calcium absorption increases to more than 50% during pregnancy. Calcium need is also increased during lactation, but some of the calcium needed to make milk appears to come from the mother's bones. After lactation stops, an increase in calcium absorption and retention of calcium by the kidney help restore bone calcium.

Bioavailability The bioavailability of calcium is decreased by the presence of tannins, fibre, phytates, and oxalates. For example, spinach is a high-calcium vegetable but only about 5% of its calcium is absorbed; the rest is bound by oxalates and excreted in the feces.[3] Vegetables such as kale, collard greens, turnip greens, mustard greens, and Chinese cabbage are low in oxalates, so their calcium is well absorbed. Chocolate also contains oxalates, but chocolate milk is still a good source of calcium because the amount of oxalates from the chocolate added to a glass of milk is small. Fibre generally causes a small reduction in calcium absorption, but there are exceptions (see Critical Thinking: Dietary Factors Can Increase Calcium Bioavailability—the Example of Inulin). However, phytates from foods such as wheat bran, and red, white, and pinto beans, can have a significant effect on the absorption of calcium. It would take almost 10 servings of red beans to provide the same amount of absorbable calcium as 1 serving of milk. When calcium intake is low, dietary components that alter absorption have a greater effect on calcium status than when calcium intake is adequate.[4]

Calcium in the Body

Calcium accounts for 1%-2% of adult body weight. Over 99% of the calcium in the body is found in the solid mineral deposit in bones and teeth.[1] The remaining 1% is present in intracellular fluid, blood, and other extracellular fluids, where it plays vital roles in nerve transmission, muscle contraction, blood pressure regulation, and the release of hormones (see **Table 11.1**).

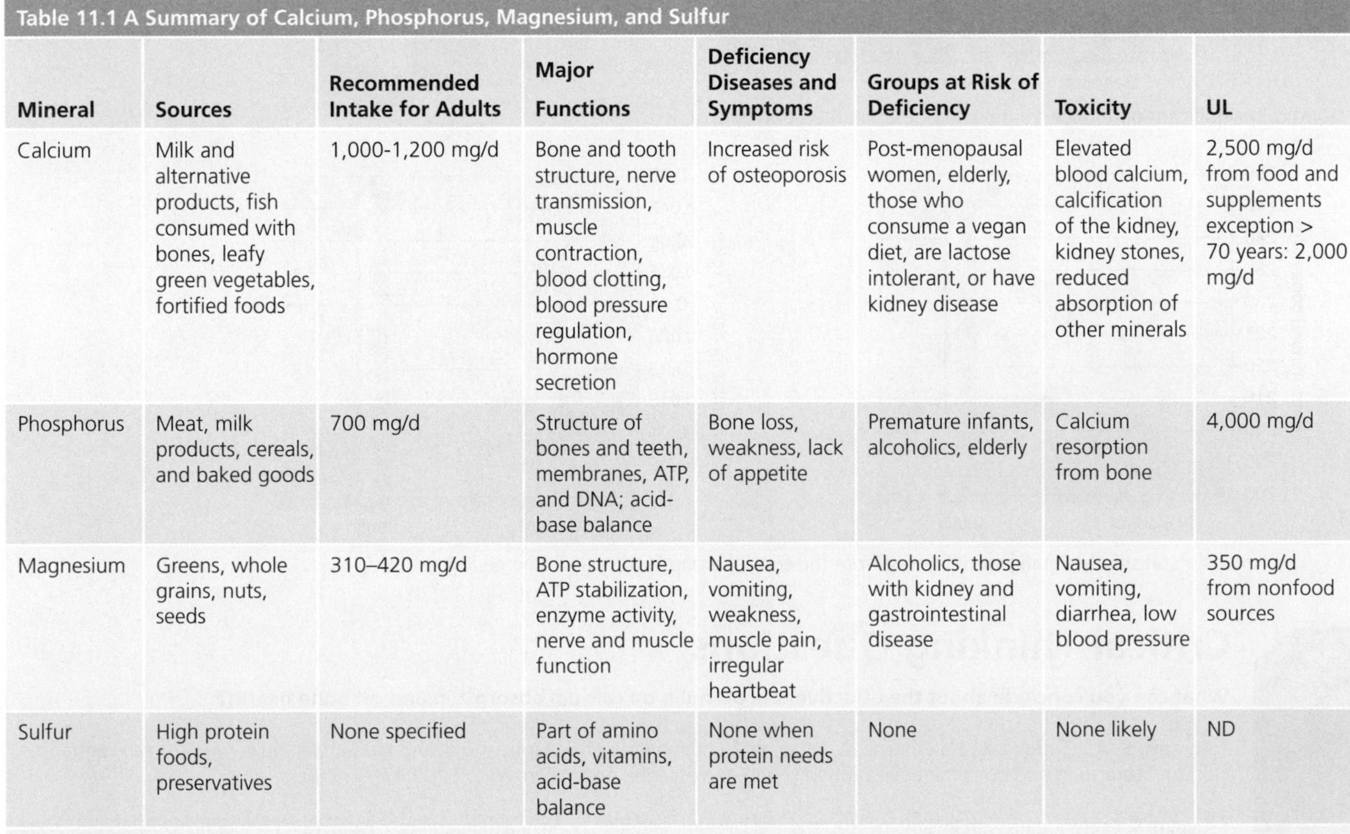

Table 11.1 A Summary of Calcium, Phosphorus, Magnesium, and Sulfur

Mineral	Sources	Recommended Intake for Adults	Major Functions	Deficiency Diseases and Symptoms	Groups at Risk of Deficiency	Toxicity	UL
Calcium	Milk and alternative products, fish consumed with bones, leafy green vegetables, fortified foods	1,000-1,200 mg/d	Bone and tooth structure, nerve transmission, muscle contraction, blood clotting, blood pressure regulation, hormone secretion	Increased risk of osteoporosis	Post-menopausal women, elderly, those who consume a vegan diet, are lactose intolerant, or have kidney disease	Elevated blood calcium, calcification of the kidney, kidney stones, reduced absorption of other minerals	2,500 mg/d from food and supplements exception > 70 years: 2,000 mg/d
Phosphorus	Meat, milk products, cereals, and baked goods	700 mg/d	Structure of bones and teeth, membranes, ATP, and DNA; acid-base balance	Bone loss, weakness, lack of appetite	Premature infants, alcoholics, elderly	Calcium resorption from bone	4,000 mg/d
Magnesium	Greens, whole grains, nuts, seeds	310–420 mg/d	Bone structure, ATP stabilization, enzyme activity, nerve and muscle function	Nausea, vomiting, weakness, muscle pain, irregular heartbeat	Alcoholics, those with kidney and gastrointestinal disease	Nausea, vomiting, diarrhea, low blood pressure	350 mg/d from nonfood sources
Sulfur	High protein foods, preservatives	None specified	Part of amino acids, vitamins, acid-base balance	None when protein needs are met	None	None likely	ND

UL = Tolerable upper intake level, ND = not determined

CRITICAL THINKING:

Dietary Factors Can Increase Calcium Bioavailability—the Example of Inulin

Calcium absorption is relatively limited, ranging from 5%-50%, depending on the food, and averaging about 25%-30% overall for a typical diet. Many dietary factors may contribute to increasing or decreasing calcium bioavailability.

In a recent study, researchers wanted to determine whether a type of fibre called inulin could enhance the calcium absorption in children and adolescents.[1] Maximizing calcium absorption during childhood would increase the amount of calcium in the bone and potentially reduce the risk of osteoporosis later in life.

Calcium is typically absorbed in the small intestine (see Figure 11.6). Any unabsorbed calcium enters the colon and is eliminated in the feces. Inulin is a fructose-containing polysaccharide used as a prebiotic (see Chapter 3: Your Choice: Should You Feed Your Flora?). Prebiotics are believed to improve calcium absorption by promoting the growth of beneficial microflora in the colon that ferment carbohydrates and produce short-chain fatty acids. These fatty acids, in turn, create an acidic environment in the colon which makes calcium more soluble and allows for additional absorption of calcium to take place in the colon.[2]

Researchers recruited 50 boys and 50 girls, aged 9-13, for their intervention trial. Subjects were randomized into 2 groups, a control group and a treatment group in which participants consumed 8 g of inulin per day. The study was a double-blind intervention trial. The treatment group was given a packet of inulin which they were advised to add to orange juice at breakfast. The control group was given a similar packet, but it contained a placebo of soluble starch polysaccharides.

Using the results of the research, these three graphs show the effect of inulin on (a) calcium absorption, (b) bone mineral content, and (c) bone mineral density. The calcium absorption of the participants was measured at 8 weeks and again after 1 year (graph a). Researchers also measured the amount of calcium in the bone, at the end of one year, using measurements such bone mineral content (BMC) and bone mineral density (BMD) (graph b and c). Both BMC and BMD increase when the amount of calcium in the bone increases (for more information on measuring bone mineral content and density, see Section 11.3 and Figure 11.10b).

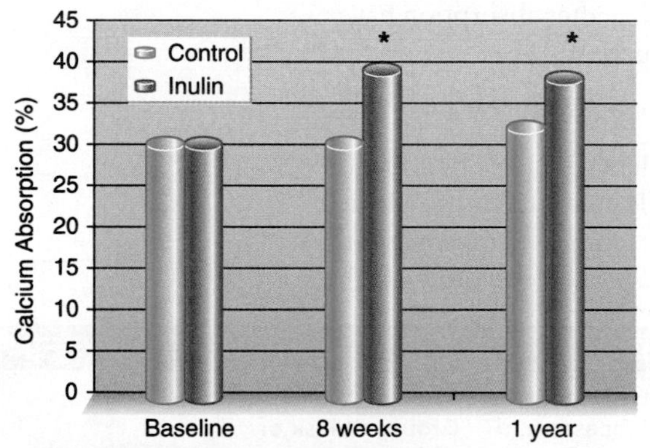

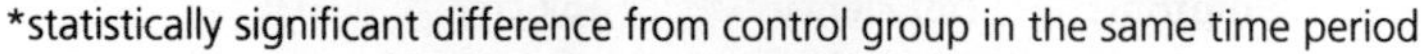

*statistically significant difference from control group in the same time period

(Artville)

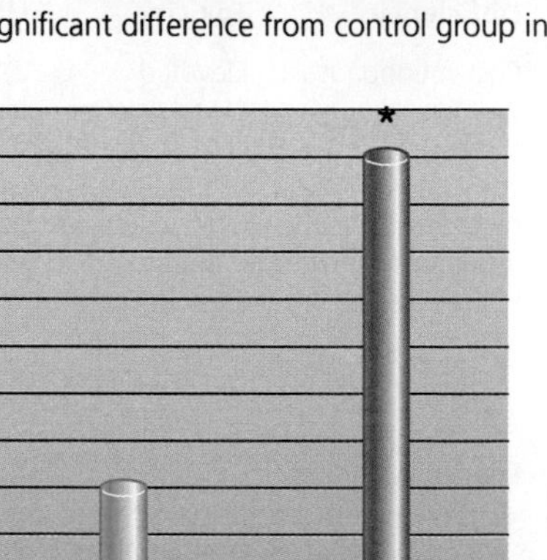

c)

Bone mineral density (g/cm²)

0, 0.005, 0.01, 0.015, 0.02, 0.025, 0.03, 0.035, 0.04, 0.045, 0.05

Control Inulin

*statistically significant difference from the control group in the same time period.

Critical Thinking Questions

What can you conclude about the effectiveness of inulin on calcium absorption and on bone health?

[1] Abrams, S. A., Griffin, I. J., Hawthorne, K. M. et al. A combination of prebiotic short- and long-chain inulin-type fructans enhances calcium absorption and bone mineralization in young adolescents. *Am. J. Clin. Nutr.* 82(2):471-6, 2005.

[2] Coxam, V. Current data with inulin-type fructans and calcium, targeting bone health in adults. *J. Nutr.* 137(11 Suppl):2527S-2533S, 2007.

Regulatory Roles of Calcium The calcium in body fluids plays critical roles in cell communication and the regulation of body processes. Calcium helps regulate enzyme activity and is necessary in blood clotting. It is involved in transmitting chemical and electrical signals in nerves and muscles. It is necessary for the release of neurotransmitters, which allow nerve impulses to pass from one nerve to another and from nerves to target tissues. Inside the muscle cells, calcium allows the two muscle proteins, actin and myosin, to interact to cause muscle contraction. The importance of calcium for proper nerve transmission and muscle contraction is illustrated by what happens when the concentration of calcium in the extracellular fluid drops too low. When this occurs, the nervous system becomes increasingly excitable and nerves fire spontaneously, triggering contractions of the muscles, a condition known as tetany. Mild tetany can cause tingling of the lips, fingers, and toes, and more serious tetany results in severe muscle contractions, tremors, cramps, and even death. Tetany is typically caused by hormonal abnormalities, not a dietary calcium deficiency.

Calcium also plays a role in blood pressure regulation, possibly by controlling the contraction of muscles in the blood vessel walls and signalling the secretion of substances that regulate blood pressure. The impact of adequate calcium on maintaining a healthy blood pressure was demonstrated by the DASH Trial (see Chapter 10, Science Applied: A Total Dietary Approach to Reducing Blood Pressure). In this trial, a diet high in potassium, magnesium, and calcium was found to be more effective at lowering blood pressure than the control diet or an experimental diet lower in calcium.[5] Following the recommendations of the DASH Eating Plan can help keep both bones and blood pressure healthy.

Regulation of Blood Calcium The roles of calcium are so vital to survival that powerful regulatory mechanisms ensure that constant intracellular and extracellular concentrations are maintained. Slight changes in blood calcium levels trigger responses that quickly raise or lower them back to normal levels. This homeostasis is maintained by the hormones **parathyroid hormone (PTH)**, which raises blood calcium, and **calcitonin**, which lowers blood calcium (**Figure 11.7**). If the level of blood calcium falls too low, it triggers the secretion of more PTH from the parathyroid glands. PTH stimulates the release of calcium from bone, reduces calcium excretion by the kidney, and activates vitamin D. Activated vitamin

parathyroid hormone (PTH) A hormone secreted by the parathyroid gland that increases blood calcium levels.

calcitonin A hormone secreted by the thyroid gland that reduces blood calcium levels.

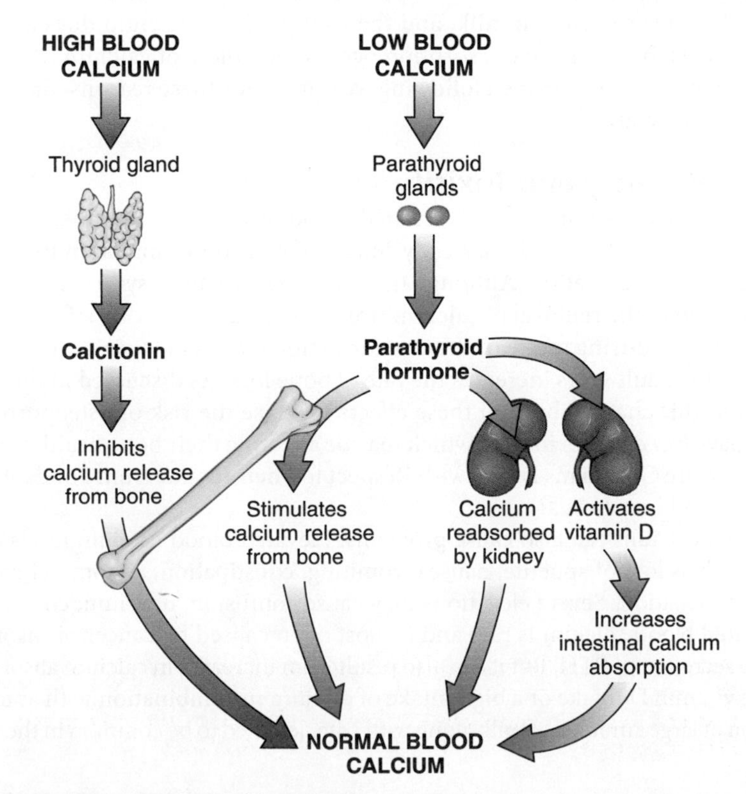

Figure 11.7 Regulation of blood calcium levels
Levels of calcium in the blood are very tightly regulated by parathyroid hormone and calcitonin.

D increases the amount of calcium absorbed from the gastrointestinal tract and, with PTH, stimulates calcium release from the bone and calcium retention by the kidney (see Section 9.3). The overall effect is to rapidly increase blood calcium levels. If blood calcium levels become too high, the secretion of PTH is reduced. This increases excretion of calcium by the kidney; decreases vitamin D activation, so less dietary calcium is absorbed; and reduces calcium release from bone. High blood calcium also stimulates the secretion of calcitonin from the thyroid gland. Calcitonin acts primarily on bone to inhibit the release of calcium. Together, low PTH levels and the presence of calcitonin cause a decrease in blood calcium levels.

Recommended Calcium Intake

accretion An accumulation by external addition; in the case of nutrition, the uptake and accumulation by the body of a nutrient.

The DRI values for calcium intake have been set at the amount that allows the maintenance of bone health. In children, this means ensuring there is sufficient calcium intake for the growth of healthy bone, that is, for calcium **accretion.** In young adults, it means maintaining sufficient calcium intake to prevent the loss of calcium from bone. In older adults, because some loss of bone is inevitably associated with aging, calcium intake should be sufficient to minimize bone loss and prevent the development of osteoporosis (see Section 11.3 for a detailed discussion of bone development and osteoporosis).

The RDA for adults age 19-50 years is 1,000 mg/day.[6] The RDA for women age 51 and older and men age 71 and older is increased to 1,200 mg/day. In children and adolescents, the RDA is set at a level that will support bone growth. For adolescents, the RDA is higher than for adults—1,300 mg/day for boys and girls age 9-18 years.

Infants thrive on the amount of calcium they obtain from human milk. For infants, an AI is set based on the mean intake of infants fed principally with human milk. Because calcium is not as well absorbed from infant formulas, formula-fed infants require more. There is no specific AI for formula-fed infants, but formulas are higher in calcium than breast milk to compensate for the reduced absorption.

Life Cycle

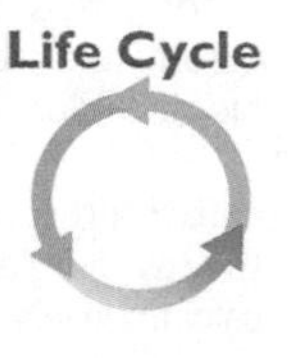

The RDA for calcium during pregnancy is not increased above nonpregnant levels. This is because there is an increase in maternal calcium absorption during pregnancy that helps to supply the calcium needed for the fetal skeleton. In addition, since there is no correlation between the number of pregnancies and bone mineral density, the maternal skeleton does not appear to be used as a supply of calcium for the fetus.[1] However, during lactation, calcium is secreted in milk, and the source of this calcium does appear to be the maternal skeleton. This bone resorption occurs regardless of calcium intake and the calcium lost appears to be regained following weaning.[7] For these reasons, the RDA is not increased during lactation.

Calcium Deficiency and Toxicity

When calcium intake is not adequate, normal blood levels are maintained by resorbing calcium from bone. This provides a steady supply of calcium to maintain its roles in cell communication and regulation. Although there are no short-term symptoms of a calcium deficiency related to the removal of calcium from bone, a deficient diet affects bone mass. Low calcium intake during the years of bone formation results in lower bone density. Low intake during the adult years increases the rate of bone loss. As discussed in the osteoporosis section in this chapter, both of these effects increase the risk of osteoporosis. Many Canadians have low calcium intakes which may be affecting their bone health (see Critical Thinking: How are Canadians Doing with Respect to Their Intake, from Food, of Calcium, Magnesium, and Phosphorus?)

Too much calcium can also cause problems. Elevated blood calcium levels can cause symptoms such as loss of appetite, nausea, vomiting, constipation, abdominal pain, thirst, and frequent urination. Severe elevations may cause confusion, delirium, coma, and even death. Elevated blood calcium is rare and is most often caused by cancer or disorders that increase the secretion of PTH. But it can also result from increases in calcium absorption due to excessive vitamin D intake or a high intake of calcium in combination with antacids. The consumption of large amounts of milk along with antacids used to be common in the treatment of peptic ulcers. This combination is associated with a condition called milk-alkali syndrome, which is characterized by high blood calcium along with calcification of the kidney that can lead to kidney failure. After the treatment of ulcers changed, the incidence of this condition declined, but it has risen again due to the increased use of calcium-containing antacids as calcium supplements and to treat heartburn.[8]

Too much calcium from supplements may also promote the formation of kidney stones. Kidney stones, which are usually composed of calcium oxalate or calcium phosphate, affect approximately 12% of the population. Although their cause is usually unknown, abnormally elevated urinary calcium, which can result from high doses of supplemental calcium, increases the risk of developing calcium stones.[1]

High calcium intake can also interfere with iron, zinc, magnesium, and phosphorus availability. Although calcium supplements inhibit iron absorption, there is no evidence that the long-term use of calcium supplements with meals affects iron status.[9] High intakes of calcium from supplements have also been found to reduce zinc absorption and thereby increase the amount of zinc needed in the diet.[10] There is no evidence of depletion of phosphorus or magnesium associated with calcium intake. Based on the occurrence of elevated blood calcium and kidney stones, as well as the potential for decreased absorption of other essential minerals, the UL for calcium in adults has been set at 2,500 mg/day from food and supplements.[1] In 2010, the IOM reduced the UL for calcium for adults over age 50 to 2,000 mg/day, because of studies suggesting that this segment of the population may be especially vulnerable to kidney stones.[6]

11.3 Calcium and Bone Health

Learning Objectives

- Explain why bone remodelling is important.
- Discuss how the rates of bone formation and breakdown change throughout life.
- Describe dietary and lifestyle factors that affect the risk of osteoporosis.

Bone is composed of a protein framework, or matrix, that is hardened by deposits of minerals. The most abundant protein in this matrix is collagen. The mineral portion of bone is composed mainly of calcium associated with phosphorus as solid mineral crystals known as **hydroxyapatite**. In addition to calcium and phosphorus, bone contains magnesium, sodium, fluoride, and a number of other trace minerals. Healthy bone requires adequate dietary protein and vitamin C to maintain collagen (see Section 8.10), adequate vitamin D (see Section 9.3) to promote calcium absorption, and a sufficient supply of minerals to ensure solidity. There is also growing evidence of the importance of vitamin K for bone health.[11] It is required for the synthesis of proteins important to the regulation of bone growth (discussed in Section 9.5).

There are two types of bone: **cortical** or **compact bone**, which makes up about 80% of the skeleton and forms the sturdy, dense outer surface layer, and **trabecular** or **spongy bone**, which forms an inner lattice that supports the cortical shell **(Figure 11.8)**. Trabecular bone is found in the knobby ends of the long bones, the pelvis, wrists, vertebrae, scapulae, and the areas of the bone that surround the bone marrow.

hydroxyapatite A crystalline compound composed of calcium and phosphorus that is deposited in the protein matrix of bone to give it strength and rigidity.

cortical or **compact bone** Dense, compact bone that makes up the sturdy outer surface layer of bones.

trabecular or **spongy bone** The type of bone that forms the inner spongy lattice that lines the bone marrow cavity and supports the cortical shell.

bone remodelling The process whereby bone is continuously broken down and re-formed to allow for growth and maintenance.

osteoblasts Cells responsible for the deposition of bone.

osteoclasts Large cells responsible for bone breakdown.

Bone: A Living Tissue

Bone is a living, metabolically active tissue that is constantly being broken down and re-formed in a process called **bone remodelling**. Bone is formed by cells called **osteoblasts** and broken down or resorbed by cells called **osteoclasts**. During bone formation, the activity of the bone-building osteoblasts exceeds that of the osteoclasts. When bone is being broken down, the osteoclasts resorb bone more rapidly than the osteoblasts can rebuild it **(Figure 11.9)**.

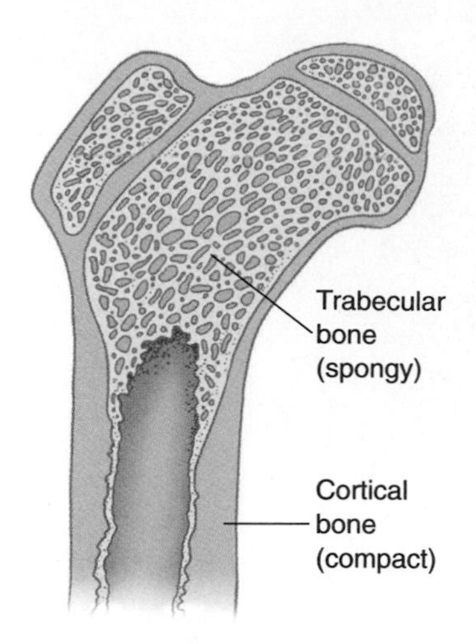

Figure 11.8 **Types of bone**
Cortical bone is compact bone that forms the dense outer layer; the spongy interior is referred to as trabecular bone.

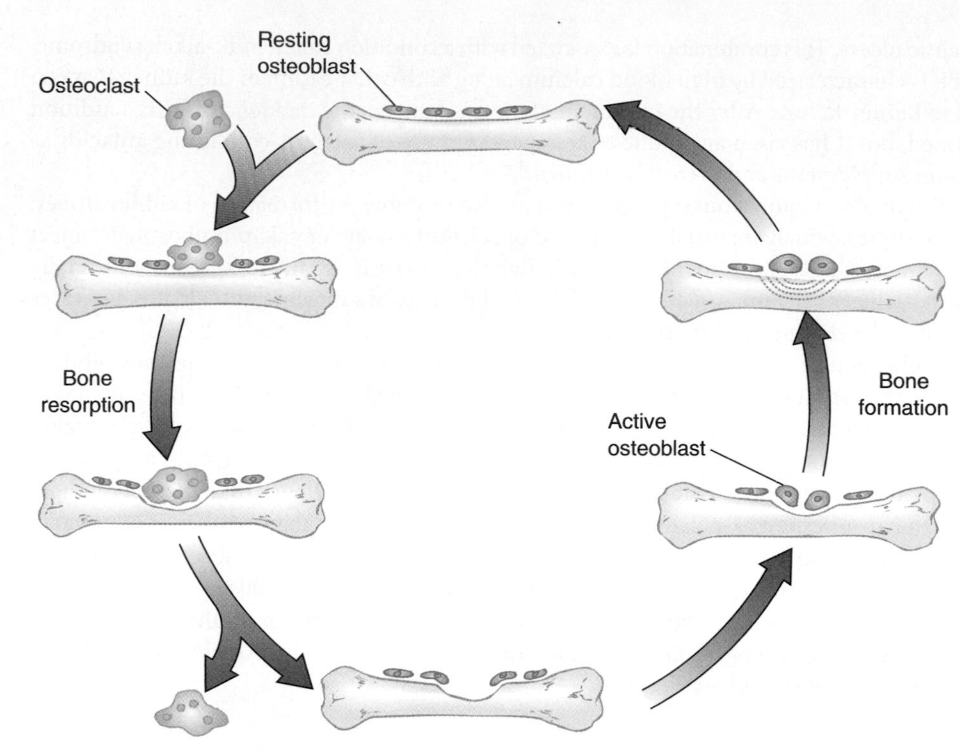

Figure 11.9 **Bone remodelling**
During bone remodelling, osteoclasts (pink) break down bone, and osteoblasts (blue) build bone.

peak bone mass The maximum bone density attained at any time in life, usually occurring in young adulthood.

Most bone is formed early in life. In the growing bones of children, bone formation occurs more rapidly than breakdown. Even after growth stops, bone mass continues to increase into young adulthood when **peak bone mass** is achieved, somewhere between the ages of 16 and 30.[12] When bone breakdown and formation are in balance, bone mass remains constant. After about age 35-45, the amount of bone broken down begins to exceed that which is formed. If enough bone is lost, the skeleton is weakened and fractures occur easily, a condition known as osteoporosis.

Osteoporosis

Osteoporosis is caused by a loss of both the protein matrix and the mineral deposits of bone, resulting in a decrease in the total amount of bone **(Figure 11.10a)**. Although both types of bone are lost with age, the greater surface area of the spongy trabecular bone gives it a higher turnover rate compared to cortical bone and it is therefore more vulnerable to bone loss. As a result, the regions in the skeleton that have higher amounts of trabecular bone, such as the spine and the upper part of the femur, are more susceptible to fracture later in life.

Detection of osteoporosis Osteoporosis is detected by using dual-energy X-ray absorptiometry (DXA) (discussed in Section 7.5). This X-ray produces an image from which the mineral content (measured in total g) and bone mineral density (measured in g mineral/cm^2-as DXA produces a flat 2-dimensional image) is determined **(Figure 11.10b)**. BMD is assessed by comparison to the density of healthy young adults and is calculated as a T-score or number of standard deviations from normal **(Table 11.2)**.

Table 11.2 A T-Score for Bone Mineral Density is a Measure of Standard Deviations from Normal

T-score	Diagnosis
–1 Standard deviation or above	Normal bone density
Between –1 and –2.5 standard deviations	Osteopenia (low bone density)
Below –2.5 standard deviations	Osteoporosis

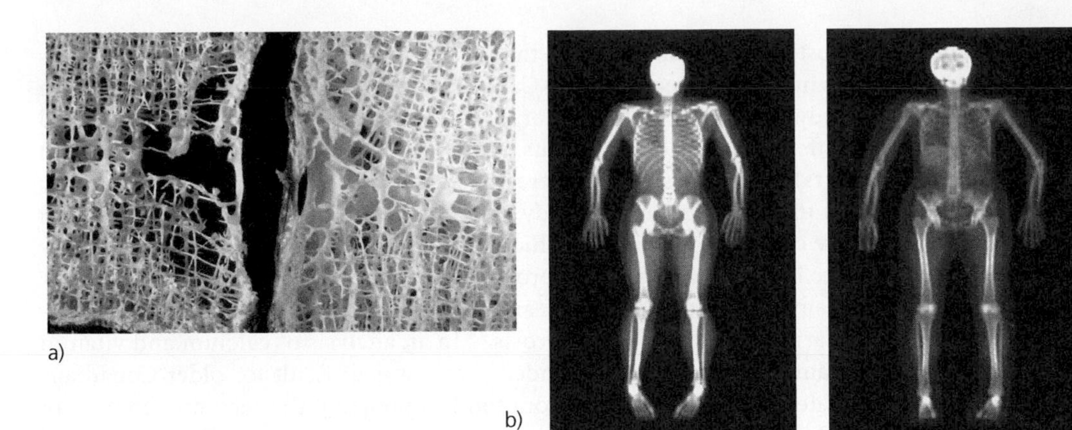

Figure 11.10 a) Healthy trabecular bone (on the right side of the photo) is denser than trabecular bone weakened by osteoporosis (left). b) The dual energy x-ray (DXA) bone mineral density scan on the left shows the skeleton of a 24-year-old woman, while the one on the right is a 72-year-old woman. The bones of the younger woman show up as bright white, indicating high bone mineral density, while the bones of the older woman appear almost transparent, indicating considerable mineral loss.
(Dr. Michael Klein/Peter Arnold, Inc./Getty; Grand Forks Human Nutrition Research Center)

Health Impact of Osteoporosis Osteoporosis is a silent disease because it initially causes no symptoms. By the fifth or sixth decade of life, the bones of individuals with this disorder have weakened enough to cause back pain and bone fractures of the spine, hip, and wrist. Spinal compression fractures may result in loss of height and a stooped posture called a dowager's hump **(Figure 11.11)**. Osteoporosis is a major public health problem in Canada and around the world. In Canada, about 2 million people over age 50 have osteoporosis; twice as many women have it than men.[13] It is also estimated that osteoporosis in responsible for at least 80% of fractures in people over 60 years of age and that osteoporosis and its associated bone fractures cost the Canadian health care system $1.9 billion annually to treat.[13]

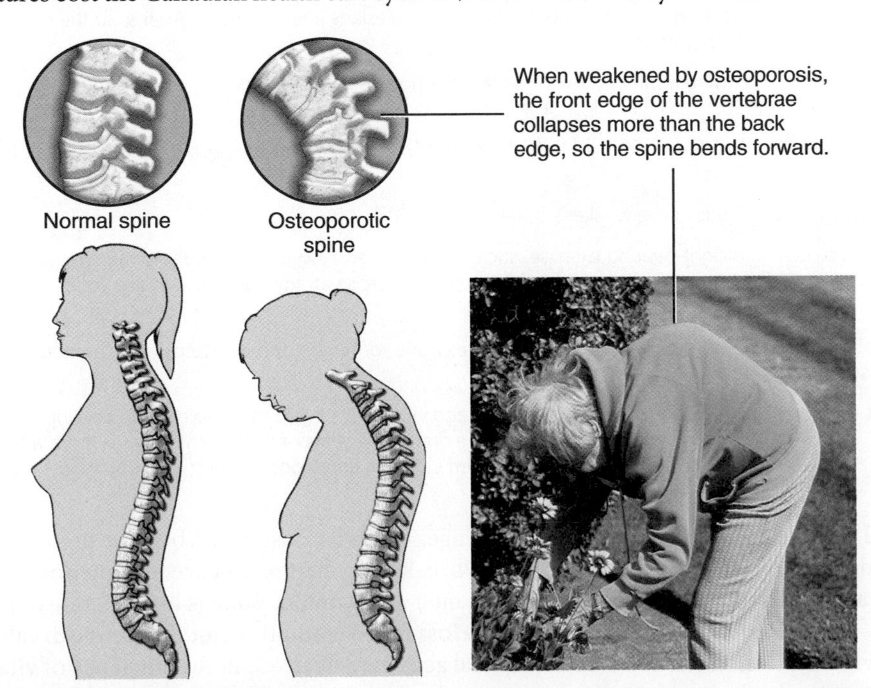

Figure 11.11 Effects of osteoporosis
Osteoporosis of the spine leads to a stooped posture and decreased stature. (Larry Mulvehill/Photo Researchers)

Understanding osteoporosis in Canada In the early 1990s, it was recognized that little was known about the impact of osteoporosis in Canada. As a result, the Canadian Multicentre Osteoporosis Study (CaMos) (www.camos.org) was formed to study how osteoporosis and bone fractures affect the lives of Canadians. Randomly selected adults (>25 years) and youth (16-24 years) from across Canada were invited to participate. Bone mineral density was measured at the beginning of the study, after 5 years, and again after 10 years, and participants were extensively surveyed, which included being asked about their dietary intake, in order to identify factors that improve the prevention, detection, and treatment of the disease. Using this information, researchers have published and continue to publish studies on various aspects of osteoporosis.[14] In an analysis of calcium and vitamin D intake, for example, researchers concluded that it was difficult for older Canadians to obtain adequate levels of vitamin D from food, supporting the recommendation of Canada's Food Guide that individuals over 50 consume a supplement. They also noted that many post-menopausal women were taking calcium supplements, likely in response to concerns about their bone health.[15]

Among the most sobering findings of CaMos was the impact of bone fractures on mortality. Compared to individuals who did not suffer a bone fracture, those who did were 3-4 times more likely to die. This is believed to be due to a decline in health resulting from immobility, loss of muscle mass, and loss of strength that accompany bone fracture and which may be especially stressful for an elderly person, who already has health problems.[16]

Factors Affecting Osteoporosis Risk The causes of osteoporosis are not fully understood, but the risk depends on the level of peak bone mass and the rate at which bone is lost. These are affected by age, genetics, gender, hormone levels, and lifestyle **(Table 11.3)**.

Table 11.3 Factors Affecting the Risk of Osteoporosis

Risk Factor	How It Affects Risk
Gender	Fractures due to osteoporosis are about twice as common in women as in men. Men are larger and heavier than women and therefore have a greater peak bone mass. Women lose more bone than men due to post-menopausal bone loss.
Age	Bone loss is a normal part of aging, and risk increases with age.
Race	Blacks have denser bones than do Caucasians and Southeast Asians, so their risk of osteoporosis is lower.
Family history	Having a family member with osteoporosis increases risk.
Body size	Individuals who are thin and light have an increased risk because they have less bone mass.
Smoking	Tobacco use weakens bones.
Exercise	Weight-bearing exercise, such as walking and jogging, throughout life strengthens bone, and increasing weight-bearing exercise at any age can increase your bone density.
Alcohol abuse	Long-term alcohol abuse reduces bone formation and interferes with the body's ability to absorb calcium.
Diet	A diet that is lacking in calcium and vitamin D plays a major role in the development of osteoporosis. Low calcium intake during the years of bone formation results in a lower peak bone mass, and low calcium intake in adulthood can accelerate bone loss.

age-related bone loss The bone loss that occurs in both cortical and trabecular bone of men and women as they advance in age.

Age The risk of osteoporosis increases with age. This is because bones become progressively less dense after about age 35 when bone breakdown begins to exceed bone formation. This **age-related bone loss** occurs in both men and women. Bone is lost at a rate of about 0.3%-0.5%/year. Factors that increase bone loss in older adults include a decline in calcium and vitamin D intake, a decrease in physical activity, a decrease in the efficiency of vitamin D activation by the kidney, and a decrease in dietary calcium absorption.[2] In addition, older adults typically spend less time in the sun and wear more clothing to cover the skin when outdoors, which may reduce calcium absorption by decreasing the amount of vitamin D synthesized in the skin.

Gender and Hormonal Factors About 80% of those affected by osteoporosis are women. Osteoporosis-related fractures occur in 1 out of every 2 women over age 50 compared to about 1 in every 8 men over 50.[17] The risk of osteoporosis is greater in women than men because men have a higher peak bone mass to begin with and because bone loss is accelerated in women for about 5 years after **menopause (Figure 11.12)**. This **post-menopausal bone loss** is related to the drop in estrogen levels that occurs around menopause. Declining estrogen affects bone cells, allowing an increase in bone breakdown. It also decreases intestinal calcium absorption.[18] During this 5-7-year period, bone loss may be increased 10-fold to a rate of 3%-5% per year. After the post-menopausal period, women continue to lose bone but at a slower rate.

menopause The physiological changes that mark the end of a woman's capacity to bear children.

post-menopausal bone loss The accelerated bone loss that occurs in women for about 5 years after estrogen production decreases.

Genetics Twin and family studies have shown that genetic factors are important determinants of bone density, bone size, bone turnover, and osteoporosis risk.[19] Hormonal and lifestyle factors such as nutrition, activity level, smoking, and alcohol consumption interact with genetic factors over time to determine actual bone density. Genetic differences among racial groups lead to differences in the risk of osteoporosis.[20] For example, American studies show that the incidence of osteoporosis in African-American women is half that of white women despite similar environmental risk factors. The reason for this lower risk of fractures is that African-American women begin menopause with higher bone density and have lower rates of post-menopausal bone loss. Bone density in Asians is generally lower than in non-Hispanic whites and that of Hispanics is similar to non-Hispanic whites, but bone density measures do not always explain racial differences in osteoporosis risk.[21,22]

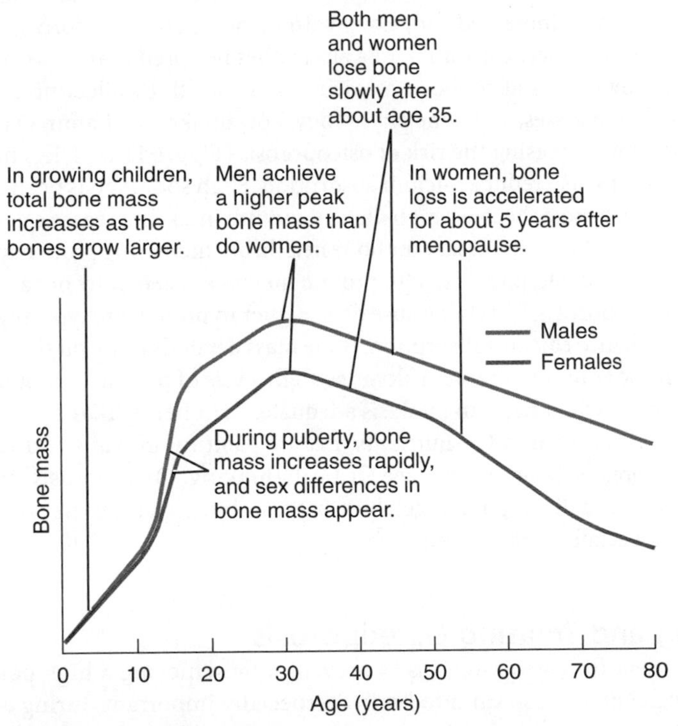

Figure 11.12 Bone mass by gender and age
Although both men and women lose bone after about age 35, women have a lower bone mineral density than men and experience accelerated bone loss after menopause.

Smoking and Alcohol Use Cigarette smoking and alcohol consumption both can decrease bone mass. Smoking affects ovarian function and calcium metabolism, leading to an increase in bone loss and the risk of osteoporosis.[23] It is estimated that smoking increases the lifetime risk of developing a vertebral fracture by 13% in women and 32% in men and that it increases the risk of a hip fracture by 31% in women and 40% in men.[24] The effect of smoking may be partially reversed by smoking cessation. Although some evidence suggests that moderate drinking may decrease the risk of fracture in post-menopausal women, long-term heavy alcohol consumption can interfere with bone growth during adolescence and affect bone turnover in adults, leading to bone loss.[25]

Exercise Weight-bearing exercise, such as walking and jogging, which puts direct weight over the skeleton, is good for bones. This mechanical stress stimulates the bones to become denser and stronger, increasing bone mass. In contrast, individuals who get no weight-bearing exercise, such as those with spinal cord injuries, those confined to bed, and astronauts in space, lose bone mass rapidly (see Science Applied: Bone: Lost in Space). Exercise during childhood and adolescence is thought to be particularly important for achieving a high peak bone mass. In adults who exercise regularly, there is a high correlation between muscle mass and bone mass, and both bone and muscle mass decrease with disuse.[26]

A greater body weight also increases bone mass.[27] Having greater body weight, whether that weight is due to an increase in muscle mass or to excess body fat, increases bone mass because it increases the amount of weight the bones must support in day-to-day activities. In post-menopausal women, excess body fat may also reduce risk because adipose tissue produces estrogen, which helps maintain bone mass and enhances calcium absorption. Therefore, the risk and severity of osteoporosis is decreased in individuals with higher body weight and fat.

Diet Diet can have a significant effect on osteoporosis risk. A low calcium intake is the most significant dietary factor contributing to osteoporosis. Calcium is necessary for bone development. Adequate calcium intake during childhood and adolescence is an important factor in maximizing bone density; low calcium intakes during the years of bone formation result in a lower peak bone mass. If calcium intake is low after peak bone mass has been achieved, the rate of bone loss may be increased and, along with it, the risk of osteoporosis.

Despite its importance, calcium intake alone does not predict the risk of osteoporosis. Other dietary components affect bone mass and bone health by affecting calcium absorption, urinary calcium losses, and bone physiology. Low intakes of vitamin D reduce calcium absorption, thereby increasing the risk of osteoporosis (Figure 11.6). Diets high in phytate, oxalates, and tannins also reduce calcium absorption. High sodium has been implicated as a risk factor for osteoporosis because high dietary sodium intake increases calcium loss in the urine.[28] Adequate protein is necessary for bone health, but increasing protein intake increases urinary calcium losses. Despite this, high protein intakes are generally not associated with a higher risk of osteoporosis. This is because diets higher in protein are typically higher in calcium and may enhance calcium absorption. Bone mass depends more on the ratio of calcium to protein than the amount of protein alone, so high levels of protein do not have a negative effect on bone mass when calcium intake is adequate.[29] Higher intakes of zinc, magnesium, potassium, fibre, and vitamin C—nutrients that are plentiful in fruits and vegetables—are associated with greater bone mass[30] (see Critical Thinking: Osteoporosis Risk). A CaMos study also found that a diet high in vegetables, fruit, and whole grains may reduce the risk of bone fracture, especially in older women.[31]

Preventing and Treating Osteoporosis

The best treatment for osteoporosis is to prevent it by achieving a high peak bone mass. Maximizing calcium deposition into bone is especially important during childhood and adolescence. Individuals with the highest peak bone mass after adolescence have a protective advantage over bone loss later in life.[17] A high peak bone mass can be achieved by maintaining an active lifestyle that includes weight-bearing exercise, limiting cigarette smoking and alcohol consumption, and consuming a diet that assures adequate calcium levels. Once osteoporosis has occurred, medications may help restore bone mass and prevent fractures.

Maximizing Dietary Calcium The risk of osteoporosis can be minimized by consuming a diet that is adequate in calcium as well as vitamin D. Limiting intake of phosphorus, protein, and sodium may also be beneficial. Milk is the major source of calcium in the diet, but teenage boys and girls today drink soft drinks rather than milk (See Your Choice: Choose Your Beverage Wisely). Canada's Food Guide recommends 3 servings of milk or milk alternatives daily to ensure adequate calcium intake. Ice cream, puddings, and

SCIENCE APPLIED

Bone: Lost in Space

In 1962, when John Glenn became the first American to orbit the Earth, there was little concern about the effect that his 5-hour flight would have on his bones. By 1997, when Shannon Lucid spent 188 days aboard the Russian Mir space station, the effect of weightlessness on bone health had become a serious concern. Weight-bearing activities such as walking, jogging, and weight training are important for the maintenance of bone health. Under the force of Earth's gravity, these activities mechanically stress the bones, which stimulate the deposition of calcium into bone. When an astronaut goes into space, weightlessness eliminates this stimulus and calcium is lost from bone.[1] This elevates calcium levels in the rest of the body and may lead to kidney stones and calcification of the soft tissues.[2] Unless bone loss in space can be prevented, prolonged space flights may not be possible. Understanding and preventing bone loss in space will also benefit those on Earth who are at risk for osteoporosis.

Information on bone loss in space has been accumulating for 40 years. Studies done back in the 1960s and early 1970s during the Gemini and Apollo space missions showed that astronauts lost calcium from their bones during space travel. Skylab missions then offered an opportunity to study the effects of more extended periods in zero gravity. In one study, 9 astronauts maintained a constant dietary intake and made continuous urine and fecal collections for 21-31 days before their flight, during their Skylab missions, and for 17-18 days after returning to Earth. The results showed that urinary calcium losses were increased during the flight. These losses occurred despite vigorous exercise regimens while in flight and were comparable to losses seen in normal adults subjected to prolonged bed rest.[3]

The study of bone metabolism in space continued with joint Russian-American studies, many conducted among astronauts who spent long periods on the Mir space station. Techniques used to study bones in space included measuring hormones and other indicators of bone metabolism in blood and urine samples before and after flights; measuring muscle strength and bone density before, during, and after return to Earth's gravity; and taking X-ray scans to determine bone mass. The most significant bone loss was found in weight-bearing parts of the skeleton; the lower vertebrae, hips, and upper femur—the same areas at risk for fracture in osteoporosis.[4] Once the astronauts returned to Earth, calcium loss slowed, but even after 6 months, bone mass had not completely recovered in most subjects.[5] Biochemical measures indicate that the bone loss that occurs during space flight is due to an increase in bone resorption and decreased intestinal calcium absorption.[1] The decrease in calcium absorption is likely due to low levels of vitamin D from insufficient dietary intake and lack of ultraviolet light exposure during space flight.[1]

(Courtesy NASA)

Shannon Lucid exercising on a treadmill aboard the Mir space station.

To try to counteract the effects of weightlessness, astronauts exercise in space on stationary bikes and treadmills and pull against bungee cords. Exercises that provide higher loads can counteract bone loss to some degree but do not prevent it.[1] In addition to weight-bearing exercise, future studies may address nutritional factors such as vitamin D and calcium intakes as well as appropriate levels of other nutrients that affect bone metabolism, such as vitamin K, sodium, and protein. Understanding how to optimize nutrient intake and exercise patterns for bone mineral retention in space will help astronauts remain healthy during long space flights and may do the same for elderly or bedridden persons on Earth who are at risk of bone loss due to inactivity.

[1] Smith, S. M., Wastney, M. E., O'Brien, K. O., et al. Bone markers, calcium metabolism, and calcium kinetics during extended-duration space flight on the Mir space station. *J. Bone Miner. Res.* 20:208-218, 2005.

[2] Whitson, P. A., Pietrzyk, R. A., Morukov, B. V., et al. The risk of renal stone formation during and after long duration space flight. *Nephron* 89:264-270, 2001.

[3] Whedon, G. D., Lutwak L., Rambaut, P., et al. Mineral and nitrogen metabolic studies on Skylab flights and comparison with effects of earth long-term recumbency. *Life Sci. Space Res.* 14:119-127, 1976.

[4] Grigoriev, A. I., Oganov, V. S., Bakulin, A. V, et al. Clinical and physiological evaluation of bone changes among astronauts after long-term space flights. *Aviakosm. Ekolog. Med.* 32:21-25, 1998.

[5] Shackelford, L. C, Oganov, V., LeBlanc, A., et al. Bone mineral loss and recovery after shuttle-Mir flights. Available online at spaceflight.nasa.gov/history/shuttle-mir/science/hls/musc/sc-hls-bone.htm. Accessed June 17, 2005.

soups made with milk are also good calcium sources but can be high in kcalories from added sugar and fat.

Individuals who are lactose intolerant can meet their calcium needs by consuming high-calcium foods that contribute little or no lactose. For those who can tolerate some lactose, fermented milk products such as yogurt and cheese may be tolerated. Lactose-free calcium sources include dark-green, leafy vegetables such as kale, broccoli, and mustard greens; soy products processed with calcium; and fish consumed with the bones (**Figure 11.13a**). Drinking milk treated with the lactose-digesting enzyme, lactase (Lactaid milk), or consuming lactase pills to help digest the lactose consumed with a meal can also help those with lactose intolerance to meet calcium needs (see Section 4.3). Calcium supplements and fortified foods such as juice products also provide calcium without lactose **(Figure 11.13b)** (see Label Literacy: Counting All Your Calcium).

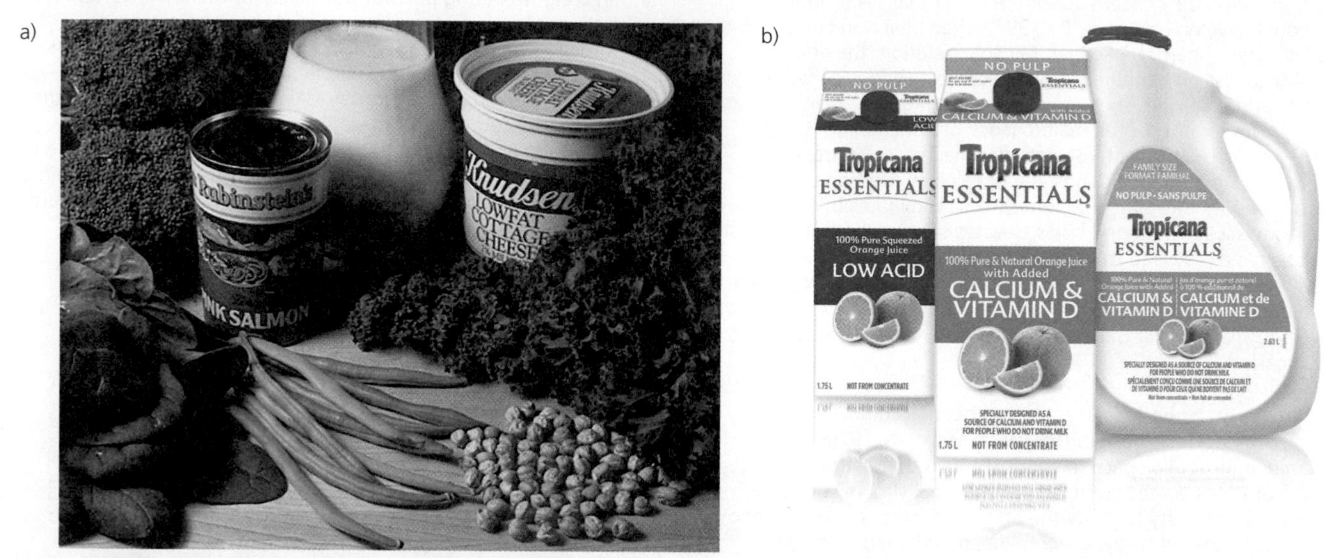

Figure 11.13 (a) Milk and alternative products, fish consumed with bones, leafy greens, and legumes are good sources of calcium. (b) A variety of calcium-fortified foods are also available to help meet needs, such as fortified orange juice. (Felicia Martinez/PhotoEdit; courtesy Pepsico)

Calcium Supplements Individuals who do not meet their calcium needs with diet alone can benefit from calcium supplementation. In young individuals, supplemental calcium can promote the development of a higher peak bone mass. In post-menopausal women, calcium supplements have a small but beneficial effect on bone mass.[32] Supplementation with 17.5-20 µg (700-800 IU) of vitamin D per day appears to reduce the risk of hip fractures in elderly persons.[33] Because calcium absorption decreases when large amounts are consumed at one time, calcium availability is better when a lower-dose calcium supplement (no more than 500 mg/dose) is taken twice a day than when a supplement that provides 100% of the RDA is taken once a day.

Osteoporosis Treatment Medications used to treat osteoporosis include hormones and drugs known as bisphosphonates. Replacing the hormones estrogen and progesterone, lost in menopause—known as hormone replacement therapy—has been shown to reduce bone loss and restore some lost bone, but this therapy carries risks and its use should be considered within the context of the individual's other health risks.[34] Administration of calcitonin, a hormone that acts by inhibiting bone resorption, by injection or nasal spray, can reduce bone loss, and its effects are enhanced by calcium supplementation.[35] Bisphosphonates are not hormones. They act by binding to the bone surface and inhibiting the activity of osteoclasts. They have been shown to prevent post-menopausal bone loss, increase bone mineral density, and reduce the risk of fractures in patients with osteoporosis.[35] Exercise can also be helpful in treating osteoporosis. Minerals other than calcium that have been used to prevent and treat bone loss include magnesium, fluoride, and boron, but results from these have been equivocal.

CRITICAL THINKING

Osteoporosis Risk

Background:

Mika is nearly 50. Although she has no symptoms, she is worried about her risk for osteoporosis. Her mother is 75 and recently suffered a fractured hip due to reduced bone density caused by osteoporosis. Her previously independent mother is now living in a nursing home and struggling to return to her former life. Mika is frightened that she will face the same future. She finds the following osteoporosis risk factor questionnaire in a health magazine and fills it out. She also records her intake for one day.

FOOD	ENERGY (KCALS)	CALCIUM (MG)
Breakfast		
Eggs (2 large)	150	5
Toast with margarine (2 slices)	200	50
Orange juice (175 ml-¾ cup)	80	15
Coffee with cream (250 ml-1 cup)	45	35
Lunch		
Bologna sandwich on white		
bread with mayonnaise	260	60
Lettuce and tomato (2 slices)	10	5
Milk (250 ml-1 cup)	120	300
Apple (1 medium)	80	10
Snack		
Chips (30 g-1 oz)	150	10
Beer (360 ml-12 fl oz)	140	20
Dinner		
Roast beef (100g-3 oz)	225	5
Mashed potatoes and gravy (250 ml-1 cup)	350	50
Green beans (250 ml-1 cup)	35	60
Iced tea (360 ml-12 oz)	4	0
Ice cream (125 ml-1/2 cup)	140	70
Total	1989	695

Data:
OSTEOPOROSIS QUESTIONNAIRE

Gender ?	__	Male
	X	Female
Age ?	__	2 to 18 years
	__	19 to 30 years
	X	31 to 50 years
	__	51 to 70 years
	__	> 70 years
Have you ever broken a bone?	__	Yes
	X	No
What is your bone density?	X	Never been measured
	__	Normal
	__	Low density
What is your Body Mass Index?	__	<18.5 kg/m^2
	__	8.5 to 24.9
	X	25.0 to 29.9
	__	30.0 to 34.9
	__	> 35
Do you smoke cigarettes?	X	Yes
	__	No
How much alcohol do you drink?	__	> 2 drinks/day
	X	1-2 drinks/day
	__	Several drinks per week
	__	None

How much milk do you drink?	X	None
	__	2 or fewer glasses a day
	__	3 or more glasses a day
How much milk did you drink as a child?	__	None
	X	2 or fewer glasses a day
	__	3 or more glasses a day
How much milk did you drink as an adolescent?	X	None
	__	2 or fewer glasses a day
	__	3 or more glasses a day
How often do you exercise?	X	Less than 3 times a week
	__	3 or more times a week
What types of activities do you participate in?	__	None
	X	Walking, jogging, tennis
	__	Swimming, bicycling
What is your exercise history?	__	Have been active all my life
	X	Was active as a child but no longer exercise often.
	__	Recently started exercising
If you are female, are you currently menstruating?	X	Yes
	__	No
If you are post-menopausal, how long ago did menopause occur?	__	Less than 5 years ago
	__	More than 5 years ago
Do you have a family history of osteoporosis?	X	Yes
	__	No

Critical Thinking Questions

Evaluate Mika's risk for osteoporosis by looking at her answers on the questionnaire. Why would her milk intake and activity level as a child affect her risk now?

How does Mika's calcium intake compare with the recommendations? Suggest changes she could make in her diet to increase her calcium without increasing her kcalorie intake.

Do you think Mika should take a calcium supplement? Why or why not?

iProfile Use iProfile to find out how much calcium is provided by the non-milk foods in your diet.

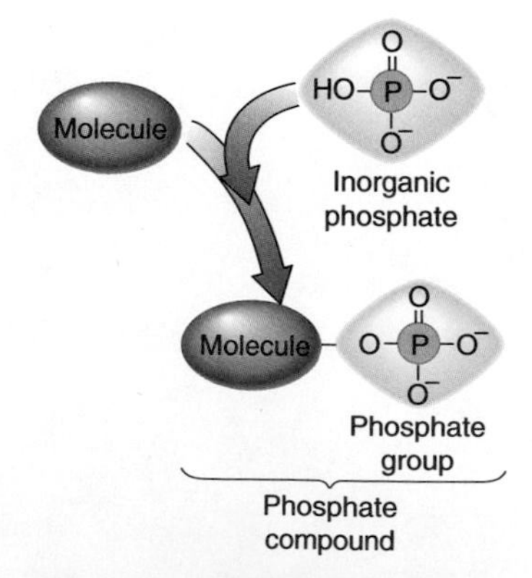

Figure 11.14 Phosphate group When inorganic phosphate (phosphorus combined with oxygen) joins with another molecule, it is called a phosphate group.

11.4 Phosphorus

Learning Objectives

- Describe the functions of phosphorus in the body.
- Plan a diet that meets the recommended intakes for calcium and phosphorus.

Phosphorus makes up about 1% of the adult body by weight, and 85% of this is found in bones and teeth.[1] The phosphorus in soft tissues has both structural and regulatory roles. In nature, phosphorus is most often found in combination with oxygen as phosphate **(Figure 11.14)**.

Phosphorus in the Diet

Phosphorus is more widely distributed in the diet than calcium. Like calcium, it is found in milk, yogurt, and cheese, but meat, cereals, bran, eggs, nuts, and fish are also good sources **(Figure 11.15)**. Food additives used in baked goods, cheese, processed meats, and soft drinks also contribute to dietary phosphorus.

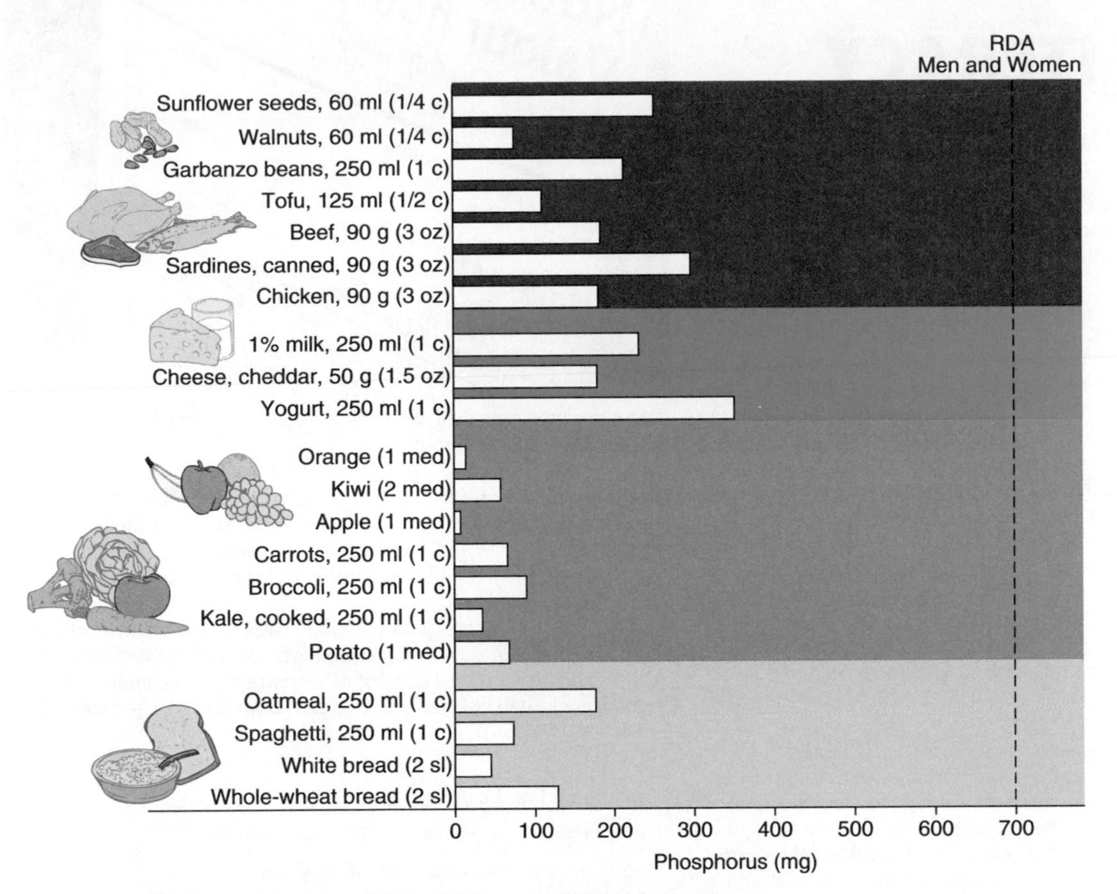

Figure 11.15 Phosphorus content of Canada's Food Guide food groups
Adults can obtain their RDA of phosphorus (dashed line) by consuming foods in all the food groups.

Phosphorus in the Digestive Tract

Phosphorus is more readily absorbed than calcium. About 60%-70% is absorbed from a typical diet. There is no evidence that the efficiency of absorption is affected by the amount in the diet. Vitamin D does aid phosphorus absorption via an active mechanism, but most absorption occurs by a mechanism that does not depend on vitamin D. Therefore, when vitamin D is deficient, phosphorus can still be absorbed, but its absorption is reduced.

Phosphorus in the Body

Phosphorus and Bone Health Phosphorus, along with calcium, forms hydroxyapatite crystals that provide rigidity to bones. Blood levels of phosphorus are not as strictly controlled as those of calcium, but levels are maintained in a ratio with calcium that allows bone mineralization. When blood levels of phosphorus are low, the active form of vitamin D is synthesized. This increases the absorption of both phosphorus and calcium from the intestine and increases their release from bone. When phosphorus intake is high, more is lost in the urine, so plasma levels rise only slightly. A rise in serum phosphorus indirectly stimulates PTH release, causing phosphorus excretion and calcium retention by the kidney as well as calcium release from bone. When PTH is not secreted (such as when calcium levels rise), phosphorus is retained by the kidney and calcium is excreted.

Other Functions of Phosphorus Phosphorus is also an important component of a number of molecules with regulatory roles (see Table 11.1). Phosphorus is a component of the water-soluble head of phospholipid molecules, which form the structure of cell membranes. Phosphorus is a major constituent of the genetic material DNA and RNA and it is essential for energy metabolism because the high-energy bonds of ATP are formed between phosphate groups **(Figure 11.16)**. Phosphorus is also a component of other high-energy compounds, including creatine phosphate, which provides energy to exercising muscles. Phosphorus-containing molecules are important in relaying signals

LABEL LITERACY

Counting All Your Calcium

(©Tim Pohl/iStockphoto)

To find out if you get enough calcium, you'll need to count all your sources. You can obtain calcium from natural sources such as milk, yogurt, and leafy greens, from foods fortified with the mineral, and from calcium supplements.

You can see if a packaged food is a good source of calcium by looking at the label. The Nutrition Facts panel lists the % Daily Value for calcium. To calculate the milligrams of calcium in that food, multiply the % Daily Value by 1,100 mg (the Daily Value for calcium). Descriptors such as "high in calcium" or "a good source of calcium" can help you identify foods that make a significant calcium contribution (see table). Foods high in calcium may include the health claim that a diet high in calcium helps reduce the risk of osteoporosis.

If you rely on natural health products such as mineral supplements to increase your calcium count, be aware that the typical multivitamin and mineral supplement provides only a small amount of the calcium you need. To get a significant amount, read the labels to find a supplement that contains a calcium compound alone or calcium with vitamin D (which aids calcium absorption).

The form of calcium in supplements is also important. Calcium carbonate is absorbed as well as the calcium from milk. Calcium citrate, calcium gluconate, calcium lactate, calcium citrate-malate, and calcium phosphate are absorbed as well as calcium from a mixed diet.[1] To optimize absorption, calcium carbonate should be taken with a meal, while calcium citrate can be taken at any time.[2]Avoid calcium preparations such as bone meal, coral calcium, powdered bone, dolomite, and oyster shell, which may contain contaminants that are dangerous if consumed routinely.[3]

Over-the-counter antacids can also be taken to supplement calcium intake. Many of these, such as Tums, which contains calcium carbonate, are safe, effective calcium supplements. However, antacids that contain aluminum and magnesium may actually increase calcium loss.

In short, to see if you are getting enough calcium, check the labels on your foods, supplements, and medications; consider the form of calcium in each; and watch for excesses of other nutrients and contaminants. Sound complicated? Maybe an extra glass of milk or fortified soy milk is easier.

Calcium on Food Labels
Excellent source of calcium: > 25% of Daily Value
Good source of calcium > 15% of Daily Value
More or added calcium > 5% of Daily Value

[1] Mortensen, L., and Charles, P. Bioavailability of calcium supplements and the effect of vitamin D: Comparisons between milk, calcium carbonate, and calcium carbonate plus vitamin D. *Am. J. Clin. Nutr*. 63:354–357, 1996.

[2] Straub, D. A. Calcium supplementation in clinical practice: a review of forms, doses, and indications. *Nutr. Clin. Pract.* 22(3):286-96, 2007.

[3] Bourgoin, B. P., Evans, D. R., Cornett, J. R., et al. Lead content in 70 brands of dietary calcium supplements. *Am. J. Public Health* 83:1155–1160, 1993.

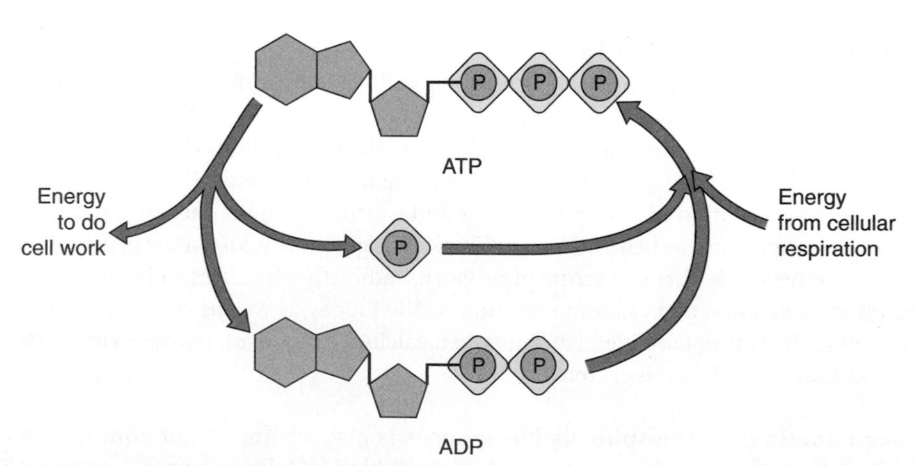

Figure 11.16 Phosphorus and ATP
Breaking the high-energy bond between the second and third phosphate groups of ATP releases energy for cellular work. The ADP (adenosine diphosphate) that is formed can be converted back to ATP by using the energy trapped by the electron transport chain of cellular respiration to add a phosphate group.

to the interior of cells to mediate hormone action and perform other metabolic activities. Phosphorus is involved in regulating enzyme activity because the addition of a phosphate group can activate or deactivate certain enzymes. It is also part of the phosphate buffer system that helps regulate the pH in the cytosol of all cells so that chemical reactions can proceed normally.

Recommended Phosphorus Intake

The RDA for phosphorus is set at 700 mg/day for men and women 19-50 years of age.[1] This is the amount needed to maintain normal blood phosphorus levels. Because neither absorption nor urinary losses change significantly with age, the RDA is the same for older adults.

For growing children and adolescents, the RDA is based on the phosphorus intake necessary to meet the needs for bone and soft-tissue growth. There is no evidence that phosphorus requirements are increased during pregnancy; intestinal absorption increases by about 10%, which is sufficient to provide the additional phosphorus needed by the mother and fetus. The RDA is not increased during lactation because the phosphorus in milk is provided by an increase in bone resorption and a decrease in urinary excretion that are independent of dietary intake of either phosphorus or calcium.

Life Cycle

Phosphorus Deficiency

Phosphorus deficiency can lead to bone loss, weakness, and loss of appetite. Deficiency is rare in healthy people because phosphorus is so widely distributed in foods. Most people in Canada easily meet their needs (see Critical Thinking: How are Canadians Doing with Respect to Their Intake, from Food, of Calcium, Phosphorus, and Magnesium?). Marginal phosphorus deficiencies are most common in premature infants, vegans, alcoholics, and the elderly. Marginal phosphorus status may also be caused by losses due to chronic diarrhea and overuse of aluminum-containing antacids, which prevent phosphorus absorption.

Phosphorus Toxicity

Toxicity from high phosphorus intake is rare in healthy adults, but excessive intakes can lead to bone resorption. Typical intakes are above recommendations. One reason is the increased use of phosphorus-containing food additives, which has lead to concern about its impact on bone health.[17] High phosphorus intake has been found to increase bone resorption, but some of this can be prevented by adequate calcium intake.[36] Therefore, levels of phosphorus intake typical in Canada are not believed to affect bone health as long as calcium intake is adequate. Based on the upper level of normal serum phosphate, a UL for phosphorus of 4 g/day has been set for adults aged 19-70.[1]

11.5 Magnesium

Learning Objectives

- Describe the functions of magnesium in the body.
- Name 3 foods that are good sources of magnesium.

There are approximately 25 g of magnesium in the adult human body. Magnesium is a mineral that affects the metabolism of calcium, sodium, and potassium.

Magnesium in the Diet

Magnesium is found in leafy greens such as spinach and kale because it is a component of chlorophyll. Nuts, seeds, bananas, and the germ and bran of whole grains are also good sources (**Figure 11.17**). Processed foods are generally poor sources. For example, removing the bran and germ of the wheat kernel reduces the magnesium content of 250 ml of white flour to only 28 mg, compared with the 166 mg in the same amount of whole-wheat flour. In areas with hard water, the water supply may provide a significant amount of magnesium.

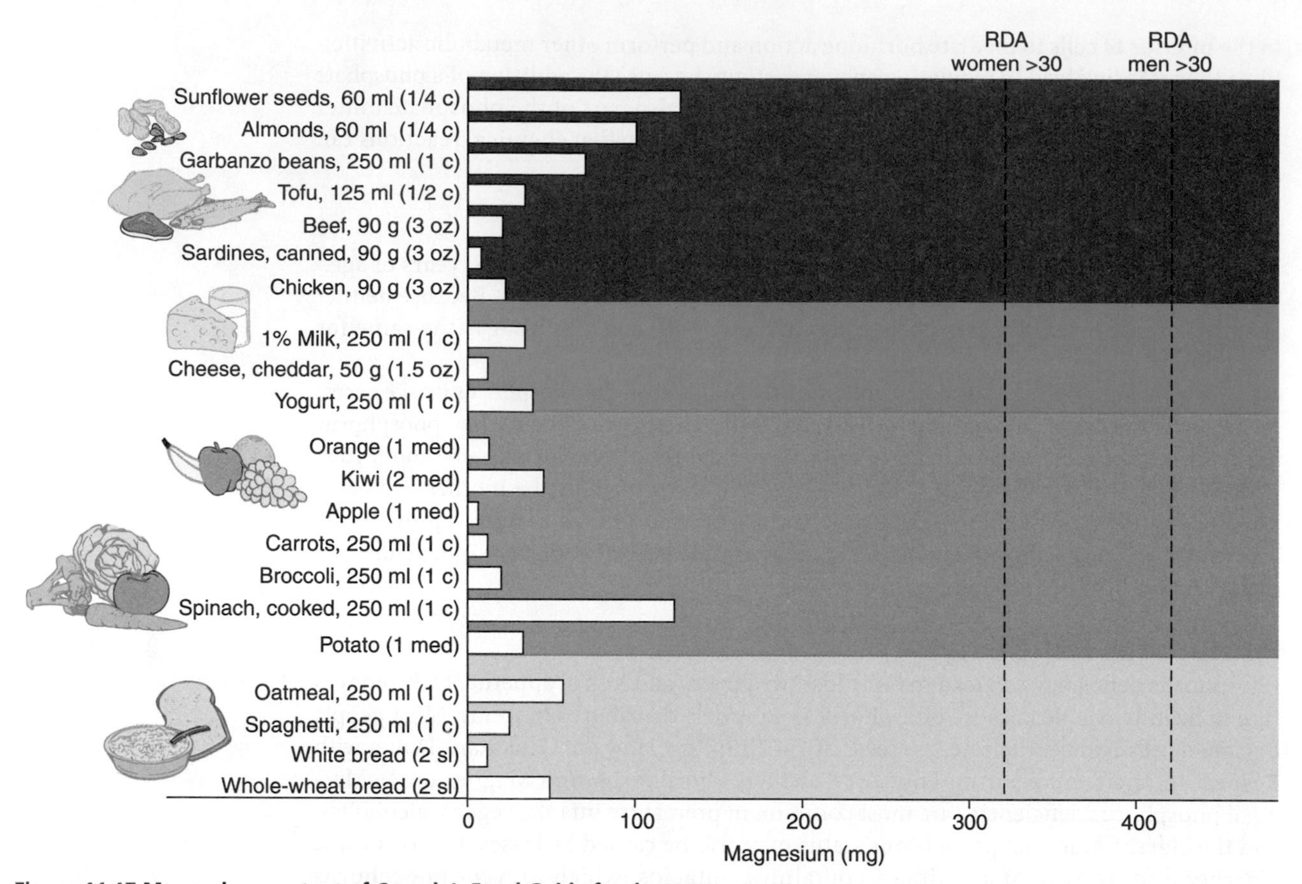

Figure 11.17 Magnesium content of Canada's Food Guide food groups
Magnesium is found in nuts, seeds, legumes, and leafy greens; the dashed lines represent the RDA for adult men and women over age 30.

Magnesium in the Digestive Tract

About 50% of the magnesium in the diet is absorbed, and the percentage decreases as intake increases. The active form of vitamin D can enhance magnesium absorption to a small extent, and the presence of phytate decreases absorption. As calcium in the diet increases, the absorption of magnesium decreases, so the use of calcium supplements can reduce the absorption of magnesium.

Magnesium in the Body

Magnesium and bone health About 50%-60% of the magnesium in the body is in bone, where it is essential for the maintenance of structure. Magnesium is also involved in regulating calcium homeostasis and is needed for the action of vitamin D and many hormones, including PTH.[37]

Other Functions of Magnesium Most of the remaining body magnesium is present inside cells, where it is the second most abundant positively charged intra-cellular ion (after potassium). Magnesium is associated with the negative charge on phosphate-containing molecules such as ATP **(Figure 11.18)**.

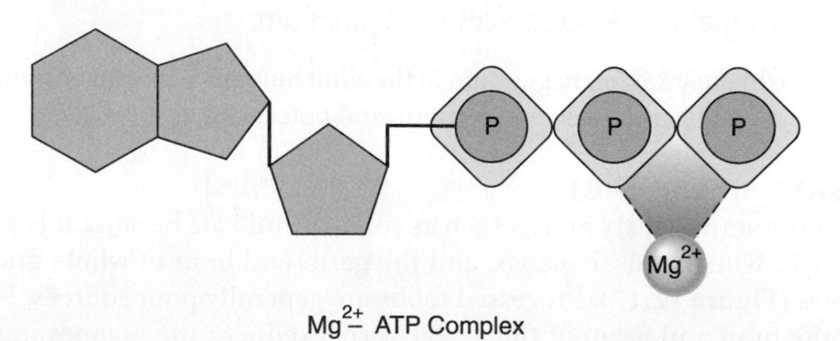

Figure 11.18 Magnesium and ATP
Magnesium stabilizes ATP structure by forming a magnesium-ATP complex, so it is important for all reactions that use or generate ATP.

Magnesium is a cofactor for more than 300 enzymes. It is necessary for the generation of ATP from carbohydrate, lipid, and protein (see Table 11.1). In some of these reactions, it is involved indirectly as a stabilizer of ATP, and in others, directly as an enzyme activator. Magnesium is needed for the activity of the sodium-potassium ATPase pump, which is responsible for active transport of sodium and potassium across membranes. It is therefore essential for maintenance of electrical potentials across cell membranes and proper functioning of the nerves and muscles, including those in the heart. It is important for DNA and RNA synthesis and for almost every step in protein synthesis. Therefore, magnesium is particularly important for dividing growing cells.

The kidneys closely regulate blood levels of magnesium. When magnesium intake is low, excretion in the urine is decreased. As intake increases, urinary excretion increases to maintain normal blood levels. This efficient regulation permits homeostasis over a wide range of dietary intakes.

Recommended Magnesium Intake

The RDA for magnesium is 400 mg/day for young men and 310 mg/day for young women.[1] This is based on the maintenance of total body magnesium balance over time. The RDA is slightly higher for men and women over age 30; 420 and 320 mg/day, respectively. A serving of whole-grain breakfast cereal, spinach, or legumes contains about 100 mg of magnesium.

The requirement for pregnancy is increased by 35 mg/day to account for the addition of lean body mass. No increase is recommended for lactation because magnesium is released when bone is resorbed and urinary excretion is decreased. An AI is set for infants based on the magnesium content of human milk.

Life Cycle

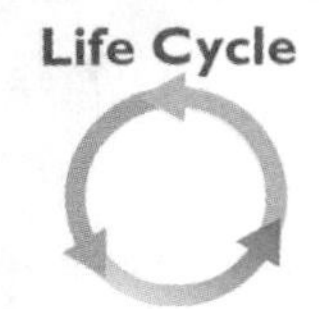

Magnesium Deficiency

Magnesium deficiency is rare in the general population. It does occur in those with alcoholism, malnutrition, kidney disease, and gastrointestinal disease, as well as in those who use diuretics that increase magnesium loss in the urine. Deficiency symptoms include nausea, muscle weakness and cramping, irritability, mental derangement, and changes in blood pressure and heartbeat. Low blood magnesium levels affect levels of blood calcium and potassium; therefore, some of these symptoms may be due to alterations in the levels of these other minerals.

Although overt deficiency is rare, a high proportion of Canadians consume less than the EAR. Low intakes of magnesium have been associated with a number of chronic diseases, including osteoporosis.[38] Low magnesium may also affect cardiovascular health. Dietary patterns higher in magnesium help reduce blood pressure and epidemiological evidence suggests that humans with good magnesium status are at a lower risk of atherosclerosis.[39] Areas with hard water, which is high in calcium and magnesium, tend to have lower rates of death from cardiovascular disease.

Magnesium Toxicity and Supplements

No adverse effects have been observed from ingestion of magnesium from food, but toxicity may occur from concentrated sources such as magnesium-containing drugs and supplements. Toxicity has been reported in elderly patients with impaired kidney function who frequently use magnesium-containing laxatives and antacids such as milk of magnesia. Magnesium toxicity is characterized by nausea, vomiting, low blood pressure, and cardiovascular changes. The UL for adults and adolescents over 9 years of age is 350 mg from nonfood sources of magnesium.

11.6 Sulfur

Learning Objective

- Discuss the role of sulfur in the body.

Dietary sulfur is found in organic molecules such as the sulfur-containing amino acids in proteins and the sulfur-containing vitamins. It is also found in some inorganic food preservatives such as sulfur dioxide, sodium sulfite, and sodium and potassium bisulfite, which are used as antioxidants.

In the body, the sulfur-containing amino acids methionine and cysteine are needed for protein synthesis. Cysteine is also part of the compound glutathione, which is important in detoxifying drugs and protecting cells from oxidative damage. The vitamins thiamin and biotin, essential for ATP production, contain sulfur. Sulfur-containing ions are important in regulating acid-base balance.

There is no recommended intake for sulfur, and no deficiencies are known when protein needs are met (see Table 11.1).

CASE STUDY OUTCOME

Margie's hip fracture is not surprising, given her medical history. She is a small woman who never got much exercise or consumed much calcium—factors that contribute to a low peak bone mass. In addition, because she has been post-menopausal for a number of years, Margie has been losing bone faster than her body has replaced it. As a result, her bones are so brittle that the impact of stepping off a curb caused a bone in her hip to break. If Margie had had regular physical exams, her doctors would have noted her low bone density and begun treatment, which might have prevented the hip fracture.

After she broke her hip, Margie spent 4 months in a rehabilitation facility. She was lucky, however, and has been able to return to her old life. Now, a year later, she continues to treat her disease with bisphosphonate drugs, which inhibit bone resorption, as well as calcium and vitamin D supplements. As her mobility improves, she is trying to increase the time she spends walking to 30 minutes on most days. She was thankful to be able to celebrate her grandson's first birthday in her own home.

APPLICATIONS

Personal Nutrition

1. **Do you get enough calcium?**

iProfile

a. Using iProfile and the food record you kept in Chapter 2, calculate your average calcium intake.

b. How does your intake compare with the RDA for calcium for someone of your age and gender?

c. If your calcium intake is below the RDA, suggest modifications to increase the amount of calcium in your diet without significantly increasing your energy intake.

2. **How much calcium do fortified foods contribute?**

a. Choose 3 foods that are fortified with calcium and use their labels to determine how much calcium is in a serving of each.

b. How many servings per day of each would you need to consume to meet your calcium needs?

c. Are there other dietary components in any of these foods that would interfere with calcium absorption or utilization?

General Nutrition Issues

1. **Many people must limit milk consumption due to lactose intolerance. How can they meet their calcium needs?**

a. Use iProfile to plan a day's diet that meets your calcium needs but does not include any milk products or calcium-fortified foods.

b. Is this a reasonable diet to follow every day? Would you recommend including fortified foods or calcium supplements? Why or why not?

2. **Imagine you are a scientist with Health Canada and you have been assigned the task of deciding on a food or group of foods that will be fortified with calcium to assure that the population meets their calcium needs.**

a. What food or group of foods would you recommend? Why?

b. Does the food or group of foods contain dietary components that interfere with calcium absorption?

c. Is this a food or group of foods that is consumed by the population groups most at risk for calcium deficiency?

d. How much calcium would you recommend adding? Would this amount meet recommendations but not put the population at risk of calcium toxicity?

SUMMARY

11.1 What Are Minerals?

- Minerals are elements needed by the body to regulate chemical reactions and provide structure. They are found in both plant and animal foods.
- Minerals are added to some foods through fortification and get into others as a result of contamination. Dietary supplements are also a source of minerals.
- Mineral bioavailability is affected by body needs as well as interactions with other minerals, vitamins, and dietary components such as fibre, phytates, oxalates, and tannins.

11.2 Calcium

- Sources of calcium in the Canadian diet include milk and alternative products, fish consumed with bones, and leafy green vegetables. Fortified foods also contribute to calcium intake.
- Adequate calcium absorption depends on adequate levels of vitamin D. The absorption of calcium is reduced by the presence of tannins, fibre, phytates, and oxalates. Calcium absorption varies with life stage and is highest during infancy and pregnancy, when needs are greatest.
- Most of the calcium in the body is in bone. Calcium not found in bone is essential for cell communication, nerve transmission, muscle contraction, blood clotting, and blood pressure regulation. Blood levels of calcium are regulated by parathyroid hormone (PTH) and calcitonin. PTH stimulates the release of calcium from bone, decreases calcium excretion by the kidney, and activates vitamin D to increase the amount of calcium absorbed from the gastrointestinal tract and released from bone. Calcitonin blocks calcium release from bone.
- The RDA for calcium ranges from 1,000-1,200 mg/day for adults and is 1,300 mg/day in adolescents.
- Calcium deficiency can reduce bone mass and increase the risk of osteoporosis. Too much calcium can contribute to the formation of kidney stones, raise blood calcium levels, and interfere with the absorption of other minerals.

11.3 Calcium and Bone Health

- Bone is a living tissue that is constantly being broken down and reformed in a process known as bone remodelling. Early in life, bone formation occurs more rapidly than bone breakdown to allow bone growth and an increase in bone mass. Peak bone mass occurs in young adulthood. With age, bone breakdown begins to outpace formation, causing a decrease in bone mass; this is accelerated in women for about 5 years after menopause.
- Osteoporosis is a condition in which loss of bone mass increases the risk of bone fractures. The risk of osteoporosis is related to the level of peak bone mass and the rate of bone loss. These are affected by age, gender, hormone levels, genetics, smoking and alcohol use, exercise, and diet.
- Osteoporosis risk can be reduced by an active lifestyle and a diet adequate in calcium and vitamin D and not excessive in phosphorus, sodium, or protein. Osteoporosis is treated with supplements of calcium and vitamin D, medications that inhibit bone breakdown, and, in some cases, hormone replacement therapy.

11.4 Phosphorus

- Phosphorus is more widely distributed in the diet than calcium. It is found in milk, yogurt, and cheese, but meat, cereals, bran, eggs, nuts, and fish are also good sources. Food additives used in baked goods, cheese, processed meats, and soft drinks also contribute to dietary phosphorus.
- About 60%-70% of the phosphorus in a typical diet is absorbed. Vitamin D aids phosphorus absorption via an active mechanism, but most absorption occurs by a mechanism that does not depend on vitamin D.

- Most of the phosphorus in the body is found in bones and teeth. In addition to its structural role in these tissues, phosphorus is an essential component of phospholipids, ATP, and DNA. Phosphorus is also part of a buffer system that helps prevent changes in pH.
- The RDA for adults is 700 mg/day.
- Phosphorus deficiency is rare in healthy people because it is so widely distributed in foods. It can lead to bone loss, weakness, and loss of appetite.
- Toxicity from high phosphorus intake is rare in healthy adults, but excessive intakes can lead to bone resorption. The levels of phosphorus intake typical in Canada are not believed to affect bone health as long as calcium intake is adequate.

11.5 Magnesium

- Magnesium is found in leafy greens such as spinach and kale because it is a component of chlorophyll. Nuts, seeds, bananas, and the germ and bran of whole grains are also good sources.
- About half of the magnesium in the diet is absorbed, and the percentage decreases as intake increases. The active form of vitamin D can enhance magnesium absorption and the presence of phytates decreases absorption.
- Magnesium is important for bone health, and it is needed as a cofactor for numerous reactions throughout the body. In reactions involved in energy metabolism, it acts as an enzyme activator and stabilizer of ATP. It is also needed to maintain membrane potentials; thus, it is essential for nerve and muscle conductivity. Homeostasis is regulated by the kidney.
- The RDA for magnesium is 400 mg/day for young men and 310 mg/day for young women.
- Magnesium deficiency is rare in the general population. It does occur in those with alcoholism, malnutrition, kidney disease, and gastrointestinal disease. Symptoms include nausea, muscle weakness and cramping, irritability, mental derangement, and changes in blood pressure and heartbeat.
- No adverse effects have been observed from ingestion of magnesium from food, but toxicity may occur from magnesium-containing drugs and supplements.

11.6 Sulfur

- Sulfur is in the diet as preformed organic molecules such as the amino acids methionine and cysteine, which are needed to synthesize proteins and glutathione; and in the vitamins thiamin and biotin, needed for energy metabolism. Sulfur is also part of a buffer system that regulates acid-base balance. A dietary deficiency is unknown in the absence of protein malnutrition.

REVIEW QUESTIONS

1. Explain the difference between major minerals and trace elements.
2. List 4 factors that can affect mineral bioavailability.
3. What is the major source of calcium in the Canadian diet?
4. What is the function of calcium in bones and teeth?
5. What are the roles of calcium in body fluids?
6. How are blood calcium levels restored when they drop too low? Rise too high?
7. What is bone remodelling?
8. How does the rate of bone formation and breakdown change throughout life?
9. How does the level of peak bone mass affect the risk of osteoporosis?
10. How is calcium intake related to the risk of osteoporosis?
11. What factors other than calcium intake are related to the risk of osteoporosis?
12. List sources of dietary calcium acceptable to those who are lactose intolerant.
13 Name some food sources of phosphorus.
14. What are the functions of phosphorus in the body?
15. Name some food sources of magnesium.
16. What is the function of magnesium in the body?
17. Where is sulfur found in the body?

REFERENCES

1. Institute of Medicine, Food and Nutrition Board. *Dietary Reference Intakes for Calcium, Phosphorus, Magnesium, Vitamin D, and Fluoride*. Washington, DC: National Academies Press, 1997.
2. Nordin, B. E. C., Need, A. G., Morris, H. A. et al. Effect of age on calcium absorption in postmenopausal women. *Am. J. Clin. Nutr.* 80:998–1002, 2004.
3. Heaney, R. P., Weaver, C. M., and Recker, R. R. Calcium absorption from spinach. *Am. J. Clin. Nutr.* 47:707–709, 1988.
4. Bronner, F., and Pansu, D. Nutritional aspects of calcium absorption. *J. Nutr.* 129:9–12, 1999.
5. Conlin, P. R., Chow, D., Miller, E. R., III, et al. The effect of dietary patterns on blood pressure control in hypertensive patients: Results from the Dietary Approaches to Stop Hypertension (DASH) trial. *Am. J. Hypertens.* 13:949–955, 2000.
6. Food and Nutrition Board, Institute of Medicine. Dietary Reference Intakes for Calcium and Vitamin D. Washington DC: National Academies Press, 2011.
7. Prentice, A. Micronutrients and the bone mineral content of the mother, fetus and newborn. *J. Nutr.* 133(5 Suppl 2): 1693S–1699S, 2003.
8. Beall, D. P., and Scofield, R. H. Milk-alkali syndrome associated with calcium carbonate consumption. Report of 7 patients with parathyroid hormone levels and an estimate of prevalence among patients hospitalized with hypercalcemia. *Medicine* (Baltimore). 74:89–96, 1995.
9. Minihane, A. M., and Fairweather-Tait, S. J. Effect of calcium supplementation on daily nonheme-iron absorption and long-term iron status. *Am. J. Clin. Nutr.* 68:96–102, 1998.
10. Wood, R. J., and Zheng, J. J. High dietary calcium intakes reduce zinc absorption and balance in humans. *Am. J. Clin. Nutr.* 65:1803–1809, 1997.

11. Shea, M. K., and Booth, S. L. Update on the role of vitamin K in skeletal health. *Nutr. Rev.* 66:549–557, 2008.

12. Wang, Q., and Seeman, E. Skeletal growth and peak bone strength. *Best Pract. Res. Clin. Endocrinol. Metab.* 22:687–700, 2008.

13. Osteoporosis Canada. Facts and statistics. Available online at http://www.osteoporosis.ca/index.php/ci_id/8867/la_id/1.htm. Accessed August 22,2010.

14. CaMos Publications. Available online at http://www.camos.org/publications.php. Accessed Sept 23, 2010.

15. Poliquin, S., Joseph, L., Gray-Donald, K. Calcium and vitamin D intakes in an adult Canadian population. *Can. J. Diet Pract. Res.* 70(1):21-7, 2009.

16. Ioannidis, G., Papaioannou, A., Hopman, W. M. et al. Relation between fractures and mortality: results from the Canadian Multicentre Osteoporosis Study. *CMAJ* 181 (5):265-71, 2009.

17. National Institutes of Health. Consensus Development Conference Statement. Osteoporosis Prevention, Diagnosis and Therapy, March 27–29, 2000. Available online at odp.od.nih.gov/consensus/cons/111/111_intro.htm/. Accessed October 8, 2009.

18. Van Cromphaut, S. J., Rummens, K., Stockmans, I. et al. Intestinal calcium transporter genes are upregulated by estrogens and the reproductive cycle through vitamin D receptor-independent mechanisms. *J. Bone Miner. Res.* 18:1725–1736, 2003.

19. Walker, M. D., Novotny, R., Bilezikian, J. P., et al. Race and diet interactions in the acquisition, maintenance, and loss of bone. *J. Nutr.* 138:1256S–1260S, 2008.

20. National Institute of Arthritis and Musculoskeletal and Skin Diseases. *Osteoporosis Overview*. Available online at www.niams.nih.gov/Health_Info/Bone/Osteoporosis/ default.asp/. Accessed January 3, 2009.

21. Ferrari, S. Human genetics of osteoporosis. *Best Pract. Res. Clin. Endocrinol. Metab.* 22:723–735, 2009.

22. Looker, A. C. The skeleton, race, and ethnicity. *J. Clin. Endocrinol. Metab*. 87:3047–3050, 2002.

23. Kapoor, D., and Jones, T. H. Smoking and hormones in health and endocrine disorders. *Eur. J. Endocrinol.* 152:491–499, 2005.

24. Ward, K. D., and Klesges, R. C. A meta-analysis of the effects of cigarette smoking on bone mineral density. *Calcif. Tissue Int.* 68:259–270, 2001.

25. Sampson, H. W. Alcohol and other factors affecting osteoporosis risk in women. *Alcohol Res. Health* 26:292–298, 2002.

26. Anderson, J. J. The important role of physical activity in skeletal development: How exercise may counter low calcium intake. *Am. J. Clin. Nutr.* 6:1384–1386, 2000.

27. Reid, I. R. Relationships between fat and bone. *Osteoporos. Int.* 19:595–606, 2008.

28. Teucher, B., and Fairweather-Tait, S. Dietary sodium as a risk factor for osteoporosis: Where is the evidence? *Proc. Nutr. Soc.* 62:859–866, 2003.

29. Kerstetter, J. E., O'Brien, K. O., Caseria, D. M. et al. The impact of dietary protein on calcium absorption and kinetic measures of bone turnover in women. *Clin. Endocrinol. Metab.* 90:26–31, 2005.

30. New, S. A., Robins, S. P., Campbell, M. K. et al. Dietary influences on bone mass and bone metabolism: Further evidence of a positive link between fruit and vegetable consumption and bone health. *Am. J. Clin. Nutr*. 71:142–151, 2000.

31. Langsetmo, L., Hanley, D. A., Prior, J. C. at al. CaMos Research Group. Dietary patterns and incident low-trauma fractures in postmenopausal women and men aged ≥ 50 y: a population-based cohort study. *Am. J. Clin. Nutr.* 93(1):192-9, 2011.

32. Shea, B., Wells, G., Cranney, A. et al. Meta-analysis of calcium supplementation for the prevention of postmenopausal osteoporosis. *Endocrine Reviews* 23:552–559, 2002.

33. Heike A., Bischoff-Ferrari, H. A., Willett, W. A. et al. Fracture prevention with vitamin D supplementation: A meta-analysis of randomized controlled trials. *JAMA*. 293:2257–2264, 2005.

34. U.S. Preventive Services Task Force. Hormone therapy for the prevention of chronic conditions in postmenopausal women: Recommendations from the U.S. Preventive Services Task Force. *Ann. Intern. Med*. 142:855–860, 2005.

35. South-Paul, J. E. Osteoporosis: Part II. Nonpharmacologic and pharmacologic treatment. *Am. Fam. Physician* 63:1121–1128, 2001.

36. Kerni,V. E., Kärkkäinen, M. U., Karp, H. J. et al. Increased calcium intake does not completely counteract the effects of increased phosphorus intake on bone: An acute dose-response study in healthy females. *Br. J. Nutr.* 99:832–839, 2008.

37. Konrad, M., and Schlingmann, K-P. Magnesium. In *Biochemical, Physiological, and Molecular Aspects of Human Nutrition*, 2nd ed. M. H. Stipanuk, ed. St. Louis: Saunders Elsevier, 2006, pp. 921–941.

38. Rude, R. K., and Gruber, H. E. Magnesium deficiency and osteoporosis: Animal and human observations. *J. Nutr. Biochem.* 12:710–716, 2004.

39. Al-Delaimy, W. K., Rimm, E. B., Willett, W. C. et al. Magnesium intake and risk of coronary heart disease among men. *J. Am. Coll. Nutr*. 23:63–70, 2004.

FOCUS ON
Nonvitamin/Nonmineral Supplements

(©iStockphoto)

Focus Outline

If you are looking for a dietary supplement, you won't have to look far. They come as tablets, capsules, powders, softgels, gelcaps, and liquids. They are sold in health food stores, supermarkets, drug stores, and national discount chain stores, as well as through mail-order catalogues, television, the Internet, and direct sales. Most of these products contain vitamins and minerals, but some contain nutrients other than vitamins or minerals and substances that are not classified as nutrients at all. Surveys have reported that more than 43% of adult Canadians use supplements of all types; 38% use nutrient supplements and 15% use herbal products.[1] They are taken to promote health and prevent and treat disease. Others contain substances such as hormones, enzymes, and coenzymes, which are made in the body but are not dietary essentials. Some contain plant-derived substances such as phytochemicals and herbs. For many, there is scientific evidence of beneficial effects. For others, the benefits are more questionable, and for a few, the risk of using them clearly outweighs any benefits they may provide. An office of Health Canada, called the Natural Health Products Directorate was formed in 2003 to regulate these supplements, which are called natural health products. Natural health products were first described in section 2.5. Their biological functions and their regulation in Canada are discussed in detail here.

F4.1 What Is a Natural Health Product?

Learning Objectives

CANADIAN CONTENT

- List the types of substances that are included in natural health products.
- Explain the importance of the natural product number (NPN).
- List the contents of natural health product label.

Health Canada defines a natural health product (NHP) as "vitamins and minerals, herbal remedies, homeopathic medicines, traditional medicines such as Traditional Chinese

Medicines, probiotics, and other products like amino acids and essential fatty acids."[2] The Natural Health Products Directorate was created to ensure that products available to Canadians meet certain quality standards. A manufacturer of a natural health product in Canada must apply for a licence, which is issued when the manufacturer has provided evidence of product safety and effectiveness, based on criteria set by the directorate. The processor must indicate the health benefit of the product, that is, its intended use, and provide evidence that the product is effective for that use. If a product has a documented history of use beyond 50 years, then this traditional use can be an acceptable level of evidence of safety and effectiveness. If a non-traditional use is proposed, then stronger evidence of effectiveness is required. This is usually a careful review of existing scientific literature, or, if this evidence is incomplete, the completion of at least one good quality intervention trial.[3] When a natural health product licence application process is completed satisfactorily, then a natural product number (NPN) is issued and manufacturers are required to put this number on their product labels. This number is an indicator to consumers that the product has undergone an assessment of safety and effectiveness. It is important to recognize that because of the acceptance of traditional uses, there are natural health products available in Canada for which strong experimental scientific evidence of effectiveness may not be available. The stated role of the Natural Health Products Directorate is "to ensure that Canadians have ready access to natural health products that are safe, effective and of high quality while respecting freedom of choice and philosophical and cultural diversity."[4]

Health Canada maintains a database of approved natural health products[5] and a compendium of monographs, which summarizes the scientific information on natural health product ingredients that have the strongest evidence for safety and effectiveness.[6]

Natural Health Product Labels

In addition to the NPN, manufacturers are required to include on the label of their product the product name, quantity, list of both medicinal and non-medicinal ingredients, the health benefit of taking the product, that is to say, its health claim, the recommended dosage, how the product should be taken (e.g., with a meal, on an empty stomach, etc.), any warnings such as contraindications and adverse reactions such as interactions with over-the-counter or prescription drugs, lot number, expiry date, name and address of manufacturer or distributor, and any special storage conditions.[7] These labels, although providing very important information, do not provide the nutritional information found in a nutrition facts table, such as the total kcal or % DV (see Section 2.5).

F4.2 Macronutrient Supplements

Learning Objective

- Explain why people might take supplements containing protein, amino acids, or fatty acids.

A varied balanced diet will meet nutrient needs for most healthy people (see Chapter 9: Label Literacy: Think Before You Supplement). However, some people choose to supplement specific nutrients because they do not think they get enough in their diets or because they are looking for additional beneficial effects, such as disease prevention or performance enhancement. For example, carbohydrate and protein supplements are frequently used by athletes. Carbohydrate pills wouldn't provide enough carbohydrate to have a significant effect, so supplemental carbohydrate usually comes in the form of sports beverages, bars, or carbohydrate gels. These are packaged in convenient containers and are absorbed from the GI tract quickly to provide glucose to exercising muscles. Protein supplements are generally in the form of powders, drinks, and bars. They do not provide any benefits beyond that obtained from the protein in an adequate diet. Individual amino acids and fatty acids are also marketed for their effect on performance and body composition as well as disease prevention.

CANADIAN CONTENT

Amino Acids

Amino acid supplements are popular among athletes. These include the amino acids arginine, ornithine, and lysine, which are often promoted as stimulating growth hormone release or enhancing muscle growth. Arginine is licensed as a natural health product for the purpose of "modest" improvement in exercise capacity.[8] Some natural health products are available that contain both arginine and ornithine but are marketed as promoting protein metabolism, not exercise performance. Lysine is also licensed, but not for any exercise-related purposes.[9] These limited and non-exercise purposes are consistent with controlled studies that indicate oral supplements of these amino acids taken before exercise do not boost growth hormone release, or increase muscle mass or strength above that obtained from training alone.[10]

Branched-chain amino acids (leucine, isoleucine, and valine), also promoted to athletes, are licensed in Canada, but their stated purpose is to promote protein synthesis; no exercise-related claims are made.[5] Intervention trials using these supplements have not demonstrated enhanced exercise performance.[11] Glutamine is another popular supplement. Unlike some other amino acids, it is licensed for an exercise-related purpose, notably for muscle cell repair and recovery from physical stress.[12]

Because some amino acids share transport systems, supplementing one amino acid can cause a deficiency of others that use that same transport system, so recommended doses should not be exceeded (see Section 6.3). All amino acid supplements come with warnings to individuals with certain medical conditions or taking certain medications that they should not use these supplements without first consulting a medical practitioner.

Fatty Acids

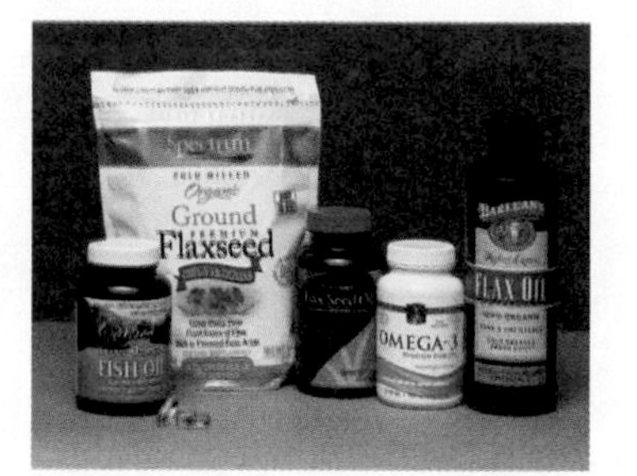

Figure F4.1 Fish oil and flaxseed oil supplements can increase your omega-3 fatty acid intake, but are not as beneficial to your health as eating fish and flaxseeds, as these foods provide additional nutrients. (Andy Washnik)

Supplements of omega-3 fatty acids from fish oils, which contain omega-3 fatty acids EPA and DHA, are licensed for the maintenance of cardiovascular health.[13,14] Supplemental flaxseed oil is also licensed, but for the maintenance of good health because it contains alpha-linolenic acid, an essential fatty acid (**Figure 4.1**).[15] This fatty acid can be converted in the body to the longer chain EPA and DHA, but this conversion is inefficient, so flaxseed oil alone may not be a sufficient source of these longer-chain fatty acids (see Section 5.5). While supplements provide important nutrients, eating whole food is more beneficial. Fish oil supplements, for example, can ensure a good intake of DHA and EPA but eating fish several times a week also provides high quality protein. Flaxseed oil supplements are an excellent source of the essential fatty acid alpha-linolenic acid, but whole flaxseed also provides dietary fibre and phytochemicals. The risks of taking supplements containing omega-3 fatty acids are minimal for healthy people, but these products are not recommended for those taking blood-thinning medications because they inhibit blood clotting.

F4.3 Substances Made in the Body

Learning Objectives

- Discuss why oral enzyme supplements, digestive enzymes excepted, are ineffective.
- Describe the function of DHEA in the body and why it is a controlled substance in Canada.
- Describe the function of melatonin.

Many molecules essential to normal metabolic and physiological function are made in the body in amounts sufficient to meet needs and are therefore not essential in the diet. No deficiency symptoms occur when they are absent from the diet. Nonetheless, supplements of many of these biological molecules are sold to enhance bodily functions. Most of these have little beneficial effect for healthy, well-nourished individuals.

Enzyme Supplements

Enzymes are proteins, and those needed for normal function are made in the body. When consumed in supplements, like other proteins, they are broken down into amino acids in the GI tract (see Section 6.3). The enzyme protein itself never reaches cells inside the body, so an oral enzyme supplements would have little effect on body function, an exception being digestive enzymes that act in the gastrointestinal tract. When used as supplements, these can perform their functions before they are broken down. For example, lactase, the enzyme that breaks down the milk sugar lactose, can benefit individuals with lactose intolerance when it is consumed along with foods containing lactose. The enzyme breaks down lactose that is in the GI tract. Eventually the lactase is also digested.

A number of other digestive enzymes, including proteases, lipases, amylases, and plant enzymes such as bromelain from pineapple and papain from papaya, are licensed natural health products and are marketed to improve digestion, although there is no strong scientific evidence that they are beneficial for healthy people.

Hormone Supplements

CANADIAN CONTENT

Hormones are another popular supplement ingredient. Protein or peptide hormones such as growth hormone have the same problem as enzymes—they are proteins, so they are broken down in the gut before they can reach their target; for this reason, injectable forms are often promoted as a way to increase muscle mass, to decrease body fat, and to boost stamina and a sense of well-being, even though studies have shown that growth hormone does not have a significant effect on muscle strength or the aging process.[16] In Canada, injectable growth hormone is a available only as a prescription drug and its use is limited to the treatment of specific medical conditions where growth hormone is not adequately produced in the body.[17]

Hormones that are not proteins can be absorbed into the body intact and supplemental doses may affect body functions. Melatonin is a hormone made by the pineal gland in the brain. Although it is synthesized from the amino acid tryptophan, it is not a protein and supplemental doses can be absorbed intact from the GI tract. Supplements of melatonin are available in Canada, and are licensed as sleep aids.[18] Melatonin has also been touted as a supplement that slows aging or acts as an antioxidant, but there is little evidence to support these claims. On the other hand, there is significant evidence that links the hormone melatonin to sleep cycles in humans. It has been suggested that in situations where the body's production of melatonin is reduced (advancing age) or the normal sleep/wake cycle is disrupted, such as with jet lag, supplemental melatonin may improve both sleep duration and quality. Melatonin appears to decrease jet lag symptoms and hasten the return to normal energy levels. Melatonin may also improve sleep quality for people whose jobs require them to work on rotating shift schedules. However, adequate long-term studies examining the efficacy, toxicity, and optimal dosage and timing of melatonin administration are still lacking.[19]

Like melatonin, steroid hormones can be absorbed in the digestive tract and reach the bloodstream intact, thereby affecting function. DHEA (dehydroepiandrosterone), a steroid hormone made by the adrenal glands, is a sold as a supplement in the United States. In Canada, however, DHEA is considered a controlled substance (that is, a substance with the potential for abuse or addiction). A health practitioner who wants to prescribe it must seek special permission from Health Canada.

DHEA is a precursor to the sex hormones estrogen and testosterone. The production of DHEA in the body peaks in one's mid-twenties, and gradually declines with age in most people. Preventing this drop has been suggested to delay the aging process by maintaining levels of estrogen and testosterone. It has been claimed that DHEA improves energy level, strength, and immunity. Athletes use it as a substitute for anabolic steroids to increase muscle mass and decrease body fat. However, supplementation has not been found to be effective for increasing muscle size or strength or for treating advancing age, male sexual dysfunction, or menopausal symptoms.[20,21] It has been banned by many athletic organizations.

Coenzyme Supplements

Many vitamins have coenzyme functions, but there are also coenzymes that are not dietary essentials. Lipoic acid and coenzyme Q are 2 such coenzymes, which also have antioxidant activity, and are licensed as natural health products. Lipoic acid is a coenzyme and an antioxidant needed for the production of ATP from glucose or fatty acids. It is licensed as an antioxidant or as a product which promotes glucose metabolism. People with diabetes are advised to consult a physician before using it, and pregnant and breastfeeding women are advised not to take the product.[5]

Ubiquinone, also called coenzyme Q, is another coenzyme important for the production of ATP from carbohydrate, fat, and protein. It is needed in the electron transport chain, in the final stage of aerobic metabolism where ATP is generated. Because, as its name implies, it is present ubiquitously in animals, plants, and micro-organisms, and it is synthesized in the human body, supplements are not necessary for most people. However, supplements may be beneficial in individuals with inherited defects in the function of their mitochondria.[22] Low levels of coenzyme Q are also a concern in those taking cholesterol-lowering statin drugs such as Crestor and Lipitor. Statins inhibit an enzyme needed for the synthesis of both cholesterol and ubiquinone, which can cause depletion of this coenzyme. Depletion of coenzyme Q in the mitochondria of muscle cells has been suggested as a cause of the muscle weakness that appears as a side effect in some individuals taking statins. Studies suggest that coenzyme Q appears to benefit statin users and may be useful in the treatment of cardiac disease.[23] It is licensed as a natural health product for use in the maintenance of cardiovascular health. Individuals are advised to consult a health care practitioner if they are taking blood pressure medication or blood thinners.[24]

Supplements Containing Structural and Regulatory Molecules

A number of structural and regulatory molecules are taken as supplements for a variety of different reasons. For instance, carnitine and creatine are both sold to enhance athletic performance in the United States, while in Canada, only creatine is a natural health product (see Section 13.7); carnitine is available only by prescription. Carnitine is a molecule that is needed to transport fatty acids into the mitochondria where they are broken down by β-oxidation and aerobic metabolism to produce ATP. It is thought to increase endurance by improving the use of fat as an energy source during exercise, while some studies focus on its potential to enhance recovery from exercise.[25] Creatine is a small molecule used to make a high-energy called creatine phosphate, which provides energy to the muscle for short bursts of activity. Creatine supplements have been shown to benefit some athletes by enhancing strength, performance, and recovery from high-intensity exercise[26] and they are licensed as a natural health product for this purpose.[27]

Glucosamine and chondroitin are molecules needed for the maintenance of healthy joints; they are licensed natural health products, and sold to alleviate the pain and progression of arthritis.[28,29] Glucosamine and chondroitin are molecules found in and around the cells of cartilage, the type of connective tissue that cushions joints. They are made in the body and consumed in the diet in meat. Supplements of both glucosamine and chondroitin are reported to reduce arthritis pain, stop cartilage degeneration, and possibly stimulate the repair of damaged joint cartilage (see Chapter 16: Your Choice: Do Glucosamine and Chondroitin Really Help Arthritis?). Supplements of these may be beneficial in relieving the symptoms of arthritis in some people[30] (**Figure F4.2**).

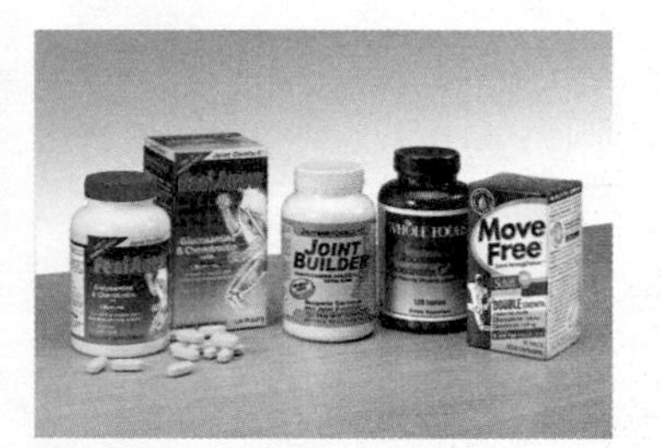

Figure F4.2 There is evidence that supplements containing glucosamine and chondroitin sulfate benefit people suffering from arthritis. The pills are large and about 3 a day are recommended to obtain benefits. (Andy Washnik)

SAM-e, chemically known as S-adenosylmethionine, is present in the body normally as an intermediate in the metabolism of the amino acid methionine. SAM-e promotes the production of cartilage and is beneficial in the treatment of arthritis. It is also claimed to be effective for treating depression and liver disease. It is licensed as a natural health product for these purposes. Although there is preliminary evidence that SAM-e may be beneficial to individuals with depression and arthritis, the results of large, well-controlled studies are not yet available and the risks of taking this supplement have not been adequately assessed.[31] Labels on these supplements contain several warnings: they advise individuals taking antidepressants, hepatoxic drugs, or monoamine oxidase inhibitors to avoid the supplement. It should not be used by anyone diagnosed with bipolar disorder. Use at night should be avoided as SAM-e may cause severe anxiety, restlessness, and insomnia. Pregnant women and nursing mothers should not take the supplement.[5]

Supplements of inositol are licensed.[5] Inositol is a component of phospholipids in cell membranes, where it plays a role in relaying messages to the inside of the cell. Inositol can be synthesized from glucose. There is no evidence that it is essential in the human diet, but it may have some clinical value in treating diseases such as diabetes and kidney failure.[32]

F4.4 Phytochemical Supplements

Learning Objective

- Explain why phytochemical supplements may not have the same benefits as foods containing these phytochemicals.

A diet rich in phytochemicals from fruits, vegetables, and whole grains has been shown to reduce the risk of heart disease and cancer (see Chapter 9: Focus on Phytochemicals).

Although we have not yet isolated and identified all of these health-promoting substances, some have been extracted, purified, and pressed into pills or capsules. These are advertised as having health-promoting properties but they do not appear to provide all the benefits obtained from whole foods that are rich sources of these as well as a host of other phytochemicals and nutrients.

Carotenoids

Carotenoids are a group of more than 600 yellow, orange, and red compounds found in living organisms. Carotenoids have antioxidant properties, and some also have vitamin A activity. The most prevalent carotenoids in the North American diet include β-carotene, α-carotene, β-cryptoxanthin, lycopene, lutein, and zeaxanthin. The major sources of carotenoids in the diet are fruits and vegetables (**Figure F4.3**).

Figure F4.3 Vegetables and fruit are the major source of dietary carotenoids. (Andy Washnik)

Lutein and zeaxanthin are carotenoids found in leafy green vegetables. They also appear in natural health products. They do not have vitamin A activity but are concentrated in the macula of the eye, where they protect against oxidative damage. Diets high in these carotenoids have been associated with a reduced incidence of age-related eye disorders (see Section16.4) such as macular degeneration, and supplements of lutein have been shown to improve vision in individuals with macular degeneration.[33]

The most common carotenoid supplement is β-carotene. It is licensed as a natural health product for its role as a source of vitamin A, which is beneficial for the maintenance of eyesight, immune function, skin membranes, and bone.[34] Beta-carotene has also been promoted for its antioxidant properties, but under some circumstances, it may promote rather than prevent oxidative damage. One study, using megadoses of β-carotene, far exceeding those recommended in natural health products, found an increase in the incidence of lung cancer among smokers supplemented with this carotenoid. This is believed to be due to β-carotene acting as a pro-oxidant rather than an antioxidant in the lungs of smokers and causing tissue damage that promoted tumour growth.[35] One must recognize that antioxidants may vary in their activity according to dose. The balance of antioxidants obtained in a nutritious diet is likely to be safer than artificially high doses of antioxidants from supplements.

Flavonoids

Like carotenoids, flavonoids, which are often called bioflavonoids, are antioxidants. Supplements containing bioflavonoids such as rutin, hesperidin, and pycnogenol are licensed in Canada. These are often included in supplements containing vitamin C, which is also an antioxidant. Although the foods containing these phytochemicals have been shown to have health-promoting properties, the heath benefits of these supplements are less clear.

Resveratrol, a compound first isolated from red wine but also found in certain seaweeds, is an antioxidant with a chemical structure similar to flavonoids; it has also appeared as an ingredient in licensed natural health products.[5] The compound may help to prevent cardiovascular disease.[36]

F4.5 Herbal Supplements

Learning Objective

- Discuss the benefits and risks of herbal supplements.

Technically, a herb is a non-woody, seed-producing plant that dies at the end of the growing season. However, the term herb is generally used to refer to any botanical or plant-derived substance. Throughout history folk medicine has used herbs to prevent and treat disease. Today, herbs and herbal supplements are still popular in treating illness and promoting health. It is estimated that more than half of all patients diagnosed with cancer explore alternative medicine—mostly herbal medicine.[37]

Herbal supplements are readily available and relatively inexpensive. They can be purchased without a trip to the doctor or a prescription. Although consumers who want to manage their

CANADIAN CONTENT

own health may view this as beneficial, it can also cause problems. Herbal products are known to be more variable in their composition than prescription drugs, and can harbour harmful bacteria or other contaminants, such as heavy metals like mercury and cadmium. To address these potential problems, Natural Health Products licensees must adhere to good manufacturing practices which help to maintain a standard of composition, purity, and absence of contamination in a natural health product.[3]

Because consumers decide what to treat, herbal remedies can be used inappropriately. To minimize adverse reactions, regulations require that known interactions with drugs or other products be indicated on the label or if individuals, such as pregnant or breastfeeding women, or those with certain medical conditions, should avoid the product. A recent study of natural health product users in Ontario found that many people who were taking prescription drugs also used natural health products for the same purpose. As about a quarter of these users had not informed their doctors,[38] warnings on labels may be an especially important source of information, particularly for this type of consumer. The use of herbal supplements can be inappropriate at certain times. For example, St. John's wort can prolong and intensify the effects of narcotic drugs and anaesthetic agents, so it should not be taken for 2-3 weeks prior to surgery. Herbal products should also be avoided during pregnancy and breastfeeding and carry this warning on their labels. Black cohosh (used to treat menstrual cramps), juniper (used for heartburn), pennyroyal or rosemary (used for digestive problems), sage (used for stomach upset), and thuja (used for respiratory infections), and even raspberry tea (used to treat morning sickness) may stimulate uterine contractions, which can increase the risk of miscarriage or premature labour.[39] Recently, Health Canada banned the herb kava because of concerns that it may cause liver damage[40] and limited the dosage of ephedra in natural health products (see Chapter 7: Your Choice: Can a Weight-Loss Supplement Help you Trim the Fat?)

Some guidelines to follow if you are considering taking herbs or herbal supplements are included in **Table F4.1**. When you have questions about natural health products it is best to consult a qualified health care professional. In a recent study of Canadian pharmacies and health food stores across Canada, researchers found that the information given to consumers about natural health products and health concerns was far more accurate in pharmacies (68% of questions answered accurately) compared to health food stores (7% of questions answered accurately).[41] The most popular herbal products used by Canadians are, in order, garlic, echinacea, ginseng, ginkgo biloba, and St. John's wort.[1]

Table F4.1 Considerations When Choosing Herbal Supplements
Do not use a herbal supplement if it does not have an NPN as this may indicate that the product is not authorized for sale in Canada.
• If you are ill or taking medications, consult your physician before taking herbs.
• Do not take herbs if you are pregnant or breastfeeding, without first consulting a physician.
• Do not give herbs to children.
• Do not assume herbal products are safe.
• Read label ingredients and the list of precautions.
• Start with low doses, and stop taking any product that causes side effects.
• Do not take combinations of herbs without consulting a health care professional.
• Do not use herbs for long periods.
• Do not choose products that claim to be a secret cure and be wary of terms such as "breakthrough," "magical," "miracle cure," and "new discovery."

Garlic

Garlic (**Figure F4.4a**) is licensed as a natural health product to maintain cardiovascular health, as recent research has shown that it may lower blood cholesterol.[42,43] Supplement manufacturers have provided a way to increase intake without eating this odiferous food at every meal; some preparations contain a deodorized form. Even though we spice our food with it, garlic in supplement form is not safe for everyone. Labels caution individuals to consult a health care practitioner if taking medications, if they have diabetes, or are taking blood thinners and protease inhibitors, which are used in the treatment of AIDS.[42]

Echinacea

Petals of the echinacea plant were used by the First Nations people as a treatment for colds, flu, and infections (**Figure F4.4b**). Based on this traditional use, the echinacea root is licensed as a natural health product, primarily as a cold remedy. Echinacea is hypothesized to act as an immune system stimulant, but there is little evidence that it is beneficial in either preventing or treating the common cold.[44] Labels advise individuals to seek the advice of a health care professional if they have tuberculosis, leukosis, collagenosis or multiple sclerosis, if they are taking an immunosuppressant, if their respiratory symptoms persist, or if they have certain allergies.[45]

Ginseng

Ginseng has a long history in traditional medicine (**Figure F4.4c**). It has been used in Asia for centuries for its energizing, stress-reducing, and aphrodisiac properties. It is licensed as a natural health product for a number of traditional uses, such as the treatment of respiratory infections, maintaining blood glucose levels, relieving nervousness, and maintaining a healthy immune system.[46]

Although there is animal and cell culture research that demonstrates potential benefits of ginseng on cardiovascular health and in the prevention of cancer, there are few clinical trials in humans to support these effects.[47,48] Despite these potential benefits and a history of use in Chinese medicine, ginseng may not be safe for everyone. Natural health product labels carry warnings to consult a health care practitioner if taking antidepressants, blood thinners, digoxin, or if you have diabetes. Side effects such as insomnia, anxiety, and headaches can occur.[49]

Ginkgo Biloba

Ginkgo biloba (**Figure F4.4d**) has been used for millennia in Chinese medicine to treat asthma and skin infections, but today it is licensed as a natural health product to enhance memory. Despite its popularity for this use, a relatively recent controlled clinical trial found ginkgo biloba supplements to have no measurable benefit on memory in older adults.[50] Ginkgo biloba also interacts with a number of medications and labels warn against its use if taking medication for diabetes, high blood pressure, and seizures and if taking medication that alters blood coagulation, such as blood thinners, clotting factor replacements, acetylsalicylic acid, ibuprofen, fish oils, or vitamin E, as spontaneous bleeding can result.[51]

Figure F4.4 Herbal supplements are derived from plants, many of which are common in fields, gardens, or kitchens. (a) Garlic is used to spice food in the kitchen but extracts are also processed into pills or tablets and marketed as herbal supplements. (b) Flowers of the echinacea plant, like the ones shown here, are native to Manitoba and Saskatchewan. (c) Ginseng capsules, extracts, and powders are made from the dried root of the ginseng plant. (d) The leaves of the ginkgo biloba plant are used to make the popular herbal supplement. (e) These yellow flowers of St. John's wort grow wild in fields and pastures and are considered a perennial weed. (Mary Ellen Bartley/StockFoodAmerica; © istockphoto.com/Ruud de Man; Lynn Johnson/NG Image Collection; Altrendo/Getty Images; Michael Gadomski/Photo Researchers)

St. John's Wort

St. John's wort (**Figure F4.4e**) is traditionally used in herbal medicine to "help relieve restlessness and/or nervousness (sedative and/or calmative)."[52] Analysis reveals that it contains low doses of the chemical found in the antidepressant drug fluoxetine (Prozac). Clinical trials on its effectiveness are inconsistent but a number do suggest that it is effective for the treatment of mild to moderate depression.[53] Side effects include nausea and allergic reactions. Natural health product labels warn against its use in combination with other drugs. It should not be used in conjunction with prescription antidepressant drugs or with anticoagulants, heart medications, birth control pills, and medications used to treat HIV infection.[52,54]

REFERENCES

1. Troppmann, L., Johns, T., and Gray-Donald, K. Natural Health Product use in Canada. *Canadian Journal of Public Health* 93(6):426-430, 2002.
2. Health Canada. Natural Health Products Frequently asked questions. Available online at http://www.hc-sc.gc.ca/dhp-mps/prodnatur/faq/question_general-eng.php. Accessed June 2, 2011.
3. Nestmann, E. R., Harwood, M. and Martyres, S. An innovative model for regulating supplement products: Natural health products in Canada. *Toxicology* 221 (2006)50-58, 2006.
4. Health Canada. Compendium of Monographs, Guidance Documents. Available online at http://www.hc-sc.gc.ca/dhp-mps/prodnatur/legislation/docs/compendium-eng.php. Accessed June 2, 2011.
5. Health Canada. Licensed Natural Health Products Database Available online at http://www.hc-sc.gc.ca/dhp-mps/prodnatur/applications/licen-prod/lnhpd-bdpsnh-eng.php. Accessed June 2, 2011.
6. Health Canada. Compendium of Monographs. Available online at http://www.hc-sc.gc.ca/dhp-mps/prodnatur/applications/licen-prod/monograph/index-eng.php. Accessed June 2, 2011.
7. Mine, Y. and Young, D. Regulation of natural health products in Canada. *Food Sci. Technol. Research* 15(5): 459-468, 2009.
8. Health Canada. Arginine-L. Compendium of Monographs. Available online at http://webprod.hc-sc.gc.ca/nhpid-bdipsn/monoReq.do?id=124&lang=eng. Accessed June 4, 2011.
9. Health Canada. Lysine. Compendium of Monographs. Available online at http://webprod.hc-sc.gc.ca/nhpid-bdipsn/monoReq.do?id=134&lang=eng. Accessed July 30, 2011.
10. Chromiak, J. A., and Antonio, J. Use of amino acids as growth hormone-releasing agents by athletes. *Nutrition* 18:657–661, 2002.
11. Gleeson, M. Interrelationship between physical activity and branched-chain amino acids. *J. Nutr.* 135:1591S–1595S, 2005.
12. Health Canada. Glutamine-L. Compendium of Monographs. Available online at http://webprod.hc-sc.gc.ca/nhpid-bdipsn/monoReq.do?id=126&lang=eng. Accessed June 2, 2011.
13. Health Canada. Fish oils. Compendium of Monographs. Available online at http://webprod.hc-sc.gc.ca/nhpid-bdipsn/monoReq.do?id=88&lang=eng. Accessed July 30, 2011.
14. Watts, G.F. and Mori, T. A. Recent advances in understanding the role and use of marine ω3 polyunsaturated fatty acids in cardiovascular protection. *Curr. Opin. Lipidol.* 22(1):70-1, 2011.
15. Health Canada. Flaxseed Oil. Compendium of Monographs. Available online at http://webprod.hc-sc.gc.ca/nhpid-bdipsn/monoReq.do?id=220&lang=eng. Accessed June 4, 2011.
16. Jenkins, P. J. Growth hormone and exercise. *Clin. Endocrinol.* 50:683–689, 2000.
17. Health Canada. Drug Product Database. Available online at http://webprod.hc-sc.gc.ca/dpd-bdpp/index-eng.jsp. Accessed June 2, 2011.
18. Health Canada. Melatonin-Oral. Compendium of Monographs. Available online at http://webprod.hc-sc.gc.ca/nhpid-bdipsn/monoReq.do?id=136&lang=eng. Accessed June 4, 2011.
19. Caldwell, J. L. The use of melatonin: An information paper. *Aviat. Space Environ. Med.* 71:238–244, 2000.
20. Cameron, D. R., and Braunstein, G. D. The use of dehydroepiandrosterone therapy in clinical practice. *Treat. Endocrinol.* 4:95–114, 2005.
21. Dayal, M., Sammel, M. D., Zhao, J. et al. Supplementation with DHEA: Effect on muscle size, strength, quality of life, and lipids. *J. Womens Health* 14:391–400, 2005.
22. Marriage, B., Clandinin, M. T., and Glerum, D. M. Nutritional cofactor treatment in mitochondrial disorders. *J. Am. Diet. Assoc.* 103:1029–1038, 2003.
23. Kumar, A., Kaur, H., Devi, P. et al. Role of coenzyme Q10 (CoQ10) in cardiac disease, hypertension and Meniere-like syndrome. *Pharmacol. Ther.* 124(3):259-68, 2009.
24. Health Canada. Coenyzme Q10. Compendium of Monographs. Available online at http://webprod.hc-sc.gc.ca/nhpid-bdipsn/monoReq.do?id=70&lang=eng. Accessed June 2, 2011.
25. Kraemer, W. J., Volek, J. S., and Dunn-Lewis, C. L-carnitine supplementation: Influence upon physiological function. *Curr. Sports Med. Rep.* 7:218–223, 2008.
26. Paddon-Jones, D., Borsheim, E., and Wolfe, R. R. Potential ergogenic effects of arginine and creatine supplementation. *J. Nutr.* 134(10 Suppl):2888S–2894S, 2004.
27. Health Canada. Creatine monohydrate. Compendium of Monographs. Available online at http://webprod.hc-sc.gc.ca/nhpid-bdipsn/atReq.do?atid=creatine.mono&lang=eng. Accessed June 2, 1011.
28. Health Canada. Glucosamine. Compendium of Monographs. Available online at http://webprod.hc-sc.gc.ca/nhpid-bdipsn/monoReq.do?id=103&lang=eng. Accessed June 2, 2011.
29. Health Canada. Chondroitin sulfate. Compendium of Monographs. Available online at http://webprod.hc-sc.gc.ca/nhpid-bdipsn/monoReq.do?id=66&lang=eng. Accessed June 2, 2011.
30. Gregory, P. J., Sperry, M., and Wilson, A. F. Dietary supplements for osteoarthritis. *Am. Fam. Physician.* 15:177–184, 2008.
31. Ramos, L. Beyond the headlines: SAM-e as a supplement. *J. Am. Diet. Assoc.* 100: 414, 2000.
32. Ooms, L. M., Horan, K. A., Rahman, P. et al. The role of the inositol polyphosphate 5-phosphatases in cellular function and human disease. *Biochem. J.* 419:29–49, 2009.
33. Richer, S., Stiles, W., Statkute, L. et al. Double-masked, placebo-controlled, randomized trial of lutein and antioxidant supplementation in the intervention of atrophic age-related macular degeneration: The Veterans LAST study (Lutein Antioxidant Supplementation Trial). *Optometry* 75:216–230, 2004.

34. Health Canada. Beta-carotene. Compendium of Monographs. Available online at http://webprod.hc-sc.gc.ca/nhpid-bdipsn/monoReq.do?id=38&lang=eng. Accessed June 4, 2011.
35. Palozza, P. Prooxidant actions of carotenoids in biologic systems. *Nutr. Rev*. 56:257–265, 1998.
36. Csiszar, A. Anti-inflammatory effects of resveratrol: possible role in prevention of age-related cardiovascular disease. *Ann. N. Y. Acad. Sci.* 1215:117–22, 2011.
37. Boon, H., and Wong, J. Botanical medicine and cancer: A review of the safety and efficacy. *Expert Opin. Pharmacother*. 5:2485–2501, 2004.
38. Levine, M. A., Xu, S., Gaebel, K. et al. Self-reported use of natural health products: a cross-sectional telephone survey in older Ontarians. *Am. J. Geriatr. Pharmacother.* 7(6):383–92, 2009.
39. Schweitzer, A. Dietary Supplements During Pregnancy. *J. Perinat. Educ.* 15(4): 44–45, 2006.
40. Health Canada. Advisories, warnings, and recalls 2002. Health Canada issues a stop-sale order for all products containing kava [2002-09-21]. Available online at http://www.hc-sc.gc.ca/ahc-asc/media/advisories-avis/2002-eng.php. Accessed July 30, 2011.
41. Temple, N.J., Eley, D., Nowrouzi, B. Advice on dietary supplements: a comparison of health food stores and pharmacies in Canada. J Am Coll Nutr. 28(6):674–7, 2009.
42. Health Canada. Garlic. Compendium of Monographs. Available online at http://www.hc-sc.gc.ca/dhp-mps/prodnatur/applications/licen-prod/monograph/mono_garlic-ail-eng.php. Accessed June 2, 2011.
43. Hermansen, K., Dinesen, B., Hoie, L. H. et al. Effects of soy and other natural products on LDL: HDL ratio and other lipid parameters: A literature review. *Adv. Ther*. 20:50–78, 2003.
44. Caruso, T. J., and Gwaltney, J. M., Jr. Treatment of the common cold with echinacea: A structured review. *Clin. Infect. Dis.* 40:807–810, 2005.
45. Health Canada. Echinacea. Compendium of Monographs. Available online at http://webprod.hc-sc.gc.ca/nhpid-bdipsn/monoReq.do?id=80&lang=eng. Accessed June 2, 2011.
46. Health Canada. Ginseng, American. Compendium of Monographs. Available online at http://webprod.hc-sc.gc.ca/nhpid-bdipsn/monoReq.do?id=29&lang=eng. Accessed July 30, 2011.
47. Zhou, W., Chai, H., Lin, P. H. et al. Molecular mechanisms and clinical applications of ginseng root for cardiovascular disease. *Med. Sci. Monit.* 10:RA187–RA192, 2004.
48. Helms, S. Cancer prevention and therapeutics: Panax ginseng. *Altern. Med. Rev*. 9:259–274, 2004.
49. Health Canada. Ginseng, Panax. Compendium of Monographs. Available online at http://webprod.hc-sc.gc.ca/nhpid-bdipsn/monoReq.do?id=146&lang=eng. Accessed June 2, 2011.
50. Snitz, B. E., O'Meara, E. S., Carlson, M. C, et al. Ginkgo Evaluation of Memory (GEM) Study Investigators. Ginkgo biloba for preventing cognitive decline in older adults: a randomized trial. *JAMA*. 302(24):2663–70, 2009.
51. Health Canada. Ginkgo Biloba. Compendium of Monographs. Available online at http://webprod.hc-sc.gc.ca/nhpid-bdipsn/monoReq.do?id=100&lang=eng. Accessed June 2, 2011.
52. Health Canada. St John's Wort—Oral. Compendium of Monographs. Available online at http://webprod.hc-sc.gc.ca/nhpid-bdipsn/monoReq.do?id=163&lang=eng. Accessed June 2, 2011.
53. Freeman, M. P., Fava, M., Lake, J., et al. Complementary and alternative medicine in major depressive disorder: the American Psychiatric Association Task Force report. *J Clin Psychiatry.* 71(6):669-81, 2010.
54. Mills, E., Montori, V., Perri, D. et al. Natural health product-HIV drug interactions: A systematic review. *Intl. J. STD AIDS* 16:181–186, 2005.

12 The Trace Elements

Case Study

Nissi has been feeling tired all the time and has noticed a lump in her neck that seems to be getting bigger. She lives in a small village in India. For generations, her family has farmed the flood plains of the Ganges River valley as subsistence farmers—that is, they use their crops primarily to feed their family. Nissi's diet consists almost entirely of these nutritious homegrown grains, pulses (seeds, beans, and lentils), and vegetables. A typical meal might be flat fried bread, called chapattis; a porridge made with lentils, called dahl; vegetables such as potatoes and cauliflower; and yogurt. Nissi's mother takes her to the clinic, where the doctor determines that her malaise and swollen neck are caused by an iodine deficiency. He gives Nissi an injection of iodine. Soon she has more energy, and the bulge at the front of her neck begins to disappear.

How could an apparently healthy, well-nourished young girl have a nutritional deficiency? Despite the small amounts of iodine needed in the diet—only about 150 μg/day—iodine deficiency is a problem not only for Nissi but for others in her village and other villages and cities throughout India, because the repeated flooding of the Ganges River valley over the centuries has washed the iodine out of the soil. Food grown there is therefore low in iodine. A diet based solely on local foods does not provide enough of this essential mineral to meet needs. Left untreated, Nissi's symptoms would have continued to worsen. If she had become pregnant, her baby would have been at risk for developmental abnormalities.

One way to prevent iodine deficiency is to include foods that are higher in iodine. However, this approach is not practical for Nissi, because all of the food grown in the region—and thus her entire diet—is deficient in iodine. The alternative for Nissi's family is to use iodized salt—salt to which iodine has been added.

In this chapter, you will learn about the role that iodine and other trace minerals play in human health.

(PhotosIndia.com/Getty Images, Inc.)

(©iStockphoto)

(© istockphoto.com/JackJelly)

Chapter Outline

12.1 Trace Elements in the Canadian Diet

Learning Objective

- Discuss why bioavailability is so important in meeting trace element needs.

The trace elements, which include iron, zinc, copper, manganese, selenium, iodine, fluoride, chromium, and molybdenum, as well as several others, are required by the body in an amount of 100 mg or less per day or present in the body in an amount of 0.01% or less of body weight. Like the major minerals, they provide a variety of essential structural and regulatory roles. Some of their functions are unique: iodine is needed to make thyroid hormones, iron is needed to carry oxygen to body cells, and fluoride is needed for strong teeth. Other functions are similar and complementary; selenium, copper, zinc, iron, and manganese are each cofactors for antioxidant enzyme systems.

Although the trace elements are distinguished from the major minerals only by the amounts present in the diet and required in the body, the presence of such small amounts makes them difficult to study. Usually, mineral needs are evaluated by feeding a diet devoid of that nutrient. However, with some trace elements, needs are so small that contamination from the environment and minerals already present in the body can obscure experimental results. Bioavailability is also more of a concern with trace elements because such small amounts are present in the diet. Phytates, tannins, oxalate, and fibre can bind minerals, reducing their absorption. For example, when the diet is based on unleavened grains, the phytate content may be high enough to decrease zinc absorption and cause a zinc deficiency. The interactions among the minerals that can affect their absorption and utilization also have a greater impact on trace element status. For example, a deficiency of copper can decrease available iron by reducing the amount of iron that can bind to iron transport proteins in the blood.

Determining the trace element content of foods is difficult and is compromised by the fact that the amounts of some are affected by the soil content where the food is grown or produced. For example, a loaf of bread made from wheat grown in one location may supply a different amount of selenium than a loaf made from wheat grown elsewhere. When modern transportation systems make foods produced in many locations available, this variation is unlikely to affect mineral status, but in countries where the diet consists predominantly of locally grown foods, individual trace element deficiencies and excesses are more likely.

12.2 Iron (Fe)

Learning Objectives

- List some dietary sources of heme and nonheme iron.
- Explain how the amount of iron in the body is regulated.
- Describe the primary function of iron and the physiological effects of iron deficiency and iron toxicity.

Iron was identified as a major constituent of blood in the eighteenth century. By 1832, iron tablets were used to treat young women in whom "colouring matter" was lacking in the blood. Today we know that the red colour in blood is due to the iron-containing protein **hemoglobin** and that a deficiency of iron decreases hemoglobin production. Despite the fact that iron is one of the best understood of the trace elements, iron deficiency remains the most common nutritional deficiency worldwide and is a problem among certain population groups in Canada (see: Critical Thinking: How are Canadians Doing with Respect to Their Intake, from Food, of Iron and Zinc?).

hemoglobin An iron-containing protein in red blood cells that binds oxygen and transports it through the bloodstream to cells.

heme iron A readily absorbed form of iron found in animal products that is chemically associated with proteins such as hemoglobin and myoglobin.

myoglobin An iron-containing protein in muscle cells that binds oxygen.

Iron in the Diet

Iron in the diet comes from both plant and animal sources (**Figure 12.1**). Much of the iron in animal products is **heme iron**—iron that is part of a chemical complex, called a heme group, found in proteins, such as hemoglobin in blood and **myoglobin** in muscle (**Figure 12.2**). Meat,

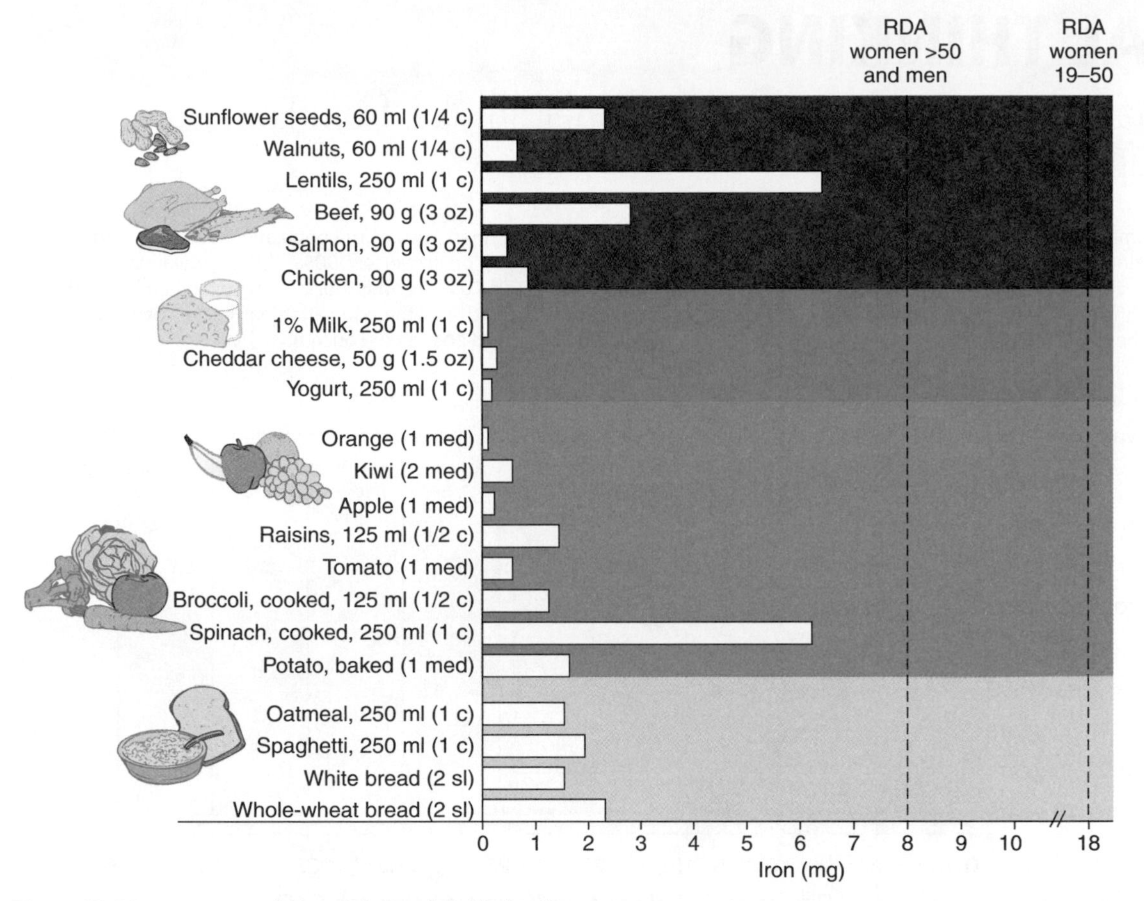

Figure 12.1 Iron content of Canada's Food Guide food groups
Both plant and animal foods are good sources of iron. The dashed lines represent the RDA for women of child-bearing age and for men and post-menopausal women.

poultry, and fish are good sources of heme iron. Heme iron accounts for about 5%-10% of the dietary iron in Western countries.[1]

Leafy green vegetables, legumes, and whole and enriched grains are good sources of **nonheme iron**. Another source of nonheme iron in the diet is iron cooking utensils, from which iron leaches into food. Leaching is enhanced by acidic foods. For example, 50 ml of spaghetti sauce cooked in a glass pan contains about 0.6 mg of iron, but the same sauce cooked in an iron skillet contains about 5.7 mg, depending on how long it is cooked.

nonheme iron A poorly absorbed form of iron found in both plant and animal foods that is not part of the iron complex found in hemoglobin and myoglobin.

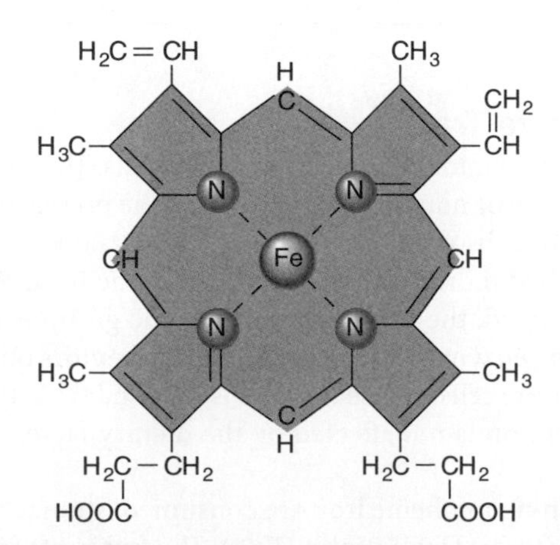

Figure 12.2 Heme group
A heme group contains four 5-membered nitrogen-containing rings that form a cage around a central iron ion. In hemoglobin and myoglobin, the iron ion is in the Fe^{2+} state.

CRITICAL THINKING

How are Canadians Doing with Respect to Their Intake, from Food, of Iron and Zinc?

The CCHS collected information on the intake of 2 nutrients discussed in this chapter, iron and zinc. The results are shown in the figure.

Health Canada considers the prevalence of inadequate intake to be low when less than 10% of the population has intakes below the EAR, as this means that less than 10% of the population is *not* meeting their nutrient requirement, or, conversely, 90% of the population *is* meeting its requirement. This 10% cutoff is indicated by the green line in the figure. If more than 10% of the population has intakes below the EAR, then this is considered an area of concern.

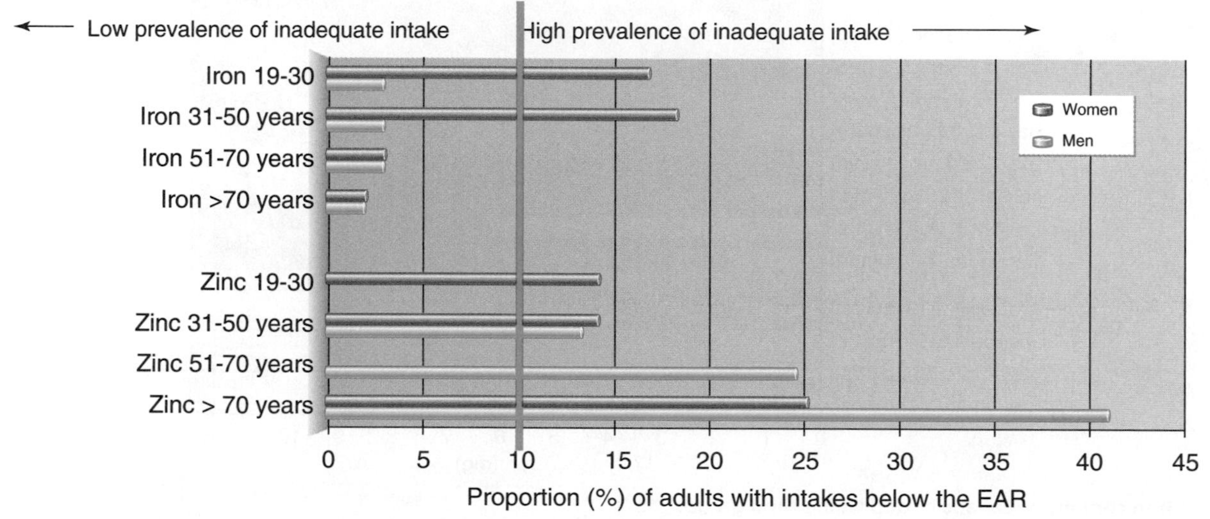

Source: Canadian Community Health Survey cycle 2.2, nutrition (2004). Available online at http://www.hc-sc.gc.ca/fn-an/surveill/nutrition/commun/index-eng.php. Accessed Sept 23, 2010.

Critical Thinking Questions

What would you conclude about the adequacy of zinc intake? How does it change with age?*

What would you conclude about the adequacy of iron intake in women? How does it change with age? Why? (Note that to account for menstrual losses of iron, the EAR for women under 50 is 8.1 mg and it drops to 5 mg in post-menopausal women, meaning those 50 and over.)

*Answers to all Critical Thinking questions can be found in Appendix J.

Iron in the Digestive Tract

Iron from the diet is absorbed into the intestinal mucosal cells. The amount absorbed depends on whether the iron is heme or nonheme iron as well as the presence of dietary components that enhance or inhibit iron absorption.

Heme iron is absorbed more efficiently than nonheme iron. When foods containing heme proteins are consumed, the iron-containing heme group is released from the proteins by protein-digesting enzymes. The heme binds to receptors on the surface of mucosal cells, allowing it to enter the cells, where the iron is released from the heme group (**Figure 12.3**). Heme iron absorption is not affected by the dietary factors that impact nonheme iron absorption.

When foods containing nonheme iron are consumed, stomach acid helps convert the ferric form (Fe^{3+}) of iron to the ferrous form (Fe^{2+}). The ferrous form of iron remains more soluble when it enters the intestine and therefore is absorbed into the mucosal cells more readily. When foods containing nonheme iron are consumed with foods containing acids, such as ascorbic acid (vitamin C), citric acid, or lactic acid, iron absorption is enhanced

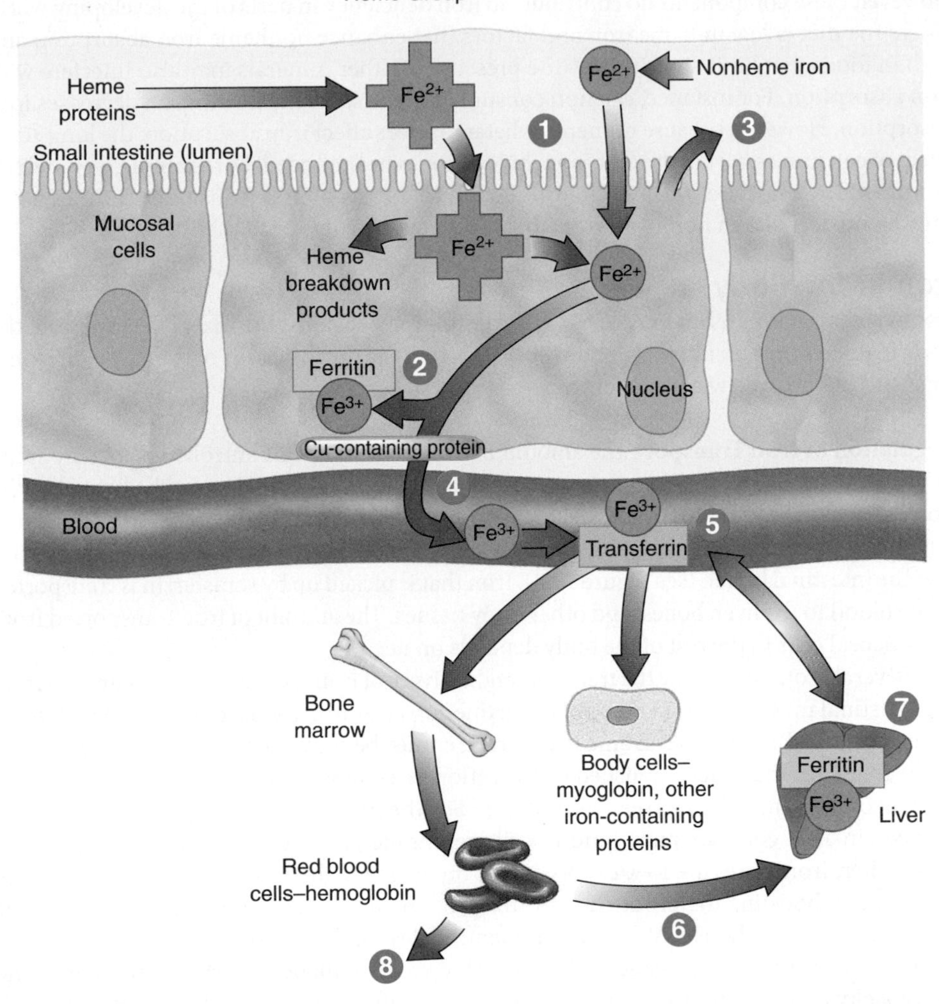

1. Heme iron is absorbed as part of the heme group. Nonheme iron is absorbed in the ferrous form (Fe^{2+}).
2. Once inside the mucosal cells, some iron may be bound to ferritin for storage.
3. When the mucosal cells die, iron that remains bound to ferritin is excreted in the feces.
4. Iron that enters the blood is converted to ferric iron (Fe^{3+}) by a copper-containing protein in the cell membrane. The Fe^{3+} binds to transferrin for transport.
5. Transferrin transports iron to liver, bone, and other body cells.
6. When red blood cells die, they are broken down by cells in the liver, spleen, or bone marrow and the iron is released for reuse.
7. Excess iron is stored primarily in the liver, bound to ferritin.
8. Most iron loss is due to blood loss.

Figure 12.3 Iron absorption, transport, storage, and loss
The amount of iron that leaves the mucosal cells for transport to liver, bone, and other tissues is carefully regulated because little iron is lost from the body.

because the acids help to keep iron in the ferrous (Fe^{2+}) form.[2] The best studied of these acids is vitamin C, which enhances nonheme iron absorption both by keeping the iron in its more absorbable form and by forming a complex with iron that remains soluble and more bioavailable.[1] Vitamin C can enhance nonheme iron absorption up to sixfold. Another dietary component that increases the absorption of nonheme iron is meat such as beef, fish, or poultry. For example, a small amount of ground beef in a pot of chili, in addition to being a source of iron on its own, will also enhance the body's absorption of nonheme iron from the beans.

Dietary factors that interfere with the absorption of nonheme iron include fibre, phytates found in cereals, tannins found in tea, and oxalates found in some leafy greens such as spinach. These prevent absorption by binding iron in the gastrointestinal tract. They are not consumed in large enough quantities in the diet of most developed countries to cause an iron deficiency.

However, these components do contribute to iron deficiency in parts of the developing world where the diet is low in heme iron and factors that enhance nonheme iron absorption and high in foods that limit aborption.[3,4] The presence of other minerals may also interfere with iron absorption. For instance, calcium consumed in the same meal with iron decreases iron absorption. However, because numerous dietary factors affect iron absorption, the long-term effect of calcium as part of the diet as a whole is not as pronounced. For example, a study that examined the impact of the amount of calcium in a glass of milk consumed 3 times a day showed no decrease in nonheme-iron absorption over a 4-day period.[5]

Iron in the Body

Iron is essential for life but in excess it is toxic. To protect against the toxic effects of iron, the body regulates the amount that enters the blood from the mucosal cells of the gastrointestinal tract and has evolved ways to safely transport and store it.

ferritin The major iron storage protein.

transferrin An iron transport protein in the blood.

Regulation of Iron Transport The amount of iron in the body is controlled primarily at the intestine. Iron that has entered the mucosal cells of the small intestine can be bound to the iron storage protein **ferritin** or picked up by the iron transport protein **transferrin**. Iron that remains bound to ferritin is excreted in the feces when mucosal cells die and are sloughed off into the intestinal lumen (see Figure 12.3). Iron that is picked up by transferrin is transported in the blood to the liver, bones, and other body tissues. The amount of iron transported from the mucosal cells to the rest of the body depends on need.

Several proteins regulate the transport and delivery of iron. Transferrin picks up iron from the intestinal mucosal cells in the small intestine as well as iron released from the breakdown of hemoglobin. Before iron can bind transferrin, it must be converted to the ferric (Fe^{3+}) form. In the intestine, this is accomplished by the action of a copper-containing protein located in the mucosal cell membrane (see Figure 12.3).[6] For the iron to be taken up by body cells, the transferrin-iron complex must bind to cell membrane proteins called **transferrin receptors**. When iron stores are low, expression of the gene for the transferrin receptor protein is increased, boosting the production of this protein and thereby allowing more iron to be transported into body cells. When iron is plentiful, less of the transferrin receptor protein is made and expression of the gene for ferritin is increased, enhancing ferritin production and the capacity to store iron. Therefore, when iron is plentiful, more is stored in the mucosal cells and, thus, is lost when these cells die.

transferrin receptor Protein found in cell membranes that binds to the iron-transferrin complex and allows it to be taken up by cells.

Iron Stores Iron that is transported out of the mucosal cell in excess of immediate needs can be stored in the protein ferritin, primarily in the liver, spleen, and bone marrow. Levels of ferritin in the blood can be used to estimate iron stores. When ferritin concentrations in the liver become high, some is converted to an insoluble storage protein called **hemosiderin**. Iron can be mobilized from body stores as needed, and deficiency signs appear only after stores are depleted.

hemosiderin An insoluble iron storage compound that stores iron when the amount of iron in the body exceeds the storage capacity of ferritin.

Iron Losses Iron is not readily excreted. Even when red blood cells die, the iron in their hemoglobin is not lost from the body. The red blood cells are removed from the blood by cells in the liver, spleen, and bone marrow and degraded; the iron is then attached to transferrin for transport back to body tissues, including the bone, where it can be incorporated into new red blood cells (see Figure 12.3). Most iron loss even in healthy individuals occurs through blood loss, including that lost during menstruation and the small amounts lost from the gastrointestinal tract. Some iron is also lost through the shedding of cells from the intestine, skin, and urinary tract.

Functions of Iron Iron in the body is essential for the delivery of oxygen to cells. It is a component of 2 oxygen-carrying proteins, hemoglobin and myoglobin. Most of the iron in the body is part of hemoglobin (**Table 12.1**). Hemoglobin in red blood cells transports oxygen to body cells and carries carbon dioxide away from cells for elimination by the lungs. Myoglobin is found in the muscle, where it enhances the amount of oxygen available for use in muscle contraction. Iron is also essential for ATP production as a part of several proteins involved in the citric acid cycle and the electron transport chain. Iron-containing proteins are involved in drug metabolism and immune function. Iron is also part of the enzyme catalase, which protects the cells from oxidative damage by destroying the reactive oxygen species hydrogen peroxide before it can form free radicals (see Section 8.10).

Table 12.1 A Summary of the Trace Elements

Mineral	Sources	Recommended Intake for Adults	Major Functions	Deficiency Diseases and Symptoms	Groups at Risk of Deficiency	Toxicity	UL
Iron	Red meats, leafy greens, dried fruit, whole and enriched grains	8-18 mg/d	Part of hemoglobin, which delivers oxygen to cells, myoglobin, which holds oxygen in muscle, and electron carriers in the electron transport chain; needed for immune function	Iron deficiency anemia: fatigue, weakness, small, pale red blood cells, low hemoglobin	Infants and preschool children, adolescents, women of child-bearing age, pregnant women, athletes	Gastrointestinal upset, liver damage	45 mg/d
Zinc	Meat, seafood, whole grains, eggs	8-11 mg/d	Regulates protein synthesis; functions in growth, development, wound healing, immunity, and antioxidant protection	Poor growth and development, skin rashes, decreased immune function	Vegetarians, low-income children, elderly	Decreased copper absorption, depressed immune function	40 mg/d
Copper	Organ meats, nuts, seeds, whole grains, seafood, cocoa	900/μg/d	A part of proteins needed for iron absorption, lipid metabolism, collagen synthesis, nerve and immune function, and antioxidant protection	Anemia, poor growth, bone abnormalities	Those who over-supplement zinc	Vomiting	10 mg/d
Manganese	Nuts, legumes, whole grains, tea	1.8–2.3 mg/d[a]	Functions in carbohydrate and lipid metabolism and antioxidant protection	Growth retardation	None	Nerve damage	11 mg/d
Selenium	Organ meats, seafood, eggs, whole grains	55/μg/d	Antioxidant protection as part of glutathione peroxidase, synthesis of thyroid hormones, spares vitamin E	Muscle pain, weakness, Keshan disease	Populations in areas with low-selenium soil	Nausea, diarrhea, vomiting, fatigue, hair changes	400 μg/d
Iodine	Iodized salt, salt water fish, seafood, dairy products	150 μg/d	Needed for synthesis of thyroid hormones	Goiter, cretinism, mental retardation, growth and developmental abnormalities	Populations in areas with low-iodine soil and where iodized salt is not used	Enlarged thyroid	1110 μg/d
Chromium	Brewers yeast, nuts, whole grains, mushrooms	25-35 μg/d	Enhances insulin action	High blood glucose	Malnourished children	None reported	ND
Fluoride	Fluoridated water, tea, fish, toothpaste	3–4 mg/d[a]	Strengthens tooth enamel, enhances remineralization of tooth enamel, reduces acid production by bacteria in the mouth	Increased risk of dental caries	Populations in areas with unfluoridated water	Mottled teeth, kidney damage, bone abnormalities	10 mg/d
Molybdenum	Milk, organ meats, grains, legumes	45 μ/d	Cofactor for a number of enzymes	Unknown in humans	None	Arthritis and joint inflammation	2 mg/d

[a]Value is an Adequate Intake (AI).

UL = Tolerable Upper Intake Level; ND = insufficient data to determine a UL.

Recommended Iron Intake

The RDA for iron is based on the amount needed to maintain normal function but only minimal iron stores. The RDA is set at 8 mg/day for adult men age 19 and older and for post-menopausal women (**Table 12.2**).[7] The RDA for menstruating women is increased to 18 mg/day to compensate for the iron lost in menstruation. Other specific recommendations have been made for each gender and life-stage group by considering the percentage of dietary iron absorbed, iron losses from the body, and conditions that increase needs, such as growth and pregnancy. A separate RDA category has been created for vegetarians because iron is poorly absorbed from plant sources (see Table 12.2).

Table 12.2 Dietary Reference Intake Values for Iron

Gender/Life Stage	Recommended Intake
Infants	
0–6 months	0.27 mg/d[a]
7–12 months	11 mg/d
Children	
1–3 years	7 mg/d
4–8 years	10 mg/d
Males	
9–13 years	8 mg/d
14–18 years	11 mg/d
≥ 19 years	8 mg/d
Females	
9–13 years	8 mg/d
14–18 years	15 mg/d
19–50 years	18 mg/d
≥ 51 years	8 mg/d
Females taking oral contraceptives	
14–18 years	11.4 mg/d
19–50 years	10.9 mg/d
Pregnant women	27 mg/d
Lactating women	
≤ 18 years	10 mg/d
19–50 years	9 mg/d
Vegetarians[b]	
Men 19 years	14 mg/d
Women ≥ 51 years	14 mg/d
Menstruating women (19–50 years)	32 mg/d
Adolescent girls	27 mg/d

[a]This value is an AI; all other values are RDAs. [b]Value is RDA X 1.8.

Life Cycle

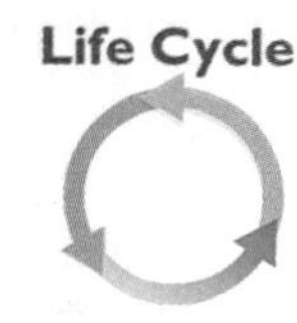

The recommended iron intake during pregnancy is increased to 27 mg/day to account for the iron deposited in the fetal and maternal tissues. The RDA during lactation is set lower than the RDA for menstruating women because little iron is lost in milk and menstruation is usually absent. The RDA for infants, children, and adolescents considers the additional iron needed for growth. An AI has been set for infants from newborn to 6 months based on the mean iron intake of infants principally fed human milk. Because the iron in human milk is more bioavailable than that in infant formula, it is recommended that infants who are not fed human milk or are only partially nourished with human milk be fed iron-fortified formula.[8]

Iron Deficiency

When iron is deficient, hemoglobin cannot be produced. When not enough hemoglobin is available, the red blood cells that are formed are small (microcytic) and pale (hypochromic) and unable to deliver adequate oxygen to the tissues. This is known as **iron deficiency anemia** (**Figure 12.4**). Anemia is the last stage of iron deficiency. Earlier stages have no symptoms because they do not affect the amount of iron in red blood cells. Iron depletion can be detected by blood tests that measure indicators of iron levels in the plasma and in body stores (**Figure 12.5**).

iron deficiency anemia An iron deficiency disease that occurs when the oxygen-carrying capacity of the blood is decreased because there is insufficient iron to make hemoglobin.

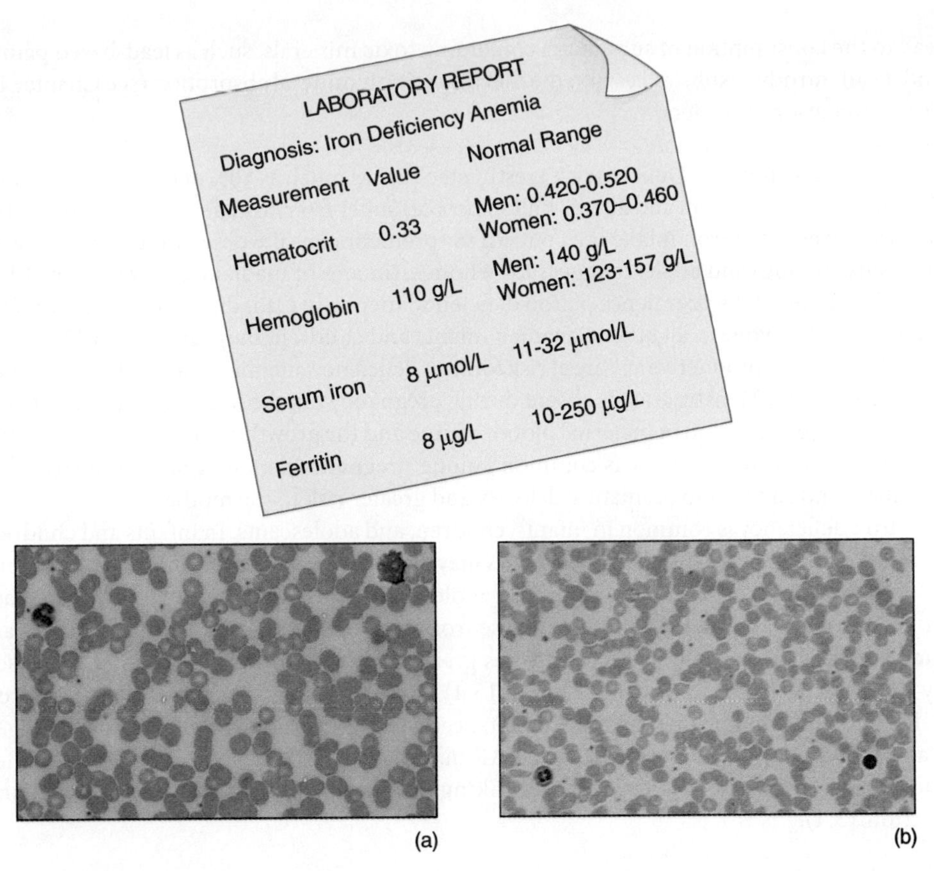
LABORATORY REPORT

Diagnosis: Iron Deficiency Anemia

Measurement	Value	Normal Range
Hematocrit	0.33	Men: 0.420-0.520 Women: 0.370–0.460
Hemoglobin	110 g/L	Men: 140 g/L Women: 123-157 g/L
Serum iron	8 µmol/L	11-32 µmol/L
Ferritin	8 µg/L	10-250 µg/L

Figure 12.4 Iron deficiency anemia
Iron deficiency anemia is diagnosed when the volume of red blood cells (hematocrit), proteins containing iron (hemoglobin and ferritin), or serum iron levels are low. Under a microscope, normal red blood cells (a) appear larger and darker in colour than red blood cells from an individual with iron deficiency anemia (b). (B & B Photos/Custom Medical Stock Photo, Inc.; Custom Medical Stock Photo, Inc.)

Symptoms The symptoms of iron deficiency anemia include fatigue, weakness, headache, decreased work capacity, an inability to maintain body temperature in a cold environment, changes in behaviour, decreased resistance to infection, adverse pregnancy outcomes, impaired development in infants, and an increased risk of lead poisoning in young children. One strange symptom, thought to be related to iron deficiency, is **pica**. This is a compulsion to eat nonfood items such as clay, ice, paste, laundry starch, paint chips, and ashes. Pica can

pica The compulsive ingestion of nonfood substances such as clay, laundry starch, and paint chips.

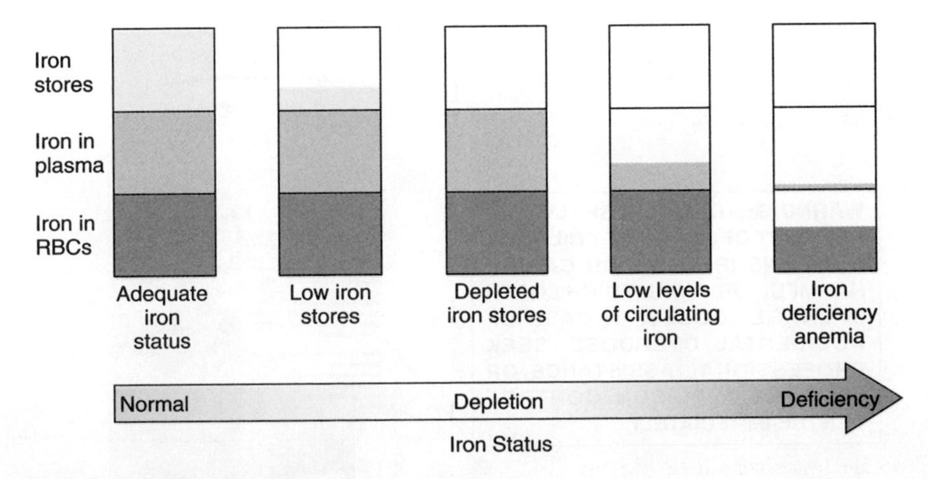

Figure 12.5 Stages of iron deficiency
Iron deficiency anemia is the final stage of iron deficiency. Inadequate iron first causes a decrease in the amount of stored iron, followed by low iron levels in the plasma. It is only after plasma levels drop that there is no longer enough iron available to maintain adequate hemoglobin in red blood cells.

lead to the consumption of substances containing toxic minerals, such as lead-based paints, and it can introduce substances into the diet that inhibit mineral absorption (see Chapter 15: Focus on Eating Disorders).

Groups at Risk for Iron Deficiency It is estimated that as much as 80% of the world's population may be iron deficient and 30% (2 billion people) suffer from iron deficiency anemia.[9] The CCHS suggests that iron intake is a concern for premenopausal women (adult women who are menstruating) and adolescent girls for whom estimates of inadequacy range from 12%-18%. Estimates of the prevalence of iron deficiency anemia in Canadian Aboriginal children are 11%-35%,[10] while in all other Canadian infants and children, they range from 3%-7%.

Women of reproductive age are at risk for iron deficiency anemia because of iron loss due to menstruation. Menstruation is absent during pregnancy but the need for iron is increased because of the expansion of maternal blood volume and the growth of other maternal tissues and the fetus. Iron deficiency is common among pregnant women even in industrialized countries and can lead to premature delivery and greater risk to the mother.

Life Cycle

Iron deficiency is common in infants, children, and adolescents. In infants and children, rapid growth increases iron needs. Toddlers may also be at risk because finicky eating habits often reduce intake (see Section 15.2). In adolescent boys, rapid growth and an increase in muscle mass and blood volume increase iron need. In adolescent girls, iron needs are increased because weight gain is almost as great as in boys and iron losses are increased by the onset of menstruation (see Section 15.4). Athletes are another group at risk for iron deficiency. This may be due to a low iron intake, as well as increased losses due to prolonged training. Based on the amount lost, the EAR may be 30%-70% higher for athletes than for the general population[7] (see Critical Thinking: Increasing Iron Intake and Uptake, and Section 13.4).

Iron Toxicity

Iron is essential for cellular metabolism, but too much can be toxic. Iron promotes the formation of free radicals and causes cell death due to excess oxidation of cellular components. Iron toxicity can be acute, resulting from ingestion of a single large dose at one time, or chronic, due to the accumulation of iron in the body over time, referred to as iron overload. A UL has been set at 45 mg/day from all sources.

Life Cycle

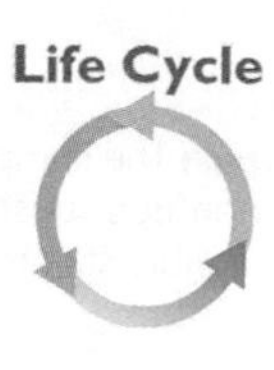

Acute Toxicity Even a single large dose of iron can be life-threatening. Iron poisoning can damage the intestinal lining and cause abnormalities in body pH, shock, and liver failure. Iron toxicity from supplements is one of the most common forms of poisoning among children under age 6 and is the leading cause of liver transplants in children. To protect children from accidental poisoning from iron-containing drugs and supplements, these products display a warning on the label (**Figure 12.6**).[11]

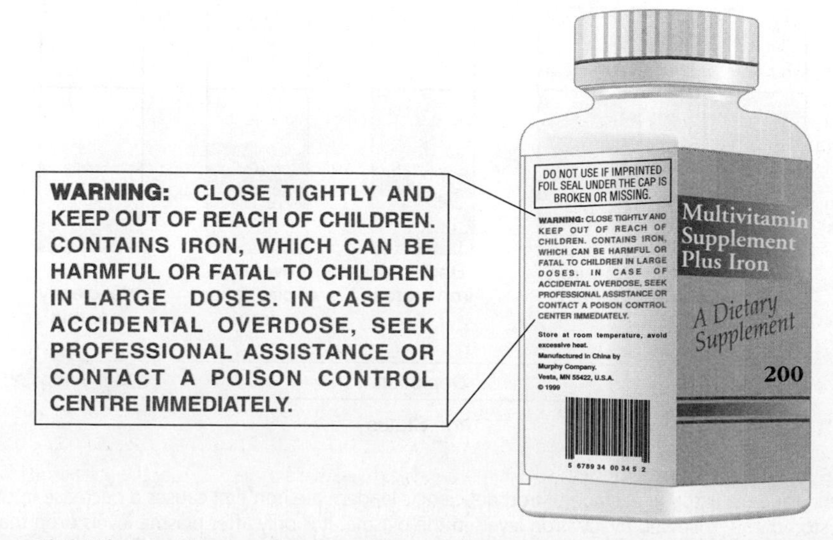

Figure 12.6 Why must labels on iron-containing supplements and medications carry a toxicity warning?

CRITICAL THINKING

Increasing Iron Intake and Uptake

Background:

Hanna is a 23-year-old graduate student from Alberta who has been feeling tired and run down all term. She recently read an article about iron deficiency in young women and has become concerned about her iron status. She decides to go to the health centre where she has blood drawn. The results of her tests indicate that she does not have iron deficiency anemia, but her iron stores are very low.

A review of her typical diet shows that Hanna's iron intake is less than the recommended amount. She decides to try to increase the amount of iron she gets from her diet before considering iron supplements.

Hanna consumes a vegetarian diet. At home, her mother prepared meals using iron cookware, but the pans in Hanna's college apartment are stainless steel. Her typical diet is shown in the table.

(© istockphoto.com/ Richard Rudisill)

Data:

TYPICAL DIET

FOOD	AMOUNT	IRON (MG)
Breakfast		
Hot cornmeal	250 ml (1 cup)	0.3
butter	5 ml (1 tsp)	0
Raisins	60 ml (¼ cup)	0.7
Whole wheat toast	1 slice	1.2
Apple juice	175ml (¾ cup)	0.7
Tea with	(250 ml) 1 cup	0
sugar	5 ml (1 tsp)	0
Lunch		
Apple	1 medium	0.2
Cornbread with	1 piece	0.5
butter	5 ml (1 tsp)	0
Yogurt	250 ml (1 cup)	0.2
Tomato	1 medium	0.5
Tea with	250 ml (1 cup)	0
sugar	5 ml (tsp)	0
Dinner		
Rice	250 ml (1 cup)	2.4
Peanuts	80 ml (⅓ cup)	0.9
Kale	250 ml (1 cup)	1.2
Yams	250 ml (1 cup)	1.1
Apple juice	175 ml (¾ cup)	0.7
Tea with	250 ml (1 cup)	0
sugar	5 ml (1 tsp)	0
Total		**10.6**

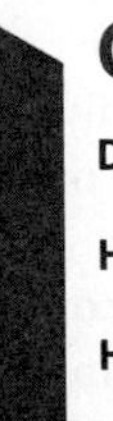

Critical Thinking Questions

Does Hanna's iron intake meet the RDA for a young female vegetarian?

How could Hanna increase her iron intake without adding iron to her diet?

How could Hanna increase the absorption of iron in her meals?

Does Hanna's diet provide good vegetarian sources of calcium? Are there other nutrient deficiencies for which she may be at risk?

Iron Overload If too much iron enters the body, over time it accumulates in tissues such as the heart and liver. Iron overload can occur in people with conditions that cause abnormal red blood cell synthesis and in those with diseases requiring frequent blood transfusions, but the most common cause of chronic iron overload is **hemochromatosis.**[12] Hemochromatosis is an inherited condition that causes increased iron absorption. It afflicts about 1 in 200 to 1 in 500 North Americans and is the most common genetic disorder in the Caucasian population.[13]

Hemochromatosis has no symptoms early in life, but in middle age nonspecific symptoms such as weight loss, fatigue, weakness, and abdominal pain typically begin. If allowed to

hemochromatosis An inherited condition that results in increased iron absorption.

progress, the accumulation of excess iron that occurs in hemochromatosis causes oxidative changes resulting in heart and liver damage, diabetes, certain types of cancer, and other chronic conditions. Iron deposits also darken the skin. To have these symptoms, an individual must inherit the hemochromatosis gene from both parents. The 1 in 10 people who inherit the gene from only a single parent don't have these serious symptoms but do absorb iron better than people who do not have the gene at all.

The rate at which iron accumulates and leads to serious symptoms in those with hemochromatosis depends on the amount of iron consumed and other dietary factors that affect iron absorption, as well as factors that cause iron loss, such as menstruation and blood donations. The public health impact of hemochromatosis is potentially significant. The availability of red meat and the prevalence of iron-fortified foods in the Canadian diet virtually assure that individuals with 2 genes for hemochromatosis will eventually accumulate damaging levels of iron. If individuals with hemochromatosis can be identified, treatment is simple: regular blood withdrawal. This will prevent the complications of iron overload, but to be effective, it must be initiated before organs are damaged. Therefore, genetic screening to identify and treat young healthy individuals is essential in preventing complications.

Overconsumption of iron supplements or a diet high in absorbable iron can also increase iron stores. Although these iron stores are not high enough to cause the serious problems that occur with hemochromatosis, it has been hypothesized that people with high iron stores are at increased risk of the same diseases that occur in hemochromatosis—heart disease, diabetes, and cancer. Iron is hypothesized to promote heart disease by increasing the formation of oxidized LDL cholesterol, which then leads to atherosclerosis. In some studies, elevated levels of the iron-storage protein ferritin are associated with an increased risk of heart attacks, but the majority of studies do not support this association.[14] A relationship between higher iron stores and an increased risk of diabetes has been found in overweight and obese individuals with impaired glucose tolerance.[15] Iron is hypothesized to contribute to diabetes because it promotes the formation of free radicals, which contribute to insulin resistance and eventually decreased insulin secretion. The increase in free radical formation may also increase cancer risk.

Meeting Iron Needs: Consider the Total Diet

Both the amount and the bioavailability of iron from the diet need to be considered when meeting iron needs. The best sources of iron are red meats and organ meats such as liver and kidney. Good nonheme sources are legumes, dried fruit, leafy greens such as spinach and kale, and fortified grain products (**Figure 12.7**). Nonheme iron absorption can be enhanced by including meat, fish, poultry, and foods rich in vitamin C in meals containing iron, while decreasing the consumption of dairy products, which are high in calcium, at these meals.[1]

Figure 12.7 The bioavailability of iron from legumes, plant sources, and fortified foods such as flour is lower than that from animal sources. (Luisa Begani)

Because iron is a nutrient at risk for deficiency in the Canadian diet, the iron content of packaged foods must be listed on food labels. It is given as a percent of the Daily Value for iron, which is 14 mg for adults. Therefore, if your breakfast cereal provides 10% of the Daily Value for iron, it contains about 1.4 mg of iron per serving.

Although diet is the ideal way to meet iron needs, supplements are often recommended for groups at risk for deficiency such as small children, women of child-bearing age, and pregnant women. Iron is commonly available as an individual supplement or as part of multivitamin and mineral supplements. These contain nonheme iron. As with nonheme iron in the diet, to enhance the absorption of iron in a supplement, it should be consumed with foods containing vitamin C, such as orange juice; taken with a meal containing meat, fish, or poultry; and not taken with dairy products, calcium supplements, or substances that bind iron. Iron from supplements that contain the ferrous form (Fe^{2+}) of iron, such as ferrous sulfate, is more readily absorbed than iron from those with the ferric form (Fe^{3+}). Iron supplements can improve iron status, but large intakes of iron from supplements can interfere with the absorption of zinc and copper (**Table 12.3**). Iron-containing supplements should be taken only as suggested on the label and stored out of the reach of children or others who may consume them in excess.

Table 12.3 Benefits and Risks of Trace Element Supplements

Supplement	Claims	Actual Benefits or Risks
Iron	Increases energy	Needed to make hemoglobin to deliver oxygen to tissues. Supplements are beneficial if iron is deficient. High doses cause constipation, liver damage, and death.
Zinc	Treats colds, prevents aging, improves immune function, enhances fertility	Needed for enzyme function, protein synthesis, and vitamin and hormone function. Supplements do not enhance these effects and there is little evidence that they help prevent or treat colds. High doses cause copper deficiency, nausea, and vomiting.
Copper	Prevents heart disease and osteoporosis, alleviates arthritis symptoms, maintains healthy skin and hair colour, treats hypoglycemia	Supplements are useful for improving bone health and improving blood lipids in those with copper deficiency but there is no evidence that intakes above recommended levels prevent heart disease or are effective for the treatment of arthritis or skin conditions. High doses can cause vomiting.
Selenium	Protects against cancer, promotes heart health, immune function	Antioxidant; evidence that it may protect against cancer in those with low levels. High doses cause loss of hair and nail changes.
Chromium	Controls diabetes, lowers cholesterol, reduces body fat, and increases lean tissue	Needed for insulin action. Supplements may improve blood sugar regulation but do not affect body composition. High doses may be related to headaches, sleep disturbances, and mood swings.
Vanadium	Aids insulin action; allows more rapid and intense muscle pumping for body builders	No evidence to support a benefit for body builders. Supplements can reduce insulin requirements but the dose required exceeds the UL.

12.3 Zinc (Zn)

Learning Objectives

- Describe how zinc absorption is regulated.
- Discuss the role of zinc in gene expression.

The essentiality of zinc in the human diet was first recognized in the early 1960s, when a syndrome of growth depression and delayed sexual development, seen in Iranian and Egyptian adolescents and young men consuming diets based on vegetable protein, was alleviated by supplemental zinc.[16] Although the diet was not low in zinc, it was high in grains containing phytates, which interfered with zinc absorption, causing a deficiency.

Zinc in the Diet

Zinc is found in foods from both plant and animal sources. Zinc from animal sources is better absorbed than that from plants because the zinc in plant foods is often bound by phytates. Zinc is abundant in red meat, liver, eggs, dairy products, vegetables, and some seafood (**Figure 12.8**). Whole grains are a good source, but refined grains are not because zinc is lost in milling and not added back in enrichment. Grain products leavened with yeast provide more zinc than unleavened products because the yeast leavening of breads reduces the phytate content.[7]

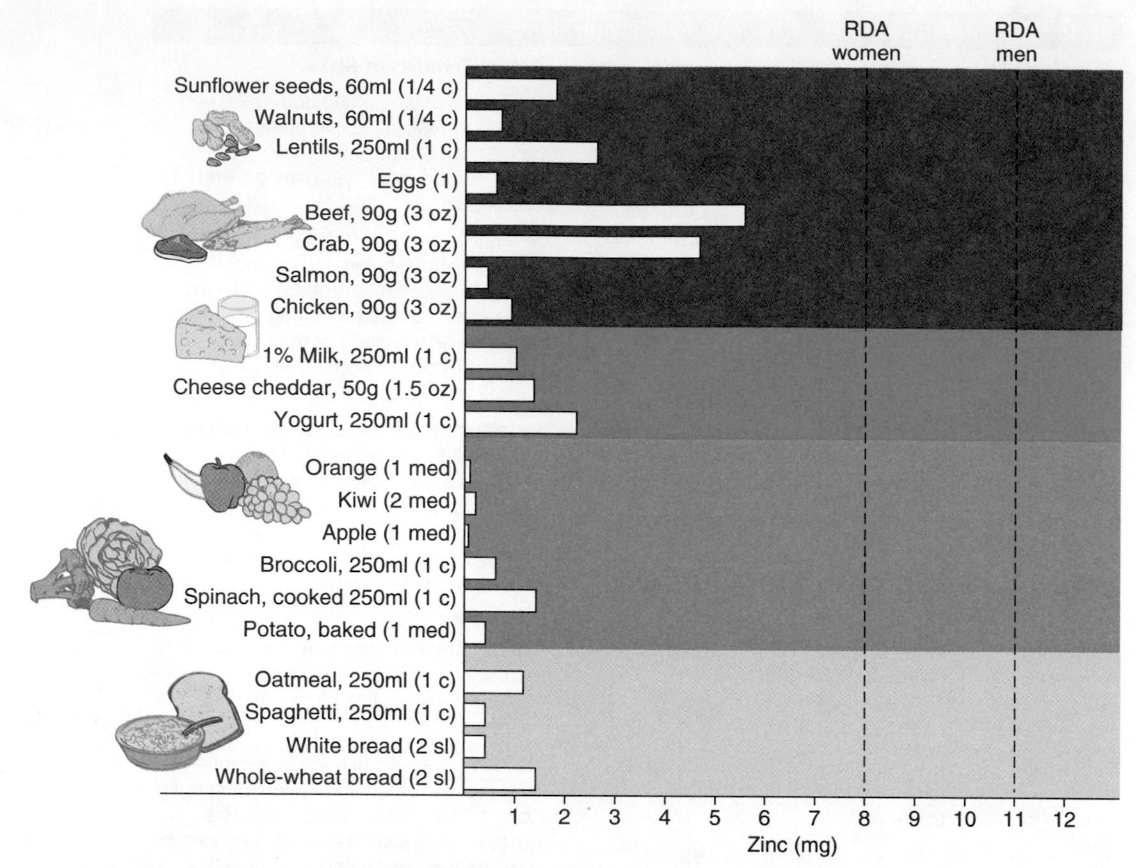

Figure 12.8 Zinc content of Canada's Food Guide food groups
Meat, seafood, dairy products, and fortified foods are good sources of zinc; the dashed lines represent the RDA for adult men and women.

Zinc in the Digestive Tract

The gastrointestinal tract is the major site for regulation of zinc homeostasis. Both the amount of zinc in the mucosal cells and the amount that leaves these cells for distribution to the rest of the body are regulated.

Zinc transport proteins regulate the amount of zinc in the cytosol of the mucosal cells. Some of these proteins increase the amount of zinc absorbed into the mucosal cell by promoting the transport of zinc from the intestinal lumen into the cell. Others reduce the amount of zinc in the cytosol of the mucosal cell by transporting zinc back into the lumen or into storage vesicles in the cell (**Figure 12.9**).[17] The amount of zinc that is in the cytosol can be regulated by increasing or decreasing the synthesis of proteins that transport zinc in versus those that transport it back out of the mucosal cells. For example, if zinc intake is low (see Figure 12.9a), expression of zinc transport proteins that move zinc from the lumen into the mucosal cells will increase relative to the expression of proteins that export zinc out of the mucosa. High zinc levels will have the opposite effect, increasing zinc export to the lumen relative to transport into the cell (see Figure 12.9b).

The amount of zinc that passes from the mucosal cell into the blood is regulated by a metal binding protein called **metallothionein**. When zinc intake is high, metallothionein is synthesized. Zinc in the mucosal cell binds to metallothionein, slowing its transfer into the blood; this provides more opportunity for export of zinc back into the lumen when zinc intakes are high or for it to be lost if the mucosal cell dies (see Figure 12.9b). Metallothionein also binds copper, and high levels can inhibit copper absorption.

metallothionein Refers to proteins that bind minerals. One such protein binds zinc and copper in intestinal cells, limiting their absorption into the blood.

Zinc in the Body

Once zinc has been absorbed, homeostasis can be maintained to some extent by regulating excretion. Zinc is secreted in pancreatic and intestinal juices, which enter the lumen of the intestine. When zinc levels in the body are low, the zinc that enters the gastrointestinal tract can be reabsorbed and recycled. When levels in the body are high, less is reabsorbed and more is therefore eliminated in the feces.

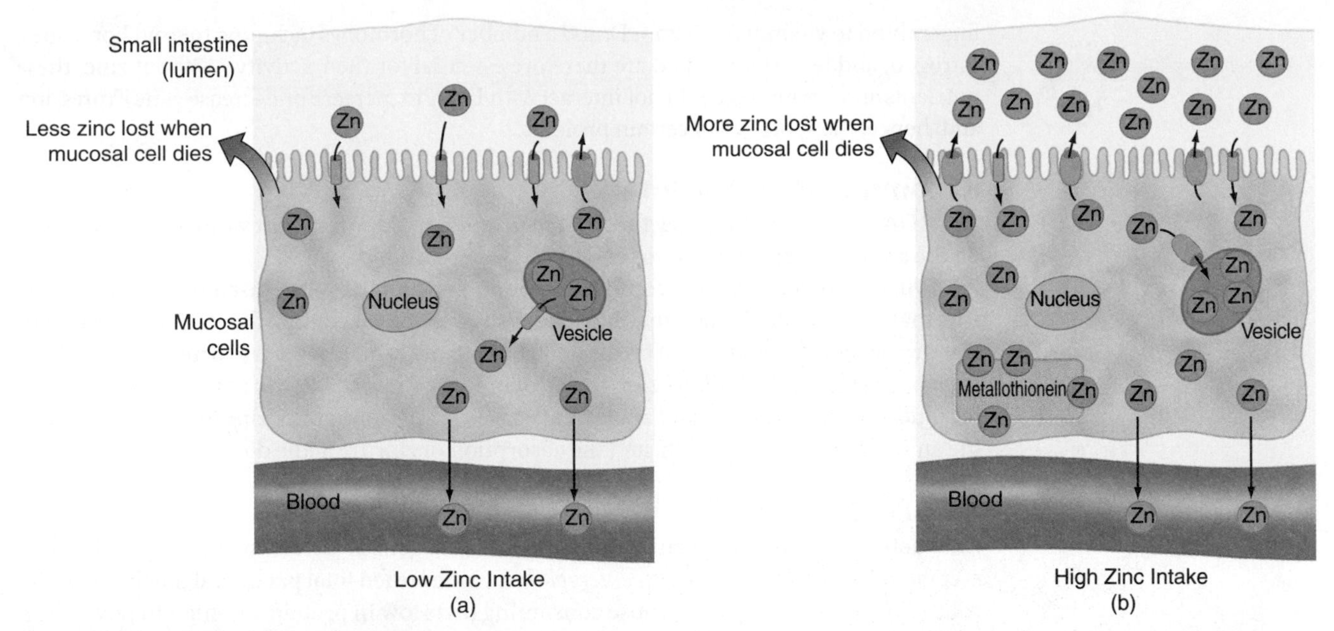

Figure 12.9 Regulation of zinc absorption
a) When zinc intake is low, more zinc moves from the lumen into the mucosal cells and from vesicles into the cytosol, and little metallothionein is synthesized. b) When zinc intake is high, little zinc is transported from the lumen into the mucosal cells and more zinc moves out of the mucosal cells into the lumen and from the cytosol into vesicles. The synthesis of metallothionein, which binds zinc and limits its uptake into the blood, increases.

Zinc Functions Zinc is the most abundant intracellular trace element. It is found in the cytosol, in cellular organelles, and in the nucleus. Zinc is involved in the functioning of more than 300 different enzymes, including a form of **superoxide dismutase (SOD)**, which is vital for protecting cells from free radical damage. It is needed to maintain adequate levels of metallothionein proteins, which also scavenge free radicals.[18] Zinc is also needed by enzymes that function in the synthesis of DNA and RNA, in carbohydrate metabolism, in acid-base balance, and in a reaction that is necessary for the absorption of folate from food. Zinc plays a role in the storage and release of insulin, the mobilization of vitamin A from the liver, and the stabilization of cell membranes. It influences hormonal regulation of cell division and is therefore needed for the growth and repair of tissues, the activity of the immune system, and the development of sex organs and bone.

superoxide dismutase (SOD) An enzyme that protects the cell from oxidative damage by neutralizing superoxide free radicals. One form of the enzyme requires zinc and copper for activity, and another form requires manganese.

Zinc and Gene Expression Some of the functions of zinc can be traced to its role in gene expression. For example, zinc stimulates the production of metallothionein by binding to a regulatory factor and activating the transcription of the gene for this protein. Zinc also plays a structural role in proteins essential for gene expression. Proteins containing zinc fold around the zinc atom to form a loop or "finger." These zinc fingers allow protein receptors to bind to regulatory regions on DNA, stimulating the transcription of specific genes and therefore the synthesis of the proteins for which they code (**Figure 12.10**). Protein receptors containing zinc

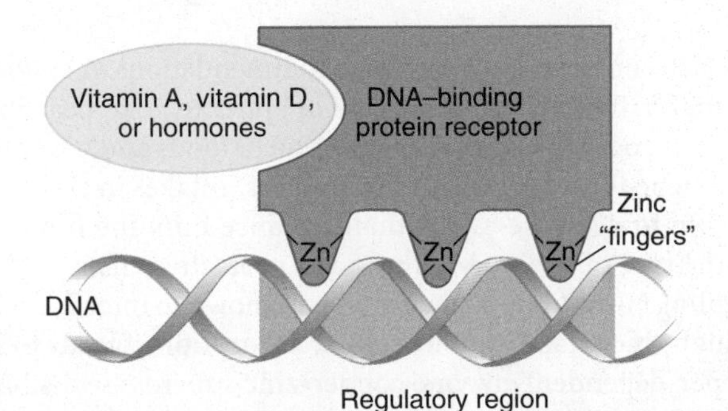

Figure 12.10 Zinc fingers and gene expression
Finger-like structures called zinc fingers allow nuclear protein receptors that bind to vitamin A, vitamin D, and hormones to interact with the regulatory region of a gene and thus affect gene expression.

fingers bind to vitamin A, vitamin D, and a number of hormones including thyroid hormones, estrogen, and testosterone, and are therefore essential for their activity. Without zinc, these nutrients and hormones could not interact with DNA to increase or decrease gene expression and, hence, the synthesis of certain proteins.

Recommended Zinc Intake

The RDA for zinc is 11 mg/day for adult men and 8 mg/day for adult women.[7] This is based on the amount of zinc needed to replace daily losses from the body.

During pregnancy, the recommendation for zinc intake is increased to account for the zinc that accumulates in maternal and fetal tissues. During lactation, the RDA is increased to compensate for zinc secreted in breast milk. For infants from newborn to 6 months, an AI has been established based on the zinc intake of breastfed infants. RDAs have been established for older infants, children, and adolescents based on the amount of zinc lost from the body, the amount needed for growth, and the absorption of zinc from the diet.

Zinc Deficiency

Zinc deficiency has been seen in individuals with a genetic defect in zinc absorption and metabolism called *acrodermatitis enteropathica*, in those fed total parenteral nutrition (TPN) solutions lacking zinc, and in those consuming diets low in protein and high in phytates. It may also occur in individuals with kidney disease, sickle cell anemia, alcoholism, cancer, and AIDS.

The symptoms of zinc deficiency include poor growth and development, skin rashes, hair loss, diarrhea, neurological changes, impaired reproduction, skeletal abnormalities, and reduced immune function.[18] Many of these symptoms reflect zinc's importance in protein synthesis and gene expression. Because it is needed for the proper functioning of vitamins A and D and the activity of numerous enzymes, some of the zinc deficiency symptoms resemble deficiencies of other essential nutrients. Decreased immune function is one of the main concerns with even moderate zinc deficiency. The impact of zinc deficiency on immune function is rapid and extensive, causing a decrease in the number and function of immune cells in the blood, which can lead to an increased incidence of infections.

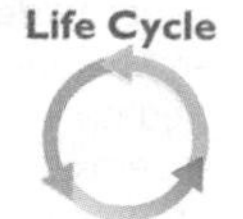

Symptomatic zinc deficiency is relatively uncommon in North America. However, a significant portion of the Canadian population does not consume adequate zinc. As indicated by CCHS (see Critical Thinking: How are Canadians Doing with Respect to Their Intake, from Food, of Iron and Zinc?), more than 10% of the adult population consumes less than the EAR for zinc. Groups especially vulnerable to inadequate intakes include adolescent females and adults 71 years of age and older. In developing countries, zinc deficiency is more likely to have important health and developmental consequences. Through its relationship with immune function, it contributes to infection and overall mortality in children.[19] The risk of zinc deficiency is greater in areas of the world where the diet is high in phytate, fibre, tannins, and oxalates. Pregnant women, the elderly, low-income children, and vegans are at particular risk. Some foods in Canada are fortified with zinc to increase intake from vegetarian sources (see Your Choice: Zinc and Vegetarians: Choosing the Right Foods to Meet Requirements).

Zinc Toxicity

Zinc can be toxic when consumed in excess of recommendations. A single dose of 1-2 g can cause gastrointestinal irritation, vomiting, loss of appetite, diarrhea, abdominal cramps, and headaches. This has occurred with consumption of foods and beverages contaminated with zinc that has leached from galvanized containers. Intakes in the range of 50-300 mg/day have been shown to decrease rather than enhance immune function and to reduce HDL cholesterol, the type of cholesterol that has a protective effect against heart disease.[7] Supplements providing 50 mg/day of zinc have been shown to interfere with the absorption of copper. When high zinc intake inhibits copper absorption, it leads to a reduction in the activity of the copper-dependent enzyme copper-zinc superoxide dismutase in red blood cells. A UL has been set at 40 mg/day from all sources, based on the adverse effect of excess zinc on copper metabolism.

YOUR CHOICE

(©iStockphoto)

Zinc and Vegetarians: Choosing the Right Foods to Meet Requirements

The CCHS suggests that a fair number of Canadians are not meeting their requirements for zinc (see Critical Thinking: How are Canadians Doing with Respect to Their Intake, from Food, of Iron and Zinc?). The richest sources of zinc are found in animal products (see table).

Zinc Content of Animal Products

Food (1 Canada's Food Guide Serving)	Zinc (mg)
Oysters	136.0
Beans, baked, canned with pork and tomato sauce	10.2
Beef, sirloin steak	8.5
Lamb shank	6.5
Crab, Alaska	5.7
Beef, brisket	5.2
Lobster	5.0
Beef, ground, lean	4.0

Source: Health Canada. Canadian Nutrient File version 2010. Available online at http://webprod3.hc-sc.gc.ca/cnf-fce/index-eng.jsp. Accessed August 21, 2011.

Vegetarians have always been more vulnerable to inadequate intakes. Furthermore, the RDA (see Section 2.20) for zinc for vegans may be 50% higher due to the lower bioavailability of zinc from vegan sources (see zinc RDA table).

Zinc RDAs

	RDA (mg)-mixed diets	RDA (mg)–vegan
Men (+19 years)	11	up to 17
Women (+ 19 years)	8	up to 12

Source: Health Canada DRI tables. Available online at http://www.hc-sc.gc.ca/fn-an/nutrition/reference/table/index-eng.php. Accessed August 21, 2011.

Fortunately, by making the right choices, vegetarians can ensure adequate intakes (see Non-Animal Products table). Zinc is a nutrient that is widely distributed in small amounts in many foods, so the key to meeting requirements is to eat a variety of foods.

Zinc Content of Non-animal Products

Food (1 Canada's Food Guide serving)	Zinc (mg)
Soy foods	
Soybeans, cooked (175 ml)	2.9
Soy milk (250 ml)	1.0-2.0
Tofu, firm (175 ml)	1.4
Legumes	
Baked beans, canned, vegetarian (175 ml)	2.5
Lentils (175 ml)	1.7
Navy beans (175 ml)	3.2
Nuts, peanuts, seeds	
Almonds (60 ml)	1.2
Cashews (60 ml)	1.9
Peanut butter (30 ml)	0.9
Pumpkin seeds (60 ml)	2.6
Sunflower seeds (60 ml)	1.8
Sesame tahini (30 ml)	1.4
Breads and grains	
Quinoa (125 ml)	0.8
Wheat germ (30 ml)	1.8
Whole wheat bread (1 slice)	0.5
Milk and eggs	
Milk, cow's (250 ml)	1.0
Yogurt (175 ml)	1.1-1.5
Egg, large (2)	1.0

Source: American Dietetic Association; Dietitians of Canada. Position of the American Dietetic Association and Dietitians of Canada: vegetarian diets. *Can. J. Diet. Pract. Res.* 64(2):62-81, 2003.

Canadian regulations also require some foods, such as simulated meat products and beverages derived from legumes, nuts, cereal grains and potatoes, to be fortified with zinc.[1] Simulated meat products often contain 60%-130% of the DV, which for zinc is 9 mg.[2] Breakfast cereals may also be fortified with zinc. When zinc is added to a food, it will be listed on the Nutrition Facts Table. With careful choices, zinc requirements can be met.

[1]Canadian Food Inspection Agency. Guide to Food Labelling and Advertising. Annex 7-1 Foods to which vitamins, mineral nutrients, and amino acids may or must be added. Available online at http://www.inspection.gc.ca/english/fssa/labeti/guide/ch7-1e.shtml. Accessed May 4, 2011.

[2]Yves Veggie Cuisine. Available online at http://www.yvesveggie.ca/. Accessed May 4, 2011.

Zinc Supplements

Zinc is often marketed as a supplement to improve immune function and enhance fertility and sexual performance. For individuals consuming adequate zinc, there is no evidence that extra is beneficial. In individuals with a mild zinc deficiency, supplementation may result in improved wound healing, immunity, and appetite; in children, it can result in improved growth and learning. In healthy older adults, supplements of zinc have been shown to improve the immune response[20] (see sections 16.3 and 16.4). Zinc supplements in lozenge form are currently popular for preventing and treating colds. Supplemental zinc is also used therapeutically to treat genetic diseases.

Zinc and the Common Cold It has been suggested that zinc lozenges, containing either zinc glyconate or zinc acetate, reduce the duration and severity of the common cold by preventing the cold virus from binding to cells of the mucous membranes in the nose and throat.[21] The zinc swallowed in a mineral supplement will not have any effect because this zinc goes to your stomach and does not contact the mucosal surfaces affected by cold viruses. Although clinical trials to assess the efficacy of zinc lozenges have been inconsistent, many do support the value of zinc in reducing the duration and severity of symptoms when administered within 24 hours of the onset of common cold symptoms.[22,23] If zinc lozenges are used as a cold remedy, they should be used only for the period of time recommended on the label, usually about a week. Too much zinc can suppress the immune system, lower HDL cholesterol levels, and impair copper absorption. Zinc lozenges each contain about 11-14 mg of elemental zinc (**Figure 12.11**). Taking 4 of these in a day will exceed the UL of 40 mg/day.

Figure 12.11 Zinc lozenges are marketed to reduce the prevalence and severity of the common cold. Do they really work? Do they pose a risk? (Stockbyte/Getty Images, Ltd.)

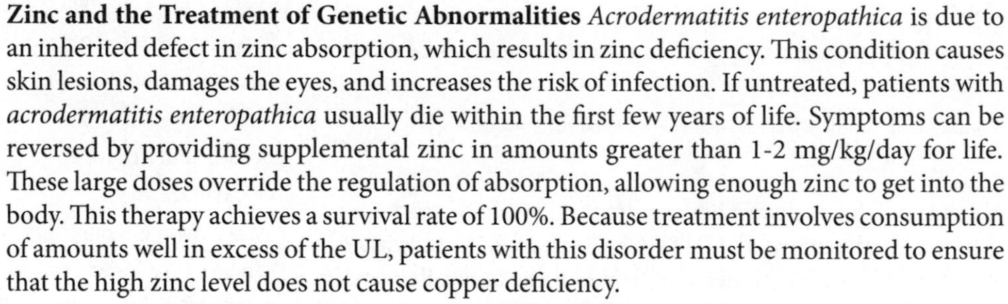

Zinc and the Treatment of Genetic Abnormalities *Acrodermatitis enteropathica* is due to an inherited defect in zinc absorption, which results in zinc deficiency. This condition causes skin lesions, damages the eyes, and increases the risk of infection. If untreated, patients with *acrodermatitis enteropathica* usually die within the first few years of life. Symptoms can be reversed by providing supplemental zinc in amounts greater than 1-2 mg/kg/day for life. These large doses override the regulation of absorption, allowing enough zinc to get into the body. This therapy achieves a survival rate of 100%. Because treatment involves consumption of amounts well in excess of the UL, patients with this disorder must be monitored to ensure that the high zinc level does not cause copper deficiency.

Zinc supplements are also used to treat Wilson's disease. Wilson's disease is due to an inherited defect in the excretion of copper and causes copper to accumulate in the body, leading to toxicity. A few drugs are available to remove copper from the body, but supplemental zinc acetate, which blocks the absorption of copper, increases copper excretion in the stool, and causes no serious side effects, is considered the best treatment.[24]

12.4 Copper (Cu)

Learning Objectives

- Explain why copper deficiency can lead to anemia.
- Describe how high intakes of zinc affect copper absorption.

The ability of copper to treat certain types of anemia helped establish the essentiality of this mineral in human nutrition.[25] Further understanding of the impact of copper deficiency in humans came from studying individuals who were inadvertently fed intravenous (TPN) solutions deficient in copper and those with a rare genetic disease called Menkes disease or kinky hair disease, in which there is a defect in intestinal copper absorption.

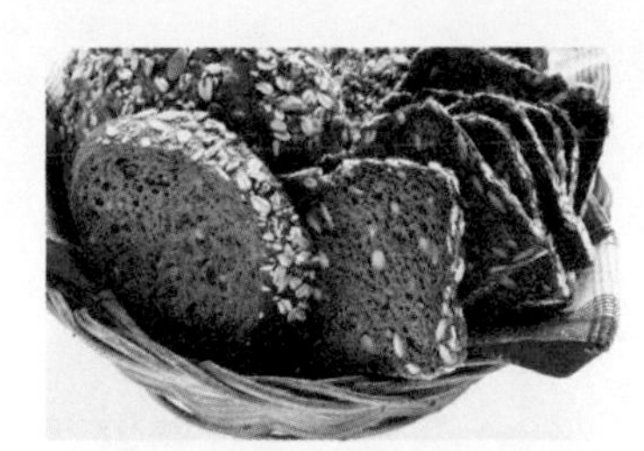

Figure 12.12 Whole grain breads, nuts, and seeds are good sources of copper. (Food Collection/StockFood America)

Copper in the Diet

The richest dietary sources of copper are organ meats such as liver and kidney. Seafood, nuts and seeds, whole-grain breads and cereals, and chocolate are also good sources (**Figure 12.12**). As with many other trace elements, soil content affects the amount of copper in plant foods.

Copper in the Digestive Tract

About 30%-40% of the copper in a typical diet is absorbed.[7] The absorption of copper is affected by the presence of other minerals in the diet. As discussed, the zinc content of the diet can have a major impact on copper absorption. When zinc intake is high, it stimulates the synthesis of the protein metallothionein in the mucosal cells. Although metallothionein binds zinc, it binds to copper more tightly. Therefore, when metallothionein is synthesized, it binds copper, preventing it from being moved out of mucosal cells into the blood (**Figure 12.13**). The antagonism between copper and zinc is so great that phytates, which inhibit zinc absorption, actually increase the absorption and utilization of copper. Copper absorption is also reduced by high intakes of iron, manganese, and molybdenum. Other factors that affect copper absorption include vitamin C, which decreases copper absorption, and large doses of antacids, which also inhibit copper absorption and, over the long term, can cause copper deficiency.

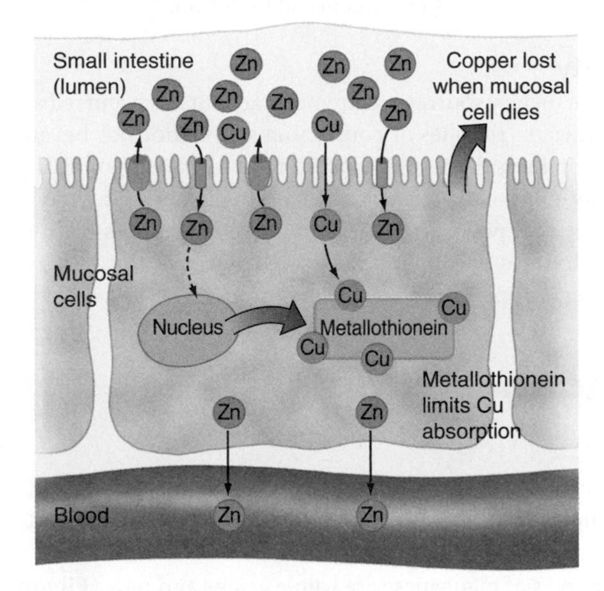

Figure 12.13 Inhibition of copper absorption by zinc
High levels of dietary zinc can inhibit copper absorption by stimulating the synthesis of metallothionein, which then preferentially binds copper and limits its absorption.

Copper in the Body

Once absorbed, copper binds to albumin, a protein in the blood, and travels to the liver, where it binds to the protein **ceruloplasmin** for delivery to other tissues. Copper must be transported bound to proteins such as albumin and ceruloplasmin because free copper ions can trigger oxidation leading to cellular damage. Copper can be removed from the body by secretion in the bile and subsequent elimination in the feces. Copper functions in a number of important proteins and enzymes that are involved in iron and lipid metabolism, connective tissue synthesis, maintenance of heart muscle, and the functioning of the immune and central nervous systems.[7] The copper-containing plasma protein ceruloplasmin converts iron into a form that can bind to transferrin for transport from body cells and an analogous copper-containing protein found in intestinal cells is essential for the transport of absorbed iron from the intestine.[6] Copper is an essential component of a form of the antioxidant enzyme superoxide dismutase. Copper also plays a role in cholesterol and glucose metabolism; elevated blood cholesterol levels have been reported in copper deficiency. Copper is needed for the synthesis of the neurotransmitters

ceruloplasmin The major copper-carrying protein in the blood.

norepinephrine and dopamine, and several blood-clotting factors. It may also be involved in the synthesis of myelin, which is necessary for transmission of nerve signals.

Recommended Copper Intake

Life Cycle

The RDA for copper for adults is 900 μg/day. This recommendation is based on the amount of copper needed to maintain normal blood levels of copper and ceruloplasmin. During pregnancy, the RDA is increased to 1000 μg/day to account for the copper that accumulates in the fetus and maternal tissues. The RDA for lactation is 1,300 μg/day to account for the copper secreted in human milk. The amount of copper in the North American diet is slightly above the RDA.

Copper Deficiency

Severe copper deficiency is relatively rare, occurring most often in pre-term infants. Marginal copper deficiency may be more prevalent but has been difficult to diagnose. The most common manifestation of copper deficiency is anemia. This is due primarily to the importance of copper-containing proteins for iron transport. In copper deficiency, even if iron is sufficient in the diet, iron cannot be transported out of the intestinal mucosa. Copper deficiency causes skeletal abnormalities similar to those seen in vitamin C deficiency (scurvy). This is because the enzyme needed for the cross-linking of connective tissue requires copper. Copper deficiency has also been associated with impaired growth, degeneration of the heart muscle, degeneration of the nervous system, and changes in hair colour and structure.[7] Because of copper's role in the development and maintenance of the immune system, a diet low in copper decreases the immune response and increases the incidence of infection.[26]

Copper Toxicity

Copper toxicity from dietary sources is extremely rare but has occurred as a result of drinking from contaminated water supplies or consuming acidic foods or beverages that have been stored in copper containers. Excessive copper intake causes abdominal pain, vomiting, and diarrhea. These symptoms may occur with copper intakes of 4.8 mg/day in some individuals, but there is evidence that people can adapt to higher exposures without experiencing any adverse effects. High doses of copper have also been shown to cause liver damage. The UL has been set at 10 mg of copper/day.[7]

12.5 Manganese (Mn)

Learning Objective

- Discuss the antioxidant function of manganese, copper, and zinc.

Figure 12.14 Legumes, nuts, and whole grains are high in manganese. (istockphoto.com/Brian Dicks)

The best dietary sources of manganese are whole grains and nuts (**Figure 12.14**). Fruits and vegetables are fair sources; meat, dairy products, and refined grains are poor sources.

Manganese homeostasis is maintained by regulating both absorption and excretion. Manganese absorption increases when intake is low and decreases when intake is high. Manganese is eliminated by secretion into the intestinal tract in bile. It is a constituent of some enzymes and an activator of others. Manganese-requiring enzymes are involved in amino acid, carbohydrate, and cholesterol metabolism; cartilage formation; urea synthesis; and antioxidant protection. Like copper and zinc, manganese is needed for the activity of a form of superoxide dismutase. The form of the enzyme requiring manganese is located inside the mitochondria.

Recommended Manganese Intake

There is not sufficient evidence to set an RDA for manganese; the AI is 2.3 mg/day for men and 1.8 mg/day for women based on the amounts consumed in the healthy population. Recommended intakes are higher during pregnancy and lactation.[7]

Manganese Deficiency and Toxicity

Manganese deficiency in animals results in growth retardation, reproductive problems, congenital malformations in the offspring, and abnormalities in brain function, bone formation, glucose regulation, and lipid metabolism.

Although a naturally occurring manganese deficiency has never been reported in humans, a man participating in a study of vitamin K was inadvertently fed a diet deficient in manganese for 6 months. He lost weight, his black hair turned a red colour, and he developed dermatitis and excessively low blood cholesterol (which is not desirable). Manganese deficiency was further studied in young male volunteers fed a manganese-deficient diet for 39 days. These men developed dermatitis and had altered blood levels of cholesterol, calcium, and phosphorus.[27]

Toxic levels of manganese result in damage to the nervous system. In humans, toxicity has been reported in mine workers exposed to high concentrations of inhaled manganese dust. The UL is 11 mg/day from all sources.[7]

12.6 Selenium (Se)

Learning Objectives

- Compare the antioxidant functions of selenium and vitamin E.
- Discuss the relationship between selenium and cancer.

Although selenium was discovered about 180 years ago, its essential role in human nutrition was not recognized until the 1970s when it was found to prevent a heart disorder in children living in regions of China with low soil selenium levels. Today, selenium is known to be an important part of the body's antioxidant defenses.

Selenium in the Diet

Seafood, kidney, liver, and eggs are excellent sources of selenium (**Figure 12.15**). Fruits, vegetables, and drinking water are generally poor sources. Grains and seeds can be good sources depending on the selenium content of the soil where they were grown. For example, wheat grown in Ontario has a different selenium content from wheat grown in Saskatchewan.[28] Soil selenium can have a significant impact on the selenium intake of populations consuming

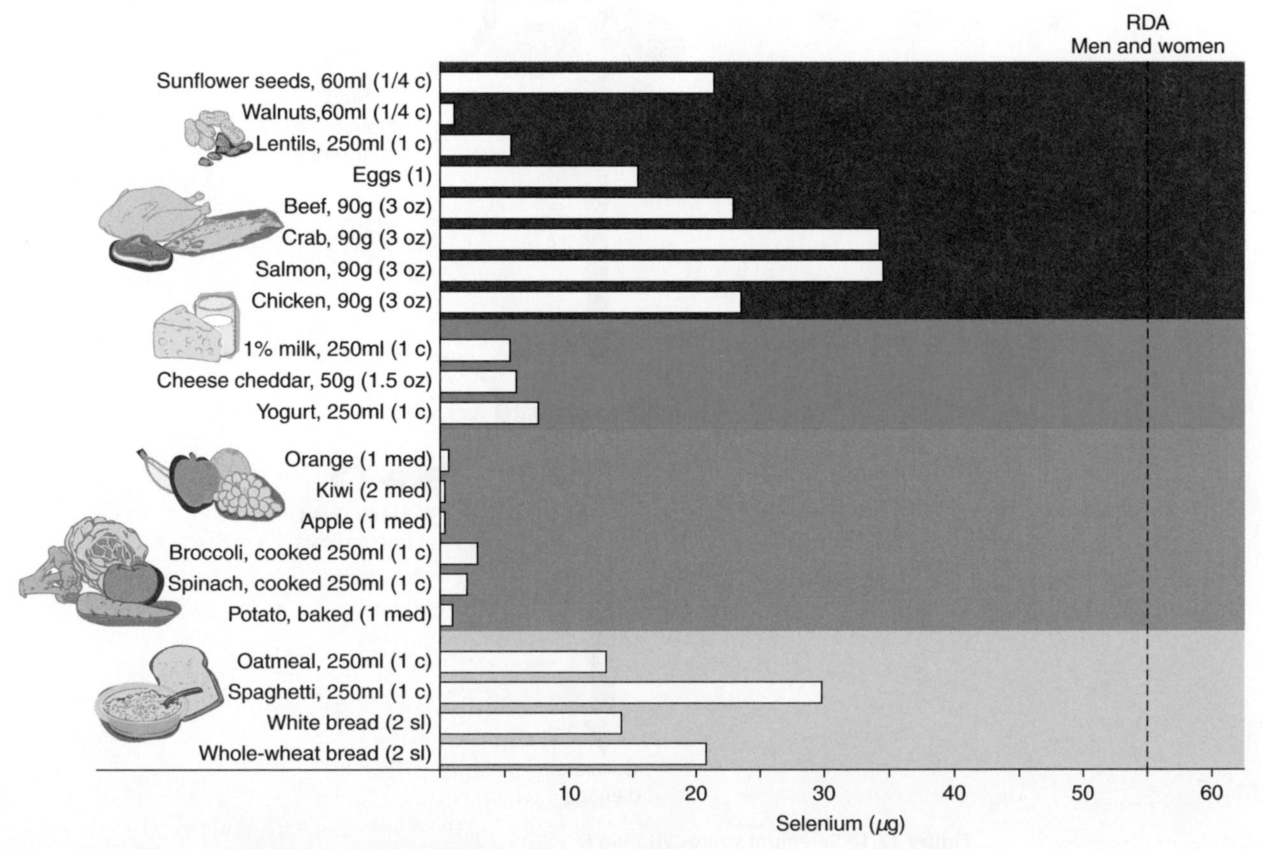

Figure 12.15 Selenium content of Canada's Food Guide food groups
Both plant and animal foods are good sources of selenium. The dashed line represents the RDA for adult men and women.

primarily locally grown food. Selenium deficiency is not likely to be a problem when the diet includes foods produced in many different locations.

Selenium in the Body

Selenium absorption is efficient and does not appear to be regulated. Once absorbed, selenium homeostasis is maintained by regulating its excretion in the urine. Selenium is a mineral that functions mostly through association with proteins called **selenoproteins**. Several of these, including **glutathione peroxidase**, are enzymes that help protect cells from oxidative damage. Glutathione peroxidase neutralizes peroxides so they no longer form free radicals, which cause oxidative damage. By reducing free radical formation, selenium can reduce the need for vitamin E because this vitamin stops the action of free radicals once they are produced (**Figure 12.16**). Selenium is also needed for the synthesis of the thyroid hormones, which regulate basal metabolic rate.

selenoproteins Proteins that contain selenium as a structural component of their amino acids. Selenium is most often found as selenocysteine, which contains an atom of selenium in place of the sulfur.

glutathione peroxidase A selenium-containing enzyme that protects cells from oxidative damage by neutralizing peroxides.

Recommended Selenium Intake

The RDA for selenium for adults is 55 μg/day.[7] This is based on the amount needed to maximize the activity of the enzyme glutathione peroxidase in the blood. Estimates of the selenium content of the Canadian diet suggest that Canadians meet this requirement.[29]

An increase in selenium intake is recommended during pregnancy based on the amount transferred to the fetus and during lactation to account for the amount of selenium secreted in milk. The AI for infants is based on the amount contained in breast milk.

Keshan disease A type of heart disease that occurs in areas of China where the soil is very low in selenium. It is believed to be caused by a combination of viral infection and selenium deficiency.

Selenium Deficiency

Symptoms of selenium deficiency include muscular discomfort and weakness. Deficiency was not identified in humans until the late 1970s, when it was observed in patients fed TPN solutions inadvertently deficient in selenium. At the same time, scientists in China described a disease of the heart muscle called **Keshan disease**, which was linked to selenium deficiency.

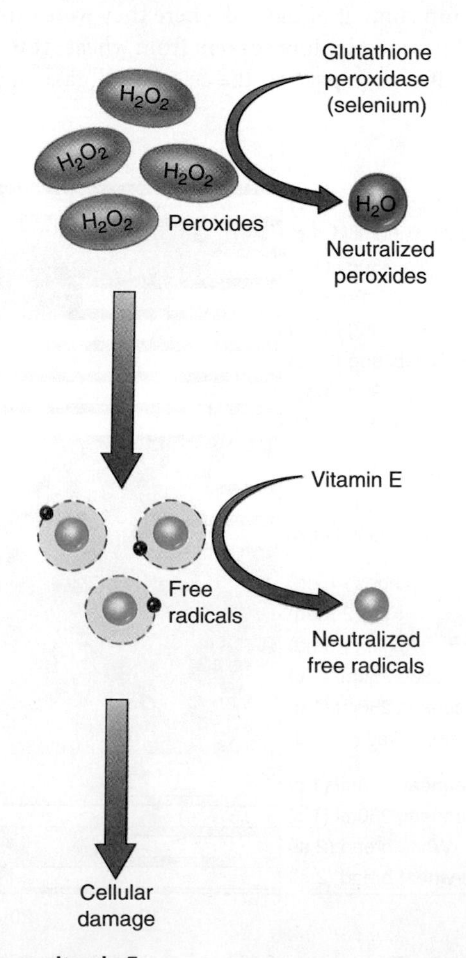

Figure 12.16 Selenium spares vitamin E
Selenium is a part of the enzyme glutathione peroxidase, which neutralizes peroxides before they form free radicals. Fewer free radicals means less vitamin E is needed to eliminate them.

Life Cycle

Selenium and Keshan Disease Keshan disease causes an enlarged heart and poor heart function. It used to be endemic in regions of China where the diet was restricted to locally grown food and the soil was deficient in selenium (**Figure 12.17**).

It affected primarily children and women of child-bearing age. Selenium supplementation was found to dramatically reduce the incidence of Keshan disease, but it could not reverse heart damage once it had occurred. Although Keshan disease is now virtually eliminated by selenium supplementation, the disease itself is believed to be due to a combination of selenium deficiency and infection with a virus. When selenium is deficient, the virus becomes more virulent, causing the symptoms of Keshan disease.[30]

Selenium and Cancer The role of selenium in cancer has been under investigation for three decades. An increased incidence of cancer has been observed in regions where selenium intake is low. In 1996 a study investigating the effect of selenium supplements on people with a history of skin cancer found that the supplement had no effect on the recurrence of skin cancer but the incidence of lung, prostate, and colon cancer all decreased in the selenium-supplemented group.[31] There was a great deal of excitement about this result and many believed selenium supplements could reduce cancer risk. Subsequent research, however, has not supported this result. Evidence now suggests that selenium supplements actually increase the incidence of certain types of skin cancer.[32] The reduction in the incidence of lung and prostate cancer seen in the 1996 study is now believed to have occurred primarily in people who began the study with low levels of selenium. So, selenium deficiency can increase the risk of cancer, but supplements of selenium have not been shown to be of additional benefit in the general population. A major intervention trial, called SELECT (selenium and vitamin E cancer trial) involving 30,000 men similarly found no effect of either selenium, vitamin E, or their combination, on prostate cancer risk.[33]

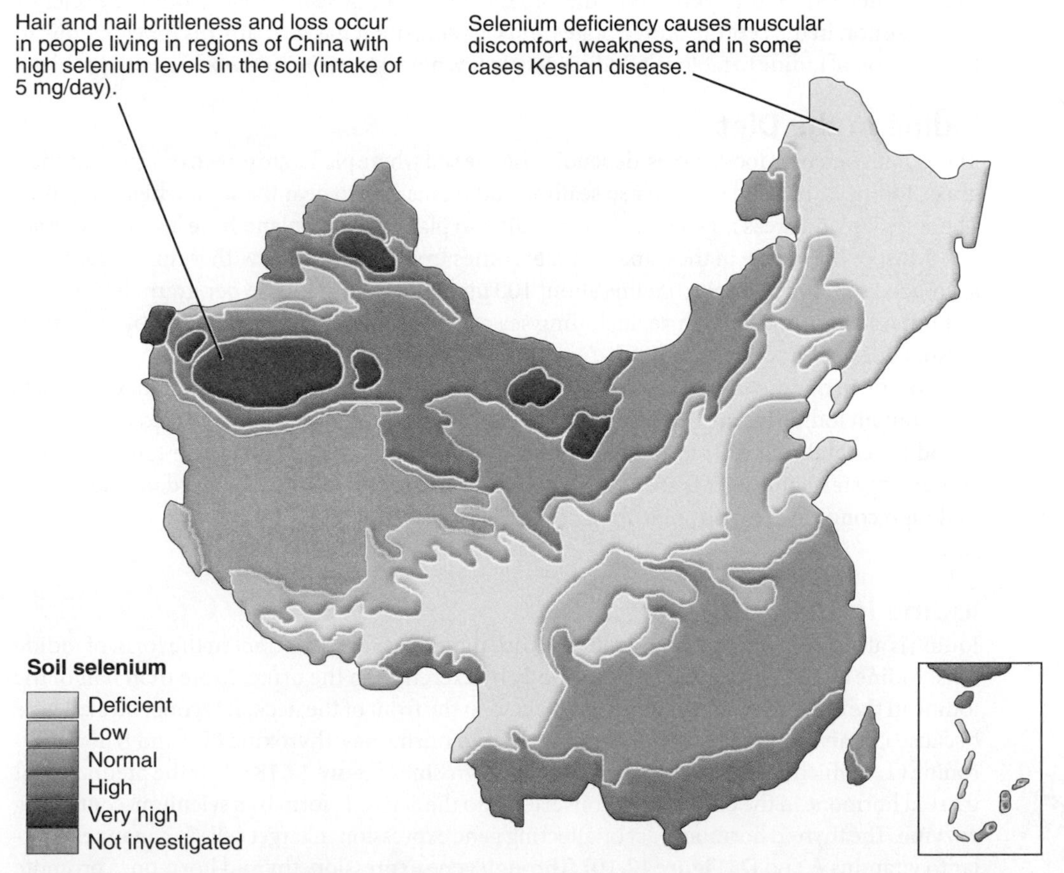

Figure 12.17 Levels of soil selenium in China
The soil content of selenium affects the selenium content of crops grown in the soil. Because the diet in rural China consists predominantly of locally grown foods, the incidence of Keshan disease corresponds to the belt of selenium deficient soil that crosses China from the northeast to the southwest.

Selenium Toxicity

In a region of China with very high selenium in the soil, an intake of 5 mg/day resulted in fingernail changes and hair loss (see Figure 12.17). Selenium toxicity has also been reported in the United States because of a manufacturing error that created mineral supplements containing a dose of 27 mg of selenium per day. The individuals who used these supplements had symptoms that included nausea, diarrhea, abdominal pain, fingernail and hair changes, nervous system abnormalities, fatigue, and irritability.[27] Hair loss, fingernail changes, and gastrointestinal upset have been reported at much lower levels. The UL for adults is 400 μg/day from diet and supplements combined.[7]

Selenium Supplements

Selenium supplements are marketed with claims that they will protect against environmental pollutants, prevent cancer and heart disease, slow the aging process, and improve immune function. Although selenium does play a role in these processes, supplements that increase intake above the RDA will not provide additional benefits.

12.7 Iodine (I)

Learning Objectives

- Describe the function of iodine.
- Explain why iodine deficiency causes the thyroid gland to enlarge.
- Discuss factors that impact the iodine status of a population.

Iodine is needed for the synthesis of thyroid hormones. In the early 1900s, iodine deficiency was common in the central United States and Canada, but it has virtually disappeared due to the addition of iodine to table salt. Iodine deficiency, however, remains a world health problem.

Iodine in the Diet

The iodine content of foods varies, depending on the soil where plants are grown or where animals graze. Iodine is found in sea water so seafood and plants grown near the sea are high in iodine. The soil in inland areas is generally low in iodine so plants grown inland have lesser amounts.

Most of the iodine in the Canadian diet comes from salt fortified with iodine, referred to as iodized salt. Iodized salt contains about 100 μg of potassium iodide per gram. In Canada, salt intended for household use, including sea salt, must be iodized.[34] Salt used by food processors does not have to be iodized.

Iodine in our diet also comes from contaminants and additives in foods. Dairy products may contain iodine because of the iodine-containing additives used in cattle feed and the use of iodine-containing disinfectants on cows, milking machines, and storage tanks. Iodine-containing sterilizing agents are also used in food service establishments, and iodine is used in dough conditioners and some food colourings.

Iodine in the Body

Iodine is absorbed completely and rapidly from the gastrointestinal tract in the form of iodide ions. Iodine can be eliminated from the body by excretion in the urine. More than half of the iodine in the body is located in the thyroid gland in the front of the neck. It is concentrated here because it is an essential component of the thyroid hormones, thyroxine (T_4) and triiodothyronine (T_3), which are made from the amino acid tyrosine (**Figure 12.18**). T_4 is the predominant thyroid hormone in the blood and is converted into the active T_3 form by a selenium-containing enzyme. The thyroid hormones act by affecting gene expression in target cells in a manner similar to vitamins A and D (**Figure 12.19**). Through gene expression, thyroid hormones promote protein synthesis and regulate basal metabolic rate, growth, and development.

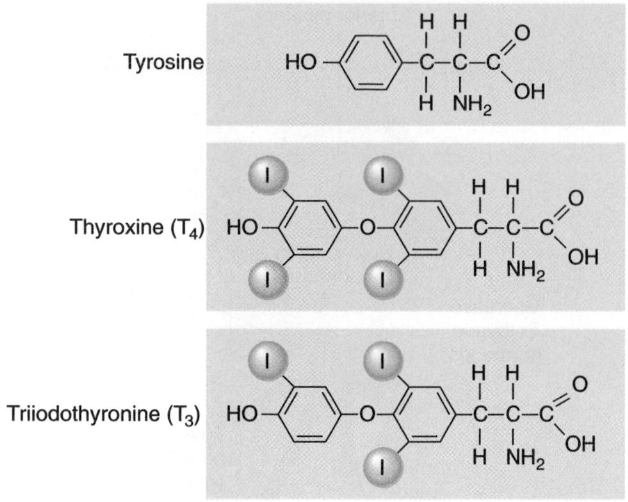

Figure 12.18 **Structure of the thyroid hormones**
The thyroid hormones thyroxine (T_4) and triiodothyronine (T_3) are made from the amino acid tyrosine.

Levels of the thyroid hormones are carefully controlled. If blood levels drop, **thyroid-stimulating hormone** is released from the anterior pituitary. This hormone signals the thyroid gland to take up iodine and synthesize thyroid hormones. When the supply of iodine is adequate, thyroid hormones can be made and their presence turns off the synthesis of thyroid-stimulating hormone (**Figure 12.20**).

thyroid-stimulating hormone A hormone that stimulates the synthesis and secretion of thyroid hormones from the thyroid gland.

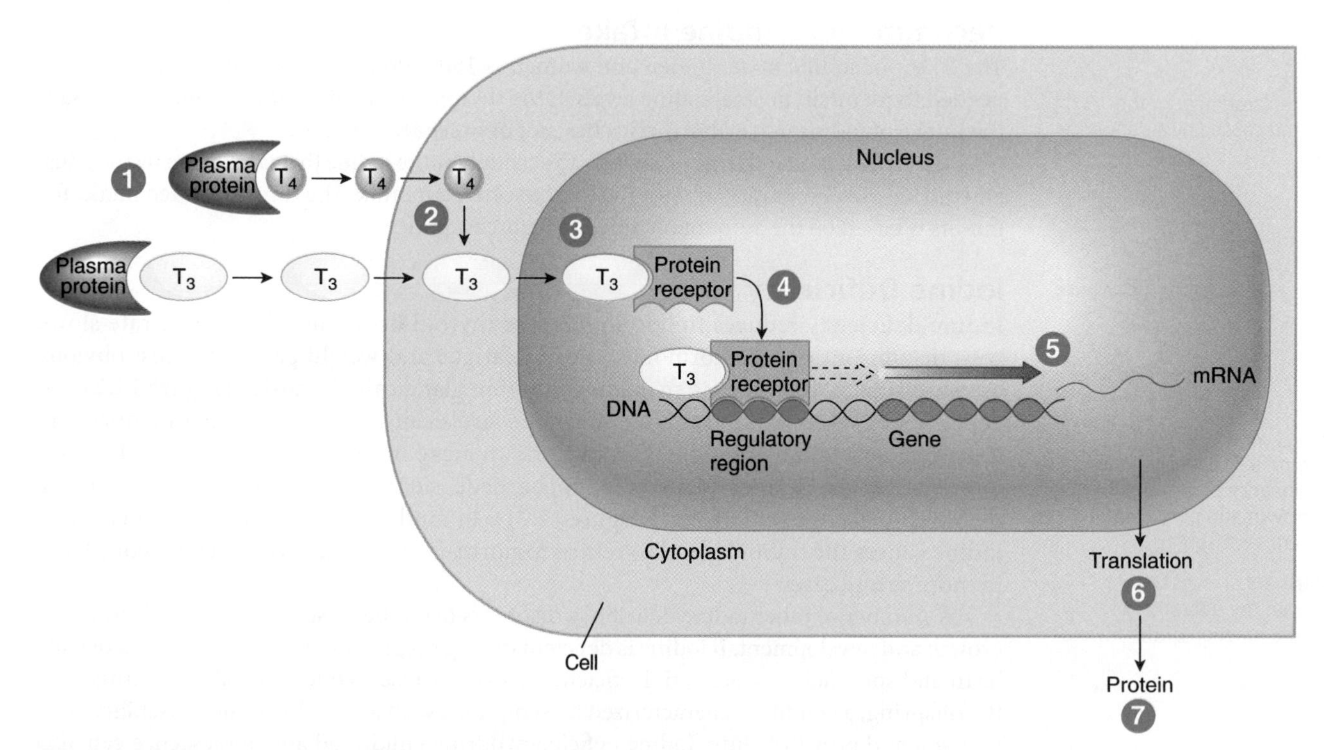

1. Thyroid hormones (T_4 and T_3) circulate in the blood bound to plasma proteins.
2. T_4 and T_3 enter the cell where a selenium-containing enzyme converts T_4 to T_3.
3. T_3 enters the nucleus and binds to a nuclear protein receptor.
4. The T_3-protein receptor complex then binds to a regulatory region of a target gene.
5. Transcription of the gene is turned on, increasing the amount of mRNA made.
6. mRNA directs translation, increasing the synthesis of the protein coded by this gene.
7. There is an increase in the amount of protein and hence the cellular functions and body processes affected by this protein.

Figure 12.19 Role of thyroid hormones in gene expression

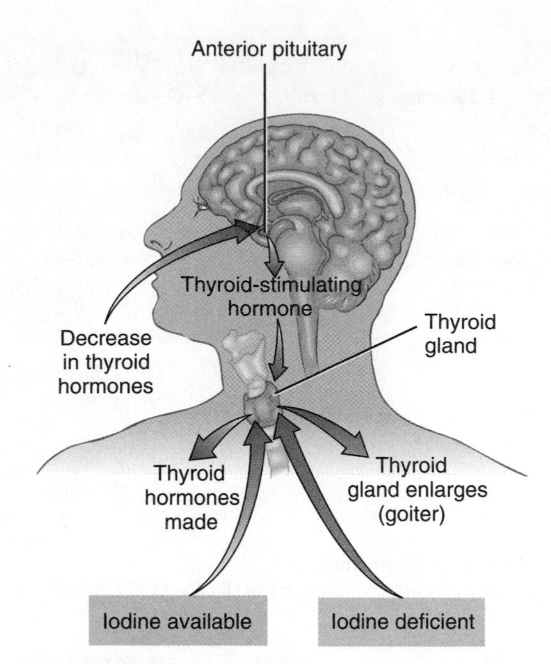

Figure 12.20 Regulating thyroid hormone levels
When thyroid hormone levels drop too low, thyroid-stimulating hormone is released and stimulates the thyroid gland to take up iodine and synthesize more hormones. If iodine is not available (brown arrows), thyroid hormones cannot be made and the stimulation continues, causing the thyroid gland to enlarge.

Recommended Iodine Intake

goiter An enlargement of the thyroid gland caused by a deficiency of iodine.

The RDA for iodine in adult men and women is 150 μg/day. This is based on the amount needed to maintain normal iodine levels in the thyroid gland. Since the iodinization of salt, the intake of iodine in North America has met or exceeded the RDA.

The RDA is higher during pregnancy to account for the iodine that is taken up by the fetus, and during lactation to account for the iodine secreted in milk. The recommended intake for infants is based on the amount obtained from breast milk.

Iodine Deficiency

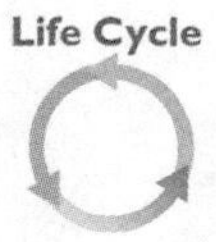

cretinism A condition resulting from poor maternal iodine intake during pregnancy that causes stunted growth and poor mental development in offspring.

goitrogens Substances that interfere with the utilization of iodine or the function of the thyroid gland.

Iodine deficiency reduces the production of thyroid hormones. Metabolic rate slows with insufficient thyroid hormones, causing fatigue and weight gain. The most obvious outward sign of deficiency is an enlarged thyroid gland called a **goiter** (**Figure 12.21**). A goiter forms when reduced thyroid hormone levels cause thyroid-stimulating hormone to be released, stimulating the thyroid gland to make more thyroid hormones. Because iodine is unavailable, the hormones cannot be made and the stimulation continues, causing the thyroid gland to enlarge (see Figure 12.21). In milder cases of goiter, treatment with iodine causes the thyroid gland to return to normal size, but this result is not consistent in more severe cases.

A number of other iodine deficiency disorders occur because of the effect of iodine on growth and development. If iodine is deficient during pregnancy, it increases the risk of stillbirth and spontaneous abortion. Deficiency also can cause a condition called **cretinism** in the offspring, a condition characterized by symptoms such as developmental disability, deaf mutism, and growth failure. Iodine deficiency during childhood and adolescence can also result in goiter and impaired mental function that lowers intellectual capacity.

Goitrogens The risk of iodine deficiency is increased by consuming **goitrogens**, substances in food that interfere with the utilization of iodine or with thyroid function. Goitrogens are found in turnips, rutabaga, cabbage, cassava, and millet. When these foods are boiled, the goitrogen content is reduced because some of these compounds leach into the cooking water. They are primarily a problem in African countries where cassava is a dietary staple. In the Canadian diet, goitrogens are not a problem because they are present in foods that are not consumed in significant quantities in the typical diet.

Iodine Fortification Since it was first used in Switzerland in the 1920s, iodized salt has been the major means of combating iodine deficiency (see Label Literacy: Why is Table Salt "Iodized"?). Because of the fortification of table salt with iodine, cretinism and goiter are now rare in North America, but worldwide, 600 million people have goiter and 1.5 billion people are at risk for iodine deficiency.[35] Dramatic improvements in iodine status are seen in areas where iodization has been in place for more than 5 years. For groups who do not have access to iodized salt or who will not use it, other forms of iodine supplementation, such as injections or oral doses of iodized oil, may be effective for control of iodine deficiency.[36]

Figure 12.21
Iodine deficiency causes goiter, an enlargement of the thyroid gland seen as a swelling in the neck. (John Paul Kay/Peter Arnold, Inc.)

Iodine Toxicity

Acute toxicity can occur with very large doses of iodine. Intakes between 200-500 μg/kg of body weight have caused death in laboratory animals.[27] Chronically high intakes of iodine can cause an enlargement of the thyroid gland that resembles goiter. The UL for adults is 1100 μg/day from all sources.[7] Goiter from excessive iodine can also occur if iodine intake changes drastically. For example, in a population with a marginal intake, a large increase in intake due to supplementation can cause thyroid enlargement even at levels that would not be toxic in a healthy population.

12.8 Chromium (Cr)

Learning Objective

- Explain the relationship between blood glucose levels and chromium.

It has been known since the 1950s that chromium is needed for normal glucose utilization, but only recently have scientists begun to understand the role of chromium in normal insulin function. The popular supplement chromium picolinate is promoted to increase lean body mass.

Chromium in the Diet

Dietary sources of chromium include liver, brewer's yeast, nuts, and whole grains. Milk, vegetables, and fruit are poor sources. Refined carbohydrates such as white breads, pasta, and white rice, are also poor sources because chromium is lost in milling and not added back in the enrichment process. Cooking in stainless steel cookware can increase chromium intake because chromium leaches from the steel into the food.

Chromium in the Body

After absorption, chromium is bound to the iron transport protein transferrin for transport in the blood. Chromium is involved in carbohydrate and lipid metabolism. When carbohydrate is consumed, insulin is released and binds to receptors in cell membranes. This binding triggers the uptake of glucose by cells, an increase in protein and lipid synthesis, and other effects. Chromium is believed to act as part of a small peptide that binds to the insulin receptor after insulin is bound, enhancing its effect[7] (**Figure 12.22**). When chromium is deficient, it takes more insulin to produce the same effect.

Recommended Chromium Intake

Based on the amount of chromium in a balanced diet, an AI has been set at 35 μg/day for men and 25 μg/day for women. The AI is increased during pregnancy and lactation. The AI for older adults is slightly lower because energy intake decreases with age.

Chromium Deficiency

Overt chromium deficiency is not a problem in the Canadian population. Deficiencies have been reported in patients receiving long-term TPN not containing chromium and in malnourished children. Deficiency symptoms include impaired glucose tolerance with diabetes-like symptoms, such as elevated blood glucose levels and increased insulin levels.[37] Chromium deficiency may also cause elevated blood cholesterol and triglyceride levels, but the role of chromium in lipid metabolism is not fully understood.

LABEL LITERACY

Why is Table Salt "Iodized"?

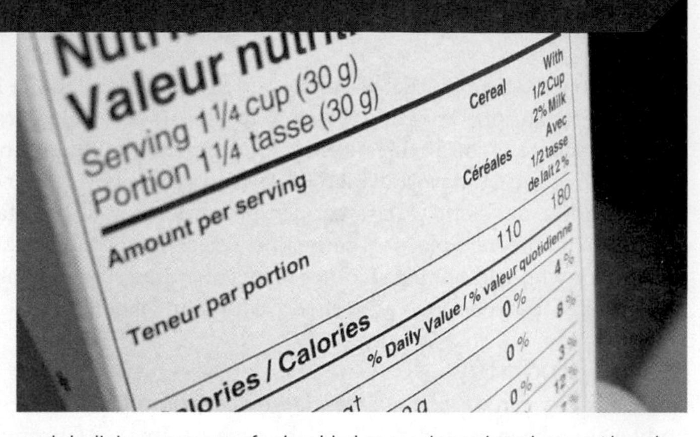

Iodized salt is salt to which the trace element iodine has been added. In Canada, all salt sold for household use, including sea salt, must be iodized; this addition is stated on the label. Why is Canadian salt iodized?

The amount of iodine you consume in your diet depends as much on where the foods are grown as on which foods you choose. Foods from the ocean or produced near it, where the soil is rich in iodine, are better sources of iodine than foods produced inland. This is because the iodine in inland areas has been washed into the oceans by glaciers, snow, rain, and flood waters. The iodine content of plants grown in iodine-deficient soil may be 100 times less than that of plants grown in iodine-rich soil.[1]

Iodine deficiency has been known for centuries in areas where the soil is depleted. In Europe, the presence of iodine deficiency was recorded in classical art, which portrayed even some of the wealthy with goiter or cretinism. Leonardo da Vinci is said to have been more knowledgeable about goiter than medical professors of his time. A century ago, goiter was endemic in the central regions of North America. However, the iodinization of salt, which began in the early twentieth century, has virtually eliminated iodine deficiency in North America, Switzerland, and other European countries.

Why fortify salt? Salt was selected as the vehicle for added iodine because it is a food item consistently consumed by the majority of the population at risk. Also, iodine can be added to salt uniformly, inexpensively, and in a form that is well utilized by the body. It can be added in amounts that eliminate deficiency with typical consumption but not cause toxicity in people consuming larger amounts of iodized salt or individuals meeting iodine needs from other sources.

Since the introduction of iodized salt, iodine intake in Canada has been adequate, and iodine deficiency has been rare. However, one must always be vigilant as nutrient deficiencies can re-emerge. Studies from the United States, for example, have shown a significant reduction in average iodine status.[2] Factors that may have contributed to this include increased consumption of processed foods, which contain non-iodized salt; reduced consumption of eggs, which are rich in iodine; and declining amounts of salt added to meals made at home. Also, the iodine content of milk has decreased due to a reduction in the amount of iodine added to cattle feed, and the baking industry has eliminated some iodine-containing dough conditioners, thus reducing the iodine content of commercially manufactured breads.[2] If you live on the coast or buy food imported from many locations, you still probably get plenty of iodine. However, if you eat little seafood, live inland where the soil is deficient in iodine, and consume primarily foods grown locally, then iodized salt is an important source of iodine.

(© istockphoto.com/John Madden)

[1]Dunn, J. T. Iodine. In *Modern Nutrition in Health and Disease*, 10th ed. M. E. Shils, M. Shike, A. C. Ross, et al. eds. Philadelphia: Lippincott Williams & Wilkins, 2006, pp. 300–311.

[2]Pearce, E. N. National trends in iodine nutrition: Is everyone getting enough? *Thyroid* 17:823–827, 2007.

Chromium Supplements and Toxicity

Chromium supplements, particularly as chromium picolinate, are marketed to reduce body fat and increase lean body tissue (**Figure 12.23**). This appeals to individuals wanting to lose weight as well as to athletes trying to build muscle. Because chromium is needed for insulin action and insulin promotes protein synthesis, it is likely that adequate chromium is necessary to increase lean body mass. However, most recent studies on the effects of chromium picolinate or other chromium supplements in healthy human subjects have found no beneficial effects on muscle strength, body composition, weight loss, or other aspects of health.[38] Chromium supplementation has been shown to have beneficial effects on blood glucose levels in individuals with type 2 diabetes.[39]

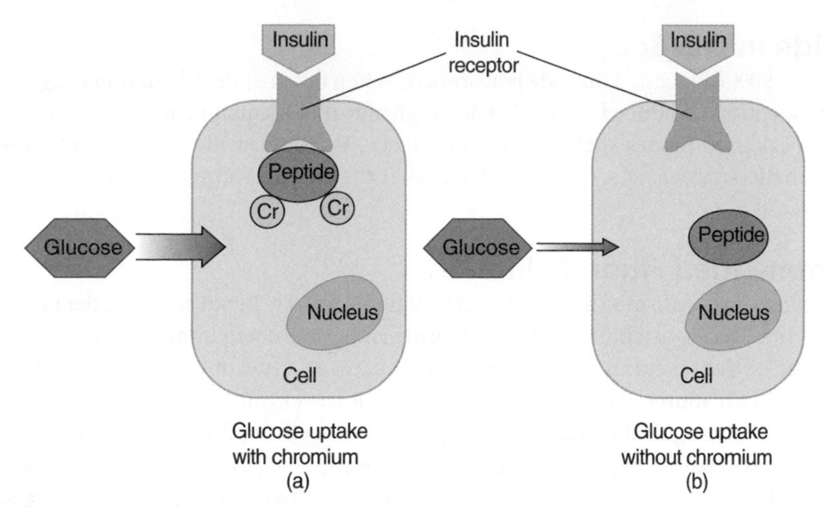

Figure 12.22 Chromium and insulin function
a) When chromium is present, a small peptide inside cells becomes active and enhances the action of insulin by binding to the insulin receptor, which increases glucose uptake. b) When chromium is deficient, the active peptide is not formed and thus cannot bind the insulin receptor. The result is that insulin is less effective and less glucose can enter the cell.

Controlled trials have reported no dietary chromium toxicity in humans.[7] Despite the apparent safety of chromium supplements, a few concerns have been raised. Two cases of renal failure have been associated with chromium picolinate supplements, but both of these individuals were taking other drugs known to cause renal toxicity, so it is unclear whether the effect was due to the chromium supplement.[40] The safety of chromium picolinate has also been questioned because of studies in cell culture that suggest it may cause DNA damage.[41] This effect is specific to the picolinate form of chromium and may be due to the ability of this form to generate DNA-damaging free radicals. Human studies using the standard supplemental doses of chromium picolinate have not detected an increase in DNA damage, but more work is needed to completely rule out any risk.[42] Despite these concerns, the DRI committee concluded that there were insufficient data to establish a UL for chromium.

Figure 12.23 Do chromium supplements help increase lean body mass and decrease body fat? (© istockphoto.com/Milos Markovic)

12.9 Fluoride (F)

Learning Objective

- Discuss the role of fluoride in maintaining dental health.

The importance of fluoride for dental health has been recognized since the 1930s, when an association between the fluoride content of drinking water and the prevalence of dental caries was noted (see Science Applied: From Colorado Brown Stain to Water Fluoridation).

Fluoride in the Diet

Fluoride is present in small amounts in almost all soil, water, plants, and animals. The richest dietary sources of fluoride are fluoridated water, tea, and marine fish consumed with their bones (**Figure 12.24**). Tea contributes significantly to total fluoride intake in countries that consume large amounts of the beverage. Brewed tea contains 1-6 mg/L of fluoride depending on the amount of dry tea used, the brewing time, and the fluoride content of the water.[43] In Canada most of the fluoride in the diet comes from toothpaste and from fluoride added to the water supply— usually 0.7-1.2 mg/L. (Water companies often report fluoride levels in parts per million [ppm], 1 mg/L = 1 ppm.) Because food readily absorbs the fluoride in cooking water, the fluoride content of food can be significantly increased when it is handled and prepared using fluoridated water. Cooking utensils also affect food fluoride content. Foods cooked with Teflon utensils can pick up fluoride from the Teflon, whereas aluminum cookware can decrease fluoride content. Fluoride is absorbed into the body in proportion to its content in the diet.

Figure 12.24 Dietary sources of fluoride include water, tea, fish eaten with bones, and toothpaste. (Luisa Begani)

Fluoride in the Body

About 80%-90% of ingested fluoride is absorbed. When taken with milk or other high-calcium foods, absorption is reduced. Fluoride has a high affinity for calcium and so is usually associated with calcified tissues such as bones and teeth. When fluoride is incorporated into the hydroxyapatite crystals of tooth enamel, it makes the enamel more resistant to the acid that causes decay.

Recommended Fluoride Intake

Epidemiology has confirmed the effectiveness of fluoridated water in reducing dental cavities.[44] The criterion used to establish an AI for fluoride is the estimated intake shown to reduce the occurrence of dental caries maximally without causing unwanted side effects. The AI for fluoride from all sources is set at 0.05 mg/kg/day for everyone 6 months of age and older, because it protects against dental caries with no adverse effects.[43] Thus, for children aged 4-8 years, the AI is set at 1.1 mg/day using a reference weight of 22 kg. For adult men age 19 and older, the AI is 3.8 mg/day based on a weight of 76 kg, for women, it is 3.1 mg/day based on a weight of 61 kg. The AI is not increased during pregnancy or lactation.

Life Cycle

Breast milk is low in fluoride, and ready-made infant formulas are prepared with unfluoridated water. Unless infant formula is prepared at home with fluoridated water, it contains little fluoride. The Canadian Paediatric Society suggests a supplement of 0.25 mg of fluoride per day for children 6 months to 3 years of age, 0.5 mg/day for ages 3-6 years, and 1.0 mg/day for ages 6-16 who are receiving less than 0.3 mg/L of fluoride in the water supply.[45] These supplements are available by prescription for children living in areas with low water fluoride concentrations. Swallowed toothpaste is estimated to contribute about 0.6 mg /day of fluoride in young children.[43]

Fluoride Deficiency

Adequate dietary fluoride is important for bone and dental health. When fluoride is deficient, tooth decay is more common. Because there are few food sources of fluoride and the fluoride content of water is variable, fluoride supplements or water fluoridation is often needed to minimize dental caries. Fluoride has its greatest effect on dental caries prevention early in life, during maximum tooth development up to the age of 13. During this time, it can be incorporated into tooth enamel, making the enamel more acid resistant. It has also been shown to protect teeth in other ways, making it beneficial for adults as well as children.[46] Fluoride in saliva reduces cavities by reducing acid produced by bacteria, inhibiting the dissolution of tooth enamel by acid, and increasing enamel remineralization after acid exposure.[43] Fluoride seems to stimulate new bone formation and has therefore been suggested to strengthen bones in adults with osteoporosis. Slow-release fluoride supplements have been shown to increase bone mass and prevent new fractures. [47]

Fluoride Toxicity

fluorosis A condition caused by chronic overconsumption of fluoride, characterized by black and brown stains and cracking and pitting of the teeth.

Fluoride can cause adverse effects in high doses. In children, fluoride intakes of 2-8 mg/day can cause stained, pitted teeth, a condition called **fluorosis** (**Figure 12.25**). A recent increase in the prevalence of this condition has occurred due to the chronic ingestion of toothpaste containing fluoride.[48] In adults, doses of 20-80 mg/day can result in changes in bone that can be crippling, as well as changes in kidney function and possibly nerve and muscle function. Death has been

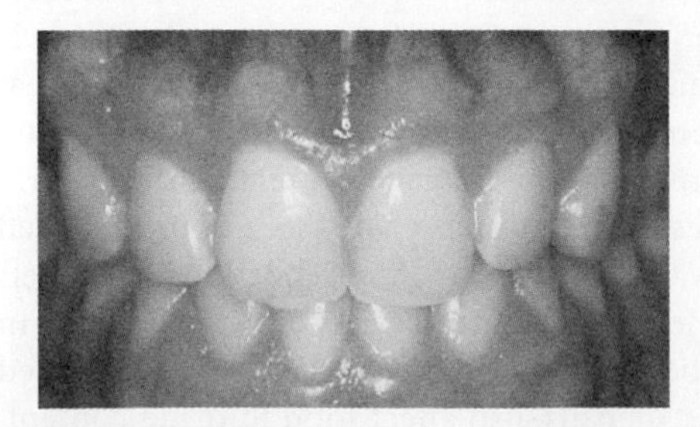

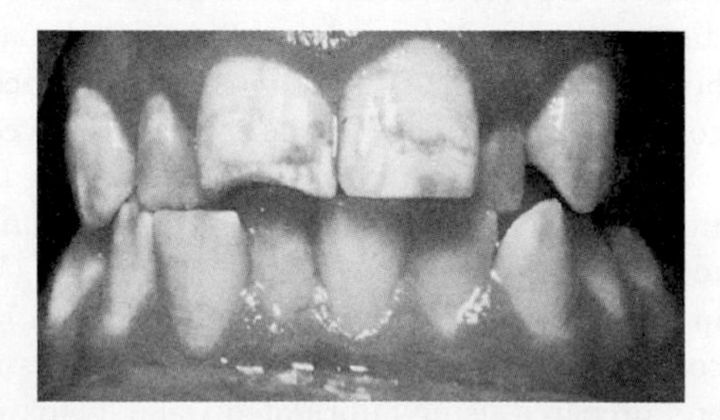

Figure 12.25
Too much dietary fluoride causes staining and pitting of the teeth. These photos compare normal teeth (left) and teeth showing enamel fluorosis (right). [©E.H. Gill/Custom Medical Stock Photo, Inc. (right) ©NIH/Custom Medical Stock Photo, Inc.]

SCIENCE APPLIED

CANADIAN CONTENT

From Colorado Brown Stain to Water Fluoridation

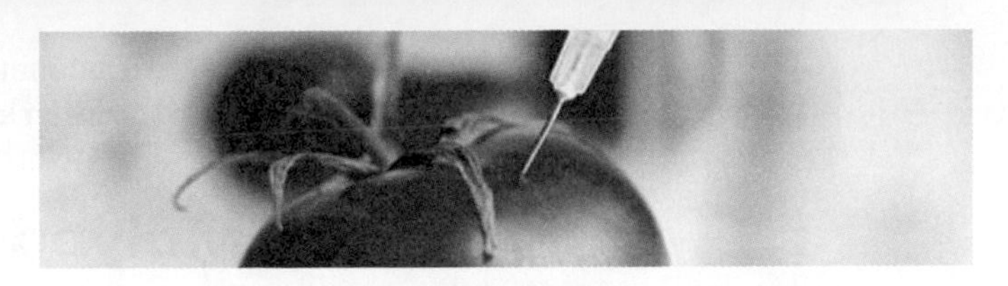

When Dr. Fredrick McKay set up his dental practice in Colorado Springs in 1901, he noticed that many of his patients had stained or mottled tooth enamel. McKay noted that those with stained teeth, a condition called "Colorado brown stain," seemed to be less susceptible to tooth decay.[1] At the time, dental caries were extremely prevalent, there was no known way to prevent the disease, and the most common way to treat it was to extract the affected teeth.

McKay believed that Colorado brown stain was due to something in the water supply. In 1930, he sent water samples to a chemist for analysis. Using a new methodology called spectrographic analysis, the chemist identified high levels of fluoride in McKay's samples. This finding led to the establishment of the Dental Hygiene Unit at the National Institutes of Health headed by Dr. H. Trendley Dean. His task was to investigate the association between fluoride intake and mottled enamel, which he termed "fluorosis."

Dean conducted epidemiological surveys to establish the prevalence of fluorosis across the country. He noted a strong inverse relationship between the prevalence of fluorosis and the prevalence of dental caries among children.[2] In other words, children with fluorosis had fewer dental caries. Further studies revealed that the protective effect of fluoride on dental caries was seen at water fluoride levels of 1 ppm, a level low enough to cause little fluorosis (see graph).[3] Thus, work designed to identify the harm caused by too much fluoride had discovered the benefits of enough fluoride.

The first intervention trial to test the effectiveness of community water fluoridation began in 1945 and included 3 pairs of American cities in Michigan, New York, and Illinois.

In Canada, Dr, W. L. Hutton of the Brant County Health unit organized what became known as the Brantford-Sarnia-Stratford fluoridation caries study in Ontario. In this study, Brantford, Ontario, became the first city in Canada to artificially fluoridate its water. The development of dental cavities was compared in Brantford, in Stratford, which had naturally fluoridated water, and Sarnia, which had no fluoride. Researchers found that only 1% of Sarnia children were cavity-free, while 16% of children in Brantford and Stratford were cavity-free. Half of Sarnia's children had decay-free permanent incisors, while the number rose to 87% and 89% in Brantford and Stratford, respectively.[4] By the mid-1960s, this study, along with others, was used to make the first recommendation for an optimal range of fluoride concentration (0.7–1.2 ppm) in the water supply, with the lower range suggested for warmer climates, where more water is consumed, and the higher range for colder climates.[3]

Some people still believe that water fluoridation represents a public health hazard and increases the risk of cancer, but these beliefs are not supported by scientific fact. Based on epidemiological data and available evidence related to the adverse effects of fluoride, the small amounts consumed in drinking water do not pose a health risk.

Today, 43% of the Canadian population receives fluoridated water, but there are substantial regional differences, ranging from a high of 75% of Albertans having access to fluoridated water to no access in the Yukon (see "fluoridated water by region" graph). Fluoride intake has also increased due to the widespread use of fluoride toothpaste and fluoridated water in foods and beverages that are distributed in nonfluoridated areas. Although dental caries remains a public health problem, increased fluoride intake, combined with advances in dental care, have dramatically improved the dental health of the public.

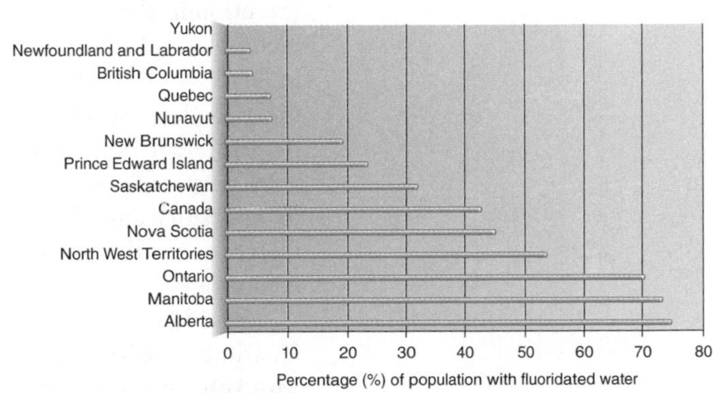

Fluoridated water by region
Source: Health Canada. Fluoride in drinking water. Available online at http://www.hc-sc.gc.ca/ewh-semt/consult/_2009/fluoride-fluorure/b-table-b-tableau-eng.php#tab1. Accessed May 4, 2011.

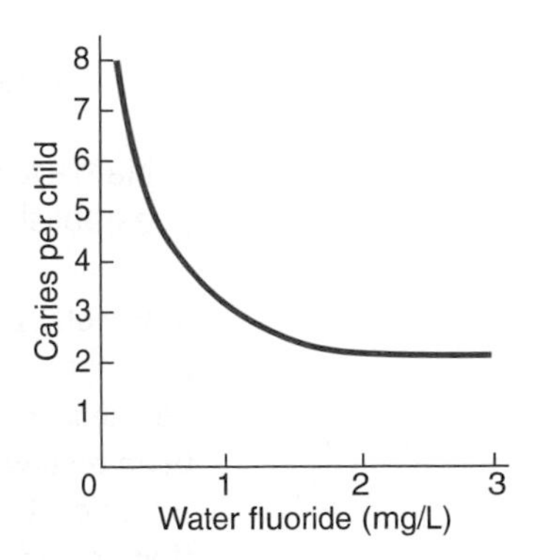

This graph illustrates the effect of water fluoridation level on dental caries in children 12-14 years of age.

[1]McKay, F. S. Relation of mottled enamel to caries. *J. Am. Dent. Assoc.* 15:1429-1437, 1928.

[2]Dean, H. T. Endemic fluorosis and its relation to dental caries. *Public Health Rep.* 53:1443-1452, 1938.

[3]Achievements in public health, 1900–1999: Fluoridation of drinking water to prevent dental caries. *Morb. Mortal. Wkly. Rep.* 48:933–940, 1999

[4]Brown, H. K. The Brantford-Sarnia-Stratford fluoridation caries study-1961 report. *Canadian Journal of Public Health* 53:401, 1962.

reported with an intake of 5-10 g/day. Due to concern over excess fluoride intake, labels of toothpaste intended for children advise parental supervision to prevent swallowing and suggest that only a pea-sized amount of toothpaste be used. The UL for fluoride is set at 0.1 mg/kg/day for infants and children less than 9 years of age and at 10 mg/day for people ages 9-70.[43]

12.10 Molybdenum (Mo)

Learning Objective

- Discuss how the molybdenum content of the soil affects the content of food.

Like many other trace elements, molybdenum is needed to activate enzymes. The molybdenum content of food varies with the molybdenum content of the soil where the food is produced. The most reliable sources include milk, milk products, organ meats, breads, cereals, and legumes.

Molybdenum is readily absorbed from foods. The amount in the body is regulated by excretion in the urine and bile. Molybdenum is a cofactor for enzymes necessary for the metabolism of sulfur-containing amino acids and nitrogen-containing compounds present in DNA and RNA, the production of uric acid (a nitrogen-containing waste product), and the oxidation and detoxification of various other compounds.

Although molybdenum deficiency in humans has been reported as a result of long-term TPN, a naturally occurring deficiency has never been reported. Deficiency has been induced in laboratory animals by feeding them high doses of the element tungsten, which inhibits molybdenum absorption. The resulting deficiency caused growth retardation, decreased food intake, impaired reproduction, and decreased life expectancy.

Based on the results of molybdenum balance studies, an RDA has been set at 45 μg/day for adults. The RDA is increased for pregnancy and lactation. An AI has been set for infants based on the amount of molybdenum in breast milk.

There are few data on adverse effects of high intakes of molybdenum in humans. A UL of 2000 μg/day was set based on impaired growth and reproduction in animals.[7]

12.11 Other Trace Elements

Learning Objective

- Name four essential trace elements for which no RDA or AI has been established.

Many other trace elements are found in minute amounts in the human body. Some of these may be essential for human health, and others, such as lead (see Section 15.2) may be present only as a result of environmental exposure. Arsenic, boron, nickel, silicon, and vanadium have been reviewed by the DRI committee and found to have a significant role in human health.[7] Arsenic is better known as a poison than an essential nutrient, but the organic forms of arsenic that occur in foods are nontoxic. It is hypothesized that arsenic is involved in the conversion of the amino acid methionine into compounds that affect heart function and cell growth. Arsenic deficiency has been correlated with nervous system disorders, blood vessel diseases, and cancer. Boron may be involved in vitamin D and estrogen metabolism. Nickel is thought to function in enzymes involved in the metabolism of certain fatty acids and amino acids, and it may play a role in folate metabolism. Silicon, the primary constituent of sand, is involved in the synthesis of collagen and the calcification of bone. Vanadium has been shown to have an insulin-like action and to stimulate cell proliferation and differentiation. There are insufficient data to establish an AI or an RDA for any of these elements, but ULs have been set for boron, nickel, and vanadium. Other trace elements that play a physiological role include aluminum, bromine, cadmium, germanium, lead, lithium, rubidium, and tin. The specific functions of these have not been defined, and they have not been evaluated by the DRI committee. All the minerals, both those known to be essential and those that are still being assessed for their role in human health, can be obtained by choosing a variety of foods from each of the food groups of Canada's Food Guide.

CASE STUDY OUTCOME

Nissi's family and many others in the village are now using iodized salt. Nissi has grown to be an energetic teenager, with no trace of goiter. Iodine deficiency was common in Nissi's village, because the people grow the food they need and cannot afford expensive imports. The iodine content of food, like the content of many other trace elements, depends on where that food is grown or produced, and the food grown near Nissi's village is low in iodine. The Canadian diet, on the other hand, made up of foods from many locations, usually contains adequate amounts of all of the trace minerals. About a year after Nissi was diagnosed with goiter, the Indian government began promoting the use of salt fortified with iodine and banned the production and sale of non-iodized salt. Some families in Nissi's village are still resistant to using the new salt, but information and education programs are spreading the message that iodized salt is safe and affordable.

UNICEF also has established a program that allows children to bring samples of their salt to school to test its iodine content.

APPLICATIONS

Personal Nutrition

1. **Are you at risk for iron deficiency?**

iProfile

   a. Use iProfile and the 3-day food intake record you kept in Chapter 2 to calculate your average daily intake of iron.

   b. How does your iron intake compare with the recommendation for someone of your age, gender, and life stage?

   c. If your intake is low, suggest modifications to your diet that would increase your iron intake enough to meet the RDA for someone your age and gender.

   d. If your diet already meets the recommendations for iron, make a list of foods you like that are good sources of iron.

   e. Identify the major food sources of iron in your diet and indicate whether they contribute heme iron.

2. **Do you consume enough zinc?**

   a. Use iProfile and your food record to calculate your zinc intake.

   b. If you eliminated meat from your diet, would you meet the RDA for zinc?

   c. What foods could you substitute for meats that are good sources of zinc?

General Nutrition Issues

1. **What promises are made about trace element supplements?**

   a. Using the Internet, search for information on a supplement discussed in this chapter—for instance, zinc lozenges or chromium picolinate.

   b. How does the information compare to the discussion in the text?

   c. Who provided the information? Does it promote the sale of a product?

   d. Is the information supported by scientific studies?

2. **Imagine the diet in a developing country is deficient in iodine. Most of the food is grown and prepared locally. To solve the problem, the government imports iodized salt. When iodine deficiency continues to be a problem, a study of the local diet finds it to be very low in added salt. Not enough of the imported iodized salt was used to have an effect on the iodine status of the population. Use the following information to suggest a way that iodine intake might be increased in this population:**

   The following foods are typically consumed:

   Corn and corn tortillas

Dried beans

White rice

Fresh tomatoes

Fresh vegetables—greens and squash

Pork

Fresh fruits

A typical breakfast is cornmeal and fruit; lunch is a hot meal, usually with some kind of vegetable soup served with tortillas; and dinner is meat, beans, rice, tortillas, and fruit.

3. **A researcher asks his new technician to prepare a diet for his laboratory animals. The technician is interrupted several times while mixing the diet and is unfamiliar with the scale he is using to weigh the diet ingredients. After the diet is fed to the animals for several months, they begin to show signs of anemia.**

An error in diet preparation is suspected. The diet ingredients are:

Starch	Potassium
Sucrose	Magnesium
Casein (protein)	Chloride
Corn oil	Zinc
Mixed plant fibres	Iron
Vitamin A	Iodine
Vitamin D	Selenium
Vitamin E	Copper
Vitamin K	Manganese
B vitamin mix	Chromium
Calcium	Molybdenum
Sodium	

a. What trace element deficiencies can cause anemia?

b. What trace element excess can cause anemia?

c. What vitamin deficiencies can cause anemia?

d. What dietary components can contribute to anemia by interfering with the absorption of trace elements?

SUMMARY

12.1 Trace Elements in the Canadian Diet

- Trace elements are needed in minute amounts but deficiencies can be as devastating as deficiencies of nutrients required in larger amounts.

12.2 Iron (Fe)

- Iron is found in both plant and animal foods. Heme iron, the easily absorbable form, is found in animal products. Nonheme iron, which is not well absorbed, comes from both animal and plant sources.
- The amount of iron that is absorbed from the diet depends on the type of iron and the presence of other dietary components. The absorption of nonheme iron can be increased by consuming it with meat or acidic foods; its absorption is decreased by consuming it with foods containing phytates, oxalates, and tannins. If iron stores are low, more iron is bound to transferrin and transported from the intestinal mucosa to body cells. When body stores are adequate, less iron is transported from the mucosa and more is bound to ferritin and lost when mucosal cells die.
- Iron functions as part of hemoglobin, which transports oxygen in the blood, and myoglobin, which enhances the amount of oxygen available during muscle contraction. Iron is also a component of proteins involved in ATP production and is needed for activity of the antioxidant enzyme catalase.
- The RDA for iron for women age 19-50 is 18 mg/day, more than double the RDA of 8 mg/day for adult men and post-menopausal women.
- When iron is deficient, adequate hemoglobin cannot be made, resulting in iron deficiency anemia—the most common nutritional deficiency worldwide.
- Iron can be toxic. Ingestion of a single large dose can be fatal. The accumulation of iron in the body over time causes heart and liver damage and contributes to diabetes and cancer. The most common cause of chronic iron overload is hemochromatosis, a genetic disorder in which too much iron is absorbed.

12.3 Zinc (Zn)

- Good sources of zinc include red meats, eggs, dairy products, and whole grains.
- Zinc absorption is regulated by zinc transport proteins that determine how much zinc is in the mucosal cell and by metallothionein, a protein that binds zinc in the mucosal cell. When zinc intake is high, more metallothionein is synthesized and zinc absorption is limited.
- Zinc is needed for the activity of many enzymes, including a form of the antioxidant enzyme superoxide dismutase. Many of the functions of zinc are related to its role in gene expression. Zinc is needed for tissue growth and repair, development of sex organs and bone, proper immune function, storage and release of insulin, mobilization of vitamin A from the liver, and stabilization of cell membranes.
- The RDA for zinc is 11 mg/day for adult men and 8 mg/day for adult women.
- Zinc deficiency results in poor growth, delayed sexual maturation, skin changes, hair loss, skeletal abnormalities, and depressed immunity.
- Since copper also binds metallothionein, an excess of zinc can stimulate its synthesis and trap copper in the mucosal cells, causing a copper deficiency.

12.4 Copper (Cu)

- Good sources of copper in the diet include organ meats, seafood, nuts, and seeds.
- The absorption of copper is affected by the presence of other minerals in the diet .The zinc content of the diet can have a major impact on copper absorption.

- Copper functions in a number of important proteins that affect iron and lipid metabolism, synthesis of connective tissue, and antioxidant protection. Copper is transported in the blood bound to proteins such as ceruloplasmin.
- The RDA for copper for adults is 900 μg/day.
- A copper deficiency can cause anemia and connective tissue abnormalities. Copper toxicity from dietary sources is extremely rare.

12.5 Manganese (Mn)

- Good dietary sources of manganese include whole grains and nuts. The AI is 2.3 mg/day for men and 1.8 mg/day for women.
- Manganese is necessary for the activity of some enzymes, including a form of the antioxidant enzyme superoxide dismutase. Manganese is involved in amino acid, carbohydrate, and lipid metabolism.

12.6 Selenium (Se)

- Dietary sources of selenium include seafood, eggs, organ meats, and plant foods grown in selenium-rich soils.
- Selenium protects against oxidative damage as an essential part of the enzyme glutathione peroxidase. Glutathione peroxidase destroys peroxides before they can form free radicals. Adequate dietary selenium reduces the need for vitamin E.
- The RDA for selenium for adults is 55 μg/day.
- Severe selenium deficiency is rare except in regions with very low soil selenium content and limited diets. In China, selenium deficiency contributes to the development of a heart condition known as Keshan disease. Low selenium intake has been linked to increased cancer risk.
- Very high selenium intake (5 mg/day) causes fingernail changes and hair loss.
- Selenium supplements are marketed with claims that they will protect against environmental pollutants, prevent cancer and heart disease, slow the aging process, and improve immune function. Supplements that increase intake above the RDA will not provide additional benefits.

12.7 Iodine (I)

- The best sources of iodine in the diet are seafood, foods grown near the sea, and iodized salt.
- Iodine is an essential component of thyroid hormones, which promote protein synthesis and regulate basal metabolic rate, growth, and development.
- The RDA for iodine in adult men and women is 150 μg/day.
- When iodine is deficient, continued release of thyroid-stimulating hormone causes the thyroid gland to enlarge, forming a goiter. Iodine deficiency during pregnancy causes a condition in the offspring known as cretinism, which is characterized by growth failure and developmental disability. Iodine deficiency during childhood and adolescence can impair mental function. Although iodine deficiency is a world health problem, it has been virtually eliminated in North America through the use of iodized salt.
- Acute toxicity can occur with very large doses of iodine. Chronically high intakes of iodine can cause an enlargement of the thyroid gland that resembles goiter.

12.8 Chromium (Cr)

- Chromium is found in liver, brewer's yeast, nuts, and whole grains.
- Chromium is needed for normal insulin action and glucose utilization.
- Recommended chromium intake is 35 μg/day for men and 25 μg/day for women.
- Overt chromium deficiency is not a problem in the Canadian population.
- Chromium supplements are marketed to control blood sugar and increase lean body mass although little scientific evidence supports these functions. Controlled trials have reported no dietary chromium toxicity in humans.

12.9 Fluoride (F)

- Most of the fluoride in the diet in Canada comes from fluoridated drinking water and toothpaste.
- Fluoride is necessary for the maintenance of bones and teeth. Adequate dietary fluoride helps prevent dental caries.
- The recommended fluoride intake from all sources is set at 0.05 mg/kg/day for everyone 6 months of age and older, because it protects against dental caries with no adverse effects.

12.10 Molybdenum (Mo)

- Molybdenum is a cofactor for enzymes involved in the metabolism of the amino acids methionine and cysteine and nitrogen-containing compounds such as DNA and RNA.

12.11 Other Trace Elements

- There is evidence that boron, arsenic, nickel, silicon, and vanadium may be essential in humans as well as animals. These elements may be necessary in small amounts but can be toxic if consumed in excess.

REVIEW QUESTIONS

1. What are the functions of iron in the body?
2. Why does iron deficiency cause red blood cells to be small and pale?
3. List 3 life-stage groups at risk for iron deficiency anemia and explain why they are at risk.
4. List several good sources of iron in the diet and indicate if they contain heme or nonheme iron.
5. Discuss 3 factors that affect iron absorption.
6. Explain the roles of ferritin, transferrin, and transferrin receptors in regulating the amount of iron in the body.
7. What is hemochromatosis?
8. How does zinc affect the synthesis of proteins?
9. How does a high zinc intake reduce the amount of zinc that enters the blood?
10. Why does excess zinc cause a deficiency of copper?
11. Explain why a deficiency of copper can contribute to anemia.
12. What is the role of selenium in the body?
13. Why does selenium decrease the need for vitamin E?
14. What is a goiter and why does iodine deficiency cause it to form?
15. What is the role of chromium in the body?
16. What do zinc, copper, and manganese have in common?
17. How does fluoride function in dental health?

REFERENCES

1. Sharp, P., and Srai, S. K. Molecular mechanisms involved in intestinal iron absorption. *World J. Gastoenterol.* 13:4716–4724, 2007.
2. Teucher, B., Olivares, M., and Cori, H. Enhancers of iron absorption: Ascorbic acid and other organic acids. *Intl. J. Vit. Nutr. Res.* 74:403–419, 2004.
3. Zijp, I. M., Korver, O., and Tijburg, L. B. Effect of tea and other dietary factors on iron absorption. *Crit. Rev. Food Sci. Nutr.* 40:371–398, 2000.
4. Hurrell, R. F., Lynch, S., Bothwell, T. et al. SUSTAIN Task Force. Enhancing the absorption of fortification iron. A SUSTAIN Task Force report. *Intl. J. Vit. Nutr. Res.* 74:387–401, 2004.
5. Grinder-Pedersen, L., Bukhave, K., Jensen, M. et al. Calcium from milk or calcium-fortified foods does not inhibit nonheme-iron absorption from a whole diet consumed over a 4-d period. *Am. J. Clin. Nutr.* 80:404–409, 2004.
6. Anderson, G. J., Frazer, D. M., McKie, A. T., et al. The ceruloplasmin homolog hephaestin and the control of intestinal iron absorption. *Blood Cells Mol. Dis.* 29:367–375, 2002.
7. Food and Nutrition Board, Institute of Medicine. Dietary Reference Intakes: Vitamin A, Vitamin K, Arsenic, Boron, Chromium, Copper, Iodine, Iron, Manganese, Molybdenum, Nickel, Silicon, Vanadium, and Zinc. Washington, DC: National Academies Press, 2001.
8. O'Connor, N. R. Infant formula. *Am. Fam. Physician* 79:565–570, 2009.
9. World Health Organization. Micronutrient Deficiencies, Iron Deficiency Anemia: The Challenge. Available online at www.who.int/nutrition/topic/ida/en/. Accessed Jan 21, 2009.
10. Verrall, T. and Gray-Donald, K. Impact of a food-based approach to improve iron nutrition of at-risk infants in northern Canada. *Preventative Medicine* 40:896-903, 2005.
11. Guidance for industry: Iron-containing supplements small entity compliance guide, October, 2003. Available online at www.fda.gov/Food/GuidanceCompliance RegulatoryInformation/GuidanceDocuments/Dietary Supplements/ucm073014.htm. Accessed September 8, 2009.
12. Siah, C. W., Trinder, D., and Olynyk, J. K. Iron overload. *Clin. Chem. Acta.* 358:24–36, 2005.
13. Rajpathak, S. N., Wylie-Rosett, J., Gunter, M. J. et al. Biomarkers of body iron stores and risk of developing type 2 diabetes. *Diabetes Obes. Metab.* 11:472–479, 2009.
14. Zegrean, M. Association of body iron stores with development of cardiovascular disease in the adult population: A systematic review of the literature. *Can. J. Cardiovasc. Nurs.* 19:26–32, 2009.
15. Fowler, C. F. Hereditary hemochromatosis: Pathophysiology, diagnosis, and management. *Crit. Care Nurs. Clin. N. Am.* 20:191–201, 2008.
16. Prasad, A. S. Discovery of human zinc deficiency and studies in an experimental human model. *Am. J. Clin. Nutr.* 53:403–412, 1991.
17. Liuzzi, J. P., and Cousins, R. J. Mammalian zinc transporters. *Annu. Rev. Nutr.* 24:151–172, 2004.
18. Tapiero, H., and Tew, K. D. Trace elements in human physiology and pathology: Zinc and metallothioneins. *Biomed. Pharmacother.* 57:399–411, 2003.
19. Black, R. E. Zinc deficiency, infectious disease, and mortality in the developing world. *J. Nutr.* 133:1485S–1489S, 2003.
20. Haase, H., Mocchegiani, E., and Rink, L. Correlation between zinc status and immune function in the elderly. *Biogerontology* 7:421–428, 2006.
21. Hulisz, D. Efficacy of zinc against common cold viruses: An overview. *J. Am. Pharm. Assoc.* 44:594–603, 2004.
22. Caruso, T. J., Prober, C. G., and Gwaltney, J. M., Jr. Treatment of naturally acquired common colds with zinc: a structured review. *Clin. Infect. Dis.* 45:569–574, 2007.
23. Prasad, A. S., Beck, F. W., Bao, B. et al. Duration and severity of symptoms and levels of plasma interleukin-1 receptor antagonist, soluble tumor necrosis factor receptor, and adhesion molecules in patients with common cold treated with zinc acetate. *J. Infect. Dis.* 197:795–802, 2008.
24. National Institute of Neurological Disorders and Stroke. NINDS Wilson's Disease Information Page. Available online at www.ninds.nih.gov/disorders/wilsons/wilsons. html/. Accessed September 8, 2009.
25. Mills, E. S. The treatment of idiopathic (hypochromic) anemia with iron and copper. *Can. Med. Assoc. J.* 22:175–178, 1930.
26. Failla, M. L. Trace elements and host defense: Recent advances and continuing challenges. *J. Nutr.* 133(5 suppl 1):1443S–1447S, 2003.
27. National Research Council, Food and Nutrition Board. *Recommended Dietary Allowances*, 10th ed. Washington, DC: National Academies Press, 1989.
28. Arthur, D. Selenium content of some feed ingredients available in Canada. *Can. J. Anim. Sci.* 51:(1) 71-74,1971
29. Thompson, J. N., Erdody, P., and Smith, D. C. Selenium content of food consumed by Canadians. *J. Nutr.* 105(3):274-277, 1975.
30. Beck, M. A., Levander, O. A., and Handy, J. Selenium deficiency and viral infection. *J. Nutr.* 133(5 suppl 1): 1463S–1467S, 2003.
31. Clark, L. C., Combs, G. F. Jr., Turnbull, B. W. et al. Effect of selenium supplementation for cancer prevention in patients with carcinoma of the skin. *JAMA* 276:1957–1968, 1996.
32. Duffield-Lillio, A. J., Slate, E. H., Reid, M.E. et al. Selenium supplentation and secondary prevention of nonmelanoma skin cancer in a randomized trial. *J. Natl. Cancer Inst.* 95:1477–1481, 2003.
33. Ledesma, M. C., Jung-Hynes, B., Schmit, T. L. et al. Selenium and vitamin E for prostate cancer: post-SELECT (Selenium and Vitamin E Cancer Prevention Trial) status. *Mol. Med.* 17(1-2):134-143, 2011.
34. Canadian Food Inspection Agency. Decisions: Salt. Available online at http://www.inspection.gc.ca/english/fssa/labeti/decisions/salsele.shtml. Accessed Sept 25, 2010.
35. World Health Organization. Micronutrient Deficiencies: Iodine Deficiency Disorders. Available online at www.who.int/nutrition/topics/idd/en/. Accessed April 8, 2009.
36. World Health Organization. *Micronutrient Deficiencies: Iodine Deficiency Disorders*. Available online at www.who.int/ nutrition/topics/idd/en/index.html. Accessed May 20, 2008.
37. Vincent, J. B. Recent advances in the nutritional biochemistry of trivalent chromium. *Proc. Nutr. Soc.* 63:41–47, 2004.
38. Di Luigi, L. Supplements and the endocrine system in athletes. *Clin. Sports Med.* 27:131–151, 2008.
39. Balk, E. M., Tatsioni, A., Lichtenstein, A. H. et al. Effect of chromium supplementation on glucose metabolism and lipids: A systematic review of randomized controlled trials. *Diabetes Care* 30:2154–2163, 2007.
40. Jeejeebhoy, K. N. The role of chromium in nutrition and therapeutics and as a potential toxin. *Nutr. Rev.* 57:329–335, 1999.
41. Stearns, D. M., Wise, J. P. Sr., Patierno, S. R., et al. Chromium (III) picolinate produces chromosome damage in Chinese hamster ovary cells. *FASEB J.* 9:1643–1648, 1995.
42. Vincent, J. B. The potential value and toxicity of chromium picolinate as a nutritional supplement, weight loss agent and muscle development agent. *Sports Med.* 33:213–230, 2003.

43. Institute of Medicine, Food and Nutrition Board. *Dietary Reference Intakes for Calcium, Phosphorus, Magnesium, Vitamin D, and Fluoride*. Washington, DC: National Academies Press, 1997.

44. Parnell, C., Whelton, H., and O'Mullane, D. Water fluoridation. *Eur. Arch. Paediatr. Dent.* 10:141–148, 2009.

45. Canadian Paediatric Society Position Statement: The use of fluoride in infants and children. *Paediatrics and Child Health* 7(8):569-72, 2002.

46. American Dental Association. Fluoridation facts. Available online at: www.ada.org/public/topics/fluoride/facts/ fluoridation_facts.pdf. Accessed October 11, 2009.

47. Rubin, C. D., Pak, C.Y., Adams-Huet, B. et al. Sustained-release sodium fluoride in the treatment of the elderly with established osteoporosis. *Arch. Intern. Med.* 161:2325–2333, 2001.

48. Beltrán-Aguilar E. D., Griffin S. O., and Lockwood S. A. Prevalence and trends in enamel fluorosis in the United States from the 1930s to the 1980s *J. Am. Dent. Assoc.* 133:157–165, 2002.

Circa/Imagezoo/Age Fotostock

FOCUS ON

Integrating Nutrient Function in the Body

Focus Outline

In chapters 4–12, the functions of many nutrients have been discussed, usually by describing the multiple roles of each nutrient. In this feature, the function of the nutrients will be reviewed, but from a different perspective. The focus will be on describing how multiple nutrients function together, in an integrated way, to maintain human health at the cellular level and at the level of several important organ systems.

F5.1 Nutrient Function at the Cellular Level

Learning objectives

- Describe how nutrients influence gene expression.
- Describe the roles of nutrients in energy metabolism.
- Describe which nutrients are most important in protecting the cell from oxidative stress.
- List the nutrients important in fluid balance.

In this first section, we will focus on the role of various nutrients at the cellular level, including a discussion of how nutrients influence gene expression, how they function in energy metabolism, protect the cell from oxidative stress, and maintain fluid balance. The functions described here take place in the majority of cells in the body.

Nutrients, DNA, and Gene Expression

Almost all cells contain a nucleus; within the nucleus is DNA **(Figure F5.1).** Folate, vitamin B_{12}, and phosphorus are 3 nutrients that have roles in the biosynthesis of DNA. Folate, because of its role of single carbon transfer, participates in the synthesis of the bases that make up DNA (see Section 8.8), while phosphorus, in the form of phosphate, is a constituent of the DNA backbone **(Figure F5.2).** Vitamin B_{12} is also required to ensure that folate is available in the proper chemical form to participate in DNA synthesis (see Section 8.9; Figure 8.31).

Nutrients influence gene expression in several ways. In addition to its role in DNA synthesis, folate participates in reactions that result in DNA methylation. This alters which proteins are synthesized, as genes in the methylated regions of DNA are not expressed (see Section 8.8; Figure 8.25).

Vitamin A, vitamin D, glucose, and fatty acids influence gene expression because they, or, more precisely, chemical compounds derived from them, act by binding to a receptor protein,

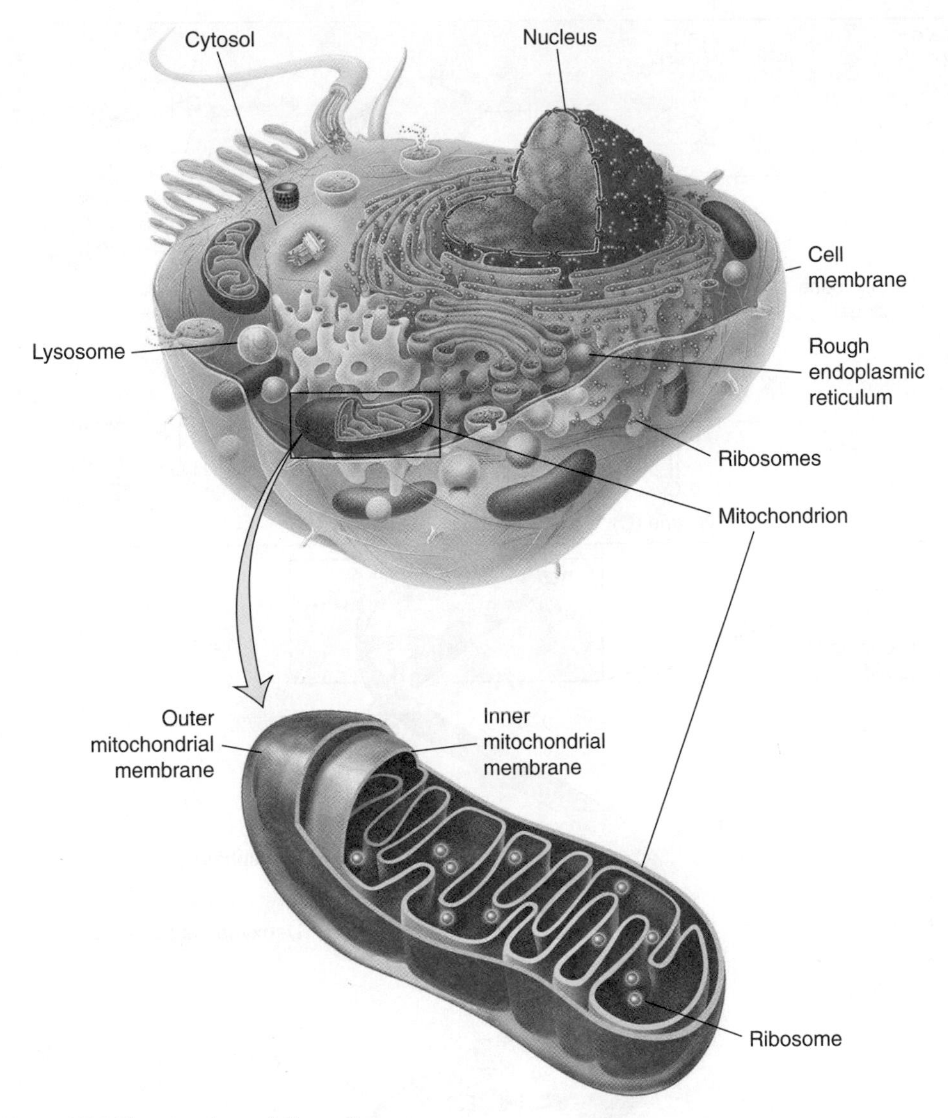

Figure F5.1 The structure of the cell.
All human cells are surrounded by a cell membrane and most contain a nucleus, mitochondria, lysosomes, endoplasmic reticulum, and ribosomes in their cytosol.

which in turn binds to DNA (See sections 9.2 and 9.3). When the receptor binds to DNA, a specific gene is expressed. This means that mRNA forms and a specific protein is synthesized **(Figure F5.3)**. If the nutrient is absent, then the nutrient-receptor complex does not form and protein synthesis does not occur. Iodine similarly influences gene expression, as a component of the thyroid hormones, which like the derivatives of vitamin A, vitamin D, glucose, and fatty acids, bind to receptor proteins (see Section 12.7 , Figure 12.19).

Zinc also plays a role in gene expression. Receptor proteins interact with zinc to form zinc fingers, which creates a structure that allows these proteins to attach to DNA (Step 2, Figure F5.3; see also Section 12.3; Figure 12.10).

Finally, the ultimate outcome of gene expression is protein synthesis and this depends on an adequate supply of essential amino acids (see Section 6.2).

Nutrients and Energy Metabolism

Many nutrients play a role in the metabolic pathways that result in the oxidation of carbohydrates, fats, and protein, and the generation of ATP. Iodine, as a constituent of thyroid hormones, influences the expression of genes involved in the regulation of basal metabolism, which is the amount of energy consumed at rest (see Section 12.7; Figure 12.19). In the pancreas, high levels of blood glucose enhance the binding of a derivative of glucose to a nuclear receptor (see Figure F5.3), which increases the expression of the insulin gene; insulin in turn promotes the uptake of glucose by cells, where it is used as a source of energy. Chromium plays a role

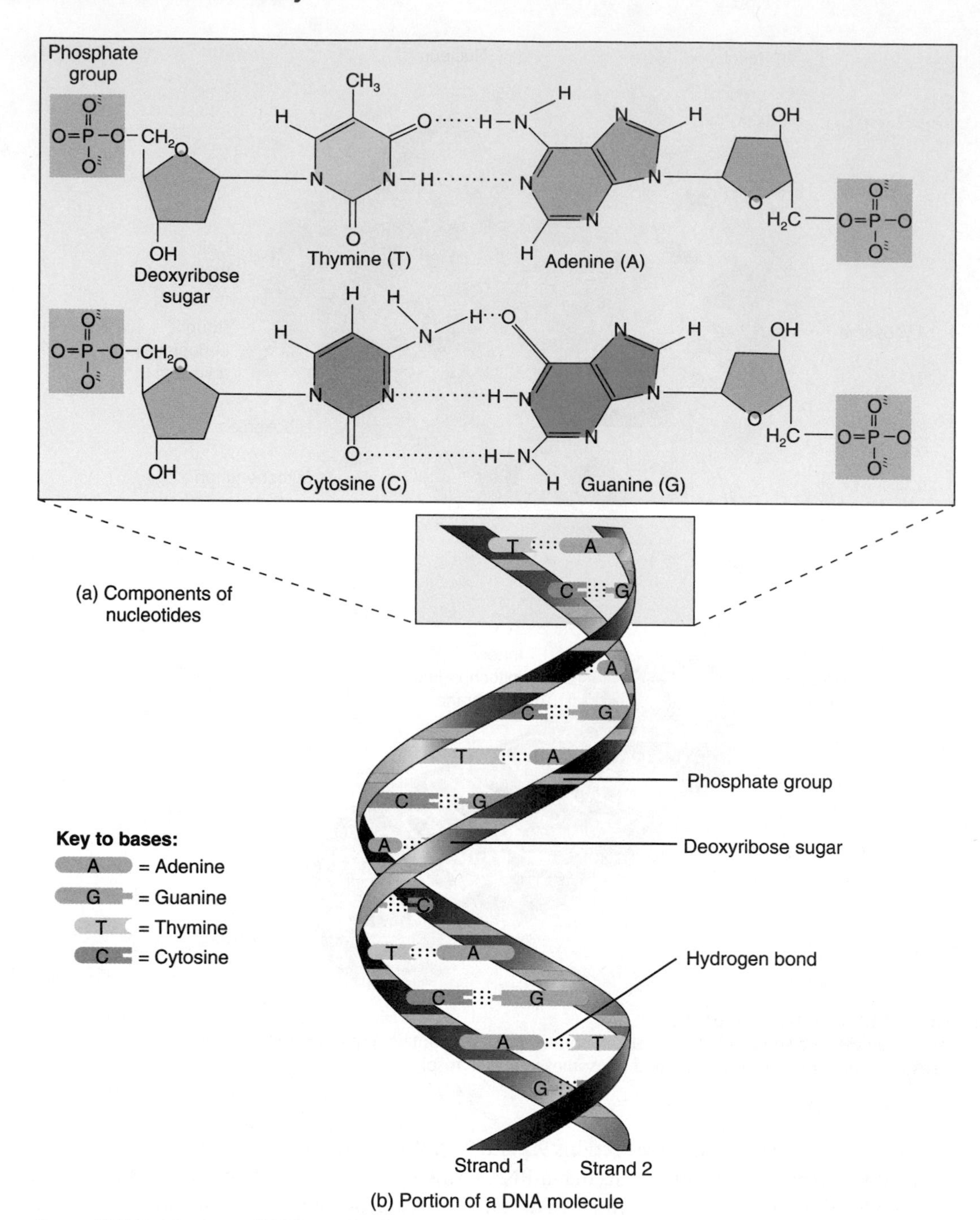

Figure F5.2 Nutrients and DNA synthesis.
Folate is required for the biosynthesis of the bases that make up DNA and phosphate is a integral part of the nucleotide.

in energy metabolism by increasing the ability of insulin to promote glucose uptake by cells (see Section 12.8; Figure 12.22).

Many B vitamins, for example thiamin, riboflavin, niacin, pantothenic acid, and biotin, because of their role as co-enzymes, are essential for the oxidation of glucose, fatty acids, and amino acids, and the formation of ATP (**Figure F5.4**). The minerals iron and copper are essential components of some of the proteins that make up the electron transport chain. Phosphorus in the form of phosphate is a component of ATP.

Nutrients and Antioxidant Activity

Energy metabolism, especially the electron transport chain, is one of several sources of reactive oxygen species in the cell (see Section 8.10 for more information about reactive oxygen species). To minimize cellular damage, cells contain several enzymes to neutralize these species

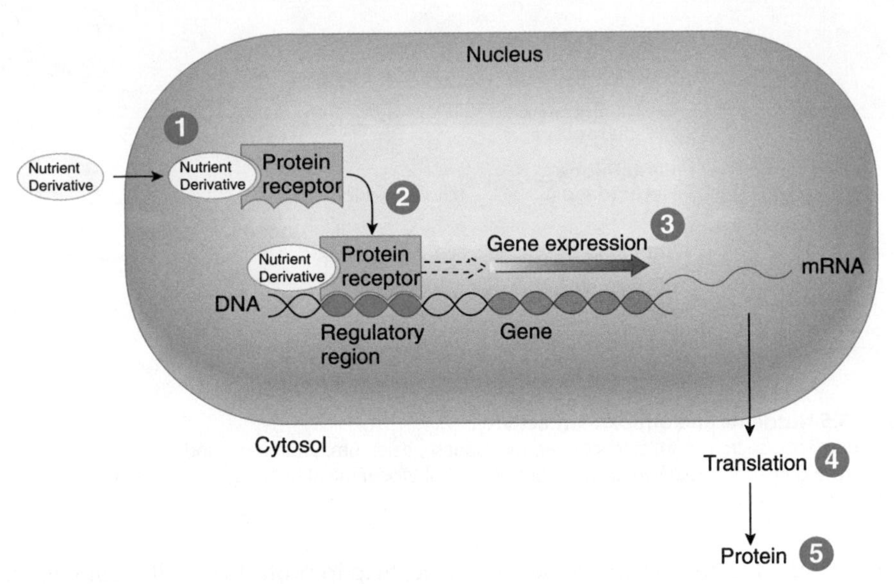

Figure F5.3 Nutrients and gene expression.
(1) Derivatives of nutrients such as vitamin A, vitamin D, glucose, and fatty acids bind to receptor protein. (2) The nutrient derivative-receptor complex binds to DNA. (3) The binding of the complex to DNA promotes gene expression and mRNA is formed. (4) The mRNA is translated into a protein (5).

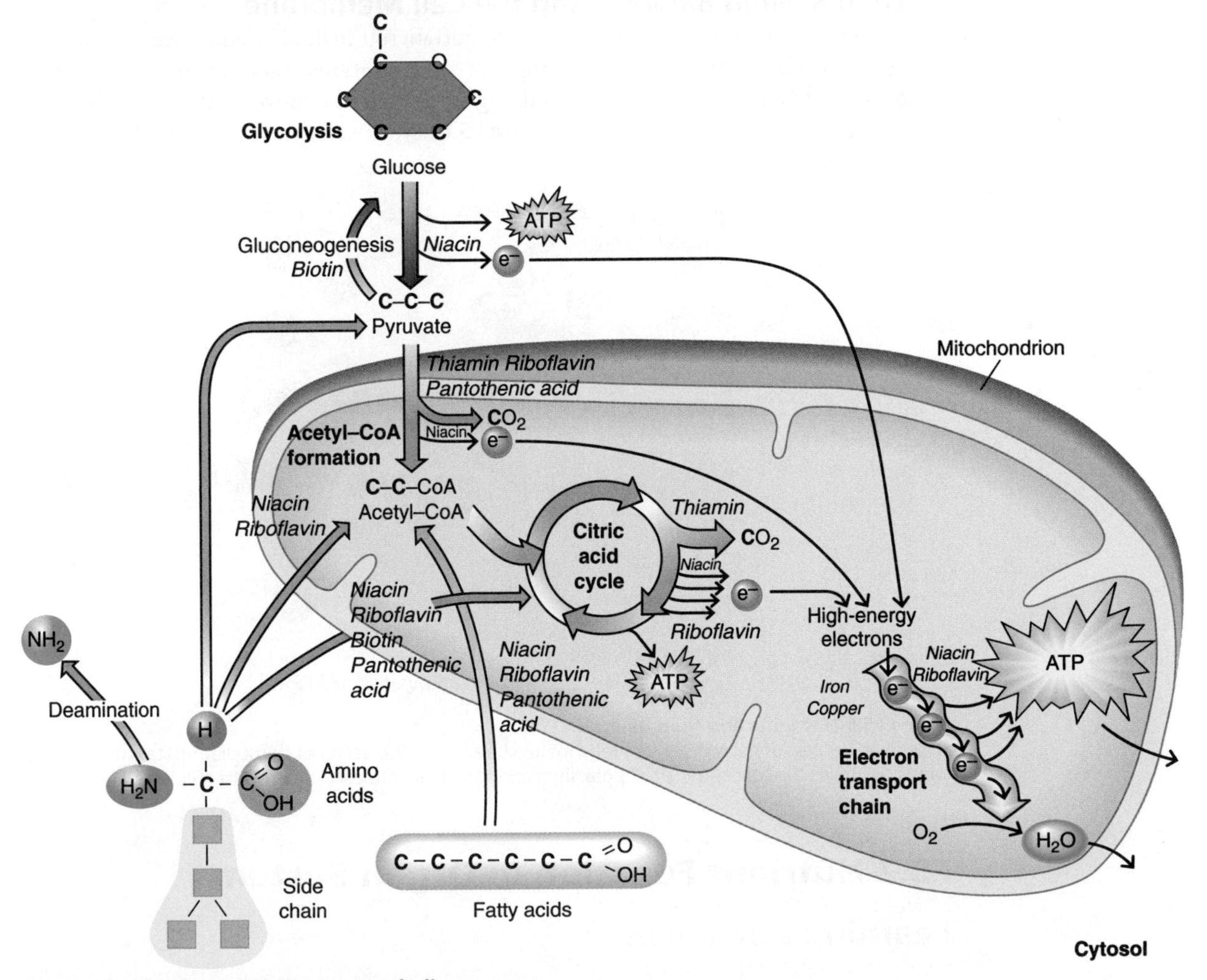

Figure F5.4 Nutrients and energy metabolism.
Several B vitamins and minerals such as iron, copper, and phosphorus play important roles in energy metabolism.

(Figure F5.5). These enzymes are dependent on several essential minerals for their activity. They include the enzymes glutathione peroxidase, which depends on selenium; cytoplasmic superoxide dismutase, which depends on zinc and copper; catalase, which depends on iron; and mitochondrial superoxide dismutase, which depends on manganese. In addition, antioxidants,

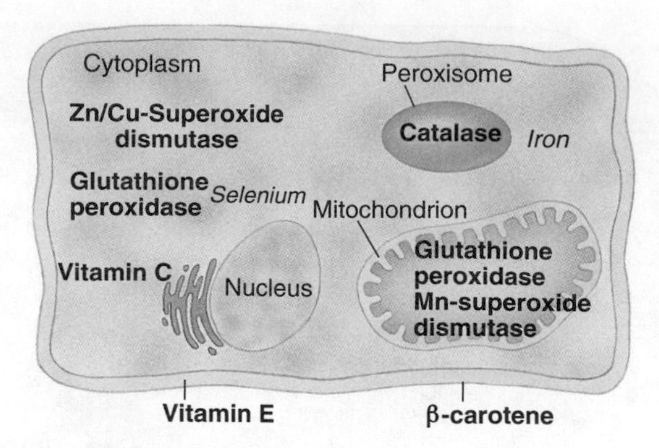

Figure F5.5 Nutrients and antioxidant activity.
Several nutrients, such as iron, zinc, copper, manganese, selenium, vitamin C, and vitamin E, protect the cell from damage by reactive oxygen species. The phytochemical beta-carotene also has antioxidant activity.

vitamin E, and the phytochemical beta-carotene, help to protect the cell membrane from damage by reactive oxygen species. Vitamin E is spared by selenium (see Section 12.6; Figure 12.16) and regenerated by vitamin C (see Section 9.4; Figure 9.14).

Nutrients, Fluid Balance, and the Cell Membrane

The nutrients sodium and potassium play an important role in fluid balance (See section 10.2). The Na-K-ATPase pump maintains appropriate concentrations of sodium and potassium on either side of the cell membrane and in doing so controls the movement of water between intracellular and extracellular spaces (**Figure F5.6**). Magnesium is also required for the AT-Pase pump to function properly.

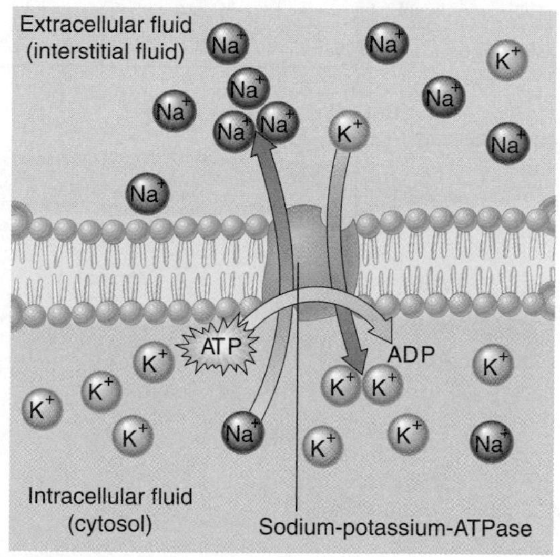

Figure F5.6 Nutrients and fluid balance.
Sodium and potassium play important roles in fluid balance via the action of the sodium-potassium ATPase which maintains the concentration of potassium and sodium in intracellular and extracellular fluid.

F5.2 Nutrient Function in Organ Systems

Learning objectives

- Describe the nutrients important to the development of the brain.
- List the nutrients that are most important to the function of the immune system.
- List the nutrients that affect the function of blood vessels.
- List the nutrients that are important to the organic and inorganic matrices of bone.

In addition to the role of nutrients at the cellular level, multiple nutrients have been identified as playing key roles in organ systems. These include the central nervous system, the immune system, the blood and circulatory system, and the skeletal system (bone).

Nutrients and the Central Nervous System

The nervous system is made up of the brain, the spinal cord, sensory organs such as the eyes and ears, and the nerves, which interconnect these organs with the rest of the body. Several nutrients play key roles in the fetal development of the nervous system. Folate is important in preventing defects during the formation of the neural tube, which occurs shortly after conception, and which ultimately develops into the nervous system. It is for this reason that Canada's food supply is fortified with folic acid and that women of reproductive age are urged to take folic acid supplements. The long-chain fatty acid DHA, found in fish oils, is important for the development of the brain and eye during the fetal stage and in early infancy. Pregnant women who are iodine deficient may have babies with cretinism, a condition resulting in poor mental development. Iron deficiency in infants has also been associated with limitations in mental development.[1]

The neuron is the basic cell of the nervous system; it communicates messages between the brain and the body. A neuron **(Figure F5.7)** contain axons, long projections that carry electrical signals from neurons to others cells in the body or from neuron to neuron. Axons are covered by a fatty myelin sheath that insulates the axon and enhances signal transmission. At the end of an axon is a region called the synapse. When the electrical signal travelling along the axon reaches the synapse, it stimulates the release of chemicals called neurotransmitters, which transmit the signal from the axon, across a small gap (the synaptic cleft), to an adjacent cell, such as another neuron or a muscle cell; neurotransmitters reach specific receptors on these adjacent cells and bind to them. Examples of neurotransmitters include acetylcholine, norepinephrine, dopamine, and serotonin, which have different functions that will not be discussed here.

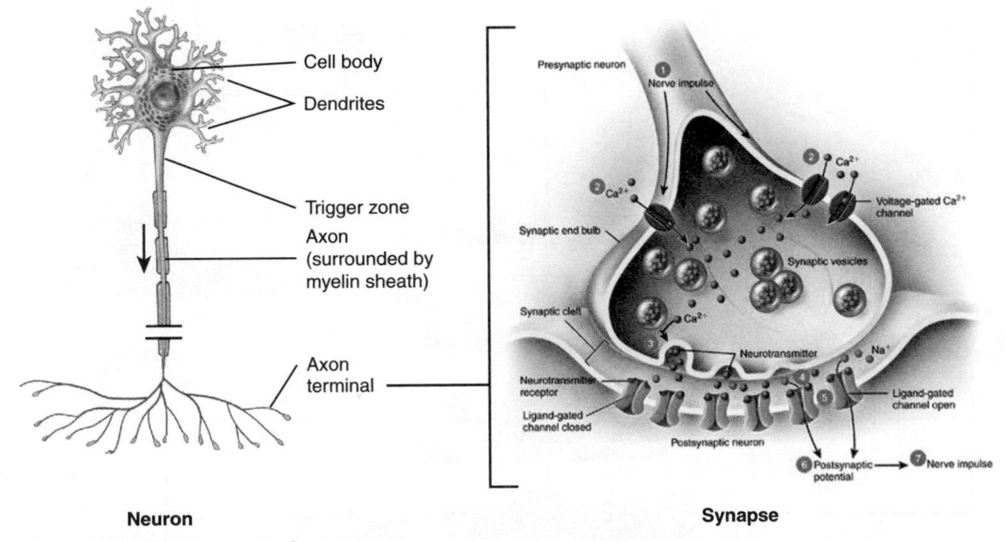

Figure F5.7 The neuron and synapse.
Neural signals travel along the axon of a neuron until they reach the synapse. Here, neurotransmitters are released, which allows the signal to be communicated to an adjacent cell, by binding to neurotransmitter receptors on this cell.

Several nutrients influence neuron function. The electrical signal transmitted along an axon results from a stimulus that increases the nerve cell membrane's permeability to sodium, allowing the signal to travel along the axon (see Section 10.2 Figure 10.14). Both vitamin B_6 and B_{12} are needed for the synthesis and maintenance of the myelin sheath. Vitamin B_{12} deficiency results in neurological symptoms such as numbness, tingling, abnormalities in gait, memory loss, and disorientation. Vitamin E helps to maintain the integrity of the cell membrane of the neuron; its deficiency results in poor muscle coordination, weakness, and impaired vision. The amino acid tryptophan is required for the synthesis of the neurotransmitter serotonin, while tyrosine is required for the synthesis of dopamine and norepinephrine. Acetylcholine is not synthesized from an amino acid, but from choline, which may be an essential nutrient at some life stages. The neurological symptoms of the thiamin-deficiency disease beriberi, such as nerve tingling, poor coordination, and paralysis, are believed to be due to the role that the vitamin thiamin plays in acetylcholine synthesis. The importance of niacin in neurological function is implied by the dementia that develops in the niacin-deficiency disease pellagra, although the functional role of niacin is not known.

Sensory organs, such as the eyes, are also part of the central nervous system. Vitamin A is essential to eye health, by preventing xeropthalmia (see Section 9.2). The phytochemicals lutein and zeaxanthin are antioxidants that accumulate in the macula, which is the central portion of the retina of the eye. High intakes of these are associated with reduced risk of macular degeneration, the leading cause of blindness in older adults (see Focus On Phytochemicals).

Nutrients and the Immune System

Two major types of cells circulate in the blood: red blood cells, which carry oxygen to tissue, and white blood cells, which form the immune system. There are 2 major types of white blood cells, phagocytes and lymphocytes, which together help to fight infection. There are 3 main ways that the immune system can destroy harmful, disease-causing, organisms **(Figure F5.8)**. These include (a) the engulfing of harmful organisms by phagocytes; (b) the enhancement of phagocytosis by the synthesis of antibodies, which requires B-cells and helper T-cells; and (c) the destruction of infected cells by the binding of cytotoxic T-cells.

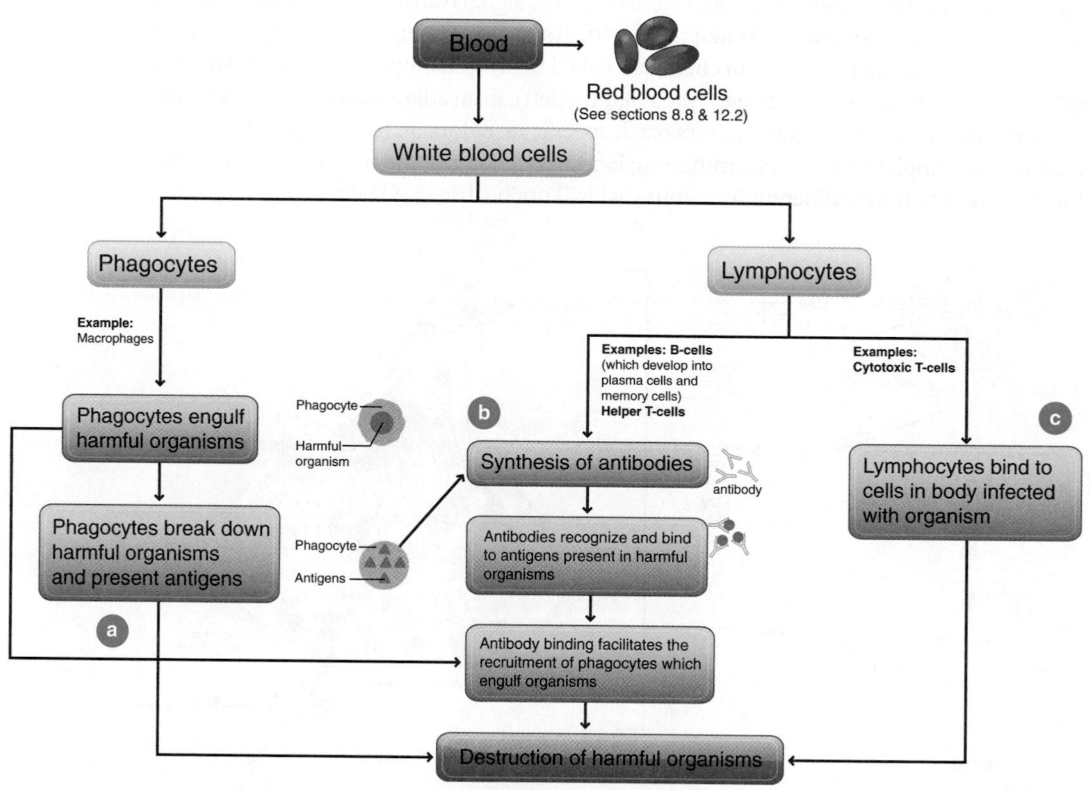

Figure F5.8 The immune system.
The immune system fights infection by (a) the engulfing of harmful organisms by phagocytes; (b) the enhancement of phagocytosis by the synthesis of antibodies, which requires B-cells and helper T-cells; and (c) the destruction of infected cells by the binding of cytotoxic T-cells.

Several nutrients have been identified as playing key roles in the immune system. The cells of the immune system must form and replicate quickly if the body is to be successful in fighting infections, so many nutrients involved in DNA synthesis and cell division are linked to the proper functioning of the immune system. Vitamin B_6 is required for the formation of lymphocytes such as B- and T-cells and vitamin B_6 deficiency has been linked to decreased antibody production.[2] Folate and vitamin B_{12} are also involved in DNA synthesis and deficiencies of both vitamins have similarly been associated with impaired immune function.[2] The role of vitamin A in the immune system is well established. Vitamin A controls the gene expression that induces the differentiation of immature immune cells into phagocytes and lymphocytes. Similarly, vitamin A regulates the gene expression that results in the formation of mucus-secreting cells in epithelial tissue. These mucus-secreting cells act as barriers to infection in various parts of the body, such as the intestine, lungs, vagina, and bladder, by trapping harmful organisms. When vitamin A intake is inadequate, this "barrier function" is compromised. Combined with impaired phagocyte and lymphocyte function, the result is a high susceptibility to infection.

Vitamin D, like vitamin A, also regulates gene expression and the vitamin D receptor is present in many cells of the immune system. One of vitamin D's functions is to increase the formation of macrophages.[2] Zinc, because of its role in the zinc finger, is required for proteins to bind to DNA. For this reason, zinc deficiency resembles the deficiency of nutrients involved in gene expression, such as vitamins A and D, and has been found to increase susceptibility to infection (see Section 12.3; Figure 12.10).

Antioxidants also play a role in the proper functioning of the immune system. Phagocytes destroy the organisms that they engulf by generating reactive oxygen species which are toxic to the organism. In order to successfully destroy the organism, phagocytes must protect their cell membranes from possible damage from these reactive oxygen species. They do this by maintaining high levels of antioxidants. For this reason, antioxidant vitamins such as vitamins C and E, as well as the minerals that are cofactors in endogenous enzymes, including selenium, iron, copper, zinc, and manganese, are important to the proper functioning of the immune system.

Finally, the cells of the immune function secrete proteins that help to regulate inflammation, which is a beneficial response to injury, as it stimulates the immune response and limits the spread of infection. The cells of the immune system also secrete anti-inflammatory proteins to carefully balance this process. Fatty acids can influence this balance (see Section 5.5; Figure 5.16). The omega-3 fatty acids, EPA and DHA, or, more specifically, eicosanoids derived from them, tend to promote the secretion of anti-inflammatory factors, while eicosanoids derived from the omega-6 fatty acid arachidonic acid tend to favour inflammatory factors.[3] EPA and DHA are present in fatty fish and fish oils and can also be synthesized in the body from the essential fatty acid alpha-linolenic acid, while arachidonic acid is present in many foods and can also be synthesized from the essential fatty acid, linoleic acid.

Nutrients, Blood, and the Circulatory System

Many of the components of the immune system described in this chapter circulate in the blood, as do erythrocytes (red blood cells). Many nutrients are essential to the development of the properly functioning red blood cells. Folate and vitamin B_{12} play a vital role in the development of the mature red blood cell, and deficiencies of these nutrients result in megaloblastic anemia (see sections 8.8 and 8.9). Iron is essential for the synthesis of hemoglobin; iron deficiency anemia is the most widespread micronutrient deficiency globally (see Section 12.2).

Nutrients also play a key role in blood clotting. Just as omega-3 and omega-6 fatty acids have counterbalancing roles in the control of inflammation, they have a similar role in blood coagulation; derivatives of omega-3 fatty acids generally inhibit coagulation, while derivatives of omega-6 fatty acids promote coagulation (see Section 5.5; Figure 5.16). Vitamin K plays a key role in the synthesis of the protein prothrombin and its conversion to thrombin during blood clotting (see Section 9.5 Figure 9.16). Calcium is also required for blood clotting, while vitamin E has an anti-coagulatory effect.

In addition to contributing to the many factors that circulate in blood, nutrients contribute to the health of blood vessels. Inadequate folate intake results in elevated levels of homocysteine in the blood. While the homocysteine hypothesis (see Section 8.7), linking low folate intakes to increased risk of cardiovascular disease, remains controversial, there is evidence that homocysteine damages blood vessels in a number of ways, such as promoting inflammation and decreasing flexibility of the vessel walls, which may contribute to the development of atherosclerosis.[4]

The relationship between dietary fat intake, LDL cholesterol levels, and atherosclerosis is well researched (see Section 5.6; Figure 5.24). Saturated fatty acids promote atherosclerosis by reducing the number of LDL receptors, while *trans* fatty acids are also harmful because they increase LDL cholesterol levels and decrease beneficial HDL cholesterol, which functions to clear cholesterol from the blood. Polyunsaturated fatty acids reduce the risk of atherosclerosis by lowering LDL-cholesterol levels. Omega-6 fatty acids do so primarily by increasing

LDL receptor levels and by promoting bile acid secretion, while omega-3 increases fatty acid oxidation, which in turn reduces the secretion of VLDL in the blood.

Nutrients and the Skeletal System

Several nutrients are important in bone health. The nutrients that support the organic components of bone include vitamin C and copper, which are needed for the proper synthesis of collagen, and vitamin K, for the proper synthesis of other bone proteins. To support the inorganic matrix of bone, the hydroxyapatite crystals, calcium is required, as is vitamin D, which promotes the absorption of calcium. Phosphorus is also an important component of hydroxyapatite. Similarly, magnesium is part of the inorganic matrix of bone and is also required for the proper functioning of vitamin D. Finally, the trace mineral fluoride, by forming fluoroapatite, strengthens tooth enamel.

The descriptions given here represent the best understood nutrient functions, because these functions are most adversely effected when nutrient intake is inadequate. Nutrients function, however, in every cell and every system in the body and there is still much to be learned about their role in human health.

REFERENCES

1. Peirano, P. D , Algarín, C. R., Chamorro, R., et al. Sleep and neurofunctions throughout child development: lasting effects of early iron deficiency. *J. Pediatr. Gastroenterol. Nutr.* 48 (Suppl) 1:S8–15, 2009.
2. Wintergerst, E. S., Maggini, S., and Hornig, D. H. Contribution of selected vitamins and trace elements to immune function. *Ann. Nutr. Metab.* 51(4):301–23, 2007.
3. Chapkin, R. S., Kim, W., Lupton, J. R. et al. Dietary docosahexaenoic and eicosapentaenoic acid: emerging mediators of inflammation. *Prostaglandins Leukot. Essent. Fatty Acids.* 81(2-3):187–91, 2009.
4. Joseph, J., Joseph, L. Hyperhomocysteinemia and cardiovascular disease: new mechanisms beyond atherosclerosis. *Metab. Syndr. Relat. Disord.* 1(2):97–104, 2003.

13 Nutrition and Physical Activity

Case Study

Jeffrey, an avid cyclist, is having a great summer, doing what he always wanted to do. He is participating in the Tour du Canada, cycling across Canada. He starts out in Vancouver in late June, and his travels end just before Labour Day at Signal Hill in St John's. The Tour is not a race but an adventure; a way for cyclists, riding in small groups, to see Canada and meet Canadians from coast to coast. Many days they ride more than 160 km (100 miles). It is also a "no-frills" adventure. The riders spend most of their evenings on campgrounds, sleeping in tents. The Tour du Canada organizers provide food for breakfast and dinner, which is rich in the carbohydrates that cyclists need to fuel the many hours they spend riding. This means a lot of pasta salads, beans, potatoes, and desserts. But during the day, riders stop for snacks and meals in the many small towns they pass along the way.

On one such day in early July, Jeffrey is making his way across Alberta with his riding partners, Steven and Jacob, when they decide to stop for lunch at a local diner near Drumheller. The ride has been especially fast, with the wind in their backs and the temperature being somewhat cooler than typical for a mid-July day. Jeffrey is feeling confident that he and his companions will make it to the next campground 50 km east of Drumheller in good time. He orders a 300g (12 oz) steak and a green salad for lunch while his riding partners dine on spaghetti and meatballs and macaroni and cheese, respectively.

(© istockphoto.com/Nicole Waring)

On the final hours of the ride that day, Jeffrey notices that his energy is flagging and that he is having trouble keeping up with his riding partners. He makes it to the campgrounds that evening, but the afternoon ride proves far, far more exhausting and more difficult than the morning's ride. Jeffrey realizes that his choice of lunch may be at the root of his problem.

In this chapter, you will learn about the benefits of physical activity on human health and how good nutrition promotes athletic performance.

(©iStockphoto)

(© istockphoto.com/Silvia Jansen)

Chapter Outline

13.1 Exercise, Fitness, and Health

Learning Objectives

- Describe the characteristics of a fit individual.
- Explain what is meant by the overload principle.
- Evaluate the impact of exercise on health.
- Discuss the role of exercise in weight management.

fitness The ability to perform routine physical activity without undue fatigue.

overload principle The concept that the body will adapt to the stresses placed on it.

cardiorespiratory system The circulatory and respiratory systems, which together deliver oxygen and nutrients to cells.

aerobic exercise Endurance exercise such as jogging, swimming, or cycling that increases heart rate and requires oxygen in metabolism.

stroke volume The volume of blood pumped by each beat of the heart.

resting heart rate The number of times that the heart beats per minute while a person is at rest.

aerobic capacity or **VO_2 max** The maximum amount of oxygen that can be consumed by the tissues during exercise. This is also called maximal oxygen consumption.

Exercise improves **fitness** and overall health. For some people, fitness means being able to easily walk around the block, mow the lawn, or play with their children. For more seasoned athletes, fitness means optimal performance of strenuous exercise. For everyone, fitness reduces the risk of chronic diseases such as cardiovascular disease, diabetes, and obesity. You don't need to run 10-km races, ride the Tour du Canada, or compete in the Olympics to be physically fit. Even a small amount of exercise is better than none, and, within reason, more exercise is better than less. Whether you are 8 or 80, fitness through regular exercise can improve your overall health.

Exercise Improves Fitness

Fitness level is defined by endurance, strength, flexibility, and body composition. Engaging in regular exercise improves these parameters. When you exercise, you breathe harder, your heart beats faster, and your muscles stretch and strain. Regular exercise causes adaptations that result in long-term physiological changes. This is known as the **overload principle:** the more you do, the more you are capable of doing. For example, if you run 3 times a week, in a few weeks, you can run farther; if you lift weights a few days a week, in a few weeks, you will have more muscle and will be able to lift more weight more easily; if you stretch to touch your toes every morning, in a few days, it becomes less of a stretch. These adaptations improve overall fitness.

Cardiorespiratory Endurance Cardiorespiratory endurance determines how long you can continue a task, whether it is climbing stairs, raking leaves, or running a race. It requires muscle strength but also involves the cardiovascular and respiratory systems, referred to jointly as the **cardiorespiratory system.** Endurance is increased by **aerobic exercise**, the type of exercise that increases heart rate and uses oxygen. To be aerobic, an activity should be performed at an intensity low enough to allow you to carry on a conversation but high enough that you cannot sing while exercising. Aerobic activities include walking, dancing, jogging, cross-country skiing, cycling, and swimming.

Regular aerobic exercise strengthens the heart muscle and increases **stroke volume**, which is the amount of blood pumped with each beat of the heart. This in turn decreases **resting heart rate**, which is the rate at which the heart must beat to supply blood to the tissues at rest. Resting heart rate can be measured by counting the number of pulses, or heartbeats, per minute while at rest (Figure 13.1). The more fit a person is, the lower their resting heart rate and the more blood their heart can pump to muscles during exercise. In addition to increasing the amount of oxygen-rich blood that is pumped to muscles, regular aerobic exercise increases the muscle's ability to use oxygen to produce ATP. The body's maximum ability to generate ATP by aerobic metabolism during exercise is called **aerobic capacity**, or **VO_2 max**. Aerobic capacity is dependent on the ability of the cardiorespiratory system to deliver oxygen to the cells and the ability of the cells to use oxygen to produce ATP. The greater a person's aerobic capacity, the more intense activity he or she can perform before a lack of oxygen affects performance.

Figure 13.1 To estimate your heart rate, place your fingers over the carotid artery, which is at either side of your neck just below the jawbone, and count the number of pulses per minute. (Michael Newman/PhotoEdit)

Aerobic capacity can be determined in an exercise laboratory by measuring oxygen uptake during exercise. To perform this measurement, an individual runs on a treadmill or rides a stationary bicycle while the gases he or she breathes are measured. Oxygen consumption or uptake is calculated by subtracting the amount of oxygen exhaled from the amount of oxygen inhaled. The workload is then increased by increasing the speed and/or grade of the treadmill or resistance on the bike until the individual can no longer continue **(Figure 13.2)**.

The amount of oxygen consumed at the highest workload achieved is the aerobic capacity. A trained athlete will have a greater aerobic capacity than an untrained individual.

Muscle Strength and Endurance Muscle strength and endurance enhance the ability to perform tasks such as pushing or lifting. In daily life, this could mean lifting heavy boxes, unscrewing the lid of a jar, or shovelling snow. Muscle strength and endurance are increased by repeatedly using muscles in activities that require moving against a resisting force. This type of exercise is called strength-training or resistance-training and includes activities such as weight lifting (**Figure 13.3**). Lifting a heavy weight stresses muscles. This stress or overload causes muscles to adapt by increasing in size and strength—a process called **hypertrophy**. The larger, stronger muscles can now easily lift the same weight that stressed them the first time. By progressively increasing the amount of weight at each exercise session, the muscle slowly hypertrophies.

When muscles are not used due to a lapse in training, an injury, or illness, they become smaller and weaker. This process is called **atrophy**. For example, when an individual is bedridden and unable to move about, their muscles atrophy. Once they are up and active again, their muscles regain strength and size. There is truth behind the expression "use it or lose it."

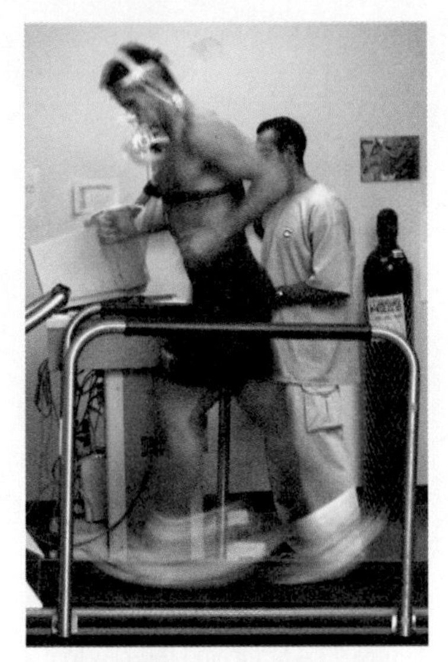

Figure 13.2 Aerobic capacity can be estimated by measuring oxygen uptake while running to exhaustion on a treadmill. (CP PHOTO/Paul Chiasson)

Flexibility Fitness is not just about bulging muscles, it also involves flexibility. Flexibility determines range of motion—how far you can bend and stretch muscles and ligaments. If flexibility is poor, a person cannot easily bend to tie their shoes or stretch to remove packages from the car. Being flexible may reduce the risk of pulled muscles and tendons. In a competitive athlete, improving flexibility can increase speed. This is because too-tight muscles, tendons, and ligaments can restrict motion at a joint and thus decrease stride length and increase the energy needed to overcome this motion-resisting stiffness. Regularly moving the limbs, neck, and torso through their full ranges of motion helps increase and maintain flexibility (**Figure 13.4**).

Figure 13.3 Building muscle requires an increase in the amount of resistance exercise, not simply an increase in protein intake. (© istockphoto.com/Will Selarep)

Body Composition Exercise builds and maintains muscle. Individuals who are physically fit have a greater proportion of lean body tissue than unfit individuals of the same body weight (**Figure 13.5**). Not everyone who is fit is thin, but in a fit person who carries extra pounds, more of the weight is from muscle. How much body fat a person has is also affected by gender and age. In general, women have more stored body fat than men. For young adult women, the desirable percent of body fat is 21%-32% of total weight; in adult men, the desirable percent is about 8%-19%.[1] With aging, lean body mass decreases in both men and women, and there is an increase in the percentage of body fat even if body weight remains the same. Some of this change may be prevented by staying physically active (Section 16.4).

hypertrophy An increase in the size of a muscle or organ.

atrophy Wasting or decrease in the size of a muscle or other tissue caused by lack of use.

Health Benefits of Exercise

In addition to making the tasks of everyday life easier, maintaining fitness through regular activity offers many health benefits. A regular exercise program makes it easier to maintain a healthy body weight; helps maintain muscles, bones, and joints; and reduces the risk of osteoporosis.[2] It can help to prevent or delay the onset of cardiovascular disease, hypertension, diabetes, and colon cancer. It can also prevent depression and improve mood, sleep patterns, and overall outlook on life.

Weight Management People who exercise regularly are more likely to be at and maintain a healthy body weight. Exercise makes weight management easier because it increases energy needs so more kcalories can be consumed without weight gain. During exercise, energy expenditure can rise well above the resting rate and some of this increase persists for many hours after activity slows.[3] Exercise can also boost energy expenditure through its effect on body composition. Exercise increases lean tissue mass and because, even at rest, lean tissue uses more energy than fat tissue, this increases basal energy needs. The combination of increased energy output during exercise, the rise in expenditure that persists after exercise, and the increase in basal needs can have a major impact on total energy expenditure (**Figure 13.6**). Besides increasing energy needs, exercise also promotes the loss of body fat and slows the loss of lean tissue that occurs with energy restriction. This makes exercise an essential component of any weight-reduction program.

Figure 13.4 Stretching muscles to increase and maintain flexibility is an important component of any exercise regimen. (© istockphoto.com/ Steven Debenport)

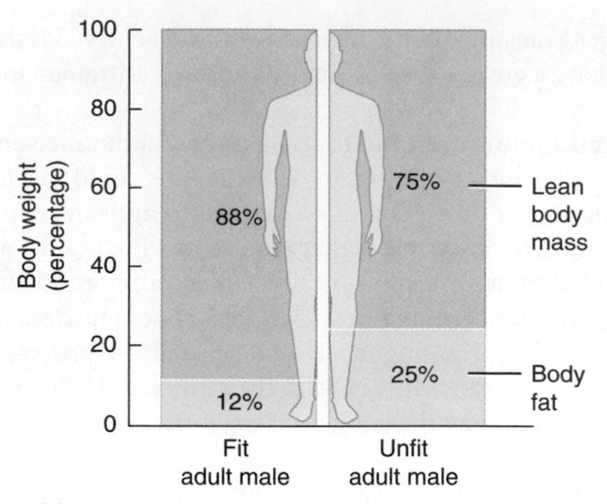

Figure 13.5 Body composition
Body composition, which refers to the percentage of fat versus non-fat or lean tissue (muscle, bones, cartilage, skin, nerves, and internal organs), is an indicator of health and fitness.

Cardiovascular Health Exercise reduces the risk of cardiovascular disease.[4] Because aerobic exercise strengthens the heart muscle, it reduces the number of times the heart must beat to deliver blood to the tissues at rest and during exercise. The changes that occur with exercise also help to lower blood pressure and increase HDL cholesterol levels in the blood. All of these effects help to reduce the risk of cardiovascular diseases such as heart attack and stroke.[5]

Diabetes Prevention and Management People with excess body fat are more likely to develop diabetes (see Section 7.4). By keeping body fat within the normal range, aerobic exercise can decrease the risk of developing type 2 diabetes. Physical activity that includes both aerobic exercise and strength training is also important in the treatment of diabetes because exercise increases the sensitivity of tissues to insulin.[6] Exercise can reduce or eliminate the need for medication to maintain normal blood glucose levels. Therefore, to prevent low blood glucose, people with diabetes should develop exercise programs with the help of their physicians and dietitians.

Bone and Joint Health Just as lifting weights helps maintain muscle size and strength, weight-bearing exercise stimulates bones to become denser and stronger. One of the causes of bone

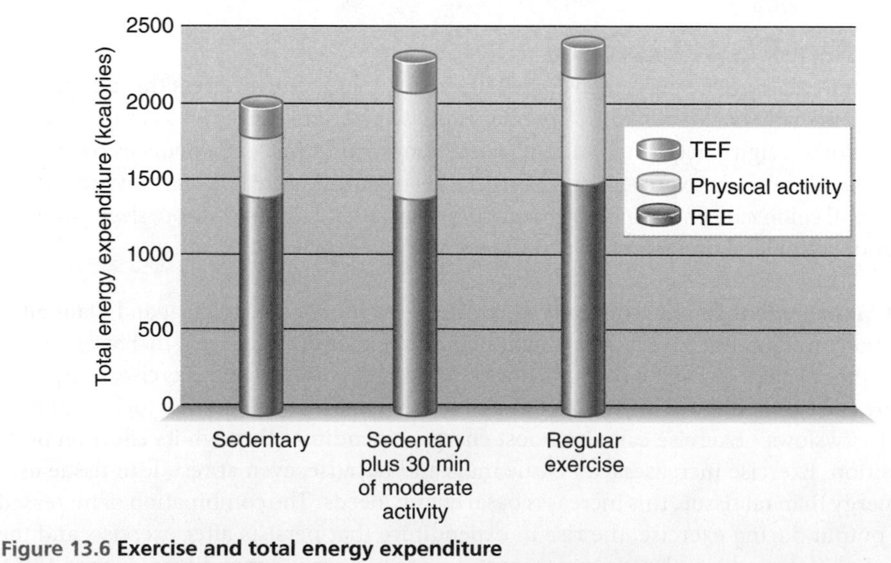

Figure 13.6 Exercise and total energy expenditure
Total energy expenditure is the sum of the energy used for resting energy expenditure (REE), physical activity, and the thermic effect of food (TEF). Exercise increases energy expenditure by increasing the amount of energy expended for physical activity, and if it is regular, it leads to an increase in muscle mass, which increases REE.

loss, like muscle loss, is lack of use; therefore, weight-bearing exercise such as walking, running, and aerobic dance can increase peak bone mass and prevent bone loss, and therefore reduce the risk of osteoporosis (see Section 11.3). Exercise can also benefit individuals with arthritis because the strength and flexibility promoted by exercise helps manage pain and allow arthritic joints to move more easily.

Cancer Risk Individuals who exercise regularly may be reducing their cancer risk by as much as 40%.[7] There is evidence that exercise reduces breast cancer risk; the risk reduction is related to exercise intensity, duration, and the age at which the exercise is performed.[8] The evidence that exercise reduces colon cancer risk is also strong; active individuals are less likely to develop colon cancer than their sedentary counterparts.[7] When evaluating the impact of exercise on cancer risk, diet and other lifestyle factors also must be considered. It is possible that some of the effect is due to the fact that people who exercise regularly are more likely to have healthier overall diets and lifestyles.

Overall Well-Being Physical activity improves mood, boosts self-esteem, and increases overall well-being.[9] It has been shown to reduce depression and anxiety, and improve the quality of life.[10] The exact mechanisms involved are not clear but one hypothesis has to do with the production of **endorphins**. Exercise stimulates the release of these chemicals, which are thought to be natural mood enhancers that play a role in triggering what athletes describe as an "exercise high." In addition to causing this state of exercise euphoria, endorphins are thought to aid in relaxation, pain tolerance, and appetite control. Exercise may also benefit mental well-being by affecting the levels of certain mood-enhancing neurotransmitters in the brain, releasing muscle tension, improving sleep patterns, and reducing levels of the stress hormone cortisol. Exercise also raises body temperature, which is believed to have a calming effect. These changes in both the brain and the body can reduce anxiety, irritability, stress, fatigue, anger, self-doubt, and hopelessness.[10]

endorphins Compounds that cause a natural euphoria and reduce the perception of pain under certain stressful conditions.

The Toll of Physical Inactivity in Canada An analysis of the role of physical activity in human disease indicates that physical inactivity contributes substantially to the development of chronic disease, as indicated in **Figure 13.7**. This figure suggests, for example, that 14% of breast cancer cases and up to 24% of stroke cases could be eliminated if Canadians increased their physical activity.

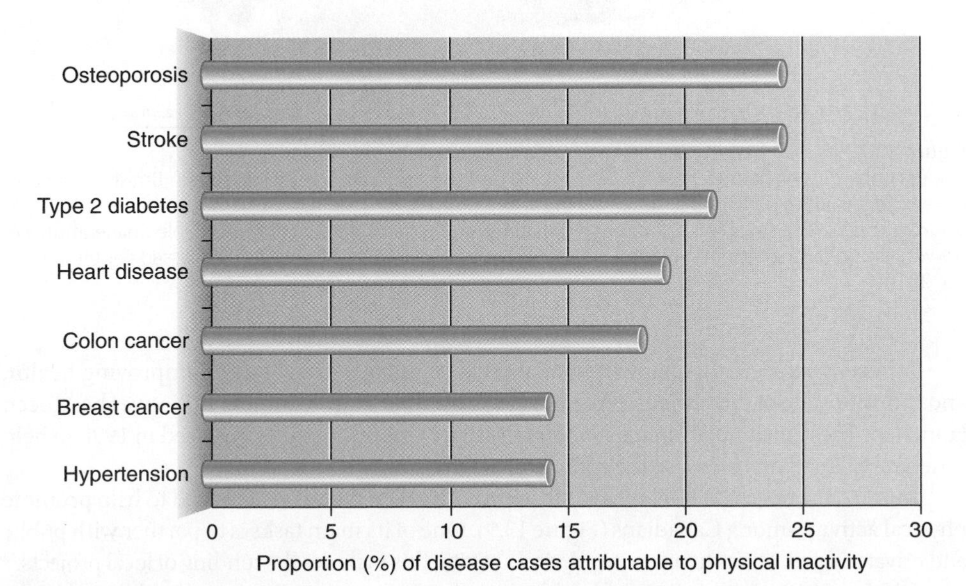

Figure 13.7 The proportion of Canadian disease cases due to physical inactivity.
Physical inactivity contributes to disease. It is estimated, for example, that 24% of stroke or 18% of colon cancer cases could be eliminated if Canadians increased their physical activity.
Source: Katzmarzyk, P. T. and Janssen, I. The economic costs associated with physical inactivity and obesity in Canada: an update. Can. J. Appli. Physiol. 29(1):90-115, 2004.

13.2 Exercise Recommendations

Learning Objectives

- Describe the amounts and types of exercise recommended to improve health.
- Classify activities as aerobic or anaerobic.
- Plan a fitness program that can be integrated into your daily routine.
- Explain overtraining syndrome.

Most Canadians do not exercise regularly; 55% of adults are physically inactive, although there is some regional variation as indicated in **Figure 13.8.** Between 2007 and 2009, a survey, called the Canadian Health Measures Survey, collected data on a number of health-related parameters; this included an examination of fitness among Canadians.[11] This study documented a decline in the fitness of Canadians between a survey conducted in 1981 and the Canadian Health Measures Survey in areas of flexibility and muscular strength.

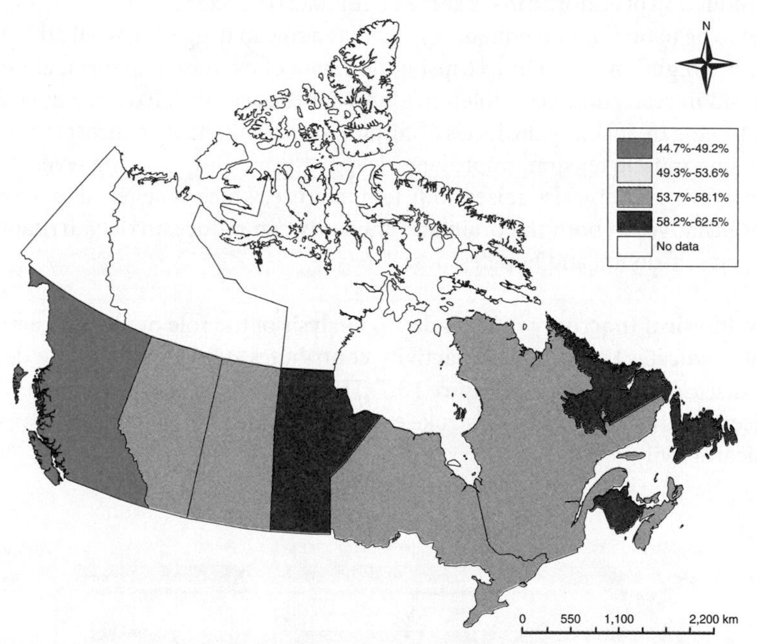

Figure 13.8 Physical activity levels among Canadian adults
The majority of Canadians (53%-62%, depending on the region) are inactive, except in British Columbia, were a significant minority (44%-49%) are inactive.
Source: Used with permission from Health Canada. Nutrition and Health Atlas. Available online at http://www.hc-sc.gc.ca/fn-an/surveill/atlas/map-carte/physic_s_a_inacti_mf-hf-eng.php. Accessed September 30, 2010.

In recent years, recognizing the impact that physical activity has on improving health, and the unfit state of the Canadian population, a number of government initiatives have been launched. These included *Canada's Physical Activity Guide* for adults, released in 1998 to help sedentary Canadians increase their physical activity.

In 2007, ParticipACTION (www.participACTION.com) was relaunched to help promote physical activity among Canadians (Figure 13.9). One of its main tasks is to partner with public and private organizations to promote physical activity; this includes the funding of local projects.[12]

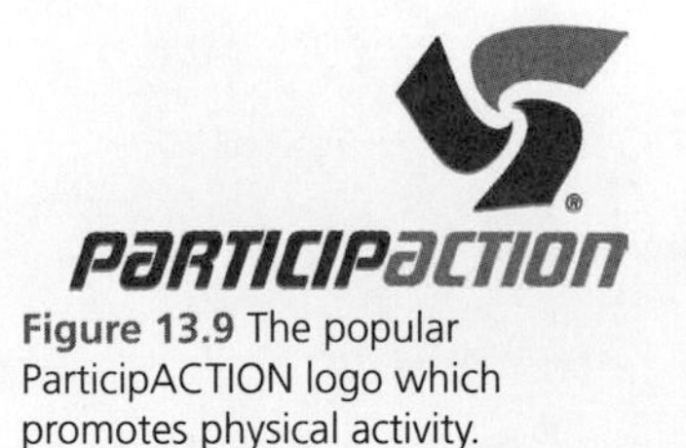

Figure 13.9 The popular ParticipACTION logo which promotes physical activity.

In May 2010, the minister of health announced that consultations had begun to update *Canada's Physical Activity Guide*, based on additional research over the last decade.[13] In January 2011, *Canada's Physical Activity Guide* was replaced with new physical activity guidelines in 4 categories: children (5 -11 years), youth (12-17 years) (both discussed in Section 15.3), adults (18-64 years), and old adults (> 65 years, discussed in Section 16.5). These new recommendations were developed by the Canadian Society of Exercise Physiology and

Canadian Physical Activity Guidelines

FOR ADULTS - 18 – 64 YEARS

Guidelines

To achieve health benefits, adults aged 18-64 years should accumulate at least 150 minutes of moderate- to vigorous-intensity aerobic physical activity per week, in bouts of 10 minutes or more.

It is also beneficial to add muscle and bone strengthening activities using major muscle groups, at least 2 days per week.

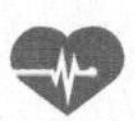

More physical activity provides greater health benefits.

Let's Talk Intensity!

Moderate-intensity physical activities will cause adults to sweat a little and to breathe harder. Activities like:

- Brisk walking
- Bike riding

Vigorous-intensity physical activities will cause adults to sweat and be 'out of breath'. Activities like:

- Jogging
- Cross-country skiing

Being active for at least **150 minutes** per week can help reduce the risk of:

- Premature death
- Heart disease
- Stroke
- High blood pressure
- Certain types of cancer
- Type 2 diabetes
- Osteoporosis
- Overweight and obesity

And can lead to improved:

- Fitness
- Strength
- Mental health (morale and self–esteem)

Pick a time. Pick a place. Make a plan and move more!

☑ Join a weekday community running or walking group.
☑ Go for a brisk walk around the block after dinner.
☑ Take a dance class after work.
☑ Bike or walk to work every day.
☑ Rake the lawn, and then offer to do the same for a neighbour.
☑ Train for and participate in a run or walk for charity!
☑ Take up a favourite sport again or try a new sport.
☑ Be active with the family on the weekend!

Now is the time. Walk, run, or wheel, and embrace life.

PARTICIPACTION

CSEP | SCPE THE GOLD STANDARD IN EXERCISE SCIENCE AND PERSONAL TRAINING

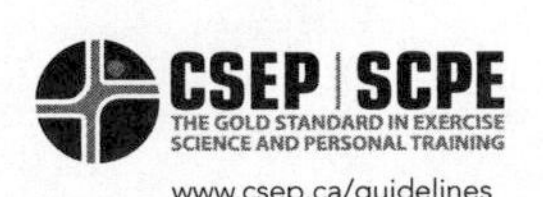

www.csep.ca/guidelines

Figure 13.10 Canadian physical activity guidelines.
These guidelines, released in January 2011, include recommendations for aerobic activity as well as muscle- and bone-strengthening activities,
Source: Canadian Physical Activity Guidelines, © 2011. Used with permission from the Canadian Society for Exercise Physiology, www.csep.ca/guidelines.

the guidelines for adults (18-64 years) are shown in **Figure 13.10**. It recommends that adults should accumulate at least 150 min. of moderate to vigorous intensity aerobic exercise weekly, in bouts of 10 min. or more. Muscle and bone strengthening activities at least 2 days/week is also advised. The guidelines also include an explanation of intensity and provides suggestions for suitable activities (see Figure 13.10).

Components of a Good Exercise Regimen

A well-planned exercise regimen includes aerobic exercise, which raises heart rate and therefore improves cardiorespiratory fitness; stretching, which promotes and maintains flexibility; and strength training, which increases the strength and endurance of specific muscles. Some aerobic and strength-training activities also help strengthen bone. Exercise should be integrated into an active lifestyle that includes a variety of everyday activities, enjoyable recreational activities, and a minimum amount of time spent in sedentary activities.

Aerobic Activity Adults should get at least 150 min. weekly of moderate physical activity. For example, exercising 5 times a week, 30 min./day would be one way to achieve this goal. If 30 min. is too long, activity can be divided into intervals at least 10 min. long.[14]

Examples of aerobic exercise include walking, bicycling, skating, swimming, or jogging. An activity is in the aerobic zone if it raises heart rate to 60%-85% of its maximum (**Figure 13.11**). **Maximum heart rate** is the maximum number of beats/min. that the heart can attain. It is dependent on age and can be estimated by subtracting your age from 220. For example, a 40-year-old individual would have a maximum heart rate of 180 beats/min. and should exercise at a pace that keeps his or her heart rate between 108-153 beats/min. (see Figure 13.11). When someone is exercising at this level, he or she will begin to feel warm and their rate of breathing will increase.

maximum heart rate The maximum number of beats/min. that the heart can attain. It declines with age and can be estimated by subtracting age in years from 220.

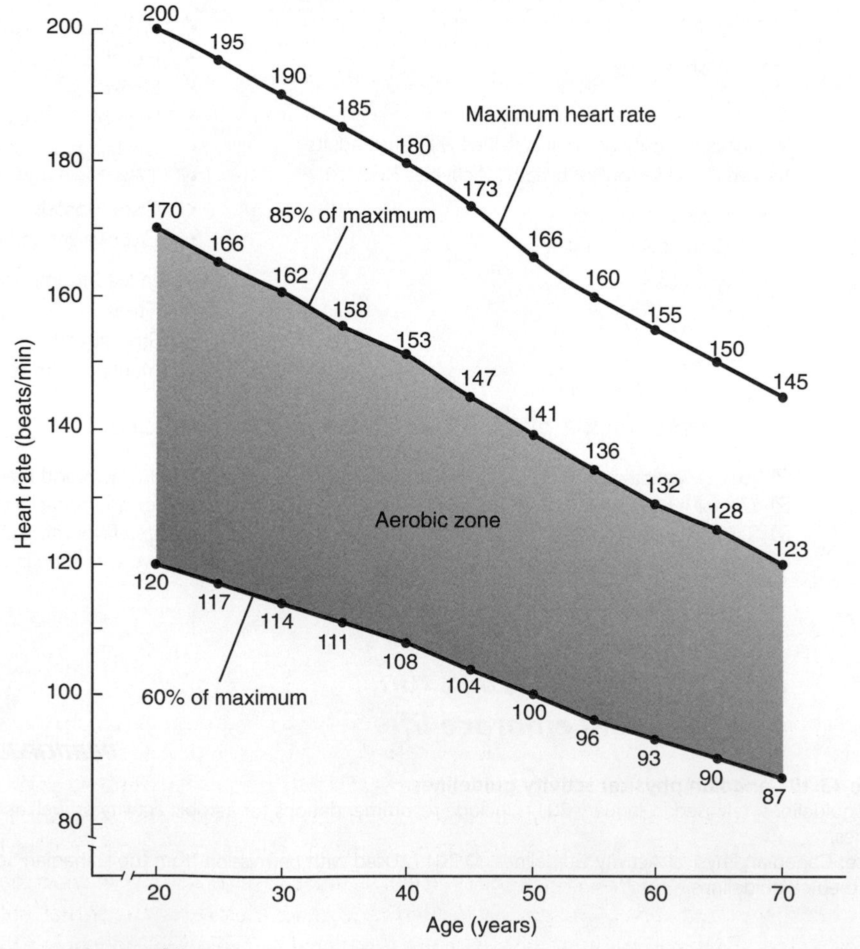

Figure 13.11 The aerobic zone
The aerobic zone (orange area) is between 60%-85% of maximum heart rate.

Each exercise session should begin with a warm-up, such as mild stretching and easy jogging, to increase blood flow to the muscles, and end with a cool-down period, such as walking or stretching, to help prevent muscle cramps and slowly reduce heart rate. Aerobic activities of different intensities can be combined to meet recommendations and achieve health benefits. The total amount of energy expended in physical activity depends on the intensity, duration, and frequency of the activity. Vigorous physical activity, such as jogging, that raises the heart rate to the high end of the aerobic zone (70%-85%) improves fitness more and burns more kcalories per unit of time than does moderate-intensity activity, such as walking, that raises heart rate only to the low end of the aerobic zone (60%-69%). For a sedentary individual beginning an exercise program, mild exercise such as walking can raise the heart rate into this range. As fitness improves, exercisers must perform more intense activity to raise their heart rates to this level.

Strength Training Adults should include training to strengthen muscle and bone at least 2 days/week.[14] A typical muscle- and bone-strengthening routine might include a minimum of 8-10 exercises that train the major muscle groups. Each exercise should be repeated 8-12 times, and the weights should be heavy enough to cause the muscle to be near exhaustion after the 8-12 repetitions. Increasing the amount of weight lifted will increase muscle strength, whereas increasing the number of repetitions will improve endurance. Running, walking, and yoga can also contribute to bone-strengthening.[15]

Flexibility Flexibility exercises should be done at least 3-4 days a week.[14] A typical regime for flexibility might include stretching muscles to a position of mild discomfort, holding for 10-30 sec. and repeating 3-5 times.

Exercise Recommendations for Children

Children and adolescents should spend at least 60 min./day engaged in moderate- to vigorous-intensity physical activity.[16] This exercise should include vigorous activities (e.g., running, hockey, soccer) at least 3 days a week and muscle- and bone-strengthening activities (e.g., skipping, jumping, or playing in the park) at least 3 times a week. There should be an emphasis on variety with total activity achieved through play, games, sport, walking, and physical education at the family, school, and community levels (**Figure 13.12**).

Life Cycle

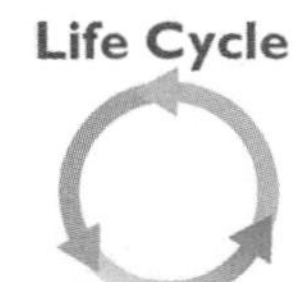

Modern lifestyles do not promote activity in children and adolescents; television, computers, video games, and cell phones are often chosen over physical activities. Reducing the amount of time spent in these sedentary activities can increase fitness, lower BMI, and improve blood pressure and cholesterol levels.[17] Children who learn to enjoy physical activity are more likely to be active adults who maintain a healthy body weight and have a lower risk of cardiovascular disease, diabetes, osteoporosis, and certain types of cancer. Learning by example is always best. Children who have physically active parents are the leanest and the fittest.

Figure 13.12 Children who participate in and enjoy exercise are more likely to have active lifestyles as adults. (© istockphoto.com/ Christopher Futcher)

Planning an Active Lifestyle

Almost everyone can participate in some form of exercise, no matter where they live, how old they are, or what physical limitations they have. Exercise classes are taught in nursing homes. People with heart disease, visual impairments, and physical disabilities compete in athletic events. You are never too old to exercise, and it is never too late to start.

Find Convenient, Enjoyable Activities Incorporating exercise into day-to-day life may require a behaviour change, and changing behaviour is not easy. The first step in beginning an exercise program is to recognize the reasons for not exercising and identify ways to overcome them. Many people avoid exercise because they do not enjoy it, feel they have to join an expensive health club, have little motivation to do it alone, or find it inconvenient and uncomfortable. Finding a type of exercise that is enjoyable, a time that is realistic and convenient, and a place that is appropriate and safe are important first steps in adopting an exercise program. Riding your bike to class or work rather than driving or taking the bus, taking a walk on your lunch break, and enjoying a game of catch or tag with your children are all effective ways to increase your everyday activity level. The goal is to gradually make lifestyle changes that increase physical activity. Behavioural strategies such as those listed in **Table 13.1** may help promote regular exercise (see Critical Thinking: Incorporating Exercise Sensibly).

Table 13.1 Suggestions for Starting and Maintaining an Exercise Program

Start slowly. Set specific attainable goals. Once you have met them, add more.

- Walk around the block after dinner.
- Get off the bus or subway one stop early.
- Use half of your lunch break to exercise.
- Do a few biceps curls each time you take the milk out of the refrigerator.

Make your exercise fun and convenient.

- Opt for activities you enjoy—bowling and dancing are more fun than a treadmill in the basement.
- Find a partner to exercise with you.
- Choose times that fit your schedule.

Stay motivated.

- Vary your routine—swim one day and mountain bike the next.
- Challenge your strength or endurance once or twice a week and do moderate workouts on other days.
- Track your progress by recording your activity.
- Use a pedometer, a device which allows you to count your steps, and work toward accumulating 10,000 steps daily.
- Reward your success with a new book, movie, or some workout clothes.

Keep your exercise safe.

- Warm up before you start and cool down when you are done.
- Wear light-coloured or reflective clothing that is appropriate for the environmental conditions.
- Don't overdo it—alternate hard days with easy days and take a day off when you need it.
- Listen to your body so you stop before an injury occurs.

Keep Exercise Safe Safety should be a consideration in planning any exercise regimen. Before beginning, everyone should check with their physician to be sure that their plans are safe, considering their medical history. Then the location and environment for exercise can be considered. Busy work schedules often force people to exercise in the dark, early morning or evening hours. Exercisers who use the street for walking or jogging should wear light-coloured, reflective clothing so they can be seen by motorists. Exercising with a partner is safer and more enjoyable.

Weather conditions can also be a safety concern. Physical activity produces heat, which normally is dissipated to the environment, partly by the evaporation of sweat. When the environmental temperature is high, heat is not efficiently transferred to the environment, and when humidity is high, sweat evaporates slowly, making it difficult to cool the body. Thus, exercise should be reduced or curtailed in hot and humid conditions. Cold environments can also pose problems for the outdoor exerciser. In general, cold does not impair exercise capacity, but the numbing of exposed flesh and the bulk of extra clothing can cause problems for joggers and bicyclists. Clothing must allow the body to dissipate heat while providing protection from the cold. For swimmers, cold water can cause performance to deteriorate.

Tailor Exercise Frequency, Intensity, and Duration Individuals should structure their fitness program based on their needs, goals, and abilities. For example, some people might prefer a short, intense workout such as 30 min. of running, while others would rather work out for a longer time, at a lower intensity, such as a 1-hour walk. Some may choose to complete all their exercise during the same session, while others may spread their exercise throughout the day, in shorter bouts. Three short bouts of 10-min. duration can be as effective as a continuous bout of 30 min. for reducing the risk of chronic disease.[2,18] A combination of intensities, such as a brisk 30-min. walk twice during the week in addition to a 20-min. jog on 2 other days, can meet recommendations. Also, what is best for a middle-aged man trying to reduce his risk of chronic disease is different from what is best for a 19-year-old college basketball player, and different still from what is best for an octogenarian trying to continue living independently. Young, healthy athletes may require very intense activity

CRITICAL THINKING:

Incorporating Exercise Sensibly

Background:

Nicole recently celebrated her forty-fifth birthday. Her promise to herself was to get back in shape. Nicole rarely gets any exercise and when she does, she suffers for the next few days with sore muscles. When the family goes on outings, she finds that she tires long before her husband and children.

Before beginning her exercise program she checks with her physician, who recommends that she do stretching and strength-training exercises as well as aerobic activities that keep her heart rate between 60%-85% of her maximum.

Data:

Nicole is 1.6 m (5′ 7″) tall and weighs 70 kg (155 lb.). Although her body mass index is still within the healthy range, she is about 2 kg (5 lb). above her usual weight. She would like to lose weight, but more importantly, she would like to increase her strength and endurance.

(© istockphoto.com/Jenny Swanson)

Nicole decides she will exercise for 90 min./day, 5 days/week. Her plan is to join a gym and stretch and lift weights for 30 min., followed by an hour of aerobic exercise outdoors, either jogging or riding a bicycle in the park.

After 3 days she realizes that she is spending a lot of time away from her family and, she is tired, sore, and ready to give up.

Critical Thinking Questions

List 3 things that are wrong with Nicole's exercise program.*

What should her heart rate be to keep it in her aerobic zone?

Calculate Nicole's EER before and after the addition of her exercise regimen. See Table 7.5. Do you think she will lose weight?

Suggest some modifications to her exercise program that will keep her from getting sore and allow her to spend more time with her family.

*Answers to all Critical Thinking questions can be found in Appendix J.

iProfile Use iProfile to calculate your energy expenditure for one day.

---

to obtain a training effect. Older adults and those who have not previously been active can increase their fitness by exercising at a lower intensity if the duration and frequency of exercise are increased.

Don't Overdo It To improve cardiorespiratory fitness and muscle strength, the body must be stressed and respond to the stress by increasing aerobic capacity and muscle size and strength. Initially, training can cause fatigue and weakness, but during rest the body rebuilds to become stronger. If not enough rest occurs between exercise sessions, there is no time to regenerate, so fitness and performance do not improve. In serious athletes, excessive training can lead to **overtraining syndrome**, which involves emotional, behavioural, and physical symptoms that persist for weeks to months. It is caused by repeatedly training without sufficient rest to allow for recovery. The most common symptom of overtraining syndrome is fatigue that limits workouts and is felt even at rest. Some athletes experience a decrease in appetite and weight loss as well as muscle soreness, increased frequency of viral illnesses, and increased incidence of injuries. They may become moody, easily irritated, depressed, have altered sleep patterns, or lose their competitive desire and enthusiasm. Overtraining syndrome occurs only in serious athletes who are training extensively, but rest is essential for anyone working to increase fitness (see Science Applied: Training: Sometimes Less Is Better).

overtraining syndrome A collection of emotional, behavioural, and physical symptoms that occurs in serious athletes when training without sufficient rest persists for weeks to months.

13.3 Fuelling Exercise

Learning Objectives

- Compare the fuels used to generate ATP by anaerobic and aerobic metabolism.
- Discuss the effect of exercise duration and intensity on the type of fuel used.
- Describe the physiological changes that occur in response to exercise.

Whether your goal is maintaining health or competing in athletic events, nutrition provides a launching pad from which physical fitness can be improved. Just as an automobile engine runs on energy from gasoline, the body machine runs on energy from the carbohydrate, fat, and protein in food and body stores. These fuels are needed whether you are writing a letter, walking around the block, or running a marathon. But before nutrients can be used to fuel activity, their energy must be converted into the high-energy compound ATP. ATP provides an immediate source of energy for all body functions, including muscle contraction. ATP can be generated both in the presence of oxygen by **aerobic metabolism** and in the absence of oxygen by **anaerobic metabolism** or **anaerobic glycolysis**. The way ATP is produced during activity depends on how long an activity is performed, the intensity of the activity, and the physical conditioning of the exerciser. This in turn affects how much carbohydrate, fat, and protein are used to produce this ATP.

aerobic metabolism Metabolism in the presence of oxygen. In aerobic metabolism, glucose, fatty acids, and amino acids are completely broken down to form carbon dioxide and water and produce ATP.

anaerobic metabolism or **anaerobic glycolysis** Metabolism in the absence of oxygen. Each molecule of glucose generates 2 molecules of ATP. Glucose is metabolized in this way when the blood cannot deliver oxygen to the tissues quickly enough to support aerobic metabolism.

The Effect of Exercise Duration

Resting muscles do not need much energy. At rest, the heart and lungs are able to deliver enough oxygen to meet energy demands using aerobic metabolism. During exercise, to increase the amount of energy provided by aerobic metabolism, the amount of oxygen available at the muscle must be increased (**Figure 13.13**). To do this, breathing and heart rate are increased, but this takes time. When exercise first begins, breathing and heart rate have not yet had enough time to increase the amount of oxygen available at the muscle.

Instant Energy: Stored ATP and Creatine Phosphate When you jump up to answer the phone or take the first steps of your morning jog, your muscles increase their activity but your heart and lungs have not had time to step up oxygen delivery to them. To get the needed energy, the muscles rely on small amounts of ATP that are stored in resting muscle. It is enough to sustain activity for a few seconds. As the ATP in muscle is used, enzymes break down another high-energy compound, called **creatine phosphate**, to replenish the ATP supply and allow activity to continue. But, like ATP, the amount of creatine phosphate stored in the muscle at any time is small. It will fuel muscle activity for about an additional 8-10 sec. before it, too, is used up. So, during the first 10-15 sec. of exercise, the muscles rely on energy from the ATP and creatine phosphate that is stored in them (**Figure 13.14**).

creatine phosphate A compound found in muscle that can be broken down quickly to make ATP.

Short-Term Energy: Anaerobic Metabolism As exercise continues beyond 10-15 sec., the ATP and creatine phosphate in the muscles are used up but the heart and lungs have still not had time to increase oxygen delivery to the muscles. Therefore, the additional ATP needed to fuel muscle contraction must be produced without oxygen. By 30 sec. into activity, anaerobic pathways are operating at full capacity. This anaerobic metabolism takes place in the cytosol. It includes glycolysis, which breaks glucose into the 3-carbon molecule pyruvate, releases electrons, and produces 2 molecules of ATP (**Figure 13.15**). At this point, if oxygen is unavailable, the pyruvate and released electrons combine to form **lactic acid**, which is transported out of the muscle for use in other tissues.

lactic acid A compound produced from the breakdown of glucose in the absence of oxygen.

Anaerobic metabolism can produce ATP very rapidly, but can only use glucose as a fuel. This glucose may come from the breakdown of glycogen inside the muscle or from glucose delivered via the bloodstream. The glucose delivered in the blood comes from the breakdown of liver glycogen, the synthesis of glucose by the liver, or the ingestion of carbohydrate during exercise. Anaerobic metabolism predominates during the first few minutes of exercise (see Figure 13.14) and is also important during periods of intense exercise, because oxygen cannot be delivered to the cells quickly enough to meet energy demands.

SCIENCE APPLIED

Training: Sometimes Less is Better

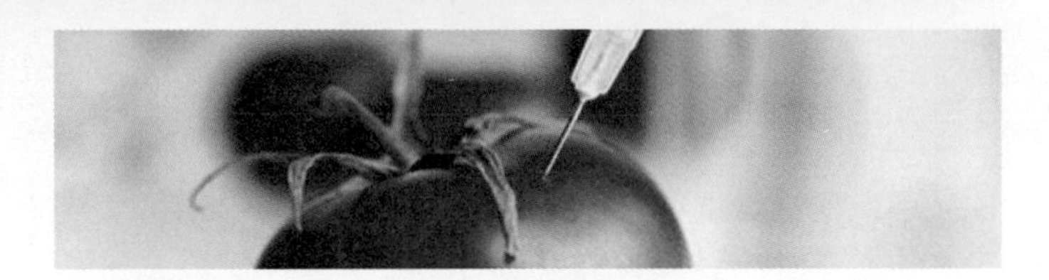

The Ball State University swim team was talented. They trained long and hard. Why, then, couldn't they perform consistently well? In search of an answer, the coach enlisted the help of Dr. David Costill, director of the school's Human Performance Laboratory. Costill began by assessing the condition of the swimmers. He attached monitors to measure heart rates and took blood samples to measure lactic acid levels. These measurements would help assess how well the athletes' bodies were responding to their training schedule. With continued training, heart rates and lactic acid accumulation were expected to decrease. However, Costill found that many of the swimmers had high lactic acid levels after practice, despite the fact that they were well-trained athletes.[1] Perhaps they were training too long and too hard. Costill conducted a series of experiments that resulted in a better understanding of "overtraining syndrome" and carbohydrate metabolism in the muscle during training.

Further studies helped to clarify the problems and suggest some solutions. Costill studied the swimmers on an increased training schedule; they swam at their normal pace but doubled the distance.[2] Some of the swimmers were unable to tolerate the heavier training schedule. Costill took muscle biopsies and found that muscle glycogen was depleted in these swimmers, indicating that they were running out of fuel. Their carbohydrate intake was not sufficient to meet the needs of the increased training, and their performance was suffering.[2] Other swimmers were able to tolerate the heavier training but it was not enhancing their performance. In addition, the extra training had other negative consequences. The swimmers' ratings of their muscle soreness, depression, anger, fatigue, and overall mood disturbances became more negative and they reported a reduction in their general sense of well-being.[3] Could cutting down on training actually improve performance?

Traditionally, it was believed that more training meant better strength and endurance. However, Costill believed that the swimmers were overtrained. To test his hypothesis, he divided the team into 2 groups. One group of swimmers trained for 1.5 hours in the morning and 1.5 hours in the afternoon. A second group participated only in the afternoon session. Costill found that the group that trained the most experienced a decline in speed, whereas the second group showed an improvement.

So, doubling the training time did not enhance performance.[4] Costill suggested that the Ball State swimming coach cut training time in half: Instead of having the team work out twice a day, the coach cut out the morning workout. The team now consisted of more rested swimmers who swam faster than before; they ended with their best season in 10 years.

Costill's work suggests that when athletes train too much, they are not able to maintain muscle glycogen, and increases in the amount of energy and carbohydrate in the diet may not be able to keep up. Performance can be optimized by a combination of proper diet and the right amount of exercise. Just as dietary excesses and deficiencies can hurt performance, too little or too much training can result in less-than-peak performance. When excessive training continues, athletes are at risk for overtraining syndrome, characterized by underperformance, persistent fatigue, altered mood, and increased rates of infection.[5] Athletes can help prevent overtraining by resting one day each week, alternating hard and easy training days, and optimizing their nutrition, especially carbohydrate and energy intake.

[1]The Champion Within. Infinite Voyage Video Series, Intellimation, Inc. Santa Barbara, C A: QED Communications, Inc., and Washington, D.C.: National Academy of Sciences, 1991.

[2]Costill, D. L, Flynn, M. G., and Kirwan, J. P. Effects of repeated days of intensified training on muscle glycogen and swimming performance. *Med. Sci. Sports Exerc.* 20:249-254, 1988.

[3]Morgan,W. P., Costill, D. L, Flynn, M. G. et al. Mood disturbances following increased training in swimmers. *Med. Sci. Sports Exerc.* 20:408–414, 1988.

[4]Costill, D. L, Thomas, R., Robergs, R. A. et al. Adaptations to swimming training: Influence of training volume. *Med. Sci. Sports Exerc.* 23:371-377, 1991.

[5]Pearce, P. Z. A practical approach to the overtraining syndrome. *Curr. Sports Med. Rep.* 1:179-183, 2002.

Anaerobic metabolism uses glucose rapidly. Because the amount of glucose available is limited, if activity is to continue, the body must use its glucose more efficiently and find a more plentiful fuel source.

Long-Term Energy: Aerobic Metabolism After 2-3 min. of exercise, breathing and heart rate have increased to supply more oxygen to the muscles. When oxygen is available, ATP can be produced by aerobic metabolism. The reactions of aerobic metabolism take place in the mitochondria (see Figure 13.15). When glucose is broken down by aerobic metabolism, the pyruvate produced by glycolysis is converted to acetyl-CoA, so lactic acid is not formed. Acetyl-CoA is broken down by the citric acid cycle, producing carbon dioxide and some ATP and releasing high-energy electrons. The electrons are shuttled to the electron transport chain, where their energy is harnessed to produce ATP and water is formed.

Aerobic metabolism produces ATP at a slower rate than anaerobic metabolism but is much more efficient, producing about 18 times more ATP for each molecule of glucose. In addition, aerobic metabolism can use fatty acids, and sometimes amino acids from protein,

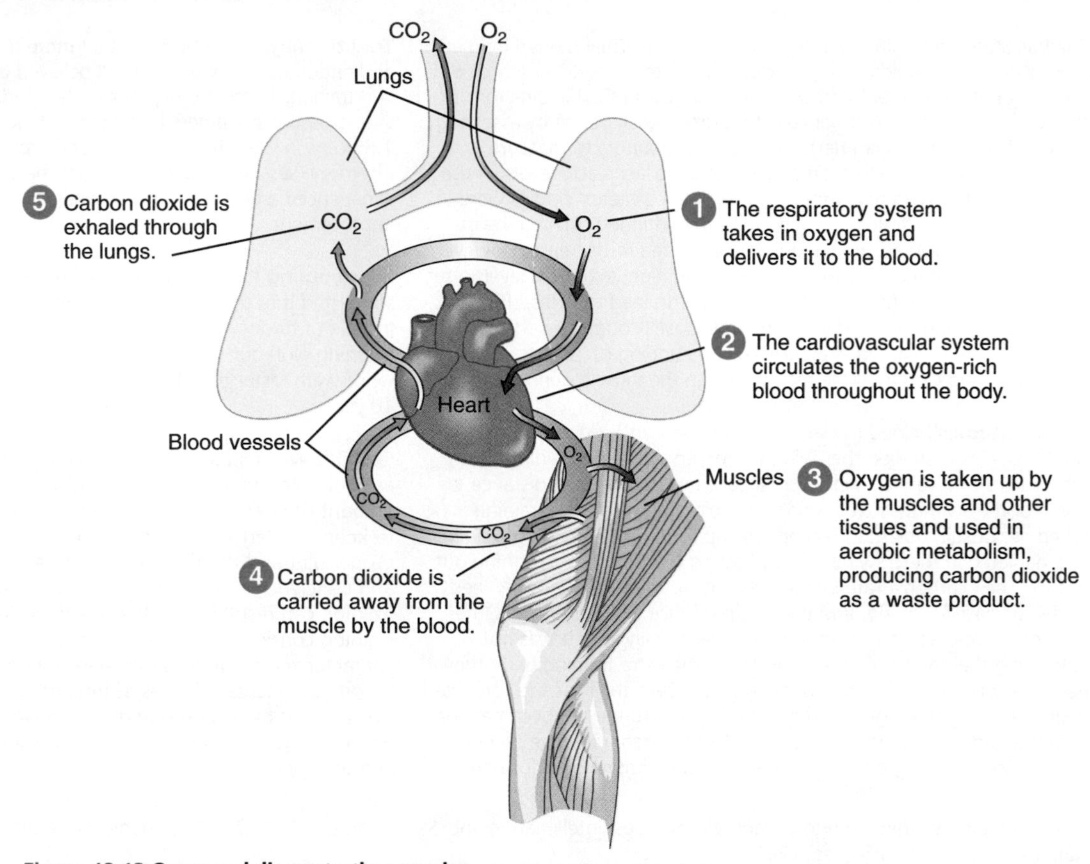

Figure 13.13 Oxygen delivery to the muscles
When you exercise, your muscles demand more energy, which requires oxygen. Your body responds by breathing faster and deeper in order to take in more oxygen and by increasing heart rate in order to deliver more oxygen to your muscles.

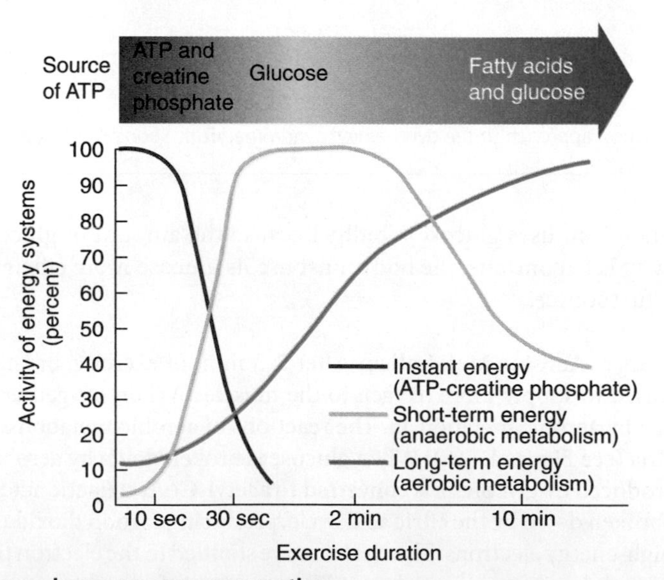

Figure 13.14 Change in energy sources over time
The ATP for muscle contraction is first derived from ATP and creatine phosphate stored in the muscle, after which anaerobic glycolysis, which breaks down glucose, becomes the predominant source of ATP. After about 2-3 min., oxygen delivery to the muscles has increased enough for aerobic metabolism, which uses fatty acids and glucose to produce ATP, to make a significant contribution to ATP production.

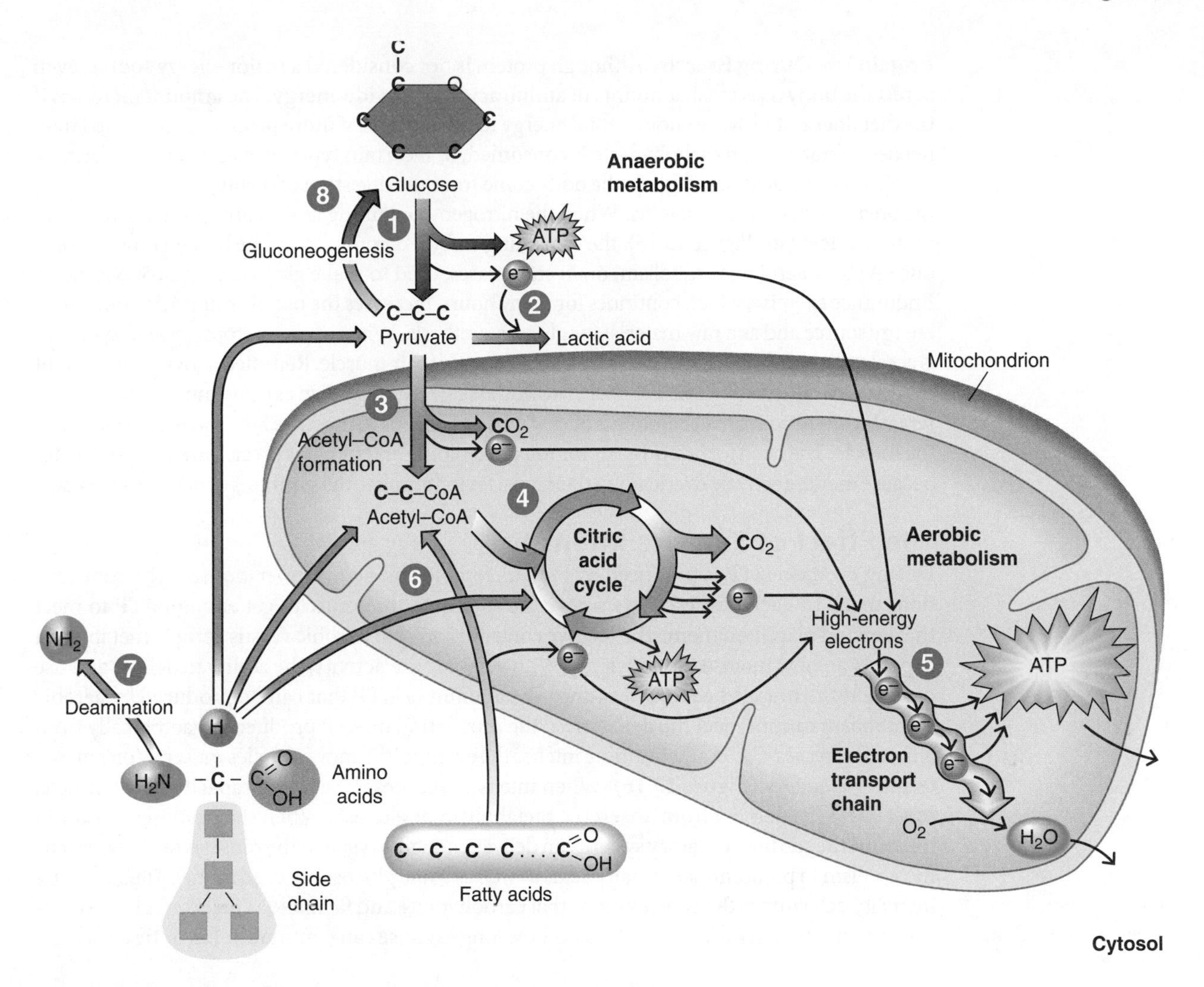

1. Glucose is split into 2 molecules of pyruvate, releasing electrons, and producing 2 molecules of ATP.
2. In the absence of oxygen, the pyruvate and released electrons combine to form lactic acid.
3. When oxygen is available, the pyruvate is converted to acetyl-CoA.
4. Acetyl-CoA is broken down by the citric acid cycle.
5. The electrons released are shuttled to the electron transport chain, where their energy is harnessed to produce ATP.
6. Fatty acids are broken into 2-carbon units that form acetyl-CoA.
7. Amino acids are deaminated and then used as an energy source.
8. After deamination, some amino acids can be used to synthesize glucose by gluconeogenesis.

Figure 13.15 Anaerobic versus aerobic metabolism

In the absence of oxygen, ATP is produced by the anaerobic glycolysis of glucose. When oxygen is present fatty acids and amino acids can also be used for energy.

to generate ATP (see Figure 13.15). Fatty acids to fuel muscle contraction come from triglycerides stored in adipose tissue as well as small amounts stored in the muscle itself. During exercise, triglycerides are broken down into fatty acids and glycerol. Fatty acids from adipose tissue are released into the blood and are then taken up by the muscle cells. Inside the muscle cell, fatty acids from triglycerides within the muscle and those delivered by the blood must be transported into the mitochondria to produce ATP. To enter the mitochondria, fatty acids must be activated with the help of **carnitine**. Inside the mitochondria, fatty acids are broken into 2-carbon units by beta-oxidation to form acetyl-CoA (see Figure 13.15). Acetyl-CoA is metabolized via the citric acid cycle and electron transport chain to produce ATP, carbon dioxide, and water.

carnitine A molecule synthesized in the body that is needed to transport fatty acids and some amino acids into the mitochondria for metabolism.

When exercise continues at a low to moderate intensity, aerobic metabolism predominates and fat becomes the principal fuel source for exercising muscles. If exercise intensity increases, the proportion of energy generated by anaerobic versus aerobic metabolism changes, as do the relative amounts of glucose and fatty acids used.

Protein Use During Exercise Although protein is not considered a major energy source, even at rest the body uses small amounts of amino acids to provide energy. The amount increases if the diet does not provide enough total energy to meet needs, if more protein is consumed than needed, if not enough carbohydrate is consumed, or if certain types of exercise are performed.

The amino acids available to the body come from the digestion of dietary proteins and from the breakdown of body proteins. When the nitrogen-containing amino group is removed from an amino acid (see Figure 13.15), the remaining carbon compound can be broken down to produce ATP by aerobic metabolism, or, in some cases, used to make glucose via gluconeogenesis. Endurance exercise, which continues for many hours, increases the use of amino acids both as an energy source and as a raw material for glucose synthesis. When exercise stops, protein synthesis is accelerated so amino acids are needed to build and repair muscle. Repeated activity with a slight overload stimulates the muscle to adapt to the stress by breaking down existing muscle proteins and replacing them with greater amounts of new muscle proteins to meet the higher demand placed on the muscle. The need for amino acids for muscle building and repair is greater in strength athletes because they are actively overloading their muscles to stimulate the synthesis of new muscle tissue.

The Effect of Exercise Intensity

During exercise, ATP is produced by both anaerobic and aerobic metabolism. The contributions made by each of these systems overlap to ensure that muscles get enough ATP to meet the demand placed on them. The relative contribution of anaerobic versus aerobic metabolism depends on how intense the activity is. With very intense activity, the ability to deliver and use oxygen at the muscle becomes limiting. The amount of ATP that can be produced by aerobic metabolism cannot meet the demand, so the proportion of ATP produced anaerobically from glucose increases. Generally, the more intense the exercise, the more muscles must rely on glucose to provide energy (**Figure 13.16**). When intensity reaches the aerobic capacity of the athlete, most energy is derived from anaerobic metabolism of glucose. When the exercise is lower in intensity, the cardiorespiratory system can deliver enough oxygen to the muscles to allow aerobic metabolism to predominate, so fatty acids as well as some glucose are used as fuel. Thus, exercise intensity determines the contributions that carbohydrate and fat make as fuels for ATP production. In turn, which fuels are used affects how long exercise can continue before **fatigue** occurs.

fatigue The inability to continue an activity at an optimal level.

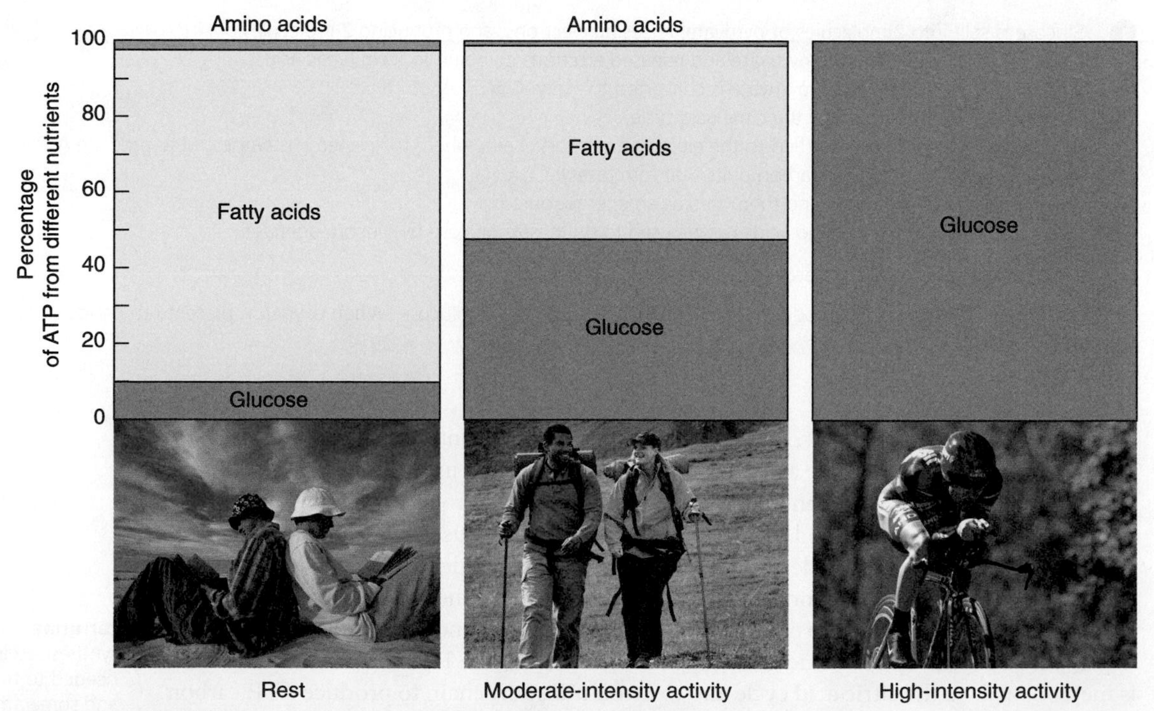

Figure 13.16 The effect of exercise intensity on fuel use
As exercise intensity increases, the proportion of energy supplied by carbohydrate increases. During exercise, the total amount of energy expended is greater than during rest.
Source: Adapted from Horton, E. S. Effects of low-energy diets on work performance. *Am. J. Clin. Nutr.* 35:1228–1233, 1982. Yva Momatiuk and John Eastcott/Minden Pictures/NG Image Collection; Alaska Stock Images/NG Image Collection; Rich Reid/NG Image Collection

High-Intensity Exercise Contributes to Fatigue If you run faster, you tire sooner. This is because more intense exercise relies more on anaerobic metabolism, which uses glucose more rapidly than aerobic metabolism and produces lactic acid. Until recently, it was assumed that lactic acid buildup was the cause of muscle fatigue, but we now know that although lactic acid buildup is associated with fatigue, it does not cause it.[19] Fatigue most likely has many causes, including glycogen depletion and changes in the muscle cells and the concentrations of molecules involved in muscle energy metabolism.

When athletes run out of glycogen, they experience a feeling of overwhelming fatigue that is sometimes referred to as "hitting the wall" or "bonking." Glycogen depletion is a concern for athletes because the amount of glycogen available to produce glucose during exercise is limited. There are between 60-120 g of glycogen stored in the liver; stores are highest just after a meal. Liver glycogen is used to maintain blood glucose between meals and during the night. Eating breakfast replenishes the liver glycogen used overnight. There are about 200-500 g of glycogen in the muscles of a 70-kg person. The glycogen in a muscle is used to fuel the activity of that muscle. Muscle glycogen levels can be increased by a combination of rest and a very high-carbohydrate diet.

Small amounts of lactic acid produced by anaerobic metabolism can be carried away from the muscle and used by other tissues as a fuel or by the liver to produce glucose. But if the amount of lactic acid produced exceeds the amount that can be used by other tissues, this by-product begins to build up in the muscles and subsequently the blood. **Anaerobic threshold** or **lactate threshold** refers to the exercise intensity at which lactic acid starts to accumulate in the blood faster than it can be metabolized. This is normally somewhere between 80%-90% of maximum heart rate. Although the cause of muscle fatigue is not fully understood, it correlates with lactic acid buildup. When exercise stops and oxygen is available again, lactic acid can be either carried away by the blood to other tissues to be broken down or metabolized aerobically in the muscle.

anaerobic threshold or **lactate threshold** The exercise intensity at which the reliance on anaerobic metabolism results in the accumulation of lactic acid.

Low-Intensity Exercise Can Continue Longer Lower-intensity exercise can be continued for longer periods because it relies on aerobic metabolism, which is more efficient than anaerobic metabolism and uses both glucose and fatty acids for energy. The body's fat reserves are almost unlimited, so if fat is the fuel, exercise can theoretically continue for a very long time. For example, it is estimated that a 60 kg (130-lb) woman has enough energy stored as body fat to run 1,000 miles.[20] However, even aerobic activity uses some glucose, so if exercise continues long enough, glycogen stores will eventually be depleted and contribute to fatigue.

The Effect of Training

Training with repeated bouts of aerobic exercise causes physiological changes that increase aerobic capacity—the amount of oxygen that can be delivered to and used by the muscle cells. This in turn affects which fuels can be used by the exercising muscle cells.

Aerobic exercise causes adaptations in the cardiorespiratory system. The heart becomes larger and stronger so that the stroke volume is increased (**Figure 13.17**). The number of

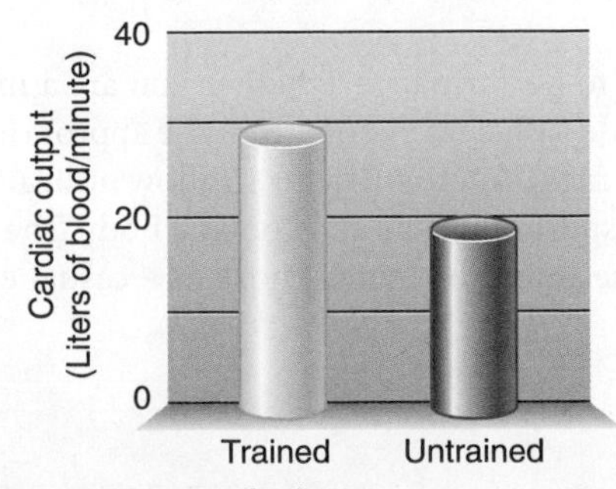

Figure 13.17 Effect of exercise training on the heart
Training increases the amount of blood pumped with each beat, so the heart of a trained athlete can pump more blood per minute than can the heart of an untrained individual.

capillary blood vessels in the muscles increases so that blood is delivered to muscles more efficiently. And the total blood volume and number of red blood cells expands, increasing the amount of hemoglobin so more oxygen can be transported to the cells.

Training also causes changes at the cellular level that affect the ability of cells to use different types of fuel to produce ATP. There is an increase in the ability to store glycogen, and there is an increase in the number and size of muscle-cell mitochondria (**Figure 13.18**). Because aerobic metabolism occurs in the mitochondria, this increases the cell's capacity to burn fatty acids to produce ATP. The use of fatty acids spares glycogen, which delays the onset of fatigue. Because trained athletes store more glycogen and use it more slowly, they can sustain aerobic exercise for longer periods at higher intensities than can untrained individuals. Conditioned athletes can also exercise at a higher percentage of their aerobic capacity before lactic acid begins to accumulate. Canadian researchers are examining ways to structure exercise sessions to maximize aerobic capacity in the shorter periods (see Critical Thinking: The Benefits of Interval Training).

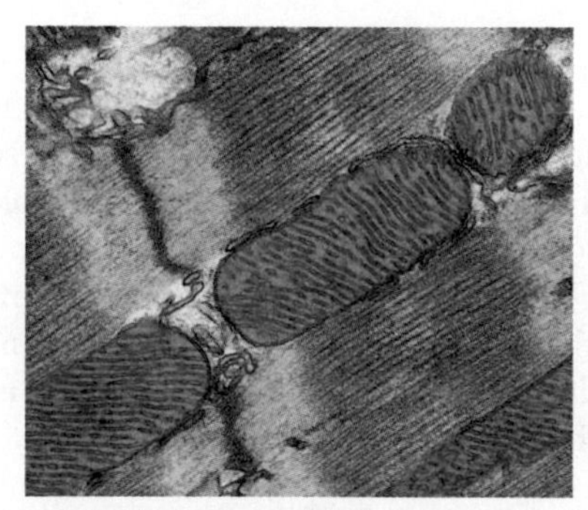

Figure 13.18 Training increases the number of mitochondria in muscle cells, which increases aerobic capacity. (Thomas Deerinck, NCMIR/Photo Researchers, Inc.)

Living and working at high altitudes, where the atmosphere contains less oxygen, also causes adaptations that improve the ability of the cardiorespiratory system to deliver oxygen. This is why endurance athletes often train at high altitudes—to enhance their aerobic capacity.

13.4 Energy and Nutrient Needs for Physical Activity

Learning Objectives

- Compare the energy and macronutrient needs of athletes and nonathletes.
- Explain why athletes are at risk for iron deficiency.
- Describe the female athlete triad.

Good nutrition is essential to performance, whether you are a marathon runner or a mall walker. The diet must provide sufficient energy from the appropriate sources to fuel activity, protein to maintain muscle mass, micronutrients to allow utilization of the energy-yielding nutrients, and water to transport nutrients and cool the body. The major difference between the nutritional needs of a serious athlete and those of a casual exerciser is the amount of energy and fluid required.

CRITICAL THINKING:

CANADIAN CONTENT

The Benefits of Interval Training

One of the main reasons people give for not exercising is that they don't have the time. The Exercise Metabolism Research Group at McMaster University in Hamilton, Ontario, was interested in determining whether the amount of time spent exercising could be reduced without compromising fitness, so they conducted an intervention trial involving 10 young men and 10 young women, all sedentary, with an average age of 23 years.[1] Two approaches to exercising were studied: sprint interval training (SIT) and traditional endurance training (ET). In interval training, subjects go "all out" on an exercise bicycle for 1min. followed by a recovery period of 4.5 min. of slower cycling for a total of 30 min. This is repeated 3 times a week for a total weekly time commitment of 1.5 hours. Traditional endurance training involved cycling 45-60 min. per session at moderate intensity, 5 days a week, for an average weekly commitment of 4.5 hours. Half the subjects were assigned to SIT and half to ET. Despite the large differences in time commitments, both groups had similar improvements in the aerobic capacity of muscle, that is, the ability of their muscles to oxidize carbohydrates and lipids as fuel. This was determined by looking at the levels of proteins, in muscle cells, involved in the citric acid cycle and electron transport chain. The proteins were found to be similarly increased in both SIT and ET groups, compared to levels at the beginning of the experiment.

The McMaster group was next interested in looking at whether interval training could be effective with older subjects, so they tested the impact of interval training on 7 subjects, 3 women and 4 men, with an average age of 43 years. The interval training pattern was modified in this experiment; subjects alternated between 1 min of high intensity exercise and 1 min of lower intensity exercise for a total of 20 min.[2] As in the previous experiment, researchers measured levels of proteins associated with aerobic capacity and found that in the older subjects, aerobic capacity improved significantly, just as it had in their younger subjects.

The researchers also measured the levels of a certain protein called the GLUT4 transporter, which is a protein on the cell membrane of muscle cells that helps transport glucose from the blood into the cell.

Critical Thinking Questions

What do you think the researchers observed when they measured GLUT4 levels? Explain why.

What are the implications of these studies?

What is one limitation of these two studies?

[1] Burgomaster, K. A., Howarth, K. R., Phillips, S. M. et al. Similar metabolic adaptations during exercise after low volume sprint interval and traditional endurance training in humans. *J. Physiol.* 1;586(1):151-60,2008.

[2] Hood, M. S., Little, J. P., Tarnopolsky, M. A. et al. Low-Volume Interval Training Improves Muscle Oxidative Capacity in Sedentary Adults. *Med. Sci. Sports Exerc.* 43(10):1849-56, 2011.

Energy Needs

The amount of energy needed for activity depends on the intensity, duration, and frequency of the activity, as well as the characteristics of the exerciser, and even his or her location. For a casual exerciser, the energy needed for activity may increase energy expenditure by only a few hundred kcal/day. For an endurance athlete, such as a marathon runner, the energy needed for training may increase expenditure by 2,000-3,000 kcal/day. Some athletes require 6,000 kcal/day to maintain body weight. In general, the more intense the activity, the more energy it requires, and the more time spent exercising, the more energy it burns (**Table 13.2** and Appendix H). For example, walking for 30 min. involves less work than running for the same amount of time and therefore requires less energy. Riding a bicycle for 60 min. requires 6 times the energy needed to ride for 10 min. The body weight of the exerciser is another factor in determining energy needs. Moving a heavier body requires more energy than moving a lighter one, so it requires less energy for a 55-kg (120-lb) woman to walk for 30 min. than it does for a 110-kg (240-lb) woman.

Table 13.2 Kcalorie Needs for Various Activities

Body Weight (kg)		**45**	**57**	**64**	**70**	**77**	**84**	**91**
Activity	**Rate**	**Energy (kcal/hr)**						
Bicycling	< 16 km/h	137	171	191	212	233	252	273
	16-19.2 km/h	228	285	318	353	388	420	455
	19.3-22.4 km/h	319	399	445	494	543	588	637
	22.5-25.6 km/h	410	513	572	635	698	756	819
	25.7-30.5 km/h	501	627	699	776	853	924	1,001
Running	7.5 min/km	319	399	445	494	543	588	637
	6.3 min/km	410	513	572	635	698	756	819
	5 min/km	523	713	730	811	891	966	1,047
	3.8 min/km	683	855	953	1,058	1,163	1,260	1,365
Skiing, cross-country	4 km/h	273	342	381	423	465	504	546
	6.4-7.8 km/h	319	399	445	494	543	588	637
	8-12.7 km/h	364	456	508	564	620	672	728
Swimming	leisurely	228	285	318	353	388	420	455
	45 m/min	319	399	445	494	543	588	637
	70 m/min	455	570	635	705	775	840	910
Walking	3.2 km/h	68	86	95	106	116	126	137
	4.8 km/h	105	131	146	162	178	193	209
	6.4 km/h	182	228	254	282	310	336	364

The DRIs have developed equations to estimate energy requirements based on an individual's age, gender, size, and physical activity (PA) level (see section 7.3 and inside cover). For the purposes of calculating estimated energy requirement (EER), an individual who performs no exercise other than the activities of daily living is in the "sedentary" PA category and one who performs less than an hour of moderate activity fits into the "low-active" PA category. Someone who engages in 60 min. of moderate exercise each day is considered to be in the "active" PA category (see Chapter 7, Table 7.4). An "active" activity level can be achieved with less than 60 min. of exercise if the exercise is more intense, for example, jogging at 8 km/h or greater or swimming at a moderate to fast pace. Individuals who perform moderate exercise for more than 2.5 hour/day or more intense exercise for more than 1 hour/day are in the "very active" PA category. Calculating EER at different activity levels can demonstrate the dramatic impact activity can have on energy needs.[3] For example, the EER for a 25-year-old, sedentary, 1.8 m (5′11″) tall, 70 kg (154-lb.) man is about 2,500 kcal. If this same person becomes a runner and trains several hours/day, his energy needs may increase to 3,500 kcal/day or more (**Figure 13.19**).

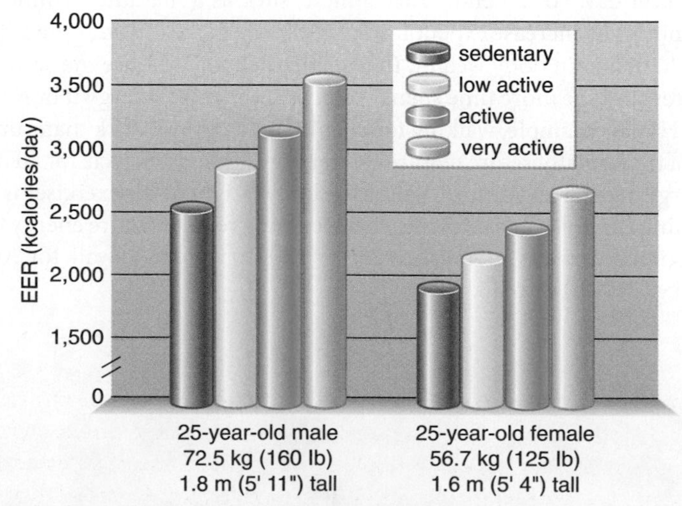

Figure 13.19 Effect of activity level on energy expenditure
Changing the amount of activity you routinely engage in can have a significant impact on your EER.

There are also some special considerations that affect the energy needed for activity. For example, because of the buoyancy of adipose tissue, the energy required for an individual with excess body fat to swim may be less than for his or her lean counterpart. If a lean individual and an obese individual were in the weightlessness of space, it would require no more energy for one to leap across the room than for the other. There are also special circumstances that affect the amount of energy an individual needs for daily activity. A paraplegic in a wheelchair may have lower energy needs because many of the major muscles in the body are always inactive. At the other extreme, a person with the form of cerebral palsy that causes uncontrolled muscle movements may have higher energy needs because the muscles never stop moving.

Weight Loss Body weight and composition can affect exercise performance. Athletes involved in activities where small, light bodies offer an advantage—for instance, ballet, gymnastics, and certain running events—may restrict energy intake to maintain a low body weight. While a slightly leaner physique may be beneficial, dieting to maintain an unrealistically low weight may threaten health and performance. The general guidelines for healthy weight loss should be followed—reduce energy intake, increase activity, and change the behaviours that led to weight gain (see Section 7.8). To preserve lean body mass and enhance fat loss, weight should be lost at a rate of about 0.25 – 1 kg/week (0.5-2 lb./week). This can be accomplished by reducing total energy intake by 200-500 kcal/day and increasing exercise. An athlete who needs to lose weight should do so in advance of the competitive season to prevent the restricted diet from affecting performance.[21]

Unhealthy Weight-Loss Practices Athletes are often under extreme pressure to achieve and maintain a body weight that optimizes their performance. Failure to meet weight-loss goals may have serious consequences such as being cut from the team or restricted from competition. This pressure may compel some athletes to use strict diets and maintain body weights that are not healthy. This, combined with the self-motivation and discipline that characterizes successful athletes, makes them vulnerable to eating disorders such as anorexia and bulimia (see Chapter 15: Focus on Eating Disorders).[22] In athletes with anorexia, restricted food intake causes a deficiency of energy and nutrients, which can affect growth and maturation and impair exercise performance. In athletes with bulimia, purging can cause dehydration and electrolyte imbalance, which affect performance and put overall health at risk. In addition to using restricted food intake or purging to keep body weight low, some athletes focus on the other side of the energy balance equation by exercising compulsively to burn kcalories (see Chapter 15 Focus on Eating Disorders).

Athletes involved in sports with weight classes, such as wrestling and boxing, are at particular risk for unhealthy weight-loss practices because they are under pressure to lose weight before a competition so they can compete in lower weight classes (**Figure 13.20**). Competing at the high end of a weight class is thought to give one an advantage over smaller opponents. To lose weight rapidly, these athletes may use sporadic diets that severely restrict energy intake or dehydrate themselves through such practices as vigorous exercise, fluid restriction, wearing plastic vapour-impermeable suits, and using hot environments such as saunas and steam rooms. They may also resort to even more extreme measures, such as self-induced vomiting and the use of diuretics and laxatives. These practices can be dangerous and even fatal. They may impair performance and can adversely affect heart and kidney function, temperature regulation, and electrolyte balance. In 1997, three young wrestlers died while trying to "make weight."[23] As a result of these deaths, wrestling weight classes were altered to eliminate the lightest weight class, plastic sweat suits were banned, a maximum wrestling room temperature of 75°F was established, weigh-ins were moved to 1 hour before competition, and mandatory weight-loss rules were put in place restricting the amount of weight that could be lost. There are minimum limits for percent body fat of 5% for college wrestlers and 7% for high school wrestlers.

Figure 13.20 Energy restriction and dehydration used by wrestlers to keep weight down can harm health and impair exercise performance. (Enigma/Alamy)

Weight Gain In sports such as football and weight lifting in which being large is advantageous, an increase in body weight may be desirable. To gain weight, 500-1000 extra kcal/day should be consumed. Strength training should accompany weight gain to promote an increase in lean tissue. To increase muscle mass, many amateur and professional athletes have been drawn to the use of hormones called anabolic steroids. Although they do stimulate muscle growth, they are dangerous and illegal (see Your Choice: Ergogenic Hormones: Anything for an Edge).

Carbohydrate, Fat, and Protein Needs

The source of dietary energy can be as important as the amount of energy in an athlete's diet. In general, the diets of physically active individuals should contain the same proportion of carbohydrate, fat, and protein as is recommended to the general public—about 45%-65% of total energy as carbohydrate, 20%-35% of energy as fat, and 10%-35% of energy as protein.

Carbohydrate Carbohydrate is needed to maintain blood glucose levels during exercise and to replace glycogen stores after exercise. The amount recommended for athletes depends on the total energy expenditure, type of sport, gender, and environmental conditions but ranges from 6-10 g/kg of body weight per day.[21] For a 70-kg (154-lb) person burning 3,000 kcal/day, this would be about 60% of kcalories or about 450 g of carbohydrate. Most of the carbohydrate in the diet should be complex carbohydrates from whole grains and starchy vegetables, with some naturally occurring simple sugars from fruit and milk. These foods provide vitamins, minerals, phytochemicals, and fibre as well as energy. Snacks and meals consumed before or during exercise should be lower in fibre to avoid cramping and gastrointestinal distress.

Fat Dietary fat supplies essential fatty acids, ensures the absorption of fat-soluble vitamins, and provides an important source of energy. Body stores of fat provide enough energy to support the needs of even the longest endurance events. For physically active individuals, diets providing 20%-25% of energy as fat have been recommended to allow adequate carbohydrate intake.[21] Diets too high in fat do not contain enough carbohydrate to maximize glycogen stores and optimize performance. Excess dietary fat is unnecessary and excess energy consumed as fat, carbohydrate, or protein can cause an increase in body fat. Diets very low in fat (less than 20% of kcalories) have not been found to benefit performance.

Protein Protein is not a significant energy source, accounting for only about 5%-10% of energy expended, but dietary protein is needed to maintain and repair lean tissues, including muscle. Enough protein is essential to maintain muscle mass and strength, but eating extra protein does not produce bigger muscles. Muscle growth is stimulated by exercise, not by increasing protein intake.

A diet that contains the RDA for protein (0.8 g/kg) provides adequate protein for most active individuals. Competitive athletes participating in endurance and strength sports may require more protein.[21] In endurance events such as marathons, protein is used for energy and to maintain blood glucose so these athletes may benefit from 1.2-1.4 g/kg/day of protein. Strength athletes who require amino acids to synthesize new muscle proteins may benefit from 1.2-1.7 g/kg /day. While this amount is greater than the RDA, it is not greater than the amount of protein habitually consumed by athletes.[24] For example, an 85-kg man consuming 3,000 kcal/day, 18% of which is from protein, would be consuming 135 g, or 1.6 g of protein per kilogram of body weight.

Vitamin and Mineral Needs

An adequate intake of vitamins and minerals is essential for optimal performance. These nutrients are needed for energy metabolism, oxygen delivery, antioxidant protection, and repair and maintenance of body structures. During exercise, the amounts of many vitamins and minerals used in energy metabolism are increased and after exercise, the amounts of those needed to repair tissue damage are increased. Exercise may also increase the losses of some nutrients. Nonetheless, most athletes can meet their needs by consuming the amounts of vitamins and minerals recommended for the general population. In addition, because athletes must eat more food to satisfy their higher energy needs, they consume extra vitamins and minerals in these foods, particularly if nutrient-dense choices are made. Athletes who restrict their intake to maintain a low body weight may be at risk for vitamin or mineral deficiencies.

B Vitamins B vitamins such as thiamin, riboflavin, and niacin are important for the production of ATP from carbohydrates and fat. Vitamin B_6, folate, and vitamin B_{12} are needed for proper synthesis of red blood cells, which deliver oxygen to body tissues. Vitamin B_6 is needed to break down glycogen to release glucose and to make the protein hemoglobin, which carries oxygen in red blood cells. Despite the importance of all of these roles during exercise, the recommended intake of B vitamins is not any greater for athletes than for the rest of the population.

YOUR CHOICE

(©iStockphoto)

Ergogenic Hormones: Anything for an Edge

Athletes are always searching for things that will help them sprint faster, jump higher, or run farther. Some of the most effective substances used are hormones; however, they are also some of the most deadly. Because of the unfair advantage offered by these products, their use is banned in most athletic competitions, and because of their risks, many are also illegal (see Table 13.6).

Anabolic Steroids have muscle-building effects. They attracted the attention of the athletic community when athletes from Eastern European nations began to dominate international strength events. These hormones accelerate protein synthesis and growth. The anabolic steroids used by athletes are synthetic versions of the human steroid hormone testosterone. Natural testosterone stimulates and maintains the male sexual organs and promotes the development of bones and muscles and the growth of skin and hair. The synthetic testosterone used by athletes has a greater effect on muscle development and bone, skin, and hair than it does on sexual organs. When synthetic testosterone is taken in conjunction with exercise and an adequate diet, muscle mass increases. However, these drugs also make the body think testosterone is being produced, and therefore the production of natural testosterone is reduced. Without natural testosterone, the sexual organs are not maintained; this leads to testicular shrinkage and a decrease in sperm production.[1] In adolescents, the use of synthetic testosterone causes bone growth to stop and height to be stunted. Use may also cause oily skin and acne, water retention, yellowing of eyes and skin, coronary artery disease, liver disease, and sometimes death. Psychological and behavioural side effects may lead to suicide. Because steroids are illegal and their manufacturing and distribution are not regulated, users can't be sure of their potency and purity.

Steroid Precursors are compounds that the body can convert into steroid hormones. The best known is androstenedione, or "andro," a precursor to testosterone marketed as an alternative to anabolic steroids to increase testosterone levels. The majority of studies have not supported that contention, but an increase in estrogen concentration has been seen.[1] In addition, no studies have shown a significant ergogenic effect. Andro has been shown to lower HDL cholesterol, and therefore increase cardiovascular disease risk. Because andro is metabolized to estrogen and testosterone in the body, it may cause hormonal imbalances that could result in symptoms such as shrinkage of the testicles, impotence, and breast enlargement in men, and male pattern baldness, deepening voice, increased facial hair, and abnormal menstrual cycles in women. In children it may cause earlier puberty and stunt bone growth. Steroid precursors, like anabolic steroids, are classified as controlled substances (substances with the potential for abuse). They have been banned by all major national and international sports federations.

Human Growth Hormone is produced by the pituitary gland. In children it is important for tissue building during growth; in adults it maintains lean tissue, stimulates fat breakdown, increases the number of red blood cells, and boosts heart function. Genetically engineered growth hormone is used to treat children with growth failure. This hormone is appealing to athletes because it increases muscle protein synthesis, but ergogenic benefits among athletes remain unproven.[2] Prolonged use of growth hormone can cause heart dysfunction and high blood pressure, as well as excessive growth of the hands, feet, and facial features.

EPO, which is short for erythropoietin, is another popular hormone among endurance athletes. Natural erythropoietin is produced by the kidneys and stimulates cells in the bone marrow to differentiate into red blood cells. Genetically engineered EPO is used to treat anemia due to kidney disease, chemotherapy, HIV infection, and blood loss. It can enhance the performance of endurance athletes by increasing the ability to transport oxygen to the muscles.[3] It therefore increases aerobic capacity and spares glycogen. However, too much EPO can cause production of too many red blood cells and lead to excessive blood clotting, heart attacks, and strokes. The International Olympic Committee banned EPO in 1990 after it was linked to the death of more than a dozen cyclists.[4]

[1]Tokish, J. M., Kocher, M. S., and Hawkins, R. J. Ergogenic aids: A review of basic science, performance, side effects, and status in sports. *Am. J. Sports Med.* 32:1543–1553, 2004.

[2]Jenkins, P. J. Growth hormone and exercise. *Clin. Endocrinol.* 50:683–689, 2000.

[3]Ritter, S. K. Faster, higher, stronger. *Chemical Engineering News* 77:42–52, 1999.

[4]Birkeland, K. I., Stray-Gundersen, J., Hemmersbach, P. et al. Effect of rhEPO administration on serum levels of sTfR and cycling performance. *Med. Sci. Sports Exerc.* 32:1238–1243, 2000.

Antioxidant Nutrients Exercise increases the amount of oxygen at the muscle and the rate of metabolic reactions that produce ATP; oxygen utilization in active muscles can rise as much as 200-fold above resting levels.[25] This increased oxygen use increases the production of free radicals that can lead to oxidative damage and contribute to muscle fatigue.[26] Antioxidants, such as vitamins C and E, β-carotene, and selenium, help protect the body from oxidative damage. Despite the increase in free radical production that occurs during exercise, there is no evidence that supplementation of antioxidants improves performance or that athletes require more of these nutrients than the general public.[27]

Iron and Anemia Iron is important for exercise because it is required for the formation of hemoglobin and myoglobin and a number of iron-containing proteins that are essential for the production of ATP by aerobic metabolism. Although a specific RDA has not been set for athletes, the DRIs acknowledge that based on iron losses, the EAR may be 30%-70% higher for athletes than for the general population.[28]

It is not uncommon for athletes, particularly female athletes, to have reduced iron stores.[29,30] If this deficiency progresses to anemia, it can impair exercise performance and reduce immune function. Low iron stores can be caused by inadequate iron intake, increased iron needs, increased iron losses, or a redistribution of iron due to exercise training. Iron intake can be a problem in athletes who are attempting to keep body weight low and in those who consume a vegetarian diet and therefore do not eat meat—an excellent source of readily absorbable heme iron. Iron needs may be increased in athletes because exercise stimulates the production of red blood cells, so more iron is needed for hemoglobin synthesis. An increase in iron losses with prolonged training, possibly because of increased fecal, urinary, and sweat losses, may also occur.[21] Iron losses may also be increased by foot-strike hemolysis, the breaking of red blood cells from impact or the contraction of large muscles in events such as running. Although most of the iron from these cells is recycled, some is lost. Despite this, foot-strike hemolysis rarely causes anemia because the rupture of red blood cells stimulates the production of new ones.[29]

sports anemia Reduced hemoglobin levels that occur as part of a beneficial adaptation to aerobic exercise in which expanded plasma volume dilutes red blood cells.

Some athletes experience a condition known as **sports anemia**, which is a temporary decrease in hemoglobin concentration that occurs during exercise training. This is an adaptation to training that does not seem to impair delivery of oxygen to tissues. It occurs when blood volume expands to increase oxygen delivery, but the synthesis of red blood cells lags behind the increase in plasma volume (**Figure 13.21**).

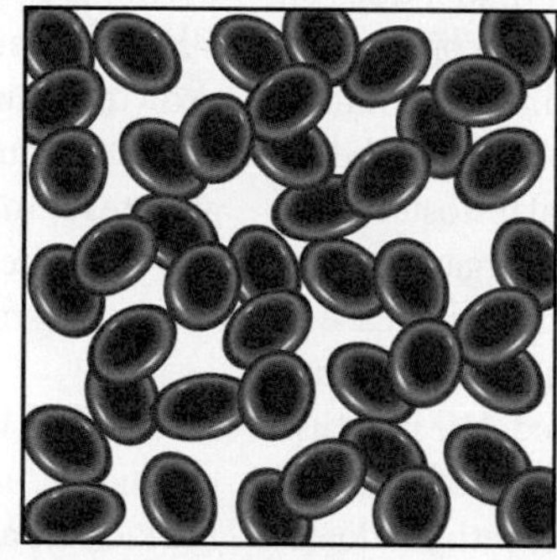

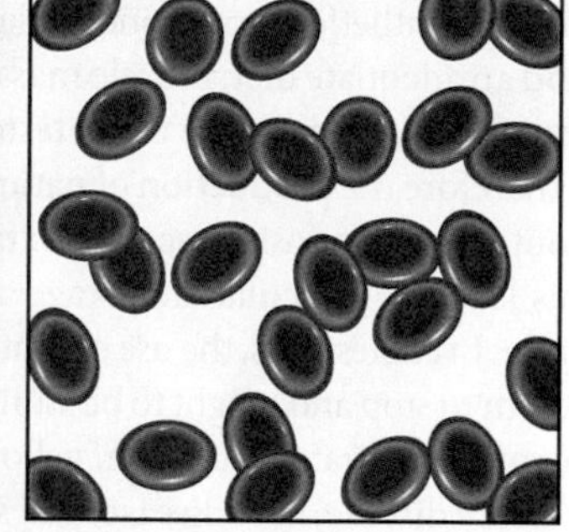

Figure 13.21 Sports anemia
Training causes a decrease in the percentage of blood volume that is red blood cells. As training progresses, the number of red blood cells increases to catch up with the increase in total blood volume.

female athlete triad The combination of disordered eating, amenorrhea, and osteoporosis that occurs in some female athletes, particularly those involved in sports in which low body weight and appearance are important.

amenorrhea Delayed onset of menstruation or the absence of 3 or more consecutive menstrual cycles.

Calcium and Bone Health Calcium is needed to maintain blood calcium levels and promote and maintain healthy bone density, which in turn reduces the risk of osteoporosis. In general, exercise—particularly weight-bearing exercise—increases bone density, thereby reducing the risk of osteoporosis. Although calcium needs are not different for athletes, in female athletes, too much exercise combined with restricted food intake can cause hormonal abnormalities that affect calcium metabolism and put bone health at risk. This **female athlete triad** is a combination of restrictive eating patterns that can lead to eating disorders, abnormalities in hormone levels that cause **amenorrhea**, and disturbances in bone formation and breakdown that contribute to osteoporosis (See chapter 15: Focus on Eating Disorders). Hormonal abnormalities occur when extreme energy restriction and exercise create a physiological condition similar to starvation. Estrogen levels drop, causing amenorrhea. Because estrogen is needed for calcium homeostasis in the bone and calcium absorption in the intestines, low levels lead to premature bone loss, low peak bone mass, and an increased risk of stress fractures. Neither adequate dietary calcium nor the increase in bone mass caused by weight-bearing exercise can compensate for bone loss due to low estrogen levels. Treatment for female athlete triad involves increasing energy intake and reducing activity so that menstrual cycles resume. This is essential for preserving long-term bone health.[31]

13.5 Fluid Needs for Physical Activity

Learning Objectives

- Discuss dehydration in relation to performance and heat-related illness.
- Describe a scenario that might lead to hyponatremia.
- Explain why the types of fluid recommended for a 30-min. workout and a 2-hour workout are different.

During exercise, water is needed to eliminate heat and to transport both oxygen and nutrients to the muscles and waste products away from the muscles. The ability to dissipate the heat generated during exercise is affected by the hydration status of the exerciser as well as by environmental conditions. At rest in a temperate environment, an individual loses about 1.2 L (about 4½ cups) of water per day, or 50 mL /hour, through evaporation from the skin and lungs. Exercise in a hot environment can increase losses more than 10-fold. Even when fluids are consumed at regular intervals during exercise, it may not be possible to drink enough to compensate for losses from sweat and evaporation through the lungs. Failure to consume adequate fluids to replace water lost can be critical to even the most casual exerciser. If heat cannot be lost from the body, body temperature rises and exercise performance as well as health may be jeopardized.

Dehydration

Dehydration occurs when water loss is great enough for blood volume to decrease, thereby reducing the ability to deliver oxygen and nutrients to exercising muscles. Dehydration hastens the onset of fatigue and makes a given exercise intensity seem more difficult. Even mild dehydration—a body water loss of 2%-3% of body weight—can impair exercise performance (**Figure 13.22**).[21] A 3% reduction in body weight can significantly reduce the amount of blood pumped with each heartbeat. This reduces the ability of the circulatory system to deliver oxygen and nutrients to cells and remove waste products. The decrease in blood volume that occurs with dehydration reduces blood flow to the skin and sweat production, which limits the body's ability to sweat and cool itself. Core body temperature can then increase and with it the risk of various **heat-related illnesses**.

heat-related illness Conditions, including heat cramps, heat exhaustion, and heat stroke, that can occur due to an unfavourable combination of exercise, hydration status, and climatic conditions.

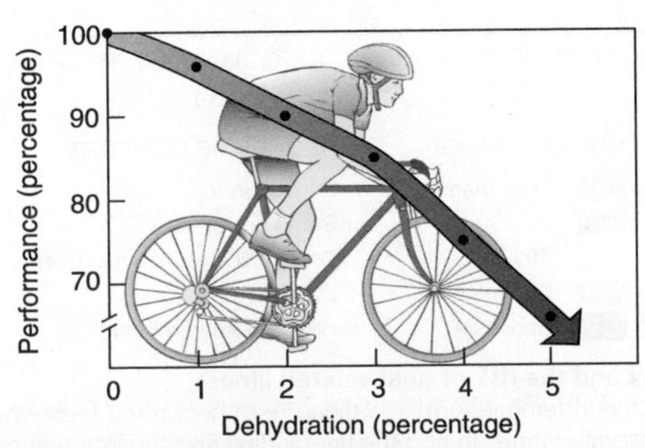

Figure 13.22 Effect of dehydration on exercise performance
As the severity of dehydration increases, exercise performance declines.
Source: Adapted from Saltin, B., and Castill, D. I. Fluid and electrolyte balance during prolonged exercise. In Exercise, Nutrition, and Energy Metabolism. E. S. Horton, and R. I. Tergung, eds. New York: Macmillan, 1988.

The risk of dehydration is greater in hot environments, but it may also occur when exercising in the cold. Cold air tends to be dry air so evaporative losses from the lungs are greater. Insulated clothing may increase sweat losses and fluid intake may be reduced because a chilled athlete may be reluctant to drink a cold beverage if warm or hot beverages are unavailable. Female athletes tend to limit fluid intake to avoid the inconvenience of removing clothing to urinate.[21]

Children, older athletes, and obese individuals are at a greater risk for dehydration and heat-related illness. Children produce more heat, are less able to transfer heat from muscles to the skin, take longer to acclimatize to heat, and sweat less than adults. To reduce risks on hot days, children should rest periodically in the shade, consume fluids frequently, and limit the intensity and duration of activities. Also, children lose more heat in cold environments than adults because they have a greater surface area per unit of body weight. Therefore, they are more prone to **hypothermia**. Older athletes are at greater risk of dehydration because the thirst sensation decreases with age, and the kidneys may be less able to concentrate urine thus increasing the amount of fluid lost in urine. Excess weight increases the risk of heat-related illness because it increases the amount of work and therefore the amount of heat produced in a given activity. The fat also acts as an insulator, retarding the conduction of heat to the body surface. Obese individuals also have a smaller surface area-to-body mass ratio than lean people so they are less efficient at dissipating heat through blood flow to the surface and the evaporation of sweat.

hypothermia A condition in which body temperature drops below normal. Hypothermia depresses the central nervous system, resulting in the inability to shiver, sleepiness, and eventually coma.

Life Cycle

Heat-Related Illness

Exercising in hot weather can lead to heat-related illness. Both the temperature and humidity greatly affect the risk. As environmental temperature rises, it becomes more difficult to dissipate heat and as humidity rises, the ability to cool the body by evaporation declines.[32] When the humidity is high, the same air temperature feels hotter than when the humidity is lower. For example, when the humidity is 85%, a temperature of 28°C feels the same as a temperature of 32°C and a humidity of only 50%. The risks associated with exercising in these conditions are similar (**Figure 13.23**). Conditioning with repeated bouts of exercise can reduce the risk of

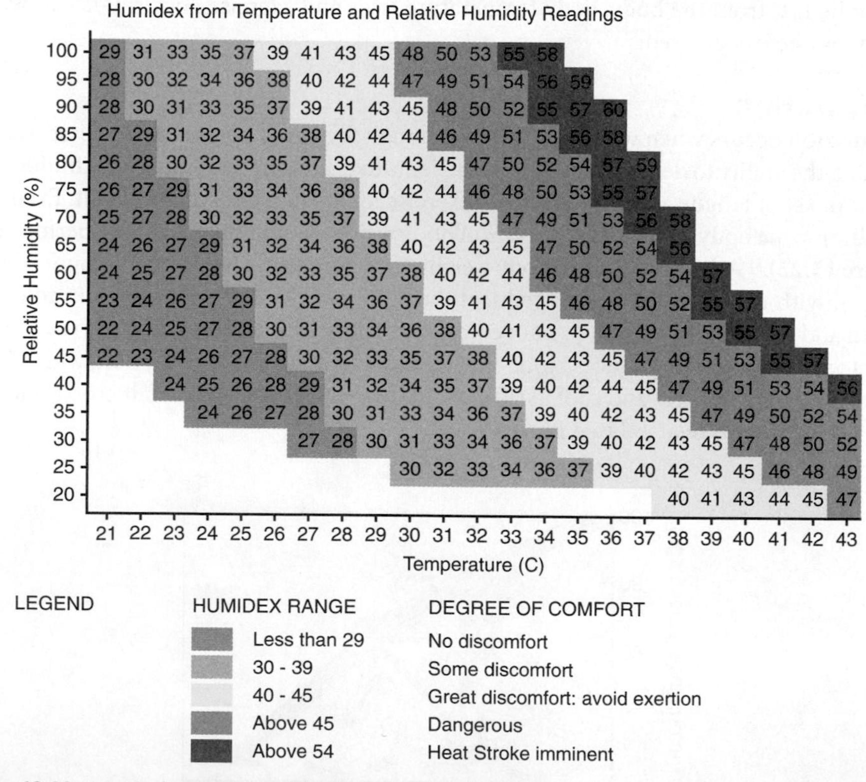

Humidex from Temperature and Relative Humidity Readings

Relative Humidity (%) \ Temperature (C)	21	22	23	24	25	26	27	28	29	30	31	32	33	34	35	36	37	38	39	40	41	42	43
100	29	31	33	35	37	39	41	43	45	48	50	53	55	58									
95	28	30	32	34	36	38	40	42	44	47	49	51	54	56	59								
90	28	30	31	33	35	37	39	41	43	45	48	50	52	55	57	60							
85	27	29	31	32	34	36	38	40	42	44	46	49	51	53	56	58							
80	26	28	30	32	33	35	37	39	41	43	45	47	50	52	54	57	59						
75	26	27	29	31	33	34	36	38	40	42	44	46	48	50	53	55	57						
70	25	27	28	30	32	33	35	37	39	41	43	45	47	49	51	53	56	58					
65	24	26	27	29	31	32	34	36	38	40	42	43	45	47	50	52	54	56					
60	24	25	27	28	30	32	33	35	37	38	40	42	44	46	48	50	52	54	57				
55	23	24	26	27	29	31	32	34	36	37	39	41	43	45	46	48	50	52	55	57			
50	22	24	25	27	28	30	31	33	34	36	38	40	41	43	45	47	49	51	53	55	57		
45	22	23	24	26	27	28	30	32	33	35	37	38	40	42	43	45	47	49	51	53	55	57	
40			24	25	26	28	29	31	32	34	35	37	39	40	42	44	45	47	49	51	53	54	56
35				24	26	27	28	30	31	33	34	36	37	39	40	42	43	45	47	49	50	52	54
30							27	28	30	31	33	34	36	37	39	40	42	43	45	47	48	50	52
25										30	32	33	34	36	37	39	40	42	43	45	46	48	49
20																		40	41	43	44	45	47

LEGEND

HUMIDEX RANGE	DEGREE OF COMFORT
Less than 29	No discomfort
30 - 39	Some discomfort
40 - 45	Great discomfort: avoid exertion
Above 45	Dangerous
Above 54	Heat Stroke imminent

Figure 13.23 Heat index and the risk of heat-related illness
The "heat index" or "apparent temperature" is a measure of how hot it feels when the relative humidity is added to the actual air temperature. To find the heat index, find the intersection of the temperature on the left side of the table and the relative humidity across the top of the table. The shaded zones indicate the relative risk of heat-related illness with continued exposure and/or physical activity.
Source: Canadian Centre for Occupational Health and Safety. Humidex Rating and Work. Available online at http://www.ccohs.ca/oshanswers/phys_agents/humidex.html. Accessed August 21, 2011.

heat-related illness but cannot compensate for a lack of water. Dehydration reduces the ability to cool the body and increases the risk of these disorders even when it is not extremely hot or humid.

Heat-related illnesses include heat cramps, heat exhaustion, and heat stroke. Heat cramps are involuntary muscle spasms that occur during or after intense exercise, usually in the muscles involved in exercise. They are caused by an imbalance of the electrolytes sodium and potassium at the muscle cell membranes and can occur when water and salt are lost during extended exercise. Heat exhaustion occurs when fluid loss causes blood volume to decrease so much that it is not possible to both cool the body and deliver oxygen to active muscles. It is characterized by a rapid weak pulse, low blood pressure, fainting, profuse sweating, and disorientation. Someone experiencing the symptoms of heat exhaustion should stop exercising and move to a cooler environment. If exercise continues, heat exhaustion may progress to heat stroke. Heat stroke, the most serious form of heat-related illness, occurs when the temperature regulating centre of the brain fails due to a very high core body temperature (greater than 105°F). Heat stroke is characterized by elevated body temperature, hot dry skin, extreme confusion, and unconsciousness. It requires immediate medical attention.

Hyponatremia

The evaporation of sweat helps cool the body during exercise. Sweat is a blood filtrate produced by sweat glands in the skin. It is made up of 99% water along with small amounts of minerals, primarily sodium and chloride, acids, and trace amounts of other substances. Because sweat is mostly water, for most activities sweat losses can be replaced with plain water. However, when sweating continues for more than 4 hours, as it may during endurance events such as marathons, enough sodium may be lost to affect electrolyte balance. A reduction in the concentration of sodium in the blood is referred to as **hyponatremia**. Hyponatremia can occur if an athlete loses large amounts of water and salt in sweat, and replaces the loss with water alone. This causes the sodium that remains in blood to be diluted so the amount of water is too great for the amount of sodium. This is analogous to taking a full glass of salt water, dumping out half, and replacing what was poured out with plain water. The sodium in the glass is now more dilute (**Figure 13.24**). It is also possible to develop hyponatremia when salt losses from sweating are not excessive. This

hyponatremia Abnormally low concentration of sodium in the blood.

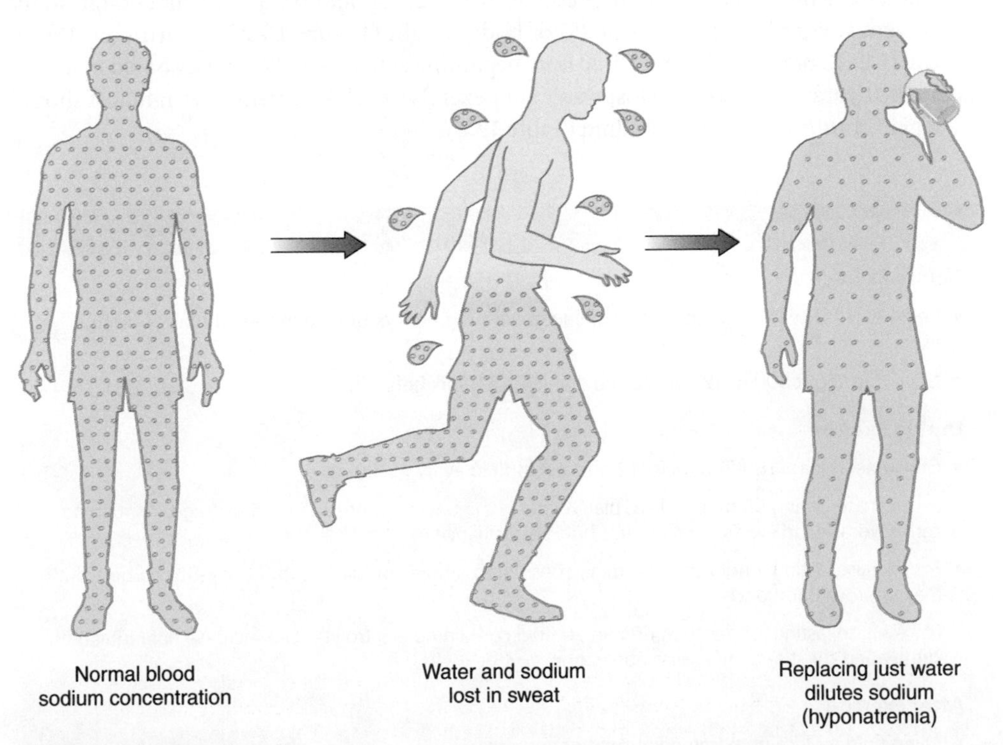

Figure 13.24 Hyponatremia
Hyponatremia can occur when an athlete loses water and sodium in sweat but replaces these losses with plain water, diluting the remaining sodium in the blood.

can occur if an athlete drinks more water than is lost in sweat, diluting the sodium in their system. A study of runners in the Boston Marathon found that 13% of those tested had hyponatremia and 0.6% had serum sodium concentrations low enough for this condition to be considered critical.[33] It is the concentration of sodium that is important, not the absolute amount.

Hyponatremia causes a number of problems. Solutes in the blood help hold fluid in the blood vessels. As sodium concentration drops, water will leave the bloodstream by osmosis and accumulate in the tissues, causing swelling. Fluid accumulation in the lungs interferes with gas exchange, and fluid accumulation in the brain causes disorientation, seizure, coma, and death. The early symptoms of hyponatremia may be similar to dehydration: nausea, muscle cramps, disorientation, slurred speech, and confusion. Drinking water alone will make the problem worse and can result in seizure, coma, or death.

For most types of exercise, hyponatremia is not a concern and lost electrolytes can be replaced during the meals following exercise. During long-distance events when hyponatremia is more likely, risk can be reduced by increasing sodium intake several days prior to competition, consuming sodium-containing sports drinks during the event, and avoiding Tylenol, aspirin, ibuprofen, and other non-steroidal anti-inflammatory agents. These medications interfere with kidney function and may contribute to the development of hyponatremia. Mild symptoms of hyponatremia can be treated by eating salty foods or drinking sodium-containing beverages such as sports drinks. More severe symptoms require medical attention.

Recommended Fluid Intake for Exercise

Anyone exercising should consume extra fluids. Good hydration is important before exercise and since thirst is not a reliable indicator of immediate fluid needs, it is important to schedule regular fluid breaks during exercise. The amount and type of fluid that is best depends on how much water you lose and how long you exercise. Because many people do not consume enough during exercise, beverages consumed after exercise must restore hydration.

Figure 13.25 Consume fluids before, during, and after exercise to maintain adequate hydration. (Mario Tama/Getty Images, Inc.)

How Much Should You Drink? To ensure hydration, adequate fluids should be consumed before, during, and after exercise. Exercisers should drink generous amounts of fluid in the 24 hours before the exercise session and about 500 ml (2 cups) of fluid at least 4 hours before exercise, to ensure that they are fully hydrated at the beginning of the exercise period. During exercise, whether casual or competitive, exercisers should try and drink enough water to prevent weight loss in excess of 2% of body weight (**Figure 13.25**).[18] Drinking 180 to 360 ml (6 to 12 oz) of fluid every 15-20 min. beginning at the start of exercise should maintain adequate hydration. To restore lost water after exercise, each kilogram of weight lost should be replaced with 1 to 1.5 litres of fluid (**Table 13.3**).

Table 13.3 Recommended Fluid Intake

Before Exercise

- Begin exercise well hydrated by consuming generous amounts of fluid in the 24 hours before exercise.
- Consume about 500 ml (2 cups) of fluid at least 4 hours before exercise.

During Exercise

- Consume at least 180-360 ml (6-12 ounces) of fluid every 15-20 min.
- For exercise lasting 60 min. or less, plain water is the only fluid needed but beverages containing carbohydrate and electrolytes will not hurt performance.
- For exercise lasting longer than 60 min., consuming a fluid containing about 6%-8% carbohydrate may improve endurance.
- For exercise lasting longer than 60 min., a fluid containing electrolytes can increase fluid intake by stimulating thirst and increasing absorption.

After Exercise

- Begin fluid replacement immediately after exercise.
- Consume 500-750 ml (16-24 ounces) of fluid for each 0.5 kg (1 lb) of weight lost.

What Should You Drink During Short Workouts? For exercise lasting an hour or less, water is the only fluid needed. For a 20-min. jog, 40 min. at the gym, or a brisk walk through the park, sports drinks offer no advantage over a water bottle. Sports drinks will not hurt your performance in a short workout, but they may be counterproductive if the goal of exercise is weight loss. A typical sports drink contains about 200 kcal/L, so drinking a 500 ml (16 oz) bottle at the gym will replace about half of the 200 kcal expended during your 40-min. ride on the stationary bicycle.

What Should You Drink During Long Workouts? For exercise lasting more than 60 min., beverages containing a small amount of carbohydrate and electrolytes are recommended. Exercise depletes body carbohydrate stores. Consuming carbohydrate in a beverage helps to maintain blood glucose levels, therefore providing a source of glucose for the muscle and delaying fatigue. A good sports drink should empty rapidly from the stomach, enhance intestinal absorption, and promote fluid retention. As the amount of carbohydrate in the beverage increases, the rate at which the solution leaves the stomach decreases. Beverages containing 60-80 g of carbohydrate/L (6%-8%) are best. This is the amount of carbohydrate found in popular sports beverages such as Gatorade and PowerAde. Beverages containing larger amounts of carbohydrate, such as fruit juices and soft drinks, are not recommended unless they are diluted with an equal volume of water. Water and carbohydrate trapped in the stomach do not benefit the athlete.

Small amounts of minerals, including sodium and chloride, are lost in sweat, but sweat consists mostly of water, so the amounts lost during exercise lasting less than 3-4 hours are usually not enough to affect health or performance, particularly if sodium was present in the previous meal. Even though there may not be a physiological need to replace sodium, a beverage containing 500-700 mg/L of sodium is recommended for exercise lasting more than an hour.[21] This is because the sodium enhances palatability and the drive to drink so it may cause an increase in fluid intake. The presence of small amounts of sodium and glucose also tend to slightly increase the rate of water absorption. A sodium-containing beverage will also help prevent hyponatremia in athletes who overhydrate and in those participating in endurance events, such as ultramarathons or iron man triathlons, in which significant amounts of sodium may be lost in sweat.

13.6 Food and Drink to Maximize Performance

Learning Objectives

- Discuss the advantages and disadvantages of carbohydrate loading.
- Explain the recommendations for food and drink during extended exercise.
- Plan pre- and post-competition meals for a marathon runner.

For most of us, a trip to the gym requires no special nutritional planning, but for competitive athletes, when they eat and what they eat before, during, and after competition is as important as a balanced overall diet. Food eaten at these times may give or take away the extra seconds that can mean victory or defeat.

Maximizing Glycogen Stores

Glycogen provides a source of stored glucose. Larger glycogen stores allow exercise to continue for longer periods. Glycogen stores and, hence, endurance are increased by increasing carbohydrate intake (**Figure 13.26**). Serious endurance athletes who want to substantially increase their glycogen stores before a competition may choose to follow a dietary regimen referred to as **glycogen supercompensation** or **carbohydrate loading**. This involves resting for 1-3 days before competition while consuming a very high-carbohydrate diet.[34,35] The diet should provide 10-12 g of carbohydrate per kilogram of body weight per day. For a 70-kg (150-lb) person, this is equivalent to about 700 g of carbohydrate per day. Having a stack of

glycogen supercompensation or **carbohydrate loading** A regimen designed to maximize muscle glycogen stores before an athletic event.

pancakes with syrup and a large glass of juice for breakfast will provide about a third of the carbohydrate recommended for a day. A number of commercial high-carbohydrate beverages, containing 200-250 g/L of carbohydrate, are available to help athletes consume the amount of carbohydrate recommended to maximize glycogen stores. These should not be confused with sports drinks designed to be consumed during competition, which contain only about 60-80 g/L of carbohydrate. Trained athletes who follow a carbohydrate-loading regimen can double their muscle glycogen content.[34]

Although glycogen supercompensation is beneficial to endurance athletes, it will provide no benefit and even has some disadvantages for people exercising for periods less than 90 min. For every gram of glycogen in the muscle, 3 g of water are also deposited. This water can cause a 0.5–3.5-kg (1–7-lb) weight gain and may cause some muscle stiffness. The extra weight is a disadvantage for those competing in events of short duration. As glycogen is used, the water is released; this can be an advantage when exercising in hot weather.

What to Eat Before Exercise

An athlete who is hungry will not perform at his or her best, but the wrong meal can hinder performance more than the right one can enhance it. The size, composition, and timing of the pre-exercise meal are all important. The goal of meals eaten before exercise is to maximize glycogen stores and provide adequate hydration while minimizing hunger and any undigested food in the stomach that can lead to gastric distress.

Ideally, a pre-exercise meal should provide enough fluid to maintain hydration and be high in carbohydrate (60%-70% of kcalories). This will help to maintain blood glucose and maximize glycogen stores. Muscle glycogen is depleted by exercise, but liver glycogen is used to supply blood glucose and is depleted even during rest if no food is ingested. So, first thing in the morning, liver glycogen stores have been reduced by the overnight fast. A high-carbohydrate meal eaten 2-4 hours before the event will fill liver glycogen stores. In addition to being high in carbohydrate, the pre-exercise meal should be moderate in protein (10%-20%) and low in fat (10%-25%) and fibre to minimize gastrointestinal distress and bloating during competition. A 250 ml (1 cup) serving of pasta with tomato sauce and a slice of bread, or a turkey sandwich and juice (250 ml) are good choices. Spicy foods that could cause heartburn, and large amounts of simple sugars that could cause diarrhea, should be avoided, unless the athlete is accustomed to eating these foods.

In addition to providing nutritional clout, a meal that includes "lucky" foods may provide some athletes with an added psychological advantage. Some athletes find that in addition to a precompetition meal, a small high-carbohydrate snack or beverage consumed shortly before an event, may enhance endurance. Because foods affect people differently, athletes should test the effect of these meals and snacks during training, not during competition.

What to Eat During Exercise

Regardless of the type or duration of exercise, maintaining adequate fluid intake is important while exercising (see Table 13.3). For exercise that lasts more than an hour, consuming about 30-60 g of carbohydrate per hour (the amount in a banana or an energy bar) can enhance endurance.[18] Consuming carbohydrate during exercise is particularly important for athletes who exercise in the morning when liver glycogen levels are low. Carbohydrate intake should begin shortly after exercise commences and regular amounts should be consumed every 15-20 min. during exercise. The carbohydrate should provide glucose, glucose polymers (chains of glucose molecules), or a combination of glucose and fructose.[21] Fructose alone is not as effective and may cause diarrhea. Some athletes may prefer to obtain this carbohydrate from a sports drink, while others prefer a high-carbohydrate snack or energy gel consumed with water. Energy gels consist of a thick carbohydrate syrup packaged in a palm-sized packet. The contents can be sucked out of the packet, providing about 20 g of carbohydrate (see Label Literacy: What Are You Getting from That Sports Bar?).

During exercise, sodium and other minerals are lost in sweat. Although the amounts lost during exercise lasting less than 3-4 hours are usually not enough to affect health or performance, a snack or beverage containing sodium is recommended for exercise lasting 1 hour or more. The sodium enhances the palatability of beverages and increases the drive to drink so even if sodium losses are small, consuming it during exercise may cause an increase in fluid intake.

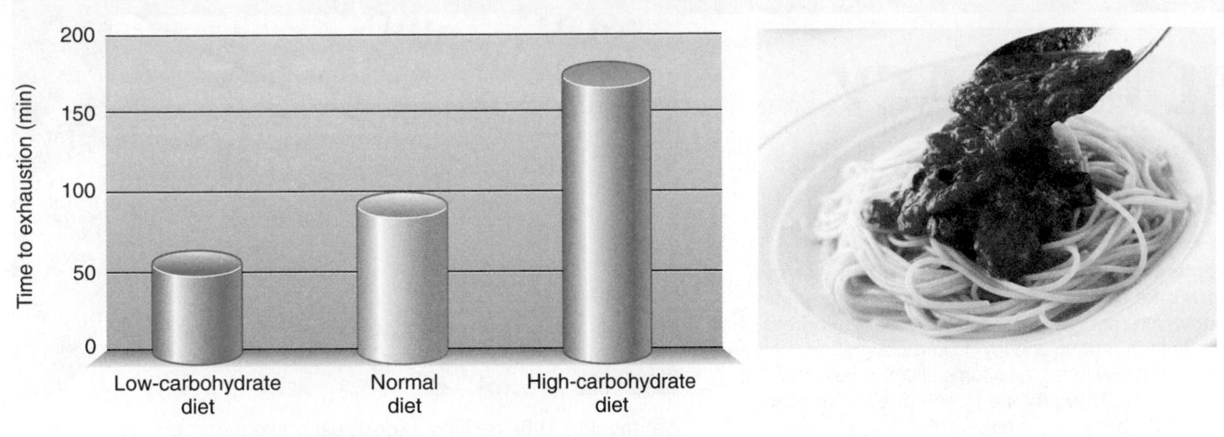

Figure 13.26 Effect of carbohydrate consumption on endurance
Endurance capacity is shown during cycling exercise after 3 days of a very-low-carbohydrate diet (less than 5% of energy from carbohydrate), a normal diet (about 55% carbohydrate), and a high-carbohydrate diet (82% carbohydrate). **Source:** From Bergstrom, J., Hermansen, L., Hultman, E., et al. Diet, muscle glycogen and physical performance. *Acta. Physiol. Scand.* 71:140–150, 1967.) (© Datacraft/Age Fotostock)

What to Eat After Exercise

When exercise ends, the body must shift from the catabolic state of breaking down glycogen, triglycerides, and muscle proteins for fuel to the anabolic state of restoring muscle and liver glycogen, depositing lipids, and synthesizing muscle proteins. The goal for meals after exercise is to replenish fluid, electrolyte, and glycogen losses and to provide amino acids for muscle protein synthesis and repair. For example, a mixed meal such as pancakes and a glass of milk consumed soon after a strenuous competition or training session will help the athlete prepare for the next exercise session.

The first priority for all exercisers is to replace fluid losses. For serious athletes it may also be important to rapidly replenish glycogen stores. Appropriate post-exercise intake can replenish muscle and liver glycogen within 24 hours of the athletic event. To maximize glycogen replacement, a high-carbohydrate meal or drink should be consumed within 30 min. of the competition and again every 2 hours for 6 hours after the event. Ideally, the meals should provide about 1.0-1.5 g of carbohydrate per kilogram of body weight, which is about 70-100 g of carbohydrate for a 70-kg (150-lb) person—the equivalent of about 750 ml (2.5 cups) of pasta.[21] This type of regimen to restore glycogen is critical for athletes who must perform again the following day, but is not necessary if the athlete has 1 or more days to replace glycogen stores before the next intense exercise. Those not competing again the next day can replace glycogen more slowly by consuming high-carbohydrate foods for the next day or so.

Most people who are not serious competitive athletes do not need a special glycogen replacement strategy to ensure glycogen stores are full by their next gym visit. Eating a typical diet that provides about 55% of kcalories as carbohydrate will quickly replace the glycogen used during a 30-60-min. workout at the gym.

13.7 Ergogenic Aids: Do Supplements Enhance Athletic Performance?

Learning Objectives

- Assess the health risks associated with anabolic steroids.
- Explain why creatine supplements affect sprint performance.
- Describe one way in which a supplement might improve endurance.

"Citius, Altius, Fortius"—faster, higher, stronger—the Olympic motto. For as long as there have been competitions, athletes have yearned for something—anything—that would give them the competitive edge. Anything designed to enhance performance can be considered an **ergogenic aid**; running shoes are mechanical aids; psychotherapy is a psychological aid; drugs are pharmacological aids. Many dietary supplements are also used as ergogenic aids.

ergogenic aid Anything designed to increase physical work or improve exercise performance.

LABEL LITERACY

What Are You Getting from That Sports Bar?

Looking for a convenient snack that can give you an energy boost during your bike ride or day of skiing? A sports bar may be the answer, but which should you choose? There are hundreds of varieties. Some are high in protein and low in carbohydrate; others are high in carbohydrate and low in fat; and some claim to have just the right balance of everything. They promise to optimize performance, build lean muscle, reduce body fat, increase strength, and speed recovery.

Carbohydrate is the fuel that becomes limiting during prolonged exercise, so if you want to have the energy to keep pedalling or skiing, choose high-carbohydrate bars, called energy or endurance bars. They have the carbohydrate needed to prevent hunger and maintain blood glucose during sports activities. Use the label to check out the amount of carbohydrate, fat, protein, and energy in different bars. A bar that provides about 45 g of carbohydrate (70% of kcalories) will help maintain your blood glucose level during exercise. Watch the fat and protein; in 300 kcal, you want no more than about 8 g of fat and 16 g of protein. Bars higher in fat or protein or lower in carbohydrate will not give you the blood glucose boost that you need to continue exercising. High protein bars are sold for use after exercise to promote recovery. The differences in macronutrient composition are evident in the table.

Is a sports bars any better for you than a chocolate bar? As shown in the table, sports bars are more nutrient-dense, being lower in fat and higher in fibre and protein. Often, energy bar manufacturers add vitamins and minerals to their products. When this happens, all the added micronutrients are listed on both the nutrition facts table as well as in the ingredients list.

Macronutrient Composition **Per bar**	**Energy bar intended for use before exercise**	**Protein bar intended for use after exercise**	**Milk Chocolate bar**
Calories	200	300	220
Total Dietary Fat (g)	2.2	6	11
Saturated Fat (g)	0.7	3.5	7
Total Carbohydrates (g)	43	38	26
Fibre (g)	3.5	1	1
Protein (g)	9	24	3
Vitamin A (% Daily Value)	20*	0	4
Vitamin C (% Daily Value)	23*	0	0
Calcium (% Daily Value)	20*	15	10
Iron (% Daily Value)	15*	15	8

*Additional vitamins and minerals, not listed here, were added to this bar.

So, should you be packing a sports bar on your next outing? It won't take the place of the whole grains, fresh vegetables and fruit, low-fat milk and alternatives, and lean meats and alternatives that make up a healthy diet. But, if having a compact, individually wrapped bar that can travel with you means the difference between consuming this snack or no food at all, sports bars can be beneficial. They may also provide a psychological edge if you believe they will enhance your performance.

If you choose to use these bars, wash them down with plenty of water. They don't provide fluid—an essential during any activity. Also remember that one sports bar provides around 200-300 kcal. Even though they are eaten to support activity, they still add to your overall energy intake and can contribute to weight gain if consumed in excess.

(© istockphoto.com/John A Meents)

Although these supplements are often expensive and most have not been shown to improve performance, athletes are vulnerable to their enticements. When considering the use of an ergogenic supplement, an athlete should first weigh the health risks against potential benefits (**Table 13.4**). The following sections discuss some of the products that are often promoted to athletes. Others are reviewed in **Table 13.5**. As indicated in the table, most of these supplements are licensed natural health products and a few have been licensed for exercise-related purposes. When products are not licensed for exercise-related purposes, it suggests that there is no evidence that these supplements benefit exercise performance (see Chapter 12: Focus on Nonvitamin/Nonmineral Supplements for more information on the licensing of natural health products).

Table 13.4 Evaluating the Benefits and Risks of Ergogenic Supplements

Does the supplement meet your needs?

- Does the product contain the nutrient or other ingredient you are looking for?
- Has it been shown to provide the benefits you want?

Are the ingredients safe for you?

- Does it contain any ingredients that have been shown to be hazardous to someone like you?
- Do you have a medical condition that would make it dangerous to take this product?
- Are you taking prescription medication that might interact with the supplement?

Is the dose safe?

- Follow the recommended dose on the label. More isn't always better and may cause side effects.

How much does it cost?

- More expensive is not always better.
- Compare costs and ingredients before you buy.

Vitamin Supplements

There are many vitamin preparations licensed as natural health products, most for the maintenance of good health (see Section 2.5 and Focus on Nonvitamin/Nonmineral Supplements for more information on natural health products).[36] Many of the promises made to athletes about the benefits of vitamin supplements are extrapolated from their biochemical functions. For example, B vitamin supplements are promoted to enhance ATP production because of their roles in muscle energy metabolism. Vitamins B_6, B_{12}, and folic acid are promoted for aerobic exercise because they are involved in the transport of oxygen to exercising muscles. These vitamins are indeed needed for energy metabolism, and a deficiency of one or more of these will interfere with ATP production and impair athletic performance; however, providing more than the recommended amount does not deliver more oxygen to the muscle, cause more ATP to be produced, or enhance athletic performance. Because athletes must consume more food to meet energy needs, they consume more vitamins as well. A reasonably well-planned diet that is based on whole grains, vegetables, and fruits and includes lean meats and low-fat dairy products will provide enough of all the B vitamins to meet an athlete's needs.

Table 13.5 Claims, Benefits, and Risks of Popular Ergogenic Aids

Ergogenic Aid	Promoter Claims	Proven Benefits	Potential Risks	Licensed as Natural Health Product?
Arginine, ornithine, and lysine	Causes the release of growth hormone, which stimulates muscle development and decreases body fat.	No increase in lean body mass observed with supplementation.	Reduced absorption of other amino acids. Diarrhea at high doses.	Yes, arginine, for exercise-related purpose

Table 13.5 *(continued)*

Ergogenic Aid	Promoter Claims	Proven Benefits	Potential Risks	Licensed as Natural Health Product?
Bee pollen	Causes faster recovery from training workouts, which enables a higher level of training.	No evidence that it improves training level.	Allergic reactions.	Yes, a minor ingredient in many multi-ingredient preparations
Bicarbonate (sodium bicarbonate, baking soda)	Helps buffer lactic acid produced during exercise and delays fatigue.	Increases blood pH and may enhance performance and strength during intense anaerobic activities.	Causes bloating, diarrhea, and high blood pH.	No, common food ingredient
Branched-chain amino acids (leucine, isoleucine, and valine)	Improves endurance and prevents fatigue.	Evidence of an effect is inconsistent.	No toxicity reported.	Yes, for the maintenance of protein metabolism
Caffeine	Increases the release of fatty acids from adipose tissue, spares glycogen, and enhances endurance.	Increases endurance in some individuals.	Dehydration, nervousness, anxiety, insomnia, digestive discomfort, abnormal heartbeat.	Yes, to promote energy
Carnitine	Enhances the utilization of fatty acids and spares glycogen.	No increase in fatty acid utilization or improvement in exercise performance found.	D,L-carnitine and D-carnitine forms can be toxic.	No, prescription required
Chromium (chromium picolinate)	Increases lean body mass, decreases body fat, delays fatigue.	No effect on protein or lipid metabolism unless a chromium deficiency exists.	No toxicity reported in humans.	Yes, to support healthy glucose metabolism
Coenzyme Q10	Increases mitochondrial ATP production, acts as an antioxidant, and may combat fatigue.	No effect on exercise performance observed.	No toxicity reported.	Yes, to maintain cardiovascular health
Creatine (creatine monohydrate)	Increases ATP production and speeds recovery after high-intensity exercise.	Increases muscle creatine and creatine phosphate synthesis after exercise. Enhances strength, performance, and recovery from high-intensity exercise.	Stomach pain.	Yes, to increase muscle mass and strength during resistance training
DHEA (dehydroepiandrosterone)	Builds muscles, burns fat, and delays chronic diseases associated with aging.	No proven benefits.	Acne, oily skin, facial hair, voice deepening, hair loss, mood changes, liver damage, and stimulation of existing cancers.	No, controlled substance; may be obtained only by prescription and the permission of Health Canada

Ergogenic Aid	Promoter Claims	Proven Benefits	Potential Risks	Licensed as Natural Health Product?
Ginseng (*Panax ginseng* or Chinese ginseng)	Enhances performance.	Little evidence of ergogenic effects.	May increase the effects and side effects of other stimulants such as caffeine.	Yes, for a variety of traditional uses including to enhance physical capacity
Glutamine	Increases muscle glycogen deposition following intense exercise, enhances immune function, and prevents the adverse effects of overtraining.	Little evidence that glutamine increases immune function, prevents the symptoms of overtraining, or increases glycogen synthesis.	No evidence of toxicity.	Yes, for muscle cell repair and recovery from physical stress
Glycerol	Improves hydration and endurance.	Evidence of an effect is equivocal.	May cause cellular dehydration, nausea, vomiting, diarrhea.	No
HMB (β-hydroxy-β-methylbutyrate)	Increases ability to build muscle and burn fat in response to exercise.	Some evidence of an increase in lean body mass and strength.	No toxicity in animals, but little information in humans.	Yes, for enhancing strength in untrained individuals during intense resistance training
Medium-chain triglycerides (MCT)	Provides energy without promoting fat deposition; reduces muscle protein breakdown during prolonged exercise.	Provides energy and must be metabolized before they can be stored as body fat. They increase endurance and fatty acid oxidation in mice, but there is no evidence of a benefit in humans.	None known.	No
Ribose	Increases cellular ATP and muscular power.	No research.	None known.	No
Vanadium (vanadyl sulfate)	Aids insulin action; allows more rapid and intense muscle pumping for body builders.	No evidence to support a benefit for body builders.	Reduces insulin production.	Yes, most commonly part of a multivitamin mineral supplement, for the maintenance of health

Supplements of vitamin E, vitamin C, and β-carotene are promoted to athletes because of their antioxidant functions. Exercise increases oxidative processes, and therefore increases the production of free radicals, which cause cellular damage and have been associated with fatigue during exercise. It has been suggested that antioxidant supplements reduce the levels of free radicals and hence delay fatigue. Research, however, has not found that supplementation of antioxidant nutrients improves performance.[37] There is also no clear evidence that antioxidant supplements reduce oxidative stress. In fact, free radical production is believed to serve as a signal to the muscle to adapt to the stress by enhancing its natural antioxidant defenses. Exercise training has been shown to increase the activity of the enzyme superoxide dismutase in muscle and a variety of protective enzymes in the blood, thus reducing the risk of oxidative damage to tissues.[25] Although antioxidant supplements do not appear to be ergogenic, as long as athletes do not consume them in amounts that exceed the ULs, there is little risk associated with their use.[28] A better way to ensure adequate antioxidant protection is a diet that includes plenty of whole grains, fruits, and vegetables, which are rich in antioxidant nutrients as well as antioxidant phytochemicals.

Mineral Supplements

Supplements of chromium, vanadium, selenium, zinc, and iron are promoted to strengthen muscles or enhance endurance. As with vitamin supplements, many of the claims made about these minerals are based on their physiological functions. And as with vitamins, there is little evidence that consuming more than the recommended amount provides any benefits.

Chromium supplements, as chromium picolinate, are purported to increase lean body mass and decrease body fat. Chromium is needed for insulin action and insulin promotes protein synthesis. Therefore, adequate chromium status is likely to be important for lean tissue synthesis. The picolinate form is typically used because it is believed to be absorbed better than other forms of chromium. Studies in humans have not consistently demonstrated an effect of supplemental chromium picolinate on muscle strength, body composition, body weight, or other aspects of health.[38] Because no adverse effects have been associated with chromium intake from food or supplements, no UL has been established.

Vanadium, usually as vanadyl sulfate, is another mineral promoted for its ability to promote the action of insulin. Vanadium supplements promise to increase lean body mass, but there is no evidence that they have an anabolic effect, and toxicity is a concern.[39] A UL of 1.8 mg/day of elemental vanadium has been set for adults age 19 and older.

Selenium is promoted for its antioxidant properties and zinc for its role in protein synthesis and tissue repair, but neither of these supplements has been found to improve athletic performance in individuals with adequate mineral status. Iron is also marketed as an ergogenic mineral because it is needed for hemoglobin synthesis. If an iron deficiency exists, as it frequently does in female athletes, supplements can be of benefit.

Most minerals in licensed natural health products are components of multivitamin and mineral preparations for the purpose of maintaining good health. Products containing only one or a limited number of minerals are also available.[36]

Protein Supplements

Hundreds of protein supplements are available—from powders you mix in your beverage to bars you put in your backpack. They are often marketed to athletes with the promise of enhancing muscle growth or improving performance. Muscles enlarge in response to exercise. Adequate protein is necessary for this to occur, but consuming extra protein either as food or supplements does not increase muscle growth or strength.[39] The protein provided by expensive supplements will not meet an athlete's needs any better than the protein found in a balanced diet. If an athlete's diet provides enough energy, it usually provides enough protein without adding a supplement.

Amino Acid Supplements

Supplements of individual amino acids are also promoted to athletes. These include ornithine, arginine, and lysine, glutamine, and the branched-chain amino acids (leucine, isoleucine, and valine). As described in Chapter 12: Focus On Nonvitamin/Nonmineral Supplements, these amino acids are licensed as natural health products in Canada. Glutamine and arginine are sold for exercise-related purposes. Arginine is licensed for modest improvement in exercise capacity and glutamine for muscle cell repair and recovery from physical stress. The other amino acids are licensed for more general purposes such as the maintenance of protein metabolism, implying there is little evidence that these amino acids have any beneficial effects (see Focus on Nonvitamin/Nonmineral Supplements for more details).

Supplements to Enhance Short, Intense Performance

Although amino acid supplements are not ergogenic, creatine and β-hydroxy-β-methylbutyrate (HMB), which are made from amino acids, do provide some benefits for athletes who seek to increase muscle mass and strength and improve performance in activities that depend on quick bursts of intense exercise (see Critical Thinking: Does it Provide an Ergogenic Boost?). Bicarbonate will not increase muscle mass or strength but it may provide some advantages for intense activities.

Creatine Supplements Creatine is a nitrogen-containing compound found primarily in muscle, where it is used to make creatine phosphate, which is a source of energy for short-term exercise. It can be synthesized in the liver, kidneys, and pancreas from the amino acids arginine, glycine, and methionine. It is also consumed in the diet in meat and milk. The more creatine in the diet, the greater the muscle creatine stores. Creatine supplements increase levels of both creatine and creatine phosphate in muscle (**Figure 13.27**). Higher levels of these provide muscles with more quick energy for short-term maximal exercise. Creatine supplementation has been shown to improve performance in high-intensity exercise lasting 30 sec. or less.[40,41] It is therefore beneficial for exercise that requires explosive bursts of energy, such as sprinting and weight lifting, but not for long-term endurance activities such as marathons.

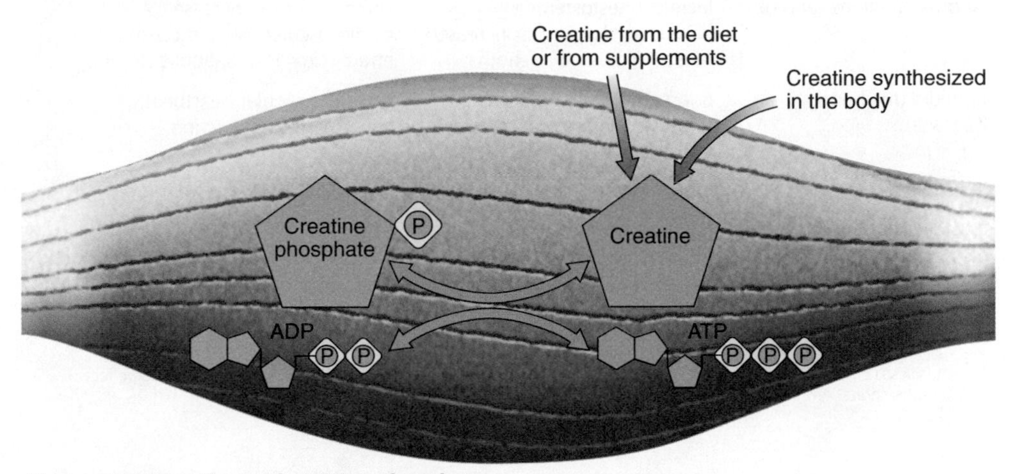

Figure 13.27 Creatine and creatine phosphate
The more creatine consumed, the greater the amount of creatine and creatine phosphate, which is made from it and stored in the muscles. During short bursts of intense activity, creatine phosphate can transfer a phosphate group to ADP, forming creatine and ATP that can be used for muscle contraction.

Creatine supplements are also taken by athletes to increase muscle mass and strength. Some of the increase in lean body mass that occurs when supplements are taken is believed to be due to water retention related to creatine uptake in the muscle. In addition to this, an increase in muscle mass and strength may occur in response to the greater amount and intensity of training that may be achieved when muscle creatine levels are higher.[41,42]

Long-term creatine supplementation appears to be safe at intakes of up to 5 g/day, but the safety of higher doses over the long term has not been established.[43] These supplements are licensed as natural health products for the purpose of increasing muscles mass and strength during resistance training, but Health Canada requires a warning that anyone with a kidney disorder, or is pregnant or breastfeeding seek the advice of a physician before use.

Beta-Hydroxy-Beta-Methylbutyrate β-hydroxy-β-methylbutyrate, known as HMB, is a metabolite of the branched-chain amino acid leucine and studies have found that it can increase muscle mass and strength, especially in untrained individuals.[44] HMB is a licensed natural health product for the purpose of enhancing "muscle strength in previously untrained individuals in combination with intense resistance training exercise."[36]

Bicarbonate Bicarbonate may provide some advantage for intense activities. Because bicarbonate acts as a buffer in the body, supplementing it is thought to neutralize acid and thus delay fatigue and allow improved performance. Sodium bicarbonate is just baking soda from the kitchen cupboard. Taking some before exercise has been found to improve performance and delay exhaustion in sports, such as sprint cycling, that entail intense exercise lasting only 1-7 min., but it is of no benefit for lower-intensity aerobic exercise.[45] Just because baking soda is an ingredient in your cookies does not mean that it is risk free. Many people experience abdominal cramps and diarrhea after taking sodium bicarbonate, and other possible side effects have not been carefully researched.

Banned Ergogenic Aids There are a number of substances used by athletes to enhance strength or endurance that have been banned by sporting organizations. Most have dangerous side effects and athletes who test positive for these drugs face sanctions (**Table 13.6**).

Table 13.6 Banned Ergogenic Aids

Ergogenic Aid	Claim/Effectiveness	Adverse Effects
Amphetamines (beans or greenies)	Decreases reaction time and increases endurance	Anxiety, irregular heartbeat, hallucinations, addiction, death
Anabolic steroids (testosterone or "T")	Increases muscle mass and strength, decreases body fat	Testicular atrophy, liver damage, heart disease, hypertension
Androstenedione (andro)	Increases testosterone production leading to increased muscle mass and strength	Unknown, may increase cardiovascular risk and cause side effects similar to anabolic steroids
Ephedra (ma huang, guarana)	Boosts metabolism, burns fat, increases alertness, increases endurance	Anxiety, irregular heartbeat, hallucinations, addiction
Erythropoietin (EPO)	Increases endurance, increases oxygen-carrying capacity of the blood	Increased blood viscosity, increased risk of heart attack, or pulmonary embolism
Human growth hormone (HGH)	Stimulates body growth to increase muscle mass	Muscle and cardiac abnormalities, carpal tunnel syndrome, abnormal growth

Source: Tokish, J. M., Kocher, M. S., and Hawkins, R. J. Ergogenic aids: A review of basic science, performance, side effects, and status in sports. *Am. J. Sports Med.* 32:1543–1553, 2004.

Supplements to Enhance Endurance

Sprinters and weight lifters can benefit from increases in creatine phosphate levels, but endurance athletes have different fuel concerns. Long-distance runners and cyclists are more concerned about running out of glycogen. Glycogen is spared when fat is used as an energy source, allowing exercise to continue for a longer time before glycogen is depleted and fatigue sets in. A number of substances, including carnitine, medium-chain triglycerides, and caffeine are taken to increase endurance by increasing the amount of fat available to the muscle cell.

Carnitine Carnitine supplements are promoted in the popular media as fat burners—substances that increase the utilization of fat during exercise. Carnitine is an amino-acid-like molecule produced in the body from the amino acids lysine and methionine. It is needed to transport fatty acids into the mitochondria where they are used to produce ATP by aerobic metabolism, but due to concerns about toxicity, it is available in Canada only by prescription.

Medium-Chain Triglycerides Most of the fatty acids used by the muscle to generate ATP are delivered in the blood, so theoretically, higher levels in the blood increase the availability of fatty acids as a fuel for exercise. Most of the fat consumed in the diet is absorbed in chylomicrons, which enter the lymphatic system before they appear in the blood. However, medium-chain triglycerides (MCT), those made up of fatty acids with medium-length carbon chains (8-10 carbons), do not need to be incorporated into chylomicrons. They are soluble in water so they can be absorbed directly into the blood. In addition to being absorbed much more quickly, the fatty acids cross cell membranes more easily and can enter the mitochondria for oxidation without the help of carnitine. When MCTs are ingested, blood fatty acid levels increase. Despite this, research has not found that supplementation with MCT increases endurance, spares glycogen, or enhances performance.[46,47]

CRITICAL THINKING:

Does it Provide an Ergogenic Boost?

Background

Paulo is on the college track team. To improve his performance, he decides to try some ergogenic supplements. Based on the articles and advertisements he's read in sports magazines, he chooses creatine to improve his sprint times and chromium to increase his lean body mass. But before he begins taking these, he wants to explore their risks and benefits.

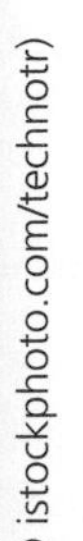
(© istockphoto.com/technotr)

Data

The ads and articles about these supplements make the following claims:

- Creatine will increase muscle creatine phosphate levels to provide quick energy and speed recovery after exercise.

- Chromium will increase lean body tissue and enhance fat loss.

Critical Thinking Questions

Use the questions in Table 13.4 to help Paulo evaluate the claims made by the articles and ads.

Based on the information given in this chapter, explain the advantages and risks of these products.

Suggest places Paulo could look to get additional information he can trust.

Would you recommend Paulo take these supplements? Why or why not?

Use iProfile to compare the amount of protein in a powdered supplement with the amount in a chicken breast.

Caffeine Caffeine is a stimulant found in coffee, tea, and some soft drinks. It has been shown to enhance performance and endurance during prolonged exhaustive exercise and to a lesser degree it also enhances short-term high-intensity exercise.[48,49] Caffeine also improves concentration, enhances alertness, and reduces fatigue. How caffeine causes all these physiological effects is not completely understood, but it may enhance endurance by increasing the release of fatty acids. When fatty acids are used as a fuel source, glycogen is spared, delaying the onset of fatigue. Your morning coffee is probably not enough to have an effect but drinking 750 ml (2.5 cups) of brewed coffee up to an hour before exercising has been shown to improve endurance. The effectiveness of caffeine varies with each person. Athletes who are unaccustomed to caffeine respond better than those who routinely consume it. Caffeine is a diuretic but it does not cause significant dehydration during exercise. In some athletes caffeine may impair performance by causing gastrointestinal upset. Regardless of its effectiveness, athletes should know that consuming excess caffeine before a competition is illegal. The International Olympic Committee prohibits athletes from competing when urine caffeine levels are 12 μg/mL or greater. For urine caffeine to reach this level, an individual would need to drink 6-8 cups of coffee within about a 2-hour period. Caffeine is widely available in energy drinks such as Red Bull (80 mg of caffeine) and 5-hour Energy (about 100 mg of caffeine) and is also found in pill form in products such as Wake Ups, which contains about 100 mg of caffeine per tablet—about the same amount as that in 250 ml (1 cup) of coffee (**Table 13.7**).

Table 13.7 The Caffeine Content of Foods and Medications

Food or Medication	Amount	Caffeine (mg)
Coffee, regular	250 ml (1 cup)	139
Coffee, decaffeinated	250 ml (1 cup)	3
Tea, brewed	250 ml (1 cup)	45
Pepsi	355 ml (12 oz-can)	37
Diet Pepsi	355 ml (12 oz-can)	50
Mountain Dew	355 ml (12 oz can)	54
Hot chocolate	250 ml (1 cup)	7
Brownie	1	14
Chocolate bar	30 g (1 oz)	15
Wake Ups	1 tablet	100
Excedrin	1 tablet	65
Anacin	1 tablet	32
Red Bull	1 can	80
5-hour Energy	1 bottle	100 (amount of 1 cup of coffee)

Other Supplements

In addition to the nutrient and non-nutrient supplements discussed thus far, there are hundreds of other products marketed to athletes. Most have no effect on performance. For example, bee pollen, which is a mixture of the pollen of flowering plants, plant nectar, and bee saliva has been promoted as an ergogenic aid even though it contains no extraordinary factors and has not been shown to have any performance-enhancing effects. In addition, ingesting or inhaling bee pollen can be hazardous to individuals allergic to various plant pollens.[50] Brewer's yeast is a source of B vitamins and some minerals but has not been demonstrated to have any ergogenic properties. Likewise, there is no evidence to support claims that wheat germ oil will aid endurance. As an oil, it is high in fat, but it is no better as an energy source than any other fat. Royal jelly is a substance produced by worker bees to feed to the queen bee. While it helps the queen bee grow to twice the size of worker bees and to live 40 times longer, royal jelly does not appear to enhance athletic capacity in humans. Supplements of DNA and RNA are marketed to aid in tissue regeneration. In the body, they carry genetic information and are needed to synthesize proteins, but DNA and RNA are not required in the diet, and supplements do not help replace damaged cells. A variety of herbal products is also marketed to athletes. Most of these have not been studied extensively for their ergogenic effects so the evidence of their benefits is only anecdotal. Ginseng is promoted to increase endurance, but human clinical trials of Chinese ginseng (*Panax ginseng*) have not yet demonstrated conclusively that it enhances physical performance.[51] Ginseng supplements are generally considered safe based on more than 2,000 years of use with few reported side effects but the same is not true of all herbal products. Many can harm health as well as performance, so athletes should consider the risks before using these products (see Focus on Nonvitamin/Nonmineral Supplements).

The Impact of Diet Versus Supplements

Supplements may garner most of the press when it comes to exercise performance, but they are only the very tip of the iceberg when it comes to the things athletes can do to enhance their performance. The foundation of good athletic performance is talent, hard work, and a healthy diet. A healthy diet is one that provides the right number of kcalories to keep weight in the desirable range; the proper balance of carbohydrate, protein, and fat to fuel activity and maintain tissues; plenty of water; and sufficient but not excessive amounts of essential vitamins and minerals. It is rich in whole grains, fruits and vegetables, high in fibre, moderate in fat and sodium, and low in saturated fat, cholesterol, *trans* fat, and added sugars. Whether you are a couch potato or an Olympic hopeful, the recommendations of Canada's Food Guide can help you choose such a diet. This diet provides the foundation from which to optimize performance. Performance can be further improved by using appropriate foods and fluids to help refuel and rehydrate during workouts and events. Where do supplements fit in? Most of them don't, but a few specific types of athletes in specific events will receive an additional small benefit from a few select supplements (**Figure 13.28**).

Figure 13.28 The impact of diet and supplements on exercise performance
Eating a healthy overall diet provides the foundation for optimization of performance. Foods and beverages used to supply energy and ensure hydration during an event can provide additional benefits but ergogenic aids provide little or no performance boost.

CASE STUDY OUTCOME

On the final hours of the ride that day, Jeffrey notices his energy is flagging and that he is having trouble keeping up with his riding partners. He makes it to the camp grounds that evening, but the afternoon ride proves far more exhausting than the morning's ride and concentrating on the road is difficult. Jeffrey realizes that he did not make the best choices for his lunchtime meal. Steak and green salad, while a nutritious meal high in protein and micronutrients, is very low in carbohydrates. By not eating carbohydrates, Jeffrey reduces the supply of glucose to his body, forcing it to use stored glycogen in his muscles and liver. By the time he makes it to the campgrounds that evening, those stores are so depleted, his blood sugar is beginning to drop and he is experiencing the fatigue and mental confusion characteristic of hypoglycemia.

Jeffrey learns from his experience, and while enjoying the local cuisine in the many diners he frequents during the summer, he makes a point of ensuring that a healthy portion of carbohydrates is part of every meal.

While you will probably not spend an entire summer cycling the 7,500 km between Vancouver and St John's, even a weekend ride or run can become more difficult if your body is unable to provide glucose to fuel your muscles and brain.

APPLICATIONS

1. How much exercise do you get?

a. Keep a log of your activity for one day.

b. Refer to Chapter 7, Table 7.3, to help you determine the number of hours you spend engaged in:

i Activities of daily living

ii Moderate intensity activity

iii Vigorous activity

c. Use Table 7.4 to determine your physical activity level and PA value.

d. Use Table 7.5 to determine your estimated energy expenditure (EER).

iProfile

e. If you increased your exercise enough to move to the "active" physical activity level, what would your new EER be? (If you are already active, what would your EER be in the "very active" level?)

f. Use iProfile to find foods that you could add to your diet to balance the added expenditure of this increase in activity.

2. What types and amounts of exercise work for you?

a. Taking into consideration your typical weekly schedule of activities and events, design a reasonable exercise program for yourself. Include the types of activities, the times during the week you will be involved in each activity, and the length of time you will engage in each activity. Choose activities you enjoy and schedule them for practical lengths of time and at reasonable frequencies.

b. What everyday changes have you made that will increase the energy expended in day-to-day activities?

c. Which activities are aerobic, which are for strength training? Are any of your activities bone-strengthening?

d. Can each of these activities be performed year-round? If not, suggest alternative activities and locations for inclement weather.

General Nutrition Issues

1. David is beginning an exercise program. He plans to run before lunch and then play racquetball every night after dinner. Once he begins his exercise program, he finds that he feels lethargic and hungry before his late-morning run. After running, he doesn't have much of an appetite, so he saves his fast-food lunch until mid-afternoon. He is still hungry enough to eat dinner at home with his family, but finds that he is getting stomach cramps and is too full when he goes to play racquetball. His typical diet is listed below:

BREAKFAST	LUNCH	DINNER
Orange juice	Big Mac	Steak
Coffee	French fries	Baked potato with sour cream and butter
	Milk shake	Green beans in butter sauce. Salad with Italian dressing. Whole milk

a. How might David change his diet so it is better suited to his exercise program?

b. Does his exercise program include both an aerobic and a strength-training component?

c. Do you think David will be able to stick with this exercise program? Why or why not?

d. Suggest some changes that would make David's exercise program more balanced.

2. Do a risk-benefit analysis of an ergogenic aid (a quick way to do this is to use the Internet to collect information). List the risks and benefits and then write a conclusion stating why you would or would not recommend this substance.

SUMMARY

13.1 Exercise, Fitness, and Health

- Regular exercise improves fitness. How fit an individual is depends on his or her cardiorespiratory endurance, muscle strength, muscle endurance, flexibility, and body composition.
- Regular exercise can reduce the risk of chronic diseases such as obesity, heart disease, diabetes, and osteoporosis. It can reduce overall mortality even in obese individuals.
- Exercise helps manage body weight by increasing energy expenditure and by increasing the proportion of body weight that is lean tissue.

13.2 Exercise Recommendations

- The current Canadian Physical Activity Guidelines for adults (18-65 years) suggest a minimum 150 min./week of moderate to vigorous intensity aerobic activity and muscle- and bone-strengthening exercise at least 2 days/week.
- An activity is moderate if it increases heart rate and breathing and requires moderate exertion. Vigorous activity will increase heart rate more and leave a person out of breath.
- The current Canadian Physical Activity Guidelines for children and youth suggest a minimum of 60 min. of moderate to vigorous-intensity activity per day, some of which is muscle-strengthening and bone-strengthening. Activities should be age appropriate.
- An exercise program should include activities that are enjoyable, convenient, and safe. Rest is important to allow the body to recover and rebuild. In serious athletes, inadequate rest can lead to overtraining syndrome.

13.3 Fuelling Exercise

- During the first 10-15 sec. of exercise, ATP and creatine phosphate stored in the muscle provide energy to fuel activity. During the next 2-3 min., the amount of oxygen at the muscle remains limited, so ATP is generated by the anaerobic metabolism of glucose. After a few minutes, the delivery of oxygen at the muscle increases, and ATP can be generated by aerobic metabolism. Aerobic metabolism is more efficient than anaerobic metabolism and can utilize glucose, fatty acids, and amino acids as energy sources. The use of protein as an energy source increases when exercise continues for many hours.

- For short-term, high-intensity activity, ATP is generated primarily from the anaerobic metabolism of glucose from muscle glycogen stores. Anaerobic metabolism uses glucose rapidly. For lower-intensity exercise of longer duration, aerobic metabolism predominates, and both glucose and fatty acids are important fuel sources.
- Fitness training causes changes in the cardiovascular system and muscles that improve oxygen delivery and utilization and increase glycogen stores, allowing aerobic exercise to be sustained for longer periods at higher intensity.

13.4 Energy and Nutrient Needs for Physical Activity

- The diet of an active individual should provide sufficient energy to fuel activity. The EERs base energy needs on activity level as well as age, gender, and body size.
- To maximize glycogen stores, optimize performance, and maintain and repair lean tissue, a diet providing about 60% of energy from carbohydrate, 20%-25% of energy from fat, and about 15%-20% of energy from protein is recommended.
- Sufficient vitamins and minerals are needed to generate ATP from macronutrients, to maintain and repair tissues, and to transport oxygen and wastes to and from the cells. Most athletes who consume a varied diet that meets their energy needs also meet their vitamin and mineral needs. Those who restrict their food intake may be at risk for deficiencies. The pressure to compete and maintain a body weight that is optimal for their sport puts some athletes at risk for eating disorders. A combination of excessive exercise and energy restriction puts female athletes at risk for the female athlete triad. Increased iron needs and greater iron losses due to fitness training put athletes, particularly female athletes, at risk of iron deficiency.

13.5 Fluid Needs for Physical Activity

- Water is needed to ensure that the body can be cooled and that nutrients and oxygen can be delivered to body tissues. Adequate fluid intake before exercise ensures that athletes begin exercise well hydrated. Fluid intake during and after exercise must replace water lost in sweat and from evaporation through the lungs.
- If water intake is inadequate, dehydration can lead to a decline in exercise performance and increase the risk of heat-related illness. This can be life-threatening in severe cases.
- Drinking plain water during extended exercise increases the risk of hyponatremia, a relative deficiency of sodium in the blood.
- Plain water is an appropriate fluid to consume for most exercise. Beverages containing carbohydrate and sodium are recommended for exercise lasting more than an hour.

13.6 Food and Drink to Maximize Performance

- Competitive endurance athletes may benefit from glycogen supercompensation (carbohydrate loading), which maximizes glycogen stores before an event.
- Meals eaten before competition should help ensure adequate hydration, provide moderate amounts of protein, be high enough in carbohydrate to maximize glycogen stores, be low in fat and fibre to speed gastric emptying, and satisfy the psychological needs of the athlete.
- During exercise, athletes need beverages and food to replace lost fluid and provide carbohydrate and sodium.
- Post-competition meals should replace lost fluids and electrolytes, provide carbohydrate to restore muscle and liver glycogen, and provide protein for muscle protein synthesis and repair.

13.7 Ergogenic Aids: Do Supplements Enhance Athletic Performance?

- Many types of ergogenic aids are marketed to improve athletic performance. An individual risk–benefit analysis should be used to determine whether a supplement is appropriate for you.
- Vitamin and mineral supplements are usually not necessary for athletes who meet their kcalorie needs.
- Protein needs can be met by diet; amino acid supplements are not recommended.
- Creatine supplementation has been shown to improve performance in short-duration, high-intensity exercise.
- Caffeine use can improve performance in endurance activities, but high doses are illegal during athletic competitions.
- Anabolic steroids combined with resistance-training exercise increase muscle size and strength, but these supplements are illegal and have dangerous side effects.
- A healthy diet is the base for successful athletic performance. Beverages and foods that supply fluids and energy can enhance performance, and there are a few supplements that benefit specific activities.

REVIEW QUESTIONS

1. What characterizes a fit individual?
2. What is aerobic exercise?
3. How does aerobic exercise affect resting heart rate?
4. What is strength training?
5. What causes muscle hypertrophy? Muscle atrophy?
6. List five of the health benefits of exercise.
7. How much of what types of exercise is recommended?
8. What is aerobic capacity? How is it affected by training?
9. From where does the ATP to fuel the first few minutes of exercise come?
10. What fuels are used to produce ATP in anaerobic metabolism?
11. Which is more efficient, aerobic or anaerobic metabolism?
12. What factors affect the availability of oxygen and the type of fuel used during exercise?
13. What fuels are used in exercise of long duration such as marathon running?
14. What are the recommendations for fluid intake before, during, and after exercise?
15. How does exercise affect protein needs?
16. What is glycogen supercompensation or carbohydrate loading?
17. Explain why creatine supplements are ergogenic. Are they safe?

REFERENCES

1. Gallagher, D., Heymsfield, S., Heo, M. et al. Healthy percentage body fat ranges: An approach for developing guidelines based on body mass index. *Am. J. Clin. Nutr.* 72:694–701, 2000.
2. United States Department of Health and human Services. 2008 Physical Activity Guidelines for Americans. Available online at www.health.gov/paguidelines/. Accessed June 16, 2009.
3. Institute of Medicine, Food and Nutrition Board. *Dietary Reference Intakes for Energy, Carbohydrates, Fiber, Fat, Protein and Amino Acids.* Washington, DC: National Academies Press, 2002.
4. The President's Council on Physical Fitness and Sports. *Physical Activity Protects against the Health Risks of Obesity.* Available online at www.fitness.gov/digest_dec2000.htm. Accessed June 12, 2009.
5. Centers for Disease Control and Prevention. *Physical Activity and Health: A Report of the Surgeon General.* Available online at www.cdc.gov/nccdphp/sgr/adults.htm/. Accessed February 3, 2009.
6. Zanuso, S., Jimenez, A., Pugliese, G. et al. Exercise for the management of type 2 diabetes: A review of the evidence. *Acta. Diabetol.* June 3, 2009 [Epub].
7. Newton, R. U., and Galvão, D. A. Exercise in prevention and management of cancer. *Curr. Treat. Options Oncol.* 9:135–146, 2008.
8. Reigle, B. S., and Wonders, K. Breast cancer and the role of exercise in women. *Methods Mol. Biol.* 472:169–189, 2009.
9. Fontaine, K. R. *Physical Activity Improves Mental Health.* Available online at www.physsportsmed.com/index.php?art=psm_10_2000?article=1256. Accessed April 7, 2009.
10. Antunes, H. K., Stella, S. G., Santos, R. F. et al.Depression, anxiety and quality of life scores in seniors after an endurance exercise program. *Rev. Bras. Psiquiatr.* 27:266–271, 2005.
11. Shields, M., Tremblay, M. S., Laviolette, M. et al. Fitness of Canadian adults: Results from the 2007-2009 Health Measures Survey. Health Reports 21(1): 1-15. Available online at http://www.statcan.gc.ca/pub/82-003-x/2010001/article/11064-eng.pdf. Accessed June 5, 2011.
12. ParticiPACTION. Participaction Initiatives. Available online at http://www.participaction.com/en-us/FindProgramsAndEvents/ParticipACTIONInitiatives.aspx. Accessed May 8, 2011.
13. Warburton, D. E. R., Katzmarzyk, P. T., Rhodes, R. E. et al. Evidence-informed physical activity guidelines for Canadian adults. *Appl. Physiolo. Nutr. Metab.* 32: S16-S68, 2007.
14. Canadian Society for Exercise Physiologists. Canadian Physical Activity Guidelines. Clinical Practice Guideline Development Report. 2011. Available online at http://www.csep.ca/CMFiles/Guidelines/CPA. Guideline_Report_JAN2011.pdf. Accessed May 8, 2011.
15. Public Health Agency of Canada. Tips to get active: Information and tips for adults (ages 18-64 years). Available online at http://www.phac-aspc.gc.ca/hp-ps/hl-mvs/pa-ap/07paap-eng.php. Accessed May 9, 2011.
16. Canadian Society of Exercise Physiology, ParticipACTION (2010): Fact Sheet—New Physical Activity Recommendations. Available online at http://63.134.208.23/CMFiles/PAMGpdfs/CSEP_PAC%20-%20Fact%20Sheet%20-%20EN.pdf. Accessed Sept 26,2010.
17. Swinburn, B., and Shelly, A. Effect of TV time and other sedentary pursuits. *Int. J. Obes.* (Lond) 32:S5132–S5136, 2008.
18. Haskell, W. L., Lee, I-M, Pate, R. R. et al. Physical Activity and Public Health Updated Recommendation for Adults: Updated Recommendation from the American College of Sports Medicine and the American Heart Association. *Circulation* 116:1081–1093, 2007.
19. Cairns, S. P. Lactic acid and exercise performance: Culprit or friend? *Sports Med.* 36:279–291, 2006.
20. Manore, M., and Thompson, J. *Sport Nutrition for Health and Performance.* Champaign, IL: Human Kinetics, 2000.
21. Rodriguez, N. R., DiMarco, N. M., and Langley, S. Position of the American Dietetic Association, Dietitians of Canada, and the American College of Sports Medicine: Nutrition and athletic performance. *J. Am. Diet Assoc.* 109:509–527, 2009.
22. Sundgot-Borgen, J., and Torstveit, M. K. Prevalence of eating disorders in elite athletes is higher than in the general population. *Clin. J. Sport Med.* 14: 25–32, 2004.
23. Remick, D., Chancellor, K., Pederson, J. et al. Hyperthermia and dehydration-related deaths associated with intentional rapid weight loss in three collegiate wrestlers—North Carolina, Wisconsin, and Michigan, November–December, 1997. MMWR Morb Mortal Wkly Rep 47:105–108, 1998. Available online at www.cdc.gov/ mmwr/preview/mmwrhtml/00051388.htm. Accessed March 25, 2009.
24. Tipton, K. D., and Wolfe, R. R. Protein and amino acids for athletes. *J. Sports. Sci.* 22:65–79, 2004.
25. Jackson, M. J., Khassaf, M., Vasilaki, F. et al. Vitamin E and the oxidative stress of exercise. *Ann. N.Y. Acad. Sci.* 1031:158–168, 2004.
26. Finaud, J., Lac, G., and Filaire, E. Oxidative stress: Relationship with exercise and training. *Sports Med.* 36(4):327–358, 2006.
27. Margaritis, I., and Rousseau, A. S. Does physical exercise modify antioxidant requirements? *Nutrition Research Reviews* 21:3–12, 2008.
28. Food and Nutrition Board, Institute of Medicine. *Dietary Reference Intakes: Vitamin A, Vitamin K, Arsenic, Boron, Chromium, Copper, Iodine, Iron, Manganese, Molybdenum, Nickel, Silicon, Vanadium, and Zinc.* Washington, DC: National Academies Press, 2001.
29. Suedekum, N. A., and Dimeff, R. J. Iron and the athlete. *Curr. Sports Med. Rep.* 4:199–202, 2005.
30. Di Santolo, M., Stel, G., Banfi, G. et al. Anemia and iron status in young fertile non-professional female athletes. *Eur. J. Appl. Physiol.* 102(6):703–709, 2008.
31. Kazis, K., Iglesias, E. The female athlete triad. *Adolescent Medicine* 14:87–95, 2003.
32. NOAA's National Weather Service: *Heat Index.* Available online at www.weather.gov/os/heat/index.shtml. Accessed March 10, 2009.
33. Almond, C. S. D., Shin, A. Y., Fortescue, E. B. et al. Hyponatremia among runners in the Boston marathon. *N. Engl. J. Med.* 352:1550–1556, 2005.
34. Bussar, V. A., Fairchild, T. J., Rao, A. et al. Carbohydrate loading in human muscle: An improved 1 day protocol. *Eur. J. Appl. Physiol.* 87:290–295, 2002.
35. Burke, L. M. Nutrition strategies for the marathon. Fuel for training and racing. *Sports Med.* 37:344–347, 2007.
36. Health Canada. Natural Health Products Database. Available online at http://webprod.hc-sc.gc.ca/lnhpd-bdpsnh/start-debuter.do?lang=eng. Accessed June 4, 2011.
37. Urso, M. L. and Clarkson, P. M. Oxidative stress, exercise, and antioxidant supplementation. *Toxicology* 189:41–54, 2003.
38. Di Luigi, L. Supplements and the endocrine system in athletes. *Clin. Sports Med.* 27:131–151, 2008.
39. Nissen S. L., and Sharp R. L. Effect of dietary supplements on lean mass and strength gains with resistance exercise: A meta-analysis. *J. Appl. Physiol.* 94:651–659, 2003.

40. Bemben, M. G., and Lamont, H.S. Creatine supplementation and exercise performance: Recent findings. *Sports Med.* 35:107–125, 2005.

41. Paddon-Jones, D., Borsheim, E., and Wolfe, R. R. Potential ergogenic effects of arginine and creatine supplementation. *J. Nutr.* 134(10 Suppl):2888S–2894S, 2004.

42. Volek, J. S., and Rawson, E. S. Scientific basis and practical aspects of creatine supplementation for athletes. *Nutrition* 20:609–614, 2004.

43. Shao, A., and Hathcock, J. N. Risk assessment for creatine monohydrate. *Regul. Toxicol. Pharmacol.* 45:242–251, 2006.

44. Portal, S., Eliakim, A., Nemet, D. et al. Effect of HMB supplementation on body composition, fitness, hormonal profile and muscle damage indices. *J. Pediatr. Endocrinol. Metab.* 23(7):641-50, 2010.

45. Raymer, G. H., Marsh, G. D., Kowalchuk, J. M., et al. Metabolic effects of induced alkalosis during progressive forearm exercise to fatigue. *J. Appl. Physiol.* 96:2050–2056, 2004.

46. Jeukendrup, A. E., and Aldred, S. Fat supplementation, health, and endurance performance. *Nutrition* 20:678–688, 2004.

47. Horowitz, J. F., Mora-Rodriguez, R., Byerley, L. O., et al. Preexercise medium-chain triglyceride ingestion does not alter muscle glycogen use during exercise. *J. Appl. Physiol.* 88:219–225, 2000.

48. Paluska, S. A. Caffeine and exercise. *Curr. Sports Med. Rep.* 2:213–219, 2003.

49. Sökmen, B., Armstrong, L. E., Kraemer, W. J. et al. Caffeine use in sports: Considerations for the athlete. *J. Strength Cond. Res.* 22:978–986, 2008.

50. Bauer, L., Kohlich, A., Hirschwehr, R. et al. Food allergy: Pollen or bee products? *J. Allergy Clin. Immunol.* 97:65–73, 1996.

51. Bahrke, M. S., Morgan, W. P., and Stenger, A. Is ginseng an ergogenic aid? *Int. J. Sport Nutr. Exerc. Metab.* 19:298–322, 2009.

14 Nutrition During Pregnancy and Lactation

Case Study

Jasmine is 26 years old and expecting her first baby. At her initial prenatal visit, the obstetrician tells her that she and the baby seem healthy except that her iron stores are low. The doctor prescribes a prenatal supplement and gives her a pamphlet on nutrition during pregnancy. Jasmine is concerned because the pamphlet recommends that she gain 11.5-16 kg (25-35 lb) over the course of her pregnancy and suggests a dietary intake that includes much more than she usually eats. Jasmine has always worried about her weight. She typically skips breakfast, has only yogurt and a diet soda for lunch, and then eats a big dinner with her husband in the evening. Jasmine's sister, who has three young children, is now 11 kg (25 lb) heavier than she was before the birth of her first child 8 years ago, and Jasmine doesn't want to end up like that. She also remembers her grandmother telling her that gaining too much weight during pregnancy makes the delivery more difficult. So Jasmine decides to continue to follow her usual eating pattern. After all, she reasons, the prenatal supplement will give her all the nutrients her baby needs.

By the fourth month of her pregnancy, Jasmine has gained only 0.5 kg (1 lb) and is feeling tired and run down. She tells the doctor that she stopped taking the supplement because it was making her nauseous and constipated, and that she has been limiting what she eats because she is afraid of gaining too much weight. The doctor explains to Jasmine that during pregnancy, her body undergoes physiological changes that support the baby's growth. New tissues in her body plus the growth of the baby cause weight gain, and she will lose the weight after the baby is born. The obstetrician warns Jasmine that her diet must supply all of the nutrients necessary for the changes in her physiology as well as for the growth and development of the infant.

A deficiency of nutrients or energy may cause birth defects or increase the risk of having a baby that is born too soon or too small.

In this chapter you will learn about the nutritional requirements of pregnancy.

(© istockphoto.com/ Mona Makela)

(PhotoAlto/Sigrid Olsson/Getty Images, Inc.)

(Joe McBride/Getty Images, Inc.)

Chapter Outline

14.1 The Physiology of Pregnancy

Learning Objectives

- Explain why the mother's nutrient intake during pregnancy is so important.
- Discuss the risks associated with gaining too little or too much weight during pregnancy.
- List some types of exercise that are appropriate for pregnant women.
- Describe some of the nutrition-related discomforts and complications of pregnancy.

conception The union of sperm and egg (ovum) that results in pregnancy.

lactation Milk production and secretion.

Pregnancy, from **conception** to birth, usually lasts 40 weeks in humans. During pregnancy, a single cell grows and develops into an infant that is ready for life outside the womb. This development requires a safe environment to which oxygen and nutrients are provided in the right amounts and from which waste products are removed. Many physiological changes take place in the mother to support prenatal development and prepare her for **lactation**.

fertilization The union of sperm and egg (ovum).

oviducts or **fallopian tubes** Narrow ducts leading from the ovaries to the uterus.

zygote The cell produced by the union of sperm and ovum during fertilization.

implantation The process by which the developing embryo embeds in the uterine lining.

Prenatal Growth and Development

Reproduction requires the **fertilization** of an egg, or ovum, from the mother by a sperm from the father. Fertilization, which occurs in the **oviduct** or **fallopian tube**, produces a single-celled **zygote**. The zygote travels down the mother's fallopian tube into the uterus. Along the way it divides many times to form a ball of smaller cells. In the uterus, the developing embryo imbeds in the uterine lining in a process known as **implantation (Figure 14.1)**.

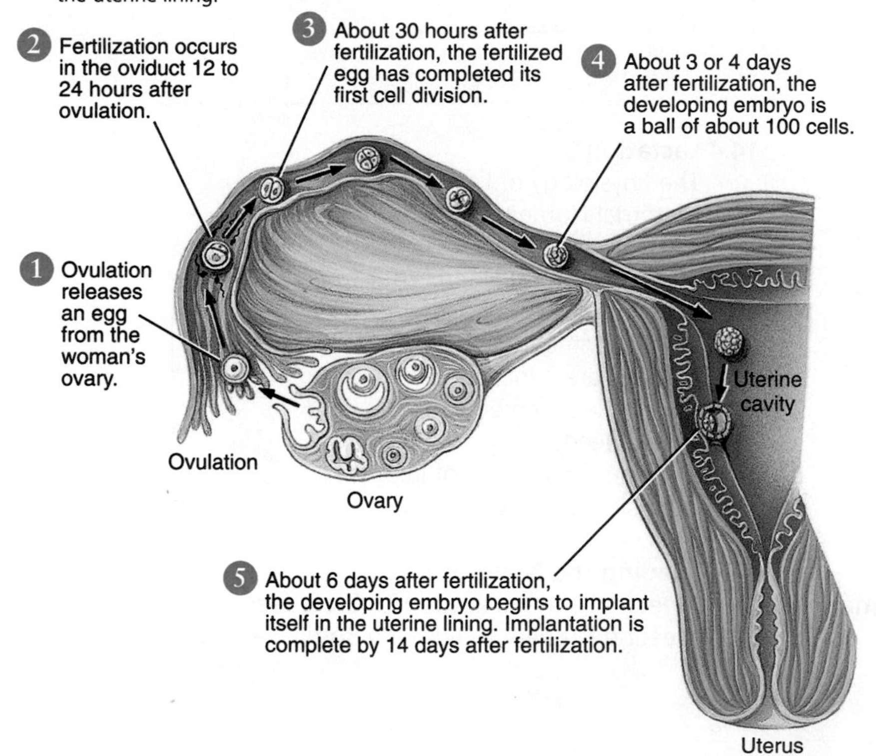

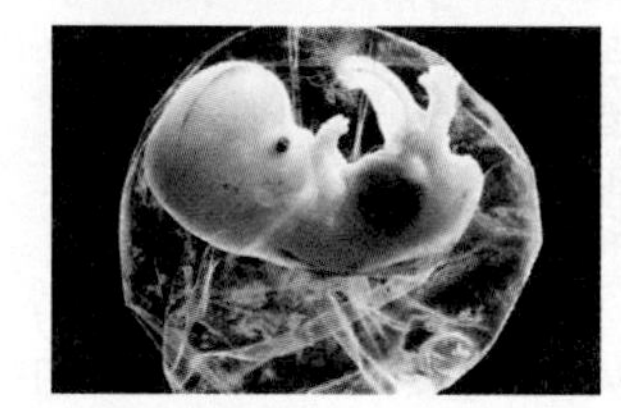

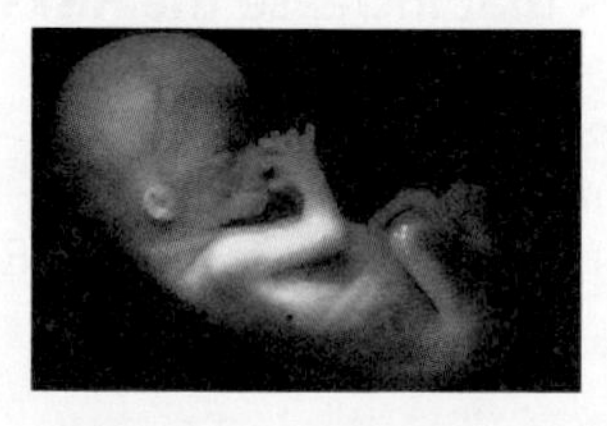

Figure 14.1 Prenatal development
This cross section shows the path of the egg and developing embryo from the ovary, where the egg is produced, through the oviduct to the uterus, where most prenatal development occurs. (centre/right photo: Biphoto Associates/Photo Researchers, Inc.; bottom right photo: Meitchik/Custom Medical Stock Photo, Inc.)

embryo The developing human from 2-8 weeks after fertilization. All organ systems are formed during this time.

Prenatal growth and development continue after implantation. The cells differentiate into the multitude of specialized cell types that make up the human body, and arrange themselves in the proper shapes and locations to form organs and other structures. About 2 weeks after fertilization, implantation is complete and the developing offspring is known as an **embryo** (see Figure 14.1). The embryonic stage of development lasts until the eighth week after fertilization, when rudimentary organ systems have been formed. The embryo at this point is approximately 3 cm long and has a beating heart. All major external and internal structures have been formed. Beginning at the ninth week of development and continuing until birth, the developing offspring is known as a **fetus** (see Figure 14.1). During the fetal period of development, structures that appeared during the embryonic period continue to grow and mature. Anything that interferes with development can cause birth defects. If the birth defects are severe, they may result in a **spontaneous abortion** or **miscarriage**.

The early embryo gets its nourishment by breaking down the lining of the uterus, but soon this source is inadequate to meet its growing needs. After about 5 weeks, the **placenta** takes over the role of nourishing the embryo (**Figure 14.2**). The placenta is a network of blood vessels and tissues that allows nutrients and oxygen to be transferred from mother to fetus and waste products to be transferred from the fetus to the mother's blood for elimination. The placenta is made up of tissue from both the mother and the fetus. The maternal portion of the placenta develops from the uterine lining. The fetal portion of the placenta develops from the outer layer of pre-embryonic cells. These cells divide to form branch-like projections that grow into the lining of the uterus where they are surrounded by pools of maternal blood. The projections contain blood vessels that supply the developing fetus. Although maternal and fetal blood do not mix, the close proximity of fetal blood vessels to maternal blood allows nutrients and oxygen to easily pass from mother to fetus, and allows carbon dioxide and other wastes to pass from fetus to mother for elimination (see Figure 14.2). The placenta also secretes hormones that are necessary to maintain pregnancy.

fetus The developing human from the ninth week to birth. Growth and refinement of structures occur during this time.

spontaneous abortion or **miscarriage** Termination of pregnancy, due to natural expulsion of fetus, prior to the seventh month.

placenta An organ produced from both maternal and embryonic tissues. It secretes hormones, transfers nutrients and oxygen from the mother's blood to the fetus, and removes wastes.

gestation The time between conception and birth, which lasts about 9 months (or about 40 weeks) in humans.

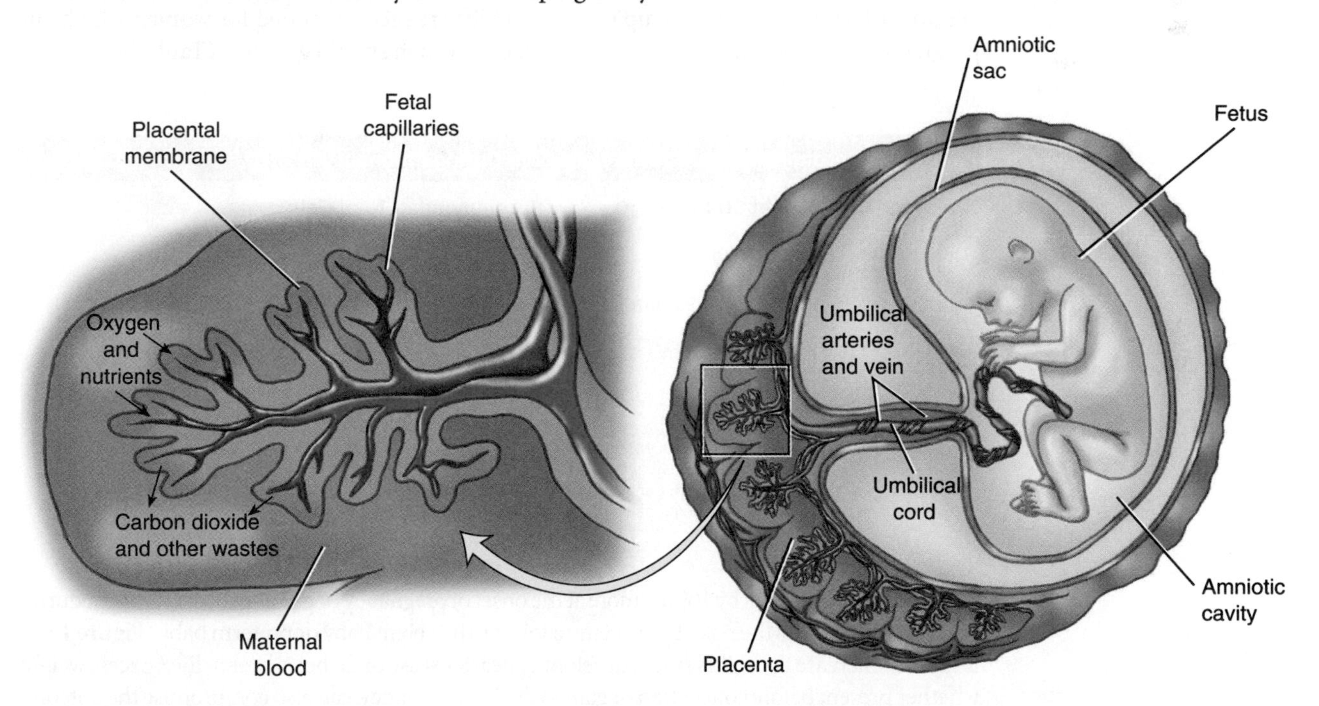

Figure 14.2 Placenta and amniotic sac
The placenta allows the transfer of nutrients and wastes between mother and developing fetus. The amniotic sac and the fluid it contains protect the fetus.

The pregnancy usually ends after 40 weeks of **gestation** with the birth of an infant weighing about 3-4 kg (6.6-8.8 lb).[1] Infants who are born on time but have failed to grow normally in the uterus are said to be **small-for-gestational-age**. Those born before 37 weeks of gestation are said to be **preterm** or **premature**. Whether born too soon or just too small, **low-birth-weight** infants (those weighing less than 2.5 kg [5.5 lb] at birth) and **very-low-birth-weight** infants (those weighing less than 1.5 kg [3.3 lb]), are at increased risk for illness and early death.[2] They often require special care and a special diet in order to successfully continue to grow and develop. Survival improves with increasing gestational age and birth weight. Today, with advances in medical and nutritional care, infants born as early as 25 weeks of gestation and those weighing as little as 1 kg (2.2 lb) can survive (**Figure 14.3**).

small-for-gestational-age An infant born at term weighing less than 2.5 kg (5.5 lb).

preterm or **premature** An infant born before 37 weeks of gestation.

low-birth-weight A birth weight less than 2.5 kg (5.5 lb).

very-low-birth-weight A birth weight less than 1.5 kg (3.3 lb).

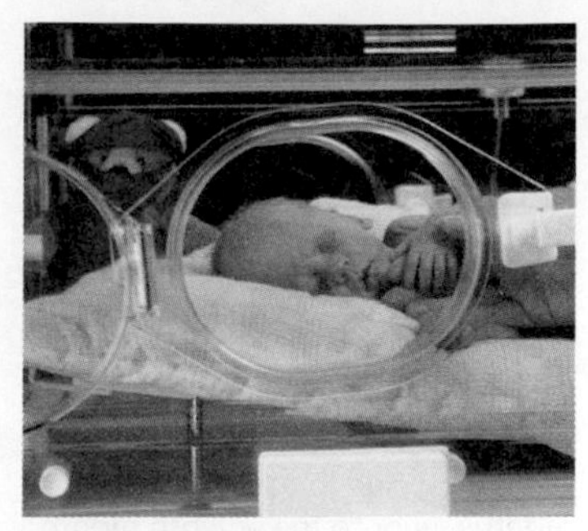

Figure 14.3 Even though incubators and other technological aids can help premature babies survive, they face greater risk for illness and death than full-term babies. (Brad Nelson/ Phototake)

Changes in the Mother

A woman's body undergoes many changes during pregnancy to support the growth and development of the embryo and fetus. These continuous physiological adjustments affect the metabolism and distribution of nutrients in her body. Maternal blood volume increases by 50%, and the heart, lungs, and kidneys work harder to deliver nutrients and oxygen and remove wastes. The placenta develops and the hormones produced by it orchestrate other changes: they promote uterine growth; they relax muscles and ligaments to accommodate the growing fetus and allow for childbirth; they promote breast development; and they increase fat deposition to provide the energy stores that will be needed during late pregnancy and lactation. These changes result in weight gain and can affect the type and level of physical activity that is safe for the pregnant woman.

Weight Gain During Pregnancy Gaining the right amount of weight during pregnancy is essential to the health of both mother and fetus. The recommended weight gain for healthy, normal-weight women is 11.5 -16 kg (25-35 lb). Typically the weight of the infant at birth is about 25% of the total pregnancy weight gain. The placenta, amniotic fluid, and changes in maternal tissues account for the rest. This includes increases in the size of the uterus and breasts, expansion of blood and extracellular fluid volume, and deposition of fat stores (**Figure 14.4**). The rate of weight gain is as important as the total weight gain. Little gain is expected in the first 3 months, or **trimester**, of pregnancy—usually about 0.9-1.8 kg (2-4 lb). In the second and third trimesters, when the fetus grows from less than 0.5 kg to 3-4 kg, the recommended maternal weight gain is about 0.5 kg (1 lb)/ week. Women who are underweight, and women who are overweight or obese at conception should still gain weight at a slow, steady rate (**Figure 14.5**). Weight gains of up to 18 kg (40 lb) are recommended for women who begin pregnancy underweight. Overweight and obese women should gain less (**Table 14.1**).[1]

trimester A term used to describe each third, or 3-month period, of a pregnancy.

Table 14.1 Recommendations for Weight Gain During Pregnancy

Prepregnancy Weight Status[a]	Recommended Total Gain
Underweight (BMI <18.5 kg/m²)	13–18 kg (28–40 lb)
Normal weight (BMI 18.5–24.9 kg/m²)	11.5–16 kg (25–35 lb)
Overweight (BMI 25.0–29.9 kg/m²)	7–11.5 kg (15–25 lb)
Obese (BMI <30.0 kg/m²)	5–9 kg (11–20 lb)

Source: From Committee to Reexamine IOM Pregnancy Weight Guidelines, Institute of Medicine, National Research Council. *Weight Gain During Pregnancy: Reexamining the Guidelines*. Washington, DC: National Academies Press, 2009.These guidelines have been adopted by Health Canada: Canadian Gestational Weight Gain recommendations: Available online at http://www.hc-sc.gc.ca/fn-an/nutrition/prenatal/qa-gest-gros-qr-eng.php. Accessed Nov 19, 2010.

Being underweight by 10% or more at the onset of pregnancy or gaining too little weight during pregnancy increases the risk of producing a low-birth-weight baby or preterm baby (**Figure 14.6**). It can also increase the child's risk of developing heart disease or diabetes later in life.[2] Excess weight, whether present before conception or gained during pregnancy, can also compromise the outcome of the pregnancy. The mother's risks for high blood pressure, diabetes, difficult delivery, and **Caesarean section** are increased by excess weight, as is the risk of having a **large-for-gestational-age** baby (see Figure 14.6). Maternal obesity may also increase the risk of neural tube defects and fetal death.[3] Despite this, dieting during pregnancy is not advised even for obese women. If possible, excess weight should be lost before the pregnancy; ideally, women should have a BMI less than 25 or, if not possible, less than 30 when entering a pregnancy. During pregnancy, women are encouraged to exercise to stay within recommended weight gains (see Table 14.1).[4]

Caesarean section The surgical removal of the fetus from the uterus.

large-for-gestational-age An infant weighing more than 4 kg (8.8 lb) at birth.

Approximately 5 kg (10 lb) are lost at birth from the weight of the baby, amniotic fluid, and placenta. In the week after delivery, another 2.5 kg (5 lb) of fluid are typically lost. Once this initial fluid and tissue weight is lost, further weight loss requires that energy intake be less than energy output. After the mother has recovered from delivery, a balanced diet, with a small deficit of kcalories, combined with moderate exercise, will promote gradual weight loss and the return of muscle tone. Gradual weight loss is important when the mother is breastfeeding, to ensure that milk production is not compromised.

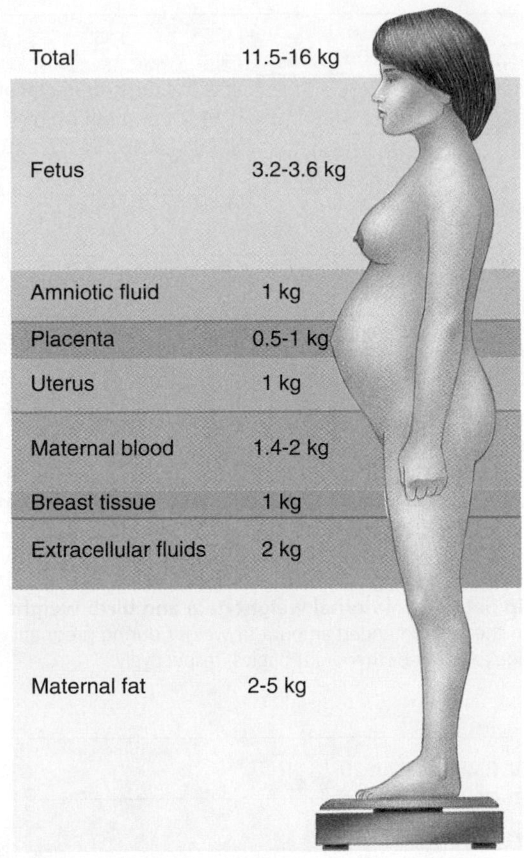

Figure 14.4 Sources of weight gain in pregnancy
Weight gained during pregnancy is due to increases in the weight of the mother's tissues, as well as the weight of the fetus, placenta, and amniotic fluid.

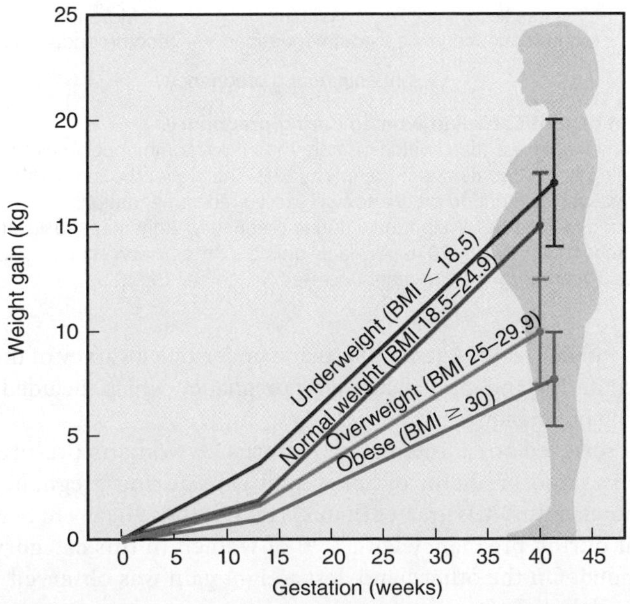

Figure 14.5 Recommended weight gain during pregnancy
The same pattern of weight gain is recommended for women who are normal weight, underweight, overweight, or obese at the start of pregnancy, but total weight gain recommendations differ.
Source: Adapted from Committee to Reexamine IOM Pregnancy Weight Guidelines, Institute of Medicine, National Research Council. *Weight Gain During Pregnancy: Reexamining the Guidelines.* Washington, DC: National Academies Press, 2009.

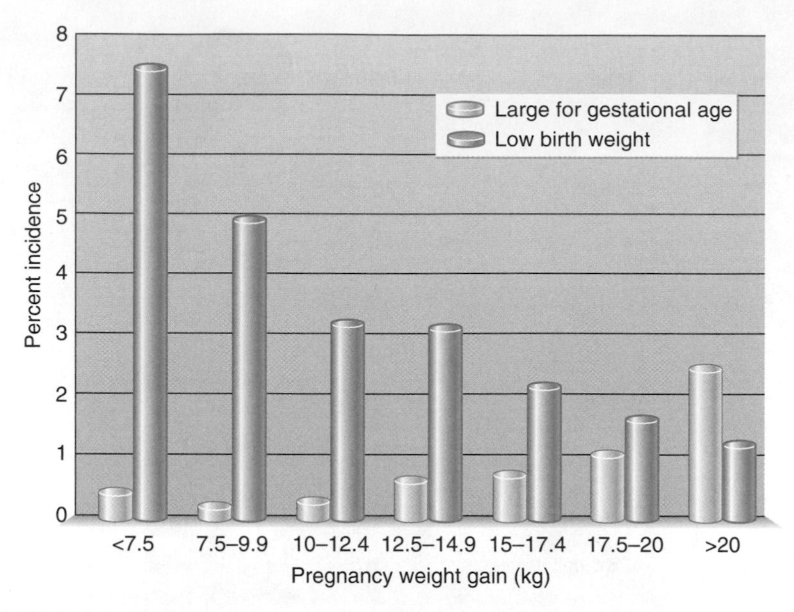

Figure 14.6 Relationship between maternal weight gain and birth weight
Gaining more or less than the recommended amount of weight during pregnancy increases the incidence of large-for-gestational-age and low-birth-weight babies, respectively.

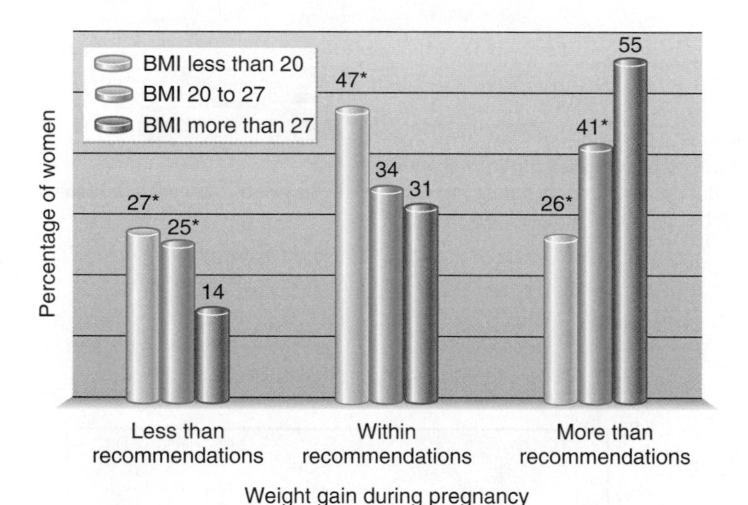

Figure 14.7 Weight gain of Canadian women during pregnancy.
Some Canadian women gain less than, within or more than the recommended weight during pregnancy, but it varies depending on the woman's pre-pregnancy BMI. The higher the mother's pre-pregnancy BMI, the more likely her weight gain during pregnancy will exceed recommendations.
Source: Lowell, H. and Miller, D. C. Weight gain during pregnancy: Adherence to Health Canada's guidelines. Health Reports 21(2), 1-5, 2010. Available online at http://www.statcan.gc.ca/bsolc/olc-cel/olc-cel?catno=82-003-X&chropg=1&lang=eng. Accessed November 19, 2010.

CANADIAN CONTENT

In 2006, the Public Health Agency of Canada undertook a survey of the experiences of Canadian women during and immediately after pregnancy, which included information on weight gain during pregnancy.[5]

The results revealed some interesting patterns. A woman's pre-pregnancy weight was found to be a good predictor of her weight gain during pregnancy. As shown in **Figure 14.7**, women with BMIs greater than 27, that is, those heavy to begin with, gained the most weight during pregnancy with 55% of women in this category gaining more than recommended. On the other hand, less weight gain was observed in women with normal or low body weights.

The survey also revealed that higher than recommended weight gain was more common among women having their first baby, among less educated women, and among Aboriginal women. Women who gained more weight during pregnancy were also more likely to retain that extra weight after birth.[5] Since this extra weight may contribute both to complications during pregnancy and the development of obesity, these results suggest that the prenatal and postpartum care of Canadian women could be improved by providing more assistance in weight management both during pregnancy and after the baby is born.

Physical Activity During Pregnancy In the early 1980s, active women began asking their doctors about the safety of exercise during pregnancy; should they exercise? How much and what kind of exercise was appropriate? This sparked research into the role of exercise during pregnancy and the development of guidelines for exercise during pregnancy by the Canadian Society for Exercise Physiology and the Society of Obstetricians and Gynaecologists of Canada.[6] Their major recommendations are listed in **Table 14.2.** Researchers also developed the Physical Activity Readiness Medical Examination form, which is available to physicians to access whether a woman can exercise during pregnancy or whether her particular pregnancy has contraindications such as medical conditions that recommend against exercise.[7]

Table 14.2 Recommendations for Exercise in Pregnancy and the Postpartum Period	
1)	All women without contraindications should be encouraged to participate in aerobic and strength-conditioning exercises as part of a healthy lifestyle during their pregnancy.
2)	Reasonable goals of aerobic conditioning in pregnancy should be to maintain a good fitness level throughout pregnancy without trying to reach peak fitness or train for an athletic competition.
3)	Women should choose activities that will minimize the risk of loss of balance and fetal trauma.
4)	Women should be advised that adverse pregnancy or neonatal outcomes are not increased for exercising women.
5)	Initiation of pelvic-floor exercises in the immediate postpartum period may reduce the risk of future urinary incontinence.
6)	Women should be advised that moderate exercise during lactation does not affect the quantity or composition of breast milk or impact infant growth.

Source: Davies, G. L. A., Wolfe, L. A., Mottola, M. F. et al. Joint SOGC/CSEP Clinical Practice Guideline: Exercise in pregnancy and the postpartum period. *Can. J. Appl. Physiol.* 28(3): 329-341, 2003.

The major research finding was that healthy women without contraindications can safely engage in aerobic and strength-conditioning exercises with appropriate precautions (**Table 14.3**). For example, pregnancy is not a time for intense athletic training but should focus on maintaining and modestly improving fitness levels.[6] The benefits of exercise include improved maternal fitness, less weight gain during pregnancy, easier labour and delivery, better posture and reduced back pain, reduced risk of developing diabetes during pregnancy (gestational diabetes), and reduced blood pressure.[8]

Table 14.3 Safety Considerations During Exercise in Pregnancy
Important safety considerations:
• Avoid exercise in warm/humid environments, especially during the first trimester.
• Avoid isometric exercise or straining while holding your breath.
• Maintain adequate nutrition and hydration—drink liquids before and after exercise.
• Avoid exercise while lying on your back past the 4th month of pregnancy.
• Avoid activities which involve physical contact or danger of falling.
• Know your limits—pregnancy is not a good time to train for athletic competition.
• Know the reasons to stop exercise and consult a qualified health care provider immediately if they occur. These are: excessive shortness of breath; chest pain; painful uterine contractions (more than 6-8/hour); vaginal bleeding; any "gush" of fluid from vagina (suggesting premature rupture of the membranes); dizziness or faintness.

Source: PARmed-X for Pregnancy: available online at from Canadian Society for Exercise Physiology: http://www.csep.ca/english/view.asp?x=698. Accessed August 22, 2011.

The second trimester is considered the best time to begin an exercise program, as the first trimester is often characterized by fatigue and morning sickness. In the third trimester, the size of the developing fetus can limit some activities.[6]

The guidelines recommend that aerobic exercise be conducted for a maximum of 30 min./day, 4 days a week. Previously sedentary mothers should begin with 15-min. sessions 3 times a week and gradually increase time and frequency. Recommended activities are those that

Figure 14.8 During pregnancy, exercising in the water can reduce stress on joints and help keep the body cool. (Tracy Frankel/The Image Bank/Getty Images, Inc.)

minimize the loss of balance and any possible trauma to the fetus. These include brisk walking, stationary cycling, cross-country skiing, and aqua-fitness programs (**Figure 14.8**). Running and jogging are not advised because of potential damage to joints, which become more flexible during pregnancy.[6]

The degree of exertion should be kept moderate and can be determined by measuring heart rate, rating perceived exertion at no more than "somewhat hard" and/or by conducting a talk test; a person should be able to carry on a conversation during exercise.[7]

Muscle conditioning should involve all major muscle groups with focus on promoting good posture, and strengthening abdominal muscles and muscles of the pelvic floor that support the weight of the uterus.

Discomforts of Pregnancy

The physiological changes that occur during pregnancy can cause uncomfortable side effects for the mother. Some are caused by changes in fluid distribution, others by hormonal changes that affect the digestive tract. Most of these problems are minor, but in some cases they may endanger the mother and the fetus.

edema Swelling due to the buildup of extracellular fluid in the tissues.

Edema During pregnancy, blood volume expands to nourish the fetus, but this expansion may also cause the accumulation of extracellular fluid in the tissues, known as **edema**. Edema is common in the feet and ankles during pregnancy because the growing uterus puts pressure on the veins that return blood from the legs to the heart. This causes blood to pool in the legs, forcing fluid from the veins into the tissues of the feet and ankles (**Figure 14.9**). Edema can be uncomfortable but does not increase medical risks unless it is accompanied by a rise in blood pressure. Reducing fluid and sodium intake below the amount recommended for the general population is not recommended.

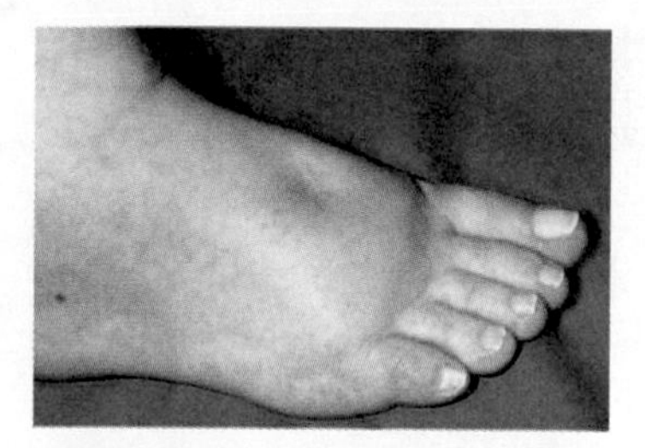

Figure 14.9 Edema in the feet and ankles is common during pregnancy. The swelling may be reduced by elevating the feet. (P. Marazzi/Science Photo Library/Photo Researchers)

morning sickness Nausea and vomiting that affect many women during the first few months of pregnancy and in some women can continue throughout the pregnancy.

Morning Sickness **Morning sickness** is a syndrome of nausea and vomiting that occurs in about 80% of women during pregnancy.[9] The incidence peaks at about 8-12 weeks of pregnancy and symptoms usually resolve by week 20. The term "morning sickness" is somewhat of a misnomer because symptoms can occur anytime during the day or night. Although the cause is unknown, this nausea and vomiting are hypothesized to be related to the hormonal changes of pregnancy. The symptoms may be alleviated to some extent by eating small, frequent snacks of dry, starchy foods, such as plain crackers or bread. In most women, symptoms are mild and do not affect maternal or fetal health. Fewer than 1% of pregnant women have a severe and intractable form of nausea and vomiting called *hyperemesis gravidarum* that may result in weight loss; nutritional deficiencies; and abnormalities in fluids, electrolyte levels, and acid-base balance. Treatment may require intravenous nutrition to assure that needs are met, and medications to reduce nausea.

Heartburn Heartburn, a burning sensation caused by stomach acid leaking up into the esophagus, is another common digestive complaint during pregnancy because the hormones produced to relax the muscles of the uterus also relax the muscles of the gastrointestinal tract. This involuntary relaxation of the gastroesophageal sphincter allows the acidic stomach contents to back up into the esophagus, causing irritation. The problem gets more severe as pregnancy progresses because the growing baby crowds the stomach (**Figure 14.10**). The fuller the stomach, the more likely that its contents will back up into the esophagus. Heartburn can be reduced by avoiding substances that are known to cause heartburn, such as caffeine and chocolate, and by consuming many small meals throughout the day rather than a few large meals. Limiting intake of high-fat foods that leave the stomach slowly, such as fried foods, rich sauces, and desserts, can also help reduce heartburn. Because a reclining position makes it easier for acidic juices to flow into the esophagus, remaining upright after eating, limiting eating in the hours before bedtime, and sleeping with extra pillows to produce a semi-reclining sleep position can also reduce heartburn.

Constipation and Hemorrhoids Constipation is a frequent complaint during pregnancy. The pregnancy-related hormones that cause muscles to relax also decrease intestinal motility and slow transit time. Constipation becomes more of a problem late in pregnancy when the enlarging uterus puts pressure on the gastrointestinal tract (see Figure 14.10). Iron supplements prescribed during pregnancy also contribute to constipation. Maintaining a moderate

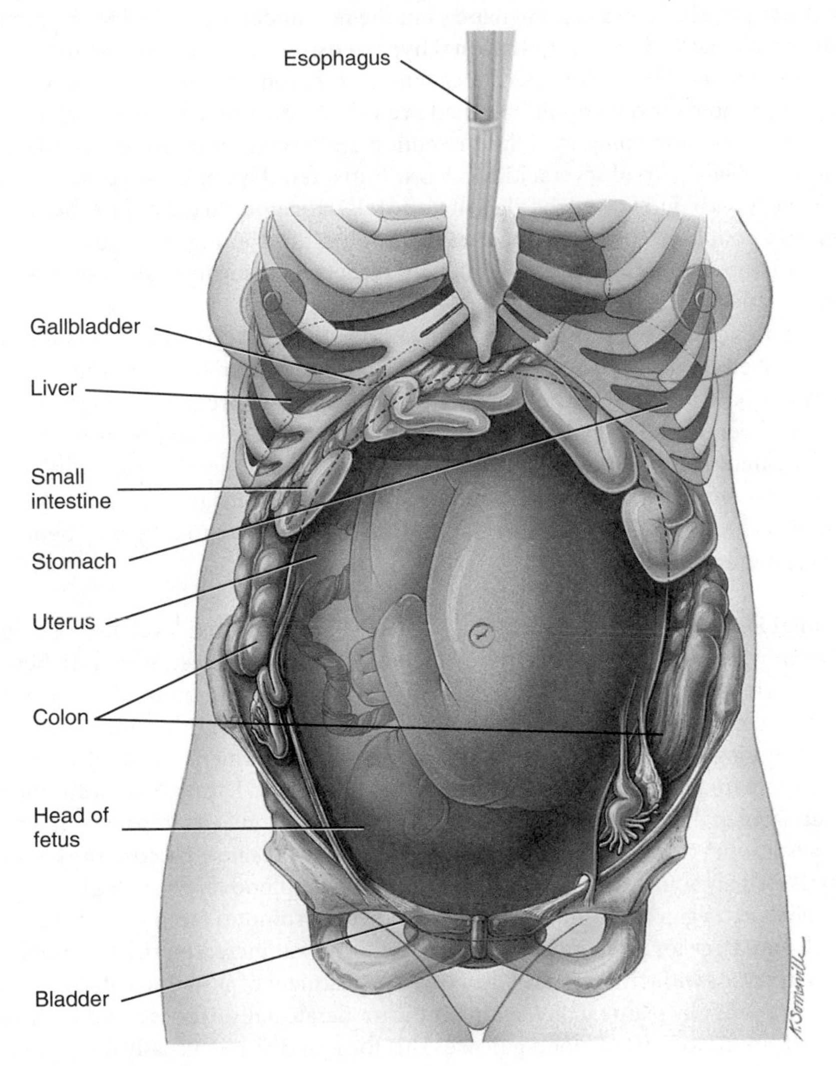

Figure 14.10 Crowding of the gastrointestinal tract
During pregnancy the uterus enlarges and pushes higher into the abdominal cavity, exerting pressure on the stomach and intestines.

level of physical activity and consuming plenty of water and other fluids along with high-fibre foods such as whole grains, vegetables, and fruits, are recommended to prevent constipation. Hemorrhoids are also more common during pregnancy, as a result of both constipation and physiological changes in blood flow.

Complications of Pregnancy

While most pregnancies are problem-free, some women do experience complications that increase risks to both mother and baby.[10,11] Some, such as anemia, can be prevented by proper nutrition and good prenatal care. For others, such as hypertension and diabetes, the cause is not as well understood, making prevention difficult. However, if caught early, these complications can usually be managed, allowing for a healthy delivery (**Figure 14.11**).

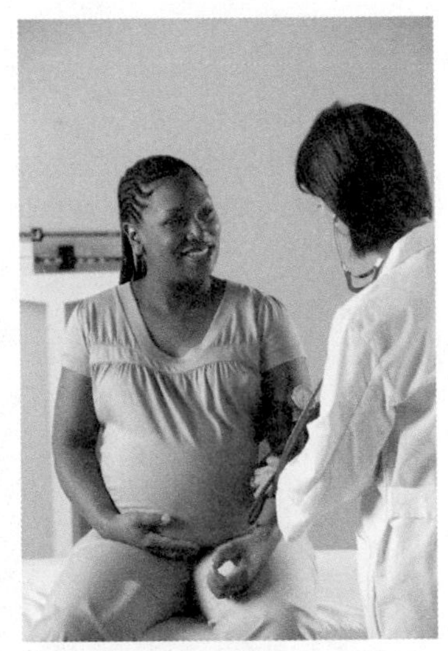

Figure 14.11 Prenatal care, which monitors the mother's blood pressure, weight, and blood sugar and the baby's size and heartbeat, can allow early identification and treatment of pregnancy complications. (Terry Vine/ © Blend Images/ Getty)

High Blood Pressure About 5%-10% of pregnant women experience high blood pressure during pregnancy. **Hypertensive disorders of pregnancy**, previously known as pregnancy-induced hypertension, refers to a spectrum of conditions involving elevated blood pressure during pregnancy and are a major cause of pregnancy-related maternal deaths.[12] It is especially common in mothers under 20 and over 35 years of age, low-income mothers, and mothers with chronic hypertension or kidney disease.

hypertensive disorders of pregnancy High blood pressure during pregnancy that is due to chronic hypertension, gestational hypertension, pre-eclampsia, eclampsia, or pre-eclampsia superimposed on chronic hypertension.

gestational hypertension The development of hypertension after the twentieth week of pregnancy.

pre-eclampsia A condition characterized by an increase in body weight, elevated blood pressure, protein in the urine, and edema. It can progress to **eclampsia**, which can be life-threatening to mother and fetus.

About one-third of the hypertensive disorders of pregnancy are due to chronic hypertension that was present before the pregnancy, but the remainder are related to the pregnancy. The least problematic of these is **gestational hypertension**, an abnormal rise in blood pressure that occurs after the 20th week of pregnancy. Gestational hypertension may signal the potential for a more serious condition called **pre-eclampsia**. Pre-eclampsia is characterized by high blood pressure along with fluid retention and excretion of protein in the urine; it can result in weight gain of several kilograms within a few days. It is dangerous to the fetus because it reduces blood flow to the placenta, and it is dangerous to the mother because it can progress to a condition called **eclampsia**, in which life-threatening seizures occur. Women with pre-eclampsia require bed rest and careful medical monitoring. The condition usually resolves after delivery.

The causes of pre-eclampsia are not fully understood. At one time, low-sodium diets were prescribed to prevent it, but studies have not found them to be effective.[13] Calcium may play a role in preventing the hypertensive disorders of pregnancy; calcium supplements have been found to reduce the risk of high blood pressure and pre-eclampsia.[14] Although calcium supplements are not routinely recommended for healthy pregnant women, pregnant teens, individuals with inadequate calcium intake, and women who are known to be at risk of developing hypertension during pregnancy may benefit from additional dietary calcium.

gestational diabetes mellitus A consistently elevated blood glucose level that develops during pregnancy and returns to normal after delivery.

Gestational Diabetes Mellitus Consistently elevated blood glucose level during pregnancy in a woman without previously diagnosed diabetes is known as **gestational diabetes mellitus**. It occurs in about 7% of all pregnancies and is most common in obese women and those with a family history of type 2 diabetes.[2,15] Canadian research suggests it occurs more frequently among Aboriginal women than non-Aboriginal women.[16] It has also been found to occurs more frequently among African American, Hispanic/Latino American, and Native American women than among Caucasian women.[17] In addition to its impact on the mother's health, gestational diabetes increases risks for the baby. Because glucose in the mother's blood passes freely across the placenta, when the mother's blood levels are high, the growing fetus receives extra glucose kcalories. This extra energy promotes rapid growth, resulting in babies who are large for gestational age and consequently at increased risk of complications during delivery. As with other types of diabetes, the treatment of gestational diabetes involves consuming a carefully planned diet, maintaining moderate daily exercise, and in some cases using medications to control blood glucose. This form of diabetes usually disappears when the pregnancy is completed, although the mother remains at higher risk for developing type 2 diabetes (see Section 4.5). Babies born to mothers with gestational diabetes are at risk of developing diabetes as adults.[18]

14.2 The Nutritional Needs of Pregnancy

Learning Objectives

- Compare the nutrient needs of pregnant women with those of nonpregnant women.
- Discuss the possible effects of too little folic acid during pregnancy.

In order to produce a healthy baby, maternal intake must supply all the nutrients needed to provide for the growth and development of the fetus while continuing to meet the mother's needs. Because the increased need for energy is proportionately smaller than the increased need for protein, vitamins, and minerals, a well-balanced, nutrient-dense diet is required.

Energy Needs During Pregnancy

Energy needs increase during pregnancy to deposit and maintain the new fetal and maternal tissues. The estimated energy requirement (EER) for pregnancy is calculated by totalling the energy needs of nonpregnant women, the increase in energy needs due to pregnancy, and the energy deposited in tissues.[19] During the first trimester, total energy expenditure changes little, so the EER is not increased above nonpregnant levels. During the second and third trimesters, an additional 350 and 450 kcal/day, respectively, are recommended (**Figure 14.12**). This is the amount of energy contained in a snack such as a sandwich, an apple, and a glass of milk.

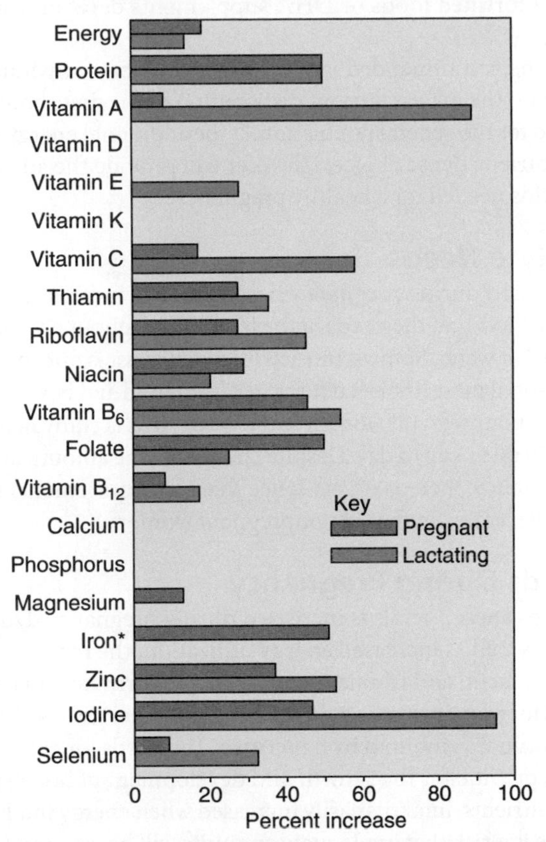

Figure 14.12 Energy and nutrient needs during pregnancy and lactation
This graph illustrates the percentage increase in recommended nutrient intakes for a 25-year-old woman during the third trimester of pregnancy and during lactation. *The RDA for iron during lactation is equal to half the RDA for nonpregnant, nonlactating women.

Protein, Carbohydrate, and Fat Recommendations

The RDA for protein is increased during pregnancy. The additional protein is needed because protein is essential for the formation and growth of new cells. During pregnancy the placenta develops and grows, the uterus and breasts enlarge, and a single cell develops into a fully formed infant. An additional 25 g of protein per day above the RDA for nonpregnant women or 1.1 g/kg/day is recommended for the second and third trimesters of pregnancy. For a woman weighing 62 kg (136 lb), this increases protein needs to about 75 g/day. This is the amount of protein in 750 ml (3 cups) of milk or yogurt plus 210 g (7 oz.) of meat. Studies suggest that vegetarian women may have lower protein intakes than non-vegetarians, but there is no evidence that this lower intake is harmful to the fetus.[19] By proper complementation of proteins, an adequate intake of essential amino acids is ensured (see Section 6.6 for more details on protein complementation).

The RDA for carbohydrate is increased by 45 g during pregnancy to provide sufficient glucose to fuel the fetal and maternal brains. Therefore the RDA for carbohydrate during pregnancy is 175 g/day. This is well below the typical intake of about 300 g/day, and therefore most women do not need to consciously increase carbohydrate intake.

Although total fat intake does not need to increase during pregnancy, more of the essential fatty acids linoleic and alpha-linolenic acid are recommended because these are incorporated into the placenta and the fetal tissues. Since there are insufficient data to determine how much is required to meet these needs, AIs for these fatty acids have been established based on the median intake of a healthy population.

DHA, a fatty acid derived from fish oil or synthesized in the body from the essential fatty acid, alpha-linolenic acid, is important in pregnancy as DHA functions in the development of the retina and brain in the fetus and infant. The best dietary source of DHA is fish.

Infants born to vegetarian mothers who do not eat fish tend to have lower levels of DHA in their blood. DHA levels in breast milk are also lower in vegetarians. Although the implications of these lower DHA levels is unclear, vegetarians and vegans should include sources of DHA in their diet from fortified foods or DHA supplements derived from microalgae, a non-animal product.[20]

Despite increases in the recommended intakes of protein, carbohydrate, and specific fatty acids during pregnancy, the macronutrient distribution of the diet should be about the same as that recommended for the general population. If the additional energy needed during pregnancy comes from nutrient-dense choices, the diet will provide the additional protein, carbohydrate, and fatty acids needed for a healthy pregnancy.

Water and Electrolyte Needs

The need for water is increased during pregnancy because of the increase in blood volume, the production of amniotic fluid, and the needs of the fetus. Throughout pregnancy, a woman will accumulate about 6-9 L of water. Some is intracellular, but most is due to increases in the volume of blood and interstitial fluid. The need for water from food and beverages is therefore increased from 2.7 L/day in nonpregnant women to 3 L/day.[20] This is equivalent to drinking a little more than an extra 250 ml (1 cup) a day. Despite changes in the amount and distribution of body water during pregnancy, there is no evidence that the requirements for potassium, sodium, or chloride are different from that of nonpregnant women.

Micronutrient Needs During Pregnancy

The need for many vitamins and minerals is increased during pregnancy. Due to growth in maternal and fetal tissues as well as increased energy utilization, the requirements for the B vitamins, such as thiamin, niacin, and riboflavin, increase. To form new maternal and fetal cells and to meet the needs for protein synthesis in fetal and maternal tissues, the requirements for folate, vitamin B_{12}, vitamin B_6, zinc, and iron increase. The needs for calcium, vitamin D, and vitamin C increase to provide for the growth and development of bone and connective tissue. For many of these nutrients, intake is easily increased when energy intake rises to meet needs, but for others, there is a risk that inadequate amounts will be consumed.

Calcium The fetus retains about 30 g of calcium over the course of gestation. Most of the calcium is deposited in the last trimester when the fetal skeleton is growing most rapidly and the teeth are forming. Many women have trouble getting enough calcium to meet their own needs, let alone enough to provide this amount for the fetus. Fortunately, they don't need to consume any more than is recommended for nonpregnant women because calcium absorption increases during pregnancy.[21] This increase is believed to be due in part to the rise in estrogen that occurs during pregnancy as well as an increase in the concentration of active vitamin D in the blood.[22] At one time, there was concern that the calcium needed by the fetus would come from maternal bones if intake was not increased. It is now known that the increased need for calcium does not increase maternal bone resorption, and studies have found no correlation between the number of pregnancies a woman has had and the density of her bones.[23] Therefore, the RDA for calcium for pregnant women age 19 and older—1,000 mg/day—is not increased above nonpregnant needs. This RDA can be met by consuming 3-4 servings of milk or alternatives daily. Women who are lactose intolerant can meet their calcium needs with yogurt, cheese, reduced-lactose milk, calcium-rich vegetables, fish consumed with bones, calcium-fortified foods, such as soy milk, or calcium supplements.

Vitamin D Adequate vitamin D is essential to ensure efficient calcium absorption, but the recommended intake for vitamin D during pregnancy is not increased above nonpregnant levels. Pregnant women who receive regular exposure to sunlight can synthesize sufficient vitamin D. If exposure to sunlight is limited, dietary sources such as milk must supply the needed amounts. The incidence of vitamin D deficiency has been increasing. One American study found that more than 40% of African-American women of child-bearing age were vitamin D deficient.[24] Inadequate vitamin D may be a problem in African-American women because their milk intake is often low due to lactose intolerance and their darker pigmentation reduces the synthesis of vitamin D in the skin. In Canada, vitamin D deficiencies have been found among pregnant Aboriginal women, especially those living in the North where sunlight is limited.[25] If sufficient vitamin D is not consumed in the diet, careful supplementation should be considered.

Vitamin C Vitamin C is important for bone and connective tissue formation because it is needed for the synthesis of collagen, which gives structure to skin, tendons, and the protein matrix of bones. Vitamin C deficiency during pregnancy increases the risk for premature birth and pre-eclampsia. The RDA is increased by 10 mg/day during pregnancy.[26] The requirement for vitamin C can easily be met by including citrus fruits and juices in the diet, and supplements are generally not necessary.

neural tube The portion of the embryo that develops into the brain and spinal cord.

Folate Folate is needed for the synthesis of DNA and thus for cell division. During pregnancy, cells multiply to form the placenta, expand maternal blood, and allow for fetal growth. Adequate folate intake is crucial even before conception because rapid cell division occurs in the first days and weeks of pregnancy.

If maternal folate intake is low, there is an increased risk of fetal abnormalities that involve the formation of the **neural tube**. During development, neural tissue forms a groove; the groove closes when the sides rise and fold together to form the neural tube (**Figure 14.13**). This neural tube closure occurs between 21 and 28 days of development. If it does not occur normally, the infant will be born with a neural tube defect. These defects include anencephaly, in which the brain and skull do not develop normally—a condition that is fatal at or shortly after birth—and spina bifida, a condition in which the vertebrae do not close completely, causing part of the spinal cord to be exposed. Spina bifida is generally not fatal, but can result in mild to severe paralysis. Although the mechanism underlying the protective effect of folate is unknown, it is likely to involve the vitamin's role in single-carbon metabolism and DNA methylation (see Section 8.8).

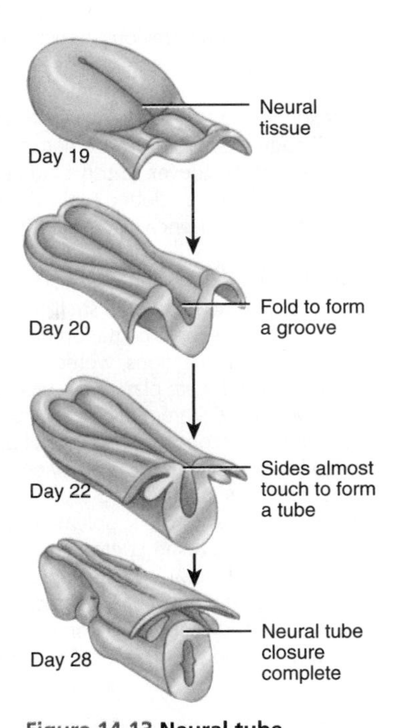

Figure 14.13 Neural tube formation
The neural tube develops from a flat plate of neural tissue.

Because the neural tube closes so early in development, often before a woman even knows she is pregnant, recommendations are that women who could become pregnant consume a multivitamin with at least 400 μg/day of synthetic folic acid, in addition to consuming a varied diet rich in natural sources of folate (see Section 8.8).[27] Since the initiation of folic acid fortification, the incidence of neural tube defects has been reduced by 25% in the U.S. and 50% in Canada[28,29] (see Science Applied: Folate: From Epidemiology to Health Policy).

Adequate folate continues to be important even after the neural tube closes. Cell division continues in both embryonic and fetal development and folate is central because of its role in DNA synthesis. Marginal folate status can impair growth in both the fetus and the placenta. If folate is inadequate during pregnancy, megaloblastic anemia—the type of anemia in which blood cells do not mature properly—may result (see Section 8.8). Low dietary folate intakes and low circulating folate levels are associated with increased risk of preterm delivery, low birth weight, and fetal growth retardation.[30] Thus, to maintain red blood cell folate levels in pregnant women, the RDA for folate is set at 600 μg dietary folate equivalents per day.[31] Natural sources of folate include orange juice, legumes, leafy green vegetables, and organ meats. A 125 ml (½ cup) serving of legumes plus 250 ml (1 cup) of raw spinach provide about a third of the RDA. Fortified sources include enriched breads, cereals, and other grain products. Folic acid supplements can also be used to meet this goal. Prenatal supplements provide at least 400 μg of folic acid.

Vitamin B_{12} Vitamin B_{12} is essential for the regeneration of active forms of folate, so a deficiency of vitamin B_{12} can also result in megaloblastic anemia. Vitamin B_{12} is transferred from the mother to the fetus during pregnancy. Based on the amount transferred and the increased efficiency of vitamin B_{12} absorption that occurs during pregnancy, the RDA for pregnancy is set at 2.6 μg/day.[31] This recommendation is easily met by a diet containing even small amounts of animal products. Vegetarian diets are generally safe for pregnant women but vegans must consume foods fortified with vitamin B_{12} or take vitamin B_{12} supplements daily to meet the needs of mother and fetus.

Zinc Zinc is involved in the synthesis of DNA, RNA, and proteins. It is therefore extremely important for growth and development. Zinc deficiency during pregnancy is associated with an increased risk of fetal malformations, prematurity, and low birth weight.[32] Because zinc absorption is inhibited by high iron intakes, iron supplements may compromise zinc status if the diet is low in zinc. The RDA is 11 mg/day for pregnant women 19 years of age and older.[33] A 90 g serving of lean ground beef provides about 2 mg of zinc.

SCIENCE APPLIED

CANADIAN CONTENT

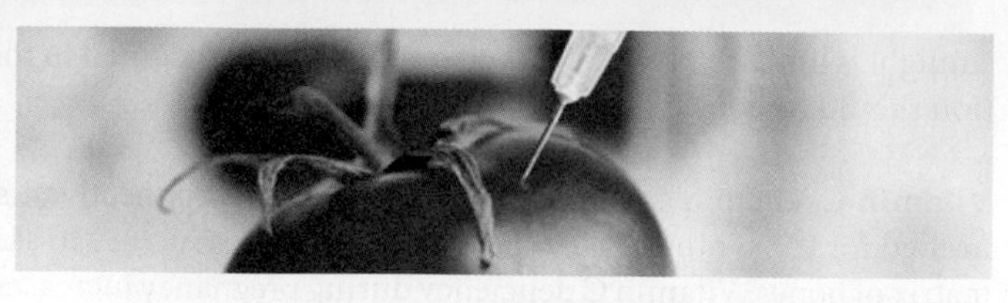

Folate: From Epidemiology to Health Policy

Until the late 1990s, neural tube defects (NTDs), such as spina bifida and anencephaly, affected approximately 1.6 births out of every 1,000 in Canada. For decades, researchers had suspected that NTDs might be related to the mother's dietary intake. Today, public health policies mandate the fortification of certain foods with folic acid, the synthetic form of folate.

The link between maternal folate status and NTDs was identified in the 1970s when lower, first-trimester, red blood cell folate concentrations were found in women who later gave birth to NTD-affected babies.[1] An intervention study completed in 1980 demonstrated that folic acid supplementation in early pregnancy reduced the incidence of NTDs in women who had previously given birth to a baby with an NTD.[2] These findings were supported by a second, large-scale study, which randomly assigned nonpregnant women to receive either folic acid, other supplemental vitamins, or a placebo.[3] This study involved women from 7 countries, including women in Halifax, Nova Scotia, Hamilton, Ontario, and Vancouver, British Columbia. This trial was stopped early when researchers concluded that folic acid supplementation alone reduced NTD recurrence by 71%.

A link between NTDs and folic acid was now clear, but early trials used 800 μg of folic acid. Could a smaller amount be as effective? Also, would folic acid supplementation prevent the first occurrence of NTDs? To answer these questions, women who had no previous NTD–affected pregnancies and were planning a pregnancy were randomly assigned to receive a multivitamin containing folic acid or a placebo for at least 1 month before conception and until at least the date of the second missed menstrual cycle.[4] The study evaluated 5,453 pregnancies. In the 2,391 women receiving the placebo, 6 babies were born with NTDs; in the 2,471 women in the supplement group, there were no NTDs.[5] This and other studies helped determine that 400 μg of supplemental folic acid dramatically reduced the incidence of NTDs.[6,7] In Canada, these results lead to the publication of recommendations, in the early 1990s, by Health Canada and other Canadian organizations. They advised that women who could become pregnant obtain intakes of folic acid, from supplementation or diet, of at least 400 μg daily, to prevent neural tube defects.[8] **In the United States**, the Centers for Disease Control and Prevention and the U.S. Public Health Service made similar recommendations.

But how could the population's folate intake be increased? Educating women to consume foods high in folate would be costly and likely have limited effectiveness. Folic acid supplements were recommended, but after 5 years, American studies revealed that only one-third of women of child-bearing age were consuming a supplement containing the recommended amount.[9] Therefore, it was concluded by both American and Canadian authorities that the most reliable way to increase folate consumption was through food fortification.

Fortification of the food supply requires a careful analysis to determine how much of the nutrient to add so it will provide the needed benefit without undue risk. The right amount of supplemental folic acid could reduce the incidence of NTD-affected pregnancies. But because a high folic acid intake can mask the symptoms of vitamin B_{12} deficiency, a level of fortification needed to be chosen that would reduce NTDs without compromising the elderly population at risk for B_{12} deficiency (see Section 8.9 for more information about the masking of vitamin B_{12} deficiency).

Folate fortification became mandatory in Canada and the United States in 1998. White flour, cornmeal, and pasta were chosen for mandatory fortification in Canada. These products are regularly consumed by the target population—women of child-bearing age of all races and cultures. Food processors can also voluntarily add folic acid to other products, including breakfast cereals, pre-cooked rice, and fruit-flavoured drinks.[10] The amounts of folic acid added varies with the food product, but are chosen to balance the need to provide enough folic acid to reduce the risk of NTDs without the possibility of masking vitamin B_{12} deficiency. The success of the folic acid fortification program can be seen in the decline in the estimated number of NTD-affected pregnancies from 1.6 to 0.86 births out of 1,000, a reduction of 46%.[11]

[1]Smithells, R. W., Sheppard, S., and Schorah, C. J. Vitamin deficiencies and neural tube defects. *Arch. Dis. Child.* 51:944-949, 1976.

[2]Smithells, R.W., Sheppard, S., Schorah, C.J. et al. Possible prevention of neural tube defects by periconceptional vitamin supplementation. *Lancet* 1:339-340, 1980.

[3]MRC Vitamin Study Research Group. Prevention of neural tube defects: Results of the MRC vitamin study. *Lancet* 338:131-137, 1991.

[4]Czeizel,A. E., and Dudás, I. Prevention of the first occurrence of neural-tube defects by periconceptional vitamin supplementation. *N. Engl. J. Med.* 327:1832-1835, 1992.

[5]Czeizel,A. E. Folic acid in the prevention of neural tube defects./ *Pediatr. Gastroenterol. Nutr.* 20:4-16, 1995.

[6]Werler, M. M., Shapiro, S., and Mitchell, A.A. Periconceptional folic acid exposure and the risk of occurrent neural tube defects. *JAMA* 269:1257-1261, 1993.

[7]Daly S., Mills, J.L., Molloy A. M., et al. Minimum effective dose of folic acid for food fortification to prevent neural-tube defects. *Lancet* 350:1666-1669, 1997.

[8] Canadian Task Force on the Periodic Health Examination, Health Canada. Periodic health examination, update: 3. Primary and secondary prevention of neural tube defects. *CMAJ.* 151(2):159-66, 1994.

[9]Use of folic acid-containing supplements among women of childbearing age—United States, 1997. *MMWR* 47:131-134, 1998. Available online at cdc.gov/mmwr/preview/mmwrhtml/00051435.htm/. Accessed February 10, 2001.

[10] CFIA. Guide to Food Labelling and Advertising. Annex 7.1 Available online at http://www.inspection.gc.ca/english/fssa/labeti/guide/ch7-1e.shtml. Accessed June 17, 2011.

[11] De Wals, P., Tairou, F., Van Allen, M.I. et al. Reduction in neural tube defects after folic acid fortification in *Canada.N. Engl. J. Med.* 357:135-142, 2007.

Iron Iron needs are high during pregnancy to allow for the synthesis of hemoglobin and other iron-containing proteins in both maternal and fetal tissues. The physiological changes of pregnancy allow for increased iron absorption and iron losses are decreased due to the cessation of menstruation. Nonetheless, iron-deficiency anemia is common during pregnancy. Part of the reason for this is that low iron stores are common among women of child-bearing age so many women start pregnancy with diminished iron stores and quickly become deficient.

The RDA for iron during pregnancy is 27 mg/day compared with 18 mg for nonpregnant women.[33] It takes an exceptionally well-planned diet to meet iron needs during pregnancy. Red meats, leafy green vegetables, and fortified cereals are good sources of iron. Foods that enhance iron absorption, such as citrus fruit and meat, should also be included in the diet. Iron supplements are typically recommended during the second and third trimesters of pregnancy (see Critical Thinking: Nutrient Needs for a Successful Pregnancy).

When iron needs are not met, iron-deficiency anemia may occur. Iron-deficiency anemia during pregnancy has been associated with an increased risk of low birth weight and preterm delivery.[34] The fetus draws iron from the mother to ensure adequate fetal hemoglobin production, mostly during the last trimester. Babies born prematurely may not have had time to accumulate sufficient iron, but babies born at term usually have adequate iron stores even if the mother is deficient.

Meeting Nutrient Needs with Food and Supplements

The energy and nutrient needs of pregnancy can be met by following the recommendations of Canada's Food Guide (**Figure 14.14**). Additional grains, vegetables, and fruits provide energy, protein, folate, vitamin C, and fibre, particularly if whole grains are chosen. An extra serving of milk provides energy, protein, calcium, vitamin D, and riboflavin. Additional lean meat provides energy, protein, vitamin B_6, vitamin B_{12}, iron, and zinc. For example, adding a snack such as a turkey sandwich on whole-grain bread, an apple, and a glass of low-fat milk, will provide all the extra energy and nutrients needed daily for a healthy pregnancy.

Supplements Even when a healthy diet based on Canada's Food Guide is consumed, it is difficult to meet all vitamin and mineral needs. Generally, supplements of folic acid are recommended before and during pregnancy, and iron supplements are recommended during the second and third trimesters.[2] A multivitamin and mineral supplement may also be necessary in those whose food choices are limited, such as vegetarians, or in those whose needs are very high, such as pregnant teenagers. A prenatal supplement, however, must be taken in conjunction with, not in place of, a carefully planned diet (see Label Literacy: What's in a Prenatal Supplement?).

Figure 14.15 This woman is eating a white clay called kaolin, which some women crave during pregnancy. Eating kaolin is also a traditional remedy for morning sickness. This example of pica may be related to cultural beliefs and traditions. (© Michael DiBari, Jr./AP/wide world photos)

Food Cravings and Aversions Most women change their diets during pregnancy. Some changes are made in an effort to improve nutrition to ensure a healthy infant, but other changes are based on cravings, aversions, or cultural or family traditions. Foods that are commonly craved include fruit and fruit juices, sweets, candy (particularly chocolate), and dairy products. Common aversions include coffee and other caffeinated drinks, alcohol, meat, fish, poultry, eggs, highly seasoned foods, or fried foods. It has been suggested that hormonal or physiological changes during pregnancy—in particular, changes in taste and smell—may be the cause of such cravings and aversions.

Usually the food cravings of pregnant women can be indulged within reason with no harmful effects. But abnormal cravings leading to consumption of nonfood substances can have serious consequences. **Pica** is an abnormal craving for and ingestion of nonfood substances having little or no nutritional value (see Chapter 15: Focus on Eating Disorders). Pica has been described since antiquity but its cause is still a mystery. It was once thought that pica was an attempt to meet micronutrient needs. It is now believed that pica may also be related to cultural or behaviourial factors (**Figure 14.15**). It has been documented in those with a family or personal history of the practice. Research on its prevalence across various ethnicities and according to socioeconomic status is limited.[35] Women with pica commonly consume clay, laundry starch, ice and freezer frost, baking soda, cornstarch, and ashes. Consuming large amounts of these can reduce the intake of nutrient-dense foods,

pica An abnormal craving for and ingestion of unusual food and nonfood substances.

CRITICAL THINKING:

Nutrient Needs for a Successful Pregnancy

Background

During the first trimester, women should follow Canada's Food Guide recommendations for their age group. During the second trimester, women need an additional 350 kcalories and an additional 450 kcalories in the third trimester. These increases in kcaloric requirement can be met by women eating 2-3 additional Canada's Food Guide servings per day.[1]

Consider how these recommendations might be applied to the case of Tina:

Tina is 26 years old and 3 months pregnant (her first trimester). From the start— before she tried to conceive—she has been careful about her nutritional health. She even took a prenatal vitamin supplement before she knew she was pregnant to be sure she got enough folic acid. Now that she is approaching her second trimester, her doctor is concerned about her intake of iron. Even though there is iron in her supplement, she needs to increase the iron in her diet.

Data

Tina's food intake for one day is shown here.

(©Mark Burstyn/Masterfile)

FOOD	IRON (mg)
Breakfast	
250 ml (1 cup) corn flakes	0.4
with 250 ml (1 cup) reduced-fat milk	0.1
125 ml (½ cup) orange juice	1.1
250 ml (1 cup) decaffeinated coffee	
with sugar and cream	0.1
Lunch	
Tuna sandwich	
75 g (2.5 oz.) tuna	1.2
10 ml (2 tsp) mayonnaise	0.5
2 slices white bread	2.5
20 french fries	1.3
1 can orange soda	0.2
3 chocolate chip cookies	0.8
1 apple	0.4
Dinner	
75 g (2.5 oz.) chicken leg	1.5
125 ml (½ cup) peas	1.2
1 piece corn bread	0.8
5 ml (1 tsp) margarine	0
250 ml (1 cup) lettuce and tomato salad	1.3
15 ml (1 Tbsp) Italian dressing	0
250 ml (1 cup) reduced-fat milk	0.1
Total	**13.5**

Critical Thinking Questions

What is Tina's intake of food guide servings for each of the 4 food groups: Vegetables & Fruit, Grain Products, Milk and Alternatives, Meat and Alternatives. How often does Tina consume "foods to limit," which are foods high in calories, fat, sugar, and salt, listed on page 6 of Canada's Food Guide (see Section 2.3)*.

How does Tina's current food intake compare with Canada's Food Guide recommendations for women in their first trimester?

Give examples of the types of foods she could add to her diet to meet the increased kcaloric requirements of the second and third trimesters.

What about iron? Does her current diet meet her needs without the supplement?

* Answers to all Critical Thinking questions can be found in Appendix J.

[1] Health Canada. Eating Well with Canada's Food Guide: A resource for educators and communicators: 5. Advice for different ages and stages. Available online at. http://www.hc-sc.gc.ca/fn-an/pubs/res-educat/res-educat_5-eng.php. Accessed December 12, 2010.

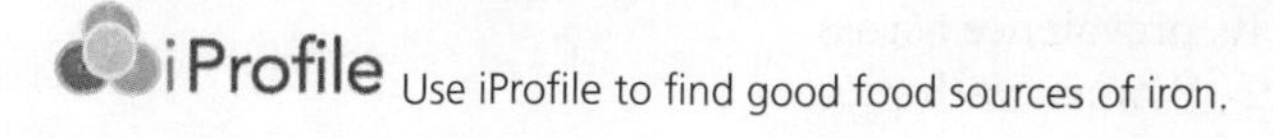

Use iProfile to find good food sources of iron.

Recommended Number of Food Guide Servings per Day

	Children			Teens		Adults			
Age in Years	2-3	4-8	9-13	14-18		19-50		51+	
Sex	Girls and Boys			Females	Males	Females	Males	Females	Males
Vegetables and Fruit	4	5	6	7	8	7-8	8-10	7	7
Grain Products	3	4	6	6	7	6-7	8	6	7
Milk and Alternatives	2	2	3-4	3-4	3-4	2	2	3	3
Meat and Alternatives	1	1	1-2	2	3	2	3	2	3

Advice for different ages and stages...

Women of childbearing age

All women who could become pregnant and those who are pregnant or breastfeeding need a multivitamin containing **folic acid** every day. Pregnant women need to ensure that their multivitamin also contains **iron**. A health care professional can help you find the multivitamin that's right for you.

Pregnant and breastfeeding women need more calories. Include an extra 2 to 3 Food Guide Servings each day.

Here are two examples:

- Have fruit and yogurt for a snack, or
- Have an extra slice of toast at breakfast and an extra glass of milk at supper.

Oils and Fats

- Include a small amount – 30 to 45 mL (2 to 3 Tbsp) – of unsaturated fat each day. This includes oil used for cooking, salad dressings, margarine and mayonnaise.
- Use vegetable oils such as canola, olive and soybean.
- Choose soft margarines that are low in saturated and trans fats.
- Limit butter, hard margarine, lard and shortening.

Figure 14.14 Using Canada's Food Guide during pregnancy
Canada's Food Guide's general recommendations, when combined with the special advice given to women of child-bearing age, can be effectively used for meal planning during pregnancy.
Source: Health Canada: Canada's Food Guide: Get Your Copy. Available online at http://www.hc-sc.gc.ca/fn-an/food-guide-aliment/order-commander/index-eng.php#1. Accessed December 12, 2010.

reduce nutrient absorption from food, increase the risk of consuming toxins and harmful micro-organisms, and even cause intestinal obstructions. Complications of pica include iron-deficiency anemia, lead poisoning, and parasitic infections.[35] Anemia and hypertensive disorders of pregnancy are more common in mothers who practice pica, but it is not clear if pica is a result of these conditions or a cause. In newborns, anemia and low birth weight are often related to pica in the mother.

There has been little research on effective treatments for pica. Nutritional treatment for any deficiencies which may cause or be the consequence of pica are suggested. Cases have been reported that have responded well when iron deficiencies were corrected. After nutritional treatment, education, counselling, and the reduction of stress is recommended.[35]

LABEL LITERACY

What's in a Prenatal Supplement?

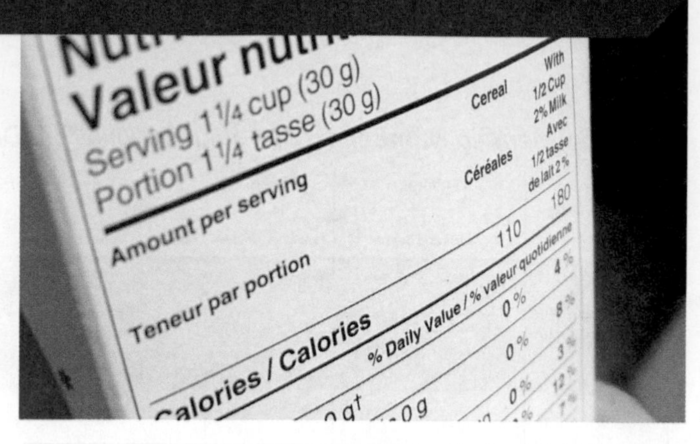

(© istockphoto.com/Tim Pohl)

Most pregnant women leave their first prenatal doctor's visit with a prescription for a prenatal vitamin and mineral supplement. What's in these supplements? Do they meet all the needs of pregnancy?

A look at a label shows that a typical prenatal supplement contains more than 15 vitamins and minerals. It supplies enough folate, iron, and many other micronutrients to meet recommendations, but taking the supplement does not mean that pregnant women can ignore their diet. Some nutrients in these supplements are present in amounts that do not meet the needs of pregnancy. For example, the tablet whose ingredients are shown in the table contains only 250 mg of calcium, which is only 25% of the recommended intake for a pregnant woman aged 19 or older. To meet her calcium needs, a pregnant woman taking this tablet would still need to consume enough milk or other high calcium foods to provide about 1,000 mg of calcium or use an additional calcium supplement with higher levels of calcium.

Because excessive intakes of vitamins can be harmful, prenatal supplements are carefully formulated. Preformed vitamin A, for example, can cause birth defects at extremely high doses of 3,000 μg RAE (10,000 IU) which is the UL for vitamin A.[1] Beta-carotene, which can be converted to vitamin A, has not been found to be similarly teratogenic. For this reason, prenatal supplements have relatively low amounts of retinol (preformed vitamin A) and vitamin A is provided as beta-carotene, as shown in the table.

Vitamin and mineral supplements can never provide everything needed in an adequate diet. Prenatal supplements do not contain the protein needed for tissue synthesis or the complex carbohydrates required for energy. They lack fibre, which helps prevent constipation, and they do not contain fluid needed for expanding blood volume and maintaining normal bowel function. They also don't contain other food components such as the phytochemicals supplied by a diet rich in whole grains, fruits, and vegetables. Thus, taking a multivitamin and mineral supplement during pregnancy can be beneficial, but remember that the recommended dosage should not be exceeded and the supplement should not take the place of a carefully planned diet.

Dose: Women 19 years and older – 1 Tablet	
VITAMINS:	
Beta-carotene (a source of vitamin A)	2,500 IU
Vitamin A (Vitamin A as acetate)	1,000 IU
Vitamin C (ascorbic acid)	85 mg
Vitamin D (cholecalciferol)	400 IU
Vitamin E (dl-alpha tocopheryl) acetate	30 IU
Folic acid	1 mg
Vitamin B1 (thiamine mononitrate)	1.4 mg
Vitamin B2 (riboflavin)	1.4 mg
Vitamin B12 (cyanocobalamin)	2.6 μg
Biotin	30 μg
Pantothenic Acid (calcium pantothenate)	6 mg
MINERALS:	
Calcium (calcium carbonate)	250 mg
Magnesium (magnesium oxide)	50 mg
Iodine	220 μg
Iron	27 mg
Copper	1 mg
Zinc (zinc oxide)	7.5 mg
Chromium (chromium oxide)	30 μg
Manganese (manganese sulfate)	2 mg
Selenium (sodium selenite)	30 μg

[1] Rothman, K. J., Moore, L. L, Singer, M. R. et al. Teratogenicity of high vitamin A intake. *N. Engl. J. Med.* 333(21):1369-73,1995.

14.3 Factors that Increase the Risks of Pregnancy

Learning Objectives

- Explain how maternal nutritional status, health status, age, and income level affect the risks of pregnancy.
- Describe the effects of maternal drug and alcohol use on pregnancy outcome.
- Describe the purpose of the Canadian Prenatal Nutrition Program.

Most of the women who give birth every year in Canada are healthy during pregnancy and produce healthy babies. However, child-bearing is not without risks. In Canada, 5.5 out of every 100,000 women die as a result of childbirth and 5.1 out of 1,000 babies born die within the first year of life.[16] The reasons for poor pregnancy outcome vary. Malnutrition is a factor in some women. Others are at increased risk because of their age and pre-existing health problems or socioeconomic factors, such as limited access to health care, lack of a supportive home environment, or insufficient resources to acquire nutritious foods (**Table 14.4**). Some women and babies are at risk because they are exposed to harmful substances from the diet or environment.

Maternal Nutritional Status

Proper nutrition is important before pregnancy to support conception and maximize the likelihood of a healthy pregnancy. At any time during pregnancy, maternal malnutrition due to a deficiency or excess of energy or individual nutrients can affect pregnancy outcome.

Table 14.4 Factors that Increase Pregnancy Risks

Maternal Factor	Maternal Risk	Infant/Fetal Risk
Prepregnant BMI < 19.8 or gaining too little weight during pregnancy	Anemia, premature rupture of the membranes, hemorrhage after delivery	Low birth weight, preterm birth
Prepregnancy BMI > 26 or gaining too much weight during pregnancy	Hypertensive disorders of pregnancy, gestational diabetes, difficult delivery, Caesarean section	Large-for-gestational-age, low Apgar scores (a score used to assess the health of a baby in the first minutes after birth), and neural tube defects
Malnutrition	Decreased ability to conceive, anemia	Fetal growth retardation, low birth weight, birth defects, preterm birth, spontaneous abortion, stillbirth, increased risk of chronic disease later in life
Phenylketonuria	High blood levels of phenylketones	Mental retardation if low phenylalanine diet is not carefully followed by mother
Hypertension	Stroke, heart attack, premature separation of the placenta from the uterine wall	Low birth weight, fetal death
Diabetes	Difficulty adjusting insulin dose, pre-eclampsia, Caesarean section	Large-for-gestational-age, congenital abnormalities, fetal death
Frequent pregnancies: 3 or more during a 2-year period	Malnutrition	Low birth weight, preterm birth
Poor obstetric history or history of poor fetal outcome	Recurrence of problem in subsequent pregnancy	Birth defects, death
Age:		
Adolescent	Malnutrition, hypertensive disorders of pregnancy	Low birth weight
Older than 35	Hypertensive disorders of pregnancy, gestational diabetes	Down syndrome and other chromosomal abnormalities
Alcohol consumption	Poor nutritional status	Alcohol-related birth defects, alcohol-related neurodevelopmental disorders, fetal alcohol syndrome
Cigarette smoking	Lung cancer and other lung diseases, miscarriage	Low birth weight, miscarriage, stillbirth, preterm birth, sudden infant death syndrome, respiratory problems
Cocaine use	Hypertension, miscarriage, premature labour and delivery	Intrauterine growth retardation, low birth weight, preterm birth, birth defects, sudden infant death syndrome

Nutritional Status Before Pregnancy A woman's nutritional status before she becomes pregnant may affect her ability to conceive and successfully complete a pregnancy. Starvation diets, anorexia nervosa, and excessive exercise, such as marathon running, can reduce body fat and affect hormone levels. If hormone levels are too low, ovulation does not occur and conception is not possible. Too much body fat can also reduce fertility by altering hormone levels. Deficiencies or excesses of nutrients can also affect pregnancy outcome. For instance, a deficiency of folate or an excess of vitamin A early in pregnancy can cause birth defects.

Nutritional status can also be affected by some birth control methods and these can therefore have an impact on a subsequent pregnancy. For example, oral contraceptives are associated with reduced blood levels of vitamins B_6 and B_{12}.[31] If conception occurs soon after oral contraceptive use stops, these levels will not have had time to return to normal before pregnancy begins.

Malnutrition During Pregnancy Maternal malnutrition can cause fetal growth retardation, low infant birth weight, birth defects, premature birth, spontaneous abortion, and stillbirth. The effect of malnutrition depends on how severe the nutrient deficiency or excess is and when during the pregnancy it occurs. In general, poor nutrition early in pregnancy affects embryonic development and the potential of the embryo to survive, and poor nutrition in the latter part of pregnancy affects fetal growth.

Immediate Effects of Maternal Malnutrition A low energy intake during early pregnancy is not likely to interfere with fetal growth because the energy demands of the embryo are small. However, if the embryo does not receive adequate amounts of the nutrients needed for cell division and differentiation, such as folate and vitamin A, malformations or death can result. Inadequate folate intake in the first few weeks of pregnancy may affect neural tube development.[31] Too much vitamin A is of particular concern because the risk of kidney problems and central nervous system abnormalities in the offspring increases even when maternal intake is not extremely high. High intakes early in pregnancy are the most damaging. A UL of 3,000 μg RAE/day has been established for pregnant women ages 19-50 years.[33] Supplements consumed during pregnancy typically contain β-carotene, which is not damaging to the fetus (see Label Literacy: What's in a Prenatal Supplement?).

Malnutrition is most devastating during the first trimester. After the first trimester, nutrient deficiencies or excesses are less likely to cause developmental defects (malformations) because most organs and structures have already formed. However, undernutrition in the mother after the first trimester can interfere with fetal growth. Even a mild energy restriction during the last trimester, when the fetus is growing rapidly, can affect birth weight. Malnutrition also interferes with the growth and function of the placenta. Then, in turn, a poorly developed placenta cannot deliver sufficient nutrients to the fetus and the result is a small infant who may also have other developmental abnormalities.

Long-Term Effects of Maternal Malnutrition It has been proposed that problems in maternal nutrition can cause adaptations that change fetal structure, physiology, and metabolism and can affect the child's risk of developing chronic diseases later in life. Evidence for this comes from epidemiological studies that suggest that individuals who were small at birth or disproportionately thin or short have higher rates of heart disease, high blood pressure, high blood cholesterol, and diabetes in middle age.[36]

Maternal Health Status

The general health as well as nutritional status of the mother affects the outcome of pregnancy. Women who begin pregnancy with chronic diseases such as hypertension, diabetes, and phenylketonuria (PKU) must manage their health carefully to assure a healthy pregnancy. The effect of hypertension depends on when it develops and how severe it is. As in nonpregnant individuals, high blood pressure in pregnant women increases the risk of stroke and heart attack, but in pregnancy it also increases the risk of low birth weight and premature separation of the placenta from the wall of the uterus, resulting in fetal death.

Women with diabetes must carefully manage diet and medication to ensure that glucose levels stay in the normal range throughout pregnancy. Uncontrolled diabetes early in pregnancy increases the risk of birth defects. When normal blood glucose is maintained throughout pregnancy, the risk of complications is greatly reduced. The need for insulin increases during the second and third trimesters, so women with pre-existing diabetes may need to adjust their medication dosage. When maternal blood glucose is elevated, it provides extra nutrients to the growing fetus, resulting in large-for-gestational-age newborns who are at increased risk.

Reproductive History Frequent pregnancies, with little time between, increase the risk of poor pregnancy outcomes. One reason is that the mother may not have replenished nutrient stores depleted in a previous pregnancy when she becomes pregnant again. An interval of less than 18 months increases the risk of delivering a full-term but small-for-gestational-age infant. An interval of only 3 months has been shown to increase the risk of a preterm or small-for-gestational-age infant as well as neonatal death.[37] Women with a history of poor pregnancy outcome are also at increased risk. For example, a woman who has had a number of miscarriages is more likely to have another, and a woman who has had one child with a birth defect has an increased risk for defects in subsequent children.

The Pregnant Adolescent Pregnancy places a stress on the body at any age, but this is compounded when the mother herself is still growing. Although the rate of teen pregnancy in Canada has been decreasing, from 6.8% of all births in 1995, to 4.8% in 2004, it remains a major public health problem.[38] Pregnant teens are at greater risk of developing hypertensive disorders of pregnancy and are more likely to deliver preterm and low-birth-weight babies. To produce a healthy baby, a pregnant adolescent needs early medical intervention and nutritional counselling.

Adolescent girls continue to grow and mature physically for about 4-7 years after menstruation begins, so the diet of a pregnant teen must provide both for her growth and that of her baby. Even teens who deliver normal-birth-weight infants may stop growing themselves.[39] Because the nutrient needs of a pregnant girl may be higher than those of a pregnant woman, the DRIs include a special set of nutrient recommendations for pregnant teens (**Figure 14.16**). Consuming a diet that meets these needs can be challenging. Even nonpregnant girls often fall short of meeting their nutrient needs. Nutrients that are commonly low in the diets of pregnant teens are calcium, iron, zinc, magnesium, vitamin D, folate, and vitamin B_6.[26,31]

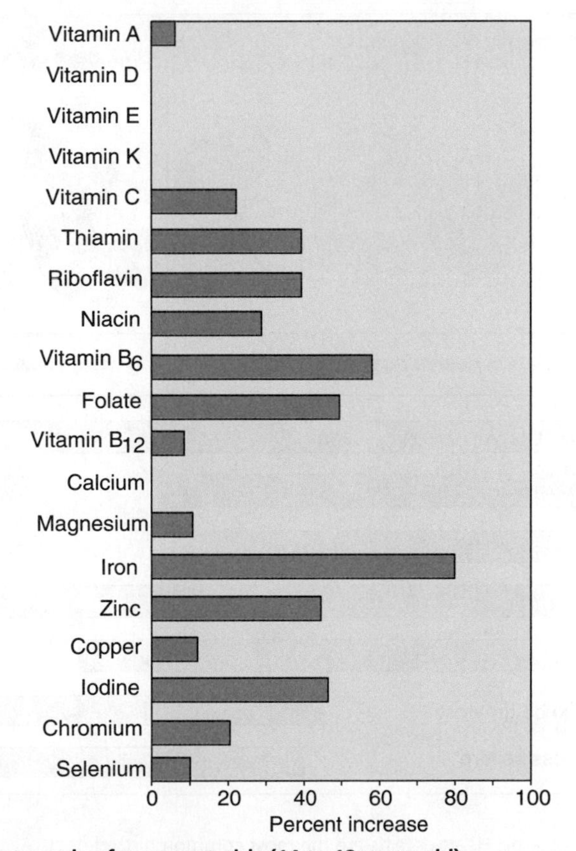

Figure 14.16 Nutrient needs of pregnant girls (14 to 18 years old)
The percentage increase in micronutrient needs above nonpregnant levels is shown here for a 14- to 18-year-old girls during pregnancy.

Down syndrome A disorder caused by extra genetic material that results in distinctive facial characteristics, mental retardation, and other abnormalities.

The Older Mother The nutritional requirements for older women during pregnancy are no different than for women in their twenties, but pregnancy after the age of 35 does carry additional risks because older women are more likely to start pregnancy with medical conditions such as cardiovascular disease, kidney disease, obesity, and diabetes. During pregnancy, they also are more likely to develop gestational diabetes, hypertensive disorders of pregnancy, and other complications. They also have a higher incidence of low-birth-weight deliveries and of chromosomal abnormalities, especially **Down syndrome**. Today, careful medical monitoring throughout pregnancy is reducing the risks to older mothers and their babies (**Figure 14.17**).

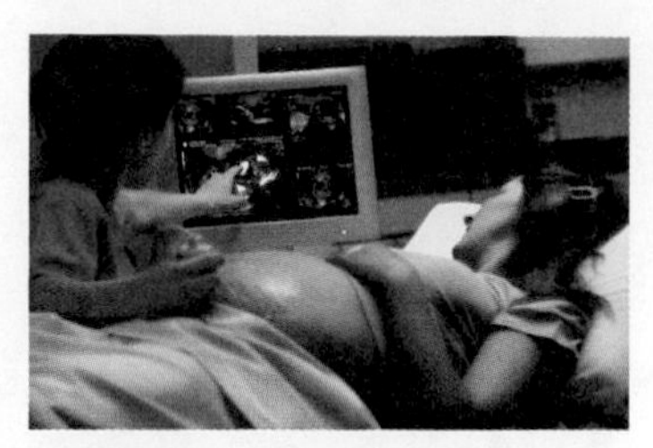

Figure 14.17 Prenatal care with careful medical monitoring can help older women have uncomplicated pregnancies and healthy babies. (Jose Luis Pelaez Inc/Blend Images/ Getty Images)

Exposure to Toxic Substances

During development, cells are particularly vulnerable to damage because they are dividing rapidly, differentiating, and moving to form organs and other structures. Anything that interferes with this development can cause a baby to be born too soon or too small or can result in birth defects. A **teratogen** is any chemical, biological, or physical agent that causes birth defects. Even some vitamins have been found to be teratogens in extremely high doses. The placenta prevents some teratogens from passing from the mother's blood to the embryonic or fetal blood, but it cannot prevent the passage of all hazardous substances.

teratogen A substance that can cause birth defects.

Each organ system develops at a different rate and time, so each has a critical period when exposure to a teratogen or other insult can disrupt development and cause irreversible damage (**Figure 14.18**). If the damage is severe, it may result in a miscarriage. Because the majority of cell differentiation occurs during the embryonic period, this is the time when exposure to teratogens can do the most damage, but vital body organs can still be affected during the fetal period. As discussed previously, deficiencies or excesses of energy or nutrients during pregnancy can affect the health of the embryo and fetus and cause developmental errors. Other substances present in the environment, consumed in the diet, or taken as medications or recreational drugs can also act as teratogens.

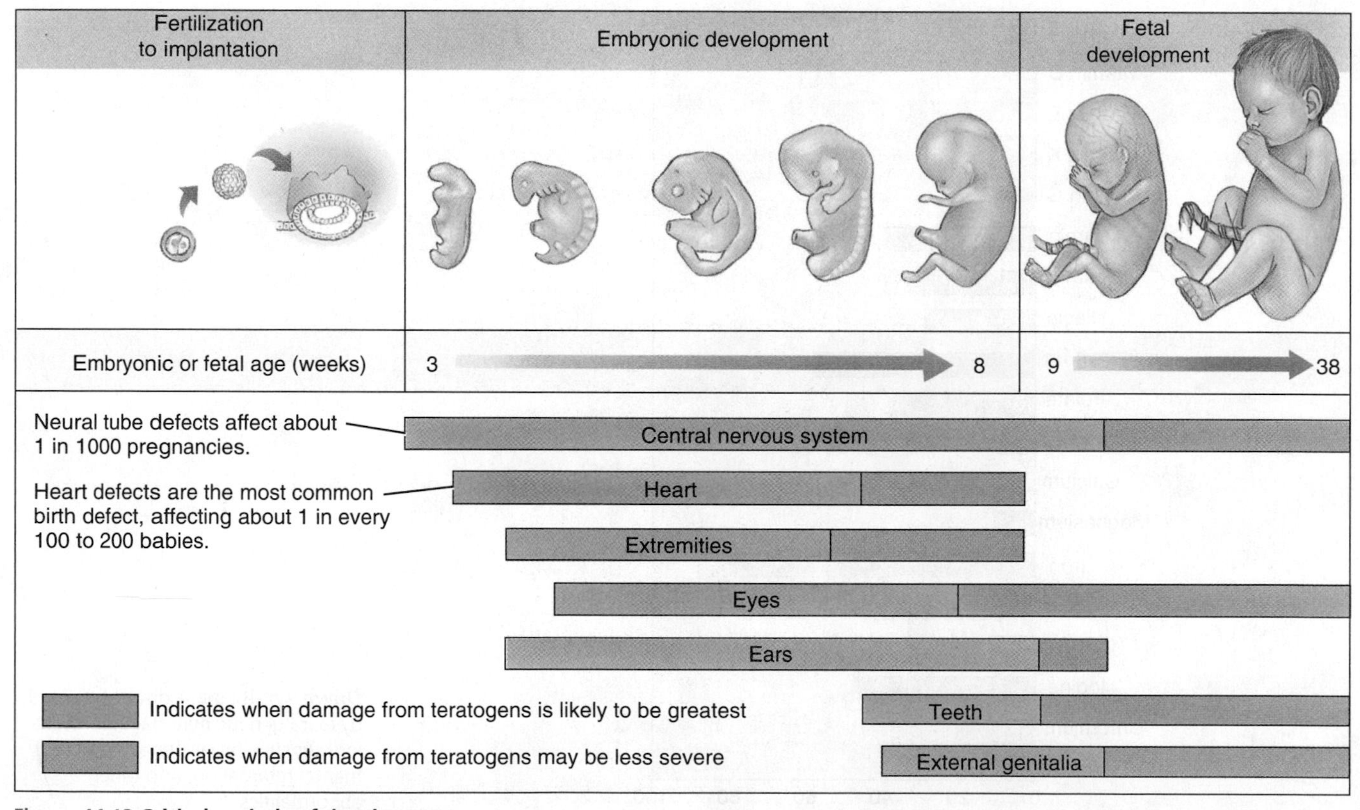

Figure 14.18 Critical periods of development
The critical periods of development are different for various body systems. Heart defects are the most common birth defect.
Source: Adapted from Moore, K., and Persaud, T. The Developing Human, 5th ed. Philadelphia: W. B. Saunders Company, 1993.

CANADIAN CONTENT

Environmental Toxins In a pregnant woman, exposure to environmental toxins—such as cleaning solvents, lead and mercury, some insecticides, or paint—can affect her developing child. Therefore, pregnant women need to be aware of the potential toxins in their food, water, and environment. Fish has both benefits and risks for pregnant women. It is a source of lean protein for tissue growth and of omega-3 fatty acids and iodine needed for brain development, but if it is contaminated with mercury, consumption during pregnancy can cause developmental delays and brain damage. Rather than avoid fish, pregnant women should be informed consumers. Health Canada reports that most fish commonly consumed in Canada have very low mercury levels. The exceptions are fresh or frozen tuna, shark, swordfish, marlin, orange roughy, and escolar, and Canadians are advised to limit their intakes to the amounts posted on the Health Canada website, which for women who are planning to get pregnant, or are pregnant or breastfeeding is 4 Canada's Food Guide servings per month.[40] Although fresh and frozen tuna are mentioned, canned light tuna, a commonly consumed fish, has low levels of mercury and is not of concern. Canned albacore tuna, a more expensive form of tuna, has sufficient mercury that Health Canada recommends limits on intake for women who are planning to get pregnant, are pregnant or are breastfeeding to no more than 4 Canada's Food Guide servings weekly.[40]

Caffeine and Herbs Caffeine-containing beverages like coffee are a part of our typical diet, but consuming too much caffeine during pregnancy has been associated with reductions in birth weight and an increased risk of miscarriage.[41] It is recommended that pregnant women avoid consuming more than 200 mg of caffeine per day. This is the amount in about 500 ml (2 cups) of regular coffee or 1.25 L (5 cups) of tea or cola beverages (see Table 13.7). Herbal teas are also popular beverages, and some are consumed to treat the discomforts of pregnancy. For example, ginger and raspberry leaves are often used to treat morning sickness. Because there is not a great deal of research on herbal teas and other herbal products, pregnant women should avoid them until they are shown to be safe during pregnancy.[2]

Artificial Sweeteners Because women are often concerned about excessive weight gain during pregnancy, they may consider using products containing artificial sweeteners. A fact sheet published by Dietitians of Canada indicates that aspartame, acesulfame-potassium, and sucralose are considered safe for pregnant women, but saccharin and cyclamate should be avoided.[42]

Food-Borne Illness The immune system is weakened during pregnancy, increasing susceptibility to and the severity of certain food-borne illnesses. *Listeria* infections are about 20 times more likely during pregnancy and are especially dangerous for pregnant women, often resulting in miscarriage, premature delivery, stillbirth, or infection of the fetus.[43] About one-quarter of babies born with *Listeria* infections do not survive. The bacteria are commonly found in unpasteurized milk, soft cheeses, and uncooked hot dogs and lunch meats.

Toxoplasmosis is an infection caused by a parasite. If a pregnant woman becomes infected, there is about a 40% chance that she will pass the infection to the fetus.[44] Some infected babies develop vision and hearing loss, intellectual disability, and/or seizures. The toxoplasmosis parasite is found in cat feces, soil, and undercooked, infected meat. Pregnant women should follow the safe food-handling recommendations discussed in Section 17.3.

Alcohol Alcohol consumption during pregnancy is one of the leading causes of preventable birth defects (see Chapter 5: Focus on Alcohol). Alcohol is a teratogen that is particularly damaging to the developing central nervous system.[45] It also indirectly affects fetal growth and development because it is a toxin that reduces blood flow to the placenta, thereby decreasing the delivery of oxygen and nutrients to the fetus. The use of alcohol can also impair maternal nutritional status, further increasing the risk to the embryo or fetus. Despite this, about 12% of women report drinking alcohol during pregnancy.[2]

Prenatal exposure to alcohol can cause a spectrum of disorders depending on the dose, timing, and duration of the exposure. One of the most severe outcomes of drinking alcohol during pregnancy is **fetal alcohol syndrome (FAS)**, which causes facial deformities, growth retardation, and permanent brain damage (**Figure 14.19**). Newborns with the syndrome may be shaky and irritable, with poor muscle tone and alcohol withdrawal symptoms. Other problems include heart and urinary tract defects, impaired vision and hearing, and delayed language development. Intellectual disability is the most common and most serious effect. Not all babies exposed to alcohol have FAS, but many have some alcohol-related problems. **Alcohol-related neurodevelopmental disorders (ARND)** are functional or mental impairments linked to prenatal alcohol exposure, and **alcohol-related birth defects (ARBD)** are malformations in the skeleton or major organ systems. These conditions are less severe than FAS, but occur about 3 times more often.

Because alcohol consumption in each trimester has been associated with abnormalities and because there is no level of alcohol consumption that is known to be safe, complete abstinence from alcohol is recommended during pregnancy. The Public Health Agency of Canada position is very clear: "There is no safe amount or safe time to drink alcohol during pregnancy."[46] In the Yukon and Northwest Territories, alcohol beverages carry warning labels about the risks

fetal alcohol syndrome A characteristic group of physical and mental abnormalities in an infant resulting from maternal alcohol consumption during pregnancy.

alcohol-related neurodevelopmental disorders A spectrum of learning and developmental disabilities and behavioural abnormalities in a child due to maternal alcohol consumption during pregnancy.

alcohol-related birth defects Malformations in the skeleton or major organ systems in a child due to maternal alcohol consumption during pregnancy.

Figure 14.19 Facial characteristics shared by children with fetal alcohol syndrome include a low nasal bridge, a short nose, distinct eyelids, and a thin upper lip. (David Young-Wolff/PhotoEdit)

CANADIAN CONTENT

of drinking during pregnancy (**Figure 14.20a**). In Ontario, establishments serving alcohol are required to post signs warning of the risks of drinking during pregnancy (**Figure 14.20b**).

a)

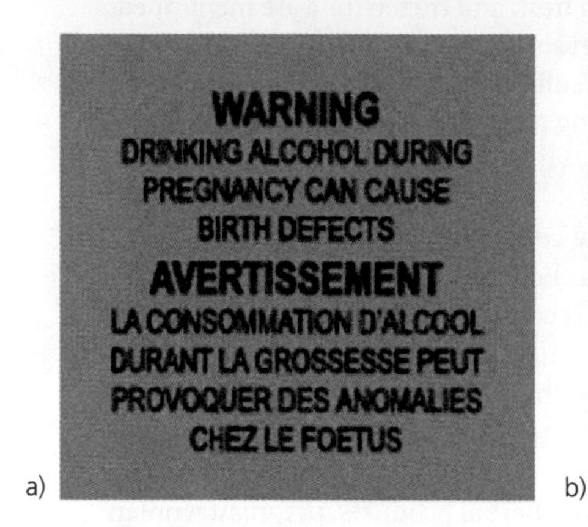

b)

Figure 14.20 Warnings about the dangers of alcohol and pregnancy
a) In the Yukon, this label appears on alcoholic beverages. b) In Ontario, this sign appears in all establishments selling alcoholic beverages.
Source: With permission from the Government of Yukon. Yukon Liquor Corporation: Social Responsibility. Available online at http://www.ylc.yk.ca/socialresp.html. Accessed June 20, 2011; with permission from the Alcohol and Gaming Commission of Ontario. Signage Requirement-Sandy's Law. Available online at http://www.agco.on.ca/en/services/igns_sandyslaw_LSL.aspx. Accessed June 20, 2011.

sudden infant death syndromes (SIDS), or **crib death** The unexplained death of an infant, usually during sleep.

Cigarette Smoke If a woman smokes cigarettes during pregnancy, her baby will be affected before birth and throughout life. The carbon monoxide in tobacco smoke binds to hemoglobin, reducing oxygen delivery to fetal tissues. The nicotine absorbed from cigarette smoke is a teratogen that can affect brain development.[47] It also constricts arteries and limits blood flow, reducing both oxygen and nutrient delivery to the fetus.[48] Cigarette smoking during pregnancy reduces birth weight, and increases the risk of preterm delivery, stillbirth, neurobehavioural problems, and early death.[2,47] Even exposure to cigarette smoke from the environment has been found to increase the risk of low birth weight. The risks of **sudden infant death syndrome (SIDS,** or **crib death)** and respiratory problems are also increased in children exposed to cigarette smoke both in the uterus and after birth. The effects of maternal smoking follow children throughout life; they are more likely to have frequent colds and develop lung problems later in life.[49] The Public Health Agency of Canada recommends that all pregnant women quit smoking completely and maintain a smoke-free environment during and after pregnancy.[50]

Legal and Illicit Drug Use The use of drugs—whether over-the-counter, prescribed, or illicit—can also affect both fertility and pregnancy outcome. For example, the acne medications Accutane and Retin-A, which are derivatives of vitamin A, can cause birth defects if taken during pregnancy. A woman who is considering pregnancy should discuss her plans with her physician in order to determine the risks associated with any medication she is taking.

Substance abuse during pregnancy is an important health issue. Marijuana and cocaine are drugs that are commonly used during pregnancy. Both cross the placenta and enter the fetal blood. There is little evidence that marijuana affects fetal outcome, but cocaine use increases the risk of complications to the mother and creates problems for the infant before, during, and after delivery. Cocaine is a central nervous system stimulant, but many of its effects during pregnancy occur because it constricts blood vessels, thereby reducing the flow of oxygen and nutrients to the rapidly dividing fetal cells. Cocaine use during pregnancy is associated with a high rate of miscarriage, intrauterine growth retardation, spontaneous abortion, premature labour and delivery, low birth weight, and birth defects.[51] Exposure to cocaine, opiates, or amphetamines has been shown to affect infant behaviour and impact learning and attention span during childhood.[52]

CANADIAN CONTENT

Reaching Canadian Mothers at Risk

Canadian women enjoy excellent prenatal care. in Canada, the maternal death rate is 5.5/100,000 women. In Africa, it is 830/100,000.[53] Despite these statistics, it is known that some Canadian women face greater risks of adverse effects during their pregnancy. The Canadian Prenatal

Nutrition Program (CPNP), launched in 1995, has targeted these women for special support to improve maternal health and nutrition during pregnancy, to reduce unhealthy birth weights (either too low or too high), and to encourage breastfeeding.[54]

The characteristics of women targeted by this program are listed in **Table 14.5** and many are linked to low socioeconomic status. Between 1998 to 2003, a modest 7% of all pregnant Canadian women accessed CPNP, but these women included 60% of low-income women and 40% of pregnant teens.[54]

Table 14.5 Characteristics of Women Participating in the Canadian Prenatal Nutrition Program

Characteristic of Women Accessing CPNP:	% of all women in CPNP (national average)
Education: less than grade 12	69
Income: less than $1,300/month	52
Single parent	34
Smoked during pregnancy	31
In Canada less than 10 years	29
Experienced abuse during pregnancy	14
Used alcohol during pregnancy	7

Source: Public Health Agency of Canada. The Canada Prenatal Nutrition Program: A decade of promoting the health of mothers, babies, and communities, 2007. Available online at http://www.phac-aspc.gc.ca/hp-ps/dca-dea/publications/pdf/mb_e.pdf. Accessed Dec 18, 2010.

The CPNP is a series of individual projects designed and implemented in communities through Canada to meet locally identified needs. There are more than 300 CPNP sites that service more than 50,000 pregnant women in 2,000 communities throughout Canada. A directory of projects is available online.[55] **Figure 14.21** illustrates the types of services offered in many CPNP projects.

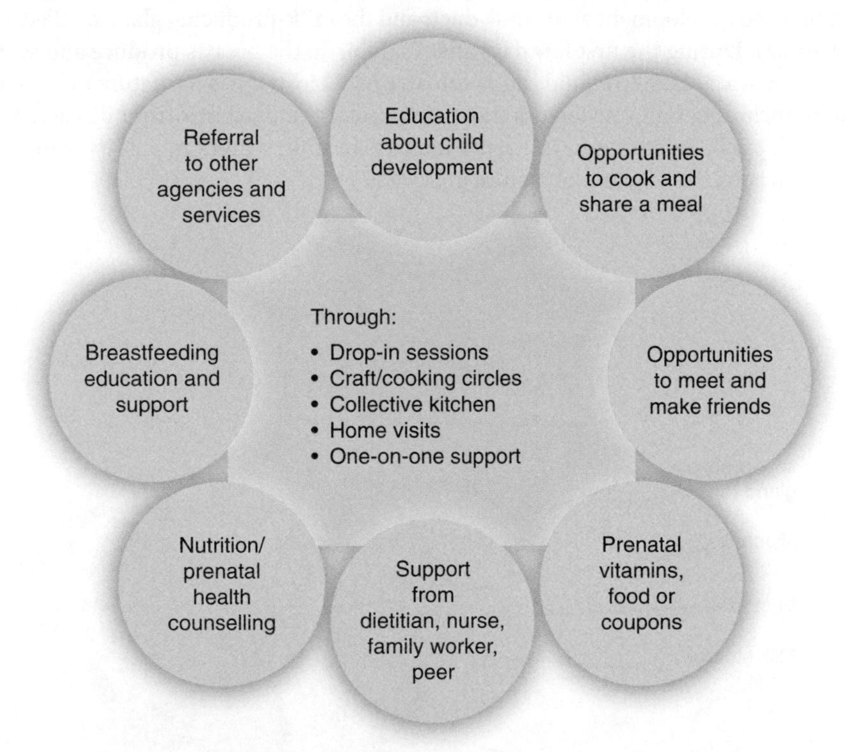

Figure 14.21 Services offered to pregnant women at CPNP projects
Source: With permission from Public Health Agency of Canada. The Canada Prenatal Nutrition Program: A decade of promoting the health of mothers, babies, and communities, 2007. Available online at www.phac-aspc.gc.ca/hp-ps/dca-dea/publications/pdf/mb_e.pdf. Accessed December 18, 2010.

Examples of projects include *Buns in the Oven* in Ottawa, which responds to the needs of young, isolated pregnant women with a program that includes cooking lessons, nutrition counselling, advice on breastfeeding, and opportunities to learn life skills and forge relationships with their peers.[56]

In Whitehorse, Yukon, *Healthy Moms, Healthy Babies* has been described as "a refuge where women can escape their troubles" as many participants are isolated and have had to deal with violence and abuse. Nutrition counselling is provided during weekly drop-in sessions and home visits by program staff.[57]

The *Alberta Prenatal Tobacco Cessation Pilot Project* was initiated in response to a provincial survey indicating that 58% of CPNP users smoked during pregnancy. Its counselling sessions successfully increased the number of women who stopped smoking.[54]

In addition to these programs, many municipal departments of public health also provide free information to pregnant women and new mothers.

14.4 Lactation

Learning Objectives

- Compare the nutritional needs of lactating women with those of nonpregnant, nonlactating women of child-bearing age.
- Explain the relationship between suckling and milk production and let-down.

The nutrient requirements of pregnancy include those needed to prepare for lactation. After childbirth, the breastfeeding mother's nutrient intake must support milk production and can influence the nutrient composition of her milk.

The Physiology of Lactation

Lactation involves both the synthesis of the milk components, such as milk proteins, lactose, and milk lipids, and the movement of these milk components through the milk ducts to the nipple. Throughout pregnancy, hormones prepare the breasts for lactation by stimulating the enlargement and development of the milk ducts and the milk-producing glands, called alveoli (**Figure 14.22**). During the first few days after childbirth, the breasts produce and secrete a small amount of a clear yellow fluid called **colostrum**. Colostrum is immature milk. It is rich in protein, including immune factors that help protect the newborn from disease. Within about a week of childbirth, there is a rapid increase in milk secretion, and its composition changes from colostrum to that of mature milk.

colostrum The first milk, which is secreted in late pregnancy and up to a week after birth. It is rich in protein and immune factors.

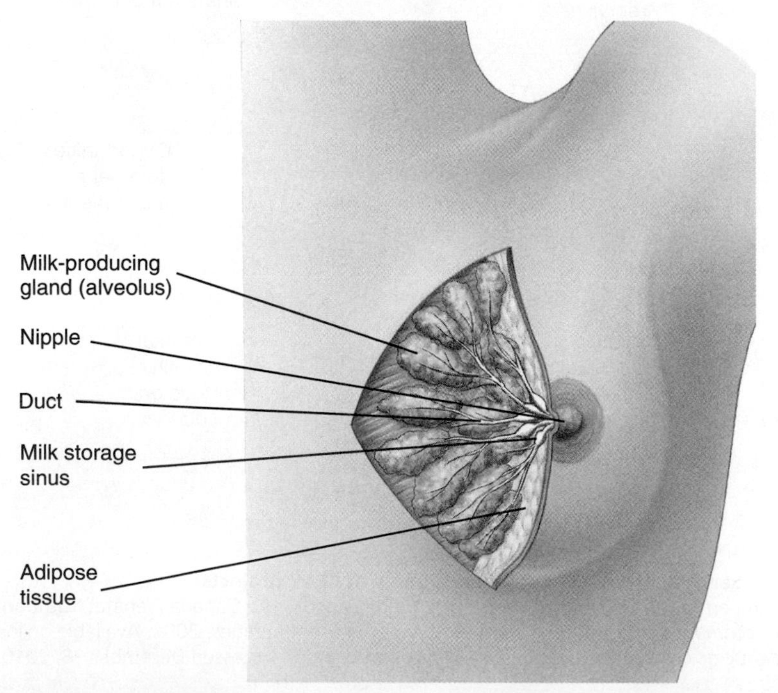

Figure 14.22 Anatomy of lactation
During lactation, milk travels from the milk-producing glands through the ducts to milk storage sinuses and then to the nipple.

prolactin A hormone released by the anterior pituitary that acts on the milk-producing glands in the breast to stimulate and sustain milk production.

Milk production and release is triggered by hormones that are released in response to the suckling of the infant. The pituitary hormone **prolactin** stimulates milk production; the more the infant suckles, the more milk is produced. The release of milk from the milk-producing glands and its movement through the ducts and storage sinuses is referred to as **let-down** (**Figure 14.23**). The let-down of milk is caused by **oxytocin**, another hormone produced by the pituitary gland that is released in response to the suckling of the infant. As nursing becomes more automatic, oxytocin release and the let-down of milk may occur in response to the sight or sound of an infant. It can be inhibited by nervous tension, fatigue, or embarrassment. The let-down response is essential for successful breastfeeding and makes suckling easier for the child. If let-down is slow, the child can become frustrated and difficult to feed.

let-down A hormonal reflex triggered by the infant's suckling that causes milk to be released from the milk glands and flow through the duct system to the nipple.

oxytocin A hormone produced by the posterior pituitary gland that acts on the uterus to cause uterine contractions and on the breast to cause the movement of milk into the secretory ducts that lead to the nipple.

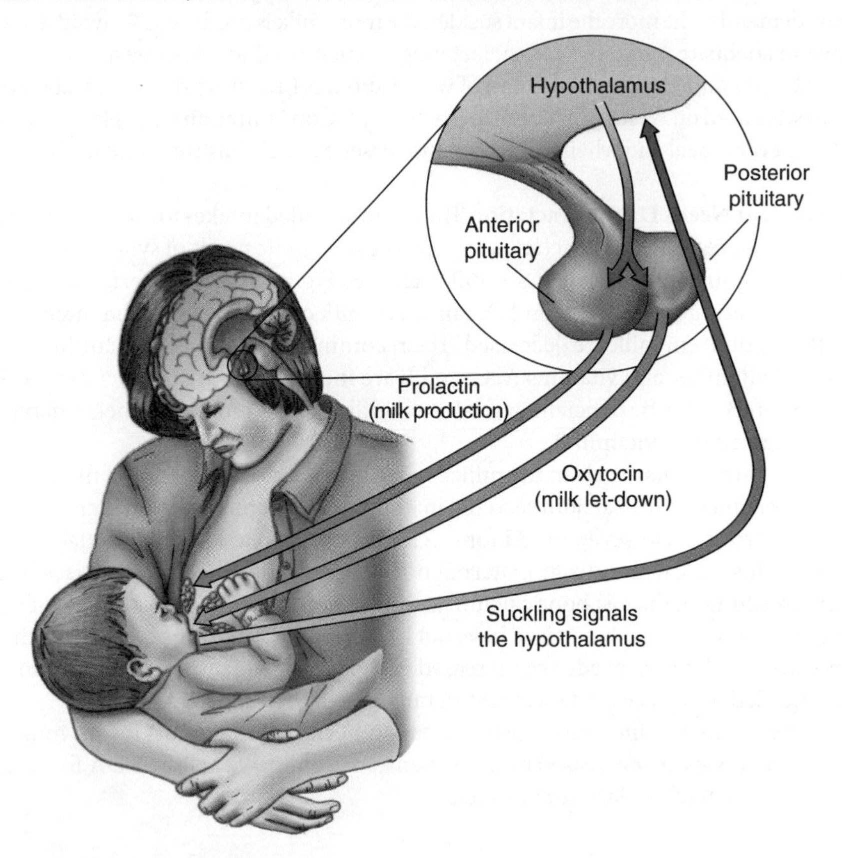

Figure 14.23 Hormones of lactation
When the infant suckles, nerve receptors in the nipple send signals to the hypothalamus, which signals the release of prolactin and oxytocin.

Maternal Nutrient Needs During Lactation

The need for many nutrients is even greater during lactation than during pregnancy. This is because the mother is still providing for all of the energy and nutrient needs of the infant, who is growing faster and is more active than the fetus. Meeting the needs of lactation requires a varied nutrient-dense diet. Most lactating women can meet all their needs without supplements.

Energy and Macronutrient Needs During the first 6 months of lactation, approximately 600-900 ml of milk is produced daily, depending on how much the infant consumes. Human milk contains about 65 kcal/100 ml so providing an infant with 750 ml of milk would require the mother to expend approximately 500 kcal. Much of this must come from the diet, but some can come from mobilization of maternal fat stores, which increase during pregnancy. The EER for lactation is estimated by adding the total energy expenditure of nonlactating women and the energy in milk and then subtracting the energy supplied by maternal fat stores.[19] This is equal to an additional 330 kcal/day during the first 6 months of lactation, and 400 kcal/day during the second 6 months (see Figure 14.12). To ensure adequate protein for milk production, the RDA for lactation is increased by 25 g/day. The RDA for carbohydrate and the AIs for linoleic and alpha-linolenic acids are also higher during lactation.[19]

Even though some of the energy for lactation comes from maternal fat stores, the impact of lactation on maternal weight loss is variable. Some studies report that breastfeeding does not affect the amount of weight lost, whereas others suggest it does so initially or if breastfeeding continues for at least 6 months.[58] Beginning 1 month after birth, most lactating women lose 0.5-1 kg (1-2 lb) per month for 6 months. Some women will lose more, and others may maintain or even gain weight regardless of whether or not they breastfeed. Rapid weight loss is not recommended during lactation because it can decrease milk production, but a modest kcalorie deficient can result in gradual weight loss. Regular exercise can also make weight loss easier and does not impair milk production.

Water Needs During Lactation The amount of milk a woman produces depends on how much her baby demands. The more the infant suckles, the more milk is produced. To avoid dehydration and ensure adequate milk production, lactating women need to consume about 1 L of additional water per day. The AI of 3.8 L/day, of which about 3.1 L is from drinking water and other beverages, is based on typical intake during lactation.[13] Consuming an extra glass of milk, juice, or water at every meal and whenever the infant nurses can help ensure adequate fluid intake.

Micronutrient Needs During Lactation The recommended intakes for several vitamins and minerals are increased during lactation to meet the metabolic needs of synthesizing milk and to replace the nutrients secreted in the milk itself (see Figure 14.12). Maternal intake of some vitamins, including C, B_6, B_{12}, A, and D, can affect milk composition. When maternal intake is low, the amounts in milk are decreased. The recommended intakes of vitamin B_6, vitamin B_{12}, other B vitamins, and vitamins A, C, and E are increased above nonlactating levels. Because vitamin B_{12} may be deficient in the breast milk of vegan mothers, their infants should be supplemented with vitamin B_{12}.

For other nutrients, levels in the milk are maintained at the expense of maternal stores. For example, much of the calcium secreted in human milk comes from an increase in maternal bone resorption. However, the AI for calcium is not increased above nonlactating levels because the loss of calcium from maternal bones is not prevented by increases in dietary calcium. In addition, the lost bone calcium is fully restored within a few months of weaning, and therefore women who breastfeed have not been found to have a long-term deficit in bone mineral density.[59] Folate needs are increased above nonpregnant levels to account for the amount needed to replace folate secreted in milk.[31]

Iron needs are not increased during lactation because little iron is lost in milk, and, in most women, losses are decreased because menstruation is absent. The RDA for lactation is 9 mg/day, half that of nonlactating women.

14.5 The Nutritional Needs of Infancy

Learning Objectives

- Compare the energy and macronutrient needs of newborns with those of adults.
- List micronutrients that are likely to be deficient in infants.
- Explain the best way to monitor the adequacy of an infant's dietary intake.

When a child is born and the umbilical cord is cut, he or she suddenly becomes actively involved in obtaining nutrients rather than being passively fed through the placenta. The energy and nutrients the infant consumes must support his or her continuing growth and development and increasing level of activity (**Table 14.6**).

Table 14.6 Energy and Nutrient Needs of Infants Compared to Adults

Nutrient/Energy	Newborn Recommendation (0–6 mo)	Adult Recommendation
Energy[a]	493-606 kcal/day (~100 kcal/kg/day)	2,403-3,067 kcal/day (~30 kcal/kg/day)
Protein	9.1 g/day 1.52 g/kg/d	46-56 g/day 0.8 g/kg/day

Nutrient/Energy	Newborn Recommendation (0–6 mo)	Adult Recommendation
Carbohydrate	at least 60 g/day	at least 130 g/day
	40% of energy intake[b]	45%–65% of energy intake
Fat	50% of energy[b]	20%–35% of energy
Linoleic acid	4.4 g/day[c]	12-17 g/day
α-linolenic acid	0.5 g/day[d]	1.1–1.6 g/day
Fluid	0.7 L	2.7–3.7 L

[a]The energy values are based on EER prediction equations for infants 0–6 months of age and for adults >19 years of age.

[b]Based on the composition of human milk.

[c]Refers to all omega-6 polyunsaturated fatty acids.

[d]Refers to all omega-3 polyunsaturated fatty acids.

Energy Needs During Infancy

During the first few months after birth, growth is more rapid than at any other time of life. As a result, newborn infants need more kcalories per kilogram of body weight than at any other time. EERs for infants are calculated from total energy expenditure plus the energy deposited in tissues due to growth[19] (see inside cover). After the first few months, growth slows some and activity increases. Differences in growth rates and activity levels are reflected in the separate EER prediction equations for infants 0-3 months, 4-6 months, and 7-12 months.

Fat Recommendations

Healthy infants consume about 55% of their energy as fat during the first 6 months of life, and 40% during the second 6 months. This is far greater than the 20%-35% of energy from fat recommended in the adult diet (**Figure 14.24**). The high proportion of fat increases the energy density of the diet, allowing the infant's small stomach to hold enough to meet energy needs. An AI for total fat has been set at 31 g/day for infants from birth to 6 months of age and at 30 g/day for infants 7-12 months of age.

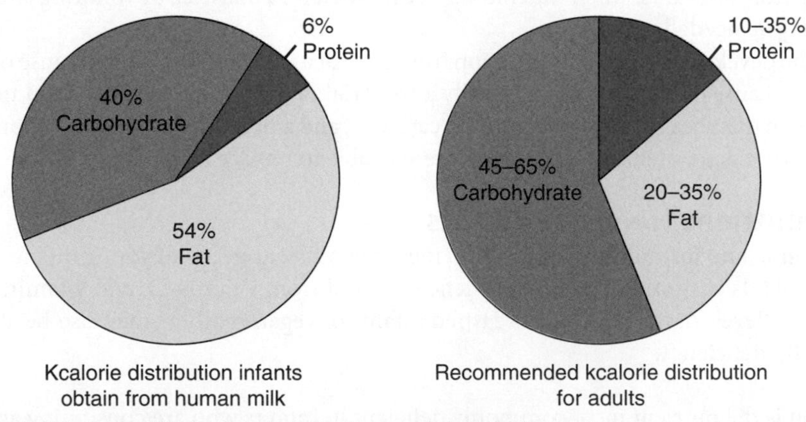

Figure 14.24 Kcalorie distribution in infant versus adult diets
A comparison of the proportions of kcalories from carbohydrate, fat, and protein in human milk with the proportions recommended for an adult illustrates how much more fat infants need.

In addition to getting enough fat, infants need the right kinds of fat. A sufficient supply of the long-chain polyunsaturated fatty acids docosahexaenoic acid (DHA, an omega-3 fatty acid) and arachidonic acid (an omega-6 fatty acid) are important for nervous system development. These fatty acids are constituents of cell membranes and are incorporated into brain tissue and the retina of the eye. Breast milk contains both of these fatty acids. Infant formulas supplemented with DHA and arachidonic acid are available but the addition of these fatty acids to infant formulas is not required in Canada.[60] Infants can synthesize these fatty acids from linoleic and alpha-linolenic acid, but since breastfed infants have higher plasma concentrations of these long-chain polyunsaturated fatty acids than infants fed non-fortified formula, it is hypothesized that the rate of conversion may not be optimal. Evidence that inclusion of these fatty acids in infant formulas benefits growth, visual function, and cognitive development is equivocal.[61] AIs for infants have been set for total omega-3 and total omega-6 fatty acids based on the amounts of these types of fatty acids in human milk.[19]

Carbohydrate Intake

Carbohydrate, like fat, is a major contributor to energy intake in the infant. The source of carbohydrate for breastfed infants and most bottle-fed infants is lactose. About 39% of the energy in breast milk is from lactose. As the infant grows and solid foods are introduced into the diet, the percentage of kcalories from carbohydrate in the diet increases and the percentage from fat decreases.

Protein Requirements

The infant's protein requirement per unit of body weight is very high compared with the adult requirement: The AI is 1.52 g/kg/day from birth to 6 months of age, compared with 0.8 g/kg/day for an adult. The ideal protein source for newborns is human milk. Infant formulas are designed to mimic its amino acid pattern. A diet too high in protein may lead to dehydration because the excretion of metabolic wastes produced when excess protein is consumed increases water loss.

Water Needs

Infants have a higher proportion of body water than adults and they lose proportionately more body water in urine and through evaporation. Urine losses are high because the infant kidneys are poorly developed and unable to reabsorb much of the water that passes through them. Infants lose proportionately more water through evaporation because they have a larger surface area relative to their total body volume. These factors, in addition to the fact that infants cannot tell us they are thirsty, put them at risk for dehydration. Despite this, infants who are exclusively breastfed do not require additional water. The AI is based on the volume of human milk consumed and the water content of the milk. It is set at 0.7 L/day for infants 0-6 months and at 0.8 L/day for older infants (7-12 months).[13] In older infants, some fluid is obtained from other beverages and foods. Although breast milk can meet fluid needs in healthy infants, when water losses are increased by diarrhea or vomiting additional fluids may be needed.

In the developing world, dehydration from diarrhea is the most common cause of infant death. The cause of the diarrhea is usually a bacterial or viral infection. The fluid intake of infants with diarrhea should be monitored carefully, and a pediatrician should be contacted. Mixtures of sugar, water, and electrolytes are available to replace lost fluids.

Micronutrient Needs of Infants

Human milk and formula are designed to meet the nutrient needs of young infants. Nonetheless, infants may still be at risk for deficiencies of iron, vitamin D, and vitamin K, and suboptimal levels of fluoride. The breastfed infants of vegan mothers may also be at risk of vitamin B_{12} deficiency.

Iron Iron is the nutrient most commonly deficient in infants who are consuming adequate energy and protein. Iron deficiency is usually not a problem during the first 4-6 months of life because infants have iron stores at birth and the iron in human milk, though not particularly abundant, is very well absorbed. The AI for iron from birth to 6 months is only 0.27 mg/day.[33] After 6 months, iron stores are depleted but iron needs remain high to provide for hemoglobin synthesis, tissue growth, and iron storage. The RDA for infants 7-12 months jumps to 11 mg/day.[33] To meet needs after 6 months, the diets of breastfed infants should contain other sources of iron, such as iron-fortified rice cereal or meat. Formula-fed infants can obtain iron from fortified formula.

Vitamin D Newborns are also potentially at risk for vitamin D deficiency. Breast milk is relatively low in vitamin D, so breastfed infants who do not receive adequate exposure to sunlight, such as those living in cold climates, may not obtain adequate vitamin D. To synthesize adequate vitamin D, about 15 min./day of sun exposure, with only the face uncovered, is needed for light-skinned babies; a longer time is required for darker-skinned babies (**Figure 14.25**). An AI of 10 μg/day of vitamin D has been set for infants 0-12 months of age. The Canadian Pediatric Society recommends that all infants receive 10 μg/day during the first year, with an increase to 20 μg/day between October and April north of the 55th parallel (approximate latitude of Edmonton) to compensate for the lack of sunshine during the winter months. Because breast milk contains little vitamin D, supplements are required for breastfed babies. Vitamin D is added to formula to meet the needs of formula-fed babies.[62]

CANADIAN CONTENT

Figure 14.25 Dark skin pigmentation reduces the amount of vitamin D that can be synthesized in the skin, putting darker-skinned babies at greater risk for deficiency. (Purestock)

Vitamin K Vitamin K, which is essential for normal blood clotting, is another nutrient for which newborns are at risk of deficiency. Little of this vitamin crosses the placenta from mother to fetus, and because the gut is sterile at birth, no microbial vitamin K synthesis occurs. Breast milk is also low in vitamin K, so breastfed infants are at risk of hemorrhage due to vitamin K deficiency. To prevent this, it is recommended that all breastfed infants receive a single intramuscular injection containing 0.5-1.0 mg of vitamin K after the first feeding is completed and within the first 6 hours of life.[63] This provides them with enough vitamin K to last until their intestines are colonized with the bacteria that synthesize it.

Fluoride Fluoride is important in the development of teeth, even before they erupt. Breast milk is low in fluoride, and formula manufacturers use unfluoridated water in preparing liquid formula. Therefore, breastfed infants, infants fed premixed formula, and those fed formula mixed with low-fluoride water are often supplemented beginning at 6 months of age. In areas where the drinking water is fluoridated, infants fed formula reconstituted with tap water should not be given fluoride supplements.

Vitamin B_{12} Breast milk and infant formula typically contain enough vitamin B_{12} to meet the infant's needs. An exception is the breast milk of a vegan mother, which may be deficient in vitamin B_{12}. Therefore, breastfed infants of vegan mothers should be supplemented with vitamin B_{12}.

Assessing Infant Growth

Although nutrient needs for infants are fairly well defined, it is difficult to calculate an infant's actual nutrient intake, particularly if they are breastfeeding. The best indicator of adequate nourishment is normal growth. Most healthy infants follow standard patterns of growth, so their growth can be monitored using growth charts (see Appendix B).

Growth Charts The World Health Organization growth charts, which are recommended for use by health professionals in Canada, plot growth patterns of healthy infants, children, and adolescents (**Figure 14.26**).[64] They can be used to monitor an infant's pattern of growth and to compare length, weight, and head circumference to standards for infants of the same age (Figure 14.27). The resulting ranking, or percentile, indicates where the infant's growth falls in relation to population standards. For example, if a 2-month-old girl is at the 55th percentile for weight, it means that 55% of newborn girls weigh less and 45% weigh more. Children usually continue at the same percentiles as they grow.

Whether an infant is 2.5 kg or 3.5 kg (5.5 lb or 8 lb) at birth, the pattern of growth should be approximately the same—rapid initially and slowing slightly as the infant approaches 1 year of age. A rule of thumb is that an infant's birth weight should double by 4 months and triple by 1 year of age. In the first year of life, most infants increase their length by 50%. Small infants and premature infants often follow a pattern parallel to but below the growth curve for a period of time and then experience catch-up growth that brings them onto the growth curve in a place compatible with their genetic growth potential.

Abnormal Growth Slight fluctuations in growth rate are normal, but a consistent pattern of not following the growth curve or a sudden change in growth pattern is cause for concern and could indicate overnutrition or undernutrition. A rapid increase in weight without an increase in height may be an indicator that the infant is being overfed. Because overweight children grow into overweight adults, this pattern of weight increase should be addressed early in life.

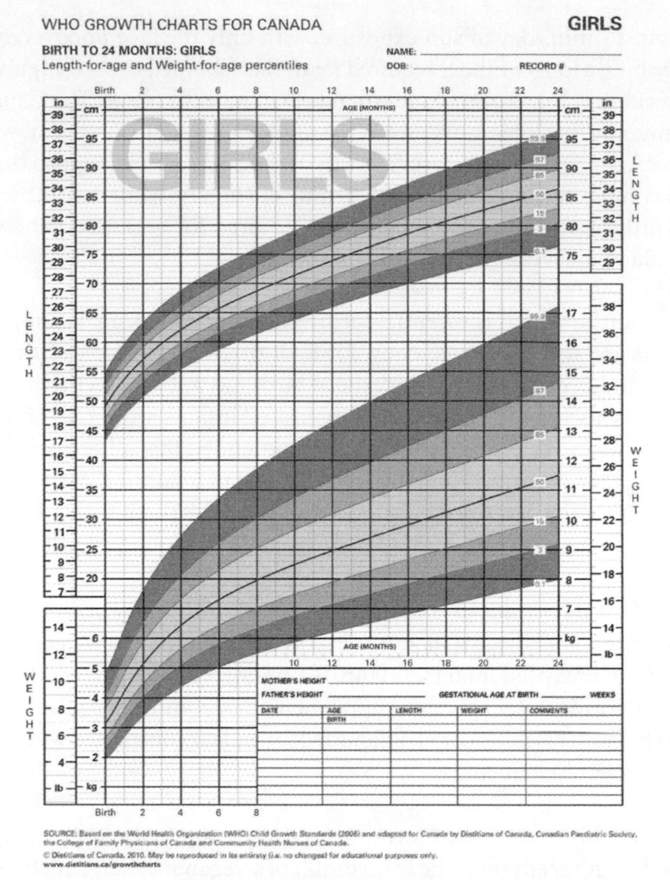

Figure 14.26 WHO Growth Charts are used to monitor infant growth
Source: © 2011. Dietitians of Canada. All rights reserved. Permission to print in its entirety. For noncommercial use only. WHO Growth Charts adapted for Canada. Available online at http://www.dietitians.ca/Secondary-Pages/Public/Who-Growth-Charts.aspx. Accessed December 18, 2010.

failure to thrive The inability of a child's growth to keep up with normal growth curves.

Growth that is slower than the predicted pattern indicates **failure to thrive**. This is a catch-all term for any type of growth failure in a young child. The cause may be a congenital condition, the presence of disease, poor nutrition, neglect, abuse, or psychosocial problems. The treatment is usually an individualized plan that includes adequate nutrition and careful monitoring by physicians, dietitians, and other health-care professionals. Just as there are critical periods in fetal life, there are critical periods for growth and development during infancy when undernutrition can permanently affect development.

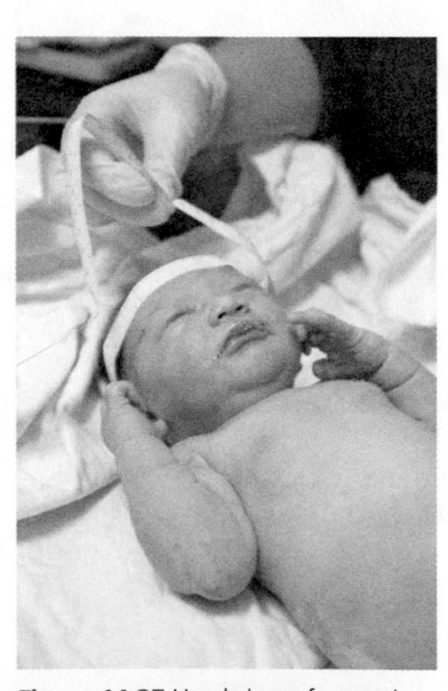

Figure 14.27 Head circumference is used as a indicator of the growth of infants, along with length and weight. (© istockphoto.com/Craig Foster)

14.6 Feeding the Newborn

Learning Objective

- List the advantages and disadvantages of breastfeeding and bottle-feeding.

Newborns have small stomachs, can consume only liquids, and have high nutrient requirements. The ideal food for the newborn is breast milk, but infant formula can also meet a newborn's needs. From birth until 6 months of age, infants don't need anything other than breast milk or formula. Solid food should not be introduced into the diet until the child is 6 months of age because the infant's feeding abilities and gastrointestinal tract are not mature enough to handle solid foods.

A relatively common problem in infants is colic. Colic involves daily periods of inconsolable crying that cannot be stopped by holding, feeding, or changing the infant. Colic usually begins at a few weeks of age and continues through the first 2-3 months, when it resolves on its own. It occurs in both breast- and bottle-fed infants. Although its cause is unknown, it may be related to the immaturity of the central nervous system. Some studies suggest that reducing the infant's exposure to cow's milk protein may also be beneficial. For formula-fed babies, this means hydrolyzing (breaking down) the cow's milk proteins typically used in formula. For breastfed babies, this means putting the mother on a hypoallergenic diet, limiting exposure to cow's milk products.[65]

Meeting Nutrient Needs with Breastfeeding

Breast milk meets the nutrient needs of the human newborn, requires no special preparation, and the amount available varies with demand. Thus, breastfeeding is the preferred form of infant nutrition and is usually the recommended choice for feeding the newborn of a healthy, well-nourished mother.[66,67]

Nutritional Advantages of Breast Milk The nutrient composition of breast milk is specifically designed for the human infant and changes over time as the infant develops, meeting the nutrient needs of the child for up to the first year of life. The first fluid that is produced by the breast after delivery is colostrum. This yellowish fluid is higher in water, protein, immune factors, minerals, and some vitamins than mature breast milk. It is produced for up to a week after delivery. While colostrum is produced, it may seem that the newborn is not receiving enough to eat; however, supplemental bottle feedings are not necessary. The nutrients in colostrum meet infant needs until mature milk production begins. Colostrum also has beneficial effects on the gastrointestinal tract, acting as a laxative that helps the baby excrete the thick, mucusy stool produced during life in the womb.

Lactalbumin is the predominant protein in human milk. In the infant's stomach, it forms a soft, easily digested curd. The amino acids methionine and phenylalanine, which are difficult for the infant to metabolize, are present in lower amounts in human milk proteins than in cow's milk proteins. Human milk is also a good source of taurine, an amino acid needed for bile salt formation and eye and brain function.

The lipids in human milk are easily digested. They are high in cholesterol and the fatty acids linoleic acid, arachidonic acid, and DHA, which are essential for normal brain development, eyesight, and growth. The fat content of breast milk changes throughout a feeding, gradually increasing during the nursing session. Thus, for the baby to attain satiety and obtain adequate energy, it is important for nursing to continue long enough for the infant to obtain the higher-fat milk.

Lactose is the primary carbohydrate in human milk. It is digested slowly so it stimulates the growth of acid-producing bacteria. It also promotes the absorption of calcium and other minerals and provides a source of the sugar galactose for nervous system development.

Breast milk is low in sodium and the zinc, iron, and calcium present are in forms that are easily absorbed.[67] About 50% of the iron in human milk is absorbed, compared with only 2%-30% from many other foods.

Other Advantages of Breastfeeding Breastfeeding can be a relaxing, emotionally enjoyable interaction for both mother and infant. In addition to its nutritional advantages, breastfeeding is convenient, inexpensive, and has immunological and physiological benefits for both mother and child (**Table 14.7**).

Table 14.7 Advantages and Disadvantages of Breast- and Formula-Feeding

Advantage/ Disadvantage	Breastfeeding	Formula-Feeding
Nutrients	Ideal food for babies. Composition changes as they eat and grow.	Modelled after human milk, but certain components cannot be duplicated. Composition does not change with time. Must be prepared carefully to supply the correct nutrient mix and ratio of nutrients to fluid.
Amount	Underfeeding can be a problem in newborns if the mother is not well versed in breastfeeding and the signs of dehydration in the infant.	Overfeeding is a risk because of the desire of caregivers to have the baby empty the bottle.
Immunity	Immune factors are transferred from mother to infant.	There are no immune factors in formula.
Allergies	Allergies to breast milk are very rare and the risk of food allergies is reduced.	There are a variety of choices if the infant is allergic to one type of formula.

Table 14.7 *(continued)*

Risk from mother	Certain contaminants such as environmental pollutants, medications, illicit drugs, and disease-causing organisms such as HIV can pass from mother to baby.	None.
Environmental contamination	Breast milk is sterile, but pumped milk can become contaminated if stored improperly.	Bacterial contamination is a risk if formula is prepared under unsanitary conditions or stored improperly.
Ease for caregivers	No equipment to wash, always available, but may require more time from the mother.	Requires more preparation and washing, but other family members can share responsibility for feeding.
Ease for baby	Suckling is harder for the baby but aids in development of teeth and facial muscles needed for speech. Weak or sick infants can easily consume pumped breast milk.	Easier for baby, which is especially important for weak or sick infants.*
Benefit to mother	Promotes uterine contractions, which help the uterus return to prepregnancy size. May promote loss of weight and body fat. May reduce risk of breast cancer.	May allow more sleep.*
Cost	Cheaper, but the mother must be well nourished.	More expensive than nursing and cost includes formula as well as equipment and energy used in preparation.

*Breast milk fed from a bottle can be used to nourish weak or sick infants and can give the mother a break from breastfeeding.

During the first few months of life, the immune factors provided first by colostrum, and later by mature milk, compensate for the infant's immature immune system. These include antibody proteins and immune system cells that pass from the mother into her milk. Breast milk also contains a number of enzymes and other proteins that prevent the growth of harmful micro-organisms. Several carbohydrates have been identified that protect against disease-causing micro-organisms, including viruses that cause diarrhea. One substance favours the growth of the beneficial bacterium *Lactobacillus bifidus* in the infant's colon, which inhibits the growth of harmful bacteria. Breastfed babies have fewer allergies, ear infections, respiratory illnesses, and urinary tract infections than formula-fed babies, and have fewer problems with constipation and diarrhea. There is also evidence that breastfeeding protects against sudden infant death syndrome, diabetes, and chronic digestive diseases.[67]

In addition to providing disease protection, the strong suckling required by breastfeeding aids in the development of facial muscles, which help in speech development and the correct formation of teeth. Breastfed babies are also less likely to be overfed, because the amount of milk consumed cannot be monitored visually. In bottle-feeding, it is often tempting to encourage the baby to finish the entire bottle whether or not he or she is hungry.

For the mother, breastfeeding has the advantage of providing a readily available and inexpensive source of nourishment for her infant. It requires no preparation or bottles and nipples that must be washed. It is more ecological because it doesn't require energy for manufacture or generate waste from discarded packaging. Physiologically, breastfeeding causes contractions that help the mother's uterus return to normal size more quickly and may promote weight loss in some women. Women who breastfeed may have a lower risk of developing osteoporosis and breast and ovarian cancer. Lactation also inhibits ovulation, lengthening the time between pregnancies; however, it does not reliably prevent ovulation and so cannot be effectively used for birth control. Oral contraceptives can be used immediately postpartum, but those containing only progestin are preferable because they do not affect milk volume or composition. Oral contraceptives containing estrogen may decrease milk volume.

Figure 14.28 By the time an infant is a week old, mother and child have usually adjusted to breastfeeding. (© istockphoto.com/Goldmund Lukic)

How Much Is Enough? A strong, healthy baby will be able to suckle shortly after birth. Within a week, milk production and breastfeeding are usually fully established (**Figure 14.28**). During the early weeks of breastfeeding, the infant should be fed about 8-12 times every 24 hours (every 2-3 hours) or whenever the infant shows early signs of hunger. A feeding should last approximately 8-12 min. at each breast. A well-fed newborn should urinate enough to soak 6-8 diapers a day and gain about 100 to 250 g/week (0.3 to 0.5 pounds/week)

How Long Should Breastfeeding Continue? Physiologically, lactation can continue as long as suckling is maintained. Breastfeeding alone is sufficient to support optimal growth for about 6 months and Health Canada and many other health organizations recommend exclusive breastfeeding for the first 6 months of life.[68] However, currently only about 14% of infants are exclusively breastfed for 6 months (See Critical Thinking: The Prevalence of Exclusive Breastfeeding in Canada).[69] Breastfeeding along with supplemental feeding of solids is recommended for at least the first year of life and beyond for as long as mutually desired by mother and child. After 12 months, the baby no longer needs breast milk to meet nutrient needs. As the infant obtains more and more of its energy from solid foods, milk production decreases due to reduced demand by the infant. However, breastfeeding beyond 12 months continues to provide nutrition, comfort, and an emotional bond between mother and child. In developing nations, it may continue to give children the nutritional advantage needed to fight infection and stay healthy. The World Health Organization recommends that infants in developing nations be breastfed for 2 years or more.[70]

Practical Aspects of Breastfeeding Breastfeeding does not always come naturally to mother or infant and can require practice and patience. Effective suckling by the infant and relaxation of the mother are essential to successful breastfeeding. Some foods and other substances in the mother's diet, such as garlic or spicy foods, contain chemicals or flavours that pass into breast milk and cause adverse reactions in some babies. These reactions seem to be individual to the mother and child. As long as a food does not affect the infant's response to feeding, it can be included in the mother's diet. Caffeine in the mother's diet can make the infant jittery and excitable, so large amounts should be avoided while breastfeeding. Alcohol, which is harmful for infants, passes into breast milk. It is most concentrated an hour to an hour and a half after consumption and is cleared from the milk at about the same rate it disappears from the bloodstream. Therefore, occasional limited alcohol consumption while breastfeeding is probably not harmful if intake is timed to minimize the amount present in milk when the infant is fed.

Breastfeeding does not mean that a mother must be available for every feeding. Milk may be pumped from the breast and stored for later feedings (**Figure 14.29**). Since pumped milk is exposed to pumps and bottles, care must be taken to avoid contamination. If pumped milk is not immediately fed to the baby, it should be refrigerated. It can be kept refrigerated for 24-48 hours, but if it will not be used within that period, it should be frozen in clean containers. Warming breast milk in a microwave is not recommended because this destroys some of its immune properties and may result in dangerously hot portions of the milk. The best way to warm milk is by running warm water over the bottle.

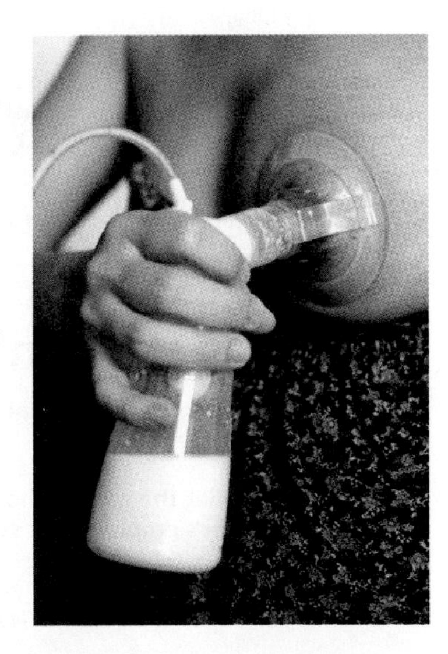

Figure 14.29 Breast pumps can be used to pump milk for bottle-feedings, relieving a mother of responsibility for all feedings. (© istockphoto.com/Joakim Leroy)

Meeting Nutrient Needs with Bottle-Feeding

A hundred years ago, a baby who could not be breastfed had little chance of survival. Today, infants who cannot breastfeed can still thrive. There are many commercially available infant formulas modelled after the nutrient content of breast milk.

When Is Bottle-Feeding Better There are some situations when breastfeeding is not the best choice. An infant who is small or weak may not have the strength to receive adequate nutrition from breastfeeding. In this case, formula, which provides almost the same nutrients as breast milk, can be used, or pumped breast milk can be offered to the infant in a bottle.

Bottle-feeding prevents the transmission of drugs and disease via breast milk. Women who are taking medications should check with their physician as to whether it is safe to breastfeed. If prescription drugs are taken by the mother for only a short time, a breast pump may be used to maintain milk production and the milk discarded, until the medication is no longer needed. Because alcohol and drugs such as cocaine and marijuana can be passed to the baby in breast milk, alcoholic and drug-addicted mothers are counselled not to breastfeed. Nicotine from cigarette smoke is also rapidly transferred from maternal blood to milk, and heavy smoking may decrease the supply of milk. Tuberculosis and certain viral infections can be transmitted to the infant in breast milk, but common illnesses such as colds, flu, and skin infections should not interfere with breastfeeding. Human immunodeficiency virus (HIV), the virus that causes AIDS, can be transmitted to the infant in breast milk. In Canada, women who are infected with HIV are advised not to breastfeed,[68] but in developing nations, the risk of malnutrition associated with not breastfeeding often outweighs the risk of passing this infection on to the infant.

CRITICAL THINKING:

The Prevalence of Exclusive Breastfeeding in Canada

Health Canada recommends that all babies be feed only breast milk during the first 6 months of life. Infant formula, and solid foods, should not be introduced until after the age of 6 months. In 2010, a study looked at the prevalence of exclusive 6-month breastfeeding among Canadian women (graph) and determined which maternal characteristics best predicted this practice.[1]

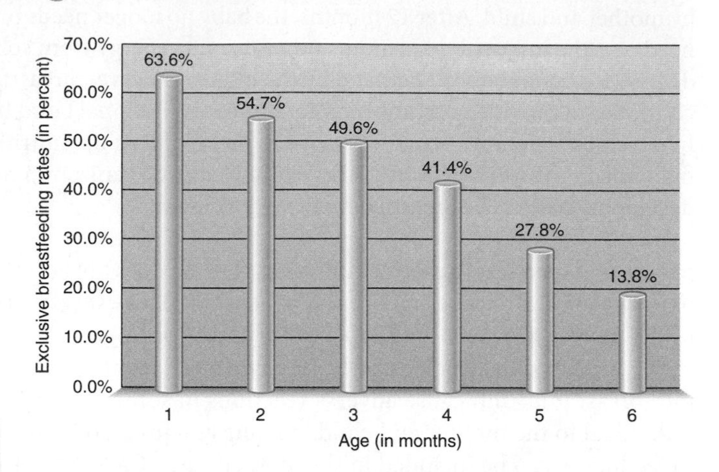

To assess the characteristics that predicted breastfeeding, researchers calculated odds ratios which are interpreted like Relative Risks (see Chapter 2 : Science Applied: Does Following a Healthy Dietary Pattern Reduce the Risk of Disease?). If the odds ratio was statistically significantly greater than 1, then the characteristic was directly associated with increased exclusive breastfeeding. For example, they considered marital status:

Marital status	Odds ratio of exclusive breastfeeding
No partner	1
Have a partner	1.61*
*statistically significant difference (compared to no partner)	

The statistically significant odds ratio of 1.61 indicates that women who have partners are more likely to breastfeed than women who do not. The odds ratios for several other maternal characteristics are shown in the following tables.

Smoking during pregnancy	Odds ratio of exclusive breastfeeding
No	2.11*
Yes	1
*statistically significant difference (compared to Smoking during pregnancy)	

Type of setting of baby's birth	Odds ratio of exclusive breastfeeding
Hospital	1
Private Home	5.29*
*statistically significant difference (compared to hospital birth)	

Mother's employment status	Odds ratio of exclusive breastfeeding
Working	1
Not working	1.55*
*statistically significant difference (compared to working)	

Critical Thinking Questions

Consider the graph that shows the results of the breastfeeding study. What does it indicate about exclusive breastfeeding in Canada?

Consider the results shown in the tables. What would you conclude about the impact of smoking, baby's birth place, and mother's employment status on exclusive breastfeeding? Suggest explanations for these results.

[1]Al-Sahab, B., Lanes, A., Feldman, M. et al. Prevalence and predictors of 6-month exclusive breastfeeding among Canadian women: a national survey. *BMC Pediatrics* 10:20, 2010. Available online at http://www.biomedcentral.com/1471-2431/10/20. Accessed August 21, 2011.

How Much Is Enough? As with breastfed infants, formula-fed infants should be fed on demand every few hours. Newborns have small stomachs, so at each feeding they may consume very small amounts. As the infant grows, the amount consumed at each feeding will increase to 125-250 ml. Caregivers should respond to cues from the infant that hunger is satisfied, even if a bottle of formula is not finished. Encouraging infants to finish every bottle can result in overfeeding and excess weight gain. As with breastfed infants, adequate intake can be judged from the amount of urine produced and the amount of weight gained.

Practical Aspects of Bottle-Feeding Infant formula must be prepared carefully in order to avoid mixing errors and contamination. If the proper measurements are not used in preparing formula, the child can receive an excess or deficiency of nutrients and an improper ratio of nutrients to fluids. If the water and all the equipment used in preparing formula are not clean or if the prepared formula is left unrefrigerated, food-borne illness may result. Because sanitation is often a problem in developing nations, infections that lead to diarrhea and dehydration occur more commonly in formula-fed than in breastfed infants. Commercially prepared formulas are sterile and powdered formulas contain no harmful micro-organisms. To avoid introducing harmful micro-organisms, the water used to mix powdered formula should be boiled for 1-2 min. and allowed to cool before mixing. Hands should be washed before preparing formula, and bottles and nipples should be washed in a dishwasher or placed in a pan of boiling water for 5 min. Formula should be prepared immediately before a feeding, and any excess should be discarded. Opened cans of ready-to-feed and liquid concentrate formula should be covered and refrigerated and used within the time indicated on the can. Formula may be fed either warm or cold, but the temperature should be consistent.

The position of the child is important during feeding. The infant's head should be higher than his or her stomach, and the bottle should be tilted so that there is no air in the nipple (**Figure 14.30**). If the hole in the nipple is too large, the infant may feel full before receiving adequate nutrition. If the hole is too small, the infant may tire before nutrient needs are met. Just as breastfed infants alternate breasts, bottle-fed infants should be held alternately between the left and right arms to promote equal development of the head and neck muscles.

Figure 14.30 Proper feeding position during bottle-feeding allows the formula to be swallowed easily and prevents air from being swallowed. (istockphoto.com/ Paul Kline)

Infants should never be put to bed with a bottle of formula. At night, while the child sleeps, the flow of saliva is decreased and the sugary formula is allowed to remain in contact with the teeth for many hours. This causes the rapid and serious decay of the upper teeth referred to as **nursing bottle syndrome**. Usually, the lower teeth are protected by the tongue and are unaffected (**Figure 14.31**).

nursing bottle syndrome Extreme tooth decay in the upper teeth resulting from putting a child to bed with a bottle containing milk or other sweet liquids.

Formula Choices Infant formulas can never duplicate the living cells, active hormones, enzymes, and immune system molecules in human milk, but formulas today try to replicate human milk as closely as possible in order to match the growth, nutrient absorption, and other parameters obtained with breastfeeding. Formula is available in ready-to-feed, liquid concentrate, or powdered forms (see Your Choice: Which Infant Formula Is Best?). Although most formulas are based on cow's milk, unmodified cow's milk should never be fed to infants. It is difficult for an infant to digest and its higher protein and mineral content taxes the kidneys and predisposes the infant to dehydration. Young infants may also become anemic if fed cow's milk because it contains little absorbable iron and can lead to iron loss by causing small amounts of gastrointestinal bleeding. Unmodified goat's milk is also not recommended for infants. Although it is less allergenic and more easily digested than cow's milk, it is low in vitamin D as well as iron, vitamin B_{12}, and folate, which can lead to an iron deficiency or megaloblastic anemia. Cow's milk and goat's milk can be introduced at about 9 months of age when organ systems are more mature and missing nutrients can be provided by other foods.

In addition to formulas for healthy infants, there are special formulas available for infants with allergies, premature infants, and those with genetic abnormalities that alter dietary needs. For infants who cannot tolerate human milk or cow's milk-based formula, soy protein formulas are available. And for those who are allergic to soy protein, formulas made from predigested proteins (elemental formulas), called protein hydrolysates, are an option.

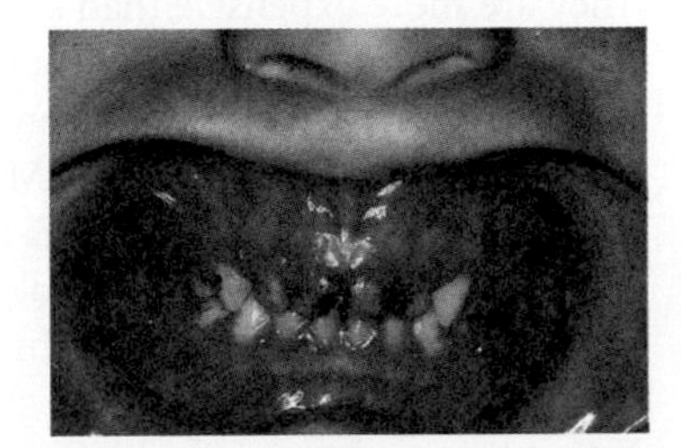

Figure 14.31 Nursing bottle syndrome causes rapid decay of the child's upper front teeth. (K. L. Boyd/ Custom Medical Stock Photo, Inc.)

Premature infants have special needs because they do not have fully developed organ systems or metabolic pathways. If they are too small or weak to nurse or take a bottle, pumped breast milk or formula can be fed through a tube. Some nutrients that are produced in the bodies of full-term infants are essential in the diets of premature infants. For example, preterm infants are less able to synthesize the amino acids tyrosine and cysteine and the fatty acid DHA. These and other substances, such as taurine and carnitine, are needed in greater amounts in the diets of premature babies. The energy, protein, and micronutrient requirements of preterm infants are also higher due to their rapid growth and development. Preterm infant formulas are available to meet the needs of premature babies.

YOUR CHOICE

(©iStockphoto)

Which Infant Formula Is Best?

CANADIAN CONTENT

While breast milk is the ideal food for human infants, there are situations when exclusive breast feeding is not possible and decisions about the use of formula have to be made. Selecting an infant formula can be confusing. Is one safer or more nutritious? Is cow's milk formula better than soy-based? Should you choose a formula supplemented with iron or fatty acids? Is premixed better than powdered?

The composition of infant formulas sold in Canada is described in the Canadian Food and Drug Regulations and these regulations are enforced by the Canadian Food Inspection Agency.[1] The food and drug regulations specify minimum amounts of carbohydrates, fat, protein, vitamins, and minerals and sets limits on the amounts of other additives or ingredients, such as sodium chloride.

The biggest difference between formulas is in the type of protein and sugar they contain. Most are based on cow's milk. The standard cow's-milk-based formulas contain heat-treated cow's-milk protein (at reduced concentrations) and the sugar lactose. Infants who cannot tolerate cow's-milk formulas can consume soy-based formulas or hydrolyzed-cow's-milk formulas, referred to as elemental formulas. Soy formulas are made with soy protein and contain sucrose or corn syrup instead of lactose. In elemental formulas, the proteins have been broken down to amino acids. They are more expensive than soy-based formulas but may be necessary for infants who are allergic to both milk and soy proteins.

Another difference between formulas is the amounts of iron and long-chain fatty acids. Most infant formulas are fortified with iron. The fats in formula come from vegetable oils and provide little of the long-chain polyunsaturated fatty acids DHA (docosahexanoic acid) and arachidonic acid, both of which are found in breast milk and are thought to help infant development. Formulas supplemented with these fatty acids are available, but they are more expensive and it is still not clear whether they improve growth or visual or neurological development.

Formulas are marketed in three basic forms: ready-to-feed, liquid concentrates, and powdered. Ready-to-feed formulas require no preparation and are packaged in many sizes.

Liquid concentrates and powdered formulas are prepared for use by mixing them with specific amounts of water. When properly prepared, all provide the needed nutrients in appropriate concentrations. Problems arise when formulas are mixed incorrectly or when the water used to prepare them is contaminated. Water for formula should come from a safe source and be boiled before use.

In short, ready-to-feed formulas are easiest to use but may cost more and are heavier and bulkier to carry home from the store. Liquid concentrates are a good compromise because they provide more formula for less weight and are easy to mix. Powders are the least expensive and the easiest to transport home, but they require more measuring and mixing. Since all of these products are nutritionally comparable, this choice depends on the needs of the caregivers.

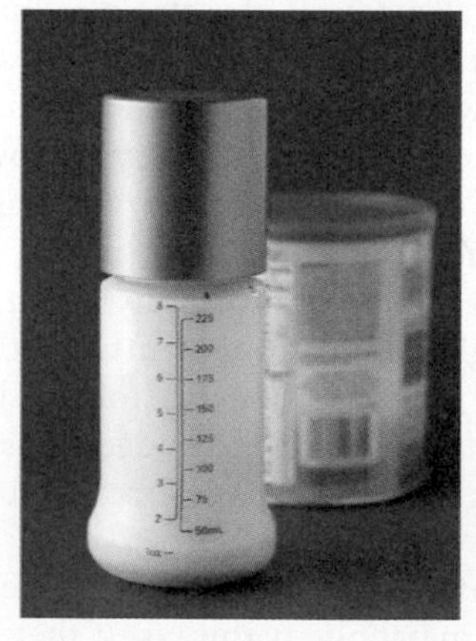

(© istockphoto.com/Edyta Anna Grzybowska)

[1] Department of Justice Canada. Food and Food Regulations. Available online at http://laws-lois.justice.gc.ca/en/F-27/C.R.C.-c.870. Accessed Dec 18, 2010.

Genetic abnormalities that prevent the normal metabolism of specific nutrients may alter dietary needs. For instance, infants with the genetic disease PKU lack an enzyme needed to metabolize the amino acid phenylalanine (see Section 6.5). If a child with PKU is fed breast milk or a formula that contains too much phenylalanine, the by-products of phenylalanine metabolism accumulate and cause brain damage. This can be prevented by feeding infants with PKU a special formula that provides only enough phenylalanine to meet the need for protein synthesis. Because this special diet must be started as soon as possible, infants born in Canada are tested for PKU at birth.

CANADIAN CONTENT

Safe Baby Bottles In 2008, Canada became the first country to ban infant bottles containing bisphenol A. This chemical can migrate from the plastics into foods, especially if the food is hot. As water and formula are typically sterilized by heating prior to use, and may be added to bottles while still hot, there was concern about the presence of bisphenol A in baby bottles. Bisphenol A's chemical structure is similar to estrogen and it may disrupt hormone function. Based on a review of scientific evidence, Health Canada concluded that exposures to bisphenol A from plastic materials is unlikely to harm human health. Despite this general conclusion, newborns and infants were identified as a vulnerable population and as a precaution, a higher margin of safety is applied to this group, resulting in the ban.[71]

CASE STUDY OUTCOME

After her doctor's visit, Jasmine realized that the health of her baby was more important to her than whether or not she was a bit overweight after the delivery. Her doctor helped her understand that she needed to eat a diet higher in energy, protein, and micronutrients than the one she consumed before she became pregnant. Jasmine took a prenatal supplement to help meet her micronutrient needs, but a pill can't provide the kcalories and protein she needs to build new tissues or fibre to prevent constipation. To construct her diet, Jasmine used Canada's Food Guide, taking note of the additional servings of food she would need in her second and third trimesters. She increased her intake of grains and vegetables, which added energy, fibre, protein, folate, and vitamin C to her diet, and extra-lean meat, which provided iron, protein, zinc, and B vitamins. After 9 months, Jasmine had gained 15 kg (32 lb) and gave birth to a healthy 3,400 g (7 lb, 8 oz.) baby boy. She breastfed him for a year. While she was lactating, her need for energy, fluid, and certain micronutrients were even higher than they were during pregnancy.

Although she felt like she was eating all the time, after 6 months she was back to her pre-pregnancy weight.

APPLICATIONS

Personal Nutrition

1. **If you were a 25-year-old pregnant woman, would your diet meet your needs?**

   a. Pick one day of the food record you kept in Chapter 2. Use iProfile to determine whether this diet meets the third trimester energy and protein recommendations of a 1.7 m (5′5″) tall, 25-year-old, sedentary pregnant woman who weighed 60 kg (130 lb) at the beginning of her pregnancy, If it doesn't, what foods would you add to the diet to meet the needs of pregnancy?

   b. Does this diet meet the iron and calcium needs of a 25-year-old pregnant woman? List 3 foods that are good sources of each.

   c. Does this diet meet the folate needs of a 25-year-old pregnant woman? What foods could you add to a diet that is low in folate to meet needs without supplements? What foods in this diet are fortified with folic acid?

2. **How do the energy and nutrient needs of nonpregnant, pregnant, and lactating women differ?**

   a. For each of the following, describe any differences between the needs of nonpregnant, pregnant, and lactating women of similar age and size.
   - Energy
   - Protein
   - Calcium
   - Iron
   - Folate

   b. For each of the above, explain why the requirements for pregnancy and lactation do or do not differ from those for the nonpregnant state.

General Nutrition

1. **Use the CPNP Project Directory of the Canadian Public Health Agency to find out about the Canadian Prenatal Nutrition Programme projects in your area: http://cpnp-pcnp.phac-aspc.gc.ca/index-eng.php.**

   a. Would it be easy for you to use this program if you were a pregnant or lactating woman or had a young child?

   b. What income levels does it serve?

2. **Marina is 16 years old and is pregnant with her first child. She is 1.6 m (5′4″) and weighs 50 kg (110 lb). She eats breakfast at home with her mother and two brothers and has lunch in the school cafeteria. After school she often has a snack with friends and then has dinner at home. A day's sample diet is listed here:**

SAMPLE DIET

FOOD	SERVING SIZE	FOOD	SERVING SIZE
Breakfast		*Snack*	
Pastry		Ice cream cone	
Fruit punch	125 ml (½ cup)	*Dinner*	
Lunch		Chicken	drumstick
Hamburger		Rice	250 ml (1 cup)
on bun		Refried beans	125 ml (½ cup)
Canned peaches	125 ml (½ cup)	Tortillas	3
Diet cola	30 ml (⅛ cup)	Fruit punch	30 ml(⅛ cup)

   a. What is the RDA or AI for folate, vitamin D, calcium, iron, and zinc for a 16-year-old pregnant woman?

   b. Does Marina's diet meet the recommendations for these nutrients?

   c. Use iProfile to determine how many kcalories her diet provides?

   d. Will Marina gain the recommended amount of weight if she consumes this diet throughout her pregnancy?

   e. What dietary changes would you suggest to meet the needs of the second trimester of pregnancy and to ensure a healthy pregnancy for both Marina and her baby?

SUMMARY

14.1 The Physiology of Pregnancy

- Pregnancy begins with the fertilization of an egg by a sperm. About 2 weeks after fertilization implantation is complete. The embryo grows, and the cells differentiate to form the organs and structures of the body. Growth and maturation continue in the fetal period, which begins at 9 weeks, and continues until birth, about 38 weeks after fertilization.
- During pregnancy, the placenta develops; maternal blood volume increases; the uterus and supporting muscles expand; body fat is deposited; the heart, lungs, and kidneys work harder; the breasts enlarge; and total body weight increases. Changes in the mother and growth of the fetus result in weight gain. Recommended weight gain during pregnancy is 11.5-16 kg (25–35 lb) for normal-weight women. Normal-weight, underweight, overweight, and obese mothers should all gain weight at a steady rate during pregnancy.
- During healthy pregnancies, a carefully planned program of moderate-intensity exercise can be beneficial and safe.
- Changes in blood volume and hormone levels and the enlargement of the uterus can result in edema, morning sickness, heartburn, constipation, and hemorrhoids during pregnancy.
- The hypertensive disorders of pregnancy include chronic hypertension, gestational hypertension, and pre-eclampsia, which can lead to life-threatening eclampsia. Gestational diabetes can result in a large-for-gestational-age baby.

14.2 The Nutritional Needs of Pregnancy

- During pregnancy, the requirements for energy, protein, water, vitamins, and minerals increase. The B vitamins are needed to support increased energy and protein metabolism; calcium, vitamin D, and vitamin C are needed for bone and connective tissue growth; protein, folate, vitamin B_{12}, and zinc are needed for cell replication; and iron is needed for red blood cell synthesis.
- Even with a nutrient-dense diet, supplements of folic acid and iron are recommended during pregnancy.

14.3 Factors that Increase the Risks of Pregnancy

- Nutritional status is important before, during, and after pregnancy. Poor nutrition before pregnancy can decrease fertility or lead to a poor pregnancy outcome. Malnutrition during pregnancy can affect fetal growth and development and the risk that the child will develop chronic disease later in life.
- Poor maternal health status, age that is under 20 or over 35 years, a short interval between pregnancies, a history of poor reproductive outcomes, and poverty all increase the risks of complications for the mother and baby.
- Because the embryo and fetus are developing and growing rapidly, they are susceptible to damage from physical, chemical, or other environmental teratogens. Mercury in food, food-borne pathogens, cigarette smoking, alcohol use, and certain prescription and illegal drugs can interfere with growth and development of the embryo and fetus.

14.4 Lactation

- Milk production and let-down are triggered by the hormones prolactin and oxytocin, respectively. They are released in response to the suckling of the infant.
- During lactation the need for protein, water, and many vitamins and minerals is even greater than during pregnancy.

14.5 The Nutritional Needs of Infancy

- Newborns grow more rapidly and require more energy and protein per kilogram of body weight than at any other time in life. Fat and water needs are also proportionately higher than in adults.
- A diet that meets energy, protein, and fat needs may not necessarily meet the needs for iron, fluoride, and vitamins D and K.
- Growth, which is assessed using growth charts, is the best indicator of adequate nutrition in an infant.

14.6 Feeding the Newborn

- Breast milk is the ideal food for new babies. It is designed specifically for the human newborn; is always available; requires no special equipment, mixing, or sterilization; and provides immunities.
- There are many infant formulas on the market that are patterned after human milk and provide adequate nutrition to the baby. Infant formulas are the best option when the mother is ill or is taking prescription or illicit drugs, or when the infant has special nutritional needs. The major disadvantages of formula-feeding are the potential for bacterial contamination, overfeeding, and the possibility of errors in mixing formula.

REVIEW QUESTIONS

1. **List 3 physiological changes that occur in the mother's body during pregnancy.**
2. **List 3 common digestive system discomforts that afflict pregnant women and explain why they occur.**
3. **Explain why the hypertensive disorders of pregnancy can be a risk to the mother and baby.**
4. **Why does gestational diabetes increase the risk of having a large-for-gestational-age baby?**
5. **How do energy and protein requirements change during pregnancy?**
6. **Why does the recommendation for iron intake increase during pregnancy?**
7. **How much weight should a woman gain during pregnancy?**
8. **How do the recommendations for weight gain differ for underweight, overweight, and obese women?**
9. **What kind of exercise is safe during pregnancy?**
10. **Why are folic acid supplements recommended even before pregnancy for women of child-bearing age?**
11. **Are vegetarian diets safe for pregnant women? Why or why not?**
12. **Why does malnutrition early in pregnancy have different effects than malnutrition during the last trimester?**
13. **How does maternal age affect nutrient requirements during pregnancy?**
14. **How does alcohol consumed by a woman during pregnancy affect the child?**
15. **Why is the need for energy and some nutrients greater during lactation than pregnancy?**
16. **What is the best indicator of adequate nutrition in an infant?**
17. **What are the advantages of breastfeeding?**
18. **When is bottle-feeding a better choice?**

REFERENCES

1. Committee to Reexamine IOM Pregnancy Weight Guidelines, Institute of Medicine, National Research Council. *Weight Gain During Pregnancy: Reexamining the Guidelines*. Washington, DC: National Academies Press, 2009.
2. Kaiser, L., and Allen, L. A. Position of the American Dietetic Association: Nutrition and lifestyle for a healthy pregnancy outcome. *J. Am. Diet. Assoc.* 108:553–561, 2008.
3. American College of Obstetricians and Gynecologists. ACOG Committee Opinion number 315, September 2005. Obesity in pregnancy. *Obstet. Gynecol.* 106(3):671–675, 2005.
4. Davies, G. A., Maxwell, C., McLeod, L. et al.; Society of Obstetricians and Gynaecologists of Canada. SOGC Clinical Practice Guidelines: Obesity in pregnancy. No. 239, *Int. J. Gynaecol. Obstet.* 110(2):167-73, 2010.
5. Lowell, H. and Miller, D. C. Weight gain during pregnancy: Adherence to Health Canada's guidelines. Health Reports 21(2):1-5, 2010. Available online at http://www.statcan.gc.ca/bsolc/olc-cel/olc-cel?catno=82-003-X&chropg=1&lang=eng. Accessed November 19, 2010.
6. Davies, G. L. A., Wolfe, L. A., Mottola, M. F. et al. Joint SOGC/CSEP Clinical Practice Guideline: Exercise in pregnancy and the postpartum period. *Can. J. Appl. Physiol.* 28(3):329-341, 2003.
7. Canadian Society for Exercise Physiology. *PARmed-X for Pregnancy*: Physical Activity Readiness Medical Examination. Available online at http://www.csep.ca/cmfiles/publications/parq/parmed-xpreg.pdf. Accessed June 22, 2011.

8. Wolfe, L.A. and Davies, G. A. L. Canadian guidelines for exercise in pregnancy. *Clinical Obstetrics and Gynecology* 46(2):488-495, 2003.
9 Quinlan, J. D., and Hill, D. A. Nausea and vomiting of pregnancy. *Am. Fam. Physician* 68:121–128, 2003.
10. National Center for Health Statistics, Centers for Disease Control and Prevention. *FastStats: Infant Health*. Available online at www.cdc.gov/nchs/fastats/infant_health.htm. Accessed July 3, 2008.
11. National Center for Health Statistics, Centers for Disease Control and Prevention. *Deaths: Final Data for 2004*. Available online at www.cdc.gov/nchs/products/pubs/_pubd/hestats/finaldeaths04/finaldeaths04.htm. Accessed July 30, 2008.
12. Chang, J., Elam-Evans, L. D., Berg, C. J. et al. Pregnancy-related mortality surveillance—United States, 1991–1999. *MMWR Morb. Mortal. Wkly. Rep.* 52:1–8, 2003.
13. Food and Nutrition Board, Institute of Medicine. Dietary Reference Intakes Water, Potassium, Sodium, Chloride, and Sulfate. Washington, DC: National Academies Press, 2004.
14. Kumar, A., Devi, S. G., Batra, S. et al. Calcium supplementation for the prevention of pre-eclampsia. *Int. J. Gynaecol. Obstet.* 104: 32–36, 2009.
15. Centers for Disease Control and Prevention. National Agenda for Public Health Action: A National Public Health Initiative on Diabetes and Women's Health. Available online at www.cdc.gov/diabetes/pubs/action/facts.htm/. Accessed April 11, 2009.
16. Dyck, R., Klomp, H., Tan, L. K. et al.A comparison of rates, risk factors, and outcomes of gestational diabetes between aboriginal and non-aboriginal women in the Saskatoon health district. *Diabetes Care.* 25(3):487-93, 2002.
17. Office of Minority Health and Health Disparities, Centers for Disease Control and Prevention. Eliminate Disparities in Diabetes. Available online at www.cdc.gov/omhd/AMH/factsheets/diabetes.htm. Accessed February 23, 2009.
18. Damm, P. Future risk of diabetes in mother and child after gestational diabetes mellitus. *Int. J. Gynaecol. Obstet.* 104(Suppl. 1):S25–S26, 2009.
19. Institute of Medicine, Food and Nutrition Board. *Dietary Reference Intakes for Energy, Carbohydrates, Fiber, Fat, Protein and Amino Acids.* Washington, DC: National Academies Press, 2002.
20. Craig, W. J. and Mangels, A. R. American Dietetic Association. Position of the American Dietetic Association: vegetarian diets. *J. Am. Diet. Assoc.* 109(7):1266-82, 2009.
21. Food and Nutrition Board, Institute of Medicine. Dietary Reference Intakes for Calcium, Phosphorus, Magnesium, Vitamin D, and Fluoride. Washington, DC: National Academies Press, 1997.
22. Prentice, A. Micronutrients and the bone mineral content of the mother, fetus and newborn. *J. Nutr.* 133(5 Suppl 2):1693S–1699S, 2003.
23. Kalkwarf, H. J., and Specker, B. L. Bone mineral changes during pregnancy and lactation. *Endocrine* 17:49–53, 2002.
24. Nesby-O'Dell, S., Scanlon, K. S., Cogswell, M. E. et. al. Hypovitaminosis D prevalence and determinants among African American and white women of reproductive age: Third National Health and Nutrition Examination Survey, 1988–1994. Am. J. Clin. Nutr. 76:187–192, 2002.
25. Schroth, R. J., Lavelle, C. L. and Moffatt, M. E. Review of vitamin D deficiency during pregnancy: who is affected? *Int. J. Circumpolar. Health.* 64(2):112-20, 2005.
26. Food and Nutrition Board, Institute of Medicine. Dietary Reference Intakes for Vitamin C, Vitamin E, Selenium, and Carotenoids. Washington, DC: National Academies Press, 2000.
27. Public Health Agency of Canada. Healthy Pregnancy. Folic acid. Available online at http://www.phac-aspc.gc.ca/hp-gs/know-savoir/folic-folique-eng.php. Accessed June 20, 2011.
28. Centers for Disease Control and Prevention. Spina Bifida and Anencephaly Before and After Folic Acid Mandate— United States, 1995–1996 and 1999–2000, MMWR Morb. Mortal. Wkly. Rep. 17: 362–365, 2004. Available online at www.cdc.gov/MMWR/preview/mmwrhtml/mm5317a3.htm. Accessed April 17, 2008.
29. De Wals, P., Tairou, F., Van Allen, M. I. et al. Reduction in neural-tube defects after folic acid fortification in Canada. *N. Engl. J. Med.* 357: 135–142, 2007.
30. Scholl, T. O., and Johnson, W. G. Folic acid: Influence on the outcome of pregnancy. *Am. J. Clin. Nutr.* 71(Suppl.):1295S–1303S, 2000.
31. Food and Nutrition Board, Institute of Medicine. Dietary Reference Intakes for Thiamin, Riboflavin, Niacin, Vitamin B-6, Folate, Vitamin B-12, Pantothenic Acid, Biotin, and Choline. Washington, DC: National Academies Press, 1998.
32. King, J. C. Determinants of maternal zinc status during pregnancy. *Am. J. Clin. Nutr.* 71(Suppl.):1334S–1343S, 2000.
33. Food and Nutrition Board, Institute of Medicine. Dietary Reference Intakes for Vitamin A, Vitamin K, Arsenic, Boron, Chromium, Copper, Iodine, Iron, Manganese, Molybdenum, Nickel, Silicon, Vanadium, and Zinc. Washington, DC: National Academies Press, 2001.
34. Scholl, T. O. Iron status during pregnancy: Setting the stage for mother and infant. *Am. J. Clin. Nutr.* 81:1218S–1222S, 2005.
35. Mills, M. E. Craving more than food: The implications of pica in pregnancy. *Nurs. Womens Health* 11:266–273, 2007.
36. Gluckman, P. D., Hanson, M. A., Cooper, C., and Thornburg, K. L. Effect of in utero and early-life conditions on adult health and disease. *N. Engl. J. Med.* 359:61–73, 2008.
37. Centers for Disease Control and Prevention. Pediatric and Pregnancy Nutrition Surveillance System, Health Indicators. Available online at www.cdc.gov/pednss/what_is/pnss_health_indicators.htm#Maternal%20Health %20Indicators/. Accessed January 12, 2006.
38. Public Health Agency of Canada. Canadian Perinatal Health Report. Available online at http://www.phac-aspc.gc.ca/rhs-ssg/phic-ispc/index-eng.php. Accessed December 12, 2010.
39. Casanueva, E., Roselló-Soberón, M. E., and De-Regil, L. M. Adolescents with adequate birth weight newborns diminish energy expenditure and cease growth. J Nutr 136:2498–2501, 2006.
40. Health Canada Consumption Advice: Making informed choices about fish. Available online at http://www.hc-sc.gc.ca/fn-an/securit/chem-chim/environ/mercur/cons-adv-etud-eng.php. Accessed March 19, 2011.
41. Weng, X., Odouli, R., and Li, D. K. Maternal caffeine consumption during pregnancy and the risk of miscarriage: A prospective cohort study. *Am. J. Obstet. Gynecol.* 198:279e1–279e8. Epub 2008.
42. Dietitians of Canada. I am concerned about eating too much sugar; are artificial sweeteners safe? Available online at http://www.dietitians.ca/Nutrition-Resources-A-Z/Fact-Sheet-Pages(HTML)/Food-Safety/Artificial-Sweeteners.aspx. Accessed June 21, 2011.
43. Delgado, A. R. Listeriosis in pregnancy. *J.Midwifery Womens Health* 53(3):255–259, 2008.
44. Center for the Evaluation of Risks to Human Reproduction (CERHR). Toxoplasmosis. Available online at http://cerhr.niehs.nih.gov/common/toxoplasmosis.html. Accessed July 16, 2008.
45. Goodlett, C. R., and Horn, K. H., Mechanisms of alcohol-induced damage to the developing nervous system. *Alcohol. Res. Health*

25:175–184, 2001. Available online at http://pubs.niaaa.nih.gov/publications/arh25-3/ 175–184.pdf/. Accessed April 20, 2009.

46. Public Health Agency of Canada. Healthy Pregnancy. Alcohol and Pregnancy. Available online at http://www.phac-aspc.gc.ca/hp-gs/know-savoir/alc-eng.php. Accessed June 20, 2011.

47. Rogers, J.M. Tobacco and pregnancy: Overview of exposures and effects. *Birth Defects Res. C. Embryo Today* 84:1–15, 2008.

48. Xiao, D., Huang, X.,Yang, S., et al. Direct effects of nicotine on contractility of the uterine artery in pregnancy. *J. Pharmacol. Exp. Ther.* 322(1): 180–185, 2007.

49. National Center for Chronic Disease Prevention and Health Promotion. Women and Smoking: A Report of the Surgeon General, March 2001. Available online at www.surgeongeneral.gov/library/womenandtobacco/. Accessed April 20, 2009.

50. Public Health Agency of Canada. Healthy Pregnancy: Smoking and Pregnancy. Available online at http://www.phac-aspc.gc.ca/hp-gs/know-savoir/smoke-fumer-eng.php. Accessed June 20, 2011.

51. Fajemirokun-Odudeyi, O., and Lindow, S. W. Obstetric implications of cocaine use in pregnancy: A literature review. *Eur. J. Obstet. Gyncol. Reprod. Biol.* 112:2–8, 2004.

52. Schiller, C., and Allen, P. J. Follow-up of infants prenatally exposed to cocaine. *Pediatr. Nurs.* 31:427–436, 2005.

53. McCourt, C., Paquette, D., Pelletier, L. et al. Make every mother and child count: Report on maternal and child health in Canada. Available online at http://www.phac-aspc.gc.ca/rhs-ssg/pdf/whd_05epi_e.pdf. Accessed Dec 18, 2010.

54. Public Health Agency of Canada. Canada Prenatal Nutrition Program: A decade of promoting the health of mothers, babies, and communities, 2007. Available online at http://www.phac-aspc.gc.ca/hp-ps/dca-dea/publications/pdf/mb_e.pdf. Accessed Dec 18, 2010

55. Public Health Agency of Canada. Canadian Prenatal Program. Projects Directory Online. Available online at http://cpnp-pcnp.phac-aspc.gc.ca/index-eng.php. Accessed Dec 18, 2010.

56. Public Health Agency of Canada. Buns in the Oven. Available online at http://www.phac-aspc.gc.ca/hp-ps/dca-dea/prog-ini/cpnp-pcnp/buns-oven_ca-mijote/index-eng.php. Accessed Dec 18, 2010.

57. Public Health Agency of Canada. Healthy Moms, Healthy Babies. Available online at http://www.phac-aspc.gc.ca/hp-ps/dca-dea/prog-ini/cpnp-pcnp/yukon-whitehorse/index-eng.php. Accessed Dec 18, 2010.

58. Haiek, L. N., Kramer, M. S., Ciampi, A., et al. Postpartum weight loss and infant feeding. *J. Am. Board Fam. Pract.* 14:85–94, 2001.

59. Kovacs, C. S. Calcium and bone metabolism during pregnancy and lactation. *J. Mammary Gland Biol.* Neoplasia 10:105–118, 2005.

60. Canadian Food Inspection Agency: Requirements related to nutrition information and health claims for infant formula. Available online at http://www.inspection.gc.ca/english/fssa/labeti/inform/20070112e.shtml. Accessed December 18, 2010.

61. Wright, K., Coverston, C.,Tiedeman, M., et al. A. Formula supplemented with docosahexaenoic acid (DHA) and arachidonic acid (ARA): A critical review of the research. *J. Spec. Pediatr. Nurs.* 11:100–112, 2006.

62. First Nations, Inuit and Métis Health Committee, Canadian Paediatric Society (CPS). Vitamin D supplementation: Recommendations for Canadian mothers and infants. *Paediatr. Child. Health.*;12(7):583-589, 2007.

63. American Academy of Pediatrics, Committee on Fetus and Newborn. Controversies concerning vitamin K and the newborn. *Pediatrics* 112:191–192, 2003.

64. Dietitians of Canada. WHO Growth Charts adapted for Canada. Available online at http://www.dietitians.ca/Secondary-Pages/Public/Who-Growth-Charts.aspx. Accessed December 18, 2010.

65. Critch, J. N. Canadian Paediatric Society, Nutrition and Gastroenterology Committee. Infantile colic: Is there a role for dietary interventions? *Paediatr. Child Health.*16(1):47-49, 2011.

66. American Academy of Pediatrics, Section on Breast-Feeding. Breast-feeding and the use of human milk. *Pediatrics* 115:496–506, 2005.

67. American Dietetic Association. Position of the American Dietetic Association: Promoting and supporting breastfeeding. *J. Am. Diet. Assoc.* 105:810–818, 2005.

68. Health Canada. Breastfeeding. Available online at http://www.hc-sc.gc.ca/fn-an/pubs/infant-nourrisson/nut_infant_nourrisson_term_3-eng.php. Accessed December 18, 2010.

69. Al-Sahab, B., Lanes, A., Feldman, M. et al. Prevalence and predictors of 6-month exclusive breastfeeding among Canadian women: a national survey. BMC Pediatrics 10:20, 2010. Available online at http://www.biomedcentral.com/1471-2431/10/20. Accessed August 21, 2011.

70. WHO Fact Sheet No. 178: Reducing mortality from major childhood killer diseases. September 1997. Available online at www.who.int/inf-fs/en/fact178.html/. Accessed March 15, 2009.

71. Health Canada. Bisphenol A Fact sheet. Available online at http://www.chemicalsubstanceschimiques.gc.ca/fact-fait/bisphenol-a-eng.php. Accessed June 21, 2011.

15 Nutrition from Infancy to Adolescence

Case Study

Thirteen-year-old Felicia has been overweight since she was 3. When she entered the eighth grade, she was 1.5 m (5 feet) tall and weighed 65 kg (142 lb) At a recent checkup, her physician noted that her blood pressure was elevated and her cholesterol levels were at the high end of normal. The physician recommended that Felicia and her mother meet with a dietitian to discuss ways to manage Felicia's weight and improve her diet.

At that meeting, Felicia explained that she knows she should eat less and exercise more but can't find a way to make those changes. She spends most of her school day sitting at a desk. Physical education class meets only twice a week. She buys her lunch at the school cafeteria, which most days is pizza and french fries. It is difficult for her to exercise after school because she goes to the library study program until her mother gets out of work at 5 p.m. Felicia gets a little more exercise on weekends by going out with friends, but they often go out for ice cream or fast food and Felicia feels left out if she doesn't join them.

The dietitian helps Felicia figure out some changes that will help reduce her kcalorie intake. She also works with Felicia's mother to plan low-kcalorie, nutritious dinners that the entire family will enjoy. Finally, Felicia and her mother decide to get some regular exercise by walking after dinner. These evening walks soon progress to activities that include the whole family such as skating and bike riding. Felicia is on the right track, as good nutrition and exercise patterns learned early in life are key to maintaining long-term health later on.

In this chapter you will learn about the nutritional needs of children and adolescents and the problem of childhood obesity.

(Rubberball/Erik Isakson/Getty Images)

(©iStockphoto)

Chapter Outline

15.1 Starting Right for a Healthy Life

Learning Objectives

- Discuss the quality of the diet of Canadian children and youth.
- Describe the trends in obesity and chronic disease among adolescents.

Figure 15.1 The foods and nutrients this child eats influence the eating habits and health of the adult she will become. (MIXA/Age Fotostock America, Inc.)

Nutrient intake during childhood helps shape the adult that a child will become. A nutritious, well-balanced eating pattern and active lifestyle allow children to grow to their potential and can prevent or delay the onset of the chronic diseases that plague Canadian adults. Therefore, teaching healthy eating and exercise habits will benefit not only today's children, but tomorrow's adults (**Figure 15.1**).

What Are Canadian Children Eating?

The Canadian Community Health Survey (CCHS) found that the diets of Canadian children and youth are not as healthy as they could be. The Canadian Healthy Eating Index, described in Section 2.6, generates a score for the overall quality of a diet based on its conformance to Canada's Food Guide. The index has a maximum score of 100 points. The more closely an individual's diet comes to meeting the recommendations of Canada's Food Guide, the higher the score. Points are lost if an inadequate number of servings is consumed in each of the 4 food groups and if high amounts of saturated fat, sodium or "foods to limit" are consumed. The Canadian Healthy Eating Index scores for the diets of Canadian children and youth show that younger children have a higher average score that declines in adolescence (**Figure 15.2a**). About one-quarter of adolescents score less than 50 (**Figure 15.2b**).[1]

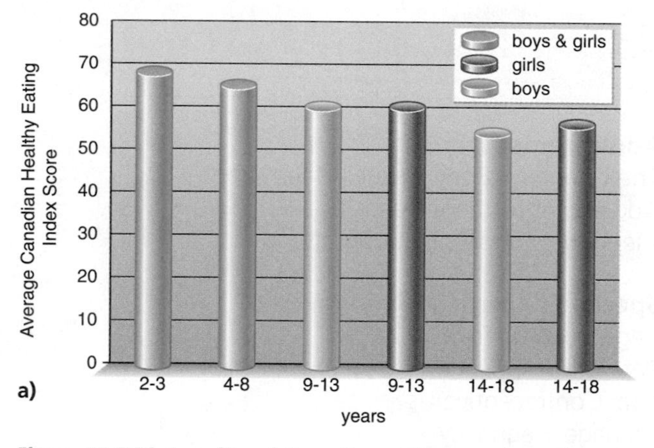

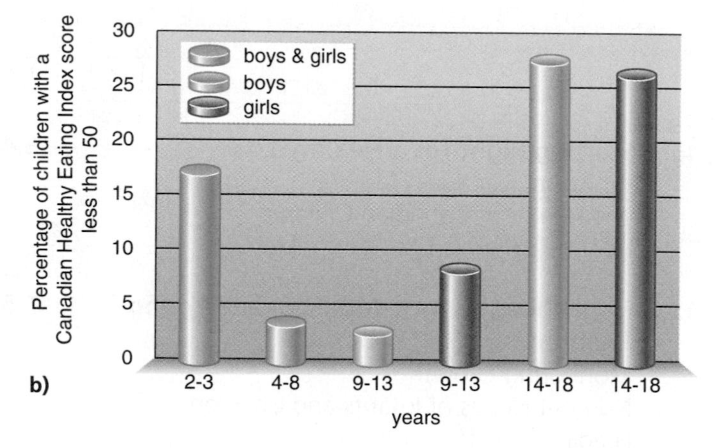

Figure 15.2 Diet quality of Canadian children and youth.
(a) The average Canadian Healthy Eating Index scores for children and youth are between 50 and 70, and decline with age. (b) The proportion of children with CHEI score less than 50 is lowest between ages 4 and 13 years and at its highest levels between ages 14 to 18 years.
Source: Garriguet, D. Diet quality in Canada. Health Reports 20(3):1-11, 2009. Available online at http://www.statcan.gc.ca/pub/82-003-x/2009003/article/10914-eng.htm. Accessed December 19, 2010.

These relatively low scores are due to low intakes of fruits and vegetables, milk and alternatives, and grain products.[2] Not surprisingly, the CCHS found that children's intakes from food alone of potassium, found in abundant amounts in fruits and vegetables, and of fibre, found in both fruits and vegetables and grain products, are of concern.[3] In adolescents, these same nutrients are also of concern, as is that of calcium. In addition, more than 10% of adolescents are not meeting their requirements for vitamin A and magnesium, and among girls, intakes of folate, iron, zinc, and vitamins B_6 and B_{12} are also low.[4]

The saturated fat intake of both children and adolescents could also be reduced, sodium intake is very high, and 20% of children and 30% of adolescents consume calories in excess of their energy needs.[3,4] The CCHS indicates that on any given day, about 20% of children (ages 4–13) and almost 40% of adolescents (ages 14–18) consume some food purchased in a fast food outlet.[2]

Today's children and youth are also more sedentary than in the past. In 2007, a survey called the Canadian Health Measures Surveys (CHMS) found that compared to children in 1981, fitness levels have declined significantly.[5] It should not be surprising that one of the consequences of excess caloric intake and reduced physical activity is an increase in childhood obesity. Compared to children in 1978, the proportion of obese or overweight children and youth aged 6–17 has doubled (**Figure 15.3**).

The Health of Canadian Children

The high-kcalorie, high-saturated-fat diet and low-activity lifestyle that contributes to obesity and chronic disease in adults is having the same effect in children and adolescents. As with adult obesity, childhood obesity increases the risks of chronic disease. Inactive, overweight children may have high blood cholesterol and glucose levels and elevated blood pressure. These factors increase the risk of developing heart disease, diabetes, and hypertension.[6,7]

In the United States, where the trends in childhood obesity are similar to those in Canada, type 2 diabetes, a disease that was, until recently, seen only in adults, is now being reported in children ages 10–19 years, with the incidence higher among minorities.[8,9] Recent Canadian studies suggest similarly disturbing trends. In a study of Ontario youth, about 6% were found to have metabolic syndrome, that is, abdominal obesity and at least 2 of the following: high serum triglycerides, low HDL cholesterol, high blood pressure, and high fasting blood glucose.[10] A second study, in 2008, of ninth graders (ages 14–15 years) found that 21% had at least one high level risk factor for cardiovascular disease, such as high blood pressure, high serum cholesterol, or obesity, a proportion that had increased significantly from 17% in 2002.[11] The Canadian Health Measures Survey found that among Canadian children and adolescents, 2.9% had either borderline or elevated blood pressure measurements.[12]

Obese children and adolescents also face social and psychological challenges. For example, childhood obesity may trigger low self-esteem,[13] and obese children and youth are often the victims of **weight stigmatization**. They are less well accepted by their peers than normal-weight children and are frequently ridiculed and teased. They may be discriminated against by adults as well as by their peers. This contributes to feelings of rejection and social isolation.

weight stigmatization Disapproval and lack of acceptance suffered by someone because he or she is obese.

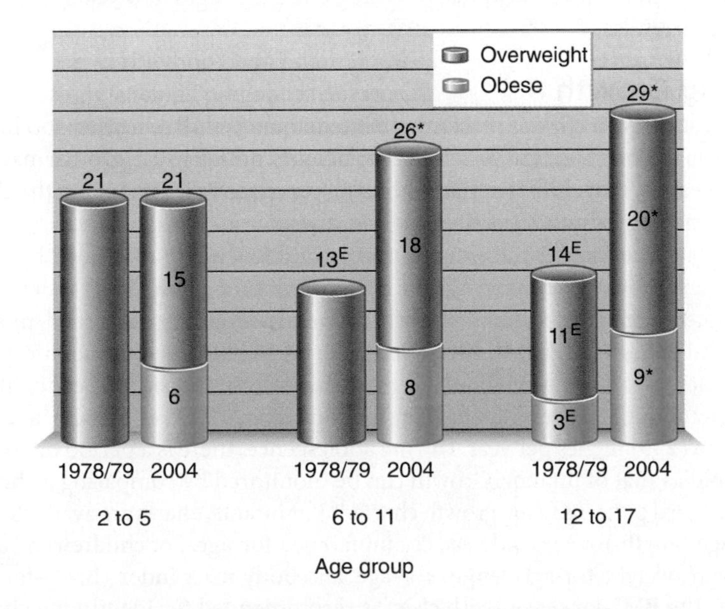

Figure 15.3 Percentage of Canadian children and adolescents who are overweight or obese
*statistically significant difference from 1978/79;

[E]co-efficient of variation 16.6% to 33.3% (interpret with caution).
Source: Shields, M. Overweight and obesity among children and youth. *Health Reports* 17(3): 27-42, 2006. Available online at http://www.statcan.gc.ca/cgi-bin/af-fdr.cgi?l=eng&loc=http://www.statcan.gc.ca/studies-etudes/82-003/archive/2006/9277-eng.pdf&t=Overweight and obesity among children and youth. Accessed December 19, 2010.

Figure 15.4 The social isolation experienced by many overweight teens contributes to inactivity and overeating, further exacerbating their weight problems. (Robert E. Daemmrich/Stone/Getty Images)

The isolation of obese adolescents from teen society, in turn, results in boredom, depression, inactivity, and withdrawal—all of which can cause an increase in eating and a decrease in energy output, worsening the problem (**Figure 15.4**).

Healthy Eating Is Learned for Life

Much of what we choose to eat as adults depends on what we learned to eat as children. Children learn by example, so the eating patterns, attitudes, and feeding styles of their caregivers influence what they learn to eat. When caregivers drink milk, choose whole grains, and eat plenty of fruits and vegetables, children follow their example. Likewise, if a child's role models eat a diet high in fat and low in fruits and vegetables, the child will follow suit. The eating and exercise habits developed during childhood and adolescence are important because they establish a pattern that may last a lifetime and affect how healthy people will be as they get older.

15.2 Nourishing Infants, Toddlers, and Young Children

Learning Objectives

- Explain why growth is the best indicator of nutrient intake.
- Compare the energy and protein requirements of infants and children with those of adults.
- Give examples of how environment influences children's nutrient intake.
- Summarize the recommendations for preventing and managing food allergies.

Nourishing a growing child is not always an easy task. The foods offered must supply enough energy and nutrients to meet the needs of maintenance, activity, and growth, as well as suit children's tastes. This can be a challenge to caregivers whether they are feeding infants (ages 4 months to 1 year) sampling solid foods for the first time, toddlers (ages 1–3 years) experimenting with new foods, or young children (ages 4–8 years) eating meals at school or with friends.

Monitoring Growth

The best indicator that a child is receiving adequate nourishment, neither too little nor too much, is a normal growth pattern. If a child does not get enough to eat, growth may be slowed. If intake is excessive, that child is at risk for becoming obese and developing the chronic diseases that are increasingly common in Canadian adults.

The ultimate size (height and weight) that a child will attain is affected by genetic, environmental, and lifestyle factors. A child whose parents are 1.5 m (5 feet) tall may not have the genetic potential to grow to 1.8 m (6 feet), but when adequately nourished, most children follow standard patterns of growth. Growth is most rapid in the first year of life, when an infant's length increases by 50%, or about 25 cm (10 inches). In the second year of life, children generally grow about 12.5 cm (5 inches); in the third year, 10 cm (4 inches); and thereafter, about 5-7.5 cm (2–3) inches per year. During adolescence, there is a period of growth that is almost as rapid as that of infancy. Growth can be monitored by comparing a child's growth pattern to standard patterns using growth charts.[14] For infants, charts are available to monitor weight-for-age, length-for age, and head circumference-for-age. For children and adolescents ages 2–19 years, weight-for-age, length-for-age, and body mass index (BMI)-for-age charts are available. The BMI-for-age growth chart is recommended for identifying children who are underweight, overweight, or obese (**Figure 15.5**).

Although growth often occurs in spurts and plateaus, overall growth patterns are predictable. If a child's overall pattern of growth changes, his or her dietary intake should be evaluated to determine the reason for the sudden change. Children who fall below the third percentile of the BMI-for-age distribution are considered underweight and their intake should be evaluated to be sure they are meeting their needs. Malnutrition during childhood can cause lasting damage for which adequate nutrition later on may not be able to compensate. Children are considered overweight when their BMI is greater than or equal to the 85th percentile and less than the 97th percentile and are considered obese when their BMI is greater than or equal to the 97th percentile (see Figure 15.5). An increase in activity and/or decrease in energy intake may be needed to keep a child's weight in the healthy range.

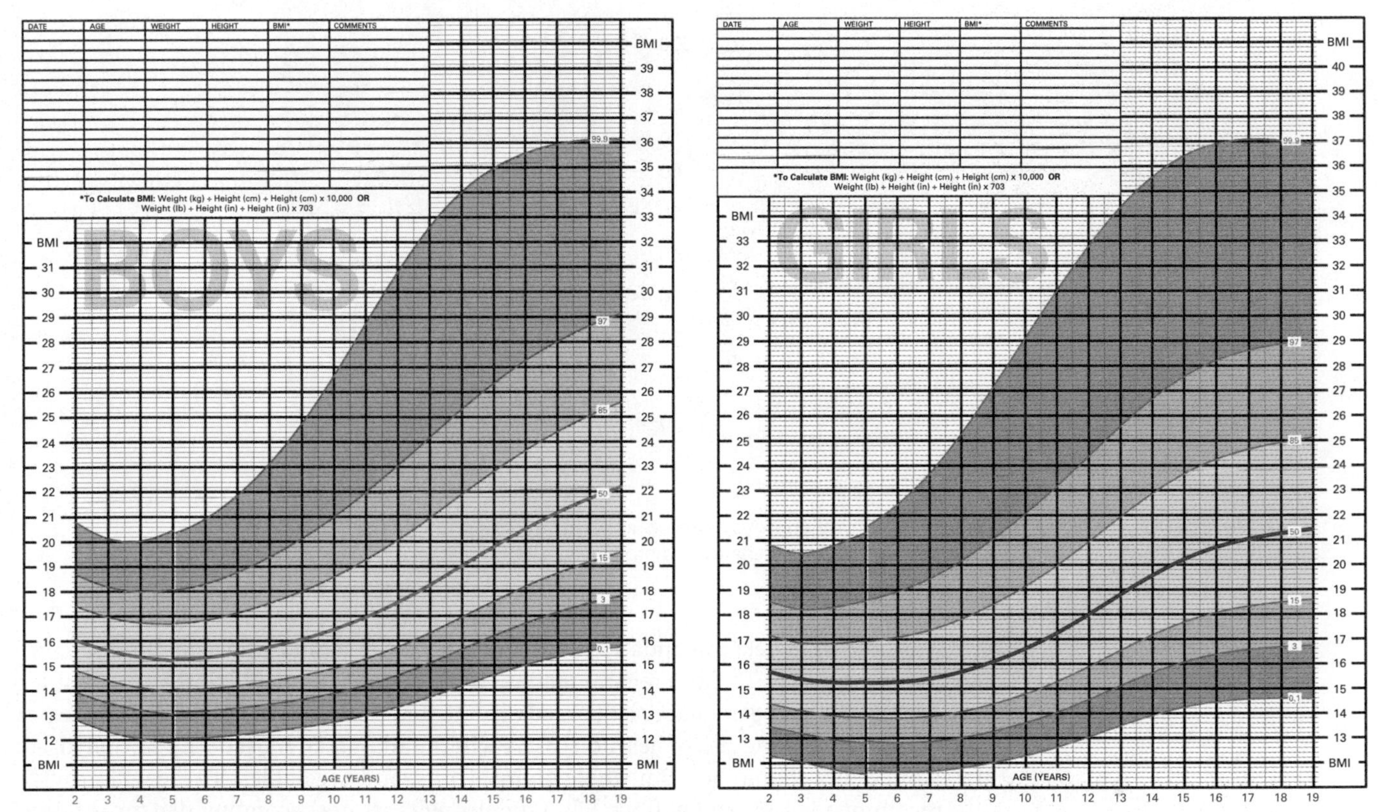

Figure 15.5. Body mass index-for-age percentiles for boys and girls ages 2 to 19 years.
Source: Based on the World Health Organization (WHO) Child Growth Standards (2006) and WHO Reference (2007) adapted for Canada by Dietitians of Canada, Canadian Paediatric Society, the College of Family Physicians of Canada and Community Health Nurses of Canada.

Nutrient Needs of Infants and Children

As children grow, their nutrient requirements per unit of body weight decrease, but total needs increase because they gain weight and become more active. Recommended intakes are not different for boys and girls until about 9 years of age, at which time sexual maturation causes differences in nutrient requirements between the sexes.

Energy and Energy-Yielding Nutrients The amount of energy and protein needed per kilogram of body weight decreases with age, but the total amount of each increases as body size increases. The average 2-year-old needs about 1,000 kcal and 13 g of protein per day. By age 6, that child will need about 1,600 kcal and 19 g of protein per day (**Figure 15.6**).[15]

Infants need a high-fat diet (40%–55% of energy intake) to support their rapid growth and development, but by age 4, the recommended proportion of kcalories from fat is reduced to provide adequate energy without increasing the risk of developing chronic disease (see Label Literacy: Reading Labels on Food for Young Children). The acceptable range for fat intake is 30%–40% of energy for children ages 1-3 years and 25%–35% of energy for those 4-18 years of age compared to 20%–35% for adults.[15] To promote health the type of fat is also important. The diets of children must provide adequate amounts of essential fatty acids; specific AIs have been set for linoleic and alpha-linolenic acid.[15] It should also be low in cholesterol, saturated fat, and *trans* fat.[15]

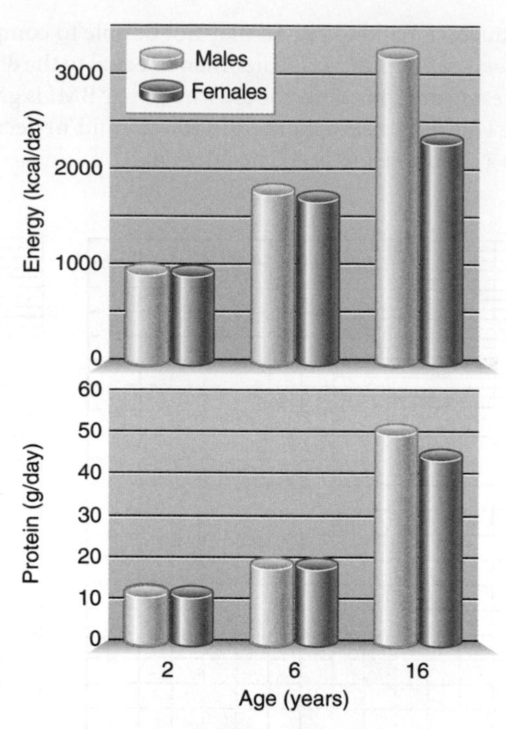

Figure 15.6 Total energy and protein needs by age and gender.
The total need for both energy and protein increases with age.

Carbohydrate recommendations for children are the same as those for adults: 45%–65% of energy. Specific fibre recommendations have not been made for infants, but for children an AI has been established based on data that show an intake of 14 g of fibre per 1,000 kcal reduces the risk of heart disease. As in the adult diet, most of the carbohydrate in a child's diet should be from whole grains, fruits, and vegetables. These will provide the recommended amount of fibre. Fibre supplements are not recommended for children since high intakes can limit the amount of food and, consequently, the nutrients that a small child can consume. Foods high in added sugars, such as cookies, candy, and pop, should be limited.[16]

Water and Electrolytes By 1 year of age, a child's kidneys have matured and the amount of fluid lost through evaporation has decreased, so total fluid losses decline. As with adults, under most situations drinking enough to satisfy thirst will provide sufficient water. In children 1–3 years of age about 1.3 L (5½ cups) of water daily will meet needs; about 1 L (4 cups) of this should be from water and other fluids. Older children, ages 4–8, need about 1.7 L (7 cups)/day.[17] Water needs increase with illness, when the environmental temperature is high, and when activity increases sweat losses.

A UL of 2.3 g of sodium per day has been set for adults and teens 14–18 years of age because a high sodium intake is associated with elevated blood pressure. The UL is somewhat lower in children and younger teens (see inside cover). The typical sodium intake in children and teens currently exceeds the UL.[17]

Micronutrients Children are smaller than adolescents and adults, and for the most part the recommended amounts of micronutrients are also smaller (see inside cover). Generally, a nutrient-dense diet that follows the Canada's Food Guide recommendations for children will meet needs. Consuming the recommended amounts of meats and whole and enriched grains helps ensure enough B vitamins. Adequate fruits and vegetables provide vitamin C and vitamin A. Milk provides calcium and vitamins A and D. Fortified breakfast cereals help compensate for poorer diets by providing the large amounts of a variety of vitamins and minerals in a single serving. However, despite the relative abundance of vitamins and minerals in the modern diet, poor food choices can put many children today at risk for deficiencies.

LABEL LITERACY

CANADIAN CONTENT

Reading Labels on Food for Young Children

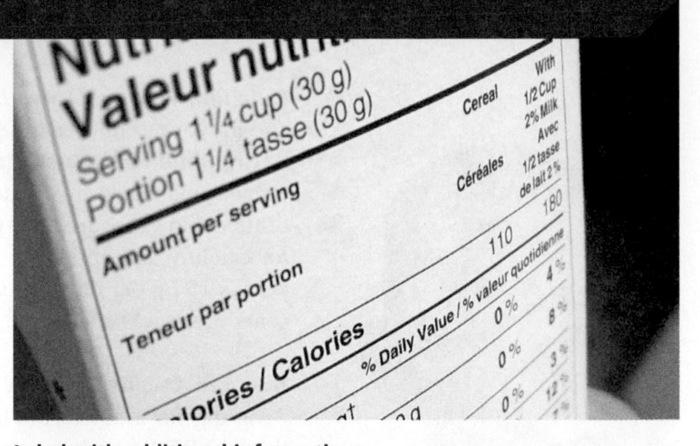

Children have different nutrient needs than adults. Therefore, the labels on foods designed for young children must follow different rules. Included here are some examples of Nutrition Facts labels for foods intended solely for children under age 2. The most obvious difference is how fat is listed in the Nutrition Facts. Labels for foods intended for children younger than 2 years list total fat, but may or may not list the amount of saturated fat, polyunsaturated fat, monounsaturated fat, cholesterol, kcalories from fat, or kcalories from saturated fat.[1] These labels also may not carry most claims about a food's nutrient content or health effects. The reason is that dietary fat is needed for brain development and meeting energy needs during the rapid growth and development of infancy and early childhood. It is hoped that excluding information about fat content on the label will prevent caregivers from restricting fats in the diets of young children. Canada's Food Guide also advises parents of young children not to restrict nutritious foods because of the amount of fat. This concern appears to be well-founded. The CCHS found that almost half of Canadian 1- to 3-year-olds obtain less than the recommended amount of fat (30%–40% kcal) in their diets. This puts children at risk for reduced intake of fat-soluble vitamins and essential fatty acids.[2]

Another difference between standard food labels and those for foods designed for children under age 2 is the absence of percent Daily Values for total fat, saturated fat, cholesterol, total carbohydrate, fibre, and sodium.[1] For children under 2 years, only Daily Values for vitamins and minerals are shown. Nutrient content claims allowed on young children's foods are limited to claims about protein, sodium, and sugar content and claims that describe the percentage of vitamins or minerals as they apply to the Daily Values for children under age 2, such as "an excellent source of vitamin C" for foods that provide 50% of the DV of vitamin C.[1] Labelling of infant formula is governed by another set of regulations that do not require a nutrition facts table. Instead, the label of infant formula must list, per 100 g or 100 ml, the number of calories and the amount (grams or milligrams) of protein, fat, digestible carbohydrate, fibre, vitamins, and minerals. Food processors may also indicate if DHA or arachidonic acid have been added to the formula.[3]

Many of the foods consumed by young children do not have special labels because they are also adult foods. When selecting these foods, keep in mind that the needs of young children, especially for fat, are different than the needs of adults.

Standard Label:

Nutrition Facts
Per 1 Jar (128 mL)

	Amount
Calories	150
Fat	0 g
Sodium	10 mg
Carbohydrate	27 g
Fibre	4 g
Sugars	18 g
Protein	0 g

% Daily value			
Vitamin A	6 %	Vitamin C	45 %
Calcium	2 %	Iron	2 %

Valeur nutritive
Pour 1 pot (128 mL)

	Teneur
Calories	150
Lipides	0 g
Sodium	10 mg
Glucides	27 g
Fibres	4 g
Sucres	18 g
Protéines	0 g

% Valeur quotidienne			
Vitamine A	6 %	Vitamine C	45 %
Calcium	2 %	Fer	2 %

Label with additional information:

Nutrition Facts
Serving Size 125 mL (26 g)
Servings per Container 8

	Amount Per Serving
Calories 100	(400 KJ)
Total Fat	1 g
Saturated	0 g
Trans	0 g
Omega-6 Polyunsaturated	0.5 g
Omega-3 Polyunsaturated	0 g
Monounsaturated	0.2 g
Cholesterol	0 mg
Sodium	15 mg
Potassium	80 mg
Total Carbohydrate	20 g
Dietary Fibre	2 g
Soluble Fibre	1 g
Insoluble Fibre	1 g
Sugars	7 g
Sugar Alcohols	0 g
Starch	16 g
Protein	3 g

% Daily value			
Vitamin A	0 %	Vitamin C	0 %
Calcium	60 %	Iron	120 %
Vitamin D	0 %	Vitamin E	0 %
Vitamin K	0 %	Thiamine	100 %
Riboflavin	100 %	Niacin	100 %
Vitamin B_6	4 %	Folate	4 %
Vitamin B_{12}	0 %	Biotin	0 %
Pantothenate	0 %	Phosphorus	60 %
Iodide	0 %	Magnesium	10 %
Zinc	4 %	Selenium	20 %
Copper	4 %	Manganese	10%
Chromium	0 %	Molybdenum	0 %
Chloride	0 %		

Valeur nutritive
Portion 125 mL (26 g)
Portions par Contenant 8

	Teneur par portion
Calories 100	(400 KJ)
Total des lipides	1 g
saturés	0 g
trans	0 g
polyinsaturés oméga-6	0.5 g
polyinsaturés oméga-3	0 g
monoinsaturés	0.2 g
Cholestérol	0 mg
Sodium	15 mg
Potassium	80 mg
Total des glucides	20 g
Fibres alimentaires	2 g
Fibres solubles	1 g
Fibres insolubles	1 g
Sucres	7 g
Polyalcools	0 g
Amidon	16 g
Protéines	3 g

% valeur quotidienne			
Vitamine A	0 %	Vitamine C	0 %
Calcium	60 %	Fer	120 %
Vitamine D	0 %	Vitamine E	0 %
Vitamine K	0 %	Thiamine	100 %
Riboflavine	100 %	Niacin	100 %
Vitamine B_6	4 %	Folate	4 %
Vitamine B_{12}	0 %	Biotine	0 %
Pantothenate	0 %	Phosphore	60 %
Iodure	0 %	Magnésium	10 %
Zinc	4 %	Sélénium	20 %
Cuivre	4 %	Manganese	10%
Chrome	0 %	Molybdène	0 %
Chlorure	0 %		

Source: CFIA: Nutrition Labelling Toolkit Templates, Available online at http://www.inspection.gc.ca/english/fssa/labeti/nutrikit/nutrikite.shtml. Accessed December 26, 2010.

© istockphoto.com/CGissemann

[1]Canadian Food Inspection Agency. Nutrition Labelling: Chapter 5.16: Foods intended solely for children under two years of age. Available online at http://www.inspection.gc.ca/english/fssa/labeti/guide/ch5be.shtml#a5_16_7. Accessed December 26, 2010.

[2]Health Canada. Do Canadian children meet their nutrient requirements through food intake alone? Cat H164-112/1-2009E-PDF, 2009. Available online at http://www.hc-sc.gc.ca/fn-an/surveill/nutrition/commun/art-nutr-child-enf-eng.php. Accessed December 19, 2010.

[3]Canadian Food Inspection Agency. Letter to Industry: Requirements related to nutrition information and nutrition and health claims for infant formula. Available online at http://www.inspection.gc.ca/english/fssa/labeti/inform/20070112e.shtml. Accessed December 26, 2010.

Figure 15.7 **Milk is an excellent source of calcium;** 750 ml (3 cups) supplies about 900 mg of calcium—enough to meet the RDA for children up to age 8. (Jamie Grill/ Age Fotostock America, Inc.)

Calcium, Vitamin D, and Bone Health The RDA for children 1-3 years of age is 700 mg/day and for young children (4–8 years) it is 1,000 mg/day (**Figure 15.7**). Adequate calcium intake during childhood is essential for achieving maximum peak bone mass, which is important in preventing osteoporosis later in life (see Section 11.3). Low intakes of milk, combined with limited sun exposure, may put children at risk of vitamin D deficiency. Vitamin D is needed for calcium absorption, so it is also essential for bone health. The RDA for children over 1 year is 15 μg per day. The CCHS found that calcium intake in Canadian children under 8 years of age was adequate.[3]

Iron and Anemia The RDA for iron for infants 7–12 months of age is 11 mg/day. This drops to 7 mg/day for toddlers, but for young children ages 4–8 years, it is 10 mg/day, which is higher than the RDA for adult men. Good sources of iron that are acceptable to small children include fortified grains and breakfast cereals, raisins, eggs, and lean meats. The CCHS found that more than 90% Canadian children (ages 1–8 years) had adequate intakes of iron.[3] This is good news as iron deficiency can lower a child's resistance to illness and slow recovery time. It can affect learning ability, intellectual performance, stamina, and mood.[18,19] If anemia is diagnosed, iron supplements are usually prescribed until iron stores are repleted. These supplements should be kept out of the reach of children. Overdoses of iron-containing supplements are the leading cause of poisoning deaths among children under 6 years of age.[20] To help protect children, products containing iron include a warning about the hazards to children of ingesting large amounts of iron.

Feeding Infants

Although breast milk or infant formula meets most nutritional needs until 1 year of age, semisolid and solid foods can be gradually introduced into the infant's diet starting at 6 months.[21] Introducing solid foods earlier provides no nutritional or developmental advantages. Despite this, some parents offer semisolid foods before this age often by adding infant cereal to the baby's bottle because they think the infant is hungry or the added food will help the child sleep through the night. Studies have shown that there is no difference in sleeping patterns based on such feeding practices and it may cause choking and increase the risk of food allergies.[22]

Before 6 months of age, the infant's feeding abilities and gastrointestinal tract are not mature enough to handle foods other than breast milk or formula. The young infant takes milk by a licking motion of the tongue called suckling, which strokes or milks the liquid from the nipple. Solid food placed in the mouth at an early age is usually pushed out as the tongue thrusts forward. By 4–6 months of age, the early reflex to bring the tongue to the front of the mouth to suckle has diminished, and the tongue is held farther back in the mouth, allowing solid food to be accepted without being expelled. By 6 months of age, the infant also can hold his or her head up steadily and is able to sit, either with or without support (**Figure 15.8**). Internally, the digestive tract has developed, and enzymes are present for starch digestion. The kidneys are more mature and better able to concentrate urine. With all of these changes, the child is ready to begin a new approach to eating.

CANADIAN CONTENT

What Foods to Introduce First? Health Canada makes several recommendations on the introduction of solid foods to infants.[21] The first foods offered should be iron-containing. Canadian parents most commonly start with iron-fortified infant rice cereal mixed with formula or breast milk. It is easily digested and rarely causes allergic reactions. Other iron-containing foods that can be offered at this stage include cooked egg yolks, meats, fish, or poultry or vegetarian options such as well-cooked legumes and tofu. Pureed vegetables or fruits can be introduced next; some suggest that vegetables be offered before fruits so that the child will learn to enjoy food that is not sweet before being introduced to sweet foods. Milk products such as cheese and yogurt can follow. Health Canada suggests that fluid milk not be introduced before 9 months of age, as it is low in iron. If it is introduced at an earlier age before a variety of iron-containing foods are part of the diet, it may increase the risk of iron-deficiency anemia. Egg whites are traditionally not given to infants until one year of age to reduce the risk of allergies from egg proteins. Once teeth have erupted, foods with more texture can be added (Figure 15.8). For the 6–12-month-old child, small pieces of soft or ground fruits, vegetables, and meats are appropriate (**Table 15.1**).

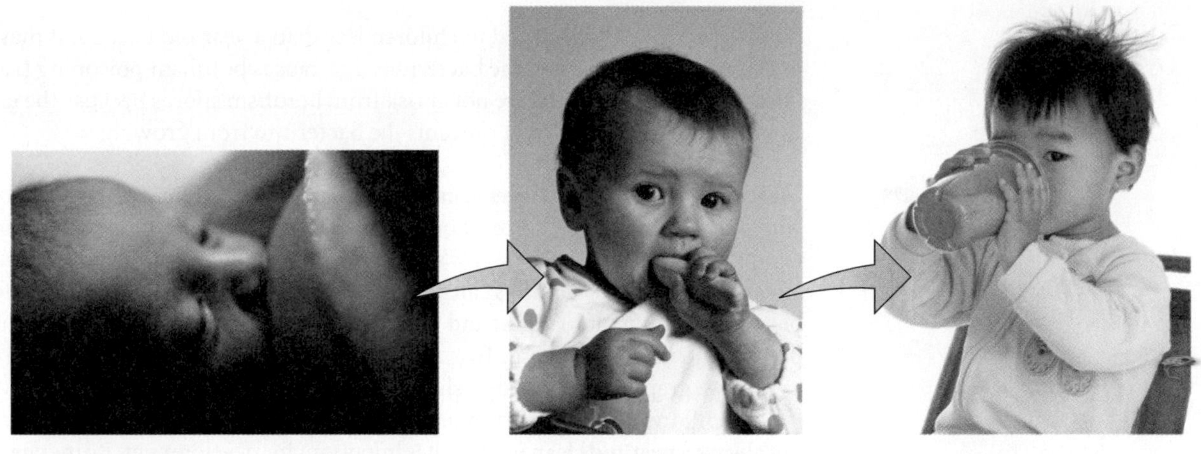

Age	Birth to 6 months	6 to 9 months	9 to 12 months
Developmental milestone	The infant takes milk by suckling. Solid food placed in the mouth is usually pushed out because the tongue is thrust forward during suckling. By age 6 months this tongue action is sufficiently diminished that baby can accept solid food.	The infant can hold his/her head up, sit, chew, hold food, and easily move hand to mouth.	The infant can drink from a cup and feed him/herself.
Foods	Breast milk or iron-fortified infant formula	Breast milk or formula, iron-fortified infant cereal, pureed or strained vegetables, fruits, meats and beans, limited finger foods.	Breast milk or formula, iron-fortified infant cereal, chopped vegetables, soft fruits, meats and beans, fruit juice, nonchoking finger foods such as dry cereal, cooked pasta, and cut well-cooked vegetables.

Figure 15.8 Nourishing a developing infant
(Raul Touzon/NG Image Collection; Laura Dwight/PhotoEdit; Alamy; iStockphoto)

Table 15.1 Typical Meal Patterns for Infants

		Servings per Day		
Food	**Serving Size**	**0–6 Months**	**6–9 Months**	**9–12 Months**
Formula or breast milk[a]	250 ml	4	4	4
Dry infant cereal	30 ml	—	4	4
Vegetables	40-45 ml	—	2	3
Fruits	30 ml	—	2	4
Fruit juice	125 ml	—	—	1 (in a cup)
Meats (or egg yolks)	15 ml	—	2–4 (strained)	4–6 (chopped)
Finger foods		—	1[b]	4[c]

[a]includes formula or breast milk added to cereal
[b]Dry toast, teething biscuits.
[c]Table foods except "choking" foods (e.g., foods in shapes and sizes that are likely to cause choking, such as large pieces of meat, whole grapes, or hot dogs or carrots cut in circular slices).

Honey should not be fed to children less than a year old because it may contain spores of *Clostridium botulinum*, the bacterium that causes botulism poisoning (see Section 17.3). Older children and adults are not at risk from botulism spores because the environment in a mature gastrointestinal tract prevents the bacterium from growing.

CANADIAN CONTENT

Increasing Variety with Developmentally Appropriate Choices As the child becomes familiar with more variety, food choices should be made from each of the food groups in Canada's Food Guide. At 1 year of age, whole cow's milk should be offered and continued until 2 years of age, after which reduced-fat milks can be used. Cow's milk should not be used before 1 year of age because it is difficult to digest and the infant's kidneys are too immature to handle its higher protein and mineral content. To avoid choking, foods that can easily lodge in the throat, such as carrots, grapes, and hot dogs, should not be offered to infants or toddlers.

As children become more independent, they will want to feed themselves. Although this is not always a neat and clean process, it is important for development. By the age of 8 or 9 months, infants can hold a bottle and self-feed finger foods such as crackers. By 10 months, most infants can drink from a cup, so water and fruit juices can be offered (Figure 15.8). The amount of juice offered to children should be limited. The Canadian Paediatric Society recommends no more than 125–250 ml/day (½-1 cup/day) of 100% juice for children.[23] Drinking too much juice can cause diarrhea, over- or undernutrition, and dental caries. It is recommended that juice not be offered to children in containers that can be carried around, encouraging continuous sipping.

A Caution about Food Allergies Although true food allergies are relatively rare, they are more common in infants.[24] Their immature digestive tracts allow incompletely digested proteins to be absorbed, causing a reaction involving the immune system. Exposure to an **allergen** for the first time causes the immune system to produce antibodies to that allergen (see sections 3.2 and 6.5 and Chapter 6 Label Literacy: Is it Safe for You?). When the allergen is encountered again by eating the same food, allergy symptoms such as vomiting, diarrhea, asthma, hives, eczema, runny nose and swelling of tissues, hay fever, and general cramps and aches may result as the immune system battles the allergen. The symptoms may occur almost immediately or take up to 24 hours to appear, and can vary from mild to severe and life-threatening. Foods that commonly cause allergies include wheat, peanuts, eggs, milk, nuts, fish, shellfish, and soy (see Your Choice: How can You Find Peanut-Free Foods?).

allergen A substance, usually a protein, that stimulates an immune response.

To monitor for food allergies when solid foods are introduced to infants, it is important to introduce new foods one at a time. Each new food should be offered for a few days without the addition of any other new foods. If an allergic reaction occurs, it is most likely due to the newly introduced food. Foods that cause allergy symptoms such as rashes, digestive upsets, or respiratory problems should be discontinued before any other new foods are added. After an infant is about 3 months of age, the risk of developing food allergies is reduced because incompletely digested proteins are less likely to be absorbed. Many children who develop food allergies before the age of 3 years will outgrow them. For example, most children allergic to eggs at 1 year of age will no longer be allergic by age 5. Allergies that appear after 3 years of age are more likely to be a problem for life.

Adverse reactions to foods are also caused by **food intolerances**. In contrast to food allergies, food intolerances do not involve antibody production by the immune system. Rather, they are caused by foods that create problems during digestion. Food intolerances can be caused by chemical components in foods, by toxins that occur naturally in foods, by substances added to foods during processing or preparation, or simply by large amounts of foods, such as onions or prunes, that cause local gastrointestinal irritation. Lactose intolerance is an example of a food intolerance caused by a reduced ability to digest milk sugar (see Section 4.3). It is not an allergy to milk proteins.

food intolerance An adverse reaction to a food that does not involve antibody production by the immune system.

Diagnosing Food Allergies Several laboratory methods are available to identify foods that are likely to cause an individual's allergic reaction, but they cannot determine the source of the problem with 100% reliability. The cause of a food allergy can be confirmed by using an **elimination diet** and **food challenge**. This involves eliminating all foods suspected of causing an allergic reaction from the diet. Once a diet that causes no symptoms has been established, it should be consumed for 2 weeks. Then in the food challenge, small amounts of a food suspected of causing a reaction are reintroduced under a doctor's

elimination diet and **food challenge** A regimen that eliminates potential allergy-causing foods from an individual's diet and then systematically adds them back to identify any foods that cause an allergic reaction.

YOUR CHOICE

(©iStockphoto)

How Can You Find Peanut-Free Foods?

An allergy to peanuts is one of the most common food allergies. It is rarely outgrown, and life-threatening reactions occur from exposure to minuscule amounts. Because reactions result from such low levels of peanut exposure, individuals with this allergy must avoid consuming any peanut protein and some people are so sensitive that they must avoid all contact with peanuts.

Shopping for safe foods for someone with peanut allergy is challenging. Individuals with peanut allergy or their caregivers need to rely on food labels to identify safe and unsafe choices. Peanuts, peanut butter, and peanut butter candy are obvious foods to avoid. Others are not so obvious; some foods, such as cookies and crackers, contain peanut flour and peanut oil. Crude peanut oil (less refined), which still contains some peanut protein, was found to cause an allergic reaction in many allergic subjects, whereas refined peanut oil did not cause a reaction in the majority of peanut-allergic individuals.[1] The ingredient list may not indicate if the oil is crude or refined, and other peanut-derived ingredients may be hidden in the list, but allergen labelling makes it easier to identify products containing peanuts and other allergens. Labels are required to indicate whether the product contains any of the top 8 food allergens; milk, eggs, fish, shellfish, peanuts, tree nuts, wheat, and soy.[2] This is most often done by listing them at the end of the ingredient list after the word "Contains," for example, "Contains milk, peanuts."

In the case of peanut allergy, even foods that do not contain peanuts or peanut products can pose a risk if they are manufactured in the same location as peanut-containing foods. For instance, the cheese "sandwich" crackers, M&Ms, and granola bars shown in the photo seem like safe foods, but a thorough reading of the label reveals that this is not the case. Because all of these products were manufactured in facilities that process peanuts for products such as peanut butter sandwich crackers, peanut M&Ms, and granola bars with peanuts, cross contamination with small amounts of peanuts is possible. To warn consumers of this, the label includes a statement such as "May contain peanuts" or "not suitable for consumption by persons with an allergy to peanuts."[3] This information protects both the company from legal action and the allergic individual from inadvertently consuming peanut-containing products.

Parents of children with allergies must read labels carefully. These products do not look like they contain peanuts, but the labels indicate that they may have inadvertently become contaminated with peanuts during manufacturing. (comstock/Getty Images)

[1]Crevel, R.W., Kerkhoff, M.A., and Koning, M. M. Allergenicity of refined vegetable oils. *Food Chem. Toxicol.* 38:385–393, 2000.

[2]FDA, CFSAN. Food Allergen and Consumer Protection Act of 2004. Available online at www.cfsan.fda.gov/~dms/alrgact.html/. Accessed June 8, 2009.

[3]Health Canada. The use of food allergen precautionary statements on prepackaged foods. Available online at http://www.hc-sc.gc.ca/fn-an/label-etiquet/allergen/precaution_label-etiquette-eng.php. Accessed December 26, 2010.

---

supervision. If no reaction to the food occurs, then increasing amounts are introduced until a normal portion is offered. If there is still no reaction, then the food can be ruled out as an allergen.

Preventing and Managing Food Allergies Preventing the development of food allergies is not always possible. Breastfeeding can reduce the risk of food allergies and is recommended for infants from families with a history of allergies.[25] Infants who are breastfed are less likely to be exposed to foreign proteins that cause food allergies. In addition, their gut matures earlier and they are protected by antibodies and other components of human milk. There is no evidence that delaying the introduction of solid food, including foods thought to be highly allergenic, past 6 months protects against the development of food allergies.[25]

Once a food allergy has developed, the best way to manage it is to avoid consuming the offending food. The information on food labels can be helpful in identifying foods that contain allergy-causing ingredients (see Chapter 6, Label Literacy: Is It Safe for You?). The Allergy and Asthma Information Association (www.aaia.ca) provides useful information to parents of children with allergies.

Feeding Children

The development of nutritious eating habits begins in infancy and childhood. Parents and caregivers influence which foods children learn to eat through the examples they set and the amounts and types of foods offered. Children are more likely to eat foods that are available and easily accessible, and they tend to eat greater quantities when larger portions are provided.[26] Caregivers are responsible for deciding what foods should be offered to a child, when they should be offered, and where they should be eaten. The child must then decide whether to eat, what foods to eat, and how much to consume.[27] As children get older, their choices are affected more by social activities, what they see at school, and what their friends are eating.

What to Offer? Children should be offered a balanced and varied diet adequate in energy and essential nutrients and appropriate to the child's developmental needs. A healthy meal plan is based on whole grains, vegetables, and fruits; is adequate in milk and other high-protein foods; and is moderate in fat and sodium. Food choices to meet these goals can be determined by following the recommendations of Canada's Food Guide and spreading these amounts throughout the meals and snacks served each day (**Figure 15.9**). For example, a 6-year-old, moderately active boy would only need 4 servings of grains. These could include serving of cereal (30 g) at breakfast, some crackers for a snack, half a sandwich at lunch, and a 125 ml (½ cup) of rice at dinner. This child would need 2 servings of milk (500 ml) a day but these may be consumed in smaller portions throughout the day (**Table 15.2**). The need for nutrient-dense choices from all food groups is just as important for children as it is for adults.

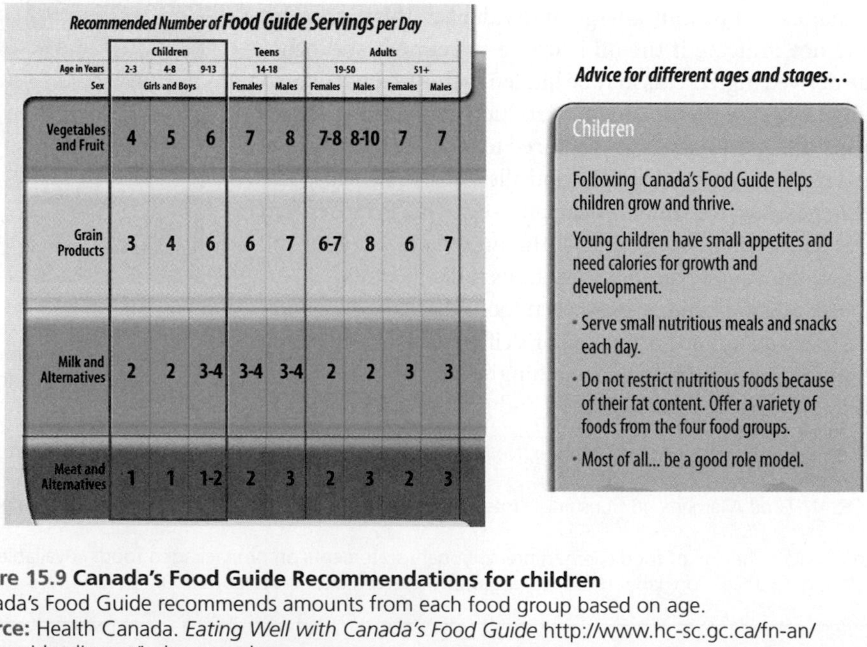

Recommended Number of ***Food Guide Servings*** *per Day*

	Children			Teens		Adults			
Age in Years	2-3	4-8	9-13	14-18		19-50		51+	
Sex	Girls and Boys			Females	Males	Females	Males	Females	Males
Vegetables and Fruit	4	5	6	7	8	7-8	8-10	7	7
Grain Products	3	4	6	6	7	6-7	8	6	7
Milk and Alternatives	2	2	3-4	3-4	3-4	2	2	3	3
Meat and Alternatives	1	1	1-2	2	3	2	3	2	3

Advice for different ages and stages...

Children

Following Canada's Food Guide helps children grow and thrive.

Young children have small appetites and need calories for growth and development.

- Serve small nutritious meals and snacks each day.
- Do not restrict nutritious foods because of their fat content. Offer a variety of foods from the four food groups.
- Most of all... be a good role model.

Figure 15.9 Canada's Food Guide Recommendations for children
Canada's Food Guide recommends amounts from each food group based on age.
Source: Health Canada. *Eating Well with Canada's Food Guide* http://www.hc-sc.gc.ca/fn-an/food-guide-aliment/index-eng.php

Getting a child to eat all the foods recommended by Canada's Food Guide may not always be easy. To increase variety, new foods should regularly be introduced into a child's diet. Children's food preferences are learned through repeated exposure to foods; they develop an increased preference for a food if it is offered a minimum of 8–10 times.[16] Children are also more likely to try a new food if it is introduced at the beginning of a meal, when the child is hungry, and if the child sees his or her parents or peers eating it. If a new food becomes associated with a bad

Table 15.2 A Typical Meal and Snack Pattern for Three- and Nine-Year-Old Children

Child's Age		**3 years**					**9 years**			
		Canada's Food Guide Servings					**Canada's Food Guide Servings**			
Food	**Amount**	**Vegetables and Fruit**	**Grain Products**	**Milk & Alternatives**	**Meat & Alternatives**	**Amount**	**Vegetables and Fruit**	**Grain Products**	**Milk & Alternatives**	**Meat & Alternatives**
Breakfast										
Cheerios	15 g		0.5			30 g		1		
Milk, 2%	125 ml			0.5		250 ml			1	
Strawberries	125 ml	1				125 ml	1			
Snack										
Peanut butter	15 ml				0.5	30 ml				1
Wheat crackers	2 (20 g)		0.5			4 (40 g)		1		
banana	1/2	0.5				1	1			
water										
Lunch										
Vegetable soup	125 ml	0.5				250 ml	1			
Tuna sandwich	25 g tuna 1 slice bread		1		0.33	50 g tuna 2 slice bread		2		0.67
Apple	1/2	0.5				1	1			
Milk, 2%	125 ml			0.5		250 ml			1	
Snack										
Yogurt	85 g			0.5		175 g			1	
Orange wedges	60 ml	0.5				125 ml	1			
Dinner										
Rice	125 ml		1			250 ml		2		
Chicken breast	40 g				0.5	75 g				1
Diced carrots	125 ml	1				125 ml	1			
Milk 2%	125 ml			0.5		250 ml			1	
Total CFG servings		4	3	2	1.33		6	6	4	2.67
Recommended CFG servings		4	3	2	1		6	6	3-4	1-2

experience, such as burning the mouth, the child will be unlikely to try it again. Incorporating refused foods into familiar dishes can also increase the variety of the diet. If vegetables are refused, they can be added to soups and casseroles. Fruit can be served on cereal or in milkshakes. Cheese can be included in recipes such as macaroni and cheese, cheese sauce, and pizza. Milk can be added to hot cereal, cream soups, puddings, and custards, and powdered milk can be used in baking. Meats can be added to spaghetti sauce, stews, casseroles, burritos, or pizza.

Children often have periods known as **food jags**, when they will eat only certain foods and nothing else. For example, a child may refuse to eat anything other than peanut butter and jelly sandwiches for breakfast, lunch, and dinner. The general guideline is to continue to offer other foods along with those the child is focused on. What a child will not touch at one meal, he or she may eat the next day or the next week. No matter how erratic children's food intake may be, caregivers should offer a variety of appropriate healthy food choices at each meal and let their children select what and how much they will eat (**Figure 15.10**). Food jags rarely last long enough to do any harm.

food jag When a child will eat only one food item meal after meal.

How Often to Offer Meals and Snacks? Children have small stomachs and high nutrient needs; therefore, they should consume nutritious meals and snacks throughout the day. For

SCIENCE APPLIED

Is Breakfast Food Really Brain Food?

Did your mother tell you that breakfast is the most important meal of the day? Research tells us that she was probably right. Breakfast provides the first dietary energy source of the day, fuelling your body and your brain. A study that evaluated the contribution of breakfast to daily nutrient intake found that children who eat breakfast have healthier diets than those who skip breakfast.[1] Breakfast eaters are more likely to meet nutrient intake recommendations, and compared with breakfast skippers, they have higher daily intakes of vitamins A and C, riboflavin, folic acid, calcium, zinc, iron, and fibre and lower intakes of fat and cholesterol.[2–4] Children and adolescents who skip breakfast do not, on average, make up the nutrient deficits later in the day.

Breakfast may also help maintain a healthy weight. Compared with breakfast eaters, breakfast skippers tend to have a lower total daily energy intake, and studies have shown that children and teens who eat breakfast are less likely to be overweight.[2,4]

The impact of eating breakfast on cognitive and academic performance, psychosocial function, and school attendance has been studied widely.[2] Children who are hungry are more likely to have academic, emotional, and behavioural problems.[5] Eating breakfast could therefore improve cognitive performance simply by alleviating hunger. This is particularly important in children whose diets barely meet nutrient requirements, but even well-nourished children benefit from the short-term effects of eating breakfast. Breakfast provides energy and nutrients to the brain. Glucose is the primary energy source for the brain, and blood glucose levels affect the performance of tasks involving recall and memory.[6] Without breakfast, the brain must rely on energy from body stores for morning activities.

Studies have found that compared with nonbreakfast eaters, children in school breakfast programs have higher nutrient intakes, and the improvements in intakes are associated with improvements in academic performance, reductions in hyperactivity, better psychosocial behaviours, and less absence and tardiness.[7,8] Over the school year, students who ate breakfast showed improved cognitive performance and reduced tardiness and absenteeism.[9] Breakfast consumption may therefore enhance learning not only by fuelling the brain, but also by improving school attendance, which provides children greater learning opportunities.[10]

Any breakfast is probably better than none, but a healthy breakfast can maximize the potential benefits. Ideally, breakfast should provide a quarter to a third of the day's nutrients without excessive kcalories. For example, a breakfast of orange juice, toast with jelly, and oatmeal with milk and raisins provides about 450 kcal as well as protein; B vitamins; vitamins C, A, and D; and the minerals calcium and iron. But if the only food a child will eat in the morning is a bowl of sweetened cereal, that is better than nothing. Despite the high-sugar, low-fibre content of many breakfast cereals marketed to children, these cereals have few other nutritional strikes against them. For example, although 40% of the energy in Cap'n Crunch comes from simple sugars, this food is still low in fat and provides 20% or more of the Daily Value for thiamin, riboflavin, niacin, vitamin B_6, folate, vitamin B_{12}, pantothenic acid, and iron. When 125 ml (½ cup) of reduced-fat milk is added to the cereal, it also provides 15% of the Daily Value for calcium. Even children who cannot or will not eat breakfast before they leave the house can snack on fruit, yogurt, a bag of dry cereal, or half a sandwich on the way to school or during recess. Having breakfast at school may be an option where such programs are available. In Canada, a non-profit organization, Breakfast for Learning, has helped many schools organize breakfast programs.[11]

[1]Basiotis, P., Lino, M., and Anand, R. Eating breakfast greatly improves schoolchildren's diet quality. *Fam. Econ. Nutr. Rev.* 12:81-83, 1999.

[2]Rampersaud, G. C, Pereira, M. A., Girard, B. L, et al. Breakfast habits, nutritional status, body weight, and academic performance in children and adolescents. *J. Am. Diet Assoc.* 105:743-760, 2005.

[3]Nicklas, T. A., O'Neil, C. E., and Berenson, G. S. Nutrient contribution of breakfast, secular trends, and the role of ready-to-eat cereals: A review of the data from the Bogalusa Heart Study. *Am. J. Clin. Nutr.* 67:757S-763S, 1998.

[4]Barton, B. A., Eldridge, A. L., Thompson, D. et al. The relationship of breakfast and cereal consumption to nutrient intake and body mass index: The national heart, lung, and blood institute growth and health study. *J. Am. Diet Assoc.* 105:1383-1389, 2005.

[5]Alaimo, K. Olson, C, and Frongillo, E. Food insufficiency and American school-aged children's cognitive, academic, and psychosocial development. *Pediatrics* 108:44-53, 2001.

[6]Benton, D., and Parker, P. Y. Breakfast, blood glucose, and cognition. *Am. J. Clin. Nutr.* 67(Suppl.)772S-778S, 1998.

[7]Kennedy E., and Davis, C. USDA School Breakfast Program. *Am. J. Clin. Nutr.* 67:798S-803S, 1998.

[8]Kleinman, R. E., Hall, S., Green, H, et al. Diet, breakfast, and academic performance in children. *Ann. Nutr. Metab.* 46 (Suppl. 1): 24-30, 2002.

[9]Meyers, A. F., Sampson, A. E., Weitzman, M., et al. School breakfast program and school performance. *Am. J. Dis. Child.* 143: 1234-1239, 1989.

[10]Pollitt, E., and Mathews, R. Breakfast and cognition: An integrative summary. *Am. J. Clin. Nutr.* 67(Suppl.):804S-813S, 1998.

[11]Breakfast for Learning. Available online at http://www.breakfastforlearning.ca/. Accessed December 26, 2010.

young children, a meal or snack should be offered every 2–3 hours and, because children thrive on routines and feel secure in knowing what to expect, a consistent pattern should be maintained from day to day. A missed meal or snack can leave a child without sufficient energy to perform optimally at school or play. A good breakfast is particularly important for ensuring optimal performance at school (see Science Applied: Is Breakfast Food Really Brain Food?). Snacks should be as nutritious as meals and should focus on fruits, vegetables, low-fat dairy products, lean meats, and whole-grain products, not soft drinks and chips (**Table 15.3**).

Table 15.3 Healthy Snacks for Young Children

Vegetables and Fruit—dip baby carrots in ranch dressing, add peanut butter to celery, serve salsa with baked tortilla chips; make popsicles from fruit juice, cut fruit into interesting shapes, try frozen grapes, make trail mix with raisins, dried cranberries, and other dried fruit.
Grain Products—try whole-grain breadsticks or pretzels, have graham crackers instead of cookies, pop some corn.
Milk and Alternatives—include milk with after-school cookies, try melted cheese on a tortilla, top some fruit with yogurt, snack on string cheese, make a yogurt and fruit shake.
Meat and Alternatives—put peanut butter on apple slices or crackers, snack on some turkey slices, roll some black beans into a tortilla, spread hummus on crackers.

Meals at Daycare or School All meals need to contribute to a child's nutrient intake, but parents may have little input into what children eat while at daycare or school. Ensuring that meals eaten away from home are nutritious is not easy because there is no guarantee that what is served or brought from home will be eaten. A packed lunch should contain foods the child likes and that do not require refrigeration (even if a refrigerator is available, the child is likely to forget to put the lunch in it). Even the most carefully planned lunch is not nutritious if it is not eaten.

Children often purchase their meals in school (**Figure 15.11**). Unfortunately, the quality of food in many schools is not very nutritious. A recent survey of Canadian high schools found that the most widely sold foods are hamburgers, pizza, french fries, cookies, muffins, soft drinks, and fruit drinks.[28] Fast food outlets are often located near high schools, so school cafeteria and vending machine operators feel they have to serve similar fast food items in order to compete with these outlets for revenues. Since the revenues from cafeterias and vending machines are often used to support school activities, there are real pressures working against providing nutritious meals.[28]

On the other hand, when nutrition programs make more nutritious foods available in schools and encourage their consumption by students, improved diet quality, lower rates of obesity, and increased physical activity are observed.[29] One example of a successful program was the Annapolis Valley Health Promoting School Project, which was begun by concerned parents in 8 schools and spread to become a province-wide program in Nova Scotia.[30]

Canada is one of the few developed countries that does not have a national school nutrition program. Some provincial governments have recently implemented school nutrition standards to restrict the types of foods that can be sold in school cafeterias and vending machines. Critics suggest that these standards are not strict enough to substantially reduce the amount of non-nutritious foods in schools, but they are a good start.[31] Given the obvious benefits of good nutrition in childhood, it is hoped that these standards are improved and more nutrition programs are implemented in schools.

Figure 15.10 Children should be allowed to select what and how much they will eat from a variety of healthy choices. (© istockphoto.com/ Nicole S. Young)

Figure 15.11 Children often purchase meals at school
Nutrition programs can help improve the quality of these meals. (Baerbel Schmidt/Stone/Getty Images, Inc.)

Vitamin and Mineral Supplements As with adults, children who consume a well-selected, varied diet can meet all their vitamin and mineral requirements with food. In fact, average intakes of most vitamins and minerals for children 2–11 years of age exceed 100% of the recommended amounts.[16] Occasional skipped meals and unfinished dinners are a normal part of most children's eating behaviour. However, children with particularly erratic eating habits, those on regimens to manage obesity, those with limited food availability, and those who consume a vegan diet may benefit from supplements that provide doses comparable to their RDA. If a children's supplement is offered, it should be monitored by caregivers and stored safely.

Eating Environment Factors such as whether families eat together, watch TV during meals, and choose food prepared at home or at a restaurant influence children's eating patterns.[26] To develop sound eating habits children need companionship, conversation, and a pleasant location at mealtimes. Caregivers should sit with children and eat what they eat (**Figure 15.12**). Children should be given plenty of time to finish eating. Slow eaters are unlikely to finish eating if they are abandoned by siblings who run off to play and adults who leave to wash dishes.

To make mealtime a nutritious, educational, and enjoyable experience, it should not be a battle zone. Threats and bribes are counterproductive and can create a problem where none had previously existed. Food is not a reward or a punishment: it is simply nutrition.

Figure 15.12 Companionship and conversation at meals help create a positive eating environment and foster good eating patterns. (©Masterfile)

15.3 Nutrition and Health Concerns in Children

Learning Objectives

- Discuss the relationship between food intake patterns and dental caries.
- Explain how lead toxicity can be avoided.
- Describe the impact of the modern lifestyle on childhood obesity.
- Describe how children can become more active.

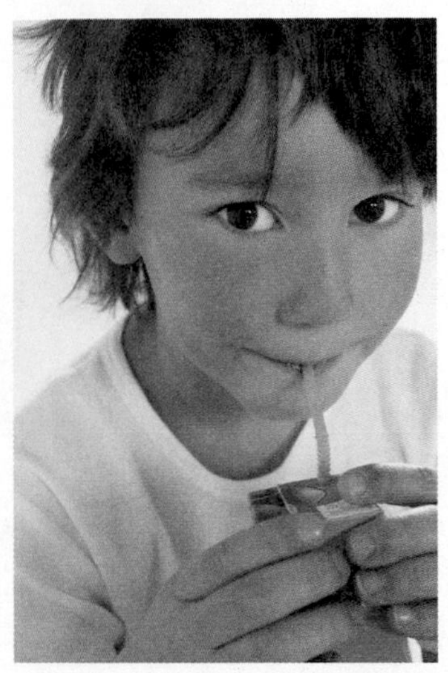

Figure 15.13 When juice is sipped slowly over a long period, it provides a continuous supply of sugar to feed cavity-causing bacteria. (Tom Merton/Age Fotostock America, Inc.)

A number of diet and lifestyle factors put children at risk for illness and malnutrition. Some are a greater risk in young children because of their size and stage of development and others are problems that may continue into adolescence.

Dental Caries

Children and teens typically eat more sugar than is recommended. Added sugars reduce the nutrient density of foods, so excessive consumption of foods high in added sugars make it difficult to meet nutrient needs. In addition, a diet high in sugary foods promotes tooth decay, causing dental caries, commonly known as cavities. Decay occurs when there is prolonged contact between sugar and bacteria on the surface of the teeth. Because the primary teeth guide the growth of the permanent teeth, maintaining healthy primary teeth is just as important as preserving permanent ones.

Much of the added sugar in children's diets come from soft drinks and other sweetened beverages; when these are sipped slowly between meals, the contact time between sugar and teeth is prolonged, hence increasing the risk of tooth decay (**Figure 15.13**). Changing the types of beverages consumed by children may impact the incidence of dental caries. A study found fewer caries in children who drank more milk and more in those who consumed pop and juice.[32] The Canadian Paediatric Society recommends that juice intake be limited to 125-250 ml (½ -1 cup). of 100% juice per day.[23] Although sugary foods are the most cavity-promoting, any carbohydrate-containing food can cause tooth decay, especially if the food sticks to the teeth (see Section 4.6).

Both diet and dental hygiene can affect the risk of dental caries. Preventing tooth decay involves limiting high carbohydrate snacks, especially those that stick to teeth; brushing teeth frequently to remove sticky sweets; and consuming adequate fluoride. Children's teeth should be brushed as soon as they erupt and children 3 years of age and older should be examined by a dentist regularly.

Diet and Hyperactivity

attention deficit hyperactivity disorder (ADHD) A condition that is characterized by a short attention span, acting without thinking, and a high level of activity, excitability, and distractibility.

Hyperactivity is a problem in 5%–10% of school-age children, occurring more frequently in boys than in girls. This syndrome involves extreme physical activity, excitability, impulsiveness, distractibility, short attention span, and a low tolerance for frustration. Hyperactive children have more difficulty learning but usually are of normal or above-average intelligence. Hyperactivity is now considered part of a larger syndrome known as **attention deficit hyperactivity disorder (ADHD)**.

One popular misconception is that hyperactivity is caused by a high sugar intake, but research on sugar intake and behaviour has failed to support this hypothesis.[33,34] Hyperactive behaviour that is observed after sugar consumption is likely the result of other circumstances in that child's life. For example, the excitement of a birthday party rather than the cake is most likely the cause of hyperactive behaviour. Other situations that might cause hyperactivity include lack of sleep, overstimulation, the desire for more attention, or lack of physical activity.

Specific foods and food additives have also been implicated as a cause of hyperactivity. In 1975, an allergist named Dr. Benjamin Feingold published a book suggesting that hyperactive behaviour and learning disabilities were caused by high intakes of artificial colours and flavours. Despite the appeal of this type of dietary management, numerous studies done over the years have failed to provide sufficient scientific evidence for the efficacy of this diet to treat children with hyperactivity. There are, however, some children with behavioural problems who are sensitive to specific additives and who may benefit from a diet that eliminates these.[34]

Another possible cause of hyperactive behaviour in children is caffeine. Caffeine is a stimulant that can cause sleeplessness, restlessness, and irregular heartbeats. Beverages, foods, and medicines containing caffeine are often a part of children's diets. For example, caffeinated beverages such as Coke are commonly included in children's fast-food meals.

Lead Toxicity

Lead is an environmental contaminant that can be toxic, especially in children under 6 years of age. Children are particularly susceptible because they absorb lead much more efficiently than do adults. It is estimated that infants and young children may absorb as much as 50% of ingested lead, whereas adults absorb only about 10%–15%.[35] Malnourished children are at particular risk because malnutrition increases lead absorption due to the fact that lead is better absorbed from an empty stomach and when other minerals such as calcium, zinc, and iron are deficient. Once absorbed from the gastrointestinal tract, lead circulates in the bloodstream and then accumulates in the bones and, to a lesser extent, the brain, teeth, and kidneys. Lead disrupts the activity of neurotransmitters and thus interferes with the functioning of the nervous system. Higher levels of lead can contribute to iron-deficiency anemia, altered kidney function, nervous system changes, and even seizures, coma, and death. In young children, lead poisoning can cause learning disabilities and behaviour problems.[35–37]

Lead is found naturally in the earth's crust, but over the years industrial activities have redistributed it in the environment. Lead is now found in soil contaminated with lead paint dust; it enters drinking water from old, corroded lead plumbing, lead solder on copper pipes, or brass faucets. It is found in polluted air, in leaded glass, and in glazes used on imported and antique pottery. Lead from these can contaminate food and beverages. Ways to avoid some of these sources of lead are described in **Table 15.4**.

Because of the risks of lead toxicity from environmental contamination, lead is no longer used in house paint, gasoline, or solder. Health Canada reports low blood levels of lead in children but suggests that the lead levels of children be tested in communities where, because of past or current industrial activities, environmental levels of lead may be high.[38] In the United States, certain groups have been found to be at higher risk of elevated blood lead levels (**Figure 15.14**). Low-income children are at particular risk. Low-income families are more likely to live in older buildings that still have lead paint and lead plumbing. In Canada, the lead levels in a small proportion of Aboriginal women of reproductive age have been found to exceed guidelines.[39] These higher lead levels may be linked to the use of lead shotshell for hunting and as a result the federal government is promoting the use of non-toxic steel shotshell.[40]

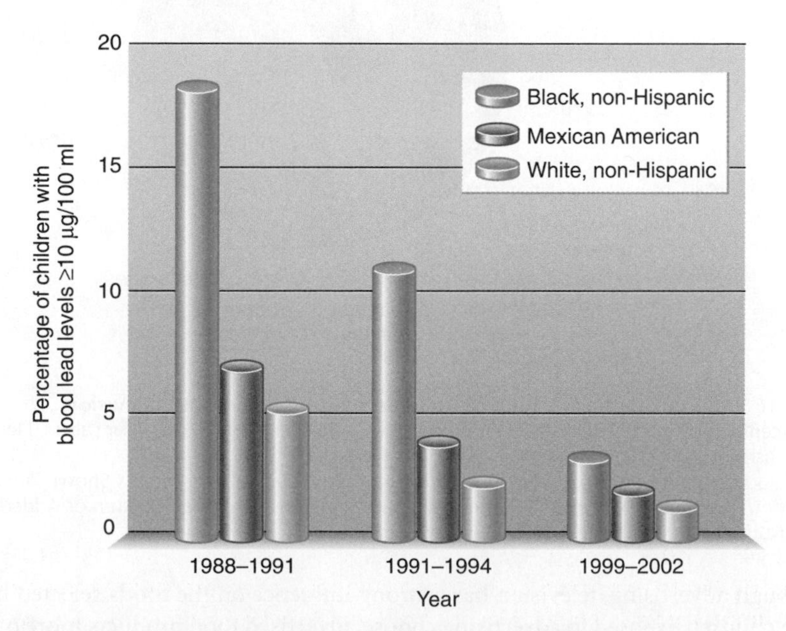

Figure 15.14 Percentage of children aged 1-5 years with elevated blood lead levels
American data indicate that blood lead levels have declined since the 1980s, but vary based on ethnic group; this may be related to differences in income levels.
Source: CDC, Blood lead levels-United States, 1999–2002, *MMWR* 54: 513–516, 2005.

Table 15.4 What Can Be Done to Reduce Lead Exposure

Lead-based paint Lead-based paint was widely used in the 1970s and earlier. If you are concerned about lead-based paint, scientific testing can determine the levels in your home. Lead-based paint is not hazardous if it is not chipping and is not in an area where children can ingest it. One way to deal with lead paint is to cover the area with wallpaper or panelling. To remove the paint, chemical strippers are recommended. Mechanical removal, by sanding, is not advised.
Reducing exposure from tap water If your home has old plumbing, lead may be leaching into your tap water. More lead leaches into hot water than cold, so use only cold water for drinking or cooking. Water that has been standing in the pipes has more lead, so allow water to run in the morning before drinking. Note that typical morning activities such as showering or flushing toilets are often sufficient to flush water pipes if done before water is used for drinking or cooking.
Reducing exposure from food container Pottery: If pottery contains lead-based glaze and is poorly fired, lead can be released into food, particularly acidic foods. While the amount of leachable or releasable lead from ceramic foodware is strictly regulated in Canada, lead exposure may occur from pottery obtained abroad. Such pottery should not be used to store foods. Leaded crystal: The amount of lead that can be released during the consumption of beverages for leaded crystal in the course of a meal is very low. Storing beverages in leaded crystal containers should be avoided as this can result in the accumulation of large amounts of lead. Do not serve pregnant women or children beverages in leaded crystal.
For additional information Search the Health Canada website for the lead information package.

Source: Health Canada. Lead information package. Available online at http://www.hc-sc.gc.ca/ewh-semt/contaminants/lead-plomb/asked_questions-questions_posees-eng.php. Accessed December 26, 2010.

Childhood Obesity

Although genes are important determinants of body size and weight, changes in lifestyle that have led to decreases in activity and increases in energy intake are the major reasons for the increase in obesity that has occurred among Canadian children.[41] Reversing this trend requires changes in the way children play and eat (see Critical Thinking: At Risk for Malnutrition) and may also require parents to make changes (see Critical Thinking: Parental Influences on Childhood Eating Habits).

Figure 15.15 Television watching influences activity level, snacking behaviour, and the kinds of foods children choose. (© Ocean/Corbis)

The Impact of Screen Time Many children today spend more time watching television or at computer screens than they do in any activity other than sleep. Television affects nutritional status in a number of ways: it introduces children to foods they might otherwise not be exposed to, it promotes snacking, and it reduces physical activity (**Figure 15.15**).

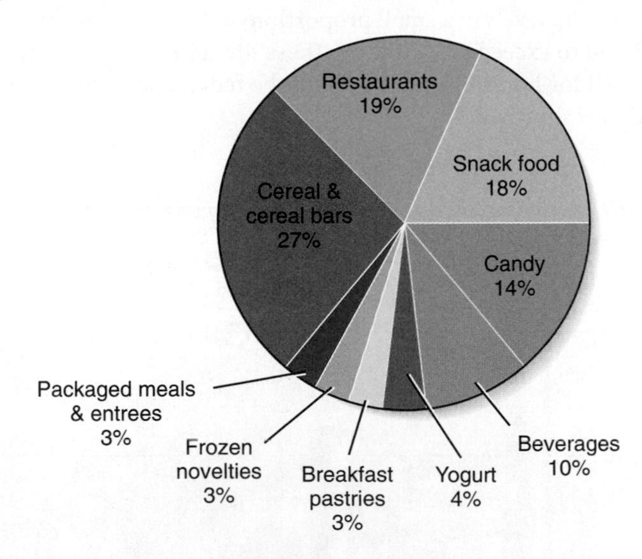

Figure 15.16 Types of foods advertised on children's Saturday morning television
A high percentage of foods advertised on children's television is high in fat, sugar, or salt and low in nutrient density.
Source: Batada, A, Seitz, M, Wotan, M, et al. Nine out of Ten Food Advertisements Shown During Saturday Morning Children's Television Programming are for Food High in Fat. Sodium or Added Sugars, or Low in Nutrients *J. Am. Diet. Assoc.* 108:673–678, 2008.

Through advertising, television has a strong influence on the foods selected by young children; children exposed to advertising choose advertised food products more often than

CRITICAL THINKING:

At Risk for Malnutrition

Background

Jamar is 8 years old and has been gaining weight. All he wants to do is lie around, watch TV, and play video games. Previously he enjoyed playing basketball with his friends and was eager to go on hikes with the family. Jamar's parents are both overweight. They are concerned that Jamar will also have a weight problem, so they take him to see their pediatrician. The nurse weighs and measures Jamar and draws a blood sample for routine analysis.

Data

Last year, Jamar's BMI was at the 50th percentile, but it is now almost at the 75th. The blood sample reveals that Jamar has iron-deficiency anemia. The pediatrician prescribes an iron supplement and refers Jamar and his parents to a dietitian. By reviewing Jamar's diet and exercise patterns, the dietitian learns that he gets 25 min. of exercise during recess at school and that he has been watching TV or playing video games for about 6 hours a day. Below are the results of a food frequency questionnaire that she recorded:

	FREQUENCY	
FOOD	**SERVINGS/DAY**	**SERVINGS/WEEK**
Milk and dairy products		
Regular fat	6	
Reduced fat		
Fat-free		
Meat and eggs		
Red meat		1
Chicken		2
Fish		1
Eggs		
Grains and cereals		
Whole grains	2	
Refined grains	4	
Fruit and juices		
Citrus	1	
Other	2	
Vegetables		
Dark green leafy		
Other	1	
Added fats	3	
Snack foods		
Chips, etc.	2	
Candy	1	

Critical Thinking Questions

Why is Jamar's energy balance "out of balance"? Does he get the amount of daily activity recommended for a child his age (see Figure 15.18)? Evaluate his diet by comparing the number of servings per day he consumes from the various food groups to Canada's Food Guide recommendations for an 8-year-old boy.*

Why is he anemic? What foods groups does he consume that are good sources of iron? How would his intake of dairy products affect his iron status?

How might Jamar's iron deficiency anemia have contributed to his weight gain?

Suggest some dietary changes that could increase his iron intake and decrease his energy intake.

*Answers to the Critical Thinking questions can be found in Appendix J.

iProfile Use iProfile to find the iron content of several fast-food meals.

children who are not exposed.[42] Food is the most frequently advertised product category on children's TV. The majority of these ads target highly sweetened products such as sweetened breakfast cereals; sweets such as candy, cookies, doughnuts, and other desserts; snacks; and beverages; and more recently, the proportion from fast-food meal promotions has been growing (**Figure 15.16**).[43] The more hours of TV a child watches, the more he or she asks for advertised food items, and the more likely these items are to be in the home. Television also promotes snacking behaviour. Although snacks are an important part of a growing child's diet, while watching TV many children snack on sweet and salty foods that are low in nutrient density. In terms of overall diet, more hours of TV watching have been associated with higher intakes of energy, fat, sweet and salty snacks, and carbonated beverages and lower intakes of fruits and vegetables.[42]

CANADIAN CONTENT

In Canada, the advertising of food and beverages to children is regulated by government legislation and by a self-regulatory organization called the Canadian Children's Food and Beverage Advertising Initiative. The principles promoted by this initiative include: a) to advertise only healthy dietary choices to children under 12; b) to include only healthy foods in interactive games directed to children under 12; c) to not place food and beverage products in children's programming, and d) to not advertise in elementary schools. Most major food companies advertising in Canada have agreed to follow the initiative's principles.[44] Dietitians of Canada, in their position paper on the subject of advertising food to children, indicated that the initiative was a good starting point, but that all food companies must participate in the initiative, that "healthy foods" must be clearly defined based on scientific evidence, and that there should be a system of monitoring compliance and reporting complaints. Finally, Dietitians of Canada believes that the advertising of healthy foods should be encouraged.[45]

In addition to influencing food choices through advertising, perhaps the most important nutritional impact of television is that it reduces activity. Hours spent watching television are hours when physical activity is at a minimum. One study showed that children who watch 4 or more hours of TV per day are 40% more likely to be overweight than those who watch an hour or less a day.[46] Generally, the more physically active a child is the less TV he or she watches. In addition to television, children and adolescents today replace time spent at more physically demanding activities with time spent at the computer or playing video games.

Figure 15.17 Consuming too much fast food and making poor fast-food choices can result in a diet high in fat and low in fruits and vegetables. (Arthur R. Hill/Visuals Unlimited)

The Impact of Fast Food Children and teens generally love fast food, and there is nothing wrong with an occasional fast-food meal. But a steady diet of burgers, fries, and soft drinks will likely contribute to an overall diet that is high in energy, fat, sugar, and salt and low in calcium, fibre, and vitamins A and C (**Figure 15.17**).

The portions served at fast-food restaurants are a major contributor to children's increased energy intake. A meal that includes a hamburger, small fries, and a small drink will provide about 600 kcal, 12 g of saturated fat, and 36 g of sugar. But a meal with a Big Mac, a large order of fries, and a large drink will increase the numbers to 1,390 kcal, 15 g of saturated fat, and 94 g of sugar. A steady diet of such meals can significantly increase kcalorie intake and consequently body weight. Choosing carefully from the old fast-food standbys such as plain, single-patty hamburgers or newer low-fat options such as grilled chicken sandwiches can keep fat and energy intake reasonable (see **Table 15.5**).

Table 15.5 Make Healthier Fast-Food Choices

Instead of . . .	Choose . . .
Double-patty hamburger with cheese, mayonnaise, special sauce, and bacon	Regular single-patty hamburger without mayonnaise, special sauce, and bacon
Breaded and fried chicken sandwich	Grilled chicken sandwich
Chicken nuggets or tenders	Grilled chicken strips
Large french fries	Baked potato, side salad, or small order of fries
Fried chicken wings	Broiled skinless wings

Crispy-shell chicken taco with extra cheese and sour cream	Grilled-chicken soft taco without sour cream
Nachos with cheese sauce	Tortilla chips with bean dip
12-in. meatball sub	6-in. turkey breast sub with lots of vegetables
Thick-crust pizza with extra cheese and meat toppings	Thin-crust pizza with extra veggies
Doughnut	Cinnamon and raisin bagel with low-fat cream cheese

Some food groups are limited in typical fast-food meals. For example, the few pieces of shredded lettuce and the slice of tomato that garnish a burger or taco make only a small contribution to the 5 Canada's Food Guide servings of vegetables and fruit that should be included in the diet of an average 8-year-old boy. Typically, fast-food meals are also lacking in milk and fruits. Many fast-food franchises now offer vegetables, salads, yogurt, fruit, and milk, which can increase calcium and vitamin intake if selected. Even if they are not, keep in mind that a fast-food meal is only one part of the total diet. If the missing milk, fruits, and vegetables are consumed at other times during the day and overall energy intake balances output, the total diet can still be a healthy one.

Preventing and Treating Obesity In 2009, the Public Health Agency of Canada published a report on the state of public health in Canada which focused on the health of children.[47] In this report, it was recognized that childhood obesity must be addressed at multiple levels, including nutrition education, the regulation of the advertising of food to children, the creation of environments that promote physical activity (e.g., the presence of recreational facilities, parks, and playgrounds) and provision of nutritious foods (by influencing restaurant menu items and ensuring that healthy foods can be purchased locally), and programs that target parents, school, and community. An example of a successful community program is Saskatoon in Motion (www.in-motion.ca/). Begun in 2000, it targets children and youth with school programs, inactive adults in the workplace, and older adults, and encourages them to get active.[48] Programs like Saskatoon in Motion are important because one of the major contributors to the increase in body weight among children is lack of physical activity. Watching television, playing video games, and surfing the Web have replaced neighbourhood games of tag and soccer for many children. Whether or not a child is overweight, he or she should be physically active.

CANADIAN CONTENT

To promote physical activity, the Canadian Public Health Agency has developed several physical activity guidelines. These include Canada's Physical Activity Guide for Children, geared to children 5–11 years of age, and Canada's Physical Activity Guide for Youth, geared to ages 12–17. The guidelines recommend that both children and youth be physically active for 60 min./day, and include both moderate and vigorous intensity activities[49] (**Figure 15.18**). Recommended activities include running, jumping, swimming, gymnastics, dance, and climbing playground equipment. Parents are encouraged to plan games, walks after dinner, bike rides, hikes, swimming, and other activities that can be enjoyed by the whole family. This sends a positive message to children to be more active. Involvement of the whole family is key.[50]

Physical activity should also be an important part of the school curriculum. The Canadian Fitness and Lifestyle Research Institute, a non-profit organization which researches how to improve Canadian fitness levels, found in a 2006 survey that only 53% of Canadian schools have implemented policies to provide a range of physical activities to students, only 46% of schools hire teachers with university training in physical education, and only 35% of schools provide daily physical activity.[51] Improving these numbers would benefit the health of all Canadian students.

In 2011, the Public Health Agency of Canada launched a website entitled "Our Health, Our Future" which invited Canadians, including Canadian youth, to post ideas on how to reduce childhood obesity. The most popular ideas, all of which would benefit Canadian children, included a) shifting the focus of health messages away from weight to more broadly based messages promoting health and physical activity for all, so as to reduce weight stigmatization; b) developing a national cycling strategy to allow children to ride bicycles safely in more places; c) restricting the advertising of "junk foods" to children; d) encouraging the participation of families in health-promoting activities; and e) including cooking classes in the high school curriculum.[52]

CRITICAL THINKING:

Parental Influences on Childhood Eating Habits

Research has indicated that the health behaviour of children is influenced by the behaviour of their parents. Shown is an example from data collected as part of CCHS. The graph compares the body weight of 12-19-year-old teens (as indicated by the percentage of teens who are overweight or obese), compared to their parents' body weight (i.e., underweight or acceptable weight, overweight, or obese).

Fuse/Getty Images

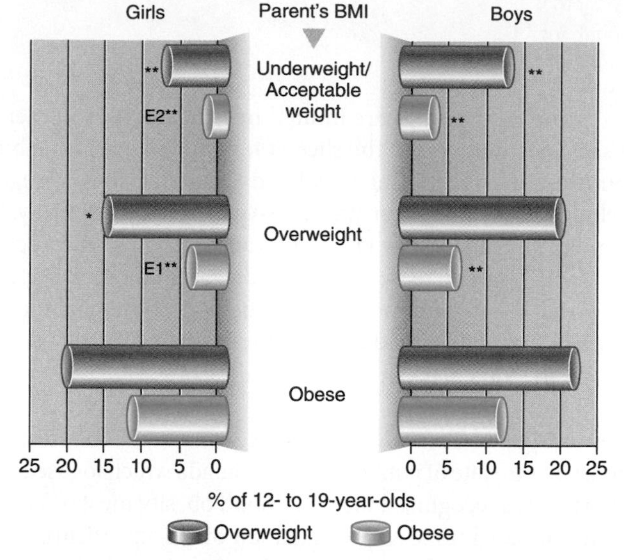

Source: 2000/01 Canadian Community Health Survey, cycle 1.1

* Significantly different from value for corresponding category in household with an obese parent ($p < 0.05$, adjusted for multiple comparisons)

** Significantly different from value for corresponding category in household with an obese parent ($p < 0.01$, adjusted for multiple comparisons)

E1 Coefficient of variation between 16.6% and 25.0%

E2 Coefficient of variation between 25.1% and 33.3%

Critical Thinking Questions

Describe the relationship between teen body weight and parental body weight.

What environmental and behavioural factors might explain this association?

What other factors might explain this association?

What are the implications of these trends in the planning of programs to reduce childhood obesity?

Source: Carriere, G. Parent and child factors associated with youth obesity. Supplement to Health Reports 2003: 29-39. Available online at www.statcan.gc.ca/bsolc/olc-cel/olc-cel?lang=eng&catno=82-003-X20031016679. Accessed December 26, 2010.

15.4 Adolescents

Learning Objectives

- Describe how growth and body composition are affected by puberty.
- Compare the energy needs of adolescents with those of children and adults.
- Explain why iron and calcium are of particular concern during the teen years.
- Use Canada's Food Guide to plan a day's diet that would appeal to an adolescent.

Once a child has reached about 9-12 years of age, the physical changes associated with sexual maturation begin to occur. These physical changes, along with the social and psychological changes that accompany them, have a significant impact on the nutritional needs and nutrient intakes of adolescents. The DRIs divide recommended intakes for adolescence into ages 9–13 and ages 14–18.

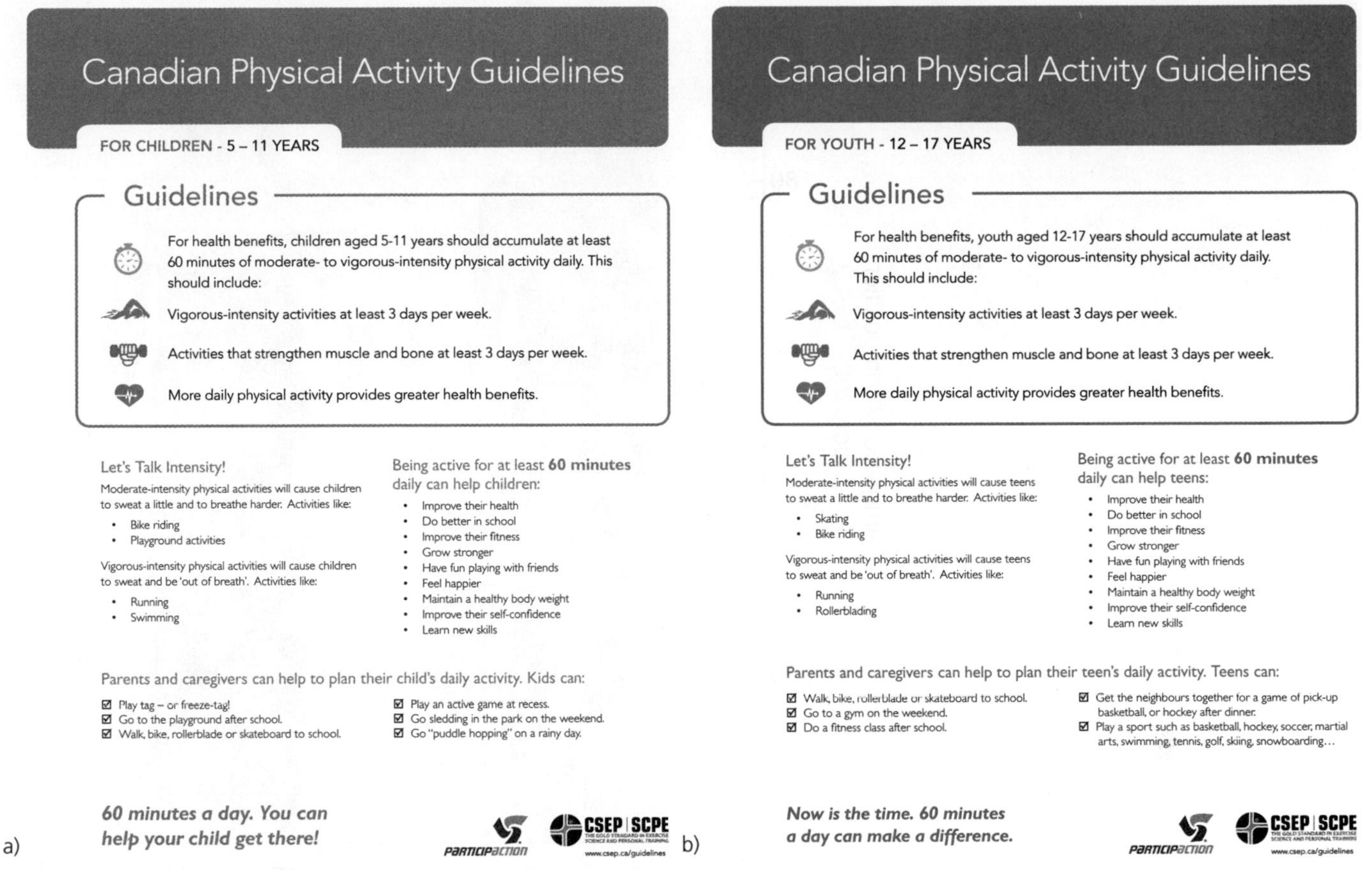

Canadian Physical Activity Guidelines
FOR CHILDREN - 5 – 11 YEARS
Guidelines
For health benefits, children aged 5-11 years should accumulate at least 60 minutes of moderate- to vigorous-intensity physical activity daily. This should include:
Vigorous-intensity activities at least 3 days per week.
Activities that strengthen muscle and bone at least 3 days per week.
More daily physical activity provides greater health benefits.
Let's Talk Intensity!
Moderate-intensity physical activities will cause children to sweat a little and to breathe harder. Activities like:
• Bike riding
• Playground activities
Vigorous-intensity physical activities will cause children to sweat and be 'out of breath'. Activities like:
• Running
• Swimming
Being active for at least 60 minutes daily can help children:
• Improve their health
• Do better in school
• Improve their fitness
• Grow stronger
• Have fun playing with friends
• Feel happier
• Maintain a healthy body weight
• Improve their self-confidence
• Learn new skills
Parents and caregivers can help to plan their child's daily activity. Kids can:
☑ Play tag – or freeze-tag!
☑ Go to the playground after school.
☑ Walk, bike, rollerblade or skateboard to school.
☑ Play an active game at recess.
☑ Go sledding in the park on the weekend.
☑ Go "puddle hopping" on a rainy day.
60 minutes a day. You can help your child get there!
PARTICIPACTION
CSEP | SCPE
www.csep.ca/guidelines
a)
Canadian Physical Activity Guidelines
FOR YOUTH - 12 – 17 YEARS
Guidelines
For health benefits, youth aged 12-17 years should accumulate at least 60 minutes of moderate- to vigorous-intensity physical activity daily. This should include:
Vigorous-intensity activities at least 3 days per week.
Activities that strengthen muscle and bone at least 3 days per week.
More daily physical activity provides greater health benefits.
Let's Talk Intensity!
Moderate-intensity physical activities will cause teens to sweat a little and to breathe harder. Activities like:
• Skating
• Bike riding
Vigorous-intensity physical activities will cause teens to sweat and be 'out of breath'. Activities like:
• Running
• Rollerblading
Being active for at least 60 minutes daily can help teens:
• Improve their health
• Do better in school
• Improve their fitness
• Grow stronger
• Have fun playing with friends
• Feel happier
• Maintain a healthy body weight
• Improve their self-confidence
• Learn new skills
Parents and caregivers can help to plan their teen's daily activity. Teens can:
☑ Walk, bike, rollerblade or skateboard to school.
☑ Go to a gym on the weekend.
☑ Do a fitness class after school.
☑ Get the neighbours together for a game of pick-up basketball, or hockey after dinner.
☑ Play a sport such as basketball, hockey, soccer, martial arts, swimming, tennis, golf, skiing, snowboarding...
Now is the time. 60 minutes a day can make a difference.
PARTICIPACTION
CSEP | SCPE
www.csep.ca/guidelines
b)

Figure 15.18 Canada's Physical Activity Guidelines for children (a) and youth (b).
Source: Canadian Physical Activity Guidelines, © 2011. Used with permission from the Canadian Society for Exercise Physiology, www.csep.ca/guidelines.

The Changing Body: Sexual Maturation

During adolescence, organ systems develop and grow, **puberty** occurs, body composition changes, and the growth rates and nutritional requirements of boys and girls diverge. During the teenage years, boys and girls grow about 28 cm (11 inches) and gain about 40% of their eventual skeletal mass.[53] From ages 10–17, girls gain about 24 kg (53 lbs) and boys about 32 kg (70 lb). During adolescence, there is an 18-24-month period of peak growth velocity, called the **adolescent growth spurt**. In girls, the growth spurt occurs between 10 and 13 years. In boys, it occurs between 12 and 15 years (**Figure 15.19**).

puberty A period of rapid growth and physical changes that ends in the attainment of sexual maturity.

adolescent growth spurt An 18-24-month period of peak growth velocity that begins at about ages 10–13 in girls and 12–15 in boys.

The hormonal changes that occur with sexual development orchestrate growth and affect body composition. During the growth spurt, boys tend to grow taller and heavier than girls and do so at a faster rate. Boys gain fat but also add so much lean mass as muscle and bone that their percentage of body fat actually decreases. In girls, **menarche**, the onset of menstruation,

menarche The onset of menstruation, which occurs normally between the ages of 10-15 years.

Adolescent growth spurt
Growth rate (cm/yr)
20
17.5
15
12.5
10
7.5
5
2.5
0
Males
Females
0
5
10
15
20
Age (years)

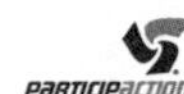

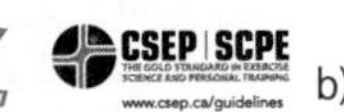

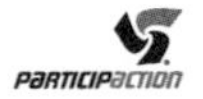

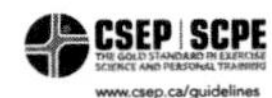

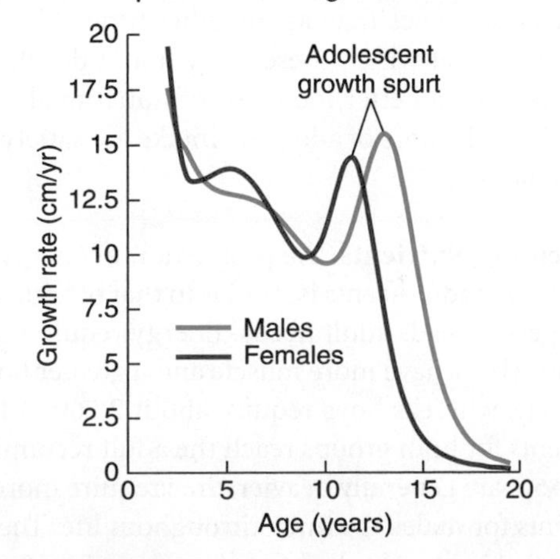

Figure 15.19 The adolescent growth spurt
During a 1-year growth spurt, boys can gain 10 cm in height and girls 9 cm.

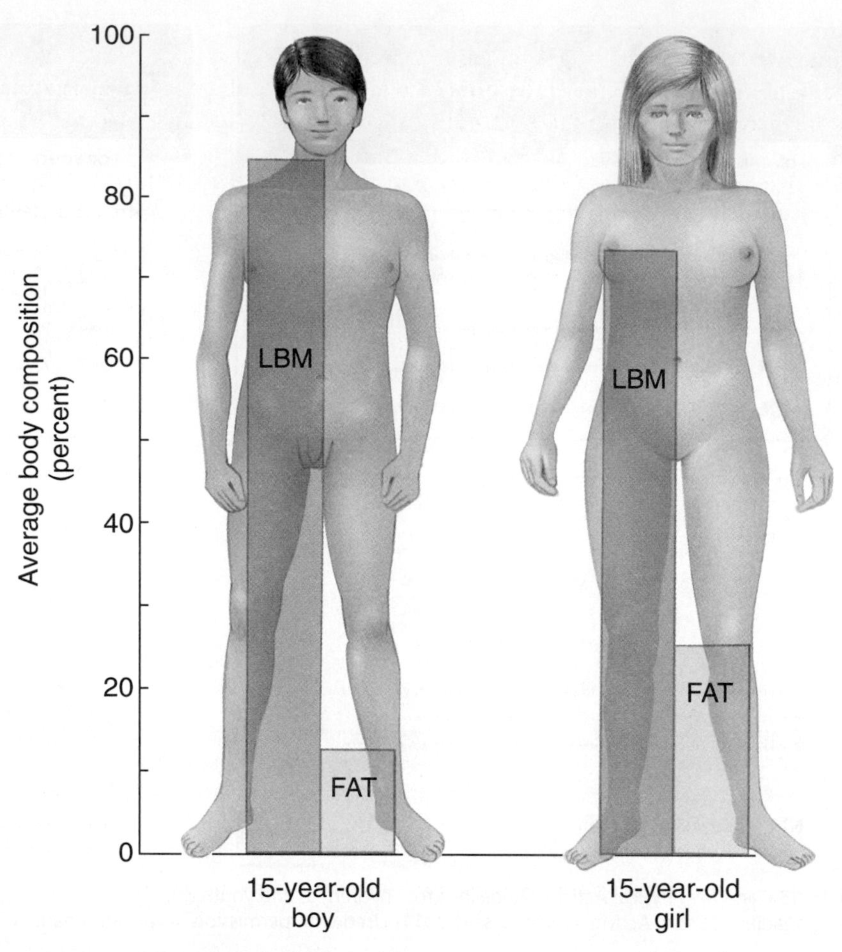

Figure 15.20 Body composition of males and females
After puberty, males have a higher percentage of lean body mass and less body fat than females.
Source: Adapted from Forbes, G. B. Body composition. In Present Knowledge in Nutrition, 6th ed. M. L. Brown, ed. Washington, DC: International Life Sciences Institute-Nutrition Foundation, 1990.

is typically followed by a deceleration in growth rate and an increase in fat deposition. By age 20, females have about twice as much adipose tissue as males and only about two-thirds as much lean tissue (**Figure 15.20**).

Nutrient intake during childhood and adolescence can affect sexual development. Nutritional deficiencies can cause poor growth and delayed sexual maturation. Taller, heavier children usually enter puberty sooner than shorter, lighter ones.[54]

Adolescent Nutrient Needs

The physiological changes that occur during adolescence affect nutrient needs. Total nutrient needs are greater during adolescence than at any other time of life. Because there is a large individual variation in the age at which these growth and developmental changes occur, the stage of maturation is often a better indicator of nutritional requirements than actual chronological age. The best indicators of adequate intake are satiety and growth that follows the curve of the growth charts.

Energy and Energy-Yielding Nutrients The proportion of energy from carbohydrate, fat, and protein recommended for adolescents is similar to that of adults, but the total amount of energy needed by teenagers exceeds adult needs. Energy requirements for boys are greater than those for girls because boys have more muscle and a greater body size. Adolescent girls need 2,100–2,400 kcal/day, whereas boys require about 2,200–3,150 kcal/day (see Figure 15.6). Protein requirements for both groups reach the adult recommendation of 0.8 g/kg by about age 19, but since boys are generally heavier, they require more total protein than girls. These higher requirements for males continue throughout life. The CCHS found that most Canadian children (ages 9–13 years) and adolescents (ages 14–18 years) consumed macronutrients within the AMDRs.[4]

Micronutrients The requirements for many vitamins and minerals are greater during adolescence than childhood. Some of these are increased because of increased energy needs and others are increased because of their roles in growth and maturation.

Vitamins The need for most of the vitamins rises to adult levels during adolescence. The requirement for B vitamins, which are involved in energy metabolism, is much higher in adolescence than in childhood because of higher energy needs. The rapid growth of adolescence further increases the need for vitamin B_6, which is important for protein synthesis, and for folate and vitamin B_{12}, which are essential for cell division. The CCHS found that Canadian boys age 9-13 had adequate intakes of most vitamins, except vitamin A, for which 12% were not meeting their requirements from food intake alone. In boys 14–18 years, this trend continues with 38% not meeting their vitamin A requirements. Intakes of vitamin A were similarly poor for adolescent girls, with 23% of 9-13-year-olds, and 42% of those aged 14–18 years, not meeting their requirements. These numbers, because they exceed 10%, are areas of concern (See Chapter 8: Critical Thinking: How are Canadians Doing with Respect to Their Intake, from Food, of Water Soluble Vitamins?). These low intakes are consistent with a low consumption of vegetables and fruits. Girls aged 14–18 years were also found to have low intakes of vitamin B_6 (11% not meeting requirements), folate (20% not meeting requirements), and vitamin B_{12} (16% not meeting requirements).[4]

Iron Iron-deficiency anemia is common in adolescence. Iron is needed to synthesize hemoglobin for the expansion of blood volume and myoglobin for the increase in muscle mass. Because blood volume expands at a faster rate in boys than in girls, boys require more iron for tissue synthesis than girls. However, the iron loss due to menstruation makes total needs greater in young women. The RDA is set at 11 mg/day for boys (this is greater than the 8 mg RDA for adult men) and 15 mg/day for girls ages 14–18.[55] Girls are more likely than boys to consume less than the recommended amount because they require more iron, tend to eat fewer iron-rich foods, and consume fewer overall kcalories. This is confirmed by the CCHS, which found that most adolescent boys were meeting their requirements, but that almost 12% of adolescent girls were not meeting their iron requirements, putting them at risk for iron-deficiency anemia.[4]

Calcium The adolescent growth spurt increases both the length and the mass of bones, and adequate calcium is essential to form healthy bone. Calcium retention varies with growth rate, with the fastest-growing adolescents retaining the most calcium. The RDA for calcium during adolescence is 1,300 mg/day for both sexes (300 mg greater than the RDA for adults ages 19–50), but intake is typically below this in both adolescent boys and girls.[56] From the CCHS data, it can be estimated that around 70% of girls (ages 9–13 years) and 30% of boys (ages 14–18 years) are not meeting their requirements for calcium.[57] This low intake in combination with a sedentary lifestyle in childhood and adolescence can impede skeletal growth and bone mineralization and increase the risk of developing osteoporosis later in life.

Foods common in the teen diet that are good sources of calcium include milk, yogurt and frozen yogurt, ice cream, and cheese added to hamburgers, nachos, and pizza. Although milk and cheese are the biggest sources of calcium in teen diets, they can be high in saturated fat, so adolescents should be encouraged to consume reduced-fat dairy products and vegetable sources of calcium. One factor that has contributed to low calcium intake among teens is the use of soft drinks as a beverage, rather than milk (**Figure 15.21**). In addition to calcium, milk is an important source of vitamin D, phosphorus, magnesium, potassium, protein, riboflavin, vitamin A, and zinc. Adolescent girls are likely to drink less milk in order to cut kcalories, favouring low-kcalorie soft drinks (see Chapter 11: Your Choice: Choose Your Beverage Wisely).

Zinc During adolescence, the increase in protein synthesis required for the growth of skeletal muscle and the development of organs increases the need for zinc. The RDA is 11 mg/day for boys and 9 mg/day for girls ages 14-18. A long-term deficiency results in growth retardation and altered sexual development. Although severe zinc deficiency is rare in developed countries, even mild deficiency can cause poor growth, affect appetite and taste, impair immune response, and interfere with vitamin A metabolism. Since adolescents are growing rapidly and maturing sexually, adequate zinc is essential for this age group.[55] Good sources of zinc include meats and whole grains. The CCHS found that most boys met their requirements, but about 15% of girls aged 9–13, and 20% of girls aged 14–18, were not meeting their requirements.[4]

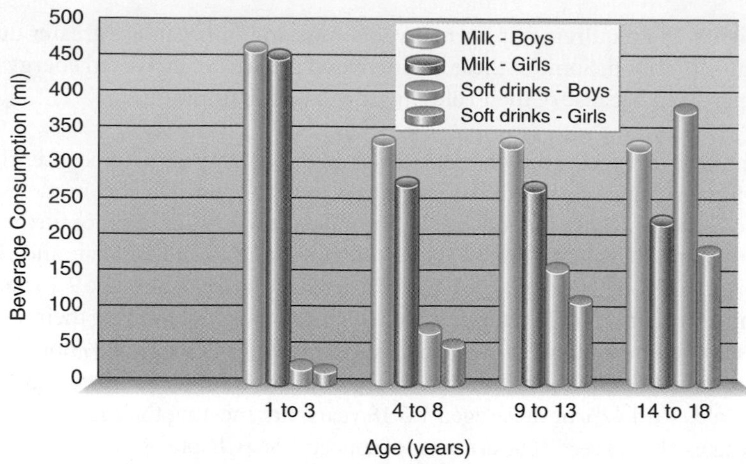

Figure 15.21 Average consumption of milk and soft drinks (ml) among Canadian children and teens. Consumption of milk declines while that of soft drinks increases with age.
Source: Garriguet, D. Beverage consumption of children and teens. *Health Reports* 19(4):1-6, 2008. Available online at www.statcan.gc.ca/pub/82-003-x/2008004/article/6500228-eng.htm. Accessed December 26, 2010.

Most of the micronutrient shortfalls in the diets of children and adolescents are linked to low intakes of vegetables and fruits, whole grains, and milk and alternatives.

Meeting Adolescent Nutrient Needs

During adolescence, physiological changes dictate nutritional needs but peer pressure may dictate food choices. Parents often have little control over what adolescents eat and skipped meals and meals away from home are common. A food is more likely to be selected because it tastes good, it is easy to grab on the go, or friends are eating it than because it is healthy. No matter when foods are consumed throughout the day, an adolescent's diet should follow the recommendations of Canada's Food Guide for the appropriate age, sex, and activity level (**Figure 15.22**). For example, an 18-year-old boy would need 7 servings of grains daily. This may seem like a lot of food, but it is not when spread over the course of a day. A large bowl of cereal and a slice of toast for breakfast is 2 servings; two tacos for lunch and crackers after school provide 3 more servings; and a dinner of spaghetti can add another 2.

As shown in Figure 15.22, the diet should also provide 8 servings of vegetables and fruit. Vegetables and fruit is the food group most likely to be lacking in the teen diet. Potatoes in the form of french fries are widely consumed but fries, which are high in fat and salt, are more properly categorized as "food to limit." Sources of fruits and vegetables acceptable to teens include fruit juice, salads, and tomato sauce and vegetables on pizza and spaghetti.

Snacks are an important part of the adolescent diet; they provide about a quarter of the kcalories for a typical teen. Unfortunately, many of these snacks are burgers, fries, potato chips, candy, cookies, and other high-fat, high-sodium, high-sugar foods. Most of these choices are low in calcium, fibre, folate, vitamins A and C, and iron, and make the typical teen diet, especially that of teenage boys, high in fat, saturated fat, cholesterol, and sodium. To improve the teen diet, meals and snacks offered at home should be low in saturated fat and sodium and high in dairy products, vegetables, and fruits.

15.5 Special Concerns of Teenagers

Learning Objectives

- Explain how a vegetarian diet could be high in saturated fat.
- Discuss how teens' concerns for appearance and performance can affect their nutritional status.
- Describe why peer pressure affects nutrition in teenagers.

As teens grow and become more independent, they are responsible for decisions about their diet and lifestyle. Choices they make regarding the foods they eat, their appearance, and their

Recommended Number of Food Guide Servings per Day

Age in Years	Children 2-3	Children 4-8	Children 9-13	Teens 14-18		Adults 19-50		Adults 51+	
Sex	Girls and Boys			Females	Males	Females	Males	Females	Males
Vegetables and Fruit	4	5	6	7	8	7-8	8-10	7	7
Grain Products	3	4	6	6	7	6-7	8	6	7
Milk and Alternatives	2	2	3-4	3-4	3-4	2	2	3	3
Meat and Alternatives	1	1	1-2	2	3	2	3	2	3

Figure 15.22 Canada's Food Guide recommendations for teens
Canada's Food Guide recommends amounts from each food group based on age and sex.
Source: with permission from Health Canada. Eating Well with Canada's Food Guide,. Available online at http://www.hc-sc.gc.ca/fn-an/alt_formats/hpfb-dgpsa/pdf/food-guide-aliment/print_eatwell_bienmang-eng.pdf. Accessed Sept 10, 2011.

physical activities, as well as the use of tobacco, alcohol, and illegal drugs can affect their nutrition as well as their overall health.

Vegetarian Diets

Some children and adolescents consume vegetarian diets (described in more detail in Section 6.7) because their families are vegetarians, but teens may also decide to consume a vegetarian diet even if the rest of the family does not. Some do it for health reasons or to lose weight, but most give up meat because they are concerned about animals and the environment. Although there are no Canadian data, American studies suggest that about 2% of children and teens ages 6–17 are vegetarian.[58]

We typically think of vegetarian diets as healthful alternatives and for many teens, they are. One American study found that teenage vegetarians were much more likely than their meat-eating counterparts to consume less than 30% of kcalories from fat and less than 10% from saturated fat, and to have 5 servings of vegetables and fruit.[59] However, as with any diet, vegetarian foods must be carefully chosen to meet needs and avoid excesses. Meatless diets can be low in iron and zinc, and vegan diets, which contain no animal products, may put teens at risk of vitamin B_{12} deficiency and inadequate calcium and vitamin D intake. If the diet includes high-fat dairy products it can be high in saturated fat and cholesterol. If only refined grains are chosen and fruit and vegetable intake goals are not met, the diet can be low in fibre. For example, a slice of cheese pizza and a can of cola provide calcium but are low in iron and fibre and high in fat and sugar. In contrast, a vegetarian lunch of whole grain pita bread stuffed with hummus (chickpea dip), tomatoes, and spinach, along with some dried fruit and a glass of reduced-fat milk, is low in saturated fat and cholesterol, high in complex carbohydrate and fibre, and contains good sources of calcium and plant sources of iron. The key to a healthy vegetarian diet, like all healthy diets, is to choose a variety of nutrient-dense vegetables, fruits, grains, legumes, and milk and alternatives.

Eating for Appearance and Performance

Appearance is probably of more concern during adolescence than at any other time of life. Many girls want to lose weight even if they are not overweight. Some boys also want to reduce their weight, but more want to gain weight to achieve a muscular, strong appearance and enhance their athletic abilities (see Critical Thinking: Less Food May Not Mean Fewer Kcalories).

Eating Disorders An eating disorder is a persistent disturbance in eating behaviour or other behaviours intended to control weight that affects physical health and psychosocial functioning (see Focus on Eating Disorders). Eating disorders typically begin in adolescence, although the excessive concern about weight and body image that characterizes these conditions may begin as early as the preschool years. As children grow, the stresses of taking on the responsibilities of adulthood, combined with pressure from peers and society to be thin, may contribute to the development of eating disorders. It is estimated that 0.5% of adolescent females have anorexia and 1%–5% have bulimia.[60] Disordered eating is often hidden by other eating patterns. For example, many women choose vegetarian diets for weight control and vegetarian college women have been found to be at greater risk of disordered eating than non-vegetarians.[61] The nutritional consequences of an eating disorder can affect growth and development and have a lifelong impact on health.

The Impact of Athletics Despite all the benefits of exercise, nutrition misinformation and the desire to excel in a sport can cause adolescent athletes to take dietary supplements, use anabolic steroids, or consume inappropriate training diets and experiment with fad diets, all of which can impact health (see Section 13.4).

Teen athletes may require more water, energy, protein, carbohydrate, and micronutrients than their less active peers, but supplements are rarely needed to meet these needs. If the extra energy needs of teen athletes are met with whole grains, fresh fruits and vegetables, and dairy products, their protein, carbohydrate, and micronutrient needs will easily be met. An exception is iron, which may need to be supplemented, particularly in female athletes. The combination of poor iron intake, iron losses from menstruation and sweat, and increased needs for building new lean tissue puts many female athletes at risk for iron-deficiency anemia.[62]

Many of the most dangerous practices associated with adolescent sports are those that attempt to control body weight. Some sports such as football demand that the athlete be large and heavy (**Figure 15.23**). In order to "bulk up," high school athletes may experiment with anabolic steroids, androstenedione, and creatine. Anabolic steroids are illegal, and although they do increase muscle mass, the risks far outweigh the benefits (see Chapter 13: Your Choice: Ergogenic Hormones: Anything for an Edge). Creatine improves exercise performance in sports requiring short bursts of activity and has not been associated with serious side effects.[63] Nonetheless, the best and safest way for young athletes to increase muscle mass is the hard way: lift weights and eat more. Lifting weights 3 times a week will stimulate the muscles to enlarge and adding snacks such as milkshakes and peanut butter sandwiches will provide the energy and protein needed to support muscle growth.

Female athletes involved in physical activities that require lean, light bodies, such as gymnastics and ballet, are likely to abuse weight-loss diets. Sexual maturation, which causes an increase in body fat and changes in weight distribution, can be disturbing to young women involved in such activities. The combination of hard training and weight restriction can lead to a syndrome known as the female athlete triad that includes disordered eating, amenorrhea, and osteoporosis[64] (see Focus on Eating Disorders and Section 13.4).

Figure 15.23 High school football players often use dietary supplements and illegal ergogenic aids to gain muscle mass and strength. (Dennis MacDonald/Age Fotostock America, Inc.)

Weight loss is also a concern for adolescents participating in sports such as wrestling that require athletes to fit into a specific weight class on the day of the event. In these athletes, dangerous methods of quick weight loss—such as severe energy intake restriction, water deprivation, self-induced vomiting, and diuretic and laxative abuse—are common practice. Low-energy diets can interfere with normal growth and may be too limited in variety to meet these athletes' needs for vitamins and minerals. Even more of a danger is the practice of restricting water intake and encouraging sweat loss to decrease body weight. This may achieve the temporary weight loss necessary to put the athlete in a lower weight class, but dehydration is dangerous and can impair athletic performance (see sections 10.1 and 13.4).

Oral Contraceptive Use

Oral contraceptive hormones may be prescribed to adolescent girls for a number of reasons and can change nutritional status because they affect nutrient metabolism. Oral contraceptives may cause a rise in fasting blood sugar and a tendency toward abnormal glucose tolerance in those with a family history of diabetes. They may also cause changes in body composition, including weight gain due to water retention and an increase in lean body mass. Oral contraceptives may reduce the need for iron by reducing menstrual flow and increasing iron absorption. Therefore, a special RDA for iron of 11.4 mg/day has been established for those taking oral contraceptives.[55]

CRITICAL THINKING:

Less Food May Not Mean Fewer Kcalories

Background

Jenny is a busy 16-year-old high school student. Until recently she hadn't paid much attention to her diet because she ate all her meals at home or at school. Now she has a part-time job and frequently eats dinner and snacks on her own. She notices that she has gained a bit of weight and decides that she should change her diet. Despite eating less food with her new diet, she continues to gain weight.

(Image Source/Getty Images, Inc.)

Data

JENNY'S DIET		JENNY'S NEW DIET	
FOOD	**ENERGY (KCAL)**	**FOOD**	**ENERGY (KCAL)**
Breakfast			
Corn flakes	97	Bagel	187
Low-fat milk	120		
Orange juice	112		
Toast	140		
Margarine	68		
Lunch			
Hamburger	260		
Apple	81		
Low-fat milk	120		
Corn chips	153		
Snack			
Slice of pizza	200	Frozen yogurt	288
Cola	185	Candy bar	300
		Potato chips	230
Dinner			
Ham and cheese sandwich	350	Double burger	576
Potato chips	150	French fries	315
Cola	185	Vanilla shake	503
Total	**2,221**		**2,319**

Critical Thinking Questions

Why is Jenny continuing to gain weight on her new diet despite eating a smaller breakfast and skipping lunch?

How could Jenny modify her new diet to reduce her kcalorie intake and still fit her busy schedule?

iProfile Use iProfile to find foods that are nutrient-dense and filling, but still low in kcalories.

Teenage Pregnancy

Because adolescent girls continue to grow and mature for several years after menstruation starts, the diet of a pregnant teenager must meet her own nutrient needs for growth and development as well as the needs of pregnancy. These elevated needs put the pregnant adolescent at nutritional risk. In order for the mother and fetus to remain healthy, special attention must be paid to all aspects of prenatal care, including nutrient intake (see Section 14.3). Due to the special nutrient needs of this group, the DRIs have included a life-stage group for pregnant girls age 18 or younger.

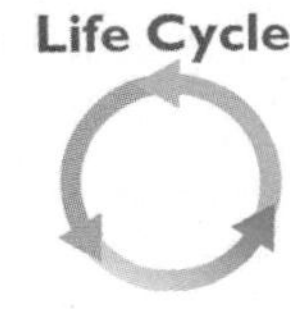

Life Cycle

Tobacco Use

In 2008, 3% of students in grades 6-9 and 13% of students in grades 10-12 reported being smokers.[65] Smoking affects hunger, body weight, and nutritional status, and increases the risk of cardiovascular disease and lung cancer. Many teens start smoking in order to promote weight loss or maintenance. Because smoking is associated with lower body weights in adult women, some teens believe smoking will curb their appetite and help them stay thin or lose weight.[66] Smokers often do not want to quit because they fear that they will gain weight. Smoking impacts nutrient intake. A comparison of the diets of smokers and nonsmokers found that smokers had higher intakes of total and saturated fat and consumed fewer fruits and vegetables leading to lower intakes of folate, vitamin C, and fibre.[67] This dietary pattern can affect nutritional status and further increase the risk of developing heart disease and cancer. In addition to being associated with lower intakes of vitamin C, smoking increases oxidative stress raising the requirement for vitamin C. The vitamin is needed to neutralize the reactive oxygen species created by cigarette smoke. As a result the DRIs recommend that smokers consume an extra 35 mg/day (see Section 8.10).[68]

Alcohol Use

Although it is illegal to sell alcohol to those under the legal drinking age, alcoholic beverages are commonly available at teen social gatherings, and the peer pressure to consume them is strong. Among Canadian youth, 91% report having consumed alcohol and 46% report drinking heavily on a monthly basis.[69] This often includes **binge drinking**. Alcohol is a drug that has short-term effects that occur soon after ingestion and long-term health consequences that are associated with overuse. It provides 7 kcal/g but no nutrients. Alcohol consumption displaces foods that are nutritious. Once ingested, alcohol alters nutrient absorption and metabolism. The metabolism of alcohol as well as its impact on nutritional and overall health are discussed in Focus on Alcohol.

binge drinking The consumption of 5 or more drinks in a row for males or 4 or more for females.

CASE STUDY OUTCOME

After a few months of consuming healthy, low-kcalorie meals and exercising regularly, Felicia has lost a little weight. More important, she has begun to feel better about herself and is even more motivated to change her old habits to improve her health and appearance. Her fitness has improved enough for her to ride her bike to school. She has begun helping her mother find healthy recipes and has learned to make lower-kcalorie choices when she eats out or snacks with friends. For example, instead of ice cream she has sorbet or low-fat frozen yogurt, and she passes on burgers and fries in favour of a grilled chicken sandwich and a salad. When she is hungry, she reaches for fruit or raw vegetables. Felicia now takes a multivitamin and mineral supplement to make sure she gets enough iron and drinks fat-free milk with her meals for a low-kcalorie source of calcium. She still eats lunch in the school cafeteria, but now she makes better choices from the foods offered. This has been made easier since her school became involved in a nutrition program to promote the consumption of fruits and vegetables.

Felicia, now 15 years old, has grown 7.5 cm (3 inches) and gained only 2.2 kg (5 pounds) over the last 2 years. Her BMI is at about the 90th percentile. Her self-confidence has increased so much that she is considering trying out for the field hockey team next year. If she grows another 5 cm (2 inches) and gains weight slowly, by the time she is 17 she will have a BMI in the healthy range.

APPLICATIONS

Personal Nutrition

1. What food groups are included in a fast-food lunch?

a. How much food from each food group of Canada's Food Guide is provided by a Big Mac, a small order of fries, and a 591 ml cola?

b. If you consumed this fast-food lunch, how much more food from each group would you need to consume to satisfy the daily recommendations of Canada's Food Guide?

c. Could you consume this fast-food lunch and meet the Canada's Food Guide requirements without going over your EER?

2. What's in your favourite fast-food meal?

iProfile

a. Use iProfile to look up the nutrient composition of your favourite fast-food meal.

b. What is the percent of kcalories from carbohydrate and fat in the meal?

c. Compare the amount of energy, fat, protein, iron, calcium, vitamin C, and vitamin A in this meal to the recommended amounts of each for a person of your age and gender.

General Nutrition Issues

1. Is this girl growing normally?

Age	Height (cm)	Weight (kg)
6	114	20
7	122	24
8	127	35
9	132	44

a. Use the height and weight measurements recorded in the table above for a girl from age 6 to age 9 to calculate her BMI at each age and plot these values on the appropriate BMI-for-age growth chart in Figure 15.5.

b. What recommendations do you have about this girl's weight?

2. What recommendations would you make for each of the following young athletes?

a. Frank is a wrestler. His usual weight is just at the low end of a weight class and he wants to lose enough weight to be in the next lower class.

b. Sam is a football player. He has been getting hit hard in the last few games and wants to bulk up. He is looking into taking supplements to speed up the process.

c. Talia is a dancer who has recently gained some weight. Her instructor notices that she no longer drinks or snacks during practice.

SUMMARY

15.1 Starting Right for a Healthy Life

- The current diet of Canadian children is low in fruits and vegetables and whole grains and high in saturated fat and sodium.
- Good nutrition during childhood sets the stage for nutrition and health in the adult years.
- Healthy eating habits learned in childhood will extend into the adult years.

15.2 Nourishing Infants, Toddlers, and Young Children

- Growth that follows standard patterns is the best indicator of adequate nutrition. BMI-for-age growth charts can be used to evaluate whether a child is normal weight, underweight, overweight, or obese.
- Energy and protein needs per kilogram of body weight decrease as children grow, but total needs increase because of the increase in total body weight and activity level. Children need a greater percentage of fat and the same percentage of carbohydrates as adults. Dietary carbohydrates should come primarily from whole grains, vegetables, fruits, and milk. Baked goods, candy, and soda should be limited.
- Under most situations, water needs in children can be met by consuming enough fluid to alleviate thirst. Activity and a hot environment increase water needs. As with adults, the typical sodium intake in children and teens currently exceeds the recommended amounts.
- Calcium and iron intakes are important in children's diets. Adequate calcium is important for preventing osteoporosis later in lifeand iron intake is essential for healthy red blood cells.
- Introducing solid foods at 6 months of age adds iron and other nutrients to the diet and aids in muscle development. Newly introduced foods should be appropriate to the child's stage of development and offered one at a time to monitor for food allergies.
- Food allergies are caused by the absorption of allergens, most of which are proteins. Food allergies involve antibody production by the immune system and are more common in infancy because the infant's immature gastrointestinal tract is more likely to absorb whole proteins. Specific foods that cause allergies can be identified by an elimination diet and food challenge. Unlike food allergies, food intolerances do not involve the immune system.
- In order to meet nutrient needs and develop nutritious habits, a variety of healthy foods should be offered at meals and snacks throughout the day. Some children can benefit from vitamin/mineral supplements, but children who consume a well-selected, varied diet can meet all their vitamin and mineral requirements with food.

15.3 Nutrition and Health Concerns in Children

- Dietary patterns high in sugars combined with poor dental hygiene put children at risk of dental caries. Sugar intake as well as food additives have been blamed for hyperactive behaviour, but there is little evidence that these are the cause of hyperactivity.

- Children's health is harmed by lead. Lead disrupts the activity of neurotransmitters and thus interferes with the functioning of the nervous system. Due to reductions in the use of lead the number of children with high blood lead levels has been declining, but the incidence remains higher in certain minorities.
- Obesity is a growing problem among children. Watching television contributes to childhood obesity by promoting high-fat, high-sugar, and high-salt foods and by reducing the amount of exercise children get. Fast food can also add excess kcalories to a child's diet unless nutrient-dense choices are made throughout the rest of the day. Solutions to the problem of childhood obesity involve action from government, industry, health-care providers, communities, schools, and parents and families. Individuals who are overweight need to change behaviours to decrease food intake and increase activity.

15.4 Adolescents

- During adolescence, body composition changes and the nutritional requirements of boys and girls diverge. Males gain more lean body tissue, while females have a greater increase in body fat.
- During adolescence, accelerated growth and sexual maturation have an impact on nutrient requirements. Total energy and protein requirements are higher than at any other time of life. Vitamin requirements increase to meet the needs of rapid growth. The minerals calcium, iron, and zinc are likely to be low in the adolescent diet. Calcium intake is often low because teens are drinking pop instead of milk. Iron-deficiency anemia is common, especially in girls as they begin losing iron through menstruation.
- Adolescent nutrient needs can be met by following the recommendations of Canada's Food Guide. Since meals are frequently missed, healthy snacks should be included in the diet.

15.5 Special Concerns of Teenagers

- Teens often turn to vegetarian diets, sometimes for environmental reasons and sometimes for weight control. Poorly planned vegetarian diets can put teens at risk of iron, zinc, calcium, vitamin D, or vitamin B_{12} deficiency. Poor food choices can result in vegetarian diets that are high in saturated fat and cholesterol and low in fibre.
- Psychosocial changes occurring during the adolescent years make physical appearance of great concern. Eating disorders are more common in adolescence than at any other time. Adolescent athletes are susceptible to nutrition misinformation, and they may experiment with dangerous practices such as using anabolic steroids to increase muscle mass or fad diets and fluid restriction to lose weight.
- During the teen years, pregnancy, the use of oral contraceptives, and tobacco may affect nutritional status. Alcohol consumption and overconsumption often start in the teen years and can have negative nutritional and social consequences.

REVIEW QUESTIONS

1. How does nutrient intake during childhood affect health later in life?
2. What improvements should be made in the diets of children?
3. How can parents and caregivers influence children's food choices?
4. What is the best way to determine if a child is eating enough?
5. What factors influence the maximum height a child will reach?
6. What nutritional problems can be signalled by sudden changes in weight patterns?
7. How do the recommendations for fat intake change as a child gets older?
8. When should solid and semisolid foods be introduced into an infant's diet?
9. How should new foods be introduced to monitor for the development of food allergies?
10. Why is anemia a problem in teenage girls?
11. Why are snacks an important part of children's diets?
12. Why is breakfast important?
13. Why are malnourished children at greater risk of lead toxicity than adults or well-nourished children?
14. How does screen time at the TV or computer impact the nutritional status of children?
15. How can fast foods be incorporated into a healthy diet?
16. How does the treatment of obesity in children differ from treatment in adults?
17. What is the adolescent growth spurt? How does it affect nutrient requirements?
18. Describe two physiological differences between males and females after puberty that affect their nutrient needs.
19. Explain why soft drink intake among teens may be contributing to osteoporosis.
20. Why are vegetarian diets not always healthier than diets that include meat?
21. Why might participation in athletics contribute to the development of eating disorders in adolescents?

REFERENCES

1. Garriguet, D. Diet quality in Canada. Health Reports 20(3):1-11, 2009. Available online at http://www.statcan.gc.ca/pub/82-003-x/2009003/article/10914-eng.htm. Accessed December 19, 2010.
2. Garriguet, D. Canadians' Eating Habits. Health Reports 18(2):17-32, 2007. Available online at http://www.statcan.gc.ca/bsolc/olc-cel/olc-cel?catno=82-003-X20060069609&lang=eng. Accessed December 19, 2010.
3. Health Canada. Do Canadian children meet their nutrient requirements through food intake alone? Cat H164-112/1-2009E-PDF, 2009. Available online at http://www.hc-sc.gc.ca/fn-an/surveill/nutrition/commun/art-nutr-child-enf-eng.php. Accessed December 19, 2010.
4. Health Canada. Do Canadian adolescents meet their nutrient requirements through food intake alone? Cat H164-112/2-2009E-PDF, 2009. Available online at http://www.hc-sc.gc.ca/fn-an/surveill/nutrition/commun/art-nutr-adol-eng.php. Accessed December 19, 2010.
5. Tremblay, M. S., Shields, M., Laviolette, M. et al. Fitness of Canadian Children and Youth. Results of the 2007 to 2009 Canadian Health Measures Survey Health Reports 21(1): 1-14, 2010. Available online at http://www.statcan.gc.ca/pub/82-003-x/2010001/article/11065-eng.htm. Accessed December 19, 2010.

6. DHHS, 2008 Physical Activity Guidelines for Americans. Available online at http://www.health.gov/paguidelines/pdf/paguide.pdf. Accessed May 13, 2009.

7. Ogden, C. L., Carroll, M. D., and Flegal, K. M. High body mass index for age among U.S. children and adolescents, 2003–2006. *JAMA* 299:2401–2405, 2008.

8. National Diabetes Information Clearinghouse. *National Diabetes Statistics, 2007*. Available online at http://diabetes.niddk.nih.gov/dm/pubs/statistics/index.htm#13. Accessed April 30, 2009.

9. Mayer-Davis, E. J. Type 2 diabetes in youth: Epidemiology and current research toward prevention and treatment. *J. Am. Diet. Assoc.* 108:S45–S51, 2008.

10. Vuksan, V., Peeva, V., Rogovik, A. et al. The metabolic syndrome in healthy, multiethnic adolescents in Toronto, Ontario: The use of fasting blood glucose as a simple indicator. *Can. J. Cardiol.* 26(3):128-132, 2010.

11. McCrindle, B. W., Manlhiot, C., Millar, K. et al. Population trends toward increasing cardiovascular risk factors in Canadian adolescents. *J. Pediatr.* 157:837-843, 2010.

12. Paradis, G., Tremblay, M. S., Janssen, I. et al. Blood pressure in Canadian children and adolescents. *Health Rep.* 21(2):15-22, 2010.

13. Wang, F., Wild, T. C., Kipp, W. et al. The influence of childhood obesity on the development of self-esteem. Health Reports 20(2):21-27, 2009. Available online at http://www.statcan.gc.ca/pub/82-003-x/2009002/article/10871-eng.htm. Assessed December 10, 2010.

14. Dietitians of Canada. WHO Growth Charts adapted for Canada Available online at http://www.dietitians.ca/Secondary-Pages/Public/Who-Growth-Charts.aspx. Accessed Sept 11, 2011.

15. Institute of Medicine, Food and Nutrition Board. Dietary Reference Intakes for Energy, Carbohydrate, Fiber, Fat, Protein and Amino Acids. Washington, DC: National Academies Press, 2002.

16. Nicklas, T. A., and Hayes, D. Position of the American Dietetic Association: Nutrition guidance for healthy children ages 2 to 11 years. *J. Am. Diet. Assoc.* 108:1038–1047, 2008.

17. Institute of Medicine, Food and Nutrition Board. Dietary Reference Intakes for Water, Potassium, Sodium, Chloride, and Sulfate. Washington, DC: National Academies Press, 2004.

18. Haas, J. D., and Brownlie, T. IV. Iron deficiency and reduced work capacity: A critical review of the research to determine a causal relationship. J. Nutr. 131:676S–690S, 2001.

19. Grantham-McGregor, S., and Ani, C. A review of studies on the effect of iron deficiency on cognitive development in children. J. Nutr. 131:649S–668S, 2001.

20. U.S. Food and Drug Administration. Preventing iron poisoning in children. FDA Backgrounder. January 15, 1997. Available online at vm.cfsan.fda.gov/~dms/bgiron.html. Accessed September 29, 2009.

21. Health Canada. Nutrition for Healthy Term Infants - Statement of the Joint Working Group: Canadian Paediatric Society, Dietitians of Canada and Health Canada. Available online at http://www.hc-sc.gc.ca/fn-an/nutrition/infant-nourisson/vita_d_supp-eng.php. Accessed June 22, 2011.

22. Macknin, M. L., Medendorp, S. V., and Maier, M. C. Infant sleep and bedtime cereal. *Am. J. Dis. Child.* 143:1066–1068, 1989.

23. Canadian Paediatric Society. Healthy active living for children and youth, Available online at http://www.cps.ca/caringforkids/growinglearning/HealthyActive.htm. Accessed December 26, 2010.

24. Formanek, R., Food allergies: When food becomes the enemy. *FDA Consumer*, July/Aug, 2001. Available online at www.fda.gov/fdac/features/2001/401_food.html/. Accessed November 17, 2003.

25. Greer, F. R., Sicherer, S. H., and Burks, A. W. Effects of early nutritional interventions on the development of atopic disease in infants and children: the role of maternal dietary restriction, breastfeeding, timing of introduction of complementary foods, and hydrolyzed formulas. *Pediatrics* 121:183–191, 2008.

26. Patrick, H., and Nicklas, T. A. A review of family and social determinants of children's eating patterns and diet quality. *J. Am. Coll. Nutr.* 24:83–92, 2005.

27. Slatter, E. *Child of Mine*. Palo Alto, CA: Bull Publishing, 1986.

28. Winson, A. School food environments and the obesity issue: content, structural determinants, and agency in Canadian high schools. *Agriculture and human values* 25(4):499-511, 2008.

29. Veugelers, P. J. and Fitzgerald, A. L. Effectiveness of school programs in preventing childhood obesity: a multilevel comparison. *Am. J. Public Health* 95(3):432-435, 2005.

30. Public Health Agency of Canada. Canadian best practices portal. Annapolis Valley Health promoting schools. Available online at http://cbpp-pcpe.phac-aspc.gc.ca/intervention/291/view-eng.html. Accessed December 26, 2010.

31. Leo A. Centre for Science in the Public Interest. Are schools making the grade?: School Nutrition Policies across Canada, 2007. Available online at http://www.cspinet.org/canada/pdf/makingthegrade_1007.pdf. Accessed December 26, 2010.

32. Marshall, T. A., Levy, S. M., Broffitt, B., et al. Dental caries and beverage consumption in young children. *Pediatrics.* 112:e184–191, 2003.

33. Cormier, E., and Elder, J. H. Diet and child behavior problems: Fact or fiction? *Pediatr. Nurs.* 33:138–143, 2007.

34. Rojas, N. L., and Chan, E. Old and new controversies in the alternative treatment of attention-deficit hyperactivity disorder. *Ment. Retard. Dev. Disabil. Res. Rev.* 11:116–130, 2005.

35. Chandramouli, K., Steer, C. D., Ellis, M., et al. Effects of early childhood lead exposure on academic performance and behavior of school age children. *Arch. Dis. Child.* September 21, 2009 [Epub].

36. Centers for Disease Control and Prevention. Blood lead levels—United States, 1999–2002. MMWR *Morb. Mortal. Wkly. Rep.* 54:513–516, 2005. Available online at www.cdc.gov/mmwr/preview/mmwrhtml/mm5420a5.htm. Accessed April 20, 2009.

37. Jusko, T. A., Henderson, C. R. Jr, Lanphear, B. P., et al. Blood lead concentrations below 10 microg/dL and child intelligence at 6 years of age. *Environ. Health. Perspect.* 116:243–248, 2008.

38. Health Canada. Lead information package. Available online at http://www.hc-sc.gc.ca/ewh-semt/contaminants/lead-plomb/exposure-exposition-eng.php#a51. Accessed December 26, 2010.

39. Walker, J. B., Houseman, J., Seddon, L. et al. Maternal and umbilical cord blood levels of mercury, lead, cadmium, and essential trace elements in Arctic Canada. *Environmental Research* 100295-318, 2006,

40. Tsuji, L. S. J., Wainman, B. C., Martin, I. D. et al. Elevated blood-lead levels in First Nations people of northern Ontario, Canada; Policy implications. *Bulletin of environmental contamination and toxicology* 80(1):14-18, 2008.

41. Shields, M. Overweight and obesity among children and youth. *Health Reports* 17(3): 27-42, 2006. Available online at http://www.statcan.gc.ca/cgi-bin/af-fdr.cgi?l=eng&loc=http://www.statcan.gc.ca/studies-etudes/82-003/archive/2006/9277-eng.pdf&t=Overweight and obesity among children and youth. Accessed December 19, 2010.

42. Coon, K. A., and Tucker, K. L. Television and children's consumption patterns. A review of the literature. *Minerva Pediatr.* 54:423–436, 2002.

43. Batada, A., Seitz, M., Woxan, M. et. al. Nine out of ten food advertisements shown during Saturday morning children's television programming are for food high in fat, sodium, or added sugars, or low in nutrients. *J. Am. Diet. Assoc.* 108:673–678, 2008.

44. Canadian Children's Food and Beverage Advertising Initiative. Available online at: http://www.adstandards.com/en/childrensinitiative/default.htm. Accessed June 24, 2011.

45. Dietitians of Canada. Advertising of Food and Beverages to Children. Position of Dietitians of Canada, 2010. Available online at http://www.dietitians.ca/Downloadable-Content/Public/Advertising-to-Children-position-paper.aspx. Accessed June 24, 2011.

46. Eisenmann, J. C., Bartee, R. T., and Wang, M. Q. Physical activity, TV viewing, and weight in U.S. youth: 1999 Youth Risk Behavior Survey. *Obes. Res.* 10:379–385, 2002.

47. Public Health Agency of Canada. The chief public health officer's report on the state of public health in Canada 2009: Growing up well: Priorities for a healthy future. Available online at http://www.phac-aspc.gc.ca/cphorsphc-respcacsp/2009/fr-rc/index-eng.php. Accessed December 26, 2010.

48. Public Health Agency of Canada. Canadian Best Practices Portal. Saskatoon in Motion, Available online at http://cbpp-pcpe.phac-aspc.gc.ca/intervention/541/view-eng.html. Accessed September 11, 2011.

49. Canadian Society of Exercise Physiologists. Canada's Physical Activity Guidelines Guide for Healthy Living. Available online at http://www.phac-aspc.gc.ca/hp-ps/hl-mvs/pag-gap/downloads-eng.php. Accessed December 26, 2010.

50. Centers for Disease Control. Making Physical Activity a Part of a Child's Life. Available online at http://www.cdc.gov/physicalactivity/everyone/getactive/children.html. Accessed May 4, 2009.

51. Canadian Fitness and Lifestyle Research Institute. Schools. 2006 Capacity Study. Available online at http://72.10.49.94/pub_page/135. Accessed June 25, 2011.

52. Public Health Agency of Canada. Our Health Our Future. A National Dialogue on Healthy Weights,. Available online at http://ourhealthourfuture.gc.ca/home/. Accessed June 27, 2011.

53. Mandel, D., Zimlichman, E., Mimouni, F. B., et al. Age at menarche and body mass index: A population study. *J. Pediatr. Endocrinol. Metab.* 17:1507–1510, 2004.

54. Weng, F. L., Shults, J., Leonard, M. B. et al. Risk factors for low serum 25-hydroxyvitamin D concentrations in otherwise healthy children and adolescents. *Am. J. Clin. Nutr:* 86:150–158, 2007.

55. Institute of Medicine, Food and Nutrition Board. *Dietary Reference Intakes for Vitamin A, Vitamin K, Arsenic, Boron, Chromium, Copper, Iodine, Iron, Manganese, Molybdenum, Nickel, Silicon, Vanadium, and Zinc.* Washington, DC: National Academies Press, 2001.

56. Food and Nutrition Board, Institute of Medicine. *Dietary Reference Intakes for Calcium and Vitamin D.* Washington DC: National Academies Press, 2011.

57. Health Canada and Statistics Canada. Canada Community Health Survey Cycle 2.2., Nutrient Intakes from Food. Data files on CD. Cat No. 978-0-662-06542-5, 2004 andinterpolation by author (DG).

58. The Vegetarian Resource Group. How Many Teens are Vegetarian? How Many Kids Don't Eat Meat. Available online at www.vrg.org/press/2000novteen.htm/. Accessed September 29, 2009.

59. Zlotkin, S. Adolescent vegetarians: How well do their dietary patterns meet the healthy people 2010 objectives? *Arch. Pediatr. Adolesc. Med.* 156:426–427, 2002.

60. American Academy of Pediatrics. Policy Statement. Identifying and Treating Eating Disorders. *Peds.* 111:204–211, 2003.

61. Klopp, S. A., Heiss, C. J., and Smith, H. S. Self-reported vegetarianism may be a marker for college women at risk for disordered eating. *J. Am. Diet. Assoc.* 103:745–747, 2003.

62. Rodriguiez, N. R., Dimarco, N. M., and Langley, S. Position of the American Dietetic Association, Dietitians of Canada, and the American College of Sports Medicine: Nutrition and athletic performance. *J. Am. Diet. Assoc.* 109:509–527, 2009.

63. Bemben, M. G., and Lamont, H. S. Creatine supplementation and exercise performance: Recent findings. *Sports Med.* 35:107–125, 2005.

64. Kazis, K., and Iglesias, E. The female athlete triad. *Adolesc. Med.* 14:87–95, 2003.

65. Health Canada. Summary of results of the 2008-2009 Youth Smoking Survey. Available online at http://www.hc-sc.gc.ca/hc-ps/tobac-tabac/research-recherche/stat/_survey-sondage_2008-2009/result-eng.php. Accessed December 26, 2010.

66. Facchini, M., Rozensztejn, R., and Gonzalez, C. Smoking and weight control behaviors. *Eat. Weight Disord.* 10:1–7, 2005.

67. Palaniappan, U., Jacobs Starkey, L., O'Loughlin, J., et al. Fruit and vegetable consumption is lower and saturated fat intake is higher among Canadians reporting smoking. *J. Nutr.* 131:1952–1958, 2001.

68. Food and Nutrition Board, Institute of Medicine. *Dietary Reference Intakes for Vitamin C, Vitamin E, Selenium and Carotenoids.* Washington, DC: National Academies Press, 2000.

69. Health Canada. Substance abuse by Canadian youth, Available online at http://www.hc-sc.gc.ca/hc-ps/pubs/adp-apd/cas_youth-etc_jeunes/chap3-eng.php#a1. Accessed September 11, 2011.

(©istockphoto.com/DNY59)

FOCUS ON
Eating Disorders

Focus Outline

Normal eating patterns are flexible. One day, you may eat twice as much as on another. On some days, your food choices are varied and nutritious, while on others, you may survive on snacks and fast food. Lunch one day may include an appetizer, entrée, and dessert and on the next it may be a carton of yogurt grabbed on the run. Normal eating patterns include eating more than we need at a party or other special occasion and less than we need when we are busy or stressed. Normal eating may also involve limiting intake in order to manage weight and meet recommendations for a healthy diet. What and how much people eat varies in response to social occasions, emotions, time limitations, hunger, and the availability of food, but generally, people eat when they are hungry, choose foods they enjoy, and stop eating when they are satisfied. Abnormal eating occurs when a person is overly concerned with food, eating, and body size and shape. When the emotional aspects of food and eating overpower the role of food as nourishment, an **eating disorder** may develop.

eating disorder A persistent disturbance in eating behaviour or other behaviours intended to control weight that affects physical health and psychosocial functioning.

Eating disorders can occur at any point in the lifecycle. The most vulnerable time is during adolescence, although concerns about body image often start in childhood. The recent Canadian Community Health Survey found that 3.8% of Canadian girls and women (aged 15 to 24 years) were at risk for eating disorders and this risk dropped to 2.9% for women 25 to 64 years of age.[1]

CANADIAN CONTENT

A survey of Canadian high school students found almost 30% of girls in grades nine and ten had tried to lose weight in the past year.[2]

F6.1 What Are Eating Disorders?

Learning Objective

- Name the three categories of eating disorders.

Eating disorders are psychological disorders that involve a persistent disturbance in eating patterns or other behaviours intended to control weight. They affect physical and nutritional health and psychosocial functioning. If untreated, eating disorders can be fatal.

According to mental health guidelines, there are three categories of eating disorders. The first, **anorexia nervosa**, is characterized by self-starvation to reduce weight or prevent weight gain. **Bulimia nervosa**, the second category, involves frequent episodes of **bingeing** or **binge eating**, during which extremely large amounts of high-kcalorie foods are consumed. These episodes are almost always followed by depression, guilt, and **purging** behaviours, such as self-induced vomiting, to rid the body of the extra energy. The final category is **eating disorders not otherwise specified (EDNOS)**, which includes abnormal eating behaviours that don't fit into the other two categories. More than 50% of all people who seek treatment for an eating disorder are categorized as EDNOS. For example, someone who has a body weight that is very low but not low enough to be classified as anorexia would fit into the EDNOS category. Someone who binges and purges, but not often enough to be considered bulimic, would also fit the EDNOS category. **Binge-eating disorder**, which involves bingeing without purging, is also included in the EDNOS category[3] (**Table F6.1**).

anorexia nervosa An eating disorder characterized by self-starvation, a distorted body image, and below normal body weight.

bulimia nervosa An eating disorder characterized by the consumption of large amounts of food at one time (binge eating), followed by purging behaviours such as vomiting or the use of laxatives to eliminate kcalories from the body.

bingeing or **binge eating** The rapid consumption of a large amount of food in a discrete period of time associated with a feeling that eating is out of control.

purging Behaviours such as self-induced vomiting and misuse of laxatives, diuretics, or enemas to rid the body of kcalories.

eating disorders not otherwise specified (EDNOS) A category of eating disorders that includes abnormal eating behaviours that don't fit into the anorexia or bulimia nervosa categories.

binge-eating disorder An eating disorder characterized by recurrent episodes of binge eating in the absence of purging behaviour.

Table F6.1 Diagnostic Criteria for Eating Disorders

Anorexia nervosa

- Refusal to maintain body weight at or above 85% of normal weight for age and height.
- Intense fear of gaining weight or becoming fat, even though underweight.
- Disturbance in the way body weight or shape is experienced, or denial of the seriousness of the current low body weight.
- In women, absence of at least three consecutive menstrual cycles without other known cause.

Restricting Type: During the current episode of anorexia nervosa, the person does not regularly engage in binge-eating or purging behaviour (i.e., self-induced vomiting or the misuse of laxatives, diuretics, or enemas).

Binge-Eating Type or **Purging Type:** During the current episode of anorexia nervosa, the person regularly engages in binge-eating or purging behaviour (i.e., self-induced vomiting or the misuse of laxatives, diuretics, or enemas).

Bulimia nervosa

- Recurrent episodes of binge eating.
- Recurrent inappropriate compensatory behaviour to prevent weight gain, such as self-induced vomiting; misuse of laxatives, diuretics, enemas, or other medications; fasting; or excessive exercise.
- Occurrence, on average, of binge eating and inappropriate compensatory behaviours at least twice a week for 3 months.
- Undue influence by body shape and weight on self-evaluation.
- Disturbance does not occur exclusively during episodes of anorexia nervosa.

Purging Type: During the current episode of bulimia nervosa, the person regularly engages in self-induced vomiting or the misuse of laxatives, diuretics, or enemas.

Nonpurging Type: During the current episode of bulimia nervosa, the person uses other inappropriate compensatory behaviours, such as fasting or excessive exercise, but does not regularly engage in self-induced vomiting or the misuse of laxatives, diuretics, or enemas.

Eating disorders not otherwise specified (EDNOS)

- Criteria for anorexia nervosa are met except the individual menstruates regularly.
- Criteria for anorexia nervosa are met except that, despite substantial weight loss, the individual's current weight is in the normal range.
- Criteria for bulimia nervosa are met except binges occur at a frequency of less than twice a week and for a duration of less than 3 months.

Table F6.1 *(Continued)*
• Inappropriate compensatory behaviour after eating small amounts of food in individuals of normal body weight.
• Regularly chewing and spitting out, without swallowing, large amounts of food.
• Recurrent episodes of binge eating in the absence of regular use of inappropriate compensatory behaviours characteristic of bulimia (binge eating disorder).

Source: From the *DSM-IV, Diagnostic and Statistical Manual of Mental Disorders*, 4th ed., TR, Washington DC: American Psychiatric Association, 2000.

F6.2 What Causes Eating Disorders?

Learning Objectives

- Discuss the genetic, psychological, and sociocultural factors that influence the development of eating disorders.
- Describe how a society's body ideal affects the incidence of eating disorders.

Genetic, psychological, and sociocultural factors all contribute to the development of eating disorders (**Figure F6.1**). Eating disorders typically begin in adolescence when physical, psychological, and social developments are occurring rapidly, but they occur in people of all ages, races, and socioeconomic backgrounds. They are more common in women than men. The Public Health Agency of Canada reports that 3% of women will be affected by eating disorders in their lifetime.[4] In an Ontario community, the prevalence of anorexia nervosa and bulimia nervosa was 2.1% in women and 0.3% in men.[5]

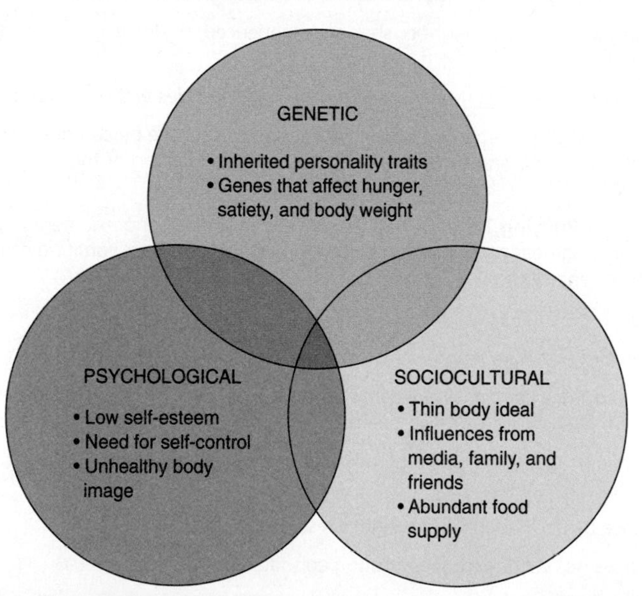

Figure F6.1 Causes of eating disorders
Medical professionals must address genetic, psychological, and sociocultural factors if treatment for eating disorders is to be effective.

Eating disorders occur in the greatest frequency in groups that are concerned about maintaining a low body weight, such as professional dancers and models.[6] They are on the rise among athletes, especially those involved in sports that require the athlete to be thin, such as gymnastics and figure skating, or to fit into a particular weight class, such as wrestling.

The Role of Genetics

Eating disorders are not necessarily passed from parent to child, but the genes you inherit contribute to personality traits and other biological characteristics that might predispose you to developing an eating disorder. For example, inherited abnormalities in the levels of neurotransmitters such as serotonin, which affects food intake, have been hypothesized to contribute to the behaviours typical of anorexia and bulimia.[7] Binge-eating disorder may be linked to a defect in a gene called the melanocortin 4 receptor gene. The protein made by this gene helps control hunger and satiety. If this gene is abnormal and makes too little protein, the body feels too much hunger. In one study, all carriers of the mutant gene were binge eaters and mutations were found in 5% of obese subjects.[8] Genes such as this contribute to eating disorders but a single gene is not likely to be the sole cause. These are complex diseases that are the result of the interaction of multiple genes with the environment. Each gene may have a small effect, but when taken together, they can increase risk severalfold. When placed in an environment conducive to eating disorders, individuals who carry such genes will be more likely to develop one.

Psychological Characteristics

Certain personality characteristics and psychological problems are common among individuals with eating disorders.[6] For example, people with eating disorders often have low **self-esteem.** Self-esteem refers to the judgements people make and maintain about themselves—a general attitude of approval or disapproval that indicates if the person thinks he or she is worthy and capable. Eating disorders are also rooted in the need for self-control. Those with eating disorders are often perfectionists who set very high standards for themselves and others. In order to be perfect, they strive to be in control of their bodies and their lives. They view everything as either a success or a failure. Being fat is seen as failure, thin as success, and thinner as even more successful. In spite of their many achievements, those with eating disorders feel inadequate, defective, and worthless.

self-esteem The general attitude of approval or disapproval that people make and maintain about themselves.

People with eating disorders often try to use their relationship with food to gain control over their lives and boost their self-esteem. They believe that controlling their food intake and weight demonstrates their ability to control other aspects of their lives and solve other problems. Their fixation with food and weight loss and their ability to control their intake and weight help them to feel better about themselves. Even if they feel insecure, helpless, or dissatisfied in other aspects of life, if they are in control of their food intake, weight, and body size, they can associate this control with success. This feeling of being in control can become addictive.

Sociocultural Messages About the Ideal Body

body image The way a person perceives and imagines his or her body.

While genetic and psychological issues may predispose individuals to eating disorders, sociocultural and economic factors are important triggers for the onset of these disorders. An important sociocultural factor is body ideal. What is viewed as an ideal body differs across cultures and has changed throughout history. Ancient drawings and figurines show women with large breasts and swollen abdomens. This plump body ideal is still prevalent today in cultures where food is not readily available. Young women in these cultures may struggle to gain weight to achieve what is viewed as the ideal female body (**Figure F6.2**). In contrast, modern Western women strive to achieve a thin, lean body. The sociocultural ideals about body size are linked to **body image** and the incidence of eating disorders.[9] Eating disorders occur in societies where food is abundant and the body ideal is thin. They do not occur where food is scarce and people must worry about where their next meal is coming from.

Figure F6.2 A fuller figure is still desirable in many cultures, such as the Zulu of South Africa. As television images of very thin Western women become more accessible, the Zulu cultural view of plumpness as desirable may be changing. (SCPhotos/Alamy)

Body Ideal in Modern Society From television and movies to magazines and advertisements and even toys, the modern culture today is a culture of thinness. Messages about what society views as a perfect body—the ideal we should strive for—are constantly delivered by the mass media. The perfect body is long, lean, and muscled. The tall, dark, muscular man gets the girl; the thin, athletic woman gets her man. Thin fashion models adorn billboards and magazine covers to show off the latest fashions (**Figure F6.3**). Being thin is associated with beauty, success, intelligence, and vitality. Being plump, on the other hand, is associated with failure, stupidity, and clumsiness. What young woman would want to be plump when exposed to these negative associations? Although men are also affected by these messages, there is much more emphasis in the media on the appearance of women's bodies. A young woman facing a future where she must be independent, have a prestigious job, maintain a successful love relationship, bear and nurture children, manage a household, and look fashionable can become overwhelmed. Unable to master all these roles, she may look for some aspect of her life that she can control. Food intake and body weight are natural choices, since being thin brings

Figure F6.3 The fashion models whom women want to emulate are thinner than 98% of women. (Andy Washnik)

the societal associations of success. These messages about how we should look are difficult to ignore and can create pressure to achieve this ideal body. But it is a standard that is very hard to meet—a standard that is contributing to disturbances in body image and eating behaviour. This is illustrated by the fact that as the body dimensions of female models, actresses, and other cultural icons have become thinner over the last several decades, the incidence of eating disorders has increased (**Figure F6.4**).

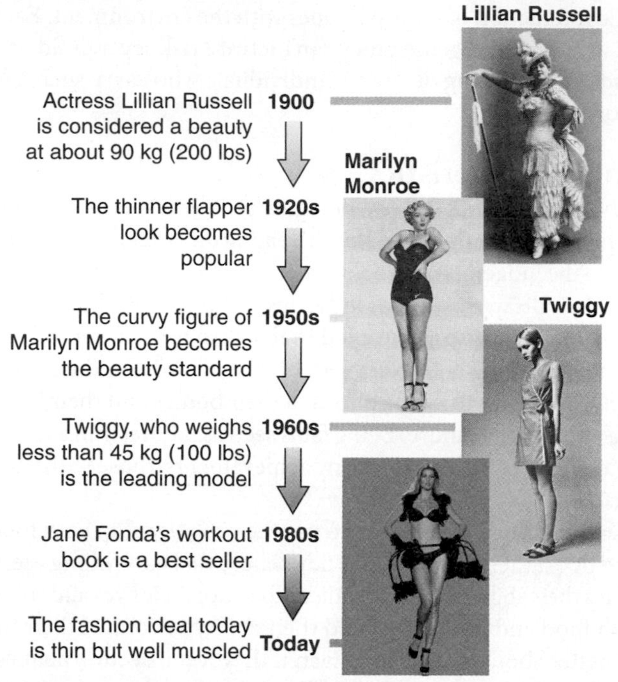

Figure F6.4 The changing female body ideal
As this timeline shows, extreme thinness has not always been the beauty standard. (Bettmann/© Corbis; Bettmann/© Corbis; Bettmann/ © Corbis; Antonio de Moraes Barros Filho/WireImage/Getty Images, Inc.

Body Image and Eating Disorders What individuals think they should look like or wish they would look like is affected by the ideals of their culture and society. Many women and girls, particularly teenage girls, are dissatisfied with their bodies because they look different than what they and their culture see as ideal. The average woman is 163 cm tall, and weighs 64 kg. The average fashion model is 180 cm and weighs 53 kg. The difference between models and average women has been increasing over the years. Forty years ago, the average weight of a model was 8% lower than that of an average woman; today, the difference is more than 25%.[10] Almost everyone has something that they would like to change about their bodies, but for some, this becomes a pathological concern with body weight and shape, and as a result, body image may become distorted. A distorted body image means that an individual is unable to judge the size of his or her own body and does not see him or herself as as he or she really is (**Figure F6.5**). Body image distortion is common with eating disorders, so even if the person achieves a body weight comparable to that of a fashion model, he or she may continue to see himself or herself as fat and strive to lose more weight (**Table F6.2**).

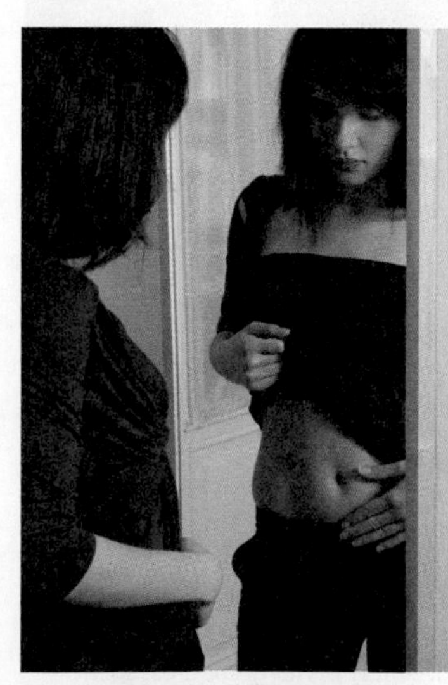

Figure F6.5 When people with a distorted body image look in the mirror, they see themselves as fat even when they are normal or underweight. (David Young Wolff/ PhotoEdit)

Table F6.2 To Maintain a Healthy Body Image

Try to...

- Accept that healthy bodies come in many shapes and sizes.
- Recognize your positive qualities.
- Remember that you can be your worst critic.
- Explore your internal self, as well as your external appearance.
- Spend your time and energy enjoying the positive things in your life.
- Be aware of your own weight prejudice. Explore how your feelings may affect your self-esteem.

And try not to...

- Let your body define who or what you are.
- Judge others on the basis of appearance, body size, or shape.
- Forget that society changes its ideals of beauty over the years.
- Forget that you are not alone in your pursuit of self-acceptance.
- Be afraid to enjoy life.

F6.3 Anorexia Nervosa

Learning Objectives

- Describe the features of anorexia nervosa.
- Explain how anorexia is treated.

"Anorexia" means lack of appetite, but in the case of the eating disorder anorexia nervosa, it is a desire to be thin, rather than a lack of appetite, that causes individuals to decrease their food intake. Anorexia nervosa was first recognized by physicians in the second half of the nineteenth century and the characteristics they described are still true of the syndrome today: severe weight loss, constipation, restlessness, and **amenorrhea** in women. Anorexia nervosa affects 1% of female adolescents in the United States.[11] The average age of onset is 17 years. There is a 5% death rate in the first 2 years, and this can reach 20% in untreated individuals.[6] Canadian trends are likely similar.

amenorrhea Delayed onset of menstruation or the absence of 3 or more consecutive menstrual cycles.

Psychological Issues

The psychological component of anorexia nervosa revolves around an overwhelming fear of gaining weight, even in those who are already underweight (**Figure F6.6**). It is not uncommon for individuals with anorexia to feel that they would rather be dead than fat. Anorexia is also characterized by disturbances in body image or perception of body size that prevent those affected from seeing themselves as underweight even when they are dangerously thin. Those with this disorder may use body weight and shape as a means of self-evaluation: many believe that if they weren't so "fat," then everyone would like and respect them and that they wouldn't have other problems, However, no matter how much weight they lose, individuals with anorexia nervosa do not gain self-respect, inner assurance, or the happiness they seek. Therefore, they continue to restrict their intake and use other behaviours in order to lose weight.

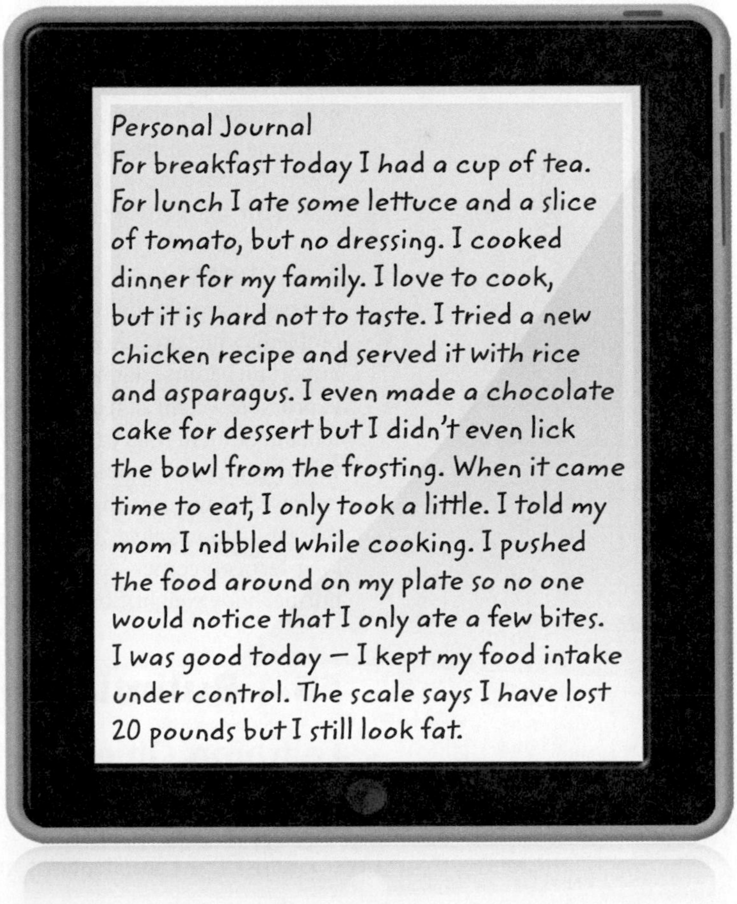

Anorexic Behaviours

The most obvious behaviours associated with anorexia are those that contribute to the maintenance of a body weight that is 15% or more below normal body weight. These behaviours include restriction of food intake, binge eating and purging episodes, eating rituals, and excessive exercise. Based on these behaviours, anorexia is subdivided into two subtypes. Those with the Restricting Type maintain their low body weight solely by restricting their food intake and increasing their activity. Those with the Binge Eating/Purging Type also typically restrict their food intake but in addition, regularly engage in binge eating and/or purging behaviours (see Table F6.1). It is estimated that about half of people with anorexia use purging as a means of weight control.[12]

Personal Journal
For breakfast today I had a cup of tea. For lunch I ate some lettuce and a slice of tomato, but no dressing. I cooked dinner for my family. I love to cook, but it is hard not to taste. I tried a new chicken recipe and served it with rice and asparagus. I even made a chocolate cake for dessert but I didn't even lick the bowl from the frosting. When it came time to eat, I only took a little. I told my mom I nibbled while cooking. I pushed the food around on my plate so no one would notice that I only ate a few bites. I was good today — I kept my food intake under control. The scale says I have lost 20 pounds but I still look fat.

Figure F6.6 **A day in the life of an anorexic**
People with anorexia nervosa carefully regulate what they eat to maintain a very low body weight.

Figure F6.7 People with anorexia nervosa often use exercise as well as food restriction to achieve and maintain a very low body weight. (Rubberball/Getty Images, Inc.)

For individuals with anorexia, food and eating become an obsession. In addition to restricting the total amount of food consumed, anorexics develop personal diet rituals, limiting certain foods and eating them in specific ways. Although they do not consume very much food, they are preoccupied with food and spend an enormous amount of time thinking about food, talking about food, and preparing food for others. Instead of eating, they move the food around the plate and cut it into tiny pieces.

Both hyperactivity and overactivity are behaviours that are also typical of anorexia. This is in contrast to the decrease in activity and fatigue characteristic of other starvation states associated with weight loss. Many anorexics exercise excessively to burn kcalories. For some, the activity is surreptitious, such as going up and down stairs repeatedly or getting off the bus a few stops too early. For others, the activity takes the form of strenuous physical exercise. They may become fanatical athletes and feel guilty if they cannot exercise. The exercise is typically done alone and is performed as a regular, rigid routine. They may link exercise and eating, so a certain amount of exercise earns them the right to eat, and if they eat too much they must pay the price by adding extra exercise. Those who use exercise to increase energy expenditure do not stop when they are tired; instead, they train compulsively beyond reasonable endurance (**Figure F6.7**).

Individuals with anorexia often access pro-anorexia, or "pro-ana" websites that tend to glorify eating disorders, provide tips on how to hide weight loss from family and doctors, and post photos of anorexics as "thinspiration." The Canadian organization National Eating Disorders Information Centre (NEDIC) considers these sites very harmful because they promote practices that are dangerous and make disordered eating seem normal and emaciated bodies seem beautiful.[13]

CANADIAN CONTENT

Physical Symptoms of Anorexia

The first obvious physical manifestation of anorexia is weight loss. As weight loss becomes severe, symptoms of starvation begin to appear. Starvation affects mental function, causing those with anorexia to become apathetic, dull, exhausted, and depressed. Physical symptoms include depletion of fat stores; wasting of muscles; inflammation and swelling of the lips; flaking and peeling of skin; growth of fine hair, called lanugo hair, on the body, and dry, thin, brittle hair on the head that may fall out. In females, estrogen levels drop, and menstruation becomes irregular or stops. This can delay sexual maturation and can have long-term effects on bone density. In males, testosterone levels decrease. In the final stages of starvation, there are abnormalities in electrolyte and fluid balance and cardiac irregularities. Ketones are typically absent because fat stores are depleted. Immune function is suppressed, leading to infections which further increase nutritional needs.

Treatment

The goal of treatment for anorexia nervosa is to help resolve the psychological and behavioural problems while providing for physical and nutritional rehabilitation. Early treatment of anorexia is important because starvation may cause irreversible damage. The goal of nutrition intervention is to promote weight gain by increasing energy intake and expanding dietary choices.[14] Nutritional rehabilitation in mild cases involves learning about nutrition and meal planning in order to develop healthy eating patterns. In more severe cases, hospitalization is required so food intake and exercise behaviours can be carefully controlled. Intravenous nutrition may be necessary to keep these individuals alive. Although some people recover fully from anorexia, about half have poor long-term outcomes—remaining irrationally concerned about weight gain and never achieving normal body weight. Some patients with anorexia also transition to bulimia nervosa.[15]

F6.4 Bulimia Nervosa

Learning Objectives

- Explain what is meant by the binge/purge cycle.
- Discuss the complications and treatment of bulimia nervosa.

"Bulimia" is from the Greek *bous* (ox) and *limos* (hunger), denoting hunger of such intensity that a person could "eat an ox." The modern concept of bulimia nervosa as an eating disorder arose in the early 1970s, when a set of symptoms was identified and distinguished from anorexia and obesity. Many different names were used for this disorder, including dysorexia, bulimarexia, thin-fat syndrome, binge/purge syndrome, and dietary chaos syndrome. The term bulimia nervosa was coined in 1979 by a British psychiatrist who suggested that bulimia consisted of powerful urges to overeat in combination with a morbid fear of becoming fat and the avoidance of the fattening effects of food by inducing vomiting or abusing purgatives, or both.[16] Today, an estimated 2%-5% of the female population will have bulimia nervosa in their lifetime.[6] A diagnosis of bulimia is based on the frequency with which episodes of binge eating and inappropriate compensatory behaviours occur.

Psychological Issues

As with anorexia, people with bulimia have an intense fear of becoming fat. They have a negative body image accompanied by a distorted perception of their body size. Since their self-esteem is highly tied to their impressions of their body shape and weight, they blame all of their problems on their appearance; this allows them to avoid facing the real problems in their lives. People with bulimia are preoccupied with the fear that once they start eating, they will not be able to stop. They may engage in continuous dieting, which leads to a preoccupation with food. They also think they are the only person in the world with this problem, and as a result, they are often socially isolated. In addition, they may avoid situations that will expose them to food, such as going to parties or out to dinner, further isolating themselves.

Bulimic Behaviours

Bulimia typically begins with food restriction motivated by the desire to be thin. Overwhelming hunger may finally cause the dieting to be interrupted by a period of overeating. Eventually, a pattern develops involving semi-starvation, interrupted by periods of gorging. During a food binge, a person with bulimia experiences a sense of lack of control. The amount of food consumed during a binge varies, but is typically on the order of 3,400 kcal, while a normal teenager may consume 2,000-3,000 kcal in an entire day. One study found that bulimics consumed an average of about 7,000 kcal in a 24-hour period.[17] Binges usually last less than 2 hours and occur in secrecy. They stop when the food runs out, or when pain, fatigue, or an interruption intervenes. The amount of food consumed in a binge may not always be enormous, but it is perceived by the bulimic individual as a binge episode. Binging and purging are then followed by intense feelings of guilt and shame (**Figure F6.8**).

After binge episodes, individuals with bulimia use various inappropriate compensatory behaviours to eliminate the extra kcalories and prevent weight gain. Bulimia is subdivided into two types based on the type of compensatory behaviour used. Nonpurging bulimia involves behaviours such as fasting or excessive exercise to prevent weight gain, whereas purging bulimia involves regularly engaging in behaviours that may include self-induced vomiting and misuse of enemas, laxatives, and diuretics, or other medications (see Table F6.1). Self-induced vomiting is the most common purging behaviour. It is used at the end of a binge but also after normal eating to eliminate food before it is absorbed and the energy it provides can cause weight gain. At first, a physical manoeuvre such as sticking a finger down the throat is needed to induce vomiting but bulimics eventually

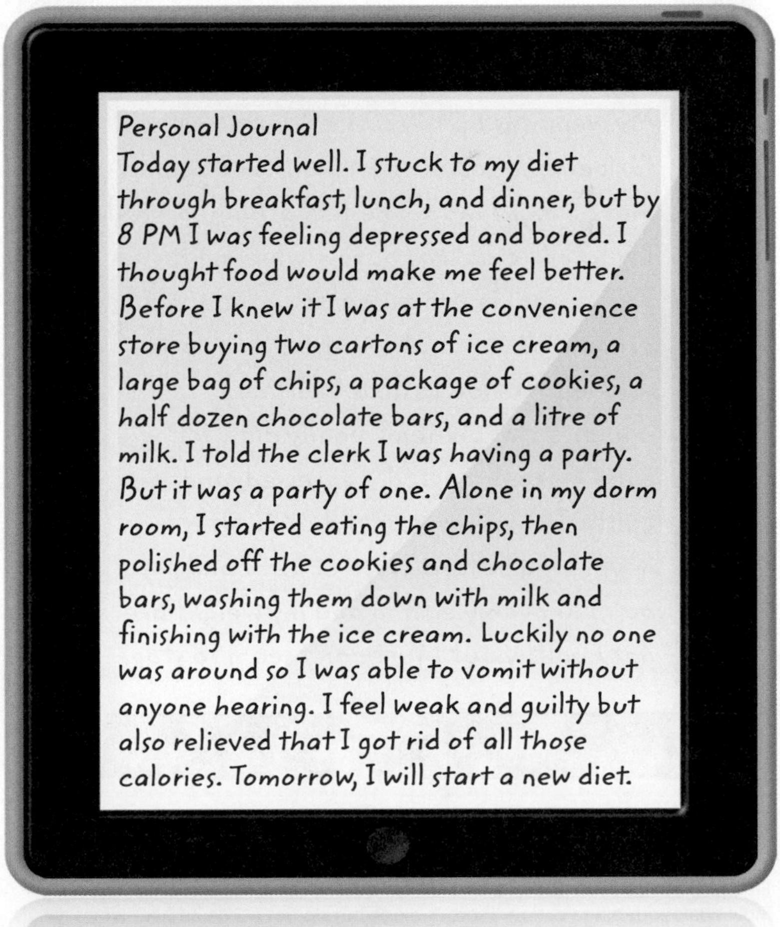

Personal Journal
Today started well. I stuck to my diet through breakfast, lunch, and dinner, but by 8 PM I was feeling depressed and bored. I thought food would make me feel better. Before I knew it I was at the convenience store buying two cartons of ice cream, a large bag of chips, a package of cookies, a half dozen chocolate bars, and a litre of milk. I told the clerk I was having a party. But it was a party of one. Alone in my dorm room, I started eating the chips, then polished off the cookies and chocolate bars, washing them down with milk and finishing with the ice cream. Luckily no one was around so I was able to vomit without anyone hearing. I feel weak and guilty but also relieved that I got rid of all those calories. Tomorrow, I will start a new diet.

Figure F6.8 **A day in the life of a bulimic**
Bulimia is characterized by binge eating followed by purging.

learn to vomit at will. Vomiting does not purge all kcalories consumed in a binge. After a binge containing 3,530 kcal, on average 1,209 kcalories were retained. Interestingly, after a smaller binge of only 1,549 kcal, almost the same amount of energy remained in the stomach, 1,128 kcal.[18] Some bulimic individuals take laxatives to induce diarrhea. Although bulimics believe the diarrhea prevents kcalories from being absorbed, in fact, nutrient absorption is almost complete before food enters the colon, where laxatives have their effect. The weight loss associated with laxative abuse is due to dehydration. Diuretics also cause water loss, but via the kidney rather than the GI tract. They do not cause fat loss. Some bulimia sufferers use a combination of purging and nonpurging methods to eliminate excess kcalories.

Physical Complications of Bulimia

It is the purging portion of the binge-purge cycle that is most hazardous to health in bulimia nervosa. Purging by vomiting brings stomach acid into the mouth. Frequent vomiting affects the GI tract by causing tooth decay, sores in the mouth and on the lips, swelling of the jaw and salivary glands, irritation of the throat, inflammation of the esophagus, and changes in stomach capacity and stomach emptying.[6] It also causes broken blood vessels in the face from the force of vomiting, electrolyte imbalance, dehydration, muscle weakness, and menstrual irregularities in women. Laxative and diuretic abuse can also cause dehydration and electrolyte imbalance. Rectal bleeding may occur from laxative overuse.

Treatment

The overall goal of therapy for people with bulimia nervosa is to separate eating from their emotions and from their perceptions of success, and to promote eating in response to hunger and satiety. Psychological counselling is needed to address issues related to body image and a sense of lack of control over eating. Nutritional therapy addresses physiological imbalances caused by purging episodes as well as provides education on nutrient needs and how to meet them. Antidepressant medications may be beneficial in reducing the frequency of binge episodes. Treatment has been found to speed recovery, especially if it is provided soon after symptoms begin, but for some individuals, this disorder may remain a chronic problem throughout life.[19]

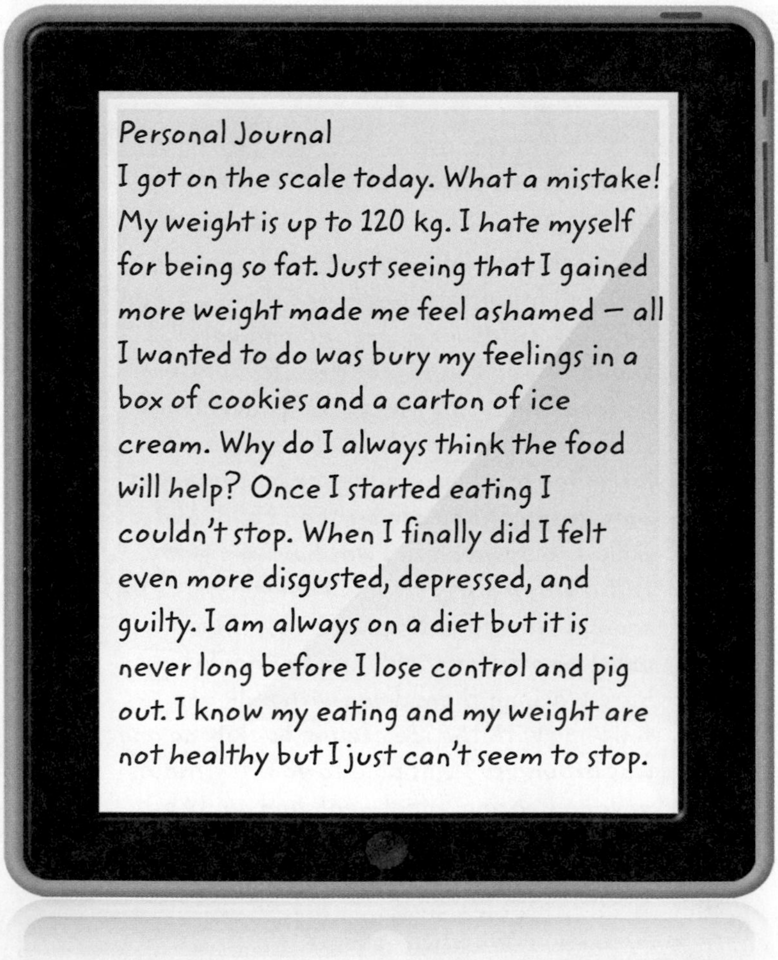

F6.5 Binge-Eating Disorder

Learning Objective

- Distinguish binge-eating disorder from anorexia and bulimia.

Binge-eating disorder, which is in the EDNOS category, is probably the most common eating disorder. It affects at least 1% of the total population.[20] Unlike anorexia and bulimia, binge-eating disorder is not uncommon in men who account for about 40% of cases.[21] It is more common in overweight individuals (**Figure F6.9**). Individuals with binge-eating disorder engage in recurrent episodes of binge eating but do not regularly engage in purging and other

Figure F6.9 A day in the life of a binge eater
Individuals with binge eating disorder binge but do not purge and are typically overweight.

inappropriate compensatory behaviours such as vomiting, fasting, or excessive exercise (**Table F6.3**). The major complications of binge-eating disorder are the conditions that accompany obesity, which include diabetes, high blood pressure, high cholesterol levels, gallbladder disease, heart disease, and certain types of cancer.[21] Treatment of binge-eating disorder involves counselling to improve body image and self-acceptance, a healthy nutritious diet, and increased exercise to promote weight loss, along with behaviour therapy to reduce bingeing.

Table F6.3 Diagnostic Criteria for Binge-Eating Disorder
Recurrent episodes of binge eating. An episode is characterized by: • Eating a larger amount of food than normal during a short period of time (within any 2-hour period). • Lack of control over eating during the binge episode (i.e., the feeling that one cannot stop eating).
Binge eating episodes are associated with 3 or more of the following: • Eating until feeling uncomfortably full. • Eating large amounts of food when not physically hungry. • Eating much more rapidly than normal. • Eating alone because you are embarrassed by how much you're eating. • Feeling disgusted, depressed, or guilty after overeating.
Marked distress regarding binge eating is present.
Binge eating occurs, on average, at least 2 days a week for 6 months.
The binge eating is not associated with the regular use of inappropriate compensatory behaviour (i.e., purging, excessive exercise, etc.) and does not occur exclusively during the course of bulimia nervosa or anorexia nervosa.

Source: From the *DSM-IV, Diagnostic and Statistical Manual of Mental Disorders*, 4th ed. Washington DC: American Psychiatric Association, 2000.

F6.6 Eating Disorders in Special Groups

Learning Objectives

- Describe how eating disorders are different in men and women.
- List some eating disorders that occur in children, pregnant women, and athletes.

Although anorexia, bulimia, and binge-eating disorder are most common in women in their teens and twenties, eating disorders occur in both genders and all age groups. They can be a complication in pregnant women and a problem for athletes and young children. They also occur in individuals with diseases that have a nutrition component, such as diabetes. In addition, there are a number of less common eating disorders that appear in special groups and in the general population. These are listed in **Table F6.4**.

Eating Disorders in Men

The incidence of anorexia, bulimia, and binge-eating disorder is much lower among men than women. One reason for the lower incidence is that the cultural pressure for males to be thin is less intense. Women are encouraged to be thin to attract friends and romantic partners and to be successful at school and work, whereas men are encouraged to be strong and powerful. The male ideal is a V-shaped upper body that is muscular, moderate in weight, and low in body fat. This difference in societal expectations is reflected in the BMI of men and women when they first "feel fat" and begin dieting. Women who develop eating disorders generally feel fat and begin dieting when their BMI is in the healthy range, whereas men who develop eating disorders usually do not start dieting until they have a BMI in the overweight range.[22] Men also often acquire an eating disorder at an older age than women.

Table F6.4 Other Eating Disorders

Eating Disorder	Characteristics	Who Is Affected	Consequences
Anorexia athletica	Engaging in compulsive exercise to lose weight or maintain a very low body weight.	Athletes	Can lead to more serious eating disorders and serious health problems including kidney failure, heart attack, and death.
Avoidance emotional disorder	Similar to anorexia nervosa in that the child avoids eating and experiences weight loss and the physical symptoms of anorexia. However, there is no distorted body image or fear of weight gain.	Children	Weight loss, reduced body fat, malnutrition.
Bigorexia (muscle dysmorphia or reverse anorexia)	Obsession with being small and underdeveloped. Individuals believe their muscles are inadequate even when they have a good muscle mass.	Bodybuilders and avid gym-goers, more common in men than women	Sufferers are at risk if they take steroids or other muscle-enhancing drugs.
Body dysmorphic disorder	An obsession with a perceived defect in the sufferer's body or appearance.	Affects males and females equally	Increased risk for depression and suicide.
Chewing and spitting	The person puts food in his/her mouth, tastes it, chews it, and then spits it out.	Those with other eating disorders	Since the food is not swallowed, this can result in the same symptoms as starvation dieting.
Female athlete triad	A triad of disordered eating, amenorrhea, and osteoporosis.	Female athletes in weight-dependent sports	Low estrogen levels, which interfere with calcium balance, eventually causing reductions in bone mass and an increased risk of bone fractures.
Insulin misuse (diabulimia)	Withholding insulin to cause weight loss or prevent weight gain.	People with type 1 diabetes	Uncontrolled blood sugar, which can lead to blindness, kidney disease, heart disease, nerve damage, and amputations.
Night-eating syndrome	Most of the day's kcalories are eaten late in the day or at night. A similar disorder, in which a person may eat while asleep and have no memory of the events, is called nocturnal sleep-related eating disorder. It is considered a sleep disorder, not an eating disorder.	Obese adults and those experiencing stress	Obesity
Orthorexia nervosa	Obsession with eating food considered to be healthy or beneficial. Focus on the quality of the food, not the quantity.	No particular group	Harmful to interpersonal relationships.
Pica	Craving and eating nonfood items such as dirt, clay, paint chips, plaster, chalk, laundry starch, coffee grounds, and ashes.	Pregnant women, children, people whose family or ethnic customs include eating certain nonfood substances.	Mineral deficiencies, perforated intestines, intestinal infections.
Rumination syndrome	Eating, swallowing, and then regurgitating food back into the mouth where it is chewed and swallowed again.	Infants and adults with mental and emotional impairment	Bad breath, indigestion, chapped lips, damage to dental enamel and tissues in the mouth, aspiration of food leading to pneumonia, weight loss and failure to grow (children), electrolyte imbalance, and dehydration.
Selective eating disorder	Eating only a few foods, mostly carbohydrate.	Children	Malnutrition

Although men currently represent a small percentage of those with eating disorders, the numbers seem to be on the rise.[23] This is likely due to increasing pressure to achieve an ideal male body. Advertisements directed at men today are showing more and more exposed skin with a focus on well-defined abdominal and chest muscles (**Figure F6.10**). Just as the Barbie doll set an unrealistic standard for young women, male action figures, superhero cartoons, and media ideals set a standard that is impossible for young men to achieve.

The physical consequences of eating disorders are similar in men and women. Both lose bone but men are more severely affected by disorders related to bone loss and tend to have lower bone mineral density than women with the same disorder.[24] Rather than amenorrhea, men experience a gradual drop in testosterone levels, which causes a loss of sexual desire.

Men with eating disorders have psychiatric conditions that are similar to those affecting women, including mood and personality disorders. Like women, men with eating disorders require professional help in order to recover and the outcome of treatment is similar in men and women. Men, however, are less likely to seek treatment because they do not want to be perceived as having a "woman's disease."

Figure F6.10 The ideal male body is as difficult for most men to achieve as the thin, athletic ideal is for women. (Ryan McVay/Getty Images, Inc.)

Eating Disorders During Pregnancy

Eating disorders are common in women in their twenties, an age when many people choose to start a family. If the eating disorder interrupts the menstrual cycle, it will cause infertility.

Some women with eating disorders are able to conceive. What and how much they eat during pregnancy influences their health and the health of the baby before and after birth. Pregnancy can also make other medical problems related to the eating disorder worse, such as liver, heart, and kidney damage. Pregnant women with eating disorders are at increased risk of caesarean delivery and have a higher rate of miscarriages, premature birth, babies who are small for their age or have congenital malformations.[25] Babies born to women with eating disorders are more likely to be slower to grow and develop and they may lag behind intellectually and emotionally and remain dependent. They may also have difficulty developing social skills and relationships with other people.

Life Cycle

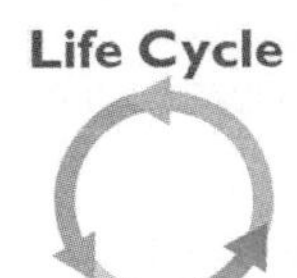

An eating disorder that is more common in pregnancy is **pica** (see Section 14.2). This is an abnormal craving for and ingestion of nonfood substances having little or no nutritional value. Commonly consumed substances include dirt, clay, chalk, paint chips, laundry starch, ice and freezer frost, baking soda, cornstarch, coffee grounds, and ashes. Pica during pregnancy is potentially dangerous. The consumption of large amounts of nonfood substances may cause micronutrient deficiencies by reducing the intake of nutrient-dense foods and by interfering with the absorption of certain minerals from food. Substances consumed could also cause intestinal obstruction or perforation and could contain toxins or harmful bacteria. The cause of pica is unknown but there is some evidence that it results from cultural beliefs related to pregnancy, along with changes in food preferences that occur during pregnancy.

pica An abnormal craving for and ingestion of nonfood items.

Eating Disorders in Children

Life Cycle

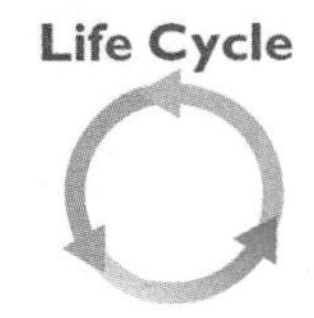

Eating disorders also occur in children. These can be difficult to diagnose because most children are finicky eaters at some point in their development. But when being finicky becomes extreme and growth is impaired, an eating disorder may be the cause. Disorders may involve consumption of a very limited number of foods or a general restriction of food intake (see Table F.2.4). Anorexia may also begin in children under the age of 13, but the incidence is much lower than it is in late adolescence and early adulthood. There is concern, however, that its prevalence is increasing in younger children as they are exposed to our cultural values about food and body weight. An American study found that 42% of girls in grades 1-3 want to be thinner. Eighty-one percent of 10-year-olds are afraid of being fat, and 46% of 9- to 11-year-olds are "sometimes" or "very often" on diets.[26] More girls than boys are affected but the proportion of boys is greater than the proportion of men in the older age groups.[27] Canadian trends are probably similar. Diagnosing anorexia in girls under 13 is more challenging than in older girls because many have not started menstruating and it is difficult to calculate expected weight because growth may have slowed. Despite these problems, there is little doubt that childhood-onset anorexia does occur and is a serious illness.

Children with anorexia, like older sufferers, are perfectionistic, conscientious, and hardworking. They also exhibit similar symptoms, including weight loss, food avoidance, preoccupation with food and kcalories, fear of fatness, excessive exercise, self-induced vomiting, and laxative abuse. Other physical changes that may accompany the weight loss include growth of lanugo hair, low blood pressure, slow heart rate, poor peripheral circulation, cold peripheries, and delayed or arrested growth. Bone density may be reduced and bone age delayed. Vitamin and mineral deficiencies are common.

As with older individuals, the prognosis for children with anorexia is variable.[27] If treatment is not effective, the resulting undernutrition is even more likely to cause physical complications in children (such as heart and circulatory failure), than it does in older people. Growth is affected, but the long-term consequences depend on the outcome of treatment. If treatment restores normal eating, these children will catch up in growth. Complications that may persist include delayed growth, impaired fertility, osteoporosis, and amenorrhea in girls.

Eating Disorders in Athletes

The relationship between body weight and performance in certain sports contributes to the higher prevalence of eating disorders in athletes than in the general population.[28] It is higher in female athletes than in male athletes, and more common among those competing in leanness-dependent and weight-dependent activities such as ballet and other dance, figure skating, gymnastics, track and field, swimming, cycling, crew, wrestling, and horse racing[25] (**Figure F6.11**). The most problematic women's sports are cross-country, gymnastics, swimming, and track and field. The male sports with the highest number of participants with eating disorders are wrestling and cross-country.[29]

Figure F6.11 In sports such as gymnastics, the advantages offered by a small, light body can motivate athletes to diet to stay thin, which can potentially contribute to the development of an eating disorder. (Pornchai Kittiwongsakul/AFP/Getty Images, Inc.)

Both anorexia nervosa and bulimia occur in athletes. The regimented schedule of an athlete makes it easy for him or her to use training diets and schedules, travel, or competition as an excuse to not eat normally and hide the eating disorder. Over time, the continued starvation characteristic of anorexia leads to serious health problems as well as a decline in athletic performance. Starvation can lead to abnormal heart rhythms, low blood pressure, and atrophy of the heart muscle. The lack of food means that neither energy nor nutrients are sufficient to support activity and growth. Bulimia nervosa is more common in athletes than anorexia. It may begin because an athlete is unable to stick with a restrictive diet or the hunger associated with a very-low-kcalorie diet leads to bingeing. As in nonathletes, most of the health complications associated with bulimia are the result of purging. It causes fluid loss and low potassium levels, which can result in extreme weakness, as well as dangerous and sometimes lethal heart rhythms.

Anorexia Athletica Compulsive exercise, which has been termed *anorexia athletica*, is a type of eating disorder that is a particular problem in athletes. People with this disorder focus on exercise rather than food, but anorexia athletica is considered an eating disorder because the goal of the behaviour is to expend kcalories to control weight. Extreme training is easy to justify because it is a common belief that serious athletes can never work too hard or too long and pain is accepted as an indicator of achievement. Compulsive exercisers will force themselves to exercise even when they don't feel well and may miss social events in order to fulfill their exercise quota. They often calculate exercise goals based on how much they eat. They believe that any break in the training schedule will cause them to gain weight and performance will suffer. Compulsive exercise can lead to more serious eating disorders, such as anorexia and bulimia, as well as serious health problems, including kidney failure, heart attack, and death.

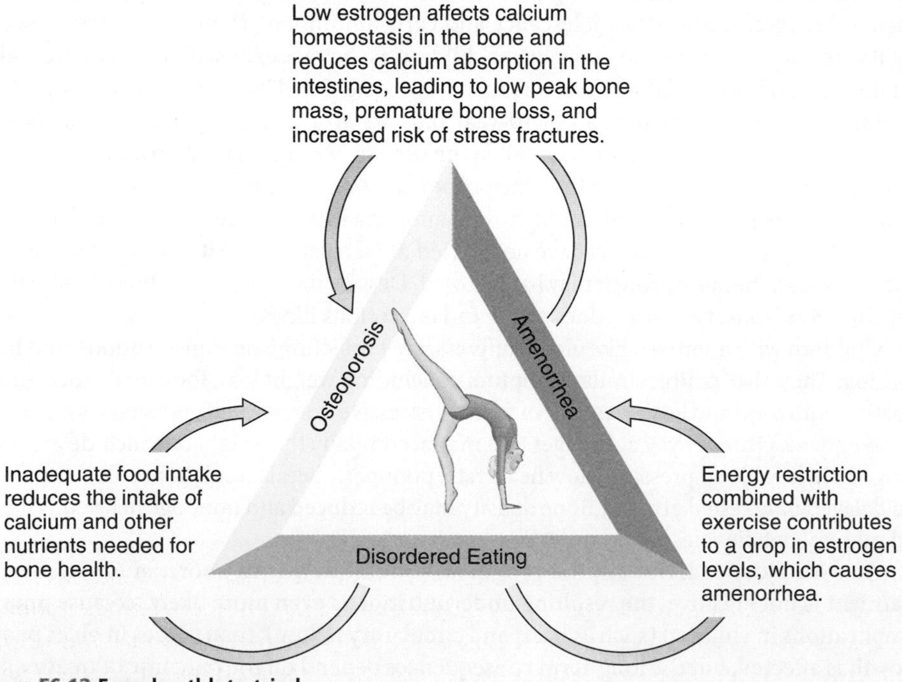

Figure F6.12 Female athlete triad
Women with female athlete triad syndrome typically have low body fat, do not menstruate regularly, and may experience multiple or recurrent stress fractures.

Female Athlete Triad Female athletes with eating disorders are at risk for a syndrome of interrelated disorders referred to as the **female athlete triad** (see Section 13.4) (**Figure F6.12**). This syndrome includes disordered eating, amenorrhea, and **osteoporosis**. The three are linked because the extreme energy restriction that occurs in eating disorders creates a physiological condition similar to starvation, which leads to menstrual irregularities. High levels of exercise can also affect the menstrual cycle by increasing energy demands and causing hormonal changes.[30] When combined, energy restriction and excessive exercise contribute to amenorrhea. The low estrogen levels associated with amenorrhea then interfere with calcium balance, leading to reductions in bone mass and bone-mineral density. Low estrogen levels also reduce calcium absorption and when combined with poor calcium intake (common in female athletes and females in general), lead to premature bone loss, failure to reach maximal peak bone mass, and an increased risk of stress fractures.

female athlete triad The combination of disordered eating, amenorrhea, and osteoporosis that occurs in some female athletes, particularly those involved in sports in which low body weight and appearance are important.

osteoporosis A bone disorder characterized by a decrease in bone mass, an increase in bone fragility, and an increased risk of fractures.

Eating Disorders and Diabetes

Diabetes does not cause eating disorders but it may set the stage for them, both physically and emotionally, and can also be used to hide them. Diabetes is a disease characterized by a chronic elevation in blood glucose that is due to abnormalities in the production or effectiveness of the hormone insulin (see Section 4.5). Treatment involves paying careful attention to diet, exercise, body weight, and blood glucose levels. The timing of exercise and timing and composition of meals are crucial to good glucose control. This regimentation, which is part of routine diabetes management, may contribute to the development of eating disorders because it places attention on food portions and body weight; a focus similar to that seen in women with eating disorders who do not have diabetes.[31]

Control is a central issue in diabetes as it is in eating disorders. A person with diabetes may feel guilty or out of control if their blood sugar is too high. Anorexics feel the same way if their weight increases. People with diabetes become consumed with strategies to control blood sugar and those with an eating disorder become consumed with ways to control weight. Both are preoccupied with weight, food, and diet. Because this is expected in diabetes, people with this disease can use their diabetes to hide anorexia or bulimia. They are expected to watch what they eat and the diabetes can be blamed for weight loss.

Those who take insulin to control their diabetes, especially those with type 1 diabetes, are at particular risk because they can misuse it to control their weight, a condition that has been termed diabulimia. Insulin is responsible for allowing glucose to enter cells. If patients cut back on the amount of insulin they take, the sugar in their blood cannot enter cells and it is excreted in the urine. This causes weight loss, but at a very high cost. The long-term complications of having high levels of blood glucose include blindness, kidney disease, cardiovascular disease, impaired circulation, and nerve death that can lead to limb amputations. Once a person with diabetes starts to control their weight by withholding insulin, they are reluctant to stop and may also begin using other inappropriate behaviours to control weight. If the weight loss continues, it can lead to organ failure and death. Diabulimia is very common among girls with type 1 diabetes.[32]

F6.7 Preventing and Getting Treatment for Eating Disorders

Learning Objectives

- Describe factors that predispose people to eating disorders.
- List the steps you could take if you had a friend with an eating disorder.

Reducing the incidence of, and morbidity from, eating disorders involves action on a number of levels. The first step is to recognize individuals who are at risk. Early intervention can help prevent those who are at risk from developing serious eating disorders, and the actions of family and friends can help those who are affected get help before their health is impaired. To reduce the overall incidence of eating disorders, changes in social attitudes that contribute to their development need to occur.

Recognizing the Risks

To prevent eating disorders, it is important to first recognize factors that increase risk. Excessive concerns about body weight, having friends who are preoccupied with weight, being teased by peers about weight, and problems with one's family all predispose people to eating disorders. There is an association between parental criticism and children's weight preoccupation. Dieting also increases risk. Girls and women who diet are more likely to develop an eating disorder than those who don't diet.[2] Those who have a mother, sister, or friend who diets are also at increased risk. Exposure to media pressure to be thin is also associated with the development of eating disorders.

Getting Help for a Friend or Family Member

Those who are at risk for eating disorders can be targeted for intervention. For teens, parents play an important role. Arranging an evaluation with a physician and a mental health specialist when the first symptoms are discovered may help prevent the disorder.

Once an eating disorder has developed, people usually do not get better by themselves. Helping them to get medical and psychological treatment can avoid severe physical consequences, but getting a friend or relative with an eating disorder to agree to seek help is not always easy. People with eating disorders are good at hiding their behaviours and denying the problem, and often do not want help.

If you suspect a friend has an eating disorder, you can alert a parent, teacher, coach, religious leader, school nurse, or other trusted adult about your concerns, or confront your friend or relative yourself and express your concern. If you approach the person yourself, you need to be firm but supportive and caring. The goal in discussing a person's eating disorder is to encourage them to seek help, but help is only effective if it is desired, and people with eating disorders are likely to refuse help initially.

When approaching someone about an eating disorder, it is important to make it clear that you are not trying to force them to do anything they don't want to do. Continued encouragement can help some people to seek professional help. The first reaction of someone confronted about an eating disorder is often to deny that they have a problem. Support your suspicions with examples of behaviours you have seen that make you believe your friend has a problem. You should be prepared for all possible reactions. People with eating disorders usually try to hide their behaviours so it is traumatic for them when someone has discovered their secret. One person may be relieved that you are concerned and willing to help, whereas another may be angry and defensive (**Table F6.5**).

Table F6.5 Eating Disorders: How to Help
• Get the person to a doctor; the sooner the illness is treated, the more likely there will be a successful outcome.
• Talk to the parents, spouse, or other family members.
• Explain your concerns and the potential hazards of the disease.
• Do not expect the person to co-operate; denial is common.
• If you work with the person, contact your employee assistance program.

Reducing the Prevalence of Eating Disorders

Eating disorders are easier to prevent than to cure. The biggest impact on prevention can be made by social interventions that target the elimination of weight-related teasing and criticism from peers and family members. Another important target for reducing the incidence of eating disorders is the media. If the unrealistically thin body ideal presented by the media could be altered, the incidence of eating disorders would likely decrease. Even with these interventions, however, eating disorders are unlikely to go away, but education through schools and communities about the symptoms and complications of eating disorders can help people identify friends and family members at risk and persuade those with early symptoms to seek help.

REFERENCES

1. Statistics Canada Table 105-110: Mental Health and Well-being profile, Canadian Community Health Survey (CCHS), by age group and sex, Canada and provinces, occasional: Available online at: http://www5.statcan.gc.ca/cansim/a16?id=1051100&lang=eng&pattern=&stByVal=3&requestID=2011070122205435182&MBR%5B%27MHWBPROF%27%5D=38&MBR%5B%27UNIT%27%5D=5&MBR%5B%27AGE%27%5D=2*3&MBR%5B%27GEO%27%5D=1&MBR%5B%27SEX%27%5D=2*2&retrLang=eng&syear=2002&eyear=2002&exporterId=TABLE_HTML_TIME_AS_COLUMN&__checkbox_accessible=true&action%3Aa23=Retrieve+now. Accessed July 1, 2011.
2. Boyce W. Young people in Canada: their health and well being. Available online at http://www.phac-aspc.gc.ca/hp-ps/dca-dea/publications/hbsc-2004/index-eng.php. Accessed July 1, 2011.
3. American Psychiatric Association. *Diagnostic and Statistical Manual of Mental Disorders*, 4th ed., TR, Washington, DC: American Psychiatric Association, 2000.
4. Public Health Agency of Canada. A report on mental illnesses in Canada. Chapter 6: Eating Disorders. Available online at http://www.phac-aspc.gc.ca/publicat/miic-mmac/index-eng.php. Accessed July 12, 2010.
5. Woodside, D. B. , Garfinkel, P. E., Lin, E. et al. Comparisons of men with full or partial eating disorders, men without eating disorders, and women with eating disorders in the community. *Am. J. Psychiatry* 158: 570-574, 2001.
6. American Dietetic Association. Position of the American Dietetic Association: Nutrition intervention in the treatment of anorexia nervosa, bulimia nervosa, and other eating disorders. *J Am Diet Assoc* 106:2073–2082, 2006.
7. Milano, W., Petrella, C., Sabatino, C., et al. Treatment of bulimia nervosa with sertraline: A randomized controlled trial. *Adv. Ther.* 21:232–237, 2004.
8. Branson, R., Potoczna, N., Kral, J. G. et al. Binge eating as a major phenotype of melanocortin 4 receptor gene mutations. *N. Engl. J. Med.* 348:1096–1103, 2003.
9. Stice, E. Sociocultural influences on body weight and eating disturbance. In: *Eating Disorders and Obesity: A Comprehensive Handbook*, 2nd ed., C. G. Fairburn, and K. D. Brownell, eds. New York: The Guilford Press 2002, pp. 103–107.
10. Mediascope. Body Image and Advertising. Available online at www.mediascope.org/pubs/ibriefs/bia.htm/. Accessed April 20, 2005.
11. ANRED. Anorexia Nervosa and Related Eating Disorders, Inc. Statistics: How Many People Have Eating Disorders? Available online at www.anred.com/stats.html/. Accessed April 2, 2002.
12. Peat, C., Mitchell, J. E., Hoek, H. W., et al. Validity and utility of subtyping anorexia nervosa. *Int. J. Eat. Disord.* 42:1–7, 2009.
13. NEDIC.Pro-anorexia websites. Available online at http://www.nedic.ca/knowthefacts/proanorexiasites.shtml. Accessed July 2, 2011.
14. Fitzgibbon, M., and Stolley, M. Minority Women: The Untold Story. NOVA Online: Dying to Be Thin. Available online at www.pbs.org/wgbh/nova/thin/minorities.html/. Accessed December 21, 2005.
15. Tozzi, F., Thornton, L. M., Klump, K. L. et al. Symptom fluctuation in eating disorders: Correlates of diagnostic crossover. *Am. J. Psychiatry* 162:732–740, 2005.
16. Vandereycken, W. History of anorexia nervosa and bulimia nervosa. In: *Eating Disorders and Obesity: A Comprehensive Handbook*, 2nd ed., C. G. Fairburn, and K. D. Brownell, eds. New York: The Guilford Press, 2002, pp. 151–154.
17. Kaye, W. H., Weltzin, T. E., McKee, M. et al. Laboratory assessment of feeding behavior in bulimia nervosa and healthy women. Methods of developing a human feeding laboratory. *Am. J. Clin. Nutr.* 55:372–380, 1992.
18. Kaye, W. H., Weltzin, T. E., Hsu, L. K. et al. Amount of calories retained after binge eating and vomiting. *Am. J. Psychiatry* 150:969–971, 1993.
19. Ebeling, H., Tapanainen, P., and Joutsenoja, A. A practice guideline for treatment of eating disorders in children and adolescents. *Ann. Med.* 35:488–501, 2003.
20. Hoek H. W., van Hoeken, D. Review of the prevalence and incidence of eating disorders. *Int. J. Eat Disord.* 34:383–396, 2003.
21. National Eating Disorders Association. Binge Eating Disorder. Available online at http://www.nationaleatingdisorders.org/nedaDir/files/documents/handouts/BED.pdf. Accessed June 13, 2009.
22. Andersen, A. E., and Holman, J. E. Males with eating disorders: Challenges for treatment and research. *Psychopharmacol. Bull.* 33:391–397, 1997.
23. Woodside, D. B., Garfinkel, P. E., Lin, E. et al. Comparisons of men with full or partial eating disorders, men without eating disorders, and women with eating disorders in the community. *Am. J. Psychiatry* 158:570–574, 2001.
24. Andersen, A. E., Watson, T., and Schlechte, J. Osteoporosis and osteopenia in men with eating disorders. *Lancet* 355:1967–1968, 2000.
25. Mitchell-Gieleghem, A., Mittelstaedt, M. E., and Bulik, C. M. Eating disorders and childbearing: Concealment and consequences. *Birth* 29:182–191, 2002.
26. University of Nebraska Medical Center, Department of Obstetrics and Gynecology. Don't Weigh Your Self Esteem—Every Body Counts. Available online at www.unmc.edu/olson/ education/ed.htm/. Accessed June 12, 2009.
27. Casper, R. C. Eating disturbances and eating disorders in childhood. Psychopharmacology—The Fourth Generation of Progress. The American College of Neuropsychopharmacology, 2000, Available online at www.acnp.org/g4/GN401000162/Default.htm/. Accessed April 15, 2005.
28. Sundgot-Borgen, J., and Torstveit, M. K. Prevalence of eating disorders in elite athletes is higher than in the general population. *Clin. J. Sports Med.* 14:25–32, 2004.
29. National Eating Disorders Information Center. Available online at www.nedic.ca/glossary.html/. Accessed June 12, 2009.
30. Warren, M. P., and Goodman, L. R. Exercise-induced endocrine pathologies. *J. Endocrinol. Invest.* 26:873–878, 2003.
31. Goebel-Fabbri, A. E., Fikkan, J., Connell, A. et al. Identification and treatment of eating disorders in women with type 1 diabetes mellitus. *Treat. Endocrinol.* 1:155–162, 2002.
32. Hasken, J., Kresl, L., Nydegger, T. et al. Diabulimia and the role of school health personnel. *J Sch Health.* 80(10):465-9, 2010.

16 Nutrition and Aging: The Adult Years

Case Study

Min is 70 years old and has always eaten well and got plenty of exercise. When her husband died last year, though, her life changed. Right after his death, her family spent a lot of time with her, bringing groceries, helping with the cooking, or sharing a meal or snack. Now they don't visit as often, and Min usually eats alone. She used to walk every morning with her friends, but during the year and a half that her husband was ill, she stopped because she wasn't comfortable leaving him alone. She has gained a little weight and is tired all the time. At her last doctor's visit, she learned that both her blood pressure and her blood sugar are slightly elevated. She tried walking with her friends a few weeks ago but found she could no longer keep up with them. She was so embarrassed that she doesn't plan to walk with them again.

Although we tend to focus on the physical changes that occur with aging, Min's case demonstrates that the psychological and social changes that are common in the elderly can also have a significant effect on lifestyle and nutritional status. Older adults and those of us who care for them need to consider all these factors when managing nutritional health.

In this chapter, you will learn about the nutritional needs of the older adult and the many factors that influence food intake in the older adult.

(Amana Images, Inc/Alamy)

(©iStockphoto)

Chapter Outline

CANADIAN CONTENT

16.1 What Is Aging?

Learning Objectives

- Define aging.
- Explain what is meant by compression of morbidity.

life span The maximum age to which members of a species can live.

Figure 16.1 Aging is a process that occurs continuously in individuals of all ages. (B2M Productions/Getty Images)

Biologically, aging is not something that begins at age 55, 65, or 75; it is a process that begins with conception and continues throughout life (**Figure 16.1**). It can be defined as the inevitable accumulation of changes with time that are associated with and responsible for an ever-increasing susceptibility to disease and death. The maximum age to which any human can live—the **life span**—is about 100 to 120 years. Life span is a characteristic of a species; dogs can live about 20 years, gorillas 39 years, and mice only about 3 years. Life span is believed to be genetically determined. The only experimental treatment that has been found to extend the life span of a variety of mammalian species is a nutritional manipulation called *kcaloric restriction*.[1] Restricting the energy intake of animals to about 70% of typical intake slows and/or delays the aging processes. Although the underlying biological mechanism responsible for its life span extending effects are still not known, kcaloric restriction is believed to act by affecting the organism's genes.[2] Despite promising data from animal studies, kcaloric restriction is not currently considered a viable option for extending life in humans (see Science Applied: Eat Less—Live Longer?).

How Long Can Canadians Expect to Live?

life expectancy The average length of life for a population of individuals.

Although humans can live to be 120 years, most do not live that long. **Life expectancy**, the average length of time that a person can be expected to live, varies between and within populations. It is affected by genetics, lifestyle, and environmental factors. Due to advances in health care and improved nutrition, life expectancy has increased in Canada in the last 50 years. Canadians born today can expect to live more than a decade longer than Canadians born in 1950. Life expectancy for men has increased from 66 to 77 years, while the increase for women has been from 71 to 82 years[3] (**Figure 16.2a**). With this increase in life expectancy, the number of older adults is increasing. The proportion of Canadians over age 65 has increased from 8% of the total population in the 1950s to 13% in 2005. (**Figure 16.2b**) By 2056, it is projected that nearly 27% of Canadians will be over age 65 and almost 6% will be over 85 years of age.[4]

How Long Can Canadians Expect To Be Healthy?

Even though average life expectancy in Canada has increased to 77 years for men and 82 years for women, the average healthy life expectancy is about 10 years less at 68 years and 71 years for men and women respectively.[5] This means that, on average, the last decade of life is restricted by disease and disability. The goal of successful aging is to increase not only life expectancy but the number of years of healthy life that an individual can expect. This is achieved by slowing the changes that accumulate over time and postponing the diseases of aging long enough to approach or reach the limits of life span before any adverse symptoms appear. This is referred to as **compression of morbidity.** When applied to the population as a whole, this term means that people are healthier and living longer (**Figure 16.3**); applied to the individual, it means staying healthy through the later years of life.

compression of morbidity The postponement of the onset of chronic disease such that disability occupies a smaller and smaller proportion of the life span.

Because the incidence of disease and disability increases with increasing age, the older population accounts for a large part of the public health budget.[6] Thus, keeping older adults healthy is beneficial not only for the aging individuals themselves, but for the health-care system as well. Postponing the changes that occur with age is an important public health goal.

SCIENCE APPLIED

Eat Less—Live Longer?

Throughout history, humans have searched for ways to stay young, avoid the ravages of old age, and extend life. Researchers have found that limiting kcaloric intake extends life and maintains health and vitality—at least in animals. Is this true in humans as well?

Kcaloric restriction, which refers to undernutrition without malnutrition, involves a diet providing 30%–40% fewer kcalories than would typically be consumed, but containing all the necessary nutrients to support life. Since the 1930s, studies have described the life-extending effect of kcalorie restriction in rats and other organisms.[1,2] A kcalorie-restricted diet has been shown to extend the maximal life span of organisms as diverse as worms, insects, and rodents by as much as 50%. It also reduces age-related chronic disease, improves immune function, increases resistance to numerous stressors and toxins, and maintains function later into life.

To explore the possibility that kcaloric restriction would lengthen life and reduce chronic disease in animals more closely related to humans, in the late 1980s researchers began to study the effect of this diet in rhesus monkeys. Adult monkeys were fed nutritionally adequate diets containing 30% fewer kcalories than a control group. After 20 years, 80% of the kcalorie-restricted animals were still alive, compared to only 50% of the control animals.[3] The kcalorie-restricted group had a lower incidence of diabetes, cancer, cardiovascular disease, and brain atrophy. This study confirms that kcalorie restriction can slow aging in a primate species.

The fact that kcaloric restriction has extended the life of every species tested suggests that it would do the same in humans. Human studies are the only way to know for sure how effective this diet would be. However, a study of kcaloric restriction in humans is fraught with problems, including the fact that it would take decades to produce results. Instead, scientists have looked at the people of Okinawa. They provide one of the best examples of how kcaloric restriction might affect health and longevity in humans.[4] Okinawans have active lives and consume a nutrient-dense diet of vegetables and fish and typically eat only until they are 80% full. Okinawans enjoy the longest life expectancy in the world (81.2 years). They maintain a low BMI and the incidence of the chronic diseases of aging, such as heart disease and cancer, is much lower among Okinawans than among people living in the United States or on mainland Japan (see graph).

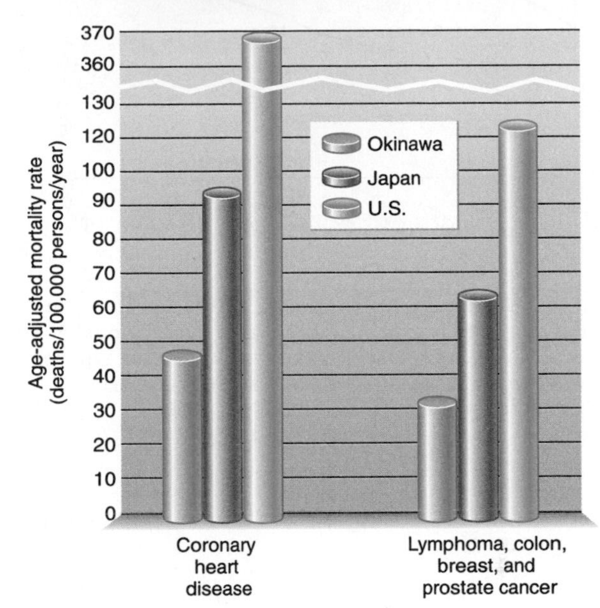

Despite the health and longevity enjoyed by the Okinawan people, kcalorie restriction is probably not a realistic means to extend life span, as most people would not be able to maintain continuing food deprivation. Furthermore, humans are not caged animals living under controlled conditions, so their longevity is affected by many factors. A mathematical model that factored in information related to the kcaloric intake and longevity of typical Japanese men, Japanese sumo wrestlers who must eat large amounts of food, and Okinawan men who eat less than is typical predicted that humans following this restrictive regimen would increase their life span by only 3%–7%—far less than the almost 50% extension in rodent species.[5] The biggest improvement in longevity was between the Sumo wrestlers, who consumed 5,500 kcal/day and lived an average of 56 years and the average Japanese man who consumed 2,300 kcal/day and lived an average of 76.7 years, a gain of 20 years. The Okinawan, on the other hand, consumed 1,900 kcal/day and gained on average an additional year of life. Perhaps the important lesson from the research into kcalorie restriction is not that we should be consider extreme food deprivation as a path to longevity, but that avoiding excessive food intake and increasing energy expenditure through physical activity is a realistic path to health.

[1]McCay, C. M., Crowell, M. F., and Maynard, LA. The effect of retarded growth upon the length of life and upon ultimate size. *J. Nutr.* 10:63-79, 1935.

[2]Masoro, E.J. Subfield history: Caloric restriction, slowing aging, and extending life. *Sci. Aging Knowl. Environ.* 2003: RE2, 2003.

[3]Colman, R. J., Anderson, R. M. Johnson, E. K. et al. Caloric restriction delays disease onset and mortality in rhesus monkeys. *Science* 325:201-204, 2009.

[4]Willcox, B. J., Willcox, D. C, Todoriki, H. et al. Caloric restriction, the traditional Okinawan diet, and healthy aging: The diet of the world's longest-lived people and its potential impact on morbidity and life span. *Ann. N.Y. Acad. Sci.* 1114:434–455, 2007.

[5]Phelan, J. P., and Rose, M. R. Why dietary restriction substantially increases longevity in animal models but won't in humans. *Aging Res. Rev.* 4: 339-350, 2005.

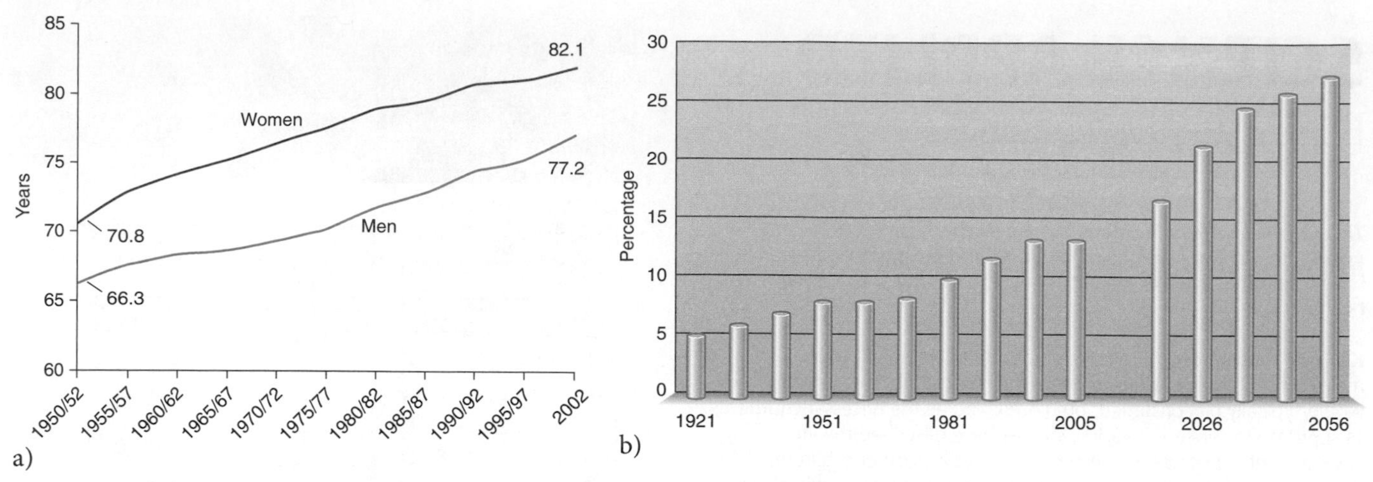

Figure 16.2 Life expectancy in Canada a) Life expectancy in Canada has steadily increased since 1950. b) The proportion of Canadians over the age of 65 is expected to increase over the next 50 years.
Source: a) St-Arnaud, J., Beaudet, M. P., and Tully, P. Life Expectancy. *Health Reports* 17 (1): 43-47, 2005. Available online at http://www.statcan.gc.ca/bsolc/olc-cel/olc-cel?catno=82-003-X20050018709&lang=eng. Accessed December 28, 2010. b) Turcotte, M. and Schellenberg, G. A Portrait of Seniors in Canada, Statistics Canada Cat no 89-519-XIE, 2006. Available online at http://www.statcan.gc.ca/bsolc/olc-cel/olc-cel?catno=89-519-X&lang=eng. Accessed December 27, 2010.

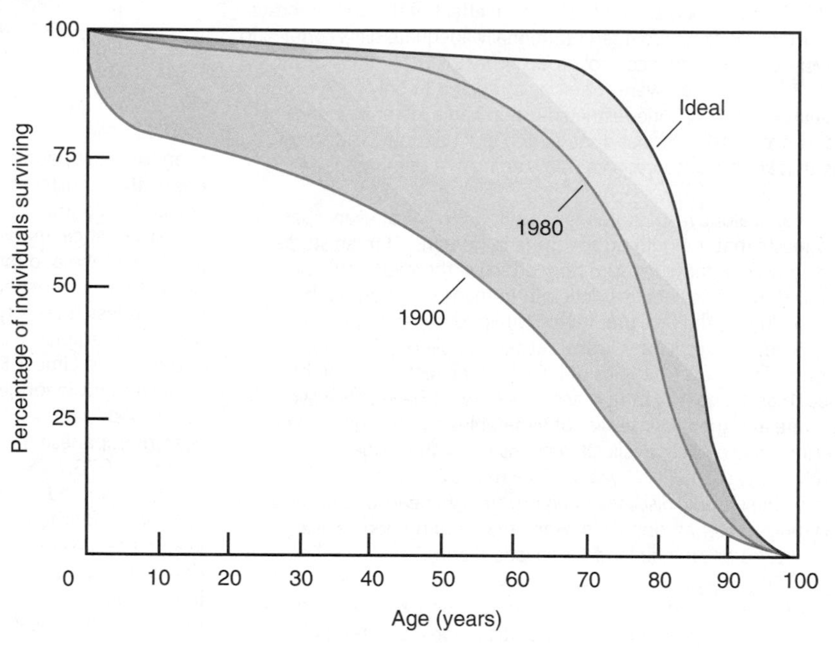

Figure 16.3 Effect of compression of morbidity on survival
The decline in deaths from infectious diseases between 1900 and 1980 allowed more people to survive into adulthood. Delaying the onset of chronic diseases will allow more people to remain healthy and survive into their seventies and eighties.
Source: Adapted from Fries, J. F. Aging, natural death, and the compression of morbidity. *N. Engl. J. Med.* 303:130–135, 1980.

16.2 What Causes Aging?

Learning Objectives

- Describe the biological causes of aging.
- Discuss genetic, environmental, and lifestyle factors that affect the aging process.

Although universal to all living things, aging is a process we don't fully understand. We do know that as organisms become older, the number of cells they contain decreases and the function of the remaining cells declines. As tissues and organs lose cells, the ability of the organism to perform the physiological functions necessary to maintain homeostasis decreases; disease becomes increasingly common and the risk of malnutrition increases. This loss of cells and cell function occurs throughout life, but the effects are not felt for many years because organisms begin life with extra functional capacity, or **reserve capacity**. Reserve capacity allows an organism to continue functioning normally despite a decrease in the number and function of cells. In young adults, the reserve capacity of organs is 4–10 times that required to sustain life. As a person ages and reserve capacity decreases, the effects of aging become evident in all body systems.

reserve capacity The amount of functional capacity that an organ has above and beyond what is needed to sustain life.

There are two major hypotheses to explain why aging occurs. One favours the idea of a genetic clock and argues that the cell death associated with aging is a genetically programmed event. The other views the events of aging as the result of cellular wear and tear. The actual cause of the cell death associated with aging is probably some combination of both of these, and the rate at which cell death occurs and at which aging proceeds is determined by the interplay among our genetic makeup, our lifestyle, and the environment in which we spend our years.

Programmed Cell Death

One hypothesis about aging proposes that cell death is triggered when genes that disrupt cell function are activated.[7] This causes the selective, orderly death of individual cells or groups of cells and is referred to as **programmed cell death**. This hypothesis is supported by the fact that cells grown in the laboratory divide only a certain number of times before they die. Cells from older individuals will divide fewer times than those from younger individuals, and cells from longer-lived species will divide more times than those from shorter-lived species. If cells in an organism stop reproducing and continue to die, the total number of cells will decline, resulting in a loss of organ function.

programmed cell death The death of cells at specific predictable times.

Wear and Tear

A second hypothesis suggests that aging is the result of an accumulation of cellular damage. This wear and tear may result from errors in DNA synthesis, increases in glucose levels, or damage caused by free radicals. Free radicals are reactive chemical substances that are generated from both normal metabolic processes and exposure to environmental factors (see Section 8.10). They cause oxidative damage to proteins, lipids, carbohydrates, and DNA, and may also indirectly harm cells by producing toxic products. For example, age spots—brown spots that appear on the skin with age—are caused by the oxidation of lipids, which produces a pigment called lipofusoin, or age pigment. The damage done to cells by free radicals is associated with aging and has been implicated in the development of a number of chronic diseases common among older adults, including cardiovascular disease and cancer.

Genetics, Environment, and Lifestyle

The rate at which the changes associated with aging accumulate depends on the genes people inherit, the environment they live in, and their lifestyle (**Figure 16.4**). Genes determine the efficiency with which cells are maintained and repaired. Individuals with less cellular repair capacity will lose cells more readily and consequently age more quickly. Likewise, genes determine susceptibility to age-related diseases such as cardiovascular disease and cancer. However, individuals who inherit a low capacity to repair cellular damage may still live long lives if they reside in an environment with few factors that damage cells and if they eat well and exercise regularly. In contrast, individuals with exceptional cellular repair ability may accumulate cellular damage rapidly and die young if they smoke cigarettes, consume a diet high in saturated fat and low in antioxidant nutrients, and live sedentary lives. No matter what individuals' genes predict about how long they will live, their actual **longevity** is also affected by lifestyle factors and the extent to which they are able to avoid accidents and disease.

longevity The duration of an individual's life.

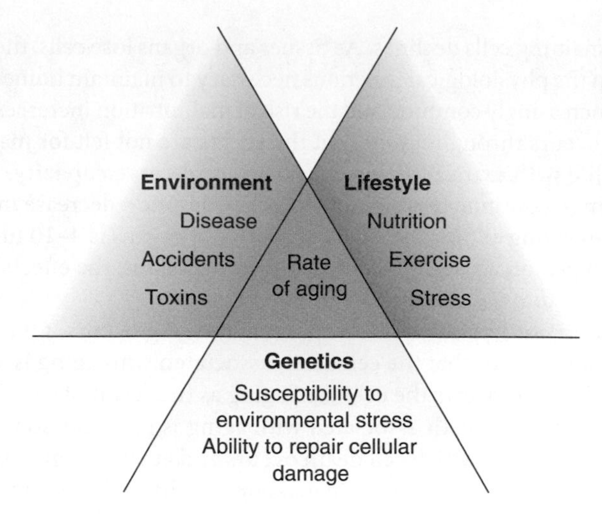

Figure 16.4 Factors that affect the rate of aging
The rate at which individuals age is affected by their genetic makeup, the environment in which they live, and the lifestyle choices they make.

16.3 Nutritional Needs and Concerns of Older Adults

Learning Objectives

- Compare the nutrient recommendations for older adults to those for young adults.
- List nutrients that may need to be supplemented in the diets of adults over the age of 50 years.

Older adults are a diverse population. There are 70-year-olds running marathons while others are confined to wheelchairs (**Figure 16.5**). The physiological and health changes that accompany aging affect the requirements for some nutrients, how nutrient needs must be met, and the risk of malnutrition. In order to best address the nutrient needs of adults, the DRIs divide adulthood into 4 age categories: young adulthood, ages 19–30; middle age, 31–50 years; adulthood, ages 51–70; and older adults, those over 70 years of age. Recommendations have been developed to meet the needs of the majority of healthy individuals in each age group. Although the incidence of chronic diseases and disabilities increases with advancing age, these conditions are not considered when making general nutrient intake recommendations.

Figure 16.5 Chronological age is not always the best indicator of a person's health. A person who is 75 may have the vigour and health of someone who is 55, or vice versa. Some older adults are healthy, independent, and active, while others are chronically ill, dependent, and at high risk for malnutrition. (Alaska Stock Images/NG Image Collection;Thinkstock/Getty Images, Inc.)

CRITICAL THINKING:

Changes in Diet Quality as Canadians Age

The Canadian Healthy Eating Index (CHEI), described in Section 2.6, generates a score for the overall quality of a diet based on its conformance to Canada's Food Guide. The index has a maximum score of 100. Points are lost if an inadequate number of servings is consumed in each of the 4 food groups and if high amounts of saturated fat, sodium or "foods to limit" are consumed (see Section 15.1). These two figures show the CHEI scores for Canadian adults.[1]

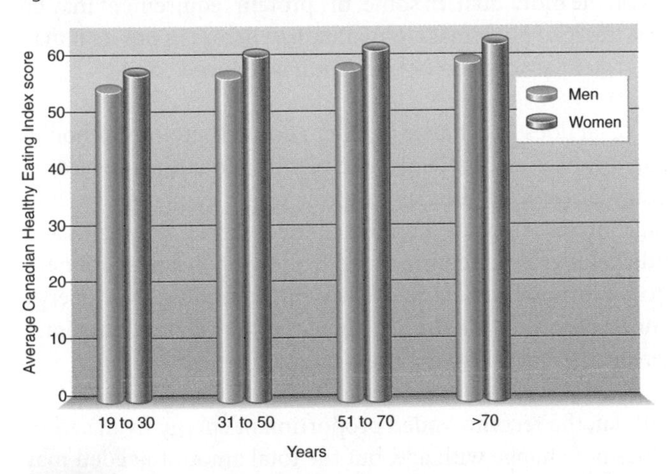

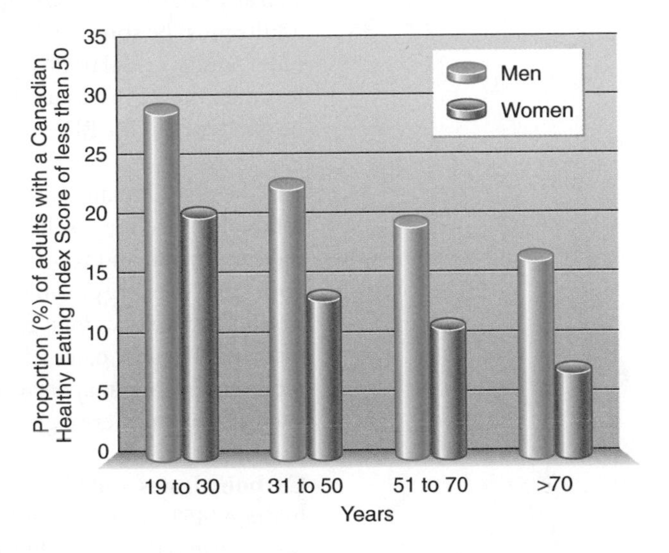

Critical Thinking Questions

The first figure shows the average scores for Canadians in different age groups. How does the average CHEI score change as Canadians age?*

The second figure shows the proportion of Canadians with a score less than 50. What would you conclude from these results?

How do men and women compare?

*Answers to all Critical Thinking questions can be found in Appendix J.

[1]Garriguet, D. Diet quality in Canada. *Health Reports* 20(3): 1-11. Available online at http://www.statcan.gc.ca/pub/82-003-x/2009003/article/10914-eng.htm. Accessed June 27, 2011.

Maintaining Health throughout Adulthood

In order to maintain health as an older adult (>70 years), good lifestyle practices have to be established in youth. Most young adults are at the healthiest stage of their lives and their challenge is to maintain this health throughout the middle and later years of adulthood, largely by reducing their risks of chronic diseases such as type 2 diabetes, cardiovascular disease, cancer, hypertension, osteoporosis, and other chronic diseases. While genetic and environmental factors beyond an individual's control play a role in disease risk, maintaining a healthy body weight, staying physically active, and eating a nutritious diet based on Canada's Food Guide are among the best ways that an individual can maintain health. By doing this well, adults will enter their later years in a position to best manage the physical changes that occur with aging. Analysis of data from the CCHS suggests that the quality of Canadians' diet does tend to improve modestly with age (See Critical Thinking: Changes in Diet Quality as Canadians Age).

Energy and Macronutrient Needs

Adult energy needs typically decline with age. This is due to a decrease in all components of total energy expenditure. Basal metabolic rate (BMR) decreases as adults get older, in part due to a decrease in lean body mass. There is a 2%–3% drop in BMR every decade after about age 20. The thermic effect of food also declines with age and is about 20% lower in older than in younger adults. These decreases are reflected in the energy needs of older adults. For example, the estimated energy requirement (EER) for an 80-year-old man is almost 600 kcal/day less than that for a 20-year-old man of the same height, weight, and physical activity level. For women, the difference in EER between an 80- and a 20-year-old of the same height, weight, and physical activity level is about 400 kcal/day.[8] The decrease in EER that occurs with age is even greater if activity level declines, which it typically does. Decreased physical activity is estimated to account for about one-half of the decrease in total energy expenditure that occurs with aging.[9] The decrease in activity also contributes to the reduction in lean body mass and BMR.

Protein Protein is needed at all ages to repair and maintain tissues. Therefore, unlike energy requirements, the requirement for protein does not decline with age. The RDA for older as well as younger adults is 0.8 g/kg of body weight per day. As a result, an adequate diet for older adults must be somewhat higher in protein relative to energy intake. Due to the diversity of older adults, actual need depends on the individual. In some, the protein requirement may be less than the RDA because there is less lean body mass to maintain, whereas in others it may be greater than the RDA because protein absorption or utilization is reduced.

Fat The digestion and absorption of fat does not change in older adults. Therefore, although total fat intake may be lower due to lower energy needs, the recommendations regarding the proportion and types of dietary fat apply to older as well as younger adults. A diet with 20%–35% of energy from fat that contains adequate amounts of the essential fatty acids and limits saturated fat, *trans* fat, and cholesterol is recommended. Following these recommendations will allow older adults to meet their nutrient needs without exceeding their energy requirements and may delay the onset of chronic disease. However, there are certain situations, such as being underweight, where greater fat intake may be warranted.

Carbohydrates and Fibre As with fat, the recommended proportion of energy from carbohydrate (45%–65% of energy) does not change with age, but the total amount needed may be lower in older adults due to lower energy needs. Dietary carbohydrate should come from whole grains, fruits, vegetables, and dairy products, and foods high in added sugars should be limited. This pattern will help assure adequate nutrients without excess energy.

The recommendations for fibre intake for older adults are slightly lower than for younger adults because the AIs for fibre are based on total energy needs.[8] For example, the fibre intake for older women is 21 g/day compared to 25 g/day for younger women and 30 g instead of 38 g for men. Fibre, from whole grains, fruits, vegetables, and legumes, when consumed with adequate fluid, helps prevent constipation, hemorrhoids, and diverticulosis—conditions that are common in older adults. High-fibre diets may also be beneficial in the prevention and management of diabetes, cardiovascular disease, and obesity.

Water The recommended water intake for older adults is the same as that for younger adults; however, changes in the homeostatic mechanisms that regulate water balance may make meeting these needs more challenging. With age, there is a reduction in the sense of thirst, which can decrease fluid intake.[10] In addition, the kidneys are no longer as efficient at conserving water, so water loss increases. Other physical and psychological changes also increase the risk of dehydration. For instance, difficulty in swallowing and restricted mobility may limit access to water even in the presence of thirst. Depression, which decreases water intake, and medications such as laxatives and diuretics, which increase water loss, also contribute to dehydration in the elderly. The elderly may also voluntarily restrict fluid intake to avoid accidents due to incontinence or because numerous trips to the bathroom increase pain from arthritis. In addition to impairing organ function, inadequate fluid intake contributes to the development of constipation.

Micronutrient Needs

Although the recommended intake for many of the micronutrients is no different for older adults than for younger adults, the decrease in energy intake that occurs with age causes a decline in the intakes of micronutrients, especially the B vitamins, calcium, iron, and zinc (**Figure 16.6**).[10] Changes in digestion, absorption, and metabolism also affect micronutrient status. In turn, inadequate levels of certain micronutrients contribute to the development of some of the disorders that are common in older adults.

B Vitamins The only B vitamins for which recommendations differ between older and younger adults are vitamins B_6 and B_{12}. The RDA for vitamin B_6 is greater in people ages 51 and older than

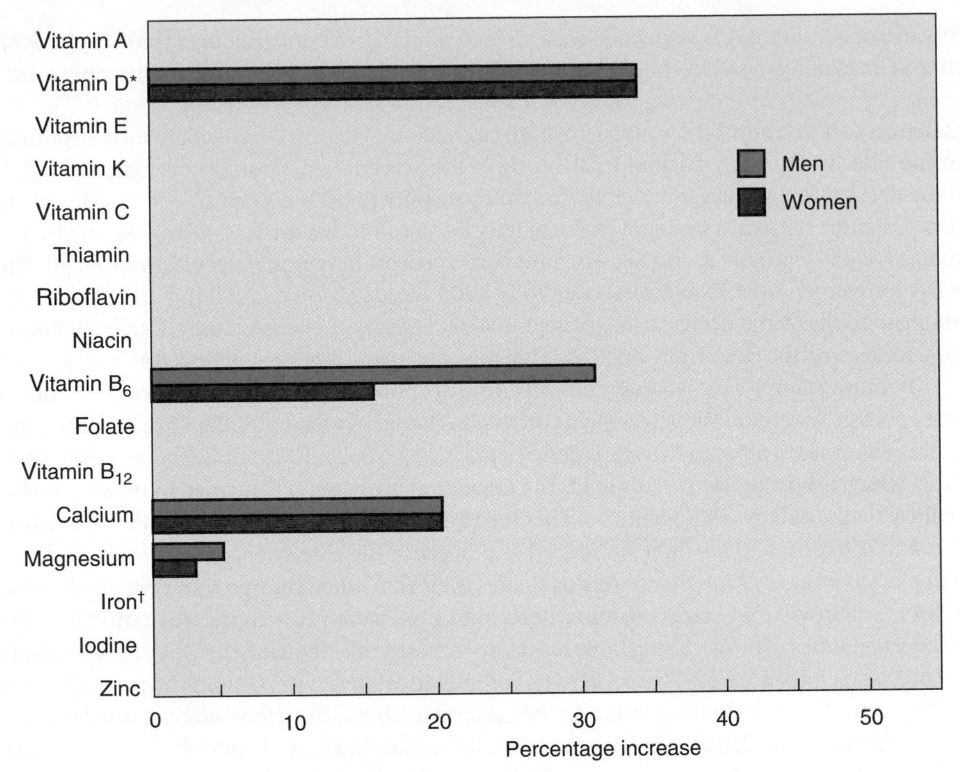

Figure 16.6 Vitamin and mineral needs of older adults
The percentage increases in micronutrient recommendations are shown here for adults aged 51 and older compared to those of young adults ages 19–30. *For vitamin D recommendations for adults > 70 years are compared to those for adults 19- 70 years. †The RDA for iron is decreased by more than 50% in women over 50.

for younger adults because higher dietary intakes are needed to maintain the same functional levels in the body. Vitamin B_{12} is a nutrient of concern for older adults because of both reduced absorption and low dietary intakes, especially among the poor. Absorption of vitamin B_{12} naturally present in food is reduced in many older adults due to an inflammation of the stomach that reduces stomach acid so food-bound vitamin B_{12} cannot be released. The RDA for vitamin B_{12} is not increased (see Figure 16.6), but it is recommended that individuals over the age of 50 meet their RDA for vitamin B_{12} by consuming foods fortified with vitamin B_{12} or by taking a supplement containing vitamin B_{12}.[11] This recommendation is made because the vitamin B_{12} in fortified foods and supplements is not bound to proteins, so it is absorbed even when stomach acid is low (see section 8.9).

The RDA for folate is the same for adults of all ages but folate intake is a concern in older adults for several reasons. Deficiencies of folate and vitamin B_{12} contribute to anemia, which is common in older adults. In addition, low folate, along with inadequate levels of vitamin B_6 and B_{12}, may result in an elevated homocysteine level, which may increase the risk of cardiovascular disease. Due to the importance of folate in DNA synthesis, low folate intake has also been hypothesized to cause DNA changes that contribute to cancer development.[12] The fortification of enriched grain products with folate, which began in 1998, has increased intake of this vitamin. However, when folate is consumed in excess it can mask the symptoms of vitamin B_{12} deficiency. Diets that include both fortified products and high-dose folate supplements may exceed the UL for folate. Levels above the UL increase the risk that vitamin B_{12} deficiency will be masked and therefore go untreated (see Section 8.9).[12]

Antioxidant Vitamins The recommended intake of antioxidant vitamins is not increased in older adults, but low dietary intakes of vitamins C and E and carotenoids are a concern due to low fruit and vegetable consumption among the elderly. Only about half of Canadians 55 years of age and older consume 5 or more servings of fruits and vegetables daily.[13] Eye disorders as well as mental impairment have been correlated with low levels of antioxidants in the elderly.

Vitamin A Compared to its action in young adults, preformed vitamin A is not as readily cleared from the blood of older adults, resulting in the possible accumulation of toxic levels of vitamin A, if intakes are very high, as might occur with the excessive use of supplements.[14]

Since excess vitamin A is associated with an increased risk of bone fractures (see Section 9.2), intakes exceeding the RDA should be avoided by older adults who are a risk for osteoporosis.[15]

Calcium and Vitamin D Low intakes of both calcium and vitamin D contribute to osteoporosis in the elderly. Although calcium intake early in life prevents osteoporosis by ensuring a high bone density, low intakes in older adults can contribute to osteoporosis by accelerating bone loss. Calcium status is a problem in the elderly because intakes are low, primarily due to low intakes of dairy products, and because intestinal absorption typically decreases with age. The RDA for women over 50 and men over 70 is 1,200 mg/day, which is 200 mg greater than for younger adults.[16] The decrease in estrogen that occurs at menopause causes accelerated bone loss. Increasing the RDA from 1,000 to 1,200 mg in women over 50 may slow this loss.

Because vitamin D is necessary for calcium absorption, a deficiency may also contribute to osteoporosis. Vitamin D deficiency is a concern in the elderly for a number of reasons. Intakes of this vitamin are often low in the elderly population, usually due to limited consumption of milk, which is fortified with vitamin D. The amount of provitamin D formed in the skin is also reduced in the elderly. This occurs both because the capacity to synthesize provitamin D when the skin is exposed to sunlight is reduced and because the elderly spend less time outdoors and tend to wear clothing that covers or shades their skin when they go out. Even if adequate amounts of provitamin D are consumed or formed, the capacity to activate provitamin D in the kidney decreases with age. Using bone loss as an indicator of adequacy, the RDA for vitamin D for men and women age 19–70 is 15 μg/day. For individuals over age 70, this is further increased to 20 μg/day. Because these amounts can be difficult to obtain from food alone, Canada's Food Guide recommends that all adults over age 50 take a vitamin D supplement (see Figure 16.13).

Iron The iron needs of women decline sharply at menopause when blood loss through menstruation stops. The RDA for iron for women over 50 years of age is less than half of that of menstruating women. The iron needs of men do not change. Nonetheless, iron-deficiency anemia does occur, especially when energy intake is low. Common causes are chronic blood loss from disease and medications and poor iron absorption due to low stomach acid and antacid use. The problem is more common in people 85 years or older. In non-institutionalized adults 65 and older, about 11% of men and 10.2% of women are anemic. In men and women 85 years or older, this increases to 25% and 20%, respectively.[10]

Zinc The RDA for zinc is not changed in older adults, but lower energy intakes as well as malabsorption, physiological stress, trauma, muscle wasting, and prescription and over-the-counter medications can all contribute to poor zinc status.[17] The consequences of poor zinc status may include loss of taste acuity and impaired immune function and wound healing. Loss of taste acuity can contribute to malnutrition by reducing food intake. Reduction in immune function and wound healing increases the risk of infection, which can also impair nutritional status. The CCHS found that a high proportion (25% of women and 41% of men) of Canadians older than 70 years were not meeting their requirements for zinc.[18]

16.4 Factors that Increase the Risk of Malnutrition

Learning Objectives

- Describe how the normal physiological changes of aging can affect nutritional status.
- Explain how changes in body composition affect energy needs.
- Explain how the nutrient needs of older adults are affected by disease and medication use.
- Discuss social and economic factors that increase the risk of malnutrition.

Low energy intakes and poor food choices put many older adults at risk of malnutrition, but these are not the only reasons malnutrition is so prevalent among this group. Many of the physiological changes associated with aging can affect nutritional status (**Table 16.1**). In addition, the elderly have a higher frequency of acute and chronic illnesses and are therefore more likely to be taking multiple medications. All of these factors can contribute to malnutrition (**Figure 16.7**). Also contributing to malnutrition are social economic factors that result in **food insecurity** which is "limited, inadequate, or insecure access of individuals and households to sufficient, safe, nutritious, and personally acceptable food to meet their dietary requirements for a productive and healthy life."[19]

food insecurity Limited, inadequate, or insecure access of individuals and households to sufficient, safe, nutritious, and personally acceptable food to meet their dietary requirements for a productive and healthy life.

Table 16.1 Factors that Increase the Risk of Malnutrition Among the Elderly

Reduced food intake due to:
Decreased appetite due to lack of exercise, depression, or social isolation
Changes in taste, smell, and vision
Dental problems
Limitations in mobility
Medications that restrict mealtimes or affect appetite
Lack of money to buy food
Lack of nutrition knowledge
Reduced nutrient absorption and utilization due to:
Gastrointestinal changes
Medications that affect absorption
Diseases such as diabetes, kidney disease, alcoholism, and gastrointestinal disease
Increased nutrient requirements due to:
Illness with fever or infection
Injury or surgery
Increased nutrient losses due to:
Medications that increase excretion of nutrients
Diseases such as gastrointestinal and kidney disease

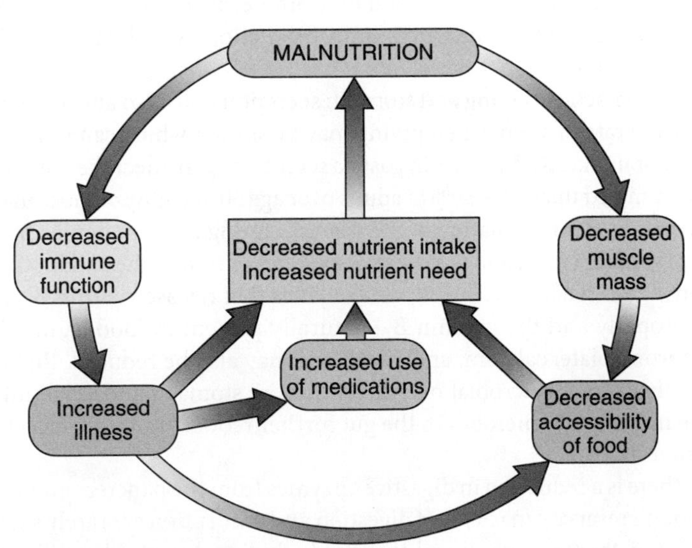

Figure 16.7 Causes and consequences of malnutrition
The decreases in muscle mass and immune function that occur with age contribute to malnutrition and, in turn, malnutrition makes these problems worse.

Physiological Changes

It is difficult to determine which of the changes that occur with aging are inevitable consequences of the aging process and which are the result of disease states. But whether caused by disease or the inescapable loss of cells and cell function, the changes that occur in organs and organ systems can affect nutritional status by altering the appeal of food, nutrient digestion and absorption, nutritional requirements, and the ability to obtain food.

Sensory Decline Beginning around age 60, there is a progressive decline in the ability to taste and smell; this becomes more severe in persons over 70. The deterioration of these senses can contribute to impaired nutritional status by decreasing the appeal and enjoyment of food.[10] Some studies suggest that the drop in taste acuity is due to a reduction in the number of taste buds on the tongue; others suggest that it is the result of changes in sensitivity to specific flavours such as salty and sweet. The appeal of food is also affected by the decline in the sense of smell. Odours provide important clues to food acceptability before food enters the mouth and once in the mouth, some molecules reach the nasal cavity where their odour is detected. It is the blending of the odour message from the nasal cavity and the taste message from the tongue that provides the overall food flavour. When the sense of smell is diminished, food is not as flavourful.

Vision also typically declines with age, making shopping for and preparation of food difficult. **Macular degeneration** is the most common cause of blindness in older Canadians. The macula is a small area of the retina of the eye that distinguishes fine detail. With age, oxidative damage reduces the number of viable cells in the macula. As the macula degenerates, visual acuity declines, ultimately resulting in blindness. **Cataracts** are another common reason for declining sight (**Figure 16.8**). Of people who live to age 85, half will have cataracts that impair vision. Oxidative damage is believed to cause both macular degeneration and cataracts, so a diet high in foods containing antioxidant nutrients might slow or prevent these eye disorders.[10] The carotenoid phytochemical, lutein, has been associated with decreased risk of macular degeneration (see Chapter 9: Focus on Phytochemicals).

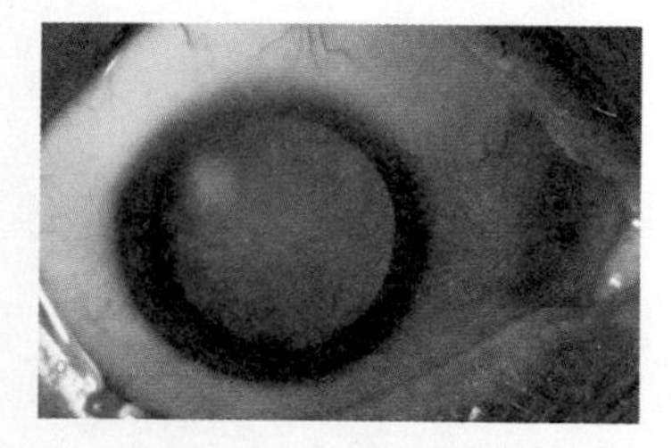

Figure 16.8 Cataracts cause the lens of the eye to become cloudy and impair vision. When cataracts obscure vision, the affected lens can be removed and replaced with an artificial plastic lens. (Custom Medical Stock Photo)

macular degeneration Degeneration of a portion of the retina that results in a loss of visual detail and blindness.

cataracts A disease of the eye that results in cloudy spots on the lens (and sometimes the cornea), which obscure vision.

periodontal disease A degeneration of the area surrounding the teeth, specifically the gum and supporting bone.

atrophic gastritis An inflammation of the stomach lining that causes a reduction in stomach acid and allows bacterial overgrowth.

Alterations in Gastrointestinal Function Aging causes changes in the gastrointestinal tract and its accessory organs that may alter the palatability as well as the digestion of food and the absorption of nutrients. One change is a decrease in the secretion of saliva into the mouth. Saliva provides lubrication for easy swallowing and mixes with food to allow it to be tasted. A reduction in saliva causes dryness which makes swallowing difficult and decreases the taste of food. Saliva is also an important defense against tooth decay because it helps wash material away from the teeth and it contains substances that kill bacteria. Thus, a dry mouth increases the likelihood of tooth decay and **periodontal disease**. Loss of teeth and improperly fitting dentures limit food choices and can contribute to poor nutrition in the elderly.

Changes in stomach emptying and stomach secretions can also affect nutritional status. In older adults, the rate of stomach emptying may be slower, which can reduce hunger and, therefore, nutrient intake. Reductions in gastric secretions can affect the absorption of some nutrients. It is estimated that 10%–30% of adults over age 50 and 40% of those in their 80s have **atrophic gastritis**. This inflammation of the stomach lining is accompanied by a decrease in the secretion of stomach acid, and in severe cases, a reduction in the production of intrinsic factor.[11] When stomach acid is reduced, the enzymes that release vitamin B_{12} from food do not function properly and the vitamin B_{12} naturally present in food cannot be absorbed. Absorption of iron, folate, calcium, and vitamin K may also be reduced. Reduced stomach acid secretion also allows microbial overgrowth in the stomach and small intestine.[11] The increase in the numbers of microbes in the gut further reduce B_{12} absorption by competing for available vitamin B_{12}.

With age, there is a reduction in digestive enzymes from the pancreas and small intestine, but there is enough reserve capacity that digestion and absorption are rarely significantly impaired. In the colon, there are functional changes, including decreased motility and elasticity, weakened abdominal and pelvic muscles, and decreased sensory perception, which can lead to constipation. Low fibre and fluid intakes and lack of activity also contribute to constipation. Constipation is a problem that occurs in fewer than 2% of persons in the nonelderly population but affects as many as 26% of men and 34% of women over 65 years of age.[20] It is estimated that more than 75% of elderly patients in hospitals and nursing homes use laxatives for bowel regulation. Maintaining regular exercise and consuming adequate fluid and fibre are safer ways to prevent constipation.

Changes in Other Organs Age-related changes in organs other than those of the gastrointestinal tract may also affect nutrient metabolism. Most absorbed nutrients travel from the intestine to the liver for metabolism or storage. The liver has a greater regenerative capacity than most organs, but with age, there is a decrease in liver size and blood flow and an increase in fat accumulation, which eventually decrease the liver's ability to metabolize nutrients and break down drugs and alcohol. With age, the pancreas may become less responsive to blood glucose levels, and the body cells may become more resistant to insulin, resulting in diabetes. Changes in the heart and blood vessels reduce blood flow to the kidneys, making waste removal less efficient. The kidneys themselves become smaller and their ability to filter blood and to excrete the products of protein breakdown declines.[21] In some individuals, blood urea levels may increase if protein intake is too high. The ability of the kidneys to concentrate urine also decreases with age, as does the sensation of thirst, increasing the risk of dehydration.

Excess Body Fat The prevalence of obesity in older age groups, as with younger adults, has increased over the past 25 years. In 2004, about 30% of Canadians age 55–64 were obese, while in 1978, the number was only 20%.[22] Although the risk of death associated with obesity is lower in older than in younger adults, obesity still increases the risk of health complications that reduce the quality of life. It contributes to a higher risk of cardiovascular disease, certain cancers, hypertension, stroke, sleep apnea, type 2 diabetes, and arthritis. Obesity also contributes to suboptimal physical functioning.[9]

As with younger adults, moderate weight loss can decrease obesity-related complications in older adults and, if combined with physical activity, improve physical function and quality of life. The approach to weight loss in the older population must place more emphasis on preventing the loss of muscle and bone mass that occurs with age because it can be further accelerated with weight loss.[9] Including exercise as part of a weight-loss regimen can improve strength, endurance, and overall well-being.

Weight Loss and Changes in Body Composition Although obesity is a problem among older adults, the prevalence declines with age. While 30% of those aged 55–64 years are obese, that number drops to 24% for those over 75 years.[22] As people get older, extreme thinness and unintentional weight loss also become important health risks and increase the risk of malnutrition. Numerous studies have demonstrated that in older adults, low BMI is associated with a higher risk of mortality.[23]

Even when body weight is in a healthy range, the changes in body composition that occur with age can affect nutritional and overall health. With age, there is an increase in body fat, especially in the abdomen, and a decrease in lean tissue, including a loss of muscle mass and strength, referred to as **sarcopenia** (**Figure 16.9**).[24,25] In older adults, the wasting of lean tissue can result in a high percentage of body fat even when body weight is low or stable. The maximum amount of fat mass occurs at ages 60–70, and after this, both lean tissue and fat mass decrease.[9] The decline in muscle size and strength affects both the skeletal muscles needed to move the body and the heart and respiratory muscles needed to deliver oxygen to the tissues (see Figure 16.9). Therefore, both strength and endurance are decreased, making the tasks of day-to-day life more difficult. The changes in muscle strength contribute not only to **physical frailty**, which is characterized by general weakness, impaired mobility and balance, and poor endurance, but also to the risk of falls and fractures. In the oldest old (those 85 years of age and older), loss of muscle strength becomes the limiting factor determining whether they can continue to live independently. Some of the reduction in muscle strength and mass is due to changes in hormone levels and in muscle protein synthesis, but a lack of exercise is also an important contributor.[24]

sarcopenia Progressive decrease in skeletal muscle mass and strength that occurs with age.

physical frailty Impairment in function and reduction in physiological reserves severe enough to cause limitations in the basic activities of daily living.

Aging is also accompanied by a decrease in bone mass, often resulting in osteoporosis, which further increases the risk of fractures. Although obesity makes many chronic conditions worse, the loss of bone and the risk of osteoporosis is reduced in individuals who weigh more. This is partly due to the added mechanical stress on the bones caused by carrying excess body weight and partly due to the release of estrogen by body fat (see Section 11.3).

Reduced Hormone Levels Some of the hormonal changes that occur with age are considered part of the normal aging process. Others may be a symptom of a disease process. For example, about 4%–8.5% of adults have thyroid levels that are below normal. Many of these individuals

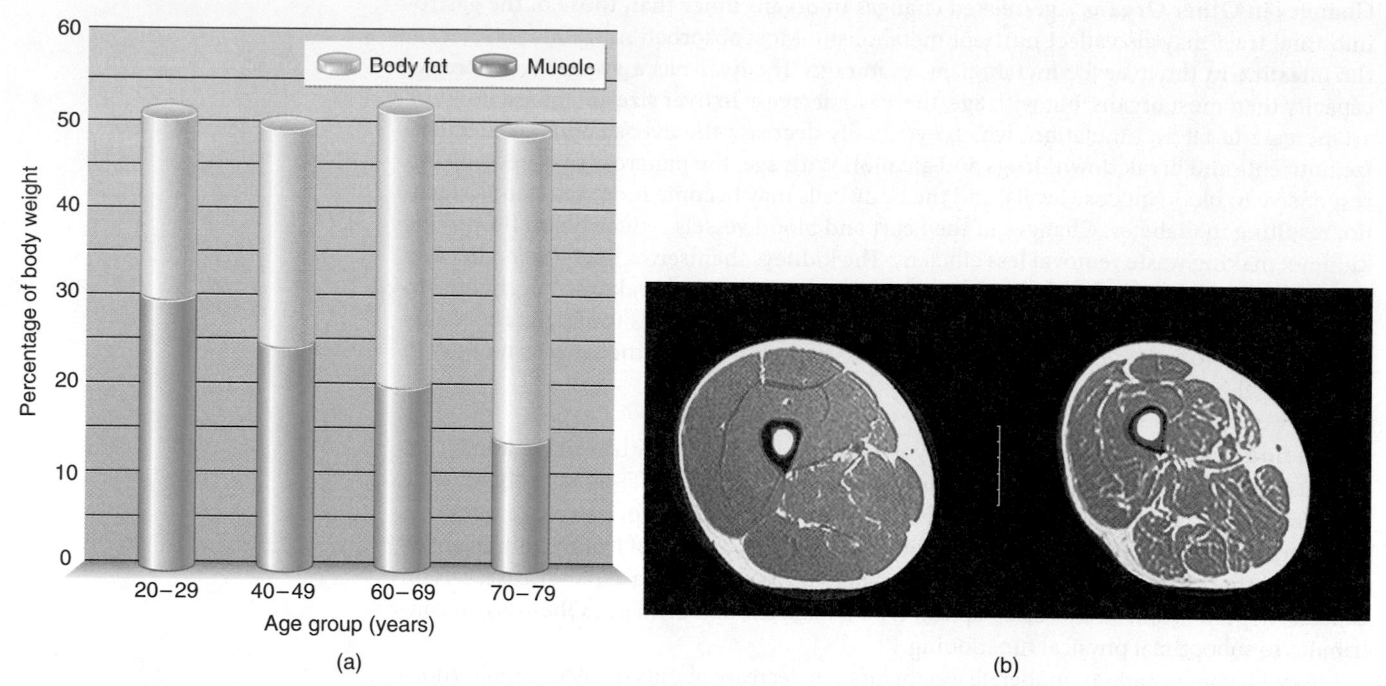

Figure 16.9 Changes in the proportion of muscle and fat with age
(a) With age, the proportion of muscle mass decreases and that of body fat increases.
Source: Adapted from Cohen, S. H., et al. Compartmental body composition based on the body nitrogen, potassium, and calcium. *Am. J. Physiol.* 239:192–200, 1980.
(b) Magnetic resonance image of a thigh cross-section from a 25-year-old man (left) and a 65-year-old man (right); note that although the thighs are of similar size, the thigh from the older man has more fat (shown in white) around and through the muscle, indicating significant muscle loss. (© Courtesy S. A. Jubias and K. E. Conley, University of Washington Medical Center)

have no symptoms of this decline. It is not clear whether administering thyroid hormones to individuals with no symptoms offers any benefits.[26] If the decrease causes symptoms, the patient is treated by administering hormones.[27]

menopause Physiological changes that mark the end of a woman's capacity to bear children.

Estrogen The most striking and rapidly occurring age-related hormonal change is **menopause.** Menopause normally occurs in women between the ages of 45 and 55. During menopause, the cyclical release of the female hormones estrogen and progesterone slows and eventually stops, causing ovulation and menstruation to cease. The period of decline in estrogen is accompanied by changes in mood, skin, and body composition (an increase in body fat and a decrease in lean tissue). The reduction in estrogen decreases the risk of breast cancer but increases the risk of heart disease to a level more similar to that in men. Reduced estrogen also increases the risk of osteoporosis by increasing the rate of bone breakdown and decreasing calcium absorption from the intestine. Estrogen used to be prescribed liberally to older women to alleviate the symptoms of menopause and reduce the risk of osteoporosis and heart disease. This "hormone replacement therapy" is no longer as common because studies have found that while it does reduce menopausal symptoms and the risk of bone fractures, it increases the risks of heart disease, blood clots, stroke, breast cancer, and problems with memory and thinking.[28]

Menopause does not occur in men, but with age, men do experience a gradual decrease in testosterone levels, which may contribute to a decrease in muscle mass and strength.

Growth Hormone Growth hormone stimulates growth and protein synthesis. Levels gradually decline with age in both men and women and may be responsible for some of the decrease in lean body mass, increase in fat mass, and bone loss that occurs with age. A few small, short-term studies have demonstrated improvements in body composition with growth hormone treatments, but there are few data on the benefits, safety, or cost effectiveness of long-term growth hormone administration.[29] When compared to a program of regular exercise, growth hormone injections did not produce any greater increases in muscle size or strength.[30] In addition, growth hormone administration has side effects, including edema, carpal tunnel syndrome, and decreases in insulin sensitivity. Despite the widespread use of growth hormone and products that supposedly increase growth hormone release in the body, until more is known about the long-term effects of growth hormone administration, the use of these products for anti-aging purposes is not recommended.[30]

DHEA DHEA (dehydroepiandrosterone) is a precursor to the sex hormones, testosterone, estrogen, and progesterone. Even though low levels of this hormone are not known to be the cause of age-associated disorders, claims have been made that this compound can strengthen bones, muscles, and the immune system, and prevent diabetes, obesity, heart disease, and cancer. Although some of these effects have been demonstrated when DHEA is administered to animals, beneficial effects of DHEA administration in humans have not been clearly established.[31]

Melatonin Melatonin is a hormone that is secreted by the pineal gland. It is involved in regulating the body's cycles of sleep and wakefulness. A decline in melatonin is hypothesized to influence aging by affecting body rhythms and triggering genetically programmed aging at a cellular level. Melatonin is also an antioxidant and may enhance immune function and reduce inflammation in the brain.[32] Its ability to extend normal longevity in humans has not been determined.[33]

Insulin One of the most common hormone-related changes that occurs with aging is elevated blood glucose. This is due to both a decrease in the amount of insulin released by the pancreas and a decrease in insulin sensitivity of the tissues. This decreased insulin sensitivity is related to poor diet, inactivity, increased abdominal fat mass, and decreased lean mass. About 20% of individuals 60 years of age or older have diabetes, and many of these cases are undiagnosed.[34] Individuals with diabetes are treated with diet and lifestyle prescriptions, medications to reduce blood glucose, and, when necessary, administration of the hormone insulin.

Changes in Immune Function The ability of the immune system to fight disease declines with age. As it does, the incidence of infections, cancers, and autoimmune diseases increases, and the effectiveness of immunizations declines (**Figure 16.10**). In turn, the increases in infections and chronic disease that occur can affect nutritional status. Some of the decrease in immune function may be due to nutritional deficiencies. The immune response depends on the ability of cells to differentiate, divide rapidly, and secrete immune factors, so nutrients that are involved in cell differentiation, cell division, and protein synthesis can influence the immune response. Deficiencies of zinc, iron, vitamin A, folic acid, and vitamins B_6, B_{12}, C, D, and E have been linked to impaired function and supplementation of some of these individual nutrients has been shown to improve immune response in the elderly.[35]

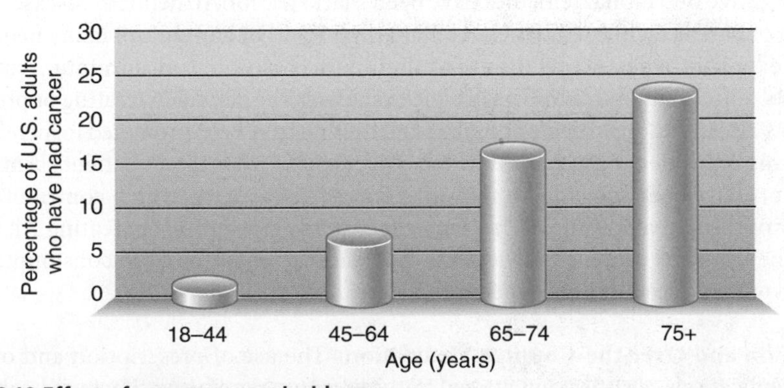

Figure 16.10 Effect of age on cancer incidence
The incidence of cancer increases with age. One reason for the higher incidence is that the immune system's ability to destroy cancer cells declines.

Acute and Chronic Illness

With age, there is an increase in the incidence of both acute and chronic illness. The reduction in reserve capacity and decline in immune function make infectious disease more frequent and more serious in the elderly. In addition, most older adults have at least one chronic medical condition.[10] The incidence of cardiovascular disease, diabetes, osteoporosis, hypertension, cancer, arthritis, and Alzheimer's disease all increase with age. Some of these diseases change nutrient requirements, some decrease the appeal of food, and some impair the ability to obtain and prepare an adequate diet by affecting mobility and mental status. All of these can increase the risk of food insecurity and malnutrition.

Conditions that Decrease Mobility More than half of the older population suffers from some form of physical disability, and the incidence increases with increasing age. disabilities often make it difficult to carry out the activities of daily life.[36] These limitations affect the ability to maintain good nutritional health by making it hard to shop, prepare food, get around the house, or go out to eat. Arthritis, a condition that causes pain upon movement, is the most common cause of disability in older individuals, affecting approximately 59% of all older adults. Fifty percent of individuals 70 years of age and older with arthritis need help with the activities of daily living, including preparing and eating meals[10] (see Your Choice: Do Glucosamine and Chondroitin Really Help Arthritis?). Osteoporosis and its associated fractures can also affect mobility, which in turn affects the ability to acquire and consume a healthy diet.

dementia A deterioration of mental state resulting in impaired memory, thinking, and/or judgement.

Alzheimer's disease A disease that results in the relentless and irreversible loss of mental function.

Conditions that Impair Mental Status Altered mental status can affect nutrition by interfering with the response to hunger and the ability to eat and to obtain and prepare food. Although many individuals maintain adequate nervous system function into old age, the incidence of dementia increases with age. **Dementia** refers to an impairment in memory, thinking, and/or judgement that is severe enough to cause personality changes and adversely affect daily activities and relationships with others. Causes of dementia include multiple strokes, alcoholism, dehydration, medication side effects and interactions, and **Alzheimer's disease**. Low levels of vitamin B_{12} and vitamin E have also been suggested to affect mental function in the elderly. With aging, there is a decrease in blood vitamin B_{12} levels and a rise in metabolites indicative of poor vitamin B_{12} status. In most cases, vitamin B_{12} supplements do not improve neurological function; however, in some elderly patients with mild dementia and low blood levels of vitamin B_{12}, supplementation does improve mental function.[8] The relationship between vitamin E and cognitive impairment is also controversial. In the elderly, lower blood levels of vitamin E have been associated with poor memory and mental functioning and those who suffer from dementia have lower plasma levels of vitamin E.[37] The effect of vitamin E and vitamin C supplements on preventing dementia is inconsistent, but some evidence suggests that vitamin E supplements may slow the progression of the disease.[37]

Over half of the cases of dementia in the elderly are due to Alzheimer's disease, a progressive, incurable loss of mental function. The brains of patients with Alzheimer's disease are characterized by the accumulation of an abnormal protein and a loss of certain types of nerve cells. Its cause is unknown, but there does appear to be a genetic component in some cases. Many ineffective nutritional remedies have been marketed for Alzheimer's disease. Because the brains of patients with Alzheimer's contain high levels of aluminum, many people tried to reduce exposure by restricting the use of aluminum cookware and aluminum-containing deodorants. Aluminum restriction has not been shown to be helpful in treating or preventing Alzheimer's disease.[38] Supplements of choline and lecithin have been promoted to increase levels of the neurotransmitter acetylcholine, which is deficient in Alzheimer's patients. Antioxidant supplements have been suggested to prevent free radical damage, which contributes to the pathology of this disease. None of these are effective for preventing or treating Alzheimer's disease. There is some evidence, however, that physical activity[39] and fish consumption[40] are associated with reduced risk of Alzheimer's disease.

Figure 16.11 It is not uncommon for older adults to take multiple medications daily. (Jose Luis Pelaez/Getty Images, Inc.)

Prescription and Over-the-Counter Medications The use of prescription and over-the-counter medications can affect nutritional status in a number of ways. Because health problems increase with increasing age, older adults are more likely to take medications; almost half of older adults take multiple medications daily (**Figure 16.11**).[41] The more medications taken, the greater the chance of side effects such as increased or decreased appetite, changes

YOUR CHOICE

(©iStockphoto)

Do Glucosamine and Chondroitin Really Help Arthritis?

Osteoarthritis is a type of arthritis that occurs when the connective tissue that cushions the joints degenerates, allowing the bones to rub together, causing pain (see figure). The treatment goals are to control pain and slow or reverse disease progression. Traditionally, arthritis has been treated with pain relievers such as acetaminophen, as well as those that also reduce inflammation, such as Aspirin and ibuprofen. These drugs can reduce pain and inflammation but do not repair the tissue. They also have side effects if taken over long periods of time. Supplements containing glucosamine and chondroitin sulfate offer an alternative.

Glucosamine and chondroitin sulfate are compounds found in and around the cells of cartilage, the connective tissue that cushions joints. These compounds are needed for the synthesis of large molecules that bind water to form a porous, gel-like material that allows cartilage to resist crushing forces and cushion the joints. Glucosamine may also inhibit inflammation and contribute to the lubricating and shock-absorbing properties of cartilage. Supplements of both glucosamine and chondroitin sulfate are said to reduce arthritis pain, stop cartilage degeneration, and possibly stimulate the repair of damaged cartilage.

What is the scientific evidence? Are these supplements beneficial? Some studies found large improvements, while others found little or no effect. A study in 2000 that integrated results from many different trials concluded that these supplements had moderate to large effects on osteoarthritis symptoms.[1] The benefits occurred after the supplements had been taken for approximately 4–6 weeks, and the effects were sustained for 4–8 weeks after the supplements were discontinued.[2] A trial to evaluate the effects of glucosamine and chondroitin sulfate on osteoarthritis of the knee found that the combination of supplements did not reduce pain in all subjects but was effective in some patients with moderate to severe knee pain.[3] On the other hand, a meta-analysis published in 2010 concluded that glucosamine and chondroitin sulfate had no effect on osteoarthritis.[4]

Should you try these supplements? The risks are low, and modest benefits have been observed. The Canadian Arthritis Society recommends a 3-month trial for those who would like to test whether the supplement may be useful to them. The supplements should be used with caution if diabetes, glucose intolerance or shellfish allergies are present.[5]

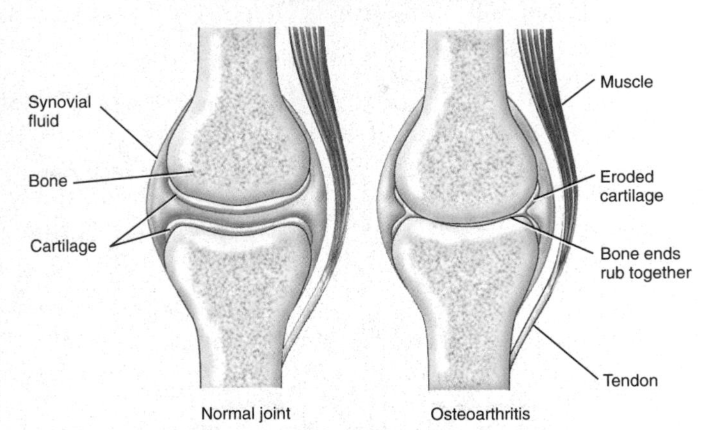

In a normal joint (left), cartilage and fluid cushion the bones; in a joint with osteoarthritis (right), the cartilage has eroded away so there is nothing to prevent the bones from rubbing together.

[1]McAlindon, T. E., LaValley, M. P., Gulin, J. P., and Felson, D. T. Glucosamine and chondroitin for treatment of osteoarthritis: A systematic quality assessment and meta-analysis. *JAMA* 283:1469–1475, 2000.

[2]Hochberg, M. C., and Dougados, M. Pharmacological therapy of osteoarthritis. *Best Pract. Res. Clin. Rheumatol.* 15:583–593, 2001.

[3]Clegg, D. O., Reda, D. I., Harris, C. L., et al. Glucosamine, chondroitin sulfate, and the two in combination for painful knee osteoarthritis. *N. Engl. J. Med.* 354:795–808, 2006.

[4] Wandel, S., Jüni, P., Tendal, B. et al. Effects of glucosamine, chondroitin, or placebo in patients with osteoarthritis of hip or knee: network meta-analysis. *BMJ.* 341:c4675, 2010.

[5] Canadian Arthritis Foundation. Glucosamine and Chondroitin. Available online at http://www.arthritis.ca/tips%20for%20living/complementary%20 therapies/types/supplements/other%20supplements/glucosamine/default.asp?s=1. Accessed January 2, 2011.

in taste, constipation, weakness, drowsiness, diarrhea, and nausea. Illness related to incorrect doses or inappropriate combinations of medications is also a significant health problem in the elderly. Both the effects of drugs on nutritional status and the effects of nutritional status on the effectiveness of the medications must be considered.

Effect of Medications on Nutritional Status Medications can affect nutritional status by altering appetite, nutrient absorption, metabolism, or nutrient excretion (**Table 16.2**). The impact is greatest in individuals who must take medications for extended periods, those who take multiple medications, and those whose nutritional status is already marginal.

Table 16.2 Commonly Used Drugs that May Cause Nutritional Deficiencies

Drug Group	Drug	Potential Deficiency
Antacids	Sodium bicarbonate	Folate, phosphorus, calcium, copper
	Aluminum hydroxide	Phosphorus
Anticonvulsants	Phenytoin, phenobarbital, primidone	Vitamins D and K
	Valproic acid	Carnitine
Antibiotics	Tetracycline	Calcium
	Gentamicin	Potassium, magnesium
Antibacterial agents	Neomycin	Fat, nitrogen
	Boric acid	Riboflavin
	Trimethoprim	Folate
	Isoniazid	Vitamins B_6 and D, niacin
Anti-inflammatory agents	Sulfasalazine	Folate
	Prednisone	Calcium
	Aspirin	Vitamin C, folate, iron
Anticancer drugs	Colchine	Fat, vitamin B_{12}
	Methotrexate	Folate, calcium
Anticoagulant drugs	Warfarin	Vitamin K
Antihypertensive drugs	Hydralazine	Vitamin B_6
Diuretics	Thiazides	Potassium
	Furosemide	Potassium, calcium, magnesium
Hypocholesterolemic agents	Cholestyramine	Fat, fat-soluble vitamins, iron, folate, vitamin B_{12}
Laxatives	Mineral oil	Fat-soluble vitamins
	Senna	Fat, calcium, vitamin B_6, folate, vitamin C
Tranquilizers	Chlorpromazine	Riboflavin

Source: Adapted from Roe, D. A. *Diet and Drug Interactions*. New York: Van Nostrand Reinhold, 1989.

Some medications directly affect the gastrointestinal tract. More than 250 drugs, including blood pressure medications, antidepressants, decongestants, and the pain reliever ibuprofen (found in Advil and Motrin), can cause mouth dryness, which can decrease interest in eating by interfering with taste, chewing, and swallowing. Aspirin is a stomach irritant and can cause small amounts of painless bleeding in the gastrointestinal tract, resulting in iron loss. Digoxin, which is a heart stimulant, can cause gastrointestinal upset, loss of appetite, and nausea. Narcotic pain medications such as codeine can lead to constipation, nausea, and vomiting.

Other drugs can decrease nutrient absorption. Cholestyramine, which is used to reduce blood cholesterol, and Orlistat, used to promote weight loss, can decrease the absorption of fat-soluble vitamins, vitamin B_{12}, iron, and folate. Metformin, a popular drug for the treatment of type 2 diabetes, can reduce the absorption of vitamin B_{12}. Antacids that contain aluminum or magnesium hydroxide (Rolaids or Maalox) combine with phosphorus in the gut to form compounds that cannot be absorbed; chronic use can result in loss of phosphorus from bone and possibly accelerate osteoporosis. Repeated use of stimulant laxatives can deplete calcium and potassium. Mineral oil laxatives prevent the absorption of fat-soluble vitamins. If it is not possible to prevent constipation by consuming a diet high in fibre and fluid, bulk-forming laxatives containing the soluble fibre psyllium (Metamucil) are a safer choice.

The metabolism of drugs can also affect nutritional status. For example, anticonvulsive drugs (used to prevent seizures) increase the liver's capacity to metabolize and eliminate vitamin D, therefore increasing the need for this vitamin.

Some drugs affect nutrient excretion. Diuretics, which are used to treat hypertension and edema, cause water loss, but some types (thiazides) also increase the excretion of potassium. People taking thiazide diuretics are advised to include several good sources of potassium in their diet each day or are prescribed supplements.

Effect of Food and Nutritional Status on the Effectiveness of Medications Food components can either enhance or retard the absorption and metabolism of drugs. Some drugs are absorbed better or faster if taken with food. Other drugs, such as Aspirin and ibuprofen, should be taken with food because they are irritating to the gastrointestinal tract. Since food can delay how quickly drugs leave the stomach, some medications are best taken with just water. Other drugs interact with specific foods. For instance, the antibiotic tetracycline should not be taken with milk because it binds with calcium, making both unavailable. The metabolism of atorvastatin (Lipitor), taken to lower cholesterol, is blocked by a compound in grapefruit so eating grapefruit or drinking grapefruit juice can result in drug toxicity.[42]

Nutritional status can also affect drug metabolism. If nutritional status is poor, the body's ability to detoxify drugs may be altered. For example, in a malnourished individual, theophylline, used to treat asthma, is metabolized slowly, resulting in high blood levels of the drug, which can cause loss of appetite, nausea, and vomiting.

Specific nutrients can also affect the metabolism of drugs. High-protein diets enhance drug metabolism in general, and low-protein diets slow it. Vitamin K hinders the action of anticoagulants taken to reduce the risk of blood clots. On the other hand, omega-3 fatty acids, such as those in fish oils, inhibit blood clotting and may intensify the effect of an anticoagulant drug and cause bleeding. It is safe to eat fish while taking anticoagulant drugs; however, the use of fish oil supplements is not recommended. Drugs can also interact with each other. For example, alcohol affects the metabolism of more than 100 medications. Drug interactions can exaggerate or, in some cases, diminish the effect of a medication. Individuals taking any medication should consult their doctor, pharmacist, or dietitian regarding how the drug could affect the action of other drugs they may be taking, how the drug could affect their nutrition, and how their nutrition could affect the action of the drug.

Economic, Social, and Psychological Factors

There are a variety of social and economic changes that often accompany aging. These factors are all interrelated and affect nutritional status by decreasing the motivation to eat and the ability to acquire and enjoy food.

Income Many older adults must live on a fixed income when they retire from their jobs, making it difficult to afford medications and a healthy diet. Food is often the most flexible expense in one's budget, so limiting the types and amounts of foods consumed may be the only option available for older adults trying to meet expenses. Substandard housing and inadequate food preparation facilities can make the situation worse because food cannot easily be prepared and eaten at home. In 2007-2008, 2.5% of Canadians over age 65 experienced food insecurity.[43]

Figure 16.12 Many older people live in assisted-living facilities and nursing homes, where nutritious meals are prepared for them. (©Ryan McVay/PhotoDisc/Getty Images, Inc.)

Dependent Living Although many older adults continue to live independently in their own homes, the physical decline and psychological issues associated with aging cause some to eventually require assistance in living (**Figure 16.12**). Poor eyesight and other physical restrictions can limit the ability to drive a car. Without help, many older adults may be unable to get to markets and food programs, restricting the types of food available to them. While a social support system consisting of family members, friends, and other caregivers can help many people stay at home, others may require assisted-living facilities, where they have their own apartments but can obtain assistance around the clock. For some, however, a nursing home is required to obtain the appropriate care.

Those in nursing homes are at increased risk for malnutrition because they are more likely to have medical conditions that increase nutrient needs or that interfere with food intake or nutrient absorption, and because they are dependent on others to provide for their care. In addition, 50% of institutionalized elderly suffer from some form of disorientation or confusion, which further increases the likelihood of decreased nutrient intake. Even when adequate meals are provided, many nursing home residents require assistance in eating and frequently do not consume all of the food served, increasing the likelihood of fluid and energy deficits.[44]

Depression Social, psychological, and physical factors all contribute to depression in the elderly.[44] Social factors such as retirement and the death or relocation of friends and family can cause social isolation and depression. Physical factors such as disability cause loss of independence. This reduces the ability to engage in normal daily activities, visit with friends and family easily, and provide for personal needs, further contributing to depression. Depression can make meals less appetizing and decrease the quantity and quality of foods consumed, thereby increasing the risk of malnutrition.

16.5 Keeping Older Adults Healthy

Learning Objectives

- Explain how exercise and a nutritious diet affect the degenerative changes of aging.
- Describe factors that may limit a senior's ability to consume a nutritious diet.

There is no secret dietary factor that will bestow immortality, but good nutrition and an active lifestyle are major determinants of successful aging. A good diet can extend an individual's healthy life span by preventing malnutrition and delaying the onset of chronic diseases. The diseases that are the major causes of disability in older adults—cardiovascular disease, hypertension, diabetes, cancer, and osteoporosis—are all nutrition-related. Exercise and a lifetime of healthy eating will not necessarily prevent these diseases, but they may slow the changes that accumulate over time, postponing the onset of disease symptoms. For example, the risk of developing cardiovascular disease can be decreased by exercise and a diet low in saturated fat, *trans* fat, and cholesterol and high in whole grains, fruits, and vegetables. The risk of osteoporosis may be reduced by adequate calcium intake and exercise throughout life. And the likelihood of developing certain types of cancer can be reduced by consuming a diet high in whole grains, vegetables, and fruits. Regular exercise can slow the loss of lean body mass, maintain fitness and independence, and allow an increase in food intake without weight gain, so micronutrient needs are more easily met.

Despite the fact that the nutrient needs of older adults are not drastically different from those of young adults, it is more challenging to meet these needs. Some of this challenge is due to changes in health and social and economic conditions that are more common in this population. Meeting needs requires consideration of each person's medical, psychological, social, and economic circumstances. For some, government aid is needed to assure adequate nutrition.

A Healthy Diet Plan for Older Adults

The first step toward meeting the nutrient needs of the elderly is to plan a healthy diet. Because older adults need less energy but the same amounts of most micronutrients, their food choices must be nutrient dense. Meals and snacks should include plenty of liquids because dehydration is a common problem. In some cases, nutrient supplements may be necessary to meet needs.

Canada's Food Guide Figure 16.13 shows the amounts of food from each group needed to meet the energy needs of sedentary Canadians over age 50.

Recommended Number of Food Guide Servings per Day

	Children			Teens		Adults			
Age in Years	2-3	4-8	9-13	14-18		19-50		51+	
Sex	Girls and Boys			Females	Males	Females	Males	Females	Males
Vegetables and Fruit	4	5	6	7	8	7-8	8-10	7	7
Grain Products	3	4	6	6	7	6-7	8	6	7
Milk and Alternatives	2	2	3-4	3-4	3-4	2	2	3	3
Meat and Alternatives	1	1	1-2	2	3	2	3	2	3

Men and women over 50

The need for **vitamin D** increases after the age of 50.

In addition to following *Canada's Food Guide*, everyone over the age of 50 should take a daily vitamin D supplement of 10 µg (400 IU).

Figure 16.13 Canada's Food Guide: Recommendations for older adults.
Source: Reprinted with permission from Health Canada. Canada's Food Guide. Available online at http://www.hc-sc.gc.ca/fn-an/alt_formats/hpfb-dgpsa/pdf/food-guide-aliment/print_eatwell_bienmang-eng.pdf. Accessed September 11, 2011.

To meet micronutrient needs without exceeding kcalorie needs, nutrient-dense choices must be made from each food group. To ensure adequate fibre, whole grains should be chosen from grain products. To maximize vitamin and phytochemical intake, a variety of fruits and orange, dark green, and starchy vegetables should be included. To maximize nutrient density, low-fat milk and alternatives and lean meats should be chosen. To ensure foods are eaten, those offered should be easy to prepare and well seasoned to enhance appeal.

Dietary Supplements Many older adults may benefit from supplementing particular nutrients. Canada's Food Guide specifically recommends Vitamin D supplements for older adults because production of this vitamin in the skin is decreased in the elderly and exposure to sunlight may be limited. A calcium supplement may be necessary to meet needs, particularly in older women, because it can be difficult to consume 1,200 mg of calcium from food without exceeding energy needs. Supplemental vitamin B_{12} from pills or fortified foods is recommended for older adults because the absorption of vitamin B_{12} often decreases with age. However, supplements should not take the place of a balanced, nutrient-dense diet high in whole grains, fruits, and vegetables. These foods also contain phytochemicals and other substances that may protect against disease. Older adults should be cautious to avoid overdoses, and the resulting toxicities, when selecting supplements.

If supplementation is required, a vitamin and mineral supplement containing no more than the RDA for the nutrient is safest; supplements containing megadoses should be avoided. Supplements of non-nutrient substances should be taken with care. Most of these provide no proven benefit, many are costly, and others can be toxic. For example, lecithin is claimed to lower cholesterol and to treat Alzheimer's disease, but there is no proof that it does either. RNA is claimed to rejuvenate old cells, improve memory, and prevent wrinkling, but there are no controlled studies to support any of these claims. Superoxide dismutase (SOD), an enzyme that protects against oxidative damage, is said to slow aging and treat Alzheimer's disease. However, SOD is a protein that is broken down to amino acids in the gastrointestinal tract, so oral supplements will not increase blood or tissue levels of this enzyme. Coenzyme Q10, a synthetic version of a compound in the electron transport chain, is marketed to older adults as a way to slow aging by enhancing the immune system. However, it does not boost immune function and may pose a risk to people with poor circulation.

The hypothesis that aging is caused by oxidative damage has contributed to the popularity of antioxidant supplements. Although antioxidant supplements will not retard the aging process, there is evidence that adequate intakes may reduce the incidence of disease. Antioxidants, including vitamin E and vitamin C, have been found to preserve the function of immune system cells and may therefore help protect the body from infectious disease.[45] There is also evidence that diets high in antioxidants from fruits and vegetables and other plant foods are associated with a reduced incidence of various chronic diseases, including cardiovascular disease and some types of cancer. Unfortunately, intervention trials have not consistently found that antioxidant supplements reduce the risk of cardiovascular disease or cancer.[46,47] So, rather than taking an antioxidant supplement, older adults, like everyone else, should consume a diet plentiful in plant foods high in these nutrients. When antioxidant nutrients are obtained from foods, they bring with them phytochemicals, some of which offer additional antioxidant protection and some of which protect us from chronic disease in other ways (**Figure 16.14**).

Figure 16.14 Older adults can increase their antioxidant intake by consuming foods that are good sources of antioxidants, which also provide fibre, energy, other micronutrients, and phytochemicals. (©istockphoto.com/Robyn Mackenzie)

Physical Activity for Older Adults

Regular physical activity can extend the years of active independent life, reduce disability, and improve the quality of life for older adults. It helps to maintain muscle mass, bone strength, and cardiorespiratory function. Exercise also increases energy expenditure, so more food can be eaten, increasing the chances that adequate amounts of all essential nutrients will be consumed and reducing the risk of weight gain. The Canadian Physical Activity Guidelines for older adults (**Figure 16.15**) recommends that adults 65 years and older should engage in at least 150 min. of moderate-to-vigorous intensity aerobic activity weekly.[48] Activity should be sustained for at least 10 min. per session. Examples include walking, biking, and swimming. Water activities such as water aerobics and swimming do not stress the joints, so they can be used to improve endurance in those with arthritis or other bone and joint disorders (**Figure 16.16**). Before starting an exercise program or increasing the level of physical activity, older adults should check with their physician. Those who have been inactive should gradually increase their activity levels. Muscle- and bone-strengthening activities that involve major muscle groups are recommended at least 2 days/week (**Figure 16.17**). This could include activities such as lifting weights or soup cans, carrying the laundry, carrying groceries, climbing stairs, wall push-ups, and standing up and sitting down several times in a row. Physical activities to enhance balance are also advised to help prevent falls.

Preventing Food Insecurity

Once a diet that will meet needs has been developed, steps must be taken to assure that the elderly individual is capable of obtaining and consuming this diet. Preventing food insecurity and ensuring adequate nutrient intake may involve providing nutrient-dense meals or instruction regarding nutrient needs, economic food choices, and food preparation. It may also require assistance with shopping and food preparation.

Overcoming Economic Limitations Being able to afford a healthy diet is a problem for many older individuals. Reduced-cost food and meals at senior centres, food banks, and soup kitchens are available to people on limited incomes. Programs that provide education about low-cost nutritious food choices can also help reduce food costs.

Overcoming Social Limitations Another problem that contributes to poor nutrient intake is loneliness (**Figure 16.18**). Living, cooking, and eating alone can decrease interest in food. This can be a problem not only for the elderly but for anyone who typically eats alone.

Canadian Physical Activity Guidelines

FOR OLDER ADULTS - 65 YEARS & OLDER

Guidelines

To achieve health benefits, and improve functional abilities, adults aged 65 years and older should accumulate at least 150 minutes of moderate- to vigorous-intensity aerobic physical activity per week, in bouts of 10 minutes or more.

It is also beneficial to add muscle and bone strengthening activities using major muscle groups, at least 2 days per week.

Those with poor mobility should perform physical activities to enhance balance and prevent falls.

More physical activity provides greater health benefits.

Let's Talk Intensity!

Moderate-intensity physical activities will cause older adults to sweat a little and to breathe harder. Activities like:

- Brisk walking
- Bicycling

Vigorous-intensity physical activities will cause older adults to sweat and be 'out of breath'. Activities like:

- Cross-country skiing
- Swimming

Being active for at least **150 minutes** per week can help reduce the risk of:

- Chronic disease (such as high blood pressure and heart disease) and,
- Premature death

And also help to:

- Maintain functional independence
- Maintain mobility
- Improve fitness
- Improve or maintain body weight
- Maintain bone health and,
- Maintain mental health and feel better

Pick a time. Pick a place. Make a plan and move more!

- ☑ Join a community urban poling or mall walking group.
- ☑ Go for a brisk walk around the block after lunch.
- ☑ Take a dance class in the afternoon.
- ☑ Train for and participate in a run or walk for charity!
- ☑ Take up a favourite sport again.
- ☑ Be active with the family! Plan to have "active reunions".
- ☑ Go for a nature hike on the weekend.
- ☑ Take the dog for a walk after dinner.

Now is the time. Walk, run, or wheel, and embrace life.

www.csep.ca/guidelines

Figure 16.15 **Canadian Physical Activity Guidelines help older adults plan physical activities.** **Source:** Canadian Physical Activity Guidelines, © 2011. Used with permission from the Canadian Society for Exercise Physiology, www.csep.ca/guidelines.

Figure 16.16 Water aerobics classes provide older adults with a low-impact aerobic activity and social interaction. (Photo and Co/Getty Images, Inc.)

Figure 16.17 Weight training at any age improves muscle strength and endurance. (© istockphoto.com/Ben Blankenburg)

Figure 16.18 The social interaction provided by congregate meals is as beneficial to older adults as the meals provided. (Thinkstock/ Getty Images, Inc.)

Buying single servings of food is an option, although it can be expensive. To avoid spoilage of perishable items, grocers can be asked to break up packages of meat, eggs, fruits, and vegetables so small amounts can be purchased. Alternatively, large packages can be purchased and shared among friends. Cooking larger portions and freezing foods in meal-sized batches can be helpful not only with cost but also to relieve the boredom of eating the same leftovers several days in a row. Creativity and flexibility in what defines a meal can also help. An easy single meal can be prepared by topping a potato with cooked vegetables and cheese, or with leftover chili or spaghetti sauce (**Figure 16.19**). Yogurt or a bowl of cereal with fruit and milk is also a nutritious dinner option.

Overcoming Physical Limitations Difficulty in cooking due to limited mobility can also reduce food intake. Precooked foods, frozen dinners, canned foods, or salad bar items, as well as instant foods such as cereals, rice and noodle dishes, and soups that just require adding water, can provide a meal with almost no preparation. Medical nutritional products such as Ensure or Boost can also be used to supplement intake. These canned, fortified products have a long shelf life and can meet nutrient needs with a small volume. Food can also be ordered by phone if it is affordable. Eating out at senior centres or low-cost restaurants or sharing shopping and cooking chores with a friend can reduce cooking demand and increase social interaction. Home health services can help with cooking and feeding, and most senior centres, health departments, and social service agencies offer meals, rides, and in-home care.

Overcoming Medical Limitations Medical conditions and the use of medications often affect food choices. Meals need to be appealing and easy to prepare and consume, as well as compatible with medical conditions. For instance, an individual with dental problems may not be able to chew fresh fruits and vegetables. Therefore, a texture modification is required. Fully cooked, canned, or soft fruit or fruit juices can be substituted for hard-to-chew fruits, and cooked vegetables can replace raw ones. Eggs and stewed meats can provide easy-to-chew protein sources. To overcome changes in the sense of taste and smell, spicy or acidic foods may be limited or emphasized, depending on individual tastes.

Special diets are typically prescribed for individuals with medical conditions such as hypertension, heart disease, diabetes, and kidney disease. These diets may contribute to malnutrition if they restrict favourite foods and if individuals prescribed the diet are not provided with enough information about how to substitute foods that will provide adequate energy, nutrients, and eating pleasure.[10] Education about what foods are appropriate and how to read food labels can help identify products that fit within dietary restrictions.

The use of prescription or over-the-counter medications to treat medical conditions can also affect eating habits. Physicians, pharmacists, and dietitians can provide information about possible effects on food intake. Purchasing all prescription medications from the same pharmacy will ensure that the pharmacist is aware of all medications taken and can advise of possible interactions. Health-care providers also need to be informed about all nonprescription medications and vitamin, mineral, or other dietary supplements used and whether or not medications are taken according to the prescription instructions.

CANADIAN CONTENT

Nutrition Programs for the Elderly

Canada does not have a comprehensive national nutrition program for older adults. A recent review of government policy at the federal and provincial levels notes that there is very limited policy on the promotion of nutrition that is specifically targeted to seniors living in the community.[49]

Programs that do exist focus on the provision of food to needy seniors and are organized in local communities. The most common programs included congregate dinners and home meal delivery. Congregate dinners, where groups of seniors get together for a meal, provide not only nutrition but social contact. The delivery of meals to isolated seniors by programs such as Meals on Wheels allows many seniors to continue to live independently at home.[50] These programs provide a valuable service, but there is a need for more nutrition education.[49]

To help identify elderly who may need nutritional assistance, Canadian researchers have developed a questionnaire called Seniors in the Community Risk Evaluation for Eating and Nutrition, or SCREEN,™ which asks seniors questions about weight changes, appetite, swallowing problems, meal preparation, and other eating related activities that impact nutritional adequacy.[51]

Using SCREEN™ (http://www.drheatherkeller.com/SCREEN.htm) and other survey questions, Statistics Canada began collecting data on the health of older Canadians in 2008 to determine the factors, including diet and physical activity, that contribute to good health (See Critical Thinking: Health-promoting Factors). Perhaps these data will provide the information needed to design programs that effectively promote healthy eating to seniors.

Figure 16.19 Selecting acceptable foods that are nutritious and easy to prepare is important in meeting the needs of the elderly. (©Tony Freeman/Photo Edit)

CRITICAL THINKING:

Health-Promoting Factors

CANADIAN CONTENT

As part of the CCHS, researchers conducted a survey of older adults (> 45 years) to determine how healthy they were and to determine what factors contribute to good health. Participants were asked to describe whether they were in good health and were then asked to indicate which of the following health-promoting factors applied to their personal situation:

- Never smoked daily/quit for 15 years or more
- Not obese (i.e., BMI < 30)
- Sleeps well
- Fruit/vegetable consumption 5 or more times/daily
- Good oral health
- Frequent walker
- Frequent social participation
- Low daily stress

Researchers compared the prevalence of self-reported good health to the total number of health-promoting factors each individual identified. The results are shown in the graph.

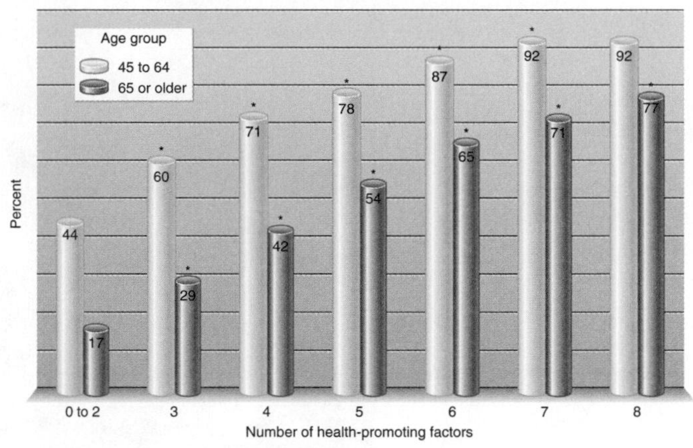

*significantly different from estimate for previous category in same age group ($p < 0.05$)

Source: Ramage-Morin, P. L., Shields, M., and Martel, L. Health-promoting factors and good health among Canadians in mid- to late life. *Health Reports.* 21(3):1-9, 2010. Available online at http://www.statcan.gc.ca/pub/82-003-x/2010003/article/11289-eng.htm. Accessed January 2, 2011.

Critical Thinking Questions

Looking at the results shown in the graph, what do you conclude about the differences between younger and older adults?

What do you conclude about the relationship between good health and the number of health-promoting factors identified?

OUTCOME

Min was lonely and became more and more depressed after her husband died. She didn't like to eat by herself, so she stopped cooking healthy meals. Instead, she ate easy-to-prepare convenience foods that were high in fat and included few vegetables or fruits. She became too depressed to seek out her old walking buddies, her fitness declined, and she rarely left the house. Her family became concerned and began visiting and phoning more often. After spending a weekend with her grandchildren, Min realized she was still loved and needed. She decided it was time to take care of herself. She began taking an exercise class, which helped her regain her fitness so she could walk with her friends. She realized that her diet of convenience foods had led to some weight gain. To improve her diet, she pulled out some of the healthy recipes she used to cook when her husband was alive, cutting the amounts in half or freezing the extra portions for another day. On Wednesday nights, she began attending church suppers. Eating there allowed her to meet people and helped her enjoy her meals again. Although Min still misses her husband, a year after his death she again has an active social life and is in good nutritional health.

APPLICATIONS

Personal Nutrition

1. **How does age affect energy needs?**
   a. How does your average energy intake from the food record you kept in Chapter 2 compare to the EER for a person who is your height, weight, and activity level but is 75 years old?
   b. Use iProfile to suggest modifications in your food choices that would allow you to meet the nutrient needs of a 75-year-old without exceeding the energy needs.
   c. Use Canada's Food Guide to determine the amounts of food you should consume from each food group if you were 75 years old, but at your current weight and height.

iProfile

2. **How do medical conditions and dietary restrictions affect food choices?**
   a. How might you modify your food choices to accommodate a low-sodium diet?
   b. How might you modify your food choices to accommodate a restriction of protein to 0.6 g/kg of body weight?
   c. How might you modify your food choices to accommodate a loss of smell and taste?
   d. How might you modify your food choices to accommodate a dry mouth and poorly fitting dentures?

General Nutrition Issues

1. **What information and resources are available for the elderly and their families?**
   a. Use the Internet to determine what kind of nutrition information is available to individuals planning for the care of their elderly parents or relatives. What are the costs?
   b. Assume that you have an elderly friend who lives and eats alone. Find resources in your area that would be able to provide meals and other services for your friend.

2. **What do seniors need to know about nutrition?**
   a. Prepare an outline for a 20-minute lecture on nutrition and aging that could be given at a senior centre in your area.
   b. Define 2 goals for the lecture.
   c. What are 5 main points that you should discuss?

SUMMARY

16.1 What Is Aging?

- Aging is the accumulation of changes over time that results in an ever-increasing susceptibility to disease and death. The longest an organism can live, or life span, is a characteristic of a species. The average age to which people in a population live, or life expectancy, is a characteristic of a population.
- As a population, we are living longer but not necessarily healthier lives. The elderly are the fastest-growing segment of

the Canadian population. Compression of morbidity, that is, increasing the number of healthy years, is an important public health goal. A healthy diet and lifestyle cannot stop aging but can postpone the onset of many of the physiological changes and diseases that are common in older adults.

16.2 What Causes Aging?

- As organisms become older, the number of cells they contain decreases and the function of the remaining cells declines. This reduces reserve capacity, lowering the organism's ability to maintain homeostasis and increasing the risk of disease.
- The loss of cells and cell function that causes aging is believed to be due to both genetic factors that limit cell life and the accumulation of cellular damage over time.
- How long we live and how long we remain healthy is determined by a combination of genetic, environmental, and lifestyle factors.

16.3 Nutritional Needs and Concerns of Older Adults

- Energy needs are lower in older adults due to a decrease in basal metabolic rate and the thermic effect of food and a reduction in physical activity, but the need for protein, water, fibre, and most micronutrients remains the same. Intakes of fibre and water are often less than recommended, increasing the risk of dehydration and constipation.
- Low intakes, reduced absorption, and changes in the metabolism of certain micronutrients, including vitamin B_{12}, vitamin D, and calcium, puts older adults at risk of deficiency. Iron requirements decrease in women after menopause but many older adults are at risk of iron deficiency due to blood loss from disease or medications.

16.4 Factors that Increase the Risk of Malnutrition

- The risk of malnutrition increases with age due to the physiological changes that accompany aging. There are changes in the sense of smell that affect the appeal of food, changes in vision that affect the ability to prepare food, changes in digestion and absorption that decrease the intake and absorption of nutrients, changes in metabolism that affect nutrient utilization, changes in weight and body composition that increase health risks and reduce independence, changes in hormonal patterns that affect body function, and changes in immune function that increase the risk of infectious and chronic disease.
- Both infectious and chronic diseases affect nutrient requirements and the ability to consume a nutritious diet. Aging increases the incidence of diseases that reduce mobility and mental capacity, limiting the ability to acquire, prepare, and consume food. The medications used to treat disease also affect nutrition, especially when the medications are taken over long periods of time and when multiple medications are taken simultaneously.
- Low income levels increase the risk of malnutrition among the elderly by limiting the ability to purchase food. Loss of independence contributes to depression, which makes meals less appetizing and decreases the quantity and quality of foods consumed.

16.5 Keeping Older Adults Healthy

- A healthy diet and regular exercise can prevent malnutrition, delay the onset of chronic conditions, and increase independence in older adults. Canada's Food Guide can be used to determine the amounts of food from each food group needed by older adults. Meals for the elderly must be nutrient-dense, provide plenty of fluid and fibre, and consider individual medical, psychological, social, and economic circumstances. Supplements of calcium, vitamin D, and vitamin B_{12} may be beneficial. In some cases, assistance with shopping and meal preparation may be needed.
- A physical activity program for older adults should include activities that improve endurance, strength, flexibility, and balance. Activities should be tailored to the individual's needs and likes. Exercise classes are advantageous for older adults because they provide both social interaction and professional support and instruction. A well-planned exercise program can reduce the risk of chronic disease, improve mobility, increase independence, and reduce the risk of falls and injuries.
- There is limited nutrition education programming in Canada specifically directed to seniors.

REVIEW QUESTIONS

1. What is life expectancy? How does it differ from healthy life expectancy? How does it differ from life span?
2. What is meant by compression of morbidity?
3. What factors determine at what age the consequences of aging become apparent?
4. Why are older adults at risk for malnutrition?
5. List 3 physiological changes that occur with aging.
6. List 3 ways in which medication use and nutrition interact.
7. What social and economic factors increase nutritional risk among the elderly?
8. Why are the energy needs of older adults reduced?
9. Why are older adults at risk for vitamin B_{12} deficiency? Vitamin D deficiency?
10. How can nutrition affect the risk of developing macular degeneration?
11. Should obese adults over 70 years of age lose weight? Why or why not?
12. Explain how physical disabilities and mental illness affect nutritional status.
13. Why is it important that elderly individuals consume a nutrient-dense diet?
14. How do energy requirements change with aging?
15. List some physical activities that are appropriate for older adults
16. What is the purpose of SCREEN™?

REFERENCES

1. Masoro, E. J. Overview of caloric restriction and ageing. *Mech. Aging Dev.* 126:913–922, 2005.
2. Lamming, D. W., Latorre-Esteves, M., Medvedik, O., et al. HST2 mediates SIR2-independent life-span extension by calorie restriction. *Science*. 309:1861–1864, 2005.
3. St-Arnaud, J., Beaudet, M. P., and Tully, Life Expectancy. *Health Reports* 17 (1): 43-47, P. 2005. Available online at http://www.statcan.gc.ca/bsolc/olc-cel/olc-cel?catno=82-003-X20050018709&lang=eng. Accessed December 28, 2010.

4. Turcotte, M., and Schellenberg, G. A Portrait of Seniors in Canada, Statistics Canada Cat no 89-519-XIE, 2006. Available online at http://www.statcan.gc.ca/bsolc/olc-cel/olc-cel?catno=89-519-X&lang=eng. Accessed December 27, 2010.
5. Statistics Canada. Health-adjusted life expectancy, by sex, 2001. Available online at http://www40.statcan.ca/l01/cst01/hlth67-eng.htm. Accessed December 28, 2010.
6. DHHS, Administration on Aging A Profile of Older Americans: 2008 Health and Health Care Available online at http://www.aoa.gov/AoARoot/Aging_Statistics/Profile/ 2008/14.aspx. Accessed May 11, 2009.
7. Troen, B. R. The biology of aging. *Mt. Sinai J. Med.* 70:3–22, 2003.
8. Institute of Medicine, Food and Nutrition Board. *Dietary Reference Intakes for Energy, Carbohydrate, Fiber, Fat, Protein, and Amino Acids*. Washington, D.C.: National Academies Press, 2002.
9. Villareal, D. T., Apovian, C. M., Kushner, R. F., et al. American Society for Nutrition; NAASO, the Obesity Society. Obesity in older adults: Technical review and position statement of the American Society for Nutrition and NAASO, the Obesity Society. *Am. J. Clin. Nutr.* 82:923–934, 2005.
10. American Dietetic Association. Position of the American Dietetic Association: Nutrition across the spectrum of aging. *J. Am. Diet. Assoc.* 105: 616–633, 2005.
11. Institute of Medicine, Food and Nutrition Board. *Dietary Reference Intakes for Thiamin, Riboflavin, Niacin, Vitamin B6, Folate, Vitamin B12, Pantothenic Acid, Biotin, and Choline*. Washington, D.C.: National Academies Press, 1998.
12. Rampersaud, G. C., Kauwell, G. P., and Bailey, L. B. Folate: A key to optimizing health and reducing disease risk in the elderly. *J. Am. Coll. Nutr.* 22:1–8, 2003.
13. Garriguet, D. Canadians' Eating Habits. *Health Reports* 18(2):17-32, 2007. Available online at http://www.statcan.gc.ca/bsolc/olc-cel/olc-cel?catno=82-003-X20060069609&lang=eng. Accessed December 19, 2010.
14. Russell, R. M. The vitamin A spectrum: from deficiency to toxicity. *Am. J. Clin. Nutr.* Apr;71(4):878-84, 2000.
15. Penniston, K. L. and Tanumihardjo, S. A. Vitamin A in dietary supplements and fortified foods: too much of a good thing? *J. Am. Diet. Assoc.* 103(9):1185-7, 2003.
16. Institute of Medicine, Food and Nutrition Board. *Dietary reference intakes for calcium and vitamin D.* Washington DC. National Academies Press. 2010.
17. Chernoff, R. Micronutrient requirements in older women. *Am. J. Clin. Nutr.* 81:1240S–1245S, 2005.
18. Health Canada. Do Canadians meet their nutrient requirements through food intake alone? Cat #H164-112/3-2009E-PDF, 2009. Available online at http://www.hc-sc.gc.ca/fn-an/alt_formats/pdf/surveill/nutrition/commun/art-nutr-adult-eng.pdf. Accessed June 27, 2011.
19. Tarasuk, V. Household food insecurity in Canada. *Topics in Clinical Nutrition* 20(4):299-312, 2005.
20. Schaefer, D. C., and Cheskin, L. J. Constipation in the elderly. *Am. Family Physician* 58:907–914, 1998.
21. Esposito, C., Plati, A., Mazzullo, T. et al. Renal function and functional reserve in healthy elderly individuals. *J. Nephrol.* 20:617–625, 2007.
22. Tjepkema, M. Adult obesity. *Health Rep.* 17(3):9-25, 2006.
23. Thomas, D. R. Weight loss in older adults. *Rev. Endocr. Metab. Disord.* 6:129–136, 2005.
24. Nikolic, M., Bajek, S., Bobinac, D. et al. Aging of human skeletal muscles. *Coll. Anthropol.* 29:67–70, 2005.
25. St-Onge, M. P. Relationship between body composition changes and changes in physical function and metabolic risk factors in aging. *Curr. Opin. Clin. Nutr. Metab.* Care 8:523–528, 2005.
26. Col, N. F., Surks, M. I., and Daniels, G. H. Subclinical thyroid disease: Clinical applications. *JAMA* 291:239–243, 2004.
27. Surks, M. I., Ortiz, E., Daniels, G. H. et al. Subclinical thyroid disease: Scientific review and guidelines for diagnosis and management. *JAMA* 291:228–238, 2004.
28. U.S. Preventive Services Task Force. Hormone therapy for the prevention of chronic conditions in postmenopausal women: Recommendations from the U.S. Preventive Services Task Force. *Ann. Intern. Med.* 142:855–860, 2005.
29. Toogood, A. A. The somatopause: An indication for growth hormone therapy? *Treat. Endocrinol.* 3:201–209, 2004.
30. Thorner, M. O. Statement by the Growth Hormone Research Society on the GH/IGF-I axis in extending health span. *J. Gerontol. A Biol. Sci. Med. Sci.* 64:1039–44, 2009.
31. Cameron, D. R., and Braunstein, G. D. The use of dehydroepiandrosterone therapy in clinical practice. *Treat. Endocrinol.* 4:95–114, 2005.
32. Bondy, S. C., Lahiri, D. K., Perreau, V. M. et al. Retardation of brain aging by chronic treatment with melatonin. *Ann. N.Y. Acad. Sci.* 1035:197–215, 2004.
33. Karasek, M. Does melatonin play a role in the aging process. *J. Physiol. Pharmacol.* 58:105–113, 2007.
34. National Institute of Diabetes and Digestive and Kidney Diseases. National Diabetes Information Clearinghouse, National Diabetes Statistics, 2007. Available online at http://diabetes.niddk.nih.gov/DM/PUBS/statistics/#allages. Accessed October 1, 2009.
35. Maggini, S., Wintergerst, E. S., Beveridge, S., et al. Selected vitamins and trace elements support immune function by strengthening epithelial barriers and cellular and humoral immune responses.*Br. J. Nutr.* 98 Suppl 1:S29-35, 2007.
36. Federal Interagency Forum on Aging-Related Statistics. *Older Americans 2008: Key Indicators of Wellbeing, Population*. Available online at www.agingstats.gov/agingstatsdotnet/Main_Site/ Data/Data_2008.aspx. Accessed September 30, 2009.
37. Cherubini, A., Martin, A., Andres-Lacueva, C. et al. Vitamin E levels, cognitive impairment and dementia in older persons: The InCHIANTI study. *Neurobiol. Aging* 26:987–994, 2005.
38. McDaniel, M. A., Maier, S. F., and Einstein, G. O. "Brain-specific" nutrients: A memory cure? *Nutrition* 19:957–975, 2003.
39. Sofi, F., Valecchi, D., Bacci, D. et al. Physical activity and risk of cognitive decline: a meta-analysis of prospective studies *J. Intern. Med.* 269(1):107-17, 2011.
40. Huang, T. L. Omega-3 fatty acids, cognitive decline, and Alzheimer's disease: a critical review and evaluation of the literature. *J. Alzheimers Dis.* 21(3):673-90, 2010.
41. Hajjar, E. R., Cafiero, A. C., and Hanlon, J. T. Polypharmacy in elderly patients. *Am. J. Geriatr. Pharmacother.* 5:314–316, 2007.
42. Karch, A. M. The grapefruit challenge: The juice inhibits a crucial enzyme, with possibly fatal consequences. *Am. J. Nurs.* 104:33–35, 2004.
43. Statistics Canada. Household food insecurity 2007-2008. Available online at http://www.statcan.gc.ca/pub/82-625-x/2010001/article/11162-eng.htm. Accessed September 23, 2011.
44. Dorner, B., Niedert, K. C., and Welch, P. K.; American Dietetic Association. Position of the American Dietetic Association: Liberalized diets for older adults in long-term care. *J. Am. Diet. Assoc.* 102:1316–1323, 2002.
45. De la Fuente, M. Effects of antioxidants on immune system aging. *Eur. J. Clin. Nutr.* 56(Suppl 3):S5–S8, 2002.

46. Stanner, S. A., Hughes, J., Kelly, C. N., et al. Review of the epidemiological evidence for the "antioxidant hypothesis." *Public Health Nutr*. 7:407–422, 2004.

47. Clarke, R., and Armitage, J. Antioxidant vitamins and risk of cardiovascular disease. Review of large-scale randomised trials. *Cardiovasc. Drugs Ther*. 16:411–415, 2002.

48. Canadian Society of Exercise Physiologists. Canadian Physical Activity Guidelines. Available online at: http://www.csep.ca/CMFiles/Guidelines/CSEP-InfoSheets-older%20adults-ENG.pdf. Accessed June 27, 2011. 33

49. More, C., and Keller, H.. Community nutrition policy for older adults in Canada. *Canadian Journal of Dietetic Practice and Research* 69(4): 198-200 2008.

50. Home and Community support: Nutrition Services. Available online at http://www.homeandcommunitysupport.ca/care_guide/nutrition.asp. Accessed January 2, 2011.

51. Keller, H. H., Goy, R., and Kane. S. L. Validity and reliability of SCREEN II (Seniors in the community: risk evaluation for eating and nutrition, Version II). *Eur. J. Clin. Nutr.* Oct;59(10):1149-57, 2005.

17 Food Safety

Case Study

One hundred children at Loghill Elementary School were absent or went home sick on Friday. Forty of them vomited at school. It might have been the stomach flu, but only a few of the teachers and parents became ill, even though they were in contact with the children and presumably would have been just as likely as the children to catch the flu. When this many people in one place become ill at the same time, foodborne illness is always suspect.

Even though almost everyone had recovered by Monday, the local health department was notified. Inspectors came to the school to investigate the cause of the illnesses and were able to trace the source to the "Welcome Back, Spring" celebration held the day before everyone became ill. For this event, four barbeques were set up to accommodate a wide range of tastes. The food served included beef burgers, chicken burgers, turkey burgers, and veggie burgers. After interviewing the children and adults who were sick, the inspectors determined that only those who ate ground turkey became ill, and they all began having symptoms within 48 hours of consuming it. Symptoms included nausea and vomiting, diarrhea, abdominal pain, and fever. Many of the sick children were seen by physicians, and *Salmonella* bacteria were isolated from their stool samples. Further investigation revealed that the Canadian Food Inspection Agency had, the day after the barbeque, issued a recall for the brand of ground turkey burger that had been served at the barbeque, because of suspected contamination with high levels of *Salmonella.* In this chapter you will learn about food safety and how to avoid foodborne illness.

(Jason Todd/Digital Vision)

(©iStockphoto)

(Chris Sattlberger/Getty Images)

Chapter Outline

17.1 How Can Food Make Us Sick?

Learning Objectives

- Name the primary cause of foodborne illness.
- Explain why a contaminated food does not cause illness in everyone who eats it.

Even though the Canadian food supply is among the safest in the world, it is not risk-free. *Salmonella* bacteria contaminate chickens, industrial waste has polluted some of our waterways, and pesticide residues are found on our fruit. Headlines announce *Escherichia coli* (*E. coli*) in apple juice; *Listeria* in lunchmeats; *Salmonella* in eggs, on vegetables, and in cereal; *Cyclospora* on fruit; *Cryptosporidium* in drinking water; and hepatitis A in frozen strawberries. It is estimated that foodborne illness attacks about 12 million Canadians annually.[1]

If given a choice, most people would elect to consume food that contains no harmful substances. However, it is nearly impossible to choose a diet that is free of all potential hazards. Food has always carried the risks of bacterial contamination and naturally occurring toxins. Today, modern agricultural technology, trade patterns, food processing, and changes in dietary habits have increased the risks associated with bacterial contamination and introduced new risks. Regulatory agencies, food manufacturers, and retailers, as well as consumers, must work together to maximize the safety of the food supply (**Figure 17.1**).

What Is Foodborne Illness?

foodborne illness An illness caused by consumption of food containing a toxin or disease-causing micro-organism.

pathogen A biological agent that causes disease.

toxins Substances that can cause harm at some level of exposure.

Foodborne illness in the broadest sense is any illness that is related to the consumption of food or contaminants or toxins in food. However, most foodborne illness is caused by the contamination of food with **pathogens**, that is, micro-organisms or microbes that can cause disease. **Toxins** produced by these micro-organisms, as well as chemical and physical contaminants from the environment and those used in the processing and packaging of food, can also cause foodborne illness. Chemical contaminants include substances such as drugs used in raising cattle and producing milk, pesticides and fertilizers used in growing crops, and wastes from industry that accumulate in the environment. Physical contaminants include substances as diverse as broken glass, packaging materials, and insect wings. Even substances we use to protect the food supply such as packaging and preservatives have the potential to cause harm if not properly used.

Figure 17.1 Choosing safe, nutritious foods is one way that consumers help control the safety of the foods they eat. (Eric Glenn/DK Stock/Getty Images)

How Does Food Get Contaminated?

Often, food is contaminated where it is grown or produced. *Salmonella enteritidis* may enter eggs directly from hens infected with the bacteria. Fish and seafood may be contaminated by agricultural runoff, sewage, and other toxins in the waters where they live. Moulds may grow on grains during unusually wet or dry growing seasons. Food can also be contaminated during processing or storage, at retail facilities, and even at home. This often occurs by **cross-contamination** from a contaminated food or piece of equipment to an uninfected one. For example, *E. coli* from a single cow can contaminate processing equipment and be transferred to thousands of kilograms of hamburger. Careful sanitation and handling can control most of these sources of foodborne illness.

cross-contamination The transfer of contaminants from one food or object to another.

When Do Contaminants Make Us Sick?

Even when a food is contaminated, it does not cause every individual who consumes it to become ill. The potential of a substance to cause harm depends on how potent it is, the amount or dose that is consumed, how frequently it is consumed, and who consumes it. Some contaminants in food cause harm even when minute amounts are consumed, and almost any substance can be toxic if a large enough amount is consumed. Many substances have a **threshold effect**; that is, they are harmless up to a certain dose or threshold, after which negative effects increase with increasing intake. Body size, nutritional status, and how a substance is metabolized by the body can also affect toxicity. Small doses are more dangerous in children and small adults because the amount of toxin per unit of body weight is greater. Poor nutritional or health status may decrease the body's natural ability to detoxify harmful substances. Substances that are stored in the body are more likely to be toxic because they accumulate over time. They are deposited in bones, adipose tissue, liver, or other tissues until toxicity symptoms occur. Substances that are easily excreted when consumed in excess are less likely to cause toxicity. The interaction of toxins with one another and with other dietary factors also affects toxicity. For example, mercury, which is extremely toxic, is not absorbed well if the diet is high in selenium, and the absorption of lead is decreased by the presence of iron and calcium in the diet.

threshold effect A reaction that occurs at a certain level of ingestion and increases as the dose increases. Below that level, there is no reaction.

17.2 Keeping Food Safe

Learning Objectives

- Explain how a HACCP system helps prevent foodborne illness.
- Discuss the roles of the federal agencies responsible for the safety of the Canadian food supply.
- Discuss the role of the consumer in keeping food safe.

Ensuring the safety of the food supply is a responsibility shared by the government, food manufacturers and retail establishments, and consumers. The steps taken to avoid foodborne illness require weighing the benefits a food provides against the potential risks it presents. This type of risk-benefit analysis is done by regulatory agencies when they evaluate the safety and sanitation of food-processing methods and food service establishments. Not every potential contaminant is harmful, nor can all be avoided, but those that have a great potential for harm need to be addressed. Consumers should also consider the risks and benefits when they choose which foods to buy, which to eat, and how to handle, store, and cook these foods.

Figure 17.2 Both provincial and municipal governments regulate the safety of food sold at restaurants. (©Gerenme/iStockPhoto)

The Government's Role

The safety of the food supply is monitored by agencies at the international, federal, provincial, and municipal levels (**Table 17.1**). At the international level, the Food and Agriculture Organization of the United Nations (FAO) and the World Health Organization (WHO) work together to share knowledge on all aspects of food quality and safety, to set food safety policy and standards, and to educate the consumer and the processor. In Canada, the Canadian Food Inspection Agency (CFIA) in co-operation with Health Canada, the Public Health Agency of Canada, and provincial and municipal public health agencies, oversees food safety in Canada (**Figure 17.2**). The functions of the CFIA are listed in **Table 17.1**.

Table 17.1 The Role of the Canadian Food Inspection Agency
• Enforcement of the standards set by Health Canada on the safety and nutritional quality of food
• Inspection of food-processing facilities to ensure that safe procedures are followed
• Education of the consumer about safe food-handling practices
• Enforcement of the nutrition-labelling regulations
• Coordination of food recalls
• Risk assessment of diseases or pests that may threaten the food supply
• Development of new regulations in response to changes in the food supply
• Consumer protection from misleading food labelling
• Certification of Canadian food exports
• Regulation of biotechnology in co-operation with Health Canada
• Prevention of the transmission of animal diseases to humans
• Management of emergencies that may threaten the Canadian food supply

Source: Canadian Food Inspection Agency. Science and regulation...working together for Canadians. Available online at http://www.inspection.gc.ca/english/agen/broch/broche.shtml. Accessed June 30, 2011.

Hazard Analysis Critical Control Point (HACCP) A food safety system that focuses on identifying and preventing hazards that could cause foodborne illness.

critical control points Possible points in food production, manufacturing, and transportation at which contamination could occur or be prevented.

Identifying Potential Problems Food safety used to be monitored by conducting spot-checks of manufacturing conditions and products. These checks often relied on visual inspection to detect contamination and typically did not find a problem until after it had already occurred. The current system for safeguarding the food supply is called **Hazard Analysis Critical Control Point (HACCP)**. This is a science-based approach designed to prevent food contamination rather than catch it after it occurs (**Figure 17.3**).

The HACCP approach to food safety involves establishing standardized procedures to prevent, control, or eliminate contamination before food reaches consumers (**Table 17.2**). It focuses on identifying points in the handling of food, called **critical control points**, where chemical, physical, or microbial contamination can be prevented, controlled, or eliminated. The HACCP system requires the food manufacturing and food service industries to anticipate where contamination might occur. It also establishes record-keeping procedures to verify that the system is working consistently. The advantages of the HACCP system over standard inspections are that it is preventative rather than punitive, it is easier to manage, and the responsibility for food safety is placed on the manufacturer, not the regulatory agencies.

Table 17.2 Seven Principles of HACCP	
1. Conduct a hazard analysis.	Analyze the processes associated with the production of a food to identify the potential biological, chemical, and physical hazards and determine what type of preventive measures, such as changes in temperature, pH, or moisture level, could be used to control or avoid these hazards.
2. Identify critical control points.	Identify steps in a food's production called critical control points, at which the potential hazard can be prevented, controlled, or eliminated—for example, cooking, cooling, packaging, and metal detection.
3. Establish critical limits.	Establish preventive procedures with measurable limits for all critical control points. For example, for a cooked food this might be a minimum cooking time and temperature required to ensure elimination of harmful microbes. If these critical limits are not met, the food safety hazards are not being prevented, eliminated, or reduced to acceptable levels.
4. Establish monitoring procedures.	Establish procedures to monitor the critical limits. For example, how and by whom will the cooking temperature be monitored? Adjustments can be made while continuing the process.
5. Establish corrective actions.	Establish plans to discard the potentially hazardous product and to correct the out-of-control process when monitoring shows that a critical limit has not been met—for example, reprocessing or discarding food if the minimum cooking temperature is not met.
6. Establish verification procedures.	Establish procedures to verify the scientific or technical validity of the hazard analysis, the adequacy of the critical control points, and the effectiveness of the HACCP plan. An example of verification is the testing of time and temperature recording devices to verify that a cooking unit is working properly.
7. Establish record-keeping and documentation procedures.	Prepare and maintain a written HACCP plan. This would include records of hazards and their control methods, the monitoring of each critical control point, and notations of corrective actions taken. Each principle must be backed by sound scientific knowledge—for example, published studies on the time and temperatures needed to control specific foodborne pathogens.

Source: U.S. Food and Drug Administration. Available online at www.cfsan.fda.gov/~lrd/bghaccp.html/.

Tracking Foodborne Illness In addition to requiring the application of HACCP principles to identify and prevent potential and actual food hazards, the government has established a system for tracking foodborne illness once it has occurred. Rapid identification of the source of the contaminant that caused an outbreak of foodborne illness can help stop its spread. This is often done by examining the DNA of the micro-organism, that is, DNA fingerprinting. The Canadian Public Health Agency maintains a computer system that can rapidly compare the DNA fingerprints of micro-organisms from across Canada and the United States.[2] For example, if outbreaks of foodborne illness in Ontario, Manitoba, and New York are caused by the same strain of an organism, epidemiologists know that the outbreaks were caused by the same food source. They can focus their search for the source of contamination on foods distributed to the three locations. To confirm the source, the DNA fingerprint isolated from the organisms found in people who became ill can be matched to the DNA fingerprint from a contaminated food source.

Figure 17.3 How does the current system for safeguarding the food supply differ from the traditional spot checks by food safety inspectors? (Thinkstock Images/ Getty Images)

The Role of Food Manufacturers and Retailers

The responsibility of providing safe food to the marketplace falls on the shoulders of food manufacturers. It is their job to establish and implement an HACCP system for their particular business. Once in place, it allows the company to anticipate where contamination might occur and to then prevent hazardous food from reaching the consumer. For example, contamination with *Salmonella* has been identified as a risk in the production of liquid egg products. To produce these products, eggs are removed from their shells, mixed together in large vats, and then heated, in a process called **pasteurization**, to kill *Salmonella* and other microbial contaminants. Pasteurization is a heating process, sufficient to kill bacteria, but mild enough to leave the eggs in liquid form. The liquid eggs are next packaged, and then refrigerated or frozen. The critical control point for preventing contaminated eggs from reaching consumers is the pasteurization process. To monitor the effectiveness of pasteurization in the liquid egg industry, bacterial tests are performed on samples of eggs following pasteurization. All the eggs are held refrigerated or frozen until the results of the bacterial tests have been obtained. If the eggs are *Salmonella*-free, they are released to the market. If they contain *Salmonella*, the entire batch of eggs is discarded and pasteurization conditions are adjusted to ensure that *Salmonella* is killed in the next batch. Extensive record keeping enables the manufacturer to trace which eggs were pasteurized when, for how long, and at what temperature, and when and where they were shipped in the event of an outbreak of foodborne illness (**Figure 17.4**).

pasteurization The process of heating food products to kill disease-causing organisms.

Food manufacturers are also responsible for proper labelling of their products. In addition to nutritional labelling, some products also contain safe-handling labels as well as some type of product dating. Canadian labelling regulations require foods that remain fresh less than 90 days, to have a "use by" or "best before" date. It refers to the last date the product is likely to be at peak flavour, freshness, and texture. Beyond this date, the product's quality may diminish, but the food may still be safe if it has been handled and stored properly. The dates can include the year (presented first) and must have a month and day. The abbreviations used for "best before" dates are shown in **Table 17.3**. Some specific foods, such as formulated liquid diets, require an expiration date. This is used to specify the last date that the food should be eaten or used.[3]

CANADIAN CONTENT

Table 17.3 Abbreviations Used on "Best Before" Dates in Canada

Best before 08 JA 30 Meilleur avant		
January: JA	May: MA	September: SE
February: FE	June: JN	October: OC
March: MR	July: JL	November: NO
April: AL	August: AU	December: DE

Source: Canadian Food Inspection Agency. Data labelling on pre-packaged foods. Available online at http://www.inspection.gc.ca/english/fssa/concen/tipcon/date.shtml. Accessed January 22, 2011.

Figure 17.4 Liquid egg products, which are used in institutions that must serve large numbers of people, are produced using HACCP guidelines to help ensure their safety. (Brian Leatart/Foodpix/Jupiter Images Corp)

Once food has left the manufacturer, it goes to restaurants and other retail establishments. These businesses are responsible for preventing contaminated food from reaching the consumer. They must monitor the food that enters the establishment and prevent infected food, utensils, and employees from cross-contaminating food served to customers. In retail establishments, food has many opportunities to be contaminated because of the large volume of food that is handled and the large number of people involved in food preparation. Although most of the foodborne illness is caused by food prepared in private homes, an outbreak in a commercial or institutional establishment usually involves more people at a time and is more likely to be reported.

Even when a restaurant uses extreme care in food preparation, customers can be a source of contamination. Because customers serve themselves at salad bars, cross-contamination from one customer to another is a risk. To limit this, salad and dessert bars in restaurants are usually equipped with "sneeze guards" (**Figure 17.5**).

Figure 17.5 Clear plastic shields placed above salad bars prevent customers from contaminating food with micro-organisms transmitted by coughs and sneezes. (© istockphoto.com/Eliza Snow)

The Role of the Consumer

Consumers should be actively involved in preventing foodborne illness. Individuals must decide what foods they will consume and evaluate the risks involved. A food that has been manufactured, packaged, and transported with the greatest care can still cause foodborne illness if it is not carefully handled at home. For example, contaminated eggs, chicken, or hamburger can cause microbial foodborne illness if they are not thoroughly cooked. Just as manufacturers are asked to identify critical control points in food handling where contamination can be prevented and monitored, consumers can take a similar approach in selecting, storing, preparing, and serving food and leftovers (see Critical Thinking: Safe Picnic Choices). Consumers can also protect themselves and others by reporting incidents involving unsanitary, unsafe, deceptive, or mislabelled food to the Canadian Food Inspection Agency by following the instructions on their website.[4]

17.3 Pathogens in Food

Learning Objectives

- Distinguish foodborne infection from foodborne intoxication.
- Discuss 3 types of bacteria that commonly cause foodborne illness.
- Explain how viruses, moulds, parasites, and prions can make us sick.
- Describe how careful food handling can prevent foodborne illness.

Most foodborne illness is caused by consuming food contaminated with pathogens (**Table 17.4**). The pathogens that most commonly affect the food supply include bacteria, viruses, moulds, and parasites. An illness caused by consuming food contaminated with pathogens

CRITICAL THINKING:

Safe Picnic Choices

Background

Tamika is organizing the annual class picnic. She decides to try to apply the HACCP food safety principles she learned in her food science class to keep the food safe. HACCP is designed to prevent or eliminate potential food hazards before they can make anyone sick. The first step in HACCP is to analyze the points in food preparation and storage where food contamination can occur.

(Comstock/Getty Images)

Data

The picnic is a potluck so the food will be made in people's homes where most contamination occurs. Also, it is summer, so the food will sit in the temperature danger zone for several hours, allowing pathogenic bacteria present in the food to multiply.

Tamika can analyze the potential for contamination in her own kitchen, but it is not possible to check every step in the preparation of each food that others are making for the picnic. She decides to find out what people are bringing and perhaps suggest some different choices if the foods seem to present a significant risk. She collects the following list of food items that her friends intend to bring:

PICNIC MENU	
Chicken salad	Cheese and crackers
Tamales	Apple pie
Fruit salad	Cookies
Raw vegetables and onion dip	Mushrooms stuffed with crab meat
Chips and salsa	Fried chicken

Critical Thinking Questions

Is the picnic food safe to eat?*

Which foods are the least risky?

Which foods carry the highest risks?

What can Tamika do to reduce the risk of foodborne spoilage during the picnic?

After the picnic is over, what foods would you consider safe to keep as leftovers and what would you throw out?

*Answers to all Critical Thinking questions can be found in Appendix J.

iProfile Use iProfile to find the number of kcalories in your favourite picnic foods.

---

that multiply in the gastrointestinal tract or other parts of the body is called a **foodborne infection**. An illness caused by consuming food containing toxins produced by a pathogen is referred to as a **foodborne intoxication**. Unlike foodborne infections, which are usually caused by ingesting large numbers of pathogens, intoxication can be caused by only a few micro-organisms that have produced a toxin. These food toxins may be difficult to destroy.

foodborne infection Illness produced by the ingestion of food containing micro-organisms that can multiply inside the body and cause injurious effects.

foodborne intoxication Illness caused by consuming a food containing a toxin.

Table 17.4 Which Bug Has You Down?

Microbe	Sources	Symptoms	Onset	Duration
Bacteria				
Salmonella	Fecal contamination, raw or undercooked eggs and meat, especially poultry	Nausea, abdominal pain, diarrhea, headache, fever	6–48 hrs.	1–2 days
Campylobacter jejuni	Unpasteurized milk, undercooked meat and poultry, untreated water	Fever, headache, diarrhea, abdominal pain	2–5 days	1-2 wks.
Listeria monocytogenes	Raw milk products, soft ripened cheeses, deli meats and cold cuts, raw and undercooked poultry and meats, raw and smoked fish, raw produce	Fever, headache, stiff neck, chills, nausea, vomiting. May cause spontaneous abortion or stillbirth in pregnant women.	Days to weeks	Days to weeks
Vibrio vulnificus	Raw seafood from contaminated water	Cramps, abdominal pain, weakness, watery diarrhea, fever, chills	15-24 hrs.	2–4 days
Staphylococcus aureus	Human contamination from coughs and sneezes, eggs, meat, potato and macaroni salads	Severe nausea, vomiting, diarrhea	2-8 hrs.	24–48 hrs.
Escherichia coli O157:H7	Fecal contamination, undercooked ground beef	Abdominal pain, bloody diarrhea, kidney failure	5–48 hrs.	3 days to 2 wks. or longer
Clostridium perfringens	Fecal contamination, deep-dish casseroles	Fever, nausea, diarrhea, abdominal pain	8–22 hrs.	6–24 hrs.
Clostridium botulinum	Canned foods, deep casseroles, honey	Lassitude, weakness, vertigo, respiratory failure, paralysis	18-36 hrs.	10 days or longer (must administer antitoxin)
Bacillus cereus	Starchy foods and food mixtures such as soups and casseroles	Diarrhea, abdominal cramps, nausea, vomiting	30 min.–15 hrs.	24 hrs.
Shigella	Fecal contamination of water or foods, especially salads such as chicken, tuna, shrimp, and potato salad	Diarrhea, abdominal pain, fever, vomiting	12–50 hrs.	5–6 days
Yersinia enterocolitica	Pork, dairy products, and produce	Diarrhea, vomiting, fever, abdominal pain; often mistaken for appendicitis	24–48 hrs.	Weeks
Viruses				
Noroviruses	Fecal contamination of water or foods, especially shellfish and salad ingredients	Diarrhea, nausea, vomiting	1–2 days	2–6 days
Hepatitis A virus	Human fecal contamination of food or water, raw shellfish	Jaundice, liver inflammation, fatigue, fever, nausea, anorexia, abdominal discomfort	10–50 days	1–2 wks. to several months
Parasites				
Giardia lamblia	Fecal contamination of water and uncooked foods	Diarrhea, abdominal pain, gas, anorexia, nausea, vomiting	5–25 days	1–2 wks. but may become chronic
Cryptosporidium parvum	Fecal contamination of food or water	Severe watery diarrhea	Hours	2–4 days but sometimes weeks
Trichinella spiralis	Undercooked pork, game meat	Muscle weakness, flu-like symptoms	Weeks	Months
Anisakis simplex	Raw fish	Severe abdominal pain	1 hr.-2 wks.	3 wks.
Toxoplasma gondii	Meat, primarily pork	Toxoplasmosis (can cause central nervous system disorders, flu-like symptoms, and birth defects in women exposed during pregnancy)	10–23 days	May become chronic carrier

Source: U.S. Food and Drug Administration, Center for Food Safety and Nutrition. Foodborne Pathogenic Microorganisms and Natural Toxins Handbook: The "Bad Bug Book."

In most cases, the symptoms of a microbial foodborne illness include abdominal pain, nausea, diarrhea, and vomiting. These relatively mild symptoms are often mistaken for the flu. Foodborne illness can also cause more severe symptoms such as spontaneous abortion; hemolytic uremic syndrome, which can lead to kidney failure and death; and long-lasting conditions like arthritis and Guillain-Barré syndrome, which is the most common cause

Figure 17.6 *Salmonella* can infect the ovaries of hens and contaminate the eggs before the shells are formed, so that the bacteria are present inside the shell when the eggs are laid. (B. Anthony Stewart/ Getty Images, Inc.)

Life Cycle

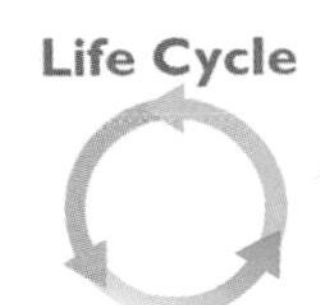

of acute paralysis. Young children, pregnant women, elderly persons, and individuals with compromised immune systems, such as people with HIV/AIDS and cancer, are most susceptible to severe reactions. Avoiding microbial foodborne illness requires a knowledge of how contamination occurs and how to handle, store, and prepare food safely.

Bacteria

Bacteria are present in the soil, on our skin and in our digestive tracts, on most surfaces in our homes, and in the food we eat. Most of the bacteria in our environment are harmless, some are beneficial, and some are pathogenic, causing disease either by growing in the body or by producing toxins in food. Pathogenic bacteria may also produce toxins within the body. Usually a large number of bacteria must be consumed to cause illness. Some common causes of bacterial infections include *Salmonella, E. coli, Campylobacter jejuni, Listeria monocytogenes*, and *Vibrio vulnificus*. *Staphylococcus aureus* and *Clostridium* are common causes of foodborne intoxication.

Salmonella *Salmonella* is a common foodborne infection. Most of the people who are infected experience only abdominal pain and diarrhea, but some infections are more serious.[5] *Salmonella* is found in animal and human feces and infects food through contaminated water or improper handling. *Salmonella* outbreaks have been caused by contaminated meat, meat products, dairy products, seafood, fresh vegetables, and cereal, but poultry and eggs are the most common food sources. Poultry products are often contaminated because poultry farms house large numbers of chickens in close proximity, allowing one infected chicken to infect thousands of others (**Figure 17.6**). One way to reduce infection is to spray chicks with beneficial bacteria. The chicks ingest the bacteria when they preen their feathers and the beneficial bacteria colonize the digestive tract, leaving no room for pathogens.

Even if food contaminated with *Salmonella* is brought into the kitchen, careful handling and cooking of the food can prevent the organisms from causing illness. Washing hands, cutting boards, and utensils can prevent cross-contamination. If a contaminated food is stored in the refrigerator, the multiplication of the *Salmonella* will be slowed. In contrast, if a contaminated food is left at room temperature, the *Salmonella* will multiply rapidly, and when the food is eaten, large numbers of bacteria will be ingested with it (**Figure 17.7**). *Salmonella* is killed by heat—so foods likely to be contaminated, such as poultry and eggs, should be cooked thoroughly.

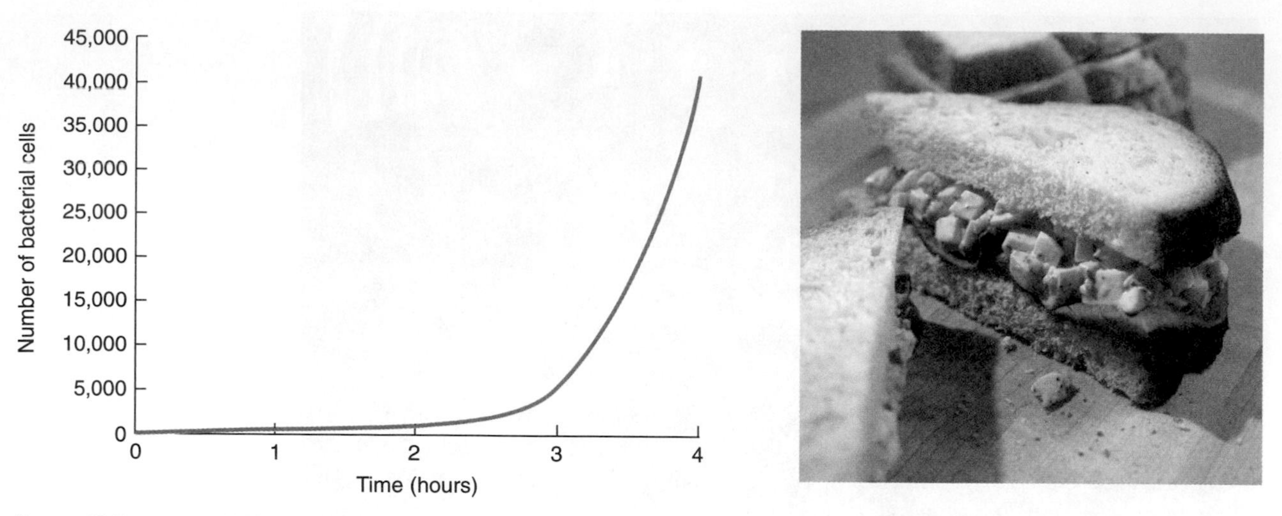

Figure 17.7 Exponential bacterial growth
The size of a population of bacterial cells doubles each time the cells divide; thus, if 10 bacterial cells contaminate an egg salad sandwich during preparation and it sits in your warm car for 4 hours, during which the cells divide every 20 minutes, there will be 40,960 bacterial cells in the sandwich by the time you eat it. (© Snowflake Studios/Stock Food America)

Escherichia coli (E. coli) *Escherichia coli* (*E. coli*) is a bacterium that inhabits the gastrointestinal tracts of humans and other animals. It comes in contact with food through fecal contamination of water or unsanitary handling of food. Transmission of *E. coli* is also a risk at daycare centres from cross-contamination if caregivers do not carefully wash their hands after diaper changes. Some strains of *E. coli* are harmless, but others can cause serious foodborne illness. One strain of *E. coli*, found in water contaminated by human or animal feces, is the cause of "travellers' diarrhea."

CANADIAN CONTENT

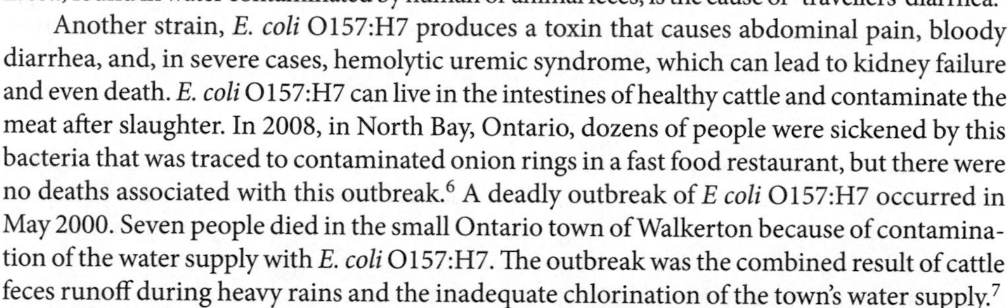

Another strain, *E. coli* O157:H7 produces a toxin that causes abdominal pain, bloody diarrhea, and, in severe cases, hemolytic uremic syndrome, which can lead to kidney failure and even death. *E. coli* O157:H7 can live in the intestines of healthy cattle and contaminate the meat after slaughter. In 2008, in North Bay, Ontario, dozens of people were sickened by this bacteria that was traced to contaminated onion rings in a fast food restaurant, but there were no deaths associated with this outbreak.[6] A deadly outbreak of *E coli* O157:H7 occurred in May 2000. Seven people died in the small Ontario town of Walkerton because of contamination of the water supply with *E. coli* O157:H7. The outbreak was the combined result of cattle feces runoff during heavy rains and the inadequate chlorination of the town's water supply.[7]

Although *E. coli* on food can multiply slowly, even at refrigerator temperatures, if a contaminated food is thoroughly cooked to 71°C, both the bacteria and the toxin are destroyed. Ground beef contaminated with *E. coli* O157:H7 is a particular risk because, unlike other meats that are likely only to be contaminated on the surface, the grinding mixes the bacteria throughout the meat (**Figure 17.8**). The *E. coli* on the outside of the meat are quickly killed during cooking, but those in the interior survive if the meat is not cooked thoroughly.

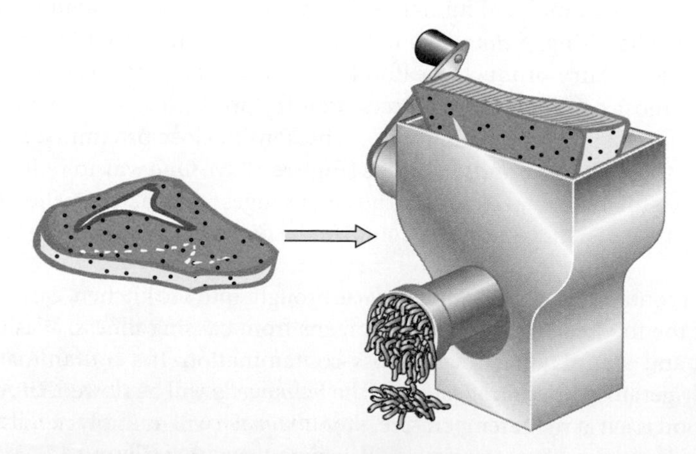

Figure 17.8 Meat grinders and *E. coli* contamination
E. coli-contaminated meat that comes into contact with a grinder may contaminate hundreds of kilograms of ground beef. During grinding, the bacteria are mixed throughout the meat.

Campylobacter There are several species of *Campylobacter* that cause foodborne infections; *Campylobacter jejuni* causes most cases of the illness. *Campylobacter* infection is a leading cause of bacterial diarrheal illness.[8] Common sources are undercooked chicken, unpasteurized milk, and untreated water. This organism grows slowly in the cold and is killed by heat, so, as with *Salmonella*, thorough cooking and careful storage help prevent infection.

Life Cycle

CANADIAN CONTENT

Listeria Another cause of bacterial foodborne infection is *Listeria monocytogenes*. Although most cases of *Listeria* infection, called listeriosis, result in flu-like symptoms, in high-risk groups such as pregnant women, children, the elderly, and the immunocompromised, it can cause meningitis and serious blood infections and has one of the highest fatality rates of all foodborne illnesses, with 20% of those infected dying.[9] *Listeria* frequently contaminates dairy products, but it is destroyed by pasteurization. It is also found in processed, ready-to-eat foods such as hot dogs and lunchmeats. In 2008, an outbreak of listeriosis in Canada killed 23 people, many of them elderly. The outbreak was linked to the contamination of ready-to-eat meats manufactured in an Ontario food plant. The meat products were heated during processing in the plant, killing the bacteria, but were re-contaminated during packaging when they contacted surfaces that harboured the bacteria. The products left the production plant with much higher levels of bacteria than was typical.[10] *Listeria* is able to survive at higher and lower temperatures than most bacteria. So while on the store shelf and in the possession of the consumer, the bacteria in these contaminated products, even when properly refrigerated, continued to increase in number. When these products were consumed without further cooking, as is often the case with ready-to-eat products, the bacteria count was so high, the stage was set for an outbreak. Properly processed and stored products would not normally contain enough bacteria to cause serious illness, but to fully prevent *Listeria* infection, ready-to-eat meats should ideally be heated to steaming by the consumer. Unpasteurized dairy products should also be avoided.

Vibrio *Vibrio vulnificus* and *Vibrio parahaemolyticus* are two species of *Vibrio* bacteria that cause vomiting, diarrhea, and abdominal pain in healthy people and can be deadly if they infect people with compromised immune systems. The most common way people become infected is by eating raw or undercooked shellfish, particularly oysters. *Vibrio* bacteria grow in warm seawater, so the incidence of *Vibrio* infection is higher during the summer months, when warm water favours growth.

Staphylococcus aureus *Staphylococcus aureus* is a common cause of microbial foodborne intoxication. These bacteria live in human nasal passages and can be transferred through coughing or sneezing when handling food. The bacteria then produce toxins as they grow on the food. When ingested, the toxin causes symptoms which include nausea, vomiting, diarrhea, abdominal cramps, and headache. Foods that are common sources of *Staphylococcus aureus* include cooked ham, salads, bakery products, and dairy products.

Clostridium perfringens The bacterium *Clostridium perfringens* may cause illness by both infection and intoxication. It is found in soil and in the intestines of animals and humans. It thrives in conditions with little oxygen (anaerobic conditions) and is difficult to kill because it forms heat-resistant **spores**. Spores are a stage of bacterial life that remains dormant until environmental conditions favour growth. *Clostridium perfringens* is often called the "cafeteria germ" because foods stored in large containers, such as those used to serve food in cafeteria lines, have anaerobic centres that provide an excellent growth environment. Sources include improperly prepared roast beef, turkey, pork, chicken, and ground beef.

spore A dormant state of some bacteria that is resistant to heat but can germinate and produce a new organism when environmental conditions are favourable.

Clostridium botulinum Another strain of *Clostridium*, *Clostridium botulinum*, produces the deadliest bacterial foodborne toxin. Although the bacteria themselves are not harmful, the toxin, produced as the bacteria begin to grow and develop, blocks nerve function, resulting in vomiting, abdominal pain, double vision, dizziness, and paralysis causing respiratory failure. If untreated, botulism poisoning is often fatal, but today, modern detection methods and rapid administration of antitoxin have reduced mortality. Low-acid foods, such as potatoes or stews, that are held in anaerobic conditions provide an optimal environment for botulism spores to germinate. Canned foods, particularly improperly home-canned foods, can also be a source of botulism. Canned foods should be discarded if the can is bulging because this indicates the presence of gas produced by bacteria as they grow. Once formed, botulism toxin can be destroyed by boiling, but if the safety of a food is in question, it should be discarded; even a taste of botulism toxin can be deadly.

Life Cycle

Infant botulism is a rare but serious type of botulism that occurs worldwide and is seen only in infants.[11] It occurs when ingested botulism spores germinate in the gastrointestinal tract, producing toxin, some of which is absorbed into the bloodstream causing weakness, paralysis, and respiratory problems. In the absence of complications, infants generally recover. Only infants get botulism from ingesting spores because in adults, competing intestinal microflora prevent spores from germinating. Because botulism spores can contaminate honey, it should never be fed to infants under 1 year of age (**Figure 17.9**).

Figure 17.9 Why are *Clostridium botulinum* spores in honey dangerous only to babies under 1 year old? (Caroline Kopp/Foodpix/Jupiter Images Corp.)

Viruses

Viruses make us ill by entering our cells and converting them into virus-making factories (**Figure 17.10**). The viruses that cause human disease cannot grow and reproduce in foods, but the virus particle can contaminate food and then infect the consumer when the food is eaten.

viruses Minute particles not visible under an ordinary microscope that depend on cells for their metabolic and reproductive needs.

Noroviruses Noroviruses are a group of viruses that cause gastroenteritis, or what we commonly think of as the "stomach flu." Symptoms usually involve more vomiting than diarrhea and resolve within 2 days. Noroviruses, which include caliciviruses and Norwalk-like viruses, are an extremely common cause of foodborne illness, but the technology to diagnose the infection was not available until the early 1990s. Noroviruses are now recognized as the most common cause of infectious gastroenteritis among persons of all ages and are believed to be responsible for about 50% of all foodborne gastroenteritis outbreaks.[12] Noroviruses spread primarily from one infected person to another, but they can also be spread by eating food that is contaminated with the virus or by touching a contaminated surface and then putting your fingers in your mouth. If kitchen workers have the virus on their hands, they can contaminate a salad or sandwich as they prepare it. Infected fishermen can contaminate oysters as they harvest them. Shellfish can also become contaminated with norovirus if the water where they live is polluted with human or animal feces. Worldwide, about 1 in 10 oysters is believed to be contaminated.[13] Because noroviruses are destroyed by cooking, water and uncooked foods such as raw shellfish and salads are the most common cause of norovirus foodborne illness.

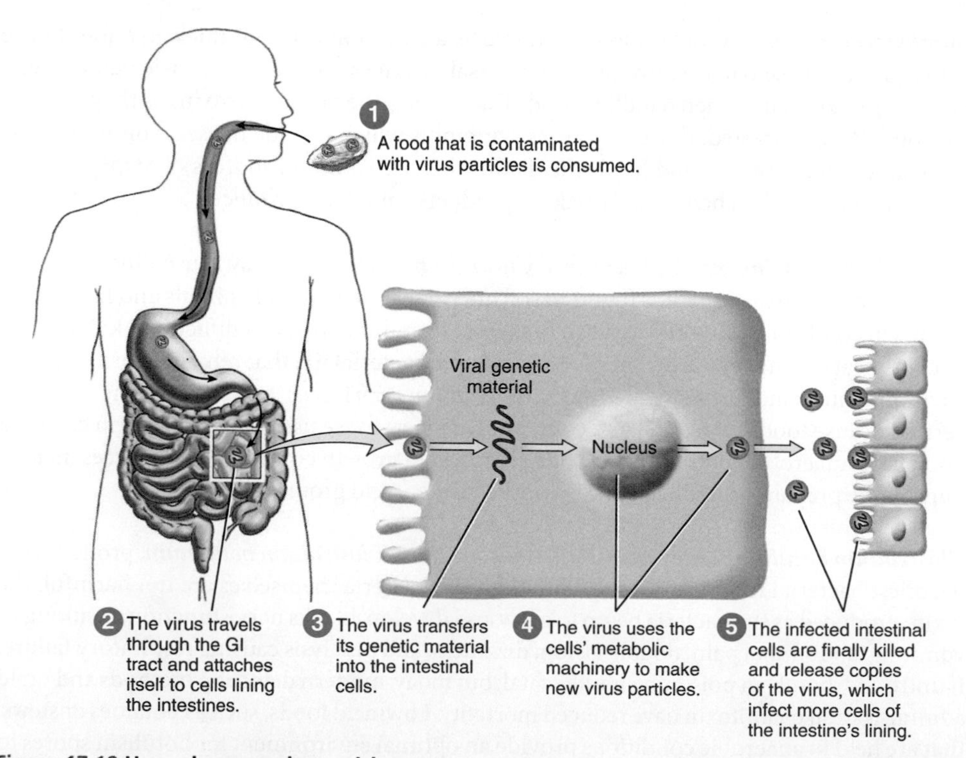

Figure 17.10 How viruses make us sick
Viruses make us sick by reproducing inside our cells. Viruses that cause foodborne illness enter the body through the gastrointestinal tract. Other types of viruses may enter the body through open cuts, the respiratory tract, or the genital tract.

Hepatitis A Hepatitis A is a highly contagious viral disorder that causes inflammation of the liver, jaundice, fever, nausea, fatigue, and abdominal pain. It can be contracted from food contaminated by unsanitary handling or from eating raw or undercooked shellfish caught in sewage-contaminated waters. Hepatitis A can require a recovery period of several months, but it usually does not require treatment and does not cause permanent liver damage. Hepatitis in drinking water is destroyed by chlorination. Cooking destroys the virus in food, and good sanitation can prevent its spread. A vaccine that protects against hepatitis A infection is available.

Moulds

CANADIAN CONTENT

Many types of **mould** grow on foods such as bread, cheese, and fruit (**Figure 17.11**). Some moulds produce toxins, called mycotoxins, that can lead to food intoxication. According to the CFIA, mycotoxins that have been detected in agricultural crops grown in Canada include trichothecenes, zaeralenone, fumonisins, ochratoxins, and ergot. They are most commonly found in cereal grains and corn, and to a lesser extent alfalfa and oilseeds. Mycotoxin contamination is also reported in foods such as coffee, cocoa, rice, beer, and wine.[14]

mould Multicellular fungi that form a filamentous branching growth.

Ergot contaminates grain, particularly rye. It causes hallucinations and is a natural source of the hallucinogenic drug LSD. Today, modern milling removes the part of the grain that harbours the mould, so the disease ergotism is rare.

Aflatoxin is a mycotoxin that is among the most potent mutagens and carcinogens known. It grows in tropical climates and can enter the Canadian food supply through products imported from those regions. It is produced by the mould *Aspergillus flavus*. This mould commonly grows on corn, rice, wheat, peanuts, almonds, walnuts, sunflower seeds, and spices such as black pepper and coriander. The level of aflatoxin that may be present in foods is regulated to prevent toxicity.[14]

Figure 17.11 Mould can grow at refrigerator temperatures and it will grow on almost anything if it remains in the refrigerator long enough. (Joern Rynio/Getty Images)

Cooking and freezing stop mould growth but do not destroy the mould toxins that have already been produced. If a food is mouldy, it should be discarded, the area where it was stored should be cleaned, and neighbouring foods should be checked to see if they have also become contaminated.

Parasites

Some **parasites** are tiny. single-celled animals, while others are worms that are easily seen with the naked eye. Parasites are killed by thorough cooking. They may be transmitted through consumption of contaminated food and water. *Giardia lamblia* (also called *Giardia duodenalis*) is a single-celled parasite that can infect the gastrointestinal tract through water or food contaminated with human or animal feces (**Figure 17.12**).

parasites Organisms that live at the expense of others.

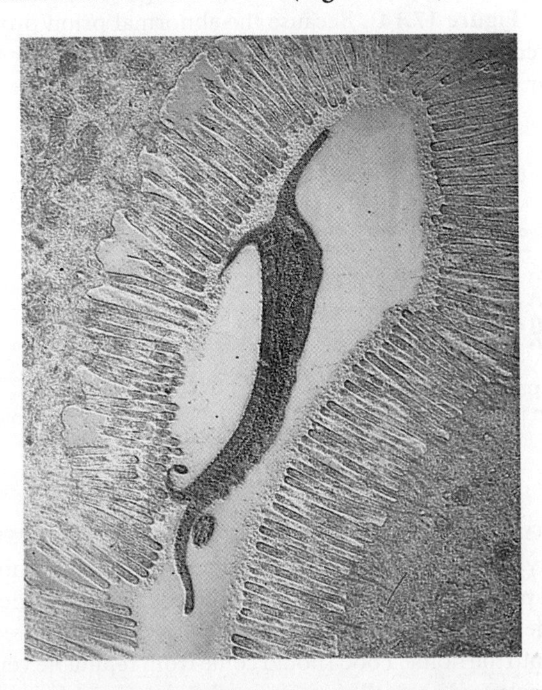

Figure 17.12 This electron micrograph shows *Giardia* (green) attached to the microvilli of the human small intestine. (CNRI/Science Photo Library/Photo Researchers)

Figure 17.13 Eating raw fish, such as this sushi, if not properly prepared, can result in parasitic infections. (© istockphoto.com/ Matteo De Stefano)

It is the most frequent cause of diarrhea not due to bacteria.[5] *Giardia* is sometimes contracted by hikers who drink untreated water from streams contaminated with animal feces, and it is becoming a problem from cross-contamination in daycare centres. *Cryptosporidium parvum* is another single-celled parasite that is commonly spread by contaminated water, but cases have also been reported from consuming unpasteurized apple cider and contaminated raw fruits and vegetables.[15,16]

Trichinella spiralis is a parasite found in raw and undercooked pork, pork products, and game meats, particularly bear. Once ingested, these small, worm-like organisms find their way to the muscles, where they grow, causing flu-like symptoms, muscle weakness, fever, and fluid retention. Trichinosis, the disease caused by *Trichinella* infection, can be prevented by thoroughly cooking meat to kill the parasite before it is ingested. The parasites are also destroyed by curing, smoking, canning, or freezing.

Fish are another common source of parasitic infections. Fish can carry the larvae (the worm-like stage of an organism's life cycle) of parasites such as roundworms, flatworms, flukes, and tapeworms. One such infection, Anisakis disease, is caused by the larval form of the small roundworm *Anisakis simplex*, or herring worm, found in raw fish.[5] Once consumed, these parasites invade the stomach and intestinal tract, causing severe abdominal pain. The fresher the fish is when it is eviscerated, the less likely it is to cause this disease because the larvae move from the fish's stomach to its flesh only after the fish dies. Parasitic infections from fish can be avoided by consuming cooked fish or freezing fish for 72 hours before consumption. If raw fish is consumed, it should be very fresh (**Figure 17.13**).

Prions

prion A pathogenic protein that is the cause of degenerative brain diseases called spongiform encephalopathies.

The strangest, scariest, yet rarest foodborne illness is caused not by a microbe but by a protein, called a **prion**, short for proteinaceous infectious particle. Abnormal prions are believed to be the cause of mad cow disease, or bovine spongiform encephalopathy (BSE), a deadly, degenerative neurological disease that affects cattle. A human form of this disease, called variant Creutzfeldt-Jakob disease (vCJD), has been identified. People are believed to contract it by eating tissue from a cow infected with BSE (see Your Choice: Should You Bypass the Beef?).[17] Symptoms of vCJD begin as mood swings and numbness and within about 14 months the nervous tissue damage progresses to dementia and death.

The abnormal prions that cause BSE and vCJD differ from normal proteins in the way they are folded—that is, in their three-dimensional structure. These rogue proteins reproduce by corrupting neighbouring proteins, essentially changing their shape, so they, too, become abnormal prions (**Figure 17.14**). Because the abnormal prion proteins are not degraded normally, they accumulate and form clumps called plaques. These plaques cause the deadly nervous tissue damage.

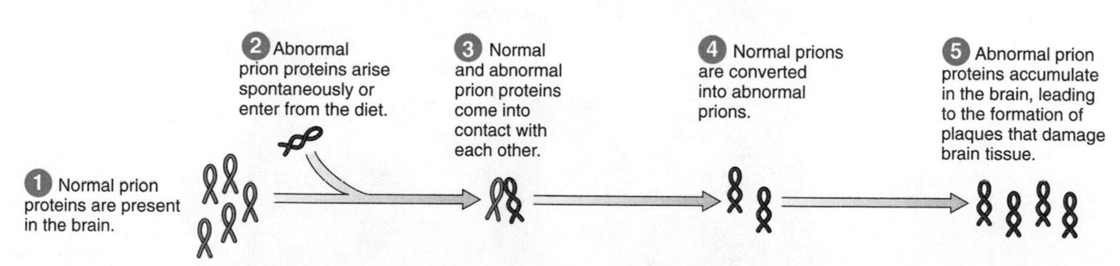

Figure 17.14 **How prions multiply**
Abnormal prions can reproduce by coming into contact with normal prion proteins and causing them to fold abnormally.

Figure 17.15 **Fight Bac!**
The Fight Bac! educational campaign recommends that consumers follow 4 steps— clean, separate, cook, and chill—to prevent foodborne illness. Reprinted with permission from the Canadian Partnership for Consumer Food Safety Education.
Source: Canadian Partnership for Consumer Food Safety Education, http://www.canfightbac.org/cpcfse/en/.

Reducing the Risk of Microbial Foodborne Illness

Despite the variety of organisms that can cause foodborne illness, most cases can be avoided if food is handled properly (**Figure 17.15, Table 17.5**). The first critical control point in preventing foodborne illness is making safe selections at the store to reduce the contaminants that are brought into the home. Food should come from reputable vendors and appear fresh. Foods that are discoloured or smell contaminated and those in damaged packages should not be purchased or consumed.

Table 17.5 Tips for Handling Food Safely

Choose foods wisely
Voluntary freshness dates should be checked and foods with expired dates avoided. Jars should be closed and seals unbroken. Cans that are rusted, dented, or bulging should be rejected. Frozen foods should not contain frost or ice crystals, and food packaging should be secure. Frozen foods should be selected from below the frost line in the freezer. When shopping, cold foods should be purchased last.
Store foods properly
Fresh or frozen foods brought from the store should be refrigerated or frozen immediately at the proper temperature. Food that has been in your refrigerator for longer than is safe should be discarded.
Wash
Hands, cooking utensils, and surfaces should be washed with warm, soapy water before each food preparation step. This will prevent cross-contamination. Hands should be washed for at least 20 seconds before handling food. To sterilize surfaces, wash with bleach.
Thaw
Foods should be thawed in the refrigerator or microwave.
Cook thoroughly
Cooking temperatures should be checked. Thorough cooking destroys most bacteria, toxins, viruses, and parasites.
Refrigerate promptly
Cooked food can be recontaminated so it should be refrigerated as soon as possible after it is served.
Reheat thoroughly
Foods should be reheated to 75°C to destroy micro-organisms that have recontaminated cooked foods and toxins that have been produced.
When in doubt, throw it out.

Store Food Properly Once selected, foods need to be stored appropriately. The goal is to keep foods from remaining at temperature that promote bacterial growth (**Figure 17.16**). Cold foods should be kept cold, at 4°C or less, and hot foods should be kept hot, at more than 60°C. Refrigerator temperature should be set be at 4°C and freezers at –18°C. Produce should be stored in the refrigerator. Fresh meat, poultry, and fish should be frozen immediately if it will not be used within 1 or 2 days. Processed meats such as hot dogs and bologna must also be kept refrigerated but can be kept longer than fresh meat (**Table 17.6**).

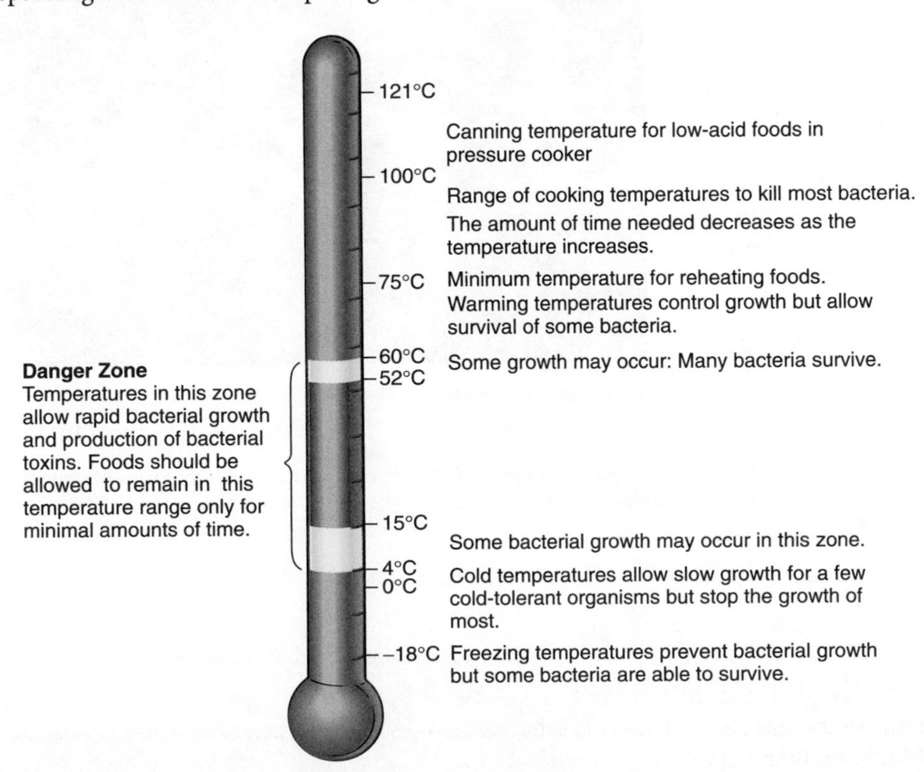

Figure 17.16 Effect of temperature on bacterial growth
Bacterial growth is most rapid between 4°C and 57°C.

YOUR CHOICE

(©iStockphoto)

Should You Bypass the Beef?

CANADIAN CONTENT

In 2003, the Canadian Food Inspection Agency announced that a case of mad cow disease, or bovine spongiform encephalopathy (BSE), had been identified in Canada. The agent that causes BSE and variant Creutzfeldt-Jakob disease (vCJD), the human form of the disease is a protein called a prion. Prions in food cannot be killed, because they aren't alive. Prions can be passed from an infected animal to an uninfected host, most likely from consumption of a food contaminated with central nervous system tissue.[1] BSE is believed to have originated from sheep that carried a similar disease, called scrapie. In the late 1980s and early 1990s, the disease spread rapidly through cattle in Britain when they were fed protein supplements containing the remains of slaughtered, diseased sheep. The disease was then passed to people when they ate products from infected cows. The first cases of vCJD were identified in Britain in 1994. The number of reported cases peaked in 2000 and has declined since, although new cases were still reported in 2010.[2]

To prevent people from getting vCJD, we need to prevent cows from getting BSE and prevent humans from eating products from infected cows. This is a challenge as the only way to determine if a cow is infected with BSE is to kill the animal and analyze its brain tissue, by which time its meat could be on the store shelf. Safeguards that have been in place for some time have helped keep Canadian beef BSE-free. Since 1997, the Canadian Food Inspection Agency has restricted the import of ruminants and most ruminant products from Europe, where BSE was first identified. The practice of including proteins from cows and other ruminants in protein supplements fed to cows also was banned. High-risk tissues such as brain, spinal cord, eyes, and intestines are also kept out of the human food supply.[3] Even though cooking does not destroy prions, the risk of contracting vCJD is extremely small; meat (if free of central nervous system tissue) and milk have not been demonstrated to transmit BSE or vCJD. The meat from cows identified with BSE in Canada had not entered the food supply and thus far, there has been no known instance of Canadian beef causing a case of vCJD. In Britain, where more than 180,000 cows were infected, about 200 definite or probable cases of vCJD have been identified from 1994–2010.[2]

The Canadian Food Inspection Agency maintains continuing surveillance of Canadian cattle to ensure that you don't have to bypass the beef.[4]

(© istockphoto.com/Dave Willman)

[1]U.S. Food and Drug Administration, Center for Food Safety and Nutrition. *Foodborne Pathogenic Microorganisms and Natural Toxins Handbook: The "Bad Bug Book."* Available online at www.fda.gov/Food/FoodSafety/Foodborneillness/FoodborneillnessFoodbornePathogensNaturalToxins/BadBugBook/default.htm. Accessed October 5, 2009.

[2] Andrew, N. J. Incidence of variant Creutzfeldt-Jakob disease diagnoses and deaths in the UK January 1994-December 2010. Available online at http://www.cjd.ed.ac.uk/cjdq68.pdf. Accessed June 30, 2011.

[3] Canadian Food Inspection Agency. Bovine Spongiform Encephalopathy. Available online at http://www.inspection.gc.ca/english/anima/disemala/bseesb/bseesbfse.shtml. Accessed September 11, 2011.

[4] Canadian Food Inspection Agency. Bovine Spongiform Encephalopathy in North America,. Available online at http://www.inspection.gc.ca/english/anima/disemala/bseesb/bseesbe.shtml. Accessed June 30, 2011.

Prevent Cross-Contamination The next critical control point for food in the home is preparation. A clean kitchen is essential for safe food preparation. Hands, countertops, cutting boards, and utensils should be washed with warm, soapy water before each food preparation step. Food should be thawed in the refrigerator, in the microwave oven, or under cold, running water—not at room temperature. Foods that are eaten raw should not be prepared on the same surfaces as foods that are going to be cooked. For example, if a chicken contaminated with *Salmonella* is cut up on a cutting board and the unwashed cutting board is then used to chop vegetables for a salad, the vegetables will become contaminated with *Salmonella.* When the chicken is cooked, the bacteria will be killed, but the contaminated vegetables are not cooked, so the bacteria can grow and cause foodborne illness. Cross-contamination can also occur when uncooked foods containing live microbes come in contact with foods that have already been cooked. Therefore, cooked meat should never be returned to the same dish that held the raw meat, and sauces used to marinate uncooked foods should never be used as a sauce on cooked food. To remind consumers how to handle meat, the packaging is labelled with safe handling guidelines (**Figure 17.17**).

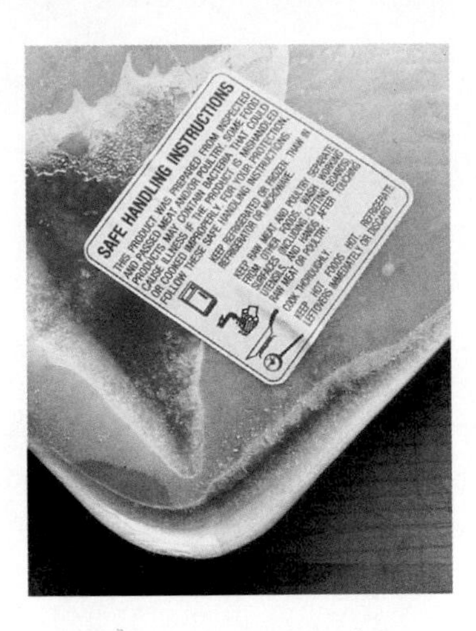

Figure 17.17 Meat labels offer safe handling guidelines. (Dennis Drenner)

Cook Food Thoroughly Cooking is one of the most important critical control points in preventing foodborne illness. Heat will destroy most harmful micro-organisms. A meat thermometer should be used when cooking meat because colour is not always a good indicator of safety (**Table 17.7**). Fish should be cooked until the flesh is opaque and separates easily with a fork. Eggs should not be eaten raw, since *Salmonella* can contaminate the inside of the shell; they should be cooked until the white is firm and the yolk is not runny.

Table 17.6 How Long Can Food Be Safely Stored?

Product	Refrigerator (4°C)	Freezer (−18°C)
Eggs		
Fresh, in shell	3 wks.	Don't freeze
Raw yolks, white	2–4 days	1 yr.
Hard cooked	1 wk.	Don't freeze
Frozen entrees and casseroles	Keep frozen until ready to serve	3–4 mos.
Soups & stews with vegetables or meat	3–4 days	2–3 mos.
Hamburger, ground, & stew meats		
Hamburger & stew meats	1–2 days	3–4 mos.
Ground turkey, veal, pork, lamb & mixtures of these	1–2 days	3–4 mos.
Hotdogs		
Opened package	1 wk.	In freezer wrap, 1–2 mos.
Unopened package	2 wks.	In freezer wrap, 1–2 mos.
Lunch meats		
Opened	3–5 days	In freezer wrap, 1–2 mos.
Unopened	2 wks.	In freezer wrap, 1–2 mos.
Bacon & sausage		
Bacon	7 days	1 mo.
Sausage, raw from pork, beef, turkey	1–2 days	1–2 mos.
Smoked breakfast links, patties	7 days	1–2 mos.
Ham		
Ham, canned—label says keep refrigerated	6–9 mos.	Don't freeze

Table 17.6 *(continued)*		
Ham, fully cooked—whole	7 days	1–2 mos.
Ham, fully cooked—slices	3–4 days	1–2 mos.
Fresh Meat and Poultry		
Steaks, beef	3–5 days	6–12 mos.
Chops, pork	3–5 days	4–6 mos.
Chops, lamb	3–5 days	6–9 mos.
Roasts, beef	3–5 days	6–12 mos.
Roasts, pork & veal	3–5 days	4–6 mos.
Chicken or turkey, whole	1–2 days	1 yr.
Chicken or turkey pieces	1–2 days	9 mos.
Meat and Poultry Leftovers		
Cooked meat & meat dishes	3–4 days	2–3 mos.
Fried chicken	3–4 days	4 mos.
Cooked poultry dishes	3–4 days	4–6 mos.
Chicken nuggets, patties	1–2 days	1–3 mos.

Source: Partnership for Food Safety Education. Available online at www.fightbac.org/doubt.cfm/.

Cooked food should be refrigerated as soon as possible after serving. The best temperatures for bacterial growth are the temperatures at which food is usually kept between service and storage. Large portions of food should be divided before refrigeration so they will cool quickly. Most leftovers should be kept for only a few days. For example, cooked pasta can be kept refrigerated for 3–5 days; cooked beef, poultry, pork, vegetables, soup, and stews for 3–4 days; and stuffing and meat in gravy for 1–2 days. When leftovers are reheated, they should be heated to 75°C to destroy any bacteria that may have grown in them.

Table 17.7 What Cooking Temperatures Are Safe?

Food Item	Internal Temperature (°C) or Description
Beef, veal, lamb	
Ground products	71
Nonground products	
Medium-rare	63
Medium	60
Well done	80
Poultry	
Ground products	74
Whole bird and poultry parts	74
Stuffing (cooked alone or in a bird)	74
Pork	
Noncured products	
Medium	71
Well done	80
Ham	
Fresh (raw)	71
Pre-cooked (to reheat)	60

Eggs & egg dishes	
Eggs	Cook until yolk and white are firm
Egg dishes	71
Fish & shellfish	
Fish	63 or until flesh is opaque and flakes easily
Shrimp, lobster, crab	Cook until flesh is pearly and opaque
Scallops	Cook until milky white or opaque and firm
Clams, mussels, oysters	Cook until shells open
Leftovers & Casseroles	74

Source: Partnership for Food Safety Education. *Cook to Safe Temperature*. Available online at www.fightbac.org/images/pdfs/cook.pdf.

Food Safety Away from Home When you prepare your own food, you can monitor the safety of the ingredients and food preparation steps, but at restaurants, picnics, and potlucks, you must rely on others to keep your food safe. Consumers should choose restaurants with safety in mind. Restaurants should be clean, and cooked foods should be served hot. Cafeteria steam tables should be kept hot enough that the water is steaming and food is kept above 60°C. Cold foods such as salad bar items should be kept refrigerated or on ice to keep food at 4°C or colder.

Picnics, potlucks, and other large events where food is served provide a prime opportunity for microbes to flourish because food is often left at room temperature or in the sun for hours before it is consumed. Foods that last well without refrigeration, such as fresh fruits and vegetables, breads, and crackers, should be selected for these occasions (see Critical Thinking: Safe Picnic Choices).

Food safety is also a concern when food must be carried out of the home. Any food that is transported should be kept cold. Lunches should be transported to and from work or school in a cooler or an insulated bag. They should be refrigerated upon arrival or kept cold with ice packs. Most foods that are brought home from work or school uneaten should be thrown out and not saved for another day.

17.4 Chemical Contaminants in Food

Learning Objectives

- Illustrate how contaminants move through the food chain and into our foods.
- Compare the risks and benefits of using pesticides with those of growing food organically.
- Describe how to minimize the risks of exposure to chemical contaminants.

Chemicals used in agricultural production and industrial wastes contaminate the environment and can find their way into the food supply. How harmful these chemicals are depends on how long they remain in the environment and whether they are stored in the organisms that consume them or can be broken down and excreted by these organisms. Some contaminants are eliminated from the environment quickly because they are broken down by micro-organisms or chemical reactions. Others remain in the environment for very long periods, and when taken up by plants and small animals, they are not metabolized or excreted. For example, fat-soluble contaminants concentrate in body fat and cannot be excreted. When these plants or small animals are consumed by larger animals that are in turn eaten by still larger animals, the contaminants accumulate, reaching higher concentrations at each level of the food chain (**Figure 17.18**). This process is called **bioaccumulation**. Because the toxins are not eliminated from the body, the greater the amount consumed, the greater the amount present in the body.

bioaccumulation The process by which compounds accumulate or build up in an organism faster than they can be broken down or excreted.

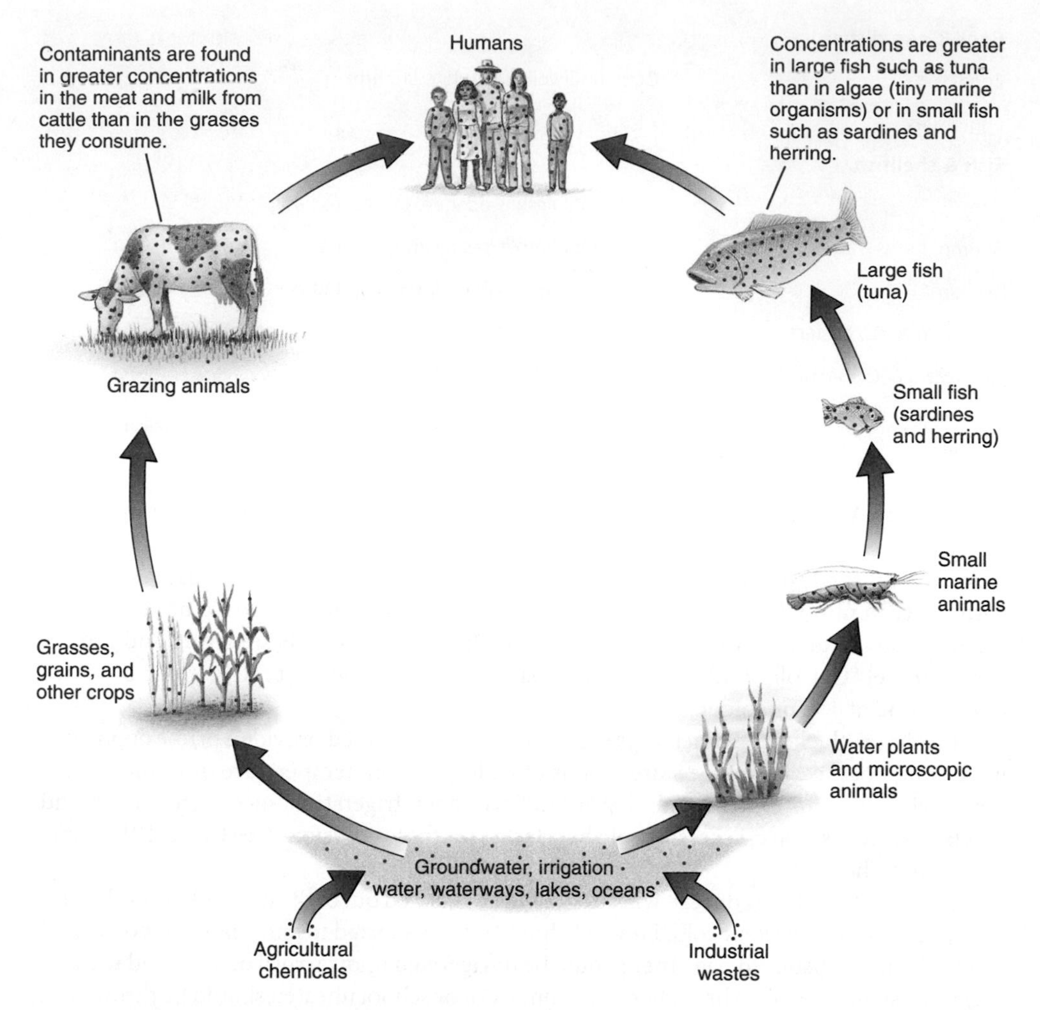

Figure 17.18 Contaminants in the food chain
Industrial pollutants and agricultural chemicals that contaminate the water supply enter the food chain and accumulate as they are passed through the chain. An animal that occupies a higher level in the food chain has higher concentrations of these contaminants because it consumes all the contaminants that have been eaten by organisms at lower feeding levels.

Risks and Benefits of Pesticides

Pesticides are applied to crops growing in the field to prevent plant diseases and insect infestations and to produce after harvesting to prevent spoilage and extend shelf life. Crops grown using pesticides generally produce higher yields and look more appealing because insect damage is limited. Residues of these chemicals remain on the fruits and vegetables that reach consumers. Pesticides can also spread, from the fields where they are applied, into water supplies, soil, and other parts of the environment, so pesticide residues are found not only on treated produce but also in meat, poultry, fish, and dairy products.[18]

Regulating Pesticides In order to protect public health and the environment, the types of pesticides that can be used on food crops, how often they may be used, and the amount of residue that can remain when foods reach consumers, are regulated. Before a pesticide can be used in food production, it must be approved and registered by Health Canada.[19] A major consideration in approving pesticides is whether they pose an unreasonable risk to humans. To determine this, Health Canada assesses the health risks associated with pesticide ingestion through foods. Maximum pesticide residue limits in food are then established. In setting the limits, Health Canada considers the known incidence of toxicity, and the predicted exposure that consumers, both adults and children, will have to the toxin. From this information, it sets a minimum level of exposure and then applies an additional margin of safety to this level to reach a **maximum residue limit**. These limits are often several hundred times lower than the level found to cause harm in test animals.[20] The Canadian Food Inspection Agency then monitors pesticide residues in foods to ensure they do not exceed these limits.

maximum residue limits The maximum amount of pesticide residues that may legally remain in food, set by Health Canada.

Life Cycle

In general, the amounts of pesticides to which people are exposed through foods are small. According to Health Canada's pesticide residue monitoring program, pesticide residue levels in foods are well below permitted limits.[21] Similar levels of pesticides were found in an American study depicted in **Figure 17.19**. Despite this, some individuals and special-interest groups are concerned that the pesticides remaining on foods do pose a risk to human health and also have concerns of potential interactions between the many different pesticides used on foods.

Reducing Pesticide Risks To reduce the risks posed by pesticides, new, more effective chemical pesticides are being developed, and the use of older, more toxic products is decreasing. One class of pesticide that is less likely to be harmful is biopesticides. These include naturally occurring substances and various micro-organisms that control pests, and pesticidal substances that are produced by plants through **genetic engineering**. In addition to developing safer pesticides, production methods are being implemented to make low-pesticide and pesticide-free produce available to the consumer.

genetic engineering A set of techniques used to manipulate DNA for the purpose of changing the characteristics of an organism or creating a new product.

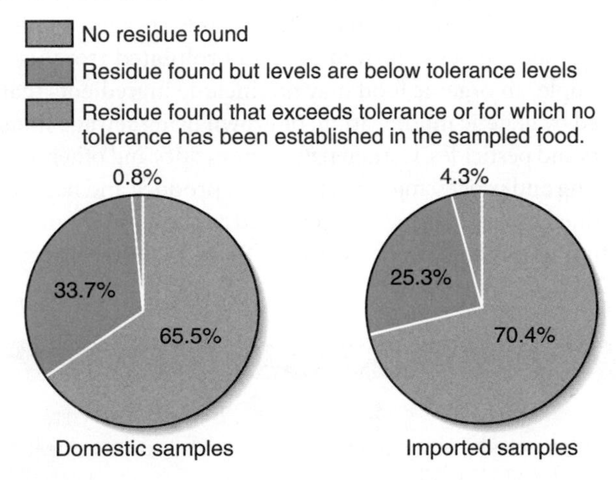

Figure 17.19 Pesticide tolerances
An analysis of samples of domestically produced food and imported food from 100 countries found no pesticide residues in 65.5% of domestic and 70.4% of import samples.
Source: U.S. Food and Drug Administration, Center for Food Safety and Applied Nutrition. Pesticide Program Residue Monitoring, 2004–2006. Available online at www.cfsan.fda.gov/~dms/pes04-06.html. Accessed August 8, 2008.; U.S. Food and Drug Administration, Center for Food Safety and Applied Nutrition. Pesticide Program Residue Monitoring, 2002. Available online at www.cfsan.fda.gov/~dms/pes02rep.html. Accessed August 8, 2008.

Natural Toxins Many toxins that occur in plants function as natural pesticides that offer protection from bacteria, moulds, and insect pests. These naturally pest-resistant crops are advantageous in developing countries because they thrive without the use of expensive added pesticides. Plants high in natural pesticides are being produced through special breeding programs as well as genetic engineering. The natural toxins in plants can also be isolated and applied to crops like synthetic pesticides.

As with all chemical toxins, natural toxins move through the food supply. For example, a cow that has foraged on toxic plants can pass the toxin into her milk and poison the consumer of the milk. The potential for toxicity, however, depends on the dose of toxin and the health of the consumer. Most natural toxins in the food supply are consumed in doses that pose little risk to the consumer.

Integrated Pest Management **Integrated pest management (IPM)** is a method of agricultural pest control that combines chemical and nonchemical methods and emphasizes the use of natural toxins and effective pesticide application. For example, increasing the use of naturally pest-resistant crop varieties that thrive without the use of added pesticides can reduce costs and do less environmental damage. Integrated pest management programs use information about the life cycles of pests and their interactions with the environment to manage pest damage economically and with the least possible hazard to people, property, and the environment.

integrated pest management (IPM) A method of agricultural pest control that integrates nonchemical and chemical techniques.

organic food Food that is produced, processed, and handled in accordance with the standards set by the Canadian General Standards Board's *Standards for Organic Agriculture* and the *Organic Products Regulations*.

Organic Techniques **Organic food** production methods emphasize the recycling of resources and the conservation of soil and water to protect the environment. Organic food is produced without using most conventional pesticides, fertilizers made with synthetic ingredients, sewage sludge, genetically modified ingredients, irradiation, antibiotics, or growth hormones.

Organic farming techniques reduce the exposure of farm workers to pesticides and decrease the quantity of pesticides introduced into the food supply and the environment. However, organic foods are not risk-free. Manure is often used for fertilizer and runoff can pollute lakes and streams. If the manure is not treated properly, the food crop can be contaminated with bacteria that cause foodborne illness. Irrigation water, rain, and a variety of other sources can introduce traces of synthetic pesticides and other agricultural chemicals not approved for organic use into organically grown foods, although producers are expected to create buffer zones or barriers to minimize this exposure.[22] In addition, organic foods are usually more expensive and available in less variety than conventionally grown foods.

CANADIAN CONTENT

The Organic Products Regulations of the Canada Agricultural Products Act and the Canadian National Standard Board govern the standards for organic foods.[22] These national standards define both substances approved for and prohibited from use in organic food production. For example, an organic food may not include ingredients that are treated with irradiation, produced by genetic modification, or grown using sewage sludge or industrially synthesized fertilizers and pesticides. Certain natural pesticides and other manufactured agents are permitted. Farming and processing operations that produce and handle foods labelled as organic must be certified by an organization accredited by the CFIA for this purpose. Products meeting the definition of "organic" as described in **Table 17.8** may display the organic logo shown in **Figure 17.20.**

Table 17.8 Labelling of Organic Foods

Labelling Term	Meaning
organic	Single ingredient or multi-ingredient products with an organic content greater than or equal to 95%. May display the organic logo.
X% organic ingredients	Multi-ingredient products with an organic content greater than 70% and less than 95%. Cannot display organic logo.
Organic product in list of ingredients	Multi-ingredient products with less than 70% organic content may not be labelled as organic but an organic product may be indicated in the ingredients list.

Source: Canada Gazette. Organic Products Regulations, 2009. Available online at http://canadagazette.gc.ca/rp-pr/p2/2009/2009-06-24/html/sor-dors176-eng.html. Accessed January13, 2011.

CANADIAN CONTENT

Antibiotics and Hormones in Food

Animals, like humans, are treated with antibiotics when they are sick. In addition, some animals are given antibiotics to prevent disease and to promote growth. This increases the amount of meat produced and reduces costs, but if used improperly, residues of these drugs can remain in the meat. To prevent passing these chemicals on to consumers, Health Canada regulates which drugs can be used to treat animals used for food production and when they can be administered. For example, withdrawal periods are established. The milk or meat from an antibiotic-treated animal cannot enter the food supply until the treatment has concluded and drug levels are not detected.[23]
The CFIA monitors tissue samples for drug residues to ensure compliance and recent reports have indicated that the detection of unsafe levels of drugs in food is very rare.[24]

Figure 17.20 The organic logo can appear on the label of agricultural products that meet the definition of "organic," as described in Table 17.8 (CFIA).

Antibiotic use in animals may also contribute to the creation of antibiotic-resistant bacteria. When bacteria are exposed to an antibiotic, those that are resistant to that antibiotic survive and produce offspring that also carry the antibiotic resistance trait. If these resistant bacteria infect humans, the resulting illness cannot be treated with that antibiotic. Since antibiotics are often used to prevent disease in animals, this use is suspected of being a major contributor to the development of antibiotic-resistant strains of bacteria.[25] Hormones are used to increase weight gain in sheep and cattle and milk production in dairy cows. Some naturally occurring hormones such as estrogen and testosterone are used in slow-release form, and levels in the treated animals are no higher than in untreated animals. Before synthetic hormones can be used, it must be demonstrated that residues in meat are within the safe limits. A synthetic hormone that has created public concern is genetically engineered bovine somatotropin (bST). Cows naturally produce somatotropin, a hormone that stimulates milk production. Genetically engineered bST is produced by bacteria and injected into cows to increase milk production. Consumer groups contend that genetically engineered bST causes health problems for the cows and for humans who consume milk or meat from the cows. An American review of the effect of bST concluded that it causes no serious long-term health effects in cows and that milk and meat from bST-treated cows are not health risks to consumers.[26] While Canadian regulators agreed that there was no risk to consumers, bST was not approved for use in Canada , because of concerns that the hormone compromised the health of dairy cows.[27]

polychlorinated biphenyls (PCBs) Carcinogenic industrial compounds that have found their way into the environment and, subsequently, the food supply. Repeated ingestion causes them to accumulate in biological tissues over time.

Contamination from Industrial Wastes

One group of industrial chemicals that pollutes the environment is **polychlorinated biphenyls (PCBs)**. These were used in the past in the manufacture of electrical capacitors and transformers, plasticizers, waxes, and paper. PCBs in runoff from manufacturing plants contaminated water, particularly near the Great Lakes. Although they are no longer produced, these compounds do not degrade and so are still found in the environment. Fish that live and feed in contaminated waters accumulate PCBs in their adipose tissue; humans who consume large quantities of contaminated fish accumulate PCBs in their adipose tissue.

PCB exposure can cause skin conditions, liver damage, and certain kinds of cancer. It is a particular problem for pregnant and lactating women because prenatal exposure and consumption of contaminated breast milk can damage the nervous system and cause learning deficits in children. Health Canada, however, concluded that levels of PCBs in breast milk in Canada did not pose a health risk.[28]

Other contaminants from manufacturing, such as chlordane (used to control termites), radioactive substances such as strontium-90, and toxic metals such as cadmium, lead, arsenic, and mercury have also found their way into fish and shellfish. Cadmium and lead can interfere with the absorption of other minerals, as well as have a direct toxic effect: cadmium can cause kidney damage; lead can impair brain development. Arsenic is believed to contribute to cancer development, and mercury damages nerve cells.[29] Health Canada reports that most fish commonly consumed in Canada have very low mercury levels. The exceptions are fresh or frozen tuna, shark, swordfish, marlin, orange roughy, and escolar, and Canadians are advised to limit their intakes to the amounts posted on the Health Canada website.[30] Although fresh and frozen tuna are mentioned, canned light tuna, a commonly consumed fish, has low levels of mercury and is not of concern. Canned albacore tuna, a more expensive form of tuna, has sufficient mercury that Health Canada recommends limits on intake for children and women who are planning to get pregnant, are pregnant, or are breastfeeding.[30]

Large fish at the top of the food chain are more likely to contain high levels of industrial contaminants, but shellfish also accumulate contaminants because they feed by passing large volumes of water through their bodies (**Figure 17.21**).

Figure 17.21 It is unsafe to consume shellfish from contaminated waters. (Paul A. Souders/© Corbis)

Choosing Wisely to Reduce Risk

Even though individual consumers cannot detect chemicals in food, care in selection and preparation can reduce the amounts that are consumed. One of the easiest ways to reduce risk is to choose a wide variety of foods, thus avoiding excessive consumption of any one food. Although some consumers are concerned about the consumption of pesticide residues, the health risk of eliminating foods from the diet that may contain pesticide residues, such as fresh fruits and vegetables, is probably greater than that of the pesticide exposure. To reduce exposure, consumers can choose locally grown produce. This is likely to have fewer pesticides because it does not contain those applied to prevent spoilage and extend the shelf life during shipping. Foods produced organically or using IPM are also likely to contain fewer pesticide residues.

Pesticides on conventionally grown produce can be removed or reduced by peeling or washing with tap water and scrubbing with a brush, if appropriate. For leafy vegetables such as lettuce and cabbage, the outer leaves can be removed and discarded. Some produce, such as cucumbers, apples, eggplant, squash, and tomatoes, is coated with wax to maintain freshness by sealing in moisture, but wax also seals in pesticides. Much of the wax can be removed

CRITICAL THINKING:

The Risks and Benefits of Food

Background

After reading a newspaper article about a child who died of foodborne illness contracted by eating an undercooked hamburger, Ron became concerned about the safety of the foods his family was eating. He thought carefully about food safety issues such as eggs and poultry contaminated with *Salmonella*, pesticide residues on fruits and vegetables, and fish contaminated with industrial pollutants, bacteria, viruses, and parasites. He started to think most of the foods his family ate were risky but then decided to see if the benefits they provide are worth the risk.

(© istockphoto.com/Robyn Mackensie)

Data

In general, Ron's family eats a healthy diet and is rarely sick, but he knows that a few of the things they like carry risks. He enjoys his meat rare, his son is an athlete who drinks protein shakes containing raw eggs, his young daughter likes to lick cookie dough (containing raw eggs) from the bowl when baking, and he and his wife enjoy eating sushi and raw oysters. Ron makes the following list of foods and records the risks and benefits of each. When he reviews his list, Ron recognizes that although the hamburger may be contaminated, it is an inexpensive source of protein and iron in his family's diet. He decides to continue to eat hamburger but to be sure to cook it to 71 °C, thereby minimizing the risks.

FOOD	RISK	BENEFIT
Hamburger	Can be contaminated with pathogenic *E. coli*.	A good source of protein and iron in the diet.
Chicken	Is often contaminated with *Salmonella*.	An economical source of protein that is low in fat.
Eggs	Can contain *Salmonella* or *Campylobacter*.	An inexpensive source of protein.
Fish	Can be contaminated with environmental pollutants such as PCBs and toxic metals.	A source of high-quality protein. Fatty fish are an excellent source of omega-3 fatty acids EPA and DHA, and have been associated with a reduced risk of cardiovascular disease.
Raw fish and shellfish	May be a source of bacterial, viral, and parasitic infections.	A low-fat source of high quality protein and omega-3 fatty acids.
Fruits and vegetables	May contain pesticide residues.	An excellent source of fibre and vitamins. They also contain health-promoting phytochemicals.

Critical Thinking Questions

Draw a similar risk-benefit conclusion for the other foods on Ron's list. Which foods do you think he should keep in his family's diet? How can he minimize the risks associated with eating them?

Are there any foods on his list that you would suggest he eliminate from the family diet? Why or why not?

iProfile Use iProfile to compare the omega-3 fatty acid content of varieties of fish and shellfish.

---

by rinsing produce in warm water and scrubbing it with a brush, but to eliminate all wax, the produce must be peeled. Although peeling fruits and vegetables eliminates some pesticides, it also eliminates some fibre and micronutrients.

The risk of ingesting chemical pollutants from fish can be minimized by choosing wisely. Smaller species of fish are safer because they are lower in the food chain, and smaller fish within a species are safer because they are younger and hence have had less time to accumulate contaminants. The safest fish are saltwater varieties caught well offshore, away from polluted waters. Freshwater fish and saltwater fish that live near shore or spend part of their life cycle in freshwater are more likely to contain contaminants. Migratory fish such as striped bass and bluefish are a problem because they may contain contaminants even when they are caught in clean water well offshore. Consuming a variety of fish rather than just one or two kinds can also reduce the risk of ingesting dangerous amounts of contaminants. Most toxins concentrate in adipose tissue, so amounts can be reduced by removing the skin, fatty material, and dark meat from fish (and trimming fat from meat and removing skin from poultry). Use cooking methods such as broiling, poaching, boiling, and baking, which allow the fatty portions of the fish to drain out. Do not eat the "tomalley" in lobster. The tomalley, a green paste inside the abdominal cavity of a cooked lobster, serves as the liver and pancreas and is the organ in which toxins accumulate. The analogous organ in blue crabs, called the "mustard" because of its yellow color, should also be avoided (see Critical Thinking: The Risks and Benefits of Food).

17.5 Food Technology

Learning Objectives

- Describe how temperature is used to prevent food spoilage.
- Discuss how irradiation preserves food.
- Explain how packaging protects food.
- Compare the risks and benefits of food additives.

Advances in food and agricultural technology have improved the safety and availability of foods. This technology includes techniques to preserve food and develop new food products. It allows food to be stored for long periods, adds nutrients lacking in the diet, and creates disease-resistant, high-yield crops. It has helped ensure that food is available even if the local growing season is not ideal. Without technology, the food supply would include only locally grown foods that must be eaten soon after harvest or slaughter. While this has some appeal, it would limit the variety of foods in the diet, particularly during the winter months, and increase the risk for malnutrition if food production were interrupted by a natural or man-made disaster. Despite the many benefits technology offers, it also creates risks.

Food spoilage occurs when the taste, texture, or nutritional value of the food changes as a result of either enzymes that are naturally present in the food or bacteria or mould that grow on the food. For thousands of years, humans have been treating foods in order to protect them from spoilage. Techniques that preserve food slow down or kill microbes or destroy enzymes present in the food. As shown in **Table 17.9**, the acronym FAT TOM reminds us of the factors that affect microbial growth. Most food preservation techniques modify one or more of these factors to stop or slow microbial growth. Many of the oldest methods of food preservation—including drying, smoking, **fermentation**, adding sugar or salt, and heating or cooling—are still used today. In addition, new methods of improving food quality and preventing spoilage and contamination, such as irradiation and specialized packaging, have been developed. The newest methods of enhancing and protecting our food supply rely on biotechnology (see Focus on Biotechnology). Food technology often involves adding substances to food. Health Canada considers a **food additive** to be a chemical substance that is added to food during preparation or storage and becomes part of the food or affects the characteristics of food. Health Canada determines the safety of food additives and approves their use in food.[31]

fermentation A process in which micro-organisms metabolize components of a food and therefore change the composition, taste, and storage properties of the food.

food additives Substances that are added to food during preparation or storage and become part of the food or affect its characteristics.

Table 17.9 FAT TOM

Food	Provides a growth media for bacteria.
Acidity	Most bacteria grow best at a pH near neutral. Some food additives, such as citric acid and ascorbic acid (vitamin C), are acids, which prevent microbial growth by lowering the pH of food.
Time	The longer a food sits at an optimum growth temperature, the more bacteria it will contain. Preservation methods such as canning and pasteurization kill microbes by heating food to an appropriate temperature for the right amount of time.
Temperature	The high temperatures of canning, cooking, and pasteurization kill microbes, and the low temperatures of freezing and refrigeration slow or stop microbial growth.
Oxygen	In order to grow, most bacteria need oxygen so packaging that eliminates oxygen prevents their growth.
Moisture	Bacteria need water to grow, so preservation methods such as drying or the use of high concentrations of salt or sugar, which draw water away by osmosis, prevent bacteria from growing.

aseptic processing A method that places sterilized food in a sterilized package using a sterile process.

High and Low Temperatures

Cooking food is one of the oldest methods of ensuring food safety. It kills disease-causing organisms and destroys most toxins. Cooling food with refrigeration or freezing protects us by slowing or stopping microbial growth. Other preservation techniques that rely on temperature include canning, pasteurization, sterilization, and **aseptic processing**. Aseptic processing heats foods to temperatures that result in sterilization. The sterilized foods are then placed in sterilized packages using sterilized packaging equipment. Aseptic processing is currently used to produce boxes of sterile milk and juices (**Figure 17.22**). These can remain free of microbial growth at room temperature for years (see Science Applied: Pasteurization: From Spoiled Wine to Safe Milk).

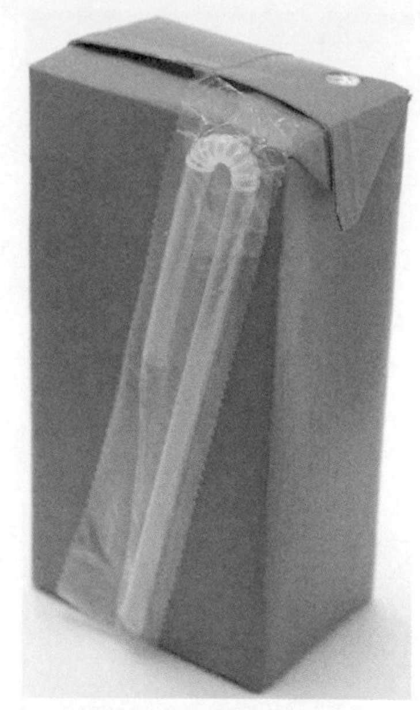

Figure 17.22 If unopened, milk and juice in aseptic packaging can be stored for long periods without refrigeration. (© istockphoto.com/ Kieran Wills)

Preservation techniques that rely on temperature benefit us by providing appealing, safe food, but they are not risk-free, particularly if used incorrectly. If foods are not heated long enough or to a high enough temperature, or if they are not kept cold enough, there is a risk of microbial foodborne illness. In addition, some types of cooking can also generate hazardous chemicals. The most familiar group of chemicals produced during cooking is the **polycyclic aromatic hydrocarbons (PAHs)**. These carcinogenic substances are formed when fat drips onto the flame of a grill and burns. They rise with the smoke and are deposited on the surface of the food. Grilled or charred meats are therefore high in PAHs. PAH formation can be minimized by selecting lower-fat meat and using a layer of aluminum foil to prevent fat from dripping on the coals.

polycyclic aromatic hydrocarbons (PAHs) A class of mutagenic substances produced during cooking when there is incomplete combustion of organic materials—such as when fat drips on a grill.

Broiled foods, which are cooked with the heat source at the top, are low in PAHs. However, broiling and other methods of high temperature cooking can result in another potential hazard—**heterocyclic amines (HCAs)**, such as benzopyrene. HCAs are formed from the burning of amino acids and other substances in meats. Well-done meat and meat cooked using hotter temperatures contain greater amounts. HCA formation can be reduced by precooking meat, marinating meat before cooking, cooking at lower temperatures, and reducing cooking time by using smaller pieces of meat and avoiding overcooking. The recommended cooking temperatures are designed to prevent microbial foodborne illness and minimize the production of PAHs and HCAs.

heterocyclic amines (HCAs) A class of mutagenic substances produced when there is incomplete combustion of amino acids during the cooking of meats—such as when meat is charred.

Another contaminant formed during food preparation is acrylamide. It forms as a result of chemical reactions during high-temperature baking or frying, particularly in carbohydrate-rich foods. The highest levels of acrylamide are found in french fries and snack chips, foods that people should already be eating less of because they are low in nutrients and high in kcalories. High doses of acrylamide have been found to cause cancer and reproductive problems in animals and to act as a neurotoxin in humans. Thus far, dietary exposure to acrylamide has not been associated with cancer in humans, and more research is needed to determine whether long-term, low-level exposure has any cumulative effects.[32] Methods for reducing the amounts and potential toxicity of acrylamide in foods are being investigated.[33]

Food Irradiation

Irradiation, also called *cold pasteurization*, exposes food to a high dose of x-rays, gamma radiation, or high-energy electrons to kill micro-organisms and insects and inactivate enzymes that cause germination and ripening of fruits and vegetables. Although irradiated foods must be labelled with the radura symbol (**Figure 17.23**) and the statement "treated with radiation" or "treated by irradiation," foods that contain irradiated spices or other irradiated ingredients do not need to display this symbol.

irradiation A process, also called cold pasteurization, that exposes foods to radiation to kill contaminating organisms and retard ripening and spoilage of fruits and vegetables.

Irradiation can be used in place of chemical treatments, so it reduces exposure to chemical pesticides and preservatives. It is a technology with great potential for improving the safety of food and reducing the incidence of foodborne illness.

CANADIAN CONTENT

Food irradiation is used in more than 40 countries to treat everything from frog legs to rice. It is, however, used relatively infrequently in Canada. Part of the reason for this is lack of irradiation facilities, but public fear and suspicion of the technology also limit its use. The word "irradiation" fosters the belief that the food itself becomes radioactive. Opponents to food irradiation claim that it introduces carcinogens, depletes the nutritional value of food, and is used to allow the sale of previously contaminated foods. In fact, irradiated food is not radioactive and scientific studies conducted over the past 50 years have found that the benefits of irradiation outweigh the potential risks.[34] It increases the safety and shelf life of foods and does not compromise nutritional quality or noticeably change food texture, taste,

CANADIAN CONTENT

SCIENCE APPLIED

Pasteurization: From Spoiled Wine to Safe Milk

In 1857, sailors in Napoleon's navy were in mutiny because wine supplies were spoiling after only a few weeks at sea. Napoleon recognized this spoilage problem as a threat to his hopes for world conquest. He turned to Louis Pasteur for help. To study the problem, Pasteur travelled to a vineyard in Arbois, France, where spoilage was causing considerable economic losses for the wine industry. By examining the spoiled wine under a microscope, Pasteur was able to demonstrate the presence of certain strains of micro-organisms. He suggested that spoilage could be prevented by heating the wine to a temperature high enough to kill the harmful microbes but low enough that it did not affect the flavour Experimentation with heating wine for various times and temperatures revealed that as Pasteur had predicted, the micro-organisms could be killed without damaging the flavour of the wine. This became the foundation for the modern treatment of bottled liquids to prevent their spoilage, a process known as pasteurization.

Pasteurization had a major public health impact, particularly when applied to the milk industry. Milk-borne illness was common at the turn of the nineteenth century. causing undulant fever, typhoid, scarlet fever, diphtheria, and tuberculosis. In 1915, the city of Toronto, Ontario, passed a by-law requiring the pasteurization of milk and after that date, the Hospital for Sick Children found that there were no cases of bovine tuberculosis among children born and raised in Toronto. All their cases were children from outside the city.[1] In 1937, the then premier of Ontario, Mitchell Hepburn, was given a tour of the Toronto hospital where he observed several children suffering from bovine tuberculosis. When he asked the cause of the cause of the ailment, doctors responded, "Drinking dirty milk." The premier was so moved that he enacted legislation making Ontario the first Canadian province to make pasteurized milk mandatory.[1]

The milk pasteurization process kills pathogenic bacteria and reduces the total number of micro-organisms present, but it allows many microbes to survive. The multiplication of the bacteria that remain eventually causes the milk to spoil. To prevent rapid multiplication after pasteurization, the milk must be cooled immediately and remain refrigerated. In addition to killing micro-organisms, the heat also destroys enzymes in the milk; the inactivation of lipase extends the shelf life of homogenized milk by preventing it from becoming rancid. The inactivation of another enzyme called phosphatase is used to gauge the adequacy of pasteurization. Lack of phosphatase activity indicates that the heat treatment has been sufficient, but if phosphatase activity remains, it indicates that the milk was not treated adequately and may not be free of pathogens.

In 1991, Health Canada made the pasteurization of milk mandatory across Canada. Pasteurized milk, of course, had been widespread and readily available in Canada well before 1991, but some unpasteurized products were still being sold and milk-related illnesses were still being reported. The legislation making pasteurization mandatory dramatically reduced the number of these illnesses in Canada.[2] Thanks to the work of Louis Pasteur, we now enjoy the nutritional benefits of milk without the risk of contamination with disease-causing organisms.

This photo depicts Louis Pasteur studying the souring of wine. (Jean Loup Charmet/SPL/Photo Researchers)

[1]Brink, G. C. How pasteurization of milk came to Ontario. *Canadian Medical Association Journal* 91:972-973, 1964.

[2]Health Canada. Tip Sheet for raw milk, Available online at http://www.hc-sc.gc.ca/fn-an/securit/kitchen-cuisine/raw-milk-lait-cru-eng.php. Accessed January 13, 2011.

or appearance as long as it is properly applied to a suitable product. Irradiation can benefit the environment by reducing the use of chemicals to kill microbes and insects. Irradiation can decrease the amounts of certain nutrients but losses are similar to those that occur with canning or cold storage.[35]

Health Canada has approved irradiation to destroy pathogens in red meat and poultry and contaminants in spices; to prevent insect infestation in flour and spices; to increase the shelf life of potatoes; to eliminate *Trichinella* in pork; to control insects in fruits, vegetables, and grains; and to slow the ripening and spoilage of some produce.[36] Because irradiation produces unique compounds in irradiated foods, it is treated as a food additive, and the level of radiation that may be used is regulated. At the allowed levels of radiation, the amounts of these unique compounds produced are almost negligible and have not been found to be a risk to consumers. Irradiated foods may cost more because of the cost of adding an extra processing step, but in the future, this may be offset by a longer shelf life (**Figure 17.24**). Irradiation should be used to complement, not replace, proper food handling by producers, processors, and consumers.

Figure 17.23 The radura symbol
This symbol is used to identify foods that have been treated with irradiation.

Figure 17.24 After 2 weeks in cold storage, the strawberries treated by irradiation remain free of mould (right), whereas the untreated strawberries picked at the same time are covered with mould (left). (Courtesy Council for Agricultural Science and Technology)

Food Packaging

An open package of cheddar cheese will grow mould in the refrigerator after only a few days. An unopened package will stay fresh for weeks. Packaging plays an important role in food preservation; it keeps moulds and bacteria out, keeps moisture in, and protects food from physical damage. Food packaging is continuously being improved.

Consumer demand for fresh, easy-to-prepare foods has led manufacturers to offer partially cooked pasta, vegetables, seafood, fresh and cured meats, and dry products such as whole-bean and ground coffee in packaging that, if unopened, will keep perishable food fresh much longer than will conventional packaging. **Modified atmosphere packaging (MAP)** uses plastics or other packaging materials that are impermeable to oxygen. The air inside the package is vacuumed out in order to remove the oxygen. The product can then remain packaged in a vacuum, or the package can be infused with another gas, such as carbon dioxide or nitrogen. The lack of oxygen prevents the growth of aerobic bacteria, slows the ripening of fruits and vegetables, and slows down oxidation reactions, which cause discoloration in fruits and vegetables and rancidity in fats.

Figure 17.25 Does your water bottle contain bisphenol A? (Jupiter Images/Getty Images)

MAP is often used to package cooked entrees such as pasta primavera or beef teriyaki. The raw ingredients are sealed in a plastic pouch, the air is vacuumed out, and the pouch and its contents are partially precooked and immediately refrigerated. This processing eliminates the need for the extreme cold of freezing or the extreme heat of canning, so flavour and nutrients are better preserved. Because these products are not heated to temperatures high enough to kill all bacteria and are not stored at temperatures low enough to prevent all bacteria from growing, they could pose a food safety risk. To ensure safety, fresh refrigerated foods should be purchased only from reputable vendors, used before the expiration date printed on the package, refrigerated until use, and heated according to the time and temperature directions on the package.

Packaging can protect food from spoilage, but even the best packaging can introduce risk if it becomes a part of the food. A variety of substances leach into foods from plastics, paper, and even dishes. One potential contaminant from plastics containers is bisphenol A, a chemical found in polycarbonate plastic used to manufacture hard, transparent water bottles, baby bottles, and food containers (**Figure 17.25**). A recent scientific review found that although there is negligible risk of adverse effects from the transfer of bisphenol A from containers to food, it, nonetheless, identified infants as a vulnerable group (see Section 14.6). For this reason, Canada became the first country to ban bisphenol A from plastic baby bottles.[37] Substances that are known to contaminate foods are indirect food additives and the amounts and types are regulated by Health Canada tolerance levels and CFIA inspections.

Food Additives

Health Canada considers a **food additive** to be a chemical substance that is added to food during preparation or storage and becomes part of the food or affects the characteristics of food. Food additives keep bread from moulding, give margarine its yellow colour, and keep parmesan cheese from clumping in the shaker (**Figure 17.26**). Other substances unintentionally get into our food, like the oil used to lubricate food-processing machinery. Health Canada regulates the amounts and types of food additives in food. The CFIA maintains a list of non-food chemicals, packaging materials, and construction materials that are acceptable for use in food processing facilities and contact with food.[38]

food additives Substances intentionally added to foods.

CANADIAN CONTENT

Additives that Prevent Spoilage Many substances are added to prevent bacteria and moulds from causing food spoilage, to extend shelf life, or to protect the natural colour and flavour of food. Sugar and salt are two of the oldest **preservatives**. They prevent microbial growth by decreasing the water availability in the product; without adequate water, microbes cannot grow. For example, the high concentration of sugar in jams and jellies draws water away from the microbial cells and prevents them from growing. Antioxidants such as sulfites and BHT (butylated hydroxytoluene) are also used as preservatives. They prevent fats and oils in baked goods and other foods from becoming rancid or developing an off-flavour and prevent cut fruits such as apples from turning brown when exposed to air. Other preservatives act by blocking the natural ripening and enzymatic processes that continue to occur in foods even after harvest.

Additives that Improve Nutrient Content, Colour, Texture, and Flavour Additives are not just used to make food safer and last longer. They are also used to enhance the nutrient content, texture, colour, and flavour of foods (see Label Literacy: Should You Avoid Food Additives?).

Additives to Maintain or Improve the Nutritional Quality Nutrients that are added to foods are considered additives. Refined grains are enriched with iron and some of the B vitamins that are lost in processing; these are considered additives. Food is also fortified with nutrients, such as calcium, that are typically lacking in the diet. Although the addition of these additives to foods benefits the population by increasing the nutrient content of the diet, it can also increase the risk of nutrient toxicities.

preservatives Compounds that extend the shelf life of a product by retarding chemical, physical, or microbiological changes.

Additives to Improve and Maintain Texture Many different types of additives are used in product processing and preparation. Emulsifiers improve the homogeneity, stability, and consistency of products such as ice cream. Stabilizers, thickeners, and texturizers, such as pectins and gums, are used to improve consistency or texture in pudding and to stabilize emulsions in foods such as salad dressing. Leavening agents are added to incorporate gas into breads and cakes, causing them to rise. Acids are added as flavour enhancers, preservatives, and antioxidants. Humectants, such as propylene glycol, cause moisture to be retained so products stay fresh. Anti-caking agents prevent crystalline products such as powdered sugar from absorbing moisture and caking or lumping.

Additives to Affect Flavour and Colour Additives are also used to enhance the flavour and colour of foods. For example, both natural and alternative sweeteners are added to sweeten foods such as yogurt and fruit drinks. Colour additives enhance the appearance of foods. A colour additive is defined as any dye, pigment, or substance that can impart colour when added or applied to a food, drug, cosmetic, or to the human body. Colour additives are used in foods for many reasons, including to balance colour loss due to storage or processing and to even out natural variations in a food's colour. Colours can be used to make foods appear more appetizing; however, they cannot be used as deception to conceal inferiority. Food colours include synthetic dyes as well as colours derived from plant, animal, and certain mineral sources. All colours must meet safety standards before they are approved for use in foods. Examples of plant-derived colours include annatto extract (yellow), dehydrated beets (bluish-red to brown), caramel (yellow to tan), β-carotene (yellow to orange), and grape skin extract (red, green). In February, 2010, Health Canada announced that it was undertaking a review of the labelling of food colours. Currently, food processors can name the individual colours used or simply list "colours" as an ingredient. The new proposal is to require the naming of each food colour used.[39]

Nitrates and nitrites are chemical substances added to cured meat products such as luncheon meats, salami, hot dogs, sausages, and so forth, to enhance the colour of meats. There have been safety concerns about these additives because they can react with amino acids in the stomach to form nitrosamines, compounds that are carcinogenic. Health Canada has not banned the use of nitrates and nitrites because they are also beneficial in preventing the growth of *Clostridium botulinum*. Instead, the levels permitted in food are strictly regulated to the minimum needed to be effective.[40] Food processors also add the antioxidant sodium erythorbate, which is similar in structure to vitamin C, to reduce the formation of nitrosamines.

Regulating Food Additives Food additives improve food quality and help protect us from disease, but if the wrong additive is used or if the wrong amount is added, it could do more harm than good. Health Canada maintains a list of acceptable food additives and their functions.[41] When a manufacturer wants to use a new food additive, a petition must be submitted to Health Canada. The petition describes the chemical composition of the additive, how it is manufactured, and how it is detected and measured in food. The manufacturer must prove that the additive will be effective for its intended purpose at the proposed levels, that it is safe for its intended use, and that its use is necessary. Additives may not be used to disguise inferior products or deceive consumers. They cannot be used if they significantly destroy nutrients or where the same effect can be achieved by sound manufacturing processes.

Figure 17.26 The additives in these foods prevent the bread from moulding and the fruit snacks from hardening; they also smooth the texture of the pudding and give colour and flavour to soft drinks and candy. (Luisa Begani)

LABEL LITERACY

Should You Avoid Food Additives?

CANADIAN CONTENT

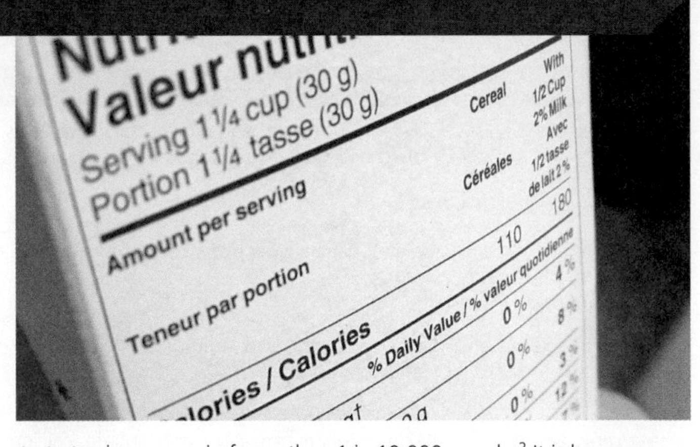

Sometimes the ingredients listed on food labels sound like a chemical soup. Calcium propionate is added to bread, disodium EDTA is added to canned kidney beans, and BHA is in potato chips. Are these chemical additives necessary in our food supply? Are they safe? Understanding what these chemicals are used for can help make the ingredient list a source of information rather than a cause for concern. The common food additives table included with this box provides some examples of additives, explains what each does, and gives examples of the types of food where they are used.

Health Canada does not approve food additives unless they are safe for most consumers, but this doesn't mean they are safe for everyone. For individuals who are sensitive or allergic to certain additives such as preservatives or colours, the ingredient list provides information that can be life-saving. For example, in sensitive individuals, sulfites can cause symptoms that range from a stomach ache and hives to severe asthmatic reactions. It has been estimated that 1 in 100 people are sulfite-sensitive.[1] Sulfites are used to preserve foods such as baked goods, canned vegetables, condiments, and maraschino cherries. Individuals sensitive to sulfites should read food labels. Foods served in restaurants are also a concern because sulfites are sometimes used in food preparation. For example, a potato dish on the menu may be prepared using potatoes that were peeled and soaked in a sulfite solution before cooking.

Food colours can also cause reactions in sensitive individuals. The colour additive tartrazine may cause itching and hives in sensitive people. It is also found in beverages, desserts, and processed vegetables. Sensitivity to tartrazine occurs in fewer than 1 in 10,000 people.[2] It is because of these potential reactions that Health Canada is re-evaluating the labelling of food colours. Currently, colours can be named separately or simply listed as "colours." In February, 2010, Health Canada proposed instituting the requirement that each colour be listed by name. Health Canada has invited commentary from interested Canadians.[3]

There are more additives than you might think in your whipped topping and maraschino cherry. (Comstock Images/Gettyone.com)

Common food additives

Type of additive	Examples of what's on the label	What they do	Where they are used
Preservatives	Ascorbic acid, citric acid, sodium benzoate, calcium propionate, sodium erythorbate, sodium nitrite, calcium sorbate, potassium sorbate, BHA, BHT, EDTA, tocopherols	Maintain freshness; prevent spoilage caused by bacteria, moulds, fungi, or yeast; slow or prevent changes in colour, flavour, or texture; and delay rancidity	Jellies, beverages, baked goods, cured meats, oils and margarines, cereals, dressings, snack foods, fruits and vegetables
Sweeteners	Sucrose, glucose, fructose, sorbitol, mannitol, corn syrup, high-fructose corn syrup, saccharin, aspartame, sucralose, acesulfame potassium (acesulfame-K), neotame	Add sweetness with or without extra kcalories	Beverages, baked goods, table-top sweeteners, many processed foods
Colour additives	Brillliant Blue FCF, Indigotine, Fast Green FCF, Erythrosine, Sunset Yellow FCF, orange B, citrus red no. 2, annatto extract, beta-carotene, grapeskin extract, cochineal extract or carmine, paprika oleoresin, caramel colour, fruit and vegetable juices, saffron, colourings or colour added	Prevent colour loss due to exposure to light, air, temperature extremes, and moisture; enhance colours; give colour to colourless and "fun" foods	Processed foods, candies, snack foods, margarine, cheese, soft drinks, jellies, puddings and pie fillings
Flavours, spices, and flavour enhancers	Natural flavouring, artificial flavour, spices, monosodium glutamate (MSG), hydrolyzed soy protein, autolyzed yeast extract, disodium guanylate or inosinate	Add specific flavours or enhance flavours already present in foods	Processed foods, puddings and pie fillings, gelatin mixes, cake mixes, salad dressings, candies, soft drinks, ice cream, BBQ sauce
Nutrients	Thiamine hydrochloride, riboflavin (vitamin B_2), niacin, niacinamide, folate or folic acid, beta-carotene, potassium iodide, iron or ferrous sulfate, alpha-tocopherols, ascorbic acid, vitamin D, amino acids (L-tryptophan, L-lysine, L-leucine, L-methionine)	Replace vitamins and minerals lost in processing; add nutrients that may be lacking in the diet	Flour, breads, cereals, rice, pasta, margarine, salt, milk, fruit beverages, energy bars, breakfast drinks

Common food additives *(continued)*			
Emulsifiers	Soy lecithin, mono- and diglycerides, egg yolks, polysorbates, sorbitan monostearate	Allow smooth mixing and prevent separation; reduce stickiness; control crystallization; keep ingredients dispersed	Salad dressings, peanut butter, chocolate, margarine, frozen desserts
Stabilizers and thickeners, binders, and texturizers	Gelatin, pectin, guar gum, carrageenan, xanthan gum, whey	Produce uniform texture, improve "mouth-feel"	Frozen desserts, dairy products, cakes, pudding and gelatin mixes, dressings, jams and jellies, sauces
pH control agents and acidulants	Lactic acid, citric acid, ammonium hydroxide, sodium carbonate	Control acidity and alkalinity, prevent spoilage	Beverages, frozen desserts, chocolate, low-acid canned foods, baking powder
Leavening agents	Baking soda, monocalcium phosphate, calcium carbonate	Promote rising of baked goods	Breads and other baked goods
Anti-caking agents	Calcium silicate, iron ammonium citrate, silicon dioxide	Keep powdered foods free-flowing, prevent moisture absorption	Salt, baking powder, confectioners' sugar
Humectants	Glycerine, sorbitol	Retain moisture	Shredded coconut, marshmallows, soft candies, confections

[1]Papazian, R. Sulfites: Safe for most, dangerous for some. *FDA Consumer* 30:11–14, December 1996. Available online at http://www.fda.gov/fdac/features/096_sulf.html. Accessed May 20, 2009.

[2]U.S. Food and Drug Administration. Food Color Facts. January 1993. Available online at www.cfsan.fda.gov/~lrd/colorfac.html/. Accessed May 20, 2009.

[3]Health Canada. Health Canada proposal to improve food colour labelling requirements, Available online at http://www.hc-sc.gc.ca/fn-an/consult/_feb2010-food-aliments-col/draft-ebauche-eng.php. Accessed January 13, 2011.

CASE STUDY OUTCOME

How could turkey burgers, which sound so wholesome, make so many people sick? Unfortunately turkey often harbours the *Salmonella* bacteria and the mixing that takes place during the making of ground burgers evenly distributes the bacteria throughout the meat product. Proper heating kills the bacteria, even in a contaminated food, but a ground burger product must reach at internal temperature of least 74°C to ensure safety. Health inspectors found that the parent volunteers, who were in charge of the turkey burger barbeque, did not use an internal thermometer to measure the temperature of the burgers. Furthermore, there were many people at the celebration and the line ups at the barbeque were very long. The parents felt rushed and as a result probably served undercooked burgers. This combined with the higher than usual levels of bacteria in the original product resulted in many people getting ill.

The recall issued by the Canadian Food Inspection Agency prevented further outbreaks by taking the tainted product out of the stores. Furthermore, the turkey burger processors were able to link the high bacterial levels in the turkey to the poor cleaning of equipment in their processing plant. The cleaning procedures were revised and made more stringent and Salmonella levels in the turkey product returned to safer levels. All of the individuals who became ill after the barbeque recovered, but *Salmonella* infection can be very serious, causing severe illness and even death in the elderly, infants, and persons with impaired immune function.

APPLICATIONS

Personal Nutrition

1. **Can your kitchen pass the food safety test? To find out, have a look at Health Canada's Safe Food Handling Interactive Guide (visit www.healthycanadians.gc.ca and search for "safe food handling interactive guide") to find safety tips for avoiding foodborne illnesses. After you complete the exercise, answer the following questions:**

   a. Based on how you answered these questions, what changes should you make in the way you store and handle foods in your kitchen?

   b. Based on how you answered these questions, are there foods that you will eliminate from your diet?

2. **What food additives are in your favourite snacks?**

   a. Choose a packaged product you like to snack on and write out all the ingredients it contains.

   b. Which of the ingredients are food additives?

   c. Use the table in the Label Literacy feature in this chapter to describe the function of each additive.

   d. If these additives could not be used, how would the product differ?

General Nutrition Issues

1. **Sixty-seven people became ill after consuming food at a company picnic. Testing of food samples revealed that the tossed salad, the egg salad, and the turkey slices were all contaminated with *Salmonella.* Invent a scenario that would explain how all three became contaminated.**

2. **What are the risks and benefits for each scenario described below?**

   a. A restaurant decides not to replace their old dishwasher even though it no longer heats the water to above 60°C.

   b. A town decides that they can improve the health of their citizens and the environment by banning the production and sale of all but organically produced foods.

3. **A train crash spills a load of industrial waste into a river that feeds into a local reservoir. How would this spill affect:**

   a. The safety of the drinking water?

   b. The milk from dairy cattle grazing nearby?

   c. The fish that swim in the river?

   d. The crops irrigated by this water?

SUMMARY

17.1 How Can Food Make Us Sick?

- Foodborne illness is any illness that is related to the consumption of food or contaminants or toxins in food.
- Food may become contaminated where it is grown or produced, during processing or storage, or even in the home. Once contaminated, food preparation surfaces can cross-contaminate other foods.
- The harm caused by contaminants in the food supply depends on the type of toxin, the dose, the length of time over which it is consumed, and the size and health status of the consumer.

17.2 Keeping Food Safe

- The safety of the food supply is monitored by agencies at the international, federal, provincial, and municipal levels. The government promotes the use of HACCP (Hazard Analysis Critical Control Point) principles to ensure food safety. HACCP systems help prevent or eliminate food contamination and track contaminated foods to prevent foodborne illness.
- It is the job of the food manufacturer to establish and implement a HACCP system for their particular business.
- Consumers play an important role in limiting the risks of developing foodborne illness through the way they store, handle, and prepare food.

17.3 Pathogens in Food

- The pathogens that affect the food supply include bacteria, viruses, moulds, parasites, and prions. Some bacteria cause foodborne infection because they are able to grow in the gastrointestinal tract when ingested. Others produce toxins in food, and consumption of the toxin causes foodborne intoxication.
- Viruses do not grow on food, but when consumed in food, they can reproduce in human cells and cause foodborne illness.
- Moulds that grow on foods produce toxins that can harm consumers.
- Parasites include microscopic single-celled animals, as well as worms that can be seen with the naked eye. They are consumed in contaminated water or food.
- Improperly folded prion proteins cause bovine spongiform encephalopathy (BSE) in cattle. The risk of acquiring the human form of this deadly degenerative neurological disease is extremely low.
- The risk of foodborne illness can be decreased through proper food selection, preparation, and storage.

17.4 Chemical Contaminants in Food

- Contaminants such as pesticides applied to crops and industrial wastes that leach into water may find their way into the food supply. To decrease the potential risk of pesticides, safer ones are being developed and farmers are reducing the amounts applied by using integrated pest management and organic methods.
- Antibiotics are given to animals to prevent disease and increase growth. The widespread use of these drugs may be contributing to the emergence of antibiotic-resistant strains of bacteria.

- Industrial pollutants such as PCBs, radioactive substances, and toxic metals have contaminated some waterways and the fish that live in them. As these contaminants move through the food chain, their concentrations increase.
- Consumers can reduce the amounts of pesticides and other environmental contaminants in food by careful selection and handling of produce; selection of low-fat saltwater varieties of fish caught well offshore in unpolluted waters; and trimming fat from meat, poultry, and fish before cooking.

17.5 Food Technology

- High and low temperatures are used to prevent food spoilage. Cold temperatures slow or prevent microbial growth. High temperatures used in canning, pasteurization, sterilization, and cooking kill micro-organisms. However, cooking can also introduce hazardous substances such as polycyclic aromatic hydrocarbons, heterocyclic amines, and acrylamide.
- Irradiation preserves food by exposing it to x-rays, gamma radiation, or high-energy electrons. It kills micro-organisms, destroys insects, and slows the germination and ripening of fruits and vegetables.
- Packaging can help preserve food, but can introduce risk if it leaches into the food. Modified atmosphere packaging reduces the oxygen available for microbial growth.
- Food additives include all substances that can reasonably be expected to find their way into a food during processing.

REVIEW QUESTIONS

1. What is the major cause of foodborne illness?
2. List three factors that determine the likelihood that a contaminant will cause foodborne illness in an individual.
3. How is the federal government involved in ensuring a safe food supply?
4. Explain what HACCP is and how it can prevent or eliminate food contamination.
5. What is the difference between a foodborne infection and a foodborne intoxication?
6. List three common bacterial food contaminants. What can be done to avoid the foodborne illnesses caused by them?
7. What temperature range allows the most rapid bacterial growth?
8. Explain how cross-contamination can occur in home kitchens.
9. How do pesticides applied to crops find their way into animal products?
10. List some food-processing and packaging techniques that reduce foodborne illnesses.
11. What is food irradiation? Is it safe?
12. List four reasons for using food additives.

REFERENCES

1. Canadian Food Inspection Agency: Causes of Foodborne Illness. Available online at http://www.inspection.gc.ca/english/fssa/concen/causee.shtml. Accessed January 2, 2011.
2. Public Health Agency of Canada. National Microbiology Laboratory. Available online at http://www.nml-lnm.gc.ca/Pulsenet/index-eng.htm. Accessed January 2, 2011.
3. Canadian Food Inspection Agency. Data labelling on pre-packaged foods. Available online at http://www.inspection.gc.ca/english/fssa/concen/tipcon/date.shtml. Accessed January 22, 2011.
4. Canadian Food Inspection Agency. Report a Food Safety or Labelling Concern. Available online at http://www.inspection.gc.ca/english/fssa/concen/reporte.shtml. Accessed June 30, 2011.
5. U.S. Food and Drug Administration, Center for Food Safety and Nutrition. *Foodborne Pathogenic Microorganisms and Natural Toxins Handbook: The "Bad Bug Book."* Available online at www.foodsafety.gov/~mow/intro.html. Accessed August 1, 2008.
6. North Bay Parry Sound District Health Unit. Investigative Summary of the Escherichia Coli Outbreak Associated with a Restaurant in North Bay, Ontario, 2009. Available online at http://www.healthunit.biz/docs/Ecoli%20Outbreak/2008%20NBPSDHU%20Ecoli%20Report_June%202009_Formatted.pdf. Accessed June 26, 2011.
7. Hrudey, S. E. and Hrudey, E. T. Walkerton and North Battleford—key lessons for public health professionals. *Canadian Journal of Public Health* 93 (5): 332, 2002.
8. Samuel, M. C., Vugia, D. J., Shallow, S. et al. Epidemiology of sporadic *Campylobacter* infection in the United States and declining trend in incidence. FoodNet 1996–1999. *Clin. Infect. Dis.* 38 (Suppl. 3):S165–S174, 2004.
9. Centers for Disease Control and Prevention, Division of Foodborne, Bacterial, and Mycotic Disease. *Listeriosis*. Available online at www.cdc.gov/nczved/dfbmd/disease_listing/listeriosis_gi.html. Accessed August 6, 2008.
10. Maple Leaf Foods Blog. Food Safety Knowledge Exchange in Bogota. Columbia, 2010. Available online at http://blog.mapleleaf.com/. Accessed June 26, 2011.
11. Koepke, R., Sobel, J., and Arnon, S. S. Global occurrence of infant botulism, 1976–2006. *Pediatrics* 122:73–82, 2008.
12. Widdowson, M. A., Sulka, A., Bulens, S. N. et al. Norovirus and foodborne disease, United States, 1991–2000. *Emerg. Infect. Dis.* 11:95–102, 2005.
13. Cheng, P. K., Wong, D. K., Chung, T. W., et al. Norovirus contamination found in oysters worldwide. *J. Med. Virol.* 76:593–597, 2005.
14. Canadian Food Inspection Agency: Fact Sheet: Mycotoxins:Available online at http://www.inspection.gc.ca/english/anima/feebet/pol/mycoe.shtml. Accessed January 13, 2011.
15. Outbreaks of *Escherichia coli* O157:H7 infection and cryptosporidosis associated with drinking unpasteurized apple cider—Connecticut and New York, October 1996. *MMWR* 46:4–8, 1997.
16. Centers for Disease Control and Prevention. *Cryptospiridium* Infection. Available online at www.cdc.gov/ncidod/dpd/parasites/cryptosporidiosis/factsht_cryptosporidiosis.htm. Accessed August 1, 2008.
17. Centers for Disease Control and Prevention. *vCJD (Variant Creutzfeldt-Jakob Disease)*. Available online at www.cdc.gov/ncidod/dvrd/vcjd/. Accessed August 7, 2008.
18. U.S. Environmental Protection Agency. *Setting Tolerances for Pesticide Residues in Foods*. www.epa.gov/pesticides/factsheets/stprf.htm. Accessed August 1, 2008.

19. Health Canada. Pest Management Regulatory Agency. Available online at http://www.hc-sc.gc.ca/ahc-asc/branch-dirgen/pmra-arla/index-eng.php. Accessed January 13, 2011.

20. Health Canada. Consultation document on the use of safety and uncertainty factors in the human health risk assessment of pesticides—regulatory proposal-PRO2007-1, 2007. Available online at http://www.hc-sc.gc.ca/cps-spc/pest/part/consultations/_pro2007-01/index-eng.php#multi. Accessed January 22, 2011.

21. Health Canada. Pesticides and Foods. Available online at http://www.hc-sc.gc.ca/cps-spc/pubs/pest/_fact-fiche/pesticide-food-alim/index-eng.php. Accessed January 13, 2011.

22. CGSB. Standards for Organic Agriculture. Available online at http://www.tpsgc-pwgsc.gc.ca/cgsb/on_the_net/organic/index-e.html. Accessed January 13, 2011.

23. Health Canada. Setting standards for maximum residue limits (mrls) of veterinary drugs used in food-producing animals. Available online at http://www.hc-sc.gc.ca/dhp-mps/vet/mrl-lmr/mrl-lmr_levels-niveaux-eng.php. Accessed January 13, 2011.

24. CFIA. Report on pesticides, agricultural chemicals, veterinary drugs, environmental pollutants and other impurities in agri-food commodities of animal origin. Available online at http://www.inspection.gc.ca/english/fssa/microchem/resid/2004-2005/anima_e.shtml. Accessed January 13, 2011.

25. Pew Commission on Industrial Farm Animal Production. Final *Report: Putting Meat on The Table: Industrial Farm Animal Production in America.* Available online at www.ncifap.org/_images/PCIFAPFin.pdf. Accessed August 11, 2008.

26. U.S. Food and Drug Administration, Center for Veterinary Medicine. *Report on the Food and Drug Administration's Review of the Safety of Recombinant Bovine Somatotropin.* Available online at www.fda.gov/cvm/RBRPTFNL.htm. Accessed August 1, 2008.

27. Health Canada. Recombinant Bovine Somatotropin. Available online at http://www.hc-sc.gc.ca/dhp-mps/vet/issues-enjeux/rbst-stbr/index-eng.php. Accessed January 13, 2011.

28. Health Canada. Breastfeeding. http://www.hc-sc.gc.ca/fn-an/pubs/infant-nourrisson/nut_infant_nourrisson_term_3-eng.php. Accessed January 13, 2011.

29. U.S. Department of Health and Human Services and U.S. Environmental Protection Agency. *What You Need to Know about Mercury in Fish and Shellfish.* EPA-823-R-04–005, March 2004. Available online at www.cfsan.fda.gov/~dms/admehg3.html. Accessed August 1, 2008.

30. Health Canada Consumption Advice: Making informed choices about fish. Available online at http://www.hc-sc.gc.ca/fn-an/securit/chem-chim/environ/mercur/cons-adv-etud-eng.php. Accessed March 19, 2011.

31. Health Canada. Food Additives. Available online at http://www.hc-sc.gc.ca/fn-an/securit/addit/index-eng.php. Accessed January 13, 2011.

32. Exon, J. H. A review of the toxicology of acrylamide. *J. Toxicol. Environ. Health B. Crit. Rev.* 9:397–412, 2006.

33. Friedman, M., and Levin, C. E. Review of methods for the reduction of dietary content and toxicity of acrylamide. *J. Agric. Food Chem.* 56:6113–6140, 2008.

34. Osterholm, M. T., and Norgan, A. P. The role of irradiation in food safety. *N. Engl. J. Med.* 350:1898–1901, 2004.

35. U.S. Food and Drug Administration. *Food Irradiation: A Safe Measure*. January 2000. Available online at www.fda.gov/opacom/catalog/irradbro.html. Accessed July 31, 2008.

36. Health Canada. Frequently asked questions about food irradiation. Available online at http://www.hc-sc.gc.ca/fn-an/securit/irridation/faq_food_irradiation_aliment01-eng.php. Accessed January 13, 2011.

37. Health Canada. Consumer information: Safety of Plastic Containers commonly found in the home. Available online at http://www.chemicalsubstanceschimiques.gc.ca/fact-fait/plastic-plastique-eng.php. Accessed January 13, 2011.

38. CFIA. Reference listing of accepted construction materials, packaging materials and non-food chemical products. Available online at http://inspection.gc.ca/english/fssa/reference/refere.shtml. Accessed January 14, 2011.

39. Health Canada. Health Canada proposal to improve food colour labelling requirements. Available online at http://www.hc-sc.gc.ca/fn-an/consult/_feb2010-food-aliments-col/draft-ebauche-eng.php. Accessed January 13, 2011.

40. Ontario Ministry of Agriculture, Food & Rural Affairs. Food Inspection Branch. Manufacture of cured product ingredients. Available online at: http://www.omafra.gov.on.ca/english/food/inspection/meatinsp/manual/p9100401.htm. Accessed June 26, 2011.

41. Health Canada. Dictionary of Food Additives. Available online at http://www.hc-sc.gc.ca/fn-an/securit/addit/diction/index-eng.php. Accessed January 14, 2011.

(© istockphoto.com/kkgas)

FOCUS ON
Biotechnology

Focus Outline

In 1909, British physician Archibald Garrod hypothesized that genes might be involved in creating proteins. By 1966, investigators had deciphered the genetic code, which links the information in DNA to the synthesis of proteins. The proteins made affect the traits that an organism exhibits. Then, in 1972, a discussion over hot pastrami and corned beef sandwiches between Dr. Stanley Cohen of Stanford University and Dr. Herbert Boyer of the University of California at San Francisco led to the birth of genetic engineering. The collaboration that resulted brought together the information needed to allow genetic instructions from one organism to be inserted into another. Cohen's laboratory had been studying how bacterial cells take up small loops of DNA. Boyer had been studying enzymes that could cut DNA at specific locations and paste it back together again. Cohen and Boyer realized that fragments of DNA, produced by cutting DNA with Boyer's enzymes, could be pasted into the small loops of DNA and introduced into bacterial cells using the procedure developed in Cohen's lab (**Figure F7.1**). As the bacteria multiplied, so would the new piece of DNA—making copies, or clones. These techniques for recombining DNA from different sources and cloning it are the basis for all genetic engineering.[1]

Figure F7.1 Chimera
Boyer and Cohen called the small loops of bacterial DNA that contain DNA from another organism chimeras, after the mythical, fire-breathing beast with the heads of a lion and a goat, the body of a goat, and a serpent for a tail.

F7.1 How Does Biotechnology Work?

Learning Objectives

- Explain how genetic engineering introduces new traits into plants.
- Compare and contrast traditional breeding methods and modern biotechnology.

Modern **biotechnology** is possible because of the emergence of techniques that allow scientists to alter the DNA of plants, animals, yeast, and bacteria. By modifying DNA, these techniques, referred to as **genetic engineering**, allow scientists to change the proteins that a cell or organism can make, introducing new traits, enhancing desirable ones, or eliminating undesirable ones. Genetic engineering often involves taking the gene for a desired trait from one organism and transferring it to another. It has allowed researchers to create bacteria that make medicines for humans, plants that are disease-resistant, and foods that provide a healthier mix of nutrients.

biotechnology The process of manipulating life forms via genetic engineering in order to provide desirable products for human use.

genetic engineering A set of techniques used to manipulate DNA for the purpose of changing the characteristics of an organism or creating a new product.

As this new technology has evolved, so has the vocabulary to describe it (**Table F7.1**). Crops and other organisms and food products produced using these techniques are often referred to as *genetically modified* or *GM*.

Table F7.1 Terms Used in Genetic Engineering

Biotechnology	Manipulating life forms via genetic engineering to provide desirable products for human use.
Bioengineered foods	Foods that have been produced using biotechnology.
Chimera	A DNA molecule composed of DNA from 2 different species, or an organism consisting of tissues of diverse genetic constitution.
Cloning	Producing an exact duplicate of a gene or an organism.
DNA	A long, thread-like molecule that carries the genetic information of an organism.
Gene	A unit of DNA that provides genetic information coding for a trait. It is the physical basis for the transmission of the characteristics of living organisms from one generation to another.
Gene splicing	The precise joining of DNA from different sources to create a new gene structure.
Genetic engineering	Technology used to selectively alter genes. It can be used to manipulate the genetic material of an organism in such a way as to allow it to produce new and different types of proteins. Other terms applicable to the same techniques are gene splicing, gene manipulation, genetic modification, or recombinant DNA technology.
Genome	The total complement of genetic information in an organism.
GM crops	Genetically modified crops.
GMO	Genetically modified organism.
Hybridization	The mating of different plants to enhance favourable characteristics.
Plasmid	An independent, stable, self-replicating piece of circular DNA in bacterial cells. It is not a part of the normal cell genome.
Recombinant DNA	DNA that has been formed by joining the DNA from different sources.
Transgenic	An organism with a gene or group of genes intentionally transferred from another species or breed.

Genetics: From Genes to Traits

The characteristics of a plant or animal are carried in its genes. These genes, which are segments of DNA, are passed from generation to generation. Genes contain the information that directs the synthesis of proteins. The specific proteins that are made then determine the traits that an individual organism displays.

DNA Structure DNA is a long, thread-like molecule consisting of 2 strands that twist around each other, forming a double helix. Each strand has a backbone made up of alternating units of the 5-carbon sugar deoxyribose and phosphate groups. Each deoxyribose sugar is attached to a molecule called a base. The 4 different bases that occur in DNA—adenine, thymine, cytosine, and guanine—are usually abbreviated as A, T, C, and G, respectively. The bases on adjacent strands bind to one another, connecting the two strands. Each base binds only to its complementary base: adenine to thymine and cytosine to guanine (**Figure F7.2**).

The DNA of all organisms—plants, bacteria, and animals, including humans—is made up of the same DNA bases. The differences in the sequence of bases in DNA are responsible for all of the genetic differences between living things, whether they are differences between species or differences between individuals of the same species. Different individuals of the same species have small differences in the base sequence of their DNA; only identical twins share the same base sequence. Organisms of different species, such as humans and corn plants, have larger differences in the sequences of DNA bases.

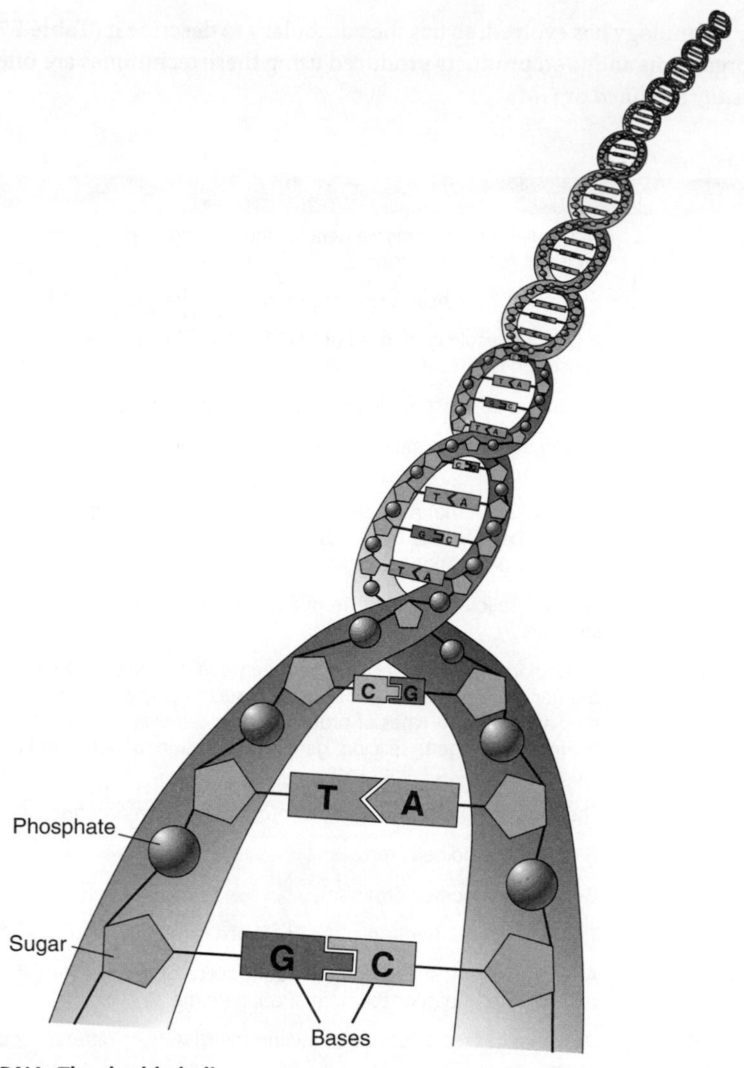

Figure F7.2 DNA: The double helix
DNA is a double-stranded, helical molecule. Each strand has a backbone made up of phosphate groups and sugar molecules. Each sugar is attached to adenine (A), thymine (T), cytosine (C), or guanine (G). The 2 strands of the DNA molecule are bonded together by these bases; A bonds to T, and C bonds to G.

Genes to Proteins to Traits The sequence of bases present in a gene specifies the sequence of amino acids that will be present in a protein (see Section 6.4). In the genetic code, 3 bases code for a single amino acid. The same 3 bases will always code for the same amino acid; for example, the DNA base sequence ACC corresponds to the amino acid tryptophan. Because the code is universal among all life on earth, a particular 3-base sequence will code for the same amino acid whether it is in a bacterium, a mosquito, a corn plant, or a human being.

When a gene is expressed, the protein it codes for is made, providing certain characteristics to the organism. For example, if a protein that stimulates growth is made, the organism will grow bigger, and if a protein pigment is made, it will affect the colour of the plant or animal. The presence or absence of specific proteins determines the traits that an individual organism displays (**Figure F7.3**).

Even though organisms of different species are very different, they may have genes with similar base sequences if they both need to make the same protein. For example, humans and pigs both rely on the protein hemoglobin to carry oxygen in the blood, so both humans and pigs have a gene with a very similar sequence of bases that codes for the protein hemoglobin. It is estimated that 25% of the genes found in plants are also present in humans, presumably because they code for proteins needed by both organisms.[2]

Passing Traits from Parent to Offspring When an organism reproduces, it passes its genes—and thus the instructions to make the proteins coded for by these genes—to the next

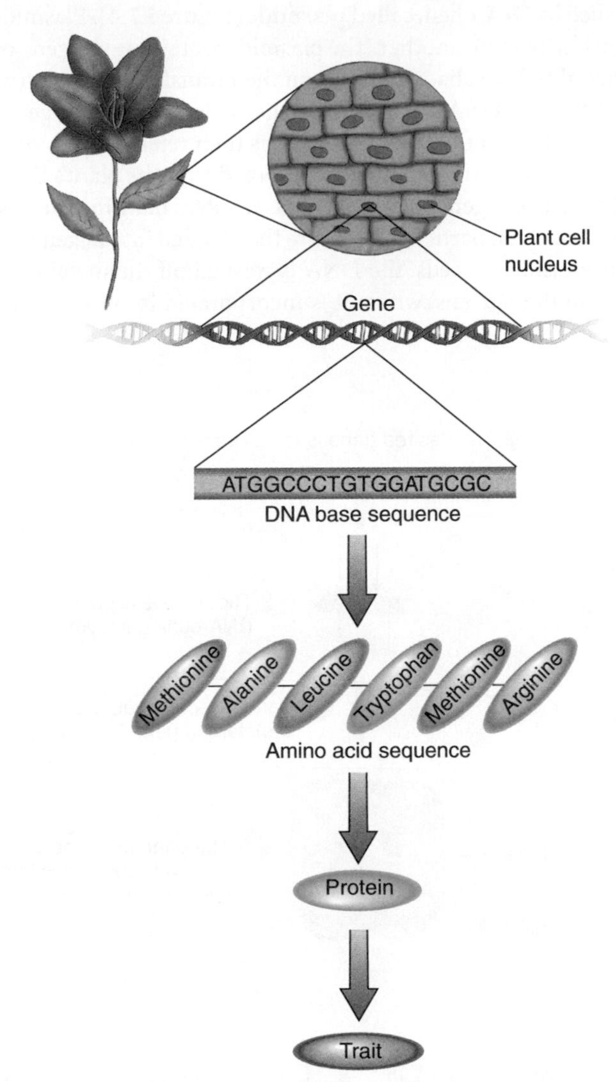

Figure F7.3 Relationships among genes, proteins, and traits
The DNA base sequence in a gene codes for the sequence of amino acids that are joined to form a protein. Which proteins are made determines an organism's traits.

generation. When 2 organisms breed, some genes from each are passed to the offspring. The result is a new combination of genes and the traits for which they code. Over time, mutations can occur in the sequence of bases that make up the genes. Some of these mutations are harmful, reducing the organism's ability to survive and reproduce. Because the organism dies or cannot reproduce, these harmful mutations are not passed on to the next generation. Other mutations are beneficial and result in traits that allow the organism to survive more easily and thus to reproduce and pass on the altered genes. Over millions of years, many genes have been changed by mutations, and those that code for beneficial traits have been passed on because the organisms carrying them thrive and reproduce. The idea behind biotechnology is to speed up the process of introducing beneficial traits that can then be passed from generation to generation.

Methods of Biotechnology

The first step in genetic modification is to identify a stretch of DNA, or gene, for a desired trait, such as resistance to a particular disease. This gene of interest could be from a plant, animal, or bacterial cell. The gene is then clipped out with the specific DNA-cutting enzymes studied by Dr. Boyer, which are called **restriction enzymes** (see Chapter 6: Science Applied: Discovering How to Manipulate Genes).

restriction enzyme A bacterial enzyme used in genetic engineering that has the ability to cut DNA in a specific location.

A number of different techniques can be used to introduce the gene into the host cell. If the host cell is a plant cell, the gene can be pasted into, or recombined with, the small loops of

plasmid A loop of bacterial DNA that is independent of the bacterial chromosome.

bacterial DNA studied by Dr. Cohen, called **plasmids** (**Figure F7.4**). Plasmids have the ability to carry genes from one place to another. The plasmid, containing the gene of interest, can be taken up by a bacterial cell. The bacterial cell can then transfer the gene to a plant cell. Once inside the plant cell, the new DNA can migrate to the nucleus, where the gene for the new trait can be integrated into the plant's DNA. The DNA is then referred to as **recombinant DNA** because the DNA from the plasmid has been combined with the plant's DNA (**Figure F7.5**). A second method used to get genes into plant cells involves painting the desired segment of DNA onto microscopic metal particles. These are then loaded into a "gene gun" and shot into the plant cells. Once inside the cells, the DNA is washed off the metal particles by cellular fluids and migrates to the nucleus, where it is incorporated into the plant's DNA, forming recombinant DNA.

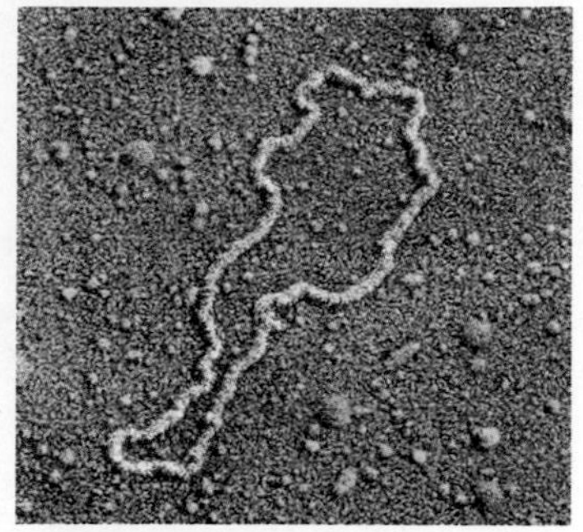

Figure F7.4 This electron micrograph shows a plasmid, which is a small loop of DNA found in bacterial cells. (Stanley N. Cohen/ Photo Researchers)

recombinant DNA DNA that has been formed by joining DNA from different sources.

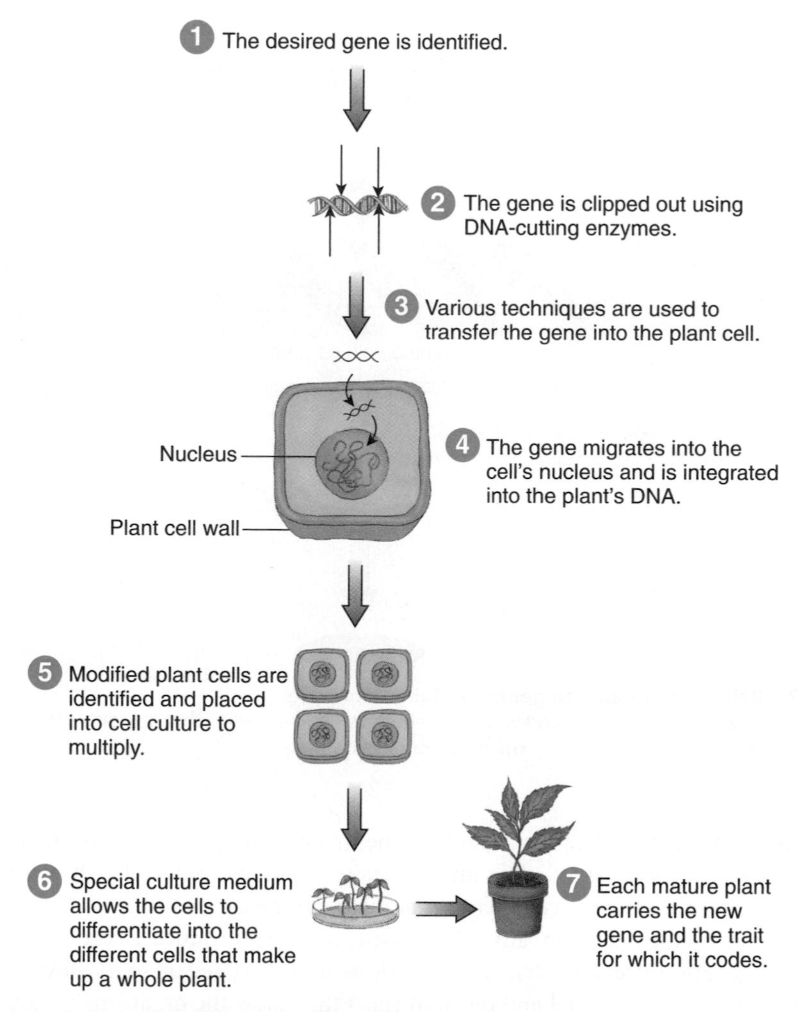

Figure F7.5 Engineering genetically modified plants
Crops developed by using the genetic engineering steps shown here are grown all over the world.

The modified plant cells produced by either technique are then allowed to multiply. As they do, the new gene is reproduced with them. The cells are then placed in a special culture medium that allows them to differentiate into the different types of cells that make up a whole plant. The new plant is a **transgenic** organism. Each cell in the new plant contains the transferred gene for the desired trait, such as disease resistance (see Figure F7.5).

transgenic An organism with a gene or group of genes intentionally transferred from another species or breed.

Genetic engineering is more difficult in animals because animal cells do not take up genes as easily as plant cells do, and making copies of these cells (clones) is also more difficult. However, these techniques have been used to produce cows that yield more milk, cattle and pigs with more meat on them, and sheep that grow more wool.[3]

Is Genetic Modification Really New?

Biotechnology methods are new, but humans have been directing the genetic modification of plants and animals for about 10,000 years. Farmers thousands of years ago didn't use gene guns or bacterial plasmids, but they selected seeds from plants with the most desirable characteristics to plant for the next year's crop, bred the animals that grew fastest or produced the most milk to improve the productivity of the next generation of animals, and cross-bred plant varieties to combine the desired traits of each. Almost every fruit, vegetable, or crop grown today has been in some way genetically modified using traditional **selective breeding** techniques. Some of these crops, such as pumpkins, potatoes, sugar beets, and varieties of corn, oats, and rice, would not have developed without human intervention. This intervention has allowed modern farmers to grow plants that produce more food that is more nutritious and can better withstand harsh environments and resist disease.

selective breeding Techniques to selectively control reproduction in plants and animals for the purpose of producing organisms that better serve human needs.

Traditional Breeding Technology Traditional breeding begins when farmers and ranchers select plants or animals with desired characteristics, such as high yield, palatability, resistance to disease and insects, or aesthetic characteristics. The traits can then be brought together by controlled mating. Plant breeders use a process called **hybridization**, in which 2 related plants are cross-pollinated or cross-fertilized (**Figure F7.6**). The resulting offspring has characteristics from both parent plants. For example, a breeder who wanted to produce a variety of wheat that was high-yielding and was resistant to cool temperatures would cross plants with these traits. The breeder would then select the offspring that acquired both of the desired traits.

hybridization The process of cross-fertilizing 2 related plants with the goal of producing an offspring that has the desirable characteristics of both parent plants.

Cross-breeding of animals has a similar goal. Animal breeders use both inbreeding and outbreeding. Inbreeding involves crosses between closely related animals. It can intensify desirable traits but may also intensify undesirable traits. Outbreeding crosses unrelated animals to reduce undesirable traits, increase variability, and introduce new traits. Today, inbreeding and outbreeding are carried out by artificial insemination.

Figure F7.6 Cross-pollination or cross-fertilization, as illustrated with these oat seedlings, is a traditional method for breeding new plant varieties. (David Woodfall/Stone/Getty Images, Inc.)

Advantages of Modern Biotechnology Traditional breeding techniques work well but they have limitations in terms of both time and outcome. Breeding generations of plants or animals to consistently produce a desired trait is time-consuming; a new trait can only be produced once in the reproductive cycle of the plant or animal. Only plants or animals of the same or closely related species can be interbred, and not every offspring will inherit the desirable traits. Because cross-breeding transfers a set of genes from the parents to the offspring, both desirable and undesirable traits are transferred. Eliminating the undesirable genes while keeping the desirable ones can require many crosses.

Biotechnology, which selects specific genes in the laboratory, has significantly sped up this process and removed certain limitations. It enables breeders to select, modify, and transfer single genes. This speeds up the process by reducing the time and cost of breeding the crosses that are needed to select out the undesirable traits (**Figure F7.7**). In addition, biotechnology is not limited by whether 2 animals are capable of cross-breeding but, rather, can select desirable genes from any species, because the way DNA codes for proteins in plants, animals, yeast, and bacteria is the same. This allows traits from different species and completely different organisms to be used.

F7.2 Applications of Modern Biotechnology

Learning Objectives

- List some ways in which genetic engineering is being used to enhance the food supply.
- Describe how genetic engineering can be used to combat human disease.

The techniques of biotechnology can be used in a variety of ways in both production and processing to alter the quantity, quality, cost, safety, and shelf life of the food supply (**Table F7.2**). This technology also has great potential for addressing the problem of world hunger and malnutrition.

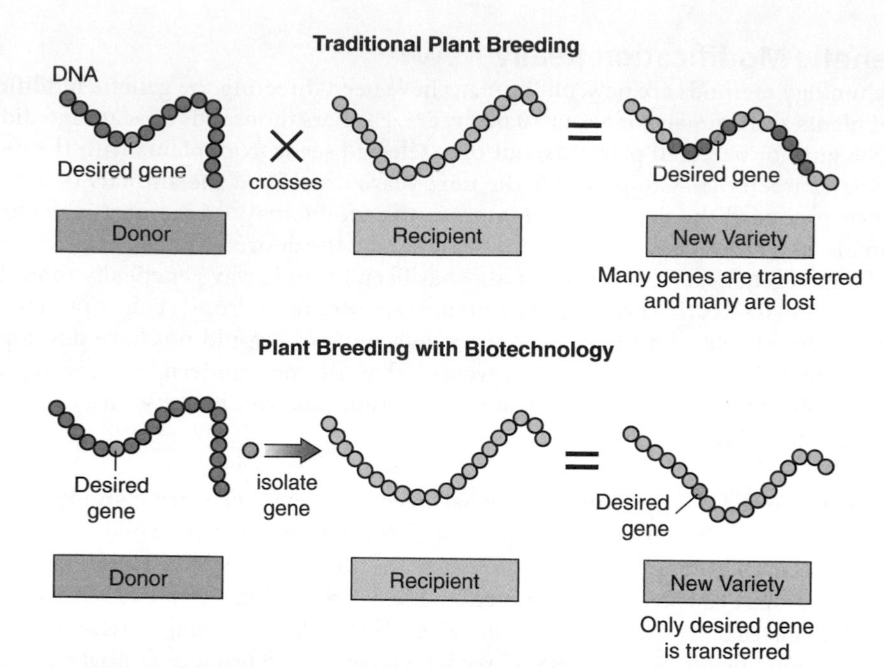

Figure F7.7 Traditional plant breeding versus biotechnology
(Top) When traditional genetic modification methods are used, thousands of genes are mixed. Not all the new varieties contain the desired gene and it may require many attempts over many years to remove the unwanted traits. (Bottom) Genetic engineering is more precise, more predictable, and faster. It allows the insertion of 1 or 2 genes into a plant without the transfer of genes coding for undesirable traits.

Table F7.2 Characteristics Introduced Using Genetic Engineering

Food	Characteristic
Cherry tomato	Better taste, colour, texture
Flavr Savr tomato	Delayed ripening
Tomato	Thicker skin, altered pectin content
Corn	Insect protection, herbicide resistance
Cotton	Insect protection, herbicide resistance
Squash	Virus resistance
Papaya	Virus resistance
Potatoes	Potato beetle resistance, virus resistance
Soybeans	Herbicide resistance, high oleic content to reduce need for hydrogenation
Sugar beets	Herbicide resistance
Sunflower	High oleic acid content to reduce need for hydrogenation

Increasing Crop Yields

Genetic modification of crops can increase yields either directly, by inserting genes that improve the efficiency with which plants convert sunlight into food, or indirectly, by creating plants that are resistant to herbicides, pesticides, and plant diseases, thus reducing crop losses. Scientists are also working to develop plants that can withstand drought, freezing, and high salt concentrations. Many attempts in the last century to increase crop yields in developing countries have failed because they required expensive machinery and chemicals. With genetically engineered crops, simply providing a new type of seed has the potential to increase food production.

Herbicide Resistance Herbicides are chemicals that are sprayed on crops in the field to control weeds. Effective herbicides kill weeds but do not harm crops. A crop such as soybeans is not harmed by a particular herbicide because it contains enzymes that inactivate the herbicide. These enzymes are crop specific so only certain herbicides can be used with certain crops. In some cases, the weeds and the crops are resistant to the same herbicide so the farmer has few or no choices for weed control. Genetic engineering allows researchers to transfer genes to plants that make them resistant to specific herbicides. Farmers then have more options for weed control as they can use these specific herbicides that are more effective and less environmentally damaging.

Insect Resistance Genetic engineering techniques can increase insect resistance in plants. For example, a gene from the bacterium *Bacillus thuringiensis* produces a protein that is toxic to certain insects but safe for humans and other animals (**Figure F7.8**). By inserting the gene for this protein into plant cells, scientists have created plants that manufacture their own insecticide. The protein produced by this gene, known as Bt, has been used as an insecticide for more than 30 years. When plants manufacture their own insecticide, Bt does not need to be sprayed on the plants, which saves the farmer money, fuel, and time. It also benefits the environment because the Bt affects only insects feeding on the crop of interest and is not spread to surrounding foliage. Corn, potatoes, and cotton have been genetically modified to produce the Bt protein. During the first 3 years of commercial availability, cotton with the Bt gene reduced chemical insecticide use by 900 metric tonnes.

Figure F7.8 Genetically engineered corn that produces the Bt protein is toxic to this European corn borer. (Courtesy Marlin E. Rice, Iowa State University)

In addition to creating insect-resistant plants, genetic engineering has been used to create environmentally friendly pesticides. These pesticides are produced by bacteria, which are killed and then sprayed on plants.

Disease Resistance Potato, squash, cucumber, watermelon, and papaya have been modified to resist viral infections. The benefits of this technology are illustrated by how it helped to preserve Hawaii's second largest crop. In the mid-1990s, the papaya ring-spot virus threatened to wipe out Hawaii's papaya crop. Traditional plant breeding was not able to produce a virus-resistant strain of papaya, but by inserting a gene that acted like a vaccine into the papaya plant DNA, scientists were able to produce papaya plants that were immune to the virus (**Figure F7.9**).[4]

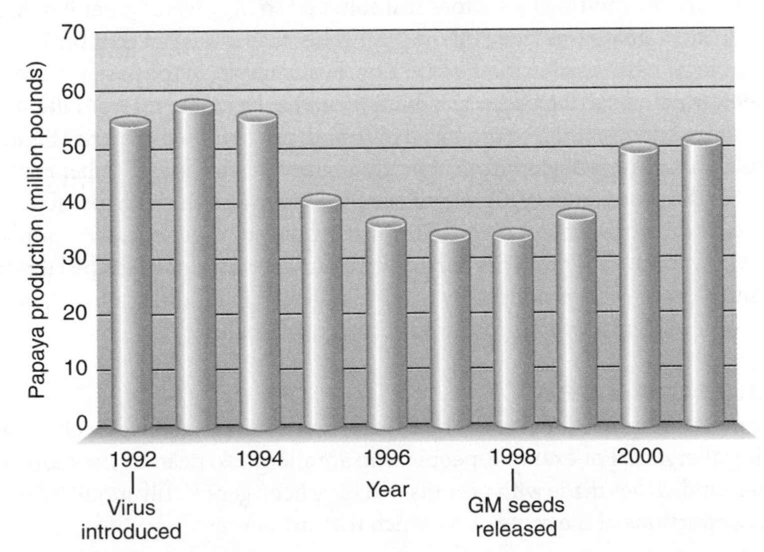

Figure F7.9 Virus-resistant papaya
Seeds for papaya that are resistant to the papaya ring-spot virus were released for commercialization in 1998, allowing Hawaiian papaya production to rebound. This was the first genetically enhanced fruit crop on the market.

Improving Nutritional Quality

Although the causes of malnutrition are rooted in political, economic, and cultural issues that cannot be resolved by agricultural technology alone, genetically modified crops that target some of the major nutritional deficiencies worldwide are being developed. To address protein deficiency, varieties of corn, soybeans, and sweet potatoes with higher levels of essential amino

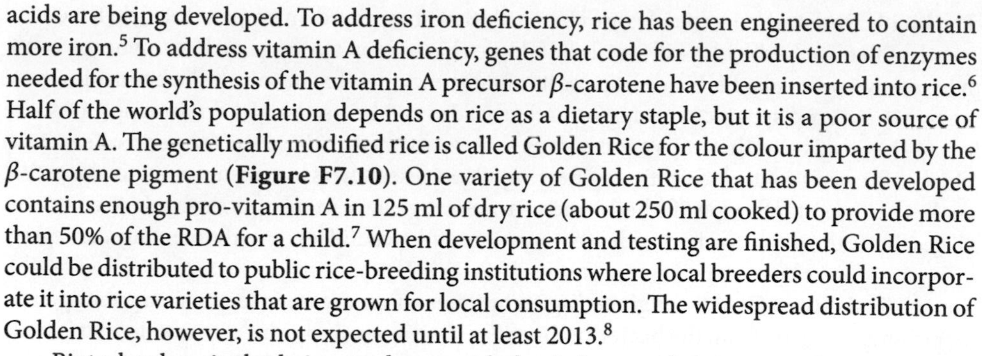

acids are being developed. To address iron deficiency, rice has been engineered to contain more iron.[5] To address vitamin A deficiency, genes that code for the production of enzymes needed for the synthesis of the vitamin A precursor β-carotene have been inserted into rice.[6] Half of the world's population depends on rice as a dietary staple, but it is a poor source of vitamin A. The genetically modified rice is called Golden Rice for the colour imparted by the β-carotene pigment (**Figure F7.10**). One variety of Golden Rice that has been developed contains enough pro-vitamin A in 125 ml of dry rice (about 250 ml cooked) to provide more than 50% of the RDA for a child.[7] When development and testing are finished, Golden Rice could be distributed to public rice-breeding institutions where local breeders could incorporate it into rice varieties that are grown for local consumption. The widespread distribution of Golden Rice, however, is not expected until at least 2013.[8]

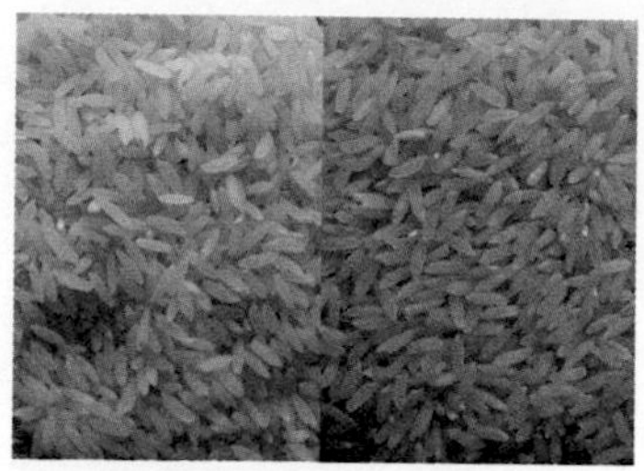

Figure F7.10 The rice on the (right), called Golden Rice, has been genetically modified to produce the vitamin A precursor β-[carotene, which gives it a yellow-orange colour. The white rice on the left does not contain this genetic modification. (Courtesy Golden Rice Humanitarian Board www.goldenrice.org)

Biotechnology is also being used to provide foods that may help prevent chronic disease. For example, genetic modification has produced a variety of soybeans that has a higher percentage of monounsaturated fat—a type of fat that may help reduce the risk of cardiovascular disease. Vegetable oils that have lower amounts of saturated fat are also being developed, as are soybeans that contain more of the antioxidant vitamin E than the traditional varieties.[9]

Other qualities that affect a food's role in the diet can also be changed by genetic engineering. For example, potatoes that are denser and have less water have been developed because they absorb less oil when fried—lowering the fat content of an order of french fries.

A transgenic animal is an animal that has been genetically modified by the insertion into its genome of a gene from another species. One reason for doing this is to improve the nutritional quality of the meat that comes from this animal. Although no such meat is currently commercially available, a transgenic pig has been developed that has high levels of omega-3 fatty acids, raising the possibility of high-omega-3 pork.[10]

Advancing Food Quality and Processing

Genetic engineering is also being used to improve the appeal and quality of food. The first genetically modified whole food available on the market was the Flavr Savr tomato, introduced in 1994. In this case, rather than adding a gene, an existing gene for an enzyme that controls ripening was inactivated to slow the ripening process and prolong the tomato's shelf life. Unfortunately, the modified tomatoes still softened so they were difficult to harvest and ship. The Flavr Savr tomato was taken off the market because it was not economically feasible.

Biotechnology is also used in food processing. For example, in the past, the enzyme preparation rennet, which is used in cheese production, had to be extracted from the stomachs of slaughtered calves, but now it can be produced by genetically modified bacteria (**Figure F7.11**). About 60% of hard cheese is made with genetically engineered enzymes.[11] Other enzymes, such as those used in the production of high-fructose corn syrup and the enzyme lactase, which is used to reduce the lactose content of milk, are also produced by genetically modified microorganisms. Many food colour and flavour additives are also produced in the laboratory. For example, vanilla can now be produced by plant cells grown in culture, rather than harvested from vanilla orchid plants, which grow only in tropical climates.

Figure F7.11 During cheese production, an enzyme preparation known as rennet is added to clot the milk. Much of the hard cheese produced today relies on rennet produced by genetically modified bacteria. (Rosenfeld Images Ltd./ Photo Researchers)

Enhancing Food Safety

Biotechnology can improve food safety by engineering foods to reduce or eliminate naturally occurring allergens. For example, people who are allergic to peanuts may someday enjoy peanut butter sandwiches made with peanuts that have been genetically modified to eliminate the proteins or portions of the proteins to which they are allergic.

Improving Animal and Seafood Production

Genetic engineering is being used to improve methods of preventing, diagnosing, and treating animal disease as well as to enhance growth efficiency and fertility in food animals. Although not used in Canada (see Section 17.4), recombinant bovine somatotropin use in the United States is an early example of the use of biotechnology in animal production. To produce recombinant bST, scientists isolated the gene from cow cells and then inserted it into bacterial cells. The bST, which is produced in large amounts by the bacterial cells, was then isolated, purified, and injected into dairy cows to increase milk production.

aquaculture The controlled cultivation and harvest of aquatic plants or animals.

Biotechnology is also being applied to **aquaculture**, which is the fastest growing sector of animal production worldwide. There is concern about the environmental impact of raising large numbers of fish in a restricted area and the safety of consuming this fish. Today, most of the salmon consumed is farm-raised. Because farm-raised salmon are fed pellets made of concentrated fish products, they are also fed the toxins that were present in the bodies of other fish. A comparison between wild salmon and farm-raised salmon found that the farm-raised salmon had higher levels of PCB (polychlorinated biphenyl), dioxin, toxaphene, and dieldrin—all of these are suspected to cause cancer in humans.[12] There is also concern that parasites and sea lice have spread from farmed salmon to wild salmon, causing a decline in the numbers of wild salmon. This observation has resulted in a call for changes in aquaculture practices to reduce the contact between farmed and wild fish.[13] Despite these problems, there is also optimism that aquaculture will help reduce the world's dependence on wild stocks of fish. Biotechnology allows scientists to identify and combine traits in fish and shellfish to increase productivity and improve quality. Scientists at Memorial University began research in the early 1990s that led to the development of an Atlantic salmon with an enhanced growth rate. This fish, which as of 2011 is not yet in commercial use, grows to market weight in about 18 months, compared to 24–30 months for unmodified fish.[14]

CANADIAN CONTENT

The EnviroPig™ is a transgenic animal developed at the University of Guelph. This animal has been modified to produce an enzyme that promotes the absorption of the essential nutrient, phosphorus. Pigs are normally inefficient in their absorption of phosphorus, which is excreted in urine and eliminated in feces. Because pig urine and feces is spread on the soil, phosphorus can become an environmental pollutant. Depending on the age and diet of the animal, the waste of the EnviroPig™ contains from 30%-70% less phosphorus than a normal pig. Research on the development and safety of this animal continues. It is not yet commercially available (as of 2011), but it has the potential to reduce the environmental impact of raising pigs.[15]

Combating Human Disease

One of the ways that biotechnology is impacting human health and the treatment of disease is through the production of genetically engineered medicines. In 1978, bacteria were engineered to produce human insulin; before this, diabetics relied on insulin extracted from pigs or cows. Other engineered proteins used to treat human disease include tissue plasminogen activator to dissolve blood clots in heart attack victims, growth factors to stimulate cell replication in bone marrow transplants, hepatitis B vaccine, and interferon to attack viruses and stimulate the immune system.

F7.3 Safety and Regulation of Genetically Modified Foods

Learning Objectives

- Explain how genetic engineering might create foods that harm consumers.
- Discuss the potential impact of genetic engineering on the environment.
- Describe how genetically modified food and crops are regulated.

The use of biotechnology is expanding rapidly, producing a variety of different products (**Figure F7.12**). Despite the potential benefits of these products, there is concern that this technology may create health problems and cause environmental damage. A number of regulations are currently in place to protect consumers and the environment. Researchers have concluded that the risks posed by agricultural products produced by modern biotechnology are the same as those for products produced by traditional plant breeding.[16] Despite this, many consumers and scientists alike believe that the conclusions regarding the health and environmental effects of these relatively new products are premature and that the impact of this booming technology has not yet become apparent.[17] They urge that this technology be used with caution to avoid introducing health or environmental risks that outweigh the benefits.

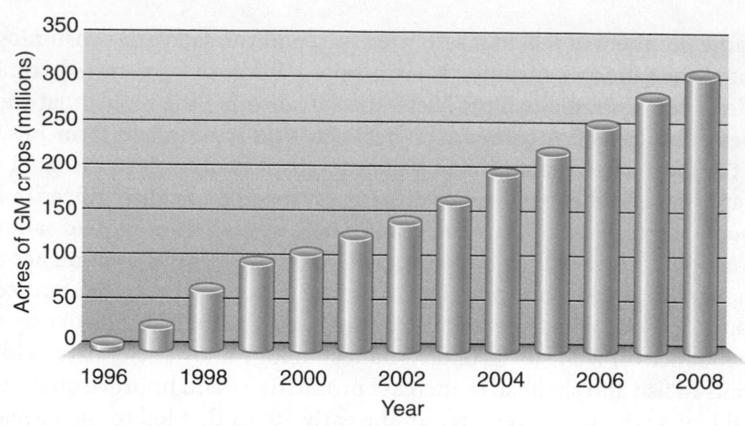

Figure F7.12 Worldwide growth of GM crops, 1996–2008
Despite concerns about the impact of GM crops, the number of acres planted with them has risen steadily. The most common GM crops are soybeans, corn, cotton, and rapeseed or canola oil.

Consumer Safety

Consumer safety concerns related to bioengineered foods include the possibility that an allergen or toxin may have inadvertently been introduced into a previously safe food, or that the nutrient content of a food has been negatively affected. Although foods containing ingredients derived from plant biotechnology are not generally required to carry special labels, those that contain potentially harmful allergens or toxins, or that are altered nutritionally, must carry labels (see Label Literacy: Should Genetically Modified Foods Carry Special Labels?). Another potential health concern is that the antibiotic resistance genes used in biotechnology may promote the development of antibiotic-resistant strains of bacteria.

Allergens and Toxins Genes code for proteins, so when a new gene is introduced into a food product, a new protein is made. These new proteins may in some cases be allergens or toxins. If a food contains a new protein that is an allergen, allergic individuals will now react to this food, which may previously have been safe for them. For example, if DNA from fish or peanuts—foods that commonly cause allergic reactions—were introduced into tomatoes or corn, these foods might be dangerous to allergic individuals. To prevent this from happening unintentionally, biotechnology companies have established systems for monitoring the allergenic potential of proteins used for plant genetic engineering.[18] Testing for the transfer of allergens has already proved to be valuable. In 1996, allergy testing successfully prevented soybeans containing a gene from a Brazil nut from entering the market.[19] If these soybeans had entered the food supply, they could have caused allergic reactions in people allergic to nuts. However, despite mandatory testing programs, individuals with food allergies cannot assume that new foods are safe.

Likewise, when a product is created using either a donor or recipient organism that is known to produce a toxin, the manufacturer must verify that the resulting product does not have high levels of the toxin. Toxins are also an issue when plants are produced by traditional breeding; toxic varieties of celery and potatoes have resulted from traditional breeding methods.

Changes in Nutrient Content Changing the proteins made by a plant or animal could also affect the nutrient content of foods. For example, tomatoes are an excellent source of vitamin C. If tomatoes were modified to have no vitamin C, people who rely on tomatoes for this vitamin might no longer get enough in their diets. As with foods containing potentially harmful allergens or toxins, foods with altered nutrient content must be labelled to disclose this information.

Antibiotic Resistance There is a concern that the use of genetic engineering will spread antibiotic resistance traits to bacteria in the environment or in the gastrointestinal tract. If pathogenic bacteria were to acquire this trait, it would make some of the antibiotics used to treat disease ineffective. The reason for this concern is that genetic engineering techniques use genes that code for antibiotic resistance as marker genes. By inserting a marker gene along with the gene they want to transfer, scientists are able to check to see that the gene transfer

LABEL LITERACY

CANADIAN CONTENT

Should Genetically Modified Foods Carry Special Labels?

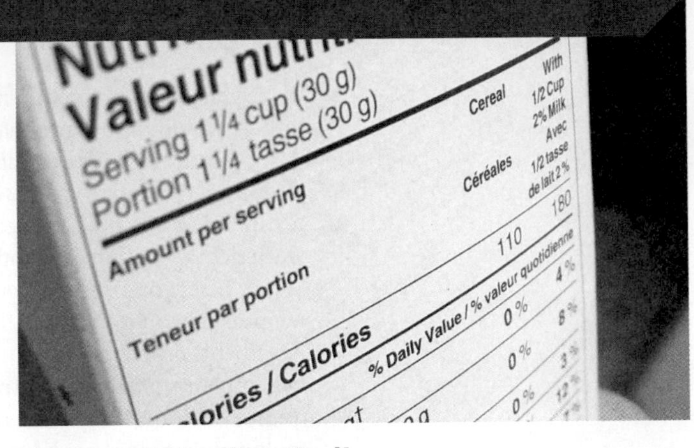

Can you identify foods containing ingredients created using genetic engineering? The answer is usually no because there is no mandatory labelling of genetically modified foods. In Canada only voluntary labelling exists.[1]

Some people feel that all foods containing GM ingredients should be labelled as such. Proponents of mandatory labelling argue that consumers have a right to this information and that it will help to ensure that the regulatory measures set up to protect consumer and environmental safety are working. Opponents say such labelling would be misleading, since even foods that are not different in quality, nutrient composition, and safety would be viewed as somehow different from the traditional foods.

Ethical, Philosophical, and Religious Issues

Can a tomato that contains some DNA from a fish be included in a vegan diet? Is corn that contains DNA from a pig appropriate for Jews and Muslims to eat? Regulators currently believe that the answer to these questions is yes, since plants and animals already share some of the same DNA segments. Jewish organizations agree and say that a single gene or even several genes transferred from an animal or shellfish source create no automatic conflict with a kosher diet. Despite this, some have suggested special labelling of GM products for kosher and vegetarian dietary requirements.

Economic Issues

Labelling increases costs at many levels. Hundreds, if not thousands, of food products contain GM ingredients and would therefore need to be labelled if regulations change. DNA-modified plants would need to be segregated during planting and harvesting, and products containing ingredients from them would require different manufacturing, transport, and storage facilities. Despite opposition, however, the need to sell to markets such as the European Union that do not want GM crops is already forcing Canadian farmers to separate and label GM crops.

Current Labelling Policy

When considering the safety and labelling of foods, there has traditionally been more focus on the characteristics of a food, not on how it is produced. Virtually all plants have been genetically modified through traditional plant breeding, and the Canadian Food Inspection Agency does not require declarations regarding those modifications. Under this precedent, labelling decisions are based on whether there are differences between the new food and the traditional food. Labelling of foods containing GM ingredients is therefore required only if the nutritional composition of the food has been altered; if it contains potentially harmful allergens, toxins, pesticides, or herbicides; if it contains ingredients that are new to the food supply; or if the food has been changed significantly enough that its traditional name no longer applies.

What Is an Appropriate Label?

Whether the label is voluntary or mandatory, designing an appropriate label is a complex issue. For example the label "Grown from genetically modified seed" may imply to some people that the product is inherently better than other products, while other people will feel this means that the food is more dangerous than others. However, a more detailed label that describes the way that a particular product was modified would be too lengthy and would not be understood by many consumers.

The Bottom Line

Food labelling allows customers to know what is in their food. They can then use this information to decide whether or not to choose the food. The question remains, are GM ingredients different enough from ingredients produced without genetic engineering to warrant the economic and logistical costs involved in providing this information on food labels?

[1]Canadian General Standards Board. Canadian General Standards Board Voluntary Labelling and Advertising of Foods That Are and Are Not Products of Genetic Engineering,. http://www.tpsgc-pwgsc.gc.ca/ongc-cgsb/internet/032-0315/fiches-facts/fs04-eng.html. Accessed January 14, 2011.

---

was successful. Antibiotic-resistant genes are used for this purpose because it is easy to verify that the transfer has occurred by exposing the bacteria to antibiotics—those that survive the antibiotic contain the transferred genes. Current techniques are unlikely to cause problems because the marker genes used are already widespread in the normal bacteria that inhabit the gut and in harmless environmental bacteria. In addition, markers are not used if they confer resistance to clinically important antibiotics. Newer techniques are under development that will remove the antibiotic resistance markers before the plants leave the laboratory.[20]

Environmental Concerns

Some of the arguments against the use of genetically modified crops are that they will reduce biodiversity, promote the evolution of pesticide-resistant insects, or create "superweeds" that will overgrow our agricultural and forest lands.

Biodiversity The ability of populations of organisms to adapt to new conditions, diseases, or other hazards depends on the presence of many different species that provide a diversity of genes. When there is this biodiversity, a harmful event, such as the emergence of a new disease strain, does not eliminate all the plants—it might kill one species, but others would have genes that protect them. Likewise, a drought might be harmful to some species, while others would have genes that make them drought-resistant.

A concern with biotechnology is that farmers will prefer new, resistant, high-yielding crop varieties and stop planting other varieties, causing certain varieties to eventually become extinct. For example, if every farmer began to use only genetically modified rice, then the varieties of rice that carry other traits would become extinct and those traits would be lost. If a new insect or virus emerged that killed the genetically modified rice, breeders and genetic engineers would not have the other varieties available to search for a gene that allows survival. This problem is not unique to genetically engineered crops. Biodiversity is also reduced when plants produced by traditional plant breeding are used to the exclusion of others. To preserve biodiversity and ensure a large supply of traits for use in future breeding, biotechnology techniques are being applied to establish gene banks and seed banks and to identify and characterize the genes in many species.[21] These precautions are designed to help prevent genes from being lost, but they will not prevent our agricultural land from being dominated by only a few varieties of crops.

Superweeds and Superbugs Other environmental concerns that have been raised with regard to genetically modified crops are that they will promote the development of superweeds or superbugs. A superweed might arise if a plant that has been modified to grow faster or to better survive begins growing in areas beyond the farmer's field. Most experts do not feel this is a major concern because domesticated crops depend on a managed agricultural environment and carry traits that make them unable to compete in the wild.

It has also been suggested that the genes inserted to produce hardy, high-yield, fast-growing crops could be transferred to wild relatives by natural cross-breeding. This could result in fast-growing superweeds. Although a possibility, this scenario is unlikely for a number of reasons. First, the probability that a weed growing near a genetically modified plant that is closely enough related to cross-breed is small. Even if it does occur, the chances that the new plant will survive and have inherited traits that enhance its survival is even smaller. As a further safeguard to prevent environmental risks, most developers avoid adding traits that could increase the competitiveness or other undesirable properties of weedy relatives.

There is also concern that crops engineered to produce pesticides will promote the evolution of pesticide-resistant insects. Although this is an important concern, the risk of it occurring is no greater when pesticides are engineered into the crops than it is when pesticides are sprayed on crops. An illustration of this problem involves insects that are resistant to the Bt toxin.[21] As more and more of the insect's food supply is made up of plants that produce this pesticide or plants that are sprayed with it, only insects that carry genes making them resistant to Bt can survive and reproduce. This increases the number of Bt-resistant insects and therefore reduces the effectiveness of Bt as a method of pest control. To address this problem, strategies are being developed to prevent the number of Bt-resistant insects from increasing. Farmers who grow pesticide-resistant crops are required to grow nonmodified plants in adjacent fields. This provides a food supply for—and encourages the continued existence of—nonresistant insect pests, thereby reducing the likelihood that the number of pesticide-resistant insects will increase.

Figure F7.13 Field tests of genetically modified crops attempt to determine the impact of the new crops on the environment and how well the plants function. There is concern, however, that the tests themselves may pose a risk to the environment. (Chris Knapton/Photo Researchers)

Regulation of Genetically Engineered Food Products

Although the government does not scrutinize every step of the development of new plant varieties, it is involved in overseeing the process. The government sets guidelines to help researchers address safety and environmental issues at all stages, from the early development of genetically engineered plants through field-testing and, eventually, commercialization. Companies that develop new plant varieties must provide data to support the safety and wholesomeness of the product. Crops created by both traditional breeding and biotechnology methods must be field-tested for several seasons to make sure only desirable changes have been made (**Figure F7.13**). Plants are examined to ensure that they look right, grow right, and produce food that is safe and tastes right. Analytical tests must be performed to determine if the levels of nutrients in the new variety are different and if the food is safe to eat. Health Canada and the Canadian Food Inspection Agency (CFIA) are both involved in the oversight of plant biotechnology.

Health Canada Health Canada classifies genetically modified plants as plants with novel traits. In addition to genetically modified foods, plants with novel traits include plants developed by traditional plant breeding techniques and which do not have a history of safe use as a food, or plants developed by new processes not previously used in food.[22]

The policy is that the safety of a food product should be determined based on the characteristics of the food or food product, not the method used to produce it. Foods developed using biotechnology are therefore evaluated to determine their equivalence to foods produced by traditional plant breeding. Emphasis is placed on whether the food creates a new or increased allergenic risk, has an increased level of a naturally occurring toxin, contains a substance not previously present in the food supply, or is nutritionally different from the traditional plant. Health Canada must give its approval before the plant can enter the food supply.

Canadian Food Inspection Agency While Health Canada evaluates the impact that a genetically modified food might have on human health, the CFIA assesses the environmental impact of cultivating this plant in Canada. For example, if there is a high probability that a new plant variety will cross-breed with a weed and that the transfer of the new trait could allow the weed plant to survive better, development of this plant might be halted. If a plant has been studied and tested and does not pose environmental risks, field-testing is allowed. CFIA continues to oversee the testing until it is determined that the plant is safe.[23]

REFERENCES

1. Cohen, S. N., Chang, A. C., Boyer, H. W., et al. Construction of biologically functional bacterial plasmids *in vitro*. *Proc. Natl. Acad. Sci. U.S.A.* 70:3240–3244, 1973.
2. Cook, J. R. Testimony before the U.S. House of Representatives Subcommittee on Basic Research hearing on "Plant Genome Research: From the Lab to the Field to the Market." October 5, 1999, Serial No. 106–60. Washington, D.C.: Government Printing Office, 1999.
3. Margawati, E. T. *Transgenic Animals: Their Benefits to Human Welfare.* Available online at www.actionbioscience.org/biotech/margawati.html. Accessed March 21, 2009.
4. Gonsalves, D., and Ferreira, S. *Transgenic Papaya: A Case for Managing Risks of Papaya Ringspot Virus in Hawaii*. Available online at www.plantmanagementnetwork.org/pub/php/review/2003/papaya/. Accessed August 21, 2008.
5. Sautter, C., Poletti, S., Zhang, P., et al. Biofortification of essential nutritional compounds and trace elements in rice and cassava. *Proc. Nutr. Soc.* 65:153–159, 2006.
6. Ye, X., Al-Babili, S., Kloti, A., et al. Engineering the provitamin A (beta-carotene) biosynthetic pathway into (carotenoid-free) rice endosperm. *Science* 287:303–305, 2000.
7. Paine, J. A., Shipton, C. A., Chaggar, S., et al. Improving the nutritional value of Golden Rice through increased pro-vitamin A content. *Nat. Biotechnol.* 23:482–487, 2005.
8. Potrykus I. Lessons from the 'Humanitarian Golden Rice' project: regulation prevents development of public good genetically engineered crop products. *N Biotechnol*. 27(5):466-72, 2010.
9. Sattler, S. E., Cheng, Z., and DellaPenna, D. From *Arabidopsis* to agriculture: Engineering improved vitamin E content in soybean. *Trends Plant Sci.* 9:365–367, 2004.
10. Lai, L., Kang, J. X., Li, R. et al. Generation of cloned transgenic pigs rich in omega-3 fatty acids. *Nat. Biotechnol.* 24(4):435-6, 2006.
11. U.S. Department of Agriculture, Agricultural Marketing Service, National Organic Program. Final Rule. Available online at www.ams.usda.gov/nop/nop2000/nop/finalrulepages/finalrulemap.htm/. Accessed March 1, 2001.
12. Hites, R. A., Foran, J. A., Carpenter, D. O., et al. Global assessment of organic contaminants in farmed salmon. *Science* 303:226–229, 2004.
13. Krkosek, M., Connors, B. M., Morton, A. et al. Effects of parasites from salmon farms on productivity of wild salmon. PNAS 108(35):14700-14074, 2011.
14. Eenennaam A. L. V., Muir, W. M. Transgenic salmon: a final leap to the grocery shelf? Nature Biotech 29:706–710, 2011.
15. University of Guelph. EnviroPig™. Environmental Benefits. Available online at http://www.uoguelph.ca/enviropig/environmental_benefits.shtml. Accessed June 25, 2011.
16. National Academy of Sciences, National Research Council. *Genetically Modified Pest-Protected Plants: Science and Regulation*. Washington, D.C.: National Academies Press, 2000.
17. Singh, O. V., Ghai, S., Paul, D., et al. Genetically modified crops: Success, safety assessment, and public concern. *Appl. Microbiol. Biotechnol.* 71:598–607, 2006.
18. Goodman, R. E., Vieths, S., Sampson, H. A. et al. Allergenicity assessment of genetically modified crops— what makes sense? *Nat. Biotechnol.* 26:73–81, 2008.
19. Nordlee, J. A., Taylor, S. L., Townsend, J. A., et al. Identification of a Brazil-nut allergen in transgenic soybeans. *N. Engl. J. Med.* 334: 688–692, 1996.
20. Darbani, B., Eimanifar, A., Stewart, C. N. Jr., et al. Methods to produce marker-free transgenic plants. *Biotechnol. J.* 2:83–90, 2007.
21. Lemaux, P. G. Genetically engineered plants and foods: A scientist's analysis of the issues (Part II). *Annu. Rev. Plant Biol.* 60:511–559, 2009.
22. Health Canada. Genetically Modified Foods and Other Novel Foods. Available online at http://www.hc-sc.gc.ca/fn-an/gmf-agm/index-eng.php. Accessed January 14, 2011.
23. Canadian Food Inspection Agency. The regulation of plants with novel traits in Canada,. Available online at http://www.inspection.gc.ca/english/plaveg/bio/pntchae.shtml. Accessed June 30, 2011.

18 World Hunger and Malnutrition

Case Study

Vincent wanted to use social media to help the university student union raise awareness of World Food Day. This event, which commemorates the founding of the Food and Agriculture Organization of the United Nations in Quebec City on October 16, 1945, is observed annually in more than 150 countries. Its goal is to heighten public awareness and understanding of the plight of the world's hungry and malnourished, as well as to promote year-round action to alleviate world hunger. Vincent arranged to create a page on the student union social networking site to illustrate the inadequate food supply and resulting hunger and malnutrition that plagues most of the world. Research on the Internet turned up information and stories from around the globe.

(Hola Images/Getty Images, Inc.)

When Vincent looked at all of this information, though, he realized that his understanding of malnutrition around the world was not as complete as he had thought. In India, where he had believed hunger was common, there was a grain surplus. In Canada, where the grocery-store shelves were always packed, some people were going hungry. Vincent began to realize that hunger and malnutrition are complicated problems and that viewing them as merely the result of a lack of food is an oversimplification.

(©iStockphoto)

(Tony Krumba/AFP/Getty Images, Inc.)

Chapter Outline

18.1 The Two Faces of Malnutrition

Learning Objectives

- Describe the two faces of malnutrition in the world today.
- Explain what is meant by nutrition transition.

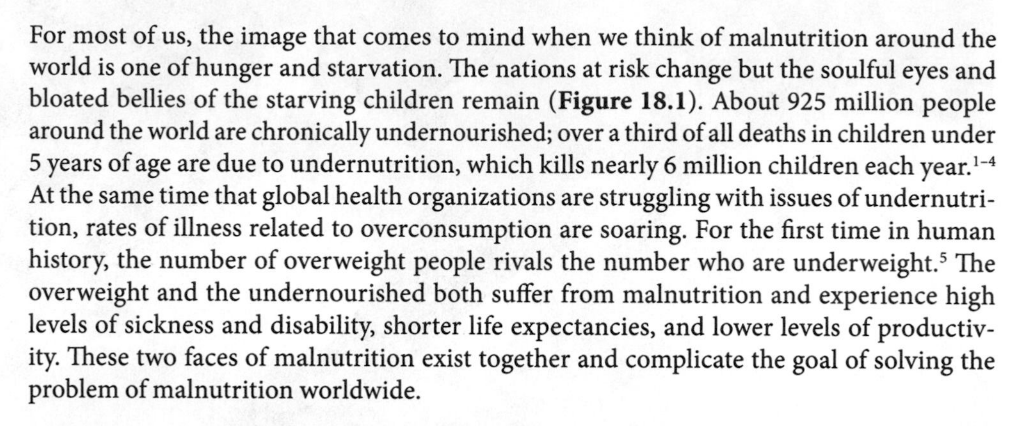

For most of us, the image that comes to mind when we think of malnutrition around the world is one of hunger and starvation. The nations at risk change but the soulful eyes and bloated bellies of the starving children remain (**Figure 18.1**). About 925 million people around the world are chronically undernourished; over a third of all deaths in children under 5 years of age are due to undernutrition, which kills nearly 6 million children each year.[1–4] At the same time that global health organizations are struggling with issues of undernutrition, rates of illness related to overconsumption are soaring. For the first time in human history, the number of overweight people rivals the number who are underweight.[5] The overweight and the undernourished both suffer from malnutrition and experience high levels of sickness and disability, shorter life expectancies, and lower levels of productivity. These two faces of malnutrition exist together and complicate the goal of solving the problem of malnutrition worldwide.

Figure 18.1 Undernutrition is more common in developing nations, especially among children, because they have high nutrient needs. (Reuters/Bettman/©Corbis)

Nutrition Transition

As economic conditions in a country improve, changes occur in the way food is grown, produced, and obtained, and traditional diets give way to more modern food intake patterns. This **nutrition transition** begins the shift from concern with undernutrition to concern with overnutrition and, at times, the two exist together in the same population (**Figure 18.2**).

nutrition transition A series of changes in diet, physical activity, health, and nutrition that occurs as poor countries become more prosperous.

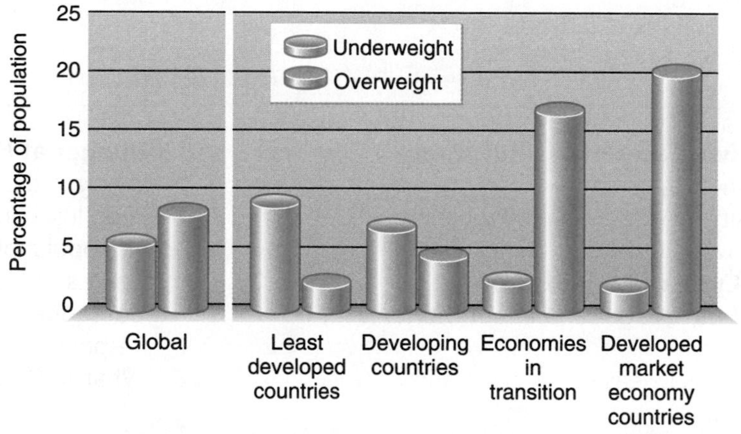

Figure 18.2 Impact of economic development on the incidence of underweight and overweight As countries develop economically, the incidence of underweight drops, and the percentage of the population that is overweight increases. Higher rates of obesity are found in cities because of greater food availability and more sedentary occupations.

Traditional diets in developing countries are based on a limited number of foods—primarily starchy root vegetables. As incomes increase and food availability improves, the diet becomes more varied and is likely to include more meat, milk, refined grains, fat, and sugar. Along with this dietary transition come changes in lifestyle that decrease activity. There is a shift toward less physically demanding occupations, an increase in the use of transportation to get to work or school, more labour-saving technology in the home, and more passive leisure time. As a result, nutrition-related chronic diseases such as cardiovascular disease, cancer, diabetes, and obesity are newly appearing, rapidly rising, or already established in every country around the world (**Figure 18.3**).

Some of the effects of this economic and nutrition transition are positive. Shifts in diet are accompanied both by increases in life expectancy and by decreases in the frequency of low birth weight, infectious disease, and nutrient deficiencies. However, at the same

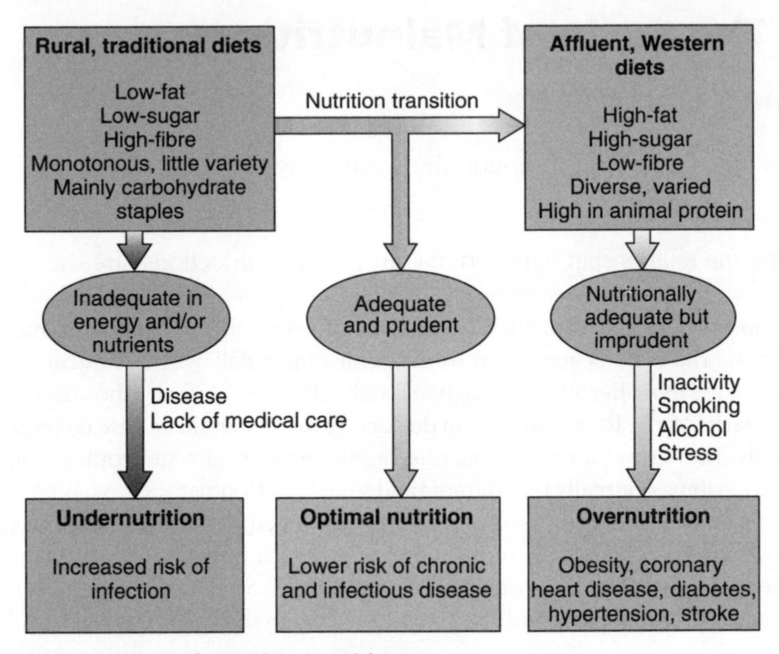

Figure 18.3 Consequences of nutrition transition
A diet that falls somewhere between the traditional rural diet that may be inadequate in energy, protein, or micronutrients and the affluent Western diet that meets nutrient needs but is high in fat and sugar and low in fibre is optimal for health.
Source: Adapted from Vorster, H. H., Bourne, L. T., Venter, C. S., and Oosthuizen, W. Contribution of nutrition to the health transition in developing countries: A framework for research and intervention. *Nutr. Rev.* 57: 341–349, 1999.

time, rates of heart disease, cancer, diabetes, and obesity increase.[6,7] The increased reliance on animal proteins as well as on refined and processed foods also increases the amount of energy and natural resources needed to produce the food, which in the long term may damage the environment.

Obesity: A World Health Problem

There are now more than 1 billion adults worldwide who are overweight and 300 million of them carry enough excess body fat to be classified as obese.[5] The prevalence of obesity around the world ranges from less than 5% in rural China, Japan, and some African countries to as high as 75% of the adult population in urban Samoa.[5] In Argentina, Colombia, Mexico, Paraguay, Peru, and Uruguay, more than half of the population is overweight, and more than 15% are obese.[8] Countries such as China and India, which have historically been plagued by undernutrition, must now also contend with overnutrition. In China, the number of adults who are overweight jumped from 9% in 1989 to 15% in 1992, and rates are almost 20% in some cities.[9] In parts of Africa, obesity is now considered a major disease along with AIDS and malnutrition. Because obesity increases the risk of cardiovascular disease, hypertension, stroke, type 2 diabetes, certain cancers, arthritis, and other conditions, it is now a major contributor to what has been called the "global burden of chronic disease and disability (see Figure 18.3)."[5]

Life Cycle

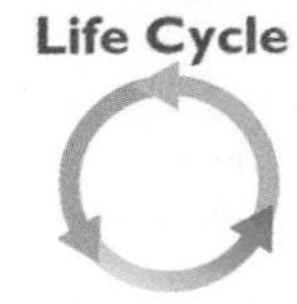

The growing prevalence of overweight and obesity among children is also a major concern. Currently, there are an estimated 43 million preschool children worldwide who are overweight and obese; of these, 35 million live in developing countries. Another 92 million preschoolers worldwide are at risk of overweight.[10] In some regions of the world, a high prevalence of overweight now exists alongside a high frequency of undernourished children. For example, in Africa, 8.5% of preschool children are overweight and another 13.8% are at risk of overweight,[10] whereas 21% of African preschoolers are underweight.[11] A similar situation exists in Asia, where 4.9% of children under 5 years are overweight and another 12.2% are at risk of overweight,[10] while 28% of Asian preschoolers are underweight.[11]

18.2 The Cycle of Malnutrition

Learning Objectives

- Discuss the impact of malnutrition throughout the life cycle.
- Define stunting.
- Describe the relationship between malnutrition and infectious disease.

cycle of malnutrition A cycle in which malnutrition is perpetuated by an inability to meet nutrient needs at all life stages.

In populations where undernutrition is a chronic problem, there is a **cycle of malnutrition** (**Figure 18.4**). The cycle begins when women consume a deficient diet during pregnancy. These women are more likely to give birth to low-birth-weight infants who are susceptible to illness and early death. The children who do survive may be small and weakened physically and mentally. They grow into undernourished adults, who are also susceptible to disease and unable to contribute optimally to economic and social development. The women in this next generation also begin their pregnancies poorly nourished and are therefore likely to give birth to low-birth-weight infants. Interruption of this cycle of malnutrition at any point can benefit the individuals and the society. Healthy children can then grow into healthy adults who produce healthy offspring and can contribute fully to society.

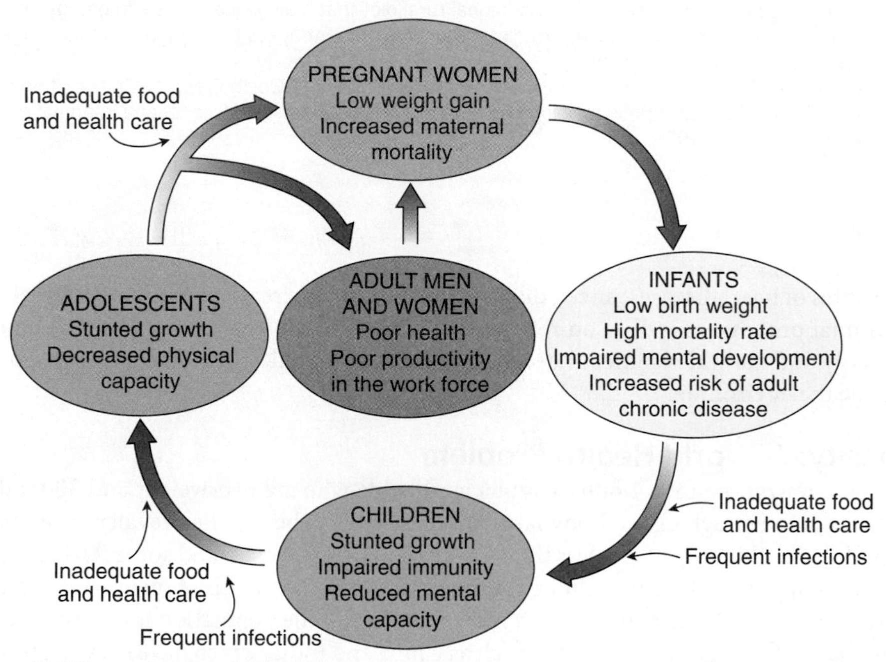

Figure 18.4 **The cycle of malnutrition**
Malnutrition affects the health and productivity of individuals at every stage of life. It often begins in the womb, continues through infancy and childhood, and extends into adolescent and adult life.

Low Birth Weight and Infant Mortality

infant mortality rate The number of deaths during the first year of life per 1,000 live births.

Low-birth-weight infants—those weighing less than 5.5 lb. (2.5 kg) at birth—are at greater risk of complications, illness, and early death. A higher number of low-birth-weight infants means a higher **infant mortality rate**, the number of deaths per 1,000 live births in a population. The infant mortality rate and the number of low-birth-weight births are indicators of the health and nutritional status of a population. In industrialized countries like Canada, Sweden, the United States, and Japan, the infant mortality rate is generally no more than 7 per 1,000 live births; in developing countries like Angola, Somalia, and Afghanistan, the rate is more than 100 per 1,000 live births (**Table 18.1**).[12] Low-birth-weight infants who do survive require extra nutrients, which are usually not available. Malnutrition in infancy and childhood has a profound effect on mental and physical growth and development as well as susceptibility to infectious disease.

Table 18.1 Indicators of Poverty and Malnutrition

	Infant Mortality (deaths per 1,000 live births)	Life Expectancy (years)	Illiteracy (percent of population)	Access to Medical Care (people per physician)
More Developed Countries	20	76	3	680
Less Developed Countries				
Sierra Leone	182	37.2	66.7	—
Central Africa	113	44.9	57.6	25,920
Ghana	67	60	33.6	22,970
Ivory Coast	81	46.7	57.4	11,739
El Salvador	31	69.6	23	848
Cuba	7	76	4.1	176
Haiti	71	54.1	54.2	4,000
India	70	62.6	46.5	2,459
Bangladesh	79	58.1	61	12,884

Source: World Bank Indicators Data: www.worldbank.org/data/wdi2000/pdfs/tab2_18.pdf/. Accessed November 12, 2011.

stunting A decrease in linear growth rate, which is an indicator of nutritional well-being in populations of children.

Stunting

Malnourished children grow poorly. The prevalence of decreased growth in height, referred to as **stunting**, is used as an indicator of the well-being of a population's children (**Figure 18.5**). It is estimated that more than 30% of children under 5 years of age in developing countries are stunted.[13] Deficiencies of energy, protein, iron, and zinc, as well as prolonged infections, have been implicated as causes. Stunting in childhood produces smaller adults who have a reduced work capacity. Stunted women are more likely to give birth to low-birth-weight babies. In addition, those who had lower birth weights and early childhood stunting are more likely to have abdominal obesity in adulthood.[14] Abdominal obesity increases the risk of morbidity from cardiovascular disease, hypertension, and diabetes.

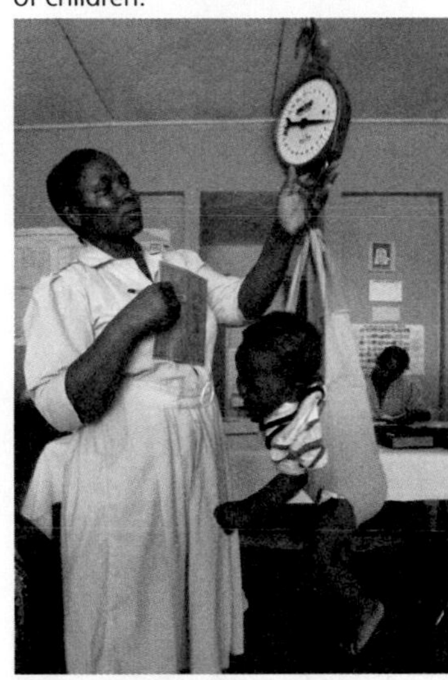

Figure 18.5 Stunted children may never regain the height lost as a result of malnutrition, and most children will never gain the corresponding body weight. Stunting also leads to premature death later in life because vital organs never fully develop. (Jorgen Schytte/Peter Arnold, Inc.)

Infectious Diseases

Infectious diseases are more common in undernourished children (**Figure 18.6**) because they suffer from depressed immune systems, which reduces their ability to resist infection. Mortality from infections is increased among malnourished children; they may die of infectious

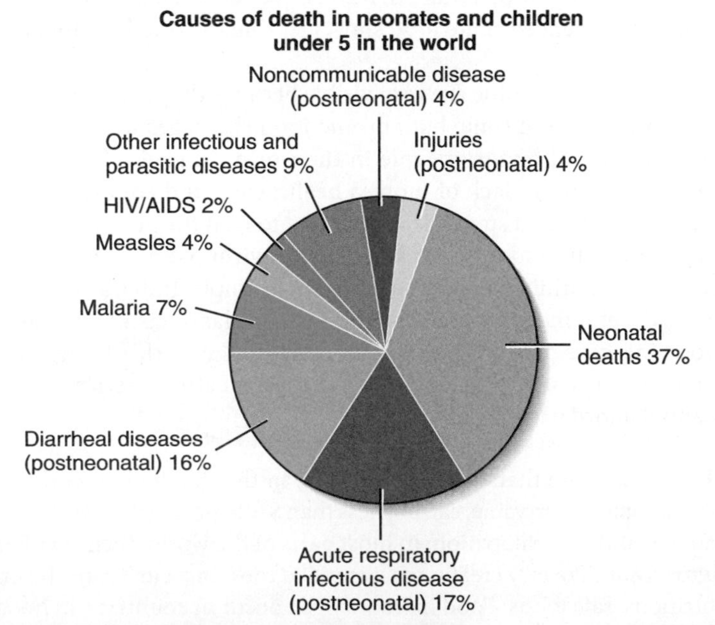

Figure 18.6 Worldwide causes of death in children under 5
Infectious diseases such as measles, diarrheal diseases, and respiratory infections, are often fatal for malnourished children. About 35% of deaths in young children might be prevented by adequate nutrition.
Source: World Health Organization. Black, R. E., Allen, L. H., Bhutta, Z. A., et al. Maternal and child undernutrition: Global and regional exposures and health consequences. *Lancet* 371:243–260, 2008.

diseases that would not be life-threatening in well-nourished children. Well over half of all deaths in children under 5 years are due to infectious disease.[4] It is estimated that 35% of deaths in children under age 5 occur due to the presence of undernutrition.[3] Mortality is increased even among children with mild to moderate malnutrition. Even immunization programs, designed to reduce the incidence of infectious disease, may be ineffective because the immune systems of undernourished individuals cannot respond normally.

18.3 Causes of Undernutrition

Learning Objectives

- Discuss the factors that cause food shortages for populations and for individuals.
- Explain the concept of food insecurity.
- List 3 common nutrient deficiencies worldwide and explain the consequences of these deficiencies.

The specific reasons for hunger vary with the time and location, but the underlying cause is that the food available in the world is not distributed equitably. This inequitable distribution results in either not enough food or the wrong combination of foods to meet nutrient needs. This in turn results in protein-energy malnutrition and micronutrient deficiencies.

Food Shortages

famine A widespread lack of access to food due to a disaster that causes a collapse in the food production and marketing systems.

The most obvious example of a food shortage is **famine**, which is a widespread failure in the food supply due to a collapse in the food production and marketing systems. Drought, floods, earthquakes, and crop destruction by diseases or pests are natural causes of famines. Man-made causes include wars and civil conflicts as well as poor preparedness on the part of governments to prevent or deal with food shortages. Regions that produce barely enough food for survival under normal conditions are vulnerable to the disaster of famine. This situation is analogous to a man standing in water up to his nostrils: if all is calm, he can breathe, but if there is a ripple, he will drown. When a ripple such as a natural or civil disaster occurs, it cuts the margin of survival and creates famine (**Figure 18.7**). For example, in recent years, sub-Saharan African countries, such as South Sudan, Malawi, and Somalia have experienced acute food shortages due to armed conflict or crop failure as a result of drought.

Figure 18.7 In October 2005, an earthquake struck northern Pakistan, killing almost 100,000 people, causing widespread damage, and interrupting the production and distribution of food. Without adequate relief efforts, widespread famine following such a disaster could kill more people than the disaster itself. (Reuters/Thierry Roge/Landov)

Food shortages due to famine are very visible because they cause many deaths in one area during a short period of time, but chronic food shortages take a greater toll when it comes to the number of hungry people in the world. Chronic shortages occur when economic inequities result in lack of money, health care, and education for individuals or populations; when the food supply is insufficient to feed the population; when cultural and religious practices limit food choices; or when environmental resources are misused, limiting the ability to continue to produce food. For example, India, a country undergoing rapid economic development with a growing middle class, has more than 300 million chronically food insecure people who represent 25% of the world's hungry poor.[15] Closer to home, Bolivia remains one of the poorest countries in Latin America, where 40% of the population cannot afford basic food needs.[16]

food insecurity A situation in which people lack adequate physical, social, or economic access to sufficient, safe, nutritious food that meets their dietary needs and food preferences for an active and healthy life.

Poverty and Hunger More than 1.4 billion people in the developing world currently live below the international poverty line, earning less than $1.25 per day.[17] Poverty is central to the problem of hunger and undernutrition; in most parts of the world, their incidence is almost identical (**Figure 18.8**). Poverty creates what is called **food insecurity**, or the limited ability to acquire nutritious, safe foods. Food insecurity can occur in countries, in households, and among individuals. In wealthy countries, social safety nets, such as food banks, soup kitchens, government food assistance programs, and job training programs, help the hungry to obtain food or money to buy food. In poor countries, a family that cannot grow enough food, or earn enough money to buy food, may have nowhere to turn for help.[18]

Figure 18.8 Ninety-six percent of the world's undernourished people live in the developing world, where poverty is most prevalent. (Per-Anders Pettersson/Getty Images, Inc.)

Disease and disability are more prevalent among the poor. Poverty reduces access to health care so disease goes untreated (see Table 18.1). When left untreated, illness increases nutrient needs and further limits the ability to obtain an adequate diet, contributing to malnutrition. Lack of immunizations and medical treatment result in an increased incidence of and morbidity from infectious disease and a decrease in survival rates from chronic diseases such as cancer. Lack of health care also increases infant mortality and the incidence of low-birth-weight infants.

The poor have less access to education, which reduces the opportunities to escape poverty, which also increases the risk of undernutrition and disease because lack of education leads to inadequate care for infants, children, and pregnant women. A lack of education about food preparation and storage can affect food safety and the health of the household—unsanitary food preparation increases the incidence of gastrointestinal diseases, which contribute to malnutrition.

Overpopulation Overpopulation exists when a region has more people than its natural resources can support. A fertile river valley can support more people per hectare than can a desert environment. But even in fertile regions of the world, if the number of people increases too much, resources are overwhelmed and food shortages occur.

The human population is currently growing at a rate of more than 82 million persons per year, and most of this growth is occurring in developing countries (**Figure 18.9**).[19] These countries cannot escape from poverty because their economy cannot keep pace with the rate of population growth. Efforts to produce enough food can damage the soil and deplete environmental resources, further reducing the capacity to produce food in the future. The problem of hunger today is due primarily to the unequal distribution of resources, but it is estimated that, worldwide, food production has begun to lag behind population growth. If this trend continues, there will be too little food in the world to feed the population.

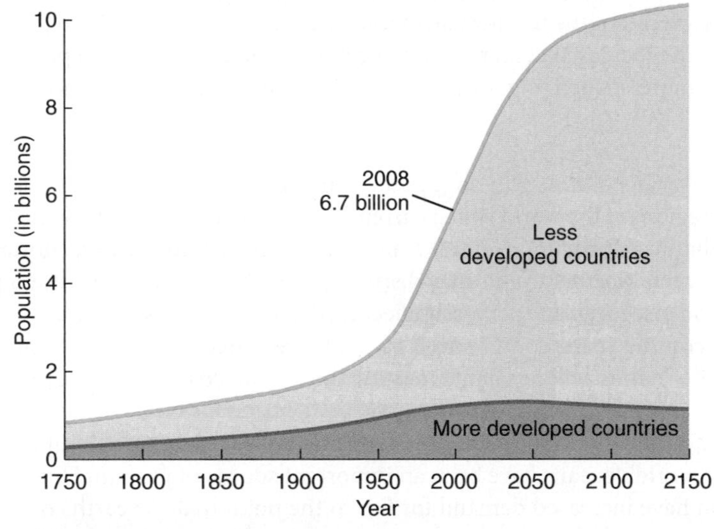

Figure 18.9 **World population growth, 1750 to 2150**
Since about 1950, most of the increase in world population has occurred in developing countries, and this trend is expected to continue.

Cultural Practices In some cultures, access to food may be limited for certain individuals within households. For example, women and girls may receive less food than men and boys, because culturally they are viewed as less important. How much food is available to an individual within a household depends on gender, control of income, education, age, birth order, and genetic endowments.

The cultural acceptability or unacceptability of foods also contributes to food shortages and malnutrition. If available foods are culturally unacceptable, a food shortage exists unless the population can be educated to use and accept new foods. For example, insects are eaten in some cultures and provide an excellent source of protein, but they are unacceptable to people in other cultures.

Limited Environmental Resources The land and resources available to produce food are limited. Some resources, such as minerals and fossil fuels, are present in the earth in finite amounts and are nonrenewable—that is, once used they cannot be replaced in a reasonable amount of time. Other resources such as soil and water are **renewable resources**, because they will be available indefinitely if they are used at a rate at which the earth can restore them. For example, when agricultural land is used wisely—crops rotated, erosion prevented, contamination limited—it can be reused almost endlessly. However, if this land is not used carefully, soil erosion, nutrient depletion, and accumulation of pollutants in soil and water may exceed the ability to restore and reuse this resource. Sometimes the methods used to increase food production have damaging long-term effects on the environment.

renewable resources Resources that are restored and replaced by natural processes and that can therefore be used forever.

Crops and Cattle Modern mechanized agricultural methods increase yields, but use more energy and resources and cause more environmental damage than more traditional labour-intensive farming. In industrialized nations, about 17% of energy used is for food production.[20]

In general, the environmental cost of producing plant-based foods is lower than that of producing animal products, but the cost may still be substantial. Modern large-scale farming can erode the soil and deplete its nutrients. Fertilizers used to restore soil and pesticides used to kill insects can contaminate groundwater and eventually waterways. And if the plant products are shipped long distances, require refrigeration or freezing, or need other types of processing, the environmental costs are increased even more (see Your Choice: Does Choosing Vegetarian Help Alleviate World Hunger?).

Modern cattle production uses rangeland, grain, water, and fossil fuels, and creates both air and water pollution. The animals themselves produce methane gas in their gastrointestinal tracts, which enters the air and contributes to the greenhouse effect. When animal sewage is stored in ponds and heaps, it decomposes, producing more methane. Livestock is responsible for about 18% of greenhouse gas emissions, a larger share than that of all the cars in the world combined.[21]

As more countries undergo nutrition transition, the demand for meat-based diets will increase, as will the use of natural resources and energy. In addition, the demand for grain is increasing as a result of the recent sharp acceleration in the use of grain to produce ethanol to fuel cars.[22] The increased demand has contributed to increases in food prices, which have made it even more challenging for low- and middle-income families worldwide to obtain enough food.

Climate Change Global climate change poses a potential threat to food security, especially in low-latitude regions of the world which currently experience food shortages including South Asia, sub-Saharan Africa, and Central America. Warmer temperatures, shortages of fresh water, and extreme weather events may disrupt and further reduce agricultural production, increasing food insecurity and adversely affecting the health of those living in these and other areas. In the coming years, conservation of natural resources and more sustainable and efficient food production will be vital to meeting the food needs of the world's population.[23,24]

Fishing It is not only the resources of the land that are at risk. Throughout human history, fish from the world's oceans have been an important source of protein. However, increases in population have increased demand for fish to the point that the earth's oceans are being depleted.[25] Because the ocean is open to fishermen from around the world, its use has been difficult to control. Many marine species have been harvested until their numbers have been severely reduced. Pollution also threatens the world's fishing grounds. Oil spills and deliberate dumping can occur offshore, and sewage, pesticides, organic pollutants, and sediments from erosion wash into coastal waters where most fish spend at least part of their lives.

Poor Quality Diets

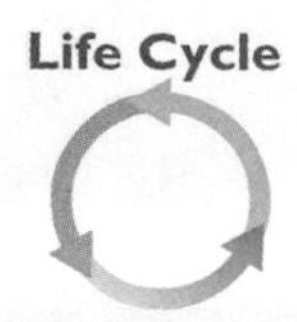

Even when there is enough food, undernutrition can occur if the quality of the diet is poor. The typical diet in developing countries is based on high-fibre grain products and has little variety. Adults who are able to consume a relatively large amount of this diet may be able to meet their nutrient needs, but those with increased needs or a limited capacity to consume these foods are at risk for nutrient deficiencies. Children, pregnant women, the elderly, and the ill may not be able to eat enough of this bulky grain diet to meet their needs. Deficiencies of protein, iron, iodine, and vitamin A are common because of poor-quality diets (**Table 18.2**).

YOUR CHOICE

(©iStockphoto)

Does Choosing Vegetarian Help Alleviate World Hunger?

While many people choose a vegetarian diet for its health benefits (see Section 6.7), others may choose to consume a vegetarian diet out of concern for the environment. They argue that producing animal products consumes large amounts of energy, destroys forests and grazing lands, and pollutes the air and water. These problems, in turn, limit the amount of food that can be produced and thus contribute to world hunger. Although these arguments are valid in some instances, the impact of food animals on the environment and nutrient intake depends on how these animals are integrated into the ecosystem.

On a traditional small farm, animals can consume crop wastes, kitchen scraps, and grasses that people cannot eat and turn them into meat, milk, and eggs that make important contributions to the human diet. But animals raised in agribusinesses are fed grain rather than grasses and kitchen scraps. Worldwide, 33% of total arable land is used to grow crops to feed animals.[1] Humans who eat these animals then get back only a fraction of the food energy they could have gotten from eating the grain. The energy from fossil fuel used in livestock production is also high, ranging from 4 kcal of fossil energy to obtain 1 kcal of chicken protein to 40 kcal of fossil fuel to produce 1 kcal of beef protein.[2] Livestock production accounts for more than 8% of global human water use and 70% of all agricultural land.[1] The world simply would not have enough energy, water, or land to produce enough food if everyone were to adopt the typical North American, meat-based diet.

Management of animal waste materials also affects the environment. On small farms, manure is used for fertilizer in local fields. On huge farms, thousands of animals are confined to a small area so manure builds up rapidly; runoff can pollute nearby rivers and lakes that supply drinking water. Animal wastes also produce gases that are released into the atmosphere, contributing to acid rain and global warming.

The sheer number of domestic animals is also destructive to the environment. Overstocking of pastures and overgrazing reduce the potential to continue to use those lands, and new grazing land is created by cutting down forests. Deforestation allows carbon dioxide to accumulate in the atmosphere and contributes to global warming.

Is the elimination of animal foods the answer to feeding the world and saving the planet? Not really. In the developing world, small amounts of meat and milk in the diet can mean the difference between survival and starvation. In Canada and other developed nations, animal foods provide important sources of vitamin B_{12}, calcium, and highly absorbable forms of iron and zinc. Eliminating animal products would reduce both the variety and nutrient content of the human diet.

By using sustainable agricultural systems, we could both nourish the world's population and preserve the environment. Fertile land could be used to grow crops for human consumption, and cattle and sheep could be fed only from grazing lands unsuitable for growing crops. To absorb the resultant drop in animal production, developed nations would have to decrease their demand for animal products. Consuming a diet that is, whenever possible, higher in locally-grown grains, vegetables, and fruits and lower in animal products is therefore a goal that is compatible not only with the recommendations for a healthy diet but also with the ecological health of the planet. Completely eliminating animal products is neither necessary nor beneficial.

Grazing cattle on land that is unfit for growing crops preserves arable land for human food production. (© istockphoto.com/Harris Shiffman)

[1]Steinfeld, H., Gerber, P., Wassenaar, T. et al. *Livestock's Long Shadow: Environmental Issues and Options* 2006. Available online at http://www.fao.org/docrep/010/a0701e/a0701e00.HTM. Accessed March 17, 2009.

[2]Pimentel, D., and Pimentel, M. Sustainability of meat-based and plant-based diets and the environment. *Am. J. Clin. Nutr.* 78 (Suppl.):660S–663S, 2003.

Table 18.2 Malnutrition at Different Life Stages

Life Stage	Common Deficiencies	Consequence
In utero	Energy	Low birth weight
	Iodine	Brain damage
	Folate	Neural tube defects
Infant/young child	Protein, energy	Growth retardation, increased risk of infection
	Iron	Anemia
	Iodine	Developmental retardation, goiter
	Vitamin A	Infection, blindness
Adolescent	Protein, energy	Stunting, delayed growth
	Iron	Anemia
	Iodine	Impaired intellectual development, goiter
	Vitamin A	Infection, blindness
	Calcium	Inadequate bone mineralization
Pregnant women	Protein, energy	Intrauterine growth retardation, increased mortality of mother and fetus
	Folate	Maternal anemia, neural tube defects in infants
	Iron	Maternal anemia
	Iodine	Cretinism in infant, goiter in mother
	Vitamin A	Infection, blindness
Adult	Energy, protein	Thinness, lethargy
	Iron	Anemia
Elderly adult	Energy	Thinness, lethargy
	Calcium and protein	Osteoporosis fractures, falls

Source: World Health Organization.

Figure 18.10 Because they are small, children are often unable to eat enough of a bulky grain diet to meet their protein needs. (Jim Sugar/ © Corbis)

Protein-Energy Malnutrition Protein and energy deficiencies usually occur together and are most common in children. When there is a general lack of food, marasmus results, but when the diet is limited to starchy grains and vegetables, protein deficiency can predominate, particularly in individuals with high protein needs—those who are growing, developing, or healing (see Section 6.5). Kwashiorkor, a deficiency of protein but not energy, occurs as a result of the wrong combination of foods rather than of a general lack of food. It is common in children over 18 months of age when the main food is a bulky cereal grain low in high-quality protein. Children have small stomachs and are not able to consume enough of this grain to meet their protein needs (**Figure 18.10**). Other factors such as metabolic changes caused by infection may also play a role in the development of kwashiorkor.

Iron Deficiency Iron deficiency is the most common nutritional deficiency worldwide. As many as 4–5 billion people, 66%–80% of the world's population, have low body iron stores, an early stage of iron deficiency; 2 billion people worldwide have iron-deficiency anemia, the most severe stage of iron deficiency. Nine out of 10 anemia sufferers live in developing countries.[26] Iron deficiency can result from too little dietary iron, an increased need for iron, or a chronic loss of iron due to blood loss (see Section 12.2). Limited meat consumption in the developing world restricts iron intake to poorly absorbed nonheme plant sources. Also, intestinal parasites, especially hookworm infections, cause gastrointestinal blood loss, which leads to iron-deficiency anemia. The greater rates of both acute and chronic infections, such as malaria, in the developing world aggravate dietary iron deficiency.

Iron deficiency can have a major impact on the health and productivity of a population. Anemia during pregnancy increases the risk of maternal and fetal mortality, premature delivery, and low birth weight. Iron deficiency in infants and children can stunt growth, retard mental development, decrease resistance to infection, and increase morbidity due to disease. In older children and adults it causes fatigue and decreases productivity.

Life Cycle

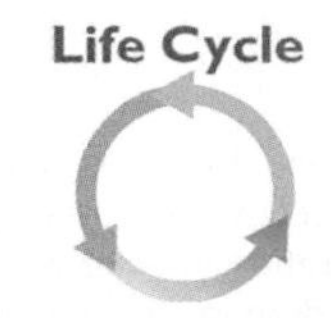

Iodine Deficiency Iodine is an essential trace element that is a constituent of the thyroid hormones (see Section 12.7). Globally, more than 1.9 billion people, including 285 million school-aged children, have inadequate iodine intake as defined by low urinary iodine excretion.[27] When iodine is deficient in the food supply, it affects virtually all members of the community. During pregnancy, iodine deficiency increases the incidence of stillbirths, spontaneous abortions, and developmental abnormalities such as cretinism in the offspring.[28] Although goiter is the most visible symptom of iodine deficiency, cretinism is the most severe manifestation (**Figure 18.11**). Cretinism is characterized by impaired cognitive and physical development. It is devastating to individuals and families, but the more subtle effects of iodine deficiency on mental performance and work capacity may have a greater impact on the population as a whole. Iodine-deficient children have lower IQs and impaired school performance. Iodine deficiency in children and adults is associated with apathy and decreased initiative and decision-making capabilities. Worldwide, iodine deficiency diseases are believed to be the greatest single cause of preventable brain damage in the fetus and infant, and of retarded psychomotor development in young children.[28]

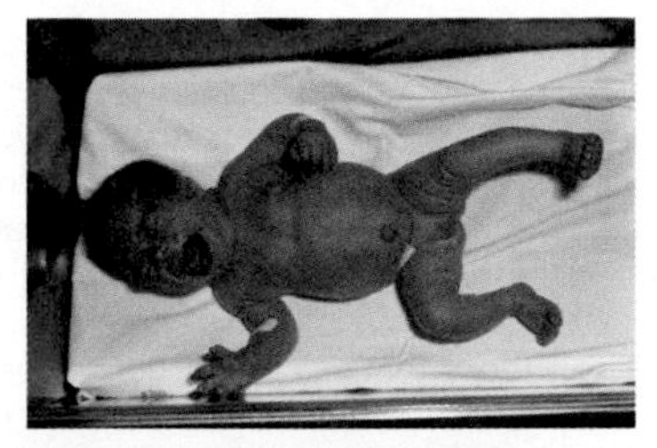

Figure 18.11 Cretinism is associated with cognitive impairment, deaf mutism and spasticity, and weakness in the limbs. (Custom Medical Stock Photo, Inc.)

Iodine deficiency occurs in regions with iodine-deficient soil that rely extensively on locally produced food. The eastern Mediterranean region and Africa have the highest incidence of iodine-deficiency disorders. Because soil iodine is low in regions where deficiency is common, the problem can be solved only by importing foods high in iodine or by adding iodine to the local diet through fortification or supplementation. In 1990, it was estimated that 1.6 billion people lived in areas considered to be at risk for iodine deficiency. Since then, as a result of campaigns to introduce iodized salt, this number has been drastically reduced. However, there are still 54 countries where iodine deficiency is a major public health problem.[27]

Vitamin A Deficiency It is estimated that more than 250 million preschool children worldwide suffer from vitamin A deficiency.[29] It causes blindness; depresses immune function, which increases the risk of infections; retards growth; and is often accompanied by anemia (see Section 9.2). It is the leading cause of preventable blindness in children. It is estimated that 250,000–500,000 children go blind from vitamin A deficiency every year; half die within a year of losing their sight.[29] In communities where vitamin A deficiency exists, supplementation has been shown to significantly reduce childhood deaths due to infection (see Science Applied: Vitamin A: The Anti-Infective Vitamin).[30]

Life Cycle

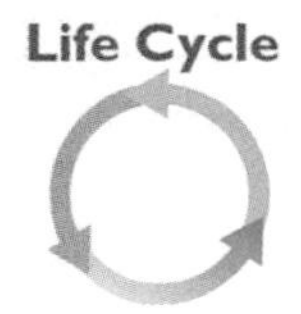

Obtaining sufficient vitamin A is a particular problem during periods of rapid growth and development, such as infancy, early childhood, pregnancy, and lactation. Need is increased by frequent infections, such as those causing diarrhea, and illnesses such as measles. Deficiencies of other nutrients, including fat, protein, and zinc, can contribute to vitamin A deficiency because they are needed to absorb and transport the vitamin in the body.

Other Nutrients of Concern In addition to deficiencies of protein, iron, iodine, and vitamin A, there are several vitamin and mineral deficiencies that have recently emerged or re-emerged as problems throughout the world. Beriberi, pellagra, and scurvy, caused by deficiencies of thiamin, niacin, and vitamin C, respectively, are rare in the developed world but still occur among the extremely poor and underprivileged and in large refugee populations. Folate deficiency is also a problem in many parts of the world. It causes megaloblastic anemia during pregnancy and often compounds existing iron-deficiency anemia. In women of child-bearing age, low folate intake increases the risk of having a baby with a neural tube defect. In Canada, United States, Chile, and Costa Rica, enriched grain products are fortified with folic acid to assure adequate intake and to reduce the incidence of neural tube defects (see Section 8.8).

Deficiencies of the minerals zinc, selenium, and calcium are also of concern.[31] Zinc deficiency affects about one-third of the world's population and is believed to cause as many

SCIENCE APPLIED

Vitamin A: The Anti-Infective Vitamin

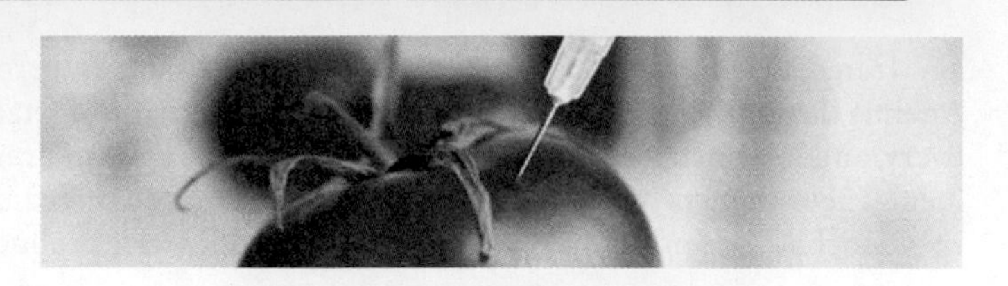

In 1910, one out of every 4 U.S. infants died before 1 year of age. The major causes were infectious diseases: epidemics of diarrhea during the summer and respiratory infections during the winter.[1] Similar problems affected children in Canada and Europe. In the early part of the twentieth century, the discovery of vitamins led to observations that nutritional deficiency, particularly vitamin A deficiency, was associated with an increased incidence and severity of infectious disease.[2]

In 1925, an epidemic of pneumonia swept through a colony of dogs in a research laboratory in England. The pneumonia occurred almost exclusively in vitamin A-deficient animals. It was hypothesized that the deficiency increased susceptibility to respiratory infections and that this might be relevant to infections in children.[3] Animal experiments confirmed this hypothesis and vitamin A was dubbed the "anti-infective vitamin."[4]

The theory that vitamin A could prevent infections triggered 20 years of clinical investigations. Clinical trials were conducted to determine if vitamin A, given as cod liver oil, could reduce morbidity and mortality from respiratory diseases, measles, and other infections. Although results were mixed, the pharmaceutical industry began promoting cod liver oil to decrease severity and recovery time in ailments such as whooping cough, measles, mumps, chicken pox, and scarlet fever.[1] The administration of cod liver oil became routine for millions of children in the United States, Canada, and Europe in the 1940s.

In 1959, the World Health Organization (WHO) published a paper that reviewed the mounting evidence of a relationship between nutritional status and infection.[5] It recognized that poor nutritional status leads to more frequent and more severe infectious illnesses, and infection triggers metabolic responses that cause nutrient losses. To a well-nourished child, common infectious diseases such as measles are usually a passing illness, whereas to a malnourished child, they can result in life-long disabilities or death.

Today, antibiotics, vaccinations, and a nutritious and varied diet have reduced infant morbidity and mortality in Canada and other developed nations. Worldwide, though, infections are still responsible for more than half of the deaths that occur in children under age 5 every year, and about 35% of these deaths are associated with malnutrition.[6,7] Improved vitamin A status can help reduce the number of child deaths.

As this 1940 ad shows, cod liver oil, which is a good source of vitamin A, was promoted to reduce the incidence and severity of infections in children. (Bettman/©Corbis)

The relationship between vitamin A and measles has been studied extensively. Before a vaccine was developed, measles claimed 7 million to 8 million lives a year. Today it remains a major problem in developing nations. A deficiency of vitamin A reduces the ability of the immune system to defend itself against infection, and the infection itself causes loss of vitamin A that could precipitate acute vitamin A deficiency and blindness.[8] Because of these interactions, the WHO and UNICEF currently advise that large doses of vitamin A be provided to children with measles and that vitamin A be supplemented at the time of measles vaccination.[9] Providing vitamin A supplements is a short-term answer that can accompany long-term solutions to vitamin A deficiency and malnutrition, such as changes in dietary intake patterns and fortification of appropriate foods with vitamin A.

[1]Semba, R. D. Vitamin A as "anti-infective" therapy, 1920-1940. *J. Nutr.* 129:783-791, 1999.

[2]Brundtland, G. H. Nutrition and infection: Malnutrition and mortality in public health. *Nutr. Rev.* 58(II):S1–S4, 2000.

[3]Mellanby, E. Diet and disease, with special reference to the teeth, lungs, and prenatal feeding. *Lancet* 1:151-519, 1926.

[4]Green, H. N., and Mellanby, E. Vitamin A as an anti-infective agent. *Br. Med. J.* 2:691-696, 1928.

[5]Scrimshaw, N. S., Taylor, C. E., and Gordon, J. E. Interaction of nutrition and infection. *Am. J. Med. Sci.* 237:367–403, 1959.

[6]World Health Organization. The Global Burden of Disease: 2004 update, 2008. Available online at www.who.int/healthinfo/global_burden_disease/2004_report_update/en/index.html. Accessed May 4, 2009.

[7]Black, R. E., Allen, L. H., Bhutta, Z. A. et al. Maternal and child undernutrition: Global and regional exposures and health consequences. *Lancet* 371:243-260, 2008.

[8]West, C. E. Vitamin A and measles. *Nutr. Rev.* 58 (II): S46-S54, 2000.

[9]WHO/UNICEF. Joint Statement Reducing Measles mortality in emergencies. Available online at www.unicef.at/fileadmin/medien/pdf/Measles_Emergencies.pdf. Accessed May 8, 2009.

CANADIAN CONTENT

SCIENCE APPLIED

Canada's Role in Addressing Global Micronutrient Deficiencies

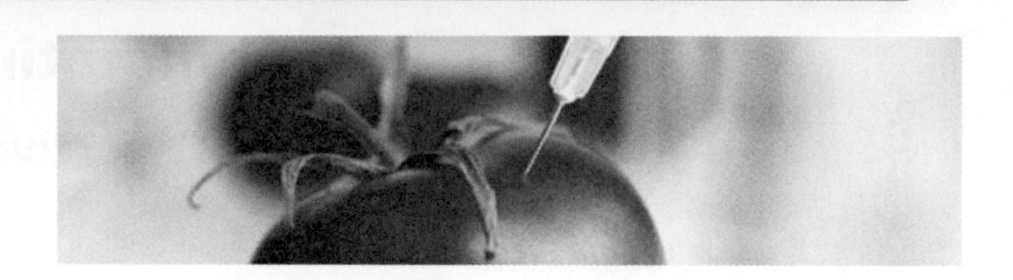

Canada has played an important role in combating global deficiencies of vitamins and minerals. The Micronutrient Initiative (MI), an independent, not-for-profit organization headquartered in Ottawa, with regional offices in New Delhi, India, and Dakar, Senegal, has been a leader in the field. Working in partnership with governments, the food industry, and other organizations, MI provides technical expertise and financial assistance to develop, implement, and evaluate interventions for hidden hunger appropriate to local communities. MI also plays an important advocacy and educational role in highlighting the need for action in addressing micronutrient deficiencies.

Examples of the Micronutrient Initiative's work include its efforts to reduce vitamin A, iodine, and iron deficiencies in developing countries. Since 1997, the organization has provided more than 75% of the world's vitamin A capsules and drops used in supplementation programs among young children (**see Figure 18.12**). It has also played an important role in decreasing iodine deficiency through its support of universal salt iodization programs and its work with small-scale salt producers in countries such as Senegal, Ghana, and Bangladesh. In collaboration with Dr. Levente Diosady in the Department of Chemical Engineering and Applied Chemistry at the University of Toronto, and with financial support from the Canadian International Development Agency and the World Bank, the Micronutrient Initiative developed a microencapsulation process to double-fortify salt with iodine and iron. This was a challenging, 15-year undertaking, given the chemical incompatibility of the two minerals. Today, more than 3.5 million school children in the state of Tamil Nadu, India, benefit from double-fortified salt as part of their daily school meal program. MI hopes to introduce double-fortified salt in 15–20 more countries within the next 5 years. To learn more about the work of the Micronutrient Initiative, go to www.micronutrient.org

Another Canadian success story in efforts to reduce hidden hunger has been the development of Sprinkles. In the 1990s, UNICEF issued a challenge to nutritional scientists to find a practical means of addressing childhood iron-deficiency anemia. Interventions for iron-deficiency anemia available at the time consisted of iron-containing drops and syrups, which were not well accepted because of poor taste, discolouration of food and teeth, and other side effects. In response, Dr. Stanley Zlotkin, a paediatrician and professor in the Departments of Nutritional Sciences and Paediatrics at the University of Toronto, and his research team at Toronto's Hospital for Sick Children, developed a novel, home fortification program for iron deficiency in infants and young children. Sprinkles are single-dose sachets which may contain iron or a combination of powdered vitamins and minerals that can be mixed into infant food at 6 months of age. Since the iron (ferrous fumarate) is encapsulated in a thin layer of lipid, there is little or no interaction with food and minimal change in its taste, texture, or appearance. Research conducted in many countries, including Ghana, Bangladesh, China, Haiti, India, and Northern Canada, has proven Sprinkles to be easy to use, highly acceptable among families and children, and effective in addressing iron-deficiency anemia. In addition, Sprinkles can be tailored to the individual nutritional needs of a population, such as including vitamin D to prevent rickets in Mongolia, and vitamin A to prevent blindness in Ghana. In recognition of his research and its application in countries around the world, Dr. Zlotkin received the Order of Canada in 2006. You can learn more about the ongoing research and work of the Sprinkles Global Health Initiative by visiting www.sghi.org.

(Used with permission from Sprinkles Global Health Initiative.)

deaths as vitamin A or iron deficiency.[32] Zinc deficiency can cause growth retardation or failure, diarrhea, immune deficiencies, skin and eye lesions, delayed sexual maturation, night blindness, and behavioural changes. It may also contribute to intrauterine growth retardation and neural tube defects in the fetus, and in the elderly it may affect taste acuity and cause dermatitis and impaired immune function. Selenium deficiency has been identified in population groups in China, New Zealand, and the Russian Federation. Selenium deficiency is associated with an increased incidence of Keshan disease (see Section 12.6), a type of heart disease that affects mainly children and young women.[31] Inadequate calcium intake is also a concern worldwide due to its association with the occurrence of osteoporosis.[31] Although factors other than low calcium intake, such as hormone levels and exercise, play a role in the development of osteoporosis, calcium supplementation has been proposed as a means of combating the high prevalence of spine and hip fractures due to osteoporosis, particularly in post-menopausal women.

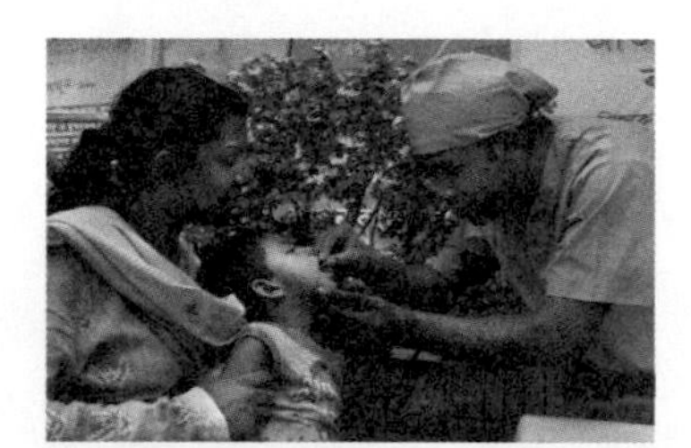

Figure 18.12 Micronutrient Initiative
Children around the world have benefited from vitamin A supplements provided by the Micronutrient Initiative, based in Ottawa, Ontario. The organization has supplied more than 75% of global vitamin A supplements provided to children aged six months to 5 years. (© Micronutrient Initiative)

18.4 Eliminating World Hunger

Learning Objectives

- Discuss 2 strategies that can help reduce population growth.
- Explain how international trade can help eliminate hunger.
- Discuss the role of sustainable agriculture in maintaining the food supply.
- List 3 considerations needed to plan food fortification.

In 1996, the World Food Summit set a goal of cutting world hunger in half by 2015. Progress was made in the late 1990s, resulting in a slow, steady decrease in the number of chronically hungry people, but the most recent numbers indicate that the number of hungry people has increased and is now higher than it was when the World Food Summit set its 2015 goal.[33]

Solving the problem of world hunger is a daunting task. It involves controlling population growth, meeting the nutritional needs of a large and diverse population with culturally acceptable foods, increasing production of nutrient-dense foods, and maintaining the global ecosystem. It requires international co-operation, commitment from national and local governments, and the involvement of local populations. The solutions involve economic policies, technical advancement, education, and legislative measures. It requires input from experts in many fields, including politicians, nutrition scientists, agricultural specialists, economists, and the food industry. Programs and policies must be in place to provide food in the short term and in the long term establish sustainable programs to allow continued production and distribution of food. During the United Nations (UN) Millennium Summit in September 2000, 189 nations adopted 8 goals called the Millennium development goals (**Table 18.3**). To be achieved by 2015, these goals correspond to the world's main development challenges. In order to achieve the first goal, stamping out hunger, most of the others must also be addressed.

Table 18.3 UN Millennium development goals
Eradicate extreme poverty and hunger
Achieve universal primary education
Promote gender equality and empower women
Reduce child mortality
Improve maternal health
Combat HIV/AIDS, malaria and other diseases
Ensure environmental sustainability
Global partnership for development

Source: With permission from United Nations. End Poverty: Millennium Development Goals 2015. Available online at www.un.org/millenniumgoals.

Short-Term Emergency Aid

When people are starving, short-term food and medical aid must be provided right away. The standard approach has been to bring food into the stricken area (**Figure 18.13**). Until recently, these foods generally consisted of agricultural surpluses from other countries and often were not well planned in terms of their nutrient content. Today, donors provide an increasing amount of food aid in the form of cash which can be used to purchase needed food in a neighbouring country or region.[34] Although this type of relief is necessary for a population to survive an immediate crisis such as famine, it does little to prevent future hunger.

Figure 18.13 Many international relief organizations provide food to hungry people throughout the world. (© AP/Wide World Photos)

There are many international, national, and private organizations working toward the goal of relieving world hunger. The United Nations World Food Programme is the world's largest humanitarian agency involved in food aid. It plays a central role in coordinating and providing for emergency food relief during times of crisis or natural disasters. The Food and Agriculture Organization (FAO) works to improve the production, intake, and distribution of food worldwide. The World Health Organization (WHO) targets community health centres and emphasizes the prevention of nutrition problems, such as micronutrient deficiencies. The World Bank finances projects such as supplementation and fortification to foster economic development. The United Nations Children's Fund (UNICEF), which relies on volunteer support, distributes food to all countries in need with a goal of assisting developing countries that occasionally suffer periods of starvation. The Red Cross, the UN Disaster Relief Organization, and the UN High Commissioner for Refugees concentrate on famine relief. More and more agencies are engaging in both development and relief. A few examples include the Canadian International Development Agency (CIDA), the U.S. Agency for International Development (USAID), Oxfam, the Hunger Project, and Catholic Relief Services (**Table 18.4**).

Table 18.4 Organizations that Work to Alleviate World Hunger

Organization	What They Do	Where to Find Them
Bread for the World Institute	Helps hungry people by engaging in research and education on policies related to hunger and development.	www.bread.org
CARE	Helps fight poverty by providing aid to poor communities.	www.care.org
Food and Agriculture Organization (FAO), Sustainable Development Department	Provides information and advice on biophysical, biological, socioeconomic, and social dimensions of sustainable development. Promotes sustainable methods and concepts and strategies.	www.fao.org
Freedom from Hunger	Provides cash, credit, and education to women in poor rural areas to help them better nourish their children, keep their families healthy, and develop profitable businesses.	www.freefromhunger.org
The International Fund for Agricultural Development (IFAD)	Finances agricultural development projects that improve nutrition and enable the rural poor to enhance food production and overcome poverty.	www.ifad.org
World Food Programme	Coordinates food relief efforts, provides technical and logistical expertise, and helps vulnerable communities rebuild and become more food secure.	www.wfp.org
Oxfam Canada	Works to eliminate the social and economic problems that prevent people from getting the skills and resources they need to be self-sufficient.	www.oxfam.ca
The United Nations Children's Fund (UNICEF)	Works to improve the health and lives of children through education and vaccination as well as responding to crisis situations.	www.unicef.org
The World Bank	Provides loans, policy advice, technical assistance, and information-sharing services to low- and middle-income countries to reduce poverty.	www.worldbank.org
World Health Organization (WHO)	Focuses on all aspects of international health. Targets community health centres and emphasizes the prevention of nutrition problems, such as micronutrient deficiencies.	www.who.int

Providing Enough Food

In the long term, solving the problem of world hunger requires balancing the number of people with the amount of food that can be produced. The rate of population growth worldwide has slowed from 5 children per woman in 1950 to 2.6 in 2009.[35] A balance between population and the amount of food available globally and locally is needed to feed the world. Inequities within populations must be addressed by eliminating poverty and providing opportunities for education—both to help people escape poverty and to teach them what constitutes a healthy diet and how to prepare food safely. Long-term solutions need to be based on the cultural and economic needs of the local population.

Controlling Population Growth Population growth can be slowed directly by controlling birth rates through family planning. To be successful, family-planning efforts must be acceptable to the population and compatible with their cultural and religious needs. A number of approaches, such as provision of contraceptives, education, and economic incentives have been used to decrease population growth. In Singapore, Thailand, Colombia, and Costa Rica, programs that provide contraceptive information, services, and supplies have been somewhat successful in slowing population growth. In some countries, population-control education is being integrated into the school curriculum, and family-planning messages are carried by popular television programs.

An indirect way to reduce population growth is to increase the general level of education and provide economic security. Birthrates decrease when the educational level and economic status of women is improved.[36] Women with more education tend to marry later and have fewer children. Education also increases the likelihood that women will have control over their fertility, provides information to improve family health, decreases infant and child mortality rates, and offers options other than having numerous children.

Changes in economic policies can help reduce population growth. In some developing countries, higher birth rates are due to the economic and societal roles of children. They are needed to work the farms, support the elders, and otherwise contribute to the economic survival of the family. Another reason for high birth rates is high infant mortality. When infant mortality rates are high, people choose to have many children to ensure that some will survive. Programs that foster economic development and ensure access to food, shelter, and medical care have been shown to cause a decline in birth rates because people feel secure having fewer children and because economic development reduces the need for children as workers.

food self-sufficiency The ability of an area to produce enough food to feed its population.

Growing and Importing Adequate Food **Food self-sufficiency** is a country's capacity to feed its population. Developing manageable systems for producing acceptable, sustainable sources of food can increase the level of food self-sufficiency. In countries with limited agricultural resources, imports can increase the food supply and help reduce hunger.

Agricultural Technology Technological advancements in agriculture can help a country boost food production. These include newer varieties of plants, better agricultural techniques, and improvements in irrigation. One type of technology being used to increase the quantity and improve the quality of food is genetic engineering (see Focus on Biotechnology). Crop yields can be increased either directly, by inserting genes that improve plant growth, or indirectly, by creating plants that are resistant to herbicides, insects, and plant diseases, thus reducing crop losses. Genes that impart insect and herbicide resistance also help the environment because they allow farmers to achieve insect-free crops and weed-free fields with fewer pesticides and herbicides. Biotechnology can also affect plant characteristics that are of benefit after harvest, such as ease-of-transport, longer shelf lives, and slower ripening. The availability of older technologies such as freezing and refrigeration and better storage facilities can reduce food losses due to insects and rodents and thus also increase the amount of food available to the population.

International Trade Some countries have the resources to grow enough food to feed their population and others do not. When a country has few natural resources, access to international trade systems can help provide for their population. The newly industrialized countries of Asia, such as Thailand and South Korea, are examples of how an increase in food imports can decrease the number of hungry people. In general, countries are becoming more interdependent on food imports and on exports to pay for this food. This interdependence can increase the availability of food for the world population (see Off the Label: What's on Food Labels Around the World?).

subsistence crops Crops grown as food for the local population.

cash crops Crops grown to be sold for monetary return rather than to be used for food locally.

Whether a country's agricultural emphasis focuses on producing **subsistence crops** for local consumption or on producing **cash crops**, which can be sold on the national and international market, has an influence on the availability of food for its people. Shifting to cash crops improves the cash flow of the country but uses local resources to produce crops for export and limits the ability of the local people to produce enough food to feed their families. For example, if a large portion of the arable land in West Africa is used to grow cash crops such as coffee and cotton, little agricultural land remains to grow grains and vegetables that nourish the local population. If, however, the cash from the crop is used to purchase nutritious foods for the local people, this decision may help alleviate undernutrition.

Maintaining the Environment The resources needed to support food production depend on the methods used. In developing nations, the resources used by a single person are small, but the number of people is large so it is difficult to produce sufficient food without depleting natural resources such as soil, forests, and water supplies. In developed nations, the population is less dense but the resource demands made by each individual are far greater because of lifestyle and the methods of food production and distribution. Solutions to the problem of providing enough food must assure that natural resources are conserved in both industrialized and developing nations to allow continued food production for future generations.

sustainable agriculture Agricultural methods that maintain soil productivity and a healthy ecological balance while having minimal long-term impacts.

Sustainable Agriculture **Sustainable agriculture** uses food production methods that prevent damage to the environment and allow the land to restore itself so food can be produced indefinitely. For example, contour plowing and terracing help prevent soil erosion, keeping the soil available for future crops. Rotating the crops grown in a specific field prevents the depletion of nutrients in the soil, reducing the need for added fertilizers. Sustainable agriculture

LABEL LITERACY

What's on Food Labels Around the World?

Did you know that the information on food labels varies from country to country? Labels reflect differences in national nutrition and food-safety guidelines, as well as economic and political agendas. Canada and the United States are two of only a few nations where nutrition labelling is mandatory and the information is presented based on common serving sizes. In most countries, nutrition labelling is voluntary unless a product makes nutrition claims, and it often takes higher math to figure out how much of a nutrient is in the portion you consume. For example, in England, nutrients are listed per 100 g of the product. So, if you want to find out how much sodium is in the blob of ketchup next to your fish and chips, you'd better find out what fraction of 100 g of ketchup is on your plate.

Canada may have some of the most comprehensive nutrition information on labels, but we don't come out on top when it comes to other types of information. For example, a can of tomatoes labelled in the European Union would show that the product is 80% tomatoes and 20% water.[1] Here, we would know only that tomatoes were the most abundant ingredient by weight. The information provided about how a food is produced also varies among countries. If you want to know if your food is organically produced, irradiated, or made using genetically modified ingredients, you need to research the labelling guidelines of the country from which you are purchasing the food.

The best labels provide consumers with the information they need to make informed choices. In Canada, consumers are concerned with overconsumption of saturated fat, cholesterol, sodium, and sugar. They can find information about these nutrients on all food labels, but the amounts of niacin, thiamin, and riboflavin are not required because deficiencies are not a concern in the population. In countries where niacin, thiamin, and riboflavin deficiencies are still prevalent, however, food labels would ideally include the content of these nutrients. In today's global economy, countries should learn from one another and incorporate the best label components from around the world to help their consumers choose wisely.

(Stefano Bianchetti/© Corbis)

[1]Food Labeling for the 21st Century: A Global Agenda for Action. A Report by the Center for Science in the Public Interest, 1998. Available online at http://www.cspinet.org/reports/labelrept.pdf. Accessed May 19, 2009.

uses environmentally friendly chemicals that degrade quickly and do not persist as residues in the environment. It also relies on diversification. This approach to farming maximizes natural methods of pest control and fertilization and protects farmers from changes in the marketplace (**Figure 18.14**).

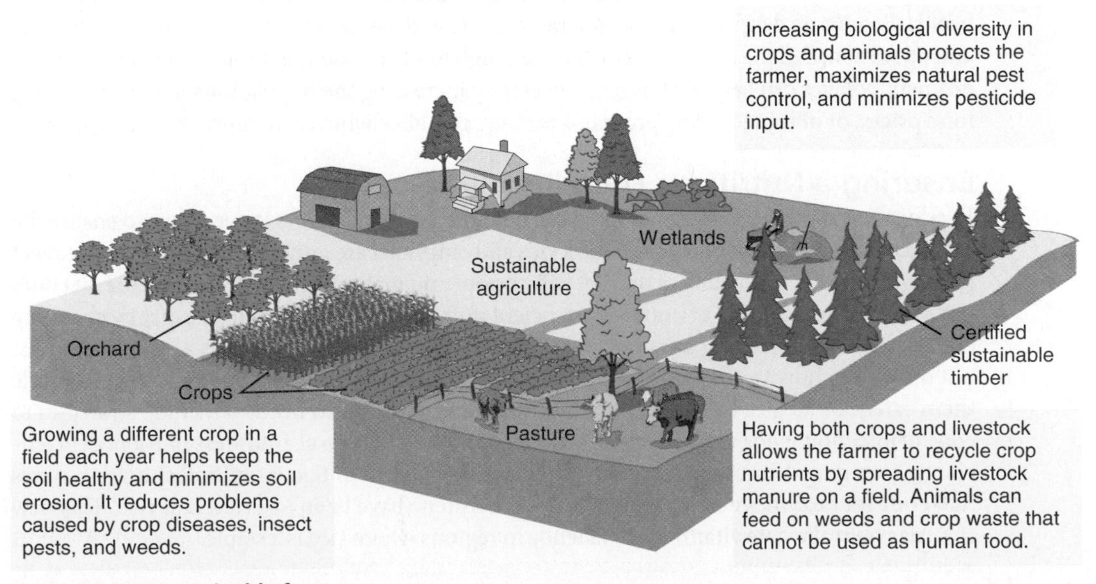

Figure 18.14 A sustainable farm
A sustainable farm consists of a total agricultural ecosystem rather than a single crop. It may include field crops, fruit- and nut-bearing trees, herds of livestock, and forests.

Sustainable agriculture is not a single program but involves choosing options that mesh well with the local soil, climate, and farming techniques. In some cases, organic farming, which does not use synthetic pesticides, herbicides, and fertilizers (see Section 17.4), may be the more sustainable option. Organic techniques have a smaller environmental impact because they reduce the use of agricultural chemicals and the release of pollutants into the environment. Organic farming is also advantageous in terms of soil quality and biodiversity, but it is a disadvantage when it comes to land use because crop yields are lower. A combination of organic and conventional techniques, as is used with integrated pest management (see Section 17.4), might be best for improving land use and protecting the environment.

Other sustainable programs include agroforestry, in which techniques from forestry and agriculture are used together to restore degraded areas; natural systems agriculture, which attempts to develop agricultural systems that include many types of plants and therefore function like natural ecosystems; and the technique of reducing fertilizer use by matching nutrient resources with the demands of the particular crop being grown.[36]

CANADIAN CONTENT

Sustainable Choices The choices individuals make can also influence the environmental impact of food production. (see Critical Thinking: What Can You Do?) Choosing a diet that is primarily plant-based, with smaller amounts of animal products, can help to minimize the ecological impact of food production (see Your Choice: Does Choosing Vegetarian Help Alleviate World Hunger?). Choosing organically grown, minimally processed, and ecologically packaged foods can further minimize the environmental impact of food production. In recent years, there has been a groundswell of interest in locavorism, that is, relying more on locally produced foods. Choosing locally grown foods in season can reduce the energy costs and pollution due to reduced costs of food transport and provide fresher, potentially more nutritious foods. Buying locally also supports local agricultural producers, an important contributor to a more food-secure community. The non-profit organization Local Food Plus (LFP) certifies food producers who practice sustainable production, including the use of few or no pesticides, reduced energy input, conservation of land, soil and wildlife habitat, and who, whenever possible, sell products locally. LFP then connects farmers with commercial buyers and consumers. In 2008, Markham, Ontario, became the first municipality in Canada to adopt LFP practices for its food services. Currently, LFP is partnering with producers, retailers, and institutions across Canada in British Columbia, the Prairies, Ontario, and Atlantic Canada. To learn more about LFP and its work, visit www.localfoodplus.ca.

Eliminating Poverty

Although controlling population growth and ensuring adequate food are essential steps in eliminating world hunger, hunger will still exist as long as there is poverty. Even when food is plentiful in a region, the poor do not have access to enough of the right foods to maintain their nutritional health. Economic development that guarantees safe and sanitary housing, access to health care and education, and the resources to acquire enough food are essential to eliminate hunger. Government policies can help to eliminate poverty by increasing the population's income, lowering food prices, or offering feeding programs for the poor, all of which can improve food security.

Ensuring a Nutritious Food Supply

In addition to sufficient energy in the diet, the right mix of nutrients is necessary to ensure the nutritional health of a population. If the foods and crops that are grown or imported do not meet all nutrient needs, the quality of the diet will be poor and malnutrition will occur. If the diet does not provide the right mix of nutrients, deficient nutrients can be added to the diet by changing dietary patterns, fortifying existing foods, or including dietary supplements. Consumers must not only learn how to choose foods that provide the needed nutrients but also how to handle them safely. Strategies to reduce micronutrient deficiencies also need to include strategies to control infectious and parasitic diseases. Genetic engineering can also address nutrient deficiencies by changing the nutrient content of foods. As discussed in Focus on Biotechnology, genes that code for enzymes needed to synthesize β-carotene have been inserted into rice. This new rice may help alleviate vitamin A deficiency in regions where rice is a staple.

Food Fortification Fortification is the process of adding one or more nutrients to commonly consumed foods with the goal of increasing the nutrient intake of a population. Food

CRITICAL THINKING:
What Can You Do?

Background

Keesha is concerned about the problems of hunger, malnutrition, and global ecology. Although she is a college student who cannot afford to make monetary contributions to relief organizations, she would like to contribute in other ways.

Data

She enjoys working with children, so she arranges to spend one afternoon a week helping with nutrition education programs for children. She also volunteers to spend one evening a week helping to prepare and serve food in a church soup kitchen near campus.

To be more ecological, Keesha buys a canvas bag to take to the grocery store. This will reduce the amount of waste she generates by eliminating the need for a new plastic bag each time she shops. She begins recycling cans, bottles, and paper goods and tries to avoid purchasing products in non-recyclable containers. This will reduce the amount of nonrecyclable, non-biodegradable waste she generates. Instead of driving the 4 km from home to campus, she rides her bike, takes the bus, or carpools with a friend. This decreases the use of fossil fuels and reduces air pollution. Keesha's efforts to reduce her ecological footprint inspire her family to contact their local utility company to arrange an energy audit of their home and make suggestions that will reduce energy usage in their home. Keesha then makes a list of other things she can do to reduce her environmental footprint.

KEESHA'S LIST

Instead of buying bottled water or coffee at school, she can drink tap water or carry a thermos from home.

She can begin composting the leftover vegetable scraps and other plant matter from her kitchen.

She can choose organically grown produce.

She can select locally grown foods when possible.

Critical Thinking Questions

Looking at Keesha's list, do you think her actions will help? For each item on her list, identify an environmental benefit.*

How will each of these actions affect her spending?

* Answers to all Critical Thinking questions can be found in Appendix J.

iProfile Use iProfile to look up the nutritional composition of some foods available at your local farmer's market.

---

fortification will not provide energy to a hungry population, but it can increase the protein quality of the diet and eliminate micronutrient deficiencies. Fortification programs have been created by partnerships among industry, academia, and government. Industry and academia can provide the technology for adding nutrients to foods, and government public health policies can promote the consumption of these fortified foods.

In order for fortification to solve a nutritional problem in a population, it must be implemented wisely. Fortification works if vulnerable groups consume foods that are centrally processed. The foods selected for fortification should be among those consistently eaten by the majority of the population so that extensive promotion and re-education are not needed to encourage their consumption. The nutrient should be added uniformly and in a form that optimizes its utilization. Fortification has been used successfully in preventing health problems in Canada. The fortification of cow's milk to increase vitamin D intake was a major factor in the elimination of infantile rickets, and the enrichment of grains with niacin helped eliminate pellagra. In 1998, the Canadian government introduced the mandatory fortification of white flour, enriched pasta, and cornmeal with folic acid to reduce neural tube defects in newborns (see sections 8.8 and 14).

Fortification has also been used successfully in developing countries. The number of countries with salt-iodization programs doubled over the last decade, rising from 46 to 93, and as a result, the global rates of goiter, cognitive disability, and cretinism are falling fast[27] (**Figure 18.15**). Likewise, the fortification of foods with other micronutrients such as vitamin A can help reduce deficiencies.[37]

Figure 18.15 Iodized salt logo The global iodized salt logo is used around the world as an indicator of iodized salt.

Supplementation Supplementing specific nutrients to at-risk segments of the population can also help reduce the prevalence of malnutrition. Of countries where vitamin A deficiency is a public health problem, 78% have policies supporting regular vitamin A supplementation in children. Many have also adopted the WHO recommendation to provide all breastfeeding women with a high-dose supplement of vitamin A within 8 weeks of delivery. This improves maternal vitamin A status and increases the amount of vitamin A that is in breast milk that is passed to the infant. Many countries have adopted programs to supplement children older than 6 months with iron, and pregnant women with iron and folate.

Providing Education

Education can help people improve their nutrient intake by teaching them what foods to eat, how to prepare them safely, and how to grow them. Education must include information about which foods are good nutrient sources so choices made when purchasing foods or growing vegetables at home can meet micronutrient needs. Education is particularly important when introducing a new crop. No matter how nutritious, a new plant variety is not beneficial unless local farmers know how to grow it and the population accepts it as part of their diet and knows how to prepare it. For instance, white yams are common in some regions but are a poor source of β-carotene, which the body can use to make vitamin A. If the yellow yam, which is rich in β-carotene, became an acceptable choice, the vitamin A available to the population would increase (**Figure 18.16**). Food safety is also a concern when changing traditional dietary practices. For example, introducing papaya to the diet as a source of vitamin A will not improve nutritional status if the fruit is washed in unsanitary water and causes dysentery among the people it is meant to nourish.

Figure 18.16 Replacing white yams, which are a poor source of β-carotene, with yellow yams, which contain enough β-carotene to meet the requirement for vitamin A in a single serving, can improve the quality of the diet. (© istockphoto.com/YinYang)

Education to encourage breastfeeding can also improve nutritional status and health. Breastfeeding reduces the risk of infectious diseases in infants. When infants are not breastfed, education about nutritious breast milk substitutes and their safe preparation is essential.

CANADIAN CONTENT

18.5 Hunger at Home

Learning Objectives

- List the population groups that are at greatest risk for food insecurity in Canada. Discuss the causes of food insecurity in Canada.
- Describe the potential consequences of food insecurity.
- Discuss Canada's response to food insecurity. How could it be improved?

In Canada and other developed countries, most of the nutritional problems are related to overnutrition. According to the 2004 Canadian Community Health Survey 2.2 (CCHS 2.2) results, more than one-third of Canadian adults were overweight and close to one-quarter were obese.[38] Heart disease, hypertension and cancer—all related to obesity—are the leading causes of death. While much of the population is concerned with consuming a diet to lower the risks for these chronic diseases, food insecure families may be choosing foods mainly on the basis of cost or seeking charitable food assistance at food banks. Problems such as poverty and unemployment lead to **food insecurity** in a land of plenty which may be moderate, resulting in compromised quality and/or quantity of food consumed; in other cases, however, food security is severe, resulting in reduced food intake and disruption of normal eating patterns. Government must develop social policy targeting improved economic and income security and at the same time develop food and nutrition policies to promote healthy diets to reduce diseases related to overconsumption.

Who are the food insecure?

The latest survey results for food insecurity in the Canadian provinces indicate that more than 9% of Canadian households, or 2.7 million Canadians, are affected (**Figure 18.17**).[39] The rate of household food insecurity across Canada varied by province and ranged from 8.1% in Saskatchewan to a high of 14.6% in Nova Scotia. Among those experiencing a higher prevalence of food insecurity were lower-income households, households with children, families led by single mothers, and Aboriginal households living off reserves. Given that the 2004 Canadian Community Health Survey did not include other populations known to experience an even higher prevalence of food insecurity, including those living in the Territories,[40] Aboriginals living on reserves,[41,42] and the homeless,[43] the actual number of food-insecure Canadians may be 3 million or more.

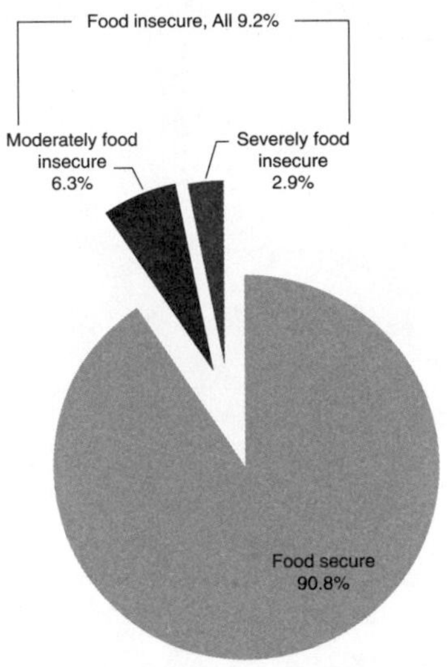

Figure 18.17 In 2003, more than 1.1 million households (9.2%) experienced food insecurity. One-third of these households reported severe food insecurity.
Source: Statistics Canada, Canadian Community Health Survey, Cycle 2.2, 2004 – Share File, Household Weights.

Food Insecurity in Canada's Aboriginal Population

Canada's Aboriginal peoples (First Nations, Inuit, and Métis) now exceed 1 million in number with an increasing number living off-reserve or in urban areas. Food insecurity is pervasive among Aboriginal groups whether they live in urban areas, off-reserve, or in more remote, isolated communities. High rates of unemployment and poverty, dependence on social assistance, low levels of education, and families headed by single mothers are associated with food insecurity.[44]

The 2004 Canadian Community Health Survey found that 33% of off-reserve Aboriginal households were food insecure and they were more likely to experience severe food security compared to non-Aboriginal households.[39]

In more remote, northern Aboriginal communities, the prevalence of food security has been reported to be much higher, ranging from 40% to 83%.[41,45,46]

A recent survey of food insecurity in 16 Nunavut communities found that almost 70% of preschoolers lived in food-insecure households, and that 25% of these children were severely food insecure.[47]

In these more isolated northern communities, a shift away from traditional ways of life and decreased use of country foods such as Arctic char, caribou, and muskox has lead to increased dependence on store-bought, market foods. Lack of year-round road access and dependence on air transport means market foods available in these communities are often significantly more expensive, at least double the cost of the same items sold in southern parts of the country. For example, on average, in 2009, $16.05 in British Columbia would buy 4 L of milk, one loaf of bread, 454 g of apples, and 4.54 kg of potatoes. In a small, remote community in northern British Columbia, the same food items would cost $34.85, or 117% more.[48] Choice of food may be limited and of poor quality; this is particularly true for perishables such as fresh fruits and vegetables and dairy products.[49,50] The high cost of perishable foods and availability of highly processed, less nutritious, store-bought foods along with other factors, such as lifestyle changes and genetic susceptibility, have contributed to the very high prevalence of diet-related chronic disease such as obesity and type 2 diabetes. Type 2 diabetes has been reported to be 3–5 times higher in Aboriginal peoples as compared to non-Aboriginals.[51] Nutrition North Canada, a new Government of Canada food subsidy program launched in April 2011, is one effort designed to improve access to healthy foods in eligible, isolated northern communities. The new subsidy program will focus on the most nutritious perishable foods consistent with Eating Well With Canada's Food Guide; these include foods such as fruits, vegetables, bread, meat, milk, and eggs.[52]

Tackling the problems of food insecurity and poor health in Canada's Aboriginal peoples will require concerted efforts to address income inadequacy, poverty alleviation efforts, and lifestyle interventions.

Causes of Food Insecurity in Canada

Canada has previously experienced hunger, for example, during the Great Depression in the 1930s, but the food insecurity seen today is a growing and relatively new problem. During the economic recession of the early 1980s, food banks appeared as a short-term solution to address concerns about hunger in the Canadian population. Instead, even as the economy improved, the demand for charitable food assistance persisted and grew. Restructuring of Canada's social programs in the 1990s reduced spending on publicly funded social programs, such as subsidized housing and social assistance, which contributed to growing income insecurity. Declines in the manufacturing, mining, forestry, agriculture, and fisheries industries have further weakened income and employment opportunities for many Canadians. Today, the income of Canada's middle class is no better than it was 30 years ago, and the lowest income group earns even less than it did 3 decades ago.[53]

The likelihood of food insecurity increases as income declines. Lower-income families who rely on part-time or low-paying jobs, and those on social assistance, employment insurance, or workers' compensation are at increased risk of food insecurity (**Figure 18.18**).[39] While older Canadian adults may experience food insecurity due to economic, social, physical, or medical limitations (see Section 16.5), overall they do so at a much lower rate than other groups. In 2004, less than 5% of households dependent on pensions and seniors' benefits were food insecure compared to 60% of households dependent on social assistance and 29% of those households dependent on workers' compensation or employment insurance.[39] Meals On Wheels, a volunteer-run, community-based program aimed primarily at seniors unable to care for their nutritional needs, helps to promote food security in this population. (See Section 16.5).

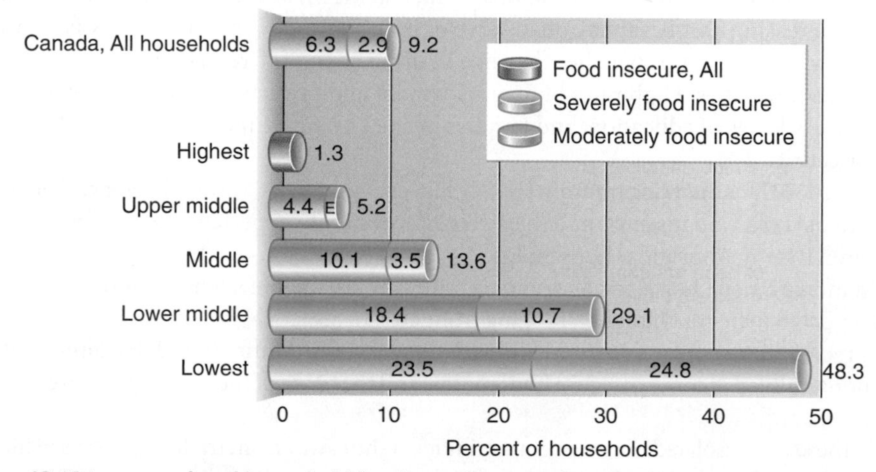

Figure 18.18 Income-related household food security status in Canada by income adequacy category, 2004.
CCHS 2.2 results clearly show a relationship between household food security and adequacy of income. Those in the lowest income category reported the highest prevalence of food insecurity and half of these experienced severe food insecurity. [E]Interpret data with caution.
Source: With permission from CCHS 2.2, Nutrition (2004) Income-Related Household Food Security in Canada, 2007.

Consequences of Food Insecurity

Whereas in developing countries, food insecurity is often severe, resulting in obvious malnutrition and poor health, the consequences of food insecurity in developed countries are less understood. Recent evidence in Canada suggests that household food insecurity in adults and adolescents compromises the diet, resulting in poorer quality diets and lower intake of nutrients. Household food insecurity has been associated with reduced consumption of more expensive foods such as fruits, vegetables, and milk products, and to a lesser degree, meat and meat alternates, and lower intakes of nutrients such as protein, zinc, magnesium, phosphorus, vitamin A, folate, riboflavin, thiamine, vitamin B_{12}, and vitamin B_6.[54] These results suggest that the food insecure may be at risk of nutrition-related health problems; the extent to which this may occur would depend on the frequency, severity, and duration of inadequate intakes. Although there is little Canadian research in this area, a large U.S. population survey among adults found lower serum levels for some nutrients in those from food-insufficient households as compared to food-sufficient households.[55] The nutritional intakes of young children in food-insecure households are less likely to be compromised, perhaps because older family members reduce the quality and quantity of their own intake to provide for younger family members.[54,56]

Increasing numbers of homeless in Canada and other developed nations raise concerns about the nutritional health of this population. Homeless youth in downtown Toronto, Ontario, have been reported to rely on varied food sources, including charitable food donations, purchased fast food, and even scavenging, resulting in a high prevalence of inadequate nutrient intakes and evidence of some chronically low energy intakes.[57,58] Nutritional deprivation in this young population raises concerns regarding their future health and ability to gain employment or seek further education.

Responses to Food Insecurity

In contrast to the United States, which provides a variety of government-run, food-based assistance programs, Canada historically has provided the majority of assistance in the form of income supports such as social insurance or welfare, employment insurance or benefits during times of unemployment, and pensions.[59] Social assistance rates in Canada are set by the provinces and Territories and vary by household type, but in most cases, social assistance income now falls below Canada's unofficial poverty lines, the Statistics Canada Low-Income Cut-Offs, and is insufficient to cover the cost of basic needs.[60] As a result, low-income families may divert money to rent and other essential needs that could otherwise be used for food.

One federally funded program which plays some role in promoting food security is Canada's Prenatal Nutrition Program (CPNP), which has existed since 1994. This federally funded program targets those women most at risk of having unhealthy babies because of poor health and nutrition and operates in communities across Canada and in Inuit and on-reserve First Nation communities. The majority of participants are pregnant women living in poverty, those in remote communities with poor access to health services, or pregnant teens. Most CPNP projects offer nutritional counselling and food supplements, as well as vitamin supplements and breastfeeding support. Given that the CPNP provides food supplements, it addresses food insecurity concerns in this vulnerable population to some extent[61] (see Section 14.3 for more details about CPNP projects).

Charitable Food Assistance in Canada With the erosion of publicly funded assistance programs, the response to food insecurity in Canada has been the growth of charitable food assistance, including food banks and prepared meals offered at soup kitchens, drop-in centres, and shelters. At present, an indirect but regular estimate of food insecurity is the use of charitable food assistance provided by Food Banks Canada's annual HungerCount survey of food bank users. In March 2010, more than 865,000 Canadians visited a food bank **(Figure 18.19)**.[62]

Figure 18.19 Food Bank Use in Canada
The annual HungerCount survey conducted by Food Banks Canada, a national organization of food banks, in March 2010, reported that 867,948 Canadians visited a food bank that month, the highest level since 1997.
Source: HungerCount 2010. A Comprehensive Report on Hunger and Food Bank Use in Canada, and Recommendations for Change. Available online at http://foodbankscanada.ca/documents/HungerCount2010_web.pdf. Accessed September 27, 2011.

These figures, however, underestimate the actual numbers of Canadians using food assistance, since they do not include those who visit other such programs offered at shelters or drop-in centres.[63]

The ability of charitable food initiatives such as food banks to enhance food security has been challenged in that participation rates in these initiatives among low-income families may be low and do little to alleviate food insecurity. Contrary to common assumptions, food-insecure households tend to be exceptionally efficient at budgeting and preparing meals with food available. Indeed, visiting a food bank is likely to be a last resort among the food insecure rather than a regular means of supplementing food in the home.[64] In addition, foods available at food banks are dependent on donations from the food industry and public and may be limited, of poor quality, and offer little choice.[65] More common strategies to alleviate household food insecurity among the very low-income food insecure include delaying bill or rent payment or eliminating household services such as telephone.[64] In addition, charitable meal programs offered at overnight shelters, soup kitchens, and other locations feed the hungry, but may be uncoordinated, intermittent, and provide only limited assistance.[63]

Another charitable response in the late 1980s to concerns about food insecurity in Canada was the emergence of child feeding programs, such as snack, breakfast, or lunch programs in schools or at community sites. These programs arose because of perceived hunger among children but recognized that children may arrive at school hungry for a variety of reasons unrelated to poverty, including lack of time, or long bus rides. Today, these programs typically receive funding from a variety of sources, among them, municipal and provincial governments, the Canadian Living Breakfast Learning Foundation, boards of education, and service clubs; they also rely to a great extent on volunteers for leadership and labour.[60] Children who are well fed may experience fewer academic and other problems at school[66] but the extent to which these child feeding programs improve food security is debatable, given that participation of children from low-income families in meal and snack programs has been reported to be low and not associated with food insecurity.[64]

Other responses to food insecurity In recent years, there has been increasing interest in an alternative approach to food insecurity which focuses on community food security. Community food security aims to provide all citizens with access to a safe and nutritious diet provided through a more sustainable food system. Examples of programs using this approach include **community gardens, community** or **collective kitchens**, promotion of farmers' markets, and alternative food distribution networks, such as **Good Food Box programs** which provide locally grown produce at a reasonable cost **(Figure 18.20)**.[67]

While these initiatives support a more sustainable food system which is accessible to all, the programs may operate on a relatively small scale and are unlikely to greatly help food-insecure households where the primary problem is poverty and inadequate income.[64,68]

community garden A plot of land on which community members can grow food, usually in an urban setting. Community gardens promote urban green spaces and foster community development.

community or **collective kitchen** A group that gathers to prepare meals together, then eats or takes food home for later use. Participants improve food preparation skills, increase knowledge about healthy food choices, and enjoy time spent socializing.

Good Food Box program An alternative means of distributing food which supplies members with affordable, fresh produce, most grown locally and bought directly from farmers; food may be delivered or picked up at a designated central location.

What To Do About Food Insecurity

Food insecurity is a serious problem in Canada, affecting a significant portion of the population, yet governments at all levels have failed to tackle the issue. Instead, charitable food assistance programs have become institutionalized. Some have argued that non-governmental and other civil society organizations providing charitable "solutions" to food insecurity are part of the problem because they divert attention away from what should be a social policy debate about how to ensure all Canadians can afford a nutritious and adequate diet.[69]

This is particularly true given that Canada has been signatory to many international agreements recognizing the right to an adequate standard of living, including the fundamental right to freedom from hunger.[69]

There are various ways Canada could address the issue of low income and associated food insecurity. These include developing a national strategy to reduce and prevent poverty, as well as a federal housing strategy to provide more affordable housing for families; investing more federal money into affordable and quality childcare so that it is possible for parents to enter into the workforce; making the Employment Insurance system more fair and in keeping with the changing labour market; introducing taxation that is more favourable to low-income individuals and families; and ensuring that provincial minimum wages and social insurance rates are increased so that workers and recipients can afford basic necessities including adequate food.

Figure 18.20 Volunteers at Toronto's Food Share pack Good Food Boxes.
(Laura Berman - GreenFuse Images.com)

In recent years, most Canadian provinces and Territories have adopted or moved toward implementing poverty-reduction strategies.[70] Several provinces have taken steps to increase minimum wage, including Manitoba, Ontario, Nova Scotia, and Newfoundland and Labrador.[71]

Tackling the problem of food insecurity requires commitment on the part of all levels of government. This would be accomplished best within the framework of a coordinated food and nutrition policy involving the federal, provincial, and territorial governments. It is important that nutrition and health professionals and civil society continue to advocate for social justice and progressive policy responses addressing food insecurity in Canada. Doing so will help to ensure the health and productivity of Canadians in the coming years.[68]

OUTCOME

Vincent hadn't realized that most hunger is caused by economic, political, sociocultural, and environmental situations that prevent enough food, or the right kinds of food, from reaching people in need. Sometimes, war or natural disaster disrupts food production and distribution, leading to famine, but poverty is one of the most common causes of hunger. To illustrate this on the page he created for the student union social networking site, Vincent tried to highlight articles that focused on the cause of malnutrition in particular areas. He was also shocked to find that obesity was a growing problem even in developing countries. Thus, he juxtaposed an article about rising worldwide obesity rates with one on malnourished children in Latin America to show that these problems often exist side by side.

Vincent is convinced that a combination of poverty reduction and nutrition education could eliminate malnutrition in Canada, because almost everyone has access to food if he or she has the money to purchase it. The problem is more challenging in other parts of the world. Vincent plans to spend next summer working for a relief organization in Honduras, where malnutrition is widespread.

APPLICATIONS

Personal Nutrition

1. **Fill in the table below to help you decide the least expensive way to buy these products:**

PRODUCT	COST		
	CORNER STORE ($)	GROCERY STORE ($)	DISCOUNT/ WAREHOUSE STORE ($)
Orange juice (250 ml)			
Orange juice (1.75 L)			
White bread (1 loaf)			
Whole-grain bread (1 loaf)			
Fresh apples (price/kg)			

   a. Calculate the cost per 125-ml serving of orange juice for the smaller and larger size orange juice at each type of store.

   b. How does the size of the container affect cost?

   c. How does the type of store affect the cost of these food items?

   d. Will following the recommendation in Eating Well with Canada's Food Guide to make half your grain products whole grain each day affect food costs?

2. **How much money do you spend on food?**

   a. Keep a record of how much money you spend on food in a day and use this to estimate your monthly food costs.

   b. Suggest 2 or 3 changes in the foods you choose that will reduce your food costs.

   c. How do these changes affect the nutrient content of your diet?

   d. What could you eat if your food budget for the day was only $3?

   e. Would the $3-a-day diet you put together meet your nutrient needs? Which nutrients are deficient? Which are excessive?

General Nutrition Issues

1. **What could you do for World Food Day (October 16)? List some ideas for campus-wide programs to increase awareness of global nutrition issues.**

2. **What do you know about hunger in various parts of the world?**

   a. Use the Internet to locate websites for organizations such as Worldwatch or Bread for the World Institute.

   b. Choose one area of the world where hunger and undernutrition are a major problem and explain the cause of undernutrition in this area.

   c. What solutions are in place or proposed to solve these problems?

SUMMARY

18.1 The Two Faces of Malnutrition

- Undernutrition remains a problem around the world.
- Due to changes in diet and lifestyle that occur as economic conditions in a country improve, overnutrition is now also a global health problem that coexists with undernutrition in both developed and developing nations around the world.

18.2 The Cycle of Malnutrition

- In poorly nourished populations, a cycle of malnutrition exists in which undernourished women give birth to low-birth-weight infants at risk of disease and early death. Children who survive grow poorly and become adults who are physically unable to fully contribute to society.
- Malnutrition causes stunting. The prevalence of stunting is used as an indicator of nutritional well-being in populations of children.
- Malnutrition depresses immune function, causing an increase in the frequency and severity of infectious diseases.

18.3 Causes of Undernutrition

- World hunger exists due to the inequitable distribution of available food. Natural and man-made disasters can temporarily disrupt food production and distribution and cause famine.
- Chronic food shortage occurs when economic inequities result in lack of money, health care, and education for individuals or populations; when overpopulation and limited natural resources create a situation in which there are more people than food; when cultural practices limit food choices; and when environmental resources are misused, limiting the ability to continue to produce food.
- Malnutrition occurs when the quality of the diet is poor. High-risk groups with special nutrient needs, such as pregnant women, children, the elderly, and the ill, may not be able to meet their nutrient needs with the available diet. Deficiencies of protein, iron, iodine, vitamin A, and zinc are common worldwide.

18.4 Eliminating World Hunger

- Short-term solutions to undernutrition provide food through relief at the local, national, and international levels.
- Long-term solutions to undernutrition must ensure the availability of food by controlling population growth, increasing the food supply through agricultural technology or importation, developing sustainable systems that will provide food without damaging the environment, and improving economic conditions to eliminate poverty.
- Long-term solutions to undernutrition must ensure a nutritionally adequate food supply. Food fortification and dietary supplementation can be used to increase protein quality and eliminate micronutrient deficiencies and improve the overall quality of the diet.
- Education about what to eat and how to prepare food safely can help eliminate malnutrition.

18.5 Hunger at Home

- Both undernutrition and overnutrition are problems in Canada. Over the last 20 years, erosion of Canada's social safety net in the form of income supports, such as a decline in social insurance rates and reduced subsidized housing, have increased poverty. Food insecurity is associated with poverty, which limits education, access to well-paying jobs, and adequate housing.

Limited access to food increases nutritional risk in certain segments of the population and may contribute to diet-related chronic disease.

- Charitable food assistance programs and community food initiatives, such as community gardens and community kitchens, provide only limited assistance to those who are food insecure. Real change in reducing food insecurity will depend on tackling the issue of poverty in Canada.

REVIEW QUESTIONS

1. What is meant by the statement, "world nutrition policies must address the two faces of malnutrition"?
2. What is meant by nutrition transition?
3. What is the cycle of malnutrition?
4. How does overpopulation contribute to food shortage?
5. How does poverty contribute to world hunger?
6. How are economic growth and population growth related?
7. What segments of the world population are at greatest risk for undernutrition?
8. List 3 micronutrient deficiencies that are world health problems.
9. Why are environmental issues important in maintaining the world's food supply?
10. How can sustainable agriculture reduce environmental damage?
11. How can food fortification be used to help eliminate malnutrition?
12. List 4 population groups in Canada who are at risk for food insecurity.
13. What are the potential consequences of food insecurity?

REFERENCES

1. Food and Agriculture Organization of the United Nations. *The State of Food Insecurity in the World 2010.* Available online at www.fao.org/docrep/013/i1683e/i1683e00.htm. Accessed November 28, 2010.
2. Caulfield, L. E., de Onis, M. Blössner, M., et al. Undernutrition as an underlying cause of child deaths associated with diarrhea, pneumonia, malaria, and measles. *Am. J. Clin. Nutr.* 80:193–198, 2004.
3. Black, R. E., Allen, L. H., Bhutta, Z. A. et al. Maternal and child undernutrition: Global and regional exposures and health consequences. *Lancet* 371:243–260, 2008.
4. World Health Organization. *The Global Burden of Disease: 2004 update, 2008.* Available online at www.who.int/ healthinf/global_burden_disease/2004_report_update/ en/index.html. Accessed May 4, 2009.
5. World Health Organization. *Global Strategy on Diet Physical Activity and Health: Obesity and Overweight.* Available online at www.who.int/dietphysicalactivity/ publications/facts/obesity/en/. Accessed August 27, 2008.
6. Kennedy, E. T. The global face of nutrition: What can governments and industry do? *J. Nutr.* 135:913–915, 2005.
7. Popkin, B. M. The nutrition transition: An overview of world patterns of change. *Nutr. Rev.* 62(7 Pt 2):S140–143, 2004.
8. Eberwine, D., Pan American Health Organization. Globesity: A crisis of growing proportions. Perspective's in Health 7(3):2002. Available online at www.paho.org/ English/DPI/Number15_article2_2.htm/. Accessed February 12, 2006.
9. Chen, C. M. Overview of obesity in mainland China. *Obes. Rev.* 9:14–21, 2008.
10. De Onis, M., Blössner, M., and Borghi, E. Global prevalence and trends of overweight and obesity among preschool children. *Am. J. Clin. Nutr.* 92:1257–1264, 2010.
11. UNICEF. Progress for Children. Achieving the Millennium Development Goals with Equity. Available online at: www.unicef.org/publications/files/Progress_for_Children-No.9_EN_081710.pdf. Accessed November 28, 2010.
12. Infant Mortality Rates. Available online at www.geographyiq.com/ranking/ranking_Infant_ Mortality_Rate_aall.htm. Accessed August 28, 2008.
13. UNICEF. Progress for Children. A World Fit for Children. Statistical Review. Available online at www.unicef.org/progressforchildren/2007n6/index_41401.htm. Accessed May 4, 2009.
14. James, P. T., Leach, R., Kalamara, E., et al. The worldwide obesity epidemic. *Obes. Res.* 9(Suppl 4):228S–233S, 2001.
15. WFP. India. Available online at www.wfp.org/countries/india. Accessed June 5, 2011.
16. WFP. Bolivia. Available online at www.wfp.org/countries/bolivia. Accessed June 5, 2011.
17. The World Bank. *Global Purchasing Power Parities and Real Expenditures: 2005 International Comparison Program.* Available online at http://siteresources.worldbank.org/ ICPINT/Resources/icp-final.pdf. Accessed April 23, 2009.
18. Are We on Track to End Hunger? Hunger Report 2004. Bread for the World Institute. Available online at www.bread.org/institute/hunger_report/index.html/. Accessed May 7, 2004.
19. Population Reference Bureau. World population highlights: Key findings from PRB's 2008 world population data sheet, *Population Bulletin* 63(3):1–16, 2008. Available online at www.prb.org/pdf08/63.3highlights.pdf. Accessed August 30, 2008.
20. Reijnders, L., and Soret, S. Quantification of the environmental impact of different dietary protein choices. *Am. J. Clin. Nutr.* 78(Suppl):664S–668S, 2003.
21. Steinfeld, H., Gerber, P., Wassenaar, T. et al. *Livestock's Long Shadow: Environmental Issues and Options.* 2006. Available online at http://www.fao.org/docrep/010/a0701e/ a0701e00.HTM. Accessed March 17, 2009.
22. Brown, L. R. *World Facing Huge New Challenge on Food Front—Business-as-Usual Not a Viable Option.* April 16, 2008. Available online at www.earthpolicy.org/Updates/2008/ Update72.htm. Accessed August 30, 2008.
23. Godfray, H. C. J., Beddington, J. R., Crute, I. R. et al. Food security: The challenge of feeding 9 billion people. *Science* 327:812–818, 2010.
24. McMichael A. J., Butler, C. J. D. The effect of environmental change on food production, human nutrition and health. *Asia Pac. J. Clin. Nutr.* 14(CD supplement):39–47, 2005.

25. Worm, B., Hilborn, R., Baum, J. K. et al. Rebuilding global fisheries. *Science* 325:578–585, 2009.

26. World Health Organization. Micronutrient Deficiencies. Iron Deficiency Anaemia. Available online at www.who.int/ nutrition/ topics/ida/en/index.html/. Accessed April 24, 2009.

27. De Benoist, B., Andersson, M., Takkouche, B., et al. Prevalence of iodine deficiency worldwide. *Lancet* 362:1859–1860, 2003.

28. World Health Organization. Micronutrient Deficiencies. Iodine Deficiency Disorders. Available online at www.who.int/nutrition/ topics/idd/en/index.html/. Accessed May 20, 2008.

29. World Health Organization. Micronutrient Deficiencies. Vitamin A Deficiency. Available online at www.who.int/nutrition/topics/vad/ en/ index.html/. Accessed April 23, 2008.

30. Jones, G., Steketee, R. W., Black, R. E. et al. How many child deaths can we prevent this year? *Lancet* 362:65–71, 2003.

31. World Health Organization. *The World Health Report, 1998: Life in the 21st Century—A Vision for All.* Geneva: World Health Organization, 1998. Available online at www.who.int/whr/1998/en/ index.html. Accessed April 24, 2009.

32. Lopez, A. Malnutrition and the burden of disease. *Asia Pac. J. Clin. Nutr.* 13(Suppl):S7, 2004.

33. Food and Agriculture Organization of the United Nations. The State of Food Insecurity in the World 2010. Available online at www.fao.org/ docrep/013/i1683e/i1683e00.htm. Accessed November 28, 2010.

34. Maxwell, D. Global factors shaping the future of food aid: the implications for WFP. *Disasters* 31(S1):S25–S39, 2007.

35. Central Intelligence Agency. *World Fact Book. Country Comparisons: Total Fertility Rate.* Available online at www.cia.gov/library/ publications/the-world-factbook/ rankorder/2127rank.html. Accessed July 23, 2009.

36. Berg, L. R., and Hager, M. C. *Visualizing Environmental Science.* Hoboken: John Wiley & Sons, 2007.

37. Bhutta, Z. A. Micronutrient needs of malnourished children. *Curr. Opin. Clin. Nutr. Metab. Care* 11:309–314, 2008.

38. Tjepkema, M. Measured obesity. Adult obesity in Canada: measured height and weight. Statistics Canada, Ottawa, Ontario, 2005.

39. Canadian Community Health Survey, Cycle 2.2, Nutrition (2004). Income-Related Household Food Security in Canada. Office of Nutrition Policy and Promotion. Health Products and Food Branch, Health Canada, Ottawa, Canada, 2007. Available online at http:// dsp-psd.pwgsc.gc.ca/collection_2007/hc-sc/H164-42-2007E.pdf. Accessed June 12, 2011.

40. Ledrou, I., and Gervais, J. Food insecurity. *Health Reports* 16(3):47–51. PMID: 15971515, 2005.

41. Indian Affairs and Northern Development. Nutrition and food security in Fort Severn, Ontario: Baseline survey for the food mail pilot project. Prepared by Lawn, J. & Harvey, D., Ottawa: Indian Affairs and Northern Development, 2004.

42. Chan, H. M., Fediuk, K., Hamilton, S. et al. Food security in Nunavut, Canada: barriers and recommendations. *Int. J. Circumpolar. Health* 65:416–431, 2006.

43. Tarasuk, V., Dachner, N., Poland, B. et al. Food deprivation is integral to the "hand to mouth" existence of homeless youths in Toronto. *Public Health Nutrition* 12:1437–1442, 2009.

44. Willows, N. D., Veugelers, P., Raine, K. et al. Prevalence and sociodemographic risk factors related to household food security in Aboriginal peoples in Canada. *Public Health Nutrition* 12(8):1150–1156, 2009.

45. Indian Affairs and Northern Development. Nutrition and food security in Kugaaruk, Nunavut: Baseline survey for the food mail pilot project. Prepared by Lawn, J. & Harvey, D. Ottawa: Indian Affairs and Northern Development, 2003.

46. Indian Affairs and Northern Development. Nutrition and food security in Kangiqsujuaq, Nunavik: Baseline survey for the food mail pilot project. Prepared by Lawn, J. & Harvey, D. Ottawa: Indian Affairs and Northern Development, 2004.

47. Egeland, G. M., Pacey, A., Cao, Z. et al. Food insecurity among Inuit preschoolers: Nunavut Inuit Child Health Survey, 2007-2008, *Can. Med. Assoc. J.* 182(3):243–248, 2010.

48. The Cost of Eating in B. C. 2009. Available online at www.dietitians. ca/Downloadable-Content/Public/BC_CostofEating_2009-(1).aspx. Accessed June 9, 2011.

49. Green, C., Blanchard, J. F., Kue et al. The epidemiology of diabetes in the Manitoba-registered First Nation population: current patterns and comparative trends. *Diabetes Care* 26:1993-1998, 2003.

50. Willows, N. Determinants of healthy eating in Aboriginal peoples in Canada. *Can. J. Pub. Health* 96 (supplement 3):S32–S36, 2005.

51. Young, T. K., Reading, J., Elias B. et al. Type 2 diabetes mellitus in Canada's First Nations: status of an epidemic in progress. *Can. Med. Assoc. J.* 163:561–566, 2000.

52. Indian and Northern Affairs Canada. Nutrition North Canada. Available online at www.ainc-inac.gc.ca/nth/fon/nn/fs-eng.asp. Accessed January 12, 2011.

53. Statistics Canada, 2008. Earnings and income of Canadians over the past quarter century, 2006 census, Government of Canada, all figures in 2005 constant dollars. Available online at http://dsp-psd. pwgsc.ca/collection_2008/statcan/97-563-X/97-563-XIE20061.pdf. Accessed January 15, 2011.

54. Kirkpatrick, S. L. and Tarasuk, V. Food insecurity is associated with nutrient inadequacies among Canadian adults and adolescents. *J. Nutr. Educ.* 138:604-612, 2007.

55. Dixon, L. B, Winkleby, M. A, and Radimer, K. L. Dietary intakes and serum nutrients differ between adults from food-insufficient and food-sufficient families: Third National Health and Nutrition Examination Survey, 1988-1994. *J. Nutr.* 131:1232–1246, 2001.

56. McIntyre, L., Glanville, N. T., Raine, K. D. et al. Do low-income lone mothers compromise their nutrition to feed their children? *Can. Med. Assoc. J.* 168:686–691, 2003.

57. Dachner, N. and Tarasuk, V. Homeless "squeegee kids": food insecurity and daily survival. *Social Science* 54:1039–1049, 2002.

58. Tarasuk, V., Dachner, N. and Li, J. Homeless youth in Toronto are nutritionally vulnerable. *J. Nutr.* 135:1926–1933, 2005.

59. Tarasuk, V. Responses to food insecurity in the changing Canadian welfare state. *J. Nutr. Educ.* 28:71–75, 1996.

60. Power, E. Individual and household food insecurity in Canada: Position of Dietitians of Canada. Available online at http://www. dietitians.ca/Downloadable-content/public/householdfoodsec-position-paper.aspx. Accessed January 9, 2011.

61. Public Health Agency of Canada. Canada Prenatal Nutrition Program (CPNP). Available online at http://www.qa.phac-aspc.gc.ca/dca-dea/ programs-mes/cpnp_goals-eng.php. Accessed January 12, 2011.

62. HungerCount 2010. A Comprehensive Report on Hunger and Food Bank Use in Canada, and Recommendations for Change. Available online at http://foodbankscanada.ca/documents/HungerCount2010_ web.pdf. Accessed September 27, 2011.

63. Tarasuk, V. and Dachner, N. The proliferation of charitable meal programs in Toronto. *Canadian Public Policy* 35(4):433–450, 2009.

64. Kirkpatrick, S. and Tarasuk, V. Food insecurity and participation in community food programs among low-income Toronto families. *Can. J. Pub. Health* 100(2):135–139, 2009.

65. Teron, A. C. and Tarasuk, V. Charitable food assistance: what are food banks receiving? *Can. J. Pub. Health* 90:382–384, 1999.

66. Alaimo, K, Olson, C., and Frongillo, E. Food insufficiency and

American school-aged children's cognitive, academic and psychosocial development. Pediatrics 108:44–53, 2001.

67. Slater, J. Community food security. Position of Dietitians of Canada. Available online at www.dietitians.ca/Dietitians-View/Community-Food-Security.aspx. Accessed September 27, 2011.

68. Tarasuk, V. Household food insecurity in Canada. *Topics. Clin. Nutr.* 20:299–312, 2005.

69. Rideout, R., Riches, G., Ostry A. et al. Bringing home the right to food in Canada: challenges and possibilities for achieving food security. *Public Health Nutrition* 10(6):566–573, 2007.

70. Provincial and Territorial Anti-Poverty Strategies and Poverty Reduction Campaigns. Available online at www.canadiansocialresearch.net/antipoverty.htm. Accessed January 15, 2011.

71. Human Resources and Skills Development Canada. Current and forthcoming minimum hourly wage rates for experienced adult workers in Canada. Available online at http://srv116.services.gc.ca/dimt-wid/sm-mw/rpt1.aspx?lang=eng. Accessed January 15, 2011.

Appendices

Appendix A Additional DRI Tables

All other DRI tables are included in the front and back covers of this text.

Dietary Reference Intakes: Recommended Intakes for Individuals: Essential Amino Acids									
Life Stage Group	**Histidine**	**Isoleucine**	**Leucine**	**Lysine**	**Methionine+ Cysteine**	**Phenylalanine+ Tyrosine**	**Threonine**	**Tryptophan**	**Valine**
	(mg/kg/ day)	**(mg/kg/ day)**	**(mg/kg/ day)**	**(mg/kg/ day)**	**(mg/kg/day)**	**(mg/kg/day)**	**(mg/kg/ day)**	**(mg/kg/ day)**	**(mg/kg/ day)**
Infants									
0–6 mo.*	23	88	156	107	59	135	73	28	87
7–12 mo.	32	43	93	89	43	84	49	13	58
Children									
1-3 yr.	21	28	63	58	28	54	32	8	37
4–8 yr.	16	22	49	46	22	41	24	6	28
Males									
9-13 yr.	17	22	49	46	22	41	24	6	28
14–18 yr.	15	21	47	43	21	38	22	6	27
19-30 yr.	14	19	42	38	19	33	20	5	24
31-50 yr.	14	19	42	38	19	33	20	5	24
51-70 yr.	14	19	42	38	19	33	20	5	24
> 70 yr.	14	19	42	38	19	33	20	5	24
Females									
9-13 yr.	15	21	47	43	21	38	22	6	27
14-18 yr.	14	19	44	40	19	35	21	5	24
19-30 yr.	14	19	42	38	19	33	20	5	24
31–50 yr.	14	19	42	38	19	33	20	5	24
51-70 yr.	14	19	42	38	19	33	20	5	24
> 70 yr.	14	19	42	38	19	33	20	5	24
Pregnancy	18	25	56	51	25	44	26	7	31
Lactation	19	30	62	52	26	51	30	9	35

*Values for this age group are AI (Adequate Intakes).
Source: Institute of Medicine, Food and Nutrition Board *Dietary Reference Intakes for Energy, Carbohydrates, Fiber, Fat, Protein and Amino Acids.* Washington, DC: National Academies Press, 2002.

Dietary Reference Intake Values for Energy: Total Energy Expenditure (TEE) Equations for Overweight and Obese Individuals		
Life Stage Group	**TEE Prediction Equation**	**PA Values**
Overweight boys aged 3–18 years	TEE =114 — (50.9 X age in yr.) + PA[(19.5 X weight in kg) + (1161.4 X height in m)]	Sedentary = 1.00 Low active =1.12 Active = 1.24 Very active = 1.45
Overweight girls aged 3–18 years	TEE = 389 — (41.2 X age in yr.) + PA[(15.0 X weight in kg) + (701.6 X height in m)]	Sedentary = 1.00 Low active = 1.18 Active = 1.35 Very active = 1.60
Overweight and obese men aged 19 years and older	TEE = 1086 — (10.1 X age in yr.) + PA[(13.7 X weight in kg) + (416 X height in m)]	Sedentary = 1.00 Low Active = 1.12 Active = 1.29 Very active = 1.59
Overweight and obese women aged 19 years and older	TEE = 448 — (7.95 X age in yr.) + PA[(11.4 X weight in kg) + (619 X height in m)]	Sedentary = 1.00 Low active = 1.16 Active = 1.27 Very active = 1.44

Source: Institute of Medicine, Food and Nutrition Board Dietary Reference Intakes for Energy, Carbohydrates, Fiber, Fat, Protein and Amino Acids, Washington, DC: National Academies Press, 2002.

Appendix B

Standards for Body Size

Body Mass Index (BMI)

Information on how to calculate your body mass index can be found at the Health Canada website (www.healthcanada.gc.ca) by searching "BMI nomogram".

The site provides three methods to determine BMI: a nomogram, a calculator where you enter your height and weight and your BMI is automatically determined, and a mathematical equation. The following classifications are used for estimating health risk, although for individuals, other factors such as age, ethnicity, fitness level, lifestyle habits, and so on, also impact risk.

Classification	BMI Range	Risk of Health Problems
Underweight	<18.5	Increased
Normal Weight	18.5-24.9	Lowest
Overweight	25.0-29.9	Increased
Obese class I	30.0-34.9	High
Obese class II	35.0-39.9	Very High
Obese class III	≥ 40	Extremely High

Source: Health Canada. Canadian Guidelines for Body Weight Classification in Adults 2003. Available online at http://www.hc-sc.gc.ca/fn-an/alt_formats/hpfb-dgpsa/pdf/nutrition/weight_book-livres_des_poids-eng.pdf. Accessed Oct 23, 2011.

Growth Charts

In 2007, Dietitians of Canada, Canadian Pediatric Society, the College of Family Physicians of Canada, and Community Health Nurses of Canada jointly recommended the use of the WHO Growth Charts to monitor and evaluate the growth of Canadian infants and children. These charts, developed by the World Health Organization, were judged to more accurately reflect healthy growth patterns for Canadian children than the previously used charts, which were developed in the United States. To access the WHO Growth Charts, visit: www.dietitians.ca/Secondary-Pages/Public/Who-Growth-Charts.aspx.

Appendix C

Normal Blood Values of Nutritional Relevance

Red blood cells	
Men	$4.5\text{-}6.5 \times 10^{12}$/L[1]
Women	$4.0\text{-}5.6 \times 10^{12}$/L[1]
White blood cells	$4.0\text{-}10.0 \times 10^{9}$/L[1]
Hematocrit	
Men	0.43-0.52[2]
Women	0.37-0.46[2]
Hemoglobin	
Men	140-174 g/L[2]
Women	123-157 g/L[2]
Ferritin	10-250 µg/L[1]
Calcium	2.15-2.55 mmol/L[1]
Iodine	0.24-0.63 µmol/L[1]
Iron	7-36 nmol/L[2]
Zinc	9.8-20.2 µmol/L[2]
Magnesium	0.65-1.05 mmol/L[2]
Potassium	3.5-5.0 mmol/L[2]
Sodium	135-145 mmol/L[2]
Chloride	98-107 mmol/L[2]
Vitamin A	1.2-2.8 µmol/L[2]
Vitamin B_{12}	200-672 pmol/L[2]
Vitamin C	≥ 25 µmol/L[2]
Carotene	0.9-4.7 µmol/L[2]
Folate (red cells)	634-1792 nmol/L[2]
pH	7.35-7.45[1]
Total protein	64-83 g/L[2]
Albumin	35-52 g/L[2]
Cholesterol	<5.2 mmol/L[1]
LDL cholesterol	< 3.5 mmol/L[3*]
HDL Cholesterol	>1.0 mmol/L[3*]
Total Cholesterol/HDL cholesterol	<5.0 mmol/L[3*]
Triglycerides	<1.7 mol/L[3*]
Glucose (serum)	3.3-5.8 mmol/L[1]

*These numbers may vary from individual to individual, depending on other risk factors (e.g., hypertension, diabetes).

Sources:

[1]Laboratory Test Information Guide. London Laboratory Services Group, London Ontario, 2010. Available online at http://www.lhsc.on.ca/cgibin/view_labtest.pl. Accessed August 5, 2011.

[2]The Medical Council of Canada 2011 Clinical Laboratory Test. Normal Values. Available online at www.mcc.ca/objectives_online/objectives.pl?lang=english&loc=values. Accessed August 5, 2011.

[3]McPherson R, Frohlich J, Fodor G, Genest J, Canadian Cardiovascular Society. Canadian Cardiovascular Society position statement--recommendations for the diagnosis and treatment of dyslipidemia and prevention of cardiovascular disease. *Can J Cardiol.* 22(11):913-27, 2006.

Major Risk Factors (Exclusive of LDL Cholesterol) that Increase Heart Disease Risk	
Risk Factors You Can Control	**Risk Factors You Cannot Control**
High Blood Pressure (BP≥ 140/90mmHg)	Age
High Blood Cholesterol	Gender
Diabetes	Family History
Being Overweight (waist circumference in men >102cm or in women >88cm; or BMI >25kg/m2)	Ethnicity
Excessive Alcohol Consumption	History of Stroke
Physical Inactivity	
Smoking	
Stress	

Source: The Heart and Stroke Foundation. Prevention of Risk Factors. March 2010. Available online at: www.heartandstroke.com/site/c.ikIQLcMWJtE/b.3483919/k.F2CA/Heart_disease__Prevention_of_Risk_Factors.htm. Accessed August 5, 2011.

High Blood Pressure (Hypertension) Guidelines	
Category	Systolic/Diastolic
Normal	120-129/ 80-84
High-normal	130-139/85-89
High blood pressure	140/90 or higher
High blood pressure for people with diabetes or kidney disease	130/80 or higher

Source: Heart and Stroke Foundation of Canada. High blood pressure. Available online at http://www.heartandstroke.com/site/c.ikIQLcMWJtE/b.3484023/k.2174/Heart_disease__High_blood_pressure.htm. Accessed August 5, 2011.

Appendix E

Beyond the Basics: Meal Planning for Healthy Eating, Diabetes Prevention and Management

This discussion of the Canadian Diabetes Association's *Beyond the Basics: Meal Planning for Healthy Eating, Diabetes Prevention and Management* consists of two parts:

- Part 1 is a detailed description of *Beyond the Basics* included as part of a case study.
- Part 2 is a comparison of Canada's Food Guide and *Beyond the Basics*.

Part 1: Case Study

Samantha is a registered dietitian and certified diabetes educator. The endocrinologist at the diabetes clinic where she works has asked her to meet with Robert to discuss a meal plan. Robert has been diagnosed with type 2 diabetes for 3 years and was previously taught *Just the Basics* for diabetes meal planning to help manage his condition (see **Section 4.5**). Samantha's dietary assessment shows that Robert has inconsistent carbohydrate intake from day to day and poor carbohydrate distribution throughout the day, which can lead to poor glycemic control (i.e., fluctuations in his blood glucose levels). He also tends to consume a lot of high glycemic index (GI) foods (see Section 4.5 for a discussion of GI). Robert has a BMI of 33, is physically active, and is motivated to make changes to his diet and lose weight, but requires a more advanced knowledge of nutrition management of diabetes. Samantha decides that the appropriate next step for education on nutrition management of diabetes for Robert is the Canadian Diabetes Association's tool *Beyond the Basics*.

After reading this appendix about *Beyond the Basics*, you will be able to understand this tool and its use for counselling individuals with type 2 diabetes.

Introduction to Beyond the Basics

Beyond the Basics: Meal Planning forHealthy Eating, Diabetes Prevention and Management was created to help primarily the adult with type 2 diabetes, but all forms of diabetes were considered when it was developed. It is a tool, in the form of a poster, to help Canadians control blood glucose levels, maintain a healthy weight, and meet daily nutritional needs, while navigating different eating situations and including ethnoculturally appropriate foods. It is based on current thinking around carbohydrate counting and reflects scientific evidence around heart health and glycemic index. Carbohydrate counting is a method of meal planning that is based on the premise that carbohydrate-containing foods increase blood glucose levels; it entails keeping track of the total carbohydrate consumed throughout the day to help control blood glucose levels. *Beyond the Basics* can be used as a progression from the Canadian Diabetes Association's *Just the Basics* tool (see Section 4.5) or as a standalone resource. It is described in detail on the Canadian Diabetes Association website, at www.diabetes.ca/for-professionals/resources/nutrition/beyond-basics.

Components of *Beyond the Basics*

Figure E.1 shows the layout of the *Beyond the Basics* poster. Foods are divided into 7 groups. Four food groups, which are rich in carbohydrates, are found on the left-hand side of the poster under the heading *Carbohydrate-Containing Foods* and include (**Figure E. 2**):

- Grains and Starches
- Fruits
- Milk and Alternatives
- Other Choices

Three food groups, containing foods with little or no carbohydrates, are found on the right-hand side of the poster (**Figure E.3**) and include:

- Vegetables
- Meat and Alternatives
- Fats

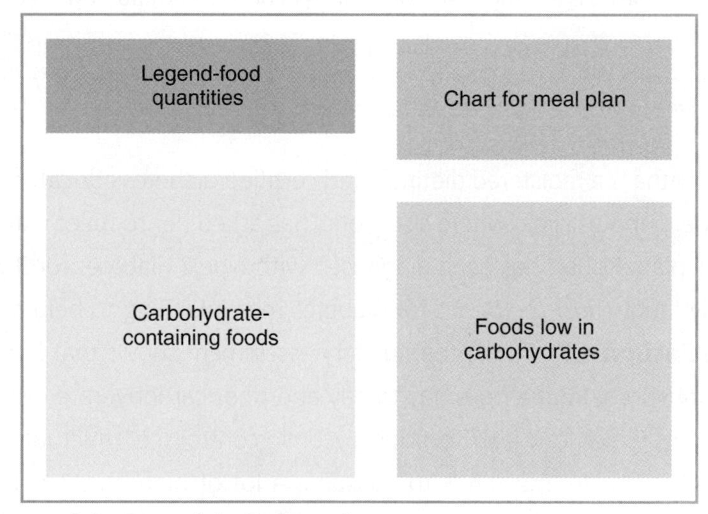

Figure E.1 Layout of the *Beyond the Basics* tool
Source: Canadian Diabetes Association: Beyond the Basics. Available online at http://www.diabetes.ca/for-professionals/resources/nutrition/beyond-basics/. Accessed April 6, 2011.

Within each of the food groups, there are several features to note:

- The Legend (**Table E.1**) provides portion sizes for each of the illustrated foods.
- Portions of carbohydrate-containing foods are based on a serving that will provide approximately 15 g of available carbohydrate. Available carbohydrate = (total carbohydrate) - (total fibre) because fibre does not increase blood glucose levels. *Beyond the Basics* uses household measures to identify portion sizes; therefore, carbohydrate content in individual servings is approximate.
- All foods have been placed in a green or yellow box, with a green or yellow triangle in the upper-left corner of the box. Foods to "choose more often" are shown in green boxes; these foods are higher in vitamins, minerals, and fibre, have lower GI values, and/or lower fat content. Foods to "choose less often" are shown in yellow boxes may have a high fat or sugar content or may be low in fibre or other essential nutrients.

- The back page of the *Beyond the Basics* poster has a space for goal setting and a *Nutrition Facts Table* label-reading exercise for people with diabetes.
- The *Beyond the Basics* poster contains a limited number of foods under each category due to space restrictions. Complete food lists can be found on the Canadian Diabetes Association website and allow for greater variety and flexibility in meal planning.

Symbol	Explanation	Symbol	Explanation
	1 cup (250 mL)		1 teaspoon (5 mL)
	½ cup (125 mL)		1 ounce (30 grams) by weight
	¼ cup (60 mL)		Measure after cooking
	1 tablespoon (15 mL)		

Table E.1 Legend Items of *Beyond the Basics* Tool
Source: Used with permission from Canadian Diabetes Association: Beyond the Basics. Available online at http://www.diabetes.ca/for-professionals/resources/nutrition/beyond-basics/. Accessed April 17, 2011.

The *Beyond the Basics* Food Groups

Food Groups that Contain Carbohydrate-rich Foods

Figure E.2 illustrates the four food groups which contain carbohydrate-rich foods: Grains and Starches, Fruits, Milk and Alternatives, and Other Choices. One serving of these foods is equal to 15 g of available carbohydrates or 1 carbohydrate choice.

Grains and Starches This category includes grains, bread, pasta, rice, and starchy vegetables such as potatoes, plantains, and corn. These foods are considered a good source of carbohydrate, fibre, vitamins, minerals, and energy. People need a minimum of 4 servings per day, depending on age, sex, activity, and weight; however, excessive amounts should not be consumed because it will result in elevated blood glucose. "Choose more often" items found in green boxes are typically higher in fibre and vitamins and minerals, and lower in total and saturated fat compared to "choose less often" items.

Additional messages that can be included in the counselling session include:

- Choose whole grains, such as brown rice, quinoa, and whole wheat pasta, more often.
- Emphasize the natural taste of grain products by limiting spreads and sauces.
- Read the ingredient lists to find whole grain starches with maximum amounts of fibre.
- Choose low or medium GI foods. These will not raise blood glucose as much as high GI foods.

Fruits Fresh, frozen, canned, and dried fruit, as well as fruit juices and applesauce, are included in this food group. Fruits are a good source of carbohydrates, fibre, vitamins, minerals, and energy. Consuming 2–3 servings of fruit per day is sufficient. Most fruits are categorized as "choose more often" items. "Choose less often" items have been categorized as such because of their dramatic impact on blood glucose levels and in order to emphasize portion control. Points to emphasize include:

- Try to choose whole fruit and fruit with skins to increase fibre content.
- Try to spread fruit consumption across meals and snacks throughout the day.
- Canned fruit: read the label to see if it is preserved in sugar syrup. Rinse the fruit from the sugary syrup or adjust the amount eaten to follow your meal plan.

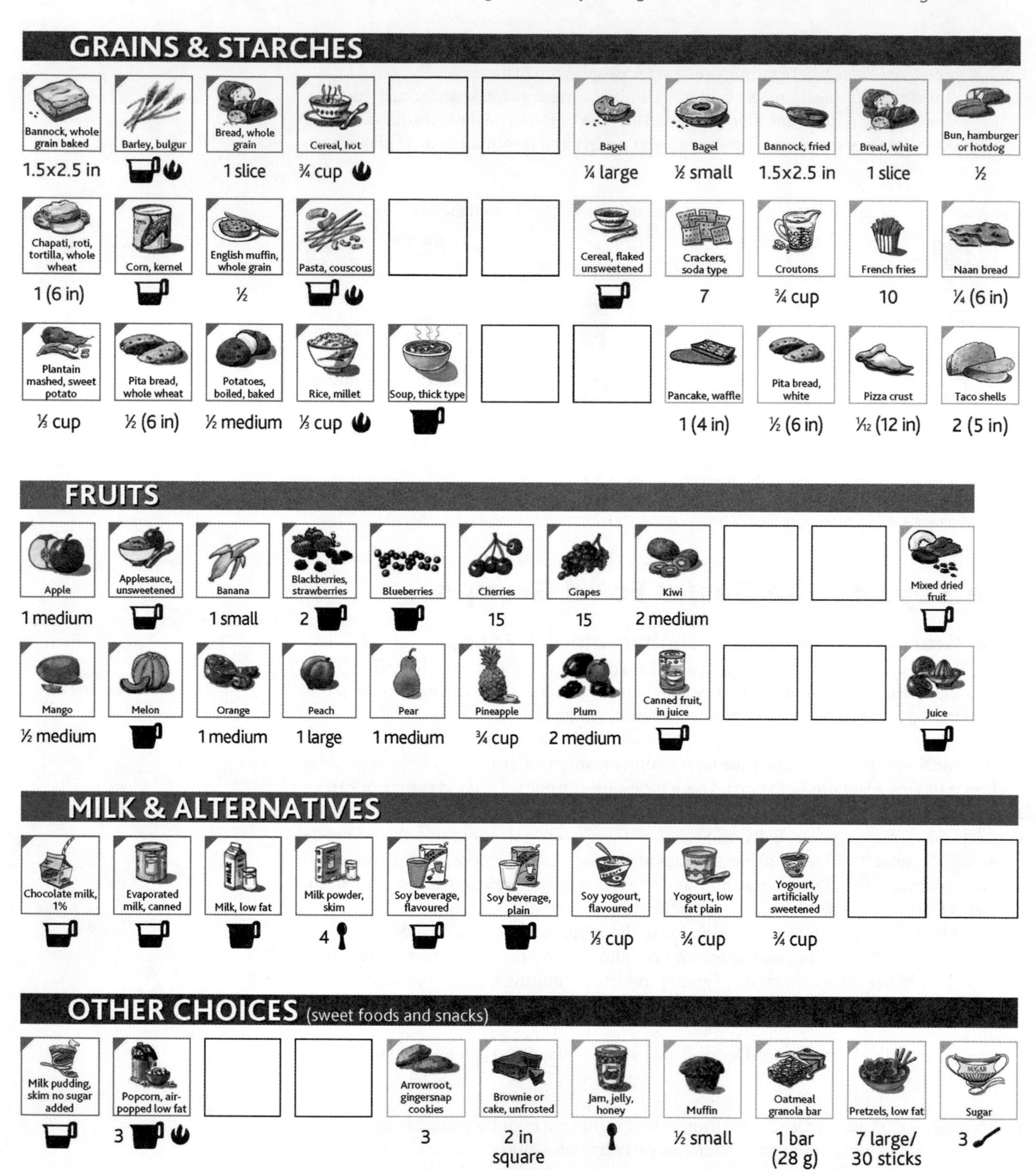

Figure E.2 Carbohydrate containing foods. These include Grains and Starches, Fruits, Milk and Alternatives, and Other Choices.
Source: With permission from Canadian Diabetes Association: Beyond the Basics. Available online at http://www.diabetes.ca/for-professionals/resources/nutrition/beyond-basics/. Accessed April 6, 2011.

Milk and Alternatives This category includes cow's milk, fortified soy beverages, flavoured yogurts, and chocolate milk. "Choose more often" items are lower in fat (skim, 1%, 2%) and contain calcium and vitamin D. Milk products and yogurts can vary considerably in their carbohydrate, fat, and protein contents so label reading is particularly important with this food group.

Other Choices (Sweet Foods and Snacks) The "Other Choices" category covers a wide array of sweet foods and snacks, including popcorn, pudding, jams, cookies, and pretzels. These foods are typically lower in nutrients and higher in fat, sugar, salt, and calories and therefore should be eaten only occasionally. Most of these products are "choose less often" items except for plain popcorn and skim, no added sugar milk pudding. The "choose less often" items do not have to be eliminated from the diet; however, portions should be limited to the serving guide suggested in *Beyond the Basics.*

Food Groups that Contain Little or No Carbohydrates

The *Beyond the Basics* food groups that have little or no carbohydrates are vegetables, meat and alternatives, and fats. These are found in Figure E.3.

Vegetables Vegetables are nutritionally dense; they provide fibre, vitamins, and minerals. Non-starchy vegetables found in this category are illustrated in Figure E.3 and are considered "free items" (that is, people with diabetes can consume as many as they would like) to encourage increased consumption of this nutritious group. It is recommended that people with

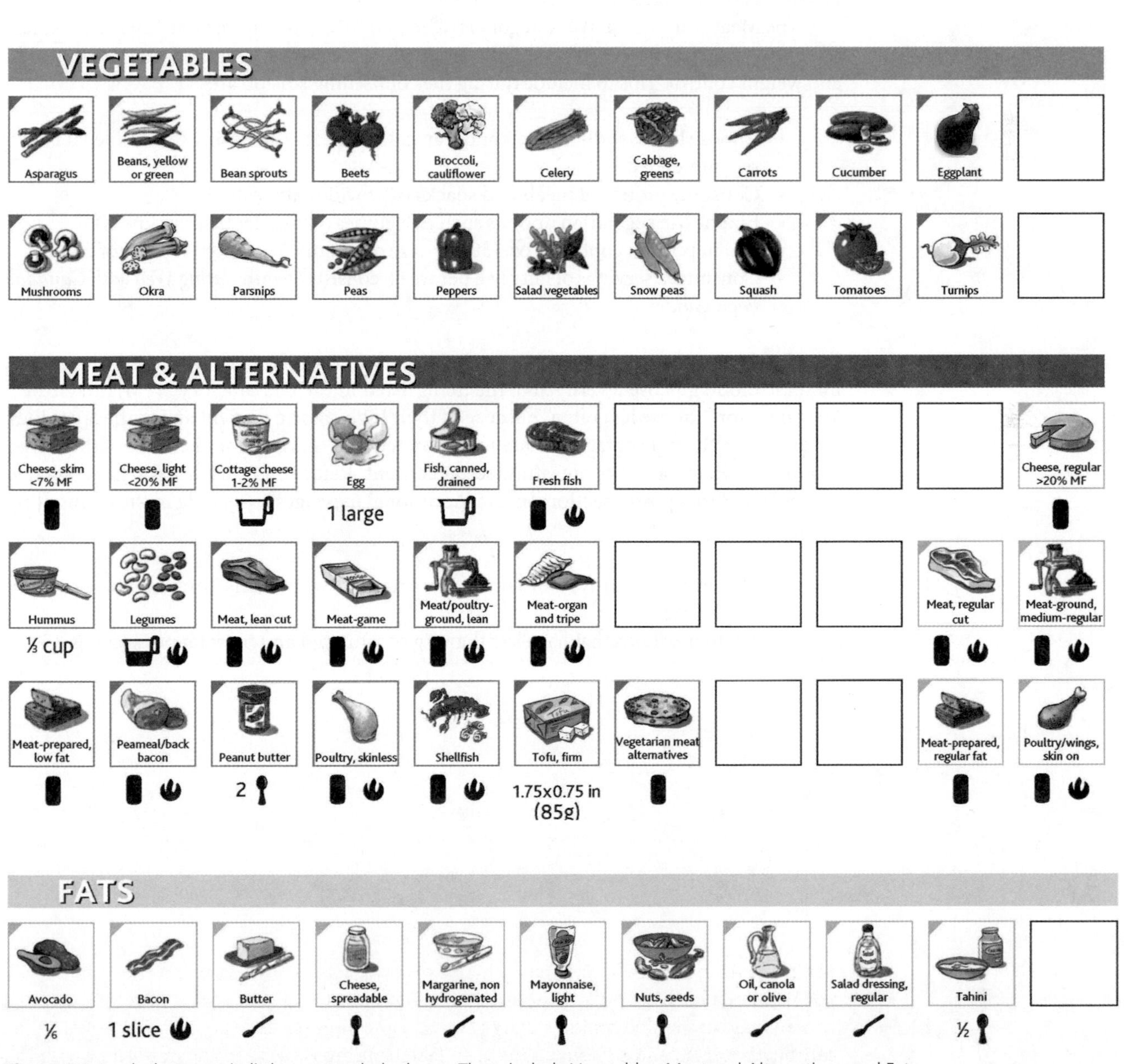

Figure E.3 Foods that contain little or no carbohydrates. These include Vegetables, Meat and Alternatives, and Fats.
Source: With permission from Canadian Diabetes Association: Beyond the Basics. Available online at http://www.diabetes.ca/for-professionals/resources/nutrition/beyond-basics/. Accessed April 6, 2011.

diabetes consume 4–5 servings a day (a serving is 125 ml (½ cup) cooked or 250 ml (1 cup) raw vegetables).

Most of the vegetables listed in this category have 2 g of protein and less than 5 g of carbohydrate per choice. However, 250 ml (1 cup) of squash, peas, and/or parsnips provides 15 g of available carbohydrate. Carrots and beets contain less carbohydrate than popularly believed due to their high fibre content (e.g., 22 miniature carrots contain 15 g of available carbohydrate). Tips to emphasize in the counselling session include:

- When preparing, limit cooking time to ensure vitamin retention.
- Choose vegetables that are prepared with little or no added fat, sugar, and salt.

Meat and Alternatives Items listed in this category have 7 g of protein and 3–5 g of fat per choice. Most of the items listed have no (0 g) carbohydrate per serving. Legumes contain carbohydrates but also have considerable amounts of fibre, which helps to decrease the absorption of glucose and therefore minimizes the blood glucose elevation. Cheese is found in the Meat and Alternatives category (rather than Milk and Alternatives) because it has 0 g of available carbohydrate and it is a source of protein. Cheese has a higher fat content, so lower-fat cheeses (< 20% milk fat) are recommended.

The Meat and Alternatives category is divided into "choose more often" and "choose less often" based on fat content (total and/or saturated) to encourage healthy eating, heart health, and weight control. Tips to include during the counselling session are:

- Choose lower-sodium items and lean cuts of meats. Trimming the visible fat is also recommended.
- Consume protein at meals and snacks to provide extra satiation.
- Emphasize vegetarian options such as legumes (beans, lentils, chickpeas).
- Watch portion control: 1 oz. (30 g) = size of thumb; 3 oz. (90g) = palm of hand.
- Consume 2 servings of fatty fish per week, consistent with *Eating Well with Canada's Food Guide.*

Fats Fat choices contain 5 g of fat per choice and no carbohydrate, and therefore do not increase blood glucose levels. All of the items listed in the fats category are written in yellow boxes for "choose less often" because of their high calorie content and/or because they are an unhealthy type of fat. Nuts are incorporated into this category because of their high monounsaturated and polyunsaturated fat content. Refer to **Figure E.3** for the complete *Beyond the Basics* poster section on fats. Additional messages to include in the counselling session are:

- The amount and type of fat are important to watch; unsaturated fats should be emphasized.
- Try to use foods that have less than 2 g of saturated and *trans* fat per serving.

Meal Planning with *Beyond the Basics*

The simplest way to conduct Meal Planning with ***Beyond the Basics*** is to fill in **Table E.2** using the guidelines detailed in this appendix.

Meal Plan							
TIME							
CARBOHYDRATES *(grams / choices)*							
GRAINS & STARCHES							
FRUITS							
MILK & ALTERNATIVES							
OTHER CHOICES							
VEGETABLES							
MEAT & ALTERNATIVES							
FATS							

Table E.2 Meal Plan Framework
Source: Used with permission from Canadian Diabetes Association: Beyond the Basics. Available online at http://www.diabetes.ca/files/Meal%20plan%20web.pdf. Accessed April 6, 2011.

Carbohydrates

The main focus of meal planning for people with diabetes is the amount and type of carbohydrate they consume from the following groups: grains and starches, fruits, milk and alternatives, and other choices.

Amount of carbohydrates As per the *Beyond the Basics* poster, 1 portion is approximately 15 g of available carbohydrate, or 1 carbohydrate choice, regardless of the type of carbohydrate-containing food consumed (fruit, grains, etc.). Typical quantities of carbohydrate choices per person per meal or snack are found in **Table E.3.**

Table E.3 Typical Quantities of Carbohydrate Choices per Person per Meal or Snack

Men/Women	Meal/Snack	Number of Carbohydrate Choices*	Grams of Carbohydrates*
Men	Meal	3 – 5	45 – 75 g
Women	Meal	2 – 4	30 – 60 g
Men and women	Snack**	1 – 2	15 – 30 g

* Depending on physical activity level, age, size, and weight loss/maintenance goals
** Optional throughout day

Type of carbohydrates People with diabetes should aim to consume a variety of foods from each food group and more of the “choose more often” items.

Vegetables

Vegetables should be consumed at each meal. These items are considered “free” as they are nutritious and do not increase blood glucose levels.

Meat and Alternatives

Meat and alternatives can be consumed throughout the day, as these items do not increase blood glucose levels. The following serving sizes will help meet protein needs: 30-60 g (1–2 oz). for smaller meals and snacks (if desired), and 90-120 g (3–4 oz.) (smaller appetites) or 150 -180 g (5–6 oz) (bigger appetites) for main meals.

Fats

Fats should be used in moderation. Consuming too much fat can lead to weight gain which makes it harder for the body to control blood glucose levels.

Case Study Outcome

In the counselling session, Samantha, the diabetes educator, has gained an understanding of Robert's typical meals through the assessment process. She has also described the components of and food groups within *Beyond the Basics* to Robert. Now it is time to develop an individualized meal plan for Robert.

As demonstrated in Robert's meal plan in **Table E. 4**, he should aim for 3–5 carbohydrate choices (45–75 g of carbohydrate) per meal, and 1–2 choices (15–30 g of carbohydrate) per snack, when he chooses to eat snacks. These carbohydrate choices can fall within the different carbohydrate-containing food groups; however, variety within groups should be emphasized. Vegetables can be consumed freely throughout the day as indicated by the check mark in Table E.4. Portion sizes of Meat and Alternatives should be watched, with a focus on "choose more often" items. Fats should be consumed in moderation.

Table E.4 Robert's Meal Plan

TIME	Breakfast	Snack	Lunch	Snack	Dinner	Snack
CARBOHYDRATES *(grams / choices)*	45 – 75 g	15 – 30 g	45 – 75 g	15 – 30 g	45 – 75 g	15 – 30 g
GRAINS & STARCHES FRUITS MILK & ALTERNATIVES OTHER CHOICES	3 – 5 choices	1 – 2 choices	3 – 5 choices	1 – 2 choices	3 – 5 choices	1 – 2 choices
VEGETABLES	✓	✓	✓	✓	✓	✓
MEAT & ALTERNATIVES	AS IS*					
FATS	USE IN MODERATION					

✓ Means "eat freely."

* Meat and Alternatives can be consumed as per usual, that is, consumed "as is," with a focus on watching portion size, meaning eating Canada's Food Guide servings of 75 g (the size of a deck of cards or the palm of the hand) and consuming "choose more often" items (e.g., lean meats).

Here is a sample meal plan that Samantha prepared for Robert:

Breakfast: 2 slices of toasted whole grain bread, 1 cup (250 ml) 1% milk, ¾ cup (175 ml) yogurt, 2 poached eggs, 3 tomato slices, 1 tablespoon (15 ml) tub margarine, and tea or coffee with artificial sweetener → **4 choices, 60 g**

Lunch: 2 cups (500 ml) thick soup, ½ small bagel, 2 oz. (60 g) skim mozzarella cheese, baby carrots, sliced peppers, 1 orange → **4 choices, 60 g**

P.M. snack: 7 whole wheat soda crackers, 1 cup (250 ml) low fat milk, 2 tablespoons (60 ml) peanut butter → **2 choices, 30 g**

Dinner: 1 medium baked potato with melted cheese, mixed vegetables (carrots, green beans) with margarine, 5 oz. (150 g) roasted chicken, ½ cup (125 ml) corn, 1 cup (250 ml) cantaloupe, ½ cup (125 ml) cranberry juice → **5 choices, 75 g**

Evening snack: 1 medium apple → **1 choice, 15 g**

A follow-up appointment indicated that Robert, with the help of *Beyond the Basics*, was able to consume a more nutritionally balanced diet, stabilize his inconsistent carbohydrate intake and improve his poor carbohydrate distribution, which led to improved blood glucose control. He also made some progress towards his weight loss goal.

Part 2: Comparison of Canada's Food Guide and *Beyond the Basics*

Eating Well with Canada's Food Guide and *Beyond the Basics* are meal planning guides to help Canadians make healthier dietary choices. Canada's Food Guide was developed with healthy Canadians in mind; however, concepts of Canada's Food Guide were incorporated in the development of *Beyond the Basics.* Although *Beyond the Basics* is targeted to people with diabetes, it also can be used by Canadians without diabetes to plan their meals. These two guides are compared in **Table E.5**.

Table E. 5 Comparison of Eating Well with Canada's Food Guide and Beyond the Basics

ITEM OF COMPARISON	BEYOND THE BASICS	CANADA'S FOOD GUIDE
Target Audience	People with Diabetes	Healthy Canadians
Item classification	Meal planning tool to assist with healthy eating	Healthy eating guidelines
Developed by	Canadian Diabetes Association	Health Canada
Aim of tool or guidelines	Using the *Beyond the Basics* meal planning tool will help Canadians: • Control blood glucose levels • Maintain a healthy weight • Meet daily nutritional needs • Navigate different eating situations	Following *Canada's Food Guide* will help Canadians: • Meet nutrient needs • Reduce risk for developing obesity, type 2 diabetes, heart disease, osteoporosis, and certain types of cancer • Contribute to overall health and vitality
How foods are categorized within the tool	Foods are classified as: Carbohydrate-containing foods: Grains and Starches, Fruits, Milk and Alternatives, Other Choices Foods that contain little or no carbohydrate: Vegetables, Meat and Alternatives, Fats Items which do not obviously belong into a specific food group are classified based on most common usage (e.g., legumes) and botanical classification (e.g., corn)	Foods are classified into the four Food Groups (Vegetables and Fruit, Grain Products, Milk and Alternatives, Meat and Alternatives) based on the following criteria: • Foods originating from the same agricultural base • How foods traditionally have been classified • How people use foods
Serving Sizes	½ cup (125 ml) cooked pasta or couscous	½ cup (125 ml) cooked pasta or couscous
	1/3 cup (80 ml) cooked rice	½ cup (125 ml) cooked rice
	1 slice bread, ½ small bagel	1 slice bread, ½ bagel (45 g)
	¾ cup (175 ml) cooked hot cereal	¾ cup (175 ml) hot cereal
	½ medium potato	½ cup (125 ml) potatoes
	1 medium apple or pear 2 medium kiwis or plums	1 fruit
	2 cups (500 ml) blackberries or strawberries 15 grapes or cherries ½ cup (125 ml) fruit juice ¼ cup (60 ml) dried fruit	½ cup (125 ml) fresh, frozen, canned fruit or fruit juice
	½ cup (125 ml) chocolate milk or flavoured soy beverage 1 cup (250 ml) low fat milk or plain soy beverage	1 cup (250 ml) chocolate or low fat milk, or fortified soy beverage
	¾ cup (175 ml) yogurt	¾ cup (175 g) yogurt
	1 oz. (30 g) cheese	1 ½ oz. (50 g) cheese
	½ cup (125 ml) cooked or 1 cup (250 ml) raw vegetables	½ cup (125 ml) fresh, frozen, canned vegetables or cooked leafy vegetables 1 cup (250 ml) raw leafy vegetables
	½ cup (125 ml) cooked legumes	¾ cup (175 ml) cooked legumes
	1 oz. (30 g) cooked fresh fish or lean meat, poultry	2.5 oz. (75 g) cooked fish, lean meat, poultry
	1 tablespoon (50 ml) nuts, seeds	¼ cup (60 ml) shelled nuts and seeds

Number of Recommended Servings/Choices	Grains and Starches	Minimum of 4 servings per day Adults should consume 1–2 servings of carbohydrate foods at snacks, and either 3–5 servings (males) or 2–4 servings (females) at meals	Females 19-50: 6–7 servings Males 19–50: 8 servings Females 51+: 6 servings Males 51+: 7 servings
	Fruit	2–3 servings, spread throughout the day Adults should consume 1–2 servings of carbohydrate foods at snacks, and either 3–5 servings (males) or 2–4 servings (females) at meals	Grouped with "Vegetables" *Vegetables and Fruit*: Females 19–50: 7–8 servings Males 19–50: 8–10 servings Females 51+: 7 servings Males 51+: 7 servings
	Milk and Alternatives	Adults should consume 1–2 servings of carbohydrate foods at snacks, and either 3–5 servings (males) or 2–4 servings (females) at meals	Females 19–50: 2 servings Males 19–50: 2 servings Females 51+: 3 servings Males 51+: 3 servings
	Other Choices	Adults should consume 1–2 servings of carbohydrate foods at snacks, and either 3–5 servings (males) or 2–4 servings (females) at meals	No Food Group for "Other Choices"
	Vegetables	"Free"	Grouped with "Fruits" *Vegetables and Fruit:* Females 19–50: 7–8 servings Males 19–50: 8–10 servings Females 51+: 7 servings Males 51+: 7 servings
	Meat and Alternatives	"As is"	Females 19–50: 2 servings Males 19–50: 3 servings Females 51+: 2 servings Males 51+: 3 servings
	Fats	"Consume in Moderation"	Referred to as "Oils and Fats" Include small amount 30 – 45 ml (2 to 3 tablespoons) of unsaturated fat each day Limit saturated and trans fats
Food item classification	Cheese	Classified under "Meat and Alternatives"	Classified under "Milk and Alternatives"
	Potato, Corn, Sweet potato, Plantains	Classified under "Grains and Starches"	Classified under "Vegetables and Fruit"
	Nuts	Classified under "Fats"	Classified under "Meat and Alternatives"
	Avocado	Classified under "Fats"	Classified under "Vegetables and Fruit"

World Health Organization Nutrition Recommendations

The Population Nutrient Goals from WHO

Dietary Factor	**Goal (% of total energy intake, unless otherwise stated)**
Total dietary fat	15-30
Saturated fatty acids (SFA)	<10
Polyunsaturated fatty acids (PUFA)	6-10
omega-6 PUFA	5-8
omega-3 PUFA	1-2
Trans fatty acids (TFA)	<1%
Monounsaturated fatty acids	(total dietary fat – [SFA +PUFA + TFA])
Total carbohydrate	55-75
Free sugars (all monosaccharides and disaccharides added during processing + monosaccharides and disaccharides present in honey, syrups, and fruit juices)	<10
Protein	10-15
Cholesterol	<300 mg/day
Sodium chloride (sodium)—salt should be iodized	<5 g/day (<2 g/day)
Vegetables and fruit	≥400 g/day
Total dietary fibre (obtained from whole grain cereals, vegetables, fruit)	>25 g/day
Non-starch polysaccharides (NSP) (obtained from whole grain cereals, vegetables, fruit)	>20 g/day

Source: World Health Organization. Diet, Nutrition, and the Prevention of Chronic Diseases. WHO Technical Report Series 916, 2003. Available online at http://whqlibdoc.who.int/trs/who_trs_916.pdf. Accessed August 5, 2011.

Appendix G

U.S. Nutrition Recommendations

The 2010 edition of the Dietary Guidelines for Americans, published by the United States Department of Agriculture and the Department of Health and Human Services, and available at http://www.cnpp.usda.gov/DGAs2010-PolicyDocument.htm, provides general recommendations for developing healthy eating and lifestyle patterns. Because of the high prevalence of obesity in the United States, the 2010 recommendations emphasize reducing the intake of calories and increasing physical activity. Americans are encouraged to eat more nutrient-dense foods such as vegetables, fruits, whole grains, fat-free and low-fat dairy products, and fish, and to reduce their intake of sodium, *trans* and saturated fats, added sugars, and refined grain products.

Applying the Dietary Guidelines with MyPlate

To help Americans apply the recommendations set out in the Dietary Guidelines to their daily lives, the USDA has developed an educational icon called *MyPlate*. The plate illustrates 5 food groups: grains, vegetables, fruits, dairy, and protein; it also helps people get a sense of the portions of each food group that should be consumed as well as the amounts of food overall that should be eaten **(Figure G.1)**.

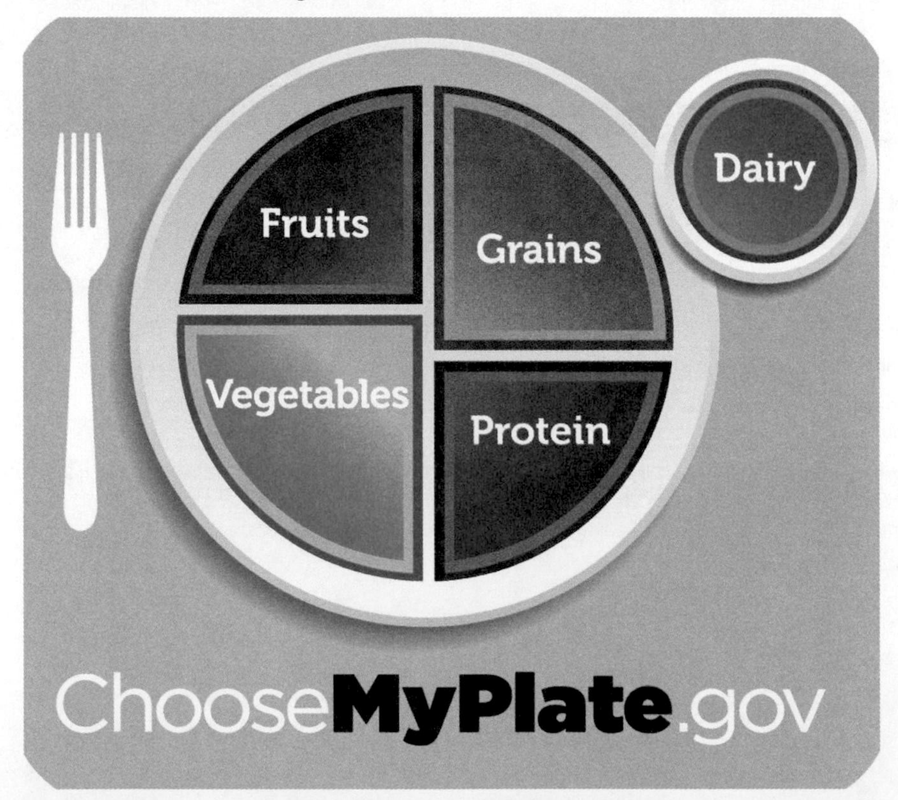

Figure G.1 The MyPlate icon. The MyPlate icon was developed to help Americans make healthier food choices. Half the plate should be covered by fruits and vegetables, while only a quarter should be taken up by protein and grains, respectively.
Source: USDA Choose MyPlate. Available online at www.ChooseMyPlate.gov. Accessed August 5, 2011.

To personalize MyPlate to each individual's needs, one can visit the www.ChooseMyPlate.gov website. After entering personal information such as sex, age, height, weight, and activity level, a person is told how much of each food group should be eaten daily. If someone is overweight, the website can also direct him or her to a weight-loss plan that reduces the number of calories consumed.

The amount of food recommended from each food group is expressed in measures such as cups, ounces, and teaspoons. For example, for an 18-year-old male who exercises 30-60 minutes a day, *MyPlate* recommends a daily diet that contains 10 oz of grains, 3.5 cups of vegetables, 2.5 cups of fruit, 7 oz of proteins, 3 cups of milk, and 8 tsp of oil, for a total of 2,800 kcal. A person can also enter the foods he or she has eaten and analyze the nutrient content of the diet using online features called the Tracker and Menu Planner, found at http://www.choosemyplate.gov/tools.html.

By choosing nutrient-dense foods, as recommended by the dietary guidelines, it is possible to meet daily nutrient needs with fewer kcalories than is required to maintain a constant weight. For example, for a 2,000-kcal diet, approximately 265 kcals remain after the specified number of choices have been made from each food group. These extra kcalories are referred to as discretionary kilocalories **(Figure G.2)**. Discretionary kilocalories can be spent on foods from within the food groups, such as cookies or cakes, or from dietary substances that are not in a food group, such as butter, added sugar, or alcohol, referred to as empty calories. Americans are advised not exceed their allotment of discretionary kilocalories.

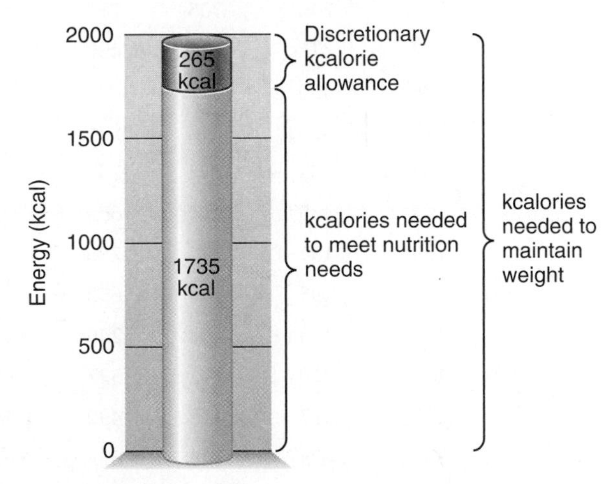

Figure G.2 Discretionary kilocalories

Appendix H Energy Expenditure for Various Activities

Type of Activity	Kcalories per Hour (by body weight)				
	45 kg/ 100 lb	55 kg/ 120 lb	68 kg/ 150 lb	82 kg/ 180 lb	90 kg/ 200 lb
Aerobics, high impact	318	381	476	572	635
Aerobics, low impact	227	272	340	408	454
Backpacking, general	318	381	476	572	635
Ballroom dancing, fast (disco, folk, square)	249	299	374	449	499
Ballroom dancing, slow (waltz, foxtrot)	136	163	204	245	272
Badminton, social singles and doubles, general	204	245	306	367	408
Baseball, playing catch	113	136	170	204	227
Basketball, game, structured	363	435	544	653	726
Basketball, shooting baskets	204	245	306	367	408
Boxing, punching bag	272	327	408	490	544
Boxing, sparring	408	490	612	735	816
Basketball, wheelchair	295	354	442	531	590
Bowling	136	163	204	245	272
Calisthenics, heavy, vigorous (pushups, pullups)	363	435	544	653	726
Calisthenics, light/moderate	159	191	238	286	318
Circuit training, general	363	435	544	653	726
Cleaning, heavy (wash car, wash windows, mop)	136	163	204	245	272
Cycling, <16 km/h (<10 mph), leisurely	181	218	272	327	363
Cycling, 16-19 km/h (10–12 mph), light	272	327	408	490	544
Cycling, 19-23 km/h (12–14 mph), moderate	363	435	544	653	726
Dancing, general	204	245	306	367	408
Fencing	272	327	408	490	544
Fishing, general	136	163	204	245	272
Football, competition	408	490	612	735	816
Football, playing catch	113	136	170	204	227
Golf, pulling clubs	227	272	340	408	454
Golf, using power cart	159	191	238	286	318
Gymnastics, general	181	218	272	327	363
Hacky Sack	181	218	272	327	363
Handball, general	544	653	816	980	1089
Hiking, cross country	272	327	408	490	544
Horseback riding, trotting	295	354	442	531	590
Ice hockey	363	435	544	653	726
Ice or in-line skating, general	318	381	476	572	635
Ice or in-line skating, speed, competition	680	816	1021	1225	1361
Jai alai	544	653	816	980	1089

Jog/walk combination	272	327	408	490	544
Jumping rope, moderate, general	454	544	680	816	907
Kayaking	227	272	340	408	454
Mowing lawn, general	249	299	374	449	499
Playing with children, heavy (walk/run)	227	272	340	408	454
Playing with children, light (standing)	127	152	191	229	254
Racquetball, casual, general	318	381	476	572	635
Rowing or canoeing, 6.4–10km/h (4.0–5.9 mph)	318	381	476	572	635
Running, 16 km/h (10 mph)	726	871	1089	1306	1451
Running, 12 km/h (7.5 mph)	567	680	850	1021	1134
Running, 14 km/h (8.6 mph)	635	762	953	1143	1270
Sailing, Sunfish/Laser/Hobby Cat, keel boat, ocean	136	163	204	245	272
Skateboarding	227	272	340	408	454
Skiing, cross country, vigorous downhill	363	435	544	653	726
Skiing, downhill, moderate effort	272	327	408	490	544
Soccer, casual, general	318	381	476	572	635
Softball, fast or slow pitch	227	272	340	408	454
Surfing, body or board	136	163	204	245	272
Swimming, laps, freestyle, fast	454	544	680	816	907
Swimming, laps, freestyle, slow	363	435	544	653	726
Table tennis (ping pong)	181	218	272	327	363
Tae kwan do, judo, juijitsu, karate, kick boxing	454	544	680	816	907
Tai chi	181	218	272	327	363
Tennis, general	318	381	476	572	635
Volleyball, noncompetitive, general	136	163	204	245	272
Walking, 4 km/h (2.5 mph), firm surface	136	163	204	245	272
Walking, 6.4 km/h (4 mph), level, firm surface	227	272	340	408	454
Weight lifting, free or machine	272	327	408	490	544
Yoga, hatha	113	136	170	204	227

Data provided by Axxya Systems—Nutritionist Pro.

Appendix I Chemistry, Metabolism, and Structures

A Review of Basic Chemistry

Chemistry is the science of the structure and interactions of matter. All living and nonliving things consist of matter, which is anything that occupies space and has mass. Mass is the amount of matter in any object, which does not change.

Chemical Elements

All forms of matter—both living and nonliving—are made up of a limited number of building blocks called chemical elements. Each element is a substance that cannot be split into a simpler substance by ordinary chemical means. Scientists now recognize 112 elements. Of these, 92 occur naturally on Earth. The rest have been produced from the natural elements using particle accelerators or nuclear reactors. Each element is designated by a chemical symbol, one or two letters of the element's name in English, Latin, or another language. Examples of chemical symbols are H for hydrogen, C for carbon, O for oxygen, N for nitrogen, Ca for calcium, and Na for sodium.

Twenty-six different elements are normally present in the human body. Just 4 elements, called the major elements, constitute about 96% of the body's mass: oxygen (O), carbon (C), hydrogen (H), and nitrogen (N). Eight others, the lesser elements, contribute 3.8% of the body's mass: calcium (Ca), phosphorus (P), potassium (K), sulfur (S), sodium (Na), chlorine (Cl), magnesium (Mg), and iron (Fe). An additional 14 elements—the trace elements—are present in tiny amounts. Together, they account for the remaining 0.2% of the body's mass.

Structure of Atoms

Each element is made up of atoms, the smallest units of matter that retain the properties and characteristics of the element. Atoms are extremely small. Hydrogen atoms, the smallest atoms, have a diameter less than 0.1 nanometer, and the largest atoms are only 5 times larger.

Dozens of different subatomic particles compose individual atoms. However, only three types of subatomic particles are important for understanding the chemical reactions in the human body: protons, neutrons, and electrons (**Figure I.1**). The dense central core of an atom is its nucleus. Within the nucleus are positively charged protons and uncharged (neutral) neutrons. The tiny, negatively charged electrons move about in a large space surrounding the nucleus. They do not follow a fixed path or orbit but instead form a negatively charged "cloud" that envelops the nucleus.

Even though their exact positions cannot be predicted, specific groups of electrons are most likely to move about within certain regions around the nucleus. These regions, called **electron shells,** are depicted as simple circles around the nucleus. Because each electron shell can hold a specific number of electrons, the electron shell model best conveys this aspect of atomic structure (see Figure I.1). The first electron shell (nearest the nucleus) never holds more than 2 electrons. The second shell holds a maximum of 8 electrons, and the third can hold up to 18 electrons. The electron shells fill with electrons in a specific order, beginning with the first shell. For example, notice in **Figure I.2** that sodium (Na), which has 11 electrons total, contains 2 electrons in the first shell, 8 electrons in the second shell, and 1 electron in the third shell. The number of electrons in an atom of an element always equals the number of protons. Because each electron and proton carries one charge, the negatively charged electrons and

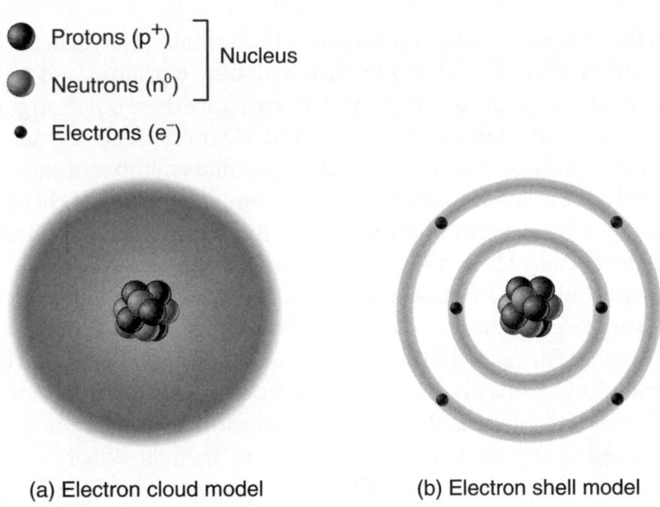

Figure I.1 Two representations of the structure of an atom. Electrons move about the nucleus, which contains neutrons and protons. (a) In the electron cloud model of an atom, the shading represents the chance of finding an electron in regions outside the nucleus. (b) In the electron shell model, filled circles represent individual electrons, which are grouped into concentric circles according to the shells they occupy. Both models depict a carbon atom, with 6 protons, 6 neutrons, and 6 electrons.

the positively charged protons balance each other. Thus, each atom is electrically neutral; its total charge is zero.

The number of protons in the nucleus of an atom is an atom's **atomic number**. For example, hydrogen has an atomic number of 1 because its nucleus has 1 proton, whereas sodium has an atomic number of 11 because its nucleus has 11 protons (see Figure I.2).

The **mass number** of an atom is the sum of its protons and neutrons. Because sodium has 11 protons and 12 neutrons, its mass number is 23 (see Figure I.2). Although all atoms of one element have the same number of protons, they may have different numbers of neutrons and thus different mass numbers. Isotopes are atoms of an element that have different numbers

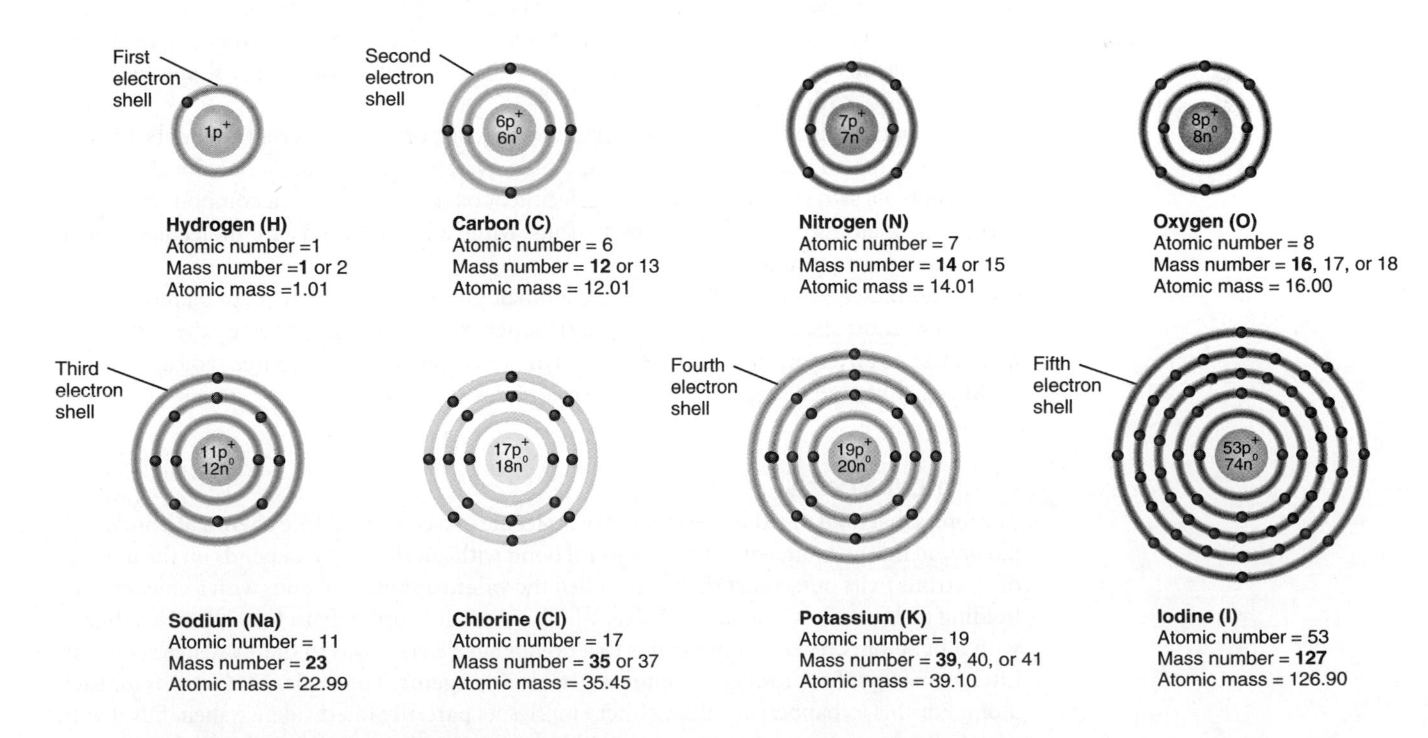

Figure I.2 Atomic structures of several stable atoms. The atoms of different elements have different atomic numbers because they have different numbers of protons.

of neutrons and therefore different mass numbers. In a sample of oxygen, for example, most atoms have 8 neutrons, and a few have 9 or 10, but all have 8 protons and 8 electrons. Most isotopes are stable, which means that their nuclear structure does not change over time. The stable isotopes of oxygen are designated ^{16}O, ^{17}O, and ^{18}O (or O-16, O-17, and O-18). As you may already have determined, the numbers indicate the mass number of each isotope. As you will discover shortly, the number of electrons of an atom determines its chemical properties. Although the isotopes of an element have different numbers of neutrons, they have identical chemical properties because they have the same number of electrons.

Certain isotopes, called radioactive isotopes, are unstable; their nuclei decay into a stable configuration. As they decay, these atoms emit radiation and in the process often transform into a different element. For example, the radioactive isotope of carbon, C-14, decays to N-14. The decay of a radioisotope may be as fast as a fraction of a second or as slow as millions of years. The half-life of an isotope is the time required for half of the radioactive atoms in a sample of that isotope to decay into a more stable form. The half-life of C-14, which is used to determine the age of organic samples, is 5,600 years, whereas the half-life of I-131, an important clinical tool, is 8 days.

Ions, Molecules, and Compounds

The atoms of each element have a characteristic way of losing, gaining, or sharing their electrons when interacting with other atoms to achieve stability. The way that electrons behave enables atoms in the body to exist in electrically charged forms called ions, or to join with each other into the complex combinations called molecules. If an atom either *gives up* or *gains* electrons, it becomes an ion. An ion is an atom that has a positive or negative charge because it has unequal numbers of protons and electrons. An ion of an atom is symbolized by writing its chemical symbol followed by the number of its positive or negative charges. Thus, Ca^{2+} stands for a calcium ion that has 2 positive charges because it has lost 2 electrons.

When 2 or more atoms *share* electrons, the resulting combination is called a molecule. A *molecular formula* indicates the elements and the number of atoms of each element that make up a molecule. A molecule may consist of 2 atoms of the same kind, such as an oxygen molecule. The molecular formula for a molecule of oxygen is O_2. The subscript 2 indicates that the molecule contains 2 atoms of oxygen. Two or more different kinds of atoms may also form a molecule, as in a water molecule (H_2O). In H_2O, 1 atom of oxygen shares electrons with 2 atoms of hydrogen.

A compound is a substance that contains atoms of 2 or more different elements. Most of the atoms in the body are joined into compounds. Water (H_2O) and sodium chloride (NaCl), common table salt, are compounds. A molecule of oxygen (O_2) is not a compound because it consists of atoms of only one element. Thus, while all compounds are molecules, not all molecules are compounds.

A free radical is an electrically charged atom or group of atoms with an unpaired electron in the outermost shell. A common example is superoxide, which is formed by the addition of an electron to an oxygen molecule. Having an unpaired electron makes a free radical unstable, highly reactive, and destructive to nearby molecules.

Chemical Bonds

The forces that hold together the atoms of a molecule or a compound are **chemical bonds**. The likelihood that an atom will form a chemical bond with another atom depends on the number of electrons in its outermost shell, also called the **valence shell**. An atom with a valence shell holding 8 electrons is *chemically stable*, which means it is unlikely to form chemical bonds with other atoms. Two or more atoms that do not have 8 electrons in their valence shells can interact in ways that produce a chemically stable arrangement of 8 valence electrons for each atom. For this to happen, an atom either empties its partially filled valence shell, fills it with donated electrons, or shares electrons with other atoms. The way that valence electrons are distributed determines what kind of chemical bond results. An ionic bond is formed when positively and negatively charged ions are attracted to one another. As shown in **Figure I.3**, sodium has 1 valence electron and chlorine has 7 valence electrons. When an atom of sodium donates its sole valence electron to an atom of chlorine, the resulting positive and negative charges pull both ions tightly together, forming an ionic bond. The resulting compound is sodium chloride, written NaCl.

A **covalent bond** forms when 2 or more atoms *share* electrons rather than gaining or losing them. Atoms form a covalently bonded molecule by sharing one, two, or three pairs of valence electrons (**Figure I.4**). They are the most common chemical bonds in the body, and the compounds that result from them form most of the body's structures. In a **polar covalent bond**, the sharing of electrons between two atoms is unequal: the nucleus of one atom attracts the shared electrons more strongly than the nucleus of the other atom. When polar covalent bonds form, the resulting molecule has a partial negative charge near the atom that attracts electrons more strongly (**Figure I.5**).

Chemical Reactions

A **chemical reaction** occurs when new bonds form or old bonds break between atoms. Chemical reactions are the foundation of all life processes. Each chemical reaction involves energy

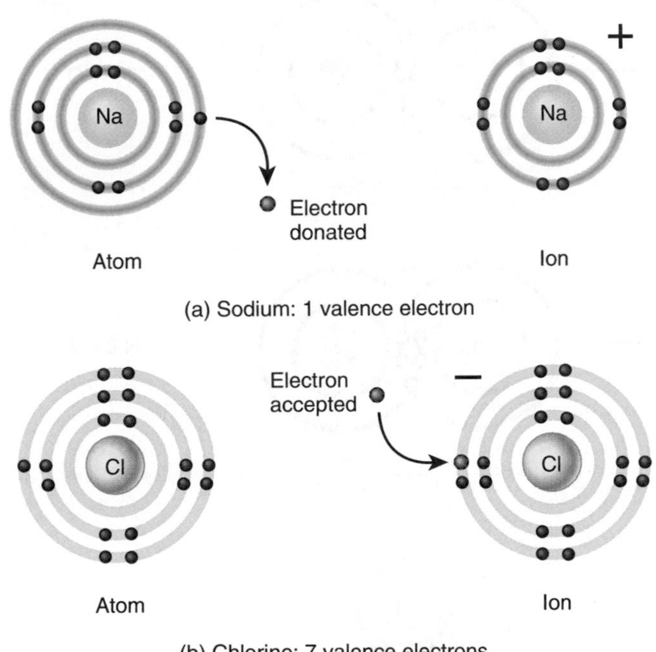

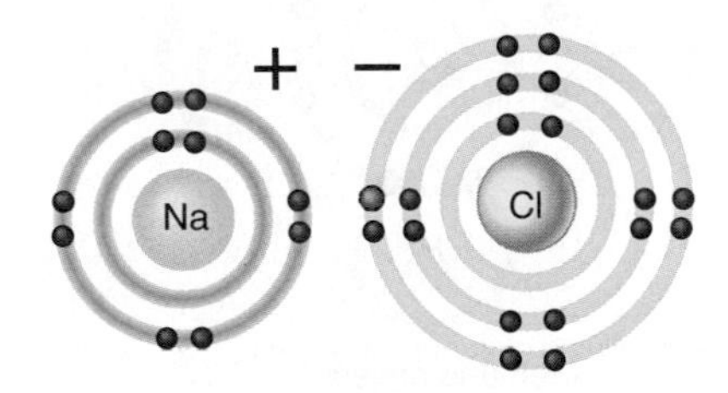

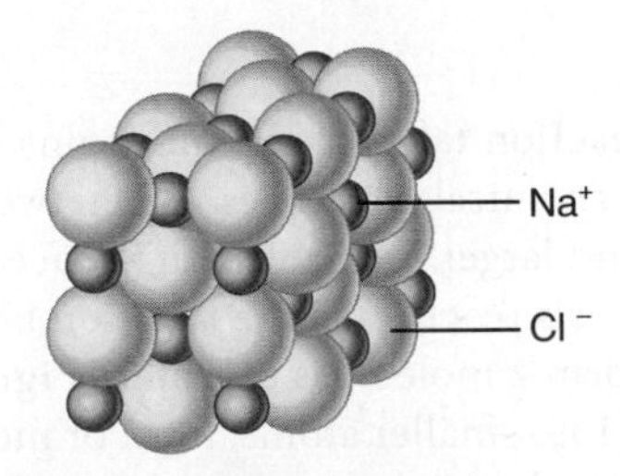

Figure I.3 Ions and ionic bond formation. (a) A sodium atom can have a complete octet of electrons in its outermost shell by losing 1 electron. (b) A chlorine atom can have a complete octet by gaining 1electron. (c) An ionic bond may form between oppositely charged ions.

changes. **Chemical energy** is a form of energy that is stored in the bonds of compounds and molecules. The total amount of energy present at the beginning and end of a chemical reaction is the same. Although energy can be neither created nor destroyed, it may be converted from one form to another.

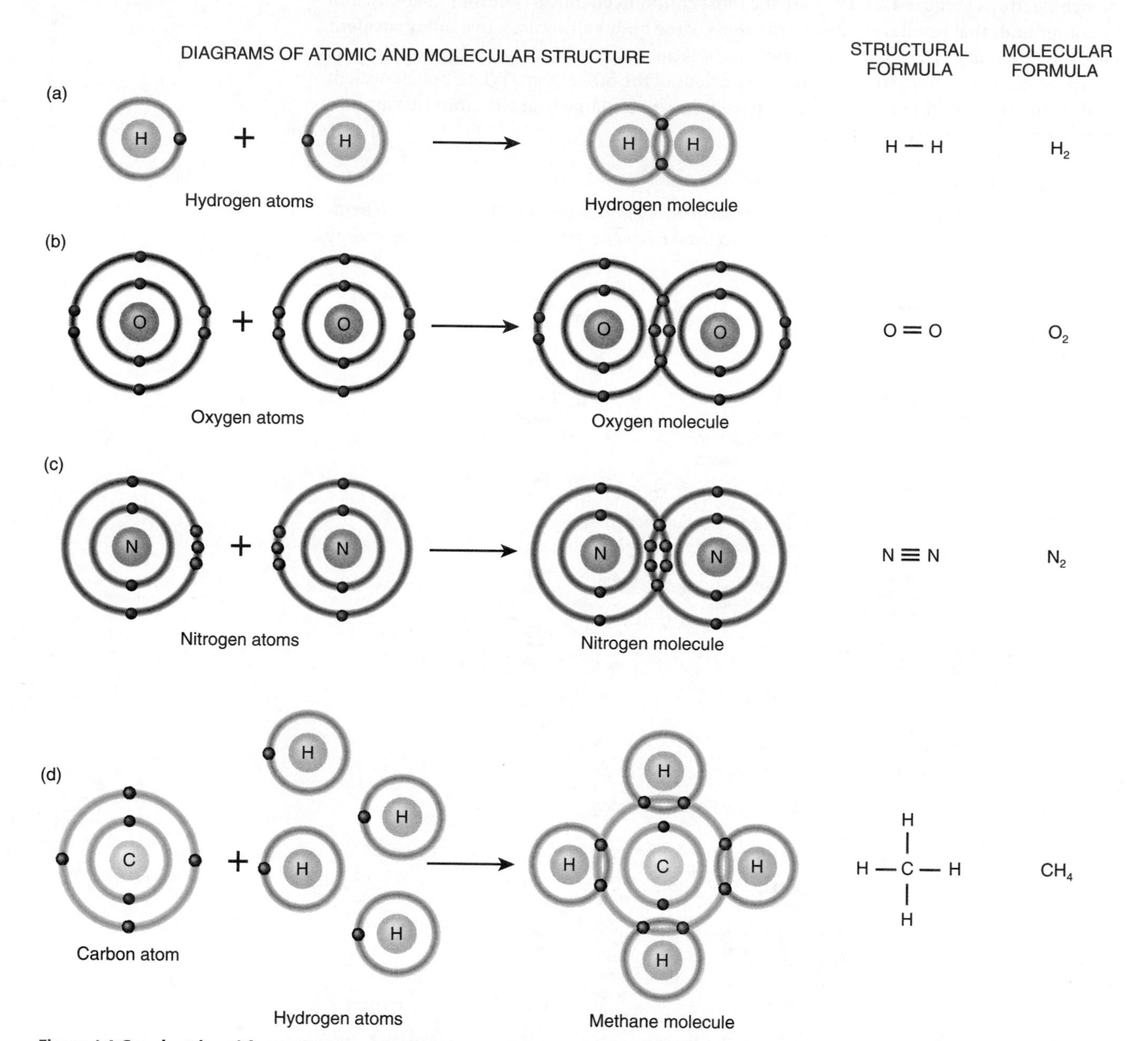

Figure I.4 Covalent bond formation. The red electrons are shared equally. In writing the structural formula of a covalently bonded molecule, each straight line between the chemical symbols for 2 atoms denotes a pair of shared electrons. In molecular formulas, the number of atoms in each molecule is noted by subscripts.

After a chemical reaction takes place, the atoms of the reactants are rearranged to yield products with new chemical properties. When two or more atoms, ions, or molecules combine to form new and larger molecules, the processes are called synthesis reactions. One example of a synthesis reaction is the reaction between 2 hydrogen molecules and 1 oxygen molecule to form 2 molecules of water (**Figure I.6**). Decomposition reactions split up large molecules into smaller atoms, ions, or molecules. For instance, the series of reactions that break down glucose to pyruvic acid, with the net production of 2 molecules of ATP, are important catabolic reactions in the body. Many reactions in the body are **exchange reactions**; they consist of both synthesis and decomposition reactions. Some chemical reactions may be reversible. In a **reversible reaction**, the products can revert to the original reactants.

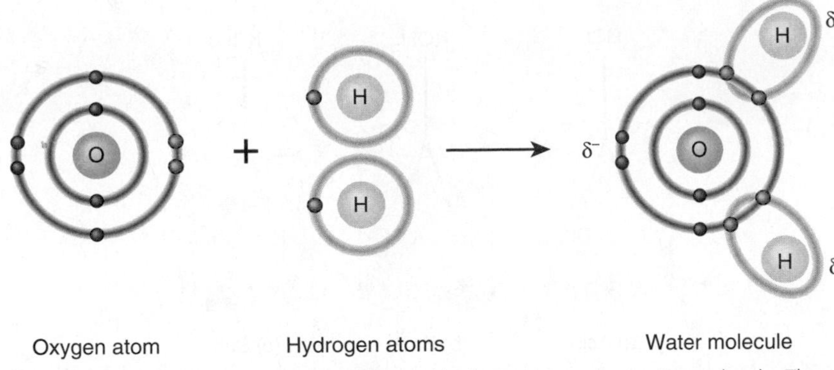

Figure I.5 Polar covalent bonds between oxygen and hydrogen atoms in a water molecule. The red electrons are shared unequally. Because the oxygen nucleus attracts the shared electrons more strongly, the oxygen end of a water molecule has a partial negative charge, written δ$^-$, and the hydrogen ends have partial positive charges, written δ$^+$.

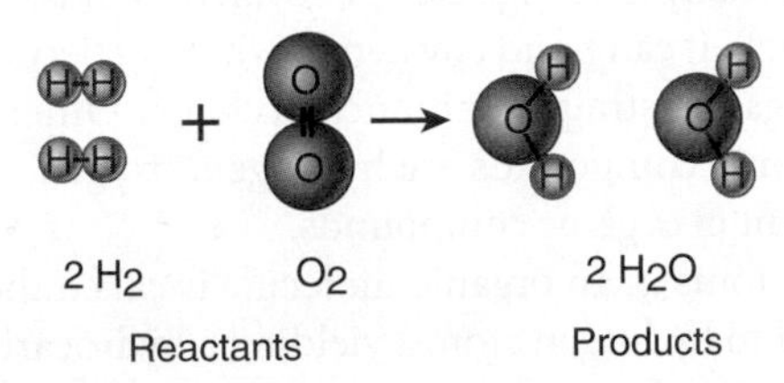

Figure I.6 The chemical reaction between 2 hydrogen molecules (H_2) and 1oxygen molecule (O_2) to form 2 molecules of water (H_2O). Note that the reaction occurs by breaking old bonds and making new bonds.

Inorganic Compounds and Solutions

Most of the chemicals in your body exist in the form of compounds. **Inorganic compounds** contain no more than 1 carbon atom. They include water and many salts, acids, and bases. Inorganic compounds may have either ionic or covalent bonds. **Organic compounds** always contain carbon and always have covalent bonds. Most are large molecules and many are made up of long chains of carbon atoms.

Inorganic Acids, Bases, and Salts When inorganic acids, bases, or salts dissolve in water, they dissociate; they separate into ions and become surrounded by water molecules. An acid (**Figure I.7a**) is a substance that dissociates into one or more **hydrogen ions** (H) and one or more negatively charged anions. Because H is a single proton with 1 positive charge, an acid is also referred to as a **proton donor**. A **base**, by contrast (**Figure I.7b**), removes H^+ from a solution and is therefore a **proton acceptor**.

To ensure homeostasis, intracellular and extracellular fluids must contain almost balanced quantities of acids and bases. The more hydrogen ions (H^+) dissolved in a solution, the more acidic the solution; the more hydroxide ions (OH^-), the more basic (alkaline) the solution. The chemical reactions that take place in the body are very sensitive to even small changes in the acidity or alkalinity of the body fluids in which they occur. Any departure from the narrow limits of normal H^+ and OH^- concentrations greatly disrupts body functions.

A solution's acidity or alkalinity is expressed on the pH scale, which extends from 0 to 14. This scale is based on the concentration of H+ in moles per litre. The midpoint of the pH scale is 7, where the concentrations of H+ and OH– are equal. A substance with a pH of 7, such as pure water, is neutral. A solution that has more H+ than OH– is an acidic solution and has a pH below 7. A solution that has more OH– than H+ is a basic (alkaline) solution and has a pH above 7. Although the pH of body fluids may differ, as we have discussed, the normal limits for each fluid are quite narrow. Homeostatic mechanisms maintain the pH of blood between 7.35 and 7.45, which is slightly more basic than pure water. If the pH of blood falls below 7.35, a condition called acidosis occurs, and if the pH rises above 7.45, it results in a condition called alkalosis; both conditions can seriously compromise homeostasis. Even though strong acids and bases are continually taken into and formed by the body, the pH of fluids inside and outside cells remains almost constant. One important reason for this is the presence of buffer systems, which function to convert strong acids or bases into weak acids or bases.

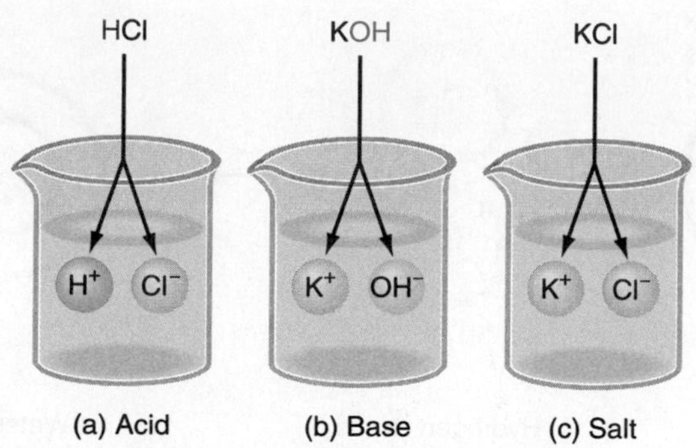

Figure I.7 Dissociation of inorganic acids, bases, and salts. Dissociation is the separation of inorganic acids, bases, and salts into ions in a solution.

Organic Compounds

Organic compounds are usually held together by covalent bonds. Carbon has 4 electrons in its outermost (valence) shell. It can bond covalently with a variety of atoms, including other carbon atoms, to form rings and straight or branched chains. Other elements that most often bond with carbon in organic compounds are hydrogen, oxygen, and nitrogen. Sulfur and phosphorus are also present in organic compounds.

The chain of carbon atoms in an organic molecule is called the carbon skeleton. Many of the carbons are bonded to hydrogen atoms, yielding a hydrocarbon. Also attached to the carbon skeleton are distinctive functional groups, other atoms or molecules bound to the hydrocarbon skeleton. Each type of functional group has a specific arrangement of atoms that confers characteristic chemical properties upon the organic molecule attached to it.

Nucleic Acids: Deoxyribonucleic Acid (DNA) and Ribonucleic Acid (RNA) Nucleic acids, so named because they were first discovered in the nuclei of cells, are huge organic molecules that contain carbon, hydrogen, oxygen, nitrogen, and phosphorus. Nucleic acids are of two varieties. The first, deoxyribonucleic acid (DNA), forms the inherited genetic material inside each human cell (**Figure I.8**). In humans, each gene is a segment of a DNA molecule. Our genes determine the traits we inherit, and by controlling protein synthesis, they regulate most of the activities that take place in body cells throughout a lifetime. When a cell divides, its hereditary information passes on to the next generation of cells. **Ribonucleic acid (RNA)**, the second type of nucleic acid, relays instructions from the genes to guide each cell's synthesis of proteins from amino acids.

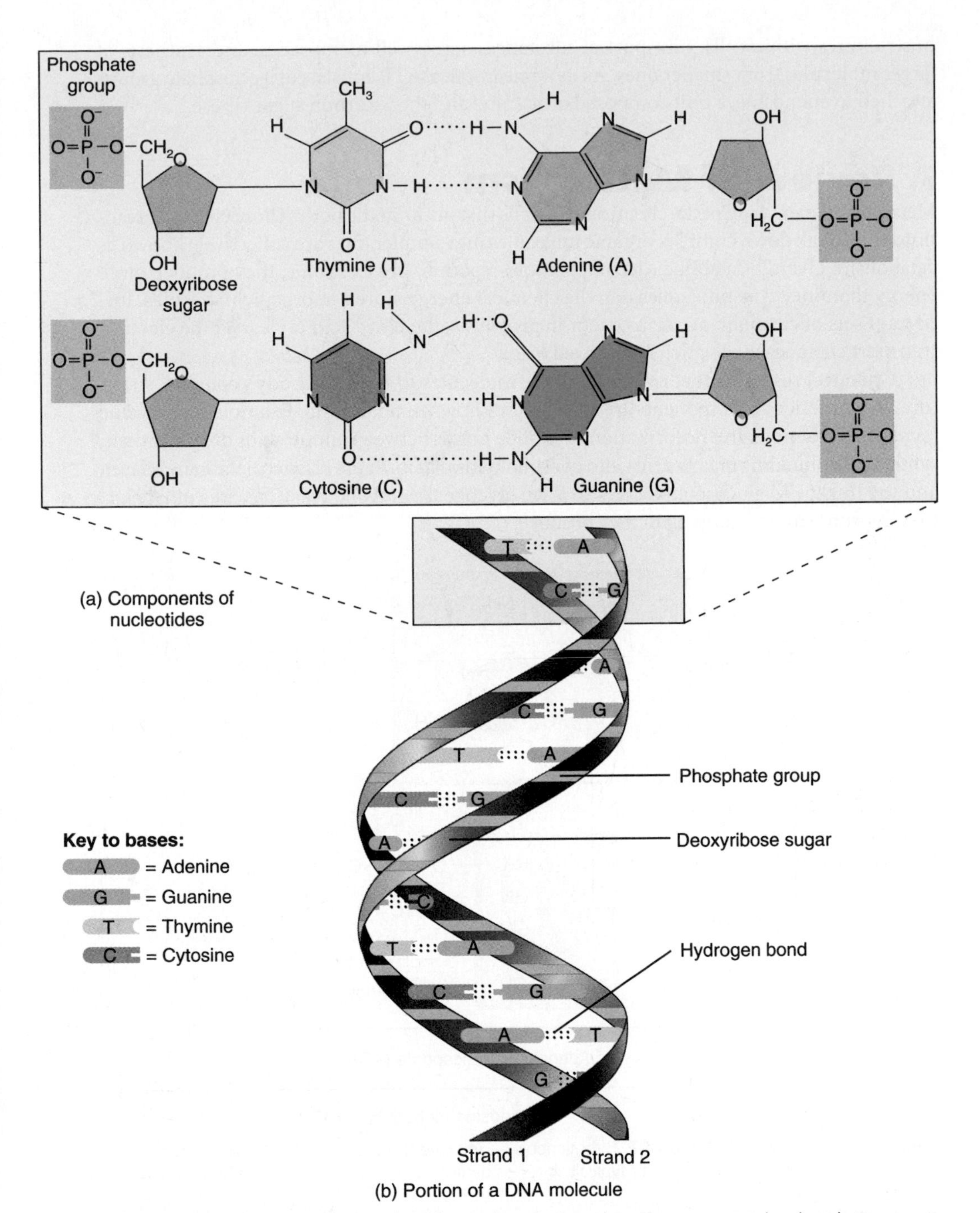

(b) Portion of a DNA molecule

Figure I.8 DNA molecule. (a) A nucleotide consists of a base, a pentose sugar, and a phosphate group. (b) The paired bases project toward the center of the double helix. The structure is stabilized by hydrogen bonds (dotted lines) between each base pair. There are 2 hydrogen bonds between adenine and thymine and 3 between cytosine and guanine. Hydrogen bonds result from the attraction of oppositely charged parts of molecules and are weak compared to ionic or covalent bonds.

DNA and RNA molecules consist of a chain of repeating nucleotides. Each nucleotide consists of 3 parts: a nitrogenous base, a pentose sugar, and a phosphate group. DNA contains the sugar deoxyribose and 4 different nitrogenous bases: adenine (A), thymine (T), cytosine (C), and guanine (G). Adenine and guanine are larger, double-ring bases called purines; thymine and cytosine are smaller, single-ring bases called pyrimidines. RNA contains the sugar ribose and instead of thymine it contains the pyrimidine base uracil (U).

Adenosine Triphosphate Adenosine triphosphate or **ATP** is the "energy currency" of living systems (**Figure I.9**). ATP transfers the energy liberated in exergonic catabolic reactions to power cellular activities that require energy (endergonic reactions). Among these cellular activities are muscular contractions, movement of chromosomes during cell division, movement of structures within cells, transport of substances across cell membranes, and synthesis of larger molecules from smaller ones. As its name implies, ATP consists of 3 phosphate groups attached to adenosine, a unit composed of adenine and the 5-carbon sugar ribose.

A Review of Metabolism

Metabolism refers to all of the chemical reactions that occur in the body. Those chemical reactions that break down complex organic molecules into simpler ones are collectively known as catabolism. Overall, catabolic (decomposition) reactions are *exergonic*; they produce more energy than they consume, releasing the chemical energy stored in organic molecules. Important sets of catabolic reactions occur in glycolysis, the citric acid cycle, and the electron transport chain, each of which is reviewed below.

Chemical reactions that combine simple molecules to form the body's complex structural and functional components are collectively known as anabolism. Examples of anabolic (synthesis) reactions are the formation of peptide bonds between amino acids during protein synthesis, the building of fatty acids into phospholipids that form the plasma membrane bilayer, and the linkage of glucose molecules to form glycogen. Anabolic reactions are *endergonic*; they consume more energy than they produce.

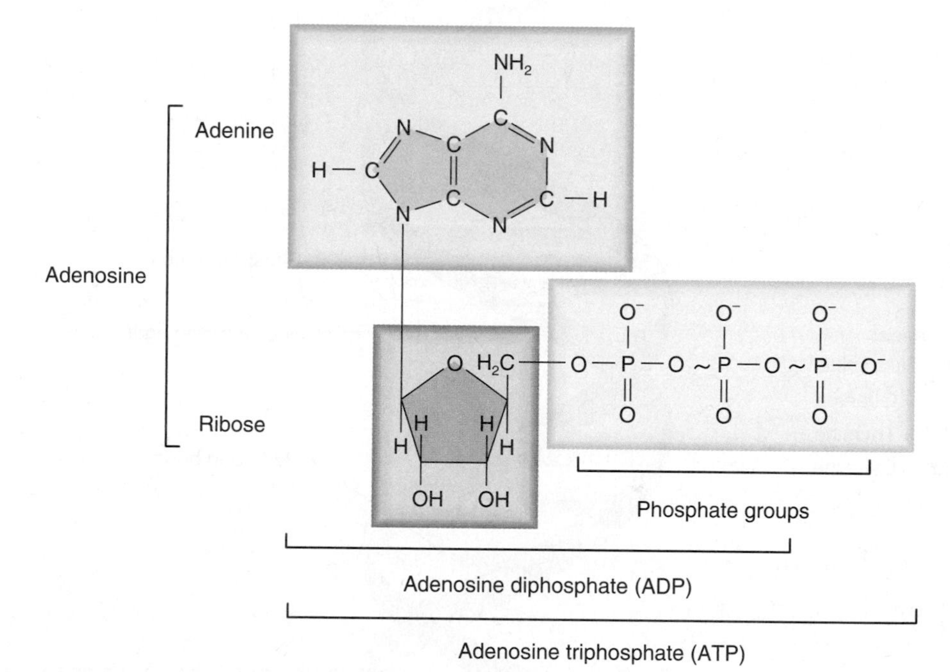

Figure I.9 Structures of ATP and ADP. Tilde symbols (~) indicate the 2 phosphate bonds that can be used to transfer energy. Energy transfer typically involves hydrolysis of the last phosphate bond of ATP.

Carbohydrate Metabolism

As discussed in Section 4.2, both polysaccharides and disaccharides are hydrolyzed into the monosaccharides glucose (about 80%), fructose, and galactose during the digestion of carbohydrates. Hepatocytes (liver cells) convert most of the remaining fructose and practically all the galactose to glucose. So the story of carbohydrate metabolism is really the story of glucose metabolism.

Glucose Catabolism The oxidation of glucose to produce ATP is also known as cellular respiration, and it involves 4 sets of reactions: glycolysis, the formation of acetyl coenzyme A, the citric cycle, and the electron transport chain (**Figure I.10**).

1. Glycolysis is a set of reactions in which 1 glucose molecule is oxidized and 2 molecules of pyruvic acid are produced (**Figure I.11**). The reactions also produce 2 molecules of ATP and 2 energy-containing NADH + H^+. Because glycolysis does not require oxygen, it is a way to produce ATP anaerobically (without oxygen) and is known as anaerobic cellular respiration.
2. Formation of acetyl coenzyme A is a transition step that prepares pyruvic acid for

entrance into the citric acid cycle. This step also produces energy-containing NADH + H^+ plus carbon dioxide (CO_2).

3. Citric acid cycle reactions oxidize acetyl coenzyme A and produce CO_2, ATP, energy-containing NADH + H^+, and $FADH_2$. (**Figure I.12, Figure I.13**)
4. Electron transport chain reactions oxidize NADH + H^+ and $FADH_2$ and transfer their electrons through a series of electron carriers. The citric acid cycle and the electron transport chain both require oxygen to produce ATP and are collectively known as aerobic cellular respiration.

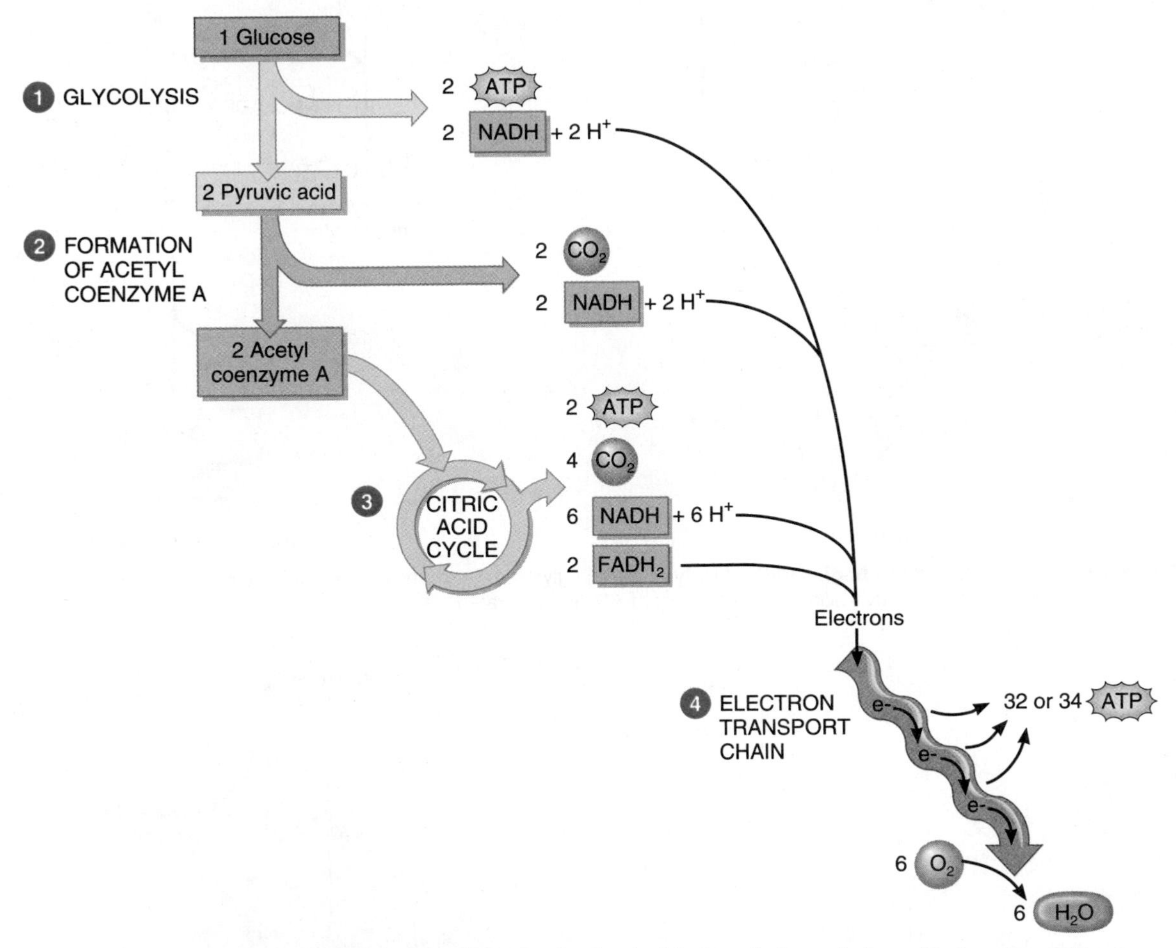

Figure I.10 Overview of cellular respiration (oxidation of glucose). A modified version of this figure appears in several places in this appendix to indicate the relationships of particular reactions to the overall process of cellular respiration. The oxidation of glucose involves glycolysis, the formation of acetyl coenzyme A, the citric acid cycle, and the electron transport chain.

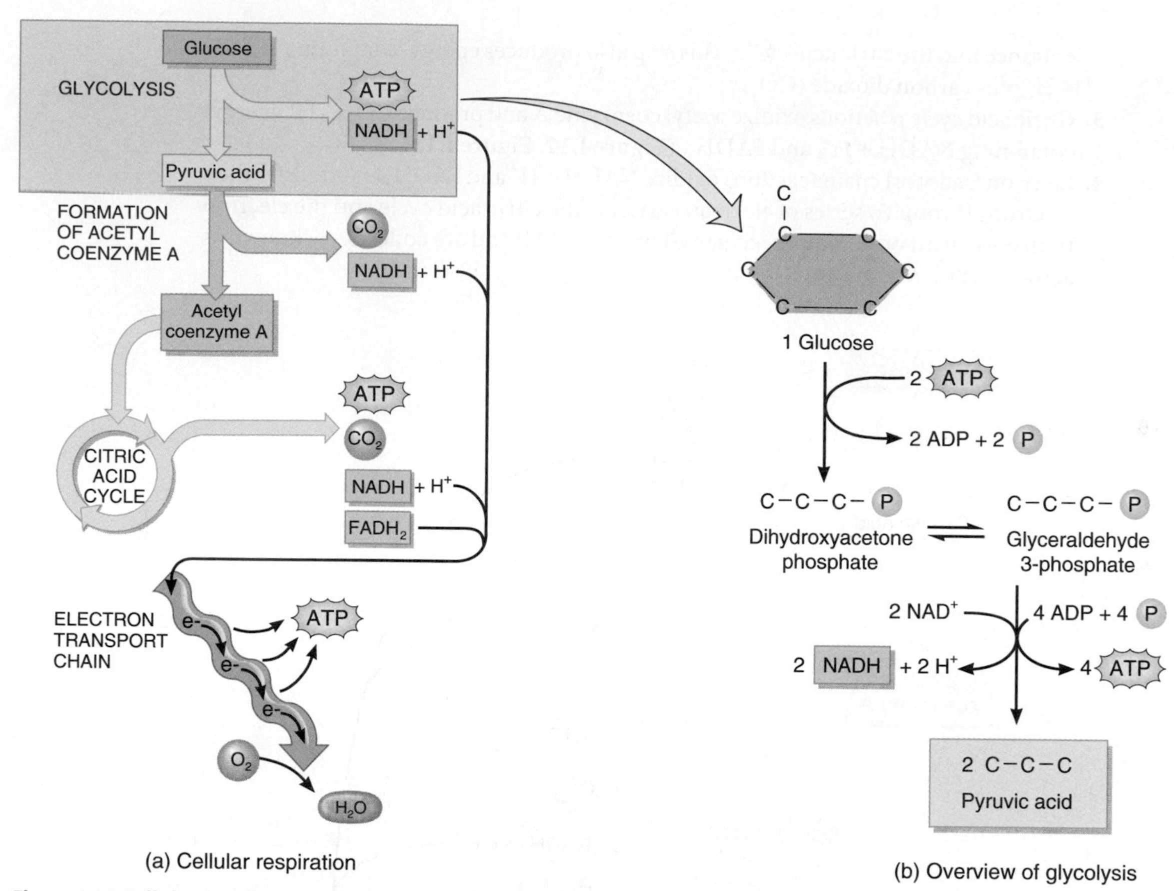

Figure I.11 Cellular respiration begins with glycolysis. During glycolysis, each molecule of glucose is converted to 2 molecules of pyruvic acid and 2 molecules of ATP are generated.

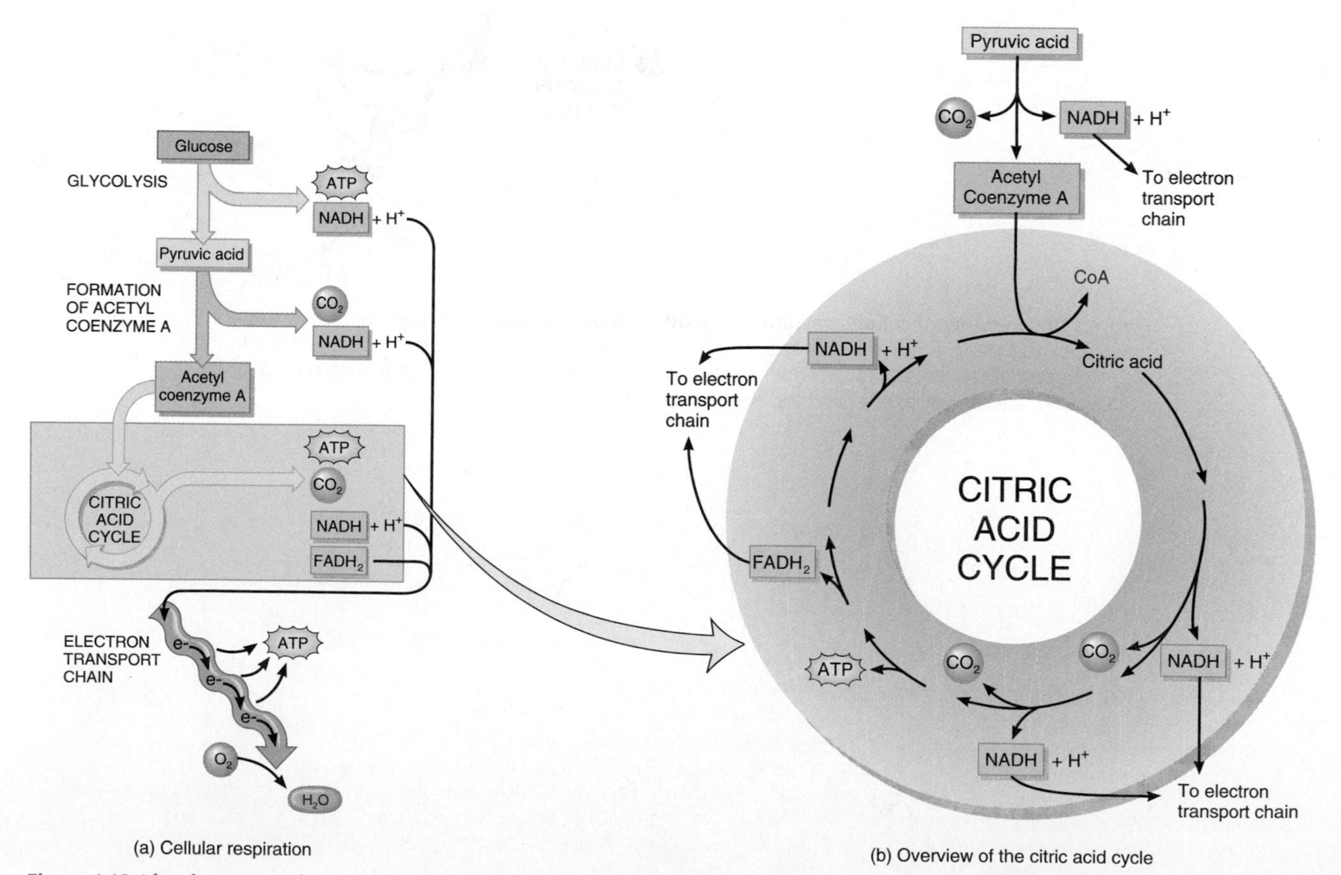

Figure I.12 After formation of acetyl coenzyme A, the next stage of cellular respiration is the citric acid cycle. Reactions of the citric acid cycle occur in the matrix of mitochondria.

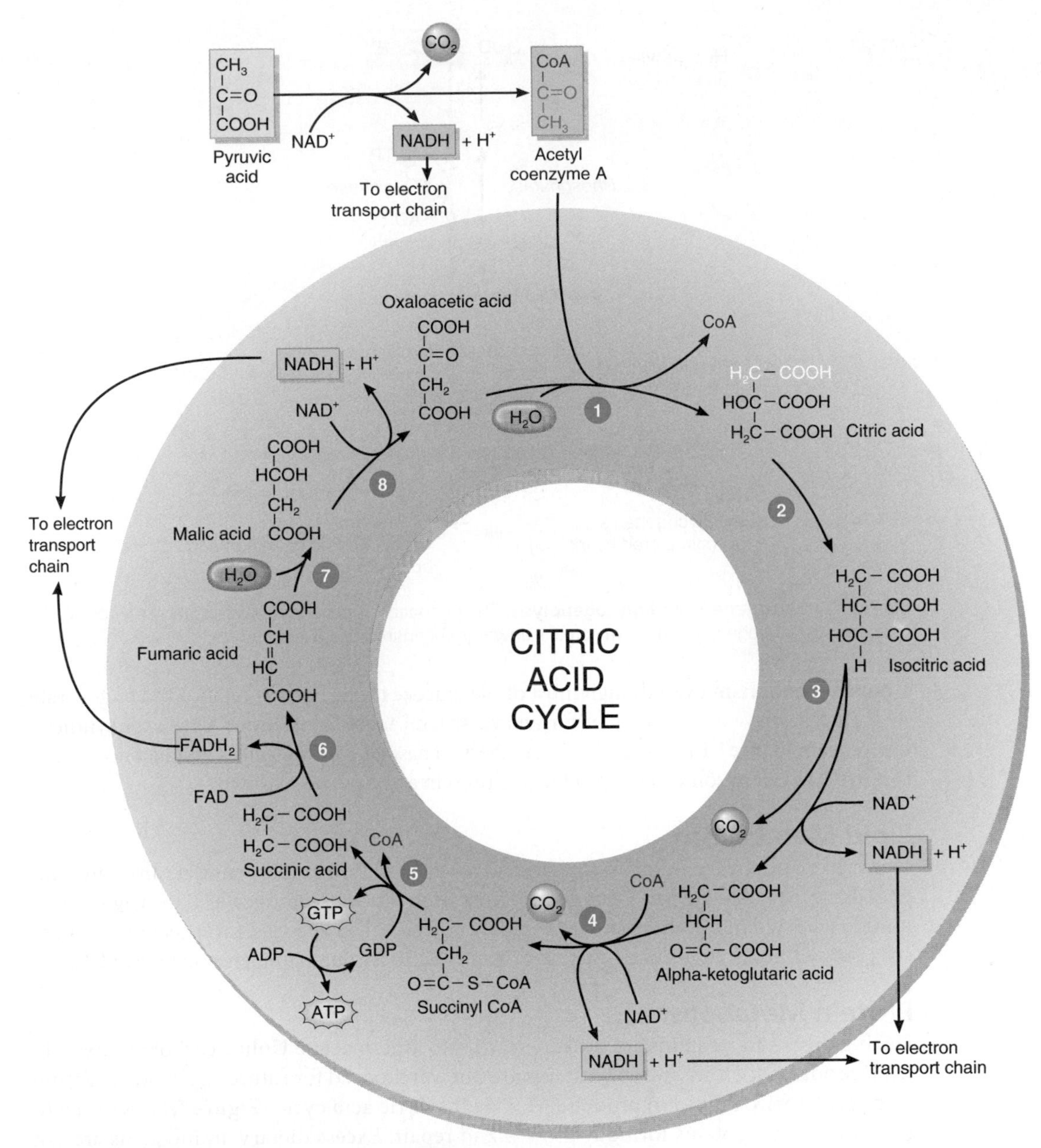

Figure I.13 The 8 reactions of the citric acid cycle.

1. Entry of the acetyl group. The chemical bond that attaches the acetyl group to coenzyme A (CoA) breaks, and the 2-carbon acetyl group attaches to a 4-carbon molecule of oxaloacetic acid to form a 6-carbon molecule called citric acid. CoA is free to combine with another acetyl group from pyruvic acid and repeat the process.
2. Isomerization. Citric acid undergoes isomerization to isocitric acid, which has the same molecular formula as citrate. Notice, however, that the hydroxyl group (-OH) is attached to a different carbon.
3. Oxidative decarboxylation. Isocitric acid is oxidized and loses a molecule of CO_2, forming alpha-ketoglutaric acid. The H^+ from the oxidation is passed on to NAD^+, which is reduced to NADH + H^+.
4. Oxidative decarboxylation. Alpha-ketoglutaric acid is oxidized, loses a molecule of CO_2, and picks up CoA to form succinyl CoA.
5. Substrate-level phosphorylation. CoA is displaced by a phosphate group, which is then transferred to guanosine diphosphate (GDP) to form guanosine triphosphate (GTP). GTP can donate a phosphate group to ADP to form ATP.
6. Dehydrogenation. Succinic acid is oxidized to fumaric acid as 2 of its hydrogen atoms are transferred to the coenzyme flavin adenine nucleotide (FAD), which is reduced to $FADH_2$.
7. Hydration. Fumaric acid is converted to malic acid by the addition of a molecule of water.
8. Dehydrogenation. In the final step in the cycle, malic acid is oxidized to re-form oxaloacetic acid. Two hydrogen atoms are removed and one is transferred to NAD^+, which is reduced to NADH + H^+. The regenerated oxaloacetic acid can combine with another molecule of acetyl CoA, beginning a new cycle. The 3 main results of the citric acid cycle are the production of reduced coenzymes (NADH + H^+ and $FADH_2$), which contain stored energy; the generation of GTP, a high-energy compound that is used to produce ATP; and the formation of CO_2, which is transported to the lungs and exhaled.

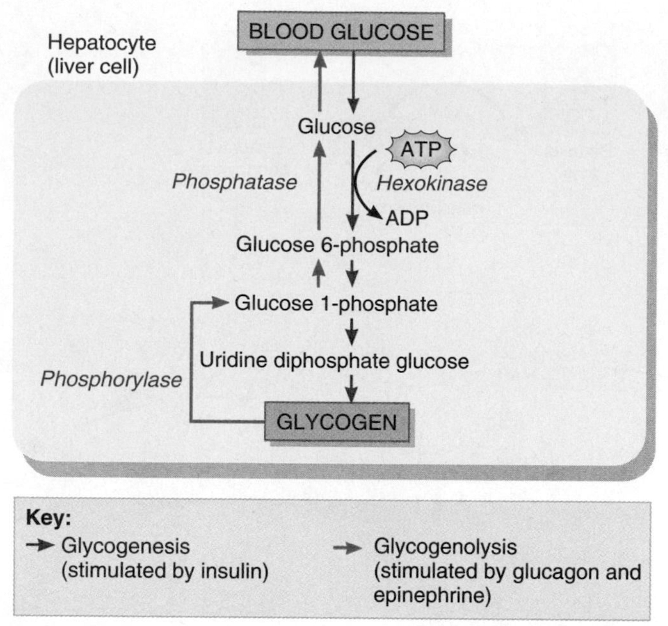

Figure I.14 Glycogenesis and glycogenolysis. The glycogenesis pathway converts glucose into glycogen; the glycogenolysis pathway breaks down glycogen into glucose.

Glucose Anabolism Even though most of the glucose in the body is catabolized to generate ATP, glucose may take part in or be formed via several anabolic reactions. One is the synthesis of glycogen (**Figure I.14**); another is the synthesis of new glucose molecules via gluconeogenesis (**Figure I.15**) from some of the products of protein and lipid breakdown.

Lipid Metabolism

Lipids, like carbohydrates, may be oxidized to produce ATP. In order for muscle, liver, and adipose tissue to oxidize the fatty acids derived from triglycerides to produce ATP, the triglycerides must first be split into glycerol and fatty acids, a process called lipolysis. Liver cells and adipose cells can synthesize lipids from glucose or amino acids through lipogenesis (**Figure I.16**).

Protein Metabolism

During digestion, proteins are broken down into amino acids. Unlike carbohydrates and triglycerides, which are stored, proteins are not warehoused for future use. Instead, amino acids are either oxidized to produce ATP via the citric acid cycle (**Figure I.17**) or used to synthesize new proteins for body growth and repair. Excess dietary amino acids are not excreted in the urine or feces but instead are converted into glucose (gluconeogenesis) or triglycerides (lipogenesis).

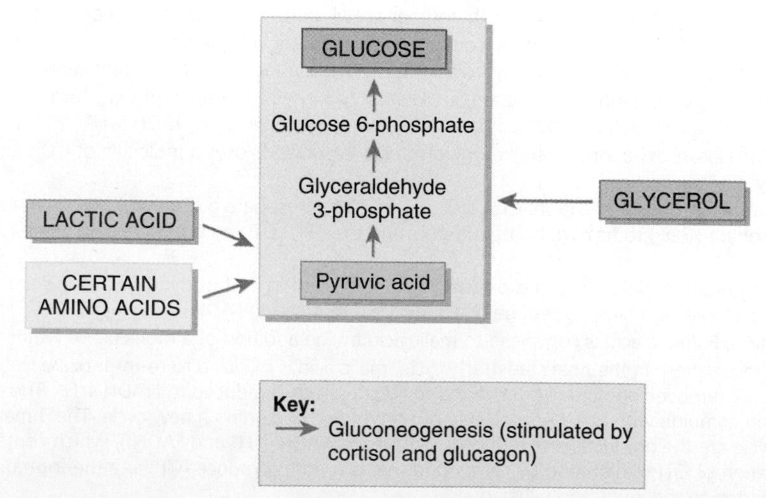

Figure I.15 Gluconeogenesis. The process of gluconeogenesis involves the conversion of noncarbohydrate molecules (amino acids, lactic acid, and glycerol) into glucose.

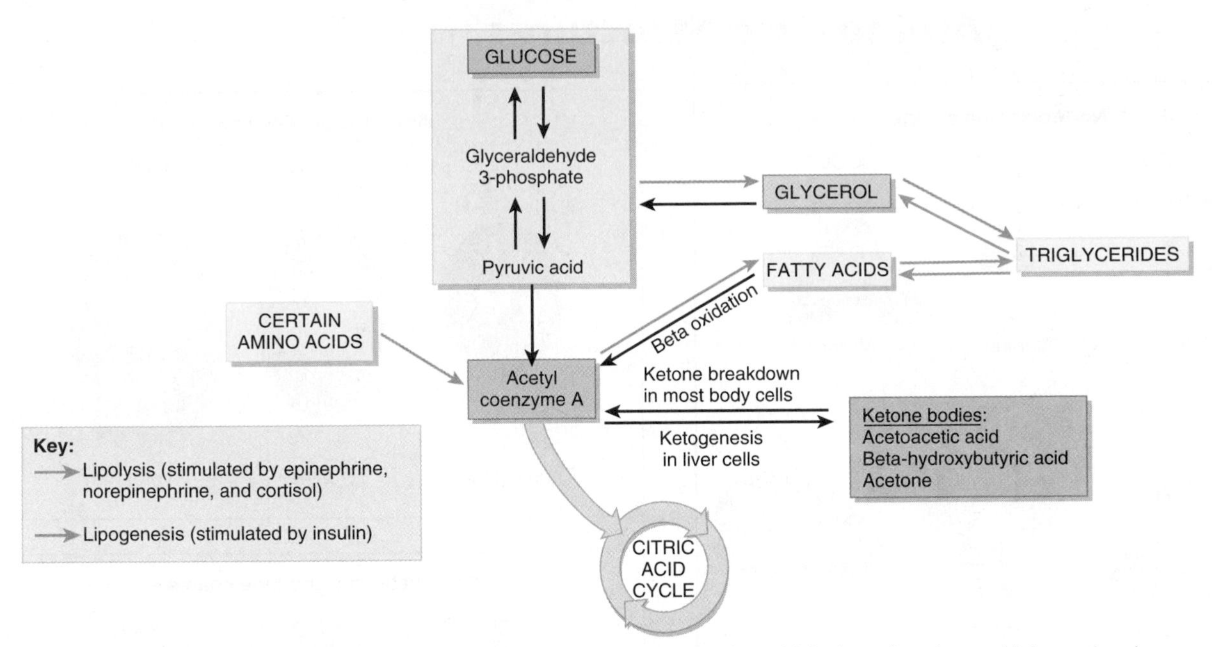

Figure I.16 Pathways of lipid metabolism. Glycerol may be converted to glyceraldehyde 3-phosphate, which can then be converted to glucose or enter the citric acid cycle for oxidation. Fatty acids undergo beta oxidation and enter the citric acid cycle via acetyl coenzyme A. The synthesis of lipids from glucose or amino acids is called lipogenesis.

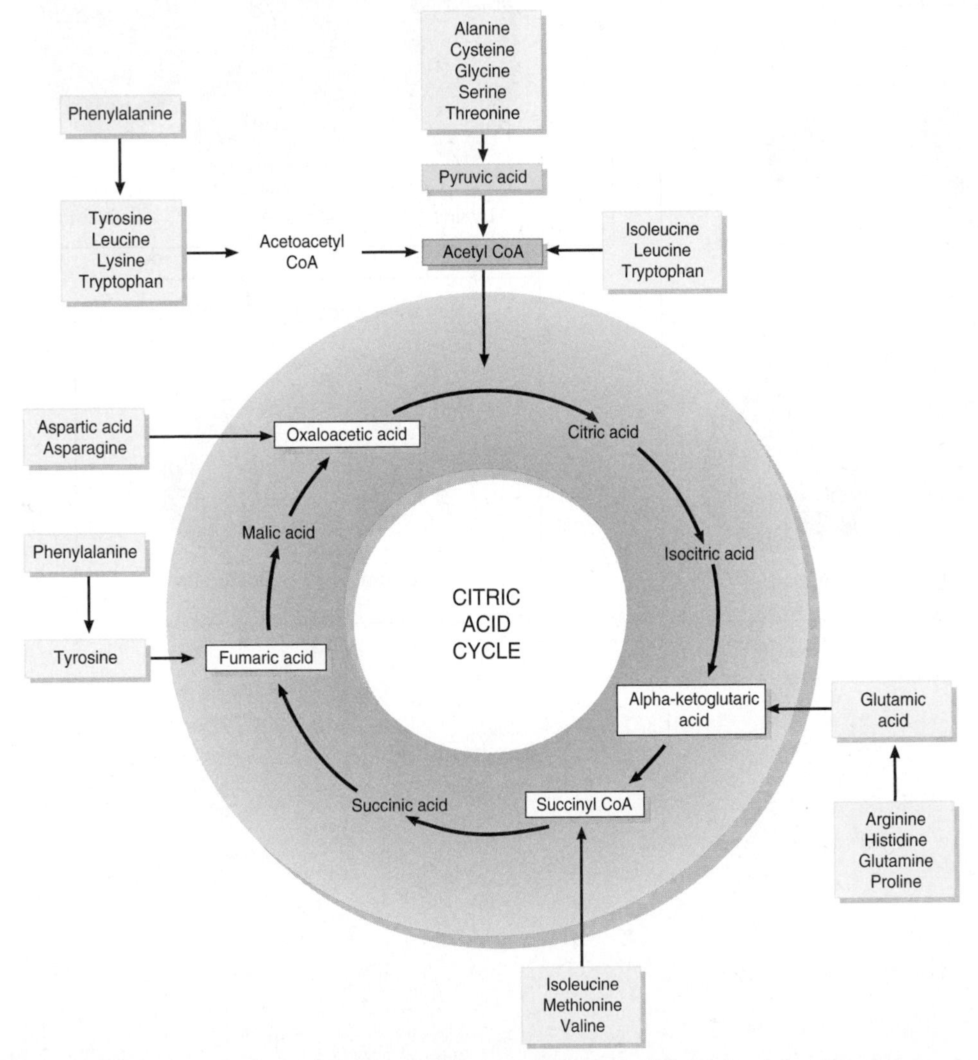

Figure I.17 Various points at which amino acids (shown in yellow boxes) enter the citric acid cycle for oxidation.

Amino Acid Structures

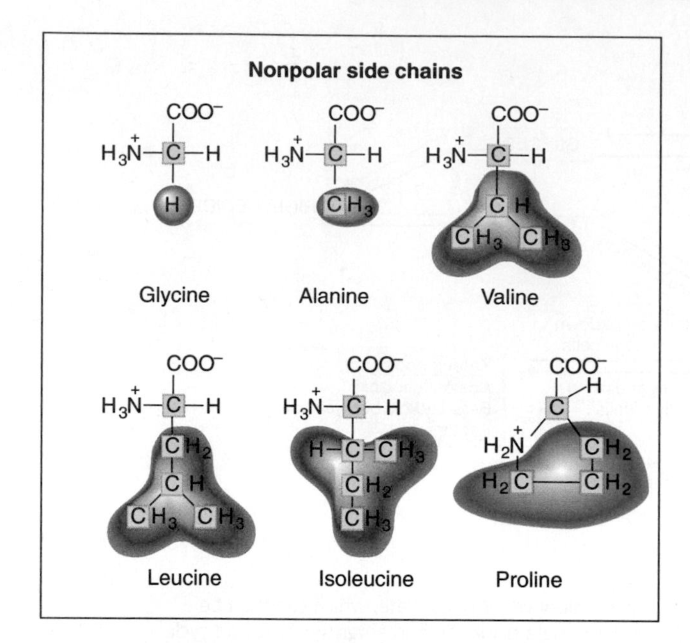

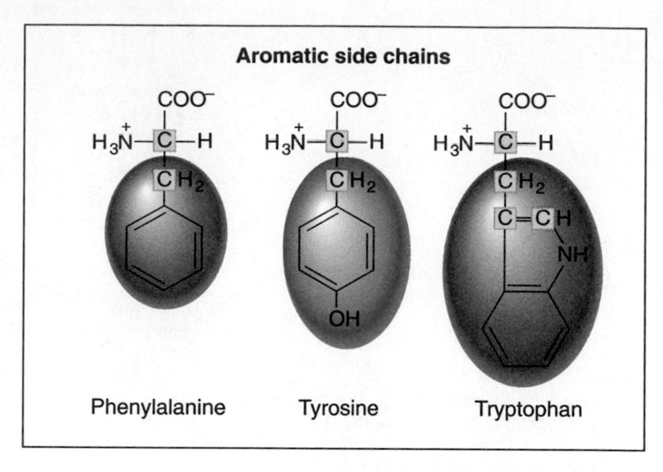

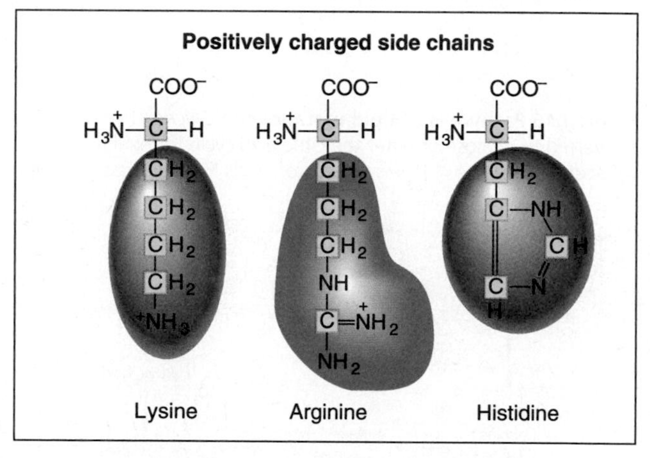

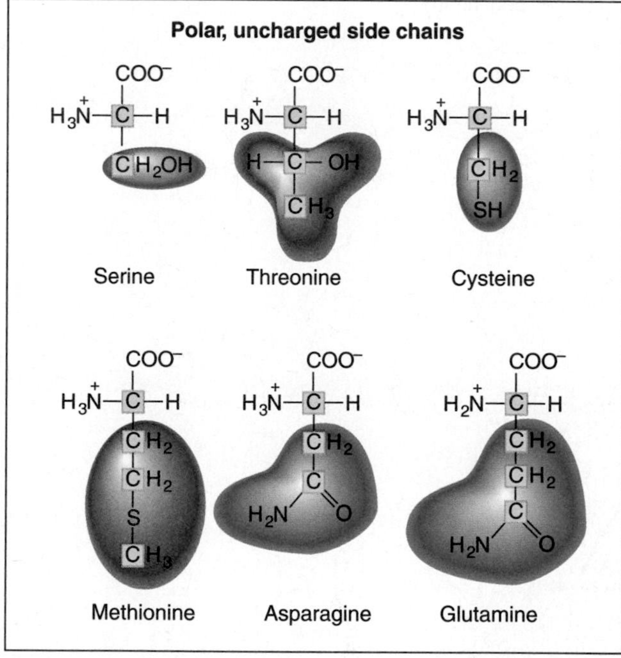

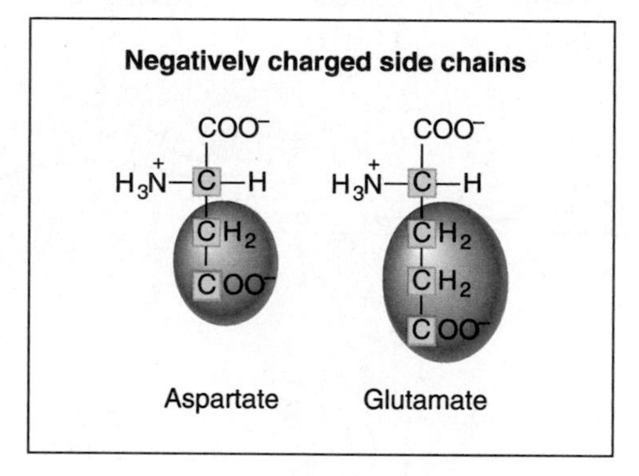

Essential amino acids side chain

Nonessential amino acids side chain

Vitamin Structures

Water-Soluble Vitamins

Thiamin

Thiamin structure

Thiamin pyrophosphate (TPP): The active coenzyme form of thiamin.

Riboflavin

Oxidized form

Reduced form

Riboflavin structure: In riboflavin coenzymes, the nitrogens can pick up hydrogen atoms.

Flavin mononucleotide (FMN): One of the active coenzyme forms of riboflavin.

Pyrophosphate

Adenine

Riboflavin

Ribose

Flavin adenine dinucleotide (FAD): One of the active coenzyme forms of riboflavin.

Niacin

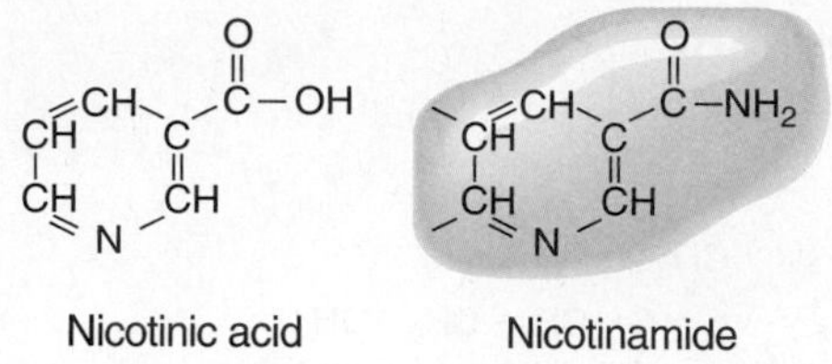

O ‖ CH C C–OH CH CH CH N

O ‖ CH C C–NH₂ CH CH CH N

Nicotinic acid Nicotinamide

Forms of niacin: Both nicotinic acid and nicotinamide can provide a source of niacin in the diet. Both can be used to make niacin coenzymes.

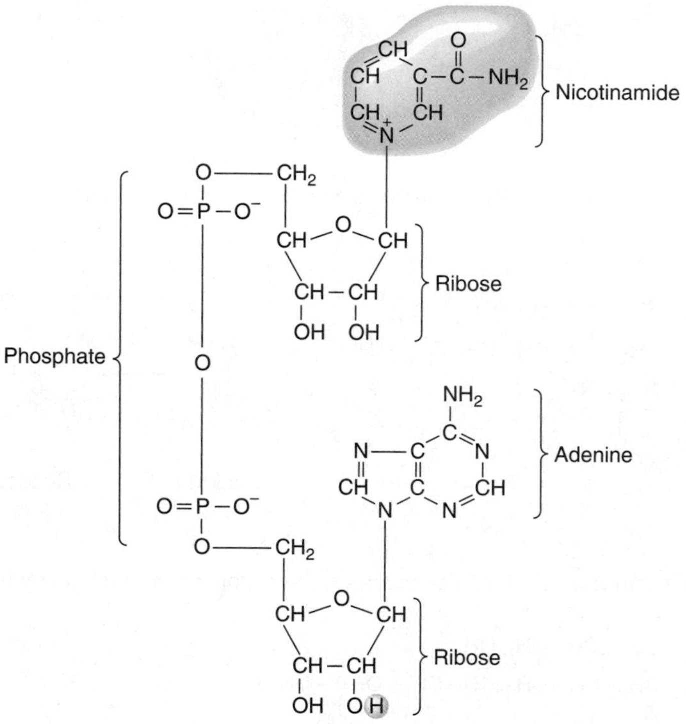

CH O CH C–C–NH₂ CH CH N⁺ — Nicotinamide

O—CH₂ O=P–O⁻ CH–O–CH CH–CH OH OH — Ribose

Phosphate O

NH₂ C N–C N CH C CH N N — Adenine

O=P–O⁻ O—CH₂ CH–O–CH CH–CH OH OH — Ribose

Nicotinamide adenine dinucleotide (NAD⁺) and nicotinamide adenine dinucleotide phosphate (NADP⁺): The active coenzyme forms of niacin. NADP has the same structure as NAD except a phosphate group is attached to the oxygen instead of the highlighted H. These niacin coenzymes can pick up a hydrogen and 2 electrons to form NADH or NADPH.

Biotin

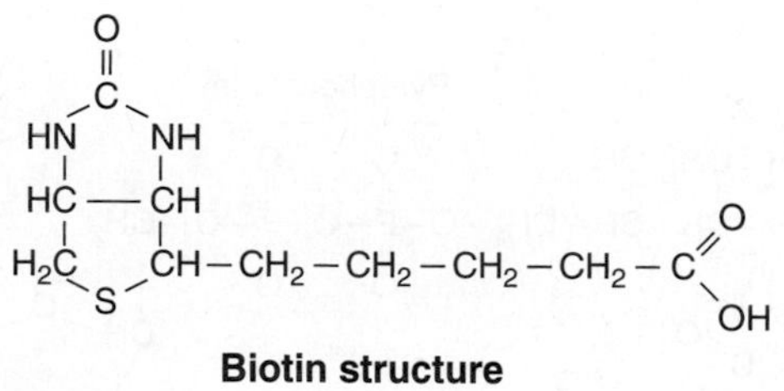

O ‖ C HN NH HC — CH H_2C S CH–CH_2–CH_2–CH_2–CH_2–C(=O)OH

Biotin structure

Pantothenic acid

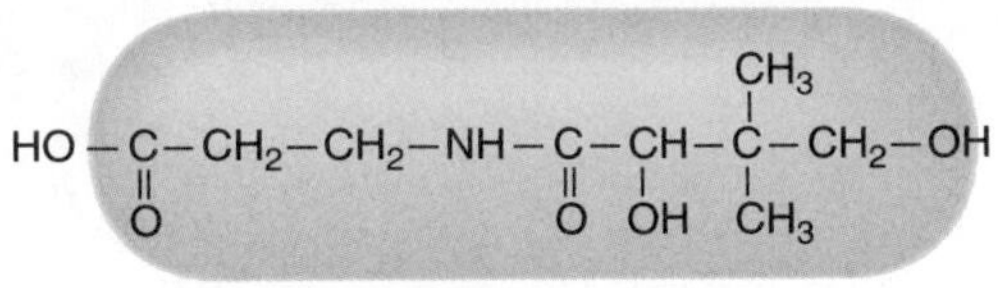

$HO-C(=O)-CH_2-CH_2-NH-C(=O)-CH(OH)-C(CH_3)_2-CH_2-OH$

Pantothenic acid: This molecule is part of the structure of coenzyme A (CoA).

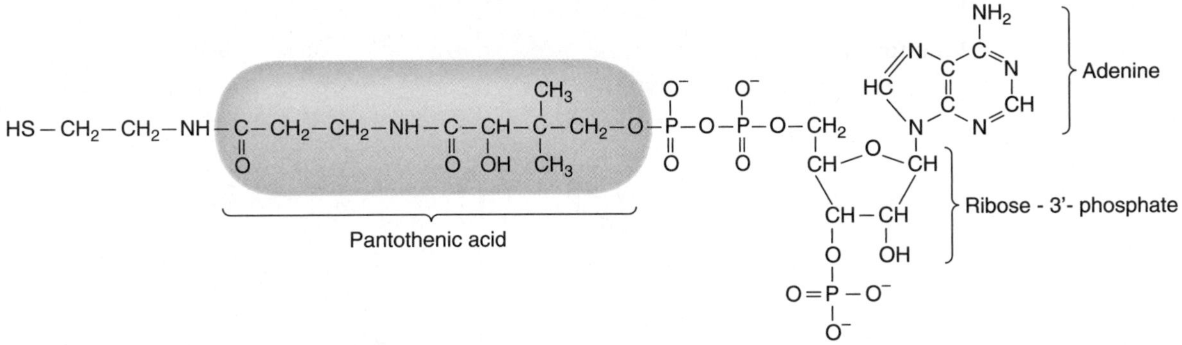

Adenine

Ribose - 3'- phosphate

Pantothenic acid

Coenzyme A: This coenzyme includes pantothenic acid as part of its structure.

Vitamin B6

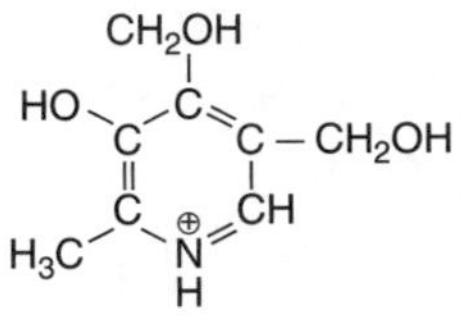

Pyridoxine

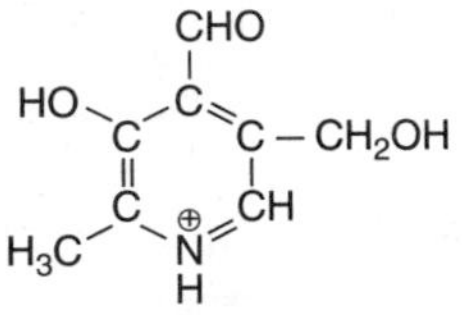

Pyridoxal

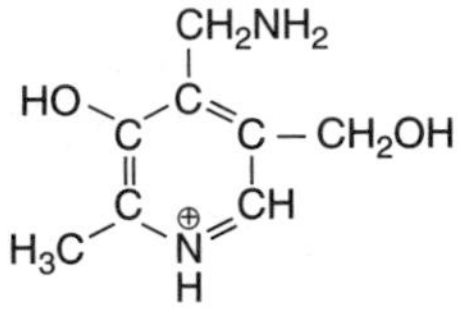

Pyridoxamine

Forms of vitamin B6: Pyridoxine, pyridoxal, and pyridoxamine are all converted into the active form of vitamin B6.

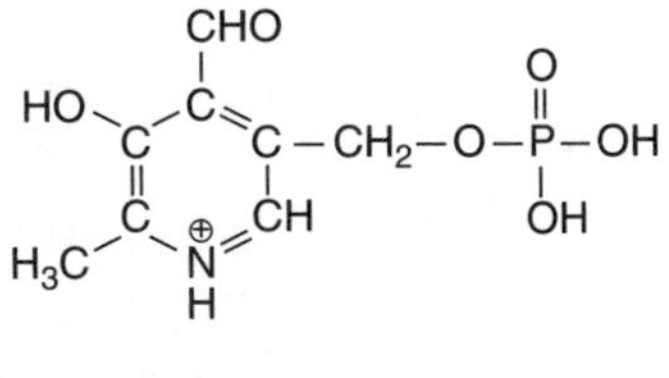

Pyridoxal Phosphate: The active coenzyme form of vitamin B_6.

Folate

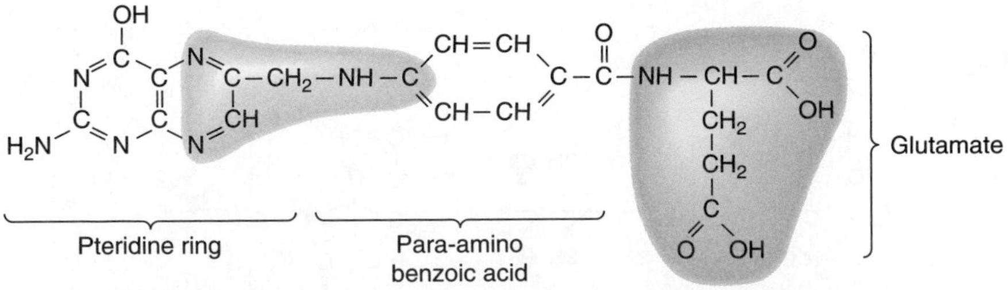

Pteridine ring

Para-amino benzoic acid

Glutamate

Folate structure: Folate consists of a pteridine ring combined with para-amino benzoic acid and at least 1 glutamate (a nonessential amino acid). The monoglutamate form is called folic acid. The folate naturally occurring in foods is the polyglutamate form.

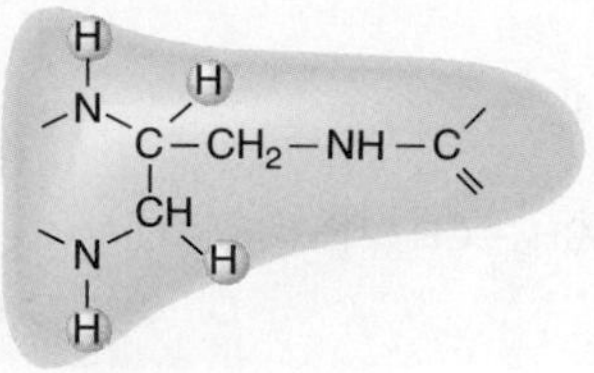

Tetrahydrofolate: The active coenzyme form of folate has 4 added hydrogens. Derivatives of this form of folate carry and transfer different types of 1-carbon units during synthetic reactions.

Vitamin B12

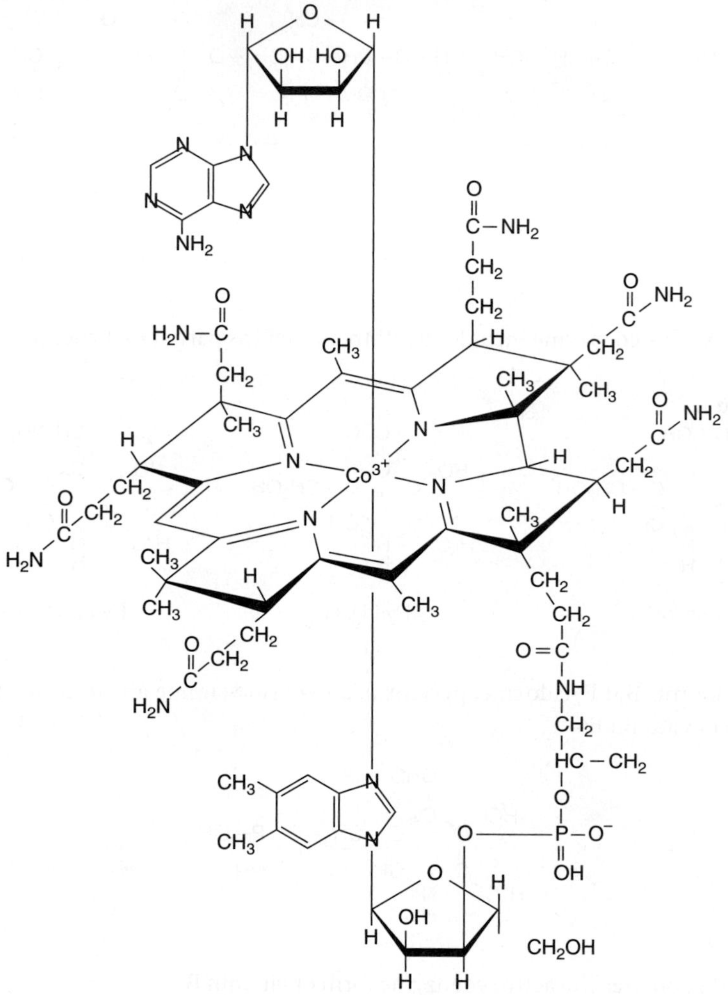

Vitamin B12 structure: Cobalamin, commonly known as vitamin B_{12}, is composed of a complex ring structure with a cobalt ion in the centre.

Vitamin C

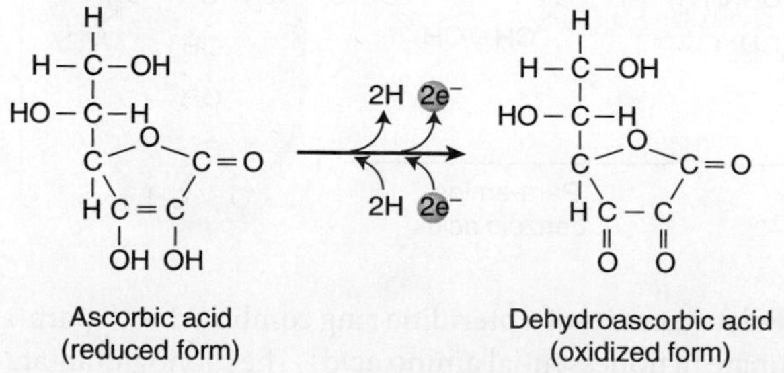

Ascorbic acid (reduced form)

Dehydroascorbic acid (oxidized form)

Vitamin C: Ascorbic acid can donate 2 hydrogen atoms with their electrons.

Fat-Soluble Vitamins

Vitamin A

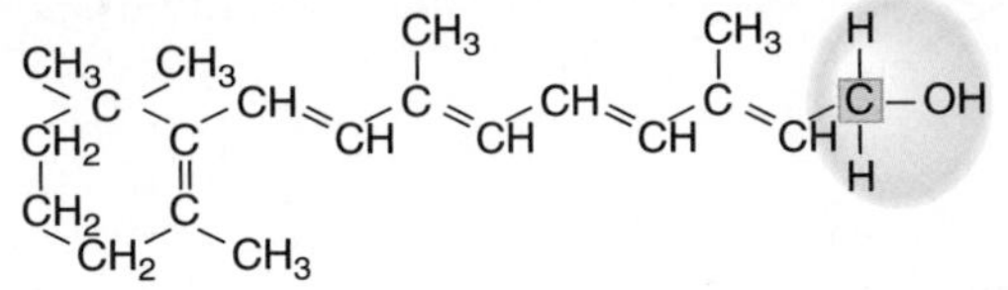

Retinol: The alcohol form of vitamin A.

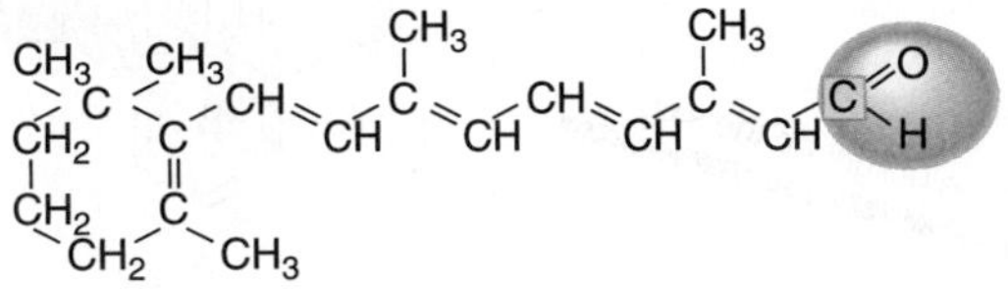

Retinal: The aldehyde form of vitamin A.

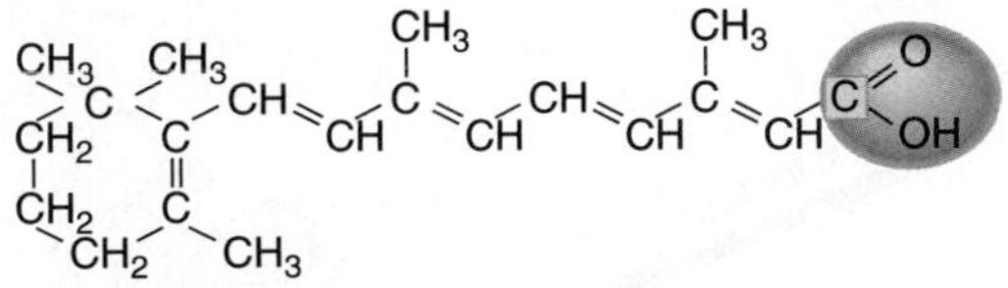

Retinoic acid: The acid form of vitamin A.

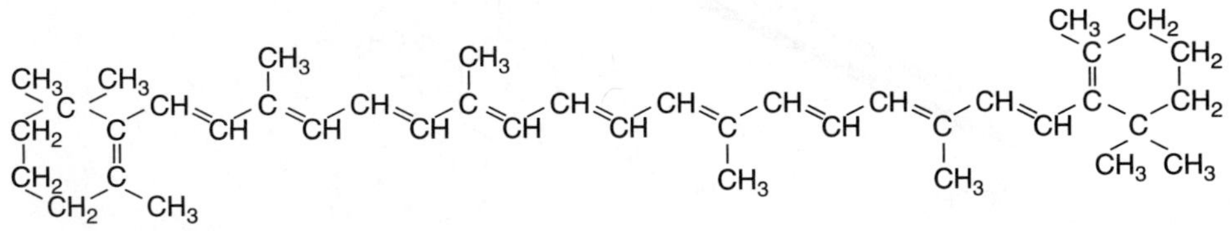

β-Carotene: A vitamin A precursor.

Vitamin D

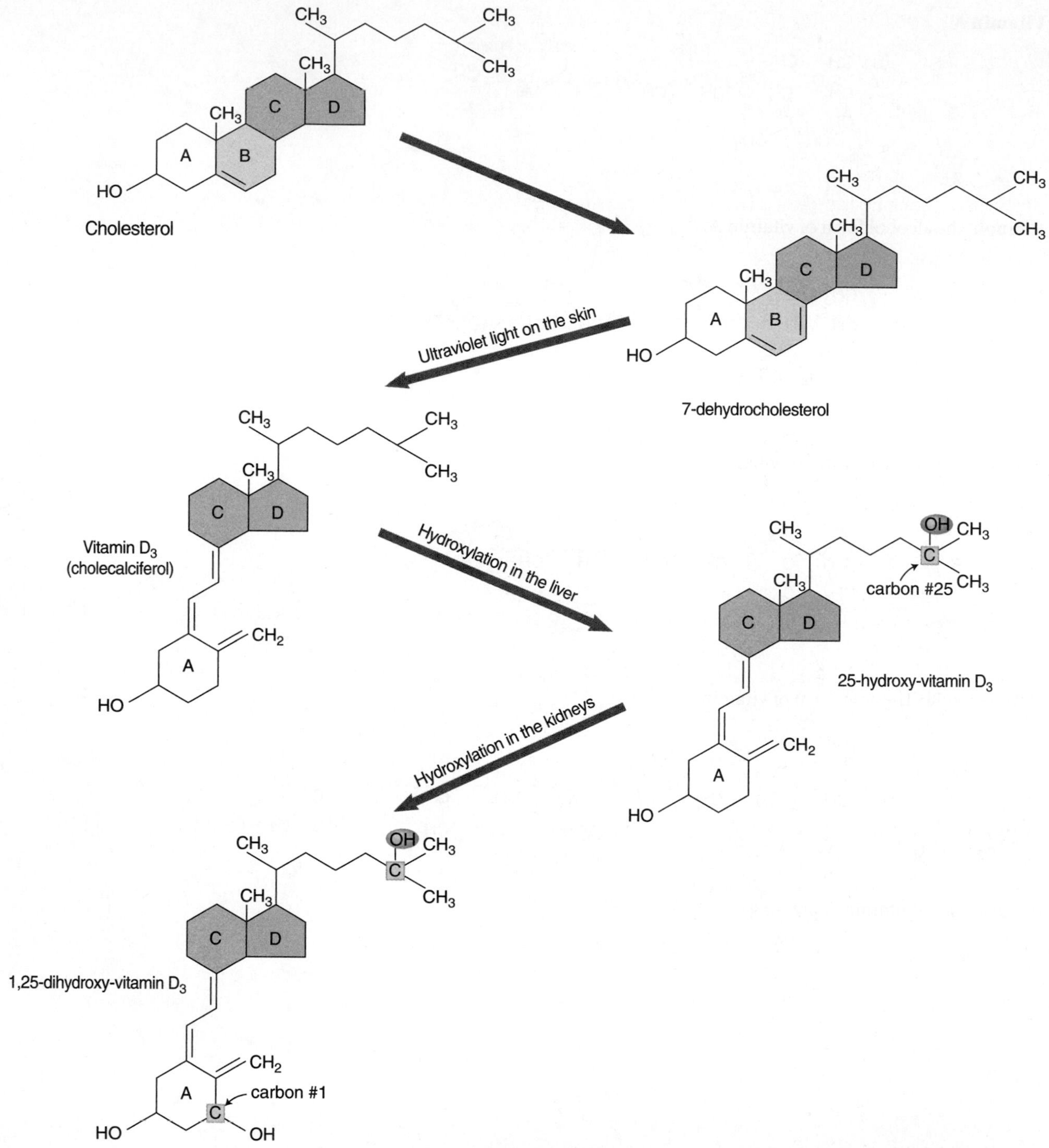

Cholesterol

7-dehydrocholesterol

Ultraviolet light on the skin

Vitamin D_3 (cholecalciferol)

Hydroxylation in the liver

25-hydroxy-vitamin D_3

carbon #25

Hydroxylation in the kidneys

1,25-dihydroxy-vitamin D_3

carbon #1

Vitamin D synthesis and metabolism: Vitamin D is synthesized in the skin during exposure to sunlight. Hydroxylation in the liver and kidneys converts it to its active form, 1,25-dihydroxy vitamin D_3.

Vitamin E

HO, H_3C, CH_3, CH_3, CH_2, CH_2, O, CH_3

$-CH_2-CH_2-CH_2-CH(CH_3)-CH_2-CH_2-CH_2-CH(CH_3)-CH_2-CH_2-CH_2-CH(CH_3)-CH_3$

Vitamin E: The alpha-tocopherol form is shown here. In other tocopherol isomers, the number and positions of the methyl groups differ. In other vitamin E isomers, called tocotrienols, the carbon chains contain double bonds.

Vitamin K

Phylloquinone

$R= -CH_2-CH=C(CH_3)-CH_2-(CH_2-CH_2-CH(CH_3)-CH_2)_3-H$

Menaquinone

$R= -(CH_2-CH=C(CH_3)-CH_2)_n-H$

Menadione

$R= -H$

Vitamin K: Phylloquinone (from plants) and menaquinones (from bacteria) are naturally occurring compounds with vitamin K activity. Menadion is a synthetic compound with vitamin K activity.

Appendix J

Critical Thinking Answers

Chapter 1

Using the Scientific Method

Which of the two hypotheses is supported by these experiments? Which is refuted? The second hypothesis has been supported because the Japanese Canadians consuming a traditional Japanese diet had a lower incidence of colon cancer than those consuming a typical North American diet. The first hypothesis has been refuted. Japanese Canadians are not genetically different from Japanese living in Japan, so the difference in colon cancer incidence between these two groups cannot be due to genetic differences.

Could you propose other experiments to obtain more information about the causes of colon cancer in Canada?

Epidemiological observations could be used to look for correlations between dietary factors in the North American diet and the incidence of colon cancer. Case controlled studies could be done to identify dietary factors that are unique to the cancer group. Finally, intervention studies could be done to test different hypotheses. For example, if epidemiology suggested that a high-fibre diet reduced colon cancer risk an intervention trial could be designed to compare people on high- and low-fibre diets. An intervention study could also be done in Japan comparing the incidence of colon cancer of Japanese people who consume a western diet with those who consume a traditional diet.

Based on the evidence presented here, propose a theory to explain the difference in the incidence of colon cancer in Canada and Japan. Some component of the typical American diet contributes to the development of colon cancer.

Chapter 2

Canada's Food Guide: Additional Statements

An elderly woman requires seven servings of vegetables and fruit daily. Is she following Canada's Food Guide if she consumes three servings of orange juice and four servings of mashed potatoes? No, Canada's Food Guide recommends a variety of foods and the consumption of an orange and green vegetable daily.

A sedentary young man consumes the appropriate number of servings of foods from all four food groups. Is he following Canada's Food Guide if he regularly enjoys his food deep-fried or served with creamy sauces? No. Deep-fried foods and creamy sauces are high in fat and Canada's Food Guide recommends foods low in fat, salt, and sugar.

A young woman consumes fortified soy milk instead of cow's milk. Is she following Canada's Food Guide, with respect to milk and alternatives? Yes. Fortified soy milk is an acceptable alternative to milk.

A young man eats a well-balanced nutritious diet but has french fries at a fast-food outlet once every three weeks. Is he following the recommendations of Canada's Food Guide? Yes. Canada's Food Guide recommends limiting the consumption of high-fat foods including french fries. Consumption once every three weeks would be considered infrequent and unlikely to negatively effect an otherwise nutrious and balanced diet.

Is a middle-aged woman who regularly skips breakfast following Canada's Food Guide? No. Canada's Food Guide recommends the consumption of breakfast.

An elderly woman, interested in increasing her intake of vegetables, does so by consuming canned soups regularly. Is she following Canada's Food Guide? (Additional information: One serving of soup contains 75% of the woman's Adequate Intake for sodium). No. Canada's Food Guide recommends the consumption of vegetables low in salt. The soup is very high in salt.

A young woman wants to increase her intake of fish, but is concerned about mercury contamination. Does Canada's Food Guide direct her to information on this subject? Yes. Canada's Food Guide recommends that consumers check the Helath Canada website for more information on the mercury contamination of fish.

A middle-aged man routinely drinks tap water because it is inexpensive. Is this the reason why Canada's Food Guide recommends drinking water? No. While tap water is cheaper than bottled water, Canada's Food Guide recommends the drinking of water because it is low in calories.

A 58-year old woman does not take any supplements because she believes that all nutrients should come from food. Has she interepreted Canada's Food Guide correctly? No. There are some situations where supplements are appropriate. Canada's Food Guide recommends a vitamin D supplement for anyone over age 50, such as the 58-year old woman in this question.

Applying My Food Guide Servings Tracker

Using the check boxes, how well does Jarad's diet compare to the recommendations for each food group for a man of his age? With the exception of milk and alternatives, Jarad does not meet the requirements of any of the food groups.

Is he following the recommendations of Canada's Food Guide's additional statements? Contrary to the additional statements, Jarad has selected many high foods high in fats and added sugars. Jarad vegetable intake is lacking orange and green vegetables and he did not consume whole grains, as recommended in the additional statements.

Why do you think he is gaining weight? Jarad has selected many high foods high in fats and added sugars, including some explicitly listed in the foods to limit category, such as whole milk, ice cream, cheeses, regular ground beef, French fries, cookies, and soft drinks. These foods add to Jarad's caloric intake and have caused his weight gain.

What changes do you think Jarad should make to his diet and activity levels? To control his weight he should consider at least 30 minutes/day of exercise and should increase his intake of whole grains, vegetables (especially orange and green), and switch to leaner meats and plant sources of protein.

Should Canadians Eat According to Canada's Food Guide or the Mediterranean Diet?

No answer provided. Opinions will vary.

Assessing Nutritional Health

Do you think Darra has iron deficiency anemia? Evaluate this by looking up the normal values for hemoglobin and hematocrit in Appendix C. Yes, both her hemoglobin and hematocrit are below normal: Darra's hemoglobin is 112 g/L blood and the normal range is 123-157g/L. Darra's hematocrit is 0.34 and the normal range is 0.37-0.46.

What about her iron intake? Compare her iron intake to the recommendations for a woman her age. She consumes only 6 mg of iron, which is 33% of the recommended 18 mg/day.

Should Darra be concerned about the nutrients she is consuming in excess of the recommended amount? Use the DRI tables to determine if they are likely to pose a risk. Although she exceeds the recommendations for vitamins A and C and calcium, she does not exceed the UL for any of these nutrients so she does not need to be concerned.

Chapter 3

Gastrointestinal Problems Can Affect Nutrition

What effect would this have on his nutrition and health?If he doesn't have enough saliva, he will have difficulty swallowing and tasting his food. This will decrease the appeal of food and may reduce his food intake. Also because saliva helps protect the teeth, the likelihood that he will develop cavities will increase.

How will this affect the digestion and absorption of nutrients in the carrots? If he can't chew the carrots well, the enzymes needed to digest them will not have access to all of the carrot. As a result, pieces will remain undigested and pass through the GI tract and the vitamins and minerals they contain will not be available for absorption.

Which person's stomach will empty faster? Why? The stomach of the student who ate just the cereal with skim milk and black coffee will empty faster because the meal is smaller and is much lower in fat. The amount of food and the high fat content of the eggs, sausage, biscuits with butter, and whole milk will cause this meal to remain in the stomach longer.

How will this affect food consumption and nutrient absorption? The surgery reduces the size of her stomach so she can only consume small amounts of food at any one time. Small meals are also important because the surgery bypasses the sphincter that regulates the entry of food into the small intestine. A large meal would cause too much material to enter the small intestine too quickly and would cause diarrhea. She does not absorb all the nutrients from her food because a section of the small intestine has been bypassed, reducing the area available for absorption. In addition, the smaller stomach may reduce the efficiency of both mechanical and chemical digestion in the stomach. Undigested food cannot be absorbed.

What effect would this have on her ability to digest and absorb proteins? Pancreatic enzymes are needed to digest carbohydrate, fat, and protein. If these enzymes are lacking, digestion will be incomplete, and nutrient absorption will be compromised. Protein-digesting enzymes in the stomach and mucosal cells can partially compensate for her reduction in pancreatic enzymes.

What type of foods should be avoided and why? She should avoid fatty foods. Bile is needed for fat absorption. Fat entering the small intestine causes the gallbladder to contract and release bile. A low-fat diet will minimize gallbladder contraction and therefore pain.

How would this affect fluid needs? The large intestine absorbs much of the water that enters with the contents of the small intestine. If most of his large intestine is removed he will lose more water and will need to consume more than before to prevent dehydration.

How might this affect the feces? The amount of intestinal gas? His feces will be larger and softer assuming he consumes additional fluid with his fibre. If he doesn't consume more fluid, this additional fibre could make him constipated. The added fibre provides a plentiful food supply for the intestinal microflora so it will cause an increase in the amount of intestinal gas produced.

Chapter 4

Dietary Fibre, Glycemic Index, and Type 2 Diabetes

What was the purpose of the study? To determine if there is an association between glycemic index, dietary fibre, and risk of type 2 diabetes.

What kind of observational study is this? Prospective cohort study

Describe the population in this study. Young and middle-aged women

How long were they observed? Eight years

What dietary components were the authors interested in? Foods of high and low glycemic index, foods high in cereal fibre

What health outcome was being evaluated? The incidence of type 2 diabetes

Interpret the relative risks reported. Note that there were statistically significant differences between relative risks. As glycemic index increased, risk of type 2 diabetes increased; as cereal fibre intake increased, risk of type 2 diabetes decreased.

What do you conclude about the relationship between diet and type 2 diabetes? The incidence of type 2 diabetes is associated with increased glycemic index and decreased cereal fibre intake.

Does the study demonstrate causation? Why or why not? No; this is an observational study; only intervention trials can determine causation.

Based on what is known about the effects of cereal fibre and low glycemic index foods on blood glucose levels, do the results of the study make sense? Yes, the results makes sense. Type 2 diabetes is characterized by elevated blood glucose levels. Low GI foods and fibre, especially soluble fibres, tend to reduce blood glucose levels. So diets which have a low GI and are high in cereal fibre may reduce blood levels slowing the development of type-2-diabetes.

How are Canadians Doing with Respect to Fibre Intake?

What would you conclude about the fibre intake of Canadians? The majority of Canadian adults, in all age categories, consume less than the AI. Therefore, we cannot conclude that fibre intake is adequate for Canadian adults. It may or may not be.

Becoming Less Refined

Determine whether Emma eats enough carbohydrate by calculating her percent of kcalories from carbohydrate in her diet (see Table 4.3).

(350 g of carbohydrate × 4 kcal/g) / (2340 kcalories) × 100 = 59.8% of kcalories

This is within the recommended range of 45 to 65 percent of kcalories.

Now consider her fibre intake. Does it meet recommendations? If not, list specific changes she could make in her diet to increase her fibre intake to the recommended level. She only consumes 19.6 grams of fibre, which is below the recommended 25 grams per day for adult women. She could increase her fibre intake by having oatmeal or another whole grain cereal for breakfast and adding a high-fibre vegetable such as broccoli to her dinner.

What about added refined sugars? Which food adds the most sugar to her diet? Identify other foods in her diet that are high in added sugar. Suggest foods she could substitute for these to reduce her added sugar intake. The soft drink adds the most sugar, but the fruit punch, chocolate bar, ice cream, and cherry syrup are also high in added sugar. The milk and apple contribute sugar, but these are unrefined sources. To reduce the amount of added sugar she can replace the soft drink and chocolate bar with low-fat milk and 2 oatmeal raisin cookies or some peanut butter and whole grain crackers and a glass of orange juice. For her evening snack she could top her ice cream with fresh berries instead of cherry syrup. These substitutions will also increase her fibre intake.

Chapter 5

Fish Consumption and Heart Disease

Describe one strength and one limitation of the study. Strength: the results pooled a large number of studies to produce results for an extremely large number of individuals, who were followed for years. Limitation: this is an analysis of observational studies and as a result, demonstrates an association between fish consumption and heart disease rather than causation (see Section 1.5 for a discussion of causation versus association).

Based on the results presented here, do you agree or disagree with the authors' conclusions? Explain your answer by interpreting the results of the table. Agree; the relative risk decreased steadily with increasing consumption of fish in an association that was statistically significant.

Lowering the Risk of Heart Disease

Is Rafael at risk for a heart attack? Review his risk factor questionnaire and list the factors that increase his risk. He has a family history of heart disease because his mother died of a heart attack before the age of 65. He is a smoker and is inactive, so his lifestyle further increases his risk. His blood pressure is elevated and his total blood cholesterol of 5.4 mmol/L is over the recommendation of less than 5.2 mmol/L. His LDL cholesterol of 4.3 mmol/L is also above the recommended level and his HDL cholesterol is less than recommended. His triglycerides are in the healthy range.

What about his diet? Based on the information from his typical intake, how does his fat intake compare with recommendations? Are there any other food choices he makes that increase his risk? The percent of energy from fat in his diet is within the AMDR, but he consumes less polyunsaturated fat and more saturated fat and cholesterol than recommended and his diet is high in *trans* fat. His choices of whole-fat dairy products and red meat increase his saturated fat and cholesterol intake. He does not consume enough whole grains and does not eat any legumes or nuts so his intake of fibre is less than that recommended. By not consuming legumes he is missing out on soluble fibre that can help lower blood cholesterol and by avoiding nuts he is missing the heart healthy monounsaturated and omega-3 fatty acids that they provide.

Suggest some diet and lifestyle changes that would reduce Rafael's risk of developing heart disease. To reduce his saturated and *trans* fat intake he can use low-fat dairy products and reduce his intake of red meat and solid fats. To increase the amounts of unsaturated fat in his diet he can use liquid oils for cooking and eat more nuts, seeds, and fish. To increase his fibre intake he can choose more whole grains and add some legumes to his diet. Lifestyle changes that will help reduce his risk of heart disease include increasing his activity to at least 30 minutes most days of the week and stopping smoking.

Eliminating *Trans* Fat from the Canadian Food Supply

In this continuing story, what do you think the government of Canada should do next?

There is no right answer to this question. The government can proceed directly to impose regulations or it can give the baking industry more time to comply voluntarily, since most of the problem foods are in this sector. Alternatively, the government can do nothing, deciding no further reductions in *trans* fats can realistically be made.

Why do you think the voluntary compliance was so low for baked foods? This may be related to the special problems that this sector faces. They may genuinely need more time to reformulate their products. Alternatively, because removing *trans* fat may compromise quality, companies may not want to voluntarily comply for fear they may lose market share to better-tasting products still high in *trans* fat. In this case, regulations to "level the playing field," that is, make all companies comply, may be necessary.

Using *trans* fat as an example, suggest general steps that a government might follow when trying to improve the nutrient composition of the food supply.

Possible approaches:

1) Implement changes in nutrition-labelling regulations.
2) Increase consumer awareness of new labelling regulations.
3) Ask for recommendations from representatives of both industry and other health experts, that is, relevant stakeholders.
4) If recommendations require industry to reformulate products, first consider a program of voluntary compliance from industry.
5) Monitor the food supply to confirm voluntary compliance.
6) If compliance is inadequate, impose regulation.

Eating Healthier Fats

Assuming Isabel is eating 2,250 kcalories per day, calculate the percent of energy from fat and saturated fat (sat fat) in her original diet. How do these percentages compare to recommendations? What foods are the biggest contributors to her saturated fat intake? To her *trans* fat intake? To her cholesterol intake? Her original diet contains 45% of kcalories from total fat and 15% of kcalories from saturated fat. It is recommended that total fat intake be 20 to 35% of kcalorie intake and that the amount of saturated fat be limited to less than 10% of kcalories. The biggest contributor to her saturated fat intake is the Big Mac. The biggest contributors to her *trans* fat intake are the french fries, tater tots, and coconut cookies. The bran muffin, whole milk, Big Mac, and fish sticks all add significant amounts of cholesterol.

Now look at the changes she has made. Assuming her kcalorie intake stayed the same, calculate the percent of energy from fat and saturated fat in her modified diet. Her modified diet contains 17% of kcalories from total fat and 4% of kcalories from saturated fat.

Fat is not the only concern in choosing a healthy diet. How does her modified diet stack up to the Canada's Food Guide recommendations in terms of grains, fruits, and vegetables?

Canada's Food Guide recommendations are 7-8 servings of vegetables and fruit and 6 servings of grain products. Her diet meets the serving number recommendations. The bran muffin (140 g) is comparable to 4 slices of bread or 4 servings, rice noodles are 2 servings, and pretzels are 1 serving for a total of seven. The muffin is made with bran, but none of her other grains are whole. Her diet contains 7 servings of vegetables and fruit from the stir fry, the baked potato, the green beans, the salad, the orange, and the apple.

Chapter 6

What Does Nitrogen Balance Tell Us?

Subject A is in negative nitrogen balance. List some possible reasons that would explain this. She is losing nitrogen and therefore losing protein. This would occur if she was losing weight, if her diet was not providing enough protein to meet her needs, or if she was experiencing a stress such as illness, injury, or surgery that caused her body to break down protein.

Calculate Subject B's nitrogen balance. What can you determine about subject B based on your answer? Nitrogen balance = 11.2 — 11.2 = 0

Subject B is in nitrogen balance and therefore he is consuming enough protein to meet body needs and the total amount of protein in his body is not changing.

What would you predict about Subject C's nitrogen balance based on the fact that she is pregnant? Calculate her nitrogen balance. Explain whether it supports your prediction. If she is pregnant she should be building new tissue. This new tissue is mostly protein so she would retain some of the nitrogen in her diet. Her nitrogen balance of 12.8—10.4 = +2.4 confirms this.

Scientific Evidence for the Benefits of a Vegetarian Diet

What do you conclude about disease risk from the first table comparing non-vegetarians and vegetarians? Vegetarian diets are associated with a reduction in risk of cardiovascular disease.

What effect does age have on disease risk? Vegetarian diets were more protective for younger individuals. There was no statistically significant difference between vegetarians and non-vegetarians at the oldest age range (80-89 years), possibly because people who live that long are generally healthier because of genetic predisposition, so diet has less of an impact.

What can you conclude from these results shown in the third table comparing frequent and non-frequent meat eaters? Reducing meat consumption, without completely eliminating it, is also associated with a health benefit.

Choosing a Healthy Vegetarian Diet

Does Ajay get enough protein? Compare his intake with the RDA for someone his age and size. The RDA is 0.8 grams per kilogram of body weight. Ajay weighs 70 kg. Therefore his recommended protein intake is 56 g per day (70 kg × 0.8 g/kg = 56 g). His diet provides 70.5 grams, more than enough to meet his RDA.

Does his diet contain complementary proteins? List the protein sources in his diet and explain how they complement each other. At breakfast the wheat protein in the toast complements the protein in the peanut butter. At lunch the legume (lentils) protein in the dahl complements the protein in the rice. At dinner the protein in the chickpeas complements the protein in the rice and the wheat protein in the poori. The milk he has at breakfast and the yogurt (and ice cream) he has at dinner provide animal proteins that further complement the plant proteins in his diet.

If Ajay decided to become a vegan, what could he substitute for his dairy foods in order to meet his protein needs? At breakfast he could put soy milk on his cereal, which would provide 2 grams per half cup and compliment the wheat protein in the Grape Nuts cereal. At dinner he could substitute soy yogurt, which would provide about 6 grams of protein. Instead of ice cream for dessert he could have a smoothie made with bananas, strawberries, and soy milk, which would add about 4 grams of protein.

How could he make sure his vegan diet meets his need for calcium and vitamin D? To increase his calcium intake he could eat more leafy green vegetables, which are a good source of calcium, and use calcium-fortified products like orange juice, soy milk, and other beverages. To meet his need for vitamin D he could spend plenty of time outdoors in the sun to ensure adequate vitamin D synthesis and use vitamin D-fortified breakfast cereal and soy milk to boost his vitamin D intake.

Chapter 7

Obesity, BMI, and Cancer: Establishing a Relationship

What do you conclude about the relationship between obesity and death from colon cancer? Obesity is associated with an increased the risk of death from colon cancer.

How does the risk between men and women differ? There is a lower risk for women than for men.

NOTE: the researchers were unable to fully explain why the risk was lower in women than men. They speculated that in women, estrogen may have a protective effect or that visceral fat, which represents a higher proportion of body fat in men than women, may be especially detrimental.

The results shown were age-adjusted. Why is this important? Age adjustment eliminates age as a potential confounding factor.

What do you conclude about the relationship between obesity and risk for breast cancer? Obesity is associated with an increased risk of breast cancer.

Balancing Energy: Genetics and Lifestyle

Is Aysha overweight? Calculate her BMI. Aysha is 1.63 m tall and weighs 70 kg. Her BMI is equal to $70/(1.63)^2 = 26.3$ kg/m^2, which is in the overweight range.

Is she in energy balance? Calculate her EER. How do her energy needs compare with her typical intake?

Aysha is 1.63 m tall and weighs 70 kg.

Aysha is in the low active activity category, so PA = 1.12

Using the equation in table 7.5 Aysha's EER is equal to:

EER = 354— (6.91 × 23 yrs) + 1.12 [(9.36 × 70.4 kg) + (726 × 1.63 m)] = 2,254 kcal

note: This answer uses the EER equations in table 7.5. Because Aysha is overweight, it is technically correct to use the equations in Appendix A but results are not shown.

She consumes about 2,450 kcal per day but is expending only 2,254 kcal per day, so she is not in energy balance. She is consuming 196 kcal per day more than she is expending.

If Aysha's weight does not change, but she increases her activity to 60 minutes per day, what will her new EER be? This will boost Aysha into the active physical activity category (PA = 1.27) and her EER will increase to:

EER = 354 — (6.91 × 23 yrs) + 1.27[(9.36 × 70.4 kg) + (726 × 1.63 m)] = 2,530 kcal

If she maintains her higher activity level, but doesn't change her intake, how long will it take her to lose 4.5 kg pounds (assume 7,700 kcalories/kg)? Her new energy balance = 2,450 kcal consumed — 2,530 kcal expended = -80 kcal/day. To lose 4.5 kg she will need to shift energy balance by almost 35,000 kcalories:

35,000/80 kcal/day = 438 days

It will take her a little over a year to lose 4.5 kg. If she reduces her intake, the weight will come off faster.

Is she destined to be overweight based on her family history? Why or why not? Ashya is not destined to be overweight, but to maintain her weight in a healthy range she probably needs to monitor her energy intake more carefully and exercise more than an individual with no genetic tendency to store excess body fat. If she makes small changes in her diet and exercise patterns that she can stick with, she is more likely to succeed.

Choosing a Weight Loss Plan that Works for You

Would Rose benefit from weight loss based on the data given about her in the table?Yes. Her BMI is in the obese category, her waist circumference is greater than 88 cm, and she has high blood pressure and elevated total and LDL blood cholesterol. Weight loss will help lower her blood pressure and blood cholesterol levels and reduce her risk of type 2 diabetes.

Evaluate the programs she is considering: Are the diets nutritionally sound? Do they recommend activity and include social support? What about cost? The low-carbohydrate plan lacks variety and may be low in certain nutrients because it limits choices from grains, vegetables, fruits, and milk. The liquid formula is adequate nutritionally because the formula is fortified with nutrients. The one meal that is allowed adds additional nutrients as well as fibre and phytochemicals if nutrient-dense foods are selected. The low-calorie diet is also nutritionally sound. Allowing varied choices from multiple food groups at each meal helps ensure that all nutrient needs are met.

The only plan that includes any exercise or social support is the low-calorie diet, which includes a walking group that meets 5 days a week. The cost of the 3 plans is similar. Low-carbohydrate diets include a lot of animal products so they tend to be expensive—about 3 times the cost of a standard diet. The low-calorie diet carries a fee of $25.00 per week in addition to food, which makes the cost comparable to the low-carbohydrate plan. Similarly the liquid formula is about $30 per week for the formula plus the costs of one meal a day.

Which plan do you think would benefit Rose in the long term? The low-kcal diet is best in the long term because it is easy to follow and will teach her diet planning skills that will help her maintain her weight loss in the long term. It will integrate well into her current lifestyle and will also address the need to increase her activity level.

Chapter 8

How are Canadians Doing with Respect to Their Intake from Food of Water-Soluble Vitamins?

Based on the data in the bar graph, identify areas of concern; that is, identify the vitamins and the population groups with low intakes. Folate intake in women; B_{12} intake in women; Vitamin C intake in both men and women, smokers and non-smokers.

What Canada's Food Guide food groups would you recommend to Canadians to improve their intake of these vitamins? Vegetables and fruits are the best food group to address intake of Vitamin C; to increase folate intake, vegetables and fruits as well as beans (from the meat and alternatives group) are good sources. Some women may have to increase their intake of products from animal sources, for instance, from the milk and alternatives group (milk, cheese, fortified soy milk), or from the meat and alternatives group (meat, fish or poultry) to ensure adequate intake of vitamin B_{12}.

Speculate on how the numbers in the bar graph might change if intake from supplements is included. The prevalence of inadequate intake might decline if individuals with a low intake from food are compensating for the low intake with supplements.

Where in Canada is the intake of vitamin C most adequate? Inadequate? Most adequate areas: British Columbia and Quebec; least adequate areas: New Brunswick and Prince Edward Island.

Why do you think these differences exist? Differences in lifestyle, socioeconomic status, access to fresh fruits and vegetables.

Meeting Folate Recommendations

Why is folate a concern for Mercedes when planning a pregnancy? Research has shown that consuming extra folic acid during the month before conception and during early pregnancy can reduce the incidence of neural tube defects in the baby.

Look at Mercedes's diet. Which foods are highest in folate? Of these, which are fortified with folic acid? The orange juice, refried beans, white rice, flour tortilla, and salad each provide over 50 µg of folate. The white rice and flour tortilla are fortified with folic acid because they are processed, enriched grains.

Mercedes has oatmeal for breakfast because it is a whole grain, but why is it low in folate? Oatmeal is a whole grain, so it has not been fortified with folic acid.

If she replaced her white rice with brown rice and her tortilla and hamburger bun with whole wheat products, how would this affect her total folate intake? The other grain products in her diet, such as the white rice, hamburger bun, and tortilla are refined and fortified so they contain added folic acid. Replacing her white rice with brown rice and her hamburger bun and tortilla with whole-wheat products would reduce her total folate intake (see table) and would reduce her intake of the folic acid form of folate. Despite this she should not pass up whole grains—they are good sources of most B vitamins, minerals, and fibre. It is recommended that at least half her grain product servings be whole grains.

Grain product	Folate (µg)
White flour tortilla	98
Whole-wheat tortilla	10
White rice	110
Brown rice	13
White hamburger bun	61
Whole-wheat hamburger bun	13

Would you recommend Mercedes take a folic acid supplement? Yes; although Mercedes's total intake of folate exceeds the RDA of 400 µg, she is planning to get pregnant so she should follow Health Canada's recommendation to take a multivitamin that contains at least 400 µg of folic acid in addition to dietary sources of folate.

Folate Intake in Canada: Too Little or Too Much?

Do you agree with the recommendation that the amount of folate added to food in Canada should be increased to reduce the number of women of child-bearing age that have folate blood levels below the optimum to reduce neural tube defects? There is no absolute right or wrong answer. More folate supplementation may be warranted as there is strong evidence that folate reduces NTD and limited evidence that it increases either cancer risk or the masking of B_{12} deficiency. Alternatively, a more cautious approach can be taken. There is some evidence of the potential harm from high intakes, perhaps enough to advise against adding more folate into the food supply until more research is conducted; instead, more targeted programs to identify and reach those 22% of women of child-bearing age who are not yet getting enough folate might be preferable. It may be useful to determine whether the 22% of women who have folate levels too low to protect against neural tube defects have any characteristics in common. Are they women who are likely to get pregnant or not?

Natural Health Product Choices

Review the ingredients in Hazel's supplements. If she takes all these products at the frequency recommended in the table, will she be exceeding the UL for any nutrients? List them.

Hazel is exceeding the UL for vitamin C.

From the vitamin C supplement, she is getting 2 caps X 1,000 mg/cap = 2,000 mg; from B-complex plus C, she is getting 2 caps X 500 mg = 1,000 mg; from the zinc lozenges, she is getting 5 X 250 mg = 1,250 mg, for a total of 4,250 mg , well above the UL of 2,000 mg/day.

She is also exceeding the UL for zinc, 5 X 15 mg = 75 mg, but this is unlikely to be harmful if the lozenges are taken as directed for a maximum of only 7 days.

What is responsible for Hazel's symptoms? Her symptoms are typical for excessive intake of vitamin C.

Chapter 9

How are Canadians Doing with Respect to their Intake of Vitamin A?

What would you conclude about the intake of vitamin A by Canadians? More than 44% of Canadian men and 35% of Canadian women are not meeting their requirements for vitamin A intake. These numbers are above 10% and are of concern.

What does Canada's Food Guide recommend to ensure adequate intake of vitamin A? The consumption of vegetables and fruit, especially the consumption of orange and green vegetables daily is recommended because these foods are rich in beta-carotene. Preformed Vitamin A is also found in milk and alternatives and meat and alternatives.

The data reflect vitamin A intake from food alone. What impact would vitamin supplements have on overall intake? Vitamin A intake would likely increase somewhat.

Consider the map. Are there regional differences in the vitamin A intake of Canadian adults? Why might these differences exist? Yes, a very wide range. Differences may reflect different access to foods, lifestyles, and socioeconomic status all of which influence food choices.

How Much Vitamin A Is In Your Fast-Food Meal?

Looking at his fast food options, John notices that pizza has lots of vitamin A. What ingredients in pizza make it so much higher in vitamin A than the other choices? The pizza is a good source of vitamin A because it has cheese, which is a source of preformed vitamin A, and tomato sauce, tomatoes, and peppers, which provide provitamin A carotenoids. Meats, such as the hamburger and chicken, as well as the bun and potatoes in the other meals are all poor sources of vitamin A.

Should John be concerned about his vitamin A intake? Use iProfile to estimate how much vitamin A he is getting in his breakfast and lunch. How much vitamin C is provided by these meals? His breakfast and lunch provide about 534 (μg (RDA = 900 (μg) of vitamin A and 22 mg of vitamin C (RDA = 90 mg).

How could John's breakfast and lunch be modified to increase his intake of vitamin A and vitamin C? He could increase his intake of preformed vitamin A by replacing the soft drink he has at lunch with milk. He could boost his provitamin A and vitamin C intake by increasing his fruit and vegetable intake, especially orange and green vegetables. Adding cantaloupe, mango, or grapefruit to his breakfast will increase his provitamin A and vitamin C intake. Strawberries, orange juice, and kiwis can add additional vitamin C to his breakfast or lunch. Adding vegetables to his lunch can also boost both provitamin A and vitamin C intake. For example, he could put some spinach on his sandwich and bring a bag of baby carrots to munch on to add provitamin A. Including raw broccoli florets with his lunch will provide a source of vitamin C and if he chooses sliced sweet peppers, it will add both vitamin A and vitamin C to his lunch.

Chapter 10

How are Canadians Doing with Respect to their Intake, from Food, of Sodium and Potassium?

Based on the graph showing the proportion of Canadian adults with sodium intakes above the UL, what would you conclude about Canadians' exposure to adverse health effects arising from their sodium intake? A large proportion of Canadians exceed the UL and may be exposing themselves to the adverse effects of high sodium intake.

What does the map suggest about regional differences in sodium intake across Canada? There are small differences with intakes, with Saskatchewan's and Ontario's intakes being somewhat lower than the rest of the country.

What does Canada's Food Guide recommend to maintain lower sodium intake? Additional statements advise Canadians to select foods that are low in sodium from each food group. The list of foods to limit include foods that are high in salt. Canadian are also advised to check nutrition labels to compare sodium levels in food.

Can anything be definitively concluded about the potassium intake of Canadian adults? Because less than 50% of Canadians have potassium intakes above the AI, and because of the limitations of the methodology in using the AI, one cannot conclude anything definitive about the adequacy of intake for these population groups. Intakes may or may not be adequate.

Which of Canada's Food Guide food groups should Canadians consume to increase their potassium intake? What is the impact of processing on potassium content of food? Vegetables and fruit are high in potassium. Processed foods have more sodium and less potassium than unprocessed foods.

A Diet for Health

How healthy is Rashamel's current diet? Does it meet the recommendations of Canada's Food Guide for grains, fruits, and vegetables? As seen in the table Rashamel consumes less than the recommended servings of vegetables and fruit, milk and alternatives, and meat and alternatives.

Food Group	Rashamel's CFG servings	Recommended CFG
Vegetable & Fruits	4	7
Grain Products	9 (3 WG*)	7
Milk and Alternatives	1	3
Meat and Alternatives	2	3
*WG = whole grain		

Now compare his diet to the recommendations of the DASH Eating Plan for someone who eats 2,600 kcal/day. Does he meet these recommendations? How many additional servings from grains, fruits, and vegetables would he need to add? As seen in the table below Rashamel's current diet provides fewer servings of grains, fruits, vegetables, and beans, nuts, and seeds than recommended by the DASH Eating Plan.

	Rashamel's DASH servings	Recommended DASH
Grains	9	10-11
Vegetables	1	5-6
Fruit	2.7	5-6
Low-fat dairy	1	2-3
Meat	1.6	2
Beans	0	1
Fats & oils	3.5	3

Rashamel's wife is concerned about her risk for osteoporosis. Use Table 10.3 and iProfile to plan a day's meals that follow a 1,600-kcalorie DASH Eating Plan and meet her calcium needs.

She could choose:

Meal	Serving	DASH serving
Breakfast		
Fortified cereal	250 ml (1 cup)	2 grains (100 mg calcium)
Low-fat milk	250 ml (1 cup)	1 milk (300 mg calcium)
Banana	1 medium	2 fruits
Calcium-fortified		
orange juice	175 ml (¾ cup)	1 fruit (400 mg calcium)
Lunch		
Chef salad	500 ml (2 cups)	2 vegetables
with egg		30 g (1 oz) meat
and chopped ham	30 g (1 oz)	30 g (1 oz) meat
and salad dressing	30 ml (2 tbsp)	1 fat and oils
Whole wheat crackers	6	1 grain
Low-fat yogurt	250 ml (1 cup)	1 milk (300 mg calcium)
Snack		
Sunflower seeds	80 ml (⅓ cup)	1 beans, nuts and seeds
Dinner		
Chicken breast	90 g (3 oz)	90 g (3 oz)
Brown rice	250 ml (1 cup)	2 grains
Steamed spinach	250 ml (1 cup)	2 vegetables (100 mg calcium)
Apple		1 fruit
Low-fat milk	250 ml (1 cup)	1 milk (300 mg calcium)

Chapter 11

How are Canadians Doing with Respect to their Intake, from Food, of Calcium, Magnesium, and Phosphorus?

What would you conclude about Canadians' intake, from food, of calcium, phosphorus, and magnesium? About 35%-90% of Canadians are not meeting their calcium requirements from food and about 35%-65% Canadians are not meeting their magnesium requirements from food. These numbers are above 10% and are of concern. Phosphorus intake is adequate in both men and women.

How and why does age affect the adequacy of intake? As age increases, generally, the proportion of the population not meeting their requirements from food increases.

With respect to calcium, it is known that calcium supplements are popular among older women. How would that affect the prevalence of inadequate intake? The prevalence of inadequate intake would decline in that population.

Dietary Factors Can Increase Calcium Bioavailability—the Example of Inulin

What can you conclude about the effectiveness of inulin on calcium absorption and on bone health? Inulin effectively increases calcium absorption and this extra calcium is taken up by bone, as reflected in increased BMC and BMD, improving bone health.

Osteoporosis Risk

Evaluate Mika's risk for osteoporosis by looking at her answers on the questionnaire. Why would her milk intake and activity level as a child affect her risk now? Her risk is increased by the fact that she is a female, drank little milk as a child, and still typically consumes less than the recommended amount of milk. Her risk is further increased because she smokes, gets little exercise, and has a family history of osteoporosis. Her intake of calcium and the amount of exercise she got as a child affect her risk now because there are factors that determine peak bone mass, which is usually established by about age 30. Less calcium and exercise in childhood lead to lower peak bone mass, which increased her risk of osteoporosis as an adult.

How does Mika's calcium intake compare with the recommendations? Suggest changes she could make in her diet to increase her calcium without increasing her kcalorie intake. Her intake of 695 mg is below the RDA of 1,000 mg for women age 31 to 50 years. For breakfast she could drink calcium-fortified orange juice. At lunch she could replace a slice of bologna with a slice of cheese on her sandwich. For a snack she could have low-fat yogurt instead of chips. For dinner she could have a leafy green vegetable such as Swiss chard instead of green beans and drink milk instead of iced tea.

Do you think Mika should take a calcium supplement? Why or why not? Yes; without major changes in her diet, Mika is unlikely to meet her calcium needs. In addition, she is approaching menopause, when the risk of osteoporosis increases, and she does not meet exercise recommendations. The benefits of a supplement outweigh the risks.

Chapter 12

How are Canadians Doing with Respect to Their Intake, from Food, of Iron and Zinc?

What would you conclude about the adequacy of zinc intake? How does it change with age? There is some concern about zinc intakes as the level of inadequacy is above 10% for all Canadian adults and increases to 25% and 41% for women and men, respectively, who are over age 70.

What would you conclude about the adequacy of iron intake in women? How does it change with age? Why? (Note that to account for menstrual losses of iron, the EAR for women under 50 is 8.1 mg and it drops to 5 mg in post-menopausal women, meaning those 50 and over.) There is some concern about iron intakes in young women, as more than 10% have intakes below the EAR. Post-menopausal women, like men, have adequate intakes.

Increasing Iron Intake and Uptake

Does Hanna's iron intake meet the RDA for a young female vegetarian? No, she consumes only 12 mg of iron. The RDA is 32 mg of iron per day for vegetarian women of child-bearing age.

How could Hanna increase her iron intake without adding kcals to her diet? She could go back to using iron cookware—the iron leaches into food. Cooking acidic foods in iron cook-ware further increases the amount of iron that leaches into the food. She can also include more iron-fortified foods in her diet, such as breakfast cereals and soy milk.

What could Hanna do to increase the absorption of the iron in her meals? Vitamin C-rich foods enhance iron absorption, so Hanna could increase her vitamin C intake by replacing her apple juice at breakfast with orange juice and substituting strawberries for the apple juice at dinner. Since tannins interfere with the absorption of iron she could have her tea between meals.

Does Hanna's diet provide good vegetarian sources of calcium? Are there other nutrient deficiencies for which she may be at risk? Good vegan sources of calcium include calcium-fortified drinks, tofu, almonds, legumes, and green leafy vegetables. For lacto and lacto-ovo vegetarians dairy products provide a good calcium source. Hanna's diet includes kale and yogurt, which are good sources of calcium. Since she does consume dairy products she is unlikely to be at risk for calcium, vitamin D, or B_{12} deficiency but the lack of meat in her diet puts her at risk of zinc deficiency.

Chapter 13

Incorporating Exercise Sensibly

List 3 things that are wrong with Nicole's exercise program. She started with too much exercise, she included strength training 5 days per week rather than the recommended 2–3 times per week, and she did not plan her exercise to fit well into her lifestyle.

What should her heart rate be to be in the aerobic zone?

Maximum heart rate	=–220 – age = 220 – 45
	= 175 beats per minute
60% of maximum	= 175 × 0.6
	= 105 beats per minute
85% of maximum	= 175 × 0.85
	= 149 beats per minute

To stay in her aerobic range she should exercise at a heart rate of at least 105 but no more than 149 beats per minute.

Calculate Nicole's EER before and after the addition of her exercise regimen. Do you think she will lose weight? Before her activity change, Nicole's physical activity level was in the sedentary category. Her estimated energy requirement (EER) can be calculated using the following equation for an adult woman (see inside cover):

EER = 354 – (6.91 × age in yrs) + PA [(9.36 × weight in kg) + (726 × height in m)]

Age = 45 yrs, PA = 1.0,weight = 70kg, height = 1.6 m

EER = 354 – (6.91 × 45) + 1.0[(9.36 × 70) + (726 × 1.6)] = 1,860 kcalories/day

Her exercise program put her in the active physical activity category so her EER increases as follows:

EER = 354 — (6.91 × 45) + 1.27[(9.36 × 70) + (726 × 1.6)] = 2,350 kcalories/day

If she could stick with this and not increase her food intake she would lose weight. Her energy expenditure has increased by almost 500 kcalories a day. This would result in a weight loss of about a 0.5 kg/week.

Suggest some modifications to her exercise program that will keep her from getting sore and allow more time with family. She should start slowly—perhaps 20 minutes of walking— and build up to 45 minutes of walking 3 days a week. She could combine this with going to the gym twice a week for 30 minutes of weight lifting and light calisthenics. She needs to plan these activities so they do not interfere with family time.

The Benefits of Interval Training

What do you think the researchers observed when they measured GLUT4 levels? Explain why. GLUT transporter increased. If aerobic capacity increased, one would expect the glucose would be taken into cells more readily.

What are the implications of these studies? These results suggest that interval training could be used to reduce the time spent exercising, with the same physiological benefits of traditional endurance training.

What is one limitation of these two studies? Very small sample sizes. Experiments should be repeated with larger groups.

Does it Provide an Ergogenic Boost?

Use the questions in Table 13.4 to help Paulo evaluate these claims. Since Paulo is a sprinter, the claims made by these supplements would be beneficial to his performance. Based on the questions in Table 13.4 he explores the scientific evidence to support these claims, whether there are any dangerous side effects, and considers how much each will cost.

Based on the information given in this chapter, explain the advantages and risks of these products. Creatine is a precursor for creatine phosphate, which is a source of ATP for short-term exercise. Supplements increase creatine phosphate levels in the muscle, which provides more energy for short bursts of activity such as sprinting. The risks of taking creatine are low, but there is no information on the long-term effects of this supplement. Creatine is approved in Canada as a natural health product.

Chromium is a mineral that is needed for insulin to perform optimally. Insulin has many essential roles, including getting glucose into cells, turning on protein synthesis, and stimulating the synthesis of fat. The claim that it will increase lean tissue makes some sense metabolically, but supplements have not been found to increase lean tissue or enhance performance. The risk of taking chromium picolinate is low.

Suggest places Paulo could look to get additional information he can trust. In order to find sound scientific studies he should search Pubmed and articles in peer-reviewed journals such as the *International Journal of Sport Nutrition* and *Medicine and Science in Sports and Exercise.* He should focus on studies that include athletes involved in the types of activities he performs. He can also check the Health Canada natural health product database.

Would you recommend Paulo take these supplements? Why or why not? If Paulo participates in sprint events requiring short bursts of activity, creatine might improve his performance. To have this benefit he needs to continue to take the supplement, and there is no good long-term studies on creatine safety. There is little risk if Paulo wants to try creatine to see if his performance is improved, but he should be cautious about taking it over the long term. The evidence supporting the benefits of chromium picolinate is questionable so, although the risks of taking it are small, they still outweigh the benefits.

Chapter 14

Nutrient Needs for a Successful Pregnancy

How does Tina's current food intake compare with Canada's Food Guide recommendations for women in their first trimester? Tina's intake of Milk and Alternatives and Meat and Alternatives is adequate; Vegetables and Fruit, and Grain Products are low (See Table J.1). She should reduce her intake of chocolate chip cookies and add more nutrient-dense grain products, including whole-grain products, which were largely absent from her diet. She should consider limiting her intake of french fries and adding more nutrient-dense vegetables, especially green and orange vegetables, which were absent from her food intake.

Give examples of the types of foods she could add to her diet to meet the increased kcaloric requirements of the second and third trimesters. She can add nutrient-dense foods such as a fruit, yogurt, beans, a slice of whole-grain bread, etc. She should add 2-3 servings each day.

What about iron? Does her current diet meet her needs without the supplement? No, the RDA for iron is 27 mg, and Tina receives only 13.5 mg. Her diet alone does not ensure that she will be meeting her requirements. She should try to include more iron-rich foods in her diet and to take a supplement to be certain her requirements are being met.

The Prevalence of Exclusive Breastfeeding in Canada

Consider the graph that shows the results of the breastfeeding study. What does it indicate about exclusive breastfeeding in Canada? At the end of the first month, only 63.6% of women are breastfeeding exclusively, and this proportion drops substantially with time. At the end of 6 months, only 13.8% of mothers were exclusively breastfeeding.

What would you conclude about the impact of smoking, baby's birthplace, and mother's employment status on exclusive breastfeeding? Suggest explanations for these results. Mothers who don't smoke, give birth at home, and are not employed are more likely to exclusively breastfeed. Mothers who don't smoke are likely to be more health-conscious and more likely to favour breastfeeding. Alternatively, mothers who smoke may want to avoid exposing their infants to tobacco-related toxins they fear may be in their breast milk. Similarly, the planning of a home birth may be indicative of a mother better able to commit to breastfeeding exclusively, as is a mother who is at home, rather than working.

Table J.1 Comparing Tina's Intake to Food Guide Servings

		Canada's Food Guide Servings (#)					
FOOD	IRON (mg)	Vegetables-Fruit	Grain Products	Milk and Alternatives	Meat and Alternatives	Oils	Foods to Limit
Breakfast							
250 ml (1 cup) corn flakes (30 g)	0.4		1				
with 250 ml (1 cup) reduced-fat milk	0.1			1			
125 ml (½ cup) orange juice	1.1	1					

Table J.1 *(Continued)*

FOOD	IRON (mg)	Vegetables-Fruit	Grain Products	Milk and Alternatives	Meat and Alternatives	Oils	Foods to Limit
250 (1 cup) decaffeinated coffee							
with sugar and cream	0.1						X
Lunch							
Tuna sandwich							
75 g (2.5 oz.) tuna	1.2				1		
10 ml (2 tsp) mayonnaise	0.5					5 ml	
2 slices white bread	2.5		2				
20 french fries	1.3						X
1 can orange soda	0.2						X
3 chocolate chip cookies	0.8						X
1 apple	0.4	1					
Dinner							
75 g (2.5 oz.) chicken leg	1.5				1		
125 ml (½ cup) peas	1.2	1					
1 piece corn bread	0.8		1				
5 ml (1 tsp) margarine	0					5 ml	
250 ml (1 cup) lettuce and tomato salad	1.3	1					
15 ml (1 Tbsp) Italian dressing	0					10 ml	
250 ml (1 cup) reduced-fat milk	0.1			1			
Total	**13.5**	4	4	2	2	20 ml	4 X
Recommended-First trimester		7-8	6-7	2	2	30-45 ml	

Chapter 15

At Risk for Malnutrition

Why is Jamar's energy balance out of balance? Does he get the amount of daily activity recommended for a child his age? Evaluate his diet by comparing the number of servings per day he consumes from the various food groups to Canada's Food Guide recommendations for an 8-year-old boy. His energy balance is out of balance because he is consuming more energy than he is expending. His high intake is due to a diet that is very high in kcalories from high-fat milk and dairy products and snack foods. His low expenditure is due to a lack of physical activity. Children should get at least an hour of moderate to vigorous intensity exercise each day. Jamar only gets 25 minutes of exercise during recess at school 5 days a week. Most of his leisure time is spent in sedentary activities.

Canada's Food Guide recommendations for an 8-year-old sedentary boy 5 servings of vegetables and fruit, 4 servings of grain, 2 servings of milk and alternatives, and one serving of meat and alternatives. Jamar consumes about 3 times more milk as recommended, but falls short on the amounts of grain products, vegetables and fruit, and meat and alternatives.

Why is he anemic? What foods does he consume that are good sources of iron? How would his intake of dairy products affect his iron status? Jamar is anemic because his diet is low in heme and nonheme iron and in foods that promote iron absorption. He consumes little red meat, which is a good source or absorbable heme iron. Chicken and fish are also good sources of heme iron, but he does not consume these everyday. He consumes both enriched and whole grains, which provide nonhemeiron, but his intake of nonheme iron from vegetables is low. Citrus fruit and other sources of vitamin C enhance absorption of nonheme iron, but his diet is low in these. His high intake of dairy products provides lots of calcium but large amounts of calcium consumed with iron can reduce iron absorption.

How might Jamar's iron deficiency anemia have contributed to his weight gain? Iron deficiency anemia causes lethargy, which would decrease his activity level and contribute to a positive energy balance.

Suggest some dietary changes that could increase his iron intake and decrease his energy intake. He could have a serving of red meat in spaghetti sauce or as a pizza topping, or he could have iron-fortified breakfast cereal and dried fruit for snacks. To further increase iron intake, Jamar's food could be cooked in iron cookware. To increase iron absorption he could consume dietary iron sources with foods that are high in vitamin C. For instance, having a glass of orange juice with his iron-fortified breakfast cereal would provide iron and enhance its absorption. To reduce his energy intake he could switch to low-fat milk and snack on fruit and whole-grain crackers, rather than candy and chips.

Parental Influences on Childhood Eating Habits

Describe the relationship between teen body weight and parental body weight. The heavier the parent, the heavier the teen.

What environmental and behavioural factors might explain this association? Parents and teens may live in the same obesogenic environment; parental habits are learned by children.

What other factors might explain this association? Children may share the same genetic predisposition to obesity as their parents.

What are the implications of these trends in the planning of programs to reduce childhood obesity? Because of the strong association between child and parent body weight, for the most successful programming, parents should be integrated into any programming and have opportunities to improve their health. Parents may have to make changes in their behaviour and environment to maximize the benefits of any programming directed to children.

Less Food May Not Mean Fewer Kcalories

Why is Jenny continuing to gain weight on her new diet despite eating a smaller breakfast and skipping lunch? Even though Jenny seems to be eating less food, the foods she has chosen are high in fat and so contain more kcalories per gram than the foods in her original diet. Her new diet provides about 100 kcalories more than her original diet.

How could Jenny modify her new diet to reduce her kcalorie intake and still fit her busy schedule? The bagel she chose for breakfast is fine, but she should have a whole-grain variety and add some fruit or juice as well as some cheese or peanut butter to make sure she is not hungry before lunch. For a quick lunch, she could have a turkey sandwich from a nearby deli. She won't be as hungry in the afternoon so she could have a carton of low-fat yogurt for an afternoon snack, rather than frozen yogurt, a chocolate bar, and potato chips. For dinner, she could have the pizza she snacked on in her original diet, and top it with vegetables such as mushrooms, peppers, broccoli, and tomatoes. Including a salad and a glass of low-fat milk with her pizza at dinner will fill her up and be much lower in kcalories than a double burger, fries, and a shake.

Chapter 16

Changes in Diet Quality as Canadians Age

How does the average CHEI score change as Canadians age? There is a small improvement in scores as Canadians age.

What would you conclude from the second graph, which shows the proportion with a score less than 50? Diet quality tends to improve with age but many have low scores.

How do men and women compare? Diet quality is better for women than men.

Health-Promoting Factors

What do you conclude about the differences between younger and older adults? Younger adults are in better health.

What do you conclude about the relationship between good health and the number of health-promoting factors identified? There is a direct association between self-reported good health and the number of health-promoting factors that apply to an individual.

Chapter 17

Safe Picnic Choices

Is this picnic food safe to eat? It is safe only if none of the foods were contaminated with pathogens that can cause foodborne illness, or if they were prepared in ways that killed any pathogens that were present, or were stored in ways that prevented or slowed the growth of pathogens.

Which foods are the least risky? The safest foods are those that involve no preparation and are intended to be served at room temperature such as the chips, cookies, and cheese and crackers. Store-bought dips are a good choice because they can remain unopened until the picnic begins. Acidic dips such as tomato-based salsa are safe because the acid inhibits bacterial growth. The apple pie is also probably safe during the picnic as long as it has been refrigerated beforehand. The raw fruits and vegetables are safe as long as they are washed before they are prepared and served.

Which foods carry the highest risks? The more food is handled, the more likely it is to be contaminated. The chicken salad, tamales, and stuffed mushrooms pose a risk because they are handled extensively in preparation. The fried chicken, tamales, and stuffed mushrooms are cooked, but when left at room temperature, the inside may stay warm enough to provide a good environment for microbial growth.

What can Tamika do to reduce the risk of foodborne spoilage during the picnic? She can bring ice and coolers to keep foods like the chicken salad, onion dip, fruit salad, and vegetables at cool temperatures below the danger zone. There is no way to keep hot foods hot but she can encourage people to eat the hot foods first so they do not cool off and then sit in the warm sun for hours.

After the picnic is over, what foods would you consider safe to keep as leftovers and what would you throw out? The fruit salad, raw vegetables, chips, crackers and cheese, and cookies would be the safest to keep. The chicken salad and cooked items that were left at room temperature for hours (fried chicken, stuffed mushrooms, and tamales) should be discarded, as should the chicken salad and onion dip.

The Risks and Benefits of Food

Draw a similar risk-benefit conclusion for the other foods on Ron's list. Which foods do you think he should keep in his family's diet? How can he minimize the risks associated with eating them? To decrease the risks associated with using raw eggs, he can advise his son that he does not need protein shakes but that if he wants to drink them, he should use nonfat dry milk rather than raw eggs to add protein. He can have his daughter wait for the cookies to bake before tasting them. To decrease the risk of consuming high levels of contaminants from fish, he can eat a wider variety of fresh and saltwater fish to avoid consuming high levels of any one contaminant. Because the risks of eating raw fish are greater, he can limit sushi to a rare treat purchased only from a restaurant that he knows buys the fish fresh daily. The pesticide risks from consuming produce are small compared with the benefits of a diet high in fruits and vegetables. To reduce the amounts of pesticides ingested, fruits and vegetables can be washed thoroughly or consumed without skins. He can also reduce pesticide consumption by purchasing organic produce, but he must consider that it is more expensive. He can also buy locally grown produce and grow some of his own vegetables in the summer.

Are there any foods on his list that you would suggest he eliminate from the family diet? Why or why not? The riskiest food on Ron's list is raw shellfish. When raw clams and oysters are eaten, so are the contents of their digestive tracts, which can include any bacteria, viruses, chemical contaminants, and other impurities that are present in the water where they feed. The risk of foodborne infection can be reduced by cooking the shellfish, but they still may contain chemical contaminants, and other impurities that are not destroyed by heat.

Chapter 18

What Can You Do?

Looking at Keesha's list, do you think her actions will help? For each item on her list, identify an environmental benefit. Bringing her juice in a thermos reduces the amount of garbage (packaging, disposable cups, etc.) she generates from her meals; composting reduces the amount of garbage sent to the landfill; buying organically grown produce supports agricultural techniques that reduce the amount of chemical fertilizers and pesticides used and therefore introduced into the environment; and buying locally grown foods reduces the energy cost of transporting and storing produce grown at distant locations.

How will each of these actions affect her spending? Bringing her juice in a thermos reduces her costs because it is more expensive to buy individual juice boxes than to buy juice in a large container. This savings will soon offset the cost of the thermos. Composting will save money because she will have less garbage to take to the landfill and she will not have to purchase fertilizer for her garden. Buying organically grown produce will cost more, but the cost of locally grown foods is usually less than the cost of foods imported from other locations.

Appendix K

Calculations and Conversions

Weights and Measures		
Measure	**Abbreviation**	**Equivalent**
1 gram	g	1,000 milligrams
1 milligram	mg	1,000 micrograms
1 microgram	g	1/1000000 of a gram
1 nanogram	ng	1/1000000000 of a gram
1 picogram	pg	1/1000000000000 of a gram
1 kilogram	kg	1,000 grams 2.2 lb.
1 pound	lb.	454 grams 16 ounces
1 teaspoon	tsp	approximately 5 grams
1 tablespoon	Tbsp	3 teaspoons
1 ounce	oz.	28.4 grams
1 cup	c	8 fluid ounces 16 tablespoons
1 pint	pt.	2 cups 16 fluid ounces
1 quart	qt.	2 pints 32 fluid ounces
1 gallon	gal.	128 fluid ounces 4 quarts
1 litre	l (sometimes L)	1.06 quarts 1,000 millilitres
1 millilitre	ml (sometimes mL)	1,000 microlitres
1 decilitre	dl	100 millilitres
1 kcalorie	kcal, Cal	1,000 calories 4.167 kilojoules
1 kilojoule	kJ	1,000 joules

Glossary

A

24-hour recall A method of assessing dietary intake in which a trained interviewer helps an individual remember what he or she ate during the previous day.

absorption The process of taking substances into the interior of the body.

acceptable macronutrient distribution ranges (AMDRs) Ranges of intake for energy-yielding nutrients, expressed as a percentage of total energy intake, that are associated with reduced risk of chronic disease while providing adequate intakes of essential nutrients.

accretion An accumulation by external addition; in the case of nutrition, the uptake and accumulation by the body of a nutrient.

acetyl-CoA A metabolic intermediate formed during the breakdown of glucose, fatty acids, and amino acids. It is a 2-carbon compound attached to a molecule of CoA.

active transport The transport of substances across a cell membrane with the aid of a carrier molecule and the expenditure of energy. This may occur against a concentration gradient.

adaptive thermogenesis The change in energy expenditure induced by factors such as changes in ambient temperature and food intake.

added sugars Sugars and syrups that have been added to foods during processing or preparation.

adequacy A state in which there is a sufficient amount of a nutrient or nutrients in the diet to maintain health.

adequate intakes (AIs) Intakes that should be used as a goal when no RDA exists. These values are an approximation of the average nutrient intake that appears to sustain a desired indicator of health.

adipocytes Fat-storing cells.

adipose tissue Tissue found under the skin and around body organs that is composed of fat-storing cells.

adolescent growth spurt An 18-24-month period of peak growth velocity that begins at about ages 10–13 in girls and 12–15 in boys.

aerobic capacity or **VO_2 max** The maximum amount of oxygen that can be consumed by the tissues during exercise. This is also called maximal oxygen consumption.

aerobic exercise Endurance exercise such as jogging, swimming, or cycling that increases heart rate and requires oxygen in metabolism.

aerobic metabolism Metabolism in the presence of oxygen, which can completely break down glucose to yield carbon dioxide, water, and as many as 38 ATP molecules.

age-related bone loss The bone loss that occurs in both cortical and trabecular bone of men and women as they advance in age.

alcohol-related birth defects Malformations in the skeleton or major organ systems in a child due to maternal alcohol consumption during pregnancy.

alcohol-related neurodevelopmental disorders A spectrum of learning and developmental disabilities and behavioural abnormalities in a child due to maternal alcohol consumption during pregnancy.

aldosterone A hormone that increases sodium reabsorption by the kidney and therefore enhances water retention.

allergen A substance that causes an allergic reaction.

allergen A substance, usually a protein, that stimulates an immune response.

alpha-tocopherol (α-tocopherol) The most common form of vitamin E in human blood.

Alzheimer's disease A disease that results in the relentless and irreversible loss of mental function.

amenorrhea Delayed onset of menstruation or the absence of 3 or more consecutive menstrual cycles.

amino acid pool All of the amino acids in body tissues and fluids that are available for use by the body.

amino acids The building blocks of proteins. Each contains a central carbon atom bound to a hydrogen atom, an amino group, an acid group, and a side chain.

anabolic Energy-requiring processes in which simpler molecules are combined to form more complex substances.

anaerobic metabolism or **anaerobic glycolysis** Metabolism in the absence of oxygen. Each molecule of glucose generates 2 molecules of ATP. Glucose is metabolized in this way when the blood cannot deliver oxygen to the tissues quickly enough to support aerobic metabolism.

anaerobic threshold or **lactate threshold** The exercise intensity at which the reliance on anaerobic metabolism results in the accumulation of lactic acid.

anaphylaxis An immediate and severe allergic reaction to a substance (e.g., food or drugs). Symptoms include breathing difficulty, loss of consciousness, and a drop in blood pressure and can be fatal.

anencephaly A birth defect due to failure of the neural tube to close that results in the absence of a major portion of the brain, skull, and scalp. This is a fatal defect.

angiotensin II A compound that causes blood vessel walls to constrict and stimulates the release of the hormone aldosterone.

anthropometric measurements External measurements of the body, such as height, weight, limb circumference, and skinfold thickness.

antibodies Proteins produced by the body's immune system that recognize foreign substances in the body and help destroy them.

antidiuretic hormone (ADH) A hormone secreted by the pituitary gland that increases the amount of water reabsorbed by the kidney and therefore retained in the body.

antigen A foreign substance (almost always a protein) that, when introduced into the body, stimulates an immune response.

antioxidant A substance that is able to neutralize reactive oxygen molecules and thereby reduce oxidative damage.

anus The outlet of the rectum through which feces are expelled.

appetite The desire to consume specific foods that is independent of hunger.

ariboflavinosis The condition resulting from a deficiency of riboflavin.

arteries Vessels that carry blood away from the heart.

ascorbic acid or **ascorbate** The chemical term for vitamin C.

aseptic processing A method that places sterilized food in a sterilized package using a sterile process.

atherosclerosis A type of cardiovascular disease that involves the buildup of fatty material in the artery walls.

atherosclerotic plaque The cholesterol-rich material that is deposited in the arteries of individuals with atherosclerosis. It consists of cholesterol, smooth-muscle cells, fibrous tissue, and eventually calcium.

atoms The smallest units of an element that still retain the properties of that element.

ATP (adenosine triphosphate) The high-energy molecule used by the body to perform energy-requiring activities.

atrophic gastritis An inflammation of the stomach lining that causes a reduction in stomach acid and allows bacterial overgrowth.

atrophy Wasting or decrease in the size of a muscle or other tissue caused by lack of use.

attention deficit hyperactivity disorder (ADHD) A condition that is characterized by a short attention span, acting without thinking, and a high level of activity, excitability, and distractibility.

B

balance study A study that compares the total amount of a nutrient that enters the body with the total amount that leaves the body.

basal energy expenditure (BEE) The energy expended to maintain an awake resting body that is not digesting food.

basal metabolic rate (BMR) The rate of energy expenditure under resting conditions. BMR measurements are performed in a warm room in the morning before the subject rises, and at least 12 hours after the last food or activity.

behaviour modification A process used to gradually and permanently change habitual behaviours.

beriberi The disease resulting from a deficiency of thiamin.

beta-carotene (β-carotene) A carotenoid that has more provitamin A activity than other carotenoids. It also acts as an antioxidant.
beta-oxidation The first step in the production of ATP from fatty acids. This pathway breaks the carbon chain of fatty acids into 2-carbon units that form acetyl-CoA and releases high-energy electrons that are passed to the electron transport chain.
bile A substance made in the liver and stored in the gallbladder, which is released into the small intestine to aid in fat digestion and absorption.
binge drinking The consumption of 5 or more drinks in a row for males or 4 or more for females.
bioaccumulation The process by which compounds accumulate or build up in an organism faster than they can be broken down or excreted.
bioavailability A general term that refers to how well a nutrient can be absorbed and used by the body.
bioelectric impedance analysis A technique for estimating body composition that measures body fat by directing a low-energy electric current through the body and calculating resistance to flow.
biological value A measure of protein quality determined by comparing the amount of nitrogen retained in the body with the amount absorbed from the diet.
biomarker A biological measurement that is an indicator of future disease development.
blood pressure The amount of force exerted by the blood against the artery walls.
blood-glucose response curve A curve that illustrates the change in blood glucose that occurs after consuming food.
body mass index (BMI) A measure of body weight in relation to height that is used to compare body size with a standard.
bomb calorimeter An instrument used to determine the energy content of food. It measures the heat energy released when a dried food is combusted.
bone remodelling The process whereby bone is continuously broken down and re-formed to allow for growth and maintenance.
bran The protective outer layers of whole grains. It is a concentrated source of dietary fibre.
brown adipose tissue A type of fat tissue that has a greater number of mitochondria than the more common white adipose tissue. It can waste energy by producing heat.

C

Caesarean section The surgical removal of the fetus from the uterus.
calcitonin A hormone secreted by the thyroid gland that reduces blood calcium levels.
Canadian Community Health Survey A comprehensive survey of health-related issues, including the eating habits of 35,000 Canadians, that was begun in 2000 and continues to collect data annually.
capillaries Small, thin-walled blood vessels where the exchange of gases and nutrients between blood and cells occurs.
carcinogens Cancer-causing substances.
cardiorespiratory system The circulatory and respiratory systems, which together deliver oxygen and nutrients to cells.
cardiovascular disease Any disease affecting the heart and blood vessels.
carnitine A molecule synthesized in the body that is needed to transport fatty acids and some amino acids into the mitochondria for metabolism.
carotenoids Natural pigments synthesized by plants and many micro-organisms. They give yellow and red-orange fruits and vegetables their colour.
case-control study A type of observational study that compares individuals with a particular condition under study with individuals of the same age, sex, and background who do not have the condition.
cash crops Crops grown to be sold for monetary return rather than to be used for food locally.
catabolic The processes by which substances are broken down into simpler molecules, releasing energy.
cataracts A disease of the eye that results in cloudy spots on the lens (and sometimes the cornea), which obscure vision.
causation A relationship between two factors where one factor causes the second factor to occur.
celiac disease A disorder that causes damage to the intestines when the protein gluten is eaten.
cell differentiation Structural and functional changes that cause cells to mature into specialized cells.
cell membrane The membrane that surrounds the cell contents.
cells The basic structural and functional units of plant and animal life.
cellular respiration The reactions that break down carbohydrates, fats, and proteins in the presence of oxygen to produce carbon dioxide, water, and energy in the form of ATP.
ceruloplasmin The major copper-carrying protein in the blood.
chemical bonds Forces that hold atoms together.
chemical or **amino acid score** A measure of protein quality determined by comparing the essential amino acid content of the protein in a food with that in a reference protein. The lowest amino acid ratio calculated is the chemical score.
cholecalciferol The chemical name for vitamin D_3. It can be formed in the skin of animals by the action of sunlight on a form of cholesterol called 7-dehydrocholesterol.
cholecystokinin (CCK) A hormone released by the duodenum that stimulates the release of pancreatic juice rich in digestive enzymes and causes the gallbladder to contract and release bile into the duodenum.
cholesterol A lipid that consists of multiple chemical rings and is made only by animal cells.
chylomicrons Lipoproteins that transport lipids from the mucosal cells of the small intestine and deliver triglycerides to other body cells.
chyme A mixture of partially digested food and stomach secretions.
citric acid cycle Also known as the Krebs cycle or the tricarboxylic acid cycle, this is the stage of cellular respiration in which two carbons from acetyl-CoA are oxidized, producing two molecules of carbon dioxide.
coagulation The process of blood clotting.
cobalamin The chemical term for vitamin B_{12}.
coenzyme A small, organic molecule (not a protein but sometimes a vitamin) that is necessary for the proper functioning of many enzymes.
coenzymes Small nonprotein organic molecules that act as carriers of electrons or atoms in metabolic reactions and are necessary for the proper functioning of many enzymes.
cofactor An inorganic ion or coenzyme required for enzyme activity.
collagen The major protein in connective tissue.
colon The largest portion of the large intestine.
colostrum The first milk, which is secreted in late pregnancy and up to a week after birth. It is rich in protein and immune factors.
community garden A plot of land on which community members can grow food, usually in an urban setting. Community gardens promote urban green spaces and foster community development.
community or **collective kitchen** A group that gathers to prepare meals together, then eats or takes food home for later use. Participants improve food preparation skills, increase knowledge about healthy food choices, and enjoy time spent socializing.
comorbidity Two disease states or health conditions that occur together, such as obesity and type 2 diabetes.
complete dietary protein Protein that provides essential amino acids in the proportions needed to support protein synthesis.
complex carbohydrates Carbohydrates composed of monosaccharide molecules linked together in straight or branching chains. They include glycogen, starches, and fibres.
compression of morbidity The postponement of the onset of chronic disease such that disability occupies a smaller and smaller proportion of the life span.
conception The union of sperm and egg (ovum) that results in pregnancy.
condensation reaction Chemical reaction that joins 2 molecules together. Hydrogen and oxygen are lost from the 2 molecules to form water.
conditionally essential amino acids Amino acids that are essential in the diet only under certain conditions or at certain times of life.
confounding factor In scientific studies, a factor that is related to both the outcome being investigated (e.g., disease) and a factor that might influence outcome (dietary intake).
control group A group of participants in an experiment that is identical to the experimental group except that no experimental treatment is used. It is used as a basis of comparison.
correlation or **association** Two or more factors occurring together. The correlation can be direct (positive) or inverse (negative). A direct or positive relationship is observed when increased nutrient intake increases disease risk; an inverse or negative relationship is observed when decreased nutrient intake increases disease risk. Correlations do not prove causation.

cortical or **compact bone** Dense, compact bone that makes up the sturdy outer surface layer of bones.
creatine phosphate A compound found in muscle that can be broken down quickly to make ATP.
cretinism A condition resulting from poor maternal iodine intake during pregnancy that causes stunted growth and poor mental development in offspring.
criterion of adequacy A functional indicator, such as the level of a nutrient in the blood, that can be measured to determine the biological effect of a level of nutrient intake.
critical control points Possible points in food production, manufacturing, and transportation at which contamination could occur or be prevented.
cross-contamination The transfer of contaminants from one food or object to another.
cycle of malnutrition A cycle in which malnutrition is perpetuated by an inability to meet nutrient needs at all life stages.
cytosol The liquid found within cells.

D

daily value A nutrient reference value used on food labels to help consumers make comparisons between foods and select more nutritious food.
deamination The removal of the amino group from an amino acid.
dehydration A condition that results when not enough water is present to meet the body's needs.
dementia A deterioration of mental state resulting in impaired memory, thinking, and/or judgement.
denaturation The alteration of a protein's three-dimensional structure.
dental caries The decay and deterioration of teeth caused by acid produced when bacteria on the teeth metabolize carbohydrate.
depletion-repletion study A study that feeds subjects a diet devoid of a nutrient until signs of deficiency appear, and then adds the nutrient back to the diet to a level at which symptoms disappear and health is restored.
diabetes mellitus A disease caused by either insufficient insulin production or decreased sensitivity of cells to insulin. It results in elevated blood-glucose levels.
diet history Information about dietary habits and patterns. It may include a 24-hour recall, a food record, or a food frequency questionnaire to provide information about current intake patterns.
dietary antioxidant A substance in food that significantly decreases the adverse effects of reactive species on normal physiological function in humans.
dietary fibre A mixture of indigestible carbohydrates and lignin that is found intact in plants.
dietary folate equivalents (DFEs) The unit used to express the amount of folate present in food. One DFE is equivalent to 1 µg of folate naturally occurring in food, 0.6 µg of synthetic folic acid from fortified food or supplements consumed with food, or 0.5 µg of synthetic folic acid consumed on an empty stomach.
dietary pattern A description of a way of eating that includes the types and amounts of recommended foods and food groups, rather than individual nutrients.
dietary reference intakes (DRIs) A set of reference values for the intake of energy, nutrients, and food components that can be used for planning and assessing the diets of healthy people in the United States and Canada.
digestion The process of breaking food into components small enough to be absorbed into the body.
dipeptide Two amino acids linked by a peptide bond.
direct calorimetry A method of determining energy use that measures the amount of heat produced.
disaccharide A sugar formed by linking two monosaccharides.
dissociate To separate two charged ions.
diverticula Sacs or pouches that protrude from the wall of the large intestine in the disease **diverticulosis**. When these become inflamed, the condition is called **diverticulitis**.
double-blind study An experiment in which neither the study participants nor the researchers know who is in the control or the experimental group.
doubly-labelled water technique A method for measuring energy expenditure based on measuring the disappearance of heavy (but not radioactive) isotopes of hydrogen and oxygen in body fluids after consumption of a defined amount of water labelled with both isotopes.
Down syndrome A disorder caused by extra genetic material that results in distinctive facial characteristics, mental retardation, and other abnormalities.

E

EAR cutpoint method A method that indicates the proportion of a population that is not meeting its requirements, indicated by the proportion of the population with intakes below the EAR.
edema Swelling due to the buildup of extracellular fluid in the tissues.
eicosanoids Regulatory molecules, including prostaglandins and related compounds, that can be synthesized from omega-3 and omega-6 fatty acids.
electrolytes Positively and negatively charged ions that conduct an electrical current in solution. Commonly refers to sodium, potassium, and chloride.
electron High-energy particle carrying a negative charge that orbits the nucleus of an atom.
electron transport chain The final stage of cellular respiration in which electrons are passed down a chain of molecules to oxygen to form water and produce ATP.
elements Substances that cannot be broken down into products with different properties.
elimination diet and **food challenge** A regimen that eliminates potential allergy-causing foods from an individual's diet and then systematically adds them back to identify any foods that cause an allergic reaction.
embryo The developing human from 2-8 weeks after fertilization. All organ systems are formed during this time.
empty kcalories Refers to foods that contribute energy but few other nutrients.
emulsifiers Substances that allow water and fat to mix by breaking large fat globules into smaller ones.
endorphins Compounds that cause a natural euphoria and reduce the perception of pain under certain stressful conditions.
endosperm The largest portion of a kernel of grain, which is primarily starch and serves as a food supply for the sprouting seed.
energy balance The amount of energy consumed in the diet compared with the amount expended by the body over a given period.
energy-yielding nutrients Nutrients that can be metabolized to provide energy in the body.
enrichment Refers to a food that has had nutrients added to restore those lost in processing to a level equal to or higher than originally present.
enteral or **tube-feeding** A method of feeding by providing a liquid diet directly to the stomach or intestine through a tube placed down the throat or through the wall of the GI tract.
enzymes Protein molecules that accelerate the rate of specific chemical reactions without being changed themselves.
epidemiology The study of the interrelationships between health and disease and other factors in the environment or lifestyle of different populations.
epigenetics The study of genetics unrelated to changes in the sequence of DNA but to its chemical modification due to reactions such as methylation.
epiglottis A piece of elastic connective tissue at the back of the throat that covers the opening of the passageway to the lungs during swallowing.
ergogenic aid Anything designed to increase physical work or improve exercise performance.
esophagus A portion of the GI tract that extends from the pharynx to the stomach.
essential fatty acid deficiency A condition characterized by dry, scaly skin and poor growth that results when the diet does not supply sufficient amounts of the essential fatty acids.
essential fatty acids Fatty acids that must be consumed in the diet because they cannot be made by the body or cannot be made in sufficient quantities to meet needs.
essential nutrients Nutrients that must be provided in the diet because the body either cannot make them or cannot make them in sufficient quantities to satisfy its needs.
essential or **indispensable amino acids** Amino acids that cannot be synthesized by the human body in sufficient amounts to meet needs and therefore must be included in the diet.
estimated average requirements (EARs) Intakes that meet the estimated nutrient needs of 50% of individuals in a gender and life-stage group.

estimated energy requirements (EER) The amount of energy recommended by the DRIs to maintain body weight in a healthy person based on age, gender, size, and activity level.

extracellular fluid The fluid located outside cells. It includes fluid found in the blood, lymph, gastrointestinal tract, spinal column, eyes, joints, and that found between cells and tissues.

F

facilitated diffusion The movement of substances across a cell membrane from an area of higher concentration to an area of lower concentration with the aid of a carrier molecule. No energy is required.

failure to thrive The inability of a child's growth to keep up with normal growth curves.

famine A widespread lack of access to food due to a disaster that causes a collapse in the food production and marketing systems.

fatigue The inability to continue an activity at an optimal level.

fat-soluble vitamins Vitamins that dissolve in fat.

fatty acids Organic molecules made up of a chain of carbons linked to hydrogen atoms with an acid group at one end.

feces Body waste, including unabsorbed food residue, bacteria, mucus, and dead cells, which is excreted from the gastrointestinal tract by passing through the anus.

female athlete triad The combination of disordered eating, amenorrhea, and osteoporosis that occurs in some female athletes, particularly those involved in sports in which low body weight and appearance are important.

fermentation A process in which micro-organisms metabolize components of a food and therefore change the composition, taste, and storage properties of the food.

ferritin The major iron storage protein.

fertilization The union of sperm and egg (ovum).

fetal alcohol syndrome A characteristic group of physical and mental abnormalities in an infant resulting from maternal alcohol consumption during pregnancy.

fetus The developing human from the ninth week to birth. Growth and refinement of structures occur during this time.

fitness The ability to perform routine physical activity without undue fatigue.

flavin adenine dinucleotide (FAD) and **flavin mononucleotide (FMN)** The active coenzyme forms of riboflavin. The structure of these molecules allows them to pick up and donate hydrogens and electrons in chemical reactions.

fluorosis A condition caused by chronic overconsumption of fluoride, characterized by black and brown stains and cracking and pitting of the teeth.

folate and **folacin** General terms for the many forms of this vitamin.

folic acid The monoglutamate form of folate, which is present in the diet in fortified foods and vitamin supplements.

food additives Substances that are added to food during preparation or storage and become part of the food or affect its characteristics.

food diary A method of assessing dietary intake that involves an individual keeping a written record of all food and drink consumed during a defined period.

food disappearance surveys Surveys that estimate the food use of a population by monitoring the amount of food that leaves the marketplace.

food frequency questionnaire A method of assessing dietary intake that gathers information from individuals about how often certain categories of food are consumed.

food insecurity Limited, inadequate, or insecure access of individuals and households to sufficient, safe, nutritious, and personally acceptable food to meet their dietary requirements for a productive and healthy life.

food intolerance An adverse reaction to a food that does not involve antibody production by the immune system.

food jag When a child will eat only one food item meal after meal.

food self-sufficiency The ability of an area to produce enough food to feed its population.

foodborne illness An illness caused by consumption of food containing a toxin or disease-causing micro-organism.

foodborne infection Illness produced by the ingestion of food containing micro-organisms that can multiply inside the body and cause injurious effects.

foodborne intoxication Illness caused by consuming a food containing a toxin.

fortification A term used generally to describe the process of adding nutrients to foods, such as the addition of vitamin D to milk.

fortified foods Foods to which one or more nutrients have been added, typically to replace nutrient losses during processing or to prevent known inadequacies in the Canadian diet.

fortified or **enriched grains** Grains to which specific amounts of thiamin, riboflavin, niacin, and iron have been added. Since 1998, folic acid has also been added to enriched grains.

free radical One type of highly reactive molecule that causes oxidative damage.

fructose A monosaccharide that is the primary form of carbohydrate found in fruit.

functional fibre Isolated indigestible carbohydrates that have been shown to have beneficial physiological effects in humans.

G

galactose A monosaccharide that combines with glucose to form lactose or milk sugar.

gallbladder An organ of the digestive system that stores bile, which is produced by the liver.

gastric banding A surgical procedure in which an adjustable band is placed around the upper portion of the stomach to limit the volume that the stomach can hold and the rate of stomach emptying.

gastric bypass A surgical procedure to treat morbid obesity that both reduces the size of the stomach and bypasses a portion of the small intestine.

gastrin A hormone secreted by the stomach mucosa that stimulates the secretion of gastric juice.

gastroesophageal reflux disease (GERD) A chronic condition in which acidic stomach contents leak back up into the esophagus, causing pain and damaging the esophagus.

gastrointestinal tract A hollow tube consisting of the mouth, pharynx, esophagus, stomach, small intestine, large intestine, and anus, in which digestion and absorption of nutrients occur.

gene A length of DNA containing the information needed to synthesize RNA or a polypeptide chain. It is responsible for inherited traits.

gene expression The events of protein synthesis in which the information coded in a gene is used to synthesize a product, either a protein or a molecule of RNA.

genetic engineering A set of techniques used to manipulate DNA for the purpose of changing the characteristics of an organism or creating a new product.

germ The embryo or sprouting portion of a kernel of grain, which contains vegetable oil, protein, fibre, and vitamins.

gestation The time between conception and birth, which lasts about 9 months (or about 40 weeks) in humans.

gestational diabetes A form of diabetes that occurs during pregnancy and resolves after the baby is born.

gestational diabetes mellitus A consistently elevated blood glucose level that develops during pregnancy and returns to normal after delivery.

gestational hypertension The development of hypertension after the twentieth week of pregnancy.

ghrelin A hormone produced by the stomach that stimulates food intake.

glomerulus A ball of capillaries in the nephron that filters blood during urine formation.

glucagon A hormone made in the pancreas that stimulates the breakdown of liver glycogen and the synthesis of glucose to increase blood sugar.

gluconeogenesis The synthesis of glucose from simple, noncarbohydrate molecules. Amino acids from protein are the primary source of carbons for glucose synthesis.

glucose A monosaccharide that is the primary form of carbohydrate used to provide energy in the body. It is the sugar referred to as blood sugar.

glucose/fatty-acid-cycle The relationship between blood glucose and free fatty acids. When blood glucose levels are high, as in the post-prandial state, free-fatty-acid levels are low. In the fasting state, when blood glucose levels decline, free-fatty acid levels increase.

glutathione peroxidase A selenium-containing enzyme that protects cells from oxidative damage by neutralizing peroxides.

glycemic index A ranking of the effect on blood glucose of a food of a certain carbohydrate content relative to an equal amount of carbohydrate from a reference food such as white bread or glucose.

glycemic load An index of the glycemic response that occurs after eating specific foods. It is calculated by multiplying a food's glycemic index by the amount of available carbohydrate in a serving of the food.

glycemic response The rate, magnitude, and duration of the rise in blood glucose that occurs after a particular food or meal is consumed.

glycogen A carbohydrate made of many glucose molecules linked together in a highly branched structure. It is the storage form of carbohydrate in animals.

glycogen supercompensation or **carbohydrate loading** A regimen designed to maximize muscle glycogen stores before an athletic event.

glycolysis (also called **anaerobic metabolism**) Metabolic reactions in the cytosol of the cell that split glucose into two, 3-carbon pyruvate molecules, yielding two ATP molecules.

goiter An enlargement of the thyroid gland caused by a deficiency of iodine.

goitrogens Substances that interfere with the utilization of iodine or the function of the thyroid gland.

Good Food Box program An alternative means of distributing food which supplies members with affordable, fresh produce, most grown locally and bought directly from farmers; food may be delivered or picked up at a designated central location.

H

Hazard Analysis Critical Control Point (HACCP) A food safety system that focuses on identifying and preventing hazards that could cause foodborne illness.

heat-related illness Conditions, including heat cramps, heat exhaustion, and heat stroke, that can occur due to an unfavourable combination of exercise, hydration status, and climatic conditions.

heme iron A readily absorbed form of iron found in animal products that is chemically associated with proteins such as hemoglobin and myoglobin.

hemochromatosis An inherited condition that results in increased iron absorption.

hemoglobin An iron-containing protein in red blood cells that binds oxygen and transports it through the bloodstream to cells.

hemolytic anemia Anemia that results when red blood cells break open.

hemorrhoids Swollen veins in the anal or rectal area.

hemosiderin An insoluble iron storage compound that stores iron when the amount of iron in the body exceeds the storage capacity of ferritin.

hepatic portal circulation The system of blood vessels that collects nutrient-laden blood from the digestive organs and delivers it to the liver.

hepatic portal vein The vein that transports blood from the GI tract to the liver.

heterocyclic amines (HCAs) A class of mutagenic substances produced when there is incomplete combustion of amino acids during the cooking of meats—such as when meat is charred.

high-density lipoproteins (HDLs) Lipoproteins that pick up cholesterol from cells and transport it to the liver so that it can be eliminated from the body. A high level of HDL decreases the risk of cardiovascular disease.

homeostasis A physiological state in which a stable internal body environment is maintained.

hormones Chemical messengers that are produced in one location, released into the blood, and elicit responses at other locations in the body.

hormone-sensitive lipase An enzyme present in adipose cells that responds to chemical signals by breaking down triglycerides into free fatty acids and glycerol for release into the bloodstream.

human intervention trial or **clinical trial** A study of a population in which there is an experimental manipulation of some members of the population; observations and measurements are made to determine the effects of this manipulation, compared to members who did not undergo the manipulation.

hunger Internal signals that stimulate one to acquire and consume food.

hydrogenation The process whereby hydrogen atoms are added to the carbon-carbon double bonds of unsaturated fatty acids, making them more saturated.

hydrolysis reaction A type of chemical reaction in which a large molecule is broken into two smaller molecules by the addition of water.

hydroxyapatite A crystalline compound composed of calcium and phosphorus that is deposited in the protein matrix of bone to give it strength and rigidity.

hypercarotenemia A condition in which carotenoids accumulate in the adipose tissue, causing the skin to appear yellow-orange, especially the palms of the hands and the soles of the feet.

hypertension Blood pressure that is consistently elevated to 140/90 mm Hg or greater.

hypertensive disorders of pregnancy High blood pressure during pregnancy that is due to chronic hypertension, gestational hypertension, pre-eclampsia, eclampsia, or pre-eclampsia superimposed on chronic hypertension.

hypertrophy An increase in the size of a muscle or organ.

hypoglycemia A low blood-glucose level, usually below 2.2 to 2.8 mmol/L of blood plasma.

hyponatremia Abnormally low concentration of sodium in the blood.

hypothermia A condition in which body temperature drops below normal. Hypothermia depresses the central nervous system, resulting in the inability to shiver, sleepiness, and eventually coma.

hypothesis An educated guess made to explain an observation or to answer a question.

I

implantation The process by which the developing embryo embeds in the uterine lining.

incomplete dietary protein Protein that is deficient in one or more essential amino acids relative to body needs.

indirect calorimetry A method of estimating energy use that compares the amount of oxygen consumed to the amount of carbon dioxide exhaled.

infant mortality rate The number of deaths during the first year of life per 1,000 live births.

inorganic molecules Those containing no carbon–hydrogen bonds.

insensible losses Fluid losses that are not perceived by the senses, such as evaporation of water through the skin and lungs.

insoluble fibre Fibre that, for the most part, does not dissolve in water and cannot be broken down by bacteria in the large intestine. It includes cellulose, some hemicelluloses, and lignin.

insulin A hormone made in the pancreas that allows the uptake of glucose by body cells and has other metabolic effects such as stimulating protein and fat synthesis and the synthesis of glycogen in liver and muscle.

insulin resistance A situation when tissues become less responsive to insulin and do not take up glucose as readily. As a result glucose levels in the blood rise.

intake distribution A plot of the intakes of a specific nutrient in a population.

integrated pest management (IPM) A method of agricultural pest control that integrates nonchemical and chemical techniques.

interstitial fluid The portion of the extracellular fluid located in the spaces between the cells of body tissues.

intestinal microflora Micro-organisms that inhabit the large intestine.

intracellular fluid The fluid located inside cells.

intrinsic factor A protein produced in the stomach that is needed for the absorption of adequate amounts of vitamin B_{12}.

ion An atom or group of atoms that carries an electrical charge.

iron deficiency anemia An iron deficiency disease that occurs when the oxygen-carrying capacity of the blood is decreased because there is insufficient iron to make hemoglobin.

irradiation A process, also called cold pasteurization, that exposes foods to radiation to kill contaminating organisms and retard ripening and spoilage of fruits and vegetables.

isomers Molecules with the same molecular formula but a different arrangement of the atoms.

isotopes Alternative forms of an element that have different atomic masses, which may or may not be radioactive.

K

keratin A hard protein that makes up hair and nails.

keratomalacia Softening and drying and ulceration of the cornea resulting from vitamin A deficiency.

Keshan disease A type of heart disease that occurs in areas of China where the soil is very low in selenium. It is believed to be caused by a combination of viral infection and selenium deficiency.

ketones or **ketone bodies** Molecules formed in the liver when there is not sufficient carbohydrate to completely metabolize the 2-carbon units produced from fat breakdown.

kilocalorie (kcalorie, kcal) The unit of heat that is used to express the amount of energy provided by foods. It is the amount of heat required to raise the temperature of 1 kilogram of water 1 degree Celsius (1 kcalorie = 4.18 kjoules).

kilojoule (kjoule, kJ) A unit of work that can be used to express energy intake and energy output. It is the amount of work required to move an object weighing one kilogram a distance of 1 metre under the force of gravity (4.18 kjoules = 1 kcalorie).

kwashiorkor A form of protein-energy malnutrition in which only protein is deficient.

L

lactase An enzyme located in the brush border of the small intestine that breaks the disaccharide lactose into glucose and galactose.

lactation Milk production and secretion.

lacteal A tubular component of the lymphatic system that carries fluid away from body tissues. Lymph vessels in the intestine are known as lacteals and can transport large particles such as the products of fat digestion.

lactic acid A compound produced from the breakdown of glucose in the absence of oxygen.

lactose A disaccharide that is formed by linking galactose and glucose. It is commonly known as milk sugar.

lactose intolerance The inability to digest lactose because of a reduction in the levels of the enzyme lactase. It causes symptoms including intestinal gas and bloating after dairy products are consumed.

large-for-gestational-age An infant weighing more than 4 kg (8.8 lb) at birth.

LDL receptor A protein on the surface of cells that binds to LDL particles and allows their contents to be taken up for use by the cell.

lean body mass Body mass attributed to nonfat body components such as bone, muscle, and internal organs. It is also called fat-free mass.

lecithin A phosphoglyceride composed of a glycerol backbone, two fatty acids, a phosphate group, and a molecule of choline.

legumes Plants in the pea or bean family, which produce an elongated pod containing large starchy seeds. Examples include green peas, lentils, kidney beans, and peanuts.

leptin A protein hormone produced by adipocytes that signals information about the amount of body fat.

leptin resistance A lack of responsiveness to the hormone leptin; characterized, in obesity, by high levels of leptin in the blood but the lack of response to the action of leptin, which is to decrease energy intake and increase energy expenditure.

let-down A hormonal reflex triggered by the infant's suckling that causes milk to be released from the milk glands and flow through the duct system to the nipple.

life expectancy The average length of life for a population of individuals.

life span The maximum age to which members of a species can live.

life-stage groups Groupings of individuals based on stages of growth and development, pregnancy, and lactation, that have similar nutrient needs.

limiting amino acid The essential amino acid that is available in the lowest concentration in relation to the body's needs.

lipases Fat-digesting enzymes.

lipid bilayer Two layers of phosphoglyceride molecules oriented so that the fat-soluble fatty acid tails are sandwiched between the water-soluble phosphate-containing heads.

lipids A group of organic molecules, most of which do not dissolve in water. They include fatty acids, triglycerides, phospholipids, and sterols.

lipoprotein lipase An enzyme that breaks down triglycerides into fatty acids and glycerol; attached to the cell membranes of cells that line the blood vessels.

lipoproteins Particles containing a core of triglycerides and cholesterol surrounded by a shell of protein, phospholipids, and cholesterol that transport lipids in blood and lymph.

liposuction A procedure that suctions out adipose tissue from under the skin; used to decrease the size of local fat deposits such as on the abdomen or hips.

longevity The duration of an individual's life.

low-birth-weight A birth weight less than 2.5 kg (5.5 lb).

low-density lipoproteins (LDLs) Lipoproteins that transport cholesterol to cells. Elevated LDL cholesterol increases the risk of cardiovascular disease.

lymphatic system The system of vessels, organs, and tissues that drains excess fluid from the spaces between cells, transports fat-soluble substances from the digestive tract, and contributes to immune function.

lysozyme An enzyme in saliva, tears, and sweat that is capable of destroying certain types of bacteria.

M

macrocytes Larger-than-normal mature red blood cells that have a shortened life span.

macronutrients Nutrients needed by the body in large amounts. These include water and the energy-yielding nutrients: carbohydrates, lipids, and proteins.

macular degeneration Degeneration of a portion of the retina that results in a loss of visual detail and blindness.

major minerals Minerals needed in the diet in amounts greater than 100 mg/day or present in the body in amounts greater than 0.01% of body weight.

malignancy or **metastasis** A mass of cells showing uncontrolled growth, a tendency to invade and damage surrounding tissues, and an ability to seed daughter growths to sites remote from the original growth.

malnutrition Any condition resulting from an energy or nutrient intake either above or below that which is optimal.

maltose A disaccharide made up of 2 molecules of glucose. It is formed in the intestines during starch digestion.

marasmus A form of protein-energy malnutrition in which a deficiency of energy in the diet causes severe body wasting.

maximum heart rate The maximum number of beats/min. that the heart can attain. It declines with age and can be estimated by subtracting age in years from 220.

maximum residue limits The maximum amount of pesticide residues that may legally remain in food, set by Health Canada.

megaloblastic or **macrocytic anemia** A condition in which there are abnormally large immature and mature red blood cells in the bloodstream and a reduction in the total number of red blood cells and the oxygen-carrying capacity of the blood.

megaloblasts Large, immature red blood cells that are formed when developing red blood cells are unable to divide normally.

menaquinones The forms of vitamin K synthesized by bacteria and found in animals.

menarche The onset of menstruation, which occurs normally between the ages of 10-15 years.

menopause Physiological changes that mark the end of a woman's capacity to bear children.

metabolic pathway A series of chemical reactions inside of a living organism that result in the transformation of one molecule into another.

metabolic syndrome A collection of health risks, including excess fat in the abdominal region, high blood pressure, elevated blood triglycerides, low high-density lipoprotein (HDL) cholesterol, and high blood glucose that increases the chance of developing heart disease, stroke, and diabetes. The condition is also known by other names including Syndrome X, insulin resistance syndrome, and dysmetabolic syndrome.

metabolism The sum of all the chemical reactions that take place in a living organism.

metallothionein Refers to proteins that bind minerals. One such protein binds zinc and copper in intestinal cells, limiting their absorption into the blood.

micelles Particles formed in the small intestine when the products of fat digestion are surrounded by bile acids. They facilitate the absorption of fat.

micronutrients Nutrients needed by the body in small amounts. These include vitamins and minerals.

microvilli or **brush border** Minute, brush-like projections on the mucosal cell membrane that increase the absorptive surface area in the small intestine.

minerals In nutrition, elements needed by the body in small amounts for structure and to regulate chemical reactions and body processes.

mitochondrion (mitochondria) Cellular organelle responsible for providing energy in the form of ATP for cellular activities.

modified atmosphere packaging (MAP) A preservation technique used to prolong the shelf life of processed or fresh food by changing the gases surrounding the food in the package.

molecules Units of two or more atoms of the same or different elements bonded together.

monosaccharide A single sugar unit, such as glucose.

monounsaturated fatty acid A fatty acid that contains 1 carbon-carbon double bond.

morning sickness Nausea and vomiting that affect many women during the first few months of pregnancy and in some women can continue throughout the pregnancy.

mould Multicellular fungi that form a filamentous branching growth.

mucosa The layer of tissue lining the GI tract and other body cavities.

mucus A viscous fluid secreted by glands in the GI tract and other parts of the body, which acts to lubricate, moisten, and protect cells from harsh environments.

mutations Changes in DNA caused by chemical or physical agents.

myoglobin An iron-containing protein in muscle cells that binds oxygen.

N

natural health products A category of products regulated by Health Canada that include vitamin and mineral supplements, amino acids, fatty acids, probiotics, herbal remedies, and homeopathic and other traditional medicines. They occupy a middle ground between food and drugs.

nephron The functional unit of the kidney which performs the job of filtering the blood and maintaining fluid balance.

net protein utilization A measure of protein quality determined by comparing the amount of nitrogen retained in the body with the amount eaten in the diet.

neural tube defects Abnormalities in the brain or spinal cord that result from errors that occur during prenatal development. Defects in the brain are fatal, while those of the spinal cord often result in paralysis.

neural tube The portion of the embryo that develops into the brain and spinal cord.

neurotransmitters Molecules that function to transfer signals between the cells of the nervous system and can stimulate or inhibit a signal.

niacin equivalents (NEs) The measure used to express the amount of niacin present in food, including that which can be made from its precursor, tryptophan. One NE is equal to 1 mg of niacin or 60 mg of tryptophan.

nicotinamide adenine dinucleotide (NAD) and nicotinamide adenine dinucleotide phosphate (NADP) The active coenzyme forms of niacin that are able to pick up and donate hydrogens and electrons. They are important in the transfer of electrons to oxygen in cellular respiration and in many synthetic reactions.

night blindness The inability of the eye to adapt to reduced light, causing poor vision in dim light.

nitrogen balance The amount of nitrogen consumed in the diet compared with the amount excreted by the body over a given period.

nonessential or **dispensable amino acids** Amino acids that can be synthesized by the human body in sufficient amounts to meet needs.

nonexercise activity thermogenesis (NEAT) The energy expended for everything we do other than sleeping, eating, or sports-like exercise.

nonheme iron A poorly absorbed form of iron found in both plant and animal foods that is not part of the iron complex found in hemoglobin and myoglobin.

nursing bottle syndrome Extreme tooth decay in the upper teeth resulting from putting a child to bed with a bottle containing milk or other sweet liquids.

nutrient density An evaluation of the nutrient content of a food in comparison to the kcalories it provides.

nutrients Chemical substances in foods that provide energy and structure and help regulate body processes.

nutrition A science that studies the interactions that occur between living organisms and food.

nutrition transition A series of changes in diet, physical activity, health, and nutrition that occurs as poor countries become more prosperous.

nutritional assessment An evaluation used to determine the nutritional status of individuals or groups for the purpose of identifying nutritional needs and planning personal healthcare or community programs to meet those needs.

nutritional genomics or **nutrigenomics** The study of how diet affects our genes and how individual genetic variation can affect the impact of nutrients or other food components on health.

nutritional status State of health as it is influenced by the intake and utilization of nutrients.

O

obesity A condition characterized by excess body fat. It is defined as a body mass index of 30 kg/m^2 or greater.

obesity genes Genes that code for proteins involved in the regulation of food intake, energy expenditure, or the deposition of body fat. When they are abnormal, the result is abnormal amounts of body fat.

obesogenic environment An environment that promotes weight gain by encouraging overeating and physical inactivity.

observational study An epidemiological study that looks for associations between health and disease and environmental or lifestyle factors. There is no intervention or attempt to alter the behaviour or lifestyle of the study participants. Information about the subjects' lifestyle and health is collected and analyzed.

oligosaccharides Short-chain carbohydrates containing 3-10 sugar units.

omega-3 (ω-3) fatty acid A fatty acid containing a carbon-carbon double bond between the third and fourth carbons from the omega end.

omega-6 (ω -6) fatty acid A fatty acid containing a carbon-carbon double bond between the sixth and seventh carbons from the omega end.

organelles Cellular organs that carry out specific metabolic functions.

organic food Food that is produced, processed, and handled in accordance with the standards set by the Canadian General Standards Board's *Standards for Organic Agriculture* and the *Organic Products Regulations.*

organic molecules Those containing carbon bonded to hydrogen.

organs Discrete structures composed of more than one tissue that perform a specialized function.

osmosis The passive movement of water across a semipermeable membrane in a direction that will equalize the concentration of dissolved substances on both sides.

osteoblasts Cells responsible for the deposition of bone.

osteoclasts Large cells responsible for bone breakdown.

osteomalacia A vitamin D deficiency disease in adults characterized by a loss of minerals from bones. It causes bone pain, muscle aches, and an increase in bone fractures.

osteoporosis A bone disorder characterized by a reduction in bone mass, increased bone fragility, and an increased risk of fractures.

overload principle The concept that the body will adapt to the stresses placed on it.

overnutrition Poor nutritional status resulting from an energy or nutrient intake in excess of that which is optimal for health.

overtraining syndrome A collection of emotional, behavioural, and physical symptoms that occurs in serious athletes when training without sufficient rest persists for weeks to months.

overweight Being too heavy for one's height. It is defined as having a body mass index (BMI—a ratio of weight to height squared) of 25-29.9 kg/m^2.

oviducts or **fallopian tubes** Narrow ducts leading from the ovaries to the uterus.

oxalates Organic acids found in spinach and other leafy green vegetables that can bind minerals and decrease their absorption.

oxidative damage Damage caused by highly reactive oxygen molecules that steal electrons from other compounds, causing changes in structure and function.

oxidative stress A condition that occurs when there are more reactive oxygen molecules than can be neutralized by available antioxidant defenses. It occurs either because excessive amounts of reactive oxygen molecules are generated or because antioxidant defenses are deficient.

oxidized LDL cholesterol A substance formed when the cholesterol in LDL particles is oxidized by reactive oxygen molecules. It is key in the development of atherosclerosis because it contributes to the inflammatory process.

oxidized Refers to a compound that has lost an electron or undergone a chemical reaction with oxygen.

oxytocin A hormone produced by the posterior pituitary gland that acts on the uterus to cause uterine contractions and on the breast to cause the movement of milk into the secretory ducts that lead to the nipple.

P

PA (physical activity) value A numeric value associated with activity level that is a variable in the EER equations used to calculate energy needs.

pancreas An organ that secretes digestive enzymes and bicarbonate ions into the small intestine during digestion.

parasites Organisms that live at the expense of others.

parathyroid hormone (PTH) A hormone secreted by the parathyroid gland that increases blood calcium levels.

parietal cells Cells in the stomach lining that make hydrochloric acid and intrinsic factor in response to nervous or hormonal stimulation.

pasteurization The process of heating food products to kill disease-causing organisms.

pathogen A biological agent that causes disease.

peak bone mass The maximum bone density attained at any time in life, usually occurring in young adulthood.

peer review A process by which the quality of a science experiment is reviewed by experts. Experts must agree that the experiment is of good quality before it can be published.

pellagra The disease resulting from a deficiency of niacin.

pepsin A protein-digesting enzyme produced by the gastric glands. It is secreted in the gastric juice in an inactive form and activated by acid in the stomach.

pepsinogen An inactive protein-digesting enzyme produced by gastric glands and activated to pepsin by acid in the stomach.

peptic ulcer An open sore in the lining of the stomach, esophagus, or small intestine.

periodontal disease A degeneration of the area surrounding the teeth, specifically the gum and supporting bone.

peristalsis Coordinated muscular contractions that move food through the GI tract.

pernicious anemia An anemia resulting from vitamin B_{12} deficiency that occurs when dietary vitamin B_{12} cannot be absorbed due to a lack of intrinsic factor.

pH A measure of the level of acidity or alkalinity of a solution.

pharynx A funnel-shaped opening that connects the nasal passages and mouth to the respiratory passages and esophagus. It is a common passageway for food and air and is responsible for swallowing.

phenylketonuria (PKU) An inherited disease in which the body cannot metabolize the amino acid phenylalanine. If the disease is untreated, toxic by-products called phenylketones accumulate in the blood and interfere with brain development.

phosphoglycerides A class of phospholipid consisting of a glycerol molecule, 2 fatty acids, and a phosphate group.

phospholipids Types of lipids containing phosphorous. The most common are the phosphoglycerides, which are composed of a glycerol backbone with two fatty acids and a phosphate group attached.

phylloquinone The form of vitamin K found in plants.

physical frailty Impairment in function and reduction in physiological reserves severe enough to cause limitations in the basic activities of daily living.

phytic acid or **phytate** A phosphorus-containing storage compound found in seeds and grains that can bind minerals and decrease their absorption.

phytochemicals Substances found in plant foods (*phyto* means plant) that are not essential nutrients but may have health-promoting properties.

pica The compulsive ingestion of nonfood substances such as clay, laundry starch, and paint chips.

placebo A fake medicine or supplement that is indistinguishable in appearance from the real thing. It is used to disguise the control from the experimental groups in an experiment.

placenta An organ produced from both maternal and embryonic tissues. It secretes hormones, transfers nutrients and oxygen from the mother's blood to the fetus, and removes wastes.

plasma The liquid portion of the blood that remains when the blood cells are removed.

polar Used to describe a molecule that has a positive charge at one end and a negative charge at the other.

polychlorinated biphenyls (PCBs) Carcinogenic industrial compounds that have found their way into the environment and, subsequently, the food supply. Repeated ingestion causes them to accumulate in biological tissues over time.

polycyclic aromatic hydrocarbons (PAHs) A class of mutagenic substances produced during cooking when there is incomplete combustion of organic materials—such as when fat drips on a grill.

polypeptide A chain of 3 or more amino acids linked by peptide bonds.

polysaccharides Carbohydrates containing many monosaccharides units linked together.

polyunsaturated fatty acid A fatty acid that contains 2 or more carbon-carbon double bonds.

portion distortion The increase in portion sizes for typical restaurant and snack foods, observed over the last 40 years.

post-menopausal bone loss The accelerated bone loss that occurs in women for about 5 years after estrogen production decreases.

post-prandial state The time following a meal when nutrients from the meal are being absorbed.

prebiotics Indigestible carbohydrates that pass into the colon, where they serve as a food supply for bacteria, stimulating the growth or activity of certain types of beneficial bacteria.

pre-diabetes or **impaired glucose tolerance** A fasting blood-glucose level above the normal range but not high enough to be classified as diabetes.

pre-eclampsia A condition characterized by an increase in body weight, elevated blood pressure, protein in the urine, and edema. It can progress to **eclampsia**, which can be life-threatening to mother and fetus.

preservatives Compounds that extend the shelf life of a product by retarding chemical, physical, or microbiological changes.

preterm or **premature** An infant born before 37 weeks of gestation.

prion A pathogenic protein that is the cause of degenerative brain diseases called spongiform encephalopathies.

probiotics Specific types of live bacteria found in foods that are believed to have beneficial effects on human health.

processed foods Foods that have been specially treated or changed from their natural state.

programmed cell death The death of cells at specific predictable times.

prolactin A hormone released by the anterior pituitary that acts on the milk-producing glands in the breast to stimulate and sustain milk production.

pro-oxidant A substance that promotes oxidative damage.

prospective cohort study An observational study in which dietary intake information is collected by researchers and the health of the study participants is observed, usually for several years. At the end of the study, scientists determine whether there are any correlations between dietary intake and the incidence of disease.

protein complementation The process of combining proteins from different sources so that they collectively provide the proportions of amino acids required to meet needs.

protein digestibility–corrected amino acid score (PDCAAS) A measure of protein quality that reflects a protein's digestibility as well as the proportions of amino acids it provides.

protein efficiency ratio A measure of protein quality determined by comparing the weight gain of a laboratory animal fed a test protein with the weight gain of an animal fed a reference protein.

protein quality A measure of how efficiently a protein in the diet can be used to make body proteins.

protein turnover The continuous synthesis and breakdown of body proteins.

protein-energy malnutrition (PEM) A condition characterized by wasting and an increased susceptibility to infection that results from the long-term consumption of insufficient amounts of energy and protein to meet needs.

protein-sparing modified fast A very-low-kcalorie diet with a high proportion of protein, designed to maximize the loss of fat and minimize the loss of protein from the body.

prothrombin A blood protein required for blood clotting.

provitamin or **vitamin precursor** A compound that can be converted into the active form of a vitamin in the body.

puberty A period of rapid growth and physical changes that ends in the attainment of sexual maturity.

pyridoxal phosphate The major coenzyme form of vitamin B_6 that functions in more than 100 enzymatic reactions, many of which involve amino acid metabolism.

pyridoxine The chemical term for vitamin B_6.

R

randomization The process by which participants in an intervention trial are assigned to either a treatment group or a control group entirely by chance.

recommended daily intakes (RDIs) Reference values established for vitamins and minerals in Canada in the 1980s and 1990s.

recommended dietary allowances (RDAs) Intakes that are sufficient to meet the nutrient needs of almost all healthy people in a specific life-stage and gender group.

rectum The portion of the large intestine that connects the colon and anus.

reduced Refers to a compound that has gained an electron.

reference standards Reference values established for other several nutrients. The values are based on dietary recommendations for reducing the risk of chronic disease.

refined Refers to foods that have undergone processes that change or remove various components of the original food.

renewable resources Resources that are restored and replaced by natural processes and that can therefore be used forever.

renin An enzyme produced by the kidneys that converts angiotensin to angiotensin I.

requirement distribution A plot of the nutrient requirements for a group of individuals in the same life stage. Typically, the plot has the shape of a bell curve (i.e., a normal or binomial distribution).

reserve capacity The amount of functional capacity that an organ has above and beyond what is needed to sustain life.

resistant starch Starch that escapes digestion in the small intestine of healthy people.

resting energy expenditure (REE) or **resting metabolic rate (RMR)** Terms used when an estimate of basal metabolism is determined by measuring energy utilization after 5-6 hours without food or exercise.

resting heart rate The number of times that the heart beats per minute while a person is at rest.

retinoids The chemical forms of preformed vitamin A: retinol, retinal, and retinoic acid.

retinol activity equivalent (RAE) The amount of retinol, β-carotene, α-carotene, or β-cryptoxanthin that provides vitamin A activity equal to 1 μg of retinol.

retinol-binding protein A protein that is necessary to transport vitamin A from the liver to other tissues.

rhodopsin A light-absorbing compound found in the retina of the eye that is composed of the protein opsin loosely bound to retinal.

rickets A vitamin D deficiency disease in children that is characterized by poor bone development because of inadequate calcium absorption.

S

saliva A watery fluid produced and secreted into the mouth by the salivary glands. It contains lubricants, enzymes, and other substances.

salivary amylase An enzyme secreted by the salivary glands that breaks down starch.

sarcopenia Progressive decrease in skeletal muscle mass and strength that occurs with age.

satiety The feeling of fullness and satisfaction, caused by food consumption, that eliminates the desire to eat.

saturated fatty acid A fatty acid in which the carbon atoms are bound to as many hydrogens as possible and which, therefore, contains no carbon-carbon double bonds.

scavenger receptors Proteins on the surface of macrophages that bind to oxidized LDL cholesterol and allow it to be taken up by the cell.

scientific method The general approach of science that is used to explain observations about the world around us.

scurvy The vitamin C deficiency disease.

secretin A hormone released by the duodenum that signals the release of pancreatic juice rich in bicarbonate ions and stimulates the liver to secrete bile into the gallbladder.

segmentation Rhythmic local constrictions of the intestine that mix food with digestive juices and speed absorption by repeatedly moving the food mass over the intestinal wall.

selectively permeable Describes a membrane or barrier that will allow some substances to pass freely but will restrict the passage of others.

selenoproteins Proteins that contain selenium as a structural component of their amino acids. Selenium is most often found as selenocysteine, which contains an atom of selenium in place of the sulfur.

set-point theory The theory that when people finish growing, their weight remains relatively stable for long periods despite periodic changes in energy intake or output.

simple carbohydrates Carbohydrates known as sugars that include monosaccharides and disaccharides.

simple diffusion The movement of substances from an area of higher concentration to an area of lower concentration. No energy is required.

single carbon metabolism The transfer of single carbon groups, such as but not limited to methyl groups ($-CH_3$), between compounds.

single-blind study An experiment in which either the study participants or the researchers are unaware of which subjects are in the control or experimental group.

skinfold thickness A measurement of subcutaneous fat used to estimate total body fat.

small-for-gestational-age An infant born at term weighing less than 2.5 kg (5.5 lb).

soluble fibre Fibre that dissolves in water or absorbs water to form viscous solutions and can be broken down by the intestinal microflora. It includes pectins, gums, and some hemicelluloses.

solutes Dissolved substances.

solvent A fluid in which one or more substances dissolve.

sphincter A muscular valve that helps control the flow of materials in the GI tract.

spina bifida A birth defect resulting from the incorrect development of the spinal cord that can leave the spinal cord exposed. This can result in paralysis.

spontaneous abortion or **miscarriage** Termination of pregnancy, due to natural expulsion of fetus, prior to the seventh month.

spore A dormant state of some bacteria that is resistant to heat but can germinate and produce a new organism when environmental conditions are favourable.

sports anemia Reduced hemoglobin levels that occur as part of a beneficial adaptation to aerobic exercise in which expanded plasma volume dilutes red blood cells.

starch A carbohydrate made of many glucose molecules linked in straight or branching chains. The bonds that hold the glucose molecules together can be broken by human digestive enzymes.

sterols Types of lipids with a structure composed of multiple chemical rings.

stroke volume The volume of blood pumped by each beat of the heart.

stunting A decrease in linear growth rate, which is an indicator of nutritional well-being in populations of children.

subcutaneous fat Adipose tissue that is located under the skin.

subsistence crops Crops grown as food for the local population.

sucrose A disaccharide that is formed by linking fructose and glucose. It is commonly known as table sugar or white sugar.

sudden infant death syndrome (SIDS), or **crib death** The unexplained death of an infant, usually during sleep.

sugar alcohols Sweeteners that are structurally related to sugars but provide less energy than monosaccharides and disaccharides because they are not well absorbed.

superoxide dismutase (SOD) An enzyme that protects the cell from oxidative damage by neutralizing superoxide free radicals. One form of the enzyme requires zinc and copper for activity, and another form requires manganese.

sustainable agriculture Agricultural methods that maintain soil productivity and a healthy ecological balance while having minimal long-term impacts.

T

tannins Substances found in tea and some grains that can bind minerals and decrease their absorption.

teratogen A substance that can cause birth defects.

theory An explanation based on scientific study and reasoning.

thermic effect of food (TEF) or **diet-induced thermogenesis** The energy required for the digestion of food and the absorption, metabolism, and storage of nutrients. It is equal to approximately 10% of daily energy intake.

thiamin pyrophosphate The active coenzyme form of thiamin. It is the predominant form found inside cells, where it aids reactions in which a carbon-containing group is lost as CO_2.

threshold effect A reaction that occurs at a certain level of ingestion and increases as the dose increases. Below that level, there is no reaction.

thyroid-stimulating hormone A hormone that stimulates the synthesis and secretion of thyroid hormones from the thyroid gland.

tocopherol The chemical name for vitamin E.

tolerable upper intake levels (ULs) Maximum daily intakes that are unlikely to pose a risk of adverse health effects to almost all individuals in the specified life-stage and gender group.

total energy expenditure (TEE) The sum of the energy used for basal metabolism, activity, processing food, deposition of new tissue, and production of milk.

total fibre The sum of dietary fibre and functional fibre.

total parenteral nutrition (TPN) A technique for nourishing an individual by providing all needed nutrients directly into the circulatory system.

toxins Substances that can cause harm at some level of exposure.

trabecular or **spongy bone** The type of bone that forms the inner spongy lattice that lines the bone marrow cavity and supports the cortical shell.

trace elements or **trace minerals** Minerals required in the diet in amounts of 100 mg or less per day or present in the body in amounts of 0.01% of body weight or less.

***trans* fatty acid** An unsaturated fatty acid in which the hydrogen atoms are on opposite sides of the double bond.

transamination The process by which an amino group from one amino acid is transferred to a carbon compound to form a new amino acid.

transcription The process of copying the information in DNA to a molecule of mRNA.

transferrin An iron transport protein in the blood.

transferrin receptor Protein found in cell membranes that binds to the iron-transferrin complex and allows it to be taken up by cells.

transit time The time between the ingestion of food and the elimination of the solid waste from that food.

translation The process of translating the mRNA code into the amino acid sequence of a polypeptide chain.

treatment group A group of participants in an experiment who are receiving an experimental treatment. The effects of the treatment are compared to the control group.

triglycerides (Triacylglycerols) The major form of lipid in food and in the body. They consist of three fatty acids attached to a glycerol molecule.

trimester A term used to describe each third, or 3-month period, of a pregnancy.

tripeptide Three amino acids linked by peptide bonds.

tropical oils A term used in the popular press to refer to the saturated oils—coconut, palm, and palm kernel oil—that are derived from plants grown in tropical regions.

type 1 diabetes A form of diabetes that is caused by the autoimmune destruction of insulin-producing cells in the pancreas, usually leading to absolute insulin deficiency; previously known as insulin-dependent diabetes mellitus or juvenile-onset diabetes.

type 2 diabetes A form of diabetes that is characterized by insulin resistance and relative insulin deficiency; previously known as noninsulin-dependent diabetes mellitus or adult-onset diabetes.

U

undernutrition Any condition resulting from an energy or nutrient intake below that which meets nutritional needs.

underwater weighing A technique that uses the difference between body weight underwater and body weight on land to estimate body density and calculate body composition.

urea A nitrogen-containing waste product formed from the breakdown of amino acids that is excreted in the urine.

V

variable A factor or condition that is changed in an experimental setting.

vegan A pattern of food intake that eliminates all animal products.

vegetarianism A pattern of food intake that eliminates some or all animal products.

veins Vessels that carry blood toward the heart.

very-low-birth-weight A birth weight less than 1.5 kg (3.3 lb).

very-low-density lipoproteins (VLDLs) Lipoproteins assembled by the liver that carry lipids from the liver and deliver triglycerides to body cells.

very-low-kcalorie diet A weight-loss diet that provides fewer than 800 kcal/day.

villi (villus) Finger-like protrusions of the lining of the small intestine that participate in the digestion and absorption of nutrients.

viruses Minute particles not visible under an ordinary microscope that depend on cells for their metabolic and reproductive needs.

visceral fat Adipose tissue that is located in the abdomen around the body's internal organs.

vitamin D receptor (VDR) A protein to which vitamin D binds. This receptor-vitamin complex is then able to bind to DNA and alter gene expression.

vitamins Organic compounds needed in the diet in small amounts to promote and regulate the chemical reactions and processes needed for growth, reproduction, and maintenance of health.

W

water intoxication A condition that occurs when a person drinks enough water to significantly lower the concentration of sodium in the blood.

water-soluble vitamins Vitamins that dissolve in water.

weight cycling or **yo-yo dieting** The repeated loss and regain of body weight.

weight stigmatization Disapproval and lack of acceptance suffered by someone because he or she is obese.

Wernicke-Korsakoff syndrome A form of thiamin deficiency associated with alcohol abuse that is characterized by mental confusion, disorientation, loss of memory, and a staggering gait.

whole-grain The entire kernel of grain, including the bran layers, the germ, and the endosperm.

X

xerophthalmia A spectrum of eye conditions resulting from vitamin A deficiency that may lead to blindness. An early symptom is night blindness, and as deficiency worsens, lack of mucus leaves the eye dry and vulnerable to cracking and infection.

Z

zoochemicals Substances found in animal foods (*zoo* means animal) that are not essential nutrients but may have health-promoting properties.

zygote The cell produced by the union of sperm and ovum during fertilization.

Index

E

J

K

L

P

Q

R

U

V

W

Dietary Reference Intake Values for Energy: Estimated Energy Requirement (EER) Equations and Values for Active Individuals by Life Stage Group			
Life Stage Group	**EER Prediction Equation**	**EER for Active Physical Activity Level (kcal/day)[a]**	
		Male	**Female**
0–3 mo	EER = (89 × weight of infant in kg – 100) + 175	538	493 (2 mo)[c]
4–6 mo	EER = (89 × weight of infant in kg – 100) + 56	606	543 (5 mo)[c]
7–12 mo	EER = (89 × weight of infant in kg – 100) + 22	743	676 (9 mo)[c]
1–2 y	EER = (89 × weight of infant in kg – 100) + 20	1046	992 (2 y)[c]
3–8 y			
Male	EER = 88.5 – (61.9 × Age in yrs) + PA[b][(26.7 × Weight in kg) + (903 × Height in m)] + 20	1742 (6 y)[c]	
Female	EER = 135.3 – (30.8 × Age in yrs) + PA[b][(10.0 × Weight in kg) + (934 × Height in m)] + 20		1642 (6 y)[c]
9–13 y			
Male	EER = 88.5 – (61.9 × Age in yrs) + PA[b][(26.7 × Weight in kg) + (903 × Height in m)] + 25	2279 (11 y)[c]	
Female	EER = 135.3 – (30.8 × Age in yrs) + PA[b][(10.0 × Weight in kg) + (934 × Height in m)] + 25		2071 (11 y)[c]
14–18 y			
Male	EER = 88.5 – (61.9 × Age in yrs) + PA[b][(26.7 × Weight in kg) + (903 × Height in m)] + 25	3152 (16 y)[c]	
Female	EER = 135.3 – (30.8 × Age in yrs) + PA[b][(10.0 × Weight in kg) + (934 × Height in m)] + 25		2368 (16 y)[c]
19 and older			
Males	EER = 662 – (9.53 × Age in yrs) + PA[b][(15.91 × Weight in kg) + (539.6 × Height in m)]	3067 (19 y)[c]	
Females	EER = 354 – (6.91 × Age in yrs) + PA[b][(9.36 × Weight in kg) + (726 × Height in m)]		2403 (19 y)[c]
Pregnancy			
14–18 y			
1st trimester	Adolescent EER + 0		2368 (16 y)[c]
2nd trimester	Adolescent EER + 340 kcal		2708 (16 y)[c]
3rd trimester	Adolescent EER + 452 kcal		2820 (16 y)[c]
19–50 y			
1st trimester	Adult EER + 0		2403 (19 y)[c]
2nd trimester	Adult EER + 340 kcal		2743 (19 y)[c]
3rd trimester	Adult EER + 452 kcal		2855 (19 y)[c]
Lactation			
14–18 y			
1st 6 mo	Adolescent EER + 330 kcal		2698 (16 y)[c]
2nd 6 mo	Adolescent EER + 400 kcal		2768 (16 y)[c]
19–50 y			
1st 6 mo	Adult EER + 330 kcal		2733 (19 y)[c]
2nd 6 mo	Adult EER + 400 kcal		2803 (19 y)[c]

[a]The intake that meets the average energy expenditure of active individuals at a reference height, weight, and age.
[b]See table entitled "Physical Activity Coefficients (PA Values) for Use in EER Equations" to determine the PA value for various ages, genders, and activity levels.
[c]Value is calculated for an individual at the age in parentheses.

Physical Activity Coefficients (PA Values) for Use in EER Equations				
Age and Gender	**Sedentary**	**Low Active**	**Active**	**Very Active**
3 to 18 y				
Boys	1.00	1.13	1.26	1.42
Girls	1.00	1.16	1.31	1.56
≥ 19 y				
Men	1.00	1.11	1.25	1.48
Women	1.00	1.12	1.27	1.45

Source: Institute of Medicine, Food and Nutrition Board. "Dietary Reference Intakes for Energy, Carbohydrates, Fiber, Fat, Protein, and Amino Acids." Washington, D.C.: National Academy Press, 2002.

Dietary Reference Intakes: Tolerable Upper Intake Levels (UL[a]): Vitamins

Life Stage Group	Vitamin A (μg/day)[b]	Vitamin C (mg/day)	Vitamin D (μg/day)	Vitamin E (mg/day)[c,d]	Vitamin K	Thiamin	Riboflavin	Niacin (mg/day)[d]	Vitamin B_6 (mg/day)	Folate (μg/day)[d]	Vitamin B_{12}	Pantothenic Acid	Biotin	Choline (g/day)	Carotenoids[e]
Infants															
0–6 mo	600	ND[f]	25	ND	ND	ND	ND	ND	ND	ND	ND	ND	ND	ND	ND
7–12 mo	600	ND	38	ND	ND	ND	ND	ND	ND	ND	ND	ND	ND	ND	ND
Children															
1–3 y	600	400	63	200	ND	ND	ND	10	30	300	ND	ND	ND	1.0	ND
4–8 y	900	650	75	300	ND	ND	ND	15	40	400	ND	ND	ND	1.0	ND
Males, Females															
9–13 y	1,700	1,200	100	600	ND	ND	ND	20	60	600	ND	ND	ND	2.0	ND
14–18 y	2,800	1,800	100	800	ND	ND	ND	30	80	800	ND	ND	ND	3.0	ND
19–70 y	3,000	2,000	100	1,000	ND	ND	ND	35	100	1,000	ND	ND	ND	3.5	ND
> 70 y	3,000	2,000	100	1,000	ND	ND	ND	35	100	1,000	ND	ND	ND	3.5	ND
Pregnancy															
≤ 18 y	2,800	1,800	100	800	ND	ND	ND	30	80	800	ND	ND	ND	3.0	ND
19–50 y	3,000	2,000	100	1,000	ND	ND	ND	35	100	1,000	ND	ND	ND	3.5	ND
Lactation															
≤ 18 y	2,800	1,800	100	800	ND	ND	ND	30	80	800	ND	ND	ND	3.0	ND
19–50 y	3,000	2,000	100	1,000	ND	ND	ND	35	100	1,000	ND	ND	ND	3.5	ND

[a]UL = The maximum level of daily nutrient intake that is likely to pose no risk of adverse effects. Unless otherwise specified, the UL represents total intake from food, water, and supplements. Due to lack of suitable data, ULs could not be established for vitamin K, thiamin, riboflavin, vitamin B_{12}, pantothenic acid, biotin, or carotenoids. In the absence of ULs, extra caution may be warranted in consuming levels above recommended intakes.

[b]As preformed vitamin A only.

[c]As α-tocopherol; applies to any form of supplemental α-tocopherol.

[d]The ULs for vitamin E, niacin, and folate apply to synthetic forms obtained from supplements, fortified foods, or a combination of the two.

[e]β-Carotene supplements are advised only to serve as a provitamin A source for individuals at risk of vitamin A deficiency.

[f]ND = Not determinable due to lack of data of adverse effects in this age group and concern with regard to lack of ability to handle excess amounts. Source of intakes should be from food only to prevent high levels of intake.

Source: Dietary Reference Intake Tables: The Complete Set. Institute of Medicine, National Academy of Sciences. Available online at www.nap.edu